Get the Most Out of
MyMathLab®

MyMathLab, Pearson's online homework, tutorial, and assessment program, creates personalized experiences for students and provides powerful tools for instructors. With a wealth of tested and proven resources, each course can be tailored to fit your specific needs.

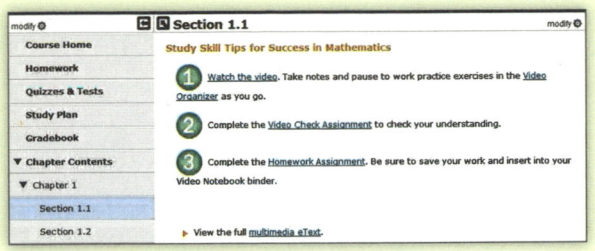

Learning in any Environment

- Because classroom formats and student needs continually change and evolve, MyMathLab has built-in flexibility to accommodate various course designs and formats.

- With an updated and streamlined design, students and instructors can access MyMathLab from most mobile devices.

Personalized Learning

Not every student learns the same way or at the same rate. Thanks to our advances in adaptive learning technology capabilities, you no longer have to teach as if they do.

- MyMathLab's **adaptive Study Plan** acts as a personal tutor, updating in real-time based on student performance throughout the course to provide personalized recommendations for practice. You can now assign the Study Plan as a prerequisite to a test or quiz with **Companion Study Plan Assignments**.

- MyMathLab can **personalize homework assignments** for students based on their performance on a test or quiz. This way, students can focus on just the topics they have not yet mastered.

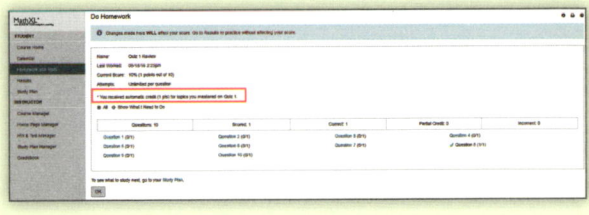

- **MyMathLab with Integrated Review courses**—available with selected titles for Developmental Mathematics through Calculus—can be used for just-in-time prerequisite review or for co-requisite courses. These courses provide videos on review topics, along with pre-made, assignable skill-review quizzes and personalized homework assignments integrated throughout your regular MyMathLab course content.

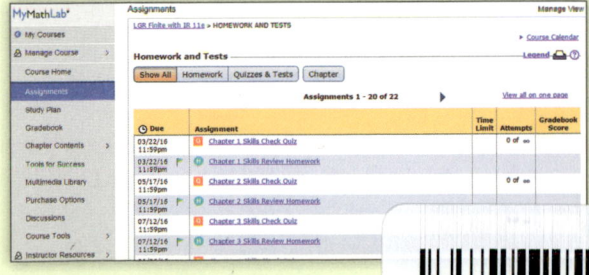

Visit www.mymathlab.com and click Get Trained to make you're getting the most out of MyMathLab.

Get the Most Out of Your Developmental Math MyMathLab Courses

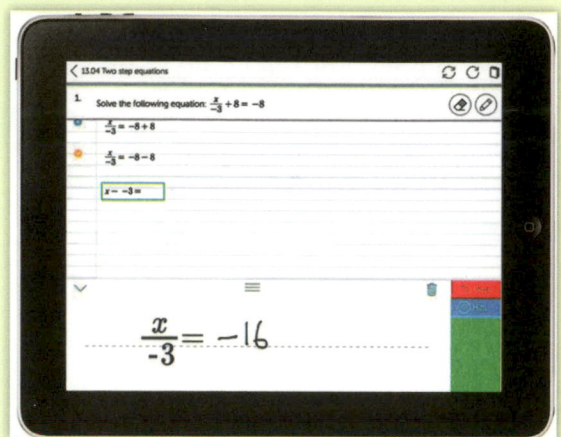

Offer Step-by-Step Exercise Support

New Workspace Assignments allow students to naturally write out their work, step-by-step, and receive instant feedback at each step. Specific, relevant hints and videos are available to guide students through the steps. Each student's work is captured in the MyMathLab gradebook so you can easily pinpoint exactly where you need to focus your instruction.

Get Students Engaged

Learning Catalytics™—a student response tool that uses students' smartphones, tablets, or laptops to engage them in more interactive tasks and thinking— is available through any MyMathLab course to foster student engagement and peer-to-peer learning. You can generate class discussion, guide your lecture, and promote peer-to-peer learning with real-time analytics.

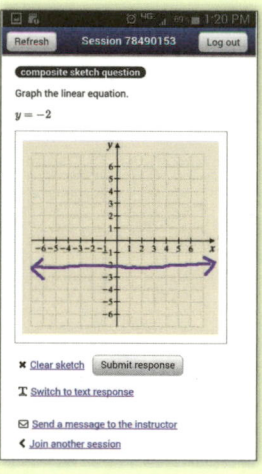

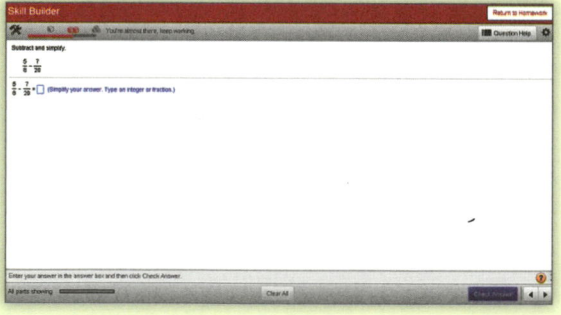

Provide Adaptive Just-in-Time Practice

When students struggle on an exercise, Skill Builder Assignments provide just-in-time, targeted support to help them build up the requisite skills needed to complete their homework assignment. As students progress, these assignments adapt to provide support exercises that are personalized to each student's activity and performance throughout the course.

Use Data to Tailor Your Course

A comprehensive gradebook with enhanced reporting functionality helps you efficiently manage your course. The new Reporting Dashboard presents student performance data at the class, section, and program levels in an accessible, visual manner. Item Analysis allows you to track class-wide understanding at the exercise level, so that you can refine your lectures or assignments to address just-in-time student needs.

Elementary & Intermediate Algebra

Fourth Edition

Michael Sullivan, III
Joliet Junior College

Katherine R. Struve
Columbus State Community College

Janet Mazzarella
Southwestern College

with contributions by

Jessica Bernards
Portland Community College

Wendy Fresh
Portland Community College

330 Hudson Street, NY NY 10013

VP, Courseware Portfolio Management: Chris Hoag
Director, Courseware Portfolio Management: Michael Hirsch
Courseware Portfolio Manager: Mary Beckwith
Courseware Portfolio Management Assistant: Alison Oehman
Content Producer: Tamela Ambush
Managing Producer: Karen Wernholm
Producer: Audra Walsh
Manager, Courseware QA: Mary Durnwald
Manager, Content Development: Rebecca Williams
Product Marketing Manager: Alicia Frankel
Field Marketing Managers: Jennifer Crum and Lauren Schur
Marketing Assistant: Hanna Lafferty
Senior Author Support/Technology Specialist: Joe Vetere
Manager, Rights and Permissions: Gina Cheselka
Manufacturing Buyer: Carol Melville, LSC Communications
Associate Director of Design: Blair Brown
Text Design, Cover Design, and Composition: Cenveo® Publisher Services
Cover Image: (grass) Val Lawless/Shutterstock; (cap) Angelo Gilardelli/Shutterstock;
(duck tape) Mexrix/Shutterstock

Library of Congress Cataloging-in-Publication Data

Names: Sullivan, Michael, III, 1967- | Struve, Katherine R. | Mazzarella,
    Janet.
Title: Elementary & intermediate algebra / Michael Sullivan, III, Joliet Junior College,
    Katherine R. Struve, Columbus State Community College, Janet Mazzarella,
    Southwestern College.
Other titles: Elementary and intermediate algebra
Description: 4th edition. | Boston: Pearson, c2018.
Identifiers: LCCN 2016020521| ISBN 9780134556079 (hardcover) | ISBN
    0134556070 (hardcover) | ISBN 9780134592350 (hardcover) | ISBN 0134592352 (hardcover)
Subjects: LCSH: Algebra—Textbooks.
Classification: LCC QA152.3 .S864 2018 | DDC 512.9—dc23
LC record available at https://lccn.loc.gov/2016020521

2 17

ISBN 13: 978-0-13-455607-9   (Student Edition)
ISBN 10: 0-13-455607-0

About the Authors

With training in mathematics, statistics, and economics, Michael Sullivan, III has a varied teaching background that includes 27 years of instruction in both high school and college-level mathematics. He is currently a full-time professor of mathematics at Joliet Junior College. Michael has numerous textbooks in publication, including an Introductory Statistics series and a Precalculus series, which he writes with his father, Michael Sullivan.

Michael believes that his experiences writing texts for college-level math and statistics courses give him a unique perspective as to where students are headed once they leave the developmental mathematics tract. This experience is reflected in the philosophy and presentation of his developmental text series. When not in the classroom or writing, Michael enjoys spending time with his three children, Michael, Kevin, and Marissa, and playing golf. Now that his two sons are getting older, he has the opportunity to do both at the same time!

Kathy Struve has been a classroom teacher for nearly 35 years, first at the high school level and, for the past 27 years, at Columbus State Community College. Kathy embraces classroom diversity: diversity of students' age, learning styles, and previous learning success. She is aware of the challenges of teaching mathematics at a large, urban community college, where students have varied mathematics backgrounds and may enter college with a high level of mathematics anxiety.

Kathy served as Lead Instructor of the Developmental Algebra sequence at Columbus State, where she developed curriculum, conducted workshops, and provided leadership to adjunct faculty in the mathematics department. She embraces the use of technology in instruction, and has taught web and hybrid classes in addition to traditional face-to-face and emporium-style classes. She is always looking for ways to more fully involve students in the learning process. In her spare time Kathy enjoys spending time with her two adult daughters, her four granddaughters, and biking, hiking, and traveling with her husband.

Born and raised in San Diego county, Janet Mazzarella spent her career teaching in culturally and economically diverse high schools before taking a position at Southwestern College 25 years ago. Janet has taught a wide range of mathematics courses from arithmetic through calculus for math/science/engineering majors and has training in mathematics, education, engineering, and accounting.

Janet has worked to incorporate technology into the curriculum by participating in the development of Interactive Math and Math Pro. At Southwestern College, she helped develop the self-paced developmental mathematics program. In addition, Janet was the Dean of the School of Mathematics, Science, and Engineering, the Chair of the Mathematics Department, the faculty union president, and the faculty coordinator for Intermediate Algebra. In the past, free time consisted of racing motorcycles off-road in the Baja 500 and rock climbing, but recently she has given up the adrenaline rush of these activities for the thrill of traveling in Europe.

To my wife, Yolanda, for her friendship, support, and encouragement.

—Michael Sullivan

To my daughters, Anne and Laura Struve, who have taught me how to be a better teacher.

—Katherine R. Struve

To my students, who have inspired me more than I can say; and to the two of whom I am most proud, my children, Kellen and Jillian.

—Janet Mazzarella

Contents

Preface

We would like to thank the reviewers, class testers, and users of the previous edition of *Elementary & Intermediate Algebra* who helped to make the book an overwhelming success. Their thoughtful comments and suggestions provided strong guidance for improvements in the fourth edition that we believe will enhance this solid, student-friendly text.

The Elementary & Intermediate Algebra course serves a diverse group of students. Some of them are new to algebra, while others were introduced to the material but have not yet grasped all the concepts. Still other students realized success in the course in the past but need a refresher. Not only do the backgrounds of students vary with respect to their mathematical abilities, but students' motivation, reading level, and study skills also range considerably.

This diversity makes teaching Elementary & Intermediate Algebra challenging. It is imperative that texts recognize the diversity of the classroom and address the array of needs of the students.

Elementary & Intermediate Algebra introduces students to the logic and precision of mathematics. We expect students to leave the course with an appreciation of this precision as well as of the power of mathematics. Our students need to understand that the concepts we teach in this course form the basis for future mathematics courses. Once they have a conceptual understanding of algebra, students recognize that the material is not merely a series of unconnected topics. Instead, they see a story in which each new chapter builds on concepts learned in previous chapters.

To reinforce this idea, we remind our students of a helpful fact—mathematics is about taking a problem and reducing it to another problem that they have already seen. Reducing a problem to its component parts makes it easier to solve and helps students to see the forest for the trees (and, to carry the metaphor further, prevent them from feeling that they are lost in the woods).

In short, to address the many needs of today's Elementary & Intermediate Algebra students, we established the following as our goals for this text:

- Provide students with a strong conceptual foundation in mathematics through a clear and thorough presentation of concepts.

- Offer comprehensive exercise sets that build students' skills, show intriguing applications of mathematics, begin to build mathematical thinking skills, and reinforce mathematical concepts.

- Provide students with ample opportunity to see the connections among the topics learned in the course.

- Present a variety of study aids and tips so students quickly come to view the text as a useful and reliable tool that can increase success in the course.

New to the Fourth Edition

The revision of this text takes advantage of MyMathLab as a tool for learning. To address the needs of students who are exposed to the material almost exclusively through MyMathLab, we have introduced the following new features to MyMathLab based on some of the hallmark features of the text.

- Discovery activities using **applets** have been developed. These explorations are carefully crafted to allow students to develop understanding of mathematical concepts through experiential learning. The applets and guided exercises that utilize the applets are found in MyMathLab. The applets may also be accessed using the QR code at the beginning of the section.

- **Guided Exercises** are now available in MyMathLab based on the popular Showcase Examples. Showcase Examples from the text are easy to recognize with the words "How To" in the example title and provide step-by-step

solutions to examples. This example structure was
MyMathLab exercises that require students to respond
steps to solving problems are developed, similar to the
feature of MyMathLab. This keeps the student comple
learning process and develops their conceptual underst
These exercises are easy to identify in the Assignme
designated "How-To-#.# Ex #-<title abbreviation>. For
Ex 6-Solve a Linear Equation.

- **Quick Response (QR) codes** now appear at each secti
  level exercises, and as part of the Chapter Tests. Students
  scanner from their smartphone for easy access to the popu
  lecture videos, select end-of-section exercise videos, the disc
  Chapter Test Prep videos.

- The authors developed a **Premade Author-Created MyMath**
  the latest MyMathLab features. Each section has two MyMat

  - The first assignment is a multimedia assignment that incor
    Action lecture videos, the new discovery applet exercises, the
    exercises, and the Quick Check exercises from the text. The
    follow many of the examples in the text. To assist student
    the Textbook learning aid for each Quick Check exercise w
    corresponding example in the text. All learning aids with th
    an Example" will be available for this portion of the homev
    as instructors has been that too many students rely on th
    doing homework, thereby reducing the effect of homewor
    mimic the View an Example content.

  - The second assignment is based on the Skill Building
    exercises from the text. Skill building exercises are tied
    the text, so the Textbook learning aid will link directly to the
    section. The idea is to reduce the amount of guidance pro
    (compared with Quick Check exercises) so they are more resp
    the problem type. The Mixed Practice exercises are based
    learned within the section or text, so the Textbook learning aid i
    The student must determine the problem type based on Q
    Building exercise experience. The "View an Example" learni
    this exercise set as well. Because this text has Skill Builder ava
    you may consider reducing the number of exercises in the s
    checking the Skill Builder box, the assignments will adap
    exercises personalized to each student's needs.

Content Change

- Systems of linear equations with dependent systems no lo
  of simply "infinitely many solutions." Rather, we express
  dependent system using set builder notation. For examp
  one of the equations in the dependent system, the solu
  $\{(x, y) \mid 3x + y = 1\}$.

Develop an Effective Text for Use In and Out of the Classroom

Given the hectic lives led by most students, coupled with the an
with which they approach this course, an outstanding develop
text must provide pedagogical support that makes the text va
they study and do assignments. Pedagogy must be presented
that teaches students how to study math; pedagogical devices m
students see as the "mystery" of mathematics—and solve that m

To encourage students and to clarify the material, we develop
features that help students develop good study skills, garner an
connections between topics, and work smarter in the process.

Elementary & Intermediate Algebra

Fourth Edition

Michael Sullivan, III
Joliet Junior College

Katherine R. Struve
Columbus State Community College

Janet Mazzarella
Southwestern College

with contributions by

Jessica Bernards
Portland Community College

Wendy Fresh
Portland Community College

 **Pearson**

330 Hudson Street, NY NY 10013

VP, Courseware Portfolio Management: Chris Hoag
Director, Courseware Portfolio Management: Michael Hirsch
Courseware Portfolio Manager: Mary Beckwith
Courseware Portfolio Management Assistant: Alison Oehman
Content Producer: Tamela Ambush
Managing Producer: Karen Wernholm
Producer: Audra Walsh
Manager, Courseware QA: Mary Durnwald
Manager, Content Development: Rebecca Williams
Product Marketing Manager: Alicia Frankel
Field Marketing Managers: Jennifer Crum and Lauren Schur
Marketing Assistant: Hanna Lafferty
Senior Author Support/Technology Specialist: Joe Vetere
Manager, Rights and Permissions: Gina Cheselka
Manufacturing Buyer: Carol Melville, LSC Communications
Associate Director of Design: Blair Brown
Text Design, Cover Design, and Composition: Cenveo® Publisher Services
Cover Image: (grass) Val Lawless/Shutterstock; (cap) Angelo Gilardelli/Shutterstock;
(duck tape) Mexrix/Shutterstock

Library of Congress Cataloging-in-Publication Data

Names: Sullivan, Michael, III, 1967- | Struve, Katherine R. | Mazzarella,
    Janet.
Title: Elementary & intermediate algebra / Michael Sullivan, III, Joliet Junior College,
    Katherine R. Struve, Columbus State Community College, Janet Mazzarella,
    Southwestern College.
Other titles: Elementary and intermediate algebra
Description: 4th edition. | Boston: Pearson, c2018.
Identifiers: LCCN 2016020521| ISBN 9780134556079 (hardcover) | ISBN
    0134556070 (hardcover) | ISBN 9780134592350 (hardcover) | ISBN 0134592352 (hardcover)
Subjects: LCSH: Algebra—Textbooks.
Classification: LCC QA152.3 .S864 2018 | DDC 512.9—dc23
LC record available at https://lccn.loc.gov/2016020521

2 17

 **Pearson**

ISBN 13: 978-0-13-455607-9   (Student Edition)
ISBN 10: 0-13-455607-0

About the Authors

With training in mathematics, statistics, and economics, Michael Sullivan, III has a varied teaching background that includes 27 years of instruction in both high school and college-level mathematics. He is currently a full-time professor of mathematics at Joliet Junior College. Michael has numerous textbooks in publication, including an Introductory Statistics series and a Precalculus series, which he writes with his father, Michael Sullivan.

Michael believes that his experiences writing texts for college-level math and statistics courses give him a unique perspective as to where students are headed once they leave the developmental mathematics tract. This experience is reflected in the philosophy and presentation of his developmental text series. When not in the classroom or writing, Michael enjoys spending time with his three children, Michael, Kevin, and Marissa, and playing golf. Now that his two sons are getting older, he has the opportunity to do both at the same time!

Kathy Struve has been a classroom teacher for nearly 35 years, first at the high school level and, for the past 27 years, at Columbus State Community College. Kathy embraces classroom diversity: diversity of students' age, learning styles, and previous learning success. She is aware of the challenges of teaching mathematics at a large, urban community college, where students have varied mathematics backgrounds and may enter college with a high level of mathematics anxiety.

Kathy served as Lead Instructor of the Developmental Algebra sequence at Columbus State, where she developed curriculum, conducted workshops, and provided leadership to adjunct faculty in the mathematics department. She embraces the use of technology in instruction, and has taught web and hybrid classes in addition to traditional face-to-face and emporium-style classes. She is always looking for ways to more fully involve students in the learning process. In her spare time Kathy enjoys spending time with her two adult daughters, her four granddaughters, and biking, hiking, and traveling with her husband.

Born and raised in San Diego county, Janet Mazzarella spent her career teaching in culturally and economically diverse high schools before taking a position at Southwestern College 25 years ago. Janet has taught a wide range of mathematics courses from arithmetic through calculus for math/science/engineering majors and has training in mathematics, education, engineering, and accounting.

Janet has worked to incorporate technology into the curriculum by participating in the development of Interactive Math and Math Pro. At Southwestern College, she helped develop the self-paced developmental mathematics program. In addition, Janet was the Dean of the School of Mathematics, Science, and Engineering, the Chair of the Mathematics Department, the faculty union president, and the faculty coordinator for Intermediate Algebra. In the past, free time consisted of racing motorcycles off-road in the Baja 500 and rock climbing, but recently she has given up the adrenaline rush of these activities for the thrill of traveling in Europe.

Contents

CHAPTER 13 **Sequences, Series, and the Binomial Theorem 896**

based upon the more than 70 years of classroom teaching experience that the authors bring to this text.

Examples are often the determining factor in how valuable a textbook is to a student. Students look to examples to provide them with guidance and instruction when they need it most—the times when they are away from the instructor and the classroom. We have developed two example formats in an attempt to provide superior guidance and instruction for the students.

Innovative Examples

The innovative *Left-to-Right Example* has a two-column format in which annotations are provided to the **left** of the algebra, rather than the right, as is the practice in most texts. Because we read from **left to right,** placing the annotation on the left will make more sense to the student. It becomes clear that the annotation describes what we are about to do instead of what was just done. The annotations may be thought of as the teacher's voice offering clarification immediately before writing the next step in the solution on the board. Consider the following:

EXAMPLE 3 **Combining Like Terms to Solve a Linear Equation**

Solve the equation: $2x - 6 + 3x = 14$

Solution

$$2x - 6 + 3x = 14$$

Combine like terms: $\qquad\qquad 5x - 6 = 14$

Add 6 to both sides of the equation: $\qquad 5x - 6 + 6 = 14 + 6$

$$5x = 20$$

Divide both sides by 5: $\qquad\qquad \dfrac{5x}{5} = \dfrac{20}{5}$

$$x = 4$$

Check

$$2x - 6 + 3x = 14$$

Substitute 4 for x in the original equation: $\quad 2(4) - 6 + 3(4) \overset{?}{=} 14$

$$8 - 6 + 12 \overset{?}{=} 14$$

$$14 = 14 \quad \text{True}$$

The solution of the equation is 4, or the solution set is $\{4\}$. ●

> **Quick ✓**
>
> *In Problems 6–9, solve each equation.*
>
> **6.** $7b - 3b + 3 = 11$ **7.** $-3a + 4 + 4a = 13 - 27$
>
> **8.** $6c - 2 + 2c = 18$ **9.** $-12 = 5x - 3x + 4$

Showcase Examples

Showcase Examples are used strategically to introduce key topics or important problem-solving techniques. These examples provide "how-to" instruction by offering a guided, step-by-step approach to solving a problem. Students can then immediately see how each of the steps is employed. We remind students that the *Showcase Example* is meant to provide "how-to" instruction by including the words "how to" in the example title. The *Showcase Example* has a three-column format in which the left column describes a step, the middle column provides a brief annotation, as needed, to explain the step, and the right column presents the algebra. With this format, students can see each step in the problem-solving process in context so that the steps make more sense. This approach is more effective than simply stating each step in the text.

EXAMPLE 6 **How to Solve a Linear Equation in One Variable**

Solve the equation: $2(z - 4) + 3z = 4 - (z + 2)$

Step-by-Step Solution

Step 1: Remove any parentheses using the Distributive Property.

$$2(z - 4) + 3z = 4 - (z + 2)$$
$$2z - 8 + 3z = 4 - z - 2$$

Step 2: Combine like terms on each side of the equation.

$$5z - 8 = 2 - z$$

Step 3: Use the Addition Property of Equality to get the terms with the variable on one side of the equation and the constants on the other side.

Add z to both sides of the equation: $5z - 8 + z = 2 - z + z$
Simplify: $6z - 8 = 2$
Add 8 to both sides of the equation: $6z - 8 + 8 = 2 + 8$
Simplify: $6z = 10$

Step 4: Use the Multiplication Property of Equality to get the coefficient of the variable to be 1.

Divide both sides of the equation by 6: $\dfrac{6z}{6} = \dfrac{10}{6}$

Simplify: $z = \dfrac{5}{3}$

Step 5: Check the solution to verify that it satisfies the original equation.

The check is left to you.

The solution of the equation is $\dfrac{5}{3}$, or the solution set is $\left\{ \dfrac{5}{3} \right\}$.

> **Quick ✓**
>
> **17.** *True or False* To solve the equation $13 - 2(7x + 1) + 8x = 12$, the first step is to subtract 2 from 13 and get $11(7x + 1) + 8x = 12$.
>
> *In Problems 18 and 19, solve each equation.*
>
> **18.** $-9x + 3(2x - 3) = -10 - 2x$ **19.** $3 - 4(p + 5) = 5(p + 2) - 12$

Quick Check Exercises

Placed at the conclusion of most examples, the *Quick Check* exercises provide students with an opportunity for immediate reinforcement. By working the problems that mirror the example just presented, students get instant feedback and gain confidence in their understanding of the concept. All *Quick Check* exercise answers are provided in the back of the text. **The *Quick Check* exercises should be assigned as homework to encourage students to read, consult, and use the text regularly.** Ideally, these exercises should be completed within one day of class.

Superior Exercise Sets: Paired with Purpose

Students learn algebra by doing algebra. The superior end-of-section exercise sets in this text provide students with ample practice of both procedures and concepts. The exercises are paired and present problem types with every possible derivative. The exercises also present a gradual increase in difficulty level. The early, basic exercises keep the student's focus on as few "levels of understanding" as possible. The later or higher-numbered exercises are "multi-task" (or Mixed Practice) exercises where students are required to utilize multiple skills, concepts, or problem-solving techniques.

Throughout the textbook, the exercise sets are grouped into eight categories—some of which appear only as needed:

1. **Are You Prepared For This Section?** problems are located at the opening of the section. They are problems that address prerequisite material for the section, along with page references, so students may remediate, if necessary. Answers to the Prepared? . . . problems appear as a footnote on the page.

2. **Quick Check** exercises, which provide the impetus to get students into the text, follow most examples and are numbered sequentially as the first problems in each section exercise set. By doing these problems as homework and the first exercises attempted, the student is directed into the material in the section. If a student gets stuck, he or she will learn that the example immediately preceding the Quick Check exercise illustrates the concepts needed to solve the problem.

3. **Building Skills** exercises are skill development problems that develop the student's understanding of the procedures and skills in working with the methods presented in the section. These exercises can be linked back to a single learning objective in the section. Notice that the Building Skills problems begin the numbering scheme where the Quick Checks leave off. For example, if the last Quick Check exercise is Problem 20, then we begin the Building Skills exercises with Problem 21. This serves as a reminder that Quick Check exercises should be assigned as homework.

4. **Mixed Practice** exercises are also skill development problems, but they offer a comprehensive assessment of the skills learned in the section by asking problems that relate to more than one concept or objective. In addition, problems from previous sections may be presented so students must first recognize the type of problem and then employ the appropriate technique to solve the problem.

5. **Applying the Concepts** exercises are problems that allow students to see the relevance of the material learned within the section. Problems in this category either are situational problems that use material learned in the section to solve "real-world" problems or are problems that ask a series of questions to enhance a student's conceptual understanding of the mathematics presented in the section.

6. **Extending the Concepts** exercises are problems that go beyond the basics. Within this block of exercises an instructor will find a variety of problems to sharpen students' critical-thinking skills.

7. **Explaining the Concepts** problems require students to think about the big picture concepts of the section and express these ideas in their own words. It is our belief that students need to improve their ability to communicate complicated ideas both orally and in writing. When they are able to explain mathematical methods or concepts to another individual, they have truly mastered the ideas. These problems can serve as a basis for classroom discussion or can be used as writing assignments.

8. Finally, we include **Technology Exercises.** Instructors' philosophies about the use of graphing technology, such as graphing calculators or Desmos, to solve problems vary considerably. Because instructors disagree about the value of these tools, we have made an effort to make graphing technology entirely optional. When appropriate, technology exercises are included at the close of a section's exercise set. Also included in the technology exercises are the new applet explorations. The applets may be found in MyMathLab or using the Quick Response (QR) code located in the section opener ribbon.

Problem Icons In addition to the carefully structured categories of exercises, selected problems are flagged with icons.

- Problems whose number is green have complete worked-out solutions found in MyMathLab.

- △ These problems focus on geometry concepts.

- 🖩 A calculator will be useful in working the problem.

Hallmark Features

Author in Action Videos

The Author in Action videos are videos of the authors presenting the content. Most of the videos are from the authors' actual classroom lectures. This makes the videos authentic and gives the viewer the sense of participating in the lecture. The videos are tied to the objectives and under 12 minutes in length. For those objectives that require more than 12 minutes, we have multiple videos. Students are alerted to the availability of a video with the ▶ icon. The videos are available in MyMathLab, the Multimedia Textbook (in MyMathLab), or through a Quick Response (QR) code ▓ located in the section opener ribbon. The videos are captioned in English and Spanish.

Video Notebook

A Video Notebook is available, which is ideal for online, emporium/redesign courses, or inverted (flipped) classrooms. This notebook assists students in taking thorough, organized, and understandable notes as they watch the Author in Action videos by asking students to complete definitions, procedures, and examples based on the content of the videos. The Video Notebook is available as an unbound, three-hole punched workbook—students can insert additional pages of notes or homework to begin a course notebook.

Quick Check Exercises: Encourage Study Skills that Lead to Independent Learning

What is one of the overarching goals of an education? We believe it is to learn to solve problems independently. In particular, we would like to see students develop the ability to pick up a text or manual and teach themselves the skills they need. In our mathematics classes, however, we are often frustrated because students rarely read the text and often struggle to understand the concepts independently.

To encourage students to use the text more effectively and to help them achieve greater success in the course, we have structured the exercises in our text differently from other mathematics textbooks. The aim of this structure is to get students "into the text" in order to increase their ability and confidence to work any math problem— particularly when they are away from the classroom and an instructor who can help.

Each section's exercise set begins with the *Quick Check* exercises. The *Quick Checks* are consecutively numbered. The end-of-section exercises begin their numbering scheme based on where the *Quick Checks* end. For example:

- Section 1.2: *Quick Checks* end at Problem 48, so the end-of-section exercise set starts with Problem 49 (see page 16).
- Section 1.3: *Quick Checks* end at Problem 24, so the end-of-section exercise set starts with Problem 25 (see page 25).

The *Quick Checks* follow most examples and provide the platform for students to get "into the text." By integrating these exercises into the exercise set, students are directed to the instructional material in that section. Our hope is that students will then become more aware of the instructional value of the text and will be more likely to succeed when studying away from the classroom and the instructor.

Answer annotations to Quick Checks and exercises have been placed directly next to each problem in the Annotated Instructor's Edition to make it easier for instructors to create assignments.

We have used the same background color for the Quick Checks and the exercise sets to reinforce the connection between them visually. The colored background will also make the Quick Checks easier to find on the page.

Answers to Selected Exercises at the back of the text integrate the answers to *every* Quick Check exercise with the answers to *every odd* problem from the section exercise sets.

Study Skills and Student Success

We have included study skills and student success as regular themes throughout this text starting with *Section 1.1 Success in Mathematics*. In addition to this dedicated section that covers many of the basics that are essential to success in any math course, we have included several recurring study aids that appear in the margin. These features are designed to anticipate the student's needs and to provide immediate help — as if the teacher were looking over his or her shoulder. These margin features include: *In Other Words; Work Smart;* and *Work Smart: Study Skills*.

Section 1.1 Success in Mathematics focuses the student on basic study skills, including what to do during the first week of the term; what to do before, during, and after class; how to use the text effectively; and how to prepare for an exam.

In Other Words helps to address the difficulty that students have in reading mathematically precise definitions and theorems by explaining them in easier to understand language.

Work Smart provides "tricks of the trade" hints, tips, reminders, and alerts. It also identifies some common errors to avoid and helps students work more efficiently.

Work Smart: Study Skills reminds students of study skills that will help them to succeed at various points in the course. Attention to these practices will help them to become better, more proficient learners.

Test Preparation and Student Success

The Chapter Tests in this text and the companion Chapter Test Prep Videos have been designed to help students make the most of their valuable study time.

Chapter Test In preparation for their classroom test, students should take the practice test to make sure they understand the key topics in the chapter. The exercises in the Chapter Tests have been crafted to reflect the level and types of exercises a student is likely to see on a classroom test.

Chapter Test Prep Videos The Chapter Test Prep Videos provide students with help at the critical juncture when they are studying for a test. The videos present step-by-step solutions to the exact exercises found in each of the book's Chapter Tests. Easy video navigation allows students instant access to the worked-out solutions to the exercises they want to study or review. These videos are available in MyMathLab or may be accessed using the QR code in the Chapter Test ribbon.

Do the Math Workbook The Do the Math Workbook is a compilation of worksheets that may be used to supplement student learning. For each section, the workbook includes Five-Minute Warm-Ups, Guided Practice (based on the Showcase Examples from the text), and Do the Math Exercises.

Seeing the Connections: The Big Picture

Another important role of the pedagogy in this text is to help students see and understand the connection among the mathematical topics presented. Several section-opening and margin features help to reinforce connections:

The Big Picture: Putting It Together (**Chapter Opener**) This feature is based on how we start each chapter in the classroom — with a quick sketch of what we plan to cover. Before tackling a chapter, we tie concepts and techniques together by summarizing material covered previously and then relate these ideas to material we are about to discuss. It is important for students to understand that content truly builds from one chapter to the next. We find that students need to be reminded that the familiar operations of addition, subtraction, multiplication, and division are being applied to different or more complex objects.

Are You Prepared for This Section? As part of this building process, we think it is important to remind students of specific skills that they will need from earlier in the course to be successful within a given section. The *Are You Prepared? . . .* feature that begins each section not only provides a list of prerequisite skills that a student should understand before tackling the content of a new section, but also acts as a short set of problems to test students' preparedness. Answers to the problems are provided in a footnote on the same page, and a cross-reference to the material in the text is provided so that the student can remediate when necessary.

Mixed Practice These problems exist within each end-of-section exercise set and draw upon material learned from multiple objectives. Sometimes, these problems simply represent a mixture of problems presented within the section, but they also may include a mixture of problems from various sections. For example, students may need to distinguish between linear and quadratic equations or students may need to distinguish between the direction *simplify* versus the direction *solve*.

Putting the Concepts Together (**Mid-Chapter Review**) Each chapter has a group of exercises at the appropriate point in the chapter, entitled *Putting the Concepts Together*. These exercises serve as a review—synthesizing material introduced up to that point in the chapter. The exercises in these mid-chapter reviews are carefully chosen to assist students in seeing the "big picture."

Cumulative Review Learning algebra is a building process, and building involves considerable reinforcement. The Cumulative Review exercises at the end of each odd-numbered chapter, starting with Chapter 3, help students to reinforce and solidify their knowledge by revisiting concepts and using them in context. This way, studying for the final exam should be fairly easy.

Getting Ready for Intermediate Algebra The transition from the Elementary Algebra portion of the course to the Intermediate Algebra part can be difficult for students. As instructors, we want our students to be aware of any deficiencies they may have in their mathematical preparation when starting Intermediate Algebra. We have written two transition quizzes for students, with one appearing after Chapter 6 and another after Chapter 7, to assess student preparedness. The answers to all problems on the quizzes are located in the back of the text. Further, a cross reference to the material presented earlier in the course is provided so that students may review important concepts, if necessary.

In Closing

When we started writing this textbook, we discussed improvements we could make in coverage; in staples such as examples and problems; and in any pedagogical features that we found truly useful. After writing and rewriting, and reading many thoughtful reviews from instructors, we focused on the following features of the text to set it apart.

- The innovative ***Left-to-Right Examples*** and ***Showcase Examples*** provide students with superior guidance and instruction when they need it most—when they are away from the instructor and the classroom. Each of the margin features ***In Other Words, Work Smart,*** and ***Work Smart: Study Skills*** are designed to improve study skills, make the textbook easier to navigate, and increase student success.

- **Exercise Sets**—The exercise sets are structured to assess student understanding of vocabulary, concepts, meaningful repetition, problem solving, and applications. The exercise sets are graded in difficulty level to build confidence and to enhance students' mathematical thinking. The ***Quick Check*** exercises provide students with immediate reinforcement and instant feedback to determine their understanding of the concepts presented in the examples.

- **The Big Picture**—Each section opens with *Are You Prepared For This Section?* problems that allows students to review material learned earlier in the course that is needed in the upcoming section. ***Mixed Practice*** problems require students to utilize material learned from multiple objectives to solve a problem. Often, these problems require students to first determine the correct approach to solving the problem prior to actually solving it. ***Putting the Concepts Together*** helps students see the big picture and provide a structure for learning each new concept and skill in the course.

Resources for Success

MyMathLab Online Course for Sullivan/Struve/ Mazzarella, *Elementary & Intermediate Algebra*, 4th edition

To give students a consistent tone, voice, and teaching method, this text's approach is tightly integrated throughout its accompanying MyMathLab course, making learning the material as seamless as possible. This course contains all of MyMathLab's powerful features, in addition to specific Sullivan/Struve/ Mazzarella tools.

Premade Author-Created Course

A premade course developed by the authors with a guided learning path for students allows instructors the ease of quick start-up, and encourages students to learn and retain the concepts in order to be more successful on their homework. The learning path guides students to first take advantage of the learning resources at their disposal, including videos and new applets, before directing them to their assignments, which are premade. The MyMathLab course is set up to help instructors get the most out of their course, but all assignments are able to be tailored to instructors' needs.

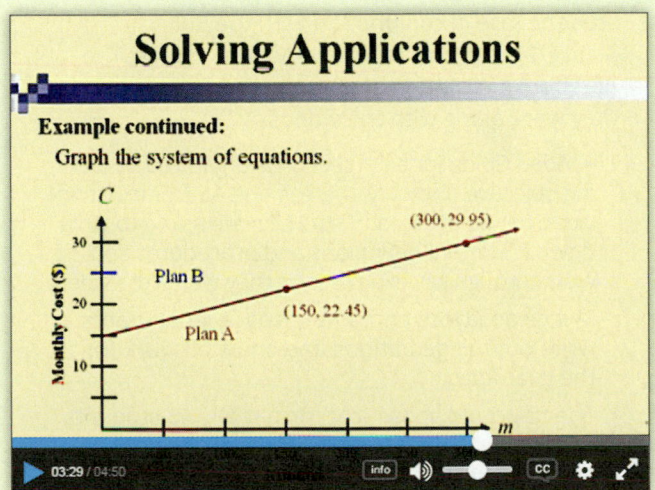

Robust Video Program

The wealth of video resources in the MyMathLab course give students just-in-time help at home, in the lab, or on the go. Video resources include:

– Author-in-Action videos featuring author Mike Sullivan's actual classroom lecture
– Example-level solution clips
– Chapter Test Prep videos

New QR codes located throughout the textbook give students instant, easy access to all the videos.

New Applets

New applets developed by the authors let students interact with the math in a visual, tangible way. These animations allow students to explore and manipulate the mathematical concepts, leading to long-lasting understanding, and corresponding exercises in MyMathLab make them truly assignable.

Guided Exercises

In addition to MyMathLab's hallmark interactive exercises, Guided Exercises walk students through each step of the problem-solving process, giving them a guided, step-by-step learning experience. These are based on the "How To" exercises from the text and were written by the authors.

www.mymathlab.com

Resources for Success

Pearson

Instructor Resources

Annotated Instructor's Edition

ISBN 10: 0134592352 **ISBN 13:** 9780134592350
The AIE provides annotations for instructors, including answers and teaching tips.

The following resources can be downloaded from www.pearsonhighered.com* or in MyMathLab.

Instructor Solutions Manual

This manual provides worked-out solutions to all exercises in the text.

Instructor's Resource Manual

This manual includes resources designed to help both new and experienced instructors with course preparation and classroom management. This includes mini-lectures for each section of the text, chapter by chapter teaching tips, sample syllabi, and more.

PowerPoints

These slides present key concepts and definitions from the text.

TestGen

TestGen® (www.pearsoned.com/testgen) enables instructors to build, edit, print, and administer tests using a computerized bank of questions developed to cover all the objectives of the text.

Student Resources

Author in Action videos

Available in MyMathLab, these videos feature each objective presented by the authors with detailed explanations and examples.

Student Solutions Manual

ISBN 10: 0134592336 **ISBN 13:** 9780134592336
This manual contains complete worked solutions to the odd-numbered problems in the end-of-section exercise sets and all of the Quick Checks and end-of-chapter exercises.

Video Notebook

ISBN 10: 0134592271 **ISBN 13:** 9780134592275
The Video Notebook is an unbound, three-hole-punched workbook/note-taking guide that students use in conjunction with the Sullivan/Struve/Mazzarella "Author in Action" videos. The notebook helps them develop organized notes as they work along with the videos.

- A Video Guide for each section is organized by learning objective. Typically, there is one Author in Action video per objective, and students are asked to write down important definitions and procedures and work through key examples as they watch the video.
- The clean layout and ample space let students write out full definitions and show all work for the examples.
- The unbound, loose-leaf format allows students to insert additional notes from class and/or homework—so they can build a course notebook and good study skills for future classes!

Do The Math Workbook

ISBN 10: 0134591941 **ISBN 13:** 9780134591940
This workbook offers a collection of 5-Minute Warm-Up exercises, Guided Practice exercises, and Do the Math exercises for each section in the text. These worksheets can be used as in-class assignments, as an in-lab study assignment, or for homework.

www.mymathlab.com

Acknowledgments

Textbooks are written by authors but evolve through the efforts of many people. We would like to extend our thanks to the following individuals for their important contributions to the project. From Pearson: Mary Beckwith, Tamela Ambush, Michael Hirsch, Lauren Morse, Melissa Parkin, Michelle Renda, Chris Hoag, Patty Bergin, Rose Kernan, and, finally, the Pearson Arts & Sciences sales team for their confidence and support of our books.

We would also like to thank Brad Davis, Jared Burch, Cindy Trimble, and John Bialas for their attention to details and consistency in accuracy checking the text and answer sections and Val Villegas for his work on the video notebook. We offer many thanks to all the instructors from across the country who participated in reviewer conferences and focus groups, reviewed or class-tested some aspect of the manuscript, and taught from the previous editions. Their insights and ideas form the backbone of this text. Hundreds of instructors contributed their time, energy, and ideas to help us shape this text. We will attempt to thank them all here. We apologize for any omissions.

The following individuals, many of whom reviewed or class-tested the previous edition, provided direction and guidance in shaping the fourth edition.

Marwan Abu-Sawwa, *Florida Community College—Jacksonville*
Darla Aguilar, *Pima State University*
Grant Alexander, *Joliet Junior College*
Philip Anderson, *South Plains College*
MaryAnne Anthony, *Santa Ana College*
Mary Lou Baker, *Columbia State Community College*
Bill Bales, *Rogers State*
Tony Barcellos, *American River College*
John Beachy, *Northern Illinois University*
Donna Beatty, *Ventura College*
David Bell, *Florida Community College—Jacksonville*
Sandy Berry, *Hinds Community College*
John Bialas, *Joliet Junior College*
Linda Blanco, *Joliet Junior College*
Kevin Bodden, *Lewis and Clark College*
Rebecca Bonk, *Joliet Junior College*
Cherie Bowers, *Santa Ana College*
Becky Bradshaw, *Lake Superior College*
Lori Braselton, *Georgia Southern University*
Tim Britt, *Jackson State Community College*
Linda Britton, *Oakland Community College*
Holly J. Broesamle, *Oakland Community College*
Beverly Broomell, *Suffolk Community College*
Joanne Brunner, *Joliet Junior College*
Hien Bui, *Hillsborough Community College—Dale Mabry*
Connie Buller, *Metropolitan Community College*
Annette Burden, *Youngstown State University*
James Butterbach, *Joliet Junior College*
Marc Campbell, *Daytona Beach Community College*
Elena Catoiu, *Joliet Junior College*
Nancy Chell, *Anne Arundel Community College*

John F. Close, *Salt Lake Community College*
Bobbi Cook, *Indian River Community College*
Carlos Corona, *San Antonio College*
Faye Dang, *Joliet Junior College*
Shirley Davis, *South Plains College*
Vivian Dennis-Monzingo, *Eastfield College*
Alvio Dominguez, *Miami Dade College—Wolfson*
Karen Driskell, *South Plains College*
Thomas Drucker, *University of Wisconsin—Whitewater*
Brenda Dugas, *McNeese State University*
Doug Dunbar, *Okaloosa-Walton Junior College*
Laura Dyer, *Southwestern Illinois State University*
Bill Echols, *Houston Community College—Northwest*
Erica Egizio, *Lewis University*
Laura Egner, *Joliet Junior College*
Jason Eltrevoog, *Joliet Junior College*
Nancy Eschen, *Florida College Jacksonville*
Mike Everett, *Santa Ana College*
Phil Everett, *Ohio State University*
Scott Fallstrom, *Shoreline Community College*
Betsy Farber, *Bucks County Community College*
Fitzroy Farquharson, *Valencia Community College—West*
Jacqueline Fowler, *South Plains College*
Dorothy French, *Community College of Philadelphia*
Randy Gallaher, *Lewis and Clark College*
Sanford Geraci, *Broward Community College*
Donna Gerken, *Miami Dade College—Kendall*
Adrienne Goldstein, *Miami Dade College—Kendall*

Marion Graziano, *Montgomery County Community College*
Susan Grody, *Broward College*
Tom Grogan, *Cincinnati State University*
Barbara Grover, *Salt Lake Community College*
Shawna Haider, *Salt Lake Community College*
Margaret Harris, *Milwaukee Area Technical College*
Sheyleah V. Harris-Plant, *South Plains College*
Teresa Hasenauer, *Indian River College*
Margy Heddens, *Highland Community College*
Mary Henderson, *Okaloosa-Walton Junior College*
Celeste Hernandez, *Richland College*
Paul Hernandez, *Palo Alto College*
Pete Herrera, *Southwestern College*
Bob Hervey, *Hillsborough College—Dale Mabry*
Teresa Hodge, *Broward College*
Sandee House, *Georgia Perimeter College*
Becky Hubiak, *Tidewater Community College—Virginia Beach*
Michelle Hurn, *Highland Community College*
Sally Jackman, *Richland College*
John Jarvis, *Utah Valley State College*
Nancy Johnson, *Broward College*
Steven Kahn, *Anne Arundel Community College*
Linda Kass, *Bergen Community College*
Donna Katula, *Joliet Junior College*
Mohammed Kazemi, *University of North Carolina—Charlotte*
Doreen Kelly, *Mesa Community College*
Mike Kirby, *Tidewater Community College—Virginia Beach*
Keith Kuchar, *College of Dupage*
Carla Kulinsky, *Salt Lake Community College*

Julie Labbiento, *Leigh Carbon Community College*

Kathy Lavelle, *Westchester Community College*

Deanna Li, *North Seattle Community College*

Heidi Lyne, *Joliet Junior College*

Brian Macon, *Valencia Community College—West*

Lynn Marecek, *Santa Ana College*

Jim Matovina, *Community College of Southern Nevada*

Jean McArthur, *Joliet Junior College*

Michael McComas, *Marshall University*

Mikal McDowell, *Cedar Valley College*

Lee McEwen, *Ohio State University*

David McGuire, *Joliet Junior College*

Angela McNulty, *Joliet Junior College*

Debbie McQueen, *Fullerton College*

Judy Meckley, *Joliet Junior College*

Lynette Meslinsky, *Erie Community College—City Campus*

Kausha Miller, *Lexington Community College*

Chris Mizell, *Okaloosa Walton Junior College*

Jim Moore, *Madison Area Technical College*

Ronald Moore, *Florida College Jacksonville*

Elizabeth Morrison, *Valencia College—West*

Roya Namavar, *Rogers State University*

Hossein Navid-Tabrizi, *Houston Community College*

Carol Nessmith, *Georgia Southern University*

Kim Neuburger, *Portland Community College*

Larry Newberry, *Glendale Community College*

Elsie Newman, *Owens Community College*

Charlotte Newsome, *Tidewater Community College*

Charles Odion, *Houston Community College*

Viann Olson, *Rochester Community and Technical College*

Linda Padilla, *Joliet Junior College*

Carol Perry, *Marshall Community and Technical College*

Faith Peters, *Miami Dade College—Wolfson*

Dr. Eugenia Peterson, *Richard J. Daley College*

Jean Pierre-Victor, *Richard J. Daley College*

Philip Pina, *Florida Atlantic University*

Carol Poos, *Southwestern Illinois University*

Elise Price, *Tarrant County College*

R.B. Pruitt, *South Plains College*

William Radulovich, *Florida College Jacksonville*

Pavlov Rameau, *Miami Dade College—Wolfson*

David Ray, *University of Tennessee—Martin*

Nancy Ressler, *Oakton Community College*

Michael Reynolds, *Valencia College—West*

George Rhys, *College of the Canyons*

Jorge Romero, *Hillsborough College—Dale Mabry*

David Ruffato, *Joliet Junior College*

Carol Rychly, *Augusta State University*

David Santos, *Community College of Philadelphia*

Togba Sapolucia, *Houston Community College*

Julia Shew, *Columbus State Community College*

Doug Smith, *Tarrant County College*

Catherine J.W. Snyder, *Alfred State College*

Gisela Spieler-Persad, *Rio Hondo College*

Raju Sriram, *Okaloosa-Walton Junior College*

Patrick Stevens, *Joliet Junior College*

Bryan Stewart, *Tarrant County College*

Jennifer Strehler, *Oakton Community College*

Elizabeth Suco, *Miami Dade College—Wolfson*

Katalin Szucs, *East Carolina University*

KD Taylor, *Utah Valley State College*

Mary Ann Teel, *University of North Texas*

Suzanne Topp, *Salt Lake Community College*

Suzanne Trabucco, *Nassau Community College*

Jo Tucker, *Tarrant County College*

Bob Tuskey, *Joliet Junior College*

Mary Vachon, *San Joaquin Delta College*

Carol Walker, *Hinds Community College*

Kim Ward, *Eastern Connecticut State University*

Richard Watkins, *Tidewater Community College*

Natalie Weaver, *Daytona Beach College*

Carol White, *Highland Community College*

Darren Wiberg, *Utah Valley State College*

Rachel Wieland, *Bergen Community College*

Christine Wilson, *Western Virginia University*

Brad Wind, *Miami Dade College—North*

Roberta Yellott, *McNeese State University*

Steve Zuro, *Joliet Junior College*

Additional Acknowledgments

We also would like to extend thanks to our colleagues at Joliet Junior College, Columbus State Community College, and Southwestern College, who provided encouragement, support, and the teaching environment where the ideas and teaching philosophies in this text were developed.

Michael Sullivan, III

Katherine R. Struve

Janet Mazzarella

1 Operations on Real Numbers and Algebraic Expressions

In the year 1202, the Italian mathematician Leonardo Fibonacci posed this problem: A certain man put a pair of rabbits in a place surrounded on all sides by a wall. How many pairs of rabbits can be produced from that pair in a year if every month each pair begets a new pair that is productive from the second month on?

The answer to Fibonacci's puzzle leads to a sequence of numbers called the *Fibonacci sequence*. See Problem 159 in Section 1.4.

The Big Picture: Putting It Together

Welcome to algebra! This course is taken by a diverse group of individuals. Some of you may never have taken an algebra course, while others may have taken algebra at some time in the past. In any case, we have written this text with both groups in mind.

The first chapter of the text reviews arithmetic. The material is presented with an eye on the future, which is algebra. This means that we will slowly build our discussion so that the shift from arithmetic to algebra is painless. Carefully study the methods used in this chapter, because these same methods will be used again in later chapters.

1.1 Success in Mathematics

Objectives

1. What to Do the First Week of the Semester
2. What to Do Before, During, and After Class
3. How to Use the Text Effectively
4. How to Prepare for an Exam

Let's start by having a discussion about the "big picture" goals of the course and how this text can help you to be successful at mathematics. Our first "big picture" goal is to develop algebraic skills and gain an appreciation for the power of algebra and mathematics. But there is also a second "big picture" goal. By studying mathematics, we develop a sense of logic and exercise the part of our brains that deals with logical thinking. The examples and problems in this text are like the crunches we do in a gym to exercise our bodies. The goal of running or walking is to get from point A to point B, so doing fifty crunches on a mat does not accomplish that goal, but crunches do make our upper bodies, backs, and hearts stronger when we need to run or walk.

Logical thinking can assist us in solving difficult everyday problems, and solving algebra problems "builds the muscles" in the part of our brain that performs logical thinking. So, when you are studying algebra and getting frustrated with the amount of work that needs to be done, and you say, "My brain hurts," remember that just like an athlete, you must practice to be successful. But, as is also true of an athlete, practice needs to be on a regular basis, not just before "the big game."

Another phrase to keep in mind is "Success breeds success." Mathematics is everywhere. You already are successful at doing some everyday mathematics. With practice, you can take your initial successes and become even more successful. Have you ever done any of the following everyday activities?

- Compare the price per ounce of different sizes of jars of peanut butter or jam.
- Leave a tip at a restaurant.
- Figure out how many calories your bowl of breakfast cereal provides.
- Compare the distances between cities as you plan a vacation.
- Order the appropriate number of gallons of paint to cover the walls of a room.
- Buy a car and take out a car loan with interest.
- Double a cookie recipe.
- Exchange American dollars for Canadian dollars.
- Find the final cost of a t-shirt after a 20%-off coupon is applied.

You may do five or ten mathematical activities in a single day! The everyday mathematics that you already know is the foundation for your success in this course.

❶ What to Do the First Week of the Semester

The first week of the semester gives you the opportunity to prepare for a successful course. Here are the things you should do:

1. **Pick a good seat.** Choose a seat that gives you a good view of the room. Sit close enough to the front so you can easily see the board and hear the professor.

2. **Read the syllabus to learn about your instructor and the course.** Take note of your instructor's name, office location, e-mail address, telephone number, and office hours. Pay attention to any additional help available, such as tutoring centers, videos, software, online tutorials, and so on. Be sure you fully understand all of the instructor's policies for the class, including the policy on absences, missed exams or quizzes, and homework. Know important dates and put them in your planner, tablet, computer, or phone. Ask questions.

3. **Learn the names of some of your classmates and exchange contact information.** One of the best ways to learn math is through group study sessions. Try to create time each week to study with your classmates. Knowing how to get in contact with classmates is also useful if you ever miss class, because you can obtain the assignment for the day.

4. **Budget your time.** Most students have a tendency to "bite off more than they can chew." To help with time management, consider the following general rule: Plan on studying *at least* two hours outside of class for each hour in class. Thus, if you enrolled in a four-hour math class, you should set aside at least eight hours each week to study for the course. You will also need to set aside time for other courses. Consider your work schedule and personal life when creating your time budget. A blank time chart is provided in the exercises for you to use to manage your time.

5. **Get a notebook or binder that is dedicated to math alone.** Don't use this notebook for any other classes, just math, so you stay organized.

❷ What to Do Before, During, and After Class

Now that the semester is under way, we present the following ideas for what to do before, during, and after each class meeting. These suggestions may sound overwhelming, but by following them, you will increase your chances to be successful in mathematics (and other courses). Also, you will find that studying for exams becomes much easier by following this plan.

Before Class

1. Read the section or sections that will be covered in the upcoming class meeting. Watch the video lectures that accompany the text. There is a QR code ▓ at the beginning of each section that will take you to the video lectures.

2. Based on your reading and watching, write down a list of questions. Your questions will probably be answered through the lecture. You can then ask any questions that are not answered completely. Also, write down any important new vocabulary and formulas in your math notebook.

3. Make sure you are mentally prepared for class. Your mind should be alert and ready to concentrate for the entire class. A good night's sleep and healthy meals high in protein can help.

During Class

1. Arrive early enough to prepare your mind and material for the lecture.

2. Avoid distractions such as cell phones, and websites that don't pertain to the class.

3. Stay alert. Do not doze off or daydream during class. If you do so, understanding the classroom discussion will be very difficult when you "return to class."

4. Take thorough notes. It is normal not to understand certain topics the first time you hear them in a lecture. However, this does not mean that you throw your hands up in despair. Rather, continue to take class notes.

5. Do not be afraid to ask questions. In fact, instructors love questions, for two reasons. First, if one student has a question, other students probably have the same question. Second, by asking questions, you teach the teacher what topics cause difficulty.

After Class

1. Reread (and possibly rewrite) your class notes. You may be amazed at how often your confusion during class disappears after studying your in-class notes after class.

2. Reread the section. This is an especially important step. Once you have heard the class discussion, the section will make more sense and you will understand much more.

Work Smart: Study Skills
The reason for homework is to build your skill and confidence. Don't skip assignments.

3. Do your homework as soon as possible. **Homework is not optional.** There is an old Chinese proverb that says,

> I hear . . . and I forget
>
> I see . . . and I remember
>
> I do . . . and I understand

This proverb applies to any situation in life in which you want to succeed. Would a pianist expect to be the best if she didn't practice? The only way you are going to learn algebra is by doing algebra.

4. When you get a problem wrong, try to figure out *why* you got the problem wrong. Once you figure out why, work a similar problem. If you can't discover your error, be sure to ask for help. If possible, connect with other students in your class to ask and answer questions about the homework.

5. If you have questions, visit your professor during office hours. You can also ask someone in your study group or go to the tutoring center on campus, if available.

❸ How to Use the Text Effectively

This text was developed so that there is more than one way to learn the material.

All of the features in the text are here to help you succeed. These features are based on techniques we use in class. The features that appear, an explanation of the purpose of each feature, and how each can be used to help you succeed in this course are outlined in the following paragraphs.

Are You Prepared for This Section?: Warming Up

Beginning with Section 1.3, each section starts with a short set of review problems. These problems ask questions about material that was presented earlier in the course and is needed for the upcoming section. Complete the problems to be sure you understand the material on which the new section is based. Answers to the problems appear in a footnote on the page where the problems appear. Check your answers. If you get a problem wrong or don't know how to do a problem, go back to the section listed and review the material.

Objectives: A "Road Map" through the Course

To the left of the review problems is a list of objectives to be covered in the section. If you follow the objectives, you will get a good idea of the section's "big picture"—the important concepts, techniques, and procedures.

The objectives are numbered. (See the numbered headline at the beginning of this section.) When we begin discussing a particular objective within the section, the objective number appears along with the stated objective.

Examples: Where to Look for Information

Examples are meant to provide you with guidance and instruction when you are away from the instructor and the classroom. With this in mind, we have developed two special example formats.

Step-by-Step Examples have a three-column format where the left column describes a step, the middle column briefly explains the step, and the right column presents the algebra. Thus the left and middle columns can be thought of as your instructor's voice during a lecture. *Step-by-Step Examples* introduce key topics or important problem-solving strategies. They provide easy-to-understand, practical instructions and can be identified by the words "how to" in the title of the example.

Annotated Examples have a two-column format with explanations to the left of the algebra. The explanation clearly describes what we are about to do. Again, annotations are like your instructor's voice as he or she writes each step of the solution on the board.

Authors in Action: Lecture Videos to Help You Learn

At the beginning of each section, there is a QR code ▦ that takes you to the accompanying video lectures. Every objective has one or more classroom lecture videos, marked with an ⏵ icon, of the authors teaching their students. These "live" classroom lectures can be used to supplement your instructor's presentations and your reading of the text. These videos can also be found in the Multimedia Library of MyMathLab.

In Other Words: Math in Everyday Language

Have you ever been given a math definition in class and said, "What in the world does that mean?" We have heard that from our students, so we added the "In Other Words" feature, which restates mathematical definitions in everyday language. This margin feature will help you understand the language of mathematics better. See page 11.

Work Smart

These "tricks of the trade" that appear in the margin can help you solve problems. They also show alternative problem-solving approaches. There is often more than one way to solve a math problem! See page 9.

Work Smart: Study Skills

These margin notes highlight the study skills required for success in this and other mathematics courses. See page 7.

Exercises: A Unique Numbering Scheme

As teachers, we know that students typically jump right to the exercises after attending class. This means they may skip all of the examples and explanations of concepts in the section. To help you use the text most effectively to learn the math, we have structured the exercises differently from other texts you have used. Our structure is designed to encourage the reading of the text, while increasing your confidence and ability to work any mathematical problem. For this reason, the exercises in each section are broken into as many as seven parts. Each exercise set will have some, or all, of the following exercise types.

1. Quick Checks
2. Building Skills
3. Mixed Practice
4. Applying the Concepts
5. Extending the Concepts
6. Explaining the Concepts
7. Technology Exercises

1. **Quick Checks: Learning to Ride a Bicycle with Training Wheels** Do you remember when you were first learning to ride a bicycle? Training wheels were placed on the bicycle to assist you in learning balance. The Quick Checks are like exercises with training wheels. These exercises appear right after the example or examples that illustrate the concept being taught. So, if you get stuck on a Quick Check problem, you can simply consult the example immediately preceding it, rather than searching through the text. These Quick Checks are intended for you to complete within one day of the class. If you do this, your retention will increase, making the rest of your work much easier! The answers to all Quick Checks are in the back of the book. For an example, see page 9 in Section 1.2.

2. **Building Skills: Learning to Ride a Bicycle with Assistance**   Once you felt ready to ride without training wheels, you probably had an adult follow closely behind you, holding the bicycle for balance and building your confidence. The Building Skills problems serve a similar purpose. They are keyed to the objectives within the section, so the directions for the problem indicate which objective is being developed. As a result, you know exactly which objective (but not exactly which example) to consult if you get stuck. For an example, see page 16 in Section 1.2.

3. **Mixed Practice: Now You Are Ready to Ride!**   After mastering training wheels and learning to balance with assistance, you are ready to ride alone. This stage corresponds to the Mixed Practice exercises. These exercises include problems that develop your ability to see the big picture of mathematics. They are not keyed to a particular objective and require you to determine the appropriate approach to solving a problem on your own. For an example, see page 17 in Section 1.2.

4. **Applying the Concepts: Where Will I Ever Use This Stuff?**   The Applying the Concepts exercises not only illustrate the application of mathematics in your life but also provide problems that test your conceptual understanding of the mathematics. For an example, see pages 17–18 in Section 1.2.

5. **Extending the Concepts: Stretching Your Mind**   Sometimes students need to be challenged further. These exercises extend your skills to a new level and provide further insight into where mathematics can be used. For an example, see page 18 in Section 1.2.

6. **Explaining the Concepts: Verbalize Your Understanding**   These problems require you to express the section's big-picture concepts in your own words. Students need to improve their ability to communicate complicated ideas (both oral and written). If you truly understand the material in the section, you should be able to articulate the concepts clearly. For an example, see page 27 in Section 1.3.

7. **Technology Exercises**   Technology can be a great way to verify answers and to help visualize results. These exercises illustrate how technology can be incorporated into the material of the section. For an example, see page 259 in Section 4.1.

Chapter Review

The chapter review is arranged section by section. For each section, we state key concepts, key terms, and objectives. We also list the examples and page numbers from the text that illustrate each objective. Also, for each objective, we list the problems in the review exercises that test your understanding. If you get a problem wrong, use this feature to determine where to look in the book to help you to work the problem.

Chapter Test

Once you think you are prepared for the exam, take the chapter test at the end of the chapter. If you do well on the chapter test, chances are you will do well on your in-class exam. Be sure to take the test under the conditions you will face in class. For example, try setting a timer to mimic the time constraints felt in class during a test, or take the test at a desk in a quiet room. If you are unsure how to solve a problem in the chapter test, watch the Chapter Test Prep Videos available in MyMathLab, on YouTube, or by using the QR code, all of which show an instructor solving each chapter test problem.

Cumulative Review: Reinforcing Your Knowledge

The building process of learning algebra involves a lot of reinforcement. Thus we provide cumulative reviews at the end of every odd-numbered chapter starting with Chapter 3. Do these cumulative reviews after each chapter test, so that you are always refreshing your memory–your brain gets its exercise. This way, studying for the final exam should be fairly easy.

❹ How to Prepare for an Exam

The following steps are time-tested suggestions to help you prepare for an exam.

Step 1: Revisit your homework and the chapter review problems. About one week before your exam, start to redo your homework assignments. If you don't understand a topic, seek out help. Work the problems in the chapter review as well. The problems are keyed to the section objectives. If you get a problem wrong, identify the objective and examples that illustrate the objective. Then review this material and try the problem in the chapter review again. If you get the problem wrong again, seek out help.

Step 2: Test yourself. A day or two before the exam, take the chapter test under test conditions. Be sure to check your answers in the back of the book or with the Chapter Test Prep videos. If you got any problems wrong, determine why you got them wrong and rework the problems correctly. Do not gamble on what topics might be on the exam; master everything.

Step 3: View the Chapter Test Prep Videos. These videos show step-by-step solutions to the problems found in each of the book's chapter tests. Follow the worked-out solutions to any of the exercises on the chapter test that you want to study or review.

Step 4: Follow these rules as you train. Be sure to arrive early at the location of the exam. Prepare your mind for the exam. Be sure you are well rested. Don't try to pull "all-nighters" because your brain can only process a certain amount of material at a time. You should be reviewing material you already know the night before an exam, not learning new material.

Work Smart: Study Skills

Do not "cram" for an exam by pulling an "all-nighter."

1.1 Exercises

1. Why do you want to be successful in mathematics? Are your goals positive or negative? If you stated your goal negatively ("Just get me out of this course!"), can you restate it positively?

2. Name three activities in your daily life that involve the use of math (for instance, playing cards, operating your computer, or reading a credit-card bill).

3. What is your instructor's name?

4. What are your instructor's office hours? Where is your instructor's office?

5. What is your instructor's e-mail address?

6. Does your class have a website? Do you know how to access it? What information is located on the website?

7. Are there tutors available for this course? If so, where are they located? When are they available?

8. Name two other students in your class. What is their contact information? When can you meet with them to study?

9. List at least two things that you should do before class begins.

10. List at least two things that you should do during class.

11. List at least two things that you should do after class.

12. What is the point of the Chinese proverb quoted on page 4 in this section?

13. Where are the "Are You Prepared for this Section" problems? How should they be used?

14. Name two features that appear in the margins. What is the purpose of each of them?

15. Name the categories of exercises that appear in this text.

16. How should the chapter review material be used?

17. How should the chapter test be used? What are the Chapter Test Prep Videos?

18. How should the cumulative review be used?

19. List the four steps that should be followed when preparing for an exam. Can you think of other methods of preparing for an exam that have worked for you?

20. Use the chart below to help manage your time. Be sure to fill in the time allocated to various activities in your life, including each of your classes, work, and leisure. Remember, for every hour you're in class, you should plan to spend two hours outside of class studying the material.

	Monday	Tuesday	Wednesday	Thursday	Friday	Saturday	Sunday
7 A.M.							
8 A.M.							
9 A.M.							
10 A.M.							
11 A.M.							
Noon							
1 P.M.							
2 P.M.							
3 P.M.							
4 P.M.							
5 P.M.							
6 P.M.							
7 P.M.							
8 P.M.							
9 P.M.							

1.2 Fractions, Decimals, and Percents

Objectives

1 Factor a Number as a Product of Prime Factors

2 Find the Least Common Multiple of Two or More Numbers

3 Write Equivalent Fractions

4 Write a Fraction in Lowest Terms

5 Round Decimals

6 Convert Between Fractions and Decimals

7 Convert Between Percents and Decimals

Work Smart: Study Skills

The QR code in the title bar of this section links to lecture videos, and the icon ▶ means a lecture video is available for this content. See page 5 for a description.

The discussion of fractions, decimals, and percents in this section is based on *natural numbers*. **Natural numbers** are the numbers 1, 2, 3, 4, and so on.

▶ **1** Factor a Number as a Product of Prime Factors

In multiplication, the numbers that are multiplied together are the **factors** and the answer is the **product**.

$$\underset{\text{factor}}{7} \quad \cdot \quad \underset{\text{factor}}{5} \quad = \quad \underset{\text{product}}{35}$$

When a number is written as a product, we say that we **factor** the number. For example, when we write 20 as the product $10 \cdot 2$, we say that we have factored 20.

Some natural numbers are *prime* numbers and others are *composite*.

Definition

A natural number is **prime** if its only factors are 1 and itself. Natural numbers that are not prime are called **composite**. The number 1 is neither prime nor composite.

The first six prime numbers are 2, 3, 5, 7, 11, and 13. When we write a composite number as the product of prime numbers, we say that we are writing the **prime factorization** of the number. A *factor tree* can be used to find the prime factorization of a number. The process begins with finding two factors of the given number. Continue to factor until all factors are prime.

EXAMPLE 1 **Finding the Prime Factorization**

Write the prime factorization of 24.

Solution

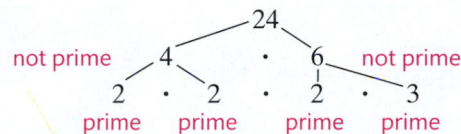

Work Smart

The factorization of 24 could have begun with the factors 8 and 3 instead of 4 and 6. Try this for yourself.

The prime factorization is complete because all the numbers in the last row are prime. The prime factorization of 24 is $2 \cdot 2 \cdot 2 \cdot 3$. Order is not important in multiplying factors. The product could also be written as $3 \cdot 2 \cdot 2 \cdot 2$ or $2 \cdot 3 \cdot 2 \cdot 2$. ●

Quick ✓

1. A natural number is _____ if its only factors are 1 and itself.

2. In the statement $6 \cdot 8 = 48$, 6 and 8 are called _____ and 48 is called the _____.

In Problems 3–6, find the prime factorization of each number. If a number is prime, state "prime."

3. 12   4. 120

5. 31   6. 117

▶ ❷ Find the Least Common Multiple of Two or More Numbers

A **multiple** of a number is the product of that number and any natural number. For example, the multiples of 2 are

$$2 \cdot 1 = 2, \quad 2 \cdot 2 = 4, \quad 2 \cdot 3 = 6, \quad 2 \cdot 4 = 8, \quad 2 \cdot 5 = 10, \quad 2 \cdot 6 = 12, \quad \text{and so on.}$$

Multiples of 3 are

$$3 \cdot 1 = 3, \quad 3 \cdot 2 = 6, \quad 3 \cdot 3 = 9, \quad 3 \cdot 4 = 12, \quad 3 \cdot 5 = 15, \quad 3 \cdot 6 = 18, \quad \text{and so on.}$$

Notice that the numbers 2 and 3 have 6 and 12 as common multiples. The *smallest common multiple*, called the *least common multiple*, of 2 and 3 is 6.

Definition

The **least common multiple (LCM)** of two or more natural numbers is the smallest number that is a multiple of each the numbers.

For example, to find the least common multiple of 6 and 15, list the multiples of each number until the smallest common multiple is found, as follows:

Multiples of 6:   6, 12, 18, 24, 30, 36, 42, . . .

Multiples of 15:   15, 30, 45, 60, . . .

The least common multiple is 30. This approach works just fine for numbers, but it does not work for algebra. For this reason, follow the steps used in Example 2 on the next page to find the least common multiple so that you will be better prepared when the LCM is discussed again later in the course.

EXAMPLE 2 **How to Find the Least Common Multiple**

Find the least common multiple of 6 and 15.

Step-by-Step Solution

Step 1: Write each number as the product of prime factors, aligning common factors vertically.

Arrange the common factor of 3 in its own column: $6 = 2 \cdot 3$
$15 = 3 \cdot 5$

Step 2: Write down the common factor(s), if any. Then write down the remaining factors.

The common factor is 3.
The remaining factors are 2 and 5.

Step 3: Multiply the factors listed in Step 2. The product is the least common multiple (LCM).

The least common multiple of 6 and 15 is $2 \cdot 3 \cdot 5 = 30$.

●

EXAMPLE 3 **Finding the Least Common Multiple**

Find the least common multiple of 18 and 15.

Solution

Write each number as the product of prime factors.

Write the common prime factors, if any. Then write down the remaining factors. Find the product of the factors.

$18 = 2 \cdot 3 \cdot 3$
$15 = 3 \cdot 5$
$ 2 \cdot 3 \cdot 3 \cdot 5$

The LCM is $2 \cdot 3 \cdot 3 \cdot 5 = 90$.

●

> **Quick ✓**
>
> 7. The ____ _____ _____ of two or more natural numbers is the smallest number that is a multiple of each of the numbers.
>
> *In Problems 8–11, find the LCM of the numbers.*
>
> **8.** 6 and 8   **9.** 45 and 72
>
> **10.** 14 and 9   **11.** 12, 18, and 30

▶ ❸ **Write Equivalent Fractions**

Figure 1

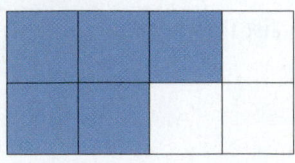

A fraction represents a part of a whole. For example, the fraction $\frac{5}{8}$ means "5 parts out of 8 parts." A fraction also indicates division: $\frac{5}{8}$ means "five divided by eight" and may be written as $8\overline{)5}$. Figure 1 shows the fraction $\frac{5}{8}$ visually.

In the fraction $\frac{5}{8}$, the number 5 is the **numerator** and the number 8 is the **denominator.** The denominator tells the number of equal parts that the whole is divided into, and the numerator tells the number of equal parts that are shaded. For example, in Figure 1 the box is divided into 8 equal parts, 5 of which are shaded.

The **whole numbers** are 0, 1, 2, 3, and so on. Whole numbers are used for the numerator of a fraction, and natural numbers are used for the denominator.

Fractions without common denominators can be rewritten in equivalent forms so they have the same denominator.

Work Smart

The denominator of a number such as 7 is 1 because $7 = \frac{7}{1}$.

Figure 2

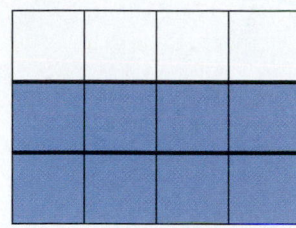

Definition

Equivalent fractions are fractions that represent the same part of a whole.

For example, $\frac{2}{3}$ and $\frac{8}{12}$ are equivalent fractions. To understand why, consider Figure 2. Divide the whole into 12 equal parts and shade 8 of these parts. Now, the shaded region represents $\frac{8}{12}$ of the rectangle. If only the 3 parts separated by the thick black lines are considered, it can be seen that 2 parts are shaded for a fraction of $\frac{2}{3}$. In each case, the same portion of the rectangle is shaded, so $\frac{2}{3}$ and $\frac{8}{12}$ are equivalent fractions.

How are equivalent fractions obtained? The answer lies in the following property.

If a, b, and c are whole numbers, then

$$\frac{a}{b} = \frac{a \cdot c}{b \cdot c} \quad \text{where } b \neq 0, c \neq 0$$

EXAMPLE 4 **Writing an Equivalent Fraction**

Write the fraction $\frac{3}{4}$ as an equivalent fraction with a denominator of 20.

Solution

Find "what over 20 equals $\frac{3}{4}$," or $\frac{3}{4} = \frac{?}{20}$. To write $\frac{3}{4}$ with a denominator of 20, multiply the numerator and denominator of $\frac{3}{4}$ by 5. Do you see why?

$$\frac{3}{4} = \frac{3 \cdot 5}{4 \cdot 5}$$
$$= \frac{15}{20}$$

Quick ✓

12. In the fraction $\frac{7}{12}$, 7 is called the _____ and 12 is called the _____.

13. Fractions that represent the same portion of a whole are called _____ _____.

In Problems 14 and 15, rewrite each fraction with the denominator indicated.

14. $\frac{1}{2}$; 10

15. $\frac{5}{8}$; 48

Sometimes it's necessary to rewrite two or more fractions so that they both have the same denominator. For example, the fractions $\frac{5}{6}$ and $\frac{3}{8}$ can be written with a common denominator of 24, 48, 96 and so on because these are common multiples of the denominators 6 and 8. Notice that 24 is the least common multiple of 6 and 8.

Definition

The **least common denominator (LCD)** is the least common multiple of the denominators of a group of fractions.

> **EXAMPLE 5**   **How to Write Two Fractions as Equivalent Fractions with the LCD**

Write $\dfrac{5}{8}$ and $\dfrac{9}{20}$ as equivalent fractions with the least common denominator.

Step-by-Step Solution

Step 1: Find the least common denominator of the fractions.

The denominators of $\dfrac{5}{8}$ and $\dfrac{9}{20}$ are 8 and 20.

Write each denominator as the product of prime factors:

$$8 = 2 \cdot 2 \cdot 2$$
$$20 = 2 \cdot 2 \cdot \;\;\; 5$$

Write the common factors; then write the remaining factors:

$$\text{LCD} = 2 \cdot 2 \cdot 2 \cdot 5$$
$$= 40$$

Step 2: Rewrite each fraction with the least common denominator.

Multiply the numerator and denominator of $\dfrac{5}{8}$ by 5:

$$\frac{5}{8} = \frac{5 \cdot 5}{8 \cdot 5}$$
$$= \frac{25}{40}$$

Multiply the numerator and denominator of $\dfrac{9}{20}$ by 2:

$$\frac{9}{20} = \frac{9 \cdot 2}{20 \cdot 2}$$
$$= \frac{18}{40}$$

> **Quick ✓**
>
> **16.** The ____ _____ _____ is the least common multiple of the denominators of a group of fractions.
>
> *In Problems 17 and 18, write the equivalent fractions with the least common denominator.*
>
> **17.** $\dfrac{1}{4}$ and $\dfrac{5}{6}$
>
> **18.** $\dfrac{9}{20}$ and $\dfrac{11}{16}$

▶ ❹ Write a Fraction in Lowest Terms

> **Definition**
>
> A fraction is written in **lowest terms** if the numerator and the denominator share no common factor other than 1.

In Other Words

To write a fraction in lowest terms, find any common factors between the numerator and denominator, and divide out the common factors.

Fractions can be written in lowest terms using the fact that

$$\frac{a \cdot c}{b \cdot c} = \frac{a}{b}$$

Thus, to write a fraction in lowest terms, write the numerator and the denominator as a product of primes and then divide out common factors.

> **EXAMPLE 6**   **Writing a Fraction in Lowest Terms**

Write $\dfrac{24}{40}$ in lowest terms.

Solution

Write the numerator and the denominator as the product of primes and divide out common factors.

Work Smart

Use different slash marks to keep track of factors that have divided out. Also, nonprime factors may be used when writing a fraction in lowest terms. In Example 6, we could write

$$\frac{24}{40} = \frac{8 \cdot 3}{8 \cdot 5}$$
$$= \frac{3}{5}$$

$$\frac{24}{40} = \frac{2 \cdot 2 \cdot 2 \cdot 3}{2 \cdot 2 \cdot 2 \cdot 5}$$

Divide out common factors:
$$= \frac{2 \cdot 2 \cdot 2 \cdot 3}{2 \cdot 2 \cdot 2 \cdot 5}$$

$$= \frac{3}{5}$$

Quick ✔

19. A fraction is written in _____ _____ if the numerator and the denominator share no common factor other than 1.

In Problems 20–22, write each fraction in lowest terms.

20. $\dfrac{45}{80}$ **21.** $\dfrac{4}{9}$ **22.** $\dfrac{16}{56}$

▶ ⑤ Round Decimals

Decimals and percentages commonly occur in everyday life. You receive a 92% on your test, there is a 10% discount on jeans, 7.75% is charged in sales tax, 45% of the people polled support a proposition. Before decimals and percents are discussed, let's review place value.

Figure 3 shows how we interpret the place value of each digit in the number 9186.347. For example, the 7 is in the thousandths position, 3 is in the tenths position, and the 8 is in the tens position.

Figure 3

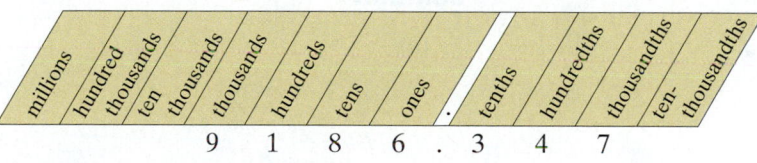

The number 9186.347 is read "nine thousand, one hundred eighty-six and three hundred forty-seven thousandths."

Quick ✔

In Problems 23–26, tell the place value of the digit in the given number.

23. 235.71; the 1   **24.** 56,701.28; the 2

25. 278,403.95; the 8   **26.** 0.189; the 9

Decimals are rounded in the same way whole numbers are rounded. First, identify the specified place value in the decimal. If the digit to the right is 5 or more, add 1 to the digit; if the digit to the right is 4 or less, leave the digit as it is. Then drop the digits to the right of the specified place value.

EXAMPLE 7 **Rounding a Decimal Number**

(a) Round 8.726 to the nearest hundredth.

(b) Round 0.9451 to the nearest thousandth.

Solution

(a) To round to the nearest hundredth, notice that 2 is in the hundredths place: 8.7$\underline{2}$6. The number to the right of 2 is 6. Because 6 is greater than 5, round 8.726 to 8.73.

(b) To round 0.9451 to the nearest thousandth, notice that 5 is in the thousandths place: 0.94$\underline{5}$1. The number to the right of 5 is 1. Because 1 is less than 5, round 0.9451 to 0.945.

Quick ✓

In Problems 27–31, round each number to the given decimal place.

27. 0.173 to the nearest tenth

28. 0.932 to the nearest hundredth

29. 1.396 to the nearest hundredth

30. 690.004 to the nearest hundredth

31. 59.98 to the nearest tenth

⑥ Convert Between Fractions and Decimals

▶ Convert a Fraction to a Decimal

To convert a fraction to a decimal, divide the numerator of the fraction by the denominator of the fraction until the remainder is 0 or the remainder repeats.

EXAMPLE 8 **Converting a Fraction to a Decimal**

Convert each number to a decimal.

(a) $\dfrac{9}{20}$

(b) $\dfrac{2}{3}$

Solution

(a)

$$\frac{9}{20} = 20)\overline{9.00} \quad \begin{array}{r} 0.45 \\ \hline \end{array}$$

$$\begin{array}{r} 8\,0 \\ \hline 100 \\ 100 \\ \hline 0 \end{array}$$

Therefore, $\dfrac{9}{20} = 0.45$.

(b)

$$\frac{2}{3} = 3)\overline{2.000} \quad \begin{array}{r} 0.666 \\ \hline \end{array}$$

$$\begin{array}{r} 1\,8 \\ \hline 20 \\ 18 \\ \hline 20 \\ 18 \\ \hline 2 \end{array}$$

Notice that the remainder, 2, repeats. So $\dfrac{2}{3} = 0.666\ldots$.

In Example 8(a), the decimal 0.45 is called a **terminating decimal** because the decimal stops after the 5. In Example 8(b), the number 0.666 . . . is called a **repeating decimal** because the 6 repeats indefinitely. The decimal 0.666 . . . can also be written as $0.\overline{6}$. The bar over the 6 means the 6 repeats.

Quick ✓

In Problems 32–35, write the fraction as a decimal.

32. $\dfrac{2}{5}$

33. $\dfrac{5}{6}$

34. $\dfrac{11}{8}$

35. $\dfrac{3}{7}$

Based on Examples 8(a) and (b) and Quick Check Problems 32–35, notice that **every fraction has a decimal representation that either terminates or repeats.**

▶ **Convert a Decimal to a Fraction**

To convert a decimal to a fraction, identify the place value of the last digit in the decimal. Write the decimal as a fraction using the place value of the last digit as the denominator, and write in lowest terms.

EXAMPLE 9   **Writing a Decimal as a Fraction**

Convert each decimal to a fraction and write in lowest terms.

    (a) 0.8   **(b)** 0.77   **(c)** 4.237

Solution

  (a) 0.8 is equivalent to 8 tenths, or $\dfrac{8}{10}$. Because $\dfrac{8}{10} = \dfrac{4 \cdot 2}{5 \cdot 2} = \dfrac{4}{5}$, $0.8 = \dfrac{4}{5}$.

  (b) 0.77 is equivalent to 77 hundredths, or $\dfrac{77}{100}$.

  (c) 4.237 is equivalent to 4237 thousandths, or $\dfrac{4237}{1000}$.

> **Quick** ✔
>
> *In Problems 36–38, convert the decimal to a fraction and write in lowest terms.*
>
> **36.** 0.6   **37.** 0.17   **38.** 0.625

❼ Convert Between Percents and Decimals

When computing with percents, it is convenient to write percents as decimals. How is a percent converted to a decimal? Let's see.

▶ **Convert a Percent to a Decimal**

Work Smart

The word "per" means "divided by." "Miles per hour" means "miles divided by hours." "Miles per gallon" means "miles divided by gallons," and so on.

> **Definition**
>
> The word **percent** means **parts per hundred** or **parts out of one hundred.**

So 25% means 25 parts out of 100 parts. Therefore, $25\% = \dfrac{25}{100} = \dfrac{1 \cdot 25}{4 \cdot 25} = \dfrac{1}{4}$.

Since the word percent means "parts per hundred," 100% means "100 parts per 100," so $100\% = 1$. Therefore, to convert from a percent to a decimal, multiply the percent by $\dfrac{1}{100\%}$.

EXAMPLE 10   **Writing a Percent as a Decimal**

Write the following percents as decimals:

    (a) 27%   **(b)** 150%

Solution

Work Smart

To convert from a percent to a decimal, move the decimal point two places to the left and drop the % symbol.

  (a) $27\% = 27\% \cdot \dfrac{1}{100\%}$

       $= \dfrac{27}{100}$

       $= 0.27$

  (b) $150\% = 150\% \cdot \dfrac{1}{100\%}$

       $= \dfrac{150}{100}$

       $= 1.5$

Quick ✔

39. The word percent means parts per _____, so 35% means __ parts out of 100 parts or $\dfrac{\overline{\quad}}{100}$.

In Problems 40–43, write the percent as a decimal.

40. 23%　　　　　　　　　　**41.** 1%

42. 72.4%　　　　　　　　　**43.** 127%

▶ Convert a Decimal to a Percent

Because $100\% = 1$, to convert a decimal to a percent, multiply the decimal by $\dfrac{100\%}{1}$.

EXAMPLE 11　**Writing a Decimal as a Percent**

Write the following decimals in percent form:

(a) 0.445　　　　　　　　　　**(b)** 1.42

Solution

Work Smart

To convert from a decimal to a percent, move the decimal point two places to the right and add the % symbol.

(a) $0.445 = 0.445 \cdot \dfrac{100\%}{1}$

$= 44.5\%$

(b) $1.42 = 1.42 \cdot \dfrac{100\%}{1}$

$= 142\%$

Quick ✔

44. To convert a decimal to a percent, multiply the decimal by _____.

In Problems 45–48, write the decimal as a percent.

45. 0.15　　　　　　　　　　**46.** 0.8

47. 1.372　　　　　　　　　　**48.** 0.004

1.2 Exercises　MyMathLab®

Exercise numbers in **green** have complete video solutions in MyMathLab or may be accessed using the QR code to the right.

*Problems **1–48** are the Quick✔s that follow the **EXAMPLES**.*

Building Skills

In Problems 49–62, find the prime factorization of each number. If a number is prime, state "prime". See Objective 1.

49. 25　　　　　**50.** 9　　　　　**51.** 28

52. 100　　　　**53.** 21　　　　**54.** 35

55. 36　　　　**56.** 54　　　　**57.** 50

58. 70　　　　**59.** 53　　　　**60.** 79

61. 252　　　**62.** 315

In Problems 63–74, find the LCM of each set of numbers. See Objective 2.

63. 6 and 21　　　　　　　**64.** 10 and 14

65. 18 and 63　　　　　　　**66.** 42 and 18

67. 15 and 14　　　　　　　**68.** 55 and 6

69. 30 and 45　　　　　　　**70.** 8 and 60

71. 5, 6, and 12　　　　　　**72.** 9, 15, and 20

73. 3, 8, and 9　　　　　　　**74.** 4, 18, and 20

In Problems 75–80, write each fraction with the given denominator. See Objective 3.

75. Write $\dfrac{2}{3}$ with denominator 12.

76. Write $\dfrac{4}{5}$ with denominator 15.

77. Write $\dfrac{3}{4}$ with denominator 24.

78. Write $\dfrac{5}{14}$ with denominator 28.

79. Write 7 with denominator 3.

80. Write 4 with denominator 10.

In Problems 81–88, write as equivalent fractions with the least common denominator. See Objective 3.

81. $\dfrac{1}{2}$ and $\dfrac{3}{8}$

82. $\dfrac{3}{4}$ and $\dfrac{5}{12}$

83. $\dfrac{3}{5}$ and $\dfrac{2}{3}$

84. $\dfrac{1}{4}$ and $\dfrac{2}{9}$

85. $\dfrac{1}{12}$ and $\dfrac{5}{18}$

86. $\dfrac{5}{12}$ and $\dfrac{7}{15}$

87. $\dfrac{2}{9}$ and $\dfrac{7}{18}$ and $\dfrac{7}{30}$

88. $\dfrac{7}{10}$ and $\dfrac{1}{4}$ and $\dfrac{5}{6}$

In Problems 89–96, write each fraction in lowest terms. See Objective 4.

89. $\dfrac{14}{21}$

90. $\dfrac{9}{15}$

91. $\dfrac{38}{18}$

92. $\dfrac{81}{36}$

93. $\dfrac{22}{66}$

94. $\dfrac{9}{27}$

95. $\dfrac{18}{3}$

96. $\dfrac{36}{4}$

In Problems 97–102, tell the place value of the indicated digit in the given number. See Objective 5.

97. 3465.902; the 0

98. 549,813.0267; the 8

99. 357.469; the 5

100. 9124.786; the 7

101. 2018.3764; the 6

102. 539.016; the 9

In Problems 103–110, round each number to the given place. See Objective 5.

103. 578.206 to the nearest tenth

104. 7298.0845 to the nearest hundredth

105. 354.678 to the nearest hundredth

106. 543.56 to the nearest whole number

107. 3682.0098 to the nearest thousandth

108. 683.098 to the nearest hundredth

109. 29.96 to the nearest whole number

110. 37.999 to the nearest tenth

In Problems 111–120, convert each fraction to a decimal. See Objective 6.

111. $\dfrac{5}{8}$

112. $\dfrac{3}{4}$

113. $\dfrac{2}{7}$

114. $\dfrac{2}{9}$

115. $\dfrac{5}{16}$

116. $\dfrac{11}{32}$

117. $\dfrac{3}{13}$

118. $\dfrac{6}{13}$

119. $\dfrac{29}{25}$

120. $\dfrac{57}{50}$

In Problems 121–126, write each decimal as a fraction in lowest terms. See Objective 6.

121. 0.75

122. 0.25

123. 0.5

124. 0.4

125. 0.982

126. 0.358

In Problems 127–132, write each percent as a decimal. See Objective 7.

127. 37%

128. 59%

129. 6.02%

130. 8.25%

131. 0.1%

132. 0.5%

In Problems 133–138, write each decimal as a percent. See Objective 7.

133. 0.2

134. 0.5

135. 0.275

136. 0.349

137. 2

138. 1

Mixed Practice

In Problems 139–144, write each fraction as a decimal, rounded to the indicated place.

139. $\dfrac{13}{6}$ to the nearest tenth

140. $\dfrac{15}{8}$ to the nearest tenth

141. $\dfrac{8}{3}$ to the nearest hundredth

142. $\dfrac{9}{7}$ to the nearest hundredth

143. $\dfrac{14}{27}$ to the nearest thousandth

144. $\dfrac{18}{31}$ to the nearest thousandth

Applying the Concepts

145. Planets in Our Solar System At a certain point, Mercury, Venus, and Earth lie on a straight line. If it takes these planets 3, 7, and 12 months, respectively, to revolve around the sun, what is the fewest number of months until they align this way again?

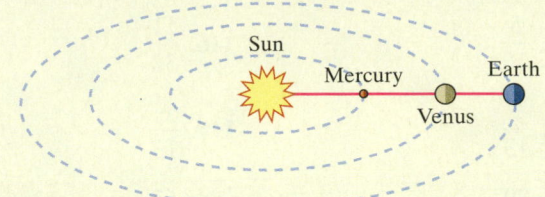

146. Talladega Raceway At Talladega, one of the crew chiefs discovered that in a given time interval, Jimmie Johnson completed 21 laps, Dale Earnhardt Jr. completed 18 laps, and Danica Patrick completed 15 laps. Suppose all three drivers begin at the same time. How many laps would need to be completed so that all three drivers were at the finish line at exactly the same time?

147. Sam's Medication Bob gives his dog Sam one type of medication every 4 days, and a second type of medication every 10 days. How often does Bob give Sam both medications on the same day?

148. Visiting Columbus Pamela and Geoff both visit Columbus on business. Pamela flies to Columbus from Atlanta every 14 days, and Geoff takes the train to Columbus from Cincinnati every 20 days. How often are both Pamela and Geoff in Columbus on business?

149. Survey Data In a survey of 500 students, 325 stated that they work at least 25 hours per week. Express the fraction of students that work at least 25 hours per week as a fraction in lowest terms.

150. Survey Data In a survey of 750 students, 450 stated that they are enrolled in 15 or more semester hours. Express the fraction of students who are enrolled in 15 or more semester hours as a fraction in lowest terms.

In Problems 151–158, express answers to the nearest hundredth of a percent, if necessary.

151. Eating Healthy In a poll of 1200 adult Americans conducted by Zogby International, 840 stated that they believe that they eat healthy food.
  (a) What fraction of adult Americans stated that they eat healthy foods? Write the fraction in lowest terms.

  (b) Express the fraction found in part (a) as a decimal.
  (c) Express the decimal found in part (b) as a percent.

152. Ghosts In a survey of 1100 adult women conducted by Harris Interactive, it was determined that 640 believe in ghosts.
  (a) What fraction of adult women stated that they believe in ghosts? Write the fraction in lowest terms.

  (b) Express the fraction found in part (a) as a decimal.
  (c) Express the decimal found in part (b) as a percent.

153. Test Score A student earns 85 points out of a total of 110 points on an exam. Express this score as a percent.

154. Test Score A student earns 80 points out of a total of 115 points on an exam. Express this score as a percent.

155. Time Utilization In a 24-hour day, Jackson sleeps for 8 hours, works for 4 hours, and goes to school and studies for 6 hours.
  (a) What percent of the time does Jackson sleep?
  (b) What percent of the time does Jackson work?
  (c) What percent of the time does Jackson go to school and study?

156. Tree Inventory An arborist counted the number of ash trees in Whetstone Park. He found that there were 48 white ash trees, 51 green ash trees, and 2 blue ash trees.
  (a) What percent of the ash trees in Whetstone Park were white ashes?
  (b) What percent of the ash trees in Whetstone Park were green ashes?
  (c) What percent of the ash trees in Whetstone Park were blue ashes?

157. Cashews A single serving of cashews contains 14 grams of fat. Of this, 3 grams is saturated fat. What percentage of fat grams is saturated fat in a single serving of cashews?

158. Cheese Pizza A single serving of cheese pizza contains 11 grams of fat. Of this, 5 grams are saturated fat. What percentage of fat grams is saturated fat in a single serving of cheese pizza?

Extending the Concepts

159. The Sieve of Eratosthenes Eratosthenes (276 B.C.–194 B.C.) was born in Cyrene, which is now in Libya in North Africa. He devised an algorithm (a series of steps that are followed to solve a problem) for identifying prime numbers. Use this algorithm to find all the prime numbers less than 100.

  Step 1: List all the natural numbers from 2 to 100. Since 1 is neither prime nor composite, do not include it on the list. It is helpful to use a table format, listing 2–10 on the first row, 11–20 on the second row, etc.

  Step 2: Circle the 2, the first prime number, and then cross out every multiple of 2. For example, cross out 4, 6, 8, . . .

  Step 3: Circle the next number on the list that has not been crossed out. This number, 3, is the next prime number. Cross out all of the multiples of this number, 6, 9, 12, . . .

Step 4: Repeat Step 3 until all of the numbers have been circled or crossed out.

The circled numbers are the prime numbers less than 100. This process can be used for any list of natural numbers, although determining whether or not 541 is a prime number by this process would be time consuming.

1.3 The Number Systems and the Real Number Line

Objectives

1. Classify Numbers
2. Plot Points on a Real Number Line
3. Use Inequalities to Order Real Numbers
4. Compute the Absolute Value of a Real Number

Work Smart

The use of the word "real" to describe numbers leads us to question, "Are there 'nonreal' numbers?" The answer is yes. We use the word "imaginary" to describe nonreal numbers. Imaginary does not mean that these numbers are made up, however. Imaginary numbers will be discussed later in the text.

Are You Prepared for This Section?

Before getting started, complete the following problems. If you get a problem wrong, go back to the section cited and review the material.

P1. Write $\dfrac{5}{8}$ as a decimal.   [Section 1.2, p. 14]

P2. Write $\dfrac{9}{11}$ as a decimal.   [Section 1.2, p. 14]

This section discusses the *real number system.* You are already familiar with real numbers because you use them every day. In short, real numbers are numbers that are used to count or measure things, such as 25 students in your class, 18.4 miles per gallon, or a $130 debt.

Various types of numbers are organized in *sets.* A **set** is a well-defined collection of objects. For example, we can identify the students enrolled in Elementary Algebra at your college as a set. The collection of numbers 0, 1, 2, 3, 4, 5, 6, 7, 8, and 9 may also be identified as a set. If A represents this set of numbers, then

$$A = \{0, 1, 2, 3, 4, 5, 6, 7, 8, 9\}$$

In this notation, braces { } are used to enclose the objects, or **elements**, in the set. A set with no elements in it is called an **empty set.** Empty sets are denoted by the symbol $\varnothing$ or { }.

EXAMPLE 1   **Writing a Set**

Write the set of letters of the English alphabet that are vowels.

Solution
The vowels are *a, e, i, o,* and *u.* If *V* represents this set, then

$$V = \{a, e, i, o, u\}$$

Quick ✔

1. Write the set that represents the first four positive, odd numbers.

2. Write the set that represents the states in the United States with names that begin with the letter A.

3. Write the set that represents the states in the United States with names that begin with the letter Z.

① Classify Numbers

It is interesting to look at the real number system in the context of the history of numbers. The first types of numbers that humans used are the *natural numbers* or *counting numbers.* Natural numbers were introduced in Section 1.2. Now a formal definition is presented using a set. The three dots in the definition are called an *ellipsis* and indicate that the pattern continues indefinitely.

Definition

The **natural numbers,** or **counting numbers**, are the numbers in the set $\{1, 2, 3, . . .\}$.

Prepared?...Answers   **P1.** 0.625
P2. $0.8181 . . .$ or $0.\overline{81}$

Figure 4
The natural numbers.

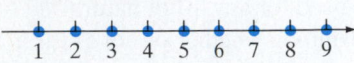

The natural numbers are often used to count things. For example, we can count the number of cars waiting at a Wendy's drive-thru. We can represent the natural numbers graphically using a number line. See Figure 4. The arrow on the right indicates the direction in which the numbers increase.

Because we do not count the number of cars waiting in the drive-thru by saying, "zero, one, two, three...," zero is not a natural, or counting, number. When the number 0 is added to the set of natural numbers, we get the set of *whole numbers*.

Definition
The **whole numbers** are the numbers in the set $\{0, 1, 2, 3, \dots\}$.

Figure 5
The whole numbers.

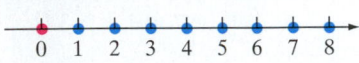

Figure 5 represents the whole numbers on the number line. Notice that the set of natural numbers is included in the set of whole numbers.

By expanding the numbers to the left of zero on the number line, the set of *integers* is obtained.

Definition
The **integers** are the numbers in the set $\{\dots, -3, -2, -1, 0, 1, 2, 3, \dots\}$.

Figure 6
The integers.

Figure 6 represents the integers on the number line. Notice that the whole numbers and natural numbers are included in the set of integers.

Integers are useful in many situations. For example, temperatures above 0°F (positive counting numbers) or below 0°F (negative counting numbers) could not be discussed without integers. A debt of 300 dollars can be represented as an integer by -300 dollars.

How can a part of a whole be represented, such as a part of last night's leftover pizza or part of a dollar? To address this problem, our number system is enlarged to include *rational numbers*.

In Other Words
A rational number is a number that can be expressed as a fraction where the numerator is any integer and the denominator is any nonzero integer.

Definition
A **rational number** is a number that can be written in the form $\dfrac{p}{q}$, where p and q are integers. However, q cannot equal zero.

Work Smart
Remember, all integers are also rational numbers. For example, $42 = \dfrac{42}{1}$.

Examples of rational numbers are $\dfrac{2}{5}, \dfrac{5}{2}, \dfrac{0}{8}, -\dfrac{7}{9}$, and $\dfrac{31}{4}$. Because $\dfrac{p}{1} = p$ for any integer p, it follows that all integers are also rational numbers. For example, 7 is an integer, but it is also a rational number because it can be written as $\dfrac{7}{1}$. This idea is illustrated below.

Here, $\frac{7}{1}$ is written as a rational number ...

$$\frac{7}{1} = 7$$

...but over here, it is written as the integer 7, which is also a natural number.

In addition to being represented as fractions, rational numbers can also be represented in decimal form as either repeating decimals or terminating decimals. Table 1 on the next page shows various rational numbers in fraction form and decimal form.

Table 1

Fraction Form of Rational Number	Decimal Form of Rational Number	Terminating or Repeating Decimal
$\dfrac{7}{2}$	3.5	Terminating
$\dfrac{1}{3}$	$0.333\ldots = 0.\overline{3}$	Repeating
$-\dfrac{3}{8}$	-0.375	Terminating
$-\dfrac{15}{11}$	$-1.3636\ldots = -1.\overline{36}$	Repeating

The repeating decimal $0.\overline{3}$ and the terminating decimal -0.375 are rational numbers because they represent fractions (see Section 1.2).

Decimals that neither terminate nor repeat are called *irrational numbers*.

> **In Other Words**
>
> Numbers that cannot be written as the ratio of two integers are irrational.

Definition

An **irrational number** is a number that has a decimal representation that neither terminates nor repeats. Therefore, irrational numbers cannot be written as the quotient (ratio) of two integers.

An example of an irrational number is 1.343343334 . . . because the decimal neither terminates nor repeats. Other examples of irrational numbers are $\sqrt{2}$, whose value is approximately 1.41421, and π, whose value is approximately 3.141593.

Here is a formal definition of the set of *real numbers*.

Definition

The set of rational numbers combined with the set of irrational numbers is called the set of **real numbers**.

Figure 7 shows the relationships among the various types of numbers. Note that the oval that represents the whole numbers surrounds the oval that represents the natural numbers. This means the set of whole numbers includes all the natural numbers.

Figure 7

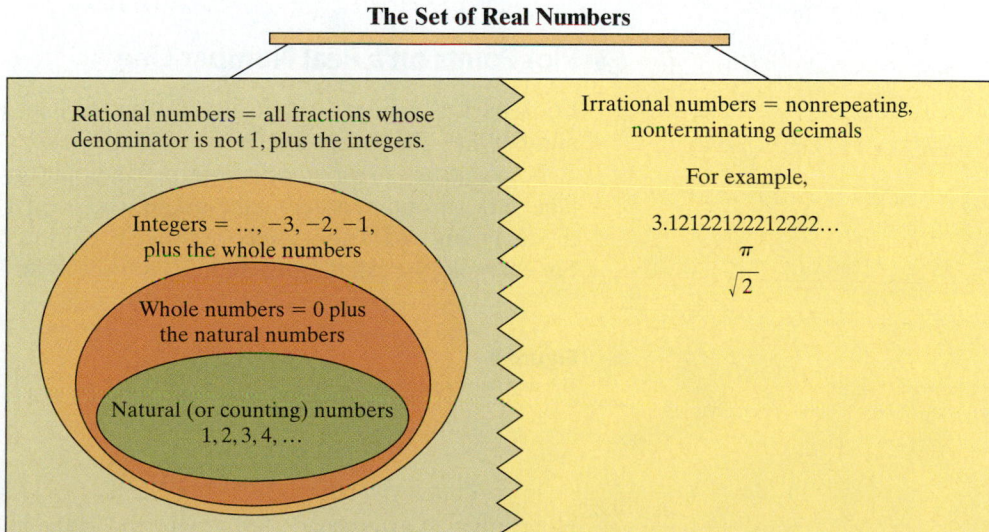

The Set of Real Numbers

Rational numbers = all fractions whose denominator is not 1, plus the integers.

Integers = ..., −3, −2, −1, plus the whole numbers

Whole numbers = 0 plus the natural numbers

Natural (or counting) numbers 1, 2, 3, 4, ...

Irrational numbers = nonrepeating, nonterminating decimals

For example,

3.12122122212222...

π

$\sqrt{2}$

The set of real numbers is composed of the set of rational numbers and the set of irrational numbers.

EXAMPLE 2 **Classifying Numbers in a Set**

List the numbers in the set

$$\left\{9, -\frac{2}{7}, -4, 0, -4.010010001\ldots, 3.\overline{632}, 18.3737\ldots\right\}$$

that are

 (a) Natural numbers   (b) Whole numbers

 (c) Integers   (d) Rational numbers

 (e) Irrational numbers   (f) Real numbers

Solution

 (a) 9 is the only natural number.

 (b) 0 and 9 are the whole numbers.

 (c) 9, −4, and 0 are the integers.

 (d) $9, -\frac{2}{7}, -4, 0, 3.\overline{632}$, and 18.3737 . . . are the rational numbers.

 (e) −4.010010001 . . . is the only irrational number because the decimal does not repeat, nor does it terminate.

 (f) All the numbers listed are real numbers. Real numbers consist of rational numbers together with irrational numbers.   ●

Quick ✓

4. *True or False* Every integer is a rational number.

5. Real numbers that can be represented with a terminating or repeating decimal are called _____ numbers.

In Problems 6–11, list the numbers in the set $\left\{\dfrac{11}{5}, -5, 12, 2.\overline{76}, 0, 2.737737773\ldots, \dfrac{18}{4}\right\}$
that are

 6. Natural numbers   **7.** Whole numbers

 8. Integers   **9.** Rational numbers

 10. Irrational numbers   **11.** Real numbers

▶ ❷ **Plot Points on a Real Number Line**

Look back at Figure 6 (page 20). Notice the gaps between the integers plotted on the number line. These gaps are filled in with the real numbers that are not integers.

 To construct a *real number line*, pick a point on a line somewhere in the center, and label it 0. This point is called the **origin**. The point 1 unit to the right of 0 corresponds to the real number 1. The distance between 0 and 1 determines the **scale** of the number line. For example, the point representing 2 is twice as far from 0 as 1 is. See Figure 8.

Figure 8
The real number line.

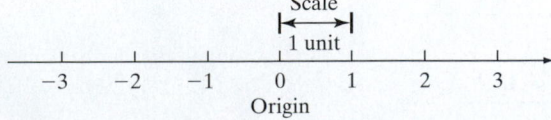

Notice that an arrowhead on the right end of the line indicates the direction in which the numbers increase. Points to the left of 0 correspond to the real numbers −1, −2, and so on.

> **Definition**
>
> The real number associated with a point P is called the **coordinate** of P. The **real number line** is the set of all points that have been assigned coordinates.

EXAMPLE 3   **Plotting Points on a Real Number Line**

On a real number line, label the points with coordinates $0, 6, -2, 2.5, -\dfrac{1}{2}$.

Solution

Draw a real number line and then plot the points. See Figure 9. Notice that 2.5 is midway between 2 and 3. Also notice that $-\dfrac{1}{2}$ is midway between -1 and 0.

Figure 9

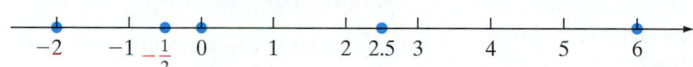

Quick ✔

12. The point on the real number line whose coordinate is 0 is called the _____.

13. On a real number line, label the points with coordinates $0, 3, -2, \dfrac{1}{2}$, and 3.5.

The real number line consists of three classes (or categories) of real numbers, as shown in Figure 10.

Figure 10

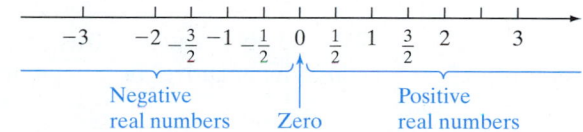

- The **negative real numbers** are the coordinates of points to the left of 0.
- The real number 0 is the coordinate of the origin.
- The **positive real numbers** are the coordinates of points to the right 0.

The **sign** of a number refers to whether the number is a positive or a negative real number. For example, the sign of -4 is negative and the sign of 100 is positive.

Work Smart

The arrowhead in an inequality always points to the smaller number—that is, the number further to the left on the number line.

Figure 11

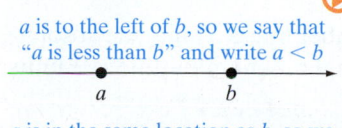

▶ ❸ **Use Inequalities to Order Real Numbers**

Given two numbers (points) a and b, a must be to the left of b (denoted $a < b$) or the same as b (denoted $a = b$) or to the right of b (denoted $a > b$). See Figure 11.

If a is less than or equal to b, write $a \leq b$. Similarly, $a \geq b$ means that a is greater than or equal to b. Collectively, the symbols $<$, $>$, $\leq$, and $\geq$ are called **inequality symbols**. The "arrowhead" in an inequality always points to the smaller number. For $3 < 5$, the "arrowhead" points to 3.

Note that $a < b$ and $b > a$ mean the same thing. For example, $2 < 3$ and $3 > 2$ mean the same thing. Do you see why?

EXAMPLE 4   **Using Inequality Symbols**

(a) We know that $3 is less than $7 and that 3 apples is fewer than 7 apples. Using the real number line, we say $3 < 7$ because the point whose coordinate is 3 lies to the left of the point whose coordinate is 7 on a real number line.

(continued)

(b) Being $2 in debt is not as bad as being $5 in debt, so $-2 > -5$. Using the real number line, $-2 > -5$ because the point whose coordinate is -2 lies to the right of the point whose coordinate is -5 on a real number line.

(c) $2.7 > \dfrac{5}{2}$ because $\dfrac{5}{2} = 2.5$ and $2.7 > 2.5$.

(d) $\dfrac{5}{6} > \dfrac{4}{5}$ because $\dfrac{5}{6} = \dfrac{25}{30}$ and $\dfrac{4}{5} = \dfrac{24}{30}$, and 25 out of 30 parts is more than 24 out of 30 parts. Also, when the two fractions are expressed as decimals, $\dfrac{5}{6} = 0.8\overline{3}$ and $\dfrac{4}{5} = 0.8$. Because $0.8\overline{3}$ is greater than 0.80, $\dfrac{5}{6} > \dfrac{4}{5}$.

Work Smart

Write fractions with a common denominator or change fractions to decimals to compare the location of the numbers on the number line.

Quick ✓

14. The symbols $<, >, \le, \ge$ are called _____ symbols.

In Problems 15–20, replace the question mark by $<$, $>$, or $=$, whichever is correct.

15. $2\ ?\ 9$

16. $-5\ ?\ -3$

17. $\dfrac{4}{5}\ ?\ \dfrac{1}{2}$

18. $\dfrac{4}{7}\ ?\ 0.5$

19. $\dfrac{4}{3}\ ?\ \dfrac{20}{15}$

20. $-\dfrac{4}{3}\ ?\ -\dfrac{5}{4}$

Based upon the discussion so far, we conclude that

$$a > 0 \qquad \text{is equivalent to} \qquad a \text{ is positive}$$

$$a < 0 \qquad \text{is equivalent to} \qquad a \text{ is negative}$$

We sometimes read $a > 0$ as "a is positive." If $a \ge 0$, then $a > 0$ or $a = 0$, so this is read as "a is nonnegative" or "a is greater than or equal to zero."

▶ ❹ Compute the Absolute Value of a Real Number

The real number line can be used to describe the concept of *absolute value*.

In Other Words

Think of absolute value as the number of units you must count to get from 0 to a number on the real number line. The absolute value of a number can never be negative because it represents a distance.

Definition

The **absolute value** of a number a, written $|a|$, is the distance from 0 to a on a real number line.

For example, because the distance from 0 to 3 on a real number line is 3 units, the absolute value of 3, $|3|$, is 3. Because the distance from 0 to -3 on a real number line is 3 units, $|-3| = 3$. See Figure 12.

Figure 12

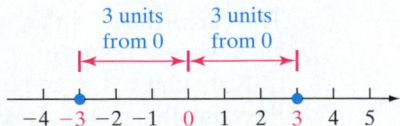

3 units from 0   3 units from 0

$-4\ -3\ -2\ -1\ \ 0\ \ 1\ \ 2\ \ 3\ \ 4\ \ 5$

EXAMPLE 5   **Computing Absolute Value**

Evaluate each of the following:

(a) $|6|$ (b) $|-7|$ (c) $|0|$ (d) $-|-1.5|$

Solution

(a) $|6| = 6$ because the distance from 0 to 6 on a real number line is 6.

(b) $|-7| = 7$ because the distance from 0 to -7 on a real number line is 7.

(c) $|0| = 0$ because the distance from 0 to 0 on a real number line is 0.

(d) $-|-1.5| = -1.5$

Quick ✔

21. The distance from zero to a point on a real number line whose coordinate is a is called the _____ ____ of a.

In Problems 22–24, evaluate each expression.

22. $|-15|$ **23.** $\left|\dfrac{3}{4}\right|$ **24.** $-|-4|$

1.3 Exercises MyMathLab®

Exercise numbers in **green** have complete video solutions in MyMathLab or may be accessed using the QR code to the right.

*Problems **1–24** are the Quick ✔s that follow the **EXAMPLES**.*

Building Skills

In Problems 25–30, write each set. See Objective 1.

25. A is the set of *whole* numbers less than 5.

26. B is the set of *natural* numbers less than 25.

27. D is the set of *natural* numbers less than 5.

28. C is the set of *integers* between -6 and 4, not including -6 or 4.

29. E is the set of even *natural* numbers less than 1.

30. F is the set of *whole* numbers less than 0.

In Problems 31–36, list the elements in the set
$$\left\{-4, 3, -\frac{13}{2}, 0, 2.303003000\ldots\right\} \text{ that are described. See Objective 1.}$$

31. natural numbers

32. whole numbers

33. integers

34. rational numbers

35. irrational numbers

36. real numbers

In Problems 37–42, list the elements in the set $\left\{-4.2, 3.\overline{5}, \pi, \dfrac{5}{5}\right\}$ *that are described. See Objective 1.*

37. real numbers

38. rational numbers

39. irrational numbers   **40.** integers

41. whole numbers   **42.** natural numbers

In Problems 43 and 44, plot the points in each set on a real number line. See Objective 2.

43. $\left\{0, \dfrac{3}{3}, -1.5, -2, \dfrac{4}{3}\right\}$

44. $\left\{\dfrac{3}{4}, \dfrac{0}{2}, -\dfrac{5}{4}, -0.5, 1.5\right\}$

In Problems 45–52, determine whether the statement is True or False. See Objective 3.

45. $-2 > -3$ **46.** $0 < -5$

47. $-6 \leq -6$ **48.** $-3 > -5$

49. $\dfrac{3}{2} = 1.5$ **50.** $4.7 = 4.\overline{7}$

51. $\pi = 3.14$ **52.** $\dfrac{1}{3} = 0.33$

In Problems 53–60, replace the ? with the correct symbol: $>$, $<$, $=$. See Objective 3.

53. $-1 \ ? \ 0$ **54.** $-8 \ ? \ -8.5$

55. $\dfrac{5}{8} \ ? \ \dfrac{6}{11}$ **56.** $\dfrac{5}{12} \ ? \ \dfrac{2}{3}$

57. $\dfrac{2}{9} \ ? \ 0.22$ **58.** $\dfrac{5}{11} \ ? \ 0.\overline{45}$

59. $\dfrac{42}{6} \ ? \ 7$ **60.** $\dfrac{3}{4} \ ? \ \dfrac{3}{5}$

In Problems 61–68, evaluate each expression. See Objective 4.

61. $|-12|$ **62.** $|-8|$

63. $|4|$

64. $|7|$

65. $\left| -\dfrac{3}{8} \right|$

66. $\left| -\dfrac{13}{9} \right|$

67. $-|-2.1|$

68. $-|-3.2|$

Mixed Practice

In Problems 69 and 70, (a) plot the points on a real number line, (b) write the numbers in ascending order, and (c) list the numbers that are (i) integers, (ii) rational numbers.

69. $\left\{ \dfrac{3}{5},\ -1,\ -\dfrac{1}{2},\ 1,\ 3.5,\ |-7|,\ -4.5 \right\}$

70. $\left\{ 8,\ -2,\ |-4|,\ -1.5,\ -\dfrac{4}{3},\ 0,\ -\dfrac{15}{3} \right\}$

Applying the Concepts

In Problems 71–78, place a ✓ in the box if the given number belongs to that set.

		Natural	Whole	Integer	Rational	Irrational	Real
71.	-100			✓	✓		✓
72.	0		✓	✓	✓		✓
73.	-10.5				✓		✓
74.	$\sqrt{2}$					✓	✓
75.	$\dfrac{75}{25}$	✓	✓	✓	✓		✓
76.	4	✓	✓	✓	✓		✓
77.	$7.56556555\ldots$					✓	✓
78.	$6.\overline{45}$				✓		✓

In Problems 79–88, determine whether the statement is True or False.

79. Every *whole* number is also an *integer*.

80. Every decimal number is a *rational* number.

81. There are numbers that are both *rational* and *irrational*.

82. 0 is a positive number.

83. Every *natural* number is also a *whole* number.

84. Every *integer* is also a *real* number.

85. Every terminating decimal is a *rational* number.

86. Some numbers in the form $\dfrac{p}{q}$, $q \neq 0$ are *integers*.

87. 0 is a nonnegative *integer*.

88. -1 is a nonpositive *integer*.

In Problems 89–94, name the set, or give the elements of the set, that matches each description.

89. nonterminating and nonrepeating decimals

90. nonnegative integers

91. the set of rational numbers combined with the set of irrational numbers

92. terminating or repeating decimals

93. numbers that are both nonnegative and nonpositive

94. numbers that are both negative and positive

Extending the Concepts

*If every element of set A is also an element of set B, we say A is a **subset** of B and we write $A \subseteq B$. In Problems 95–98, use this definition and the following sets to answer True or False to each statement.*

$$X = \{a, b, c, d, e\} \quad Y = \{c, e\} \quad Z = \{c, e, f\}$$

95. $Y \subseteq X$

96. $Z \subseteq Y$

97. $Y \subseteq Z$

98. $Z \subseteq X$

The **intersection** of two sets A and B is the set that contains the elements common to both A and B and is written $A \cap B$. The **union** of two sets A and B is the set of all elements that are in either A or B or both and is written $A \cup B$. In Problems 99–104, write the elements of each set, using sets A, B, and C below.

$$A = \{7, 8, 9, 10, 11, 12\} \quad B = \{10, 11, 12, 13, 14, 15\}$$

$$C = \{11, 12, 13, 14, 15\}$$

99. $A \cup B$

100. $A \cap B$

101. $B \cup C$

102. $B \cap C$

103. $A \cap C$

104. $A \cup C$

105. If $A = \{\text{even integers}\}$ and $B = \{\text{whole numbers less than 11}\}$, find $A \cap B$.

106. If $X = \{48, 49, 50, \dots\}$ and $Y = \{60, 62, 64, \dots, 80\}$, find $X \cap Y$.

When listing the possible subsets of a set, it is important to be orderly. Think of a pattern and then answer Problems 107 and 108.

107. Use the set $Z = \{1, 2, 3, 4\}$.

(a) List all possible subsets of set Z. *Hint:* The empty set is a subset of every set.

(b) How many subsets did you find?

108. Use the set $M = \{a, b, c\}$.

(a) List all possible subsets of M. *Hint:* The empty set is a subset of every set.

(b) How many subsets did you find?

Explaining the Concepts

109. Write a definition of "rational number" in your own words. Describe the characteristics to look for when deciding whether a number is in this set.

110. Write a definition of "irrational number" in your own words. Describe the characteristics to look for when deciding whether a number is in this set.

1.4 Adding, Subtracting, Multiplying, and Dividing Integers

Objectives

1 Add Integers

2 Determine the Additive Inverse of a Number

3 Subtract Integers

4 Multiply Integers

5 Divide Integers

Are You Prepared for This Section?

Before getting started, complete the following problems. If you get a problem wrong, go back to the section cited and review the material.

P1. Write $\dfrac{16}{36}$ as a fraction in lowest terms. [Section 1.2, pp. 12–13]

P2. Evaluate: $|-5|$ [Section 1.3, pp. 24–25]

In this section, addition, subtraction, multiplication, and division, called **operations**, are performed on integers. The symbols used in algebra for these operations are $+$, $-$, $\cdot$, and /, respectively. The results of these four operations are called the **sum, difference, product**, and **quotient**, respectively. Table 2 summarizes these ideas.

Work Smart

Don't use x for multiplication to avoid confusion with the variable x.

Table 2

Operation	Symbols	Words
Addition	$a + b$	Sum: a plus b
Subtraction	$a - b$	Difference: a minus b
Multiplication	$a \cdot b, ab, (a)b, a(b), (a)(b)$	Product: a times b
Division	$a/b, \dfrac{a}{b}, a \div b$	Quotient: a divided by b

Prepared?...Answers

P1. $\dfrac{4}{9}$ **P2.** 5

In algebra, avoid using the multiplication sign $\times$ used in arithmetic. Instead, multiply two expressions that are placed next to each other without an operation symbol, as in ab or that are in parentheses, as in $(a)(b)$, or use $\cdot$, as in $a \cdot b$.

Work Smart

The fraction $\frac{b}{c}$ is *improper* if $b \geq c$ and *proper* if $b < c$. In general, write the mixed number $a\frac{b}{c}$ as an improper fraction as follows:

$$a\frac{b}{c} = \frac{ac + b}{c}$$

A **mixed number** is a whole number followed by a fraction. We prefer not to use mixed numbers in algebra. When mixed numbers are used, addition is understood—for example, $3\frac{2}{5}$ means $3 + \frac{2}{5}$. But in algebra, use of a mixed number may be confusing because the absence of an operation symbol between two numbers is taken to mean multiplication. To avoid this confusion, $3\frac{2}{5}$ is written as an improper fraction.

$$
\begin{aligned}
3\frac{2}{5} &= 3 + \frac{2}{5} \\
&= \frac{3}{1} \cdot \frac{5}{5} + \frac{2}{5} \\
&= \frac{15}{5} + \frac{2}{5} \\
&= \frac{17}{5}
\end{aligned}
$$

Thus, $3\frac{2}{5}$ is written as $\frac{17}{5}$. Equivalently, $3\frac{2}{5}$ may be written as 3.4. You may, however, see mixed numbers used to represent measurements, such as $2\frac{3}{4}$ gallons or $5\frac{1}{2}$ hours.

▶ ❶ Add Integers

Adding Integers with the Same Sign Using a Number Line

We will use a real number line to discover a pattern for adding integers. When adding a positive integer to a given integer, move to the right on the number line, and when adding a negative integer to a given integer, move to the left on the number line.

Remember, the *sign* of a number indicates whether the number is positive or negative. For example, the sign of 4 is positive, while the sign of -12 is negative. We will first consider adding integers with the same sign.

EXAMPLE 1 | **Adding Two Positive Integers Using a Number Line**

Find the sum: $5 + 3$

Solution
Begin at 5 on the number line and move 3 spaces to the right, so $5 + 3 = 8$. See Figure 13.

Figure 13

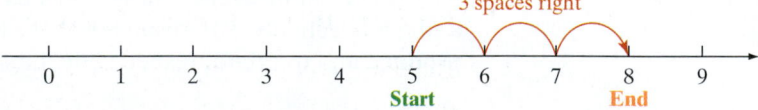

EXAMPLE 2 | **Adding Two Negative Integers Using a Number Line**

Find the sum: $-7 + (-4)$

Solution
Begin at -7 on the number line and move 4 spaces to the left, so $-7 + (-4) = -11$. See Figure 14.

Figure 14

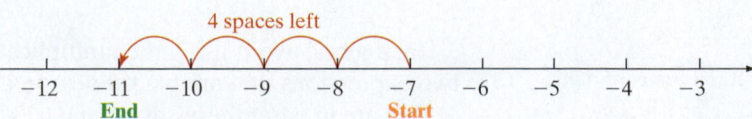

Adding Integers with Different Signs Using a Number Line

Consider the sum of two integers with different signs.

EXAMPLE 3 **Adding Integers with Different Signs Using a Number Line**

Find the sum: $-5 + 3$

Solution
Begin at -5 and then move 3 spaces to the right. From Figure 15 it can be seen that $-5 + 3 = -2$.

Figure 15

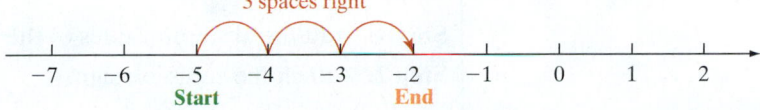

EXAMPLE 4 **Adding Integers with Different Signs Using a Number Line**

Find the sum: $7 + (-4)$

Solution
Begin at 7 and move 4 spaces to the left. So $7 + (-4) = 3$. See Figure 16.

Figure 16

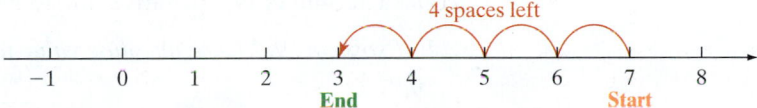

Adding Integers Using Absolute Value

Did you discover a pattern for adding integers from Examples 1–4? To add integers with the same sign (both positive or both negative), add the absolute values of the integers and attach the common sign. To add integers with different signs (one positive and one negative), subtract the smaller absolute value from the larger absolute value and attach the sign of the integer having the larger absolute value.

EXAMPLE 5 **How to Add Integers with the Same Sign Using Absolute Value**

Find $-16 + (-24)$ using absolute value.

Step-by-Step Solution

Step 1: Add the absolute values of the two integers.

$$|-16| = 16 \text{ and } |-24| = 24$$
$$16 + 24 = 40$$

Step 2: Attach the common sign, either positive or negative.

Both integers are negative in the original problem, so
$$-16 + (-24) = -40$$

EXAMPLE 6   How to Add Integers with Different Signs Using Absolute Value

Find $-31 + 16$ using absolute value.

Step-by-Step Solution

Step 1: Subtract the smaller absolute value from the larger absolute value.

$|-31| = 31$ and $|16| = 16$. The smaller absolute value is 16, so compute

$$31 - 16 = 15$$

Step 2: Attach the sign of the integer with the larger absolute value.

The larger absolute value is 31, which was negative in the original problem, so the sum is negative. Therefore,

$$-31 + 16 = -15$$ ●

Adding Two Integers Using Absolute Value

To add integers with the same sign (both positive or both negative),

Step 1: Add the absolute values of the two integers.

Step 2: Attach the common sign.

To add integers with different signs (one positive and one negative),

Step 1: Subtract the smaller absolute value from the larger absolute value.

Step 2: Attach the sign of the integer with the larger absolute value.

Quick ✓

8. The sum of two negative integers will be _____.

In Problems 9–12, use absolute value to find each sum.

9. $-11 + 7$ **10.** $5 + (-8)$

11. $-8 + (-16)$ **12.** $94 + (-38)$

▶ ❷ **Determine the Additive Inverse of a Number**

What is $3 + (-3)$? What is $10 + (-10)$? What is $-143 + 143$? The answer to all three of these questions is 0. These results are true in general.

Work Smart

The additive inverse of a, $-a$, should not be called *negative a* because it suggests that the opposite is a negative number, which may not be true! For example, the additive inverse of -11 is 11, a positive number.

Additive Inverse Property

For any real number a other than 0, there is a real number $-a$, called the **additive inverse**, or **opposite**, of a, having the following property.

$$a + (-a) = -a + a = 0$$

Any two numbers whose sum is zero are additive inverses, or opposites, of each other.

EXAMPLE 7   Finding an Additive Inverse

    (a) The additive inverse of 9 is -9 because $9 + (-9) = 0$.

    (b) The additive inverse of -12 is $-(-12) = 12$ because $-12 + 12 = 0$. ●

Notice from Example 7(b) that $-(-a) = a$ for any real number a.

Quick ✓

13. For any real number a other than 0, there is a real number $-a$, called the
_____ _____, or _____, of a such that $a + (-a) = -a + a = \_$.

In Problems 14–17, determine the additive inverse of the given real number.

14. 7   **15.** -21 **16.** $-\dfrac{8}{5}$ **17.** 5.75

▶ ❸ Subtract Integers

To subtract integers, rewrite the subtraction problem as an addition problem and use the addition rules.

From arithmetic, we write $10 - 6 = 4$. Using the additive inverse, $10 - 6 = 4$ can be written as the addition problem $10 + (-6) = 4$.

> **In Other Words**
> To subtract b from a, add the opposite of b to a.

> **Definition**
> The **difference** $a - b$, read "a minus b" or "a less b," is defined as
> $$a - b = a + (-b)$$

EXAMPLE 8   **How to Subtract Integers**

Compute the difference: $-18 - (-40)$

Step-by-Step Solution

Step 1: Change the subtraction problem to an equivalent addition problem.

$$-18 - (-40) = -18 + 40$$

Step 2: Find the sum.

$$= 22 \quad ●$$

> **In Other Words**
> The problem in Example 8 is read "negative eighteen minus negative 40" not "negative 18 minus minus 40."

> **Subtracting Nonzero Integers**
> 1. Change the subtraction problem to an equivalent addition problem using $a - b = a + (-b)$.
> 2. Find the sum.

Quick ✓

18. The answer to a subtraction problem is called the _____.

19. The subtraction problem $-3 - 10$ is equivalent to $-3 +$ _____.

In Problems 20–22, find the value of each expression.

20. $59 - (-21)$ **21.** $-32 - 146$ **22.** $-19 - (-40)$

> **Work Smart**
> Do you see most of the word **value** in **evaluate**?

For the remainder of this text, to **evaluate** will mean to find the numerical value of an expression. To evaluate an expression that has both addition and subtraction, change all subtraction to addition. Then add from left to right.

EXAMPLE 9   **Evaluating an Expression with Three Integers**

Evaluate: $10 - 18 + 25$

Solution

> **Work Smart**
> When adding and subtracting more than two numbers, change all subtraction to addition, and then add.

$$10 - 18 + 25 = 10 + (-18) + 25$$

Add from left to right: $= -8 + 25$

$$= 17 \quad ●$$

EXAMPLE 10 **Evaluating an Expression with Four Integers**

Evaluate: $162 - (-46) + 80 - 274$

Solution

$$162 - (-46) + 80 - 274 = 162 + 46 + 80 + (-274)$$

$$\text{Add from left to right: } = 208 + 80 + (-274)$$

$$= 288 + (-274)$$

$$= 14$$

●

> **Quick ✓**
>
> *In Problems 23–26, evaluate each expression.*
>
> **23.** $8 - 13 + 5$ **24.** $-27 - 49 + 18$
>
> **25.** $3 - (-14) - 8 + 3$ **26.** $-825 + 375 - (-735) + 265$

▶ ④ **Multiply Integers**

In the statement $9 \cdot 2 = 18$, 9 and 2 are the **factors** and 18 is the **product.**

$$\underset{\text{factor}}{9} \quad \cdot \quad \underset{\text{factor}}{2} \quad = \underset{\text{product}}{18}$$

Recall that multiplication is repeated addition.

$$3 \cdot 4 = \underbrace{4 + 4 + 4}_{\text{Add 4 three times}} = 12$$

Notice that the product of two positive factors is positive. We knew that from arithmetic! But what is $3 \cdot (-4)$?

$$3 \cdot (-4) = \underbrace{-4 + (-4) + (-4)}_{\text{Add } -4 \text{ three times}} = -12$$

We conclude that the product of two real numbers with *different signs* is negative.

What about the product of two negative numbers? Consider the pattern below.

$$-4 \cdot 3 = -12$$
$$-4 \cdot 2 = -8$$
$$-4 \cdot 1 = -4$$
$$-4 \cdot 0 = 0$$

Each time the second factor decreases by 1, the product increases by 4. Assuming this pattern continues,

$$-4 \cdot (-1) = 4$$
$$-4 \cdot (-2) = 8$$

The pattern suggests that the product of two negative numbers is a positive number.

In Other Words

The product is *positive* if the signs of the two factors are the *same* and *negative* if the signs of the two factors are *different.*

> **Rules of Signs for Multiplying Two Integers**
>
> **1.** The product of two positive integers is positive.
>
> **2.** The product of one positive integer and one negative integer is negative.
>
> **3.** The product of two negative integers is positive.

EXAMPLE 11 **Multiplying Integers**

(a) $2(-4) = -8$ (b) $-6(5) = -30$

(c) $(-7)(-8) = 56$ (d) $-25(18) = -450$ ●

Quick ✓

27. The product of two integers with the same sign is _____.

In Problems 28–32, find the product.

28. $-3(7)$ 29. $13(-4)$ 30. $5 \cdot 16$

31. $-9(-12)$ 32. $(-13)(-25)$

Find the Product of Several Integers

To find the product of more than two integers, multiply in order, from left to right.

EXAMPLE 12 **Multiplying Three or More Integers**

Find the product:

(a) $3(-4)(-7)$ (b) $-8(-1)(4)(-5)$

Solution

To find each product, multiply from left to right.

Work Smart

The product of an *even* number of negative factors is *positive*. The product of an *odd* number of negative factors is *negative*.

(a) $3(-4)(-7) = -12(-7)$
$= 84$

(b) $-8(-1)(4)(-5) = 8(4)(-5)$
$= 32(-5)$
$= -160$ ●

Quick ✓

33. *True or False* The product of thirteen negative factors is negative.

In Problems 34 and 35, find the product.

34. $-3 \cdot 9(-4)$ 35. $3(-4)(-5)(-6)$

▶ ❺ Divide Integers

In division, the numerator is the **dividend**, the denominator is the **divisor**, and the answer is the **quotient.**

$$\text{dividend} \rightarrow \underset{\text{divisor} \rightarrow}{\frac{28}{7}} = 4 \leftarrow \text{quotient} \quad \text{or} \quad \underset{\text{dividend}}{28} \quad \div \quad \underset{\text{divisor}}{7} \quad = \quad \underset{\text{quotient}}{4}$$

To discuss division of integers, we need to introduce the idea of a *multiplicative inverse*.

Multiplicative Inverse (Reciprocal) Property

For each *nonzero* real number a, there is a real number $\frac{1}{a}$, called the **multiplicative inverse**, or **reciprocal**, of a, having the following property:

$$a \cdot \frac{1}{a} = \frac{1}{a} \cdot a = 1 \quad \text{where } a \neq 0$$

In Other Words

Any two numbers whose product is 1 are called multiplicative inverses, or reciprocals, of each other.

EXAMPLE 13 ## Finding the Multiplicative Inverse (or Reciprocal) of an Integer

(a) The multiplicative inverse, or reciprocal, of 5 is $\dfrac{1}{5}$.

(b) The multiplicative inverse, or reciprocal, of -8 is $-\dfrac{1}{8}$. ●

Quick ✓

In Problems 36 and 37, find the multiplicative inverse, or reciprocal, of each integer.

36. 6   **37.** -2

Work Smart

Don't confuse the multiplicative inverse (reciprocal) with the additive inverse (opposite). The reciprocal of a number does not change the sign of the number. It only reverses the numbers in the numerator and denominator.

The multiplicative inverse is used to define division.

Definition

If b is a nonzero integer, the **quotient** $\dfrac{a}{b}$, read as "a divided by b" or "the **ratio** of a to b," is defined as

$$\frac{a}{b} = a \div b = a \cdot \frac{1}{b}$$

For example, $\dfrac{40}{8} = 40 \div 8 = 40 \cdot \dfrac{1}{8}$ and $\dfrac{12}{7} = 12 \div 7 = 12 \cdot \dfrac{1}{7}$. Because division can be represented as multiplication, the same rules of signs that apply to multiplication also apply to division.

In Other Words

These are the same rules as those used to multiply integers. A positive divided by a positive is positive, a positive divided by a negative is negative, and so on.

Rules of Signs for Dividing Two Integers

1. The quotient of two positive integers is positive.

2. The quotient of one positive integer and one negative integer is negative.

3. The quotient of two negative integers is positive.

Finding the quotient of two integers is the same as writing a fraction in lowest terms.

EXAMPLE 14 ## Finding the Quotient of Two Integers

Find each quotient:

(a) $\dfrac{-180}{20}$ (b) $200 \div (-5)$

Solution

(a)

$$\frac{-180}{20} = \frac{20 \cdot -9}{20 \cdot 1}$$

Divide out the common factor:
$$= \frac{\cancel{20} \cdot -9}{\cancel{20} \cdot 1}$$

$$= \frac{-9}{1}$$

$$\frac{-a}{b} = -\frac{a}{b}. \qquad\qquad = -9$$

(b)

$$200 \div (-5) = \frac{200}{-5}$$

$$= \frac{40 \cdot 5}{-1 \cdot 5}$$

Divide out the common factor: $\qquad = \frac{40 \cdot \cancel{5}}{-1 \cdot \cancel{5}}$

$$= -40$$

Quick ✓

38. In the division problem $\frac{18}{3} = 6$, 18 is called the _____, 3 is called the _____, and 6 is called the _____.

39. *True or False:* The quotient of two negative numbers is positive.

In Problems 40–42, find the quotient.

40. $\dfrac{20}{-4}$ **41.** $\dfrac{-72}{20}$ **42.** $-63 \div -7$

1.4 Exercises MyMathLab®

Exercise numbers in green have complete video solutions in MyMathLab or may be accessed using the QR code to the right.

*Problems **1–42** are the Quick ✓s that follow the **EXAMPLES**.*

Building Skills

In Problems 43–58, find the sum. See Objective 1.

43. $8 + 7$ **44.** $6 + 4$

45. $-5 + 9$ **46.** $-4 + 12$

47. $8 + (-12)$ **48.** $7 + (-13)$

49. $-11 + (-8)$ **50.** $-13 + (-5)$

51. $-16 + 37$ **52.** $-32 + 49$

53. $-119 + (-209)$ **54.** $-145 + (-68)$

55. $-14 + 21 + (-18)$

56. $(-13) + 37 + (-22)$

57. $74 + (-13) + (-23) + 5$

58. $-34 + 46 + (-12) + 72$

In Problems 59–62, determine the additive inverse of each real number. See Objective 2.

59. -325 **60.** -34

61. 125 **62.** 7

In Problems 63–78, find the difference. See Objective 3.

63. $23 - 12$ **64.** $35 - 23$

65. $9 - 17$ **66.** $12 - 19$

67. $-20 - 8$ **68.** $-15 - 9$

69. $13 - (-41)$ **70.** $14 - (-18)$

71. $-36 - (-36)$ **72.** $-15 - (-15)$

73. $0 - 41$ **74.** $0 - 18$

75. $-93 - (-62)$ **76.** $46 - (-25)$

77. $86 - (-86)$ **78.** $49 - (-49)$

In Problems 79–94, find the product. See Objective 4.

79. $5 \cdot 8$ **80.** $7 \cdot 9$

81. $8(-7)$ **82.** $9(-7)$

83. $(0)(-21)$ **84.** $-21 \cdot 0$

85. $(-48)(-3)$ **86.** $(-22)(-5)$

87. $(-42)3$ **88.** $(-128)7$

89. $-5 \cdot 6 \cdot 3$ **90.** $-6 \cdot 4 \cdot 8$

91. $-10(3)(-7)$ **92.** $-8(2)(-9)$

93. $(-2)(4)(-1)(3)(5)$

94. $(-3)(-4)(6)(-1)$

In Problems 95–100, find the multiplicative inverse (or reciprocal) of each number. See Objective 5.

95. 8 **96.** 10 **97.** -4

98. -3 **99.** 1 **100.** 2

In Problems 101–112, find the quotient. See Objective 5.

101. $10 \div 2$

102. $36 \div 9$

103. $\dfrac{-56}{-8}$

104. $\dfrac{-63}{-7}$

105. $\dfrac{-45}{3}$

106. $\dfrac{-144}{6}$

107. $\dfrac{35}{10}$

108. $\dfrac{20}{16}$

109. $\dfrac{60}{-42}$

110. $\dfrac{120}{-66}$

111. $\dfrac{-105}{-12}$

112. $\dfrac{-80}{-12}$

Mixed Practice

In Problems 113–128, evaluate the expression.

113. $-4 \cdot 18$

114. $7(-15)$

115. $-16 - (-76)$

116. $87 - 19$

117. $-9(-19)$

118. $7 \cdot 209$

119. $\dfrac{120}{-8}$

120. $\dfrac{-156}{-26}$

121. $-98 + 56$

122. $103 + (-66)$

123. $\dfrac{75}{|-20|}$

124. $\dfrac{|-42|}{12}$

125. $|-14| + |-26|$

126. $|-10| + (-62)$

127. $|-389| - 627$

128. $|-193| - (-20)$

In Problems 129–140, write each expression using mathematical symbols. Then evaluate the expression.

129. the sum of 28 and -21

130. the sum of 32 and -64

131. -21 minus 47

132. -85 subtracted from -16

133. -12 multiplied by 18

134. 32 multiplied by -8

135. -36 divided by -108

136. -40 divided by 100

137. The difference of -35 and 12.

138. The product of -11 and -5.

139. The quotient of -125 and 5.

140. The difference of -16 and -42.

Applying the Concepts

In Problems 141–146, write the positive or negative number for each amount or measurement.

141. Stock The price of IBM stock fell by 3.25 dollars.

142. Temperature The current temperature in Juneau, Alaska, is 14° below zero.

143. Chargers Football The Chargers lost 6 yards on the play.

144. Profit Mark's auto dealership showed a profit of $125,000 this quarter.

145. Checking Account Leila's checking account is now overdrawn by $48.

146. Census The number of people in Brian's hometown grew by 12,368.

147. Hiking Loren and Richard went on a hiking trip. They walked 8 miles to the base of Snow Creek Falls, where they set up camp. They went another 3 miles to see the Falls and then returned to their campsite. How many miles did they walk that day?

148. Football The Cary High School Eagles took over possession of the ball on their own 15-yard line. The following plays occurred: QB sack, loss of 7 yards; Jon Anderson ran for 14 yards; Juan Ramirez caught a pass for 26 yards. What yard marker is the ball on now?

149. Bank Balance When Martha balanced her checkbook, she had $563 in the account. Then the following transactions occurred: She wrote a check to Home Depot for $46, deposited $233, wrote a check to Vons for $63, and wrote a check to Petco for $32. What is Martha's new balance?

150. Checkbook Balance Josie began the month with $399 in her bank account. She deposited her paycheck of $839. She paid her $69 telephone bill, the electric bill for $78, and rent of $739. How much does she have left for spending money?

151. Warehouse Inventory The warehouse began the month with 725 cases of soda. During the month the following transactions occurred: 120 cases were shipped out, 590 cases were shipped out, and 310 cases were delivered to the warehouse. Does the warehouse have enough stock on hand to fill an order for 450 cases of soda? What is the difference between what it has and what has been requested?

152. Altitude of an Airplane A pilot leveled off his airplane at 35,000 feet at the beginning of the flight. The following adjustments were made during the trip: gained 4290 feet, dropped 10,400 feet, and then dropped 2605 feet. At what altitude is the plane currently flying?

153. Distance An airplane flying at 25,350 feet is directly over a submarine that is 375 feet below sea level. What is the distance between a person in the plane and a person in the submarine?

154. Elevation The highest point in California is Mt. Whitney at an elevation of 14,495 feet, and the lowest point is in Death Valley at 280 feet below sea level. What is the difference in elevation between the two points in California?

Extending the Concepts

155. Use trial and error to find two integers whose sum is -8 and whose product is 15.

156. Use trial and error to find two integers whose sum is 2 and whose product is -24.

157. Use trial and error to find two integers whose sum is -10 and whose product is -24.

158. Use trial and error to find two integers whose sum is -18 and whose product is 45.

159. The Fibonacci Sequence
The numbers in the Fibonacci sequence are $1, 1, 2, 3, 5, 8, 13, 21, 34, 55, \ldots$, where each term after the second term is the sum of the two preceding terms. This famous sequence of numbers can be used to model many phenomena in nature.

(a) Form fractions of consecutive terms in the sequence. Find the decimal approximations to $\dfrac{1}{1}, \dfrac{2}{1}, \dfrac{3}{2}, \dfrac{5}{3}$, and so on.

(b) To the nearest thousandth, what number do the ratios get close to? This number is called the **golden ratio** and has application in many different areas.

(c) Research Fibonacci numbers and cite three different applications.

Explaining the Concepts

160. Write a sentence or two that justifies the fact that the product of a positive number and a negative number is a negative number. You may use an example.

161. Explain how $42 \div 4$ may be written as a multiplication problem.

1.5 Adding, Subtracting, Multiplying, and Dividing Rational Numbers

Objectives

1 Multiply Rational Numbers in Fraction Form

2 Divide Rational Numbers in Fraction Form

3 Add or Subtract Rational Numbers in Fraction Form

4 Add, Subtract, Multiply, or Divide Rational Numbers in Decimal Form

Are You Prepared for This Section?

Before getting started, complete the following problems. If you get a problem wrong, go back to the section cited and review the material.

P1. Find the least common denominator of $\dfrac{5}{12}$ and $\dfrac{3}{16}$.   [Section 1.2, pp. 11–12]

P2. Rewrite $\dfrac{4}{5}$ as an equivalent fraction with a denominator of 30.   [Section 1.2, pp. 10–11]

P3. Write each rational number in lowest terms.   [Section 1.4, pp. 34–35]

   (a) $-\dfrac{18}{30}$ **(b)** $-\dfrac{24}{4}$

Operations on rational numbers are performed using the same properties as those used for integers. In fact, these properties apply to all real numbers. In this section, you will learn how to perform operations on rational numbers expressed as fractions and rational numbers expressed as decimals.

▶ **1** Multiply Rational Numbers in Fraction Form

The following property is used to multiply two rational numbers in fraction form:

> **In Other Words**
> To find the product of two or more fractions, multiply the numerators together. Then multiply the denominators together. Write the fraction in lowest terms, if necessary.

Multiplying Fractions

$$\frac{a}{b} \cdot \frac{c}{d} = \frac{a \cdot c}{b \cdot d} \quad \text{where } b, d \neq 0$$

The rules of signs that apply to integers also apply to rational numbers: The product of two positive rational numbers is positive; the product of a positive rational number and a negative rational number is negative; and the product of two negative rational numbers is positive.

Prepared?...Answers P1. LCD $= 48$
P2. $\dfrac{4}{5} = \dfrac{24}{30}$ **P3. (a)** $-\dfrac{3}{5}$ **(b)** -6

EXAMPLE 1 **Multiplying Rational Numbers (Fractions)**

Find the product: $\dfrac{2}{9}\left(-\dfrac{15}{19}\right)$

Solution

Begin by rewriting the rational number $-\dfrac{15}{19}$ as $\dfrac{-15}{19}$. Then multiply the numerators and multiply the denominators.

$$\dfrac{2}{9}\left(\dfrac{-15}{19}\right) = \dfrac{2(-15)}{9\cdot 19}$$

Work Smart

Notice that multiplication is not performed in the numerator or denominator until common factors have been divided out.

Write the numerator and the denominator as products of prime factors: $= \dfrac{2\cdot 3(-5)}{3\cdot 3\cdot 19}$

Divide out common factors: $= \dfrac{2\cdot \cancel{3}(-5)}{\cancel{3}\cdot 3\cdot 19}$

$$= \dfrac{2(-5)}{3\cdot 19}$$

Multiply; $\dfrac{-a}{b} = -\dfrac{a}{b}$: $= -\dfrac{10}{57}$

●

Quick ✓

In Problems 1–5, find each product, and write in lowest terms.

1. $\dfrac{3}{4}\cdot \dfrac{9}{8}$

2. $\dfrac{-5}{7}\cdot \dfrac{56}{15}$

3. $\dfrac{12}{45}\left(-\dfrac{18}{20}\right)$

4. $-\dfrac{25}{75}\left(-\dfrac{9}{4}\right)$

5. $\dfrac{7}{3}\cdot \dfrac{1}{14}\left(-\dfrac{9}{11}\right)$

▶ ❷ **Divide Rational Numbers in Fraction Form**

To divide rational numbers, we must know how to find the reciprocal of a rational number. In Section 1.4, we saw that two numbers are *reciprocals*, or multiplicative inverses, if their product is 1. This definition applies to any nonzero real number. Thus, $\dfrac{3}{2}$ and $\dfrac{2}{3}$ are reciprocals because $\dfrac{3}{2}\cdot \dfrac{2}{3} = 1$; 9 and $\dfrac{1}{9}$ are reciprocals because $9\cdot \dfrac{1}{9} = 1$; $-\dfrac{4}{7}$ and $-\dfrac{7}{4}$ are reciprocals because $-\dfrac{4}{7}\left(-\dfrac{7}{4}\right) = 1$.

Quick ✓

6. Two numbers are called multiplicative inverses, or reciprocals, if their product is equal to _.

In Problems 7–10, find the reciprocal of each number.

7. 12

8. $\dfrac{7}{5}$

9. $-\dfrac{1}{4}$

10. $-\dfrac{31}{20}$

Divide rational numbers by rewriting the division as an equivalent multiplication problem.

Dividing Rational Numbers Expressed as Fractions

$$\dfrac{a}{b} \div \dfrac{c}{d} = \dfrac{a}{b}\cdot \dfrac{d}{c} = \dfrac{a\cdot d}{b\cdot c} \quad \text{where } b, c, d \neq 0$$

EXAMPLE 2   **How to Divide Rational Numbers (Fractions)**

Find the quotient: $\dfrac{3}{10} \div \dfrac{12}{25}$

Step-by-Step Solution

Step 1: Write the equivalent multiplication problem.

$$\dfrac{3}{10} \div \dfrac{12}{25} = \dfrac{3}{10} \cdot \dfrac{25}{12}$$

Step 2: Write the product in factored form and divide out common factors.

$$= \dfrac{3 \cdot 25}{10 \cdot 12}$$

$$= \dfrac{3 \cdot 5 \cdot 5}{5 \cdot 2 \cdot 4 \cdot 3}$$

$$= \dfrac{\cancel{3} \cdot \cancel{5} \cdot 5}{\cancel{5} \cdot 2 \cdot 4 \cdot \cancel{3}}$$

$$= \dfrac{5}{2 \cdot 4}$$

Step 3: Multiply the remaining factors.

$$= \dfrac{5}{8}$$

Quick ✔

In Problems 11–14, find the quotient.

11. $\dfrac{5}{7} \div \dfrac{7}{10}$

12. $-\dfrac{9}{12} \div \dfrac{14}{7}$

13. $\dfrac{8}{35} \div \left(-\dfrac{1}{10}\right)$

14. $-\dfrac{18}{63} \div \left(-\dfrac{54}{35}\right)$

 ❸ **Add or Subtract Rational Numbers in Fraction Form**

Add or Subtract Rational Numbers (Fractions) with Like Denominators

Figure 17

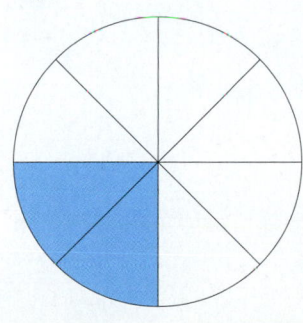

In Figure 17, two of the eight equal regions are shaded. Each of the regions represents the fraction $\dfrac{1}{8}$. Together, the two shaded regions make up $\dfrac{1}{4}$ of the circle. This implies that

$$\dfrac{1}{8} + \dfrac{1}{8} = \dfrac{1 + 1}{8}$$

$$= \dfrac{2}{8}$$

$$= \dfrac{1}{4}$$

Or, suppose Bobby has $0.25 and his grandma gives him $0.50. He now has $0.25 + $0.50 = $0.75, or $\dfrac{3}{4}$ of a dollar. Because $0.25 = \dfrac{1}{4}$ and $0.50 = \dfrac{1}{2}$, we can determine Bobby's good fortune using fractions:

$$\dfrac{1}{4} + \dfrac{1}{2} = \dfrac{1}{4} + \dfrac{2}{4}$$

$$= \dfrac{3}{4}$$

Based on these results, we conclude that to add fractions with the same denominators, add the numerators and write the result over the common denominator. Also, because any subtraction problem can be written as an equivalent addition problem, the following methods for adding or subtracting rational numbers are given.

In Other Words
To add or subtract fractions with a common denominator, add or subtract the numerators and retain the denominator.

Adding or Subtracting Rational Numbers (Fractions) with the Same Denominator

$$\frac{a}{c} + \frac{b}{c} = \frac{a+b}{c} \quad \text{where } c \neq 0 \qquad \frac{a}{c} - \frac{b}{c} = \frac{a-b}{c} = \frac{a+(-b)}{c} \quad \text{where } c \neq 0$$

EXAMPLE 3   **Adding Rational Numbers (Fractions) with the Same Denominator**

Find the sum and write in lowest terms: $-\dfrac{1}{8} + \dfrac{3}{8}$

Solution

$$-\frac{1}{8} + \frac{3}{8} = \frac{-1}{8} + \frac{3}{8}$$

Write the numerators as a sum over the common denominator: $\quad = \dfrac{-1+3}{8}$

Add the numerators: $\quad = \dfrac{2}{8}$

Factor 8 and divide out the 2s: $\quad = \dfrac{1 \cdot \cancel{2}}{4 \cdot \cancel{2}}$

$$= \frac{1}{4}$$

EXAMPLE 4   **Subtracting Rational Numbers (Fractions) with the Same Denominator**

Find the difference and write in lowest terms: $\dfrac{9}{16} - \dfrac{3}{16}$

Solution

$$\frac{9}{16} - \frac{3}{16} = \frac{9-3}{16}$$

$$= \frac{6}{16}$$

Factor 6, factor 16, and divide out the 2s: $\quad = \dfrac{3 \cdot \cancel{2}}{8 \cdot \cancel{2}}$

$$= \frac{3}{8}$$

Quick ✓

15. $-\dfrac{5}{7} + \dfrac{3}{7} = \dfrac{\underline{} + 3}{\underline{}}$

In Problems 16–19, find the sum or difference, and write in lowest terms.

16. $\dfrac{8}{11} + \dfrac{2}{11}$ **17.** $-\dfrac{18}{35} + \dfrac{3}{35}$

18. $\dfrac{19}{63} - \dfrac{10}{63}$ **19.** $-\dfrac{9}{10} - \dfrac{3}{10}$

Work Smart: Study Skills

Note the title of Example 5: "How to Add Rational Numbers with Unlike Denominators." This three-column example provides a guided, step-by-step approach to solving a problem so you can see each of the steps. Cover the third column and try to work the example yourself. Then look at the entries in the third column and check your solution. Was it correct?

▶ Adding or Subtracting Rational Numbers (Fractions) with Unlike Denominators

How do we add rational numbers with different denominators? First find the least common denominator of the two rational numbers. Recall from Section 1.2 that the **least common denominator (LCD)** is the smallest number that each denominator has as a common multiple.

EXAMPLE 5   **How to Add Rational Numbers (Fractions) with Unlike Denominators**

Find the sum: $\dfrac{5}{6} + \dfrac{3}{8}$

Step-by-Step Solution

Step 1: Find the least common denominator of the fractions.

Write each denominator as the product of prime factors, arranging common factors vertically:

$$6 = 2 \qquad \cdot 3$$
$$8 = 2 \cdot 2 \cdot 2$$

List the common factors; list the remaining factors; multiply:

$$\text{LCD} = 2 \cdot 2 \cdot 2 \cdot 3$$
$$= 24$$

Step 2: Write each rational number with the denominator found in Step 1.

Use $1 = \dfrac{4}{4}$ to change the denominator 6 to 24, and use $1 = \dfrac{3}{3}$ to change the denominator 8 to 24:

$$\frac{5}{6} + \frac{3}{8} = \frac{5}{6} \cdot \frac{4}{4} + \frac{3}{8} \cdot \frac{3}{3}$$
$$= \frac{20}{24} + \frac{9}{24}$$

Step 3: Add the numerators and write the result over the common denominator.

$$= \frac{20 + 9}{24}$$
$$= \frac{29}{24}$$

Step 4: Write in lowest terms.

The rational number is in lowest terms, so $\dfrac{5}{6} + \dfrac{3}{8} = \dfrac{29}{24}$. ●

EXAMPLE 6   **How to Subtract Rational Numbers (Fractions) with Unlike Denominators**

Find the difference: $-\dfrac{9}{14} - \dfrac{1}{6}$

Step-by-Step Solution

Step 1: Find the least common denominator of the fractions.

Write each denominator as the product of prime factors, aligning common factors vertically:

$$14 = 2 \cdot 7$$
$$6 = 2 \quad \cdot 3$$

List the common factors; list the remaining factors; multiply:

$$\text{LCD} = 2 \cdot 7 \cdot 3$$
$$= 42$$

Step 2: Write each rational number with the denominator found in Step 1.

Use $1 = \dfrac{3}{3}$ to change the denominator 14 to 42, and use $1 = \dfrac{7}{7}$ to change the denominator 6 to 42:

$$-\frac{9}{14} - \frac{1}{6} = \frac{-9}{14} \cdot \frac{3}{3} - \frac{1}{6} \cdot \frac{7}{7}$$
$$= \frac{-27}{42} - \frac{7}{42}$$

(continued)

Step 3: Subtract the numerators and write the result over the common denominator.

$$= \frac{-27 - 7}{42}$$

$$= \frac{-34}{42}$$

Step 4: Write in lowest terms.

Factor -34 and 42 and divide out common factors:

$$= \frac{2 \cdot (-17)}{2 \cdot 21}$$

$$= -\frac{17}{21}$$ ●

Adding or Subtracting Rational Numbers (Fractions) with Unlike Denominators

Step 1: Find the LCD of the rational numbers.

Step 2: Write each rational number with the LCD.

Step 3: Add or subtract the numerators and write the result over the common denominator.

Step 4: Write the result in lowest terms.

Quick ✔

In Problems 20–23, find each sum or difference, and write in lowest terms.

20. $\dfrac{3}{14} + \dfrac{10}{21}$

21. $\dfrac{5}{12} - \dfrac{5}{18}$

22. $-\dfrac{23}{6} + \dfrac{7}{12}$

23. $-\dfrac{1}{15} - \dfrac{1}{12}$

Remember, the direction "evaluate" means to find the value of the expression.

EXAMPLE 7 **Evaluating an Expression Containing Rational Numbers**

Evaluate and write in lowest terms: $4 - \dfrac{2}{3}$

Solution

The key is to remember that $4 = \dfrac{4}{1}$.

$$4 - \frac{2}{3} = \frac{4}{1} - \frac{2}{3}$$

Multiply by $1 = \dfrac{3}{3}$ to change the denominator 1 to 3:

$$= \frac{4}{1} \cdot \frac{3}{3} - \frac{2}{3}$$

$$= \frac{12}{3} - \frac{2}{3}$$

$$= \frac{10}{3}$$ ●

Quick ✔

In Problems 24 and 25, evaluate and write in lowest terms.

24. $-2 + \dfrac{7}{16}$

25. $6 - \dfrac{9}{4}$

④ Add, Subtract, Multiply, or Divide Rational Numbers in Decimal Form

▶ Adding or Subtracting Decimals

To add or subtract decimals, arrange the numbers in a column with the decimal points aligned. Then add or subtract the digits in the like place values. Put the decimal point in the answer directly below the decimal points in the problem.

EXAMPLE 8   **Adding or Subtracting Decimals That Have the Same Sign**

Evaluate each expression:

   (a) $2.93 + 7.2 + 3.026$ **(b)** $76.4 - 4.95$

Solution

  (a) Use zeros as placeholders:

$$\begin{array}{r} 1 \\ 2.930 \\ 7.200 \\ +3.026 \\ \hline 13.156 \end{array}$$

Work Smart

A whole number has an implied decimal point. For example,

$$74 = 74.000$$

  (b) Use zero as a placeholder:

$$\begin{array}{r} {}^{5\,13\,10} \\ 76.4\cancel{0} \\ -4.95 \\ \hline 71.45 \end{array}$$

EXAMPLE 9   **Adding or Subtracting Decimals That Have Different Signs**

Evaluate each expression:

   (a) $100.32 - (-32.015)$ **(b)** $-23.03 + 18.49$

Solution

  (a) Remember, $a - (-b) = a + b$, so $100.32 - (-32.015) = 100.32 + 32.015$.

$$\begin{array}{r} 100.320 \\ +32.015 \\ \hline 132.335 \end{array}$$

  (b) Recall that to add real numbers with different signs, subtract the smaller absolute value from the larger absolute value and attach the sign of the number with the larger absolute value. Because $|-23.03| = 23.03$ and $|18.49| = 18.49$, compute $23.03 - 18.49$ and attach a negative sign to the difference.

$$\begin{array}{r} 23.03 \\ -18.49 \\ \hline 4.54 \end{array}$$

So, $-23.03 + 18.49 = -4.54$.

Quick ✔

In Problems 26–29, find the sum or difference.

26. $9.67 - (-11.344)$ **27.** $-17.39 + 81.96$

28. $-74.28 + 14.832$ **29.** $-180.782 + 100.3 + 9.07$

▶ Multiplying Decimals

The rules for multiplying decimals come from the rules for multiplying rational numbers in fraction form. For example,

$$\underbrace{-0.7}_{\substack{\text{1 decimal} \\ \text{place}}} \cdot \underbrace{0.03}_{\substack{\text{2 decimal} \\ \text{places}}} = \frac{-7}{10} \cdot \frac{3}{100} = \frac{-21}{1000} = \underbrace{-0.021}_{\substack{\text{3 decimal} \\ \text{places}}}$$

Notice the number of digits to the right of the decimal point in the answer is equal to the sum of the numbers of digits to the right of each decimal point in the factors.

The rules of signs that were presented in Section 1.4 apply to decimals as well.

EXAMPLE 10 **Multiplying Decimals**

Find the product:

    (a) $3.43 \cdot 2.6$ **(b)** $-3.17 \cdot 0.02$

Solution

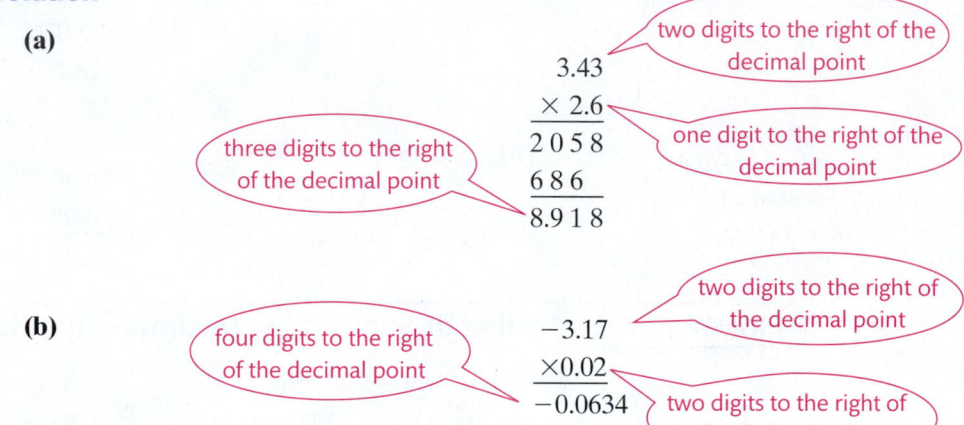

(a)

three digits to the right of the decimal point

$$\begin{array}{r} 3.43 \quad \text{two digits to the right of the decimal point} \\ \times\, 2.6 \quad \text{one digit to the right of the decimal point} \\ \hline 2\,0\,5\,8 \\ 6\,8\,6 \\ \hline 8.9\,1\,8 \end{array}$$

(b)

four digits to the right of the decimal point

$$\begin{array}{r} -3.17 \quad \text{two digits to the right of the decimal point} \\ \times\,0.02 \\ \hline -0.0634 \quad \text{two digits to the right of the decimal point} \end{array}$$

Notice that one zero was added to the right of the decimal point to get the required number of decimal places.

> **Multiplying Decimals**
>
> **Step 1:** Multiply the factors as if they were whole numbers.
>
> **Step 2:** Place the decimal point so the number of digits to the right of the decimal point in the product equals the *sum* of the numbers of digits to the right of each decimal point in the factors.

Quick ✓

In Problems 30–33, find the product.

30. $23.9 \cdot 0.2$ **31.** $257 \cdot (-3.5)$

32. $-3.45 \cdot 0.03$ **33.** $-0.03 \cdot (-0.45)$

▶ Dividing Decimals

In the division problem $2\overline{)7.94}$ (with quotient 3.97), the number 2 is the *divisor*, 7.94 is the *dividend*, and 3.97 is the *quotient*. Notice that when dividing by a whole number, the decimal points in the quotient and the dividend are lined up. In algebra, this division problem is typically written as $\dfrac{7.94}{2} = 3.97$.

In order for us to divide decimals, the divisor must be a whole number. Therefore, multiply the dividend and the divisor by a power of 10 that will make the divisor a whole number. Then divide.

EXAMPLE 11   **Dividing Decimals**

(a) Divide: $\dfrac{22.26}{15.9}$

(b) Divide: $\dfrac{0.03724}{-0.38}$

Solution

(a) Since the divisor 15.9 is fifteen and nine-tenths, we multiply $\dfrac{22.26}{15.9}$ by $\dfrac{10}{10}$ to make the divisor a whole number and obtain $\dfrac{22.26}{15.9} \cdot \dfrac{10}{10} = \dfrac{222.6}{159}$. Now divide.

$$
\begin{array}{r}
1.4 \\
159\overline{)222.6} \\
\underline{159} \\
63\,6 \\
\underline{63\,6} \\
0
\end{array}
$$

So $\dfrac{22.26}{15.9} = 1.4$.

> **In Other Words**
>
> To divide decimals, convert the divisor to a whole number and divide.

(b) Because we have a positive number divided by a negative number, the quotient will be negative. The divisor 0.38 is thirty-eight hundredths, so

multiply $\dfrac{0.03724}{0.38}$ by $\dfrac{100}{100}$ to make the divisor a whole number and obtain

$\dfrac{0.03724}{0.38} \cdot \dfrac{100}{100} = \dfrac{3.724}{38}$. Now divide.

$$
\begin{array}{r}
0.098 \\
38\overline{)3.724} \\
\underline{3\,42} \\
304 \\
\underline{304} \\
0
\end{array}
$$

Therefore, $\dfrac{0.03724}{-0.38} = -0.098$.

Dividing Decimals

Step 1: Multiply the dividend and divisor by a power of 10 that will make the divisor a whole number.

Step 2: Divide as though working with whole numbers, but be sure to keep track of the decimal point.

Quick ✓

In Problems 34–36, find the quotient.

34. $\dfrac{18.25}{73}$

35. $\dfrac{1.0032}{0.12}$

36. $\dfrac{-4.2958}{45.7}$

1.5 Exercises MyMathLab® Exercise numbers in **green** have complete video solutions in MyMathLab or may be accessed using the QR code to the right.

*Problems **1–36** are the Quick ✔s that follow the EXAMPLES.*

Building Skills

In Problems 37–42, write each rational number in lowest terms.

37. $\dfrac{14}{21}$ **38.** $\dfrac{9}{15}$ **39.** $-\dfrac{38}{18}$

40. $-\dfrac{81}{36}$ **41.** $-\dfrac{22}{44}$ **42.** $-\dfrac{24}{27}$

In Problems 43–52, find the product and write in lowest terms. See Objective 1.

43. $\dfrac{6}{5} \cdot \dfrac{2}{5}$ **44.** $\dfrac{7}{8} \cdot \dfrac{10}{21}$ **45.** $-\dfrac{5}{2} \cdot 10$

46. $-\dfrac{3}{7} \cdot 63$ **47.** $-\dfrac{3}{2} \cdot \dfrac{4}{9}$ **48.** $-\dfrac{5}{2} \cdot \dfrac{16}{25}$

49. $-\dfrac{22}{3}\left(-\dfrac{12}{11}\right)$ **50.** $-\dfrac{60}{75}\left(-\dfrac{25}{4}\right)$

51. $\dfrac{3}{16} \cdot \dfrac{8}{9}$ **52.** $\dfrac{4}{27} \cdot \dfrac{9}{16}$

In Problems 53–56, find the reciprocal of each number. See Objective 2.

53. $\dfrac{3}{5}$ **54.** $\dfrac{9}{4}$

55. -5 **56.** -8

In Problems 57–66, find the quotient and write in lowest terms. See Objective 2.

57. $\dfrac{4}{9} \div \dfrac{8}{15}$ **58.** $\dfrac{3}{2} \div \dfrac{9}{8}$

59. $-\dfrac{1}{3} \div 3$ **60.** $-\dfrac{1}{4} \div 4$

61. $\dfrac{5}{6} \div \left(-\dfrac{5}{4}\right)$ **62.** $\dfrac{4}{3} \div \left(-\dfrac{9}{10}\right)$

63. $\dfrac{2}{5} \div \dfrac{22}{5}$ **64.** $\dfrac{44}{63} \div \dfrac{88}{21}$

65. $-8 \div \left(-\dfrac{1}{4}\right)$ **66.** $-3 \div \left(-\dfrac{1}{6}\right)$

In Problems 67–84, find the sum or difference and write in lowest terms. See Objective 3.

67. $\dfrac{3}{4} + \dfrac{3}{4}$ **68.** $\dfrac{6}{11} + \dfrac{16}{11}$

69. $\dfrac{9}{8} - \dfrac{5}{8}$ **70.** $\dfrac{12}{5} - \dfrac{2}{5}$

71. $\dfrac{6}{7} - \left(-\dfrac{8}{7}\right)$ **72.** $\dfrac{2}{3} - \left(-\dfrac{7}{3}\right)$

73. $-\dfrac{5}{3} + 2$ **74.** $-\dfrac{7}{8} + 4$

75. $6 - \dfrac{7}{2}$ **76.** $3 - \dfrac{5}{3}$

77. $-\dfrac{4}{3} + \dfrac{1}{4}$ **78.** $-\dfrac{2}{5} + \left(-\dfrac{2}{3}\right)$

79. $\dfrac{7}{5} + \left(-\dfrac{23}{20}\right)$ **80.** $\dfrac{7}{15} - \left(-\dfrac{4}{3}\right)$

81. $\dfrac{8}{15} - \dfrac{7}{10}$ **82.** $\dfrac{17}{6} - \dfrac{13}{9}$

83. $-\dfrac{33}{10} - \left(-\dfrac{33}{8}\right)$ **84.** $-\dfrac{29}{6} - \left(-\dfrac{29}{20}\right)$

In Problems 85–102, perform the indicated operation(s). See Objective 4.

85. $-10.5 + 4$ **86.** $-13.2 + 7$

87. $-(-3.5) + 4.9$ **88.** $-(-32.9) + 10.3$

89. $39.1 - (-16.82)$ **90.** $29.23 - (-12.98)$

91. $-5.21 - (-6.7)$ **92.** $-4.94 - (-3.87)$

93. $45 - 2.45$ **94.** $32 - 5.68$

95. $4.3 \cdot 5.8$ **96.** $3.1 \cdot 10.9$

97. $0.075(-120)$ **98.** $0.065(-340)$

99. $\dfrac{136.08}{5.6}$ **100.** $\dfrac{332.59}{7.9}$

101. $\dfrac{-25.48}{0.052}$ **102.** $\dfrac{-48}{0.03}$

Mixed Practice

In Problems 103–134, evaluate and write in lowest terms.

103. $-\dfrac{5}{6} + \dfrac{7}{15}$ **104.** $-\dfrac{8}{9} + \left(-\dfrac{16}{21}\right)$

105. $-\dfrac{10}{21} \cdot \dfrac{14}{5}$ **106.** $\dfrac{24}{5}\left(-\dfrac{35}{4}\right)$

107. $\dfrac{3}{64} \div \left(-\dfrac{9}{16}\right)$ **108.** $-\dfrac{12}{7} \div \left(-\dfrac{4}{21}\right)$

109. $-\dfrac{5}{12} + \dfrac{2}{12}$ **110.** $-\dfrac{4}{9} + \dfrac{1}{9}$

111. $-\dfrac{2}{7} - \dfrac{17}{5}$ **112.** $-\dfrac{3}{4} - \dfrac{1}{5}$

113. $-8.7 - (-10.3)$ **114.** $-4.63 - (-12.9)$

115. $\dfrac{1}{12} + \left(-\dfrac{5}{28}\right)$ **116.** $\dfrac{3}{16} + \left(-\dfrac{7}{40}\right)$

117. $-12.03 \cdot 4.2$ **118.** $34.2(-8.43)$

119. $36\left(-\dfrac{4}{9}\right)$ **120.** $-\dfrac{8}{3} \cdot 15$

121. $-27 \div \dfrac{9}{5}$ **122.** $-24 \div \dfrac{8}{7}$

123. $3.62 - 10.2$ **124.** $4.75 - 6.2$

125. $\dfrac{-145.518}{18.42}$

126. $\dfrac{-297.078}{22.17}$

127. $\dfrac{12}{7} - \dfrac{17}{14} - \dfrac{48}{21}$

128. $\dfrac{9}{4} - \dfrac{21}{6} - \dfrac{11}{8}$

129. $54.2 - 18.78 - (-2.5) + 20.47$

130. $90.3 - 100.9 - (-34.26) + 32.95$

131. $(400)(-25.8)(0.003)$

132. $(500)(-12.4)(-0.02)$

133. $-\dfrac{11}{12} - \left(-\dfrac{1}{6}\right) + \dfrac{7}{8}$

134. $\dfrac{8}{15} - \left(-\dfrac{7}{9}\right) + \dfrac{2}{3}$

Applying the Concepts

135. **Watching TV** If Rachel spends $\dfrac{1}{8}$ of her life watching TV, how many hours of TV does she watch in one week?

136. **Halloween Candy** Henry decided to make $\dfrac{2}{3}$-oz bags of candy for treats at Halloween. If he bought 16 oz of candy, how many bags will he have to give away?

137. **Biology Class** Susan's biology class begins with 36 students. If $\dfrac{2}{3}$ will finish the course and $\dfrac{3}{4}$ of those get a passing grade, how many students will pass Susan's biology class this term?

138. **Pizza Time** Joyce and Ramie bought a pizza. Joyce ate $\dfrac{2}{5}$ of the pizza and Ramie ate $\dfrac{1}{9}$ of what was left. What fraction of the pizza remains uneaten?

139. **Hourly Pay** Last week, Jonathon received a paycheck for $442.80. The withholding for federal, state, and FICA (social security and Medicare) taxes was $97.20. If Jonathon worked 30 hours last week, what is his hourly pay rate?

140. **Average Revenue** Eric and Amelia started an online store selling beauty products. Last year, the total revenue of the company was $29,409.12. What was the average revenue per month?

141. **Stock Prices** The price per share of Intel stock has been up and down lately. On Monday it rose 2.75; on Tuesday it rose 0.87; on Wednesday it dropped 1.12; on Thursday it rose 0.52; and on Friday it fell 0.62. What was the net change in Intel's stock price per share for the week?

142. **Bank Balance** Henry started the month with $43.68 in his checking account. During the month the following transactions occurred: He deposited his paycheck of $929.30; and he wrote checks for rent $650, phone $33.49, credit card $229.50, cable service $75.50, and groceries $159.30. How much does he have in his account now?

Extending the Concepts

Problems 143–146 use the following definition.

*If P and Q are two points on a real number line with coordinates a and b, respectively, then the **distance between P and Q**, denoted by d(P, Q), is*

$$d(P, Q) = |b - a|$$

143. Find the distance between the points P and Q on the real number line if $P = -9.7$ and $Q = 3.5$.

144. Find the distance between the points P and Q on the real number line if $P = -12.5$ and $Q = 2.6$.

145. Find the distance between the points P and Q on the real number line if $P = -\dfrac{13}{3}$ and $Q = \dfrac{7}{5}$.

146. Find the distance between the points P and Q on the real number line if $P = -\dfrac{5}{6}$ and $Q = 4$.

Explaining the Concepts

147. We know that 6 divided by 2 is 3. Explain why 6 divided by $\dfrac{1}{2}$ is 12.

148. Use a figure like Figure 17 on page 39 to explain why $\dfrac{1}{6} + \dfrac{2}{6} = \dfrac{1}{2}$.

Putting the Concepts Together (Sections 1.2–1.5)

We designed these problems so that you can review Sections 1.2–1.5 and show your mastery of the concepts. Take time to work these problems before proceeding with the next section. The answers to these problems are located at the back of the text on page AN-2.

1. Write $\dfrac{7}{8}$ and $\dfrac{9}{20}$ as equivalent fractions with the least common denominator.

2. Write $\dfrac{21}{63}$ as a fraction in lowest terms.

3. Convert $\dfrac{2}{7}$ to a decimal.

4. Write 0.375 as a fraction in lowest terms.

5. Write 12.3% as a decimal.

6. Write 0.0625 as a percent.

7. Use the set $\left\{-12, -\dfrac{14}{7}, -1.25, 0, \sqrt{2}, 3, 11.2\right\}$ to list all of the elements that are:

   (a) integers

   (b) rational numbers

   (c) irrational numbers

   (d) real numbers

8. Replace the ? with the correct symbol $>, <, =: \dfrac{1}{8}$? 0.5

In Problems 9–30, perform the indicated operation and write in lowest terms.

9. $17 + (-28)$

10. $-23 + (-42)$

11. $18 - 45$

12. $3 - (-24)$

13. $-18 - (-12.5)$

14. $(-5)(2)$

15. $25(-4)$

16. $(-8)(-9)$

17. $\dfrac{-35}{7}$

18. $\dfrac{-32}{-2}$

19. $27 \div -3$

20. $-\dfrac{4}{5} - \dfrac{11}{5}$

21. $7 - \dfrac{4}{5}$

22. $\dfrac{7}{12} + \dfrac{5}{18}$

23. $-\dfrac{5}{12} - \dfrac{1}{18}$

24. $\dfrac{6}{25} \cdot 15 \cdot \dfrac{1}{2}$

25. $\dfrac{2}{7} \div (-8)$

26. $\dfrac{0}{-8}$

27. $3.56 - (-7.2)$

28. $18.946 - 11.3$

29. $62.488 \div 42.8$

30. $(7.94)(2.8)$

1.6 Properties of Real Numbers

Objectives

1. Use the Identity Properties of Addition and Multiplication

2. Use the Commutative Properties of Addition and Multiplication

3. Use the Associative Properties of Addition and Multiplication

4. Understand the Multiplication and Division Properties of 0

Are You Prepared for This Section?

Before getting started, complete the following problems. If you get a problem wrong, go back to the section cited and review the material.

P1. Find the sum: $12 + 3 + (-12)$ [Section 1.4, pp. 29–30]

P2. Find the product: $\dfrac{3}{4} \cdot 11 \cdot \dfrac{4}{3}$ [Section 1.5, pp. 37–38]

This section presents properties of real numbers. A property in mathematics is a rule that is always true. These properties will be used throughout this text and in future math courses, so it is extremely important that you understand these properties and know how to use them.

▶ 1 Use the Identity Properties of Addition and Multiplication

The real number 0 is the only number that when added to any real number a results in the same real number a.

Prepared?...Answers
P1. 3   **P2.** 11

In Other Words
The word "identity" comes from a Latin word that means "the same" or "alike." The Identity Property of Addition means adding zero to a real number keeps the number the same.

Identity Property of Addition

For any real number a,

$$0 + a = a + 0 = a$$

That is, the sum of any number and 0 is that number. The number 0 is called the **additive identity**.

Recall that $3 \cdot 5$ is equivalent to adding 5 three times, so $3 \cdot 5 = 5 + 5 + 5$. Therefore, $1 \cdot 5$ means to add 5 once, so $1 \cdot 5 = 5$. This property about the real number 1 is true in general.

Identity Property of Multiplication

For any real number a,

$$a \cdot 1 = 1 \cdot a = a$$

That is, the product of any number and 1 is that number. The number 1 is called the **multiplicative identity**.

Work Smart
$\dfrac{a}{a} = 1$ for $a \neq 0$, and $a \cdot 1 = a$. These properties were used in Sections 1.2 and 1.5 to find equivalent fractions.

The multiplicative identity, used in the last section to change a rational number to one with a different denominator, lets us create expressions that are equivalent to other expressions. For example, the expressions $\dfrac{4}{5}$ and $\dfrac{4}{5} \cdot \dfrac{3}{3}$ are equivalent because $\dfrac{3}{3} = 1$.

Conversion

One use of the multiplicative identity is *conversion*. **Conversion** is changing the units of measure (such as inches or pounds). For example, we might change a length from inches to feet or a weight from pounds to ounces.

EXAMPLE 1   ### Converting from Inches to Feet

Janice measures her family room and finds that its length is 184 inches. How many feet long is Janice's family room? (*Note*: 12 inches = 1 foot)

Solution
When doing conversions, make sure the units of measure you are trying to remove get divided out and the new units of measure remain. In this problem, we want the inches to divide out and feet to remain. Because 12 inches equals 1 foot, multiplying 184 inches by $\dfrac{1 \text{ foot}}{12 \text{ inches}}$ is the same as multiplying by 1.

Work Smart
Just as dividing out identical numbers produces a quotient of 1, dividing out identical units of measure also produces a quotient of 1.

Multiplying by 1

$$184 \text{ inches} = 184 \text{ inches} \cdot \frac{1 \text{ foot}}{12 \text{ inches}}$$

$$= \frac{184}{12} \text{ feet}$$

$184 = 2 \cdot 2 \cdot 2 \cdot 23;\ 12 = 2 \cdot 2 \cdot 3;$
Divide out common factors:
$$= \frac{2 \cdot 2 \cdot 2 \cdot 23}{2 \cdot 2 \cdot 3} \text{ feet}$$

$$= \frac{46}{3} \text{ feet}$$

(continued)

Work Smart

$$
\begin{array}{r}
15 \\
3\overline{)46} \\
-3 \\
\hline
16 \\
-15 \\
\hline
1
\end{array}
$$

So 184 inches equals $\dfrac{46}{3}$ feet. Because 46 divided by 3 is 15 with a remainder of 1 (Why? See the

Work Smart), $\dfrac{46}{3}$ feet is equivalent to $15\dfrac{1}{3}$ feet. Because $\dfrac{1}{3}$ foot $\cdot \dfrac{12 \text{ inches}}{1 \text{ foot}} = 4$ inches,

$\dfrac{46}{3}$ feet is equivalent to $15\dfrac{1}{3}$ feet, or 15 feet, 4 inches. ●

Quick ✓

1. Because $a \cdot 1 = 1 \cdot a = a$ for any real number a, 1 is called the _____ _____.

In Problems 2–4, convert each measurement to the indicated unit of measurement.

2. 96 inches = ? feet [1 foot = 12 inches]

3. 500 minutes = ? hours [60 minutes = 1 hour]

4. 88 ounces = ? pounds [16 ounces = 1 pound]

▶ ❷ **Use the Commutative Properties of Addition and Multiplication**

EXAMPLE 2 **Illustrating the Commutative Properties**

(a) $4 + 7 = 11$ and
$7 + 4 = 11$ so
$4 + 7 = 7 + 4$

(b) $3 \cdot 8 = 24$ and
$8 \cdot 3 = 24$ so
$3 \cdot 8 = 8 \cdot 3$ ●

The results of Example 2 are true in general.

In Other Words

The Commutative Property of real numbers states that the **order** in which we add or multiply real numbers does not affect the final result.

Commutative Properties of Addition and Multiplication

If a and b are real numbers, then

$$a + b = b + a \quad \text{and} \quad a \cdot b = b \cdot a$$

Real numbers are added from left to right. Real numbers are multiplied from left to right. The **Commutative Property** allows us to write $3 + 5$ as $5 + 3$ and write $3 \cdot 5$ as $5 \cdot 3$ without affecting the value of the expression. Why is this important? Rearranging addition or multiplication problems makes some expressions easier to evaluate.

EXAMPLE 3 **Using the Commutative Property of Addition**

Evaluate the expression: $18 + 3 + (-18)$

Solution

$$
\begin{aligned}
18 + 3 + (-18) &= 18 + (-18) + 3 \\
&= 0 + 3 \\
&= 3
\end{aligned}
$$

Adding from left to right results in $18 + 3 + (-18) = 21 + (-18) = 3$. Rearranging the numbers made the problem easier! ●

Does subtraction obey the Commutative Property? In other words, does $3 - 14 = 14 - 3$? Because $3 - 14 = -11$, but $14 - 3 = 11$, notice that **subtraction is not commutative.**

Quick ✔

5. The Commutative Property of Addition states that for any real numbers a and b, $a + b = \_ + \_$.

6. The Commutative Property of Multiplication states that for any real numbers a and b, $\_ \cdot \_ = \_ \cdot \_$.

In Problems 7–9, use the Commutative Property of Addition and the Additive Inverse Property to find the sum of the real numbers.

7. $(-8) + 22 + 8$

8. $\dfrac{8}{15} + \dfrac{3}{20} + \left(-\dfrac{8}{15}\right)$

9. $2.1 + 11.98 + (-2.1)$

EXAMPLE 4   **Using the Commutative Property of Multiplication**

Find each product.

(a) $-27 \cdot 7\left(-\dfrac{2}{9}\right)$ **(b)** $100 \cdot 307.5 \cdot 0.01$

Solution

(a) Rearrange the factors 7 and $\left(-\dfrac{2}{9}\right)$.

$$-27 \cdot 7\left(-\dfrac{2}{9}\right) = -27\left(-\dfrac{2}{9}\right) \cdot 7$$

Divide:
$$= -\overset{3}{27}\left(-\dfrac{2}{\underset{1}{9}}\right) \cdot 7$$

$$= -3(-2) \cdot 7$$

$$= 6 \cdot 7$$

$$= 42$$

(b) Rearrange the factors 307.5 and 0.01.

$$100 \cdot 307.5 \cdot 0.01 = 100 \cdot 0.01 \cdot 307.5$$

$$= 1 \cdot 307.5$$

$$= 307.5$$

Quick ✔

In Problems 10–12, find the product of the real numbers.

10. $\left(-\dfrac{4}{3}\right)(-13)\left(-\dfrac{3}{4}\right)$

11. $\dfrac{5}{22} \cdot \dfrac{18}{331}\left(-\dfrac{44}{5}\right)$

12. $100{,}000 \cdot 349 \cdot 0.00001$

Work Smart

Neither division nor subtraction is commutative.

Does division obey the Commutative Property? That is, does $a \div b = b \div a$? For example, does $8 \div 2 = 2 \div 8$? Because $8 \div 2 = 4$, but $2 \div 8 = \dfrac{2}{8} = \dfrac{1}{4}$, **division is not commutative.**

▶ ❸ **Use the Associative Properties of Addition and Multiplication**

Sometimes **grouping symbols** such as parentheses ( ), brackets [ ], or braces { } are used to indicate that the operation within the grouping symbols is to be performed first. For example, $5(8 + 3)$ means that we should first add 8 and 3 and then multiply this sum by 5.

Earlier we mentioned that addition is performed from left to right. We also stated that multiplication is performed from left to right. But does the order in which we add (or multiply) three or more numbers matter? Let's see.

EXAMPLE 5 **Illustrating the Associative Properties**

(a) $2 + (8 + 6) = 2 + 14 = 16$ and $(2 + 8) + 6 = 10 + 6 = 16$

so $2 + (8 + 6) = (2 + 8) + 6$

(b) $-4 \cdot (9 \cdot 2) = -4 \cdot (18) = -72$ and $(-4 \cdot 9) \cdot 2 = -36 \cdot 2 = -72$

so $-4 \cdot (9 \cdot 2) = (-4 \cdot 9) \cdot 2$ ●

Example 5 illustrates the **Associative Properties of Addition and Multiplication.**

Associative Properties of Addition and Multiplication

If a, b, and c are real numbers, then

$$a + (b + c) = (a + b) + c = a + b + c$$

$$a \cdot (b \cdot c) = (a \cdot b) \cdot c = a \cdot b \cdot c$$

EXAMPLE 6 **Using the Associative Property of Addition**

Use the Associative Property of Addition to evaluate $23 + 453 + (-453)$.

Solution
Because 453 and -453 are additive inverses, use the Associative Property of Addition to insert parentheses around these numbers.

$$23 + 453 + (-453) = 23 + (453 + (-453))$$

$$= 23 + 0$$

$$= 23$$ ●

EXAMPLE 7 **Using the Associative Property of Multiplication**

Use the Associative Property of Multiplication to evaluate $-\dfrac{3}{11} \cdot \dfrac{9}{4} \cdot \dfrac{8}{3}$.

Solution
Because the second and third factors have common factors that can be divided out, use the Associative Property of Multiplication to insert parentheses around $\dfrac{9}{4} \cdot \dfrac{8}{3}$ and perform this operation first.

$$-\frac{3}{11} \cdot \frac{9}{4} \cdot \frac{8}{3} = -\frac{3}{11} \cdot \left(\frac{9}{4} \cdot \frac{8}{3}\right)$$

$$= -\frac{3}{11} \cdot \left(\frac{\overset{3}{9}}{\underset{1}{4}} \cdot \frac{\overset{2}{8}}{\underset{1}{3}}\right)$$

$$= -\frac{3}{11} \cdot (3 \cdot 2)$$

$$= -\frac{3}{11} \cdot 6$$

$$= -\frac{18}{11}$$ ●

▶ ❹ **Understand the Multiplication and Division Properties of 0**

Now let's look at multiplication by zero.

Multiplication Property of Zero

For any real number a, the product of a and 0 is always 0; that is,

$$a \cdot 0 = 0 \cdot a = 0$$

Division properties of zero are now introduced.

Work Smart

Division by zero is not allowed. That is, 0 cannot be used as a divisor.

Division Properties of Zero

For any nonzero real number a,

1. The quotient of 0 and a is 0. That is, $\dfrac{0}{a} = 0$.

2. The quotient of a and 0 is **undefined**. That is, $\dfrac{a}{0}$ is undefined.

Why are these statements true? When we divide, we can check the quotient by multiplication. For example, $\dfrac{12}{4} = 3$ because $4 \cdot 3 = 12$. In the same way, $\dfrac{0}{4} = 0$ because $4 \cdot 0 = 0$. But what is the value of $\dfrac{12}{0}$? To determine this quotient, we should be able to determine a real number such that $0 \cdot \square = 12$. But since the product of 0 and any real number is 0, there is no value for $\square$.

EXAMPLE 8   **Using Zero as a Divisor and a Dividend**

Find the quotient:

(a) $\dfrac{23}{0}$

(b) $\dfrac{0}{17}$

Solution

(a) $\dfrac{23}{0}$ is undefined because 0 is the divisor.

(b) $\dfrac{0}{17} = 0$ because 0 is the dividend.

The properties of addition, multiplication, and division are summarized below.

Summary Properties of Addition

Identity Property of Addition   For any real number $a, 0 + a = a + 0 = a.$

Commutative Property of Addition   If a and b are real numbers, then $a + b = b + a.$

Additive Inverse Property   For any real number $a, a + (-a) = -a + a = 0.$

Associative Property of Addition   If $a, b,$ and c are real numbers, then $a + (b + c) = (a + b) + c.$

Summary Properties of Multiplication and Division

Identity Property of Multiplication $a \cdot 1 = 1 \cdot a = a$ for any real number $a.$

Commutative Property of Multiplication   If a and b are real numbers, then $a \cdot b = b \cdot a.$

Multiplicative Inverse Property $a \cdot \dfrac{1}{a} = \dfrac{1}{a} \cdot a = 1$ provided that $a \neq 0.$

Associative Property of Multiplication   If $a, b,$ and c are real numbers, then $a \cdot (b \cdot c) = (a \cdot b) \cdot c.$

Multiplication Property of Zero   For any real number $a, a \cdot 0 = 0 \cdot a = 0.$

Division Properties of Zero   For any nonzero number $a, \dfrac{0}{a} = 0$ and $\dfrac{a}{0}$ is undefined.

1.6 Exercises MyMathLab®

Exercise numbers in **green** have complete video solutions in MyMathLab or may be accessed using the QR code to the right.

*Problems **1–18** are the Quick ✓s that follow the **EXAMPLES**.*

Building Skills

In Problems 19–34, convert each measurement to the indicated unit of measurement. See Objective 1. Use the following conversions:

1 foot = 12 inches	3 feet = 1 yard
1 gallon = 4 quarts	100 centimeters = 1 meter
16 ounces = 1 pound	60 seconds = 1 minute
	60 minutes = 1 hour

19. 13 feet to inches

20. 13 yards to feet

21. 4500 centimeters to meters

22. 5900 centimeters to meters

23. 42 quarts to gallons

24. 58 quarts to gallons

25. 180 ounces to pounds

26. 120 ounces to pounds

27. 1620 seconds to minutes

28. 2250 minutes to hours

29. 258 inches to feet

30. 52 feet to yards

31. 8 hours to minutes

32. 17.5 hours to minutes

33. 6.25 gallons to quarts

34. 7.75 gallons to quarts

In Problems 35–50, state the property of real numbers that is being illustrated. See Objectives 2, 3, and 4.

35. $16 + (-16) = 0$

36. $4 \cdot 63 \cdot \dfrac{1}{4} = 4 \cdot \dfrac{1}{4} \cdot 63$

37. $\dfrac{3}{4}$ is equivalent to $\dfrac{3}{4} \cdot \dfrac{5}{5}$

38. $4 + 5 + (-4)$ is equivalent to $4 + (-4) + 5$

39. $12 \cdot \dfrac{1}{12} = 1$

40. $-236 + 236 = 0$

41. $34.2 + (-34.2) = 0$

42. $(4 \cdot 5)7 = 4(5 \cdot 7)$

43. $\dfrac{0}{17}$

44. $\dfrac{-8}{0}$

45. $\dfrac{2}{3}\left(-\dfrac{12}{43}\right) \cdot \dfrac{3}{2} = \dfrac{2}{3} \cdot \dfrac{3}{2}\left(-\dfrac{12}{43}\right)$

46. $\dfrac{5}{12} \cdot \dfrac{12}{5} = 1$

47. $5.23 + 4.98 + (-5.23) = 5.23 + (-5.23) + 4.98$

48. $16.4 \cdot 0 = 0$

49. $\dfrac{21}{0}$

50. $\dfrac{0}{106}$

Mixed Practice

In Problems 51–72, evaluate each expression by using the properties of real numbers.

51. $54 + 29 + (-54)$

52. $46 + 59 + (-46)$

53. $\dfrac{9}{5} \cdot \dfrac{5}{9} \cdot 18$

54. $\dfrac{4}{9} \cdot \dfrac{9}{4} \cdot 28$

55. $\dfrac{0}{42}$

56. $\dfrac{-12}{0}$

57. $-25 \cdot 13 \cdot \dfrac{1}{5}$

58. $36(-12)\dfrac{1}{6}$

59. $347 + 456 + (-456)$

60. $593 + 306 + (-306)$

61. $\dfrac{9}{2}\left(-\dfrac{10}{3}\right)6$

62. $\dfrac{13}{2} \cdot \dfrac{8}{39} \cdot \dfrac{39}{4}$

63. $\dfrac{7}{0}$

64. $\dfrac{0}{100}$

65. $100(-34)(0.01)$

66. $4000(0.5)(0.001)$

67. $569.003 \cdot 0$

68. $104 \cdot \dfrac{1}{104}$

69. $\dfrac{45}{3902} + \left(-\dfrac{45}{3902}\right)$

70. $30 \cdot \dfrac{4}{4}$

71. $-\dfrac{5}{44} \cdot \dfrac{80}{3} \cdot \dfrac{11}{5}$

72. $\dfrac{7}{48} \cdot \left(-\dfrac{21}{4}\right) \cdot \dfrac{12}{7}$

Applying the Concepts

73. Balancing the Checkbook Alberto's checking account balance at the start of the month was $321.03. During the month, he wrote checks for $32.84, $85.03, and $120.56. He also deposited a check for $120.56. What is Alberto's balance at the end of the month?

74. Stock Price Before the opening bell on Monday, a certain stock was priced at $32.04. On Monday the

stock was up $0.54, on Tuesday it was down $0.32, and on Wednesday it was down $0.54. What was the closing price of the stock on Wednesday?

Extending the Concepts

In Problems 75–78, insert parentheses to make the statement true.

75. $-3 - 4 - 10 = 3$

76. $-6 - 4 + 10 = -20$

77. $-15 + 10 - 4 - 8 = -1$

78. $25 - 6 - 10 - 1 = 28$

79. 16,200 seconds to hours

80. 22,500 seconds to hours

81. Convert 30 miles per hour to feet per second. (*Note*: 1 mile = 5280 feet)

82. Convert 40 miles per hour to feet per second. (*Note*: 1 mile = 5280 feet)

83. The space station orbits Earth every 90 minutes. How many sunrises do the astronauts in the space station see each day?

84. The planet Mercury makes one orbit around the sun every 88 days. That is, Mercury's "year" is 88 days long. How many minutes are there in a Mercury "year"?

Explaining the Concepts

85. In your own words, explain why 0 does not have a multiplicative inverse.

86. Why does $(24 \div 12) \div 2$ not equal $24 \div (12 \div 2)$?

87. Why does $\dfrac{0}{4} = 0$? Why is $\dfrac{4}{0}$ undefined?

88. How is the Identity Property of Addition related to the Additive Inverse Property?

89. How is the Identity Property of Multiplication related to the Multiplicative Inverse Property?

90. Let a be a real number such that $a > 0$, and let b represent a real number such that $b < 0$. Indicate whether each of the following statements is true or false, and explain your decision using actual values for a and b.

(a) $a > -a$

(b) $b > -b$

(c) $a + (-a) = 0$

(d) $-b + b = 0$

(e) $a - (-b) > 0$

(f) $ab < 0$

(g) $-ab > 0$

(h) $\dfrac{-a}{-b} > 0$

1.7 Exponents and the Order of Operations

Objectives

1. Evaluate Exponential Expressions
2. Apply the Rules for Order of Operations

Are You Prepared for This Section?

Before getting started, complete the following problems. If you get a problem wrong, go back to the section cited and review the material.

P1. Find the sum: $9 + (-19)$ [Section 1.4, pp. 29–30]

P2. Find the difference: $28 - (-7)$ [Section 1.4, pp. 31–32]

P3. Find the product: $-7 \cdot \dfrac{8}{3} \cdot 36$ [Section 1.5, pp. 37–38]

P4. Find the quotient: $\dfrac{100}{-15}$ [Section 1.4, pp. 34–35]

▶ ❶ Evaluate Exponential Expressions

If we want to multiply 2 eight times, we write $2 \cdot 2 \cdot 2 \cdot 2 \cdot 2 \cdot 2 \cdot 2 \cdot 2$. That's a lot of writing! To reduce the amount of writing needed to show repeated multiplication, **exponential notation** is used, where we write $2 \cdot 2 \cdot 2 \cdot 2 \cdot 2 \cdot 2 \cdot 2 \cdot 2$ as 2^8. In 2^8, 2 is called the **base** and 8 is called the **exponent.**

Exponential Notation

If n is a natural number and a is a real number, then

$$a^n = \underbrace{a \cdot a \cdot a \cdot \ldots \cdot a}_{n \text{ factors}}$$

where a is the **base** and n is the **exponent** or **power.** The exponent tells the number of times the base is used as a factor.

An expression written in the form a^n is said to be in **exponential form.** The expression 6^2 is read "six squared," 8^3 is read "eight cubed," and the expression 11^4 is read "eleven to the fourth power." In general, a^n is read as "a to the nth power."

EXAMPLE 1   **Writing a Numerical Expression in Exponential Form**

Write each expression in exponential form:

  (a) $5 \cdot 5 \cdot 5$ **(b)** $(-4)(-4)(-4)(-4)(-4)(-4)$

Solution

  (a) The expression $5 \cdot 5 \cdot 5$ contains three factors of 5, so $5 \cdot 5 \cdot 5 = 5^3$.

  (b) The expression $(-4)(-4)(-4)(-4)(-4)(-4)$ contains six factors of -4, so $(-4)(-4)(-4)(-4)(-4)(-4) = (-4)^6$.

Quick ✓

1. In the expression 3^6, 3 is the ____ and 6 is the _____ or _____.

In Problems 2 and 3, write each expression in exponential form.

2. $11 \cdot 11 \cdot 11 \cdot 11 \cdot 11$ **3.** $(-7)(-7)(-7)(-7)$

Prepared?...Answers   **P1.** -10 **P2.** 35   **P3.** -672 **P4.** $-\dfrac{20}{3}$

To evaluate an exponential expression, write the expression in **expanded form.** For example, 2^8 in expanded form is $2 \cdot 2 \cdot 2 \cdot 2 \cdot 2 \cdot 2 \cdot 2 \cdot 2$.

EXAMPLE 2   **Evaluating an Exponential Expression**

Evaluate each exponential expression:

(a) 6^4

(b) $\left(\dfrac{5}{3}\right)^5$

Solution

(a) $6^4 = 6 \cdot 6 \cdot 6 \cdot 6$
$= 1296$

(b) $\left(\dfrac{5}{3}\right)^5 = \left(\dfrac{5}{3}\right)\left(\dfrac{5}{3}\right)\left(\dfrac{5}{3}\right)\left(\dfrac{5}{3}\right)\left(\dfrac{5}{3}\right)$

$= \dfrac{5 \cdot 5 \cdot 5 \cdot 5 \cdot 5}{3 \cdot 3 \cdot 3 \cdot 3 \cdot 3}$

$= \dfrac{3125}{243}$

EXAMPLE 3   **Evaluating an Exponential Expression—Odd Exponent**

Evaluate each exponential expression:

(a) $(-5)^3$

(b) -5^3

Solution

(a) The base is -5. When the base is negative, it is important to use parentheses around the base as shown below.

$$(-5)^3 = (-5)(-5)(-5)$$
$$= -125$$

(b) The base is 5.

$$-5^3 = -(5 \cdot 5 \cdot 5)$$
$$= -125$$

EXAMPLE 4   **Evaluating an Exponential Expression—Even Exponent**

Evaluate each exponential expression:

(a) $(-5)^4$

(b) -5^4

Solution

Work Smart

There is a difference between $(-5)^4$ and -5^4. The parentheses in $(-5)^4$ says to use four factors of -5. However, in the expression -5^4, 5 is used as a factor four times and the result is multiplied by -1. -5^4 could also be read as "Take the opposite of the quantity 5^4."

(a) The base is -5.

$$(-5)^4 = (-5)(-5)(-5)(-5)$$
$$= 625$$

(b) The base is 5.

$$-5^4 = -(5 \cdot 5 \cdot 5 \cdot 5)$$
$$= -625$$

Quick ✓

In Problems 4–9, evaluate each exponential expression.

4. 2^4

5. $(-7)^2$

6. $\left(-\dfrac{1}{6}\right)^3$

7. $(0.9)^2$

8. -2^4

9. $(-2)^4$

▶ ❷ **Apply the Rules for Order of Operations**

To evaluate $3 \cdot 5 + 4$, do you multiply first and then add to get $15 + 4 = 19$ *or* do you add first and then multiply to get $3 \cdot 9 = 27$?

Because $3 \cdot 5$ is equivalent to $5 + 5 + 5$,

$$3 \cdot 5 + 4 = 5 + 5 + 5 + 4$$
$$= 19$$

Based on this, **whenever addition and multiplication appear in the same expression, always multiply first and then add.**

In Other Words
Multiply first then add.

Because any division problem can be written as a multiplication problem, divide before adding as well. Also, because any subtraction problem can be written as an addition problem, **always multiply and divide before adding and subtracting.**

EXAMPLE 5

Evaluating an Expression Containing Multiplication, Division, and Addition

Evaluate each expression:

(a) $11 + 2(-6)$

(b) $7 + 12 \div 3 \cdot 5$

Solution

(a)

Multiply first: $\quad 11 + 2(-6) = 11 + (-12)$

Add: $\qquad\qquad\qquad\quad = -1$

(b) Multiply/divide left to right: $\quad 7 + 12 \div 3 \cdot 5 = 7 + 4 \cdot 5$

Multiply: $\qquad\qquad\qquad\qquad\quad = 7 + 20$

$\qquad\qquad\qquad\qquad\qquad\qquad\quad = 27$ ●

Quick ✔

In Problems 10–13, evaluate each expression.

10. $1 + 7 \cdot 2$

11. $-3 \div \left(-\dfrac{1}{2}\right) + 18$

12. $9 \cdot 4 \div 2 + 5$

13. $\dfrac{15}{2} \div (-5)(8) - 7$

▶ **Parentheses**

If we want to add two numbers first and then multiply, use parentheses and write $(3 + 5)4$. In other words, **always evaluate the expression in parentheses first.**

EXAMPLE 6

Finding the Value of an Expression Containing Parentheses

Evaluate each expression:

(a) $(5 + 3)2$

(b) $\left(\dfrac{3}{2} - \dfrac{5}{2}\right)\left(\dfrac{7}{3} + \dfrac{2}{3}\right)$

Solution

(a) $(5 + 3)2 = 8 \cdot 2$
$\qquad\qquad\quad = 16$

(b) $\left(\dfrac{3}{2} - \dfrac{5}{2}\right)\left(\dfrac{7}{3} + \dfrac{2}{3}\right) = \left(-\dfrac{2}{2}\right)\left(\dfrac{9}{3}\right)$
$\qquad\qquad\qquad\qquad\qquad\qquad = (-1)(3)$
$\qquad\qquad\qquad\qquad\qquad\qquad = -3$ ●

Quick ✔

In Problems 14–16, evaluate each expression.

14. $8(2 + 3)$

15. $(2 - 9)(5 + 4)$

16. $\left(\dfrac{6}{7} + \dfrac{8}{7}\right)\left(\dfrac{11}{8} + \dfrac{5}{8}\right)$

The Division Bar

If an expression contains a division bar, treat the terms above and below the division bar as if they were in parentheses. For example,

Work Smart

The division bar acts like two sets of parentheses.

$$\frac{3 + 5}{9 + 7} = \frac{(3 + 5)}{(9 + 7)} = \frac{8}{16} = \frac{8 \cdot 1}{8 \cdot 2} = \frac{1}{2}$$

EXAMPLE 7 **Finding the Value of an Expression That Contains a Division Bar**

Evaluate each expression:

(a) $\dfrac{7 \cdot 3}{3 + 9 \cdot 2}$

(b) $\dfrac{1 + 7 \div \frac{1}{5}}{-6 \cdot 2 + 8}$

Solution

(a) Multiply: $\dfrac{7 \cdot 3}{3 + 9 \cdot 2} = \dfrac{21}{3 + 18}$

Add: $= \dfrac{21}{21}$

$= 1$

(b) Write division as multiplication: $\dfrac{1 + 7 \div \frac{1}{5}}{-6 \cdot 2 + 8} = \dfrac{1 + 7 \cdot 5}{-6 \cdot 2 + 8}$

Multiply: $= \dfrac{1 + 35}{-12 + 8}$

Add: $= \dfrac{36}{-4}$

$= \dfrac{9 \cdot 4}{-1 \cdot 4}$

$= -9$ ●

Quick ✓

In Problems 17–19, evaluate each expression.

17. $\dfrac{2 + 5 \cdot 6}{-3 \cdot 8 - 4}$

18. $\dfrac{(12 + 14)2}{13 \cdot 2 + 13 \cdot 5}$

19. $\dfrac{4 + 3 \div \frac{1}{7}}{2 \cdot 9 - 3}$

▶ **Multiple Grouping Symbols**

Grouping symbols include parentheses ( ), brackets [ ], braces { }, and absolute value symbols, | |. Operations within the grouping symbols are performed first. **When multiple grouping symbols occur, evaluate the expression in the innermost grouping symbols first and work outward.**

EXAMPLE 8 **Finding the Value of an Expression Containing Grouping Symbols**

Evaluate each expression:

(a) $2[3(6 + 3) - 7]$

(b) $3\left[4 + \left(\dfrac{2}{3}(-9)\right)\right]$

(continued)

Solution

(a)

Perform the operation in parentheses first
$\downarrow$

$$2[3(6 + 3) - 7] = 2[3 \cdot 9 - 7]$$

Perform the operations in brackets, multiply first: $= 2[27 - 7]$

$$= 2[20]$$

$$= 40$$

(b)

Perform the operation in parentheses first
$\downarrow$

$$3\left[4 + \left(\frac{2}{3}(-9)\right)\right] = 3[4 + (-6)]$$

Perform the operation in brackets: $= 3(-2)$

$$= -6$$ ●

> **Quick ✓**
>
> *In Problems 20 and 21, evaluate each expression.*
>
> **20.** $4[2(3 + 7) - 15]$ **21.** $2\{4[26 - (9 + 7)] - 15\} - 10$

⊙   When are exponents evaluated in the order of operations? In $2 \cdot 4^3$, do we multiply first and then evaluate the exponent to obtain $2 \cdot 4^3 = 8^3 = 512$, or do we evaluate the exponent first and then multiply to obtain $2 \cdot 4^3 = 2 \cdot 64 = 128$? Because $2 \cdot 4^3 = 2 \cdot 4 \cdot 4 \cdot 4 = 128$, **evaluate exponents before multiplication.** Don't forget to evaluate any expressions in the grouping symbols first.

EXAMPLE 9   **Finding the Value of an Expression Containing Exponents**

Evaluate each of the following:

(a) $2 + 7(-4)^2$ **(b)** $\dfrac{2 \cdot 3^2 + 4}{3(2 - 6)}$

Solution

(a)

Evaluate the exponent:
$\downarrow$

$$2 + 7(-4)^2 = 2 + 7 \cdot 16$$

Multiply: $= 2 + 112$

Add: $= 114$

(b)

Parentheses first:
$\downarrow$

$$\frac{2 \cdot 3^2 + 4}{3(2 - 6)} = \frac{2 \cdot 3^2 + 4}{3(-4)}$$

Evaluate the exponent: $= \dfrac{2 \cdot 9 + 4}{3(-4)}$

Find products: $= \dfrac{18 + 4}{-12}$

Add terms in numerator: $= \dfrac{22}{-12}$

Write in lowest terms: $= \dfrac{2 \cdot 11}{2 \cdot -6}$

$$= -\frac{11}{6}$$ ●

Quick ✔

In Problems 22–24, evaluate each of the following:

22. $\dfrac{7 - 5^2}{2}$

23. $3(7 - 3)^2$

24. $\dfrac{(-3)^2 + 7(1 - 3)}{3 \cdot 2^3 + 6}$

Order of Operations

Step 1: Perform all operations within *grouping symbols* first. When an expression has more than one set of grouping symbols, begin within the innermost grouping symbols and work outward.

Step 2: Evaluate expressions containing exponents.

Step 3: Perform *multiplication and division*, working from *left to right*.

Step 4: Perform *addition and subtraction*, working from *left to right*.

EXAMPLE 10   **How to Evaluate an Expression Using Order of Operations**

Evaluate: $18 + 7(2^3 - 26) + 5^2$

Step-by-Step Solution

Steps 1 and 2: Evaluate the expression in the parentheses first. Evaluate the expressions containing exponents, if present.

$$18 + 7(2^3 - 26) + 5^2 = 18 + 7(8 - 26) + 25$$

Evaluate $8 - 26$ in the parentheses: $= 18 + 7(-18) + 25$

Step 3: Perform multiplication and division, working from left to right.

Multiply $7(-18)$: $= 18 - 126 + 25$

Step 4: Perform addition and subtraction, working from left to right.

$$= -108 + 25$$
$$= -83$$ ●

EXAMPLE 11   **Evaluating a Numerical Expression Using Order of Operations**

Evaluate: $\left(\dfrac{2^3 - 6}{10 - 2 \cdot 3}\right)^2$

Solution

Evaluate the exponential expression inside the parentheses: $\left(\dfrac{2^3 - 6}{10 - 2 \cdot 3}\right)^2 = \left(\dfrac{8 - 6}{10 - 2 \cdot 3}\right)^2$

Multiply inside the parentheses: $= \left(\dfrac{8 - 6}{10 - 6}\right)^2$

Add/subtract inside the parentheses: $= \left(\dfrac{2}{4}\right)^2$

Simplify: $= \left(\dfrac{1}{2}\right)^2$

Evaluate the exponential expression: $= \dfrac{1}{4}$ ●

Quick ✔

In Problems 25–28, evaluate each expression.

25. $\dfrac{(4 - 10)^2}{2^3 - 5}$

26 $-3[(-4)^2 - 5(8 - 6)]^2$

27. $\dfrac{(2.9 + 7.1)^2}{5^2 - 15}$

28. $\left(\dfrac{4^2 - 4(-3)(1)}{7 \cdot 2}\right)^2$

1.7 Exercises MyMathLab®

*Problems **1–28** are the Quick ✓s that follow the **EXAMPLES**.*

Building Skills

In Problems 29–32, write in exponential form. See Objective 1.

29. $5 \cdot 5$

30. $4 \cdot 4 \cdot 4 \cdot 4 \cdot 4$

31. $\left(-\dfrac{3}{5}\right)\left(-\dfrac{3}{5}\right)\left(-\dfrac{3}{5}\right)$

32. $(-8)(-8)(-8)$

In Problems 33–54, evaluate each exponential expression. See Objective 1.

33. 8^2

34. 4^2

35. $(-8)^2$

36. $(-4)^2$

37. 10^3

38. 2^5

39. -10^3

40. -2^5

41. $(-10)^3$

42. $(-2)^5$

43. $(1.5)^2$

44. $(0.04)^2$

45. -8^2

46. -4^2

47. -1^{20}

48. $(-1)^{19}$

49. 0^4

50. 1^6

51. $\left(-\dfrac{1}{2}\right)^6$

52. $\left(-\dfrac{3}{2}\right)^5$

53. $\left(-\dfrac{1}{3}\right)^3$

54. $\left(-\dfrac{3}{4}\right)^2$

In Problems 55–86, evaluate each expression. See Objective 2.

55. $2 + 3 \cdot 4$

56. $12 + 8 \cdot 3$

57. $-5 \cdot 3 + 12$

58. $-3 \cdot 12 + 9$

59. $100 \div 2 \cdot 50$

60. $50 \div 5 \cdot 4$

61. $156 - 3 \cdot 2 + 10$

62. $86 - 4 \cdot 3 + 6$

63. $(2 + 3)4$

64. $(7 - 5)\dfrac{5}{2}$

65. $8 \div 4 \cdot 2$

66. $4 \div 7 \cdot 21$

67. $\dfrac{4 + 2}{2 + 8}$

68. $\dfrac{5 + 3}{3 + 15}$

69. $\dfrac{14 - 6}{6 - 14}$

70. $\dfrac{15 - 7}{7 - 15}$

71. $13 - [3 + (-8)4]$

72. $12 - [7 + (-6)3]$

73. $(-8.75 - 1.25) \div (-2)$

74. $(-11.8 - 15.2) \div (-2)$

75. $4 - 2^3$

76. $10 - 4^2$

77. $15 + 4 \cdot 5^2$

78. $10 + 3 \cdot 2^4$

79. $-2^3 + 3^2 \div (2^2 - 1)$

80. $-5^2 + 3^2 \div (3^2 + 9)$

81. $\left(\dfrac{4^2 - 3}{12 - 2 \cdot 5}\right)^2$

82. $\left(\dfrac{7 - 5^2}{8 + 4 \cdot 2}\right)^2$

83. $-2[5(9 - 3) - 3 \cdot 6]$

84. $3[6(5 - 2) - 2 \cdot 5]$

85. $\left(\dfrac{4}{3} + \dfrac{5}{6}\right)\left(\dfrac{2}{5} - \dfrac{9}{10}\right)$

86. $\left(\dfrac{3}{4} + \dfrac{1}{2}\right)\left(\dfrac{2}{3} - \dfrac{1}{2}\right)$

Mixed Practice

In Problems 87–110, evaluate each expression.

87. $4^2 - 3 \cdot 4 + 7$

88. $(-2)^2 + 4(-2) + 11$

89. $4 + 2(6 - 2)$

90. $3 + 6(9 - 5)$

91. $\dfrac{12 - 16 \div 4 + (-24)}{16 \cdot 2 - 4 \cdot 0}$

92. $\dfrac{6 + 15 \div 3 + 16}{6 + 10 \cdot 0}$

93. $\left(\dfrac{2 - (-4)^3}{5^2 - 7 \cdot 2}\right)^2$

94. $\left(\dfrac{9 \cdot 2 - (-2)^3}{4^2 + 3(-1)^5}\right)^2$

95. $\dfrac{5^2 - 10}{3^2 + 6}$

96. $\dfrac{12(2)^3}{4^2 + 4 \cdot 5}$

97. $\left|6(5 - 3^2)\right|$

98. $-6(2 + \left|2 \cdot 3 - 4^2\right|)$

99. $\dfrac{4 - (-6)}{1 - (-1)}$

100. $\dfrac{-2 - 12}{3 - (-4)}$

101. $-\dfrac{7}{20} + \dfrac{3}{8} \div \dfrac{1}{2}$

102. $-\dfrac{4}{5} + \dfrac{3}{10} \div \dfrac{2}{9}$

103. $\dfrac{21 - 3^2}{1 + 3}$

104. $\dfrac{5 + 3^2}{2 + 5}$

105. $\dfrac{3}{4}\left[\dfrac{5}{4} \div \left(\dfrac{3}{8} - \dfrac{1}{8}\right) - 3\right]$

106. $\left[\dfrac{9}{10} \div \left(\dfrac{2}{5} + \dfrac{1}{5}\right) + \dfrac{7}{2}\right]\dfrac{1}{10}$

107. $\left(\dfrac{4}{3}\right)^3 - \left(\dfrac{1}{2}\right)^2\left(\dfrac{8}{3}\right) + 2 \div 3$

108. $\dfrac{1}{18} \cdot \dfrac{46}{5} - \left(\dfrac{2}{3}\right)^2$

109. $\dfrac{5^2 - 3^3}{\left|4 - 4^2\right|}$

110. $\dfrac{3 \cdot 2^3 - 2^2 \cdot 12}{3 + 3^2}$

Applying the Concepts

In Problems 111–114, express each number as the product of prime factors. Write the answer in exponential form.

111. 72   **112.** 675

113. 48   **114.** 200

In Problems 115–120, insert grouping symbols so that the expression has the desired value.

115. $4 \cdot 3 + 6 \cdot 2$ results in 36

116. $4 \cdot 7 - 4^2$ results in -36

117. $4 + 3 \cdot 4 + 2$ results in 42

118. $6 - 4 + 3 - 1$ results in 0

119. $6 - 4 + 3 - 1$ results in 4

120. $4 + 3 \cdot 2 - 1 \cdot 6$ results in 42

121. Cost of a TV The total amount paid for a flat-screen television that costs $479, plus sales tax of 7.5%, is found by evaluating the expression $479 + 0.075(479)$. Evaluate this expression rounded to the nearest cent.

122. Manufacturing Cost Evaluate the expression $3000 + 6(100) - \dfrac{100^2}{1000}$ to find the weekly production cost of manufacturing 100 calculators.

△**123. Surface Area** The surface area of a right circular cylinder whose radius is 6 inches and height is 10 inches is given approximately by $2 \cdot 3.1416 \cdot 6^2 + 2 \cdot 3.1416 \cdot 6 \cdot 10$. Evaluate this expression. Round the answer to two decimal places.

△**124. Volume of a Cone** The volume of a cone whose radius is 3 centimeters and whose height is 12 centimeters is given approximately by $\dfrac{1}{3} \cdot 3.1416 \cdot 3^2 \cdot 12$. Evaluate this expression. Round the answer to two decimal places.

125. Investing If $1000 is invested at 3% annual interest and remains untouched for 2 years, the amount of money that is in the account after 2 years is given by the expression $1000(1 + 0.03)^2$. Evaluate this expression, rounded to the nearest cent.

126. Investing If $5000 is invested at 4.5% annual interest and remains untouched for 5 years, the amount of money that is in the account after 5 years is given by the expression $5000(1 + 0.045)^5$. Evaluate this expression, rounded to the nearest cent.

Extending the Concepts

The Angle Addition Postulate from geometry states that the measure of an angle is equal to the sum of the measures of its parts. Refer to the figure. Use the Angle Addition Postulate to answer Problems 127 and 128.

△**127.** If the measure of $\angle XYQ = 46.5°$ and the measure of $\angle QYZ = 69.25°$, find the measure of the measure of $\angle XYZ$.

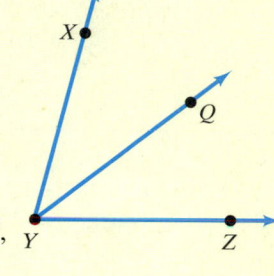

△**128.** If the measure of $\angle QYZ = 18°$ and the measure of $\angle XYZ = 57°$, find the measure of $\angle XYQ$.

Explaining the Concepts

129. Explain the difference between -3^2 and $(-3)^2$. Identify the distinguishing characteristics of the two problems, and explain how to evaluate each expression.

1.8 Simplifying Algebraic Expressions

Objectives

1 Evaluate Algebraic Expressions

2 Identify Like Terms and Unlike Terms

3 Use the Distributive Property

4 Simplify Algebraic Expressions by Combining Like Terms

Prepared?...Answers   **P1.** 5
P2. −15   **P3.** −36

Are You Prepared for This Section?

Before getting started, complete the following problems. If you get a problem wrong, go back to the section cited and review the material.

P1. Find the sum: $-3 + 8$ [Section 1.4, pp. 29–30]

P2. Find the difference: $-7 - 8$ [Section 1.4, pp. 31–32]

P3. Find the product: $-\dfrac{4}{3}(27)$ [Section 1.5, pp. 37–38]

What is algebra? The word "algebra" is derived from the Arabic word *al-jabr*, which means "restoration." Today algebra means more. **Algebra** uses symbols to represent quantities and to express general relationships that hold for all members of the set.

▶ ❶ Evaluate Algebraic Expressions

In arithmetic, we work with numbers. In algebra, letters such as $x, y, a, b,$ and c are used to represent numbers.

> **Definition**
>
> When a letter represents any number from a set of numbers, it is called a **variable**.

In this course, the set of numbers referred to in the definition will usually be the set of real numbers.

> **Definition**
>
> A **constant** is either a fixed number, such as 5, or a letter or symbol that represents a fixed number.

For example, in Einstein's Theory of Relativity, $E = mc^2$, the E and m are variables that represent total energy and mass, respectively, and c is a constant that represents the speed of light (299,792,458 meters per second).

> **Definition**
>
> An **algebraic expression** is any combination of variables, constants, grouping symbols, and mathematical operations such as addition, subtraction, multiplication, division, and exponents.

Some examples of algebraic expressions are

$$x - 5 \qquad \frac{1}{2}x \qquad 2y - 7 \qquad z^2 + 3 \qquad \text{and} \qquad \frac{b - 1}{b + 1}$$

One of the procedures performed on algebraic expressions is *evaluating an algebraic expression*.

> **Definition**
>
> To **evaluate an algebraic expression,** substitute a numerical value for each variable into the expression and simplify the result.

EXAMPLE 1   **Evaluating an Algebraic Expression**

Evaluate each expression for the given value of the variable.

    (a) $2x + 5$ for $x = 8$ **(b)** $a^2 - 2a + 4$ for $a = -3$

Solution

    (a) Substitute 8 for x in the expression $2x + 5$:

$$2(8) + 5 = 16 + 5$$
$$= 21$$

    (b) Substitute -3 for a in $a^2 - 2a + 4$:

$$(-3)^2 - 2(-3) + 4 = 9 + 6 + 4$$
$$= 19$$

EXAMPLE 2 **An Algebraic Expression for Revenue**

The expression $15.79x + 12.79y$ represents the total amount of money, in dollars, received at an IMAX theater, where x represents the number of adult tickets sold and y represents the number of children's tickets sold. Evaluate $15.79x + 12.79y$ for $x = 50$ and $y = 82$. Interpret the result.

Solution

Substitute 50 for x and 82 for y in the expression $15.79x + 12.79y$.

$$15.79(50) + 12.79(82) = 789.50 + 1048.78 = 1838.28$$

So $1838.28 was collected by selling 50 adult tickets and 82 children's tickets.

Quick ✓

1. When a letter represents any number from a set of numbers, it is called a _____.

2. To _____ an algebraic expression, substitute a numerical value for each variable into the expression and simplify the result.

In Problems 3 and 4, evaluate each expression for the given value of the variable.

3. $-3k + 5$ for $k = 4$ 4. $-2y^2 - y + 8$ for $y = -2$

5. The Amadeus Coffee Shop creates a breakfast blend of two types of coffee. They mix x pounds of a mild coffee that sells for $7.00 per pound with y pounds of a robust coffee that sells for $10.00 per pound. An algebraic expression that represents the value of the breakfast blend, in dollars, is $7x + 10y$. Evaluate this expression for $x = 8$ pounds and $y = 16$ pounds

---

▶ ❷ **Identify Like Terms and Unlike Terms**

Algebraic expressions consist of *terms*.

Definition

A **term** is a constant or the product of a constant and one or more variables raised to a power.

In algebraic expressions, the terms are separated by addition signs.

EXAMPLE 3 **Identifying the Terms in an Algebraic Expression**

Identify the terms in the following algebraic expressions.

(a) $4a^3 + 5b^2 - 8c + 12$ (b) $\dfrac{x}{4} - 7y + 8z$

Solution

(a) Rewrite $4a^3 + 5b^2 - 8c + 12$ so it contains only addition signs.

$$4a^3 + 5b^2 + (-8c) + 12$$

The four terms are $4a^3$, $5b^2$, $-8c$, and 12.

(b) The algebraic expression $\dfrac{x}{4} - 7y + 8z$ has three terms: $\dfrac{x}{4}$, $-7y$, and $8z$.

Quick ✔

6. *True or False* A constant by itself is a term.

In Problems 7–9, identify the terms in each algebraic expression.

7. $5x^2 + 3xy$

8. $9ab - 3bc + 5ac - ac^2$

9. $\dfrac{2mn}{5} - \dfrac{3n}{7}$

Definition

The **coefficient** of a term is the numerical factor of the term.

For example, the coefficient of $7x$ is 7; the coefficient of $-2x^2y$ is -2. Terms that have no number as a factor, such as mn, have a coefficient of 1 since $mn = 1 \cdot mn$. The coefficient of $-y$ is -1 since $-y = -1 \cdot y$. If a term consists of just a constant, the coefficient is the number itself. For example, the coefficient of 14 is 14.

EXAMPLE 4 **Determining the Coefficient of a Term**

Determine the coefficient of each term:

(a) $\dfrac{1}{2}xy^2$ (b) $-\dfrac{t}{12}$ (c) ab^3 (d) 12

Solution

(a) The coefficient of $\dfrac{1}{2}xy^2$ is $\dfrac{1}{2}$.

(b) The coefficient of $-\dfrac{t}{12}$ is $-\dfrac{1}{12}$ because $-\dfrac{t}{12}$ can be written as $-\dfrac{1}{12} \cdot t$.

(c) The coefficient of ab^3 is 1 because ab^3 can be written as $1 \cdot ab^3$.

(d) The coefficient of 12 is 12 because the coefficient of a constant is the number itself.

Quick ✔

In Problems 10–14, determine the coefficient of each term.

10. $2z^2$ **11.** xy **12.** $-b$

13. 5   **14.** $-\dfrac{2}{3}z$

Sometimes algebraic expressions can be simplified by combining *like terms.*

Work Smart

Like terms can have different coefficients, but they cannot have different variables or different exponents on those variables.

Definition

Terms that have the same variable factor(s) with the same exponent(s) are called **like terms.**

For example, $3x^2$ and $-7x^2$ are like terms because both contain x^2, but $3x^2$ and $-7x^3$ are not like terms because the variable x is raised to different powers. Constant terms such as -9 and 6 are like terms.

EXAMPLE 5 **Classifying Terms as Like or Unlike**

Classify the following pairs of terms as *like* or *unlike*.

(a) $2p^3$ and $-5p^3$ **(b)** $7kr$ and $\frac{1}{4}k^2r$ **(c)** 5 and 8

Solution

(a) $2p^3$ and $-5p^3$ are *like* terms because both contain the factor p^3.

(b) $7kr$ and $\frac{1}{4}k^2r$ are *unlike* terms because the factor k is raised to different powers.

(c) 5 and 8 are *like* terms because both are constants. ●

Quick ✓

15. *True or False* Like terms can have different coefficients or different exponents on the same variable.

In Problems 16–20, tell whether the terms are like or unlike.

16. $-\frac{2}{3}p^2$ and $\frac{4}{5}p^2$ **17.** $\frac{m}{6}$ and $4m$ **18.** $3a^2b$ and $-2ab^2$

19. $8a$ and 11   **20.** -7 and 12

▶ ❸ Use the Distributive Property

The *Distributive Property* will be used throughout this course and in future courses.

Work Smart

The long name for the Distributive Property is the Distributive Property of Multiplication over Addition. This name is a reminder not to distribute across multiplication. For example,

$$6x(5xy) \neq 6x \cdot 5x \cdot 6x \cdot y$$

The Distributive Property

If $a, b,$ and c are real numbers, then

$$a(b + c) = a \cdot b + a \cdot c$$
$$(a + b)c = a \cdot c + b \cdot c$$

That is, multiply each of the terms inside the parentheses by the factor on the outside.

Because $b - c = b + (-c)$, it is also true that $a(b - c) = a \cdot b - a \cdot c$.

EXAMPLE 6 **Using the Distributive Property to Remove Parentheses**

Use the Distributive Property to remove the parentheses.

(a) $3(x + 5)$ **(b)** $-\frac{1}{3}(6x - 12)$

Solution

(a) To use the Distributive Property, multiply each term in the parentheses by 3:

$$3(x + 5) = 3 \cdot x + 3 \cdot 5$$
$$= 3x + 15$$

(continued)

(b) Multiply each term in the parentheses by $-\dfrac{1}{3}$:

$$-\frac{1}{3}(6x - 12) = -\frac{1}{3} \cdot 6x - \left(-\frac{1}{3}\right) \cdot 12$$

$$= -2x + 4$$

Quick ✓

21. $a(b + c) = a \cdot \underline{} + a \cdot \underline{}$

In Problems 22–25, use the Distributive Property to remove the parentheses.

22. $6(x + 2)$

23. $-5(x + 2)$

24. $-2(k - 7)$

25. $(8x + 12)\dfrac{3}{4}$

▶ ❹ Simplify Algebraic Expressions by Combining Like Terms

An algebraic expression that contains the sum or difference of like terms may be simplified using the Distributive Property "in reverse." When the Distributive Property is used to add coefficients of like terms, we say that we are **combining like terms.**

EXAMPLE 7   **Using the Distributive Property to Combine Like Terms**

Combine like terms:

(a) $2x + 7x$

(b) $x^2 - 5x^2$

Solution

(a) Use the Distributive Property "in reverse": $2x + 7x = (2 + 7)x$
$$= 9x$$

(b) Use the Distributive Property "in reverse": $x^2 - 5x^2 = (1 - 5)x^2$
$$= -4x^2$$

Look carefully at the results of Example 7. Notice that when combining like terms, the coefficients of the like terms are added, and the variables and exponents remain the same.

Work Smart

When combining like terms, add or subtract the coefficients of the like terms and keep the variables and exponents the same.

Quick ✓

26. $4x^2 + 9x^2 = (\underline{} + \underline{})x^2$

In Problems 27–29, combine like terms.

27. $3x - 8x$

28. $-5x^2 + x^2$

29. $-7x - x + 6 - 3$

Sometimes terms must be rearranged using the Commutative Property of Addition to combine like terms.

EXAMPLE 8   **Combining Like Terms Using the Commutative Property**

Combine like terms: $4x + 5y + 12x - 7y$

Solution

Use the Commutative Property to rearrange the terms.

$$4x + 5y + 12x - 7y = 4x + 12x + 5y - 7y$$

Use the Distributive Property "in reverse": $= (4 + 12)x + (5 - 7)y$
$$= 16x + (-2y)$$

Write the answer in simplest form: $= 16x - 2y$

Quick ✔

In Problems 30–33, combine like terms.

30. $3a + 2b - 5a + 7b - 4$ **31.** $5ac + 2b + 7ac - 5a - b$

32. $5ab^2 + 7a^2b + 3ab^2 - 8a^2b$ **33.** $\dfrac{4}{3}rs - \dfrac{3}{2}r^2 + \dfrac{2}{3}rs - 5$

Parentheses may also need to be removed using the Distributive Property before combining like terms. Recall that the rules for order of operations on real numbers place multiplication before addition or subtraction. In this section, the direction **simplify** will mean to remove all parentheses and combine like terms.

EXAMPLE 9   **Combining Like Terms Using the Distributive Property**

Simplify the algebraic expression: $3 - 4(2x + 3) - (5x + 1)$

Solution

First, use the Distributive Property to remove parentheses.

$$3 - 4(2x + 3) - (5x + 1) = 3 - 8x - 12 - 5x - 1$$

Rearrange terms using the Commutative Property of Addition: $= -8x - 5x + 3 - 12 - 1$

Combine like terms: $= -13x - 10$ ●

Work Smart

Remember that multiplication comes before subtraction. In the first step of Example 9, do not compute $3 - 4$ first to obtain $-1(2x + 3)$.

Simplifying an Algebraic Expression

Step 1: Remove any parentheses using the Distributive Property.

Step 2: Combine any like terms.

Quick ✔

34. Explain what it means to simplify an algebraic expression.

In Problems 35–38, simplify each expression.

35. $3x + 2(x - 1) - 7x + 1$

36. $m + 2n - 3(m + 2n) - (7 - 3n)$

37. $2(a - 4b) - (a + 4b) + b$

38. $\dfrac{1}{2}(6x + 4) - \dfrac{1}{3}(12 - 9x)$

1.8 Exercises   **MyMathLab®**   Exercise numbers in green have complete video solutions in MyMathLab or may be accessed using the QR code to the right.

*Problems **1–38** are the Quick ✔s that follow the EXAMPLES.*

Building Skills

In Problems 39–50, evaluate each expression using the given values of the variables. See Objective 1.

39. $2x + 5$ for $x = 4$

40. $3x + 7$ for $x = 2$

41. $x^2 + 3x - 1$ for $x = 3$

42. $n^2 - 4n + 3$ for $n = 2$

43. $4 - k^2$ for $k = -5$

44. $-2p^2 + 5p + 1$ for $p = -3$

45. $\dfrac{9x - 5y}{x + y}$ for $x = 3, y = 5$

46. $\dfrac{3y + 2z}{y - z}$ for $y = 4, z = -2$

47. $(x + 3y)^2$ for $x = 3, y = -4$

48. $(a - 2b)^2$ for $a = 1, b = 2$

49. $b^2 - 4ac$ for $a = 1, b = 4, c = 3$

50. $\dfrac{a^2 - 4}{a^2 + 5a - 14}$ for $a = -3$

In Problems 51–54, for each expression, identify the terms and then name the coefficient of each term. See Objective 2.

51. $2x^3 + 3x^2 - x + 6$

52. $3m^4 - m^3n^2 + 4n - 1$

53. $z^2 + \dfrac{2y}{3}$ **54.** $t^3 - \dfrac{t}{4}$

In Problems 55–62, determine whether the terms are like or unlike. See Objective 2.

55. $8x$ and 8 **56.** $11p$ and 11

57. 54 and -21 **58.** -13 and 38

59. $12b$ and $-b$ **60.** $6a^2$ and $-3a^2$

61. r^2s and rs^2 **62.** x^2y^3 and y^2x^3

In Problems 63–70, use the Distributive Property to remove the parentheses. See Objective 3.

63. $3(m + 2)$ **64.** $3(4s + 2)$

65. $(3n^2 + 2n - 1)6$ **66.** $(6a^4 - 4a^2 + 2)3$

67. $-(x - y)$ **68.** $-5(k - n)$

69. $(8x - 6y)\left(-\dfrac{1}{2}\right)$ **70.** $(20a - 15b)\left(-\dfrac{2}{5}\right)$

In Problems 71–98, simplify each expression by using the Distributive Property to remove parentheses and combining like terms. See Objective 4.

71. $5x - 2x$ **72.** $14k - 11k$

73. $4z - 6z + 8z$ **74.** $9m - 8m + 2m$

75. $2m + 3n + 8m + 7n$ **76.** $x + 2y + 5x + 7y$

77. $0.3x^7 + x^7 + 0.9x^7$ **78.** $1.7n^4 - n^2 + 2.1n^4$

79. $-3y^6 + 13y^6$ **80.** $-7p^5 + 2p^5$

81. $-(6w + 12y - 13z)$ **82.** $-(-6m + 9n - 8p)$

83. $5(k + 3) - 8k$ **84.** $3(7 - z) - z$

85. $7n - (3n + 8)$ **86.** $18m - (6 + 9m)$

87. $(7 - 2x) - (x + 4)$ **88.** $(3k + 1) - (4 - k)$

89. $(7n - 8) - (3n - 6)$ **90.** $(5y - 6) - (11y + 8)$

91. $-6(n - 3) + 2(n + 1)$

92. $-9(7r - 6) + 9(10r + 3)$

93. $\dfrac{2}{3}x + \dfrac{1}{6}x$ **94.** $\dfrac{3}{5}y + \dfrac{7}{10}y$

95. $\dfrac{1}{2}(8x + 5) - \dfrac{2}{3}(6x + 12)$

96. $\dfrac{1}{5}(60 - 15x) + \dfrac{3}{4}(12 - 4x)$

97. $2(0.5x + 9) - 3(1.5x + 8)$

98. $3(0.2x + 6) - 5(1.6x + 1)$

Mixed Practice

In Problems 99–114, (a) evaluate the expression for the given value(s) of the variable(s) before combining like terms, (b) simplify the expression by combining like terms and then evaluate the expression for the given value(s) of the variable(s). Compare your results.

99. $5x + 3x; x = 4$ **100.** $8y + 2y; y = -3$

101. $-2a^2 + 5a^2; a = -3$ **102.** $4b^2 - 7b^2; b = 5$

103. $4z - 3(z + 2); z = 6$

104. $8p - 3(p - 4); p = 3$

105. $5y^2 + 6y - 2y^2 + 5y - 3; y = -2$

106. $3x^2 + 8x - x^2 - 6x; x = 5$

107. $\dfrac{1}{2}(4x - 2) - \dfrac{2}{3}(3x + 9); x = 3$

108. $\dfrac{1}{5}(5x - 10) - \dfrac{1}{6}(6x + 12); x = -2$

109. $3a + 4b - 7a + 3(a - 2b); a = 2, b = 5$

110. $-4x - y + 2(x - 3y); x = 3, y = -2$

111. $3p^2 - 4p + 5 - p^2 + 6p; p = -4$

112. $8k - (k^2 - 2k); k = 3$

113. $\dfrac{1}{3}(6a - 9) + \dfrac{3}{2}(8a + 2); a = -5$

114. $\dfrac{1}{6}(12w + 24) - \dfrac{1}{4}(12w + 24); w = 8$

Applying the Concepts

In Problems 115–120, evaluate each expression using the given values of the variables.

115. Area of a Trapezoid

$\dfrac{1}{2}h(b + B); h = 4, b = 5, B = 17$

116. Area of a Trapezoid

$\dfrac{1}{2}h(b + B); h = 9, b = 3, B = 12$

117. Slope of a Line Given Two Points

$\dfrac{a - b}{c - d}; a = 6, b = 3, c = -4, d = -2$

118. Slope of a Line Given Two Points

$\dfrac{a - b}{c - d}$; $a = -5, b = -2, c = 7, d = 1$

119. $b^2 - 4ac$; $a = 7, b = 8, c = 1$

120. $b^2 - 4ac$; $a = 2, b = 5, c = 3$

121. Renting a Truck The cost of renting a truck from Hamilton Truck Rental is $59.95 per day plus $0.15 per mile. The expression $59.95 + 0.15m$ represents the cost of renting a truck for one day and driving it m miles. Evaluate $59.95 + 0.15m$ for $m = 125$.

122. Renting a Car The cost of renting a compact car for one day from CMH Auto is $29.95 plus $0.17 per mile. The expression $29.95 + 0.17m$ represents the total daily cost. Evaluate the expression $29.95 + 0.17m$ for $m = 245$.

123. Ticket Sales The Center for Science and Industry sells adult tickets for $12 and children's tickets for $7. The expression $12a + 7c$ represents the total revenue from selling a adult tickets and c children's tickets. Evaluate the algebraic expression $12a + 7c$ for $a = 156$ and $c = 421$.

124. Ticket Sales A community college theatre group sold tickets to a recent production. Student tickets cost $5, and nonstudent tickets cost $8. The algebraic expression $5s + 8n$ represents the total revenue from selling s student tickets and n nonstudent tickets. Evaluate $5s + 8n$ for $s = 76$ and $n = 63$.

△**125. Rectangle** The width of a rectangle is w yards, and the length of the rectangle is $(3w - 4)$ yards. The perimeter of the rectangle is given by the algebraic expression $2w + 2(3w - 4)$.

(a) Simplify the algebraic expression $2w + 2(3w - 4)$.

(b) Determine the perimeter of a rectangle whose width w is 5 yards.

△**126. Rectangle** The length of a rectangle is l meters, and the width of the rectangle is $(l - 11)$ meters. The perimeter of the rectangle is given by the algebraic expression $2l + 2(l - 11)$.

(a) Simplify the expression $2l + 2(l - 11)$.

(b) Determine the perimeter of a rectangle whose length l is 15 meters.

127. Finance Novella invested some money in two investment funds. She placed s dollars in stocks that yield 5.5% annual interest and b dollars in corporate bonds that yield 2.35% annual interest. Evaluate the expression $0.055s + 0.0235b$ for $s =$ $2950 and $b =$ $2050. Round your answer to the nearest cent.

128. Finance Jonathan received an inheritance from his grandparents. He invested x dollars in a Certificate of Deposit that pays 1.05% and y dollars in a global pharmaceutical company that is expected to pay 5.08%. Evaluate the algebraic expression $0.0105x + 0.0508y$ for $x =$ $2500 and $y =$ $1000.

Extending the Concepts

129. Simplify the algebraic expression (cleverly!!) using the Distributive Property—in reverse! $2.75(-3x^2 + 7x - 3) - 1.75(-3x^2 + 7x - 3)$

130. Simplify the algebraic expression using the Distributive Property in reverse. $11.23(7.695x + 81.34) + 8.77(7.695x + 81.34)$

Explaining the Concepts

131. Explain why the sum $2x^2 + 4x^2$ is *not* equivalent to $6x^4$. What is the correct answer?

132. Use $x = 4$ and $y = 5$ to answer parts (a), (b), and (c).

(a) Evaluate $x^2 + y^2$.

(b) Evaluate $(x + y)^2$.

(c) Are the results the same? Is $(x + y)^2$ equal to $x^2 + y^2$? Explain your response.

Chapter 1 Activity: The Math Game

Focus: Applying order of operations and simplifying expressions

Time: 10 minutes

Group size: 2–4

- The instructor will announce when the groups may begin solving the problems to the right.
- When your group has completed all of the problems, ask the instructor to check the answers. The instructor will tell you how many answers are correct, but not which ones.
- The first group to complete all of the problems correctly will win a prize, as determined by the instructor.

1. Evaluate: $-8 \div 2^2 \cdot 6 + (-2)^3$

2. Evaluate: $\dfrac{6(-3) + 4^2}{25 + 4(-9 + 4)}$

3. Evaluate: $x^3 - x^2$ for $x = -3$

4. Evaluate: $\dfrac{(x + 2y)^2}{xy}$ for $x = 1, y = -2$

5. Simplify: $-2(4x + 3) - (5x - 1)$

6. Simplify: $\dfrac{3}{4}(8x^2 + 16) - 2x^2 + 3x$

Chapter 1 Review

Section 1.2   Fractions, Decimals, and Percents

KEY CONCEPTS

- To find the least common multiple (LCM) of two numbers, (1) factor each number as the product of prime factors; (2) write the factor(s) that the numbers share, if any; (3) write down the remaining factors the greatest number of times that the factors appear in any number. The product of the factors is the LCM.

- To write a fraction in lowest terms, find the common factors between the numerator and denominator, and use the fact that $\dfrac{a \cdot c}{b \cdot c} = \dfrac{a}{b}$ to divide out the common factors.

- To round a decimal, identify the specified place value in the decimal. If the digit to the right is 5 or more, add 1 to the digit; if the digit to the right is 4 or less, leave the digit as it is. Then drop the digits to the right of the specified place value.

- To convert a fraction to a decimal, divide the numerator of the fraction by the denominator of the fraction until the remainder is 0 or the remainder repeats.

- To convert a decimal to a fraction, identify the place value of the last digit in the decimal. Write the decimal as a fraction using the place value of the last digit as the denominator, and write in lowest terms.

- To convert a percent to a decimal, move the decimal point two places to the left and drop the % symbol.

- To convert a decimal to a percent, move the decimal point two places to the right and add the % symbol.

KEY TERMS

Factor
Product
Prime
Composite
Prime factorization
Multiple
Least common multiple (LCM)
Numerator
Denominator
Whole numbers
Equivalent fractions
Least common denominator (LCD)
Lowest terms
Terminating decimal
Repeating decimal
Percent

You Should Be Able To...	EXAMPLE	Review Exercises
❶ Factor a number as a product of prime factors (p. 8)	Example 1	1–4
❷ Find the least common multiple of two or more numbers (p. 9)	Examples 2 and 3	5, 6
❸ Write equivalent fractions (p. 10)	Examples 4 and 5	7–10
❹ Write a fraction in lowest terms (p. 12)	Example 6	11–13
❺ Round decimals (p. 13)	Example 7	14, 15
❻ Convert between fractions and decimals (p. 14)	Examples 8 and 9	16–22
❼ Convert between percents and decimals (p. 15)	Examples 10 and 11	23–31

In Problems 1–4, factor each number as a product of primes, if possible. If a number is prime, state "prime".

1. 75

2. 87

3. 81

4. 17

In Problems 5 and 6, find the LCM of the numbers.

5. 18 and 24

6. 4, 8, and 18

In Problems 7 and 8, write each number with the given denominator.

7. Write $\dfrac{7}{15}$ with the denominator 30.

8. Write 3 with the denominator 4.

In Problems 9 and 10, write equivalent fractions with the least common denominator.

9. $\dfrac{1}{6}$ and $\dfrac{3}{8}$

10. $\dfrac{9}{16}$ and $\dfrac{7}{24}$

In Problems 11–13, write each fraction in lowest terms.

11. $\dfrac{25}{60}$ **12.** $\dfrac{125}{250}$ **13.** $\dfrac{96}{120}$

In Problems 14 and 15, round each number to the given place.

14. 21.7648 to the neared hundredth

15. 14.91 to the nearest one (unit)

16. Write $\dfrac{8}{9}$ as a repeating decimal.

17. Write $\dfrac{9}{32}$ as a terminating decimal.

In Problems 18 and 19, write each fraction as a decimal rounded to the indicated place.

18. $\dfrac{11}{6}$ to the nearest hundredth.

19. $\dfrac{19}{8}$ to the nearest tenth.

In Problems 20–22, write each decimal as a fraction in lowest terms.

20. 0.6   **21.** 0.375   **22.** 0.864

In Problems 23–26, write each percent as a decimal.

23. 41%   **24.** 760%

25. 9.03%   **26.** 0.35%

In Problems 27–30, write each decimal as a percent.

27. 0.23   **28.** 1.17

29. 0.045   **30.** 3

31. A student earns 12 points out of a total of 20 points on a quiz.

(a) Express this score as a fraction in lowest terms.

(b) Express this score as a percentage.

Section 1.3   The Number Systems and the Real Number Line

KEY CONCEPTS

- $a < b$ means a is to the left of b on a real number line.
- $a = b$ means a and b are in the same position on a real number line.
- $a > b$ means a is to the right of b on a real number line.
- $|a|$ is the distance from 0 to a on a real number line.

KEY TERMS

Set	Real number line
Elements	Origin
Empty set	Scale
Natural numbers	Coordinate
Counting numbers	Negative real numbers
Whole numbers	Zero
Integers	Positive real numbers
Rational number	Sign
Irrational number	Inequality symbols
Real numbers	Absolute value

You Should Be Able To...	EXAMPLE	Review Exercises
❶ Classify numbers (p. 19)	Example 2	32–41, 57, 58
❷ Plot points on a real number line (p. 22)	Example 3	42
❸ Use inequalities to order real numbers (p. 23)	Example 4	43–47, 51–56
❹ Compute the absolute value of a real number (p. 24)	Example 5	48–50, 55, 56

In Problems 32–35, write each set.

32. A is the set of whole numbers less than 7.

33. B is the set of natural numbers less than or equal to 3.

34. C is the set of integers greater than -3 and less than or equal to 5

35. D is the set of whole numbers less than 0.

In Problems 36–41, use the set

$$\left\{-6, -3.25, 0, 5.030030003\ldots, \frac{9}{3}, 11, \frac{5}{7}\right\}.$$

List all the elements that are

36. natural numbers

37. whole numbers

38. integers

39. rational numbers

40. irrational numbers

41. real numbers

42. Plot the points $\left\{-3, -\frac{4}{3}, 0, 2, 3.5\right\}$ on a real number line.

In Problems 43–47, determine whether the statement is True or False.

43. $-3 > -1$

44. $5 \le 5$

45. $-5 \le -3$

46. $\dfrac{1}{2} = 0.5$

47. $\dfrac{2}{3} > \dfrac{5}{6}$

In Problems 48–50, evaluate each expression.

48. $-\left|\dfrac{1}{2}\right|$

49. $|-7|$

50. $-|-6|$

In Problems 51–56, replace the ? with the correct symbol: $>$, $<$, $=$.

51. $\dfrac{1}{4}$? 0.25

52. -6 ? 0

53. 0.83 ? $\dfrac{3}{4}$

54. $\dfrac{-2}{|-2|}$? $-|-1|$

55. $|-4|$? $|-3|$

56. $\dfrac{4}{5}$? $\left|-\dfrac{5}{6}\right|$

57. Explain the difference between a rational number and an irrational number. Be sure that your explanation includes a discussion of terminating decimals and nonterminating decimals.

58. What do we call the set of positive integers?

Section 1.4 Adding, Subtracting, Multiplying, and Dividing Integers

KEY CONCEPTS

- **Rules of Signs for Multiplying Two Integers**
  1. The product of two positive integers is positive.
  2. The product of one positive integer and one negative integer is negative.
  3. The product of two negative integers is positive.

- **Rules of Signs for Dividing Two Integers**
  1. The quotient of two positive integers is positive.
  2. The quotient of one positive integer and one negative integer is negative.
  3. The quotient of two negative integers is positive.

KEY TERMS

Operations
Sum
Difference
Product
Quotient
Mixed number
Additive inverse
Opposite
Evaluate
Factors
Dividend
Divisor
Multiplicative inverse
Reciprocal
Ratio

You Should Be Able To...	EXAMPLE	Review Exercises
❶ Add integers (p. 28)	Examples 1 through 6	59–66, 89, 90, 93, 99, 100, 102
❷ Determine the additive inverse of a number (p. 30)	Example 7	87, 88
❸ Subtract integers (p. 31)	Examples 8 through 10	67–72, 91, 92, 94, 101
❹ Multiply integers (p. 32)	Examples 11 and 12	73–78, 95, 96, 102
❺ Divide integers (p. 33)	Examples 13 and 14	79–86, 97, 98

In Problems 59–86, perform the indicated operation.

59. $-2 + 9$

60. $6 + (-10)$

61. $-23 + (-11)$

62. $-120 + 25$

63. $-|-2 + 6|$

64. $-|-15| + |-62|$

65. $-110 + 50 + (-18) + 25$

66. $-28 + (-35) + (-52)$

67. $-10 - 12$

68. $18 - 25$

69. $-11 - (-32)$

70. $0 - (-67)$

71. $34 - 18 + 10$

72. $-49 - 8 + 21$

73. $-6(-2)$

74. $4(-10)$

75. $13(-86)$

76. $-19(423)$

77. $(11)(13)(-5)$

78. $(-53)(-21)(-10)$

79. $\dfrac{-20}{-4}$

80. $\dfrac{60}{-5}$

81. $\dfrac{|-55|}{11}$

82. $-\left|\dfrac{-100}{4}\right|$

83. $\dfrac{120}{-15}$

84. $\dfrac{64}{-20}$

85. $\dfrac{-180}{54}$

86. $\dfrac{-450}{105}$

In Problems 87 and 88, determine the additive inverse of each number.

87. 13

88. -45

In Problems 89–98, write the expression using mathematical symbols, and then evaluate the expression.

89. -43 plus 101

90. 45 plus -28

91. -10 minus -116

92. 74 minus 56

93. the sum of 13 and -8

94. the difference between -60 and -10

95. -21 multiplied by -3

96. 54 multiplied by -18

97. -34 divided by -2

98. -49 divided by 14

99. Football Matt Forte had three possessions of the football within the first few minutes of the game. On his first possession he gained 20 yards, on his second possession he lost 6 yards, and on his third possession he gained 12 yards. What was his total yardage?

100. Temperature On a winter day in Detroit, Michigan, the temperature was 10°F in the morning. The temperature rose 12°F in the afternoon and then fell 25°F by midnight. What was the temperature at midnight in Detroit?

101. Temperature One day in Bismarck, North Dakota, the high temperature was 6°F above zero and the low temperature was 18°F below zero. What was the difference between the high and low temperatures on that day in Bismarck?

102. Test Score Ms. Rosen awards 5 points for each correct multiple-choice question and awards 8 points for each correct free-response question. On one of Ms. Rosen's tests, Sarah got 11 multiple-choice questions correct and 4 free-response questions correct. What was Sarah's test score?

Section 1.5 Adding, Subtracting, Multiplying, and Dividing Rational Numbers

KEY CONCEPTS

KEY TERMS

Least common denominator

- **Multiplying Fractions**

$$\frac{a}{b} \cdot \frac{c}{d} = \frac{a \cdot c}{b \cdot d} \quad \text{where } b, d \neq 0$$

- **Dividing Fractions**

$$\frac{a}{b} \div \frac{c}{d} = \frac{a}{b} \cdot \frac{d}{c} = \frac{a \cdot d}{b \cdot c} \quad \text{where } b, c, d \neq 0$$

- **Adding or Subtracting Fractions with the Same Denominator**

$$\frac{a}{c} + \frac{b}{c} = \frac{a + b}{c} \quad \text{where } c \neq 0$$

$$\frac{a}{c} - \frac{b}{c} = \frac{a - b}{c} = \frac{a + (-b)}{c} \quad \text{where } c \neq 0$$

- **Adding or Subtracting Fractions with Unlike Denominators**

Step 1: Find the LCD of the fractions.

Step 2: Find equivalent fractions with the LCD by multiplying by a fraction equivalent to 1.

Step 3: Add or subtract the numerators, and write the result over the common denominator.

Step 4: Simplify the result.

You Should Be Able To...	EXAMPLE	Review Exercises
❶ Multiply rational numbers in fraction form (p. 37)	Example 1	103–106, 136
❷ Divide rational numbers in fraction form (p. 38)	Example 2	107–110
❸ Add or subtract rational numbers in fraction form (p. 39)	Examples 3 through 7	111–122, 137
❹ Add, subtract, multiply, or divide rational numbers in decimal form (p. 43)	Examples 8 through 11	123–135, 138

In Problems 103–122, perform the indicated operation. Write in lowest terms.

103. $\dfrac{2}{3} \cdot \dfrac{15}{8}$

104. $-\dfrac{3}{8} \cdot \dfrac{10}{21}$

105. $\dfrac{5}{8}\left(-\dfrac{2}{25}\right)$

106. $5\left(-\dfrac{3}{10}\right)$

107. $\dfrac{24}{17} \div \dfrac{18}{3}$

108. $-\dfrac{5}{12} \div \dfrac{10}{16}$

109. $-\dfrac{27}{10} \div 9$

110. $20 \div \left(-\dfrac{5}{8}\right)$

111. $\dfrac{2}{9} + \dfrac{1}{9}$

112. $-\dfrac{6}{5} + \dfrac{4}{5}$

113. $\dfrac{5}{7} - \dfrac{2}{7}$

114. $\dfrac{7}{5} - \left(-\dfrac{8}{5}\right)$

115. $\dfrac{3}{10} + \dfrac{1}{20}$

116. $\dfrac{5}{12} + \dfrac{4}{9}$

117. $-\dfrac{7}{35} - \dfrac{2}{49}$

118. $\dfrac{5}{6} - \left(-\dfrac{1}{4}\right)$

119. $-2 - \left(-\dfrac{5}{12}\right)$

120. $-5 + \dfrac{9}{4}$

121. $-\dfrac{1}{10} + \left(-\dfrac{2}{5}\right) + \dfrac{1}{2}$

122. $-\dfrac{5}{6} - \dfrac{1}{4} + \dfrac{3}{24}$

In Problems 123–134, perform the indicated operation.

123. $30.3 + 18.2$

124. $-43.02 + 18.36$

125. $201.37 - 118.39$

126. $-35.1 - 18.64$

127. $(-0.04)(-2.01)$

128. $(87.3)(-2.98)$

129. $\dfrac{69.92}{3.8}$

130. $-\dfrac{1.08318}{0.042}$

131. $12.5 - 18.6 + 8.4$

132. $-13.5 + 10.8 - 20.2$

133. $(12.9)(1.4)(-0.3)$

134. $(2.4)(6.1)(-0.05)$

135. Checking Account Lee had a balance of $256.75 in her checking account. Lee wrote a check for $175.68 on Wednesday and wrote a check for $180.00 on Thursday. What is her checking account balance now? Is Lee's account overdrawn?

136. Super Bowl Party Jarred had 36 friends at his Super Bowl party. Two-thirds of his friends wanted the NFC team to win. How many of Jarred's friends wanted the NFC team to win the Super Bowl?

137. Ribbon Cutting Tara has a piece of ribbon that is 15 inches long. If she cuts off a $3\frac{1}{2}$ - inch piece of the ribbon, what is the length of the piece that remains?

138. Buying Clothes While shopping at her favorite store, Sierra bought 5 sweaters. If the sweaters cost $35 each and sales tax is 6.75% of the net price (net price = price × quantity), how much did Sierra spend on the clothes?

Section 1.6   Properties of Real Numbers

KEY CONCEPTS

- **Identity Property of Addition**
  For any real number a, $0 + a = a + 0 = a$.

- **Commutative Property of Addition**
  If a and b are real numbers, then $a + b = b + a$.

- **Additive Inverse Property**
  For any nonzero real number a, $a + (-a) = -a + a = 0$.

- **Associative Property of Addition**
  If a, b, and c are real numbers, then $a + (b + c) = (a + b) + c$.

- **Commutative Property of Multiplication**
  If a and b are real numbers, then $a \cdot b = b \cdot a$.

- **Multiplication Property of Zero**
  For any real number a, the product of a and 0 is always 0; that is, $a \cdot 0 = 0 \cdot a = 0$.

- **Multiplicative Identity**
  $a \cdot 1 = 1 \cdot a = a$ for any real number a.

- **Associative Property of Multiplication**
  If a, b, and c are real numbers, then $a \cdot (b \cdot c) = (a \cdot b) \cdot c$.

- **Multiplicative Inverse Property**
  $a \cdot \dfrac{1}{a} = \dfrac{1}{a} \cdot a = 1$ provided that $a \neq 0$

- **Division Properties of Zero**
  For any nonzero number a,

  1. The quotient of 0 and a is 0. That is, $\dfrac{0}{a} = 0$.

  2. The quotient of a and 0 is undefined. That is, $\dfrac{a}{0}$ is undefined.

KEY TERMS

Additive Identity
Multiplicative Identity
Conversion
Commutative Property
Grouping Symbols
Associative Property
Undefined

You Should Be Able To...	EXAMPLE	Review Exercises
❶ Use the Identity Properties of Addition and Multiplication (p. 48)	Example 1	140, 141, 144–146, 148, 151–156, 164–166
❷ Use the Commutative Properties of Addition and Multiplication (p. 50)	Examples 2 through 4	142, 143, 147, 151–154, 157, 158, 161, 162, 167, 168
❸ Use the Associative Properties of Addition and Multiplication (p. 51)	Examples 5 through 7	139, 150, 155, 156
❹ Understand the Multiplication and Division Properties of 0 (p. 53)	Example 8	149, 159, 160, 163

In Problems 139–150, state the property of real numbers that is being illustrated.

139. $(5 \cdot 12) \cdot 10 = 5 \cdot (12 \cdot 10)$

140. $20 \cdot \dfrac{1}{20} = 1$

141. $\dfrac{8}{3} \cdot \dfrac{3}{8} = 1$

142. $\dfrac{5}{3} \cdot \left(-\dfrac{18}{61}\right) \cdot \dfrac{3}{5} = \dfrac{5}{3} \cdot \dfrac{3}{5} \cdot \left(-\dfrac{18}{61}\right)$

143. $9 \cdot 73 \cdot \dfrac{1}{9} = 9 \cdot \dfrac{1}{9} \cdot 73$

144. $23.9 + (-23.9) = 0$

145. $36 + 0 = 36$

146. $-49 + 0 = -49$

147. $23 + 5 + (-23)$ is equivalent to $23 + (-23) + 5$

148. $\dfrac{7}{8}$ is equivalent to $\dfrac{7}{8} \cdot \dfrac{3}{3}$

149. $14 \cdot 0 = 0$

150. $-5.3 + (5.3 + 2.8) = (-5.3 + 5.3) + 2.8$

In Problems 151–168, evaluate each expression, if possible, by using the properties of real numbers.

151. $144 + 29 + (-144)$

152. $76 + 99 + (-76)$

153. $\dfrac{19}{3} \cdot 18 \cdot \dfrac{3}{19}$

154. $\dfrac{14}{9} \cdot 121 \cdot \dfrac{9}{14}$

155. $3.4 + 42.56 + (-42.56)$

156. $5.3 + 3.6 + (-3.6)$

157. $\dfrac{9}{7} \cdot \left(-\dfrac{11}{3}\right) \cdot 7$

158. $\dfrac{13}{5} \cdot \dfrac{18}{39} \cdot 5$

159. $\dfrac{7}{0}$

160. $\dfrac{0}{100}$

161. $1000(-334)(0.001)$

162. $400(0.5)(0.01)$

163. $43{,}569{,}003 \cdot 0$

164. $154 \cdot \dfrac{1}{154}$

165. $\dfrac{3445}{302} + \left(-\dfrac{3445}{302}\right)$

166. $130 \cdot \dfrac{42}{42}$

167. $-\dfrac{7}{48} \cdot \dfrac{20}{3} \cdot \dfrac{12}{7}$

168. $\dfrac{9}{8} \cdot \left(-\dfrac{25}{13}\right) \cdot \dfrac{48}{9}$

Section 1.7 Exponents and the Order of Operations

KEY CONCEPTS

- **Exponential Notation**

  If n is a natural number and a is a real number, then

  $$a^n = \underbrace{a \cdot a \cdot a \cdot \ldots \cdot a}_{n \text{ factors}}$$

  where a is called the base and the natural number n is called the exponent or power.

- **Rules for Order of Operations**

  Step 1: Perform all operations within *grouping symbols* first. When an expression has multiple grouping symbols, begin with the innermost pair of grouping symbols and work outward.

  Step 2: Evaluate expressions containing *exponents*.

  Step 3: Perform *multiplication and division* in the order in which they occur, working from *left to right*.

  Step 4: Perform *addition and subtraction* in the order in which they occur, working from *left to right*.

KEY TERMS

Exponential notation
Base
Exponent
Power
Exponential form
Expanded form

You Should Be Able To...	EXAMPLE	Review Exercises
❶ Evaluate exponential expressions (p. 56)	Examples 1 through 4	169–178
❷ Apply the rules for order of operations (p. 57)	Examples 5 through 11	179–186

In Problems 169–172, write in exponential form.

169. $3 \cdot 3 \cdot 3 \cdot 3$

170. $\dfrac{2}{3} \cdot \dfrac{2}{3} \cdot \dfrac{2}{3}$

171. $(-4)(-4)$

172. $(-3)(-3)(-3)$

In Problems 173–178, evaluate each expression.

173. 5^3

174. -5^3

175. $(-3)^4$

176. $(-5)^3$

177. -3^4

178. $\left(\dfrac{1}{2}\right)^6$

In Problems 179–186, evaluate each expression.

179. $-2 + 16 \div 4 \cdot 2 - 10$

180. $-4 + 3[2^3 + 4(2 - 10)]$

181. $(12 - 7)^3 + (19 - 10)^2$

182. $5 - (-12 \div 2 \cdot 3) + (-3)^2$

183. $\dfrac{2(4 + 8)}{3 + 3^2}$

184. $\dfrac{3(5 + 2^2)}{2 \cdot 3^3}$

185. $\dfrac{6[12 - 3(5 - 2)]}{5[21 - 2(4 + 5)]}$

186. $\dfrac{4[3 + 2(8 - 6)]}{5[14 - 2(2 + 3)]}$

Section 1.8   Simplifying Algebraic Expressions

KEY CONCEPTS

- **Distributive Property**
  If $a, b,$ and c are real numbers, then $a(b + c) = a \cdot b + a \cdot c$ and $(a + b)c = a \cdot c + b \cdot c$.

KEY TERMS

Algebra	Term
Variable	Coefficient
Constant	Like terms
Algebraic expression	Combining like
Evaluate an algebraic	terms
expression	Simplify

You Should Be Able To...	EXAMPLE	Review Exercises
1 Evaluate algebraic expressions (p. 64)	Examples 1 and 2	187–190, 205
2 Identify like terms and unlike terms (p. 65)	Examples 3 through 5	191–196
3 Use the Distributive Property (p. 67)	Example 6	200–204
4 Simplify algebraic expressions by combining like terms (p. 68)	Examples 7 through 9	197–204

In Problems 187–190, evaluate each expression using the given values of the variables.

187. $x^2 - y^2$ for $x = 5, y = -2$

188. $x^2 - 3y^2$ for $x = 3, y = -3$

189. $(x + 2y)^3$ for $x = -1, y = -4$

190. $\dfrac{a - b}{x - y}$ for $a = 5, b = -10, x = -3, y = 2$

In Problems 191 and 192, identify the terms and then name the coefficient of each term.

191. $3x^2 - x + 6$

192. $2x^2y^3 - \dfrac{y}{5}$

In Problems 193–196, determine whether the terms are like or unlike.

193. $4xy^2, -6xy^2$

194. $-3x, 4x^2$

195. $-6y, -6$

196. $-10, 4$

In Problems 197–204, simplify each algebraic expression.

197. $4x - 6x - x$

198. $6x - 10 - 10x - 5$

199. $0.2x^4 + 0.3x^3 - 4.3x^4$

200. $-3(x^4 - 2x^2 - 4)$

201. $20 - (x + 2)$

202. $-6(2x + 5) + 4(4x + 3)$

203. $5 - (3x - 1) + 2(6x - 5)$

204. $\dfrac{1}{6}(12x + 18) - \dfrac{2}{5}(5x + 10)$

205. **Moving Van** The cost of renting a moving van for one day is $19.95 plus $0.25 per mile. The expression $19.95 + 0.25m$ represents the total cost of renting the truck for one day and driving m miles. Evaluate the expression $19.95 + 0.25m$ for $m = 315$.

Chapter 1 Test

Step-by-step test solutions are found on the Chapter Test Prep Videos available in MyMathLab®, on You Tube™, or may be accessed using the QR code to the right.

1. Find the LCM of 2, 6, and 14.

2. Write $\dfrac{21}{66}$ in lowest terms.

3. Write $\dfrac{13}{9}$ as a decimal rounded to the nearest hundredth.

4. Write 0.425 as a fraction in lowest terms.

5. Write 0.6% as a decimal.

6. Write 0.183 as a percent.

In Problems 7–15, perform the indicated operation. Write in lowest terms.

7. $\dfrac{4}{15} - \left(-\dfrac{2}{30}\right)$

8. $\dfrac{21}{4} \cdot \dfrac{3}{7}$

9. $-16 \div \dfrac{3}{20}$

10. $14 - 110 - (-15) + (-21)$

11. $-14.5 + 2.34$

12. $(-4)(-1)(-5)$

13. $16 \div 0$

14. -6 subtracted from -20

15. -110 divided by -2

16. Use the set $\left\{-2, -\dfrac{1}{2}, 0, 2.5, 6\right\}$. List all of the elements that are:

    (a) natural numbers

    (b) whole numbers

    (c) integers

    (d) rational numbers

    (e) irrational numbers

    (f) real numbers

In Problems 17 and 18, replace the ? with the correct symbol $>, <,$ or $=$.

17. $-|-14|$? -12

18. $\left|-\dfrac{2}{5}\right|$? 0.4

In Problems 19–21, evaluate each expression.

19. $-16 \div 2^2 \cdot 4 + (-3)^2$

20. $\dfrac{4(-9) - 3^2}{25 + 4(-6 - 1)}$

21. $8 - 10[6^2 - 5(2 + 3)]$

22. Evaluate $(x - 2y)^3$ for $x = -1$ and $y = 3$.

In Problems 23 and 24, simplify each algebraic expression.

23. $-6(2x + 5) - (4x - 2)$

24. $\dfrac{1}{2}(4x^2 + 8) - 6x^2 + 5x$

25. **Bank Account** Latoya started with $675.15 in her bank account. She wrote a check for $175.50, withdrew $78.00 in cash, and made a deposit of $110.20. How much money does Latoya have in her bank account now?

26. **Perimeter** The length of a rectangle is 5 feet more than its width. The algebraic expression $2(x + 5) + 2x$ represents the perimeter of the rectangle. Simplify the expression $2(x + 5) + 2x$.

CHAPTER

2 Equations and Inequalities in One Variable

You and your friends want to buy pizza to eat while watching the Super Bowl. Your local grocery store sells medium (12″) frozen pizzas for $9.99 and small (8″) frozen pizzas that are on special for $4.49 each. Which should you buy to get the best deal?

Understanding mathematical formulas and models allows you to solve everyday problems such as this one. See Problem 83 in Section 2.4.

The Big Picture: Putting It Together

In Chapter 1, the arithmetic skills needed throughout the course were reviewed. Algebraic expressions were introduced, and were simplified and evaluated.

In this chapter, we begin the study of algebra. The word "algebra" comes from the Arabic word *al-jabr*. The word *al-jabr* means "restoration." This is a reference to the fact that if a number is added to one side of an equation, then it must also be added to the other side in order to "restore" the equality. While algebra now means a whole lot more than "restoration," we will concentrate on the "restoration" part of algebra in this chapter.

Outline

2.1 Linear Equations: The Addition and Multiplication Properties of Equality

Objectives

1. Determine Whether a Number Is a Solution of an Equation
2. Use the Addition Property of Equality to Solve Linear Equations
3. Use the Multiplication Property of Equality to Solve Linear Equations

Are You Prepared for This Section?

Before getting started, complete the following problems. If you get a problem wrong, go back to the section cited and review the material.

P1. Determine the additive inverse of 3. What is the sum of a real number and its additive inverse? [Section 1.4, p. 30]

P2. Determine the multiplicative inverse of $-\dfrac{4}{3}$. What is the product of a nonzero number and its multiplicative inverse? [Section 1.4, pp. 33–34]

P3. Evaluate: $\dfrac{2}{3}\left(\dfrac{3}{2}\right)$ [Section 1.5, pp. 37–38]

P4. Use the Distributive Property to simplify: $-4(2x + 3)$ [Section 1.8, pp. 67–68]

P5. Simplify: $11 - (x + 6)$ [Section 1.8, pp. 68–69]

Work Smart

Although there is no exponent written on x in the equation $ax + b = 0$, it is important to remember that $x = x^1$.

▶ ① Determine Whether a Number Is a Solution of an Equation

We begin with a definition.

> **Definition**
>
> A **linear equation in one variable** is an equation that can be written in the form $ax + b = 0$, where a and b are real numbers and a does not equal 0.

Examples of linear equations (in the variable x) are

$$x - 5 = 8 \qquad \frac{1}{2}x - 7 = \frac{3}{2}x + 8 \qquad 0.2(x + 5) - 1.5 = 4.25 - (x + 3)$$

The algebraic expressions in the equation are called the **sides** of the equation. For example, in the equation $\frac{1}{2}x - 7 = \frac{3}{2}x + 8$, the algebraic expression $\frac{1}{2}x - 7$ is the **left side** of the equation, and $\frac{3}{2}x + 8$ is the **right side** of the equation.

An equation may be true or false. For example, the equation $x - 5 = 8$ is true if the variable x is replaced by 13, but false if x is replaced by 2.

Because replacing x by 13 in the equation $x - 5 = 8$ results in a true statement, 13 is a *solution* of the equation $x - 5 = 8$. We could also say that $x = 13$ *satisfies* the equation.

> **Definition**
>
> The **solution** of a linear equation is the value or values of the variable that make the equation a true statement. The set of all solutions of an equation is called the **solution set**. We sometimes say that the solution **satisfies** the equation.

Set notation is used to indicate the solution set of an equation. For example, the solution set of $x - 5 = 8$ is $\{13\}$.

To determine whether a number satisfies an equation, replace the variable with the number and see whether the left side of the equation equals the right side of the equation. If it does, we have a true statement, and the number is a solution of the equation.

EXAMPLE 1   **Determine Whether a Number is a Solution of a Linear Equation**

Determine whether the given value of the variable is a solution of the equation $4x + 7 = 19$.

(a) $x = -2$ (b) $x = 3$

Solution

(a)

$$4x + 7 = 19$$

Replace x with -2: $$4(-2) + 7 \overset{?}{=} 19$$

Simplify: $$-8 + 7 \overset{?}{=} 19$$

$$-1 = 19 \quad \text{False}$$

> **In Other Words**
> The symbol $\overset{?}{=}$ is used to indicate that we are unsure whether the left side of the equation is equal to the right side of the equation.

Because the left side of the equation does not equal the right side when x is replaced by -2, $x = -2$ is *not* a solution of the equation.

(b)

$$4x + 7 = 19$$

Replace x with 3: $$4(3) + 7 \overset{?}{=} 19$$

Simplify: $$12 + 7 \overset{?}{=} 19$$

$$19 = 19 \quad \text{True}$$

Because both sides of the equation are equal when x is replaced by 3, $x = 3$ is a solution. ●

Quick ✓

1. The _____ of a linear equation is the value or values of the variable that make the equation a true statement.

In Problems 2–5, determine whether the given value of the variable is a solution of the linear equation. Answer Yes or No.

2. $a - 4 = -7; a = -3$

3. $2x + 1 = 11; x = \dfrac{21}{2}$

4. $3x - (x + 4) = 8; x = 6$

5. $-9b + 3 + 7b = -3b + 8; b = -3$

▶ ❷ **Use the Addition Property of Equality to Solve Linear Equations**

Linear equations are solved by writing a series of *equivalent equations* that result in the equation

$$x = a \ number$$

Equivalent equations are formed using mathematical properties that transform the original equation into a new equation that has the same solution. The first property presented is called the *Addition Property of Equality*.

> **In Other Words**
> The Addition Property of Equality says that whatever you add to one side of the equation, you must also add to the other side.

> **Addition Property of Equality**
> The **Addition Property of Equality** states that for real numbers $a, b,$ and $c,$
> $$\text{if } a = b, \text{ then } a + c = b + c$$

Remember, since $a - b$ is equivalent to $a + (-b)$, the Addition Property of Equality can also be used to subtract a real number from each side of the equation.

Because the goal in solving a linear equation is to get the variable by itself with a coefficient of 1, we say that we want to **isolate the variable.**

EXAMPLE 2   **How to Use the Addition Property of Equality to Solve a Linear Equation**

Solve the linear equation: $x - 6 = 11$

Step-by-Step Solution

Because the coefficient of the variable x in this equation is 1, it is only necessary to get x by itself.

Step 1: Isolate the variable x on the left side of the equation.

$$x - 6 = 11$$

Add 6 to both sides of the equation:

$$x - 6 + 6 = 11 + 6$$

Step 2: Simplify the left and right sides of the equation.

Use the Additive Inverse Property, $a + (-a) = 0$:

$$x + 0 = 17$$

Use the Additive Identity Property, $a + 0 = a$:

$$x = 17$$

Step 3: Check Verify that $x = 17$ is the solution.

$$x - 6 = 11$$

Replace x by 17 in the original equation to see whether a true statement results:

$$17 - 6 \overset{?}{=} 11$$

$$11 = 11 \quad \text{True}$$

Because $x = 17$ satisfies the original equation, the solution is 17, or the solution set is $\{17\}$. ●

EXAMPLE 3   **Using the Addition Property of Equality**

Solve the linear equation: $x + \dfrac{5}{2} = \dfrac{1}{4}$

Solution

$$x + \frac{5}{2} = \frac{1}{4}$$

Subtract $\dfrac{5}{2}$ from both sides of the equation:

$$x + \frac{5}{2} - \frac{5}{2} = \frac{1}{4} - \frac{5}{2}$$

$a + (-a) = 0$; LCD = 4:

$$x + 0 = \frac{1}{4} - \frac{5}{2} \cdot \frac{2}{2}$$

$a + 0 = a$:

$$x = \frac{1}{4} - \frac{10}{4}$$

$$x = -\frac{9}{4}$$

Check   Verify that $x = -\dfrac{9}{4}$ is the solution.

$$x + \frac{5}{2} = \frac{1}{4}$$

$$-\frac{9}{4} + \frac{5}{2} \overset{?}{=} \frac{1}{4}$$

$$-\frac{9}{4} + \frac{10}{4} \overset{?}{=} \frac{1}{4}$$

$$\frac{1}{4} = \frac{1}{4} \quad \text{True}$$

The solution is $-\dfrac{9}{4}$, or the solution set is $\left\{-\dfrac{9}{4}\right\}$. ●

Quick ✓

6. *True or False* The Addition Property of Equality says that whatever you add to one side of the equation, you must also add to the other side.

In Problems 7–12, solve each linear equation using the Addition Property of Equality.

7. $x - 11 = 21$

8. $y + 7 = 21$

9. $-8 + a = 4$

10. $-3 = 12 + c$

11. $z - \dfrac{2}{3} = \dfrac{5}{3}$

12. $\dfrac{5}{4} + x = \dfrac{1}{6}$

EXAMPLE 4 **How Much Is the Fitness Tracker?**

The total cost to buy a fitness tracker, including $6.87 in sales tax, was $144.36. To find the price p of the fitness tracker before tax, solve $p + 6.87 = 144.36$ for p.

Solution

$$p + 6.87 = 144.36$$

Subtract 6.87 from both sides of the equation: $\quad p + 6.87 - 6.87 = 144.36 - 6.87$

Use the Additive Inverse Property, $a + (-a) = 0$: $\quad\quad\quad\quad p + 0 = 137.49$

Use the Additive Identity Property, $a + 0 = a$: $\quad\quad\quad\quad\quad p = 137.49$

Because $p = 137.49$, the fitness tracker costs $137.49 before taxes. ●

Quick ✓

13. The total cost for a used car, including tax, title, and dealer preparation charges of $1472.25, is $13,927.25. To find the price p of the car before the extra charges, solve the equation $p + 1472.25 = 13,927.25$ for p.

▶ ❸ Use the Multiplication Property of Equality to Solve Linear Equations

A second property used to create equivalent equations is called the *Multiplication Property of Equality*.

In Other Words

The Multiplication Property of Equality says that when you multiply one side of an equation by a nonzero quantity, you must also multiply the other side by the same nonzero quantity.

Multiplication Property of Equality

The **Multiplication Property of Equality** states that for real numbers a, b, and c, where c does not equal 0,

$$\text{if } a = b, \quad \text{then} \quad ac = bc$$

EXAMPLE 5 **How to Solve a Linear Equation Using the Multiplication Property of Equality**

Solve the linear equation: $5x = 30$

Step-by-Step Solution

Step 1: Get the coefficient of the variable x to be 1.

$$5x = 30$$

Multiply both sides of the equation by $\dfrac{1}{5}$: $\quad \dfrac{1}{5}(5x) = \dfrac{1}{5}(30)$

(continued)

Step 2: Simplify the left and right sides of the equation.

Use the Associative Property of Multiplication: $\left(\frac{1}{5} \cdot 5\right)x = \frac{1}{5}(30)$

Use the Multiplicative Inverse Property, $a \cdot \frac{1}{a} = 1$: $1 \cdot x = 6$

Use the Multiplicative Identity Property, $1 \cdot a = a$: $x = 6$

Step 3: Check Verify that $x = 6$ is the solution.

$5x = 30$

Replace x by 6 in the original equation: $5(6) \overset{?}{=} 30$

$30 = 30$ True

The solution is 6, or the solution set is $\{6\}$.   ●

In Step 1 of Example 5, both sides of the equation were multiplied by $\frac{1}{5}$ to make the coefficient of x to equal 1. We could also divide both sides of the equation by 5, because dividing by 5 is the same as multiplying by the reciprocal of 5, $\frac{1}{5}$.

$5x = 30$

Divide both sides of the equation by 5: $\dfrac{5x}{5} = \dfrac{30}{5}$

Simplify: $x = 6$

Which approach do you prefer?

EXAMPLE 6   **Using the Multiplication Property of Equality**

Solve the linear equation: $-4n = 18$

Solution

$-4n = 18$

Multiply both sides of the equation by $-\frac{1}{4}$: $-\dfrac{1}{4}(-4n) = -\dfrac{1}{4} \cdot 18$

Use the Associative Property of Multiplication: $\left(-\dfrac{1}{4}(-4)\right)n = -\dfrac{18}{4}$

Use the Multiplicative Inverse Property; factor: $1 \cdot n = -\dfrac{9 \cdot 2}{2 \cdot 2}$

Use the Multiplicative Identity; divide out common factors: $n = -\dfrac{9}{2}$

Check   Verify the solution by replacing $n = -\dfrac{9}{2}$ in the original equation to see whether a true statement results.

$-4n = 18$

$-4\left(-\dfrac{9}{2}\right) \overset{?}{=} 18$

Divide out the common factor: $\overset{-2}{\cancel{-4}}\left(-\dfrac{9}{\cancel{2}_1}\right) \overset{?}{=} 18$

$18 = 18$ True

The solution is $-\dfrac{9}{2}$, or the solution set is $\left\{-\dfrac{9}{2}\right\}$.   ●

EXAMPLE 7   **Solving a Linear Equation with a Fraction as a Coefficient**

Solve the linear equation: $\dfrac{x}{3} = -7$

Solution

The left side of the equation, $\dfrac{x}{3}$, is equivalent to $\dfrac{1}{3}x$. To eliminate the fraction, multiply both sides of the equation by 3, the reciprocal of $\dfrac{1}{3}$:

$$\frac{x}{3} = -7$$

$$\frac{1}{3} \cdot x = -7$$

Multiply both sides of the equation by 3, the reciprocal of $\dfrac{1}{3}$: $3 \cdot \left(\dfrac{1}{3}x\right) = 3 \cdot (-7)$

Use the Associative Property of Multiplication: $\left(3 \cdot \dfrac{1}{3}\right)x = -21$

Use the Multiplicative Inverse Property, $a \cdot \dfrac{1}{a} = 1$: $1 \cdot x = -21$

Use the Multiplicative Identity, $1 \cdot a = a$: $x = -21$

Check   Let $x = -21$ in the original equation to see whether a true statement results.

$$\frac{x}{3} = -7$$

Let $x = -21$: $\dfrac{-21}{3} \overset{?}{=} -7$

$$-7 = -7 \quad \text{True}$$

The solution is -21, or the solution set is $\{-21\}$.   ●

EXAMPLE 8   **Solving a Linear Equation with a Fraction as a Coefficient**

Solve the linear equation: $12 = \dfrac{2}{3}x$

Solution

$$12 = \frac{2}{3}x$$

Work Smart

When the variable is on the right side of an equation, isolate the variable in the same way as when it was on the left side of the equation.

Multiply both sides of the equation by $\dfrac{3}{2}$, the reciprocal of $\dfrac{2}{3}$: $\dfrac{3}{2}(12) = \dfrac{3}{2}\left(\dfrac{2}{3}x\right)$

Use the Associative Property of Multiplication: $18 = \left(\dfrac{3}{2} \cdot \dfrac{2}{3}\right)x$

(continued)

$$\text{Use the Multiplicative Inverse Property, } a \cdot \frac{1}{a} = 1: \quad 18 = 1 \cdot x$$

$$\text{Use the Multiplicative Identity Property, } 1 \cdot a = a: \quad 18 = x$$

$$\text{Symmetric Property: if } b = a, \text{ then } a = b: \quad x = 18$$

Check Let $x = 18$ in the original equation to see whether a true statement results.

$$12 = \frac{2}{3}x$$

$$\text{Let } x = 18: \quad 12 \overset{?}{=} \frac{2}{3}(18)$$

$$12 \overset{?}{=} \frac{2}{\cancel{3}}(\overset{6}{\cancel{18}})$$

$$12 = 12 \quad \text{True}$$

The solution is 18, or the solution set is $\{18\}$.

> ### Quick ✓
>
> In Problems 19–21, solve each linear equation using the Multiplication Property of Equality.
>
> **19.** $\dfrac{4}{3}n = 12$ **20.** $-21 = \dfrac{7}{3}k$ **21.** $15 = -\dfrac{z}{2}$

EXAMPLE 9 **Solving a Linear Equation with Fractions**

Solve the linear equation: $\dfrac{4}{5} = -\dfrac{2}{15}p$

Solution

$$\frac{4}{5} = -\frac{2}{15}p$$

Multiply both sides of the equation by $-\dfrac{15}{2}$:

$$-\frac{15}{2} \cdot \frac{4}{5} = -\frac{15}{2}\left(-\frac{2}{15}p\right)$$

Simplify; use the Associative Property of Multiplication:

$$-\frac{\overset{3}{\cancel{15}}}{\underset{1}{\cancel{2}}} \cdot \frac{\overset{2}{\cancel{4}}}{\underset{1}{\cancel{5}}} = \left(-\frac{15}{2}\left(-\frac{2}{15}\right)\right)p$$

$$-6 = p$$

Check Let $p = -6$ in the original equation to see whether a true statement results.

$$\frac{4}{5} = -\frac{2}{15}p$$

$$\text{Let } p = -6: \quad \frac{4}{5} \overset{?}{=} -\frac{2}{15}(-6)$$

$$\text{Simplify:} \quad \frac{4}{5} \overset{?}{=} -\frac{2}{\underset{5}{\cancel{15}}}(\overset{-2}{\cancel{-6}})$$

$$\frac{4}{5} = \frac{4}{5} \quad \text{True}$$

The solution is -6, or the solution set is $\{-6\}$.

Working with fractional coefficients can be tricky. The three equations $\dfrac{-x}{3} = 7$, $-\dfrac{1}{3}x = 7$, and $\dfrac{x}{-3} = 7$ are all equivalent. Do you see why?

Work Smart

To solve an equation of the form $ax = b$, where a is an integer, either multiply by the reciprocal of a or divide by a.

To solve an equation of the form $ax = b$, where a is a fraction, multiply by the reciprocal of a.

Multiply by the reciprocal of a	Divide by a	Multiply by the reciprocal of a
$2x = 7 \qquad 5x = -\dfrac{15}{2}$	$-3x = 72$	$\dfrac{4}{5}x = -16$
$\dfrac{1}{2}(2x) = \dfrac{1}{2}\cdot 7 \quad \dfrac{1}{5}(5x) = \dfrac{1}{5}\left(-\dfrac{15}{2}\right)$	$\dfrac{-3x}{-3} = \dfrac{72}{-3}$	$\dfrac{5}{4}\left(\dfrac{4}{5}x\right) = \dfrac{5}{4}(-16)$
$x = \dfrac{7}{2} \qquad\quad x = -\dfrac{3}{2}$	$x = -24$	$x = \dfrac{5}{4}\cdot\dfrac{-16}{1}$
		$x = -20$

Quick ✓

In Problems 22–24, solve each linear equation using the Multiplication Property of Equality.

22. $\dfrac{3}{8}b = \dfrac{9}{4}$

23. $-\dfrac{4}{9} = \dfrac{-t}{6}$

24. $\dfrac{1}{4} = -\dfrac{7}{10}m$

2.1 Exercises MyMathLab®

Exercise numbers in green have complete video solutions in MyMathLab or may be accessed using the QR code to the right.

Problems 1–24 are the Quick ✓s that follow the EXAMPLES.

Building Skills

In Problems 25–32, determine whether the given value is a solution to the equation. Answer Yes or No. See Objective 1.

25. $3x - 1 = 5; x = 2$

26. $4t + 2 = 16; t = 3$

27. $4 - (m + 2) = 3(2m - 1); m = 1$

28. $3(x + 1) - x = 5x - 9; x = -3$

29. $8k - 2 = 4; k = \dfrac{3}{4}$

30. $-15 = 3x - 16; x = \dfrac{1}{3}$

31. $r + 1.6 = 2r + 1; r = 0.6$

32. $3s - 6 = 6s - 3.4; s = -1.2$

In Problems 33–48, solve the linear equation using the Addition Property of Equality. Be sure to check your solution. See Objective 2.

33. $x - 9 = 11$

34. $y - 8 = 2$

35. $x + 4 = -8$

36. $r + 3 = -1$

37. $12 = n - 7$

38. $13 = u - 6$

39. $-8 = x + 5$

40. $-2 = y + 13$

41. $x - \dfrac{2}{3} = \dfrac{4}{3}$

42. $x - \dfrac{1}{8} = \dfrac{3}{8}$

43. $z + \dfrac{1}{2} = \dfrac{3}{4}$

44. $n + \dfrac{3}{5} = \dfrac{7}{10}$

45. $\dfrac{5}{12} = x - \dfrac{3}{8}$

46. $\dfrac{3}{8} = y - \dfrac{1}{6}$

47. $w + 3.5 = -2.6$

48. $z + 4.9 = -2.6$

In Problems 49–72, solve the linear equation using the Multiplication Property of Equality. Be sure to check your solution. See Objective 3.

49. $5c = 25$

50. $8b = 48$

51. $-7n = 28$

52. $-8s = 40$

53. $4k = 14$

54. $4z = 30$

55. $-6w = 15$

56. $-8p = 20$

57. $\dfrac{5}{3}a = 35$

58. $\dfrac{4}{3}b = 16$

59. $-\dfrac{3}{11}p = -33$

60. $-\dfrac{6}{5}n = -36$

61. $\dfrac{n}{5} = 8$

62. $-\dfrac{y}{6} = 3$

63. $\dfrac{6}{5} = 2x$

64. $\dfrac{9}{2} = 3b$

65. $5y = -\dfrac{5}{3}$

66. $4r = -\dfrac{12}{5}$

67. $\dfrac{1}{2}m = \dfrac{9}{2}$

68. $\dfrac{1}{4}w = \dfrac{7}{2}$

69. $-\dfrac{3}{8}t = \dfrac{1}{6}$

70. $\dfrac{3}{10}q = -\dfrac{1}{6}$

71. $\dfrac{5}{24} = -\dfrac{y}{8}$

72. $\dfrac{11}{36} = -\dfrac{t}{9}$

Mixed Practice

In Problems 73–100, solve the linear equation. Be sure to check your solution.

73. $n - 4 = -2$

74. $m - 6 = -9$

75. $b + 12 = 9$

76. $c + 4 = 1$

77. $2 = 3x$

78. $9 = 5y$

79. $-4q = 24$

80. $-6m = 54$

81. $-39 = x - 58$

82. $-637 = c - 142$

83. $-18 = -301 + x$

84. $-46 = -51 + q$

85. $-\dfrac{x}{5} = -10$

86. $\dfrac{z}{3} = -12$

87. $m - 56.3 = -15.2$

88. $p - 26.4 = -471.3$

89. $-40 = -6c$

90. $-45 = 12x$

91. $14 = -\dfrac{7}{2}c$

92. $12 = -\dfrac{3}{2}n$

93. $\dfrac{3}{4} = -\dfrac{x}{16}$

94. $\dfrac{5}{9} = -\dfrac{h}{36}$

95. $x - \dfrac{5}{16} = \dfrac{3}{16}$

96. $w - \dfrac{7}{20} = \dfrac{9}{20}$

97. $-\dfrac{3}{16} = -\dfrac{3}{8} + z$

98. $\dfrac{5}{2} = -\dfrac{13}{4} + y$

99. $\dfrac{5}{6} = -\dfrac{2}{3}z$

100. $-\dfrac{4}{9} = \dfrac{8}{3}b$

Applying the Concepts

101. New Car The total cost for a new car, including tax, title, and dealer preparation charges of $2722.50, is $27,685.77. To find the price of the car without the extra charges, solve the equation $y + 2722.50 = 27,685.77$, where y represents the price of the car without the extra charges.

102. New Kayak The total cost for a new kayak is $862.92, including sales tax of $63.92. To find the cost of the kayak without tax, solve the equation $k + 63.92 = 862.92$, where k represents the cost of the kayak.

103. Discount The cost of a sleeping bag has been discounted by $17, so that the sale price of the bag is $51. Find the original price of the sleeping bag, p, by solving the equation $p - 17 = 51$.

104. Discount The cost of a computer has been discounted by $239, so that the sale price is $1230. Find the original price of the computer, c, by solving the equation $c - 239 = 1230$.

105. Eating Out Rebecca purchased several 20-piece boxes of Chicken McNuggets for her child's end-of-year preschool celebration. Each 20-piece box costs $5 and she spent a total of $60 on the chicken. To determine the number of boxes of Chicken McNuggets, b, that she purchased, solve the equation $5b = 60$.

106. Paperback Books Anne bought 3 paperback books to read on the flight to Europe. She paid $36 for the books (without sales tax). To find the average price, p, of each book, solve the equation $3p = 36$.

107. Interest Suppose you have a credit card debt of $3000. Last month, the bank charged you $45 interest on the debt. The solution to the equation $45 = \dfrac{3000}{12}r$ represents the annual interest rate, r, on the credit card. Find the annual interest rate on the credit card.

108. Interest Suppose you have a credit card debt of $4000. Last month, the bank charged you $40 interest on the debt. The solution to the equation $40 = \dfrac{4000}{12}r$ represents the annual interest rate, r, on the credit card. Find the annual interest rate on the credit card.

Extending the Concepts

109. Let λ, the Greek letter lambda, represent some real number. Solve the equation $x + \lambda = 48$ for x.

110. Let β, the Greek letter beta, represent some real number. Solve the equation $x - \beta = 25$ for x.

111. Let θ, the Greek letter theta, represent some real number other than 0. Solve the equation $14 = \theta x$ for x.

112. Let ψ, the Greek letter psi, represent some real number other than 0. Solve the equation $\dfrac{2}{5} = \psi x$ for x.

113. Find the value of λ in the equation $x + \lambda = \dfrac{16}{3}$ so that the solution is $x = -\dfrac{2}{9}$.

114. Find the value of β in the equation $x - \beta = -13.6$ so that the solution is $x = -4.79$.

115. Find the value of θ in the equation $-\dfrac{3}{4} = \theta x$ so that the solution is $x = \dfrac{7}{8}$.

116. Find the value of ψ in the equation $-11.92 = \psi x$ so that the solution is $x = 2.98$.

Explaining the Concepts

117. Explain what is meant by finding the *solution* of an equation.

118. Consider the equation $3z = \dfrac{15}{4}$. You could either multiply both sides of the equation by $\dfrac{1}{3}$ or divide

both sides by 3. Explain which operation you would choose and why you think it is easier.

119. Explain the difference between an algebraic expression and an algebraic equation. Write an equation involving the algebraic expression $x - 10$. Then solve the equation.

120. A classmate suggests that to solve the equation $12x = 4$, you must divide by 12 and get the answer $x = 3$. Explain why this reasoning is or is not incorrect.

121. Explain the Addition Property of Equality. How is it used to solve $x - 5 = 12$? What role does the Identity Property of Addition play in solving this equation?

2.2 Linear Equations: Using the Properties Together

Objectives

① Use the Addition and the Multiplication Properties of Equality to Solve Linear Equations

② Combine Like Terms and Use the Distributive Property to Solve Linear Equations

③ Solve a Linear Equation with the Variable on Both Sides of the Equation

④ Use Linear Equations to Solve Problems

Prepared?...Answers
P1. $10 - 3x$ **P2.** -3

Are You Prepared for This Section?

Before getting started, complete the following problems. If you get a problem wrong, go back to the section cited and review the material.

P1. Simplify by combining like terms: $6 - (4 + 3x) + 8$ [Section 1.8, pp. 68–69]

P2. Evaluate the expression $2(3x + 4) - 5$ for $x = -1$. [Section 1.8, pp. 64–65]

① Use the Addition and the Multiplication Properties of Equality to Solve Linear Equations

In the last section, we solved equations such as $x + 3 = 7$ and $\dfrac{1}{2}z = 8$ using the Addition Property or the Multiplication Property of Equality, but not both. In this section, we use both the Addition and Multiplication Properties of Equality to solve a linear equation. For example, the equation $2x + 3 = 7$ can be read as "Two *times* a number x *plus* three equals seven." We must "undo" both the multiplication and the addition to find the solution of the equation.

EXAMPLE 1 **How to Solve a Linear Equation Using the Addition and Multiplication Properties of Equality**

Solve the equation: $2z - 3 = 9$

Step-by-Step Solution

Step 1: Isolate the term containing the variable.

Use the Addition Property of Equality; add 3 to both sides of the equation:

$$2z - 3 = 9$$
$$2z - 3 + 3 = 9 + 3$$
$$2z = 12$$

Step 2: Get the coefficient of the variable to be 1.

Use the Multiplication Property of Equality; divide both sides of the equation by 2 $\left(\text{or multiply both sides by } \dfrac{1}{2}\right)$:

$$\dfrac{2z}{2} = \dfrac{12}{2}$$
$$z = 6$$

(continued)

Step 3: Check Verify that $z = 6$ is the solution of the equation.

$$2z - 3 = 9$$

Substitute 6 for z in the original equation:
$$2(6) - 3 \overset{?}{=} 9$$
$$12 - 3 \overset{?}{=} 9$$
$$9 = 9 \quad \text{True}$$

Because $z = 6$ satisfies the equation, the solution is 6, or the solution set is $\{6\}$. ●

EXAMPLE 2 **Solving a Linear Equation Using the Addition and Multiplication Properties of Equality**

Solve the equation: $\dfrac{3}{2}p + 3 = 12$

Solution

$$\frac{3}{2}p + 3 = 12$$

Subtract 3 from both sides of the equation:
$$\frac{3}{2}p + 3 - 3 = 12 - 3$$

Simplify:
$$\frac{3}{2}p = 9$$

Multiply both sides of the equation by $\dfrac{2}{3}$:
$$\frac{2}{3}\left(\frac{3}{2}p\right) = \frac{2}{3}(9)$$

Simplify:
$$p = 6$$

Check Let $p = 6$ in the original equation to see whether a true statement results.

$$\frac{3}{2}p + 3 = 12$$

Substitute 6 for p in the original equation:
$$\frac{3}{2}(6) + 3 \overset{?}{=} 12$$

Simplify:
$$9 + 3 \overset{?}{=} 12$$
$$12 = 12 \quad \text{True}$$

The solution of the equation is $p = 6$, or the solution set is $\{6\}$. ●

Quick ✓
1. *True or False* To solve the equation $2x - 11 = 40$, the first step is to add 11 to both sides of the equation.

In Problems 2–5, solve each equation.

2. $5x - 4 = 11$

3. $8 = \dfrac{2}{3}k - 4$

4. $8 - 5r = -2$

5. $-\dfrac{3}{2}n + 2 = -\dfrac{1}{4}$

▶ ❷ **Combine Like Terms and Use the Distributive Property to Solve Linear Equations**

Often we must combine like terms before using the Addition or Multiplication Property of Equality.

EXAMPLE 3　**Combining Like Terms to Solve a Linear Equation**

Solve the equation: $2x - 6 + 3x = 14$

Solution

$$2x - 6 + 3x = 14$$

Combine like terms:　　　　　　　$5x - 6 = 14$

Add 6 to both sides of the equation:　$5x - 6 + 6 = 14 + 6$

$$5x = 20$$

Divide both sides by 5:　　　　　$\dfrac{5x}{5} = \dfrac{20}{5}$

$$x = 4$$

Check

$$2x - 6 + 3x = 14$$

Substitute 4 for x in the original equation:　$2(4) - 6 + 3(4) \overset{?}{=} 14$

$$8 - 6 + 12 \overset{?}{=} 14$$

$$14 = 14 \quad \text{True}$$

The solution of the equation is 4, or the solution set is $\{4\}$.　　　　　　●

> **Quick ✔**
>
> *In Problems 6–9, solve each equation.*
>
> **6.** $7b - 3b + 3 = 11$　　　　　　**7.** $-3a + 4 + 4a = 13 - 27$
>
> **8.** $6c - 2 + 2c = 18$　　　　　　**9.** $-12 = 5x - 3x + 4$

When an equation contains parentheses, use the Distributive Property to remove the parentheses before using the Addition or Multiplication Property of Equality.

EXAMPLE 4　**Solving a Linear Equation Using the Distributive Property**

Solve the equation: $4(2x + 3) - 7 = -11$

Solution

$$4(2x + 3) - 7 = -11$$

Use the Distributive Property to remove parentheses:　$8x + 12 - 7 = -11$

Combine like terms:　　　　　　　　　$8x + 5 = -11$

Subtract 5 from both sides of the equation:　$8x + 5 - 5 = -11 - 5$

$$8x = -16$$

Divide both sides by 8:　　　　　　　$\dfrac{8x}{8} = \dfrac{-16}{8}$

$$x = -2$$

Check

$$4(2x + 3) - 7 = -11$$

Substitute -2 for x in the original equation:　$4[2(-2) + 3] - 7 \overset{?}{=} -11$

$$4(-4 + 3) - 7 \overset{?}{=} -11$$

$$4(-1) - 7 \overset{?}{=} -11$$

$$-4 - 7 \overset{?}{=} -11$$

$$-11 = -11 \quad \text{True}$$

The solution of the equation is -2, or the solution set is $\{-2\}$.　　　　　●

▶ ❸ Solve a Linear Equation with the Variable on Both Sides of the Equation

When solving a linear equation, the goal is to get the terms that contain the variable on one side of the equation and the constant terms on the other side.

EXAMPLE 5 | **Solving a Linear Equation with the Variable on Both Sides of the Equation**

Solve the equation: $9y - 5 = 5y + 9$

Solution

$$9y - 5 = 5y + 9$$

Subtract 5y from both sides of the equation: $9y - 5 - 5y = 5y + 9 - 5y$

$$4y - 5 = 9$$

Add 5 to both sides of the equation: $4y - 5 + 5 = 9 + 5$

$$4y = 14$$

Divide both sides by 4: $\dfrac{4y}{4} = \dfrac{14}{4}$

Simplify: $y = \dfrac{7}{2}$

Check

$$9y - 5 = 5y + 9$$

Substitute $\dfrac{7}{2}$ for y in the original equation: $9\left(\dfrac{7}{2}\right) - 5 \overset{?}{=} 5\left(\dfrac{7}{2}\right) + 9$

$$\dfrac{63}{2} - 5 \overset{?}{=} \dfrac{35}{2} + 9$$

LCD = 2: $\dfrac{63}{2} - \dfrac{10}{2} \overset{?}{=} \dfrac{35}{2} + \dfrac{18}{2}$

$$\dfrac{53}{2} = \dfrac{53}{2} \quad \text{True}$$

The solution of the equation is $\dfrac{7}{2}$, or the solution set is $\left\{\dfrac{7}{2}\right\}$.   ●

EXAMPLE 6 **How to Solve a Linear Equation in One Variable**

Solve the equation: $2(z - 4) + 3z = 4 - (z + 2)$

Step-by-Step Solution

Step 1: Remove any parentheses using the Distributive Property.

$$2(z - 4) + 3z = 4 - (z + 2)$$
$$2z - 8 + 3z = 4 - z - 2$$

Step 2: Combine like terms on each side of the equation.

$$5z - 8 = 2 - z$$

Step 3: Use the Addition Property of Equality to get the terms with the variable on one side of the equation and the constants on the other side.

Add z to both sides of the equation: $5z - 8 + z = 2 - z + z$

Simplify: $6z - 8 = 2$

Add 8 to both sides of the equation: $6z - 8 + 8 = 2 + 8$

Simplify: $6z = 10$

Step 4: Use the Multiplication Property of Equality to get the coefficient of the variable to be 1.

Divide both sides of the equation by 6: $\dfrac{6z}{6} = \dfrac{10}{6}$

Simplify: $z = \dfrac{5}{3}$

Step 5: Check Verify that $z = \dfrac{5}{3}$ satisfies the original equation.

The check is left to you.

The solution of the equation is $\dfrac{5}{3}$, or the solution set is $\left\{\dfrac{5}{3}\right\}$.

●

> **Quick ✓**
>
> **17.** *True or False* To solve the equation $13 - 2(7x + 1) + 8x = 12$, the first step is to subtract 2 from 13 and get $11(7x + 1) + 8x = 12$.
>
> *In Problems 18 and 19, solve each equation.*
>
> **18.** $-9x + 3(2x - 3) = -10 - 2x$ **19.** $3 - 4(p + 5) = 5(p + 2) - 12$

Here is a summary of the steps that you should follow to solve a linear equation in one variable. Not all the steps may be necessary to solve every equation, but you should use this summary as a guide.

> **Solving a Linear Equation in One Variable**
>
> **Step 1:** Remove any parentheses using the Distributive Property.
>
> **Step 2:** Combine like terms on each side of the equation.
>
> **Step 3:** Use the Addition Property of Equality to get the terms with the variable on one side of the equation and the constants on the other side.
>
> **Step 4:** Use the Multiplication Property of Equality to get the coefficient of the variable to be 1.
>
> **Step 5:** Check the solution to verify that it satisfies the original equation.

▶ ④ Use Linear Equations to Solve Problems

Real world problems, such as finding an hourly wage, can be modeled and solved using a linear equation.

EXAMPLE 7 **How Much Does Alejandro Make in an Hour?**

Alejandro, a hospital engineer, earns double time on all hours worked in excess of 40 hours each week. One week, Alejandro worked 46 hours and earned $1326 before taxes. To determine Alejandro's hourly wage, w, solve the equation $40w + 6(2w) = 1326$.

Solution

$$40w + 6(2w) = 1326$$

Simplify: $\qquad 40w + 12w = 1326$

Combine like terms: $\qquad 52w = 1326$

Divide both sides of the equation by 52: $\qquad \dfrac{52w}{52} = \dfrac{1326}{52}$

Simplify: $\qquad w = 25.5$

Check If Alejandro earns $25.50 per hour, then 40 hours × $25.50 per hour = $1020; 6 hours working at double time is 6 hours × (2 × $25.50 per hour) = $306. Alejendro's regular salary, $1020, plus his overtime wages, $306, equals $1326, so the answer $25.50 per hour is correct.

Alejandro makes $25.50 per hour.

Quick ✔

20. Marcella works at a clothing store. Whenever she works more than 40 hours in a week, she gets paid twice her regular hourly wage of $15 per hour. One week, Marcella earned $960 before taxes. To determine how many hours, h, Marcella worked, solve the equation $600 + 30(h - 40) = 960$ for h.

2.2 Exercises MyMathLab® Exercise numbers in green have complete video solutions in MyMathLab or may be accessed using the QR code to the right.

*Problems **1–20** are the* **Quick ✔** *s that follow the* **EXAMPLES**.

Building Skills

In Problems 21–34, solve the equation. Check your solution. See Objective 1.

21. $3x + 4 = 7$

22. $5t + 1 = 11$

23. $2y - 1 = -5$

24. $6z - 2 = -8$

25. $-3p + 1 = 10$

26. $-4x + 3 = 15$

27. $8y + 3 = 15$

28. $6z - 7 = 3$

29. $5 - 2z = 11$

30. $1 - 3k = 4$

31. $\dfrac{2}{3}x + 1 = 9$

32. $\dfrac{5}{4}a + 3 = 13$

33. $\dfrac{7}{2}y - 1 = 13$

34. $\dfrac{1}{5}p - 3 = 2$

In Problems 35–46, solve the equation. Check your solution. See Objective 2.

35. $3x - 7 + 2x = -17$

36. $5r + 2 - 3r = -14$

37. $2k - 7k - 8 = 17$

38. $2b + 5 - 8b = 23$

39. $2(x + 1) + 5 = -9$

40. $3(t - 4) + 7 = -11$

41. $-3(2 + r) = 9$

42. $-5(6 + z) = -20$

43. $17 = 2 - (n + 6)$

44. $21 = 5 - (2a - 1)$

45. $-8 = 5 - (7 - z)$

46. $-9 = 3 - (6 + 4y)$

In Problems 47–60, solve the equation. Check your solution. See Objective 3.

47. $2x + 9 = x + 1$

48. $7z + 13 = 6z + 8$

49. $2t - 6 = 3 - t$

50. $3 + 8x = 21 - x$

51. $14 - 2n = -4n + 7$

52. $6 - 12m = -3m + 3$

53. $-3(5 - 3k) = 6k + 6$

54. $-4(10 - 7x) = 3x + 10$

55. $2(2x + 3) = 3(x - 4)$

56. $3(5 + x) = 2(2x + 11)$

57. $13 + 3(x - 1) = 9x$

58. $-8 + 4(p + 6) = 10p$

59. $9(6 + a) + 33a = 10a$

60. $5(12 - 3w) + 25w = 2w$

Mixed Practice

In Problems 61–76, solve the equation. Check your solution.

61. $-5x + 11 = 1$

62. $-6n + 14 = -10$

63. $4m + 5 = 2$

64. $7x + 1 = -9$

65. $-2(3n - 2) = 2$

66. $-5(2n - 3) = 10$

67. $4k - (3 + k) = -(2k + 3)$

68. $11w - (2 - 4w) = 13 + 2(w - 1)$

69. $2y + 36 = 6 + 6y$

70. $7a - 26 = 13a + 2$

71. $\frac{1}{2}(-4k + 28) = 6 + 14k$

72. $\frac{2}{3}(9a - 12) = -6a - 11$

73. $-\frac{5}{2}(x + 6) + \frac{3}{2}x = -8$

74. $\frac{4}{3}a - \left(\frac{7}{3}a + 6\right) = -15$

75. $-3(2y + 3) - 1 = -4(y + 6) + 2y$

76. $-5(b + 2) + 3b = -2(1 + 5b) + 6$

Applying the Concepts

77. Burger King A Burger King Tendercrisp Chicken garden salad contains 6 more grams of fat than a McDonald's Southwestern salad with crispy chicken. If there are 38 grams of fat in the two salads, find the number of grams of fat in each salad by solving the equation $x + (x + 6) = 38$, where x represents the number of fat grams in the McDonald's Southwestern salad and $x + 6$ represents the number of fat grams in the Burger King Tendercrisp salad. (SOURCE: *Burger King and McDonald's websites*)

78. Whataburger A Whataburger Apple and Cranberry Salad with grilled chicken contains 13 grams of fat less than a Wendy's Apple-Pecan Chicken Salad. If there are 41 grams of fat in the two salads, find the number of grams of fat in each salad by solving the equation $x + (x - 13) = 41$, where x represents the number of grams of fat in a Wendy's Apple-Pecan Chicken Salad and $x - 13$ represents the number of grams of fat in the Whataburger Apple and Cranberry Salad with grilled chicken. (SOURCE: *Whataburger and Wendy's websites*)

△ **79. Dog Run** The length of a rectangular dog run is 2 feet more than twice the width, w, and the perimeter of the dog run is 30 feet. Solve the equation $2w + 2(2w + 2) = 30$ to find the width, w, of the dog run. Then find the length, $2w + 2$, of the dog run.

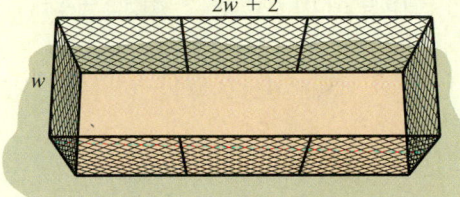

△ **80. Garden** The width of a rectangular garden is 1 yard more than one-half the length, L. The perimeter of the garden is 26 yards. Find the length, L, of the garden by solving the equation $2L + 2\left(\frac{1}{2}L + 1\right) = 26$. Then find the width, $\frac{1}{2}L + 1$, of the garden.

81. Overtime Pay Jennifer worked 44 hours last week, including 4 hours of overtime, and earned $736. She is paid at a rate of 1.5 times her regular hourly rate for overtime hours. Solve the equation $40r + 4(1.5r) = 736$ to find her regular hourly pay rate, r.

82. Overtime Pay Juan worked a total of 50 hours last week and earned $855. He earned 1.5 times his regular hourly rate for 6 hours, and double his hourly rate for 4 holiday hours. Solve the equation $40x + 6(1.5x) + 4(2x) = \$855$, where x is Juan's regular hourly rate.

△ **83. Hanging Wallpaper** Becky purchased a remnant of 42 feet of a wallpaper border to hang in a rectangular bedroom. She knows that one wall of the bedroom is 5 feet longer than the other wall. Let x represent the length of the shorter wall and $x + 5$ represent the length of the longer wall. Use the equation $2x + 2(x + 5) = 42$ to find the dimension of the bedroom.

⚠ **84. Perimeter of a Triangle** The perimeter of a triangle is 210 inches. If the lengths of the sides are 3 consecutive even integers, find the lengths of all 3 sides by solving the equation $x + (x + 2) + (x + 4) = 210$, where x represents the length of the shortest side.

Extending the Concepts

In Problems 85 and 86, solve the equation. Check your solution.

85. $8[4 - 6(x - 1)] + 5[(2x + 3) - 5] = 18x - 338$

86. $3[10 - 4(x - 3)] + 2[(3x + 6) - 2] = 2x + 360$

In Problems 87–90, use a calculator to solve the equation, and round your answer to the indicated place

87. $3(36.7 - 4.3x) - 10 = 4(10 - 2.5x) - 8(3.5 - 4.1x)$ to the nearest hundredth

88. $12(2.3 - 1.5x) - 6 = -3(18.4 - 3.5x) - 6.1(4x + 3)$ to the nearest tenth

89. $3.5\{4 - [6 - (2x + 3)] + 5\} = -18.4$ to the nearest tenth

90. $9\{3 - [4(2.3z - 1)] + 6.5\} = -406.3$ to the nearest hundredth

In Problems 91–94, determine the value of d to make the statement true.

91. In the equation $3d + 2x = 12$, the solution is -4.

92. In the equation $5d - 2x = -2$, the solution is -6.

93. In the equation $\frac{2}{3}x - d = 1$, the solution is $-\frac{3}{8}$.

94. In the equation $\frac{2}{5}x + 3d = 0$, the solution is $\frac{15}{8}$.

Explaining the Concepts

95. Explain the difference between $6x - 2(x + 1)$ and $6x - 2(x + 1) = 6$. In general, what is the difference between an algebraic expression and an algebraic equation?

96. Explain the Addition and Multiplication Properties of Equality. In the Multiplication Property of Equality, why do you think we cannot multiply both sides of the equation by 0?

97. A classmate begins to solve the equation $7x + 3 - 2x = 9x - 5$ by adding $2x$ in the following manner:

$$7x + 3 - 2x = 9x - 5$$
$$\underline{ + 2x + 2x}$$

Will this lead to the correct solution? Why or why not? Write an explanation telling the steps you would use to solve this equation.

98. A *corollary* is a rule or theorem that is closely related to a previous rule. Write a corollary to the Addition Property of Equality and title it the Subtraction Property of Equality. Write a corollary to the Multiplication Property of Equality, called the Division Property of Equality. What restrictions would you place on the Division Property of Equality? Why do you think these properties were not included in the text?

2.3 Solving Linear Equations Involving Fractions and Decimals; Classifying Equations

Objectives

1 Use the Least Common Denominator to Solve a Linear Equation Containing Fractions

2 Solve a Linear Equation Containing Decimals

3 Classify an Equation as an Identity, a Conditional Equation, or a Contradiction

4 Use Linear Equations to Solve Problems

Are You Prepared for This Section?

Before getting started, complete the following problems. If you get a problem wrong, go back to the section cited and review the material.

P1. Find the LCD of $\frac{3}{5}$ and $\frac{3}{4}$.   [Section 1.2, pp. 11–12]

P2. Find the LCD of $\frac{3}{8}$ and $-\frac{7}{12}$.   [Section 1.2, pp. 11–12]

1 Use the Least Common Denominator to Solve a Linear Equation Containing Fractions

Sometimes equations are easier to solve if they do not contain fractions. To solve a linear equation containing fractions, multiply each side of the equation by the least common denominator (LCD) to rewrite the equation without fractions. Recall that the LCD is the smallest number that each denominator has as a common multiple. The Multiplication Property of Equality allows us to multiply both sides of the equation by the LCD.

Prepared?...Answers   **P1.** 20   **P2.** 24

EXAMPLE 1   **How to Solve a Linear Equation That Contains Fractions**

Solve the equation: $\dfrac{1}{2}x + \dfrac{2}{3}x = \dfrac{14}{3}$

Step-by-Step Solution

Before following the steps in Section 2.2 on page 95, rewrite the equation as an equivalent equation by multiplying both sides of the equation by 6, the LCD of $\dfrac{1}{2}, \dfrac{2}{3},$ and $\dfrac{14}{3}$.

$$6\left(\dfrac{1}{2}x + \dfrac{2}{3}x\right) = 6\left(\dfrac{14}{3}\right)$$

Now follow Steps 1–5 from the summary in Section 2.2 to solve the equation.

Step 1: Apply the Distributive Property to remove parentheses.

$$6\left(\dfrac{1}{2}x + \dfrac{2}{3}x\right) = 6\left(\dfrac{14}{3}\right)$$

Use the Distributive Property: $\ 6\left(\dfrac{1}{2}x\right) + 6\left(\dfrac{2}{3}x\right) = 6\left(\dfrac{14}{3}\right)$

$$3x + 4x = 28$$

Step 2: Combine like terms.

$$7x = 28$$

Step 3: Use the Addition Property of Equality to get the terms with the variable on one side of the equation and the constants on the other side.

This step is not necessary because in this problem, the only variable term is already on one side of the equation, and the constant is on the other side of the equation.

Step 4: Get the coefficient of the variable to be 1.

Divide both sides of the equation by 7:

$$\dfrac{7x}{7} = \dfrac{28}{7}$$

$$x = 4$$

Step 5: Check Verify that $x = 4$ is the solution of the equation.

Substitute 4 for x in the original equation:

$$\dfrac{1}{2}x + \dfrac{2}{3}x = \dfrac{14}{3}$$

$$\dfrac{1}{2}(4) + \dfrac{2}{3}(4) \overset{?}{=} \dfrac{14}{3}$$

$$2 + \dfrac{8}{3} \overset{?}{=} \dfrac{14}{3}$$

$$\dfrac{6}{3} + \dfrac{8}{3} \overset{?}{=} \dfrac{14}{3}$$

$$\dfrac{14}{3} = \dfrac{14}{3} \quad \text{True}$$

Work Smart

Although removing fractions is not required to solve an equation, it usually makes the arithmetic easier.

The solution of the equation is 4, or the solution set is $\{4\}$.

Quick ✔

1. To solve a linear equation containing fractions, multiply each side of the equation by the ____ _____ _____ to rewrite the equation without fractions.

In Problems 2 and 3, solve each equation by multiplying by the LCD.

2. $\dfrac{2}{5}x - \dfrac{1}{4}x = \dfrac{3}{2}$

3. $\dfrac{5}{6}x + \dfrac{2}{9} = -\dfrac{1}{3}x - \dfrac{5}{18}$

EXAMPLE 2 **Solving a Linear Equation Using the LCD**

Solve the equation: $\dfrac{7n + 5}{8} = 2 + \dfrac{3n + 15}{10}$

Solution

Because the equation contains fractions, multiply both sides of the equation by 40, the LCD of $\dfrac{7n + 5}{8}$ and $\dfrac{3n + 15}{10}$.

$$\frac{7n + 5}{8} = 2 + \frac{3n + 15}{10}$$

Multiply both sides by the LCD, 40: $\quad 40\left(\dfrac{7n + 5}{8}\right) = 40\left(2 + \dfrac{3n + 15}{10}\right)$

Use the Distributive Property: $\quad 40\left(\dfrac{7n + 5}{8}\right) = 40(2) + 40\left(\dfrac{3n + 15}{10}\right)$

Divide out common factors: $\quad \overset{5}{\cancel{40}}\left(\dfrac{7n + 5}{\underset{1}{\cancel{8}}}\right) = 40(2) + \overset{4}{\cancel{40}}\left(\dfrac{3n + 15}{\underset{1}{\cancel{10}}}\right)$

Use the Distributive Property to remove parentheses: $\quad 5(7n + 5) = 40(2) + 4(3n + 15)$

Combine like terms: $\quad 35n + 25 = 80 + 12n + 60$

$\quad 35n + 25 = 140 + 12n$

Isolate the term containing n: $\quad 35n + 25 - 12n = 140 + 12n - 12n$

$\quad 23n + 25 = 140$

Isolate the constant: $\quad 23n + 25 - 25 = 140 - 25$

$\quad 23n = 115$

Divide both sides by 23: $\quad \dfrac{23n}{23} = \dfrac{115}{23}$

$\quad n = 5$

Work Smart

Multiply *all* terms in an equation by the LCD.

Check

$$\frac{7n + 5}{8} = 2 + \frac{3n + 15}{10}$$

Substitute 5 for n in the original equation: $\quad \dfrac{7(5) + 5}{8} \overset{?}{=} 2 + \dfrac{3(5) + 15}{10}$

$\quad \dfrac{35 + 5}{8} \overset{?}{=} 2 + \dfrac{15 + 15}{10}$

$\quad \dfrac{40}{8} \overset{?}{=} 2 + \dfrac{30}{10}$

$\quad 5 \overset{?}{=} 2 + 3$

$\quad 5 = 5 \quad$ True

The solution is 5, or the solution set is $\{5\}$.

Quick ✓

4. The result of multiplying the equation $\dfrac{1}{5}x + 7 = \dfrac{3}{10}$ by 10, the LCD of $\dfrac{1}{5}$ and $\dfrac{3}{10}$, is $2x +$ __ $= 3$.

In Problems 5 and 6, solve each equation by multiplying by the LCD.

5. $\dfrac{2a + 3}{4} + 3 = \dfrac{3a + 19}{8}$

6. $\dfrac{3x - 2}{4} - 1 = \dfrac{6}{5}x$

▶ ❷ Solve a Linear Equation Containing Decimals

When decimals occur in a linear equation, the equation can be rewritten as an equivalent equation without decimals by multiplying both sides of the equation by a power of 10 so that the decimal is cleared. For example, $0.8x$ is equivalent to $\frac{8}{10}x$, so multiplying by 10 clears the decimal from the term. Because 0.34 is equivalent to $\frac{34}{100}$, multiplying by 100 clears the decimal from the term.

EXAMPLE 3 **Solving a Linear Equation with a Decimal Coefficient**

Solve the equation: $0.3x - 4 = 11$

Solution

To rewrite the equation as an equivalent equation without decimals, multiply both sides of the equation by 10. Do you see why? Multiplying $0.3x = \frac{3}{10}x$ by 10 clears the decimal from the term $0.3x$.

$$0.3x - 4 = 11$$

Multiply both sides of the equation by 10: $\quad 10(0.3x - 4) = 10 \cdot 11$

Distribute: $\quad 10(0.3x) - 10 \cdot 4 = 110$

$$3x - 40 = 110$$

Add 40 to both sides of the equation: $\quad 3x = 150$

Divide both sides of the equation by 3: $\quad x = 50$

Check

$$0.3x - 4 = 11$$

Substitute 50 for x in the original equation: $\quad 0.3(50) - 4 \overset{?}{=} 11$

$$15 - 4 \overset{?}{=} 11$$

$$11 = 11 \quad \text{True}$$

The solution is 50, or the solution set is $\{50\}$. ●

It's not a requirement to clear fractions or decimals before solving an equation. Here is another way to solve $0.3x - 4 = 11$.

Work Smart

$$\frac{15}{0.3} = \frac{15}{\frac{3}{10}}$$

$$= 15 \cdot \frac{10}{3}$$

$$= \frac{150}{3}$$

$$= 50$$

$$0.3x - 4 = 11$$

Add 4 to both sides of the equation: $\quad 0.3x - 4 + 4 = 11 + 4$

$$0.3x = 15$$

Divide both sides of the equation by 0.3: $\quad \dfrac{0.3x}{0.3} = \dfrac{15}{0.3}$

$$x = 50$$

Which method do you prefer?

Quick ✓

7. To clear the decimals in the equation $0.25x + 5 = 7 - 0.3x$, multiply both sides of the equation by ___ .

In Problems 8 and 9, solve each equation.

8. $0.2z = 20$

9. $0.15p - 2.5 = 5$

EXAMPLE 4   **Solving a Linear Equation Containing Decimals**

Solve the equation: $p + 0.08p = 129.6$

Solution

Because $0.08 = \dfrac{8}{100}$ and $129.6 = \dfrac{1296}{10}$, multiplying by the LCD, 100, will clear the decimals. However, first combine like terms on the left-hand side of the equation.

$$p + 0.08p = 129.6$$

$p = 1 \cdot p$: $\quad 1p + 0.08p = 129.6$

Combine like terms: $\quad 1.08p = 129.6$

Multiply both sides of the equation by 100: $\quad 100 \cdot 1.08p = 100 \cdot 129.6$

Simplify: $\quad 108p = 12{,}960$

Divide both sides of the equation by 108: $\quad \dfrac{108p}{108} = \dfrac{12{,}960}{108}$

Simplify: $\quad p = 120$

The check is left to you. The solution is 120, or the solution set is {120}. ●

> **Quick ✓**
>
> **10.** Simplify: $n + 0.25n = $ ___ n.
>
> *In Problems 11 and 12, solve each equation.*
>
> **11.** $p + 0.05p = 52.5$ **12.** $c - 0.25c = 120$

EXAMPLE 5   **Solving a Linear Equation That Contains Decimals**

Solve the equation: $0.05x + 0.08(10{,}000 - x) = 680$

Solution

Before the equation is cleared of decimals, distribute the 0.08 and combine like terms.

$$0.05x + 0.08(10{,}000 - x) = 680$$

$$0.05x + 800 - 0.08x = 680$$

Combine like terms: $\quad -0.03x + 800 = 680$

Subtract 800 from both sides of the equation: $\quad -0.03x = -120$

Multiply both sides of the equation by 100 to clear the decimal: $\quad 100(-0.03x) = 100(-120)$

$$-3x = -12{,}000$$

Divide both sides of the equation by -3: $\quad x = 4000$

The check is left to you. The solution is 4000, or the solution set is $\{4000\}$. ●

> **Quick ✓**
>
> *In Problems 13 and 14, solve each equation.*
>
> **13.** $0.36y - 0.5 = 0.16y + 0.3$ **14.** $0.12x + 0.05(5000 - x) = 460$

▶ ❸ Classify an Equation as an Identity, a Conditional Equation, or a Contradiction

All of the equations we have solved so far have had a single solution. All other values of the variable made the equation false. These types of equations have a special name.

> **Definition**
>
> A **conditional equation** is an equation that is true for some values of the variable and false for other values of the variable.

For example, $x + 5 = 11$ is a conditional equation because it is true when $x = 6$ and false for every other real number x. The solution set of $x + 5 = 11$ is $\{6\}$. All the equations we have studied to this point have been conditional equations.

Some equations are false for all values of the variable.

> **Definition**
>
> A **contradiction** is an equation that is false for every value of the variable.

Work Smart

Do not write the empty set as $\{\varnothing\}$.

For example, the equation $2x + 3 = 7 + 2x$ is a contradiction because it is false for all values of x. Contradictions are identified by creating equivalent equations. For example, subtracting $2x$ from both sides of $2x + 3 = 7 + 2x$ gives $3 = 7$, which is clearly false. **Contradictions have no solution, so the solution set is the empty set, written as $\{\ \}$ or $\varnothing$.**

EXAMPLE 6 **Solving an Equation That Is a Contradiction**

Solve the equation: $3y - (5y + 4) = 12y - 7(2y - 1)$

Solution

$$3y - (5y + 4) = 12y - 7(2y - 1)$$

Use the Distributive Property to remove parentheses: $\quad 3y - 5y - 4 = 12y - 14y + 7$

Combine like terms: $\quad -2y - 4 = -2y + 7$

Isolate the terms containing y: $\quad -2y + 2y - 4 = -2y + 2y + 7$

$$-4 = 7 \quad \text{False}$$

The statement $-4 = 7$ is false, so the solution set is $\varnothing$ or $\{\ \}$. The equation is a contradiction. ●

Some equations are true for all real numbers for which the equation is defined.

> **Definition**
>
> An **identity** is an equation that is satisfied for all values of the variable for which both sides of the equation are defined.

For example, the equation $2(x - 5) + 1 = 5x - (9 + 3x)$ is an identity because any real number x makes the equation true. Just as with contradictions, identities are found by forming equivalent equations. For example, if the Distributive Property is used and like terms are combined in the equation $2(x - 5) + 1 = 5x - (9 + 3x)$, we obtain $2x - 9 = 2x - 9$. After subtracting $2x$ from both sides the result is $-9 = -9$, which is true. **The solution set of an equation that is an identity, where each side of the equation is a linear expression, is the set of all real numbers.**

EXAMPLE 7 **Solving an Equation That Is an Identity**

Solve the equation: $2(x + 5) = 4x - (2x - 10)$

Solution

$$2(x + 5) = 4x - (2x - 10)$$

Use the Distributive Property to remove parentheses: $\quad 2x + 10 = 4x - 2x + 10$

Combine like terms: $\quad 2x + 10 = 2x + 10$

Isolate the terms containing x: $\quad 2x - 2x + 10 = 2x - 2x + 10$

$$10 = 10$$

The statement $10 = 10$ is true for all real numbers x. Therefore, the solution set is the set of all real numbers, and the equation is an identity. ●

Quick ✓

15. A _____ _____ is an equation that is true for some values of the variable and false for other values of the variable.

16. A(n) _____ is an equation that is false for every value of the variable. A(n) _____ is an equation that is satisfied for all values of the variable for which both sides of the equation are defined.

17. *True or False* The solution of the equation $4x + 1 = 4x + 7$ is the empty set.

In Problems 18–21, solve each equation and state the solution set.

18. $3(x + 4) = 4 + 3x + 18$

19. $\frac{1}{3}(6x - 9) - 1 = 6x - [4x - (-4)]$

20. $-5 - (9x + 8) + 23 = 7 + x - (10x - 3)$

21. $\frac{3}{2}x - 8 = x + 7 + \frac{1}{2}x$

Work Smart

A conditional equation results in $x = a$ number. When the variable is eliminated and a false statement results, the solution set is the empty set. When the variable is eliminated and a true statement results, the solution set is all real numbers.

Summary Categories of Equations in One Variable

A *conditional equation* is true for some values of the variable and false for others.	A *contradiction* is false for all values of the variable.	An *identity* is true for all of the permitted values of the variable.
$3x + 2 = 11$	$4(x - 2) - x = 2(x + 1) + x$	$-2(x + 3) + 4x = 2(x - 3)$
$3x = 9$	$4x - 8 - x = 2x + 2 + x$	$-2x - 6 + 4x = 2x - 6$
$x = 3$	$3x - 8 = 3x + 2$	$2x - 6 = 2x - 6$
	$-8 = 2$	$-6 = -6$
solution: $x = 3$	no solution: $\varnothing$ or $\{\ \}$	solution: all real numbers

EXAMPLE 8

Classifying an Equation

Solve the equation $2(2 + a) - 12a = -a + 5 - 9a$. State whether the equation is a contradiction, an identity, or a conditional equation.

Solution

$$2(2 + a) - 12a = -a + 5 - 9a$$

Use the Distributive Property to remove parentheses:

$$4 + 2a - 12a = -a + 5 - 9a$$

Combine like terms:

$$4 - 10a = 5 - 10a$$

Isolate the term containing a:

$$4 - 10a + 10a = 5 - 10a + 10a$$

$$4 = 5$$

Because $4 = 5$ is false, the solution set is $\varnothing$ or $\{\ \}$. The equation is a contradiction. ●

Quick ✓

In Problems 22–25, solve each equation and state whether each equation is a contradiction, an identity, or a conditional equation.

22. $2(x - 7) + 8 = 6x - (4x + 2) - 4$

23. $\frac{4(7 - x)}{3} = x$

24. $4(5x - 4) + 1 = 20x$

25. $\frac{1}{2}(4x - 6) = 6\left(\frac{1}{3}x - \frac{1}{2}\right) + 4$

▶ ❹ **Use Linear Equations to Solve Problems**

Problems from finance can be modeled by algebraic equations, as is shown in the next example.

EXAMPLE 9   **Solving a Problem from Finance**

You invest part of $5000 in a certificate of deposit (CD) that earns 2% simple interest compounded annually, and the rest in bonds that earn 3% simple interest compounded annually. To determine the amount, c, that you should invest in the CD to earn $135 interest at the end of one year, solve the equation $0.02c + 0.03(5000 - c) = 135$.

Solution

$$0.02c + 0.03(5000 - c) = 135$$

Use the Distributive Property: $0.02c + 150 - 0.03c = 135$

Combine like terms: $-0.01c + 150 = 135$

Isolate the term containing c: $-0.01c + 150 - 150 = 135 - 150$

$$-0.01c = -15$$

Multiply both sides of the equation by 100 to eliminate the decimal: $-c = -1500$

Multiply both sides of the equation by -1: $c = 1500$

Because $c = 1500$, you must invest $1500 in the CD to earn $135 in interest at the end of one year.

Quick ✔

26. Janet Majors invested part of her lottery winnings in a long-term bond fund that has averaged earnings of 3% over the past 5 years. She invested $250 more than that in a large-cap growth fund that has averaged 6% earnings over the past 5 years. The total interest earned last year was $60. To find the amount, b, that she invested in the bond account, solve the equation $0.03b + 0.06(b + 250) = 60$.

2.3 Exercises MyMathLab® Exercise numbers in green have complete video solutions in MyMathLab or may be accessed using the QR code to the right.

*Problems **1–26** are the Quick ✔s that follow the **EXAMPLES**.*

Building Skills

In Problems 27–42, solve the equation. Check your solution. See Objective 1.

27. $\dfrac{1}{5}x + \dfrac{3}{2} = \dfrac{3}{10}$

28. $\dfrac{3}{2}n - \dfrac{4}{11} = \dfrac{91}{22}$

29. $\dfrac{a}{4} - \dfrac{a}{3} = -\dfrac{1}{2}$

30. $\dfrac{3}{2}b - \dfrac{4}{5}b = \dfrac{28}{5}$

31. $\dfrac{3x + 2}{4} = \dfrac{x}{2}$

32. $\dfrac{2x - 3}{5} = \dfrac{3}{10}x$

33. $\dfrac{2k - 1}{4} = 2$

34. $\dfrac{3a + 2}{5} = -1$

35. $-\dfrac{2x}{3} + 1 = \dfrac{5}{9}$

36. $\dfrac{4}{3}m - 1 = \dfrac{1}{9}$

37. $\dfrac{2}{3}(x + 1) = \dfrac{1}{9}(x + 4)$

38. $\dfrac{2}{3}(6 - x) = \dfrac{5}{6}x$

39. $\dfrac{y}{10} + 3 = \dfrac{y}{4} + 6$

40. $\dfrac{p}{8} - 1 = \dfrac{7p}{6} + 2$

41. $\dfrac{4x - 9}{3} + \dfrac{x}{6} = \dfrac{x}{2} - 2$

42. $\dfrac{3x + 2}{4} - \dfrac{x}{12} = \dfrac{x}{3} - 1$

In Problems 43–62, solve the equation. Check your solution. See Objective 2.

43. $0.4w = 12$

44. $0.3z = 6$

45. $-1.3c = 5.2$

46. $-1.7q = -8.5$

47. $1.05p = 52.5$

48. $1.06z = 31.8$

49. $p + 1.5p = 12$

50. $2.5a + a = 7$

51. $p + 0.05p = 157.5$

52. $p + 0.04p = 260$

53. $0.3x + 2.3 = 0.2x + 1.1$

54. $0.7y - 4.6 = 0.4y - 2.2$

55. $0.65x + 0.3x = x - 3$

56. $0.5n - 0.35n = 2.5n + 9.4$

57. $3 + 1.5(z + 2) = 3.5z - 4$

58. $5 - 0.2(m - 2) = 3.6m + 1.6$

59. $0.02(2c - 24) = -0.4(c - 1)$

60. $0.3(6a - 4) = -0.10(2a - 8)$

61. $0.15x + 0.10(250 - x) = 28.75$

62. $0.03t + 0.025(1000 - t) = 27.25$

In Problems 63–74, solve the equation. State whether the equation is a contradiction, an identity, or a conditional equation. See Objective 3.

63. $3(z + 1) = 2(z - 3) + z$

64. $4(y - 2) = 6(y + 1) - 2y$

65. $6q - (q - 3) = 2q + 3(q + 1)$

66. $-3x + 2 + 5x = 2(x + 1)$

67. $9a - 5(a + 1) = 2(a - 3)$

68. $7b + 2(b - 4) = 8b - (3b + 2)$

69. $\dfrac{4x - 9}{6} - \dfrac{x}{2} = \dfrac{x}{6} + 3$

70. $\dfrac{2m + 1}{4} - \dfrac{m}{6} = \dfrac{m}{3} - 1$

71. $\dfrac{5z + 1}{5} = \dfrac{2z - 3}{2}$

72. $\dfrac{2y - 7}{4} = \dfrac{3y - 13}{6}$

73. $\dfrac{q}{3} + \dfrac{4}{5} = \dfrac{5q + 12}{15}$

74. $\dfrac{2x}{3} + \dfrac{x + 3}{12} = \dfrac{3x + 1}{4}$

Mixed Practice

In Problems 75–100, solve the equation.

75. $-3(2n + 4) = 10n$

76. $-15(z - 3) = 25z$

77. $-2x + 5x = 4(x + 2) - (x + 8)$

78. $5m - 3(m + 1) = 2(m + 1) - 5$

79. $-6(x - 2) + 8x = -x + 10 - 3x$

80. $3 - (x + 10) = 3x + 7$

81. $\dfrac{3}{4}x = \dfrac{1}{2}x - 5$

82. $\dfrac{1}{3}x = 2 + \dfrac{5}{6}x$

83. $\dfrac{1}{2}x + 2 = \dfrac{4x + 1}{4}$

84. $\dfrac{x}{2} + 4 = \dfrac{x + 7}{3}$

85. $0.3p + 2 = 0.1(p + 5) + 0.2(p + 1)$

86. $1.6z - 4 = 2(z - 1) - 0.4z$

87. $-0.7x = 1.4$

88. $0.2a = -6$

89. $\dfrac{3(2y - 1)}{5} = 2y - 3$

90. $\dfrac{4(2n + 1)}{3} = 2n - 6$

91. $0.6x - 0.2(x - 4) = 0.4(x - 2)$

92. $0.3x - 1 = 0.5(x + 2) - 0.2x$

93. $\dfrac{3x - 2}{4} = \dfrac{5x - 1}{6}$

94. $\dfrac{x - 1}{4} = \dfrac{x - 4}{6}$

95. $0.3x + 2.6x = 5.7 - 1.8 + 2.8x$

96. $0.3(z - 10) - 0.5z = -6$

97. $\dfrac{3}{2}x - 6 = \dfrac{2(x - 9)}{3} + \dfrac{1}{6}x$

98. $\dfrac{2x - 3}{4} + 5 = \dfrac{3(x + 3)}{4} - \dfrac{x}{2} + 2$

99. $\dfrac{2}{3}\left[4 - \left(\dfrac{x}{2} + 6\right) - 2x\right] + 3 = \dfrac{5x}{6}$

100. $\dfrac{1}{2}\left[3 - \left(\dfrac{2x}{3} - 1\right) + 3x\right] = \dfrac{-4x + 1}{3} + 1$

In Problems 101–104, solve the equation and round to the indicated place.

101. $2.8x + 13.754 = 4 - 2.95x$ to the nearest hundredth

102. $-4.88x - 5.7 = 2(-3.41x) + 1.2$ to the nearest whole number

103. $x - \{1.5x - 2[x - 3.1(x + 10)]\} = 0$ to the nearest tenth

104. $-3x - 2\{4 + 3[x - (1 + x)]\} = 12$ to the nearest hundredth

Applying the Concepts

105. Sales Tax The price of a pair of jeans, including sales tax of 6%, is $53. To find the price, p, of the jeans, solve the equation $1.06p = 53$.

106. Gardening Bob Adams rented a rototiller at a cost of $7.50 per hour. Bob paid a bill of $37.50. Find the number of hours, h, he rented the tiller by solving the equation $7.50h = 37.50$.

107. Purchasing a Car The total cost (including 6% sales tax) for the purchase of an automobile was $19,080. To determine the cost of the auto before the sales tax was added, solve the equation $x + 0.06x = 19,080$, where x represents the cost of the car before taxes.

108. Purchasing a Kayak The total cost, including 5.5% sales tax, for the purchase of a kayak was $1266. Find the price of the kayak, k, before sales tax, by solving the equation $k + 0.055k = 1266$.

109. Hourly Pay Mark recently received a 4% pay increase. His hourly wage is now $15.08. To determine his hourly wage, w, before the 4% pay raise, solve the equation $w + 0.04w = 15.08$.

110. Hourly Pay A union representing airline employees recently agreed to a 6% cut in hourly wage in order to help the airline avoid filing for bankruptcy. A baggage handler will now earn $26.32 per hour. To find the baggage handler's hourly wage, w, before the pay cut, solve the equation $w - 0.06w = 26.32$.

111. Bluetooth Speakers Tamara purchased a Bluetooth speaker at a "25% off" sale for $60. To determine the original price, p, of the speakers, solve the equation $p - 0.25p = 60$.

112. Team Sweatshirt Your favorite college has logo sweatshirts on sale for $33.60. The sweatshirt has been marked down by 30%. To find the original price of the sweatshirt, x, solve the equation $x - 0.30x = 33.60$.

113. Piggy Bank Celeste saves dimes and quarters in a piggy bank. She opened the bank and discovered that she had $7.05 and that the number of dimes was 3 more than twice the number of quarters. To find the number of quarters, q, solve the equation $0.25q + 0.10(2q + 3) = 7.05$.

114. Clean Car Pablo cleaned out his car and found nickels and quarters in the car seats. He found $4.25 in change and noticed that the number of quarters was 5 less than twice the number of nickels. Solve the equation $0.05n + 0.25(2n - 5) = 4.25$ to find n, the number of nickels Pablo found.

△**115. Comparing Perimeters** The two rectangles shown in the figure have the same perimeter. Solve the equation

$$2x + 2(x + 3) = 2\left(\frac{1}{2}x\right) + 2(x + 6)$$ for x, the

width of the first rectangle

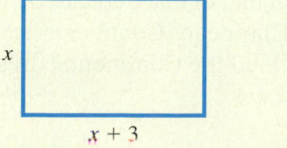
x

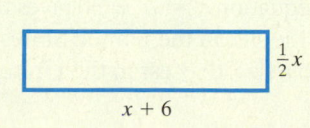
$\frac{1}{2}x$

$x + 3$ $x + 6$

△**116. Comparing Perimeters** The square and the rectangle shown have the same perimeter. Solve the equation

$$4x = 2\left(\frac{1}{2}x\right) + 2(x + 5)$$ for x, the length of the

side of the square.

x

$\frac{1}{2}x$

$x + 5$

117. Paying Your Taxes You are single and just determined that you paid $1346.25 in federal income taxes for 2015. The solution to the equation $1346.25 = 0.15(x - 9225) + 922.50$ represents your adjusted gross income in 2015. Determine your adjusted gross income in 2015.
(SOURCE: *Internal Revenue Service*)

118. Paying Your Taxes You are married and just determined that you paid $12,837.50 in federal income taxes in 2015. The solution to the equation $12,837.50 = 0.25(x - 74,900) + 10,312.50$ represents the adjusted gross income of you and your spouse in 2015. Determine the adjusted gross income of you and your spouse in 2015.
(SOURCE: *Internal Revenue Service*)

Explaining the Concepts

119. Make up an equation that has one solution. Make up an equation that has no solution. Make up an equation that is an identity. Discuss the differences and similarities in making up each equation.

120. A student solved the equation $3(x + 8) = \frac{1}{2}(6x + 4)$ and wrote the answer $24 = 2$. The instructor did not give full credit for this answer. Explain the student's error and determine the correct solution.

121. When solving the equation $\frac{2}{3}x - 5 = \frac{1}{2}x$, a student decided to multiply both sides by the LCD and wrote $6 \cdot \frac{2}{3}x - 5 = \frac{1}{2}x \cdot 6$. This resulted in the next line $4x - 5 = 3x$. Is this correct? Explain the steps necessary to finish using this technique and then suggest another list of steps that would arrive at the correct solution.

122. Explain how you can tell from the last step of the solution process whether the solution of an equation is the set of all real numbers or the empty set. For example, if the last line of the solution of an equation is $-6 = 8$, is the solution the set of all real numbers or the empty set? If the last line of the solution of an equation is $9 = 9$, is the solution the set of all real numbers or the empty set?

2.4 Evaluating Formulas and Solving Formulas for a Variable

Objectives

❶ Evaluate a Formula

❷ Solve a Formula for a Variable

Are You Prepared for This Section?

Before getting started, complete the following problems. If you get a problem wrong, go back to the section cited and review the material.

P1. Evaluate the expression $2L + 2W$ for $L = 7$ and $W = 5$.   [Section 1.8, pp. 64–65]

P2. Round the expression 0.5873 to the hundredths place.   [Section 1.2, pp. 13–14]

▶ ❶ Evaluate a Formula

In this section, *formulas* are used to solve mathematical problems.

> **Definition**
>
> A mathematical **formula** is an equation that describes how two or more variables are related.

For example, a formula for the area of a rectangle is *area = length · width*, or $A = l \cdot w$. You use mathematical formulas every day. For instance, to paint your bedroom, you must find the surface area of the walls to compute the amount of paint you will need. To determine the number of gallons of gasoline you can afford to put in your car, you may also use a formula, as shown in the next example.

EXAMPLE 1   **Evaluating a Formula**

The number of gallons of gasoline added to the tank of a car is found by using the formula $\dfrac{C}{p} = n$, where C is the total cost, p is the price per gallon, and n is the number of gallons. How many gallons of gas can be purchased for $33.00 if gas costs $2.75 per gallon?

Solution

$$\frac{C}{p} = n$$

Replace C with cost, $33, and p with price per gallon, $2.75: $\dfrac{\$33}{\$2.75} = n$

Evaluate the numerical expression: $12 = n$

When gas costs $2.75 per gallon, 12 gallons of gasoline can be purchased for $33.   ●

> **Quick ✔**
>
> 1. A _____ is an equation that describes how two or more variables are related.
>
> 2. Your best friend, who is on spring break in Europe, e-mailed you that the temperature in Paris today is 15° Celsius. Use the formula $F = \dfrac{9}{5}C + 32$, where F is degrees Fahrenheit and C is degrees Celsius, to find the Paris temperature in degrees Fahrenheit.
>
> 3. The size of a dress purchased in Europe is different from that of one purchased in the United States. The equation $c = a + 30$ gives the European (Continental) dress size c in terms of the size in the United States, a. Find the Continental dress size that corresponds to a size 10 dress in the United States.

EXAMPLE 2 **Evaluating a Formula Containing a Percent**

The formula $S = P - 0.25P$ gives the sale price S of an item originally costing P dollars that was reduced by 25%. Find the sale price of a pair of jeans that originally cost $40.00.

Solution

$$S = P - 0.25P$$

Replace P with the original price of the jeans, $40: $$S = 40 - 0.25(40)$$

Evaluate the numerical expression: $$S = 40 - 10$$

$$S = 30$$

The sale price of the jeans is $30. ●

Quick ✓

4. The formula $E = 250 + 0.05S$ is a formula for the earnings, E, of a salesman who receives $250 per week plus 5% commission on all weekly sales, S. Find the earnings of a salesman who had weekly sales of $1250.

5. The formula $N = p + 0.06p$ models the new population, N, of a town with current population p if the town is expecting population growth of 6% next year. Find the new population of a town whose current population is 5600 persons.

The Simple Interest Formula

Interest is money paid for the use of money. The total amount borrowed is the **principal.** The principal can be in the form of a loan (an individual borrows from the bank) or a deposit (the bank borrows from the individual). The **rate of interest**, expressed as a percent, is the amount charged for the use of the principal for a given period of time. The time frame is usually a year or some fraction of a year (such as 1 month, or $\frac{1}{12}$ of a year).

Work Smart

Be sure to express t in years. For computations, the interest rate r must be expressed as a decimal.

Simple Interest Formula

If an amount of money, P, called the **principal** is invested for a period of t years at an annual interest rate r, the amount of interest I earned is

$$I = Prt$$

Interest earned according to this formula is called **simple interest.**

EXAMPLE 3 **Evaluating Using the Simple Interest Formula**

Janice invested $500 in a two-year certificate of deposit (CD) at the simple interest rate of 2% per year. Find the amount of interest Janice will earn in 6 months and the total amount of money she will have at the end of 6 months.

Solution

The amount of money Janice invested, P, is $500. The interest rate r is 2% $= 0.02$. Because 6 months is half a year, $t = \dfrac{1}{2}$.

$$I = Prt$$

Let $P = \$500, r = 0.02$, and $t = \dfrac{1}{2}$: $$I = 500 \cdot 0.02 \cdot \dfrac{1}{2}$$

Evaluate the numerical expression: $$I = 5$$

At the end of 6 months Janice will have earned $5 and will have a total of $500 + $5 = $505. ●

In Other Words

"Per annum" is another way of saying an annual or yearly rate.

▶ Geometry Formulas

Let's review a few common terms from geometry.

Definitions

The **perimeter** is the sum of the lengths of all the sides of a figure, measured in linear units such as feet, meters, and so on.

The **area** is the amount of space enclosed by a two-dimensional figure, measured in square units.

The **surface area** of a solid is the sum of the areas of the surfaces of a three-dimensional figure, measured in square units.

The **volume** is the amount of space occupied by a three-dimensional figure, measured in cubic units.

The **radius** r of a circle is any line segment that extends from the center of the circle to any point on the circle.

The **diameter** of a circle is any line segment that extends from one point on the circle through the center to a second point on the circle. The length of a diameter is two times the length of the radius, $d = 2r$.

In circles, the term **circumference** is used to mean the perimeter.

Formulas from geometry are useful in solving many types of problems. Some of these formulas are listed in Table 1.

Table 1

Plane Figures	Formulas	Plane Figures	Formulas
Square	**Area:** $A = s^2$ **Perimeter:** $P = 4s$	Trapezoid	**Area:** $A = \dfrac{1}{2}h(B + b)$ **Perimeter:** $P = a + b + c + B$
Rectangle	**Area:** $A = lw$ **Perimeter:** $P = 2l + 2w$	Parallelogram	**Area:** $A = bh$ **Perimeter:** $P = 2a + 2b$
Triangle	**Area:** $A = \dfrac{1}{2}bh$ **Perimeter:** $P = a + b + c$	Circle	**Area:** $A = \pi r^2$ **Circumference:** $C = 2\pi r$ $\quad\quad\quad\quad\quad\quad = \pi d$

Table 1 (*continued*)

Solids	Formulas	Solids	Formulas
Cube 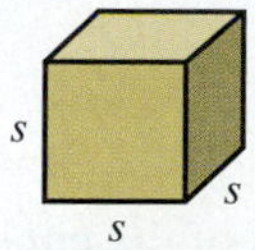	**Volume:** $V = s^3$ **Surface Area:** $S = 6s^2$	Right Circular Cylinder 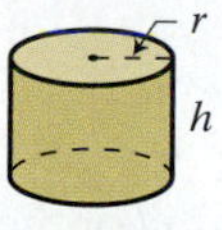	**Volume:** $V = \pi r^2 h$ **Surface Area:** $S = 2\pi r^2 + 2\pi rh$
Rectangular Solid 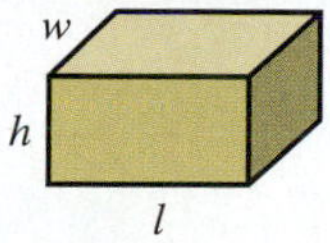	**Volume:** $V = lwh$ **Surface Area:** $S = 2lw + 2lh + 2wh$	Cone 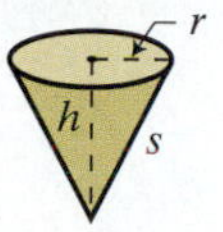	**Volume:** $V = \dfrac{1}{3}\pi r^2 h$ **Surface Area:** $S = \pi r^2 + \pi rs$
Sphere 	**Volume:** $V = \dfrac{4}{3}\pi r^3$ **Surface Area:** $S = 4\pi r^2$		

EXAMPLE 4 **Evaluating the Formula for the Perimeter of a Rectangle**

Find the number of yards of fencing needed to enclose a garden that is 10.5 yards long and 4.25 yards wide, as illustrated in Figure 1.

Solution

The garden is rectangular, so use the formula for the perimeter of a rectangle, $P = 2l + 2w$, where $l = 10.5$ yards and $w = 4.25$ yards.

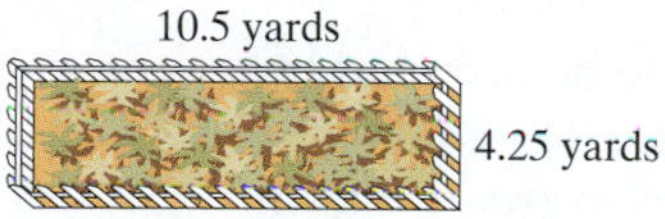

$$P = 2l + 2w$$

Let l = 10.5 and w = 4.25: $P = 2(10.5) + 2(4.25)$

Evaluate the numerical expression: $P = 21 + 8.5$

$$P = 29.5$$

A total of 29.5 yards of fencing is needed to enclose the garden.

Figure 1

10.5 yards

4.25 yards

Quick ✓

8. The _____ is the sum of the lengths of all the sides of a figure.

9. The ____ is the amount of space enclosed by a two-dimensional figure, measured in square units.

10. The _____ is the amount of space occupied by a three-dimensional figure, measured in cubic units.

11. The _____ of a circle is any line segment that extends from the center of the circle to any point on the circle.

12. *True or False* The area A of a circle whose radius is r is found using the formula $A = \pi r^2$.

13. Find the area of a trapezoid with $h = 4.5$ inches, $B = 9$ inches, and $b = 7$ inches.

Sometimes we need to use more than one geometry formula to solve a problem.

EXAMPLE 5 **Finding the Area of a Lawn and the Cost of Sod**

A circular swimming pool with a 30-foot diameter is to be installed in a 100-foot by 60-foot rectangular yard. Sod (a surface of ground with grass growing on it) will then be installed on the remaining land to create a lawn around the pool. See Figure 2 on the following page.

(a) How much sod, to the nearest square foot, is required for the lawn?

(b) If sod costs $0.25 per square foot to install, how much will the sod for the lawn cost?

(*continued*)

Figure 2

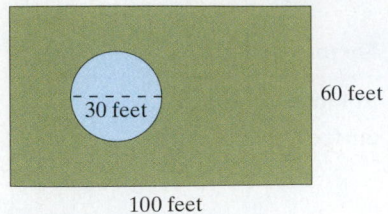

Solution

(a) The area of a rectangle is $A = lw$ and the area of a circle is $A = \pi r^2$. Find the area of the yard that will receive sod by subtracting the area of the circular swimming pool from the area of the rectangular yard.

$$\text{Area of lawn} = \text{Area of rectangle} - \text{Area of circle}$$
$$= \qquad lw \qquad - \qquad \pi r^2$$

The length l of the yard is 100 feet and the width is 60 feet. The radius of the circle is one-half its diameter, so the radius is 15 feet. Substituting into the formulas,

$$\text{Area of lawn} = (100 \text{ feet})(60 \text{ feet}) - \pi(15 \text{ feet})^2$$
$$= 6000 \text{ feet}^2 - 225\pi \text{ feet}^2$$
$$\pi \approx 3.14159: \quad \approx 5293 \text{ feet}^2$$

Approximately 5293 square feet of sod is needed for the lawn.

(b) The cost of the sod is the product of the square footage and the cost per square foot.

$$\text{Cost for sod} = 5293 \text{ square feet} \cdot \frac{\$0.25}{1 \text{ square foot}}$$
$$= \$1323.25$$

The sod for the lawn will cost \$1323.25.   ●

In Example 5(b), notice that the units "square feet" divide out, giving an answer in dollars. Getting the appropriate units in the answer gives us more confidence that the answer is correct.

Work Smart

Area is expressed in square units because you multiply units (i.e., feet · feet = feet²). Perimeter is expressed as a linear quantity.

In Other Words

The word "per" means for each one. For example, the sod costs \$0.25 for each square foot. When you see the word "per," think fraction. Thus \$0.25 per square foot is represented as

$$\frac{\$0.25}{1 \text{ square foot}}$$

Quick ✓

14. A circular water feature whose diameter is 4 feet is to be installed in a rectangular garden that is 20 feet by 10 feet. Once the water feature is installed, sod is to be installed on the remaining land to create a lawn.

(a) How much sod, to the nearest square foot, is required for the lawn?

(b) If sod costs \$0.25 per square foot to install, how much will the sod for the lawn cost?

Geometry formulas can even be used to make us informed consumers.

 EXAMPLE 6   **Determining the Better Buy**

At Mamma da Vinci's Pizza Parlor, a 16-inch pizza costs \$14.99 and a 12-inch pizza costs \$9.99. Which is the "better" buy?

Solution

The "better" buy is the pizza that costs less per square inch. The plan is: (1) find the area of each pizza and (2) find the price per square inch of each pizza. Keep in mind that a 16″ pizza has a 16″ *diameter* and an 8″ *radius*. A 12″ pizza has a 6″ radius.

1. Find the area of each pizza. Since pizzas are circular, use the formula for the area of a circle.

	Area of 16″ pizza	Area of 12″ pizza
	$A = \pi r^2$	$A = \pi r^2$
Replace r with its given value:	$A = \pi \cdot 8^2$	$A = \pi \cdot 6^2$
Exact area:	$A = 64\pi$	$A = 36\pi$
Approximate area:	$A \approx 201.06 \text{ in.}^2$	$A \approx 113.10 \text{ in.}^2$

2. Find the price per square inch of each pizza. The price per square inch is found by dividing the price of the pizza by the number of square inches of area.

price per square inch of 16" pizza: $\dfrac{\$14.99}{201.06 \text{ in.}^2} \approx \$0.075$ per square inch

price per square inch of 12" pizza: $\dfrac{\$9.99}{113.10 \text{ in.}^2} \approx \$0.088$ per square inch

The price per square inch of the 16-inch pizza is about 7.5¢. The price per square inch of the 12-inch pizza is about 8.8¢. The 16-inch pizza is the "better" buy. ●

Quick ✓

15. A homeowner plans to construct a circular brick pad for his barbeque grill. The diameter of the brick pad is to be 6 feet. Find, to the nearest hundredth of a square foot, the area of the barbeque pad.

16. An extra large (18") pizza at Dante's Pizza costs $16.99 and a small (9") pizza costs $8.99. Which is the better buy?

▶ ❷ Solve a Formula for a Variable

To "solve for a variable" means to isolate the variable with a coefficient of 1 on one side of the equation and all other variables and constants, if any, on the other side. For example, the formula for the area of a rectangle, $A = lw$, is solved for A because A is by itself with a coefficient of 1 on one side of the equation while all other variables are on the other side.

The steps used to solve formulas for a certain variable are identical to those used to solve equations.

Solve for x: $\qquad 15 = \dfrac{3}{2}x$

Multiply both sides of the equation by 2 to clear the fraction. $\qquad 2(15) = 2\left(\dfrac{3}{2}x\right)$

$$30 = 3x$$

Divide both sides by 3 to isolate the variable x. $\qquad \dfrac{30}{3} = \dfrac{3x}{3}$

$$10 = x$$

Solve for h: $\qquad A = \dfrac{1}{2}bh$

Multiply both sides of the equation by 2 to clear the fraction. $\qquad 2(A) = 2\left(\dfrac{1}{2}bh\right)$

$$2A = bh$$

Divide both sides by b to isolate the variable h. $\qquad \dfrac{2A}{b} = \dfrac{bh}{b}$

$$\dfrac{2A}{b} = h$$

EXAMPLE 7 **Solve a Formula for a Variable**

Solve the simple interest formula $I = Prt$ for t.

Solution
To solve for t, isolate t on one side of the equation.

$$I = Prt$$

Divide both sides of the equation by Pr: $\qquad \dfrac{I}{Pr} = \dfrac{Prt}{Pr}$

Simplify: $\qquad \dfrac{I}{Pr} = t$

Use the Symmetric Property: $\qquad t = \dfrac{I}{Pr}$ ●

Work Smart

The Symmetric Property states that if $a = b$, then $b = a$.

Quick ✓

17. Solve the formula $d = rt$ for t, where d represents distance, r represents an average speed, and t represents time.

18. Solve the formula $V = lwh$ for l, where V represents the volume, l represents the length, w represents the width, and h represents the height of a rectangular solid.

EXAMPLE 8   **Solve a Formula for a Variable**

Solve the formula $P = 2l + 2w$ for w.

Solution

$$P = 2l + 2w$$

Subtract $2l$ from both sides of the equation: $P - 2l = 2l + 2w - 2l$

Simplify: $P - 2l = 2w$

Divide both sides of the equation by 2: $\dfrac{P - 2l}{2} = \dfrac{2w}{2}$

$$\dfrac{P - 2l}{2} = w$$

Use the Symmetric Property: $w = \dfrac{P - 2l}{2}$

Work Smart

Both $w = \dfrac{P - 2l}{2}$ and $\dfrac{P - 2l}{2} = w$ are correct. Another correct answer is $w = \dfrac{P}{2} - l$. Do you see why?

Quick ✓

19. To convert from degrees Celsius to degrees Fahrenheit, use the formula $F = \dfrac{9}{5}C + 32$. Solve the formula for C.

20. The formula $S = 2\pi rh + 2\pi r^2$ represents the surface area S of a right circular cylinder whose radius is r and height is h. Solve the formula for h.

The next example illustrates a skill that is extremely important in the study of graphing lines.

EXAMPLE 9   **Solve a Formula for a Variable**

Solve the equation $3x + 2y = 6$ for y.

Solution

To solve for y, isolate y on one side of the equation and the variable x and constants on the other side.

$$3x + 2y = 6$$

Subtract $3x$ from both sides of the equation to isolate the term containing y: $3x + 2y - 3x = 6 - 3x$

Simplify: $2y = 6 - 3x$

Divide both sides of the equation by 2: $\dfrac{2y}{2} = \dfrac{6 - 3x}{2}$

Simplify: $y = \dfrac{6 - 3x}{2}$

Work Smart

$\dfrac{A + B}{C} = \dfrac{A}{C} + \dfrac{B}{C}$, so

$\dfrac{6 - 3x}{2} = \dfrac{6}{2} - \dfrac{3x}{2}$.

The equation $y = \dfrac{6 - 3x}{2}$ may also be written as $y = 3 - \dfrac{3}{2}x$, or $y = -\dfrac{3}{2}x + 3$, by dividing each term in the numerator by 2. Do you see why these three equations are equivalent? See the Work Smart in the margin to the left.

Quick ✓

In Problems 21 and 22, solve each equation for the indicated variable.

21. $6x - 3y = 18$; for y

22. $4x + 6y = 15$; for y

2.4 Exercises **MyMathLab®** Exercise numbers in green have complete video solutions in MyMathLab or may be accessed using the QR code to the right.

*Problems **1–22** are the **Quick ✓**s that follow the **EXAMPLES**.*

Building Skills

In Problems 23–32, substitute the given values into the formula and then evaluate to find the unknown quantity. Label units in the answer. If the answer is not exact, round your answer to the nearest hundredth. See Objective 1.

23. How Far Can I Go? The Chevrolet Volt is an electric car that also has a gas engine. The distance D, in miles, the car can travel with a fully charged battery on g gallons of gasoline is given by the formula $D = 38 + 37g$. Determine the distance the Volt can travel using 4 gallons of gasoline.

24. Maximum Heart Rate Maximum heart rate, in beats per minute, is the highest heart rate a person can safely achieve while exercising. One formula for the maximum heart rate H of an individual who is a years of age is $H = 208 - 0.7a$. Determine the maximum heart rate of an individual who is 40 years old.

25. Buying a Digital Media Player The formula $S = P - 0.20P$ gives the sale price S of an item whose original cost P was reduced by 20%. Find the sale price of a digital media player that originally cost $130.00.

26. Buying a Computer The formula $S = P - 0.35P$ gives the sale price S of an item whose original cost P was reduced by 35%. Find the sale price of a computer that originally cost $950.00.

27. Salesperson's Earnings The formula $E = 500 + 0.15S$ is a formula for the earnings, E, of a salesperson who receives $500 per week plus 15% commission on all sales, S. Find the earnings of a salesperson who had weekly sales of $1000.

28. Salesperson's Earnings The formula $E = 750 + 0.07S$ is a formula for the earnings, E, of a salesperson who receives $750 per week plus 7% commission on all sales, S. Find the earnings of a salesperson who had weekly sales of $1200.

29. Planning a Trip A businessperson is planning a trip from Tokyo to San Diego, California, in March. She learns that the average high temperature in San Diego in March is 68° Fahrenheit. Use the formula $C = \dfrac{5}{9}(F - 32)$ to convert 68° Fahrenheit to degrees Celsius.

30. Planning a Trip An English literature professor plans to take twenty students on a study-abroad trip to London in March. He learns that the average daytime temperature in London in March is 10° Celsius. Use the formula $F = \dfrac{9}{5}C + 32$ to convert 10° Celsius to degrees Fahrenheit.

31. Lottery Earnings Therese invested her $2000 West Virginia Lottery winnings in a 6-month certificate of deposit (CD) that earns 2% simple interest per annum. Find the amount of interest Therese's investment will earn using the formula $I = Prt$.

32. Investing an Inheritance Christopher invested his $5000 inheritance from his grandmother in a 9-month certificate of deposit that earns 2.5% simple interest per annum. Find the amount of interest Christopher's investment will earn using the formula $I = Prt$.

In Problems 33–42, use formulas from geometry to find each of the following. See Objective 1.

△ **33.** Find
  (a) the perimeter and
  (b) the area of the rectangle.

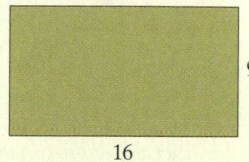

△ **34.** Find
  (a) the perimeter and
  (b) the area of the rectangle.

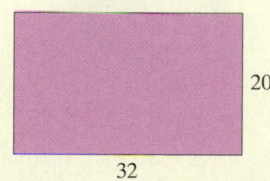

△ **35.** Find

(a) the perimeter and
(b) the area of the rectangle.

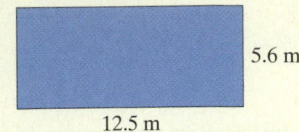

5.6 m

12.5 m

△ **36.** Find

(a) the perimeter and
(b) the area of the rectangle.

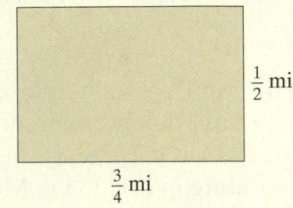

$\frac{1}{2}$ mi

$\frac{3}{4}$ mi

△ **37.** Find

(a) the perimeter and
(b) the area of the square.

9

△ **38.** Find

(a) the perimeter and
(b) the area of the square.

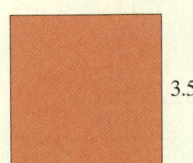

3.5

In Problems 39 and 40, find the (a) exact and approximate circumference, and (b) the exact and approximate area of each circle. For approximate measures use $\pi \approx 3.14$ and round to the nearest tenth.

△ **39.**

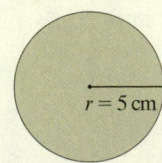

$r = 5$ cm

△ **40.**

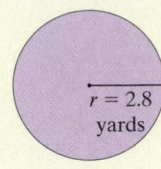

$r = 2.8$ yards

△ **41. Area of a Circle** Find the area of a circle when $\pi \approx \dfrac{22}{7}$ and $r = \dfrac{14}{3}$ inches.

△ **42. Area of a Circle** Find the area, to the nearest hundredth, of a circle when $\pi \approx 3.14$ and $r = 2.5$ km.

In Problems 43–56, solve each formula for the stated variable. See Objective 2.

43. $d = rt$; solve for r

44. $A = lw$; solve for w

45. $C = \pi d$; solve for d

46. $F = mv^2$; solve for m

47. $I = Prt$; solve for t

48. $V = LWH$; solve for W

49. $A = \dfrac{1}{2}bh$; solve for b

50. $V = \dfrac{1}{3}Bh$; solve for B

51. $P = a + b + c$; solve for a

52. $S = a + b + c$; solve for b

53. $A = P + Prt$; solve for r

54. $P = 2l + 2w$: solve for l

55. $A = \dfrac{1}{2}h(B + b)$; solve for b

56. $S = 2\pi r(r + h)$; solve for h

In Problems 57–64, solve for y. See Objective 2.

57. $3x + y = 12$

58. $-2x + y = 18$

59. $10x - 5y = 25$

60. $12x - 6y = 18$

61. $4x + 3y = 13$

62. $5x + 6y = 18$

63. $\dfrac{1}{2}x - \dfrac{1}{6}y = 2$

64. $\dfrac{2}{3}x - \dfrac{5}{2}y = 5$

Applying the Concepts

In Problems 65–78, (a) solve for the indicated variable, and then (b) find the value of the unknown quantity. When units are given, label units in the answer.

65. Profit = Revenue − Cost: $P = R - C$

(a) Solve for C.
(b) Find C when $P = \$1200$ and $R = \$1650$.

66. Profit = Revenue − Cost: $P = R - C$

(a) Solve for R.
(b) Find R when $P = \$4525$ and $C = \$1475$.

67. Simple Interest: $I = Prt$

  (a) Solve for r.

  (b) Find r when $I = \$225$, $P = \$5000$, and $t = 1.5$ years.

68. Simple Interest: $I = Prt$

  (a) Solve for t.

  (b) Find t when $I = \$42$, $P = \$525$, and $r = 4\%$.

69. Statistics Formula: $Z = \dfrac{x - \mu}{\sigma}$

  (a) Solve for x.

  (b) Find x when $Z = 2$, $\mu = 100$, and $\sigma = 15$.

70. Physics Formula: $P = mgh$

  (a) Solve for m.

  (b) Find m when $P = 8192$, $g = 32$, and $h = 1$.

71. Algebra: $y = mx + 5$

  (a) Solve for m.

  (b) Find m when $x = 3$ and $y = -1$.

72. Fahrenheit/Celsius Temperature Conversion:
$$F = \frac{9}{5}C + 32$$

  (a) Solve for C.

  (b) Find C when $F = 59°$.

73. Finance: $A = P + Prt$

  (a) Solve for r.

  (b) Find r when $A = \$540$, $P = \$500$, and $t = 2$.

74. Finance: $A = P + Prt$

  (a) Solve for t.

  (b) Find t when $A = \$249$, $P = \$240$, and $r = 2.5\% = 0.025$.

△ **75. Volume of a Right Circular Cylinder:** $V = \pi r^2 h$

  (a) Solve for h.

  (b) Find h when $V = 320\pi$ mm^3 and $r = 8$ mm.

△ **76. Volume of a Cone:** $V = \dfrac{1}{3}\pi r^2 h$

  (a) Solve for h.

  (b) Find h when $V = 75\pi$ in.^{3} and $r = 5$ in.

△ **77. Area of a Triangle:** $A = \dfrac{1}{2}bh$

  (a) Solve for b.

  (b) Find b when $A = 45$ ft and $h = 5$ ft.

△ **78. Area of a Trapezoid:** $A = \dfrac{1}{2}h(b + B)$

  (a) Solve for h.

  (b) Find h when $A = 99$ cm^2, $b = 19$ cm, and $B = 3$ cm.

79. Energy Expenditure Basal energy expenditure (E) is the amount of energy, measured in calories, required to maintain the body's normal metabolic activity such as respiration, maintenance of body temperature, and so on. For males, the basal energy expenditure is given by the formula

$$E = 66.67 + 13.75W + 5H - 6.76A$$

where W is the weight of the male (in kilograms), H is the height of the male (in centimeters), and A is the age of the male. Determine the basal energy expenditure of a 37-year-old male who is 178 cm (5 feet, 10 inches) tall and weighs 82 kg (180 pounds).

80. Energy Expenditure See Problem 79. The basal energy expenditure for females is given by

$$E = 665.1 + 9.56W + 1.85H - 4.68A$$

Compute the basal energy expenditure of a 40-year-old female who is 168 cm tall (5 feet, 6 inches) and weighs 57 kg (125 pounds).

△ **81. Cylinders** The volume V of a right circular cylinder is given by the formula $V = \pi r^2 h$, where r is the radius and h is the height.

  (a) Solve the formula for h.

  (b) Find the height of a right circular cylinder whose volume is 90π cubic inches and whose radius is 3 inches.

△ **82. Soup Can** The formula $S = 2\pi rh + 2\pi r^2$ gives the surface area S of a right circular cylinder whose radius is r and height is h.

  (a) Solve the formula for h.

  (b) Find the height of a right circular cylinder whose surface area is 19.5π square inches and radius is 1.5 inches.

△ **83. The Better Buy** You and your friends want to buy pizza to eat while watching the Super Bowl. Your local grocery store sells medium (12″) frozen pizzas for $9.99 and small (8″) frozen pizzas that are on special for $4.49 each. Which should you buy to get a better deal?

△ **84. Pizza for Dinner** Mama Mimi's Take and Bake Pizzeria is running a special on southwestern-style pizzas: a large 16″ pizza for $13.99 or two small 8″ pizzas for $12.99. Which should you choose to get the best deal?

85. Taking a Trip Jason drives a truck as an independent contractor. He bills himself out at $28 per hour. Suppose Jason has a contract that calls for him to leave a dock at 9:00 A.M. and travel 600 miles to a warehouse. Jason has driven this route many times and figures that he can travel at an average speed of 50 miles per hour.

(a) Using the formula $d = rt$, where d is the distance traveled, r is the average speed, and t is the time spent traveling, determine how long Jason expects the trip to take.

(b) How much money does Jason expect to earn from this contract?

86. **Taking a Trip** Messai drives a truck as an independent contractor. He bills himself out at $32 per hour. Messai has a contract that calls for him to leave a dock at 8:00 A.M. and travel 145 miles to a warehouse. At the warehouse, he will wait while the truck is loaded (this takes 2 hours) and then return to his original dock. Messai has driven this route many times and figures that he can travel at an average speed of 58 miles per hour.

(a) Using the formula $d = rt$, where d is the distance traveled, r is the average speed, and t is the time spent traveling, determine how long Messai expects the round-trip to take. Exclude the time Messai waits for the truck to be loaded.

(b) How much money does Messai expect to earn from this contract driving his truck?

△ 87. **Area of a Region** Find the area of the figure below.

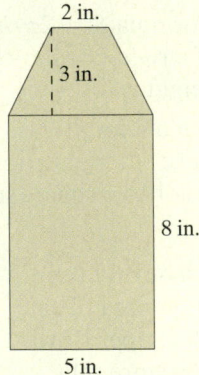

2 in.

3 in.

8 in.

5 in.

△ 88. **Area of a Region** Find the area of the figure below.

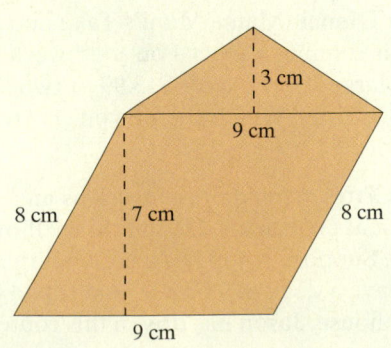

3 cm

9 cm

8 cm 7 cm 8 cm

9 cm

△ 89. **Ice Cream Cone** Find the amount of ice cream in a cone if the radius of the cone is 4 cm and its height is 10 cm. The ice cream fully fills the cone, and the hemisphere of ice cream on the top has a radius of 4 cm. Give both an exact answer and an approximate answer, to the nearest hundredth.

△ 90. **Window** Find the area of the window in the figure below given that the upper portion is a semicircle. Give both an exact answer and an approximate answer, to the nearest hundredth.

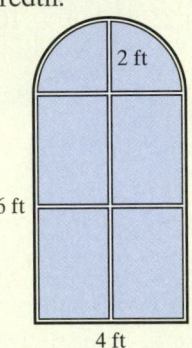

2 ft

6 ft

4 ft

91. **Federal Income Taxes** For a single filer with an annual adjusted income in 2015 over $36,900, but less than $89,350, the federal income tax T for an annual adjusted income I is found using the formula $T = 0.25(I - 36,900) + 5081.25$.

(SOURCE: *Internal Revenue Service*)

(a) Solve the formula for I.

(b) Determine the adjusted income of a single filer whose tax bill is $14,607.

92. **Computing a Grade** Jose's art history instructor uses the formula $G = \dfrac{a + b + 2c + 2d}{6}$ to compute her students' semester grade. The variables a and b represent the grades on two tests, c represents the grade on a research paper, d represents the final exam grade, and G represents the student's average.

(a) Solve the equation for d, Jose's final exam grade.

(b) A final average of 84 will earn Jose a B in the course. Compute the grade Jose must make on his final exam in order to earn a B for the semester, if he scored 78 and 74 on his tests, and 84 on his research paper.

△ 93. **Remodel a Bathroom** You plan to remodel your bathroom and you've chosen 1-foot-by-1-foot ceramic tiles for the floor. The bathroom is 7 feet 6 inches long and 8 feet 3 inches wide.

(a) How many tiles do you need to cover the floor of your bathroom?

(b) Each tile costs $6. How much will it cost to tile your floor?

(c) The store from which you purchase the tile offers a discount of 10% on orders over $350. Does your order qualify for the discount?

94. Painting a Room A gallon of paint can cover about 500 square feet. Find the number of gallon containers of paint that must be purchased to paint two coats on each wall of a rectangular room measuring 8 feet by 12 feet, with a 10-foot ceiling. *Note:* You cannot purchase a partial can of paint!

△ **95. Landscaping a Back Yard** A circular swimming pool whose diameter is 24 feet is to be installed in a rectangular yard that is 60 feet by 90 feet. Once the pool is installed, sod is to be installed on the remaining land to create a lawn around the pool.

  (a) How much sod is required for the lawn? Round your answer to the nearest square foot. Use $\pi \approx 3.14159$.

  (b) If sod costs $0.25 per square foot to install, how much will the sod for the lawn cost?

△ **96. Landscaping a Back Yard** A rectangular swimming pool whose dimensions are 12 feet by 24 feet is to be installed in a rectangular yard that is 80 feet by 40 feet. Once the pool is installed, sod is to be installed on the remaining land to create a lawn around the pool.

  (a) How much sod is required for the lawn?

  (b) A pallet of sod covers 500 square feet. How many pallets of sod are required?

  (c) Each pallet of sod costs $96. How much will the sod for the lawn cost?

Extending the Concepts

△ **97. Conversion** A rectangle has length 5 feet and width 18 inches.

  (a) What is the area in square inches?

  (b) What is the area in square feet?

△ **98. Conversion** A rectangle has length 9 yards and width 8 feet.

  (a) What is the area in square feet?

  (b) What is the area in square yards?

△ **99. Conversion** Determine a formula for converting square inches to square feet.

△ **100. Conversion** Determine a formula for converting square yards to square feet.

Explaining the Concepts

101. A student solved the equation $x + 2y = 6$ for y and obtained the result $y = \dfrac{-x + 6}{2}$. Another student solved the same equation for y and obtained the result $y = -\dfrac{1}{2}x + 3$. Are both solutions correct? Explain why or why not.

102. Make up an example of a linear equation whose coefficients are integers. Make up a similar example where the coefficients are constants (letters). Solve both and then explain which steps in the solution are alike and which are different.

Putting the Concepts Together (Sections 2.1–2.4)

We designed these problems so that you can review Sections 2.1–2.4 and show your mastery of the concepts. Take time to work these problems before proceeding with the next section. The answers are at the back of the text on page AN-4.

  1. Determine whether the given value of the variable is a solution of the equation. Answer Yes or No.

$$4 - (6 - x) = 5x - 8$$

  (a) $x = \dfrac{3}{2}$ **(b)** $x = -\dfrac{5}{2}$

  2. Determine whether the given value of the variable is a solution of the equation. Answer Yes or No.

$$\frac{1}{2}(x - 4) + 3x = x + \frac{1}{2}$$

  (a) $x = -4$ **(b)** $x = 1$

In Problems 3–14, solve the equation and check the solution.

  3. $x + \dfrac{1}{2} = -\dfrac{1}{6}$ **4.** $-0.4m = 16$

  5. $14 = -\dfrac{7}{3}p$ **6.** $8n - 11 = 13$

  7. $\dfrac{5}{2}n - 4 = -19$ **8.** $-(5 - x) = 2(5x + 8)$

  9. $7(x + 6) = 2x + 3x - 15$

  10. $-7a + 5 + 8a = 2a + 8 - 28$

  11. $-\dfrac{1}{2}(x - 6) + \dfrac{1}{6}(x + 6) = 2$

  12. $0.3x - 1.4 = -0.2x + 6$

  13. $5 + 3(2x + 1) = 5x + x - 10$

14. $3 - 2(x + 5) = -2(x + 2) - 3$

15. Investment You have $7500 to invest, and your financial advisor suggests that you put part of the money in a certificate of deposit (CD) that earns 2.4% simple interest, and the remainder in bonds that earn 4% simple interest. Determine the amount you should invest in the CD to earn $220 interest at the end of one year, by solving the equation $0.024x + 0.04(7500 - x) = 220$, where x represents the amount of money invested in CDs.

16. Area of a trapezoid: $A = \dfrac{1}{2}h(B + b)$

    (a) Solve for b.

    (a) Find b when $A = 76$ in.^{2}, $h = 8$ in., and $B = 13$ in.

17. Volume of a right circular cylinder: $V = \pi r^2 h$

    (a) Solve for h.

    (b) Find h when $V = 117\pi$ in.^{2} and $r = 3$ in.

18. Solve the equation $3x + 2y = 14$ for y.

2.5 Problem Solving: Direct Translation

Objectives

1 Translate English Phrases into Algebraic Expressions

2 Translate English Sentences into Equations

3 Build Models for Solving Direct Translation Problems

Are You Prepared for This Section?

Before getting started, complete the following problems. If you get a problem wrong, go back to the section cited and review the material.

P1. Solve the equation: $x + 34.95 = 60.03$ [Section 2.1, pp. 83–85]

P2. Solve the equation: $x + 0.25x = 60$ [Section 2.3, pp. 101–102]

▶ **1** Translate English Phrases into Algebraic Expressions

One of the neat features of mathematics is that its symbols let us express English phrases briefly and consistently as algebraic expressions. For example, the English phrase "5 more than a number x" is represented algebraically as $x + 5$.

Some words or phrases easily translate into mathematical symbols, as shown in Table 2.

Table 2 Math Symbols and the Words They Represent

Add (+)	Subtract (−)	Multiply (·)	Divide (/)
sum	difference	product	quotient
plus	minus	times	divided by
greater than	subtracted from	of	per
more than	less	twice	ratio
exceeds by	less than	double	
in excess of	decreased by	half	
added to	fewer		
increased by			
combined			
altogether			

EXAMPLE 1 **Writing English Phrases Using Math Symbols**

Express each English phrase using mathematical symbols.

    (a) The sum of 2 and 5

    (b) The difference of 12 and 7

    (c) The product of -3 and 8

Prepared? ...Answers
P1. {25.08}   **P2.** {48}

(d) The quotient of 10 and 2

(e) 9 less than 15

(f) A number z decreased by 11

(g) Three times the sum of a number x and 8

Solution

(a) The word sum indicates addition, so "The sum of 2 and 5" is represented mathematically as $2 + 5$.

(b) The word difference indicates subtraction, so "The difference of 12 and 7" is represented mathematically as $12 - 7$.

(c) "The product of -3 and 8" is represented as $-3 \cdot 8$.

(d) "The quotient of 10 and 2" is represented mathematically as $\dfrac{10}{2}$.

(e) "9 less than 15" is represented mathematically as $15 - 9$.

(f) "A number z decreased by 11" is represented algebraically as $z - 11$.

(g) "Three times the sum of a number x and 8" is represented algebraically as $3(x + 8)$.

In Example 1(g), the mathematical representation is $3(x + 8)$ rather than $3x + 8$ because the phrase "three times the sum" means to multiply the sum of the two numbers by 3. The phrase that would result in $3x + 8$ might be "the sum of three times a number and 8." Do you see the difference?

<aside>
Work Smart

When translating from English to math, try some specific examples. In Example 1(e), think of money when translating. Bob owes Mary $15. Mary says, "No, you owe me $9 less than $15," or $6. In Example 1(f), to translate "a number z decreased by 11," pick specific values of z, as in "16 decreased by 11," which would be $16 - 11$, or 5. So "z decreased by 11" is $z - 11$.
</aside>

> ### Quick ✔
>
> *In Problems 1–6, express each phrase using mathematical symbols.*
>
> **1.** The sum of 5 and 17   **2.** The difference of 7 and 4
>
> **3.** The quotient of 25 and 3   **4.** The product of -2 and 6
>
> **5.** Twice a less 2   **6.** Five times the difference of m and 6.

EXAMPLE 2   ### Translating an English Phrase into an Algebraic Expression

Write an algebraic expression for each problem.

(a) The Raiders scored p points in a football game. The Packers scored 12 more points than the Raiders. Write an algebraic expression for the number of points the Packers scored.

(b) A lumberman cuts a 50-foot log into two pieces. One piece is t feet long. Express the length of the second piece as an algebraic expression in t.

(c) The number of quarters in a vending machine is two less than the number of dimes, d, in the machine. Write an expression for the number of quarters as an algebraic expression in d.

Solution

(a) The phrase "more than" implies addition. The Packers scored $p + 12$ points in the football game.

(b) The log is 50 feet long. If the lumberman cuts one piece 20 feet long, then the other piece must be $50 - 20 = 30$ feet. In general, if the lumberman cuts one piece t feet long, then the remaining piece must be $(50 - t)$ feet.

(c) The phrase "less than" implies subtraction. Two less than the number of dimes is represented algebraically as $d - 2$.

EXAMPLE 3   **Translating an English Phrase into an Algebraic Expression**

Write an algebraic expression for each problem.

(a) The number of student tickets sold for a play is five fewer than four times the number n of nonstudent tickets sold. Write an algebraic expression for the number of student tickets sold in terms of n.

(b) The height of a full-grown maple tree is ten feet more than three times the height h of a sapling. Write an algebraic expression for the height of the full-grown tree in terms of h.

(c) Como believes he owes Abigail d dollars. Abigail says Como owes her 5 dollars less than three times the amount Como thought was owed. Write an algebraic expression for the amount Abigail says Como owes her in terms of d.

Solution

(a) The phrase "fewer than" implies subtraction. The number of student tickets sold is five fewer than four times the number of nonstudent tickets sold. So the algebraic expression representing the number of student tickets sold is 4 times the number of nonstudent tickets sold minus 5, or $4n - 5$.

(b) The full-grown tree is ten feet more than three times the height of the sapling, so $(3h + 10)$ feet represents the height of the full-grown tree.

(c) The phrase "less than" implies subtraction. Como owes Abigail 5 dollars less than 3 times d, so he owes her $3d - 5$ dollars.   ●

▶ ❷ Translate English Sentences into Equations

Nearly every word problem in algebra requires some type of translation. Learning to speak the language of math is the same as learning to speak any language. Now that we know how to translate English phrases to algebraic expressions, we can learn to translate English sentences to algebraic equations.

Work Smart

An equation is a statement in which two algebraic expressions are equal. See Section 2.1.

In English, a complete sentence must contain a subject and a verb, so expressions or "phrases" are not complete sentences. For example, the expression "Beats me!" does not contain a subject, and the expression "5 more than a number x" does not contain a verb. Therefore neither expression is a complete sentence. The statement "5 more than a number x is 18" is a complete sentence because it contains a subject and a verb, so it can be translated into a mathematical statement. In mathematics, statements can be represented symbolically as equations.

In English, statements can be true or false. For example, "The moon is made of green cheese" is a false statement, while "The sky is blue" is a true statement. Mathematical statements can be true or false as well.

Table 3 lists words that typically translate into an equal sign.

Table 3 Words That Translate into an Equal Sign			
is	yields	are	equals
was	gives	results in	is equal to
is equivalent to			

Notice that these words are verbs. The equal sign in an equation acts like a verb in a sentence.

EXAMPLE 4 **Translating English Sentences into Equations**

Translate each sentence into an equation. Do not solve the equation.

(a) Five more than a number x is 20.

(b) Four times the sum of a number z and 3 is 15.

(c) The difference of x and 5 equals the quotient of x and 2.

Solution

(a)

$$\underbrace{\text{5 more than a number } x}\quad \underbrace{\text{is}}\quad \underbrace{\text{20}}$$
$$x + 5 \qquad\quad = \qquad 20$$

(b) First determine the sum and then multiply this result by 4.

$$\underbrace{\text{Four times the sum of a number } z \text{ and } 3}\quad \underbrace{\text{is}}\quad \underbrace{15}$$
$$4(z + 3) \qquad\qquad\qquad = \qquad 15$$

(c)

$$\underbrace{\text{The difference of } x \text{ and } 5}\quad \underbrace{\text{equals}}\quad \underbrace{\text{the quotient of } x \text{ and } 2}$$
$$x - 5 \qquad\qquad = \qquad \frac{x}{2}$$

Work Smart

5 more than a number x can be written as $5 + x$ or $x + 5$. Be careful! $5 - x$ is not the same as $x - 5$. Can you explain why?

Work Smart

The English sentence "The sum of four times a number z and 3 is 15" is expressed mathematically as $4z + 3 = 15$. Do you see how this differs from Example 4(b)?

Quick ✓

13. In mathematics, English statements can be represented symbolically as _____.

In Problems 14–17, translate each English statement into an equation. Do not solve the equation.

14. The product of 3 and y is equal to 21.

15. The difference of x and 10 equals the quotient of x and 2.

16. Three times the sum of n and 2 is 15.

17. The sum of three times n and 2 is 15.

▶ **An Introduction to Problem Solving and Mathematical Models**

Every day we encounter various types of problems that must be solved. **Problem solving** is the ability to use information, tools, and our own skills to achieve a goal. For example, suppose 4-year-old Kevin wants a glass of water, but he is too short to reach the sink.

Kevin has a problem. To solve the problem, he finds a step stool and pulls it over to the sink. He uses the step stool to climb on the counter, opens the kitchen cabinet, and pulls out a cup. He then crawls along the counter top, turns on the faucet, fills the cup, and proceeds to drink the water. Problem solved!

Of course, Kevin could solve the problem other ways. Just as there are various ways to solve life's everyday problems, there are many ways to solve problems using mathematics. However, regardless of the approach, there are always some common aspects in solving any problem. For example, regardless of how Kevin ultimately ends up with his cup of water, someone must get a cup from the cabinet and someone must turn on the faucet.

One of the purposes of learning algebra is to be able to solve certain types of problems. To solve these problems, the verbal descriptions in the problems are translated into equations that can be solved. The process of turning a verbal description into a mathematical equation is known as **mathematical modeling.** The equation that is developed is called the **mathematical model.**

Not all models are mathematical. In general, a **model** is a way of using graphs, pictures, equations, or even verbal descriptions to represent a real-life situation. Because the world is a complex place, we need to simplify information when we develop a model. For example, a map is a model of our road system. Maps don't show all the details of the system (such as trees, buildings, or potholes), but they do a good job of describing how to get from point A to point B. Mathematical models are similar in that assumptions are often made regarding the world in order to make the mathematics less complicated.

Every problem is unique in some way. However, because many problems are similar, problems can be categorized. In this text, we will solve five categories of problems.

Five Categories of Problems

1. **Direct Translation**—problems in which we translate from English directly to an equation by using key words in the verbal description.

2. **Geometry**—problems in which the unknown quantities are related through geometric formulas.

3. **Mixtures**—problems in which two or more quantities are combined in some fashion.

4. **Uniform Motion**—problems in which an object travels at a constant speed.

5. **Work Problems**—problems in which two or more entities join forces to complete a job.

In this section, direct translation problems are presented. However, the following guidelines will help you solve any category of problem.

Solving Problems with Mathematical Models

Step 1: Identify What You Are Looking For Read the problem very carefully, perhaps two or three times. Identify the type of problem and the information that you wish to find. It is fairly typical that the last sentence in the problem indicates what we need to solve for in the problem.

Step 2: Give Names to the Unknowns Assign variables to the unknown quantities. Choose a variable that is representative of the unknown quantity. For example, use t for time. Be sure to include units.

Step 3: Translate the Problem into Mathematics Read the problem again. This time, after each sentence is read, determine whether the sentence can be translated into a mathematical statement or expression in terms of the variables identified in Step 2. It is often helpful to create a table, chart, or figure. If necessary, combine the mathematical statements or expressions into an equation that can be solved.

Step 4: Solve the Equation(s) Found in Step 3 Solve the equation for the variable.

Step 5: Check the Reasonableness of Your Answer Check your answer with the wording of the original problem to be sure that it makes sense. If it does not, go back and try again.

Step 6: Answer the Question Write your answer in a complete sentence.

Let's review each of these steps, one at a time.

- **Identify** Carefully read the problem. Reading a verbal description of a problem is not like reading a spy novel. You may not know how to solve the problem while reading it, but you should get a sense of which of the five categories the problem falls into, what information you are given, and what you are being asked to do.

- **Name** Assign variables to the unknowns. Write down the name of each variable and what it represents. Use this to check your final answer.

- **Translate** In this step, you develop a model (an equation) that mathematically describes the problem.

- **Solve the equation** This is generally the easy part. Most students say, "I could solve the problem if I could find the right equation."

- **Check** Checking your answer can be difficult because you can make two types of errors: 1. You correctly translate the problem into a model but then make an error solving the equation. 2. You misinterpret the problem and develop an incorrect model. Your solution will satisfy your model, but it probably will not be the solution to the original problem. We can check for this type of error by determining whether the solution is reasonable.

- **Answer the question** Always be sure to answer the question being asked.

Quick ✓

18. Letting variables represent unknown quantities and then expressing relationships among the variables in the form of equations is called _____ _____.

❸ Build Models for Solving Direct Translation Problems

Let's look at a "direct translation" problem. Remember, these problems can be set up by reading the problem and translating the verbal description into an equation.

EXAMPLE 5 **Solving a Direct Translation Problem**

The price of a "premium" ticket for a Broadway show in New York is $69 more than twice the price of a premium ticket for the same show in Chicago. If you buy a ticket for this show in Chicago, and your friend buys a ticket for the same show in New York, and the total cost of the tickets is $436.50, what does each ticket cost?

Solution

Step 1: Identify This is a direct translation problem. Find the price of a premium ticket for a Broadway show in New York and for the same show in Chicago.

Step 2: Name The price of a ticket in New York is $69 more than twice the price of a ticket for the same show in Chicago. Let c represent the ticket price in dollars in Chicago. Then $2c + 69$ represents the ticket price in New York.

Step 3: Translate Because the total cost for the tickets to both performances is $436.50, use the equation

price of ticket in Chicago plus price of ticket in New York equals total cost

$$c \quad + \quad (2c + 69) \quad = \quad 436.50 \quad \text{The Model}$$

Step 4: Solve Solve the equation.

$$c + (2c + 69) = 436.50$$

Combine like terms: $\qquad 3c + 69 = 436.50$

Subtract 69 from both sides of the equation: $\quad 3c + 69 - 69 = 436.50 - 69$

$$3c = 367.50$$

Divide both sides of the equation by 3: $\qquad \dfrac{3c}{3} = \dfrac{367.50}{3}$

$$c = 122.50$$

(continued)

Work Smart

Remember that you can use any letter to represent the unknown(s) when you make your model. Choose a letter that reminds you of what it represents. For example, use t for time.

Work Smart

It is helpful to assign the variable to the quantity that you know the **least** about.

So, the price of a ticket for the show in Chicago is $c = \$122.50$. The price of a ticket for the same Broadway show in New York is $2c + 69 = 2(122.50) + 69 = 314$. It costs $314 to see the show in New York.

Step 5: Check Is the total cost of both tickets $436.50? Because $122.50 + \$314 = \$436.50$, the answers are correct.

Step 6: Answer The price of a premium ticket to a Broadway show in Chicago is $122.50, and the price of a premium ticket to the same Broadway show in New York is $314.

> **Quick** ✔
>
> *Translate the problem into an algebraic equation and solve the equation for the unknowns.*
>
> **19.** Sean and Connor decide to buy a pizza. The pizza costs $15 and they decide to split the cost based upon how much pizza each eats. Connor eats two-thirds of the amount that Sean eats, so Connor pays two-thirds of the amount that Sean pays. How much does each pay?

Work Smart

Examples of consecutive even integers are

16, 18, 20

−10, −8, −6

752, 754, 756

Examples of consecutive odd integers are

9, 11, 13

−35, −33, −31

623, 625, 627

▶ **Consecutive Integer Problems**

Recall that an *integer* is a member of the set $\{\ldots, -3, -2, -1, 0, 1, 2, 3, \ldots\}$. An *even integer* is an integer that is divisible by 2, such as, 16 and 24. An *odd integer* is an integer that is not even, such as 9 and 17. Consecutive integers are integers such as 9 and 10, or 31 and 32, and always differ by 1. So if n represents the first integer, then $n + 1$ represents the second integer, $n + 2$ represents the third integer, and so on. Consecutive *even* integers, such as 14 and 16, differ by 2, so if n represents the first even integer, then $n + 2$ represents the second even integer and $n + 4$ represents the third even integer. Consecutive odd integers also differ by 2 (41 and 43, for example), so if n represents the first odd integer, then $n + 2$ represents the second odd integer, and $n + 4$ represents the third odd integer. See Figure 3.

Figure 3

Consecutive integers

n $n + 1$ $n + 2$

Consecutive **even** integers, if n is even

n $n + 2$ $n + 4$

Consecutive **odd** integers, if n is odd

n $n + 2$ $n + 4$

(**EXAMPLE 6**) **Solving a Direct Translation Problem: Consecutive Integers**

The sum of three consecutive even integers is 324. Find the integers.

Solution

Step 1: Identify This is a direct translation problem. Find three consecutive even integers whose sum is 324.

Step 2: Name Let n represent the first even integer, so $n + 2$ is the second even integer, and $n + 4$ is the third even integer.

Step 3: Translate The sum of the three consecutive even integers is 324, so the equation is

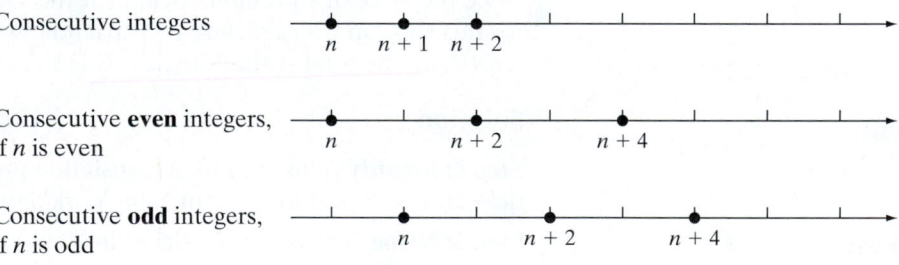

first even integer   second even integer   third even integer   sum

$$n \quad + \quad (n + 2) \quad + \quad (n + 4) \quad = 324 \quad \text{The Model}$$

Step 4: Solve

$$n + (n + 2) + (n + 4) = 324$$

Combine like terms:
$$3n + 6 = 324$$

Subtract 6 from both sides of the equation:
$$3n + 6 - 6 = 324 - 6$$

$$3n = 318$$

Divide both sides of the equation by 3:
$$\frac{3n}{3} = \frac{318}{3}$$

$$n = 106$$

Because n, the first even integer, is 106, the remaining even integers are 108 and 110.

Step 5: Check The numbers are all even integers and $106 + 108 + 110 = 324$. The answer is correct.

Step 6: Answer The three consecutive even integers are 106, 108, and 110. ●

> **Quick ✓**
>
> **20.** *True or False* If n represents the first of three consecutive odd integers, then $n + 1$ and $n + 3$ represent the next two odd integers.
>
> *In Problems 21 and 22, translate the problem into an algebraic equation and solve the equation for the unknowns.*
>
> **21.** The sum of three consecutive even integers is 270. Find the integers.
>
> **22.** The sum of 4 consecutive odd integers is 72. Find the integers.

EXAMPLE 7   Solving a Direct Translation Problem: Piece Lengths

A carpenter is building a shelving system and cuts a 14-foot length of cherry shelving into three pieces. The second piece is twice as long as the first, and the third piece is 2 feet longer than the first. Find the length of each piece of cherry shelving.

Solution

Step 1: Identify This is a direct translation problem. Find the length of each piece of shelving. We know that the board is 14 feet long.

Step 2: Name Because the lengths of the second and third pieces are described in terms of the length of the first piece, let x represent the length of the first piece in feet. The second piece is $2x$ because it is twice as long as the first piece. The third piece is $x + 2$ because it is 2 feet longer than the first piece. See Figure 4.

Figure 4

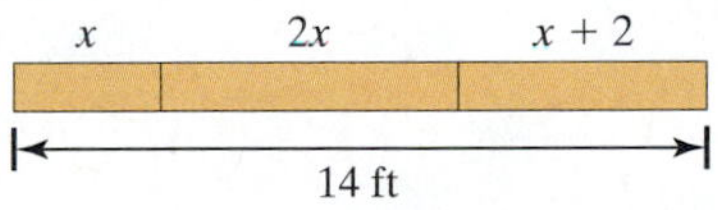

$$x \qquad 2x \qquad x + 2$$

|← 14 ft →|

Step 3: Translate The lengths of the three pieces must have a sum of 14 feet, so the equation is

length of first piece		length of second piece	length of third piece		total length
x	$+$	$2x$	$+ \ (x + 2)$	$=$	14 The Model

Step 4: Solve

$$x + 2x + (x + 2) = 14$$

Combine like terms:
$$4x + 2 = 14$$

Subtract 2 from both sides of the equation:
$$4x + 2 - 2 = 14 - 2$$

$$4x = 12$$

Divide both sides of the equation by 4:
$$\frac{4x}{4} = \frac{12}{4}$$

$$x = 3$$

Work Smart

Sometimes these problems are called "the whole equals the sum of the parts" problems because the values of the "parts" must sum to the value of the "whole."

Since the length of the first piece is 3 feet, the second piece is $2x = 2 \cdot 3 = 6$ feet and the third piece is $x + 2 = 3 + 2 = 5$ feet.

(continued)

Step 5: Check The sum of the lengths of the three pieces of cherry shelving is $3 + 6 + 5 = 14$, so we have the correct answer.

Step 6: Answer The three pieces of shelving are 3 feet, 6 feet, and 5 feet long. ●

> **Quick ✓**
>
> *Translate the problem into an algebraic equation and solve.*
>
> **23.** A 76-inch length of ribbon is to be cut into three pieces. The longest piece is to be 24 inches longer than the shortest piece, and the third piece is to be half the length of the longest piece. Find the length of each piece of ribbon.

▶ **EXAMPLE 8** **Investment Decisions**

A total of $25,000 is to be invested in bonds and certificates of deposit (CDs). The amount in CDs is to be $8000 less than the amount in bonds. How much is to be invested in each type of investment?

Solution

Step 1: Identify This is a direct translation problem. Find the amount invested in CDs and the amount invested in bonds.

Step 2: Name Let b represent the amount invested in bonds, in dollars.

Step 3: Translate Suppose $18,000 was invested in bonds, then the amount in CDs will be $8000 less than this amount, or $10,000. In general, if b is the amount, in dollars, invested in bonds, then $b - 8000$ represents the amount invested in CDs. Also, our total investment in bonds and CDs is $25,000, so

$$\underset{\text{Amount invested in bonds}}{b} + \underset{\text{Amount invested in CDs}}{(b - 8000)} = \underset{\text{Total investment}}{25,000} \quad \text{The Model}$$

Step 4: Solve
$$b + (b - 8000) = 25{,}000$$

Combine like terms: $\quad 2b - 8000 = 25{,}000$

Add 8000 to both sides of the equation: $\quad 2b = 33{,}000$

Divide both sides of the equation by 2: $\quad b = 16{,}500$

Step 5: Check If $16,500 is invested in bonds, then the amount invested in CDs should be $8000 less than this amount, or $8500. The total investment is $16,500 + $8500 = $25,000, so the answer checks.

Step 6: Answer Invest $16,500 in bonds and $8500 in CDs. ●

> **Quick ✓**
>
> *Translate the problem into an algebraic equation and solve.*
>
> **24.** A total of $18,000 is to be invested in stocks and bonds. If the amount invested in bonds is twice that invested in stocks, how much is invested in each category?

▶ **EXAMPLE 9** **Mulch Delivery**

Central Indiana Landscapers has undertaken a landscaping job that requires a truckload of mulch. Lots-O-Mulch will deliver up to 20 cubic yards of mulch for $104 plus $28.99 per cubic yard of mulch. Mulch Man will deliver up to 20 cubic yards of the same mulch for $68 plus $32.99 per cubic yard of mulch. For how many cubic yards of mulch will the charges be the same?

Solution

Step 1: Identify This is a direct translation problem. Find the number of cubic yards of mulch for which the two charges will be the same.

Step 2: Name Let m represent the number of cubic yards of mulch for which the charges will be the same.

Step 3: Translate The charge for Lots-O-Mulch is $104 plus $28.99 for each cubic yard of mulch delivered. So if one cubic yard of mulch is delivered, the charge is $104 + (28.99)(1) = 132.99$ dollars. For two cubic yards, the charge is $104 + (28.99)(2) = 161.98$ dollars. In general, for m cubic yards of mulch, the charge is $(104 + 28.99m)$ dollars. Similar logic reveals that the charge for Mulch Man is $(68 + 32.99m)$ dollars. To find the number of cubic yards for which the two charges are the same, solve

$$\underbrace{104 + 28.99m}_{\text{Cost for Lots-O-Mulch}} \underset{\text{same as}}{=} \underbrace{68 + 32.99m}_{\text{Cost for Mulch Man}} \qquad \text{The Model}$$

Step 4: Solve

$$104 + 28.99m = 68 + 32.99m$$

Subtract $28.99m$ from both sides of the equation: $\qquad 104 = 68 + 4m$

Subtract 68 from both sides of the equation: $\qquad 36 = 4m$

Divide both sides of the equation by 4: $\qquad 9 = m$

Step 5: Check The charge of Lots-O-Mulch will be $104 + 28.99(9) = \$364.91$. The charge of Mulch Man is $68 + 32.99(9) = \$364.91$. They are the same!

Step 6: Answer the Question The delivery charges will be the same for 9 cubic yards of mulch. ●

Quick ✓

25. You need to rent a moving truck. EZ-Rental charges $30 per day plus $0.15 per mile. Do It Yourself Rental charges $15 per day plus $0.25 per mile. For how many miles will the daily cost of renting from these companies be the same?

26. Carl's Appliance Repair Shop charges $69.99 for a repair call plus $30 per hour. Terry's Appliance Repair Shop charges $54.99 for a repair call plus $40 per hour. For how many hours will the repair charges be the same?

2.5 Exercises   MyMathLab®

Exercise numbers in green have complete video solutions in MyMathLab or may be accessed using the QR code to the right.

*Problems **1–26** are the Quick ✓s that follow the EXAMPLES.*

Building Skills

Problems 27–44, translate each phrase into an algebraic expression. Let x represent the unknown number. See Objective 1.

27. the sum of 5 and a number

28. a number increased by 32.3

29. the product of a number and $\dfrac{2}{3}$

30. the product of -2 and a number

31. half of a number

32. double a number

33. a number less -25

34. 8 less than a number

35. the quotient of a number and 3

36. the quotient of -14 and a number

37. $\dfrac{1}{2}$ more than a number

38. $\dfrac{4}{5}$ of a number

39. 9 more than 6 times a number

40. 21 more than 4 times a number

41. twice the sum of 13.7 and a number

42. 50 less than half of a number

43. the sum of twice a number and 31

44. the sum of twice a number and 45

In Problems 45–52, choose a variable to represent one quantity. State what that quantity represents and then express the second quantity in terms of the first. See Objective 1.

45. The Columbus Clippers scored 5 more runs than the Charlotte Knights.

46. The Toronto Blue Jays scored 3 fewer runs than the Tampa Bay Rays.

47. Jan has $0.55 more in her piggy bank than Bill.

48. Beryl has $0.25 more than 3 times the amount Ralph has.

49. Janet and Kathy will share the $200 grant.

50. Juan and Emilio will share the $1500 lottery winnings.

51. There were 1433 visitors to the Arts Center Spring show. Some were adults and some were children.

52. There were 12,765 fans at a recent NBA game. Some held paid admission tickets, and some held special promotion tickets.

In Problems 53–60, translate each statement into an equation. Let x represent the unknown number. DO NOT SOLVE. See Objective 2.

53. The sum of a number and 15 is −34.

54. The sum of 43 and a number is −72.

55. 35 is 7 less than triple a number.

56. 49 is 3 less than twice a number.

57. The quotient of a number and −4, increased by 5, is 36.

58. The quotient of a number and −6, decreased by 15, is 30.

59. Twice the sum of a number and 6 is the same as 3 more than the number.

60. Twice the sum of a number and 5 is the same as 7 more than the number.

For Problems 61–66, (a) name the unknown quantity or quantities and (b) translate the problem into a mathematical model. Do not solve the equation. See Objective 2.

61. Number of Twitter Users In 2015 there were 2.6 million more Twitter users in the 25–34 age category than there were in the 35–44 age category. There were 20 million Twitter users in those two categories in 2015. How many users were there in the 25–34 age category?

62. Number of Facebook Users There are 98 million more monthly Facebook users in Europe than there are in Canada and the United States combined. If there are a total of 532 million monthly Facebook users in Europe, Canada, and the United States, find the number of monthly Facebook users in Europe.

63. Phone Apps Jessica has 15 fewer apps on her phone than her friend Wendy. If the total number of apps on both phones is 31, find the number of apps on Jessica's phone.

64. Nursery Crops The *QRST* Nursery grows oak trees and magnolia trees for landscapers. This year, the number of oak trees is ten less than twice the number of magnolia trees. If *QRST* Nursery has 110 of these trees, find the number of oak trees in the nursery.

65. Art Display A 46-inch. piece of wire for an art exhibit is to be cut into three pieces. The shortest piece is 4 inches shorter than the middle piece, and the longest piece is 2 inches longer than the middle piece. How long is each piece?

66. Education Grant Professors Crawford, MacLean, and Betzel will share a $10,000 educational grant. Professor Crawford will receive $1000 less than Professor Betzel, and Professor MacLean will receive $2000 more than Professor Betzel. How much money will each professor receive?

Applying the Concepts

67. Number Sense The sum of a number and −12 is 71. Find the number.

68. Number Sense The difference between a number and 13 is −29. Find the number.

69. Consecutive Integers The sum of three consecutive integers is 165. Find the numbers.

70. Consecutive Integers The sum of three consecutive odd integers is 81. Find the numbers.

71. Bridges The longest suspension bridge in the United States is the Verrazano-Narrows Bridge. The second-longest suspension bridge in the United States is the Golden Gate Bridge, which is 60 feet shorter than the Verrazano-Narrows Bridge. The combined length of the two bridges is 8460 feet. Find the length of each bridge.

72. Towers The tallest buildings in the world (those having the most floors) are Burj Khalifa in Dubai, UAE, and the Shanghai Tower in China. The Shanghai Tower has 35 fewer floors than Burj Khalifa. The two building together have 291 floors. Find the number of floors in each building.

73. Buying a Motorcycle Scott buys a new motorcycle for a total price of $12,336.50. The tax, title, and dealer preparation charges amount to one-tenth of the purchase price. Find the price of the motorcycle before the extra charges.

74. Buying a Desk The total price for a new desk and chair is $336.25. The cost of the chair is one-quarter of the cost of the desk. Find the cost of the desk.

75. Finance A total of $20,000 is to be invested, some in bonds and some in certificates of deposit (CDs). The amount invested in bonds is to be $3000 greater than the amount invested in CDs. How much is to be invested in each type of investment?

76. Finance A total of $10,000 is to be divided between Sean and George. George is to receive $3000 less than Sean. How much will each receive?

77. Investments Suppose that your Aunt May has left you an unexpected inheritance of $32,000. You have decided to invest the money rather than blow it on frivolous purchases. Your financial advisor has recommended that you diversify by placing some of the money in stocks and some in bonds. Based upon current market conditions, she has recommended that the amount invested in bonds should equal three-fifths of the amount invested in stocks. How much should be invested in stocks? How much should be invested in bonds?

78. Investments Jack and Diane have $40,000 to invest. Their financial advisor has recommended that they diversify by placing some of the money in stocks and some in bonds. Based upon current market conditions, he has recommended that the amount invested in bonds should equal two-thirds of the amount invested in stocks. How much should be invested in stocks? How much should be invested in bonds?

79. Cereal A serving of Kellogg's Smart Start cereal contains 5 more grams of carbohydrates than Kashi Go Lean Crunch cereal. If you eat a serving of each cereal, you will consume 81 grams (g) of carbohydrates. Find the amount of carbohydrates in each cereal.

80. Books A paperback edition of a book costs $12.50 less than the hardback edition of the book. If you purchase one of each version, you will pay $37.40. Find the cost of the paperback edition of the book.

81. Income On a joint income tax return, Elizabeth Morrell's adjusted gross income was $2549 more than her husband Dan's adjusted gross income. Their combined adjusted gross income was $55,731. Find Elizabeth Morrell's adjusted gross income.

82. Spring Break Allison went shopping to prepare for her Spring Break trip. Her bathing suit cost $8 more than a pair of shorts, and a T-shirt cost $2 less than the shorts. Find the cost of the bathing suit if Allison spent $60 on the items, before sales tax.

83. Truck Rentals You need to rent a moving truck. You have identified two companies that rent trucks. EZ-Rental charges $35 per day plus $0.15 per mile. Do It Yourself Rental charges $20 per day plus $0.25 per mile. For how many miles will the daily cost of renting be the same?

84. Cellular Telephones You need a new cell phone for emergencies only. Company A charges $12 per month plus $0.10 per minute, while Company B charges $0.15 per minute with no monthly service charge. For how many minutes will the monthly cost be the same?

85. Comparing Printers Samuel is trying to decide between two laser printers, one manufactured by Hewlett-Packard, the other by Brother. Both have similar features and warranties, so price is the determining factor. The Hewlett-Packard costs $200, and printing costs are approximately $0.03 per page. The Brother costs $240, and printing costs are approximately $0.01 per page. How many pages need to be printed for the cost of the two printers to be the same?

86. Comparing Job Offers Hans has just been offered two sales jobs selling vacuums. The first job offer is a base monthly salary of $2000 plus a commission of $50 for each vacuum sold. The second job offer is a base monthly salary of $1200 plus a commission of $60 for each vacuum sold. How many vacuums must be sold for the two jobs to pay the same monthly salary?

87. Adjusted Gross Income On a joint income tax return, Jensen Beck's adjusted gross income was $249 more than his wife Maureen's adjusted gross income. Their combined adjusted gross income was $72,193. Find Jensen and Maureen Beck's adjusted gross income.

88. Camping Trip Jaime Juarez purchased some new camping equipment. He spent $199 on a cookware set, a lantern, and a cook stove. The cookware set cost $30 more than the lantern, and the cook stove cost $34 more than the lantern. Find the cost of each item.

89. Computing Grades Going into the final, which will count as three tests, Monica has test scores of 76, 90, 94, and 88. What score (out of 100) does Monica need on the final exam in order to have an average score of 90?

90. Computing Grades Going into the final exam, which will count as two tests, Brooke has test scores of 80, 83, 71, 61, and 95. What score (out of 100) does Brooke need on the final exam in order to have an average score of 80?

Extending the Concepts

In Problems 91–94, write a problem that would translate into the given equation.

91. $10x = 370$ **92.** $5n + 10 = 170$

93. $\dfrac{x + 74}{2} = 80$ **94.** $n + n + 2 = 98$

△ **95. Angles** The sum of the measures of the three angles in any triangle is 180 degrees. The measure of the smallest angle of a certain triangle is half the measure of the second angle. The measure of the largest angle is 40° more than 4 times the measure of the smallest. Find the measure of each angle.

△ **96. Angles** The sum of the measures of the three angles in any triangle is 180 degrees. The measure of one angle of a certain triangle is 1° more than three times the measure of the smallest angle. The measure of the third angle is 13° less than twice the measure of the second angle. Find the measure of each angle.

Explaining the Concepts

97. How is mathematical modeling related to problem solving? Why do we make assumptions when creating mathematical models?

98. What is the difference between an algebraic expression and an equation? How is each related to English phrases and English statements?

99. Two students write an equation to solve a word problem with consecutive odd integers. One student assigns the variables as $n - 1, n + 1, n + 3$, where n is an even integer. A second student uses $n, n + 2, n + 4$, where n is an odd integer. Which student is correct? Will the value for n be the same for both? Make up a problem that can be solved in more than one way, and explain how the variables were assigned.

100. Using the algebraic expression $3x + 5$, make up a problem that uses the direction "evaluate." Using the same algebraic expression, make up a problem that uses the direction "solve."

2.6 Problem Solving: Problems Involving Percent

Objectives

❶ Solve Problems Involving Percent

❷ Solve Business Problems That Involve Percent

Are You Prepared for This Section?

Before getting started, complete the following problems. If you get a problem wrong, go back to the section cited and review the material.

P1. Write 45% as a decimal. [Section 1.2, p. 15]

P2. Write 0.2875 as a percent. [Section 1.2, p. 16]

P3. Simplify: **(a)** $0.4 \cdot 50$ **(b)** $\dfrac{15}{0.3}$ [Section 1.5, pp. 44–45]

P4. Simplify: $p + 0.05p$ [Section 1.8, pp. 68–69]

▶ ❶ **Solve Problems Involving Percent**

Percent means "divided by 100" or "per hundred." The symbol % denotes percent, so 45% means 45 out of 100 or $\dfrac{45}{100}$ or 0.45. In applications involving percents, we often see the word "of," as in 20% of 60. The word "of" translates into multiplication in mathematics, so 20% of 60 means $0.20 \cdot 60$.

(**EXAMPLE 1**) **Solving an Equation Involving Percent**

A number is 35% of 40. Find the number.

Solution

Step 1: Identify This is a direct translation problem. Find the unknown number.

Step 2: Name Let n represent the number.

Step 3: Translate

$$\underset{n}{\underline{\text{a number}}} \quad \underset{=}{\underline{\text{is}}} \quad \underset{0.35}{\underline{35\%}} \quad \underset{\cdot}{\underline{\text{of}}} \quad \underset{40}{\underline{40}}$$

Step 4: Solve

$$n = 0.35(40)$$

Multiply: $n = 14$

Step 5: Check Check the multiplication: $0.35(40) = 14$.

Step 6: Answer 14 is 35% of 40.

Quick ✔

1. Percent means "divided by ___."
2. *True or False* 40% of 120 means $0.4 \cdot 120$.
3. A number is 60% of 90. Find the number.
4. A number is 3% of 80. Find the number.
5. A number is 150% of 24. Find the number.
6. A number is 8% of 40. Find the number.

EXAMPLE 2   Solving an Equation Involving Percent

The number 240 is what percent of 800?

Solution

Step 1: Identify This is a direct translation problem. Find the percent.

Step 2: Name Let x represent the percent.

Step 3: Translate

$$\underset{240}{\underline{240}} \ \underset{=}{\underline{\text{is}}} \ \underset{x}{\underline{\text{what percent}}} \ \underset{\cdot \ 800}{\underline{\text{of 800?}}}$$

Work Smart

When using "percent" in an equation, the percent will always be expressed in decimal form. To change a percent to a decimal, move the decimal point two places to the left. To change a decimal to a percent, move the decimal two places to the right.

Step 4: Solve

$$240 = 800x$$

Divide both sides by 800: $\dfrac{240}{800} = \dfrac{800x}{800}$

$$0.3 = x$$

Because we are finding a percent and our answer is a decimal, change 0.3 to a percent by moving the decimal point two places to the right: $0.30 = 30\%$.

Step 5: Check Is 240 equal to 30% of 800? Does $(0.30)(800) = 240$? Yes!

Step 6: Answer The number 240 is 30% of 800.   ●

Quick ✔

7. The number 8 is what percent of 20?
8. The number 15 is what percent of 40?
9. The number 44 is what percent of 40?

EXAMPLE 3   Solving an Equation Involving Percent

42 is 35% of what number?

Solution

Step 1: Identify This is a direct translation problem. Find the unknown number.

Step 2: Name Let x represent the number.

Step 3: Translate

$$\underset{42}{\underline{42}} \ \underset{=}{\underline{\text{is}}} \ \underset{0.35}{\underline{35\%}} \ \underset{\cdot}{\underline{\text{of}}} \ \underset{x}{\underline{\text{what number?}}}$$

Step 4: Solve

$$42 = 0.35x$$

Divide both sides by 0.35: $\dfrac{42}{0.35} = \dfrac{0.35x}{0.35}$

$$120 = x$$

Step 5: Check Is 35% of 120 equal to 42? Because $(0.35)(120) = 42$, the answer is correct.

Step 6: Answer 42 is 35% of 120.   ●

EXAMPLE 4

Millennial Population in the United States

In 2015, the population of the United States was approximately 320,000,000. If about 26% of the U.S. population in 2015 were millennials, or individuals born between 1982 and 2000, determine the number of millennials in 2015. (SOURCE: *U.S. Census Bureau*)

Solution

Step 1: Identify Use direct translation to find the number of millennials in the United States in 2015.

Step 2: Name Let m represent the number of millennials.

Step 3: Translate In the year 2015, 26% of the United States population were millennials, and in 2015 the population of the United States was approximately 320,000,000. Translate the words of the problem:

$$\underset{0.26}{\underline{26\%}} \quad \underset{\cdot}{\underline{\text{of}}} \quad \underset{320{,}000{,}000}{\underline{\text{U.S. population}}} \quad \underset{=}{\underline{\text{were}}} \quad \underset{m}{\underline{\text{millennials}}} \quad \text{The Model}$$

Step 4: Solve

$$(0.26)(320{,}000{,}000) = m$$
$$83{,}200{,}000 = m$$

Step 5: Check Recheck the arithmetic. Because $(0.26)(320{,}000{,}000) = 83{,}200{,}000$, the answer is correct.

Step 6: Answer In the year 2015, there were about 83,200,000 millennials in the United States.

▶ ② Solve Business Problems That Involve Percent

Now let's look at percent problems from business. Typically, these problems involve discounts or mark-ups that businesses use in determining their prices. They may also include finding the cost of an item excluding sales tax, as shown in Example 5.

EXAMPLE 5

Finding the Cost of an Item Excluding Sales Tax

You just purchased a new pair of jeans for $41.34, including 6% sales tax. How much did the jeans cost before sales tax?

Solution

Step 1: Identify Use direct translation to find the price of the jeans before sales tax.

Step 2: Name Let p represent the price of the jeans in dollars before sales tax.

Step 3: Translate The total cost is the sum of the price of the jeans, p, and the amount of tax. The amount of tax is 6% of the price of the jeans. Thus the total price is

Work Smart

To change a percent to a decimal, move the decimal point two places to the left.

$$\underset{p}{\underline{\text{original price}}} \quad + \quad \underset{0.06p}{\underline{\text{amount of tax}}} \quad = \quad \underset{41.34}{\underline{\text{total price}}} \quad \text{The Model}$$

Step 4: Solve

$$p + 0.06p = 41.34$$

Combine like terms: $$1.06p = 41.34$$

Divide both sides by 1.06: $$\frac{1.06p}{1.06} = \frac{41.34}{1.06}$$

$$p = 39$$

Step 5: Check If the jeans cost $39, then the jeans plus the 6% sales tax on $39 equals $39 + (0.06)($39) = $39 + $2.34 = $41.34, so the answer is correct.

Step 6: Answer The jeans cost $39 before the sales tax. ●

Quick ✓

14. Suppose you just purchased a used car. The price of the car, including 7% sales tax, was $7811. What was the price of the car before sales tax?

15. As a reward for being named "Teacher of the Year," Janet will receive a 2.5% pay raise. If Janet's current salary is $59,000, determine her new salary.

Another type of percent problem involves the discounts or mark-ups that businesses use in determining their prices. For these problems, it is helpful to remember the following:

Original Price − Discount = Sale Price

Wholesale Price + Markup = Selling Price

EXAMPLE 6   **Discount**

A local clothing store is going out of business and all merchandise is marked down (discounted) by 40%. The sale price of a jacket is $108. What was the original price?

Solution

Step 1: Identify Use direct translation to find the original price of a jacket that was marked down by 40%. The sale price of the jacket is $108.

Step 2: Name Let p represent the original price in dollars of the jacket.

Step 3: Translate The original price minus the discount is the sale price, which is $108, so

$$p - \text{discount} = 108$$

"Marked down by 40%" means that the discount is 40% of the original price, so the discount is represented by $0.40p$. Substitute this into the equation $p - \text{discount} = 108$:

$$p - 0.40p = 108 \quad \text{The Model}$$

Step 4: Solve

$$p - 0.40p = 108$$

Combine like terms: $$0.60p = 108$$

Divide both sides by 0.60: $$\frac{0.60p}{0.60} = \frac{108}{0.60}$$

$$p = 180$$

Work Smart

To combine like terms in the expression $p - 0.40p$, remember $p = 1p$ or $p = 1.00p$.

(continued)

Step 5: Check If p, the original price of the jacket, was $180, then the discount would be $0.40\,(\$180) = \$72$. Subtracting $72 from the original price of $180 gives $108, the sale price. Also notice that the answer is reasonable — the sale price is less than the original price.

Step 6: Answer The original price of the jacket was $180. ●

Quick ✓

16. Suppose a gas station marks its gasoline up 10%. If the gas station charges $2.64 per gallon of 87-octane gasoline, what does it pay for the gasoline?

17. A furniture store marks recliners down by 25%. The sale price, excluding the sales tax, is $494.25. Find the original price of each recliner.

18. Albert's house lost 2% of its value last year. The value of the house is now $148,000. To the nearest dollar, what was the value of the house one year ago?

2.6 Exercises   MyMathLab®

Exercise numbers in green have complete video solutions in MyMathLab or may be accessed using the QR code to the right.

Problems 1–18 are the Quick ✓ s that follow the EXAMPLES.

Building Skills

In Problems 19–36, find the unknown in each percent question. See Objective 1.

19. What is 50% of 160?

20. What is 80% of 50?

21. 7% of 200 is what number?

22. 75% of 20 is what number?

23. What number is 16% of 30?

24. What number is 150% of 9?

25. 18 is 15% of what number?

26. 40% of what number is 122?

27. 60% of 120 is what number?

28. 45% of what number is 900?

29. 24 is 120% of what number?

30. 11 is 5.5% of what number?

31. What percent of 60 is 24?

32. 15 is what percent of 75?

33. 1.5 is what percent of 20?

34. 4 is what percent of 25?

35. What percent of 300 is 600?

36. What percent of 16 is 12?

Applying the Concepts

37. **Sales Tax** The sales tax in Delaware County, Ohio, is 7%. The total cost of purchasing a tennis racket,
including tax, is $57.78. Find the cost of the tennis racket before sales tax.

38. **Sales Tax** The sales tax in Franklin County, Ohio, is 7.5%. The total cost of purchasing a used Honda Civic, including sales tax, is $8600. Find the cost of the car before sales tax.

39. **Pay Cut** Todd works for a computer firm, and last year he earned a salary of $120,000. Recently, Todd was required to take a 15% pay cut. Find Todd's salary after the pay cut.

40. **Pay Raise** MaryBeth works from home as a graphic designer. Recently she raised her hourly rate by 5% to cover increased costs. Her new hourly rate is $29.40. Find MaryBeth's previous hourly rate.

41. **Bad Investment** After Mrs. Fisher lost 9% of her investment, she had $22,750. What was Mrs. Fisher's original investment?

42. **Good Investment** Perry just learned that his house increased in value by 4% over the past year. The value of the house is now $208,000. What was the value of the home one year ago?

43. **Marking Up Furniture** Darvin Furniture marks up the price of a dining room set 60%. What will be the selling price of a dining room set that Darvin buys for $1400?

44. **Hailstorm** Toyota Town had a 15%-off sale on vehicles that had been damaged in a hailstorm. The original price of a truck was $28,000. What is the sale price of the truck?

45. **Discount Pricing** A wool suit that was originally priced at $700 is on the 30%-off clearance rack. What is the sale price of the suit?

46. Business: Marking Up the Price of Books A college bookstore marks up the price that it pays the publisher for a book by 30%. If the selling price of a book is $117, how much did the bookstore pay for the book?

47. Furniture Sale A furniture store discounted a dining room table by 40%. The discounted price of the dining room table was $240. Determine the original price of the dining room table.

48. Vacation Package Booking.com advertised a 7-night stay at the Halifax, Nova Scotia, Prince George Hotel for 13% off the regular price. The sale price of the package is $1007. To the nearest dollar, how much was the 7-night stay at the hotel before the 13%-off sale?   (SOURCE: *Booking.com, 2/9/2016*)

49. Voting In an election for school president, the loser received 60% of the number of votes that the winner received. If 848 votes were cast, how many did each receive?

50. Voting On a committee consisting of Republicans and Democrats, there are twice as many Republicans as Democrats. If 30% of the Republicans and 20% of the Democrats voted in favor of a bill and there were 8 yes votes, how many people are on the committee?

51. Commission Melanie receives a 3% commission on every house she sells. If she received a commission of $8571, what was the value of the house she sold?

52. Commission Mario collects a commission for bringing in advertisers to his magazine company. He receives 8% on $450 full-page ads and 5% for $300 half-page ads. If he sells twice as many half-page as full-page ads and his commission was $5610, how many of each type did he bring in?

53. Job Growth According to the Bureau of Labor Statistics, the demand for physical therapists is expected to grow by 34% over the next 10 years. If there are currently 210,900 physical therapists in the United States, how many are there expected to be in 10 years?

54. Grades If 15% of Grant's astronomy class received an A, how many students were in his astronomy class if 6 students earned A's this term?

55. Bachelors Based on data from the U.S. Census Bureau and obtained from the Pew Research Center, in 2012, 23% of the 118 million U.S. males aged 18 years or older had never married. To the nearest million, in 2012 how many males 18 years or older had never married?

56. Census Data Based on data from the U.S. Census Bureau and obtained from the Pew Research Center, in 2012, 17% of the 125 million U.S. females aged 18 years or older had never married. To the nearest million, in 2012 how many females 18 years or older had never married?

According to the U.S. Census Bureau, the level of education of the U.S. population is changing. The table below shows the number of persons (in thousands) aged 25 years and older and their educational attainment in 2004 and 2015. In Problems 57 and 58, use the information in the table to answer each question.

	2004	2015
Population 25 years and older	186,534	212,132
Associate's degree	13,244	20,867
Bachelor's degree	32,084	43,500

(SOURCE: *United States Census Bureau*)

57. Find, to the nearest tenth of a percent, the percent of the U.S. population aged 25 years and older that held a bachelor's degree in 2004 and in 2015.

58. Find, to the nearest tenth of a percent, the percent of the U.S. population aged 25 years and older that held an associate's degree in 2004 and in 2015.

Extending the Concepts

59. Discount Pricing Suppose that you are the manager of a clothing store and have just purchased 100 shirts for $12 each. After 1 month of selling the shirts at the regular price, you plan to have a sale giving 25% off the original selling price. However, you still want to make a profit of $6 on each shirt at the sale price. What should you price the shirts at initially to ensure this?

In Problems 60–65, find the percent increase or decrease. The percent increase or percent decrease is defined as:

$$\frac{\text{amount of change}}{\text{original amount}} \times 100\%$$

60. Population Growth The population in a small fishing town grew from 2500 to 2825. Find the percent increase.

61. Gas Mileage The gas mileage on your van decreases due to the heavy weight of extra passengers and a boat on a trailer. With the extra weight your van gets 15 mpg, and without the weight it gets 21 mpg. To the nearest tenth, what is your percent decrease in gas mileage with extra weight?

62. Salaries Your current consulting position pays you $77,000 each year. A competing firm has offered you $86,000 to join it. To the nearest tenth, what percent increase in salary would you get if you joined the competing firm?

63. Car Depreciation A new car decreases in value by about 20% each year. At *www.bmwusa.com* Misha priced his 428*i* at $53,545.

   (a) After 2 years, what will the car be worth?
   (b) To the nearest percent, after 2 years, what is the overall percentage decrease in the value of the car?

64. Stock Prices One day Buffalo Wild Wings stock price went from $157.37 to $160.52. To the nearest tenth, what was the percent increase in the stock price?

65. Stock Prices One day Cisco Systems stock went from $26.41 to $25.86. To the nearest tenth, what is the percent decrease in the stock price?

Explaining the Concepts

66. The sales tax rate is 6%. Explain why $1.06p$ will correctly calculate the total purchase price of any item that sells for p dollars.

67. A problem on a quiz stated: Write an equation to solve the following problem. "Jack received a 5% raise in hourly wage, effective on his birthday. Jack's new hourly wage will be $12.81. Find Jack's current hourly wage." You wrote the equation $x + 0.05 = 12.81$, and your instructor counted your answer wrong. Explain why your equation is incorrect, and then write the correct equation and solve it for Jack's current hourly wage.

68. An item is reduced by 10% and then this is reduced by another 20%. Is this the same as reducing the item by 30%? Explain why or why not.

2.7 Problem Solving: Geometry and Uniform Motion

Objectives

1 Set Up and Solve Complementary and Supplementary Angle Problems

2 Set Up and Solve "Angles of a Triangle" Problems

3 Use Geometry Formulas to Solve Problems

4 Set Up and Solve Uniform Motion Problems

Are You Prepared for This Section?

Before getting started, complete the following problems. If you get a problem wrong, go back to the section cited and review the material.

P1. Solve: $q + 2q - 30 = 180$ [Section 2.2, pp. 92–94]

P2. Solve: $30w + 20(w + 5) = 300$ [Section 2.2, pp. 92–94]

In this section, we continue using the six-step method introduced in Section 2.5.

▶ **1 Set Up and Solve Complementary and Supplementary Angle Problems**

We begin by defining *complementary* and *supplementary angles.*

Figure 5

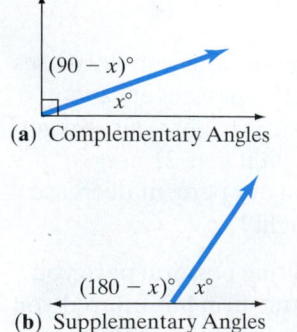

(a) Complementary Angles

(b) Supplementary Angles

> **Definitions**
>
> Two angles whose measures have a sum of 90° are called **complementary angles.** Each angle is called the *complement* of the other. Two angles whose measures have a sum of 180° are called **supplementary angles.** Each angle is called the *supplement* of the other.

For example, the angles shown in Figure 5(a) are complements because their sum is 90°. Notice the use of the symbol ⌐ to show the 90° angle. The angles shown in Figure 5(b) are supplementary because their sum is 180°.

Prepared?...Answers **P1.** {70} **P2.** {4}

EXAMPLE 1   **Solving a Complementary Angle Problem**

Find the measure of two complementary angles if the measure of the larger angle is 6° greater than twice the measure of the smaller angle.

Solution

Step 1: Identify Use a definition from geometry to find the measure of two complementary angles.

Step 2: Name Less is known about the measure of the smaller angle, so call it x.

Step 3: Translate Let the measure of the larger angle be $(2x + 6)$ because it is 6° more than twice the measure of the smaller angle.

The angles are complementary, so the sum of the measures of the two angles must equal 90°.

$$\underbrace{x}_{\text{measure of smaller angle}} + \underbrace{(2x + 6)}_{\text{measure of larger angle}} = 90 \quad \text{The Model}$$

Step 4: Solve

$$x + (2x + 6) = 90$$

$$\text{Combine like terms:} \quad 3x + 6 = 90$$

$$\text{Subtract 6 from both sides:} \quad 3x = 84$$

$$\text{Divide both sides by 3:} \quad x = 28$$

Step 5: Check The measure of the smaller angle, x, is 28° and the measure of the larger angle is $(2x + 6)$ degrees $= 2(28) + 6 = 56 + 6 = 62°$. Because the sum of the measures of these angles is $28° + 62° = 90°$, the answer is correct.

Step 6: Answer The two complementary angles measure 28° and 62°.   ●

> ### Quick ✓
>
> 1. Complementary angles are angles whose measures have a sum of __ degrees.
> 2. Find two complementary angles if the measure of the larger angle is 12° more than the measure of the smaller angle.
> 3. Find two supplementary angles if the measure of the larger angle is 30° less than twice the measure of the smaller angle.

Figure 6

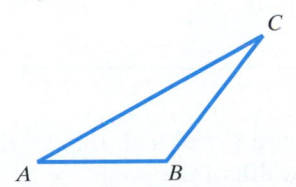

▶ ❷ **Set Up and Solve "Angles of a Triangle" Problems**

The sum of the measures of the interior angles of a triangle is 180°. In the triangle in Figure 6, the sum of the measures of angles A, B, and C must equal 180°, which is written

$$m \angle A + m \angle B + m \angle C = 180°$$

EXAMPLE 2   **Solving an "Angles of a Triangle" Problem**

The measure of the largest angle of a triangle is 20° more than twice the measure of the smallest angle, and the measure of the middle angle is 10° more than twice the measure of the smallest angle. Find the measure of each angle.

Solution

Step 1: Identify Use a relationship from geometry to find the measure of each of the three interior angles of a triangle.

(continued)

Step 2: Name We know the least about the measure of the smallest angle, so call its measure x and call the angle A.

Step 3: Translate The largest angle measures 20° more than twice the smallest angle, so $(2x + 20)$ is its measure. Call the largest angle C. The middle angle measures 10° more than twice the smallest angle so $(2x + 10)$ is its measure. Call the middle angle B. Since $m \angle A + m \angle B + m \angle C = 180°$,

$$\underbrace{x}_{m \angle A} + \underbrace{(2x + 10)}_{m \angle B} + \underbrace{(2x + 20)}_{m \angle C} = 180 \quad \text{The Model}$$

Step 4: Solve

$$x + (2x + 10) + (2x + 20) = 180$$

Combine like terms: $\qquad 5x + 30 = 180$

Subtract 30 from both sides: $\qquad 5x = 150$

Divide both sides by 5: $\qquad x = 30$

Step 5: Check Since $x = 30$ degrees $= m \angle A$, we know that $m \angle C = (2x + 20) = 2(30) + 20 = 60 + 20 = 80$ degrees and that $m \angle B = (2x + 10) = 2(30) + 10 = 60 + 10 = 70$ degrees.

The sum of the measures of these angles is $30° + 70° + 80° = 180°$, so the answer is correct.

Step 6: Answer The measures of the angles of the triangle are 30°, 70°, and 80°. ●

Quick ✓

4. The sum of the measures of the angles of a triangle is ___ degrees.

5. The measure of the smallest angle of a triangle is one-third the measure of the largest angle. The measure of the middle angle is 65° less than the measure of the largest angle. Find the measure of the angles of the triangle.

▶ ❸ Use Geometry Formulas to Solve Problems

Recall from Section 2.4 that the perimeter of a figure is the sum of the lengths of its sides. Perimeter is measured in linear units, such as feet, yards, and meters.

EXAMPLE 3 **Solving a Perimeter Problem**

Figure 7

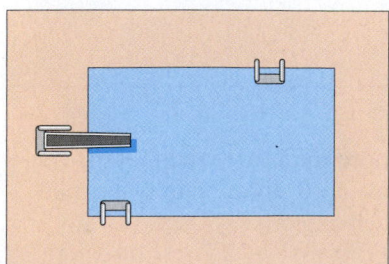

The perimeter of the rectangular swimming pool shown in Figure 7 is 80 feet. If the length is 10 feet more than the width, find the length and the width of the pool.

Solution

Step 1: Identify This is a geometry problem. Find the length and the width of the rectangular pool, given the perimeter.

Step 2: Name Because the length is given in terms of the width, let w represent the width of the pool in feet.

Step 3: Translate The length of the pool is 10 feet more than the width, so the length is $w + 10$. The formula for perimeter of a rectangle is $P = 2l + 2w$, where l is the length and w is the width. So

$$\underbrace{2(w + 10)}_{2l} + \underbrace{2w}_{2w} = \underbrace{80}_{P} \quad \text{The Model}$$

Step 4: Solve

$$2(w + 10) + 2w = 80$$

Use the Distributive Property:	$2w + 20 + 2w = 80$
Combine like terms:	$4w + 20 = 80$
Subtract 20 from both sides:	$4w = 60$
Divide both sides by 4:	$w = 15$

Step 5: Check The width of the pool is $w = 15$ feet, so the length is $w + 10 = 15 + 10 = 25$ feet. The perimeter is $2(25) + 2(15) = 80$ feet, so the answer is correct.

Step 6: Answer The length of the pool is 25 feet, and the width of the pool is 15 feet.   ●

Quick ✓

6. *True or False* The perimeter of a square is 4s, where *s* is the length of a side of the square.

7. *True or False* The perimeter of a rectangle can be found by multiplying the length of the rectangle by the width of the rectangle.

8. The perimeter of a small rectangular garden is 9 feet. If the length is twice the width, find the width and length of the garden.

Recall that the area of a plane (that is, a two-dimensional) figure is the number of square units the figure contains, such as square feet, square inches, or square centimeters.

EXAMPLE 4   **Solving an Area Problem**

A garden in the shape of a trapezoid has an area of 18 square feet. The height is 3 feet, and the length of the shorter base is 2 feet less than the length of the longer base. Find the length of each base of the trapezoid. See Figure 8.

Figure 8

Solution

Step 1: Identify This problem from geometry is about the area of a trapezoid. The formula for the area of a trapezoid is $A = \dfrac{1}{2}h(B + b)$, where *h* is the height, *B* is the length of the longer base, and *b* is the length of the shorter base. We are given the area and the height of the trapezoid.

Step 2: Name Let *B* represent the length of the longer base in feet.

Step 3: Translate Because one base is 2 feet shorter than the other base, and *B* represents the length of the longer base, $B - 2$ represents the length of the shorter

(continued)

base. Use the formula for the area of a trapezoid, $A = \frac{1}{2}h(B + b)$, to find the model.

area   height   long base   short base

$18 = \frac{1}{2} \cdot \quad 3 \quad (B + B - 2)$ **The Model**

Step 4: Solve

$$18 = \frac{1}{2} \cdot 3(B + B - 2)$$

Combine like terms in the parentheses:
$$18 = \frac{1}{2} \cdot 3(2B - 2)$$

Multiply both sides by 2 to clear fractions:
$$2[18] = 2\left[\frac{1}{2} \cdot 3(2B - 2)\right]$$

$$36 = 3(2B - 2)$$

Work Smart

Instead of distributing the 3 in

$$36 = 3(2B - 2)$$

you could divide both sides by 3. Try it! Which approach do you prefer?

Use the Distributive Property:
$$36 = 6B - 6$$

Add 6 to both sides:
$$42 = 6B$$

Divide both sides by 6:
$$7 = B$$

Step 5: Check The longer base, B, is 7 feet. The smaller base is $B - 2 = 7 - 2 = 5$ feet. The area of the trapezoidal garden is $\frac{1}{2} \cdot 3(7 + 5) = \frac{1}{2} \cdot 3(12) = 18$ square feet. The answers 5 feet and 7 feet are correct.

Step 6: Answer The lengths of the two bases are 5 feet and 7 feet. ●

> **Quick** ✔
>
> **9.** The surface area of a rectangular box is 62 square feet. If the length of the box is 3 feet and the width is 2 feet, find the height of the box.

▶ ❹ Set Up and Solve Uniform Motion Problems

Objects that move at a constant velocity (speed) are said to be in **uniform motion.** We treat the average speed of an object as its constant velocity. For example, a car traveling at an average speed of 45 miles per hour is in uniform motion. An object traveling down an assembly line at a constant speed is also in uniform motion.

In Other Words

The uniform motion formula states that distance equals rate times time.

> **Uniform Motion Formula**
>
> If an object moves at an average speed r, the distance d covered in time t is given by the formula
>
> $$d = r \cdot t$$

We use r in the uniform motion formula because "rate" is another term for speed. In solving uniform motion problems, $d = rt$ can also be written as $rt = d$. A chart can be used to set up uniform motion problems, as shown in Table 4.

Table 4

	Rate	·	Time	=	Distance
Object 1	rate 1		time 1		distance 1
Object 2	rate 2		time 2		distance 2

Rate, time, and distance must be expressed in corresponding units. For example, if rate (speed) is stated in miles per hour, then distance must be in miles, and time must be in hours. If rate is measured in kilometers per minute, then distance is in kilometers and time is in minutes.

EXAMPLE 5

Solving a Uniform Motion Problem for Time

Bob and Karen drove from Atlanta, Georgia, to Durham, North Carolina, a distance of 390 miles, to attend a family reunion. Their average speed for the first part of the trip was 60 miles per hour (mph). Due to road construction, their average speed for the remainder of the trip was 45 mph. How long did they travel at 45 mph if they drove 3 hours longer at 60 mph than at 45 mph?

Solution

Step 1: Identify This is a uniform motion problem. Find the number of hours Bob and Karen drove at 45 mph.

Step 2: Name Let t represent the number of hours Bob and Karen drove at 45 mph. They traveled 3 hours longer at 60 mph, so $t + 3$ represents the number of hours driven at 60 mph.

Step 3: Translate Set up Table 5. For the first part of the trip, the rate is 60 mph and the time is $t + 3$ hours, so the distance is $60(t + 3)$. For the second part of the trip, the rate is 45 mph and the time is t, so the distance is $45t$.

Table 5

	Rate (in mph) $\cdot$	Time $=$	Distance
First part of trip	60	$t + 3$	$60(t + 3)$
Second part of trip	45	t	$45t$
Total			390

The total distance that Bob and Karen traveled is 390 miles, so

$$\underbrace{60(t + 3)}_{\substack{\text{distance traveled} \\ \text{at 60 mph}}} + \underbrace{45t}_{\substack{\text{distance traveled} \\ \text{at 45 mph}}} = \underbrace{390}_{\substack{\text{total} \\ \text{distance}}} \qquad \text{The Model}$$

Step 4: Solve

$$60(t + 3) + 45t = 390$$

Use the Distributive Property: $\quad 60t + 180 + 45t = 390$

Combine like terms: $\quad\quad\quad 105t + 180 = 390$

Subtract 180 from both sides: $\quad 105t + 180 - 180 = 390 - 180$

$$105t = 210$$

Divide both sides by 105: $\quad\quad \dfrac{105t}{105} = \dfrac{210}{105}$

$$t = 2$$

Step 5: Check Bob and Karen drove for $t = 2$ hours at 45 mph. So they drove $t + 3 = 2 + 3 = 5$ hours at 60 mph. Does 2 hours driven at 45 mph plus 5 hours at 60 mph equal a distance of 390 miles? Because 2 hours $\cdot$ 45 mi/hr + 5 hours $\cdot$ 60 mi/hr = 90 miles + 300 miles = 390 miles, the answer checks.

Step 6: Answer the Question Bob and Karen drove for 2 hours at 45 mph.

EXAMPLE 6 | **Solving a Uniform Motion Problem for Rate**

Two groups of friends took a canoe trip down Big Darby Creek. The first group left at 12 noon. One-half hour later, the second group left the same location, paddling at an average speed that was 0.75 mph faster than the first group. At 2:30 P.M. the second group caught up to the first group. How fast was each group paddling?

Solution

Step 1: Identify This is a uniform motion problem. Find the rate at which each group paddled.

Step 2: Name Let r represent the rate at which the first group paddled in mph. The second group's paddling rate is $r + 0.75$, because its rate is 0.75 mph greater than that of the first group.

Step 3: Translate We have a specific time that the first group left, 12 noon. The second group left $\frac{1}{2}$ hour later. In the formula $d = rt$, time is measured in hours. So the first group traveled for two and a half (or 2.5) hours (12 noon until 2:30 P.M.), and the second group traveled for 2 hours. Table 6 summarizes the given information.

Table 6

	Rate (in mph)	·	Time	=	Distance
First group	r		2.5		$2.5r$
Second group	$r + 0.75$		2		$2(r + 0.75)$

Because the second group caught up to the first group, the distances the two groups traveled were the same.

<u>distance traveled by first group</u>   <u>distance traveled by second group</u>
$$2.5r \qquad = \qquad 2(r + 0.75) \qquad \text{The Model}$$

Step 4: Solve

$$2.5r = 2(r + 0.75)$$

Use the Distributive Property: $\qquad 2.5r = 2r + 1.5$

Subtract $2r$ from both sides: $\quad 2.5r - 2r = 2r - 2r + 1.5$

$$0.5r = 1.5$$

Divide both sides by 0.5: $\qquad \dfrac{0.5r}{0.5} = \dfrac{1.5}{0.5}$

$$r = 3$$

Work Smart

In solving $0.5r = 1.5$, we could also multiply both sides by 2. Why?

$$2(0.5\,r) = 2 \cdot \frac{1}{2}r = r$$

Step 5: Check The first group paddled at $r = 3$ mph, and the second group paddled at $r + 0.75 = 3 + 0.75 = 3.75$ mph. Did both groups travel the same distance? Because $(3)(2.5) = 7.5$ miles and $(2)(3.75)$ also equals 7.5 miles, the answers are correct.

Step 6: Answer the Question The first group paddled at 3 miles per hour and the second group paddled at 3.75 miles per hour. ●

Quick ✓

10. *True or False* When using $d = rt$ to calculate the distance traveled, it is not necessary to assume that the object travels at a constant speed.

11. Two bikers, Mariko and Luis, start at the same point at the same time and travel in opposite directions. Mariko's average speed is 5 miles per hour more than that of Luis, and after 3 hours the bikers are 63 miles apart. Find the average speed of each biker.

12. Tanya, a long-distance runner, runs at an average speed of 8 miles per hour. Two hours after Tanya leaves your house, you leave in your car and follow the same route. If your average speed is 40 miles per hour, how long will it be before you catch up to Tanya? How far will each of you be from your home?

2.7 Exercises   MyMathLab®

*Problems **1–12** are the **Quick ✓**s that follow the **EXAMPLES**.*

Building Skills

In Problems 13–16, identify the measure of each of the angles. See Objective 1.

△ **13.** Find two supplementary angles if the measure of the first angle is three times the measure of the second.

△ **14.** Find two supplementary angles if the measure of the first angle is four times the measure of the second.

△ **15.** The measures of two complementary angles are consecutive even integers. Find the measure of each angle.

△ **16.** Find two complementary angles if the measure of the first angle is 25° less than the measure of the second.

In Problems 17–20, find the measures of the angles of the triangle. See Objective 2.

△ **17.**

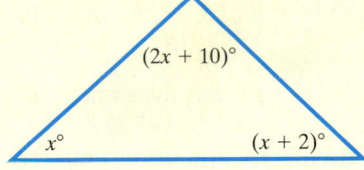

△ **18.**

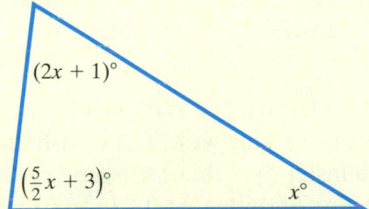

△ **19.** In a triangle, the second angle measures four times the first. The measure of the third angle is 18° more than the measure of the second. Find the measures of the three angles.

△ **20.** The measures of the angles of a triangle are consecutive even integers. Find the measure of each angle.

In Problems 21–26, use a formula from geometry to solve for the unknown quantity. See Objective 3.

△ **21.** The length of a rectangle is 8 feet longer than the width. If the perimeter is 88 feet, find the length and width of the rectangle.

△ **22.** The width of a rectangle is 10 meters less than the length. If the perimeter is 56 meters, find the length and width of the rectangle.

△ **23.** A rectangular field is divided into 2 squares of the same size and shape. If it takes 294 yards of fencing

to enclose the field and divide the field into the two parcels, find the dimensions of the field. See the figure.

△ **24.** A rectangular plot has been divided into two fields so that the length of one of the fields is twice the length of the other. The smaller field is a square, and the larger field is a rectangle. If it takes 279 meters (m) of fencing to enclose the entire plot and to divide it into two fields, find the dimensions of the original plot.

△ **25.** A trapezoid has an area of 900 square meters. The height of the trapezoid is 40 meters (m), and the length of the longer base is twice that of the shorter base. Find the length of each base of the trapezoid.

△ **26.** A parallelogram has a perimeter of 120 inches. If one side of the parallelogram is 10 inches longer than the other, find the dimensions of the figure.

In Problems 27 and 28, set up the model to solve the uniform motion problems by answering parts (a)–(d). See Objective 4.

27. Two cars leave Chicago, one traveling north and the other south. The car going north is traveling at 62 mph, and the car going south is traveling at 68 mph. How long before they are 585 miles apart?

   (a) Write an algebraic expression for the distance traveled by the car going north.

   (b) Write an algebraic expression for the distance traveled by the car going south.

   (c) Write an algebraic expression for the total distance traveled by the two cars.

   (d) Write an equation to answer the question.

28. Two trains leave Albuquerque, traveling in the same direction on parallel tracks. One train is traveling at 72 mph, and the other is traveling at 66 mph. How long before they are 45 miles apart?

   (a) Write an algebraic expression for the distance traveled by the faster train.

(b) Write an algebraic expression for the distance traveled by the slower train.

(c) Write an algebraic expression for the difference in distance between the two.

(d) Write an equation to answer the question.

In Problems 29 and 30, fill in the table from the information given. Then write the equation that will solve the problem. DO NOT SOLVE. See Objective 4.

29. Martha is running in her first marathon. She can run at a rate of 528 ft per min. Martha's mom starts the same course 10 minutes after Martha, running at a rate of 704 ft per min. How long before Mom catches up to Martha?

	Rate	·	Time	=	Distance
Martha	?		?		?
Mom	?		?		?

30. A 580-mile trip in a small plane took a total of 5 hours. The first two hours were flown at one rate, and then the plane encountered a headwind and was slowed by 10 mph. Find the rate for each portion of the trip.

	Rate	·	Time	=	Distance
Beginning of trip	?		?		?
Rest of the trip	?		?		?
Total			?		?

Applying the Concepts

△ **31. Isosceles Triangle** An isosceles triangle has exactly two sides that are equal in length (*congruent*). If the base (the third side) measures 45 inches and the perimeter is 98 inches, find the length of the two congruent sides, called *legs*.

△ **32. Isosceles Triangle** See Problem 31. An isosceles triangle has a base of 17 cm. If the perimeter is 95 cm, find the length of each of the legs.

△ **33. Billboard** A billboard along a highway has a perimeter of 110 feet. Find the length of the billboard if its height is 15 feet.

△ **34. Buying Wallpaper** Erika is buying wallpaper for her bedroom. She remembers that the perimeter of the room is 54 ft and that the room is twice as long as it is wide.

(a) Find the dimensions of the room.

(b) If the walls are 8 ft high, how many square feet of wallpaper does she need to buy?

(c) Erika arrives at the decorating store and finds that wallpaper is sold by the square yard. How many square yards of wallpaper does Erika need to buy?

△ **35. Back Yard** Bob's backyard is in the shape of a trapezoid with height of 60 feet. The shorter base is 8 feet shorter than the longer base, and the area of the backyard is 2160 square feet. Find the length of each base of the trapezoidal yard.

△ **36. Buying Fertilizer** Melinda has to buy fertilizer for a flower garden in the shape of a right triangle. If the area of the garden is 54 square feet and the base of the garden measures 9 feet, find the height of the triangular garden.

△ **37. Garden** The perimeter of a rectangular garden is 60 yards. The width of the garden is three yards less than twice the length.

(a) Find the length and width of the garden.

(b) What is the area of the garden?

△ **38. Table** The Jacksons are having a custom rectangular table made. The length of the table is 18 inches more than the width, and the perimeter is 180 inches. Find the length and the width of the table.

39. Boats Two boats leave a port at the same time, one going north and the other traveling south. The northbound boat travels 16 mph faster than the southbound boat. If the southbound boat is traveling at 47 mph, how long will it be before they are 1430 miles apart?

40. Cyclists Two cyclists leave a city at the same time, one going east and the other going west. The west-bound cyclist bikes 4 mph faster than the east-bound cyclist. After 5 hours they are 200 miles apart. How fast is the east-bound cyclist riding?

41. Road Trip Two cars leave a city on the same road, one driving 12 mph faster than the other. After 4 hours, the car traveling faster stops for lunch. After 4 hours and 30 minutes, the car traveling slower stops for lunch. Assuming that the person in the faster car is still eating lunch, the cars are now 24 miles apart. How fast is each car driving?

42. Passenger and Freight Trains Two trains leave a city on parallel tracks, traveling the same direction. The passenger train is going twice as fast as the freight train. After 45 minutes, the trains are 30 miles apart. Find the speed of each train.

43. Down the Highway A 360-mile trip began on a freeway in a car traveling at 62 mph. Once the road became a 2-lane highway, the car slowed to 54 mph. If the total trip took 6 hours, find the time spent on each type of road.

44. Tuna Fishing Max likes to fish for tuna at the Coronado Islands, located in the Pacific Ocean west of San Diego . His total trip from his dock in San Diego to his fishing spot and back is 41 miles. On his last fishing trip, it took 2 hours to get to the Islands and he increased his speed 5 mph so that he could get home in one hour. What was the speed of the boat traveling out to the Coronado Islands and what was the speed of the boat on the return trip?

45. Walking and Jogging Carol can complete her neighborhood jog in 10 minutes. It takes 30 minutes to cover the same distance when she walks. If her jogging rate is 4 mph faster than her walking rate, find the speed at which she jogs.

46. Trip to School Dien drives to school at 40 mph. Five minutes $\left(\frac{1}{12}\text{hour}\right)$ after he leaves home, his mother sees that he forgot his homework and leaves to take it to him, driving 48 mph. If they arrive at school at the same time, how far away is the school?

△ **47. Isosceles Triangle** In an isosceles triangle, the base angles (angles opposite the two congruent legs) are equal in measure (*congruent*). Find the measures of the angles of an isosceles triangle in which the third angle (called the *vertex angle*) has a measure that is 16° less than twice the measure of a base angle.

△ **48. Isosceles Triangle** See Problem 47. In an isosceles triangle, the measure of the vertex angle is 4 degrees less than twice the measure of each base angle. Find the measure of each of the angles of the triangle.

Extending the Concepts

Parallel lines are lines in the same plane that never intersect (think of railroad tracks going infinitely far out into space). A line that cuts two parallel lines is called a *transversal*. The transversal forms eight different angles that are related in following ways:

Corresponding angles are equal in measure.

Alternate interior angles are equal in measure.

Interior angles on the same side of the transversal are supplementary.

In the figure shown, lines L_1 and L_2 are parallel $(L_1 \| L_2)$ and the transversal is labeled t. In this figure, there are four pairs of corresponding angles:

$\angle 1$ and $\angle 5$, $\angle 2$ and $\angle 6$, $\angle 3$ and $\angle 7$, $\angle 4$ and $\angle 8$

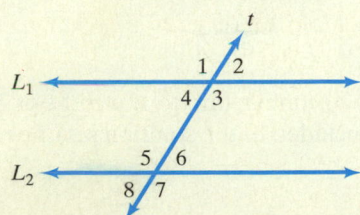

There are two pairs of alternate interior angles:
$\angle 3$ and $\angle 5$, $\angle 4$ and $\angle 6$.

There are two pairs of interior angles on the same side of the transversal: $\angle 3$ and $\angle 6$, $\angle 4$ and $\angle 5$.

In Problems 49–54, given $L_1 \| L_2$, use the appropriate properties from geometry to solve for x.

△ **49.**

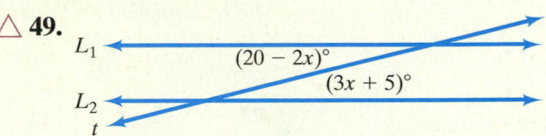

△ **50.**

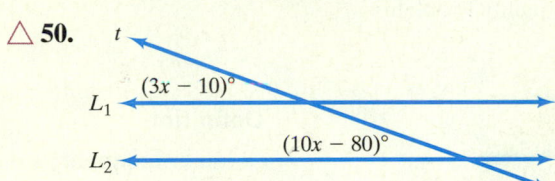

△ **51.** △ **52.**

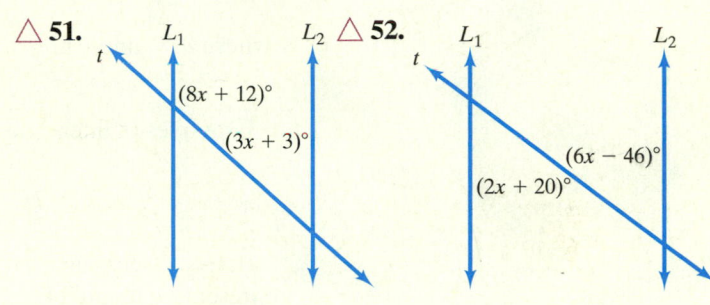

△ **53.**

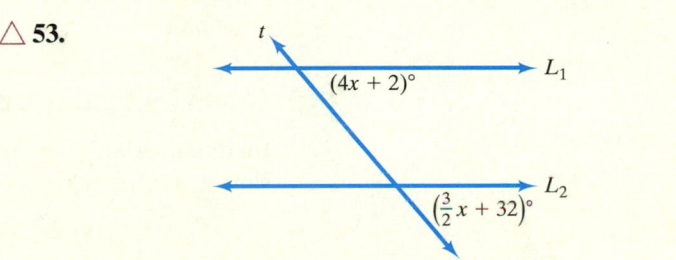

△ **54.**

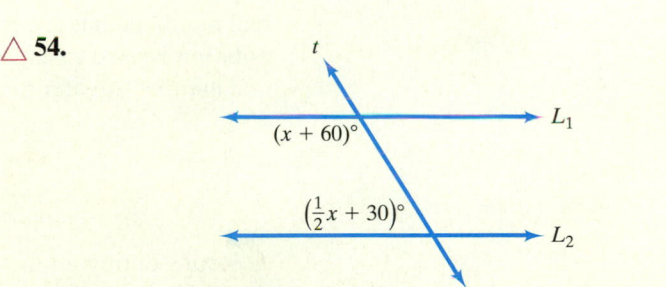

Explaining the Concepts

55. Explain the difference between complementary angles and supplementary angles.

56. Explain the difference between the area of a rectangle and the perimeter of a rectangle.

57. When setting up a uniform motion problem, you wrote $65t + 40t = 115$. Your classmate wrote $65t - 40t = 115$. Write a word problem for each of these equations, and explain the keys to recognizing the difference between the two types.

2.8 Solving Linear Inequalities in One Variable

Objectives

1. Graph Inequalities on a Real Number Line
2. Use Interval Notation
3. Solve Linear Inequalities Using Properties of Inequalities
4. Model Inequality Problems

Are You Prepared for This Section?

Before getting started, complete the following problems. If you get a problem wrong, go back to the section cited and review the material.

In Problems P1–P4, replace the question mark with $<$, $>$, or $=$ to make the statement true.

P1. $4 \, ? \, 19$ **P2.** $-11 \, ? \, -24$ **P3.** $\dfrac{1}{4} \, ? \, 0.25$ **P4.** $\dfrac{5}{6} \, ? \, \dfrac{4}{5}$ [Section 1.3, pp. 23–24]

> ### Definition
>
> A **linear inequality in one variable** is an inequality that can be written in the form
>
> $$ax + b < c \quad \text{or} \quad ax + b \leq c \quad \text{or} \quad ax + b > c \quad \text{or} \quad ax + b \geq c$$
>
> where a, b, and c are real numbers and $a \neq 0$.

Examples of linear inequalities in one variable, x, are

$$x + 2 > 7 \qquad \frac{1}{2}x + 4 \leq 9 \qquad 5x > 0 \qquad 3x + 7 \geq 8x - 3$$

Before discussing methods for solving linear inequalities, we will present three ways of representing inequalities: *set-builder notation*, graphing on a real number line, and *interval notation*.

① Graph Inequalities on a Real Number Line

Inequalities with one inequality symbol are called **simple inequalities.** For example, the simple inequality

$$x > 2 \quad \text{means} \quad \text{the set of all real numbers } x \text{ greater than 2}$$

Expressions such as $x > 2$ are in **inequality notation.** Because there are infinitely many real numbers that are greater than 2, we cannot list them all in a set. Instead **set-builder notation** is used to express the set formed by this inequality. For example, the set of all real numbers greater than 2 is represented in set-builder notation as

Representing an inequality on a number line is called graphing the inequality, and the picture is called the **graph of the inequality.** For example, graph $\{x \mid x > 2\}$ on a real number line by shading the portion of the number line that is to the right of 2. The number 2 is called an **endpoint.** Use a parenthesis to indicate that 2 is *not* included in the solution set. See Figure 9.

Figure 9
$x > 2$

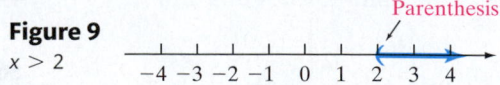

To graph the inequality $x \geq 2$, also shade the number line to the right of 2, but use a bracket on the endpoint (to indicate that 2 is included in the solution set). See Figure 10.

Figure 10
$x \geq 2$

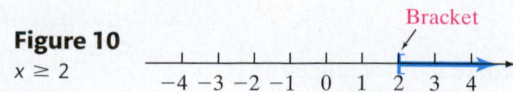

In a similar way, the inequality $x < 4$ is graphed in Figure 11(a) and $x \leq 4$ is graphed in Figure 11(b).

Figure 11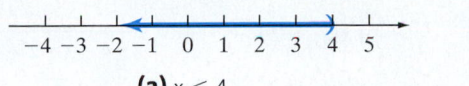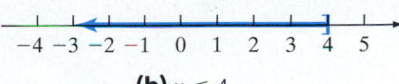

(a) $x < 4$ (b) $x \leq 4$

EXAMPLE 1

Graphing Simple Inequalities on a Real Number Line

Graph each inequality on a real number line.

(a) $x > 5$ (b) $x \leq -4$

Solution

(a) The inequality $x > 5$ represents all real numbers greater than five. Because 5 is not included in this set, place a parenthesis at 5 and shade to the right. See Figure 12.

Figure 12
$x > 5$

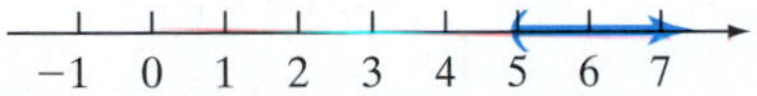

(b) The inequality $x \leq -4$ represents all real numbers less than or equal to negative four. Because -4 is included in the set, place a bracket at -4 and shade to the left. See Figure 13.

Work Smart

The inequalities $x > 5$ and $5 < x$ have the same graph. Do you see why?

Figure 13
$x \leq -4$

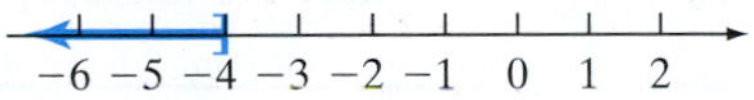

Quick ✓

1. *True or False* To graph an inequality that contains $>$ or $<$, use a parenthesis.

In Problems 2–5, graph each inequality on a real number line.

2. $n \geq 8$ 3. $a < -6$

4. $x > -1$ 5. $p \leq 0$

▶ ❷ **Use Interval Notation**

A third way to represent inequalities uses *interval notation*. To use interval notation, we need a new symbol, ∞.

The symbol ∞ (which is read as "infinity") is not a real number; it is used to indicate that there is no upper bound on an inequality. The symbol $-\infty$ (which is read as "negative infinity") also is not a real number and means that there is no lower bound on an inequality.

Work Smart

Use a parenthesis when either $-\infty$ or ∞ is an endpoint because these symbols are not real numbers.

Intervals Involving Infinity

$[a, \infty)$	consists of all real numbers x for which $x \geq a$
(a, ∞)	consists of all real numbers x for which $x > a$
$(-\infty, a]$	consists of all real numbers x for which $x \leq a$
$(-\infty, a)$	consists of all real numbers x for which $x < a$
$(-\infty, \infty)$	consists of all real numbers x (or $-\infty < x < \infty$)

There are three ways to represent all real numbers greater than 2.

SET-BUILDER NOTATION	NUMBER LINE GRAPH	INTERVAL NOTATION
$\{x \mid x > 2\}$		$(2, \infty)$

Table 7 summarizes set builder notation, interval notation, and the graphs.

Work Smart

Intervals are always written from left to right using the number line. For example $(5, \infty)$ represents all real numbers greater than 5 because 5 is the left endpoint. The interval $(-\infty, -2]$ represents all real numbers less than or equal to -2.

Table 7

Set-Builder Notation	Interval Notation	Graph
$\{x \mid x \geq a\}$	$[a, \infty)$	
$\{x \mid x > a\}$	(a, ∞)	
$\{x \mid x \leq a\}$	$(-\infty, a]$	
$\{x \mid x < a\}$	$(-\infty, a)$	
$\{x \mid x$ is a real number$\}$	$(-\infty, \infty)$	

EXAMPLE 2 **Writing an Inequality in Interval Notation**

Write each inequality using interval notation.

(a) $x > -4$ (b) $x \leq 8$

(c)   (d)

Solution

Remember, if the inequality contains $<$ or $>$, use a parenthesis. If the inequality contains $\leq$ or $\geq$, use a bracket.

(a) $(-4, \infty)$ (b) $(-\infty, 8]$ (c) $[1.75, \infty)$ (d) $(-\infty, 19)$ ●

Quick ✓

6. *True or False* The inequality $x \geq 3$ is written in interval notation as $[3, \infty]$.

7. *True or False* The inequality $x < -4$ is written in interval notation as $(-4, -\infty)$.

In Problems 8–11, write the inequality in interval notation.

8. $x \geq -3$ 9. $x < 12$

10.   11.

▶ ❸ Solve Linear Inequalities Using Properties of Inequalities

To **solve an inequality** means to find all values of the variable for which the statement is true. These values are the **solutions** of the inequality. The set of all solutions is the **solution set.** As with equations, one method for solving a linear inequality is to replace it by a series of *equivalent inequalities* until an inequality with an obvious solution, such as $x > 2$, is obtained.

Two inequalities having the same solution set are called **equivalent inequalities.** Equivalent inequalities are obtained by using some of the same operations used to find equivalent equations.

Consider the inequality $3 < 8$. If we add 2 to both sides of the inequality, the left side becomes 5 and the right side becomes 10. Since $5 < 10$, we see that adding the same quantity to both sides of an inequality does not change the sense, or direction, of the inequality. This result is called the **Addition Property of Inequality.**

In Other Words
The Addition Property of Inequality states that the direction of the inequality does not change when the same quantity is added to each side of the inequality.

> **Addition Property of Inequality**
>
> The **Addition Property of Inequality** states that for real numbers a, b, and c,
>
> $$\text{if} \quad a < b, \quad \text{then} \quad a + c < b + c$$
> $$\text{if} \quad a > b, \quad \text{then} \quad a + c > b + c$$

In Other Words
Subtracting a quantity from both sides of an inequality also does not change the direction of the inequality.

The Addition Property of Inequality also holds true for subtracting a real number from both sides of an inequality, since $a - b$ is equivalent to $a + (-b)$.

EXAMPLE 3   **How to Solve an Inequality Using the Addition Property of Inequality**

Solve the linear inequality $4y - 5 \le 3y - 2$. Express the solution set using set-builder notation and interval notation. Graph the solution set.

Step-by-Step Solution

		$4y - 5 \le 3y - 2$
Step 1: Get the terms containing variables on the left side of the inequality.	Subtract $3y$ from both sides of the inequality:	$4y - 5 - 3y \le 3y - 2 - 3y$
		$y - 5 \le -2$
Step 2: Isolate the variable y on the left side.	Add 5 to both sides of the inequality:	$y - 5 + 5 \le -2 + 5$
		$y \le 3$

Figure 14

The solution set using set-builder notation is $\{y \,|\, y \le 3\}$. The solution set using interval notation is $(-\infty, 3\,]$. The solution set is graphed in Figure 14.   ●

> **Quick ✓**
>
> 12. To ____ an inequality means to find the set of all values of the variable for which the statement is true.
>
> 13. Write the inequality that results from adding 7 to each side of $3 < 8$. What property of inequality does this illustrate?
>
> *In Problems 14–17, find the solution of the linear inequality. Express the solution set in set-builder notation and interval notation. Graph the solution set.*
>
> 14. $n - 2 > 1$ 15. $-2x + 3 < 7 - 3x$
>
> 16. $5n + 8 \le 4n + 4$ 17. $3(4x - 8) + 12 > 11x - 13$

We've seen what happens when we add a real number to both sides of an inequality. Let's look at two examples from arithmetic to see what happens when both sides of an inequality are multiplied or divided by a nonzero constant.

EXAMPLE 4 **Multiplying or Dividing an Inequality by a Positive Number**

(a) Write the inequality that results from multiplying both sides of the inequality $-2 < 5$ by 3.

(b) Write the inequality that results from dividing both sides of the inequality $18 > 14$ by 2.

Solution

(a) Multiplying both sides of $-2 < 5$ by 3 results in the numbers -6 and 15 on each side of the inequality, so we have $-6 < 15$.

(b) Dividing both sides of $18 > 14$ by 2 results in the numbers 9 and 7 on each side of the inequality, so we have $9 > 7$. ●

From Example 4 it would seem that multiplying (or dividing) both sides of an inequality by a positive real number does not change the direction of the inequality.

EXAMPLE 5 **Multiplying or Dividing an Inequality by a Negative Number**

(a) Write the inequality that results from multiplying both sides of the inequality $-2 < 5$ by -3.

(b) Write the inequality that results from dividing both sides of the inequality $18 > 14$ by -2.

Solution

(a) Multiplying both sides of $-2 < 5$ by -3 results in the numbers 6 and -15 on each side of the inequality, so we have $6 > -15$.

(b) Dividing both sides of $18 > 14$ by -2 results in the numbers -9 and -7 on each side of the inequality, so we have $-9 < -7$. ●

Example 5 suggests that multiplying (or dividing) both sides of an inequality by a negative real number results in an inequality that changes the direction of the original inequality. The results of Examples 4 and 5 lead us to the *Multiplication Properties of Inequality*.

In Other Words

The Multiplication Properties of Inequality state that if you multiply (or divide) both sides of an inequality by a positive number, the inequality symbol remains the same, but if you multiply (or divide) both sides by a negative number, the inequality symbol reverses.

Multiplication Properties of Inequality

Let a, b, and c be real numbers.

If $a < b$ and c is positive, then $ac < bc$.

If $a > b$ and c is positive, then $ac > bc$.

If $a < b$ and c is negative, then $ac > bc$.

If $a > b$ and c is negative, then $ac < bc$.

The Multiplication Properties of Inequality also holds for dividing both sides of an inequality by a nonzero real number.

EXAMPLE 6 **Solving a Linear Inequality Using the Multiplication Properties of Inequality**

Solve each linear inequality. Express the solution set using set-builder notation and interval notation. Graph the solution set.

(a) $\frac{1}{3}x > -2$

(b) $-4x \geq 24$

Solution

(a)
$$\frac{1}{3}x > -2$$

Multiply both sides of the inequality by 3: $3 \cdot \frac{1}{3}x > 3 \cdot (-2)$

$$x > -6$$

The solution set using set-builder notation is $\{x | x > -6\}$. The solution set using interval notation is $(-6, \infty)$. The graph of the solution set is shown in Figure 15.

(b)
$$-4x \geq 24$$

Divide both sides of the inequality by -4.
Remember to reverse the inequality symbol! $\dfrac{-4x}{-4} \leq \dfrac{24}{-4}$

$$x \leq -6$$

The solution set using set-builder notation is $\{x | x \leq -6\}$. The solution set using interval notation is $(-\infty, -6]$. The graph of the solution set is shown in Figure 16.

Figure 15

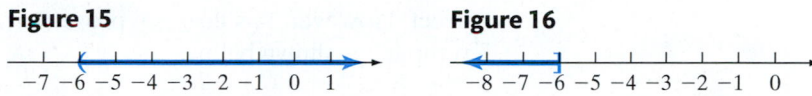

Figure 16

Work Smart

Note the difference:
$3x > -15$ is equivalent to $x > -5$ but
$-3x > -15$ is equivalent to $x < 5$.

Quick ✓

18. Write the inequality that results from multiplying both sides of the inequality $3 < 12$ by $\frac{1}{3}$. What property of inequalities does this illustrate?

19. Write the inequality that results from dividing both sides of $6 < 10$ by -2. What property of inequalities does this illustrate?

In Problems 20–23, find the solution of the linear inequality and express the solution set in set-builder notation and interval notation. Graph the solution set.

20. $\frac{1}{6}k < 2$

21. $2n \geq -6$

22. $-\frac{3}{2}k > 12$

23. $-\frac{4}{3}p \leq -\frac{4}{5}$

▶ It may be necessary to use both the Addition and Multiplication Properties of Inequality to solve an inequality. In each solution, isolate the variable on the left side of the inequality, so the inequality is easier to read. If the variable ends up on the right side, remember that

$$a < x \text{ is equivalent to } x > a \quad \text{and}$$
$$a > x \text{ is equivalent to } x < a$$

EXAMPLE 7 **How to Solve an Inequality Using Both the Addition and Multiplication Properties of Inequality**

Solve the inequality $4(x + 1) - 2 < 8x - 26$. Express the solution set using set-builder notation and interval notation. Graph the solution set.

Step-by-Step Solution

Step 1: Remove parentheses. $4(x + 1) - 2 < 8x - 26$

Use the Distributive Property: $4x + 4 - 2 < 8x - 26$

(continued)

Step 2: Combine like terms on
each side of the inequality.

$$4x + 2 < 8x - 26$$

Step 3: Get the terms containing
variables on the left side of the
inequality and the constants on the
right side.

Subtract 8x from both sides of the inequality:

$$4x + 2 - 8x < 8x - 26 - 8x$$
$$-4x + 2 < -26$$

Subtract 2 from both sides of the inequality:

$$-4x + 2 - 2 < -26 - 2$$
$$-4x < -28$$

Step 4: Get the coefficient of the
variable term to be 1.

Divide both sides of the inequality by −4:
Remember to reverse the inequality symbol!

$$\frac{-4x}{-4} > \frac{-28}{-4}$$
$$x > 7$$

Figure 17

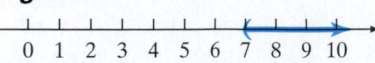

The solution set using set-builder notation is $\{x \mid x > 7\}$, or, using interval notation, $(7, \infty)$. The graph of the solution is shown in Figure 17. ●

After solving an inequality, substitute a value for the variable that is in the solution set into the original inequality and see whether a true statement is obtained. If a true statement is obtained, then there is some evidence that the solution is correct. However, this does *not* prove that the solution is correct. The check for Example 7 is shown below.

Check Because the solution of the inequality $4(x + 1) - 2 < 8x - 26$ is any real number greater than 7, let's replace x by 10.

$$4(x + 1) - 2 < 8x - 26$$

Replace x with 10: $4(10 + 1) - 2 \overset{?}{<} 8(10) - 26$

Perform the arithmetic: $4(11) - 2 \overset{?}{<} 80 - 26$

$$42 < 54 \qquad \text{True}$$

$42 < 54$ is a true statement, so $x = 10$ is in the solution set. Also, substituting 6 for x, which is less than 7, yields $26 < 22$, which is false. We have some evidence that our solution set, $(7, \infty)$, is correct.

> **Quick ✔**
>
> *In Problems 24–27, find the solution of the linear inequality. Express the solution set using set-builder notation and interval notation. Graph the solution set.*
>
> **24.** $3x - 7 > 14$ **25.** $-4n - 3 < 9$
>
> **26.** $2x - 6 < 3(x + 1) - 5$ **27.** $-4(x + 6) + 18 \geq -2x + 6$

EXAMPLE 8 **Solving a Linear Inequality Containing Fractions**

Solve the inequality $\frac{1}{2}(x - 4) \geq \frac{3}{4}(2x + 1)$. Express the solution using set-builder notation and interval notation. Graph the solution set.

Solution

Clear the fractions by multiplying both sides of the inequality by 4, the least common denominator of $\frac{1}{2}$ and $\frac{3}{4}$.

$$\frac{1}{2}(x - 4) \geq \frac{3}{4}(2x + 1)$$

Multiply both sides of the inequality by 4: $\qquad 4 \cdot \frac{1}{2}(x - 4) \geq 4 \cdot \frac{3}{4}(2x + 1)$

$$2(x - 4) \geq 3(2x + 1)$$

Use the Distributive Property: $\qquad 2x - 8 \geq 6x + 3$

Subtract $6x$ from both sides of the inequality: $\qquad 2x - 8 - 6x \geq 6x + 3 - 6x$

$$-4x - 8 \geq 3$$

Add 8 to both sides of the inequality: $\qquad -4x - 8 + 8 \geq 3 + 8$

$$-4x \geq 11$$

Divide both sides of the inequality by -4 and remember to reverse the inequality symbol: $\qquad \dfrac{-4x}{-4} \leq \dfrac{11}{-4}$

$$x \leq -\frac{11}{4}$$

The solution is $\left\{ x \middle| x \leq -\dfrac{11}{4} \right\}$, or, using interval notation, $\left(-\infty, -\dfrac{11}{4} \right]$. The graph of the solution is given in Figure 18.

Figure 18

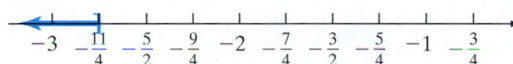

$$\begin{array}{ccccccccc} -3 & -\frac{11}{4} & -\frac{5}{2} & -\frac{9}{4} & -2 & -\frac{7}{4} & -\frac{3}{2} & -\frac{5}{4} & -1 & -\frac{3}{4} \end{array}$$

Quick ✓

In Problems 28 and 29, find the solution of the linear inequality and express the solution set using set-builder notation and interval notation. Graph the solution set.

28. $\dfrac{1}{2}(x + 2) > \dfrac{1}{5}(x + 17)$ **29.** $\dfrac{4}{3}x - \dfrac{2}{3} \leq \dfrac{4}{5}x + \dfrac{3}{5}$

Some inequalities are true for all values of the variable, and some are false for all values of the variable. These special cases are presented now.

EXAMPLE 9   **Solving an Inequality Whose Solution Set Is All Real Numbers**

Solve the inequality $3(x + 4) - 5 > 7x - (4x + 2)$. Express the solution using set-builder notation and interval notation, and graph the solution set.

Solution

$$3(x + 4) - 5 > 7x - (4x + 2)$$

Use the Distributive Property: $\qquad 3x + 12 - 5 > 7x - 4x - 2$

$$3x + 7 > 3x - 2$$

Subtract $3x$ from both sides of the inequality: $\qquad 3x + 7 - 3x > 3x - 2 - 3x$

$$7 > -2$$

Figure 19

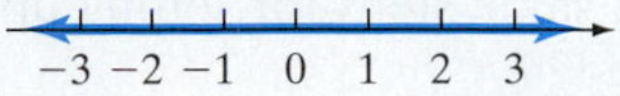

$$\begin{array}{ccccccc} -3 & -2 & -1 & 0 & 1 & 2 & 3 \end{array}$$

The statement $7 > -2$ is true, so the solution is all real numbers. The solution set is $\{ x | x \text{ is any real number} \}$, or $(-\infty, \infty)$ in interval notation. Figure 19 shows the graph of the solution set.

EXAMPLE 10 **Solving an Inequality Whose Solution Set Is the Empty Set**

Solve the inequality $8\left(\dfrac{1}{2}x - 1\right) + 2x \le 6x - 10$. Express the solution using set-builder notation and interval notation, if possible. Graph the solution set.

Solution

$$8\left(\frac{1}{2}x - 1\right) + 2x \le 6x - 10$$

Use the Distributive Property: $\qquad 4x - 8 + 2x \le 6x - 10$

$$6x - 8 \le 6x - 10$$

Subtract $6x$ from both sides of the inequality: $\qquad 6x - 8 - 6x \le 6x - 10 - 6x$

$$-8 \le -10$$

The statement $-8 \le -10$ is false, so this inequality has no solution. The solution set is the empty set, $\varnothing$ or $\{\ \}$. Figure 20 shows the graph of the solution set on a real number line.

Figure 20

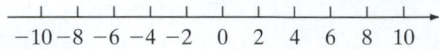

-10 -8 -6 -4 -2 0 2 4 6 8 10

Quick ✓

30. *True or False* When you are solving an inequality, if the variable is eliminated and the result is a true statement, the solution is all real numbers.

31. *True or False* When you are solving an inequality, if the variable is eliminated and the result is a false statement, the solution is the empty set or $\varnothing$.

In Problems 32–35, find the solution of the inequality. Express the solution set using set-builder notation and interval notation, if possible. Graph the solution set.

32. $-2x + 7(x + 5) \le 6x + 32$

33. $-x + 7 - 8x \ge 2(8 - 5x) + x$

34. $\dfrac{3}{2}x + 5 - \dfrac{5}{2}x < 4x - 3(x + 1)$

35. $x + 3(x + 4) \ge 2x + 5 + 3x - x$

▶ ❹ Model Inequality Problems

When solving word problems modeled by inequalities, look for key words that indicate the type of inequality symbol to use. Some key words and phrases are listed in Table 8.

Table 8

Word or Phrase	Inequality Symbol	Word or Phrase	Inequality Symbol
at least	$\ge$	at most	$\le$
no less than	$\ge$	no more than	$\le$
more than	$>$	fewer than	$<$
greater than	$>$	less than	$<$

Work Smart

The context of the words "less than" is important! "3 **less than** x" is $x - 3$, but "3 **is less than** x" is $3 < x$. Do you see the difference?

To solve applications involving linear inequalities, use the same steps for setting up applied problems that were introduced in Section 2.5, on page 124.

EXAMPLE 11 **A Handyman's Fee**

A handyman charges a flat fee of $60 plus $22 per hour for a job. How many hours does this handyman need to work at a job to make at least $500?

Solution

Step 1: Identify This is a direct translation problem. Find the number of hours the handyman must work at a job to make at least $500.

Step 2: Name Let h represent the number of hours that the handyman must work.

Step 3: Translate The sum of the flat fee that the handyman charges and the hourly charge must be greater than or equal to $500. So

$$\underbrace{\text{flat fee}}_{60} \text{ plus } \underbrace{\text{hourly rate}}_{22} \text{ times } \underbrace{\text{hours worked}}_{h} \underbrace{\text{is at least}}_{\geq} \underbrace{500}_{500} \quad \color{red}{\text{The Model}}$$

Step 4: Solve

$$60 + 22h \geq 500$$

Subtract 60 from both sides of the inequality: $\quad 60 - 60 + 22h \geq 500 - 60$

Simplify: $\quad 22h \geq 440$

Divide both sides of the inequality by 22: $\quad \dfrac{22h}{22} \geq \dfrac{440}{22}$

$$h \geq 20$$

Step 5: Check If the handyman works 23 hours (for example), will he earn more than $500? Since $60 + 22(23) = 566$ is greater than 500, there is evidence that our answer is correct.

Step 6: Answer The handyman must work 20 hours or more to earn at least $500. ●

Quick ✓

36. A worker in a large apartment complex uses an elevator to move supplies. The elevator has a weight limit of 2000 pounds. The worker weighs 180 pounds, and each box of supplies weighs 91 pounds. Find the maximum number of boxes of supplies the worker can move on one trip in the elevator.

2.8 Exercises   MyMathLab®

Exercise numbers in green have complete video solutions in MyMathLab or may be accessed using the QR code to the right.

*Problems **1–36** are the Quick ✓s that follow the **EXAMPLES**.*

Building Skills

In Problems 37–44, graph each inequality on a number line, and write each inequality in interval notation. See Objectives 1 and 2.

37. $x > 4$ **38.** $n > 8$

39. $x \leq -1$ **40.** $x \leq 6$

41. $z \geq -3$ **42.** $x \geq -2$

43. $x < 4$ **44.** $y < -3$

In Problems 45–50, use interval notation to express the inequality shown in each graph. See Objective 2.

45.
46.

47.
48.
49.
50.

In Problems 51–58, fill in the blank with the correct symbol. State which property of inequality is being used. See Objective 3.

51. If $x - 7 < 11$, then x____18

52. If $3x - 2 > 7$, then $3x$____9

53. If $\dfrac{1}{3}x > -2$, then x____-6

54. If $\dfrac{5}{3}x \geq -10$, then x____-6

55. If $3x + 2 \leq 13$, then $3x$___11

56. If $\dfrac{3}{4}x + 1 \leq 9$, then $\dfrac{3}{4}x$___8

57. If $-3x \geq 15$, then x___-5

58. If $-4x < 28$, then x___-7

In Problems 59–88, solve the inequality. Express the solution set in set-builder notation and interval notation. Graph the solution set on a real number line. See Objective 3.

59. $x + 1 < 5$

60. $x + 4 \leq 3$

61. $x - 6 \geq -4$

62. $x - 2 < 1$

63. $3x \leq 15$

64. $4x > 12$

65. $-5x < 35$

66. $-7x \geq 28$

67. $3x - 7 > 2$

68. $2x + 5 > 1$

69. $3x - 1 \geq 3 + x$

70. $2x - 2 \geq 3 + x$

71. $1 - 2x \leq 3$

72. $2 - 3x \leq 5$

73. $-2(x + 3) < 8$

74. $-3(1 - x) > x + 8$

75. $4 - 3(1 - x) \leq 1$

76. $8 - 4(2 - x) \leq -2x$

77. $\dfrac{1}{2}(x - 4) > x + 8$

78. $3x + 4 > \dfrac{1}{3}(x - 2)$

79. $4(x - 1) > 3(x - 1) + x$

80. $2y - 5 + y < 3(y - 2)$

81. $5(n + 2) - 2n \leq 3(n + 4)$

82. $3(p + 1) - p \geq 2(p + 1)$

83. $2n - 3(n - 2) < n - 4$

84. $4x - 5(x + 1) \leq x - 3$

85. $4(2w - 1) \geq 3(w + 2) + 5(w - 2)$

86. $3q - (q + 2) > 2(q - 1)$

87. $3y - (5y + 2) > 4(y + 1) - 2y$

88. $8x - 3(x - 2) \geq x + 4(x + 1)$

In Problems 89–98, write the given statement using inequality symbols. Let x represent the unknown quantity. See Objective 4.

89. Karen's salary this year will be at least $16,000.

90. Bob's salary this year will be at most $120,000.

91. There will be at most 20,000 fans at the San Diego Padres game today.

92. The cost of a new lawnmower is at least $250.

93. The cost to remodel a kitchen is more than $12,000.

94. There are no more than 25 students in your math class on any given day.

95. x is a positive number

96. x is a nonnegative number

97. x is a nonpositive number

98. x is a negative number

Mixed Practice

In Problems 99–116, solve the inequality and express the solution set in set-builder notation and interval notation, if possible. Graph the solution set on a real number line.

99. $-1 < x - 5$

100. $6 \geq x + 15$

101. $-\dfrac{3}{4}x > -\dfrac{9}{16}$

102. $-\dfrac{5}{8}x > \dfrac{25}{48}$

103. $3(x + 1) > 2(x + 1) + x$

104. $5(x - 2) < 3(x + 1) + 2x$

105. $-4a + 1 > 9 + 3(2a + 1) + a$

106. $-5b + 2(b - 1) \leq 6 - (3b - 1) + 2b$

107. $n + 3(2n + 3) > 7n - 3$

108. $2k - (k - 4) \geq 3k + 10 - 2k$

109. $\dfrac{x}{2} \geq 1 - \dfrac{x}{4}$

110. $\dfrac{x}{3} \geq 2 + \dfrac{x}{6}$

111. $\dfrac{x + 5}{2} + 4 > \dfrac{2x + 1}{3} + 2$

112. $\dfrac{3z - 1}{4} + 1 \leq \dfrac{6z + 5}{2} + 2$

113. $-5z - (3 + 2z) > 3 - 7z$

114. $2(4a - 3) \le 5a - (2 - 3a)$

115. $1.3x + 3.1 < 4.5x - 15.9$

116. $4.9 + 2.6x < 4.2x - 4.7$

Applying the Concepts

117. Auto Rental A car can be rented from Certified Auto Rental for $55 per week plus $0.18 per mile. How many miles can you drive if you have at most $280 to spend for weekly transportation?

118. Truck Rental A truck can be rented from Acme Truck Rental for $80 per week plus $0.28 per mile. How many miles can you drive if you have at most $100 to spend on truck rental?

119. Final Grade Yvette has earned 72, 78, 66, and 81 points on her algebra tests. Her final exam counts as two test grades. How many points (out of 100) does she need on the final exam to earn at least 361 points for the class?

120. Final Grade To earn an A in Mrs. Smith's elementary statistics class, Elizabeth must earn at least 540 points. Thus far, Elizabeth has earned 85, 83, 90, and 96 points on her tests. The final exam counts as two test grades. How many points (out of 100) does Elizabeth need on the final exam to earn an A?

121. Mulch Delivery Clintonville Landscapers (CL) charges a delivery fee of $30 regardless of the amount of mulch delivered within a 20-mile radius of the store. The cost of Absolute Black Ultra mulch is $39.99 per cubic yard at CL. Oregonia Nursery is having a special and charges $28.49 for Absolute Black Ultra mulch, but their delivery charge is $50 for any quantity of mulch. For how many cubic yards is buying mulch from Oregonia Nursery cheaper? Round your answer to the nearest tenth.

122. Commission A recent college graduate had an offer of a sales position that pays $15,000 per year plus 1% of all sales.

   (a) Write an expression for the total annual salary based on sales of S dollars.

   (b) For what total sales amount will the college graduate earn in excess of $150,000 annually?

123. Borrowing Money The amount of money that a lending institution will allow you to borrow mainly depends on the interest rate and your annual income. The equation $L = 2.98I - 76.11$ describes the amount of money, L, that a bank will lend at an interest rate of 7.5% for 30 years, based upon annual income, I. For what annual income, I, will a bank lend at least $150,000? (SOURCE: *Information Please Almanac*)

124. Advertising A marketing firm found that the equation $S = 2.1A + 224$ describes how the amount

of sales S of a product depends on A, the amount spent on advertising the product. Both S and A are measured in thousands of dollars. For what amount, A, is the sales of a product at least 350 thousand dollars?

Extending the Concepts

125. Grades In your Economics 101 class, you have scores of 68, 82, 87, and 89 on the first four of five tests. To earn a grade of B or higher, the average of the first five test scores must be greater than or equal to 80. Find the minimum score (out of 100) that you can make on the last test and earn a B.

126. Delivery Service A messenger service charges $10 to make a delivery to an address. In addition, each letter delivered costs $3 and each package delivered costs $8. If there are 15 more letters than packages delivered to this address, what is the maximum number of items that can be delivered for $85?

Compound inequalities are solved using the properties of inequality. Study the example below, and then solve Problems 127–134. State the solution in set-builder notation.

$$-15 \;\le\; 2x - 1 \;<\; 37$$
$$-15 + 1 \le 2x - 1 + 1 \;<\; 37 + 1$$
$$\frac{-14}{2} \le \frac{2x}{2} < \frac{38}{2}$$
$$-7 \le \quad x \quad < 19$$

127. $-3 < x + 30 < 16$ **128.** $-41 \le x - 37 \le 26$

129. $-6 \le \dfrac{3x}{2} \le 9$ **130.** $-4 < \dfrac{8x}{9} < 12$

131. $-7 \le 2x - 3 < 15$ **132.** $4 < 3x + 7 \le 28$

133. $4 < 6 - \dfrac{x}{2} \le 10$ **134.** $-1 \le 5 - \dfrac{x}{3} < 1$

Explaining the Concepts

135. In graphing an inequality, when is a left parenthesis used? When is a left bracket used?

136. Explain the circumstances in which the direction of the inequality symbol is reversed when one is solving a simple inequality.

137. Explain how you recognize when the solution of an inequality is all real numbers. Explain how you recognize when the solution of an inequality is the empty set.

138. Why is the interval notation $[-7, -\infty)$ for $x > -7$ incorrect? What is the correct notation?

Chapter 2 Activity: Pass to the Right

Focus: Solving linear equations and inequalities as a group

Time: 20–30 minutes

Group size: 3–4

1. Each member of the group should choose one of the following four equations and write it down on a piece of paper. Do not begin to solve.

(a) $\dfrac{x+2}{2} - \dfrac{5x-12}{6} = 1$

(b) $-\dfrac{1}{9}(x+27) + \dfrac{1}{3}(x+3) = x+6$

(c) $\dfrac{x+3}{3} - \dfrac{2x-12}{9} = 1$

(d) $-\dfrac{1}{2}(x+6) + \dfrac{1}{7}(x+7) = x+3$

2. Each member of the group should pass her or his equation to the person on the right. This member should perform the first step in solving the equation and, when finished with that

step, should pass the paper to the next group member on her or his right.

3. Upon receipt of the equation, each group member should check the previous member's work and then perform the next step. If an error is found, discuss and correct the error.

4. Continue passing the problems until all equations have been solved.

5. As a group, discuss the results.

6. If time permits, repeat this activity except choose one of the following inequalities. Express your final answer using interval notation.

(a) $\dfrac{1}{4}(2x+12) > \dfrac{3}{8}(x-1)$

(b) $\dfrac{3x+1}{10} - \dfrac{1+6x}{5} \le -\dfrac{1}{2}$

(c) $\dfrac{1}{2}(2x+14) > \dfrac{3}{4}(x-1)$

(d) $\dfrac{3x+1}{21} - \dfrac{1+4x}{7} \le -\dfrac{1}{3}$

Chapter 2 Review

Section 2.1 Linear Equations: The Addition and Multiplication Properties of Equality

KEY CONCEPTS

- **Linear Equation in One Variable**

  An equation equivalent to one of the form $ax + b = c$, where $a, b,$ and c are real numbers and $a \ne 0$.

- **Addition Property of Equality**

  For real numbers $a, b,$ and c, if $a = b$, then $a + c = b + c$.

- **Multiplication Property of Equality**

  For real numbers $a, b,$ and c, where $c \ne 0$, if $a = b$, then $ac = bc$.

KEY TERMS

Linear equation
Sides
Left side
Right side
Solution
Solution set
Satisfies
Equivalent equations

You Should Be Able To...	EXAMPLE	Review Exercises
❶ Determine whether a number is a solution of an equation (p. 82)	Example 1	1–4
❷ Use the Addition Property of Equality to solve linear equations (p. 83)	Examples 2 through 4	5–10, 15, 16, 19
❸ Use the Multiplication Property of Equality to solve linear equations (p. 85)	Examples 5 through 9	11–14, 17, 18, 20

In Problems 1–4, determine whether the given value is a solution to the equation. Answer Yes or No.

1. $3x + 2 = 7; x = 5$

2. $5m - 1 = 17; m = 4$

3. $6x + 6 = 12; x = \dfrac{1}{2}$

4. $9k + 3 = 9; k = \dfrac{2}{3}$

In Problems 5–18, solve the equation. Check your solution.

5. $n - 6 = 10$

6. $n - 8 = 12$

7. $x + 6 = -10$

8. $x + 2 = -5$

9. $-100 = m - 5$

10. $-26 = m - 76$

11. $\frac{2}{3}y = 16$

12. $\frac{x}{4} = 20$

13. $-6x = 36$

14. $-4x = -20$

15. $z + \frac{5}{6} = \frac{1}{2}$

16. $m - \frac{1}{8} = \frac{1}{4}$

17. $1.6x = 6.4$

18. $1.8m = 9$

19. **Discount** The cost of a used Honda Accord has been discounted by $1200, so that the sale price is $22,275. Find the original price of the Honda Accord, p, by solving the equation $p - 1200 = 22,275$.

20. **Coffee** While studying for an exam, Randi drank coffee at a local coffeehouse. She bought 3 cups of coffee for a total of $7.65. Find the cost of each cup of coffee, c, by solving the equation $3c = 7.65$.

Section 2.2   Linear Equations: Using the Properties Together

KEY CONCEPT

KEY TERM

- The steps for solving an equation in one variable are given on page 95.

Distributive Property

You Should Be Able To...	EXAMPLE	Review Exercises
❶ Use the Addition and Multiplication Properties of Equality to solve linear equations (p. 91)	Examples 1 and 2	21–24
❷ Combine like terms and use the Distributive Property to solve linear equations (p. 92)	Examples 3 and 4	25–30
❸ Solve a linear equation with the variable on both sides of the equation (p. 94)	Examples 5 and 6	31–34
❹ Use linear equations to solve problems (p. 96)	Example 7	35, 36

In Problems 21–34, solve the equation. Check your solution.

21. $5x - 1 = -21$

22. $-3x + 7 = -5$

23. $\frac{2}{3}x + 5 = 11$

24. $\frac{5}{7}x - 2 = -17$

25. $-2x + 5 + 6x = -11$

26. $3x - 5x + 6 = 18$

27. $2m + 0.5m = 10$

28. $1.4m + m = -12$

29. $-2(x + 5) = -22$

30. $3(2x + 5) = -21$

31. $5x + 4 = -7x + 20$

32. $-3x + 5 = x - 15$

33. $4(x - 5) = -3x + 5x - 16$

34. $4(m + 1) = m + 5m - 10$

35. **Ages** Skye is 4 years older than Beth. The sum of their ages is 24. Find Skye's age by solving the equation $x + x + 4 = 24$, where x represents Beth's age and $x + 4$ represents Skye's age.

36. **Parking Lot** The length of a rectangular parking lot is 10 yards longer than the width, w, and the perimeter of the parking lot is 96 yards. Solve the equation $2w + 2(w + 10) = 96$ to find the width, w, of the parking lot. Then find the length of the parking lot, $w + 10$.

Section 2.3   Solving Linear Equations Involving Fractions and Decimals; Classifying Equations

KEY CONCEPTS

KEY TERMS

- A **conditional equation** is an equation that is true for some values of the variable and false for other values of the variable.

- An equation that is false for every replacement value of the variable is called a **contradiction.**

- An equation that is satisfied for every choice of the variable for which both sides of the equation are defined is called an **identity.**

Least common denominator (LCD)
Identity
Conditional equation
Contradiction

You Should Be Able To...	EXAMPLE	Review Exercises
1 Use the least common denominator to solve a linear equation containing fractions (p. 98)	Examples 1 and 2	37–40, 45, 46
2 Solve a linear equation containing decimals (p. 101)	Examples 3 through 5	41–44, 47, 48
3 Classify an equation as an identity, a conditional equation, or a contradiction (p. 102)	Examples 6 through 8	49–54
4 Use linear equations to solve problems (p. 104)	Example 9	55, 56

In Problems 37–48, solve the equation. Check your solution.

37. $\dfrac{6}{7}x + 3 = \dfrac{1}{2}$　　　**38.** $\dfrac{1}{4}x + 6 = \dfrac{5}{6}$

39. $\dfrac{n}{2} + \dfrac{2}{3} = \dfrac{n}{6}$　　　**40.** $\dfrac{m}{8} + \dfrac{m}{2} = \dfrac{3}{4}$

41. $1.2r = -1 + 2.8$

42. $0.2x + 0.5x = 2.1$

43. $1.2m - 3.2 = 0.8m - 1.6$

44. $0.3m + 0.8 = 0.5m + 1$

45. $\dfrac{1}{2}(x + 5) = \dfrac{3}{4}$

46. $-\dfrac{1}{6}(x - 1) = \dfrac{2}{3}$

47. $0.1(x + 80) = -0.2 + 14$

48. $0.35(x + 6) = 0.45(x + 7)$

In Problems 49–54, solve each equation. State whether the equation is a contradiction, an identity, or a conditional equation.

49. $4x + 2x - 10 = 6x + 5$

50. $-2(x + 5) = -5x + 3x + 2$

51. $-5(2n + 10) = 6n - 50$

52. $8m + 10 = -2(7m - 5)$

53. $10x - 2x + 18 = 2(4x + 9)$

54. $-3(2x - 8) = -3x - 3x + 24$

55. T-Shirt Al purchased a tie-dyed T-shirt at a "20% off" sale for $12.60. What was the original price, p, of the shirt? Solve the equation $p - 0.20p = 12.60$.

56. Couch Cushions Juanita was cleaning under her couch cushions and found some nickels and dimes. She found $0.55 in change, with the number of nickels one less than twice the number of dimes. Solve the equation $0.10d + 0.05(2d - 1) = 0.55$ to find d, the number of dimes Juanita found.

Section 2.4　Evaluating Formulas and Solving Formulas for a Variable

KEY CONCEPTS

- **Simple interest formula**

  $I = Prt$, I represents the amount of interest, P is the principal, r is the rate of interest (expressed as a decimal), and t is time (expressed in years).

- **Geometry Formulas (pages 110–111)**

KEY TERMS

Formula
Interest
Principal
Rate of interest

You Should Be Able To...	EXAMPLE	Review Exercises
1 Evaluate a formula (p. 108)	Examples 1 through 6	57–60, 67–70
2 Solve a formula for a variable (p. 113)	Examples 7 through 9	61–66, 67, 68

In Problems 57–60, substitute the given values into the formula and then simplify to find the unknown quantity. Include units in the answer.

57. Area of a rectangle: $A = lw$; Find A when $l = 8$ inches and $w = 6$ inches.

58. Perimeter of a square: $P = 4s$; Find P when $s = 16$ cm.

59. Perimeter of a rectangle: $P = 2l + 2w$; Find w when $P = 16$ yards and $l = \dfrac{13}{2}$ yards.

60. Circumference of a circle: $C = \pi d$; Find C when $d = \dfrac{15}{\pi}$ mm.

In Problems 61–66, solve each formula for the stated variable.

61. $V = LWH$; solve for H.

62. $I = Prt$; solve for P.

63. $S = 2LW + 2LH + 2WH$; solve for W.

64. $\rho = mv + MV$; solve for M.

65. $2x + 3y = 10$; solve for y.

66. $6x - 3y = 15$; solve for y.

67. Finance The formula $A = P(1 + r)^t$ can be used to find the future value A of a deposit of P dollars in an account that earns an annual interest rate r (expressed as a decimal) after t years.

(a) Solve the formula for P.

(b) How much would you have to deposit today in order to have $3000 in 6 years in a bank account that pays 5% annual interest? Round your answer to the nearest cent.

68. Cylinders The surface area A of a right circular cylinder is given by the formula $A = 2\pi rh + 2\pi r^2$, where r is the radius and h is the height.

(a) Solve the formula for h.

(b) Determine the height of a right circular cylinder whose surface area is 72π square centimeters and whose radius is 4 centimeters.

69. Christmas Bonus Samuel invested his $500 Christmas bonus in a 9-month certificate of deposit that earns 2% simple interest. Use the formula $I = Prt$ to find the amount of interest Samuel's investment will earn.

70. Coffee Table Find the exact and approximate (to the nearest tenth) area of a circular coffee table top with a diameter of 3 feet.

Section 2.5   Problem Solving: Direct Translation

KEY CONCEPTS

- **Six steps for solving problems with mathematical models**
  Identify the problem type and what you are looking for
  Name the unknown(s)
  Translate to a mathematical equation
  Solve the equation
  Check the answer
  Answer the question in a complete sentence

KEY TERMS

Problem solving
Mathematical modeling
Mathematical model
Model

You Should Be Able To...	EXAMPLE	Review Exercises
1 Translate English phrases into algebraic expressions (p. 120)	Examples 1 through 3	71–76, 83–86
2 Translate English sentences into equations (p. 122)	Example 4	77–82
3 Build models for solving direct translation problems (p. 125)	Examples 5 through 9	87–90

In Problems 71–76, translate each phrase into an algebraic expression. Let x represent the unknown number.

71. the difference between a number and 6

72. eight subtracted from a number

73. the product of -8 and a number

74. the quotient of a number and 10

75. twice the sum of 6 and a number

76. four times the difference of 5 and a number

In Problems 77–82, translate each statement into an equation. Let x represent the unknown number. DO NOT SOLVE.

77. The sum of 6 and a number is equal to twice the number increased by 5.

78. The product of 6 and a number decreased by 10 is one more than double the number.

79. Eight less than a number is the same as half of the number.

80. The ratio of 6 to a number is the same as the number added to 10.

81. Four times the sum of twice a number and 8 is 16.

82. Five times the difference between double a number and 8 is -24.

In Problems 83–86, choose a variable to represent one quantity. State what the quantity represents, and then express the second quantity in terms of the first.

83. Jacob is seven years older than Sarah.

84. Maria runs twice as fast as Consuelo.

85. Irene has $6 less than Max.

86. Victor and Larry will share $350.

87. Losing Weight Over the past year Lee Lai lost 28 pounds. Find Lee Lai's weight one year ago if her current weight is 125 pounds.

88. Consecutive Integers The sum of three consecutive integers is 39. Find the integers.

89. Finance A total of $20,000 is to be divided between Roberto and Juan, with Roberto to receive $2000 less than Juan. How much will each receive?

90. Truck Rentals You need to rent a moving truck. You identified two companies that rent trucks. ABC-Rental charges $30 per day plus $0.15 per mile. U-Do-It Rental charges $15 per day plus $0.30 per mile. For how many miles will the daily cost of renting a moving truck be the same?

Section 2.6 Problem Solving: Problems Involving Percent

KEY CONCEPTS

- Percent means divided by 100 or per hundred.
- Total Cost = Original Cost + Sales Tax
- Original Price − Discount = Sale Price
- Wholesale Price + Markup = Selling Price

You Should be Able To...	EXAMPLE	Review Exercises
1 Solve problems involving percent (p. 132)	Examples 1 through 4	91–94
2 Solve business problems that involve percent (p. 134)	Examples 5 and 6	95–100

In Problems 91–94, find the unknown in each percent question.

91. What is 6.5% of 80?

92. 18 is 30% of what number?

93. 15.6 is what percent of 120?

94. 110% of what number is 55?

95. Sales Tax The base sales tax rate in Florida is 6%. The total cost of purchasing a leotard, including tax, was $19.61. Find the cost of the leotard before sales tax.

96. Tutoring Mei Ling is a private tutor. She raised her hourly fee by 8.5% to cover her traveling expenses. Her new hourly fee is $32.55. Find Mei Ling's previous hourly fee.

97. Business: Discount Pricing A sweater, discounted by 70% for an end-of-the-year clearance sale, has a price tag of $12. What was the sweater's original price?

98. Business: Mark-Up A clothing store marks up the price that it pays for a suit 80%. If the selling price of a suit is $360, how much did the store pay for the suit?

99. Salary Tanya earns $500 each week plus 2% of the value of the computers she sells each week. If Tanya wishes to earn $3000 this week, what must be the total value of the computers she sells?

100. Voting In an election for school president, the loser received 80% of the winner's votes. If 900 votes were cast, how many did each receive?

Section 2.7 Problem Solving: Geometry and Uniform Motion

KEY CONCEPTS

- Complementary angles are two angles whose measures sum to 90°.
- Supplementary angles are two angles whose measures sum to 180°.
- The sum of the measures of the interior angles of a triangle is 180°.

- **Uniform Motion**
  If an object moves at an average speed r, the distance d covered in time t is given by the formula $d = rt$.

KEY TERMS

Complementary angles
Supplementary angles
Uniform motion

You Should be Able To...	EXAMPLE	Review Exercises
1 Set up and solve complementary and supplementary angle problems (p. 138)	Example 1	101, 102
2 Set up and solve "angles of a triangle" problems (p. 139)	Example 2	103, 104
3 Use geometry formulas to solve problems (p. 140)	Examples 3 and 4	105–108
4 Set up and solve uniform motion problems (p. 142)	Examples 5 and 6	109, 110

△**101. Complementary Angles** Find two complementary angles such that the measure of the first angle is 20° more than six times the second.

△**102. Supplementary Angles** Find two supplementary angles such that the measure of the first angle is 60° less than twice the second.

△**103. Triangles** In a triangle, the measure of the second angle is twice the first. The measure of the third angle is 30° more than the second. Find the measures of the three angles.

△**104. Triangles** In a triangle, the measure of the second angle is 5° less than the first. The measure of the third angle is 5° less than twice the second. Find the measures of the three angles.

△**105. Rectangle** The length of a rectangle is 15 inches longer than twice its width. If the perimeter is 78 inches, find the length and width of the rectangle.

△**106. Rectangle** The length of a rectangle is four times its width. If the perimeter is 70 cm, find the length and width of the rectangle.

△**107. Garden** The perimeter of a rectangular garden is 120 feet. The width of the garden is twice the length.

(a) Find the length and width of the garden.

(b) What is the area of the garden

△**108. Back Yard** Yvonne's back yard is in the shape of a trapezoid with height of 80 feet. The shorter base is 10 feet shorter than the longer base, and the area of the back yard is 3600 square feet. Find the length of each base of the trapezoidal yard.

109. Boating Two motorboats leave the same dock at the same time, traveling in the same direction. One boat travels at 18 miles per hour, and the other travels at 25 miles per hour. In how many hours will the motorboats be 35 miles apart?

110. Train Station Two trains leave a station at the same time. One train is traveling east at 10 miles per hour faster than the other train, which is traveling west. After 6 hours, the two trains are 720 miles apart. At what speed did the faster train travel?

Section 2.8   Solving Linear Inequalities in One Variable

KEY CONCEPTS

- A linear inequality is of the form $ax + b < c, ax + b \le c, ax + b > c,$ or $ax + b \ge c$, where $a, b,$ and c are real numbers and $a \ne 0$.
- **Addition Property of Inequality**
  For real numbers $a, b,$ and c,
  If $a < b$, then $a + c < b + c$.
  If $a > b$, then $a + c > b + c$.
- **Multiplication Properties of Inequality**
  For real numbers $a, b,$ and c,
  If $a < b$ and $c > 0$, then $ac < bc$.
  If $a > b$ and $c > 0$, then $ac > bc$.
  If $a < b$ and $c < 0$, then $ac > bc$.
  If $a > b$ and $c < 0$, then $ac < bc$.

KEY TERMS

Linear inequality
Simple inequality
Set-builder notation
Graph of an inequality
Endpoint
Interval
Solve an inequality
Equivalent inequalities

SET-BUILDER NOTATION	INTERVAL NOTATION	GRAPH
$\{x \mid x \ge a\}$	$[a, \infty)$	
$\{x \mid x > a\}$	(a, ∞)	
$\{x \mid x \le a\}$	$(-\infty, a]$	
$\{x \mid x < a\}$	$(-\infty, a)$	
$\{x \mid x$ is any real number $\}$	$(-\infty, \infty)$	

You Should be Able To...	EXAMPLE	Review Exercises
❶ Graph inequalities on a real number line (p. 148)	Example 1	111–116
❷ Use interval notation (p. 149)	Example 2	117–120
❸ Solve linear inequalities using properties of inequalities (p. 150)	Examples 3 through 10	121–128
❹ Model inequality problems (p. 156)	Example 11	129, 130

In Problems 111–116, graph each inequality on a number line.

111. $x \leq -3$ **112.** $x > 4$

113. $m < 2$ **114.** $m \geq -5$

115. $0 < n$ **116.** $-3 \leq n$

In Problems 117–120, write the inequality in interval notation.

117. $x < -4$ **118.** $x \geq 7$

119. 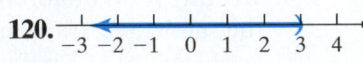 **120.**

In Problems 121–128, solve the inequality and express the solution in set-builder notation and interval notation, if possible. Graph the solution set on a real number line.

121. $4x + 3 < 2x - 10$

122. $3x - 5 \geq -12$

123. $-4(x - 1) \leq x + 8$

124. $6x - 10 < 7x + 2$

125. $-3(x + 7) > -x - 2x$

126. $4x + 10 \leq 2(2x + 7)$

127. $\frac{1}{2}(3x - 1) > \frac{2}{3}(x + 3)$

128. $\frac{5}{4}x + 2 < \frac{5}{6}x - \frac{7}{6}$

129. Moving The Rent-A-Moving-Van Company charges a flat rate of $19.95 per day plus $0.20 for each mile driven. How many miles can be driven by a customer who can afford to spend at most $32.95 for one day?

130. Bowling Travis is bowling three games in a tournament. In the first game, his score was 148. In the second game, his score was 155. What score must Travis get in the third game for his tournament average to be greater than 151?

Chapter 2 Test

Step-by-step test solutions are found on the Chapter Test Prep Videos available in MyMathLab®, on You Tube®, or may be accessed using the QR code to the right.

In Problems 1–8, solve the equation. Check your solution.

1. $x + 3 = -14$

2. $-\frac{2}{3}m = \frac{8}{27}$

3. $5(2x - 4) = 5x$

4. $-2(x - 5) = 5(-3x + 4)$

5. $-\frac{2}{3}x + \frac{3}{4} = \frac{1}{3}$

6. $-0.6 + 0.4y = 1.4$

7. $8x + 3(2 - x) = 5(x + 2)$

8. $2(x + 7) = 2x - 2 + 16$

In Problems 9 and 10, (a) solve for the indicated variable, and then (b) find the value of the unknown quantity. Include units in the answer.

9. Volume of a rectangular solid: $V = lwh$

   (a) Solve for l.

   (b) Find l when $V = 540$ in.³, $w = 6$ in., and $h = 10$ in.

10. Equation of a line: $2x + 3y = 12$

   (a) Solve for y.

   (b) Find y when $x = 8$.

11. Translate the following statement into an equation: Six times the difference between a number and 8 is equal to 5 less than twice the number. DO NOT SOLVE.

12. 18 is 30% of what number?

13. Consecutive Integers The sum of three consecutive integers is 48. Find the integers.

14. Fixer Uppers? On the show *Fixer Upper*, designer Joanna Gaines constructs a triangular art piece for a bedroom. The longest side is two inches longer than the middle side. The shortest side is 14 inches shorter than the middle side. If the perimeter of the art piece is 60 inches, what is the length of each side?

15. Buses Kimberly and Clay leave a concert hall at the same time, traveling in buses going in opposite directions. Kimberly's bus travels at 40 mph and Clay's bus travels at 60 mph. In how many hours will Kimberly and Clay be 350 miles apart?

16. Construction A carpenter cuts an oak board 21 feet long into two pieces. The longer board is 1 foot longer than three times the shorter one. Find the length of each board.

17. New Backpack Sherry purchased a new backpack for her daughter. The backpack was on sale for 20% off the regular price. If Sherry paid $28.80 for the backpack without sales tax, what was the original price?

In Problems 18 and 19, solve the inequality and express the solution in set-builder notation and interval notation. Graph the solution set on a real number line.

18. $3(2x - 5) \leq x + 15$

19. $-6x - 4 < 2(x - 7)$

20. Car Expenses Danielle sticks to her budget to ensure that she doesn't run out of money at the end of the month. Danielle pays $260 per month on her car loan and estimates $0.60 per mile driven for gas and maintenance. If her total car expenses can be at most $500, how many miles can Danielle drive each month? Round to the nearest mile.

3 Introduction to Graphing and Equations of Lines

Fall River Road was completed in 1920 and was the first road built through the Rocky Mountains in Colorado. It was so steep that sometimes early model cars had to drive up the hill in reverse to maximize the output of their weak engines and fuel systems. We can use mathematics to describe the steepness of a hill. See Problem 73 in Section 3.3.

The Big Picture: Putting It Together

It is now time to switch gears. In Chapter 2, linear equations and inequalities involving one unknown were solved. In this chapter, the focus is on linear equations and inequalities involving two unknowns.

In dealing with a single unknown, solutions to equations or inequalities can be represented graphically using real number lines. In that case, only one dimension is needed to represent the solution. When an equation has two unknowns, two dimensions are required. Representing solutions to equations or inequalities in two unknowns is accomplished through the *rectangular coordinate system*.

3.1 The Rectangular Coordinate System and Equations in Two Variables

Objectives

❶ Plot Points in the Rectangular Coordinate System

❷ Determine Whether an Ordered Pair Satisfies an Equation

❸ Create a Table of Values That Satisfy an Equation

▶ ❶ Plot Points in the Rectangular Coordinate System

Figure 1

Because pictures allow individuals to visualize ideas, they are typically more powerful than other forms of printed communication. Figure 1, which shows the results of the Manhattan Project from the test conducted on July 16, 1945, illustrates the power of the atom in a way that words never could.

Although the pictures used in mathematics might not deliver as powerful a message as Figure 1, they are powerful nonetheless. To draw pictures of mathematical relationships, we need a "canvas." The "canvas" used in this chapter is the *rectangular coordinate system.*

In Section 1.3, we learned how to plot points on a real number line. Plotting points on the real number line is plotting in one dimension. In this chapter, the *rectangular coordinate system* is used to plot points in two dimensions.

Begin by drawing two real number lines, one horizontal and one vertical, that intersect at right (90°) angles. The horizontal real number line is called the **x-axis** and the vertical real number line the **y-axis.** The point where the x-axis and y-axis intersect is called the **origin, O.** See Figure 2.

Figure 2
The rectangular coordinate system.

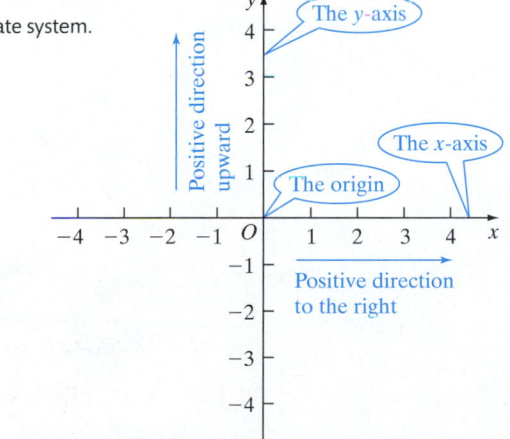

The origin O has a value of 0 on each axis. Points on the x-axis to the right of O represent positive real numbers, and points on the x-axis to the left of O represent negative real numbers. On the y-axis, points above O represent positive real numbers, and points below

O represent negative real numbers. Notice in Figure 2 that the x-axis is labeled "x" and the y-axis is labeled "y." An arrow at the end of each axis shows the positive direction.

The coordinate system in Figure 2 is called a **rectangular** or **Cartesian coordinate system,** named after René Descartes (1596–1650), a French mathematician, philosopher, and theologian. The plane formed by the x-axis and y-axis is often called the **xy-plane,** and the x-axis and y-axis are called the **coordinate axes.**

Any point P in the rectangular coordinate system can be represented by an **ordered pair (x, y)** of real numbers. In an ordered pair (x, y), x represents the distance that P is from the y-axis. If $x > 0$ (that is, if x is positive), then P is x units to the right of the y-axis. If $x < 0$, then P is $|x|$ units to the left of the y-axis. In (x, y), y represents the distance that P is from the x-axis. If $y > 0$, then P is y units above the x-axis. If $y < 0$, then P is $|y|$ units below the x-axis. The ordered pair (x, y) is also called the **coordinates** of P. To plot the point with coordinates $(2, 5)$, begin at the origin, move 2 units to the right, and then move 5 units up, as shown in Figure 3(a).

Work Smart

Be careful: The order in which numbers appear in the ordered pairs matters. For example, (3, 2) represents a different point from (2, 3).

Figure 3

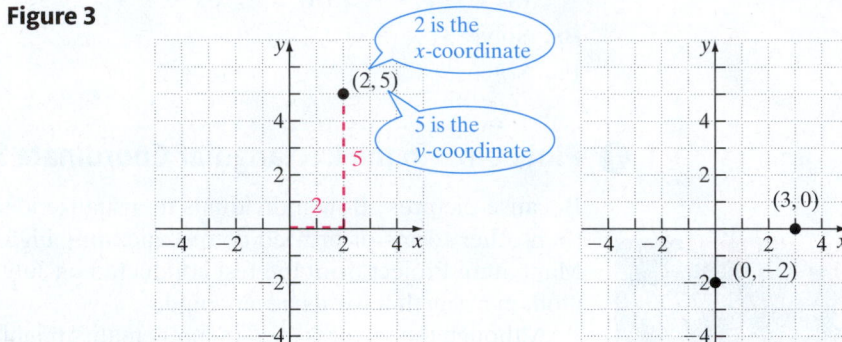

(a)   (b)

The origin O has coordinates $(0, 0)$. Any point on the x-axis has coordinates of the form $(x, 0)$, and any point on the y-axis has coordinates of the form $(0, y)$. For example, the points with coordinates $(3, 0)$ and $(0, -2)$, respectively, are shown in Figure 3(b).

If point P has coordinates (x, y), then x is called the **x-coordinate** of P, and y is called the **y-coordinate** of P.

The x- and y-axes divide the plane into four separate regions or **quadrants.** In quadrant I, both the x-coordinate and the y-coordinate are positive. In quadrant II, x is negative and y is positive. In quadrant III, both x and y are negative. In quadrant IV, x is positive and y is negative. Points on the coordinate axes do not belong to a quadrant. See Figure 4.

Figure 4

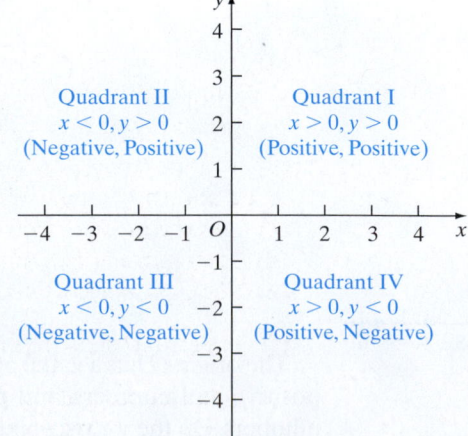

Let's summarize what we've learned so far.

> **The Rectangular Coordinate System**
>
> - Composed of two real number lines—one horizontal (the x-axis) and one vertical (the y-axis). The x- and y-axes intersect at the origin.
> - Also called the Cartesian coordinate system or xy-plane.
> - Points in the rectangular coordinate system are denoted (x, y) and are called the coordinates of the point. The notation (x, y) is referred to as an ordered pair. We call x the x-coordinate and y the y-coordinate.
> - If both x and y are positive, the point lies in quadrant I; if x is negative and y is positive, the point lies in quadrant II; if x is negative and y is negative, the point lies in quadrant III; if x is positive and y is negative, the point lies in quadrant IV.
> - Points on the x-axis have a y-coordinate of 0; points on the y-axis have an x-coordinate of 0. Points on the x- or y-axis do not lie in a quadrant.

EXAMPLE 1

Plotting Points in the Rectangular Coordinate System

Plot the following ordered pairs in the rectangular coordinate system. Tell in which quadrant each point lies or state that the point lies on the x- or y-axis.

(a) $A(3, 1)$ **(b)** $B(-4, 2)$ **(c)** $C(3, -5)$

(d) $D(4, 0)$ **(e)** $E(0, -3)$ **(f)** $F\left(-\dfrac{5}{2}, -\dfrac{7}{2}\right)$

Solution

Before plotting the points, first draw a rectangular or Cartesian coordinate system. See Figure 5(a). Then plot the points.

Figure 5

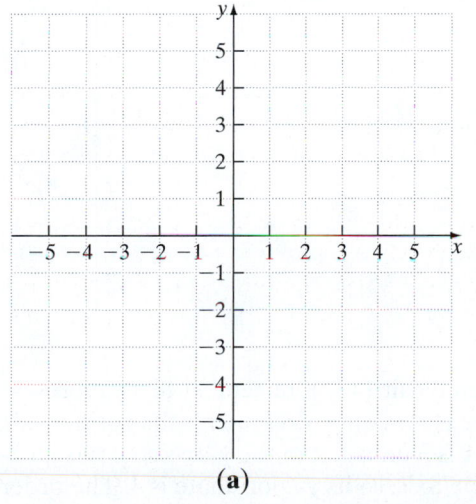

(a)

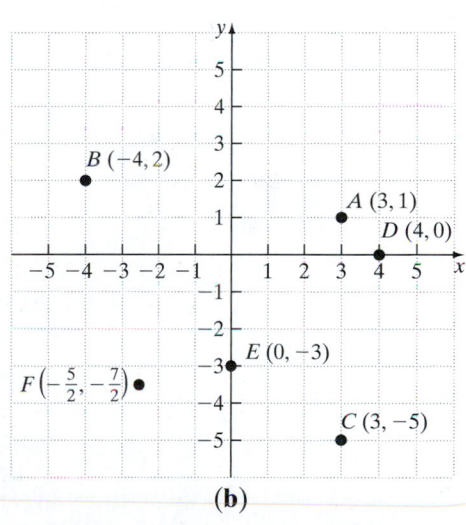

(b)

(a) To plot the point whose coordinates are $A(3, 1)$, begin at the origin O and travel 3 units to the right and then 1 unit up. Label the point A. Point A lies in quadrant I because both x and y are positive. See Figure 5(b).

(b) To plot $B(-4, 2)$, begin at the origin O and travel 4 units to the left and 2 units up. Label the point B. See Figure 5(b). Point B lies in quadrant II.

(c) See Figure 5(b). Point C lies in quadrant IV.

(d) See Figure 5(b). Point D lies on the x-axis, not in a quadrant.

(e) See Figure 5(b). Point E lies on the y-axis, not in a quadrant.

(f) It helps to convert the fractions into decimals, so $F\left(-\dfrac{5}{2}, -\dfrac{7}{2}\right) = F(-2.5, -3.5)$.

The x-coordinate of the point is halfway between -3 and -2; the y-coordinate is halfway between -4 and -3. See Figure 5(b). Point F lies in quadrant III. ●

Quick ✓

1. In the rectangular coordinate system, the horizontal real number line is called the _____ and the vertical real number line is called the _____. The point where these two axes intersect is called the _____.

2. If $(5, -1)$ are the coordinates of a point P, then 5 is called the _____ of P and -1 is called the _____ of P.

3. *True or False* The point whose ordered pair is $(-2, 4)$ is located in quadrant IV.

4. *True or False* The ordered pairs $(7, 4)$ and $(4, 7)$ represent the same point in the Cartesian plane.

In Problems 5 and 6, plot the ordered pairs in the rectangular coordinate system. Tell in which quadrant each point lies, or state that the point lies on the x-axis or y-axis.

5. **(a)** $(5, 2)$ **(b)** $(-4, -3)$ **(c)** $(1, -3)$ **(d)** $(-2, 0)$ **(e)** $(0, 6)$ **(f)** $\left(-\dfrac{3}{2}, \dfrac{5}{2}\right)$

6. **(a)** $(-6, 2)$ **(b)** $(1, 7)$ **(c)** $(-3, -2)$ **(d)** $(4, 0)$ **(e)** $(0, -1)$ **(f)** $\left(\dfrac{3}{2}, -\dfrac{7}{2}\right)$

EXAMPLE 2 **Identifying Points in the Rectangular Coordinate System**

Identify the coordinates of each point labeled in Figure 6.

Figure 6

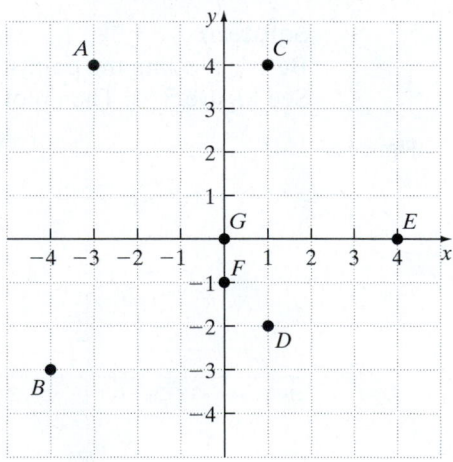

Solution

In an ordered pair (x, y), remember that x represents the position of the point left or right of the y-axis, while y represents the position of the point above or below the x-axis.

Point A is 3 units left of the y-axis so it has an x-coordinate of -3; point A is 4 units above the x-axis, so its y-coordinate is 4. The ordered pair $(-3, 4)$ corresponds to point A. Find the remaining coordinates in a similar fashion.

Point	Position	Ordered Pair
A	3 units left of the y-axis, 4 units above the x-axis	$(-3, 4)$
B	4 units left of the y-axis, 3 units below the x-axis	$(-4, -3)$
C	1 unit right of the y-axis, 4 units above the x-axis	$(1, 4)$
D	1 unit right of the y-axis, 2 units below the x-axis	$(1, -2)$
E	4 units right of the y-axis, on the x-axis	$(4, 0)$
F	On the y-axis, 1 unit below the x-axis	$(0, -1)$
G	On the y-axis, on the x-axis; origin	$(0, 0)$

Quick ✓

7. Identify the coordinates of each point labeled in the figure below.

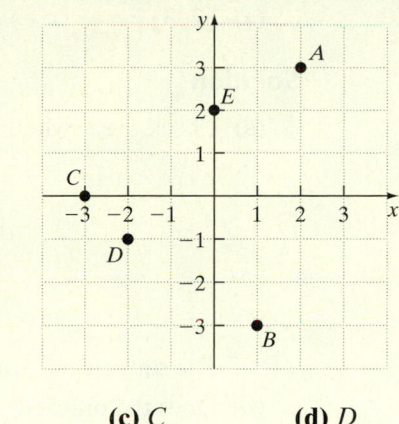

(a) A (b) B (c) C (d) D (e) E

▶ ❷ Determine Whether an Ordered Pair Satisfies an Equation

Recall from Sections 2.1–2.3, that the solution of an equation in one variable is either a single value of the variable (conditional equation), the empty set (contradiction), or all real numbers (identity). Now a method for representing the solution to an *equation in two variables* will be presented.

> **Definition**
>
> An **equation in two variables,** x and y, is a statement in which the algebraic expressions involving x and y are equal. The expressions are called **sides** of the equation.

For example, the following statements are all equations in two variables.

$$x + 2y = 6 \qquad y = -3x + 7 \qquad y = x^2 + 3$$

Since an equation is a statement, it may be true or false, depending on the values of the variables. Any values of the variables that make the equation a true statement **satisfy** the equation.

The first equation, $x + 2y = 6$, is satisfied when $x = 4$ and $y = 1$.

$$x + 2y = 6$$

Let $x = 4$ and $y = 1$: $\quad 4 + 2(1) \stackrel{?}{=} 6$

$$6 = 6 \quad \text{True}$$

We can also say that the ordered pair $(4, 1)$ satisfies the equation. Does $x = 8$ and $y = -1$ satisfy the equation $x + 2y = 6$?

$$x + 2y = 6$$

Let $x = 8$ and $y = -1$: $\quad 8 + 2(-1) \stackrel{?}{=} 6$

$$6 = 6 \quad \text{True}$$

So the ordered pair $(8, -1)$ satisfies the equation as well. In fact, infinitely many choices of x and y satisfy the equation $x + 2y = 6$, but some choices of x and y do not satisfy the equation. For example, $x = 3$ and $y = 4$ does not satisfy the equation $x + 2y = 6$.

$$x + 2y = 6$$

Let $x = 3$ and $y = 4$: $\quad 3 + 2(4) \stackrel{?}{=} 6$

$$11 = 6 \quad \text{False}$$

EXAMPLE 3 **Determining Whether an Ordered Pair Satisfies an Equation**

Determine whether the following ordered pairs satisfy the equation $x + 2y = 8$.

(a) $(2, 3)$ (b) $(6, -1)$ (c) $(-2, 5)$

Solution

(a) Check to see whether $x = 2$, $y = 3$ satisfies the equation $x + 2y = 8$.

$$x + 2y = 8$$
$$\text{Let } x = 2 \text{ and } y = 3: \quad 2 + 2(3) \overset{?}{=} 8$$
$$2 + 6 \overset{?}{=} 8$$
$$8 = 8 \quad \text{True}$$

The statement is true, so $(2, 3)$ satisfies the equation $x + 2y = 8$.

(b) Does the ordered pair $(6, -1)$, satisfy the equation?

$$x + 2y = 8$$
$$\text{Let } x = 6 \text{ and } y = -1: \quad 6 + 2(-1) \overset{?}{=} 8$$
$$6 - 2 \overset{?}{=} 8$$
$$4 = 8 \quad \text{False}$$

The statement $4 = 8$ is false, so $(6, -1)$ does not satisfy the equation.

(c) Does the ordered pair $(-2, 5)$, satisfy the equation?

$$x + 2y = 8$$
$$\text{Let } x = -2 \text{ and } y = 5: \quad -2 + 2(5) \overset{?}{=} 8$$
$$-2 + 10 \overset{?}{=} 8$$
$$8 = 8 \quad \text{True}$$

The statement is true, so $(-2, 5)$ satisfies the equation. ●

Quick ✓

8. *True or False* An equation in two variables can have more than one solution.

9. Determine whether the following ordered pairs satisfy the equation $x + 4y = 12$. Answer Yes or No.

(a) $(4, 2)$ (b) $(-2, 4)$ (c) $(1, 8)$

10. Determine whether the following ordered pairs satisfy the equation $y = 4x + 3$. Answer Yes or No.

(a) $(1, 3)$ (b) $(-2, -5)$ (c) $\left(-\dfrac{3}{2}, -3\right)$

▶ ❸ **Create a Table of Values That Satisfy an Equation**

In Example 3 we learned how to determine whether a given ordered pair satisfies an equation. Now, we learn how to find an ordered pair that satisfies an equation.

EXAMPLE 4 **How to Find an Ordered Pair That Satisfies an Equation**

Find an ordered pair that satisfies the equation $3x + y = 5$.

Step-by-Step Solution

Step 1: Choose any value for one of the variables in the equation.

Choose any value of x or y that you wish. For example, let $x = 2$.

Step 2: Substitute the chosen value of the variable into the equation. Solve for the remaining variable.

Substitute 2 for x in $3x + y = 5$ and then solve for y.

$$3x + y = 5$$

Let $x = 2$: $\quad 3(2) + y = 5$

Simplify: $\quad 6 + y = 5$

Subtract 6 from both sides: $\quad 6 - 6 + y = 5 - 6$

$$y = -1$$

One ordered pair that satisfies the equation is $(2, -1)$.   ●

Quick ✓

In Problems 11–13, determine an ordered pair that satisfies the given equation by substituting the given value of the variable into the equation.

11. $2x + y = 10; x = 3$ **12.** $-3x + 2y = 11; y = 1$ **13.** $4x + 3y = 0; x = \dfrac{1}{2}$

Look again at Example 3. Did you notice that two different ordered pairs satisfy the equation? In fact, an infinite number of ordered pairs satisfy the equation $x + 2y = 8$ because for any real number y, we can find a value of x that makes the equation a true statement. One way to find some of the solutions of an equation in two variables is to create a table of values that satisfy the equation. The table is created by choosing values of x and using the equation to find the corresponding value of y (as in Example 4) or by choosing a value of y and using the equation to find the corresponding value of x.

EXAMPLE 5   **Creating a Table of Values That Satisfy an Equation**

Use the equation $y = -2x + 5$ to complete Table 1, and then use the table to list some of the ordered pairs that satisfy the equation.

Solution

The first entry in the table is $x = -2$. Substitute -2 for x and use the equation $y = -2x + 5$ to find y.

$$y = -2x + 5$$

Let $x = -2$: $\quad y = -2(-2) + 5$

$$y = 4 + 5$$

$$y = 9$$

Now substitute 0 for x in the equation $y = -2x + 5$.

$$y = -2x + 5$$

Let $x = 0$: $\quad y = -2(0) + 5$

$$y = 0 + 5$$

$$y = 5$$

Finally, substitute 1 for x in the equation $y = -2x + 5$.

$$y = -2x + 5$$

Let $x = 1$: $\quad y = -2(1) + 5$

$$y = -2 + 5$$

$$y = 3$$

The completed table is shown as Table 2. Three ordered pairs that satisfy the equation are $(-2, 9)$, $(0, 5)$, and $(1, 3)$.   ●

Table 1

x	y = −2x + 5	(x, y)
−2		
0		
1		

Table 2

x	y = −2x + 5	(x, y)
−2	9	(−2, 9)
0	5	(0, 5)
1	3	(1, 3)

Quick ✔

In Problems 14 and 15, use the equation to complete the table. Use the table to list some of the ordered pairs that satisfy the equation.

14. $y = 5x - 2$

x	y = 5x − 2	(x, y)
−2		
0		
1		

15. $y = -3x + 4$

x	y = −3x + 4	(x, y)
−1		
2		
5		

EXAMPLE 6

Creating a Table of Values That Satisfy an Equation

Use the equation $2x - 3y = 12$ to complete Table 3, and then use the table to list some of the ordered pairs that satisfy the equation.

Table 3

x	y	(x, y)
−3		
	−4	
6		

Solution

The first entry in the table is $x = -3$. Substitute -3 for x and use the equation $2x - 3y = 12$ to find y.

$$2x - 3y = 12$$

Let $x = -3$: $\quad 2(-3) - 3y = 12$

$$-6 - 3y = 12$$

Add 6 to both sides: $\quad -3y = 18$

Divide both sides by −3: $\quad y = -6$

Substitute -4 for y in the equation $2x - 3y = 12$.

$$2x - 3y = 12$$

Let $y = -4$: $\quad 2x - 3(-4) = 12$

$$2x + 12 = 12$$

Subtract 12 from both sides: $\quad 2x = 0$

Divide both sides by 2: $\quad x = 0$

Substitute 6 for x in the equation $2x - 3y = 12$.

$$2x - 3y = 12$$

Let $x = 6$: $\quad 2(6) - 3y = 12$

$$12 - 3y = 12$$

Subtract 12 from both sides: $\quad -3y = 0$

Divide both sides by −3: $\quad y = 0$

Table 4 shows that three ordered pairs that satisfy the equation are $(-3, -6)$, $(0, -4)$, and $(6, 0)$. ●

Table 4

x	y	(x, y)
−3	−6	(−3, −6)
0	−4	(0, −4)
6	0	(6, 0)

Quick ✔

In Problems 16 and 17, use the equation to complete the table. Use the table to list some of the ordered pairs that satisfy the equation.

16. $2x + y = -8$

x	y	(x, y)
−5		
	−4	
2		

17. $2x - 5y = 18$

x	y	(x, y)
−6		
	−4	
2		

In modeling a situation, variables other than x and y can be used. For example, the equation $C = 1.20m + 3$ might be used to represent the cost of taking a taxi, where C represents the cost (in dollars) and m represents the number of miles driven.

Recall from Section 1.3 that the scale of a number line refers to the distance between tick marks on the number line. In Figures 2 through 6 of this section, a scale of 1 was used on both the x-axis and the y-axis. In applications, a different scale is often used on each axis.

 EXAMPLE 7   **An Electric Bill**

In North Carolina, Duke Power determines that the monthly electric bill for a household will be C dollars for using x kilowatt-hours (kWh) of electricity using the formula

$$C = 11.80 + 0.093581x$$

(SOURCE: *Duke Power*)

(a) Complete Table 5, and use the table to list ordered pairs that satisfy the equation. Express answers rounded to the nearest penny.

(b) Plot the ordered pairs (x, C) found in part (a) in a rectangular coordinate system.

Solution

(a) The first entry in the table is $x = 50$. Substitute 50 for x into the equation $C = 11.80 + 0.093581x$ to find C. Round C to two decimal places because C represents the cost, so the answer is rounded to the nearest penny.

$$C = 11.80 + 0.093581x$$

Let $x = 50$: $C = 11.80 + 0.093581(50)$

Use a calculator: $C = 16.48$

Now substitute 100 for x into the equation to find C.

$$C = 11.80 + 0.093581x$$

Let $x = 100$: $C = 11.80 + 0.093581(100)$

Use a calculator: $C = 21.16$

Now substitute 200 for x into the equation to find C.

$$C = 11.80 + 0.093581x$$

Let $x = 200$: $C = 11.80 + 0.093581(200)$

Use a calculator: $C = 30.52$

Table 6 shows the completed table. Three ordered pairs that satisfy the equation are $(50, 16.48)$, $(100, 21.16)$, and $(200, 30.52)$.

Table 5

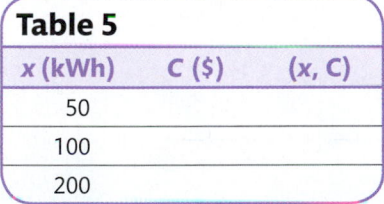

x (kWh)	C ($)	(x, C)
50		
100		
200		

Table 6

x (kWh)	C ($)	(x, C)
50	16.48	(50, 16.48)
100	21.16	(100, 21.16)
200	30.52	(200, 30.52)

(b) Because x represents the number of kilowatt-hours used, the horizontal axis is labeled x. Because C represents the bill, the vertical axis is labeled C. The ordered pairs found in part (a) are plotted in Figure 7. Different scales are used on the horizontal (scale = 25) and vertical (scale = 2) axes. ●

Figure 7

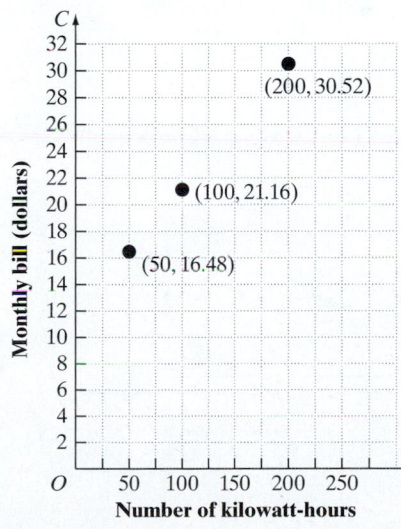

Notice in Figure 7 that the horizontal axis is labeled "Number of kilowatt-hours" and the vertical axis is labeled "Monthly bill (dollars)" so that it is clear what they represent. Labeling the axes is a good practice to follow whenever you draw a graph that models a problem, or when the coordinates represent data.

Quick ✓

18. Piedmont Natural Gas charges its customers in North Carolina a monthly fee for November through March of C dollars for using x therms of natural gas using the formula

$$C = 10 + 0.83692x$$

(SOURCE: *Piedmont Natural Gas*)

(a) Complete the table, and use the results to list ordered pairs (x, C) that satisfy the equation. Express answers rounded to the nearest penny.

x (kWh)	C ($)	(x, C)
50		
100		
150		

(b) Plot the ordered pairs found in part (a) in a rectangular coordinate system.

3.1 Exercises MyMathLab®

Exercise numbers in **green** have complete video solutions in MyMathLab or may be accessed using the QR code to the right.

*Problems **1–18** are the Quick ✓s that follow the **EXAMPLES**.*

Building Skills

In Problems 19–22, plot the following ordered pairs in the rectangular coordinate system. Tell in which quadrant each point lies or state that the point lies on the x-axis or y-axis. See Objective 1.

19. $A(-3, 2)$; $B(4, 1)$; $C(-2, -4)$; $D(5, -4)$; $E(-1, 3)$; $F(2, -4)$

20. $P(-3, -2)$; $Q(2, -4)$; $R(4, 3)$; $S(-1, 4)$; $T(-2, -4)$; $U(3, -3)$

21. $A\left(\frac{1}{2}, 0\right)$; $B\left(\frac{3}{2}, -\frac{1}{2}\right)$; $C\left(4, \frac{7}{2}\right)$; $D\left(0, -\frac{5}{2}\right)$; $E\left(\frac{9}{2}, 2\right)$; $F\left(-\frac{5}{2}, -\frac{3}{2}\right)$; $G(0, 0)$

22. $P\left(\frac{3}{2}, -2\right)$; $Q\left(0, \frac{5}{2}\right)$; $R\left(-\frac{9}{2}, 0\right)$; $S(0, 0)$; $T\left(-\frac{3}{2}, -\frac{9}{2}\right)$; $U\left(3, \frac{1}{2}\right)$; $V\left(\frac{5}{2}, -\frac{7}{2}\right)$

In Problems 23 and 24, plot the following ordered pairs in the rectangular coordinate system. Tell the location of each point: positive x-axis, negative x-axis, positive y-axis, or negative y-axis. See Objective 1.

23. $A(3, 0)$; $B(0, -1)$; $C(0, 3)$; $D(-4, 0)$

24. $P(0, -1)$; $Q(-2, 0)$; $R(0, 3)$; $S(1, 0)$

In Problems 25 and 26, identify the coordinates of each point labeled in the figure. Name the quadrant in which each point lies or state that the point lies on the x- or y-axis. See Objective 1.

25.

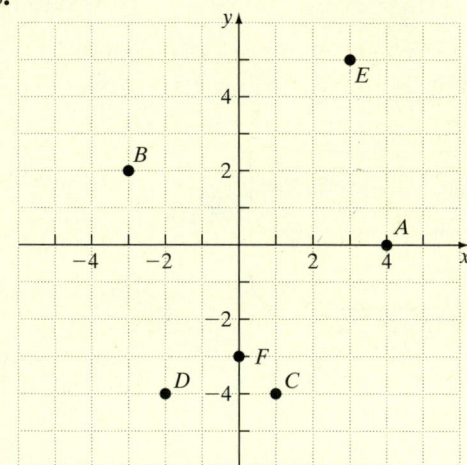

26.

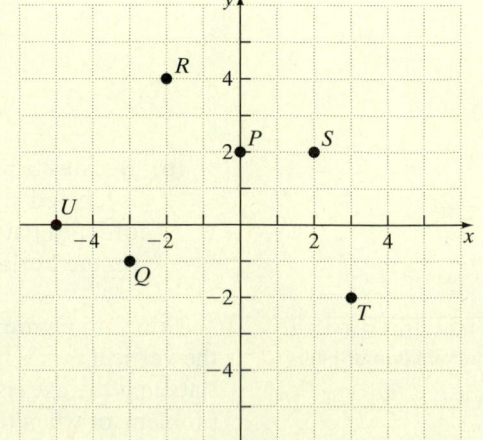

In Problems 27–32, determine whether or not the ordered pair satisfies the equation. Answer Yes or No. See Objective 2.

27. $y = -3x + 5$; $A(-2, -1)$ $B(2, -1)$ $C\left(\dfrac{1}{3}, 4\right)$

28. $y = 2x - 3$; $A(-1, -5)$ $B(4, -5)$ $C(-2, -7)$

29. $3x + 2y = 4$; $A(0, 2)$ $B(1, 0)$ $C(4, -4)$

30. $5x - y = 12$; $A(2, 0)$ $B(0, 12)$ $C(-2, -22)$

31. $\dfrac{4}{3}x + y - 1 = 0$; $A(3, -3)$ $B(-6, -9)$ $C\left(\dfrac{3}{4}, 0\right)$

32. $\dfrac{3}{4}x + 2y = 0$; $A\left(-4, \dfrac{3}{2}\right)$ $B(0, 0)$ $C\left(1, -\dfrac{3}{2}\right)$

For Problems 33–38, see Objective 3.

33. Find an ordered pair that satisfies the equation $x + y = 5$ by letting $x = 4$.

34. Find an ordered pair that satisfies the equation $x + y = 7$ by letting $x = 2$.

35. Find an ordered pair that satisfies the equation $2x + y = 9$ by letting $y = -1$.

36. Find an ordered pair that satisfies the equation $-4x - y = 5$ by letting $y = 7$.

37. Find an ordered pair that satisfies the equation $-3x + 2y = 15$ by letting $x = -3$.

38. Find an ordered pair that satisfies the equation $5x - 3y = 11$ by letting $y = 3$.

In Problems 39–52, use the equation to complete the table. Use the table to list some of the ordered pairs that satisfy the equation. See Objective 3.

39. $y = -x$

x	y	(x, y)
-3		
0		
1		

40. $y = x$

x	y	(x, y)
-4		
0		
2		

41. $y = -3x + 1$

x	y	(x, y)
-2		
-1		
4		

42. $y = 4x - 5$

x	y	(x, y)
-3		
1		
2		

43. $2x + y = 6$

x	y	(x, y)
-1		
2		
3		

44. $3x + 4y = 2$

x	y	(x, y)
-2		
2		
4		

45. $y = 6$

x	y	(x, y)
-4		
1		
12		

46. $x = 2$

x	y	(x, y)
	-4	
	0	
	8	

47. $x - 2y + 6 = 0$

x	y	(x, y)
1		
	1	
-2		

48. $2x + y - 4 = 0$

x	y	(x, y)
-1		
	-4	
	2	

49. $y = 5 + \dfrac{1}{2}x$

x	y	(x, y)
	7	
-4		
	2	

50. $y = 8 - \dfrac{1}{3}x$

x	y	(x, y)
	10	
9		
	27	

51. $\dfrac{x}{2} + \dfrac{y}{3} = -1$

x	y	(x, y)
0		
	0	
	-6	

52. $\dfrac{x}{5} - \dfrac{y}{2} = 1$

x	y	(x, y)
5	0	
	0	
	-5	

In Problems 53–64, for each equation find the missing value in the ordered pair. See Objective 3.

53. $y = -3x - 10$; $A(\underline{}, -16)$ $B(-3, \underline{})$ 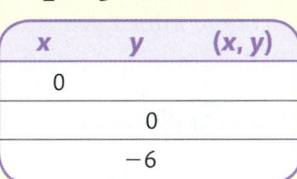 $C\left(\underline{}, -9\right)$

54. $y = 5x - 4$; $A(-1, \underline{})$ $B(\underline{}, 31)$ $C\left(-\dfrac{2}{5}, \underline{}\right)$

55. $x = -\dfrac{1}{3}y$; $A(2, \underline{})$ $B(\underline{}, 0)$ $C\left(\underline{}, -\dfrac{1}{2}\right)$

56. $y = \dfrac{2}{3}x$; $A\left(\underline{}, -\dfrac{8}{3}\right)$ $B(0, \underline{})$ $C\left(-\dfrac{5}{6}, \underline{}\right)$

57. $x = 4$; $A(\underline{}, -8)$ $B(\underline{}, -19)$ $C(\underline{}, 5)$

58. $y = -1$; $A(6, \underline{})$ $B(-1, \underline{})$ $C(0, \underline{})$

59. $y = \dfrac{2}{3}x + 2$; $A(\underline{}, 4)$ $B(-6, \underline{})$ $C\left(\dfrac{1}{2}, \underline{}\right)$

60. $y = -\dfrac{5}{4}x - 1$; $A(\underline{}, -6)$ $B(-8, \underline{})$ $C\left(\underline{}, -\dfrac{11}{6}\right)$

61. $\frac{1}{2}x - 3y = 2$; $A\left(-4, \underline{\quad}\right)$ $B\left(\underline{\quad}, -1\right)$

$C\left(-\frac{2}{3}, \underline{\quad}\right)$

62. $\frac{1}{3}x + 2y = -1$; $A\left(-4, \underline{\quad}\right)$ $B\left(\underline{\quad}, -\frac{3}{4}\right)$

$C\left(0, \underline{\quad}\right)$

63. $0.5x - 0.3y = 3.1$; $A(20, \underline{\quad})$ $B(\underline{\quad}, -17)$

$C(2.6, \underline{\quad})$

64. $-1.7x + 0.2y = -5$; $A(\underline{\quad}, -110)$ $B(40, \underline{\quad})$

$C(2.4, \underline{\quad})$

Applying the Concepts

65. Book Value Residential investment property such as apartment buildings may be depreciated over 28.5 years, according to the Internal Revenue Service (IRS). The IRS allows an investor to depreciate an apartment building whose value (excluding the land) is $285,000 by $10,000 per year.

The equation $V = -10,000x + 285,000$ represents the book value, V, of an apartment after x years.

(a) What will be the book value of the apartment building after 2 years?

(b) What will be the book value of the apartment building after 5 years?

(c) After how many years will the book value of the apartment building be $205,000?

(d) If (x, V) represents any ordered pair that satisfies $V = -10,000x + 285,000$, interpret the meaning of $(3, 255,000)$.

(e) Use the results from parts (a)–(c) to list ordered pairs (x, V) that satisfy $V = -10,000x + 285,000$. Plot the points in a rectangular coordinate system.

66. Taxi Ride The cost to take a taxi is $1.70 plus $2.00 per mile for each mile driven. The total cost in dollars, C, is given by the equation $C = 1.7 + 2m$, where m represents the total miles driven.

(a) How much will it cost to take a taxi 5 miles?

(b) How much will it cost to take a taxi 20 miles?

(c) If you spent $32.70 on cab fare, how far was your trip?

(d) If (m, C) represents any ordered pair that satisfies $C = 1.7 + 2m$, interpret the meaning of $(14, 29.7)$ in the context of this problem.

(e) Use the results from parts (a)–(c) to list ordered pairs (m, C) that satisfy $C = 1.7 + 2m$. Plot the points in a rectangular coordinate system.

67. College Graduates An equation to approximate the percentage of the U.S. population ages 25–34 with a bachelor's degree or higher can be given by the model $P = 0.38n + 30.5$, where n is the number of years after 2009.

(a) According to the model, what percentage of the U.S. population ages 25–34 had a bachelor's degree or higher in 2009 $(n = 0)$?

(b) According to the model, what percentage of the U.S. population ages 25–34 had a bachelor's degree or higher in 2015 $(n = 6)$?

(c) According to the model, what percentage of the U.S. population ages 25–34 will have a bachelor's degree in 2020 $(n = 11)$?

(d) In which year will 50% of the U.S. population ages 25–34 have a bachelor's degree or higher? Round your answer to the nearest year.

(e) According to the model, 100% of the U.S. population ages 25–34 will have a bachelor's degree or higher in 2192. Do you think this is reasonable? Why or why not?

 68. Life Expectancy The model $A = 0.183n + 67.895$ is used to estimate the life expectancy A of residents of the United States born n years after 1950.

(a) According to the model, what is the life expectancy, to the nearest year, for a person born in 1950?

(b) According to the model, what is the life expectancy, to the nearest year, for a person born in 1980 $(n = 30)$?

(c) If the model holds true for future generations, what is the life expectancy, to the nearest year, for a person born in 2020?

(d) If a person has a life expectancy of 77 years, to the nearest year, when was the person born?

(e) Do you think life expectancy will continue to increase in the future? What could happen that would change this model?

In Problems 69–72, use the equation to complete the table.
Use the table to list some of the ordered pairs that satisfy the equation.

69. $4a + 2b = -8$

a	b	(a, b)
2		
	−4	
	6	

70. $2r - 3s = 3$

r	s	(r, s)
	3	
	−1	
−3		

71. $\frac{2p}{5} + \frac{3q}{10} = 1$

p	q	(p, q)
0		
	0	
−10		

72. $\frac{4a}{3} + \frac{2b}{5} = -1$

a	b	(a, b)
	0	
0		
	10	

Extending the Concepts

In Problems 73–78, determine the value of k so that the given ordered pair is a solution to the equation.

73. Find the value of k for which $(1, 2)$ satisfies $y = -2x + k$.

74. Find the value of k for which $(-1, 10)$ satisfies $y = 3x + k$.

75. Find the value of k for which $(2, 9)$ satisfies $7x - ky = -4$.

76. Find the value of k for which $(3, -1)$ satisfies $4x + ky = 9$.

77. Find the value of k for which $\left(-8, -\dfrac{5}{2}\right)$ satisfies $kx - 4y = 6$.

78. Find the value of k for which $\left(-9, \dfrac{1}{2}\right)$ satisfies $kx + 2y = -2$.

In Problems 79–80, use the equation to complete the table. Choose any value for x and solve the resulting equation to find the corresponding value for y. Then plot these ordered pairs in a rectangular coordinate system. Connect the points and describe the figure.

79. $3x - 2y = -6$

x	y	(x, y)

80. $-x + y = 4$

x	y	(x, y)

In Problems 81–84, use the equation to complete the table. Then plot the points in a rectangular coordinate system.

81. $y = x^2 - 4$

x	y	(x, y)
−2		
−1		
0		
1		
2		

82. $y = -x^2 + 3$

x	y	(x, y)
−2		
−1		
0		
1		
2		

83. $y = -x^3 + 2$

x	y	(x, y)
−2		
−1		
0		
1		
2		

84. $y = 2x^3 - 1$

x	y	(x, y)
−2		
−1		
0		
1		
2		

Explaining the Concepts

85. Describe how the quadrants in the rectangular coordinate system are labeled and how you can determine the quadrant in which a point lies. Describe the characteristics of a point that lies on either the *x*- or *y*-axis.

86. Describe how to plot the point whose coordinates are $(3, -5)$.

Technology Exercises

Graphing calculators can also create tables of values that satisfy an equation. To do this, first solve the equation for *y*. For example, to obtain a table of values that satisfy the equation $2x - 3y = 12$ (Example 6), solve for *y* as follows:

$$2x - 3y = 12$$

Subtract 2x from both sides: $-3y = -2x + 12$

Divide both sides by -3: $y = \dfrac{-2x + 12}{-3}$

Divide each term in the numerator by -3: $y = \dfrac{-2x}{-3} + \dfrac{12}{-3}$

Simplify: $y = \dfrac{2}{3}x - 4$

Now enter the equation $y = \dfrac{2}{3}x - 4$ into the calculator and create the table shown.

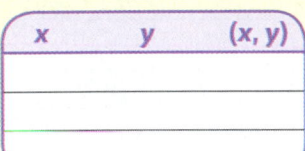

In Problems 87–94, use a graphing calculator to create a table of values that satisfy each equation. Have the table begin at −3 and increase by 1.

87. $y = 2x - 9$

88. $y = -3x + 8$

89. $x + y = 8$

90. $-2x + y = -4$

91. $y + 2x = 13$

92. $y - x = -15$

93. $y = -6x^2 + 1$

94. $y = -x^2 + 3x$

3.2 Graphing Equations in Two Variables

Objectives

1 Graph a Line by Plotting Points
2 Graph a Line Using Intercepts
3 Graph Vertical and Horizontal Lines

Are You Prepared for This Section?

Before getting started, complete the following problems. If you get a problem wrong, go back to the section cited and review the material.

P1. Solve: $4x = 24$ [Section 2.1, pp. 85–87]
P2. Solve: $-3y = 18$ [Section 2.1, pp. 85–87]
P3. Solve: $2x + 5 = 13$ [Section 2.2, pp. 91–92]

▶ 1 Graph a Line by Plotting Points

In the previous section, we found values of x and y that satisfy an equation. What does this mean? Well, it means that the ordered pair (x, y) is a point on the graph of the equation.

In Other Words

The graph of an equation is a geometric way of representing the set of all ordered pairs that make the equation a true statement. Think of the graph as a picture of the solution set.

Definition

The **graph of an equation in two variables** x and y is the set of points whose coordinates, (x, y), in the xy-plane satisfy the equation.

But how do we obtain the graph of an equation? One method for graphing an equation is the **point-plotting method.**

EXAMPLE 1 **How to Graph an Equation Using the Point-Plotting Method**

Graph the equation $y = 2x - 3$ using the point-plotting method.

Step-by-Step Solution

Step 1: Find ordered pairs that satisfy the equation by choosing some values of x and using the equation to find the corresponding values of y. See Table 7.

Table 7

x	$y = 2x - 3$	(x, y)
-2	$y = 2(-2) - 3$ $= -4 - 3$ $= -7$	$(-2, -7)$
-1	$y = 2(-1) - 3$ $= -5$	$(-1, -5)$
0	$y = 2(0) - 3$ $= -3$	$(0, -3)$
1	$y = 2(1) - 3$ $= -1$	$(1, -1)$
2	$y = 2(2) - 3$ $= 1$	$(2, 1)$

Step 2: Plot the points whose coordinates, (x, y), were found in Step 1 in a rectangular coordinate system. See Figure 8.

Figure 8

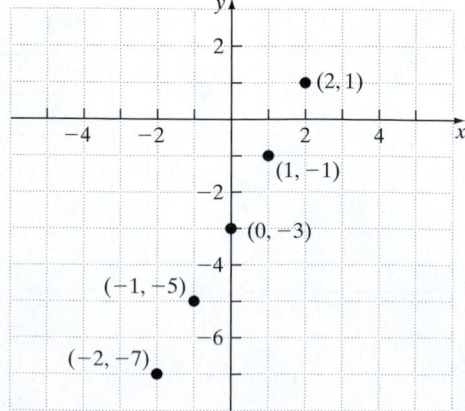

Step 3: Connect the points with a straight line. Put arrows on both ends of the graph to indicate the graph continues in the pattern shown. See Figure 9.

Figure 9
$y = 2x - 3$

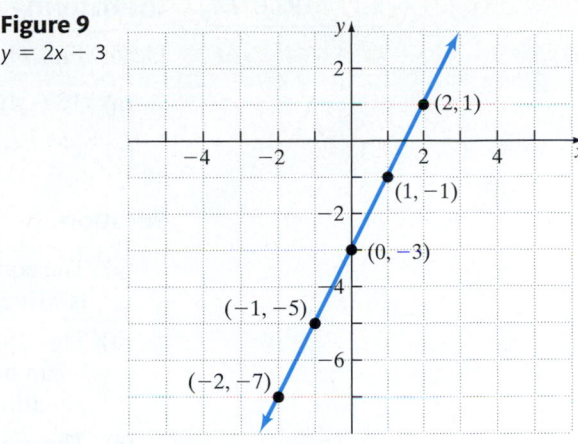

Work Smart

Remember, all coordinates of the points on the graph shown in Figure 9 satisfy the equation $y = 2x - 3$.

The graph of the equation in Figure 9 does not show all the points that satisfy the equation. For example, the point $(5, 7)$ is a part of the graph of $y = 2x - 3$ but is not shown in Figure 9. Because the graph of $y = 2x - 3$ can be extended, arrows are used to indicate that the pattern shown continues. It is important to show enough of the graph so that anyone can "see" how it continues. This is called a **complete graph.**

For the remainder of the text, we will say "the point (x, y)" for short rather than "the point whose coordinates are (x, y)."

Graphing an Equation Using the Point-Plotting Method

Step 1: Find several ordered pairs that satisfy the equation.

Step 2: Plot the points found in Step 1 in a rectangular coordinate system.

Step 3: Connect the points with a smooth curve or line.

Quick ✓

1. The _____ of an equation in two variables is the set of points whose coordinates, (x, y) satisfy the equation.

In Problems 2 and 3, draw a complete graph of each equation using point plotting.

2. $y = 3x - 2$ 3. $y = -4x + 8$

A question you may be asking yourself is "How many points do I need to find before I can be sure that I have a complete graph?" It depends on the type of equation you are graphing.

Definition

A **linear equation in two variables** is an equation that can be written in the form

$$Ax + By = C$$

where A, B, and C are real numbers. A and B cannot both be 0. A linear equation written in the form $Ax + By = C$ is said to be in **standard form.***

The equation from Example 1, $y = 2x - 3$, is a linear equation in two variables because it can be written as $-2x + y = -3$ by subtracting $2x$ from both sides of the equation.

---
* Some texts call $Ax + By = C$ the **general form** of a line.

EXAMPLE 2

Identifying Linear Equations in Two Variables

Determine whether the equation is a linear equation in two variables.

(a) $3x - 4y = 9$ (b) $3x + 2y - 9 = 0$

(c) $x^2 + 5y = 10$ (d) $-2y + 5 = 0$

Solution

(a) The equation $3x - 4y = 9$ is a linear equation in two variables because it is written in the form $Ax + By = C$ with $A = 3$, $B = -4$, and $C = 9$.

(b) The equation $3x + 2y - 9 = 0$ is a linear equation in two variables because it can be written in the form $Ax + By = C$ by adding 9 to both sides of the equation to obtain $3x + 2y = 9$. So $A = 3$, $B = 2$, and $C = 9$.

(c) The equation $x^2 + 5y = 10$ is not a linear equation because x is squared.

(d) The equation $-2y + 5 = 0$ is a linear equation in two variables because it can be written in the form $Ax + By = C$ by subtracting 5 to both sides of the equation, obtaining $-2y = -5$. Then $A = 0$, $B = -2$, and $C = -5$. ●

Quick ✓

4. A(n) _____ equation in two variables is an equation that can be written in the form $Ax + By = C$, where A, B, and C are real numbers, and A and B are not both zero. Equations written in this form are said to be in _____ ____.

In Problems 5–7, determine whether or not the equation is a linear equation in two variables. Answer Linear or Not linear.

5. $4x - y = 12$ 6. $5x - y^2 = 10$

7. $5x - 20 = 0$

Work Smart

When graphing a line, we recommend that you find at least three points to be sure your graph is correct.

For the remainder of the text, we will refer to linear equations in two variables as **linear equations.** The graph of a linear equation is a **line.** To graph a linear equation requires only two points; however, finding a third point as a check is recommended.

 EXAMPLE 3

Graphing a Linear Equation Using the Point-Plotting Method

Graph the linear equation $2x + y = 4$.

Solution

Because the coefficient of y is 1, it is easier to choose values of x and find the corresponding values of y. For example, determine the value of y for $x = -2, 0$, and 2. There is nothing magical about these choices. Any three different values of x will give us the desired results.

Work Smart

Choose values of x (or y) that make the algebra easy.

$x = -2$:	$x = 0$:	$x = 2$:
$2x + y = 4$	$2x + y = 4$	$2x + y = 4$
Let $x = -2$: $2(-2) + y = 4$	Let $x = 0$: $2(0) + y = 4$	Let $x = 2$: $2(2) + y = 4$
$-4 + y = 4$	$0 + y = 4$	$4 + y = 4$
$y = 8$	$y = 4$	$y = 0$

Plot the three points from Table 8 and connect them with a straight line. See Figure 10.

Table 8

x	y	(x, y)
−2	8	(−2, 8)
0	4	(0, 4)
2	0	(2, 0)

Figure 10

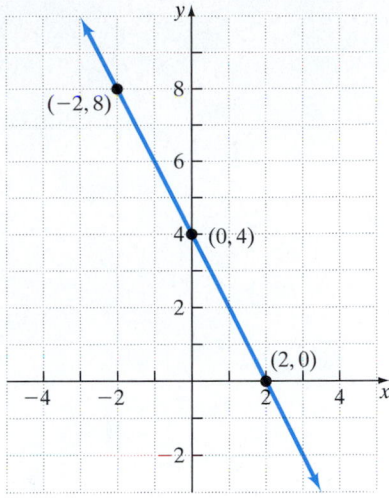

EXAMPLE 4

Cost of Renting a Car

Table 9

m	C	(m, C)
0		
50		
100		

A car-rental agency quotes you the cost of renting a car in Washington, D.C., as $30 per day plus $0.20 per mile. The linear equation $C = 0.20m + 30$ models the cost, where C represents total cost, in dollars, and m represents the number of miles that are traveled in one day.

(a) Complete Table 9 to find ordered pairs that satisfy the equation.

(b) Graph the linear equation $C = 0.20m + 30$ using the points obtained in part (a).

Solution

(a) The first entry in the table is $m = 0$. Substitute 0 for m in the equation $C = 0.20m + 30$ to find C.

$m = 0$:
$$C = 0.20m + 30$$
$$C = 0.20(0) + 30$$
$$C = 30$$

Now substitute 50 for m in the equation $C = 0.20m + 30$ to find C.

$m = 50$:
$$C = 0.20m + 30$$
$$C = 0.20(50) + 30$$
$$C = 40$$

Now substitute 100 for m in the equation to find C.

$m = 100$:
$$C = 0.20m + 30$$
$$C = 0.20(100) + 30$$
$$C = 50$$

Table 10

m	C	(m, C)
0	30	(0, 30)
50	40	(50, 40)
100	50	(100, 50)

Table 10 shows the completed table. The ordered pair $(50, 40)$ means that if a car is driven 50 miles in a day, the rental cost will be $40.

(b) Because m represents the number of miles driven, label the horizontal axis m, "Number of miles." Label the vertical axis C, "Cost (dollars)," to be clear as to what they represent. Let the scale of the horizontal axis be 10, so

(continued)

each tick mark represents 10 miles. Scale the vertical axis to 5, so each tick mark represents 5 dollars. Scaling in this way makes it easier to plot the ordered pairs. In Figure 11, plot the points found in part (a) and then draw the line. The line is drawn only for $m \geq 0$ because m represents the number of miles driven, which cannot be negative.

Figure 11

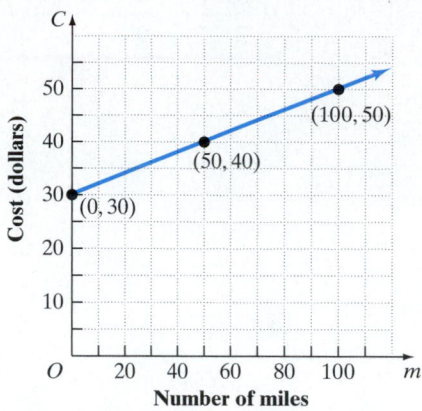

Quick ✔

11. Michelle sells computers. Her monthly salary is $3000 plus 8% of total sales. The linear equation $S = 0.08x + 3000$ models Michelle's monthly salary, S, where x represents her total sales in the month.

(a) Complete the table and use the results to list ordered pairs that satisfy the equation.

x	S	(x, S)
0		
10,000		
25,000		

(b) Graph the linear equation $S = 0.08x + 3000$ using the points obtained in part (a).

▶ ❷ Graph a Line Using Intercepts

Intercepts should always be displayed in a complete graph.

Figure 12

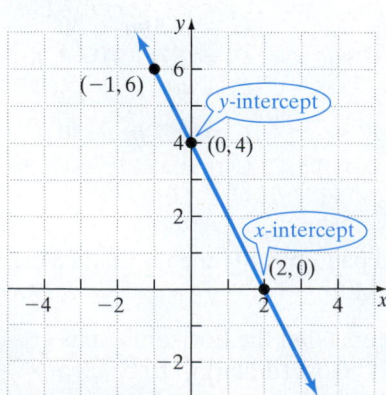

Definitions

The **intercepts** are the coordinates of the points, if any, where a graph crosses or touches a coordinate axis. A point at which the graph crosses or touches the x-axis is an **x-intercept,** and a point at which the graph crosses or touches the y-axis is a **y-intercept.**

Figure 12 illustrates the definition. The graph in Figure 12 is the graph of $2x + y = 4$, found in Example 3.

EXAMPLE 5 | **Finding Intercepts from a Graph**

Find the intercepts of the graphs shown in Figures 13(a) and 13(b). What are the x-intercept(s)? What are the y-intercept(s)?

> **In Other Words**
> An x-intercept exists when $y = 0$.
> A y-intercept exists when $x = 0$.

Figure 13

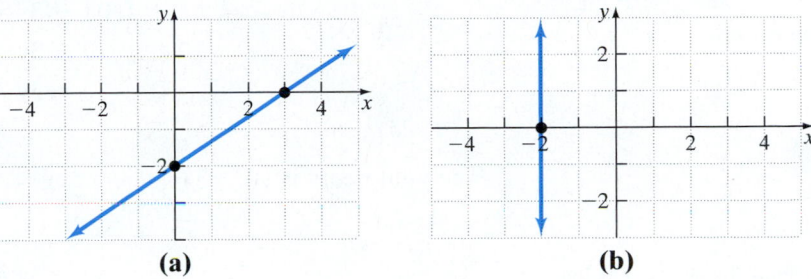

(a)   (b)

Solution

(a) The intercepts of the graph in Figure 13(a) are the points $(0, -2)$ and $(3, 0)$. The x-intercept is $(3, 0)$. The y-intercept is $(0, -2)$.

(b) The intercept of the graph in Figure 13(b) is the point $(-2, 0)$. The x-intercept is $(-2, 0)$. There are no y-intercepts. ●

> **Quick ✓**
>
> 12. The _____ are the points, if any, where a graph crosses or touches a coordinate axis.
>
> *In Problems 13 and 14, find the intercepts of the graphs shown in the figures. What are the x-intercept(s)? What are the y-intercept(s)?*
>
> **13.** 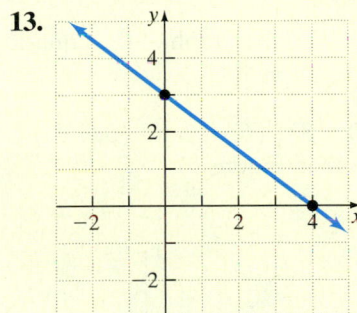   **14.**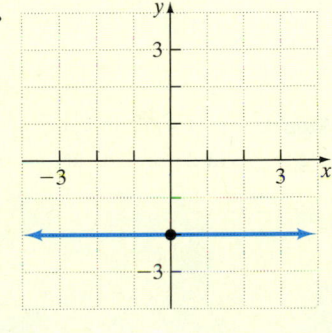

▶ From Figure 12, it should be clear that an x-intercept exists when the value of y is 0 and that a y-intercept exists when the value of x is 0. This leads to the following procedure for finding the intercepts algebraically.

Work Smart

Every point on the x-axis has a y-coordinate of 0. That's why we set $y = 0$ to find an x-intercept. Likewise, every point on the y-axis has an x-coordinate of 0. That's why we set $x = 0$ to find a y-intercept.

> **Procedure for Finding Intercepts**
>
> 1. To find the x-intercept(s), if any, of the graph of an equation, let $y = 0$ in the equation and solve for x.
>
> 2. To find the y-intercept(s), if any, of the graph of an equation, let $x = 0$ in the equation and solve for y.

The intercepts can be used to graph a line. Because the intercepts represent only two points, we find a third point so that we can check our work.

EXAMPLE 6   **How to Graph a Linear Equation by Finding Its Intercepts**

Graph the linear equation $4x - 3y = 24$ by finding its intercepts.

Step-by-Step Solution

Step 1: Find the y-intercept by letting $x = 0$ and solving the equation for y.

$$4x - 3y = 24$$
$$\text{Let } x = 0: \quad 4(0) - 3y = 24$$
$$0 - 3y = 24$$
$$-3y = 24$$
$$\text{Divide both sides by } -3: \quad y = -8$$

The y-intercept is $(0, -8)$.

Step 2: Find the x-intercept by letting $y = 0$ and solving the equation for x.

$$4x - 3y = 24$$
$$\text{Let } y = 0: \quad 4x - 3(0) = 24$$
$$4x - 0 = 24$$
$$4x = 24$$
$$\text{Divide both sides by } 4: \quad x = 6$$

The x-intercept is $(6, 0)$.

Step 3: Find one additional point on the graph by choosing any value of x that is convenient and solving the equation for y.

Let $x = 3$ and solve the equation $4x - 3y = 24$ for y.

$$\text{Let } x = 3: \quad 4(3) - 3y = 24$$
$$12 - 3y = 24$$
$$\text{Subtract 12 from both sides:} \quad -3y = 12$$
$$\text{Divide both sides by } -3: \quad y = -4$$

The point $(3, -4)$ is on the graph of the equation.

Step 4: Plot the points found in Steps 1–3 and draw the line.

Plot the points $(0, -8)$, $(6, 0)$, and $(3, -4)$. Connect the points with a straight line and obtain the graph in Figure 14.

Figure 14
$4x - 3y = 24$

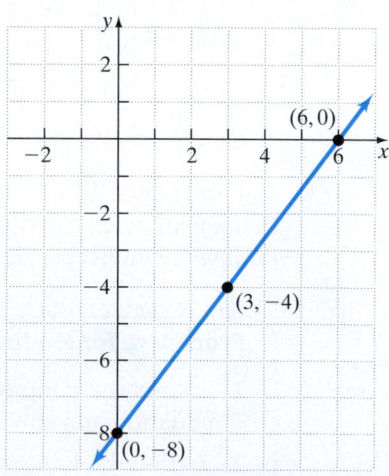

EXAMPLE 7   **Graphing a Linear Equation by Finding Its Intercepts**

Graph the linear equation $\frac{1}{2}x - 2y = 3$ by finding its intercepts.

Solution

y-intercept:

$$\frac{1}{2}x - 2y = 3$$

Let $x = 0$: $\frac{1}{2}(0) - 2y = 3$

$$-2y = 3$$

Divide both sides by -2: $y = -\frac{3}{2}$

The y-intercept is $\left(0, -\frac{3}{2}\right)$.

x-intercept:

$$\frac{1}{2}x - 2y = 3$$

Let $y = 0$: $\frac{1}{2}x - 2(0) = 3$

$$\frac{1}{2}x = 3$$

Multiply both sides by 2: $x = 6$

The x-intercept is $(6, 0)$.

Additional point (choose $x = 2$):

$$\frac{1}{2}x - 2y = 3$$

Let $x = 2$: $\frac{1}{2}(2) - 2y = 3$

$$1 - 2y = 3$$

Subtract 1 from both sides: $-2y = 2$

Divide both sides by -2: $y = -1$

The additional point is $(2, -1)$.

The points $(6, 0)$, $\left(0, -\frac{3}{2}\right)$, and $(2, -1)$ are on the graph of the equation. Plot the points $(6, 0)$, $\left(0, -\frac{3}{2}\right)$, and $(2, -1)$ and connect them with a straight line. See Figure 15.

Figure 15

$\frac{1}{2}x - 2y = 3$

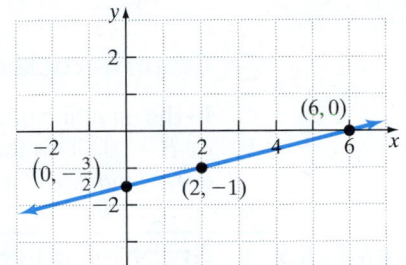

Quick ✔

15. *True or False* To find the y-intercept(s), if any, of the graph of an equation, let $y = 0$ in the equation and solve for x.

In Problems 16–18, graph each linear equation by finding its intercepts.

16. $x + y = 3$

17. $2x - 5y = 20$

18. $\frac{3}{2}x - 2y = 9$

EXAMPLE 8   **Graphing a Linear Equation of the Form $Ax + By = 0$**

Graph the linear equation $2x + 3y = 0$ by finding its intercepts.

Solution

y-intercept:

$$2x + 3y = 0$$

Let $x = 0$: $2(0) + 3y = 0$

$$3y = 0$$

Divide both sides by 3: $y = 0$

The y-intercept is $(0, 0)$.

x-intercept:

$$2x + 3y = 0$$

Let $y = 0$: $2x + 3(0) = 0$

$$2x + 0 = 0$$

Divide both sides by 2: $x = 0$

The x-intercept is $(0, 0)$.

(continued)

Work Smart
Linear equations of the form
$Ax + By = 0$, where $A \neq 0$ and
$B \neq 0$, have only one intercept at
$(0, 0)$, the origin, so two additional
points should be plotted to obtain
the graph.

Figure 16
$2x + 3y = 0$

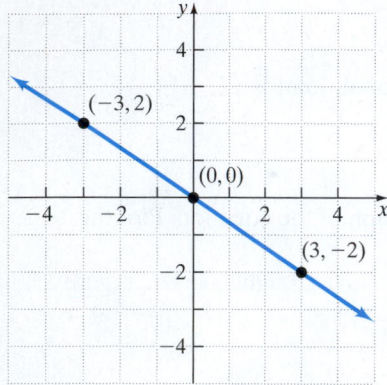

Because both the x- and y-intercepts are $(0, 0)$, find *two* additional points.

Additional point (choose $x = 3$):

$$2x + 3y = 0$$
Let $x = 3$: $2(3) + 3y = 0$
$$6 + 3y = 0$$
Subtract 6 from both sides: $\qquad 3y = -6$
Divide both sides by 3: $\qquad y = -2$

Additional point (choose $x = -3$):

$$2x + 3y = 0$$
Let $x = -3$: $2(-3) + 3y = 0$
$$-6 + 3y = 0$$
Add 6 to both sides: $\qquad 3y = 6$
Divide both sides by 3: $\qquad y = 2$

The points $(3, -2)$ and $(-3, 2)$ are on the graph of the equation. Plot the points $(0, 0)$, $(3, -2)$, and $(-3, 2)$ and connect them with a straight line. See Figure 16. ●

Quick ✓

In Problems 19 and 20, graph the equation by finding its intercepts.

19. $x - 2y = 0$ **20.** $4x + y = 0$

▶ ❸ Graph Vertical and Horizontal Lines

In the equation of a line, $Ax + By = C$, A and B cannot both be zero. But what if $A = 0$ or $B = 0$? This leads to special types of lines called *vertical lines* (when $B = 0$) and *horizontal lines* (when $A = 0$).

EXAMPLE 9   **Graphing a Vertical Line**

Graph the equation $x = 3$ using the point-plotting method.

Solution
Because the equation $x = 3$ can be written as $1x + 0y = 3$, the graph is a line. For the equation $x = 3$, any value of y has a corresponding x-value of 3. For example, if $y = -1$, then

$$1x + 0(-1) = 3$$
$$x = 3$$

See Table 11 for other choices for y. The points $(3, -2)$, $(3, -1)$, $(3, 0)$, $(3, 1)$, and $(3, 2)$ are all on the line. See Figure 17.

Table 11

x	y	(x, y)
3	-2	(3, -2)
3	-1	(3, -1)
3	0	(3, 0)
3	1	(3, 1)
3	2	(3, 2)

Figure 17
$x = 3$

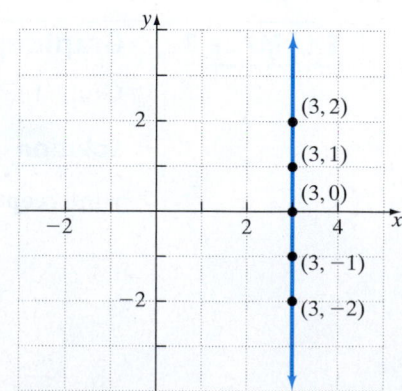

Work Smart

Graphing an equation in the form $x = a$ always gives a vertical line. The graphs of $x = 5$, $x = -4$, and $x = \dfrac{11}{4}$ are all vertical lines.

The results of Example 9 lead to a definition of a *vertical line*.

> **Definition**
>
> A **vertical line** is given by an equation of the form
> $$x = a$$
> where $(a, 0)$ is the x-intercept.

EXAMPLE 10 **Graphing a Horizontal Line**

Graph the equation $y = -2$ using the point-plotting method.

Solution

Because the equation $y = -2$ can be written as $0x + 1y = -2$, the graph is a line. For $y = -2$, any value of x has a corresponding y-value of -2. For example, if $x = -2$, then

$$0(-2) + 1y = -2$$
$$y = -2$$

See Table 12 for other choices of x. The points $(-2, -2)$, $(-1, -2)$, $(0, -2)$, $(1, -2)$, and $(2, -2)$ are all on the line. See Figure 18.

Table 12

x	y	(x, y)
-2	-2	(-2, -2)
-1	-2	(-1, -2)
0	-2	(0, -2)
1	-2	(1, -2)
2	-2	(2, -2)

Figure 18
$y = -2$

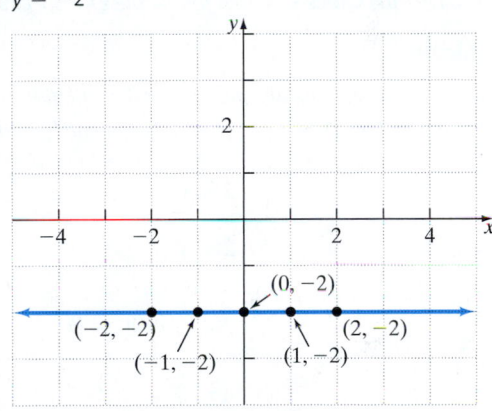

The results of Example 10 lead to a definition of a *horizontal line*.

Work Smart

Graphing an equation in the form $y = b$ always gives a horizontal line. The graphs of $y = 7$, $y = -3$, and $y = \dfrac{13}{5}$ are all horizontal lines.

> **Definition**
>
> A **horizontal line** is given by an equation of the form
> $$y = b$$
> where $(0, b)$ is the y-intercept.

Quick ✓

21. A _____ line is given by an equation of the form $x = a$, where _____ is the x-intercept.

22. A _____ line is given by an equation of the form $y = b$, where _____ is the y-intercept.

In Problems 23 and 24, graph each equation.

23. $x = -5$ 24. $y = -4$

Below is a summary to help organize the information presented.

Summary Intercepts and Equations of Lines

Topic	Comments
Intercepts: Points where the graph crosses or touches a coordinate axis.	Intercepts need to be shown for a graph to be complete.
x-intercept: Point where the graph crosses or touches the x-axis. Found by letting $y = 0$ in the equation.	
y-intercept: Point where the graph crosses or touches the y-axis. Found by letting $x = 0$ in the equation.	
Standard Form of an Equation of a Line: $Ax + By = C$, where A and B are not both zero	Can be graphed using point-plotting or intercepts.
Equation of a Vertical Line: $x = a$	Graph is a vertical line whose x-intercept is $(a, 0)$.
Equation of a Horizontal Line: $y = b$	Graph is a horizontal line whose y-intercept is $(0, b)$.

3.2 Exercises MyMathLab®

Exercise numbers in **green** have complete video solutions in MyMathLab or may be accessed using the QR code to the right.

*Problems **1–24** are the **Quick ✔**s that follow the **EXAMPLES**.*

Building Skills

In Problems 25–32, determine whether or not the equation is a linear equation in two variables. Answer Linear or Not linear. See Objective 1.

25. $2x - 5y = 10$ **26.** $y^2 = 2x + 3$

27. $x^2 + y = 1$ **28.** $y - 2x = 9$

29. $y = \dfrac{4}{x}$ **30.** $x - 8 = 0$

31. $y - 1 = 0$ **32.** $y = -\dfrac{2}{x}$

In Problems 33–50, graph each linear equation using the point-plotting method. See Objective 1.

33. $y = 2x$ **34.** $y = 3x$

35. $y = 4x - 2$ **36.** $y = -3x - 1$

37. $y = -2x + 5$ **38.** $y = x - 6$

39. $x + y = 5$ **40.** $x - y = 6$

41. $-2x + y = 6$ **42.** $5x - 2y = -10$

43. $4x - 2y = -8$ **44.** $x + 3y = 6$

45. $x = -4y$ **46.** $x = \dfrac{1}{2}y$

47. $y + 7 = 0$ **48.** $x - 6 = 0$

49. $y - 2 = 3(x + 1)$ **50.** $y + 3 = -2(x - 2)$

In Problems 51–58, find the intercepts of each graph. See Objective 2.

51.

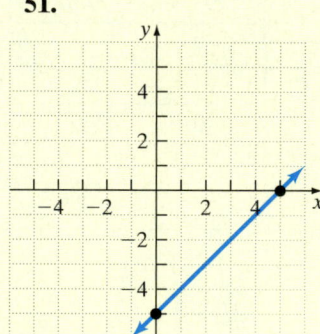

52.

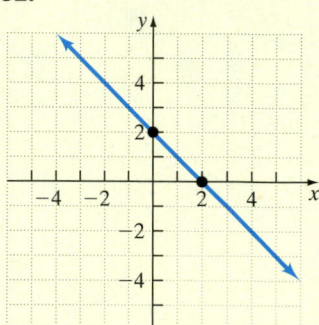

53.

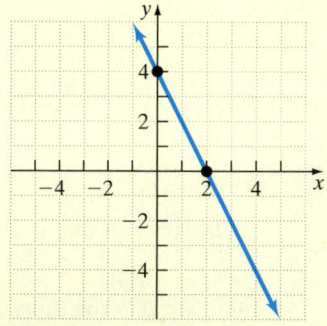

54.

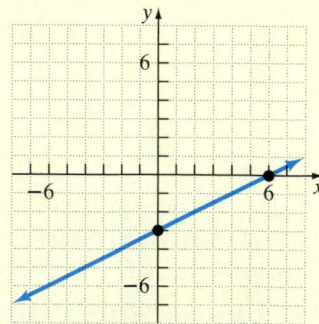

55.

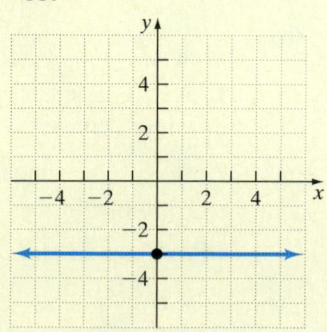

56.

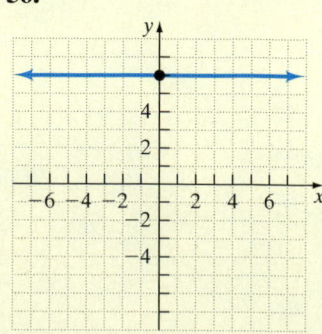

57.

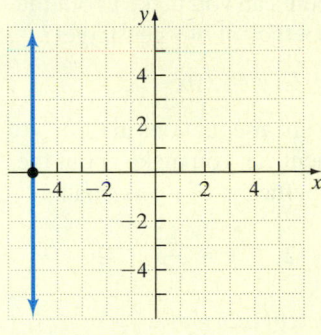

58.

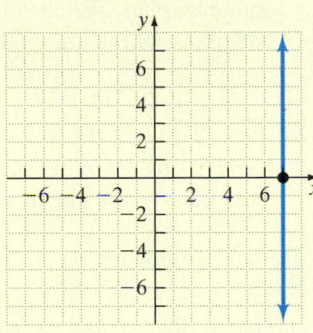

In Problems 59–70, find the intercepts of each equation. See Objective 2.

59. $2x + 3y = -12$

60. $3x - 5y = 30$

61. $x = -6y$

62. $y = 10x$

63. $y = x - 5$

64. $y = -x + 7$

65. $\dfrac{x}{6} + \dfrac{y}{8} = 1$

66. $\dfrac{x}{2} - \dfrac{y}{8} = 1$

67. $x = 4$

68. $y = 6$

69. $y + 2 = 0$

70. $x + 8 = 0$

In Problems 71–86, graph each linear equation by finding its intercepts. See Objective 2.

71. $3x + 6y = 18$

72. $3x - 5y = 15$

73. $-x + 5y = 15$

74. $-2x + y = 14$

75. $\dfrac{1}{2}x - y = 3$

76. $\dfrac{4}{3}x + y = 1$

77. $9x - 2y = 0$

78. $\dfrac{1}{3}x - y = 0$

79. $y = -\dfrac{1}{2}x + 3$

80. $y = \dfrac{2}{3}x - 3$

81. $\dfrac{1}{3}y + 2 = 2x$

82. $\dfrac{1}{2}x - 3 = 3y$

83. $\dfrac{x}{2} + \dfrac{y}{3} = 1$

84. $\dfrac{y}{4} - \dfrac{x}{3} = 1$

85. $4y - 2x + 1 = 0$

86. $2y - 3x + 2 = 0$

In Problems 87–94, graph each horizontal or vertical line. See Objective 3.

87. $x = 5$

88. $x = -7$

89. $y = -6$

90. $y = 2$

91. $y - 12 = 0$

92. $y + 3 = 0$

93. $3x - 5 = 0$

94. $2x - 7 = 0$

Mixed Practice

In Problems 95–106, graph each linear equation by the point-plotting method or by finding intercepts.

95. $y = 2x - 5$

96. $y = -3x + 2$

97. $y = -5$

98. $x = 2$

99. $2x + 5y = -20$

100. $3x - 4y = 12$

101. $2x = -6y + 4$

102. $5x = 3y - 10$

103. $x - 3 = 0$

104. $y + 4 = 0$

105. $3y - 12 = 0$

106. $-4x + 8 = 0$

107. If $(3, y)$ is a point on the graph of $4x + 3y = 18$, find y.

108. If $(-4, y)$ is a point on the graph of $3x - 2y = 10$, find y.

109. If $(x, -2)$ is a point on the graph of $3x + 5y = 11$, find x.

110. If $(x, -3)$ is a point on the graph of $4x - 7y = 19$, find x.

111. Plot the points $(3, 5)$ and $(-2, 5)$ and draw a line through the points. What is the equation of this line?

112. Plot the points $(-1, 2)$ and $(5, 2)$ and draw a line through the points. What is the equation of this line?

113. Plot the points $(-2, -4)$ and $(-2, 1)$ and draw a line through the points. What is the equation of this line?

114. Plot the points $(3, -1)$ and $(3, 2)$ and draw a line through the points. What is the equation of this line?

Applying the Concepts

In Problems 115–118, find the equation of each line.

115.

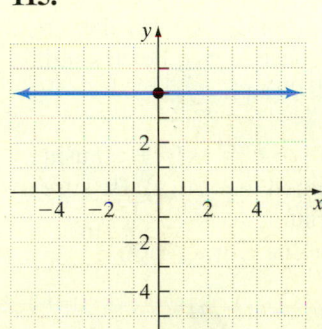

116.

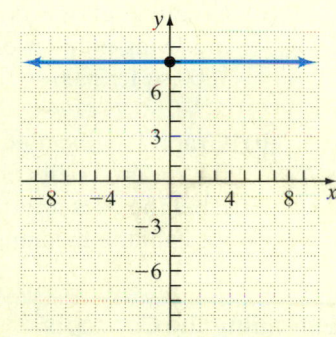

117.

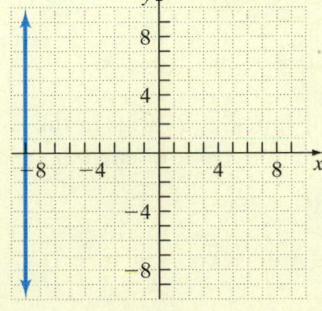

118.

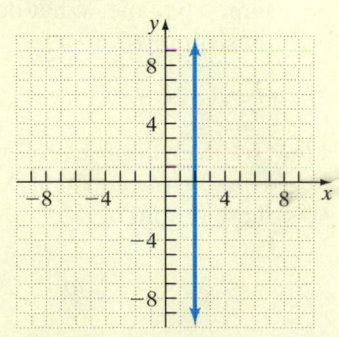

119. Create a set of ordered pairs in which the *x*-coordinates are twice the *y*-coordinates. What is the equation of this line?

120. Create a set of ordered pairs in which the *y*-coordinates are twice the *x*-coordinates. What is the equation of this line?

121. Create a set of ordered pairs in which the *y*-coordinates are 2 more than the *x*-coordinates. What is the equation of this line?

122. Create a set of ordered pairs in which the *x*-coordinates are 3 less than the *y*-coordinates. What is the equation of this line?

123. Calculating Wages Marta earns $500 per week plus $100 in commission for every car she sells. The linear equation that calculates her weekly earnings is $E = 100n + 500$, where *E* represents her weekly earnings in dollars and *n* represents the number of cars she sold during the week.

   (a) Create a set of ordered pairs (n, E) if, in three consecutive weeks, she sold 0 cars, 4 cars, and 10 cars.

   (b) Graph the linear equation $E = 100n + 500$ using the ordered pairs obtained in part (a). Be sure to label the axes appropriately.

   (c) Explain the meaning of the *E*-intercept.

124. Carpet Cleaning Harry's Carpet Cleaning charges a $50 service charge plus $0.10 for each square foot of carpet to be cleaned. The linear equation that calculates the total cost to clean a carpet is $C = 0.1f + 50$, where *C* is the total cost in dollars and *f* is the number of square feet of carpet.

   (a) Create a set of ordered pairs (f, C) for the following number of square feet to be cleaned: 1000 sq ft, 2000 sq ft, 2500 sq ft.

   (b) Graph the linear equation $C = 0.1f + 50$ using the ordered pairs obtained in part (a). Be sure to label the axes appropriately.

   (c) Explain the meaning of the *C*-intercept.

Extending the Concepts

125. Graph each of the following linear equations in the same *xy*-plane. What do you notice about each of these graphs?

$$y = 2x - 1 \qquad y = 2x + 3 \qquad 2x - y = 5$$

126. Graph each of the following linear equations in the same *xy*-plane. What do you notice about each of these graphs?

$$y = 3x + 2 \qquad 6x - 2y = -4 \qquad x = \frac{1}{3}y - \frac{2}{3}$$

127. Graph each of the following linear equations in the same *xy*-plane. What statement can you make about the steepness of the line as the coefficient of *x* gets larger?

$$y = x \qquad y = 2x \qquad y = 10x$$

128. Graph each of the following linear equations in the same *xy*-plane. What statement can you make about the steepness of the line as the coefficient of *x* gets smaller?

$$y = x + 2 \qquad y = \frac{1}{2}x + 2 \qquad y = \frac{1}{8}x + 2$$

In Problems 129–132, find the intercepts of each graph.

129.

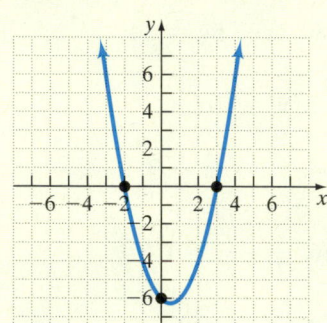

130.

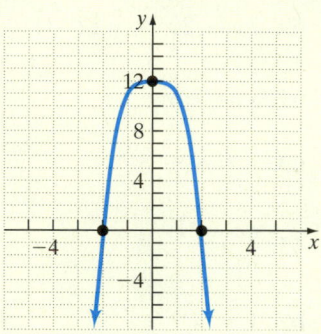

131.

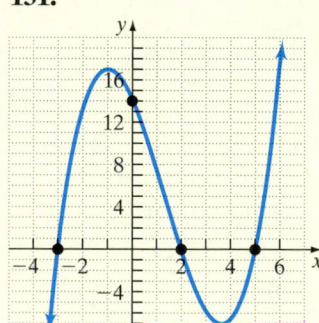

132.

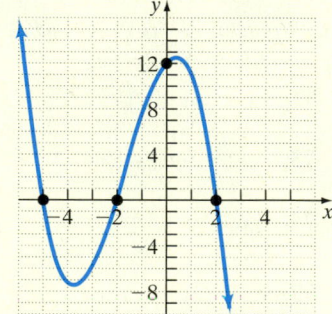

Explaining the Concepts

133. Explain what the graph of an equation represents.

134. What is meant by a complete graph?

135. How many points are required to graph a line? Explain your reasoning and why you might include additional point(s) when graphing a line.

136. Explain how to use the intercepts to graph the equation $Ax + By = C$, where *A*, *B*, and *C* are not equal to zero. Explain how to graph the same equation when *C* is equal to zero. Can you use the same techniques for both equations? Why or why not?

Technology Exercises

Graphing calculators and software such as Desmos use the point-plotting method to obtain the graph. As with creating tables, first solve the equation for *y*. For example, to graph the equation $2x - 3y = 12$ (Example 6 from Section 3.1), solve for y to obtain $y = \frac{2}{3}x - 4$. Enter the equation $y = \frac{2}{3}x - 4$ into the technology and create the graph shown below.

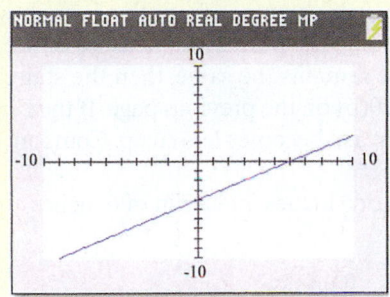

In Problems 137–142, use technology to graph each equation.

137. $2x - y = 9$

138. $3x + y = 8$

139. $y + 2x = 13$

140. $y - x = -15$

141. $y = -6x^2 + 1$

142. $y = -x^2 + 3x$

3.3 Slope

Objectives

❶ Find the Slope of a Line Given Two Points

❷ Find the Slope of Vertical and Horizontal Lines

❸ Graph a Line Using Its Slope and a Point on the Line

❹ Work with Applications of Slope

Are You Prepared for This Section?

Before getting started, complete the following problems. If you get a problem wrong, go back to the section cited and review the material.

P1. Evaluate: $\dfrac{5 - 2}{8 - 7}$ [Section 1.7, p. 59]

P2. Evaluate: $\dfrac{3 - 7}{9 - 3}$ [Section 1.7, p. 59]

P3. Evaluate: $\dfrac{-3 - 4}{6 - (-1)}$ [Section 1.7, p. 59]

Figure 19

Hill   Hill

(a)   (b)

Pretend you are on snow skis for the first time in your life. The ski resort that you are visiting has two hills, as shown in Figure 19. Which hill would you prefer to go down? Why?

It is clear that the hill in Figure 19(a) is not as steep as the hill in Figure 19(b). Mathematicians like to numerically describe situations such as the steepness of a hill. Measuring the steepness of each hill allows for an easy comparison. The numerical measure that describes the steepness of a hill is its *slope*.

▶ ❶ Find the Slope of a Line Given Two Points

Consider the staircases drawn in Figure 20. If we draw a line through the top of each riser on the staircase (in blue), we can see that every step of each staircase contains exactly the same horizontal change (or **run**) and the same vertical change (or **rise**). *Slope* is defined in terms of the rise and run.

Figure 20

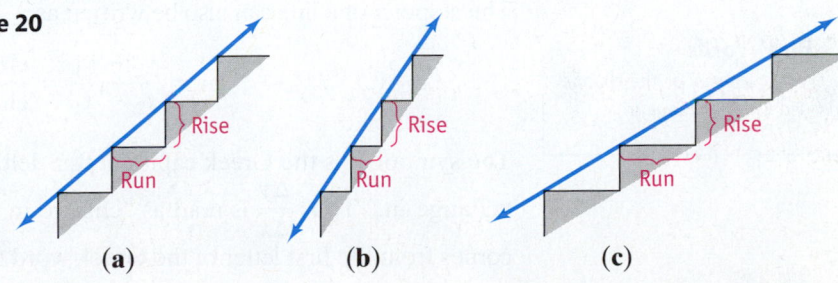

(a)   (b)   (c)

Definition

The **slope** of a line, denoted by the letter m, is the ratio of the rise to the run.

$$\text{Slope} = m = \frac{\text{rise}}{\text{run}}$$

The symbol for the slope of a line is m, which comes from the French word *monter*, which means "to go up, ascend, or climb." Slope is a numerical measure of the steepness of the line. If the run is decreased and the rise remains the same, then the staircase becomes steeper. Compare Figure 20(a) with Figure 20(b) on the previous page. If the run is increased and the rise remains the same, then the staircase becomes less steep. Compare Figure 20(a) with Figure 20(c).

If the staircase in Figure 20(a) has a rise of 6 inches and a run of 6 inches, then the slope of the line is

$$m = \frac{\text{rise}}{\text{run}} = \frac{6 \text{ inches}}{6 \text{ inches}} = 1$$

If the rise of the staircase is increased to 9 inches, then the slope of the line is

$$m = \frac{\text{rise}}{\text{run}} = \frac{9 \text{ inches}}{6 \text{ inches}} = \frac{3}{2}$$

The main idea is the steeper the line, the larger the slope. The slope of a line can be defined using rectangular coordinates.

Definition

If $x_1 \neq x_2$, the **slope** m of the line containing the points (x_1, y_1) and (x_2, y_2) is defined by the formula

$$m = \frac{y_2 - y_1}{x_2 - x_1}$$

Figure 21 illustrates the slope of a line.

Figure 21

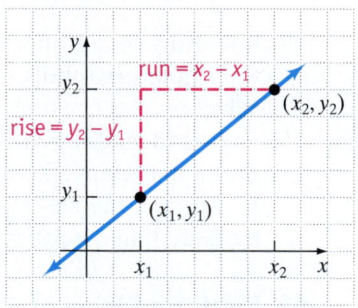

Notice that the slope m of a line is calculated as

$$m = \frac{\text{rise}}{\text{run}} = \frac{y_2 - y_1}{x_2 - x_1}.$$

The slope m of a line can also be written as

$$m = \frac{y_2 - y_1}{x_2 - x_1} = \frac{\text{change in } y}{\text{change in } x} = \frac{\Delta y}{\Delta x}.$$

The symbol Δ is the Greek capital letter delta. In mathematics, the symbol Δ is read as "change in." Thus $\dfrac{\Delta y}{\Delta x}$ is read as "change in y divided by change in x." The symbol Δ comes from the first letter of the Greek word for "difference," *diaphora*.

EXAMPLE 1 **How to Find the Slope of a Line**

Find the slope of the line drawn in Figure 22.

Figure 22

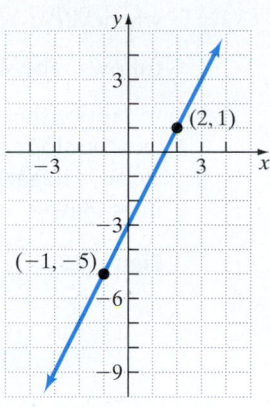

Step-by-Step Solution

Step 1: Let one of the points be (x_1, y_1) and the other point be (x_2, y_2).

Let's say that $(x_1, y_1) = (-1, -5)$ and $(x_2, y_2) = (2, 1)$.

Step 2: Find the slope by evaluating

$$m = \frac{y_2 - y_1}{x_2 - x_1}$$

$$m = \frac{y_2 - y_1}{x_2 - x_1}$$

Let $x_1 = -1, y_1 = -5$:
Let $x_2 = 2, y_2 = 1$:

$$= \frac{1 - (-5)}{2 - (-1)}$$

$$= \frac{6}{3}$$

$$= \frac{2}{1}$$

$$m = 2$$

Figure 23

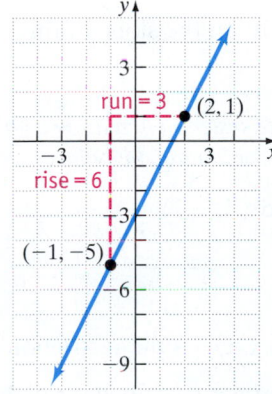

Work Smart

It doesn't matter which point is called (x_1, y_1) and which is called (x_2, y_2). The answer will be the same. In Example 1, if we let $(x_1, y_1) = (2, 1)$ and $(x_2, y_2) = (-1, -5)$, we get

$$m = \frac{y_2 - y_1}{x_2 - x_1} = \frac{-5 - 1}{-1 - 2} = \frac{-6}{-3} = \frac{2}{1} = 2$$

Remember that slope can be thought of as "rise divided by run." This description of the slope of a line is illustrated in Figure 23. The slope of the line drawn in Figure 23 can be interpreted as follows: "The value of y will increase by 6 units whenever x increases by 3 units." Or, because $\frac{6}{3} = \frac{2}{1}$, "the value of y will increase by 2 units whenever x increases by 1 unit." Both interpretations are acceptable.

EXAMPLE 2 **Finding and Interpreting the Slope of a Line**

Plot the points $(-1, 3)$ and $(2, -2)$ and draw a line through the two points. Find and interpret the slope of the line.

Solution

Plot the points $(x_1, y_1) = (-1, 3)$ and $(x_2, y_2) = (2, -2)$ in the rectangular coordinate system and draw a line through the two points. See Figure 24 on the next page. The slope of the line drawn in Figure 24 is

$$m = \frac{y_2 - y_1}{x_2 - x_1} = \frac{-2 - 3}{2 - (-1)}$$

$$= \frac{-5}{3}$$

$$= -\frac{5}{3}$$

(continued)

Work Smart

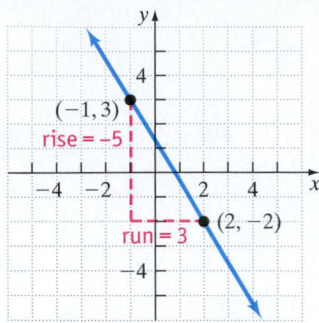

You can interpret a slope of $-\dfrac{5}{3} = \dfrac{-5}{3}$ this way: The value of y will go down 5 units whenever x increases by 3 units. Because $-\dfrac{5}{3} = \dfrac{5}{-3} = \dfrac{\text{rise}}{\text{run}}$, a second interpretation is as follows: The value of y will increase by 5 units whenever x decreases by 3 units. ●

Notice that the line drawn in Figure 23 on the previous page goes to the right and up and the slope is positive, while the line drawn in Figure 24 goes to the right and down and slope is negative. In general, lines that go to the right and up have positive slopes, and lines that go to the right and down have negative slopes. See Figure 25.

Figure 24

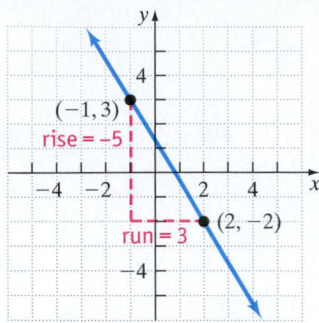

Figure 25

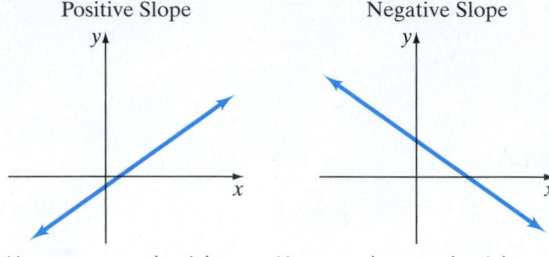

Positive Slope

Negative Slope

Line goes up to the right   Line goes down to the right

Quick ✓

1. If the run of a line is 10 and its rise is 6, then its slope is ___.

2. *True or False* If $P = (x_1, y_1)$ and $Q = (x_2, y_2)$ are two distinct points with $y_1 \neq y_2$, the slope m of the line that contains points P and Q is defined by the formula $m = \dfrac{x_2 - x_1}{y_2 - y_1}$.

3. *True or False* If the slope of a line is $\dfrac{3}{2}$, then y will increase by 3 units when x increases by 2 units.

4. If the graph of a line goes up as you move to the right, then the slope of this line must be _____.

In Problems 5 and 6, plot the points in a rectangular coordinate system. Then draw a line through the two points. Find and interpret the slope of the line containing the points.

5. $(0, 2)$ and $(4, 10)$ 6. $(-2, 2)$ and $(3, -7)$

▶ ❷ Find the Slope of Vertical and Horizontal Lines

Did you notice that in the definition of slope, $m = \dfrac{y_2 - y_1}{x_2 - x_1}$, there is the restriction that $x_1 \neq x_2$? This means the formula does not apply if the x-coordinates of the two points are the same. Why? See Example 3.

EXAMPLE 3 **The Slope of a Vertical Line**

Plot the points $(2, -1)$ and $(2, 3)$ and draw a line through the two points. Find and interpret the slope of the line.

Solution

Plot the points $(x_1, y_1) = (2, -1)$ and $(x_2, y_2) = (2, 3)$ in the rectangular coordinate system and draw a line through the two points. See Figure 26. The slope of the line drawn in Figure 26 is

$$m = \frac{y_2 - y_1}{x_2 - x_1} = \frac{3 - (-1)}{2 - 2}$$

$$= \frac{4}{0}$$

Figure 26

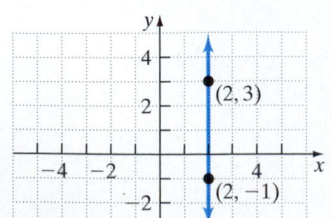

Work Smart

Remember that if a represents any nonzero number, then $\frac{0}{a} = 0$ and $\frac{a}{0}$ is undefined.

Because division by 0 is undefined, the slope of the line is undefined. When y increases by 4, there is no change in x. ●

Let's generalize the results of Example 3. Let (x_1, y_1) and (x_2, y_2) be two distinct points. If $x_1 = x_2$, then we have a **vertical line** whose slope m is **undefined** (since this results in division by 0). See Figure 27.

Figure 27
Vertical line

$$m = \frac{y_2 - y_1}{x_1 - x_1} = \frac{y_2 - y_1}{0}$$ Slope is undefined

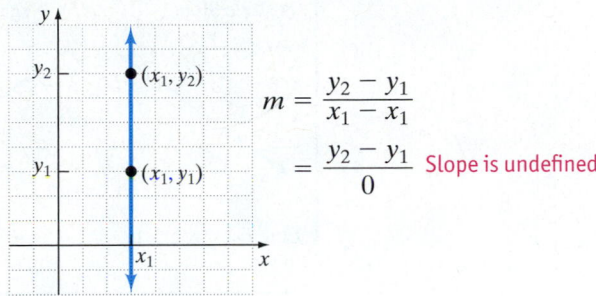

Okay, but what if $y_1 = y_2$?

EXAMPLE 4 The Slope of a Horizontal Line

Plot the points $(-2, 4)$ and $(3, 4)$ and draw a line through the two points. Find and interpret the slope of the line.

Solution

Plot the points $(x_1, y_1) = (-2, 4)$ and $(x_2, y_2) = (3, 4)$ in the rectangular coordinate system and draw a line through the two points. See Figure 28. The slope of the line drawn in Figure 28 is

Figure 28

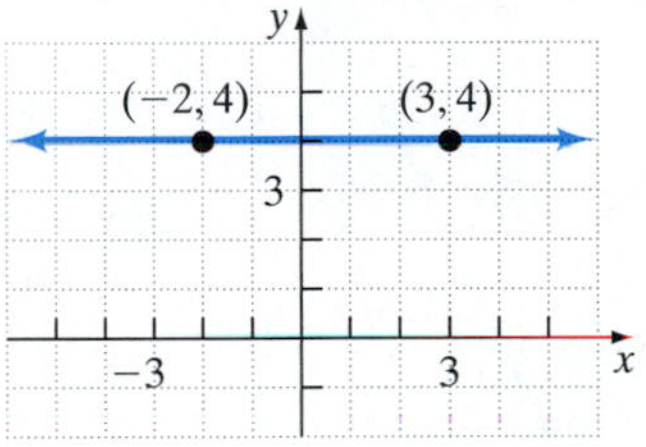

$$m = \frac{y_2 - y_1}{x_2 - x_1} = \frac{4 - 4}{3 - (-2)}$$
$$= \frac{0}{5}$$
$$= 0$$

The slope of the line is 0. A slope of 0 can be interpreted as: There is no change in y when x increases by 1 unit. ●

Let's generalize the results of Example 4. Let (x_1, y_1) and (x_2, y_2) be two distinct points. If $y_1 = y_2$, then we have a **horizontal line** whose slope m is 0. See Figure 29.

Figure 29
Horizontal line

$$m = \frac{y_1 - y_1}{x_2 - x_1} = \frac{0}{x_2 - x_1}$$ Slope is zero

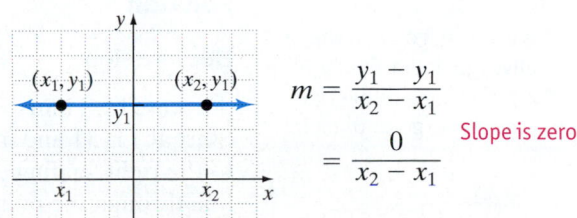

Quick ✓

7. The slope of a horizontal line is ____, while the slope of a vertical line is ________.

In Problems 8 and 9, plot the given points in a rectangular coordinate system. Then draw a line through the two points. Find the slope of the line and state whether the line is horizontal or vertical.

8. $(2, 5)$ and $(2, -1)$

9. $(2, 5)$ and $(6, 5)$

Summary The Slope of a Line

Figure 30 illustrates the four possibilities for the slope of a line. Remember, just as a text is read from left to right, graphs are also read from left to right.

Figure 30

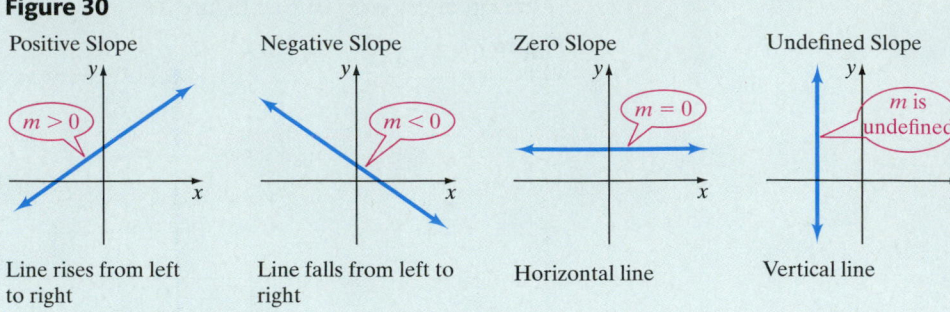

Positive Slope — $m > 0$ — Line rises from left to right

Negative Slope — $m < 0$ — Line falls from left to right

Zero Slope — $m = 0$ — Horizontal line

Undefined Slope — m is undefined — Vertical line

▶ ❸ Graph a Line Using Its Slope and a Point on the Line

We now illustrate how to use slope to graph lines. Slope may be thought of as "driving directions" from one point to another point.

EXAMPLE 5 **Graphing a Line Given a Point and Its Slope**

Draw a graph of the line that contains the point $(1, 3)$ and has a slope of 2.

Solution

The slope $= m = 2 = \dfrac{2}{1} = \dfrac{\text{rise}}{\text{run}}$ means that y will increase by 2 units (the rise), when x increases by 1 unit (the run). So start at $(1, 3)$ and move 2 units up and then 1 unit to the right, ending at the point $(2, 5)$. Then draw a line through the points $(1, 3)$ and $(2, 5)$ to obtain the graph of the line. See Figure 31.

Figure 31

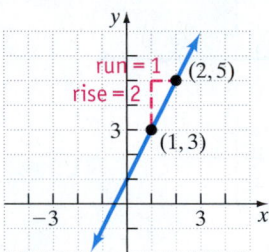

EXAMPLE 6 **Graphing a Line Given a Point and Its Slope**

Draw a graph of the line that contains the point $(1, 3)$ and has a slope of $-\dfrac{2}{3}$.

Work Smart

If the "rise" is positive, go up. If the "rise" is negative, go down. Similarly, if the "run" is positive, go to the right. If the "run" is negative, go to the left.

Solution

Because slope $= -\dfrac{2}{3} = \dfrac{-2}{3} = \dfrac{\text{rise}}{\text{run}}$, y will decrease by 2 units when x increases by 3 units. Start at $(1, 3)$ and move 2 units down and then 3 units to the right, ending at the point $(4, 1)$. Then draw a line through the points $(1, 3)$ and $(4, 1)$ to obtain the graph of the line. See Figure 32.

Figure 32

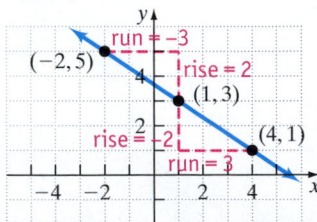

A second option is to set $\dfrac{\text{rise}}{\text{run}} = -\dfrac{2}{3} = \dfrac{2}{-3}$. In this case, move 2 units up from $(1, 3)$ and then 3 units to the left, and end up at $(-2, 5)$, which is also on the graph of the line, as indicated in Figure 32.

Quick ✔

10. Draw a graph of the line that contains the point $(1, 2)$ and has a slope of

(a) $\dfrac{1}{2}$ **(b)** -3 **(c)** 0

▶ ④ Work with Applications of Slope

In its simplest form, slope is a ratio of rise over run. For example, a hill whose grade is $5\% = 0.05 = \frac{5}{100}$ goes up 5 feet (the rise) for every 100 feet it goes horizontally (the run). See Figure 33.

Figure 33

100 feet · 5 feet

Work Smart

The pitch of a roof or grade of a road is always represented as a positive number.

Consider the pitch of a roof. If a roof's pitch is $\frac{5}{12}$, then every 5-foot measurement upward results in a horizontal measurement of 12 feet. See Figure 34.

Figure 34

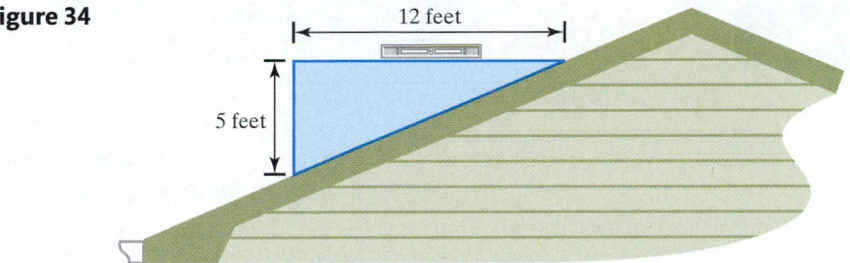
12 feet · 5 feet

EXAMPLE 7 **Finding the Grade of a Road**

In Heckman Pass, British Columbia, there is a road that rises 9 feet for every 50 feet of horizontal distance covered. What is the grade of the road as a percent?

Solution

The grade of the road is given by $\frac{\text{rise}}{\text{run}}$. Since a rise of 9 feet is accompanied by a run of 50 feet, the grade of the road is $\frac{9 \text{ feet}}{50 \text{ feet}} = 0.18 = 18\%$. ●

Work Smart

$$\begin{array}{r} 0.18 \\ 50)\overline{9.00} \\ \underline{-50} \\ 400 \\ \underline{-400} \\ 0 \end{array}$$

The slope m of a line measures the amount that y changes as x changes from x_1 to x_2. The slope of a line is also called the **average rate of change** of y with respect to x.

In applications, we are often interested in knowing how the change in one variable might affect some other variable. For example, if your income increases by $1000, how much will your spending (on average) change? Or, if the speed of your car increases by 10 miles per hour, how much (on average) will your car's gas mileage change?

EXAMPLE 8 **Slope as an Average Rate of Change**

In Austin, Texas, the price of a new two-bedroom condominium with an area of 1400 square feet is $295,000. The price of a new two-bedroom condominium with an area of 1625 square feet is $331,900. Find and interpret the slope of the line joining the points $(1400, 295{,}000)$ and $(1625, 331{,}900)$.
(SOURCE: *Edgewick.com*)

Solution

Let x represent the square footage of the condominium and y represent the price. Let $(x_1, y_1) = (1400, 295{,}000)$ and $(x_2, y_2) = (1625, 331{,}900)$ and compute the slope as

$$m = \frac{y_2 - y_1}{x_2 - x_1} = \frac{331{,}900 - 295{,}000}{1625 - 1400}$$

$$= \frac{36{,}900}{225}$$

$$= 164$$

(continued)

The unit of measure for y is dollars, and the unit of measure for x is square feet. So, the slope can be interpreted as follows: Between 1400 and 1625 square feet, the price increases by \$164 per square foot, on average.

> ### Quick ✓
>
> **11.** A road rises 4 feet for every 50 feet of horizontal distance covered. What is the grade of the road?
>
> **12.** The average annual cost of operating a Chevy Cruze is \$1370 when it is driven 10,000 miles. The average annual cost of operating a Chevy Cruze is \$1850 when it is driven 14,000 miles. Find and interpret the slope of the line joining $(10,000, 1370)$ and $(14,000, 1850)$.

3.3 Exercises MyMathLab®

Exercise numbers in **green** have complete video solutions in MyMathLab or may be accessed using the QR code to the right.

*Problems **1–12** are the Quick ✓ s that follow the* **EXAMPLES**.

Building Skills

In Problems 13–18, find the slope of the line whose graph is given. See Objective 1.

13.

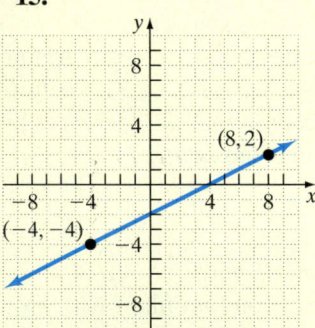

14.

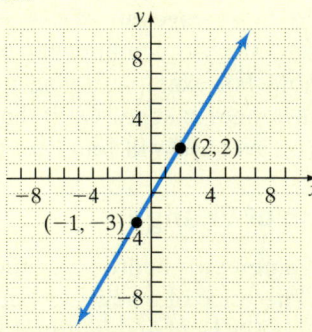

15.

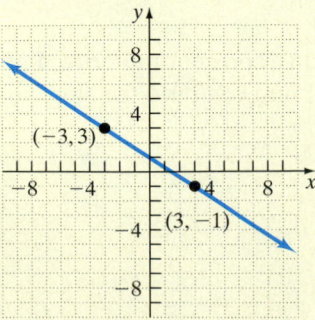

16.

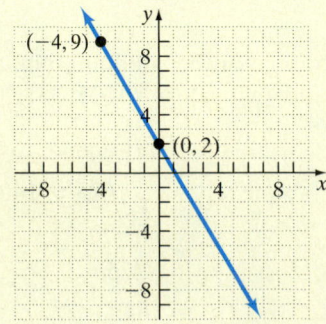

17.

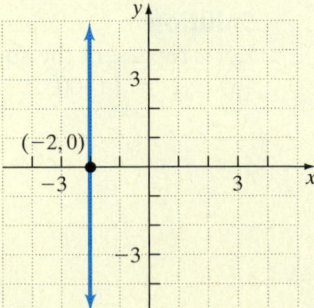

18.

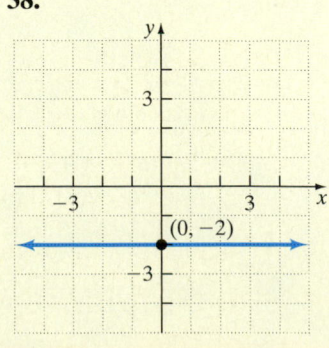

*In Problems 19–22, **(a)** plot the points in a rectangular coordinate system, **(b)** draw a line through the points, and **(c)** find and interpret the slope of the line. See Objective 1.*

19. $(-3, 2)$ and $(3, 5)$ **20.** $(2, 6)$ and $(-2, -4)$

21. $(2, -9)$ and $(-2, -1)$ **22.** $(4, -5)$ and $(-2, -4)$

In Problems 23–36, find and interpret the slope of the line containing the given points. See Objective 1.

23. $(10, 4)$ and $(6, 12)$ **24.** $(7, 3)$ and $(0, -11)$

25. $(4, -4)$ and $(12, -12)$ **26.** $(-3, 2)$ and $(2, -3)$

27. $(7, -2)$ and $(4, 3)$ **28.** $(-8, -1)$ and $(2, 3)$

29. $(0, 6)$ and $(-4, 0)$ **30.** $(-5, 0)$ and $(0, 3)$

31. $(-4, -1)$ and $(2, 3)$ **32.** $(5, 1)$ and $(-1, -1)$

33. $\left(\dfrac{1}{2}, \dfrac{3}{4}\right)$ and $\left(-\dfrac{5}{2}, -\dfrac{1}{4}\right)$ **34.** $\left(-\dfrac{1}{3}, \dfrac{2}{5}\right)$ and $\left(\dfrac{2}{3}, -\dfrac{3}{5}\right)$

35. $\left(\dfrac{1}{2}, \dfrac{1}{3}\right)$ and $\left(\dfrac{3}{4}, \dfrac{5}{6}\right)$ **36.** $\left(\dfrac{1}{4}, -\dfrac{4}{3}\right)$ and $\left(-\dfrac{5}{4}, \dfrac{1}{3}\right)$

In Problems 37–40, find the slope of the line whose graph is given. See Objective 2.

37.

38.

39.   **40.**

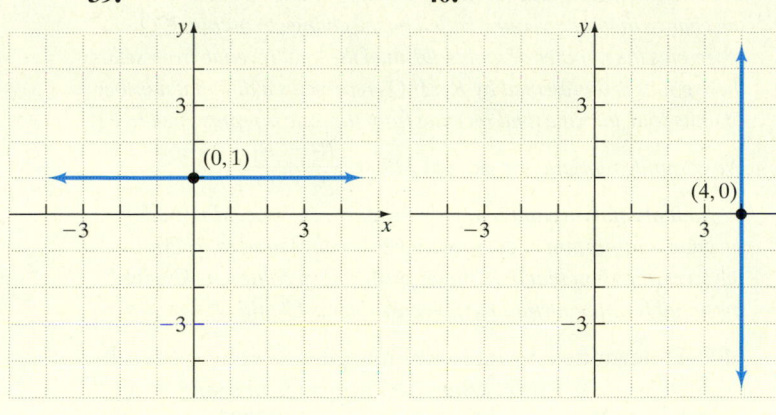

In Problems 41–44, find and interpret the slope of the line containing the given points. See Objective 2.

41. $(4, -6)$ and $(-1, -6)$ **42.** $(-1, -3)$ and $(-1, 2)$

43. $(3, 9)$ and $(3, -2)$ **44.** $(5, 1)$ and $(-2, 1)$

In Problems 45–62, draw a graph of the line that contains the given point and has the given slope. See Objective 3.

45. $(4, 2); m = 1$ **46.** $(3, -1); m = -1$

47. $(0, 6); m = -2$ **48.** $(-1, 3); m = 3$

49. $(-1, 0); m = \dfrac{1}{4}$ **50.** $(5, 2); m = -\dfrac{1}{4}$

51. $(2, -3); m = 0$ **52.** $(1, 4); m$ is undefined

53. $(2, 1); m = \dfrac{2}{3}$ **54.** $(-2, -3); m = \dfrac{5}{2}$

55. $(-1, 4); m = -\dfrac{5}{4}$ **56.** $(0, -2); m = -\dfrac{3}{2}$

57. $(0, 0); m$ is undefined   **58.** $(3, -1); m = 0$

59. $(0, 2); m = -4$ **60.** $(0, 0); m = \dfrac{1}{5}$

61. $(2, -3); m = \dfrac{3}{4}$ **62.** $(-3, 0); m = -3$

Applying the Concepts

In Problems 63–68, draw the graph of the two lines with the given properties on the same rectangular coordinate system.

63. Both lines pass through the point $(2, -1)$. One has slope of 2 and the other has slope of $-\dfrac{1}{2}$.

64. Both lines pass through the point $(3, 0)$. One has slope of $\dfrac{2}{3}$ and the other has slope of $-\dfrac{3}{2}$.

65. Both lines have a slope of $\dfrac{3}{4}$. One passes through the point $(-1, -2)$, and the other passes through the point $(2, 1)$.

66. Both lines have a slope of -1. One passes through the point $(0, -3)$, and the other passes through the point $(2, -1)$.

67. Both lines have a slope of $-\dfrac{4}{3}$. One passes through the point $(2, 1)$, and the other passes through the point $(-1, -1)$.

68. Both lines have a slope of 4. One passes through the point $(-2, -3)$, and the other passes through the point $(3, -1)$.

69. Roof Pitch A carpenter who was installing a new roof on a garage noticed that for every 1-foot horizontal run, the roof was elevated by 4 inches. What is the pitch of this roof?

70. Roof Pitch A canopy is set up on the football field. On the 45-yard line, the height of the canopy is 68 inches. The peak of the canopy is at the 50-yard marker, where the height is 84 inches. What is the pitch of the canopy?

71. Building a Doghouse To build a doghouse in his backyard, Moises has decided to use a pitch of $\dfrac{2}{5}$ for his roof. The doghouse measures 30 inches from the side to the center. Find h, the height of the doghouse roof in the figure below?

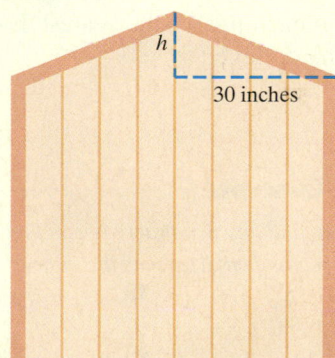

72. Building a Shed The design for a shed requires a roof pitch of $\dfrac{7}{20}$. If the shed measures 5 feet from the wall to the center, find the height h, of the roof of the shed in the figure below.

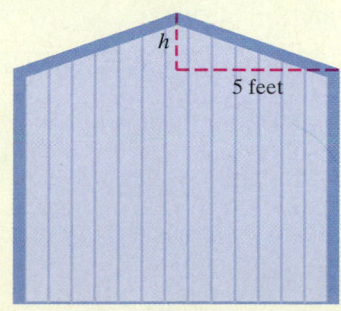

73. **Road Grade** Fall River Road was completed in 1920 and was the first road built through the Rocky Mountains in Colorado. It was so steep that sometimes the early model cars had to  drive up the hill in reverse to maximize the output of their weak engines and fuel systems. If the road rises 200 feet for every 1250 feet of horizontal change, what is the grade of this road written as a percent?

74. **Road Grade** Barbara decided to take a bicycle trip up to the observatory on Mauna Kea on the island of Hawaii. The road has a vertical rise of 120 feet for every 800 feet of horizontal change. What is the percent grade of this road?

75. **Population Growth** The population of the United States was 123,202,624 in 1930 and 321,418,820 in 2015. Use the ordered pairs $(0, 123 \text{ million})$ and $(85, 321 \text{ million})$ to find and interpret the average rate of change, to the nearest thousandth, in the population of the United States.

76. **Earning Potential** On average, a person who graduates from high school can expect to have lifetime earnings of 1.2 million dollars. It takes four years to earn a bachelor's degree, but the lifetime earnings will increase to 2.1 million dollars. Use the ordered pairs $(0, 1.2 \text{ million})$ and $(4, 2.1 \text{ million})$ to find and interpret the increase in earnings, to the nearest thousandth, due to finishing college.

Extending the Concepts

In Problems 77–80, find any two ordered pairs that lie on the given line. Graph the line and then determine the slope of the line.

77. $3x + y = -5$ 78. $2x + 5y = 12$

79. $y = 3x + 4$ 80. $y = -x - 6$

In Problems 81–86, find the slope of the line containing the given points. Assume all the lines are non-vertical.

81. $(2a, a)$ and $(3a, -a)$

82. $(4p, 2p)$ and $(-2p, 5p)$

83. $(2p + 1, q - 4)$ and $(3p + 1, 2q - 4)$

84. $(3p + 1, 4q - 7)$ and $(5p + 1, 2q - 7)$

85. $(a + 1, b - 1)$ and $(2a - 5, b + 5)$

86. $(2a - 3, b + 4)$ and $(4a + 7, 5b - 1)$

*In economics, **marginal revenue** is a rate of change defined as the change in total revenue divided by the change in output. If Q_1 represents the number of units sold, then the total revenue from selling these goods is represented by R_1. If Q_2 represents a different number of units sold, then the total revenue from this sale is represented by R_2. We compute marginal revenue as $MR = \dfrac{R_2 - R_1}{Q_2 - Q_1}$.*

So marginal revenue is a rate of change, or slope. Marginal revenue is important in economics because it is used to determine the level of output that maximizes profits for a company. Use the marginal revenue formula to solve Problems 87 and 88.

87. Determine and interpret marginal revenue if total revenue is $1000 when 400 hot dogs are sold at a baseball game and total revenue is $1200 when 500 hot dogs are sold.

88. Determine and interpret marginal revenue if total revenue is $300 when 30 compact disks are sold and total revenue is $400 when 50 compact disks are sold.

Explaining the Concepts

89. Describe a line that has one *x*-intercept but no *y*-intercept. Give two ordered pairs that could lie on this line and then describe how to find its slope.

90. Describe a line that has one *y*-intercept but no *x*-intercept. Give two ordered pairs that could lie on this line and then describe how to find its slope.

Technology Exercises

Open the "Slope" Geogebra applet for Problems 91 and 92. The applet may be found using the QR code at the beginning of this section or through the Multimedia Library in MyMathLab. If you are using a keyboard, it is best to use the arrow keys on the keypad to move the slider, instead of the mouse.

The grade of a mountain comes from the *rise*, or vertical distance between two points on the mountain, and the *run*, or horizontal distance between two points on the mountain. See the figure. The grade of the mountain is the *slope* of a line connecting two points on the line and is defined as

$$m = \frac{\text{rise}}{\text{run}} = \frac{\text{vertical change}}{\text{horizontal change}}$$

Because slope is defined in terms of rise and run, the value of the slope will change as the rise and run change.

91. **Exploration: Slope** Given the graph of a line, one way to find the slope is to choose two points on the line and find the rise and run between these two points. The ratio of the rise to the run is the slope of the line. If possible, express the slope in simplest form.

    (a) Move point A to $(2, 1)$ and point $B (6, 4)$. What is the rise? What is the run? What is the slope?

(b) Move point A to $(-1, 3)$ and point B to $(5, 7)$. What is the rise? What is the run? What is the slope?

(c) Move point A to $(1, 5)$ and point B to $(4, 1)$. What is the rise? What is the run? What is the slope?

(d) Move point A to $(8, 3)$ and point B to $(4, 5)$. What is the rise? What is the run? What is the slope?

(e) Move point A to $(-2, 4)$ and point B to $(3, 4)$. What is the rise? What is the run? What is the slope?

(f) Move point A to $(3, -2)$ and point B to $(3, 5)$. What is the rise? What is the run? What is the slope?

(g) Move point A to $(-2, -2)$ and point B to $(5, 5)$. What is the slope?

(h) Move point A to $(-4, 4)$ and point B to $(3, -3)$. What is the slope?

92. Exploration: Slope Move the points A and B to points of your choice on the line in order to see the position of the skier. Use the words positive, negative, zero, and undefined to fill in the blanks.

(a) When the skier is going "uphill" when moving left to right, the slope of the mountain is _____.

(b) When the skier is going "downhill" when moving left to right, the slope of the mountain is _____.

(c) When the skier is on flat land the slope of the mountain is _____.

(d) When the skier is going vertical the slope of the mountain is _____.

3.4 Slope-Intercept Form of a Line

Objectives

① Use the Slope-Intercept Form to Identify the Slope and y-Intercept of a Line

② Graph a Line Whose Equation Is in Slope-Intercept Form

③ Graph a Line Whose Equation Is in the Form $Ax + By = C$

④ Find the Equation of a Line Given Its Slope and y-Intercept

⑤ Work with Linear Models in Slope-Intercept Form

Are You Prepared for This Section?

Before getting started, complete the following problems. If you get a problem wrong, go back to the section cited and review the material.

P1. Solve $4x + 2y = 10$ for y. [Section 2.4, pp. 114–115]

P2. Solve: $10 = 2x - 8$ [Section 2.2, pp. 91–92]

▶ ① Use the Slope-Intercept Form to Identify the Slope and y-Intercept of a Line

In this section, the slope and y-intercept are used to graph a line. This method for graphing will be more efficient than plotting points. Why? Well, suppose we want to graph the equation $-2x + y = 5$ by plotting points. To do this, we might first solve the equation for y (get y by itself) by adding $2x$ to both sides of the equation.

$$-2x + y = 5$$

Add $2x$ to both sides: $$y = 2x + 5$$

Now create Table 13, which gives some points on the graph of the equation. Figure 35 shows the graph of the line.

Notice two things about the line in Figure 35. First, the slope is $m = 2$. Second, the y-intercept is $(0, 5)$. Notice in the equation $y = 2x + 5$ the coefficient of the variable x is 2 and the constant is 5. This is no coincidence!

Table 13

x	$y = 2x + 5$	(x, y)
-2	$2(-2) + 5 = 1$	$(-2, 1)$
-1	$2(-1) + 5 = 3$	$(-1, 3)$
0	$2(0) + 5 = 5$	$(0, 5)$

Figure 35
$-2x + y = 5$

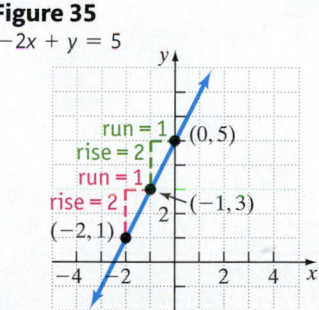

Slope-Intercept Form of an Equation of a Line

An equation of a line with slope m and y-intercept $(0, b)$ is

$$y = mx + b$$

Prepared?...Answers **P1.** $y = -2x + 5$
P2. $\{9\}$

EXAMPLE 1 **Finding the Slope and y-Intercept of a Line**

Find the slope and y-intercept of the line whose equation is $y = -3x + 1$.

Solution

Compare the equation $y = -3x + 1$ to the slope-intercept form of a line $y = mx + b$. The coefficient of x, -3, is the slope. The constant is 1, so the y-intercept is $(0, 1)$. ●

EXAMPLE 2 **Finding the Slope and y-Intercept of a Line Whose Equation Is in Standard Form**

Find the slope and y-intercept of the line whose equation is $3x + 2y = 6$.

Solution

Rewrite the equation $3x + 2y = 6$ so that it is in the form $y = mx + b$.

$$3x + 2y = 6$$

Subtract 3x from both sides: $\quad 2y = -3x + 6$

Divide both sides by 2: $\quad y = \dfrac{-3x + 6}{2}$

$\dfrac{a + b}{c} = \dfrac{a}{c} + \dfrac{b}{c}: \quad y = \dfrac{-3}{2}x + \dfrac{6}{2}$

Simplify: $\quad y = -\dfrac{3}{2}x + 3$

Work Smart

Notice that after subtracting 3x from both sides, we wrote the equation as $2y = -3x + 6$ rather than $2y = 6 - 3x$. This is because we want to get the equation in the form $y = mx + b$, so the term involving x should be first.

Now compare the equation $y = -\dfrac{3}{2}x + 3$ to the slope-intercept form of a line $y = mx + b$. The coefficient of x, $-\dfrac{3}{2}$, is the slope. The constant is 3, so the y-intercept is $(0, 3)$. ●

Any equation of the form $y = b$ can be written $y = 0x + b$. Thus, the slope of the line whose equation is $y = b$ is 0, and the y-intercept is $(0, b)$.

No equation of the form $x = a$ can be written in the form $y = mx + b$. The line whose equation is $x = a$ has an undefined slope and no y-intercept.

Work Smart

The equation of a horizontal line can be written in slope-intercept form, but the equation of a vertical line cannot.

Quick ✓

1. An equation of a line with slope m and y-intercept $(0, b)$ is _____.

In Problems 2–6, find the slope and y-intercept of the line whose equation is given.

2. $y = 4x - 3$ **3.** $3x + y = 7$ **4.** $2x + 5y = 15$

5. $y = 8$ **6.** $x = 3$

▶ ❷ **Graph a Line Whose Equation Is in Slope-Intercept Form**

If an equation is in slope-intercept form, the line can be graphed by plotting the y-intercept and using the slope to find another point on the line.

EXAMPLE 3   **How to Graph a Line Whose Equation Is in Slope-Intercept Form**

Graph the line $y = 3x - 1$ using the slope and y-intercept.

Step-by-Step Solution

Step 1: Identify the slope and y-intercept of the line.

$$y = 3x - 1$$

$$y = 3x + (-1)$$

$\boxed{m = 3}$ $\boxed{b = -1}$

The slope is $m = 3$ and the y-intercept is $(0, -1)$.

Step 2: Plot the y-intercept and then use the slope to find a second point on the graph. Draw a line through the points.

Plot the y-intercept at $(0, -1)$. Use the slope $m = \dfrac{3}{1} = \dfrac{\text{rise}}{\text{run}}$ to find a second point on the graph by counting up 3 units, then counting 1 unit to the right. Draw a line through these points. See Figure 36.

Figure 36
$y = 3x - 1$

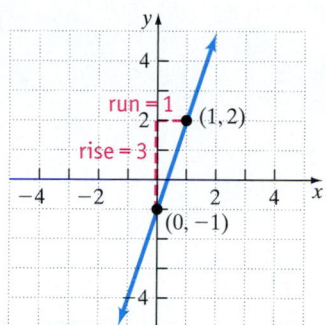

Quick ✔

In Problems 7 and 8, graph the line using the slope and y-intercept.

7. $y = 2x - 5$

8. $y = \dfrac{1}{2}x + 3$

EXAMPLE 4   **Graphing a Line Whose Equation Is in Slope-Intercept Form**

Graph the line $y = -\dfrac{4}{3}x + 2$ using the slope and y-intercept.

Solution

First, determine the slope and y-intercept.

$$y = -\dfrac{4}{3}x + 2$$

$\boxed{b = 2}$

$\boxed{m = -\dfrac{4}{3}}$

Figure 37

$y = -\dfrac{4}{3}x + 2$

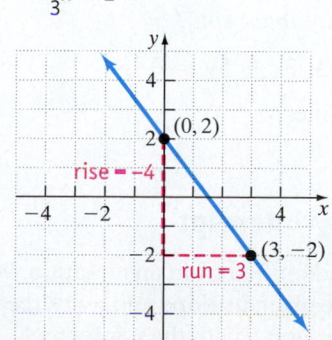

Plot the y-intercept at $(0, 2)$. Now use the slope $m = -\dfrac{4}{3} = \dfrac{-4}{3} = \dfrac{\text{rise}}{\text{run}}$ to find a second point on the graph by counting down 4 units, then counting 3 units to the right. Draw a line through these two points. See Figure 37.

Quick ✓

In Problems 9 and 10, graph the line using the slope and y-intercept.

9. $y = -3x + 1$ **10.** $y = -\dfrac{3}{2}x + 4$

▶ ❸ Graph a Line Whose Equation Is in the Form $Ax + By = C$

If a linear equation is written in standard form $Ax + By = C$, the slope and y-intercept can still be used to obtain the graph of the equation.

EXAMPLE 5 **How to Graph a Line Whose Equation Is in the Form $Ax + By = C$**

Graph the line $8x + 2y = 10$ using the slope and y-intercept.

Step-by-Step Solution

Step 1: Solve the equation for y to write it in slope-intercept form, $y = mx + b$.

$$8x + 2y = 10$$

Subtract 8x from both sides: $2y = -8x + 10$

Divide both sides by 2: $y = \dfrac{-8x + 10}{2}$

Simplify: $y = -4x + 5$

Step 2: Identify the slope and y-intercept of the line.

The slope is $m = -4$ and the y-intercept is $(0, 5)$.

Step 3: Plot the y-intercept and then use the slope to find a second point on the graph. Draw a line through the points.

Plot the point $(0, 5)$ and use the slope $m = -4 = \dfrac{-4}{1} = \dfrac{\text{rise}}{\text{run}}$ to find a second point on the graph. Draw a line through these points. See Figure 38.

Figure 38
$8x + 2y = 10$

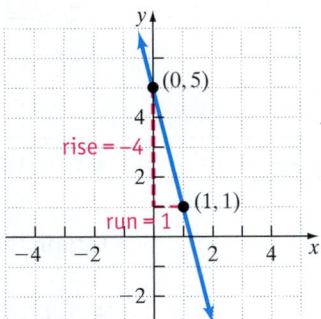

Work Smart

An alternative to graphing a linear equation using the slope and y-intercept is to graph the line using intercepts, if the intercepts are integers.

Quick ✓

In Problems 11–13, graph each line using the slope and y-intercept.

11. $-2x + y = -3$ **12.** $6x - 2y = 2$ **13.** $3x + 5y = 0$

14. List three techniques that can be used to graph a line.

▶ ❹ Find the Equation of a Line Given Its Slope and y-Intercept

Up to now, the slope and y-intercept of a line have been identified from an equation. We will now reverse the process and find the equation of a line given its slope and y-intercept. This is a straightforward process—replace m with the slope and b with the y-intercept.

EXAMPLE 6   **Finding the Equation of a Line Given Its Slope and y-Intercept**

Find the equation of a line whose slope is $\frac{3}{8}$ and whose y-intercept is $(0, -4)$. Graph the line.

Solution

The slope is $m = \frac{3}{8}$ and the y-intercept is $(0, b) = (0, -4)$. Substitute $\frac{3}{8}$ for m and -4 for b in the slope-intercept form of a line, $y = mx + b$, to obtain

$$y = \frac{3}{8}x - 4$$

Figure 39 shows the graph of the equation.

Figure 39

$y = \frac{3}{8}x - 4$

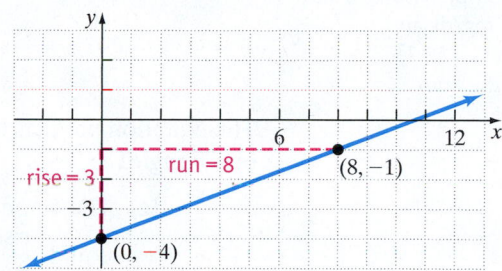

Quick ✓

In Problems 15–18, find the equation of the line whose slope and y-intercept are given. Graph the line.

15. $m = 3$, $(0, -2)$ **16.** $m = -\frac{1}{4}$, $(0, 3)$ **17.** $m = 0$, $(0, -1)$ **18.** $m = 1$, $(0, 0)$

▶ ⑤ Work with Linear Models in Slope-Intercept Form

In some situations, a linear equation can be used to describe the relationship between two variables. For example, the cost of renting a moving truck depends linearly on the number of miles driven, and a person's total cholesterol depends linearly on the person's age.

EXAMPLE 7   **A Model for Total Cholesterol**

When you have a physical exam, your doctor draws blood for your cholesterol test. Your total cholesterol count is measured in milligrams per deciliter (mg/dL). It is the sum of low-density lipoprotein cholesterol (LDL)—sometimes called "bad cholesterol"—and high-density lipoprotein cholesterol (HDL)—sometimes called "good cholesterol." Based on data from the National Center for Health Statistics, a woman's total cholesterol y is related to her age x, for $x \geq 18$, by the following linear equation:

$$y = 1.1x + 157$$

(a) Use the equation to predict the total cholesterol of a 40-year-old woman.

(b) Use the equation to predict the told cholesterol to the nearest mg/dl, of an 18-year old women.

(c) Determine and interpret the slope of the equation.

(d) Explain why it does not make sense to interpret the y-intercept of the equation.

(e) Graph the equation in a rectangular coordinate system.

(continued)

Solution

(a) Substitute 40 for x, the woman's age, in the equation $y = 1.1x + 157$ to find the total cholesterol y.

$$y = 1.1x + 157$$
$$x = 40: \quad y = 1.1(40) + 157$$
$$= 201$$

The equation predicts that a 40-year-old woman will have a total cholesterol of 201 mg/dL.

(b) Substitute 18 for x, the woman's age, in the equation $y = 1.1x + 157$ to find the total cholesterol y, to the nearest mg/dL.

$$y = 1.1x + 157$$
$$x = 18: \quad y = 1.1(18) + 157$$
$$= 176.8$$

The equation predicts that a 18-year-old woman will have a total cholesterol of 177 mg/dL.

(c) The slope of the equation $y = 1.1x + 157$ is 1.1. Because slope equals $\dfrac{\text{rise}}{\text{run}} = \dfrac{1.1 \text{ mg/dL}}{1 \text{ year}}$, the slope can be interpreted as follows: "The total cholesterol of a female increases by 1.1 mg/dL as age increases by 1 year."

(d) The y-intercept of the equation $y = 1.1x + 157$ is $(0, 157)$. The y-intercept is the value of total cholesterol, y, when $x = 0$. Since x represents age, the y-intercept can be interpreted as follows: "The total cholesterol of a newborn girl is 157 mg/dL."

(e) Figure 40 shows the graph of the equation for $x \geq 18$. Because it does not make sense for x to be less than 0, graph the equation only in quadrant I. The horizontal axis is labeled x, Age (years), and the vertical axis is labeled y, Total cholesterol (mg/dL).   ●

Work Smart

Notice that the slope was not used to obtain an additional point on the graph of the equation. It would be difficult to find an additional point with a slope of 1.1. For example, from the y-intercept, go up 1.1 mg/dL and right 1 year and end up at (1, 158.1). It would be hard to draw the line through these two points!

Figure 40
$y = 1.1x + 157$

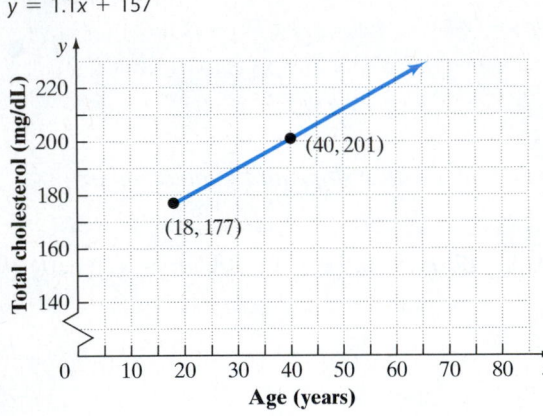

In Figure 40, notice the "broken line" (⌇) on the y-axis near the origin. This symbol indicates that a portion of the graph has been removed. This is done to avoid a lot of white space in the graph. Whenever you are reading a graph, always look carefully at how the axes are labeled and at the units on each axis.

Quick ✓

19. Based on data obtained from the National Center for Health Statistics, the birth weight y of a baby, measured in grams, is linearly related to gestation period x (in weeks) according to the equation

$$y = 143x - 2215$$

(a) Use the equation to predict the birth weight of a baby if the gestation period is 30 weeks.

(b) Use the equation to predict the birth weight of a baby if the gestation period is 36 weeks.

(c) Determine and interpret the slope of the equation.

(d) Explain why it does not make sense to interpret the y-intercept of the equation.

(e) Graph the equation in a rectangular coordinate system for $28 \leq x \leq 43$.

Slope can be interpreted as a rate of change. For this reason, when information in a problem is given as a rate of change (as in miles per gallon or dollars per pound), the rate of change will represent the slope in a linear model.

EXAMPLE 8 **Cost of Owning and Operating a Car**

Some costs involved in owning a car are affected by the number of miles driven (gas and maintenance), while others are not (comprehensive insurance, license plates, depreciation). Suppose the annual cost of operating a Chevy Cruze is $0.25 per mile plus $3000.

(a) Write a linear equation that relates the annual cost of operating the car y to the number of miles driven in a year x.

(b) What is the annual cost of driving 11,000 miles?

(c) If it cost Bernice $5375 to operate her Chevy Cruz last year, how many miles did she drive?

(d) Graph the equation in a rectangular coordinate system.

Solution

(a) It costs $0.25 per mile $= \dfrac{\$0.25}{1 \text{ mile}}$ to operate the car, so the rate of change, or slope m, of the linear equation is $0.25 per mile. The cost of $3000 is a cost that does not change with the number of miles driven. Put another way, if 0 miles are driven, the cost will be $3000, so this value represents the y-intercept, $(0, b)$. The linear equation that relates cost y to number of miles driven x is

$$y = 0.25x + 3000$$

(b) Let $x = 11{,}000$ in the equation $y = 0.25x + 3000$.

$$y = 0.25(11{,}000) + 3000$$
$$= 2750 + 3000$$
$$= \$5750$$

The cost of driving 11,000 miles in a year is $5750. This cost includes gas, insurance, maintenance, and depreciation in the value of the vehicle.

(c) Bernice spent $5375 driving her Chevy Cruz last year, so let $y = 5375$ in the equation $y = 0.25x + 3000$.

$$y = 0.25x + 3000$$
$$5375 = 0.25x + 3000$$
$$2375 = 0.25x$$

Divide both sides by 0.25: $\dfrac{2375}{0.25} = \dfrac{0.25x}{0.25}$

$$9500 = x$$

Bernice drove 9500 miles last year.

(d) See Figure 41. Notice that the graph is only drawn for $x \geq 0$. Do you know why?

Work Smart

The rate of change represents the slope of the linear equation.

Figure 41
$y = 0.25x + 3000$

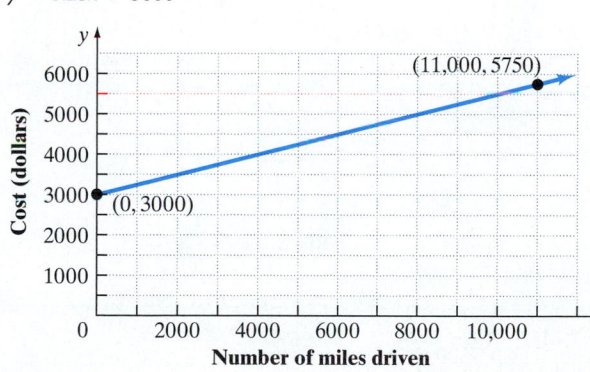
(11,000, 5750)
(0, 3000)
Cost (dollars)
Number of miles driven

Quick ✓

20. The daily cost in dollars, y, of renting a 16-foot moving truck for a day is $50 plus $0.38 per mile driven, x.

(a) Write a linear equation relating the daily cost in dollars, y, to the number of miles driven, x.

(b) Determine the cost of renting the truck if the truck is driven 75 miles.

(c) If the cost of renting the truck is $84.20, how many miles were driven?

(d) Graph the linear equation.

3.4 Exercises MyMathLab®

Problems **1–20** *are the* Quick ✔ *s that follow the* **EXAMPLES**.

Building Skills

In Problems 21–40, find the slope and y-intercept of the line whose equation is given. See Objective 1.

21. $y = 5x + 2$ **22.** $y = 7x + 1$

23. $y = x - 9$ **24.** $y = x - 7$

25. $y = -10x + 7$ **26.** $y = -6x + 2$

27. $y = -x - 9$ **28.** $y = -x - 12$

29. $2x + y = 4$ **30.** $3x + y = 9$

31. $2x + 3y = 24$ **32.** $6x - 8y = -24$

33. $5x - 3y = 9$ **34.** $10x + 6y = 24$

35. $x - 2y = 5$ **36.** $-x - 5y = 3$

37. $y = -5$ **38.** $y = 3$

39. $x = 6$ **40.** $x = -2$

In Problems 41–48, use the slope and y-intercept to graph each line whose equation is given. See Objective 2.

41. $y = x + 3$ **42.** $y = x + 4$

43. $y = -2x - 3$ **44.** $y = -4x - 1$

45. $y = \frac{2}{3}x + 2$ **46.** $y = \frac{4}{3}x - 3$

47. $y = -\frac{5}{2}x - 2$ **48.** $y = -\frac{2}{5}x + 3$

In Problems 49–56, graph each line using the slope and y-intercept. See Objective 3.

49. $4x + y = 5$ **50.** $3x + y = 2$

51. $x + 2y = -6$ **52.** $x - 2y = -4$

53. $3x - 2y = 10$ **54.** $4x + 3y = -6$

55. $6x + 3y = -15$ **56.** $5x - 2y = 6$

In Problems 57–68, find the equation of the line with the given slope and intercept. See Objective 4.

57. slope is -1; y-intercept is $(0, 8)$

58. slope is 1; y-intercept is $(0, 10)$

59. slope is $\frac{6}{7}$; y-intercept is $(0, -6)$

60. slope is $\frac{4}{7}$; y-intercept is $(0, -9)$

61. slope is $-\frac{1}{3}$; y-intercept is $\left(0, \frac{2}{3}\right)$

62. slope is $\frac{1}{4}$; y-intercept is $\left(0, \frac{3}{8}\right)$

63. slope is undefined; x-intercept is $(-5, 0)$

64. slope is 0; y-intercept is $(0, -2)$

65. slope is 0; y-intercept is $(0, 3)$

66. slope is undefined; x-intercept is $(4, 0)$

67. slope is 5; y-intercept is $(0, 0)$

68. slope is -3; y-intercept is $(0, 0)$

Mixed Practice

In Problems 69–92, graph each equation using any method you wish.

69. $y = 2x - 7$ **70.** $y = -4x + 1$

71. $3x - 2y = 24$ **72.** $2x + 5y = 30$

73. $y = -5$ **74.** $y = 4$

75. $x = -6$ **76.** $x = -3$

77. $6x - 4y = 0$ **78.** $3x + 8y = 0$

79. $y = -\frac{5}{3}x + 6$ **80.** $y = -\frac{3}{5}x + 4$

81. $2y = x + 4$ **82.** $3y = x - 9$

83. $y = \frac{x}{3}$ **84.** $y = -\frac{x}{4}$

85. $2x = -8y$ **86.** $-3x = 5y$

87. $y = -\frac{2}{3}x + 1$ **88.** $y = \frac{3}{2}x - 4$

89. $x + 2 = -7$ **90.** $y - 4 = -1$

91. $5x + y + 1 = 0$ **92.** $2x - y + 4 = 0$

Applying the Concepts

93. Mobile Device Usage In 2008, Americans spent an average of 19 minutes daily on their mobile devices. By 2015 that figure was up to 171 minutes daily. Assuming the increase in usage was constant, the number of minutes used can be calculated by the equation $y = 21.7x + 19$, where x represents the number of years after 2008 and y represents the number of daily minutes spent on mobile devices. (SOURCE: *CTIA, The Wireless Association*)

(a) To the nearest whole minute, how many minutes daily did Americans spend on their mobile devices in 2010?

(b) In which year did an average mobile device user spend approximately 106 minutes on her or his mobile devices?

(c) Interpret the slope of the equation $y = 21.7x + 19$.

(d) Can this trend continue indefinitely?

(e) Graph the equation in a rectangular coordinate system. Label the axes appropriately.

94. Counting Calories According to a National Academy of Sciences report, the recommended daily intake of calories for males between the ages of 7 and 15 can be calculated by the equation $y = 125x + 1125$, where x represents the boy's age and y represents the recommended caloric intake.

(a) What is the recommended caloric intake for a 12-year-old boy?

(b) What is the age of a boy whose recommended caloric intake is 2250 calories?

(c) Interpret the slope of $y = 125x + 1125$.

(d) Why would this equation not be accurate for a 3-year-old male?

(e) Graph the equation in a rectangular coordinate system. Label the axes appropriately.

95. Weekly Salary Dien is paid a salary of $400 per week plus an 8% commission on all sales he makes during the week.

(a) Write a linear equation that calculates his weekly income, where y represents his income and x represents the amount of sales.

(b) What is Dien's weekly income if he sold $1200 worth of merchandise?

(c) Graph the equation in a rectangular coordinate system. Label the axes appropriately.

96. Car Rental To rent a car for a day, Gloria pays $75 plus $0.10 per mile.

(a) Write a linear equation that calculates the daily cost, y, to rent a car that will be driven x miles.

(b) What is the cost to drive this car for 200 miles?

(c) If Gloria paid $87.50, how many miles did she drive?

(d) Graph the equation in a rectangular coordinate system. Label the axes appropriately.

Extending the Concepts

In Problems 97–102, find the value of the missing coefficient so that the line will have the given property.

97. $2x + By = 12$; slope is $\dfrac{1}{2}$

98. $Ax + 2y = 5$; slope is $\dfrac{3}{2}$

99. $Ax - 2y = 10$; slope is -2

100. $12x + By = -1$; slope is -4

101. $x + By = \dfrac{1}{2}$; y-intercept is $-\dfrac{1}{6}$

102. $4x + By = \dfrac{4}{3}$; y-intercept is $\dfrac{2}{3}$

In Problems 103 and 104, use the following information. In business, a cost equation relates the total cost of producing a product or good such as a refrigerator, rug, or blender to the number of goods produced. The simplest cost model is the linear cost model. In the linear cost model, the slope of the linear equation represents the cost of producing one additional unit of a product. Variable cost is reported as a rate of change, such as $40 per calculator. Examples of variable costs include labor costs and materials. The y-intercept of the linear equation represents the fixed costs of production — these are costs that exist regardless of the level of production. Fixed costs would include the cost of the manufacturing facility and insurance.

103. Manufacturing Costs Suppose the variable cost of manufacturing a graphing calculator is $40 per calculator, and the daily fixed cost is $4000.

(a) Write a linear equation that relates cost y to the number of calculators manufactured x.

(b) What is the daily cost of manufacturing 500 calculators?

(c) One day, the total cost was $19,000. How many calculators were manufactured?

(d) Graph the equation relating cost and number of calculators manufactured.

104. Cost Equations Manufacturing Costs Suppose the variable cost of manufacturing a cell phone is $35 per phone, and the daily fixed cost is $3600.

(a) Write a linear equation that relates the daily cost y to the number of cell phones manufactured x.

(b) What is the daily cost of manufacturing 400 cell phones?

(c) One day, the total cost was $13,225. How many cell phones were manufactured?

(d) Graph the equation relating cost and number of cell phones manufactured.

Explaining the Concepts

105. Describe the line whose graph is shown. Which of the following equations could have the graph that is shown?

(a) $y = 3x - 2$

(b) $y = -2x + 5$

(c) $y = 3$

(d) $2x + 3y = 6$

(e) $3x - 2y = 8$

(f) $4x - y = -4$

(g) $-5x + 2y = 12$

(h) $x - y = -3$

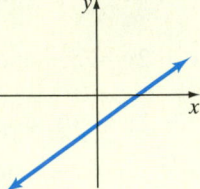

106. Without graphing, describe the orientation of each line (rises from left to right, and so on). Explain how you came to this conclusion.

(a) $y = 4x - 3$

(b) $y = -2x + 5$

(c) $y = x$

(d) $y = 4$

Technology Exercises

107. Exploration: Slope and y-intercept *Open the "Slope and y-intercept" Geogebra applet. The applet may be found using the QR code at the beginning of this section or through the Multimedia Library in MyMathLab. If you are using a keyboard, it is best to use the arrow keys on the keypad to move the slider, instead of the mouse.*

In the upper left corner are two "sliders." The top one, labeled m, controls the slope of the line, and the bottom one, labeled b, controls the y-intercept of the line. The slider changes the slope in increments of $\dfrac{1}{4}$ and changes the y-intercept in integer increments.

(a) Use the sliders to obtain the following graphs:

(1) $y = 3x$ (2) $y = 3x + 2$ (3) $y = 3x - 5$

What do all the graphs have in common? What is different about all the graphs? In general, describe the graph of $y = 3x + b$.

(b) Use the sliders to obtain the following graphs:

(1) $y = 2x + 1$

(2) $y = 5x + 1$

(3) $y = \dfrac{11}{4}x + 1$

What do all the graphs have in common? What is different about all the graphs? In general, describe the graph of $y = mx + 1$ for $m > 0$.

(c) Use the sliders to obtain the following graphs:

(1) $y = -2x + 1$

(2) $y = -5x + 1$

(3) $y = -\dfrac{11}{4}x + 1$

What do all the graphs have in common? What is different about all the graphs? In general, describe the graph of $y = mx + 1$ for $m < 0$.

(d) Consider each of the following slopes:

$4, \dfrac{1}{4}, \dfrac{1}{2}$ and 2. Rank the slopes in terms of "steepness" from least "steep" to most "steep." What conclusion can you draw from this observation?

(e) Consider each of the following slopes:

$-4, -\dfrac{1}{4}, -\dfrac{1}{2}$ and -2. Rank the slopes in terms of "steepness" from least "steep" to most "steep." What conclusion can you draw from this observation?

(f) What can you conclude about the appearance of a line when its slope is close to zero?

(g) What can you conclude about the appearance of a line when its slope approaches infinity or negative infinity (as $|m|$ becomes large)?

3.5 Point-Slope Form of a Line

Objectives

❶ Find the Equation of a Line Given a Point and a Slope

❷ Find the Equation of a Line Given Two Points

❸ Build Linear Models Using the Point-Slope Form of a Line

Are You Prepared for This Section?

Before getting started, complete the following problems. If you get a problem wrong, go back to the section cited and review the material.

P1. Solve $y - 3 = 2(x + 1)$ for y. [Section 2.4, pp. 114–115]

P2. Evaluate: $\dfrac{7 - 3}{4 - 2}$ [Section 1.7, p. 59]

 ❶ Find the Equation of a Line Given a Point and a Slope

Two forms for the equation of a line have been discussed: the standard form of a line, $Ax + By = C$, where A and B are not both zero, and the slope-intercept form of a line,

Figure 42

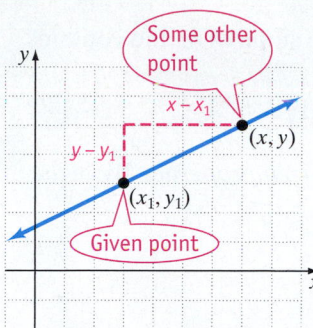

$y = mx + b$, where m is the slope and b is the y-intercept. We now introduce another form for the equation of a line.

Suppose we have a nonvertical line with slope m containing the point (x_1, y_1). See Figure 42. For any other point (x, y) on the line, the slope of the line is

$$m = \frac{y - y_1}{x - x_1}$$

Multiplying both sides by $x - x_1$ gives us

$$m(x - x_1) = y - y_1 \quad \text{or} \quad y - y_1 = m(x - x_1)$$

> **Point-Slope Form of an Equation of a Line**
>
> An equation of a nonvertical line with slope m containing the point (x_1, y_1) is
>
> $$\overset{\text{Slope}}{\underset{\uparrow \text{ Given point } \uparrow}{y - y_1 = m(x - x_1)}}$$

The point-slope form of a line can be used to write an equation in either slope-intercept form $(y = mx + b)$ or standard form $(Ax + By = C)$.

EXAMPLE 1   Using the Point-Slope Form of an Equation of a Line—Positive Slope

Figure 43
$y = 3x + 7$

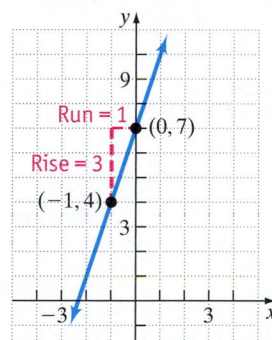

Find the equation of a line that has a slope of 3 and contains the point $(-1, 4)$. Write the equation in slope-intercept form. Graph the line.

Solution
Because the slope and a point on the line are given, use the point-slope form of a line with $m = 3$ and $(x_1, y_1) = (-1, 4)$.

$$y - y_1 = m(x - x_1)$$
$m = 3, x_1 = -1, y_1 = 4{:} \quad y - 4 = 3(x - (-1))$
$$y - 4 = 3(x + 1)$$

To write the equation in slope-intercept form, $y = mx + b$, solve the equation for y.

Distribute: $y - 4 = 3x + 3$
Add 4 to both sides: $y = 3x + 7$

See Figure 43 for a graph of the line. ●

EXAMPLE 2   Using the Point-Slope Form of an Equation of a Line—Negative Slope

Find the equation of a line that has a slope of $-\dfrac{3}{4}$ and contains the point $(-4, 2)$. Write the equation in slope-intercept form. Graph the line.

Solution
Because the slope and a point on the line are given, use the point-slope form of a line with $m = -\dfrac{3}{4}$ and $(x_1, y_1) = (-4, 2)$.

Figure 44
$y = -\dfrac{3}{4}x - 1$

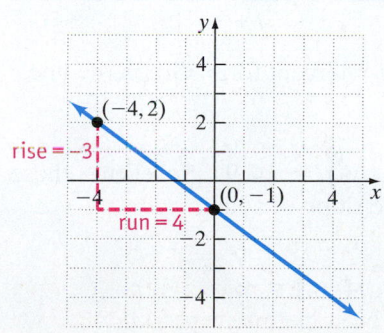

$$y - y_1 = m(x - x_1)$$
$m = -\dfrac{3}{4}, x_1 = -4, y_1 = 2{:} \quad y - 2 = -\dfrac{3}{4}(x - (-4))$

Simplify: $y - 2 = -\dfrac{3}{4}(x + 4)$

Distribute: $y - 2 = -\dfrac{3}{4}x - 3$

Add 2 to both sides: $y = -\dfrac{3}{4}x - 1$

See Figure 44 for a graph of the line. ●

EXAMPLE 3 **Finding the Equation of a Horizontal Line**

Figure 45
$y = 2$

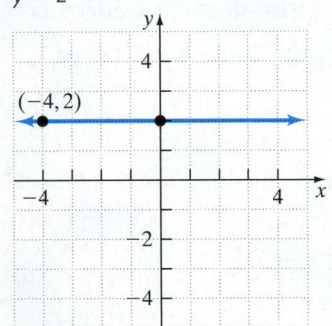

Find the equation of a horizontal line that contains the point $(-4, 2)$. Write the equation in slope-intercept form. Graph the line.

Solution
The line is horizontal, so its slope is 0. Because the slope and a point on the line are known, use the point-slope form with $m = 0$, $x_1 = -4$, and $y_1 = 2$.

$$y - y_1 = m(x - x_1)$$

Let $m = 0, x_1 = -4, y_1 = 2$: $y - 2 = 0(x - (-4))$

$$y - 2 = 0$$

Add 2 to both sides: $y = 2$

See Figure 45 for a graph of the line. ●

Work Smart

When the slope of a line is 0, the equation of the line will always be in the form "$y = $ some number."

▶ ❷ Find the Equation of a Line Given Two Points

Given two points, the equation of the line through the points can be found by first finding the slope of the line and then using the point-slope form of a line.

EXAMPLE 4 **How to Find the Equation of a Line from Two Points**

Find the equation of a line through the points $(1, 3)$ and $(4, 9)$. Write the equation in slope-intercept form. Graph the line.

Step-by-Step Solution

Step 1: Find the slope of the line containing the points.

Let $(x_1, y_1) = (1, 3)$ and $(x_2, y_2) = (4, 9)$. Substitute these values into the formula for the slope of a line.

$$m = \frac{y_2 - y_1}{x_2 - x_1} = \frac{9 - 3}{4 - 1} = \frac{6}{3} = 2$$

Step 2: Substitute the slope found in Step 1 and either point into the point-slope form of a line to find the equation.

$$y - y_1 = m(x - x_1)$$

Let $m = 2, x_1 = 1, y_1 = 3$: $\quad y - 3 = 2(x - 1)$

Step 3: Solve the equation for y.

Distribute the 2: $\quad y - 3 = 2x - 2$
Add 3 to both sides: $\quad y = 2x + 1$

The slope-intercept form of the equation is $y = 2x + 1$. See Figure 46 for the graph.

Figure 46
$y = 2x + 1$

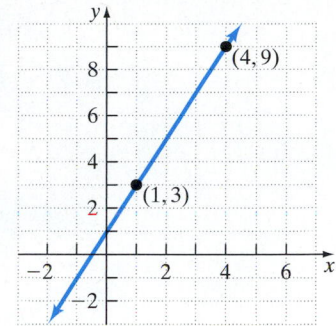

Work Smart

In example 4, the point $(1, 3)$ was used to find the equation of the line. Find the equation of the line using the point $(4, 9)$. You should get the same answer. Choose the coordinate that makes the algebra easiest.

> **Quick ✔**
>
> *In Problems 8 and 9, find the equation of the line containing the given points. Write the equation in slope-intercept form. Graph the line.*
>
> **8.** $(-3, -1); (3, 5)$ **9.** $(-1, 4); (1, -2)$

Work Smart: Study Skills

To write the equation of a nonvertical line, we must know either the slope of the line and a point on the line or two points on the line.

- If the **slope** and the **y-intercept** are known, use the **slope-intercept** form, $y = mx + b$.
- If the **slope** and a **point** that is not the y-intercept are known, use the **point-slope** form, $y - y_1 = m(x - x_1)$.
- If **two points** are known, first find the **slope** and then use that slope and one of the **points** in the **point-slope** form, $y - y_1 = m(x - x_1)$.

EXAMPLE 5 **Finding the Equation of a Vertical Line from Two Points**

Find the equation of a line through the points $(-3, 2)$ and $(-3, -4)$. Write the equation in slope-intercept form, if possible. Graph the line.

Solution

Let $(x_1, y_1) = (-3, 2)$ and $(x_2, y_2) = (-3, -4)$ in the formula for the slope of a line.

$$m = \frac{y_2 - y_1}{x_2 - x_1} = \frac{-4 - 2}{-3 - (-3)} = \frac{-6}{0}$$

The slope is undefined, so the line is vertical. No matter what value of y is chosen, the x-coordinate of the point on the line will be -3.

The equation of the line is $x = -3$. See Figure 47 for the graph.

Figure 47
$x = -3$

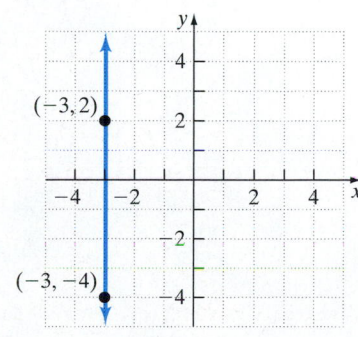

Work Smart

The equation of a vertical line cannot be written in slope-intercept form.

> **Quick ✔**
>
> **10.** Find the equation of the line containing the points $(3, 2)$ and $(3, -4)$. Write the equation in slope-intercept form, if possible. Graph the line.

Summary   Equations of Lines

Form of Line	Formula	Comments
Horizontal line	$y = b$	Graph is a horizontal line (slope is 0) with y-intercept $(0, b)$.
Vertical line	$x = a$	Graph is a vertical line (undefined slope) with x-intercept $(a, 0)$.
Point-slope	$y - y_1 = m(x - x_1)$	Useful for finding the equation of a line, given a point and the slope, or two points.
Slope-intercept	$y = mx + b$	Useful for finding the equation of a line, given the slope and y-intercept, or for quickly determining the slope and y-intercept of the line, given the equation of the line.
Standard form	$Ax + By = C$	Useful for finding the x- and y-intercepts.

Quick ✓

11. List the five forms of the equation of a line: Horizontal line: _____; Vertical line: _____; Point-slope: _____; Slope-intercept: _____; Standard form: _____.

▶ ❸ Build Linear Models Using the Point-Slope Form of a Line

The point-slope form of a line can be used to build linear models from data.

EXAMPLE 6   **Building a Linear Model from Data**

You purchased a phone card to use while traveling abroad. The remaining credit after 15 minutes of calls is $33.75, and the remaining credit after 60 minutes is $27. Assume the amount of credit remaining decreases linearly with the number of minutes used.

(a) Plot the points $(15, 33.75)$ and $(60, 27)$ in a rectangular coordinate system and graph the line.

(b) Find the linear equation in slope-intercept form that relates the credit remaining to the number of minutes used.

(c) Use the equation found in part (b) to predict the amount of credit remaining after 90 minutes of calls.

(d) Interpret the slope.

(e) Find the maximum number of minutes you can talk by solving the equation from part (b) for $y = 0$.

Solution

(a) Plot the ordered pairs $(15, 33.75)$ and $(60, 27)$ and draw a line through the points. See Figure 48 on the next page.

(b) Because two points on the line are known, find the slope of the line and then use the point-slope form of a line to find the equation of the line.

$$m = \frac{y_2 - y_1}{x_2 - x_1} = \frac{27 - 33.75}{60 - 15}$$

$$= \frac{-6.75}{45}$$

$$= -0.15$$

Figure 48

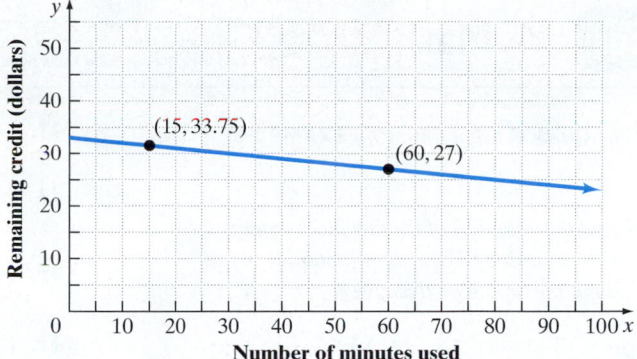

Now use the point-slope form of a line with $m = -0.15$, $x_1 = 60$, and $y_1 = 27$.

$$y - y_1 = m(x - x_1)$$

Let $m = -0.15$, $x_1 = 60$, $y_1 = 27$: $\quad y - 27 = -0.15(x - 60)$

$$y - 27 = -0.15x + 9$$

Add 27 to both sides: $\quad y = -0.15x + 36$

(c) To predict the amount of credit remaining after 90 minutes of calls, substitute 90 for x in the equation found in part (b).

$$y = -0.15x + 36$$

Let $x = 90$: $\quad = -0.15(90) + 36$

$$= 22.5$$

After 90 minutes of calls, \$22.50 credit remains on the phone card.

(d) The slope is -0.15. The amount of credit on the card is reduced by \$0.15 for every minute of calling.

(e) To find the maximum number of minutes you can talk, let $y = 0$ in the equation from part (b) and solve for x. That is, find the number of minutes used when the remaining credit on the card is \$0.

$$y = -0.15x + 36$$

Let $y = 0$: $\quad 0 = -0.15x + 36$

$$-36 = -0.15x$$

Divide both sides by -0.15: $\quad \dfrac{-36}{-0.15} = \dfrac{-0.15x}{-0.15}$

$$240 = x$$

When you have no credit left on the card, you have talked for 240 minutes. ●

Quick ✓

12. Armando owns a gas station. He has found that when the price of regular unleaded gasoline is \$2.75, he sells 400 gallons of gasoline between the hours of 7:00 A.M. and 8:00 A.M. When the price of regular unleaded gasoline is \$2.95, he sells 380 gallons of gasoline between the hours of 7:00 A.M. and 8:00 A.M. Suppose that the relation between quantity of gasoline sold and price is linear.

  (a) Plot the points $(2.75, 400)$ and $(2.95, 380)$ in a rectangular coordinate system and graph the line. Find the linear equation in slope-intercept form that relates quantity of gasoline sold y to price x.

  (b) Use the equation found in part (a) to predict the number of gallons of gasoline sold when the price is \$4.00.

  (c) Interpret the slope.

3.5 Exercises   MyMathLab®

*Problems **1–12** are the Quick ✔s that follow the EXAMPLES.*

Building Skills

In Problems 13–28, find the equation of the line that contains the given point and has the given slope. Write the equation in slope-intercept form and graph the line. See Objective 1.

13. $(2, 5)$; slope $= 3$ **14.** $(4, 1)$; slope $= 6$

15. $(-1, 2)$; slope $= -2$ **16.** $(6, -3)$; slope $= -5$

17. $(8, -1)$; slope $= \dfrac{1}{4}$ **18.** $(-8, 2)$; slope $= -\dfrac{1}{2}$

19. $(0, 13)$; slope $= -6$ **20.** $(0, -4)$; slope $= 9$

21. $(5, -7)$; slope $= 0$ **22.** $(3, 12)$; undefined slope

23. $(-4, 5)$; undefined slope   **24.** $(-7, -1)$; slope $= 0$

25. $(-3, 0)$; slope $= \dfrac{2}{3}$ **26.** $(-10, 0)$; slope $= -\dfrac{4}{5}$

27. $(-8, 6)$; slope $= -\dfrac{3}{4}$ **28.** $(-4, -6)$; slope $= \dfrac{3}{2}$

In Problems 29–36, find the equation of the line that contains the given point and satisfies the given information. Write the equation in slope-intercept form, if possible. See Objectives 1 and 2.

29. Vertical line that contains $(-3, 10)$

30. Horizontal line that contains $(-6, -1)$

31. Horizontal line that contains $(-1, -5)$

32. Vertical line that contains $(4, -3)$

33. Horizontal line that contains $(0.2, -4.3)$

34. Vertical line that contains $(3.5, 2.4)$

35. Vertical line that contains $\left(\dfrac{1}{2}, \dfrac{7}{4}\right)$

36. Horizontal line that contains $\left(\dfrac{3}{2}, \dfrac{9}{4}\right)$

In Problems 37–52, find the equation of the line that contains the given points. Write the equation in slope-intercept form, if possible. See Objective 2.

37. $(0, 4)$ and $(-2, 0)$ **38.** $(0, 3)$ and $(6, 0)$

39. $(1, 2)$ and $(0, 6)$ **40.** $(2, 4)$ and $(0, 8)$

41. $(-3, 2)$ and $(1, -4)$ **42.** $(-2, 4)$ and $(2, -2)$

43. $(-3, -11)$ and $(2, -1)$ **44.** $(4, 18)$ and $(-1, 3)$

45. $(4, -3)$ and $(-3, -3)$ **46.** $(-6, 5)$ and $(7, 5)$

47. $(2, -1)$ and $(2, -9)$ **48.** $(-3, 8)$ and $(-3, 1)$

49. $(0.1, 0.6)$ and $(0.5, 0.7)$ **50.** $(0.7, 0.8)$ and $(0.2, 0.4)$

51. $\left(\dfrac{1}{2}, -\dfrac{9}{4}\right)$ and $\left(\dfrac{5}{2}, -\dfrac{1}{4}\right)$ **52.** $\left(\dfrac{1}{3}, \dfrac{12}{5}\right)$ and $\left(\dfrac{4}{3}, \dfrac{2}{5}\right)$

Mixed Practice

In Problems 53–70, find the equation of the line described. Write the equation in slope-intercept form, if possible. Graph the line.

53. Contains $(4, -2)$ with slope $= 5$

54. Contains $(3, 2)$ with slope $= 4$

55. Horizontal line that contains $(-3, 5)$

56. Vertical line that contains $(-4, 2)$

57. Contains $(1, 3)$ and $(-4, -2)$

58. Contains $(-2, -8)$ and $(2, -6)$

59. Contains $(-2, 3)$ with slope $= \dfrac{1}{2}$

60. Contains $(-8, 3)$ with slope $= \dfrac{1}{4}$

61. Vertical line that contains $(5, 2)$

62. Horizontal line that contains $(-2, -6)$

63. Contains $(3, -19)$ and $(-1, 9)$

64. Contains $(-3, 13)$ and $(4, -22)$

65. Contains $(6, 3)$ with slope $= -\dfrac{2}{3}$

66. Contains $(-6, 3)$ with slope $= -\dfrac{2}{3}$

67. Contains $(-2, 3)$ and $(4, -6)$

68. Contains $(5, -3)$ and $(-3, 3)$

69. x-intercept: $(5, 0)$; y-intercept: $(0, -2)$

70. x-intercept: $(-6, 0)$; y-intercept: $(0, 4)$

Applying the Concepts

71. Shipping Packages The shipping department for a warehouse has noted that if 60 packages are shipped during a month, the total expenses for the department are $1635. If 120 packages are shipped during a month, the total expenses for the shipping department are $1770. Let x represent the number of packages and y represent the total expenses for the shipping department.

(a) Interpret the meaning of the point $(60, 1635)$ in the context of this problem.

(b) Plot the ordered pairs $(60, 1635)$ and $(120, 1770)$ in a rectangular coordinate system and graph the line through the points.

(c) Find the linear equation, in slope-intercept form, that relates the total expenses for the shipping department, y, to the number of packages sent, x.

(d) Use the equation found in part (c) to find the total expenses during a month when 200 packages were sent.

(e) Interpret the slope.

72. Retirement Plans Based on the retirement plan available through his employer, Kei knows that if he retires after 20 years, his monthly retirement income will be $3150. If he retires after 30 years, his monthly income increases to $3600. Let x represent the number of years of service and y represent the monthly retirement income.

(a) Interpret the meaning of the point $(30, 3600)$ in the context of this problem.

(b) Plot the ordered pairs $(20, 3150)$ and $(30, 3600)$ in a rectangular coordinate system and graph the line through the points.

(c) Find the linear equation, in slope-intercept form, that relates the monthly retirement income, y, to the number of years of service, x.

(d) Use the equation found in part (c) to find the monthly income for 15 years of service.

(e) Interpret the slope.

73. Credit Scores Your Fair Isaacs Corporation (FICO) credit score is used to determine your ability to get credit (such as a car loan or a credit card). FICO scores have a range of 300 to 850, with a higher score indicating a better credit history. Suppose a bank offers a person with a credit score of 600 a 15% interest rate on a 3-year car loan, while a person with a credit score of 750 is offered a 6% interest rate. Let x represent a person's credit score and y represent the interest rate.

(a) Fill in the ordered pairs: $(\_\_\_, 15)$; $(\_\_\_, 6)$

(b) Plot the ordered pairs from part (a) in a rectangular coordinate system, and graph the line through the points.

(c) Find the linear equation, in slope-intercept form, that relates the interest rate (in percent), y, to the credit score, x.

(d) Use the equation found in part (c) to predict the interest rate for a person with a credit score of 700.

(e) Interpret the slope.

74. U.S. Traffic Fatalities Nationwide, the statistics for traffic fatalities show a decline. In 2005, the United States had 43,510 fatal crashes, and in 2014 the number dropped to 32,675. Let y represent the number of traffic fatalities and x represent the number of years since 2005.

(SOURCE: *National Highway Traffic Safety Administration*)

(a) Fill in the ordered pairs: $(\_, 43{,}510)$; $(\_, 32{,}675)$

(b) Plot the ordered pairs from part (a) in a rectangular coordinate system, and graph the line through the points.

(c) Find the linear equation, in slope-intercept form, that relates the number of traffic fatalities, y, to the number of years since 2005, x.

(d) Assuming this trend continues, use the equation found in part (c) to find the number of traffic fatalities in 2019.

(e) Interpret the slope.

Extending the Concepts

Up to this point, when we knew the slope of a line and a point (not the y-intercept) on the line, we found the equation of the line using point-slope form. We could also use the slope-intercept form to find this equation.

For example, suppose that we were asked to find the equation of the line that has slope 6 and passes through the point $(2, -5)$. Use the slope-intercept form, $y = mx + b$, to write the equation of this line. We know $m = 6$, so we have $y = 6x + b$. We also know that $y = -5$ when $x = 2$, so substitute 2 for x and -5 for y into the equation and solve for b.

$$y = 6x + b$$

Let $x = 2$ and $y = -5$: $-5 = 6(2) + b$

Multiply: $-5 = 12 + b$

Subtract 12 from each side: $-5 - 12 = 12 - 12 + b$

$$-17 = b$$

We now know that $b = -17$, and because we also know that $m = 6$, the equation of the line is $y = 6x - 17$. Use this technique to find the slope-intercept form of the line for Problems 75–80.

75. $(-4, 2)$; slope $= 3$ **76.** $(5, -2)$; slope $= 4$

77. $(3, -8)$; slope $= -2$ **78.** $(-1, 7)$; slope $= -5$

79. $\left(\dfrac{2}{3}, \dfrac{1}{2}\right)$; slope $= 6$ **80.** $\left(\dfrac{4}{3}, -\dfrac{3}{2}\right)$; slope $= -9$

For Problems 81–86, first find the slope of the line containing the two points, then use the method just described to find the slope-intercept form of the line.

81. $(6, -13)$ and $(-2, -5)$ **82.** $(-10, -5)$ and $(2, 7)$

83. $(5, -1)$ and $(-10, -4)$ **84.** $(-6, -3)$ and $(9, 2)$

85. $(-4, 8)$ and $(2, -1)$ **86.** $(4, -9)$ and $(8, -19)$

Explaining the Concepts

87. You are asked to write the equation of the line through the points $(3, 1)$ and $(4, 7)$ in slope-intercept form. After calculating the slope of the line, you choose the point-slope form and assign $x_1 = 3$ and $y_1 = 1$. Your friend calculates the slope, and lets $x_1 = 4$ and $y_1 = 7$. Will you and your friend obtain the same answer? Explain why or why not.

88. You are asked to write the equation of the line through $(-1, 3)$ and $(0, 4)$. Which form of a line would you choose to find the equation? Explain why you chose this form. Could you also use one of the other forms?

3.6 Parallel and Perpendicular Lines

Objectives

1 Determine Whether Two Lines Are Parallel

2 Find the Equation of a Line Parallel to a Given Line

3 Determine Whether Two Lines Are Perpendicular

4 Find the Equation of a Line Perpendicular to a Given Line

▶ **1** Determine Whether Two Lines Are Parallel

Two lines in the rectangular coordinate system that do not intersect (that is, have no points in common) are said to be *parallel*. The equations of lines can be used to determine whether the lines are parallel.

> **Definition**
>
> Two nonvertical lines are **parallel** *if and only if* their slopes are equal and they have different y-intercepts. Vertical lines are parallel if they have different x-intercepts.

Work Smart

The use of the words "if and only if" in the definition of parallel lines means that there are two statements being made:

1. If two nonvertical lines are parallel, then their slopes are equal and they have different y-intercepts.

2. If two nonvertical lines have equal slopes and different y-intercepts, then they are parallel.

Figure 49(a) shows nonvertical parallel lines. Figure 49(b) shows vertical parallel lines.

Figure 49
Parallel lines

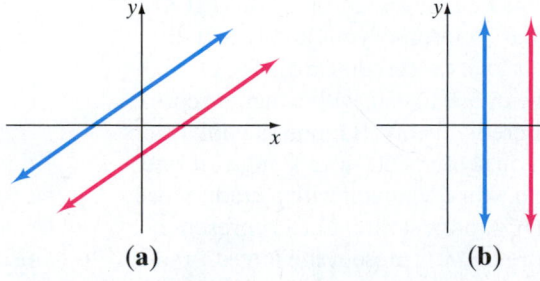

(a) (b)

Prepared?...Answers

P1. $\dfrac{1}{3}$ **P2.** $-\dfrac{5}{3}$

To determine whether two lines are parallel, find the slope and y-intercept of each line by writing the equations of the lines in slope-intercept form. If the slopes are the same but the y-intercepts are different, then the lines are parallel.

EXAMPLE 1   **Determining Whether Two Lines Are Parallel**

Determine whether the line $y = 4x - 5$ is parallel to $y = 3x - 2$. Graph the lines to confirm the results.

Solution

The line $y = 4x - 5$ has slope 4 and the y-intercept $(0, -5)$. The line $y = 3x - 2$ has slope 3 and y-intercept $(0, -2)$. Because the lines have different slopes, they are not parallel. Figure 50 shows that the lines intersect at $(3, 7)$. Therefore, the lines are not parallel.

Figure 50

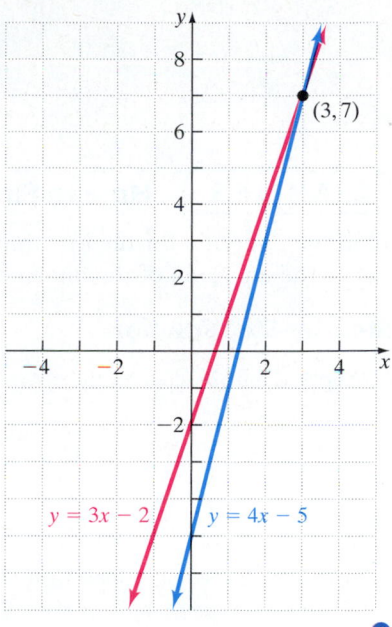

EXAMPLE 2   **Determining Whether Two Lines Are Parallel**

Determine whether the line $-3x + y = 5$ is parallel to $6x - 2y = -2$. Graph the lines to confirm the results.

Solution

Solve each equation for y so that each is in slope-intercept form.

$$-3x + y = 5$$

Add $3x$ to both sides: $\qquad y = 3x + 5$

The slope of the line $-3x + y = 5$ is 3 and the y-intercept is $(0, 5)$.

$$6x - 2y = -2$$

Subtract $6x$ from both sides: $\qquad -2y = -6x - 2$

Divide both sides by -2: $\qquad y = \dfrac{-6x - 2}{-2}$

Divide each term in the numerator by -2: $\qquad y = 3x + 1$

The slope of $6x - 2y = -2$ is 3 and the y-intercept is $(0, 1)$.
   The lines have the same slope, 3, but different y-intercepts, so they are parallel. Figure 51 shows the graph of the two lines.

Work Smart

Make sure both criteria for parallel lines are satisfied.

1. Same slope

2. Different y-intercepts

Lines with the same slope and same y-intercept are called *coincident lines*.

Figure 51

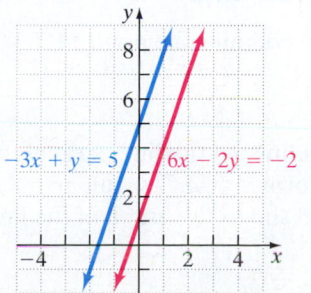

Quick ✔

1. Two nonvertical lines are parallel if and only if their _____ are equal and they have different _____. Vertical lines are parallel if they have different _____.

In Problems 2–4, determine whether the two lines are parallel. Graph the lines to confirm the results.

2. $y = 2x + 1$

$\qquad y = -2x - 3$

3. $6x + 3y = 3$

$\qquad 10x + 5y = 10$

4. $4x + 5y = 10$

$\qquad 8x + 10y = 20$

▶ ❷ Find the Equation of a Line Parallel to a Given Line

Now that we know how to identify parallel lines, we can find the equation of a line that is parallel to a given line.

EXAMPLE 3 **How to Find the Equation of a Line Parallel to a Given Line**

Find the equation for the line that is parallel to $2x + y = 5$ and contains the point $(-1, 3)$.
Write the equation of the line in slope-intercept form. Graph the lines.

Step-by-Step Solution

Step 1: Find the slope of the given line.

$$2x + y = 5$$
Subtract $2x$ from both sides: $\qquad y = -2x + 5$

The slope of the line is -2, so the slope of the parallel line is also -2.

Step 2: Use the point-slope form of a line with the given point and the slope found in Step 1 to find the equation of the parallel line.

$$y - y_1 = m(x - x_1)$$
$m = -2, x_1 = -1, y_1 = 3$: $\qquad y - 3 = -2(x - (-1))$

Step 3: Write the equation in slope-intercept form by solving for y.

$$y - 3 = -2(x + 1)$$
Distribute the -2: $\qquad y - 3 = -2x - 2$
Add 3 to both sides: $\qquad y = -2x + 1$

Figure 52

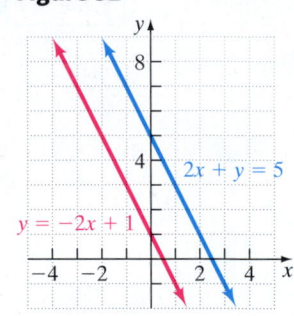

The equation of the line parallel to $2x + y = 5$ is $y = -2x + 1$. Figure 52 shows the graph of the parallel lines.

> **Quick ✓**
>
> In Problems 5 and 6, find the equation of the line that contains the given point and is parallel to the given line. Write the line in slope-intercept form. Graph the lines.
>
> **5.** $y = 2x + 1$ containing $(2, 3)$ **6.** $3x + 2y = 4$ containing $(-2, 3)$

EXAMPLE 4 **Finding the Equation of a Line Parallel to a Given Line**

Figure 53

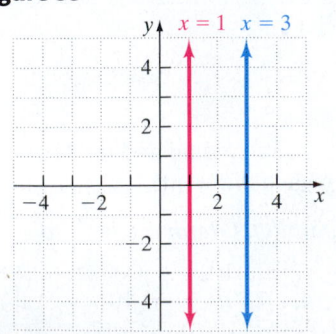

Find the equation for the line that is parallel to $x = 3$ and contains the point $(1, 5)$.
Graph the lines.

Solution

The equation of the given line is $x = 3$. This is a vertical line, so the line parallel to it will also be vertical. Vertical lines have equations of the form $x = a$. The line parallel to $x = 3$ that contains the point $(1, 5)$ is $x = 1$. Figure 53 shows the graph of the lines $x = 3$ and $x = 1$.

> **Quick ✓**
>
> In Problems 7 and 8, find the equation of the line that contains the given point and is parallel to the given line. Write the line in slope-intercept form, if possible. Graph the lines.
>
> **7.** $x = -2$ containing $(3, 1)$ **8.** $y + 3 = 0$ containing $(-2, 5)$

Figure 54
Perpendicular lines.

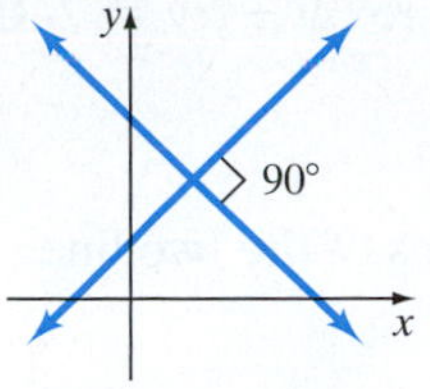

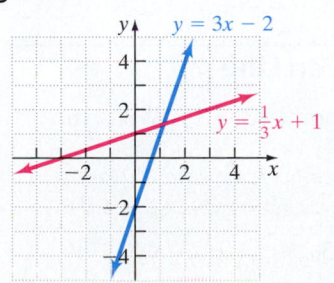

3 Determine Whether Two Lines Are Perpendicular

Two lines that intersect at a right (90°) angle are *perpendicular*. See Figure 54.

Just as slopes can tell us whether two lines are parallel, slopes can also tell us whether two lines are perpendicular.

> **Definition**
>
> Two nonvertical lines are **perpendicular** if and only if the product of their slopes is -1. Put another way, two nonvertical lines are perpendicular if their slopes are negative reciprocals of each other. Any vertical line is perpendicular to any horizontal line.

If m_1 and m_2 are negative reciprocals of each other, then $m_1 = \dfrac{-1}{m_2}$. For example, the numbers 4 and $-\dfrac{1}{4}$ are negative reciprocals. Watch out, though, because the use of the word *negative* does not mean that the slope of the perpendicular line must be negative—it means that the nonvertical lines have slopes that are opposite in sign. One is positive and one is negative.

EXAMPLE 5 **Finding the Slope of a Line Perpendicular to a Given Line**

Find the slope of a line perpendicular to a line whose slope is:

(a) 5 **(b)** $-\dfrac{2}{3}$.

Solution

(a) To find the slope of a line perpendicular to a given line, determine the negative reciprocal of the slope of the given line. The negative reciprocal of 5 is $\dfrac{-1}{5} = -\dfrac{1}{5}$. Any line whose slope is $-\dfrac{1}{5}$ will be perpendicular to the line whose slope is 5 because $5\left(-\dfrac{1}{5}\right) = -1$.

(b) The negative reciprocal of $-\dfrac{2}{3}$ is $\dfrac{-1}{-\frac{2}{3}} = (-1)\left(-\dfrac{3}{2}\right) = \dfrac{3}{2}$. Any line whose slope is $\dfrac{3}{2}$ will be perpendicular to the line whose slope is $-\dfrac{2}{3}$ because $-\dfrac{2}{3} \cdot \dfrac{3}{2} = -1$.

> **Quick ✓**
>
> **9.** Given any two nonvertical lines, if the product of their slopes is -1, then the lines are _____.
>
> **10.** *True or False* Two different lines L_1 and L_2 have slopes $m_1 = 4$ and $m_2 = -4$, so L_1 is perpendicular to L_2.
>
> *In Problems 11–13, find the slope of a line perpendicular to the line whose slope is given.*
>
> **11.** -4 **12.** $\dfrac{5}{4}$ **13.** $-\dfrac{1}{8}$

EXAMPLE 6 **Determining Whether Two Lines Are Perpendicular**

Determine whether the line $y = 3x - 2$ is perpendicular to $y = \dfrac{1}{3}x + 1$. Graph the lines to confirm the results.

Figure 55

Solution

First find the slope of each line. If the product of the slopes is -1 (or the slopes are negative reciprocals of each other), then the lines are perpendicular. The slope of $y = 3x - 2$ is $m_1 = 3$. The slope of $y = \dfrac{1}{3}x + 1$ is $m_2 = \dfrac{1}{3}$. Because the product of the slopes, $m_1 \cdot m_2 = 3 \cdot \dfrac{1}{3} = 1 \neq -1$, the lines are not perpendicular. Notice that the slopes are reciprocals of each other, but not *negative* reciprocals of each other. Figure 55 shows the graph of the two lines.

EXAMPLE 7 **Determining Whether Two Lines Are Perpendicular**

Determine whether the line $2x + 3y = -6$ is perpendicular to $3x - 2y = 2$. Graph the lines to confirm the results.

Solution

Write each equation in slope-intercept form to find the slopes of the two lines.

$$2x + 3y = -6$$

Subtract 2x from both sides: $\qquad 3y = -2x - 6$

Divide both sides by 3: $\qquad y = \dfrac{-2x - 6}{3}$

Divide each term in the numerator by 3: $\qquad y = -\dfrac{2}{3}x - 2$

The slope of $2x + 3y = -6$ is $m_1 = -\dfrac{2}{3}$.

$$3x - 2y = 2$$

Subtract 3x from both sides: $\qquad -2y = -3x + 2$

Divide both sides by -2: $\qquad y = \dfrac{-3x + 2}{-2}$

Divide each term in the numerator by -2: $\qquad y = \dfrac{3}{2}x - 1$

The slope of $3x - 2y = 2$ is $m_2 = \dfrac{3}{2}$.

The product of the slopes is $m_1 \cdot m_2 = -\dfrac{2}{3} \cdot \dfrac{3}{2} = -1$, so the lines are perpendicular.

Put another way, because the slopes are negative reciprocals of each other, the lines are perpendicular. See Figure 56 for the graph of the two lines. ●

Figure 56

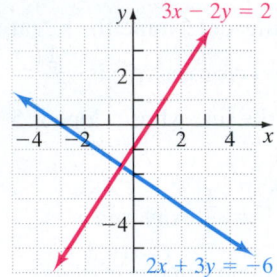

Quick ✔

In Problems 14–16, determine whether the given lines are perpendicular. Graph the lines.

14. $y = 4x - 3$

$\qquad y = -\dfrac{1}{4}x - 4$

15. $2x - y = 3$

$\qquad x - 2y = 2$

16. $5x + 2y = 8$

$\qquad 2x - 5y = 10$

▶ ④ **Find the Equation of a Line Perpendicular to a Given Line**

Now that we know how to find the slope of a line perpendicular to a second line, we can find the equation of a line that is perpendicular to a given line.

EXAMPLE 8 **How to Find the Equation of a Line Perpendicular to a Given Line**

Find the equation of the line that is perpendicular to the line $y = 3x - 2$ and contains the point $(3, -1)$. Write the equation of the line in slope-intercept form, if possible. Graph the two lines.

Step-by-Step Solution

Step 1: Find the slope of the given line. $\qquad$ The slope of the line $y = 3x - 2$ is 3.

Step 2: Find the slope of the perpendicular line.

The slope of the perpendicular line is the negative reciprocal of 3, which is $\dfrac{-1}{3} = -\dfrac{1}{3}$.

Step 3: Use the point-slope form of a line with the given point and the slope found in Step 2 to find the equation of the perpendicular line.

$$y - y_1 = m(x - x_1)$$

$m = -\dfrac{1}{3}, x_1 = 3, y_1 = -1: \quad y - (-1) = -\dfrac{1}{3}(x - 3)$

Step 4: Write the equation in slope-intercept form by solving for y.

$$y + 1 = -\dfrac{1}{3}(x - 3)$$

Distribute the $-\dfrac{1}{3}$: $\quad y + 1 = -\dfrac{1}{3}x + 1$

Subtract 1 from both sides: $\quad y = -\dfrac{1}{3}x$

Figure 57

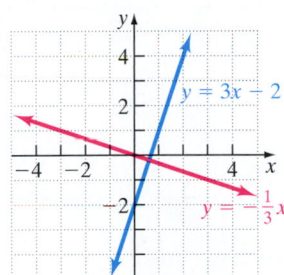

The equation of the line perpendicular to $y = 3x - 2$ through $(3, -1)$ is $y = -\dfrac{1}{3}x$. Figure 57 shows the graph of the two lines. ●

Quick ✓

In Problems 17 and 18, find the equation of the line that contains the given point and is perpendicular to the given line. Write the line in slope-intercept form, if possible. Graph the lines.

17. $(-4, 2); y = 2x + 1$

18. $(-2, -1); 2x + 3y = 3$

EXAMPLE 9

Finding the Equation of a Line Perpendicular to a Given Line

Find the equation of the line that is perpendicular to the line $x = 2$ and contains the point $(-4, 3)$. Write the equation of the line in slope-intercept form, if possible. Graph the two lines.

Solution

The equation $x = 2$ is the equation of a vertical line. Therefore, the line perpendicular to $x = 2$ will be horizontal. Horizontal lines have slopes equal to 0. To find the equation of the perpendicular line, we use the point-slope formula.

$$y - y_1 = m(x - x_1)$$

$m = 0, x_1 = -4, y_1 = 3: \quad y - 3 = 0(x - (-4))$

$$y - 3 = 0$$

Add 3 to both sides: $\quad y = 3$

Figure 58

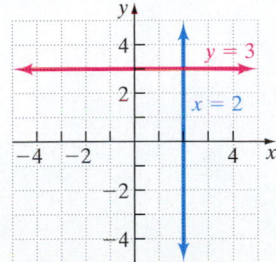

The equation of the line perpendicular to $x = 2$ through the point $(-4, 3)$ is $y = 3$. See Figure 58 for the graph of the two lines. ●

Quick ✓

In Problems 19 and 20, find the equation of the line that contains the given point and is perpendicular to the given line. Write the line in slope-intercept form, if possible. Graph the lines.

19. $x = -4$ containing $(-1, -5)$

20. $y + 2 = 0$ containing $(3, -2)$

3.6 Exercises MyMathLab®   Exercise numbers in green have complete video solutions in MyMathLab or may be accessed using the QR code to the right.

*Problems **1–20** are the Quick ✔s that follow the **EXAMPLES**.*

Building Skills

In Problems 21–28, fill in the chart with the missing slopes. See Objectives 1 and 3.

	Slope of the Given Line	Slope of a Line Parallel to the Given Line	Slope of a Line Perpendicular to the Given Line
21.	$m = -3$		
23.	$m = \dfrac{1}{2}$		
25.	$m = -\dfrac{4}{9}$		
27.	$m = 0$		

	Slope of the Given Line	Slope of a Line Parallel to the Given Line	Slope of a Line Perpendicular to the Given Line
22.	$m = 4$		
24.	$m = -\dfrac{1}{8}$		
26.	$m = \dfrac{5}{2}$		
28.	$m = $ undefined		

In Problems 29–42, determine whether the lines are parallel, perpendicular, or neither. See Objectives 1 and 3.

29. $L_1: y = 2x - 3$
  $L_2: y = -\dfrac{1}{2}x + 1$

30. $L_1: y = -4x + 3$
  $L_2: y = 4x - 1$

31. $L_1: y = \dfrac{3}{4}x + 2$
  $L_2: y = 0.75x - 1$

32. $L_1: y = 0.8x + 6$
  $L_2: y = \dfrac{4}{5}x + \dfrac{19}{3}$

33. $L_1: y = -\dfrac{5}{3}x - 6$
  $L_2: y = \dfrac{3}{5}x - 1$

34. $L_1: y = 3x - 1$
  $L_2: y = 6 - \dfrac{x}{3}$

35. $L_1: x + y = -3$
  $L_2: y - x = 1$

36. $L_1: x - 4y = 24$
  $L_2: 2x - 8y = -8$

37. $L_1: 2x + 5y = 5$
  $L_2: 5x + 2y = 4$

38. $L_1: x - 2y = -8$
  $L_2: x + 2y = 2$

39. $L_1: 4x - 5y - 15 = 0$
  $L_2: 8x - 10y + 5 = 0$

40. $L_1: x + y = 6$
  $L_2: x - y = -2$

41. $L_1: 4x = 3y + 3$
  $L_2: 6y = 8x + 36$

42. $L_1: 2x - 5y - 45 = 0$
  $L_2: 5x + 2y - 8 = 0$

In Problems 43–54, find the equation of the line that contains the given point and is parallel to the given line. Write the equation in slope-intercept form, if possible. See Objective 2.

43. $(4, -2); y = 3x - 1$ **44.** $(7, -5); y = 2x + 6$

45. $(-3, 8); y = -4x + 5$ **46.** $(-2, 6); y = -5x - 2$

47. $(3, -7); y = 4$ **48.** $(-4, 5); x = -3$

49. $(-1, 10); x = 10$ **50.** $(4, -8); y = -1$

51. $(10, 2); 3x - 2y = 5$ **52.** $(6, 7); 2x + 3y = 9$

53. $(-1, -10); x + 2y = 4$ **54.** $(-3, -5); 2x - 5y = 6$

In Problems 55–66, find the equation of the line that contains the given point and is perpendicular to the given line. Write the equation in slope-intercept form, if possible. See Objective 4.

55. $(3, 5); y = \dfrac{1}{2}x - 2$

56. $(4, 7); y = \dfrac{1}{3}x - 3$

57. $(-4, -1); y = -4x + 1$

58. $(-2, -5); y = -2x + 5$

59. $(-2, 1); x$-axis

60. $(3, -6); y$-axis

61. $(7, 5); y$-axis

62. $(11, -6); x$-axis

63. $(0, 0); 2x + 5y = 7$

64. $(0, 0)$; $6x + 4y = 3$

65. $(-10, -3)$; $5x - 3y = 4$

66. $(-6, 10)$; $3x - 5y = 2$

Mixed Practice

In Problems 67–82, find the equation of the line that has the given properties. Write the equation in slope-intercept form, if possible. Graph each line.

67. Contains $(3, -5)$; slope $= 7$

68. Contains $(-2, 10)$; slope $= 3$

69. Contains $(2, 9)$; perpendicular to the line $y = -5x + 3$

70. Contains $(8, 3)$; parallel to the line $y = -4x + 2$

71. Contains $(6, -1)$; parallel to the line $y = -7x + 2$

72. Contains $(5, -2)$; perpendicular to the line $y = 4x + 3$

73. Contains $(-6, 2)$ and $(-1, -8)$

74. Contains $(-4, -1)$ and $(-3, -5)$

75. Slope $= 3$, y-intercept $= (0, -2)$

76. Slope $= -2$, y-intercept $= (0, 7)$

77. Contains $(5, 1)$; parallel to the line $x = -6$

78. Contains $(2, 7)$; perpendicular to the line $x = -8$

79. Contains $(3, -2)$; parallel to the line $4x + 3y = 9$

80. Contains $(-3, -2)$; perpendicular to the line $6x - 2y = 1$

81. Contains $(-1, -3)$; perpendicular to the line $x - 2y = -10$

82. Contains $(-7, 2)$; parallel to the line $x + 4y = 2$

In Problems 83–90, each line contains the given points. (a) Find the slope of each line. (b) Determine whether the lines are parallel, perpendicular, or neither.

83. L_1: $(0, -1)$ and $(-2, -7)$
L_2: $(-1, 5)$ and $(2, -4)$

84. L_1: $(-3, -14)$ and $(1, 2)$
L_2: $(0, 2)$ and $(-3, -10)$

85. L_1: $(2, 8)$ and $(7, 18)$
L_2: $(-2, -3)$ and $(6, 13)$

86. L_1: $(6, 0)$ and $(-2, 8)$
L_2: $(4, 1)$ and $(-6, -9)$

87. L_1: $(-2, -5)$ and $(4, -2)$
L_2: $(-8, -5)$ and $(0, -1)$

88. L_1: $(1, 6)$ and $(-1, -10)$
L_2: $(0, 1)$ and $(-2, 17)$

89. L_1: $(-6, -9)$ and $(3, 6)$
L_2: $(10, -8)$ and $(-5, 1)$

90. L_1: $(-8, -8)$ and $(4, 1)$
L_2: $(12, 8)$ and $(-4, -4)$

Applying the Concepts

A parallelogram is a quadrilateral in which both pairs of opposite sides are parallel. In Problems 91 and 92, plot the given points, draw the figure, and then use slope to determine whether the figure is a parallelogram.

91. $A(-1, 1)$; $B(3, 5)$; $C(6, 4)$; $D(2, 0)$

92. $A(-1, -3)$; $B(1, -1)$; $C(5, 1)$; $D(3, -2)$

A rectangle is a parallelogram that contains one right angle. That is, one pair of sides are perpendicular. In Problems 93 and 94, plot the given points, draw the figure, and then use slope to determine whether the figure is a rectangle.

93. $A(6, -1)$; $B(1, -6)$; $C(-3, -2)$; $D(2, 3)$

94. $A(1, 1)$; $B(-1, 5)$; $C(5, 8)$; $D(7, 4)$

A right triangle is a triangle that contains one right angle. In Problems 95–98, plot each point and form the triangle ABC. Verify using slope that the triangle is a right triangle.

95. $A(-2, 5)$; $B(1, 3)$; $C(3, 6)$

96. $A(-5, 3)$; $B(6, 0)$; $C(5, 5)$

97. $A(4, -3)$; $B(0, -3)$; $C(4, 2)$

98. $A(-2, 5)$; $B(12, 3)$; $C(10, -11)$

Extending the Concepts

In Problems 99 and 100, find the missing coefficient so that the lines are parallel.

99. $-3y = 6x - 12$ and $4x + By = -2$

100. $Ax + 2y = 4$ and $15x = 5y + 20$

In Problems 101 and 102, find the missing coefficient so that the lines are perpendicular.

101. $Ax + 6y = -6$ and $12 - 6y = -9x$

102. $x - By = 10$ and $3y = -6x + 9$

△ **103.** An altitude of a triangle is a line segment drawn from a vertex of the triangle perpendicular to the opposite side. Plot the following points, draw triangle ABC and the segment joining points B and D, and then determine whether $\overline{BD}$ is an altitude of the triangle. $A(-6, -3); B(-4, 7); C(-1, 2); D(-3, 0)$

△ **104.** The coordinates of the vertices of a quadrilateral are $A(2, 1), B(4, 6), C(6, 6),$ and $D(9, 2)$. Use slopes to show that the diagonals of the quadrilateral, $\overline{AC}$ and $\overline{BD}$, are perpendicular to each other.

Explaining the Concepts

105. Pamela thinks that line l_1, with slope $\dfrac{3}{2}$, and line l_2, with slope $-\dfrac{3}{2}$ are perpendicular. Is Pamela correct? Explain why or why not.

106. You are asked to determine if the following lines are parallel, perpendicular or neither: L_1 contains the points $(1, 1)$ and $(3, 5)$ and L_2 contains the points $(-1, -1)$ and $(4, 9)$. List the steps you would follow to make this determination and describe the criteria you would use to answer the question.

Technology Exercises

107. Exploration: Parallel Lines *Open the "Parallel and Perpendicular Lines" Geogebra applet. The applet may be found using the QR code at the beginning of this section or through the Multimedia Library in MyMathLab. If you are using a keyboard, it is best to use the arrow keys on the keypad to move the slider, instead of the mouse.*

In the upper left corner are two pairs of "sliders." The top one in each pair, labeled slope$_1$ and slope$_2$, controls the slope of the line, and the bottom one, labeled yintercept$_1$ and yintercept$_2$, controls the y-intercept of the line. The slider changes the slope in increments of $\dfrac{1}{4}$ and the y-intercept in integer increments.

(a) Use the sliders for y_1 to obtain the graph of $y_1 = 3x + 2$. Now, use the sliders for y_2 to find an equation that is parallel to y_1. The applet will show the words "Parallel Lines!" once you have successfully found a pair. Write the equation of the parallel line.

(b) Use the sliders for y_1 to obtain the graph of $y_1 = \dfrac{1}{2}x + 4$. Now, use the sliders for y_2 to find an equation that is parallel to y_1. The applet will show the words "Parallel Lines!" once you have successfully found a pair. Write the equation of the parallel line.

(c) Use the sliders for y_1 to obtain the graph of $y_1 = -2x + 3$. Now, use the sliders for y_2 to find an equation that is parallel to y_1. The applet will show the words "Parallel Lines!" once you have successfully found a pair. Write the equation of the parallel line.

(d) What relationship did you notice exists between a pair of parallel lines? Specifically, what is the relationship between the slopes (if any) and what is the relationship between the y-intercepts (if any)?

108. Exploration: Perpendicular Lines *Open the "Parallel and Perpendicular Lines" Geogebra applet. The applet may be found using the QR code at the beginning of this section or through the Multimedia Library in MyMathLab. If you are using a keyboard, it is best to use the arrow keys on the keypad to move the slider, instead of the mouse.*

(a) Use the sliders for y_1 to create the graph of $y_1 = 4x + 2$. Now, use the sliders for y_2 to find an equation that is perpendicular to y_1. The applet will show the words "Perpendicular Lines!" once you have successfully found a pair. Write the equation of the perpendicular line.

(b) Use the sliders for y_1 to create the graph of $y_1 = \dfrac{1}{2}x + 4$. Now, use the sliders for y_2 to find an equation that is perpendicular to y_1. The applet will show the words "Perpendicular Lines!" once you have successfully found a pair. Write the equation of the perpendicular line.

(c) Use the sliders for y_1 to create the graph of $y_1 = -2x + 3$. Now, use the sliders for y_2 to find an equation that is perpendicular to y_1. The applet will show the words "Perpendicular Lines!" once you have successfully found a pair. Write the equation of the perpendicular line.

(d) What relationship did you notice exists between a pair of perpendicular lines? Specifically, what is the relationship between the slopes (if any) and what is the relationship between the y-intercepts (if any)?

Putting the Concepts Together (Sections 3.1–3.6)

We designed these problems so that you can review Sections 3.1–3.6 and show your mastery of the concepts. Take time to work these problems before proceeding with the next section. The answers are at the back of the text on page AN-13.

1. Given the linear equation $4x - 3y = 10$, determine whether the ordered pair $(1, -2)$ is a solution to the equation. Answer Yes or No.

In Problems 2 and 3, graph each equation using the point-plotting method.

2. $y = \dfrac{2}{3}x - 1$ **3.** $-5x + 2y = 10$

4. Given the equation $-8x + 2y = 6$, determine
   (a) the *x*-intercept
   (b) the *y*-intercept

5. Graph the equation $4x + 3y = 6$ by finding the *x*-intercept and the *y*-intercept.

6. Given the equation $6x + 9y = -12$, determine
   (a) the slope
   (b) the *y*-intercept

7. Find the slope of the line that contains the points $(3, -5)$ and $(-6, -2)$.

8. Given the linear equation $2y = -5x - 4$, find
   (a) the slope of a line perpendicular to the given line
   (b) the slope of a line parallel to the given line

9. Determine whether the lines are parallel, perpendicular, or neither. Explain how you came to your conclusion.

$$L_1: 10x + 5y = 2 \quad L_2: y = -2x + 3$$

In Problems 10–16, write the equation of the line that satisfies the given conditions. Write the equation in slope-intercept form, if possible.

10. slope $= 3$ and *y*-intercept is $(0, 1)$

11. slope $= -6$ and passes through $(-1, 4)$

12. through $(4, -1)$ and $(-2, 11)$

13. through $(-8, 0)$ and perpendicular to $y = \dfrac{2}{5}x - 5$

14. through $(-8, 3)$ and parallel to $-8y + 2x = -1$

15. horizontal line through $(-6, -8)$

16. through $(2, 6)$ with undefined slope

17. Shipping Expenses The shipping department records indicate that during a week when 80 packages were shipped, the total expenses recorded for the shipping department were \$1180. During a different week, 50 packages were shipped, and the expenses recorded were \$850. Use the ordered pairs $(80, 1180)$ and $(50, 850)$ to determine the average rate to ship an additional package.

18. Diamonds Suppose the relation between the cost of a diamond and its weight is linear. In looking at two diamonds, we find that one of the diamonds weighs 0.7 carat and costs \$3543, while the other diamond weighs 0.8 carat and costs \$4378. (SOURCE: *diamonds.com*)
   (a) Use the ordered pairs $(0.7, 3543)$ and $(0.8, 4378)$ to find a linear equation that relates the price of a diamond to its weight.
   (b) Interpret the slope.
   (c) Predict the price of a diamond that weighs 0.76 carat.

3.7 Linear Inequalities in Two Variables

Objectives

1 Determine Whether an Ordered Pair Is a Solution to a Linear Inequality
2 Graph Linear Inequalities
3 Solve Problems Involving Linear Inequalities

Are You Prepared for This Section?

Before getting started, complete the following problems. If you get a problem wrong, go back to the section cited and review the material.

P1. Solve: $x - 4 > 5$ [Section 2.8, pp. 150–151]
P2. Solve: $3x + 1 \le 10$ [Section 2.8, pp. 152–154]
P3. Solve: $2(x + 1) - 6x > 18$ [Section 2.8, pp. 153–154]

▶ 1 Determine Whether an Ordered Pair Is a Solution to a Linear Inequality

In Chapter 2, inequalities in one variable were solved. In this section, linear inequalities in two variables will be discussed.

> **Definition**
>
> **Linear inequalities in two variables** are inequalities equivalent to one of the forms
>
> $$Ax + By < C \quad Ax + By > C \quad Ax + By \le C \quad Ax + By \ge C$$
>
> where A and B are not both zero. A linear inequality in two variables x and y is **satisifed** by an ordered pair (a, b) if a true statement results when x is replaced by a and y is replaced by b.

EXAMPLE 1 **Determining Whether an Ordered Pair Is a Solution to a Linear Inequality in Two Variables**

Determine which of the following ordered pairs are solutions to the linear inequality $2x + y \leq 9$.

(a) $(3, 5)$ (b) $(1, 3)$ (c) $(3, -2)$

Solution

(a) Let $x = 3$ and $y = 5$ in the inequality. If a true statement results, then $(3, 5)$ is a solution to the inequality.

$$2x + y \leq 9$$
$$x = 3, y = 5: \quad 2(3) + 5 \overset{?}{\leq} 9$$
$$6 + 5 \overset{?}{\leq} 9$$
$$11 \leq 9 \quad \text{False}$$

The statement $11 \leq 9$ is false, so $(3, 5)$ is not a solution to the inequality.

(b) Let $x = 1$ and $y = 3$ in the inequality. If a true statement results, then $(1, 3)$ is a solution to the inequality.

$$2x + y \leq 9$$
$$x = 1, y = 3: \quad 2(1) + 3 \overset{?}{\leq} 9$$
$$2 + 3 \overset{?}{\leq} 9$$
$$5 \leq 9 \quad \text{True}$$

The statement $5 \leq 9$ is true, so $(1, 3)$ is a solution to the inequality.

(c) Let $x = 3$ and $y = -2$ in the inequality. If a true statement results, then $(3, -2)$ is a solution to the inequality.

$$2x + y \leq 9$$
$$x = 3, y = -2: \quad 2(3) + (-2) \overset{?}{\leq} 9$$
$$6 - 2 \overset{?}{\leq} 9$$
$$4 \leq 9 \quad \text{True}$$

The statement $4 \leq 9$ is true, so $(3, -2)$ is a solution to the inequality.   ●

Quick ✓

Determine which of the following ordered pairs are solutions to the given linear inequality. Answer Yes or No.

(a) $(2, 1)$ (b) $(3, 4)$ (c) $(-1, 10)$

1. $2x + y > 7$ **2.** $-3x + 2y \leq 8$

▶ ❷ Graph Linear Inequalities

The **graph of a linear inequality in two variables** x and y consists of all points whose coordinates (x, y) satisfy the inequality.

If we replace the inequality symbol in a linear inequality of the form

$$Ax + By < C \qquad Ax + By > C \qquad Ax + By \leq C \qquad Ax + By \geq C$$

with an equal sign, we obtain the equation of a line, $Ax + By = C$. This **boundary line** separates the xy-plane into two regions called **half-planes**. See Figure 59 on the next page.

When graphing the boundary line, use dashes if the inequality is strict ($<$ or $>$) to indicate points on the line do not satisfy the inequality, and use a solid line if the inequality is nonstrict ($\leq$ or $\geq$) to indicate points on the line do satisfy the inequality.

Figure 59

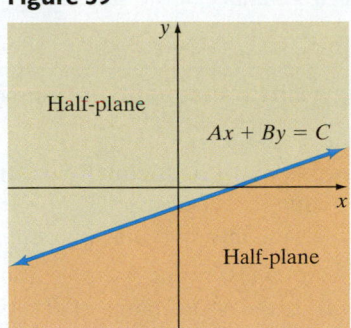

Figure 60

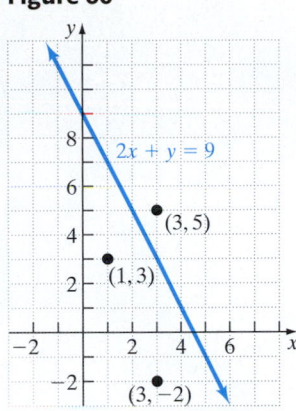

Let's consider the linear inequality $2x + y \le 9$ in Example 1. Figure 60 shows the graph of the equation $2x + y = 9$ and the three points $(3, 5)$, $(1, 3)$, and $(3, -2)$ examined in Example 1.

In Example 1, the two points below the line, $(1, 3)$ and $(3, -2)$, satisfy the inequality $2x + y \le 9$, while $(3, 5)$ does not. In fact, all points below the line satisfy the inequality, and all points above the inequality do not. What can we learn from this? If a point satisfies a linear inequality in two variables, then *all* points in the half-plane containing that point satisfy the inequality. If a point does not satisfy a linear inequality in two variables, then all points in the *opposite* half-plane satisfy the inequality. For this reason, a single **test point** is all that is required to obtain the graph of a linear inequality in two variables. Even so, it is not a bad idea to check a few points to verify your algebra.

EXAMPLE 2   How to Graph a Linear Inequality in Two Variables

Graph the linear inequality: $y > 2x - 5$

Figure 61

Step-by-Step Solution

Step 1: Replace the inequality symbol with an equal sign and graph the resulting equation.

Replace $>$ with $=$ to obtain $y = 2x - 5$. Graph the line $y = 2x - 5$. The y-intercept of the line is $(0, -5)$, and the slope of the line is $m = 2$. Starting at the y-intercept $(0, -5)$, use slope $= \dfrac{2}{1} = \dfrac{\text{rise}}{\text{run}}$ to count up 2 units and 1 unit to the right. Graph the line through $(0, -5)$ and $(1, -3)$ using a dashed line because the inequality is strict ($>$). See Figure 61.

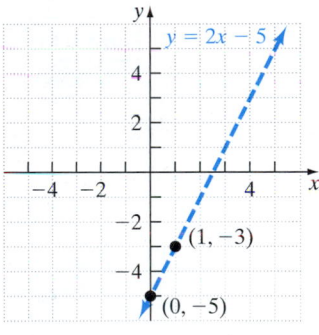

Step 2: Select any test point that is not on the line, and determine whether the test point satisfies the inequality. If the test point satisfies the inequality, shade the half-plane containing the test point; otherwise, shade the half-plane opposite it.

Use $(0, 0)$ as the test point.

$$y > 2x - 5$$

Test point $(0, 0)$: $0 \overset{?}{>} 2(0) - 5$

$0 > -5$ True

Because 0 is greater than -5, the point $(0, 0)$ satisfies the inequality $y > 2x - 5$. Therefore, shade the half-plane containing the point $(0, 0)$. See Figure 62.

Figure 62
$y > 2x - 5$

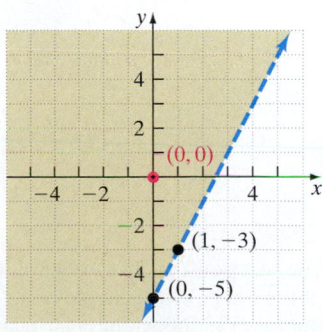

Work Smart

When the line does not contain the origin, it is usually easiest to choose the origin, $(0, 0)$, as the test point. To double-check your results, choose a point, such as $(0, 4)$, in the shaded region. Does $(0, 4)$ satisfy $y > 2x - 5$? Yes!

The shaded region represents the solution to the linear inequality. Because the inequality is strict, points on the line $y = 2x - 5$ do not satisfy the inequality.

Below is a summary of the steps for graphing a linear inequality in two variables.

> **Graphing a Linear Inequality in Two Variables**
>
> **Step 1:** Replace the inequality symbol with an equal sign, and graph the resulting equation. If the inequality is strict ($<$ or $>$), use a dashed line; if the inequality is nonstrict ($\leq$ or $\geq$), use a solid line. The graph separates the xy-plane into two half-planes.
>
> **Step 2:** Select a test point P that is not on the line.
>   **(a)** If the coordinates of P satisfy the inequality, shade the half-plane containing P.
>   **(b)** If the coordinates of P do not satisfy the inequality, shade the half-plane that does not contain P.
>
> **Note:** If the inequality is nonstrict, all points on the line also satisfy the inequality. If the inequality is strict, all points on the line *do not* satisfy the inequality.

Quick ✓

3. When drawing the boundary line for the graph of $Ax + By \geq C$, use a _____ line. When drawing the boundary line for the graph of $Ax + By < C$, use a _____ line.

4. The boundary line separates the xy-plane into two regions called _____.

In Problems 5 and 6, graph each linear inequality.

5. $y < -2x + 1$ 6. $y \geq 3x + 2$

EXAMPLE 3

Graphing a Linear Inequality in Two Variables

Graph the linear inequality: $5x + 2y \geq 10$

Solution

Graph the equation $5x + 2y = 10$ by finding the intercepts of the equation.

Find the *y*-intercept:

$$5x + 2y = 10$$
Let $x = 0$: $\quad 5(0) + 2y = 10$
$$2y = 10$$
Divide both sides by 2: $\quad\quad y = 5$

The y-intercept is $(0, 5)$.

Find the *x*-intercept:

$$5x + 2y = 10$$
Let $y = 0$: $\quad 5x + 2(0) = 10$
$$5x = 10$$
Divide both sides by 5: $\quad\quad x = 2$

The x-intercept is $(2, 0)$.

Plot the points $(0, 5)$ and $(2, 0)$ and draw the equation of the line as a solid line because the inequality is nonstrict ($\geq$). See Figure 63(a) on the next page.

Because the graph of the equation does not contain the origin, select the origin $(0, 0)$ as the test point.

$$5x + 2y \geq 10$$

Test point $(0, 0)$: $\quad 5(0) + 2(0) \overset{?}{\geq} 10$

$$0 \geq 10 \quad \text{False}$$

Work Smart

The test point (0, 0) is not a solution of the linear inequality, so none of the other points in that half-plane are solutions. That's why we shade the half-plane opposite the half-plane that contains the origin.

Because the test point $(0, 0)$ results in a false statement, shade the half-plane opposite the half-plane that contains $(0, 0)$. See Figure 63(b). The shaded region and all points on the line $5x + 2y = 10$ represent the solution set.

Figure 63

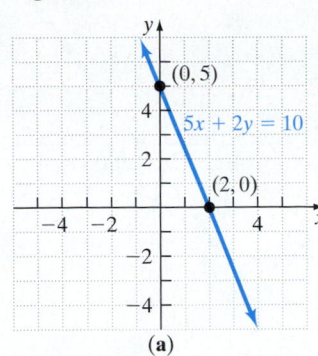

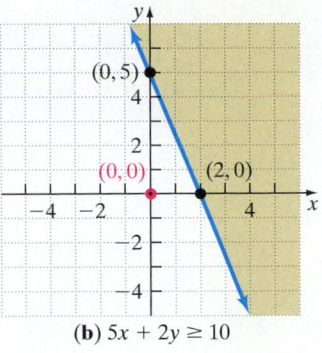

(a)

(b) $5x + 2y \geq 10$

Quick ✓

In Problems 7 and 8, graph each linear inequality.

7. $2x + 3y \leq 6$ **8.** $4x - 6y > 12$

EXAMPLE 4 **Graphing a Linear Inequality Where the Boundary Line Goes Through the Origin**

Graph the linear inequality: $-4x + 3y > 0$

Solution

Graph the equation $-4x + 3y = 0$ using a dashed line because the inequality is strict. Because this is an equation of the form $Ax + By = 0$, the graph will pass through the origin. For this reason, write the equation in slope-intercept form.

$$-4x + 3y = 0$$

Add $4x$ to both sides: $3y = 4x$

Divide both sides by 3: $y = \dfrac{4}{3}x$

Work Smart

The line $y = \dfrac{4}{3}x$ could also be graphed by plotting a point at the y-intercept, $(0, 0)$, and then going up 4 units and to the right 3 units to find another point on the line.

The line has slope $m = \dfrac{4}{3}$ and y-intercept $(0, 0)$. To find two additional points on the line, let $x = 3$ and let $x = -3$, and compute the corresponding values of y:

$$y = \frac{4}{3}x \qquad\qquad y = \frac{4}{3}x$$

$$x = 3:\ \ y = \frac{4}{3}(3) \qquad\qquad x = -3:\ \ y = \frac{4}{3}(-3)$$

$$y = 4 \qquad\qquad\qquad\qquad y = -4$$

So three points on the boundary line are $(0, 0)$, $(3, 4)$, and $(-3, -4)$. See Figure 64(a) on the next page.

The graph of the equation contains the origin, so use $(1, 3)$ as the test point.

Work Smart

Because the boundary line contains the origin, the test point cannot be the origin.

$$-4x + 3y > 0$$

Test point (1, 3): $-4(1) + 3(3) \overset{?}{>} 0$

$$-4 + 9 \overset{?}{>} 0$$

$$5 > 0 \quad \text{True}$$

Because the test point $(1, 3)$ results in a true statement, shade the half-plane that contains $(1, 3)$. See Figure 64(b) on the next page. The shaded region represents the solution set.

(continued)

Figure 64

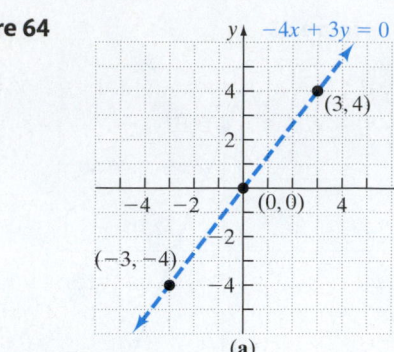

(a)

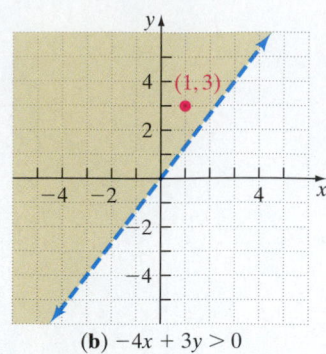

(b) $-4x + 3y > 0$

Quick ✓

In Problems 9 and 10, graph each linear inequality.

9. $3x + y < 0$ **10.** $2x - 5y \leq 0$

EXAMPLE 5   **Graphing a Linear Inequality Involving a Horizontal Line**

Graph the linear inequality $y < 2$.

Solution

Begin by graphing $y = 2$ using a dashed line (because the inequality is strict). See Figure 65(a). Next select a test point. The origin $(0, 0)$ can be selected as the test point because the line does not contain $(0, 0)$.

$$y < 2$$

Test point $(0, 0)$: $0 < 2$ True

Because the test point satisfies the inequality, shade the half-plane that contains $(0, 0)$ as shown in Figure 65(b). The shaded region represents the solution to the linear inequality. Doesn't it seem intuitive that we shade below the line since these are the y-values that are less than 2?

Work Smart: Study Skills

Have you noticed the patterns?
If $y > mx + b$, shade above.
If $y < mx + b$, shade below.
If $y > b$, shade above.
If $y < b$, shade below.
If $x > a$, shade to the right.
If $x < a$, shade to the left.

Figure 65

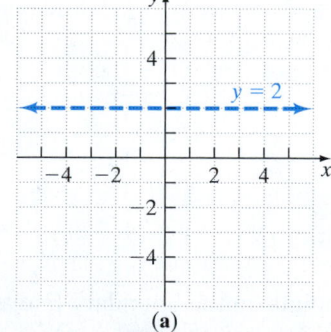

(a)

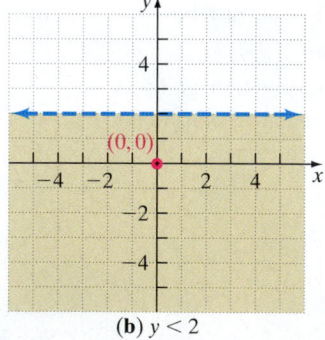

(b) $y < 2$

Graphing a linear inequality involving a vertical line is similar to graphing a linear inequality involving a horizontal line. That is, graph the vertical boundary line, choose a test point, and shade the half-plane containing the points that satisfy the inequality.

Quick ✓

In Problems 11–13, graph each linear inequality.

11. $y < -1$ **12.** $3y - 9 \geq 0$ **13.** $x > 6$

▶ ❸ Solve Problems Involving Linear Inequalities

Many applications of linear inequalities that involve two variables occur in areas such as nutrition, manufacturing, and sales. They are even used to describe weight loads in elevators!

EXAMPLE 6 **Elevator Capacity**

An elevator designed and manufactured by Otis Elevator Company has a capacity of 3000 pounds. According to the National Center for Health Statistics, the average man 20 years or older weighs 196 pounds, and the average woman 20 years or older weighs 166 pounds. (SOURCE: *otis.com*)

(a) Write a linear inequality that describes the various combinations of men and women who can ride this elevator.

(b) Can 10 men and 9 women ride in this elevator safely?

(c) Can 7 men and 9 women ride in this elevator safely?

Solution

(a) Use the first three steps in the problem-solving strategy given in Chapter 2 on page 124 to help develop the linear inequality.

Step 1: Identify Determine the number of men and women who can ride in the elevator without going over 3000 pounds.

Step 2: Name Let m represent the number of males and f represent the number of females on the elevator.

Step 3: Translate To keep things simple, assume that all the people on the elevator are "average." If one male is on the elevator, the weight on the elevator is 196 pounds. If two males are on the elevator, the weight is $2(196) = 392$ pounds. In general, if m males are on the elevator, the weight is $196m$. Similarly, if f females are on the elevator, the weight is $166f$. Because the capacity of the elevator is 3000 pounds, the weight on the elevator must be no more than 3000 pounds, so use a "less than or equal to" ($\leq$) inequality. A linear inequality that describes the weight limitations on the elevator is

$$196m + 166f \leq 3000 \quad \text{The Model}$$

(b) Letting $m = 10$ and $f = 9$, we obtain

$$196(10) + 166(9) \overset{?}{\leq} 3000$$
$$3454 \leq 3000 \quad \text{False}$$

The inequality is false, so 10 men and 9 women cannot ride the elevator safely.

(c) Letting $m = 7$ and $f = 9$, we obtain

$$196(7) + 166(9) \overset{?}{\leq} 3000$$
$$2866 \leq 3000 \quad \text{True}$$

The inequality is true, so 7 men and 9 women can ride the elevator safely. ●

Quick ✓

14. Kevin just received $2 from his grandma. He goes to a candy store where each sucker sells for $0.20 and each taffy stick sells for $0.25.

(a) Write a linear inequality that describes the various combinations of suckers, s, and taffy sticks, t, Kevin can buy.

(b) Can Kevin buy 6 suckers and 3 taffy sticks?

(c) Can Kevin buy 5 suckers and 5 taffy sticks?

*Problems **1–14** are the **Quick ✓**s that follow the **EXAMPLES**.*

Building Skills

In Problems 15–26, determine which of the following ordered pairs, if any, are solutions to the given linear inequality in two variables. Answer Yes or No. See Objective 1.

15. $y > -x + 2$ $A(2, 4)$ $B(3, -6)$ $C(0, 0)$

16. $y \leq x - 5$ $A(-4, -6)$ $B(0, -8)$ $C(3, 10)$

17. $y \leq 3x - 1$ $A(-6, -15)$ $B(0, 0)$ $C(-1, -6)$

18. $y > -2x + 1$ $A(0, 0)$ $B(-2, 3)$ $C(2, -1)$

19. $3x \geq 2y$ $A(-8, -12)$ $B(3, 5)$ $C(-5, -8)$

20. $4y < 5x$ $A(-4, -6)$ $B(5, 6)$ $C(-12, -15)$

21. $2x - 3y < -6$ $A(2, -1)$ $B(4, 8)$ $C(-3, 0)$

22. $3x + 5y \geq 4$ $A(2, 0)$ $B(-3, 4)$ $C(6, -2)$

23. $x \leq 2$ $A(7, 2)$ $B(2, 5)$ $C(4, 2)$

24. $y \leq -3$ $A(-3, -3)$ $B(-1, -4)$ $C(-6, -1)$

25. $y > -1$ $A(-1, 1)$ $B(3, -1)$ $C(4, -2)$

26. $x > 10$ $A(3, 12)$ $B(11, -6)$ $C(10, 16)$

In Problems 27–56, graph each linear inequality. See Objective 2.

27. $y > 3x - 2$ **28.** $y \leq 4x + 2$ **29.** $y \leq -x + 1$

30. $y > x - 3$ **31.** $y < \dfrac{x}{2}$ **32.** $y > \dfrac{x}{3}$

33. $y > 5$ **34.** $y < 4$ **35.** $y \leq \dfrac{2}{5}x + 3$

36. $y \geq \dfrac{3}{2}x - 2$ **37.** $y \geq -\dfrac{4}{3}x + 2$ **38.** $y \leq -\dfrac{3}{4}x + 1$

39. $x < 2$ **40.** $x > -4$ **41.** $3x - 4y < 12$

42. $2x + 6y < -6$ **43.** $2x + y \geq -4$ **44.** $6x - 8y \geq 24$

45. $x + y > 0$ **46.** $x - y > 0$ **47.** $5x - 2y < -8$

48. $3x + 2y \leq -9$ **49.** $x > -1$ **50.** $y < -3$

51. $y \leq 4$ **52.** $x \geq 2$ **53.** $\dfrac{x}{3} - \dfrac{y}{5} \geq 1$

54. $\dfrac{x}{4} + \dfrac{y}{2} < 1$ **55.** $-3 \geq x - y$ **56.** $-2 > x + y$

Applying the Concepts

In Problems 57–68, translate each statement into a linear inequality. Then graph the inequality.

57. The sum of two numbers, x and y, is at least 26.

58. One number, x, is at least 12 more than a second number, y.

59. The difference between two numbers, x and y, is at most 5.

60. The sum of twice a number x and a second number y is more than 11.

61. One number, x, is no more than 3 less than a second number, y.

62. The difference of a number, x, and half a second number, y, is positive.

63. The sum of a number, x, and 3 times a second number, y, is negative.

64. The sum of two numbers, x and y, is at least -3.

65. The difference of twice a number, x, and half a second number, y, is at least 5.

66. The difference between three times a number x and twice y is less than 1.

67. The product of a number x and 5 is at least 2 more than twice a second number, y.

68. The sum of half a number, x, and twice a second number, y, is at least 12.

69. Trip to the Aquarium A kindergarten class has a maximum of $120 to spend on a trip to the aquarium. The cost of admission for students is $3, and adults must pay $5.

(a) Write a linear inequality that describes the various combinations of the number of students s and adults a that can go on the field trip.

(b) Is there enough money to pay for 32 students and 6 adults?

(c) Is there enough money to pay for 29 students and 4 adults?

70. Fishing Trip Patrick's rowboat can hold a maximum of 500 pounds. In Patrick's circle of friends, the average adult weighs 160 pounds and the average child weighs 75 pounds.

(a) Write a linear inequality that describes the various combinations of the number of adults a and children c that can go on a fishing trip in the rowboat.

(b) Will the boat sink with 2 adults and 4 children?

(c) Will the boat sink with 1 adult and 5 children?

71. Salesperson Leon sells two different brands of bicycles. For each ten-speed bicycle sold, he makes $50, and for each three-speed bicycle sold, he makes $40. Leon has a monthly goal of earning at least $4000.

(a) Write a linear inequality that describes Leon's options for making his sales goal.

(b) Will Leon make his sales goal if he sells 30 ten-speed bicycles and 70 three-speed bicycles next month?

(c) Will Leon make his sales goal if he sells 50 ten-speed bicycles and 20 three-speed bicycles next month?

72. School Fundraiser For the Clinton Elementary School fundraiser, Guillermo earns 5 points for each magazine subscription he sells and 2 points for each novelty. It takes at least 40 points to earn the MP3 player that Guillermo wants.

(a) Write a linear inequality that describes the various combinations of subscriptions m and novelties n that Guillermo can sell to earn enough points for the MP3 player.

(b) Will he earn the MP3 player if he sells 8 novelties and 5 subscriptions?

(c) Will he earn the MP3 player if he sells 20 novelties and 2 subscriptions?

Extending the Concepts

Two inequalities joined by the word "and" or "or" form a **compound inequality.** *The directions "Solve $3x - 2y > 6$ and $x + y < 2$" mean to find the ordered pairs that satisfy both inequalities at the same time. The solution of a compound inequality of this type is found by graphing the two linear inequalities in the same Cartesian coordinate plane and then shading the area where the graphs overlap. Graph the compound inequalities in Problems 73–78.*

73. $3x - 2y > 6$ and $x + y < 2$

74. $2x - 4y \leq 4$ and $3x + 2y \geq 6$

75. $y > \dfrac{3}{4}x - 1$ and $x \geq 0$ **76.** $y > \dfrac{2}{3}x + 3$ and $y \geq 0$

77. $x < -3$ and $y \leq 4$ **78.** $y < -1$ and $x \geq -2$

Explaining the Concepts

79. Describe how you can tell whether a point (a, b) is a solution to a linear inequality in two variables. How can you decide whether (a, b) lies on the boundary line?

80. Explain why $(0, 0)$ is generally a good test point. Describe how you can decide when this is a good test point to use and when you should choose a different test point.

Technology Exercises

Technology can be used to graph linear inequalities in two variables. The figure shows the graph of $3x + y < 7$ using Desmos.

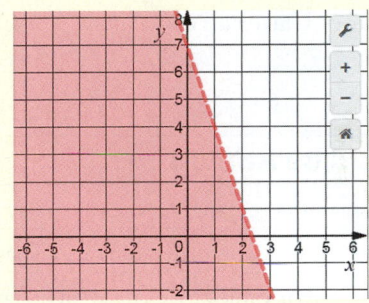

In Problems 81–92, graph the inequalities using technology.

81. $y > 3$ **82.** $y < -2$

83. $y < 5x$ **84.** $y \geq \dfrac{2}{3}x$

85. $y > 2x + 3$ **86.** $y < -3x + 1$

87. $y \leq \dfrac{1}{2}x - 5$ **88.** $y \geq -\dfrac{4}{3}x + 5$

89. $3x + y \leq 4$ **90.** $-4x + y \geq -5$

91. $2x + 5y \leq -10$ **92.** $3x + 4y \geq 12$

Chapter 3 Activity: Graphing Practice

Focus: Graphing and identifying the graphs of linear equations and inequalities.

Time: 15–20 minutes

Group size: 3–4

Materials needed: Blank piece of paper and graph paper for each group member.

1. On each of four pieces of paper, write one of the following equations or inequalities.

(a) $x + 3 \geq -4(y - 1)$

(b) $4(y - 2) + 5 = 5(x + 1)$

(c) $3(y - 7) + 4 = -2(x - 3) - 2$

(d) $\dfrac{2}{3}(x + 2y) + \dfrac{8}{3} \geq 0$

2. Place the four papers face down on the desk and mix them up. One by one, each group member should choose a piece of paper. Do not show your choice to the other members of the group.

3. Carefully graph the chosen equation/inequality on your graph paper. Do not label your graph.

4. When finished, place the unlabeled graphs in a pile.

5. As a group, with no help from the member who drew the graph, write each equation or inequality that was drawn.

6. As a group, discuss the outcome.

Chapter 3 Review

Section 3.1 The Rectangular Coordinate System and Equations in Two Variables

KEY TERMS

x-axis	xy-plane	y-coordinate
y-axis	Coordinate axes	Quadrants
Origin	Ordered pair (x, y)	Equation in two variables
Rectangular or Cartesian coordinate system	Coordinates	Sides
	x-coordinate	Satisfy

You Should Be Able To...	EXAMPLE	Review Exercises
❶ Plot points in the rectangular coordinate system (p. 169)	Examples 1 and 2	1–6
❷ Determine whether an ordered pair satisfies an equation (p. 173)	Example 3	7, 8
❸ Create a table of values that satisfy an equation (p. 174)	Examples 4 through 7	9–16

In Problems 1–4, plot the following points in the rectangular coordinate system. Tell to which quadrant each point belongs (or on which axis the point lies).

1. $A(3, -2)$
2. $B(-1, -3)$
3. $C(-4, 0)$
4. $D(0, 2)$

In Problems 5 and 6, identify the coordinates of each point labeled in the figure. Tell to which quadrant each point belongs (or on which axis the point lies).

5.

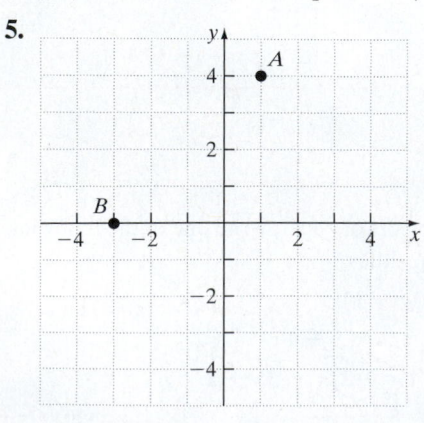

6.

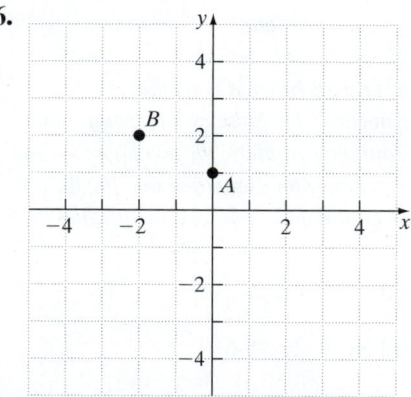

In Problems 7 and 8, determine whether the ordered pairs satisfy the given equation. Answer Yes or No.

7. $y = 3x - 7$
   $A(-1, -10)$
   $B(-7, 0)$

8. $4x - 3y = 2$
   $A(2, 2)$
   $B(6, 5)$

9. Find an ordered pair that satisfies the equation $-x = 3y$ when

   (a) $x = 4$ (b) $y = 2$

10. Find an ordered pair that satisfies the equation $x - y = 0$ when

(a) $x = 4$ **(b)** $y = -2$

In Problems 11–14, use the equation to complete the table. Use the table to list some of the ordered pairs that satisfy the equation.

11. $3x - 2y = 10$

x	y	(x, y)
−2		
0		
4		

12. $y = -x + 2$

x	y	(x, y)
−3		
2		
4		

13. $y = -\dfrac{1}{3}x - 4$

x	y	(x, y)
	2	
−6		
	−12	

14. $3x - 2y = 7$

x	y	(x, y)
	−8	
3		
	4	

15. Mail Order Shipping Martha purchases some clothes from a mail order catalog. The shipping costs for her order will be $5.00 plus $2.00 per item purchased. The equation that calculates her total shipping cost, C, is $C = 5 + 2x$, where x is the number of items purchased. Complete the table below, and then plot the ordered pairs in a rectangular coordinate system.

x	C	(x, C)
1		
2		
3		

16. Department Store Wages Isabel works in a department store where her monthly earnings are $1000 plus 10% commission on her sales. Her earnings, E, are given by the equation $E = 1000 + 0.10x$, where x is Isabel's sales for the month. Complete the table below, and then plot the ordered pairs in a rectangular coordinate system.

x	E	(x, E)
500		
1000		
2000		

Section 3.2   Graphing Equations in Two Variables

KEY CONCEPTS

- **Linear Equation**

  A linear equation in two variables is an equation of the form $Ax + By = C$, where A, B, and C are real numbers. A and B cannot both be zero.

- **Procedure for Finding Intercepts**
  1. To find the x-intercept(s), if any, of the graph of an equation, let $y = 0$ in the equation and solve for x.
  2. To find the y-intercept(s), if any, of the graph of an equation, let $x = 0$ in the equation and solve for y.

- **Vertical Line**

  A vertical line is given by the equation $x = a$, where a is the x-intercept.

- **Horizontal Line**

  A horizontal line is given by the equation $y = b$, where b is the y-intercept.

KEY TERMS

Graph of an equation in two variables
Point-plotting method
Complete graph
Linear equation in two variables
Standard form of an equation of a line
Line
Intercepts
x-intercept
y-intercept

You Should Be Able To...	EXAMPLE	Review Exercises
❶ Graph a line by plotting points (p. 182)	Examples 1, 3, and 4	17–22
❷ Graph a line using intercepts (p. 186)	Examples 5 through 8	23–32
❸ Graph vertical and horizontal lines (p. 190)	Examples 9 and 10	33–36

In Problems 17–20, graph each linear equation using the point-plotting method.

17. $y = -2x$ **18.** $y = x$

19. $4x + y = -2$ **20.** $3x - y = -1$

21. Printing Cost The cost to print pamphlets for a new healthcare clinic is a $40 set-up fee plus $2.00 per pamphlet that will be printed. The equation that calculates the total cost in dollars for printing, C, is $C = 40 + 2p$, where p is the number of pamphlets to be printed. Complete the table below, and then graph the equation.

p	C	(p, C)
20		
50		
80		

22. Performance Fees The Crickets are the newest band in town and have a gig performing at a local concert. They agree that their total fee will be $500 plus an additional $3.00 per person who attends the concert. The equation that calculates the total fee in dollars for performing, F, is $F = 500 + 3p$, where p is the number of people in attendance. Complete the table below, and then graph the equation.

p	F	(p, F)
100		
200		
500		

In Problems 23 and 24, find the intercepts of each graph.

23.

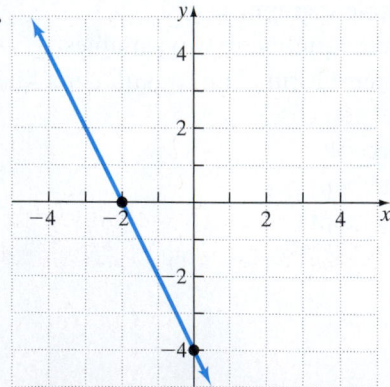

24.

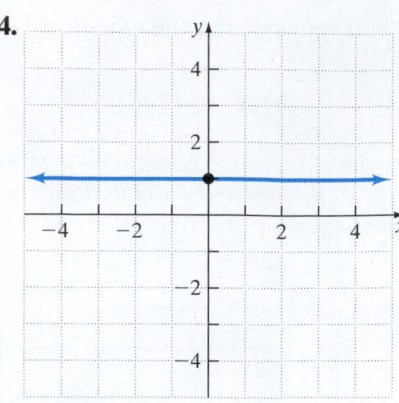

In Problems 25–28, find the x-intercept and y-intercept of each equation, if possible.

25. $-3x + y = 9$ **26.** $y = 2x - 6$

27. $x = 3$ **28.** $2x - 5y = 2$

In Problems 29–32, graph each linear equation by finding its intercepts.

29. $y - 3x = 3$ **30.** $2x + 5y = 0$

31. $\dfrac{x}{3} + \dfrac{y}{2} = 1$ **32.** $y = -\dfrac{3}{4}x + 3$

In Problems 33–36, graph each vertical or horizontal line.

33. $x = -2$ **34.** $y = 3$

35. $y = -4$ **36.** $x = 1$

Section 3.3   Slope

KEY CONCEPTS

- **Slope**

  If $x_1 \neq x_2$, the slope m of the line containing (x_1, y_1) and (x_2, y_2) is defined by the formula $m = \dfrac{y_2 - y_1}{x_2 - x_1}$.

  The slope of a vertical line is undefined.

  The slope of a horizontal line is 0.

KEY TERMS

Run
Rise
Slope
Average rate of change

You Should Be Able To...	EXAMPLE	Review Exercises
1 Find the slope of a line given two points (p. 195)	Examples 1 and 2	37–42
2 Find the slope of vertical and horizontal lines (p. 198)	Examples 3 and 4	43–46
3 Graph a line using its slope and a point on the line (p. 200)	Examples 5 and 6	47–50
4 Work with applications of slope (p. 201)	Examples 7 and 8	51, 52

In Problems 37 and 38, find the slope of the line whose graph is shown.

37.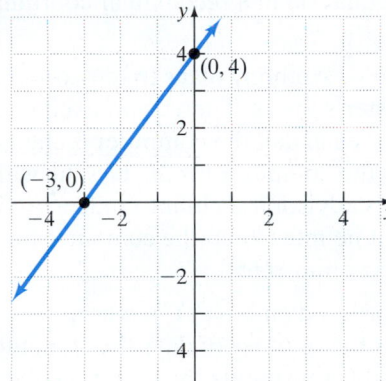

38.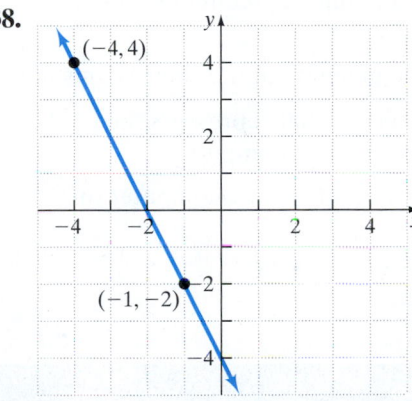

In Problems 39–46, find the slope of the line containing the given points.

39. $(-4, 6)$ and $(-3, -2)$ **40.** $(4, 1)$ and $(0, -7)$

41. $(-2, 3)$ and $(2, 13)$ **42.** $\left(-\dfrac{1}{2}, \dfrac{2}{3}\right)$ and $\left(\dfrac{3}{2}, \dfrac{1}{3}\right)$

43. $(-3, -6)$ and $(-3, -10)$ **44.** $(-5, -1)$ and $(-1, -1)$

45. $\left(\dfrac{3}{4}, \dfrac{1}{2}\right)$ and $\left(-\dfrac{1}{4}, \dfrac{1}{2}\right)$ **46.** $\left(\dfrac{1}{3}, -\dfrac{3}{5}\right)$ and $\left(\dfrac{3}{9}, -\dfrac{1}{5}\right)$

In Problems 47–50, graph the line that contains the given point and has the given slope.

47. $(-2, -3); m = 4$ **48.** $(1, -3); m = -2$

49. $(0, 1); m = -\dfrac{2}{3}$ **50.** $(2, 3); m = 0$

51. Production Cost The total cost to produce 20 bicycles is $1400, and the total cost to produce 50 bicycles is $2750. Find and interpret the slope of the line containing the points $(20, 1400)$ and $(50, 2750)$.

52. Road Grade Scott is driving to the river on a road that falls 5 feet for every 100 feet of horizontal distance. What is the grade of this road? Express your answer as a percent.

Section 3.4   Slope-Intercept Form of a Line

KEY CONCEPT

- **Slope-intercept form of an equation of a line**
  An equation of a line with slope m and y-intercept b is $y = mx + b$.

You Should Be Able To...	EXAMPLE	Review Exercises
1 Use the slope-intercept form to identify the slope and y-intercept of a line (p. 205)	Examples 1 and 2	53–56
2 Graph a line whose equation is in slope-intercept form (p. 206)	Examples 3 and 4	57–62
3 Graph a line whose equation is in the form $Ax + By = C$ (p. 208)	Example 5	63, 64
4 Find the equation of a line given its slope and y-intercept (p. 208)	Example 6	65–70
5 Work with linear models in slope-intercept form. (p. 209)	Examples 7 and 8	71, 72

In Problems 53–56, find the slope and the y-intercept of the line whose equation is given.

53. $y = -x + \dfrac{1}{2}$ **54.** $y = x - \dfrac{3}{2}$

55. $3x - 4y = -4$ **56.** $2x + 5y = 8$

In Problems 57–64, use the slope and y-intercept to graph each line whose equation is given.

57. $y = \dfrac{1}{3}x + 1$ **58.** $y = -\dfrac{1}{2}x - 1$

59. $y = -\dfrac{2}{3}x - 2$ **60.** $y = \dfrac{3}{4}x + 3$

61. $y = x$ **62.** $y = -2x$

63. $2x - y = -4$ **64.** $-4x + 2y = 2$

In Problems 65–70, find the equation of the line with the given slope and intercept.

65. slope is -3; y-intercept is $(0, 5)$

66. slope is $\dfrac{1}{5}$; y-intercept is $(0, 10)$

67. slope is undefined; x-intercept is $(-12, 0)$

68. slope is 0; y-intercept is $(0, -4)$

69. slope is 1; y-intercept is $(0, -20)$

70. slope is -1; y-intercept is $(0, -8)$

71. Car Rentals Hot Rod Car Rentals rents sports cars for $120 plus $80 per day. The equation that calculates the total cost C, in dollars, to rent a sports car is $C = 120 + 80d$, where d is the number of days that the car is used.

(a) How much will it cost to rent the car for 3 days?

(b) If the bill came to $680, how many days was the sports car rented out?

(c) Graph the equation in a rectangular coordinate system.

72. Computer Rentals Anthony plans to rent a computer to finish his master's thesis. There is no fixed fee, only a charge for each day the computer is checked out. If he keeps the computer for 22 days, he will pay $418. If his thesis advisor asks him to redo a section and he finds that he must keep the computer for 35 days, it will cost him $665.

(a) Write two ordered pairs (d, C), where d represents the number of days the computer is rented and C represents the cost. Calculate the slope of the line that contains these two points.

(b) Interpret the meaning of the slope of this line.

(c) Write an equation that represents the total cost for Anthony to rent a computer.

(d) Calculate the cost to rent a computer for 8 days.

Section 3.5   Point-Slope Form of a Line

KEY CONCEPT

- **Point-Slope Form of an Equation of a Line**
  An equation of a nonvertical line with slope m that contains the point (x_1, y_1) is $y - y_1 = m(x - x_1)$.

You Should Be Able To...	EXAMPLE	Review Exercises
1 Find the equation of a line given a point and a slope (p. 214)	Examples 1 through 3	73–78
2 Find the equation of a line given two points (p. 216)	Examples 4 and 5	79–84
3 Build linear models using the point-slope form of a line (p. 218)	Example 6	85, 86

In Problems 73–78, find the equation of the line that contains the given point and the given slope. Write the equation in slope-intercept form, if possible.

73. $(0, -3)$; slope $= 6$ **74.** $(4, 0)$; slope $= -2$

75. $(3, -1)$; slope $= -\dfrac{1}{2}$ **76.** $(-1, -3)$; slope $= \dfrac{2}{3}$

77. $\left(-\dfrac{4}{3}, -\dfrac{1}{2}\right)$; slope $= 0$

78. $\left(-\dfrac{4}{7}, \dfrac{8}{5}\right)$; slope is undefined

In Problems 79–84, find the equation of the line that contains the given points. Write the equation of the line in slope-intercept form, if possible.

79. $(-7, 0)$ and $(0, 8)$

80. $(0, -6)$ and $(4, 0)$

81. $(3, 5)$ and $(-2, -10)$

82. $(-15, 1)$ and $(-5, -3)$

83. $(-2, 7)$ and $(5, 7)$

84. $(4, -3)$ and $(4, 0)$

85. Harvesting Hay Farmer Myers noted that at the end of 3 days of harvesting, 12 acres of hay remained in his field and at the end of 5 days, 4 acres remained. Use the ordered pairs $(3, 12)$ and $(5, 4)$ to write an equation that calculates how many acres of hay, A, are left to be harvested after d days.

86. Fish Population Pat works at Scripps Institute of Oceanography and is responsible for monitoring the fish population in various waters around the world. In one bay, the sunfish population was declining at a constant rate. His initial sample netted 15 sunfish and six months later, the same location netted 13 sunfish. Use the ordered pairs $(0, 15)$ and $(6, 13)$ to write an equation that will predict how many sunfish, F, the sample will yield after m months.

Section 3.6   Parallel and Perpendicular Lines

KEY TERMS

Parallel

Perpendicular

You Should Be Able To...	EXAMPLE	Review Exercises
❶ Determine whether two lines are parallel (p. 222)	Examples 1 and 2	87, 88
❷ Find the equation of a line parallel to a given line (p. 224)	Examples 3 and 4	89–94
❸ Determine whether two lines are perpendicular (p. 225)	Examples 5 through 7	95–98
❹ Find the equation of a line perpendicular to a given line (p. 226)	Examples 8 and 9	99–102

In Problems 87 and 88, determine whether the two lines are parallel. Answer Yes or No.

87. $y = -\dfrac{1}{3}x + 2$
$x - 3y = 3$

88. $y = \dfrac{1}{2}x - 4$
$x - 2y = 6$

In Problems 89–94, find the equation of the line that contains the given point and is parallel to the given line. Write the equation in slope-intercept form, if possible.

89. $(3, -1); y = -x + 5$

90. $(-2, 4); y = 2x - 1$

91. $(-1, 10); 3x + y = -7$

92. $(4, -5); 6x + 4y = 8$

93. $(5, 19); y$-axis

94. $(-1, -12); x$-axis

In Problems 95 and 96, determine the slope of the line perpendicular to the given line.

95. $x - 2y = 5$

96. $4x - 9y = 1$

In Problems 97 and 98, determine whether the two lines are perpendicular. Answer Yes or No.

97. $x + 3y = 3$
$y = 3x + 1$

98. $5x - 2y = 2$
$y = \dfrac{2}{5}x + 12$

In Problems 99–102, write the equation of the line that contains the given point and is perpendicular to the given line. Write the equation in slope-intercept form, if possible.

99. $(-3, 4); y = -3x + 1$

100. $(4, -1); y = 2x - 1$

101. $(1, -3); 2x - 3y = 6$

102. $\left(-\dfrac{3}{5}, \dfrac{2}{5}\right); x + y = -7$

Section 3.7   Linear Inequalities in Two Variables

KEY TERMS

Satisfied   Half-planes
Graph of a linear inequality in two variables   Test point

You Should Be Able To...	EXAMPLE	Review Exercises
1 Determine whether an ordered pair is a solution to a linear inequality (p. 231)	Example 1	103, 104
2 Graph linear inequalities (p. 232)	Examples 2 through 5	105–112
3 Solve problems involving linear inequalities (p. 237)	Example 6	113, 114

In Problems 103 and 104, determine which of the following points, if any, are solutions to the given linear inequality in two variables. Answer Yes or No.

103. $y \leq 3x + 4$ $A(2, 0)$ $B(-4, -8)$ $C(7, 26)$

104. $y > \dfrac{1}{3}x + 4$ $A(6, -2)$ $B(0, 4)$ $C(-18, -1)$

In Problems 105–112, graph each linear inequality.

105. $y < -\dfrac{1}{4}x + 2$ **106.** $y > 2x - 1$

107. $3x + 2y \geq -6$ **108.** $-2x + y \geq 4$

109. $x - 3y \leq 0$ **110.** $x - 4y \geq 4$

111. $x < -3$ **112.** $y > 2$

113. Coin Jar A jar holding only quarters and dimes contains at least \$12. If there are x quarters and y dimes in the jar, write a linear inequality in two variables that describes how many of each type of coin are in the jar.

114. Number The difference between twice a number, x, and half of a second number, y, is at most 10. Write a linear inequality in two variables that describes these two numbers.

Chapter 3 Test

Step-by-step test solutions are found on the Chapter Test Prep Videos available in MyMathLab®, *on* YouTube™, *or may be accessed using the QR code to the right.*

1. Determine whether the ordered pair $(-3, -2)$ is a solution to the equation $3x - 4y = -17$.

2. Given the equation $3x - 9y = 12$, determine
   (a) the x-intercept
   (b) the y-intercept

3. Given the equation $4x - 3y = -24$, determine
   (a) the slope
   (b) the y-intercept

In Problems 4 and 5, graph each linear equation.

4. $y = -\dfrac{3}{4}x + 2$ 5. $3x - 6y = -12$

6. Find the slope of the line that contains the points $(2, -2)$ and $(-4, -1)$.

7. Given the linear equation $3y = 2x - 1$, find
   (a) the slope of a line perpendicular to the given line.
   (b) the slope of a line parallel to the given line.

8. Determine whether the lines are parallel, perpendicular, or neither. Explain how you came to your conclusion.

$$L_1: 3x - 7y = 2$$
$$L_2: y = \dfrac{7}{3}x + 4$$

In Problems 9–15, write the equation of the line that satisfies the given conditions. Write the equation in slope-intercept form, if possible.

9. slope $= -4$ and y-intercept is $(0, -15)$

10. slope $= 2$ and contains $(-3, 8)$

11. contains $(-3, -2)$ and $(-4, 1)$

12. contains $(4, 0)$ and parallel to $y = \frac{1}{2}x + 2$

13. contains $(4, 2)$ and perpendicular to $4x - 6y = 5$

14. horizontal line through $(3, 5)$

15. contains $(-2, -1)$ with undefined slope

16. Women in the U.S. Congress In 1993, 30 members of the U.S. House of Representatives were women. By 2016 that number had grown to 88. Use the ordered pairs (1993, 30) and (2016, 88) to find the average yearly increase in the number of women in the U.S. House of Representatives. Round your answer to the nearest tenth.

In Problems 17–19, graph each linear inequality.

17. $y \geq x - 3$ **18.** $-2x - 4y < 8$ **19.** $x \leq -4$

Cumulative Review Chapters 1–3

In Problems 1–3, evaluate each expression.

1. $200 \div 25 \cdot (-2)$

2. $\frac{3}{4} + \frac{1}{6} - \frac{2}{3}$

3. $\frac{8 - 3(5 - 3^2)}{7 - 2 \cdot 6}$

4. Evaluate $x^3 + 3x^2 - 5x - 7$ for $x = -3$.

5. Simplify: $8m - 5m^2 - 3 + 9m^2 - 3m - 6$

In Problems 6 and 7, solve each equation.

6. $8(n + 2) - 7 = 6n - 5$

7. $\frac{2}{5}x + \frac{1}{6} = -\frac{2}{3}$

8. Solve $A = \frac{1}{2}h(b + B)$ for B.

In Problems 9 and 10, solve each linear inequality. Express the solution using set-builder notation and interval notation. Graph the solution set.

9. $6x - 7 > -31$ **10.** $5(x - 3) \geq 7(x - 4) + 3$

11. Plot the following ordered pairs in the same Cartesian plane.

$A(-3, 0)$ $B(4, -2)$ $C(1, 5)$

$D(0, 3)$ $E(-4, -5)$ $F(-5, 2)$

In Problems 12 and 13, graph the linear equation using the method you prefer.

12. $y = -\frac{1}{2}x + 4$ **13.** $4x - 5y = 15$

In Problems 14 and 15, find the equation of the line with the given properties. Express your answer in slope-intercept form, if possible.

14. Through the points $(3, -2)$ and $(-6, 10)$

15. Parallel to $y = -3x + 10$ and through the point $(-5, 7)$

16. Graph $x - 3y > 12$.

17. Computing Grades Shawn really wants an A in his elementary algebra class. His four exam scores are 94, 95, 90, and 97. The final exam is worth two exam scores. To have an A, his average must be at least 93. What scores, out of 100, does Shawn need on the final exam to earn an A in the course?

18. Body Mass Index The body mass index (BMI) of a person 62 inches tall and weighing x pounds is given by $0.2x - 2$. A person with a BMI of 30 or more is considered to be obese. For what weights would a person 62 inches tall be considered obese?

19. Supplementary Angles Two angles are supplementary. The measure of the larger angle is 15 degrees more than twice the measure of the smaller angle. Find the angle measures.

20. Cylinders Max has 100 square inches of aluminum with which to make a closed cylinder. If the radius of the cylinder must be 2 inches, how tall will the cylinder be? Use the formula $S = 2\pi r^2 + 2\pi rh$. (Round to the nearest hundredth of an inch.)

21. Consecutive Integers Find three consecutive even integers such that the sum of the first two is 22 more than the third.

Systems of Linear Equations and Inequalities in Two Variables

A laboratory technician needs 60 milliliters (ml) of a 50% saline solution. The stockroom only has 30% saline solution and 60% saline solution available? How many ml of 30% saline solution should be added to a 60% saline solution to obtain the required mixture? Problems like this occur every day when blends are created to meet our needs. The number of milliliters of each solution required for this mixture can be found algebraically using a system of equations. See Problem 39 in Section 4.5.

The Big Picture: Putting It Together

At this point in the course, you are familiar with solving linear equations and inequalities in one variable. Recall that linear equations in one variable can have no solution, one solution, or infinitely many solutions. You also know how to graph linear equations and linear inequalities in two variables. Remember, the graph of a linear equation in two variables represents the set of all ordered pairs (x, y) that satisfy the equation.

In this chapter, methods for finding solutions that simultaneously satisfy two linear equations in two variables, called systems of equations, will be discussed. Three techniques for solving these systems will be considered. The first is a graphical method for finding the solution. Then, two algebraic methods called substitution and elimination are discussed. These systems can have no solution, one solution, or infinitely many solutions. The chapter will conclude by looking at systems of linear inequalities. The solution of these systems is a region that satisfies two or more linear inequalities simultaneously.

Outline

4.1 Solving Systems of Linear Equations by Graphing

Objectives

1. Determine Whether an Ordered Pair Is a Solution of a System of Linear Equations
2. Solve a System of Linear Equations by Graphing
3. Classify Systems of Linear Equations as Consistent or Inconsistent
4. Solve Applied Problems Involving Systems of Linear Equations

Are You Prepared for This Section?

Before getting started, complete the following problems. If you get a problem wrong, go back to the section cited and review the material.

P1. Graph: $y = 2x - 3$ [Section 3.4, pp. 206–207]

P2. Graph: $3x + 4y = 12$ [Section 3.2, pp. 186–190]

P3. Determine whether $2x + 6y = 12$ is parallel to $-3x - 9y = 18$.   [Section 3.6, pp. 222–223]

In Section 3.2, it was shown that an equation in two variables is linear provided that it can be written in the form $Ax + By = C$, where A and B cannot both be zero. Methods for finding ordered pairs that satisfy two linear equations at the same time are now presented.

Definition

A **system of linear equations** is a grouping of two or more linear equations where each equation contains one or more variables.

Example 1 gives some examples of systems of linear equations containing two equations in two variables. Throughout this chapter we consider only systems of linear equations in two variables. Each of these systems will be called a system of linear equations, for short.

EXAMPLE 1   Examples of Systems of Linear Equations

(a) $\begin{cases} 2x + y = 5 \\ x - 5y = -10 \end{cases}$ Two linear equations containing two variables, x and y

(b) $\begin{cases} a + b = 5 \\ 2a - 3b = -3 \end{cases}$ Two linear equations containing two variables, a and b ●

A brace, as shown in the systems in Example 1, reminds us that we have a system of equations.

1   Determine Whether an Ordered Pair Is a Solution of a System of Linear Equations

Definition

A **solution** of a system of equations consists of values of the variables that satisfy each equation in the system. Represent the solution of two linear equations containing two unknowns as an ordered pair, (x, y).

To determine whether an ordered pair is a solution of a system of equations, replace each variable with its value in each equation. If *both* equations are satisfied, the ordered pair is a solution of the system.

Prepared?...Answers

P1.

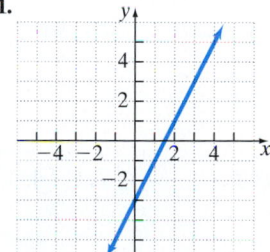

P2.

P3. Parallel

EXAMPLE 2   Determining Whether an Ordered Pair Is a Solution of a System of Linear Equations

Determine whether each ordered pair is a solution of the system of equations.

$$\begin{cases} x + 3y = 9 \\ 4x - 2y = 8 \end{cases}$$

(a) $(-3, 4)$ (b) $(3, 2)$

(continued)

Solution

(a) Let $x = -3$ and $y = 4$ in both equations. If *both* equations are true, then $(-3, 4)$ is a solution.

First equation: $\quad\quad x + 3y = 9 \quad\quad$ Second equation: $\quad\quad 4x - 2y = 8$

$$-3 + 3(4) \overset{?}{=} 9 \quad\quad\quad\quad\quad 4(-3) - 2(4) \overset{?}{=} 8$$

$$-3 + 12 \overset{?}{=} 9 \quad\quad\quad\quad\quad\quad\quad -12 - 8 \overset{?}{=} 8$$

$$9 = 9 \quad \text{True} \quad\quad\quad\quad\quad\quad -20 = 8 \quad \text{False}$$

Although $(-3, 4)$ satisfies the first equation, it does not satisfy the second equation. Because the values $x = -3$ and $y = 4$ do not satisfy both equations, $(-3, 4)$ is not a solution of the system.

(b) Let $x = 3$ and $y = 2$ in both equations. If *both* equations are true, then $(3, 2)$ is a solution.

First equation: $\quad\quad x + 3y = 9 \quad\quad$ Second equation: $\quad\quad 4x - 2y = 8$

$$3 + 3(2) \overset{?}{=} 9 \quad\quad\quad\quad\quad\quad 4(3) - 2(2) \overset{?}{=} 8$$

$$3 + 6 \overset{?}{=} 9 \quad\quad\quad\quad\quad\quad\quad\quad 12 - 4 \overset{?}{=} 8$$

$$9 = 9 \quad \text{True} \quad\quad\quad\quad\quad\quad\quad 8 = 8 \quad \text{True}$$

Because $(3, 2)$ satisfies both equations, it is a solution of the system. ●

Quick ✓

1. A _____ __ _____ _____ is a grouping of two or more linear equations, each of which contains one or more variables.

2. A _____ of a system of equations consists of values of the variables that satisfy each equation in the system.

In Problems 3 and 4, determine whether each ordered pair is a solution of the system of equations. Answer Yes or No.

3. $\begin{cases} 2x + 3y = 7 \\ 3x + y = -7 \end{cases}$
   (a) $(3, 1)$ (b) $(-4, 5)$ (c) $(-2, -1)$

4. $\begin{cases} 3x - 6y = 6 \\ -2x + 4y = -4 \end{cases}$
   (a) $(2, 0)$ (b) $(0, -1)$ (c) $(4, 1)$

▶ ❷ Solve a System of Linear Equations by Graphing

The graph of each equation in a system of linear equations in two variables is a line. So a system of two linear equations containing two variables represents a pair of lines. We know from Chapter 3 that the graph of a linear equation in two variables represents the set of all ordered pairs that make the equation a true statement. Therefore, **the point, or points, at which two lines intersect, if any, represents the solution of the system of equations because both equations are satisfied at this point.**

Let's look at an example to learn how to solve a system of linear equations by graphing.

EXAMPLE 3 **How to Solve a System of Linear Equations by Graphing**

Solve the system by graphing: $\begin{cases} 3x + y = 7 \\ -2x + 3y = -12 \end{cases}$

Step-by-Step Solution

Step 1: Graph the first equation in the system.

Graph the equation $3x + y = 7$ by writing the equation in slope-intercept form, $y = mx + b$.

$$3x + y = 7$$
$$y = -3x + 7$$

Graph the line whose slope is -3 and y-intercept is $(0, 7)$ by first plotting the y-intercept $(0, 7)$ and then using the slope

$$m = -3 = \frac{-3}{1} = \frac{\text{rise}}{\text{run}}$$

to locate a second point on the line. See Figure 1(a).

Figure 1

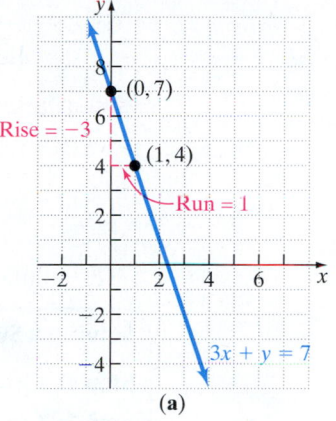

(a)

Step 2: Graph the second equation in the system in the same rectangular coordinate system as the equation in Step 1.

Graph the line $-2x + 3y = -12$ using intercepts because the equation is in standard form and the intercepts can easily be found.

x-intercept: Let $y = 0$ and solve for x.

$$-2x + 3y = -12$$
$$-2x + 3(0) = -12$$
$$-2x = -12$$
$$x = 6$$

y-intercept: Let $x = 0$ and solve for y.

$$-2x + 3y = -12$$
$$-2(0) + 3y = -12$$
$$3y = -12$$
$$y = -4$$

In the same coordinate system, plot the points $(6, 0)$ and $(0, -4)$ and draw a line through them. See Figure 1(b).

Figure 1

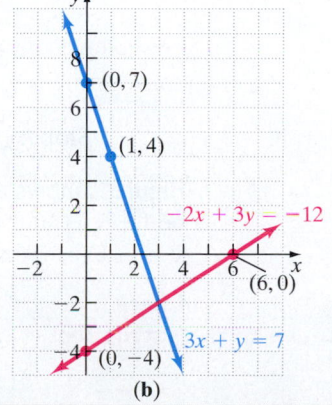

(b)

Step 3: Determine the point of intersection of the two lines, if any.

From the graphs in Figure 1(b), the lines appear to intersect at $(3, -2)$. This point is labeled in Figure 1(c) on the next page.

(continued)

Figure 1

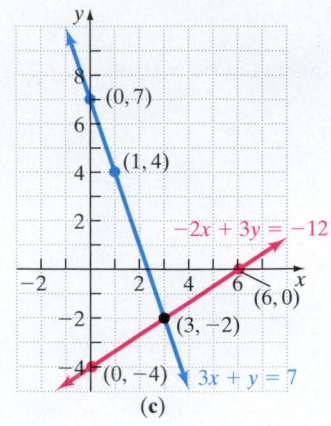

(c)

Step 4: Verify that the point of intersection found in Step 3 is a solution of the system.

Let $x = 3$ and $y = -2$ in both equations.

First equation: $3x + y = 7$

$3(3) + (-2) \stackrel{?}{=} 7$

$9 - 2 \stackrel{?}{=} 7$

$7 = 7$ True

Second equation: $-2x + 3y = -12$

$-2(3) + 3(-2) \stackrel{?}{=} -12$

$-6 - 6 \stackrel{?}{=} -12$

$-12 = -12$ True

The solution of the system is $(3, -2)$.

Solving a System of Linear Equations by Graphing

Step 1: Graph the first equation in the system.

Step 2: Graph the second equation in the system in the same rectangular coordinate system.

Step 3: Determine the point of intersection, if any. Write this point as an ordered pair (x, y), if possible. The point of intersection is the solution of the system.

Step 4: Verify that the point of intersection found in Step 3 is the solution of the system by substituting the ordered pair in both equations.

EXAMPLE 4   **Solving a System of Linear Equations Graphically**

Work Smart

We did not graph $3x - 2y = -16$ using intercepts because the x-intercept is a fraction, which is difficult to plot accurately.

Solve the system by graphing: $\begin{cases} 3x - 2y = -16 \\ 5x + 2y = 0 \end{cases}$

Solution

Graph both equations by writing each equation in slope-intercept form.

First equation: $3x - 2y = -16$

Subtract 3x from both sides: $-2y = -3x - 16$

Divide both sides by -2: $y = \dfrac{3}{2}x + 8$

Slope: $\dfrac{3}{2}$; y-intercept: $(0, 8)$

Second equation: $5x + 2y = 0$

Subtract 5x from both sides: $2y = -5x$

Divide both sides by 2: $y = -\dfrac{5}{2}x$

Slope: $-\dfrac{5}{2}$; y-intercept: $(0, 0)$

Figure 2

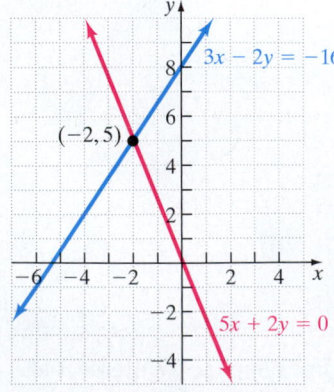

Figure 2 shows the graph of both equations. The point of intersection appears to be $(-2, 5)$, but it is important to check that it is the solution to the system. Does $x = -2$, $y = 5$ satisfy both equations?

Check

First equation: $3x - 2y = -16$

$3(-2) - 2(5) \stackrel{?}{=} -16$

$-6 - 10 \stackrel{?}{=} -16$

$-16 = -16$ True

Second equation: $5x + 2y = 0$

$5(-2) + 2(5) \stackrel{?}{=} 0$

$-10 + 10 \stackrel{?}{=} 0$

$0 = 0$ True

The point $(-2, 5)$ satisfies both equations, so the solution is $(-2, 5)$.

Quick ✓

In Problems 5 and 6, solve each system by graphing.

5. $\begin{cases} y = -2x + 9 \\ y = 3x - 11 \end{cases}$

6. $\begin{cases} 4x + y = -3 \\ 5x - 2y = -20 \end{cases}$

Sometimes the lines that make up a system of linear equations do not intersect at all.

EXAMPLE 5

Solving a System of Linear Equations That Has No Point of Intersection

Solve the system by graphing: $\begin{cases} 3x - y = -4 \\ -6x + 2y = 2 \end{cases}$

Figure 3

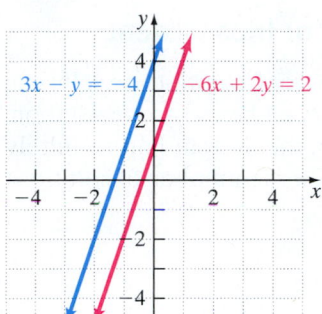

Solution

Graph both equations by writing each equation in slope-intercept form.

First equation: $3x - y = -4$	Second equation: $-6x + 2y = 2$
Subtract 3x from both sides: $\quad -y = -3x - 4$	Add 6x to both sides: $\quad 2y = 6x + 2$
Multiply both sides by -1: $\quad y = 3x + 4$	Divide both sides by 2: $\quad y = 3x + 1$
Slope: 3; y-intercept: $(0, 4)$	Slope: 3; y-intercept: $(0, 1)$

Figure 3 shows the graph of both equations. The lines are parallel and therefore do not intersect. The system of equations has no solution. Therefore, the solution set is the empty set, { } or ∅. ●

Quick ✓

In Problem 7, solve the system by graphing.

7. $\begin{cases} y = 2x + 4 \\ 2x - y = 1 \end{cases}$

Another special case with systems of linear equations in two variables occurs when the graph of each equation in the system is the same line.

EXAMPLE 6

Solving a System of Linear Equations Where the Graph of Each Equation Is the Same Line

Solve the system by graphing: $\begin{cases} 5x + 2y = -4 \\ 10x + 4y = -8 \end{cases}$

Solution

Graph both equations in the system by writing each equation in slope-intercept form.

First equation: $5x + 2y = -4$	Second equation: $10x + 4y = -8$
Subtract 5x from both sides: $\quad 2y = -5x - 4$	Subtract 10x from both sides: $\quad 4y = -10x - 8$
Divide both sides by 2: $\quad y = -\dfrac{5}{2}x - 2$	Divide both sides by 4: $\quad y = -\dfrac{5}{2}x - 2$
Slope: $-\dfrac{5}{2}$; y-intercept: $(0, -2)$	Slope: $-\dfrac{5}{2}$; y-intercept: $(0, -2)$

(continued)

Figure 4

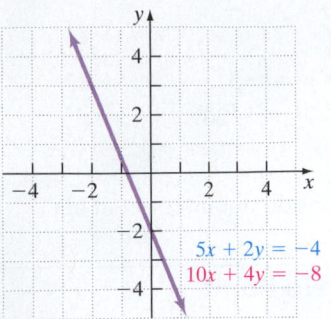

$5x + 2y = -4$
$10x + 4y = -8$

Work Smart

Either equation in a system may be used to write the solution set. So the solution to Example 6 could also be written as $\{(x, y) \mid 10x + 4y = -8\}$.

Notice that the slopes and the y-intercepts of the two lines are the same. Figure 4 shows the graph of each equation. The equations are the same, so the lines coincide. Any point on the graph of $5x + 2y = -4$ is also a point on the graph of $10x + 4y = -8$. For example, the ordered pairs $(-2, 3)$, $(0, -2)$, $(2, -7)$, and $(4, -12)$ all satisfy both equations in the system. In fact, the system of equations has infinitely many solutions, which are represented by all points on the line $5x + 2y = -4$. Therefore, the solution is written as $\{(x, y) \mid 5x + 2y = -4\}$. ●

Quick ✓

In Problem 8, solve the system by graphing.

8. $\begin{cases} 3x - y = -2 \\ -6x + 2y = 4 \end{cases}$

▶ ❸ Classify Systems of Linear Equations as Consistent or Inconsistent

Examples 3–6 show that systems of linear equations can have one solution, no solution, or infinitely many solutions. In Examples 3 and 4, the system had exactly one solution. In Example 5, the system had no solution. In Example 6, the system had infinitely many solutions. Systems of equations are classified based on the number of solutions they have.

Definitions

A **consistent system** of equations has at least one solution.

An **inconsistent system** of equations has no solution.

These possibilities are summarized below.

Classifying a System of Equations Graphically

Intersect: If the lines intersect, then the system of equations has one solution given by the point of intersection. The system is **consistent** and the equations are **independent**. See Figure 5(a).

Parallel: If the lines are parallel, then the system of equations has no solution because the lines do not intersect. The system is **inconsistent.** See Figure 5(b).

Coincident: If the lines lie on top of each other (coincide), then the system of equations has infinitely many solutions. The solution set is the set of all points on the line. The system is **consistent** and the equations are **dependent.** See Figure 5(c).

Figure 5

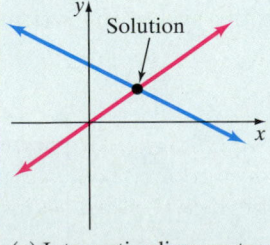

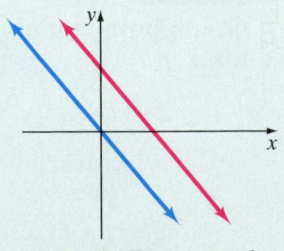

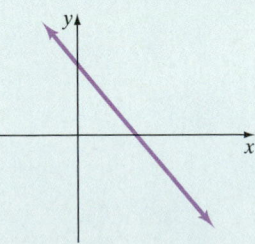

(a) Intersecting lines; system has exactly one solution

(b) Parallel lines; system has no solution

(c) Coincident lines; system has infinitely many solutions

In addition to graphing the lines to classify the system, a system can also be classified algebraically.

> **Classifying a System of Equations Algebraically**
>
> **Step 1:** Write each equation in the system in slope-intercept form.
>
> **Step 2: (a)** If the equations of the lines in the system have different slopes, the lines will intersect. The point of intersection represents the solution. The system is consistent and the equations are independent. See Examples 3 and 4.
>
> **(b)** If the equations of the lines have the same slope but different y-intercepts, the lines are parallel. The system has no solution. The system is inconsistent. See Example 5.
>
> **(c)** If the equations of the lines have the same slope and the same y-intercept, the lines coincide. The system has infinitely many solutions. The system is consistent and the equations are dependent. See Example 6.

EXAMPLE 7 **Determining the Number of Solutions in a System**

Without graphing, determine the number of solutions of the system:

$$\begin{cases} 12x + 4y = -16 \\ -9x - 3y = 3 \end{cases}$$

State whether the system is consistent or inconsistent. If the system is consistent, state whether the equations are dependent or independent.

Solution

Write the equations in slope-intercept form. Then find the slope and y-intercept to determine the number of solutions.

First equation:	$12x + 4y = -16$	Second equation:	$-9x - 3y = 3$
Subtract $12x$ from both sides:	$4y = -12x - 16$	Add $9x$ to both sides:	$-3y = 9x + 3$
Divide both sides by 4:	$y = -3x - 4$	Divide both sides by -3:	$y = -3x - 1$

The slope of both equations is -3. The y-intercept of the first equation is $(0, -4)$ and the y-intercept of the second equation is $(0, -1)$. The equations have the same slope but different y-intercepts, so the lines are parallel. Therefore, the system has no solution and is inconsistent. ●

Quick ✓

9. If a system of linear equations in two variables has at least one solution, the system is _____.

10. If a system of linear equations in two variables has exactly one solution, the equations are _____.

11. A system of linear equations in two variables that has no solution is _____.

12. *True or False* The visual representation of a consistent system consisting of dependent equations is one line.

In Problems 13–15, without graphing, determine the number of solutions of each system. State whether the system is consistent or inconsistent. If the system is consistent, state whether the equations are dependent or independent.

13. $\begin{cases} 7x - 2y = 4 \\ 2x + 7y = 7 \end{cases}$ 14. $\begin{cases} 6x + 4y = 4 \\ -12x - 8y = -8 \end{cases}$ 15. $\begin{cases} 3x - 4y = 8 \\ -6x + 8y = 8 \end{cases}$

▶ ❹ Solve Applied Problems Involving Systems of Linear Equations

Many times a system of two linear equations in two variables can be used to answer a question such as "Which truck rental company should I choose?"

EXAMPLE 8

Truck Rental

Darren wants to rent a 15-foot truck for four hours to move some sound equipment to another city for his next show. Haul-Away charges $30 plus $0.45 per mile, but Darren's friend Andy says he'll let Darren borrow his 15-foot truck if Darren will pay him a flat fee of $0.60 per mile. Find the number of miles Darren can drive in order for the two plans to cost the same amount. What is the cost for that number of miles?

Solution

Step 1: Identify Find the number of miles Darren can drive so that the cost for renting a 15-foot truck from Haul-Away is the same as the cost for using Andy's 15-foot truck.

Step 2: Name Let m represent the number of miles Darren can drive, and let C represent the cost, in dollars, of renting a 15-foot truck or borrowing Andy's truck.

Step 3: Translate Haul-Away charges $30 plus $0.45 per mile, so just renting the truck costs $30. Renting the truck and driving 1 mile costs $30 plus $1 \times \$0.45$, or $30.45; renting the truck and driving 2 miles costs $30 plus $2 \times \$0.45$, or $30 + \$0.90 = \$30.90$. To generalize, the cost for renting the truck from Haul-Away is $30 plus $0.45 times m, the number of miles driven. This leads to the equation

$$C = 0.45m + 30 \quad \text{Equation (1)}$$

Andy will let Darren borrow his truck if he pays Andy $0.60 per mile, but there are no additional charges. This leads to the equation

$$C = 0.60m \quad \text{Equation (2)}$$

Use equations (1) and (2) to form the system

$$\begin{cases} C = 0.45m + 30 \\ C = 0.60m \end{cases} \quad \text{The Model}$$

Step 4: Solve To draw the graph of each equation in the system, create a table that shows the cost for various numbers of miles driven for each company. See Table 1. Figure 6 shows the graph of the two equations in the system with the horizontal axis labeled m and the vertical axis labeled C.

Table 1

Number of miles driven, m	Cost for Renting from Haul-Away	Cost for Using Andy's Truck
0	$ 30	$ 0
100	$ 75	$ 60
250	$142.50	$150

Figure 6

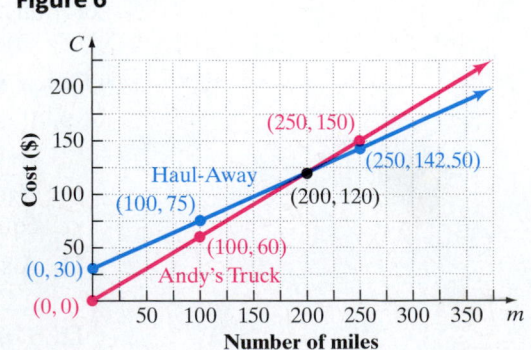

The graph shows that the charge will be the same for 200 miles. Also, $C = 120$ when $m = 200$, so driving 200 miles costs $120.

Step 5: Check To rent from Haul-Away, the cost for 200 miles is $30 + \$0.45 \times 200 = \$30 + \$90 = \$120$. To borrow Andy's truck, the cost for 200 miles is $\$0.60 \times 200 = \$120$. The cost for driving 200 miles is the same whether renting from Haul-Away or using Andy's truck.

Work Smart

Choose values for m that will yield "nice" numbers for C, such as $m = 100$ and $m = 250$, as shown in Table 1.

Step 6: Answer The cost of renting a 15-foot truck is the same, $120, when Darren drives 200 miles.

> **Quick ✓**
>
> **16.** John needs to rent a 12-foot moving truck for a day. He calls two rental companies to determine their fee structure. EZ-Rental charges $20 plus $0.40 per mile, and U-Move-It charges $35 plus $0.25 per mile. After how many miles will the cost of renting the truck be the same? What is the cost at this mileage?

4.1 Exercises MyMathLab®   Exercise numbers in **green** have complete video solutions in MyMathLab or may be accessed using the QR code to the right.

*Problems **1–16** are the **Quick ✓** s that follow the **EXAMPLES**.*

Building Skills

In Problems 17–22, determine whether the ordered pair is a solution of the system of equations. Answer Yes or No. See Objective 1.

17. $\begin{cases} x - y = -4 \\ 3x + y = -4 \end{cases}$

   (a) $(2, 6)$
   (b) $(-2, 2)$
   (c) $(2, -2)$

18. $\begin{cases} 2x + 5y = 0 \\ x - 3y = 11 \end{cases}$

   (a) $(5, -2)$
   (b) $(-5, 2)$
   (c) $(2, -3)$

19. $\begin{cases} 3x - y = 2 \\ -15x + 5y = -10 \end{cases}$

   (a) $(1, -1)$
   (b) $(-2, -8)$
   (c) $(0, -2)$

20. $\begin{cases} x - 4y = 2 \\ 5x - 8y = -6 \end{cases}$

   (a) $(-8, -2)$
   (b) $(2, 2)$
   (c) $\left(-\dfrac{10}{3}, -\dfrac{4}{3}\right)$

21. $\begin{cases} 6x - 2y = 1 \\ y = 3x + 2 \end{cases}$

   (a) $\left(0, -\dfrac{1}{2}\right)$
   (b) $(-2, -4)$
   (c) $(0, 2)$

22. $\begin{cases} y = -4x - 1 \\ 2y + 8x = 2 \end{cases}$

   (a) $\left(-\dfrac{1}{2}, 0\right)$
   (b) $(0, 1)$
   (c) $(-1, 0)$

In Problems 23–38, solve each system of equations by graphing. See Objective 2.

23. $\begin{cases} y = x + 5 \\ y = -\dfrac{1}{5}x - 1 \end{cases}$

24. $\begin{cases} y = -\dfrac{2}{3}x - 2 \\ y = \dfrac{1}{2}x + 5 \end{cases}$

25. $\begin{cases} y = \dfrac{3}{4}x - 4 \\ y = -\dfrac{1}{2}x + 1 \end{cases}$

26. $\begin{cases} y = -4x + 6 \\ y = -2x \end{cases}$

27. $\begin{cases} 2x - y = -1 \\ 3x + 2y = -5 \end{cases}$

28. $\begin{cases} x + 3y = 0 \\ x - y = 8 \end{cases}$

29. $\begin{cases} x - 4 = 0 \\ 3x - 5y = 22 \end{cases}$

30. $\begin{cases} y + 1 = 0 \\ x + 4y = -6 \end{cases}$

31. $\begin{cases} 3x - y = -1 \\ -6x + 2y = -4 \end{cases}$

32. $\begin{cases} -2x + 5y = -20 \\ 4x - 10y = 10 \end{cases}$

33. $\begin{cases} x + y = -2 \\ 3x - 4y = 8 \end{cases}$

34. $\begin{cases} 2x + 5y = 2 \\ -x + 2y = -1 \end{cases}$

35. $\begin{cases} 4x + 2y = 6 \\ 6x + 3y = 9 \end{cases}$

36. $\begin{cases} 4x + 3y = -4 \\ 8x + 2 = -6y - 6 \end{cases}$

37. $\begin{cases} y = -x + 3 \\ 3y = 2x + 9 \end{cases}$

38. $\begin{cases} 3y = 2x + 1 \\ y = 5x - 4 \end{cases}$

In Problems 39–46, determine the number of solutions of each system. State whether the system is consistent or inconsistent. For those systems that are consistent, state whether the equations are dependent or independent. State the solution of each system. See Objective 3.

39.

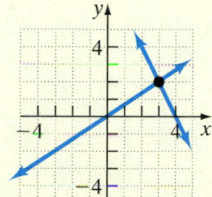

40.

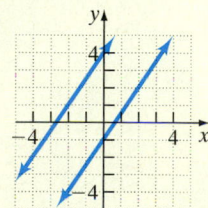

41.

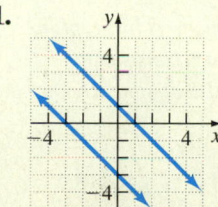

42.

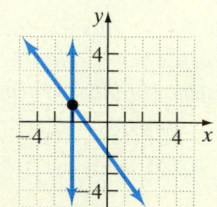

43.

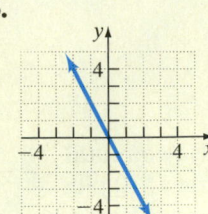

44.

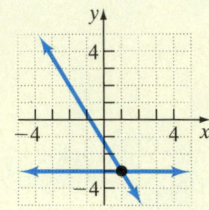

45.

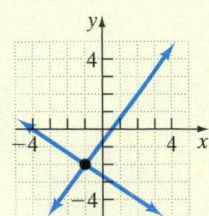

46.

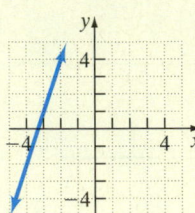

61. $\begin{cases} x - 2y = 6 \\ 2x - 4y = 0 \end{cases}$ **62.** $\begin{cases} 2x + 3y = -1 \\ 4x + 6y = -6 \end{cases}$

63. $\begin{cases} 3x - 2y = -2 \\ 2x + y = 8 \end{cases}$ **64.** $\begin{cases} 3x - y = -11 \\ -2x + y = 9 \end{cases}$

65. $\begin{cases} y = x + 2 \\ 3x - 3y = -6 \end{cases}$ **66.** $\begin{cases} -2x + 6y = -6 \\ 3x - 9y = 9 \end{cases}$

67. $\begin{cases} -5x - 2y = -2 \\ 10x + 4y = 4 \end{cases}$ **68.** $\begin{cases} 3x + y = -2 \\ 6x + 2y = -4 \end{cases}$

69. $\begin{cases} 3x = 4y - 12 \\ 2x = -4y - 8 \end{cases}$ **70.** $\begin{cases} 2x = -5y - 25 \\ -3x = 10y + 50 \end{cases}$

In Problems 47–58, without graphing, determine the number of solutions of each system of equations. State whether the system is consistent or inconsistent. For those systems that are consistent, state whether the equations are dependent or independent. See Objective 3.

47. $\begin{cases} y = 2x + 3 \\ y = -2x + 3 \end{cases}$ **48.** $\begin{cases} y = -x + 5 \\ y = x - 5 \end{cases}$

49. $\begin{cases} 6x + 2y = 12 \\ 3x + y = 12 \end{cases}$ **50.** $\begin{cases} 2x + y = -3 \\ -2x - y = 6 \end{cases}$

51. $\begin{cases} x + 2y = 2 \\ 2x + 4y = 4 \end{cases}$ **52.** $\begin{cases} -x + y = 5 \\ x - y = -5 \end{cases}$

53. $\begin{cases} y = 4 \\ x = 4 \end{cases}$ **54.** $\begin{cases} y - 3 = 0 \\ x - 3 = 0 \end{cases}$

55. $\begin{cases} x - 2y = -4 \\ -x + 2y = -4 \end{cases}$ **56.** $\begin{cases} 5x + y = -1 \\ 10x + 2y = -2 \end{cases}$

57. $\begin{cases} x + 2y = 4 \\ x + 1 = 5 - 2y \end{cases}$ **58.** $\begin{cases} 2y + 6 = 2x \\ x + 4 = y - 7 \end{cases}$

Mixed Practice

In Problems 59–70, solve each system of equations by graphing. Based on the graph, state whether the system is consistent or inconsistent. For those systems that are consistent, state whether the equations are dependent or independent.

59. $\begin{cases} y = 2x + 9 \\ y = -3x - 6 \end{cases}$ **60.** $\begin{cases} y = 3x - 5 \\ y = -x + 11 \end{cases}$

Applying the Concepts

*In business, a company's **break-even point** is the point at which the costs of production of a product will be equal to the amount of revenue from the sale of the product. In Problems 71–74, solve the system of equations by graphing to find the number of units that need to be produced for the company to break even.*

71. Producing T-Shirts The San Diego T-Shirt Company designs and sells T-shirts for promoters to sell at music concerts. The cost y (in dollars) to produce x T-shirts is given by the equation $y = 15x + 1000$, and the revenue y (in dollars) from the sale of these T-shirts is $y = 25x$. Find the break-even point: the point where the line that represents the cost, $y = 15x + 1000$, intersects the revenue line, $y = 25x$. Interpret the x-coordinate and the y-coordinate of the break-even point.

72. School Newspaper School Printers, Inc. prints and sells the school newspaper. The cost y (in dollars) to produce x newspapers is given by the equation $y = 0.05x + 8$, and the revenue y (in dollars) from the sale of these newspapers is given by the equation $y = 0.25x$. Find the break-even point: the point where the line that represents the cost, $y = 0.05x + 8$, intersects the revenue line, $y = 0.25x$. Interpret the x-coordinate and the y-coordinate of the break-even point.

73. Hockey Skates The Ice Blades Manufacturing Company produces and sells hockey skates to discount stores. The skates come in boxes of 10 pairs of skates. The cost y (in thousands of dollars) to produce x boxes of skates is given by the equation $y = 0.8x + 3.5$, and the revenue y (in thousands of dollars) from the sale of these skates is given by the equation $y = 1.5x$. Find the break-even point.

74. Patio Chairs Maumee Lumber Company produces wood patio chairs and sells them to local hardware and home improvement stores. The cost y (in hundreds of dollars) to produce x patio chairs is given by the equation $y = 0.5x + 6$. The revenue y (in hundreds of dollars) from the sale of x patio chairs is given by the equation $y = 1.5x$. Find the break-even point.

In Problems 75–78, set up a system of linear equations in two variables that models the problem. Then solve the system of linear equations.

75. Car Rental The Wheels-to-Go car rental agency rents cars for $50 daily plus $0.20 per mile. Acme Rental agency will rent the same car for $62 daily plus $0.12 per mile. On Liza's trip to Houston, she decides to rent a car. Determine the number of miles for which the cost of the car rental will be the same for both companies. If Liza plans to drive the car for 200 miles, which company should she use?

76. Car Rental The Speedy Car Rental agency charges $60 per day plus $0.14 per mile while the Slow-but-Cheap Car Rental agency charges $30 per day plus $0.20 per mile. Determine the number of miles for which the cost of the car rental will be the same for both companies. If you plan to drive the car for 400 miles, which company should you use?

77. Construction Costs You plan to build a cabin at the lake and want to get bids from two contractors to determine its cost. The first contractor gives you a quote of $60,000 plus $120 per square foot for the cabin. The second contractor gives you a quote of $80,000 plus $100 per square foot. For how many square feet will the two contractors charge the same amount? What is that cost?

78. Political Flyers A politician running for city council has found that PrintQuick can print pamphlets for her campaign for $0.04 per copy plus a one-time setup fee of $10. She has also learned that Print-A-Lot will print the same pamphlet for a flat fee of $20. Determine the number of political pamphlets that can be printed for the cost at PrintQuick to be the same as at Print-A-Lot.

Extending the Concepts

In Problems 79–82, determine the value of c so that the given system is consistent, but the equations are dependent.

79. $\begin{cases} 3x - y = -4 \\ y = cx + 4 \end{cases}$

80. $\begin{cases} 2(x + 4) = 4y + 4 \\ y = cx + 1 \end{cases}$

81. $\begin{cases} x + 3 = 3(x - y) \\ 2y + 3 = 2cx - y \end{cases}$

82. $\begin{cases} 5x + 2y = 3 \\ 4y = -10x + c \end{cases}$

Technology Exercises

Technology may be used to approximate a point of intersection of two equations. For graphing calculators, use the INTERSECT command after graphing each equation in the system. In Desmos, graph each equation in the system and click on each line. Then hover the mouse pointer over the point of intersection. Consider the following system of equations:

$$\begin{cases} x + y = -1 \\ -2x + y = -7 \end{cases}$$

Figure 7(a) shows the point of intersection using a TI-84 Plus C graphing calculator and Figure 7(b) shows the point of intersection using Desmos. The solution of the system is the ordered pair $(2, -3)$.

Figure 7

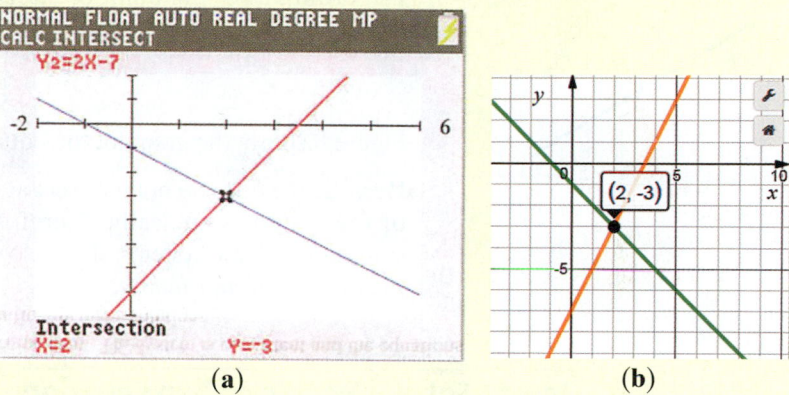

(a)   (b)

In Problems 83–88, use technology to solve each system of equations.

83. $\begin{cases} y = 3x - 1 \\ y = -2x + 5 \end{cases}$

84. $\begin{cases} y = \dfrac{3}{2}x - 4 \\ y = -\dfrac{1}{4}x + 3 \end{cases}$

85. $\begin{cases} 3x - y = -1 \\ -4x + y = -3 \end{cases}$

86. $\begin{cases} -6x - 2y = 4 \\ 5x + 3y = -2 \end{cases}$

87. $\begin{cases} 4x - 3y = 1 \\ -8x + 6y = -2 \end{cases}$

88. $\begin{cases} -2x + 5y = -2 \\ 4x - 10y = -1 \end{cases}$

Explaining the Concepts

89. Describe graphically the three possibilities for a solution of a system of two linear equations containing two variables.

90. When solving a system of linear equations using the graphing method, you find that the solution appears to be $(3, 4)$. How do you verify that $(3, 4)$ is the solution? Graphically, what does this solution represent?

4.2 Solving Systems of Linear Equations Using Substitution

Objectives

1. Solve a System of Linear Equations Using the Substitution Method
2. Solve Applied Problems Involving Systems of Linear Equations

Are You Prepared for This Section?

Before getting started, complete the following problems. If you get a problem wrong, go back to the section cited and review the material.

P1. Solve $3x - y = 2$ for y.   [Section 2.4, pp. 113–114]

P2. Solve $2x + 5y = 8$ for x.   [Section 2.4, pp. 113–114]

P3. Solve: $3x - 2(5x + 1) = 12$ [Section 2.2, pp. 92–93]

Figure 8

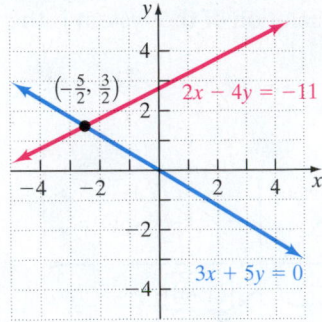

▶ 1 Solve a System of Linear Equations Using the Substitution Method

Obtaining an exact result using the graphing method can be difficult if the x- and y-coordinates of the point of intersection of two lines are not integers. Consider the following system of equations:

$$\begin{cases} 3x + 5y = 0 \\ 2x - 4y = -11 \end{cases}$$

Figure 8 shows the graph of the equations in the system. The lines intersect at $\left(-\dfrac{5}{2}, \dfrac{3}{2}\right)$.

Because the lines do not intersect at integer values, it is difficult to determine the solution of the system graphically. Therefore, rather than using graphical methods to obtain solutions of some systems, algebraic methods may be preferable. One algebraic method is the *method of substitution*.

EXAMPLE 1   **How to Solve a System of Two Equations Using Substitution**

Solve the system using substitution: $\begin{cases} y = -2x + 10 \\ 3x + 5y = 8 \end{cases}$

Step-by-Step Solution

First, label the equations (1) and (2) as shown.

$$\begin{cases} y = -2x + 10 & \text{(1)} \\ 3x + 5y = 8 & \text{(2)} \end{cases}$$

Step 1: Solve one of the equations for one of the unknowns.

Equation (1) is already solved for y: $y = -2x + 10$

Step 2: Substitute $y = -2x + 10$ into equation (2).

Equation (2): $3x + 5y = 8$

$3x + 5(-2x + 10) = 8$

Step 3: Solve the equation for x.

Distribute the 5: $3x - 10x + 50 = 8$

Combine like terms: $-7x + 50 = 8$

Subtract 50 from both sides: $-7x = -42$

Divide both sides by -7: $x = 6$

The x-coordinate of the solution is 6.

Step 4: Let $x = 6$ in equation (1) to find the value of y.

Equation (1): $y = -2x + 10$

$y = -2(6) + 10$

$y = -12 + 10$

$y = -2$

The y-coordinate of the solution is -2.

Prepared?...Answers P1. $y = 3x - 2$

P2. $x = -\dfrac{5}{2}y + 4$ **P3.** $\{-2\}$

Step 5: Verify that $x = 6$ and $y = -2$. Satisfy each equation in the original system.

Equation (1):
$$y = -2x + 10$$
$$-2 \overset{?}{=} -2(6) + 10$$
$$-2 \overset{?}{=} -12 + 10$$
$$-2 = -2 \quad \text{True}$$

Equation (2):
$$3x + 5y = 8$$
$$3(6) + 5(-2) \overset{?}{=} 8$$
$$18 - 10 \overset{?}{=} 8$$
$$8 = 8 \quad \text{True}$$

Both equations are satisfied, so the solution of the system is the ordered pair $(6, -2)$.

Figure 9 shows the graph of the equations in the system from Example 1. The graph shows the point of intersection, $(6, -2)$. ●

Figure 9

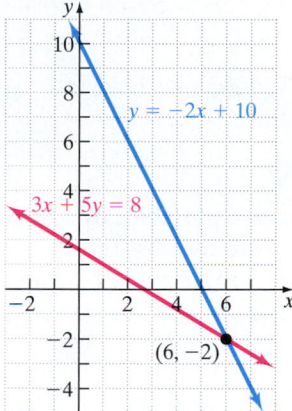

Below we summarize the steps for solving a system using substitution.

Solving a System of Linear Equations Using Substitution

Step 1: Solve one of the equations for one of the unknowns.

Step 2: Substitute the expression found in Step 1 into the *other* equation. The result will be a single linear equation in one unknown.

Step 3: Solve the linear equation in one unknown found in Step 2.

Step 4: Substitute the value of the variable found in Step 3 into one of the *original* equations to find the value of the other variable. Write the solution as an ordered pair, if possible.

Step 5: Check your answer by substituting the ordered pair into both of the original equations.

EXAMPLE 2   **Solving a System of Equations Using Substitution**

Solve the system using substitution: $\begin{cases} 2x - y = -15 & (1) \\ 4x + 3y = 5 & (2) \end{cases}$

Work Smart

Name each equation in the system as we did in Example 1 before going through the steps for solving the system.

Solution

When using substitution, solve for a variable whose coefficient is 1 or -1, if possible, to simplify the algebra. Solve equation (1) for y because the coefficient of y is -1.

Equation (1):	$2x - y = -15$
Subtract $2x$ from both sides:	$-y = -2x - 15$
Multiply both sides by -1:	$y = 2x + 15$

Substitute $2x + 15$ for y in equation (2).

Equation (2):	$4x + 3y = 5$
Let $y = 2x + 15$:	$4x + 3(2x + 15) = 5$

Work Smart

If you solve for a variable in equation (1), be sure to substitute into equation (2). If you solve for a variable in equation (2), be sure to substitute into equation (1).

Solve for x.

Distribute the 3:	$4x + 6x + 45 = 5$
Combine like terms:	$10x + 45 = 5$
Subtract 45 from both sides:	$10x = -40$
Divide both sides by 10:	$x = -4$

Let $x = -4$ in equation (1) and solve for y.

Equation (1):	$2x - y = -15$
	$2(-4) - y = -15$
	$-8 - y = -15$
Add 8 to both sides:	$-y = -7$
Multiply both sides by -1:	$y = 7$

The proposed solution is $(-4, 7)$.

(continued)

Check Let $x = -4$ and $y = 7$ in both equations in the system:

Equation (1): $\qquad 2x - y = -15$

$$2(-4) - 7 \stackrel{?}{=} -15$$

$$-8 - 7 \stackrel{?}{=} -15$$

$$-15 = -15 \quad \text{True}$$

Equation (2): $\qquad 4x + 3y = 5$

$$4(-4) + 3(7) \stackrel{?}{=} 5$$

$$-16 + 21 \stackrel{?}{=} 5$$

$$5 = 5 \quad \text{True}$$

The ordered pair $(-4, 7)$ satisfies both equations. The solution of the system is $(-4, 7)$. ●

> **Quick ✓**
>
> **1.** When solving a system of two linear equations containing two variables by substitution, if you solve for a variable in equation (1), be sure to substitute into equation ___
>
> *In Problems 2 and 3, solve the system using substitution.*
>
> **2.** $\begin{cases} y = 3x - 2 \\ 2x - 3y = -8 \end{cases}$ **3.** $\begin{cases} 2x + y = -1 \\ 4x + 3y = 3 \end{cases}$

EXAMPLE 3 **Solving a System of Equations Using Substitution**

Solve the system using substitution: $\begin{cases} 2x + 3y = 9 & (1) \\ x + 2y = \dfrac{13}{2} & (2) \end{cases}$

Solution

Solve equation (2) for x because the coefficient of x in equation (2) is 1.

Equation (2): $\qquad x + 2y = \dfrac{13}{2}$

Subtract $2y$ from both sides: $\qquad x = -2y + \dfrac{13}{2}$

Substitute $-2y + \dfrac{13}{2}$ for x in equation (1).

Equation (1): $\qquad 2x + 3y = 9$

$$2\left(-2y + \dfrac{13}{2}\right) + 3y = 9$$

Now solve this equation for y.

Distribute the 2: $\qquad -4y + 13 + 3y = 9$

Combine like terms: $\qquad -y + 13 = 9$

Subtract 13 from both sides: $\qquad -y = -4$

Multiply both sides by -1: $\qquad y = 4$

Let $y = 4$ in equation (1) and solve for x.

Equation (1): $\qquad 2x + 3y = 9$

$$2x + 3(4) = 9$$

$$2x + 12 = 9$$

Subtract 12 from both sides: $\qquad 2x = -3$

Divide both sides by 2: $\qquad \dfrac{2x}{2} = -\dfrac{3}{2}$

$$x = -\dfrac{3}{2}$$

The proposed solution is $\left(-\dfrac{3}{2}, 4\right)$.

Check Let $x = -\dfrac{3}{2}$ and $y = 4$ in both equations in the system:

Equation (1):

$$2x + 3y = 9$$

$$2\left(-\dfrac{3}{2}\right) + 3(4) \overset{?}{=} 9$$

$$-3 + 12 \overset{?}{=} 9$$

$$9 = 9 \quad \text{True}$$

Equation (2):

$$x + 2y = \dfrac{13}{2}$$

$$-\dfrac{3}{2} + 2(4) \overset{?}{=} \dfrac{13}{2}$$

$$-\dfrac{3}{2} + 8 \overset{?}{=} \dfrac{13}{2}$$

$$\dfrac{13}{2} = \dfrac{13}{2} \quad \text{True}$$

Because $\left(-\dfrac{3}{2}, 4\right)$ satisfies both equations, the solution of the system is $\left(-\dfrac{3}{2}, 4\right)$. ●

Work Smart

Solving for x or for y in equation (1) in Example 3 is possible, but it will lead to fractions, making substitution more difficult. Solving equation (1) for y yields

$$2x + 3y = 9$$

$$3y = -2x + 9$$

$$y = -\dfrac{2}{3}x + 3$$

Letting $y = -\dfrac{2}{3}x + 3$ in equation (2), $x + 2y = \dfrac{13}{2}$ becomes

$$x + 2\left(-\dfrac{2}{3}x + 3\right) = \dfrac{13}{2}$$

$$x - \dfrac{4}{3}x + 6 = \dfrac{13}{2}$$

$$-\dfrac{1}{3}x + 6 = \dfrac{13}{2}$$

$$-\dfrac{1}{3}x = \dfrac{1}{2}$$

$$x = -\dfrac{3}{2}$$

Whew! And we still need to substitute $x = -\dfrac{3}{2}$ into equation (1) or (2) to find y!

Quick ✔

In Problems 4 and 5, solve the system using substitution.

4. $\begin{cases} -4x + y = 1 \\ 8x - y = 5 \end{cases}$

5. $\begin{cases} 3x + 2y = -3 \\ -x + y = \dfrac{11}{6} \end{cases}$

⊙ Recall that a system of two linear equations containing two variables can be inconsistent. This means that the two lines in the system are parallel. What happens when the substitution method is used on an inconsistent system? Let's see.

EXAMPLE 4   **Solving an Inconsistent System Using Substitution**

Solve the system using substitution: $\begin{cases} x - 3y = 5 & (1) \\ -2x + 6y = 3 & (2) \end{cases}$

(continued)

Solution

The coefficient of x in equation (1) is 1, so solve that equation for x.

Equation (1):	$x - 3y = 5$
Add $3y$ to both sides:	$x = 3y + 5$
Equation (2):	$-2x + 6y = 3$
In equation (2), replace x by $3y + 5$:	$-2(3y + 5) + 6y = 3$
Distribute the -2:	$-6y - 10 + 6y = 3$
Combine like terms:	$-10 = 3$

Figure 10

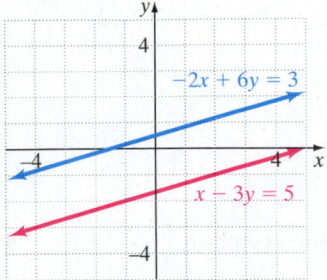

Notice that the variable y was eliminated and that the last equation, $-10 = 3$, is a false statement. Thus, the system is inconsistent. The solution is $\{\ \}$ or $\varnothing$. Figure 10 shows the graph of the equations in the system. Note that the lines are parallel. ●

So if the algebraic solution ends with a false statement such as $-10 = 3$ or $-13 = 0$, the system is inconsistent.

Now consider a system that is consistent but has dependent equations.

EXAMPLE 5 **Solving a System with Infinitely Many Solutions Using Substitution**

Solve the system using substitution: $\begin{cases} 6x - 2y = -4 & (1) \\ -3x + y = 2 & (2) \end{cases}$

Solution

It is easiest to solve equation (2) for y.

Equation (2):	$-3x + y = 2$
Add $3x$ to both sides:	$y = 3x + 2$
Equation (1):	$6x - 2y = -4$
Let $y = 3x + 2$ in equation (1):	$6x - 2(3x + 2) = -4$
Distribute the -2:	$6x - 6x - 4 = -4$
Combine like terms:	$-4 = -4$

Figure 11

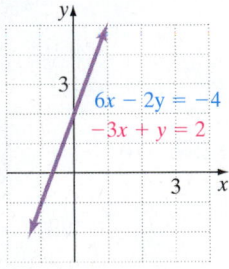

The variable x has been eliminated, but the equation $-4 = -4$ is true. Figure 11 shows the graph of the system of equations in Example 5. The graphs of the two linear equations both have slope 3 and y-intercept $(0, 2)$. The lines are coincident. This means that any values of x and y that satisfy $6x - 2y = -4$ or $-3x + y = 2$ are solutions. The system is consistent and the equations are dependent, so there are infinitely many solutions. Therefore, the solution is $\{(x, y)\,|\,-3x + y = 2\}$. ●

Work Smart

If a true statement with no variables occurs after the substitution (such as $-4 = -4$), the system has infinitely many solutions, as in Example 5. If a false statement with no variables occurs after the substitution (such as $-10 = 3$), the system has no solution, as in Example 4.

Quick ✓

6. In the process of solving a system of equations by substitution, the variables have been eliminated, and a *false* statement such as $-3 = 0$ results. This means that the solution of the system is _____

7. In the process of solving a system of equations by substitution, the variables have been eliminated, and a *true* statement such as $0 = 0$ results. This means that the system has _____ _____ _____

8. *True or False* When solving a system of equations by substitution, you solve equation (1) for x, and then substitute this expression back into equation (1), resulting in the statement $5 = 5$. This means that there are infinitely many solutions of the system.

In Problems 9–11, solve the system using substitution.

9. $\begin{cases} y = 5x + 2 \\ -10x + 2y = 4 \end{cases}$

10. $\begin{cases} 3x - 2y = 0 \\ -9x + 6y = 5 \end{cases}$

11. $\begin{cases} 2x - 6y = 2 \\ -3x + 9y = 4 \end{cases}$

▶ ❷ **Solve Applied Problems Involving Systems of Linear Equations**

EXAMPLE 6 **Supply and Demand**

The number of hot dogs h that a street vendor is willing to sell each day is given by the equation $h = 300p - 100$, where p is the price of the hot dog, in dollars. The number of hot dogs h that individuals are willing to purchase each day is given by the equation $h = -150p + 1250$, where p is the price of the hot dog. The price at which supply equals demand is the equilibrium price. To find the equilibrium price of the hot dogs, solve the system of equations

$$\begin{cases} h = 300p - 100 \\ h = -150p + 1250 \end{cases}$$

What is the equilibrium price? How many hot dogs will the vendor sell at this price?

Solution

Name the equations (1) and (2).

$$\begin{cases} h = 300p - 100 & \text{(1)} \\ h = -150p + 1250 & \text{(2)} \end{cases}$$

Solve this system of equations by substituting $300p - 100$ for h from equation (1) into equation (2). This leads to the following equation:

$$300p - 100 = -150p + 1250$$

Add 150p to both sides: $\quad 450p - 100 = 1250$

Add 100 to both sides: $\quad 450p = 1350$

Divide both sides by 450: $\quad p = 3$

The check is left to you.

The equilibrium price of the hot dogs is $3. The vendor will sell
$h = -150(3) + 1250 = 800$ hot dogs. ●

Quick ✓

12. The number of sodas s that a street vendor is willing to sell each day is given by the equation $s = 200p + 800$, where p is the price, in dollars, of the soda. The number of sodas s that individuals are willing to purchase each day is given by the equation $s = -600p + 2400$, where p is the price of the soda. The price at which supply equals demand is the equilibrium price. To find the equilibrium price of the soda, solve the system of equations

$$\begin{cases} s = 200p + 800 \\ s = -600p + 2400 \end{cases}$$

What is the equilibrium price? How many sodas will the vendor sell at this price?

4.2 Exercises **MyMathLab®**

*Problems **1–12** are the **Quick ✓** s that follow the **EXAMPLES**.*

Building Skills

In Problems 13–26, solve each system of equations using substitution. See Objective 1.

13. $\begin{cases} x + 2y = 2 \\ y = 2x - 9 \end{cases}$

14. $\begin{cases} y = 3x - 11 \\ -x + 4y = -11 \end{cases}$

15. $\begin{cases} -2x + 5y = 7 \\ x \qquad = 3y - 4 \end{cases}$

16. $\begin{cases} -4x - y = -3 \\ x \qquad = y + 7 \end{cases}$

17. $\begin{cases} x + y = -7 \\ 2x - y = -2 \end{cases}$

18. $\begin{cases} 5x + 2y = -5 \\ 3x - y = -14 \end{cases}$

19. $\begin{cases} y = \dfrac{1}{2}x - 5 \\ y = -\dfrac{3}{4}x - 10 \end{cases}$

20. $\begin{cases} y = \dfrac{2}{3}x + 1 \\ y = -\dfrac{3}{2}x + 40 \end{cases}$

21. $\begin{cases} y = 3x + 4 \\ y = -\dfrac{1}{2}x + \dfrac{5}{3} \end{cases}$

22. $\begin{cases} y = 5x - 3 \\ y = 2x - \dfrac{21}{5} \end{cases}$

23. $\begin{cases} x = -6y \\ x - 3y = 3 \end{cases}$

24. $\begin{cases} y = 4x \\ 2x - 3y = 5 \end{cases}$

25. $\begin{cases} 2x - 3y = 0 \\ 8x + 6y = 3 \end{cases}$

26. $\begin{cases} 2x + 3y = -1 \\ 2x - 9y = -9 \end{cases}$

In Problems 27–34, solve each system of equations using substitution. State whether the system is inconsistent, or consistent and with dependent equations. See Objective 1.

27. $\begin{cases} x + 3y = -12 \\ x + 3y = 6 \end{cases}$

28. $\begin{cases} 2x - y = 3 \\ y - 2x = 3 \end{cases}$

29. $\begin{cases} y = 4x - 1 \\ 8x - 2y = 2 \end{cases}$

30. $\begin{cases} y = -x + 1 \\ x + y = 1 \end{cases}$

31. $\begin{cases} x + 2y = 6 \\ x = 3 - 2y \end{cases}$

32. $\begin{cases} 4 + y = 3x \\ 6x - 2y = -2 \end{cases}$

33. $\begin{cases} 5x + 6 = 2 - y \\ y = -5x - 4 \end{cases}$

34. $\begin{cases} x + y = 2 \\ 2x + 2y = 2 \end{cases}$

Mixed Practice

In Problems 35–52, solve each system of equations using substitution.

35. $\begin{cases} x = 2y \\ x - 6y = 4 \end{cases}$

36. $\begin{cases} 2x + y = 5 \\ y = 3x \end{cases}$

37. $\begin{cases} y = \dfrac{1}{2}x \\ x = 2(y + 1) \end{cases}$

38. $\begin{cases} y = \dfrac{3}{4}x \\ x = 4(y - 1) \end{cases}$

39. $\begin{cases} y = 2x - 8 \\ x - \dfrac{1}{2}y = 4 \end{cases}$

40. $\begin{cases} x = 3y + 6 \\ y - \dfrac{1}{3}x = -2 \end{cases}$

41. $\begin{cases} 3x + 2y = 1 \\ -2x - y = 1 \end{cases}$

42. $\begin{cases} 2x + 5y = 7 \\ -x + 6y = -12 \end{cases}$

43. $\begin{cases} x - 5y = 3 \\ -2x + 10y = 8 \end{cases}$

44. $\begin{cases} -4x + y = 3 \\ 8x - 2y = 1 \end{cases}$

45. $\begin{cases} 3x - y = 1 \\ -6x + 2y = -2 \end{cases}$

46. $\begin{cases} -x + 3y = 4 \\ 2x - 6y = -8 \end{cases}$

47. $\begin{cases} 4x + 8y = -9 \\ 2x + y = \dfrac{3}{4} \end{cases}$

48. $\begin{cases} 18x - 6y = -7 \\ x + 2y = 0 \end{cases}$

49. $\begin{cases} \dfrac{3}{2}x - y = 1 \\ 3x - 2y = 2 \end{cases}$

50. $\begin{cases} \dfrac{2}{3}x + \dfrac{1}{3}y = 1 \\ \dfrac{1}{2}x + \dfrac{1}{4}y = \dfrac{3}{4} \end{cases}$

51. $\begin{cases} \dfrac{x}{2} + \dfrac{y}{3} = \dfrac{1}{12} \\ \dfrac{2x}{3} + \dfrac{y}{3} = -\dfrac{1}{3} \end{cases}$

52. $\begin{cases} \dfrac{x}{4} + \dfrac{y}{2} = \dfrac{3}{8} \\ x - \dfrac{y}{3} = \dfrac{1}{3} \end{cases}$

Applying the Concepts

In Problems 53–60, use substitution to solve each system of linear equations in two variables.

△ **53. Dimensions of a Garden** The perimeter of a rectangular garden is 34 feet. The length of the garden is 3 feet more than the width. Determine the dimensions of the garden by solving the following system of equations, where *l* and *w* represent the length and width of the garden.

$$\begin{cases} 2l + 2w = 34 & (1) \\ l = w + 3 & (2) \end{cases}$$

△ **54. Dimensions of a Rectangle** The perimeter of a rectangle is 52 inches. The length of the rectangle is 4 inches more than the width. Determine the dimensions of the rectangle by solving the following system of equations, where *l* and *w* represent the length and width of the rectangle.

$$\begin{cases} 2l + 2w = 52 & (1) \\ l = w + 4 & (2) \end{cases}$$

55. Investment Paul wants to invest part of his Ohio Lottery winnings in a safe money market fund that earns 2.5% annual interest and the rest in a risky international fund that is expected to yield 9% annual interest. The amount of money invested in the money market fund, *x*, is to be exactly twice the amount invested in the international fund, *y*. Use the following system of equations to determine the amount to be invested in each fund if a total of $560 is to be earned at the end of one year.

$$\begin{cases} x = 2y & \text{(1)} \\ 0.025x + 0.09y = 560 & \text{(2)} \end{cases}$$

56. Investment Elaine wants to invest part of her $24,000 Virginia Lottery winnings in an international stock fund that yields 4% annual interest and the remainder in a domestic growth fund that yields 6.5% annually. Use the following system of equations to determine the amount Elaine should invest in each account to earn $1360 interest at the end of one year, where x represents the amount invested in the international stock fund and y represents the amount invested in the domestic growth fund.

$$\begin{cases} x + y = 24{,}000 & \text{(1)} \\ 0.04x + 0.065y = 1360 & \text{(2)} \end{cases}$$

57. Salary Suppose that you are offered a sales position for a pharmaceutical company. They offer you two salary options. Option A would pay you an annual base salary of $15,000 plus a commission of 2% on sales. The equation $y = 15{,}000 + 0.02x$ models salary Option A, where x represents the annual sales amount and y represents the annual salary. Option B would pay you an annual base salary of $25,000 plus a commission of 1% on sales. The equation $y = 25{,}000 + 0.01x$ models salary Option B. Determine the annual sales required for the options to result in the same annual salary. What is that salary?

58. Salary Melanie has been offered a sales position for a major textbook company. Melanie was offered two options. Option A would pay her an annual base salary of $20,000 plus a commission of 4% on sales. The equation $y = 20{,}000 + 0.04x$ models salary Option A, where x represents the annual sales amount and y represents the annual salary. Option B would pay Melanie an annual base salary of $25,000 plus a commission of 2% on sales. The equation $y = 25{,}000 + 0.02x$ models salary Option B. Determine the annual sales required for the two options to result in the same annual salary. What is that salary?

59. Fun with Numbers The sum of two numbers is 17, and their difference is 7. Determine the numbers by

solving the following system of equations, using x and y to represent the unknown numbers.

$$\begin{cases} x + y = 17 & \text{(1)} \\ x - y = 7 & \text{(2)} \end{cases}$$

60. Fun with Numbers The sum of two numbers is 25, and their difference is 3. Determine the numbers by solving the following system of equations, using x and y to represent the unknown numbers.

$$\begin{cases} x + y = 25 & \text{(1)} \\ x - y = 3 & \text{(2)} \end{cases}$$

Extending the Concepts

61. For the system $\begin{cases} Ax + 3By = 2 \\ -3Ax + By = -11 \end{cases}$, find A and B such that $x = 3$, $y = 1$ is a solution.

62. For the system $\begin{cases} y = x - 3a \\ x + y = 7a \end{cases}$, consider x and y as the variables of the system of equations. Use substitution to solve for x and y in terms of a.

63. Write a system of equations that has $(3, 5)$ as a solution.

64. Write a system of equations that has $(-1, 4)$ as a solution.

65. Write a system of equations that has infinitely many solutions.

66. Write a system of equations that has no solution.

Explaining the Concepts

67. A test question asks you to solve the following system of equations by substitution. What would be a reasonable first step? Explain which variable you would solve for and why.

$$\begin{cases} \dfrac{1}{3}x - \dfrac{1}{6}y = \dfrac{2}{3} \\ \dfrac{3}{2}x + \dfrac{1}{2}y = -\dfrac{5}{4} \end{cases}$$

68. List two advantages of solving a system of linear equations by substitution rather than by the graphing method.

4.3 Solving Systems of Linear Equations Using Elimination

Objectives

1 Solve a System of Linear Equations Using the Elimination Method

2 Solve Applied Problems Involving Systems of Linear Equations

Are You Prepared for This Section?

Before getting started, complete the following problems. If you get a problem wrong, go back to the section cited and review the material.

P1. What is the additive inverse of 5? [Section 1.4, p. 30]

P2. What is the additive inverse of -8? [Section 1.4, p. 30]

P3. Distribute: $\dfrac{2}{3}(3x - 9y)$ [Section 1.8, pp. 67–68]

P4. Solve: $2y - 5y = 12$ [Section 2.2, pp. 92–93]

P5. Find the LCM of 4 and 5. [Section 1.2, pp. 9–10]

▶ 1 Solve a System of Linear Equations Using the Elimination Method

Work Smart

Use the method of substitution if it is easy to solve for one of the variables in the system. Use elimination if substitution will lead to fractions or if solving for a variable is not easy.

So far, two methods to solve a system of linear equations containing two unknowns have been used. The graphing method lets us visualize solutions, but finding the exact coordinates of the point of intersection can be difficult. The substitution method lets us find an exact solution because it is an algebraic method, but it may require the use of fractions. We would rather avoid fractions, so we introduce a second algebraic method that can be used to solve a system of equations: the *elimination method*. **The elimination method is usually preferred over the substitution method when substitution leads to fractions or solving for a variable is not straightforward.**

Remember that the Addition Property of Equality states that if the same quantity is added to both sides of an equation, an equivalent equation is obtained. The elimination method takes this principle a step further.

EXAMPLE 1 **Solving a System of Linear Equations Using Elimination**

Solve the system using elimination: $\begin{cases} x + y = 13 & (1) \\ 2x - y = -7 & (2) \end{cases}$

Solution

Equation (2) says that $2x - y$ equals -7. Thus, if we add $2x - y$ to the left side of equation (1) and -7 to the right side of equation (1), we are adding the same value to each side of equation (1). Perform this addition vertically.

$$
\begin{array}{rl}
x + y = 13 & (1) \\
\underline{2x - y = -7} & (2) \\
3x + 0 = 6 &
\end{array}
$$

Work Smart

This method is called "elimination" because one of the variables is eliminated through the process of addition. The elimination method is sometimes called the *addition method*.

Notice that the variable y has been eliminated, and we can solve for x.

$$3x + 0 = 6$$
$$3x = 6$$

Divide both sides by 3: $\quad x = 2$

Now substitute $x = 2$ into either equation (1) or equation (2) to find the value of y. In this case, substitute $x = 2$ into equation (1).

Equation (1): $\quad x + y = 13$

Substitute 2 for x: $\quad 2 + y = 13$

Subtract 2 from both sides: $\quad y = 11$

It appears that $x = 2$ and $y = 11$. Check this solution in both equations.

Check

Equation (1):	$x + y = 13$	Equation (2):	$2x - y = -7$
	$2 + 11 \stackrel{?}{=} 13$		$2(2) - 11 \stackrel{?}{=} -7$
	$13 = 13$ True		$4 - 11 \stackrel{?}{=} -7$
			$-7 = -7$ True

The solution of the system is $(2, 11)$. ●

What allows us to add two equations and use the result to replace an equation? Remember, an equation is a statement that the left side equals the right side. When equation (2) is added to equation (1), the same quantity is added to both sides of equation (1).

The idea in using the elimination method is to get the coefficients of one of the variables to be additive inverses (or opposites). In Example 1, the coefficients of the variable y are opposites: 1 and -1. Having a system of linear equations in which one of the variables has opposite coefficients is uncommon, so we need a strategy for solving systems in which there is no variable that has opposite coefficients.

EXAMPLE 2   **How to Solve a System of Linear Equations Using Elimination**

Solve the system using elimination: $\begin{cases} 2x + 3y = -6 & (1) \\ -4x + 5y = -21 & (2) \end{cases}$

Step-by-Step Solution

Step 1: Write each equation in standard form, $Ax + By = C$.

Both equations are already in standard form.

Step 2: The goal is to get the coefficients of one of the variables to be additive inverses. Because the coefficient of x in equation (1) is 2, and the coefficient of x in equation (2) is -4, this can be accomplished by multiplying both sides of equation (1) by 2.

$$\begin{cases} 2x + 3y = -6 & (1) \\ -4x + 5y = -21 & (2) \end{cases}$$

Multiply both sides of equation (1) by 2:

$$\begin{cases} 2(2x + 3y) = 2(-6) & (1) \\ -4x + 5y = -21 & (2) \end{cases}$$

Distribute the 2 in equation (1):

$$\begin{cases} 4x + 6y = -12 & (1) \\ -4x + 5y = -21 & (2) \end{cases}$$

Step 3: Notice that the coefficients of the variable x are additive inverses. Add equations (1) and (2) to eliminate x and then solve for y.

$$\begin{cases} 4x + 6y = -12 & (1) \\ -4x + 5y = -21 & (2) \end{cases}$$

Add equations (1) and (2):

$$11y = -33$$

Divide both sides by 11:

$$y = -3$$

The y-value of the solution is -3.

Step 4: Let $y = -3$ in either equation (1) or (2). Because equation (1) looks a little easier to work with, substitute $y = -3$ into equation (1) and solve for x.

Equation (1):

$$2x + 3y = -6$$
$$2x + 3(-3) = -6$$
$$2x - 9 = -6$$

Add 9 to both sides:

$$2x = 3$$

Divide both sides by 2:

$$x = \frac{3}{2}$$

The proposed solution is $\left(\frac{3}{2}, -3\right)$.

(continued)

Step 5: Check Verify that $x = \dfrac{3}{2}$ and $y = -3$ satisfy each equation in the original system.

Equation (1): $2x + 3y = -6$

$$2\left(\frac{3}{2}\right) + 3(-3) \overset{?}{=} -6$$

$$3 + (-9) \overset{?}{=} -6$$

$$-6 = -6 \quad \text{True}$$

Equation (2): $-4x + 5y = -21$

$$-4\left(\frac{3}{2}\right) + 5(-3) \overset{?}{=} -21$$

$$-6 + (-15) \overset{?}{=} -21$$

$$-21 = -21 \quad \text{True}$$

Both equations are satisfied, so the solution of the system is $\left(\dfrac{3}{2}, -3\right)$. ●

Solving a System of Linear Equations by Elimination

Step 1: Write each equation in standard form, $Ax + By = C$.

Step 2: If necessary, multiply or divide both sides of one equation (or both equations) by a nonzero constant so that the coefficients of one of the variables are opposites (additive inverses).

Step 3: Add the equations to eliminate the variable whose coefficients are now additive inverses. Solve the resulting equation for the remaining unknown.

Step 4: Substitute the value of the variable found in Step 3 into one of the *original* equations to find the value of the remaining variable. Write the solution as an ordered pair, if possible.

Step 5: Check the answer by substituting the ordered pair into both of the original equations.

Quick ✓

1. The basic idea in using the elimination method is to get the coefficients of one of the variables to be _____ _____, such as 3 and -3.

In Problems 2 and 3, solve the system using elimination.

2. $\begin{cases} x - 3y = 2 \\ 2x + 3y = -14 \end{cases}$

3. $\begin{cases} x - 2y = 2 \\ -2x + 5y = -1 \end{cases}$

EXAMPLE 3

Solve a System of Linear Equations Using Elimination—Multiply Both Equations by a Number to Create Additive Inverses

Solve the system using elimination: $\begin{cases} 3y = -2x + 3 & (1) \\ 3x + 5y = 7 & (2) \end{cases}$

Solution

Write equation (1), $3y = -2x + 3$, in standard form, $Ax + By = C$, by adding $2x$ to both sides of the equation to obtain $2x + 3y = 3$.

   The goal is to get the coefficients of one of the variables to be additive inverses. This cannot be accomplished by multiplying a single equation by a nonzero constant without introducing fractions, so *both* equations must be multiplied by a nonzero constant. For example, multiply both sides of equation (1) by 3 and multiply both sides of equation (2) by -2 so that the coefficients of x are additive inverses.

Work Smart

There is no single right way to multiply the equations in a system by nonzero constants to get coefficients to be additive inverses. For example, we could also multiply equation (1) by 5 and equation (2) by -3. Do you see why this also works?

$$\begin{cases} 2x + 3y = 3 & (1) \\ 3x + 5y = 7 & (2) \end{cases}$$

Multiply both sides of (1) by 3:
Multiply both sides of (2) by -2:
$$\begin{cases} 3(2x + 3y) = 3(3) & (1) \\ -2(3x + 5y) = -2(7) & (2) \end{cases}$$

$$\text{Distribute:} \quad \begin{cases} 6x + 9y = 9 & (1) \\ -6x - 10y = -14 & (2) \end{cases}$$

Add equations (1) and (2): $\qquad -y = -5$

Multiply both sides by -1: $\qquad y = 5$

Now substitute $y = 5$ into either original equation. In this case, substitute $y = 5$ into equation (1) to determine the value of x.

Equation (1): $\qquad 3y = -2x + 3$

$$3(5) = -2x + 3$$

Simplify: $\qquad 15 = -2x + 3$

Subtract 3 from both sides: $\qquad 12 = -2x$

Divide both sides by -2: $\qquad -6 = x$

The proposed solution is $(-6, 5)$.

Check Let $x = -6$ and $y = 5$ in both equations in the system:

Equation (1): $\quad 3y = -2x + 3$ $\qquad$ Equation (2): $\quad 3x + 5y = 7$

$$3(5) \stackrel{?}{=} -2(-6) + 3 \qquad\qquad 3(-6) + 5(5) \stackrel{?}{=} 7$$

$$15 \stackrel{?}{=} 12 + 3 \qquad\qquad\qquad -18 + 25 \stackrel{?}{=} 7$$

$$15 = 15 \quad \text{True} \qquad\qquad\qquad\qquad 7 = 7 \quad \text{True}$$

The solution to the system is $(-6, 5)$. ●

> **Quick ✓**
>
> *In Problems 4 and 5, solve the system using elimination.*
>
> **4.** $\begin{cases} 5x + 4y = 10 \\ -2x + 3y = -27 \end{cases}$ $\qquad$ **5.** $\begin{cases} 4y = -6x + 30 \\ 7x + 10y = 35 \end{cases}$

Now let's look at the results of systems that have no solution or infinitely many solutions.

EXAMPLE 4 **Solving a System of Equations with No Solution Using Elimination**

Solve the system using elimination: $\begin{cases} 4x - 6y = -5 & (1) \\ 6x - 9y = 2 & (2) \end{cases}$

Solution
Let's eliminate x.

$$\begin{cases} 4x - 6y = -5 & (1) \\ 6x - 9y = 2 & (2) \end{cases}$$

Multiply both sides of (1) by -3: $\begin{cases} -3(4x - 6y) = -3(-5) & (1) \\ 2(6x - 9y) = 2(2) & (2) \end{cases}$
Multiply both sides of (2) by 2:

$$\text{Distribute:} \quad \begin{cases} -12x + 18y = 15 & (1) \\ 12x - 18y = 4 & (2) \end{cases}$$

Add equations (1) and (2): $\qquad 0 = 19$

The statement $0 = 19$ is false. Therefore, the system is inconsistent and has no solution. The solution set is { } or $\varnothing$. If graphed, the lines would be parallel. ●

EXAMPLE 5   **Solving a System of Equations with Infinitely Many Solutions Using Elimination**

Solve the system using elimination: $\begin{cases} \dfrac{3}{2}x + \dfrac{2}{3}y = -4 & (1) \\ -\dfrac{9}{8}x - \dfrac{1}{2}y = 3 & (2) \end{cases}$

Solution

To clear each equation of fractions, multiply both sides of equation (1) by 6, the LCD of $\dfrac{3}{2}$ and $-\dfrac{2}{3}$, and multiply both sides of equation (2) by 8, the LCD of $-\dfrac{9}{8}$ and $-\dfrac{1}{2}$.

$$\begin{cases} 6\left(\dfrac{3}{2}x + \dfrac{2}{3}y\right) = 6(-4) & (1) \\ 8\left(-\dfrac{9}{8}x - \dfrac{1}{2}y\right) = 8(3) & (2) \end{cases}$$

Distribute: $\begin{cases} 9x + 4y = -24 & (1) \\ -9x - 4y = 24 & (2) \end{cases}$

Add equations (1) and (2): $0 = 0$

Notice that both variables were eliminated and that the statement $0 = 0$ is true. Therefore, the system is consistent, with dependent equations. The system has infinitely many solutions. A graph of the equations would show one line in the coordinate plane.

The solution to the system is $\left\{ (x, y) \,\middle|\, \dfrac{3}{2}x + \dfrac{2}{3}y = -4 \right\}$.   ●

Quick ✓

6. When using the elimination method to solve a system of equations, adding equation (1) and equation (2) results in the statement $-50 = -50$. This means that the equations are _____ and that the system has _____ many solutions.

In Problems 7–9, use elimination to determine whether the system has no solution or infinitely many solutions.

7. $\begin{cases} 2x - 6y = 10 \\ 5x - 15y = 4 \end{cases}$ 8. $\begin{cases} -x + 3y = 2 \\ 3x - 9y = -6 \end{cases}$ 9. $\begin{cases} -\dfrac{1}{4}x + \dfrac{1}{2}y = 1 \\ \dfrac{1}{2}x - y = -2 \end{cases}$

Summary   Which Method Should I Use?

Ask yourself the following questions in order to determine the most appropriate method for solving a system of linear equations.

Question	Method to Use	Example	Advantages/Disadvantages
Would it be beneficial to see the solutions visually?	Graphical	Find the break-even point of the cost equation $y = 20 + 4x$ and the revenue equation $y = 9x$, where x represents the number of calculators produced and sold.	This method allows the answer to be seen, but if the coordinates of the point of intersection are not integers, it can be difficult to determine the answer.
Is one of the coefficients of the variables 1 or −1?	Substitution	$\begin{cases} -x + 2y = 7 \\ 2x - 3y = -15 \end{cases}$	This method gives exact solutions. The algebra can be easy if the coefficient of one of the variables is 1 or −1. If none of the coefficients is 1 or −1, the algebra can be messy.
Are the coefficients of a variable additive inverses? If not, is it easy to get the coefficients to be additive inverses?	Elimination	$\begin{cases} 3x - 2y = -5 \\ -3x + 4y = 8 \end{cases}$	This method gives exact solutions. It is easy to use when neither variable has a coefficient of 1 or −1.

▶ ❷ Solve Applied Problems Involving Systems of Linear Equations

Applied problems are easily solved using substitution when at least one of the variables is isolated (that is, by itself). But many mathematical models that involve two equations containing two unknowns do not have one of the equations solved for one of the unknowns. If this occurs, it is better to use elimination to solve the problem.

EXAMPLE 6 **Hot Dogs and Soda at the Game**

Bill and Roger attend a baseball game with their kids. They get to the game early to watch batting practice and have dinner. Roger says that he will buy dinner. He gets 7 hot dogs and 5 Pepsis for $38.25. After the sixth inning everyone is hungry again, so Bill buys 5 hot dogs and 4 Pepsis for $28.50. Find the price of a hot dog and the price of a Pepsi by solving the system of equations

$$\begin{cases} 7h + 5p = 38.25 \\ 5h + 4p = 28.50 \end{cases}$$

where h represents the price of a hot dog and p represents the price of a Pepsi.

Solution
First, label the equations.

$$\begin{cases} 7h + 5p = 38.25 \quad (1) \\ 5h + 4p = 28.50 \quad (2) \end{cases}$$

Let's eliminate p.

Multiply both sides of (1) by -4:
Multiply both sides of (2) by 5:
$$\begin{cases} -4(7h + 5p) = -4(38.25) \quad (1) \\ 5(5h + 4p) = 5(28.50) \quad (2) \end{cases}$$

Distribute:
$$\begin{cases} -28h - 20p = -153 \quad (1) \\ 25h + 20p = 142.50 \quad (2) \end{cases}$$

Add equations (1) and (2): $\quad -3h = -10.50$
Divide both sides by -3: $\quad h = 3.50$
Equation (2): $\quad 5h + 4p = 28.50$
Let $h = 3.50$ in equation (2): $\quad 5(3.50) + 4p = 28.50$
$$17.50 + 4p = 28.50$$
Subtract 17.50 from both sides: $\quad 4p = 11$
Divide both sides by 4: $\quad p = 2.75$

We leave it to you to verify that 7 hot dogs and 5 Pepsis cost $38.25, and that 5 hot dogs and 4 Pepsis cost $28.50. A hot dog costs $3.50, and a Pepsi costs $2.75. ●

Quick ✓

10. At a fast-food joint, 5 cheeseburgers and 3 shakes cost $15.50. At the same fast-food joint, 3 cheeseburgers and 2 shakes cost $9.75. The price of a cheeseburger and the price of a shake can be determined by solving the system of equations

$$\begin{cases} 5c + 3s = 15.50 \\ 3c + 2s = 9.75 \end{cases}$$

where c represents the price of a cheeseburger and s represents the price of a shake. How much does a cheeseburger cost at the fast-food joint? How much does a shake cost?

4.3 Exercises **MyMathLab®** Exercise numbers in green have complete video solutions in MyMathLab or may be accessed using the QR code to the right.

Problems **1–10** *are the* **Quick ✔** *s that follow the* **EXAMPLES**.

Building Skills

In Problems 11–18, solve each system of equations using elimination. See Objective 1.

11. $\begin{cases} 2x + y = 3 \\ 5x - y = 11 \end{cases}$ **12.** $\begin{cases} x + y = -20 \\ x - y = 10 \end{cases}$

13. $\begin{cases} 3x - 2y = 10 \\ -3x + 12y = 30 \end{cases}$ **14.** $\begin{cases} 2x + 6y = 6 \\ -2x + y = 8 \end{cases}$

15. $\begin{cases} 2x + 3y = -4 \\ -2x + y = 6 \end{cases}$ **16.** $\begin{cases} 3x + 2y = 7 \\ -3x + 4y = 2 \end{cases}$

17. $\begin{cases} 6x - 2y = 0 \\ -9x - 4y = 21 \end{cases}$ **18.** $\begin{cases} 2x + 5y = -2 \\ -3x + 10y = -32 \end{cases}$

In Problems 19–26, solve each system using elimination. State whether the system is inconsistent, or consistent with dependent equations. See Objective 1.

19. $\begin{cases} 2x + 2y = 1 \\ -2x - 2y = 1 \end{cases}$ **20.** $\begin{cases} 3x - 2y = 4 \\ -3x + 2y = 4 \end{cases}$

21. $\begin{cases} 3x + y = -1 \\ 6x + 2y = -2 \end{cases}$ **22.** $\begin{cases} x - y = -4 \\ -2x + 2y = -8 \end{cases}$

23. $\begin{cases} 2x - 3y = 10 \\ -4x + 6y = -20 \end{cases}$ **24.** $\begin{cases} 2x + 5y = 15 \\ -6x - 15y = -45 \end{cases}$

25. $\begin{cases} -4x + 8y = 1 \\ 3x - 6y = 1 \end{cases}$ **26.** $\begin{cases} 4x + 6y = -10 \\ 9x + \dfrac{27}{2}y = 3 \end{cases}$

In Problems 27–46, solve each system of equations using elimination. See Objective 1.

27. $\begin{cases} 2x + 3y = 14 \\ -3x + y = 23 \end{cases}$ **28.** $\begin{cases} 2x + y = -4 \\ 3x + 5y = 29 \end{cases}$

29. $\begin{cases} 2x + 4y = 0 \\ 5x + 2y = 6 \end{cases}$ **30.** $\begin{cases} -2x - 2y = 3 \\ x - 2y = 1 \end{cases}$

31. $\begin{cases} x - 3y = 4 \\ -2x + 6y = 3 \end{cases}$ **32.** $\begin{cases} 5x - y = 3 \\ -10x + 2y = 2 \end{cases}$

33. $\begin{cases} 2x + 3y = -3 \\ 3x + 5y = -9 \end{cases}$ **34.** $\begin{cases} 2x + 3y = 2 \\ 5x + 7y = 0 \end{cases}$

35. $\begin{cases} 10y = 4x - 2 \\ 2x - 5y = 1 \end{cases}$ **36.** $\begin{cases} x + 3y = 6 \\ 9y = -3x + 18 \end{cases}$

37. $\begin{cases} 4x + 3y = 0 \\ 3x - 5y = 2 \end{cases}$ **38.** $\begin{cases} 5x + 7y = 6 \\ 2x - 3y = 11 \end{cases}$

39. $\begin{cases} 4x - 3y = -10 \\ -\dfrac{2}{3}x + y = \dfrac{11}{3} \end{cases}$ **40.** $\begin{cases} 12x + 15y = 55 \\ \dfrac{1}{2}x + 3y = \dfrac{3}{2} \end{cases}$

41. $\begin{cases} 1.5x + 0.5y = -0.45 \\ -0.3x - 0.4y = -0.54 \end{cases}$ **42.** $\begin{cases} -2.4x - 0.4y = 0.32 \\ 4.2x + 0.6y = -0.54 \end{cases}$

43. $\begin{cases} \dfrac{1}{2}x + \dfrac{2}{3}y = -5 \\ \dfrac{5}{2}x + \dfrac{5}{6}y = -10 \end{cases}$ **44.** $\begin{cases} x + \dfrac{5}{3}y = -1 \\ \dfrac{1}{2}x + \dfrac{1}{4}y = \dfrac{1}{4} \end{cases}$

45. $\begin{cases} 0.05x + 0.10y = 5.50 \\ x + y = 80 \end{cases}$ **46.** $\begin{cases} x + y = 1000 \\ 0.05x + 0.02y = 32 \end{cases}$

Mixed Practice

In Problems 47–68, solve by any method: graphing, substitution, or elimination.

47. $\begin{cases} x - y = -4 \\ 3x + y = 8 \end{cases}$ **48.** $\begin{cases} x - 2y = 0 \\ 3x + 5y = -11 \end{cases}$

49. $\begin{cases} 3x - 10y = -5 \\ 6x - 8y = 14 \end{cases}$ **50.** $\begin{cases} 3x - 5y = 14 \\ -2x + 6y = -16 \end{cases}$

51. $\begin{cases} -x + 3y = 6 \\ 4x + 5y = 7 \end{cases}$ **52.** $\begin{cases} 2x - 3y = -10 \\ -3x + y = 1 \end{cases}$

53. $\begin{cases} 0.3x - 0.7y = 1.2 \\ 1.2x + 2.1y = 2 \end{cases}$ **54.** $\begin{cases} 0.25x + 0.10y = 3.70 \\ x + y = 25 \end{cases}$

55. $\begin{cases} 3x - 2y = 6 \\ \dfrac{3}{2}x - y = 3 \end{cases}$ **56.** $\begin{cases} 4x + 3y = 9 \\ \dfrac{4}{3}x + y = 3 \end{cases}$

57. $\begin{cases} y = -\dfrac{2}{3}x - \dfrac{7}{3} \\ y = \dfrac{3}{4}x - \dfrac{15}{4} \end{cases}$

58. $\begin{cases} y = \dfrac{1}{7}x - 4 \\ y = -\dfrac{3}{2}x + 19 \end{cases}$

59. $\begin{cases} \dfrac{x}{2} + \dfrac{y}{4} = -2 \\ \dfrac{3x}{2} + \dfrac{y}{5} = -6 \end{cases}$

60. $\begin{cases} \dfrac{x}{3} + \dfrac{y}{5} = 2 \\ \dfrac{x}{3} - \dfrac{2y}{5} = -1 \end{cases}$

61. $\begin{cases} x - 2y = -7 \\ 3x + 4y = 6 \end{cases}$

62. $\begin{cases} -5x + y = 3 \\ 10x + 3y = 14 \end{cases}$

63. $\begin{cases} 6x - 5y = 1 \\ 8x - 2y = -22 \end{cases}$

64. $\begin{cases} -3x - 2y = -19 \\ 4x + 5y = 30 \end{cases}$

65. $\begin{cases} y = 2x - 4y \\ 4x + 1 = 10y + 3 \end{cases}$

66. $\begin{cases} 12y = 8x + 3 \\ -2 + 4(3y - 2x) = 5 \end{cases}$

67. $\begin{cases} \dfrac{x}{2} + \dfrac{3y}{4} = \dfrac{1}{2} \\ -\dfrac{3x}{5} + \dfrac{3y}{4} = -\dfrac{1}{20} \end{cases}$

68. $\begin{cases} \dfrac{x}{2} - y = 1 \\ \dfrac{x}{5} + \dfrac{5y}{6} = \dfrac{14}{15} \end{cases}$

Applying the Concepts

69. Counting Calories Suppose that Kristin ate two hamburgers and drank one medium Coke, for a total of 770 calories. Kristin's friend Jack ate three hamburgers and drank two medium Cokes (Jack takes advantage of free refills) for a total of 1260 calories. How many calories are in a hamburger? How many calories are in a medium Coke? To find the answers, solve the system

$$\begin{cases} 2h + c = 770 & (1) \\ 3h + 2c = 1260 & (2) \end{cases}$$

where h represents the number of calories in a hamburger and c represents the number of calories in a medium Coke.

70. Carbs Yvette and José go to McDonald's for breakfast. Yvette orders two sausage biscuits and one 16-ounce orange juice. The entire meal had 114 grams of carbohydrates. José orders three sausage biscuits and two 16-ounce orange juices, and his meal had 188 grams of carbohydrates. How many grams of carbohydrates are in a sausage biscuit? How many grams of carbohydrates are in a 16-ounce orange juice? To find the answers, solve the system

$$\begin{cases} 2b + u = 114 & (1) \\ 3b + 2u = 188 & (2) \end{cases}$$

where b represents the number of grams of carbohydrates in a sausage biscuit and u represents the number of grams of carbohydrates in an orange juice.

71. Planting Crops Farmer Green runs an organic farm and is planting his fields. He remembers from previous years that he planted 2 acres of tomatoes and 3 acres of zucchini in 65 hours. A different year he planted 3 acres of tomatoes and 4 acres of zucchini in 90 hours. If t represents the number of hours it takes to plant an acre of tomatoes, and z represents the number of hours it takes to plant one acre of zucchini, determine how long it takes Farmer Green to plant one acre of each crop by solving the following system of equations.

$$\begin{cases} 2t + 3z = 65 & (1) \\ 3t + 4z = 90 & (2) \end{cases}$$

If he wants to plant 5 acres of tomatoes and 2 acres of zucchini, will he get the crops in if the meteorologist predicts rain in 72 hours?

72. Farmer's Daughter Farmer Green has talked his daughter, Haydee, into helping with some of the planting in the scenario in Problem 71. Haydee's specialty is planting corn and lettuce, which she has done for the past several years. One year it took her 48 hours to plant 3 acres of corn and 2 acres of lettuce. Another year it took her 68 hours to plant 4 acres of corn and 3 acres of lettuce. If c represents the number of hours to plant one acre of corn, and l represents the number of hours required to plant one acre of lettuce, determine how long it takes Haydee to plant one acre of each crop by solving the following system of equations.

$$\begin{cases} 3c + 2l = 48 & (1) \\ 4c + 3l = 68 & (2) \end{cases}$$

Haydee has 72 hours to get the corn and lettuce crops planted. If 4 acres of corn are already planted, how many acres of lettuce can Haydee plant in the remaining time?

73. Blending Coffee Suppose that you want to blend two coffees in order to obtain a new blend. The blend will be made with Arabica beans that sell for $9.00 per pound and select African Robusta beans that sell for $11.50 per pound to obtain 100 pounds of the new blend that will sell for $10.00 per pound. How many pounds of the Arabica and the Robusta beans are required? To determine the answer, solve the system

$$\begin{cases} a + r = 100 & (1) \\ 9a + 11.50r = 1000 & (2) \end{cases}$$

where a represents the number of pounds of Arabica beans and r represents the number of pounds of African Robusta beans.

74. Candy A candy store sells chocolate-covered almonds for $6.50 per pound and chocolate-covered peanuts for $4.00 per pound. The manager decides to make a bridge mix that combines the almonds with the peanuts. She wants the bridge mix to sell for $6.00 per pound. How many pounds of chocolate-covered almonds and chocolate-covered peanuts are required to create 50 pounds of bridge mix? To determine the answer, solve the system

$$\begin{cases} a + p = 50 & (1) \\ 6.50a + 4.00p = 300 & (2) \end{cases}$$

where a represents the number of pounds of chocolate-covered almonds and p represents the number of pounds of chocolate-covered peanuts.

△ **75. Complementary Angles** Two angles are complementary if the sum of their measures is 90°. The measure of one angle is 10° more than three times the measure of its complement. If A and B represent the measures of the two complementary angles, determine the measure of each of the angles by solving the following system of equations.

$$\begin{cases} A + B = 90 & (1) \\ A = 10 + 3B & (2) \end{cases}$$

△ **76. Supplementary Angles** Two angles are supplementary if the sum of their measures is 180°. One-half the measure of one angle is 45° more than the measure of its supplement. If A and B represent the measures of the two supplementary angles,

determine the measure of each of the angles by solving the following system of equations.

$$\begin{cases} A + B = 180 & (1) \\ \dfrac{A}{2} = 45 + B & (2) \end{cases}$$

Extending the Concepts

In Problems 77 and 78, solve the system of equations for x and y using the elimination method.

77. $\begin{cases} ax + 4y = 1 \\ -2ax - 3y = 3 \end{cases}$

78. $\begin{cases} 4x + 2by = 4 \\ 5x + 3by = 7 \end{cases}$

In Problems 79 and 80, solve the system of equations for x and y using the elimination method.

79. $\begin{cases} -3x + 2y = 6a \\ x - 2y = 2b \end{cases}$

80. $\begin{cases} -7x - 3y = 4b \\ 3x + y = -2a \end{cases}$

Explaining the Concepts

81. Suppose you are given the system of equations
$\begin{cases} x - 3y = 6 \\ 2x + y = 5 \end{cases}$. Which variable would you choose to eliminate? Why? List the steps you would use to solve this system.

82. In the process of solving a system of linear equations by elimination, what tips you off that the system is consistent with dependent equations? What tips you off that the system is inconsistent?

83. The system of equations $\begin{cases} 6x + 3y = 3 & (1) \\ -4x - y = 1 & (2) \end{cases}$ appeared on a test. You solve the system by multiplying equation (2) by 3 and adding, to eliminate the variable y. Your friend multiplies the first equation by 2 and the second equation by 3 to eliminate the variable x. Assuming that neither of you makes any algebraic errors, which of you will obtain the correct answer? Explain your response. In addition to these two strategies, are there other strategies that may be used to solve the system?

84. List the three methods presented in this chapter for solving a system of linear equations in two variables. Explain the advantages of each one.

Putting the Concepts Together (Sections 4.1–4.3)

We designed these problems so that you can review Sections 4.1–4.3 and show your mastery of the concepts. Take time to work these problems before proceeding with the next section. The answers are located at the back of the text on page AN-18.

1. Determine whether the given ordered pairs are solutions of the given system of equations. Answer Yes or No.

$$\begin{cases} 4x + y = -20 \\ \quad y = -\dfrac{1}{6}x + 3 \end{cases}$$

(a) $\left(3, \dfrac{5}{2}\right)$ (b) $(-6, 4)$ (c) $(-4, -4)$

2. Suppose that you begin to solve a system of linear equations where each equation is written in slope-intercept form. You notice that the slopes and y-intercepts of the two lines are equal.

(a) How many solutions exist?
(b) State whether the system is consistent or inconsistent.
(c) If the system is consistent, state whether the equations in the system are dependent or independent.

3. Suppose that you begin to solve a system of linear equations where each equation is written in slope-intercept form. The slopes of the two lines are not equal, but the y-intercepts are the same.

(a) How many solutions exist?
(b) State whether the system is consistent or inconsistent.
(c) If the system is consistent, state whether the equations in the system are dependent or independent.

In Problems 4–13, solve each system of equations using any method: graphing, substitution, or elimination.

4. $\begin{cases} 4x + y = 5 \\ -x + y = 0 \end{cases}$

5. $\begin{cases} y = -\dfrac{2}{5}x + 1 \\ y = -x + 4 \end{cases}$

6. $\begin{cases} x = 2y + 11 \\ 3x - y = 8 \end{cases}$

7. $\begin{cases} 4x + 3y = -4 \\ x + 5y = -1 \end{cases}$

8. $\begin{cases} y = -2x - 3 \\ y = \dfrac{1}{2}x + 7 \end{cases}$

9. $\begin{cases} -2x + 4y = 2 \\ 3x + 5y = -14 \end{cases}$

10. $\begin{cases} -\dfrac{3}{4}x + \dfrac{2}{3}y = \dfrac{9}{4} \\ 3x - \dfrac{1}{2}y = -\dfrac{5}{2} \end{cases}$

11. $\begin{cases} 3(y + 3) = 1 + 4(3x - 1) \\ x = \dfrac{y}{4} + 1 \end{cases}$

12. $\begin{cases} 0.4x - 2.5y = -6.5 \\ x + y = 5.5 \end{cases}$

13. $\begin{cases} 2(y + 1) = 3x + 4 \\ x = \dfrac{2}{3}y - 3 \end{cases}$

4.4 Solving Direct Translation, Geometry, and Uniform Motion Problems Using Systems of Linear Equations

Objectives

❶ Model and Solve Direct Translation Problems

❷ Model and Solve Geometry Problems

❸ Model and Solve Uniform Motion Problems

Are You Prepared for This Section?

Before getting started, complete the following problems. If you get a problem wrong, go back to the section cited and review the material.

P1. If you travel at an average speed of 45 miles per hour for 3 hours, how far will you travel?   [Section 2.7, pp. 142–144]

P2. A rectangular garden is to be enclosed with a fence. If the garden measures $6\dfrac{1}{2}$ feet by $3\dfrac{3}{4}$ feet, how much fencing is needed?   [Section 2.4, pp. 110–112]

Prepared?...Answers   **P1.** 135 miles

P2. $20\dfrac{1}{2}$ feet

In Sections 2.5–2.7 we modeled and solved problems using a linear equation with a single variable. In this section, we learn how to model problems with two unknowns using a system of equations.

If you haven't done so already, review the problem-solving strategy given on pages 123–125 in Section 2.5.

▶ ❶ Model and Solve Direct Translation Problems

Remember, direct translation problems are problems where the words describing the problem are translated to the language of mathematics.

EXAMPLE 1   **Fun with Numbers**

The sum of two numbers is 45. Twice the first number minus the second number is 27. Find the numbers.

Solution

Step 1: Identify   Use direct translation to find two unknown numbers.

Step 2: Name   Let x represent the first number and y represent the second number.

Step 3: Translate   The sum of the two numbers is 45. The word "sum" implies addition, and the word "is" implies equal.

$$x + y = 45 \quad \text{Equation (1)}$$

Twice the first number minus the second number is 27. So,

$$2x - y = 27 \quad \text{Equation (2)}$$

Combine equations (1) and (2) to form the following system:

$$\begin{cases} x + y = 45 & (1) \\ 2x - y = 27 & (2) \end{cases} \quad \text{The Model}$$

Step 4: Solve   Use the method of elimination since the coefficients of y are opposites.

$$\begin{cases} x + y = 45 & (1) \\ 2x - y = 27 & (2) \end{cases}$$

Add equations (1) and (2): $3x \quad\;\; = 72$

Divide both sides by 3: $x \quad\;\; = 24$

Let $x = 24$ in equation (1) and solve for y.

Equation (1): $x + y = 45$

$24 + y = 45$

Subtract 24 from both sides: $y = 21$

Step 5: Check   The sum of 24 and 21 is 45. Twice 24 minus 21 is 48 minus 21, which equals 27.

Step 6: Answer   The two numbers are 24 and 21.   ●

Quick ✓

1. The sum of two numbers is 104. The second number is 25 less than twice the first number. Find the numbers.

▶ ❷ Model and Solve Geometry Problems

Formulas from geometry are needed to solve certain types of problems. It would be helpful to review the geometry formulas on pages 110–111 in Table 1 from Section 2.4.

EXAMPLE 2 **Enclosing a Yard with a Fence**

The Freese family owns a wooded lot that is on a lake. They want to enclose the lot with a fence but do not want to put up a fence along the lake. Dave Freese determined that he will need 240 feet of fence. He also knows that the width of the lot is 30 feet less than the length. See Figure 12. What are the dimensions of the lot?

Solution

Step 1: Identify Use a formula from geometry to find the length and the width of the lot.

Figure 12

Step 2: Name Let l represent the length and w represent the width of the lot.

Step 3: Translate The perimeter P of a rectangle, excluding the waterfront, is $P = l + 2w$, where l is the length and w is the width.

$$l + 2w = 240 \quad \text{Equation (1)}$$

It is also known that the width is 30 feet less than the length.

$$w = l - 30 \quad \text{Equation (2)}$$

Equations (1) and (2) form the following system:

$$\begin{cases} l + 2w = 240 & (1) \\ w = l - 30 & (2) \end{cases} \quad \text{The Model}$$

Step 4: Solve Since equation (2) is already solved for w, use the method of substitution and let $w = l - 30$ in equation (1).

Equation (1):	$l + 2w = 240$
	$l + 2(l - 30) = 240$
Distribute:	$l + 2l - 60 = 240$
Combine like terms:	$3l - 60 = 240$
Add 60 to both sides:	$3l = 300$
Divide both sides by 3:	$l = 100$

Let $l = 100$ in equation (2) and solve for w.

Equation (2):	$w = l - 30$
	$w = 100 - 30$
Simplify:	$w = 70$

Step 5: Check With $l = 100$ and $w = 70$, the perimeter (excluding the waterfront) would be $100 + 2(70) = 240$ feet. The width (70 feet) is 30 feet less than the length (100 feet).

Step 6: Answer The length of the wooded lot is 100 feet and the width is 70 feet. ●

Quick ✓

2. A rectangular field has a perimeter of 400 yards. The length of the field is three times the width. What is the length of the field? What is the width of the field?

▶ Recall from Section 2.7 that complementary angles are two angles whose measures sum to 90°. Each angle is called the *complement* of the other. For example, the angles shown in Figure 13(a) on the following page are complements because their measures sum to 90°. Supplementary angles are two angles whose measures sum to 180°. Each angle is called the *supplement* of the other. The angles shown in Figure 13(b) are supplementary.

Figure 13

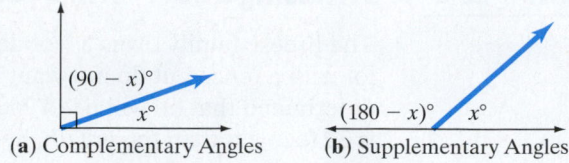

(a) Complementary Angles   (b) Supplementary Angles

The next example was first presented in Section 2.7. In Section 2.7, the problem was solved by developing a model that involved only one unknown. The same problem can be solved by developing a model of two equations containing two unknowns, as shown in Example 3.

EXAMPLE 3   ## Solve a Complementary Angle Problem

Find the measure of two complementary angles such that the measure of the larger angle is 6° greater than twice the measure of the smaller angle.

Solution

Step 1: Identify Use a definition from geometry to find the measures of two complementary angles.

Step 2: Name Let x represent the measure of the smaller angle and y represent the measure of the larger angle.

Step 3: Translate Because these are complementary angles, the sum of the measures of the angles must be 90°. So

$$x + y = 90 \quad \text{Equation (1)}$$

The measure of the larger angle y is 6° more than twice the measure of the smaller angle, x. This leads to the following equation.

$$y = 2x + 6 \quad \text{Equation (2)}$$

Equations (1) and (2) form the following system.

$$\begin{cases} x + y = 90 & (1) \\ y = 2x + 6 & (2) \end{cases} \quad \text{The Model}$$

Step 4: Solve Equation (2) is already solved for y, so let $y = 2x + 6$ in equation (1).

$$\begin{aligned}
\text{Equation (1):} && x + y &= 90 \\
&& x + (2x + 6) &= 90 \\
\text{Combine like terms:} && 3x + 6 &= 90 \\
\text{Subtract 6 from both sides:} && 3x &= 84 \\
\text{Divide both sides by 3:} && x &= 28
\end{aligned}$$

Now let $x = 28$ in equation (2) to find the measure of the larger angle, y.

$$\begin{aligned}
\text{Equation (2):} \quad y &= 2x + 6 \\
y &= 2(28) + 6 \\
y &= 62
\end{aligned}$$

Step 5: Check The measure of the smaller angle, x, is 28°. The measure of the larger angle, y, is 62°. The sum of the measures of these angles is 90° since 28° + 62° = 90°. The measure of the larger angle is 6° more than twice the measure of the smaller angle since 2(28°) + 6° = 62°. The answers check!

Step 6: Answer The two complementary angles measure 28° and 62°.   ●

▶ ❸ Model and Solve Uniform Motion Problems

Let's now look at a problem involving uniform motion. Remember, these problems use the fact that distance equals rate times time $(d = rt)$.

EXAMPLE 4   **Uniform Motion–Flying a Piper Aircraft**

The airspeed of a plane is its speed through the air. This speed is different from the plane's groundspeed—its speed relative to the ground. The groundspeed of an airplane is affected by the wind. Suppose that a Piper aircraft flying west a distance of 500 miles takes 5 hours. The return trip takes 4 hours. Find the airspeed of the plane and the effect that wind resistance has on the plane.

Solution

Step 1: Identify This is a uniform motion problem. Determine the airspeed of the plane and the effect of wind resistance.

Step 2: Name There are two unknowns in the problem—the airspeed of the plane and the effect of wind resistance on the plane. Let a represent the airspeed of the plane and w represent the effect of wind resistance.

Step 3: Translate Going west, the plane is flying into the jet stream, so the plane is slowed down by the wind. Therefore, the groundspeed of the plane will be $a - w$. Going east, the wind (jet stream) is helping the plane, so the groundspeed of the plane will be $a + w$. See Table 2.

Table 2

	Distance	Rate	Time
With Wind (East)	500	$a + w$	4
Against Wind (West)	500	$a - w$	5

Going with the wind, use distance $=$ rate $\cdot$ time and obtain the equation

$$500 = 4(a + w) \quad \text{or} \quad 4a + 4w = 500 \quad \text{Equation (1)}$$

Going against the wind, we have the equation

$$500 = 5(a - w) \quad \text{or} \quad 5a - 5w = 500 \quad \text{Equation (2)}$$

Equations (1) and (2) form a system of two linear equations containing two unknowns.

$$\begin{cases} 4a + 4w = 500 & (1) \\ 5a - 5w = 500 & (2) \end{cases} \quad \text{The Model}$$

Work Smart

In equation (1), we could have divided both sides by 4 rather than distributing. What could we have done with equation (2)?

(continued)

Step 4: Solve Use the elimination method by multiplying both sides of equation (1) by 5 and both sides of equation (2) by 4 and then adding equations (1) and (2).

$$\begin{cases} 5(4a + 4w) = 5(500) & (1) \\ 4(5a - 5w) = 4(500) & (2) \end{cases}$$

Distribute: $\begin{cases} 20a + 20w = 2500 & (1) \\ 20a - 20w = 2000 & (2) \end{cases}$

Add: $\qquad 40a \qquad\quad = 4500$

Divide both sides by 40: $\qquad a \qquad\quad = 112.5$

Use equation (1) with $a = 112.5$ to find the effect of wind resistance.

Equation (1): $\qquad\qquad 4a + 4w = 500$

$$4(112.5) + 4w = 500$$

$$450 + 4w = 500$$

Subtract 450 from both sides: $\qquad\qquad 4w = 50$

Divide both sides by 4: $\qquad\qquad w = 12.5$

Step 5: Check Flying west, the groundspeed of the plane is $112.5 - 12.5 = 100$ miles per hour, which agrees with the plane flying 500 miles in 5 hours for an average speed of 100 miles per hour. Flying east, the groundspeed of the plane is $112.5 + 12.5 = 125$ miles per hour, which agrees with the plane flying 500 miles in 4 hours for an average speed of 125 miles per hour. Everything checks!

Step 6: Answer The airspeed of the plane is 112.5 miles per hour. The impact of wind resistance on the plane is 12.5 miles per hour. ●

> **Quick ✓**
>
> 7. *True or False* In uniform motion problems use the equation $d = rt$, where d is distance, r is rate, and t is time.
>
> 8. Suppose that a plane flying 1200 miles west requires 4 hours and flying 1200 miles east requires 3 hours. Find the airspeed of the plane and the effect that wind resistance has on the plane.

4.4 Exercises MyMathLab®

Exercise numbers in **green** have complete video solutions in MyMathLab or may be accessed using the QR code to the right.

*Problems **1–8** are the **Quick ✓**s that follow the **EXAMPLES**.*

Building Skills

In Problems 9 and 10, complete the system of equations. Do not solve the system. See Objective 1.

9. The sum of two numbers is 56. Two times the smaller number is 12 more than one-half the larger number. Let x represent the smaller number and y represent the larger number.

$$\begin{cases} x + y = 56 \\ \underline{\qquad} = \underline{\qquad} \end{cases}$$

10. The sum of two numbers is 90. If 20 is added to 3 times the smaller number, the result exceeds twice

the larger number by 50. Let a represent the smaller number and b represent the larger number.

$$\begin{cases} a + b = 90 \\ \underline{\qquad} = \underline{\qquad} \end{cases}$$

In Problems 11 and 12, complete the system of equations. Do not solve the system. See Objective 2.

△ 11. The perimeter of a rectangle is 59 inches. The length is 5 inches less than twice the width. Let l represent the length of the rectangle and w represent the width of the rectangle.

$$\begin{cases} l = 2w - 5 \\ \underline{\qquad} = 59 \end{cases}$$

△ 12. The perimeter of a rectangle is 212 centimeters. The length is 8 centimeters less than three times the width. Let l represent the length of the rectangle and w represent the width of the rectangle.

$$\begin{cases} 2w + 2l = 212 \\ \underline{\hspace{2cm}} = \underline{\hspace{1.5cm}} \end{cases}$$

In Problems 13 and 14, complete the system of equations. Do not solve the system. See Objective 3.

13. A boat is rowed down a river a distance of 16 miles in 2 hours, and it is rowed upstream the same distance in 8 hours. Let r represent the rate of the boat in still water in miles per hour and c represent the rate of the stream in miles per hour.

$$\begin{cases} 2(r + c) = 16 \\ \underline{\hspace{2cm}} = \underline{\hspace{0.5cm}} \end{cases}$$

14. A plane flew with the wind for 3 hours, covering 1200 miles. It then returned over the same route to the airport against the wind in 4 hours. Let a represent the airspeed of the plane and w represent the effect of wind resistance.

$$\begin{cases} 3(a + w) = 1200 \\ \underline{\hspace{2cm}} = \underline{\hspace{0.8cm}} \end{cases}$$

Applying the Concepts

15. **Fun with Numbers** Find two numbers whose sum is 82 and whose difference is 16.

16. **Fun with Numbers** Find two numbers whose sum is 55 and whose difference is 17.

17. **Fun with Numbers** Two numbers sum to 51. Twice the first subtracted from the second is 9. Find the numbers.

18. **Fun with Numbers** Two numbers sum to 32. Twice the larger subtracted from the smaller is −22. Find the numbers.

19. **Oakland Baseball** The attendance at the games on two successive nights of Oakland A's baseball was 77,000. The attendance on Thursday's game was 7000 more than two-thirds of the attendance at Friday night's game. How many people attended the baseball game each night?

20. **Winning Baseball** The number of games the Oakland A's are expected to win this year is 8 fewer than two-thirds of the number that they are expected to lose. If there are 162 games in a season, how many games are the A's expected to win?

21. **Investments** Suppose that you received an unexpected inheritance of $21,000. You have decided to invest the money by placing some of the money in stocks and the remainder in bonds. To diversify, you decide that four times the amount invested in bonds should equal three times the amount invested in stocks. How much should be invested in stocks? How much should be invested in bonds?

22. **Investments** Marge and Homer have $40,000 to invest. Their financial advisor has recommended that they diversify by placing some of the money in stocks and the remainder in bonds. Based upon current market conditions, he has recommended that two times the amount in bonds should equal three times the amount invested in stocks. How much should be invested in stocks? How much should be invested in bonds?

△ 23. **Fencing a Garden** Melody Jackson wishes to enclose a rectangular garden with fencing, using the side of her garage as one side of the rectangle. A neighbor gave her 30 feet of fencing, and Melody wants the length of the garden along the garage to be 3 feet more than the width. What are the dimensions of the garden?

△ 24. **Perimeter of a Parking Lot** A rectangular parking lot has a perimeter of 125 feet. The length of the parking lot is 10 feet more than the width. What is the length of the parking lot? What is the width?

△ 25. **Perimeter** The perimeter of a rectangle is 70 meters. If the width is 40% of the length, find the dimensions of the rectangle.

△ 26. **Window Dimensions** The perimeter of a rectangular window is 162 inches. If the height of the window is 80% of the width, find the dimensions of the window.

△ 27. **Working with Complements** The measure of one angle is 15° more than half the measure of its complement. Find the measures of the two angles.

△ 28. **Working with Complements** The measure of one angle is 10° less than the measure of three times its complement. Find the measures of the two angles.

△ 29. **Finding Supplements** The measure of one angle is 30° less than one-third the measure of its supplement. Find the measures of the two angles.

△ 30. **Finding Supplements** The measure of one angle is 20° more than two-thirds the measure of its supplement. Find the measures of the two angles.

31. **Biking** Suppose that José bikes into the wind for 60 miles and it takes him 6 hours. After a long rest, he returns (with the wind at his back) in 5 hours. Determine the speed at which José can ride his bike in still air and the effect that the wind had on his speed.

32. Rowing On Monday afternoon, Andrew rowed his boat with the current for 4.5 hours and covered 27 miles, stopping in the evening at a campground. On Tuesday morning he returned to his starting point against the current in 6.75 hours. Find the speed of the current and the rate at which Andrew rowed in still water.

33. Kayaking Michael is kayaking on the Kankakee River. His overall speed is 3.5 mph against the current and 4.3 mph with the current. Find the speed of the current and the speed Michael can paddle in still water.

34. Southwest Airlines Plane A Southwest Airlines plane can fly 455 mph against the wind and 515 mph when it flies with the wind. Find the effect of the wind and the groundspeed of the airplane.

35. Outbound from Chicago Two trains leave Chicago going opposite directions, one going north and the other going south. The northbound train is traveling 12 mph slower than the southbound train. After 4 hours the trains are 528 miles apart. Find the speed of each train.

36. Horseback Riding Monica and Gabriella enjoy riding horses at a dude ranch in Colorado. They decide to go down different trails, which travel in opposite directions, agreeing to meet back at the ranch later in the day. Monica's horse is going 4 mph faster than the Gabriella's, and after 2.5 hours, they are 20 miles apart. Find the speed of each horse.

37. Riding Bikes Vanessa and Richie are riding their bikes down a trail to the next campground. Vanessa rides at 10 mph, and Richie rides at 8 mph. Because Vanessa is a little speedier, she stays behind and

cleans up camp for 30 minutes before leaving. How long has Richie been riding when Vanessa catches up to Richie?

38. Running a Marathon Rafael and Edith are running in a marathon to raise money for breast cancer research. Rafael can run at 12 mph, and Edith runs at 10 mph. Unfortunately, Rafael lost his car keys and went back to look for them while Edith started down the course. If Rafael was back at the starting line 15 minutes after Edith left, how long will it take him to catch up to her?

39. Wind Speed With a tailwind, a small Piper aircraft can fly 600 miles in 3 hours. Against this same wind, the Piper can fly the same distance in 4 hours. Find the effect of the wind and the average airspeed of the Piper.

40. Wind Speed The average airspeed of a single-engine aircraft is 150 miles per hour. If the aircraft flew the same distance in 2 hours with the wind as it flew in 3 hours against the wind, what was the effect of the wind on the plane?

Extending the Concepts

Problems 41 and 42 are a popular type of problem that appeared in mathematics textbooks in the 1970s and 1980s. Can you find the answers?

41. Digits The sum of the digits of a two-digit number is 6. If the digits are reversed, the difference between the new number and the original number is 18. Find the original number.

42. Digits The sum of the digits of a two-digit number is 7. If the digits are reversed, the difference between the new number and the original number is 27. Find the original number.

4.5 Solving Mixture Problems Using Systems of Linear Equations

Objectives

1. Draw Up a Plan for Solving Mixture Problems
2. Set Up and Solve Money Mixture Problems
3. Set Up and Solve Dry Mixture and Percent Mixture Problems

Are You Prepared for This Section?

Before getting started, complete the following problems. If you get a problem wrong, go back to the section cited and review the material.

P1. Suppose that Roberta has a credit card balance of $1200. Each month, the credit card charges 14% annual simple interest on any outstanding balances. What is the interest that Roberta will be charged on this loan after one month? What is Roberta's credit card balance after one month?   [Section 2.4, p. 109]

P2. Solve: $0.25x = 80$ [Section 2.3, pp. 101–102]

Prepared?...Answers **P1.** $14; $1214
P2. 320

▶ ❶ Draw Up a Plan for Solving Mixture Problems

Problems that involve mixing two or more items are called **mixture problems.** They can be solved using the formula *number of units of the same kind · rate = amount.* The *rates* used in mixture problems include cost per person, interest rates, and cost per pound. A table like Table 3 below can help organize the information given in the problem.

Table 3

	Number of Units	·	Rate	=	Amount
Item 1					
Item 2					
Total					

Work Smart

Use a chart to organize your thoughts and keep track of information.

Because each problem will be slightly different, the titles of the categories will be adjusted, but the formula *number of units of the same kind · rate = amount* will remain the same. Let's begin by filling in the chart.

EXAMPLE 1   Set Up a Table for a Mixture Problem

A class of schoolchildren and their adult chaperones took a field trip to the zoo. The total cost of admission was $230, and 40 children and adults went on the field trip. If the children paid $5 each and the adults paid $8 each, how many adults and how many children went on the trip? Fill in a table that summarizes the information in the problem. Do not solve the problem.

Solution

Both the number of children and the number of adults are unknown. Let a represent the number of adults on the field trip and c represent the number of children. Fill in Table 4 with the given quantities.

Table 4

	Number	·	Cost per Person	=	Amount
Adults	a		8		$8a$
Children	c		5		$5c$
Total	40				230

Work Smart

Not every entry in the chart must be filled.

Work Smart

Mixtures can include interest (money), solids (nuts), liquids (chocolate milk), and even gases (Earth's atmosphere).

Quick ✓

1. The formula used to solve mixture problems is number of units of the same kind · ___ = _____.

2. A single-game ticket to a Cincinnati Reds game costs $50, and a single-game ticket with $20 in concession credit costs $66. A group of 12 friends purchased some single-game tickets and some single-game tickets with $20 in concession credit, and the total cost for the tickets was $728. How many single-game tickets were purchased? Fill in the chart that summarizes the information in the problem. Let *s* represent the number of single-game tickets and *c* represent the number of single-game tickets with concession credit. Do not solve the problem.

	Number	·	Cost per ticket	=	Amount
Single-game ticket					
Single-game ticket with concession credit					
Total					

⏵ ❷ Set Up and Solve Money Mixture Problems

To solve money mixture problems, set up a table to organize the given information and then use the table to develop a system of equations (the model). We continue to use the six-step procedure that was introduced in Section 2.5.

EXAMPLE 2   **Solve a Coin Problem**

A third-grade class contributed $9.20 in dimes and quarters to the Red Cross. In all there were 56 coins. Find the number of dimes and quarters that they contributed.

Solution

Step 1: Identify This is a money mixture problem. We want to know the number of dimes and quarters the children contributed, and we know that a total of $9.20 was collected from 56 coins.

Step 2: Name Let q represent the number of quarters and d represent the number of dimes.

Step 3: Translate Fill in Table 5 with the information that is known.

Work Smart

It's always a good idea to name your variable so that it reminds you of what it represents, as in q for the number of quarters.

Table 5

	Number of Coins ·	Value per Coin in Dollars =	Total Value
Quarters	q	0.25	$0.25q$
Dimes	d	0.10	$0.10d$
Total	56		9.20

The total value of the coins is $9.20. Based on the information in Table 5,

value of quarters   value of dimes   total value of coins

$$0.25q \quad + \quad 0.10d \quad = \quad 9.20 \qquad \text{Equation (1)}$$

There are a total of 56 coins, so

$$q + d = 56 \qquad \text{Equation (2)}$$

Work Smart

The system in Example 2 could also be solved by first multiplying equation (1) by 100 in Step 4 to clear the decimals. The system would then be

$$\begin{cases} 25q + 10d = 920 & (1) \\ q + d = 56 & (2) \end{cases}$$

Which method do you prefer?

Use equations (1) and (2) to form a system of equations.

$$\begin{cases} 0.25q + 0.1d = 9.20 & (1) \\ q + d = 56 & (2) \end{cases} \quad \text{The Model}$$

Step 4: Solve Use the elimination method.

Multiply both sides of (2) by -0.1: $\begin{cases} 0.25q + 0.1d = 9.20 & (1) \\ -0.1(q + d) = -0.1(56) & (2) \end{cases}$

Distribute -0.1 in equation (2): $\begin{cases} 0.25q + 0.1d = 9.20 & (1) \\ -0.1q - 0.1d = -5.6 & (2) \end{cases}$

Add equations (1) and (2): $0.15q = 3.6$

Divide both sides by 0.15: $q = 24$

To determine the value of d, let $q = 24$ in equation (2) and solve for d.

Equation (2): $q + d = 56$

$24 + d = 56$

Subtract 24 from both sides: $d = 32$

Step 5: Check Because 24 quarters and 32 dimes add up to 56 coins and have a value of $24\,(\$0.25) + 32\,(\$0.10) = \$6.00 + \$3.20 = \$9.20$, the answer is correct.

Step 6: Answer The children collected 24 quarters and 32 dimes for the Red Cross. ●

> **Quick ✓**
>
> **3.** You have a piggy bank containing a total of 85 coins in dimes and quarters. If the piggy bank contains $14.50, how many dimes and quarters are there in the piggy bank?

Recall that Section 2.4 introduced the simple interest formula $I = Prt$, where I is interest, P is principal (an amount borrowed or deposited), r is the annual interest rate (expressed as a decimal), and t is time (in years). When solving money mixture problems involving interest, time is 1 year, so the simple interest formula reduces to $interest = principal \cdot rate \cdot 1$, or $interest = principal \cdot rate$.

EXAMPLE 3 **Solve a Money Mixture Problem Involving Interest**

You have $12,000 invested in two accounts: a money market account that paid 2% annual interest and a stock fund that paid 8% annual interest. Suppose that you earned $480 in interest at the end of one year. How much was invested in each account?

Solution

Step 1: Identify This is a mixture problem involving simple interest. Find the amount invested in the money market account and the amount invested in the stock fund so that $480 in interest is earned.

Step 2: Name Let m represent the amount invested in the money market account and s represent the amount invested in stocks.

Step 3: Translate Organize the given information in Table 6.

Table 6

	Principal $	·	Rate %	=	Interest $
Money Market	m		0.02		$0.02m$
Stock Fund	s		0.08		$0.08s$
Total	12,000				480

Work Smart

Notice that Equation (1) used the information from the column labeled "Interest $", and Equation (2) used the information from the column labeled "Principal $".

The total interest is the sum of the interest from both investments:

Interest earned from money market		Interest earned from stock fund		Total interest earned	
$0.02m$	$+$	$0.08s$	$=$	480	Equation (1)

The total investment was $12,000, so the amount invested in the money market account plus the amount invested in the stock fund equals $12,000:

$$m + s = 12{,}000 \quad \text{Equation (2)}$$

Use equations (1) and (2) to form a system of equations.

$$\begin{cases} 0.02m + 0.08s = 480 & (1) \\ m + s = 12{,}000 & (2) \end{cases} \quad \text{The Model}$$

(continued)

Step 4: Solve Use the elimination method.

Multiply both sides of (2) by −0.02:
$$\begin{cases} 0.02m + 0.08s = 480 & (1) \\ -0.02\,(m+s) = -0.02\,(12{,}000) & (2) \end{cases}$$

Distribute −0.02 in equation (2):
$$\begin{cases} 0.02m + 0.08s = 480 & (1) \\ -0.02m - 0.02s = -240 & (2) \end{cases}$$

Add equations (1) and (2): $0.06s = 240$

Divide both sides by 0.06: $s = 4000$

To determine the value of m, let $s = 4000$ in equation (2) and solve for m.

Equation (2): $m + s = 12{,}000$

$$m + 4000 = 12{,}000$$

Subtract 4000 from both sides: $m = 8000$

Step 5: Check The simple interest earned each year on the money market account is $(\$8000)(0.02)(1) = \$160$ and on the stock fund is $(\$4000)(0.08)(1) = 320$. The total interest earned is $\$160 + \$320 = \$480$. Further, the total amount invested is $\$8000 + \$4000 = \$12{,}000$. The solution checks.

Step 6: Answer You invested $8000 in the money market account and $4000 in the stock fund.   ●

Quick ✓

4. The simple interest formula states that interest $=$ _____ · ___ · ___.

5. *True or False* When solving money mixture problems involving interest, time is 1 year, so the simple interest formula reduces to interst $=$ principal · rate.

6. Faye has recently retired and requires an extra $3700 per year in income. She has $90,000 to invest and can invest in either low risk AA-rated bonds that pay 3% per annum or high risk B-rated junk bonds paying 5% annually. How much should be placed in each investment for Faye to achieve her goal exactly?

▶ ❸ Set Up and Solve Dry Mixture and Percent Mixture Problems

Mixture of Two Substances–Dry Mixture Problems

Often, new blends are created by mixing two quantities. For example, a chef might mix buckwheat flour with wheat flour to make buckwheat pancakes. Or a market might mix nuts and candy to create a trail mix.

EXAMPLE 4   **Solve a Dry Mixture Problem Using the Mixture Model**

A store manager combined nuts and M&M's to make a trail mix. She created 10 pounds of the mix and sold it for $5.10 per pound. If the price of the nuts was $3 per pound and the price of the M&M's was $6 per pound, how many pounds of each type did she use?

Solution

Step 1: Identify This is a dry mixture problem. Find the number of pounds of nuts and the number of pounds of M&M's that were required in the trail mix.

Step 2: Name Let n represent the number of pounds of nuts and m represent the number of pounds of M&M's in the mix.

Step 3: Translate   Set up Table 7.

Table 7

	Number of Pounds	·	Price \$/Pound	=	Revenue
Nuts	n		3		$3n$
M&M's	m		6		$6m$
Trail mix (blend)	10		5.10		$5.10(10) = 51$

If the mixture contains n pounds of nuts, the revenue from the nuts is \$$3n$. If the mixture contains m pounds of M&M's, the revenue from the M&M's is \$$6m$. Ten pounds of the blend selling for \$5.10 per pound yields $\$5.10(10) = \$51$ in revenue.

$$
\underset{\$3}{\underbrace{\text{Price per pound}\atop\text{of nuts}}} \cdot \underset{n}{\underbrace{\text{Pounds of}\atop\text{nuts}}} + \underset{\$6}{\underbrace{\text{Price per pound}\atop\text{of M\&M's}}} \cdot \underset{m}{\underbrace{\text{Pounds of}\atop\text{M\&M's}}} = \underset{\$5.10}{\underbrace{\text{Price per pound}\atop\text{of blend}}} \cdot \underset{10}{\underbrace{\text{Pounds of}\atop\text{blend}}}
$$

This gives the equation

$$3n + 6m = 51 \quad \text{Equation (1)}$$

The number of pounds of nuts plus the number of pounds of M&M's equals 10 pounds:

$$n + m = 10 \quad \text{Equation (2)}$$

Use equations (1) and (2) to form a system of equations.

$$
\begin{cases} 3n + 6m = 51 & (1) \\ n + m = 10 & (2) \end{cases} \quad \text{The Model}
$$

Work Smart

The system $\begin{cases} 3n + 6m = 51 \\ n + m = 10 \end{cases}$ could also have been solved by elimination. Which method do you prefer?

Step 4: Solve   Solve the system using substitution by solving equation (2) for n.

Equation (2):	$n + m = 10$
Subtract m from both sides:	$n = 10 - m$
Equation (1):	$3n + 6m = 51$
Let $n = 10 - m$ in equation (1):	$3(10 - m) + 6m = 51$
Distribute the 3:	$30 - 3m + 6m = 51$
Combine like terms:	$30 + 3m = 51$
Subtract 30 from both sides:	$3m = 21$
Divide both sides by 3:	$m = 7$

If $m = 7$ pounds, then $n = 10 - m = 10 - 7 = 3$ pounds.

Step 5: Check   A mixture of 3 pounds of nuts and 7 pounds of M&M's would sell for $\$3(3) + \$6(7) = \$9 + \$42 = \$51$, which equals the revenue obtained from selling the blend. In addition, the trail mix weighs 7 pounds + 3 pounds = 10 pounds. The solution checks.

Step 6: Answer   The trail mix has a blend of 3 pounds of nuts and 7 pounds of M&M's. ●

Quick ✓

7. A coffeehouse has Brazilian coffee that sells for \$6 per pound and Colombian coffee that sells for \$10 per pound. How many pounds of each coffee should be mixed to obtain 20 pounds of a blend that costs \$9 per pound?

▶ Mixture of Two Substances—Percent Mixture Problems

The last example dealt with mixing two dry substances. We now turn our attention to mixing two liquids. These problems require the use of percents.

EXAMPLE 5 **Solve a Percent Mixture Problem**

You work in the chemistry stockroom and your instructor asks you to prepare 4 liters of 15% hydrochloric acid (HCl). The supply room has a bottle of 12% HCl and a bottle of 20% HCl. How much of each should you mix so that your instructor has the required solution?

Solution

Step 1: Identify This is a percent mixture problem. Find the number of liters of a 12% HCl solution that must be mixed with a 20% HCl solution to prepare 4 liters of a 15% HCl solution.

Step 2: Name Let x represent the number of liters of the 12% HCl solution and y represent the number of liters of the 20% HCl solution.

Step 3: Translate Fill in Table 8 with the information that is known.

Table 8

	Number of Liters	$\cdot$	Concentration (part of solution that is pure HCl per liter)	$=$	Amount of Pure HCl
12% HCl Solution	x		0.12		$0.12x$
20% HCl Solution	y		0.20		$0.20y$
15% HCl Solution	4		0.15		$(0.15)(4)$

The amount of pure HCl from the 12% solution plus the amount of pure HCl from the 20% solution will yield the amount of pure HCl in 4 liters of 15% solution.

$$\underbrace{0.12x}_{\text{Pure HCl from 12\%}} + \underbrace{0.20y}_{\text{Pure HCl from 20\%}} = \underbrace{(0.15)(4)}_{\text{Pure HCl in 15\% solution}} \quad \text{Equation (1)}$$

The amount of 15% HCl solution will equal 4 liters.

$$x + y = 4 \quad \text{Equation (2)}$$

Use equations (1) and (2) to form a system of equations.

$$\begin{cases} 0.12x + 0.20y = 0.6 & (1) \\ x + y = 4 & (2) \end{cases} \quad \text{The Model}$$

Step 4: Solve Solve the system using substitution by solving equation (2) for y.

Equation (2):	$x + y = 4$
Subtract x from both sides:	$y = 4 - x$
Equation (1):	$0.12x + 0.20y = 0.6$
Substitute $4 - x$ for y in equation (1):	$0.12x + 0.20(4 - x) = 0.6$
Use the Distributive Property:	$0.12x + 0.8 - 0.20x = 0.6$
Combine like terms:	$-0.08x + 0.8 = 0.6$
Subtract 0.8 from both sides:	$-0.08x = -0.2$
Divide both sides by -0.08:	$x = 2.5$

Because x represents the number of liters of 12% HCl solution, you need 2.5 liters of 12% HCl solution. You need $4 - x = 4 - 2.5 = 1.5$ liters of 20% HCl solution.

Step 5: Check The amount of HCl in 2.5 liters of 12% HCl is $0.12\,(2.5) = 0.3$ liter. The amount of HCl in 1.5 liters of 20% HCl is $0.20\,(1.5) = 0.3$ liter. Combined, these solutions give us 0.6 liter of pure HCl. The amount of HCl in 4 liters of 15% HCl is $0.15\,(4) = 0.6$ liter. The answer is correct.

Step 6: Answer 2.5 liters of 12% HCl solution must be mixed with 1.5 liters of 20% HCl solution to form the 4 liters of 15% HCl solution.

Quick ✔

8. You make and sell ice cream, and decide that ice cream with 9% butterfat is the best to sell. How many gallons of ice cream with 5% butterfat should be mixed with ice cream that is 15% butterfat to make 200 gallons of the desired blend?

4.5 Exercises MyMathLab®   Exercise numbers in green have complete video solutions in MyMathLab or may be accessed using the QR code to the right.

Problems 1–8 are the Quick ✔s that follow the EXAMPLES.

Building Skills

In Problems 9–14, fill in the table from the information given. Then write the system that models the problem. Do not solve the system. See Objective 1.

9. The PTA had an ice cream social and sold adult tickets for $4 and student tickets for $1.50. The PTA treasurer found that 215 tickets had been sold and the receipts were $580.

	Number ·	Cost per Person =	Total Value
Adult Tickets	?	?	?
Student Tickets	?	?	?
Total	?		?

10. John Murphy sells jewelry at art shows. He sells bracelets for $10 and necklaces for $15. At the end of one day, John found that he had sold 69 pieces of jewelry and had receipts of $895.

	Number ·	Cost per Item =	Total Value
Bracelets	?	?	?
Necklaces	?	?	?
Total	?		?

11. Maurice has a CD (certificate of deposit) that earns 1.5% simple interest per year and a Treasury note that earns 2% simple interest. At the end of one year, Maurice received $27 in

interest on a total investment of $1600 in the two accounts.

	Principal ·	Rate =	Interest
CD	?	?	?
Treasury Note	?	?	?
Total	?		?

12. Sherry has a savings account that earns 0.5% simple interest per year and a certificate of deposit (CD) that earns 2% simple interest annually. At the end of one year, Sherry received $16 in interest on a total investment of $1700 in the two accounts.

	Principal ·	Rate =	Interest
Savings Account	?	?	?
Certificate of Deposit	?	?	?
Total	?		?

13. A coffee shop wishes to blend two types of coffee to create a breakfast blend. It mixes a mild coffee that sells for $7.50 per pound with a robust coffee that sells for $10.00 per pound. The owner wants to make 12 pounds of breakfast blend that will sell for $8.75 per pound.

	Number of Pounds ·	Price per Pound =	Total Value
Mild Coffee	?	?	?
Robust Coffee	?	?	?
Total	?	?	?

14. A merchant wishes to mix peanuts worth $5 per pound and trail mix worth $2 per pound to yield 40 pounds of a nutty mixture that will sell for $3 per pound.

Number of Pounds · Price per Pound		= Total Value
Peanuts	?	?	?
Trail Mix	?	?	?
Total	?	?	?

In Problems 15–20, complete the system of linear equations to solve the problem. Do not solve. See Objective 1.

15. A family of 11 decides to take a trip to Water World Water Park. The total cost of admission for the family is $296. If adult tickets cost $32 and children's tickets cost $24, how many adults and how many children went to Water World? Let a represent the number of adult tickets purchased, and let c represent the number of children's tickets purchased.

$$\begin{cases} a + \quad c = 11 \\ \underline{\quad} + \underline{\quad} = 296 \end{cases}$$

16. On a trip to New York City, 5 people attended the Broadway musical *The Lion King*. They paid $1008 for the tickets, which cost $154 for mezzanine seats and $273 for premium orchestra seats. How many mezzanine seats and how many premium orchestra seats for *The Lion King* were purchased? Let m represent the number of mezzanine seats purchased, and let p represent the number of premium orchestra tickets purchased.

$$\begin{cases} m + \quad p = \quad 5 \\ \underline{\quad} + \underline{\quad} = 1008 \end{cases}$$

17. You have a total of $2250 to invest. Account A pays 5% annual interest, and account B pays 7% annual interest. How much should you invest in each account if you would like the investment to earn $137.50 at the end of one year? Let A represent the amount of money invested in the account that earns 5% annual interest, and let B represent the amount of money invested in the account that earns 7% annual interest.

$$\begin{cases} A + \quad B = 2250 \\ \underline{\quad} + \underline{\quad} = 137.50 \end{cases}$$

18. You have a total of $2650 to invest. Account A pays 5% annual interest and account B pays 6.5% annual interest. How much should you invest in each account if you would like the investment to earn $155 at the end of one year? Let A represent the amount of money invested in the account that earns 5% annual interest, and let B represent the amount

of money invested in the account that earns 6.5% annual interest.

$$\begin{cases} A + \quad B = \underline{\quad} \\ \underline{\quad} + \underline{\quad} = 155 \end{cases}$$

19. Flowers on High sells flower bouquets at different prices depending on color: red for $5.85 per bouquet or yellow for $4.20 per bouquet. One day the number of red bouquets sold was 3 more than twice the number of yellow bouquets sold, and the revenue from selling the bouquets was $128.85. How many bouquets of each color were sold? Let r represent the number of red bouquets sold, and let y represent the number of yellow bouquets sold.

$$\begin{cases} r = \quad 3 \quad + 2y \\ \underline{\quad} + \underline{\quad} = \underline{\quad} \end{cases}$$

20. The Latte Shoppe sells Bold Breakfast coffee for $8.60 per pound and Wake-Up coffee for $5.75 per pound. One day, the amount of Bold Breakfast coffee, sold was 2 pounds less than twice the amount of Wake-Up coffee, and the revenue received from selling both types of coffee was $143.45. How many pounds of each type of coffee were sold that day? Let B represent the number of pounds of Bold Breakfast coffee sold, and let W represent the number of pounds of Wake-Up coffee sold.

$$\begin{cases} 8.60B + 5.75W = \underline{\quad} \\ B = \underline{\quad} - \underline{\quad} \end{cases}$$

Applying the Concepts

21. Theater Tickets to a student theater production cost $8 for students and $10 for nonstudents. The receipts for opening night came to $3270 from selling 390 tickets. How many student and nonstudent tickets were sold?

22. Ticket Pricing A ticket on the roller coaster is priced differently for adults and children. One day there were 5 adults and 8 children in a group and the cost of their tickets was $48.50. Another group of 4 adults and 12 children paid $57. What is the price of each type of ticket?

23. Bedding Plants A girls' softball team is raising money for uniforms and travel expenses. They are selling flats of bedding plants for $13.00 and hanging baskets for $18.00. They hope to sell twice as many flats of bedding plants as hanging baskets. If they do, their total revenue will be $8800. How many flats of bedding plants do they hope to sell?

24. Field Trip An elementary school group is planning a year-end day trip to Cedar Point Amusement park. The daily group rate for children is $35, and the daily group rate for chaperones is $39. A group of 23 adults and children purchased tickets for the day trip and paid $825. How many tickets for adults were purchased?

25. Tip Jar The tip jar next to the cash register has 150 coins, all in nickels and dimes. If the total value of the tips is $12, how many of each coin are in the jar?

26. Bigger Tips Christa is a waitress and collects her tips at the table. At the end of the shift she has 68 bills in her tip wallet, all ones and fives. If the total value of her tips is $172, how many of each bill does she have?

27. Buying Stamps One day Rosemarie bought 20 first-class stamps and 10 postcard stamps for $13.20. Another day she spent $56.20 on 80 first-class stamps and 50 postcard stamps. Find the cost of each type of stamp.

28. TV Commercials Advertising on television can be expensive, depending on the expected number of viewers. The Fox network sells 30-second spots for $175,000 and one-minute spots for $250,000 during the basketball playoffs. If Fox sold 13 commercial spots and earned $2,650,000, how many of each type did it sell?

29. Investments Harry has $10,000 to invest. He invests in two different accounts, one expected to return 5% and the other expected to return 8%. If he wants to earn $575 for the year, how much should he invest at each rate?

30. Investments Ann has $5000 to invest. She invests in two different accounts, one expected to return 4.5% and the other expected to return 7%. In order to earn $312.50 for the year, how much should she invest at each rate?

31. Stock Return Esmeralda invested $5000 in two stock plans. On one, the risky plan, she earned 8% annual interest, and on the other, a safer plan, she earned 4% annual interest. If Esmeralda saw a return on her investment of $328 last year, how much did she invest in each of the stock plans?

32. Real Estate Return Victor, Esmeralda's wealthy brother, also invested his money. He thought he would do better if he invested his money in real estate partnerships. He invested $12,000 in two different real estate groups. One of the groups yielded a 4% annual return on a downtown mall and the other a 6.5% annual return on a suburban office center. If Victor saw a profit of $692.50 on this investment last year, how much did he invest with each group?

33. Olive Blend Juana works at a delicatessen, and it is her turn to make the zesty olive blend. If arbequina olives are $9 per pound and green olives are $4 per pound, how much of each kind of olive should she mix to get five pounds of zesty olive blend that will sell for $6 per pound?

34. Churrascaria Platter Churrascaria is a name that comes from the Brazilian gauchos of the 1800s. To celebrate, the cowboys would take a large variety of different meats and barbeque them. Today, some specialty restaurants still serve Brazilian barbeque in the Churrascaria style. Elis da Silva's Restaurant wants to offer a one-pound Churrascaria platter for a profit of $12. If the restaurant makes a profit of $8 per pound on grilled pork and $14 per pound on barbecue flank steak, how many pounds of each meat can be served on the platter to earn $12?

35. Blending Coffee A coffee manufacturer wants to market a new blend of coffee that will sell for $3.90 per pound by mixing two coffees that sell for $2.75 per pound and $5 per pound, respectively. What amounts of each coffee should be blended to obtain 100 pounds of the desired mixture?

36. Blending Nuts The Nutty Professor store wishes to mix peanuts that sell for $2 per pound with cashews that sell for $6 per pound. The store plans to use five pounds more of the peanuts than cashews in the blend and will sell the mixture for a total of $26. How many pounds of each nut should be used in the mixture?

37. Grass Seed A nursery decides to make a grass-seed blend by mixing rye seed that sells for $4.20 per pound with bluegrass seed that sells for $3.75 per pound. The final mix is 180 pounds and sells for $3.95 per pound. How much of each type was used?

38. Stock Account A stock portfolio is currently valued at $5872.44. HiTech is currently selling at $82.01 per share, and IntRNet is currently selling at $26.52 per share. If the number of HiTech shares is 25 less than the number of IntRNet shares, find the number of shares of each.

39. Saline Solutions A lab technician needs 60 ml of a 50% saline solution. How many ml of 30% saline solution should she add to a 60% saline solution to obtain the required mixture?

40. Alcohol A laboratory assistant is asked to mix a 30% alcohol solution with 21 liters of an 80% alcohol solution to make a 60% alcohol solution. How many liters of the 30% alcohol solution should be used?

41. Silver Alloy How many liters of 10% silver must be added to 70 liters of 50% silver to make an alloy that is 20% silver?

42. Paint Four gallons of paint contain 2.5% pigment. How many gallons of paint that is 6% pigment must be added to make a paint that is 4% pigment?

Extending the Concepts

43. Antifreeze A radiator holds 3 gallons. How much of the 25% antifreeze solution should be drained and replaced with pure water to reduce the solution to 15% antifreeze?

44. Antifreeze A radiator holds 6 liters. How much of the 20% antifreeze solution should be drained and replaced with pure antifreeze to bring the solution up to 50% antifreeze?

45. Finance Jim Davidson invested $10,000 in two businesses. One business earned a profit of 5% for the year, while the other lost 7.5%. Find the amount he invested in each business if he earned $25 for the year.

46. Finance Marco invested money in two stocks. The stock that made a profit of 12% had $5000 less invested than the stock that lost 5.5%. Find the amount he invested in each if there was a net loss of $80 on his portfolio.

Explaining the Concepts

47. Without solving, explain what is wrong with the following mixture problem: How many liters of 25% ethanol should be added to 20 liters of 48% ethanol to obtain a solution of 58% ethanol?

48. Explain why the system $\begin{cases} a + b = 700 \\ 0.05a + 0.1b = 10{,}000 \end{cases}$ does not correctly model the following problem.

Molly invests $10,000 in two accounts, one paying 5% annual interest and the other paying 10% annual interest, and she earns $700 in interest at the end of the year on the two accounts. Let a represent the amount of money she invests in the account that pays 5% annual interest, and let b represent the amount of money she invests in the account that pays 10%. How much is invested in each account?

4.6 Systems of Linear Inequalities

Objectives

① Determine Whether an Ordered Pair Is a Solution of a System of Linear Inequalities

② Graph a System of Linear Inequalities

③ Solve Applied Problems Involving Systems of Linear Inequalities

Are You Prepared for This Section?

Before getting started, complete the following problems. If you get a problem wrong, go back to the section cited and review the material.

P1. Solve: $3x - 2 \geq 7$ [Section 2.8, pp. 150–154]

P2. Solve: $4(x - 1) < 6x + 4$ [Section 2.8, pp. 150–154]

P3. Graph: $y > 2x - 5$ [Section 3.7, pp. 232–235]

P4. Graph: $2x + 3y \leq 9$ [Section 3.7, pp. 232–235]

In Section 3.7, a single linear inequality in two variables was graphed. In this section, graphing a system of linear inequalities in two variables will be discussed.

Prepared?...Answers **P1.** $\{x \mid x \geq 3\}$
P2. $\{x \mid x > -4\}$
P3.

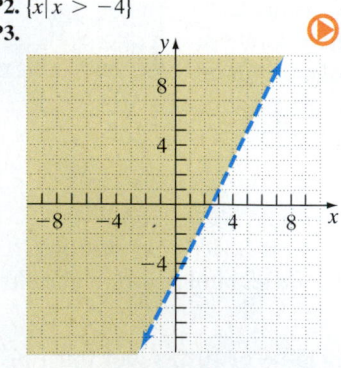

P4.

① **Determine Whether an Ordered Pair Is a Solution of a System of Linear Inequalities**

An ordered pair **satisfies** a system of linear inequalities if it makes each inequality in the system a true statement.

EXAMPLE 1 **Determining Whether an Ordered Pair Is a Solution of a System of Linear Inequalities**

Which of the following points, if any, satisfies the system of linear inequalities?

$$\begin{cases} 2x + y \leq 7 \\ 7x - 2y \geq 4 \end{cases}$$

(a) $(2, 1)$ **(b)** $(-2, 5)$

Solution

(a) Let $x = 2$ and $y = 1$ in each inequality in the system. If each statement is true, then $(2, 1)$ is a solution of the system.

$$2x + y \leq 7 \qquad\qquad 7x - 2y \geq 4$$
$$2(2) + 1 \overset{?}{\leq} 7 \qquad\qquad 7(2) - 2(1) \overset{?}{\geq} 4$$

$$4 + 1 \overset{?}{\leq} 7 \qquad \qquad 14 - 2 \overset{?}{\geq} 4$$

$$5 \leq 7 \quad \text{True} \qquad \qquad 12 \geq 4 \quad \text{True}$$

Both inequalities are true when $x = 2$ and $y = 1$, so $(2, 1)$ is a solution of the system of inequalities.

(b) Let $x = -2$ and $y = 5$ in each inequality in the system. If each statement is true, then $(-2, 5)$ is a solution of the system.

$$2x + y \leq 7 \qquad \qquad 7x - 2y \geq 4$$

$$2(-2) + 5 \overset{?}{\leq} 7 \qquad \qquad 7(-2) - 2(5) \overset{?}{\geq} 4$$

$$-4 + 5 \overset{?}{\leq} 7 \qquad \qquad -14 - 10 \overset{?}{\geq} 4$$

$$1 \leq 7 \quad \text{True} \qquad \qquad -24 \geq 4 \quad \text{False}$$

The inequality $7x - 2y \geq 4$ is not true when $x = -2$ and $y = 5$, so $(-2, 5)$ is not a solution of the system of inequalities.   ●

> ## Quick ✓
>
> 1. An ordered pair is a _____ of a system of linear inequalities if it makes each inequality in the system a true statement.
>
> 2. Determine which of the following points, if any, is a solution of the system of linear inequalities. State "solution" or "not a solution".
>
> $$\begin{cases} 4x + y \leq 6 \\ 2x - 5y < 10 \end{cases}$$
>
> **(a)** $(1, 2)$ **(b)** $(-1, -3)$

▶ ❷ Graph a System of Linear Inequalities

Work Smart

Don't forget that we use a solid line when the inequality is nonstrict ($\leq$ or $\geq$) and a dashed line when the inequality is strict ($<$ or $>$). See Section 3.7.

The graph of a system of inequalities in two variables x and y is the set of all points (x, y) that simultaneously satisfy *each* of the inequalities in the system. To graph a system of linear inequalities, graph each linear inequality individually. Then determine where, if at all, they intersect. The ONLY way we show the solution of a system of linear inequalities is by its graphical representation.

EXAMPLE 2   **How to Graph a System of Linear Inequalities**

Graph the system: $\begin{cases} y \geq 3x - 5 \\ y \leq -2x + 5 \end{cases}$

Step-by-Step Solution

Step 1: Graph the first inequality in the system, $y \geq 3x - 5$.

Graph $y \geq 3x - 5$ by graphing the line $y = 3x - 5$ (slope $= 3$, y-intercept $= (0, -5)$) with a solid line because the inequality is nonstrict ($\geq$). A test point $(0, 0)$ makes the inequality true ($0 \geq 3(0) - 5$), so shade the half-plane containing $(0, 0)$. See Figure 14.

Figure 14
$y \geq 3x - 5$

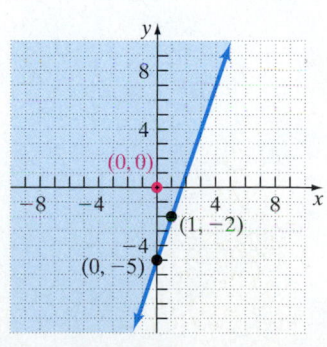

(continued)

Step 2: Graph the second inequality in the system, $y \le -2x + 5$.

Graph $y \le -2x + 5$ by graphing the line $y = -2x + 5$ (slope $= -2$, y-intercept $= (0, 5)$) with a solid line because the inequality is nonstrict ($\le$). A test point $(0, 0)$ makes the inequality true ($0 \le -2(0) + 5$), so shade the half-plane containing $(0, 0)$. See Figure 15.

Figure 15
$y \le -2x + 5$

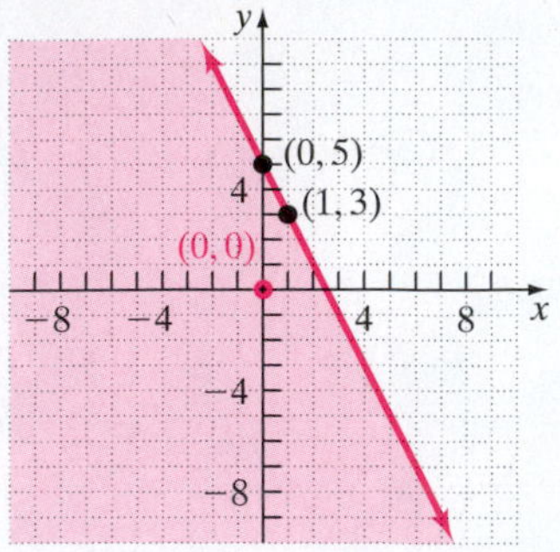

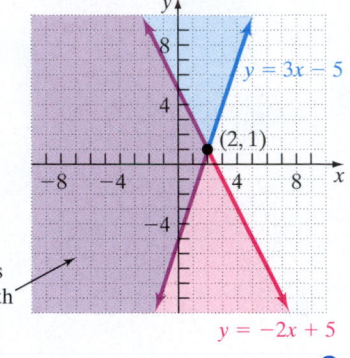

Step 3: Combine the graphs in Steps 1 and 2. The overlapping shaded region is the solution of the system of linear inequalities.

Work Smart

Inequalities can also be graphed by solving the inequality for y. Then, if the inequality is of the form $y >$ or $y \ge$, shade above the line. If the inequality is of the form $y <$ or $y \le$, shade below the line.

The graph shows an overlapping region in purple, which is the solution of the system of linear inequalities. See Figure 16.

Note: The ordered pair $(2, 1)$ represents the solution to the system

$$\begin{cases} y = 3x - 5 \\ y = -2x + 5 \end{cases}$$

Figure 16

The points in this region satisfy both inequalities.

EXAMPLE 3   **Graphing a System of Linear Inequalities**

Graph the system: $\begin{cases} 2x + y < 7 \\ 7x - 2y > 4 \end{cases}$

Solution

Graph the inequality $2x + y < 7$ ($y < -2x + 7$) in Figure 17(a). Then graph the inequality $7x - 2y > 4$ $\left(y < \dfrac{7}{2}x - 2 \right)$ in Figure 17(b). Don't forget to use dashed lines!

Combine the graphs from Figures 17(a) and (b). The overlapping shaded region in purple is the solution to the system of linear inequalities. See Figure 17(c). The point at which the two boundary lines intersect is not part of the solution of the system since the inequalities are strict.

Figure 17

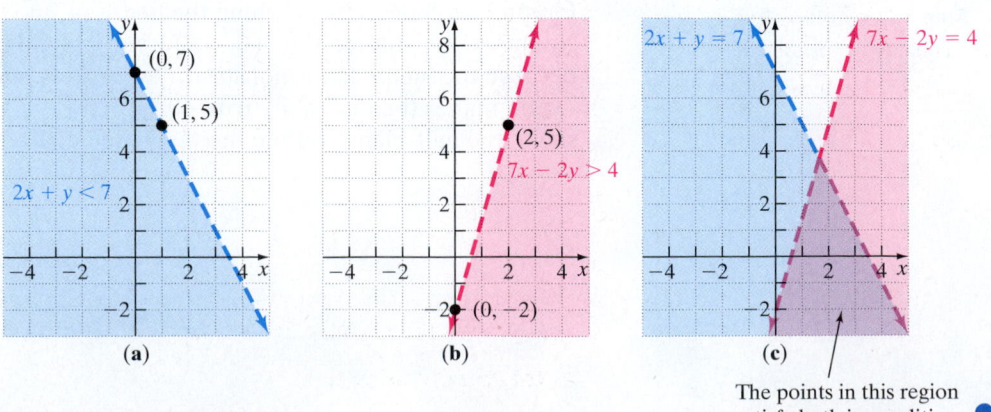

(a)   (b)   (c)

The points in this region satisfy both inequalities.

Rather than using multiple graphs to find the solution of a system of linear inequalities, we can use a single graph.

EXAMPLE 4 **Graphing a System of Linear Inequalities**

Graph the system: $\begin{cases} 2x + y \geq 3 \\ 3x - 2y < 8 \end{cases}$

Figure 18

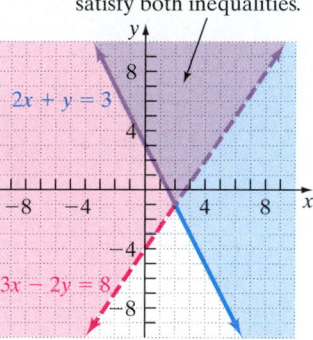

The points in this region satisfy both inequalities.

Solution
Graph the inequality $2x + y \geq 3$ ($y \geq -2x + 3$) in blue.

In the same rectangular coordinate system, graph the inequality $3x - 2y < 8$ $\left(y > \dfrac{3}{2}x - 4 \right)$ in pink.

In Figure 18, the overlapping shaded region in purple represents the solution. The intersection point of the two boundary lines is not a solution of the system because the inequality $3x - 2y < 8$ is strict.

Work Smart

$3x - 2y < 8$

$ -2y < -3x + 8$

$ y > \dfrac{3}{2}x - 4$

Remember to reverse the inequality symbol! Now graph $y = \dfrac{3}{2}x - 4$ using a dashed line and shade above.

▶ ❸ **Solve Applied Problems Involving Systems of Linear Inequalities**

Now let's look at some problems that lead to systems of linear inequalities.

EXAMPLE 5 **Financial Planning**

Maurice can invest up to $50,000. His financial advisor recommends that he place at least $10,000 in Treasury notes and no more than $35,000 in corporate bonds. A system of linear inequalities that models this situation is

$$\begin{cases} c + t \leq 50 \\ c \leq 35 \\ t \geq 10 \end{cases}$$

where c represents the amount in corporate bonds, in thousands of dollars, and t represents the amount in Treasury notes, in thousands of dollars.

  (a) Graph the system.

  (b) Can Maurice put $20,000 in corporate bonds and $30,000 in Treasury notes?

(continued)

(c) Can Maurice put $40,000 in corporate bonds and $10,000 in Treasury notes?

Solution

(a) Draw a rectangular coordinate system with the horizontal axis labeled c (for the amount in corporate bonds) and the vertical axis labeled t (for the amount in Treasury notes). Each axis will be in thousands, so we draw the line $t + c = 50$ and shade below. Draw the line $t = 10$ and shade above. Draw the line $c = 35$ and shade to the left. Figure 19 shows the graph of the system of linear inequalities.

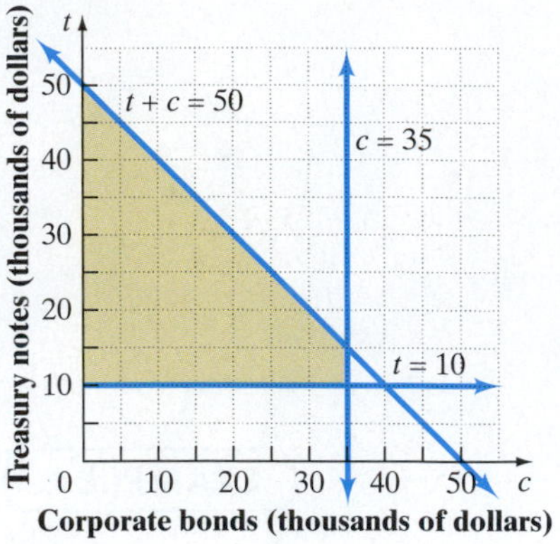

(b) Yes, Maurice can put $20,000 in corporate bonds and $30,000 in Treasury notes, because these values lie within the shaded region. In other words, $c = 20$ and $t = 30$ satisfies all three inequalities.

(c) No, Maurice cannot put $40,000 in corporate bonds and $10,000 in Treasury notes because these values do not lie within the shaded region. Put another way, $c = 40$ and $t = 10$ does not satisfy the inequality $c \le 35$.

Figure 19

Treasury notes (thousands of dollars)

$t + c = 50$ $c = 35$ $t = 10$

Corporate bonds (thousands of dollars)

Quick ✓

8. Jack and Mary can invest up to $75,000. Their financial advisor recommends that they place no more than $40,000 in corporate bonds and at least $25,000 in Treasury notes. A system of linear inequalities that models this situation is

$$\begin{cases} c + t \le 75 \\ c \quad\;\; \le 40 \\ \quad\; t \ge 25 \end{cases}$$

where c represents the amount in corporate bonds, in thousands of dollars, and t represents the amount in Treasury notes, in thousands of dollars.

(a) Graph the system.

(b) Can Jack and Mary invest $30,000 in corporate bonds and $35,000 in Treasury notes?

(c) Can Jack and Mary invest $60,000 in corporate bonds and $15,000 in Treasury notes?

4.6 Exercises   **MyMathLab**®   Exercise numbers in **green** have complete video solutions in MyMathLab or may be accessed using the QR code to the right.

Problems 1–8 are the Quick ✓ s that follow the EXAMPLES.

Building Skills

In Problems 9–12, determine which point(s), if any, is a solution of the system of linear inequalities. Answer Yes or No. See Objective 1.

9. $\begin{cases} 3x + y > 10 \\ -x + 5y \le -10 \end{cases}$

(a) $(5, -2)$
(b) $(4, -1)$
(c) $(6, -1)$

10. $\begin{cases} 2x - 3y < 3 \\ 2x + y < -5 \end{cases}$

(a) $(-4, 1)$
(b) $\left(-\dfrac{3}{2}, -2\right)$
(c) $(-1, -2)$

11. $\begin{cases} 2x + y > -4 \\ x - y \le 1 \end{cases}$

(a) $(-2, 1)$
(b) $(-1, -2)$
(c) $(2, -3)$

12. $\begin{cases} x - 2y > 2 \\ -3x - 2y \le 6 \end{cases}$

(a) $\left(-1, -\dfrac{3}{2}\right)$
(b) $(0, -4)$
(c) $(4, -1)$

In Problems 13–38, graph each system of linear inequalities. See Objective 2.

13. $\begin{cases} x > 2 \\ y \le -1 \end{cases}$

14. $\begin{cases} y > 4 \\ x < -3 \end{cases}$

15. $\begin{cases} y > -2 \\ x > -3 \end{cases}$

16. $\begin{cases} x \leq 3 \\ y < 1 \end{cases}$

17. $\begin{cases} x + y < 3 \\ x - y > 5 \end{cases}$

18. $\begin{cases} x - y > 2 \\ x + y \geq -2 \end{cases}$

19. $\begin{cases} x + y > 3 \\ 2x - y > 4 \end{cases}$

20. $\begin{cases} x - y \geq -1 \\ x + 2y > 4 \end{cases}$

21. $\begin{cases} y \geq -\dfrac{1}{2}x + 1 \\ y \leq \dfrac{1}{2}x + 3 \end{cases}$

22. $\begin{cases} y \geq -\dfrac{1}{3}x + 2 \\ y \geq \dfrac{2}{3}x - 1 \end{cases}$

23. $\begin{cases} y \geq -2 \\ y < 2x + 3 \end{cases}$

24. $\begin{cases} y > 2 \\ 2x - y \leq 1 \end{cases}$

25. $\begin{cases} x > 0 \\ y \leq \dfrac{2}{5}x - 1 \end{cases}$

26. $\begin{cases} y < -2 \\ y > 2x - 1 \end{cases}$

27. $\begin{cases} y \geq -x \\ 3x - y \geq -5 \end{cases}$

28. $\begin{cases} x + y < 2 \\ 3x - 5y \geq 0 \end{cases}$

29. $\begin{cases} x + y \leq -2 \\ y \geq x + 3 \end{cases}$

30. $\begin{cases} 3x + y > 3 \\ y > 4x - 2 \end{cases}$

31. $\begin{cases} x + 3y \geq 0 \\ 2y < x + 1 \end{cases}$

32. $\begin{cases} -y < \dfrac{2}{3}x + 1 \\ -3x + y \leq 2 \end{cases}$

33. $\begin{cases} x + y \geq 0 \\ x < 2y + 4 \end{cases}$

34. $\begin{cases} 2x - 3y \leq 9 \\ y < -2x + 3 \end{cases}$

35. $\begin{cases} x + 3y > 6 \\ 2x - y \leq 4 \end{cases}$

36. $\begin{cases} x - y \leq 3 \\ 2x + 3y \leq -9 \end{cases}$

37. $\begin{cases} -y \leq 3x - 4 \\ 2x + 3y \geq -3 \end{cases}$

38. $\begin{cases} x + 4y \leq 4 \\ 2x + 3y \geq 6 \end{cases}$

Applying the Concepts

39. House Blend The Coffee Cup coffee shop is experimenting with blending the "house" coffee. It will be made up of two varieties of coffee, French Roast and Hazelnut. The managers have decided that they will make at most 30 pounds of "house" coffee in a day. The tasters have determined that the blend should be mixed so that the amount of French Roast coffee is at least twice the amount of Hazelnut coffee. A system of linear inequalities that models this situation is

$$\begin{cases} f + h \leq 30 \\ f \geq 2h \\ f \geq 0 \\ h \geq 0 \end{cases}$$

where f represents the number of pounds of French Roast coffee and h represents the number of pounds of Hazelnut coffee.

(a) Graph the system of linear inequalities.

(b) Is it possible to use 18 pounds of French Roast coffee and 11 pounds of Hazelnut coffee in the house blend?

(c) Is it possible to use 8 pounds of French Roast coffee and 4 pounds of Hazelnut coffee in the house blend?

40. Party Food Steven and Christopher are planning a party. They plan to buy bratwurst for $4.00 per pound and hamburger patties that cost $3.00 per pound. They can spend at most $70 and think they should have no more than 20 pounds of bratwurst and hamburger patties. A system of inequalities that models the situation is

$$\begin{cases} 4b + 3h \leq 70 \\ b + h \leq 20 \\ b \geq 0 \\ h \geq 0 \end{cases}$$

where b represents the number of pounds of bratwurst and h represents the number of pounds of hamburger patties.

(a) Graph the system of linear inequalities.

(b) Is it possible to purchase 10 pounds of bratwurst and 10 pounds of hamburger patties and stay within their $70 budget?

(c) Is it possible to purchase 15 pounds of bratwurst and 5 pounds of hamburger patties and stay within their $70 budget?

41. Auto Manufacturing A manufacturing plant has 360 hours of machine time budgeted for the production of cars and trucks on a given day. It takes 12 hours to build a car and 18 hours to build a truck. Based on experience, the plant manager knows that the number of trucks is fewer than 20 less than twice the number of cars produced. Due to limitations in the machinery, the maximum number of cars that can be produced in a day is 24. A system of linear inequalities that models this situation is

$$\begin{cases} 12c + 18t \leq 360 \\ t < 2c - 20 \\ c \leq 24 \\ c \geq 0 \\ t \geq 0 \end{cases}$$

where c represents the number of cars produced and t represents the number of trucks built.

(a) Graph the system of linear inequalities.

(b) Is it possible to build 17 cars and 9 trucks at this plant?

(c) Is it possible to build 12 cars and 11 trucks at this plant?

42. Breakfast at Burger King Aman decides to eat breakfast at Burger King. He is a big fan of their French toast sticks, and he likes orange juice. He wants to eat no more than 500 calories and to consume no more than 425 mg of sodium. Each 3-piece serving of French toast contains 230 calories and 260 mg of sodium. Each small orange juice has 140 calories and 20 mg of sodium. The system of linear inequalities that represents the possible combination of French toast sticks and orange juice that Aman can consume is

$$\begin{cases} 230x + 140y \le 500 \\ 260x + 20y \le 425 \\ x \ge 0 \\ y \ge 0 \end{cases}$$

where x represents the number of 3-piece servings of Burger King French toast sticks and y represents the number of small orange juice drinks.

(a) Graph the system of linear inequalities.
(b) Is it possible for Aman to consume a 3-piece serving of French toast sticks and 1 small orange juice and stay within the allowance for calories and sodium?
(c) Is it possible for Aman to consume a 3-piece serving of French toast sticks and 3 small orange juices and stay within the allowance for calories and sodium?

Extending the Concepts

In Problems 43–46, the shaded region for the solution to each system of linear inequalities differs from what we have seen previously. Graph the solution of each system.

43. $\begin{cases} \dfrac{y}{2} - \dfrac{x}{6} \ge 1 \\ \dfrac{x}{3} - \dfrac{y}{1} \ge 1 \end{cases}$ **44.** $\begin{cases} \dfrac{x}{2} + \dfrac{y}{4} > 1 \\ -\dfrac{x}{4} - \dfrac{y}{8} > 1 \end{cases}$

45. $\begin{cases} x < \dfrac{3}{2}y + \dfrac{9}{2} \\ -2x < -3(y + 2) \end{cases}$ **46.** $\begin{cases} x > \dfrac{5}{4}y - \dfrac{1}{2} \\ 4x < 5(y + 3) \end{cases}$

In Problems 47 and 48, write a system of linear inequalities that will produce each shaded region as its solution.

47.
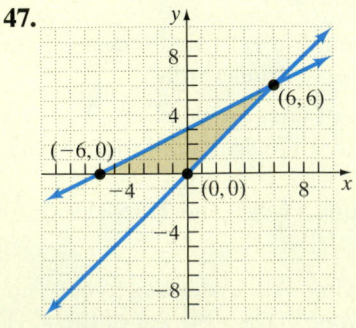

48.
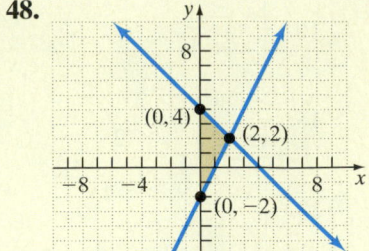

Explaining the Concepts

49. List the steps used to solve a system of linear inequalities in two variables.

50. If one inequality in a system of two linear inequalities contains a strict inequality and the other inequality contains a nonstrict inequality, assuming the lines intersect, is the intersection point of the boundary lines a solution of the system? Explain why or why not.

51. Each of the graphs below uses two of the following inequalities:

$$L_1: y \ge -2x \quad L_2: y \le -2x \quad L_3: y \ge x \quad L_4: y \le x$$

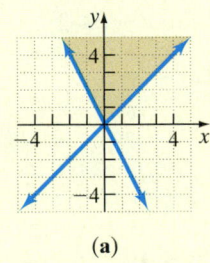

(a)

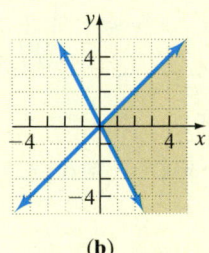

(b)

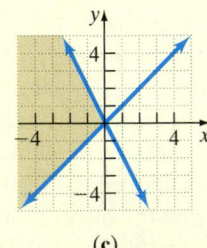

(c)

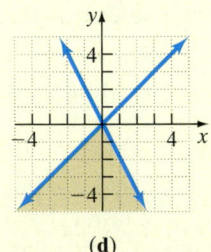

(d)

Write the system of linear inequalities for each graph and explain how you came to your conclusions.

52. When solving application problems that require solving a system of linear inequalities, typically the graph appears only in the first quadrant. Explain why this is so. Can you think of a situation in which that would not be the case?

Chapter 4 Activity: Find the Numbers

Focus: Solving systems of equations

Time: 15 minutes

Group size: 2

Consider the following dialogue between two students:

Ryan: Think of two numbers between 1 and 10, and don't tell me what they are.

Melissa: OK, I've thought of two numbers.

Ryan: Now tell me the sum of the two numbers and the difference of the two numbers, and I'll tell you what your two numbers are.

Melissa: Their sum is 14 and their difference is 6.

Ryan: Your numbers are 10 and 4.

Melissa: That's right! How did you do that?

Ryan: I set up a system of equations using the sum and difference that you gave me.

1. Each group member should set up and solve the system of equations described by Ryan. Discuss your results, and be sure that you both arrive at the solutions 10 and 4.

2. Now each of you will think of two new numbers, and the other will try to find the numbers by solving a system of equations. But this time the system will be a bit trickier! Give each other the following information about your two numbers, and then figure out each other's numbers.

   Six more than three times the sum of the numbers is ___?
   Two less than four times the difference of the numbers is ___?

3. Would the systems in this activity work if negative numbers were used? Try it and see!

Chapter 4 Review

Section 4.1   Solving Systems of Linear Equations by Graphing

KEY CONCEPTS

- **Recognizing Solutions of Systems of Two Linear Equations with Two Unknowns**
  - If the lines in a system of two linear equations containing two unknowns intersect, then the point of intersection is the solution and the system is consistent; the equations are independent.
  - If the lines in a system of two linear equations containing two unknowns are parallel, then the system has no solution and the system is inconsistent.
  - If the lines in a system of two linear equations containing two unknowns lie on top of each other (coincident), then the system has infinitely many solutions. The solution set is the set of all points on the line and the system is consistent; the equations are dependent.

KEY TERMS

System of linear equations
Solution
Consistent and independent
Inconsistent
Consistent and dependent

You Should Be Able To...	EXAMPLE	Review Exercises
❶ Determine whether an ordered pair is a solution of a system of linear equations (p. 249)	Example 2	1–4
❷ Solve a system of linear equations by graphing (p. 250)	Examples 3 through 6	5–12
❸ Classify systems of linear equations as consistent or inconsistent (p. 254)	Example 7	13–18
❹ Solve applied problems involving systems of linear equations (p. 256)	Example 8	19, 20

In Problems 1–4, determine whether the ordered pair is a solution to the system of equations. Answer Yes or No.

1. $\begin{cases} x + 2y = 6 \\ 3x - y = -10 \end{cases}$ **(a)** $(3, -1)$ **(b)** $(-2, 4)$ **(c)** $(4, 1)$

2. $\begin{cases} y = 3x - 5 \\ 3y = 6x - 5 \end{cases}$ **(a)** $(2, 1)$ **(b)** $\left(0, -\dfrac{5}{3}\right)$ **(c)** $\left(\dfrac{10}{3}, 5\right)$

3. $\begin{cases} 3x - 4y = 2 \\ 20y = 15x - 10 \end{cases}$ **(a)** $\left(\dfrac{1}{2}, -\dfrac{1}{8}\right)$ **(b)** $(6, 4)$ **(c)** $(0.4, -0.2)$

4. $\begin{cases} x = -4y + 2 \\ 2x + 8y = 12 \end{cases}$

   (a) $(10, -1)$ (b) $\left(-\dfrac{1}{2}, \dfrac{1}{2}\right)$ (c) $(2, 1)$

In Problems 5–12, solve each system of equations by graphing.

5. $\begin{cases} 2x - 4y = 8 \\ x + y = 7 \end{cases}$

6. $\begin{cases} x - y = -3 \\ 3x + 2y = 6 \end{cases}$

7. $\begin{cases} y = -\dfrac{x}{2} + 2 \\ y = x + 8 \end{cases}$

8. $\begin{cases} y = -x - 5 \\ y = \dfrac{3x}{4} + 2 \end{cases}$

9. $\begin{cases} 4x - 8 = 0 \\ 3y + 9 = 0 \end{cases}$

10. $\begin{cases} x = y \\ x + y = 0 \end{cases}$

11. $\begin{cases} 0.6x + 0.5y = 2 \\ 10y = -12x + 20 \end{cases}$

12. $\begin{cases} \dfrac{1}{4}x - \dfrac{1}{2}y = 1 \\ 3x - 6y = 12 \end{cases}$

In Problems 13–18, without graphing, determine the number of solutions to each system of equations. State whether the system is consistent or inconsistent. If the system is consistent, state whether the equations are dependent or independent.

13. $\begin{cases} y = 3x - 4 \\ y = 3x + 4 \end{cases}$

14. $\begin{cases} -2x + 4y = -8 \\ x - 2y = -4 \end{cases}$

15. $\begin{cases} -3x + 3y = -3 \\ \dfrac{1}{2}x - \dfrac{1}{2}y = 0.5 \end{cases}$

16. $\begin{cases} x - 2 = -\dfrac{2}{3}y - \dfrac{2}{3} \\ 3x = 4 - 2y \end{cases}$

17. $\begin{cases} 2x + y = -3 \\ y = 2x + 3 \end{cases}$

18. $\begin{cases} \dfrac{y}{2} = \dfrac{x}{4} + 2 \\ \dfrac{x}{8} + \dfrac{y}{4} = -1 \end{cases}$

19. **Printing Costs** Monique is creating a flier to give to local businesses to advertise her new vintage clothing store. She is trying to decide between two quotes for the printing. Printer A has given her a quote of $70 setup fee plus $0.10 per flier printed. Printer B has given her a quote of $100 plus $0.04 per flier.

   (a) Write a system of linear equations that models the problem.
   (b) Graph the system of equations to determine how many fliers she needs to have printed in order for the cost to be the same at both printers.
   (c) If 400 fliers are printed, which printer should she choose to have the lower cost?

20. **Flooring Installation** To install carpeting in Juan's bedroom, it costs $50 for installation plus $40 per square yard of carpet. In the same room, Juan can install ceramic tile for a cost of $350 for installation plus $10 per square yard of tile.

   (a) Write a system of linear equations that models the problem.
   (b) Graph the system of equations to determine how many square yards of flooring Juan needs to have installed for the cost of carpeting and the cost of tile to be the same.
   (c) If the area of Juan's bedroom is 15 square yards, which material would be cheaper?

Section 4.2   Solving Systems of Linear Equations Using Substitution

KEY CONCEPT

- **Solving Systems of Linear Equations Using Substitution**

   Step 1: Solve one of the equations for one of the unknowns. For example, we might solve equation (1) for y in terms of x.

   Step 2: Substitute the expression found in Step 1 into the *other* equation. The result will be a single linear equation in one unknown. For example, if we solved equation (1) for y in terms of x in Step 1, then we would replace y in equation (2) with the expression in x.

   Step 3: Solve the linear equation in one unknown found in Step 2.

   Step 4: Substitute the value of the variable found in Step 3 into one of the *original* equations to find the value of the other variable.

   Step 5: Check your answer by substituting the ordered pair in both of the original equations.

You Should Be Able To...	EXAMPLE	Review Exercises
❶ Solve a system of linear equations using the substitution method (p. 260)	Examples 1 through 5	21–34
❷ Solve applied problems involving systems of linear equations (p. 265)	Example 6	35, 36

In Problems 21–34, solve each system of equations using substitution.

21. $\begin{cases} x + 4y = 6 \\ y = 2x - 3 \end{cases}$

22. $\begin{cases} 7x - 3y = 10 \\ y = 3x - 4 \end{cases}$

23. $\begin{cases} 2x + 5y = 4 \\ x = 3 - 2y \end{cases}$

24. $\begin{cases} 3x + y = 10 \\ x = 8 + 2y \end{cases}$

25. $\begin{cases} y = \dfrac{2}{3}x - 1 \\ y = \dfrac{1}{2}x + 2 \end{cases}$

26. $\begin{cases} y = -\dfrac{5}{6}x + 3 \\ y = -\dfrac{4}{3}x \end{cases}$

27. $\begin{cases} 2x - y = 6 \\ 4x + 3y = 2 \end{cases}$

28. $\begin{cases} 5x + 2y = 13 \\ x + 4y = -1 \end{cases}$

29. $\begin{cases} 6x + 3y = 12 \\ y = -2x + 4 \end{cases}$

30. $\begin{cases} x = 4y - 1 \\ 8y - 2x = 4 \end{cases}$

31. $\begin{cases} -6x + 12y = 6 \\ 12x - 24y = 8 \end{cases}$

32. $\begin{cases} 12x + 9y = 9 \\ 8x + 6y = 6 \end{cases}$

33. $\begin{cases} \dfrac{1}{2}x - \dfrac{1}{4}y = \dfrac{1}{2} \\ \dfrac{1}{3}x - \dfrac{3}{4}y = -\dfrac{1}{4} \end{cases}$

34. $\begin{cases} -\dfrac{5x}{2} + \dfrac{y}{4} = -\dfrac{7}{12} \\ -\dfrac{3x}{2} - \dfrac{y}{10} = -\dfrac{3}{5} \end{cases}$

△ **35. Walk around the Park** A rectangular park is surrounded by a walking path. The park is 75 meters longer than it is wide. If a person walking completely around the park covers a distance of 650 meters, find the length and width of the park by solving the following system of equations, where *l* is the length and *w* is the width.

$$\begin{cases} 2l + 2w = 650 & \textcolor{red}{(1)} \\ l = w + 75 & \textcolor{red}{(2)} \end{cases}$$

36. Fun with Numbers The sum of two numbers is 12. If twice the smaller number is subtracted from the larger number, the difference is 21. Determine the smaller of the two numbers by solving the following system of equations, where *x* and *y* represent the unknown numbers.

$$\begin{cases} x + y = 12 & \textcolor{red}{(1)} \\ x - 2y = 21 & \textcolor{red}{(2)} \end{cases}$$

Section 4.3   Solving Systems of Linear Equations Using Elimination

KEY CONCEPT

- **Solving Systems of Linear Equations Using Elimination**

  Step 1: Write each equation in standard form, $Ax + By = C$.

  Step 2: If necessary, multiply or divide both sides of one (or both) equations by a nonzero constant so that the coefficients of one of the variables are opposites (additive inverses).

  Step 3: Add the equations to eliminate the variable whose coefficients are now additive inverses. Solve the resulting equation for the remaining unknown.

  Step 4: Substitute the value of the variable found in Step 3 into one of the original equations to find the value of the remaining variable.

  Step 5: Check your answer by substituting the ordered pair in both of the original equations.

You Should Be Able To...	EXAMPLE	Review Exercises
❶ Solve a system of linear equations using the elimination method (p. 268)	Examples 1 through 5	37–52
❷ Solve applied problems involving systems of linear equations (p. 273)	Example 6	53, 54

In Problems 37–44, solve each system of equations using elimination.

37. $\begin{cases} 4x - y = 12 \\ 2x + y = -12 \end{cases}$

38. $\begin{cases} -2x + 3y = 27 \\ 2x - 5y = -41 \end{cases}$

39. $\begin{cases} -3x + 4y = 25 \\ x - 5y = -23 \end{cases}$

40. $\begin{cases} 5x + 8y = -15 \\ -2x + y = 6 \end{cases}$

41. $\begin{cases} 4x - 3y = -1 \\ 2x - \dfrac{3}{2}y = 3 \end{cases}$

42. $\begin{cases} 2x + 5y = 3 \\ -4x - 10y = -6 \end{cases}$

43. $\begin{cases} 1.3x - 0.2y = -3 \\ -0.1x + 0.5y = 1.2 \end{cases}$ **44.** $\begin{cases} 2.5x + 0.5y = 6.25 \\ -0.5x - 1.2y = 1.5 \end{cases}$

In Problems 45–52, solve each system of equations using graphing, substitution, or elimination.

45. $\begin{cases} 2x + y = -1 \\ -6x - 8y = 13 \end{cases}$ **46.** $\begin{cases} \dfrac{1}{6}x + y = \dfrac{3}{4} \\ \dfrac{2y - 8}{3} = 4x + 4 \end{cases}$

47. $\begin{cases} y + 5 = \dfrac{2}{3}x + 3 \\ \dfrac{1}{3}x - \dfrac{1}{2}y = 1 \end{cases}$ **48.** $\begin{cases} \dfrac{1}{14} - \dfrac{x}{2} = -\dfrac{y}{7} \\ y = \dfrac{1}{2} + \dfrac{7x}{3} \end{cases}$

49. $\begin{cases} -x + y = 7 \\ -3x + 4y = 8 \end{cases}$ **50.** $\begin{cases} 4x + y = 2 \\ 9y - 3x = 5 \end{cases}$

51. $\begin{cases} 4x + 6 = 3y + 5 \\ 4(-2x - 4) = 6(-y - 3) \end{cases}$

52. $\begin{cases} 3y + 2x = 16x + 2 \\ x = -\dfrac{3}{14}y - \dfrac{1}{7} \end{cases}$

53. Valentine Gift The specialty bakery around the corner is putting together a Valentine's Day gift box to sell. The gift box is going to contain two items, heart-shaped cookies and chocolate candies. If the cookies sell for $1.50 per pound and the chocolates sell for $7.75 per pound, how many pounds of each type should be included in a 4-pound box that sells for $4.00 per pound? To find the answers, solve the system

$$\begin{cases} h + c = 4 & \text{(1)} \\ 1.50h + 7.75c = 16 & \text{(2)} \end{cases}$$

where h represents the number of pounds of heart-shaped cookies and c represents the number of pounds of candies.

54. Raffle Tickets The parent booster club is selling 50–50 raffle tickets to raise money for the upcoming hockey banquet. Raffle tickets are sold in two ways: Individual chances are sold for $1.00 each, or a ticket with a block of 6 chances can be purchased for $5.00. On one particular night, the booster club brought in $2025 from the sale of 725 raffle tickets. How many of each type of raffle ticket was sold? To find the answers. Solve the system

$$\begin{cases} t + b = 725 & \text{(1)} \\ t + 5b = 2025 & \text{(2)} \end{cases}$$

where t is the number of individual tickets and b is the number of block tickets.

Section 4.4   Solving Direct Translation, Geometry, and Uniform Motion Problems Using Systems of Linear Equations

You Should Be Able To...	EXAMPLE	Review Exercises
❶ Model and solve direct translation problems (p. 278)	Example 1	55–58
❷ Model and solve geometry problems (p. 278)	Examples 2 and 3	59–62
❸ Model and solve uniform motion problems (p. 281)	Example 4	63–66

55. Fun with Numbers Find two numbers whose difference is $\dfrac{1}{24}$ and whose sum is $\dfrac{17}{24}$.

56. Fun with Numbers The sum of two numbers is 58. If twice the smaller number is subtracted from the larger number, the difference is -20. Find the numbers.

57. Investments Aaron has $50,000 to invest. His financial advisor recommends he invest in stocks and bonds, with the amount invested in stocks equaling $4000 less than twice the amount invested in bonds. How much should be invested in stocks? How much should be invested in bonds?

58. Bookstore Sales A register at the bookstore sold only notebooks and scientific calculators. The cost of a notebook is $\dfrac{1}{3}$ of the cost of the calculator. If the calculator sells for $7.50 and the register tape shows that 24 items resulted in revenue of $110, how many of each were sold?

△ **59. Geometry** The measure of one angle is 25° less than the measure of its supplement. Find the measures of the two angles.

△ **60. Geometry** The measure of one angle is 15° more than half the measure of its complement. Find the measures of the two angles.

△ **61. Horse Corral** The O'Connells are planning to build a rectangular corral for their horses, using the river on their property as one of the sides. On the other three sides they are installing 52 meters of fencing. The longer side of the corral is opposite the water, and its length is 8 meters less than twice the shorter side. Find the dimensions of the corral.

△ **62. Geometry** In a triangle, the measure of the first angle is 10° less than three times the measure of the second. If the measure of the third angle is 30°, find the measure of the two unknown angles of the triangle.

63. Wind Resistance A plane can fly 2000 miles with a tailwind in 4 hours. It can make the return trip in 5 hours. Find the speed of the plane in still air and the effect of the wind.

64. Fishing Spot Dayle is paddling his canoe upstream, against the current, to a fishing spot 10 miles away. If he

paddles upstream for 2.5 hours and his return trip takes 1.25 hours, find the speed of the current and his paddling speed in still water.

65. Cycling in the Wind A cyclist can go 36 miles with the wind blowing at her back in 3 hours. On the return trip, after 4 hours, the cyclist still has 4 miles remaining to return to the starting point. Find the speed of the cyclist and the effect of the wind.

66. Out for a Run One speedy jogger can run 2 mph faster than his running buddy. In the last race they entered, the faster runner covered 12 miles in the same time that it took for the slower jogger to cover 9 miles. Find the speed of each of the runners.

Section 4.5	**Solving Mixture Problems Using Systems of Linear Equations**

You Should Be Able To...	EXAMPLE	Review Exercises
❶ Draw up a plan for solving mixture problems (p. 285)	Example 1	67, 68
❷ Set up and solve money mixture problems (p. 286)	Examples 2 and 3	69–72
❸ Set up and solve dry mixture and percent mixture problems (p. 288)	Examples 4 and 5	73–76

In Problems 67 and 68, fill in the table from the information given. DO NOT SOLVE.

67. Brendan has 35 coins, all nickels and dimes. He has a total of $2.25 in change.

	Number of Coins ·	Value of Each Coin =	Total Value
Dimes			
Nickels			
Total			

68. Belinda invested part of her $15,000 inheritance in a savings account with an annual rate of 6.5% simple interest and the rest of her inheritance in a mutual fund with a rate of 8% simple interest. Her total yearly interest income from both accounts was $1050.

	Principal	Rate	Interest
Savings Account			
Mutual Fund			
Total			

69. Movie Tickets One day, the Beechwold Movie Theater had sales of $9929 from selling $8 regular-priced tickets and $5.50 matinee tickets to a movie. The theater sold a total of 1498 tickets. How many regular-priced tickets were sold that day?

70. Coins Sharona has $1.70 in change consisting of three more dimes than quarters. Find the number of quarters she has.

71. Investment Carlos Singer has some money invested at 5% and $5000 more than that invested at 3%. His total annual interest income is $710. Find the amount Carlos has invested at each rate.

72. Investment Hilda invested part of her $25,000 advance in savings bonds at 2.5% annual simple interest and the rest in a stock portfolio yielding 5% annual simple interest. If her total yearly interest income is $1000, find the amount Hilda has invested at each rate.

73. Sugar A baker wants to mix a 60% sugar solution with a 30% sugar solution to obtain 10 quarts of a 51% sugar solution. How much of the 30% sugar solution will the baker use?

74. Peroxide How many pints of a 25% peroxide solution should be added to 10 pints of a 60% peroxide solution to obtain a 30% peroxide solution?

75. Almond-Peanut Butter Joni works at a health food store and is planning to grind a special blend of butter by mixing peanuts, which sell for $4.00 per pound, and almonds, which sell for $6.50 per pound.

How many pounds of each type should she use if she needs 5 pounds of almond-peanut butter, which sells for $5.00 per pound?

76. Mixing Chemicals Harold needs 20 liters of 55% acid solution. If he has 35% acid and 60% acid solutions that he can mix together, how many liters of each should he use to obtain the desired mixture?

Section 4.6 Solving Systems of Linear Inequalities

You Should Be Able To...	EXAMPLE	Review Exercises
① Determine whether an ordered pair is a solution of a system of linear inequalities (p. 294)	Example 1	77–82
② Graph a system of linear inequalities (p. 295)	Examples 2 through 4	83–94
③ Solve applied problems involving systems of linear inequalities (p. 297)	Example 5	95

In Problems 77–82, determine which point(s), if any, is a solution to the system of linear inequalities. Answer Yes or No.

77. $\begin{cases} x + y \leq 2 \\ 3x - 2y > 6 \end{cases}$
  (a) $(-1, -5)$
  (b) $(3, -1)$
  (c) $(4, 1)$

78. $\begin{cases} y \geq 3x + 5 \\ y \geq -2x \end{cases}$
  (a) $(-1, 0)$
  (b) $(-2, 4)$
  (c) $(1, 8)$

79. $\begin{cases} x > 5 \\ y < -2 \end{cases}$
  (a) $(10, -10)$
  (b) $(5, -3)$
  (c) $(7, -2)$

80. $\begin{cases} y > x \\ 2x - y \leq 3 \end{cases}$
  (a) $(4, 3)$
  (b) $(-3, -2)$
  (c) $(-1, 4)$

81. $\begin{cases} x + 2y < 6 \\ 4y - 2x > 16 \end{cases}$
  (a) $(-4, 2)$
  (b) $(-8, 6)$
  (c) $(-2, -3)$

82. $\begin{cases} 2x - y \geq 3 \\ y \leq 2x + 1 \end{cases}$
  (a) $(1, 2)$
  (b) $(3, -2)$
  (c) $(-1, 4)$

In Problems 83–94, graph the solution set of each system of linear inequalities.

83. $\begin{cases} x > -2 \\ y > 1 \end{cases}$

84. $\begin{cases} x \leq 3 \\ y > -1 \end{cases}$

85. $\begin{cases} x + y \geq -2 \\ 2x - y \leq -4 \end{cases}$

86. $\begin{cases} 3x + 2y < -6 \\ x - y < 2 \end{cases}$

87. $\begin{cases} x > 0 \\ y \leq \dfrac{3}{4}x + 1 \end{cases}$

88. $\begin{cases} y \leq 0 \\ y \leq -\dfrac{1}{2}x - 3 \end{cases}$

89. $\begin{cases} -y \geq x \\ 4x - 3y \geq -12 \end{cases}$

90. $\begin{cases} y \geq -x + 2 \\ x - y \geq -4 \end{cases}$

91. $\begin{cases} 2x + 3y \geq -3 \\ y > \dfrac{3}{2}x + 1 \end{cases}$

92. $\begin{cases} x + 4y \leq -4 \\ y \geq \dfrac{1}{4}x + 3 \end{cases}$

93. $\begin{cases} y > 2x - 5 \\ y - 2x \leq 0 \end{cases}$

94. $\begin{cases} y < x + 2 \\ y > -\dfrac{5}{2}x + 2 \end{cases}$

95. Party Planning Alexis and Sarah are planning a barbeque for their friends. They plan to serve grilled fish and carne asada and want to spend at most $40 on the meal. The fish sells for $8 per pound, and the carne asada sells for $5 per pound. They plan to buy at least twice as much fish as carne asada. A system of linear inequalities that models this situation is

$$\begin{cases} 8x + 5y \leq 40 \\ x \geq 2y \\ x \geq 0 \\ y \geq 0 \end{cases}$$

where x represents the number of pounds of fish and y represents the number of pounds of carne asada. Graph the system of linear inequalities that shows how many pounds of each they can buy and stay within their budget.

In Problems 1 and 2, determine which of the following is a solution to the system of linear equations or inequalities. Answer Yes or No.

1. $\begin{cases} 3x - y = -5 \\ \quad y = \dfrac{2}{3}x - 2 \end{cases}$

(a) $(-1, 2)$
(b) $(-9, -8)$
(c) $(-3, -4)$

2. $\begin{cases} 2x + y \geq 10 \\ \quad y < \dfrac{x}{2} + 1 \end{cases}$

(a) $(5, 3)$
(b) $(-2, 14)$
(c) $(-3, -1)$

3. The slopes of the two lines are negative reciprocals, and the y-intercepts are the same.

(a) Tell how many solutions exist.

(b) State whether the system is consistent or inconsistent.

(c) If the system is consistent, state whether the equations are dependent or independent.

4. The slopes of the two lines are the same, and the y-intercepts are different.

(a) Tell how many solutions exist.

(b) State whether the system is consistent or inconsistent.

(c) If the system is consistent, state whether the equations are dependent or independent.

In Problems 5 and 6, solve each system of equations by graphing.

5. $\begin{cases} 2x + 3y = 0 \\ x + 4y = 5 \end{cases}$

6. $\begin{cases} y = 2x - 6 \\ y = -\dfrac{1}{4}x + 3 \end{cases}$

In Problems 7 and 8, solve each system of equations using substitution.

7. $\begin{cases} 3x - y = 3 \\ 4x + 5y = -15 \end{cases}$

8. $\begin{cases} y = \dfrac{2}{3}x + 7 \\ y = \dfrac{1}{4}x + 2 \end{cases}$

In Problems 9 and 10, solve each system of equations using elimination.

9. $\begin{cases} 3x + 2y = -3 \\ 5x - y = -18 \end{cases}$

10. $\begin{cases} 4x - 5y = 12 \\ 3x + 4y = -22 \end{cases}$

In Problems 11–13, solve by graphing, substitution, or elimination.

11. $\begin{cases} \dfrac{2}{3}x - \dfrac{1}{6}y = \dfrac{25}{12} \\ -\dfrac{1}{4}x \quad\quad = \dfrac{3}{2}y \end{cases}$

12. $\begin{cases} 0.4x - 2.5y = -6.5 \\ x + \quad y = 5.5 \end{cases}$

13. $\begin{cases} 4 + 3(y - 3) = -2(x + 1) \\ x + \dfrac{3y}{2} = \dfrac{3}{2} \end{cases}$

14. Plane Trip A small plane can fly 1575 miles in 7 hours with a tailwind. On the return trip, with a headwind, the same trip takes 9 hours. Find the speed of the plane in still air and the effect of the wind.

15. Ice Cream Manny is blending a peach swirl ice cream. He mixes together vanilla ice cream, which sells for $6 per container, and peach ice cream, which sells for $11.50 per container. How many containers of each should he use if he wants to have 11 containers of swirl ice cream that he can sell for $8 each?

△ **16. Supplementary Angles** The measure of one angle is 30° less than three times the measure of its supplement. Find the measures of the two angles.

17. Playground Equipment Moises is distributing new playground equipment to several elementary schools. Each school will receive some basketballs and some volleyballs. The cost of a basketball is $25 and the cost of a volleyball is $15. If each school needs 40 balls and he can spend $750 per school, how many of each type will a school receive?

In Problems 18–20, graph the solution set of each system of inequalities.

18. $\begin{cases} 4x - 2y < 8 \\ x + 3y < 6 \end{cases}$

19. $\begin{cases} x \leq -2 \\ -2x - 4y \leq 8 \end{cases}$

20. $\begin{cases} y \leq \dfrac{2}{3}x + 4 \\ x + 4y < 0 \end{cases}$

CHAPTER

5 Exponents and Polynomials

It is very important to understand the distance that it takes a car to stop in various driving conditions, such as wet or dry pavement. This information is used to develop speed limits in school zones and busy highways with lots of traffic lights. The polynomial $0.04v^2 + 0.7v$ models the stopping distance of a car traveling on dry pavement at a speed of v miles per hour. See Problems 121 and 122 in Section 5.1.

The Big Picture: Putting It Together

In the first two chapters of the text, we reviewed and developed fundamental skills with real numbers, algebraic expressions, and equations. These skills will be used throughout the course. In particular, we will describe how new topics and skills relate to those that we already know.

This chapter introduces polynomials. Polynomial expressions, such as the one given in the chapter opener, are used to describe many situations in business, consumer affairs, the physical and biological sciences, and even the entertainment industry. In this chapter, addition, subtraction, multiplication, and division of polynomial expressions are introduced. Pay attention to these operations and how they are related to the techniques for adding, subtracting, multiplying, and dividing real numbers. Why? Algebra is easier to understand if we think of it as an extension of arithmetic.

Outline

5.1 Adding and Subtracting Polynomials

Objectives

❶ Define Monomial and Determine the Degree of a Monomial

❷ Define Polynomial and Determine the Degree of a Polynomial

❸ Simplify Polynomials by Combining Like Terms

❹ Evaluate Polynomials

Are You Prepared for This Section?

Before getting started, complete the following problems. If you get a problem wrong, go back to the section cited and review the material.

P1. What is the coefficient of $-4x^5$? [Section 1.8, p. 66]

P2. Combine like terms: $-3x + 2 - 2x - 6x - 7$ [Section 1.8, pp. 68–69]

P3. Use the Distributive Property to remove the parentheses: $-4(x - 3)$ [Section 1.8, pp. 67–68]

P4. Evaluate the expression $5 - 3x$ for $x = -2$. [Section 1.8, pp. 64–65]

Recall from Section 1.8 that a term is a number or the product of a number and one or more variables raised to a power. The numerical factor of a term is the coefficient. A constant is a single number, such as 2 or $-\dfrac{3}{2}$. Consider Table 1, where three algebraic expressions are given along with their terms and coefficients.

Table 1

Algebraic Expression	Terms	Coefficients
$3x + 2$	$3x, 2$	$3, 2$
$4x^2 - x + 5 = 4x^2 + (-x) + 5$	$4x^2, -x, 5$	$4, -1, 5$
$9x^2 + y^3$	$9x^2, y^3$	$9, 1$

Notice that $4x^2 - x + 5$ given in Table 1 is rewritten as a sum, $4x^2 + (-x) + 5$, to identify the terms and coefficients.

Work Smart

The prefix "mono" means "one." For example, a monorail has one rail. So a monomial in one variable is either a constant or a single term containing a variable raised to a whole number power.

▶ ❶ **Define Monomial and Determine the Degree of a Monomial**

The focus of this chapter is *polynomials*. Polynomials are made up of terms that are *monomials*.

> **Definition**
>
> A **monomial** in one variable is the product of a constant and a variable raised to a whole number $(0, 1, 2, \ldots)$ power. A monomial in one variable is of the form
>
> $$ax^k$$
>
> where a is a constant that is any real number, x is a variable, and k is a whole number. The constant a is called the **coefficient** of the monomial. If $a \neq 0$, then k is the **degree** of the monomial.

The degree of a nonzero constant, such as 5 or -7, is zero. What if $a = 0$? Because $0 = 0x = 0x^2 = 0x^3 = \ldots$, a degree cannot be assigned to 0. Therefore, 0 has no degree.

EXAMPLE 1 **Monomials**

Monomial	Coefficient	Degree
(a) $2x^4$	2	4
(b) $-\dfrac{7}{4}x^2$	$-\dfrac{7}{4}$	2
(c) $-x = -1 \cdot x^1$	-1	1
(d) $x^3 = 1 \cdot x^3$	1	3
(e) 8	8	0
(f) 0	0	No degree

EXAMPLE 2 **Examples of Expressions That Are Not Monomials**

(a) $5x^{\frac{1}{2}}$ is not a monomial because the exponent of the variable is $\frac{1}{2}$, and $\frac{1}{2}$ is not a whole number.

(b) $2x^{-4}$ is not a monomial because the exponent of the variable is -4, and -4 is not a whole number.

●

In Other Words

The degree of a monomial can be thought of as the number of times the variable occurs as a factor. For example, because $2x^4 = 2 \cdot x \cdot x \cdot x \cdot x$, the degree of $2x^4$ is 4.

Quick ✓

1. A _____ in one variable is the product of a number and a variable raised to a whole number power.

2. The coefficient of a monomial such as x^2 or z is _.

In Problems 3–6, determine whether the expression is a monomial. Answer Yes or No. For those that are monomials, determine the coefficient and degree.

3. $12x^6$ 4. $3x^{-3}$

5. 10 6. $n^{\frac{1}{3}}$

A monomial may contain more than one variable factor, such as $ax^m y^n$, where a is the coefficient, x and y are variables, and m and n are whole numbers. The **degree of the monomial** $ax^m y^n$ is the sum of the exponents, $m + n$.

EXAMPLE 3 **Monomials in More Than One Variable**

(a) $-4x^3 y^4$ is a monomial in x and y of degree $3 + 4 = 7$. The coefficient is -4.

(b) $10ab^5$ is a monomial in a and b of degree $1 + 5 = 6$. The coefficient is 10.   ●

Quick ✓

7. The degree of a monomial in the form $ax^m y^n$ is _____.

In Problems 8–10, determine whether the expression is a monomial. Answer Yes or No. For those that are monomials, determine the coefficient and degree.

8. $3x^5 y^2$ 9. $4ab^{\frac{1}{2}}$

10. $-x^2 y$

▶ ❷ **Define Polynomial and Determine the Degree of a Polynomial**

Definition

A **polynomial** is a monomial or the sum of monomials.

Certain polynomials have special names. A polynomial with exactly one term is a monomial; a polynomial that contains two different monomials is called a **binomial;** and a polynomial that contains three different monomials is called a **trinomial.** So

Work Smart

The prefix "bi" means "two," as in bicycle. The prefix "tri" means "three," as in tricycle. The prefix "poly" means "many."

$-14x$ is a polynomial	but more specifically	$-14x$ is a monomial
$2x^3 - 5x$ is a polynomial	but more specifically	$2x^3 - 5x$ is a binomial
$-x^3 - 4x + 11$ is a polynomial	but more specifically	$-x^3 - 4x + 11$ is a trinomial
$3x^2 + 6xy - 2y^2$ is a polynomial	but more specifically	$3x^2 + 6xy - 2y^2$ is a trinomial

Polynomials that have four or more terms, such as $5x^4 - 3x^3 + x + 7$, are simply referred to as polynomials.

A polynomial is in **standard form** if it is written with the terms in descending order according to degree. The **degree of a polynomial** is the highest degree of all the terms of the polynomial. Remember, the degree of a nonzero constant is 0, and the number 0 has no degree.

EXAMPLE 4 **Examples of Polynomials**

Polynomial	Degree
(a) $7x^3 - 2x^2 + 6x + 4$	3
(b) $3 - 8x + x^2 = x^2 - 8x + 3$	2
(c) $-7x^4 + 24$	4
(d) $x^3y^4 - 3x^3y^2 + 2x^3y$	7
(e) $p^2q - 8p^3q^2 + 3 = -8p^3q^2 + p^2q + 3$	5
(f) 6	0
(g) 0	No degree

Letters other than x may be used to represent the variable in a polynomial.

$$7t^4 + 14t^2 + 8 \text{ is a polynomial (in } t) \text{ of degree 4}$$

$$-5z^2 + 3z - 6 \text{ is a polynomial (in } z) \text{ of degree 2}$$

$$8y^3 - y^2 + 2y - 12 \text{ is a polynomial (in } y) \text{ of degree 3}$$

Because a polynomial is the sum of monomials, if any term in an algebraic expression is not a monomial, then the expression is not a polynomial.

EXAMPLE 5 **Algebraic Expressions That Are Not Polynomials**

(a) $4x^{-2} - 5x + 1$ is not a polynomial because the exponent on the first term, -2, is not a whole number.

(b) $\dfrac{4}{x^3}$ is not a polynomial because a variable expression is in the denominator.

(c) $4z^2 + 9z - 3z^{\frac{1}{2}}$ is not a polynomial because the exponent on the third term, $\dfrac{1}{2}$, is not a whole number.

Quick ✔

11. *True or False* The degree of the polynomial $3ab^2 - 4a^2b^3 + \dfrac{6}{5}a^2b^2$ is 5.

12. A polynomial is said to be written in _____ _____ if it is written with the terms in descending order according to degree.

In Problems 13–16, determine whether the algebraic expression is a polynomial. Answer Yes or No. For those that are polynomials, determine the degree.

13. $-4x^3 + 2x^2 - 5x + 3$ **14.** $2m^{-1} + 7$

15. $\dfrac{-1}{x^2 + 1}$ **16.** $5p^3q - 8pq^2 + pq$

▶ ❸ Simplify Polynomials by Combining Like Terms

To simplify a polynomial means to perform all indicated operations and combine like terms.

EXAMPLE 6 **Simplifying Polynomials: Addition**

Find the sum: $(-5x^3 + 6x^2 + 2x - 7) + (3x^3 + 4x + 1)$

Solution

The sum can be found using horizontal or vertical addition.

Horizontal Addition: The first step is to remove parentheses.

$$(-5x^3 + 6x^2 + 2x - 7) + (3x^3 + 4x + 1) = -5x^3 + 6x^2 + 2x - 7 + 3x^3 + 4x + 1$$

$$\text{Rearrange terms: } = -5x^3 + 3x^3 + 6x^2 + 2x + 4x - 7 + 1$$

$$\text{Simplify: } = -2x^3 + 6x^2 + 6x - 6$$

Work Smart

Remember, like terms have the same variable, have the same exponent on the variable, and are separated by addition symbols.

Vertical Addition: Here, line up like terms in each polynomial vertically and then add the coefficients.

$$\begin{array}{r} -5x^3 + 6x^2 + 2x - 7 \\ \underline{3x^3 \qquad\quad + 4x + 1} \\ -2x^3 + 6x^2 + 6x - 6 \end{array}$$

$\bullet$

> **Quick ✓**
>
> **17.** *True or False* $4x^3 + 7x^3 = 11x^6$
>
> *In Problems 18 and 19, add the polynomials using either horizontal or vertical addition.*
>
> **18.** $(9x^2 - x + 5) + (3x^2 + 4x - 2)$
> **19.** $(4z^4 - 2z^3 + z - 5) + (-2z^4 + 6z^3 - z^2 + 4)$

Adding polynomials in two variables is handled the same way as adding polynomials in a single variable—that is, by adding like terms.

EXAMPLE 7 **Simplifying Polynomials in Two Variables: Addition**

Find the sum: $(6a^2b + 11ab^2 - 4ab) + (a^2b - 3ab^2 + 9ab)$

Solution

Although you may add horizontally or vertically, only the horizontal format is presented here. The first step is to remove parentheses.

$$(6a^2b + 11ab^2 - 4ab) + (a^2b - 3ab^2 + 9ab) = 6a^2b + 11ab^2 - 4ab + a^2b - 3ab^2 + 9ab$$

$$\text{Rearrange terms: } = 6a^2b + a^2b + 11ab^2 - 3ab^2 - 4ab + 9ab$$

$$\text{Simplify: } = 7a^2b + 8ab^2 + 5ab$$

$\bullet$

> **Quick ✓**
>
> *In Problem 20, find the sum.*
>
> **20.** $(7x^2y + x^2y^2 - 5xy^2) + (-2x^2y + 5x^2y^2 + 4xy^2)$

▶ Polynomials can be subtracted using either the horizontal or the vertical approach. The first step in subtracting polynomials uses the method for subtracting two real numbers. Remember,

$$a - b = a + (-b)$$

So, to subtract one polynomial from another, add the opposite of *each term* in the polynomial following the subtraction sign and then combine like terms.

EXAMPLE 8 **Simplifying Polynomials: Subtraction**

Find the difference: $(6z^3 + 2z^2 - 5) - (-3z^3 + 9z^2 - z + 1)$

Solution

Horizontal Subtraction: First, write the problem as an addition problem by determining the opposite of each term following the subtraction sign.

Work Smart

Taking the opposite of each term following the subtraction symbol is the same as multiplying each term of the polynomial by −1.

$(6z^3 + 2z^2 - 5) - (-3z^3 + 9z^2 - z + 1) = (6z^3 + 2z^2 - 5) + (3z^3 - 9z^2 + z - 1)$

Remove parentheses: $= 6z^3 + 2z^2 - 5 + 3z^3 - 9z^2 + z - 1$

Rearrange terms: $= 6z^3 + 3z^3 + 2z^2 - 9z^2 + z - 5 - 1$

Simplify: $= 9z^3 - 7z^2 + z - 6$

Vertical Subtraction: First, line up like terms vertically.

$$6z^3 + 2z^2 \qquad - 5$$
$$-(-3z^3 + 9z^2 - z + 1)$$

Then change the sign of each coefficient of the second polynomial and add.

$$6z^3 + 2z^2 \qquad - 5$$
$$\underline{+3z^3 - 9z^2 + z - 1}$$
$$9z^3 - 7z^2 + z - 6$$

Quick ✓

In Problems 21 and 22, find the difference.

21. $(7x^3 - 3x^2 + 2x + 8) - (2x^3 + 12x^2 - x + 1)$

22. $(6y^3 - 3y^2 + 2y + 4) - (-2y^3 + 5y + 10)$

EXAMPLE 9 **Simplifying Polynomials in Two Variables: Subtraction**

Find the difference: $(2p^2q + 5pq^2 + pq) - (3p^2q + 9pq^2 - 6pq)$

Solution

$a - b = a + (-b)$

$(2p^2q + 5pq^2 + pq) - (3p^2q + 9pq^2 - 6pq) = (2p^2q + 5pq^2 + pq) + (-3p^2q - 9pq^2 + 6pq)$

Remove parentheses: $= 2p^2q + 5pq^2 + pq - 3p^2q - 9pq^2 + 6pq$

Rearrange terms: $= 2p^2q - 3p^2q + 5pq^2 - 9pq^2 + pq + 6pq$

Simplify: $= -p^2q - 4pq^2 + 7pq$

Quick ✓

In Problems 23 and 24, find the difference.

23. $(9x^2y + 6x^2y^2 - 3xy^2) - (-2x^2y + 4x^2y^2 + 6xy^2)$

24. $(4a^2b - 3a^2b^3 + 2ab^2) - (3a^2b - 4a^2b^3 + 5ab)$

EXAMPLE 10 **Simplifying Polynomials**

Perform the indicated operations: $(3x^2 - 2xy + y^2) - (7x^2 - y^2) + (3xy - 5y^2)$

Solution

$(3x^2 - 2xy + y^2) - (7x^2 - y^2) + (3xy - 5y^2) = 3x^2 - 2xy + y^2 - 7x^2 + y^2 + 3xy - 5y^2$

Rearrange terms: $= 3x^2 - 7x^2 - 2xy + 3xy + y^2 + y^2 - 5y^2$

Simplify: $= -4x^2 + xy - 3y^2$

▶ ❹ Evaluate Polynomials

To **evaluate** a polynomial, substitute the given number for the value of the variable and simplify, just as we did in Section 1.8.

EXAMPLE 11 **Evaluating a Polynomial**

Evaluate the polynomial $2x^3 - 5x^2 + x - 3$ for

(a) $x = 2$ (b) $x = -3$

Solution

Work Smart

Remember the order of operations: evaluate the exponent before you multiply.

(a) Let $x = 2$ in $2x^3 - 5x^2 + x - 3$.

$$2x^3 - 5x^2 + x - 3 = 2(2)^3 - 5(2)^2 + 2 - 3$$
$$= 16 - 20 + 2 - 3$$
$$= -5$$

(b) Let $x = -3$ in $2x^3 - 5x^2 + x - 3$.

$$2x^3 - 5x^2 + x - 3 = 2(-3)^3 - 5(-3)^2 + (-3) - 3$$
$$= -54 - 45 - 3 - 3$$
$$= -105$$

EXAMPLE 12 **Evaluating a Polynomial in Two Variables**

Evaluate the polynomial $2a^2b + 3a^2b^2 - 4ab$ for $a = -3$ and $b = -1$.

Solution

Let $a = -3$ and $b = -1$ in $2a^2b + 3a^2b^2 - 4ab$.

$$2a^2b + 3a^2b^2 - 4ab = 2(-3)^2(-1) + 3(-3)^2(-1)^2 - 4(-3)(-1)$$
$$= 2(9)(-1) + 3(9)(1) - 4(-3)(-1)$$
$$= -18 + 27 - 12$$
$$= -3$$

EXAMPLE 13 **How Much Revenue?**

The monthly revenue (in dollars) from selling x clocks is given by the polynomial $-0.3x^2 + 90x$. Evaluate the polynomial for $x = 200$ and explain the result.

Solution

The variable x represents the number of clocks sold in a month. The value of the polynomial $-0.3x^2 + 90x$ for $x = 200$ is the amount of revenue (in dollars) when 200 clocks are sold in a month. To find the revenue, replace x by 200 in the expression $-0.3x^2 + 90x$.

$$
\begin{aligned}
-0.3x^2 + 90x &= -0.3(200)^2 + 90(200) \\
&= -0.3(40{,}000) + 18{,}000 \\
&= -12{,}000 + 18{,}000 \\
&= 6000
\end{aligned}
$$

The revenue from selling 200 clocks in a month is \$6000.

> **Quick ✓**
>
> **28.** The polynomial $40x - 0.2x^2$ represents the monthly revenue (in dollars) realized by selling x wristwatches. Find the monthly revenue for selling 75 wristwatches by evaluating the polynomial $40x - 0.2x^2$ for $x = 75$.

5.1 Exercises

Exercise numbers in **green** have complete video solutions in MyMathLab or may be accessed using the QR code to the right.

*Problems **1–28** are the Quick ✓s that follow the **EXAMPLES**.*

Building Skills

In Problems 29–40, determine whether the given expression is a monomial. Answer Yes or No. For those that are monomials, state the coefficient and degree. See Objective 1.

29. $8y^3$

30. $-3x^2$

31. $\dfrac{x^2}{7}$

32. $4m^{101}$

33. z^{-6}

34. $\dfrac{1}{y^7}$

35. $12mn^4$

36. $-x^6y$

37. $\dfrac{3}{n^2}$

38. y^{-1}

39. 4

40. $\dfrac{2}{3}$

In Problems 41–56, determine whether the algebraic expression is a polynomial. Answer Yes or No. If it is a polynomial, write the polynomial in standard form, determine its degree, and state whether it is a monomial, a binomial, or a trinomial. If it is a polynomial with more than three terms, state the expression is a polynomial. See Objective 2.

41. $6x^2 - 10$

42. $4x + 1$

43. $\dfrac{-20}{n}$

44. $\dfrac{1}{x}$

45. $3y^{\frac{1}{3}} + 2$

46. $8m - 4m^{\frac{1}{2}}$

47. $\dfrac{1}{8}$

48. 32

49. $5z^3 - 10z^2 + z + 12$

50. $p^5 - 3p^4 + 7p + 8$

51. $7x^{-1} + 4$

52. $4y^{-2} + 6y - 1$

53. $3t^2 - \dfrac{1}{2}t^4 + 6t$

54. $-3x^3 + 4x^7 - 1$

55. $3x^2y^2 + 2xy^4 + 4$

56. $4mn^3 - 2m^2n^3 + mn^8$

In Problems 57–70, add the polynomials. Write the answer in standard form. See Objective 3.

57. $(4x - 3) + (3x - 7)$

58. $(-13z + 4) + (9z - 10)$

59. $(-4m^2 + 2m - 1) + (2m^2 - 2m + 6)$

60. $(x^2 - 2) + (6x^2 - x - 1)$

61. $(p - p^3 + 2) + (6 - 2p^2 + p^3)$

62. $(4r^4 + 3r - 1) + (2r^4 - 7r^3 + r^2 - 10r)$

63. $(2y - 10) + (-3y^2 - 4y + 6)$

64. $(3 - 12w^2) + (2w^2 - 5 + 6w)$

65. $\left(\dfrac{1}{2}p^2 - \dfrac{2}{3}p + 2\right) + \left(\dfrac{3}{4}p^2 + \dfrac{5}{6}p - 5\right)$

66. $\left(\dfrac{3}{8}b^2 - \dfrac{3}{5}b + 1\right) + \left(\dfrac{5}{6}b^2 + \dfrac{2}{15}b - 1\right)$

67. $(5m^2 - 6mn + 2n^2) + (m^2 + 2mn - 3n^2)$

68. $(4a^2 + ab - 9b^2) + (-6a^2 - 4ab + b^2)$

69.
$$4n^2 - 2n + 1$$
$$\underline{+(-6n + 4)}$$

70.
$$8x^3 \qquad + 2x + 2$$
$$\underline{+(-3x^2 - 4x - 2)}$$

In Problems 71–84, subtract the polynomials. Write the answer in standard form. See Objective 3.

71. $(7x - 10) - (4x + 6)$

72. $(7t + 8) - (2t + 4)$

73. $(12x^2 - 2x - 4) - (-2x^2 + x + 1)$

74. $(3x^2 + x - 3) - (x^2 - 2x + 4)$

75. $(y^3 - 2y + 1) - (-3y^3 + y + 5)$

76. $(m^4 - 3m^2 + 5) - (3m^4 - 5m^2 - 2)$

77. $(3y^3 - 2y) - (2y + y^2 + y^3)$

78. $(2x - 4x^3) - (-3 - 2x + x^3)$

79. $\left(\dfrac{5}{3}q^2 - \dfrac{5}{2}q + 4\right) - \left(\dfrac{1}{9}q^2 + \dfrac{3}{8}q + 2\right)$

80. $\left(\dfrac{7}{4}x^2 - \dfrac{5}{8}x - 1\right) - \left(\dfrac{7}{6}x^2 + \dfrac{5}{12}x + 5\right)$

81. $(-4m^2n^2 - 2mn + 3) - (4m^2n^2 + 2mn + 10)$

82. $(4m^2n - 2mn - 4) - (10m^2n - 6mn - 3)$

83.
$$6x - 3$$
$$\underline{-(10x + 2)}$$

84.
$$6n^2 - 2n - 3$$
$$\underline{-(4n + 7)}$$

In Problems 85–92, evaluate the polynomial for each of the given value(s). See Objective 4.

85. $2x^2 - x + 3$
  (a) $x = 0$
  (b) $x = 5$
  (c) $x = -2$

86. $-x^2 + 10$
  (a) $x = 0$
  (b) $x = -1$
  (c) $x = 1$

87. $7 - x^2$
  (a) $x = 3$
  (b) $x = -\dfrac{5}{2}$
  (c) $x = -1.5$

88. $2 + \dfrac{1}{2}n^2$
  (a) $n = 4$
  (b) $n = 0.5$
  (c) $n = -\dfrac{1}{4}$

89. $-x^2y + 2xy^2 - 3$ for $x = 2$ and $y = -3$

90. $-2ab^2 - 2a^2b - b^3$ for $a = 1$ and $b = -2$

91. $st + 2s^2t + 3st^2 - t^4$ for $s = -2$ and $t = 4$

92. $m^2n^2 - mn^2 + 3m^2 - 2$ for $m = \dfrac{1}{2}$ and $n = -1$

Mixed Practice

In Problems 93–110, perform the indicated operations. Write the answer in standard form.

93. $4t - (7t - 3)$

94. $6x^2 - (18x^2 + 2)$

95. $(5x^2 + x - 4) + (-2x^2 - 4x + 1)$

96. $(4m^2 - m + 6) + (3m^2 - 4m - 10)$

97. $(2xy^2 - 3) + (7xy^2 + 4)$

98. $(-14xy + 3) - (-xy + 10)$

99. $(4 + 8y - 2y^2) - (3 - 7y - y^2)$

100. $(9 + 2z - 6z^2) - (-2 + z + 5z^2)$

101. $\left(\dfrac{5}{6}q^2 - \dfrac{1}{3}\right) + \left(\dfrac{3}{2}q^2 + 2\right)$

102. $\left(\dfrac{7}{10}t^2 - \dfrac{5}{12}t\right) + \left(\dfrac{3}{15}t^2 + \dfrac{3}{20}t - 3\right)$

103. $14d^2 - (2d - 10) - (d^2 - 3d)$

104. $3x - (5x + 1) - 4$

105. $(4a^2 - 1) + (a^2 + 5a + 2) - (-a^2 + 4)$

106. $(2b^2 + 3b - 5) - (b^2 - 4b + 1) + (b^2 + 1)$

107. $(p^2 + 25) - (3p^2 - p + 4) - (-2p^2 - 9p + 21)$

108. $(n^2 + 4n - 5) - (2n^2 - 5n + 1) + (n^2 - 8n + 6)$

109. $(x^2 - 2xy - y^2) - (3x^2 + xy - y^2) + (xy + y^2)$

110. $(2st^4 - 3s^2t^2 + t^4) + (7s^2t^2 - 3t^4 + 8st^4)$

111. Find the sum of $3x + 10$ and $-8x + 2$.

112. Find the sum of $4x^2 - 2x - 3$ and $-5x^2 - 2x + 7$.

113. Find the difference of $-x^2 + 2x + 3$ and $-4x^2 - 2x + 6$.

114. Find the difference of $4x - 3$ and $8x - 7$.

115. Subtract $14x^2 - 2x + 3$ from $2x - 10$.

116. Subtract $-4n - 8$ from $3n^2 + 8n - 9$.

117. What polynomial should be added to $3x - 5$ so that the sum is zero?

118. What polynomial should be added to $2a + 7b$ so that the sum is $a - b$?

Applying the Concepts

119. Height of a Ball The height above the ground (in feet) of a ball dropped from the top of a 30-foot-tall building after t seconds is given by the polynomial $-16t^2 + 30$. What is the height of the ball after $\frac{1}{2}$ second?

120. Height of a Ball The height above the ground (in feet) of a ball tossed upward after t seconds is given by the polynomial $-16t^2 + 32t + 100$. What is the height of the ball after 2.5 seconds?

121. Stopping Distance on Dry Pavement The polynomial $0.04v^2 + 0.7v$ models the number of feet it takes for a car traveling v miles per hour to stop on dry, level asphalt.

   (a) How many feet will it take a car traveling 50 miles per hour to stop on dry, level asphalt.

   (b) How many feet will it take a car traveling 20 miles per hour through a school zone to stop on dry, level asphalt?

122. Stopping Distance on Wet Pavement The polynomial $0.07v^2 + 0.7v$ models the number of feet it takes for a car traveling v miles per hour to stop on wet, level asphalt.

   (a) How many feet will it take a car traveling 50 miles per hour to stop on wet, level asphalt?

   (b) How many feet will it take a car traveling 20 miles per hour through a school zone to stop on wet, level asphalt?

123. Manufacturing Calculators The revenue (in dollars) from manufacturing and selling x calculators in a day is given by the polynomial $-2x^2 + 120x$. The cost of manufacturing and selling x calculators in a day is given by the polynomial $0.125x^2 + 15x$.

   (a) Profit is equal to revenue minus cost. Write a polynomial that represents the profit from manufacturing and selling x calculators in a day.

   (b) Find the profit if 20 calculators are produced and sold each day.

124. Manufacturing DVDs Profit is equal to revenue minus cost. The revenue (in dollars) from manufacturing and selling x DVDs each day is given by the polynomial $-0.001x^2 + 6x$. The cost of manufacturing and selling x DVDs each day is given by the polynomial $0.8x + 3000$.

   (a) Write a polynomial that expresses the profit from manufacturing and selling x DVDs each day.

   (b) Find the profit if 1600 DVDs are produced and sold in a day.

125. Lawn Service Marissa started a lawn-care business in Tampa, Florida, in which she fertilizes lawns. Her costs are $5 per lawn for fertilizer and $10 per lawn for labor.

   (a) If she charges $25 per lawn, write a polynomial that represents her weekly profit for fertilizing x lawns.

   (b) If Marissa fertilizes 50 lawns in a week, what is her weekly profit?

   (c) If Marissa fertilizes 50 lawns each week for 30 weeks, what is her profit?

126. Skateboard Business Kyle owns a skateboard shop. Kyle charges $40 for each skateboard he sells. The cost for manufacturing each skateboard is $12. In addition, Kyle has weekly costs of $504, regardless of how many skateboards he manufactures.

   (a) If Kyle manufactures and sells x skateboards in a week, write a polynomial that represents his weekly profit.

   (b) What is Kyle's weekly profit if he sells 45 skateboards per week?

   (c) How many skateboards does Kyle need to sell weekly in order to break even (profit = $0)?

△ *In Problems 127–130, write the polynomial that represents the perimeter of each figure.*

127.   **128.**

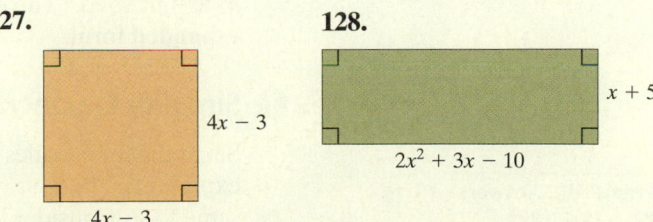

129.   **130.**

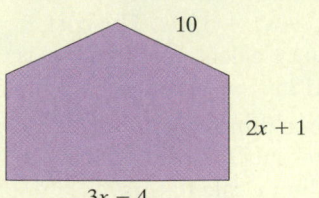

10

$2x + 1$

$3x - 4$

$x^2 - 2$

4

In Problems 131 and 132, write the polynomial that represents the unknown length.

131.

$3x - 10$

$x - 5$?

132.

?

$2x^2 + 5x$ $10x - 3$ $4x^2 - 3x + 7$

Explaining the Concepts

133. When adding polynomials, you may use either a horizontal or a vertical format. Explain which format you prefer and what you view as its advantages.

134. Use the two polynomials $3x - 5$ and $2x + 3$ to make up three different problems. In the directions to the first problem, use the word "evaluate"; in the second problem use the direction "solve"; and in the third use "simplify."

135. Explain how you find the degree of a polynomial in one variable. Explain how you find the degree of a polynomial in more than one variable. How do these processes differ?

136. Suppose you are adding two polynomials of equal degree. Must the degree of the sum equal the degree of each polynomial? Explain.

137. What is the degree of a polynomial that is linear?

138. Give a definition of "polynomial" using your own words. Provide examples of polynomials that are monomials, that are binomials, and that are trinomials.

5.2 Multiplying Monomials: The Product and Power Rules

Objectives

1 Simplify Exponential Expressions Using the Product Rule

2 Simplify Exponential Expressions Using the Power Rule

3 Simplify Exponential Expressions Containing Products

4 Multiply a Monomial by a Monomial

Work Smart

The natural numbers are 1, 2, 3, ...

Are You Prepared for This Section?

Before getting started, complete the following problems. If you get a problem wrong, go back to the section cited and review the material.

P1. Evaluate: 4^3 [Section 1.7, pp. 56–57]

P2. Use the Distributive Property to simplify: $3(4 - 5x)$ [Section 1.8, pp. 67–68]

Recall from Section 1.7, that if a is a real number and n is a natural number, then the symbol a^n means to use a as a factor n times:

$$a^n = \underbrace{a \cdot a \cdot \ldots \cdot a}_{n \text{ factors}}$$

exponent

base

For example,

$$4^3 = \underbrace{4 \cdot 4 \cdot 4}_{3 \text{ factors}} \quad \text{or} \quad y^4 = \underbrace{y \cdot y \cdot y \cdot y}_{4 \text{ factors}}$$

In the notation a^n, a is the base and n is the power or exponent. The expression a^n is read "a raised to the power n" or "a raised to the nth power." We read a^2 as "a squared" and a^3 as "a cubed." The expression 4^3 is in **exponential form** and the expression $4 \cdot 4 \cdot 4$ is in **expanded form.**

▶ **1** Simplify Exponential Expressions Using the Product Rule

Several general rules can be discovered for simplifying expressions with natural number exponents. The first rule is used when multiplying two exponential expressions with the same base. Consider the following:

$$\overbrace{}^{\text{2 factors}} \overbrace{}^{\text{4 factors}}$$
$$x^2 \cdot x^4 = (x \cdot x) \cdot (x \cdot x \cdot x \cdot x)$$
$$\text{6 factors:} \quad = x \cdot x \cdot x \cdot x \cdot x \cdot x$$
$$= x^6 \leftarrow \text{Sum of powers 2 and 4}$$

The following rule generalizes this result.

In Other Words

When multiplying two exponential expressions with the same base, add the exponents. Then write the common base to the power of this sum.

> **Product Rule for Exponents**
>
> If a is a real number, and m and n are natural numbers, then
> $$a^m \cdot a^n = a^{m+n}$$

EXAMPLE 1 | **Using the Product Rule to Evaluate Exponential Expressions**

Evaluate each expression:

    (a) $2^2 \cdot 2^4$ **(b)** $(-3)^2(-3)$

Solution

(a) $\begin{aligned} 2^2 \cdot 2^4 &= 2^{2+4} \\ &= 2^6 \\ &= 64 \end{aligned}$

(b) $\begin{aligned} (-3)^2(-3) &= (-3)^2(-3)^1 \\ &= (-3)^{2+1} \\ &= (-3)^3 \\ &= -27 \end{aligned}$ •

EXAMPLE 2 | **Using the Product Rule to Simplify Exponential Expressions**

Simplify the following expressions:

    (a) $t^4 \cdot t^7$ **(b)** $a^6 \cdot a \cdot a^4$

Solution

(a) $\begin{aligned} t^4 \cdot t^7 &= t^{4+7} \\ &= t^{11} \end{aligned}$

(b) $\begin{aligned} a^6 \cdot a \cdot a^4 &= (a^6 \cdot a^1) \cdot a^4 \\ &= a^{6+1} \cdot a^4 \\ &= a^7 \cdot a^4 \\ &= a^{7+4} \\ &= a^{11} \end{aligned}$ •

EXAMPLE 3 | **Using the Product Rule to Simplify Exponential Expressions**

Simplify the expression: $m^3 \cdot m^5 \cdot n^9$

Solution

To use the Product Rule for Exponents, the bases must be the same.

$$\begin{aligned} m^3 \cdot m^5 \cdot n^9 &= m^{3+5} \cdot n^9 \\ &= m^8 n^9 \end{aligned}$$

The expression $m^8 n^9$ is in simplest form because the bases are different. •

> **Quick ✓**
>
> **1.** The expression 12^3 is written in _____ form.
>
> **2.** $a^m \cdot a^n = a^{\underline{\quad}}$.   **3.** *True or False* $3^4 \cdot 3^2 = 9^6$
>
> *In Problems 4–8, evaluate or simplify each expression.*
>
> **4.** $3^2 \cdot 3$ **5.** $(-5)^2(-5)^3$ **6.** $c^6 \cdot c^2$
>
> **7.** $y^3 \cdot y \cdot y^5$ **8.** $a^4 \cdot a^5 \cdot b^6$

▶ ❷ Simplify Exponential Expressions Using the Power Rule

What happens when an exponential expression containing a power is raised to a power?

$$(3^2)^4 = \underbrace{3^2 \cdot 3^2 \cdot 3^2 \cdot 3^2}_{\text{4 factors}} = \underbrace{(3 \cdot 3)}_{\text{2 factors}} \cdot \underbrace{(3 \cdot 3)}_{\text{2 factors}} \cdot \underbrace{(3 \cdot 3)}_{\text{2 factors}} \cdot \underbrace{(3 \cdot 3)}_{\text{2 factors}} = 3^8$$

$$2 \cdot 4 = 8 \text{ factors}$$

The following rule results:

> **In Other Words**
>
> If an exponential expression contains a power raised to a power, keep the base and multiply the powers.

Power Rule for Exponents

If a is a real number, and m and n are natural numbers, then

$$(a^m)^n = a^{m \cdot n}$$

EXAMPLE 4 | **Using the Power Rule to Simplify Exponential Expressions**

Simplify each expression. Write the answer in exponential form.

(a) $(2^3)^2$ (b) $[(-4)^2]^5$

Solution

(a) $(2^3)^2 = 2^{3 \cdot 2}$

       $= 2^6$

(b) $[(-4)^2]^5 = (-4)^{2 \cdot 5}$

       $= (-4)^{10}$ ●

EXAMPLE 5 | **Using the Power Rule to Simplify Exponential Expressions**

Simplify each expression. Write the answer in exponential form.

(a) $(z^4)^3$ (b) $[(-n)^3]^5$

Solution

(a) $(z^4)^3 = z^{4 \cdot 3}$

       $= z^{12}$

(b) $[(-n)^3]^5 = (-n)^{3 \cdot 5}$

       $= (-n)^{15}$ ●

> **Quick ✓**
>
> **9.** If a is a real number, and m and n are natural numbers, then $(a^m)^n = a^{\underline{\hphantom{mn}}}$.
>
> *In Problems 10–13, simplify each expression. Write the answer in exponential form.*
>
> **10.** $(2^2)^4$ **11.** $[(-3)^3]^2$ **12.** $(b^2)^5$ **13.** $[(-y)^2]^4$

▶ ❸ Simplify Exponential Expressions Containing Products

What happens when a product is raised to a power?

$$\overset{\text{3 factors}}{\overbrace{(x \cdot y)^3 = (x \cdot y) \cdot (x \cdot y) \cdot (x \cdot y)}}$$

Rearrange factors: $= (x \cdot x \cdot x) \cdot (y \cdot y \cdot y)$

               $= x^3 \cdot y^3$

The following rule generalizes this result.

> **In Other Words**
>
> When raising a product to a power, raise each factor to the power. Remember, each of the numbers or variables in a multiplication problem is a factor.

Product to a Power Rule for Exponents

If a, b are real numbers and n is a natural number, then

$$(a \cdot b)^n = a^n \cdot b^n$$

EXAMPLE 6 **Using the Product to a Power Rule to Simplify Exponential Expressions**

Simplify each expression:

(a) $(3b)^4$ (b) $(-4m^3)^2$ (c) $(-2a^2b)^3$

Solution
Each expression contains the product of factors, so use the Product to a Power Rule, $(a \cdot b)^n = a^n \cdot b^n$.

(a)
$$(3b)^4 = (3)^4(b)^4$$
Evaluate 3^4: $= 81b^4$

(b)
$$(-4m^3)^2 = (-4)^2(m^3)^2$$
Evaluate $(-4)^2$; $(a^m)^n = a^{m \cdot n}$: $= 16m^{3 \cdot 2}$
$$= 16m^6$$

(c)
$$(-2a^2b)^3 = (-2)^3(a^2b)^3$$
$$= (-2)^3(a^2)^3(b)^3$$
Evaluate $(-2)^3$; $(a^m)^n = a^{m \cdot n}$: $= -8a^{2 \cdot 3}b^3$
$$= -8a^6b^3$$

Quick ✔

14. *True or False* $(ab^3)^2 = ab^6$

In Problems 15–17, simplify each expression.

15. $(2n)^3$ **16.** $(-5x^4)^3$ **17.** $(-7a^3b)^2$

Here is a summary of the three rules for exponents covered so far.

Work Smart: Study Skills

Are you having trouble remembering the rules for multiplying exponents? Try making flash cards with the property on the front and an example or two on the back. Also, write in words what each rule means and when it should be applied. There is a free applet called Quizlet that allows you to make electronic flash cards.

Rules for Exponents

If a and b are real numbers and m and n are natural numbers, then

- **Product Rule for Exponents** $a^m \cdot a^n = a^{m+n}$
- **Power Rule for Exponents** $(a^m)^n = a^{m \cdot n}$
- **Product to a Power Rule for Exponents** $(a \cdot b)^n = a^n \cdot b^n$

▶ ④ **Multiply a Monomial by a Monomial**

To multiply two monomials with the same variable base, multiply the coefficients and use the Product Rule for Exponents to multiply the variable expressions.

EXAMPLE 7 **Multiplying a Monomial by a Monomial: One Variable**

Multiply and simplify:

(a) $(5x^2)(6x^4)$ (b) $(2p^3)(-5p^2)$

Solution
In each problem, use the Commutative Property to rearrange factors.

(a) $(5x^2)(6x^4) = 5 \cdot 6 \cdot x^2 \cdot x^4$
$a^m \cdot a^n = a^{m+n}$: $= 30x^{2+4}$
$$= 30x^6$$

(b) $(2p^3)(-5p^2) = 2 \cdot (-5) \cdot p^3 \cdot p^2$
$a^m \cdot a^n = a^{m+n}$: $= -10p^{3+2}$
$$= -10p^5$$

EXAMPLE 8 **Multiplying a Monomial by a Monomial: Two Variables**

Multiply and simplify:

(a) $(-3x^2y)(-7x^5y^3)$

(b) $\left(\frac{4}{9}ab^3\right)(-a^2b)\left(\frac{27}{2}ab^2\right)$

Solution

(a) $(-3x^2y)(-7x^5y^3) = -3 \cdot (-7) \cdot x^2 \cdot x^5 \cdot y \cdot y^3$

$a^m \cdot a^n = a^{m+n}$: $= 21 \cdot x^{2+5} \cdot y^{1+3}$

$= 21x^7y^4$

(b) $\left(\frac{4}{9}ab^3\right)(-a^2b)\left(\frac{27}{2}ab^2\right) = \frac{4}{9} \cdot (-1) \cdot \frac{27}{2} \cdot a \cdot a^2 \cdot a \cdot b^3 \cdot b \cdot b^2$

$a^m \cdot a^n = a^{m+n}$: $= -6 \cdot a^{1+2+1} \cdot b^{3+1+2}$

$= -6a^4b^6$ ●

Quick ✓

In Problems 18–20, multiply and simplify.

18. $(2a^6)(4a^5)$

19. $(3m^2n^4)(-6mn^5)$

20. $\left(\frac{8}{3}xy^2\right)\left(\frac{1}{2}x^2y\right)(-12xy)$

5.2 Exercises MyMathLab®

Exercise numbers in green have complete video solutions in MyMathLab or may be accessed using the QR code to the right.

Problems **1–20** are the Quick ✓s that follow the **EXAMPLES**.

Building Skills

In Problems 21–34, simplify each expression. See Objective 1.

21. $4^2 \cdot 4^3$

22. $3 \cdot 3^3$

23. $(-2)^3(-2)^4$

24. $(-3)^2(-3)^3$

25. $m^4 \cdot m^5$

26. $a^3 \cdot a^7$

27. $b^9 \cdot b^{11}$

28. $z^8 \cdot z^{23}$

29. $x^7 \cdot x$

30. $y^{13} \cdot y$

31. $p \cdot p^2 \cdot p^6$

32. $b^5 \cdot b \cdot b^7$

33. $(-n)^3(-n)^4$

34. $(-z)(-z)^5$

In Problems 35–42, simplify each expression. See Objective 2.

35. $(2^3)^2$

36. $(3^2)^2$

37. $[(-2)^2]^3$

38. $[(-2)^3]^3$

39. $(m^2)^7$

40. $(k^8)^3$

41. $[(-b)^4]^5$

42. $[(-a)^6]^3$

In Problems 43–54, simplify each expression. See Objective 3.

43. $(3x^2)^3$

44. $(4y^3)^2$

45. $(-5z^2)^2$

46. $(-6x^3)^2$

47. $\left(\frac{1}{2}a\right)^2$

48. $\left(\frac{3}{4}n^2\right)^3$

49. $(-3p^7q^2)^4$

50. $(-2m^2n^3)^4$

51. $(-2m^2n)^3$

52. $(-4ab^2)^3$

53. $(-5xy^2z^3)^2$

54. $(-3a^6bc^4)^4$

In Problems 55–68, multiply the monomials. See Objective 4.

55. $(4x^2)(3x^3)$

56. $(5y^6)(3y^2)$

57. $(10a^3)(-4a^7)$

58. $(7b^5)(-2b^4)$

59. $(-m^3)(7m)$

60. $(-n^6)(5n)$

61. $\left(\frac{4}{5}x^4\right)\left(\frac{15}{2}x^3\right)$

62. $\left(\frac{3}{8}y^5\right)\left(\frac{4}{9}y^6\right)$

63. $(2x^2y^3)(3x^4y)$

64. $(6a^2b^3)(2a^5b)$

65. $\left(\frac{1}{4}mn^3\right)(-20mn)$

66. $\left(\frac{2}{3}s^2t^3\right)(-21st)$

67. $(4x^2y)(-5xy^3)(2x^2y^2)$

68. $\left(\frac{1}{2}b\right)(-20a^2b)\left(-\frac{2}{3}a\right)$

Mixed Practice

In Problems 69–94, simplify each expression.

69. $x^2 \cdot 4x$

70. $z^3 \cdot 5z$

71. $(-x)^2(5x^2) - (2x^2)^2$

72. $(3a^3)(-a) + (3a^2)^2$

73. $(b^3)^2 + (3b^2)^3$

74. $(m^2)^4 - (2m^4)^2$

75. $(-3b)^4 + (4b)^2$

76. $(-4a)^2 - (5a)^3$

77. $(-6p)^2\left(\frac{1}{4}p^3\right)$

78. $(-8w)^2\left(\frac{3}{16}w^5\right)$

79. $(-xy)(x^2y)(-3x)^3$

80. $(-4a)^3(bc^2)(-b)$

81. $\left(\frac{3}{5}\right)(5c)(-10d^2)$

82. $\left(-\frac{7}{3}n\right)(9n^2)\left(\frac{5}{42}n^5\right)$

83. $(5x)^2(x^2)^3$

84. $(4y)^2(y^3)^5$

85. $(x^2y)^3(-2xy^3)$

86. $(s^2t^3)^2(3st)$

87. $(-3x)^2(2x^4)^3 + (3x^{12})^2$

88. $(-2p^3)^2(3p)^3 - (2p^6)^3$

89. $\left(\frac{4}{5}q\right)^2(-5q)^2\left(\frac{1}{4}q^2\right)$

90. $\left(\frac{2}{3}m\right)^2(-3m)^3\left(\frac{3}{4}m^3\right)$

91. $(3x)^2(-2x)^4\left(\frac{1}{3}x^4\right)^2$

92. $(5y)^3(-3y)^2\left(\frac{1}{5}y^4\right)^2$

93. $-3(-5mn^3)^2\left(\frac{2}{5}m^5n\right)^3$

94. $-2(-4a^3b^2)^2\left(\frac{3}{2}ab^5\right)^3$

Applying the Concepts

△ **95. Cubes** Suppose the length of a side of a cube is x^2. Find the volume of the cube in terms of x.

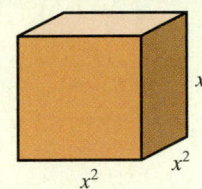

△ **96. Squares** Suppose that the length of a side of a square is $3s$. Find the area of the square in terms of s.

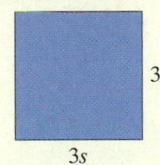

△ **97. Rectangle** Suppose that the width of a rectangle is $4x$ and its length is $12x$. Write an algebraic expression for the area of the rectangle in terms of x.

△ **98. Rectangle** Suppose that the width of a rectangle is $6a$ and its length is $8a$. Write an algebraic expression for the area of the rectangle in terms of a.

Explaining the Concepts

99. Provide a justification for the Product Rule for Exponents.

100. Provide a justification for the Power Rule for Exponents.

101. Provide a justification for the Product to a Power Rule for Exponents.

102. You simplified the expression $(-3a^2b^3)^4$ as $(-3a^2b^3)^4 = -3^4(a^2)^4(b^3)^4 = -81a^8b^{12}$. Your instructor marked your answer wrong. Find and explain the error, and then rework the problem correctly.

103. Explain the difference between simplifying the expression $4x^2 + 3x^2$ and simplifying the expression $(4x^2)(3x^2)$.

104. Explain why it is incorrect to simplify the product $\left(\frac{1}{2}x^2y\right)\left(\frac{4}{3}xy^4\right)$ by multiplying by the least common denominator.

5.3 Multiplying Polynomials

Objectives

❶ Multiply a Polynomial by a Monomial

❷ Multiply Two Binomials Using the Distributive Property

❸ Multiply Two Binomials Using the FOIL Method

❹ Multiply the Sum and Difference of Two Terms

❺ Square a Binomial

❻ Multiply a Polynomial by a Polynomial

Are You Prepared for This Section?

Before getting started, complete the following problems. If you get a problem wrong, go back to the section cited and review the material.

P1. Use the Distributive Property to simplify: $2(4x - 3)$ [Section 1.8, pp. 67–68]

P2. Find the product: $(3x^2)(-5x^3)$ [Section 5.2, pp. 321–322]

P3. Simplify: $(9a)^2$ [Section 5.2, pp. 320–321]

▶ ❶ **Multiply a Polynomial by a Monomial**

To simplify $3(2x + 1)$, multiply using the Distributive Property to obtain $3(2x) + 3(1) = 6x + 3$. In general, when multiplying a polynomial by a monomial, use the following property.

> **Extended Form of the Distributive Property**
> $$a(b + c + \cdots + z) = a \cdot b + a \cdot c + \cdots + a \cdot z$$
> where $a, b, c, \ldots, z$ are monomials.

Prepared?... Answers **P1.** $8x - 6$ **P2.** $-15x^5$ **P3.** $81a^2$

EXAMPLE 1 **Multiplying a Monomial and a Trinomial**

Multiply and simplify: $2x^2(x^2 + 3x + 5)$

Solution

Use the Extended Form of the Distributive Property and multiply each term in parentheses by $2x^2$.

$$2x^2(x^2 + 3x + 5) = 2x^2(x^2) + 2x^2(3x) + 2x^2(5)$$
$$= 2x^4 + 6x^3 + 10x^2$$

EXAMPLE 2 **Multiplying a Monomial and a Trinomial**

Multiply and simplify: $-2xy(3x^2 + 5xy + 2y^2)$

Solution

Use the Extended Form of the Distributive Property.

$$-2xy(3x^2 + 5xy + 2y^2) = -2xy(3x^2) + (-2xy)(5xy) + (-2xy)(2y^2)$$
$$= -6x^3y - 10x^2y^2 - 4xy^3$$

EXAMPLE 3 **Multiplying a Trinomial and a Monomial**

Multiply and simplify: $\left(\frac{4}{3}z^2 + 8z + \frac{1}{4}\right)\left(\frac{1}{2}z^3\right)$

Solution

$$\left(\frac{4}{3}z^2 + 8z + \frac{1}{4}\right)\left(\frac{1}{2}z^3\right) = \frac{4}{3}z^2\left(\frac{1}{2}z^3\right) + 8z\left(\frac{1}{2}z^3\right) + \frac{1}{4}\left(\frac{1}{2}z^3\right)$$
$$= \frac{2}{3}z^5 + 4z^4 + \frac{1}{8}z^3$$

Work Smart

Example 3 is an algebraic *expression*, so you do not clear fractions.

> **Quick ✓**
>
> *In Problems 1–3, multiply and simplify.*
>
> **1.** $3x(x^2 - 2x + 4)$ **2.** $-a^2b(2a^2b^2 - 4ab^2 + 3ab)$
>
> **3.** $\left(5n^3 - \frac{15}{8}n^2 - \frac{10}{7}n\right)\left(\frac{3}{5}n^2\right)$

▶ ❷ **Multiply Two Binomials Using the Distributive Property**

To understand how to multiply two binomials, we review an example of multiplication of two-digit numbers.

$$
\begin{array}{r}
32 \\
\times\, 14 \quad 10 + 4 \\
\hline
128 \leftarrow\text{This row represents } 32 \times 4 = 128. \\
320 \leftarrow\text{This row represents } 32 \times 10 = 320. \\
\hline
\text{Add vertically} \rightarrow \quad 448
\end{array}
$$

Now suppose we want to multiply $(3x + 2)$ and $(x + 4)$.

Work Smart

When multiplying polynomials vertically, align the columns of like degree.

To find the product of two binomials, proceed in exactly the same way as when multiplying two-digit numbers.

$$
\begin{array}{r}
3x + 2 \\
\times\ x + 4 \\
\hline
\end{array}
$$

$4 \cdot 3x \longrightarrow \quad 12x + 8 \quad \longleftarrow 4 \cdot 2;\ \text{this row represents } 4(3x + 2)$

$x \cdot 3x \longrightarrow \ \underline{3x^2 +\ 2x} \longleftarrow \quad x \cdot 2;\ \text{this row represents } 4(3x + 2)$

Add vertically: $\ 3x^2 + 14x + 8$

The multiplication that was performed above is known as *vertical multiplication*. *Horizontal multiplication*, which requires applying the Distributive Property, may also be used.

$$(3x + 2)(x + 4) = (3x + 2)x + (3x + 2) \cdot 4$$

Distribute x, distribute 4: $\quad = 3x \cdot x + 2 \cdot x + 3x \cdot 4 + 2 \cdot 4$

$\quad = 3x^2 + 2x + 12x + 8$

Combine like terms: $\quad = 3x^2 + 14x + 8$

This is the same result that was obtained using vertical multiplication. Here are two more examples using both horizontal and vertical multiplication.

EXAMPLE 4   **Multiplying Two Binomials**

Find the product: $(x + 3)(x - 8)$

Solution

Vertical Multiplication

$$
\begin{array}{r}
x + 3 \\
\times\ x - 8 \\
\hline
\end{array}
$$

$-8x - 24 \longleftarrow -8(x + 3)$

$\underline{x^2 + 3x} \qquad \longleftarrow\ x(x + 3)$

Add vertically: $\ x^2 - 5x - 24$

Horizontal Multiplication

Distribute $x + 3$ to each term in $x - 8$.

$(x + 3)(x - 8) = (x + 3)x + (x + 3)(-8)$

$\qquad = x^2 + 3x - 8x - 24$

Combine like terms: $\ = x^2 - 5x - 24$

In either case, $(x + 3)(x - 8) = x^2 - 5x - 24$.   ●

EXAMPLE 5   **Multiplying Two Binomials**

Find the product: $(3x + 4)(2x - 5)$

Solution

Vertical Multiplication

$$
\begin{array}{r}
3x + 4 \\
\times\ 2x - 5 \\
\hline
\end{array}
$$

$-15x - 20 \longleftarrow -5(3x + 4)$

$\underline{6x^2 + 8x} \qquad \longleftarrow 2x(3x + 4)$

Add vertically: $\ 6x^2 - 7x - 20$

Horizontal Multiplication

Distribute $3x + 4$ to each term in $2x - 5$.

$(3x + 4)(2x - 5) = (3x + 4)(2x) + (3x + 4)(-5)$

$\qquad = 3x \cdot 2x + 4 \cdot 2x + 3x \cdot (-5) + 4 \cdot (-5)$

$\qquad = 6x^2 + 8x - 15x - 20$

Combine like terms: $\quad = 6x^2 - 7x - 20$

In either case, $(3x + 4)(2x - 5) = 6x^2 - 7x - 20$.   ●

Quick ✔

In Problems 4–6, find the product.

4. $(x + 7)(x + 2)$ **5.** $(2n + 1)(n - 3)$ **6.** $(3p - 2)(2p - 1)$

▶ ③ Multiply Two Binomials Using the FOIL Method

Work Smart

The FOIL method is a form of the Distributive Property, and FOIL can be used *only* when we multiply two binomials.

Another way to multiply two binomials is known as the **FOIL method.** FOIL stands for First, Outer, Inner, Last. "First" means to multiply the *first* terms in each binomial, "Outer" means to multiply the *outside* terms of each binomial, "Inner" means to multiply the *innermost* terms of each binomial, and "Last" means to multiply the *last* terms of each binomial. See below. The horizontal method is shown for comparison.

The FOIL Method	**Horizontal Multiplication Using the Distributive Property**
First Last $(ax + b)(cx + d) = ax \cdot cx + ax \cdot d + b \cdot cx + b \cdot d$ Inner Outer	$(ax + b)(cx + d) = (ax + b)cx + (ax + b)d$ $= ax \cdot cx + b \cdot cx + ax \cdot d + b \cdot d$ Rearrange terms: $= ax \cdot cx + ax \cdot d + b \cdot cx + b \cdot d$

EXAMPLE 6 **Using the FOIL Method to Multiply Two Binomials**

Find the product:

(a) $(y + 3)(y + 5)$ (b) $(a + 1)(a - 3)$

Solution

(a)
$$(y + 3)(y + 5) = (y)(y) + (y)(5) + (3)(y) + (3)(5)$$
$$= y^2 + 5y + 3y + 15$$
$$= y^2 + 8y + 15$$

(b)
$$(a + 1)(a - 3) = (a)(a) + (a)(-3) + (1)(a) + (1)(-3)$$
$$= a^2 - 3a + a - 3$$
$$= a^2 - 2a - 3$$

●

Quick ✓

7. FOIL stands for ____, ____, ____, ____.

In Problems 8–10, find the product using the FOIL method.

8. $(x + 3)(x + 4)$ **9.** $(y + 5)(y - 3)$ **10.** $(a - 1)(a - 5)$

EXAMPLE 7 **Using the FOIL Method to Multiply Two Binomials**

Find the product:

(a) $(3a + 1)(a - 4)$ (b) $(3a - 4b)(5a - 2b)$

Solution

(a)
$$(3a + 1)(a - 4) = (3a)(a) + (3a)(-4) + (1)(a) + (1)(-4)$$
$$= 3a^2 - 12a + a - 4$$
$$= 3a^2 - 11a - 4$$

(b)

$$
\begin{array}{cccc}
 & \text{F} & \text{O} & \text{I} & \text{L} \\
(3a - 4b)(5a - 2b) = (3a)(5a) + (3a)(-2b) - (4b)(5a) - (4b)(-2b)
\end{array}
$$
$$
= 15a^2 - 6ab - 20ab + 8b^2
$$
$$
= 15a^2 - 26ab + 8b^2 \qquad \bullet
$$

Quick ✓

In Problems 11–14, find the product using the FOIL method.

11. $(2x + 3)(x + 4)$ **12.** $(3y + 5)(2y - 3)$

13. $(2a - 1)(3a - 4)$ **14.** $(5x + 2y)(3x - 4y)$

▶ ❹ Multiply the Sum and Difference of Two Terms

Certain binomials have products that result in patterns. For this reason, these products are called **special products**.

EXAMPLE 8   ### Finding a Product of the Form $(A - B)(A + B)$

Find the product: $(x - 5)(x + 5)$

Solution

Use FOIL to multiply the binomials:

$$
\begin{array}{cccc}
 & \text{F} & \text{O} & \text{I} & \text{L} \\
(x - 5)(x + 5) = (x)(x) + (x)(5) - (5)(x) - (5)(5)
\end{array}
$$
$$
= x^2 + 5x - 5x - 25
$$
$$
= x^2 - 25 \qquad \bullet
$$

Products of the form $(A - B)(A + B)$ are called "the sum and difference of two terms." Do you see why?

In Example 8, did you notice that the outer product, $5x$, and the inner product, $-5x$, were opposites? The sum of the outer and inner products of two binomials in the form $(A - B)(A + B)$ is *always* zero, so the product $(A - B)(A + B)$ is the *difference* $A^2 - B^2$.

> **Product of the Sum and Difference of Two Terms**
>
> $$(A - B)(A + B) = A^2 - B^2$$

EXAMPLE 9   ### Finding the Product of the Sum and Difference of Two Terms

Find each product:

    (a) $(2x + 5)(2x - 5)$ **(b)** $(4x - 3y)(4x + 3y)$

Solution

$$
\begin{array}{c}
(A + B) \quad (A - B) \ = \ A^2 \ - B^2 \\
\textbf{(a)} \ (2x + 5)(2x - 5) = (2x)^2 - 5^2 \\
= 4x^2 - 25
\end{array}
\qquad
\begin{array}{c}
(A - B) \quad (A + B) \ = \ A^2 \ - B^2 \\
\textbf{(b)} \ (4x - 3y)(4x + 3y) = (4x)^2 - (3y)^2 \\
= 16x^2 - 9y^2 \quad \bullet
\end{array}
$$

Quick ✓

15. $(A + B)(A - B) = \underline{\hspace{2cm}}.$

In Problems 16–18, find each product.

16. $(a - 4)(a + 4)$ **17.** $(3w + 7)(3w - 7)$ **18.** $(x - 2y^3)(x + 2y^3)$

▶ ⑤ **Square a Binomial**

Another special product is called the square of a binomial.

$$\overset{\color{magenta}{\text{F} \quad \text{O} \quad \text{I} \quad \text{L}}}{(x + 3)^2 = (x + 3)(x + 3) = x^2 + 3x + 3x + 9}$$
$$= x^2 + 6x + 9$$

Did you notice that the outer product and the inner product are the same, namely $3x$? Study Table 2 to discover the pattern.

Table 2

$(n + 5)^2 =$	$(n + 5)(n + 5) =$	$n^2 + 5n + 5n + 25 =$	$n^2 + 2(5n) + 25 =$	$n^2 + 10n + 25$
$(2a - 3)^2 =$	$(2a - 3)(2a - 3) =$	$4a^2 - 6a - 6a + 9 =$	$4a^2 - 2(6a) + 9 =$	$4a^2 - 12a + 9$
$(n + 5p)^2 =$		$n^2 + 5np + 5np + 25p^2 =$	$n^2 + 2(5np) + 25p^2 =$	$n^2 + 10np + 25p^2$
$(4b - 9)^2 =$			$16b^2 - 2(4b)(9) + 81 =$	$16b^2 - 72b + 81$

The results of Table 2 lead to the following.

$$(A + B)^2 = \underbrace{A^2}_{\color{red}{\text{(first term)}^2}} + \underbrace{2AB}_{\color{red}{\text{2(product of terms)}}} + \underbrace{B^2}_{\color{red}{\text{(second term)}^2}}$$

$$(A - B)^2 = \underbrace{A^2}_{\color{red}{\text{(first term)}^2}} - \underbrace{2AB}_{\color{red}{\text{2(product of terms)}}} + \underbrace{B^2}_{\color{red}{\text{(second term)}^2}}$$

Work Smart

$(A + B)^2 \neq A^2 + B^2$

$(A - B)^2 \neq A^2 - B^2$

Whenever you feel the urge to perform an operation that you're not quite sure about, try it with actual numbers. For example, does

$(3 + 2)^2 = 3^2 + 2^2$?

NO! So

$(x + y)^2 \neq x^2 + y^2$

Squares of Binomials

$$(A + B)^2 = A^2 + 2AB + B^2$$
$$(A - B)^2 = A^2 - 2AB + B^2$$

We call $A^2 + 2AB + B^2$ and $A^2 - 2AB + B^2$ **perfect square trinomials.**

EXAMPLE 10   **How to Find a Product of the Form $(A + B)^2$**

Find the product: $(y + 7)^2$

Step-by-Step Solution

Step 1: Because $(y + 7)^2$ is in the form $(A + B)^2$, use $(A + B)^2 = A^2 + 2AB + B^2$ with $A = y$ and $B = 7$.

$$(A + B)^2 = A^2 + 2AB + B^2$$
$$(y + 7)^2 = (y)^2 + 2(y)(7) + 7^2$$

Step 2: Simplify.

$$= y^2 + 14y + 49 \qquad \bullet$$

EXAMPLE 11   **Finding a Product of the Form $(A + B)^2$**

Find each product:

(a) $(3n + 7)^2$ **(b)** $(4 + 5k)^2$ **(c)** $(5a + 3b)^2$

Solution

(a) The first term, A, is $3n$ and the second term, B, is 7.

$$(A + B)^2 = A^2 + 2AB + B^2$$
$$(3n + 7)^2 = (3n)^2 + 2(3n)(7) + 7^2$$

Square the first term   2 times product of first and second terms   Square the second term

$$= 9n^2 + 42n + 49$$

Thus, $(3n + 7)^2 = 9n^2 + 42n + 49$.

(b)
$$(A + B)^2 = A^2 + 2AB + B^2$$
$$(4 + 5k)^2 = 4^2 + 2(4)(5k) + (5k)^2$$
$$= 16 + 40k + 25k^2$$

(c)
$$(A + B)^2 = A^2 + 2AB + B^2$$
$$(5a + 3b)^2 = (5a)^2 + 2(5a)(3b) + (3b)^2$$
$$= 25a^2 + 30ab + 9b^2$$

EXAMPLE 12   **Finding a Product of the Form $(A - B)^2$**

Find each product:

(a) $(9p - 1)^2$ (b) $(3 - r)^2$ (c) $(2x - 5y)^2$

Solution

(a) The first term, A, is $9p$ and the second term, B, is 1.

$$(A - B)^2 = A^2 - 2AB + B^2$$
$$(9p - 1)^2 = (9p)^2 - 2(9p)(1) + 1^2$$

Square the first term   2 times product of first and second terms   Square the second term

$$= 81p^2 - 18p + 1$$

So, $(9p - 1)^2 = 81p^2 - 18p + 1$.

(b)
$$(A - B)^2 = A^2 - 2AB + B^2$$
$$(3 - r)^2 = 3^2 - 2(3)(r) + r^2$$
$$= 9 - 6r + r^2$$

Work Smart

If you can't remember the formulas for a perfect square, don't panic! Simply use the fact that $(x + a)^2 = (x + a)(x + a)$ and then use FOIL. The same logic applies to perfect squares of the form $(x - a)^2$.

(c)
$$(A - B)^2 = A^2 - 2AB + B^2$$
$$(2x - 5y)^2 = (2x)^2 - 2(2x)(5y) + (5y)^2$$
$$= 4x^2 - 20xy + 25y^2$$

Quick ✓

19. $(A - B)^2 = $ _____ ; $(A + B)^2 = $ _____.

20. $x^2 + 2xy + y^2$ is referred to as a _____ _____ _____.

21. *True or False* The product of a binomial and a binomial is always a trinomial.

In Problems 22–27, find each product. Write the answer in standard form.

22. $(z - 9)^2$ **23.** $(p + 1)^2$

24. $(4 - a)^2$ **25.** $(3z - 4)^2$

26. $(5p + 1)^2$ **27.** $(2w + 7y)^2$

For convenience, a summary of binomial products is shown below.

Summary of Binomial Products

- When multiplying two binomials, the Distributive Property can be used.

$$(3x + 5)(4x - 1) = (3x + 5)(4x) + (3x + 5)(-1)$$
$$= 12x^2 + 20x - 3x - 5$$
$$= 12x^2 + 17x - 5$$

- When multiplying two *binomials*, the FOIL pattern can be used.

$$(3x + 5)(4x - 1) = 3x(4x) + 3x(-1) + 5(4x) + 5(-1)$$
$$= 12x^2 - 3x + 20x - 5$$
$$= 12x^2 + 17x - 5$$

- The product of the sum and difference of two terms is $(A - B)(A + B) = A^2 - B^2$.

$$(7a + 2b)(7a - 2b) = (7a)^2 - (2b)^2$$
$$= 49a^2 - 4b^2$$

- The square of a binomial is
$(A + B)^2 = A^2 + 2AB + B^2$ or
$(A - B)^2 = A^2 - 2AB + B^2$.
The product is called a perfect square trinomial.

$$(3x + 5)^2 = (3x)^2 + 2(3x)(5) + (5)^2$$
$$= 9x^2 + 30x + 25$$
$$(4y - 3)^2 = (4y)^2 - 2(4y)(3) + (3)^2$$
$$= 16y^2 - 24y + 9$$

▶ ❻ Multiply a Polynomial by a Polynomial

To find the product of two polynomials, make repeated use of the Extended Form of the Distributive Property. Either a horizontal or a vertical format can be used.

EXAMPLE 13 **Multiplying Polynomials Using Horizontal Multiplication**

Find the product $(2x + 3)(x^2 + 5x - 1)$ using

(a) horizontal multiplication and **(b)** vertical multiplication.

Solution

(a) Distribute $2x + 3$ to each term in the trinomial.

$$(2x + 3)(x^2 + 5x - 1) = (2x + 3) \cdot x^2 + (2x + 3) \cdot 5x + (2x + 3) \cdot (-1)$$
$$= 2x(x^2) + 3(x^2) + 2x(5x) + 3(5x) + 2x(-1) + 3(-1)$$

Simplify: $= 2x^3 + 3x^2 + 10x^2 + 15x - 2x - 3$

Combine like terms: $= 2x^3 + 13x^2 + 13x - 3$

(b) Place the polynomial with more terms on top, align terms of the same degree, and then multiply. Make sure both polynomials are written in standard form.
Now multiply the trinomial by the 3 in the binomial, and then multiply the trinomial by the $2x$ in the binomial.

$$
\begin{array}{r}
x^2 + 5x - 1 \\
\times \quad 2x + 3 \\
\hline
\end{array}
$$

$3(x^2 + 5x - 1)$: $\quad 3x^2 + 15x - 3$

$2x(x^2 + 5x - 1)$: $\quad 2x^3 + 10x^2 - 2x$

Add vertically: $\quad 2x^3 + 13x^2 + 13x - 3$

Quick ✓

28. *True or False* The product $(x - y)(x^2 + 2xy + y^2)$ can be found using the FOIL method.

In Problems 29 and 30, find the product using horizontal multiplication.

29. $(x - 2)(x^2 + 2x + 4)$

30. $(3y - 2)(y^2 + 2y + 4)$

In Problems 31 and 32, find the product using vertical multiplication.

31. $(x - 2)(x^2 + 2x + 4)$

32. $(3y - 1)(y^2 + 2y - 5)$

EXAMPLE 14 **Multiplying Three Polynomials**

Find the product: $2x(x - 4)(3x - 5)$

Solution

To find the product of three polynomials, multiply any two factors, and then multiply that product by the remaining factor. Start by using the Distributive Property to multiply $2x$ by $(x - 4)$.

$$2x(x - 4)(3x - 5) = (2x^2 - 8x)(3x - 5)$$

FOIL: $= 2x^2 \cdot 3x + 2x^2(-5) - 8x \cdot 3x - 8x(-5)$

Multiply: $= 6x^3 - 10x^2 - 24x^2 + 40x$

Combine like terms: $= 6x^3 - 34x^2 + 40x$ ●

Quick ✓

In Problems 33 and 34, find the product.

33. $-2a(4a - 1)(3a + 5)$

34. $(x + 2)(x - 1)(x + 3)$

5.3 Exercises MyMathLab®

Exercise numbers in green have complete video solutions in MyMathLab or may be accessed using the QR code to the right.

Problems 1–34 are the Quick ✓ s that follow the EXAMPLES.

Building Skills

In Problems 35–42, use the Distributive Property to find each product. See Objective 1.

35. $2x(3x - 5)$

36. $3m(2m - 7)$

37. $\frac{1}{2}n(4n - 6)$

38. $\frac{3}{5}b(15b - 5)$

39. $3n^2(4n^2 + 2n - 5)$

40. $4w(2w^2 + 3w - 5)$

41. $(4x^2y - 3xy^2)(x^2y)$

42. $(7r + 3s^2)(2r^2s)$

In Problems 43–48, use the Distributive Property to find each product. See Objective 2.

43. $(x + 5)(x + 7)$

44. $(x + 4)(x + 10)$

45. $(y - 5)(y + 7)$

46. $(n - 7)(n + 4)$

47. $(3m - 2y)(2m + 5y)$

48. $(5n - 2y)(2n - 3y)$

In Problems 49–62, find the product using the FOIL method. See Objective 3.

49. $(x + 2)(x + 3)$

50. $(x + 3)(x + 7)$

51. $(q - 6)(q - 7)$

52. $(n - 4)(n - 5)$

53. $(2x + 3)(3x - 1)$

54. $(3z - 2)(4z + 1)$

55. $(x^2 + 3)(x^2 + 1)$

56. $(x^2 - 5)(x^2 - 2)$

57. $(7 - x)(6 - x)$

58. $(2 - y)(4 - y)$

59. $(5u + 6v)(2u + v)$

60. $(3a - 2b)(a - 3b)$

61. $(2a - b)(5a + 2b)$

62. $(3r + 5s)(6r + 7s)$

In Problems 63–72, find the product of the sum and difference of two terms. See Objective 4.

63. $(x - 3)(x + 3)$

64. $(y + 7)(y - 7)$

65. $(2z + 5)(2z - 5)$

66. $(6r - 1)(6r + 1)$

67. $(4x^2 + 1)(4x^2 - 1)$

68. $(3a^2 + 2)(3a^2 - 2)$

69. $(2x - 3y)(2x + 3y)$

70. $(8a - 5b)(8a + 5b)$

71. $\left(x - \dfrac{1}{3}\right)\left(x + \dfrac{1}{3}\right)$

72. $\left(y + \dfrac{2}{9}\right)\left(y - \dfrac{2}{9}\right)$

In Problems 73–82, find the product. See Objective 5.

73. $(x - 2)^2$

74. $(x + 4)^2$

75. $(5k - 3)^2$

76. $(6b - 5)^2$

77. $(x + 2y)^2$

78. $(3x - 2y)^2$

79. $(2a - 3b)^2$

80. $(5x + 2y)^2$

81. $\left(x + \dfrac{1}{2}\right)^2$

82. $\left(y - \dfrac{1}{3}\right)^2$

In Problems 83–96, find the product. See Objective 6.

83. $(x - 2)(x^2 + 3x + 1)$

84. $(3a - 1)(2a^2 - 5a - 3)$

85. $(2y^2 - 6y + 1)(y - 3)$

86. $(2m^2 - m + 2)(2m + 1)$

87. $(2x - 3)(x^2 - 2x - 1)$

88. $(4y + 1)(2y^2 - y - 3)$

89. $2b(b - 3)(b + 4)$

90. $5a(a + 6)(a - 1)$

91. $-\dfrac{1}{2}x(2x + 6)(x - 3)$

92. $-\dfrac{4}{3}k(k + 7)(3k - 9)$

93. $(5y^3 - y^2 + 2)(2y^2 + y + 1)$

94. $(2m^2 - m + 4)(-m^3 - 2m - 1)$

95. $(b + 1)(b - 2)(b + 3)$

96. $(2a - 1)(a + 4)(a + 1)$

Mixed Practice

In Problems 97–128, perform the indicated operation(s). Write the answer as a polynomial in standard form.

97. $(x^2 - 5)(x^2 + 5)$

98. $(b^3 - 2)(b^3 + 2)$

99. $(z^2 + 9)(z + 3)(z - 3)$

100. $(y^2 + 4)(y + 2)(y - 2)$

101. $4x\left(\dfrac{1}{2}x + 3\right)\left(\dfrac{1}{2}x - 3\right) - x(x + 1)^2$

102. $25p\left(\dfrac{2}{5}p - 1\right)\left(\dfrac{2}{5}p + 1\right) - 4p(p + 2)^2$

103. $(2x^3 + 3)^2$

104. $(3a^4 - 2)^2$

105. $(x^2 + 1)(x^4 - 3)$

106. $(4y^2 - 5)(y^3 + 2)$

107. $(x + 2)(2x^2 - 3x - 1)$

108. $(x - 3)(3x^2 - x - 4)$

109. $-\dfrac{1}{2}x^2(10x^5 - 6x^4 + 12x^3) + (x^3)^2$

110. $-\dfrac{1}{3}(27y^2 - 9y + 6) - (3y)^2$

111. $7x^2(x + 3) - 2x(x^2 - 1)$

112. $2(3a^4 + 2b^4) - 3(b^4 - 2a^4)$

113. $-3w(w - 4)(w + 3)$

114. $5y(y + 5)(y - 3)$

115. $3a(a + 4)^2$

116. $2m(m - 3)^2$

117. $(n + 3)(n - 3) + (n + 3)^2$

118. $(s + 6)(s - 6) + (s - 6)^2$

119. $(a + 6b)^2 - (a - 6b)^2$

120. $(2a + 5b)^2 - (2a - 5b)^2$

121. $(x + 1)^2 - (2x + 1)(x - 1)$

122. $(a + 3)^2 - (a + 4)(3a - 1)$

123. Square $2x + 1$.   **124.** Square $3x - 2y$.

125. Find the cube of $x - 1$.

126. Find the cube of $2a + b$.

127. Subtract $x - 6$ from the product of $x + 3$ and $2x - 5$.

128. Add $2x + 3$ to the product of $x + 3$ and $2x - 3$.

Applying the Concepts

In Problems 129–132, find an algebraic expression that represents the area of the shaded region.

△ **129.**

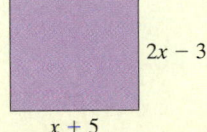

$2x - 3$

$x + 5$

△ **130.**

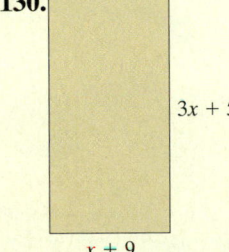

$3x + 5$

$x + 9$

△ **131.**

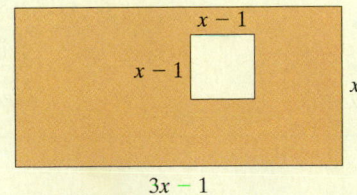

$x - 1$

$x - 1$

x

$3x - 1$

△ **132.**

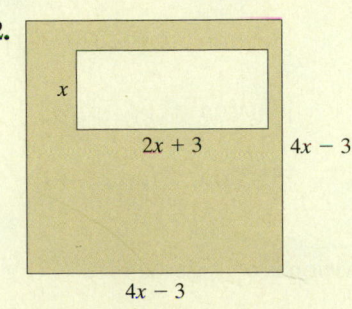

x

$2x + 3$

$4x - 3$

$4x - 3$

In Problems 133 and 134, find an algebraic expression that represents the volume of the figure.

△ **133.**

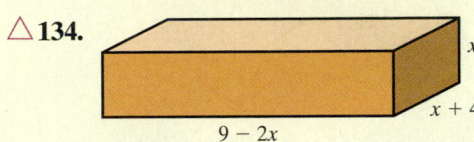

x

$x + 1$

$4x - 3$

△ **134.**

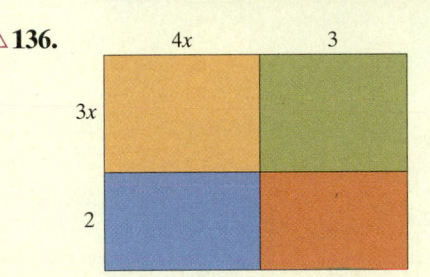

x

$x + 4$

$9 - 2x$

In Problems 135 and 136, find an algebraic expression for the area of the rectangle by finding the sum of the areas of the four interior rectangles. Then find the area of the rectangle by multiplying the width and length. Compare the two expressions. How is multiplying two binomials related to finding the area of the rectangle?

△ **135.**

x 10

x

2

△ **136.**

$4x$ 3

$3x$

2

137. Consecutive Integers If x represents the first of three consecutive integers, write a polynomial that represents the product of the next two consecutive integers.

138. Consecutive Odd Integers If x represents the first of three consecutive odd integers, write a polynomial that represents the product of the first and the third integers.

△ **139. Area of a Triangle** The length of the base of a triangle is 2 inches (in.) shorter than twice the length of its altitude, x. Write a polynomial that represents the area of the triangle.

△ **140. Area of a Triangle** The length of the base of a triangle is 3 feet (ft) greater in length than its altitude, x. Write a polynomial that represents the area of the triangle.

△ **141. Area of a Circle** Write a polynomial for the area of a circle with radius $(x + 2)$ feet.

△ **142. Area of a Circle** Write a polynomial for the area of a circle with radius $(2y - 3)$ meters.

△143. **Perfect Square** Why is the expression $(a + b)^2$ called a perfect square? Consider the figure below.

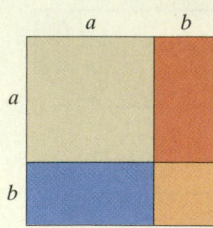

(a) Find the area of each of the four quadrilaterals.

(b) Use the result from part (a) to find the area of the entire region.

(c) Find the length and width of the entire region in terms of a and b. Use this result to find the area of the entire region. What do you notice?

△144. **Picture Frame** A photo is 13″ by 15″ is to be framed with a mat border that is x inches wide. Write the polynomial that calculates the area of the entire framed picture.

Extending the Concepts

In Problems 145–152, find the product. Write each result as a polynomial in standard form.

145. $(x - 2)(x + 2)^2$ **146.** $(x + 3)^2(x - 3)$

147. $[3 - (x + y)][3 + (x + y)]$

148. $[(x - y) + 3][(x - y) - 3]$

149. $(x + 2)^3$ **150.** $(y - 3)^3$

151. $(z + 3)^4$ **152.** $(m - 2)^4$

Explaining the Concepts

153. Explain the steps that should be followed to find the product $(3a - 5)^2$ to someone who missed the class session on squaring a binomial.

154. Explain how the product of the sum and difference of two terms can be used to calculate $19 \cdot 21$.

155. Explain when the FOIL method can be used to multiply polynomials.

156. Find and explain the error in this student's work: $2a(a^2 + 3a) + 5a = 2a(a^2) + 2a(3a) + 2a(5a)$.

5.4 Dividing Monomials: The Quotient Rule and Integer Exponents

Objectives

❶ Simplify Exponential Expressions Using the Quotient Rule

❷ Simplify Exponential Expressions Using the Quotient to a Power Rule

❸ Simplify Exponential Expressions Using Zero as an Exponent

❹ Simplify Exponential Expressions Involving Negative Exponents

❺ Simplify Exponential Expressions Using the Laws of Exponents

Are You Prepared for This Section?

Before getting started, complete the following problems. If you get a problem wrong, go back to the section cited and review the material.

P1. Evaluate: $(3a^2)^3$ [Section 5.2, pp. 320–321]

P2. Evaluate: $\left(\dfrac{2}{3}\right)^2$ [Section 1.7, pp. 56–57]

P3. Find the reciprocal: **(a)** 5   **(b)** $-\dfrac{6}{7}$ [Section 1.4, pp. 33–34]

The next two sections deal with division. Monomial division is considered first.

▶ ❶ **Simplify Exponential Expressions Using the Quotient Rule**

Observe the following:

$$\frac{y^6}{y^2} = \frac{\overbrace{y \cdot y \cdot y \cdot y \cdot y \cdot y}^{6 \text{ factors}}}{\underbrace{y \cdot y}_{2 \text{ factors}}} = \underbrace{y \cdot y \cdot y \cdot y}_{4 \text{ factors}} = y^4$$

This result suggests that $\frac{y^6}{y^2} = y^{6-2} = y^4$, which can be generalized in the rule below.

> **In Other Words**
> When dividing two exponential expressions with a common base, subtract the exponent in the denominator from the exponent in the numerator. Then write the base to that power.

Quotient Rule for Exponents

If a is a real number, and if m and n are natural numbers, then

$$\frac{a^m}{a^n} = a^{m-n} \quad \text{if } a \neq 0$$

EXAMPLE 1 **Using the Quotient Rule to Simplify Expressions**

Simplify each expression.

(a) $\dfrac{6^5}{6^3}$ **(b)** $\dfrac{n^6}{n^2}, n \neq 0$ **(c)** $\dfrac{(-7)^8}{(-7)^6}$

Solution

(a)
$$\frac{6^5}{6^3} = 6^{5-3}$$
$$= 6^2$$
Evaluate 6^2: $= 36$

(b)
$$\frac{n^6}{n^2} = n^{6-2}$$
$$= n^4$$

(c)
$$\frac{(-7)^8}{(-7)^6} = (-7)^{8-6}$$
$$= (-7)^2$$
$$= 49 \quad ●$$

EXAMPLE 2 **Using the Quotient Rule to Simplify Expressions**

Simplify each expression. All variables are nonzero.

(a) $\dfrac{25b^9}{15b^6}$ **(b)** $\dfrac{-24x^6y^4}{10x^4y}$

Solution

(a)
$$\frac{25b^9}{15b^6} = \frac{25}{15} \cdot \frac{b^9}{b^6}$$

Divide out common factors;

$$\frac{a^m}{a^n} = a^{m-n}: \quad = \frac{\cancel{5} \cdot 5}{\cancel{5} \cdot 3} \cdot b^{9-6}$$

$$= \frac{5}{3}b^3$$

(b)
$$\frac{-24x^6y^4}{10x^4y} = \frac{-24}{10} \cdot \frac{x^6}{x^4} \cdot \frac{y^4}{y}$$

Divide out common factors;

$$\frac{a^m}{a^n} = a^{m-n}: \quad = \frac{-12 \cdot \cancel{2}}{5 \cdot \cancel{2}} \cdot x^{6-4} \cdot y^{4-1}$$

$$= -\frac{12}{5}x^2y^3 \quad ●$$

1. $\dfrac{a^m}{a^n} = $ _____ provided that $a \neq 0$. 　　2. *True or False* $\dfrac{6^{10}}{6^4} = 1^6$.

In Problems 3–5, simplify each expression. All variables are nonzero.

3. $\dfrac{3^7}{3^5}$ 　　　　　 4. $\dfrac{14c^6}{10c^5}$ 　　　　　 5. $\dfrac{-21w^4z^8}{14w^3z}$

▶ ❷ Simplify Exponential Expressions Using the Quotient to a Power Rule

Now consider a quotient raised to a power:

$$\left(\frac{3}{2}\right)^4 = \left(\frac{3}{2}\right) \cdot \left(\frac{3}{2}\right) \cdot \left(\frac{3}{2}\right) \cdot \left(\frac{3}{2}\right) = \overbrace{\frac{3 \cdot 3 \cdot 3 \cdot 3}{2 \cdot 2 \cdot 2 \cdot 2}}^{\text{4 factors}} = \frac{3^4}{2^4}$$

The following property reflects this result.

In Other Words

When a quotient is raised to a power, apply the Exponent Rule to both the numerator.

> **Quotient to a Power Rule for Exponents**
>
> If a and b are real numbers and n is a natural number, then
>
> $$\left(\frac{a}{b}\right)^n = \frac{a^n}{b^n} \quad \text{if } b \neq 0$$

EXAMPLE 3 　**Using the Quotient to a Power Rule to Simplify an Expression**

Simplify the expression: $\left(\dfrac{z}{3}\right)^2$

Solution

This expression is a quotient, so apply the Quotient to a Power Rule. That is, apply the exponent to both the numerator and the denominator.

$$\left(\frac{z}{3}\right)^2 = \frac{z^2}{3^2}$$

Evaluate 3^2: 　$= \dfrac{z^2}{9}$ 　●

EXAMPLE 4 　**Using the Quotient to a Power Rule to Simplify Expressions**

Simplify each expression. All variables are nonzero.

(a) $\left(-\dfrac{2y}{z}\right)^3$ 　　　　　　　　(b) $\left(\dfrac{3a^2}{b^3}\right)^4$

Solution　$-\dfrac{a}{b} = \dfrac{-a}{b}$ 　　　　　　　　　$\left(\dfrac{a}{b}\right)^n = \dfrac{a^n}{b^n}$

(a) 　$\left(-\dfrac{2y}{z}\right)^3 \overset{\downarrow}{=} \left(\dfrac{-2y}{z}\right)^3$ 　　　(b) 　$\left(\dfrac{3a^2}{b^3}\right)^4 \overset{\downarrow}{=} \dfrac{(3a^2)^4}{(b^3)^4}$

$\left(\dfrac{a}{b}\right)^n = \dfrac{a^n}{b^n}$: 　$= \dfrac{(-2y)^3}{z^3}$ 　　　　　$(ab)^n = a^n b^n$: 　$= \dfrac{3^4(a^2)^4}{(b^3)^4}$

$(ab)^n = a^n b^n$: 　$= \dfrac{(-2)^3(y)^3}{z^3}$ 　　　Evaluate 3^4; 　$\dfrac{81a^{2 \cdot 4}}{b^{3 \cdot 4}}$

　　　　　　　　　　　　　　　　$(a^m)^n = a^{m \cdot n}$: 　$= \dfrac{81a^{2 \cdot 4}}{b^{3 \cdot 4}}$

Evaluate $(-2)^3$: 　$= -\dfrac{8y^3}{z^3}$ 　　　　　　　　$= \dfrac{81a^8}{b^{12}}$ 　●

Quick ✔

6. $\left(\dfrac{a}{b}\right)^n = \underline{\quad}$ provided that $\underline{\quad\quad}$.

In Problems 7–9, simplify each expression. All variables are nonzero.

7. $\left(\dfrac{p}{2}\right)^4$ **8.** $\left(-\dfrac{2a^2}{b^4}\right)^3$ **9.** $\left(\dfrac{-3m}{n^3}\right)^4$

▶ ❸ **Simplify Exponential Expressions Using Zero as an Exponent**

The definition of exponential expressions is now extended to integer exponents, beginning with raising a real number to the 0 power.

> **Definition of Zero as an Exponent**
>
> If a is a nonzero real number (that is, $a \neq 0$),
>
> $$a^0 = 1$$

The reason why $a^0 = 1$ is based on the Quotient Rule for Exponents.

$$\dfrac{a^n}{a^n} = a^{n-n} \quad \text{and} \quad \dfrac{a^n}{a^n} = \dfrac{a \cdot a \cdot a \cdot a \ldots \cdot a}{a \cdot a \cdot a \cdot a \ldots \cdot a}$$

$$= a^0$$

$$= \dfrac{\overbrace{\acute{a} \cdot \acute{a} \cdot \acute{a} \cdot \acute{a} \ldots \cdot \acute{a}}^{n \text{ factors}}}{\underbrace{\acute{a} \cdot \acute{a} \cdot \acute{a} \cdot \acute{a} \ldots \cdot \acute{a}}_{n \text{ factors}}}$$

$$= 1$$

Because $\dfrac{a^n}{a^n} = a^0$ and $\dfrac{a^n}{a^n} = 1$, therefore $a^0 = 1$.

EXAMPLE 5 **Using Zero as an Exponent**

In the following examples, all variables are nonzero.

(a) $3^0 = 1$ **(b)** $x^0 = 1$ **(c)** $(4y)^0 = 1$

(d) $8w^0 = 8 \cdot 1$ **(e)** $-5^0 = -1 \cdot 5^0$ **(f)** $(-4y)^0 = 1$
$\qquad\quad = 8$ $\qquad\qquad = -1$ ●

Work Smart

In Example 5(d), the exponent 0 applies *only* to the factor w because $8w^0 = 8 \cdot w^0$.

Quick ✔

10. If a is a nonzero real number, then $a^0 = \underline{\quad}$.

In Problems 11 and 12, simplify each expression. Assume variables are nonzero.

11. (a) $10^0 = \underline{\quad}$ **(b)** $-10^0 = \underline{\quad\quad}$ **(c)** $(-10)^0 = \underline{\quad}$

12. (a) $(2b)^0 = \underline{\quad}$ **(b)** $2b^0 = \underline{\quad}$ **(c)** $(-2b)^0 = \underline{\quad}$

▶ ❹ **Simplify Exponential Expressions Involving Negative Exponents**

Now let's look at exponents that are negative integers. Suppose that we wanted to simplify $\dfrac{z^3}{z^5}$. If the Quotient Rule for Exponents is used, then

$$\dfrac{z^3}{z^5} = z^{3-5} = z^{-2}$$

This expression could also be simplified directly by dividing common factors:

$$\frac{z^3}{z^5} = \frac{\cancel{z} \cdot \cancel{z} \cdot \cancel{z}}{\cancel{z} \cdot \cancel{z} \cdot \cancel{z} \cdot z \cdot z} = \frac{1}{z^2}$$

This implies that $z^{-2} = \dfrac{1}{z^2}$, which leads to the following result:

> **Definition of Negative Exponent**
>
> If n is a positive integer and if a is a nonzero real number (that is, $a \neq 0$), then
>
> $$a^{-n} = \frac{1}{a^n}$$

Remember, the reciprocal of a nonzero number a is $\dfrac{1}{a}$. For example, the reciprocal of 7 is $\dfrac{1}{7}$ and the reciprocal of $-\dfrac{5}{8}$ is $-\dfrac{8}{5}$. Whenever a negative exponent is encountered, think, "take the reciprocal of the base."

EXAMPLE 6 **Simplifying Exponential Expressions Containing Negative Integer Exponents**

Simplify:

(a) $2^{-3} = \dfrac{1}{2^3}$

$\qquad\; = \dfrac{1}{8}$

(b) $(-5)^{-2} = \dfrac{1}{(-5)^2}$

$\qquad\qquad\;\; = \dfrac{1}{25}$

(c) $-4^{-2} = -1 \cdot 4^{-2}$

$\qquad\quad = \dfrac{-1}{4^2}$

$\qquad\quad = \dfrac{-1}{16}$

$\qquad\quad = -\dfrac{1}{16}$ ●

EXAMPLE 7 **Simplifying the Sum of Exponential Expressions Containing Negative Integer Exponents**

Simplify: $2^{-2} + 4^{-1}$

Solution

$$2^{-2} + 4^{-1} = \frac{1}{2^2} + \frac{1}{4^1}$$

Evaluate: $\qquad\qquad = \dfrac{1}{4} + \dfrac{1}{4}$

Find the sum: $\qquad\quad = \dfrac{2}{4}$

Write in lowest terms: $\quad = \dfrac{1}{2}$ ●

Work Smart

Remember, $2^{-3} = \dfrac{1}{2^3}$, not -2^3.

The negative exponent means "take the reciprocal," not "take the opposite."

> **Quick ✓**
>
> **13.** If n is a positive integer and if a is a nonzero real number, then $a^{-n} = \underline{\quad}$.
>
> *In Problems 14 and 15, simplify each expression.*
>
> **14. (a)** 2^{-4} $\qquad$ **(b)** $(-2)^{-4}$ $\qquad$ **(c)** -2^{-4} $\qquad\qquad$ **15.** $4^{-1} - 2^{-3}$

EXAMPLE 8 **Simplifying Exponential Expressions Containing a Negative Integer Exponent**

Simplify each expression so that all exponents are positive integers. All variables are nonzero.

(a) x^{-4} **(b)** $(-x)^{-4}$ **(c)** $-x^{-4}$

Solution

(a) $x^{-4} = \dfrac{1}{x^4}$

(b) $(-x)^{-4} = \dfrac{1}{(-x)^4}$

$= \dfrac{1}{(-1 \cdot x)^4}$

$(ab)^n = a^n b^n$: $= \dfrac{1}{(-1)^4 x^4}$

$(-1)^4 = 1$: $= \dfrac{1}{x^4}$

(c) $-x^{-4} = -1 \cdot x^{-4}$

$= -1 \cdot \dfrac{1}{x^4}$

$= -\dfrac{1}{x^4}$

Quick ✓

In Problem 16, simplify each expression so that all exponents are positive integers. All variables are nonzero.

16. (a) y^{-8} **(b)** $(-y)^{-8}$ **(c)** $-y^{-8}$

EXAMPLE 9 **Simplifying Expressions Containing a Negative Integer Exponent**

Simplify each expression so that all exponents are positive integers. All variables are nonzero.

(a) $2a^{-3}$ **(b)** $(2a)^{-3}$ **(c)** $(-2a)^{-3}$

Work Smart

In Example 9(a), $2a^{-3} = 2 \cdot \dfrac{1}{a^3}$.

The negative exponent applies only to the base a, not to the coefficient 2. Compare Examples 9(a) and 9(b). Can you see the difference?

Solution

(a) $2a^{-3} = 2 \cdot \dfrac{1}{a^3}$

$= \dfrac{2}{a^3}$

(b) $(2a)^{-3} = \dfrac{1}{(2a)^3}$

$(ab)^n = a^n \cdot b^n$: $= \dfrac{1}{2^3 \cdot a^3}$

$= \dfrac{1}{8a^3}$

(c) $(-2a)^{-3} = \dfrac{1}{(-2a)^3}$

$(ab)^n = a^n \cdot b^n$: $= \dfrac{1}{(-2)^3 \cdot a^3}$

$= \dfrac{1}{-8a^3}$

$\dfrac{a}{-b} = -\dfrac{a}{b}$: $= -\dfrac{1}{8a^3}$

Quick ✓

In Problem 17, simplify each expression so that all exponents are positive integers. All variables are nonzero.

17. (a) $2m^{-5}$ **(b)** $(2m)^{-5}$ **(c)** $(-2m)^{-5}$

Because $\dfrac{1}{a^{-n}} = \dfrac{1}{\dfrac{1}{a^n}} = 1 \div \dfrac{1}{a^n} = 1 \cdot \dfrac{a^n}{1} = a^n$, we have the following result.

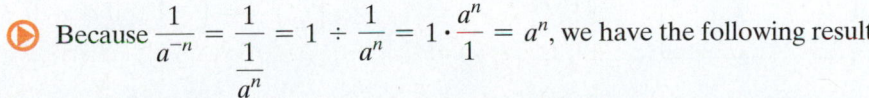

If a is a real number and n is an integer, then

$$\dfrac{1}{a^{-n}} = a^n \quad \text{if } a \neq 0$$

EXAMPLE 10 **Simplifying Quotients Containing Negative Integer Exponents**

Simplify each expression so that all exponents are positive integers. All variables are nonzero.

(a) $\dfrac{1}{2^{-3}}$ $\qquad$ (b) $\dfrac{7}{n^{-2}}$ $\qquad$ (c) $\dfrac{8}{3n^{-4}}$

Solution

(a) $\dfrac{1}{2^{-3}} = 2^3$ $\qquad$ (b) $\dfrac{7}{n^{-2}} = \dfrac{7n^2}{1}$ $\qquad$ (c) $\dfrac{8}{3n^{-4}} = \dfrac{8n^4}{3}$

$\quad = 8$ $\qquad\qquad\qquad = 7n^2$ $\qquad\qquad\qquad = \dfrac{8}{3}n^4$

Quick ✓

In Problems 18–20, simplify each expression so that all exponents are positive integers. Assume all variables are nonzero.

18. $\dfrac{1}{3^{-2}}$ $\qquad\qquad$ **19.** $\dfrac{1}{-10^{-2}}$ $\qquad\qquad$ **20.** $\dfrac{5}{2z^{-2}}$

EXAMPLE 11 **Simplifying Quotients Containing Negative Integer Exponents**

Simplify: $\left(\dfrac{3}{2}\right)^{-3}$

Solution $\qquad\qquad\qquad\qquad \left(\dfrac{3}{2}\right)^{-3} = \dfrac{1}{\left(\dfrac{3}{2}\right)^3}$

$\left(\dfrac{a}{b}\right)^n = \dfrac{a^n}{b^n}; \quad = \dfrac{1}{\dfrac{3^3}{2^3}}$

$\qquad\qquad\qquad\qquad\qquad = \dfrac{1}{\dfrac{27}{8}}$

$\qquad\qquad\qquad\qquad\qquad = 1 \cdot \dfrac{8}{27}$

$\qquad\qquad\qquad\qquad\qquad = \dfrac{8}{27}$

In Other Words

This shortcut says, "To simplify a quotient to a negative exponent, first take the reciprocal of the base, and then raise that expression to the positive exponent."

The following shortcut is based on the result of Example 11:

Quotient to a Negative Power

If a and b are real numbers and n is an integer, then

$$\left(\dfrac{a}{b}\right)^{-n} = \left(\dfrac{b}{a}\right)^n \quad \text{if } a \neq 0, b \neq 0$$

Using this rule on the problem in Example 11, we obtain $\left(\dfrac{3}{2}\right)^{-3} = \left(\dfrac{2}{3}\right)^3 = \dfrac{8}{27}$.

EXAMPLE 12 **Simplifying Quotients Containing Negative Exponents**

Simplify $\left(-\dfrac{x^3}{4}\right)^{-2}, x \neq 0$.

Solution

First, use the fact that $\left(\dfrac{a}{b}\right)^{-n} = \left(\dfrac{b}{a}\right)^{n}$ to rewrite $\left(\dfrac{-x^3}{4}\right)^{-2}$.

$$\left(-\frac{x^3}{4}\right)^{-2} = \left(-\frac{4}{x^3}\right)^{2}$$

$$\left(\frac{a}{b}\right)^{n} = \frac{a^n}{b^n}: \quad = \frac{(-4)^2}{(x^3)^2}$$

$$(a^m)^n = a^{m \cdot n}: \quad = \frac{16}{x^6}$$

Quick ✓

21. If a and b are real numbers and n is an integer, then $\left(\dfrac{a}{b}\right)^{-n} = $ _____.

In Problems 22–24, simplify each expression so that all exponents are positive integers. All variables are nonzero.

22. $\left(\dfrac{7}{8}\right)^{-1}$ **23.** $\left(\dfrac{3a}{5}\right)^{-3}$ **24.** $\left(-\dfrac{2}{3n^4}\right)^{-2}$

▶ ❺ Simplify Exponential Expressions Using the Laws of Exponents

Here is a summary of the Laws of Exponents where the exponents are integers.

The Laws of Exponents

If a and b are real numbers and if m and n are integers, then, assuming the expression is defined, the following rules apply.

Rule		Examples
Product Rule	$a^m \cdot a^n = a^{m+n}$	$7^3 \cdot 7 = 7^4; x^2 \cdot x^5 = x^7$
Power Rule	$(a^m)^n = a^{m \cdot n}$	$(3^4)^2 = 3^8; (x^3)^2 = x^6$
Product to Power Rule	$(a \cdot b)^n = a^n \cdot b^n$	$(x^3 y)^4 = x^{12} y^4$
Quotient Rule	$\dfrac{a^m}{a^n} = a^{m-n}$, if $a \neq 0$	$\dfrac{9^5}{9^3} = 9^2; \dfrac{y^{11}}{y^8} = y^3$
Quotient to Power Rule	$\left(\dfrac{a}{b}\right)^n = \dfrac{a^n}{b^n}$, if $b \neq 0$	$\left(\dfrac{3}{4}\right)^2 = \dfrac{9}{16}; \left(\dfrac{2x}{y^2}\right)^3 = \dfrac{8x^3}{y^6}$
Zero Exponent Rule	$a^0 = 1$, if $a \neq 0$	$5^0 = 1; (5m)^0 = 1$ $5m^0 = 5 \cdot 1 = 5$
Negative Exponent Rules	$a^{-n} = \dfrac{1}{a^n}$, if $a \neq 0$	$2^{-3} = \dfrac{1}{8}; b^{-4} = \dfrac{1}{b^4}$
	$\dfrac{1}{a^{-n}} = a^n$, if $a \neq 0$	$\dfrac{1}{t^{-5}} = t^5$
Quotient to a Negative Power Rule	$\left(\dfrac{a}{b}\right)^{-n} = \left(\dfrac{b}{a}\right)^{n}$, if $a \neq 0, b \neq 0$	$\left(\dfrac{5}{k}\right)^{-2} = \left(\dfrac{k}{5}\right)^{2} = \dfrac{k^2}{25}$

To evaluate or simplify exponential expressions in which one or more of the rules listed are used, ask yourself the following questions.

> ### Using the Laws of Exponents
>
> - Does the exponential expression contain numerical expressions? If so, *evaluate* them. ("Evaluate" means "find the value of.")
> - Is the expression a product of monomials? If so, use the Product Rule:
>
> $$a^m \cdot a^n = a^{m+n}$$
>
> - Is the expression a quotient (division problem)? If so, use the Quotient Rule:
>
> $$\frac{a^m}{a^n} = a^{m-n}$$
>
> - Do you see an expression raised to a power? If so, use one of the Power Rules:
>
> $$(a^m)^n = a^{m \cdot n}, \; (a \cdot b)^n = a^n \cdot b^n, \; \text{or} \; \left(\frac{a}{b}\right)^n = \frac{a^n}{b^n}.$$
>
> - Is there a negative exponent in the expression? A negative exponent means "take the reciprocal of the base." An expression is not simplified if it contains negative exponents.
> - Is an expression raised to the zero power? Remember, $a^0 = 1$ for $a \neq 0$.

EXAMPLE 13 **How to Simplify Exponential Expressions Using Exponent Rules**

Simplify $\left(\frac{3}{2}a^3b^{-2}\right)(-12a^{-4}b^5)$. Write the answer with only positive integer exponents. All variables are nonzero.

Step-by-Step Solution

Step 1: Rearrange factors.

$$\left(\frac{3}{2}a^3b^{-2}\right)(-12a^{-4}b^5) = \left(\frac{3}{2} \cdot (-12)\right)(a^3 \cdot a^{-4})(b^{-2} \cdot b^5)$$

Step 2: Find each product.

Evaluate $\left(\frac{3}{2}\right)(-12)$; $a^m \cdot a^n = a^{m+n}$: $\quad = -18a^{3+(-4)}b^{-2+5}$

$$= -18a^{-1}b^3$$

Step 3: Simplify. Write the product so that the exponents are positive integers.

$a^{-n} = \frac{1}{a^n}$: $\quad = -18 \cdot \frac{1}{a} \cdot b^3$

$$= -\frac{18b^3}{a}$$

EXAMPLE 14 **How to Simplify Expressions Using the Quotient Rule and the Negative Exponent Rule**

Simplify $-\frac{27a^7b^{-6}}{18a^{-2}b^2}$. Write the answer with only positive integer exponents. All variables are nonzero.

Step-by-Step Solution

Step 1: Write the quotient as the product of factors.

$$\frac{a}{b} = \frac{-a}{b}: \quad -\frac{27a^7b^{-6}}{18a^{-2}b^2} = \frac{-27}{18} \cdot \frac{a^7}{a^{-2}} \cdot \frac{b^{-6}}{b^2}$$

Step 2: Find each quotient.

$$\text{Simplify } \frac{-27}{18}; \frac{a^m}{a^n} = a^{m-n}: \quad = \frac{-3}{2} \cdot a^{7-(-2)} \cdot b^{-6-2}$$

$$= -\frac{3}{2} \cdot a^9 \cdot b^{-8}$$

Step 3: Simplify. Write the quotient so that the exponents are positive integers.

$$a^{-n} = \frac{1}{a^n}: \quad = -\frac{3}{2} \cdot a^9 \cdot \frac{1}{b^8}$$

$$= -\frac{3a^9}{2b^8}$$

So, $-\dfrac{27a^7b^{-6}}{18a^{-2}b^2} = -\dfrac{3a^9}{2b^8}$.

●

Quick ✔

In Problems 25–28, simplify each expression. Write the answers with only positive integer exponents. All variables are nonzero.

25. $(-4a^{-3})(5a)$

26. $\left(-\dfrac{2}{5}m^{-2}n^{-1}\right)\left(-\dfrac{15}{2}mn^0\right)$

27. $-\dfrac{16a^4b^{-1}}{12ab^{-4}}$

28. $\dfrac{45x^{-2}y^{-2}}{35x^{-4}y}$

▶ **EXAMPLE 15** **Simplifying Exponential Expressions Using Exponent Rules**

Simplify each expression. Write the answer with only positive integer exponents. All variables are nonzero.

(a) $\left(\dfrac{25a^{-2}b}{10ab^{-1}}\right)^{-2}$

(b) $(6c^{-2}d)\left(\dfrac{3c^{-3}d^4}{2d}\right)^2$

Solution

(a) Because there are common factors inside the parentheses, simplify within parentheses before applying the Power Rule.

$$a^{-n} = \frac{1}{a^n}; \frac{1}{a^{-n}} = a^n$$

$$\left(\frac{25a^{-2}b}{10ab^{-1}}\right)^{-2} \stackrel{\downarrow}{=} \left(\frac{25 \cdot b \cdot b}{10 \cdot a^2 \cdot a}\right)^{-2}$$

$$\frac{25}{10} = \frac{5 \cdot 5}{5 \cdot 2} = \frac{5}{2}; a^m a^n = a^{m+n}: \quad = \left(\frac{5b^2}{2a^3}\right)^{-2}$$

$$\left(\frac{a}{b}\right)^{-n} = \left(\frac{b}{a}\right)^n: \quad = \left(\frac{2a^3}{5b^2}\right)^2$$

$$\left(\frac{a}{b}\right)^n = \frac{a^n}{b^n}: \quad = \frac{(2a^3)^2}{(5b^2)^2}$$

$$(ab)^n = a^n b^n: \quad = \frac{2^2(a^3)^2}{5^2(b^2)^2}$$

$$(a^m)^n = a^{m \cdot n}: \quad = \frac{4a^{3 \cdot 2}}{25b^{2 \cdot 2}}$$

$$= \frac{4a^6}{25b^4}$$

Work Smart

Please note that there are many ways to simplify exponential expressions. For each problem you may use a different approach from the one we illustrate. However, be sure you are using the rules in the correct manner.

Thus, $\left(\dfrac{25a^{-2}b}{10ab^{-1}}\right)^{-2} = \dfrac{4a^6}{25b^4}$.

(continued)

(b) Because the quotient in the parentheses is messy, simplify it first.

$$(6c^{-2}d)\left(\frac{3c^{-3}d^4}{2d}\right)^2 \overset{\frac{a^n}{a^m}=a^{n-m}}{=} \frac{6c^{-2}d}{1}\left(\frac{3c^{-3}d^{4-1}}{2}\right)^2$$

$$a^{-n}=\frac{1}{a^n}: \quad = \frac{6d}{c^2}\left(\frac{3d^3}{2c^3}\right)^2$$

$$\left(\frac{a}{b}\right)^n=\frac{a^n}{b^n}; (ab)^n=a^nb^n: \quad = \frac{6d}{c^2}\left(\frac{3^2(d^3)^2}{2^2(c^3)^2}\right)$$

$$(a^m)^n=a^{m\cdot n}: \quad = \frac{6d}{c^2}\left(\frac{9d^6}{4c^6}\right)$$

$$\text{Rearrange factors:} \quad = 6\cdot\frac{9}{4}\cdot\frac{d\cdot d^6}{c^2\cdot c^6}$$

$$6\cdot\frac{9}{4}=\overset{3}{6}\cdot\frac{9}{\underset{2}{4}}=\frac{27}{2}; a^ma^n=a^{m+n}: \quad = \frac{27}{2}\cdot\frac{d^{1+6}}{c^{2+6}}$$

$$= \frac{27d^7}{2c^8}$$

So, $(6c^{-2}d)\left(\frac{3c^{-3}d^4}{2d}\right)^2=\frac{27d^7}{2c^8}$.

Quick ✓

In Problems 29–32, simplify each expression. Write the answers with only positive integer exponents. All variables are nonzero.

29. $(3y^2z^{-3})^{-2}$

30. $\left(\frac{2wz^{-3}}{7w^{-1}}\right)^2$

31. $\left(\frac{-6p^{-2}}{p}\right)(3p^8)^{-1}$

32. $(-25k^5r^{-2})\left(\frac{2}{5}k^{-3}r\right)^2$

5.4 Exercises MyMathLab®

Exercise numbers in green have complete video solutions in MyMathLab or may be accessed using the QR code to the right.

Problems **1–32** are the **Quick ✓** s that follow the **EXAMPLES**.

Building Skills

In Problems 33–42, use the Quotient Rule to simplify. All variables are nonzero. See Objective 1.

33. $\frac{2^{23}}{2^{19}}$

34. $\frac{10^5}{10^2}$

35. $\frac{x^{15}}{x^6}$

36. $\frac{x^{20}}{x^{14}}$

37. $\frac{16y^4}{4y}$

38. $\frac{9a^4}{27a}$

39. $\frac{-16m^{10}}{24m^3}$

40. $\frac{-36x^2y^5}{24xy^4}$

41. $\frac{-12m^9n^3}{-6mn}$

42. $\frac{-15r^9s^2}{-5r^8s}$

In Problems 43–50, use the Quotient to a Power Rule to simplify. All variables are nonzero. See Objective 2.

43. $\left(\frac{3}{2}\right)^3$

44. $\left(\frac{4}{9}\right)^2$

45. $\left(\frac{x^5}{3}\right)^3$

46. $\left(\frac{7}{y^2}\right)^2$

47. $\left(-\frac{x^5}{y^7}\right)^4$

48. $\left(-\frac{a^3}{b^{10}}\right)^5$

49. $\left(\frac{7a^2b}{c^3}\right)^2$

50. $\left(\frac{2mn^2}{q^3}\right)^4$

In Problems 51–62, use the Zero Exponent Rule to simplify. All variables are nonzero. See Objective 3.

51. 3^0

52. -100^0

53. $-\left(\frac{1}{2}\right)^0$

54. $\left(\frac{2}{5}\right)^0$

55. $18\cdot2^0$

56. $14^0\cdot3^0\cdot10$

57. $(-10)^0$

58. $(-8)^0$

59. $(24ab)^0$

60. $(-11xy)^0$

61. $24ab^0$

62. $-11xy^0$

In Problems 63–86, use the Negative Exponent Rules to simplify. Write answers with only positive integer exponents. All variables are nonzero. See Objective 4.

63. 10^{-3}

64. 5^{-1}

65. m^{-2}

66. k^{-2}

67. $-a^{-2}$

68. $-b^{-3}$

69. $-4y^{-3}$

70. $-7z^{-5}$

71. $2^{-1} + 3^{-2}$

72. $4^{-2} - 2^{-3}$

73. $\left(\dfrac{2}{5}\right)^{-2}$

74. $\left(\dfrac{5}{4}\right)^{-3}$

75. $\left(\dfrac{3}{z^2}\right)^{-1}$

76. $\left(\dfrac{4}{p^2}\right)^{-2}$

77. $\left(-\dfrac{2n}{m^2}\right)^{-3}$

78. $\left(-\dfrac{5}{3b^2}\right)^{-3}$

79. $\dfrac{1}{4^{-2}}$

80. $\dfrac{1}{6^{-2}}$

81. $\dfrac{6}{x^{-4}}$

82. $\dfrac{4}{b^{-3}}$

83. $\dfrac{5}{2m^{-3}}$

84. $\dfrac{9}{4t^{-1}}$

85. $\dfrac{5}{(2m)^{-3}}$

86. $\dfrac{9}{(4t)^{-1}}$

In Problems 87–96, use the Laws of Exponents to simplify. Write answers with only positive integer exponents. All variables are nonzero. See Objective 5.

87. $\left(\dfrac{4}{3}y^{-2}z\right)\left(\dfrac{5}{8}y^{-2}z^4\right)$

88. $\left(\dfrac{3}{5}ab^{-3}\right)\left(\dfrac{5}{3}a^{-1}b^3\right)$

89. $\dfrac{-7m^7n^6}{3m^{-3}n^0}$

90. $\dfrac{13r^8t^3}{-5r^0t^{-5}}$

91. $\dfrac{21y^2z^{-3}}{3y^{-2}z^{-1}}$

92. $\dfrac{30ab^{-4}}{15a^{-1}b^{-2}}$

93. $(4x^2y^{-2})^{-2}$

94. $(5x^{-4}y^3)^{-2}$

95. $(3a^2b^{-1})\left(\dfrac{4a^{-1}b^2}{b^3}\right)^2$

96. $(5a^3b^{-4})\left(\dfrac{2ab^{-1}}{b^{-3}}\right)^2$

Mixed Practice

In Problems 97–130, simplify. Write answers with only positive integer exponents. All variables are nonzero.

97. $2^5 \cdot 2^{-3}$

98. $3^8 \cdot 3^{-6}$

99. $2^{-7} \cdot 2^4$

100. $10^{-4} \cdot 10^3$

101. $\dfrac{3}{3^{-3}}$

102. $\dfrac{5^4}{5^{-1}}$

103. $\dfrac{x^6}{x^{15}}$

104. $\dfrac{b^{24}}{b^{42}}$

105. $\dfrac{8x^2}{2x^{-1}} + x^2(3x + 5x^{-1})$

106. $\dfrac{24m^2}{4m^{-3}} - m^3(2m^2 - 5m^{-1})$

107. $\dfrac{-27xy^3z^4}{18x^4y^3z}$

108. $\dfrac{8a^{15}b^2}{-18b^{20}}$

109. $(3x^2y^{-3})(12^{-1}x^{-5}y^{-6})$

110. $(-14c^2d^4)(3c^{-6}d^{-4})$

111. $(-a^4)^{-3}$

112. $(-x^{-2})^5$

113. $(3m^{-2})^3$

114. $(2z^{-1})^3$

115. $(-3x^{-2}y^{-3})^{-2}$

116. $(-4a^{-3}b^{-2}c^{-1})^{-1}$

117. $2p^{-4} \cdot p^{-3} \cdot 3p^0$

118. $7a^{-2} \cdot 2a^0 \cdot a^{-3}$

119. $(-16a^3)(-3a^4)\left(\dfrac{1}{4}a^{-7}\right)$

120. $(-3x^{-4})(2x^{-3})\left(\dfrac{1}{36}x^{10}\right)$

121. $\dfrac{8x^2 \cdot x^5}{12x^{-3} \cdot x^4} + \dfrac{(-3x)^{-1}(2x^3)}{(x^{-2})^2}$

122. $\dfrac{32x^{-3} \cdot x^2}{24x^2 \cdot x^{-5}} + \dfrac{(-3x^2)^{-1}}{(2x^2)^{-2}}$

123. $(2x^{-3}y^{-2})^4(3x^2y^{-3})^{-3}$

124. $(4a^{-2}b)^3(2a^2b^{-4})^{-2}$

125. $\left(\dfrac{y}{2z^2}\right)^{-3}\left(\dfrac{y^2}{4z^3}\right)^2 + \left(\dfrac{2}{y}\right)^{-1}$

126. $\left(\dfrac{s^2}{4t^2}\right)^3\left(\dfrac{s^2}{2t^4}\right)^{-3} + \left(\dfrac{2}{t^3}\right)^{-2}$

127. $(4x^2y)^3\left(\dfrac{2x}{3y}\right)^{-3}$

128. $(2a^5b^2)^3\left(\dfrac{4a^{-2}b^3}{3a}\right)^{-2}$

129. $\dfrac{(5a^{-3}b^2)^2}{a^{-4}b^{-4}}(15a^{-3}b)^{-1}$

130. $\left(\dfrac{2x^{-4}y^{-3}}{4xy^3}\right)^{-2}(4x^2y^{-1})^{-2}$

Applying the Concepts

△ **131. Volume of a Box** The volume of a rectangular solid is given by the equation $V = lwh$. Find the volume of a shipping box whose dimensions, measured in meters, are $l = x$, $w = x$, and $h = 3x$.

△ **132. Volume of a Box** The volume of a rectangular solid is given by the equation $V = lwh$. Find the volume of a box whose dimensions, measured in yards, are $l = 3n$, $w = 2n$, and $h = 6n$.

△ **133. Volume of a Cylinder** The volume of a right circular cylinder is given by the equation $V = \pi r^2 h$, where r is the radius of the circular base of the cylinder and h is its height. Rewrite this equation in terms of the diameter, d, of the base of the cylinder.

△ **134. Volume of a Cylinder** The volume of a right circular cylinder is given by the equation $V = \pi r^2 h$, where r is the radius of the circular base of the cylinder and h is its height. Write an algebraic expression for the volume of a cylinder in which the radius of the circular base is equal to the height of the cylinder.

△ **135. Making Buttons** How much fabric is required to cover x buttons in the shape of a circle whose radius is $3x$? Give the exact answer.

△ **136. Making a Tablecloth** How much fabric is required to make a round tablecloth if the diameter of the tablecloth is $\left(\dfrac{1}{2}x\right)$ meters? Give the exact answer.

Extending the Concepts

In Problems 137–144, simplify. Write answers with only positive exponents. All variables are nonzero.

137. $\dfrac{x^{2n}}{x^{3n}}$

138. $\dfrac{(x^{a-2})^3}{(x^{2a})^5}$

139. $\left(\dfrac{x^n y^m}{x^{4n-1} y^{m+1}}\right)^{-2}$

140. $\left(\dfrac{a^n b^{2m}}{ab^2}\right)^{-3}$

141. $(x^{2a} y^b z^{-c})^{3a}$

142. $(x^{-a} y^{-2b} z^{4c})^{2a}$

143. $\dfrac{(3a^n)^2}{(2a^{4n})^3}$

144. $\dfrac{y^a}{y^{4a}}$

Explaining the Concepts

145. Explain why $\dfrac{11^6}{11^4} \neq 1^2$.

146. Use the Quotient Rule and the expression $\left(\dfrac{a^n}{a^n}\right)$, $a \neq 0$, to explain why $a^0 = 1$, $a \neq 0$.

147. Explain the difference in simplifying the expressions $-12x^0$ and $(-12x)^0$.

148. Explain two different approaches to simplify $\left(\dfrac{x^8}{x^2}\right)^3$. Which do you prefer?

149. A friend of yours has a homework problem in which he must simplify $(x^3)^4$. He tells you that he thinks the answer is x^7. Is he right? If not, explain where he went wrong.

150. Explain why $-4x^{-3}$ is equal to $-\dfrac{4}{x^3}$. That is, why doesn't the factor -4 "move" to the denominator?

Putting the Concepts Together (Sections 5.1–5.4)

We designed these problems so that you can review Sections 5.1–5.4 and show your mastery of the concepts. Take time to work these problems before proceeding with the next section. The answers are located at the back of the text on page AN-22.

1. Determine whether the following algebraic expression is a polynomial. Answer Yes or No. If it is a polynomial, state the degree and then state whether it is a monomial, a binomial, or a trinomial: $6x^2 y^4 - 8x^5 + 3$.

2. Evaluate the polynomial $-x^2 + 3x$ for the given values: **(a)** 0 **(b)** -1 **(c)** 2.

In Problems 3–11, perform the indicated operation.

3. $(6x^4 - 2x^2 + 7) - (-2x^4 - 7 + 2x^2)$

4. $(2x^2 y - xy + 3y^2) + (4xy - y^2 + 3x^2 y)$

5. $-2mn(3m^2 n - mn^3)$

6. $(5x + 3)(x - 4)$

7. $(2x - 3y)(4x - 7y)$

8. $(5x + 8)(5x - 8)$

9. $(2x + 3y)^2$

10. $2a(3a - 4)(a + 5)$

11. $(4m + 3)(4m^3 - 2m^2 + 4m - 8)$

In Problems 12–20, simplify each expression. Write answers with only positive integer exponents. All variables are nonzero.

12. $(-2x^7 y^0 z)(4xz^8)$

13. $(5m^3 n^{-2})(-3m^{-4} n)$

14. $\dfrac{-18a^8b^3}{6a^5b}$ **15.** $\dfrac{16x^7y}{32x^9y^3}$ **18.** $\left(\dfrac{3}{2}r^2\right)^{-3}$ **19.** $(4y^{-2}z^3)^{-2}$

16. $\dfrac{7}{2ab^{-2}}$ **17.** $\dfrac{q^{-6}rt^5}{qr^{-4}t^7}$ **20.** $\left(\dfrac{3x^3y^{-3}}{2x^{-1}y^0}\right)^{-4}(2x^{-4}y^{-2})^2$

5.5 Dividing Polynomials

Objectives

1 Divide a Polynomial by a Monomial

2 Divide a Polynomial by a Binomial

Are You Prepared for This Section?

Before getting started, complete the following problems. If you get a problem wrong, go back to the section cited and review the material.

P1. Find the quotient: $\dfrac{24a^3}{6a}$ [Section 5.4, pp. 335–336]

P2. Find the quotient: $\dfrac{-49x^3}{21x^4}$ [Section 5.4, pp. 341–344]

P3. Find the product: $3x(7x-2)$ [Section 5.3, pp. 323–324]

P4. State the degree of $4x^2 - 2x + 1$.   [Section 5.1, pp. 310–311]

Work Smart

A polynomial is a monomial or the sum of monomials.

We begin polynomial division by dividing a polynomial by a monomial. Recall that in long division, such as $15\overline{)345}$ with 23 on top, the number 15 is called the *divisor,* the number 345 is called the *dividend,* and the number 23 is called the *quotient.* The same language is used with polynomial division.

▶ **1** Divide a Polynomial by a Monomial

Dividing a polynomial by a monomial requires use of the Quotient Rule for Exponents, which states that if a is a nonzero real number and m and n are integers, then $\dfrac{a^m}{a^n} = a^{m-n}$.

In Other Words

To divide a polynomial by a monomial, divide each of the terms of the polynomial in the numerator (dividend) by the monomial in the denominator (divisor).

Remember that for two rational numbers to be added, the denominators must be the same. When the denominators are the same, add the numerators and write the result over the common denominator. For example, $\dfrac{3}{11} + \dfrac{4}{11} = \dfrac{3+4}{11} = \dfrac{7}{11}$. When dividing a polynomial by a monomial, reverse this process. So if a, b, and c are monomials, $\dfrac{a+b}{c}$ can be written as $\dfrac{a}{c} + \dfrac{b}{c}$. This result can be extended to polynomials with three or more terms.

For the remainder of the chapter, assume variables in the denominator are nonzero.

EXAMPLE 1   **Dividing a Binomial by a Monomial**

Divide and simplify: $\dfrac{9x^3 - 21x^2}{3x}$

Solution

Divide each term in the numerator by the monomial in the denominator.

$$\frac{9x^3 - 21x^2}{3x} = \frac{9x^3}{3x} - \frac{21x^2}{3x}$$

$$\frac{a^m}{a^n} = a^{m-n}: \quad = \frac{9}{3}x^{3-1} - \frac{21}{3}x^{2-1}$$

$$\text{Simplify:} \quad = 3x^2 - 7x$$

Prepared?...Answers **P1.** $4a^2$

P2. $-\dfrac{7}{3x}$ **P3.** $21x^2 - 6x$ **P4.** 2

EXAMPLE 2 **Dividing a Trinomial by a Monomial**

Divide and simplify: $\dfrac{12p^4 + 24p^3 + 4p^2}{4p^2}$

Solution

Divide each term in the numerator by the monomial in the denominator.

$$\dfrac{12p^4 + 24p^3 + 4p^2}{4p^2} = \dfrac{12p^4}{4p^2} + \dfrac{24p^3}{4p^2} + \dfrac{4p^2}{4p^2}$$

$\dfrac{a^m}{a^n} = a^{m-n}: \quad = \dfrac{12}{4}p^{4-2} + \dfrac{24}{4}p^{3-2} + \dfrac{4}{4} \cdot \dfrac{p^2}{p^2}$

Simplify: $= 3p^2 + 6p + 1$ ●

EXAMPLE 3 **Dividing a Trinomial by a Monomial**

Divide and simplify: $\dfrac{8a^2b^2 - 6a^2b + 5ab^2}{2a^2b^2}$

Solution

$$\dfrac{8a^2b^2 - 6a^2b + 5ab^2}{2a^2b^2} = \dfrac{8a^2b^2}{2a^2b^2} - \dfrac{6a^2b}{2a^2b^2} + \dfrac{5ab^2}{2a^2b^2}$$

$\dfrac{a^m}{a^n} = a^{m-n};$ simplify: $= 4a^0b^0 - 3a^0b^{-1} + \dfrac{5}{2}a^{-1}b^0$

$a^0 = 1; a^{-n} = \dfrac{1}{a^n}: \quad = 4 \cdot 1 \cdot 1 - 3 \cdot 1 \cdot \dfrac{1}{b} + \dfrac{5}{2} \cdot \dfrac{1}{a} \cdot 1$

$$= 4 - \dfrac{3}{b} + \dfrac{5}{2a}$$ ●

---

1. The first step to simplify $\dfrac{4x^4 + 8x^2}{2x}$ is to rewrite $\dfrac{4x^4 + 8x^2}{2x}$ as $\dfrac{\quad}{2x} + \dfrac{\quad}{2x}$.

In Problems 2–4, find the quotient.

2. $\dfrac{10n^4 - 20n^3 + 5n^2}{5n^2}$ 3. $\dfrac{12k^4 - 18k^2 + 5}{2k^2}$

4. $\dfrac{x^4y^4 + 8x^2y^2 - 4xy}{4x^3y}$

---

❷ Divide a Polynomial by a Binomial

Dividing a polynomial by a binomial is like dividing two integers. Although this procedure should be familiar to you, it is reviewed below.

EXAMPLE 4 **Dividing an Integer by an Integer Using Long Division**

Divide 579 by 16.

Solution

$$
\begin{array}{r}
36 \quad \longleftarrow \text{quotient} \\
\text{divisor} \longrightarrow 16\overline{)579} \quad \longleftarrow \text{dividend} \\
48 \quad \longleftarrow 3 \cdot 16 = 48 \\
57 - 48 = 9 \rightarrow \quad 99 \quad \longleftarrow \text{bring down the 9} \\
96 \quad \longleftarrow 6 \cdot 16 = 96 \\
99 - 96 = 3 \rightarrow \quad 3 \quad \longleftarrow \text{remainder}
\end{array}
$$

So 579 divided by 16 equals 36 with a remainder of 3. This can be written as

$$\frac{579}{16} = 36\frac{3}{16}.$$

A long-division problem can always be checked by multiplying the quotient by the divisor and adding this product to the remainder. The result should be the dividend. That is,

$$(\text{Quotient})(\text{Divisor}) + \text{Remainder} = \text{Dividend}$$

To check Example 4, note that

$$(36)(16) + 3 = 576 + 3 = 579$$

In Example 4, the solution was written as $\frac{579}{16} = 36\frac{3}{16}$. Remember, the mixed number $36\frac{3}{16}$ means $36 + \frac{3}{16}$. So the answer is written in the form

$$\text{Quotient} + \frac{\text{Remainder}}{\text{Divisor}}$$

Dividing a polynomial by a binomial using long division involves the same process as dividing integers.

▶ EXAMPLE 5   How to Divide a Polynomial by a Binomial Using Long Division

Find the quotient when $x^2 + 10x + 21$ is divided by $x + 7$.

Step-by-Step Solution

To divide a polynomial by a binomial, first write each in standard form (descending order of degree). The dividend is $x^2 + 10x + 21$ and the divisor is $x + 7$.

Step 1: Divide the highest-degree term of the dividend, x^2, by the highest-degree term of the divisor, x. Enter the result over the term x^2.

$$\frac{x^2}{x} = x$$

$$x + 7 \overline{)x^2 + 10x + 21}$$

Step 2: Multiply x by $x + 7$. Vertically align like terms.

$$x + 7 \overline{)x^2 + 10x + 21}$$
$$\underline{x^2 + 7x} \qquad x(x + 7) = x^2 + 7x$$

Step 3: Subtract $x^2 + 7x$ from $x^2 + 10x + 21$.

$$x + 7 \overline{)x^2 + 10x + 21}$$
$$\underline{-(x^2 + 7x)}$$
$$3x + 21 \quad (x^2 + 10x + 21) - (x^2 + 7x) = 3x + 21$$

Step 4: Repeat Steps 1–3, treating $3x + 21$ as the dividend.

$$\frac{3x}{x} = 3$$

$$x + 3$$
$$x + 7 \overline{)x^2 + 10x + 21}$$
$$\underline{-(x^2 + 7x)}$$
$$3x + 21$$
$$\underline{-(3x + 21)} \quad 3(x + 7) = 3x + 21$$
$$0 \quad (3x + 21) - (3x + 21) = 0$$

The quotient is $x + 3$ and the remainder is 0.

(continued)

Step 5: Check the result by showing that
(Quotient)(Divisor) + Remainder = Dividend.

$$(x + 3)(x + 7) + 0 = x^2 + 10x + 21$$

The product checks, so $x^2 + 10x + 21$ divided by $x + 7$ is $x + 3$. This division is expressed using fractions, as follows:

$$\frac{x^2 + 10x + 21}{x + 7} = x + 3$$

▶ **EXAMPLE 6** | **How to Divide a Polynomial by a Binomial Using Long Division**

Find the quotient using long division: $\dfrac{6x^2 + 9x - 10}{2x - 1}$

Solution

Each polynomial is in standard form. The dividend is $6x^2 + 9x - 10$ and the divisor is $2x - 1$.

$$
\begin{array}{r}
3x \longleftarrow \text{Step1: } \frac{6x^2}{2x} = 3x \\
2x - 1\overline{)6x^2 + 9x - 10} \\
-\underline{(6x^2 - 3x)} \longleftarrow \text{Step 2: } 3x(2x - 1) = 6x^2 - 3x \\
12x - 10 \longleftarrow \text{Step 3: } 6x^2 + 9x - 10 - (6x^2 - 3x) = 12x - 10
\end{array}
$$

$$
\begin{array}{r}
3x + 6 \longleftarrow \text{Repeat Step 1: } \frac{12x}{2x} = 6 \\
2x - 1\overline{)6x^2 + 9x - 10} \\
-\underline{(6x^2 - 3x)} \\
12x - 10 \\
-\underline{(12x - 6)} \longleftarrow \text{Repeat Step 2: } 6(2x - 1) = 12x - 6 \\
-4 \longleftarrow \text{Repeat Step 3: } (12x - 10) - (12x - 6) = -4
\end{array}
$$

Work Smart

You know that you're finished dividing when the degree of the remainder is less than the degree of the divisor.

Because the degree of -4 is less than the degree of the divisor, $2x - 1$, the process ends. The quotient is $3x + 6$ and the remainder is -4.

Check (Quotient)(Divisor) + Remainder = Dividend.

$$(3x + 6)(2x - 1) + (-4) = 6x^2 - 3x + 12x - 6 + (-4)$$

Combine like terms: $= 6x^2 + 9x - 10$

The resulting polynomial is the same as the dividend,

so $\dfrac{6x^2 + 9x - 10}{2x - 1} = 3x + 6 - \dfrac{4}{2x - 1}$.

Quick ✔

5. To begin a polynomial division problem, write the divisor and the dividend in _____ form.

6. To check the result of long division, multiply the _____ and the divisor and add this result to the _____. If correct, this result will be equal to the _____.

In Problems 7–9, find the quotient using long division.

7. $\dfrac{x^2 - 3x - 40}{x + 5}$

8. $\dfrac{2x^2 - 5x - 12}{2x + 3}$

9. $\dfrac{4x^2 + 17x + 21}{x + 3}$

⏵ **EXAMPLE 7**   **Dividing Two Polynomials Using Long Division**

Find the quotient using long division: $\dfrac{8 - 9x + 2x^2 + 12x^3 + 5x^5}{x^2 + 3}$

Solution

The dividend needs to be written in standard form. Also, when the dividend is in standard form, notice that there is no x^4 term. When a term is missing, its coefficient is 0, so rewrite the division problem as follows:

$$\frac{5x^5 + 0x^4 + 12x^3 + 2x^2 - 9x + 8}{x^2 + 3}$$

$$
\begin{array}{r}
5x^3 \qquad\quad - 3x \ + 2 \\
x^2 + 3 \overline{)5x^5 + 0x^4 + 12x^3 + 2x^2 - 9x + 8} \\
\end{array}
$$

$-(5x^5 \qquad + 15x^3)$ ← $5x^3(x^2 + 3)$

$-3x^3 + 2x^2 - 9x + 8$ ← $5x^5 + 12x^3 + 2x^2 - 9x + 8 - (5x^5 + 15x^3) = -3x^3 + 2x^2 - 9x + 8$

$-(-3x^3 \qquad - 9x)$ ← $-3x(x^2 + 3)$

$2x^2 \qquad + 8$ ← $(-3x^3 + 2x^2 - 9x + 8) - (-3x^3 - 9x) = 2x^2 + 8$

$-(2x^2 \qquad + 6)$ ← $2(x^2 + 3)$

2 ← Remainder

The quotient is $5x^3 - 3x + 2$ and the remainder is 2.

Check $(\text{Quotient})(\text{Divisor}) + \text{Remainder} = \text{Dividend}$

$$(5x^3 - 3x + 2)(x^2 + 3) + 2 = 5x^5 + 15x^3 - 3x^3 - 9x + 2x^2 + 6 + 2$$

$$= 5x^5 + 12x^3 + 2x^2 - 9x + 8$$

The answer checks, so $\dfrac{8 - 9x + 2x^2 + 12x^3 + 5x^5}{x^2 + 3} = 5x^3 - 3x + 2 + \dfrac{2}{x^2 + 3}$. ●

Quick ✔

In Problems 10–12, find the quotient using long division.

10. $\dfrac{x + 1 - 3x^2 + 4x^3}{x + 2}$ **11.** $\dfrac{2x^3 + 3x^2 + 10}{2x - 5}$ **12.** $\dfrac{4x^3 - 3x^2 + x + 1}{x^2 + 2}$

5.5 Exercises MyMathLab®

Exercise numbers in **green** have complete video solutions in MyMathLab or may be accessed using the QR code to the right.

*Problems **1–12** are the Quick ✔s that follow the **EXAMPLES**.*

Building Skills

In Problems 13–30, divide and simplify. See Objective 1.

13. $\dfrac{4x^2 - 2x}{2x}$ **14.** $\dfrac{3x^3 - 6x^2}{3x^2}$

15. $\dfrac{9a^3 + 27a^2 - 3}{3a^2}$ **16.** $\dfrac{16m^3 + 8m^2 - 4}{8m^2}$

17. $\dfrac{5n^5 - 10n^3 - 25n}{25n}$ **18.** $\dfrac{5x^3 - 15x^2 + 10x}{5x^2}$

19. $\dfrac{15r^5 - 27r^3}{9r^3}$ **20.** $\dfrac{16a^5 - 12a^4 + 8a^3}{20a^3}$

21. $\dfrac{3x^7 - 9x^6 + 27x^3}{-3x^5}$

22. $\dfrac{7p^4 + 21p^3 - 3p^2}{-6p^4}$

46. $\dfrac{3x^4 + 7x^3 - 5x^2 + 8x + 12}{x + 3}$

23. $\dfrac{3z + 4z^3 - 2z^2}{8z}$

24. $\dfrac{-5y^2 + 15y^4 - 16y^5}{5y^2}$

47. $\dfrac{2x^3 + 7x^2 - 10x + 5}{2x - 1}$

48. $\dfrac{12a^3 + 11a^2 + 18a + 9}{4a + 1}$

25. $\dfrac{14xy - 10y}{-2y}$

26. $\dfrac{35xy + 20y}{-5y}$

49. $\dfrac{-24 + x^2 + x}{5 + x}$

50. $\dfrac{-16x + 70 + x^2}{-9 + x}$

27. $\dfrac{12y - 30x}{-2x}$

28. $\dfrac{21y^2 + 35x^2}{-7x^2}$

51. $\dfrac{4x^2 + 5}{1 + 2x}$

52. $\dfrac{9x^2 - 14}{2 + 3x}$

29. $\dfrac{25a^3b^2c + 10a^2bc^3}{-5a^4b^2c}$

30. $\dfrac{16m^2n^3 - 24m^4n^3}{-8m^3n^4}$

53. $\dfrac{x^4 + 2x^2 - 8}{x^2 - 2}$

54. $\dfrac{64x^6 - 27}{4x^2 - 3}$

In Problems 31–54, find the quotient using long division.
See Objective 2.

Mixed Practice

In Problems 55–80, perform the indicated operation.

31. $\dfrac{x^2 - 4x - 21}{x + 3}$

32. $\dfrac{x^2 + 18x + 72}{x + 6}$

55. $(a - 5)(a + 6)$

56. $(7n + 3m^2 + 4m) + (-6m^2 - 7n + 4m)$

33. $\dfrac{x^2 - 9x + 20}{x - 4}$

34. $\dfrac{x^2 + 4x - 32}{x - 4}$

57. $(2x - 8) - (3x + x^2 - 2)$

58. $3x^2(7x^2 + 2x - 3)$

35. $\dfrac{x^3 + 4x^2 - 15x + 6}{x - 2}$

36. $\dfrac{x^3 - x^2 - 40x + 12}{x + 6}$

59. $(2ab + b^2 - a^2) + (b^2 - 4ab + a^2)$

60. $\dfrac{3a^4b^{-2}}{6a^{-4}b^{-3}}$

61. $\dfrac{4 + 7x^2 - 3x^4 + 6x^3}{2x^2}$

37. $\dfrac{x^4 - x^3 + 10x - 4}{x + 2}$

38. $\dfrac{x^4 - 2x^3 + x^2 + x - 1}{x - 1}$

39. $\dfrac{x^3 - x^2 + x + 8}{x + 1}$

40. $\dfrac{x^3 - 7x^2 + 15x - 11}{x - 3}$

62. $(b - 3)(b + 2)$

63. $\dfrac{18x^3y^{-4}z^6}{27x^{-4}y^{-12}z^{-6}}$

64. $(4ab + 6ab^2) - (12a^2b - 2ab - 3ab^2)$

41. $\dfrac{2x^2 - 7x - 15}{x - 5}$

42. $\dfrac{2x^2 + 5x - 42}{x + 6}$

65. $\dfrac{6x^2 - 28x + 30}{3x - 5}$

66. $\dfrac{12x^2 - 25x - 50}{3x - 10}$

43. $\dfrac{x^3 + 4x^2 - 5x + 2}{x - 2}$

44. $\dfrac{x^3 - 2x^2 + x + 6}{x + 1}$

67. $(n - 3)^2$

68. $\dfrac{-9mn^2 - 8m^2n + 12mn}{3mn}$

45. $\dfrac{2x^4 - 3x^3 - 11x^2 - 40x - 1}{x - 4}$

69. $(x^3 + x - 4x^4)(-10x^2)$

70. $(2x^3 + 6x) + (12x - x^2 - x^3)$

71. $(2pq - q^2) + (4pq - p^2 - q^2)$

72. $(4x^2 - 6x + 3) - (x - 7)$

73. $(x^2 + x - 1)(x + 5)$

74. $(x^2 - 2x + 3)(x - 4)$

75. $(7rs^2 - 2r^2s) - (2r^2s - 8rs^2)$

76. $(x + 6)^2$

77. $(x^4 - 2x^2 + x)(-3x)$

78. $2x^4(x^2 - 2x + 3)$

79. $\dfrac{3 + 6x^2 - 11x}{2x - 3} + (x + 1)$

80. $\dfrac{7x - 5 + 6x^2}{3x + 5} - (2x + 1)$

Applying the Concepts

81. Find the quotient of $(3x^4 - 6x + 12x^2)$ and $-3x^3$.

82. Divide $x - 3$ by $x + 2$.

83. Divide the sum of $x^2 + 3x - 1$ and $x - 1$ by $-2x^3$.

84. Divide the square of the difference of $x - 9$ and $x + 3$ by x^2.

△**85. Volume of a Box** The volume of a rectangular solid is $(x^3 - 5x^2 + 6x)$ cubic feet. One side measures $(x - 2)$ feet and another measures $(x - 3)$ feet. What is the measure of the third side?

△**86. Area of a Rectangle** A rectangle has area $(x^2 + 2x - 48)$ square inches. If one side measures $(x + 8)$ inches, what is the measurement of the other side?

△**87. Area of a Rectangle** If the area of a rectangle is $(z^2 + 6z + 9)$ square inches and the length is $(z + 3)$ inches, what is the width?

△**88. Area of a Triangle** If the area of a triangle is $(6a^2 - 5a - 6)$ square yards and the length of the base is $(2a - 3)$ yards, what is the height?

△**89. Area of a Triangle** If the area of a triangle is $(6x^3 - 2x^2 - 8x)$ square feet and the height is $(3x^2 - 4x)$ feet, what is the length of the base?

△**90. Volume of a Box** The volume of a rectangular solid is $(x^3 + 2x^2 - x - 2)$ cubic feet. One side measures $(x + 2)$ feet and another measures $(x - 1)$ feet. What is the measure of the third side?

91. Average Cost The average cost of manufacturing x computers per day is given by
$$\frac{0.004x^3 - 0.8x^2 + 180x + 5000}{x}.$$

(a) Simplify the quotient by dividing each term in the numerator by the denominator.

(b) Use this result to determine the average cost of manufacturing $x = 140$ computers in one day.

92. Average Cost The average cost of manufacturing x digital cameras per day is given by
$$\frac{0.0024x^3 - 0.4x^2 + 46x + 4000}{x}.$$

(a) Simplify the quotient by dividing each term in the numerator by the denominator.

(b) Use this result to determine the average cost of manufacturing $x = 90$ cameras in one day.

Extending the Concepts

In Problems 93 and 94, determine the value of the missing term so that the remainder is zero.

93. $\dfrac{6x^2 - 13x + ?}{2x - 3}$

94. $\dfrac{3x^2 + ? - 5}{x - 1}$

Explaining the Concepts

95. Explain how to divide polynomials when the divisor is a monomial and then when the divisor is a binomial. Which procedures are the same and which are different?

96. The first steps of a division problem are written below. Describe what has occurred. Are there any potential errors with this presentation? What would you recommend this student do to improve his or her chances of obtaining the correct answer?

$$x^2 - 3 \overline{\smash{)}\, 2x^4 - 3x^3 + 2x + 1}$$
$$\underline{-(2x^4 - 6x^2)}$$

with $2x^2$ above.

5.6 Applying Exponent Rules: Scientific Notation

Objectives

❶ Convert Decimal Notation to Scientific Notation

❷ Convert Scientific Notation to Decimal Notation

❸ Use Scientific Notation to Multiply and Divide

Are You Prepared for This Section?

Before getting started, complete the following problems. If you get a problem wrong, go back to the section cited and review the material.

P1. Find the product: $(3a^6)(4.5a^4)$ [Section 5.2, pp. 321–322]

P2. Find the product: $(7n^3)(2n^{-2})$ [Section 5.4, pp. 341–344]

P3. Find the quotient: $\dfrac{3.6b^9}{0.9b^{-2}}$ [Section 5.4, pp. 341–344]

Did you know that the mass of the Sun is 1,989,000,000,000,000,000,000,000,000,000 kg? Did you know that the mass of a dust particle is 0.000000000753 kg? These numbers are difficult to write and difficult to read, so exponents are used to rewrite them.

▶ ❶ Convert Decimal Notation to Scientific Notation

The numbers 1,989,000,000,000,000,000,000,000,000,000 kg and 0.000000000753 kg are written in decimal notation. *Scientific notation* expresses a number as the product of two factors: One factor is a number between 1 and 10, including 1 but not including 10, and the other factor is an integer power of 10.

> **Definition**
>
> A number written as the product of a number x, where $1 \le x < 10$, and a power of 10 is said to be written in **scientific notation.** That is, a number is written in scientific notation when it is in the form
>
> $$x \times 10^N$$
>
> where
>
> $$1 \le x < 10 \text{ and } N \text{ is an integer}$$

Notice in the definition, $x < 10$. That's because when $x = 10$, we have 10^1, a power of 10. For example, in scientific notation,

$$\text{Mass of Sun} = 1.989 \times 10^{30} \text{ kilograms}$$

$$\text{Mass of a dust particle} = 7.53 \times 10^{-10} \text{ kilograms}$$

Prepared?...Answers **P1.** $13.5a^{10}$
P2. $14n$ **P3.** $4b^{11}$

EXAMPLE 1 **How to Convert from Decimal Notation to Scientific Notation**

Write 5283 in scientific notation.

Step-by-Step Solution

For a number to be in scientific notation, the decimal must be moved so that there is a single nonzero digit to the left of the decimal point. All remaining digits must appear to the right of the decimal point.

Step 1: The "understood" decimal point in 5283 follows the 3. Therefore, move the decimal to the left by $N = 3$ places until it is between the 5 and the 2. Do you see why?

$$5\,2\,8\,3.$$
$$3\,2\,1$$

Step 2: The original number is greater than 1, so write 5283 in scientific notation as

$$5.283 \times 10^3$$

EXAMPLE 2   **How to Convert from Decimal Notation to Scientific Notation**

Write 0.054 in scientific notation.

Step-by-Step Solution

Step 1: Because 0.054 is less than 1, move the decimal point to the right $N = 2$ places until it is between the 5 and the 4.

$$0.05\,4$$
$$1\ 2$$

Step 2: The original number is between 0 and 1, so write 0.054 in scientific notation as

$$5.4 \times 10^{-2}$$

In Other Words

For a number greater than or equal to 1, use $10^{positive\ exponent}$.

For a number between 0 and 1, use $10^{negative\ exponent}$.

Converting from Decimal Notation to Scientific Notation

To change a positive number to scientific notation:

Step 1: Count the number N of decimal places that the decimal point must be moved in order to arrive at a number x, where $1 \le x < 10$.

Step 2: If the original number is greater than or equal to 1, the scientific notation is $x \times 10^{N}$. If the original number is between 0 and 1, the scientific notation is $x \times 10^{-N}$.

Quick ✓

1. A number written as 3.2×10^{-6} is said to be written in _____ notation.

2. When 47,000,000 is written in scientific notation, the power of 10 will be _____ (positive or negative).

3. *True or False* When a number is expressed in scientific notation, it is expressed as the product of a number x, $0 \le x < 1$, and a power of 10.

In Problems 4–9, write each number in scientific notation.

4. 432   5. 10,302   6. 5,432,000

7. 0.093   8. 0.0000459   9. 0.00000008

▶ ❷ **Convert Scientific Notation to Decimal Notation**

Now we are going to convert a number from scientific notation to decimal notation. Study Table 3 to discover the pattern.

Table 3

Scientific Notation	Product	Decimal Notation	Location of Decimal Point
3.69×10^{2}	3.69×100	369	moved 2 places to the right
3.69×10^{1}	3.69×10	36.9	moved 1 place to the right
3.69×10^{0}	3.69×1	3.69	didn't move
3.69×10^{-1}	3.69×0.1	0.369	moved 1 place to the left
3.69×10^{-2}	3.69×0.01	0.0369	moved 2 places to the left

The pattern in Table 3 leads to the steps on the following page for converting a number from scientific notation to decimal notation.

> **Converting a Number from Scientific Notation to Decimal Notation**
>
> **Step 1:** Determine the exponent, N, on the number 10.
>
> **Step 2:** If the exponent is positive, then move the decimal point N decimal places to the right. If the exponent is negative, then move the decimal point $|N|$ decimal places to the left. Add zeros, as needed.

EXAMPLE 3 **How to Convert from Scientific Notation to Decimal Notation**

Write 2.3×10^3 in decimal notation.

Step-by-Step Solution

Step 1: Determine the exponent on the number 10.   The exponent on the 10 is 3.

Step 2: Because the exponent is positive, move the decimal point three places to the right. Notice zeros are added to the right of 3, as needed.

$$2.3\,0\,0.$$

So, $2.3 \times 10^3 = 2300$.   ●

EXAMPLE 4 **How to Convert from Scientific Notation to Decimal Notation**

Write 4.57×10^{-5} in decimal notation.

Step-by-Step Solution

Step 1: Determine the exponent on the number 10.   The exponent on the 10 is -5.

Step 2: Because the exponent is negative, move the decimal point five places to the left.

Add zeros to the left of the original decimal point.

$$0.0\,0\,0\,0\,4.5\,7$$

So, $4.57 \times 10^{-5} = 0.0000457$.   ●

Work Smart

In Example 3, notice that $2.3 \times 10^3 = 2.3 \times 1000$, which equals 2300.

In Example 4, notice that $4.57 \times 10^{-5} = 4.57 \times 0.00001$, which equals 0.0000457.

Quick ✓

10. *True or False* To write 3.2×10^{-6} in decimal notation, move the decimal point six places to the left.

11. *True or False* To convert 2.4×10^3 to decimal notation, move the decimal point three places to the right.

In Problems 12–16, write each number in decimal notation.

12. 3.1×10^2 **13.** 9.01×10^{-1} **14.** 1.7×10^5

15. 7×10^0 **16.** 8.9×10^{-4}

▶ **❸ Use Scientific Notation to Multiply and Divide**

To multiply and divide numbers written in scientific notation, use the Product Rule, $a^m \cdot a^n = a^{m+n}$ and the Quotient Rule, $\dfrac{a^m}{a^n} = a^{m-n}$. Use these laws where the base is 10 as follows:

$$10^m \cdot 10^n = 10^{m+n} \quad \text{and} \quad \frac{10^m}{10^n} = 10^{m-n}$$

EXAMPLE 5   **Multiplying Using Scientific Notation**

Perform the indicated operation. Express the answer in scientific notation.
$$(3 \times 10^2)(2.5 \times 10^5)$$

Solution
First, rearrange factors.
$$(3 \times 10^2)(2.5 \times 10^5) = (3 \cdot 2.5) \times (10^2 \cdot 10^5)$$

Multiply $3 \cdot 2.5$; $a^m \cdot a^n = a^{m+n}$: $= 7.5 \times 10^7$ ●

EXAMPLE 6   **Multiplying Using Scientific Notation**

Perform the indicated operation. Express the answer in scientific notation.

(a) $(4 \times 10^{-2})(6 \times 10^8)$ **(b)** $(3.2 \times 10^{-3})(4.8 \times 10^{-4})$

Solution

(a) $(4 \times 10^{-2})(6 \times 10^8) = (4 \cdot 6) \times (10^{-2} \cdot 10^8)$

Multiply $4 \cdot 6$; $a^m \cdot a^n = a^{m+n}$: $= 24 \times 10^6$

Convert 24 to scientific notation: $= (2.4 \times 10^1) \times 10^6$

$(a \cdot b) \cdot c = a \cdot (b \cdot c)$: $= 2.4 \times (10^1 \times 10^6)$

$a^m \cdot a^n = a^{m+n}$: $= 2.4 \times 10^7$

Work Smart

The result from multiplication using scientific notation initially may not be in scientific notation. See Example 6 as an illustration.

(b) $(3.2 \times 10^{-3})(4.8 \times 10^{-4}) = (3.2 \cdot 4.8) \times (10^{-3} \cdot 10^{-4})$

Multiply $3.2 \cdot 4.8$; $a^m \cdot a^n = a^{m+n}$: $= 15.36 \times 10^{-7}$

Convert 15.36 to scientific notation: $= (1.536 \times 10^1) \times 10^{-7}$

$(a \cdot b) \cdot c = a \cdot (b \cdot c)$: $= 1.536 \times (10^1 \times 10^{-7})$

$a^m \cdot a^n = a^{m+n}$: $= 1.536 \times 10^{-6}$ ●

Quick ✓

In Problems 17–20, perform the indicated operation. Express the answer in scientific notation.

17. $(3 \times 10^4)(2 \times 10^3)$ **18.** $(2 \times 10^{-2})(4 \times 10^{-1})$

19. $(5 \times 10^{-4})(3 \times 10^7)$ **20.** $(8 \times 10^{-4})(3.5 \times 10^{-2})$

EXAMPLE 7   **Dividing Using Scientific Notation**

Perform the indicated operation. Express the answer in scientific notation.

(a) $\dfrac{6 \times 10^6}{2 \times 10^2}$ **(b)** $\dfrac{2.4 \times 10^4}{3 \times 10^{-2}}$

Solution

(a) $\dfrac{6 \times 10^6}{2 \times 10^2} = \dfrac{6}{2} \times \dfrac{10^6}{10^2}$

Divide $\dfrac{6}{2}$; $\dfrac{a^m}{a^n} = a^{m-n}$: $= 3 \times 10^4$

(b) $\dfrac{2.4 \times 10^4}{3 \times 10^{-2}} = \dfrac{2.4}{3} \times \dfrac{10^4}{10^{-2}}$

Divide $\dfrac{2.4}{3}$; $\dfrac{a^m}{a^n} = a^{m-n}$: $= 0.8 \times 10^{4-(-2)}$

Convert 0.8 to scientific notation: $= (8 \times 10^{-1}) \times 10^6$

$a^m \cdot a^n = a^{m+n}$: $= 8 \times 10^5$ ●

Quick ✓

In Problems 21–24, perform the indicated operation. Express the answer in scientific notation.

21. $\dfrac{8 \times 10^6}{2 \times 10^1}$

22. $\dfrac{2.8 \times 10^{-7}}{1.4 \times 10^{-3}}$

23. $\dfrac{3.6 \times 10^3}{7.2 \times 10^{-1}}$

24. $\dfrac{5 \times 10^{-2}}{8 \times 10^2}$

EXAMPLE 8 **Visits to Facebook**

In 2015, Facebook had 1.04×10^9 visitors each day. How many visitors to Facebook were there in April 2015? Express the answer in scientific and decimal notation. (SOURCE: *Facebook*)

Solution

To find the number of visitors to Facebook for April 2015, multiply the number of visitors each day by the number of days in April, 30.

In scientific notation, $30 = 3 \times 10^1$. Thus

$$(1.04 \times 10^9)(3 \times 10^1) = (1.04 \cdot 3) \times (10^9 \cdot 10^1)$$

$$= 3.12 \times 10^{10} \qquad \text{scientific notation}$$

$$= 31{,}200{,}000{,}000 \text{ visitors} \qquad \text{decimal notation}$$

In April 2015, there were 3.12×10^{10} or 31,200,000,000 visitors to Facebook. ●

Quick ✓

25. The United States consumes 3.75×10^8 gallons of gasoline each day. How many gallons of gasoline does the United States consume in a 30-day month? Express the answer in scientific and decimal notation.

26. Attendance at Walt Disney World in Orlando, Florida, is nearly 5.3×10^4 people daily. How many visitors does Walt Disney World get in a 365-day year?

5.6 Exercises MyMathLab® Exercise numbers in green have complete video solutions in MyMathLab or may be accessed using the QR code to the right.

Problems 1–26 are the Quick ✓s that follow the EXAMPLES.

Building Skills

In Problems 27–42, write each number in scientific notation. See Objective 1.

27. 300,000

28. 421,000,000

29. 64,000,000

30. 8,000,000,000

31. 0.00051

32. 0.0000001

33. 0.000000001

34. 0.0000283

35. 8,007,000,000

36. 401,000,000

37. 0.0000309

38. 0.000201

39. 620

40. 8

41. 4

42. 120

In Problems 43–54, a number is given in decimal notation. Write the number in scientific notation. See Objective 1.

43. **World Population** According to the U.S. Census Bureau, the population of the world on February 4, 2016, was approximately 7,303,000,000 persons.

44. **United States Population** According to the U.S. Census Bureau, the population of the United States on February 4, 2016, was approximately 319,000,000 persons.

45. **Federal Debt** According to the United States Treasury, the federal debt of the United States as of February 2016 was $18,900,000,000,000.

46. Interest Payments on Debt In 2015, total interest payments on the federal debt (Problem 45) totaled $200,000,000,000.

47. Distance to the Sun The average distance from Earth to the Sun is 93,000,000 miles.

48. Distance to the Moon The average distance from Earth to the moon (of Earth) is 384,000 kilometers.

49. Smallpox Virus The diameter of a smallpox virus is 0.00003 mm.

50. Human Blood Cell The diameter of a human blood cell is 0.0075 mm.

51. Dust Mites Dust mites are microscopic bugs that are a major cause of allergies and asthma. They are approximately 0.00000025 m in length.

52. DNA The length of a DNA molecule can exceed 0.025 inch in some organisms.

53. Mass of a Penny The mass of the United States' Lincoln penny is approximately 0.00311 kg.

54. Physics The radius of a typical atom is approximately 0.0000000001 m.

In Problems 55–70, write each number in decimal notation. See Objective 2.

55. 4.2×10^5 **56.** 3.75×10^2

57. 1×10^8 **58.** 6×10^6

59. 3.9×10^{-3} **60.** 6.1×10^{-6}

61. 4×10^{-1} **62.** 5×10^{-4}

63. 3.76×10^3 **64.** 4.9×10^{-1}

65. 8.2×10^{-3} **66.** 5.4×10^5

67. 6×10^{-5} **68.** 5.123×10^{-3}

69. 7.05×10^6 **70.** 7×10^8

In Problems 71–76, a number is given in scientific notation. Write the number in decimal notation. See Objective 2.

71. Time A femtosecond is equal to 1×10^{-15} second.

72. Zika Virus The diameter of the zika virus is 5.0×10^{-8} m.

73. Vitamin A One-A-Day vitamin pill contains 2.25×10^{-3} gram of zinc.

74. Water Molecule The diameter of a water molecule is 3.85×10^{-7} m.

75. Coffee Drinkers In 2015, there were 1.945×10^8 coffee drinkers in the United States.

76. Fatal Accidents In 2014, there were 3.3×10^4 drivers in fatal auto accidents in the United States. (SOURCE: *U.S. Department of Transportation*)

In Problems 77–88, perform the indicated operations. Express your answer in scientific notation. See Objective 3.

77. $(2 \times 10^6)(1.5 \times 10^3)$

78. $(3 \times 10^{-4})(8 \times 10^{-5})$

79. $(1.2 \times 10^0)(7 \times 10^{-3})$

80. $(4 \times 10^7)(2.5 \times 10^{-4})$

81. $\dfrac{9 \times 10^4}{3 \times 10^{-4}}$ **82.** $\dfrac{6 \times 10^3}{1.2 \times 10^5}$

83. $\dfrac{2 \times 10^{-3}}{8 \times 10^{-5}}$ **84.** $\dfrac{4.8 \times 10^7}{1.2 \times 10^2}$

85. $\dfrac{56,000}{0.00007}$ **86.** $\dfrac{0.000275}{2500}$

87. $\dfrac{300,000 \times 15,000,000}{0.0005}$ **88.** $\dfrac{24,000,000,000}{0.00006 \times 2000}$

Applying the Concepts

89. Speed of Light Light travels at the rate of 1.86×10^5 miles per second. How far does light travel in 1 minute $(6.0 \times 10^1 \text{ seconds})$?

90. Speed of Sound Sound travels at the rate of 1.127×10^3 feet per second. How far does sound travel in 1 minute $(6.0 \times 10^1 \text{ seconds})$?

91. Hair Growth Human hair grows at a rate of 1×10^{-8} miles per hour. How many miles will human hair grow after 100 hours? After one week? Express each answer in decimal notation.

92. Disneyland The average number of visitors to Disneyland each day is 5.3×10^4. How many visitors visit Disneyland in a 30-day month? Express your answer in decimal notation.

93. Ice Cream Consumption Total U.S. consumption of ice cream in 2014 amounted to about 4.1 billion pounds. (SOURCE: *U.S. Department of Agriculture*)

(a) Write 4.1 billion pounds in scientific notation.

(b) The population of the United States in 2014 was approximately 319,000,000 persons. Express this number in scientific notation.

(c) Use your answers to parts (a) and (b) to find, to the nearest tenth of a pound, the average number of pounds of ice cream consumed per person in the United States in 2014.

94. M&M'S® Candy Over 400 million M&M'S® candies are produced in the United States every day. (SOURCE: *Mars.com*)

(a) Write 400 million in scientific notation.

(b) Assuming a 30-day month, how many M&M'S® candies are produced in a month in the United States? Write this answer in scientific notation.

(c) The population of the United States is expected to be approximately 323,000,000 persons in 2016. Express this number in scientific notation.

(d) Use your answers to parts (b) and (c) to find, to the nearest whole number, the average number of M&M'S candies eaten per person per month in the United States.

Extending the Concepts

Scientists often need to measure very small things, such as cells. They use the following units of measure:

$$millimeter~(mm) = 1 \times 10^{-3}~meter$$
$$micron~(\mu m) = 1 \times 10^{-6}~meter$$
$$nanometer~(nm) = 1 \times 10^{-9}~meter$$
$$picometer~(pm) = 1 \times 10^{-12}~meter$$

Write the following measurements in meters using scientific notation:

95. 250 μm **96.** 60.4 nm

97. 800 pm   **98.** 40 mm

99. 71.5 nm   **100.** 200 μm

Assume a cell is in the shape of a sphere. Given that the volume of a sphere is $V = \frac{4}{3}\pi r^3$, find the volume of each cell whose radius is given. Express the answer as a multiple of π in cubic meters.

101. 21 μm **102.** 0.75 nm

103. 6 nm   **104.** 108 μm

Explaining the Concepts

105. Explain how to convert a number written in decimal notation to scientific notation.

106. Explain how to convert a number written in scientific notation to decimal notation.

107. Dana thinks that the number 34.5×10^4 is correctly written in scientific notation. Is Dana correct? If so, explain why. If not, explain why the answer is wrong and write the correct answer.

108. Explain why scientific notation is used to perform calculations that involve multiplying and dividing but not adding and subtracting.

Chapter 5 Activity: What Is the Question?

Focus: Using exponent rules, using scientific notation, and performing operations with polynomials

Time: 15–20 minutes

Group size: 2 or 4

In this activity, you will work as a team to solve eight multiple-choice questions. However, these questions are different from most multiple-choice questions. You are given the answer to a problem and must determine which of the multiple-choice options has the correct question for the given answer.

Before beginning the activity, decide how you will approach this task as a team. For example:

- If there are two members on your team, one member will always examine choices (a) and (b), and the other will always examine choices (c) and (d).

- If there are four members on your team, one member will always examine choice (a), another member will always examine choice (b), and so on.

1. The answer is $-3x^2 - 10x$. What is the question?

(a) Simplify: $2x - 3x(x^2 + 4)$

(b) Find the quotient: $\dfrac{6x^3 - 20x^2}{-2x}$

(c) Simplify: $-3x(x^2 + 3) + 1$

(d) Find the quotient: $\dfrac{-6x^4 - 20x^3}{2x^2}$

2. The answer is $-10x^4y^8$. What is the question?

(a) Simplify: $(-5x^2y^4)^2$

(b) Simplify: $(5x^3y^3)(-2xy^5)$

(c) Simplify: $\dfrac{-30x^{-2}y^6}{3x^2y^{-2}}$

(d) Simplify: $(5x^2y^4)(-2x^2y^2)$

3. The answer is $12x^2 - 16x - 3$. What is the question?

(a) Multiply: $(6x - 1)(2x + 3)$

(b) Divide: $\dfrac{24x^3 - 32x^2 + 6x}{2x}$

(c) Multiply: $(6x + 1)(2x - 3)$

(d) Simplify: $(14x^2 - x + 2) - (2x^2 + 16x + 5)$

4. The answer is $x^2 + 5x + 6$. What is the question?

   (a) Find the product: $(x + 6)(x - 1)$

   (b) Simplify: $2x^2 + 7x + 9 - (x^2 - 2x - 3)$

   (c) Find the product: $(x + 2)(x + 3)$

   (d) Simplify: $(x + 6)^2$

5. The answer is 3. What is the question?

   (a) What is the name of the variable in $16z^2 + 3z - 5$?

   (b) What is the degree of the polynomial $2mn + 6m - 3$?

   (c) How many terms are in the polynomial $2mn + 6m - 3$?

   (d) What is the coefficient of b in the polynomial $3a^2b - 9a + 5b$?

6. The answer is 43. What is the question?

   (a) Evaluate: $4x^2 - 2x + 1$ for $x = -3$

   (b) Evaluate: $2x^2 + 3x - 4$ for $x = -2$

   (c) Evaluate: $-x^2 - 5x + 1$ for $x = -1$

   (d) Evaluate: $x^2 + 4x - 8$ for $x = -5$

7. The answer is $\dfrac{2x^5}{y^5}$. What is the question?

   (a) Simplify: $(6x^{-4}y)^0 \left(\dfrac{12x^{-1}y^{-2}}{x^{-4}y^3} \right)$

   (b) Simplify: $(6x^{-3}y^0)^{-1} \left(\dfrac{12xy^{-3}}{x^{-2}y^2} \right)$

   (c) Simplify: $(6x^2y^0)^{-1} \left(\dfrac{12x^{-2}y^3}{xy^{-4}} \right)$

   (d) Simplify: $(6x^{-2}y^0)^{-1} \left(\dfrac{12x^0y^{-2}}{x^{-3}y^3} \right)$

8. The answer is 2.5×10^{-8}. What is the question?

   (a) Find the quotient: $\dfrac{2 \times 10^3}{8 \times 10^{-4}}$

   (b) Find the product: $(5 \times 10^{-4})(5 \times 10^{-4})$

   (c) Find the quotient: $\dfrac{2 \times 10^{-3}}{8 \times 10^4}$

   (d) Find the product: $(5 \times 10^{12})(5 \times 10^{-4})$

Chapter 5 Review

Section 5.1   Adding and Subtracting Polynomials

KEY CONCEPTS

- In a monomial in the form ax^k, k is the degree of the monomial and k is a whole number.
- The degree of a polynomial is the highest degree of all the terms of the polynomial.

KEY TERMS

Monomial coefficient
Degree of a monomial
Polynomial
Binomial
Trinomial
Standard form
Degree of a polynomial

You Should Be Able To ...	EXAMPLE	Review Exercises
❶ Define monomial and determine the degree of a monomial (p. 309)	Examples 1 through 3	1–4
❷ Define polynomial and determine the degree of a polynomial (p. 310)	Examples 4 and 5	5–10
❸ Simplify polynomials by combining like terms (p. 311)	Examples 6 through 10	11–16
❹ Evaluate polynomials (p. 314)	Examples 11 through 13	17–20

In Problems 1–4, determine whether the given expression is a monomial. Answer Yes or No. For those that are monomials, state the coefficient and the degree.

1. $4x^3$ **2.** $6x^{-3}$

3. $m^{\frac{1}{2}}$ **4.** mn^2

In Problems 5–10, determine whether the algebraic expression is a polynomial. Answer Yes or No. If it is a polynomial, state the degree and then state whether it is a monomial, a binomial, or a trinomial.

5. $4x^6 - 4x^{\frac{1}{2}}$ **6.** $\dfrac{3}{x} - \dfrac{1}{x^2}$

7. 6   **8.** $3x^3 - 4xy^4$

9. $-2x^5y - 7x^4y + 7$ **10.** $\frac{1}{2}x^3 + 2x^{10} - 5$

In Problems 11–14, perform the indicated operation.

11. $(6x^2 - 2x + 1) + (3x^2 + 10x - 3)$

12. $(-7m^3 - 2mn) + (8m^3 - 5m + 3mn)$

13. $(4x^2y + 10x) - (5x^2y - 2x)$

14. $(3y^2 - yz + 3z^2) - (10y^2 + 5yz - 6z^2)$

15. Find the sum of $-6x^2 + 5$ and $4x^2 - 7$.

16. Subtract $20y^2 - 10y + 5$ from $-18y + 10$.

In Problems 17–20, evaluate the polynomial for the given value(s).

17. $3x^2 - 5x$

    (a) $x = 0$

    (b) $x = -1$

    (c) $x = 2$

18. $-x^2 + 3$

    (a) $x = 0$

    (b) $x = -1$

    (c) $x = \frac{1}{2}$

19. $x^2y + 2xy^2$ for $x = -2$ and $y = 1$

20. $4a^2b^2 - 3ab + 2$ for $a = -1$ and $b = -3$

Section 5.2 Multiplying Monomials: The Product and Power Rules

KEY CONCEPTS

- **Product Rule for Exponents**

  If a is a real number, and m and n are natural numbers, then $a^m \cdot a^n = a^{m+n}$.

- **Power Rule for Exponents**

  If a is a real number, and m and n are natural numbers, then $(a^m)^n = a^{m \cdot n}$.

- **Product to a Power Rule for Exponents**

  If a and b are real numbers and n is a natural number, then $(a \cdot b)^n = a^n \cdot b^n$.

KEY TERMS

Base

Power

Exponent

You Should Be Able To...	EXAMPLE	Review Exercises
❶ Simplify exponential expressions using the Product Rule (p. 318)	Examples 1 through 3	21, 22, 25, 26
❷ Simplify exponential expressions using the Power Rule (p. 320)	Examples 4 and 5	23, 24, 27, 28
❸ Simplify exponential expressions containing products (p. 320)	Example 6	29–32
❹ Multiply a monomial by a monomial (p. 321)	Examples 7 and 8	33–40

In Problems 21–32, simplify each expression.

21. $6^2 \cdot 6^5$

22. $\left(-\frac{1}{3}\right)^2 \left(-\frac{1}{3}\right)^3$

23. $(4^2)^6$

24. $[(-1)^4]^3$

25. $x^4 \cdot x^8 \cdot x$

26. $m^4 \cdot m^2$

27. $(r^3)^4$

28. $(m^8)^3$

29. $(4x)^3(4x)^2$

30. $(-2n)^3(-2n)^3$

31. $(-3x^2y)^4$

32. $(2x^3y^4)^2$

In Problems 33–40, multiply.

33. $3x^2 \cdot 5x^4$

34. $-4a \cdot 9a^3$

35. $-8y^4(-2y)$

36. $12p(-p^5)$

37. $\frac{8}{3}w^3 \cdot \frac{9}{2}w$

38. $\frac{1}{3}z^2\left(-\frac{9}{4}z\right)$

39. $(3x^2)^3(2x)^2$

40. $(-4a)^2(5a^4)$

Section 5.3   Multiplying Polynomials

KEY CONCEPTS

- **Extended form of the Distributive Property**
  $a(b + c + \ldots + z) = a \cdot b + a \cdot c + \ldots + a \cdot z$ where
  $a, b, c, \ldots, z$ are real numbers.
- **FOIL Method for multiplying two binomials**

  $$(ax + b)(cx + d) = \overset{F}{ax \cdot cx} + \overset{O}{ax \cdot d} + \overset{I}{b \cdot cx} + \overset{L}{b \cdot d}$$

- **Product of the Sum and Difference of Two Terms**
  $(A - B)(A + B) = A^2 - B^2$
- **Squares of Binomials**
  $(A + B)^2 = A^2 + 2AB + B^2$
  $(A - B)^2 = A^2 - 2AB + B^2$

KEY TERMS

FOIL method
Special products
Sum and difference of two terms
Squares of binomials
Perfect square trinomial

You Should be Able To...	EXAMPLE	Review Exercises
1 Multiply a polynomial by a monomial (p. 323)	Examples 1 through 3	41, 42
2 Multiply two binomials using the Distributive Property (p. 324)	Examples 4 and 5	43–46
3 Multiply two binomials using the FOIL method (p. 326)	Examples 6 and 7	49–54
4 Multiply the sum and difference of two terms (p. 327)	Examples 8 and 9	55, 56, 59, 60, 63, 64
5 Square a binomial (p. 328)	Examples 10 through 12	57, 58, 61, 62, 65, 66
6 Multiply a polynomial by a polynomial (p. 330)	Examples 13 and 14	47, 48

In Problems 41–48, multiply using the Distributive Property.

41. $-2x^3(4x^2 - 3x + 1)$ **42.** $\dfrac{1}{2}x^4(4x^3 + 8x^2 - 2)$

43. $(3x - 5)(2x + 1)$ **44.** $(4x + 3)(x - 2)$

45. $(x + 5)(x - 8)$ **46.** $(w - 1)(w + 10)$

47. $(4m - 3)(6m^2 - m + 1)$

48. $(2y + 3)(4y^4 + 2y^2 - 3)$

In Problems 49–54, use the FOIL method to find each product.

49. $(x + 5)(x + 3)$ **50.** $(2x - 1)(x - 8)$

51. $(2m + 7)(3m - 2)$ **52.** $(6m - 4)(8m + 1)$

53. $(3x + 2y)(7x - 3y)$ **54.** $(4x - y)(5x + 3y)$

In Problems 55–66, find the special products.

55. $(x - 4)(x + 4)$ **56.** $(2x + 5)(2x - 5)$

57. $(2x + 3)^2$ **58.** $(7x - 2)^2$

59. $(3x + 4y)(3x - 4y)$ **60.** $(8m - 6n)(8m + 6n)$

61. $(5x - 2y)^2$ **62.** $(2a + 3b)^2$

63. $(x - 0.5)(x + 0.5)$ **64.** $(r + 0.25)(r - 0.25)$

65. $\left(y + \dfrac{2}{3}\right)^2$

66. $2x(x + 3)(x - 3) - x(x + 2)^2$

Section 5.4 Dividing Monomials: The Quotient Rule and Integer Exponents

KEY CONCEPTS

- **Quotient Rule for Exponents**

  If a is a nonzero real number, and if m and n are integers, then $\dfrac{a^m}{a^n} = a^{m-n}$.

- **Definition of Zero as an Exponent**

  If a is a nonzero real number, then $a^0 = 1$.

- **Quotient to a Power Rule for Exponents**

  If a and b are real numbers and n is an integer, then $\left(\dfrac{a}{b}\right)^n = \dfrac{a^n}{b^n}$ if $b \neq 0$.

  If n is negative or 0, then a cannot be 0.

- **Definition of a Negative Exponent**

  If n is a positive integer and if a is a nonzero real number, then $a^{-n} = \dfrac{1}{a^n}$ and $\dfrac{1}{a^{-n}} = a^n$.

- **Quotient to a Negative Power**

  If a and b are nonzero real numbers and n is an integer, then $\left(\dfrac{a}{b}\right)^{-n} = \left(\dfrac{b}{a}\right)^n$.

You Should be Able To...	EXAMPLE	Review Exercises
❶ Simplify exponential expressions using the Quotient Rule (p. 335)	Examples 1 and 2	67–70, 75, 76
❷ Simplify exponential expressions using the Quotient to a Power Rule (p. 336)	Examples 3 and 4	77–80
❸ Simplify exponential expressions using zero as an exponent (p. 337)	Example 5	71–74
❹ Simplify exponential expressions involving negative exponents (p. 337)	Examples 6 through 12	81–86
❺ Simplify exponential expressions using the Laws of Exponents (p. 341)	Examples 13 through 15	87–92

In Problems 67–92, simplify. Write answers with only positive integer exponents. All variables are nonzero.

67. $\dfrac{6^5}{6^3}$

68. $\dfrac{7}{7^4}$

69. $\dfrac{x^{16}}{x^{12}}$

70. $\dfrac{x^3}{x^{11}}$

71. 5^0

72. -5^0

73. $m^0, \; m \neq 0$

74. $-m^0, \; m \neq 0$

75. $\dfrac{25x^3y^7}{10xy^{10}}$

76. $\dfrac{3x^4y^2}{9x^2y^{10}}$

77. $\left(\dfrac{x^3}{y^2}\right)^5$

78. $\left(\dfrac{7}{x^2}\right)^3$

79. $\left(\dfrac{2m^2n}{p^4}\right)^3$

80. $\left(\dfrac{3mn^2}{p^5}\right)^4$

81. -5^{-2}

82. $\dfrac{1}{4^{-3}}$

83. $\left(\dfrac{2}{3}\right)^{-4}$

84. $\left(\dfrac{1}{3}\right)^{-3}$

85. $2^{-2} + 3^{-1}$

86. $4^{-1} - 2^{-3}$

87. $\dfrac{16x^{-3}y^4}{24x^{-6}y^{-1}}$

88. $\dfrac{15x^0y^{-6}}{35xy^4}$

89. $(2m^{-3}n)^{-4}(3m^{-4}n^2)^2$

90. $(4m^{-6}n^0)^3(3m^{-6}n^3)^{-2}$

91. $\left(\dfrac{3rs^{-1}}{4s^2}\right)^{-2}(2r^{-6}t^0)^{-1}$

92. $\dfrac{9a}{3a^{-2}} + a(4a^2 + 7a^{-1})$

Section 5.5   Dividing Polynomials

KEY CONCEPTS

- If a, b, and c are monomials, then $\dfrac{a+b}{c} = \dfrac{a}{c} + \dfrac{b}{c}$.

- (Quotient)(Divisor) + Remainder = Dividend

KEY TERMS

Divisor
Dividend
Quotient
Remainder

You Should Be Able To...	EXAMPLE	Review Exercises
❶ Divide a polynomial by a monomial (p. 347)	Examples 1 through 3	93–98
❷ Divide a polynomial by a binomial (p. 348)	Examples 4 through 7	99–104

In Problems 93–104, perform the indicated division. All variables are nonzero.

93. $\dfrac{36x^7 - 24x^6 + 30x^2}{6x^2}$

94. $\dfrac{15x^5 + 25x^3 - 30x^2}{5x}$

95. $\dfrac{16n^8 + 4n^5 - 10n}{4n^5}$

96. $\dfrac{30n^6 - 20n^5 - 16n^3}{5n^5}$

97. $\dfrac{2p^8 + 4p^5 - 8p^3}{-16p^5}$

98. $\dfrac{3p^4 - 6p^2 + 9}{-6p^2}$

99. $\dfrac{8x^2 - 2x - 21}{2x + 3}$

100. $\dfrac{3x^2 + 17x - 6}{3x - 1}$

101. $\dfrac{6x^2 + x^3 - 2x + 1}{x - 1}$

102. $\dfrac{-6x + 2x^3 - 7x^2 + 8}{x - 2}$

103. $\dfrac{x^3 + 8}{x + 2}$

104. $\dfrac{3x^3 + 2x - 7}{x - 5}$

Section 5.6   Applying Exponent Rules: Scientific Notation

KEY CONCEPTS

- **Definition of Scientific Notation**
  A number is written in scientific notation when it is in the form $x \times 10^N$, where $1 \le x < 10$ and N is an integer.

- **Convert from Decimal Notation to Scientific Notation**
  To change a positive number into scientific notation:

  Step 1: Count the number N of decimal places that the decimal point must be moved to arrive at a number x, where $1 \le x < 10$.

  Step 2: If the original number is greater than or equal to 1, the scientific notation is $x \times 10^N$. If the original number is between 0 and 1, the scientific notation is $x \times 10^{-N}$.

- **Convert from Scientific Notation to Decimal Notation**

  Step 1: Determine the exponent, N, on the number 10.

  Step 2: If the exponent is positive, then move the decimal point N decimal places to the right. If the exponent is negative, then move the decimal point $|N|$ decimal places to the left. Add zeros, as needed.

KEY TERMS

Decimal notation
Scientific notation

You Should Be Able To...	EXAMPLE	Review Exercises
❶ Convert decimal notation to scientific notation (p. 354)	Examples 1 and 2	105–110
❷ Convert scientific notation to decimal notation (p. 355)	Examples 3 and 4	111–116
❸ Use scientific notation to multiply and divide (p. 356)	Examples 5 through 8	117–122

In Problems 105–110, write each number in scientific notation.

105. 27,000,000

106. 1,230,000,000

107. 0.00006

108. 0.00000305

109. 3

110. 8

In Problems 111–116, write each number in decimal notation.

111. 6×10^{-4}

112. 1.25×10^{-3}

113. 6.13×10^5

114. 8×10^4

115. 3.7×10^{-1}

116. 5.4×10^7

In Problems 117–122, perform the indicated operations. Express your answer in scientific notation.

117. $(1.2 \times 10^{-5})(5 \times 10^8)$

118. $(1.4 \times 10^{-10})(3 \times 10^2)$

119. $\dfrac{2.4 \times 10^{-6}}{1.2 \times 10^{-8}}$

120. $\dfrac{5 \times 10^6}{25 \times 10^{-3}}$

121. $\dfrac{200{,}000 \times 4{,}000{,}000}{0.0002}$

122. $\dfrac{1{,}200{,}000}{0.003 \times 2{,}000{,}000}$

Chapter 5 Test

Step-by-step test solutions are found on the Chapter Test Prep Videos available in MyMathLab®, on YouTube™, or may be accessed using the QR code to the right.

1. Determine whether the algebraic expression $6x^5 - 2x^4$ is a polynomial. Answer Yes or No. If it is a polynomial, state the degree and then state whether it is a monomial, a binomial, or a trinomial.

2. Evaluate the polynomial $3x^2 - 2x + 5$ for the given values:

   (a) $x = 0$ **(b)** $x = -2$ **(c)** $x = 3$

In Problems 3–11, perform the indicated operation.

3. $(3x^2y^2 - 2x + 3y) + (-4x - 6y + 4x^2y^2)$

4. $(8m^3 + 6m^2 - 4) - (5m^2 - 2m^3 + 2)$

5. $-3x^3(2x^2 - 6x + 5)$

6. $(x - 5)(2x + 7)$

7. $(2x - 7)^2$

8. $(4x - 3y)(4x + 3y)$

9. $(3x - 1)(2x^2 + x - 8)$

10. $\dfrac{6x^4 - 8x^3 + 9}{3x^3}$

11. $\dfrac{3x^3 - 2x^2 + 5}{x + 3}$

In Problems 12–16, simplify each expression. Write answers with only positive integer exponents. All variables are nonzero.

12. $(4x^3y^2)(-3xy^4)$

13. $\dfrac{18m^5n}{27m^2n^6}$

14. $\left(\dfrac{m^{-2}n^0}{m^{-7}n^4}\right)^{-6}$

15. $(4x^{-3}y)^{-2}(2x^4y^{-3})^4$

16. $(2m^{-4}n^2)^{-1}\left(\dfrac{16m^0n^{-3}}{m^{-3}n^2}\right)$

17. Write 0.000012 in scientific notation.

18. Write 2.101×10^5 in decimal notation.

In Problems 19 and 20, perform the indicated operation. Express your answer in scientific notation.

19. $(2.1 \times 10^{-6})(1.7 \times 10^{10})$

20. $\dfrac{3 \times 10^{-4}}{15 \times 10^2}$

Cumulative Review Chapters 1–5

1. Use the set $\left\{-6, -\dfrac{4}{2}, 0, 1.4, \sqrt{7}, \sqrt{25}\right\}$. List all of the elements that are
   (a) natural   (b) whole   (c) integer
   (d) rational   (e) irrational   (f) real

In Problems 2 and 3, evaluate each expression.

2. $-\dfrac{1}{2} + \dfrac{2}{3} \div 4 \cdot \dfrac{1}{3}$

3. $2 + 3[3 + 10(-1)]$

In Problems 4 and 5, simplify each algebraic expression.

4. $6x^3 - (-2x^2 + 3x) + 3x^2$

5. $-4(6x - 1) + 2(3x + 2)$

6. Solve: $-2(3x - 4) + 6 = 4x - 6x + 10$

7. Translate the following statement into an equation. DO NOT SOLVE. Four times the difference of a number and 5 is equal to 10 more than twice the number.

8. **Paycheck** Kathy's earnings this month from her part-time job working at an electronics store totaled $659.20. This amount included a 3% raise over her previous month's earnings. What were Kathy's monthly earnings before the 3% raise?

9. **Driving** Cheyenne and Amber live 306 miles apart. They start driving toward each other and meet in 3 hours. If Cheyenne drives 12 miles per hour faster than Amber, find Amber's driving speed.

10. Solve and graph the following inequality: $-5x + 2 > 17$

11. Evaluate $\dfrac{x^2 - y^2}{z}$ when $x = 3, y = -2$, and $z = -10$.

12. Find the slope of the line through $(-3, 8)$ and $(1, 2)$.

13. Graph the line $2x + 3y = 24$ by finding the intercepts.

14. Graph the line $y = -3x + 8$.

15. Solve the system of linear equations:
$$\begin{cases} 2x - 3y = 27 \\ -4x + 2y = -26 \end{cases}$$

16. Solve the system of linear equations:
$$\begin{cases} 3x - 2y = 8 \\ -6x + 4y = 8 \end{cases}$$

In Problems 17–23, perform the indicated operation.

17. $(4x^2 + 6x) - (-x + 5x^2) + (6x^3 - 2x^2)$

18. $(4m - 3)(7m + 2)$

19. $(3m - 2n)(3m + 2n)$

20. $(7x + y)^2$

21. $(2m + 5)(2m^2 - 5m + 3)$

22. $\dfrac{14xy^2 + 7x^2y}{7x^2y^2}$

23. $\dfrac{x^3 + 27}{x + 3}$

In Problems 24–27, simplify the expression. Write your answers with only positive integer exponents. All variables are nonzero.

24. $(4m^0 n^3)(-6n)$

25. $\dfrac{25m^{-6}n^{-2}}{-10m^{-4}n^{-10}}$

26. $\left(\dfrac{2xy^4}{z^{-2}}\right)^{-6}$

27. $(x^4 y^{-2})^{-4}\left(\dfrac{6x^{-4}y^3}{3y^{-8}}\right)^{-1}$

28. Write 0.0000605 in scientific notation.

29. Write 2.175×10^6 in decimal notation.

30. Perform the indicated operation. Write your answer in scientific notation.
$$(3.4 \times 10^8)(2.1 \times 10^{-3})$$

6 Factoring Polynomials

David is installing new stone tiles around his fireplace, but he wants an opening above the fireplace to install a television. The TV he purchased is 53 inches on the diagonal, excluding the casing on the sides of the TV screen. The length of the TV is 17 inches more than the width. What should the dimensions of the opening above the fireplace be if the casing on the outside of the television is 1.5 inches? See Problem 37 in Section 6.7.

The Big Picture: Putting It Together

One of the big ideas in mathematics is to be able to "undo" any operation. We have already learned how to multiply polynomials. In this chapter, we write a polynomial as a product of two or more polynomials. This process of "undoing" multiplication is called *factoring*.

Our earlier work included solving linear (first-degree) equations such as $2x + 5 = 8$. In this chapter, second-degree equations such as $2x^2 + 7x + 3 = 0$ will be solved by factoring. The approach requires that $2x^2 + 7x + 3$ be rewritten as the product of two polynomials of degree 1. This leads to solving two linear equations, which we already know how to do! This is one of the goals of algebra: Simplify a problem until it becomes a problem you already know how to solve.

Factoring is important for solving equations that contain polynomial expressions, and it plays a major role throughout the rest of the course. Be sure to master the factoring strategies presented in this chapter.

6.1 Greatest Common Factor and Factoring by Grouping

Objectives

1 Find the Greatest Common Factor of Two or More Expressions

2 Factor Out the Greatest Common Factor in Polynomials

3 Factor Polynomials by Grouping

Are You Prepared for This Section?

Before getting started, complete the following problems. If you get a problem wrong, go back to the section cited and review the material.

P1. Write 48 as the product of prime numbers.   [Section 1.2, pp. 8–9]

P2. Distribute: $2(5x - 3)$ [Section 1.8, pp. 67–68]

P3. Find the product: $(2x + 5)(x - 3)$ [Section 5.3, pp. 324–327]

▶ Consider the following products:

$$5 \cdot 3 = 15$$
$$5(y + 5) = 5y + 25$$
$$(3x - 1)(x + 5) = 3x^2 + 14x - 5$$

The expressions on the left side of the equal signs are called **factors** of the expressions on the right side. For instance, $3x - 1$ and $x + 5$ are factors of $3x^2 + 14x - 5$.

In the last chapter, we learned how to multiply expressions such as $3x - 1$ and $x + 5$ to obtain the polynomial $3x^2 + 14x - 5$. In this chapter, we learn how to obtain the factors of a polynomial. That is, we learn how to write $3x^2 + 14x - 5$ as $(3x - 1)(x + 5)$.

> **Definition**
>
> To **factor** a polynomial means to write the polynomial as a product of two or more polynomials.

The process of factoring reverses the process of multiplying, as shown below.

> **In Other Words**
> Factoring is "undoing" multiplication.

$$\text{Multiplication} \rightarrow$$

Factored form → $3x(x - 7) = 3x^2 - 21x$ ← Product

$$\leftarrow \text{Factoring}$$

▶ **1** Find the Greatest Common Factor of Two or More Expressions

Work Smart

Remember, all monomials and binomials are polynomials. So $4, -6, x^3, 5x^4$, and $x + 1$ are all examples of polynomials.

In the illustration above, notice that $3x$ is the largest polynomial that divides evenly into both $3x^2$ and $21x$ in the expression $3x^2 - 21x$. So $3x$ is the *greatest common factor* of $3x^2 - 21x$.

> **Definition**
>
> The **greatest common factor (GCF)** is the largest polynomial that divides evenly into all the polynomials.

Prepared?...Answers P1. $3 \cdot 2^4$
P2. $10x - 6$ **P3.** $2x^2 - x - 15$

Writing $3x^2 - 21x$ as $3x(x - 7)$ is referred to as factoring out the *greatest common factor*. But how can we find the GCF? The following example shows us.

EXAMPLE 1   **How to Find the Greatest Common Factor (GCF) of a List of Numbers**

Find the GCF of 12 and 18.

Step-by-Step Solution

Step 1: Write each number as the product of prime factors.

$$12 = 2 \cdot 2 \cdot 3$$
$$18 = 2 \quad \cdot 3 \cdot 3$$

Step 2: Determine the common prime factors.

The common prime factors are 2 and 3.

Step 3: Find the product of the common prime factors from Step 2. This number is the GCF.

The GCF is $2 \cdot 3 = 6$.

EXAMPLE 2 **Finding the Greatest Common Factor (GCF) of a List of Numbers**

Find the GCF of 24, 40, and 72.

Solution

Write each number as the product of prime factors. It is helpful to align the common factors vertically.

$$24 = 4 \cdot \ 6 = 2 \cdot 2 \cdot 2 \cdot 3$$
$$40 = 4 \cdot 10 = 2 \cdot 2 \cdot 2 \cdot \quad 5$$
$$72 = 8 \cdot \ 9 = 2 \cdot 2 \cdot 2 \cdot 3 \cdot 3$$

All three numbers have three factors of 2, so the GCF is $2 \cdot 2 \cdot 2 = 8$.

The greatest common factor in Example 2 could be written as 2^3. The exponent 3 shows the number of times the factor 2 appears in the factorization of each number.

Work Smart

Remember, a prime number is a number greater than 1 that has no factors other than itself and 1. For example, 3, 7, and 13 are prime numbers, whereas $4 (= 2 \cdot 2)$, $12 (= 2 \cdot 2 \cdot 3)$, and $35 (= 5 \cdot 7)$ are not prime. For more explanation on finding prime factorization, see Section 1.2.

> **Quick ✓**
>
> 1. The largest polynomial that divides evenly into a list of polynomials is called the _____ _____ _____.
> 2. In the product $(3x - 2)(x + 4) = 3x^2 + 10x - 8$, the polynomials $(3x - 2)$ and $(x + 4)$ are called _____ of the polynomial $3x^2 + 10x - 8$.
>
> *In Problems 3–5, find the greatest common factor of each list of numbers.*
>
> **3.** 32, 40   **4.** 12, 45   **5.** 21, 35, 84

▶ The approach to finding the greatest common factor of two or more expressions that contain variables is the same as it is for numbers. For example, x^3, x^5, and x^6 can be written as the product of factors of x as follows:

$$x^3 = x \cdot x \cdot x$$
$$x^5 = x \cdot x \cdot x \cdot x \cdot x$$
$$x^6 = x \cdot x \cdot x \cdot x \cdot x \cdot x$$

Work Smart

The GCF of a common variable factor is the smallest power of that variable.

Each of the monomials (x^3, x^5, x^6) contains three factors of x, so the greatest common factor is x^3. Notice that the exponent of the GCF is 3, the smallest exponent of x^3, x^5, and x^6. This approach to finding the GCF for variable expressions will work in general.

EXAMPLE 3 **Finding the Greatest Common Factor (GCF)**

Find the GCF of $8y^4$ and $12y^2$.

Solution

Step 1: Find the GCF of the coefficients, 8 and 12.

$$8 = 2 \cdot 2 \cdot 2$$
$$12 = 2 \cdot 2 \ \cdot 3$$

The GCF of the coefficients is $2 \cdot 2 = 4$.

Step 2: The variable factors are y^4 and y^2. For each variable, determine the smallest exponent to which the common variable is raised. The GCF of y^4 and y^2 is y^2.

Step 3: Multiply the common factors from Steps 1 and 2 to get the GCF, $4y^2$.

> **Finding the Greatest Common Factor (GCF)**
>
> **Step 1:** Find the GCF of the coefficients of each variable expression.
>
> **Step 2:** For each variable factor common to all the expressions, determine the smallest exponent to which the variable factor is raised.
>
> **Step 3:** Find the product of the common factors found in Steps 1 and 2. This expression is the GCF.

EXAMPLE 4   **Finding the Greatest Common Factor**

Find the GCF:

(a) $3x^3, 9x^2, 21x$

(b) $10x^5y^4, 15x^2y^3, 25x^3y^5$

Solution

Find the GCF of the coefficients, and then find the common variable factor(s) with the smallest exponent. The product of these factors is the GCF.

(a) The coefficients are 3, 9, and 21.

Factor the coefficients as products of primes:
$$3 = 3$$
$$9 = 3 \cdot 3$$
$$21 = 3 \cdot \quad 7$$

The GCF of the coefficients is 3.
  The smallest exponent of x^3, x^2, and x is 1, so the GCF of the variable factors is x. Therefore, the GCF of $3x^3, 9x^2$, and $21x$ is $3x$.

(b) The coefficients are 10, 15, and 25.

Factor the coefficients as products of primes:
$$10 = 5 \cdot 2$$
$$15 = 5 \cdot \quad 3$$
$$25 = 5 \cdot \quad\quad 5$$

The GCF of the coefficients is 5.
  The GCF of x^5, x^2, and x^3 is x^2. The GCF of y^4, y^3, and y^5 is y^3. Therefore, the GCF of $10x^5y^4, 15x^2y^3$, and $25x^3y^5$ is $5x^2y^3$. ●

Quick ✓

In Problems 6–8, find the greatest common factor (GCF).

6. $14y^3, 35y^2$

7. $6z^3, 8z^2, 12z$

8. $4x^3y^5, 8x^2y^3, 24xy^4$

The greatest common factor can be a binomial, as illustrated in the following example.

EXAMPLE 5   **The Greatest Common Factor as a Binomial**

Find the GCF of each pair of expressions.

(a) $3(x - 1)$ and $8(x - 1)$

(b) $2(z + 3)(z + 5)$ and $4(z + 5)^2$

Solution

(a) The coefficients, 3 and 8, have no common factor. However, each expression has $x - 1$ as a factor, so the GCF of $3(x - 1)$ and $8(x - 1)$ is $x - 1$.

(b) The GCF of 2 and 4 is 2. The GCF of $(z + 3)(z + 5)$ and $(z + 5)^2$ is $z + 5$. The GCF of $2(z + 3)(z + 5)$ and $4(z + 5)^2$ is $2(z + 5)$. ●

Quick ✓

In Problems 9 and 10, find the greatest common factor (GCF).

9. $7(2x + 3)$ and $-4(2x + 3)$

10. $9(k + 8)(3k - 2)$ and $12(k - 1)(k + 8)^2$

▶ ❷ **Factor Out the Greatest Common Factor in Polynomials**

The first step in factoring any polynomial is to look for the greatest common factor of the terms of the polynomial. Then use the Distributive Property "in reverse" to factor the polynomial as shown below.

$$ab + ac = a(b + c) \quad \text{or} \quad ab - ac = a(b - c)$$

The process of using the Distributive Property "in reverse" is called "factoring out" the greatest common factor.

EXAMPLE 6 **How to Factor Out the Greatest Common Factor in a Polynomial**

Factor $2x - 10$ by factoring out the greatest common factor.

Step-by-Step Solution

Step 1: Find the GCF of the terms of the polynomial.

$$2x = 2 \cdot x$$
$$-10 = 2 \; \cdot (-5)$$
$$\text{GCF} = 2$$

Step 2: Rewrite each term as the product of the GCF and the remaining factor.

$$2x - 10 = 2(x) - 2(5)$$

Step 3: Factor out the GCF.

$$= 2(x - 5)$$

Step 4: Check Use the Distributive Property to multiply.

$$2(x - 5) = 2(x) - 2(5)$$
$$= 2x - 10$$

Therefore, $2x - 10 = 2(x - 5)$.

Work Smart

The Distributive Property comes in handy for factoring out the GCF. Note how it is used in the check step, too.

> **Factoring a Polynomial Using the Greatest Common Factor**
>
> **Step 1:** Identify the greatest common factor (GCF) of the terms of the polynomial.
>
> **Step 2:** Rewrite each term as the product of the GCF and the remaining factor.
>
> **Step 3:** Use the Distributive Property "in reverse" to factor out the GCF.
>
> **Step 4:** Check using the Distributive Property.

EXAMPLE 7 **Factoring Out the Greatest Common Factor in a Binomial**

Factor the binomial $9z^3 + 36z^2$ by factoring out the greatest common factor.

Solution

The greatest common factor of 9 and 36 is 9. The greatest common factor of z^3 and z^2 is z^2. Therefore, the GCF is $9z^2$.

Rewrite each term as the product of the GCF and the remaining factor: $\quad 9z^3 + 36z^2 = 9z^2(z) + 9z^2(4)$
Factor out the GCF: $\qquad\qquad = 9z^2(z + 4)$

Check $\quad 9z^2(z + 4) = 9z^2(z) + 9z^2(4)$
$$= 9z^3 + 36z^2$$

So, $9z^3 + 36z^2 = 9z^2(z + 4)$.

EXAMPLE 8 **Factoring Out the Greatest Common Factor in a Trinomial**

Factor the trinomial $6a^2b^2 - 8ab^3 + 18a^3b^4$ by factoring out the greatest common factor.

Solution

The GCF of $6a^2b^2 - 8ab^3 + 18a^3b^4$ is $2ab^2$. Now rewrite each term as the product of the GCF and the remaining factor.

$$6a^2b^2 - 8ab^3 + 18a^3b^4 = 2ab^2(3a) - 2ab^2(4b) + 2ab^2(9a^2b^2)$$

Factor out the GCF: $= 2ab^2(3a - 4b + 9a^2b^2)$

Check $\quad 2ab^2(3a - 4b + 9a^2b^2) = 2ab^2(3a) - 2ab^2(4b) + 2ab^2(9a^2b^2)$

$$= 6a^2b^2 - 8ab^3 + 18a^3b^4$$

Therefore, $6a^2b^2 - 8ab^3 + 18a^3b^4 = 2ab^2(3a - 4b + 9a^2b^2)$. ●

Quick ✓

11. To _____ a polynomial means to write the polynomial as a product of two or more polynomials.

12. When factoring a polynomial using the GCF, use the _____ Property "in reverse."

In Problems 13–16, factor each polynomial by factoring out the greatest common factor.

13. $5z^2 + 30z$

14. $12p^2 - 12p$

15. $16y^3 - 12y^2 + 4y$

16. $6m^4n^2 + 18m^3n^4 - 22m^2n^5$

▶ When the coefficient of the term of highest degree is negative, factor the negative out of the polynomial.

EXAMPLE 9 **Factoring Out a Negative Greatest Common Factor**

Factor $-7a^3 + 14a$ by factoring out the greatest common factor.

Solution

This binomial is in standard form. Because the coefficient of the highest-degree term, $-7a^3$, is negative, factor the negative out. Thus the GCF is $-7a$.

$$-7a^3 + 14a = -7a(a^2) + (-7a)(-2)$$

Factor out the GCF: $= -7a(a^2 - 2)$

Check $\quad -7a(a^2 - 2) = -7a(a^2) + (-7a)(-2)$

$$= -7a^3 + 14a$$

Therefore, $-7a^3 + 14a = -7a(a^2 - 2)$. ●

Quick ✓

In Problems 17 and 18, factor out the greatest common factor.

17. $-4y^2 + 8y$

18. $-6a^3 + 12a^2 - 3a$

Recall from Example 7 that the greatest common factor may be a binomial.

EXAMPLE 10   **Factoring Out a Binomial as the Greatest Common Factor**

Factor out the greatest common factor: $5x(x - 2) + 3(x - 2)$

Solution

Do you see that $x - 2$ is common to both terms? The GCF is the binomial $(x - 2)$.

$$5x(x - 2) + 3(x - 2) = 5x(x - 2) + 3(x - 2)$$

Factor out $(x - 2)$: $= (x - 2)(5x + 3)$

Check $(x - 2)(5x + 3) = (x - 2)5x + (x - 2)3$

$$= 5x(x - 2) + 3(x - 2)$$

Therefore, $5x(x - 2) + 3(x - 2) = (x - 2)(5x + 3)$.   ●

> **Quick ✓**
>
> **19.** *True or False* The polynomial $(2x + 1)(x - 3) + (2x + 1)(2x + 7)$ is factored as $(2x + 1)^2(3x + 4)$.
>
> *In Problems 20 and 21, factor out the greatest common factor.*
>
> **20.** $2a(a - 5) + 3(a - 5)$ **21.** $7z(z + 5) - 4(z + 5)$

▶ ❸ **Factor Polynomials by Grouping**

Work Smart

Try factoring by grouping when a polynomial contains four terms.

Sometimes a common factor does not occur in every term of the polynomial. If a polynomial contains four terms, it may be possible to find a GCF of the first two terms and a different GCF of the second two terms and have the same remaining factor from each pair. When this happens, the common factor can be factored out of each group of terms. This technique is called **factoring by grouping.**

EXAMPLE 11   **How to Factor by Grouping**

Factor by grouping: $3x - 3y + ax - ay$

Step-by-Step Solution

Step 1: Group terms with common factors. $3x - 3y + ax - ay = (3x - 3y) + (ax - ay)$

Step 2: In each grouping, factor out the GCF. $= 3(x - y) + a(x - y)$

Step 3: Factor out the common factor that remains.   Factor out $(x - y)$: $= (x - y)(3 + a)$

Step 4: Check

Multiply: $(x - y)(3 + a) = 3x + ax - 3y - ay$ F O I L

Rearrange terms: $= 3x - 3y + ax - ay$

Therefore, $3x - 3y + ax - ay = (x - y)(3 + a)$.   ●

Based upon Example 11, use the following steps for factoring by grouping.

Work Smart

The answer to Example 11 could have been written as $(3 + a)(x - y)$. Do you know why?

> **Factoring a Polynomial By Grouping**
>
> **Step 1:** Group the terms with common factors.
>
> **Step 2:** In each grouping, factor out the greatest common factor (GCF).
>
> **Step 3:** If the remaining factor in each grouping is the same, factor it out.
>
> **Step 4:** Check your work by finding the product of the factors.

► EXAMPLE 12

Factoring by Grouping

Factor by grouping: $5x - 5y - 4bx + 4by$

Work Smart

Be careful when factoring a negative factor from a polynomial. In Example 12, notice that $-4b$ is factored out in the second grouping. If $4b$ is factored out instead, we do not end up with the common factor, $x - y$.

Solution

First, group terms with common factors. Be careful when grouping the third and fourth terms because of the minus sign with $-4bx$. It is helpful to rewrite subtraction as addition using the definition $a - b = a + (-b)$ so that the signs in the factored form are correct.

$$5x - 5y - 4bx + 4by = 5x - 5y + (-4bx) + 4by$$
$$= (5x - 5y) + (-4bx + 4by)$$

Factor out the common factor in each group: $= 5(x - y) + (-4b)(x - y)$

Factor out the common factor that remains: $= (x - y)(5 - 4b)$

Check   Multiply: $(x - y)(5 - 4b) = 5x - 4bx - 5y + 4by$

Rearrange terms: $= 5x - 5y - 4bx + 4by$

Therefore, $5x - 5y - 4bx + 4by = (x - y)(5 - 4b)$.   ●

In Example 12, we could have used the Commutative Property of Addition to rewrite $5x - 5y - 4bx + 4by$ as $5x - 4bx - 5y + 4by$ and factored as follows:

$$5x - 4bx - 5y + 4by = x(5 - 4b) - y(5 - 4b)$$
$$= (5 - 4b)(x - y)$$

Because $(5 - 4b)(x - y) = (x - y)(5 - 4b)$ by the Commutative Property of Multiplication, the answer is equivalent to the result of Example 12.

Work Smart

Sometimes terms need to be rearranged before factoring by grouping.

In any factoring problem, always look for the greatest common factor first.

EXAMPLE 13

Factoring by Grouping

Factor: $3x^3 + 12x^2 - 6x - 24$

Solution

Do all four terms contain a common factor? Yes!! There is a GCF of 3, so factor it out first.

Work Smart

Whenever factoring, always look for a GCF first.

$$3x^3 + 12x^2 - 6x - 24 = 3(x^3 + 4x^2 - 2x - 8)$$

Group terms with common factors: $= 3[(x^3 + 4x^2) + (-2x - 8)]$

Factor out the common factor in each group: $= 3[(x^2(x + 4) + (-2)(x + 4)]$

Factor out the common factor that remains: $= 3[(x + 4)(x^2 - 2)]$ or $3(x + 4)(x^2 - 2)$

Check $3(x + 4)(x^2 - 2) = 3(x^3 - 2x + 4x^2 - 8)$

Distribute the 3: $= 3x^3 - 6x + 12x^2 - 24$

Rearrange terms: $= 3x^3 + 12x^2 - 6x - 24$

Therefore, $3x^3 + 12x^2 - 6x - 24 = 3(x + 4)(x^2 - 2)$.   ●

Quick ✔

In Problems 25 and 26, factor by grouping.

25. $3z^3 + 12z^2 + 6z + 24$

26. $2n^4 + 2n^3 - 4n^2 - 4n$

6.1 Exercises MyMathLab®

Exercise numbers in green have complete video solutions in MyMathLab or may be accessed using the QR code to the right.

Problems **1–26** are the **Quick ✔** s that follow the **EXAMPLES**.

Building Skills

In Problems 27–46, find the greatest common factor, GCF, of each group of expressions. See Objective 1.

27. 8, 12

28. 49, 35

29. 15, 14

30. 6, 55

31. 12, 36, 54

32. 16, 40, 64

33. x^{10}, x^2, x^8

34. y^3, y^5, y

35. $7x, 14x^3$

36. $8a^4, 20a^2$

37. $45a^2b^3, 75ab^2c$

38. $26xy^2, 39x^2y$

39. $4a^2bc^3, 6ab^2c^2, 8a^2b^2c^4$

40. $2x^2yz, xyz^2, 5x^3yz^2$

41. $3(x - 1)$ and $10(x - 1)$

42. $8(x + y)$ and $9(x + y)$

43. $2(x - 4)^2$ and $4(x - 4)^3$

44. $6(a - b)$ and $15(a - b)^3$

45. $12(x + 2)(x - 3)^2$ and $18(x - 3)^2(x - 2)$

46. $15(2a - 1)^2(2a + 1)$ and $18(2a - 1)^3(2a + 3)^2$

In Problems 47–70, factor the GCF from the polynomial. See Objective 2.

47. $12x - 18$

48. $3a + 6$

49. $x^2 + 12x$

50. $b^2 - 6b$

51. $5x^2y - 15x^3y^2$

52. $8a^3b^2 + 12a^5b^2$

53. $3x^3 + 6x^2 - 3x$

54. $5x^4 + 10x^3 - 25x^2$

55. $-5x^3 + 10x^2 - 15x$

56. $-2y^2 + 10y - 14$

57. $9m^5 - 18m^3 - 12m^2 + 81$

58. $5z^2 + 10z^4 - 15z^3 - 45z^5$

59. $-12z^3 + 16z^2 - 8z$

60. $-22n^4 + 18n^2 + 14n$

61. $10 - 5b - 15b^3$

62. $14m^2 - 16m^3 - 24m^4$

63. $15a^2b^4 - 60ab^3 + 45a^3b^2$

64. $12r^3s^2 + 3rs - 6rs^4$

65. $x(x - 3) - 5(x - 3)$

66. $a(a - 5) + 6(a - 5)$

67. $x^2(x - 1) + y^2(x - 1)$

68. $(b + 2)a^2 - (b + 2)b$

69. $x^2(4x + 1) + 2x(4x + 1) + 5(4x + 1)$

70. $s^2(s^2 + 1) + 4s(s^2 + 1) + 7(s^2 + 1)$

In Problems 71–80, factor by grouping. See Objective 3.

71. $xy + 3y + 4x + 12$

72. $x^2 + ax + 2a + 2x$

73. $yz + z - y - 1$

74. $mn - 3n + 2m - 6$

75. $x^3 - x^2 + 2x - 2$

76. $z^3 + 4z^2 + 3z + 12$

77. $2t^3 - t^2 - 4t + 2$

78. $x^3 - x^2 - 5x + 5$

79. $2t^4 - t^3 - 6t + 3$

80. $6yz - 8y - 9z + 12$

Mixed Practice

In Problems 81–102, factor each polynomial.

81. $4y - 20$

82. $3z - 21$

83. $28m^3 + 7m^2 + 63m$

84. $10x^3 - 15x^2 - 5x$

85. $12m^3n^2p - 18m^2n$

86. $4s^2t^3 - 24st$

87. $(2p - 1)(p + 3) + (7p + 4)(p + 3)$

88. $(3x - 2)(4x + 1) + (3x - 2)(x - 10)$

89. $18ax - 9ay - 12bx + 6by$

90. $30xm + 15xn - 20ym - 10yn$

91. $(x - 2)(x - 3) + (x - 2)$

92. $(a + 2)(2b - 1) - (2b - 1)$

93. $15x^4 - 6x^3 + 30x^2 - 12x$

94. $2a^3 - 4a^2 + 8a - 16$

95. $-3x^3 + 6x^2 - 9x$

96. $-8z^4 - 12z^3 + 28z^2$

97. $-12b + 16b^2$

98. $-8a + 20a^2$

99. $12xy + 9x - 8y - 6$

100. $-12mxz - 3xz + 24mx + 6x$

101. $\frac{1}{3}x^3 - \frac{2}{9}x^2$

102. $\frac{3}{4}p^4 - \frac{1}{4}p^3$

Applying the Concepts

103. Height of a Toy Rocket The height of a toy rocket after t seconds, when it is fired straight up from the ground with an initial speed of 150 feet per second, is given by the polynomial $-16t^2 + 150t$. Write the polynomial $-16t^2 + 150t$ in factored form.

104. Height of a Ball The height of a ball after t seconds, when it is thrown straight up from a height of 48 feet with an initial speed of 80 feet per second, is given by the polynomial $-16t^2 + 80t + 48$. Write the polynomial $-16t^2 + 80t + 48$ in factored form.

105. Selling Calculators A manufacturer of calculators found that the number of calculators sold at a price of p dollars is given by the polynomial $21{,}000 - 150p$. Write $21{,}000 - 150p$ in factored form.

106. Revenue A manufacturer of a gas clothes dryer has found that the revenue (in dollars) from selling clothes dryers at a price of p dollars is given by the expression $-4p^2 + 4000p$. Write $-4p^2 + 4000p$ in factored form.

△ **107. Area of a Rectangle** A rectangle has an area of $(8x^5 - 28x^3)$ square feet. Write $8x^5 - 28x^3$ in factored form.

△ **108. Area of a Parallelogram** A parallelogram has an area of $(18n^4 - 15n^3 + 6n)$ square cm. Write $18n^4 - 15n^3 + 6n$ in factored form.

△ **109. Surface Area** The surface area of a cylindrical can whose radius is r inches and height is 4 inches is given by $S = (2\pi r^2 + 8\pi r)$ square inches. Express the surface area in factored form.

△ **110. Surface Area** The surface area of a cylindrical can whose radius is r inches and height is 8 inches is given by $S = (2\pi r^2 + 16\pi r)$ square inches. Express the surface area in factored form.

In Problems 111 and 112, the area of the polygon is given. Write a polynomial that represents the missing length.

△ **111.**

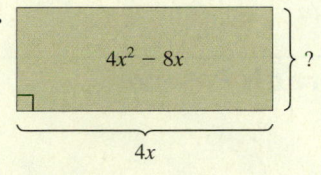

△ **112.**

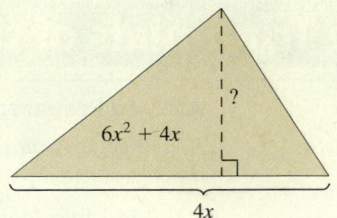

Extending the Concepts

In Problems 113–120, find the missing factor.

113. $8x^{3n} + 10x^n = 2x^n \cdot ?$

114. $3x^{2n+2} + 6x^{6n} = 3x^{2n} \cdot ?$

115. $3 - 4x^{-1} + 2x^{-3} = x^{-3} \cdot ?$

116. $1 - 3x^{-1} + 2x^{-2} = x^{-2} \cdot ?$

117. $x^2 - 3x^{-1} + 2x^{-2} = x^{-2} \cdot ?$

118. $2x^2 + x^{-3} - x^{-4} = x^{-4} \cdot ?$

119. $\dfrac{6}{35}x^4 - \dfrac{1}{7}x^2 + \dfrac{2}{7}x = \dfrac{2}{7}x \cdot ?$

120. $\dfrac{2}{15}x^3 - \dfrac{1}{9}x^2 + \dfrac{1}{3}x = \dfrac{1}{3}x \cdot ?$

Explaining the Concepts

121. Write a list of steps for finding the greatest common factor, and then write a second list of steps for factoring the GCF from a polynomial.

122. Describe how to factor a negative from a polynomial. What types of errors might happen during this process?

123. Explain the error in the following student's work:

$$3a(x + y) - 4b(x + y) = (3a - 4b)(x + y)^2$$

124. On a test, your answer for factoring

$$5z(x - 4) - 6(x - 4)$$

was $(x - 4)(5z - 6)$. Your friend Jack's answer was $(5z - 6)(x - 4)$. Explain why both answers are correct.

6.2 Factoring Trinomials of the Form $x^2 + bx + c$

Objectives

① Factor Trinomials of the Form $x^2 + bx + c$

② Factor Out the GCF, Then Factor $x^2 + bx + c$

Are you Prepared for This Section?

Before getting started, complete the following problems. If you get a problem wrong, go back to the section cited and review the material.

P1. Find two factors of 18 whose sum is 11.   [Section 1.2, pp. 8–9]

P2. Find two factors of -24 whose sum is -2.   [Section 1.4, pp. 29–30]

P3. Find two factors of -12 whose sum is 1.   [Section 1.4, pp. 29–30]

P4. Find two factors of 35 whose sum is -12.   [Section 1.4, pp. 29–30]

P5. Determine the coefficients of $3x^2 - x - 4$.   [Section 1.8, p. 66]

P6. Find the product: $-5p(p + 4)$ [Section 5.3, pp. 323–324]

P7. Find the product: $(z - 1)(z + 4)$ [Section 5.3, pp. 324–327]

P8. Find the degree: **(a)** $3x^2$ **(b)** $-z$ [Section 5.1, pp. 309–310]

▶ In this section, we factor trinomials of degree 2. Because the word **quadratic** means "relating to a square," these trinomials are called *quadratic trinomials*.

> **Definition**
>
> A **quadratic trinomial** is a polynomial of the form $ax^2 + bx + c, a \neq 0$, where a represents the coefficient of the squared (second-degree) term, b represents the coefficient of the linear (first-degree) term, and c represents the constant.

In the trinomial $ax^2 + bx + c$, a is called the **leading coefficient.** We begin by looking at quadratic trinomials where the leading coefficient, a, is 1. Examples of quadratic trinomials whose leading coefficient is 1 are

$$x^2 + 4x + 3 \qquad a^2 + 4a - 21 \qquad p^2 - 11p + 18$$

▶ ① Factor Trinomials of the Form $x^2 + bx + c$

The goal in factoring a quadratic (or second-degree) trinomial is to write it as the product of two first-degree polynomials.

For example,

Multiplication →

Factored form → $(x + 3)(x - 7) = x^2 - 4x - 21$ ← Product

← Factoring

The factors of $x^2 - 4x - 21$ are $x + 3$ and $x - 7$. Notice the following:

$$x^2 - 4x - 21 = (x + 3)(x - 7)$$

The sum of -7 and 3 is -4.

The product of -7 and 3 is -21.

In general, if $x^2 + bx + c = (x + m)(x + n)$, then $mn = c$ and $m + n = b$.

EXAMPLE 1   **How to Factor a Trinomial of the Form $x^2 + bx + c$, Where c Is Positive**

Factor: $x^2 + 7x + 12$

Step-by-Step Solution

Step 1: In comparing $x^2 + 7x + 12$ with $x^2 + bx + c$, notice that $b = 7$ and $c = 12$. Find factors of $c = 12$ whose sum is $b = 7$. Begin by listing all factors of 12 and computing the sum of these factors.

Factors of 12	1, 12	2, 6	3, 4	$-1, -12$	$-2, -6$	$-3, -4$
Sum	13	8	7	-13	-8	-7

Because $3 \cdot 4 = 12$ and $3 + 4 = 7$, let $m = 3$ and $n = 4$.

Step 2: Write the trinomial in the form $(x + m)(x + n)$.

$$x^2 + 7x + 12 = (x + 3)(x + 4)$$

Step 3: Check Multiply the binomials.

$$F \quad O \quad I \quad L$$
$$(x + 3)(x + 4) = x^2 + 4x + 3x + 3(4)$$
$$= x^2 + 7x + 12$$

Therefore, $x^2 + 7x + 12 = (x + 3)(x + 4)$. ●

Work Smart

Because multiplication is commutative, the order in which the factors are listed does not matter.

> **Factoring a Trinomial of the Form $x^2 + bx + c$**
>
> **Step 1:** Find the pair of integers whose product is c and whose sum is b. That is, determine m and n such that $mn = c$ and $m + n = b$.
>
> **Step 2:** Write $x^2 + bx + c = (x + m)(x + n)$.
>
> **Step 3:** Check your work by multiplying the binomials.

In Example 1, notice that the coefficient of the middle term is positive and the constant is positive. **If the coefficient of the middle term and the constant are both positive, then both factors of the constant must be positive.**

EXAMPLE 2   **Factoring a Trinomial of the Form $x^2 + bx + c$, Where c Is Positive**

Factor: $p^2 - 11p + 24$

Solution

Because $b = -11$ and $c = 24$, find factors of 24 whose sum is -11. Begin by listing all factors of 24 and computing the sum of these factors.

Factors of 24	1, 24	2, 12	3, 8	4, 6	$-1, -24$	$-2, -12$	$-3, -8$	$-4, -6$
Sum	25	14	11	10	-25	-14	-11	-10

Because $(-3)(-8) = 24$ and $-3 + (-8) = -11$, let $m = -3$ and $n = -8$. Now write the trinomial in the form $(p + m)(p + n)$.

$$p^2 - 11p + 24 = (p + (-3))(p + (-8))$$
$$= (p - 3)(p - 8)$$

Check $(p - 3)(p - 8) = p^2 - 8p - 3p + (-3)(-8)$
$$= p^2 - 11p + 24$$

So, $p^2 - 11p + 24 = (p - 3)(p - 8)$. ●

Work Smart

The sum of two even numbers is even, the sum of two odd numbers is even, the sum of an odd and an even number is odd. In Example 2, the coefficient of the middle term is odd (-11), so one of the factors of 24 will be odd, and the other will be even.

Use this result on the remaining examples to reduce the list of possible factors.

In Example 2, notice that the coefficient of the middle term is negative and the constant is positive. **If the coefficient of the middle term is negative and the constant is positive, then both factors of the constant must be negative.**

Quick ✓

1. A _____ _____ is a polynomial of the form $ax^2 + bx + c, a \neq 0$.

2. When factoring $x^2 - 10x + 24$, look for two numbers whose product is __ and whose sum is ___.

3. *True or False* $4 + 4x + x^2$ has a leading coefficient of 4.

In Problems 4 and 5, factor each trinomial.

4. $y^2 + 9y + 20$ 5. $z^2 - 9z + 14$

⊙ **EXAMPLE 3** | **Factoring a Trinomial of the Form $x^2 + bx + c$, Where c Is Negative**

Factor: $z^2 + 3z - 28$

Solution
Because $b = 3$ and $c = -28$, look for factors of -28 whose sum is 3. Because c is negative, one of the factors must be positive and the other negative. And because b is odd, one factor must be odd and the other even.

Factors of -28	$-1, 28$	$-4, 7$	$1, -28$	$4, -7$
Sum	27	3	-27	-3

Notice that $-4 \cdot 7 = -28$ and $-4 + 7 = 3$, so let $m = -4$ and $n = 7$. Write the trinomial in the form $(z + m)(z + n)$.

$$z^2 + 3z - 28 = (z + (-4))(z + 7)$$
$$= (z - 4)(z + 7)$$

Check $(z - 4)(z + 7) = z^2 + 7z - 4z + (-4)(7)$
$$= z^2 + 3z - 28$$

Therefore, $z^2 + 3z - 28 = (z - 4)(z + 7)$.   ●

In Example 3, notice that the coefficient of the middle term is positive and the constant is negative. **If the constant is negative, then the factors of the constant must have opposite signs. In addition, if the coefficient of the middle term is positive, then the factor with the larger absolute value must be positive.**

EXAMPLE 4 | **Factoring a Trinomial of the Form $x^2 + bx + c$, Where c Is Negative**

Factor: $a^2 - 7a - 18$

Solution
Find factors of $c = -18$ whose sum is $b = -7$. Begin by listing all factors of -18 and computing their sum.

Factors of -18	$-1, 18$	$-2, 9$	$-3, 6$	$1, -18$	$2, -9$	$3, -6$
Sum	17	7	3	-17	-7	-3

Because $(2)(-9) = -18$ and $2 + (-9) = -7$, let $m = 2$ and $n = -9$. Write the trinomial in the form $(a + m)(a + n)$.

$$a^2 - 7a - 18 = (a + 2)(a + (-9))$$
$$= (a + 2)(a - 9)$$

Check $(a + 2)(a - 9) = a^2 - 9a + 2a + 2(-9)$
$$= a^2 - 7a - 18$$

Therefore, $a^2 - 7a - 18 = (a + 2)(a - 9)$.   ●

In Example 4, notice that the coefficient of the middle term is negative and the constant is also negative. **If the coefficient of the middle term is negative and the constant is negative, then the factors of the constant must have opposite signs, and the factor with the larger absolute value must be negative.**

Quick ✔

6. *True or False* When factoring $x^2 - 10x - 24$, the factors of 24 must have opposite signs.

In Problems 7 and 8, factor each trinomial.

7. $y^2 - 2y - 15$ **8.** $w^2 + w - 12$

Table 1 summarizes the four possibilities for factoring a quadratic trinomial in the form $x^2 + bx + c = (x + m)(x + n)$.

Table 1 Factoring $x^2 + bx + c$

Signs of b and c	Signs of m and n	Example
b and c are both positive	Both factors are positive.	$x^2 + 3x + 2 = (x + 2)(x + 1)$
b is negative and c is positive	Both factors are negative.	$a^2 - 7a + 12 = (a - 4)(a - 3)$
b is positive and c is negative	Factors have opposite signs; the factor with the larger absolute value is positive.	$y^2 + 2y - 24 = (y + 6)(y - 4)$
b and c are both negative	Factors have opposite signs; the factor with the larger absolute value is negative.	$b^2 - 4b - 21 = (b - 7)(b + 3)$

Definition

A polynomial that cannot be written as the product of two other polynomials (other than 1 or -1) is a **prime polynomial.**

▶ **EXAMPLE 5**   **Identifying a Prime Trinomial**

Show that $y^2 + 10y + 12$ is prime.

Solution

Look for factors of $c = 12$ whose sum is $b = 10$. Because both b and c are positive, the factors of 12 must both be positive, so consider only positive factors of 12 and the sum of these factors.

Factors of 12	1, 12	2, 6	3, 4
Sum	13	8	7

There are no factors of 12 whose sum is 10. Therefore, $y^2 + 10y + 12$ is prime.   ●

Quick ✓

9. A polynomial that cannot be written as the product of two other polynomials (other than 1 or -1) is a _____ polynomial.

In Problems 10 and 11, factor the trinomial. If the trinomial cannot be factored, state that it is prime.

10. $z^2 - 5z + 8$ **11.** $q^2 + 4q - 45$

If a trinomial has more than one variable, take the same approach that was used for trinomials in one variable. Therefore, trinomials of the form

$$x^2 + bxy + cy^2$$

factor as

$$(x + my)(x + ny)$$

where

$$mn = c \quad \text{and} \quad m + n = b$$

EXAMPLE 6 **Factoring Trinomials in Two Variables**

Factor: $p^2 + 4pq - 21q^2$

Solution

The trinomial $p^2 + 4pq - 21q^2$ factors as $(p + mq)(p + nq)$, where $mn = -21$ and $m + n = 4$. In other words, find factors of -21 whose sum is 4. Also, because c is negative and b is positive, the factors of -21 have opposite signs, and the factor with the larger absolute value is positive.

Factors of -21	$-1, 21$	$-3, 7$
Sum	20	4

Because $-3 \cdot 7 = -21$ and $-3 + 7 = 4$, let $m = -3$ and $n = 7$. Write the trinomial in the form $(p + mq)(p + nq)$.

$$p^2 + 4pq - 21q^2 = (p + (-3)q)(p + 7q)$$
$$= (p - 3q)(p + 7q)$$

Check $(p - 3q)(p + 7q) = p^2 + 7pq - 3pq - 21q^2$
$$= p^2 + 4pq - 21q^2$$

So, $p^2 + 4pq - 21q^2 = (p - 3q)(p + 7q)$.

Quick ✔

In Problems 12 and 13, factor each trinomial.

12. $x^2 + 9xy + 20y^2$ **13.** $m^2 + mn - 42n^2$

EXAMPLE 7 **Factoring a Trinomial Not Written in Standard Form**

Factor: $2w + w^2 - 8$

Solution

First write the trinomial in standard form (descending order of degree) and then factor.

$$2w + w^2 - 8 = w^2 + 2w - 8$$

Look for factors of $c = -8$ whose sum is $b = 2$. Because $c = -8$, the factors of -8 have opposite signs. Because $b = 2$, the factor with the larger absolute value is positive.

Factors of -8	$-1, 8$	$-2, 4$
Sum	7	2

Because $-2 \cdot 4 = -8$ and $-2 + 4 = 2$, let $m = -2$ and $n = 4$. Therefore,

$$2w + w^2 - 8 = w^2 + 2w - 8$$
$$= (w - 2)(w + 4)$$

Check $(w - 2)(w + 4) = w^2 + 4w - 2w + (-2)(4)$
$$= w^2 + 2w - 8$$

So, $2w + w^2 - 8 = (w - 2)(w + 4)$.

▶ ❷ **Factor Out the GCF, Then Factor $x^2 + bx + c$**

Some algebraic expressions can be factored as trinomials in the form $x^2 + bx + c$ after we factor out a greatest common factor.

EXAMPLE 8   **Factoring Trinomials with a Common Factor**

Factor: $3u^3 - 21u^2 - 90u$

Solution

Notice that the trinomial has a GCF of $3u$. First, factor out $3u$:

$$3u^3 - 21u^2 - 90u = 3u(u^2 - 7u - 30)$$

Now factor the trinomial in parentheses, $u^2 - 7u - 30$. Look for factors of $c = -30$ whose sum is $b = -7$. Because c is negative and b is negative, the factors have opposite signs, and the factor with the larger absolute value is negative.

Factors of -30	$1, -30$	$2, -15$	$3, -10$	$5, -6$
Sum	-29	-13	-7	-1

Because $(3)(-10) = -30$ and $3 + (-10) = -7$, let $m = 3$ and $n = -10$. Write the trinomial in the form $(u + m)(u + n)$.

$$u^2 - 7u - 30 = (u + 3)(u - 10)$$

Don't forget to include the GCF in the final answer:

$$3u^3 - 21u^2 - 90u = 3u(u^2 - 7u - 30)$$
$$= 3u(u + 3)(u - 10)$$

Work Smart

The result of Example 8 could have been checked as follows:

$3u(u + 3)(u - 10)$

$= (3u^2 + 9u)(u - 10)$

$= 3u^3 - 30u^2 + 9u^2 - 90u$

$= 3u^3 - 21u^2 - 90u$

$$\text{Check} \quad 3u(u + 3)(u - 10) \overset{\text{F O I L}}{=} 3u(u^2 - 10u + 3u - 30)$$

$$\text{Combine like terms:} = 3u(u^2 - 7u - 30)$$

$$\text{Distribute:} = 3u^3 - 21u^2 - 90u$$

So, $3u^3 - 21u^2 - 90u = 3u(u + 3)(u - 10)$. ●

A polynomial is **factored completely** if each factor in the final factorization is prime. For example, $3x^2 - 6x - 45 = (x - 5)(3x + 9)$ is not factored completely because $3x + 9$ has a common factor of 3. However, $3x^2 - 6x - 45 = 3(x - 5)(x + 3)$ is factored completely.

Sometimes the leading coefficient of a trinomial is negative. If this is the case, factor out the negative to make factoring easier.

EXAMPLE 9 **Factoring Trinomials with a Negative Leading Coefficient**

Factor completely: $-w^2 - 5w + 24$

Solution
Notice that the leading coefficient is -1. It is easier to factor a trinomial when the leading coefficient is positive, so use -1 as the GCF and rewrite $-w^2 - 5w + 24$ as

$$-w^2 - 5w + 24 = -1(w^2 + 5w - 24)$$

Work Smart

It's easier to factor a trinomial in standard form when the leading coefficient is positive.

To factor $w^2 + 5w - 24$, look for two integers whose product is -24 and whose sum is 5. One factor will be positive and the other negative. Because the coefficient of the middle term is odd, one of the factors of -24 is odd and the other is even. Lastly, the factor of -24 with the larger absolute value is positive.

Factors of -24	$-1, 24$	$-3, 8$
Sum	23	5

Factor $w^2 + 5w - 24$ as $(w - 3)(w + 8)$. But remember that the greatest common factor, -1, was factored out.

$$\begin{aligned} -w^2 - 5w + 24 &= -1(w^2 + 5w - 24) \\ &= -1(w - 3)(w + 8) \\ &= -(w - 3)(w + 8) \end{aligned}$$

The check is left to you. The trinomial $-w^2 - 5w + 24$ factors completely as $-(w - 3)(w + 8)$.

Quick ✔

In Problems 21 and 22, factor each trinomial completely.

21. $-w^2 - 3w + 10$ **22.** $-2a^3 - 8a^2 + 24a$

6.2 Exercises MyMathLab®   Exercise numbers in **green** have complete video solutions in MyMathLab or may be accessed using the QR code to the right.

Problems 1–22 are the Quick ✔s that follow the EXAMPLES.

Building Skills

In Problems 23–44, factor each trinomial completely. If the trinomial cannot be factored, state that it is prime. See Objective 1.

23. $x^2 + 5x + 6$ **24.** $p^2 + 7p + 6$

25. $m^2 + 9m + 18$ **26.** $n^2 + 12n + 20$

27. $x^2 - 15x + 36$ **28.** $z^2 - 13z + 36$

29. $p^2 - 8p + 12$ **30.** $z^2 - 7z + 12$

31. $-x - 12 + x^2$ **32.** $-8y - 9 + y^2$

33. $x^2 + 6x - 20$ **34.** $t^2 + 2t - 38$

35. $z^2 + 12z - 45$ **36.** $y^2 + 6y - 40$

37. $x^2 - 5xy + 6y^2$ **38.** $x^2 - 14xy + 24y^2$

39. $r^2 + rs - 6s^2$ **40.** $p^2 + 5pq - 14q^2$

41. $x^2 - 3xy - 4y^2$

42. $x^2 - 9xy - 36y^2$

43. $z^2 + 7zy + 8y^2$

44. $m^2 + 16mn + 18n^2$

In Problems 45–58, factor each trinomial completely by factoring out the GCF first and then factoring the resulting trinomial. See Objective 2.

45. $3x^2 + 3x - 6$

46. $5x^2 + 30x + 40$

47. $3n^3 - 24n^2 + 45n$

48. $4p^4 - 4p^3 - 8p^2$

49. $5x^2z - 20xz - 160z$

50. $8x^2z^2 - 56xz^2 + 80z^2$

51. $-3x^2 + x^3 - 18x$

52. $30x - 2x^2 - 100$

53. $-2y^2 + 8y - 8$

54. $-3x^2 - 24x - 48$

55. $4x^3 - 32x^2 + x^4$

56. $-75x + x^3 + 10x^2$

57. $2x^2 + x^3 - 15x$

58. $x^3 - 20x - 8x^2$

Mixed Practice

In Problems 59–92, factor each polynomial completely. If the polynomial cannot be factored, state that it is prime.

59. $x^2 - 3xy - 28y^2$

60. $x^2 - 9xy - 36y^2$

61. $x^2 + x + 6$

62. $t^2 - 8t - 7$

63. $k^2 - k - 20$

64. $x^2 - 2x - 35$

65. $2w^3 - 8w^2 - 12w$

66. $4d^4 + 28d^3 + 32d^2$

67. $s^2t^2 - 8st + 15$

68. $x^2y^2 + 3xy + 2$

69. $-3p^3 + 3p^2 + 6p$

70. $-2z^3 - 2z^2 + 24z$

71. $-x^2 - 6x - 9$

72. $-r^2 - 12r - 36$

73. $g^2 - 4g + 21$

74. $x^2 - x + 6$

75. $n^4 - 30n^2 - n^3$

76. $-16x + x^3 - 6x^2$

77. $2x^3 - 10x^2 + 6x - 30$

78. $3a^4 - 12a^3 - 6a^2 + 24a$

79. $35 + 12s + s^2$

80. $25 + 10x + x^2$

81. $n^2 - 9n - 45$

82. $x^2 - 6x - 42$

83. $m^2n^2 - 8mn + 12$

84. $x^2y^2 - 3xy - 18$

85. $-x^3 + 12x^2 + 28x$

86. $-3r^3 + 6r^2 - 3r$

87. $y^2 - 12y + 36$

88. $x^2 - 14x + 49$

89. $-36x + 20x^2 - 2x^3$

90. $-20mn^2 + 30m^2n - 5m^3$

91. $-21x^3y - 14xy^2$

92. $-12x^2y + 8xy^3$

Applying the Concepts

93. Punkin Chunkin In a recent Punkin Chunkin contest, in which a pumpkin is shot in the air with a homemade cannon, the height of a pumpkin after t seconds was given by the trinomial $(-16t^2 + 64t + 80)$ feet. Write this polynomial in factored form.

94. Punkin Chunkin In a recent Punkin Chunkin contest, in which a pumpkin is shot into the air with a catapult, the height of a pumpkin after t seconds was given by the trinomial $(-16t^2 + 16t + 32)$ feet. Write this polynomial in factored form.

△ **95. A Rectangular Field** The trinomial $(x^2 + 9x + 18)$ square meters represents the area of a rectangular field. Find two binomials that might represent the length and width of the field.

△ **96. Another Rectangular Field** The trinomial $(x^2 + 6x + 8)$ square yards represents the area of a rectangular field. Find two binomials that might represent the length and width of the field.

△ **97. Area of a Triangle** The area of a triangle is given by the trinomial $\left(\frac{1}{2}x^2 + x - \frac{15}{2}\right)$ square inches. Find algebraic expressions that might represent the base and height of the triangle. (*Hint:* Factor out $\frac{1}{2}$ as a common factor.)

△ **98. Area of a Triangle** The area of a triangle is given by the trinomial $\left(\frac{1}{2}x^2 + 5x + 12\right)$ square kilometers. Find algebraic expressions that might represent the base and height of the triangle. (*Hint:* Factor out $\frac{1}{2}$ as a common factor.)

Extending the Concepts

In Problems 99–102, find all possible values of b so that the trinomial is factorable.

99. $x^2 + bx + 6$

100. $x^2 + bx - 10$

101. $x^2 + bx - 21$

102. $x^2 + bx + 12$

In Problems 103-106, find all possible positive values of c so that the trinomial is factorable.

103. $x^2 - 2x + c$

104. $x^2 - 3x + c$

105. $x^2 - 7x + c$

106. $x^2 + 6x + c$

Explaining the Concepts

107. The answer key to a homework assignment in which you were asked to factor $10 - 3x - x^2$ states the factored form is $-(x + 5)(x - 2)$, but you found the factored form to be $(5 + x)(2 - x)$. Is your answer correct? Explain.

108. The answer key to your algebra exam said the factored form of $2 - 3x + x^2$ is $(x - 1)(x - 2)$. You have $(1 - x)(2 - x)$ on your paper. Is your answer correct or incorrect? Explain your reasoning.

109. In the trinomial $x^2 + bx + c$, both b and c are negative. Explain how to factor the trinomial. Make up a trinomial and then use your rules to factor it.

6.3 Factoring Trinomials of the Form $ax^2 + bx + c, a \neq 1$

Objectives

❶ Factor $ax^2 + bx + c, a \neq 1$, by Grouping

❷ Factor $ax^2 + bx + c, a \neq 1$, Using Trial and Error

Are you Prepared for This Section?

Before getting started, complete the following problems. If you get a problem wrong, go back to the section cited and review the material.

P1. List the prime factorization of 24.   [Section 1.2, pp. 8–9]

P2. Determine the coefficients of $5x^2 - 3x + 7$.   [Section 1.8, p. 66]

P3. Find the product: $(2x + 7)(3x - 1)$ [Section 5.3, pp. 324–327]

▶ In this section, we factor trinomials of the form $ax^2 + bx + c$ in which the leading coefficient, a, is not 1. Examples of trinomials of the form $ax^2 + bx + c, a \neq 1$, are

$$2x^2 + 3x + 1 \qquad 5y^2 - y - 4 \qquad 10z^2 - 7z + 6$$

Let's begin by reviewing multiplication of binomials using FOIL.

Factored Form	F O I L	Polynomial
$(2x + 3)(x + 4) =$	$2x^2 + 8x + 3x + 12 =$	$2x^2 + 11x + 12$
$(3x - 4)(x + 7) =$	$3x^2 + 21x - 4x - 28 =$	$3x^2 + 17x - 28$
$(5m - 2n)(3m - n) =$	$15m^2 - 5mn - 6mn + 2n^2 =$	$15m^2 - 11mn + 2n^2$

Work Smart

You may have learned another acceptable method to factor trinomials. There are more techniques than just the two that are presented here.

To factor a trinomial of the form $ax^2 + bx + c$, reverse the FOIL multiplication process. For example, a trinomial $2x^2 + 11x + 12$ is written in factored form as $(2x + 3)(x + 4)$.

Two methods to factor trinomials of the form $ax^2 + bx + c, a \neq 1$, are presented.

  1. Factoring by grouping

  2. Trial and error

Each method has pros and cons, which will be pointed out as we proceed.

▶ ❶ Factor $ax^2 + bx + c, a \neq 1$, by Grouping

First, let's consider how to factor trinomials of the form $ax^2 + bx + c, a \neq 1$, by grouping.

Prepared?...Answers **P1.** $2^3 \cdot 3$
P2. $5, -3, 7$ **P3.** $6x^2 + 19x - 7$

EXAMPLE 1   How to Factor $ax^2 + bx + c, a \neq 1$, by Grouping

Factor: $3x^2 + 14x + 15$

Step-by-Step Solution

First, notice that $3x^2 + 14x + 15$ has no common factors. In this trinomial, $a = 3$, $b = 14$, and $c = 15$.

Step 1: Find the value of ac.

The value of $ac = 3 \cdot 15 = 45$.

Step 2: Find the pair of integers, m and n, whose product is $ac = 45$ and whose sum is $b = 14$.

Find integer factors of 45 whose sum is 14. Because both 14 and 45 are positive, list only the positive factors of 45.

Factors of 45	1, 45	3, 15	5, 9
Sum	46	18	14

The integers whose product is 45 and whose sum is 14 are 5 and 9. Let $m = 9$ and $n = 5$.

Step 3: Write $ax^2 + bx + c$ as $ax^2 + mx + nx + c$.

Write $3x^2 + 14x + 15$ as $3x^2 + 9x + 5x + 15$.

$14x = 9x + 5x$

Step 4: Factor the expression in Step 3 by grouping.

$$3x^2 + 9x + 5x + 15 = (3x^2 + 9x) + (5x + 15)$$

Common factor in 1st group: $3x$;
common factor in 2nd group: 5: $= 3x(x + 3) + 5(x + 3)$

Factor out $(x + 3)$: $= (x + 3)(3x + 5)$

Step 5: Check Multiply the factors.

$$(x + 3)(3x + 5) = 3x^2 + 5x + 9x + 15$$
$$= 3x^2 + 14x + 15$$

Therefore, $3x^2 + 14x + 15 = (x + 3)(3x + 5)$.   ●

The steps used in Example 1 are summarized below.

Work Smart

Notice the title in the steps to factor $ax^2 + bx + c, a \neq 1$, by grouping. It specifies *no common factors*. If a polynomial has a common factor, the first step is to factor the GCF out!

> **Factoring $ax^2 + bx + c, a \neq 1$, by Grouping, Where a, b, and c Have No Common Factors**
>
> **Step 1:** Find the value of ac.
> **Step 2:** Find the pair of integers, m and n, whose product is ac and whose sum is b.
> **Step 3:** Write $ax^2 + bx + c = ax^2 + mx + nx + c$.
> **Step 4:** Factor the expression in Step 3 by grouping.
> **Step 5:** Check by multiplying the factors.

(**EXAMPLE 2**)   **Factoring $ax^2 + bx + c, a \neq 1$, by Grouping**

Factor: $12x^2 - x - 1$

Solution

There is no common factor in $12x^2 - x - 1$ and $a = 12$, $b = -1$, and $c = -1$.
    The value of ac is $(12)(-1) = -12$. Find the factors of -12 whose sum is -1. Because the product ac is -12, one factor must be positive and the other negative, and because $b = -1$, the factor of -12 with the larger absolute value will be negative.

Factors of -12	1, -12	2, -6	-4
Sum	-11	-4	-1

The factors of -12 whose sum is -1 are 3 and -4. Let $m = 3$ and $n = -4$.

(continued)

Write $12x^2 - x - 1$ as $12x^2 + 3x - 4x - 1$, and factor by grouping.

$$-x = 3x - 4x$$

$$12x^2 + 3x - 4x - 1 = (12x^2 + 3x) + (-4x - 1)$$

Common factor in 1st group: $3x$;
common factor in 2nd group: -1: $= 3x(4x + 1) - 1(4x + 1)$

Factor out $(4x + 1)$: $= (4x + 1)(3x - 1)$

Check $(4x + 1)(3x - 1) = 12x^2 - 4x + 3x - 1$

$$= 12x^2 - x - 1$$

Thus, $12x^2 - x - 1 = (4x + 1)(3x - 1)$.

Work Smart

In Example 2, we could have written $12x^2 - x - 1$ as $12x^2 - 4x + 3x - 1$ and obtained the same result:

$12x^2 - 4x + 3x - 1$

$= (12x^2 - 4x) + (3x - 1)$

$= 4x(3x - 1) + 1(3x - 1)$

$= (3x - 1)(4x + 1)$

> **Quick** ✔
>
> 1. When factoring $6x^2 + x - 1$ by grouping, $ac = $ ____ and $b = $ ___.
>
> 2. To factor $2x^2 - 13x + 6$ by grouping, begin by finding factors whose product is ___ and sum is ___.
>
> *In Problems 3 and 4, factor each trinomial completely by grouping.*
>
> 3. $3x^2 - 2x - 8$ 4. $10z^2 + 21z + 9$

EXAMPLE 3

Factoring a Trinomial with a Negative Leading Coefficient by Grouping

Factor: $-18x^2 + 33x + 30$

Solution

First ask, "Is there a greatest common factor in the expression?" Remember, if the leading coefficient is negative, factor the negative out as part of the GCF. So the GCF, -3, is factored out.

$$-18x^2 + 33x + 30 = -3(6x^2 - 11x - 10)$$

Now factor $6x^2 - 11x - 10$ by grouping. Notice that $a = 6$, $b = -11$, and $c = -10$. The value of $ac = (6)(-10) = -60$.

What factors of -60 have a sum of -11?

Factors of -60	1, −60	2, −30	3, −20	4, −15	5, −12	6, −10
Sum	−59	−28	−17	−11	−7	−4

The factors of -60 whose sum is -11 are 4 and -15. So let $m = 4$ and $n = -15$. Write $6x^2 - 11x - 10$ as $6x^2 + 4x - 15x - 10$.

$$-11x = 4x - 15x$$

Now, factor by grouping.

$$6x^2 + 4x - 15x - 10 = (6x^2 + 4x) + (-15x - 10)$$

Common factor in 1st group: $2x$;
common factor in 2nd group: -5: $= 2x(3x + 2) - 5(3x + 2)$

Factor out $(3x + 2)$: $= (3x + 2)(2x - 5)$

Check Don't forget to include the GCF of -3.

$$-3(3x + 2)(2x - 5) = -3(6x^2 - 15x + 4x - 10)$$

$$= -3(6x^2 - 11x - 10)$$

$$= -18x^2 + 33x + 30$$

So, $-18x^2 + 33x + 30 = -3(3x + 2)(2x - 5)$.

The advantage of factoring trinomials of the form $ax^2 + bx + c, a \neq 1$, by grouping is that it is algorithmic (that is, step by step). However, if the product ac is large, then there are many factors of ac whose sum must be determined. This can get overwhelming, as Example 3 illustrates. Under these circumstances, it may be better to employ trial and error, which is discussed in Objective 2 of this section.

Quick ✓

In Problems 5 and 6, factor the trinomial completely by grouping.

5. $24x^2 + 6x - 9$ **6.** $-10n^2 + 17n - 3$

EXAMPLE 4   ### Factoring Trinomials with Two Variables by Grouping

Factor: $12x^2 + xy - 6y^2$

Solution

There is no common factor in this trinomial. Factor $12x^2 + xy - 6y^2$ by grouping with $a = 12$, $b = 1$, and $c = -6$. The value of $ac = (12)(-6) = -72$.
Which factors of -72 have a sum of 1?

Factors of -72	72, −1	36, −2	24, −3	18, −4	12, −6	9, −8
Sum	71	34	21	14	6	1

The factors of -72 whose sum is 1 are 9 and -8. Write

$$12x^2 + xy - 6y^2 \qquad \text{as} \qquad 12x^2 + 9xy - 8xy - 6y^2$$

$$xy = 9xy - 8xy$$

Now factor by grouping.

$$12x^2 + 9xy - 8xy - 6y^2 = (12x^2 + 9xy) + (-8xy - 6y^2)$$

Common factor in 1st group is $3x$;
common factor in 2nd group is $-2y$: $= 3x(4x + 3y) - 2y(4x + 3y)$

Factor out $(4x + 3y)$: $= (4x + 3y)(3x - 2y)$

Check $(4x + 3y)(3x - 2y) = 12x^2 - 8xy + 9xy - 6y^2$
$$= 12x^2 + xy - 6y^2$$

So, $12x^2 + xy - 6y^2 = (4x + 3y)(3x - 2y)$.   ●

Quick ✓

In Problems 7 and 8, factor the trinomial completely by grouping.

7. $6x^2 + 23xy + 21y^2$ **8.** $-36a^2 + 21ab + 30b^2$

▶ ❷ Factor $ax^2 + bx + c, a \neq 1$, Using Trial and Error

In the trial and error method, list various binomials, and multiply to find their product until you find the binomials whose product is the original trinomial. This method may sound haphazard, but experience and logic can help to minimize the number of possibilities you must try before the factored form is found.

EXAMPLE 5 **How to Factor $ax^2 + bx + c$, $a \neq 1$, Using Trial and Error**

Factor: $2x^2 + 7x + 5$

Step-by-Step Solution

Step 1: List the possibilities for the first terms of each binomial whose product is ax^2.

There is only one way to represent the first term, $2x^2$, because 2 is a prime number.

$$(2x + \_\_)(x + \_\_)$$

Step 2: List the possibilities for the last terms of each binomial whose product is c.

The last term, 5, is also prime. The pairs of integers whose product is 5 are 1, 5 and -1, -5. Notice that the coefficient of x, 7, is positive. Thus, to produce a positive sum, $7x$, there must be two positive factors. Therefore, exclude the integer pair -1, -5 in the trials.

Step 3: Write out all the combinations of factors from Steps 1 and 2. Multiply the binomials until you find a product that equals the trinomial.

Possible Factorization of $2x^2 + 7x + 5$	Product
$(2x + 1)(x + 5)$	$2x^2 + 11x + 5$
$(2x + 5)(x + 1)$	$2x^2 + 7x + 5$

The second row has the correct product, so $2x^2 + 7x + 5 = (2x + 5)(x + 1)$. ●

The steps used in Example 5 are summarized below.

> **Factoring $ax^2 + bx + c$, $a \neq 1$, Using Trial and Error, Where a, b, and c Have No Common Factors**
>
> **Step 1:** List the possibilities for the first terms of each binomial whose product is ax^2.
>
> $$(\square x + \_\_)(\square x + \_\_) = ax^2 + bx + c$$
>
> **Step 2:** List the possibilities for the last terms of each binomial whose product is c.
>
> $$(\_\_x + \square)(\_\_x + \square) = ax^2 + bx + c$$
>
> **Step 3:** Write out all the combinations of factors found in Steps 1 and 2. Multiply the binomials until a product that equals the trinomial is found.

> **Quick ✔**
>
> In Problems 9 and 10, factor each trinomial using trial and error.
>
> **9.** $3x^2 + 5x + 2$　　　　　　　**10.** $7y^2 + 22y + 3$

EXAMPLE 6 **Factoring $ax^2 + bx + c$, $a \neq 1$, Using Trial and Error**

Factor completely: $7x^2 - 18x + 8$

Solution

List possible ways of representing the first term, $7x^2$. Because 7 is a prime number, the only possibility is

$$(7x + \_\_)(x + \_\_)$$

Look at the last term, 8, and list its factors:

Factors of 8			
1, 8	2, 4	$-1, -8$	$-2, -4$

Now concentrate on the middle term, $-18x$. To produce a negative sum, $-18x$, from a positive product, 8, there must be two *negative* factors. Therefore, do not include the

Work Smart

When factoring using trial and error, it is only necessary to find the sum of the outer and inner products to see which factorization works.

factor pairs 1, 8 or 2, 4 in the list of possible factors. The sums of the "outer" and "inner" products are shown in blue.

Possible Factorization	Product
$(7x - 1)(x - 8)$	$7x^2 - 57x + 8$
$(7x - 8)(x - 1)$	$7x^2 - 15x + 8$
$(7x - 2)(x - 4)$	$7x^2 - 30x + 8$
$(7x - 4)(x - 2)$	$7x^2 - 18x + 8$

The highlighted row shows that $7x^2 - 18x + 8 = (7x - 4)(x - 2)$.

Quick ✓

In Problems 11 and 12, factor each trinomial completely using trial and error.

11. $3x^2 - 13x + 12$ **12.** $5p^2 - 21p + 4$

EXAMPLE 7   **Factoring $ax^2 + bx + c, a \neq 1$, Using Trial and Error**

Factor completely: $10x^2 - 13x - 3$

Solution

List possible ways of representing the first term, $10x^2$.

$$(10x + \underline{})(x + \underline{})$$
$$(5x + \underline{})(2x + \underline{})$$

The last term, -3, has factor pairs $-1, 3$ or $1, -3$.

List the possible combinations of factors. The sums of the "outer" and "inner" products are shown in blue.

Work Smart

As the list of possibilities gets longer, it becomes more important to write the possible factorizations in an organized fashion.

Possible Factorization	Product
$(10x - 1)(x + 3)$	$10x^2 + 29x - 3$
$(10x + 3)(x - 1)$	$10x^2 - 7x - 3$
$(10x - 3)(x + 1)$	$10x^2 + 7x - 3$
$(10x + 1)(x - 3)$	$10x^2 - 29x - 3$
$(5x - 1)(2x + 3)$	$10x^2 + 13x - 3$
$(5x + 3)(2x - 1)$	$10x^2 + x - 3$
$(5x - 3)(2x + 1)$	$10x^2 - x - 3$
$(5x + 1)(2x - 3)$	$10x^2 - 13x - 3$

The highlighted row shows that $10x^2 - 13x - 3 = (5x + 1)(2x - 3)$.

Quick ✓

In Problems 13 and 14, factor each trinomial completely using trial and error.

13. $2n^2 - 17n - 9$ **14.** $4w^2 - 5w - 6$

Factoring trinomials of the form $ax^2 + bx + c, a \neq 1$ using trial and error can at first seem overwhelming. There are so many possibilities! Plus, the technique seems haphazard.

Calling the technique "trial and error," however, is not quite truth in advertising because some thought is necessary to reduce the list of possible factors. To keep the list of possibilities small, ask yourself the following questions before listing the possible factorizations.

> **Hints for Using Trial and Error to Factor $ax^2 + bx + c$, $a \neq 1$**
>
> - Are there any common factors? If so, then factor out the GCF. Is the leading coefficient negative? If so, then factor it out as part of the GCF.
> - Is the constant, c, positive? If so, then the factors of the constant, c, must be the same sign as the coefficient of the middle term, b. For example,
>
> $$2a^2 + 11a + 5 = (2a + 1)(a + 5)$$
> $$2a^2 - 11a + 5 = (2a - 1)(a - 5)$$
>
> - Is the constant, c, negative? If so, then the factors of c must have opposite signs. For example,
>
> $$10b^2 + 19b - 15 = (5b - 3)(2b + 5)$$
> $$10b^2 - 19b - 15 = (5b + 3)(2b - 5)$$
>
> - If $ax^2 + bx + c$ has no common factor, then the binomials in the factored form cannot have common factors.
> - Is the value of b small? If so, then choose factors of a and factors of c that are close in value. If the value of b is large, then choose factors of a and factors of c that are not close in value.
> - Is the value of the middle term correct, but it has the wrong sign? Then interchange the signs in the binomial factors.

EXAMPLE 8   **Factoring $ax^2 + bx + c$, $a \neq 1$, Using Trial and Error**

Factor completely: $18x^2 + 3x - 10$

Solution
Remember, the first step in any factoring problem is to look for common factors. There are no common factors in $18x^2 + 3x - 10$. Now, list all possible ways of representing $18x^2$.

$$(18x + \underline{})(x + \underline{})$$
$$(9x + \underline{})(2x + \underline{})$$
$$(6x + \underline{})(3x + \underline{})$$

Look at the last term, -10, and list its factors:

Factors of -10	$-10, 1$	$10, -1$	$-5, 2$	$5, -2$

Before listing the possible combinations of factors, ask some questions. Is the coefficient of the middle term of the polynomial $18x^2 + 3x - 10$ small? Yes, it is $+3$. Because the coefficient of the middle term is positive and small, the binomial factors listed should have outer and inner products that sum to a positive, small number. Therefore, start with $(6x + \underline{})(3x + \underline{})$ and the factor pairs $2, -5$ and $-2, 5$. Do not use $6x + 2$ or $6x - 2$ as possible factors because these binomials have a common factor of 2 but $18x^2 + 3x - 10$ has no common factors.

Try $(6x - 5)(3x + 2)$.

$$(6x - 5)(3x + 2) = 18x^2 + 12x - 15x - 10$$
$$= 18x^2 - 3x - 10$$

Close! The only problem is that the middle term has the sign opposite to the one needed. Therefore, interchange the signs of -5 and 2 in the binomials.

$$(6x + 5)(3x - 2) = 18x^2 - 12x + 15x - 10$$
$$= 18x^2 + 3x - 10$$

It works! Therefore, $18x^2 + 3x - 10 = (6x + 5)(3x - 2)$.   ●

The moral of the story in Example 8 is that the name *trial and error* is a bit misleading. With some thought, you won't have to choose binomial factors haphazardly or for very long, provided that you use the helpful hints given and stay alert.

Quick ✓

In Problems 15 and 16, factor each trinomial completely using trial and error.

15. $12x^2 + 17x + 6$ **16.** $12y^2 + 32y - 35$

EXAMPLE 9   **Factoring Trinomials with Two Variables Using Trial and Error**

Factor completely: $48x^2 + 4xy - 30y^2$

Solution

Notice that there is a greatest common factor of 2. First factor out this GCF and obtain the polynomial $2(24x^2 + 2xy - 15y^2)$. The trinomial in parentheses factors in the form $24x^2 + 2xy - 15y^2 = (\_x + \_y)(\_x + \_y)$.

List all possible ways of representing $24x^2$.

$$(24x + \_\_y)(x + \_\_y)$$
$$(12x + \_\_y)(2x + \_\_y)$$
$$(8x + \_\_y)(3x + \_\_y)$$
$$(6x + \_\_y)(4x + \_\_y)$$

Now list the factors of the coefficient of the last term, -15.

Factors of -15	$15, -1$	$-15, 1$	$5, -3$	$-5, 3$

Do not use $24x + 3y$, $12x + 3y$, $3x + 3y$, or $6x + 3y$ as possible factors because there is a common factor in these binomials. Also, because the middle term has a small coefficient, start with $(6x + \_\_y)(4x + \_\_y)$ and the factor pairs $3, -5$ and $-3, 5$.
Try $(6x - 5y)(4x + 3y)$.

$$(6x - 5y)(4x + 3y) = 24x^2 - 2xy - 15y^2$$

Close! The only problem is that the middle term has the sign opposite to the one needed. Therefore, interchange the signs of -5 and 3 in the binomials.

$$(6x + 5y)(4x - 3y) = 24x^2 + 2xy - 15y^2$$

Thus, $48x^2 + 4xy - 30y^2 = 2(6x + 5y)(4x - 3y)$.   ●

Work Smart

Don't forget to include the GCF in the factored form!

Quick ✓

17. What is the first step in factoring any polynomial?
18. *True or False* The trinomial $12x^2 + 22x + 6$ is completely factored as $(4x + 6)(3x + 1)$.

In Problems 19 and 20, factor each trinomial completely using trial and error.

19. $-8x^2 + 28xy + 60y^2$ **20.** $90x^2 + 21xy - 6y^2$

EXAMPLE 10 **Factoring a Trinomial with a Negative Leading Coefficient**

Factor: $-14x^2 + 29x + 15$

Solution

There are no common factors in $-14x^2 + 29x + 15$, but notice that the coefficient of the squared term is negative. Factor -1 out of the trinomial to obtain

$$-14x^2 + 29x + 15 = -1(14x^2 - 29x - 15)$$

Now factor the expression in parentheses and obtain

$$\begin{aligned}-14x^2 + 29x + 15 &= -1(14x^2 - 29x - 15)\\ &= -1(7x + 3)(2x - 5)\\ &= -(7x + 3)(2x - 5)\end{aligned}$$

Work Smart

It's easier to factor a trinomial when the leading coefficient is positive.

Quick ✓

In Problems 21 and 22, factor each trinomial completely using trial and error.

21. $-6y^2 + 23y + 4$ **22.** $-6x^2 - 3x + 45$

Work Smart

Let's compare the two methods presented in this section, factoring by grouping and using trial and error to factor $3x^2 + 10x + 8$. The first question to ask is: Is there a GCF? No, there's not, so let's continue.

Grouping	**Trial and Error**
Step 1: For the polynomial $3x^2 + 10x + 8$, $a = 3$ and $c = 8$; the value of ac is $3 \cdot 8 = 24$.	**Step 1:** The coefficient of x^2, 3, is prime, so list the possibilities for the binomial factors: $(3x + \_)(x + \_)$.
Step 2: The two integers whose product is $ac = 24$ and whose sum is $b = 10$ are 6 and 4.	**Step 2:** The last term, 8, is not prime. Its factor pairs are 1, 8 or 2, 4 or $-1, -8$ or $-2, -4$. Because $c = 8$ is positive and the coefficient of the middle term is also positive, consider only 1, 8 and 2, 4.
Step 3: Rewrite $3x^2 + 10x + 8$ as $3x^2 + 10x + 8 = 3x^2 + 6x + 4x + 8$.	
Step 4: Factor the expression $3x^2 + 6x + 4x + 8$ by grouping. $3x^2 + 6x + 4x + 8$ $= 3x(x + 2) + 4(x + 2)$ $= (x + 2)(3x + 4)$ So, $3x^2 + 10x + 8 = (x + 2)(3x + 4)$.	**Step 3:** The coefficient of the middle term is not large, so start by trying $(3x + 4)(x + 2)$. $(3x + 4)(x + 2) = 3x^2 + 10x + 8$ It works! Therefore, $3x^2 + 10x + 8 = (3x + 4)(x + 2)$.

Because $(3x + 4)(x + 2) = (x + 2)(3x + 4)$, both methods give the same result. Which method do you prefer?

6.3 Exercises MyMathLab®

Exercise numbers in green have complete video solutions in MyMathLab or may be accessed using the QR code to the right.

Problems 1–22 are the **Quick ✓** *s that follow the* **EXAMPLES.**

Building Skills

In Problems 23–46, factor each polynomial completely using the grouping method. Hint: None of the polynomials are prime. See Objective 1.

23. $2x^2 + 13x + 15$ **24.** $3x^2 + 22x + 7$

25. $5w^2 + 13w - 6$ **26.** $7n^2 - 27n - 4$

27. $4w^2 - 8w - 5$ **28.** $4x^2 + 4x - 3$

29. $27z^2 + 3z - 2$ **30.** $25t^2 + 5t - 2$

31. $6y^2 - 5y - 6$ **32.** $20t^2 + 21t + 4$

33. $4m^2 + 8m - 5$ **34.** $6m^2 - 5m - 4$

35. $12n^2 + 19n + 5$ **36.** $12p^2 - 23p + 5$

37. $-5 - 9x + 18x^2$ **38.** $-4 - 3x + 10x^2$

39. $12x^2 + 2xy - 4y^2$

40. $18x^2 + 6xy - 4y^2$

41. $8x^2 + 28x + 12$

42. $30m^2 - 85m + 60$

43. $7x - 5 + 6x^2$

44. $15 + 8x - 12x^2$

45. $-8p^2 + 6p + 9$

46. $-10y^2 + 47y + 15$

In Problems 47–70, factor each polynomial completely using the trial and error method. Hint: None of the polynomials are prime. See Objective 2.

47. $2x^2 + 5x + 3$

48. $3x^2 + 16x + 5$

49. $5n^2 + 7n + 2$

50. $7z^2 + 22z + 3$

51. $5y^2 + 2y - 3$

52. $11z^2 + 32z - 3$

53. $-4p^2 + 11p + 3$

54. $-6w^2 + 11w + 2$

55. $5w^2 + 13w - 6$

56. $3x^2 + 16x - 12$

57. $7t^2 + 37t + 10$

58. $5x^2 + 16x + 3$

59. $6n^2 - 17n + 10$

60. $11p^2 - 46p + 8$

61. $2 - 11x + 5x^2$

62. $4 - 17x + 4x^2$

63. $2x^2 + 3xy + y^2$

64. $3x^2 + 7xy + 2y^2$

65. $2m^2 - 3mn - 2n^2$

66. $3y^2 + 7yz - 6z^2$

67. $6x^2 + 2xy - 4y^2$

68. $6x^2 - 14xy - 12y^2$

69. $-2x^2 - 7x + 15$

70. $-6x^2 - 3x + 45$

Mixed Practice

In Problems 71–96, factor completely. If a polynomial cannot be factored, state that it is prime.

71. $15x^2 - 23x + 4$

72. $12x^3 - 11x^2 - 15x$

73. $-13y + 12 - 4y^2$

74. $2x + 12x^2 - 24$

75. $10x^2 - 8xy - 24y^2$

76. $12n^2 + 7n - 10$

77. $9x^2 - 18x + 10$

78. $15x^2 - 4x + 8$

79. $-12x + 9 - 24x^2$

80. $-24z^2 + 18z + 2$

81. $4x^3y^2 - 8x^2y^3 - 4x^2y^2$

82. $9x^3y + 6x^2y + 3xy$

83. $4m^2 + 13mn + 3n^2$

84. $4m^2 - 19mn + 12n^2$

85. $6x^2 - 17x - 12$

86. $8x^2 + 14x - 7$

87. $48xy + 24x^2 - 30y^2$

88. $-20y^2 + 6x^2 + 7xy$

89. $18m^2 + 39mn - 24n^2$

90. $63y^3 + 60xy^3 + 12x^2y^3$

91. $-6x^3 + 10x^2 - 4x^4$

92. $21n^2 - 18n^3 + 9n$

93. $30x + 22x^2 - 24x^3$

94. $48x - 74x^2 + 28x^3$

95. $6x^2(x^2 + 1) - 25x(x^2 + 1) + 14(x^2 + 1)$

96. $10x^2(x - 1) - x(x - 1) - 2(x - 1)$

Applying the Concepts

△ **97. Area of a Triangle** A triangle has area described by the polynomial $\left(3x^2 + \dfrac{13}{2}x - 14\right)$ square meters (m²). Find the base and height of the triangle. *Hint:* Factor out $\dfrac{1}{2}$ as a common factor.

△ **98. Area of a Rectangle** A rectangle has area described by the polynomial $6x^2 + x - 1$ square centimeters (cm²). Find the length and width of the rectangle.

99. Suppose we know that one factor of $6x^2 - 11x - 10$ is $3x + 2$. What is the other factor?

100. Suppose we know that one factor of $8x^2 + 22x - 21$ is $2x + 7$. What is the other factor?

Extending the Concepts

In Problems 101–104, factor completely.

101. $27z^4 + 42z^2 + 16$

102. $15n^6 + 7n^3 - 2$

103. $3x^{2n} + 19x^n + 6$

104. $2x^{2n} - 3x^n - 5$

In Problems 105 and 106, find all possible integer values of b so that the polynomial is factorable.

105. $3x^2 + bx - 5$

106. $6x^2 + bx + 7$

Explaining the Concepts

107. Describe when you would use trial and error and when you would use grouping to factor a trinomial. Make up two examples that demonstrate your reasoning.

108. How can you tell if a trinomial is not factorable? Make up an example to demonstrate your reasoning.

6.4 Factoring Special Products

Objectives

1. Factor Perfect Square Trinomials
2. Factor the Difference of Two Squares
3. Factor the Sum or Difference of Two Cubes

Are You Prepared for This Section?

Before getting started, complete the following problems. If you get a problem wrong, go back to the section cited and review the material.

P1. Evaluate: 5^2 [Section 1.7, pp. 56–57]
P2. Evaluate: $(-2)^3$ [Section 1.7, pp. 56–57]
P3. Find the product: $(5p^2)^3$ [Section 5.2, pp. 320–321]
P4. Find the product: $(3z + 2)^2$ [Section 5.3, pp. 328–329]
P5. Find the product: $(4m + 5)(4m - 5)$ [Section 5.3, p. 327]

Let's review two "special products" that we saw in Section 5.3.

Product	Example
$(A + B)^2 = A^2 + 2AB + B^2$	$(3a + 5)^2 = (3a)^2 + 2(3a)(5) + 5^2$ $= 9a^2 + 30a + 25$
$(A - B)^2 = A^2 - 2AB + B^2$	$(2p - 3)^2 = (2p)^2 - 2(2p)(3) + 3^2$ $= 4p^2 - 12p + 9$
$(A + B)(A - B) = A^2 - B^2$	$(4z + 7)(4z - 7) = (4z)^2 - 7^2$ $= 16z^2 - 49$

In Other Words

A perfect square trinomial is a trinomial in which the first term and the third term are perfect squares and the second term is either 2 times or -2 times the product of the expressions being squared in the first and third terms.

Recall that $A^2 + 2AB + B^2$ and $A^2 - 2AB + B^2$ are called **perfect square trinomials,** and $A^2 - B^2$ is called the **difference of two squares**.

In this section, we factor polynomials that are special products: perfect square trinomials and the difference of two squares. By recognizing these polynomials, you'll be able to factor them quickly without using the grouping method or trial and error.

▶ ❶ Factor Perfect Square Trinomials

Using the formulas $(A + B)^2 = A^2 + 2AB + B^2$ and $(A - B)^2 = A^2 - 2AB + B^2$ in reverse results in a method for factoring perfect square trinomials.

Perfect Square Trinomials

$$A^2 + 2AB + B^2 = (A + B)^2$$
$$A^2 - 2AB + B^2 = (A - B)^2$$

For a polynomial to be a perfect square trinomial, two conditions must be satisfied.

1. The first and last terms must be perfect squares. Any variable raised to an even exponent is a perfect square. Thus x^2, $x^4 = (x^2)^2$, and $x^6 = (x^3)^2$ are all perfect squares. Other examples of perfect squares are $49 = 7^2$, $9x^2 = (3x)^2$, and $25a^4 = (5a^2)^2$.

2. The middle term must equal 2 times or -2 times the product of the expressions being squared in the first and last term.

Work Smart

The first five perfect squares are 1, 4, 9, 16, and 25.

Prepared?...Answers P1. 25 **P2.** -8
P3. $125p^6$ **P4.** $9z^2 + 12z + 4$
P5. $16m^2 - 25$

EXAMPLE 1 **How to Factor Perfect Square Trinomials**

Factor completely: $z^2 + 6z + 9$

Step-by-Step Solution

Step 1: Determine whether the first term and the third term are perfect squares.

The first term, z^2, is the square of z and the third term, 9, is the square of 3.

Step 2: Determine whether the middle term is 2 times or -2 times the product of the expressions being squared in the first and last term.

The middle term, $6z$, is 2 times the product of z and 3.

Step 3: Use $A^2 + 2AB + B^2 = (A + B)^2$ to factor the expression.

$$
\begin{array}{c}
A^2 \quad + 2 \cdot A \cdot B + B^2 \\
\downarrow \qquad \downarrow \ \downarrow \quad \downarrow \\
z^2 + 6z + 9 = z^2 \ + 2 \cdot z \cdot 3 + 3^2
\end{array}
$$

Factor as $(A + B)^2$ with $A = z$ and $B = 3$: $= (z + 3)^2$

Step 4: Check

$$
\begin{aligned}
(z + 3)^2 &= (z + 3)(z + 3) \\
&= z^2 + 3z + 3z + 9 \\
&= z^2 + 6z + 9
\end{aligned}
$$

Therefore, $z^2 + 6z + 9 = (z + 3)^2$.   ●

EXAMPLE 2 **Factoring Perfect Square Trinomials**

Factor completely:

   (a) $4x^2 - 20x + 25$ **(b)** $9x^2 + 42xy + 49y^2$

Solution

(a) The first term is $4x^2 = (2x)^2$. The third term is $25 = 5^2$. The middle term, $-20x$, equals $-2 \cdot (2x \cdot 5)$. Factor the perfect square trinomial using $A = 2x$ and $B = 5$.

$$
\begin{array}{c}
A^2 \quad - 2 \cdot A \ \cdot B + B^2 \\
\downarrow \qquad \downarrow \ \downarrow \quad \downarrow \\
4x^2 - 20x + 25 = (2x)^2 - 2 \cdot 2x \cdot 5 + 5^2 \\
= (2x - 5)^2
\end{array}
$$

The check is left to you. So, $4x^2 - 20x + 25 = (2x - 5)^2$.

(b) The first term is $9x^2 = (3x)^2$. The third term is $49y^2 = (7y)^2$. The middle term, $42xy$, equals $2 \cdot (3x \cdot 7y)$. Factor the perfect square trinomial using $A = 3x$ and $B = 7y$.

Work Smart

Perfect square trinomials can also be factored using trial and error or grouping, but if you recognize the perfect square pattern, you won't have to use either of these approaches.

$$
\begin{array}{c}
A^2 \quad + 2 \cdot A \ \cdot B \ + B^2 \\
\downarrow \qquad \downarrow \ \downarrow \qquad \downarrow \\
9x^2 + 42xy + 49y^2 = (3x)^2 + 2 \cdot 3x \cdot 7y + (7y)^2 \\
= (3x + 7y)^2
\end{array}
$$

The check is left to you. Thus, $9x^2 + 42xy + 49y^2 = (3x + 7y)^2$.   ●

Quick ✔

1. The expression $A^2 + 2AB + B^2$ is called a _____ _____ _____.

2. $A^2 - 2AB + B^2 =$ _____.

In Problems 3-5, factor each trinomial completely.

3. $x^2 - 12x + 36$ **4.** $16x^2 + 40x + 25$ **5.** $9a^2 - 60ab + 100b^2$

EXAMPLE 3 **Factoring a Trinomial**

Factor completely, if possible: $b^2 - 10b + 36$

Solution

The first term, b^2, is the square of b. The third term, 36, equals 6^2. Does the middle term, $-10b$, equal $-2 \cdot (b \cdot 6)$?

$$-2 \cdot (b \cdot 6) = -12b \neq -10b$$

(continued)

Therefore, $b^2 - 10b + 36$ is not a perfect square trinomial. Can $b^2 - 10b + 36$ be factored using another strategy? Because $b^2 - 10b + 36$ is in the form $x^2 + bx + c$, find two factors of 36 whose sum is -10. Choose negative factors of 36 because the coefficient of the middle term is negative. The possible factor pairs are $-1, -36$ or $-2, -18$ or $-3, -12$ or $-9, -4$. None of these factor pairs sum to -10. Therefore, $b^2 - 10b + 36$ is prime.

> **Quick ✓**
>
> In Problems 6–8, factor each trinomial completely, if possible.
>
> **6.** $z^2 - 8z + 16$ **7.** $4n^2 + 12n + 9$ **8.** $y^2 + 10y + 36$

EXAMPLE 4 **Factoring a Perfect Square Trinomial Having a GCF**

Factor completely: $32m^4 - 48m^2 + 18$

Solution

What is the first step in any factoring problem? It is to look for the GCF. There is a common factor of 2 in the trinomial, so factor it out.

$$32m^4 - 48m^2 + 18 = 2(16m^4 - 24m^2 + 9)$$

Now attempt to factor the trinomial $16m^4 - 24m^2 + 9$. The first term is $16m^4 = (4m^2)^2$. The third term is $9 = 3^2$. The middle term, $-24m^2$, equals $-2 \cdot (4m^2 \cdot 3)$.

$$A^2 \quad - 2 \cdot A \cdot B + B^2$$
$$16m^4 - 24m^2 + 9 = (4m^2)^2 - 2 \cdot 4m^2 \cdot 3 + 3^2$$
$$\text{Factor as } (A - B)^2: \quad = (4m^2 - 3)^2$$

Work Smart

Don't forget the GCF that was factored out in the first step.

Therefore,

$$32m^4 - 48m^2 + 18 = 2(16m^4 - 24m^2 + 9)$$
$$= 2(4m^2 - 3)^2$$

The check is left to you.

> **Quick ✓**
>
> In Problems 9 and 10, factor each trinomial completely.
>
> **9.** $4z^2 + 24z + 36$ **10.** $50a^3 + 80a^2 + 32a$

❷ Factor the Difference of Two Squares

In Chapter 5, we found products such as $(x + 2y)(x - 2y) = x^2 - (2y)^2 = x^2 - 4y^2$. In general,

$$(A - B)(A + B) = A^2 - B^2$$

In Other Words

The difference of two squares is just that! A perfect square minus a perfect square. This pattern is easy to recognize: Just look for two perfect squares that have been subtracted.

The multiplication process in reverse will result in a method to find the factors of the difference of two squares.

Difference of Two Squares

$$A^2 - B^2 = (A - B)(A + B)$$

EXAMPLE 5 **Factoring the Difference of Two Squares**

Factor completely:

(a) $n^2 - 81$ (b) $16x^2 - 9y^2$

Solution

(a) Notice that $n^2 - 81$ is the difference of two squares, n^2 and $81 = 9^2$. Thus

$$\overset{\overset{\displaystyle A^2 \quad - \quad B^2}{\downarrow \qquad \downarrow}}{n^2 - 81 = (n)^2 - (9)^2}$$

$$A^2 - B^2 = (A - B)(A + B): \qquad = (n - 9)(n + 9)$$

Check You can check the answer to any of these differences of two squares by using FOIL.

$$(n - 9)(n + 9) = n^2 + 9n - 9n - 81$$
$$= n^2 - 81$$

So, $n^2 - 81 = (n - 9)(n + 9)$.

(b) Notice that $16x^2 - 9y^2$ is the difference of two squares, $16x^2 = (4x)^2$ and $9y^2 = (3y)^2$.

$$\overset{\overset{\displaystyle A^2 \quad - \quad B^2}{\downarrow \qquad \downarrow}}{16x^2 - 9y^2 = (4x)^2 - (3y)^2}$$

$$A^2 - B^2 = (A - B)(A + B): \qquad = (4x - 3y)(4x + 3y)$$

Check $\qquad\qquad (4x - 3y)(4x + 3y) = 16x^2 + 12xy - 12xy - 9y^2$
$$= 16x^2 - 9y^2$$

So, $16x^2 - 9y^2 = (4x - 3y)(4x + 3y)$. ●

Work Smart

Remember that the Commutative Property of Multiplication tells us that the order of the factors in the answer doesn't matter:

$(a - b)(a + b) = (a + b)(a - b)$

Quick ✓

11. $4m^2 - 81n^2$ is called the _____ of ___ _____ and factors into two binomials.

12. $P^2 - Q^2 = (\_ - \_)(\_ + \_)$

In Problems 13-15, factor completely.

13. $z^2 - 25$ 14. $81m^2 - 16n^2$ 15. $16a^2 - \dfrac{4}{9}b^2$

For the remainder of the section, we will leave the check of the factorization to you.

EXAMPLE 6 **Factoring the Difference of Two Squares**

Factor completely:

(a) $49k^4 - 100$ (b) $p^4 - 1$

Solution

(a) Notice that $49k^4 = (7k^2)^2$ and $100 = 10^2$ are both perfect squares.

$$\overset{\overset{\displaystyle A^2 \quad - \quad B^2}{\downarrow \qquad \downarrow}}{49k^4 - 100 = (7k^2)^2 - 10^2}$$

$$A^2 - B^2 = (A - B)(A + B): \qquad = (7k^2 - 10)(7k^2 + 10)$$

So, $49k^4 - 100 = (7k^2 - 10)(7k^2 + 10)$.

(b) The expressions p^4 and 1 are both perfect squares; $p^4 = (p^2)^2$ and $1 = 1^2$.

$$\overset{\overset{\displaystyle A^2 \quad - \quad B^2}{\downarrow \qquad \downarrow}}{p^4 - 1 = (p^2)^2 - 1^2}$$

$$A^2 - B^2 = (A - B)(A + B): \qquad = (p^2 - 1)(p^2 + 1)$$

(continued)

But $p^2 - 1$ is a difference of two squares, so factor again!

$$(p^2 - 1)(p^2 + 1) = (p - 1)(p + 1)(p^2 + 1)$$

Therefore, $p^4 - 1 = (p - 1)(p + 1)(p^2 + 1)$. ●

You may be asking, "What about the sum of two squares—how does it factor?" If a and b are real numbers, **the sum of two squares, $a^2 + b^2$, is prime and does not factor.** Thus, binomials such as $a^2 + 1$ and $4y^2 + 81$ are prime.

> **Quick ✓**
>
> **16.** *True or False* The sum of two squares, such as $x^2 + 9$, is prime.
>
> *In Problems 17 and 18, factor each polynomial completely.*
>
> **17.** $100k^4 - 81w^2$ **18.** $x^4 - 16$

EXAMPLE 7   **Factoring the Difference of Two Squares Containing a GCF**

Factor completely: $50x^2 - 72y^2$

Solution
What is the first step when we factor? Look for a greatest common factor! Factor out 2, the GCF of 50 and 72, first.

$$50x^2 - 72y^2 = 2(25x^2 - 36y^2)$$

Difference of two squares with $A = 5x$ and $B = 6y$: $= 2[(5x)^2 - (6y)^2]$

Factor as $(A - B)(A + B)$: $= 2(5x - 6y)(5x + 6y)$

Thus, $50x^2 - 72y^2 = 2(5x - 6y)(5x + 6y)$. ●

> **Quick ✓**
>
> **19.** *True or False* $4x^2 - 16y^2$ factors completely as $(2x - 4y)(2x + 4y)$.
>
> *In Problems 20 and 21, factor each polynomial completely.*
>
> **20.** $147x^2 - 48$ **21.** $-2x^4y + 8y^7$

▶ ❸ Factor the Sum or Difference of Two Cubes

Consider the following products:

$$(A + B)(A^2 - AB + B^2) = A^3 - A^2B + AB^2 + A^2B - AB^2 + B^3$$
$$= A^3 + B^3$$
$$(A - B)(A^2 + AB + B^2) = A^3 + A^2B + AB^2 - A^2B - AB^2 - B^3$$
$$= A^3 - B^3$$

These products show that the sum or difference of two cubes can be factored as follows:

Work Smart

To remember the formulas for the sum and the difference of two cubes, think of "SOAP," which stands for Same (as the operation in the original polynomial), Opposite, Always Positive.

$$\underset{\text{S}}{A^3} + B^3 = (A + \underset{\text{O}}{B})(A^2 - AB + \underset{\text{AP}}{B^2})$$
$$A^3 - B^3 = (A - B)(A^2 + AB + B^2)$$

The Sum of Two Cubes

$$A^3 + B^3 = (A + B)(A^2 - AB + B^2)$$

The Difference of Two Cubes

$$A^3 - B^3 = (A - B)(A^2 + AB + B^2)$$

Notice in the formulas that the sign between the cubes matches the sign in the binomial factor and is the opposite of the sign of the middle term of the trinomial factor. Remember that the perfect cubes are $1^3 = 1, 2^3 = 8, 3^3 = 27$, and so on. Any variable raised to a multiple of 3 is a perfect cube. So $x^3, x^6 = (x^2)^3, x^9 = (x^3)^3$, and so on are all perfect cubes.

EXAMPLE 8 **Factoring the Sum or Difference of Two Cubes**

Factor completely:

(a) $x^3 - 8$ (b) $8m^3 + 125n^6$

Solution

(a) This binomial is the difference of two cubes, x^3 and $8 = 2^3$. Let $A = x$ and $B = 2$.

$$A^3 - B^3 = (A - B)(A^2 + A \cdot B + B^2)$$
$$x^3 - 8 = x^3 - 2^3 = (x - 2)(x^2 + x(2) + 2^2)$$
$$= (x - 2)(x^2 + 2x + 4)$$

So, $x^3 - 8 = (x - 2)(x^2 + 2x + 4)$.

(b) Because $8m^3 = (2m)^3$ and $125n^6 = (5n^2)^3$, the expression $8m^3 + 125n^6$ is the sum of two cubes. Let $A = 2m$ and $B = 5n^2$.

$$8m^3 + 125n^6 = (2m)^3 + (5n^2)^3$$
$$A^3 + B^3 = (A + B)(A^2 - AB + B^2): \quad = (2m + 5n^2)[(2m)^2 - (2m)(5n^2) + (5n^2)^2]$$
$$= (2m + 5n^2)(4m^2 - 10mn^2 + 25n^4)$$

Therefore, $8m^3 + 125n^6 = (2m + 5n^2)(4m^2 - 10mn^2 + 25n^4)$.   ●

Quick ✓

22. The binomial $27x^3 + 64y^3$ is called the ___ of ___ ___.

23. $A^3 - B^3 = (\_ - \_)(\_ + \_ + \_)$

24. *True or False* $(x - 1)(x^2 - x + 1)$ can be factored further.

In Problems 25 and 26, factor each polynomial completely.

25. $z^3 + 125$ **26.** $8p^3 - 27q^6$

EXAMPLE 9 **Factoring the Sum or Difference of Two Cubes Having a GCF**

Factor completely: $250x^4 + 54x$

Solution

Always look for a common factor first. Here, $2x$ is a common factor.

$$250x^4 + 54x = 2x(125x^3 + 27)$$
$$(5x)^3 = 125x^3 \text{ and } 3^3 = 27: \quad = 2x[(5x)^3 + 3^3]$$
$$A^3 + B^3 = (A + B)(A^2 - AB + B^2): \quad = 2x[(5x + 3)((5x)^2 - (5x)(3) + 3^2)]$$
$$= 2x(5x + 3)(25x^2 - 15x + 9)$$

Thus, $25x^4 - 54x = 2x(5x + 3)(25x^2 - 15x + 9)$.   ●

Quick ✓

In Problems 27 and 28, factor each polynomial completely.

27. $-16a^4 - 54a$ **28.** $-375b^3 + 3$

6.4 Exercises MyMathLab®

Exercise numbers in green have complete video solutions in MyMathLab or may be accessed using the QR code to the right.

Problems 1–28 are the Quick ✓s that follow the EXAMPLES.

Building Skills

In Problems 29–38, factor each perfect square trinomial completely. See Objective 1.

29. $x^2 + 10x + 25$

30. $m^2 + 12m + 36$

31. $4p^2 - 4p + 1$

32. $9a^2 - 12a + 4$

33. $16x^2 + 24x + 9$

34. $16y^2 - 72y + 81$

35. $x^2 - 4xy + 4y^2$

36. $4a^2 + 20ab + 25b^2$

37. $4z^2 - 12z + 9$

38. $25k^2 - 70k + 49$

In Problems 39–48, factor each difference of two squares completely. See Objective 2.

39. $x^2 - 49$

40. $m^2 - 121$

41. $4x^2 - 25$

42. $25m^2 - 9$

43. $100n^8 - 81p^4$

44. $36s^6 - 49t^4$

45. $k^8 - 256$

46. $a^4 - 16$

47. $25p^4 - 49q^2$

48. $36b^4 - 121a^2$

In Problems 49–58, factor each sum or difference of two cubes completely. See Objective 3.

49. $27 + x^3$

50. $8v^3 + 1$

51. $8x^3 - 27y^3$

52. $64r^3 - 125s^3$

53. $x^6 - 8y^3$

54. $m^9 - 27n^6$

55. $27c^3 + 64d^9$

56. $125y^3 + 27z^6$

57. $16x^3 + 250y^3$

58. $40x^3 + 135y^6$

Mixed Practice

In Problems 59–94, factor completely. If a polynomial cannot be factored, state that it is prime.

59. $16m^2 + 40mn + 25n^2$

60. $25p^2q^2 - 80pq + 64$

61. $18 - 12x + 2x^2$

62. $50 + 20x + 2x^2$

63. $-6a^2 - a + 40$

64. $-15a^2 - 4a + 4$

65. $2x^3 + 10x^2 + 16x$

66. $-5y^3 + 20y^2 - 80y$

67. $x^4y^2 - x^2y^4$

68. $16a^2b^2 - 25a^4b^2$

69. $x^8 - 25y^{10}$

70. $32a^3 + 4b^6$

71. $b^2 + 13b + 36$

72. $p^2 + 4p - 96$

73. $3s^7 + 24s$

74. $4x - 4x^3$

75. $2x^5 - 162x$

76. $48z^4 - 3$

77. $6x^2 + 19x + 10$

78. $8x^2 + 30x - 27$

79. $4x^2 - 12x - 72$

80. $8x^2 + 16x - 64$

81. $3n^2 + 14n + 36$

82. $12m^2 - 14m + 21$

83. $2x^2 - 8x + 8$

84. $3x^2 + 18x + 27$

85. $9x^2 + y^2$

86. $16a^2 + 49b^2$

87. $x^4y^3 + 216xy^3$

88. $54b^5 - 16b^2$

89. $48n^4 - 24n^3 + 3n^2$

90. $18x^2 - 24x^3 + 8x^4$

91. $2x(x^2 - 4) + 5(x^2 - 4)$

92. $4z(z^2 - 9) + 3(z^2 - 9)$

93. $2y^3 + 5y^2 - 32y - 80$

94. $m^3 + 2m^2 - 25m - 50$

Applying the Concepts

△ **95. Area of a Square** The area of a square is given by the polynomial $(4x^2 + 20x + 25)$ square meters. What is an algebraic expression for the length of one side of the square

△ **96. Area of a Square** The area of a square is given by the polynomial $(9x^2 - 6x + 1)$ square feet. What is an algebraic expression for the length of one side of the square

Extending the Concepts

In Problems 97–106, factor completely and then simplify, if possible.

97. $(x - 2)^2 - (x + 1)^2$

98. $(m - p)^2 - (m + p)^2$

99. $(x - y)^3 + y^3$

100. $(z + 1)^3 - z^6$

101. $(x + 1)^2 - 9$

102. $(x + 3)^2 - 25$

103. $2a^2(x + 1) - 17a(x + 1) + 30(x + 1)$

104. $25a^2(x - 1)^2 - 5a(x - 1)^2 - 2(x - 1)^2$

105. $5(x + 2)^2 - 7(x + 2) - 6$

106. $9(2a + b)^2 + 6(2a + b) - 8$

108. Explain why the polynomial $4x^2 - 9$ is factorable but the polynomial $4x^2 + 9$ is prime.

109. Explain the error in factoring the polynomial $x^3 + y^3$ as $(x + y)(x^2 - xy + y^2) = (x + y)(x - y)^2$.

Explaining the Concepts

107. Create a list of steps for factoring the sum or difference of two cubes.

110. Factor $x^6 - 1$ first as a difference of two squares and then as a difference of two cubes. Are the factors the same? Explain your results.

6.5 Summary of Factoring Techniques

Objective

1 Factor Polynomials Completely

▶ **1** **Factor Polynomials Completely**

The objective of this section is to put together all the various factoring techniques that have been discussed. Recall that a polynomial is factored completely if each factor in the final factorization is prime.

Steps for Factoring

Step 1: Is there a greatest common factor (GCF)? If so, factor it out.

Step 2: Count the number of terms.

Step 3: **(a)** Two terms (binomials)

- Is it the difference of two squares? If so,

$$A^2 - B^2 = (A - B)(A + B)$$

- Is it the sum of two squares? If so, stop! The expression is prime.

- Is it the difference of two cubes? If so,

$$A^3 - B^3 = (A - B)(A^2 + AB + B^2)$$

- Is it the sum of two cubes? If so,

$$A^3 + B^3 = (A + B)(A^2 - AB + B^2)$$

(b) Three terms (trinomials)

- Is it a perfect square trinomial? If so,

$$A^2 + 2AB + B^2 = (A + B)^2 \quad \text{or} \quad A^2 - 2AB + B^2 = (A - B)^2$$

- Is the coefficient of the squared term 1? If so,

$$x^2 + bx + c = (x + m)(x + n), \text{ where } mn = c \text{ and } m + n = b$$

- Is the coefficient of the squared term not 1? If so, factor by grouping or use trial and error.

(c) Four terms

- Try factoring by grouping.

Step 4: Check your work by multiplying the factors.

⊙ **EXAMPLE 1** **How to Factor Completely**

Factor completely: $2x^2 - 4x - 48$

Step-by-Step Solution

Step 1: Is there a GCF?

Yes. Factor out the GCF, 2.
$$2x^2 - 4x - 48 = 2(x^2 - 2x - 24)$$

Step 2: Count the number of terms in the polynomial in parentheses.

The polynomial in parentheses has three terms.

Step 3: The trinomial in parentheses, $x^2 - 2x - 24$, is not a perfect square trinomial. The leading coefficient is 1, so try $(x + m)(x + n)$, where $mn = c$ and $m + n = b$.

Find two factors of -24 whose sum is -2. Because $(-6)(4) = -24$ and $-6 + 4 = -2$, $m = -6$ and $n = 4$.
$$2(x^2 - 2x - 24) = 2(x - 6)(x + 4)$$

Step 4: Check

$$2(x - 6)(x + 4) = 2(x^2 + 4x - 6x - 24)$$
$$= 2(x^2 - 2x - 24)$$
Distribute: $= 2x^2 - 4x - 48$

Therefore, $2x^2 - 4x - 48 = 2(x - 6)(x + 4)$. ●

> **Quick ✓**
>
> **1.** The first step in any factoring problem is to look for the _____ _____ ____.
>
> *In Problems 2 and 3, factor each polynomial completely.*
>
> **2.** $2p^2 + 8p - 90$
>
> **3.** $-45x^3 + 3x^2 + 6x$

⊙ **EXAMPLE 2** **How to Factor Completely**

Factor completely: $12p^2 + 5p - 3$

Step-by-Step Solution

Step 1: Is there a GCF?

There is no GCF.

Step 2: Count the number of terms.

This polynomial has three terms.

Step 3: The trinomial is not a perfect square trinomial because the first and last terms are not perfect squares. Use trial and error because the leading coefficient is not equal to 1.

Notice that the coefficient of the middle term is fairly small. For this reason, the factors of the first term are probably close in value. Thus start by trying to factor $12p^2 + 5p - 3$ as
$$(4p + \_\_)(3p + \_\_)$$

The last term, -3, has only the factor pairs 1, -3 or -1, 3. The factor 3 (or -3) cannot be in the binomial factor $(3p + \_\_)$ because this would result in a common factor. Try the factor 3 with $(4p + \_\_)$ as follows:
$$(4p + 3)(3p - 1)$$

Step 4: Check

$$(4p + 3)(3p - 1) = 12p^2 - 4p + 9p - 3$$
$$= 12p^2 + 5p - 3$$

Therefore, $12p^2 + 5p - 3 = (4p + 3)(3p - 1)$. ●

EXAMPLE 3   **How to Factor Completely**

Factor completely: $12x^2 + 36xy + 27y^2$

Step-by-Step Solution

Step 1: Is there a GCF?

Notice that the coefficients 12, 36, and 27 are all multiples of 3, so factor out the GCF, 3.

$$12x^2 + 36xy + 27y^2 = 3(4x^2 + 12xy + 9y^2)$$

Step 2: Count the number of terms in the polynomial in parentheses.

The polynomial in parentheses has three terms.

Step 3: Is the trinomial in parentheses a perfect square trinomial?

Yes. The first and third terms are perfect squares, $4x^2 = (2x)^2$ and $9y^2 = (3y)^2$. And, the middle term equals $2 \cdot (2x \cdot 3y)$.

$$3(4x^2 + 12xy + 9y^2) = 3[(2x)^2 + 2 \cdot 2x \cdot 3y + (3y)^2]$$

$$A^2 + 2AB + B^2 = (A + B)^2\text{:} = 3(2x + 3y)^2$$

Step 4: Check

Multiply: $3(2x + 3y)^2 = 3(2x + 3y)(2x + 3y)$

$$= 3(4x^2 + 6xy + 6xy + 9y^2)$$
$$= 3(4x^2 + 12xy + 9y^2)$$
$$= 12x^2 + 36xy + 27y^2$$

So, $12x^2 + 36xy + 27y^2 = 3(2x + 3y)^2$.   ●

▶ **EXAMPLE 4**   **Factoring Completely**

Factor completely: $16m^3 + 54$

Solution

Factor out the greatest common factor of 2.
The polynomial in the parentheses has two terms. Both terms are perfect cubes, $8m^3 = (2m)^3$ and $27 = 3^3$.

$$16m^3 + 54 = 2(8m^3 + 27)$$

Sum of two cubes with $A = 2m$ and $B = 3$: $= 2[(2m)^3 + 3^3]$

$A^3 + B^3 = (A + B)(A^2 - AB + B^2)$: $= 2[(2m + 3)((2m)^2 - (2m)(3) + (3)^2)]$

$$= 2(2m + 3)(4m^2 - 6m + 9)$$

Check   Multiply the polynomials and then distribute the 2.

$$2(2m + 3)(4m^2 - 6m + 9) = 2[8m^3 - 12m^2 + 18m + 12m^2 - 18m + 27]$$
$$= 2[8m^3 + 27]$$
$$= 16m^3 + 54$$

Therefore, $16m^3 + 54 = 2(2m + 3)(4m^2 - 6m + 9)$.   ●

▶ **EXAMPLE 5**   **Factoring Completely**

Factor completely: $-5x^5 + 80x$

Solution

Because the leading coefficient is negative, factor out a negative as part of the GCF, $-5x$.

$$-5x^5 + 80x = -5x(x^4 - 16)$$

The factor in parentheses has two terms. The terms $x^4 = (x^2)^2$ and $16 = 4^2$ are perfect squares. Thus $x^4 - 16$ is a difference of two squares with $A = x^2$ and $B = 4$.

$$-5x(x^4 - 16) = -5x(x^2 - 4)(x^2 + 4).$$

Difference of two squares with $A = x$ and $B = 2$: $= -5x(x - 2)(x + 2)(x^2 + 4)$

Work Smart

Check each factor to determine if any factor can be factored again.

Check $-5x(x - 2)(x + 2)(x^2 + 4) = -5x(x^2 - 4)(x^2 + 4)$

$$= -5x(x^4 - 16)$$
$$= -5x^5 + 80x$$

So, $-5x^2 + 80x = -5x^2(x - 2)(x + 2)(x^2 + 4)$.   ●

EXAMPLE 6   **Factoring Completely**

Factor completely: $4x^3 - 6x^2 - 36x + 54$

Solution

Because there are four terms, try to factor by grouping. Factor out the GCF, 2, first.

Factor out the GCF of 2: $4x^3 - 6x^2 - 36x + 54 = 2(2x^3 - 3x^2 - 18x + 27)$

Factor by grouping: $= 2[(2x^3 - 3x^2) + (-18x + 27)]$

Factor out the common factor in each group: $= 2[x^2(2x - 3) - 9(2x - 3)]$

Factor out $(2x - 3)$: $= 2(2x - 3)(x^2 - 9)$

$x^2 - 9$ is the difference of two squares: $= 2(2x - 3)(x - 3)(x + 3)$

Work Smart: Study Skills

You know how to begin factoring a polynomial—look for the GCF. But do you know when a polynomial is factored completely? That is, do you know when to stop? Which of the following is not completely factored?

(a) $(x + 5)(2x + 6)$
(b) $(z - 3)(z + 3)(z^2 + 9)$
(c) $(2n - 5)(5n + 7)$
(d) $(4y - 3yz)(2 + 7z)$

Did you recognize that (a) and (d) are not completely factored because each has a common factor in one of the binomials?

Check $2(2x - 3)(x - 3)(x + 3) = 2(2x - 3)(x^2 - 9)$

$$= 2(2x^3 - 18x - 3x^2 + 27)$$
$$= 4x^3 - 6x^2 - 36x + 54$$

Therefore, $4x^3 - 6x^2 - 36x + 54 = 2(2x - 3)(x - 3)(x + 3)$.   ●

EXAMPLE 7   **Factoring Completely**

Factor completely: $-4xy^2 + 4xy + 12x$

Solution

Each term has a common factor of $-4x$, so factor it out.

$$-4xy^2 + 4xy + 12x = -4x(y^2 - y - 3)$$

The polynomial in parentheses has three terms, $y^2 - y - 3$, where $a = 1$, $b = -1$, and $c = -3$. It is not a perfect square trinomial (because 3 is not a perfect square). Look for two factors of -3 whose sum is -1. There are no such factors, so $y^2 - y - 3$ is prime.

Check $-4x(y^2 - y - 3) = -4xy^2 + 4xy + 12x$

Therefore, $-4xy^2 + 4xy + 12x = -4x(y^2 - y - 3)$.   ●

Work Smart

$ax^2 + bx + c$ is factorable only when the value of $b^2 - 4ac$ results in a perfect square. Notice that in Example 7,

$$(-1)^2 - 4(1)(-3) = 13$$

so the trinomial $y^2 - y - 3$ is not factorable.

Quick ✔

In Problems 15 and 16, factor each polynomial completely.

15. $-3z^2 + 9z - 21$ 16. $6xy^2 + 15x^3$

6.5 Exercises   MyMathLab®

Exercise numbers in green have complete video solutions in MyMathLab or may be accessed using the QR code to the right.

*Problems **1–16** are the Quick ✔s that follow the **EXAMPLES**.*

Mixed Practice

In Problems 17–90, factor completely. If a polynomial cannot be factored, state that it is prime.

17. $x^2 - 100$
18. $x^2 - 256$
19. $t^2 + t - 6$
20. $n^2 + 5n + 6$
21. $x + y + 2ax + 2ay$
22. $xy - 2ay + 3bx - 6ab$
23. $a^3 - 8$
24. $1 - y^9$
25. $a^2 - ab - 6b^2$
26. $x^2 - xy - 30y^2$
27. $2x^2 - 5x - 7$
28. $12n^2 - 23n + 5$
29. $2x^2 - 6xy - 20y^2$
30. $2x^2 + 4xy - 6y^2$
31. $2m^2 + 7m + 4$
32. $3y^2 - 8y + 3$
33. $9 - a^2$
34. $25 - m^2$
35. $u^2 - 14u + 33$
36. $x^2 + 5x - 6$
37. $xy - ay - bx + ab$
38. $2x^3 - x^2 - 18x + 9$
39. $w^2 + 6w + 8$
40. $x^2 - 8x + 15$
41. $36a^2 - 49b^4$
42. $100x^2 - 25y^2$
43. $x^2 + 2xm - 8m^2$
44. $s^2 - 11st + 24t^2$
45. $6x^2y^2 - 13xy + 6$
46. $6s^2t^2 + st - 1$
47. $x^3 + x^2 + x + 1$
48. $x^3 - 3x^2 + 2x - 6$

49. $12z^2 + 12z + 18$
50. $3y^2 + 6y + 3$
51. $14c^2 + 19c - 3$
52. $24x^2 + 66x + 45$
53. $27m^3 + 64n^6$
54. $8x^6 + 125y^3$
55. $2j^6 - 2j^2$
56. $48m - 3m^9$
57. $8a^2 + 18ab - 5b^2$
58. $10p^2 - 15q^2 + 19pq$
59. $2a^3 + 8a$
60. $4t^4 + 64t^2$
61. $12z^2 - 3$
62. $8n^3 - 18n$
63. $x^2 - x + 6$
64. $n^2 + 2n + 8$
65. $16a^4 + 2ab^3$
66. $24p^3q + 81q^4$
67. $p^2q^2 + 6pq - 7$
68. $x^2y^2 + 20xy + 96$
69. $s^2(s + 2) - 4(s + 2)$
70. $x^2(x + y) - 16(x + y)$
71. $-12x^3 + 2x^2 + 2x$
72. $-3x^3 - 13x^2 + 10x$
73. $10v^2 - 2 - v$
74. $27h - 5 + 18h^2$
75. $4n^2 - n^4 + 3n^3$
76. $4x - 2x^2 - 2x^3$
77. $-4a^3b + 2a^2b - 2ab$
78. $-4x^3y + 4x^2y - 4xy$
79. $12p - p^3 + p^2$
80. $-9x - 3x^3 - 12x^2$
81. $-32x^3 + 72xy^2$
82. $-48p^4 + 75p^2$
83. $2n^3 - 10n^2 - 6n + 30$
84. $6x^4 + 3x^3 - 24x^2 - 12x$
85. $16x^2 + 4x - 12$
86. $36x^2 + 30x - 24$

87. $14x^2 + 3x^4 + 8$

88. $14 + 53r^3 + 14r^6$

89. $-2x^3y + x^2y^2 + 3xy^3$

90. $6a^3 - 5a^2b + ab^2$

Applying the Concepts

91. Profit on Newspapers The revenue for selling x newspapers is given by the expression $x^3 + 8x$. The cost to produce x newspapers is $8x^2 - 7x$. If profit is calculated as revenue minus cost, write an expression, in factored form, that calculates the profit from selling x newspapers.

92. Profit on T-shirts Candy and her boyfriend produce silk-screened T-shirts. The cost to produce n T-shirts is given by the expression $12n - 2n^2$, and the revenue from selling the same number of T-shirts is $n^3 + 2n^2$. Write an expression, in factored form, that calculates the profit from selling n T-shirts.

△ **93. Volume of a Box** The volume of a box is given by the formula $V = lwh$. The volume of the box is represented by the expression $(4x^3 - 10x^2 - 6x)$ cubic feet. Factor this expression to determine algebraic expressions that may represent the dimensions of the box.

△ **94. Volume of a Box** The volume of a box is given by the formula $V = lwh$. The volume of the box is represented by the expression $(18x^3 + 3x^2 - 6x)$ cubic centimeters. Factor this expression to determine algebraic expressions that may represent the dimensions of the box.

Extending the Concepts

Sometimes it is possible to factor a complicated-looking polynomial by substituting one variable for another. This approach is called **factoring by substitution.**

Example: Factoring by Substitution

Factor: $2n^6 - n^3 - 15$

Solution

Notice that $2n^6 - n^3 - 15$ can be written $2(n^3)^2 - n^3 - 15$ so that the trinomial is in the form $au^2 + bu + c$, where $u = n^3$.

$$2n^6 - n^3 - 15 = 2(n^3)^2 - n^3 - 15$$

$$\text{Let } n^3 = u: = 2u^2 - u - 15$$

$$\text{Factor: } = (2u + 5)(u - 3)$$

$$\text{Let } u = n^3: = (2n^3 + 5)(n^3 - 3)$$

In Problems 95–102, factor each trinomial by substitution.

95. $y^4 - 2y^2 - 24$

96. $x^4 + 3x^2 + 2$

97. $4z^6 - 13z^3 + 10$

98. $2m^6 + 11m^3 + 12$

99. $(3r - 1)^2 - 9(3r - 1) + 20$

100. $(5z - 3)^2 - 12(5z - 3) + 32$

101. $2(y - 3)^2 + 13(y - 3) + 15$

102. $3(z + 3)^2 + 14(z + 3) + 8$

In Problems 103–108, expressions that occur in calculus are given. Factor completely each expression.

103. $2(3x + 4)^2 + (2x + 3) \cdot 2(3x + 4) \cdot 3$

104. $5(2x + 1)^2 + (5x - 6) \cdot 2(2x + 1) \cdot 2$

105. $2x(2x + 5) + x^2 \cdot 2$

106. $3x^2(8x - 3) + x^3 \cdot 8$

107. $(4x - 3)^2 + x \cdot 2(4x - 3) \cdot 4$

108. $3x^2(3x + 4)^2 + x^3 \cdot 2(3x + 4) \cdot 3$

Although the most common pattern for factoring a polynomial with four terms is to group the first two terms and group the second two terms, it is not the only possibility. In the following example, the polynomial has been written as three terms in the first group and a single term in the second group. Study the example and then try to find a grouping that will factor each polynomial.

$$x^2 + 2xy + y^2 - z^2 = (x^2 + 2xy + y^2) - z^2$$
$$= (x + y)^2 - z^2$$
$$= (x + y + z)(x + y - z)$$

In Problems 109–112, factor completely.

109. $4m^2 - 4mn + n^2 - p^2$

110. $b^2 + 2bc + c^2 - a^2$

111. $x^2 - y^2 + 2yz - z^2$

112. $16x^2 - y^2 - 8y - 16$

Explaining the Concepts

113. You are grading the paper of a student who factors the polynomial $x^2 + 4x - 3x - 12$ by grouping as follows:

$$(x^2 + 4x)(-3x - 12) = x(x + 4) - 3(x - 4)$$
$$= (x - 3)(x + 4)$$

Did the student correctly factor the polynomial? What comments would you write on the student's paper?

114. Describe the method of factoring you find most difficult. Write a polynomial and factor it using this method.

Putting the Concepts Together (Sections 6.1–6.5)

We designed these problems so that you can review Sections 6.1 to 6.5 and show your mastery of the concepts. Take time to work these problems before proceeding with the next section. The answers are located at the back of the text on page AN-25.

1. Find the GCF of $10x^3y^4z$, $15x^5y$, and $25x^2y^7z^3$.

In Problems 2–16, factor each polynomial completely. If a polynomial cannot be factored, state that it is prime.

2. $x^2 - 3x - 4$ **3.** $x^6 - 27$

4. $6x(2x + 1) + 5z(2x + 1)$

5. $x^2 + 5xy - 6y^2$ **6.** $x^3 + 64$

7. $4x^2 + 49y^2$ **8.** $3x^2 + 12xy - 36y^2$

9. $12z^5 - 44z^3 - 24z^2$ **10.** $x^2 + 6x - 5$

11. $4m^4 + 5m^3 - 6m^2$ **12.** $5p^2 - 17p + 6$

13. $10m^2 + 25m - 6m - 15$ **14.** $36m^2 + 6m - 6$

15. $4m^2 - 20m + 25$ **16.** $5x^2 - xy - 4y^2$

17. Surface Area The surface area of a right circular cylinder is given by the formula $S = 2\pi rh + 2\pi r^2$. Write the right side of this formula in factored form.

18. Rocket Height A toy rocket is launched upward from the ground with an initial velocity of 48 feet per second. Its height, h, in feet, after t seconds, is given by the formula $h = 48t - 16t^2$. Write the right side of this formula in factored form.

6.6 Solving Polynomial Equations by Factoring

Objectives

1 Solve Quadratic Equations Using the Zero-Product Property

2 Solve Polynomial Equations of Degree 3 or Higher Using the Zero-Product Property

Are You Prepared for This Section?

Before getting started, complete the following problems. If you get a problem wrong, go back to the section cited and review the material.

P1. Solve: $x + 5 = 0$ [Section 2.1, pp. 83–85]

P2. Solve: $2(x - 4) - 10 = 0$ [Section 2.2, pp. 92–94]

P3. Evaluate $2x^2 + 3x - 4$ when
  (a) $x = 2$ **(b)** $x = -1$.   [Section 1.8, pp. 64–65]

Work Smart

Remember, the degree of a polynomial in one variable is the value of the *largest exponent* on the variable. For example, the degree of $4x^3 - 9x^2 + 1$ is 3.

▶ Questions that you may have been asking yourself are "Why do I care about factoring? What good is it?" It turns out that there are many uses of factoring. For example, factoring is essential for solving *polynomial equations*.

Definitions

A **polynomial equation** is an equation that can be written in the form "polynomial expression equals zero." The **degree of a polynomial equation** is the degree of the polynomial expression in the equation.

Some examples of polynomial equations are

$$4x + 5 = 17 \qquad 2x^2 - 5x - 3 = 0 \qquad y^3 + 4y^2 = 3y + 18$$

Polynomial equation of degree 1   Polynomial equation of degree 2   Polynomial equation of degree 3

We solved polynomial equations of degree 1 back in Chapter 2. We now learn how to solve polynomial equations of degree 2.

▶ **1 Solve Quadratic Equations Using the Zero-Product Property**

The property on the next page will be used in this section.

The Zero-Product Property

If the product of two factors is zero, then at least one of the factors is 0. That is,

if $ab = 0$, then $a = 0$ or $b = 0$ or both a and b are 0.

EXAMPLE 1 **Using the Zero-Product Property**

Solve: $(x + 4)(2x - 5) = 0$

Solution

The product of two factors, $x + 4$ and $2x - 5$, is equal to 0. According to the Zero-Product Property, at least one of the factors must equal 0. Therefore, set each factor equal to 0 and solve each equation separately.

$$x + 4 = 0 \quad \text{or} \qquad\qquad 2x - 5 = 0$$

Subtract 4 from both sides: $\quad x = -4 \qquad$ Add 5 to both sides: $\quad 2x = 5$

Divide both sides by 2: $\quad \dfrac{2x}{2} = \dfrac{5}{2}$

$$x = \dfrac{5}{2}$$

Check $\quad (x + 4)(2x - 5) = 0 \qquad\qquad (x + 4)(2x - 5) = 0$

Let $x = -4$: $(-4 + 4)(2(-4) - 5) \overset{?}{=} 0 \qquad$ Let $x = \dfrac{5}{2}$: $\left(\dfrac{5}{2} + 4\right)\left(2\left(\dfrac{5}{2}\right) - 5\right) \overset{?}{=} 0$

$$0(-13) \overset{?}{=} 0 \qquad\qquad\qquad \left(\dfrac{13}{2}\right)(5 - 5) \overset{?}{=} 0$$

$$0 = 0 \quad \text{True} \qquad\qquad\qquad \dfrac{13}{2} \cdot 0 \overset{?}{=} 0$$

$$0 = 0 \quad \text{True}$$

The solution set is $\left\{-4, \dfrac{5}{2}\right\}$. ●

Quick ✓

1. The equation $5x^2 + 3x - 7 = 0$ is a polynomial equation of degree __.

2. The Zero-Product Property states that if $ab = 0$, then either _____ or _____.

In Problems 3 and 4, use the Zero-Product Property to solve the equation.

3. $x(x + 3) = 0$ 4. $(x - 2)(4x + 5) = 0$

 The Zero-Product Property comes in handy when we need to solve *quadratic equations*.

Work Smart

In a quadratic equation, why can't a equal 0? Because if a were equal to zero, the equation would be a linear equation.

Definition

A **quadratic equation** is a polynomial equation that can be written in the form

$$ax^2 + bx + c = 0$$

where a, b, and c are real numbers and $a \neq 0$.

The following are examples of quadratic equations.

$$2x^2 + 5x - 3 = 0 \qquad -6z^2 + 12z = 0 \qquad y^2 - 25 = 0 \qquad p^2 + 12p = -36$$

Notice that $y^2 - 25 = 0$ is a quadratic equation even though it is missing the "y" term and that $-6z^2 + 12z = 0$ is a quadratic equation even though it is missing a constant term.

The term "quadratic" means "of, or relating to, a square." There are many real-world situations that are modeled by quadratic equations, such as problems that involve revenue for selling x units of a good or describing the height of a projectile over time.

Sometimes a quadratic equation is called a **second-degree equation** because the polynomial in the equation is of degree 2.

In Other Words

A quadratic equation is in standard form if the polynomial is written in descending order of exponents and is set equal to zero.

> **Definition**
>
> A quadratic equation is said to be in **standard form** if it is written in the form $ax^2 + bx + c = 0$.

For example, $2x^2 + x - 5 = 0$ is in standard form, while $p^2 + 12p = -36$ is not. To write $p^2 + 12p = -36$ in standard form, add 36 to both sides of the equation to obtain $p^2 + 12p + 36 = 0$.

When a quadratic equation is written in standard form, $ax^2 + bx + c = 0$, it may be possible to factor the expression $ax^2 + bx + c$ as the product of two first-degree polynomials. If it is possible to factor the trinomial, use the Zero-Product Property to solve the quadratic equation.

EXAMPLE 2   How to Solve a Quadratic Equation by Factoring

Solve: $x^2 - 4x = 21$

Step-by-Step Solution

Step 1: Write the equation in standard form.

$$x^2 - 4x = 21$$
Subtract 21 from both sides: $x^2 - 4x - 21 = 21 - 21$
$$x^2 - 4x - 21 = 0$$

Step 2: Factor the expression on the left side of the equation.

Two integers whose product is -21 and whose sum is -4 are 3 and -7: $(x + 3)(x - 7) = 0$

Step 3: Set each factor equal to 0.

$$x + 3 = 0 \quad \text{or} \quad x - 7 = 0$$

Step 4: Solve each first-degree equation.

$$x = -3 \quad \text{or} \quad x = 7$$

Step 5: Check

$$x^2 - 4x = 21$$
$x = -3$: $(-3)^2 - 4(-3) \overset{?}{=} 21$
$$9 + 12 \overset{?}{=} 21$$
$$21 = 21 \quad \text{True}$$

$$x^2 - 4x = 21$$
$x = 7$: $(7)^2 - 4(7) \overset{?}{=} 21$
$$49 - 28 \overset{?}{=} 21$$
$$21 = 21 \quad \text{True}$$

The solution set is $\{-3, 7\}$.

Look at Step 3 in the solution to Example 2. To solve a quadratic equation put the equation into a form that we already know how to solve. That is, "transform" the quadratic equation $x^2 - 4x - 21 = 0$ into two linear equations, $x + 3 = 0$ and $x - 7 = 0$. This is an important idea to understand about math. We use procedures to reduce a problem to a simpler problem. In this case, a quadratic equation was reduced to two linear equations.

Other methods for solving $ax^2 + bx + c = 0$ when the expression $ax^2 + bx + c$ is not factorable will be presented later in the text.

> **Solving a Quadratic Equation by Factoring**
>
> **Step 1:** Write the quadratic equation in standard form, $ax^2 + bx + c = 0$.
>
> **Step 2:** Factor the polynomial.
>
> **Step 3:** Set each factor found in Step 2 equal to zero using the Zero-Product Property.
>
> **Step 4:** Solve each first-degree equation for the variable.
>
> **Step 5:** Check your answers by substituting the values of the variable into the *original* equation.

EXAMPLE 3 Solving a Quadratic Equation

Solve: $3x^2 + 5x = 14x$

Solution

First write the equation in standard form, $ax^2 + bx + c = 0$.

$$3x^2 + 5x = 14x$$

Subtract 14x from both sides of the equation: $\quad 3x^2 + 5x - 14x = 14x - 14x$

$$3x^2 - 9x = 0$$

Factor: $\quad 3x(x - 3) = 0$

Set each factor equal to 0: $\quad 3x = 0 \quad$ or $\quad x - 3 = 0$

Solve each first-degree equation: $\quad x = 0 \quad$ or $\quad x = 3$

Work Smart

Notice that a quadratic equation is a second-degree polynomial equation and has two solutions.

Check Substitute $x = 0$ and $x = 3$ into the original equation.

$$3x^2 + 5x = 14x \qquad\qquad\qquad 3x^2 + 5x = 14x$$

$x = 0: \quad 3(0)^2 + 5(0) \stackrel{?}{=} 14(0) \qquad\qquad x = 3: \quad 3(3)^2 + 5(3) \stackrel{?}{=} 14(3)$

$$0 = 0 \quad \text{True} \qquad\qquad\qquad\qquad 27 + 15 \stackrel{?}{=} 42$$

$$42 = 42 \quad \text{True}$$

The solution set is $\{0, 3\}$. ●

> **Quick ✓**
>
> 5. A _____ equation is an equation that can be written in the form $ax^2 + bx + c = 0$, where $a, b,$ and c are real numbers and $a \neq 0$.
>
> 6. Quadratic equations are also known as _____ -degree equations.
>
> 7. *True or False* $3x + x^2 = 6$ is written in standard form.
>
> *In Problems 8–11, use the Zero-Product Property to solve the equation.*
>
> 8. $p^2 - 6p + 8 = 0$ $\qquad\qquad$ 9. $2t^2 - 5t = 3$
>
> 10. $2x^2 + 3x = 5$ $\qquad\qquad$ 11. $z^2 + 20 = -9z$

EXAMPLE 4 Solving a Quadratic Equation

Solve: $3m^2 - 3m = 2 - 2m$

Solution

First write the quadratic equation in standard form, $ax^2 + bx + c = 0$.

$$3m^2 - 3m = 2 - 2m$$

Add 2m to both sides; subtract 2 from both sides: $\quad 3m^2 - 3m + 2m - 2 = 2 - 2 - 2m + 2m$

$$3m^2 - m - 2 = 0$$

Factor the polynomial: $\quad (3m + 2)(m - 1) = 0$

Set each factor equal to 0: $\quad 3m + 2 = 0 \quad$ or $\quad m - 1 = 0$

Solve each first-degree equation: $\quad 3m = -2 \quad$ or $\quad m = 1$

$$m = -\frac{2}{3}$$

Check The check is left to you. Substitute $m = -\frac{2}{3}$ and $m = 1$ in the original equation to check the solutions.

The solution set is $\left\{ -\frac{2}{3}, 1 \right\}$.

Quick ✓

In Problems 12 and 13, solve each quadratic equation by factoring.

12. $5k^2 + 3k - 1 = 3 - 5k$ $\qquad$ **13.** $3x^2 + 9x = 4 - 2x$

▶ **EXAMPLE 5** **Solving a Quadratic Equation**

Solve: $(2x + 5)(x - 3) = 6x$

Solution

$$(2x + 5)(x - 3) = 6x$$

Multiply using FOIL: $\qquad 2x^2 - x - 15 = 6x$

Write in standard form: $\qquad 2x^2 - 7x - 15 = 0$

Factor the polynomial: $\quad (2x + 3)(x - 5) = 0$

Set each factor equal to 0: $\qquad 2x + 3 = 0 \quad$ or $\quad x - 5 = 0$

Solve each first-degree equation: $\qquad 2x = -3 \quad$ or $\qquad x = 5$

$$x = -\frac{3}{2}$$

Work Smart

Do not attempt to solve $(2x + 5)(x - 3) = 6x$ by setting each factor equal to $6x$ as in $2x + 5 = 6x$ and $x - 3 = 6x$. The Zero-Product Property can be applied only when the product equals zero.

Check The check is left to you. Substitute $x = -\frac{3}{2}$ and $x = 5$ into the original equation to verify the answer.

The solution set is $\left\{ -\frac{3}{2}, 5 \right\}$.

Quick ✓

14. *True or False* $x(x - 3) = 4$ means that $x = 4$ or $x - 3 = 4$.

In Problems 15 and 16, solve each quadratic equation by factoring.

15. $(x - 3)(x + 5) = 9$ $\qquad$ **16.** $(x + 3)(2x - 1) = 7x - 3x^2$

▶ **EXAMPLE 6** **Solving a Quadratic Equation**

Solve: $4k^2 + 9 = -12k$

Solution

$$4k^2 + 9 = -12k$$

Write in standard form: $\qquad 4k^2 + 12k + 9 = 0$

Factor the polynomial: $\quad (2k + 3)(2k + 3) = 0$

Set each factor equal to 0: $\qquad 2k + 3 = 0 \quad$ or $\quad 2k + 3 = 0$

Solve each first-degree equation: $\qquad 2k = -3 \qquad\qquad 2k = -3$

$$k = -\frac{3}{2} \qquad\qquad k = -\frac{3}{2}$$

(continued)

Check Substitute $k = -\dfrac{3}{2}$ into the original equation to check the solution.

The solution set is $\left\{-\dfrac{3}{2}\right\}$.

Notice that in Example 6 the solution $k = -\dfrac{3}{2}$ occurred twice. When this occurs, the solution is called a **double root.**

> **Quick ✔**
>
> *In Problems 17 and 18, solve the quadratic equation by factoring.*
>
> **17.** $9p^2 + 16 = 24p$ **18.** $x^2 + 11x + 24 = x - 1$

EXAMPLE 7   **Solving a Quadratic Equation Containing a GCF**

Solve: $-3x^2 + 6x + 72 = 0$

Work Smart

Another technique that could have been used in Example 7 is to use the Multiplication Property of Equality and divide both sides of the equation by -3. Then solve the quadratic equation $x^2 - 2x - 24 = 0$ to find the solutions $x = -4$ and $x = 6$.

Solution

$$-3x^2 + 6x + 72 = 0$$

Factor out the GCF, -3: $-3(x^2 - 2x - 24) = 0$

Factor the trinomial: $-3(x - 6)(x + 4) = 0$

Set each factor equal to 0: $-3 = 0$ or $x - 6 = 0$ or $x + 4 = 0$

Solve each first-degree equation: $x = 6$ $x = -4$

The statement $-3 = 0$ is false, so the solutions are 6 and -4.

Check Substitute $x = 6$ and $x = -4$ into the original equation to check.

The solution set is $\{-4, 6\}$.

> **Quick ✔**
>
> *In Problems 19 and 20, solve the quadratic equation by factoring.*
>
> **19.** $4x^2 + 12x - 72 = 0$ **20.** $-2x^2 + 2x = -12$

EXAMPLE 8   **Throwing a Ball from the Top of a Building**

A ball is thrown upward from the top of a building 96 feet tall with an initial velocity of 80 feet per second. Solve the equation $-16t^2 + 80t + 96 = 192$ to find the time t (in seconds) at which the ball will be 192 feet from the ground. See Figure 1.

Figure 1

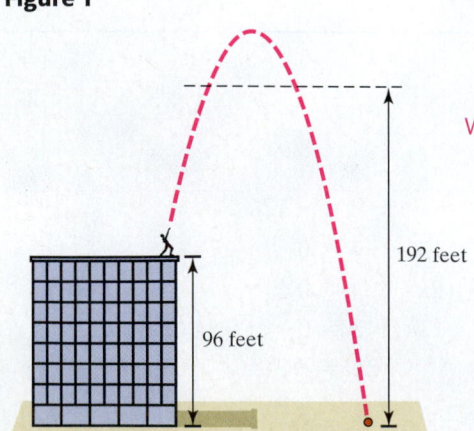

192 feet

96 feet

Solution

$$-16t^2 + 80t + 96 = 192$$

Write the quadratic equation in standard form: $-16t^2 + 80t + 96 - 192 = 192 - 192$

$$-16t^2 + 80t - 96 = 0$$

Factor out the GCF, -16: $-16(t^2 - 5t + 6) = 0$

Factor the trinomial: $-16(t - 3)(t - 2) = 0$

Set each factor equal to 0: $-16 = 0$ or $t - 3 = 0$ or $t - 2 = 0$

Solve each first-degree equation: $t = 3$ or $t = 2$

The statement $-16 = 0$ is false. The solution is $t = 3$ or $t = 2$. After 2 seconds and after 3 seconds, the ball will be 192 feet from the ground. Do you see why? ●

Quick ✔

21. A toy rocket is shot directly up from the ground with an initial velocity of 80 feet per second. Solve the equation $-16t^2 + 80t = 64$ to find the time (in seconds) at which the toy rocket is 64 feet from the ground.

▶ ❷ Solve Polynomial Equations of Degree 3 or Higher Using the Zero-Product Property

The Zero-Product Property is also used to solve higher-degree polynomial equations.

EXAMPLE 9 **How to Solve a Polynomial Equation of Degree 3 or Higher**

Solve: $2x^3 + 3x^2 = 18x + 27$

Step-by-Step Solution

Step 1: Write the equation in standard form.

$$2x^3 + 3x^2 = 18x + 27$$
$$2x^3 + 3x^2 - 18x - 27 = 0$$

Step 2: Factor the polynomial.

Factor by grouping: $(2x^3 + 3x^2) + (-18x - 27) = 0$

Factor out the common factor in each group: $x^2(2x + 3) - 9(2x + 3) = 0$

Factor out $(2x + 3)$: $(2x + 3)(x^2 - 9) = 0$

$x^2 - 9$ is the difference of two squares: $(2x + 3)(x + 3)(x - 3) = 0$

Step 3: Set each factor equal to 0.

$2x + 3 = 0$ or $x + 3 = 0$ or $x - 3 = 0$

Step 4: Solve each first-degree equation.

$2x = -3$ $\qquad$ $x = -3$ $\qquad$ $x = 3$
$x = -\dfrac{3}{2}$

Step 5: Check each solution.

The check is left to you.

The solution set is $\left\{ -\dfrac{3}{2}, -3, 3 \right\}$

Quick ✔

22. *True or False* $x^3 - 4x^2 - 12x = 0$ can be solved using the Zero-Product Property.

In Problems 23 and 24, solve each polynomial equation.

23. $3x^3 + 9x^2 + 6x = 0$ $\qquad$ **24.** $3x^3 - 4x^2 - 3x + 4 = 0$

6.6 Exercises MyMathLab®

Exercise numbers in green have complete video solutions in MyMathLab or may be accessed using the QR code to the right.

Problems 1–24 are the Quick ✔s that follow the EXAMPLES.

Building Skills

In Problems 25–30, solve each equation using the Zero-Product Property. See Objective 1.

25. $2x(x + 4) = 0$ $\qquad$ **26.** $3x(x + 9) = 0$

27. $(n + 3)(n - 9) = 0$ $\qquad$ **28.** $(a + 8)(a - 4) = 0$

29. $(3p + 1)(p - 5) = 0$ $\qquad$ **30.** $(4z - 3)(z + 4) = 0$

In Problems 31–34, identify each equation as a linear equation or a quadratic equation. See Objective 1.

31. $3(x + 4) - 1 = 5x + 2$

32. $2x + 1 - (x + 7) = 3x + 1$

33. $x^2 - 2x = 8$

34. $(x + 2)(x - 2) = 14$

In Problems 35–58, solve each quadratic equation by factoring. See Objective 1.

35. $x^2 - 3x - 4 = 0$ **36.** $x^2 + 2x - 63 = 0$

37. $n^2 + 9n + 14 = 0$ **38.** $p^2 - 5p - 24 = 0$

39. $4x^2 + 2x = 0$ **40.** $14x - 49x^2 = 0$

41. $2x^2 - 3x - 2 = 0$ **42.** $3x^2 + x - 14 = 0$

43. $a^2 - 6a + 9 = 0$ **44.** $k^2 + 12k + 36 = 0$

45. $6x^2 = 36x$ **46.** $2x^2 = 5x$

47. $n^2 - n = 6$ **48.** $a^2 - 6a = 16$

49. $x^2 + 5x + 4 = x$ **50.** $x^2 - 5x + 6 = x - 3$

51. $1 - 5m = -4m^2$ **52.** $4p - 3 = -4p^2$

53. $n(n - 2) = 24$ **54.** $p(p + 1) = 2$

55. $(x - 2)(x - 3) = 56$ **56.** $(x + 5)(x - 3) = 9$

57. $(c + 2)^2 = 9$ **58.** $(2a - 1)^2 = 16$

In Problems 59–64, solve each polynomial equation by factoring. See Objective 2.

59. $2x^3 + 2x^2 - 12x = 0$ **60.** $3x^3 + x^2 - 14x = 0$

61. $y^3 + 3y^2 - 4y - 12 = 0$

62. $m^3 + 2m^2 - 9m - 18 = 0$

63. $2x^3 + 3x^2 = 8x + 12$ **64.** $-2x + 3 = 3x^2 - 2x^3$

Mixed Practice

In Problems 65–90, solve each equation. Be careful; the problems represent a mix of linear, quadratic, and third-degree polynomial equations.

65. $(5x + 3)(x - 4) = 0$ **66.** $(3x - 2)(x + 5) = 0$

67. $p^2 - p - 20 = 0$ **68.** $z^2 - 13z + 40 = 0$

69. $4w + 3 = 2w - 7$ **70.** $7y + 3 = 2y - 12$

71. $4a^2 - 25a = 21$ **72.** $5m^2 = 18m + 8$

73. $2a(a + 1) = a^2 + 8$ **74.** $2y(y + 5) = y^2 + 11$

75. $2x^3 + x^2 = 32x + 16$

76. $2n^3 + 4 = n^2 + 8n$

77. $4(b - 3) - 3b = 8$ **78.** $2(p - 3) = p + 1$

79. $y^2 + 5y = 5(y + 20)$ **80.** $3z^2 + 7z = 7(z + 21)$

81. $(x - 2)(x + 7) = x - 2$ **82.** $(x + 3)(x + 1) = x + 3$

83. $(2k - 3)(2k^2 - 9k - 5) = 0$

84. $(7m - 11)(3m^2 - m - 2) = 0$

85. $(w - 3)^2 = 9 + 2w$

86. $3(k + 2)^2 = 5k + 8$

87. $\frac{1}{2}x^2 + \frac{5}{4}x = 3$ **88.** $z^2 + \frac{29}{4}z = 6$

89. $8x^2 + 44x = 24$ **90.** $9q^2 = 3q + 6$

Applying the Concepts

91. Tossing a Ball A ball is thrown upward from the top of a building 80 feet tall with an initial velocity of 64 feet per second. Solve the equation $-16t^2 + 64t + 80 = 128$ to find the time t (in seconds) at which the ball is 128 feet from the ground.

92. Tossing a Ball A ball is thrown upward from the ground with an initial velocity of 64 feet per second. Solve the equation $-16t^2 + 64t = 48$ to find the time t (in seconds) at which the ball is 48 feet from the ground.

93. Convex Polygon A **convex polygon** is a polygon whose interior angles are between 0° and 180°. The number of diagonals D in a convex polygon with n sides is given by the formula $D = \frac{1}{2}n(n - 3)$. Determine the number of sides n in a convex polygon that has 27 diagonals by solving the equation $27 = \frac{1}{2}n(n - 3)$.

94. Consecutive Integers The sum S of the consecutive integers $1, 2, 3, \ldots, n$ is given by the formula $S = \frac{1}{2}n(n+1)$.

That is, $1 + 2 + 3 + \cdots + n = \frac{1}{2}n(n+1)$. Determine the number of consecutive integers that must be added to obtain a sum of 55 by solving the equation $55 = \frac{1}{2}n(n+1)$.

95. Consecutive Integers The product of two consecutive integers is 12. Find the integers.

96. Consecutive Odd Integers The product of two consecutive odd integers is 143. Find the integers.

97. Consecutive Even Integers Find three consecutive even integers such that the product of the first and the third is 96.

98. Consecutive Odd Integers Find three consecutive odd integers such that the product of the second and the third is 99.

△ **99. Rectangle** The length and width of two sides of a rectangle are consecutive odd integers. The area of the rectangle is 255 square units. Find the dimensions of the rectangle.

△ **100. Garden Area** The State University Landscape Club wants to establish a horticultural garden near the administration building. The length and width of the space that is available are consecutive even integers, and the area of the garden is 440 square feet. Find the dimensions of the garden.

The equation $N = \dfrac{t^2 - t}{2}$ models the number of soccer games that must be scheduled in a league with t teams, when each team plays every other team exactly once. Use this equation to solve Problems 101 and 102.

101. If a league has 28 games scheduled, how many teams are in the league?

102. If a league has 36 games scheduled, how many teams are in the league?

Extending the Concepts

103. Write a polynomial equation in factored form with integer coefficients that has $x = 3$ and $x = -5$ as solutions. What is the degree of this polynomial equation?

104. Write a polynomial equation in factored form with integer coefficients that has $x = 0$ and $x = -8$ as solutions. What is the degree of this polynomial equation?

105. Write a polynomial equation in factored form with integer coefficients that has $z = 6$ as a double root. What is the degree of this polynomial equation?

106. Write a polynomial equation in factored form with integer coefficients that has $x = -2$ as a double root. What is the degree of this polynomial equation?

107. Write a polynomial equation in factored form with integer coefficients that has $x = -3$, $x = 1$ and $x = 5$ as solutions. What is the degree of this polynomial equation?

108. Write a polynomial equation in factored form with integer coefficients that has $a = \frac{1}{2}$, $a = \frac{2}{3}$, and $a = 1$ as solutions. What is the degree of this polynomial equation?

In Problems 109–112, solve for x in each equation.

109. $x^2 - ax + bx - ab = 0$

110. $x^2 + ax - 6a^2 = 0$

111. $2x^3 - 4ax^2 = 0$

112. $4x^2 - 6ax + 10bx - 15ab = 0$

Explaining the Concepts

113. A student solved a quadratic equation using the following procedure. Explain the student's error, and then work the problem correctly.

$$15x^2 = 5x$$
$$\frac{15x^2}{x} = \frac{5x}{x}$$
$$15x = 5$$
$$x = \frac{5}{15} = \frac{1}{3}$$

114. A student solved a quadratic equation using the following procedure. Explain the student's error, and then work the problem correctly.

$$(x - 4)(x + 3) = -6$$
$$x - 4 = -6 \quad \text{or} \quad x + 3 = -6$$
$$x = -2 \qquad\qquad x = -9$$

115. When solving polynomial equations, we always begin by writing the equation in standard form. Explain why this is important.

116. Explain the difference between a quadratic polynomial and a quadratic equation.

6.7 Modeling and Solving Problems with Quadratic Equations

Objectives

1 Model and Solve Problems Involving Quadratic Equations

2 Model and Solve Problems Using the Pythagorean Theorem

Are You Prepared for This Section?

Before getting started, complete the following problems. If you get a problem wrong, go back to the section cited and review the material.

P1. Evaluate: 15^2 [Section 1.7, pp. 56–57]

P2. Solve: $x^2 - 5x - 14 = 0$ [Section 6.6, pp. 409–412]

The solutions to many applied problems require solving polynomial equations by factoring.

▶ **1** Model and Solve Problems Involving Quadratic Equations

Let's begin by solving a quadratic equation to determine the time at which a projectile is at a certain height.

EXAMPLE 1 **Projectile Motion**

A child throws a ball upward off a cliff from a height of 240 feet above sea level. See Figure 2. The height h of the ball above the water (in feet) at any time t (in seconds) can be modeled by the equation

$$h = -16t^2 + 32t + 240$$

(a) When will the ball be 240 feet above sea level?

(b) When will the ball strike the water?

Solution

(a) To determine when the ball will be 240 feet above sea level, let $h = 240$ and solve the resulting equation.

$$-16t^2 + 32t + 240 = 240$$

Write the equation in standard form: $-16t^2 + 32t = 0$

Factor out $-16t$: $-16t(t - 2) = 0$

Set each factor equal to 0: $-16t = 0$ or $t - 2 = 0$

Solve each first degree equation: $t = 0$ or $t = 2$

The ball will be at a height of 240 feet the instant it leaves the child's hand and after 2 seconds of flight.

(b) The ball strikes the water when its height is 0, so let $h = 0$ and solve the resulting equation.

$$-16t^2 + 32t + 240 = 0$$

Factor out -16: $-16(t^2 - 2t - 15) = 0$

Factor the trinomial: $-16(t - 5)(t + 3) = 0$

Set each factor equal to 0: $-16 = 0$ or $t - 5 = 0$ or $t + 3 = 0$

Solve each first-degree equation: $t = 5$ or $t = -3$

The equation $-16 = 0$ is false, and because t represents time, discard the solution $t = -3$ because time, t, must be nonnegative. Therefore, the ball strikes the water after 5 seconds. ●

Figure 2

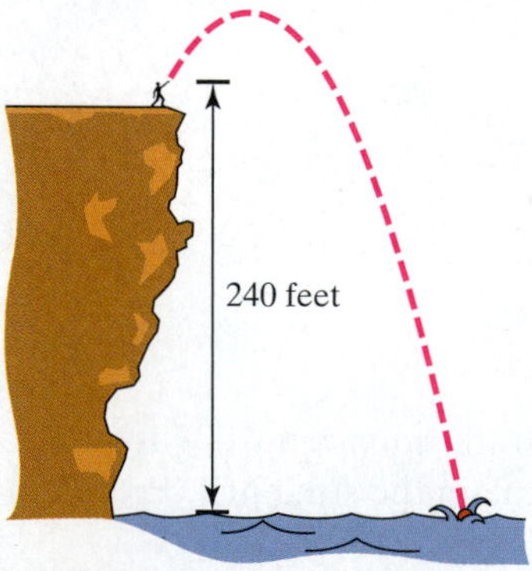

240 feet

Work Smart

Remember, nonnegative numbers are numbers greater than or equal to 0.

Quick ✔

1. A model rocket is fired straight up from the ground. The height h of the rocket (in feet) at any time t (in seconds) can be modeled by the equation $h = -16t^2 + 160t$.

(a) When will the rocket be 384 feet above the ground?

(b) When will the rocket strike the ground?

The next two examples use the problem-solving strategy first presented in Section 2.5.

EXAMPLE 2   ## Geometry: Area of a Rectangle

A carpet installer finds that the length of a rectangular hallway is 3 feet more than twice the width. If the area of the hallway is 44 square feet, what are the dimensions of the hallway? See Figure 3.

Solution

Step 1: Identify This is a geometry problem involving the area of a rectangle.

Step 2: Name Because we know less about the width of the hallway, let w represent the width. The length of the hallway is 3 feet more than twice the width, so let $2w + 3$ represent the length.

Figure 3

Step 3: Translate The area of the hallway is 44 square feet, and the area of a rectangle $=$ (length)(width), so

$$\text{area} = (\text{length})(\text{width})$$

$$44 = (2w + 3)(w) \quad \text{The Model}$$

Step 4: Solve Now solve the equation. Do you see that this equation is a quadratic equation? The first step is to put it in standard form, $ax^2 + bx + c = 0$.

$$w(2w + 3) = 44$$

Distribute:
$$2w^2 + 3w = 44$$

Subtract 44 from both sides:
$$2w^2 + 3w - 44 = 0$$

Factor:
$$(2w + 11)(w - 4) = 0$$

Set each factor equal to 0:
$$2w + 11 = 0 \quad \text{or} \quad w - 4 = 0$$

Solve each first-degree equation:
$$2w = -11 \quad \text{or} \quad w = 4$$

$$\frac{2w}{2} = -\frac{11}{2}$$

$$w = -\frac{11}{2}$$

Step 5: Check Because w represents the width of the hallway, and a measurement must be nonnegative, discard the solution $w = -\dfrac{11}{2}$. If the width of the hallway is 4 feet, then the length would be $2w + 3 = 2(4) + 3 = 11$ feet. The area of a hallway that is 4 feet by 11 feet is $4(11) = 44$ square feet.

Step 6: Answer The dimensions of the hallway are 4 feet by 11 feet. ●

> **Quick ✓**
> **2.** A rectangular plot of land has length that is 3 kilometers less than twice its width. If the area of the land is 104 square kilometers, what are the dimensions of the land?

EXAMPLE 3   ## Geometry: Area of a Triangle

The height of a triangle is 5 inches less than the length of the base, and the area of the triangle is 42 square inches. Find the height of the triangle.

Solution

Step 1: Identify This is a geometry problem involving the area of a triangle.

Step 2: Name The height of the triangle is 5 inches less than its base. Let b represent the base and $b - 5$ represent the height. See Figure 4 on the next page.

(continued)

Figure 4

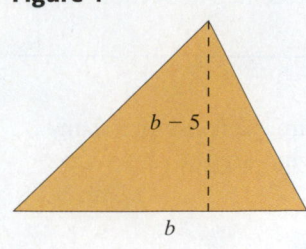

$b - 5$

b

Step 3: Translate The area of a triangle is given by the formula

Area $= \dfrac{1}{2}$ (base)(height), and the triangle's area is 42 square feet, so

$$\text{Area} = \frac{1}{2}(\text{base})(\text{height})$$

$$42 = \frac{1}{2}(b)(b - 5) \quad \text{The Model}$$

Step 4: Solve The model is a quadratic equation. Put the equation in standard form, $ax^2 + bx + c = 0$.

$$42 = \frac{1}{2}(b)(b - 5)$$

Work Smart

Multiply only the $\dfrac{1}{2}$ by 2—don't multiply the factors b and $(b - 5)$ by 2 also.

Multiply by 2 to clear fractions: $2(42) = 2\left[\dfrac{1}{2}(b)(b - 5)\right]$

$$84 = b(b - 5)$$

Distribute: $84 = b^2 - 5b$

Subtract 84 from both sides: $0 = b^2 - 5b - 84$

Factor: $0 = (b - 12)(b + 7)$

Set each factor equal to 0: $b - 12 = 0$ or $b + 7 = 0$

Solve each first-degree equation: $b = 12$ or $b = -7$

Work Smart

In application problems, the solution must make sense. If a variable represents a measurement, the value of the variable cannot be negative.

Step 5: Check Because b represents the base of the triangle, discard the solution $b = -7$. The base of the triangle is 12 inches, so the height is $b - 5 = 12 - 5 = 7$ inches. The area of a triangle with a 12-inch base and a 7-inch height is $\dfrac{1}{2} \cdot 12 \cdot 7 = 42$ square inches.

Step 6: Answer The height of the triangle is 7 inches.   ●

> **Quick** ✓
>
> **3.** The base of a triangular garden is 4 yards longer than the height, and the area of the garden is 48 square yards. Find the dimensions of the triangle.

▶ ❷ **Model and Solve Problems Using the Pythagorean Theorem**

The Pythagorean Theorem is a statement about the lengths of the sides of *right triangles*.

> **Definitions**
>
> A **right triangle** is one that contains a **right angle**—that is, an angle that measures 90°. The side of the triangle opposite the 90° angle is the **hypotenuse**; the remaining two sides are the **legs**.

In Figure 5, c represents the length of the hypotenuse and a and b represent the lengths of the legs. Notice the use of the symbol ⌐ to show the 90° angle.

Figure 5

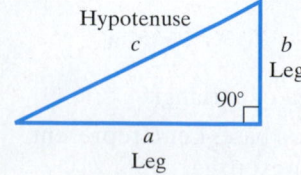

Hypotenuse
c
b
Leg
90°
a
Leg

> **The Pythagorean Theorem**
>
> In a right triangle, the square of the length of the hypotenuse equals the sum of the squares of the lengths of the legs. That is, in the right triangle shown in Figure 5,
>
> $$a^2 + b^2 = c^2 \quad \text{or} \quad \text{leg}^2 + \text{leg}^2 = \text{hypotenuse}^2$$

EXAMPLE 4 ## Using the Pythagorean Theorem

Find the lengths of the sides of the right triangle in Figure 6.

Solution

Figure 6 shows a right triangle, so the lengths of the sides of the triangle must satisfy the Pythagorean Theorem. The legs have lengths x and $x + 7$, and the hypotenuse has length 13.

Pythagorean Theorem:	$a^2 + b^2 = c^2$
Substitute $a = x, b = x + 7, c = 13$:	$x^2 + (x + 7)^2 = 13^2$
$(A + B)^2 = A^2 + 2AB + B^2$:	$x^2 + x^2 + 14x + 49 = 169$
Combine like terms:	$2x^2 + 14x + 49 = 169$
Write the equation in standard form:	$2x^2 + 14x - 120 = 0$
Factor out the GCF, 2:	$2(x^2 + 7x - 60) = 0$
Factor:	$2(x + 12)(x - 5) = 0$
Set each factor equal to 0:	$2 = 0$ or $x + 12 = 0$ or $x - 5 = 0$
Solve each first-degree equation:	$x = -12$ or $x = 5$

The equation $2 = 0$ is false, and the solution $x = -12$ makes no sense because x is a length. Thus, $x = 5$ is the length of one leg of the triangle. The other leg is $x + 7 = 5 + 7 = 12$.

To check this solution, replace a by 5 and b by 12 in the Pythagorean Theorem. Because $5^2 + 12^2 = 13^2$, or $25 + 144 = 169$, the lengths of the legs are correct. The lengths of the sides of the triangle are 5, 12, and 13. ●

Figure 6

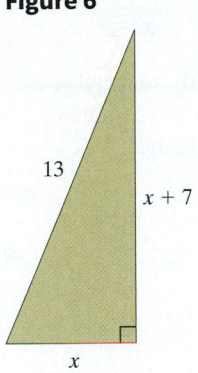

Quick ✓

4. Find the length of each leg of the right triangle pictured below.

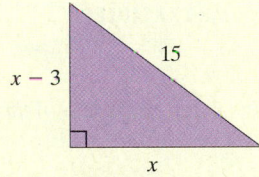

5. One leg of a right triangle is 17 inches longer than the other leg. The hypotenuse has a length of 25 inches. Find the lengths of the two legs of the triangle.

EXAMPLE 5 ## Will the Television Fit in the Media Cabinet?

A rectangular 20-inch television screen (measured diagonally) is 4 inches wider than it is tall. Will this TV fit in your new media cabinet that is 20 inches wide?

Solution

Step 1: Identify This is a geometry problem involving the lengths of the sides of a right triangle.

Step 2: Name The diagonal of the television screen (hypotenuse) is 20 inches. Let x represent the height of the screen. The screen is 4 inches wider than it is tall, so $x + 4$ represents the width.

Step 3: Translate Use the Pythagorean Theorem, which states the relationship between the lengths of the sides of a right triangle.

Pythagorean Theorem:	$a^2 + b^2 = c^2$
Substitute $a = x, b = x + 4, c = 20$:	$x^2 + (x + 4)^2 = 20^2$

(continued)

Work Smart

Remember

$(x + 4)^2 \neq x^2 + 16$

$(x + 4)^2 = x^2 + 8x + 16$

Work Smart

Rather than factoring out 2, we could have divided both sides of the equation by 2.

Step 4: Solve

$$x^2 + (x + 4)^2 = 20^2$$

$(A + B)^2 = A^2 + 2AB + B^2$: $x^2 + x^2 + 8x + 16 = 400$

Combine like terms: $2x^2 + 8x + 16 = 400$

Subtract 400 from both sides: $2x^2 + 8x - 384 = 0$

Factor out the GCF, 2: $2(x^2 + 4x - 192) = 0$

Factor: $2(x + 16)(x - 12) = 0$

Set each factor equal to 0: $2 = 0$ or $x + 16 = 0$ or $x - 12 = 0$

Solve: $x = -16$ or $x = 12$

Step 5: Check The statement $2 = 0$ is false. Since x represents a length, discard the solution $x = -16$. If the shorter leg of the triangle (height of the TV) is 12 inches, then the longer leg (width of the television) is $x + 4 = 12 + 4 = 16$ inches. Because $12^2 + 16^2 = 20^2$, the answer is correct!

Step 6: Answer The television screen is 16 inches wide, so it will fit in the new cabinet.

> **Quick ✓**
>
> **6.** A rectangular 10-inch television screen (measured diagonally) is 2 inches wider than it is tall. What are the dimensions of the TV screen?

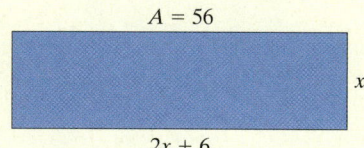 **6.7 Exercises** MyMathLab®
Exercise numbers in **green** have complete video solutions in MyMathLab or may be accessed using the QR code to the right.

*Problems **1–6** are the **Quick ✓**s that follow the **EXAMPLES**.*

Building Skills

In Problems 7–10, use the given area to find the missing sides of the rectangle. See Objective 1.

△ **7.**
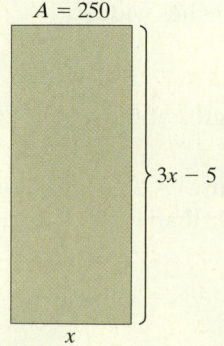
$A = 56$
$2x + 6$
x

△ **8.**
$A = 250$
$3x - 5$
x

△ **9.** $A = 18$
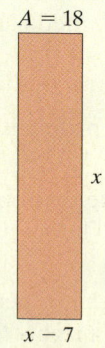
x
$x - 7$

△ **10.**
$A = 364$
x
$x + 12$

In Problems 11–14, use the given area to find the height and base of the triangle. See Objective 1.

△ **11.**
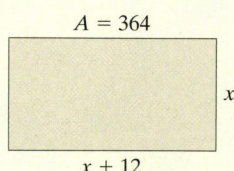
$A = 104$
x
$3x + 2$

△ **12.**
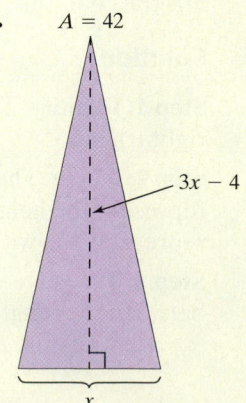
$A = 42$
$3x - 4$
x

△ **13.**

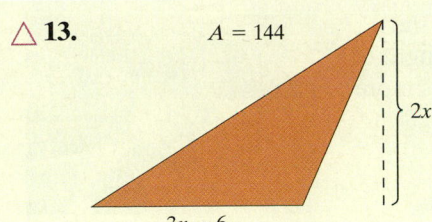

△ **14.**

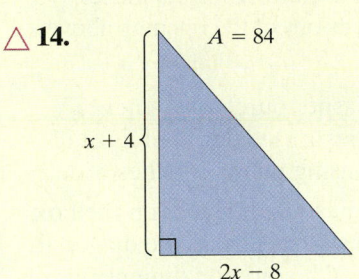

In Problems 15–18, use the given area to find the dimensions of the quadrilateral. See Objective 1.

△ **15.**

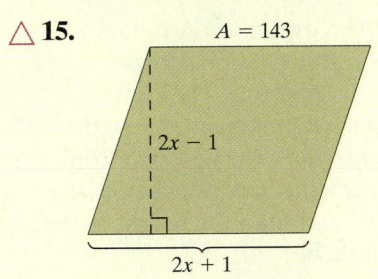

△ **16.**

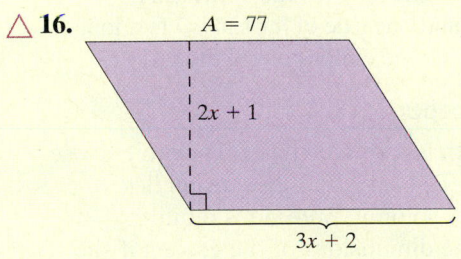

△ **17.**

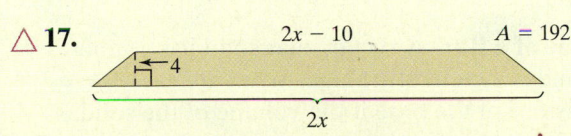

△ **18.**

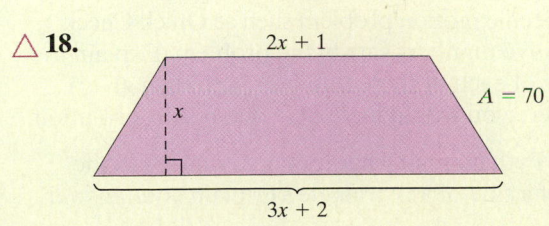

In Problems 19–22, use the Pythagorean Theorem to find the lengths of the sides of the triangle. See Objective 2.

△ **19.**

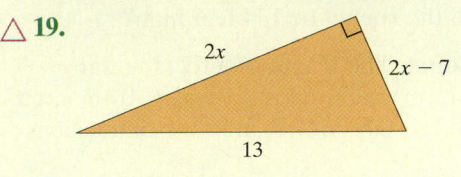

△ **20.**

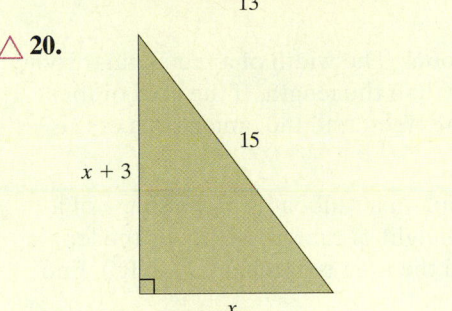

△ **21.**

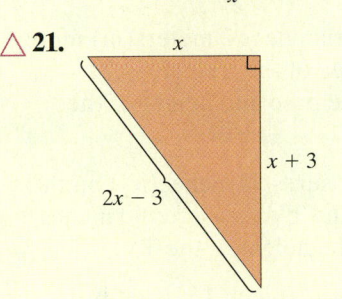

△ **22.**

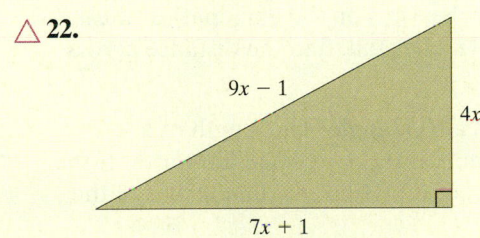

Applying the Concepts

23. Projectile Motion The height, h in feet, of an object t seconds after it is dropped from a cliff 256 feet tall is given by the equation $h = -16t^2 + 256$. Suppose you dropped your glasses off the cliff. Fill in the table below to find the height of your glasses at each time, t.

Time, in Seconds	0	0.5	1	1.5	2	2.5	3	3.5	4
Height, in Feet									

24. Projectile Motion The height, h in feet, of an object t seconds after it is propelled from ground level is given by the equation $h = -16t^2 + 64t$. Fill in the table below to find the height of the object at each time, t.

Time, in Seconds	0	0.5	1	1.5	2	2.5	3	3.5	4
Height, in Feet									

25. Projectile Motion If $h = -16t^2 + 96t$ represents the height of a rocket, in feet, t seconds after it was fired, when will the rocket hit the ground? (*Hint:* The rocket is on the ground when $h = 0$.)

26. Projectile Motion If $h = -16t^2 + 96t$ represents the height of a rocket, in feet, t seconds after it was fired, when will the rocket be 144 feet high?

△ **27. Rectangular Room** The length of a rectangular room is 8 meters(m) more than the width. If the area of the room is 48 square meters, find the dimensions of the room.

△ **28. Rectangular Room** The width of a rectangular room is 4 feet(ft) less than the length. If the area of the room is 21 square feet, find the dimensions of the room.

△ **29. Sailboat** The sail on a sailboat is in the shape of a triangle. If the height of the sail is 3 times the length of the base, and the area is 54 square feet(ft²), find the dimensions of the sail.

△ **30. Triangle** The base of a triangle is 2 meters(m) more than the height. If the area of the triangle is 24 square meters, find the base and height of the triangle.

△ **31. Television** Your TV measures 50 inches(in.) on the diagonal. If the front of the TV measures 40 inches across the bottom, find the height of the TV.

△ **32. Television** Hannah owns a 29-inch TV (that is, it measures 29 inches(in.) on the diagonal). If the television is 20 inches tall, find the distance across the bottom.

△ **33. Dimensions of a Rectangle** The length of a rectangle is 1 mm more than twice the width. If the area is 300 square mm, find the dimensions of the rectangle.

△ **34. Playing Field Dimensions** The length of a rectangular playing field is 3 yards(yd) less than twice the width. If the area of the field is 104 square yards, what are its dimensions?

△ **35. Rectangle and Square** A rectangle and a square have the same area. The width of the rectangle is 6 cm less than the side of the square, and the length of the rectangle is 5 cm more than twice the side of the square. What are the dimensions of the rectangle?

△ **36. Square and Rectangle** A rectangle and a square have the same area. The width of the rectangle is 2 in. less than the side of the square, and the length of the rectangle is 3 in. less than twice the side of the square. What are the dimensions of the rectangle?

△ **37. Hanging a TV** David is installing new stone tiles around his fireplace, but he wants an opening above the fireplace to install a television. The TV he purchased is 53 inches(in.) on the diagonal, excluding the casing around the TV screen. The length of the TV is 17 inches more than the width.

(a) What are the dimensions of the television screen?

(b) If the casing around the entire TV is 1.5 inches, what should the dimensions of the opening above the fireplace be?

△ **38. Encasing a Television** Jasper purchased a new TV. The TV screen is 10 inches(in.) wider than it is tall and is surrounded by a casing that is 2 inches wide.

(a) Jasper lost his tape measure but sees on the box that the TV measures 50 inches on the diagonal, including the casing. What are the dimensions of the TV screen?

(b) Jasper knows that the size of the opening where he wants the TV installed is 29 by 42 inches. Will this TV fit into his space?

△ **39. How Tall Is the Pole?** The owners of A & L Auto Sales want to hang decorative banners to draw attention to their used car offerings. They have 10 feet(ft) of wire to attach banners and know that, according to zoning regulations, the distance from the pole to the point where the wire attaches to the ground must be 2 feet greater than the height of the pole. How tall can the pole be?

△ **40. Sailing on Lake Erie** A sail on a sailboat is in the shape of a right triangle. The longest side of the sail is 13 feet(ft) long, and one side of the sail is 7 feet longer than the other. Find the dimensions of the sail.

Extending the Concepts

△ **41. Gardening** Beth has 28 feet(ft) of fence to enclose a small garden. The length of the garden lies along her house, so only three sides require fencing. Find the dimensions of the garden if she encloses 98 square feet of the garden with the fence.

△ **42. Volume of a Box** A rectangular solid has a square base and is 8 meters(m) high. What are the dimensions of the base if the volume of the solid is 128 cubic meters?

Explaining the Concepts

43. In a projectile motion problem such as Quick Check 1, two positive numbers satisfy the problem. Explain the meaning of each of these numbers, and then give a case where you would have only one positive solution.

44. Can the Pythagorean Theorem be used to find the length of a side of any triangle? Explain your answer.

Chapter 6 Activity: Which One Does Not Belong?

Focus: Factoring polynomials.

Time: 20–30 minutes

Group size: 3–4

Each group member should decide which polynomial from each row does not belong. Answers will vary. As a group, discuss each

member's results. Each group member should be prepared to explain WHY that particular polynomial does not belong with the other three. Be creative!

	A	B	C	D
1.	$15x^2 - 10ax - 15xy + 10ay$	$xr - 3xs - ry + 3sy$	$2ax - 14bx - 2ay + 14by$	$3ab - 3ay - 3bx + 3xy$
2.	$-x^2 - x + 6$	$3x^2 + 39x + 120$	$2x^2 - 12xy - 54y^2$	$7x^2 + 7x - 140$
3.	$3x^2 - 16x + 21$	$6x^2 - 3x + 15$	$3x^2 - 13x - 56$	$3x^2 + x + 1$
4.	$9a^2 + 30a + 25$	$16x^2 - 24x + 9$	$4x^2 - 12xy + 9y^2$	$18s^2 - 60s + 9$
5.	$4x^2 + 13x + 10$	$16x^2 + 40xy + 25y^2$	$16x^2 - 25y^2$	$8x^2 - 2xy - 15y^2$

Chapter 6 Review

Section 6.1   Greatest Common Factor and Factoring by Grouping

KEY CONCEPTS

To find the greatest common factor (GCF) of two or more expressions,

Step 1: Find the GCF of the coefficients of each variable expression.

Step 2: For each common variable, determine the smallest exponent to which the variable is raised.

Step 3: Find the product of the common factors found in Steps 1 and 2. This expression is the GCF.

To factor a polynomial using the greatest common factor,

Step 1: Identify the greatest common factor (GCF) of the terms of the polynomial.

Step 2: Rewrite each term as the product of the GCF and the remaining factor.

Step 3: Use the Distributive Property "in reverse" to factor out the GCF.

Step 4: Check using the Distributive Property.

KEY TERMS

Factors
Greatest common factor
Factoring by grouping

You Should Be Able To...	EXAMPLE	Review Exercises
1 Find the greatest common factor of two or more expressions (p. 369)	Examples 1 through 5	1–10
2 Factor out the greatest common factor in polynomials (p. 372)	Examples 6 through 10	11–16
3 Factor polynomials by grouping (p. 374)	Examples 11 through 13	17–20

In Problems 1–10, find the greatest common factor (GCF) of each group of expressions.

1. 24, 36

2. 27, 54

3. 10, 20, 30

4. 8, 16, 28

5. x^4, x^2, x^8

6. m^3, m, m^5

7. $30a^2b^4, 45ab^2$

8. $18x^4y^2z^3, 24x^3y^5z$

9. $4(2a + 1)^2$ and $6(2a + 1)^3$

10. $9(x - y)$ and $18(x - y)$

In Problems 11–16, factor the GCF from the polynomial.

11. $-18a^3 - 24a^2$

12. $-9x^2 + 12x$

13. $15y^2z + 5y^7z + 20y^3z$

14. $7x^3y - 21x^2y^2 + 14xy^3$

15. $x(5 - y) + 2(5 - y)$ **16.** $z(a + b) + y(a + b)$

18. $2xy + y^2 + 2x^2 + xy$

19. $8x + 16 - xy - 2y$

In Problems 17–20, factor by grouping.

20. $xy^2 + x - 3y^2 - 3$

17. $5m^2 + 2mn + 15mn + 6n^2$

Section 6.2   Factoring Trinomials of the Form $x^2 + bx + c$

KEY CONCEPT

KEY TERMS

- **To factor a trinomial of the form $x^2 + bx + c$, use the following steps.**

  Step 1: Find the pair of integers whose product is c and whose sum is b. That is, determine m and n such that $mn = c$ and $m + n = b$.

  Step 2: Write $x^2 + bx + c = (x + m)(x + n)$.

  Step 3: Check your work by multiplying the binomials.

Quadratic trinomial
Leading coefficient
Prime polynomial

You Should Be Able To...	EXAMPLE	Review Exercises
❶ Factor trinomials of the form $x^2 + bx + c$ (p. 378)	Examples 1 through 7	21–28
❷ Factor out the GCF, then factor $x^2 + bx + c$ (p. 383)	Examples 8 and 9	29–34

In Problems 21–34, factor completely. If the polynomial cannot be factored, say it is prime.

21. $x^2 + 5x + 6$

22. $x^2 + 6x + 8$

23. $x^2 - 21 - 4x$

24. $3x + x^2 - 10$

25. $m^2 + m + 20$

26. $m^2 - 6m - 5$

27. $x^2 - 8xy + 15y^2$

28. $m^2 + 4mn - 5n^2$

29. $-p^2 - 11p - 30$

30. $-y^2 + 2y + 15$

31. $3x^3 + 33x^2 + 36x$

32. $4x^2 + 36x + 32$

33. $2x^2 - 2xy - 84y^2$

34. $4y^3 + 12y^2 - 40y$

Section 6.3   Factoring Trinomials of the Form $ax^2 + bx + c, a \neq 1$

KEY CONCEPTS

- **Factoring $ax^2 + bx + c, a \neq 1$, by grouping, where a, b, and c have no common factors**

  Step 1: Find the value of ac.

  Step 2: Find the pair of integers, m and n, whose product is ac and whose sum is b.

  Step 3: Write $ax^2 + bx + c = ax^2 + mx + nx + c$.

  Step 4: Factor the expression in Step 3 by grouping.

  Step 5: Check by multiplying the factors.

- **Factoring $ax^2 + bx + c, a \neq 1$, using trial and error, where a, b, and c have no common factors**

  Step 1: List the possibilities for the first terms of each binomial whose product is ax^2.

  $$(\Box x + \_\_)(\Box x + \_\_) = ax^2 + bx + c$$

  Step 2: List the possibilities for the last terms of each binomial whose product is c.

  $$(\_\_x + \Box)(\_\_x + \Box) = ax^2 + bx + c$$

  Step 3: Write out all the combinations of factors found in Steps 1 and 2. Multiply the binomials until a product that equals the trinomial is found.

You Should Be Able To...	EXAMPLE	Review Exercises
❶ Factor $ax^2 + bx + c, a \neq 1$, by grouping (p. 386)	Examples 1 through 4	35–44
❷ Factor $ax^2 + bx + c, a \neq 1$, using trial and error (p. 389)	Examples 5 through 10	35–44

In Problems 35–44, factor completely using any method you wish. If a polynomial cannot be factored, state that it is prime.

35. $5y^2 + 14y - 24$

36. $6y^2 - 41y - 7$

37. $-5x + 2x^2 + 3$

38. $23x + 6x^2 + 7$

39. $2x^2 - 7x - 6$

40. $8m^2 + 18m + 9$

41. $9m^3 + 30m^2n + 21mn^2$

42. $14m^2 + 16mn + 2n^2$

43. $-15x^3 - x^2 + 2x$

44. $-12p^4 - 2p^3 + 2p^2$

Section 6.4 Factoring Special Products

KEY CONCEPTS

- **Perfect Square Trinomials**

  $A^2 + 2AB + B^2 = (A + B)^2$; $A^2 - 2AB + B^2 = (A - B)^2$

- **Difference of Two Squares**

  $A^2 - B^2 = (A - B)(A + B)$

- **Sum or Difference of Two Cubes**

  $A^3 + B^3 = (A + B)(A^2 - AB + B^2)$; $A^3 - B^3 = (A - B)(A^2 + AB + B^2)$

KEY TERMS

Perfect square trinomial
Difference of two squares
Sum of two cubes
Difference of two cubes

You Should Be Able To...	EXAMPLE	Review Exercises
1 Factor perfect square trinomials (p. 396)	Examples 1 through 4	45–50
2 Factor the difference of two squares (p. 398)	Examples 5 through 7	51–56
3 Factor the sum or difference of two cubes (p. 400)	Examples 8 and 9	57–62

In Problems 45–62, factor completely. If a polynomial cannot be factored, state that it is prime.

45. $4x^2 - 12x + 9$

46. $x^2 - 10x + 25$

47. $x^2 + 6xy + 9y^2$

48. $9x^2 + 24xy + 4y^2$

49. $8m^2 + 8m + 2$

50. $2m^2 - 24m + 72$

51. $4x^2 - 25y^2$

52. $49x^2 - 36y^2$

53. $x^2 + 25$

54. $x^2 + 100$

55. $x^4 - 81$

56. $x^4 - 625$

57. $m^3 + 27$

58. $m^3 + 125$

59. $27p^3 - 8$

60. $64p^3 - 1$

61. $y^9 + 64z^6$

62. $8y^3 + 27z^6$

Section 6.5 Summary of Factoring Techniques

Steps for Factoring

Step 1: Is there a greatest common factor (GCF)? If so, factor out the GCF.

Step 2: Count the number of terms.

Step 3:

(a) Two terms (binomials)

- Is it the difference of two squares? If so,

  $A^2 - B^2 = (A - B)(A + B)$

- Is it the sum of two squares? If so, stop! The expression is prime.

- Is it the difference of two cubes? If so,

  $A^3 - B^3 = (A - B)(A^2 + AB + B^2)$

- Is it the sum of two cubes? If so,

  $A^3 + B^3 = (A + B)(A^2 - AB + B^2)$

(b) Three terms (trinomials)

- Is it a perfect square trinomial? If so,

  $A^2 + 2AB + B^2 = (A + B)^2$

  or $A^2 - 2AB + B^2 = (A - B)^2$

- Is the coefficient of the squared term 1? If so,

  $x^2 + bx + c = (x + m)(x + n)$

  where $mn = c$ and $m + n = b$

- Is the coefficient of the squared term not 1? If so, factor by grouping or using trial and error.

(c) Four terms

- Try factoring by grouping

Step 4: Check your work by multiplying the factors.

You Should Be Able To...	EXAMPLE	Review Exercises
❶ Factor polynomials completely (p. 403)	Examples 1 through 7	63–78

In Problems 63–78, factor completely. If a polynomial cannot be factored, state that it is prime.

63. $15a^3 - 6a^2b - 25ab^2 + 10b^3$

64. $12a^2 - 9ab + 4ab - 3b^2$

65. $x^2 - xy - 48y^2$

66. $x^2 - 10xy - 24y^2$

67. $x^3 - x^2 - 42x$

68. $3x^6 - 30x^5 + 63x^4$

69. $6x^2 + 11x + 3$

70. $10z^2 + 9z - 9$

71. $27x^3 + 8$

72. $8z^3 - 1$

73. $4y^2 + 18y - 10$

74. $5x^3y^2 - 8x^2y^2 + 3xy^2$

75. $25k^2 - 81m^2$

76. $x^4 - 9$

77. $16m^2 + 1$

78. $m^4 + 25$

Section 6.6   Solving Polynomial Equations by Factoring

KEY CONCEPT

- **The Zero-Product Property**
  If the product of two factors is zero, then at least one of the factors is 0.
  That is, if $ab = 0$, then $a = 0$ or $b = 0$ or both a and b are 0.

KEY TERMS

Polynomial equation
Degree of a polynomial
  equation
Quadratic equation
Second-degree equation
Standard form

You Should Be Able To...	EXAMPLE	Review Exercises
❶ Solve quadratic equations using the Zero-Product Property (p. 409)	Examples 1 through 8	79–88
❷ Solve polynomial equations of degree 3 or higher using the Zero-Product Property (p. 415)	Example 9	89–92

In Problems 79–92, solve each equation by factoring.

79. $(x - 4)(2x - 3) = 0$

80. $(2x + 1)(x + 7) = 0$

81. $x^2 - 12x - 45 = 0$

82. $x^2 - 7x + 10 = 0$

83. $3x^2 + 6x = 0$

84. $4x^2 + 18x = 0$

85. $3x(x + 1) = 2x^2 + 5x + 3$

86. $2x^2 + 6x = (3x + 1)(x + 3)$

87. $5x^2 + 7x + 16 = 3x^2 - 5x$

88. $8x^2 - 10x = -2x + 6$

89. $x^3 = -11x^2 + 42x$

90. $-3x^2 = -x^3 + 18x$

91. $3x^3 - 4x^2 - 27x + 36 = 0$

92. $2x^3 + 5x^2 - 8x = 20$

Section 6.7   Modeling and Solving Problems with Quadratic Equations

KEY CONCEPT

- **The Pythagorean Theorem**
  In a right triangle, the square of the length of the hypotenuse is equal to the sum of the squares of the lengths of the legs. That is, if a and b are the lengths of the legs and c is the length of the hypotenuse, then $a^2 + b^2 = c^2$ or $\text{leg}^2 + \text{leg}^2 = \text{hypotenuse}^2$.

KEY TERMS

Right triangle
Right angle
Hypotenuse
Legs

You Should Be Able To...	EXAMPLE	Review Exercies
❶ Model and solve problems involving quadratic equations (p. 418)	Examples 1 through 3	93–96
❷ Model and solve problems using the Pythagorean Theorem (p. 420)	Examples 4 and 5	97, 98

93. Geysers If $h = -16t^2 + 80t$ represents the height of a jet of water from a geyser t seconds after the geyser erupts, when will the water hit the ground?

94. More Geysers If $h = -16t^2 + 80t$ represents the height of the jet of water from a geyser at time t, when will the jet of water be 96 feet high?

△ **95. Tabletop** The length of a rectangular tabletop is 3 feet(ft) shorter than twice its width. If the area of the tabletop is 54 square feet, what are the dimensions of the tabletop?

△ **96. Tarp** The length of a rectangular tarp is 1 yard(yd) shorter than twice its width. If the area of the tarp is 15 square yards, what are the dimensions of the tarp?

△ **97. Right Triangle** The shorter leg of a right triangle is 2 feet(ft) shorter than the longer leg. The hypotenuse is 10 feet. How long is each leg?

△ **98. Right Triangle** The shorter leg of a right triangle is 14 feet(ft) shorter than the longer leg. The hypotenuse is 26 feet. How long is each leg?

Chapter 6 Test   *Step-by-step test solutions are found on the Chapter Test Prep Videos available in* MyMathLab®, *on* YouTube, *or may be accessed using the QR code to the right.*

1. Find the GCF of $16x^5y^2$, $20x^4y^6$, and $24x^6y^8$.

In Problems 2–16, factor each polynomial completely. If the polynomial cannot be factored, state that it is prime.

2. $x^4 - 256$

3. $18x^3 - 9x^2 - 27x$

4. $xy - 7y - 4x + 28$

5. $27x^3 + 125$

6. $y^2 - 8y - 48$

7. $6m^2 - m - 5$

8. $4x^2 + 25$

9. $4(x - 5) + y(x - 5)$

10. $3x^2y - 15xy - 42y$

11. $x^2 + 4x + 12$

12. $2x^6 - 54y^3$

13. $9x^3 + 39x^2 + 12x$

14. $6m^2 + 7m + 2$

15. $4m^2 - 6mn + 4$

16. $25x^2 + 70xy + 49y^2$

In Problems 17 and 18, solve the equation by factoring.

17. $5x^2 = -16x - 3$

18. $5x^3 - 20x^2 + 20x = 0$

19. The length of a rectangle is 8 inches(in.) shorter than three times its width. If the area of the rectangle is 35 square inches, find the dimensions of the rectangle.

20. The hypotenuse of a right triangle is 1 inch(in.) longer than the longer leg. The shorter leg is 7 inches shorter than the longer leg. Find the length of all three sides of the triangle.

Getting Ready for Intermediate Algebra:
A Review of Chapters 1–6

The following problems cover important concepts from Chapters 1–6. We designed these problems so that you can review material that will be needed for the rest of the course. Take time to work these problems before proceeding to the next chapter. The answers to these problems are located at the back of the text on page AN-26. If you get any problems wrong, go to the section cited and review the material.

1. Using the set

$$\left\{ 0, 1, -6, \frac{2}{5}, -0.83, 0.5454\ldots, 1.010010001\ldots \right\},$$

   list all the numbers that are ...   Section 1.3, pp. 19–22

   (a) Natural numbers

   (b) Integers

   (c) Rational numbers

   (d) Irrational numbers

   (e) Real numbers

In Problems 2–8, evaluate each expression.

2. $4(-5)$ Section 1.4, pp. 32–33

3. $\dfrac{-72}{60}$ Section 1.4, pp. 33–35

4. $\dfrac{12}{5} \cdot \left(-\dfrac{35}{8} \right)$ Section 1.5, pp. 37–38

5. $\dfrac{5}{12} + \dfrac{11}{12}$ Section 1.5, pp. 39–40

6. $\dfrac{5}{6} - \dfrac{7}{30}$ Section 1.5, pp. 41–42

7. $5 - 2(1 - 4)^3 + 5 \cdot 3$ Section 1.7, pp. 57–61

8. $\dfrac{|3 - 3 \cdot 5|}{-2^3}$ Section 1.7, pp. 57–61

9. Solve: $5(x + 1) - (2x + 1) = x + 10$ Section 2.2, pp. 94–95

10. Solve: $\dfrac{1}{6}(x + 1) = 3 - \dfrac{1}{2}(x + 4)$ Section 2.3, pp. 98–100

11. Two angles are supplementary. The measure of the larger angle is 15 degrees more than twice the measure of the smaller angle. Find the measure of each angle.   Section 2.7, pp. 138–139

12. Solve $2(x + 3) \le 3x - 1$. Express your answer using set-builder notation and interval notation. Graph the inequality.   Section 2.8, pp. 150–156

13. The body mass index (BMI) of a person 62 inches tall and weighing x pounds is given by $0.2x - 2$. A BMI of 30 or more is considered to be obese. For what weights would a person 62 inches tall be considered obese?   Section 2.8, pp. 156–157

14. What is the degree of $3x^5 - 4x^3 + 2x^2 - 6x - 1$?　　Section 5.1, pp. 310–311

15. Evaluate $3x^2 + 2x - 7$ when $x = -2$.　　Section 5.1, pp. 314–315

16. Add: $(4y^3 + 8y^2 - y + 3) + (y^3 - 3y^2 - 9)$　　Section 5.1, pp. 311–312

17. Simplify: $(-3x^4)^2$　　Section 5.2, pp. 320–321

In Problems 18–21, perform the indicated operation. Express your answer as a polynomial in standard form.

18. $-4x(x^2 - 5x + 3)$　　Section 5.3, pp. 323–324

19. $(4x + 1)(3x - 5)$　　Section 5.3, pp. 324–327

20. $(2x - 7)^2$　　Section 5.3, pp. 328–329

21. $(5x + 3y)(5x - 3y)$　　Section 5.3, p. 327

22. Simplify: $\dfrac{(4y^2z)^{-1}}{y^{-3}z^2}$　　Section 5.4, pp. 341–344

In Problems 23–24, divide.

23. $\dfrac{4a^3 - 12a^2 + 2a}{2a}$　　Section 5.5, pp. 347–348

24. $\dfrac{3x^3 + 10x^2 - 23x + 1}{x + 5}$　　Section 5.5, pp. 348–351

In Problems 25–32, factor each polynomial completely.

25. $-4x^4 + 16x^2 - 20x$　　Section 6.1, pp. 372–374

26. $8z^3 - 4z^2 + 6z - 3$　　Section 6.1, pp. 374–376

27. $w^2 - 2w - 35$　　Section 6.2, pp. 378–383

28. $-3c^3 + 15c^2 + 72c$　　Section 6.2, pp. 383–384

29. $m^2 - 4m - 28$　　Section 6.2, p. 381

30. $3y^2 + 8y - 16$　　Section 6.3, pp. 386–394

31. $p^4 - 16$　　Section 6.4, pp. 398–400

32. $4a^2 + 20ab + 25b^2$　　Section 6.4, pp. 396–398

33. Solve: $(b - 3)(2b + 5) = 0$　　Section 6.6, pp. 409–410

34. Solve: $x^2 + 20 = 9x$　　Section 6.6, pp. 410–415

35. Solve: $6k^2 + 17k = 14 = 0$　　Section 6.6, pp. 410–415

7 Rational Expressions and Equations

Environmental scientists often use cost-benefit models to estimate the cost of removing a certain percentage of a pollutant from the environment. An equation such as $C = \dfrac{25x}{100 - x}$ may be used by environmental protection authorities to determine the cost C, in millions of dollars, to remove x percent of the pollutants from a lake or stream. See Problem 91 in Section 7.7.

The Big Picture: Putting It Together

In Section 1.5, we discussed how to simplify, add, subtract, multiply, and divide rational numbers expressed as fractions. Recall that a rational number is a quotient of two integers, where the denominator is not zero. In this chapter, we discuss rational expressions. A *rational expression* is the quotient of two polynomials where the denominator is not zero.

The skills we discussed in Section 1.5 apply to simplifying, adding, subtracting, multiplying, and dividing rational expressions. Performing these operations also requires the skills learned in Chapter 5 for polynomial operations plus the factoring skills acquired in Chapter 6. Being able to factor a polynomial plays a *huge* role in being able to simplify a rational expression, so make sure you are good at factoring. Because the methods used to perform operations on rational expressions are identical to those used on rational numbers, we really aren't learning any new methods here—just new ways to apply skills we already have!

7.1 Simplifying Rational Expressions

Objectives

1. Evaluate a Rational Expression
2. Determine Values for Which a Rational Expression Is Undefined
3. Simplify Rational Expressions

Are You Prepared for This Section?

Before getting started, complete the following problems. If you get a problem wrong, go back to the section cited and review the material.

P1. Evaluate $\dfrac{3x + y}{2}$ for $x = -1$ and $y = 7$. [Section 1.8, pp. 64–65]

P2. Factor: $2x^2 + x - 3$ [Section 6.3, pp. 386–394]

P3. Solve: $3x^2 - 5x - 2 = 0$ [Section 6.6, pp. 409–414]

P4. Write $\dfrac{21}{70}$ in lowest terms. [Section 1.2, pp. 12–13]

P5. Divide: $\dfrac{x^3 y^4}{xy^2}$ [Section 5.4, pp. 335–336]

Recall that a rational number is the quotient of two integers, where the denominator is not zero. The numbers $\dfrac{4}{3}$, $-\dfrac{5}{2}$, and 18 are examples of rational numbers.

> ### Definition
>
> A **rational expression** is the quotient of two polynomials. That is, a rational expression is written in the form $\dfrac{p}{q}$, where p and q are polynomials and $q \neq 0$.

Some examples of rational expressions are:

(a) $\dfrac{x - 3}{4x + 1}$ **(b)** $\dfrac{x^2 - 4x - 12}{x^2 - 5}$ **(c)** $\dfrac{3a^2 + 7b + 2b^2}{a^2 - 2ab + 8b^2}$

Expressions (a) and (b) are rational expressions in one variable, x, and expression (c) is a rational expression in two variables, a and b.

▶ 1 Evaluate a Rational Expression

To **evaluate** a rational expression, replace the variable with its assigned numerical value and perform the arithmetic, using the order of operations.

EXAMPLE 1   Evaluating a Rational Expression

Evaluate $\dfrac{-2}{x + 5}$ for **(a)** $x = -3$ **(b)** $x = 11$.

Solution

(a) Substitute -3 for x and simplify.

$$\frac{-2}{x + 5} = \frac{-2}{-3 + 5}$$

$$= \frac{-2}{2}$$

$$= -1$$

Work Smart

When evaluating a rational expression, always express the final answer in lowest terms.

(b) Substitute 11 for x and simplify.

$$\frac{-2}{x + 5} = \frac{-2}{11 + 5}$$

$$= \frac{-2}{16}$$

Divide out common factors: $= \dfrac{-1 \cdot 2}{8 \cdot 2}$

$\dfrac{-a}{b} = -\dfrac{a}{b}$: $= -\dfrac{1}{8}$

Prepared?...Answers

P1. 2 **P2.** $(2x + 3)(x - 1)$

P3. $\left\{ -\dfrac{1}{3}, 2 \right\}$ **P4.** $\dfrac{3}{10}$ **P5.** $x^2 y^2$

EXAMPLE 2 **Evaluating a Rational Expression**

Evaluate $\dfrac{p^2 - 9}{2p^2 + p - 10}$ for **(a)** $p = 1$ **(b)** $p = -2$.

Solution

(a) Substitute 1 for p and simplify.

$$\frac{p^2 - 9}{2p^2 + p - 10} = \frac{(1)^2 - 9}{2(1)^2 + 1 - 10}$$

$$= \frac{1 - 9}{2(1) + 1 - 10}$$

$$= \frac{-8}{2 + 1 - 10}$$

$$= \frac{-8}{-7}$$

$\dfrac{-a}{-b} = \dfrac{a}{b}. \qquad = \dfrac{8}{7}$

Work Smart

Remember to use parentheses when substituting a number for a variable with an exponent (as in Example 2(b)).

(b) Substitute -2 for p and simplify.

$$\frac{p^2 - 9}{2p^2 + p - 10} = \frac{(-2)^2 - 9}{2(-2)^2 + (-2) - 10}$$

$$= \frac{4 - 9}{2(4) - 2 - 10}$$

$$= \frac{-5}{8 - 2 - 10}$$

$$= \frac{-5}{-4}$$

$\dfrac{-a}{-b} = \dfrac{a}{b}. \qquad = \dfrac{5}{4}$

Quick ✓

1. The quotient of two polynomials is called a _____ _____.

In Problems 2 and 3, evaluate the rational expression for **(a)** $x = -5$ **(b)** $x = 3$.

2. $\dfrac{3}{5x + 1}$ **3.** $\dfrac{x^2 + 6x + 9}{x + 1}$

EXAMPLE 3 **Evaluating a Rational Expression with More Than One Variable**

Evaluate: **(a)** $\dfrac{w - 3v}{v + 3w}$ for $w = 2$ and $v = -4$ **(b)** $\dfrac{2a + 4b}{3c - 1}$ for $a = 1, b = -2, c = 3$

Solution

(a) Substitute 2 for w and -4 for v.

$$\frac{w - 3v}{v + 3w} = \frac{2 - 3(-4)}{-4 + 3(2)}$$

$$= \frac{2 + 12}{-4 + 6}$$

$$= \frac{14}{2}$$

$$= 7$$

(b) Substitute 1 for a, -2 for b, and 3 for c.

$$\frac{2a + 4b}{3c - 1} = \frac{2(1) + 4(-2)}{3(3) - 1}$$

$$= \frac{2 + (-8)}{9 - 1}$$

$$= \frac{-6}{8}$$

Divide out common factors: $\quad = \dfrac{\cancel{2} \cdot (-3)}{\cancel{2} \cdot 4}$

$$= \frac{-3}{4}$$

$\dfrac{-a}{b} = -\dfrac{a}{b}: \quad = -\dfrac{3}{4}$

Quick ✓

In Problems 4 and 5, evaluate the rational expression for the given values of each variable.

4. $\dfrac{5y + 2}{2y - z}$ for $y = 2, z = 1$ $\qquad$ **5.** $\dfrac{3m - 5n}{p - 4}$ for $m = 2, n = -1, p = 6$

▶ ❷ Determine Values for Which a Rational Expression Is Undefined

Work Smart

A rational expression is undefined if the *denominator* equals 0. It is okay for the *numerator* to equal 0.

A rational expression is **undefined** for values of the variable(s) that make the denominator zero. **To find the values that make a rational expression undefined, set the denominator equal to zero and solve the resulting equation.** Finding the values that make a rational expression undefined is important because it will play a role in solving rational equations, presented later in this chapter.

EXAMPLE 4 $\quad$ **Determining the Values for Which a Rational Expression Is Undefined**

Find the value of x for which the rational expression $\dfrac{2}{x + 3}$ is undefined.

Solution

Find all values of x in the rational expression $\dfrac{2}{x + 3}$ that make the denominator, $x + 3$, zero.

Set the denominator equal to 0: $\quad x + 3 = 0$

Subtract 3 from both sides: $\quad x = -3$

Because -3 makes the denominator, $x + 3$, equal 0, $\dfrac{2}{x + 3}$ is undefined for $x = -3$. ●

EXAMPLE 5 $\quad$ **Determining the Values for Which a Rational Expression Is Undefined**

Find the values of x for which the rational expression $\dfrac{8x}{x^2 - 2x - 3}$ is undefined.

Solution

Find all values of x in the rational expression $\dfrac{8x}{x^2 - 2x - 3}$ that make the denominator, $x^2 - 2x - 3$, zero.

Work Smart

Do not divide out common factors when finding values of the variable for which the rational expression is undefined.

Set the denominator equal to 0: $\quad x^2 - 2x - 3 = 0$

Factor: $\quad (x - 3)(x + 1) = 0$

Set each factor equal to 0: $\quad x - 3 = 0 \quad$ or $\quad x + 1 = 0$

Solve each equation for x: $\quad x = 3 \quad$ or $\quad x = -1$

(continued)

Work Smart

In Example 5, the expression $\dfrac{8x}{x^2-2x-3}$ is *not* undefined at $x=0$ because 0 causes the *numerator* to equal 0, which is okay.

Check Verify the answer by evaluating $\dfrac{8x}{x^2-2x-3}$ for each of these values.

If $x=3$: $\quad \dfrac{8(3)}{(3)^2-2(3)-3}=\dfrac{24}{9-6-3}=\dfrac{24}{0}$ undefined

If $x=-1$: $\quad \dfrac{8(-1)}{(-1)^2-2(-1)-3}=\dfrac{-8}{1+2-3}=\dfrac{-8}{0}$ undefined

The rational expression $\dfrac{8x}{x^2-2x-3}$ is undefined for $x=3$ or $x=-1$.

> **Quick ✓**
>
> 6. A rational expression is _____ for those values of the variable(s) that make the denominator zero.
>
> *In Problems 7–9, find the value(s) for which the rational expression is undefined.*
>
> 7. $\dfrac{3}{x+7}$ 8. $\dfrac{n-3}{3n+5}$ 9. $\dfrac{x}{x^2-6x-40}$

▶ ③ Simplify Rational Expressions

Remember that fractions are written in lowest terms by using the fact that $\dfrac{a \cdot c}{b \cdot c}=\dfrac{a}{b}$.

Thus, to write $\dfrac{18}{45}$ in lowest terms, factor both 18 and 45 and divide out common factors.

$$\frac{18}{45}=\frac{2\cdot 3\cdot 3}{3\cdot 3\cdot 5}=\frac{2}{5}$$

To **simplify** a rational expression means to write the rational expression in the form $\dfrac{p}{q}$, where p and q are polynomials that have no common factors. Use the same approach to simplify rational expressions as was used to write fractions in lowest terms.

In Other Words

A rational expression is simplified if the numerator and the denominator share no common factor other than 1.

> **Simplifying Rational Expressions**
>
> If p, q, and r are polynomials, then
>
> $$\frac{p\cdot r}{q\cdot r}=\frac{p}{q} \quad \text{where } q\neq 0 \text{ and } r\neq 0$$

To simplify a rational expression, factor the numerator, factor the denominator, and divide out the common factors.

(**EXAMPLE 6**)   **How to Simplify a Rational Expression**

Simplify: $\dfrac{7x^2+14x}{49x}$

Step-by-Step Solution

Step 1: Completely factor the numerator and the denominator.
$$\frac{7x^2+14x}{49x}=\frac{7x(x+2)}{7\cdot 7\cdot x}$$

Step 2: Divide out common factors.
$$=\frac{\cancel{7}\cancel{x}(x+2)}{\cancel{7}\cdot 7\cdot \cancel{x}}$$
$$=\frac{x+2}{7}$$

In simplifying a rational expression by dividing out common factors, the replacement values allowed for the variable may change. In Example 6, we found that $\dfrac{7x^2+14x}{49x}$ equals $\dfrac{x+2}{7}$. This is not quite true since x cannot equal 0 in $\dfrac{7x^2+14x}{49x}$, but x can equal 0 in $\dfrac{x+2}{7}$. Thus the values of the variable that make the original expression undefined should be stated. For example, write $\dfrac{7x^2+14x}{49x}=\dfrac{x+2}{7}$, $x\neq 0$. For the remainder of the text, the restrictions on the variable needed to maintain equality with the original expression will not be stated.

> **Simplifying a Rational Expression**
>
> **Step 1:** Completely factor the numerator and denominator of the rational expression.
>
> **Step 2:** Divide out common factors.

EXAMPLE 7 **Simplifying a Rational Expression**

Simplify: **(a)** $\dfrac{2x-10}{4x^2-20x}$ **(b)** $\dfrac{x^2-9}{2x^2-3x-9}$

Solution

Factor the numerator and denominator, and then simplify.

(a)
$$\frac{2x-10}{4x^2-20x}=\frac{2(x-5)}{4x(x-5)}$$

Divide out common factors:
$$=\frac{2\cancel{(x-5)}}{2\cdot 2x\cancel{(x-5)}}$$

$$=\frac{1}{2x}$$

Work Smart

Notice in Example 7(a) that when all the factors in the numerator divide out, we are left with a factor of 1, not 0.

(b)
$$\frac{x^2-9}{2x^2-3x-9}=\frac{(x+3)(x-3)}{(2x+3)(x-3)}$$

Divide out common factors:
$$=\frac{(x+3)\cancel{(x-3)}}{(2x+3)\cancel{(x-3)}}$$

$$=\frac{x+3}{2x+3}$$

Work Smart

When simplifying, divide out only common factors. Remember, an expression can be written as the product of two or more *factors*, and an expression can be written as the sum of two or more *terms*.

WRONG! $\dfrac{x+3}{x}=\dfrac{\cancel{x}+3}{\cancel{x}}=3$ WRONG! $\dfrac{x^2-4x+3}{x+3}=\dfrac{x^2-4\cancel{x}+\cancel{3}}{\cancel{x}+\cancel{3}}=x^2-4$

If you aren't sure what you can divide out, try the computation with numbers to see whether it works. For example, does $\dfrac{2+3}{2}=\dfrac{\cancel{2}+3}{\cancel{2}}=3$? No!

Quick ✓

10. To _____ a rational expression means to write the rational expression in the form $\dfrac{p}{q}$, where p and q are polynomials that have no common factors.

11. *True or False* $\dfrac{2n+3}{2n+6}=\dfrac{\cancel{2n}+3}{\cancel{2n}+6}=\dfrac{3}{6}=\dfrac{1}{2}$

In Problems 12–14, simplify the rational expression.

12. $\dfrac{3n^2+12n}{6n}$

13. $\dfrac{2z^2+6z+4}{-4z-8}$

14. $\dfrac{a^2+3a-28}{2a^2-a-28}$

EXAMPLE 8   **Simplifying a Rational Expression**

Simplify: $\dfrac{a^3 + 3a^2 + 2a + 6}{a^2 + 6a + 9}$

Solution

Because the numerator contains four terms, factor by grouping.

Work Smart

Remember: You can simplify only after the numerator and denominator have been completely factored!

Factor the numerator
and the denominator

$$\dfrac{a^3 + 3a^2 + 2a + 6}{a^2 + 6a + 9} = \dfrac{a^2(a + 3) + 2(a + 3)}{(a + 3)^2}$$

$$= \dfrac{(a + 3)(a^2 + 2)}{(a + 3)^2}$$

Divide out common factors: $= \dfrac{\cancel{(a + 3)}(a^2 + 2)}{\cancel{(a + 3)}(a + 3)}$

$$= \dfrac{a^2 + 2}{a + 3}$$

> **Quick ✔**
>
> In Problems 15 and 16, simplify the rational expression.
>
> **15.** $\dfrac{z^3 + z^2 - 3z - 3}{z^2 - z - 2}$ **16.** $\dfrac{4k^2 + 4k + 1}{4k^2 - 1}$

Notice what happens when a factor in the numerator is the *opposite* of the factor in the denominator.

3 and -3 are opposites, so $\dfrac{3}{-3} = -1$

$$\dfrac{5 - x}{x - 5} = \dfrac{-1(-5 + x)}{x - 5} = \dfrac{-1(x - 5)}{x - 5} = \dfrac{-1\cancel{(x - 5)}}{\cancel{(x - 5)}} = -1$$

$$\dfrac{4x - 7}{7 - 4x} = \dfrac{4x - 7}{-1(-7 + 4x)} = \dfrac{4x - 7}{-1(4x - 7)} = \dfrac{\cancel{(4x - 7)}}{-1\cancel{(4x - 7)}} = -1$$

EXAMPLE 9   **Simplifying a Rational Expression Containing Opposite Factors**

Simplify: $\dfrac{4 - x^2}{2x^2 - x - 6}$

Solution

Factor the numerator and denominator

$$\dfrac{4 - x^2}{2x^2 - x - 6} = \dfrac{(2 + x)(2 - x)}{(2x + 3)(x - 2)}$$

Factor -1 from $(2 - x)$: $= \dfrac{(2 + x)(-1)(-2 + x)}{(2x + 3)(x - 2)}$

Divide out common factors: $= \dfrac{(x + 2)(-1)\cancel{(x - 2)}}{(2x + 3)\cancel{(x - 2)}}$

$$= \dfrac{-(x + 2)}{2x + 3}$$

$\dfrac{-a}{b} = -\dfrac{a}{b}$: $= -\dfrac{x + 2}{2x + 3}$

Quick ✔

17. *True or False* $\dfrac{a + b}{a - b} = -1$

In Problems 18–20, simplify the rational expression.

18. $\dfrac{7a - 7b}{b - a}$

19. $\dfrac{12 - 4x}{4x^2 - 13x + 3}$

20. $\dfrac{25z^2 - 1}{3 - 15z}$

7.1 Exercises · MyMathLab®

Exercise numbers in green have complete video solutions in MyMathLab or may be accessed using the QR code to the right.

*Problems **1–20** are the Quick ✔s that follow the EXAMPLES.*

Building Skills

In Problems 21–28, evaluate each expression for the given values. See Objective 1.

21. $\dfrac{x}{x - 5}$
 (a) $x = 10$
 (b) $x = -5$
 (c) $x = 0$

22. $\dfrac{x}{x + 4}$
 (a) $x = 8$
 (b) $x = -4$
 (c) $x = 0$

23. $\dfrac{2a - 3}{a}$
 (a) $a = 3$
 (b) $a = -3$
 (c) $a = 9$

24. $\dfrac{2m - 1}{m}$
 (a) $m = 4$
 (b) $m = 1$
 (c) $m = -1$

25. $\dfrac{x^2 - 2x}{x - 4}$
 (a) $x = 3$
 (b) $x = 2$
 (c) $x = -3$

26. $\dfrac{a^2 - 2a}{a - 4}$
 (a) $a = 5$
 (b) $a = -1$
 (c) $a = -4$

27. $\dfrac{x^2 - y^2}{2x - y}$
 (a) $x = 2, y = 2$
 (b) $x = 3, y = 4$
 (c) $x = 1, y = -2$

28. $\dfrac{b^2 - a^2}{(a - b)^2}$
 (a) $a = 3, b = 2$
 (b) $a = 5, b = 4$
 (c) $a = -2, b = 2$

In Problems 29–40, find the value(s) of the variable for which the rational expression is undefined. See Objective 2.

29. $\dfrac{2 - 4x}{3x}$

30. $\dfrac{-x + 1}{5x}$

31. $\dfrac{3p}{p - 5}$

32. $\dfrac{5m^3}{m + 8}$

33. $\dfrac{8}{3 - 2x}$

34. $\dfrac{12}{4a - 3}$

35. $\dfrac{6z}{z^2 - 36}$

36. $\dfrac{5x}{25 - x^2}$

37. $\dfrac{x}{x^2 - 7x + 10}$

38. $\dfrac{2x^2}{x^2 + x - 2}$

39. $\dfrac{12x + 5}{x^3 - x^2 - 6x}$

40. $\dfrac{3h + 2}{h^3 + 5h^2 + 4h}$

In Problems 41–52, simplify each rational expression. Assume that no variable has a value that results in a denominator with a value of zero. See Objective 3.

41. $\dfrac{5x - 10}{15}$

42. $\dfrac{3n + 9}{3}$

43. $\dfrac{3z^2 + 6z}{3z}$

44. $\dfrac{6p^2 + 3p}{3p}$

45. $\dfrac{p - 3}{p^2 - p - 6}$

46. $\dfrac{x - 3}{x^2 - 4x + 3}$

47. $\dfrac{2 - x}{x - 2}$

48. $\dfrac{4 - z}{z - 4}$

49. $\dfrac{2k^2 - 14k}{7 - k}$

50. $\dfrac{2v^2 - 6v}{3 - v}$

51. $\dfrac{x^2 - 1}{2x^2 + 5x + 3}$

52. $\dfrac{x^2 - 9}{x^2 + 5x + 6}$

Mixed Practice

In Problems 53–76, simplify each rational expression. Assume that no variable has a value that results in a denominator with a value of zero.

53. $\dfrac{6x + 30}{x^2 - 25}$

54. $\dfrac{n^2 - 49}{2n + 14}$

55. $\dfrac{-3x - 3y}{x + y}$

56. $\dfrac{-2a - 2b}{a + b}$

57. $\dfrac{b^2 - 25}{4b + 20}$

58. $\dfrac{3p - 12}{p^2 - 16}$

59. $\dfrac{x^2 - 2x - 15}{x^2 - 8x + 15}$

60. $\dfrac{x^2 + x - 2}{x^2 + 5x + 6}$

61. $\dfrac{3x^3 + 8x^2 - 16x}{x^3 - x^2 - 20x}$

62. $\dfrac{4x^3 + 5x^2 - 6x}{x^3 - x^2 - 6x}$

63. $\dfrac{x^2 - y^2}{y - x}$

64. $\dfrac{a - b}{b^2 - a^2}$

65. $\dfrac{x^3 + 4x^2 + x + 4}{x^2 + 5x + 4}$

66. $\dfrac{z^3 + 3z^2 + 2z + 6}{z^2 + 5z + 6}$

67. $\dfrac{16 - c^2}{(c - 4)^2}$

68. $\dfrac{9 - x^2}{(x - 3)^2}$

69. $\dfrac{4x^2 - 20x + 24}{6x^2 - 48x + 90}$

70. $\dfrac{2x^2 + 5x - 3}{4x^2 - 8x + 3}$

71. $\dfrac{6 + x - x^2}{6x^2 + 7x - 10}$

72. $\dfrac{5 + 4x - x^2}{4x^2 - 17x - 15}$

73. $\dfrac{2t^2 - 18}{t^4 - 81}$

74. $\dfrac{x^3 + 4x}{x^4 - 16}$

75. $\dfrac{12w - 3w^2}{w^3 - 5w^2 + 4w}$

76. $\dfrac{7a - a^2}{a^3 - 5a^2 - 14a}$

Applying the Concepts

77. Drug Concentration The concentration, C, in mg/mL, of a drug in a patient's bloodstream t minutes after an injection is given by the formula

$$C = \dfrac{50t}{t^2 + 25}.$$

   (a) Find the concentration in a patient 5 minutes after an injection.

   (b) Find the concentration in a patient 10 minutes after an injection.

78. Drug Concentration The concentration, D, in mg/mL, of a drug in a patient's bloodstream t minutes after an injection is given by the formula

$$D = \dfrac{t}{2t^2 + 1}.$$

   (a) Find the concentration in a patient 30 minutes after an injection.

   (b) Find the concentration in a patient 60 minutes after an injection.

79. Body Mass Index Body Mass Index, or BMI, is used to determine if a person's weight is in a healthy range. The formula used to determine Body Mass Index is given by BMI $= \dfrac{k}{m^2}$, where k is the person's weight in kilograms and m is the person's height in meters. If a BMI range of 19 to 24.9 is considered healthy and 25 to 29.9 is considered overweight, should a person 2 meters tall weighing 110 kilograms start looking for a weight-loss program? What is this person's BMI?

80. Body Mass Index A formula to calculate the BMI when the weight, w, is given in pounds and the height, h, is given in inches is BMI $= \dfrac{705w}{h^2}$. What is

the BMI of a person weighing 120 pounds who is 5 feet tall?

81. Cost of a Car The average cost, in thousands of dollars, to produce Ford Fiestas is given by the rational expression $\dfrac{0.2x^3 - 2.3x^2 + 14.3x + 10.2}{x}$, where x is the number of cars produced. If two Fiestas are produced, what is the average cost per car?

82. Cost of a Car Use the rational expression from Problem 81 to find the average cost of producing 10 Fiestas. Do you believe that the cost will continue to decrease as more cars are made?

Extending the Concepts

In Problems 83–88, simplify each rational expression.

83. $\dfrac{c^8 - 1}{(c^4 - 1)(c^6 + 1)}$

84. $\dfrac{(1 - x)(x^6 + 1)}{x^4 - 1}$

85. $\dfrac{x^4 - x^2 - 12}{x^4 + 2x^3 - 9x - 18}$

86. $\dfrac{n^3 + 3n^2 - 16n - 48}{n^3 - 4n^2 + 3n - 12}$

87. $\dfrac{(t + 2)^3(t^4 - 16)}{(t^3 + 8)(t + 2)(t^2 - 4)}$

88. $\dfrac{24x^4 - 16x^3 + 27x - 18}{18 - 27x + 16x^2 - 24x^3}$

Explaining the Concepts

89. Explain the error in simplifying each of the following:

   (a) $\dfrac{5x^2 - 5}{x^2 - 1} = \dfrac{5\cancel{x^2} - 5}{\cancel{x^2} - 1} = \dfrac{5 - 5}{-1} = \dfrac{0}{-1} = 0$

   (b) $\dfrac{x - 2}{3x - 6} = \dfrac{x - 2}{3(x - 2)} = \dfrac{\cancel{x - 2}}{3(\cancel{x - 2})} = 3$

90. Explain why we can divide out only common *factors*, not common *terms*.

91. What does it mean for a rational expression to be undefined? Explain why the expression $\dfrac{x^2 - 9}{3x - 6}$ is undefined for $x = 2$.

92. Compare the process for simplifying a rational number such as $-\dfrac{21}{35}$ with simplifying a rational algebraic expression such as $\dfrac{x^2 - 2x - 3}{1 - x^2}$.

93. Why is $\dfrac{x - 7}{7 - x} = -1$ whereas $\dfrac{x - 7}{x + 7}$ is in simplest form?

94. Rational expressions that represent the same quantity are said to be *equivalent*. Are the rational expressions $\dfrac{x - y}{y - x}$ and $\dfrac{y - x}{x - y}$ equivalent? Explain how to determine whether two rational expressions are equivalent.

7.2 Multiplying and Dividing Rational Expressions

Objectives

1 Multiply Rational Expressions

2 Divide Rational Expressions

Are You Prepared for This Section?

Before getting started, complete the following problems. If you get a problem wrong, go back to the section cited and review the material.

P1. Find the product: $\dfrac{3}{14} \cdot \dfrac{28}{9}$ [Section 1.5, pp. 37–38]

P2. Find the reciprocal of $\dfrac{5}{8}$ [Section 1.4, pp. 33–34]

P3. Find the quotient: $\dfrac{12}{25} \div \dfrac{12}{5}$ [Section 1.5, pp. 38–39]

P4. Factor: $3x^3 - 27x$ [Section 6.4, pp. 398–400]

P5. Simplify the rational expression: $\dfrac{3x + 12}{5x^2 + 20x}$ [Section 7.1, pp. 436–439]

▶ **1** Multiply Rational Expressions

Multiplying rational expressions is similar to multiplying rational numbers. Recall that,

$$\frac{2}{3} \cdot \frac{12}{5} = \frac{2 \cdot 12}{3 \cdot 5} = \frac{2 \cdot 3 \cdot 4}{3 \cdot 5} = \frac{2 \cdot 4}{5} = \frac{8}{5}$$

EXAMPLE 1   **How to Multiply Rational Expressions**

Multiply: $\dfrac{x - 3}{5} \cdot \dfrac{5x + 35}{x^2 - 9}$

Step-by-Step Solution

Step 1: Completely factor the polynomials in each numerator and denominator.

$$\frac{x - 3}{5} \cdot \frac{5x + 35}{x^2 - 9} = \frac{x - 3}{5} \cdot \frac{5(x + 7)}{(x + 3)(x - 3)}$$

Step 2: Multiply.

$$= \frac{5(x - 3)(x + 7)}{5(x + 3)(x - 3)}$$

Step 3: Divide out common factors in the numerator and denominator.

The common factors are $(x - 3)$ and 5:

$$= \frac{\cancel{5}\,\cancel{(x - 3)}\,(x + 7)}{\cancel{5}\,(x + 3)\,\cancel{(x - 3)}}$$

$$= \frac{x + 7}{x + 3}$$

Remember that $\dfrac{x + 7}{x + 3}$ cannot be simplified further. Divide out *factors*, not terms.

So, $\dfrac{x - 3}{5} \cdot \dfrac{5x + 35}{x^2 - 9} = \dfrac{x + 7}{x + 3}$.

In Other Words

To multiply two rational expressions, factor each polynomial, write the expression as a single fraction in factored form, and then divide out common factors.

Multiplying Rational Expressions

Step 1: Factor the polynomials in each numerator and denominator.

Step 2: Use the fact that if $\dfrac{a}{b}$ and $\dfrac{c}{d}$, where $b \neq 0$ and $d \neq 0$, are two rational expressions, then $\dfrac{a}{b} \cdot \dfrac{c}{d} = \dfrac{ac}{bd}$.

Step 3: Divide out common factors in the numerator and denominator. Leave the remaining polynomials in factored form.

Prepared?...Answers   **P1.** $\dfrac{2}{3}$ **P2.** $\dfrac{8}{5}$

P3. $\dfrac{1}{5}$ **P4.** $3x(x + 3)(x - 3)$ **P5.** $\dfrac{3}{5x}$

EXAMPLE 2 **Multiplying Rational Expressions**

Multiply: $\dfrac{7y + 28}{4y} \cdot \dfrac{8}{y^2 + 2y - 8}$

Solution

Begin by factoring each polynomial in the numerator and denominator.

$$\dfrac{7y + 28}{4y} \cdot \dfrac{8}{y^2 + 2y - 8} = \dfrac{7(y + 4)}{4y} \cdot \dfrac{4 \cdot 2}{(y + 4)(y - 2)}$$

$$\dfrac{a}{b} \cdot \dfrac{c}{d} = \dfrac{a \cdot c}{b \cdot d} = \dfrac{7(y + 4) \cdot 4 \cdot 2}{4y(y + 4)(y - 2)}$$

Divide out common factors 4 and $(y + 4)$: $= \dfrac{7\cancel{(y + 4)} \cdot \cancel{4} \cdot 2}{\cancel{4} \cdot y \cancel{(y + 4)}(y - 2)}$

$$= \dfrac{7 \cdot 2}{y(y - 2)}$$

Multiply: $= \dfrac{14}{y(y - 2)}$ ●

Work Smart

Notice that we did not multiply 4 and 2 in the numerator. This is because we anticipated dividing out the 4.

EXAMPLE 3 **Multiplying Rational Expressions**

Multiply: $\dfrac{p^2 - 9}{p^2 + 5p + 6} \cdot \dfrac{p + 2}{6 - 2p}$.

Work Smart

$6 - 2p = 2(3 - p) = -2(p - 3)$

Solution

Factor the numerator and denominator ↓

$$\dfrac{p^2 - 9}{p^2 + 5p + 6} \cdot \dfrac{p + 2}{6 - 2p} = \dfrac{(p + 3)(p - 3)}{(p + 3)(p + 2)} \cdot \dfrac{p + 2}{(-2)(p - 3)}$$

$$\dfrac{a}{b} \cdot \dfrac{c}{d} = \dfrac{a \cdot c}{b \cdot d} = \dfrac{(p + 3)(p - 3)(p + 2)}{(p + 3)(p + 2)(-2)(p - 3)}$$

Divide out common factors $(p + 3)$, $(p - 3)$, and $(p + 2)$: $= \dfrac{\cancel{(p + 3)}\cancel{(p - 3)}\cancel{(p + 2)}}{\cancel{(p + 3)}\cancel{(p + 2)}(-2)\cancel{(p - 3)}}$

$$= \dfrac{1}{-2}$$

$$\dfrac{a}{-b} = -\dfrac{a}{b} = -\dfrac{1}{2}$$ ●

Work Smart

When all the factors in the numerator divide out, the numerator is left with a factor of 1, *not* 0.

Quick ✓

1. *True or False* $\dfrac{x + 4}{24} \cdot \dfrac{24}{(x + 4)(x - 4)} = \dfrac{(x + 4) \cdot 24}{24 \cdot (x + 4)(x - 4)} = x - 4$

2. *True or False* $\dfrac{8}{x - 7} \cdot \dfrac{7 - x}{16x} = \dfrac{1}{2x}$

In Problems 3–6, multiply and simplify the rational expressions.

3. $\dfrac{p^2 - 9}{5} \cdot \dfrac{25p}{2p - 6}$

4. $\dfrac{2x + 8}{x^2 + 3x - 4} \cdot \dfrac{7x - 7}{6x + 30}$

5. $\dfrac{x - 7}{x^2 - 9} \cdot \dfrac{x + 3}{7 - x}$

6. $\dfrac{15 - 3a}{7} \cdot \dfrac{3 + 2a}{2a^2 - 7a - 15}$

EXAMPLE 4   **Multiplying Rational Expressions with Two Variables**

Multiply: $\dfrac{m^2 - n^2}{10m^2 - 10mn} \cdot \dfrac{10m + 5n}{2m^2 + 3mn + n^2}$

Solution

Factor the numerator and denominator

$$\dfrac{m^2 - n^2}{10m^2 - 10mn} \cdot \dfrac{10m + 5n}{2m^2 + 3mn + n^2} = \dfrac{(m + n)(m - n)}{10m(m - n)} \cdot \dfrac{5(2m + n)}{(2m + n)(m + n)}$$

$\dfrac{a}{b} \cdot \dfrac{c}{d} = \dfrac{a \cdot c}{b \cdot d}, \ 10 = 5 \cdot 2:$ $= \dfrac{(m + n)(m - n) \cdot 5(2m + n)}{5 \cdot 2m(m - n)(2m + n)(m + n)}$

Divide out common factors $(m + n)$, $(m - n)$, $(2m + n)$, and 5: $= \dfrac{\cancel{(m + n)} \cancel{(m - n)} \cdot \cancel{5} \cancel{(2m + n)}}{\cancel{5} \cdot 2m \cancel{(m - n)} \cancel{(2m + n)} \cancel{(m + n)}}$

$$= \dfrac{1}{2m}$$

> **Quick ✓**
>
> In Problem 7, multiply the rational expressions.
>
> 7. $\dfrac{a^2 + 2ab + b^2}{3a^2 + 3ab} \cdot \dfrac{a - b}{a^2 - b^2}$

▶ ❷ **Divide Rational Expressions**

Recall that in the expression $\dfrac{2}{3} \div \dfrac{7}{9}$, $\dfrac{2}{3}$ is called the *dividend* and $\dfrac{7}{9}$ is called the *divisor*. The answer is called the *quotient*.

Finding the quotient of rational expressions is similar to finding the quotient of rational numbers. For example:

reciprocal of divisor
↓
$$\dfrac{2}{3} \div \dfrac{7}{9} = \dfrac{2}{3} \cdot \dfrac{9}{7} = \dfrac{2 \cdot 3 \cdot 3}{3 \cdot 7} = \dfrac{2 \cdot 3}{7} = \dfrac{6}{7}$$
↑
dividend

EXAMPLE 5   **How to Divide Rational Expressions**

Divide: $\dfrac{x + 3}{2x - 8} \div \dfrac{9}{4x}$

Step-by-Step Solution

Step 1: Multiply the dividend by the reciprocal of the divisor.

The reciprocal of $\dfrac{9}{4x}$ is $\dfrac{4x}{9}$
↓
$$\dfrac{x + 3}{2x - 8} \div \dfrac{9}{4x} = \dfrac{x + 3}{2x - 8} \cdot \dfrac{4x}{9}$$

Step 2: Completely factor the polynomials in each numerator and denominator.

$$= \dfrac{x + 3}{2(x - 4)} \cdot \dfrac{2 \cdot 2 \cdot x}{9}$$

(continued)

Step 3: Multiply.

$$= \frac{(x+3) \cdot 2 \cdot 2 \cdot x}{2(x-4) \cdot 9}$$

Step 4: Divide out common factors in the numerator and denominator.

Divide out common factors: $= \dfrac{(x+3) \cdot \cancel{2} \cdot 2 \cdot x}{\cancel{2}(x-4) \cdot 9}$

$$= \frac{2x(x+3)}{9(x-4)}$$

Work Smart

In Step 2 of Example 5, we factored 4 as $2 \cdot 2$, but we did not factor 9. Why? Because we anticipated dividing out a common factor of 2. We could see there were no factors of 9 in the numerator, so we did not factor 9.

Thus, $\dfrac{x+3}{2x-8} \div \dfrac{9}{4x} = \dfrac{2x(x+3)}{9(x-4)}$.

Dividing Rational Expressions

Step 1: Multiply the dividend by the reciprocal of the divisor. That is, $\dfrac{a}{b} \div \dfrac{c}{d} = \dfrac{a}{b} \cdot \dfrac{d}{c}$, where $b \neq 0, c \neq 0,$ and $d \neq 0$.

Step 2: Factor each polynomial in the numerator and denominator.

Step 3: Multiply.

Step 4: Divide out common factors in the numerator and denominator. Leave the remaining polynomials in factored form.

EXAMPLE 6 **Dividing Rational Expressions**

Find the quotient: $\dfrac{y^2-9}{2y^2-y-15} \div \dfrac{3y^2+10y+3}{2y^2+y-10}$

Solution

Multiply the dividend by the reciprocal of the divisor.

$$\frac{y^2-9}{2y^2-y-15} \div \frac{3y^2+10y+3}{2y^2+y-10} = \frac{y^2-9}{2y^2-y-15} \cdot \frac{2y^2+y-10}{3y^2+10y+3}$$

Factor the numerator and denominator: $= \dfrac{(y-3)(y+3)}{(2y+5)(y-3)} \cdot \dfrac{(2y+5)(y-2)}{(3y+1)(y+3)}$

$\dfrac{a}{b} \cdot \dfrac{c}{d} = \dfrac{a \cdot c}{b \cdot d}$: $= \dfrac{(y-3)(y+3)(2y+5)(y-2)}{(2y+5)(y-3)(3y+1)(y+3)}$

Divide out common factors: $= \dfrac{\cancel{(y-3)}\cancel{(y+3)}\cancel{(2y+5)}(y-2)}{\cancel{(2y+5)}\cancel{(y-3)}(3y+1)\cancel{(y+3)}}$

$$= \frac{y-2}{3y+1}$$

Quick ✔

8. $\dfrac{8}{9} \div \dfrac{2}{3} = \dfrac{8}{9} \cdot \underline{} = \underline{}$

In Problems 9 and 10, find the quotient.

9. $\dfrac{12}{x^2-x} \div \dfrac{4x-2}{x^2-1}$

10. $\dfrac{x^2-9}{x^2-16} \div \dfrac{x^2-x-12}{x^2+x-12}$

EXAMPLE 7 **Dividing Rational Expressions**

Find the quotient: $\dfrac{a^2+3a+2}{a^2-9} \div (a+2)$

Solution

Write the divisor $(a + 2)$ as $\dfrac{a + 2}{1}$

$$\frac{a^2 + 3a + 2}{a^2 - 9} \div (a + 2) = \frac{a^2 + 3a + 2}{a^2 - 9} \div \frac{a + 2}{1}$$

Multiply by the reciprocal of the divisor: $\quad = \dfrac{a^2 + 3a + 2}{a^2 - 9} \cdot \dfrac{1}{a + 2}$

Factor the numerator and denominator: $\quad = \dfrac{(a + 2)(a + 1)}{(a + 3)(a - 3)} \cdot \dfrac{1}{a + 2}$

$\dfrac{a}{b} \cdot \dfrac{c}{d} = \dfrac{a \cdot c}{b \cdot d} \quad = \dfrac{(a + 2)(a + 1)}{(a + 3)(a - 3)(a + 2)}$

Divide out common factors: $\quad = \dfrac{\cancel{(a + 2)}(a + 1)}{(a + 3)(a - 3)\cancel{(a + 2)}}$

$$= \frac{a + 1}{(a + 3)(a - 3)}$$

> **Quick ✓**
>
> In Problem 11, find the quotient.
>
> **11.** $\dfrac{q^2 - 6q - 7}{q^2 - 25} \div (q - 7)$

EXAMPLE 8   **Dividing Rational Expressions Written Vertically**

Find the quotient: $\dfrac{\dfrac{12x}{5x + 20}}{\dfrac{4x^2}{x^2 - 16}}$

Work Smart

$$\frac{\dfrac{12x}{5x + 20}}{\dfrac{4x^2}{x^2 - 16}} = \frac{12x}{5x + 20} \div \frac{4x^2}{x^2 - 16}$$

Solution

Rewrite division as a multiplication problem by multiplying by the reciprocal of the denominator

$$\frac{\dfrac{12x}{5x + 20}}{\dfrac{4x^2}{x^2 - 16}} = \frac{12x}{5x + 20} \cdot \frac{x^2 - 16}{4x^2}$$

Factor the numerator and denominator: $\quad = \dfrac{4 \cdot 3 \cdot x}{5(x + 4)} \cdot \dfrac{(x - 4)(x + 4)}{4 \cdot x \cdot x}$

$\dfrac{a}{b} \cdot \dfrac{c}{d} = \dfrac{a \cdot c}{b \cdot d} \quad = \dfrac{4 \cdot 3 \cdot x(x - 4)(x + 4)}{5(x + 4) \cdot 4 \cdot x \cdot x}$

Divide out common factors: $\quad = \dfrac{\cancel{4} \cdot 3 \cdot \cancel{x}(x - 4)\cancel{(x + 4)}}{5\cancel{(x + 4)} \cdot \cancel{4} \cdot \cancel{x} \cdot x}$

$$= \frac{3(x - 4)}{5x}$$

> **Quick ✓**
>
> In Problems 12 and 13, find the quotient.
>
> **12.** $\dfrac{\dfrac{2}{x + 1}}{\dfrac{8}{x^2 - 1}}$ **13.** $\dfrac{\dfrac{x + 3}{x^2 - 4}}{\dfrac{4x + 12}{7x^2 + 14x}}$

EXAMPLE 9 **Dividing Rational Expressions with More Than One Variable**

Find the quotient: $\dfrac{5a - 5b}{7c^2} \div \dfrac{10b - 10a}{21c}$

Solution

Multiply by the reciprocal of the divisor
↓

$$\frac{5a - 5b}{7c^2} \div \frac{10b - 10a}{21c} = \frac{5a - 5b}{7c^2} \cdot \frac{21c}{10b - 10a}$$

Factor the numerator and denominator: $= \dfrac{5(a - b)}{7 \cdot c \cdot c} \cdot \dfrac{7 \cdot 3 \cdot c}{5 \cdot 2 \cdot (-1)(a - b)}$

$\dfrac{a}{b} \cdot \dfrac{c}{d} = \dfrac{a \cdot c}{b \cdot d}$: $= \dfrac{5(a - b) \cdot 7 \cdot 3 \cdot c}{7 \cdot c \cdot c \cdot 5 \cdot 2 \cdot (-1)(a - b)}$

Divide out common factors: $= \dfrac{\cancel{5}\cancel{(a - b)} \cdot \cancel{7} \cdot 3 \cdot \cancel{c}}{\cancel{7} \cdot c \cdot \cancel{c} \cdot \cancel{5} \cdot 2 \cdot (-1)\cancel{(a - b)}}$

$= \dfrac{3}{c \cdot 2 \cdot (-1)}$

Multiply: $= -\dfrac{3}{2c}$

Quick ✔

In Problem 14, find the quotient and simplify.

14. $\dfrac{3m - 6n}{5n} \div \dfrac{m^2 - 4n^2}{10mn}$

7.2 Exercises MyMathLab®   Exercise numbers in **green** have complete video solutions in MyMathLab or may be accessed using the QR code to the right.

Problems **1–14** are the **Quick ✔**s that follow the **EXAMPLES**.

Building Skills

In Problems 15–22, multiply. See Objective 1.

15. $\dfrac{x + 5}{7} \cdot \dfrac{14x}{x^2 - 25}$

16. $\dfrac{x - 4}{4} \cdot \dfrac{12x}{x^2 - 16}$

17. $\dfrac{x^2 - x}{x^2 - x - 2} \cdot \dfrac{x - 2}{x^2 - 1}$

18. $\dfrac{8n - 8}{n^2 - 3n + 2} \cdot \dfrac{n + 2}{12}$

19. $\dfrac{p^2 - 1}{2p - 3} \cdot \dfrac{2p^2 + p - 6}{p^2 + 3p + 2}$

20. $\dfrac{z^2 - 4}{3z - 2} \cdot \dfrac{3z^2 + 7z - 6}{z^2 + z - 6}$

21. $(x + 1) \cdot \dfrac{x - 6}{x^2 - 5x - 6}$

22. $(x - 3) \cdot \dfrac{x - 2}{x^2 - 5x + 6}$

In Problems 23–30, divide. See Objective 2.

23. $\dfrac{x - 4}{2x - 8} \div \dfrac{3x}{2}$

24. $\dfrac{p + 7}{3p + 21} \div \dfrac{p}{6}$

25. $\dfrac{m^2 - 16}{6m} \div \dfrac{m^2 + 8m + 16}{12m}$

26. $\dfrac{z^2 - 25}{10z} \div \dfrac{z^2 - 10z + 25}{5z}$

27. $\dfrac{x^2 - x}{x^2 - 1} \div \dfrac{x + 2}{x^2 + 3x + 2}$

28. $\dfrac{3y^2 + 6y}{8} \div \dfrac{y + 2}{12y - 12}$

29. $\dfrac{\dfrac{x^2 + 4x + 4}{x^2 - 4}}{\dfrac{x^2 - x - 6}{x^2 - 5x + 6}}$

30. $\dfrac{\dfrac{4z^2 + 12z + 9}{8z + 16}}{\dfrac{4z^2 + 12z + 9}{12z + 24}}$

Mixed Practice

In Problems 31–66, perform the indicated operation.

31. $\dfrac{14}{9} \cdot \dfrac{15}{7}$

32. $\dfrac{7}{52} \cdot \dfrac{77}{13}$

33. $-\dfrac{20}{16} \div \left(-\dfrac{30}{24}\right)$

34. $-\dfrac{60}{35} \div \dfrac{15}{77}$

35. $\dfrac{8}{11} \div (-2)$

36. $\dfrac{16}{3} \div (-4)$

37. $\dfrac{3y}{y^2 - y - 6} \cdot \dfrac{4y + 8}{9y^2}$

38. $\dfrac{x - 4}{12x - 18} \cdot \dfrac{6}{x^2 - 16}$

39. $\dfrac{4a + 8b}{a^2 + 2ab} \cdot \dfrac{a}{12}$

40. $\dfrac{m^2 - n^2}{3m - 3n} \cdot \dfrac{6}{2m + 2n}$

41. $\dfrac{3x^2 - 6x}{x^2 - 2x - 8} \div \dfrac{x - 2}{x + 2}$

42. $\dfrac{x + 5}{3} \div \dfrac{30x}{4x + 20}$

43. $\dfrac{\dfrac{2w^2 - 16w + 32}{4w^2 - 17w + 4}}{\dfrac{2w^2 - 11w + 12}{8w^2 - 10w - 3}}$

44. $\dfrac{\dfrac{-3w^2 + 12w - 12}{5w^2 - 7w - 6}}{\dfrac{2w^2 - 7w + 6}{10w^2 - 9w - 9}}$

45. $\dfrac{3xy - 2y^2 - x^2}{x + y} \cdot \dfrac{x^2 - y^2}{x^2 - 2xy}$

46. $\dfrac{2r^2 + rs - 3s^2}{r^2 - s^2} \cdot \dfrac{r^2 - 2rs - 3s^2}{2r + 3s}$

47. $\dfrac{\dfrac{2c - 4}{8}}{\dfrac{2 - c}{2}}$

48. $\dfrac{\dfrac{2x - 3}{3}}{\dfrac{3}{6x - 9}}$

49. $\dfrac{8n^2 - 6n - 9}{6n + 18} \cdot \dfrac{9n^2 - 81}{2n^2 + 5n - 12}$

50. $\dfrac{2x^2 - 5x + 3}{x^2 - 1} \cdot \dfrac{x^2 + 1}{2x^2 - x - 3}$

51. $\dfrac{x^2 y}{2x^2 - 5xy + 2y^2} \div \dfrac{(2xy^2)^2}{2x^2 y - xy^2}$

52. $\dfrac{(x - 3)^2}{8xy^2} \div \dfrac{x^2 - 9}{(4x^2 y)^2}$

53. $\dfrac{2a^2 + 3ab - 2b^2}{a^2 - b^2} \cdot \dfrac{a^2 - ab}{2a^3 + 4a^2 b}$

54. $\dfrac{3y^2 - 3x^2}{2x^2 + xy - y^2} \cdot \dfrac{3x - 6y}{6x - 6y}$

55. $\dfrac{6x^2 + 17x + 5}{2x^2 - 2x - 24} \cdot \dfrac{4x^2 - 13x - 12}{12x^2 + 13x + 3}$

56. $\dfrac{6x^2 + 21x - 45}{3x^2 - 11x - 20} \cdot \dfrac{6x^2 + 11x + 4}{4x^2 - 4x - 3}$

57. $\dfrac{3x^2 + 12x + 12}{x^2 - 4} \div \dfrac{-x^2 + x + 6}{x^2 - 5x + 6}$

58. $\dfrac{4z^2 + 12z + 9}{8z + 16} \div \dfrac{(2z + 3)^3}{12z + 24}$

59. $\dfrac{9 - x^2}{x^2 + 5x + 4} \div \dfrac{x^2 - 2x - 3}{x^2 + 4x}$

60. $\dfrac{1}{b^2 + b - 12} \div \dfrac{1}{b^2 - 5b - 36}$

61. $\dfrac{\dfrac{a^2 - b^2}{a^2 + b^2}}{\dfrac{4a - 4b}{2a^2 + 2b^2}}$

62. $\dfrac{\dfrac{t^2}{t^2 - 16}}{\dfrac{t^2 - 3t}{t^2 - t - 12}}$

63. $\dfrac{\dfrac{x^3 - 1}{x^4 - 1}}{\dfrac{3x^2 + 3x + 3}{x^3 + x^2 + x + 1}}$

64. $\dfrac{\dfrac{p^3 - 8q^3}{p^2 - 4q^2}}{\dfrac{p^2 + 4pq + 4q^2}{(p + 2q)^2}}$

65. $\dfrac{xy - ay + xb - ab}{xy + ay - xb - ab} \cdot \dfrac{2xy - 2ay - 2xb + 2ab}{4b + 4y}$

66. $\dfrac{x^3 - x^2 + x - 1}{x + 1} \cdot \dfrac{x^2 + 2x + 1}{1 - x^2}$

Applying the Concepts

67. Find the product of $\dfrac{x^2 + 3xy + 2y^2}{x^2 - y^2}$ and $\dfrac{3x - 3y}{9x^2 + 9xy - 18y^2}$.

68. Find the product of $\dfrac{x}{6y^2}$ and $\dfrac{21x^2 y}{(7x)^2}$.

69. Find the quotient of $\dfrac{x}{2y}$ and $\dfrac{(2xy)^2}{9xy^3}$.

70. Find the quotient of $\dfrac{x - 3}{2x + 6}$ and $\dfrac{x - 9}{4x^2 - 36}$.

71. Find $\dfrac{x - y}{x + y}$ squared divided by $x^2 - y^2$.

72. Find $(a - b)$ squared divided by $a^2 - b^2$.

73. What is $\dfrac{3x - 9}{2x + 4}$ divided into $\dfrac{6x}{x^2 - 4}$?

74. What is $\dfrac{2x^2}{3}$ divided into $\dfrac{x^3 - 3x}{6x}$?

△ **75. Area of a Rectangle** Write an algebraic expression for the area of the rectangle.

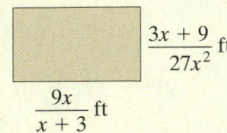

$\dfrac{3x + 9}{27x^2}$ ft

$\dfrac{9x}{x + 3}$ ft

△ **76. Area of a Rectangle** Write an algebraic expression for the area of the rectangle.

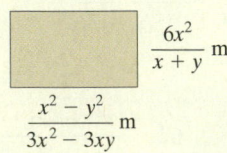

$\dfrac{6x^2}{x + y}$ m

$\dfrac{x^2 - y^2}{3x^2 - 3xy}$ m

△ **77. Area of a Triangle** Write an algebraic expression for the area of the triangle.

$h = (4x^2 + 20x + 24)$ in.

$b = \dfrac{1}{x^2 - 9}$ in.

△ **78. Area of a Triangle** Write an algebraic expression for the area of the triangle.

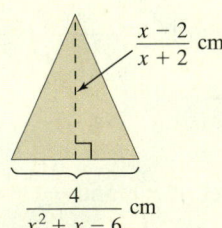

$\dfrac{x - 2}{x + 2}$ cm

$\dfrac{4}{x^2 + x - 6}$ cm

Extending the Concepts

In Problems 79–86, perform the indicated operation.

79. $\dfrac{x^2 + x - 12}{x^2 - 2x - 35} \div \dfrac{x + 4}{x^2 + 4x - 5} \div \dfrac{12 - 4x}{x - 7}$

80. $\dfrac{x^2 - 6xy + 9y^2}{x^2 - 4y^2} \div \dfrac{x - 3y}{x^2 - 5xy + 6y^2} \div \dfrac{x^2 - 9y^2}{x^2 - xy - 6y^2}$

81. $\dfrac{a^2 - 2ab}{2b - 3a} \div \dfrac{3a^2 - 4ab - 4b^2}{16a^2b^2 - 36a^4} \div (6a)$

82. $\dfrac{(2v - 1)^6}{(1 - 2v)^5} \cdot \dfrac{1 - 8v^3}{4v^2 - 4v + 1} \div (2v - 1)$

83. $\dfrac{x^2 + xy - 3x - 3y}{x^3 + y^3} \cdot \dfrac{x^2 + 2xy + y^2}{x^2 - x - 6}$

84. $\dfrac{a^2 + 4a - 12}{a^2 + 6a - 27} \cdot \dfrac{a^2 - 9}{2a^2 + 12a} \div (4a^2)$

85. $\dfrac{p^3 - 27q^3}{9pq} \cdot \dfrac{(3p^2q)^3}{p^2 - 9q^2} \div \dfrac{1}{p^2 + 2pq - 3q^2}$

86. $\dfrac{x^2 - 1}{x^4 - 81} \div \dfrac{x^2 + 1}{(x - 3)^2} \cdot \dfrac{x^3 + x}{x^2 - 2x - 3}$

In Problems 87 and 88, find the missing expression.

87. $\dfrac{2x}{x^3 - 3x^2} \cdot \dfrac{x^2 - x - 6}{?} = \dfrac{1}{3x}$

88. $\dfrac{x^2 + 12x + 36}{?} \div \dfrac{x^2 + 11x + 30}{x^2 + 3x - 10} = \dfrac{x + 6}{x - 2}$

Explaining the Concepts

89. List the steps for multiplying two rational algebraic expressions.

90. List the steps for dividing two rational algebraic expressions.

91. Find and describe the error in the following simplification, and then work the problem correctly.

$$\dfrac{n^2 - 2n - 24}{6n - n^2} = \dfrac{(n - 6)(n + 4)}{n(6 - n)}$$

$$= \dfrac{n + 4}{n}$$

92. A unit fraction is any representation of 1. For instance, both $\dfrac{x}{x}$ and $\dfrac{1 \text{ foot}}{12 \text{ inches}}$ are unit fractions. Because multiplying by 1 does not change the value of any real number, use this concept to explain how to convert 56 inches into feet.

7.3 Adding and Subtracting Rational Expressions with a Common Denominator

Objectives

① Add Rational Expressions with a Common Denominator

② Subtract Rational Expressions with a Common Denominator

③ Add or Subtract Rational Expressions with Opposite Denominators

Are You Prepared for This Section?

Before getting started, complete the following problems. If you get a problem wrong, go back to the section cited and review the material.

P1. Write $\dfrac{12}{15}$ in lowest terms. [Section 1.2, pp. 12–13]

P2. Find each sum and write in lowest terms:

(a) $\dfrac{7}{5} + \dfrac{2}{5}$ (b) $\dfrac{5}{6} + \dfrac{11}{6}$ [Section 1.5, pp. 39–40]

P3. Find each difference and write in lowest terms:

(a) $\dfrac{7}{9} - \dfrac{5}{9}$ (b) $\dfrac{7}{8} - \dfrac{5}{8}$ [Section 1.5, pp. 39–40]

P4. Determine the additive inverse of 5. [Section 1.4, pp. 30–31]

P5. Simplify: $-(x - 2)$ [Section 1.8, pp. 67–68]

▶ ① Add Rational Expressions with a Common Denominator

Rational expressions are added in the same way rational numbers are added. If the denominators of two rational expressions are the same, add the numerators, write the result over the common denominator, and simplify. That is, $\dfrac{a}{c} + \dfrac{b}{c} = \dfrac{a+b}{c}$, where $c \neq 0$. See the examples below.

ADDING RATIONAL NUMBERS	ADDING RATIONAL EXPRESSIONS
$\dfrac{7}{12} + \dfrac{11}{12} = \dfrac{18}{12} = \dfrac{6 \cdot 3}{6 \cdot 2} = \dfrac{3}{2}$	$\dfrac{7}{12z} + \dfrac{11}{12z} = \dfrac{18}{12z} = \dfrac{6 \cdot 3}{6 \cdot 2 \cdot z} = \dfrac{3}{2z}, z \neq 0$

Prepared?...Answers P1. $\dfrac{4}{5}$

P2. (a) $\dfrac{9}{5}$ **(b)** $\dfrac{8}{3}$ **P3. (a)** $\dfrac{2}{9}$ **(b)** $\dfrac{1}{4}$

P4. -5 **P5.** $2 - x$

Throughout the section, we assume that a variable cannot take on any values that result in division by zero.

EXAMPLE 1 **How to Add Rational Expressions with a Common Denominator**

Find the sum: $\dfrac{5}{x+1} + \dfrac{3}{x+1}$

Step-by-Step Solution

The rational expressions in the sum $\dfrac{5}{x+1} + \dfrac{3}{x+1}$ have a common denominator, $x + 1$.

Step 1: Add the numerators and write the result over the common denominator.

$$\dfrac{a}{c} + \dfrac{b}{c} = \dfrac{a+b}{c}: \qquad \dfrac{5}{x+1} + \dfrac{3}{x+1} = \dfrac{5+3}{x+1}$$

Combine like terms in the numerator:

$$= \dfrac{8}{x+1}$$

Step 2: Simplify the rational expression by dividing out common factors.

The sum is already simplified.

So, $\dfrac{5}{x+1} + \dfrac{3}{x+1} = \dfrac{8}{x+1}$.

Adding Rational Expressions with a Common Denominator

Step 1: Use the fact that if $\dfrac{a}{c}$ and $\dfrac{b}{c}$, where $c \neq 0$, are two rational expressions, then

$$\frac{a}{c} + \frac{b}{c} = \frac{a + b}{c}$$

Step 2: Simplify the sum by dividing out common factors.

EXAMPLE 2 **Adding Rational Expressions with a Common Denominator**

Find the sum:

(a) $\dfrac{x^2}{x + 7} + \dfrac{7x}{x + 7}$

(b) $\dfrac{2x^2 + x}{x^2 - 4} + \dfrac{x - x^2}{x^2 - 4}$

Solution

Because the denominators are the same, add the numerators and write the sum over the common denominator.

(a)
$$\frac{x^2}{x + 7} + \frac{7x}{x + 7} = \frac{x^2 + 7x}{x + 7}$$

Factor the numerator: $= \dfrac{x(x + 7)}{x + 7}$

Divide out common factors: $= \dfrac{x\cancel{(x + 7)}}{\cancel{(x + 7)}}$

$= \dfrac{x}{1}$

$\dfrac{a}{1} = a$: $= x$

(b)
$$\frac{2x^2 + x}{x^2 - 4} + \frac{x - x^2}{x^2 - 4} = \frac{2x^2 + x + x - x^2}{x^2 - 4}$$

Combine like terms in the numerator: $= \dfrac{x^2 + 2x}{x^2 - 4}$

Factor the numerator and denominator: $= \dfrac{x(x + 2)}{(x + 2)(x - 2)}$

Divide out common factors: $= \dfrac{x\cancel{(x + 2)}}{\cancel{(x + 2)}(x - 2)}$

$= \dfrac{x}{x - 2}$

Quick ✓

1. If $\dfrac{a}{c}$ and $\dfrac{b}{c}$, where $c \neq 0$, are two rational expressions, then $\dfrac{a}{c} + \dfrac{b}{c} = $ _____.

2. *True or False* $\dfrac{2x}{a} + \dfrac{4x}{a} = \dfrac{6x}{2a} = \dfrac{3x}{a}$

In Problems 3–6, find the sum.

3. $\dfrac{1}{x - 2} + \dfrac{3}{x - 2}$

4. $\dfrac{2x + 1}{x + 1} + \dfrac{x^2}{x + 1}$

5. $\dfrac{9x}{6x - 5} + \dfrac{2x - 3}{6x - 5}$

6. $\dfrac{2x - 2}{2x^2 - 7x - 15} + \dfrac{5}{2x^2 - 7x - 15}$

▶ ② Subtract Rational Expressions with a Common Denominator

The steps for subtracting rational expressions are the same as the steps for subtracting rational numbers. If the denominators of two rational expressions are the same, subtract the numerators and write the result over the common denominator. That is, $\dfrac{a}{c} - \dfrac{b}{c} = \dfrac{a - b}{c}$, where $c \neq 0$.

SUBTRACTING RATIONAL NUMBERS	**SUBTRACTING RATIONAL EXPRESSIONS**
$\dfrac{3}{10} - \dfrac{7}{10} = \dfrac{3 - 7}{10} = \dfrac{-4}{10} = \dfrac{-2 \cdot 2}{5 \cdot 2} = -\dfrac{2}{5}$	$\dfrac{3}{10b} - \dfrac{7}{10b} = \dfrac{3 - 7}{10b} = \dfrac{-4}{10b} = \dfrac{-2 \cdot 2}{5 \cdot 2 \cdot b} = -\dfrac{2}{5b}$

Once again, we exclude values of the variable that result in division by zero.

EXAMPLE 3   **How to Subtract Rational Expressions with a Common Denominator**

Find the difference: $\dfrac{2n^2}{n^2 - 1} - \dfrac{2n}{n^2 - 1}$

Step-by-Step Solution

Step 1: Subtract the numerators and write the result over the common denominator.

$$\dfrac{a}{c} - \dfrac{b}{c} = \dfrac{a - b}{c}$$
$$\downarrow$$
$$\dfrac{2n^2}{n^2 - 1} - \dfrac{2n}{n^2 - 1} = \dfrac{2n^2 - 2n}{n^2 - 1}$$

Step 2: Simplify the rational expression by dividing out common factors.

Factor the numerator and denominator: $= \dfrac{2n(n - 1)}{(n - 1)(n + 1)}$

Divide out common factors: $= \dfrac{2n\cancel{(n - 1)}}{\cancel{(n - 1)}(n + 1)}$

$$= \dfrac{2n}{n + 1}$$ ●

Subtracting Rational Expressions with a Common Denominator

Step 1: Use the fact that if $\dfrac{a}{c}$ and $\dfrac{b}{c}$, where $c \neq 0$, are two rational expressions, then

$$\dfrac{a}{c} - \dfrac{b}{c} = \dfrac{a - b}{c}$$

Step 2: Simplify the difference by dividing out common factors.

In Other Words
To subtract rational expressions with a common denominator, subtract the numerators and write the result over the common denominator. Then simplify.

Quick ✓
In Problems 7 and 8, find the difference.

7. $\dfrac{8y}{2y - 5} - \dfrac{6}{2y - 5}$ 8. $\dfrac{10 + 3z}{6z} - \dfrac{7}{6z}$

If the numerator of the rational expression being subtracted contains more than one term, enclose the terms in parentheses. This will remind you to distribute the minus sign across all the terms.

EXAMPLE 4   **Subtracting Rational Expressions in Which a Numerator Contains More Than One Term**

Find the difference: $\dfrac{9x + 1}{x + 1} - \dfrac{6x - 2}{x + 1}$

(continued)

Solution

Subtract the numerators and write the result over the common denominator, $x + 1$.

$$\frac{a}{c} - \frac{b}{c} = \frac{a - b}{c};$$

notice the use of parentheses

$$\frac{9x + 1}{x + 1} - \frac{6x - 2}{x + 1} = \frac{9x + 1 - (6x - 2)}{x + 1}$$

Subtract the binomial: $= \dfrac{9x + 1 - 6x + 2}{x + 1}$

Combine like terms: $= \dfrac{3x + 3}{x + 1}$

Factor the numerator: $= \dfrac{3(x + 1)}{x + 1}$

Divide out common factors: $= \dfrac{3(\cancel{x + 1})}{(\cancel{x + 1})}$

$$= 3$$

Work Smart

Remember, $\dfrac{a}{1} = a$.

Quick ✔

9. $\dfrac{3x + 1}{x - 4} - \dfrac{x + 3}{x - 4} = \dfrac{3x + 1 - (\underline{\hspace{1cm}})}{x - 4}$

In Problems 10 and 11, find the difference.

10. $\dfrac{2x^2 - 5x}{3x} - \dfrac{x^2 - 13x}{3x}$

11. $\dfrac{3x^2 + 8x - 1}{x^2 - 3x - 28} - \dfrac{2x^2 + 2x - 9}{x^2 - 3x - 28}$

▶ ❸ Add or Subtract Rational Expressions with Opposite Denominators

Suppose you were asked to find the sum $\dfrac{w^2 - 3w}{w - 4} + \dfrac{4}{4 - w}$. Notice the denominators are additive inverses (opposites) of each other. Recall that $4 - w = -w + 4 = -1(w - 4)$. Use this idea to rewrite the sum so that each rational expression has the same denominator.

EXAMPLE 5 **Adding Rational Expressions with Opposite Denominators**

Find the sum: $\dfrac{w^2 - 3w}{w - 4} + \dfrac{4}{4 - w}$

Solution

To rewrite the denominators so they are the same, factor -1 from $4 - w$.

$$\frac{w^2 - 3w}{w - 4} + \frac{4}{4 - w} = \frac{w^2 - 3w}{w - 4} + \frac{4}{-1(w - 4)}$$

$\dfrac{a}{-b} = \dfrac{-a}{b}$: $= \dfrac{w^2 - 3w}{w - 4} + \dfrac{-4}{w - 4}$

$\dfrac{a}{c} + \dfrac{b}{c} = \dfrac{a + b}{c}$: $= \dfrac{w^2 - 3w - 4}{w - 4}$

Factor the numerator: $= \dfrac{(w - 4)(w + 1)}{w - 4}$

Divide out common factors: $= \dfrac{(\cancel{w - 4})(w + 1)}{(\cancel{w - 4})}$

$$= \frac{(w + 1)}{1}$$

$$= w + 1$$

Quick ✓

12. *True or False* $8 - x = -1(x - 8)$

In Problems 13 and 14, find the sum.

13. $\dfrac{3x}{x - 5} + \dfrac{1}{5 - x}$

14. $\dfrac{a^2 + 2a}{a - 7} + \dfrac{a^2 + 14}{7 - a}$

EXAMPLE 6   **Subtracting Rational Expressions with Opposite Denominators**

Find the difference: $\dfrac{b^2 - 11}{b^2 - 25} - \dfrac{3b + 1}{25 - b^2}$

Solution

To rewrite the denominators so they are the same, factor -1 from $25 - b^2$.

$$\frac{b^2 - 11}{b^2 - 25} - \frac{3b + 1}{25 - b^2} = \frac{b^2 - 11}{b^2 - 25} - \frac{3b + 1}{-(b^2 - 25)}$$

$$-\frac{a}{-b} = +\frac{a}{b}: \quad = \frac{b^2 - 11}{b^2 - 25} + \frac{3b + 1}{b^2 - 25}$$

$$\frac{a}{c} + \frac{b}{c} = \frac{a + b}{c}: \quad = \frac{b^2 - 11 + 3b + 1}{b^2 - 25}$$

$$\text{Combine like terms:} \quad = \frac{b^2 + 3b - 10}{b^2 - 25}$$

$$\text{Factor the numerator and denominator:} \quad = \frac{(b + 5)(b - 2)}{(b + 5)(b - 5)}$$

$$\text{Divide out common factors:} \quad = \frac{(\cancel{b + 5})(b - 2)}{(\cancel{b + 5})(b - 5)}$$

$$= \frac{b - 2}{b - 5}$$

●

Quick ✓

In Problems 15 and 16, find the difference.

15. $\dfrac{2n}{n^2 - 9} - \dfrac{6}{9 - n^2}$

16. $\dfrac{2k}{6k - 6} - \dfrac{9 + 4k}{6 - 6k}$

7.3 Exercises   **MyMathLab®**   Exercise numbers in **green** have complete video solutions in MyMathLab or may be accessed using the QR code to the right.

*Problems **1–16** are the **Quick ✓**s that follow the **EXAMPLES**.*

Building Skills

In Problems 17–32, add the rational expressions. See Objective 1.

17. $\dfrac{3p}{8} + \dfrac{11p}{8}$

18. $\dfrac{2m}{9} + \dfrac{4m}{9}$

19. $\dfrac{n}{2} + \dfrac{3n}{2}$

20. $\dfrac{3}{x} + \dfrac{y}{x}$

21. $\dfrac{4a - 1}{3a} + \dfrac{2a - 2}{3a}$

22. $\dfrac{4n + 1}{8} + \dfrac{12n - 1}{8}$

23. $\dfrac{8c - 3}{c - 1} + \dfrac{2c - 1}{c - 1}$

24. $\dfrac{4n}{6n + 27} + \dfrac{18}{6n + 27}$

25. $\dfrac{2x}{x + y} + \dfrac{2y}{x + y}$

26. $\dfrac{4}{2x + 1} + \dfrac{8x}{2x + 1}$

27. $\dfrac{14x}{x+2} + \dfrac{7x^2}{x+2}$

28. $\dfrac{4p-1}{p-1} + \dfrac{p+1}{p-1}$

29. $\dfrac{12a-1}{2a+6} + \dfrac{13-8a}{2a+6}$

30. $\dfrac{3x^2+4}{3x-6} + \dfrac{3x^2-28}{3x-6}$

31. $\dfrac{a}{a^2-3a-10} + \dfrac{2}{a^2-3a-10}$

32. $\dfrac{2x}{2x^2-x-15} + \dfrac{5}{2x^2-x-15}$

In Problems 33–44, subtract the rational expressions. See Objective 2.

33. $\dfrac{x^2}{x^2-9} - \dfrac{3x}{x^2-9}$

34. $\dfrac{4a^2}{3a-1} - \dfrac{a^2+a}{3a-1}$

35. $\dfrac{x^2-x}{2x} - \dfrac{2x^2-x}{2x}$

36. $\dfrac{3x+3x^2}{x} - \dfrac{x^2-x}{x}$

37. $\dfrac{2c-3}{4c} - \dfrac{6c+9}{4c}$

38. $\dfrac{7x+13}{3x} - \dfrac{x+1}{3x}$

39. $\dfrac{x^2-6}{x-2} - \dfrac{x^2-3x}{x-2}$

40. $\dfrac{2x-3}{x-1} - \dfrac{x+1}{x-1}$

41. $\dfrac{2}{c^2-4} - \dfrac{c^2-2}{c^2-4}$

42. $\dfrac{2z^2+7z}{z^2-1} - \dfrac{z^2-6}{z^2-1}$

43. $\dfrac{2x^2+5x}{2x+3} - \dfrac{4x+3}{2x+3}$

44. $\dfrac{2n^2+n}{n^2-n-2} - \dfrac{n^2+6}{n^2-n-2}$

In Problems 45–56, add or subtract the rational expressions. See Objective 3.

45. $\dfrac{n}{n-3} + \dfrac{3}{3-n}$

46. $\dfrac{4}{x-1} + \dfrac{8x}{1-x}$

47. $\dfrac{12q}{2p-2q} - \dfrac{8p}{2q-2p}$

48. $\dfrac{2a}{a-b} - \dfrac{4b}{b-a}$

49. $\dfrac{2p^2-1}{p-1} - \dfrac{2p+2}{1-p}$

50. $\dfrac{x^2+3}{x-1} - \dfrac{x+3}{1-x}$

51. $\dfrac{2a}{a-b} + \dfrac{2a-4b}{b-a}$

52. $\dfrac{4k^2}{2k-1} + \dfrac{2k}{1-2k}$

53. $\dfrac{3}{p^2-3} + \dfrac{p^2}{3-p^2}$

54. $\dfrac{x^2+6x}{x^2-1} + \dfrac{4x+3}{1-x^2}$

55. $\dfrac{2x}{x^2-y^2} - \dfrac{2y}{y^2-x^2}$

56. $\dfrac{m^2+2mn}{n^2-m^2} - \dfrac{n^2}{m^2-n^2}$

Mixed Practice

In Problems 57–76, perform the indicated operation.

57. $\dfrac{4}{7x} - \dfrac{11}{7x}$

58. $\dfrac{3}{2x} - \left(\dfrac{-5}{2x}\right)$

59. $\dfrac{5}{2x-5} - \dfrac{2x}{2x-5}$

60. $\dfrac{2b+1}{b+1} - \dfrac{b}{b+1}$

61. $\dfrac{2n^3}{n-1} - \dfrac{2n^3}{n-1}$

62. $\dfrac{3n^2}{2n-1} - \dfrac{3n^2}{2n-1}$

63. $\dfrac{n^2-3}{2n+3} + \dfrac{n^2+n}{2n+3}$

64. $\dfrac{5x}{x^2+1} + \dfrac{x^2-3x}{x^2+1}$

65. $\dfrac{3v-1}{v^2-9} - \dfrac{8}{v^2-9}$

66. $\dfrac{x^2-x}{x^2-1} - \dfrac{2}{x^2-1}$

67. $\dfrac{x^2}{x^2-1} + \dfrac{2x+1}{x^2-1}$

68. $\dfrac{x^2}{x^2-9} + \dfrac{6x+9}{x^2-9}$

69. $\dfrac{2x^2+3}{x^2-3x} - \dfrac{x^2}{x^2-3x}$

70. $\dfrac{4x}{x^2-4} + \dfrac{8}{x^2-4}$

71. $\dfrac{3x^2}{x+1} - \dfrac{x^2+2}{x+1}$

72. $\dfrac{x^2}{x^2-1} - \dfrac{3x+2}{x^2-1}$

73. $\dfrac{3x-1}{x-y} + \dfrac{1-3y}{y-x}$

74. $\dfrac{2s-3t}{s-t} + \dfrac{6s-4t}{t-s}$

75. $\dfrac{2p-3q}{3p^2-18q^2} - \dfrac{9q+p}{18q^2-3p^2}$

76. $\dfrac{n+3}{2n^2-n-3} - \dfrac{-n^2-3n}{2n^2-n-3}$

Applying the Concepts

77. Find the sum of $\dfrac{7a^2}{a}$ and $\dfrac{5a^2}{a}$.

78. Find the sum of $\dfrac{3n}{3n+2}$ and $\dfrac{n^2+2}{3n+2}$.

79. Find the difference of $\dfrac{2n}{n+1}$ and $\dfrac{n+3}{n+1}$.

80. Find the difference of $\dfrac{p^2+4}{p-4}$ and $\dfrac{4p}{p-4}$.

81. Subtract $\dfrac{3}{2n}$ from $\dfrac{11}{2n}$.

82. Subtract $\dfrac{3x+4}{4}$ from $\dfrac{x-4}{4}$.

83. Find a rational expression that, when subtracted from $\dfrac{x^2}{x^2-9}$, gives a difference of $\dfrac{3x}{x^2-9}$.

84. Find a rational expression that, when subtracted from $\dfrac{6x}{x-3}$, gives a difference of 1.

85. Find a rational expression that, when added to $\dfrac{-3x+4}{x+2}$, gives a sum of 1.

86. Find a rational expression that, when added to $\dfrac{x-3}{x+1}$, gives a sum of $\dfrac{2x-5}{x+1}$.

△ **87. Perimeter of a Rectangle** Find the perimeter of the rectangle.

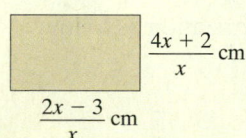

$\dfrac{4x+2}{x}$ cm

$\dfrac{2x-3}{x}$ cm

△ **88. Perimeter of a Rectangle** Find the perimeter of the rectangle.

$\dfrac{2n+3}{2n+1}$ yd

$\dfrac{n-3}{2n+1}$ yd

Extending the Concepts

In Problems 89–94, perform the indicated operations.

89. $\dfrac{7x-3y}{x^2-y^2} - \left(\dfrac{2x-3y}{x^2-y^2} - \dfrac{x+7y}{x^2-y^2} \right)$

90. $\dfrac{2g}{g-3} - \left(\dfrac{g+3}{g-3} - \dfrac{g-4}{g-3} \right)$

91. $\dfrac{x^2}{3x^2+5x-2} - \dfrac{x}{3x-1} \cdot \dfrac{2x-1}{x+2}$

92. $\dfrac{4}{x-2} \cdot \dfrac{x-2}{x+2} + \dfrac{3x-1}{x^2-4}$

93. $\dfrac{5x}{x-2} + \dfrac{2(x+3)}{x-2} - \dfrac{6(2x-1)}{x-2}$

94. $\dfrac{2a-b}{a-b} - \dfrac{6(a-2b)}{a-b} + \dfrac{3(2b+a)}{b-a}$

In Problems 95 and 96, find the missing expression.

95. $\dfrac{2n+1}{n-3} - \dfrac{?}{n-3} = \dfrac{6n+7}{n-3}$

96. $\dfrac{?}{n^2-1} - \dfrac{n-3}{n^2-1} = \dfrac{1}{n+1}$

Explaining the Concepts

97. Is the following student work correct or incorrect? Explain your reasoning.

$$\dfrac{x-2}{x} - \dfrac{x+4}{x} = \dfrac{x-2-x+4}{x} = \dfrac{2}{x}$$

98. The denominators in the expression $\dfrac{4x}{x-2} - \dfrac{2x}{2-x}$ are opposites in sign. Explain how to rewrite this difference so that the terms have a common denominator, and then describe the steps to simplify the result.

99. On a multiple-choice quiz, you see the following problem.

Find the sum and simplify: $\dfrac{5}{x} + \dfrac{2}{x}$

The possible answers are **(a)** $\dfrac{7}{2x}$ **(b)** $\dfrac{7}{x^2}$ **(c)** $\dfrac{7}{x}$

Determine the correct answer, and explain why the others are incorrect.

100. Find and explain the error in the following problem. Then work the problem correctly.

$$\dfrac{a}{a^2-b^2} - \dfrac{b}{a^2-b^2} = \dfrac{a-b}{a^2-b^2} = \dfrac{a-b}{a^2-b^2} = \dfrac{1}{a-b}$$

7.4 Finding the Least Common Denominator and Forming Equivalent Rational Expressions

Objectives

1 Find the Least Common Denominator of Two or More Rational Expressions

2 Write a Rational Expression Equivalent to a Given Rational Expression

3 Use the LCD to Write Equivalent Rational Expressions

Are You Prepared for This Section?

Before getting started, complete the following problems. If you get a problem wrong, go back to the section cited and review the material.

P1. Write $\dfrac{5}{12}$ as a fraction with 24 as its denominator.   [Section 1.2, pp. 10–11]

P2. Find the least common denominator (LCD) of $\dfrac{4}{15}$ and $\dfrac{7}{25}$.   [Section 1.2, pp. 11–12]

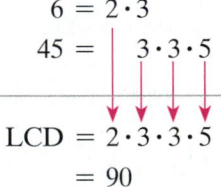

1 Find the Least Common Denominator of Two or More Rational Expressions

Work Smart

Don't confuse the LCD (least common denominator) with the GCF (greatest common factor)!

For rational expressions to be added or subtracted, the denominators must be the same. What if the denominators are different? Use the same idea we used for adding or subtracting rational numbers with unlike denominators: Rewrite each rational expression as an equivalent rational expression over the least common denominator (LCD).

Let's review how to find the LCD of rational numbers.

EXAMPLE 1 **How to Find the Least Common Denominator of Rational Numbers**

Find the least common denominator (LCD) of $\dfrac{5}{6}$ and $\dfrac{11}{45}$.

Step-by-Step Solution

Step 1: Write each denominator as the product of prime factors.

$$6 = 2 \cdot 3$$
$$45 = 3 \cdot 3 \cdot 5$$

Step 2: Find the product of each prime factor the greatest number of times it appears in any factorization.

$$\text{LCD} = 2 \cdot 3 \cdot 3 \cdot 5$$
$$= 90$$

So, the least common denominator (LCD) of $\dfrac{5}{6}$ and $\dfrac{11}{45}$ is 90. ●

In Example 1, we could have written the factorization as follows:

$$6 = 2 \cdot 3$$
$$45 = 3^2 \cdot 5$$
$$\text{LCD} = 2 \cdot 3^2 \cdot 5$$
$$= 90$$

We write $3 \cdot 3$ as 3^2 because the greatest number of times that 3 appears in any factorization is two.

We will follow the same approach for finding the LCD of rational expressions.

Definition

The **least common denominator (LCD)** of two or more rational expressions is the polynomial of least degree that is a multiple of each denominator in the rational expressions.

EXAMPLE 2 **How to Find the Least Common Denominator of Rational Expressions**

Find the least common denominator (LCD) of the rational expressions $\dfrac{3}{4x^3}$ and $\dfrac{5}{6x}$.

Step-by-Step Solution

Step 1: Write each denominator as the product of prime factors.

$$4x^3 = 2^2 \quad \cdot x^3$$
$$6x = 2 \cdot 3 \cdot x$$

Step 2: Find the product of each prime factor the greatest number of times it appears in any factorization.

$$LCD = 2^2 \cdot 3 \cdot x^3$$
$$= 12x^3$$

So, the LCD of $\dfrac{3}{4x^3}$ and $\dfrac{5}{6x}$ is $12x^3$.

Finding the Least Common Denominator of Rational Expressions

Step 1: Factor each denominator completely. Write the factored form using powers. For example, write $x^2 + 4x + 4$ as $(x + 2)^2$.

Step 2: The LCD is the product of each prime factor the greatest number of times it appears in any factorization.

EXAMPLE 3 **Finding the Least Common Denominator of Rational Expressions with Monomial Denominators**

Find the least common denominator (LCD) of the rational expressions $\dfrac{1}{15xy^2}$ and $\dfrac{7}{18x^3y}$.

Solution

Write denominators as the product of prime factors:
$$15xy^2 = \quad 3 \cdot 5 \cdot x \cdot y^2$$
$$18x^3y = 2 \cdot 3^2 \cdot x^3 \cdot y$$

List each factor the greatest number of times it appears:
$$LCD = 2 \cdot 3^2 \cdot 5 \cdot x^3 \cdot y^2$$

Multiply:
$$= 90\, x^3 y^2$$

Work Smart

To find the LCD, list every factor, using the highest power that appears.

Quick ✓

1. The ___ _____ _____ of two or more rational expressions is the polynomial of least degree that is a multiple of each denominator in the rational expressions.

2. *True or False* The least common denominator of the expressions $\dfrac{3}{4a^2b}$ and $\dfrac{5}{4ab}$ is $16a^3b^2$.

In Problems 3 and 4, find the least common denominator of the rational expressions.

3. $\dfrac{5}{8x^2y}$ and $\dfrac{1}{12xy^3}$

4. $\dfrac{1}{6a^3b^2}$ and $\dfrac{5}{21ab^3}$

EXAMPLE 4 **Finding the Least Common Denominator of Rational Expressions with Polynomial Denominators**

Find the LCD of the rational expressions $\dfrac{7}{3a}$ and $\dfrac{5}{6a + 6}$.

(continued)

Solution

Factor the denominators: $\quad 3a = \quad 3 \cdot a$

$$6a + 6 = 6(a + 1) = 2 \cdot 3 \cdot \quad (a + 1)$$

Write the LCD: $\quad$ LCD $= 2 \cdot 3 \cdot a \cdot (a + 1)$

Multiply: $\qquad = 6a(a + 1)$ ●

Work Smart

The a in the denominator $3a$ is a *factor*; the a in $(a + 1)$ is a *term*.

EXAMPLE 5 **Finding the Least Common Denominator of Rational Expressions with Polynomial Denominators**

Find the LCD of the rational expressions $\dfrac{7}{x^2 - x - 2}$ and $\dfrac{3}{x^2 + 2x + 1}$.

Solution

Factor the denominators: $\quad x^2 - x - 2 = (x - 2)(x + 1)$

$$x^2 + 2x + 1 = \qquad (x + 1)^2$$

Write the LCD: $\quad$ LCD $= (x - 2)(x + 1)^2$ ●

> **Quick ✓**
>
> **5.** *True or False* The least common denominator of two rational expressions is unique.
>
> *In Problems 6 and 7, find the least common denominator of the rational expressions.*
>
> **6.** $\dfrac{2}{15z}$ and $\dfrac{7}{5z^2 + 5z}$ $\qquad$ **7.** $\dfrac{3}{x^2 + 4x - 5}$ and $\dfrac{1}{x^2 + 10x + 25}$

EXAMPLE 6 **Finding the Least Common Denominator with Opposite Factors**

Find the LCD of the rational expressions $\dfrac{1}{x^2 - 25}$ and $\dfrac{9}{10 - 2x}$.

Solution

Factor each denominator: $\quad x^2 - 25 = (x + 5)(x - 5)$

$$10 - 2x = 2(5 - x)$$

It looks as though there are no common factors. But notice that the factors $x - 5$ and $5 - x$ are opposites. Write the factorization of $10 - 2x$ as follows:

$$10 - 2x = 2(5 - x)$$

Factor out -1: $\quad = 2(-1)(x - 5)$

$$= -2(x - 5)$$

So, to find the LCD of $\dfrac{1}{x^2 - 25}$ and $\dfrac{9}{10 - 2x}$:

Work Smart

In Example 6 we could also have factored $10 - 2x$ as $-2(-5 + x) = -2(x - 5)$.

Factor each denominator: $\quad x^2 - 25 = \quad (x - 5)(x + 5)$

$$10 - 2x = -2(x - 5)$$

Write the LCD: $\quad$ LCD $= -2(x - 5)(x + 5)$ ●

Quick ✔

In Problem 8, find the least common denominator of the rational expressions.

8. $\dfrac{3}{21 - 3x}$ and $\dfrac{8}{x^2 - 49}$

▶ ❷ **Write a Rational Expression Equivalent to a Given Rational Expression**

Rational expressions that represent the same quantity are *equivalent*. If we multiply the numerator and denominator of a rational expression by the same quantity, we get an equivalent rational expression.

$$\frac{p}{q} = \frac{p}{q} \cdot 1 = \frac{p}{q} \cdot \frac{r}{r} = \frac{pr}{qr}$$

In other words, we are multiplying the rational expression by 1, the multiplicative identity. Let's review how to do this using rational numbers.

EXAMPLE 7 **Forming Equivalent Rational Numbers**

Write $\dfrac{2}{5}$ as an equivalent fraction with a denominator of 20.

Solution
To change from a denominator of 5 to a denominator of 20:

$$\frac{2}{5} \cdot \frac{?}{?} = \frac{\square}{20}$$

Because $5 \cdot 4 = 20$, form the factor of $1 = \dfrac{4}{4}$.

$$\frac{2}{5} \cdot \frac{4}{4} = \frac{8}{20}$$

The equivalent fraction is $\dfrac{8}{20}$.

The approach used to form equivalent fractions is the same as the approach used to form equivalent rational expressions.

EXAMPLE 8 **How to Form an Equivalent Rational Expression**

Write the rational expression $\dfrac{4}{5x^2 + x}$ as an equivalent rational expression with denominator $10x^2 + 2x$.

Step-by-Step Solution

Step 1: Write each denominator in factored form.

Original denominator: $5x^2 + x = \quad x(5x + 1)$

New denominator: $10x^2 + 2x = 2 \cdot x(5x + 1)$

Step 2: Determine the factor(s) missing in the original denominator compared to the new denominator.

The original denominator is missing a factor of 2 compared to the new denominator.

Step 3: Multiply the original rational expression by $1 = \dfrac{2}{2}$.

$$\frac{4}{5x^2 + x} = \frac{4}{x(5x + 1)} \cdot \frac{2}{2}$$

Step 4: Find the product. Leave the denominator in factored form.

$$= \frac{8}{2x(5x + 1)}$$

So, $\dfrac{4}{5x^2 + x} = \dfrac{8}{2x(5x + 1)}$. Notice that $2x(5x + 1) = 10x^2 + 2x$, as required.

Forming Equivalent Rational Expressions

Step 1: Write each denominator in factored form.

Step 2: Determine the factor(s) the new denominator has that is (are) missing from the original denominator.

Step 3: Multiply the original rational expression by $1 = \dfrac{\text{missing factor(s)}}{\text{missing factor(s)}}$.

Step 4: Find the product. Leave the denominator in factored form.

Quick ✓

9. To rewrite the rational expression $\dfrac{2x + 1}{x - 1}$ with a denominator of $(x - 1)(x + 3)$, multiply $\dfrac{2x + 1}{x - 1}$ by $1 = \underline{\hspace{2cm}}$.

In Problem 10, write the rational expression as an equivalent rational expression with the given denominator.

10. $\dfrac{3}{4p^2 - 8p}$ with denominator $16p^3(p - 2)$.

▶ ❸ Use the LCD to Write Equivalent Rational Expressions

There is one more skill we need before we add or subtract rational expressions with unlike denominators. It is to find the LCD of two or more rational expressions and then write them as equivalent rational expressions.

EXAMPLE 9 **Using the LCD to Write Equivalent Rational Expressions**

Find the LCD of the rational expressions $\dfrac{4}{x^2 + 3x + 2}$ and $\dfrac{9}{x^2 - 4}$. Then rewrite each rational expression with the LCD.

Solution
Factor each denominator to find the LCD.

$$x^2 + 3x + 2 = (x + 1)(x + 2)$$
$$x^2 - 4 = \qquad (x + 2)(x - 2)$$
$$\text{LCD} = (x + 1)(x + 2)(x - 2)$$

Rewrite each rational expression with a denominator of $(x + 1)(x + 2)(x - 2)$. Multiply the numerators, but leave the denominator in factored form.

$$\dfrac{4}{x^2 + 3x + 2} = \dfrac{4}{(x + 2)(x + 1)} \cdot \dfrac{(x - 2)}{(x - 2)} \qquad\qquad \dfrac{9}{x^2 - 4} = \dfrac{9}{(x - 2)(x + 2)} \cdot \dfrac{(x + 1)}{(x + 1)}$$

$$= \dfrac{4x - 8}{(x + 2)(x + 1)(x - 2)} \qquad\qquad\qquad = \dfrac{9x + 9}{(x - 2)(x + 2)(x + 1)}$$

Use the Commutative Property to rearrange the factors in the denominator:
$$= \dfrac{4x - 8}{(x + 1)(x + 2)(x - 2)} \qquad\qquad\qquad = \dfrac{9x + 9}{(x + 1)(x + 2)(x - 2)} \quad ●$$

Quick ✓

In Problems 11 and 12, find the least common denominator of the rational expressions. Then rewrite each rational expression with the LCD.

11. $\dfrac{3}{6a^2b}$ and $\dfrac{-5}{20ab^3}$

12. $\dfrac{5}{x^2 - 4x - 5}$ and $\dfrac{-3}{x^2 - 7x + 10}$

7.4 Exercises MyMathLab®

Problems 1–12 are the Quick ✓s that follow the EXAMPLES.

Building Skills

In Problems 13–32, find the LCD of the given rational expressions. See Objective 1.

13. $\dfrac{14}{5x^2}; \dfrac{4}{25x}$

14. $\dfrac{3}{7y^2}; \dfrac{4}{49y}$

15. $\dfrac{7}{12xy^2}; \dfrac{4}{15x^3y}$

16. $\dfrac{5}{36xy^2}; \dfrac{1}{24x^2y}$

17. $\dfrac{3}{x}; \dfrac{4}{x+1}$

18. $\dfrac{2x-1}{2x+1}; \dfrac{3x}{2}$

19. $\dfrac{4}{2x+1}; 2$

20. $\dfrac{7}{y-1}; 3$

21. $\dfrac{7}{2b-6}; \dfrac{3b}{4b-12}$

22. $\dfrac{2}{6x-18}; \dfrac{3}{8x-24}$

23. $\dfrac{6}{p^2+p}; \dfrac{7}{p^2-p-2}$

24. $\dfrac{5}{2x^2-12x+18}; \dfrac{4}{4x^2-36}$

25. $\dfrac{3}{r^2+4r+4}; \dfrac{4}{r^2-r-2}$

26. $\dfrac{8}{x^2-1}; \dfrac{3}{x^2-2x+1}$

27. $\dfrac{2}{x-4}; \dfrac{3}{4-x}$

28. $\dfrac{7}{3-a}; \dfrac{1}{a-3}$

29. $\dfrac{8}{x^2-9}; \dfrac{2}{6-2x}$

30. $\dfrac{-1}{c^2-49}; \dfrac{5}{21-3c}$

31. $\dfrac{1}{(x-1)(x+2)}; \dfrac{11x}{(1-x)(2+x)}$

32. $\dfrac{3z}{(z+5)(z-4)}; \dfrac{-z}{(4-z)(5+z)}$

In Problems 33–42, write an equivalent rational expression with the given denominator. See Objective 2.

33. $\dfrac{4}{x}$ with denominator $3x^3$

34. $\dfrac{7}{x}$ with denominator $2x^2$

35. $\dfrac{3+c}{a^2b^2c}$ with denominator $a^2b^2c^2$

36. $\dfrac{7a+1}{abc}$ with denominator a^2b^2c

37. $\dfrac{x+2}{x+4}$ with denominator x^2-16

38. $\dfrac{x-5}{x+2}$ with denominator x^2+5x+6

39. $\dfrac{3n}{2n+2}$ with denominator $6n^2-6$

40. $\dfrac{7a}{3a^2+9a+6}$ with denominator $6a^2+18a+12$

41. $4t$ with denomin.ator $t-1$

42. 7 with denominator t^2+1

In Problems 43–60, find the LCD of the rational expressions. Then rewrite each as an equivalent rational expression with the LCD. See Objective 3.

43. $\dfrac{2x}{3y}; \dfrac{4}{9y^2}$

44. $\dfrac{4}{5n^2}; \dfrac{3}{7n}$

45. $\dfrac{3a+1}{2a^3}; \dfrac{4a-1}{4a}$

46. $\dfrac{2p^2-1}{6p}; \dfrac{3p^3+2}{8p^2}$

47. $\dfrac{2}{m}; \dfrac{3}{m+1}$

48. $\dfrac{x+2}{x}; \dfrac{x}{x+2}$

49. $\dfrac{y-2}{4y}; \dfrac{y}{8y-4}$

50. $\dfrac{3b}{7a}; \dfrac{2a}{7a+14b}$

51. $\dfrac{1}{x-1}; \dfrac{2x}{1-x^2}$

52. $\dfrac{2a}{a-2}; \dfrac{a}{4-a^2}$

53. $\dfrac{4x}{x^2 - 4}; \dfrac{2}{x + 2}$ **54.** $\dfrac{4a}{a^2 - 1}; \dfrac{7}{a + 1}$

55. $\dfrac{3x}{x^2 - 6x - 7}; \dfrac{5}{x^2 - 4x - 21}$

56. $\dfrac{4}{m^2 - 5m - 6}; \dfrac{2m}{m^2 - 12m + 36}$

57. $\dfrac{x + 1}{x^2 - 9}; \dfrac{x + 2}{x^2 - 6x + 9}$ **58.** $\dfrac{4n}{n^2 - n - 6}; \dfrac{2}{n^2 + 4n + 4}$

59. $\dfrac{3}{x + 4}; \dfrac{2x - 1}{2x^2 + 7x - 4}$ **60.** $\dfrac{3}{2n^2 - 7n + 3}; \dfrac{2n}{n - 3}$

Applying the Concepts

61. Painting a Room It takes an experienced painter x hours to paint a room alone; assuming that he works at a constant rate, he completes $\dfrac{1}{x}$ of the job per hour. It takes an apprentice 3 times as long to paint the same room as it takes the experienced painter, so the apprentice completes $\dfrac{1}{3x}$ of the job per hour. Find the LCD of the two rational expressions that represent the rate of work of the experienced painter and the apprentice, and then write each as an equivalent rational expression using this denominator.

62. Mowing the Lawn It takes Mario 2 hours longer to mow the lawn than it takes Marco. If Marco takes x hours, he completes $\dfrac{1}{x}$ of the job per hour. Since Mario requires 2 hours more, he completes $\dfrac{1}{x + 2}$ of the job per hour. Find the LCD of the two rational expressions, and then write each as an equivalent rational expression using this denominator.

63. Traveling on a River The time to complete a journey can be calculated by the formula $t = \dfrac{d}{r}$, where t is the time, d is the distance traveled, and r is the rate. Suppose the time for a boat to travel a distance of 12 miles upstream on a river whose current is 4 miles per hour is given by $\dfrac{12}{r - 4}$. The time for the boat to travel 12 miles downstream on the same river is $\dfrac{12}{r + 4}$. Find the LCD of the two rational expressions, and then write each as an equivalent rational expression using this denominator.

64. Traveling by Plane The time to complete a journey can be calculated by the formula $t = \dfrac{d}{r}$, where t is the time, d is the distance traveled, and r is the rate. Suppose a plane flies 500 miles west into a headwind of 50 miles per hour in a time of $\dfrac{500}{r - 50}$ hours, where r is the speed of the plane in still air. The plane flies 500 miles east with a tailwind of 50 miles per hour in a time of $\dfrac{500}{r + 50}$ hours. Find the LCD of the two rational expressions, and then write each as an equivalent rational expression using this denominator.

Extending the Concepts

In Problems 65–70, identify the LCD of the rational expressions.

65. $\dfrac{4}{x^3 + 8}; \dfrac{1}{x^2 - 4}; \dfrac{5}{x^3 - 8}$

66. $\dfrac{x}{2x^3 - 2}; \dfrac{1}{3x^2 - 3}; \dfrac{2x + 1}{4x - 4}$

67. $\dfrac{11}{p^2 - p}; \dfrac{-2}{p^3 - p^2}; \dfrac{8}{p^2 - 4p + 3}$

68. $\dfrac{7}{4n^3 - 2n^2}; \dfrac{9}{8n^2 - 4n}; \dfrac{6}{4n^2 - 1}$

69. $\dfrac{a}{2a^4 - 2a^2b^2}; \dfrac{b}{4ab^2 + 4b^3}; \dfrac{ab}{a^3b - b^4}$

70. $\dfrac{5}{3x^2 + xy - 2y^2}; \dfrac{2}{x^2 - xy - 2y^2}$

Explaining the Concepts

71. If you were teaching the class how to write an equivalent rational expression with a given denominator, what steps would you list on the board for the class?

72. A member of your class says that $\dfrac{1}{(x + 3)(x + 2)}$ and $\dfrac{4}{x + 2}$ have an LCD of $(x + 3)(x + 2)^2$. This class member then rewrites the rational expressions as $\dfrac{x + 2}{(x + 3)(x + 2)^2}$ and $\dfrac{4(x + 3)(x + 2)}{(x + 3)(x + 2)^2}$. Use this result to explain why $(x + 3)(x + 2)^2$ is not the least common denominator.

7.5 Adding and Subtracting Rational Expressions with Unlike Denominators

Objective

1. Add and Subtract Rational Expressions with Unlike Denominators

Are You Prepared for This Section?

Before getting started, complete the following problems. If you get a problem wrong, go back to the section cited and review the material.

P1. Find the least common denominator (LCD) of $\frac{2}{15}$ and $\frac{7}{24}$.　　　　[Section 1.2, pp. 11–12]

P2. Factor: $2x^2 + 3x + 1$　　　　[Section 6.3, pp. 386–394]

P3. Simplify: $\frac{3x + 9}{6x}$　　　　[Section 7.1, pp. 436–437]

P4. Simplify: $\frac{1 - x^2}{2x - 2}$　　　　[Section 7.1, pp. 436–438]

P5. Find the sum: $\frac{5}{2x} + \frac{7}{2x}$　　　　[Section 7.3, pp. 449–450]

▶ ① Add and Subtract Rational Expressions with Unlike Denominators

Before we add and subtract rational expressions with unlike denominators, let's review how to add rational numbers with unlike denominators.

EXAMPLE 1　**How to Add Rational Numbers with Unlike Denominators**

Evaluate $\frac{1}{6} + \frac{9}{14}$.

Step-by-Step Solution

Step 1: Find the least common denominator.

$$6 = 2 \cdot 3$$
$$14 = 2 \cdot \ \ \ 7$$
$$\text{LCD} = 2 \cdot 3 \cdot 7$$
$$= 42$$

Step 2: Rewrite each fraction as an equivalent fraction with the least common denominator.

Because $6 \cdot 7 = 42$, use $1 = \frac{7}{7}$ to rewrite $\frac{1}{6}$ using the LCD.

Because $14 \cdot 3 = 42$, use $1 = \frac{3}{3}$ to rewrite $\frac{9}{14}$ using the LCD: $\quad \frac{1}{6} \cdot \frac{7}{7} + \frac{9}{14} \cdot \frac{3}{3} = \frac{7}{42} + \frac{27}{42}$

Step 3: Find the sum of the numerators, and write the sum over the common denominator.

$$= \frac{7 + 27}{42}$$
$$= \frac{34}{42}$$

Step 4: Simplify.

Factor 34 and 42: $= \frac{2 \cdot 17}{2 \cdot 21}$

Divide out common factors: $= \frac{\cancel{2} \cdot 17}{\cancel{2} \cdot 21}$

$$= \frac{17}{21}$$

So, $\frac{1}{6} + \frac{9}{14} = \frac{17}{21}$.

Quick ✓

1. The least common denominator of $\dfrac{1}{8}$ and $\dfrac{5}{18}$ is ___.

In Problems 2 and 3, find each sum or difference. Simplify the result.

2. $\dfrac{5}{12} + \dfrac{5}{18}$ 3. $\dfrac{8}{15} - \dfrac{3}{10}$

Now that you've practiced adding and subtracting rational numbers, let's explore adding and subtracting rational expressions.

EXAMPLE 2   **How to Add Rational Expressions with Unlike Monomial Denominators**

Find the sum: $\dfrac{5}{6x^2} + \dfrac{4}{15x}$

Step-by-Step Solution

Step 1: Find the least common denominator.

$$6x^2 = 2 \cdot 3 \cdot\ \ x^2$$
$$15x = \ \ \ 3 \cdot 5 \cdot x$$
$$\text{LCD} = 2 \cdot 3 \cdot 5 \cdot x^2$$
$$= 30x^2$$

Step 2: Rewrite each rational expression as an equivalent rational expression with the common denominator.

We need the denominator of $\dfrac{5}{6x^2}$ to be $30x^2$, so multiply $\dfrac{5}{6x^2}$ by $1 = \dfrac{5}{5}$. Similarly, multiply $\dfrac{4}{15x}$ by $1 = \dfrac{2x}{2x}$. Do you see why?

$$\frac{5}{6x^2} = \frac{5}{6x^2} \cdot \frac{5}{5} = \frac{25}{30x^2} \qquad\qquad \frac{4}{15x} = \frac{4}{15x} \cdot \frac{2x}{2x} = \frac{8x}{30x^2}$$

Step 3: Add the rational expressions found in Step 2.

$$\frac{5}{6x^2} + \frac{4}{15x} = \frac{25}{30x^2} + \frac{8x}{30x^2}$$

$$\frac{a}{c} + \frac{b}{c} = \frac{a+b}{c}: \ \ = \frac{25 + 8x}{30x^2}$$

Step 4: Simplify.

The rational expression is already simplified.

Therefore, $\dfrac{5}{6x^2} + \dfrac{4}{15x} = \dfrac{25 + 8x}{30x^2}$. ●

Adding or Subtracting Rational Expressions with Unlike Denominators

Step 1: Find the least common denominator.

Step 2: Rewrite each rational expression as an equivalent rational expression with the common denominator.

Step 3: Add or subtract the rational expressions found in Step 2.

Step 4: Simplify the result.

Quick ✓

4. The first step in adding rational expressions with unlike denominators is to determine the ____ _____ _____.

5. *True or False* To rewrite a rational expression with the least common denominator, multiply by $1 = \dfrac{\text{missing factor(s)}}{\text{missing factor(s)}}$.

In Problems 6 and 7, find each sum.

6. $\dfrac{1}{8a^3b} + \dfrac{5}{12ab^2}$

7. $\dfrac{1}{15xy^2} + \dfrac{7}{18x^3y}$

EXAMPLE 3 **Adding Rational Expressions with Unlike Polynomial Denominators**

Find the sum: $\dfrac{-1}{x+3} + \dfrac{3}{x+2}$

Solution

The least common denominator is $(x+3)(x+2)$ because the two denominators have no common factors. To rewrite each rational expression as an equivalent rational expression with the LCD, multiply $\dfrac{-1}{x+3}$ by $1 = \dfrac{x+2}{x+2}$ and multiply $\dfrac{3}{x+2}$ by $1 = \dfrac{x+3}{x+3}$.

$$\frac{-1}{x+3} + \frac{3}{x+2} = \frac{-1}{x+3} \cdot \frac{x+2}{x+2} + \frac{3}{x+2} \cdot \frac{x+3}{x+3}$$

Multiply the numerators; leave the denominators in factored form:
$$= \frac{-1(x+2)}{(x+3)(x+2)} + \frac{3(x+3)}{(x+2)(x+3)}$$

Distribute:
$$= \frac{-x-2}{(x+3)(x+2)} + \frac{3x+9}{(x+2)(x+3)}$$

$\dfrac{a}{c} + \dfrac{b}{c} = \dfrac{a+b}{c}$:
$$= \frac{-x-2+3x+9}{(x+3)(x+2)}$$

Combine like terms:
$$= \frac{2x+7}{(x+3)(x+2)}$$

Because there are no common factors in the numerator and denominator, the expression is fully simplified. So, $\dfrac{-1}{x+3} + \dfrac{3}{x+2} = \dfrac{2x+7}{(x+3)(x+2)}$ ●

Quick ✓

In Problems 8 and 9, find each sum.

8. $\dfrac{5}{x-4} + \dfrac{3}{x+2}$

9. $\dfrac{-1}{n-3} + \dfrac{4}{n+1}$

EXAMPLE 4 **Adding Rational Expressions with Unlike Polynomial Denominators**

Find the sum: $\dfrac{4}{x+1} + \dfrac{16}{x^2-2x-3}$

Solution

The first step is to find the least common denominator. Begin by factoring each denominator.

$$x+1 = \qquad (x+1)$$

$$x^2 - 2x - 3 = (x-3)(x+1)$$

(continued)

The LCD is $(x - 3)(x + 1)$. Now rewrite each rational expression as an equivalent rational expression with the least common denominator.

$$\frac{4}{x + 1} + \frac{16}{x^2 - 2x - 3} = \frac{4}{x + 1} + \frac{16}{(x - 3)(x + 1)}$$

Rewrite each rational expression with the common denominator:
$$= \frac{4}{x + 1} \cdot \frac{x - 3}{x - 3} + \frac{16}{(x + 1)(x - 3)}$$

Work Smart

Don't simplify $\dfrac{4(x - 3)}{(x + 1)(x - 3)}$ or you'll be right back where you started!

$$= \frac{4(x - 3)}{(x + 1)(x - 3)} + \frac{16}{(x + 1)(x - 3)}$$

Distribute the 4: $\quad = \dfrac{4x - 12}{(x + 1)(x - 3)} + \dfrac{16}{(x + 1)(x - 3)}$

$\dfrac{a}{c} + \dfrac{b}{c} = \dfrac{a + b}{c}$: $\quad = \dfrac{4x - 12 + 16}{(x + 1)(x - 3)}$

Add: $\quad = \dfrac{4x + 4}{(x + 1)(x - 3)}$

Factor the numerator: $\quad = \dfrac{4(x + 1)}{(x + 1)(x - 3)}$

Divide out common factors: $\quad = \dfrac{4\cancel{(x + 1)}}{\cancel{(x + 1)}(x - 3)}$

$$= \frac{4}{x - 3}$$ ●

Quick ✓

In Problems 10 and 11, find each sum.

10. $\dfrac{-z + 1}{z^2 + 7z + 10} + \dfrac{2}{z + 5}$

11. $\dfrac{1}{x^2 + 5x} + \dfrac{1}{x^2 - 5x}$

⊙ **EXAMPLE 5** **How to Subtract Rational Expressions with Unlike Denominators**

Find the difference: $\dfrac{8}{x} - \dfrac{2}{x + 1}$

Step-by-Step Solution

Step 1: Find the least common denominator.

The LCD is $x(x + 1)$.

Step 2: Rewrite each rational expression as an equivalent rational expression with the common denominator. Multiply the numerators, but leave the denominators in factored form.

Multiply $\dfrac{8}{x}$ by $1 = \dfrac{x + 1}{x + 1}$; multiply $\dfrac{2}{x + 1}$ by $1 = \dfrac{x}{x}$:

$$\frac{8}{x} - \frac{2}{x + 1} = \frac{8}{x} \cdot \frac{x + 1}{x + 1} - \frac{2}{x + 1} \cdot \frac{x}{x}$$

$$= \frac{8(x + 1)}{x(x + 1)} - \frac{2x}{x(x + 1)}$$

$$= \frac{8x + 8}{x(x + 1)} - \frac{2x}{x(x + 1)}$$

Step 3: Subtract the rational expressions found in Step 2.

$\dfrac{a}{c} - \dfrac{b}{c} = \dfrac{a - b}{c}$: $\quad = \dfrac{8x + 8 - 2x}{x(x + 1)}$

Combine like terms: $\quad = \dfrac{6x + 8}{x(x + 1)}$

Step 4: Simplify the result.

Factor the numerator: $= \dfrac{2(3x + 4)}{x(x + 1)}$

Because there are no common factors, the expression is fully simplified.

Therefore, $\dfrac{8}{x} - \dfrac{2}{x + 1} = \dfrac{2(3x + 4)}{x(x + 1)}$.

> **Quick ✓**
>
> In Problems 12 and 13, find each difference.
>
> **12.** $\dfrac{-4}{5ab^2} - \dfrac{3}{4a^2b^3}$ **13.** $\dfrac{5}{x} - \dfrac{3}{x - 4}$

EXAMPLE 6   **Adding Rational Expressions with Unlike Denominators—Both Denominators Factor**

Perform the indicated operation: $\dfrac{-2}{a^2 - 4a} + \dfrac{2}{4a - 16}$

Solution
Factor each denominator to find the LCD.

$$a^2 - 4a = \quad a \cdot (a - 4)$$

$$4a - 16 = 4 \cdot \quad (a - 4)$$

The LCD is $4a(a - 4)$. Rewrite each rational expression as an equivalent expression using the LCD $4a(a - 4)$. Multiply $\dfrac{-2}{a(a - 4)}$ by $1 = \dfrac{4}{4}$ and multiply $\dfrac{2}{4(a - 4)}$ by $1 = \dfrac{a}{a}$.

$$\dfrac{-2}{a(a - 4)} + \dfrac{2}{4(a - 4)} = \dfrac{-2}{a(a - 4)} \cdot \dfrac{4}{4} + \dfrac{2}{4(a - 4)} \cdot \dfrac{a}{a}$$

Multiply the numerators: $= \dfrac{-8}{4a(a - 4)} + \dfrac{2a}{4a(a - 4)}$

$\dfrac{a}{c} + \dfrac{b}{c} = \dfrac{a + b}{c}$: $\quad = \dfrac{2a - 8}{4a(a - 4)}$

Factor the numerator and denominator: $= \dfrac{2(a - 4)}{2 \cdot 2a(a - 4)}$

Divide out common factors: $= \dfrac{2(a - 4) \cdot 1}{2 \cdot 2a(a - 4)}$

$$= \dfrac{1}{2a}$$

EXAMPLE 7   **Subtracting Rational Expressions with Unlike Denominators—Both Denominators Factor**

Perform the indicated operation: $\dfrac{a - 2}{a^2 + 5a + 6} - \dfrac{2a - 3}{3a^2 + 9a}$

Solution
First, factor each denominator.

$$a^2 + 5a + 6 = \quad (a + 3)(a + 2)$$

$$3a^2 + 9a = 3a(a + 3)$$

(continued)

The LCD is $3a(a + 3)(a + 2)$. Now rewrite each expression as an equivalent rational expression using the LCD $3a(a + 3)(a + 2)$.

$$\frac{a - 2}{a^2 + 5a + 6} - \frac{2a - 3}{3a^2 + 9a} = \frac{a - 2}{(a + 3)(a + 2)} \cdot \frac{3a}{3a} - \frac{2a - 3}{3a(a + 3)} \cdot \frac{a + 2}{a + 2}$$

$\frac{a}{b} \cdot \frac{c}{d} = \frac{a \cdot c}{b \cdot d}:$
$$= \frac{(a - 2) \cdot 3a}{3a(a + 3)(a + 2)} - \frac{(2a - 3)(a + 2)}{3a(a + 3)(a + 2)}$$

Distribute; FOIL:
$$= \frac{3a^2 - 6a}{3a(a + 3)(a + 2)} - \frac{2a^2 + a - 6}{3a(a + 3)(a + 2)}$$

$\frac{a}{c} - \frac{b}{c} = \frac{a - b}{c};$ don't forget the parentheses:
$$= \frac{3a^2 - 6a - (2a^2 + a - 6)}{3a(a + 3)(a + 2)}$$

Subtract the trinomial:
$$= \frac{3a^2 - 6a - 2a^2 - a + 6}{3a(a + 3)(a + 2)}$$

Combine like terms:
$$= \frac{a^2 - 7a + 6}{3a(a + 3)(a + 2)}$$

Factor the numerator:
$$= \frac{(a - 6)(a - 1)}{3a(a + 3)(a + 2)}$$

Quick ✓

In Problems 14 and 15, perform the indicated operation.

14. $\dfrac{3}{(x - 5)(x + 4)} - \dfrac{2}{(x - 5)(x - 1)}$

15. $\dfrac{x - 2}{x^2 - 3x} + \dfrac{x + 3}{4x - 12}$

EXAMPLE 8 **Adding Rational Expressions Containing Opposite Factors**

Find the sum: $\dfrac{1}{1 - n} + \dfrac{2n}{n^2 - 1}$

Solution
First, factor each denominator:

$$1 - n = 1 - n$$
$$n^2 - 1 = (n + 1)(n - 1)$$

Do you see that the factors $n - 1$ and $1 - n$ are opposites? Factor -1 from $1 - n$ so that $1 - n = -1(-1 + n) = -1(n - 1)$. Rewrite the sum as

$$\frac{1}{1 - n} + \frac{2n}{n^2 - 1} = \frac{1}{-1(n - 1)} + \frac{2n}{(n + 1)(n - 1)}$$

$\frac{a}{-b} = \frac{-a}{b}:$
$$= \frac{-1}{n - 1} + \frac{2n}{(n + 1)(n - 1)}$$

In this form, the LCD is $(n - 1)(n + 1)$. Now rewrite each expression as an equivalent expression with the common denominator.

$$\text{Multiply } \frac{-1}{n-1} \text{ by } 1 = \frac{n+1}{n+1}: \quad = \frac{-1}{n-1} \cdot \frac{n+1}{n+1} + \frac{2n}{(n+1)(n-1)}$$

$$\frac{a}{b} \cdot \frac{c}{d} = \frac{a \cdot c}{b \cdot d}: \quad = \frac{-1(n+1)}{(n+1)(n-1)} + \frac{2n}{(n+1)(n-1)}$$

$$\text{Distribute the } -1: \quad = \frac{-n-1}{(n+1)(n-1)} + \frac{2n}{(n+1)(n-1)}$$

$$\frac{a}{c} + \frac{b}{c} = \frac{a+b}{c}: \quad = \frac{-n-1+2n}{(n+1)(n-1)}$$

$$\text{Combine like terms:} \quad = \frac{n-1}{(n+1)(n-1)}$$

$$\text{Divide out common factors:} \quad = \frac{1}{n+1}$$

Work Smart

Remember that when common factors are divided out, they form a quotient of 1. That's why

$$\frac{n-1}{(n+1)(n-1)}$$

simplifies to

$$\frac{(n-1)}{(n+1)(n-1)} = \frac{1}{n+1}$$

Quick ✔

16. *True or False* $\dfrac{3}{x} + \dfrac{6}{-x} = \dfrac{3-6}{x}$

In Problems 17 and 18, perform the indicated operation.

17. $\dfrac{2p^2 - 8}{p^2 + p - 2} + \dfrac{p + 1}{1 - p}$ **18.** $\dfrac{y^2 - 13}{y^2 - 2y - 3} - \dfrac{1}{3 - y}$

EXAMPLE 9   **Adding an Integer and a Rational Expression**

Find the sum: $5 + \dfrac{2}{3x - 4}$

Solution

Because $5 = \dfrac{5}{1}$, rewrite the expression as $\dfrac{5}{1} + \dfrac{2}{3x - 4}$. The LCD is $3x - 4$.

$$\frac{5}{1} + \frac{2}{3x - 4} = \frac{5}{1} \cdot \frac{3x - 4}{3x - 4} + \frac{2}{3x - 4}$$

$$\frac{a}{b} \cdot \frac{c}{d} = \frac{a \cdot c}{b \cdot d}: \quad = \frac{5(3x - 4)}{3x - 4} + \frac{2}{3x - 4}$$

$$\text{Distribute the 5:} \quad = \frac{15x - 20}{3x - 4} + \frac{2}{3x - 4}$$

$$\frac{a}{c} + \frac{b}{c} = \frac{a+b}{c}: \quad = \frac{15x - 20 + 2}{3x - 4}$$

$$\text{Combine like terms:} \quad = \frac{15x - 18}{3x - 4}$$

$$\text{Factor the numerator:} \quad = \frac{3(5x - 6)}{3x - 4}$$

Quick ✓

In Problems 19 and 20, perform the indicated operation.

19. $2 + \dfrac{3}{x - 1}$

20. $\dfrac{6}{x + 3} - 3$

EXAMPLE 10 **Adding and Subtracting Rational Expressions**

Perform the indicated operations.

$$\frac{5}{n - 2} + \frac{5}{n + 2} - \frac{20}{n^2 - 4}$$

Solution

Factor each denominator to find the LCD. Because $n^2 - 4 = (n - 2)(n + 2)$, the LCD is $(n - 2)(n + 2)$.

$$\frac{5}{n - 2} + \frac{5}{n + 2} - \frac{20}{n^2 - 4} = \frac{5}{n - 2} + \frac{5}{n + 2} - \frac{20}{(n - 2)(n + 2)}$$

Rewrite each rational expression with the common denominator:
$$= \frac{5}{n - 2} \cdot \frac{n + 2}{n + 2} + \frac{5}{n + 2} \cdot \frac{n - 2}{n - 2} - \frac{20}{(n - 2)(n + 2)}$$

$\dfrac{a}{b} \cdot \dfrac{c}{d} = \dfrac{a \cdot c}{b \cdot d}$; distribute:
$$= \frac{5n + 10}{(n - 2)(n + 2)} + \frac{5n - 10}{(n - 2)(n + 2)} - \frac{20}{(n - 2)(n + 2)}$$

$\dfrac{a}{c} + \dfrac{b}{c} = \dfrac{a + b}{c}$:
$$= \frac{5n + 10 + 5n - 10 - 20}{(n - 2)(n + 2)}$$

Combine like terms:
$$= \frac{10n - 20}{(n - 2)(n + 2)}$$

Factor the numerator:
$$= \frac{10(n - 2)}{(n - 2)(n + 2)}$$

Divide out common factors:
$$= \frac{10\cancel{(n - 2)}}{\cancel{(n - 2)}(n + 2)}$$

$$= \frac{10}{n + 2}$$

Quick ✓

In Problems 21 and 22, perform the indicated operations.

21. $\dfrac{1}{x + 1} - \dfrac{2}{x^2 - 1} + \dfrac{3}{x - 1}$

22. $\dfrac{3}{x} - \left(\dfrac{1}{x - 2} - \dfrac{6}{x^2 - 2x} \right)$

7.5 Exercises **MyMathLab**® Exercise numbers in **green** have complete video solutions in MyMathLab or may be accessed using the QR code to the right.

Problems **1–22** are the **Quick ✓**s that follow the **EXAMPLES**.

Building Skills

In Problems 23–34, find each sum and simplify. See Objective 1.

23. $-\dfrac{4}{3} + \dfrac{1}{2}$

24. $\dfrac{7}{12} + \dfrac{3}{4}$

25. $\dfrac{2}{3x} + \dfrac{1}{x}$

26. $\dfrac{5}{2x} + \dfrac{6}{5}$

27. $\dfrac{a}{2a - 1} + \dfrac{3}{2a + 1}$

28. $\dfrac{2}{x - 1} + \dfrac{x - 1}{x + 1}$

29. $\dfrac{3}{x - 4} + \dfrac{5}{4 - x}$

30. $\dfrac{7}{n - 4} + \dfrac{8}{4 - n}$

31. $\dfrac{a + 5}{5a - a^2} + \dfrac{a + 3}{4a - 20}$

32. $\dfrac{2x - 6}{x^2 - x - 6} + \dfrac{x + 4}{x + 2}$

33. $\dfrac{2x+4}{x^2+2x}+\dfrac{3}{x}$

34. $\dfrac{3}{x-4}+\dfrac{x+4}{x^2-16}$

61. $\dfrac{3x-1}{x}-\dfrac{9}{x^2-9x}$

In Problems 35–44, find each difference and simplify. See Objective 1.

62. $\dfrac{2x}{8x-12}-\dfrac{3}{2x+2}$

35. $\dfrac{7}{15}-\dfrac{9}{25}$

36. $\dfrac{8}{21}-\dfrac{6}{35}$

63. $\dfrac{3}{x^2-4x-5}-\dfrac{2}{x^2-6x+5}$

37. $m-\dfrac{16}{m}$

38. $\dfrac{9}{x}-x$

64. $\dfrac{3}{x^2+7x+10}-\dfrac{4}{x^2+6x+5}$

39. $\dfrac{x}{x-3}-\dfrac{x-2}{x+3}$

40. $\dfrac{x-2}{x+2}-\dfrac{x+2}{x-2}$

65. $\dfrac{5}{2a-a^2}-\dfrac{3}{2a^2-4a}$

41. $\dfrac{x+2}{x+3}-\dfrac{x^2+3x}{x^2+6x+9}$

42. $\dfrac{x}{x+4}-\dfrac{-4}{x^2+8x+16}$

66. $\dfrac{-1}{3n-n^2}+\dfrac{1}{3n^2-9n}$

43. $\dfrac{-3x-9}{x^2+x-6}-\dfrac{x+3}{2-x}$

44. $\dfrac{p+1}{p^2-2p}-\dfrac{2p+3}{2-p}$

67. $\dfrac{2n+1}{n^2-4}+\dfrac{3n}{6-n^2-n}$

68. $\dfrac{a-5}{2a^2-6a}-\dfrac{6}{12a^2-4a^3}$

Mixed Practice

In Problems 45–78, perform the indicated operation.

69. $\dfrac{m+3}{m^2+2m-8}-\dfrac{m+2}{m^2-4}$

45. $\dfrac{5}{3y^2}-\dfrac{3}{4y}$

46. $\dfrac{5}{4n}-\dfrac{3}{n^2}$

70. $\dfrac{n+3}{n^2-8n+15}+\dfrac{n+3}{n^2-9}$

47. $\dfrac{9}{5x}-\dfrac{6}{10x}$

48. $\dfrac{7}{4b^2}-\dfrac{3b^2}{2b}$

71. $\dfrac{2n+1}{n+3}+\dfrac{7-2n^2}{n^2+n-6}$

49. $\dfrac{2x}{2x+3}-1$

50. $a+\dfrac{a}{a-2}$

72. $\dfrac{4}{x^2+2x-15}+\dfrac{3}{x^2-x-6}$

51. $\dfrac{n}{n-2}+\dfrac{n+2}{n}$

52. $\dfrac{6}{x+2}+\dfrac{4}{x-3}$

73. $\dfrac{-3x-9}{x^2+x-6}-\dfrac{x+3}{2-x}$

53. $\dfrac{2x}{x-3}-\dfrac{5}{x}$

54. $\dfrac{x}{x-2}-\dfrac{2}{x-1}$

74. $\dfrac{2x-1}{1-x}-\dfrac{-x^2-2x}{x^2+x-2}$

55. $\dfrac{y+8}{y-3}-\dfrac{10y+30}{y^2-9}$

56. $\dfrac{y+3}{y+2}-\dfrac{4y+4}{y^2-4}$

75. $\dfrac{4}{x+2}+\dfrac{-5x-2}{x^2+2x}-\dfrac{3-x}{x}$

57. $\dfrac{4}{2n+1}+1$

58. $\dfrac{x-2}{x+3}+2$

76. $\dfrac{x+1}{x-3}+\dfrac{x+2}{x-2}-\dfrac{x^2+3}{x^2-5x+6}$

77. $\dfrac{2}{m+2}-\dfrac{3}{m}+\dfrac{m+10}{m^2-4}$

59. $\dfrac{-12}{3x-6}+\dfrac{4x-1}{x-2}$

60. $\dfrac{2}{4x+4}+\dfrac{8}{3x+3}$

78. $\dfrac{7}{w-3}-\dfrac{5}{w}-\dfrac{2w+6}{w^2-9}$

Applying the Concepts

In Problems 79–86, perform the indicated operation.

79. Find the quotient of $\dfrac{x-3}{x+2}$ and $\dfrac{x^2-9}{x^2+4}$.

80. Find the quotient of $\dfrac{x}{x+1}$ and $\dfrac{5}{3x+3}$.

81. Find the difference of $\dfrac{4x+6}{2x^2+x-3}$ and $\dfrac{x-1}{x^2-1}$.

82. Find the difference of $\dfrac{x+4}{2x^2-8}$ and $\dfrac{3}{4x-8}$.

83. Find the sum of $\dfrac{x+2}{x^2+x-6}$ and $\dfrac{x-3}{x^2+5x+6}$.

84. Find the sum of $\dfrac{x^2}{x-4}$ and $\dfrac{16}{4-x}$.

85. Find the product of $\dfrac{2x-3}{x+6}$ and $\dfrac{x-1}{x-7}$.

86. Find the product of $\dfrac{x^2-3x+2}{x^2+2x-3}$ and $\dfrac{x^2+x-6}{x^2-4}$.

△ **87. Perimeter of a Rectangle** Find the perimeter of the rectangle.

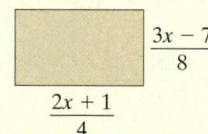

$\dfrac{3x-7}{8}$

$\dfrac{2x+1}{4}$

△ **88. Area of a Rectangle** Find the area of the rectangle.

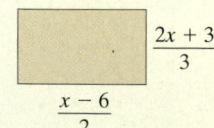

$\dfrac{2x+3}{3}$

$\dfrac{x-6}{2}$

Extending the Concepts

In Problems 89–98, perform the indicated operation.

89. $\dfrac{2}{a} - \left(\dfrac{2}{a-1} - \dfrac{3}{(a-1)^2} \right)$

90. $\dfrac{2}{b} - \left(\dfrac{2}{b+2} - \dfrac{2}{(b+2)^2} \right)$

91. $\dfrac{1}{x} - \dfrac{2}{x^2+x} + \dfrac{3}{x^3-x^2}$

92. $\dfrac{x}{(x-1)^2} + \dfrac{2}{x} - \dfrac{x+1}{x^3-x^2}$

93. $\dfrac{2a+b}{a-b} \cdot \dfrac{2}{a+b} - \dfrac{3a+3b}{a^2-b^2}$

94. $\dfrac{2a+b}{a-b} \div \dfrac{a+b}{2} + \dfrac{3}{a+b}$

95. $\dfrac{x-3}{x-4} + \dfrac{x+2}{x-4} \cdot \dfrac{4}{x+1}$

96. $\dfrac{3}{m+n} - \dfrac{m-2n}{m-n} \cdot \dfrac{2}{m+n}$

97. $\dfrac{x+2}{x-2} - \dfrac{x-2}{x+2} \div \dfrac{1}{x^2-4}$

98. $\dfrac{a+3}{a-3} - \dfrac{a+3}{a-3} \cdot \dfrac{a^2-4a+3}{a^2+5a+6}$

Explaining the Concepts

99. Explain the steps that should be used to add or subtract rational expressions with unlike denominators.

100. There are two ways that you could add $\dfrac{1}{x-7} + \dfrac{3}{7-x}$. Describe these two ways. Which method do you prefer?

101. Suppose you are asked to simplify $\dfrac{3x+1}{(2x+5)(x-1)} - \dfrac{x-4}{(2x+5)(x-1)}$. Explain the steps you should use to simplify the expression.

102. Explain the error in the following problem, and then rework the problem correctly.

$$\dfrac{2}{x-1} - \dfrac{x+2}{x} = \dfrac{2}{x-1} \cdot \dfrac{x}{x} - \dfrac{x+2}{x} \cdot \dfrac{x-1}{x-1}$$

$$= \dfrac{2x}{x(x-1)} - \dfrac{(x+2)(x-1)}{x(x-1)}$$

$$= \dfrac{2x-(x+2)(x-1)}{x(x-1)}$$

$$= \dfrac{2x-(x+2)(x\cancel{-1})}{x(x\cancel{-1})}$$

$$= \dfrac{2x-x-2}{x}$$

$$= \dfrac{x-2}{x}$$

7.6 Complex Rational Expressions

Objectives

1 Simplify a Complex Rational Expression by Simplifying the Numerator and Denominator Separately (Method I)

2 Simplify a Complex Rational Expression Using the Least Common Denominator (Method II)

Are You Prepared for This Section?

Before getting started, complete the following problems. If you get a problem wrong, go back to the section cited and review the material.

P1. Factor: $6y^2 - 5y - 6$ [Section 6.3, pp. 386–394]

P2. Find the quotient: $\dfrac{x + 3}{12} \div \dfrac{x^2 - 9}{15}$ [Section 7.2, pp. 443–446]

> **Definition**
>
> A **complex rational expression** is a quotient (fraction) whose numerator and/or denominator contains a rational expression.

The following are examples of complex rational expressions.

$$\frac{3}{\dfrac{4}{x} - \dfrac{x}{3}} \qquad \frac{\dfrac{x + 1}{x + 4} - 1}{16} \qquad \frac{3 - \dfrac{1}{x}}{1 + \dfrac{1}{x}}$$

To **simplify** a complex rational expression means to write the rational expression in the form $\dfrac{p}{q}$, where p and q are polynomials that have no common factors.

In this section, the division techniques introduced in Section 7.2 will be used. Review Example 1 in that section to refresh your memory.

▶ **EXAMPLE 1** **Dividing Rational Expressions**

Find the quotient: $\dfrac{\dfrac{3}{x + 4}}{\dfrac{9}{x^2 - 16}}$

Solution

To find the quotient, multiply the rational expression in the numerator by the reciprocal of the rational expression in the denominator.

$$\frac{\dfrac{3}{x + 4}}{\dfrac{9}{x^2 - 16}} = \frac{3}{x + 4} \cdot \frac{x^2 - 16}{9}$$

Factor the numerator and denominator: $= \dfrac{3}{x + 4} \cdot \dfrac{(x + 4)(x - 4)}{3 \cdot 3}$

$\dfrac{a}{b} \cdot \dfrac{c}{d} = \dfrac{a \cdot c}{b \cdot d}: \quad = \dfrac{3(x + 4)(x - 4)}{(x + 4) \cdot 3 \cdot 3}$

Divide out common factors: $= \dfrac{\cancel{3}\,\cancel{(x + 4)}(x - 4)}{\cancel{(x + 4)} \cdot \cancel{3} \cdot 3}$

$= \dfrac{x - 4}{3}$

Prepared?...Answers

P1. $(3y + 2)(2y - 3)$ **P2.** $\dfrac{5}{4(x - 3)}$

Quick ✓

1. An expression such as $\dfrac{\dfrac{x}{2} + \dfrac{5}{x}}{\dfrac{2x-1}{3}}$ is called a _____ _____ _____.

2. To _____ a complex rational expression means to write the rational expression in the form $\dfrac{p}{q}$, where p and q are polynomials that have no common factors.

In Problems 3 and 4, simplify each expression.

3. $\dfrac{\dfrac{2}{k+3}}{\dfrac{4}{k^2+4k+3}}$

4. $\dfrac{\dfrac{2}{n+3}}{\dfrac{8n}{2n+6}}$

▶ ❶ Simplify a Complex Rational Expression by Simplifying the Numerator and Denominator Separately (Method I)

There are two ways to simplify a complex rational expression. We'll start with Method I, simplifying the numerator and denominator separately.

EXAMPLE 2 **How to Simplify a Complex Rational Expression (Method I)**

Simplify: $\dfrac{\dfrac{1}{5} + \dfrac{1}{x}}{\dfrac{x+5}{2}}$

Step-by-Step Solution

Step 1: Write the numerator of the complex fraction as a single rational expression.

The numerator is $\dfrac{1}{5} + \dfrac{1}{x}$.

$$\text{LCD} = 5x: \quad \frac{1}{5} + \frac{1}{x} = \frac{1}{5} \cdot \frac{x}{x} + \frac{1}{x} \cdot \frac{5}{5}$$

$$= \frac{x}{5x} + \frac{5}{5x}$$

$$\frac{a}{c} + \frac{b}{c} = \frac{a+b}{c}: \quad = \frac{x+5}{5x}$$

Step 2: Write the denominator of the complex rational expression as a single rational expression.

The denominator is $\dfrac{x+5}{2}$. It is already written as a single rational expression.

Step 3: Rewrite the complex rational expression using the rational expressions found in Steps 1 and 2.

$$\frac{\dfrac{1}{5} + \dfrac{1}{x}}{\dfrac{x+5}{2}} = \frac{\dfrac{x+5}{5x}}{\dfrac{x+5}{2}}$$

Step 4: Simplify the rational expression by multiplying the numerator by the reciprocal of the denominator.

Multiply the numerator by the reciprocal of the denominator: $= \dfrac{x+5}{5x} \cdot \dfrac{2}{x+5}$

$$\frac{a}{b} \cdot \frac{c}{d} = \frac{a \cdot c}{b \cdot d}, \text{ divide out common factors: } = \frac{(x+5) \cdot 2}{5x \,(x+5)}$$

$$= \frac{2}{5x}$$

●

The steps used in Method I are summarized below.

> **Simplifying a Complex Rational Expression by Simplifying the Numerator and Denominator Separately (Method I)**
>
> **Step 1:** Write the numerator of the complex rational expression as a single rational expression.
>
> **Step 2:** Write the denominator of the complex rational expression as a single rational expression.
>
> **Step 3:** Rewrite the complex rational expression using the rational expressions found in Steps 1 and 2.
>
> **Step 4:** Simplify the rational expression by multiplying the numerator by the reciprocal of the denominator.

> **▶ EXAMPLE 3** **Simplifying a Complex Rational Expression (Method I)**
>
> Simplify: $\dfrac{\dfrac{2}{y} - \dfrac{8}{y^3}}{\dfrac{2}{y^2} + \dfrac{1}{y}}$

Solution

Write the numerator and the denominator of the complex rational expression as single rational expressions.

Work Smart

First simplify the numerator and the denominator into a single rational expression before you take the reciprocal and multiply.

NUMERATOR:

LCD is y^3:
$$\frac{2}{y} - \frac{8}{y^3} = \frac{2}{y} \cdot \frac{y^2}{y^2} - \frac{8}{y^3}$$
$$= \frac{2y^2 - 8}{y^3}$$
$$= \frac{2(y^2 - 4)}{y^3}$$
$$= \frac{2(y + 2)(y - 2)}{y^3}$$

DENOMINATOR:

LCD is y^2:
$$\frac{2}{y^2} + \frac{1}{y} = \frac{2}{y^2} + \frac{1}{y} \cdot \frac{y}{y}$$
$$= \frac{2}{y^2} + \frac{y}{y^2}$$
$$= \frac{2 + y}{y^2}$$
$$= \frac{y + 2}{y^2}$$

Now rewrite the complex rational expression with the new numerator and denominator.

$$\frac{\dfrac{2}{y} - \dfrac{8}{y^3}}{\dfrac{2}{y^2} + \dfrac{1}{y}} = \frac{\dfrac{2(y + 2)(y - 2)}{y^3}}{\dfrac{y + 2}{y^2}}$$

Multiply the numerator by the reciprocal of the denominator:
$$= \frac{2(y + 2)(y - 2)}{y^3} \cdot \frac{y^2}{y + 2}$$

$\dfrac{a}{b} \cdot \dfrac{c}{d} = \dfrac{a \cdot c}{b \cdot d}$; divide out common factors:
$$= \frac{2(y + 2)(y - 2)\,y^2}{y^2 \cdot y(y + 2)}$$

$$= \frac{2(y - 2)}{y}$$

EXAMPLE 4 **Simplifying a Complex Rational Expression (Method I)**

Simplify: $\dfrac{\dfrac{1}{x+2}+1}{x-\dfrac{3}{x+2}}$

Solution

Write the numerator and denominator of the complex rational expression as a single rational expression.

NUMERATOR:

$$\dfrac{1}{x+2}+1 = \dfrac{1}{x+2}+1\cdot\dfrac{x+2}{x+2}$$

$$= \dfrac{1}{x+2}+\dfrac{x+2}{x+2}$$

$\dfrac{a}{c}+\dfrac{b}{c}=\dfrac{a+b}{c}$: $\qquad = \dfrac{1+x+2}{x+2}$

Combine like terms: $\quad = \dfrac{x+3}{x+2}$

DENOMINATOR:

$$x-\dfrac{3}{x+2} = \dfrac{x}{1}\cdot\dfrac{x+2}{x+2}-\dfrac{3}{x+2}$$

Distribute: $\quad = \dfrac{x^2+2x}{x+2}-\dfrac{3}{x+2}$

$\dfrac{a}{c}-\dfrac{b}{c}=\dfrac{a-b}{c}$: $\quad = \dfrac{x^2+2x-3}{x+2}$

Factor: $\quad = \dfrac{(x+3)(x-1)}{x+2}$

Rewrite the complex rational expression using the numerator and denominator just found, and then simplify.

$$\dfrac{\dfrac{1}{x+2}+1}{x-\dfrac{3}{x+2}} = \dfrac{\dfrac{x+3}{x+2}}{\dfrac{(x+3)(x-1)}{x+2}}$$

Multiply the numerator by the reciprocal of the denominator: $\quad = \dfrac{x+3}{x+2}\cdot\dfrac{x+2}{(x+3)(x-1)}$

$\dfrac{a}{b}\cdot\dfrac{c}{d}=\dfrac{a\cdot c}{b\cdot d}$; divide out common factors: $\quad = \dfrac{\cancel{(x+3)}\,\cancel{(x+2)}}{\cancel{(x+2)}\,\cancel{(x+3)}(x-1)}$

$$= \dfrac{1}{x-1}$$

▶ ② **Simplify a Complex Rational Expression Using the Least Common Denominator (Method II)**

We now introduce a second method for simplifying complex rational expressions.

EXAMPLE 5 **How to Simplify a Complex Rational Expression Using the Least Common Denominator (Method II)**

Simplify: $\dfrac{\dfrac{2}{y} - \dfrac{8}{y^3}}{\dfrac{2}{y^2} + \dfrac{1}{y}}$

Step-by-Step Solution

Step 1: Find the least common denominator among all the denominators in the complex rational expression.

The denominators of the complex rational expression are y, y^3, and y^2. The least common denominator is y^3.

Step 2: Multiply both the numerator and the denominator of the complex rational expression by the least common denominator found in Step 1.

Multiply by $1 = \dfrac{y^3}{y^3}$: $\left(\dfrac{\dfrac{2}{y} - \dfrac{8}{y^3}}{\dfrac{2}{y^2} + \dfrac{1}{y}} \right) \cdot \left(\dfrac{y^3}{y^3} \right) = \dfrac{\dfrac{2}{y} \cdot y^3 - \dfrac{8}{y^3} \cdot y^3}{\dfrac{2}{y^2} \cdot y^3 + \dfrac{1}{y} \cdot y^3}$

Divide out common factors: $= \dfrac{\dfrac{2}{\cancel{y}} \cdot \cancel{y^3}y^2 - \dfrac{8}{\cancel{y^3}} \cdot \cancel{y^3}1}{\dfrac{2}{\cancel{y^2}} \cdot \cancel{y^3}y + \dfrac{1}{\cancel{y}} \cdot \cancel{y^3}y^2}$

Step 3: Simplify.

Multiply: $= \dfrac{2y^2 - 8}{2y + y^2}$

Factor out the GCF: $= \dfrac{2(y^2 - 4)}{y(2 + y)}$

Factor the difference of squares: $= \dfrac{2(y + 2)(y - 2)}{y(y + 2)}$

Divide out common factors: $= \dfrac{2\cancel{(y + 2)}(y - 2)}{y\cancel{(y + 2)}}$

$= \dfrac{2(y - 2)}{y}$

Therefore, $\dfrac{\dfrac{2}{y} - \dfrac{8}{y^3}}{\dfrac{2}{y^2} + \dfrac{1}{y}} = \dfrac{2(y - 2)}{y}$.

Compare the approach used in Example 5 with that used in Example 3.

> **Simplifying a Complex Rational Expression Using the Least Common Denominator (Method II)**
>
> **Step 1:** Find the least common denominator among all the denominators in the complex rational expression.
>
> **Step 2:** Multiply both the numerator and the denominator of the complex rational expression by the least common denominator found in Step 1.
>
> **Step 3:** Simplify the rational expression.

▶ **EXAMPLE 6** **Simplifying a Complex Rational Expression Using the Least Common Denominator (Method II)**

Simplify: $\dfrac{\dfrac{1}{x+2}+1}{x-\dfrac{3}{x+2}}$

Solution

The least common denominator among all denominators is $x+2$, so multiply the complex rational expression by $1 = \dfrac{x+2}{x+2}$.

$$\frac{\dfrac{1}{x+2}+1}{x-\dfrac{3}{x+2}} = \frac{\dfrac{1}{x+2}+1}{x-\dfrac{3}{x+2}}\cdot\frac{x+2}{x+2}$$

Distribute the LCD to each term: $= \dfrac{\dfrac{1}{x+2}(x+2)+1(x+2)}{x(x+2)-\dfrac{3}{x+2}(x+2)}$

Divide out common factors: $= \dfrac{\dfrac{1}{\cancel{x+2}}(\cancel{x+2})+x+2}{x(x+2)-\dfrac{3}{\cancel{x+2}}(\cancel{x+2})}$

$= \dfrac{1+x+2}{x^2+2x-3}$

Combine like terms; factor: $= \dfrac{x+3}{(x+3)(x-1)}$

Divide out common factors: $= \dfrac{1}{x-1}$

Work Smart

$\dfrac{1}{x+2}\cdot(x+2) = \dfrac{1}{x+2}\cdot\dfrac{x+2}{1} = 1$

Compare the approach used in Example 6 with that used in Example 4.

Quick ✓

In Problems 9 and 10, simplify the complex rational expression using Method II.

9. $\dfrac{\dfrac{3}{2}+\dfrac{2}{3}}{\dfrac{1}{2}-\dfrac{1}{3}}$

10. $\dfrac{\dfrac{1}{x}+\dfrac{2}{y}}{\dfrac{2}{x}-\dfrac{1}{y}}$

7.6 Exercises MyMathLab®

Exercise numbers in green have complete video solutions in MyMathLab or may be accessed using the QR code to the right.

Problems 1–10 are the Quick ✓s that follow the EXAMPLES.

In Problems 11–24, simplify the complex rational expression using Method I. See Objective 1.

11. $\dfrac{1-\dfrac{3}{4}}{\dfrac{1}{8}+2}$

12. $\dfrac{\dfrac{2}{3}-\dfrac{3}{4}}{\dfrac{1}{6}+\dfrac{1}{2}}$

13. $\dfrac{\dfrac{x^2}{12}-\dfrac{1}{3}}{\dfrac{x+2}{18}}$

14. $\dfrac{\dfrac{4}{t^2}-1}{\dfrac{t+2}{t^3}}$

15. $\dfrac{\dfrac{x+3}{x^2}}{\dfrac{x^2}{9}-1}$

16. $\dfrac{\dfrac{n+2}{4}}{\dfrac{4}{n^2}-1}$

17. $\dfrac{\dfrac{m}{2} + n}{\dfrac{m}{n}}$

18. $\dfrac{\dfrac{1}{x} - \dfrac{1}{y}}{\dfrac{2}{xy}}$

19. $\dfrac{\dfrac{5}{a} + \dfrac{4}{b^2}}{\dfrac{5b + 4}{b^2}}$

20. $\dfrac{\dfrac{2}{x} + \dfrac{3}{x^2}}{\dfrac{2x + 3}{x}}$

21. $\dfrac{\dfrac{8}{y + 3} - 2}{y - \dfrac{4}{y + 3}}$

22. $\dfrac{\dfrac{5}{n - 1} + 3}{n - \dfrac{2}{n - 1}}$

23. $\dfrac{\dfrac{1}{4} - \dfrac{6}{y}}{\dfrac{5}{6y} - y}$

24. $\dfrac{\dfrac{2}{a + b}}{\dfrac{1}{a} + \dfrac{1}{b}}$

In Problems 25–38, simplify the complex rational expression using Method II. See Objective 2.

25. $\dfrac{\dfrac{3}{2} - \dfrac{1}{4}}{\dfrac{5}{6} + \dfrac{1}{2}}$

26. $\dfrac{\dfrac{5}{4} - \dfrac{1}{8}}{\dfrac{3}{8} + \dfrac{9}{2}}$

27. $\dfrac{\dfrac{3}{m} + \dfrac{2}{m^2}}{\dfrac{6}{m} + \dfrac{4}{m^2}}$

28. $\dfrac{\dfrac{3c}{4} + \dfrac{3d}{10}}{\dfrac{3c}{2} - \dfrac{6d}{5}}$

29. $\dfrac{1 - \dfrac{49}{b^2}}{1 + \dfrac{7}{b}}$

30. $\dfrac{\dfrac{a}{2} + 4}{2 - \dfrac{a}{2}}$

31. $\dfrac{1 + \dfrac{5}{x}}{1 + \dfrac{1}{x + 4}}$

32. $\dfrac{\dfrac{x}{x + 1}}{1 + \dfrac{1}{x - 1}}$

33. $\dfrac{\dfrac{1}{x^2} - \dfrac{1}{y^2}}{x - y}$

34. $\dfrac{6x + \dfrac{3}{y}}{\dfrac{9x + 3}{y}}$

35. $\dfrac{a + \dfrac{1}{b}}{a + \dfrac{2}{b}}$

36. $\dfrac{4 + \dfrac{1}{x}}{8 + \dfrac{2}{x}}$

37. $\dfrac{12}{\dfrac{4}{n} - \dfrac{2}{3n}}$

38. $\dfrac{7}{\dfrac{2}{a} + \dfrac{3}{4a}}$

Mixed Practice

In Problems 39–52, simplify the complex rational expression using either Method I or Method II.

39. $\dfrac{\dfrac{x}{x - y} + \dfrac{y}{x + y}}{\dfrac{xy}{x^2 - y^2}}$

40. $\dfrac{\dfrac{b}{b + 1} - 1}{\dfrac{b + 3}{b} - 2}$

41. $\dfrac{\dfrac{2}{x + 4}}{\dfrac{2}{x + 4} - 4}$

42. $\dfrac{\dfrac{4}{x + 1} - 1}{\dfrac{4}{x + 1} - 3}$

43. $\dfrac{\dfrac{b^2}{b^2 - 16} - \dfrac{b}{b + 4}}{\dfrac{b}{b^2 - 16} - \dfrac{1}{b - 4}}$

44. $\dfrac{\dfrac{-6}{y^2 + 5y + 6}}{\dfrac{2}{y + 3} - \dfrac{3}{y + 2}}$

45. $\dfrac{1 - \dfrac{2}{x} - \dfrac{3}{x^2}}{1 - \dfrac{9}{x^2}}$

46. $\dfrac{2 - \dfrac{3}{x} - \dfrac{2}{x^2}}{1 - \dfrac{5}{x} + \dfrac{6}{x^2}}$

47. $\dfrac{1 - \dfrac{a^2}{4b^2}}{1 + \dfrac{a}{2b}}$

48. $\dfrac{1 - \dfrac{n^2}{25m^2}}{1 - \dfrac{n}{5m}}$

49. $\dfrac{\dfrac{x + 2}{x - 3} - 5}{\dfrac{x + 2}{x - 3} + 1}$

50. $\dfrac{\dfrac{x + 4}{2x - 5} - 1}{\dfrac{x + 4}{2x - 5} - 2}$

51. $\dfrac{\dfrac{3}{n - 2} + 1}{5 + \dfrac{1}{n - 2}}$

52. $\dfrac{\dfrac{5}{2b + 3} + \dfrac{1}{2b - 3}}{\dfrac{6b}{8b^2 - 18}}$

Applying the Concepts

53. **Finding the Mean** The arithmetic mean of a set of numbers is found by adding the numbers and then dividing by the number of entries on the list. Write a complex rational expression to find the arithmetic mean of the expressions $\dfrac{n}{6}, \dfrac{n + 3}{2}$, and $\dfrac{2n - 1}{8}$, and then simplify the complex rational expression.

54. **Finding the Mean** See Problem 53. Write a complex rational expression to find the arithmetic mean of the expressions $\dfrac{z}{2}, \dfrac{z - 3}{4}$, and $\dfrac{2z - 1}{6}$, and then simplify the complex rational expression.

△ **55. Area of a Rectangle** In a rectangle, the length can be found by dividing the area by the width. If the area of a rectangle is $\left(\dfrac{2x+3}{x^2} - \dfrac{3}{x}\right)$ square feet and the width is $\dfrac{x^2-9}{x^5}$ feet, write a complex rational expression to find the length and then simplify the complex rational expression.

△ **56. Area of a Rectangle** In a rectangle, the length can be found by dividing the area by the width. If the area of a rectangle is $\left(\dfrac{x-4}{x^3} - \dfrac{2}{x^2}\right)$ square feet and the width is $\dfrac{x^2-16}{x}$ feet, write a complex rational expression to find the length and then simplify the complex rational expression.

57. Electric Circuits An electric circuit contains two resistors connected in parallel, as shown in the figure. If they have resistance R_1 and R_2 ohms, respectively, then their combined resistance R is given by the formula

$$R = \dfrac{1}{\dfrac{1}{R_1} + \dfrac{1}{R_2}}$$

(a) Express R as a simplified rational expression.

(b) Evaluate the rational expression if $R_1 = 6$ ohms and $R_2 = 10$ ohms.

58. Electric Circuits An electric circuit contains three resistors connected in parallel. If the circuits have resistance R_1, R_2, and R_3 ohms, respectively, then their combined resistance is given by the formula

$$R = \dfrac{1}{\dfrac{1}{R_1} + \dfrac{1}{R_2} + \dfrac{1}{R_3}}$$

(a) Express R as a simplified rational expression.

(b) Evaluate the rational expression if $R_1 = 4$ ohms, $R_2 = 6$ ohms, and $R_3 = 10$ ohms.

Extending the Concepts

In Problems 59–62, use the definition of a negative exponent to simplify each complex rational expression.

59. $\dfrac{x^{-1}-3}{x^{-2}-9}$

60. $\dfrac{1-16x^{-2}}{1-3x^{-1}-4x^{-2}}$

61. $\dfrac{2x^{-1}+5}{4x^{-2}-25}$

62. $\dfrac{x^{-2}-3x^{-1}}{x^{-2}-9}$

In Problems 63–66, simplify each expression.

63. $1 + \dfrac{1}{1+\dfrac{1}{x}}$

64. $1 - \dfrac{1}{1-\dfrac{1}{x-2}}$

65. $\dfrac{1}{1-\dfrac{1}{2-\dfrac{1}{3-x}}}$

66. $1 + \dfrac{2}{1+\dfrac{2}{x+\dfrac{2}{x}}}$

Explaining the Concepts

67. Is the expression $\dfrac{\dfrac{2}{x+1}}{\dfrac{1}{x-5}}$ in simplified form? Explain why or why not.

68. Describe the steps to simplify a complex rational expression using Method I.

69. State which method you prefer to use when simplifying complex rational expressions. State your reasons for choosing this method, and then explain how to use this method to simplify

$$\dfrac{\dfrac{1}{x}+\dfrac{1}{y}}{\dfrac{1}{x^2}-\dfrac{1}{y^2}}.$$

70. Explain the property of real numbers that is being applied in Method II. Make up a complex rational expression, and explain how to use Method II to simplify the expression.

Putting the Concepts Together (Sections 7.1–7.6)

We designed these problems so that you can review Sections 7.1–7.6 and show your mastery of the concepts. Take time to work these problems before proceeding with the next section. The answers are located at the back of the text on page AN-28.

1. Evaluate $\dfrac{a^3-b^3}{a+b}$ when $a=2$ and $b=-3$.

2. Find the values for which the following rational expressions are undefined:

(a) $\dfrac{-3a}{a-6}$

(b) $\dfrac{y+2}{y^2+4y}$

3. Simplify:

(a) $\dfrac{ax + ay - 4bx - 4by}{2x + 2y}$ (b) $\dfrac{x^2 + x - 2}{1 - x^2}$

4. Find the least common denominator (LCD) of the rational expressions:

$$\frac{5}{2a + 4b}, \frac{-1}{4a + 8b}, \frac{3ab}{8a - 32b}$$

5. Write $\dfrac{7}{3x^2 - x}$ as an equivalent rational expression with denominator $5x^2(3x - 1)$.

In Problems 6–13, perform the indicated operations.

6. $\dfrac{y^2 - y}{3y} \cdot \dfrac{6y^2}{1 - y^2}$

7. $\dfrac{m^2 + m - 2}{m^3 - 6m^2} \cdot \dfrac{2m^2 - 14m + 12}{m + 2}$

8. $\dfrac{4y + 12}{5y - 5} \div \dfrac{2y^2 - 18}{y^2 - 2y + 1}$

9. $\dfrac{4x^2 + x}{x + 3} + \dfrac{12x + 3}{x + 3}$

10. $\dfrac{x^2}{x^3 + 1} - \dfrac{x - 1}{x^3 + 1}$

11. $\dfrac{-8}{2x - 1} - \dfrac{9}{1 - 2x}$

12. $\dfrac{4}{m + 3} + \dfrac{3}{3m + 2}$

13. $\dfrac{3m}{m^2 + 7m + 10} - \dfrac{2m}{m^2 + 6m + 8}$

14. Simplify the complex rational expression:

$$\frac{\dfrac{-1}{m + 1} - 1}{m - \dfrac{2}{m + 1}}$$

15. Simplify the complex rational expression: $\dfrac{\dfrac{4}{a} + \dfrac{1}{6}}{\dfrac{3}{a^2} + \dfrac{1}{2}}$

7.7 Rational Equations

Objectives

1 Solve Equations Containing Rational Expressions

2 Solve for a Variable in a Rational Equation

Are You Prepared for This Section?

Before getting started, complete the following problems. If you get a problem wrong, go back to the section cited and review the material.

P1. Solve: $3k - 2(k + 1) = 6$ [Section 2.2, pp. 92–94]

P2. Factor: $3p^2 - 7p - 6$ [Section 6.3, pp. 386–394]

P3. Solve: $8z^2 - 10z - 3 = 0$ [Section 6.6, pp. 409–414]

P4. Find the values for which the expression $\dfrac{x + 4}{x^2 - 2x - 24}$ is undefined. [Section 7.1, pp. 435–436]

P5. Solve for y: $4x - 2y = 10$ [Section 2.4, pp. 113–115]

▶ 1 Solve Equations Containing Rational Expressions

So far, we have solved linear equations (Sections 2.1–2.3), quadratic equations (Section 6.6), and equations that contain polynomial expressions that can be factored (Section 6.6). Now, we look at another type of equation.

Definition

A **rational equation** is an equation that contains a rational expression.

Examples of rational equations are

$$\frac{3}{x + 4} = \frac{1}{x + 2} \quad \text{and} \quad \frac{4}{3} + \frac{7}{x - 4} = \frac{x - 1}{3x - 12}$$

The idea in solving a rational equation is to use algebraic techniques to rewrite it as an equation you already know how to solve, such as a linear or quadratic equation. In Chapter 2, you learned how to solve equations such as $\dfrac{3x}{4} - \dfrac{x}{2} = \dfrac{1}{2}$. Remember that one approach to solving an equation whose coefficients are fractions is to multiply each side of the equation by the least common denominator of the fractions, which, when simplified, will result in an equivalent equation without fractions.

To solve an equation such as $\dfrac{8}{p} + \dfrac{1}{4p} = \dfrac{11}{8}$ use the same process. Let's compare a problem we already know how to solve from Chapter 2 to the approach used to solve this new type of equation.

EXAMPLE 1   **Solving a Rational Equation**

Solve:

(a) $\dfrac{3x}{4} - \dfrac{x}{2} = \dfrac{1}{2}$

(b) $\dfrac{8}{p} + \dfrac{1}{4p} = \dfrac{11}{8}, \quad p \neq 0$

Solution

Notice that Example 1(a) is a linear equation with fractional coefficients. Example 1(b) is a rational equation because it contains a rational expression. Notice that the steps for solving the two equations are identical and are the same steps that were used in Chapter 2.

(a)
$$\frac{3x}{4} - \frac{x}{2} = \frac{1}{2}$$

(b)
$$\frac{8}{p} + \frac{1}{4p} = \frac{11}{8}, \quad p \neq 0$$

Find the LCD: The LCD is 4.

The LCD is $8p$.

Multiply each side of the equation by the LCD:
$$4\left(\frac{3x}{4} - \frac{x}{2}\right) = 4\left(\frac{1}{2}\right)$$
$$8p\left(\frac{8}{p} + \frac{1}{4p}\right) = 8p\left(\frac{11}{8}\right)$$

Distribute and divide out common factors:
$$4\left(\frac{3x}{4}\right) - 4\left(\frac{x}{2}\right) = 4\left(\frac{1}{2}\right)$$
$$8p\left(\frac{8}{p}\right) + 8p\left(\frac{1}{4p}\right) = 8p\left(\frac{11}{8}\right)$$

Solve the linear equation:
$$3x - 2x = 2$$
$$x = 2$$

$$64 + 2 = 11p$$
$$66 = 11p$$
$$6 = p$$

The check is left to you. The solution set is $\{2\}$.

The check is left to you. The solution set is $\{6\}$.   ●

Did you notice the restriction that $p \neq 0$ for the equation $\dfrac{8}{p} + \dfrac{1}{4p} = \dfrac{11}{8}$ in Example 1(b)?

That condition is included because if p is replaced by 0, both $\dfrac{8}{p}$ and $\dfrac{1}{4p}$ are undefined.

When solving a rational equation, always find the value(s) of the variable that will cause the rational expression(s) in the rational equation to be undefined.

Also notice in Example 1(b) that by multiplying both sides by the LCD, a linear equation was obtained—an equation we already know how to solve. Again, this is a key to mathematics: Solve a problem by rewriting it as one you already know how to solve.

EXAMPLE 2   **How to Solve a Rational Equation**

Solve: $\dfrac{3}{2b - 1} = \dfrac{4}{3b}$

Step-by-Step Solution

Step 1: Determine the value(s) of the variable that cause the rational expression(s) in the rational equation to be undefined.

Find the values of the variable that cause either denominator to equal 0. If $2b - 1 = 0$, then $b = \dfrac{1}{2}$; if $3b = 0$, then $b = 0$. Therefore, $b \neq 0$, $b \neq \dfrac{1}{2}$.

Step 2: Determine the least common denominator (LCD) of all the denominators.

The LCD is $3b(2b - 1)$.

Step 3: Multiply both sides of the equation by the LCD, and simplify the expressions on each side.

$$\frac{3}{2b - 1} = \frac{4}{3b}$$

Multiply both sides by $3b(2b - 1)$: $3b(2b - 1)\left(\frac{3}{2b - 1}\right) = 3b(2b - 1)\left(\frac{4}{3b}\right)$

Divide out common factors: $3b\cancel{(2b - 1)}\left(\frac{3}{\cancel{2b - 1}}\right) = \cancel{3b}(2b - 1)\left(\frac{4}{\cancel{3b}}\right)$

Step 4: Solve the resulting equation.

$$3b(3) = (2b - 1)4$$

Multiply and distribute: $9b = 8b - 4$

Subtract $8b$ from both sides: $b = -4$

Step 5: Check Verify the solution using the original equation.

Let $b = -4$: $\dfrac{3}{2(-4) - 1} \overset{?}{=} \dfrac{4}{3(-4)}$

$$\frac{3}{-8 - 1} \overset{?}{=} \frac{4}{-12}$$

$$\frac{3}{-9} \overset{?}{=} \frac{4}{-12}$$

$$-\frac{1}{3} = -\frac{1}{3} \quad \text{True}$$

The solution checks, so the solution set is $\{-4\}$.

Solving a Rational Equation

Step 1: Determine the value(s) of the variable that result in an undefined rational expression in the rational equation.

Step 2: Determine the least common denominator (LCD) of all the denominators.

Step 3: Multiply both sides of the equation by the LCD, and simplify the expressions on each side of the equation.

Step 4: Solve the resulting equation.

Step 5: Verify the solution using the original equation.

Quick ✓

1. A _____ _____ is an equation that contains a rational expression.

In Problems 2–4, solve the rational equation.

2. $\dfrac{5}{2} + \dfrac{1}{z} = 4$

3. $\dfrac{8}{x + 4} = \dfrac{12}{x - 3}$

4. $\dfrac{4}{3b} + \dfrac{1}{6b} = \dfrac{7}{2b} + \dfrac{1}{3}$

EXAMPLE 3   **Solving a Rational Equation**

Solve: $\dfrac{4}{3} + \dfrac{7}{x - 4} = \dfrac{x - 1}{3x - 12}$

(continued)

Solution

First factor the denominator $3x - 12$. Factoring the denominator helps find the undefined values of the variable and the LCD. Because $3x - 12 = 3(x - 4)$, write the equation as

$$\frac{4}{3} + \frac{7}{x - 4} = \frac{x - 1}{3(x - 4)}$$

Now solve $x - 4 = 0$ to find the values of x that result in an undefined rational expression.

$$x - 4 = 0$$
$$x = 4$$

The rational expression is undefined when $x = 4$, so $x \neq 4$. The factored form of the denominators also shows that the LCD is $3(x - 4)$.

$$\frac{4}{3} + \frac{7}{x - 4} = \frac{x - 1}{3(x - 4)}$$

Multiply both sides of the equation
by the LCD, $3(x - 4)$: $\quad 3(x - 4)\left(\dfrac{4}{3} + \dfrac{7}{x - 4}\right) = 3(x - 4)\left[\dfrac{x - 1}{3(x - 4)}\right]$

Distribute: $\quad 3(x - 4)\left(\dfrac{4}{3}\right) + 3(x - 4)\left(\dfrac{7}{x - 4}\right) = 3(x - 4)\left[\dfrac{x - 1}{3(x - 4)}\right]$

Divide out common factors: $\quad 3(x - 4)\left(\dfrac{4}{3}\right) + 3(x - 4)\left(\dfrac{7}{x - 4}\right) = 3(x - 4)\left[\dfrac{x - 1}{3(x - 4)}\right]$

$$4(x - 4) + 3(7) = x - 1$$

Distribute the 4; multiply: $\quad 4x - 16 + 21 = x - 1$

Combine like terms: $\quad 4x + 5 = x - 1$

Subtract x from both sides: $\quad 3x + 5 = -1$

Subtract 5 from both sides: $\quad 3x = -6$

Divide both sides by 3: $\quad x = -2$

Check Let $x = -2$ in the *original* equation.

$$\frac{4}{3} + \frac{7}{x - 4} = \frac{x - 1}{3x - 12}$$

Let $x = -2$: $\quad \dfrac{4}{3} + \dfrac{7}{-2 - 4} \overset{?}{=} \dfrac{-2 - 1}{3(-2) - 12}$

$$\frac{4}{3} + \frac{7}{-6} \overset{?}{=} \frac{-3}{-6 - 12}$$

$$\frac{4}{3} - \frac{7}{6} \overset{?}{=} \frac{-3}{-18}$$

$$\frac{8}{6} - \frac{7}{6} \overset{?}{=} \frac{1}{6}$$

$$\frac{1}{6} = \frac{1}{6} \quad \text{True}$$

The solution checks, so the solution set is $\{-2\}$.

Quick ✓

In Problem 5, solve the rational equation.

5. $\dfrac{3}{2} + \dfrac{5}{x - 3} = \dfrac{x + 9}{2x - 6}$

EXAMPLE 4 Solving a Rational Equation in Which the LCD Is the Product of the Denominators

Solve: $\dfrac{5}{x + 2} + \dfrac{3}{x - 2} = 1$

Solution

The rational expression $\dfrac{5}{x + 2} + \dfrac{3}{x - 2}$ is undefined when $x = -2$ and $x = 2$. The LCD is $(x + 2)(x - 2)$.

$$\frac{5}{x + 2} + \frac{3}{x - 2} = 1$$

Multiply both sides of the equation by the LCD: $\quad (x + 2)(x - 2)\left(\dfrac{5}{x + 2} + \dfrac{3}{x - 2}\right) = (x + 2)(x - 2)(1)$

Distribute: $\quad (x + 2)(x - 2)\left(\dfrac{5}{x + 2}\right) + (x + 2)(x - 2)\left(\dfrac{3}{x - 2}\right) = (x + 2)(x - 2)(1)$

Divide out common factors: $\quad \cancel{(x + 2)}(x - 2)\left(\dfrac{5}{\cancel{x + 2}}\right) + (x + 2)\cancel{(x - 2)}\left(\dfrac{3}{\cancel{x - 2}}\right) = (x + 2)(x - 2)(1)$

$$(x - 2)(5) + (x + 2)(3) = (x + 2)(x - 2)$$

Distribute: $\quad 5x - 10 + 3x + 6 = x^2 - 4$

Combine like terms: $\quad 8x - 4 = x^2 - 4$

Write the quadratic equation in standard form: $\quad 0 = x^2 - 8x$

Factor the polynomial: $\quad 0 = x(x - 8)$

Use the Zero-Product Property: $\quad x = 0 \quad \text{or} \quad x - 8 = 0$

Solve for x: $\quad x = 8$

Work Smart

Recall that a quadratic equation is a second-degree equation. It is solved by writing the equation in standard form, $ax^2 + bx + c = 0$, factoring, and then using the Zero-Product Property.

Check Substitute $x = 0$ and $x = 8$ in the *original* equation. We leave it to you to confirm that the solution set is $\{0, 8\}$. ●

Quick ✓

In Problem 6, solve the rational equation.

6. $\dfrac{4}{x - 3} - \dfrac{3}{x + 3} = 1$

▶ Did you wonder why we determine the values of the variable that make a rational expression in the rational equation undefined? Well, in some instances the solution process gives results that do not satisfy the original equation.

Definition

An **extraneous solution** is a solution found through the solving process that does not satisfy the original equation.

EXAMPLE 5 Solving a Rational Equation with an Extraneous Solution

Solve: $\dfrac{x^2 + 8x + 6}{x^2 + 3x - 4} = \dfrac{3}{x - 1} - \dfrac{2}{x + 4}$

(continued)

Solution
Rewrite the equation so that each denominator is in factored form. Because
$x^2 + 3x - 4 = (x - 1)(x + 4)$, write the equation as

$$\frac{x^2 + 8x + 6}{(x - 1)(x + 4)} = \frac{3}{x - 1} - \frac{2}{x + 4}$$

Notice that $x \neq 1$ and $x \neq -4$ because these values cause the denominator to equal 0.
Also notice that the LCD is $(x - 1)(x + 4)$.
    Now multiply both sides of the equation by the LCD and divide out common factors.

$$(x - 1)(x + 4)\left[\frac{x^2 + 8x + 6}{(x - 1)(x + 4)}\right] = (x - 1)(x + 4)\left(\frac{3}{x - 1} - \frac{2}{x + 4}\right)$$

$$(x - 1)(x + 4)\left[\frac{x^2 + 8x + 6}{(x - 1)(x + 4)}\right] = (x - 1)(x + 4)\left(\frac{3}{x - 1}\right) - (x - 1)(x + 4)\left(\frac{2}{x + 4}\right)$$

$$\cancel{(x - 1)(x + 4)}\left[\frac{x^2 + 8x + 6}{\cancel{(x - 1)(x + 4)}}\right] = \cancel{(x - 1)}(x + 4)\left(\frac{3}{\cancel{x - 1}}\right) - (x - 1)\cancel{(x + 4)}\left(\frac{2}{\cancel{x + 4}}\right)$$

$$x^2 + 8x + 6 = 3(x + 4) - 2(x - 1)$$

Distribute: $\qquad x^2 + 8x + 6 = 3x + 12 - 2x + 2$

Combine like terms: $\qquad x^2 + 8x + 6 = x + 14$

Write the quadratic equation in standard form: $\qquad x^2 + 7x - 8 = 0$

Factor the polynomial: $\quad (x + 8)(x - 1) = 0$

Use the Zero-Product Property: $\qquad\qquad x + 8 = 0 \quad$ or $\quad x - 1 = 0$

Solve each equation for x: $\qquad\qquad\qquad x = -8 \quad$ or $\qquad x = 1$

Because $x = 1$ is one of the values of x that results in division by zero, it is extraneous and is rejected as a solution. We leave it to you to verify that the solution set is $\{-8\}$. ●

Quick ✓

7. Solutions obtained through the solving process that do not satisfy the original equation are called _____ _____.

In Problem 8, solve the rational equation.

8. $\dfrac{5}{z - 4} + \dfrac{3}{z - 2} = \dfrac{z^2 - z - 2}{z^2 - 6z + 8}$

▶ (**EXAMPLE 6**) **Solving a Rational Equation with No Solution**

Solve: $\dfrac{6}{z^2 - 1} = \dfrac{5}{z - 1} - \dfrac{3}{z + 1}$

Solution

$$\frac{6}{z^2 - 1} = \frac{5}{z - 1} - \frac{3}{z + 1}$$

Factor $z^2 - 1$: $\qquad \dfrac{6}{(z - 1)(z + 1)} = \dfrac{5}{z - 1} - \dfrac{3}{z + 1}$

Notice that $z \neq 1$ and $z \neq -1$. Also, the LCD is $(z - 1)(z + 1)$. Multiply both sides of the equation by the LCD and divide out common factors.

$$(z - 1)(z + 1)\left[\frac{6}{(z - 1)(z + 1)}\right] = (z - 1)(z + 1)\left(\frac{5}{z - 1} - \frac{3}{z + 1}\right)$$

$$\cancel{(z - 1)}\cancel{(z + 1)}\left[\frac{6}{\cancel{(z - 1)}\cancel{(z + 1)}}\right] = \cancel{(z - 1)}(z + 1)\left(\frac{5}{\cancel{z - 1}}\right) - (z - 1)\cancel{(z + 1)}\left(\frac{3}{\cancel{z + 1}}\right)$$

$$6 = 5(z + 1) - 3(z - 1)$$

Distribute:	$6 = 5z + 5 - 3z + 3$
Combine like terms:	$6 = 2z + 8$
Subtract 8 from both sides:	$-2 = 2z$
Divide both sides by 2:	$-1 = z$

Work Smart

The solution set { } is not the same as the solution set {0}.

Because z cannot equal -1 (it results in division by 0), it is extraneous, and $z = -1$ is rejected as a solution. Therefore, the equation has no solution. The solution set is { } or $\varnothing$.

Work Smart: Study Skills

Showing your work makes it easier for you to see how you get the extraneous solution. Don't skip steps to get the answer. Double-check every step as you proceed through the problem to catch possible errors.

Quick ✓

9. *True or False* A rational equation can have no solution.

In Problem 10, solve the rational equation.

10. $\dfrac{4}{y + 1} = \dfrac{7}{y - 1} - \dfrac{8}{y^2 - 1}$

Work Smart

When a rational equation is solved, multiply both sides of the equation by the LCD, but when rational expressions are added or subtracted, retain the LCD. Notice the difference in the directions: *simplify* expressions and *solve* equations.

Simplify: $\dfrac{2}{x} - \dfrac{1}{6} + \dfrac{5}{2x} - \dfrac{1}{3}$

$\dfrac{2}{x} - \dfrac{1}{6} + \dfrac{5}{2x} - \dfrac{1}{3} = \dfrac{2}{x} \cdot \dfrac{6}{6} - \dfrac{1}{6} \cdot \dfrac{x}{x} + \dfrac{5}{2x} \cdot \dfrac{3}{3} - \dfrac{1}{3} \cdot \dfrac{2x}{2x}$

$= \dfrac{12}{6x} - \dfrac{x}{6x} + \dfrac{15}{6x} - \dfrac{2x}{6x}$

$= \dfrac{12 - x + 15 - 2x}{6x}$

$= \dfrac{27 - 3x}{6x}$

$= \dfrac{9 - x}{2x}$

Solve: $\dfrac{2}{x} - \dfrac{1}{6} = \dfrac{5}{2x} - \dfrac{1}{3}$

$6x\left(\dfrac{2}{x} - \dfrac{1}{6}\right) = 6x\left(\dfrac{5}{2x} - \dfrac{1}{3}\right)$

$6x\left(\dfrac{2}{x}\right) - 6x\left(\dfrac{1}{6}\right) = 6x\left(\dfrac{5}{2x}\right) - 6x\left(\dfrac{1}{3}\right)$

$6(2) - x = 3(5) - 2x$

$12 - x = 15 - 2x$

$12 + x = 15$

$x = 3$

EXAMPLE 7 **An Application of Rational Equations: Average Daily Cost**

Suppose that the average daily cost $\overline{C}$, in dollars, of manufacturing x bicycles is given by the equation

$$\overline{C} = \frac{x^2 + 75x + 5000}{x}$$

Determine the number of bicycles for which the average daily cost will be $225.

(continued)

Solution

First, notice that x must be greater than 0. Do you know why? Because we want to know the level of production to obtain an average daily cost of \$225, solve the equation where $\overline{C} = 225$.

$$\overline{C} = \frac{x^2 + 75x + 5000}{x}$$

$$225 = \frac{x^2 + 75x + 5000}{x}$$

Multiply both sides by x: $\quad x \cdot 225 = \frac{x^2 + 75x + 5000}{x} \cdot x$

$$225x = x^2 + 75x + 5000$$

Write the quadratic equation in standard form: $\quad 0 = x^2 - 150x + 5000$

Factor the polynomial: $\quad 0 = (x - 50)(x - 100)$

Use the Zero-Product Property: $\quad x - 50 = 0 \quad$ or $\quad x - 100 = 0$

Solve each equation: $\quad x = 50 \quad$ or $\quad x = 100$

The average daily cost is \$225 when the level of production is at 50 or 100 bicycles. ●

Quick ✓

11. The concentration C in milligrams per liter of a certain drug in a patient's bloodstream t hours after injection is given by $C = \dfrac{50t}{t^2 + 25}$. When will the concentration of the drug be 4 milligrams per liter?

Work Smart

The steps that are followed when solving formulas for a variable are identical to those that are followed when solving rational equations.

▶ ❷ **Solve for a Variable in a Rational Equation**

Recall from Section 2.4 that "solve for the variable" means getting the variable by itself on one side of the equation, with all other variable and constant terms, if any, on the other side. Use the same steps to solve formulas for a variable that are used to solve rational equations.

EXAMPLE 8 **How to Solve for a Variable in a Rational Equation**

Solve $\dfrac{x}{1 + y} = z$ for y.

Step-by-Step Solution

Isolate the variable y. That is, get y by itself on one side of the equation and the other terms on the other side.

Step 1: Determine the value(s) of the variable that cause any rational expression in the rational equation to be undefined.	Because $y = -1$ results in division by zero, $y \neq -1$.
Step 2: Find the least common denominator (LCD) of all the denominators.	The LCD is $1 + y$.
Step 3: Multiply both sides of the equation by the LCD, and simplify the expressions on each side of the equation.	$\dfrac{x}{1 + y} = z$ $(1 + y)\left(\dfrac{x}{1 + y}\right) = z(1 + y)$ $x = z(1 + y)$

Step 4: Solve the resulting equation for y.

Divide both sides by z:
$$\frac{x}{z} = \frac{z(1+y)}{z}$$

$$\frac{x}{z} = 1 + y$$

Subtract 1 from both sides:
$$\frac{x}{z} - 1 = 1 - 1 + y$$

$$\frac{x}{z} - 1 = y \qquad \bullet$$

Look back at Step in Example 8, where both sides of the equation were divided by z. Did you think about using the Distributive Property to simplify $x = z(1+y)$? If you did, that is perfectly correct! This approach in solving $\dfrac{x}{1+y} = z$ for y is shown below:

$$x = z(1+y)$$

Use the Distributive Property:
$$x = z + zy$$

Isolate the expression containing the variable y:
$$x - z = z - z + zy$$

$$x - z = zy$$

Divide both sides by z:
$$\frac{x-z}{z} = \frac{zy}{z}$$

$$\frac{x-z}{z} = y$$

The result using this method, $y = \dfrac{x-z}{z}$, is equivalent to the equation $y = \dfrac{x}{z} - 1$, because $\dfrac{x-z}{z} = \dfrac{x}{z} - \dfrac{z}{z} = \dfrac{x}{z} - 1$.

Quick ✓

In Problems 12 and 13, solve the equation for the indicated variable.

12. Solve $R = \dfrac{4g}{x}$ for x **13.** Solve $S = \dfrac{a}{1-r}$ for r

---

EXAMPLE 9 **Solving for a Variable in a Rational Equation**

Solve $\dfrac{1}{a} + \dfrac{1}{b} = \dfrac{1}{c}$ for b.

Solution

Neither a, b, nor c can be equal to zero. Further, the LCD is abc.

$$\frac{1}{a} + \frac{1}{b} = \frac{1}{c}$$

Multiply both sides by the LCD:
$$abc\left(\frac{1}{a} + \frac{1}{b}\right) = abc\left(\frac{1}{c}\right)$$

Distribute:
$$abc\left(\frac{1}{a}\right) + abc\left(\frac{1}{b}\right) = abc\left(\frac{1}{c}\right)$$

Divide out common factors:
$$\not{a}bc\left(\frac{1}{\not{a}}\right) + a\not{b}c\left(\frac{1}{\not{b}}\right) = ab\not{c}\left(\frac{1}{\not{c}}\right)$$

$$bc + ac = ab$$

(continued)

Work Smart

To solve for a variable that occurs on both sides of the equation, get all terms containing that variable on the same side of the equation. Then factor out the variable you wish to solve for.

To solve for b, both terms containing b need to be on the same side of the equation.

Subtract bc from both sides: $\qquad$ $ac = ab - bc$

Factor out b as a common factor: $\qquad$ $ac = b(a - c)$

Divide each side by $(a - c)$: $\qquad$ $\dfrac{ac}{(a - c)} = \dfrac{b(a - c)}{(a - c)}$

$$\dfrac{ac}{a - c} = b$$

In solving $\dfrac{1}{a} + \dfrac{1}{b} = \dfrac{1}{c}$ for b, the equation is rewritten as $b = \dfrac{ac}{a - c}$.

> **Quick ✓**
>
> In Problem 14, solve the equation for the indicated variable.
>
> **14.** $\dfrac{1}{f} = \dfrac{1}{p} + \dfrac{1}{q}$ for p

7.7 Exercises MyMathLab®

Exercise numbers in **green** have complete video solutions in MyMathLab or may be accessed using the QR code to the right.

*Problems **1–14** are the Quick ✓s that follow the EXAMPLES.*

Building Skills

In Problems 15–46, solve each equation and state the solution set. Remember to identify the values of the variable for which the expressions in each rational equation are undefined. See Objective 1.

15. $\dfrac{5}{3y} - \dfrac{1}{2} = \dfrac{5}{6y} - \dfrac{1}{12}$

16. $\dfrac{4}{x} - \dfrac{11}{5} = \dfrac{3}{2x} - \dfrac{6}{5}$

17. $\dfrac{6}{x} + \dfrac{2}{3} = \dfrac{4}{2x} - \dfrac{14}{3}$

18. $\dfrac{4}{p} - \dfrac{5}{4} = \dfrac{5}{2p} + \dfrac{3}{8}$

19. $\dfrac{4}{x - 1} = \dfrac{3}{x + 1}$

20. $\dfrac{4}{x - 4} = \dfrac{5}{x + 4}$

21. $\dfrac{2}{x + 2} + 2 = \dfrac{7}{x + 2}$

22. $\dfrac{4}{x + 1} + 2 = \dfrac{3}{x + 1}$

23. $\dfrac{r - 4}{3r} + \dfrac{2}{5r} = \dfrac{1}{5}$

24. $\dfrac{x - 2}{4x} - \dfrac{x + 2}{3x} = \dfrac{1}{2x} - \dfrac{1}{2}$

25. $\dfrac{2}{a - 1} + \dfrac{3}{a + 1} = \dfrac{-6}{a^2 - 1}$

26. $\dfrac{6}{t} - \dfrac{2}{t - 1} = \dfrac{2 - 4t}{t^2 - t}$

27. $\dfrac{1}{4 - x} + \dfrac{2}{x^2 - 16} = \dfrac{1}{x - 4}$

28. $\dfrac{4}{x - 3} - \dfrac{3}{x - 2} = \dfrac{2x + 1}{x^2 - 5x + 6}$

29. $\dfrac{3}{2t - 2} + 4 = \dfrac{2t}{3t - 3}$

30. $\dfrac{2x + 3}{x - 1} - 2 = \dfrac{3x - 1}{4x - 4}$

31. $\dfrac{2}{5} + \dfrac{3 - 2a}{10a - 20} = \dfrac{2a + 1}{a - 2}$

32. $\dfrac{1}{x - 2} + \dfrac{2}{2 - x} = \dfrac{5}{x + 1}$

33. $\dfrac{6}{j^2 - 1} - \dfrac{4j}{j^2 - 5j + 4} = -\dfrac{4}{j - 1}$

34. $\dfrac{3}{2x + 2} - \dfrac{5}{4x - 4} = \dfrac{2x}{x^2 - 1}$

35. $\dfrac{x}{x-2} = \dfrac{3}{x+8}$

36. $\dfrac{x+5}{x-7} = \dfrac{x-3}{x+7}$

37. $\dfrac{x}{x+3} = \dfrac{6}{x-3} + 1$

38. $\dfrac{3x^2}{x+1} = 2 + \dfrac{3x}{x+1}$

39. $x = \dfrac{2-x}{6x}$

40. $x = \dfrac{6-5x}{6x}$

41. $\dfrac{2x+3}{x-1} = \dfrac{x-2}{x+1} + \dfrac{6x}{x^2-1}$

42. $\dfrac{2x}{x+4} = \dfrac{x+1}{x+2} - \dfrac{7x+12}{x^2+6x+8}$

43. $\dfrac{2}{b+2} - \dfrac{5b+6}{b^2-b-6} = \dfrac{-b}{b-3}$

44. $\dfrac{2x}{x+3} - \dfrac{2x^2+2}{x^2-9} = \dfrac{-6}{x-3} + 1$

45. $\dfrac{1}{n-3} = \dfrac{3n-1}{9-n^2}$

46. $\dfrac{5}{x-2} - \dfrac{2}{2-x} = \dfrac{4}{x+1}$

In Problems 47–64, solve the equation for the indicated variable.
See Objective 2.

47. $x = \dfrac{2}{y}$ for y

48. $4 = \dfrac{k}{m}$ for m

49. $I = \dfrac{E}{R}$ for R

50. $T = \dfrac{D}{R}$ for R

51. $h = \dfrac{2A}{B+b}$ for b

52. $m = \dfrac{y-z}{x-w}$ for y

53. $\dfrac{x}{3+y} = z$ for y

54. $\dfrac{a}{b+2} = c$ for b

55. $\dfrac{1}{R} = \dfrac{1}{S} + \dfrac{1}{T}$ for S

56. $\dfrac{3}{i} - \dfrac{4}{j} = \dfrac{8}{k}$ for j

57. $m = \dfrac{n}{y} - \dfrac{p}{ay}$ for y

58. $\dfrac{1}{a} = \dfrac{1}{b} + \dfrac{1}{c}$ for c

59. $A = \dfrac{xy}{x+y}$ for x

60. $B = \dfrac{k}{x} - \dfrac{m}{cx}$ for x

61. $\dfrac{2}{x} - \dfrac{1}{y} = \dfrac{6}{z}$ for y

62. $y = \dfrac{a}{a+b}$ for b

63. $y = \dfrac{x}{x-c}$ for x

64. $X = \dfrac{ab}{a-b}$ for a

Mixed Practice

In Problems 65–88, simplify the expression or solve the equation.

65. $\dfrac{1}{x} + \dfrac{3}{x+5}$

66. $\dfrac{2}{x-5} + \dfrac{5}{x-5}$

67. $x - \dfrac{6}{x} = 1$

68. $z + \dfrac{3}{z} = 4$

69. $\dfrac{3}{x-1} \cdot \dfrac{x^2-1}{6} + 3$

70. $5 - \dfrac{n^2-3n-4}{n^2-4}$

71. $2b - \dfrac{5}{3} = \dfrac{1}{3b}$

72. $3a + \dfrac{7}{3} = \dfrac{2}{3a}$

73. $\dfrac{5x}{2x-3} = \dfrac{3x}{x-1} - \dfrac{5}{2x^2-5x+3}$

74. $\dfrac{2x+1}{x^2+2x-3} = \dfrac{x-1}{x^2+5x+6} + \dfrac{x+1}{x^2+x-2}$

75. $\dfrac{x^2}{x^2-4} - \dfrac{1}{x}$

76. $\dfrac{x}{x+1} + \dfrac{2x-3}{x-1}$

77. $\dfrac{x^2-9}{x+1} \div \dfrac{3x^2+9x}{x^2-1}$

78. $\dfrac{2t+1}{t^2-16} \cdot \dfrac{t^2+t-12}{4t+2}$

79. $\dfrac{\dfrac{1}{a}+3}{\dfrac{3a+1}{5}}$

80. $\dfrac{3n+4}{5n} - \dfrac{2}{n^2+2n}$

81. $\dfrac{a}{2a - 2} - \dfrac{2}{3a + 3} = \dfrac{5a^2 - 2a + 9}{12a^2 - 12}$

82. $\dfrac{x}{12} + \dfrac{1}{2} = \dfrac{1}{3x} + \dfrac{2}{x^2}$

83. $\dfrac{\dfrac{3}{b + c} + \dfrac{5}{b - c}}{\dfrac{4b + c}{b^2 - c^2}}$

84. $\dfrac{\dfrac{1}{2a} + \dfrac{3}{a^2}}{\dfrac{1}{3a} - \dfrac{4}{a^2}}$

85. $\dfrac{2x^2 + 5x + 2}{x^2 + 6} = 2$

86. $\dfrac{x^2 + x - 12}{x^2 - 4} = 1$

87. $\dfrac{x^3 + 12}{x^2 - 4} = x$

88. $\dfrac{x^3 + 8}{x^2 + 2} = x$

Applying the Concepts

89. Drug Concentration The concentration C of a drug in a patient's bloodstream, in milligrams per liter, t hours after ingestion is modeled by $C = \dfrac{40t}{t^2 + 9}$. When will the concentration of the drug be 4 milligrams per liter?

90. Drug Concentration The concentration C of a drug in a patient's bloodstream, in milligrams per liter, t hours after ingestion is modeled by $C = \dfrac{40t}{t^2 + 3}$. When will the concentration of the drug be 10 milligrams per liter?

91. Cost-Benefit Model Environmental scientists often use cost-benefit models to estimate the cost of removing a certain percentage of a pollutant from the environment. Suppose a cost-benefit model for the cost C (in millions of dollars) of removing x percent of the pollutants from Maple Lake is given by $C = \dfrac{25x}{100 - x}$. If the federal government budgets $100 million to clean up the lake, what percent of the pollutants can be removed?

92. Average Cost Suppose that the average daily cost in dollars, $\overline{C}$, of manufacturing x bicycles is given by the equation $\overline{C} = \dfrac{x^2 + 75x + 5000}{x}$. Determine the level of production for which the average daily cost will be $240.

Extending the Concepts

93. For what value of k will the solution set of $\dfrac{4x + 3}{k} = \dfrac{x - 1}{3}$ be $\{2\}$?

94. For what value of k will the solution set of $\dfrac{1}{2} + \dfrac{3x}{k} = 1 + \dfrac{x}{3}$ be $\{3\}$?

Explaining the Concepts

95. On a test, a student wrote the following steps to solve the equation $\dfrac{2}{x - 1} - \dfrac{4}{x^2 - 1} = -2$ but then got stuck and quit. Find and explain the error in the following *incorrect* solution.

$$\frac{2}{x - 1} - \frac{4}{x^2 - 1} = -2$$

$$\frac{2}{x - 1} - \frac{4}{(x - 1)(x + 1)} = -2$$

$$[(x - 1)(x + 1)]\left[\frac{2}{x - 1} - \frac{4}{(x - 1)(x + 1)}\right] = -2[(x - 1)(x + 1)]$$

$$2(x + 1) - 4 = -2(x - 1)(x + 1)$$

$$2x + 2 - 4 = (-2x + 2)(-2x - 2)$$

$$2x - 2 = 4x^2 - 4$$

$$0 = 4x^2 + 2x - 2$$

96. Explain why it is important to find the values of the variable that make the denominator zero when solving a rational equation.

97. Explain the difference between the directions *simplify* and *solve*. Make up a problem using each of these directions with the rational expressions $\dfrac{5}{x + 1}$ and $\dfrac{6}{x - 1}$.

98. Explain how using the LCD to solve an equation that contains rational expressions is different from using the LCD in a problem that requires adding and subtracting rational expressions with unlike denominators.

7.8 Models Involving Rational Equations

Objectives

1 Model and Solve Ratio and Proportion Problems

2 Model and Solve Problems with Similar Figures

3 Model and Solve Work Problems

4 Model and Solve Uniform Motion Problems

Are You Prepared for This Section?

Before getting started, complete the following problem. If you get the problem wrong, go back to the section cited and review the material.

P1. Solve: $\dfrac{150}{r} = \dfrac{250}{r + 20}$ [Section 7.7, pp. 481–488]

1 Model and Solve Ratio and Proportion Problems

We begin with a definition.

> **Definition**
>
> A **ratio** is the quotient of two numbers or two quantities. The ratio of two numbers a and b can be written as
>
> $$a \text{ to } b \quad \text{or} \quad a : b \quad \text{or} \quad \frac{a}{b}$$

In algebraic problems, ratios are written as $\dfrac{a}{b}$. For example, the probability of winning the "Pick Three" Instant Ohio Lottery is 1 in 1000, which is written as the ratio $\dfrac{1}{1000}$. Ratios are written as fractions reduced to lowest terms. If 8 oz of a cleaner is needed for every 4 gallons of water, the ratio is $\dfrac{8 \text{ oz}}{4 \text{ gal}} = \dfrac{2 \text{ oz}}{1 \text{ gal}}$.

A rational equation that involves two ratios is called a *proportion*.

Work Smart

The equation $\dfrac{3}{8} = \dfrac{7}{x} + 1$ is NOT a proportion. Do you see why?

> **Definition**
>
> A **proportion** is an equation of the form $\dfrac{a}{b} = \dfrac{c}{d}$, where $b \neq 0$ and $d \neq 0$. The **terms** of the proportion are a, b, c, and d. The **means** of the proportion are b and c, and the **extremes** of the proportion are a and d.

A proportion can be solved the same way we solved a rational equation.

EXAMPLE 1   **Solving a Proportion**

Solve the proportion: $\dfrac{x}{5} = \dfrac{63}{105}$

Solution

Because $105 = 21 \cdot 5$, the LCD is 105.

$$\frac{x}{5} = \frac{63}{105}$$

Multiply both sides of the equation by 105: $\quad 105\left(\dfrac{x}{5}\right) = 105\left(\dfrac{63}{105}\right)$

Divide out common factors: $\quad 21x = 63$

Divide both sides by 21: $\quad x = 3$

Prepared?...Answer **P1.** $\{30\}$

The solution set is $\{3\}$.

Another method used to solve a proportion is **cross multiplication**, which is based on the *Means-Extremes Theorem*.

> **Means-Extremes Theorem**
>
> If $\dfrac{a}{b} = \dfrac{c}{d}$, where $b \neq 0$ and $d \neq 0$, then $ad = bc$.

Cross multiplication can be used to solve a proportion because the LCD for the proportion $\dfrac{a}{b} = \dfrac{c}{d}$, is bd. Multiply both sides of the equation $\dfrac{a}{b} = \dfrac{c}{d}$ by the LCD.

$$bd\left(\frac{a}{b}\right) = \left(\frac{c}{d}\right)bd$$
$$ad = bc$$

Work Smart

Use this visual for cross multiplication.

$$\frac{x}{5} \diagdown\!\!\!\!\diagup \frac{63}{105}$$

$$105x = 5 \cdot 63$$

Solving the proportion in Example 1, $\dfrac{x}{5} = \dfrac{63}{105}$, using cross multiplication yields

$$\frac{x}{5} = \frac{63}{105}$$

Cross multiply: $\quad 105x = 5(63)$

$$105x = 315$$

Divide both sides by 105: $\quad x = 3$

> **Quick ✓**
>
> 1. A _____ is an equation of the form $\dfrac{a}{b} = \dfrac{c}{d}$, where $b \neq 0$ and $d \neq 0$.
>
> 2. If $\dfrac{a}{b} = \dfrac{c}{d}$, then $a \cdot \underline{} = b \cdot \underline{}$.
>
> *In Problems 3 and 4, solve the proportion for the indicated variable.*
>
> 3. $\dfrac{2p + 1}{4} = \dfrac{p}{8}$ 4. $\dfrac{6}{x^2} = \dfrac{2}{x}$

EXAMPLE 2   **Solving an Exchange Rate Problem**

Last summer Laura took a Study Abroad trip to Italy. She spent 92 euros (€) on a romantic gondola ride touring the canals of Venice. If \$1 U.S. is equal to 0.89€, then to the nearest cent, how much did the ride cost in U.S. dollars?

Solution

Step 1: Identify Use a proportion to find the cost of the canal ride in U.S. dollars.

Step 2: Name Let c represent the cost of the canal ride in U.S. dollars.

Step 3: Translate Because 1 U.S. dollar is equivalent to 0.89 euro, set up the proportion using $\dfrac{\text{dollars}}{\text{euros}}$. The ratio on the left side of the proportion contains the equivalence, and the ratio on the right side has the cost of the canal ride, where c is the unknown cost (in dollars) for a ride costing 92€. Note that the units of measure in the model are included. This is done to keep track of the units and tells us the answer is in dollars.

Work Smart

You could also set up the model using $\dfrac{\text{euros}}{\text{dollars}}$.

$$\frac{0.89 €}{\$1} = \frac{92 €}{c}$$

$$\frac{\$1}{0.89\,€} = \frac{c}{92\,€} \qquad \text{The Model}$$

Step 4: Solve

Cross multiply: $\$1(92€) = (0.89€)c$

Divide both sides by 0.89€: $\dfrac{\$1(92€)}{0.89€} = \dfrac{(0.89€)c}{0.89€}$

Simplify using a calculator: $\$103.37 \approx c$

Step 5: Check Is the answer reasonable? First, the answer is in dollars, which is what we were looking for in this question. Next, the ratio $1 to 0.89€ should equal the ratio $103.37 to 92€. Using a calculator to divide reveals that both ratios are approximately equal to 1.124, so the answer checks.

Step 6: Answer The canal ride that Laura took in Italy cost $103.37.   ●

Quick ✓

5. An automobile manufacturer is advertising special financing on its autos. A buyer will have a monthly payment of $16.67 for every $1000 borrowed. If Clem borrows $14,000, find his monthly payment.

Sometimes a proportion can be used as a general model for economic behavior.

EXAMPLE 3   **Model and Solve a Problem from Business**

A real estate agent knows that an 1800-square-foot house in a particular neighborhood sold for $150,000. For how much should the agent appraise a 2100-square-foot house in the same neighborhood?

Solution

Step 1: Identify Use a proportion to find the appraisal price of a 2100-square-foot house in the same neighborhood as an 1800-square-foot house that sold for $150,000.

Step 2: Name Let p represent the appraisal price of the 2100-square-foot house.

Step 3: Translate The selling price of an 1800-square-foot house is $150,000. We need the price p of a 2100-square-foot house. Use the ratio $\dfrac{\text{square feet}}{\text{price}}$ to set up the proportion

$$\dfrac{1800 \text{ ft}^2}{\$150{,}000} = \dfrac{2100 \text{ ft}^2}{p} \quad \text{The Model}$$

Step 4: Solve

Work Smart

You could also cross multiply to solve the equation in Example 3.

Multiply by the LCD $150,000p: $\$150{,}000p\left(\dfrac{1800 \text{ ft}^2}{\$150{,}000}\right) = \$150{,}000p\left(\dfrac{2100 \text{ ft}^2}{p}\right)$

Divide out common factors: $(1800 \text{ ft}^2)p = (\$150{,}000)(2100 \text{ ft}^2)$

Divide both sides by 1800 ft²: $p = \dfrac{(\$150{,}000)(2100 \text{ ft}^2)}{1800 \text{ ft}^2}$

$p = \$175{,}000$

(continued)

Step 5: Check The answer is in dollars, and the bigger house is appraised at a higher price. Plus, the ratio $\dfrac{1800}{150{,}000} = 0.012$ equals the ratio $\dfrac{2100}{175{,}000} = 0.012$, so the answer checks.

Step 6: Answer The appraisal price should be $175,000.   ●

Quick ✓

6. On a map, $\dfrac{1}{4}$ inch represents a distance of 15 miles. According to the map, Springfield and Brookhaven are $3\dfrac{1}{2} = \dfrac{7}{2}$ inches apart. Find the number of miles between the two cities.

▶ ❷ Model and Solve Problems with Similar Figures

Proportions may be used to solve problems involving *similar figures* from geometry.

Definition

Two figures are **similar** if their corresponding angle measures are equal and their corresponding sides are in the same ratio.

Figure 1

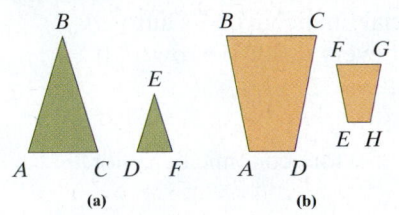

(a)   (b)

Figure 1 shows examples of similar figures.

In Figure 1(a), $\triangle ABC$ is similar to $\triangle DEF$, and in Figure 1(b), quadrilateral $ABCD$ is similar to quadrilateral $EFGH$. Do you see that similar figures have the same shape but differ in size? Because $\triangle ABC$ is similar to $\triangle DEF$, the ratio of AB to DE equals the ratio of AC to DF. That is,

$$\frac{AB}{DE} = \frac{AC}{DF}$$

The principle that the ratios of corresponding sides are equal can be used to find unknown lengths in similar figures.

EXAMPLE 4   **Solving a Problem with Similar Triangles**

Find the length of side DE in triangle $\triangle DEF$, given that $\triangle ABC$ is similar to $\triangle DEF$, as shown in Figure 2. All measurements are in inches.

Solution

Because $\triangle ABC$ is similar to $\triangle DEF$, the corresponding sides have the same ratio, so $\dfrac{AB}{DE} = \dfrac{AC}{DF}$. Substitute the known values and solve for the unknown value, letting $x = DE$.

Figure 2

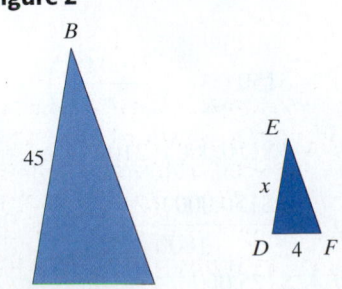

$$\frac{AB}{DE} = \frac{AC}{DF}$$

$AB = 45, AC = 10, DF = 4$: $\dfrac{45}{x} = \dfrac{10}{4}$ The Model

Cross multiply: $10x = 45 \cdot 4$

Multiply: $10x = 180$

Divide both sides by 10: $x = 18$

The length of side DE is 18 inches.   ●

In Example 4, the proportion $\dfrac{AB}{AC} = \dfrac{DE}{DF}$ could have been used and solving the proportion $\dfrac{45}{10} = \dfrac{x}{4}$ would yield the same result.

Quick ✓

7. In geometry, two figures are _____ if their corresponding angle measures are equal and their corresponding sides have the same ratio.

In Problem 8, find the length of side XY given that △MNP is similar to △XYZ.

8.

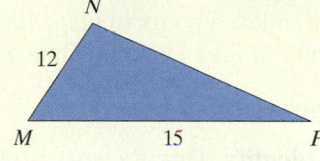
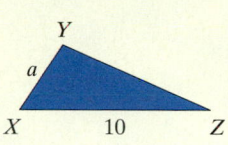

EXAMPLE 5   **Finding the Height of a Tree**

A fifth-grade student is conducting an experiment to find the height of a tree in the schoolyard. The student measures the length of the tree's shadow and then immediately measures the length of the shadow that a yardstick forms. The tree's shadow is 24 feet long, and the yardstick's shadow is 4 feet long. Find the height of the tree.

Solution

See Figure 3. The ratio of the length of the tree's shadow to the height of the tree, h, equals the ratio of the length of the yardstick's shadow to the height of the yardstick, 3 feet. This is because the tree and its shadow form a triangle that is similar to that formed by the yardstick and its shadow.

The proportion problem is set up as

Figure 3

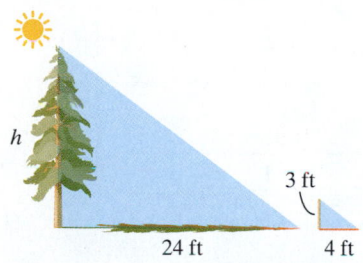

$$\frac{24}{h} = \frac{4}{3} \qquad \text{The Model}$$

Cross multiply: $4h = 24 \cdot 3$

Multiply: $4h = 72$

Divide both sides by 4: $h = 18$

Work Smart

We could have also solved the proportion $\dfrac{h}{3} = \dfrac{24}{4}$. Do you see why?

The height of the tree is 18 feet.   ●

Quick ✓

9. A tree casts a shadow 25 feet long. At the same time of day, a 6-foot-tall man casts a shadow 2.5 feet long. How tall is the tree?

▶ ❸ **Model and Solve Work Problems**

Work problems assume that jobs are performed at a **constant rate**, which means an individual works at the same pace throughout the entire job, no matter how many other people are working on the same job. Although this assumption is reasonable for machines, it is not reasonable for people because of the old phrase "too many chefs spoil the broth." Think of it this way—if you continually add more people to paint a room, the time to complete the job may decrease initially, but eventually the painters get in each other's way, and the time to completion actually increases.

While we could take this into account when modeling situations such as this, we will make the "constant-rate" assumption for humans to keep the mathematics manageable.

The constant-rate assumption states that if it takes t units of time to complete a job, then $\frac{1}{t}$ of the job is done in 1 unit of time. For example, if it takes 5 hours to paint a room, then $\frac{1}{5}$ of the room should be painted in 1 hour.

EXAMPLE 6 **Planting Seedlings**

Dan, an experienced horticulturist, can plant 1000 seedlings in 2 hours. Hala, a student assistant, takes 5 hours to plant 1000 seedlings. To the nearest tenth of an hour, how long would it take Dan and Hala working together to plant 1000 seedlings?

Solution

Step 1: Identify This is a work problem. Find how long it will take for Dan and Hala, working together, to plant 1000 seedlings.

Step 2: Name Let t represent the time (in hours) it takes them to plant the seedlings working together. Then in 1 hour they will complete $\frac{1}{t}$ of the job.

Step 3: Translate Because Dan can finish the job in 2 hours, Dan will finish $\frac{1}{2}$ of the job in 1 hour. Because Hala can finish the job in 5 hours, Hala will finish $\frac{1}{5}$ of the job in 1 hour. Using this information, set up Table 1.

Table 1

	Number of Hours to Complete the Job	Part of the Job Completed per Hour
Dan	2	$\frac{1}{2}$
Hala	5	$\frac{1}{5}$
Together	t	$\frac{1}{t}$

Work Smart

The part of the job completed per hour is the reciprocal of the number of hours to complete the job.

The model is created using the following logic:

$$\begin{pmatrix} \text{Part done by Dan} \\ \text{in 1 hour} \end{pmatrix} + \begin{pmatrix} \text{Part done by Hala} \\ \text{in 1 hour} \end{pmatrix} = \begin{pmatrix} \text{Part done together} \\ \text{in 1 hour} \end{pmatrix}$$

$$\frac{1}{2} \qquad + \qquad \frac{1}{5} \qquad = \qquad \frac{1}{t} \qquad \text{The Model}$$

Step 4: Solve $$\frac{1}{2} + \frac{1}{5} = \frac{1}{t}$$

Multiply both sides by the LCD, 10t: $$10t\left(\frac{1}{2} + \frac{1}{5}\right) = 10t\left(\frac{1}{t}\right)$$

Distribute: $$10t\left(\frac{1}{2}\right) + 10t\left(\frac{1}{5}\right) = 10t\left(\frac{1}{t}\right)$$

Divide out common factors: $$5t + 2t = 10$$

Combine like terms: $$7t = 10$$

Divide both sides by 7: $$t = \frac{10}{7}$$

$$t \approx 1.4$$

Work Smart

To convert 0.4 hour to minutes, multiply 0.4 by 60 minutes to obtain 24 minutes.

Step 5: Check Check to make sure the answer is reasonable. The answer should be greater than 0 but less than 2 (because the job takes Dan 2 hours working by himself). The answer of 1.4 hours, or 1 hour 24 minutes, seems reasonable.

Step 6: Answer It will take Dan and Hala about 1.4 hours to plant 1000 seedlings. ●

Quick ✔

10. When solving work problems, we assume jobs are performed at a _____ rate. Thus, if a job can be completed in 4 hours, ___ of the job is completed in 1 hour.

11. It takes Molly 6 hours to assemble notebooks for attendees at an educational technology conference. It takes Jacqueline 3 hours to do the same job. Working together, how long does it take Molly and Jacqueline to assemble the conference notebooks?

EXAMPLE 7 **Mowing the Lawn**

Sara can cut the grass in 3 hours working by herself. When Sara cuts the grass with her younger brother Brian, it takes 2 hours. How long would it take Brian to cut the grass if he worked by himself?

Solution

Step 1: Identify In this work problem, find how long it will take Brian to cut the grass by himself.

Step 2: Name Let t represent the time (in hours) that it takes Brian to cut the grass by himself. Then, in 1 hour he will complete $\frac{1}{t}$ of the job.

Step 3: Translate Because Sara can finish the job in 3 hours, she will finish $\frac{1}{3}$ of the job in 1 hour. Together the job is completed in 2 hours, so $\frac{1}{2}$ of the job is completed in 1 hour. To set up the model, use the following logic:

$$\begin{pmatrix} \text{Part done by} \\ \text{Sara in 1 hour} \end{pmatrix} + \begin{pmatrix} \text{Part done by} \\ \text{Brian in 1 hour} \end{pmatrix} = \begin{pmatrix} \text{Part done} \\ \text{together in 1 hour} \end{pmatrix}$$

$$\frac{1}{3} \qquad + \qquad \frac{1}{t} \qquad = \qquad \frac{1}{2} \qquad \text{The Model}$$

Step 4: Solve
$$\frac{1}{3} + \frac{1}{t} = \frac{1}{2}$$

Multiply both sides by the LCD, $6t$:
$$6t\left(\frac{1}{3} + \frac{1}{t}\right) = 6t\left(\frac{1}{2}\right)$$

Distribute:
$$6t\left(\frac{1}{3}\right) + 6t\left(\frac{1}{t}\right) = 6t\left(\frac{1}{2}\right)$$

Divide out common factors:
$$2t + 6 = 3t$$

Subtract $2t$ from each side:
$$6 = t$$

Step 5: Check The answer of 6 hours is reasonable since this time is greater than the time required when working together. The mathematics is correct because $\frac{2}{6} + \frac{1}{6} = \frac{3}{6} = \frac{1}{2}$.

Step 6: Answer It will take Brian 6 hours to cut the grass, working by himself. ●

▶ ④ Model and Solve Uniform Motion Problems

Uniform motion problems were introduced in Section 2.7. Recall that these problems use the fact that distance equals rate times time—that is, $d = rt$. When modeling uniform motion problems that lead to rational equations, we often use the model, $t = \dfrac{d}{r}$. These problems use the idea of constant rate, just as work problems do. For example, a bicyclist may not ride for several hours at exactly the same speed. To make the mathematics easier, however, assume that speed is constant throughout the time period.

EXAMPLE 8 **A Boat Trip on the Mighty Olentangy**

The Olentangy River has a current of 2 miles per hour. A motorboat goes 48 miles downstream in the same time it takes to go 36 miles upstream. What is the speed of the boat in still water?

Solution

Step 1: Identify This is a uniform motion problem. Find the speed of the boat in still water.

Step 2: Name Let r represent the speed of the boat in still water.

Step 3: Translate The current pushes the boat when it is going downstream, so the total speed of the boat is the speed of the boat in still water plus the speed of the current. If the speed of the current is 2 mph, $r + 2$ represents the speed of the boat downstream. Going upstream, the current slows the boat, so the total speed of the boat upstream is $r - 2$. Using this information, set up Table 2.

Work Smart

Remember, $d = rt$, so $\dfrac{d}{r} = t$.

Table 2

	Distance (miles)	Rate (miles per hour)	Time $= \dfrac{\text{Distance}}{\text{Rate}}$ (hours)
Downstream	48	$r + 2$	$\dfrac{48}{r + 2}$
Upstream	36	$r - 2$	$\dfrac{36}{r - 2}$

The amount of time traveling downstream is the same as the amount of time spent traveling upstream, so

$$\frac{48}{r + 2} = \frac{36}{r - 2} \quad \text{\color{red}{The Model}}$$

Step 4: Solve

$$\frac{48}{r + 2} = \frac{36}{r - 2}$$

Cross multiply: $48(r - 2) = 36(r + 2)$

Distribute: $48r - 96 = 36r + 72$

Subtract 36r from both sides: $\quad 12r - 96 = 72$

Add 96 to both sides: $\quad\quad\quad 12r = 168$

Divide each side by 12: $\quad\quad \dfrac{12r}{12} = \dfrac{168}{12}$

$$r = 14$$

Step 5: Check Are the downstream and upstream travel times the same? Going downstream, the boat travels at $14 + 2 = 16$ miles per hour, so it takes $\dfrac{48}{16} = 3$ hours.

Going upstream, the boat travels at $14 - 2 = 12$ miles per hour, and it takes $\dfrac{36}{12} = 3$ hours. The answer checks!

Step 6: Answer The boat travels 14 miles per hour in still water. ●

Quick ✓

13. A small airplane can travel 120 mph in still air. The plane can fly 700 miles with the wind in the same time that it can travel 500 miles against the wind. Find the speed of the wind.

EXAMPLE 9   **The Ride Back Through the Forest**

Every weekend, you ride your bicycle on a forest preserve path. The path is 12 miles long and ends at a waterfall, where you relax before making the trip back to the starting point. One weekend, you notice that because the ride to the waterfall is partially uphill, your cycling speed returning is 2 miles per hour faster than the speed riding to the waterfall. If the round trip takes 5 hours (excluding resting time), find the average speed cycling back from the waterfall.

Solution

Step 1: Identify This is a uniform motion problem. Find the average speed returning from the waterfall.

Step 2: Name Let r represent your cycling speed when you are cycling to the waterfall. Then $r + 2$ represents the speed cycling back from the waterfall.

Step 3: Translate The distance to the waterfall is 12 miles. Set up Table 3.

Table 3

	Distance (miles)	Rate (miles per hour)	Time $=\dfrac{\text{Distance}}{\text{Rate}}$ (hours)
Going to the Waterfall	12	r	$\dfrac{12}{r}$
Returning Home	12	$r + 2$	$\dfrac{12}{r + 2}$

Given that the total time spent cycling was 5 hours, set up the equation

$$\left(\begin{array}{c}\text{Time cycling to}\\ \text{waterfall}\end{array}\right) + \left(\begin{array}{c}\text{Time cycling to}\\ \text{starting point}\end{array}\right) = 5 \text{ hours}$$

$$\dfrac{12}{r} \quad + \quad \dfrac{12}{r + 2} \quad = 5 \quad \text{The Model}$$

(continued)

Step 4: Solve Solve for r.

$$\frac{12}{r} + \frac{12}{r+2} = 5$$

Multiply both sides by the LCD, $r(r+2)$:

$$r(r+2)\left(\frac{12}{r} + \frac{12}{r+2}\right) = r(r+2)(5)$$

Distribute; use the Commutative Property:

$$r(r+2)\left(\frac{12}{r}\right) + r(r+2)\left(\frac{12}{r+2}\right) = 5r(r+2)$$

Divide out common factors:

$$(r+2)12 + r(12) = 5r(r+2)$$

Multiply and distribute:

$$12r + 24 + 12r = 5r^2 + 10r$$

Combine like terms:

$$24r + 24 = 5r^2 + 10r$$

Write the equation in standard form:

$$0 = 5r^2 - 14r - 24$$

Factor the polynomial:

$$0 = (5r + 6)(r - 4)$$

Use the Zero-Product Property:

$$5r + 6 = 0 \quad \text{or} \quad r - 4 = 0$$

Solve each equation:

$$5r = -6 \qquad\qquad r = 4$$

$$r = -\frac{6}{5}$$

Discard the solution $r = -\frac{6}{5}$ because the cycling speed, r, cannot be negative. Thus $r = 4$ miles per hour is the speed cycling to the waterfall. The speed cycling back to the starting point is $r + 2 = 4 + 2 = 6$ mph.

Step 5: Check The time cycling to the waterfall and back is 5 hours. Because $\frac{12 \text{ mi}}{4 \text{ mph}} + \frac{12 \text{ mi}}{6 \text{ mph}} = 3 \text{ hr} + 2 \text{ hr} = 5 \text{ hr}$, the answer is correct.

Step 6: Answer Your average speed cycling from the waterfall is 6 mph.

Quick ✔

14. To prepare for a marathon run, Sue ran 18 miles before she blistered her heel, and then she walked 1 additional mile. Her running speed was 12 times as fast as her walking speed. She was running and walking for 5 hours. Find Sue's running speed.

7.8 Exercises   MyMathLab®

Exercise numbers in **green** have complete video solutions in MyMathLab or may be accessed using the QR code to the right.

*Problems **1–14** are the **Quick ✔** s that follow the **EXAMPLES**.*

Building Skills

In Problems 15–34, solve the proportion. See Objective 1.

15. $\dfrac{9}{x} = \dfrac{3}{4}$

16. $\dfrac{x}{5} = \dfrac{12}{4}$

17. $\dfrac{4}{7} = \dfrac{2x}{9}$

18. $\dfrac{6}{5} = \dfrac{8}{3x}$

19. $\dfrac{6}{5} = \dfrac{x+2}{15}$

20. $\dfrac{5}{9} = \dfrac{2x+3}{6}$

21. $\dfrac{b}{b+6} = \dfrac{4}{9}$

22. $\dfrac{k}{k+3} = \dfrac{6}{15}$

23. $\dfrac{y}{y - 10} = \dfrac{2}{27}$

24. $\dfrac{y}{y - 3} = \dfrac{3}{10}$

25. $\dfrac{p + 2}{4} = \dfrac{2p + 4}{5}$

26. $\dfrac{n + 6}{3} = \dfrac{n + 4}{5}$

27. $\dfrac{2z - 1}{z} = \dfrac{3}{5}$

28. $\dfrac{2k - 3}{k} = \dfrac{4}{3}$

29. $\dfrac{2}{v^2 - v} = \dfrac{1}{3 - v}$

30. $\dfrac{n - 2}{4n + 7} = \dfrac{-2}{n + 1}$

31. $\dfrac{10 - x}{4x} = \dfrac{1}{x - 1}$

32. $\dfrac{4}{x^2} = \dfrac{1}{x + 3}$

33. $\dfrac{2p - 3}{p^2 + 12p + 32} = \dfrac{1}{p + 4}$

34. $\dfrac{1}{z + 3} = \dfrac{2z + 1}{z^2 + 10z + 21}$

In Problems 35–38, △ABC is similar to △XYZ. See Objective 2.

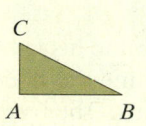 

△ **35.** If $AB = 6$, $XY = 9$, $AC = 8$, find XZ.

△ **36.** If $XY = 8$, $YZ = 7$, $AB = 24$, find BC.

△ **37.** If $XY = n$, $ZY = 2n - 1$, $BC = 5$, and $AB = 3$, find n and ZY.

△ **38.** If $AB = n$, $BC = n + 2$, $XY = 5$, and $YZ = 15$, find n and BC.

In Problems 39 and 40, rectangle ABCD is similar to rectangle EFGH. See Objective 2.

△ **39.** If $AB = x - 2$, $EF = 4$, $BC = 2x + 3$, and $FG = 9$, find x and AB.

△ **40.** If $DC = 6$, $HG = x - 1$, $DA = 8$, and $HE = 2x - 3$, find x and HE.

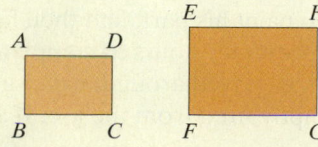

In Problems 41–44, write an algebraic expression that represents each phrase. See Objective 3.

41. If Mariko worked x hours and Natalie worked twice as many hours as Mariko, write an algebraic expression that represents the number of hours Natalie worked.

42. If Brian worked for n days and Kellen worked for half as many days, write an algebraic expression that represents the number of days Kellen worked.

43. If David painted for 3 days less than Pierre and Pierre painted for t days, write an algebraic expression that represents the number of days David painted.

44. If Valerie danced for p minutes and Melody danced for 45 minutes less than Valerie, write an algebraic expression that represents the number of minutes Melody danced.

In Problems 45–48, write an equation that could be used to model each of the following. SET UP BUT DO NOT SOLVE THE EQUATION. See Objective 3.

45. Painting Chairs Christina can paint a chair in 5 hours, and Victoria can paint a chair in 3 hours. How many hours will it take to paint the chair when the two girls work together?

46. Weeding the Yard Bill can weed the backyard in 6 hours. When he worked with Tamra, it took only 2 hours. How long would it take Tamra to weed the backyard if she worked alone?

47. Filling a Tank There are two different inlet pipes that can be used to fill a 3000-gallon tank. Pipe A takes 4 hours longer than Pipe B to fill the tank. With both pipes open, it takes 7 hours to fill the tank. How long would it take Pipe B alone to fill the tank?

48. Canning Peaches A factory that was producing canned peaches had an old canning machine. They decided to add a newer machine that could work twice as fast. With both machines on line, the plant could produce 10,000 cans in 6 hours. How long would it take to produce the same number of cans using only the newer machine?

In Problems 49–52, write an algebraic expression that represents each phrase. See Objective 4.

49. If the rate of the current of a stream is 2 mph and Joe can swim r mph in still water, write an algebraic expression that represents Joe's rate when he swims upstream.

50. If the wind is blowing at 25 mph and a plane flies at r mph in still air, write an algebraic expression that represents the rate the plane is traveling when it flies with the wind.

51. The wind is blowing at 48 mph, and a plane flies at r mph in still air. Write an algebraic expression that represents the rate the plane is traveling when it flies against the wind.

52. If the rate of the current in a stream is 3 mph and Betsy can paddle her canoe at a rate of r mph in still water, write an algebraic expression that represents Betsy's rate when she paddles downstream.

In Problems 53–56, write an equation that could be used to model each of the following. SET UP BUT DO NOT SOLVE THE EQUATION. See Objective 4.

53. Bob's River Trip Bob's boat travels at 14 mph in still water. Find the speed of the current if he can go 4 miles upstream in the same time that it takes to go 7 miles downstream.

54. Marty's River Trip A stream has a current of 3 mph. Find the speed of Marty's boat in still water if she can go 10 miles downstream in the same time it takes to go 4 miles upstream.

55. Assen's River Trip Assen can travel 12 miles downstream in his motorboat in the same time he travels 8 miles upstream. If the current of the river is 3 mph, what is the boat's rate in still water?

56. Airplane Travel A small airplane takes the same amount of time to fly 200 miles against a 30-mph wind as it does to fly 300 miles with a 30-mph wind. Find the speed of the airplane in still air.

Applying the Concepts

57. Yasmine's Map On Yasmine's map, $\frac{1}{4}$ inch represents 10 miles. According to the map, Yellow Springs and Hillsboro are $3\frac{5}{8}$ inches apart. Find the number of miles between the two towns.

58. Beth's Map On Beth's map, 0.5 cm represents 60 miles. If it is 1840 miles from San Diego to New Orleans, how many centimeters is this distance on her map?

59. Making Bread If it takes 5 lb of flour to make 3 loaves of bread, how much flour is needed to make 5 loaves of bread?

60. Buying Candy If 2 lb of candy costs $3.50, how much candy can be purchased for $8.75?

61. Comparing Rubles and Dollars. If $5 U.S. is worth approximately 325.75 Russian rubles, to the nearest whole, how many rubles can Hortencia purchase for $1300 U.S.?

62. Comparing Pesos and Pounds If 50 Mexican pesos are worth approximately 2.1 British pounds and if Sean takes 80 pounds as spending money in Mexico City, to the nearest whole peso, how many pesos will he have?

63. Tree Shadow If a 6-ft man casts a shadow that is 2.5 ft long, how tall is a tree that has a shadow 10 ft long?

64. Telephone Pole Shadow A bush that is 4 m tall casts a shadow that is 1.75 m long. How tall is a telephone pole if its shadow is 8.75 m?

65. Cleaning the Math Building Josh can clean the math building on his campus in 3 hours. Ken takes 5 hours to clean the same building. If they work together, how long will it take them to clean this building?

66. Pruning Trees Martin can prune his fruit trees in 4 hours. His neighbor can prune the same trees for him in 7 hours. If they work together on this job, how long, to the nearest tenth of an hour, will it take to prune the trees?

67. Retrieving Volleyballs After hitting practice for the Long Beach State volleyball team, Dyanne can retrieve all of the balls in the gym in 8 minutes. It takes Makini 6 minutes to retrieve all the balls. If they work together, how long, to the nearest tenth of a minute, will it take these two players to return the volleyballs and be ready to start the next round of hitting practice?

68. Stuffing Envelopes It took Kirsten 6 hours to stuff 3000 envelopes. When she worked with Brittany, it took only 4 hours to stuff the next 3000 envelopes. How long would it take Brittany working alone to stuff the 3000 envelopes?

69. Painting City Hall Three people were given the job of painting city hall. They divided the job into equal portions and decided they would each go home after completing their assigned portion. It took José 9 hours to paint his part, and then he went home. It took Joaquín 12 hours to complete his part. The third person played around, never got started, and was promptly fired from the job. If José and Joaquín went back the second day and worked together to paint the undone portion, how long, to the nearest tenth of an hour, did it take them?

70. Mowing the Grass It takes a mower 18 minutes to cut the grass in the outfield of the stadium. A new mower was purchased, and with both mowers working, the same task takes 8 minutes. If only the new mover is used, how long would it take the new mower to cut the outfield grass?

71. Replacing Pipes It takes an apprentice twice as long as the experienced plumber to replace the pipes under an old house. If it takes them 5 hours when they work together, how long would it take the apprentice alone?

72. Making Care Packages It takes Rocco 3 times as long as Traci to make 100 Red Cross care packages for refugees. If it takes them 10 hours when they work together to make 100 care packages, how long would it take Rocco working alone to make the same number?

73. Filling the Pool Using a single hose, Janet can fill a pool in 6 hours. The same pool can be drained in 8 hours by opening a drainpipe. If Janet forgets to close the drainpipe, how long will it take her to fill the pool?

74. Jill's Bucket Jill has a bucket with a hole in it. When the hole did not exist, Jill could fill the bucket in 30 seconds. With the hole, the bucket now empties in 210 seconds (3.5 minutes). How long will it take Jill to fill the bucket now that it has a hole in it?

75. Travel by Boat A boat can travel 12 km down the river in the same time it can go 4 km up the river. If the current in the river is 2 km per hour, how fast can the boat travel in still water?

76. Travel by Plane A small plane can travel 1000 miles with the wind in the same time it can go 600 miles against the wind. If the speed of the plane in still air is 180 mph, what is the speed of the wind?

77. Iron Man Training While training for an iron man competition, Tony bikes for 60 miles and runs for 15 miles. If his biking speed is 8 times his running speed and it takes 5 hours to complete the training, how long did he spend on his bike?

78. Robin's Workout Robin can run twice as fast as she walks. If she runs for 12 miles and walks for 8 miles, the total time to complete the trip is 5 hours. To the nearest tenth of an hour, how many hours did she run?

79. Driving During a Snowstorm Claire and Chris were driving to their home in Milwaukee when they came upon a snowstorm. They drove at an average rate of 20 miles per hour slower for the last 60 miles during the snowstorm than they drove for the first 90 miles. Find their average rate of speed during the last 60 miles if the trip took 3 hours.

80. Driving During Road Construction David and Lisa drove 25 miles through road construction at a certain average rate. By increasing their speed by 20 miles per hour, they traveled the next 35 miles without road construction in the same time they spent on the 25-mile leg of their trip. Find their average rate during the part of the trip that had road construction.

81. Tough Commute You have a 20-mile commute into work. Since you leave very early, the trip going to work is easier than the trip home. You can travel to work in the same time that it takes you to travel 16 miles on the trip back home. Your average speed coming home is 7 miles per hour slower than your average speed going to work. What is your average speed going to work?

82. A Bike Trip A bicyclist rides his bicycle 12 miles up a hill and then 16 miles on level terrain. His speed on level ground is 3 miles per hour faster than his speed going uphill. The cyclist rides for the same amount of time going uphill and peddling on level ground. Find his speed going uphill.

Extending the Concepts

In geometry, when a line is parallel to one side of a triangle, the intersection of the parallel line with the other two sides of the triangle will form a new triangle that is similar to the original triangle. In the figure below, $\overline{XY}$ is parallel to $\overline{BC}$, so $\triangle XAY$ is similar to $\triangle BAC$.

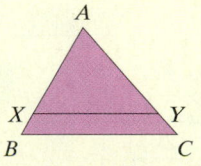 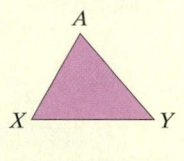 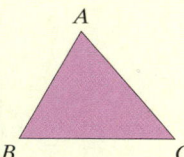

In Problems 83–88, solve for x.

△ **83.** $XA = 4$, $BA = 12$, $CA = 18$, and $YA = x$.

△ **84.** $XA = 4$, $BA = 10$, $CA = 30$, and $YA = x$.

△ **85.** $AX = 9$; $XB = 5$; $XY = 7$; $BC = x$.

△ **86.** $AX = 15$; $XB = 3$; $XY = 12$, $BC = x$.

△ **87.** $YC = 5$; $AY = 12$; $AX = 10$, $XB = x$.

△ **88.** $XB = 9$; $YC = 10$; $AY = 15$, $AX = x$.

Explaining the Concepts

89. A student reads the following sentence from a word problem: *Barbara rows her boat in a stream that has a current of 1 mph. She travels 10 miles upstream and returns to her starting point 5 hours later.* She makes the accompanying chart. Is the information correctly placed for this part of the problem?

d	r	t
10	$r + 1$	5
10	$r - 1$	5

90. When modeling work problems, we assume that individuals work at a constant rate. When modeling uniform motion problems, we assume that individuals travel at a constant rate. Explain what these assumptions mean.

91. You are teaching the class how to solve the following problem: *Marcos can write a budget analysis in 8 hours, and Alonso can complete the same analysis in 6 hours. If they work together, how long will it take to finish the analysis?* Write an equation that could be used to find the answer to the question, and then explain your justification for setting up the equation your way. If you can think of another way to do the problem, explain how to set it up differently, and then explain why this logic will also solve the problem correctly.

92. Explain why the following proportions all yield the same result.

(a) $\dfrac{x}{2} = \dfrac{4}{9}$ **(b)** $\dfrac{2}{x} = \dfrac{9}{4}$

(c) $\dfrac{x}{4} = \dfrac{2}{9}$ **(d)** $\dfrac{9}{2} = \dfrac{4}{x}$

(e) $\dfrac{4}{9} = \dfrac{x}{2}$

Chapter 7 Activity: Correct the Quiz

Focus: Performing operations with rational expressions and solving rational equations

Time: 15–20 minutes

Group size: 2

In this activity you will work as a team to grade the student quiz shown below. If an answer is correct, mark it correct. If an answer is wrong, mark it wrong and show the correct answer.

   Once all of the quiz questions are graded, compute the final score for the quiz. Be prepared to discuss your results with the rest of the class.

Student Quiz

Name: *Ima Student*   Quiz Score:_____

(1) Multiply: $\dfrac{6x - 12}{4x + 8} \cdot \dfrac{3x + 6}{2x - 4}$ Answer: $\dfrac{9}{4}$

(2) Subtract: $\dfrac{xy}{x^2 - y^2} - \dfrac{y}{x + y}$ Answer: $\dfrac{y^2}{x^2 - y^2}$

(3) Solve: $\dfrac{5x}{x + 1} = 2 + \dfrac{2x}{x + 1}$ Answer: $\{2\}$

(4) Divide: $\dfrac{x^2 + x - 6}{5x^2 - 7x - 6} \div \dfrac{3x^2 + 13x + 12}{6x^2 + 17x + 12}$ Answer: $\dfrac{2}{5}$

(5) Solve: Joe can mow his lawn in 2 hr. Mike can mow the same lawn in 3 hr. If they work together, how long will it take them to mow the lawn?   Answer: $\dfrac{6}{5}$ hour or 1 hr, 12 min

(6) Simplify: $\dfrac{\dfrac{1}{x + 5} - \dfrac{2}{x - 7}}{\dfrac{4}{x - 7} + \dfrac{1}{x + 5}}$ Answer: $\dfrac{x + 17}{5x + 13}$

(7) Add: $\dfrac{x}{x^2 + 10x + 25} + \dfrac{4}{x^2 + 6x + 5}$ Answer: $\dfrac{x^2 + 5x + 20}{(x + 5)(x + 1)}$

Chapter 7 Review

Section 7.1   Simplifying Rational Expressions

KEY CONCEPTS

- To find the values that make a rational expression undefined, set the denominator equal to zero, and solve the resulting equation.
- **Simplifying a Rational Expression**
  First, completely factor the numerator and the denominator of the rational expression; then divide out common factors using the fact that if p, q, and r are polynomials, then $\dfrac{p \cdot r}{q \cdot r} = \dfrac{p}{q}$ where $q \neq 0$ and $r \neq 0$.

KEY TERMS

Rational expression
Evaluate
Undefined
Simplify

You Should Be Able To...	EXAMPLE	Review Exercises
1 Evaluate a rational expression (p. 433)	Examples 1 through 3	1–4
2 Determine values for which a rational expression is undefined (p. 435)	Examples 4 through 5	5–10
3 Simplify rational expressions (p. 436)	Examples 6 through 9	11–16

In Problems 1–4, evaluate each expression for the given values.

1. $\dfrac{2x}{x+3}$
  (a) $x = 1$
  (b) $x = -2$
  (c) $x = 3$

2. $\dfrac{3x}{x-4}$
  (a) $x = 0$
  (b) $x = 1$
  (c) $x = -2$

3. $\dfrac{x^2 + 2xy + y^2}{x - y}$
  (a) $x = 1$, $y = -1$
  (b) $x = 2$, $y = 1$
  (c) $x = -3$, $y = -4$

4. $\dfrac{2x^2 - x - 3}{x + z}$
  (a) $x = 1$, $z = 1$
  (b) $x = 1$, $z = 2$
  (c) $x = 5$, $z = -4$

In Problems 5–10, find the value(s) of the variable for which the rational expression is undefined.

5. $\dfrac{3x}{3x-7}$

6. $\dfrac{x+1}{4x-2}$

7. $\dfrac{5}{x^2-25}$

8. $\dfrac{17}{4x^2+49}$

9. $\dfrac{5x+2}{x^2+12x+20}$

10. $\dfrac{x+1}{x^2-3x-4}$

In Problems 11–16, simplify each rational expression, if possible. Assume that no variable has a value that results in a denominator with a value of zero.

11. $\dfrac{5y^2+10y}{25y}$

12. $\dfrac{2x^3-8x^2}{10x}$

13. $\dfrac{3k-21}{k^2-5k-14}$

14. $\dfrac{-2x-2}{x^2-2x-3}$

15. $\dfrac{x^2+8x+15}{2x^2+5x-3}$

16. $\dfrac{3x^2+5x-2}{2x^3+16}$

Section 7.2 Multiplying and Dividing Rational Expressions

KEY CONCEPTS

* **Multiplying Rational Expressions**

  If $\dfrac{a}{b}$ and $\dfrac{c}{d}$, where $b \neq 0$ and $d \neq 0$, are two rational expressions, then $\dfrac{a}{b} \cdot \dfrac{c}{d} = \dfrac{ac}{bd}$.

* **Dividing Rational Expressions**

  If $\dfrac{a}{b}$ and $\dfrac{c}{d}$, where $b \neq 0$, $c \neq 0$, and $d \neq 0$, are two rational expressions, then $\dfrac{\frac{a}{b}}{\frac{c}{d}} = \dfrac{a}{b} \div \dfrac{c}{d} = \dfrac{a}{b} \cdot \dfrac{d}{c} = \dfrac{ad}{bc}$.

You Should Be Able To...	EXAMPLE	Review Exercises
1 Multiply rational expressions (p. 441)	Examples 1 through 4	17, 18, 21, 22, 25, 26
2 Divide rational expressions (p. 443)	Examples 5 through 9	19, 20, 23, 24, 27, 28

In Problems 17–28, perform the indicated operations.

17. $\dfrac{12m^4n^3}{7} \cdot \dfrac{21}{18m^2n^5}$

18. $\dfrac{3x^2y^4}{4} \cdot \dfrac{2}{9x^3y}$

19. $\dfrac{10m^2n^4}{9m^3n} \div \dfrac{15mn^6}{21m^2n}$

20. $\dfrac{5ab^3}{3b^4} \div \dfrac{10a^2b^8}{6b^2}$

21. $\dfrac{5x - 15}{x^2 - x - 12} \cdot \dfrac{x^2 - 6x + 8}{3 - x}$

22. $\dfrac{4x - 24}{x^2 - 18x + 81} \cdot \dfrac{x^2 - 9}{6 - x}$

23. $\dfrac{\frac{x^2 - 4}{x^2 - 8x + 15}}{\frac{12x + 24}{3x - 15}}$

24. $\dfrac{\frac{5x^3 + 10x^2}{3x}}{\frac{2x + 4}{18x^2}}$

25. $\dfrac{3x^2 + 14x - 5}{x^2 + x - 30} \cdot \dfrac{x^2 - 2x - 15}{3x^2 + 8x - 3}$

26. $\dfrac{y^2 - 5y - 14}{y^2 - 2y - 35} \cdot \dfrac{y^2 + 6y + 5}{y^2 - y - 6}$

27. $\dfrac{x^2 - 9x}{x^2 + 3x + 2} \div \dfrac{x^2 - 81}{x^2 + 2x}$

28. $\dfrac{y^2 - 9}{2y^2 - y - 15} \div \dfrac{3y^2 + 10y + 3}{2y^2 + y - 10}$

Section 7.3 Adding and Subtracting Rational Expressions with a Common Denominator

KEY CONCEPT

* **Adding/Subtracting Rational Expressions**

  If $\dfrac{a}{c}$ and $\dfrac{b}{c}$, where $c \neq 0$ are two rational expressions, then $\dfrac{a}{c} + \dfrac{b}{c} = \dfrac{a + b}{c}$ and $\dfrac{a}{c} - \dfrac{b}{c} = \dfrac{a - b}{c}$. Simplify the sum or difference by dividing out common factors.

You Should Be Able To...	EXAMPLE	Review Exercises
1 Add rational expressions with a common denominator (p. 449)	Examples 1 and 2	29–34
2 Subtract rational expressions with a common denominator (p. 451)	Examples 3 and 4	35–38
3 Add or subtract rational expressions with opposite denominators (p. 452)	Examples 5 and 6	39–42

In Problems 29–42, perform the indicated operation.

29. $\dfrac{4}{x-3} + \dfrac{5}{x-3}$

30. $\dfrac{3}{x+4} + \dfrac{7}{x+4}$

31. $\dfrac{m^2}{m+3} + \dfrac{3m}{m+3}$

32. $\dfrac{1}{6m} + \dfrac{5}{6m}$

33. $\dfrac{-m+1}{m^2-4} + \dfrac{2m+1}{m^2-4}$

34. $\dfrac{2m^2}{m+1} + \dfrac{5m+3}{m+1}$

35. $\dfrac{11}{15m} - \dfrac{6}{15m}$

36. $\dfrac{15b}{2b^2} - \dfrac{11b}{2b^2}$

37. $\dfrac{2y^2}{y-7} - \dfrac{y^2+49}{y-7}$

38. $\dfrac{6m+5}{m^2-36} - \dfrac{5m-1}{m^2-36}$

39. $\dfrac{3}{x-4} + \dfrac{10}{4-x}$

40. $\dfrac{7}{a-b} + \dfrac{3}{b-a}$

41. $\dfrac{2x}{x^2-25} - \dfrac{x-5}{25-x^2}$

42. $\dfrac{x+5}{2x-6} - \dfrac{x+3}{6-2x}$

Section 7.4 Finding the Least Common Denominator and Forming Equivalent Rational Expressions

KEY CONCEPTS

- The steps to find the LCD of rational expressions are given on page 457.
- The steps to form equivalent rational expressions are given on page 460.

KEY TERMS

Least common denominator (LCD)
Equivalent rational expressions

You Should Be Able To...	EXAMPLE	Review Exercises
1 Find the least common denominator of two or more rational expressions (p. 456)	Examples 1 through 6	43–48
2 Write a rational expression equivalent to a given rational expression (p. 459)	Examples 7 and 8	49–52
3 Use the LCD to write equivalent rational expressions (p. 460)	Example 9	53–58

In Problems 43–48, identify the LCD of the given rational expressions.

43. $\dfrac{6}{4x^2y^7}; \dfrac{8}{6x^4y}$

44. $\dfrac{3}{20a^3bc^4}; \dfrac{7}{30ab^4c^7}$

45. $\dfrac{11}{4a}; \dfrac{7a}{8a+16}$

46. $\dfrac{6}{5a}; \dfrac{17}{a^2+6a}$

47. $\dfrac{x+1}{4x-12}; \dfrac{3x}{x^2-2x-3}$

48. $\dfrac{11}{x^2-7x}; \dfrac{x+2}{x^2-49}$

In Problems 49–52, write an equivalent rational expression with the given denominator.

49. $\dfrac{6}{x^3y}$ with denominator x^4y^7

50. $\dfrac{11}{a^3b^2}$ with denominator a^7b^5

51. $\dfrac{x-1}{x-2}$ with denominator x^2-4

52. $\dfrac{m+2}{m+7}$ with denominator $m^2+5m-14$

In Problems 53–58, identify the LCD and then write each rational expression as an equivalent rational expression with that denominator.

53. $\dfrac{5y}{6x^3}; \dfrac{7}{8x^5}$

54. $\dfrac{6}{5a^3}; \dfrac{11b}{10a}$

55. $\dfrac{4x}{x-2}; \dfrac{6}{4-x^2}$

56. $\dfrac{3}{m-5}; \dfrac{-2m}{15+2m-m^2}$

57. $\dfrac{2}{m^2+5m-14}; \dfrac{m+1}{m^2+9m+14}$

58. $\dfrac{n-2}{n^2-5n}; \dfrac{n}{n^2-25}$

Section 7.5   Adding and Subtracting Rational Expressions with Unlike Denominators

KEY CONCEPT

- The steps for adding or subtracting rational expressions with unlike denominators are given on page 464.

You Should Be Able To...	EXAMPLE	Review Exercises
❶ Add and subtract rational expressions with unlike denominators (p. 463)	Examples 2 through 10	59–72

In Problems 59–72, perform the indicated operations.

59. $\dfrac{4}{3x^2} + \dfrac{8}{9x}$

60. $\dfrac{x}{2x^3y} + \dfrac{y}{10xy^3}$

61. $\dfrac{x}{x+7} + \dfrac{2}{x-7}$

62. $\dfrac{4}{2x+3} + \dfrac{x+1}{2x-3}$

63. $\dfrac{x+5}{x} - \dfrac{x+7}{x-2}$

64. $\dfrac{p+1}{p+3} - \dfrac{p+17}{p^2-p-12}$

65. $\dfrac{3x-4}{4x+1} + \dfrac{3x+6}{4x^2+9x+2}$

66. $\dfrac{7}{3x^2+x-4} + \dfrac{9x+2}{3x^2-2x-8}$

67. $\dfrac{m}{m-3} - \dfrac{4m+12}{m^2-9}$

68. $\dfrac{3}{y^2+7y+10} - \dfrac{4}{y^2+6y+5}$

69. $\dfrac{3}{m-2} + \dfrac{12}{4-m^2}$

70. $\dfrac{m}{m-n} - \dfrac{n}{m+n}$

71. $4 + \dfrac{x}{x+3}$

72. $7 - \dfrac{1}{x+2}$

Section 7.6   Complex Rational Expressions

KEY CONCEPT

- There are two methods that can be used to simplify a complex rational expression. The steps for Method I are presented on page 475, and the steps for Method II are presented on page 477.

KEY TERMS

Complex rational expression
Simplify

You Should Be Able To...	EXAMPLE	Review Exercises
❶ Simplify a complex rational expression by simplifying the numerator and denominator separately (Method I) (p. 474)	Examples 2 through 4	73–80
❷ Simplify a complex rational expression using the least common denominator (Method II) (p. 477)	Examples 5 and 6	73–80

In Problems 73–80, simplify the complex rational expressions.

73. $\dfrac{\dfrac{1}{2} - \dfrac{2}{3}}{\dfrac{4}{9} + \dfrac{5}{6}}$

74. $\dfrac{\dfrac{1}{4} + \dfrac{1}{2}}{\dfrac{5}{8} - \dfrac{1}{6}}$

75. $\dfrac{\dfrac{1}{5} - \dfrac{1}{m}}{\dfrac{1}{10} + \dfrac{1}{m^2}}$

76. $\dfrac{\dfrac{1}{m^2} + \dfrac{2}{3}}{\dfrac{1}{m} - \dfrac{5}{6}}$

77. $\dfrac{\dfrac{x}{4} - \dfrac{1}{2}}{\dfrac{3x}{2} - 3}$

78. $\dfrac{\dfrac{8}{y+4} + 2}{\dfrac{12}{y+4} - 2}$

79. $\dfrac{\dfrac{m-4}{m+4} + \dfrac{m-4}{m-2}}{1 - \dfrac{2}{m-2}}$

80. $\dfrac{\dfrac{a^2}{a^2-36} - \dfrac{a}{a+6}}{\dfrac{a}{a^2-36} - \dfrac{1}{a-6}}$

Section 7.7   Rational Equations

KEY CONCEPTS

- The steps for solving a rational equation are given on page 483.
- To solve for a variable in a rational equation, get the variable by itself on one side of the equation with all other variable and constant terms, if any, on the other side.

KEY TERMS

Rational equation
Extraneous solution

You Should Be Able To...	EXAMPLE	Review Exercises
❶ Solve equations containing rational expressions (p. 481)	Examples 1 through 7	83–92
❷ Solve for a variable in a rational equation (p. 488)	Examples 8 and 9	93–96

In Problems 81 and 82, state the values of the variable for which the expressions in the rational equation are undefined.

81. $\dfrac{4}{x + 5} + \dfrac{1}{x} = 3$

82. $\dfrac{2}{x^2 - 36} = \dfrac{1}{x + 6}$

87. $\dfrac{m + 4}{m - 3} = \dfrac{m + 10}{m + 2}$

88. $\dfrac{m - 6}{m + 5} = \dfrac{m - 3}{m + 1}$

89. $\dfrac{2x}{x - 1} - 5 = \dfrac{2}{x - 1}$

In Problems 83–92, solve each equation and state the solution set. Be sure to check for extraneous solutions.

83. $\dfrac{2}{x} - \dfrac{3}{4} = \dfrac{5}{x}$

84. $\dfrac{4}{x} + \dfrac{3}{4} = \dfrac{2}{3x} + \dfrac{23}{4}$

85. $\dfrac{4}{m} - \dfrac{3}{2m} = \dfrac{1}{2}$

86. $\dfrac{2}{m} + \dfrac{5}{m + 2} = -3$

90. $\dfrac{1}{x + 3} + \dfrac{1}{x - 3} = \dfrac{-5}{x^2 - 9}$

91. $\dfrac{2x}{x - 2} - 1 = \dfrac{8x - 24}{x^2 - 5x + 6}$

92. $\dfrac{3}{5 - x} - \dfrac{11}{x^2 - 2x - 15} = \dfrac{4}{x + 3}$

In Problems 93–96, solve the equation for the indicated variable.

93. $y = \dfrac{4}{k}$ for k

94. $6 = \dfrac{x}{y}$ for y

95. $\dfrac{1}{x} + \dfrac{1}{y} = \dfrac{1}{z}$ for y

96. $\dfrac{1}{x} + \dfrac{1}{y} = \dfrac{1}{z}$ for z

Section 7.8   Models Involving Rational Equations

KEY CONCEPT

- **Means-Extremes Theorem**
  If $\dfrac{a}{b} = \dfrac{c}{d}$, where $b \neq 0$ and $d \neq 0$, then $ad = bc$.

KEY TERMS

Ratio
Proportion
Terms
Means

Extremes
Cross multiplication
Similar figures
Constant rate

You Should Be Able To...	EXAMPLE	Review Exercises
❶ Model and solve ratio and proportion problems (p. 493)	Examples 1 through 3	97–100, 103, 104
❷ Model and solve problems with similar figures (p. 496)	Examples 4 and 5	101, 102
❸ Model and solve work problems (p. 497)	Examples 6 and 7	105–108
❹ Model and solve uniform motion problems (p. 500)	Examples 8 and 9	109–112

In Problems 97–100, solve the proportion.

97. $\dfrac{6}{4y + 5} = \dfrac{2}{7}$

98. $\dfrac{2}{y - 3} = \dfrac{5}{y}$

99. $\dfrac{y + 1}{8} = \dfrac{1}{4}$

100. $\dfrac{6y + 7}{10} = \dfrac{2y + 9}{6}$

In Problems 101 and 102, use the similar triangles to solve for x.

101.
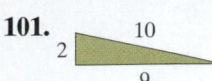 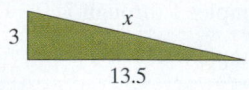

102.
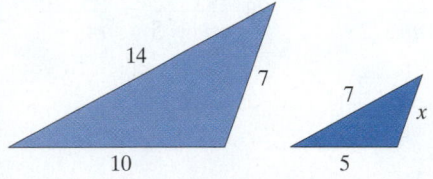

103. Cement A cement mixer uses 4 tanks of water to mix 15 bags of cement. How many tanks of water are needed to mix 30 bags of cement?

104. Pizza If 4 small pizzas cost $15.00, find the cost of 7 small pizzas.

105. Washing Lucille can wash the walls in 3 hours working alone, and Teresa can wash the walls in 2 hours. To the nearest tenth of an hour, how long would it take them to wash the walls working together?

106. Carpet Fred can install the carpet in a room in 3 hours, but Barney needs 5 hours to install the carpet. To the nearest tenth of an hour, how long would it take them to complete the carpet installation if they worked together?

107. Dishes Working alone, it takes Jake 5 minutes longer to wash the dishes than it takes Adrienne when she washes the dishes alone. Washing the dishes together, Jake and Adrienne can finish the job in 6 minutes. How long does it take Jake to wash the dishes by himself?

108. Painting Working together, Donovan and Ben can paint a room in 4 hours. Working alone, it takes Donovan 7 hours to paint the room. To the nearest tenth of an hour, how long would it take Ben to paint the room working alone?

109. Paddling Paul can paddle a kayak 15 miles upstream in the same amount of time it takes him to paddle a kayak 27 miles downstream. If the current is 2 mph, what is Paul's speed as he paddles downstream?

110. Cruising A cruise ship traveled for 145 miles with the current in the same amount of time it traveled 105 miles against the current. The speed of the current was 4 mph. What was the speed of the cruise ship as it traveled with the current?

111. Vacation On their vacation, a family traveled 135 miles by train and then traveled 855 miles by plane. The speed of the plane was three times the speed of the train. If the total time of the trip was 6 hours, what was the speed of the train?

112. Driving Tamika drove for 90 miles in the city. When she got on the highway, she increased her speed by 20 mph and drove for 130 miles. If Tamika drove a total of 4 hours, how fast did she drive in the city?

Chapter 7 Test

Step-by-step test solutions are found on the Chapter Test Prep Videos available in MyMathLab®, on YouTube™, or may be accessed using the QR code to the right.

1. Evaluate $\dfrac{3x - 2y^2}{6z}$ when $x = 2$, $y = -3$, and $z = -1$.

2. Find the values for which the following rational expression is undefined: $\dfrac{x + 5}{x^2 - 3x - 10}$

3. Simplify: $\dfrac{x^2 - 4x - 21}{14 - 2x}$

In Problems 4–11, perform the indicated operation.

4. $\dfrac{\dfrac{35x^6}{9x^4}}{\dfrac{25x^5}{18x}}$

5. $\dfrac{5x - 15}{3x + 9} \cdot \dfrac{5x + 15}{3x - 9}$

6. $\dfrac{\dfrac{2x^2 - 5xy - 12y^2}{x^2 + xy - 20y^2}}{\dfrac{4x^2 - 9y^2}{x^2 + 4xy - 5y^2}}$

7. $\dfrac{y^2}{y + 3} + \dfrac{3y}{y + 3}$

8. $\dfrac{x^2}{x^2 - 9} - \dfrac{8x - 15}{x^2 - 9}$

9. $\dfrac{6}{y - z} + \dfrac{7}{z - y}$

10. $\dfrac{x}{x - 2} + \dfrac{3}{2x + 1}$

11. $\dfrac{2x}{x^2 + 5x + 6} - \dfrac{x + 1}{x^2 + 2x - 3}$

12. Simplify the following complex rational expression:

$$\dfrac{\dfrac{1}{9} - \dfrac{1}{y^2}}{\dfrac{1}{3} + \dfrac{1}{y}}$$

In Problems 13 and 14, solve the rational equations. Check for extraneous solutions.

13. $\dfrac{m}{5} + \dfrac{5}{m} = \dfrac{m + 3}{4}$

14. $\dfrac{4}{x + 3} + \dfrac{5}{x - 6} = \dfrac{4x + 1}{x^2 - 3x - 18}$

15. Solve $\dfrac{1}{x} + \dfrac{1}{y} = \dfrac{1}{z}$ for y.

16. Solve the proportion: $\dfrac{2}{y + 1} = \dfrac{1}{y - 2}$

17. Use the similar triangles to solve for x.

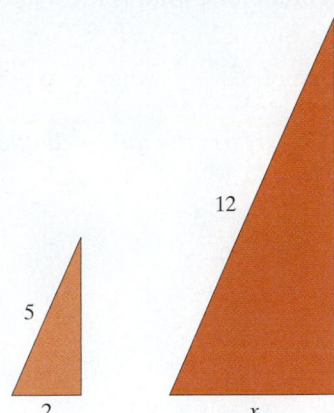

18. Barrettes If 8 hair barrettes cost \$22.00, how much would 12 hair barrettes cost?

19. Car Washing It takes Frank 18 more minutes than Juan to wash a car. If they can wash the car together in 12 minutes, how long does it take Frank to wash the car by himself?

20. Excursion On a vacation excursion, some tourists walked 7 miles on a nature path and then hiked 12 miles up a mountainside. The tourists walked 3 mph faster than they hiked. The total time of the excursion was 4 hours. At what rate did the tourists hike?

Cumulative Review Chapters 1–7

In Problems 1 and 2, simplify each expression.

1. $-6^2 + 4(-5 + 2)^3$

2. $3(4x - 2) - (3x + 5)$

In Problems 3–5, solve each equation.

3. $-3(x - 5) + 2x = 5x - 4$

4. $3(2x - 1) + 5 = 6x + 2$

5. $0.25x + 0.10(x - 3) = 0.05(22)$

6. Movie Poster In the entranceway of the Sawgrass Cinema, there is a large rectangular poster advertising a movie. If the poster's length is 3 feet less than twice its width, and its perimeter is 24 feet, what is the length of the poster?

7. Integers The sum of three consecutive even integers is 138. Find the integers.

8. Solve and graph the following inequality:

$$2(x - 3) - 5 \le 3(x + 2) - 18$$

9. Multiply: $(3x - 2y)^2$

In Problems 10 and 11, simplify each expression. Write your answers with positive exponents only.

10. $\left(\dfrac{2a^5 b}{4ab^{-2}}\right)^{-4}$

11. $(3x^0 y^{-4} z^3)^3$

12. Factor completely: $8a^2 b + 34ab - 84b$

13. Solve the equation using the Zero-Product Property:

$$6x^3 - 31x^2 = -5x$$

14. Rectangle The length of a rectangle is 2 cm longer than twice its width. If the area of the rectangle is 40 square centimeters, what is the length of the rectangle?

15. Find the value of $\dfrac{x + 5}{x^2 + 25}$ when $x = -5$.

In Problems 16–19, perform the indicated operation and simplify, if possible.

16. $\dfrac{3x + 3}{5x - 5x^2} \cdot \dfrac{2x^2 + x - 3}{4x^2 - 9}$

17. $\dfrac{x^2 - x - 2}{10} \div \dfrac{2x + 4}{5}$

18. $\dfrac{2x + 3}{x^2 - x - 30} - \dfrac{x - 2}{x^2 - x - 30}$

19. $\dfrac{15}{2x - 4} + \dfrac{x}{x^2 - 4}$

20. Simplify the following complex rational expression:

$$\dfrac{\dfrac{2}{x^2} - \dfrac{3}{5x}}{\dfrac{4}{x} + \dfrac{1}{4x}}$$

In Problems 21 and 22, solve each equation.

21. $\dfrac{3}{x - 4} = \dfrac{5x + 4}{x^2 - 16} - \dfrac{4}{x + 4}$

22. $\dfrac{x - 5}{3} = \dfrac{x + 2}{2}$

23. Map On a city map, 4 inches represents 50 miles. How many inches would represent 125 miles?

24. Inventory It takes Trent 9 hours longer than Sharona to do a store's inventory. If they can finish the inventory in 6 hours working together, how long does it take Sharona to do the inventory working alone?

25. Racing Francisco ran for 35 miles and then walked for 6 miles. Francisco runs 4 mph faster than he walks. If it took him 7 hours to finish the race, how fast was he running?

26. Find the slope of the line joining the points $(-1, 3)$ and $(5, 11)$ in the Cartesian plane.

27. Find the equation of the line that has slope $-\dfrac{3}{2}$ and contains the point $(4, -1)$.

28. Graph the line $2x - 3y = -12$ by finding the intercepts of the graph.

29. Find the slope of the line parallel to $3x + 5y = 15$. What is the slope of the line perpendicular to $3x + 5y = 15$?

30. Solve: $\begin{cases} 2x + 3y = 1 \\ -3x + 2y = 18 \end{cases}$

Getting Ready for Intermediate Algebra:
A Review of Chapters 1–7

The following problems cover important concepts from Chapters 1 –7. We designed these problems so that you can review important concepts that will be needed for the rest of the course. Take time to work these problems before proceeding to the next chapter. The answers to these problems are located at the back of the text on page AN-28. If you get any problems wrong, go back to the -section cited and review the material.

1. Evaluate: $\dfrac{4 - (-6)}{8 - 2}$

Section 1.7, p. 59

2. Evaluate $-\dfrac{2}{3}x + 5$ for $x = 6$.

Section 1.8, pp. 64–65

In Problems 3–6, solve each equation. State whether the equation is a contradiction, an identity, or a conditional equation.

3. $4x + 3 = 17$

Section 2.2, pp. 91–104

4. $5(x - 2) - 2x = 3x + 4$

Section 2.3, pp. 102–104

5. $\dfrac{1}{2}(x - 4) + \dfrac{2}{3}x = \dfrac{1}{6}(x - 4)$

Section 2.3, pp. 98–104

6. $0.3(x + 1) - 0.1(x - 7) = 0.4x - 0.2(x - 5)$

Section 2.3, pp. 101–104

In Problems 7 and 8, solve each linear inequality. Express the solution using set-builder notation and interval notation. Graph the solution set.

7. $6x - 7 > -31$

Section 2.8, pp. 150–154

8. $5(x - 3) \geq 7(x - 4) + 3$

Section 2.8, pp. 150–156

9. Plot the following ordered pairs in the same Cartesian plane.

Section 3.1, pp. 169–173

$A(-3, 0), B(4, -2), C(1, 5), D(0, 3), E(-4, -5), F(-5, 2)$

10. Graph $-4x + 3y = 24$ by finding its intercepts.

Section 3.2, pp. 187–190

11. Graph: $x = 5$

Section 3.2, pp. 190–191

12. Find the slope of the line joining $(3, 6)$ and $(-1, -4)$.

Section 3.3, pp. 195–198

13. Graph: $y = 3x + 1$

Section 3.4, pp. 206–208

14. Find the equation of the line with slope $-\dfrac{4}{3}$ through $(-3, 1)$. Express your answer in slope-intercept form.

Section 3.5, pp. 214–216

15. Find an equation of the line through $(-2, 5)$ and $(2, 3)$. Express your answer in slope-intercept form.

Section 3.5, pp. 216–218

16. Find the equation of the line parallel to $y = -3x + 10$ through the point $(-5, 7)$. Express your answer in slope-intercept form.

Section 3.6, pp. 223–224

17. Graph: $x - 3y > 12$

Section 3.7, pp. 232–236

18. Solve: $\begin{cases} y = 4x - 3 \\ 4x - 3y = 5 \end{cases}$

Section 4.2, pp. 260–264

19. Solve: $\begin{cases} x + y = 3 \\ 3x + 2y = 2 \end{cases}$

Section 4.3, pp. 268–272

20. Graph the system: $\begin{cases} x + y \geq 2 \\ -3x + y \leq 10 \end{cases}$

Section 4.6, pp. 295–297

In Problems 21–23, perform the indicated operation.

21. $(12x^3 + 5x^2 - 3x + 1) - (2x^3 - 4x + 8)$

Section 5.1, pp. 311–314

22. $(2x - 5)(x + 3)$

Section 5.3, pp. 324–327

23. $\dfrac{x^3 - 2x^2 - 5x - 3}{x - 4}$

Section 5.5, pp. 348–351

In Problems 24–28, factor each polynomial completely.

24. $-2x^3 + 6x^2 - 8x + 24$

Section 6.1, pp. 374–376

25. $5p^3 + 50p^2 + 80p$

Section 6.2, pp. 383–384

26. $8y^2 - 2y - 3$

Section 6.3, pp. 386–394

27. $9a^2 + 24a + 16$

Section 6.4, pp. 396–398

28. $3k^4 - 27k^2$

Section 6.4, pp. 398–400

29. Solve: $8b^2 + 10b = 3$

Section 6.6, pp. 410–415

30. Solve: $(2x + 3)(x - 1) = 6x$

Section 6.6, pp. 410–415

31. Simplify: $\dfrac{2x^2 + x - 21}{x^2 + 6x - 27}$

Section 7.1, pp. 436–439

32. Simplify: $\dfrac{3x^2 + 14x - 5}{x^2 + x - 30} \cdot \dfrac{x^2 - 2x - 15}{3x^2 + 8x - 3}$

Section 7.2, pp. 441–443

33. Simplify: $\dfrac{\dfrac{y^2 - 9}{2y^2 - y - 15}}{\dfrac{3y^2 + 10y + 3}{2y^2 + y - 10}}$

Section 7.2, pp. 443–446

34. Subtract: $\dfrac{2x}{x - 3} - \dfrac{x + 1}{x + 2}$

Section 7.5, pp. 463–467

35. Solve: $\dfrac{9}{k - 2} = \dfrac{6}{k} + 3$

Section 7.7, pp. 481–488

36. Solve: $\dfrac{7}{y^2 + y - 12} - \dfrac{4y}{y^2 + 7y + 12} = \dfrac{6}{y^2 - 9}$

Section 7.7, pp. 481–488

Getting Ready for Chapter 8: Interval Notation

Objective

1 Represent Inequalities Using the Real Number Line and Interval Notation

1 Represent Inequalities Using the Real Number Line and Interval Notation

Suppose that a and b are two real numbers and $a < b$. The notation

$$a < x < b$$

means that x is a number *between* a and b. So the expression $a < x < b$ is equivalent to the two inequalities $a < x$ and $x < b$. Similarly, the expression $a \leq x \leq b$ is equivalent to the two inequalities $a \leq x$ and $x \leq b$. In addition, $a \leq x < b$ and $a < x \leq b$ are defined similarly. Expressions such as $-2 < x < 5$ and $x \geq 5$ are in **inequality notation.**

Work Smart

Remember that the inequalities $<$ and $>$ are called strict inequalities, while $\leq$ and $\geq$ are called nonstrict inequalities.

Even though the expression $3 \geq x \geq 2$ is technically correct, we prefer that the numbers in an inequality go from smaller values to larger values. So write $3 \geq x \geq 2$ as $2 \leq x \leq 3$.

A statement such as $3 \leq x \leq 1$ is false because there is no number x for which $3 \leq x$ and $x \leq 1$. Also never mix inequalities directions, as in $2 \leq x \geq 3$.

Now let's see how to use **interval notation** to represent the solution set of an inequality.

Definitions

Let a and b represent two real numbers with $a < b$.

A **closed interval,** denoted by $[a, b]$, consists of all real numbers x for which $a \leq x \leq b$.

An **open interval,** denoted by (a, b), consists of all real numbers x for which $a < x < b$.

The **half-open, or half-closed,** intervals are $(a, b]$, consisting of all real numbers x for which $a < x \leq b$, and $[a, b)$, consisting of all real numbers x for which $a \leq x < b$.

In each of these definitions, a is the **left endpoint** and b is the **right endpoint** of the interval.

The symbol ∞ (which is read "infinity") is not a real number but a notational device that indicates unboundedness in the positive direction. In other words, the symbol ∞ means the inequality has no right endpoint. The symbol $-\infty$ (which is read "minus infinity" or "negative infinity") is also not a number but a notational device that indicates unboundedness in the negative direction. The symbol $-\infty$ means the inequality has no left endpoint. The symbols ∞ and $-\infty$ allow us to define five other kinds of intervals.

Work Smart

The symbols ∞ and $-\infty$ are never included as endpoints because they are not real numbers. Therefore, we use parentheses when $-\infty$ or ∞ is an endpoint.

Interval Notation

$[a, \infty)$	consists of all real numbers x where $x \geq a$
(a, ∞)	consists of all real numbers x where $x > a$
$(-\infty, a]$	consists of all real numbers x where $x \leq a$
$(-\infty, a)$	consists of all real numbers x where $x < a$
$(-\infty, \infty)$	consists of all real numbers x (that is, $-\infty < x < \infty$)

Another way to represent inequalities is to draw a graph on the real number line. The inequality $x > 3$ or the interval $(3, \infty)$ consists of all numbers x that lie to the right of 3 on the real number line. These values are graphed by shading the real number line to the right of 3. Use a *parenthesis* on the endpoint to indicate that 3 is not included in the set. See Figure 1.

Figure 1

$x > 3$

To graph the inequality $x \geq 3$ or the interval $[3, \infty)$, shade to the right of 3 but use a *bracket* on the endpoint to indicate that 3 is included in the set. See Figure 2.

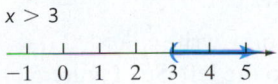

Figure 2

$x \geq 3$

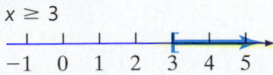

Table 1 summarizes interval notation, inequality notation, and their graphs.

Table 1

Interval Notation	Inequality Notation	Graph
The open interval (a, b)	$\{x \mid a < x < b\}$	
The closed interval $[a, b]$	$\{x \mid a \leq x \leq b\}$	
The half-open interval $[a, b)$	$\{x \mid a \leq x < b\}$	
The half-open interval $(a, b]$	$\{x \mid a < x \leq b\}$	
The interval $[a, \infty)$	$\{x \mid x \geq a\}$	
The interval (a, ∞)	$\{x \mid x > a\}$	
The interval $(-\infty, a]$	$\{x \mid x \leq a\}$	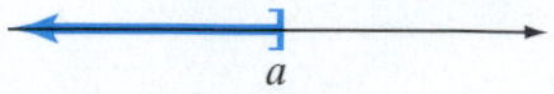
The interval $(-\infty, a)$	$\{x \mid x < a\}$	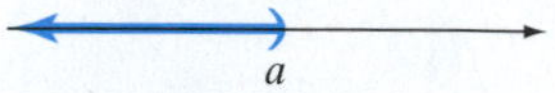
The interval $(-\infty, \infty)$	$\{x \mid x \text{ is a real number}\}$	

EXAMPLE 1

Using Interval Notation and Graphing Inequalities

Write each inequality using interval notation. Graph the inequality.

(a) $-2 \leq x \leq 4$ (b) $1 < x \leq 5$

Solution

(a) $-2 \leq x \leq 4$ describes all numbers x between -2 and 4, inclusive. In interval notation, write $[-2, 4]$. To graph $-2 \leq x \leq 4$, place brackets at -2 and 4 and shade in between. See Figure 3.

Figure 3

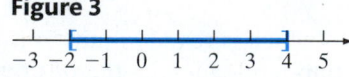

(b) $1 < x \leq 5$ describes all numbers x greater than 1 and less than or equal to 5. In interval notation, write $(1, 5]$. To graph $1 < x \leq 5$, place a parenthesis at 1 and a bracket at 5 and shade in between. See Figure 4.

Figure 4

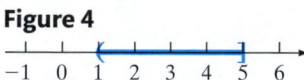

EXAMPLE 2

Using Interval Notation and Graphing Inequalities

Write each inequality using interval notation. Graph the inequality.

(a) $x < 2$ (b) $x \geq -3$

Solution

(a) $x < 2$ describes all numbers x less than 2. In interval notation, write $(-\infty, 2)$. To graph $x < 2$, place a parenthesis at 2 and then shade to the left. See Figure 5.

Figure 5

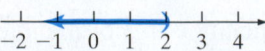

(b) $x \geq -3$ describes all numbers x greater than or equal to -3. In interval notation, write $[-3, \infty)$. To graph $x \geq -3$, place a bracket at -3 and then shade to the right. See Figure 6.

Figure 6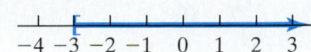

Quick ✔

1. A(n) _____ _____, denoted $[a, b]$, consists of all real numbers x for which $a \leq x \leq b$.

2. In the interval (a, b), a is called the _____ _____ and b is called the _____ _____ of the interval.

In Problems 3–6, write each inequality in interval notation. Graph the inequality.

3. $-3 \leq x \leq 2$

4. $3 \leq x < 6$

5. $x \leq 3$

6. $\dfrac{1}{2} < x < \dfrac{7}{2}$

EXAMPLE 3

Using Inequality Notation and Graphing Inequalities

Write each interval in inequality notation involving x. Graph the inequality.

(a) $[-2, 4)$

(b) $(1, 5)$

Solution

(a) The interval $[-2, 4)$ consists of all numbers x for which $-2 \leq x < 4$. See Figure 7 for the graph.

(b) The interval $(1, 5)$ consists of all numbers x for which $1 < x < 5$. See Figure 8 for the graph.

Figure 7

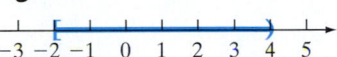

Figure 8

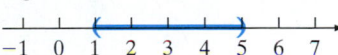

EXAMPLE 4

Using Inequality Notation and Graphing Inequalities

Write each interval in inequality notation involving x. Graph the inequality.

(a) $\left[\dfrac{3}{2}, \infty\right)$

(b) $(-\infty, 1)$

Solution

(a) The interval $\left[\dfrac{3}{2}, \infty\right)$ consists of all numbers x for which $x \geq \dfrac{3}{2}$. See Figure 9 for the graph.

(b) The interval $(-\infty, 1)$ consists of all numbers x for which $x < 1$. See Figure 10 for the graph.

Figure 9

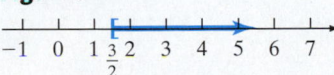

Figure 10

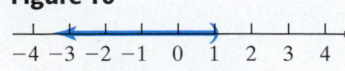

Quick ✔

In Problems 7–10, write each interval as an inequality. Graph the inequality.

7. $(0, 5]$

8. $(-6, 0)$

9. $(5, \infty)$

10. $\left(-\infty, \dfrac{8}{3}\right]$

Exercises   MyMathLab®

*Problems **1–10** are the Quick ✔s that follow the EXAMPLES.*

Building Skills

In Problems 11–18, write each inequality using interval notation. Graph the inequality. See Objective 1.

11. $2 \le x \le 10$

12. $1 < x < 7$

13. $-4 \le x < 0$

14. $-8 < x \le 1$

15. $x \ge 6$

16. $x < 0$

17. $x < \dfrac{3}{2}$

18. $x \ge -\dfrac{5}{2}$

In Problems 19–26, write each interval as an inequality involving x. Graph each inequality. See Objective 1.

19. $(1, 8)$

20. $[-2, 3]$

21. $(-5, 1]$

22. $[1, 4)$

23. $(-\infty, 5)$

24. $(2, \infty)$

25. $[3, \infty)$

26. $(-\infty, 8]$

8 Graphs, Relations, and Functions

Your cell phone bill depends on the cost of the data plan you choose. The revenue a Brownie troop earns from selling Girl Scout cookies depends on the number of boxes of cookies they sell. The area of a circle depends on the length of its radius. These three relations are examples of *functions*. Did you know that your weekly earnings are a function of the number of hours you work? See Problem 87 in Section 8.3.

The Big Picture: Putting It Together

In Chapter 3, linear equations in two variables were reviewed. The solution of a linear equation in two variables was graphed using the rectangular coordinate system. This chapter begins with a brief review of the rectangular coordinate system and graphs of equations. Then the concept of a function is introduced, probably the single most important concept of algebra. Next it discusses compound inequalities, absolute value equations and inequalities, and the graphs of solutions to these equations and inequalities.

8.1 Graphs of Equations

Objectives

❶ Graph an Equation Using the Point-Plotting Method

❷ Identify the Intercepts from the Graph of an Equation

❸ Interpret Graphs

Are You Prepared for This Section?

Before getting started, complete the following problems. If you get a problem wrong, go back to the section cited and review the material.

P1. Plot the following points on the real

number line: $-2, 4, 0, \dfrac{1}{2}$. [Section 1.3, pp. 22–23]

P2. Determine which of the following are solutions to the equation $3x - 5(x + 2) = 4$.
  (a) $x = 0$ **(b)** $x = -3$ **(c)** $x = -7$ [Section 2.1, pp. 82–83]

P3. Evaluate the expression $2x^2 - 3x + 1$ for the given values of the variable.
  (a) $x = 0$ **(b)** $x = 2$ **(c)** $x = -3$ [Section 1.8, pp. 64–65]

P4. Solve the equation $3x + 2y = 8$ for y. [Section 2.4, pp. 113–115]

P5. Evaluate $|-4|$. [Section 1.3, pp. 24–25]

P6. Plot the following ordered pairs in a rectangular coordinate system: $(3, 5)$, $(-2, 3)$, $(-1, -2)$, $(5, -3)$, $(0, 4)$. Tell which quadrant each point lies in or which coordinate axis the point lies on. [Section 3.1, pp. 169–173]

P7. Which of the following points satisfies $3x - 2y = 7$? for the given values of the variable.
  (a) $(1, -2)$ **(b)** $(3, 1)$ **(c)** $(-2, -5)$ [Section 3.1, pp. 173–174]

P8. Graph the equation: $y = -4x + 5$. [Section 3.4, pp. 206–208]

P9. Graph the line $3x - 5y = 15$ using intercepts. [Section 3.2, pp. 186–190]

▶ In Section 1.3 a point on the real number line was located by assigning it a single real number, called the *coordinate of the point*. See *Are You Prepared for This Section?* Problem P1. When a point is graphed on the real number line, we are working in one dimension. When we wish to work in two dimensions, we locate a point using two real numbers.

We begin by drawing two real number lines—one horizontal and one vertical—that intersect at right (90°) angles. The horizontal line is called the **x-axis,** and the vertical line is called the **y-axis.** The point where the x-axis and y-axis intersect is called the **origin, O.** See Figure 1.

Prepared?...Answers

P1.

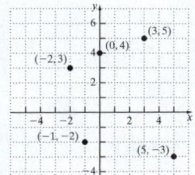

P2. (a) No   **(b)** No   **(c)** Yes   **P3. (a)** 1

(b) 3   **(c)** 28   **P4.** $y = -\dfrac{3}{2}x + 4$ **P5.** 4

P6.

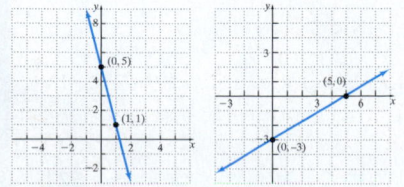

(3, 5): Quadrant I
(−2, 3): Quadrant II
(−1, −2): Quadrant III
(5, −3): Quadrant IV
(0, 4): y-axis

P7. (a) Yes   **(b)** Yes   **(c)** No

P8.   **P9.**

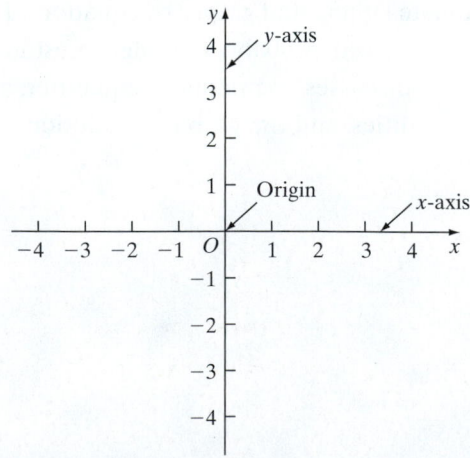

Figure 1

The origin O has a value of 0 on each axis. On the x-axis, points to the right of O are positive real numbers, and points to the left are negative. On the y-axis, points above O are positive real numbers, and points below are negative. In Figure 1 we label the x-axis "x" and the y-axis "y." An arrow at the end of each axis denotes the positive direction.

The coordinate system presented in Figure 1 is called a **rectangular** or **Cartesian coordinate system,** named after René Descartes (1596–1650), a French mathematician, philosopher, and theologian. The plane formed by the x-axis and y-axis is often called the **xy-plane,** and the x-axis and y-axis are called the **coordinate axes.**

Any point P in the rectangular coordinate system can be represented by using an **ordered pair (x, y)** of real numbers. If $x > 0$, then P is x units to the right of the y-axis; if $x < 0$, then P is $|x|$ units to the left of the y-axis. If $y > 0$, then P is y units above the x-axis; if $y < 0$, then P is $|y|$ units below the x axis. The ordered pair (x, y) is also called the **coordinates** of P.

The origin O has coordinates $(0, 0)$. Any point on the x-axis has coordinates of the form $(x, 0)$, and any point on the y-axis has coordinates of the form $(0, y)$.

If P has coordinates (x, y), then x is called the **x-coordinate,** or **abscissa,** of P and y is called the **y-coordinate,** or **ordinate,** of P.

Look back at Figure 1. Notice that the x- and y-axes divide the plane into four separate regions or **quadrants.** In quadrant I, both the x- and y-coordinates are positive; in quadrant II, x is negative and y is positive; in quadrant III, both x and y are negative; and in quadrant IV, x is positive and y is negative. Points on the coordinate axes do not belong to a quadrant. See Figure 2.

Figure 3 shows the ordered pairs $A(3, 2)$, $B(-2, 4)$, $C(-1, -3)$, $D(3, -4)$, and $E(-2, 0)$. Point A lies in quadrant I, point B lies in quadrant II, point C lies in quadrant III, point D lies in quadrant IV, and point E lies on the negative x-axis.

Figure 2

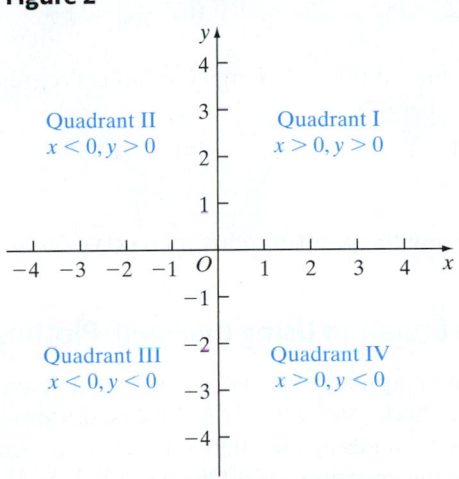

Figure 3

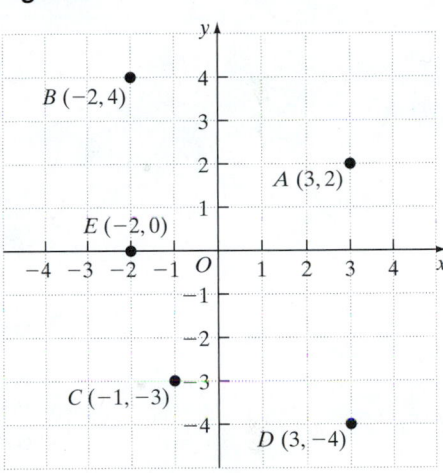

Quick ✔

1. The point where the x-axis and the y-axis intersect in the Cartesian coordinate system is called the _____.

2. *True or False* If a point lies in quadrant III of the Cartesian coordinate system, then both x and y are negative.

In Problems 3 and 4, plot each point in the xy-plane. Tell in which quadrant or on which coordinate axis each point lies. If you are struggling with these problems, go back and review pages 169–173 in Section 3.1.

3. **(a)** $A(5, 2)$

   (b) $B(4, -2)$

   (c) $C(0, -3)$

   (d) $D(-4, -3)$

4. **(a)** $A(-3, 2)$

   (b) $B(-4, 0)$

   (c) $C(3, -2)$

   (d) $D(6, 1)$

▶ Recall from Chapter 3 that an **equation in two variables,** x and y, is a statement in which the algebraic expressions involving x and y are equal. The expressions are called **sides** of the equation. Because an equation is a statement, it may be true or false, depending upon the values of the variables.

For example,

$$x^2 = y + 2 \qquad 3x + 2y = 6 \qquad y = -4x + 5$$

are all equations in two variables. The equation $x^2 = y + 2$ is satisfied when $x = 3$ and $y = 7$ since $3^2 = 7 + 2$. It is also satisfied when $x = -2$ and $y = 2$. In fact, there are infinitely many choices of x and y that satisfy the equation $x^2 = y + 2$. Some choices of x and y do not satisfy the equation $x^2 = y + 2$. For example, $x = 3$ and $y = 4$ do not satisfy the equation because $3^2 \neq 4 + 2$ (that is, $9 \neq 6$).

The **graph of an equation in two variables** x and y is the set of all ordered pairs (x, y) in the xy-plane that satisfy the equation.

> **In Other Words**
>
> The graph of an equation is a geometric way of representing the set of all ordered pairs that make the equation a true statement.

Quick ✔

5. *True or False* The graph of an equation in two variables x and y is the set of all points whose coordinates, (x, y), in the Cartesian plane satisfy the equation.

If you are struggling with Problems 6 and 7, go back and review pages 173–174 in Section 3.1.

6. Determine if the following coordinates represent points that are on the graph of $2x - 4y = 12$.

(a) $(2, -3)$ (b) $(2, -2)$ (c) $\left(\dfrac{3}{2}, -\dfrac{9}{4} \right)$

7. Determine if the following coordinates represent points that are on the graph of $y = x^2 + 3$.

(a) $(1, 4)$ (b) $(-2, -1)$ (c) $(-3, 12)$

Let's review the point-plotting method of obtaining the graph of an equation in two variables.

▶ ❶ Graph an Equation Using the Point-Plotting Method

Recall from Chapter 3 that one way to graph an equation is the **point-plotting method.** With this method, choose values for one of the variables and use the equation to find the corresponding values of the other variable. It does not matter whether we choose values of x and use the equation to find the corresponding values of y or choose y and find x. Let convenience determine which variable to begin with when using the point-plotting method.

EXAMPLE 1 **How to Graph an Equation by Plotting Points**

Graph the equation $y = -2x + 4$ by plotting points.

Step-by-Step Solution

Step 1: Find points (x, y) that satisfy the equation. To determine these points, choose values of x (do you see why?) and use the equation to determine the corresponding values of y. See Table 1.

Table 1

x	$y = -2x + 4$	(x, y)
-3	$y = -2(-3) + 4 = 10$	$(-3, 10)$
-2	$y = -2(-2) + 4 = 8$	$(-2, 8)$
-1	$y = -2(-1) + 4 = 6$	$(-1, 6)$
0	$y = -2(0) + 4 = 4$	$(0, 4)$
1	$y = -2(1) + 4 = 2$	$(1, 2)$
2	$y = -2(2) + 4 = 0$	$(2, 0)$
3	$y = -2(3) + 4 = -2$	$(3, -2)$

Step 2: Plot the ordered pairs listed in the third column of Table 1 as shown in Figure 4(a). Now connect the points to obtain the graph of the equation (*a line*), as shown in Figure 4(b).

Figure 4

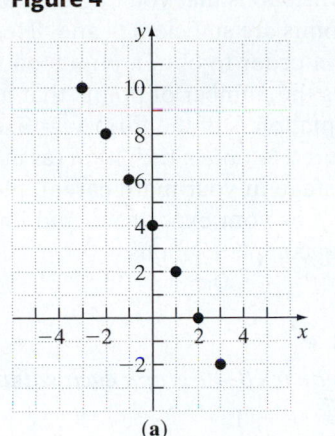

(a)

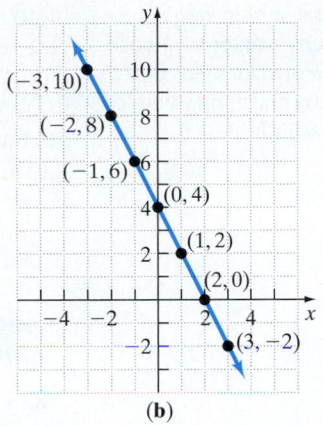

(b)

The graph of the equation in Figure 4(b) does not show all of the equation's points such as the point $(8, -12)$. Use arrows on the ends of the graph to indicate that the pattern continues. It is important to show enough of the graph so that the rest of the graph is obvious. Such a graph is called a **complete graph.**

Now let's look at the graph of a *nonlinear equation*.

EXAMPLE 2   **Graphing an Equation by Plotting Points**

Graph the equation $y = x^2$ by plotting points.

Solution

Table 2 shows several points on the graph.

Table 2

x	$y = x^2$	(x, y)
-4	$y = (-4)^2 = 16$	$(-4, 16)$
-3	$y = (-3)^2 = 9$	$(-3, 9)$
-2	$y = (-2)^2 = 4$	$(-2, 4)$
-1	$y = (-1)^2 = 1$	$(-1, 1)$
0	$y = (0)^2 = 0$	$(0, 0)$
1	$y = (1)^2 = 1$	$(1, 1)$
2	$y = (2)^2 = 4$	$(2, 4)$
3	$y = (3)^2 = 9$	$(3, 9)$
4	$y = (4)^2 = 16$	$(4, 16)$

In Figure 5(a), plot the ordered pairs listed in Table 2. In Figure 5(b), connect the points in a smooth curve and put arrows on either end to show the graph will continue indefinitely in that direction.

Figure 5

Work Smart

Note that in Figure 5, different scales are used on the x- and y-axes.

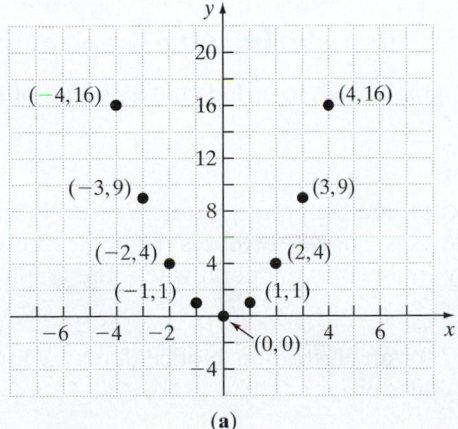

(a)

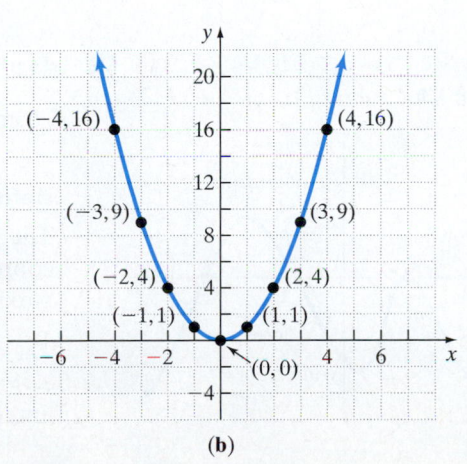

(b)

Two questions that you might be asking yourself right now are "How do I know how many points are sufficient?" and "How do I know which x-values (or y-values) I should choose in order to obtain points on the graph?" Often, the type of equation to graph indicates the number of points that are necessary. For example, as shown in Chapter 3, if the equation is of the form $y = mx + b$, then its graph is a line, and only two points are required to obtain the graph (as in Example 1). Other times, more points are required. At this stage in your math career, plot quite a few points to obtain a complete graph. However, as your experience and knowledge grow, you will become more efficient in obtaining complete graphs.

Quick ✔

In Problems 8–10, graph each equation using the point-plotting method.

8. $y = 3x + 1$ **9.** $2x + 3y = 8$ **10.** $y = x^2 + 3$

EXAMPLE 3   **Graphing the Equation $x = y^2$**

Graph the equation $x = y^2$ by plotting points.

Solution

Because the equation is solved for x, choose values of y and use the equation to find the corresponding values of x. See Table 3. Plot the ordered pairs listed in Table 3 and connect the points in a smooth curve. See Figure 6.

Table 3

y	$x = y^2$	(x, y)
-3	$x = (-3)^2 = 9$	$(9, -3)$
-2	$x = (-2)^2 = 4$	$(4, -2)$
-1	$x = (-1)^2 = 1$	$(1, -1)$
0	$x = 0^2 = 0$	$(0, 0)$
1	$x = 1^2 = 1$	$(1, 1)$
2	$x = 2^2 = 4$	$(4, 2)$
3	$x = 3^2 = 9$	$(9, 3)$

Figure 6

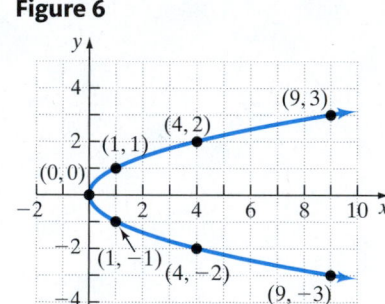

Quick ✔

In Problems 11 and 12, graph each equation using the point-plotting method.

11. $x = y^2 + 2$ **12.** $x = (y - 1)^2$

② Identify the Intercepts from the Graph of an Equation

A complete graph needs to include the *intercepts* of the graph.

Figure 7

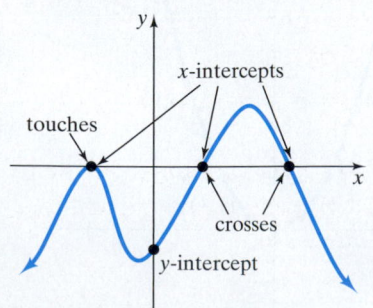

Definition

The **intercepts** are the points, if any, where a graph crosses or touches the coordinate axes. An **x-intercept** is the point where the graph crosses or touches the x-axis. A **y-intercept** is the point where the graph crosses or touches the y-axis.

See Figure 7 for an illustration. Notice that an x-intercept exists when $y = 0$ and a y-intercept exists when $x = 0$.

EXAMPLE 4   **Finding Intercepts from a Graph**

Find the intercepts of the graph shown in Figure 8. What are the x-intercepts? What are the y-intercepts?

Figure 8

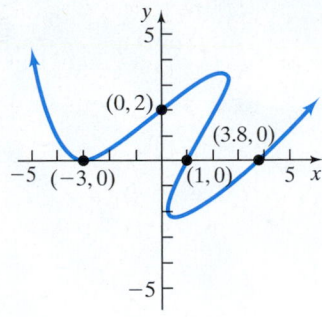

Work Smart

The x-intercept will *always* have coordinates $(a, 0)$ because *every* point on the x-axis has a y-coordinate of 0.

Solution

The intercepts of the graph are the points $(-3, 0)$, $(0, 2)$, $(1, 0)$, and $(3.8, 0)$. The x-intercepts are $(-3, 0)$, $(1, 0)$, and $(3.8, 0)$; the y-intercept is $(0, 2)$.   ●

Quick ✓

13. The points, if any, at which a graph crosses or touches a coordinate axis are called _____.

14. *True or False* An x-intercept exists when $x = 0$.

15. Find the intercepts of the graph shown in the figure. What are the x-intercepts? What are the y-intercepts?

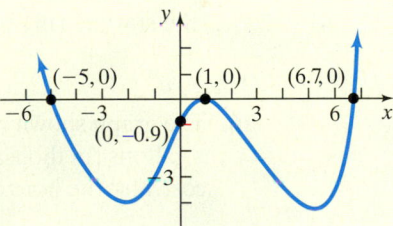

 ❸ Interpret Graphs

Graphs help us to visualize a relationship between two variables or quantities. A graph is a "picture" that illustrates the relationship between two variables. Visualizing allows important information to be seen and conclusions drawn about the relationship between the two variables.

EXAMPLE 5   **Interpreting a Graph**

The graph in Figure 9 on the following page shows the profit P for selling x gallons of gasoline in an hour at a gas station. The vertical axis represents the profit, and the horizontal axis represents the number of gallons of gasoline sold.

(continued)

Figure 9

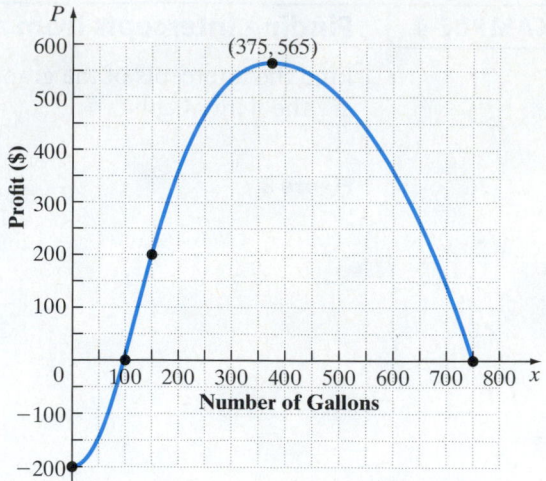

(a) What is the profit if 150 gallons of gasoline are sold?

(b) How many gallons of gasoline are sold when profit is highest? What is the highest profit?

(c) Identify and interpret the intercepts.

Solution

(a) Locate the point on the graph with x-coordinate 150. The y-coordinate at this point is 200. The profit from selling 150 gallons of gasoline is $200.

(b) The profit is highest when 375 gallons of gasoline are sold. The highest profit is $565.

(c) The intercepts are $(0, -200)$, $(100, 0)$, and $(750, 0)$. For $(0, -200)$: If 0 gallons of gasoline are sold, the profit is $-$200. The negative profit is due to the fact that the company has $0 in revenue but has hourly expenses of $200. For $(100, 0)$: When 100 gallons of gasoline are sold, there is $0 profit. The company sells just enough gas to pay its bills. For $(750, 0)$: If 750 gallons of gasoline are sold, the profit is $0. The 750 gallons sold represents the maximum number of gallons that the station can pump and still break even. ●

Quick ✓

16. The graph shown represents the cost C (in thousands of dollars) of refining x gallons (in thousands) of gasoline per hour. The vertical axis represents the cost, and the horizontal axis represents the number of gallons of gasoline refined.

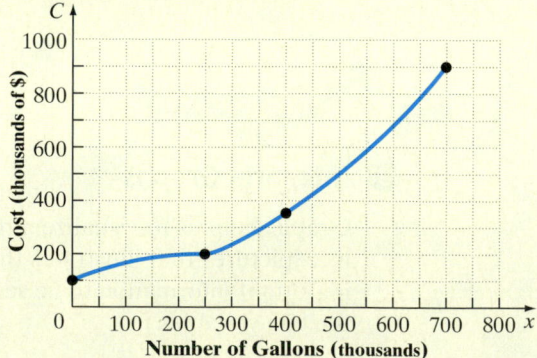

(a) What is the cost of refining 250 thousand gallons of gasoline per hour?

(b) What is the cost of refining 400 thousand gallons of gasoline per hour?

(c) In the context of the problem, explain the meaning of the graph ending at 700 thousand gallons of gasoline.

(d) Identify and interpret the intercept.

*Problems **1–16** are the **Quick ✓**s that follow the **EXAMPLES**.*

Building Skills

17. Determine the coordinates of each of the points plotted. Tell in which quadrant or on what coordinate axis each point lies. See Objective 1.

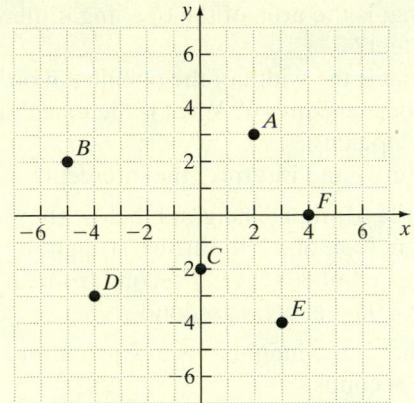

18. Determine the coordinates of each of the points plotted. Tell in which quadrant or on what coordinate axis each point lies. See Objective 1.

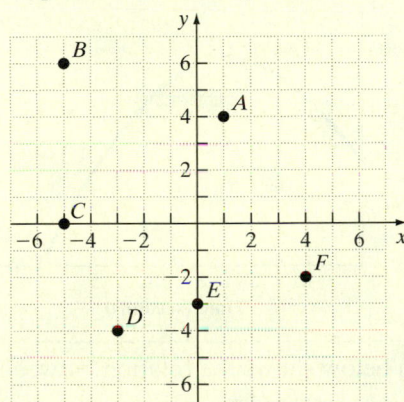

In Problems 19 and 20, plot each point in the xy-plane. Tell in which quadrant or on what coordinate axis each point lies. See Objective 1.

19. $A(3, 5)$
$B(-2, -6)$
$C(5, 0)$
$D(1, -6)$
$E(0, 3)$
$F(-4, 1)$

20. $A(-3, 1)$
$B(-6, 0)$
$C(2, -5)$
$D(-6, -2)$
$E(1, 2)$
$F(0, -5)$

In Problems 21–26, determine whether the given points are on the graph of the equation. See Objective 2.

21. $2x + 5y = 12$
(a) $(1, 2)$
(b) $(-2, 3)$
(c) $(-4, 4)$
(d) $\left(-\dfrac{3}{2}, 3\right)$

22. $-4x + 3y = 18$
(a) $(1, 7)$
(b) $(0, 6)$
(c) $(-3, 10)$
(d) $\left(\dfrac{3}{2}, 4\right)$

23. $y = -2x^2 + 3x - 1$
(a) $(-2, -15)$
(b) $(3, 10)$
(c) $(0, 1)$
(d) $(2, -3)$

24. $y = x^3 - 3x$
(a) $(2, 2)$
(b) $(3, 8)$
(c) $(-3, -18)$
(d) $(0, 0)$

25. $y = |x - 3|$
(a) $(1, 4)$
(b) $(4, 1)$
(c) $(-6, 9)$
(d) $(0, 3)$

26. $x^2 + y^2 = 1$
(a) $(0, 1)$
(b) $(1, 1)$
(c) $\left(\dfrac{1}{2}, \dfrac{1}{2}\right)$
(d) $\left(\dfrac{\sqrt{3}}{2}, \dfrac{1}{2}\right)$

In Problems 27–54, graph each equation by plotting points. See Objective 3.

27. $y = 4x$

28. $y = 2x$

29. $y = -\dfrac{1}{2}x$

30. $y = -\dfrac{1}{3}x$

31. $y = x + 3$

32. $y = x - 2$

33. $y = -3x + 1$

34. $y = -4x + 2$

35. $y = \dfrac{1}{2}x - 4$

36. $y = -\dfrac{1}{2}x + 2$

37. $2x + y = 7$

38. $3x + y = 9$

39. $y = -x^2$

40. $y = x^2 - 2$

41. $y = 2x^2 - 8$

42. $y = -2x^2 + 8$

43. $y = |x|$

44. $y = |x| - 2$

45. $y = |x - 1|$

46. $y = -|x|$

47. $y = x^3$

48. $y = -x^3$

49. $y = x^3 + 1$

50. $y = x^3 - 2$

51. $x^2 - y = 4$

52. $x^2 + y = 5$

53. $x = y^2 - 1$

54. $x = y^2 + 2$

In Problems 55–58, the graph of an equation is given. List the intercepts of the graph. See Objective 4.

55.

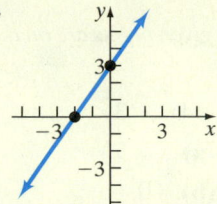

56.

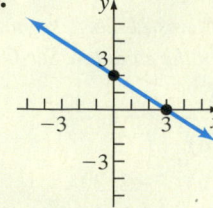

57.

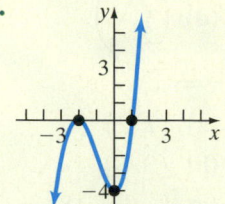

58.

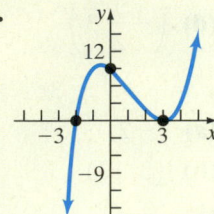

Applying the Concepts

59. If $(a, 4)$ is a point on the graph of $y = 4x - 3$, what is a?

60. If $(a, -2)$ is a point on the graph of $y = -3x + 5$, what is a?

61. If $(3, b)$ is a point on the graph of $y = x^2 - 2x + 1$, what is b?

62. If $(-2, b)$ is a point on the graph of $y = -2x^2 + 3x + 1$, what is b?

63. Area of a Window The Property Brothers wish to put a new window in a home. They want the perimeter of the window to be 100 feet. The graph on the top of right column shows the relation between the width, x, of the opening and the area of the opening.

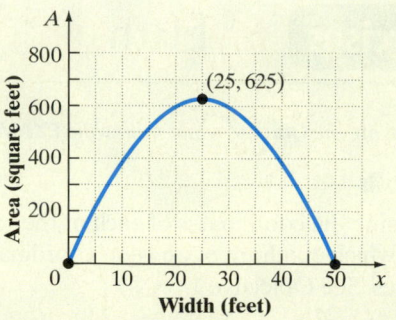

(a) What is the area of the opening if the width is 10 feet?

(b) What is the width of the opening in order for area to be a maximum? What is the maximum area of the opening?

(c) Identify and interpret the intercepts.

64. Projectile Motion The graph below shows the height, in feet, of a ball thrown straight up with an initial speed of 80 feet per second from an initial height of 96 feet after t seconds.

(a) What is the height of the object after 1.5 seconds?

(b) At what time is the height a maximum? What is the maximum height?

(c) Identify and interpret the intercepts.

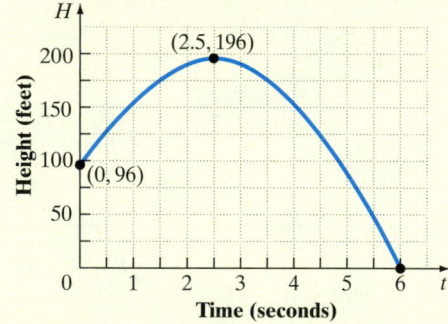

65. Cell Phones We all struggle with selecting a cell phone plan. The graph below shows the relation between the monthly cost of a cell phone and the number of minutes of talk used, m. (SOURCE: *Verizon.com*)

(a) What is the cost of talking for 200 minutes in a month? 400 minutes?

(b) What is the cost of talking 8000 minutes in a month?

(c) Identify and interpret the intercept.

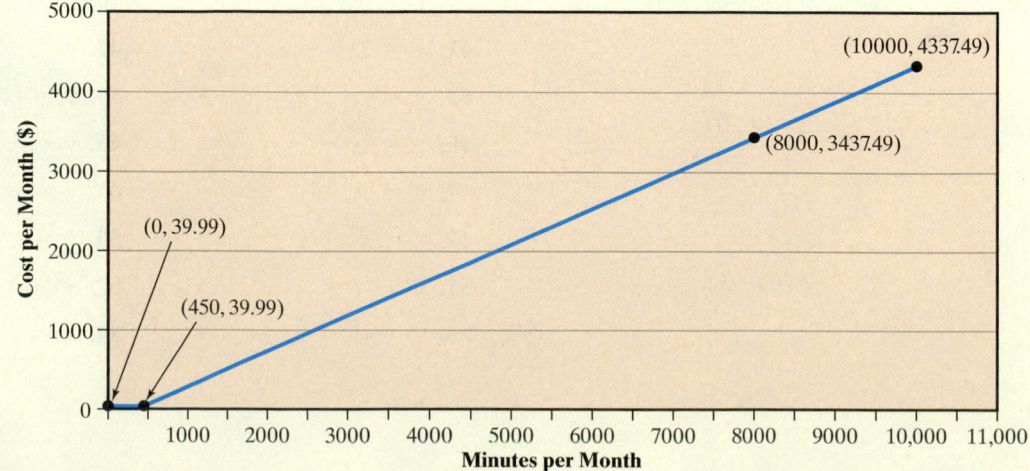

66. Wind Chill It is 10° Celsius outside. The wind is calm but then gusts up to 20 meters per second. You feel the chill go right through your bones. The following graph shows the relation between the wind chill temperature (in degrees Celsius) and wind speed (in meters per second).

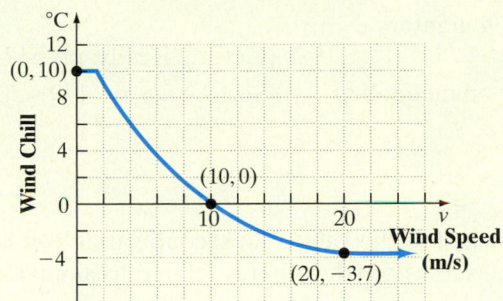

(a) What is the wind chill if the wind is blowing 4 meters per second?

(b) What is the wind chill if the wind is blowing 20 meters per second?

(c) Identify and interpret the intercepts.

67. Plot the points $(4, 0)$, $(4, 2)$, $(4, -3)$, and $(4, -6)$. Describe the set of all points of the form $(4, y)$ where y is a real number.

68. Plot the points $(4, 2)$, $(1, 2)$, $(0, 2)$, and $(-3, 2)$. Describe the set of all points of the form $(x, 2)$ where x is a real number.

Extending the Concepts

69. Draw a graph of an equation that contains two x-intercepts, $(-2, 0)$ and $(3, 0)$. At the x-intercept $(-2, 0)$ the graph crosses the x-axis; at the x-intercept $(3, 0)$, the graph touches the x-axis. Compare your graph with those of your classmates. How are they similar? How are they different?

70. Draw a graph that contains the points $(-3, -1)$, $(-1, 1)$, $(0, 3)$, and $(1, 5)$. Compare your graph with those of your classmates. How many of the graphs are straight lines? How many are "curved"?

71. Make up an equation that is satisfied by the points $(2, 0)$, $(4, 0)$, and $(1, 0)$. Compare your equation with those of your classmates. How are they similar? How are they different?

72. Make up an equation that is satisfied by the points $(0, 3)$, $(1, 3)$, and $(-4, 3)$. Compare your equation with those of your classmates. How many are the same?

Explaining the Concepts

73. Explain what is meant by a complete graph.

74. Explain what the graph of an equation represents.

75. What is the point-plotting method for graphing an equation?

76. What is the y-coordinate of a point that is an x-intercept? What is the x-coordinate of a point that is a y-intercept?

Technology Exercises

Just as we have graphed equations using point plotting, the graphing calculator also graphs equations by plotting points. Figure 10 shows the graph of $y = x^2$, and Table 4 shows points on the graph of $y = x^2$, using a graphing calculator.

Figure 10

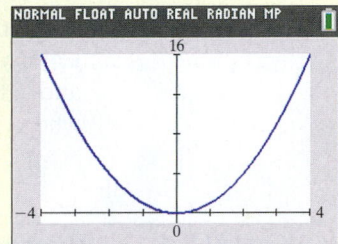

Table 4

X	Y1
-5	25
-4	16
-3	9
-2	4
-1	1
0	0
1	1
2	4
3	9
4	16
5	25

X= -5

In Problems 77–84, use technology to draw a complete graph of each equation. Remember, a complete graph shows all the interesting features of a graph, including intercepts.

77. $y = 3x - 9$

78. $y = -5x + 8$

79. $y = -x^2 + 8$

80. $y = 2x^2 - 4$

81. $y + 2x^2 = 13$

82. $y - x^2 = -15$

83. $y = x^3 - 6x + 1$

84. $y = -x^3 + 3x$

8.2 Relations

Objectives

① Understand Relations
② Find the Domain and the Range of a Relation
③ Graph a Relation Defined by an Equation

Are You Prepared for This Section?

Before getting started, complete the following problems. If you get a problem wrong, go back to the section cited and review the material.

P1. Write the inequality $-4 \leq x \leq 4$ in interval notation.   [Getting Ready pp. 517–520]

P2. Write the interval $[\,2, \infty\,)$ using an inequality.   [Section 2.8, pp. 149–150]

▶ ① Understand Relations

When the value of one variable is related to the value of a second variable, it is called a *relation*. For example, an individual's level of education is related to annual income. Engine size is related to gas mileage.

> **Definition**
>
> A **relation** exists when the elements in one set are associated with elements in a second set. If x and y are two distinct elements in these sets, and if a relation exists between x and y, then x **corresponds** to y, or y **depends on** x, which is written as $x \rightarrow y$. A relation where y depends on x may also be written as an ordered pair (x, y).

EXAMPLE 1   **Illustrating a Relation**

The data presented in Figure 11 show a correspondence between states and senators in 2016 for randomly selected senators. A possible name for the relation might be "is represented in the U.S. Senate by." Thus the relation reveals that "Indiana is represented by Dan Coats," for example. In Figure 11, **mapping** is used to represent the relation by drawing an arrow from an element in the set "state" to an element in the set "senator."

Figure 11

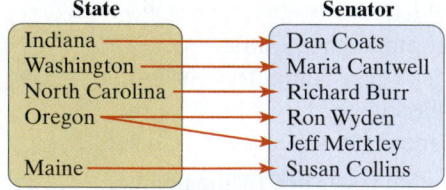

This relation could also be represented by using ordered pairs in the form (state, senator) as follows:

$$\{\, (\text{Indiana, Dan Coats}), (\text{Washington, Maria Cantwell}),$$
$$(\text{North Carolina, Richard Burr}), (\text{Oregon, Ron Wyden}),$$
$$(\text{Oregon, Jeff Merkley}), (\text{Maine, Susan Collins}) \,\}$$

●

Quick ✔

1. If a relation exists between x and y, then say that x _____ to y or that y _____ on x, and write $x \rightarrow y$.

Prepared?...Answers   **P1.** $[-4, 4]$ **P2.** $x \geq 2$

2. Use the map to represent the relation as a set of ordered pairs.

Friend

Birthday

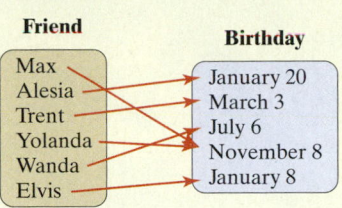

Max
Alesia
Trent
Yolanda
Wanda
Elvis

January 20
March 3
July 6
November 8
January 8

3. Use the set of ordered pairs to represent the relation as a map.

$$\{(1,3),(5,4),(8,4),(10,13)\}$$

❷ Find the Domain and the Range of a Relation

▶ In a relation we say that y depends on x and can write the relation as a set of ordered pairs (x,y). The set of all x can be thought of as the **inputs** of the relation. The set of all y can be thought of as the **outputs** of the relation. This interpretation of a relation can be used to define *domain* and *range*.

Definition

The **domain** of a relation is the set of all inputs of the relation. The **range** is the set of all outputs of the relation.

EXAMPLE 2   **Finding the Domain and the Range of a Relation**

Find the domain and the range of the relation presented in Figure 11 from Example 1.

Solution

The domain of the relation is the set of all inputs, therefore, the domain of the relation is

{ Indiana, Washington, North Carolina, Oregon, Maine }

The range of the relation is the set of all outputs. Therefore, the range of the relation is

{ Dan Coats, Maria Cantwell, Richard Burr, Ron Wyden,
Jeff Merkley, Susan Collins }

Work Smart

Never list elements in the domain or range more than once.

Notice that Oregon was not listed twice in the domain. The domain and the range are sets, and elements in a set should *not* be listed more than once. Also, the order in which the elements in the domain or range are listed does not matter.

Quick ✓

4. The _____ of a relation is the set of all inputs of the relation. The _____ is the set of all outputs of the relation.

5. State the domain and the range of the relation.

Friend

Birthday

Max
Alesia
Trent
Yolanda
Wanda
Elvis

January 20
March 3
July 6
November 8
January 8

6. State the domain and the range of the relation.

$$\{(1,3),(5,4),(8,4),(10,13)\}$$

Relations can also be represented by plotting a set of ordered pairs. The set of all x-coordinates represents the domain of the relation, and the set of all y-coordinates represents the range of the relation.

<table>
</table>

(EXAMPLE 3)

Finding the Domain and the Range of a Relation

Figure 12 shows the graph of a relation. Identify the domain and the range of the relation.

Figure 12

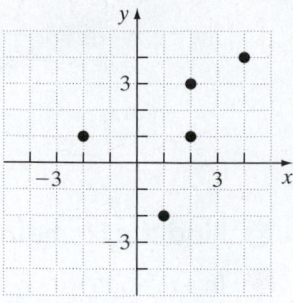

Work Smart

Write the points as ordered pairs to assist in finding the domain and range.

Solution

The ordered pairs in the graph are $(-2, 1)$, $(1, -2)$, $(2, 1)$, $(2, 3)$, and $(4, 4)$, so the domain is the set of all x-coordinates: $\{-2, 1, 2, 4\}$. The range is the set of all y-coordinates: $\{-2, 1, 3, 4\}$. •

Quick ✓

7. Identify the domain and the range of the relation shown in the figure.

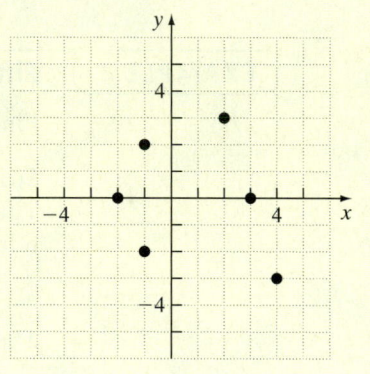

A third way to define a relation is by a graph. Remember that the graph of an equation is the set of all ordered pairs (x, y) that make the equation a true statement. If a graph exists for some ordered pair (x, y), then the x-coordinate is in the domain and the y-coordinate is in the range.

(EXAMPLE 4)

Identifying the Domain and the Range of a Relation from Its Graph

Figure 13 shows the graph of a relation. Determine the domain and the range of the relation.

Figure 13

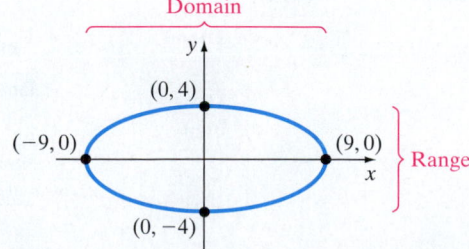

Work Smart

You can think of the domain as the shadow created by the graph on the x-axis by vertical beams of light. The range can be thought of as the shadow created by the graph on the y-axis by horizontal beams of light.

Solution

The domain of the relation consists of all x-coordinates at which the graph exists. The graph exists at all x-values between −9 and 9, inclusive. Therefore, the domain is $\{x \mid -9 \le x \le 9\}$ or, using interval notation, $[-9, 9]$.

The range of the relation consists of all y-coordinates at which the graph exists. The graph exists at all y-values between −4 and 4, inclusive. Therefore, the range is $\{y \mid -4 \le y \le 4\}$ or, using interval notation, $[-4, 4]$. ●

Quick ✓

8. *True or False* If the graph of a relation does not exist at $x = 3$, then 3 is not in the domain of the relation.

9. *True or False* The range of a relation is always the set of all real numbers.

In Problems 10 and 11, identify the domain and range of the relation from its graph.

10.

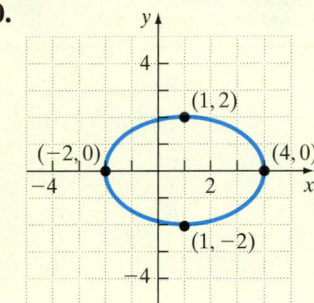

11.

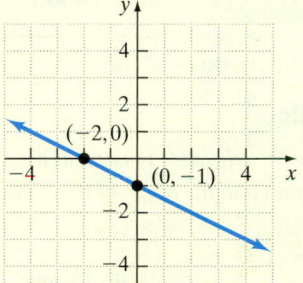

Work Smart

Relations can be defined by

1. Mapping
2. Sets of ordered pairs
3. Graphs
4. Equations

▶ ❸ Graph a Relation Defined by an Equation

Another way to define a relation (instead of a map, a set of ordered pairs, or a graph) is to use equations such as $x + y = 4$ or $x = y^2$. When relations are defined by equations, graph the relation in order to visualize how the variables are related. As seen in Example 4, the graph of the relation also helps identify its domain and range.

EXAMPLE 5 **Relations Defined by Equations**

Graph the relation $y = -x^2 + 4$. Find its domain and range using the graph.

Solution

The relation says to take the input x, square it, multiply this result by −1, and then add 4 to get the output y. Use the point-plotting method to graph the relation. Table 5 shows some points on the graph. Figure 14 shows a graph of the relation.

Figure 14

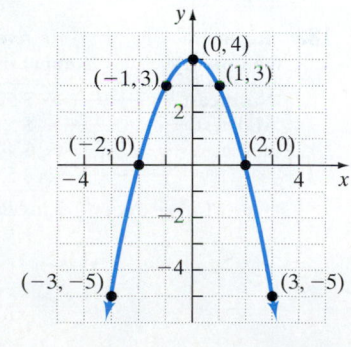

Table 5

x	$y = -x^2 + 4$	(x, y)
−3	$-(-3)^2 + 4 = -5$	(−3, −5)
−2	$-(-2)^2 + 4 = 0$	(−2, 0)
−1	3	(−1, 3)
0	4	(0, 4)
1	3	(1, 3)
2	0	(2, 0)
3	−5	(3, −5)

(continued)

The graph extends indefinitely to the left and to the right (that is, the graph exists for all x-values). Therefore, the domain of the relation is the set of all real numbers, or $\{x \mid x \text{ is any real number}\}$, or, using interval notation, $(-\infty, \infty)$. Notice that there are no y-values greater than 4, but the graph exists everywhere for y-values less than or equal to 4. The range of the relation is $\{y \mid y \leq 4\}$ or, using interval notation, $(-\infty, 4]$. •

Quick ✓

In Problems 12–14, graph each relation. Use the graph to identify the domain and range.

12. $y = 3x - 8$ **13.** $y = x^2 - 8$ **14.** $x = y^2 + 1$

8.2 Exercises MyMathLab®

*Problems **1–14** are the Quick ✓ s that follow the EXAMPLES.*

Building Skills

In Problems 15–18, write each relation as a set of ordered pairs. Then identify the domain and the range of the relation. See Objectives 1 and 2.

15.

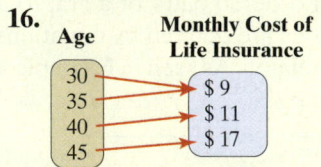

Newspaper	Daily Circulation (in millions)
USA Today	2.28
Wall Street Journal	2.06
New York Times	1.1
Los Angeles Times	0.82
Washington Post	0.70

SOURCE: *Information Please Almanac*

16.

Age	Monthly Cost of Life Insurance
30	$ 9
35	$ 11
40	$ 17
45	

SOURCE: *wholesaleinsurance.net*

17.

Level of Education	Average Annual Income, 2015
Less than 9th Grade	$ 20,791
9th – 12th Grade — No Diploma	$ 23,234
High School Graduate	$ 32,456
Associate's Degree	$ 42,508
Bachelor's Degree	$ 62,240

SOURCE: *United States Census Bureau*

18.

Region of the Country	Average Annual Income, 2014
Northeast	$ 59,210
Midwest	$ 54,267
South	$ 49,655
West	$ 57,688

SOURCE: *United States Census Bureau*

In Problems 19–24, write each relation as a map. Then identify the domain and the range of the relation. See Objectives 1 and 2.

19. $\{(-3, 4), (-2, 6), (-1, 8), (0, 10), (1, 12)\}$

20. $\{(-2, 6), (-1, 3), (0, 0), (1, -3), (2, 6)\}$

21. $\{(-2, 4), (-1, 2), (0, 0), (1, 2), (2, 4)\}$

22. $\{(-2, -8), (-1, -1), (0, 0), (1, 1), (2, 8)\}$

23. $\{(0, -4), (-1, -1), (-2, 0), (-1, 1), (0, 4)\}$

24. $\{(-3, 0), (0, 3), (3, 0), (0, -3)\}$

In Problems 25–32, identify the domain and the range of the relation from the graph. See Objective 2.

25.

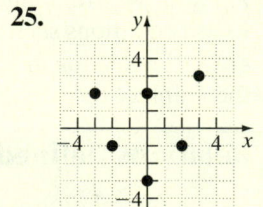

26.

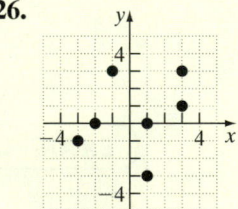

27.

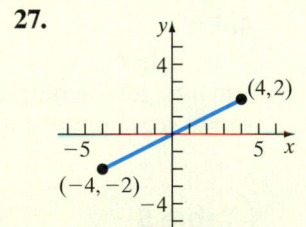

28.

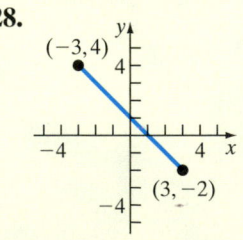

29.

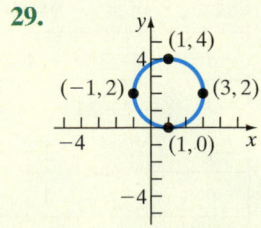

30.

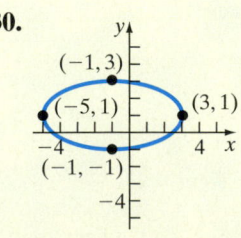

31. 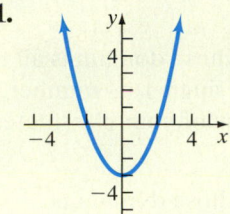   **32.**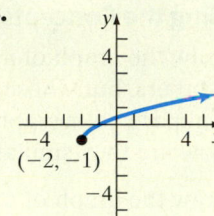

In Problems 33–54, graph the relation to identify the domain and the range of the relation. Note: These relations were graphed in Problems 33–54 in Section 8.1. See Objective 3.

33. $y = -3x + 1$ **34.** $y = -4x + 2$

35. $y = \dfrac{1}{2}x - 4$ **36.** $y = -\dfrac{1}{2}x + 2$

37. $2x + y = 7$ **38.** $3x + y = 9$

39. $y = -x^2$ **40.** $y = x^2 - 2$

41. $y = 2x^2 - 8$ **42.** $y = -2x^2 + 8$

43. $y = |x|$ **44.** $y = |x| - 2$

45. $y = |x - 1|$ **46.** $y = -|x|$

47. $y = x^3$ **48.** $y = -x^3$

49. $y = x^3 + 1$ **50.** $y = x^3 - 2$

51. $x^2 - y = 4$ **52.** $x^2 + y = 5$

53. $x = y^2 - 1$ **54.** $x = y^2 + 2$

Applying the Concepts

55. Area of a Window Chip Gaines wishes to put a new window in his home. He wants the perimeter of the window to be 100 feet. The graph in the next column shows the relation between the width, x, of the opening (in feet) and the area of the opening.

(a) Determine the domain and the range of the relation.

(b) Explain why the domain obtained in part (a) makes sense.

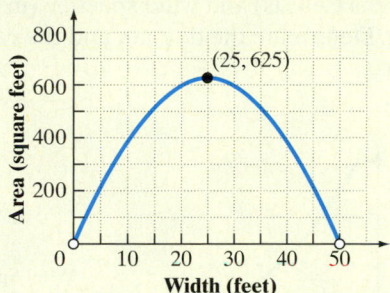

56. Projectile Motion The graph below shows the height, in feet, of a ball thrown straight up with an initial speed of 80 feet per second from an initial height of 96 feet after t seconds. Determine the domain and the range of the relation.

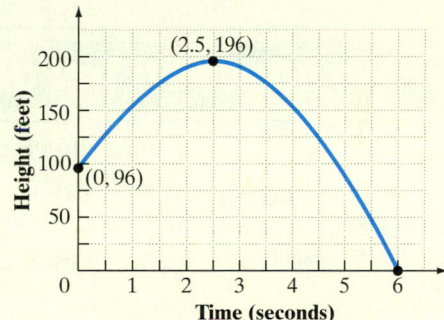

57. Cell Phones The graph below shows the relation between the monthly cost, C, of a cell phone and the number of anytime minutes used, m.

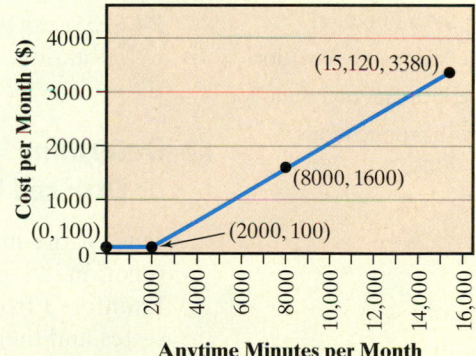

(a) Determine the domain and the range of the relation.

(b) If anytime minutes are from 7:00 A.M. to 7:00 P.M. Monday through Friday, explain why the domain obtained in part (a) makes sense, assuming there are 21 nonweekend days per month.

58. Wind Chill It is 10° Celsius outside. The wind is calm but then gusts up to 20 meters per second. You feel the chill go right through your bones. The graph below shows the relation between the wind chill temperature (in degrees Celsius) and wind speed, v (in meters per second). Determine the domain and the range of the relation.

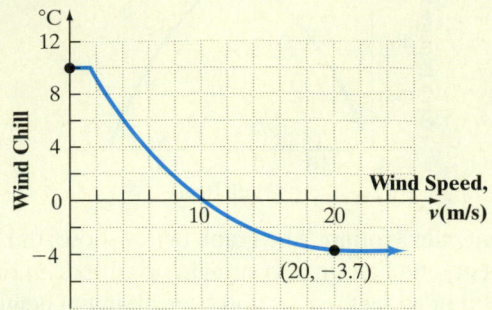

(20, −3.7)

Extending the Concepts

59. Draw the graph of a relation whose domain is all real numbers, but whose range is a single real number. Compare your graph with those of your classmates. How are they similar?

60. Draw the graph of a relation whose domain is a single real number, but whose range is all real numbers. Compare your graph with those of your classmates. How are they similar?

Explaining the Concepts

61. Explain what a relation is. Be sure to include an explanation of domain and range.

62. State the four methods for describing a relation that are presented in this section. When is using ordered pairs most appropriate? When is using a graph most appropriate? Support your opinion.

8.3 An Introduction to Functions

Objectives

❶ Determine Whether a Relation Expressed as a Map or Ordered Pairs Represents a Function

❷ Determine Whether a Relation Expressed as an Equation Represents a Function

❸ Determine Whether a Relation Expressed as a Graph Represents a Function

❹ Find the Value of a Function

❺ Find the Domain of a Function

❻ Work with Applications of Functions

Are You Prepared for This Section?

Before getting started, complete the following problems. If you get a problem wrong, go back to the section cited and review the material.

P1. Evaluate the expression $2x^2 - 5x$ for

    (a) $x = 1$ **(b)** $x = 4$ **(c)** $x = -3$ [Section 1.8, pp. 64–65]

P2. Evaluate $\dfrac{3}{2x + 1}$ for $x = -\dfrac{1}{2}$.   [Section 1.8, pp. 64–65]

P3. Express the inequality $x \le 5$ using interval notation.   [Section 2.8, pp. 149–150]

P4. Express the interval $(2, \infty)$ using set-builder notation.   [Section 2.8, pp. 148–149]

▶ ❶ **Determine Whether a Relation Expressed as a Map or Ordered Pairs Represents a Function**

One of the most important concepts in algebra is now presented—the *function*. A function is a special type of relation. To understand functions, let's revisit the relation in Example 1 from Section 8.2, shown again in Figure 15. In this correspondence between states and their senators, we named the relation "is represented in the U.S. Senate by."

Figure 15

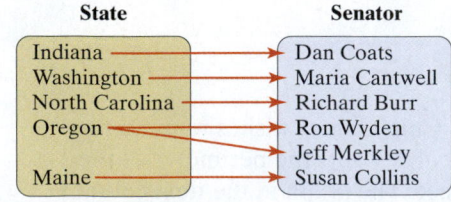

In this relation, if you were asked to name the senator who represents Oregon, you could respond, "Ron Wyden" or "Jeff Merkley." In other words, the input "state" does not correspond to a single output "senator."

Let's consider the relation in Figure 16(a) on the next page, a correspondence between states and their populations. If asked for the population that corresponds

Prepared?...Answers **P1. (a)** −3
(b) 12 **(c)** 33 **P2.** undefined
P3. $(-\infty, 5]$ **P4.** $\{x \mid x > 2\}$

to North Carolina, you could only respond, "10,043 thousand." In other words, each input "state" corresponds to exactly one output "population."

Figure 16(b) is a relation that shows a correspondence between animals and life expectancies. If asked to determine the life expectancy of a dog you could only respond, "11 years." You would have the same answer about the life expectancy of a cat.

Figure 16

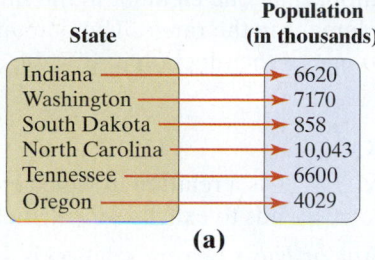
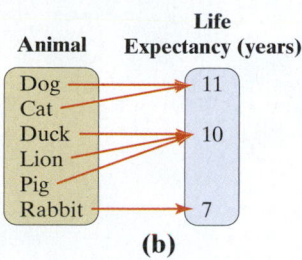

(a)   (b)

What do the relations in Figures 16(a) and 16(b) have in common that is missing from the relation in Figure 15? In the relations in Figure 16(a) and 16(b), each input corresponds to only one output. In Figure 15, however, the input Oregon corresponds to two outputs—Ron Wyden and Jeff Merkley. This leads to the definition of a *function*.

> **In Other Words**
>
> For a relation to be a function, each input may have only one output.

Definition

A **function** is a relation in which each input, or element in the domain of the relation, corresponds to exactly one output, or element in the range of the relation.

The idea behind functions is predictability. If an input is known, a function can be used to determine the output with 100% certainty, as seen in Figures 16(a) and (b). Nonfunctions do not have this predictability (Figure 15).

EXAMPLE 1   **Determining Whether a Relation Represents a Function**

Determine whether the relation is a function. If so, state its domain and range.

(a) See Figure 17(a). The domain represents the length (mm) of the right humerus, and the range represents the length (mm) of the right tibia for each of five rats that had been sent to space.

(b) See Figure 17(b). The domain represents the age of six people, and the range represents their height.

(c) See Figure 17(c). The domain represents the age of five males, and the range represents their HDL (or "good") cholesterol (mg/dL).

Figure 17

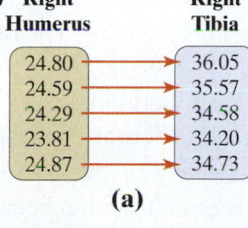
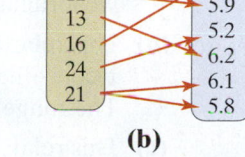
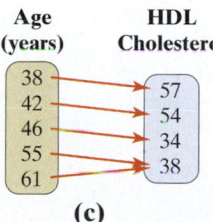

(a)   (b)   (c)

SOURCE: *NASA Life Sciences Data Archive*   SOURCE: *diamonds.com*

Solution

(a) The relation in Figure 17(a) is a function because each element in the domain corresponds to exactly one element in the range. The domain is { 24.80, 24.59, 24.29, 23.81, 24.87 }. The range is { 36.05, 35.57, 34.58, 34.20, 34.73 }.

(continued)

(b) The relation in Figure 17(b) is not a function because an element in the domain, 0.86, corresponds to two elements in the range. A single price cannot be determined for the 0.86-carat diamonds.

(c) The relation in Figure 17(c) is a function because each element in the domain corresponds to exactly one element in the range. Notice that it is okay for more than one element in the function's domain to correspond to the same element in the range. The domain of the function is $\{38, 42, 46, 55, 61\}$. The range of the function is $\{57, 54, 34, 38\}$. ●

Quick ✓

1. A _____ is a relation in which each element in the domain of the relation corresponds to exactly one element in the range of the relation.

2. *True or False* Every relation is a function.

In Problems 3 and 4, determine whether the relation is a function. If so, state its domain and range.

3.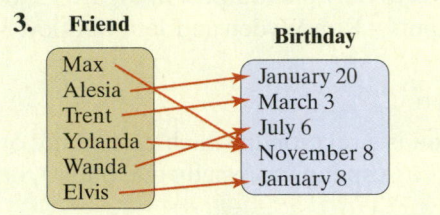

4.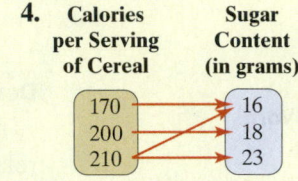

Work Smart

All functions are relations, but not all relations are functions!

A function can also be thought of as a set of ordered pairs (x, y) in which no ordered pairs have the same x-coordinate and different y-coordinates.

<hr>

EXAMPLE 2 **Determining Whether a Relation Represents a Function**

Determine whether the relation is a function. If so, state its domain and range.

(a) $\{(1, 3), (-1, 4), (0, 6), (2, 8)\}$
(b) $\{(-2, 6), (-1, 3), (0, 2), (1, 3), (2, 6)\}$
(c) $\{(0, 3), (1, 4), (4, 5), (9, 5), (4, 1)\}$

Solution

(a) This relation is a function because no ordered pairs have the same x-coordinate and different y-coordinates. The domain of the function is the set of all x-coordinates, $\{-1, 0, 1, 2\}$. The range of the function is the set of all y-coordinates, $\{3, 4, 6, 8\}$.

(b) This relation is a function because no ordered pairs have the same x-coordinate and different y-coordinates. The domain is, $\{-2, -1, 0, 1, 2\}$. The range is, $\{2, 3, 6\}$.

(c) This relation is not a function because two ordered pairs, $(4, 5)$ and $(4, 1)$, have the same x-coordinate and different y-coordinates. ●

In Example 2(b), notice that inputs -2 and 2 each correspond to output 6. This does not violate the definition of a function – two different x-coordinates can have the same y-coordinate. A violation of the definition occurs when two ordered pairs have the same x-coordinate and different y-coordinates, as in Example 2(c).

Work Smart

In a function, two different inputs can correspond to the same output, but two different outputs cannot be the result of a single input.

Quick ✔

In Problems 5 and 6, determine whether each relation is a function. If so, state its domain and range.

5. $\{(-3, 3), (-2, 2), (-1, 1), (0, 0), (1, 1)\}$

6. $\{(-3, 2), (-2, 5), (-1, 8), (-3, 6)\}$

▶ ❷ Determine Whether a Relation Expressed as an Equation Represents a Function

We now know how to identify when a relation defined by a map or ordered pairs is a function. In Section 8.2, relations were also expressed as equations. We will now address the circumstances under which equations are functions.

To determine whether an equation, where y depends on x, is a function, it is often easiest to solve the equation for y. If each value of x corresponds to exactly one value of y, the equation is a function; otherwise, it is not.

EXAMPLE 3 **Determining Whether an Equation Is a Function**

Determine whether the equation $y = 3x + 5$ shows y as a function of x.

Solution

The rule for getting from x to y is to multiply x by 3 and then add 5. Since only one output y can result by performing these operations on any input x, the equation is a function. ●

EXAMPLE 4 **Determining Whether an Equation Is a Function**

Determine whether the equation $y = \pm x^2$ shows y as a function of x.

In Other Words
The symbol $\pm$ is a shorthand device and is read "plus or minus." For example, ± 4 means "negative four or positive four."

Solution

Notice that for any single value of x (other than 0), two values of y result. For example, if $x = 2$, then $y = \pm 4$ (-4 or $+4$). Since a single x corresponds to more than one y, the equation is not a function. ●

Quick ✔

In Problems 7–9, determine whether each equation shows y as a function of x.

7. $y = -2x + 5$ **8.** $y = \pm 3x$ **9.** $y = x^2 + 5x$

▶ ❸ Determine Whether a Relation Expressed as a Graph Represents a Function

Remember that the graph of an equation is the set of all ordered pairs (x, y) that satisfy the equation. For a relation to be a function, each number x in the domain can correspond to only one number y in the range. This means that a graph is *not* a function if two points with the same x-coordinate have different y-coordinates. This leads to the following test.

Vertical Line Test

A set of points in the xy-plane is the graph of a function if and only if every vertical line intersects the graph in at most one point.

EXAMPLE 5 **Using the Vertical Line Test to Identify Graphs of Functions**

Which of the graphs in Figure 18 are graphs of functions?

Figure 18

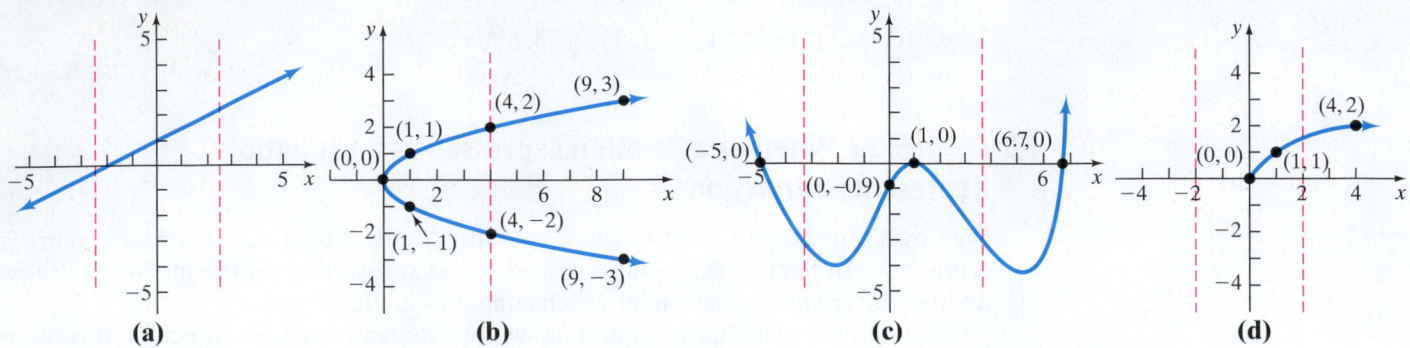

(a)   (b)   (c)   (d)

Solution

The graphs in Figures 18(a), (c), and (d) are functions because every vertical line intersects these graphs in at most one point. The graph in Figure 18(b) is not a function, because a vertical line intersects the graph in more than one point.   ●

Does Example 5 show you why the vertical line test works? If a vertical line intersects the graph of an equation in two or more points, then the same x-coordinate corresponds to two or more different y-coordinates, and we have violated the definition of a function.

> **Quick ✓**
>
> **10.** *True or False* For a graph to be a function, any vertical line can intersect the graph in at most one point.
>
> *In Problems 11 and 12, use the vertical line test to determine whether the graph is a function.*
>
> **11.** 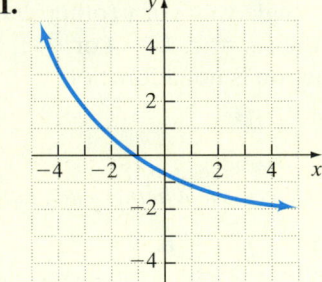   **12.**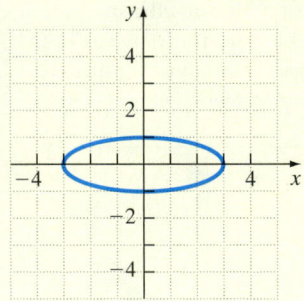

④ Find the Value of a Function

▶ Functions are often denoted by letters such as f, F, g, G, and so on. If f is a function, use special notation, called *function notation*, to represent the function. The following table shows some equations where y is a function of x. In each case, denote the function by the letter f.

Equation in Two Variables	Function Notation
$y = \dfrac{1}{2}x + 5$	$f(x) = \dfrac{1}{2}x + 5$
$y = -3x + 1$	$f(x) = -3x + 1$
$y = 2x^2 - 3x$	$f(x) = 2x^2 - 3x$

Work Smart

Be careful with function notation. In the expression $y = f(x)$, y is the dependent variable, x is the independent variable, and f is the name given to a rule that relates the input x to the output y.

Notice that $f(x)$ replaces the variable y. If f is a function, then for each number x in its domain, the corresponding value in the range is denoted by $f(x)$, which is read as "f of x" or as "f at x." Understand that $f(x)$ does not mean "f times x."

We call $f(x)$ the **value of the function f at the number x**; $f(x)$ is the number that results when the function is applied to x. Finding the value of a function parallels finding the value of y in an equation in two variables when y is a function of x. For example, "Given the equation $y = \frac{1}{2}x + 5$, find the value of y when $x = 4$" means the same as "Given the function $f(x) = \frac{1}{2}x + 5$, evaluate $f(4)$." In both cases, replace the value of x with 4 and evaluate.

EXAMPLE 6 **Finding Values of a Function**

If $f(x) = \frac{3}{2}x - 6$ and $g(x) = x^2 + 6x$, evaluate:

(a) $f(10)$ **(b)** $f(-8)$ **(c)** $g(3)$ **(d)** $g(-2)$

Solution

(a) Wherever an x is in the equation defining the function f, substitute 10.

$$f(10) = \frac{3}{2}(10) - 6$$
$$= 15 - 6$$
$$= 9$$

So $f(10) = 9$.

(b) Substitute -8 for x in the expression $\frac{3}{2}x - 6$ to get

$$f(-8) = \frac{3}{2}(-8) - 6$$
$$= -12 - 6$$
$$= -18$$

So $f(-8) = -18$.

(c) Substitute 3 for x in the expression $x^2 + bx$ to get.

$$g(3) = (3)^2 + 6(3)$$
$$= 9 + 18$$
$$= 27$$

Work Smart

Notice the use of parentheses when -2 is squared in Example 6(d) Remember, $-2^2 \neq (-2)^2$.

(d) Substitute -2 for x in the expression $x^2 + 6x$ to get

$$g(-2) = (-2)^2 + 6(-2)$$
$$= 4 + (-12)$$
$$= -8$$ ●

Quick ✓

In Problems 13–16, let $f(x) = 3x + 2$ and $g(x) = -2x^2 + x + 3$. Evaluate each function.

13. $f(4)$ **14.** $f(-2)$ **15.** $g(-3)$ **16.** $g(1)$

For a function $y = f(x)$, the variable x is called the **independent variable**, because it can be assigned any of the numbers in the domain. The variable y is called the **dependent variable**, because its value depends on x.

The independent variable is also called the **argument** of the function. For example, if f is the function defined by $f(x) = x^2$, then f tells us to square the argument. Thus, $f(2)$ means to square 2, $f(a)$ means to square a, and $f(x + h)$ means to square the quantity $x + h$.

The notation $f(x)$ plays a dual role—it represents the rule for getting from the input to the output and it represents the output y of the function. In Example 6(c), the rule for getting from the input to the output is given by $g(x) = x^2 + 6x$. In other words, the function says to "take some input x, square it, and add the result to six times the input x." If the input is 3, then $g(3)$ represents the output, 27.

▶ **EXAMPLE 7** **Evaluate a Function**

For the function $F(z) = 4z + 7$, evaluate:

(a) $F(z + 3)$ **(b)** $F(z) + F(3)$

Solution

(a) Wherever a z is in the equation defining F, substitute $z + 3$ to get

$$F(z + 3) = 4(z + 3) + 7$$
$$= 4z + 12 + 7$$
$$= 4z + 19$$

(b) $F(z) + F(3) = \underbrace{4z + 7}_{F(z)} + \underbrace{4 \cdot 3 + 7}_{F(3)}$
$$= 4z + 7 + 12 + 7$$
$$= 4z + 26$$ ●

Quick ✓

17. In the function $H(q) = 2q^2 - 5q + 1$, H is called the _____ variable, and q is called the _____ variable or _____.

In Problems 18 and 19, let $f(x) = 2x - 5$. Evaluate each function.

18. $f(x - 2)$ **19.** $f(x) - f(2)$

Summary   Important Facts About Functions

1. Each x in the domain has exactly one corresponding y in the range.

2. Letters such as f are used to denote the function. It represents the rule used to get from an x in the domain to $f(x)$ in the range.

3. If $y = f(x)$, then x is the independent variable or argument of f, and y is the dependent variable or the value of f at x.

▶ ⑤ **Find the Domain of a Function**

In working with functions, it is important to know which inputs are possible and produce an output.

In Other Words

The domain of a function is the set of all inputs for which the function gives an output that is a real number.

Definition

When only the equation of a function is given, the **domain of f** is the set of real numbers x for which $f(x)$ is a real number.

When identifying the domain of a function, do not forget that division by zero is undefined. Exclude values of the variable that result in division by zero.

EXAMPLE 8 **Finding the Domain of a Function**

Find the domain of each of the following functions:

(a) $G(x) = x^2 + 1$ **(b)** $g(z) = \dfrac{z-3}{z+1}$

Solution

(a) The function G squares a number x and then adds 1 to the result. These operations can be performed on any real number, so the domain of G is the set of all real numbers, which we can express as $\{x \mid x \text{ is a real number}\}$ or, using interval notation, $(-\infty, \infty)$.

(b) The function g involves division. Since division by 0 is not defined, the denominator $z + 1$ cannot be 0. Therefore, z cannot equal -1. The domain of g is $\{z \mid z \neq -1\}$.

Work Smart

In Example 8(b), it is understood that $\{z \mid z \neq -1\}$ includes all real numbers except -1.

Quick ✓

20. When only the equation of a function f is given, the _____ of f is the set of real numbers x for which $f(x)$ is a real number.

In Problems 21 and 22, find the domain of each function.

21. $f(x) = 3x^2 + 2$ **22.** $h(x) = \dfrac{x+1}{x-3}$

In an application, a function's domain may be restricted by physical or geometric considerations, in addition to mathematical restrictions. For example, the domain of $f(x) = x^2$ is the set of all real numbers. However, if f is used to find the area of a square given side x, then restrict the domain of f to the positive real numbers, since the length of a side cannot be 0 or negative.

EXAMPLE 9 **Finding the Domain of a Function**

The number N of computers produced at a Dell Computers' manufacturing facility in one day after t hours is given by the function, $N(t) = 336t - 7t^2$. What is the domain of this function?

Solution

The independent variable is t, the number of hours in the day. Therefore the function's domain is $\{t \mid 0 \leq t \leq 24\}$, or the interval $[0, 24]$.

Quick ✓

23. The function $A(r) = \pi r^2$ gives the area of a circle A as a function of the radius r. What is the domain of the function?

▶ ❻ Work with Applications of Functions

In practice, the symbols used for the independent and dependent variables should remind us of what they represent. For example, in economics, C is used for cost and q for quantity, so that $C(q)$ represents the cost of manufacturing q units of a good. Here C is the dependent variable, q is the independent variable, and $C(q)$ is the rule that tells us how to get the output C from the input q.

EXAMPLE 10 **Life-Cycle Hypothesis**

The Life-Cycle Hypothesis from economics was presented by Franco Modigliani in 1954. It states that income is a function of age. The function

$$I(a) = -55a^2 + 5119a - 54{,}448$$

represents the relation between average annual income I and age a.

(a) Identify the dependent and independent variables.

(b) Evaluate $I(20)$ and explain what $I(20)$ represents.

Solution

(a) Because income depends on age, the dependent variable is income, I, and the independent variable is age, a.

(b) Let $a = 20$ in the function.

$$I(20) = -55(20)^2 + 5119(20) - 54{,}448$$
$$= 25{,}932$$

The average annual income of a 20-year-old is $25,932.

Quick ✓

24. In 2010, the *Deepwater Horizon* oil explosion spilled millions of gallons of oil in the Gulf of Mexico. The oil slick takes the shape of a circle. Suppose that the area A (in square miles) of the circle contaminated with oil can be determined using the function $A(t) = 0.25\pi t^2$, where t represents the number of days since the rig exploded.

(a) Identify the dependent and independent variables.

(b) Evaluate $A(30)$ and explain what $A(30)$ means.

8.3 Exercises MyMathLab® Exercise numbers in green have complete video solutions in MyMathLab or may be accessed using the QR code to the right.

Problems 1–24 are the Quick ✓s that follow the EXAMPLES.

Building Skills

In Problems 25–34, determine whether each relation represents a function. State the domain and the range of each relation. See Objective 1.

25.

State   Number of US House Seats

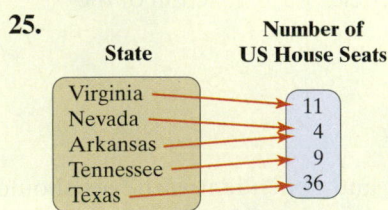

26.

Animal   Gestation Period (days)

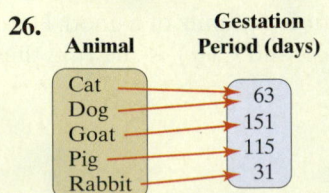

27. Horse-power   Top Speed

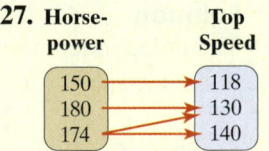

28. Grade on Exam   Study Time (hours)

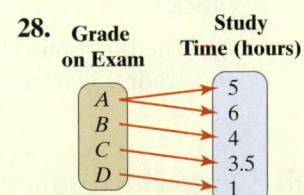

29. $\{(0, 3), (1, 4), (2, 5), (3, 6)\}$

30. $\{(-1, 4), (0, 1), (1, -2), (2, -5)\}$

31. $\{(-3, 5), (1, 5), (4, 5), (7, 5)\}$

32. $\{(-2, 3), (-2, 1), (-2, -3), (-2, 9)\}$

33. $\{(-10, 1), (-5, 4), (0, 3), (-5, 2)\}$

34. $\{(-5, 3), (-2, 1), (5, 1), (7, -3)\}$

In Problems 35–44, determine whether each equation shows y as a function of x. See Objective 2.

35. $y = 2x + 9$

36. $y = -6x + 3$

37. $2x + y = 10$

38. $6x - 3y = 12$

39. $y = \pm 5x$

40. $y = \pm 2x^2$

41. $y = x^2 + 2$

42. $y = x^3 - 3$

43. $x + y^2 = 10$

44. $y^2 = x$

In Problems 45–52, determine whether the graph is that of a function. See Objective 3.

45.
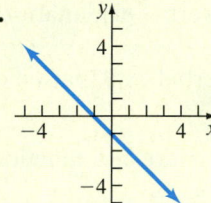

46.
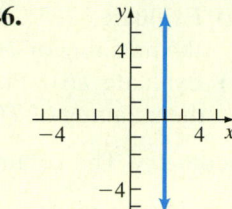

47.
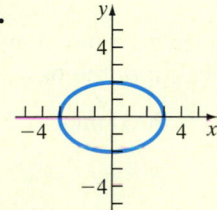

48.
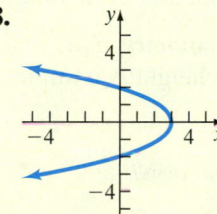

49.
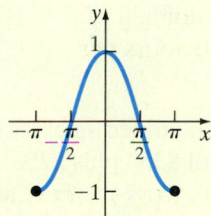

50.
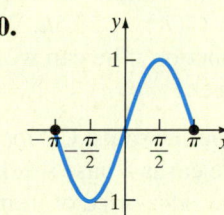

51.
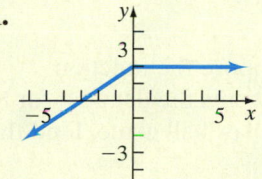

52.
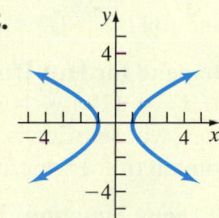

In Problems 53–60, find the indicated value of each function. See Objective 4.

   (a) $f(0)$ **(b)** $f(3)$ **(c)** $f(-2)$

53. $f(x) = 2x + 3$

54. $f(x) = 3x + 1$

55. $f(x) = -5x + 2$

56. $f(x) = -2x - 3$

57. $f(x) = x^2 - 3x$

58. $f(x) = 2x^2 + 5x$

59. $f(x) = -x^2 + x + 3$

60. $f(x) = -x^2 + 2x - 5$

In Problems 61–64, find the indicated value of each function. See Objective 4.

   (a) $f(-x)$ **(b)** $f(x + 2)$ **(c)** $f(2x)$
   (d) $-f(x)$ **(e)** $f(x + h)$

61. $f(x) = 2x - 5$ **62.** $f(x) = 4x + 3$

63. $f(x) = 7 - 5x$ **64.** $f(x) = 8 - 3x$

In Problems 65–72, evaluate each function. See Objective 4.

65. $f(x) = x^2 + 3; f(2)$

66. $f(x) = -2x^2 + x + 1; f(-3)$

67. $s(t) = -t^3 - 4t; s(-2)$

68. $g(h) = -h^2 + 5h - 1; g(4)$

69. $F(x) = |x - 2|; F(-3)$

70. $G(z) = 2|z + 5|; G(-6)$

71. $F(z) = \dfrac{z + 2}{z - 5}; F(4)$

72. $h(q) = \dfrac{3q^2}{q + 2}; h(2)$

In Problems 73–80, find the domain of each function. See Objective 5.

73. $f(x) = 4x + 7$ **74.** $G(x) = -8x + 3$

75. $F(z) = \dfrac{2z + 1}{z - 5}$ **76.** $H(x) = \dfrac{x + 5}{2x + 1}$

77. $f(x) = 3x^4 - 2x^2$ **78.** $s(t) = 2t^2 - 5t + 1$

79. $G(x) = \dfrac{3x - 5}{3x + 1}$ **80.** $H(q) = \dfrac{1}{6q + 5}$

Applying the Concepts

81. If $f(x) = 3x^2 - x + C$ and $f(3) = 18$, what is the value of C?

82. If $f(x) = -2x^2 + 5x + C$ and $f(-2) = -15$, what is the value of C?

83. If $f(x) = \dfrac{2x + 5}{x - A}$ and $f(0) = -1$, what is the value of A?

84. If $f(x) = \dfrac{-x + B}{x - 5}$ and $f(3) = -1$, what is the value of B?

△ **85. Geometry** Express the area A of a circle as a function of its radius, r. Determine the area of a circle whose radius is 4 inches. That is, find $A(4)$.

△ **86. Geometry** Express the area A of a triangle as a function of its height h, assuming that the length of the base is 8 centimeters. Determine the area of this triangle if its height is 5 centimeters. That is, find $A(5)$.

87. Salary Express the gross salary G of Jackie, who earns \$15 per hour, as a function of the number of hours worked, h. Then determine the gross salary of Jackie if she works 25 hours. That is, find $G(25)$.

88. Commissions Roberta is a commissioned salesperson. She earns a base weekly salary of \$250 per week plus 15% of the sales price of items sold. Express her gross salary G as a function of the price p of items sold. Determine the weekly gross salary of Roberta if the value of items sold is \$10,000. That is, find $G(10,000)$.

89. Population as a Function of Age The function $P(a) = -1.02a^2 + 57.36a + 3694.1$ represents the population (in thousands) of U.S. residents in 2014, P, that are a years of age.
SOURCE: *United States Census Bureau*

(a) Identify the dependent and independent variables.

(b) Evaluate $P(20)$. Provide a verbal explanation of the meaning of $P(20)$.

(c) Evaluate $P(0)$. Provide a verbal explanation of the meaning of $P(0)$.

90. Number of Rooms The function $N(r) = -1.33r^2 + 14.68r - 17.09$ represents the number of housing units (in millions), N, in 2015 that have r rooms, where $1 \leq r \leq 9$.
SOURCE: *United States Census Bureau*

(a) Identify the dependent and independent variables.

(b) Evaluate $N(3)$. Provide a verbal explanation of the meaning of $N(3)$.

(c) Why is it unreasonable to evaluate $N(0)$?

91. Revenue Function The function $R(p) = -p^2 + 200p$ represents the daily revenue R earned from selling MP3 players at p dollars for $0 \leq p \leq 200$.

(a) Identify the dependent and independent variables.

(b) Evaluate $R(50)$. Provide a verbal explanation of the meaning of $R(50)$.

(c) Evaluate $R(120)$. Provide a verbal explanation of the meaning of $R(120)$.

92. Average Trip Length The function $T(x) = 0.01x^2 - 0.12x + 8.89$ represents the average vehicle trip length T (in miles) x years since 1969.

(a) Identify the dependent and independent variables.

(b) Evaluate $T(35)$. Provide a verbal explanation of the meaning of $T(35)$.

(c) Evaluate $T(0)$. Provide a verbal explanation of the meaning of $T(0)$.

△ **93. Geometry** The volume V of a sphere as a function of its radius r is given by $V(r) = \dfrac{4}{3}\pi r^3$. What is the domain of this function?

△ **94. Geometry** The area A of a triangle as a function of its height h, assuming that the length of the base is 5 centimeters, is $A = \dfrac{5}{2}h$. What is the domain of the function?

95. Salary The gross salary G of Kevin as a function of the number of hours worked, h, is given by $G(h) = 22.5h$. What is the domain of the function if he can work up to 60 hours per week?

96. Commissions Carlos is a commissioned salesperson. He earns a base weekly salary of \$350 plus 12% of the sales price of items sold. His gross salary G as a function of the price p of items sold is given by $G(p) = 350 + 0.12p$. What is the domain of the function?

97. Demand for Hot Dogs Suppose the function $D(p) = 1200 - 10p$ represents the demand for hot dogs, whose price is p, at a baseball game. Find the domain of the function.

98. Revenue Function The function $R(p) = -p^2 + 200p$ represents the daily revenue earned from selling MP3 players at p dollars for $0 \leq p \leq 200$. Explain why any p greater than \$200 is not in the domain of the function.

Extending the Concepts

99. Math for the Future: College Algebra A **piecewise-defined function** is a function defined by more than one equation. For example, the absolute value function $f(x) = |x|$ is actually defined by two equations: $f(x) = x$ if $x \geq 0$ and $f(x) = -x$ if $x < 0$. We can combine these equations into one expression as

$$f(x) = \begin{cases} -x & x < 0 \\ x & x \geq 0 \end{cases}$$

To evaluate $f(3)$, we recognize that $3 \geq 0$, so we use the rule $f(x) = x$ and obtain $f(3) = 3$. To evaluate $f(-4)$, we recognize that $-4 < 0$, so we use the rule $f(x) = -x$ and obtain $f(-4) = -(-4) = 4$.

(a) $f(x) = \begin{cases} x + 3 & x < 0 \\ -2x + 1 & x \geq 0 \end{cases}$

(i) Find $f(3)$.   **(ii)** Find $f(-2)$.   **(iii)** Find $f(0)$.

(b) $f(x) = \begin{cases} -3x + 1 & x < -2 \\ x^2 & x \geq -2 \end{cases}$

(i) Find $f(-4)$.   **(ii)** Find $f(2)$.   **(iii)** Find $f(-2)$.

100. Math for the Future: Calculus

(a) If $f(x) = 3x + 7$, find $\dfrac{f(x + h) - f(x)}{h}$.

(b) If $f(x) = -2x + 1$, find $\dfrac{f(x + h) - f(x)}{h}$.

Explaining the Concepts

101. Investigate when the use of function notation $y = f(x)$ first appeared. Start by researching Lejeune Dirichlet.

102. Are all relations functions? Are all functions relations? Explain your answers.

103. Explain what a function is. Be sure to include the terms *domain* and *range* in your explanation.

104. Explain why the vertical line test can be used to identify the graph of a function.

105. What are the four forms of a function presented in this section?

106. Explain why the terms *independent variable* for x and *dependent variable* for y make sense in the function $y = f(x)$.

Technology Exercises

Graphing calculators have the ability to evaluate any function you wish. Figure 19 shows the results obtained in Example 6(c) and 6(d) using a graphing calculator.

Figure 19

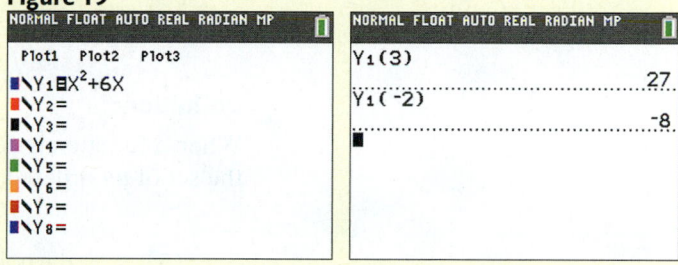

In Problems 107–114, use technology to find the value of each function.

107. $f(x) = x^2 + 3; f(2)$

108. $f(x) = -2x^2 + x + 1; f(-3)$

109. $F(x) = |x - 2|; F(-3)$

110. $G(z) = 2|z + 5|; G(-6)$

111. $H(x) = \sqrt{4x - 3}; H(7)$

112. $g(h) = \sqrt{2h + 1}; g(4)$

113. $F(z) = \dfrac{0.1z + 2.2}{0.2z - 4.8}; F(4)$

114. $h(q) = \dfrac{1.2q^2}{q + 2.8}; h(2)$

8.4 Functions and Their Graphs

Objectives

1 Graph a Function

2 Obtain Information from the Graph of a Function

3 Know Properties and Graphs of Basic Functions

4 Interpret Graphs of Functions

Prepared?...Answers   P1. $\{4\}$

P2.

Are You Prepared for This Section?

Before getting started, complete the following problems. If you get a problem wrong, go back to the section cited and review the material.

P1. Solve: $3x - 12 = 0$ [Section 2.2, pp. 91–92]

P2. Graph $y = x^2$ by point-plotting.   [Section 8.1, pp. 524–526]

▶ **1 Graph a Function**

The graph of the linear equation $y = -2x + 4$ (see Figure 20 on the next page) passes the vertical line test, so the equation is a function. This can be written as $f(x) = -2x + 4$. The graph of a function is the same as the graph of the equation that defines the function. The horizontal axis represents the independent variable and the vertical axis represents the dependent variable. When graphing functions, label the horizontal axis x and the vertical axis either y or by the name of the function.

Figure 20

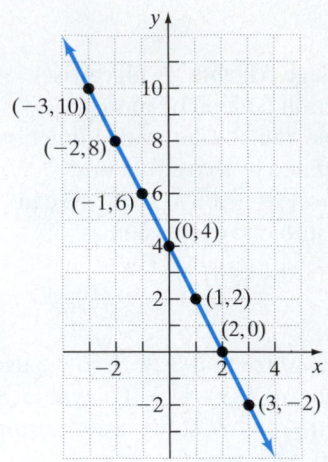

Definition

When a function is defined by an equation in x and y, the **graph of the function** is the set of *all* ordered pairs (x, y) such that $y = f(x)$.

So, if $f(3) = 8$, the point whose ordered pair is $(3, 8)$ is on the graph of $y = f(x)$.

EXAMPLE 1 **Graphing a Function**

Graph the function $f(x) = |x|$.

Solution
To graph $f(x) = |x|$, first determine some ordered pairs $(x, f(x)) = (x, y)$ such that $y = |x|$. See Table 6. Now plot the ordered pairs (x, y) from Table 6 and connect the points as shown in Figure 21.

Table 6

x	$f(x) = |x|$	$(x, f(x))$
-3	$|-3| = 3$	$(-3, 3)$
-2	$|-2| = 2$	$(-2, 2)$
-1	$|-1| = 1$	$(-1, 1)$
0	$|0| = 0$	$(0, 0)$
1	$|1| = 1$	$(1, 1)$
2	$|2| = 2$	$(2, 2)$
3	$|3| = 3$	$(3, 3)$

Figure 21

$f(x) = |x|$

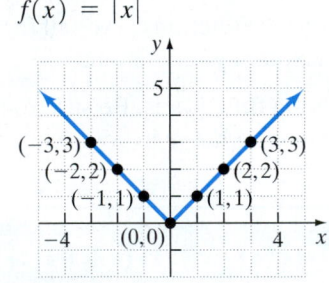

Quick ✓

1. When a function is defined by an equation in x and y, the _____ of the _____ is the set of all ordered pairs (x, y) such that $y = f(x)$.

2. If $f(4) = -7$, then the point whose ordered pair is $(\_, \_)$ is on the graph of $y = f(x)$.

In Problems 3–5, graph each function.

3. $f(x) = -2x + 9$ 4. $f(x) = x^2 + 2$ 5. $f(x) = |x - 2|$

❷ Obtain Information from the Graph of a Function

▶ Remember, the domain of a function is the set of all inputs, and the range is the set of all outputs of the function. The domain and the range of a function can be found from its graph.

EXAMPLE 2 **Finding the Domain and Range of a Function from Its Graph**

Figure 22 shows the graph of a function.

    (a) Determine the function's domain and range.

    (b) Identify the intercepts.

Figure 22

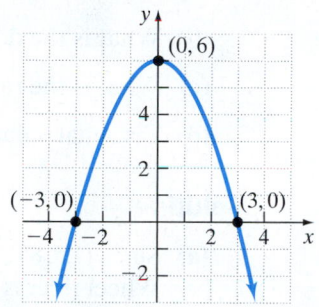

Solution

    (a) The domain of the function consists of all of the graph's *x*-coordinates. Because the graph exists for all real numbers *x*, the domain is $\{x \mid x \text{ is any real number}\}$, or the interval, $(-\infty, \infty)$.

        The range of the function consists of all of the graph's *y*-coordinates. Because the graph exists for all real numbers *y* less than or equal to 6, the range is $\{y \mid y \le 6\}$, or the interval, $(-\infty, 6]$.

    (b) The *x*-intercepts are $(-3, 0)$ and $(3, 0)$. The *y*-intercept is $(0, 6)$.   ●

Quick ✓

6. Use the graph of the function to answer parts (a) and (b).

    (a) Determine the domain and range of the function.

    (b) Identify the intercepts.

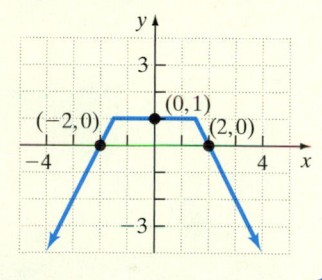

The next example illustrates how to obtain information about a function from its graph. Remember, if $(1, 5)$ is a point on the graph of *f*, then $f(1) = 5$.

EXAMPLE 3 **Obtaining Information from the Graph of a Function**

The Wonder Wheel is a Ferris wheel in Coney Island. See Figure 23 on the next page. Let *f* be the distance (in feet) above the ground of a person riding in a car on the Wonder Wheel as a function of time *x* (in minutes). Use the graph of *f* in Figure 24 to answer the following questions.

(continued)

Figure 23

Figure 24

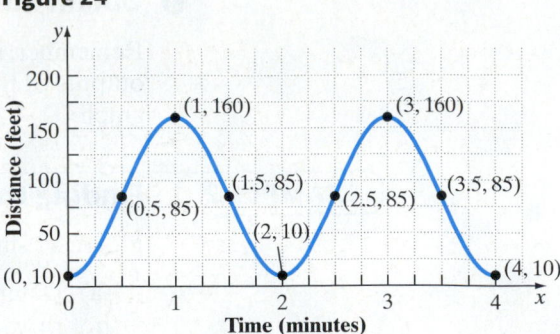

(a) Find $f(1.5)$ and $f(3)$. Interpret these values.

(b) What is the domain of f?

(c) What is the range of f?

(d) For what values of x does $f(x) = 85$? That is, solve $f(x) = 85$.

Solution

(a) Since $(1.5, 85)$ is on the graph of f, then $f(1.5) = 85$. After 1.5 minutes, a Wonder Wheel rider is 85 feet above the ground. Similarly, since $(3, 160)$ is on the graph, $f(3) = 160$. After 3 minutes, a Wonder Wheel rider is 160 feet above the ground.

(b) To determine the domain of f, notice that the graph exists for each number x between 0 and 4, inclusive. Therefore, the domain of f is $\{x \mid 0 \le x \le 4\}$, or the interval $[0, 4]$.

(c) The points on the graph have y-coordinates between 10 and 160, inclusive. Therefore, the range of f is $\{y \mid 10 \le y \le 160\}$, or the interval $[10, 160]$.

(d) Since $(0.5, 85)$, $(1.5, 85)$, $(2.5, 85)$, and $(3.5, 85)$ are the only points on the graph for which $y = f(x) = 85$, the solution set to the equation $f(x) = 85$ is $\{0.5, 1.5, 2.5, 3.5\}$. ●

Quick ✓

7. If the point $(3, 8)$ is on the graph of a function f, then $f(\_)=\_$. If $g(-2) = 4$, then $(\_,\_)$ is a point on the graph of g.

8. Use the graph of $y = f(x)$ to answer the following questions.

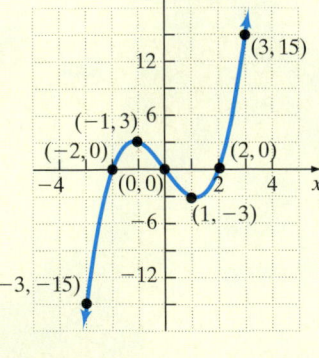

(a) Find $f(-3)$ and $f(1)$.

(b) What is the domain of f?

(c) What is the range of f?

(d) Identify the intercepts.

(e) For what value of x does $f(x) = 15$? That is, solve $f(x) = 15$.

▶ **EXAMPLE 4**   **Obtaining Information about the Graph of a Function**

Consider the function $f(x) = 2x - 5$.

(a) Is the point $(3, -1)$ on the graph of the function?

(b) If $x = 1$, what is $f(x)$? Based on this result, what point is on the graph of the function?

(c) If $f(x) = 3$, what is x? Based on this result, what point is on the graph of f?

Solution

(a) If $x = 3$, then

$$f(x) = 2x - 5$$
$$f(3) = 2(3) - 5 = 6 - 5 = 1$$

Since $f(3) = 1$, the point $(3, 1)$ is on the graph but $(3, -1)$ is not.

(b) If $x = 1$, then

$$f(1) = 2(1) - 5 = 2 - 5 = -3$$

Therefore, the point $(1, -3)$ is on the graph of f.

(c) If $f(x) = 3$, then

$$f(x) = 3$$
$$2x - 5 = 3$$

Add 5 to both sides: $\qquad 2x = 8$

Divide both sides by 2: $\qquad x = 4$

If $f(x) = 3$, then $x = 4$. Therefore, the point $(4, 3)$ is on the graph of f. ●

> **Work Smart: Study Skills**
>
> Do not confuse the directions "Find $f(3)$" with "If $f(x) = 3$, what is x?" Write down and study errors that you commonly make so that you can avoid them.

Quick ✔

9. Consider the function $f(x) = -3x + 7$.

(a) Is the point $(-2, 1)$ on the graph of the function?

(b) If $x = 3$, what is $f(x)$? Based on this result, what point is on the graph of the function?

(c) If $f(x) = -8$, what is x? What point is on the graph of f?

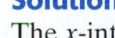 The Zero of a Function

> **In Other Words**
>
> A zero of a function is an input that makes the function equal 0.

If $f(r) = 0$ for some number r, then r is a **zero** of f. For example, if $f(x) = x^2 - 4$, then -2 and 2 are zeros of f because $f(-2) = 0$ and $f(2) = 0$. The zeros of a function can be identified from its graph by identifying the x-intercepts of the graph. Why? If $f(r) = 0$, then the point $(r, 0)$ is on the graph of f, and any point with coordinates $(r, 0)$ is an x-intercept of the graph.

EXAMPLE 5 **Finding the Zeros of a Function from Its Graph**

Find the zeros of the function f whose graph is shown in Figure 25.

Figure 25

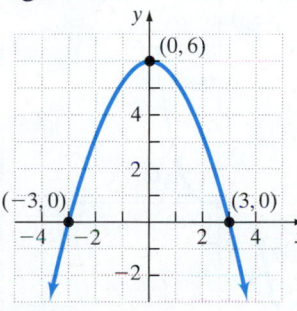

Solution

The x-intercepts of the graph are $(-3, 0)$ and $(3, 0)$. Therefore, the zeros of f are -3 and 3. ●

Quick ✔

In Problems 10–12, determine whether the value is a zero of the function.

10. $f(x) = 2x + 6; -3$ $\qquad$ **11.** $g(x) = x^2 - 2x - 3; 1$

12. $h(z) = -z^3 + 4z; 2$

13. Find the zeros of the function f whose graph is shown.

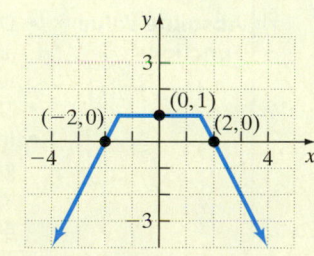

▶ ❸ Know Properties and Graphs of Basic Functions

In Table 7, we list a number of basic functions, their properties, and their graphs.

Table 7

Function	Properties	Graph
Linear Function $f(x) = mx + b$ m and b are real numbers	• Domain and range are all real numbers. • Graph is nonvertical line with slope $= m$ and y-intercept $= (0, b)$.	y, $f(x) = mx + b, m > 0$ $(0, b)$ x
Identity Function (special type of linear function) $f(x) = x$	• Domain and range are all real numbers. • Graph is a line with slope of $m = 1$ and y-intercept $= (0, 0)$. • The line consists of all points for which the x-coordinate equals the y-coordinate.	y, $f(x) = x$ $(1, 1)$ $(0, 0)$ $(-1, -1)$ x
Constant Function (special type of linear function) $f(x) = b$ b is a real number	• Domain is the set of all real numbers, and range is the set consisting of a single number b. • Graph is a horizontal line with slope $m = 0$ and y-intercept of $(0, b)$.	y $f(x) = b$ $(0, b)$ x
Square Function $f(x) = x^2$	• Domain is the set of all real numbers, and range is the set of nonnegative real numbers. • The intercept of the graph is $(0, 0)$.	y, $f(x) = x^2$ $(-2, 4)$ $(2, 4)$ $(-1, 1)$ $(1, 1)$ $(0, 0)$ x
Cube Function $f(x) = x^3$	• Domain and range are the set of all real numbers. • The intercept of the graph is $(0, 0)$.	y, $f(x) = x^3$ $(1, 1)$ $(0, 0)$ $(-1, -1)$ x
Absolute Value Function $f(x) = \lvert x \rvert$	• Domain is the set of all real numbers, and range is the set of nonnegative real numbers. • The intercept of the graph is $(0, 0)$. • If $x \geq 0$, then $f(x) = x$, and the graph of f is part of the line $y = x$; if $x < 0$, then $f(x) = -x$, and the graph of f is part of the line $y = -x$.	y, $f(x) = \lvert x \rvert$ $(-2, 2)$ $(2, 2)$ $(-1, 1)$ $(1, 1)$ $(0, 0)$ x 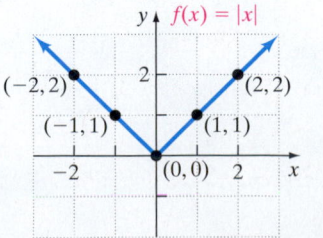

❹ Interpret Graphs of Functions

The graph of a function can be used to give a visual description of many different scenarios. Consider the following example.

EXAMPLE 6   **Graphing a Verbal Description**

Maria decides to take a walk. She leaves her house and walks 3 blocks in 2 minutes at a constant speed. She realizes that she left her front door unlocked, so she runs home in 1 minute. It takes Maria 1 minute to find her keys and lock the door. She next runs 10 blocks in 3 minutes and then rests for 1 minute. She walks 4 more blocks in 10 minutes, and finally hitches a ride home with her neighbor, who happens to drive by, and gets home in 2 minutes. Draw a graph of Maria's distance from home (in blocks) as a function of time.

Solution

Because distance from home is a function of time, draw a Cartesian plane with the horizontal axis representing the independent variable, time, and the vertical axis representing the dependent variable, distance from home.

The ordered pair $(0, 0)$ corresponds to starting the walk. The ordered pair $(2, 3)$ represents being 3 blocks from home after 2 minutes. Start the graph at $(0, 0)$ and then draw a straight line from $(0, 0)$ to $(2, 3)$. Next, draw a straight line to $(3, 0)$, which represents the return trip home to lock the door. Draw a line segment from $(3, 0)$ to $(4, 0)$ to represent the time it takes to lock the door. Draw a line segment from $(4, 0)$ to $(7, 10)$, which represents the 10 block run in 3 minutes. Now draw a horizontal line from $(7, 10)$ to $(8, 10)$. This represents the resting period. Draw a line from $(8, 10)$ to $(18, 14)$ to represent the 4-block walk in 10 minutes. Finally, draw a line segment from $(18, 14)$ to $(20, 0)$ to represent the ride home. See Figure 26.

Figure 26

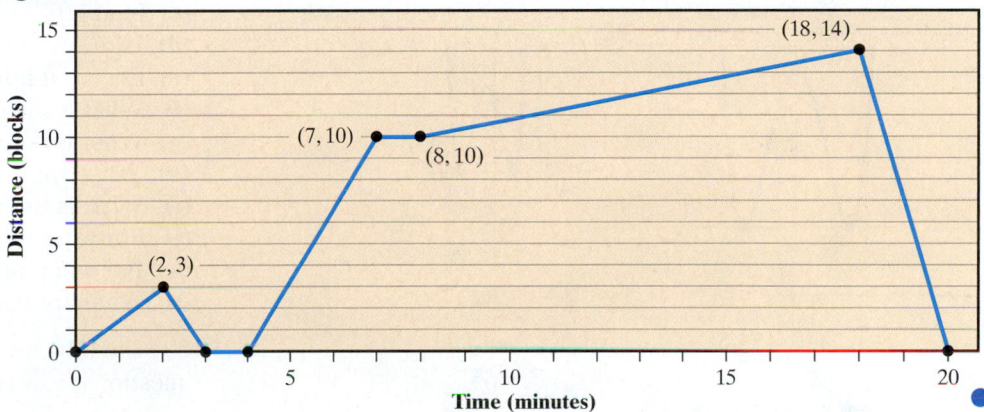

Quick ✔

14. Clara decides to take a walk. She leaves her house and walks 5 blocks in 5 minutes at a constant speed. She realizes that she left her front door unlocked, so she runs home in 2 minutes. It takes her 1 minute to find her keys and lock the door. She next jogs 8 blocks in 5 minutes, and then runs 3 blocks in 1 minute. After resting for 2 minutes she walks home in 10 minutes. Draw a graph of Clara's distance from home (in blocks) as a function of time (in minutes).

8.4 Exercises   MyMathLab®   Exercise numbers in **green** have complete video solutions in MyMathLab or may be accessed using the QR code to the right.

*Problems **1–14** are the **Quick ✔** s that follow the **EXAMPLES**.*

Building Skills

In Problems 15–22, graph each function. See Objective 1.

15. $f(x) = 4x - 6$ **16.** $g(x) = -3x + 5$

17. $h(x) = x^2 - 2$ **18.** $F(x) = x^2 + 1$

19. $G(x) = |x - 1|$ **20.** $H(x) = |x + 1|$

21. $g(x) = x^3$ **22.** $h(x) = x^3 - 3$

In Problems 23–32, for each graph of a function, find (a) the domain and the range, (b) the intercepts, if any, and (c) the zeros, if any. See Objective 2.

23.

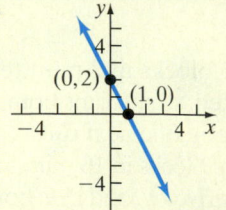

24.

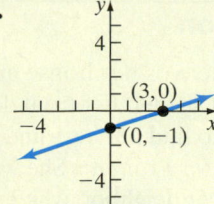

25.

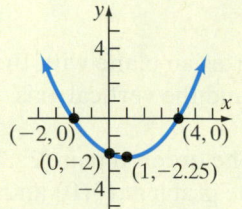

26.

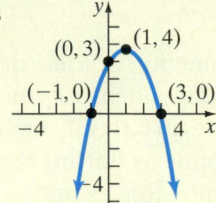

27.

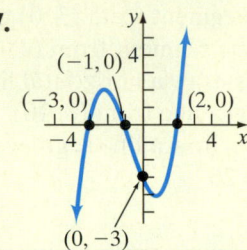

28.

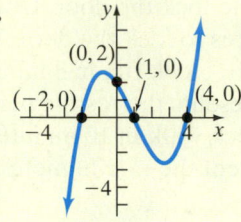

29.

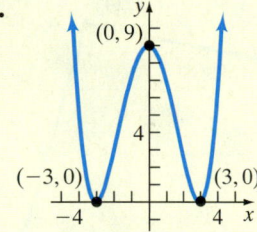

30.

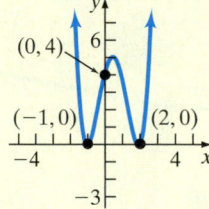

31.

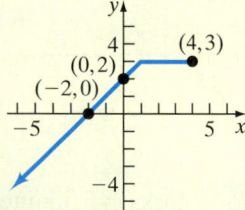

32.

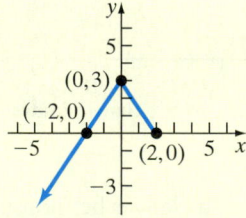

33. Use the graph of the function *f* shown to answer parts (a) – (l).

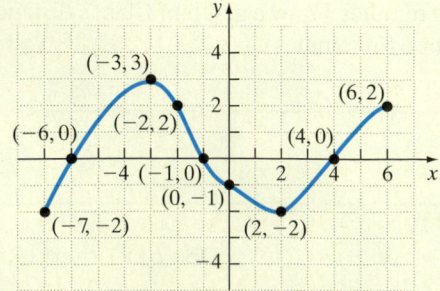

(a) Find $f(-7)$.
(b) Find $f(-3)$.

(c) Find $f(6)$.
(d) Is $f(2)$ positive or negative?
(e) For what numbers x is $f(x) = 0$?
(f) What is the domain of f?
(g) What is the range of f?
(h) What are the x-intercepts?
(i) What is the y-intercept?
(j) For what numbers x is $f(x) = -2$?
(k) For what number x is $f(x) = 3$?
(l) What are the zeros of f?

34. Use the graph of the function g shown to answer parts (a) – (l).

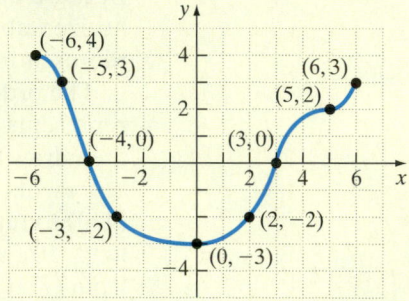

(a) Find $g(-3)$.
(b) Find $g(5)$.
(c) Find $g(6)$.
(d) Is $g(-5)$ positive or negative?
(e) For what numbers x is $g(x) = 0$?
(f) What is the domain of g?
(g) What is the range of g?
(h) What are the x-intercepts?
(i) What is the y-intercept?
(j) For what numbers x is $g(x) = -2$?
(k) For what number x is $g(x) = 3$?
(l) What are the zeros of g?

35. Use the table of values for the function F to answer questions (a) – (e).

x	F(x)
−4	0
−2	3
−1	5
0	2
3	−6

(a) What is $F(-2)$?
(b) What is $F(3)$?
(c) For what number(s) x is $F(x) = 5$?
(d) What is the x-intercept of the graph of F?
(e) What is the y-intercept of the graph of F?

36. Use the table of values for the function G to answer questions (a) – (e).

x	$G(x)$
−7	−3
−4	0
0	5
3	8
7	5

(a) What is $G(3)$?   (b) What is $G(7)$?
(c) For what number(s) x is $G(x) = 5$?
(d) What is the x-intercept of the graph of G?
(e) What is the y-intercept of the graph of G?

In Problems 37–40, answer the questions about the given function. See Objective 2.

37. $f(x) = 4x - 9$

(a) Is the point $(2, 1)$ on the graph of the function?
(b) If $x = 3$, what is $f(x)$? What point is on the graph of the function?
(c) If $f(x) = 7$, what is x? What point is on the graph of f?
(d) Is 2 a zero of f?

38. $f(x) = 3x + 5$

(a) Is the point $(-2, 1)$ on the graph of the function?
(b) If $x = 4$, what is $f(x)$? What point is on the graph of the function?
(c) If $f(x) = -4$, what is x? What point is on the graph of f?
(d) Is −2 a zero of f?

39. $g(x) = -\dfrac{1}{2}x + 4$

(a) Is the point $(4, 2)$ on the graph of the function?
(b) If $x = 6$, what is $g(x)$? What point is on the graph of the function?
(c) If $g(x) = 10$, what is x? What point is on the graph of g?
(d) Is 8 a zero of g?

40. $H(x) = \dfrac{2}{3}x - 4$

(a) Is the point $(3, -2)$ on the graph of the function?
(b) If $x = 6$, what is $H(x)$? What point is on the graph of the function?
(c) If $H(x) = -4$, what is x? What point is on the graph of H?
(d) Is 6 a zero of H?

In Problems 41–46, match each graph to the function listed whose graph most resembles the one given. See Objective 3.

(a) Constant function   (b) Linear function
(c) Square function   (d) Cube function
(e) Absolute value function   (f) Identity function

41.

42.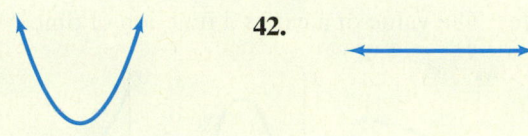

43. 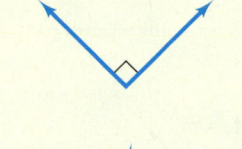 **44.**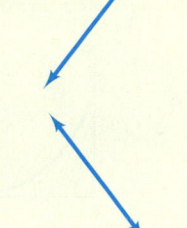

45. 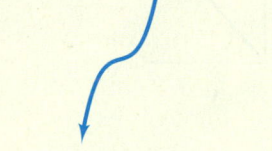 **46.**

In Problems 47–50, sketch the graph of each function. Label at least three points. See Objective 3.

47. $f(x) = x^2$ **48.** $f(x) = x^3$

49. $f(x) = |x|$ **50.** $f(x) = 4$

Applying the Concepts

51. Match each of the following functions with the graph that best describes the situation.

(a) The distance from ground level of a person who is jumping on a trampoline as a function of time
(b) The cost of a telephone call as a function of time
(c) The height of a human as a function of time
(d) The height of a rocket as a function of time
(e) The book value of a machine that is depreciated by equal amounts each year as a function of the year

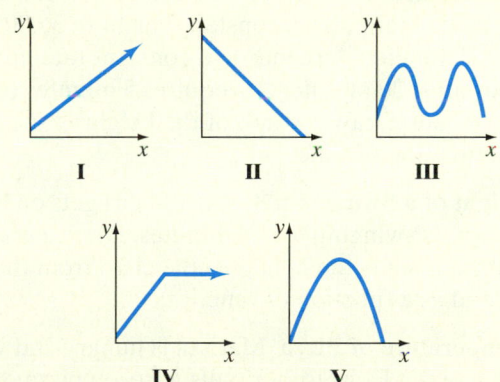

52. Match each of the following functions with the graph on the following page that best describes the situation.

(a) The average high temperature each day as a function of the day of the year
(b) The height of a human as a function of their age
(c) The distance that a person rides her bicycle at a constant speed as a function of time
(d) The temperature of a pizza after it is removed from the oven as a function of time

(e) The value of a car as a function of time

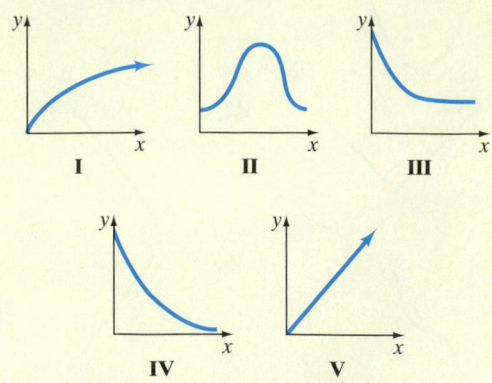

53. **Pulse Rate** Consider the following scenario: Zach starts jogging on a treadmill. His resting pulse rate is 70 beats per minute (b.p.m.). As he continues to jog on the treadmill, his pulse increases at a constant rate until, after 10 minutes, his pulse is 120 b.p.m. He then starts jogging faster and his pulse increases at a constant rate for 2 minutes, at which time his pulse is up to 150 b.p.m. He then begins a cooling-off period for 7 minutes until his pulse goes down to 110 b.p.m. He then gets off the treadmill and his pulse returns to 70 b.p.m. after 12 minutes. Draw a graph of Zach's pulse as a function of time.

54. **Altitude of an Airplane** Suppose that a plane is flying from Chicago to New Orleans. The plane leaves the gate and taxis for 5 minutes. The plane takes off and gets up to 10,000 feet after 5 minutes. The plane continues to ascend at a constant rate until it reaches its cruising altitude of 35,000 feet after another 25 minutes. For the next 80 minutes, the plane maintains a constant height of 35,000 feet. The plane then descends at a constant rate until it lands after 20 minutes. It requires 5 minutes to taxi to the gate. Draw a graph of the height of the plane as a function of time.

55. **Height of a Swing** An 8-year-old girl gets on a swing and starts swinging for 10 minutes. Draw a graph that represents the height of the child from the ground as a function of time.

56. **Temperature of Pizza** Marissa is hungry and would like a pizza. Her mother pulls a frozen pizza out of the freezer and puts it in the oven. After 12 minutes the pizza is done, but Mom lets the pizza cool for 5 minutes before serving it to Marissa. Draw a graph that represents the temperature of the pizza as a function of time.

57. The graph below shows the weight in pounds of a person as a function of his age in years. Describe the weight of the individual over the course of his life.

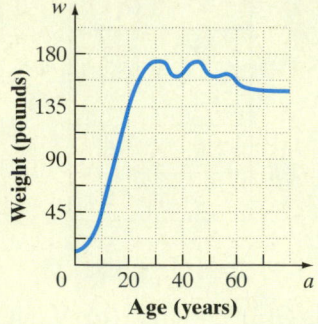

58. The following graph shows the depth of a lake (in feet) as a function of time (in days). Describe the depth of the lake over the course of the 400 days.

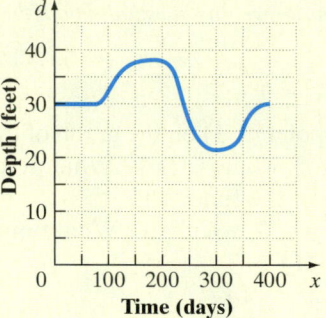

Extending the Concepts

59. Draw a graph of a function f with the following characteristics: x-intercepts: $(-4, 0)$, $(-1, 0)$, and $(2, 0)$; y-intercept: $(0, -2)$; $f(-3) = 7$ and $f(3) = 8$.

60. Draw a graph of a function f with the following characteristics: x-intercepts: $(-3, 0)$, $(2, 0)$, and $(5, 0)$; y-intercept: $(0, 3)$; $f(3) = -2$.

Explaining the Concepts

61. Using the definition of a function, explain why the graph of a function can have at most one y-intercept.

62. Explain what the domain of a function is. In your explanation, discuss how domains are determined in applications.

63. Explain what the range of a function is.

64. Explain the relationship between the x-intercepts of a function and its zeros.

65. Why are y-intercepts not considered zeros of a function?

Putting the Concepts Together (Sections 8.1–8.4)

We designed these problems so that you can review Sections 8.1 through 8.4 and show your mastery of the concepts. Take time to work these problems before proceeding with the next section. The answers are located at the back of the text on page AN-33.

1. Plot the following ordered pairs in the same xy-plane. Tell in which quadrant or on what coordinate axis each point lies.

$$A(7, 0), B(-2, 6), C(8, 4), D(0, -9),$$
$$E(-5, -10), F(6, -3)$$

2. Determine whether the ordered pair is a point on the graph of the equation $y = 4x - \dfrac{3}{2}$.

   (a) $\left(1, \dfrac{5}{2}\right)$ (b) $\left(\dfrac{1}{2}, \dfrac{1}{2}\right)$ (c) $\left(\dfrac{1}{4}, \dfrac{1}{4}\right)$

In Problems 3 and 4, graph the equations by plotting points.

3. $y = |x| + 3$ 4. $y = \dfrac{1}{2}x^2 - 1$

5. Identify the intercepts from the graph below.

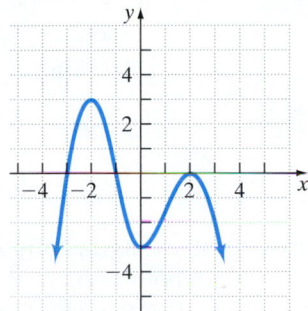

6. Explain why the following relation is a function. Then express the function as a set of ordered pairs.

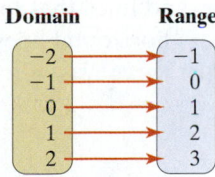

7. Determine which of the following relations represent functions.

   (a) $y = x^3 - 4x$ (b) $y = \pm 4x + 3$

8. Is the following relation a function? If so, state the domain and range.

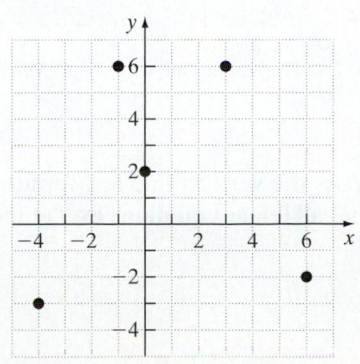

9. Explain why the relation whose graph below is a function. If the name of the function is f, find $f(5)$.

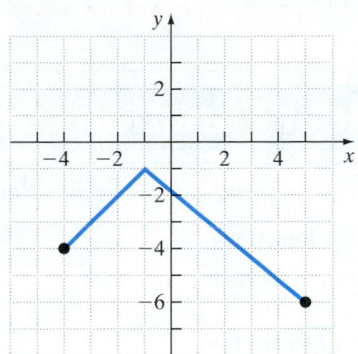

10. What is the zero of the function whose graph is shown?

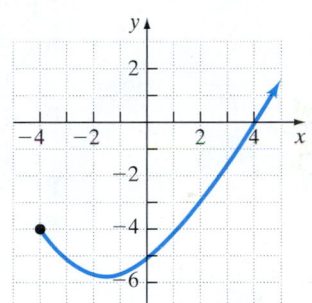

11. Let $f(x) = -5x + 3$ and $g(x) = -2x^2 + 5x - 1$. Find the value of each of the following.

   (a) $f(4)$ (b) $g(-3)$

   (c) $f(x) - f(4)$ (d) $f(x - 4)$

12. Find the domain of each of the following functions.

   (a) $G(h) = h^2 + 4$ (b) $F(w) = \dfrac{w - 4}{3w + 1}$

13. Graph $f(x) = |x| - 2$. Use the graph to state the domain and the range of f.

14. **Vertical Motion** The graph below shows the height, h, in feet, of a ball thrown straight up with an initial speed of 40 feet per second from an initial height of 80 feet after t seconds.

   (a) Find and interpret $h(2.5)$.

   (b) What is the domain of h?

   (c) What is the range of h?

   (d) For what value of t does $h(t) = 105$?

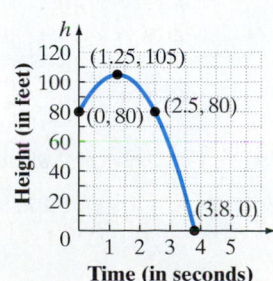

15. Consider the function $f(x) = 5x - 2$.

(a) Is the point $(3, 12)$ on the graph of the function?

(b) If $x = -2$, what is $f(x)$? What point is on the graph of the function?

(c) If $f(x) = -22$, what is x? What point is on the graph of f?

(d) Is $\dfrac{2}{5}$ a zero of f?

8.5 Linear Functions and Models

Objectives

1 Graph Linear Functions

2 Find the Zero of a Linear Function

3 Build Linear Models from Verbal Descriptions

4 Build Linear Models from Data

Are You Prepared for This Section?

Before getting started, complete the following problems. If you get a problem wrong, go back to the section cited and review the material.

P1. Graph: $y = 2x - 3$ [Section 3.4, pp. 206–207]

P2. Graph: $\dfrac{1}{2}x + y = 2$ [Section 3.2, pp. 186–190]

P3. Graph: $y = -4$ [Section 3.2, pp. 190–191]

P4. Graph: $x = 5$ [Section 3.2, pp. 190–191]

P5. Find and interpret the slope of the line through $(-1, 3)$ and $(3, -4)$.   [Section 3.3, pp. 195–198]

P6. Find the equation of the line through $(1, 3)$ and $(4, 9)$.   [Section 3.5, pp. 216–217]

P7. Solve: $0.5(x - 40) + 100 = 84$ [Section 2.3, pp. 101–102]

P8. Solve: $4x + 20 \geq 32$ [Section 2.8, pp. 150–156]

▶ **1** Graph Linear Functions

Recall from Chapter 3 that a linear equation in two variables has the form $Ax + By = C$, where A, B, and C are real numbers, and A and B are not both zero.

Consider the graphs of the four lines in Figure 27. Notice the lines that rise from left to right have positive slope, and lines that fall from left to right have negative slope. Lines that have zero slope are horizontal lines, and lines that have undefined slope are vertical lines. Remember, just as we read a text from left to right, we also read graphs from left to right.

Figure 27

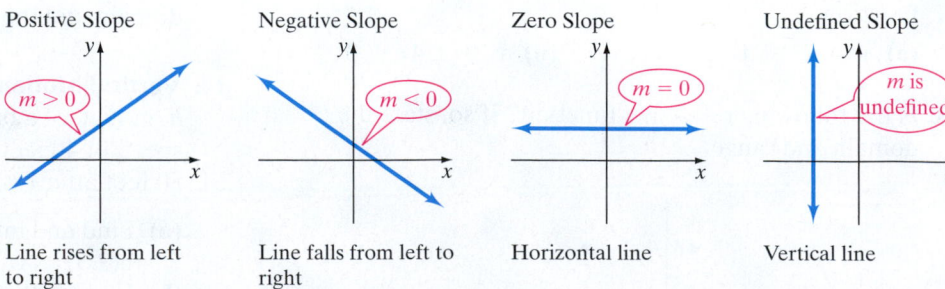

Positive Slope — Line rises from left to right

Negative Slope — Line falls from left to right

Zero Slope — Horizontal line

Undefined Slope — Vertical line

Prepared?...Answers

P1. **P2.** **P3.** **P4.**

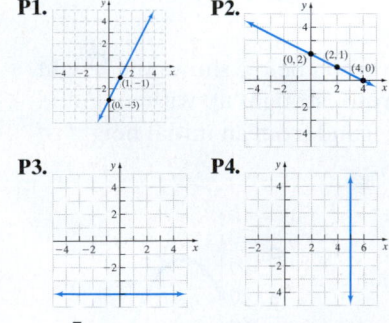

P5. $-\dfrac{7}{4}$; y decreases by 7 when x increases by 4. **P6.** $y = 2x + 1$ **P7.** $\{8\}$ **P8.** $\{x \mid x \geq 3\}$; $[3, \infty)$

In Figure 27, all the graphs except one pass the vertical line test for identifying graphs of functions. This leads to the conclusion that **all linear equations except those of the form $x = a$, vertical lines, are functions.**

Therefore any linear equation that is in the form $Ax + By = C$ can be written using function notation, provided $B \neq 0$, as follows:

$$Ax + By = C \quad B \neq 0$$

Subtract Ax from both sides: $\quad By = -Ax + C$

Divide both sides by B: $\quad \dfrac{By}{B} = \dfrac{-Ax + C}{B}$

Simplify: $\quad y = -\dfrac{A}{B}x + \dfrac{C}{B}$

$$\updownarrow \qquad \updownarrow \quad \updownarrow$$

$$f(x) = mx + b$$

This leads to the following definition:

> **Definition**
>
> A **linear function** is a function of the form
> $$f(x) = mx + b$$
> where m is the slope and $(0, b)$ is the y-intercept. The graph of a linear function is a line.

Linear functions can be graphed using the same techniques used to graph linear equations written in slope-intercept form, $y = mx + b$ (See Section 3.4).

EXAMPLE 1

Graphing a Linear Function

Graph the linear function: $f(x) = 3x - 5$

Figure 28

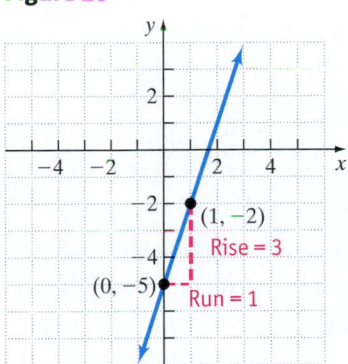

Solution
Since $f(x) = 3x - 5$ is in the form $f(x) = mx + b$, the y-intercept is $(0, -5)$, so plot that point in the coordinate plane. Because the slope $m = 3 = \dfrac{3}{1} = \dfrac{\Delta y}{\Delta x} = \dfrac{\text{rise}}{\text{run}}$, go to the right 1 unit and up 3 units and end up at $(1, -2)$. The line through these two points is the graph of $f(x) = 3x - 5$, shown in Figure 28. ●

Quick ✓

1. For the graph of a linear function $f(x) = mx + b$, m is the _____ and $(0, b)$ is the _____.

2. The graph of a linear function is called a ___.

3. *True or False* All linear equations are functions.

4. For the linear function $G(x) = -2x + 3$, the slope is ___ and the y-intercept is _____.

In Problems 5–8, graph each linear function.

5. $f(x) = 2x - 3$ 6. $G(x) = -5x + 4$ 7. $h(x) = \dfrac{3}{2}x + 1$ 8. $f(x) = 4$

Work Smart

Because $\dfrac{3}{1}$ is equivalent to $\dfrac{-3}{-1}$, we could go to the left 1 and down 3 and still remain on the line.

② Find the Zero of a Linear Function

In Section 8.4 we stated that if r is a zero of a function f, then $f(r) = 0$. To find the zero of any function f, solve the equation $f(x) = 0$.

EXAMPLE 2 **Finding the Zero of a Linear Function**

Find the zero of $f(x) = -4x + 12$.

Solution

Find the zero by solving $f(x) = 0$.

$$f(x) = 0$$
$$-4x + 12 = 0$$

Subtract 12 from both sides of the equation: $\qquad -4x = -12$

Divide both sides of the equation by -4: $\qquad x = 3$

Check: Since $f(3) = -4(3) + 12 = 0$, the zero of f is 3. ●

> **Quick ✓**
>
> In Problems 9–11, find the zero of each linear function.
>
> **9.** $f(x) = 3x - 15$ **10.** $G(x) = \dfrac{1}{2}x + 4$ **11.** $F(p) = -\dfrac{2}{3}p + 8$

▶ Applications of Linear Functions

Linear functions have many applications. For example, the cost of cab fare, sales commissions, and the cost of breakfast as a function of the number of eggs ordered can be modeled by linear functions.

EXAMPLE 3 **Sales Commissions**

Tony's weekly salary at Apple Chevrolet is 0.75% of his weekly sales plus $450, so $S(x) = 0.0075x + 450$ describes Tony's weekly salary S as a linear function of his weekly sales x.

(a) What is the implied domain of the function?

(b) If Tony sells cars worth a total of $50,000 one week, what is his salary?

(c) If Tony earned $600 one week, what was the value of the cars that he sold?

(d) Draw a graph of the function.

(e) For what value of cars sold will Tony's weekly salary exceed $1200?

Solution

(a) The independent variable is weekly sales, x. Because negative weekly sales do not make sense, the function's domain is $\{x \mid x \geq 0\}$ or, using interval notation, $[0, \infty)$.

(b) If Tony's weekly sales are $x = \$50,000$, then he earns

$$S(50,000) = 0.0075(50,000) + 450$$
$$= \$825$$

Tony will earn $825 for the week, if he sells $50,000 worth of cars.

(c) Here, solve the equation $S(x) = 600$.

$$0.0075x + 450 = 600$$

Subtract 450 from both sides: $\qquad 0.0075x = 150$

Divide both sides by 0.0075: $\qquad x = \$20,000$

If Tony earned $600, then he sold $20,000 worth of cars in a week.

(d) Plot the independent variable, weekly sales, along the horizontal axis and the dependent variable, salary, along the vertical axis. Graph the equation by plotting points. Part (b) shows that (50,000, 825) is on the graph. Part (c) indicates that (20,000, 600) is on the graph. Why do we also know that (0, 450) is on the graph? Plot these points to get the graph in Figure 29.

Figure 29

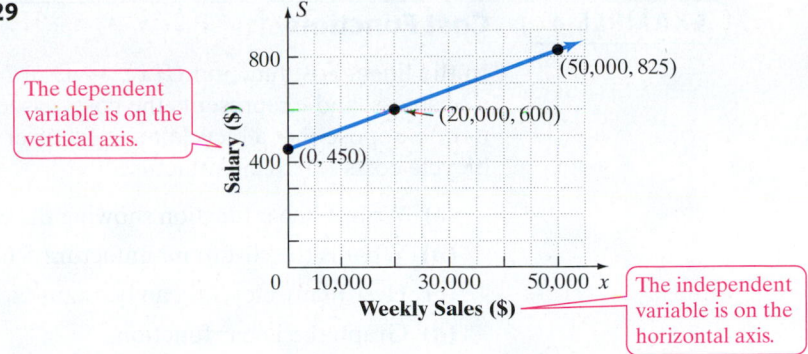

(e) Solve the inequality $S(x) > 1200$.

$$0.0075x + 450 > 1200$$

Subtract 450 from both sides: $0.0075x > 750$

Divide both sides of the inequality by 0.0075: $x > 100,000$

If Tony sells more than \$100,000 worth of cars for the week, his salary will exceed \$1200.   ●

Notice that in Figure 29 the function was graphed only over its domain, $[0, \infty)$—that is, only in quadrant I. Also notice that the horizontal axis is labeled x for the independent variable, weekly sales, and the vertical axis is labeled S for the dependent variable, salary. For this reason, the intercept on the vertical axis is the S-intercept, not the y-intercept. It is also indicated on the axes what x and S represent. Labeling your axes is always a good practice.

Quick ✔

12. The cost, C, of renting a 12-foot moving truck for a day is \$40 plus \$0.35 times the number of miles driven. The linear function $C(x) = 0.35x + 40$ describes the cost C of driving the truck x miles.

(a) What is the domain of this linear function?

(b) Determine the C-intercept of the graph of the linear function.

(c) What is the rental cost if the truck is driven 80 miles?

(d) How many miles was the truck driven if the rental cost is \$85.50?

(e) Graph the linear function.

(f) How many miles can you drive if you can spend up to \$127.50?

❸ Build Linear Models from Verbal Descriptions

A linear function has the form $f(x) = mx + b$, where m is the slope of the linear function and $(0, b)$ is its y-intercept. In Section 3.3, we learned that slope can be thought of as an average rate of change. Slope describes how much a dependent variable changes for a given change in the independent variable. For example, in the linear function $f(x) = 4x + 3$, the slope is $4 = \dfrac{4}{1} = \dfrac{\Delta y}{\Delta x}$, so the dependent variable y increases by 4 units for every 1-unit increase in x (the independent variable). When the average rate of change of a function is constant, a linear function is used to model the situation. For example, if your phone company charges \$0.05 per minute to talk regardless of the number of minutes on the phone, then a linear function can be used to model the cost of talking, with slope $m = \dfrac{0.05 \text{ dollar}}{1 \text{ minute}}$.

▶ **EXAMPLE 4** **Cost Function**

In the linear cost function $C(x) = ax + b$, b represents the fixed costs of operating a business, and a represents the costs associated with manufacturing one additional item. Suppose that a bicycle manufacturer has daily fixed costs of $2000 and each bicycle costs $80 to manufacture.

(a) Write a linear function showing the cost to manufacture x bicycles in a day.

(b) What is the cost to manufacture 5 bicycles in a day?

(c) How many bicycles can be manufactured for $2800?

(d) Graph the linear function.

Solution

(a) Because each bicycle costs $80 to manufacture, $a = 80$. The fixed costs are $2000, so $b = 2000$. Therefore, the cost function is

$$C(x) = 80x + 2000$$

(b) Evaluate the function for $x = 5$.

$$C(5) = 80(5) + 2000$$
$$= \$2400$$

It will cost $2400 to manufacture 5 bicycles.

(c) Solve $C(x) = 2800$.

$$C(x) = 2800$$
$$80x + 2000 = 2800$$

Subtract 2000 from both sides: $\qquad 80x = 800$

Divide both sides by 80: $\qquad x = 10$

Ten bicycles can be manufactured for $2800.

(d) Label the horizontal axis x and the vertical axis C. See Figure 30 for the graph of the cost function.

Figure 30

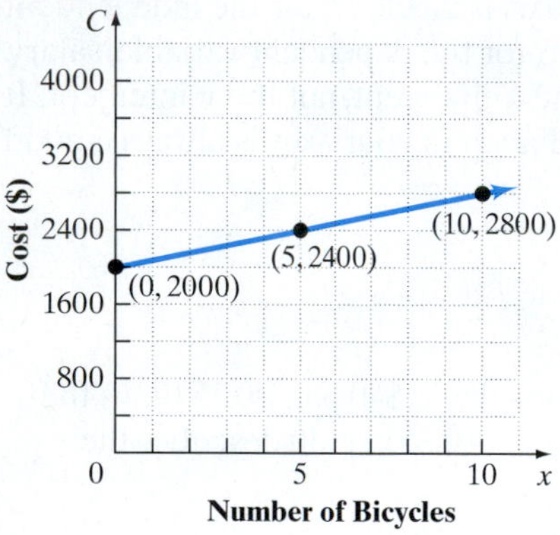

Number of Bicycles

Quick ✓

13. Suppose the business presented in Example 4 must pay a tax of $1 per bicycle produced.

(a) Write a linear function that expresses the cost C of producing x bicycles in a day.

(b) What is the cost of manufacturing 5 bicycles in a day?

(c) How many bicycles can be manufactured for $2810?

(d) Graph the linear function.

▶ **EXAMPLE 5** **Straight-Line Depreciation**

The *book value* of an asset such as a building or piece of machinery is the value that the company uses to create its balance sheet. Some companies use *straight-line depreciation* so that the book value of the asset declines by a constant amount each year. The amount of the decline depends on the useful life that the company places on the asset. Suppose Pearson Publishing Company just purchased a new fleet of cars for its sales force at a cost of $29,400 per car. The company uses the straight-line depreciation method for 7 years.

(a) Write a linear function that expresses the book value V of each car as a function of its age, x.

(b) What is the domain of this linear function?

(c) What is the book value of each car after 3 years?

(d) When will the book value of each car be $12,600?

(e) Graph the linear function.

Solution

(a) Let the linear function $V(x) = mx + b$ represent the book value of each car after x years. The original value of each car is $29,400$, so $V(0) = 29,400$. Thus the V-intercept of the function is $(0, 29,400)$. After 7 years, the book value of the car is 0. Use the ordered pairs $(0, 29,400)$ and $(7, 0)$ to find the slope, or amount of yearly (annual) depreciation, of V:

$$m = \frac{0 - 29,400}{7 - 0} = \frac{-29,400}{7} = -4200$$

So each car depreciates by $4200 per year. The linear function that represents the book value of each car after x years is

$$V(x) = -4200x + 29,400$$

(b) A car cannot have a negative age, so the age x must be greater than or equal to zero. In addition, each car is depreciated over 7 years, at which time $V(7) = 0$. Therefore, the domain of the function is $\{x \mid 0 \le x \le 7\}$, or $[0, 7]$ using interval notation.

(c) The book value of each car after $x = 3$ years is given by $V(3)$.

$$V(3) = -4200(3) + 29,400$$
$$= \$16,800$$

(d) To find when the book value is $12,600$, solve the equation

$$V(x) = 12,600$$
$$-4200x + 29,400 = 12,600$$

Subtract 29,400 from both sides: $-4200x = -16,800$

Divide both sides by -4200: $x = 4$

Each car will have a book value of $12,600 in 4 years.

(e) Label the horizontal axis x and the vertical axis V. Since $V(0) = 29,400$, the point $(0, 29,400)$ is on the graph. Since $V(7) = 0$, the point $(7, 0)$ is on the graph. To graph the function, we use these intercepts, along with the points $(3, 16,800)$ and $(4, 12,600)$ from parts (c) and (d). See Figure 31. ●

Figure 31

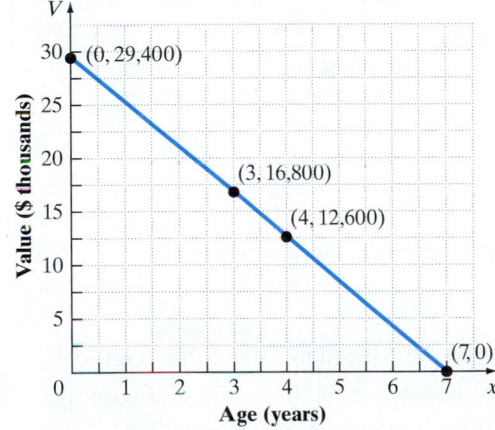

Quick ✓

14. Roberta's monthly payments for her new car are $250 per month. She estimates that maintenance and gas cost her $0.18 per mile.

(a) Write a linear function that expresses the monthly cost C of operating the car as a function of miles driven, x.

(b) What is the domain of this linear function?

(c) What is the monthly cost of driving 320 miles?

(d) How many miles can Roberta drive each month if she can afford the monthly cost to be $282.40?

(e) Graph the linear function.

▶ ❹ Build Linear Models from Data

If we have a set of data with more than two points, how can we tell whether the data (the two variables) are related linearly? There are some methods (beyond the scope of this course) for determining whether two variables are linearly related, but we can also draw a picture of the data, a *scatter diagram*, and learn whether the variables *might* be linearly related.

Scatter Diagrams

The graph consisting of the ordered pairs that make up a relation is called a **scatter diagram**.

EXAMPLE 6 **Drawing a Scatter Diagram**

The on-base percentage for a baseball team is the percent of time that the team safely reaches base. Table 8 shows the number of runs scored and the on-base percentage for various teams in the 2015 season.

Table 8

Team	On-Base Percentage, x	Runs Scored, y	(x, y)
NY Yankees	32.3	764	(32.3, 764)
Los Angeles Angels	30.7	661	(30.7, 661)
Texas Rangers	32.5	751	(32.5, 751)
Toronto Blue Jays	34.0	891	(34.0, 891)
Minnesota Twins	30.5	696	(30.5, 696)
Oakland A's	31.2	694	(31.2, 694)
Kansas City Royals	32.2	724	(32.2, 724)
Baltimore Orioles	30.7	713	(30.7, 713)

SOURCE: *espn.com*

(a) Draw a scatter diagram of the data, treating on-base percentage as the independent variable.

(b) Describe what happens as the on-base percentage increases.

Solution

(a) To draw a scatter diagram, plot the ordered pairs listed in Table 8. See Figure 32.

(b) The scatter diagram reveals that for the most part, as the on-base percentage increases, the number of runs scored also increases. While this relation is not perfectly linear (because the points don't all fall on a straight line), the *pattern* of the data is linear.

Figure 32

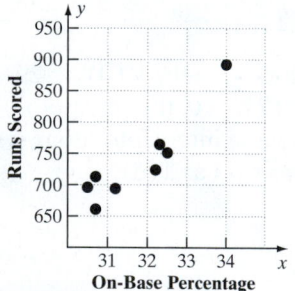

Quick ✔

15. The data listed below represent the total cholesterol (in mg/dL) and age of males.

Age, x	Total Cholesterol, y	Age, x	Total Cholesterol, y
25	180	38	239
25	195	48	204
28	186	51	243
32	180	62	228
32	197	65	269

(a) Draw a scatter diagram, treating age as the independent variable.

(b) Describe the relation between age and total cholesterol.

Recognizing the Type of Relation Between Two Variables

Scatter diagrams help us see the type of relation between two variables. This section will focus on distinguishing linear relations from nonlinear relations. See Figure 33.

Figure 33

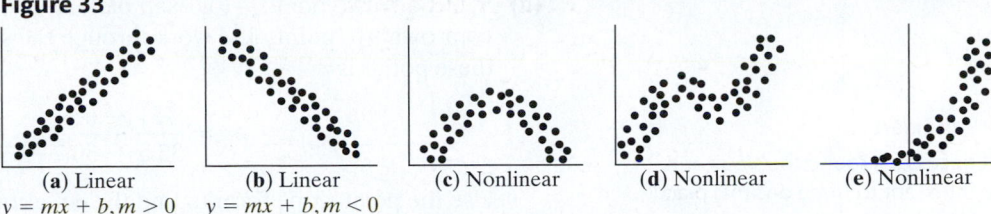

(a) Linear
$y = mx + b, m > 0$

(b) Linear
$y = mx + b, m < 0$

(c) Nonlinear

(d) Nonlinear

(e) Nonlinear

EXAMPLE 7

Distinguishing Between Linear and Nonlinear Relations

Determine whether each relation in Figure 34 is linear or nonlinear. If it is linear, state whether the slope is positive or negative.

Figure 34

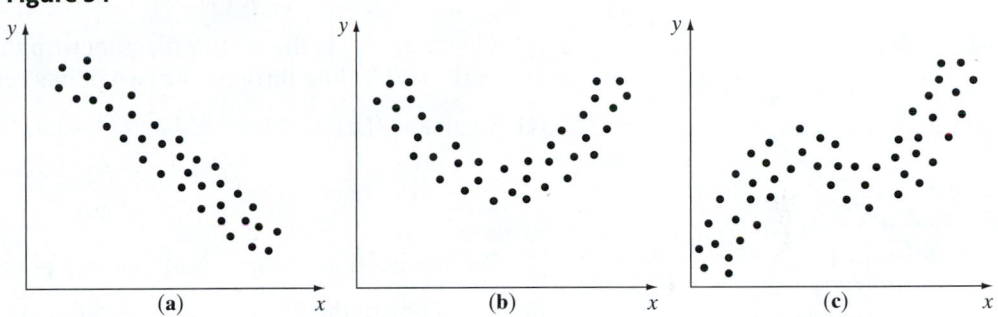

(a)   (b)   (c)

Solution

(a) Linear with negative slope   **(b)** Nonlinear   **(c)** Nonlinear   ●

> ### Quick ✔
>
> *In Problems 16 and 17, determine whether each relation is linear or nonlinear. If it is linear, state whether the slope is positive or negative.*
>
> **16.** 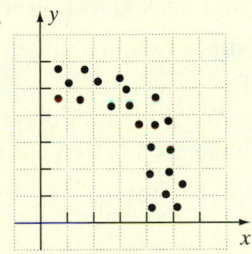   **17.**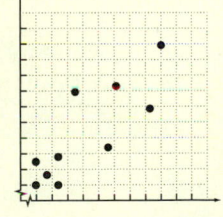

Fitting a Line to Data

Suppose a scatter diagram indicates a linear relationship, as in Figure 33(a) or (b). To find a linear equation for such data, draw a line through two points on the scatter diagram. Then determine the equation of that line by using the point-slope form of a line, $y - y_1 = m(x - x_1)$. To review using this formula, work Problem P6 in "*Are You Prepared for This Section?*" on page 560, if you haven't already done so.

EXAMPLE 8

Finding a Model for Linearly Related Data

Using the data in Table 8 from Example 6,

(a) Select two points and find a linear function for the line connecting the points.

(b) Graph the line on the scatter diagram from Example 6(a).

(c) Use the function found in part (a) to predict the number of runs scored by a team whose on-base percentage is 33.5%.

(continued)

(d) Interpret the slope as a rate of change in the context of the problem. Does it make sense to interpret the *y*-intercept? Explain.

Solution

(a) Select any two points—for example, $(30.7, 661)$ and $(32.5, 751)$. (You should select your own two points and work through the solution.) The slope of the line joining these points is

Work Smart

If necessary, round the slope and *y*-intercept to three decimal places.

$$m = \frac{751 - 661}{32.5 - 30.7} = \frac{90}{1.8} = 50$$

Use the point-slope form to find the equation of the line that has slope 50 passing through $(30.7, 661)$:

$$\begin{aligned}
\text{Point-slope form:} \quad & y - y_1 = m(x - x_1) \\
m = 50; x_1 = 30.7, y_1 = 661: \quad & y - 661 = 50(x - 30.7) \\
\text{Distribute:} \quad & y - 661 = 50x - 1535 \\
\text{Add 661 to both sides:} \quad & y = f(x) = 50x - 874
\end{aligned}$$

Figure 35

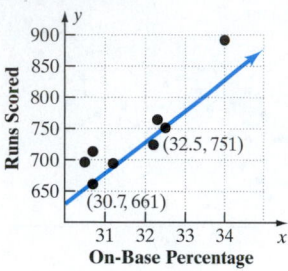

On-Base Percentage

(b) Figure 35 shows the scatter diagram with the graph of the line found in part (a). The line through the two points selected in part (a) is graphed.

(c) Evaluate $f(x) = 50x - 874$ at $x = 33.5$.

$$\begin{aligned}
f(33.5) &= 50(33.5) - 874 \\
&= 801
\end{aligned}$$

We predict that a team whose on-base percentage is 33.5% will score 801 runs.

(d) The slope of the linear function is 50, so if the on-base percentage increases by 1, then the number of runs scored will increase by 50. The *y*-intercept, -874, represents the runs scored when the on-base percentage is 0. Since negative runs scored does not make sense, interpreting the *y*-intercept does not make sense.

Quick ✓

18. Using the data from Quick Check Problem 15 on page 566:

(a) Select two points and find a linear model that describes the relation between the points.

(b) Graph the line on the scatter diagram obtained in Quick Check Problem 15 (page 566).

(c) Predict the total cholesterol of a 39-year-old male. Round to the nearest whole number.

(d) Interpret the slope as a rate of change in the context of the problem. Does it make sense to interpret the *y*-intercept?

8.5 Exercises MyMathLab®

Exercise numbers in **green** have complete video solutions in MyMathLab or may be accessed using the QR code to the right.

Problems **1–18.** *are the* **Quick ✓** *s that follow the* **EXAMPLES.**

Building Skills

For Problems 19–30, graph each linear function. See Objective 1.

19. $F(x) = 5x - 2$

20. $F(x) = 4x + 1$

21. $G(x) = -3x + 7$

22. $G(x) = -2x + 5$

23. $H(x) = -2$

24. $P(x) = 5$

25. $f(x) = \frac{1}{2}x - 4$

26. $f(x) = \frac{1}{3}x - 3$

27. $F(x) = -\frac{5}{2}x + 5$

28. $P(x) = -\frac{3}{5}x - 1$

29. $G(x) = -\frac{3}{2}x$

30. $f(x) = \frac{4}{5}x$

In Problems 31–38, find the zero of the linear function.
See Objective 2.

31. $f(x) = 2x + 10$

32. $f(x) = 3x + 18$

33. $G(x) = -5x + 40$

34. $H(x) = -4x + 36$

35. $s(t) = \dfrac{1}{2}t - 3$

36. $p(q) = \dfrac{1}{4}q + 2$

37. $P(z) = -\dfrac{4}{3}z + 12$

38. $F(t) = -\dfrac{3}{2}t + 6$

In Problems 39–42, determine whether the scatter diagram indicates that a linear relation may exist between the two variables. If a linear relation does exist, indicate whether the slope is positive or negative. See Objective 4.

39.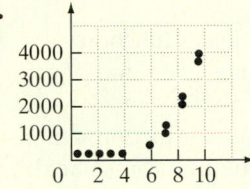

40.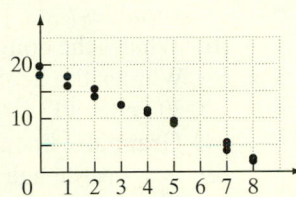

41.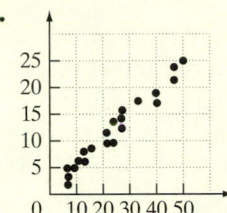

42.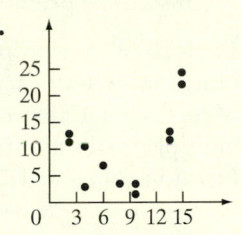

In Problems 43–46,

(a) *Draw a scatter diagram of the data.*

(b) *Select two points from the scatter diagram and find the equation of the line containing the points selected.*

(c) *Graph the line found in part (b) on the scatter diagram. See Objective 4.*

43.

x	2	4	5	8	9
y	1.4	1.8	2.1	2.3	2.6

44.

x	2	3	5	6	7
y	5.7	5.2	2.8	1.9	1.8

45.

x	1.2	1.8	2.3	3.5	4.1
y	8.4	7.0	7.3	4.5	2.4

46.

x	0	0.5	1.4	2.1	3.9
y	0.8	1.3	1.9	2.5	5.0

Mixed Practice

47. Suppose that $f(x) = 3x + 2$.

(a) What is the slope?
(b) What is the y-intercept?
(c) What is the zero of f?
(d) Solve $f(x) = 5$. What point is on the graph of f?
(e) Solve $f(x) \le -1$.
(f) Graph f.

48. Suppose that $g(x) = 8x + 3$.

(a) What is the slope?
(b) What is the y-intercept?
(c) What is the zero of g?
(d) Solve $g(x) = 19$. What point is on the graph of g?
(e) Solve $g(x) > -5$.
(f) Graph g.

49. Suppose that $f(x) = x - 5$ and $g(x) = -3x + 7$.

(a) Solve $f(x) = g(x)$. What is the value of f at the solution? What point is on the graph of f? What point is on the graph of g?
(b) Solve $f(x) > g(x)$.
(c) Graph f and g in the same Cartesian plane. Label the intersection point.

50. Suppose that $f(x) = \dfrac{4}{3}x + 5$ and $g(x) = \dfrac{1}{3}x + 1$.

(a) Solve $f(x) = g(x)$. What is the value of f at the solution? What point is on the graph of f? What point is on the graph of g?
(b) Solve $f(x) \le g(x)$.
(c) Graph f and g in the same Cartesian plane. Label the intersection point.

51. Find a linear function f such that $f(2) = 6$ and $f(5) = 12$. What is $f(-2)$?

52. Find a linear function g such that $g(1) = 5$ and $g(5) = 17$. What is $g(-3)$?

53. Find a linear function h such that $h(3) = 7$ and $h(-1) = 14$. What is $h\left(\dfrac{1}{2}\right)$?

54. Find a linear function F such that $F(2) = 5$ and $F(-3) = 9$. What is $F\left(-\dfrac{3}{2}\right)$?

55. In parts (a)–(e), use the figure shown below.

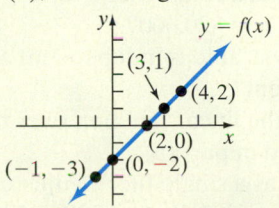

(a) Solve $f(x) = 1$.   (b) Solve $f(x) = -3$.
(c) What is $f(4)$?
(d) What are the intercepts of the function $y = f(x)$?
(e) Write the equation of the function whose graph is given in the form $f(x) = mx + b$.

56. In parts (a)–(e), use the figure shown below.

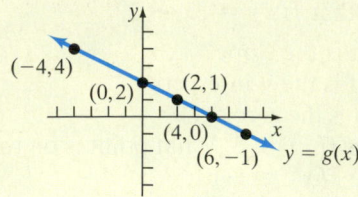

(a) Solve $g(x) = 1$.
(b) Solve $g(x) = -1$.
(c) What is $g(-4)$?
(d) What are the intercepts of the function $y = g(x)$?
(e) Write the equation of the function whose graph is given in the form $g(x) = mx + b$.

Applying the Concepts

57. Cab Fare The linear function $C(m) = 1.5m + 2$ describes the cab fare C for a ride of m miles.

(a) What is the domain of this linear function?
(b) What is $C(0)$? Explain what this result means.
(c) What is cab fare for a 5-mile ride?
(d) Graph the linear function.
(e) How many miles can you ride in a cab if you have $13.25?
(f) Over what range of miles can you ride if you can spend no more than $39.50?

58. Sales Commissions Tanya works for Pearson Education as a book representative. The linear function $I(s) = 0.01s + 20,000$ describes the annual income I of Tanya when she has total sales s.

(a) What is the domain of this linear function?
(b) What is $I(0)$? Explain what this result means.
(c) What is Tanya's salary if she sells $500,000 in books for the year?
(d) Graph the linear function.
(e) At what level of sales will Tanya's income be $45,000?

59. Taxes The function $T(x) = 0.15(x - 9226) + 922.50$ represents the federal income tax bill T of a single person whose adjusted gross income in 2015 was x dollars for income between $9226 and $37,450, inclusive. SOURCE: *Internal Revenue Service*

(a) What is the domain of this linear function?
(b) What was a single filer's tax bill if adjusted gross income was $20,000?
(c) Which variable is independent and which is dependent?
(d) Graph the linear function over the domain specified in part (a).
(e) What was a single filer's adjusted gross income if his or her tax bill was $2996.25?

60. Luxury Tax In 2002, Major League Baseball signed a labor agreement with the players. In this agreement, any team whose payroll exceeds

$189 million in 2015 will have to pay a luxury tax of 17.5% (for first offenses). The linear function $T(p) = 0.175(p - 189)$ describes the luxury tax T of a team whose payroll was p (in millions).

(a) What is the domain of this linear function?
(b) What was the luxury tax for the Boston Red Sox whose payroll was $200 million in 2015?
(c) Graph the linear function.
(d) What was the payroll of the San Francisco Giants, who paid a luxury tax of $1.3 million in 2015?

61. Health Costs The annual cost of health insurance H as a function of age a is given by the function $H(a) = 22.8a - 117.5$ for $15 \le a \le 90$.
SOURCE: *Statistical Abstract*

(a) What are the independent and dependent variables?
(b) What is the domain of this linear function?
(c) What is the health insurance premium of a 30-year-old?
(d) Graph the linear function over its domain.
(e) What is the age of an individual whose health insurance premium is $976.90?

62. Birth Rate A multiple birth is any birth with 2 or more children born. The birth rate is the number of births per 1000 women. The birth rate B of multiple births as a function of age a is given by the function $B(a) = 1.73a - 14.56$ for $15 \le a \le 44$.
SOURCE: *Centers for Disease Control*

(a) What are the independent and dependent variables?
(b) What is the domain of this linear function?
(c) What is the multiple birth rate of women who are 22 years of age, according to the model?
(d) Graph the linear function over its domain.
(e) What is the age of women whose multiple birth rate is 49.45?

63. Phone Charges Sprint has a long-distance phone plan that charges a monthly fee of $5.95 plus $0.05 per minute. SOURCE: *Sprint.com*

(a) Find a linear function that expresses the monthly bill B as a function of minutes used m.
(b) What are the independent and dependent variables?
(c) What is the domain of this linear function?
(d) What is the monthly bill if 300 minutes are used for long-distance phone calls?
(e) How many minutes were used for long distance if the long-distance phone bill was $17.95?
(f) Graph the linear function.
(g) Over what range of minutes can you talk each month if you don't want to spend more than $18.45?

64. RV Rental The weekly rental cost R of a class C 20-foot recreational vehicle is $129 plus $0.32 per mile, up to a maximum of 500 miles. SOURCE: *westernrv.com*

(a) Find a linear function that expresses the cost R as a function of miles driven m.

(b) What are the independent and dependent variables?

(c) What is the domain of this linear function?

(d) What is the rental cost if 360 miles are driven?

(e) How many miles were driven if the rental cost is $275.56?

(f) Graph the linear function.

(g) Over what range of miles can you drive if you have a budget of $273?

65. Depreciation Suppose that a company has just purchased a new computer for $2700. The company chooses to depreciate the computer using the straight-line method over 3 years.

(a) Find a linear function that expresses the book value V of the computer as a function of its age x.

(b) What is the domain of this linear function?

(c) What is the book value of the computer after the first year?

(d) What are the intercepts of the graph of the linear function?

(e) When will the book value of the computer be $900?

(f) Graph the linear function.

66. Depreciation Suppose that a company just purchased a new machine for its manufacturing facility for $1,200,000. The company chooses to depreciate the machine using the straight-line method over 20 years.

(a) Find a linear function that expresses the book value V of the machine as a function of its age x.

(b) What is the domain of this linear function?

(c) What is the book value of the machine after three years?

(d) What are the intercepts of the graph of the linear function?

(e) When will the book value of the machine be $480,000?

(f) Graph the linear function.

67. Diamonds The relation between the cost of a diamond and its weight is linear. In looking at two diamonds, we find that one of the diamonds weighs 0.7 carat and costs $3340, while the other diamond weighs 0.8 carat and costs $4065. source: *diamonds.com*

(a) Find a linear function that relates the price of a diamond, C, to its weight, x, treating weight as the independent variable.

(b) Predict the price of a diamond that weighs 0.77 carat.

(c) Interpret the slope as a rate of change.

(d) If a diamond costs $5300, what do you think it should weigh?

68. Apartments In the North Chicago area, an 820-square-foot apartment rents for $1507 per month. A 970-square-foot apartment rents for $1660. Suppose that the relation between area and rent is linear. source: *apartments.com*

(a) Find a linear function that relates the rent of a North Chicago apartment, R, to its area, x, treating area as the independent variable.

(b) Predict the rent of a 900-square-foot apartment in North Chicago.

(c) Interpret the slope as a rate of change in the context of the problem.

(d) If the rent of a North Chicago apartment is $1300 per month, how big would you expect it to be?

69. The Consumption Function A famous theory in economics developed by John Maynard Keynes states that personal consumption expenditures are a linear function of disposable income. An economist wishes to develop a model that relates income and consumption and obtains the following information from the United States Bureau of Economic Analysis. In 2010, personal disposable income was $11,380 billion and personal consumption expenditures were $10,349 billion. In 2015, personal disposable income was $15,605 billion and personal consumption expenditures were $13,618 billion.

(a) Find a linear function that relates personal consumption expenditures, C, to disposable income, x, treating disposable income as the independent variable.

(b) In 2013, personal disposable income was $14,221 billion. Use this information to find personal consumption expenditures in 2013 to the nearest billion.

(c) Interpret the slope as a rate of change in the context of the problem. In economics, this slope is called the **marginal propensity to consume.**

(d) If personal consumption expenditures were $13,117 billion, what do you think that disposable income was to the nearest billion?

70. Birth Weight According to the National Center for Health Statistics, the average birth weight of babies born to 22-year-old mothers is 3280 grams. The average birth weight of babies born to 32-year-old mothers is 3370 grams. Suppose that the relation between age of mother and birth weight is linear.

(a) Find a linear function that relates age of mother a to birth weight W, treating age of mother as the independent variable.

(b) Predict the birth weight of a baby born to a mother who is 30 years old.

(c) Interpret the slope as a rate of change in the context of the problem.

(d) If a baby weighs 3310 grams, how old do you expect the mother to be?

71. Concrete As concrete cures, it gains strength. The data represents the 7-day and 28-day strength (in pounds per square inch) of a certain type of concrete.

7-day Strength, x	28-day Strength, y
2300	4070
3390	5220
2430	4640
2890	4620
3330	4850
2480	4120
3380	5020
2660	4890
2620	4190
3340	4630

(a) Draw a scatter diagram of the data, treating 7-day strength as the independent variable.
(b) What type of relation appears to exist between 7-day strength and 28-day strength?
(c) Select two points and find an equation of the line containing the points.
(d) Graph the line on the scatter diagram drawn in part (a).
(e) Predict the 28-day strength of a slab of concrete if its 7-day strength is 3000 psi.
(f) Interpret the slope of the line as a rate of change in the context of the problem found in part (c).

72. Candy The following data represent the weight (in grams) of various candy bars and the corresponding number of calories.

Candy Bar	Weight, x	Calories, y
Hershey's Milk Chocolate	44.28	230
Nestle Crunch	44.84	230
Butterfinger	61.30	270
Baby Ruth	66.45	280
Almond Joy	47.33	220
Twix (with Caramel)	58.00	280
Snickers	61.12	280
Heath	39.52	210

SOURCE: *Megan Pocius, student at Joliet Junior College*

(a) Draw a scatter diagram of the data, treating weight as the independent variable.
(b) What type of relation appears to exist between the weight of a candy bar and the number of calories?
(c) Select two points and find an equation of the line containing the points.
(d) Graph the line on the scatter diagram drawn in part (a).

(e) Predict the number of calories in a candy bar that weighs 62.3 grams.
(f) Interpret the slope of the line as a rate of change in the context of the problem found in part (c).

73. Raisins The following data represent the weight (in grams) of a box of raisins and the number of raisins in the box.

Weight, w	Number of Raisins, N
42.3	87
42.7	91
42.8	93
42.4	87
42.6	89
42.4	90
42.3	82
42.5	86
42.7	86
42.5	86

SOURCE: *Jennifer Maxwell, student at Joliet Junior College*

(a) Does the relation defined by the set of ordered pairs (w, N) represent a function?
(b) Draw a scatter diagram of the data, treating weight as the independent variable.
(c) Select two points and find the equation of the line containing the points.
(d) Graph the line on the scatter diagram drawn in part (b).
(e) Express the relationship found in part (c) using function notation.
(f) Predict the number of raisins in a box that weighs 42.5 grams.
(g) Interpret the slope of the line as a rate of change in the context of the problem found in part (c).

74. Height versus Head Circumference The following data represent the height (in inches) and head circumference (in inches) of 9 randomly selected children.

Height, h	Head Circumference, C
25.25	16.4
25.75	16.9
25	16.9
27.75	17.6
26.50	17.3
27.00	17.5
26.75	17.3
26.75	17.5
27.5	17.5

SOURCE: *Denise Slucki, student at Joliet Junior College*

(a) Does the relation defined by the set of ordered pairs (h, C) represent a function?

(b) Draw a scatter diagram of the data, treating height as the independent variable.

(c) Select two points and find the equation of the line containing the points.

(d) Graph the line on the scatter diagram drawn in part (b).

(e) Express the relationship found in part (c) using function notation.

(f) Predict the head circumference of a child who is 26.5 inches tall.

(g) Interpret the slope of the line as a rate of change in the context of the problem found in part (c).

Extending the Concepts

75. Math for the Future: Calculus The **average rate of change** of a function $y = f(x)$ from c to x is defined as

$$\text{Average rate of change} = \frac{\Delta y}{\Delta x} = \frac{f(x) - f(c)}{x - c}, \quad x \neq c$$

provided that c is in the domain of f. The average rate of change of a function is simply the slope of the line joining the points $(c, f(c))$ and $(x, f(x))$. The line joining these points is called a **secant line.** The slope of the secant line is

$$m_{\text{sec}} = \frac{f(x) - f(c)}{x - c}$$

The following figure illustrates the idea.

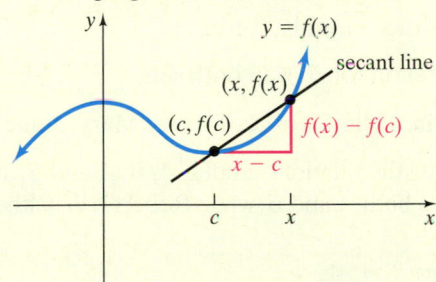

Below, we show the graph of the function $f(x) = 2x^2 - 4x + 1$.

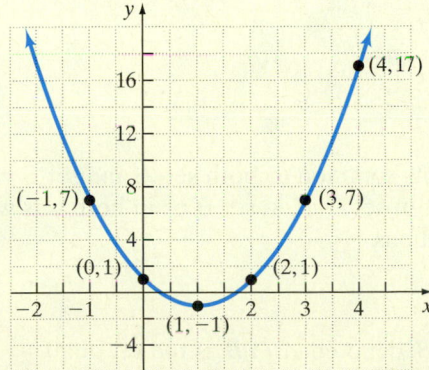

(a) On the graph of $f(x) = 2x^2 - 4x + 1$, draw a line through the points $(1, f(1))$ and $(x, f(x))$, where $x = 4$.

(b) Find the slope of the secant line through $(1, f(1))$ and $(x, f(x))$, where $x = 4$.

(c) Find the equation of the secant line through $(1, f(1))$ and $(x, f(x))$, where $x = 4$.

(d) Repeat parts (a)–(c) for $x = 3$, $x = 2$, $x = 1.5$, and $x = 1.1$.

(e) What happens to the slope of the secant line as x gets closer to 1?

76. A strain of *E. coli* Beu 397-recA441 is placed into a Petri dish at 30° Celsius and allowed to grow. The population is estimated by means of an optical device in which the amount of light that passes through the Petri dish is measured. The data below are collected. Do you think that a linear function could be used to describe the relation between the two variables? Why or why not?

Time, x	Population, y
0	0.09
2.5	0.18
3.5	0.26
4.5	0.35
6	0.50

SOURCE: *Dr. Polly Lavery, Joliet Junior College*

Technology Exercises

The equation of the line obtained in Example 8 depends on the points selected, which will vary from person to person. So the line we found might be different from the line that you found. Although the line that we found in Example 8 fits the data well, there may be a line that "fits better." Do you think that your line fits the data better? Is there a line of *best fit*? As it turns out, there is a method for finding the line that best fits linearly related data (called the *line of best fit*).*

Graphing utilities can be used to draw scatter diagrams and find the line of best fit. Figure 36(a) shows a scatter diagram of the data presented in Table 8 from Example 6, drawn on a graphing calculator. Figure 36(b) shows the line of best fit from a graphing calculator.

The line of best fit is $y = 53.529x - 963.453$.

Figure 36 (a)   **(b)**

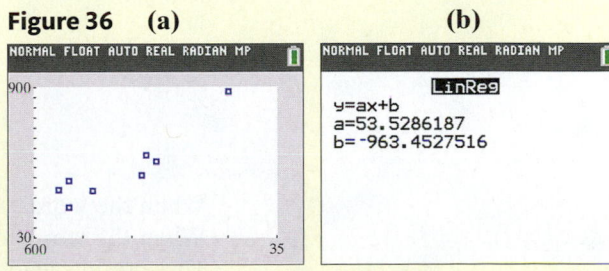

In Problems 77–80,

(a) *Draw a scatter diagram using a graphing calculator.*

(b) *Find the line of best fit for the data in the problem specified using technology.*

77. Problem 71   **78.** Problem 72

79. Problem 73   **80.** Problem 74

---

* This text will not discuss the underlying mathematics of lines of best fit. Books on elementary statistics discuss this topic.

8.6 Compound Inequalities

Objectives

① Determine the Intersection or Union of Two Sets

② Solve Compound Inequalities Involving "and"

③ Solve Compound Inequalities Involving "or"

④ Solve Problems Using Compound Inequalities

Are You Prepared for This Section?

Before getting started, complete the following problems. If you get a problem wrong, go back to the section cited and review the material.

P1. Use set-builder notation and interval notation to name the set of all real numbers x such that $-2 \le x \le 5$. [Section 2.8, pp. 148–150]

P2. Graph the inequality $x \ge 4$. [Section 2.8, pp. 148–149]

P3. Use interval notation to express the inequality shown in the graph. [Section 2.8, pp. 148–149]

P4. Solve: $2(x + 3) - 5x = 15$ [Section 2.2, pp. 91–94]

P5. Solve: $2x + 3 > 11$ [Section 2.8, pp. 150–154]

P6. Solve: $x + 8 \ge 4(x - 1) - x$ [Section 2.8, pp. 150–154]

▶ ① Determine the Intersection or Union of Two Sets

Table 9

Student	Age	Gender
Grace	19	Female
Sophia	23	Female
Kevin	20	Male
Robert	32	Male
Jack	19	Male
Mary	35	Female
Nancy	40	Female
George	22	Male
Teresa	20	Female

Table 9 contains information about students in an Intermediate Algebra course. These people can be classified in various sets. For example, if set A is defined as the set of all students whose age is less than 25. Then

$$A = \{ \text{Grace, Sophia, Kevin, Jack, George, Teresa} \}$$

If set B is defined as the set of all students who are female, then

$$B = \{ \text{Grace, Sophia, Mary, Nancy, Teresa} \}$$

Now list all the students who are in set A and set B, that is, students who are under 25 years of age and female.

$$A \text{ and } B = \{ \text{Grace, Sophia, Teresa} \}$$

Now list all the students who are in set A or in set B or in both sets.

$$A \text{ or } B = \{ \text{Grace, Sophia, Kevin, Jack, George, Teresa, Mary, Nancy} \}$$

Figure 37 shows a Venn diagram illustrating the relations among A, B, A and B, and A or B. Notice that Grace, Sophia, and Teresa are in both A and B, while Robert is in neither A nor B.

Figure 37

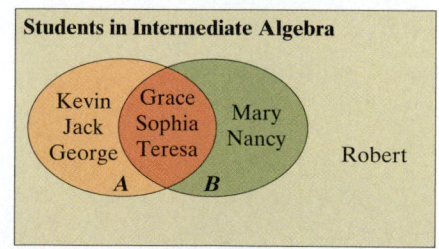

When the word "and" was used, elements common to both set A and set B were listed. When the word "or" was used elements in set A or in set B or in both sets were listed. These results lead to the following definitions.

Prepared?...Answers

P1. $\{x | -2 \le x \le 5\}$; $[-2, 5]$

P2.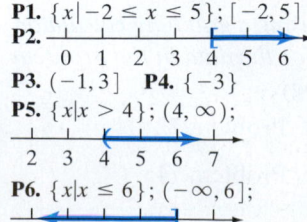

P3. $(-1, 3]$ **P4.** $\{-3\}$

P5. $\{x | x > 4\}$; $(4, \infty)$;

P6. $\{x | x \le 6\}$; $(-\infty, 6]$;

Definitions

• The **intersection** of two sets A and B, denoted $A \cap B$, is the set of all elements in both set A and set B.

• The **union** of two sets A and B, denoted $A \cup B$, is the set of all elements in set A or in set B or in both set A and set B.

• The word **and** implies intersection, while the word **or** implies union.

EXAMPLE 1 **Finding the Intersection and Union of Sets**

Let $A = \{1, 3, 5, 7, 9\}$, $B = \{1, 2, 3, 4, 5\}$, and $C = \{2, 4\}$. Find

(a) $A \cap B$ (b) $A \cup B$ (c) $A \cup C$

Solution

(a) $A \cap B$ is the set of all elements that are in both A and B. So,

$$A \cap B = \{1, 3, 5\}$$

Work Smart

When finding the union of two sets, list each element only once, even if it occurs in both sets.

(b) $A \cup B$ is the set of all elements that are in A or B, or both. So,

$$A \cup B = \{1, 2, 3, 4, 5, 7, 9\}$$

(c) $A \cup C$ is the set of all elements that are in A or C, or both. So,
$A \cup C$ is the empty set which can be notated so $\varnothing$ or $\{\ \}$.

●

Quick ✓

1. The _____ of two sets A and B, denoted $A \cap B$, is the set of all elements that belong to both set A and set B.

2. The word ___ implies intersection. The word ___ implies union.

3. *True or False* The intersection of two sets can be the empty set.

4. *True or False* The symbol for the union of two sets is $\cap$.

In Problems 5–10, let $A = \{1, 2, 3, 4, 5, 6\}$, $B = \{1, 3, 5, 7\}$, *and* $C = \{2, 4, 6, 8\}$.

5. Find $A \cap B$.   6. Find $A \cap C$.

7. Find $A \cup B$.   8. Find $A \cup C$.

9. Find $B \cap C$.   10. Find $B \cup C$.

EXAMPLE 2 **Finding the Intersection and Union of Two Sets**

Suppose $A = \{x \mid x \le 5\}$, $B = \{x \mid x \ge 1\}$, and $C = \{x \mid x < -2\}$.

(a) Determine $A \cap B$. Graph the set on a real number line and write it in set-builder notation and interval notation.

(b) Determine $B \cup C$. Graph the set on a real number line and write it in set-builder notation and interval notation.

Solution

(a) $A \cap B$ is the set of all real numbers less than or equal to 5 and greater than or equal to 1. Identify this set by determining where the graphs of the inequalities overlap. See Figure 38.

Figure 38

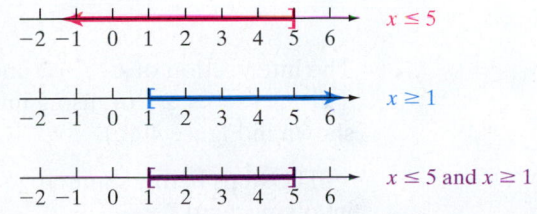

$A \cap B$ is $\{x \mid 1 \le x \le 5\}$ in set-builder notation, and $[1, 5]$ in interval notation.

Figure 39

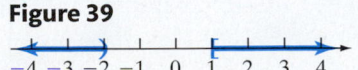

(b) $B \cup C$ is the set of all real numbers greater than or equal to 1 or less than -2. See Figure 39. Therefore, $B \cup C$ is $\{x \mid x < -2 \text{ or } x \ge 1\}$ in set-builder notation, and $(-\infty, -2) \cup [1, \infty)$ in interval notation.

●

Work Smart

Throughout the text when a solution contains more than one interval, and is written in set-builder notation, use the word "or." When a solution is written in interval notation and contains more than one interval, use the union symbol, $\cup$. See the solution to Quick ✓ 12.

Quick ✓

Let $A = \{x \mid x > 2\}$, $B = \{x \mid x < 7\}$, and $C = \{x \mid x \le -3\}$.

11. Determine $A \cap B$. Graph the set on a real number line and write it in set-builder notation and interval notation.

12. Determine $A \cup C$. Graph the set on a real number line and write it in set-builder notation and interval notation.

▶ ❷ Solve Compound Inequalities Involving "and"

A **compound inequality** is formed by joining two inequalities with the word "and" or "or." For example,

$$3x + 1 > 4 \quad \text{and} \quad 2x - 3 < 7$$
$$5x - 2 \le 13 \quad \text{or} \quad 2x - 5 > 3$$

are compound inequalities. To **solve a compound inequality** means to find all possible values of the variable such that the compound inequality results in a true statement. For example, the compound inequality

$$3x + 1 > 4 \quad \text{and} \quad 2x - 3 < 7$$

is true for $x = 2$ but false for $x = 0$. It is false for $x = 0$ because the solution is only true for the inequality $2x - 3 < 7$. It is not true for both inequalities.

Look at an example that illustrates how to solve compound inequalities involving the word "and."

EXAMPLE 3 **How to Solve a Compound Inequality Involving "and"**

Solve $3x + 2 > -7$ and $4x + 1 \le 9$. Express your solution using set-builder and interval notation. Graph the solution set.

Step-by-Step Solution

Step 1: Solve each inequality separately.

	$3x + 2 > -7$		$4x + 1 \le 9$
Subtract 2 from both sides:	$3x > -9$	Subtract 1 from both sides:	$4x \le 8$
Divide both sides by 3:	$x > -3$	Divide both sides by 4:	$x \le 2$

Step 2: Find the intersection of the solution sets, which will represent the solution set to the compound inequality.

To find the intersection of the two solution sets, graph each inequality separately. See Figures 40(a) and (b).

Figure 40

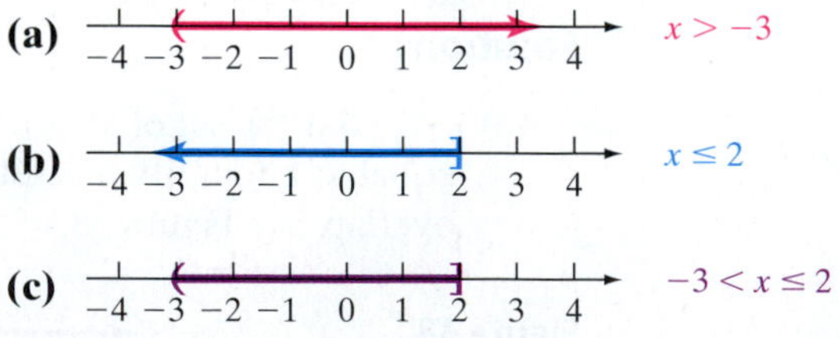

(a) $x > -3$

(b) $x \le 2$

(c) $-3 < x \le 2$

The intersection of $x > -3$ and $x \le 2$ is $-3 < x \le 2$, so the solution set is $\{x \mid -3 < x \le 2\}$ or, using interval notation, $(-3, 2]$. The graph of the solution set is shown in Figure 40(c). ●

The steps below summarize the procedure for solving compound inequalities involving "and."

Work Smart

The words "and" and "intersection" suggest overlap. When solving these types of problems, look for the overlap of the graphs.

Solving Compound Inequalities Involving "and"

Step 1: Solve each inequality separately.

Step 2: Find the INTERSECTION of the solution sets of the respective inequalities.

EXAMPLE 4 | **Solving a Compound Inequality with "and"**

Solve $-2x + 5 > -1$ and $5x + 6 \le -4$. Express the solution using set-builder notation and interval notation. Graph the solution set.

Solution

Solve each inequality separately:

$$-2x + 5 > -1 \qquad\qquad\qquad 5x + 6 \le -4$$

Subtract 5 from both sides: $\quad -2x > -6 \qquad$ Subtract 6 from both sides: $\quad 5x \le -10$

Divide both sides by -2 and reverse the direction of the inequality since you are dividing $\qquad x < 3 \qquad$ Divide both sides by 5: $\qquad x \le -2$
by a negative number:

The intersection of the solution sets is the solution set to the compound inequality. See Figures 41(a) and (b).

Figure 41

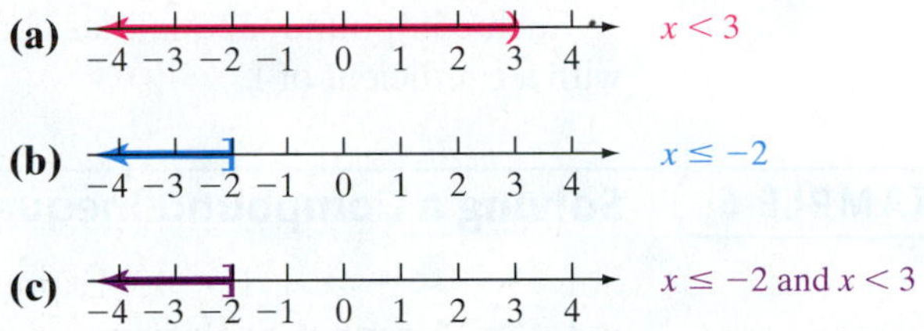

(a)

$x < 3$

(b)

$x \le -2$

(c)

$x \le -2$ and $x < 3$

The intersection of $x < 3$ and $x \le -2$ is $x \le -2$. The solution set is $\{x \mid x \le -2\}$ or, using interval notation, $(-\infty, -2\,]$. The graph of the solution set is shown in Figure 41(c). ●

> **Quick ✓**
>
> In Problems 13–15, solve each compound inequality. Express the solution using set-builder notation and interval notation. Graph the solution set.
>
> **13.** $2x + 1 \ge 5$ and $-3x + 2 < 5$
>
> **14.** $4x - 5 < 7$ and $3x - 1 > -10$
>
> **15.** $-8x + 3 < -5$ and $\dfrac{2}{3}x + 1 < 3$

EXAMPLE 5 | **Solving a Compound Inequality with "and"**

Solve $x - 5 > -1$ and $2x - 3 \le -5$. Express the solution using set-builder and interval notation. Graph the solution set.

Solution

Solve each inequality separately:

$$x - 5 > -1 \qquad\qquad\qquad 2x - 3 \le -5$$

Add 5 to both sides: $\qquad x > 4 \qquad$ Add 3 to both sides: $\qquad 2x \le -2$

Figure 42

$x > 4$

$x \le -1$

Divide both sides by 2: $\qquad x \le -1$

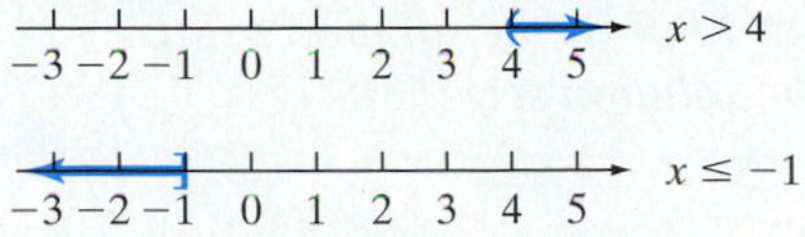

The intersection of the solution sets is the solution set to the compound inequality. See Figure 42. The intersection is the empty set, so the solution set is $\{\ \}$ or $\varnothing$. ●

> **Quick ✓**
>
> In Problems 16 and 17, solve each compound inequality. Express the solution using set-builder notation and interval notation. Graph the solution set.
>
> **16.** $3x - 5 < -8$ and $2x + 1 > 5$
>
> **17.** $5x + 1 \le 6$ and $3x + 2 \ge 5$

Sometimes, "and" inequalities can be combined into a more streamlined notation.

> **Writing Inequalities Involving "and" Compactly**
>
> If $a < b$, then we can write
>
> $$a < x \quad \text{and} \quad x < b$$
>
> more compactly as
>
> $$a < x < b$$

For example,

$$-3 < -4x + 1 \quad \text{and} \quad -4x + 1 < 13$$

can be written as

$$-3 < -4x + 1 < 13$$

· Such compound inequalities can be solved by isolating the variable "in the middle" with a coefficient of 1.

▶ **EXAMPLE 6**   **Solving a Compound Inequality**

Solve $-3 < -4x + 1 < 13$. Express the solution using set-builder notation and interval notation. Graph the solution set.

Solution

Isolate the variable "in the middle" with a coefficient of 1:

$$-3 < -4x + 1 < 13$$

Subtract 1 from all three parts (Addition Property): $\quad -3 - 1 < -4x + 1 - 1 < 13 - 1$

$$-4 < -4x < 12$$

Divide all three parts by –4 and reverse the direction of both inequalities: $\quad \dfrac{-4}{-4} > \dfrac{-4x}{-4} > \dfrac{12}{-4}$

$$1 > x > -3$$

If $b > x > a$, then $a < x < b$: $\quad -3 < x < 1$

The solution set is $\{x \mid -3 < x < 1\}$ or, in interval notation, $(-3, 1)$. Figure 43 shows the graph of the solution set.

Figure 43

To visualize the results of Example 6, look at Figure 44, which shows the graph of $f(x) = -3$, $g(x) = -4x + 1$, and $h(x) = 13$. Note that the graph of $g(x) = -4x + 1$ is between the graphs of $f(x) = -3$ and $h(x) = 13$ for $-3 < x < 1$. Thus the solution set of $-3 < -4x + 1 < 13$ is $\{x \mid -3 < x < 1\}$.

> **Quick ✓**
>
> In Problems 18–20, solve each compound inequality. Express the solution using set-builder notation and interval notation. Graph the solution set.
>
> **18.** $-2 < 3x + 1 < 10$
>
> **19.** $0 < 4x - 5 \leq 3$
>
> **20.** $3 \leq -2x - 1 \leq 11$

Figure 44

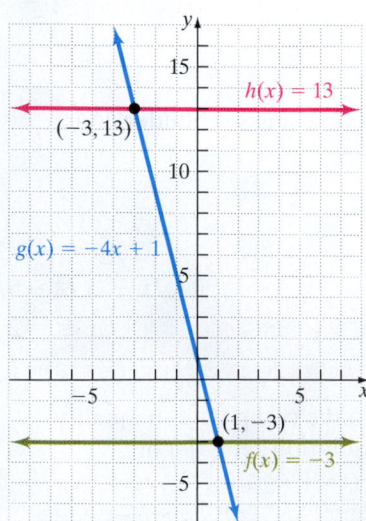

▶ ❸ **Solve Compound Inequalities Involving "or"**

The solution to compound inequalities involving the word "or" is the union of the solutions to each inequality.

EXAMPLE 7 **How to Solve a Compound Inequality Involving "or"**

Solve $3x - 5 < -2$ or $4 - 5x \le -16$. Express the solution using set-builder notation and interval notation. Graph the solution set.

Step-by-Step Solution

Step 1: Solve each inequality separately.

$$3x - 5 < -2 \qquad\qquad 4 - 5x \le -16$$

Add 5 to each side: $\quad 3x < 3$ $\qquad$ Subtract 4 from both sides: $\quad -5x \le -20$

Divide both sides by 3: $\quad x < 1$ $\qquad$ Divide both sides by –5;
Reverse the direction of the
inequality: $\quad x \ge 4$

Step 2: Find the union of the solution sets, which will represent the solution set to the compound inequality.

The union of the two solution sets is $x < 1$ or $x \ge 4$. The solution set using set-builder notation is $\{x \mid x < 1 \text{ or } x \ge 4\}$. The solution set using interval notation is $(-\infty, 1) \cup [4, \infty)$. Figure 45 shows the graph of the solution set.

Figure 45

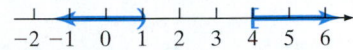

●

Steps for Solving Compound Inequalities Involving "or"

Step 1: Solve each inequality separately.

Step 2: Find the UNION of the solution sets of each inequality.

Quick ✓

In Problems 21–24, solve each compound inequality. Express the solution using set-builder notation and interval notation. Graph the solution set.

21. $x + 3 < 1$ or $x - 2 > 3$

22. $3x + 1 \le 7$ or $2x - 3 > 9$

23. $2x - 3 \ge 1$ or $6x - 5 \ge 1$

24. $\dfrac{3}{4}(x + 4) < 6$ or $\dfrac{3}{2}(x + 1) > 15$

Work Smart

A common error is to write the solution $x < 1$ or $x \ge 4$ as $1 > x \ge 4$, which is incorrect. There are no real numbers that are less than 1 *and* greater than or equal to 4. Another common error is to "mix" symbols, as in $1 < x \ge 4$. This notation does not make sense!

EXAMPLE 8 **Solving Compound Inequalities Involving "or"**

Solve $\dfrac{1}{2}x - 1 < 1$ or $\dfrac{2x - 1}{3} \ge -1$. Express the solution using set-builder notation and interval notation. Graph the solution set.

Solution

Solve each inequality separately: $\quad \dfrac{1}{2}x - 1 < 1$ $\qquad\qquad\qquad \dfrac{2x - 1}{3} \ge -1$

Add 1 to each side: $\quad \dfrac{1}{2}x < 2$ $\qquad$ Multiply both sides by 3: $\quad 2x - 1 \ge -3$

Multiply both sides by 2: $\quad x < 4$ $\qquad$ Add 1 to both sides: $\quad 2x \ge -2$

Divide both sides by 2: $\quad x \ge -1$

Find the union of these solution sets. The graphs in Figure 46 on the following page shows that the union of the two solution sets is the set of all real numbers.

(continued)

Figure 46

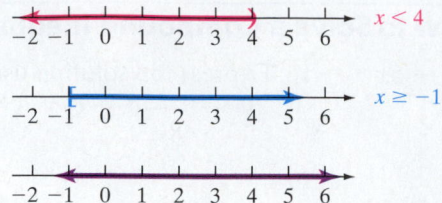

The solution set is $\{x \mid x$ is any real number $\}$ in set-builder notation, and $(-\infty, \infty)$ in interval notation.

> **Quick ✓**
>
> *In Problems 25 and 26, solve each compound inequality. Express the solution using set-builder notation and interval notation. Graph the solution set.*
>
> **25.** $3x - 2 > -5$ or $2x - 5 \le 1$
>
> **26.** $-5x - 2 \le 3$ or $7x - 9 > 5$

▶ ❹ Solve Problems Using Compound Inequalities

We now look at an application involving compound inequalities.

EXAMPLE 9 | **Federal Income Taxes**

In 2015, married couples filing a joint federal tax return who were in the 25% tax bracket paid between $10,313 and $29,388 in federal income taxes. These couples' federal income taxes equal $10,313 plus 25% of their taxable income over $74,900. Find the range of taxable incomes for which a married couple would have been in the 25% tax bracket.
SOURCE: *Internal Revenue Service*

Solution

Step 1: Identify We want to find the range of taxable incomes for married couples in the 25% tax bracket. This direct translation problem involves an inequality.

Step 2: Name Let t represent the taxable income.

Step 3: Translate The federal tax bill equals $10,313 plus 25% of the taxable income over $74,900. If the couple had taxable income equal to $75,900, their tax bill was $10,313 plus 25% of $1000 ($1000 is the amount over $74,900). In general, if the couple has taxable income t, then their tax bill will be

Work Smart

The word "range" tells us that an inequality is to be solved, since it indicates more than one number.

$$\underbrace{\$10,313}_{\$10,313} \quad \underbrace{\text{plus}}_{+} \quad \underbrace{25\%}_{0.25} \cdot \underbrace{\text{of the amount over } \$74,900}_{(t - \$74,900)}$$

Because the tax bill was between $10,313 and $29,388, we have

$$10,313 \le 10,313 + 0.25(t - 74,900) \le 29,388 \quad \text{The Model}$$

Step 4: Solve

$$10,313 \le 10,313 + 0.25(t - 74,900) \le 29,388$$

Distribute 0.25: $\qquad 10,313 \le 10,313 + 0.25t - 18,725 \le 29,388$

Combine like terms: $\qquad 10,313 \le -8412 + 0.25t \le 29,388$

Add 8412 to all three parts: $\qquad 18,725 \le 0.25t \le 37,800$

Divide all three parts by 0.25: $\qquad 74,900 \le t \le 151,200$

Step 5: Check If a married couple had taxable income of $74,900, then their tax bill was $10,313 + 0.25(\$74,900 - \$74,900) = \$10,313$. If a married couple had taxable income of $151,200, then their tax bill was $10,313 + 0.25(\$151,200 - \$74,900) = \$29,388$.

Step 6: Answer the Question A married couple who filed a joint federal income tax return with a tax bill between $10,313 and $29,388 had taxable income between $74,900 and $151,200.

Quick ✔

27. In 2015, an individual filing a federal tax return whose income placed him or her in the 25% tax bracket paid federal income taxes between $5156.25 and $18,481.25. The individual had to pay federal income taxes equal to $5156.25 plus 25% of the amount over $37,450. Find the range of taxable income in order for an individual to have been in the 25% tax bracket. (SOURCE: *Internal Revenue Service*)

28. Juan's monthly payments for his new car are $260 per month and he estimates that maintenance and gas cost him $0.20 per mile. During the course of a year, Juan's total monthly cost for his car ranges between $420 and $560. What is the range of miles he drives per month?

8.6 Exercises **MyMathLab®** Exercise numbers in **green** have complete video solutions in MyMathLab or may be accessed using the QR code to the right.

*Problems 1–28 are the **Quick ✔** s that follow the **EXAMPLES**.*

Building Skills

In Problems 29–34, use $A = \{4, 5, 6, 7, 8, 9\}$, $B = \{1, 5, 7, 9\}$, and $C = \{2, 3, 4, 6\}$ to find each set. See Objective 1.

29. $A \cup B$ **30.** $A \cup C$

31. $A \cap B$ **32.** $A \cap C$

33. $B \cap C$ **34.** $B \cup C$

In Problems 35–38, use the graph of the inequality to find each set. See Objective 1.

35. $A = \{x \mid x \le 5\}$; $B = \{x \mid x > -2\}$
Find (a) $A \cap B$ and (b) $A \cup B$.

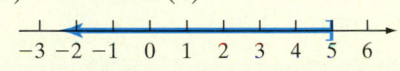

36. $A = \{x \mid x \ge 4\}$; $B = \{x \mid x < 1\}$
Find (a) $A \cap B$ and (b) $A \cup B$.

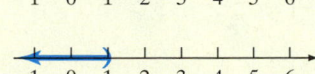

37. $E = \{x \mid x > 3\}$; $F = \{x \mid x < -1\}$
Find (a) $E \cap F$ and (b) $E \cup F$.

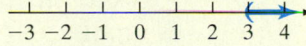

38. $E = \{x \mid x \le 2\}$; $F = \{x \mid x \ge -2\}$
Find (a) $E \cap F$ and (b) $E \cup F$.

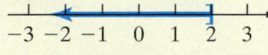

In Problems 39–42, use the graph to solve the compound inequality. Graph the solution set. See Objectives 2 and 3.

39. (a) $-5 \le 2x - 1 \le 3$
    (b) $2x - 1 < -5$ or $2x - 1 > 3$

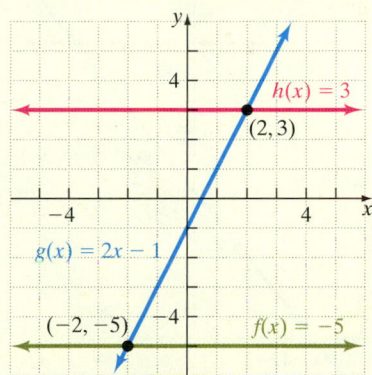

40. (a) $-1 \le \frac{1}{2}x + 1 \le 3$

    (b) $\frac{1}{2}x + 1 < -1$ or $\frac{1}{2}x + 1 > 3$

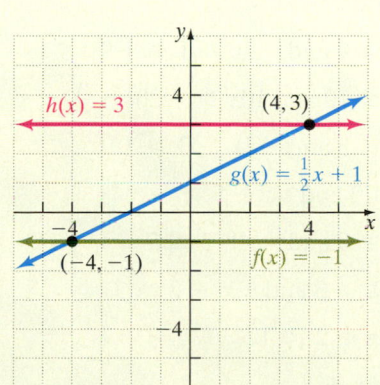

41. (a) $-4 < -\dfrac{5}{3}x + 1 < 6$

 (b) $-\dfrac{5}{3}x + 1 \le -4$ or $-\dfrac{5}{3}x + 1 \ge 6$

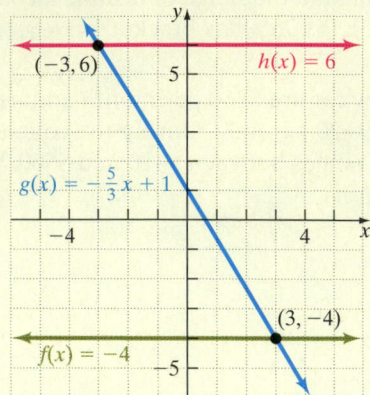

42. (a) $-3 < \dfrac{5}{4}x + 2 < 7$

 (b) $\dfrac{5}{4}x + 2 \le -3$ or $\dfrac{5}{4}x + 2 \ge 7$

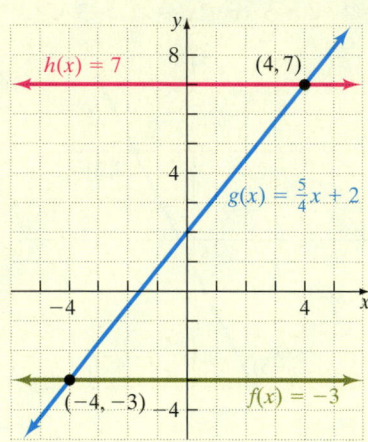

In Problems 43–66, solve each compound inequality. Graph the solution set. See Objective 2.

43. $x < 3$ and $x \ge -2$

44. $x \le 5$ and $x > 0$

45. $4x - 4 < 0$ and $-5x + 1 \le -9$

46. $6x - 2 \le 10$ and $10x > -20$

47. $4x - 3 < 5$ and $-5x + 3 > 13$

48. $x - 3 \le 2$ and $6x + 5 \ge -1$

49. $7x + 2 \ge 9$ and $4x + 3 \le 7$

50. $-4x - 1 < 3$ and $-x - 2 > 3$

51. $-3 \le 5x + 2 < 17$

52. $-10 < 6x + 8 \le -4$

53. $-3 \le 6x + 1 \le 10$

54. $-12 < 7x + 2 \le 6$

55. $3 \le -5x + 7 < 12$

56. $-6 < -3x + 6 \le 4$

57. $-1 \le \dfrac{1}{2}x - 1 \le 3$

58. $0 < \dfrac{3}{2}x - 3 \le 3$

59. $3 \le -2x - 1 \le 11$

60. $-3 < -4x + 1 < 17$

61. $\dfrac{2}{3}x + \dfrac{1}{2} < \dfrac{5}{6}$ and $-\dfrac{1}{5}x + 1 < \dfrac{3}{10}$

62. $x - \dfrac{3}{2} \le \dfrac{5}{4}$ and $-\dfrac{2}{3}x - \dfrac{2}{9} < \dfrac{8}{9}$

63. $-2 < \dfrac{3x + 1}{2} \le 8$

64. $-4 \le \dfrac{4x - 3}{3} < 3$

65. $-8 \le -2(x + 1) < 6$

66. $-6 < -3(x - 2) < 15$

In Problems 67–80, solve each compound inequality. Graph the solution set. See Objective 3.

67. $x < -2$ or $x > 3$

68. $x < 0$ or $x \ge 6$

69. $x - 2 < -4$ or $x + 3 > 8$

70. $x + 3 \le 5$ or $x - 2 \ge 3$

71. $6(x - 2) < 12$ or $4(x + 3) > 12$

72. $4x + 3 > -5$ or $8x - 5 < 3$

73. $-8x + 6x - 2 > 0$ or $5x > 3x + 8$

74. $3x \ge 7x + 8$ or $x < 4x - 9$

75. $2x + 5 \le -1$ or $\dfrac{4}{3}x - 3 > 5$

76. $-\dfrac{4}{5}x - 5 > 3$ or $7x - 3 > 4$

77. $\dfrac{1}{2}x < 3$ or $\dfrac{3x - 1}{2} > 4$

78. $\dfrac{2}{3}x + 2 \le 4$ or $\dfrac{5x - 3}{3} \ge 4$

79. $3(x - 1) + 5 < 2$ or $-2(x - 3) < 1$

80. $2(x + 1) - 5 \le 4$ or $-(x + 3) \le -2$

Mixed Practice

In Problems 81–94, solve each compound inequality. Graph the solution set.

81. $3a + 5 < 5$ and $-2a + 1 \le 7$

82. $5x - 1 < 9$ and $5x > -20$

83. $5(x + 2) < 20$ or $4(x - 4) > -20$

84. $3(x + 7) < 24$ or $6(x - 4) > -30$

85. $-4 \le 3x + 2 \le 10$

86. $-8 \le 5x - 3 \le 4$

87. $2x + 7 < -13$ or $5x - 3 > 7$

88. $3x - 8 < -14$ or $4x - 5 > 7$

89. $5 < 3x - 1 < 14$

90. $-5 < 2x + 7 \le 5$

91. $\dfrac{x}{3} \le -1$ or $\dfrac{4x - 1}{2} > 7$

92. $\dfrac{x}{2} \le -4$ or $\dfrac{2x - 1}{3} \ge 2$

93. $-3 \le -2(x + 1) < 8$

94. $-15 < -3(x + 2) \le 1$

Applying the Concepts

In Problems 95–100, use the Addition Property and/or Multiplication Properties to find a and b.

95. If $-3 < x < 4$, then $a < x + 4 < b$.

96. If $-2 < x < 3$, then $a < x - 3 < b$.

97. If $4 < x < 10$, then $a < 3x < b$.

98. If $2 < x < 12$, then $a < \dfrac{1}{2}x < b$.

99. If $-2 < x < 6$, then $a < 3x + 5 < b$.

100. If $-4 < x < 3$, then $a < 2x - 7 < b$.

101. Systolic Blood Pressure Blood pressure is measured using two numbers. One of the numbers measures systolic blood pressure. The systolic blood pressure represents the pressure while the heart is beating. In a healthy person, the systolic blood pressure should be greater than 90 and less than 140. If we let the variable x represent a person's systolic blood pressure, express the systolic blood pressure of a healthy person using a compound inequality.

102. Diastolic Blood Pressure Blood pressure is measured using two numbers. One of the numbers measures diastolic blood pressure. The diastolic blood pressure represents the pressure while the

heart is resting between beats. In a healthy person, the diastolic blood pressure should be greater than 60 and less than 90. If we let the variable x represent a person's diastolic blood pressure, express the diastolic blood pressure of a healthy person using a compound inequality.

103. Computing Grades Joanna desperately wants to earn a B in her history class. Her current test scores are 74, 86, 77, and 89. Her final exam is worth two test scores. In order to earn a B, Joanna's average must lie between 80 and 89, inclusive. What range of scores can Joanna receive on the final and earn a B in the course?

104. Computing Grades Jack needs to earn a C in his sociology class. His current test scores are 67, 72, 81, and 75. His final exam is worth three test scores. In order to earn a C, Jack's average must lie between 70 and 79, inclusive. What range of scores can Jack receive on the final exam and earn a C in the course?

105. Federal Tax Withholding The percentage method of withholding for federal income tax (2014) states that a single person whose weekly wages, after subtracting withholding allowances, are over \$436, but not over \$1506, shall have \$34.90 plus 15% of the excess over \$436 withheld. Over what range does the amount withheld vary if the weekly wages vary from \$800 to \$900, inclusive? SOURCE: *Internal Revenue Service*

106. Federal Tax Withholding Rework Problem 105 if the weekly wages vary from \$1000 to \$1100, inclusive.

107. Commission Gerard had an offer for a medical equipment sales position that pays \$2500 per month plus 1% of all sales. What total sales is required to earn between \$3000 and \$5000 per month?

108. Commission Juanita had a job offer to be an automobile sales position that pays \$1500 per month plus 2.5% of all sales. What total sales is required to earn between \$4000 and \$6000 per month?

109. Electric Bill In North Carolina, Duke Energy charges \$42.41 plus \$0.092897 for each additional kilowatt hour (kwh) used during the months from July through October for usage in excess of 350 kwh. Suppose one homeowner's electric bill ranged from a low of \$88.86 to a high of \$137.16 during this time period. Over what range (in kwh) did the usage vary?

110. Electric Bill In North Carolina, Duke Energy charges \$42.41 plus \$0.084192 for each additional kilowatt hour (kwh) used during the months from November through June for usage in excess of

350 kwh. Suppose one homeowner's electric bill ranged from a low of $55.04 to a high of $89.56 during this time period. Over what range (in kwh) did the usage vary?

Extending the Concepts

111. The Arithmetic Mean If $a < b$, show that $a < \dfrac{a+b}{2} < b$. We call $\dfrac{a+b}{2}$ the **arithmetic mean** of a and b.

⚠ **112. Identifying Triangles** A triangle has the property that the length of the longest side is greater than the difference of the other sides, and the length of the longest side is less than the sum of the other sides. That is, if a, b, and c are sides such that $a \le b \le c$, then $b - a < c < b + a$. Determine which of the following could be lengths of the sides of a triangle.

  (a) 3, 4, 5   **(b)** 4, 7, 12
  (c) 3, 3, 5   **(d)** 1, 9, 10

113. Solve $2x + 1 \le 5x + 7 \le x - 5$.

114. Solve $x - 3 \le 3x + 1 \le x + 11$.

115. Solve $4x + 1 > 2(2x + 1)$. Provide an explanation that generalizes the result.

116. Solve $4x - 2 \ge 2(2x - 1)$. Provide an explanation that generalizes the result.

117. Consider the following analysis, assuming that $x < 2$.

$$5 > 2$$
$$5(x - 2) > 2(x - 2)$$
$$5x - 10 > 2x - 4$$
$$3x > 6$$
$$x > 2$$

How can it be that the final line in the analysis states that $x > 2$, when the original assumption stated that $x < 2$?

8.7 Absolute Value Equations and Inequalities

Objectives

1 Solve Absolute Value Equations

2 Solve Absolute Value Inequalities Involving $<$ or $\le$

3 Solve Absolute Value Inequalities Involving $>$ or $\ge$

4 Solve Applied Problems Involving Absolute Value Inequalities

Figure 47
$|-5| = 5$

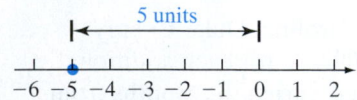

Are You Prepared for This Section?

Before getting started, complete the following problems. If you get a problem wrong, go back to the section cited and review the material.

In Problems P1–P4, evaluate each expression.

P1. $|3|$ **P2.** $|-4|$ **P3.** $|-1.6|$ **P4.** $|0|$ [Section 1.3, pp. 24–25]

P5. Express the distance between the origin, 0, and 5 as an absolute value.   [Section 1.3, pp. 24–25]

P6. Express the distance between the origin, 0, and −8 as an absolute value.   [Section 1.3, pp. 24–25]

P7. Solve: $4x + 5 = -9$ [Section 2.2, pp. 91–92]

P8. Solve: $-2x + 1 > 5$ [Section 2.8, pp. 150–155]

Recall that the absolute value of a number is its distance from the origin on the real number line. For example, $|-5| = 5$ because the distance on the real number line from 0 to −5 is 5 units. See Figure 47.

▶ **1** Solve Absolute Value Equations

EXAMPLE 1 | **Solving an Absolute Value Equation**

Solve the equation $|x| = 4$.

Solution
Two geometric solutions and one algebraic solution are presented.

Geometric Solution 1: The equation $|x| = 4$ asks, "Which real numbers x are 4 units from 0 on the number line?" Figure 48 on the next page shows the two numbers, −4 and 4. The solution set is $\{-4, 4\}$.

Prepared?...Answers **P1.** 3 **P2.** 4
P3. 1.6 **P4.** 0 **P5.** $|5|$ **P6.** $|-8|$
P7. $\left\{-\dfrac{7}{2}\right\}$ **P8.** $\{x | x < -2\}; (-\infty, -2)$

Figure 48

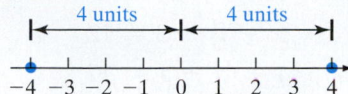

Figure 49

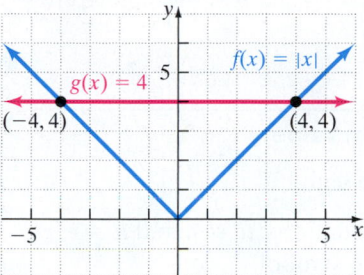

Geometric Solution 2: $|x| = 4$ can be solved by graphing $f(x) = |x|$ and $g(x) = 4$ on the same xy-plane, as in Figure 49. The x-coordinates of the points of intersection are -4 and 4, which are the solutions to the equation $f(x) = g(x)$. Again, the solution set is $\{-4, 4\}$.

Algebraic Solution: Recall that $|a| = a$ if $a \geq 0$ and $|a| = -a$ if $a < 0$. Because it is not known whether x is positive or negative in the equation $|x| = 4$, solve the problem for $x \geq 0$ or $x < 0$.

If $x \geq 0$	If $x < 0$
$|x| = 4$	$|x| = 4$
$|x| = x$ since $x \geq 0$: $x = 4$	$|x| = -x$ since $x < 0$: $-x = 4$
	Multiply both sides by -1: $x = -4$

The solution set is $\{-4, 4\}$.

> **Quick ✔**
>
> *In Problems 1 and 2, solve the equation geometrically and algebraically.*
>
> **1.** $|x| = 7$ **2.** $|z| = 1$

The results of Example 1 and Quick Checks 1 and 2 lead us to the following result.

Figure 50

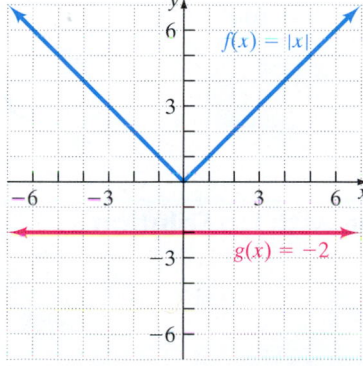

> **Equations Involving Absolute Value**
>
> If a is a positive real number and u is any algebraic expression, then
>
> $$|u| = a \quad \text{is equivalent to} \quad u = a \quad \text{or} \quad u = -a$$
>
> **Note:** If $a = 0$, the equation $|u| = 0$ is equivalent to $u = 0$. If $a < 0$, the equation $|u| = a$ has no solution, as explained below.

In the equation $|u| = a$, it is required that a be nonnegative (greater than or equal to 0). If a is negative, the equation has no real solution. To see why, consider the equation $|x| = -2$. Figure 50 shows the graph of $f(x) = |x|$ and $g(x) = -2$. Notice that the graphs do not intersect, which shows that the equation $|x| = -2$ has no real solution. The solution set is the empty set, $\varnothing$ or $\{\ \}$.

EXAMPLE 2   **How to Solve an Equation Involving Absolute Value**

Solve the equation: $|2x - 1| + 3 = 12$

Step-by-Step Solution

Step 1: Isolate the expression containing the absolute value.

$$|2x - 1| + 3 = 12$$

Subtract 3 from both sides: $|2x - 1| = 9$

Step 2: Rewrite the equation $|u| = a$ as: $u = a$ or $u = -a$, where u is the algebraic expression in the absolute value symbol.

In the equation $|2x - 1| = 9$, $u = 2x - 1$ and $a = 9$.

$$2x - 1 = 9 \qquad \text{or} \qquad 2x - 1 = -9$$

Step 3: Solve each equation.

Add 1 to each side: $2x = 10$ Add 1 to each side: $2x = -8$

Divide both sides by 2: $x = 5$ Divide both sides by 2: $x = -4$

(continued)

Step 4: Check: Verify each solution.

$$|2x - 1| + 3 = 12 \qquad\qquad |2x - 1| + 3 = 12$$

Let $x = 5$: $\quad |2(5) - 1| + 3 \overset{?}{=} 12 \qquad$ Let $x = -4$: $\quad |2(-4) - 1| + 3 \overset{?}{=} 12$

$$|10 - 1| + 3 \overset{?}{=} 12 \qquad\qquad |-8 - 1| + 3 \overset{?}{=} 12$$

$$9 + 3 \overset{?}{=} 12 \qquad\qquad\qquad 9 + 3 \overset{?}{=} 12$$

$$12 = 12 \quad \text{True} \qquad\qquad\qquad 12 = 12 \quad \text{True}$$

Both solutions check, so the solution set is $\{-4, 5\}$. ●

Solving Absolute Value Equations with One Absolute Value

Step 1: Isolate the expression containing the absolute value.

Step 2: Rewrite the absolute value equation as two equations: $u = a$ and $u = -a$, where u is the expression in the absolute value symbol.

Step 3: Solve each equation.

Step 4: Verify your solution.

Quick ✓

3. $|u| = a$ is equivalent to $u = \underline{}$ or $u = \underline{}$

4. $|2x + 3| = 5$ is equivalent to $2x + 3 = 5$ or $\underline{}$.

In Problems 5–8, solve each equation.

5. $|2x - 3| = 7$

6. $|3x - 2| + 3 = 10$

7. $|-5x + 2| - 2 = 5$

8. $3|x + 2| - 4 = 5$

▶ (**EXAMPLE 3**) **Solving an Equation Involving Absolute Value with No Solution**

Solve the equation: $|x + 5| + 7 = 5$

Solution

$$|x + 5| + 7 = 5$$

Subtract 7 from both sides: $\qquad |x + 5| = -2$

Work Smart

The equation $|u| = a$, where a is a negative real number, has no real solution. Why? See Figure 40 on the previous page.

Since the absolute value of any real number is always nonnegative (greater than or equal to zero), the equation has no real solution. The solution set is $\{\ \}$ or $\varnothing$. ●

Quick ✓

9. *True or False* $|x| = -4$ has no real solution.

In Problems 10–12, solve each equation.

10. $|5x + 3| = -2$

11. $|2x + 5| + 7 = 3$

12. $|x + 1| + 3 = 3$

What if an absolute value equation has two absolute values, as in $|3x - 1| = |x + 5|$? The signs of the algebraic expressions in the absolute value symbol have four possibilities:

1. both algebraic expressions are nonnegative,

2. both are negative,

3. the left is positive, and the right is negative, or

4. the left is negative and the right is nonnegative.

To see how the solution works, use the fact that $|a| = a$, if $a \geq 0$ and $|a| = -a$, if $a < 0$.

Thus, if $3x - 1 \geq 0$, then $|3x - 1| = 3x - 1$. However, if $3x - 1 < 0$, then $|3x - 1| = -(3x - 1)$. This leads to a method for solving absolute value equations with two absolute values.

Case 1: Both Expressions Are Nonnegative	Case 2: Both Expressions Are Negative	Case 3: The Expression on the Left Is Nonnegative, and That on the Right Is Negative	Case 4: The Expression on the Left Is Negative, and That on the Right Is Nonnegative
$\|3x - 1\| = \|x + 5\|$ $3x - 1 = x + 5$	$\|3x - 1\| = \|x + 5\|$ $-(3x - 1) = -(x + 5)$ $3x - 1 = x + 5$	$\|3x - 1\| = \|x + 5\|$ $3x - 1 = -(x + 5)$	$\|3x - 1\| = \|x + 5\|$ $-(3x - 1) = x + 5$

When both algebraic expressions are nonnegative, or both negative, we end up with equivalent equations. Therefore, Case 1 and Case 2 result in equivalent equations. Also, if one side is nonnegative and the other is negative, we end up with equivalent equations. Therefore, Case 3 and Case 4 result in equivalent equations. The four possibilities reduce to two possibilities.

Equations Involving Two Absolute Values

If u and v are any algebraic expressions, then

$$|u| = |v| \quad \text{is equivalent to} \quad u = v \quad \text{or} \quad u = -v$$

EXAMPLE 4 **Solving an Absolute Value Equation Involving Two Absolute Values**

Solve the equation: $|2x - 3| = |x + 6|$

Solution
The equation is in the form $|u| = |v|$, where $u = 2x - 3$ and $v = x + 6$. Rewrite the equation in the form $u = v$ or $u = -v$ and then solve each equation.

$u = v$: $2x - 3 = x + 6$
Add 3 to both sides: $2x = x + 9$
Subtract x from both sides: $x = 9$

$u = -v$: $2x - 3 = -(x + 6)$
Distribute the -1: $2x - 3 = -x - 6$
Add 3 to both sides: $2x = -x - 3$
Add x to both sides: $3x = -3$
Divide both sides by 3: $x = -1$

Check
$x = 9$: $|2(9) - 3| \stackrel{?}{=} |9 + 6|$
$|18 - 3| \stackrel{?}{=} |15|$
$|15| \stackrel{?}{=} 15$
$15 = 15$ True

$x = -1$: $|2(-1) - 3| \stackrel{?}{=} |-1 + 6|$
$|-2 - 3| \stackrel{?}{=} |5|$
$|-5| \stackrel{?}{=} 5$
$5 = 5$ True

Both solutions check, so the solution set is $\{-1, 9\}$.

Quick ✓

13. $|u| = |v|$ is equivalent to __ = __ or __ = __.

In Problems 14–17, solve each equation.

14. $|x - 3| = |2x + 5|$

15. $|8z + 11| = |6z + 17|$

16. $|3 - 2y| = |4y + 3|$

17. $|2x - 3| = |5 - 2x|$

⊳ **❷ Solve Absolute Value Inequalities Involving $<$ or $\leq$**

The method for solving absolute value equations relies on the geometric interpretation of absolute value. That is, the absolute value of a real number x is its distance from the origin on the real number line. The geometric interpretation of absolute value can also be used to solve absolute value inequalities.

EXAMPLE 5 **Solving an Absolute Value Inequality**

Solve the inequality $|x| < 4$. Graph the solution set.

Solution
The inequality $|x| < 4$ asks for all real numbers x that are less than 4 units from the origin on the real number line. See Figure 51. Notice from the figure that any number between -4 and 4 satisfies the inequality. The solution set is $\{x|-4 < x < 4\}$ or, using interval notation, $(-4, 4)$.

These results could also be visualized by graphing $f(x) = |x|$ and $g(x) = 4$. See Figure 52. To solve $f(x) < g(x)$, look for all x-coordinates such that the graph of $f(x)$ is below the graph of $g(x)$. Notice that this is true for all x between -4 and 4, as found above. ●

The results of Example 5 lead to the following.

Figure 51

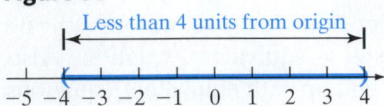
Less than 4 units from origin

Figure 52

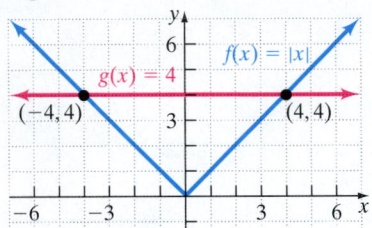

Inequalities of the Form $<$ or $\leq$ Involving Absolute Value

If a is a positive real number and if u is an algebraic expression, then

$$|u| < a \quad \text{is equivalent to} \quad -a < u < a$$
$$|u| \leq a \quad \text{is equivalent to} \quad -a \leq u \leq a$$

Note: If $a = 0$, $|u| < 0$ has no solution; $|u| \leq 0$ is equivalent to $u = 0$. If $a < 0$, the inequality has no solution.

Quick ✓

18. If $a > 0$, then $|u| < a$ is equivalent to _____.

19. To solve $|3x + 4| < 10$, solve ___ $< 3x + 10 <$ __.

In Problems 20 and 21, solve each inequality. Graph the solution set.

20. $|x| \leq 5$ **21.** $|x| < \dfrac{3}{2}$

EXAMPLE 6 **How to Solve an Absolute Value Inequality Involving $\leq$**

Solve the inequality $|2x + 3| \leq 5$. Express the solution set in set-builder notation and interval notation. Graph the solution set.

Step-by-Step Solution

Step 1: The inequality is in the form $|u| \leq a$, where $u = 2x + 3$ and $a = 5$. Rewrite the inequality as a compound inequality that does not involve absolute value.

$$|2x + 3| \leq 5$$
$|u| \leq a$ means $-a \leq u \leq a$: $\quad -5 \leq 2x + 3 \leq 5$

Step 2: Solve the resulting compound inequality.

Subtract 3 from all three parts: $\quad -5 - 3 \leq 2x + 3 - 3 \leq 5 - 3$
$$-8 \leq 2x \leq 2$$

Divide all three parts by 2: $\quad \dfrac{-8}{2} \leq \dfrac{2x}{2} \leq \dfrac{2}{2}$

$$-4 \leq x \leq 1$$

Figure 53

The solution is $\{x|-4 \leq x \leq 1\}$ or, in interval notation, $[-4, 1]$. Figure 53 shows the graph of the solution set. ●

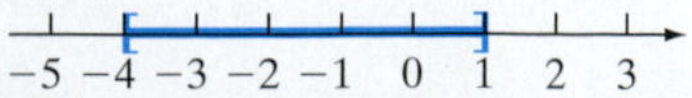

Work Smart

As a partial check of the solution of Example 6, we can check a number in the interval. Let's try $x = -3$.

$$|2(-3) + 3| \overset{?}{\le} 5$$
$$|-6 + 3| \overset{?}{\le} 5$$
$$|-3| \overset{?}{\le} 5$$
$$3 \le 5 \quad \text{True}$$

Quick ✓

In Problems 22–24, solve each inequality. Express the solution set in set-builder notation and interval notation. Graph the solution set.

22. $|x + 3| < 5$

23. $|2x - 3| \le 7$

24. $|7x + 2| < -3$

EXAMPLE 7   **Solving an Absolute Value Inequality Involving $<$**

Solve the inequality $|-3x + 2| + 4 < 14$. Express the solution set in set-builder notation and interval notation. Graph the solution set.

Solution

First, isolate the absolute value.

$$|-3x + 2| + 4 < 14$$

Subtract 4 from both sides: $\quad |-3x + 2| < 10$

$|u| < a$ means $-a < u < a$: $\quad -10 < -3x + 2 < 10$

Subtract 2 from all three parts: $\quad -10 - 2 < -3x + 2 - 2 < 10 - 2$

$$-12 < -3x < 8$$

Divide all three parts by -3.
Reverse the direction of the inequalities: $\quad \dfrac{-12}{-3} > \dfrac{-3x}{-3} > \dfrac{8}{-3}$

$$4 > x > -\dfrac{8}{3}$$

$b > x > a$ is equivalent to $a < x < b$: $\quad -\dfrac{8}{3} < x < 4$

Figure 54

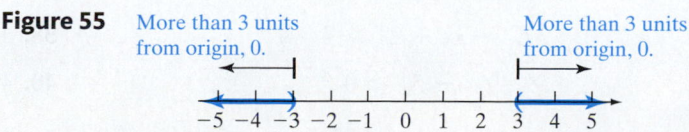

The solution is $\left\{ x \,\middle|\, -\dfrac{8}{3} < x < 4 \right\}$, using set-builder notation, and $\left(-\dfrac{8}{3}, 4 \right)$, using interval notation. Figure 54 shows the graph of the solution set.

Quick ✓

25. *True or False* $|x + 1| + 2 < 7$ is equivalent to $-7 < x + 1 + 2 < 7$.

In Problems 26–29, solve each inequality. Express the solution set in set-builder notation and interval notation. Graph the solution set.

26. $|x| + 4 < 6$

27. $|x - 3| + 4 \le 8$

28. $3|2x + 1| \le 9$

29. $|-3x + 1| - 5 < 3$

▶ ❸ **Solve Absolute Value Inequalities Involving $>$ or $\ge$**

EXAMPLE 8   **Solving an Absolute Value Inequality Involving $>$**

Solve the inequality $|x| > 3$. Graph the solution set.

Solution

The inequality $|x| > 3$ asks for all real numbers x that are more than 3 units from the origin on the real number line. See Figure 55.

Figure 55   More than 3 units from origin, 0.   More than 3 units from origin, 0.

Figure 56

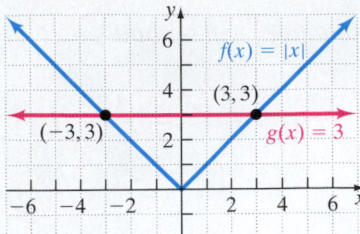

Notice that any number less than -3 or greater than 3 satisfies the inequality. The solution set is $\{x \mid x < -3 \text{ or } x > 3\}$ or, using interval notation, $(-\infty, -3) \cup (3, \infty)$.

These results could also be visualized by graphing $f(x) = |x|$ and $g(x) = 3$. See Figure 56. To solve $f(x) > g(x)$, look for all x-coordinates such that the graph of $f(x)$ is above the graph of $g(x)$. Notice that this is true for all x less than -3 or all x greater than 3, as we found above.

Based on Example 8, we have the following results.

Inequalities of the Form $>$ or $\geq$ Involving Absolute Value

If a is a positive real number and u is an algebraic expression, then

$$|u| > a \qquad \text{is equivalent to} \qquad u < -a \quad \text{or} \quad u > a$$
$$|u| \geq a \qquad \text{is equivalent to} \qquad u \leq -a \quad \text{or} \quad u \geq a$$

Quick ✓

30. $|u| > a$ is equivalent to _____ or _____.

31. $|5x - 2| \geq 7$ is equivalent to $5x - 2 \leq$ ___ or $5x - 2 \geq$ _.

In Problems 32 and 33, solve each inequality. Graph the solution set.

32. $|x| \geq 6$

33. $|x| > \dfrac{5}{2}$

EXAMPLE 9   How to Solve an Inequality Involving $>$

Solve the inequality $|2x - 5| > 3$. Express the solution set in set-builder notation and interval notation. Graph the solution set.

Step-by-Step Solution

Step 1: The inequality is in the form $|u| > a$, where $u = 2x - 5$ and $a = 3$. Rewrite the inequality as a compound inequality that does not involve absolute value.

$$|2x - 5| > 3$$

$|u| > a$ means $u < -a$ or $u > a$: $2x - 5 < -3$ or $2x - 5 > 3$

Step 2: Solve each inequality separately.

	$2x - 5 < -3$		$2x - 5 > 3$
Add 5 to both sides:	$2x < 2$	Add 5 to both sides:	$2x > 8$
Divide both sides by 2:	$x < 1$	Divide both sides by 2:	$x > 4$

Step 3: Find the union of the solution sets of each inequality.

The solution set is $\{x \mid x < 1 \text{ or } x > 4\}$ or, using interval notation, $(-\infty, 1) \cup (4, \infty)$. See Figure 57 for the graph of the solution set.

Figure 57

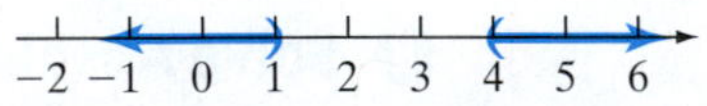

$$\overset{\qquad -2 \;\; -1 \;\;\; 0 \;\;\; 1 \;\;\; 2 \;\;\; 3 \;\;\; 4 \;\;\; 5 \;\;\; 6}{}$$

Work Smart

$$|u| > a$$

CANNOT be written as

$$-a > u > a$$

Quick ✓

34. $|x - 9| > 6$ is equivalent to $x - 9 > 6$ or $x - 9$___-6.

In Problems 35–40, solve each inequality. Express the solution set in set-builder notation and interval notation. Graph the solution set.

35. $|x + 3| > 4$

36. $|4x - 3| \geq 5$

37. $|-3x + 2| > 7$

38. $|2x + 5| - 2 > -2$

39. $|6x - 5| \geq 0$

40. $|2x + 1| > -3$

Summary   Solving Absolute Value Equations and Inequalities

Absolute Value Form	Equation/Inequality Form	Example	
		Algebraic Solution	**Graphical Solution**
$\lvert u \rvert = a$	$u = a$ or $u = -a$	$\lvert x + 3 \rvert = 5$ $x + 3 = 5$ or $x + 3 = -5$ $x = 2$ or $x = -8$ Solution set: $\{-8, 2\}$	
$\lvert u \rvert = \lvert v \rvert$	$u = v$ or $u = -v$	$\lvert x - 2 \rvert = \lvert 2x \rvert$ $x - 2 = 2x$ or $x - 2 = -2x$ $-2 = x$ or $3x = 2$ $x = -2$ or $x = \dfrac{2}{3}$ Solution set: $\left\{ -2, \dfrac{2}{3} \right\}$	
$\lvert u \rvert < a$ $\lvert u \rvert \le a$	$-a < u < a$ $-a \le u \le a$	$\lvert x - 1 \rvert \le 4$ $-4 \le x - 1 \le 4$ $-3 \le x \le 5$ Solution set: $\{ x \mid -3 \le x \le 5 \}$ or $[-3, 5]$	
$\lvert u \rvert > a$ $\lvert u \rvert \ge a$	$u < -a$ or $u > a$ $u \le -a$ or $u \ge a$	$\lvert x + 1 \rvert > 3$ $x + 1 < -3$ or $x + 1 > 3$ $x < -4$ or $x > 2$ Solution set: $\{ x \mid x < -4$ or $x > 2 \}$ or $(-\infty, -4) \cup (2, \infty)$	

❹ Solve Applied Problems Involving Absolute Value Inequalities

You may have read phrases such as "margin of error" and "tolerance" in the newspaper. For example, according to a recent Gallup poll, 67% of those polled said that recent increases in gas prices have caused financial hardship for their household. The poll had a margin of error of 4%. The 67% is an estimate of the true percentage of Americans who have experienced financial hardship as a consequence of increases in gas prices. If p represents the true percentage of Americans who have experienced financial hardship due to the increase in gas prices, then the poll's margin of error can be represented mathematically as

$$\lvert p - 67 \rvert \le 4$$

As another example, the tolerance of a belt used in a pulley system whose width is 6 inches is $\dfrac{1}{16}$ inch. If x represents the actual width of the belt, then the acceptable belt widths can be represented as

$$\lvert x - 6 \rvert \le \dfrac{1}{16}$$

EXAMPLE 10   **Analyzing the Margin of Error in a Poll**

The inequality

$$\lvert p - 67 \rvert \le 4$$

(continued)

represents the percentage of Americans who said they experienced financial hardship due to increase in gas prices. Solve the inequality and interpret the results.

Solution

$$|p - 67| \le 4$$

$|u| \le a$ means $-a \le u \le a$: $\quad -4 \le p - 67 \le 4$

Add 67 to all three parts: $\quad 63 \le p \le 71$

The percentage of Americans who have experienced financial hardship due to increases in gas prices is between 63% and 71%, inclusive. ●

Quick ✓

41. The inequality $|x - 4| \le \dfrac{1}{32}$ represents the acceptable belt width x (in inches) for a belt that is manufactured for a pulley system. Determine the acceptable belt width.

42. In a poll conducted by ABC News, 9% of Americans stated that they have been shot at. The margin of error in the poll was 1.7%. If we let p represent the true percentage of people who have been shot at, we can represent the margin of error as

$$|p - 9| \le 1.7$$

Solve the inequality and interpret the results.

8.7 Exercises MyMathLab® Exercise numbers in **green** have complete video solutions in MyMathLab or may be accessed using the QR code to the right.

Problems **1–42** *are the* **Quick ✓** *s that follow the* **EXAMPLES**.

Building Skills

In Problems 43–68, solve each absolute value equation. See Objective 1.

43. $|x| = 10$

44. $|z| = 9$

45. $|p| = -2$

46. $|4| = -1$

47. $|y - 3| = 4$

48. $|x + 3| = 5$

49. $|-3x + 5| = 8$

50. $|-4y + 3| = 9$

51. $|y| - 7 = -2$

52. $|x| + 3 = 5$

53. $|2x + 3| - 5 = 3$

54. $|3y + 1| - 5 = -3$

55. $-2|x - 3| + 10 = -4$

56. $3|y - 4| + 4 = 16$

57. $|-3x| - 5 = -5$

58. $|-2x| + 9 = 9$

59. $\left| \dfrac{3x - 1}{4} \right| = 2$

60. $\left| \dfrac{2x - 3}{5} \right| = 2$

61. $|3x + 2| = |2x - 5|$

62. $|5y - 2| = |4y + 7|$

63. $|8 - 3x| = |2x - 7|$

64. $|5x + 3| = |12 - 4x|$

65. $|4y - 7| = |9 - 4y|$

66. $|5x - 1| = |9 - 5x|$

67. $|2x - 1| = -|x|$

68. $-|x + 1| = |3x - 2|$

In Problems 69–82, solve each absolute value inequality. Graph the solution set on a real number line. See Objective 2.

69. $|x| < 9$

70. $|x| \le \dfrac{5}{4}$

71. $|x - 4| \le 7$

72. $|y + 4| < 6$

73. $|3x + 1| < 8$

74. $|4x - 3| \le 9$

75. $|6x + 5| < -1$ **76.** $|4x + 3| \le 0$

77. $2|x - 3| + 3 < 9$ **78.** $3|y + 2| - 2 < 7$

79. $|2 - 5x| + 3 < 10$ **80.** $|-3x + 2| - 7 \le -2$

81. $|(2x - 3) - 1| < 0.01$ **82.** $|(3x + 2) - 8| < 0.01$

In Problems 83–94, solve each absolute value inequality. Graph the solution set on the real number line. See Objective 3.

83. $|y - 5| > 2$ **84.** $|x + 4| \ge 7$

85. $|-4x - 3| \ge 5$ **86.** $|-5y + 3| > 7$

87. $2|y| + 3 > 1$ **88.** $3|z| + 8 > 2$

89. $|-5x - 3| - 7 > 0$ **90.** $|-9x + 2| - 11 \ge 0$

91. $4|-2x + 1| > 4$ **92.** $3|8x + 3| \ge 9$

93. $|1 - 2x| \ge |-5|$ **94.** $|3 - 5x| > |-7|$

Mixed Practice

In Problems 95–98, use the graphs of the functions given to solve each problem.

95. $f(x) = |x|, g(x) = 5$
   (a) $f(x) = g(x)$
   (b) $f(x) \le g(x)$
   (c) $f(x) > g(x)$

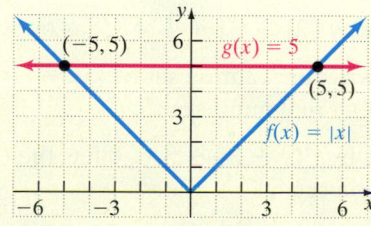

96. $f(x) = |x|, g(x) = 6$
   (a) $f(x) = g(x)$

   (b) $f(x) \le g(x)$
   (c) $f(x) > g(x)$

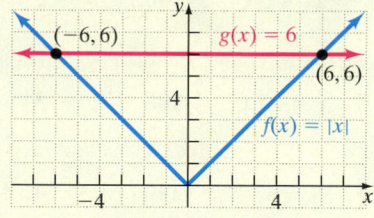

97. $f(x) = |x + 2|, g(x) = 3$
   (a) $f(x) = g(x)$
   (b) $f(x) < g(x)$
   (c) $f(x) \ge g(x)$

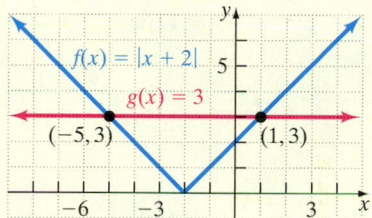

98. $f(x) = |2x|, g(x) = 10$
   (a) $f(x) = g(x)$
   (b) $f(x) < g(x)$
   (c) $f(x) \ge g(x)$

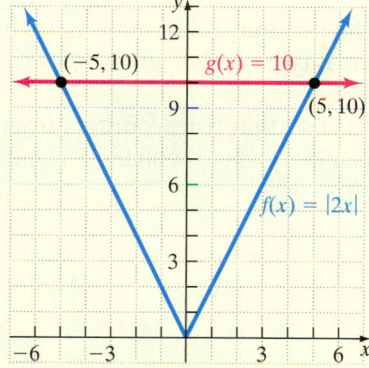

In Problems 99–120, solve each absolute value equation or inequality. For absolute value inequalities, graph the solution set on a real number line.

99. $|x| > 5$ **100.** $|x| \ge \dfrac{8}{3}$

101. $|2x + 5| = 3$ **102.** $|4x + 3| = 1$

103. $7|x| = 35$

104. $8|y| = 32$

105. $|2x + 1| - 2 = 5$

106. $|3x - 2| + 1 = 8$

107. $|-2x + 3| + 4 = 0$

108. $|7y - 3| < 11$

109. $|-2x + 3| = -4$

110. $|3x - 4| = -9$

111. $|3x + 2| \geq 5$

112. $|5y + 3| > 2$

113. $|3x - 2| + 7 > 9$

114. $|4y + 3| - 8 \geq -3$

115. $|5x + 3| = |3x + 5|$

116. $|3z - 2| = |z + 6|$

117. $|4x + 7| + 6 < 5$

118. $|4x + 1| > 0$

119. $\left|\dfrac{x - 2}{4}\right| = \left|\dfrac{2x + 1}{6}\right|$

120. $\left|\dfrac{1}{2}x - 3\right| = \left|\dfrac{2}{3}x + 1\right|$

Applying the Concepts

121. Express the fact that x differs from 5 by less than 3 as an inequality involving absolute value. Solve for x.

122. Express the fact that x differs from -4 by less than 2 as an inequality involving absolute value. Solve for x.

123. Express the fact that twice x differs from -6 by more than 3 as an inequality involving absolute value. Solve for x.

124. Express the fact that twice x differs from 7 by more than 3 as an inequality involving absolute value. Solve for x.

125. Tolerance A certain rod in an internal combustion engine is supposed to be 5.7 inches long. The tolerance on the rod is 0.0005 inch. If x represents the length of a rod, the acceptable lengths of a rod can be expressed as $|x - 5.7| \leq 0.0005$. Determine the acceptable lengths of the rod.
SOURCE: *WiseCo Piston*

126. Tolerance A certain rod in an internal combustion engine is supposed to be 6.125 inches long. The tolerance on the rod is 0.0005 inch. If x represents the length of a rod, the acceptable lengths of a rod can be expressed as $|x - 6.125| \leq 0.0005$. Determine the acceptable lengths of the rod.

127. IQ Scores According to the Stanford-Binet IQ test, a normal IQ score is 100. It can be shown that anyone with an IQ x that satisfies the inequality $\left|\dfrac{x - 100}{15}\right| > 1.96$ has an unusual IQ score. Determine the IQ scores that would be considered unusual.

128. Gestation Period The length of human pregnancy is about 266 days. It can be shown that a mother whose gestation period x satisfies the inequality $\left|\dfrac{x - 266}{16}\right| > 1.96$ has an unusual length of pregnancy. Determine the length of pregnancy that would be considered unusual.

Extending the Concepts

In Problems 129–136, solve each equation.

129. $|x| - x = 5$

130. $|y| + y = 3$

131. $z + |-z| = 4$

132. $y - |-y| = 12$

133. $|4x + 1| = x - 2$

134. $|2x + 1| = x - 3$

135. $|x + 5| = -(x + 5)$

136. $|y - 4| = y - 4$

Explaining the Concepts

137. Explain why $|2x - 3| + 1 = 0$ has no solution.

138. Explain why the solution set of $|5x - 3| > -5$ is the set of all real numbers.

139. Explain why $|4x + 3| + 3 < 0$ has the empty set as the solution set.

140. Solve $|x - 5| = |5 - x|$. Explain why the result is reasonable. What do we call this type of equation?

8.8 Variation

Objectives

❶ Model and Solve Direct Variation Problems

❷ Model and Solve Inverse Variation Problems

❸ Model and Solve Joint Variation and Combined Variation Problems

Are You Prepared for This Section?

Before getting started, complete the following problems. If you get a problem wrong, go back to the section cited and review the material.

P1. Solve: $30 = 5x$ [Section 2.1, pp. 85–89]

P2. Solve: $4 = \dfrac{k}{3}$ [Section 2.1, pp. 85–89]

P3. Graph: $y = 3x$ [Section 3.4, pp. 206–208]

▶ ❶ **Model and Solve Direct Variation Problems**

Often two variables are related in terms of proportions. Examples include "Revenue is proportional to sales" and "Force is proportional to acceleration." When one variable is proportional to another variable, it is called *variation*. **Variation** indicates how one quantity varies in relation to some other quantity. Quantities may vary *directly*, *inversely*, or *jointly*. Direct variation is discussed first.

> **In Other Words**
>
> If y is directly proportional to x, then y and x are related through a linear equation whose y-intercept is $(0, 0)$ and whose slope is k, the constant of proportionality.

Definition

If x and y represent two quantities, then y **varies directly** with x, or y is **directly proportional to** x, if there is a nonzero number k such that

$$y = kx$$

The number k is called the **constant of proportionality** or the **constant of variation**.

Figure 58
$y = kx, k > 0, x \geq 0$

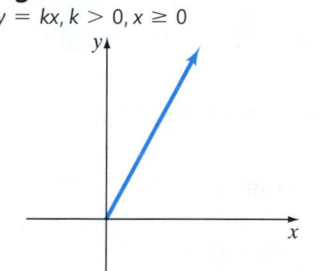

If y varies directly with x, then y is a linear function of x. The graph in Figure 58 shows the relationship between y and x if y varies directly with x and $k > 0, x \geq 0$. The constant of proportionality is the slope of the line, and the y-intercept is $(0, 0)$.

If two quantities vary directly, then knowing the value of each quantity in one instance enables us to write a formula that is true in all cases.

EXAMPLE 1   **Hooke's Law**

Hooke's Law states that the force (or weight) on a spring is directly proportional to the length that the spring stretches from its "at rest" position. That is, $F = kx$, where F is the force exerted, x is the extension of the spring, and k is the proportionality constant that varies from spring to spring.

(a) Suppose a 20-pound weight causes a spring to stretch 10 inches. See Figure 59. Find the constant of proportionality, k.

Figure 59

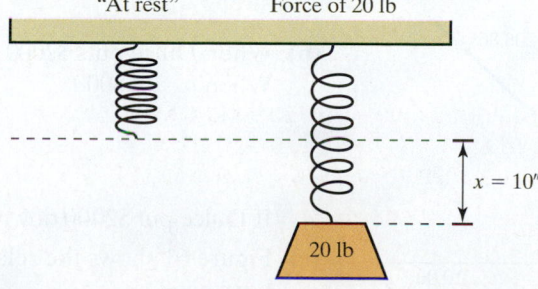

Prepared?...Answers   **P1.** $\{6\}$

P2. $\{12\}$

P3.

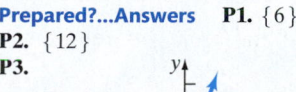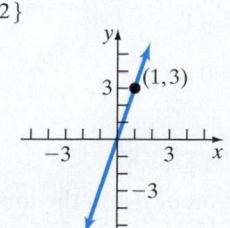

(b) Write the relation between force and stretch length using function notation.

(c) What amount of weight would make the spring stretch 8 inches?

(d) Graph the relation between force and the length the spring stretches.

Solution

(a) It is known that $F = kx$ and that $F = 20$ pounds when $x = 10$ inches. Use the given information to find k, the constant of proportionality.

$$F = kx$$
$$20 = k(10)$$
$$k = 2$$

Figure 60

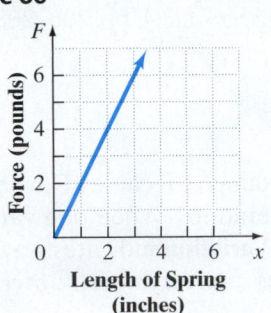

Length of Spring (inches)

(b) Because $k = 2$, then $F = 2x$. Using function notation, write

$$F(x) = 2x$$

(c) If the spring stretches $x = 8$ inches, then the force exerted is

$$F(8) = 2(8) = 16 \text{ pounds}$$

(d) Figure 60 shows the relation between force and the length of the spring. ●

EXAMPLE 2 **Car Payments**

Suppose Dulce just purchased a used car for $10,000. She decides to put $1000 down on the car and borrows the remaining $9000. The bank lends Dulce $9000 at 4.9% interest for 48 months. Her monthly payment is $206.86. The monthly payment p on a car varies directly with the amount borrowed b.

(a) Find a function that relates the monthly payment p to the amount borrowed b for any car loan with the same terms.

(b) Suppose that Dulce put $2000 down on the car instead. What would her monthly payment be?

(c) Graph the relation between monthly payment and amount borrowed.

Solution

(a) Because p varies directly with b, it is known that for some constant k,

$$p = kb$$

Work Smart

If possible, find the exact value of k. If you cannot avoid approximate values, do not round the value of k to fewer than 5 decimal places.

Because $p = \$206.86$ when $b = \$9000$, it follows that

$$206.86 = k(9000)$$

Divide by 9000: $k = 0.022984$

Rewrite the equation as

$$p = 0.022984b$$

Write this as a linear function:

$$p(b) = 0.022984b$$

(b) When Dulce puts $2000 down, she will need to borrow $8000. When $b = \$8000$,

$$p(8000) = 0.022984(8000)$$
$$\approx \$183.87$$

If Dulce put $2000 down, her monthly payment would be $183.87.

Figure 61

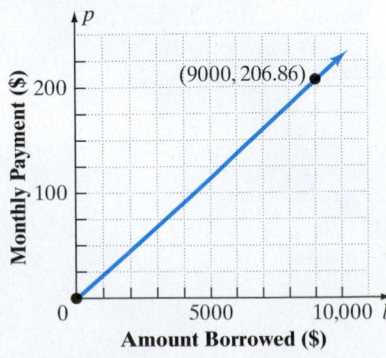

Amount Borrowed ($)

(c) Figure 61 shows the relation between the monthly payment p and the amount borrowed b. ●

Quick ✓

1. _____ indicates how one quantity varies in relation to some other quantity.

2. If x and y are two quantities, then y is directly proportional to x if there is a nonzero number k such that _____.

3. The cost of gas C varies directly with the number of gallons pumped, g. Suppose that the cost of pumping 8 gallons of gas is $22.00.

   (a) Find a function that relates the cost of gas C to the number of gallons pumped g.

   (b) Suppose that 4.6 gallons are pumped into your car. What will the cost be?

   (c) Graph the relation between cost and number of gallons pumped.

❷ Model and Solve Inverse Variation Problems

Figure 62

$y = \dfrac{k}{x}, k > 0, x > 0$

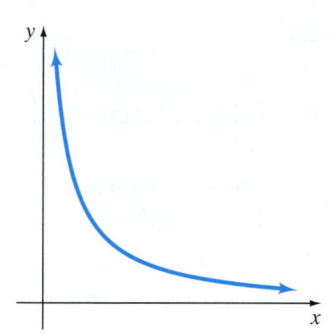

Another kind of variation is now discussed.

Definition

If x and y represent two quantities, then y **varies inversely** with x, or y is **inversely proportional to** x, if there is a nonzero number k such that

$$y = \frac{k}{x}$$

where k is the constant of variation (or constant of proportionality).

The graph in Figure 62 illustrates the relationship between y and x if y varies inversely with x, where $k > 0$ and $x > 0$. Notice from the graph that as x increases, the value of y decreases.

EXAMPLE 3 **Weight That Can Be Supported by a Beam**

The weight W that can be safely supported by a 2-inch by 4-inch (2-by-4) piece of lumber varies inversely with its length l. See Figure 63.

Figure 63

(a) Experiments indicate that the maximum weight a 12-foot pine 2-by-4 can support is 400 pounds. Find a function that relates the maximum weight to the length l for any pine 2-by-4.

(b) Determine the maximum weight a 15-foot pine 2-by-4 can sustain.

Solution

(a) Because W varies inversely with l, it is known that for some constant k,

$$W = \frac{k}{l}$$

Because $W = 400$ when $l = 12$,

$$400 = \frac{k}{12}$$

Multiply both sides by 12: $\quad k = 4800$

Substitute into $W = \dfrac{k}{l}$: $\quad W = \dfrac{4800}{l}$

Write as a function: $\quad W(l) = \dfrac{4800}{l}$

(b) When $l = 15$ feet,

$$W(15) = \dfrac{4800}{15}$$

$$= 320 \text{ pounds}$$

The maximum weight that a 15-foot pine 2-by-4 can sustain is 320 pounds. ●

Quick ✔

4. If x and y represent two quantities, then y varies inversely with x, or y is inversely proportional to x, if there is a nonzero number k such that _____.

5. Suppose that y varies inversely with x for $x > 0$.

   (a) Find an equation that relates x and y if $y = 2$ when $x = 3$.

   (b) Use the equation in part (a) to find y when $x = 4$.

6. The rate of vibration (in oscillations per second) V of a string under constant tension varies inversely with the length l.

   (a) If a string is 30 inches long and vibrates 500 times per second, find a function that relates the rate of vibration of a string to its length.

   (b) What is the rate of vibration of a string that is 50 inches long?

▶ ❸ Model and Solve Joint Variation and Combined Variation Problems

When a variable quantity Q is proportional to the product of two or more other variables, Q **varies jointly** with these quantities. For example, $y = kxz$ is read as "y varies jointly with x and z." When direct and inverse variation occur at the same time, it is called **combined variation**. For example, $y = \dfrac{kx}{z}$, is read as "y varies directly with x and inversely with z." Similarly, $y = \dfrac{kmn}{p}$ can be read as "y varies jointly with m and n and inversely with p."

EXAMPLE 4 | **Force of the Wind—Joint Variation**

The force F of the wind on a flat surface positioned at a right angle to the direction of the wind varies jointly with the area A of the surface and the square of the speed v of the wind. A wind of 30 miles per hour blowing on a window measuring 4 feet by 5 feet has a force of 150 pounds. What is the force on a window measuring 2 feet by 6 feet caused by a hurricane-force wind of 100 miles per hour?

Solution
Because F varies jointly with A and the square of v, it is known that for some constant k,

$$F = kAv^2$$

The area A of the window is $(4 \text{ feet})(5 \text{ feet}) = 20$ square feet. Substitute $F = 150$, $A = 20$, and $v = 30$ to get

$$150 = k(20)(30^2)$$

Solve this equation for k.

$$150 = 18{,}000k$$

Divide by 18,000: $k = \dfrac{1}{120}$

Substitute into $F = kAv^2$: $F = \dfrac{1}{120}Av^2$

Thus, for a 100-mile-per-hour wind blowing on a window whose area is $A = (2 \text{ feet})(6 \text{ feet}) = 12$ square feet, the force F is

$$F = \frac{1}{120}(12)(100)^2$$

$$= 1000 \text{ pounds}$$

Quick ✔

7. The equation $t = ksp$ is an example of ____ variation.

8. The kinetic energy K of an object varies jointly with its mass and the square of its velocity. The kinetic energy of a 110-kg linebacker running at 9 meters per second is 4455 joules. Determine the kinetic energy of a 140-kg linebacker running at 5 meters per second.

EXAMPLE 5   ## Centripetal Force–Combined Variation

Figure 64

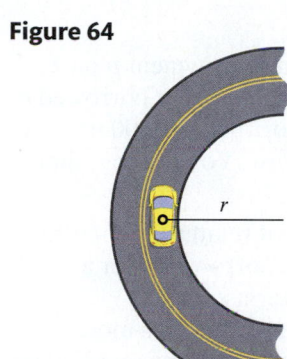

The force required to keep an object traveling in a circular motion is called centripetal force. The centripetal force F required to keep an object of a fixed mass in circular motion varies directly with the square of the velocity v of the object and inversely with the radius r of the circle. See Figure 64.

The force required to keep a car traveling 30 meters per second on a circular road with radius 50 meters is 21,600 newtons. Determine the force required to keep the same car on a circular road if it is traveling 40 meters per second on a road whose radius is 30 meters.

Solution

The force varies directly with the square of the velocity of the object and inversely with the radius of the circle. Then for some constant k, combined variation is

$$F = \frac{kv^2}{r}$$

Because $F = 21,600$ when $v = 30$ and $r = 50$,

$$21,600 = \frac{k \cdot 30^2}{50}$$

Solving for k yields $k = 1200$. Therefore,

$$F = \frac{1200v^2}{r}$$

For a car traveling 40 meters per second on a circular road with radius 30 meters, the force required to keep the car on the road is

$$F = \frac{1200 \cdot 40^2}{30}$$

$$= 64,000 \text{ newtons}$$

Quick ✔

9. When direct and inverse variation occur at the same time, it is called

_____.

10. The electrical resistance R of a wire varies directly with the length l of the wire and inversely with the square of the diameter d of the wire. If a wire 432 feet long and 4 millimeters in diameter has a resistance of 1.24 ohms, find the resistance in a wire that is 282 feet long with a diameter of 3 millimeters. Round your answer to two decimal places.

8.8 Exercises MyMathLab®

Exercise numbers in **green** have complete video solutions in MyMathLab or may be accessed using the QR code to the right.

*Problems **1–10** are the **Quick ✓**s that follow the **EXAMPLES**.*

Building Skills

In Problems 11–16, (a) find the constant of proportionality k, (b) write the linear function relating the two variables, and (c) find the quantity indicated. See Objective 1.

11. Suppose that y varies directly with x. When $x = 5$, $y = 30$. Find y when $x = 7$.

12. Suppose that y varies directly with x. When $x = 3$, $y = 15$. Find y when $x = 5$.

13. Suppose that y is directly proportional to x. When $x = 7$, $y = 3$. Find y when $x = 28$.

14. Suppose that y is directly proportional to x. When $x = 20$, $y = 4$. Find y when $x = 35$.

15. Suppose that y is directly proportional to x. When $x = 8$, $y = 4$. Find y when $x = 30$.

16. Suppose that y is directly proportional to x. When $x = 12$, $y = 8$. Find y when $x = 20$.

In Problems 17–20, (a) find the constant of proportionality k, (b) write the function relating the two variables, and (c) find the quantity indicated. See Objective 2.

17. Suppose that y varies inversely with x. When $x = 10$, $y = 2$. Find y if $x = 5$.

18. Suppose that y varies inversely with x. When $x = 3$, $y = 15$. Find y if $x = 5$.

19. Suppose that y is inversely proportional to x. When $x = 7$, $y = 3$. Find y if $x = 28$.

20. Suppose that y is inversely proportional to x. When $x = 20$, $y = 4$. Find y if $x = 35$.

In Problems 21–24 (a) find the constant of proportionality k, (b) write the function relating the variables, and (c) find the quantity indicated. See Objective 3.

21. Suppose that y varies jointly with x and z. When $x = 8$ and $z = 5$, $y = 10$. Find y if $x = 12$ and $z = 9$.

22. Suppose that y varies jointly with x and z. When $x = 6$ and $z = 10$, $y = 20$. Find y if $x = 8$ and $z = 15$.

23. Suppose that Q varies directly with x and inversely with y. When $x = 5$ and $y = 6$, $Q = \dfrac{13}{12}$. Find Q if $x = 9$ and $y = 4$.

24. Suppose that Q varies directly with x and inversely with y. When $x = 4$ and $y = 3$, $Q = \dfrac{14}{5}$. Find Q if $x = 8$ and $y = 3$.

Applying the Concepts

25. Mortgage Payments The monthly payment p on a mortgage varies directly with the amount borrowed b. Suppose that you decide to borrow \$190,000 using a 30-year mortgage at 4% interest. You are told that your monthly payment is \$907.09.

(a) Write a linear function that relates the monthly payment p to the amount borrowed b for a mortgage with the same terms.

(b) Assume that you have decided to buy a more expensive home that requires you borrow \$220,000. What will your approximate monthly payment be?

(c) Graph the relation between monthly payment and amount borrowed.

26. Mortgage Payments The monthly payment p on a mortgage varies directly with the amount borrowed b. Suppose that you decide to borrow \$190,000 using a 15-year mortgage at 3% interest. You are told that your payment is \$1312.11.

(a) Write a linear function that relates the monthly payment p to the amount borrowed b for a mortgage with the same terms.

(b) Assume that you have decided to buy a more expensive home that requires you borrow \$220,000. What will your approximate monthly payment be?

(c) Graph the relation between monthly payment and amount borrowed.

27. Cost Function The cost C of purchasing chocolate-covered almonds varies directly with the weight w in pounds. Suppose that the cost of purchasing 5 pounds of chocolate-covered almonds is \$28.

(a) Write a linear function that relates the cost C to the number of pounds of chocolate-covered almonds purchased w.

(b) What would it cost to purchase 3.5 pounds of chocolate-covered almonds?

(c) Graph the relation between cost and weight.

28. Conversion Suppose that you are planning a trip to Europe, so you need to obtain some euros. The amount received in euros varies directly with the amount in U.S. dollars. Your friend just converted \$600 into 548 euros.

(a) Write a linear function that relates the number of euros E to the number of U.S. dollars d.

(b) If you convert $700 into euros, how many euros will you receive?

(c) Graph the relation between euros and U.S. dollars.

29. Falling Objects The velocity of a falling object (ignoring air resistance) v is directly proportional to the time t of the fall. If, after 2 seconds, the velocity of the object is 64 feet per second, what will its velocity be after 3 seconds?

30. Circumference of a Circle The circumference of a circle C is directly proportional to its radius r. If the circumference of a circle whose radius is 5 inches is 10π inches, what is the circumference of a circle whose radius is 8 inches?

31. Demand Suppose that the demand D for candy at the movie theater is inversely related to the price p.

(a) When the price of candy is $2.50 per bag, the theater sells 150 bags of candy. Express the demand for candy as a function of its price.

(b) Determine the number of bags of candy that will be sold if the price is raised to $3 a bag.

32. Driving to School The time t that it takes you to get to school varies inversely with your average speed s.

(a) Suppose that it takes you 30 minutes to drive to school when your average speed is 35 miles per hour. Express your driving time to school as a function of your average speed.

(b) Suppose that your average speed driving to school is 30 miles per hour. How long will it take you to get to school?

33. Pressure The volume of a gas V held at a constant temperature in a closed container varies inversely with its pressure P. If the volume of a gas is 600 cubic centimeters (cc) when the pressure is 150 millimeters of mercury (mm Hg), find the volume when the pressure is 200 mm Hg.

34. Resistance The current I in a circuit is inversely proportional to its resistance R measured in ohms. Suppose that when the current in a circuit is 30 amperes, the resistance is 8 ohms. Find the current in the same circuit when the resistance is 10 ohms.

35. Weight The weight of an object above the surface of Earth varies inversely with the square of the distance from the center of Earth. Maria weighs 120 pounds when she is on the surface of Earth (3960 miles from the center). Determine Maria's weight if she is at the top of Mount McKinley (3.8 miles from the surface of Earth).

36. Intensity of Light The intensity I of light (measured in foot-candles) varies inversely with the square of the distance from the bulb. Suppose the intensity of a 100-watt light bulb at a distance of 2 meters is 0.075 foot-candle. Determine the intensity of the bulb at a distance of 3 meters.

37. Drag Force When an object moves through air, a frictionlike drag force tends to slow the object down. The drag force D on a free-falling parachutist varies jointly with the surface area of the parachutist and the square of his velocity. The drag force on a parachutist with surface area 2 square meters falling at 40 meters per second is 1152 newtons. Find the drag force on a parachutist whose surface area is 2.5 square meters falling at 50 meters per second.

38. Kinetic Energy The kinetic energy K (measured in joules) of a moving object varies jointly with the mass of the object and the square of its velocity v. The kinetic energy of a linebacker weighing 110 kilograms and running at a speed of 8 meters per second is 3520 joules. Find the kinetic energy of a wide receiver weighing 90 kilograms and running at a speed of 10 meters per second.

39. Newton's Law of Gravitation According to Newton's law of universal gravitation, the force F of gravity between any two objects varies jointly with the masses of the objects m_1 and m_2 and inversely with the square of the distance between the objects r. The force of gravity between a 105-kg man and his 80-kg wife when they are separated by a distance of 5 meters is 2.24112×10^{-8} newton. Find the force of gravity between the man and his wife when they are 2 meters apart.

40. Electrical Resistance The electrical resistance of a wire varies directly with the length of the wire and inversely with the square of the diameter of the wire. If a wire 50 feet long and 3 millimeters in diameter has a resistance of 0.255 ohm, find the length of a wire of the same material whose resistance is 0.147 ohm and whose diameter is 2.5 millimeters.

41. Stress of Material The stress in the material of a pipe subject to internal pressure varies jointly with the internal pressure and internal diameter of the pipe and inversely with the thickness of the pipe. The stress is 100 pounds per square inch when the diameter is 5 inches, the thickness is 0.75 inch, and the internal pressure is 25 pounds per square inch. Find the stress when the internal pressure is 50 pounds per square inch, the diameter is 6 inches, and the thickness is 0.5 inch.

42. Gas Laws The volume V of an ideal gas varies directly with the temperature T and inversely with the pressure P. If a cylinder contains oxygen at a temperature of 300 kelvin (K) and a pressure of 15 atmospheres in a volume of 100 liters, what is the constant of proportionality k? If a piston is lowered into the cylinder, decreasing the volume occupied by the gas to 70 liters and raising the temperature to 315 K, what is the pressure?

Extending the Concepts

43. David and Goliath The force F (in newtons) required to maintain an object in a circular path varies jointly with the mass m (in kilograms) of the object and the square of its speed (measured in meters per second) and inversely with the radius r (in meters) of the circular path. Suppose that David has a rope that is 3 meters long. On the end of the rope he has attached a pouch that holds a 0.5-kilogram stone. Suppose that David is able to spin the rope in a circular motion at the rate of 50 revolutions per minute.

(a) The spinning rate of 50 revolutions per minute can be converted into an *angular velocity* ω (lowercase Greek letter omega) using the formula $\omega = 2\pi$ (revolutions per minute). Write the spinning rate as an angular velocity rounded to two decimal places.

(b) Use the fact that $v = \omega r$, where r is the radius of the circle, to find the linear velocity (in meters per minute) of the stone if it were released. Now convert the linear velocity to meters per second.

(c) Suppose that the force on the rope required to keep the rock in a circular motion is 2.3 newtons. Use the result from part (b) to find the constant of proportionality k.

(d) David is fairly certain that he will require more force than this to beat Goliath in a battle, so he increases the circular motion to 80 revolutions per minute and increases the length of the rope to 4 meters. What is the force required to keep the stone in a circular motion?

44. Suppose that y is directly proportional to x^2. If x is doubled, what happens to the value of y?

Chapter 8 Activity: Shifting Discovery

Focus: Using graphing skills, discover the possible "rules" for graphing functions.

Time: 30–35 minutes

Group size: 4

Materials Needed: Graph paper (2–3 pieces)

Each member of the group needs to:

1. Draw a coordinate plane and label the x-axis and y-axis.

2. By plotting points, graph the primary function: $f(x) = x^2$.

3. In the same coordinate plane, each group member graphs *one* of the following functions by plotting points. Be sure to graph the primary function and one of the functions (a)–(d) in the same coordinate plane.

   (a) $f(x) = x^2 + 3$ (b) $f(x) = (x - 3)^2$

   (c) $f(x) = x^2 - 3$ (d) $f(x) = (x + 3)^2$

As a group, discuss the following:

4. What shape are the graphs?

5. Each member of the group should share the difference between the graph of your primary function and the other function you chose.

6. As a group, can you develop rules for these differences?

With this possible rule in mind, each member of the group needs to:

7. Draw a coordinate plane and label the x-axis, the y-axis, and -10 to 10 on each axis.

8. Graph the primary function by plotting points: $f(x) = |x|$.

9. In the same coordinate plane, each group member graphs *one* of the following functions by plotting points. Be sure to graph the primary function and one of the functions (a)–(d) on the same coordinate plane.

   (a) $f(x) = |x| + 4$ (b) $f(x) = |x - 4|$

   (c) $f(x) = |x| - 4$ (d) $f(x) = |x + 4|$

10. Did the rules you developed in Problem 6 hold true? Discuss.

Chapter 8 Review

Section 8.1   Graphs of Equations

KEY CONCEPTS

- **Graph of an Equation in Two Variables**
  The set of all ordered pairs (x, y) in the xy-plane that satisfy the equation

- **Intercepts**
  The points, if any, where a graph crosses or touches the coordinate axes

KEY TERMS

x-axis	Quadrants
y-axis	Equation in two
Origin	variables
Rectangular or Cartesian	Sides
coordinate system	Satisfy
xy-plane	Graph of an equation
Coordinate axes	in two variables
Ordered pair	Point-plotting method
Coordinates	Complete graph
x-coordinate	Intercept
y-coordinate	x-intercept
Abscissa	y-intercept
Ordinate	

You Should Be Able To...	EXAMPLE	Review Exercises
❶ Graph an equation using the point-plotting method (p. 524)	Examples 1 through 3	5–10
❷ Identify the intercepts from the graph of an equation (p. 526)	Example 4	11
❸ Interpret graphs (p. 527)	Example 5	12

In Problems 1 and 2, plot each point in the same xy-plane. Tell in which quadrant or on what coordinate axis each point lies.

1. $A(2, -4)$
   $B(-1, -3)$
   $C(0, 4)$
   $D(-5, 1)$
   $E(1, 0)$
   $F(4, 3)$

2. $A(3, 0)$
   $B(1, 5)$
   $C(-3, -5)$
   $D(-1, 4)$
   $E(5, -2)$
   $F(0, -5)$

In Problems 3 and 4, determine whether the given points are on the graph of the equation.

3. $3x - 2y = 7$
   (a) $(3, 1)$
   (b) $(2, -1)$
   (c) $(4, 0)$
   (d) $\left(\dfrac{1}{3}, -3\right)$

4. $y = 2x^2 - 3x + 2$
   (a) $(-1, 3)$
   (b) $(1, 1)$
   (c) $(-2, 16)$
   (d) $\left(\dfrac{1}{2}, \dfrac{3}{2}\right)$

In Problems 5–10, graph each equation by plotting points.

5. $y = x + 2$

6. $2x + y = 3$

7. $y = -x^2 + 4$

8. $y = |x + 2| - 1$

9. $y = x^3 + 2$

10. $x = y^2 + 1$

In Problem 11, the graph of an equation is given. List the intercepts of the graph.

11.

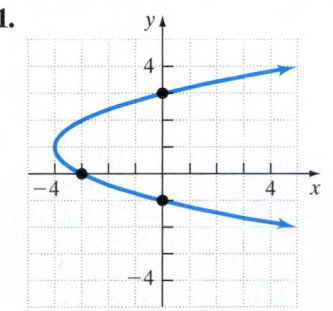

12. **Cell Phones** A cell phone company offers a plan for $59.99 per month for 900 minutes with additional minutes costing $0.40 per minute. The graph below shows the monthly cost, in dollars, when x minutes are used.

   (a) If you talk for 2250 minutes in a month, how much is your monthly bill?
   (b) Use the graph to estimate your monthly bill if you talk for 9500 minutes.

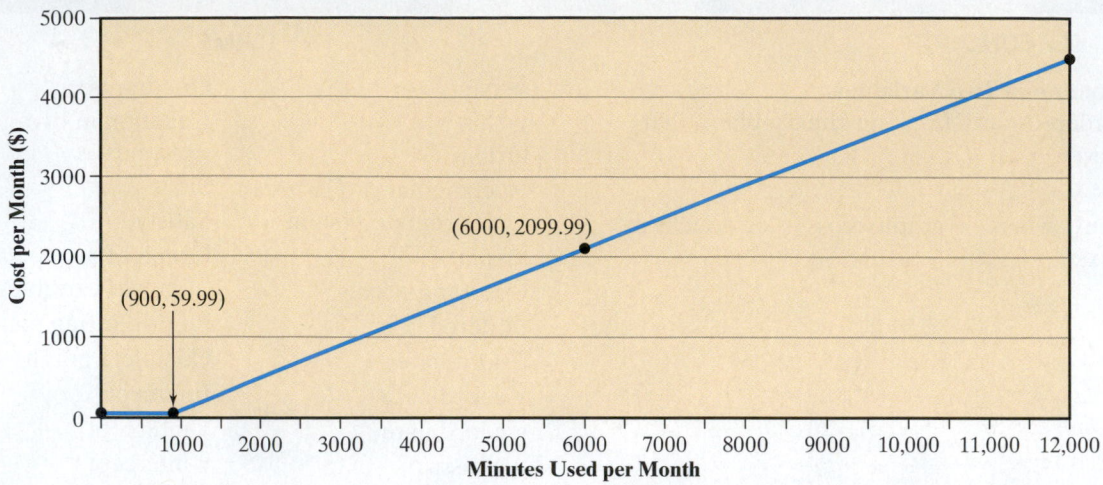

Section 8.2 Relations

KEY CONCEPT

- **Relation**
  A correspondence between two variables x and y, where y depends on x. Relations can be represented through maps, sets of ordered pairs, equations, or graphs.

KEY TERMS

Relation	Inputs
Corresponds	Outputs
Depends on	Domain
Mapping	Range

You Should Be Able To...	EXAMPLE	Review Exercises
1 Understand relations (p. 532)	Example 1	13–16
2 Find the domain and the range of a relation (p. 533)	Examples 2 through 4	13–28
3 Graph a relation defined by an equation (p. 535)	Example 5	21–26

In Problems 13 and 14, write each relation as a set of ordered pairs. Then identify the domain and range of the relation.

13.

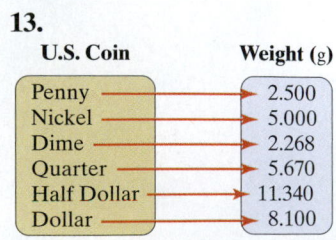

SOURCE: *U.S. Mint website*

14.

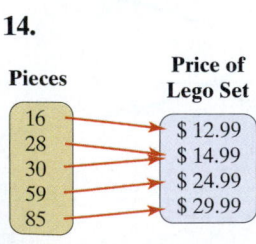

SOURCE: *Lego website*

In Problems 15 and 16, write each relation as a map. Then identify the domain and range of the relation.

15. $\{(2,7), (-4,8), (3,5), (6,-1), (-2,-9)\}$

16. $\{(3,1), (3,7), (5,1), (-2,8), (1,4)\}$

In Problems 17–20, identify the domain and range of the relation from the graph.

17.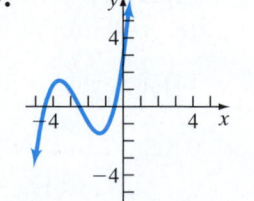

18.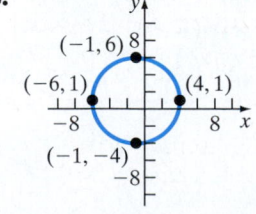

19.
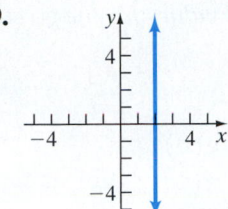

20.
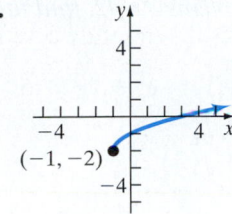

In Problems 21–26, graph the relation. Use the graph of the relation to identify the domain and range of the relation.

21. $y = x + 2$

22. $2x + y = 3$

23. $y = -x^2 + 4$

24. $y = |x + 2| - 1$

25. $y = x^3 + 2$

26. $x = y^2 + 1$

27. Cell Phones A cell phone company offers a plan for $40 per month for 3000 minutes with additional minutes costing $0.05 per minute. The graph to the right shows the monthly cost, in dollars, when x minutes are used.

(a) What are the domain and range of the relation?
(b) Explain why the domain obtained in part (a) is reasonable.

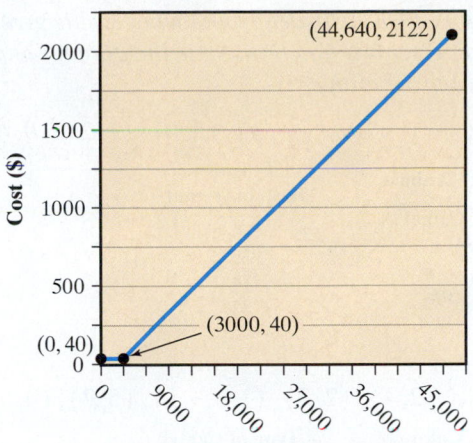

28. Vertical Motion The graph below shows the height, in feet, of a ball thrown straight up with an initial speed of 40 feet per second from an initial height of 96 feet after t seconds. What are the domain and the range of the relation?

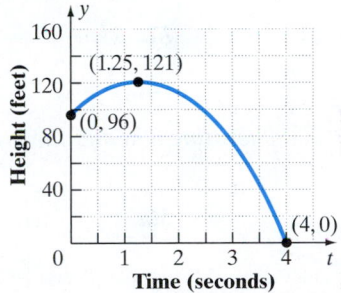

Section 8.3   An Introduction to Functions

KEY CONCEPTS

KEY TERMS

- **Functions**
  A special type of relation where any given input, x, corresponds to only one output y. Functions can be represented through maps, sets of ordered pairs, equations, or graphs.

- **Vertical Line Test**
  A set of points in the xy-plane is the graph of a function if and only if every vertical line intersects the graph in at most one point.

- **Domain of a Function**
  When only an equation of a function is given, the domain of the function is the set of real numbers x for which $f(x)$ is a real number. However, in applications, the domain of a function is the largest set of real numbers for which the output of the function is reasonable.

Function
Vertical Line Test
Value of f at the number x
Independent variable
Dependent variable
Argument
Domain of f
Graph of the function

You Should Be Able To...	EXAMPLE	Review Exercises
❶ Determine whether a relation expressed as a map or ordered pairs represents a function (p. 538)	Examples 1 and 2	29, 30
❷ Determine whether a relation expressed as an equation represents a function (p. 541)	Examples 3 and 4	31–34
❸ Determine whether a relation expressed as a graph represents a function (p. 541)	Example 5	35–38
❹ Find the value of a function (p. 542)	Examples 6 and 7	39–42
❺ Find the domain of a function (p. 544)	Examples 8 and 9	43–46
❻ Work with applications of functions (p. 545)	Example 10	47, 48

In Problems 29 and 30, determine whether the given relation represents a function. State the domain and the range of each relation.

29. (a) $\{(-1,-2),(-1,3),(5,0),(7,2),(9,4)\}$
(b)

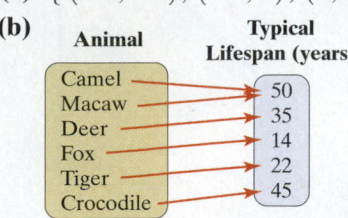

30. (a) $\{(-2,4),(2,3),(-3,-1),(5,7),(4,7)\}$
(b)

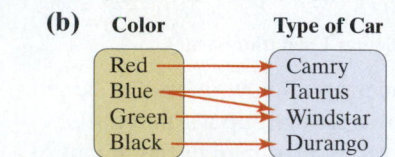

In Problems 31–34, determine whether each relation shows y as a function of x.

31. $3x - 5y = 18$

32. $x^2 + y^2 = 81$

33. $y = \pm 10x$

34. $y = x^2 - 14$

In Problems 35–38, determine whether the graph is that of a function.

35.

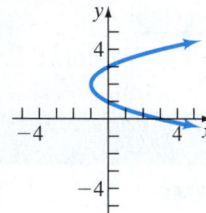

36.

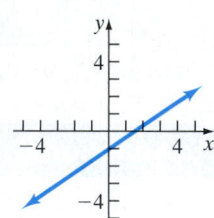

37.

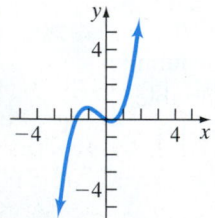

38.

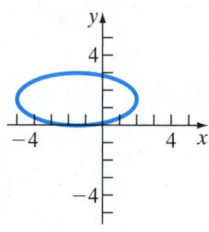

In Problems 39–42, find the indicated values for the given functions.

39. $f(x) = x^2 + 2x - 5$
(a) $f(-2)$
(b) $f(3)$

40. $g(z) = \dfrac{2z + 1}{z - 3}$
(a) $g(0)$
(b) $g(2)$

41. $F(x) = -2x + 7$
(a) $F(5)$
(b) $F(-x)$

42. $G(x) = 2x + 1$
(a) $G(7)$
(b) $G(x + h)$

For Problems 43–46, find the domain of each function.

43. $f(x) = -\dfrac{3}{2}x + 5$

44. $g(w) = \dfrac{w - 9}{2w + 5}$

45. $h(t) = \dfrac{t + 2}{t - 5}$

46. $G(t) = 3t^2 + 4t - 9$

47. Population Using census data from 1900 to 2014, the function

$$P(t) = 0.136t^2 - 5.043t + 46.927$$

represents the population, P, of Orange County in Florida (in thousands) t years after 1900.

(a) Identify the dependent and independent variables.
(b) Evaluate $P(120)$ and explain what it represents.
(c) Evaluate $P(-70)$ and explain what it represents. Is the result reasonable? Explain.

48. Education The function

$$P(a) = -0.0064a^2 + 0.6826a - 6.82$$

represents the percent of the population a years of age with an advanced degree, where $a \geq 25$.
SOURCE: *Current Population Survey*

(a) Identify the dependent and independent variables.
(b) Evaluate $P(30)$ and explain what it represents.

Section 8.4 Functions and Their Graphs

KEY CONCEPTS

- **Graph of a Function**
  The graph of a function, f, is the set of all ordered pairs $(x, f(x))$.

- **Zero of a Function**
  If $f(r) = 0$ for some number r, then r is a zero of f.

KEY TERMS

Zero of a function

You Should Be Able To...	EXAMPLE	Review Exercises
1 Graph a function (p. 549)	Example 1	49–52
2 Obtain information from the graph of a function (p. 551)	Examples 2 through 5	53–57, 59, 60
3 Know properties and graphs of basic functions (p. 554)	page 554	58
4 Interpret graphs of functions (p. 555)	Example 6	61, 62

In Problems 49–52, graph each function.

49. $f(x) = 2x - 5$ **50.** $g(x) = x^2 - 3x + 2$

51. $h(x) = (x - 1)^3 - 3$ **52.** $f(x) = |x + 1| - 4$

In Problems 53–56, for each function whose graph is shown, find (a) the domain and the range, and (b) the intercepts, if any.

53.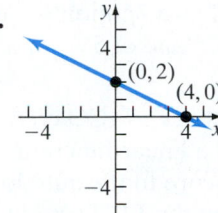

54.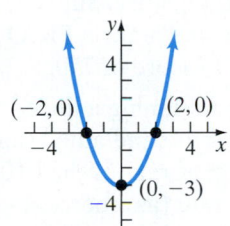

55.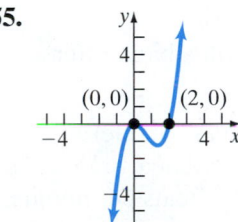

56.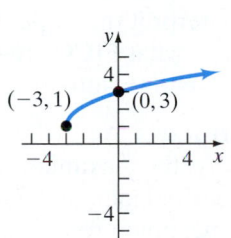

57. The graph of $y = f(x)$ is shown.

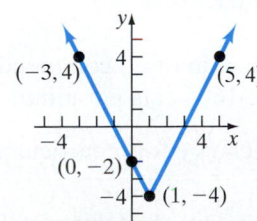

   (a) What is $f(-3)$?
   (b) For what value of x does $f(x) = -4$?
   (c) What are the zeros of f?

58. Graph each of the following functions.
   (a) $f(x) = x^2$ **(b)** $f(x) = x$

In Problems 59 and 60, answer the questions about the given function.

59. $h(x) = 2x - 7$
   (a) Is the point $(3, -1)$ on the graph of the function?
   (b) If $x = -2$, what is $h(x)$? What point is on the graph of the function?
   (c) If $h(x) = 4$, what is x? What point is on the graph of h?

60. $g(x) = \dfrac{3}{5}x + 4$

   (a) Is the point $(-5, 2)$ on the graph of the function?
   (b) If $x = 3$, what is $g(x)$? What point is on the graph of the function?
   (c) If $g(x) = -2$, what is x? What point is on the graph of g?

61. Travel by Train A Metrolink train leaves L.A. Union Station and travels 6 miles at a constant speed for 10 minutes, arriving at the Glendale station, where it waits 1 minute for passengers to board and depart. The train continues traveling at the same speed for 5 more minutes to reach downtown Burbank, which is 3 miles from Glendale. Sketch a graph that represents the distance of the train as a function of time until it reaches downtown Burbank.

62. Filling a Tub With the faucet running at a constant rate, it takes Angie 7 minutes to fill her bathtub. She turns off the faucet when the tub is full and realizes the water is too hot. She then opens the drain, letting water out at a constant rate that is half the rate of the faucet. After draining for 2 minutes, she stops the drain and turns on the faucet at the same rate as before (but at a cooler temperature) until the tub is full. Sketch a graph that represents the amount of water in the tub as a function of time.

Section 8.5   Linear Functions and Models

KEY CONCEPT

- **Linear Function**
  A linear function is a function of the form $f(x) = mx + b$, where m is the slope and $(0, b)$ is the y-intercept of the graph.

KEY TERMS

Linear function
Line
Scatter diagram

You Should Be Able To...	EXAMPLE	Review Exercises
1 Graph linear functions (p. 560)	Example 1	63–66
2 Find the zero of a linear function (p. 561)	Example 2	63–66
3 Build linear models from verbal descriptions (p. 563)	Examples 4 and 5	69–72
4 Build linear models from data (p. 565)	Examples 6 through 8	73–76

In Problems 63–66, graph each linear function. Find the zero of each function.

63. $g(x) = 2x - 6$

64. $H(x) = -\dfrac{4}{3}x + 5$

65. $F(x) = -x - 3$

66. $f(x) = \dfrac{3}{4}x - 3$

67. Monthly Car Cost Sarah's monthly payments for her new car are $260 per month. She estimates that maintenance and gas cost her $0.20 per mile. The equation for her monthly cost is $C(m) = 0.20m + 260$, where C is the monthly cost in dollars and m is the number of miles driven that month.

(a) What is the domain of this function?
(b) What is $C(0)$? Explain what this result means in the context of the problem.
(c) What is Sarah's monthly cost if she drives 1000 miles in a month?
(d) Graph the linear function.
(e) Over what range of miles can she drive in a month if she wants to keep her monthly cost to no more than $550?

68. Straight-Line Depreciation Using straight-line depreciation, the value V of a particular computer x years after purchase is given by the linear function $V(x) = 1800 - 360x$ for $0 \le x \le 5$.

(a) What are the independent and dependent variables?
(b) What is the domain of this linear function?
(c) What is the initial value of the computer?
(d) What is the value of the computer 2 years after purchase?
(e) Graph the linear function over its domain.
(f) After how long will the value of the computer be $0?

69. Credit Scores Your Fair Isaacs Corporation (FICO) credit score is used to determine your ability to get a loan. The higher your credit score, the better your credit history. Suppose a bank offers a 36-month auto loan at 9% for a FICO score of 675 and at 5% for a FICO score of 750.

(a) Assuming a linear relation between FICO credit score and auto loan rate, find a linear function that relates the FICO credit score to the auto loan rate (as a percentage), treating the FICO credit score as the independent variable.
(b) Predict the auto loan rate for a FICO credit score of 710, to the nearest whole percent.
(c) Interpret the slope.
(d) For what FICO credit score will a bank offer a 36-month auto loan at 6.5%?

70. Heart Rates According to the American Geriatric Society, the maximum recommended heart rate for a 20-year-old man under stress is 200 beats per minute. The maximum recommended heart rate for a 60-year-old man under stress is 160 beats per minute.

(a) Find a linear function that relates the maximum recommended heart rate for men to age.
(b) Predict the maximum recommended heart rate for a 45-year-old man under stress.
(c) Interpret the slope.
(d) For what age would the maximum recommended heart rate under stress be 168 beats per minute?

71. Car Rental The daily rental charge for a particular car is $35 plus $0.12 per mile.

(a) Find a linear function that expresses the rental cost C as a function of the miles driven m.
(b) What are the independent and dependent variables?
(c) What is the domain of this linear function?
(d) For a one-day rental, what is the rental cost if 124 miles are driven?
(e) For a one-day rental, how many miles were driven if the rental cost was $67.16?
(f) Graph the linear function.

72. Satellite Television Bill A satellite television company charges $33.99 per month for a 100-channel package, plus $3.50 for each pay-per-view movie watched that month.

    (a) Find a linear function that expresses the monthly bill B as a function of x, the number of pay-per-view movies watched that month.

    (b) What are the independent and dependent variables?

    (c) What is the domain of this linear function?

    (d) What is the monthly bill if 5 pay-per-view movies are watched that month?

    (e) For one month, how many pay-per-view movies were watched if the bill was $58.49?

    (f) Graph the linear function.

In Problems 73 and 74,

    (a) *Draw a scatter diagram of the data.*

    (b) *Select two points from the scatter diagram and find the equation of the line containing the points selected.*

    (c) *Graph the line found in part (b) on the scatter diagram.*

73.

x	2	5	8	11	14
y	13.3	11.6	8.4	7.2	4.6

74.

x	0	0.4	1.5	2.3	4.2
y	0.6	1.1	1.3	1.8	3.0

75. The table below gives the number of calories and the total carbohydrates (in grams) for a one-cup serving of seven name-brand cereals (not including milk).

Cereal	Calories, x	Total Carbohydrates (in grams), y
Rice Krispies®	96	23.2
Life®	160	33.3
Lucky Charms®	120	25.0
Kellogg's Complete®	120	30.7
Wheaties®	110	24.0
Cheerios®	110	22.0
Honey Nut Chex®	160	34.7

SOURCE: *Quaker Oats, General Mills, and Kellogg*

    (a) Draw a scatter diagram of the data, treating calories as the independent variable.

    (b) What type of relation appears to exist between calories and total carbohydrates in a one-cup serving of cereal?

    (c) Select two points and find an equation of the line containing the points.

    (d) Graph the line on the scatter diagram drawn in part (a).

    (e) Predict the total carbohydrates in a one-cup serving of cereal that has 140 calories.

    (f) Interpret the slope of the line found in part (c).

76. Second-Day Delivery Costs The table below lists some selected prices charged by Federal Express for FedEx 2Day delivery, depending on the weight of the package.

Weight (in pounds), x	FedEx 2Day® Delivery Charge, y
1	$11.60
3	$12.20
6	$13.55
8	$16.20
9	$16.90
11	$18.90

SOURCE: *Federal Express Corporation*

    (a) Draw a scatter diagram of the data, treating weight as the independent variable.

    (b) What type of relation appears to exist between the weight of the package and the FedEx 2Day delivery charge?

    (c) Select two points and find an equation of the line containing the points.

    (d) Graph the line on the scatter diagram drawn in part (a).

    (e) Predict the FedEx 2Day delivery charge for shipping a 5-pound package.

    (f) Interpret the slope of the line found in part (c).

Section 8.6   Compound Inequalities

KEY CONCEPT

- If $a < b$, then we can write $a < x$ and $x < b$ as $a < x < b$.

KEY TERMS

Intersection   Compound inequality

Union   Solve a compound inequality

You Should Be Able To...	EXAMPLE	Review Exercises
1 Determine the intersection or union of two sets (p. 574)	Examples 1 and 2	77–82
2 Solve compound inequalities involving "and" (p. 576)	Examples 3 through 6	83, 84, 87, 88, 92
3 Solve compound inequalities involving "or" (p. 578)	Examples 7 and 8	85, 86, 89, 90, 91
4 Solve problems using compound inequalities (p. 580)	Example 9	93, 94

In Problems 77–80, use $A = \{2, 4, 6, 8\}$,
$B = \{-1, 0, 1, 2, 3, 4\}$, *and* $C = \{1, 2, 3, 4\}$ *to find
each set.*

77. $A \cup B$ **78.** $A \cap C$

79. $B \cap C$ **80.** $A \cup C$

*In Problems 81 and 82, use the graph of the inequality to
find each set.*

81. $A = \{x \mid x \le 4\}; B = \{x \mid x > 2\}$
Find **(a)** $A \cap B$ and **(b)** $A \cup B$.

82. $E = \{x \mid x \ge 3\}; F = \{x \mid x < -2\}$
Find **(a)** $E \cap F$ and **(b)** $E \cup F$.

*In Problems 83–92, solve each compound inequality.
Graph the solution set.*

83. $x < 4$ and $x + 3 > 2$

84. $3 < 2 - x < 7$

85. $x + 3 < 1$ or $x > 2$

86. $x + 6 \ge 10$ or $x \le 0$

87. $3x + 2 \le 5$ and $-4x + 2 \le -10$

88. $1 \le 2x + 5 < 13$

89. $x - 3 \le -5$ or $2x + 1 > 7$

90. $3x + 4 > -2$ or $4 - 2x \ge -6$

91. $\dfrac{1}{3}x > 2$ or $\dfrac{2}{5}x < -4$

92. $x + \dfrac{3}{2} \ge 0$ and $-2x + \dfrac{3}{2} > \dfrac{1}{4}$

93. Heart Rates The normal heart rate for healthy adults between the ages of 21 and 60 should be between 70 and 75 beats per minute (inclusive). If we let x represent the heart rate of an adult between the ages of 21 and 60, express the normal range of values using a compound inequality.

94. Heating Bills For usage above 800 kilowatt hours, the non-space heat energy charge for American Electric Power residential service was $43.56 plus $0.038752 per kilowatt hour. During one winter, a customer's charge ranged from a low of $52.62 to a high of $88.22. Over what range of values did electrical usage vary (in kilowatt hours)? Express answers rounded to the nearest tenth of a kilowatt hour.

Section 8.7   Absolute Value Equations and Inequalities

KEY CONCEPTS

- **Equations Involving Absolute Value**
  If a is a positive real number and if u is any algebraic expression, then $|u| = a$ is equivalent to $u = a$ or $u = -a$.
- **Equations Involving Two Absolute Values**
  If u and v are any algebraic expressions, then $|u| = |v|$ is equivalent to $u = v$ or $u = -v$.
- **Inequalities of the Form $<$ or $\le$ Involving Absolute Value**
  If a is a positive real number and if u is any algebraic expression, then $|u| < a$ is equivalent to $-a < u < a$, and $|u| \le a$ is equivalent to $-a \le u \le a$.
- **Inequalities of the Form $>$ or $\ge$ Involving Absolute Value**
  If a is a positive real number and if u is any algebraic expression, then $|u| > a$ is equivalent to $u < -a$ or $u > a$, and $|u| \ge a$ is equivalent to $u \le -a$ or $u \ge a$.

You Should Be Able To...	EXAMPLE	Review Exercises
1 Solve absolute value equations (p. 584)	Examples 1 through 4	95–100
2 Solve absolute value inequalities involving $<$ or $\leq$ (p. 588)	Examples 5 through 7	101, 103, 106, 107
3 Solve absolute value inequalities involving $>$ or $\geq$ (p. 589)	Examples 8 and 9	102, 104, 105, 108
4 Solve applied problems involving absolute value inequalities (p. 591)	Example 10	109, 110

In Problems 95–100, solve the absolute value equation.

95. $|x| = 4$

96. $|3x - 5| = 4$

97. $|-y + 4| = 9$

98. $-3|x + 2| - 5 = -8$

99. $|2w - 7| = -3$

100. $|x + 3| = |3x - 1|$

In Problems 101–108, solve each absolute value inequality. Graph the solution set on a real number line.

101. $|x| < 2$

102. $|x| \geq \dfrac{7}{2}$

103. $|x + 2| \leq 3$

104. $|4x - 3| \geq 1$

105. $3|x| + 6 \geq 1$

106. $|7x + 5| + 4 < 3$

107. $|(x - 3) - 2| \leq 0.01$ **108.** $\left|\dfrac{2x - 3}{4}\right| > 1$

109. Tolerance The diameter of a certain ball bearing is required to be 0.503 inch. The tolerance on the bearing is 0.001 inch. If x represents the diameter of a bearing, the acceptable diameters of the bearing can be expressed as $|x - 0.503| \leq 0.001$. Determine the acceptable diameters of the bearing.

110. Tensile Strength The tensile strength of paper used to make grocery bags is about 40 lb/in.² A paper grocery bag whose tensile strength satisfies the inequality $\left|\dfrac{x - 40}{2}\right| > 1.96$ has an unusual tensile strength. Determine the tensile strengths that would be considered unusual.

Section 8.8   Variation

You Should Be Able To...	EXAMPLE	Review Exercises
1 Model and solve direct variation problems (p. 595)	Examples 1 and 2	111, 112, 117, 118
2 Model and solve inverse variation problems (p. 597)	Example 3	113, 115, 119, 120
3 Model and solve joint variation and combined variation problems (p. 598)	Examples 4 and 5	114, 116, 121, 122

In Problems 111–116, (a) find the constant of proportionality k, (b) write the function relating the variables, and (c) find the quantity indicated.

111. Suppose that y varies directly with x. If $y = 30$ when $x = 6$, find y when $x = 10$.

112. Suppose that y varies directly with x. If $y = 18$ when $x = -3$, find y when $x = 8$.

113. Suppose that y varies inversely with x. If $y = 15$ when $x = 4$, find y when $x = 5$.

114. Suppose that y varies jointly with x and z. If $y = 45$ when $x = 6$ and $z = 10$, find y when $x = 8$ and $z = 7$.

115. Suppose that s varies inversely with the square of t. If $s = 18$ when $t = 2$, find s when $t = 3$.

116. Suppose that w varies directly with x and inversely with z. If $w = \dfrac{4}{3}$ when $x = 10$ and $z = 12$, find w when $x = 9$ and $z = 16$.

117. Snow-Water Equivalent The amount of water in snow is directly proportional to the depth of the snow. Suppose the amount of water in 40 inches of snow is 4.8 inches. How much water is contained in 50 inches of snow?

118. Car Payments Roberta is buying a car. The monthly payment p for the car varies directly with the amount borrowed b. Suppose the dealership tells her that if she borrows $15,000, her monthly payment will be $293.49. What will be Roberta's monthly payment if she borrows $18,000?

119. **Radio Signals** The frequency of a radio signal varies inversely with the wavelength. A signal of 800 kilohertz has a wavelength of 375 meters. What frequency has a signal of wavelength 250 meters?

120. **Ohm's Law** The electrical current flowing through a wire varies inversely with the resistance of the wire. If the current is 8 amperes when the resistance is 15 ohms, for what resistance will the current be 10 amperes?

121. **Volume of a Cylinder** The volume V of a right circular cylinder varies jointly with the height h and

the square of the diameter d. If the volume of a cylinder is 231 cubic centimeters when the diameter is 7 centimeters and the height is 6 centimeters, find the volume when the diameter is 8 centimeters and the height is 14 centimeters.

122. **Volume of a Pyramid** The volume V of a pyramid varies jointly with the base area B and the height h. If the volume of a pyramid is 270 cubic inches when the base area is 81 square inches and the height is 10 inches, find the volume of a pyramid with base area 125 square inches and height 9 inches.

Chapter 8 Test

Step-by-step test solutions are found on the Chapter Test Prep Videos available in MyMathLab®, *on* You Tube*, or may be accessed using the QR code to the right.*

1. Plot the following ordered pairs in the same xy-plane. Tell in which quadrant or on what coordinate axis each point lies.

$A(3, -4), B(0, 2), C(3, 0), D(2, 1), E(-1, -4), F(-3, 5)$

2. Determine whether the ordered pair is a point on the graph of the equation $y = 3x^2 + x - 5$.

   (a) $(-2, 4)$ (b) $(-1, -3)$ (c) $(2, 9)$

3. Identify the intercepts from the graph below.

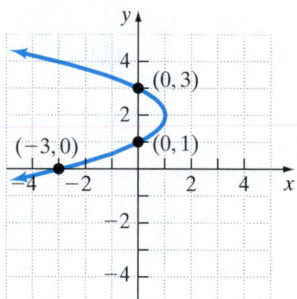

4. The following graph represents the speed of a car over time.

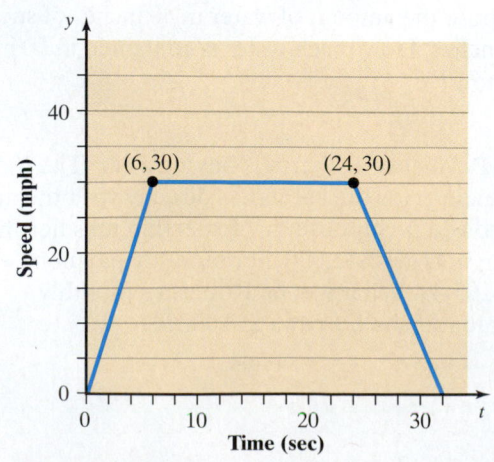

(a) When does the car stop accelerating?

(b) For how long does the car maintain a constant speed?

5. Write the relation as a map. Then identify the domain and the range of the relation.

$\{(2, 8), (5, -2), (7, 12), (-4, -7), (7, 3), (5, -1)\}$

6. Identify the domain and range of the relation from the graph.

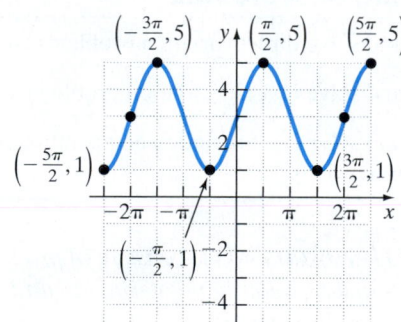

7. Graph the relation $y = x^2 - 3$ by plotting points. Use the graph of the relation to identify the domain and range.

In Problems 8 and 9, determine whether each relation represents a function. Identify the domain and the range of each relation.

8.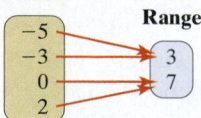

9.

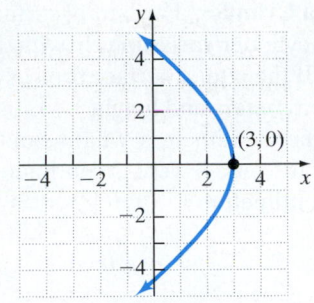

$(3, 0)$

10. Does the equation $y = \pm 5x$ represent a function? Why or why not?

11. For $f(x) = -3x + 11$, find $f(x + h)$.

12. For $g(x) = 2x^2 + x - 1$, find the indicated values.

(a) $g(-2)$ **(b)** $g(0)$ **(c)** $g(3)$

13. Sketch the graph of $f(x) = x^2 + 3$.

14. Using data from 1996 to 2015, the function $P(x) = 0.22x + 4.44$ approximates the average movie ticket price (in dollars) x years after 1996.
SOURCE: *National Association of Theater Owners*

(a) Identify the dependent and independent variables.

(b) Evaluate $P(25)$ and explain what it represents.

(c) In what year will the average movie ticket be $12.00?

15. Find the domain of $f(x) = \dfrac{-15}{x + 2}$.

16. $h(x) = -5x + 12$

(a) Is the point $(2, 2)$ on the graph of the function?
(b) If $x = 3$, what is $h(x)$? What point is on the graph of the function?
(c) If $h(x) = 27$, what is x? What point is on the graph of h?
(d) What is the zero of h?

17. Crafts Fair Sales Henry plans to sell small wooden shelves at a crafts fair for $30 each. A booth at the fair costs $100 to rent. Henry estimates his expenses for producing the shelves to be $12 each, so his profit will be $18 per shelf.

(a) Write a function that expresses Henry's profit P as a function of the number of shelves sold x.
(b) What is the implied domain of this linear function?
(c) What is the profit if Henry sells 34 shelves?
(d) Graph the linear function.
(e) If Henry's profit is $764, how many shelves did he sell?

18. Shetland Pony Weights The table below lists the average weight of a Shetland pony, depending on the age of the pony.

Age (months), x	Average Weight (kilograms), y
3	60
6	95
12	140
18	170
24	185

SOURCE: *The Merck Veterinary Manual*

(a) Draw a scatter diagram of the data, treating age as the independent variable.
(b) What type of relation appears to exist between the age and the weight of a Shetland pony?
(c) Select two points and find an equation of the line containing the points.
(d) Graph the line on the scatter diagram drawn in part (a).
(e) Predict the weight of a 9-month-old Shetland pony.
(f) Interpret the slope of the line found in part (c).

19. Solve: $|2x + 5| - 3 = 0$

In Problems 20–23, solve each inequality and graph the solution set on a real number line.

20. $x + 2 < 8$ and $2x + 5 \geq 1$

21. $x > 4$ or $2(x - 1) + 3 < -2$

22. $2|x - 5| + 1 < 7$

23. $|-2x + 1| \geq 5$

24. Using a Lever Using a lever, the force F required to lift a weight is inversely proportional to the length l of the force arm of the lever (assuming all other factors are constant). If a force of 50 pounds is required to lift a granite boulder when the force arm length is 4 feet, how much force will be required to lift the boulder if the force arm length is 10 feet?

25. Lateral Surface Area of a Cylinder The lateral surface area L of a right circular cylinder varies jointly with its radius r and height h. If the lateral surface area of a cylinder with radius 7 centimeters and height 12 centimeters is 528 square centimeters, what would be the lateral surface area if the radius were 9 centimeters and the height were 14 centimeters?

9 Radicals and Rational Exponents

Wind chill chart (Fahrenheit)								temperature (°F)									
calm	40	35	30	25	20	15	10	5	0	–5	–10	–15	–20	–25	–30	–35	–40
5	36	31	25	19	13	7	1	–5	–11	–16	–22	–28	–34	–40	–46	–52	–57
10	34	27	21	15	9	3	–4	–10	–16	–22	–28	–35	–41	–47	–53	–59	–66
15	32	25	19	13	6	0	–7	–13	–19	–26	–32	–39	–45	–51	–58	–64	–71
20	30	24	17	11	4	–2	–9	–15	–22	–29	–35	–42	–48	–55	–61	–68	–74
25	29	23	16	9	3	–4	–11	–17	–24	–31	–37	–44	–51	–58	–64	–71	–78
30	28	22	15	8	1	–5	–12	–19	–26	–33	–39	–46	–53	–60	–67	–73	–80
35	28	21	14	7	0	–7	–14	–21	–27	–34	–41	–48	–55	–62	–69	–76	–82
40	27	20	13	6	–1	–8	–15	–22	–29	–36	–43	–50	–57	–64	–71	–78	–84
45	26	19	12	5	–2	–9	–16	–23	–30	–37	–44	–51	–58	–65	–72	–79	–86
50	26	19	12	4	–3	–10	–17	–24	–31	–38	–45	–52	–60	–67	–74	–81	–88
55	25	18	11	4	–3	–11	–18	–25	–32	–39	–46	–54	–61	–68	–75	–82	–89
60	25	17	10	3	–4	–11	–19	–26	–33	–40	–48	–55	–62	–69	–76	–84	–91

wind speed (mph)

wind chill (°F) = $35.74 + 0.6215T - 35.75(V^{0.16}) + 0.4275T(V^{0.16})$ T = air temperature (°F)

frostbite times   ■ 30 minutes   ■ 10 minutes   ■ 5 minutes V = wind speed (mph)

Source: U.S. National Weather Service; Meteorological Services of Canada

According to the National Weather Service, the wind chill temperature (WCT) index uses science, technology, and computer modeling to provide an accurate, understandable, and useful formula for calculating the dangers from winter winds and freezing temperatures. The wind chill temperature is defined only for temperatures at or below 50 degrees Fahrenheit (50°F) and wind speeds above 3 miles per hour.

What is the wind chill temperature if the temperature is 0°F and the wind speed is 10 miles per hour? See Problem 137 in Section 9.2.

The Big Picture: Putting It Together

In Chapter 5 we simplified polynomial expressions by adding, subtracting, multiplying, and dividing. In Chapter 6 we factored polynomials. In Chapter 7 we used the skills learned in Chapters 5 and 6 to simplify rational expressions and perform operations on rational expressions.

We now present a similar discussion with radical expressions. We will learn how to add, subtract, multiply, and divide radical expressions. In addition, we will use factoring to simplify radical expressions. Throughout this discussion, keep in mind that radicals perform the "inverse" of raising a real number to a positive integer exponent. For example, a square root undoes the squaring operation.

Outline

9.1 Square Roots

Objectives

❶ Evaluate Square Roots of Perfect Squares

❷ Determine Whether a Square Root Is Rational, Irrational, or Not a Real Number

❸ Find Square Roots of Variable Expressions

Are You Prepared for This Section?

Before getting started, complete the following problems. If you get a problem wrong, go back to the section cited and review the material.

In Problems P1–P3, use the set $\left\{-4, \frac{5}{3}, 0, \sqrt{2}, 6.95, 13, \pi\right\}$.

P1. Which of the numbers are integers?   [Section 1.3, pp. 19–22]

P2. Which of the numbers are rational numbers?   [Section 1.3, pp. 19–22]

P3. Which of the numbers are irrational numbers?   [Section 1.3, pp. 19–22]

P4. Evaluate: **(a)** $\left(\frac{3}{2}\right)^2$ **(b)** $(0.4)^2$ [Section 1.7, pp. 56–57]

In Section 1.7, exponents were introduced. Exponents indicate repeated multiplication. For example, 4^2 means $4 \cdot 4$, so $4^2 = 16; (-6)^2$ means $(-6)(-6) = 36$. This section will show how to "undo" the process of raising a number to the second power and ask questions such as, "What number, or numbers, when squared, give me 16?"

▶ ❶ Evaluate Square Roots of Perfect Squares

A real number is squared when it is raised to the power 2. The inverse of squaring a number is finding the **square root.** For example, since $5^2 = 25$ and $(-5)^2 = 25$, the square roots of 25 are -5 and 5. The square roots of $\frac{16}{49}$ are $-\frac{4}{7}$ and $\frac{4}{7}$.

If only the positive square root of a number is wanted, use the symbol $\sqrt{}$, called a **radical sign,** to denote the **principal square root,** or nonnegative (zero or positive) square root.

> **In Other Words**
>
> Taking the square root of a number "undoes" squaring a number.

> **In Other Words**
>
> The notation $b = \sqrt{a}$ means "produce the number b greater than or equal to 0 whose square is a."

> **Definition**
>
> If a is a nonnegative real number, the nonnegative real number b such that $b^2 = a$ is the **principal square root** of a and is denoted by $b = \sqrt{a}$.

For example, the positive square root of 25 is written $\sqrt{25} = 5$. We read $\sqrt{25} = 5$ as "the principal (or positive) square root of 25 is 5." If we want the negative square root of 25, use the expression $-\sqrt{25} = -5$.

> **Properties of Square Roots**
>
> - Every positive real number has two square roots, one positive and one negative.
> - The square root of 0 is 0. That is, $\sqrt{0} = 0$.
> - Use the symbol $\sqrt{}$, called a radical sign, to denote the nonnegative square root of a real number. The nonnegative square root is called the principal square root.
> - The number under the radical sign is called the **radicand.** For example, the radicand in $\sqrt{25}$ is 25.
> - For any real number c, such that $c \geq 0, (\sqrt{c})^2 = c$. For example, $(\sqrt{4})^2 = 4$ and $(\sqrt{8.3})^2 = 8.3$.

Prepared?...Answers **P1.** $-4, 0, 13$

P2. $-4, \frac{5}{3}, 0, 6.95, 13$ **P3.** $\sqrt{2}, \pi$

P4. (a) $\frac{9}{4}$ **(b)** 0.16

To **evaluate** a square root, ask, "What is the nonnegative number whose square is equal to the radicand?"

EXAMPLE 1 **Evaluating Square Roots**

Evaluate each square root.

(a) $\sqrt{36}$ (b) $\sqrt{\dfrac{1}{9}}$ (c) $\sqrt{0.01}$ (d) $(\sqrt{2.3})^2$

Solution

(a) Is there a positive number whose square is 36? Because $6^2 = 36$, $\sqrt{36} = 6$.

(b) $\sqrt{\dfrac{1}{9}} = \dfrac{1}{3}$ because $\left(\dfrac{1}{3}\right)^2 = \dfrac{1}{9}$.

(c) $\sqrt{0.01} = 0.1$ because $0.1^2 = 0.01$.

(d) $(\sqrt{2.3})^2 = 2.3$ because $(\sqrt{c})^2 = c$ when $c \geq 0$.   ●

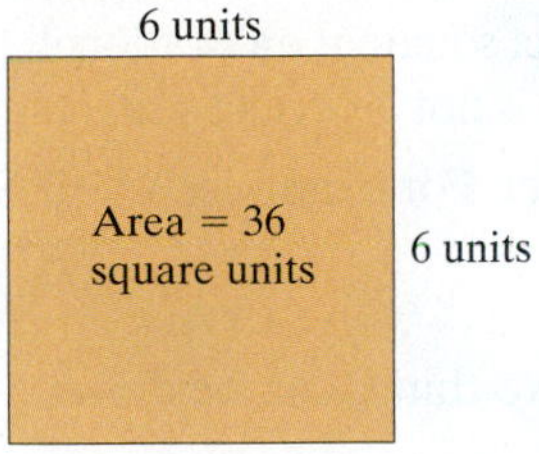

A rational number is a **perfect square** if it is the square of a rational number. Examples 1(a), (b), and (c) are square roots of perfect squares since $6^2 = 36$, $\left(\dfrac{1}{3}\right)^2 = \dfrac{1}{9}$,

and $0.1^2 = 0.01$. Perfect squares can be thought of geometrically as shown in Figure 1, where there is a square whose area is 36 square units. The square root of the area, $\sqrt{36}$, gives the length of each side of the square, 6 units.

Figure 1

6 units

Area = 36 square units

6 units

Quick ✓

1. The symbol $\sqrt{\ }$ is called a _____ ___.

2. If a is a nonnegative real number, the nonnegative number b such that $b^2 = a$ is the _____ _____ ___ of a and is denoted by $b = \sqrt{a}$.

3. The square roots of 16 are __ and __.

In Problems 4–8, evaluate each square root.

4. $\sqrt{81}$ 5. $\sqrt{900}$ 6. $\sqrt{\dfrac{9}{4}}$ 7. $\sqrt{0.16}$ 8. $(\sqrt{13})^2$

EXAMPLE 2 **Evaluating an Expression Containing Square Roots**

Evaluate each expression:

(a) $-4\sqrt{36}$ (b) $\sqrt{9} + \sqrt{16}$ (c) $\sqrt{9 + 16}$ (d) $\sqrt{64 - 4 \cdot 7 \cdot 1}$

Solution

(a) The expression $-4\sqrt{36}$ represents -4 times the positive square root of 36. To simplify, first find the positive square root of 36 and then multiply this result by -4.

$$-4\sqrt{36} = -4 \cdot 6$$
$$= -24$$

(b) $\sqrt{9} + \sqrt{16} = 3 + 4$
$$= 7$$

(c) $\sqrt{9 + 16} = \sqrt{25}$
$$= 5$$

(d) $\sqrt{64 - 4 \cdot 7 \cdot 1} = \sqrt{64 - 28}$
$$= \sqrt{36}$$
$$= 6$$ ●

Work Smart

In Examples 2(b) and (c), notice that

$$\sqrt{9} + \sqrt{16} \neq \sqrt{9 + 16}$$

In general,

$$\sqrt{a} + \sqrt{b} \neq \sqrt{a + b}$$

The radical sign acts as a grouping symbol, just like parentheses, so always simplify the radicand before taking the square root.

Quick ✓

In Problems 9–12, evaluate each expression.

9. $5\sqrt{9}$ 10. $\sqrt{36 + 64}$ 11. $\sqrt{36} + \sqrt{64}$ 12. $\sqrt{25 - 4 \cdot 3 \cdot (-2)}$

▶ ❷ Determine Whether a Square Root Is Rational, Irrational, or Not a Real Number

Work Smart

Recall from Section 1.3 that a rational number is a number that can be written as a quotient of two integers.

Not all radical expressions will simplify to a rational number. For example, because there is no rational number whose square is 5, $\sqrt{5}$ is not a rational number. In fact, $\sqrt{5}$ is an *irrational* number. Remember, an irrational number is a number that cannot be written as the quotient of two integers.

Work Smart

The square roots of negative real numbers are not real numbers.

What about evaluating $\sqrt{-16}$? Because any positive real number squared is positive, any negative real number squared is also positive, and 0 squared is 0, there is no real number whose square is -16. **Therefore, negative real numbers do not have square roots that are real numbers!**

These points are summarized below.

> **More Properties of Square Roots**
>
> - The square root of a perfect square is a rational number.
> - The square root of a positive rational number that is not a perfect square is an irrational number. For example, $\sqrt{20}$ is an irrational number because 20 is not a perfect square.
> - The square root of a negative real number is not a real number. For example, $\sqrt{-2}$ is not a real number.

If a radical has a radicand that is not a perfect square, one of two things can be done:

1. Write a decimal approximation of the radical.
2. Simplify the radical using properties of radicals, if possible (Section 9.4).

Figure 2

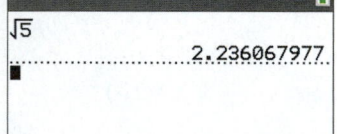

EXAMPLE 3 | Writing a Radical as a Decimal Using Technology

Write $\sqrt{5}$ as a decimal rounded to two decimal places.

Solution

We know that $\sqrt{4} = 2$ and $\sqrt{9} = 3$, so it seems reasonable to expect $\sqrt{5}$ to be between 2 and 3 because 5 is between 4 and 9. We can use technology or Appendix D to approximate $\sqrt{5}$. Figure 2 shows the results from a TI-84 Plus graphing calculator. From the display, we see that $\sqrt{5} \approx 2.24$. ●

EXAMPLE 4 | Determining Whether a Square Root of an Integer Is Rational, Irrational, or Not a Real Number

Determine whether each square root is rational, irrational, or not a real number. Evaluate each real square root that is rational. For each square root that is irrational, express the square root as a decimal rounded to two decimal places.

   (a) $\sqrt{51}$ **(b)** $\sqrt{169}$ **(c)** $\sqrt{-81}$

Solution

Work Smart

$\sqrt{-81}$ is not a real number, but $-\sqrt{81}$ is a real number because $-\sqrt{81} = -9$. Note the placement of the negative sign.

   (a) $\sqrt{51}$ is irrational because 51 is not a perfect square. That is, there is no rational number whose square is 51. Using technology, $\sqrt{51} \approx 7.14$.

   (b) $\sqrt{169}$ is a rational number because $13^2 = 169$, so $\sqrt{169} = 13$.

   (c) $\sqrt{-81}$ is not a real number. There is no real number whose square is -81. ●

> **Quick ✓**
>
> **13.** *True or False* Negative numbers do not have square roots that are real numbers.
>
> *In Problems 14–17, determine whether each square root is rational, irrational, or not a real number. Evaluate each square root that is rational. For each square root that is irrational, approximate the square root rounded to two decimal places.*
>
> **14.** $\sqrt{400}$ **15.** $\sqrt{40}$ **16.** $\sqrt{-25}$ **17.** $-\sqrt{196}$

▶ ❸ Find Square Roots of Variable Expressions

What is $\sqrt{4^2}$? Because $4^2 = 16$, $\sqrt{4^2} = \sqrt{16} = 4$. This result suggests that $\sqrt{a^2} = a$ for any real number a. But wait. Does $\sqrt{(-4)^2} = -4$? No! $\sqrt{(-4)^2} = \sqrt{16} = 4$, so both $\sqrt{4^2}$ and $\sqrt{(-4)^2}$ equal 4. Regardless of whether the "a" in $\sqrt{a^2}$ is positive or negative, the result is positive. Therefore, $\sqrt{a^2} = a$ is incorrect, because the sign of a is unknown. How can this "formula" be fixed? Section 1.3 demonstrated that $|a|$ is a positive number if a is nonzero. From this, the following results:

> **In Other Words**
> The square root of a nonzero number squared will always be positive. The absolute value ensures this.

For any **real number** a,

$$\sqrt{a^2} = |a|$$

The bottom line is this—the square root of a variable expression raised to the second power is the absolute value of the variable expression.

EXAMPLE 5 **Evaluating Square Roots**

Evaluate each square root.

(a) $\sqrt{7^2}$ (b) $\sqrt{(-15)^2}$ (c) $\sqrt{x^2}$ (d) $\sqrt{(3x-1)^2}$ (e) $\sqrt{x^2 + 6x + 9}$

Solution

(a) $\sqrt{7^2} = 7$

(b) $\sqrt{(-15)^2} = |-15| = 15$

(c) It is unknown whether the real number x is positive, negative, or zero. To ensure that the result is positive or zero, write $\sqrt{x^2} = |x|$.

(d) $\sqrt{(3x-1)^2} = |3x-1|$

(e) Notice that the radicand, $x^2 + 6x + 9$, factors to $(x+3)^2$. Therefore,

$$\sqrt{x^2 + 6x + 9} = \sqrt{(x+3)^2}$$
$$= |x+3|$$

●

> **Quick ✓**
>
> 18. $\sqrt{a^2} = \underline{\quad}$.
>
> *In Problems 19–22, evaluate each square root.*
>
> 19. $\sqrt{(-14)^2}$ 20. $\sqrt{z^2}$ 21. $\sqrt{(2x+3)^2}$ 22. $\sqrt{p^2 - 12p + 36}$

9.1 Exercises   MyMathLab®

Exercise numbers in **green** have complete video solutions in MyMathLab or may be accessed using the QR code to the right.

Problems 1–22 are the Quick ✓s that follow the EXAMPLES.

Building Skills

In Problems 23–32, evaluate each square root. See Objective 1.

23. $\sqrt{1}$ **24.** $\sqrt{9}$

25. $-\sqrt{100}$ **26.** $-\sqrt{144}$

27. $\sqrt{\dfrac{1}{4}}$ **28.** $\sqrt{\dfrac{4}{81}}$

29. $\sqrt{0.36}$ **30.** $\sqrt{0.25}$

31. $(\sqrt{1.6})^2$ **32.** $(\sqrt{3.7})^2$

In Problems 33–44, tell whether the square root is rational, irrational, or not a real number. If the square root is rational, find the exact value; if the square root is irrational, write the approximate value rounded to two decimal places. See Objective 2.

33. $\sqrt{-14}$ **34.** $\sqrt{-50}$

35. $\sqrt{64}$ **36.** $\sqrt{121}$

37. $\sqrt{\dfrac{1}{16}}$ **38.** $\sqrt{\dfrac{49}{100}}$

39. $\sqrt{44}$ **40.** $\sqrt{24}$

41. $\sqrt{50}$ **42.** $\sqrt{12}$

43. $\sqrt{-16}$ **44.** $\sqrt{-64}$

In Problems 45–56, simplify each square root. See Objective 3.

45. $\sqrt{8^2}$

46. $\sqrt{5^2}$

47. $\sqrt{(-19)^2}$

48. $\sqrt{(-13)^2}$

49. $\sqrt{r^2}$

50. $\sqrt{w^2}$

51. $\sqrt{(x+4)^2}$

52. $\sqrt{(x-8)^2}$

53. $\sqrt{(4x-3)^2}$

54. $\sqrt{(5x+2)^2}$

55. $\sqrt{4y^2+12y+9}$

56. $\sqrt{9z^2-24z+16}$

Mixed Practice

In Problems 57–74, simplify each expression.

57. $\sqrt{25+144}$

58. $\sqrt{9+16}$

59. $\sqrt{25}+\sqrt{144}$

60. $\sqrt{9}+\sqrt{16}$

61. $\sqrt{-144}$

62. $\sqrt{-36}$

63. $3\sqrt{25}$

64. $-10\sqrt{16}$

65. $5\sqrt{\dfrac{16}{25}}-\sqrt{144}$

66. $2\sqrt{\dfrac{9}{4}}-\sqrt{4}$

67. $\sqrt{8^2-4\cdot1\cdot7}$

68. $\sqrt{9^2-4\cdot1\cdot20}$

69. $\sqrt{(-5)^2-4\cdot2\cdot5}$

70. $\sqrt{(-3)^2-4\cdot3\cdot2}$

71. $\dfrac{-(-1)+\sqrt{(-1)^2-4\cdot6(-2)}}{2(-1)}$

72. $\dfrac{-7+\sqrt{7^2-4\cdot2\cdot6}}{2\cdot2}$

73. $\sqrt{(6-1)^2+(15-3)^2}$

74. $\sqrt{(2-(-1))^2+(6-2)^2}$

75. What are the square roots of 36? What is $\sqrt{36}$?

76. What are the square roots of 64? What is $\sqrt{64}$?

Math for the Future: Statistics *For Problems 77 and 78, use the formula* $Z=\dfrac{X-\mu}{\dfrac{\sigma}{\sqrt{n}}}$ *from statistics (a formula used to determine the value of one observation relative to that of another) to evaluate the expression for the given values. Write the exact value and then write your answer rounded to two decimal places.*

77. $X=120,\mu=100,\sigma=15,n=13$

78. $X=40,\mu=50,\sigma=10,n=5$

Explaining the Concepts

79. Explain why $\sqrt{a^2}=|a|$. Provide examples to support your explanation.

9.2 *n*th Roots and Rational Exponents

Objectives

1 Evaluate *n*th Roots

2 Simplify Expressions of the Form $\sqrt[n]{a^n}$

3 Evaluate Expressions of the Form $a^{\frac{1}{n}}$

4 Evaluate Expressions of the Form $a^{\frac{m}{n}}$

Are You Prepared for This Section?

Before getting started, complete the following problems. If you get a problem wrong, go back to the section cited and review the material.

P1. Simplify: $\left(\dfrac{x^2y}{xy^{-2}}\right)^{-3}$ [Section 5.4, pp. 341–344]

P2. Simplify: $(\sqrt{7})^2$ [Section 9.1, pp. 616–617]

P3. Evaluate: $\sqrt{64}$ [Section 9.1, pp. 616–617]

P4. Evaluate: $\sqrt{(x+1)^2}$ [Section 9.1, pp. 618–619]

P5. Simplify: **(a)** 3^{-2} **(b)** x^{-4} [Section 5.4, pp. 337–339]

In the last section, skills for evaluating square roots were reviewed. This section will extend this skill to other types of roots.

▶ **1** Evaluate *n*th Roots

A real number is cubed when it is raised to the power 3. The inverse of cubing a number is finding the *cube root*. For example, because $2^3=8$, the cube root of 8 is 2; because $(-2)^3=-8$, the cube root of -8 is -2. In general, *n*th roots of numbers can be found.

In Other Words

The notation $\sqrt[n]{a} = b$ means "a number b such that raising that number to the *n*th power gives *a*." Thus $\sqrt[3]{125} = x$ means "a number x such that raising x to the third power gives 125."

Definition

The **principal *n*th root of a number *a*,** symbolized by $\sqrt[n]{a}$, where $n \geq 2$ is an integer, is defined as follows:

$$\sqrt[n]{a} = b \qquad \text{means} \qquad a = b^n$$

- If $n \geq 2$ and even, then *a* and *b* must be greater than or equal to 0.
- If $n \geq 3$ and odd, then *a* and *b* can be any real number.

Work Smart

If the index is even, then the radicand must be greater than or equal to zero in order for a radical to simplify to a real number. If the index is odd, the radicand can be any real number.

In the notation $\sqrt[n]{a}$, the integer n, $n \geq 2$, is called the **index.** A radical written without the index means the square root, so $\sqrt{a}$ represents the square root of *a*. If the index is 3, $\sqrt[3]{a}$ is called the **cube root** of *a*.

If the index is even, then the radicand must be greater than or equal to 0. If the index is odd, then the radicand can be any real number. Do you know why? Since $\sqrt[n]{a} = b$ means $b^n = a$, if the index n is even, then $b^n \geq 0$ so $a \geq 0$. If there is an odd index n, then b^n can be any real number, so a can be any real number.

Before *n*th roots are evaluated, some "perfect" powers of 2, 3, 4, and 5 are listed. This list will help you find roots in the following examples. Notice in the display that follows that perfect cubes and perfect fifths can be negative, but perfect squares and perfect fourths cannot. Can you determine why?

Prepared?...Answers **P1.** $\dfrac{1}{x^3 y^9}$ **P2.** 7

P3. 8 **P4.** $|x + 1|$ **P5. (a)** $\dfrac{1}{9}$ **(b)** $\dfrac{1}{x^4}$

Perfect Squares	Perfect Cubes		Perfect Fourths	Perfect Fifths	
$(\pm 1)^2 = 1$	$1^3 = 1$	$(-1)^3 = -1$	$(\pm 1)^4 = 1$	$1^5 = 1$	$(-1)^5 = -1$
$(\pm 2)^2 = 4$	$2^3 = 8$	$(-2)^3 = -8$	$(\pm 2)^4 = 16$	$2^5 = 32$	$(-2)^5 = -32$
$(\pm 3)^2 = 9$	$3^3 = 27$	$(-3)^3 = -27$	$(\pm 3)^4 = 81$	$3^5 = 243$	$(-3)^5 = -243$
$(\pm 4)^2 = 16$	$4^3 = 64$	$(-4)^3 = -64$	$(\pm 4)^4 = 256$	$4^5 = 1024$	$(-4)^5 = -1024$
$(\pm 5)^2 = 25$	$5^3 = 125$	$(-5)^3 = -125$	$(\pm 5)^4 = 625$	$5^5 = 3125$	$(-5)^5 = -3125$
and so on	and so on		and so on	and so on	

EXAMPLE 1 **Evaluating *n*th Roots of Real Numbers**

Evaluate:

(a) $\sqrt[3]{1000}$ **(b)** $\sqrt[4]{16}$ **(c)** $\sqrt[3]{-8}$ **(d)** $\sqrt[4]{-81}$

Solution

(a) Look for a number whose cube is 1000. Since $10^3 = 1000$, $\sqrt[3]{1000} = 10$.

(b) $\sqrt[4]{16} = 2$ beacause $2^4 = 16$.

(c) $\sqrt[3]{-8} = -2$ because $(-2)^3 = -8$.

(d) Because there is no real number b such that $b^4 = -81$, $\sqrt[4]{-81}$ is not a real number.

Quick ✓

1. In the notation $\sqrt[n]{a}$, the integer n, $n \geq 2$, is called the _____.

In Problems 2–6, evaluate each root.

2. $\sqrt[3]{64}$ **3.** $\sqrt[4]{81}$ **4.** $\sqrt[3]{-216}$

5. $\sqrt[4]{-32}$ **6.** $\sqrt[5]{\dfrac{1}{32}}$

The *n*th roots in Examples 1(a)–(c) are all rational numbers, but some *n*th roots are not. Just as square roots can be approximated using technology, *n*th roots can also be approximated.

EXAMPLE 2 **Approximating an *n*th Root Using a Calculator**

(a) Write $\sqrt[3]{25}$ as a decimal rounded to two decimal places.

(b) Write $\sqrt[4]{18}$ as a decimal rounded to two decimal places.

Solution

A graphing calculator can be used to approximate both values.

(a) Because $\sqrt[3]{8} = 2$ and $\sqrt[3]{27} = 3$, $\sqrt[3]{25}$ is between 2 and 3 (closer to 3). Figure 3(a) shows that $\sqrt[3]{25} \approx 2.92$.

(b) Because $\sqrt[4]{16} = 2$ and $\sqrt[4]{81} = 3$, $\sqrt[4]{18}$ is between 2 and 3 (closer to 2). Figure 3(b) shows that $\sqrt[4]{18} \approx 2.06$.

Figure 3

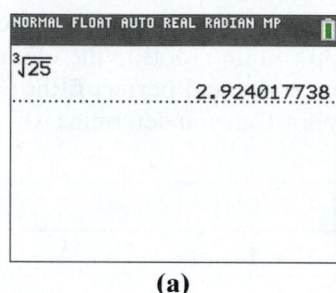

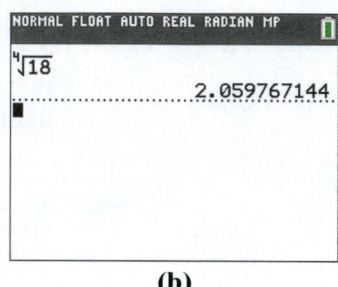

(a)   (b)

Quick ✓

In Problems 7 and 8, use technology to write the approximate value of each radical rounded to two decimal places.

7. $\sqrt[3]{50}$ **8.** $\sqrt[4]{80}$

▶ ❷ Simplify Expressions of the Form $\sqrt[n]{a^n}$

Remember that $\sqrt{a^2} = |a|$. But what about $\sqrt[3]{a^3}$ or $\sqrt[4]{a^4}$? Recall that the definition of the principal *n*th root, $\sqrt[n]{a}$, requires that *a* be nonnegative (greater than or equal to zero) when *n* is even, but *a* can be any real number when *n* is odd.

Simplifying $\sqrt[n]{a^n}$

If $n \geq 2$ is a positive integer and *a* is a real number, then

$$\sqrt[n]{a^n} = |a| \quad \text{if } n \geq 2 \text{ is even}$$
$$\sqrt[n]{a^n} = a \quad \text{if } n \geq 3 \text{ is odd}$$

EXAMPLE 3 **Simplifying Radicals**

Simplify:

(a) $\sqrt[3]{x^3}$ (b) $\sqrt[4]{(x-7)^4}$ (c) $-\sqrt[6]{(-3)^6}$ (d) $\sqrt[3]{-\dfrac{8}{125}}$

Solution

(a) Because the index, 3, is odd, $\sqrt[3]{x^3} = x$.

(b) Because the index, 4, is even, $\sqrt[4]{(x-7)^4} = |x - 7|$.

(c) $-\sqrt[6]{(-3)^6} = -|-3| = -3$

(d) $\sqrt[3]{-\dfrac{8}{125}} = \sqrt[3]{\dfrac{-8}{125}} = \sqrt[3]{\dfrac{(-2)^3}{5^3}} = \sqrt[3]{\left(\dfrac{-2}{5}\right)^3} = -\dfrac{2}{5}$

Work Smart

Note in Example 3(d) that the negative sign could be applied to 125 instead of 8 and the answer would be the same.

Work Smart

Remember, a rational number is a number of the form $\frac{p}{q}$, where p and q are integers and $q \neq 0$.

Quick ✓

In Problems 9–13, simplify each radical.

9. $\sqrt[4]{5^4}$ **10.** $\sqrt[6]{z^6}$ **11.** $\sqrt[7]{(3x-2)^7}$

12. $\sqrt[8]{(-2)^8}$ **13.** $\sqrt[5]{-\dfrac{32}{243}}$

▶ ❸ Evaluate Expressions of the Form $a^{\frac{1}{n}}$

In Chapter 5, methods for simplifying algebraic expressions that contain integer exponents were developed. Now the rules that apply to integer exponents will be extended to rational exponents.

Start by defining "a raised to the power $\dfrac{1}{n}$," where a is a real number and n is a positive integer. This definition needs to obey all the laws of exponents presented earlier. For example, because $a^2 = a \cdot a$,

$$\left(5^{\frac{1}{2}}\right)^2 = 5^{\frac{1}{2}} \cdot 5^{\frac{1}{2}}$$

$$a^m \cdot a^n = a^{m+n}: \quad = 5^{\frac{1}{2}+\frac{1}{2}}$$

$$= 5^1$$

$$= 5$$

Also, because $\left(\sqrt{5}\right)^2 = 5$, it is reasonable to conclude that

$$5^{\frac{1}{2}} = \sqrt{5}$$

This suggests the following definition:

> **Definition of $a^{\frac{1}{n}}$**
>
> If a is a real number and n is an integer with $n \geq 2$, then
> $$a^{\frac{1}{n}} = \sqrt[n]{a}$$
>
> provided that $\sqrt[n]{a}$ exists.

EXAMPLE 4 **Evaluating Expressions Containing Exponents of the Form $a^{\frac{1}{n}}$**

Write each of the following expressions as a radical and simplify, if possible.

(a) $9^{\frac{1}{2}}$ **(b)** $(-64)^{\frac{1}{3}}$ **(c)** $-100^{\frac{1}{2}}$ **(d)** $(-100)^{\frac{1}{2}}$ **(e)** $z^{\frac{1}{2}}$

Solution

(a) $9^{\frac{1}{2}} = \sqrt{9}$
$= 3$

(b) $(-64)^{\frac{1}{3}} = \sqrt[3]{-64}$
$= -4$

(c) $-100^{\frac{1}{2}} = -1 \cdot 100^{\frac{1}{2}}$
$= -1 \cdot \sqrt{100}$
$= -1 \cdot 10$
$= -10$

(d) $(-100)^{\frac{1}{2}} = \sqrt{-100}$ is not a real number because there is no real number whose square is -100.

(e) $z^{\frac{1}{2}} = \sqrt{z}$

Work Smart

Notice how parentheses are used or not used in Examples 4(c) and (d). In Example 4(c), $-100^{\frac{1}{2}}$ means evaluate $100^{\frac{1}{2}}$ first and then multiply the result by -1. Remember the order of operations: Evaluate exponents before multiplying.

Quick ✓

14. If a is a nonnegative real number and $n \geq 2$ is an integer, then $a^{\frac{1}{n}} =$ ___.

In Problems 15–19, write each of the following expressions as a radical, and simplify, if possible.

15. $25^{\frac{1}{2}}$ **16.** $(-27)^{\frac{1}{3}}$ **17.** $-64^{\frac{1}{2}}$ **18.** $(-64)^{\frac{1}{2}}$ **19.** $b^{\frac{1}{2}}$

EXAMPLE 5 **Writing Radicals with Rational Exponents**

Rewrite each of the following radicals with a rational exponent.

(a) $\sqrt[4]{7a}$ **(b)** $\sqrt[5]{\dfrac{xy^3}{4}}$

Solution

(a) The index on the radical is 4, so 4 becomes the denominator of the rational exponent. Use parentheses because the radicand is $7a$, and the exponent $\dfrac{1}{4}$ is applied to both 7 and a.

$$\sqrt[4]{7a} = (7a)^{\frac{1}{4}}$$

(b) The index on the radical is 5, so the rational exponent is $\dfrac{1}{5}$.

$$\sqrt[5]{\frac{xy^3}{4}} = \left(\frac{xy^3}{4}\right)^{\frac{1}{5}}$$

> **Quick ✓**
>
> In Problems 20–22, rewrite each of the following radicals with a rational exponent.
>
> **20.** $\sqrt[5]{8b}$ **21.** $\sqrt[8]{\dfrac{mn^5}{3}}$ **22.** $4\sqrt[3]{y}$

▶ ❹ **Evaluate Expressions of the Form $a^{\frac{m}{n}}$**

Now $a^{\frac{m}{n}}$ is defined, where m and n are integers.

> **Definition of $a^{\frac{m}{n}}$**
>
> If a is a real number and $\dfrac{m}{n}$ is a rational number in lowest terms with $n \geq 2$, then
>
> $$a^{\frac{m}{n}} = \sqrt[n]{a^m} = \left(\sqrt[n]{a}\right)^m$$
>
> provided that $\sqrt[n]{a}$ exists.

In Other Words

The expression $a^{\frac{m}{n}} = \sqrt[n]{a^m}$ means raise a to the mth power first, and then take the nth root of the result. The expression $a^{\frac{m}{n}} = (\sqrt[n]{a})^m$ means take the nth root of a first, and then raise that result to the power of m.

In the expression $a^{\frac{m}{n}}$, $\dfrac{m}{n}$ is expressed in lowest terms, and $n \geq 2$. The definition obeys all the laws of exponents presented earlier. For example,

$$a^{\frac{m}{n}} = a^{m \cdot \frac{1}{n}} = (a^m)^{\frac{1}{n}} = \sqrt[n]{a^m}$$

and

$$a^{\frac{m}{n}} = a^{\frac{1}{n} \cdot m} = \left(a^{\frac{1}{n}}\right)^m = \left(\sqrt[n]{a}\right)^m$$

When evaluating or simplifying $a^{\frac{m}{n}}$, use either $\sqrt[n]{a^m}$ or $(\sqrt[n]{a})^m$, whichever makes simplifying the expression easier. Generally, taking the root first, as in $(\sqrt[n]{a})^m$, is easier.

EXAMPLE 6 **Evaluating Expressions of the Form $a^{\frac{m}{n}}$**

Evaluate each of the following expressions, if possible.

(a) $25^{\frac{3}{2}}$ **(b)** $64^{\frac{2}{3}}$ **(c)** $-9^{\frac{5}{2}}$ **(d)** $(-8)^{\frac{4}{3}}$ **(e)** $(-81)^{\frac{7}{2}}$

Solution

(a) $25^{\frac{3}{2}} = \left(\sqrt{25}\right)^3 = 5^3 = 125$ **(b)** $64^{\frac{2}{3}} = \left(\sqrt[3]{64}\right)^2 = 4^2 = 16$

(c) $-9^{\frac{5}{2}} = -1 \cdot 9^{\frac{5}{2}} = -1\left(\sqrt{9}\right)^5 = -1 \cdot 3^5 = -1 \cdot 243 = -243$

(d) $(-8)^{\frac{4}{3}} = \left(\sqrt[3]{-8}\right)^4 = (-2)^4 = 16$

(e) $(-81)^{\frac{7}{2}}$ is not a real number because $(-81)^{\frac{7}{2}} = \left(\sqrt{-81}\right)^7$, and $\sqrt{-81}$ is not a real number.

Quick ✓

23. If a is a real number and $\dfrac{m}{n}$ is a rational number in lowest terms with $n \geq 2$, then $a^{\frac{m}{n}} = $ _____ or _____, provided that $\sqrt[n]{a}$ exists.

In Problems 24–28, evaluate each expression, if possible.

24. $16^{\frac{3}{2}}$ **25.** $27^{\frac{2}{3}}$ **26.** $-16^{\frac{3}{4}}$

27. $(-64)^{\frac{2}{3}}$ **28.** $(-25)^{\frac{5}{2}}$

The expressions in Examples 6(a)–(d) all simplified to rational numbers. Not all expressions involving rational exponents simplify to rational numbers.

EXAMPLE 7   **Approximating Expressions Involving Rational Exponents**

Write $34^{\frac{3}{4}}$ as a decimal rounded to two decimal places.

Figure 4

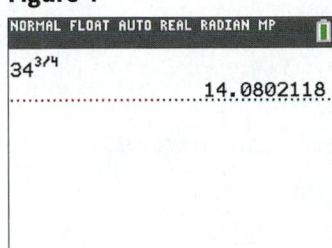

Solution

Figure 4 shows the results obtained from a graphing calculator. So $34^{\frac{3}{4}} \approx 14.08$.   ●

Quick ✓

In Problems 29 and 30, approximate the expression rounded to two decimal places.

29. $50^{\frac{2}{3}}$ **30.** $40^{\frac{3}{20}}$

EXAMPLE 8   **Writing Radicals with Rational Exponents**

Rewrite each of the following radicals with a rational exponent.

   (a) $\sqrt[3]{x^2}$ **(b)** $\left(\sqrt[5]{10a^2b}\right)^4$

Solution

   (a) The index, 3, is the denominator of the rational exponent, and the power on the radicand, 2, is the numerator of the rational exponent.

$$\sqrt[3]{x^2} = x^{\frac{2}{3}}$$

   (b) The index, 5, is the denominator of the rational exponent, and the power, 4, is the numerator of the rational exponent.

$$\left(\sqrt[5]{10a^2b}\right)^4 = (10a^2b)^{\frac{4}{5}}$$ ●

Quick ✓

In Problems 31–33, write each radical with a rational exponent.

31. $\sqrt[8]{a^3}$ **32.** $\sqrt[4]{t^{12}}$ **33.** $\left(\sqrt[4]{12ab^3}\right)^9$

If a rational exponent is negative, then use the rule for negative rational exponents given below.

Negative-Exponent Rule

If $\dfrac{m}{n}$ is a rational number, and if a is a nonzero real number (that is, if $a \neq 0$), then define

$$a^{-\frac{m}{n}} = \frac{1}{a^{\frac{m}{n}}} \quad \text{and} \quad \frac{1}{a^{-\frac{m}{n}}} = a^{\frac{m}{n}} \quad \text{if } a \neq 0, \quad \text{and } \sqrt[n]{a} \text{ is defined.}$$

EXAMPLE 9 **Evaluating Expressions with Negative Rational Exponents**

Rewrite each of the following with positive exponents, and completely simplify, if possible.

(a) $36^{-\frac{1}{2}}$ **(b)** $\dfrac{1}{27^{-\frac{2}{3}}}$ **(c)** $(6a)^{-\frac{5}{4}}$

Solution

(a) $36^{-\frac{1}{2}} = \dfrac{1}{36^{\frac{1}{2}}} = \dfrac{1}{\sqrt{36}} = \dfrac{1}{6}$

(b) Because the negative exponent is in the denominator, use $\dfrac{1}{a^{-\frac{m}{n}}} = a^{\frac{m}{n}}$ to

simplify:

$$\dfrac{1}{27^{-\frac{2}{3}}} = 27^{\frac{2}{3}} = (\sqrt[3]{27})^2 = 3^2 = 9$$

(c) $(6a)^{-\frac{5}{4}} = \dfrac{1}{(6a)^{\frac{5}{4}}}$

Quick ✓

In Problems 34–36, rewrite each of the following with positive exponents, and completely simplify, if possible.

34. $81^{-\frac{1}{2}}$ **35.** $\dfrac{1}{8^{-\frac{2}{3}}}$ **36.** $(13x)^{-\frac{3}{2}}$

Exercises 9.2 MyMathLab® Exercise numbers in green have complete video solutions in MyMathLab or may be accessed using the QR code to the right.

Problems **1–36** are the **Quick ✓**s that follow the **EXAMPLES**.

Building Skills

In Problems 37–46, evaluate each root. See Objective 1.

37. $\sqrt[3]{125}$ **38.** $\sqrt[3]{216}$

39. $\sqrt[3]{-27}$ **40.** $\sqrt[3]{-64}$

41. $-\sqrt[4]{625}$ **42.** $-\sqrt[4]{256}$

43. $\sqrt[3]{-\dfrac{1}{8}}$ **44.** $\sqrt[3]{\dfrac{8}{125}}$

45. $-\sqrt[5]{-243}$ **46.** $-\sqrt[5]{-1024}$

In Problems 47–50, use technology to write each expression as a decimal rounded to two decimal places. See Objective 1.

47. $\sqrt[3]{25}$ **48.** $\sqrt[3]{85}$

49. $\sqrt[4]{12}$ **50.** $\sqrt[4]{2}$

In Problems 51–58, simplify each radical. See Objective 2.

51. $\sqrt[3]{5^3}$ **52.** $\sqrt[4]{6^4}$

53. $\sqrt[4]{m^4}$ **54.** $\sqrt[5]{n^5}$

55. $\sqrt[9]{(x-3)^9}$ **56.** $\sqrt[6]{(2x-3)^6}$

57. $-\sqrt[4]{(3p+1)^4}$ **58.** $-\sqrt[3]{(6z-5)^3}$

In Problems 59–72, evaluate each expression, if possible. See Objective 3.

59. $4^{\frac{1}{2}}$ **60.** $16^{\frac{1}{2}}$

61. $-36^{\frac{1}{2}}$ **62.** $-25^{\frac{1}{2}}$

63. $8^{\frac{1}{3}}$ **64.** $27^{\frac{1}{3}}$

65. $-16^{\frac{1}{4}}$ **66.** $-81^{\frac{1}{4}}$

67. $\left(\dfrac{4}{25}\right)^{\frac{1}{2}}$ **68.** $\left(\dfrac{8}{27}\right)^{\frac{1}{3}}$

69. $(-125)^{\frac{1}{3}}$ **70.** $(-216)^{\frac{1}{3}}$

71. $(-4)^{\frac{1}{2}}$ **72.** $(-81)^{\frac{1}{2}}$

In Problems 73–76, rewrite each of the following radicals with a rational exponent. See Objective 3.

73. $\sqrt[3]{3x}$ **74.** $\sqrt[5]{2y}$

75. $\sqrt[4]{\dfrac{x}{3}}$

76. $\sqrt{\dfrac{w}{2}}$

In Problems 77–92, evaluate each expression. See Objective 4.

77. $4^{\frac{5}{2}}$

78. $25^{\frac{3}{2}}$

79. $-16^{\frac{3}{2}}$

80. $-100^{\frac{5}{2}}$

81. $8^{\frac{4}{3}}$

82. $27^{\frac{4}{3}}$

83. $(-64)^{\frac{2}{3}}$

84. $(-125)^{\frac{2}{3}}$

85. $-(-32)^{\frac{3}{5}}$

86. $-(-216)^{\frac{2}{3}}$

87. $144^{-\frac{1}{2}}$

88. $121^{-\frac{1}{2}}$

89. $\dfrac{1}{25^{-\frac{3}{2}}}$

90. $\dfrac{1}{49^{-\frac{3}{2}}}$

91. $\dfrac{1}{8^{-\frac{5}{3}}}$

92. $27^{-\frac{4}{3}}$

In Problems 93–100, rewrite each of the following radicals with a rational exponent. See Objective 4.

93. $\sqrt[4]{x^3}$

94. $\sqrt[3]{p^5}$

95. $(\sqrt[5]{3x})^2$

96. $(\sqrt[4]{6z})^3$

97. $\sqrt{\left(\dfrac{5x}{y}\right)^3}$

98. $\sqrt[6]{\left(\dfrac{2a}{b}\right)^5}$

99. $\sqrt[3]{(9ab)^4}$

100. $\sqrt[4]{(3pq)^7}$

In Problems 101–106, use technology to write each expression as a decimal rounded to two decimal places. See Objective 4.

101. $20^{\frac{1}{2}}$

102. $5^{\frac{1}{2}}$

103. $4^{\frac{5}{3}}$

104. $100^{\frac{3}{4}}$

105. $10^{0.1}$

106. $100^{0.25}$

Mixed Practice

In Problems 107–130, simplify each expression, if possible.

107. $\sqrt[3]{x^3} + 4\sqrt[5]{x^5}$

108. $\sqrt[3]{(x-1)^3} + \sqrt[5]{32}$

109. $-16^{\frac{3}{4}} - \sqrt[5]{32}$

110. $\dfrac{5\sqrt[3]{x^7}}{\sqrt[3]{x^3}}, x \neq 0$

111. $\sqrt[3]{512}$

112. $\sqrt[3]{-125}$

113. $9^{2.5}$

114. $100^{1.5}$

115. $\sqrt[4]{-16}$

116. $\sqrt[4]{-1}$

117. $144^{-\frac{1}{2}} \cdot 3^2$

118. $125^{-\frac{1}{3}}$

119. $\sqrt[3]{0.008}$

120. $\sqrt[4]{0.0081}$

121. $4^{0.5} + 25^{1.5}$

122. $100^{0.5} - 4^{1.5}$

123. $(-25)^{\frac{5}{2}}$

124. $(-125)^{-\frac{1}{3}}$

125. $\sqrt[3]{(3p-5)^3}$

126. $\sqrt[5]{(6b-1)^5}$

127. $-9^{\frac{3}{2}} + \dfrac{1}{27^{-\frac{2}{3}}}$

128. $\dfrac{1}{64^{\frac{1}{2}}} - 4^{-\frac{3}{2}}$

129. $\sqrt[6]{(-2)^6}$

130. $\sqrt[4]{(-10)^4}$

In Problems 131–134, evaluate each function.

131. $f(x) = x^{\frac{3}{2}}$; find $f(4)$

132. $g(x) = x^{-\frac{3}{2}}$; find $g(16)$

133. $F(z) = z^{\frac{4}{3}}$; find $F(-8)$

134. $G(a) = a^{\frac{5}{3}}$; find $G(-8)$

Applying the Concepts

135. What is the cube root of 1000? What is $\sqrt[3]{1000}$?

136. What is the cube root of 729? What is $\sqrt[3]{729}$?

137. **Wind Chill** According to the National Weather Service, the wind chill temperature is how cold people and animals feel when outside. Wind chill is based on the rate of heat loss from exposed skin caused by wind and cold. The formula for computing wind chill W is

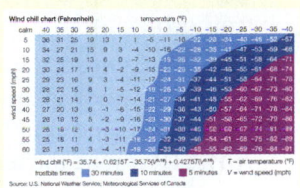

$$W = 35.74 + 0.6215T - 35.75v^{0.16} + 0.4275Tv^{0.16}$$

where T is the air temperature in degrees Fahrenheit and v is the wind speed in miles per hour.

(a) What is the wind chill if it is 30°F and the wind speed is 10 miles per hour?

(b) What is the wind chill if it is 30°F and the wind speed is 20 miles per hour?

(c) What is the wind chill if it is 0°F and the wind speed is 10 miles per hour?

138. **Money** The annual rate of interest r (expressed as a decimal) required to have A dollars after t years from an initial deposit of P dollars is given by

$$r = \left(\dfrac{A}{P}\right)^{\frac{1}{t}} - 1$$

(a) If you deposit $100 in a mutual fund today and have $144 in the account in 2 years, what was your annual rate of interest earned?

(b) If you deposit $100 in a mutual fund today and have $337.50 in 3 years, what was your annual rate of interest earned?

(c) The Rule of 72 states that your money will double in $\dfrac{72}{100r}$ years, where r is the rate of interest earned (expressed as a decimal). Suppose that you deposit $1000 in a mutual fund today and have $2000 in 8 years. What rate of interest did you earn? Compute $\dfrac{72}{100r}$ for this rate of interest. Is it close?

139. Terminal Velocity Terminal velocity is the maximum speed that a body falling through air can reach (because of air resistance). Terminal velocity is given by the formula $v_t = \sqrt{\dfrac{2mg}{C\rho A}}$, where m is the mass of the falling object, g is acceleration due to gravity (≈ 9.81 meters per second^2), C is a drag coefficient with $0.5 \le C \le 1.0$, ρ is the density of air (≈ 1.2 kg/m^3), and A is the cross-sectional area of the object. Suppose that a raindrop whose radius is 1.5 mm falls from the sky. The mass of the raindrop is given by $m = \dfrac{4}{3}\pi r^3 \rho_w$, where r is its radius and $\rho_w = 1000$ kg/m^3. The cross-sectional area of the raindrop is $A = \pi r^2$.

(a) Substitute the formulas for the mass and area of a raindrop into the formula for terminal speed, and simplify the expression.

(b) Determine the terminal velocity of a raindrop whose radius is 0.0015 m with $C = 0.6$.

140. Kepler's Law Early in the seventeenth century, Johannes Kepler (1571–1630) discovered that the square of the period T of a planet to rotate around the Sun varies directly with the cube of its mean distance r from the Sun. The period of a planet is the amount of time (in Earth years) for the planet to complete one orbit around the Sun. Kepler's Law can be expressed using rational exponents as $T = kr^{\frac{3}{2}}$, where k is the constant of proportionality.

(a) The period of Mercury is 0.241 year, and its mean distance from the Sun is 5.79×10^{10} meters. Use this information to state Kepler's Law (find the value of k).

(b) The mean distance of Mars from the Sun is 2.28×10^{11} m. Use this information, along with the result of part (a), to find the amount of time it takes Mars to complete one orbit around the Sun.

Explaining the Concepts

141. Explain why $(-9)^{\frac{1}{2}}$ is not a real number, but $-9^{\frac{1}{2}}$ is a real number.

142. In your own words, provide a justification for why $a^{\frac{1}{n}} = \sqrt[n]{a}$.

143. Under what conditions is $a^{\frac{m}{n}}$ a real number?

Synthesis Review

In Problems 144–147, simplify each expression completely.

144. $\left(\dfrac{x^2 y}{y^{-2}}\right)^3$

145. $\dfrac{(x + 2)^2 (x - 1)^4}{(x + 2)(x - 1)}$

146. $\dfrac{(3a^2 + 5a - 3) - (a^2 - 2a - 9)}{4a^2 + 12a + 9}$

147. $\dfrac{(4z^2 - 7z + 3) + (-3z^2 - z + 9)}{(4z^2 - 2z - 7) + (-3z^2 - z + 9)}$

9.3 Simplifying Expressions Using the Laws of Exponents

Objectives

④ Simplify Expressions Involving Rational Exponents

⑤ Simplify Radical Expressions

⑥ Factor Expressions Containing Rational Exponents

Are You Prepared for This Section?

Before getting started, complete the following problems. If you get a problem wrong, go back to the section cited and review the material.

P1. Simplify: z^{-3} [Section 5.4, pp. 337–339]

P2. Simplify: $x^{-2} \cdot x^5$ [Section 5.4, pp. 341–344]

P3. Simplify: $\left(\dfrac{2a^2}{b^{-1}}\right)^3$ [Section 5.4, pp. 341–344]

P4. Evaluate: $\sqrt{144}$ [Section 9.1, pp. 616–617]

▶ ① Simplify Expressions Involving Rational Exponents

The Laws of Exponents presented in the Chapter 5 applied to integer exponents. These laws apply to rational exponents as well.

Work Smart: Study Skills

When learning new rules, it is often helpful to write each rule on a notecard that you can study regularly. For example, you could put a^0 on one side of a notecard and its value of 1 on the other side. This will allow you to quiz yourself. There is also a free app called Quizlet that can be used.

The Laws of Exponents

If a and b are real numbers and if r and s are rational numbers, then, assuming the expression is defined, the following rules apply.

Zero-Exponent Rule:	$a^0 = 1$	if $a \neq 0$
Negative-Exponent Rule:	$a^{-r} = \dfrac{1}{a^r}$	if $a \neq 0$
Product Rule:	$a^r \cdot a^s = a^{r+s}$	
Quotient Rule:	$\dfrac{a^r}{a^s} = a^{r-s} = \dfrac{1}{a^{s-r}}$	if $a \neq 0$
Power Rule:	$(a^r)^s = a^{r \cdot s}$	
Product-to-a-Power Rule:	$(a \cdot b)^r = a^r \cdot b^r$	
Quotient-to-a-Power Rule:	$\left(\dfrac{a}{b}\right)^r = \dfrac{a^r}{b^r}$	if $b \neq 0$
Quotient-to-a-Negative-Power Rule:	$\left(\dfrac{a}{b}\right)^{-r} = \left(\dfrac{b}{a}\right)^r$	if $a \neq 0, b \neq 0$

Work Smart

We *simplify* expressions (no equal sign) and *solve* equations (these have equal signs).

Prepared?...Answers **P1.** $\dfrac{1}{z^3}, z \neq 0$

P2. x^3 **P3.** $8a^6b^3$ **P4.** 12

The direction **simplify** shall mean the following:

- All the exponents are positive. For example, x^{-3} is written as $\frac{1}{x^3}$ when simplified.
- Each base occurs only once. For example, $x^3y^2x^2$ is simplified as x^5y^2.
- There are no powers written to powers. For example, $(x^3)^2$ is simplified as x^6.

EXAMPLE 1   **Simplifying Expressions Involving Rational Exponents**

Simplify each of the following:

(a) $27^{\frac{1}{2}} \cdot 27^{\frac{5}{6}}$

(b) $\dfrac{8^{\frac{1}{3}}}{8^{\frac{5}{3}}}$

Solution

(a)

$$a^r \cdot a^s = a^{r+s}$$
$$\downarrow$$
$$27^{\frac{1}{2}} \cdot 27^{\frac{5}{6}} = 27^{\frac{1}{2}+\frac{5}{6}}$$
$$= 27^{\frac{3}{6}+\frac{5}{6}}$$
$$= 27^{\frac{8}{6}}$$
$$= 27^{\frac{4}{3}}$$
$$a^{\frac{m}{n}} = (\sqrt[n]{a})^m: \quad = (\sqrt[3]{27})^4$$
$$= 3^4$$
$$= 81$$

Work Smart

Example 1(b) could also be simplified as follows:

$$\dfrac{8^{\frac{1}{3}}}{8^{\frac{5}{3}}} = \dfrac{1}{8^{\frac{5}{3}-\frac{1}{3}}}$$
$$= \dfrac{1}{8^{\frac{4}{3}}}$$
$$= \dfrac{1}{(\sqrt[3]{8})^4}$$
$$= \dfrac{1}{16}$$

(b)

$$\dfrac{a^r}{a^s} = a^{r-s}$$
$$\downarrow$$
$$\dfrac{8^{\frac{1}{3}}}{8^{\frac{5}{3}}} = 8^{\frac{1}{3}-\frac{5}{3}}$$
$$= 8^{-\frac{4}{3}}$$
$$= \dfrac{1}{8^{\frac{4}{3}}}$$
$$a^{\frac{m}{n}} = (\sqrt[n]{a})^m: \quad = \dfrac{1}{(\sqrt[3]{8})^4}$$
$$= \dfrac{1}{2^4}$$
$$= \dfrac{1}{16}$$

EXAMPLE 2 **Simplifying Expressions Involving Rational Exponents**

Simplify each of the following. Assume all variables are positive.

(a) $\left(36^{\frac{2}{5}}\right)^{\frac{5}{4}}$

(b) $\left(x^{\frac{1}{2}} \cdot y^{\frac{2}{3}}\right)^{\frac{3}{2}}$

Solution

(a)
$$(a^r)^s = a^{r \cdot s}$$
$$\downarrow$$
$$\left(36^{\frac{2}{5}}\right)^{\frac{5}{4}} = 36^{\frac{2}{5} \cdot \frac{5}{4}}$$
$$= 36^{\frac{10}{20}}$$
$$= 36^{\frac{1}{2}}$$
$$a^{1/2} = \sqrt{a}: \quad = \sqrt{36}$$
$$= 6$$

(b)
$$(ab)^r = a^r \cdot b^r$$
$$\downarrow$$
$$\left(x^{\frac{1}{2}} \cdot y^{\frac{2}{3}}\right)^{\frac{3}{2}} = \left(x^{\frac{1}{2}}\right)^{\frac{3}{2}} \cdot \left(y^{\frac{2}{3}}\right)^{\frac{3}{2}}$$
$$(a^r)^s = a^{r \cdot s}: \quad = x^{\frac{1 \cdot 3}{2 \cdot 2}} \cdot y^{\frac{2 \cdot 3}{3 \cdot 2}}$$
$$= x^{\frac{3}{4}} y$$

> **Quick ✓**
>
> 1. If a and b are real numbers and if r is a rational number, then assuming the expression is defined, $(ab)^r = $ ___.
>
> 2. If a is a real number and if r and s are rational numbers, then assuming the expression is defined, $a^r \cdot a^s = $ ___.
>
> *In Problems 3–7, simplify each expression. Assume all variables are positive.*
>
> **3.** $5^{\frac{3}{4}} \cdot 5^{\frac{1}{6}}$
>
> **4.** $\dfrac{32^{\frac{6}{5}}}{32^{\frac{3}{5}}}$
>
> **5.** $\left(100^{\frac{3}{8}}\right)^{\frac{4}{3}}$
>
> **6.** $\left(a^{\frac{3}{2}} \cdot b^{\frac{5}{4}}\right)^{\frac{2}{3}}$
>
> **7.** $\dfrac{x^{\frac{1}{2}} \cdot x^{\frac{1}{3}}}{\left(x^{\frac{1}{12}}\right)^2}$

EXAMPLE 3 **Simplifying Expressions Involving Rational Exponents**

Simplify each of the following. Assume all variables are positive.

(a) $\left(x^{\frac{2}{3}} y^{-1}\right)\left(x^{-1} y^{\frac{1}{2}}\right)^{\frac{2}{3}}$

(b) $\left(\dfrac{9xy^{\frac{4}{3}}}{x^{\frac{5}{6}} y^{-\frac{2}{3}}}\right)^{\frac{1}{2}}$

Solution

(a)
Product-to-a-Power Rule: $(ab)^r = a^r b^r$
$$\downarrow$$
$$\left(x^{\frac{2}{3}} y^{-1}\right)\left(x^{-1} y^{\frac{1}{2}}\right)^{\frac{2}{3}} = x^{\frac{2}{3}} y^{-1} \left(x^{-1}\right)^{\frac{2}{3}} \left(y^{\frac{1}{2}}\right)^{\frac{2}{3}}$$
Power Rule: $(a^r)^s = a^{rs}: \quad = x^{\frac{2}{3}} y^{-1} x^{-\frac{2}{3}} y^{\frac{1}{3}}$
Product Rule: $a^r \cdot a^s = a^{r+s}: \quad = x^{\frac{2}{3} + \left(-\frac{2}{3}\right)} y^{-1 + \frac{1}{3}}$
$$= x^0 y^{-\frac{2}{3}}$$
$a^0 = 1$; Negative-Exponent Rule: $a^{-r} = \dfrac{1}{a^r}: \quad = \dfrac{1}{y^{\frac{2}{3}}}$

(b)
Quotient Rule: $\dfrac{a^r}{a^s} = a^{r-s}$
$$\downarrow$$
$$\left(\dfrac{9xy^{\frac{4}{3}}}{x^{\frac{5}{6}} y^{-\frac{2}{3}}}\right)^{\frac{1}{2}} = \left(9x^{1-\frac{5}{6}} y^{\frac{4}{3} - \left(-\frac{2}{3}\right)}\right)^{\frac{1}{2}}$$
$x^{1-\frac{5}{6}} = x^{\frac{6}{6} - \frac{5}{6}} = x^{\frac{1}{6}};$
$y^{\frac{4}{3} - \left(-\frac{2}{3}\right)} = y^{\frac{4}{3} + \frac{2}{3}} = y^{\frac{6}{3}} = y^2: \quad = \left(9x^{\frac{1}{6}} y^2\right)^{\frac{1}{2}}$
Product-to-a-Power Rule: $(a \cdot b)^r = a^r \cdot b^r: \quad = 9^{\frac{1}{2}} \cdot \left(x^{\frac{1}{6}}\right)^{\frac{1}{2}} \left(y^2\right)^{\frac{1}{2}}$
$9^{\frac{1}{2}} = \sqrt{9} = 3$; Power Rule: $(a^r)^s = a^{rs}: \quad = 3x^{\frac{1}{12}} y$

> **Quick ✔**
>
> *In Problems 8–10, simplify each expression. Assume all variables are positive.*
>
> **8.** $\left(8x^{\frac{3}{4}}y^{-1}\right)^{\frac{2}{3}}$
>
> **9.** $\left(\dfrac{25x^{\frac{1}{2}}y^{\frac{3}{4}}}{x^{-\frac{3}{4}}y}\right)^{\frac{1}{2}}$
>
> **10.** $8\left(125a^{\frac{3}{4}}b^{-1}\right)^{\frac{2}{3}}$

▶ ❷ Simplify Radical Expressions

Rational exponents can be used to simplify radicals.

EXAMPLE 4　**Simplifying Radicals Using Rational Exponents**

Use rational exponents to simplify the radicals. Assume all variables are positive.

(a) $\sqrt[8]{16^4}$　　**(b)** $\sqrt[3]{-64x^6y^3}$　　**(c)** $\dfrac{\sqrt{x}}{\sqrt[3]{x^2}}$　　**(d)** $\sqrt{\sqrt[3]{z}}$

Solution

The idea in all these problems is to rewrite the radical as an expression involving a rational exponent. Then use the Laws of Exponents to simplify the expression. Finally, write the simplified expression as a radical.

(a)　　　　Write radical as a rational exponent using $\sqrt[n]{a^m} = a^{\frac{m}{n}}$.

$$\sqrt[8]{16^4} = 16^{\frac{4}{8}}$$

Simplify the exponent:　$= 16^{\frac{1}{2}}$

Write rational exponent as a radical:　$= \sqrt{16}$

$$= 4$$

(b)　　　　$\sqrt[n]{a} = a^{\frac{1}{n}}$

$$\sqrt[3]{-64x^6y^3} = \left(-64x^6y^3\right)^{\frac{1}{3}}$$

$(ab)^r = a^r \cdot b^r$:　$= (-64)^{\frac{1}{3}}\left(x^6\right)^{\frac{1}{3}}\left(y^3\right)^{\frac{1}{3}}$

$(a^r)^s = a^{rs}$:　$= (-64)^{\frac{1}{3}} \cdot x^{6 \cdot \frac{1}{3}} \cdot y^{3 \cdot \frac{1}{3}}$

$(-64)^{\frac{1}{3}} = \sqrt[3]{-64} = -4$:　$= -4x^2y$

(c)　　　　$\sqrt{a} = a^{\frac{1}{2}}; \ \sqrt[n]{a^m} = a^{\frac{m}{n}}$

$$\frac{\sqrt{x}}{\sqrt[3]{x^2}} = \frac{x^{\frac{1}{2}}}{x^{\frac{2}{3}}}$$

$\dfrac{a^r}{a^s} = a^{r-s}$:　$= x^{\frac{1}{2}-\frac{2}{3}}$

$\dfrac{1}{2} - \dfrac{2}{3} = \dfrac{3}{6} - \dfrac{4}{6} = -\dfrac{1}{6}$:　$= x^{-\frac{1}{6}}$

$a^{-r} = \dfrac{1}{a^r}$:　$= \dfrac{1}{x^{\frac{1}{6}}}$

Write rational exponent as a radical:　$= \dfrac{1}{\sqrt[6]{x}}$

(d)　　　　Write radicand with a rational exponent.

$$\sqrt{\sqrt[3]{z}} = \sqrt{z^{\frac{1}{3}}}$$

$\sqrt{a} = a^{\frac{1}{2}}$:　$= \left(z^{\frac{1}{3}}\right)^{\frac{1}{2}}$

$(a^r)^s = a^{rs}$:　$= z^{\frac{1}{3} \cdot \frac{1}{2}}$

$$= z^{\frac{1}{6}}$$

Write rational exponent as a radical:　$= \sqrt[6]{z}$

●

Quick ✓

In Problems 11–14, use rational exponents to simplify each radical. Assume all variables are positive.

11. $\sqrt[10]{36^5}$

12. $\sqrt[4]{16a^8b^{12}}$

13. $\dfrac{\sqrt[3]{x^2}}{\sqrt[4]{x}}$

14. $\sqrt[4]{\sqrt[3]{a^2}}$

▶ ❸ Factor Expressions Containing Rational Exponents

Often, expressions involving rational exponents contain a common factor. When this occurs, factor out the common factor to write the expression in simplified form. The goal of this type of problem is to write the expression as either a single product or a single quotient. The following examples illustrate this idea.

EXAMPLE 5 **Writing an Expression Containing Rational Exponents as a Single Product**

Simplify $9x^{\frac{4}{3}} + 4x^{\frac{1}{3}}(3x + 5)$ by factoring out the common factor of both terms, $x^{\frac{1}{3}}$.

Solution

$x^{\frac{1}{3}}$ is a factor of the second term, $4x^{\frac{1}{3}}(3x + 5)$. It is also a factor of the first term, $9x^{\frac{4}{3}}$. This can be seen by rewriting $9x^{\frac{4}{3}}$ as $9x^{\frac{3}{3}+\frac{1}{3}} = 9x^{\frac{3}{3}} \cdot x^{\frac{1}{3}} = 9x \cdot x^{\frac{1}{3}}$. Now factor out $x^{\frac{1}{3}}$.

$$9x^{\frac{4}{3}} + 4x^{\frac{1}{3}}(3x + 5) = 9x \cdot x^{\frac{1}{3}} + 4x^{\frac{1}{3}}(3x + 5)$$

Factor out $x^{1/3}$: $= x^{\frac{1}{3}}(9x + 4(3x + 5))$

Distribute the 4: $= x^{\frac{1}{3}}(9x + 12x + 20)$

Combine like terms: $= x^{\frac{1}{3}}(21x + 20)$ ●

Work Smart

When factoring out the greatest common factor, factor out the variable expression raised to the largest exponent that the expressions have in common. For example,

$$3x^5 + 12x^2 = 3x^2(x^3 + 4)$$

Quick ✓

15. Simplify $8x^{\frac{3}{2}} + 3x^{\frac{1}{2}}(4x + 3)$ by factoring out the common factor, $x^{\frac{1}{2}}$.

EXAMPLE 6 **Writing an Expression Containing Rational Exponents as a Single Quotient**

Simplify $4x^{\frac{1}{2}} + x^{-\frac{1}{2}}(2x + 1)$ by factoring out the common factor, $x^{-\frac{1}{2}}$.

Solution

$x^{-\frac{1}{2}}$ is a factor of the second term, $x^{-\frac{1}{2}}(2x + 1)$. It is also a factor of the first term, $4x^{\frac{1}{2}}$. This can be seen by rewriting $4x^{\frac{1}{2}}$ as $4x^{\frac{2}{2}-\frac{1}{2}} = 4x^{\frac{2}{2}} \cdot x^{-\frac{1}{2}} = 4x \cdot x^{-\frac{1}{2}}$. Factor out $x^{-\frac{1}{2}}$ as the greatest common factor (GCF).

$$4x^{\frac{1}{2}} + x^{-\frac{1}{2}}(2x + 1) = 4x \cdot x^{-\frac{1}{2}} + x^{-\frac{1}{2}}(2x + 1)$$

Factor out $x^{-1/2}$: $= x^{-\frac{1}{2}}(4x + (2x + 1))$

Combine like terms: $= x^{-\frac{1}{2}}(6x + 1)$

Rewrite without negative exponents: $= \dfrac{6x + 1}{x^{\frac{1}{2}}}$ ●

Quick ✓

16. Simplify $9x^{\frac{1}{3}} + x^{-\frac{2}{3}}(3x + 1)$ by factoring out the common factor, $x^{-\frac{2}{3}}$.

9.3 Exercises MyMathLab®

Exercise numbers in green have complete video solutions in MyMathLab or may be accessed using the QR code to the right.

Problems 1–16 are the Quick ✔s that follow the EXAMPLES.

Building Skills

In Problems 17–38, simplify each of the following expressions. Assume all variables are positive. See Objective 1.

17. $5^{\frac{1}{2}} \cdot 5^{\frac{3}{2}}$

18. $3^{\frac{1}{3}} \cdot 3^{\frac{5}{3}}$

19. $\dfrac{8^{\frac{5}{4}}}{8^{\frac{1}{4}}}$

20. $\dfrac{10^{\frac{7}{5}}}{10^{\frac{2}{5}}}$

21. $2^{\frac{1}{3}} \cdot 2^{-\frac{3}{2}}$

22. $9^{-\frac{5}{4}} \cdot 9^{\frac{1}{3}}$

23. $\dfrac{x^{\frac{1}{4}}}{x^{\frac{5}{6}}}$

24. $\dfrac{y^{\frac{1}{5}}}{y^{\frac{9}{10}}}$

25. $(4^{\frac{4}{3}})^{\frac{3}{8}}$

26. $(9^{\frac{3}{5}})^{\frac{5}{6}}$

27. $(25^{\frac{3}{4}} \cdot 4^{-\frac{3}{4}})^2$

28. $(36^{-\frac{1}{4}} \cdot 9^{\frac{3}{4}})^{-2}$

29. $(x^{\frac{3}{4}} \cdot y^{\frac{1}{3}})^{\frac{2}{3}}$

30. $(a^{\frac{5}{4}} \cdot b^{\frac{3}{2}})^{\frac{2}{5}}$

31. $(x^{-\frac{1}{3}} \cdot y)(x^{\frac{1}{2}} \cdot y^{-\frac{4}{3}})$

32. $(a^{\frac{4}{3}} \cdot b^{-\frac{1}{2}})(a^{-2} \cdot b^{\frac{5}{2}})$

33. $(4a^2b^{-\frac{3}{2}})^{\frac{1}{2}}$

34. $(25p^{\frac{2}{5}}q^{-1})^{\frac{1}{2}}$

35. $\left(\dfrac{x^{\frac{2}{3}}y^{-\frac{1}{3}}}{8x^{\frac{1}{2}}y}\right)^{\frac{1}{3}}$

36. $\left(\dfrac{64m^{\frac{1}{2}}n}{m^{-2}n^{\frac{4}{3}}}\right)^{\frac{1}{2}}$

37. $\left(\dfrac{50x^{\frac{3}{4}}y}{2x^{\frac{1}{2}}}\right)^{\frac{1}{2}} + \left(\dfrac{x^{\frac{1}{2}}y^{\frac{1}{2}}}{9x^{\frac{3}{4}}y^{\frac{3}{2}}}\right)^{-\frac{1}{2}}$

38. $\left(\dfrac{27x^{\frac{1}{2}}y^{-1}}{y^{-\frac{2}{3}}x^{-\frac{1}{2}}}\right)^{\frac{1}{3}} - \left(\dfrac{4x^{\frac{1}{3}}y^{\frac{4}{9}}}{x^{-\frac{1}{3}}y^{\frac{2}{3}}}\right)^{\frac{1}{2}}$

In Problems 39–54, use rational exponents to simplify each radical. Assume all variables are positive. See Objective 2.

39. $\sqrt{x^8}$

40. $\sqrt[3]{x^6}$

41. $\sqrt[12]{8^4}$

42. $\sqrt[9]{125^6}$

43. $\sqrt[3]{-8a^3b^{12}}$

44. $\sqrt{25x^4y^6}$

45. $\dfrac{\sqrt{x}}{\sqrt[4]{x}}$

46. $\dfrac{\sqrt[3]{y^2}}{\sqrt[4]{y}}$

47. $\sqrt{x} \cdot \sqrt[3]{x}$

48. $\sqrt[4]{p^3} \cdot \sqrt[3]{p}$

49. $\sqrt{\sqrt[4]{x^3}}$

50. $\sqrt[3]{\sqrt{x^3}}$

51. $\sqrt{3} \cdot \sqrt[3]{9}$

52. $\sqrt{5} \cdot \sqrt[3]{25}$

53. $\dfrac{\sqrt{6}}{\sqrt[4]{36}}$

54. $\dfrac{\sqrt[4]{49}}{\sqrt{7}}$

For Problems 55–64, see Objective 3.

55. Simplify $5x^{\frac{3}{2}} + 3x^{\frac{1}{2}}(x+5)$ by factoring out the common factor, $x^{\frac{1}{2}}$.

56. Simplify $5x^{\frac{4}{3}} + 4x^{\frac{1}{3}}(2x-3)$ by factoring out the common factor, $x^{\frac{1}{3}}$.

57. Simplify $5(x+2)^{\frac{2}{3}}(3x-2) + 9(x+2)^{\frac{5}{3}}$ by factoring out the common factor, $(x+2)^{\frac{2}{3}}$.

58. Simplify $3(x-5)^{\frac{1}{2}}(3x+1) + 6(x-5)^{\frac{3}{2}}$ by factoring out the common factor, $(x-5)^{\frac{1}{2}}$.

59. Simplify $x^{-\frac{1}{2}}(2x+5) + 4x^{\frac{1}{2}}$ by factoring out the common factor, $x^{-\frac{1}{2}}$.

60. Simplify $x^{-\frac{2}{3}}(3x+2) + 9x^{\frac{1}{3}}$ by factoring out the common factor, $x^{-\frac{2}{3}}$.

61. Simplify $2(x-4)^{-\frac{1}{3}}(4x-3) + 12(x-4)^{\frac{2}{3}}$ by factoring out the common factor, $2(x-4)^{-\frac{1}{3}}$.

62. Simplify $4(x+3)^{\frac{1}{2}} + (x+3)^{-\frac{1}{2}}(2x+1)$ by factoring out the common factor, $(x+3)^{-\frac{1}{2}}$.

63. Simplify $15x(x^2+4)^{\frac{1}{2}} + 5(x^2+4)^{\frac{3}{2}}$.

64. Simplify $24x(x^2-1)^{\frac{1}{3}} + 9(x^2-1)^{\frac{4}{3}}$.

Mixed Practice

In Problems 65–76, simplify each expression. Assume all variables are positive.

65. $\sqrt[8]{4^4}$

66. $\sqrt[6]{27^2}$

67. $(-2)^{\frac{1}{2}} \cdot (-2)^{\frac{3}{2}}$

68. $25^{\frac{3}{4}} \cdot 25^{\frac{3}{4}}$

69. $(100^{\frac{1}{3}})^{\frac{3}{2}}$

70. $(8^4)^{\frac{5}{12}}$

71. $(\sqrt[4]{25})^2$

72. $(\sqrt[6]{27})^2$

73. $\sqrt[4]{x^2} - \dfrac{\sqrt[4]{x^6}}{x}$

74. $\sqrt[9]{a^6} - \dfrac{\sqrt[6]{a^5}}{\sqrt[6]{a}}$

75. $(4 \cdot 9^{\frac{1}{4}})^{-2}$

76. $(4^{-1} \cdot 81^{\frac{1}{2}})^{\frac{1}{2}}$

In Problems 77–82, distribute and simplify.

77. $x^{\frac{1}{2}}(x^{\frac{3}{2}} - 2)$

78. $x^{\frac{1}{3}}(x^{\frac{5}{3}} + 4)$

79. $2y^{-\frac{1}{3}}(1 + 3y)$

80. $3a^{-\frac{1}{2}}(2 - a)$

81. $4z^{\frac{3}{2}}(z^{\frac{3}{2}} - 8z^{-\frac{3}{2}})$

82. $8p^{\frac{2}{3}}(p^{\frac{4}{3}} - 4p^{-\frac{2}{3}})$

Applying the Concepts

83. If $3^x = 25$, what does $3^{\frac{x}{2}}$ equal?

84. If $5^x = 64$, what does $5^{\frac{x}{3}}$ equal?

85. If $7^x = 9$, what does $\sqrt{7^x}$ equal?

86. If $5^x = 27$, what does $\sqrt[3]{5^x}$ equal?

Extending the Concepts

In Problems 87 and 88, simplify the expression using rational exponents.

87. $\sqrt[4]{\sqrt[3]{\sqrt{x}}}$

88. $\sqrt[5]{\sqrt[3]{\sqrt{x^2}}}$

89. Without using technology, determine the value of $(6^{\sqrt{2}})^{\sqrt{2}}$.

90. Determine the domain of $g(x) = (x - 3)^{\frac{1}{2}}(x - 1)^{-\frac{1}{2}}$.

91. Determine the domain of $f(x) = (x + 3)^{\frac{1}{2}}(x + 1)^{-\frac{1}{2}}$.

Synthesis Review

In Problems 92–95, simplify each epression.

92. $(2x - 1)(x + 4) - (x + 1)(x - 1)$

93. $3a(a - 3) + (a + 3)(a - 2)$

94. $\dfrac{\sqrt{x^2 + 4x + 4}}{x + 2}, x + 2 > 0$

95. $\dfrac{x^2 - 4}{x + 2}(x + 5) - (x + 4)(x - 1)$

9.4 Simplifying Radical Expressions Using Properties of Radicals

Objectives

1 Use the Product Property to Multiply Radical Expressions

2 Use the Product Property to Simplify Radical Expressions

3 Use the Quotient Property to Simplify Radical Expressions

4 Multiply Radicals with Unlike Indices

Are You Prepared for This Section?

Before getting started, complete the following problems. If you get a problem wrong, go back to the section cited and review the material.

P1. List the perfect squares that are less than 200.

P2. List the perfect cubes that are less than 200.

P3. Simplify: **(a)** $\sqrt{16}$ **(b)** $\sqrt{p^2}$ [Section 9.1, pp. 616–619]

Prepared?...Answers P1. 1, 4, 9, 16, 25, 36, 49, 64, 81, 100, 121, 144, 169, 196 **P2.** 1, 8, 27, 64, 125 **P3. (a)** 4 **(b)** $|p|$

1 Use the Product Property to Multiply Radical Expressions

Perhaps you are noticing a trend: After a new algebraic expression is introduced, how to multiply, divide, add, and subtract the algebraic expression is explained. Well, here we go again! We begin with multiplying radical expressions when they have the same index.

Consider that

$$\sqrt{4 \cdot 25} = \sqrt{100} = 10 \quad \text{and} \quad \sqrt{4} \cdot \sqrt{25} = 2 \cdot 5 = 10$$

This suggests the following result:

In Other Words

$\sqrt[n]{a} \cdot \sqrt[n]{b} = \sqrt[n]{ab}$ means the product of the roots equals the root of the product, provided the index is the same.

Product Property Of Radicals

If $\sqrt[n]{a}$ and $\sqrt[n]{b}$ are real numbers and $n \geq 2$ is an integer, then

$$\sqrt[n]{a} \cdot \sqrt[n]{b} = \sqrt[n]{ab}$$

This formula can be justified using rational exponents.

$$\sqrt[n]{a} \cdot \sqrt[n]{b} = a^{\frac{1}{n}} \cdot b^{\frac{1}{n}}$$

Product-to-a-Power Rule: $= (a \cdot b)^{\frac{1}{n}}$

$a^{\frac{1}{n}} = \sqrt[n]{a}: = \sqrt[n]{a \cdot b}$

EXAMPLE 1 **Using the Product Property to Multiply Radicals**

Multiply.

(a) $\sqrt{5} \cdot \sqrt{3}$ **(b)** $\sqrt[3]{2} \cdot \sqrt[3]{13}$ **(c)** $\sqrt{x - 3} \cdot \sqrt{x + 3}$ **(d)** $\sqrt[5]{6c} \cdot \sqrt[5]{7c^2}$

Solution

(a) $\sqrt{5} \cdot \sqrt{3} = \sqrt{5 \cdot 3} = \sqrt{15}$

(b) $\sqrt[3]{2} \cdot \sqrt[3]{13} = \sqrt[3]{2 \cdot 13} = \sqrt[3]{26}$

(c) $\sqrt{x - 3} \cdot \sqrt{x + 3} = \sqrt{(x - 3)(x + 3)} = \sqrt{x^2 - 9}$ if $x \geq 3$

(d) $\sqrt[5]{6c} \cdot \sqrt[5]{7c^2} = \sqrt[5]{6c \cdot 7c^2} = \sqrt[5]{42c^3}$

Work Smart

In Example 1(c), notice that $\sqrt{x^2 - 9}$ does *not* equal $\sqrt{x^2} - \sqrt{9}$.

Quick ✓

1. If $\sqrt[n]{a}$ and $\sqrt[n]{b}$ are real numbers and $n \geq 2$ is an integer, then $\sqrt[n]{a} \cdot \sqrt[n]{b} =$ _____.

In Problems 2–5, multiply each radical expression.

2. $\sqrt{11} \cdot \sqrt{7}$ 3. $\sqrt[4]{6} \cdot \sqrt[4]{7}$ 4. $\sqrt{x-5} \cdot \sqrt{x+5}, x \geq 5$ 5. $\sqrt[7]{5p} \cdot \sqrt[7]{4p^3}$

▶ ❷ Use the Product Property to Simplify Radical Expressions

So far, radicals have been simplified only when the radicand was a perfect square, such as $\sqrt{81} = 9$, or a perfect cube, such as $\sqrt[3]{\dfrac{1}{8}} = \dfrac{1}{2}$. When a radical does not simplify to a rational number, one of two things can be done:

1. Write a decimal approximation of the radical.

2. Simplify the radical using properties of radicals, if possible.

Using technology to find the decimal approximation of radicals was illustrated in Sections 9.1 and 9.2. Now how to use properties of radicals to write the radical in simplified form will be presented.

The advantage of simplifying over approximating is that simplifying maintains an *exact* radical rather than an approximate value.

Recall that a number that is the square of a rational number is called a perfect square. Thus $1^2 = 1, 2^2 = 4, 3^2 = 9$, and so on are perfect squares. A number that is the cube of a rational number is called a perfect cube. Thus $1^3 = 1, 2^3 = 8, 3^3 = 27$, $(-1)^3 = -1, (-2)^3 = -8$, and so on are perfect cubes. In general, if n is the index of a radical, then a^n is a perfect power of a, where a is a rational number.

Work Smart

index $\rightarrow \sqrt[n]{a} \leftarrow$ radicand

A radical expression with index n is **simplified** if the radicand does not contain any factors that are perfect nth powers. For example, $\sqrt{50}$ is not simplified because 25 is a factor of 50, and 25 is a perfect square; $\sqrt[3]{16}$ is not simplified because 8 is a factor of 16, and 8 is a perfect cube; $\sqrt[3]{x^5}$ is not simplified because $x^5 = x^3 \cdot x^2$, and x^3 is a perfect cube.

To simplify radicals, use the Product Property of Radicals "in reverse." That is, use $\sqrt[n]{ab} = \sqrt[n]{a} \cdot \sqrt[n]{b}$.

EXAMPLE 2   **How to Use the Product Property to Simplify a Radical**

Simplify: $\sqrt{18}$

Step-by-Step Solution

Step 1: What is the index on the radical? Because the index is 2, write each factor of the radicand as the product of two factors, one of which is a perfect square.

The perfect squares are 1, 4, 9, 16, 25, ...
Because 9 is a factor of 18, and 9 is a perfect square, write 18 as $9 \cdot 2$: $\sqrt{18} = \sqrt{9 \cdot 2}$

Step 2: Write the radical as the product of two radicals, one of which contains a perfect square.

$\sqrt{ab} = \sqrt{a} \cdot \sqrt{b}$: $= \sqrt{9} \cdot \sqrt{2}$

Step 3: Take the square root of each perfect square. $= 3\sqrt{2}$ ●

The steps used in Example 2 are summarized below.

Work Smart

When performing Step 1 with real numbers, look for the *largest* factor of the radicand that is a perfect power of the index.

> **Simplifying a Radical Expression of the Form $\sqrt[n]{a}$**
>
> **Step 1:** Write each factor of the radicand as the product of two factors, one of which is a perfect nth power.
>
> **Step 2:** Using the Product Property of Radicals, write the radicand as the product of two radicals, one of which contains perfect nth powers.
>
> **Step 3:** Take the nth root of each perfect nth power.

EXAMPLE 3 **Using the Product Property to Simplify a Radical**

Simplify each of the following:

(a) $5\sqrt[3]{24}$　　　　　(b) $\sqrt{128x^2}$　　　　　(c) $\sqrt[4]{20}$

Solution

(a) First, notice that the index is 3. Therefore, find the largest factor of 24 that is a perfect cube. The positive perfect cubes are $1, 8, 27, \dots$. Because 8 is a factor of 24 and is a perfect cube, write 24 as $8 \cdot 3$.

$$5\sqrt[3]{24} = 5 \cdot \sqrt[3]{8 \cdot 3}$$
$$\sqrt[n]{ab} = \sqrt[n]{a} \cdot \sqrt[n]{b}: \quad = 5 \cdot \sqrt[3]{8} \cdot \sqrt[3]{3}$$
$$= 5 \cdot 2 \cdot \sqrt[3]{3}$$
$$= 10\sqrt[3]{3}$$

(b) The index is 2. Because 64 is a factor of 128, and 64 is a perfect square, write 128 as $64 \cdot 2$; x^2 is a perfect square.

$$\sqrt{128x^2} = \sqrt{64x^2 \cdot 2}$$
$$\sqrt[n]{ab} = \sqrt[n]{a} \cdot \sqrt[n]{b}: \quad = \sqrt{64x^2} \cdot \sqrt{2}$$
$$\sqrt{64x^2} = \sqrt{64} \cdot \sqrt{x^2} = 8|x|: \quad = 8|x|\sqrt{2}$$

(c) In $\sqrt[4]{20}$, the index is 4. The fourth powers (or perfect fourths) are $1, 16, 81, \dots$. There are no factors of 20 that are fourth powers, so the radical $\sqrt[4]{20}$ cannot be simplified.　●

> **Quick ✔**
>
> **6.** List the first six positive integers that are perfect squares.
>
> **7.** List the first six positive integers that are perfect cubes.
>
> *In Problems 8–11, simplify each of the radical expressions.*
>
> **8.** $\sqrt{48}$　　　　**9.** $4\sqrt[3]{54}$　　　　**10.** $\sqrt{200a^2}$　　　　**11.** $\sqrt[4]{40}$

EXAMPLE 4 **Simplifying an Expression Involving a Square Root**

Simplify: $\dfrac{4 - \sqrt{20}}{2}$

Solution

This expression can be simplified using two different approaches.

Method 1:

4 is the largest perfect square factor of 20
$$\frac{4 - \sqrt{20}}{2} = \frac{4 - \sqrt{4 \cdot 5}}{2}$$
$$\text{Use } \sqrt{a \cdot b} = \sqrt{a} \cdot \sqrt{b}: \quad = \frac{4 - \sqrt{4} \cdot \sqrt{5}}{2}$$
$$= \frac{4 - 2 \cdot \sqrt{5}}{2}$$
$$\text{Factor out the 2 in the numerator:} \quad = \frac{2(2 - \sqrt{5})}{2}$$
$$\text{Divide out common factor:} \quad = 2 - \sqrt{5}$$

Method 2:

4 is the largest perfect square factor of 20
$$\frac{4 - \sqrt{20}}{2} = \frac{4 - \sqrt{4 \cdot 5}}{2}$$
$$\text{Use } \sqrt{a \cdot b} = \sqrt{a} \cdot \sqrt{b}: \quad = \frac{4 - \sqrt{4} \cdot \sqrt{5}}{2}$$
$$= \frac{4 - 2 \cdot \sqrt{5}}{2}$$
$$\text{Use } \frac{A + B}{C} = \frac{A}{C} + \frac{B}{C}: \quad = \frac{4}{2} - \frac{2 \cdot \sqrt{5}}{2}$$
$$\text{Divide out common factor:} \quad = 2 - \sqrt{5}$$

●

Quick ✓

In Problems 12 and 13, simplify the expression.

12. $\dfrac{6 + \sqrt{45}}{3}$

13. $\dfrac{-2 + \sqrt{32}}{4}$

Recall how to simplify $\sqrt[n]{a^n}$ from Section 9.2.

Simplifying $\sqrt[n]{a^n}$

If $n \geq 2$ is a positive integer and a is a real number, then

$$\sqrt[n]{a^n} = |a| \quad \text{if } n \geq 2 \text{ is even}$$
$$\sqrt[n]{a^n} = a \quad \text{if } n \geq 3 \text{ is odd}$$

This means that

$$\sqrt{a^2} = |a| \quad \sqrt[3]{a^3} = a \quad \sqrt[4]{a^4} = |a| \quad \sqrt[5]{a^5} = a \quad \text{and so on.}$$

In order to make our mathematical lives a little easier, assume throughout this section that all variables in the radicand are greater than or equal to zero (nonnegative) unless stated otherwise. Thus

$$\sqrt{a^2} = a \quad \sqrt[3]{a^3} = a \quad \sqrt[4]{a^4} = a \quad \sqrt[5]{a^5} = a \quad \text{and so on.}$$

What if the exponent on the radicand is greater than the index, as in $\sqrt{x^3}$ or $\sqrt[3]{x^6}$? The Laws of Exponents along with the rule for simplifying $\sqrt[n]{a^n}$ could be used, or rational exponents could be used.

$$\sqrt[3]{x^6} = \sqrt[3]{(x^2)^3} = x^2 \quad \text{or} \quad \sqrt[3]{x^6} = x^{\frac{6}{3}} = x^2$$
$$\sqrt[3]{x^{12}} = \sqrt[3]{(x^4)^3} = x^4 \quad \text{or} \quad \sqrt[3]{x^{12}} = x^{\frac{12}{3}} = x^4$$

EXAMPLE 5 **Simplifying a Radical with a Variable Radicand**

Simplify $\sqrt{20x^{10}}$.

Solution

Because 4 is the largest perfect square factor of 20, write 20 as $4 \cdot 5$.

$$\sqrt{20x^{10}} = \sqrt{4 \cdot 5 \cdot x^{10}}$$
$$= \sqrt{4x^{10}} \cdot \sqrt{5}$$

$\sqrt{4} = 2, \sqrt{x^{10}} = \sqrt{(x^5)^2} = x^5$

or $\sqrt{x^{10}} = x^{\frac{10}{2}} = x^5$: $\quad = 2x^5\sqrt{5}$ ●

Quick ✓

14. Simplify $\sqrt{75a^6}$.

What if the index does not divide evenly into the exponent on the variable in the radicand, as in $\sqrt[3]{x^8}$? In this case, rewrite the variable expression as the product of two variable expressions where one of the factors has an exponent that is a multiple of the index:

3 divides evenly into 6: $\quad \sqrt[3]{x^8} = \sqrt[3]{x^6 \cdot x^2}$

So

$$\sqrt[3]{x^8} = \sqrt[3]{x^6 \cdot x^2}$$
$$= \sqrt[3]{x^6} \cdot \sqrt[3]{x^2}$$
$$= x^2\sqrt[3]{x^2}$$

EXAMPLE 6   **Simplifying Radicals**

Simplify:

(a) $\sqrt{80a^3}$

(b) $\sqrt[3]{27m^4n^{14}}$

Solution

(a)

$$80 = 16 \cdot 5; \, a^3 = a^2 \cdot a$$

$$\sqrt{80a^3} = \sqrt{16 \cdot 5 \cdot a^2 \cdot a}$$
$$= \sqrt{16a^2 \cdot 5a}$$
$\sqrt[n]{ab} = \sqrt[n]{a} \cdot \sqrt[n]{b}$: $= \sqrt{16a^2} \cdot \sqrt{5a}$
$\sqrt{a^2} = a$ assuming $a \geq 0$: $= 4a\sqrt{5a}$

(b)

$$m^4 = m^3 \cdot m; \quad n^{14} = n^{12} \cdot n^2$$

$$\sqrt[3]{27m^4n^{14}} = \sqrt[3]{27 \cdot m^3 \cdot m \cdot n^{12} \cdot n^2}$$
$$= \sqrt[3]{27 \cdot m^3 \cdot n^{12} \cdot m \cdot n^2}$$
$\sqrt[n]{ab} = \sqrt[n]{a} \cdot \sqrt[n]{b}$: $= \sqrt[3]{27m^3n^{12}} \cdot \sqrt[3]{mn^2}$
$\sqrt[3]{27} = 3; \, \sqrt[3]{m^3} = m; \, \sqrt[3]{n^{12}} = n^{\frac{12}{3}} = n^4$: $= 3mn^4\sqrt[3]{mn^2}$

Quick ✓

In Problems 15–17, simplify each radical.

15. $\sqrt{18a^5}$ **16.** $\sqrt[3]{-128x^6y^{10}}$ **17.** $\sqrt[4]{16a^5b^{11}}$

▶ **EXAMPLE 7**   **Multiplying and Simplifying Radicals**

Multiply and simplify:

(a) $\sqrt{3} \cdot \sqrt{15}$ (b) $3\sqrt[3]{4x} \cdot \sqrt[3]{2x^4}$ (c) $\sqrt[4]{27a^2b^5} \cdot \sqrt[4]{6a^3b^6}$

Solution

Remember, to multiply two radicals, the index must be the same. When radicals have the same index, multiply the radicands and then simplify the product.

(a) Notice that neither radical can be simplified. However, the index on both radicals is 2, so multiply the radicands.

$$\sqrt{3} \cdot \sqrt{15} = \sqrt{3 \cdot 15}$$
$$= \sqrt{45}$$
9 is the largest perfect square factor of 45: $= \sqrt{9 \cdot 5}$
Product Property of Radicals: $= \sqrt{9} \cdot \sqrt{5}$
$$= 3\sqrt{5}$$

(b) The index on both radicals is 3, so multiply the radicands.

$$3\sqrt[3]{4x} \cdot \sqrt[3]{2x^4} = 3\sqrt[3]{4x \cdot 2x^4}$$
$$= 3\sqrt[3]{8x^5}$$
x^3 is a perfect cube; 8 is a perfect cube: $= 3\sqrt[3]{8x^3 \cdot x^2}$
Product Property of Radicals: $= 3\sqrt[3]{8x^3} \cdot \sqrt[3]{x^2}$
$\sqrt[3]{8} = 2; \, \sqrt[3]{x^3} = x$: $= 3 \cdot 2 \cdot x \cdot \sqrt[3]{x^2}$
$$= 6x\sqrt[3]{x^2}$$

(c) The index on both radicals is 4, so multiply the radicands.

$$\sqrt[4]{27a^2b^5} \cdot \sqrt[4]{6a^3b^6} = \sqrt[4]{3^3 \cdot a^2b^5 \cdot 3 \cdot 2 \cdot a^3b^6}$$
$$= \sqrt[4]{3^4 \cdot 2 \cdot a^5b^{11}}$$

3^4, a^4, and b^8 are perfect fourths: $= \sqrt[4]{3^4 \cdot 2 \cdot a^4 \cdot a \cdot b^8 \cdot b^3}$

Product Property of Radicals: $= \sqrt[4]{3^4a^4b^8} \cdot \sqrt[4]{2ab^3}$

$$= 3ab^2\sqrt[4]{2ab^3}$$

Work Smart

Notice that $27 = 3^3$ and that $6 = 3 \cdot 2$, so that $27 \cdot 6 = 3^3 \cdot 3 \cdot 2 = 3^4 \cdot 2$. This makes finding the perfect fourth powers a lot easier!

Quick ✓

In Problems 18–20, multiply and simplify the radicals.

18. $\sqrt{6} \cdot \sqrt{8}$ **19.** $\sqrt[3]{12a^2} \cdot \sqrt[3]{10a^4}$ **20.** $4\sqrt[3]{8a^2b^5} \cdot \sqrt[4]{6a^2b^4}$

▶ ❸ Use the Quotient Property to Simplify Radical Expressions

Now consider that

$$\sqrt{\frac{64}{4}} = \sqrt{\frac{4 \cdot 16}{4}} = \sqrt{16} = 4 \quad \text{and} \quad \frac{\sqrt{64}}{\sqrt{4}} = \frac{8}{2} = 4$$

This suggests the following result:

In Other Words

$$\sqrt[n]{\frac{a}{b}} = \frac{\sqrt[n]{a}}{\sqrt[n]{b}}$$

means "the root of the quotient equals the quotient of the roots" provided that the radicals have the same index.

> **Quotient Property of Radicals**
>
> If $\sqrt[n]{a}$ and $\sqrt[n]{b}$ are real numbers, $b \neq 0$, and $n \geq 2$ is an integer, then
>
> $$\frac{\sqrt[n]{a}}{\sqrt[n]{b}} = \sqrt[n]{\frac{a}{b}}$$

This formula can be justified using rational exponents.

$$\frac{\sqrt[n]{a}}{\sqrt[n]{b}} = \frac{a^{\frac{1}{n}}}{b^{\frac{1}{n}}} = \left(\frac{a}{b}\right)^{\frac{1}{n}} = \sqrt[n]{\frac{a}{b}}$$

EXAMPLE 8   **Using the Quotient Property to Simplify Radicals**

Simplify:

(a) $\sqrt{\dfrac{18}{25}}$ **(b)** $\sqrt[3]{\dfrac{6z^3}{125}}$ **(c)** $\sqrt[4]{\dfrac{10a^2}{81b^4}}, b \neq 0$

Solution

In each of these problems, the expression in the denominator is a perfect nth power, and the index is n. Therefore, simplify the expression using the Quotient Rule "in reverse": $\sqrt[n]{\dfrac{a}{b}} = \dfrac{\sqrt[n]{a}}{\sqrt[n]{b}}$.

(a) $\sqrt{\dfrac{18}{25}} = \dfrac{\sqrt{18}}{\sqrt{25}}$

$= \dfrac{3\sqrt{2}}{5}$

(b) $\sqrt[3]{\dfrac{6z^3}{125}} = \dfrac{\sqrt[3]{6z^3}}{\sqrt[3]{125}}$

$= \dfrac{z\sqrt[3]{6}}{5}$

(c) $\sqrt[4]{\dfrac{10a^2}{81b^4}} = \dfrac{\sqrt[4]{10a^2}}{\sqrt[4]{81b^4}}$

$= \dfrac{\sqrt[4]{10a^2}}{3b}$

Quick ✓

In Problems 21–23, simplify the radicals.

21. $\sqrt{\dfrac{13}{49}}$ **22.** $\sqrt[3]{\dfrac{27p^3}{8}}$ **23.** $\sqrt[4]{\dfrac{3q^4}{16}}$

EXAMPLE 9 **Using the Quotient Property to Simplify Radicals**

Simplify:

(a) $\dfrac{\sqrt{24a^3}}{\sqrt{6a}}$ (b) $\dfrac{-2\sqrt[3]{54a}}{\sqrt[3]{2a^4}}$ (c) $\dfrac{\sqrt[3]{-375x^2y}}{\sqrt[3]{3x^{-1}y^7}}$

Solution

In these problems, the radical expression in the denominator cannot be simplified. However, the index on the numerator and denominator of each expression is the same, so write each expression as a single radical.

(a) $\dfrac{\sqrt{24a^3}}{\sqrt{6a}} = \sqrt{\dfrac{24a^3}{6a}}$

$= \sqrt{4a^2}$

$= 2a$

(b) $\dfrac{-2\sqrt[3]{54a}}{\sqrt[3]{2a^4}} = -2 \cdot \sqrt[3]{\dfrac{54a}{2a^4}}$

$= -2 \cdot \sqrt[3]{\dfrac{27}{a^3}}$

$= -2 \cdot \dfrac{3}{a}$

$= -\dfrac{6}{a}$

(c) $\dfrac{\sqrt[3]{-375x^2y}}{\sqrt[3]{3x^{-1}y^7}} = \sqrt[3]{\dfrac{-375x^2y}{3x^{-1}y^7}}$

$= \sqrt[3]{\dfrac{-125x^{2-(-1)}}{y^{7-1}}}$

$= \sqrt[3]{\dfrac{-125x^3}{y^6}}$

$= -\dfrac{5x}{y^2}$

Quick ✓

In Problems 24–26, simplify the radicals.

24. $\dfrac{\sqrt{12a^5}}{\sqrt{3a}}$ 25. $\dfrac{\sqrt[3]{-24x^2}}{\sqrt[3]{3x^{-1}}}$ 26. $\dfrac{\sqrt[3]{250a^5b^{-2}}}{\sqrt[3]{2ab}}$

❹ Multiply Radicals with Unlike Indices

The rule $\sqrt[n]{a} \cdot \sqrt[n]{b} = \sqrt[n]{ab}$ only works when the index on each radical is the same. What if the indices on the radicals are different? Can the product still be simplified? The answer is yes! Use the fact that $\sqrt[n]{a} = a^{\frac{1}{n}}$.

EXAMPLE 10 **Multiplying Radicals with Unlike Indices**

Multiply and simplify: $\sqrt[4]{8} \cdot \sqrt[3]{5}$

Solution

Notice that the indices are not the same, so $\sqrt[n]{a} \cdot \sqrt[n]{b} = \sqrt[n]{ab}$ cannot be used. Use rational exponents along with $\sqrt[n]{a} = a^{\frac{1}{n}}$ instead.

$\sqrt[4]{8} \cdot \sqrt[3]{5} = 8^{\frac{1}{4}} \cdot 5^{\frac{1}{3}}$

LCD = 12: $\qquad = 8^{\frac{3}{12}} \cdot 5^{\frac{4}{12}}$

$a^{\frac{r}{s}} = (a^r)^{\frac{1}{s}}$: $\qquad = \left[(8^3)^{\frac{1}{12}} \cdot (5^4)^{\frac{1}{12}} \right]$

$a^r \cdot b^r = (ab)^r$: $\qquad = \left[(8^3)(5^4) \right]^{\frac{1}{12}}$

$\qquad = (320{,}000)^{\frac{1}{12}}$

$a^{\frac{1}{n}} = \sqrt[n]{a}$: $\qquad = \sqrt[12]{320{,}000}$

Quick ✓

In Problems 27 and 28, multiply and simplify.

27. $\sqrt[4]{5} \cdot \sqrt[3]{3}$

28. $\sqrt{10} \cdot \sqrt[3]{12}$

9.4 Exercises   MyMathLab®

Exercise numbers in green have complete video solutions in MyMathLab or may be accessed using the QR code to the right.

Problems **1–28** *are the* **Quick ✓** *s that follow the* **EXAMPLES**.

Building Skills

In Problems 29–36, use the Product Property to multiply. Assume that all variables can be any real number unless stated otherwise. See Objective 1.

29. $\sqrt[3]{6} \cdot \sqrt[3]{10}$

30. $\sqrt[3]{-5} \cdot \sqrt[3]{7}$

31. $\sqrt{3a} \cdot \sqrt{5b}$

32. $\sqrt[4]{6a^2} \cdot \sqrt[4]{7b^2}$

33. $\sqrt{x-7} \cdot \sqrt{x+7}, x \geq 7$

34. $\sqrt{p-5} \cdot \sqrt{p+5}, p \geq 5$

35. $\sqrt{\dfrac{5x}{3}} \cdot \sqrt{\dfrac{3}{x}}, x > 0$

36. $\sqrt[3]{\dfrac{-9x^2}{4}} \cdot \sqrt[3]{\dfrac{4}{3x}}, x \neq 0$

In Problems 37–64, simplify each radical using the Product Property. Assume that all variables can be any real number. See Objective 2.

37. $\sqrt{50}$

38. $\sqrt{32}$

39. $\sqrt[3]{54}$

40. $\sqrt[4]{162}$

41. $\sqrt{48x^2}$

42. $\sqrt{20a^2}$

43. $\sqrt[3]{-27x^3}$

44. $\sqrt[3]{-64p^3}$

45. $\sqrt[4]{32m^4}$

46. $\sqrt[4]{48z^4}$

47. $\sqrt{12p^2q}$

48. $\sqrt{45m^2n}$

49. $\sqrt{162m^4}$

50. $\sqrt{98w^8}$

51. $\sqrt{y^{13}}$

52. $\sqrt{s^9}$

53. $\sqrt[3]{c^8}$

54. $\sqrt[5]{x^{12}}$

55. $\sqrt{125p^3q^4}$

56. $\sqrt{243ab^5}$

57. $\sqrt[3]{-16x^9}$

58. $\sqrt[3]{-54q^{12}}$

59. $\sqrt[5]{-16m^8n^2}$

60. $\sqrt{75x^6y}$

61. $\sqrt[4]{(x-y)^5}, x > y$

62. $\sqrt[3]{(a+b)^5}$

63. $\sqrt[3]{8x^3 - 8y^3}$

64. $\sqrt[3]{8a^3 + 8b^3}$

In Problems 65–70, simplify each expression. See Objective 2.

65. $\dfrac{4 + \sqrt{36}}{2}$

66. $\dfrac{5 - \sqrt{100}}{5}$

67. $\dfrac{9 + \sqrt{18}}{3}$

68. $\dfrac{10 - \sqrt{75}}{5}$

69. $\dfrac{7 - \sqrt{98}}{14}$

70. $\dfrac{-6 + \sqrt{108}}{6}$

In Problems 71–88, multiply and simplify. Assume that all variables are greater than or equal to zero. See Objective 2.

71. $\sqrt{5} \cdot \sqrt{5}$

72. $\sqrt{6} \cdot \sqrt{6}$

73. $\sqrt{2} \cdot \sqrt{8}$

74. $\sqrt{3} \cdot \sqrt{12}$

75. $\sqrt[3]{4} \cdot \sqrt[3]{2}$

76. $\sqrt[3]{9} \cdot \sqrt[3]{3}$

77. $\sqrt{5x} \cdot \sqrt{15x}$

78. $\sqrt{6x} \cdot \sqrt{30x}$

79. $\sqrt[3]{4b^2} \cdot \sqrt[3]{6b^2}$

80. $\sqrt[3]{9a} \cdot \sqrt[3]{6a^2}$

81. $2\sqrt{6ab} \cdot 3\sqrt{15ab^3}$

82. $3\sqrt{14pq^3} \cdot 2\sqrt{7pq}$

83. $\sqrt[4]{27p^3q^2} \cdot \sqrt[4]{12p^2q^2}$

84. $\sqrt[3]{16m^2n} \cdot \sqrt[3]{27m^2n}$

85. $\sqrt[5]{-8a^3b^4} \cdot \sqrt[5]{12a^3b}$

86. $\sqrt[5]{-27x^4y^2} \cdot \sqrt[5]{18x^3y^4}$

87. $\sqrt[4]{8(x-y)^2} \cdot \sqrt[4]{6(x-y)^3}, x > y$

88. $\sqrt[3]{9(a+b)^2} \cdot \sqrt[3]{6(a+b)^5}$

In Problems 89–96, simplify each expression. Assume that all variables are greater than zero. See Objective 3.

89. $\sqrt{\dfrac{3}{16}}$

90. $\sqrt{\dfrac{5}{36}}$

91. $\sqrt[4]{\dfrac{5x^4}{16}}$

92. $\sqrt[4]{\dfrac{2a^8}{81}}$

93. $\sqrt{\dfrac{9y^2}{25x^2}}$

94. $\sqrt{\dfrac{4a^4}{81b^2}}$

95. $\sqrt[3]{\dfrac{-27x^9}{64y^{12}}}$

96. $\sqrt[5]{\dfrac{-32a^{15}}{243b^{10}}}$

In Problems 97–110, divide and simplify. Assume that all variables are greater than zero. See Objective 3.

97. $\dfrac{\sqrt{8}}{\sqrt{2}}$

98. $\dfrac{\sqrt{27}}{\sqrt{3}}$

99. $\dfrac{\sqrt[3]{128}}{\sqrt[3]{2}}$

100. $\dfrac{\sqrt[4]{64}}{\sqrt[4]{4}}$

101. $\dfrac{\sqrt{48a^3}}{\sqrt{6a}}$

102. $\dfrac{\sqrt{54y^5}}{\sqrt{3y}}$

103. $\dfrac{\sqrt{24a^5b}}{\sqrt{3ab^3}}$

104. $\dfrac{\sqrt{360m^7n^3}}{\sqrt{5mn^5}}$

105. $\dfrac{\sqrt{512a^7b}}{3\sqrt{2ab^3}}$

106. $\dfrac{\sqrt{375x^2y^7}}{10\sqrt{3y}}$

107. $\dfrac{\sqrt[3]{104a^5}}{\sqrt[3]{4a^{-1}}}$

108. $\dfrac{\sqrt[3]{-128x^8}}{\sqrt[3]{2x^{-1}}}$

109. $\dfrac{\sqrt{90x^3y^{-1}}}{\sqrt{2x^{-3}y}}$

110. $\dfrac{\sqrt{96a^5b^{-3}}}{\sqrt{3a^{-5}b}}$

In Problems 111–118, multiply and simplify. See Objective 4.

111. $\sqrt[3]{3} \cdot \sqrt[3]{4}$

112. $\sqrt[3]{2} \cdot \sqrt[3]{7}$

113. $\sqrt[3]{2} \cdot \sqrt[6]{3}$

114. $\sqrt[4]{3} \cdot \sqrt[8]{5}$

115. $\sqrt[3]{3} \cdot \sqrt[3]{18}$

116. $\sqrt[3]{6} \cdot \sqrt[3]{9}$

117. $\sqrt[4]{9} \cdot \sqrt[6]{12}$

118. $\sqrt[3]{8} \cdot \sqrt[10]{16}$

Mixed Practice

In Problems 119–134, perform the indicated operation and simplify. Assume all variables are greater than zero.

119. $\sqrt[3]{\dfrac{5x}{8}}$

120. $\sqrt[3]{\dfrac{7a^2}{64}}$

121. $\sqrt[3]{5a} \cdot \sqrt[3]{9a}$

122. $\sqrt[5]{8b^2} \cdot \sqrt[5]{3b}$

123. $\sqrt{72a^4}$

124. $\sqrt{24b^6}$

125. $\sqrt[3]{6a^2b} \cdot \sqrt[3]{9ab}$

126. $\sqrt[4]{8x^3y^2} \cdot \sqrt[4]{4x^2y^3}$

127. $\dfrac{\sqrt[3]{-32a}}{\sqrt[3]{2a^4}}$

128. $\dfrac{\sqrt[3]{-250p^2}}{\sqrt[3]{2p^5}}$

129. $-5\sqrt[3]{32m^3}$

130. $-7\sqrt[3]{250p^3}$

131. $\sqrt[3]{81a^4b^7}$

132. $\sqrt[5]{32p^7q^{11}}$

133. $\sqrt[3]{12} \cdot \sqrt[3]{18}$

134. $\sqrt[4]{8} \cdot \sqrt[4]{18}$

Applying the Concepts

△ **135. Length of a Line Segment** The length of the line segment joining the points $(2, 5)$ and $(-1, -1)$ is given by

$$\sqrt{(5 - (-1))^2 + (2 - (-1))^2}$$

(a) Plot the points in the Cartesian plane, and draw a line segment connecting the points.

(b) Express the length of the line segment as a radical in simplified form.

△ **136. Length of a Line Segment** The length of the line segment joining the points $(4, 2)$ and $(-2, 4)$ is given by

$$\sqrt{(4 - 2)^2 + (-2 - 4)^2}$$

(a) Plot the points in the Cartesian plane, and draw a line segment connecting the points.

(b) Express the length of the line segment as a radical in simplified form.

137. Revenue Growth Suppose that the annual revenue R (in millions of dollars) of a company after t years of operating is modeled by the function

$$R(t) = \sqrt[3]{\dfrac{t}{2}}$$

(a) Predict the revenue of the company after 8 years of operation.

(b) Predict the revenue of the company after 27 years of operation.

△ **138. Sphere** The radius r of a sphere whose volume is V is given by

$$r = \sqrt[3]{\dfrac{3V}{4\pi}}$$

(a) Write the radius of a sphere whose volume is 9 cubic centimeters as a radical in simplified form.

(b) Write the radius of a sphere whose volume is 32π cubic meters as a radical in simplified form.

Extending the Concepts

139. Suppose that $f(x) = \sqrt{2x}$ and $g(x) = \sqrt{8x^3}$.

(a) Find $(f \cdot g)(x)$.

(b) Evaluate $(f \cdot g)(3)$.

In Problems 140–143, evaluate the formula

$$x = \dfrac{-b \pm \sqrt{b^2 - 4ac}}{2a}$$

for the given values of a, b, and c. Note that the symbol $\pm$ is shorthand notation to indicate that there are two solutions. One solution is obtained when you add the quantity after the $\pm$ symbol, and another is obtained when you subtract. This formula can be used to solve any equation of the form $ax^2 + bx + c = 0, a \neq 0$.

140. $a = 1, b = 4, c = 1$

141. $a = 1, b = 6, c = 3$

142. $a = 2, b = 1, c = -1$

143. $a = 3, b = 4, c = -1$

Explaining the Concepts

144. Explain how you would simplify $\sqrt[3]{16a^5}$.

145. To use the Product Property to multiply radicals, what must be true about the index in each radical?

146. In Example 1(c) the restriction $x \geq 3$ was included.

(a) Find the domain of each factor in $(\sqrt{x + 3})(\sqrt{x - 3})$.

(b) The domain of the product, $\sqrt{x^2 - 9}$, is $x \leq -3$ or $x \geq 3$. Show that the product is a real number if $x = -4$ or $x = 4$ but is not a real number if $x = 0$. What conclusion can you draw about the domain of the product of two square roots?

Synthesis Review

In Problems 147–150, solve the following.

147. $4x + 3 = 13$

148. $2|7x - 1| + 4 = 16$

149. $\frac{3}{5}x + 2 \le 28$ **150.** $\frac{5}{2}|x + 1| + 1 \le 11$

151. How are the solutions in Problems 147 and 149 similar? How are the solutions in Problems 148 and 150 similar?

Technology Exercises

152. Exploration To understand the circumstances under which absolute value symbols are required when simplifying radicals, use technology to complete the following.

(a) Graph $Y_1 = \sqrt{x^2}$ and $Y_2 = x$. Do you think that $\sqrt{x^2} = x$? Now graph $Y_1 = \sqrt{x^2}$ and $Y_2 = |x|$. Do you think that $\sqrt{x^2} = |x|$?

(b) Graph $Y_1 = \sqrt[3]{x^3}$ and $Y_2 = x$. Do you think that $\sqrt[3]{x^3} = x$? Now graph $Y_1 = \sqrt[3]{x^3}$ and $Y_2 = |x|$. Do you think that $\sqrt[3]{x^3} = |x|$?

(c) Graph $Y_1 = \sqrt[4]{x^4}$ and $Y_2 = x$. Do you think that $\sqrt[4]{x^4} = x$? Now graph $Y_1 = \sqrt[4]{x^4}$ and $Y_2 = |x|$. Do you think that $\sqrt[4]{x^4} = |x|$?

(d) In your own words, make a generalization about $\sqrt[n]{x^n}$.

9.5 Adding, Subtracting, and Multiplying Radical Expressions

Objectives

1 Add or Subtract Radical Expressions

2 Multiply Radical Expressions

Are You Prepared for This Section?

Before getting started, complete the following problems. If you get a problem wrong, go back to the section cited and review the material.

P1. Add: $4y^3 - 2y^2 + 8y - 1 + (-2y^3 + 7y^2 - 3y + 9)$ [Section 5.1, p. 312]

P2. Subtract: $5z^2 + 6 - (3z^2 - 8z - 3)$ [Section 5.1, p. 313]

P3. Multiply: $(4x + 3)(x - 5)$ [Section 5.3, pp. 325–327]

P4. Multiply: $(2y - 3)(2y + 3)$ [Section 5.3, p. 327]

▶ 1 Add or Subtract Radical Expressions

Recall that a radical expression is an algebraic expression that contains a radical. **Like radicals** have the same index and the same radicand. For example, the expressions $3\sqrt{5}$ and $8\sqrt{5}$ contain like radicals because each radical has the same index, 2, and the same radicand, 5; $4\sqrt[3]{x - 4}$ and $10\sqrt[3]{x - 4}$ also contain like radicals because each radical has the same index, 3, and the same radicand, $x - 4$.

Just as like terms are added or subtracted using the Distributive Property, radical expressions can be added or subtracted by combining like radicals using the Distributive Property.

EXAMPLE 1   **Adding and Subtracting Radical Expressions**

Add or subtract, as indicated. Assume all variables are greater than or equal to zero.

(a) $5\sqrt{2x} + 9\sqrt{2x}$

(b) $3\sqrt[3]{10} + 7\sqrt[3]{10} - 5\sqrt[3]{10}$

Work Smart

To add or subtract radicals, *both* the index and the radicand must be the same.

Solution

(a) Both radicals have the same index, 2, and the same radicand, $2x$, so they are like radicals.

$$5\sqrt{2x} + 9\sqrt{2x} = (5 + 9)\sqrt{2x}$$

$$5 + 9 = 14: = 14\sqrt{2x}$$

(continued)

Prepared?...Answers

P1. $2y^3 + 5y^2 + 5y + 8$

P2. $2z^2 + 8z + 9$ **P3.** $4x^2 - 17x - 15$

P4. $4y^2 - 9$

(b) All three radicals have the same index, 3, and the same radicand, 10, so they are like radicals.

$$3\sqrt[3]{10} + 7\sqrt[3]{10} - 2\sqrt[3]{10} = (3 + 7 - 2)\sqrt[3]{10}$$
$$3 + 7 - 2 = 8: \quad = 8\sqrt[3]{10}$$

Quick ✓

1. Two radicals are ___________ if each radical has the same index and the same radicand.
2. *True or False* $\sqrt{7} + \sqrt{5} = \sqrt{7 + 5}$

In Problems 3 and 4, add or subtract, as indicated.

3. $9\sqrt{13y} + 4\sqrt{13y}$ 4. $\sqrt[4]{5} + 9\sqrt[4]{5} - 3\sqrt[4]{5}$

Sometimes, before adding or subtracting, the radicals must be simplified so that the radicands are the same.

EXAMPLE 2 Adding and Subtracting Radical Expressions

Add or subtract, as indicated. Assume all variables are greater than or equal to zero.

(a) $3\sqrt{12} + 7\sqrt{3}$ **(b)** $3x\sqrt{20x} - 7\sqrt{5x^3}$ **(c)** $3\sqrt{5} + 7\sqrt{13}$

Solution

(a) The index on each radical is the same, but the radicands are different. However, the radicals can be simplified to make the radicands the same.

$$\sqrt{12} = \sqrt{4}\cdot\sqrt{3}$$

$$\begin{aligned} 3\sqrt{12} + 7\sqrt{3} &= 3\sqrt{4}\cdot\sqrt{3} + 7\sqrt{3} \\ \sqrt{4} = 2: \quad &= 3\cdot 2\sqrt{3} + 7\sqrt{3} \\ &= 6\sqrt{3} + 7\sqrt{3} \\ \text{Factor out } \sqrt{3}: \quad &= (6 + 7)\sqrt{3} \\ &= 13\sqrt{3} \end{aligned}$$

(b) The index on each radical is the same, but the radicands are different. However, each square root can be simplified.

$$\sqrt{20x} = \sqrt{4}\cdot\sqrt{5x} \qquad \sqrt{5x^3} = \sqrt{x^2}\cdot\sqrt{5x}$$

$$\begin{aligned} 3x\sqrt{20x} - 7\sqrt{5x^3} &= 3x\cdot\sqrt{4}\cdot\sqrt{5x} - 7\cdot\sqrt{x^2}\cdot\sqrt{5x} \\ \text{Simplify radicals:} \quad &= 3x\cdot 2\cdot\sqrt{5x} - 7\cdot x\cdot\sqrt{5x} \\ \text{Multiply:} \quad &= 6x\sqrt{5x} - 7x\sqrt{5x} \\ \text{Factor out } \sqrt{5x}: \quad &= (6x - 7x)\sqrt{5x} \\ &= -x\sqrt{5x} \end{aligned}$$

(c) For $3\sqrt{5} + 7\sqrt{13}$, the index on each radical is the same, but the radicands are different and cannot be simplified to make the radicands the same. Therefore, the terms cannot be added.

Quick ✓

In Problems 5–7, add or subtract, as indicated. Assume all variables are greater than or equal to zero.

5. $4\sqrt{18} - 3\sqrt{8}$ 6. $-5x\sqrt[3]{54x} + 7\sqrt[3]{2x^4}$ 7. $7\sqrt{10} - 6\sqrt{3}$

EXAMPLE 3 **Adding or Subtracting Radical Expressions**

Add or subtract, as indicated.

(a) $\sqrt[3]{16x^4} - 7x\sqrt[3]{-2x} + \sqrt[3]{54x}$ (b) $3\sqrt[4]{m^4n} - 5m\sqrt[8]{n^2}, m \geq 0$

Solution

(a) The index on each radical is the same. The radicands are different but can be simplified to make the radicands the same.

$$\boxed{\sqrt[3]{16x^4} = \sqrt[3]{8x^3} \cdot \sqrt[3]{2x}} \quad \boxed{\sqrt[3]{-2x} = \sqrt[3]{-1} \cdot \sqrt[3]{2x}} \quad \boxed{\sqrt[3]{54x} = \sqrt[3]{27} \cdot \sqrt[3]{2x}}$$

$$\sqrt[3]{16x^4} - 7x\sqrt[3]{-2x} + \sqrt[3]{54x} = \sqrt[3]{8x^3} \cdot \sqrt[3]{2x} - 7x\sqrt[3]{-1} \cdot \sqrt[3]{2x} + \sqrt[3]{27} \cdot \sqrt[3]{2x}$$

$$\sqrt[3]{8x^3} = 2x; \sqrt[3]{-1} = -1; \sqrt[3]{27} = 3: \quad = 2x\sqrt[3]{2x} - 7x(-1)\sqrt[3]{2x} + 3\sqrt[3]{2x}$$

$$-7x(-1) = 7x: \quad = 2x\sqrt[3]{2x} + 7x\sqrt[3]{2x} + 3\sqrt[3]{2x}$$

$$\text{Factor out } \sqrt[3]{2x}: \quad = (2x + 7x + 3)\sqrt[3]{2x}$$

$$\text{Simplify:} \quad = (9x + 3)\sqrt[3]{2x}$$

$$\text{Factor out 3:} \quad = 3(3x + 1)\sqrt[3]{2x}$$

Work Smart: Study Skills

Contrast adding radicals with multiplying radicals:

Add: $3\sqrt{5} + 8\sqrt{5} = (3 + 8)\sqrt{5}$
$= 11\sqrt{5}$

Multiply:
$3\sqrt{5} \cdot 8\sqrt{5} = 3 \cdot 8 \cdot \sqrt{5} \cdot \sqrt{5}$
$= 24\sqrt{25}$
$= 24 \cdot 5$
$= 120$

Ask yourself these questions:
How must the radicals be "like" to be added?
How must the radicals be "like" to be multiplied?

(b) Here, the index on the radicals *and* the radicands are different. Start by dealing with the index using rational exponents to try to get the same index on both radicals.

$$\sqrt[n]{a^m} = a^{\frac{m}{n}}$$

$$3\sqrt[4]{m^4n} - 5m\sqrt[8]{n^2} = 3\sqrt[4]{m^4n} - 5m \cdot n^{\frac{2}{8}}$$

$$\text{Simplify the rational exponent:} \quad = 3\sqrt[4]{m^4n} - 5m \cdot n^{\frac{1}{4}}$$

$$\text{Rewrite as radical:} \quad = 3\sqrt[4]{m^4n} - 5m \cdot \sqrt[4]{n}$$

Now the index is the same, but the radicands are still different.

$$\sqrt[4]{m^4n} = \sqrt[4]{m^4} \cdot \sqrt[4]{n}: \quad = 3\sqrt[4]{m^4} \cdot \sqrt[4]{n} - 5m \cdot \sqrt[4]{n}$$

$$\text{Simplify:} \quad = 3m \cdot \sqrt[4]{n} - 5m \cdot \sqrt[4]{n}$$

$$\text{Factor out } \sqrt[4]{n}: \quad = (3m - 5m) \cdot \sqrt[4]{n}$$

$$\text{Simplify:} \quad = -2m\sqrt[4]{n}$$

Quick ✓

In Problems 8 and 9, add or subtract, as indicated. Assume all variables are greater than or equal to zero.

8. $\sqrt[3]{8z^4} - 2z\sqrt[3]{-27z} + \sqrt[3]{125z}$ **9.** $\sqrt{25m} - 3\sqrt[4]{m^2}$

---

❷ Multiply Radical Expressions

Now let's look at multiplying radical expressions involving more than one radical. Multiply these expressions the same way polynomials are multiplied.

EXAMPLE 4 **Multiplying Radical Expressions**

Multiply and simplify:

(a) $\sqrt{5}(3 - 4\sqrt{5})$ (b) $\sqrt[3]{2}(3 + \sqrt[3]{4})$ (c) $(3 + 2\sqrt{7})(2 - 3\sqrt{7})$

(continued)

Solution

(a) Use the Distributive Property and multiply each term in the parentheses by $\sqrt{5}$.

$$\sqrt{5}(3 - 4\sqrt{5}) = \sqrt{5}\cdot 3 - \sqrt{5}\cdot 4\sqrt{5}$$

Multiply radicals: $\quad = 3\sqrt{5} - 4\cdot\sqrt{25}$

$\sqrt{25} = 5$: $\quad = 3\sqrt{5} - 4\cdot 5$

Simplify: $\quad = 3\sqrt{5} - 20$

(b) Distribute $\sqrt[3]{2}$ to each term in the parentheses.

$$\sqrt[3]{2}(3 + \sqrt[3]{4}) = \sqrt[3]{2}\cdot 3 + \sqrt[3]{2}\cdot\sqrt[3]{4}$$

Multiply radicals: $\quad = 3\sqrt[3]{2} + \sqrt[3]{8}$

$\sqrt[3]{8} = 2$: $\quad = 3\sqrt[3]{2} + 2$

(c) Treat this just like the product of two binomials and multiply using FOIL.

$$\underset{\text{First}}{} \quad \underset{\text{Last}}{} \qquad \text{F} \qquad \text{O} \qquad \text{I} \qquad \text{L}$$

$$(3 + 2\sqrt{7})(2 - 3\sqrt{7}) = 3\cdot 2 - 3\cdot 3\sqrt{7} + 2\sqrt{7}\cdot 2 - 2\sqrt{7}\cdot 3\sqrt{7}$$

Inner / Outer

Multiply: $\quad = 6 - 9\sqrt{7} + 4\sqrt{7} - 6\sqrt{49}$

$\sqrt{49} = 7$: $\quad = 6 - 9\sqrt{7} + 4\sqrt{7} - 6\cdot 7$

$\quad = 6 - 9\sqrt{7} + 4\sqrt{7} - 42$

Simplify: $\quad = -36 - 5\sqrt{7}$

Work Smart

$-36 - 5\sqrt{7} \neq -41\sqrt{7}$

Do you know why?

> **Quick ✓**
>
> In Problems 10–12, multiply and simplify.
>
> **10.** $\sqrt{6}(3 - 5\sqrt{6})$ **11.** $\sqrt[3]{12}(3 - \sqrt[3]{2})$ **12.** $(2 - 7\sqrt{3})(5 + 4\sqrt{3})$

The formulas for perfect squares, $(A + B)^2 = A^2 + 2AB + B^2$ and $(A - B)^2 = A^2 - 2AB + B^2$, as well as the formula for the difference of two squares, $(A + B)(A - B) = A^2 - B^2$, can be used to multiply radicals.

▶ **EXAMPLE 5** **Multiplying Radical Expressions Involving Special Products**

Multiply and simplify.

(a) $(2\sqrt{3} + \sqrt{5})^2$ (b) $(3 + \sqrt{7})(3 - \sqrt{7})$

Solution

(a) Notice that $(2\sqrt{3} + \sqrt{5})^2$ is in the form $(A + B)^2$, so

$$(A + B)^2 = A^2 + 2\cdot A\cdot B + B^2$$
$$(2\sqrt{3} + \sqrt{5})^2 = (2\sqrt{3})^2 + 2\cdot 2\sqrt{3}\cdot\sqrt{5} + (\sqrt{5})^2$$

Multiply: $\quad = 4\sqrt{9} + 4\sqrt{15} + \sqrt{25}$

Simplify: $\quad = 4\cdot 3 + 4\sqrt{15} + 5$

Combine like terms: $\quad = 17 + 4\sqrt{15}$

Work Smart

Notice, in Example 5(a), that

$(2\sqrt{3} + \sqrt{5})^2 \neq (2\sqrt{3})^2 + (\sqrt{5})^2$

(b) Notice that $(3 + \sqrt{7})(3 - \sqrt{7})$ is in the form $(A + B)(A - B)$, so

$$(A + B)(A - B) = A^2 - B^2$$
$$(3 + \sqrt{7})(3 - \sqrt{7}) = 3^2 - (\sqrt{7})^2$$
$$= 9 - 7$$
$$= 2$$

Notice that the product found in Example 5(b) is an integer. There are no radicals in the product. Radical expressions such as $3 + \sqrt{7}$ and $3 - \sqrt{7}$ are called **conjugates** of each other. When square roots that are conjugates are multiplied, the result never contains a radical. This result plays a huge role in the next section.

> **Quick ✓**
>
> **13.** *True or False* $(\sqrt{a} + \sqrt{b})^2 = (\sqrt{a})^2 + (\sqrt{b})^2$
>
> **14.** The radical expressions $4 + \sqrt{5}$ and $4 - \sqrt{5}$ are examples of _____.
>
> **15.** *True or False* The conjugate of $-5 + \sqrt{2}$ is $5 - \sqrt{2}$.
>
> *In Problems 16–18, multiply and simplify.*
>
> **16.** $(2 - \sqrt{5})^2$ **17.** $(\sqrt{7} - 3\sqrt{2})^2$ **18.** $(\sqrt{3} + \sqrt{2})(\sqrt{3} - \sqrt{2})$

9.5 Exercises MyMathLab®

Exercise numbers in **green** have complete video solutions in MyMathLab or may be accessed using the QR code to the right.

*Problems **1–18** are the Quick ✓s that follow the **EXAMPLES**.*

Building Skills

In Problems 19–26, add or subtract as indicated. Assume all variables are greater than or equal to zero. See Objective 1.

19. $3\sqrt{2} + 7\sqrt{2}$ **20.** $6\sqrt{3} + 8\sqrt{3}$

21. $5\sqrt[3]{x} - 3\sqrt[3]{x}$ **22.** $12\sqrt[4]{z} - 5\sqrt[4]{z}$

23. $8\sqrt{5x} - 3\sqrt{5x} + 9\sqrt{5x}$

24. $4\sqrt[3]{3y} + 8\sqrt[3]{3y} - 10\sqrt[3]{3y}$

25. $4\sqrt[3]{5} - 3\sqrt{5} + 7\sqrt[3]{5} - 8\sqrt{5}$

26. $12\sqrt{7} + 5\sqrt[4]{7} - 5\sqrt{7} + 6\sqrt[4]{7}$

In Problems 27–48, add or subtract as indicated. Assume all variables are greater than or equal to zero. See Objective 1.

27. $\sqrt{8} + 6\sqrt{2}$ **28.** $6\sqrt{3} + \sqrt{12}$

29. $\sqrt[3]{24} - 4\sqrt[3]{3}$ **30.** $\sqrt[3]{32} - 5\sqrt[3]{4}$

31. $\sqrt[3]{54} - 7\sqrt[3]{128}$ **32.** $7\sqrt[4]{48} - 4\sqrt[4]{243}$

33. $5\sqrt{54x} - 3\sqrt{24x}$ **34.** $2\sqrt{48z} - \sqrt{75z}$

35. $2\sqrt{8} + 3\sqrt{10}$ **36.** $4\sqrt{12} + 2\sqrt{20}$

37. $\sqrt{12x^3} + 5x\sqrt{108x}$ **38.** $3\sqrt{63z^3} + 2z\sqrt{28z}$

39. $\sqrt{12x^2} + 3x\sqrt{2} - 2\sqrt{98x^2}$

40. $\sqrt{48y^2} - 4y\sqrt{12} + \sqrt{108y^2}$

41. $\sqrt[3]{-54x^3} + 3x\sqrt[3]{16} - 2\sqrt[3]{128}$

42. $2\sqrt[3]{-5x^3} + 4x\sqrt[3]{40} - \sqrt[3]{135}$

43. $\sqrt{9x - 9} + \sqrt{4x - 4}$

44. $\sqrt{4x + 12} - \sqrt{9x + 27}$

45. $\sqrt{16x} - \sqrt[6]{x^3}$ **46.** $\sqrt{25x} - \sqrt[4]{x^2}$

47. $\sqrt[3]{27x} + 2\sqrt[9]{x^3}$ **48.** $\sqrt[4]{16y} + \sqrt[8]{y^2}$

In Problems 49–82, multiply and simplify. Assume all variables are greater than or equal to zero. See Objective 2.

49. $\sqrt{3}(2 - 3\sqrt{2})$ **50.** $\sqrt{5}(5 + 3\sqrt{3})$

51. $\sqrt{3}(\sqrt{2} + \sqrt{6})$ **52.** $\sqrt{2}(\sqrt{5} - 2\sqrt{10})$

53. $\sqrt[3]{4}(\sqrt[3]{3} - \sqrt[3]{6})$ **54.** $\sqrt[3]{6}(\sqrt[3]{2} + \sqrt[3]{12})$

55. $\sqrt{2x}(3 - \sqrt{10x})$ **56.** $\sqrt{5x}(6 + \sqrt{15x})$

57. $(3 + \sqrt{2})(4 + \sqrt{3})$ **58.** $(5 + \sqrt{5})(3 + \sqrt{6})$

59. $(6 + \sqrt{3})(2 - \sqrt{7})$ **60.** $(7 - \sqrt{3})(6 + \sqrt{5})$

61. $(4 - 2\sqrt{7})(3 + 3\sqrt{7})$

62. $(9 + 5\sqrt{10})(1 - 3\sqrt{10})$

63. $(\sqrt{2} + 3\sqrt{6})(\sqrt{3} - 2\sqrt{2})$

64. $(2\sqrt{3} + \sqrt{10})(\sqrt{5} - 2\sqrt{2})$

65. $(2\sqrt{5} + \sqrt{3})(4\sqrt{5} - 3\sqrt{3})$

66. $(\sqrt{6} - 2\sqrt{2})(2\sqrt{6} + 3\sqrt{2})$

67. $(1 + \sqrt{3})^2$ **68.** $(2 - \sqrt{3})^2$

69. $(\sqrt{2} - \sqrt{5})^2$ **70.** $(\sqrt{7} - \sqrt{3})^2$

71. $(\sqrt{x} - \sqrt{2})^2$ **72.** $(\sqrt{z} + \sqrt{5})^2$

73. $(\sqrt{2} - 1)(\sqrt{2} + 1)$

74. $(\sqrt{3} - 1)(\sqrt{3} + 1)$

75. $(3 - 2\sqrt{5})(3 + 2\sqrt{5})$

76. $(6 + 3\sqrt{2})(6 - 3\sqrt{2})$

77. $(\sqrt{2x} + \sqrt{3y})(\sqrt{2x} - \sqrt{3y})$

78. $(\sqrt{5a} + \sqrt{7b})(\sqrt{5a} - \sqrt{7b})$

79. $(\sqrt[3]{x} + 4)(\sqrt[3]{x} - 3)$

80. $(\sqrt[3]{y} - 6)(\sqrt[3]{y} + 3)$

81. $(\sqrt[3]{2a} - 5)(\sqrt[3]{2a} + 5)$

82. $(\sqrt[3]{4p} - 1)(\sqrt[3]{4p} + 3)$

Mixed Practice

In Problems 83–106, perform the indicated operation and simplify. Assume all variables are greater than or equal to zero.

83. $\sqrt{5}(\sqrt{3} + \sqrt{10})$

84. $\sqrt{7}(\sqrt{14} + \sqrt{3})$

85. $\sqrt{28x^5} - x\sqrt{7x^3} + 5\sqrt{175x^5}$

86. $\sqrt{180a^5} + a^2\sqrt{20} - a\sqrt{80a^3}$

87. $(2\sqrt{3} + 5)(2\sqrt{3} - 5)$

88. $(4\sqrt{2} - 2)(4\sqrt{2} + 2)$

89. $\sqrt[3]{7}(2 + \sqrt[3]{4})$

90. $\sqrt[3]{9}(5 + 2\sqrt[3]{2})$

91. $(2\sqrt{2} + 5)(4\sqrt{2} - 4)$

92. $(5\sqrt{5} - 3)(3\sqrt{5} - 4)$

93. $4\sqrt{18} + 2\sqrt{32}$

94. $5\sqrt{20} + 2\sqrt{80}$

95. $(\sqrt{5} - \sqrt{3})^2 - \sqrt{60}$

96. $(\sqrt{2} - \sqrt{7})^2 - \sqrt{56}$

97. $3\sqrt[3]{5x^3y} + \sqrt[3]{40y}$

98. $5\sqrt[3]{3m^3n} + \sqrt[3]{81n}$

99. $(\sqrt{2x} - \sqrt{7y})(\sqrt{2x} + \sqrt{7y}) - 2\sqrt{x^2}$

100. $(\sqrt{3a} - \sqrt{4b})(\sqrt{3a} + \sqrt{4b}) + 4\sqrt{b^2}$

101. $(2 + \sqrt{x + 1})^2$

102. $(4 + \sqrt{2x + 3})^2$

103. $(5 - \sqrt{x + 2})^2$

104. $(1 - \sqrt{3x - 4})^2$

105. $-\dfrac{3}{5} \cdot \left(-\dfrac{\sqrt{5}}{5}\right) - \dfrac{4}{5} \cdot \left(-\dfrac{2\sqrt{5}}{5}\right)$

106. $\dfrac{4}{5} \cdot \left(-\dfrac{\sqrt{5}}{5}\right) + \left(-\dfrac{3}{5}\right)\left(-\dfrac{2\sqrt{5}}{5}\right)$

107. Suppose that $f(x) = \sqrt{3x}$ and $g(x) = \sqrt{12x}$; find
(a) $(f + g)(x)$ **(b)** $(f + g)(4)$
(c) $(f \cdot g)(x)$

108. Suppose that $f(x) = \sqrt{4x - 4}$ and $g(x) = \sqrt{25x - 25}$; find
(a) $(f + g)(x)$ **(b)** $(f + g)(10)$
(c) $(f \cdot g)(x)$

109. Show that $-2 + \sqrt{5}$ is a solution to the equation $x^2 + 4x - 1 = 0$. Show that $-2 - \sqrt{5}$ is also a solution.

110. Show that $3 + \sqrt{7}$ is a solution to the equation $x^2 - 6x + 2 = 0$. Show that $3 - \sqrt{7}$ is also a solution.

Applying the Concepts

In Problems 111 and 112, find the perimeter and area of the figures shown. Express your answer as a radical in simplified form.

111.

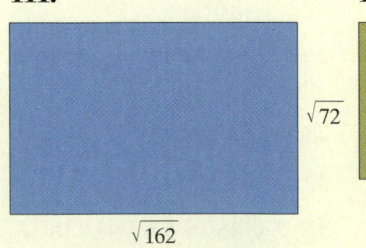

$\sqrt{72}$

$\sqrt{162}$

112.

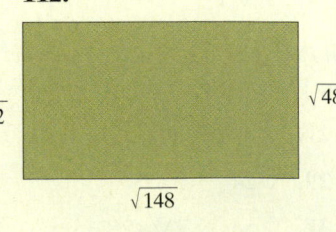

$\sqrt{48}$

$\sqrt{148}$

Problems 113 and 114, use **Heron's Formula** *for finding the area of a triangle whose sides are known. Heron's Formula states that the area A of a triangle with sides a, b, and c is*

$$A = \sqrt{s(s-a)(s-b)(s-c)}$$

where

$$s = \frac{1}{2}(a+b+c)$$

Find the area of the shaded region by computing the difference in the areas of the triangles. That is, compute "area of larger triangle minus area of smaller triangle." Write your answer as a radical in simplified form.

113.   **114.**

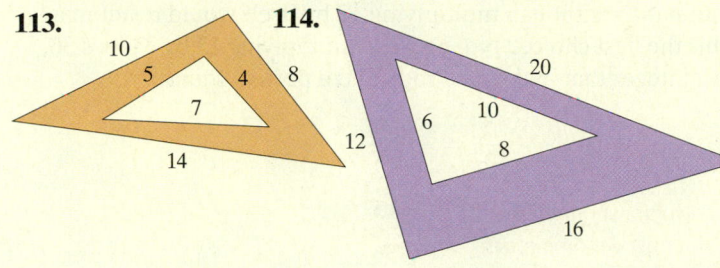

Explaining the Concepts

115. Explain how to add or subtract radicals.

116. Multiply $(\sqrt{a} - \sqrt{b})(\sqrt{a} + \sqrt{b})$ and provide a general result regarding the product of conjugates involving square roots.

Synthesis Review

In Problems 117–122, multiply each of the following.

117. $(3a^3b)(4a^2b^4)$ **118.** $(3p-1)(2p+3)$

119. $(3y+2)(2y-1)$ **120.** $(m-4)(m+4)$

121. $(5w+2)(5w-2)$ **122.** $(\sqrt{x}+2)(\sqrt{x}-2)$

9.6 Rationalizing Radical Expressions

Objectives

1 Rationalize a Denominator Containing One Term

2 Rationalize a Denominator Containing Two Terms

> **In Other Words**
> The process of "rationalizing the denominator" is called this because the denominator becomes a rational number (no radicals).

Are You Prepared for This Section?

Before getting started, complete the following problems. If you get a problem wrong, go back to the section cited and review the material.

P1. What would we need to multiply 12 by in order to make it the smallest perfect square that is a multiple of 12?

P2. Simplify: $\sqrt{25x^2}$, $x > 0$ [Section 9.4, pp. 635–639]

When radical expressions appear in the denominator of a quotient, it is customary to rewrite the quotient without radicals in the denominator. This process is called **rationalizing the denominator**. This section discusses how to rationalize denominators that contain one or two terms.

▶ **1** Rationalize a Denominator Containing One Term

To rationalize a denominator containing a single square root, multiply the numerator and denominator of the quotient by a square root so that the radicand in the denominator becomes a perfect square.

For example, if the denominator of a quotient contains $\sqrt{5}$, multiply the numerator and denominator by $\sqrt{5}$ because $5 \cdot 5 = 25$, a perfect square. If the denominator contains $\sqrt{8}$, multiply the numerator and denominator by $\sqrt{2}$ because $8 \cdot 2 = 16$, a perfect square. Because both the numerator and the denominator are being multiplied by the same value, the radical expression is being multiplied by 1, so the value of the radical expression is unchanged.

EXAMPLE 1   **Rationalizing a Denominator Containing a Square Root**

Rationalize the denominator of each expression:

(a) $\dfrac{1}{\sqrt{7}}$ (b) $\dfrac{\sqrt{5}}{\sqrt{12}}$ (c) $\dfrac{2}{3\sqrt{2x}}$, $x > 0$

Prepared?...Answers   P1. 3   **P2.** $5x$

(continued)

Solution

(a) The denominator contains $\sqrt{7}$. Ask, "What can I multiply 7 by to obtain a perfect square?" Because $7 \cdot 7 = 49$, a perfect square, multiply the numerator and denominator by $\sqrt{7}$.

$$\frac{1}{\sqrt{7}} = \frac{1}{\sqrt{7}} \cdot \frac{\sqrt{7}}{\sqrt{7}}$$

Multiply numerators;
multiply denominators: $= \dfrac{\sqrt{7}}{\sqrt{49}}$

$\sqrt{49} = 7$: $= \dfrac{\sqrt{7}}{7}$

> **Work Smart**
>
> Remember that $a = 1 \cdot a$ and that 1 can take many forms. In Example 1(a), $1 = \dfrac{\sqrt{7}}{\sqrt{7}}$.

(b) $\sqrt{12}$ is in the denominator. Although multiplying 12 by itself would result in a perfect square, is this the best choice? No, because multiplying 12 by 3 gives 36, which is the smallest integer that makes the radicand a perfect square.

$$\frac{\sqrt{5}}{\sqrt{12}} = \frac{\sqrt{5}}{\sqrt{12}} \cdot \frac{\sqrt{3}}{\sqrt{3}}$$

Multiply numerators;
multiply denominators: $= \dfrac{\sqrt{15}}{\sqrt{36}}$

$\sqrt{36} = 6$: $= \dfrac{\sqrt{15}}{6}$

> **Work Smart**
>
> An alternative approach to Example 1(b) would be to simplify $\sqrt{12}$ first, as follows:
>
> $$\frac{\sqrt{5}}{\sqrt{12}} = \frac{\sqrt{5}}{\sqrt{4 \cdot 3}}$$
> $$= \frac{\sqrt{5}}{2\sqrt{3}}$$
> $$= \frac{\sqrt{5}}{2\sqrt{3}} \cdot \frac{\sqrt{3}}{\sqrt{3}}$$
> $$= \frac{\sqrt{15}}{6}$$

(c)

Multiply by $1 = \dfrac{\sqrt{2x}}{\sqrt{2x}}$

$$\frac{2}{3\sqrt{2x}} = \frac{2}{3\sqrt{2x}} \cdot \frac{\sqrt{2x}}{\sqrt{2x}}$$

Multiply numerators;
multiply denominators: $= \dfrac{2\sqrt{2x}}{3\sqrt{4x^2}}$

$\sqrt{4x^2} = 2x$ because $x > 0$: $= \dfrac{2\sqrt{2x}}{3 \cdot 2x}$

Divide out the common factor, 2: $= \dfrac{\sqrt{2x}}{3x}$

Quick ✓

1. Rewriting a quotient to remove radicals from the denominator is called _____ ___ _____.

2. To rationalize the denominator of $\dfrac{\sqrt{5}}{\sqrt{11}}$, multiply the numerator and denominator by ____.

In Problems 3–5, rationalize each denominator.

3. $\dfrac{1}{\sqrt{3}}$ **4.** $\dfrac{\sqrt{5}}{\sqrt{8}}$ **5.** $\dfrac{5}{\sqrt{10x}}$

In general, to rationalize a denominator containing a single radical with index n, multiply the numerator and denominator of the quotient by a radical so that the product in the denominator has a radicand that is a perfect nth power. So, if the denominator contains a radical whose index is 3, multiply the numerator and denominator by a cube root so that the radicand in the denominator becomes a perfect cube. For example, if the denominator of a quotient contains $\sqrt[3]{4}$, multiply the numerator and denominator of the quotient by $\sqrt[3]{2}$ because $4 \cdot 2 = 8$, a perfect cube.

EXAMPLE 2 **Rationalizing a Denominator Containing Cube Roots and Fourth Roots**

Rationalize the denominator of each expression:

(a) $\dfrac{1}{\sqrt[3]{6}}$ **(b)** $\sqrt[3]{\dfrac{5}{18}}$ **(c)** $\dfrac{6}{\sqrt[4]{4z^3}}, z > 0$

Solution

(a) Notice that $\sqrt[3]{6}$ is in the denominator. To make the radicand in the denominator a perfect cube (because the index is 3), multiply the numerator and denominator by $\sqrt[3]{6^2} = \sqrt[3]{36}$ because $6 \cdot 36 = 6 \cdot 6^2 = 6^3 = 216$, a perfect cube.

$$\frac{1}{\sqrt[3]{6}} = \frac{1}{\sqrt[3]{6}} \cdot \frac{\sqrt[3]{6^2}}{\sqrt[3]{6^2}}$$

$$\text{Multiply numerators;} \qquad = \frac{\sqrt[3]{36}}{\sqrt[3]{6^3}}$$
$$\text{multiply denominators:}$$

$$\sqrt[3]{6^3} = 6: \qquad = \frac{\sqrt[3]{36}}{6}$$

Work Smart

The radicand in $\sqrt[3]{6}$ is 6, or 6^1. To make it a perfect cube, 6^3, multiply by 6^2. Then $6^1 \cdot 6^2 = 6^3$. The cube root of 6^3, or 216, is 6.

(b) First, use the Quotient Property $\left(\sqrt[n]{\dfrac{a}{b}} = \dfrac{\sqrt[n]{a}}{\sqrt[n]{b}} \right)$ to rewrite the radical as the quotient of two radicals.

$$\sqrt[3]{\frac{5}{18}} = \frac{\sqrt[3]{5}}{\sqrt[3]{18}}$$

What does $\sqrt[3]{18}$ need to be multiplied by to make the radicand a perfect cube? First, write 18 as $9 \cdot 2 = 3^2 \cdot 2^1$. If $3^2 \cdot 2^1$ is multiplied by $3^1 \cdot 2^2 = 12$, then the radicand will be a perfect cube in the denominator.

$$\sqrt[3]{\frac{5}{18}} = \frac{\sqrt[3]{5}}{\sqrt[3]{18}} = \frac{\sqrt[3]{5}}{\sqrt[3]{3^2 \cdot 2}} \cdot \frac{\sqrt[3]{3 \cdot 2^2}}{\sqrt[3]{3 \cdot 2^2}}$$

$$\text{Multiply numerators;} \qquad = \frac{\sqrt[3]{60}}{\sqrt[3]{3^3 \cdot 2^3}}$$
$$\text{multiply denominators:}$$

$$\sqrt[3]{3^3 \cdot 2^3} = 3 \cdot 2 = 6: \qquad = \frac{\sqrt[3]{60}}{6}$$

(c) Rewrite the denominator as $\sqrt[4]{2^2 \cdot z^3}$. To make the radicand a perfect fourth power, multiply $\sqrt[4]{2^2 \cdot z^3}$ by $\sqrt[4]{2^2 \cdot z}$ to obtain $\sqrt[4]{2^4 z^4}$ in the denominator.

$$\frac{6}{\sqrt[4]{4z^3}} = \frac{6}{\sqrt[4]{2^2 \cdot z^3}} \cdot \frac{\sqrt[4]{2^2 \cdot z}}{\sqrt[4]{2^2 \cdot z}}$$

$$\text{Multiply numerators;} \qquad = \frac{6\sqrt[4]{4z}}{\sqrt[4]{2^4 \cdot z^4}}$$
$$\text{multiply denominators:}$$

$$\sqrt[4]{2^4 \cdot z^4} = 2z: \qquad = \frac{6\sqrt[4]{4z}}{2z}$$

$$\text{Simplify:} \qquad = \frac{3\sqrt[4]{4z}}{z}$$

▶ ❷ **Rationalize a Denominator Containing Two Terms**

To rationalize a denominator containing two terms involving square roots, use the fact that

$$(A + B)(A - B) = A^2 - B^2$$

and multiply both the numerator and the denominator of the quotient by the conjugate of the denominator. For example, if the quotient is $\dfrac{3}{\sqrt{3} + 2}$, multiply both the numerator and the denominator by the conjugate of $\sqrt{3} + 2$, which is $\sqrt{3} - 2$. Recall from the last section that the product $(\sqrt{3} + 2)(\sqrt{3} - 2)$ will not contain a radical.

EXAMPLE 3 **Rationalizing a Denominator Containing Two Terms**

Rationalize the denominator: $\dfrac{\sqrt{2}}{\sqrt{6} + 2}$

Solution
Because $\sqrt{6} + 2$ is in the denominator of the quotient, multiply the numerator and denominator by the conjugate of $\sqrt{6} + 2$, $\sqrt{6} - 2$.

$$\frac{\sqrt{2}}{\sqrt{6} + 2} = \frac{\sqrt{2}}{\sqrt{6} + 2} \cdot \frac{\sqrt{6} - 2}{\sqrt{6} - 2}$$

Multiply the numerators and denominators:
$$= \frac{\sqrt{2}(\sqrt{6} - 2)}{(\sqrt{6} + 2)(\sqrt{6} - 2)}$$

Distribute $\sqrt{2}$ in the numerator; $(A + B)(A - B) = A^2 - B^2$ in the denominator:
$$= \frac{\sqrt{12} - 2\sqrt{2}}{(\sqrt{6})^2 - 2^2}$$

$\sqrt{12} = 2\sqrt{3}$:
$$= \frac{2\sqrt{3} - 2\sqrt{2}}{6 - 4}$$

Factor out the GCF of 2 in numerator:
$$= \frac{2(\sqrt{3} - \sqrt{2})}{2}$$

Divide out common factor, 2:
$$= \sqrt{3} - \sqrt{2}$$ ●

EXAMPLE 4   **Rationalizing a Denominator Containing Two Terms**

Rationalize the denominator: $\dfrac{\sqrt{6}-3}{\sqrt{10}-\sqrt{6}}$

Solution

Multiply the numerator and denominator by the conjugate of the denominator, $\sqrt{10}+\sqrt{6}$.

$$\frac{\sqrt{6}-3}{\sqrt{10}-\sqrt{6}}=\frac{\sqrt{6}-3}{\sqrt{10}-\sqrt{6}}\cdot\frac{\sqrt{10}+\sqrt{6}}{\sqrt{10}+\sqrt{6}}$$

Multiply the numerators and denominators: $=\dfrac{(\sqrt{6}-3)(\sqrt{10}+\sqrt{6})}{(\sqrt{10}-\sqrt{6})(\sqrt{10}+\sqrt{6})}$

Use FOIL in the numerator;
$(A+B)(A-B)=A^2-B^2$ in the denominator: $=\dfrac{\sqrt{60}+\sqrt{36}-3\sqrt{10}-3\sqrt{6}}{(\sqrt{10})^2-(\sqrt{6})^2}$

Simplify radicals: $=\dfrac{2\sqrt{15}+6-3\sqrt{10}-3\sqrt{6}}{10-6}$

$$=\dfrac{2\sqrt{15}+6-3\sqrt{10}-3\sqrt{6}}{4}$$

Quick ✔

In Problem 12, rationalize the denominator.

12. $\dfrac{\sqrt{5}+4}{\sqrt{5}-\sqrt{2}}$

9.6 Exercises   MyMathLab®   Exercise numbers in **green** have complete video solutions in MyMathLab or may be accessed using the QR code to the right.

*Problems 1–12 are the Quick ✔ s that follow the **EXAMPLES**.*

Building Skills

In Problems 13–36, rationalize each denominator. Assume all variables are positive. See Objective 1.

13. $\dfrac{1}{\sqrt{2}}$

14. $\dfrac{2}{\sqrt{3}}$

15. $-\dfrac{6}{5\sqrt{3}}$

16. $-\dfrac{3}{2\sqrt{3}}$

17. $\dfrac{3}{\sqrt{12}}$

18. $\dfrac{5}{\sqrt{20}}$

19. $\dfrac{\sqrt{2}}{\sqrt{6}}$

20. $\dfrac{\sqrt{3}}{\sqrt{11}}$

21. $\sqrt{\dfrac{2}{p}}$

22. $\sqrt{\dfrac{5}{z}}$

23. $\dfrac{\sqrt{8}}{\sqrt{y^3}}$

24. $\dfrac{\sqrt{32}}{\sqrt{a^5}}$

25. $\dfrac{2}{\sqrt[3]{2}}$

26. $\dfrac{5}{\sqrt[3]{3}}$

27. $\sqrt[3]{\dfrac{7}{q}}$

28. $\sqrt[3]{\dfrac{-4}{p}}$

29. $\sqrt[3]{\dfrac{-3}{50}}$

30. $\sqrt[3]{\dfrac{-5}{72}}$

31. $\dfrac{2}{\sqrt[3]{20y}}$

32. $\dfrac{8}{\sqrt[3]{36z^2}}$

33. $\dfrac{-4}{\sqrt[4]{3x^3}}$

34. $\dfrac{6}{\sqrt[4]{9b^2}}$

35. $\dfrac{12}{\sqrt[5]{m^3n^2}}$

36. $\dfrac{-3}{\sqrt[5]{ab^3}}$

In Problems 37–56, rationalize each denominator. Assume all variables are positive. See Objective 2.

37. $\dfrac{4}{\sqrt{6}-2}$

38. $\dfrac{6}{\sqrt{7}-2}$

39. $\dfrac{5}{\sqrt{5}+2}$

40. $\dfrac{10}{\sqrt{10}+3}$

41. $\dfrac{8}{\sqrt{7}-\sqrt{3}}$

42. $\dfrac{12}{\sqrt{11}-\sqrt{7}}$

43. $\dfrac{\sqrt{2}}{\sqrt{10}-\sqrt{6}}$

44. $\dfrac{\sqrt{3}}{\sqrt{15}-\sqrt{6}}$

45. $\dfrac{\sqrt{p}}{\sqrt{p}+\sqrt{q}}$

46. $\dfrac{\sqrt{a}}{\sqrt{a}+\sqrt{b}}$

47. $\dfrac{18}{2\sqrt{3}+3\sqrt{2}}$

48. $\dfrac{15}{3\sqrt{5}+4\sqrt{3}}$

49. $\dfrac{\sqrt{7}+3}{\sqrt{7}-3}$ **50.** $\dfrac{\sqrt{5}+3}{\sqrt{5}-3}$ **51.** $\dfrac{\sqrt{3}-4\sqrt{2}}{2\sqrt{3}+5\sqrt{2}}$

52. $\dfrac{3\sqrt{6}+5\sqrt{7}}{2\sqrt{6}-3\sqrt{7}}$ **53.** $\dfrac{\sqrt{p}+2}{\sqrt{p}-2}$ **54.** $\dfrac{\sqrt{x}-4}{\sqrt{x}+4}$

55. $\dfrac{\sqrt{2}-3}{\sqrt{8}-\sqrt{2}}$ **56.** $\dfrac{2\sqrt{3}+3}{\sqrt{12}-\sqrt{3}}$

Mixed Practice

In Problems 57–64, perform the indicated operation and simplify.

57. $\sqrt{3}+\dfrac{1}{\sqrt{3}}$ **58.** $\sqrt{5}-\dfrac{1}{\sqrt{5}}$

59. $\dfrac{\sqrt{10}}{2}-\dfrac{1}{\sqrt{2}}$ **60.** $\dfrac{\sqrt{5}}{2}+\dfrac{3}{\sqrt{5}}$

61. $\sqrt{\dfrac{1}{3}}+\sqrt{12}+\sqrt{75}$ **62.** $\sqrt{\dfrac{2}{5}}+\sqrt{20}-\sqrt{45}$

63. $\dfrac{3}{\sqrt{18}}-\sqrt{\dfrac{1}{2}}$ **64.** $\sqrt{\dfrac{4}{3}}+\dfrac{4}{\sqrt{48}}$

In Problems 65–76, simplify each expression so that the denominator does not contain a radical. Work smart because in some of the problems, it will be easier if you divide the radicands before attempting to rationalize the denominator.

65. $\dfrac{\sqrt{3}}{\sqrt{12}}$ **66.** $\dfrac{\sqrt{2}}{\sqrt{18}}$ **67.** $\dfrac{3}{\sqrt{72}}$

68. $\dfrac{7}{\sqrt{98}}$ **69.** $\sqrt{\dfrac{4}{3}}$ **70.** $\sqrt{\dfrac{9}{5}}$

71. $\dfrac{\sqrt{3}-3}{\sqrt{3}+3}$ **72.** $\dfrac{\sqrt{2}-5}{\sqrt{2}+5}$ **73.** $\dfrac{2}{\sqrt{5}+2}$

74. $\dfrac{5}{\sqrt{6}+4}$ **75.** $\dfrac{\sqrt{8}}{\sqrt{2}}$ **76.** $\dfrac{\sqrt{75}}{\sqrt{3}}$

In Problems 77–82, find the reciprocal of the given number. Be sure to rationalize the denominator.

77. $\sqrt{3}$ **78.** $\sqrt{7}$ **79.** $\sqrt[3]{12}$

80. $\sqrt[3]{18}$ **81.** $\sqrt{3}+5$ **82.** $7-\sqrt{2}$

Applying the Concepts

Math for the Future: Trigonometry *In Problems 83 and 84, simplify each expression.*

83. $\dfrac{1}{\sqrt{2}}\cdot\dfrac{\sqrt{3}}{2}-\dfrac{1}{\sqrt{2}}\cdot\dfrac{1}{2}$

84. $-\sqrt{\dfrac{2}{3}}\cdot\left(-\dfrac{2}{\sqrt{5}}\right)+\dfrac{1}{\sqrt{3}}\cdot\dfrac{1}{\sqrt{5}}$

Sometimes it may be useful to rationalize a numerator. In Problems 85–88, rationalize each expression by multiplying the numerator and denominator by the conjugate of the numerator.

85. $\dfrac{\sqrt{2}+1}{3}$ **86.** $\dfrac{\sqrt{3}+2}{2}$

87. $\dfrac{\sqrt{x}-\sqrt{h}}{\sqrt{x}}$ **88.** $\dfrac{\sqrt{a}-\sqrt{b}}{\sqrt{2}}$

Extending the Concepts

89. When two quantities a and b are positive, you can verify that $a=b$ by showing that $a^2=b^2$. Verify that $\dfrac{\sqrt{6}+\sqrt{2}}{4}=\dfrac{\sqrt{2}+\sqrt{3}}{2}$ by squaring each side.

90. Rationalize the denominator: $\dfrac{2}{\sqrt{2}+\sqrt{3}-\sqrt{9}}$

91. Math for the Future: Calculus The following problem comes up in calculus. Consider the rational expression:

$$\dfrac{\sqrt{x+h}-\sqrt{x}}{h}$$

(a) Rationalize the numerator by multiplying the numerator and denominator by $\sqrt{x+h}+\sqrt{x}$. Be sure to simplify the expression completely.

(b) Evaluate the expression found in part (a) at $h=0$.

(c) The expression found in part (b) represents the formula for the slope of the line tangent to the graph of $f(x)=\sqrt{x}$ at any value of $x\geq 0$. Find the slope of the line tangent to the graph of $f(x)=\sqrt{x}$ at $x=4$.

(d) If $f(x)=\sqrt{x}$, what is $f(4)$? What point is on the graph of $f(x)=\sqrt{x}$?

(e) Find the equation of the line tangent to $f(x)=\sqrt{x}$ at $x=4$, using the slope found in part (c) and the point found in part (d).

(f) Graph the function $f(x)=\sqrt{x}$ and the tangent line in the same Cartesian plane.

Explaining the Concepts

92. Explain why it is necessary to multiply the numerator and denominator by the conjugate of the denominator when rationalizing a denominator containing two terms.

93. Explain why removing irrational numbers from the denominator is called "rationalization."

Synthesis Review

In Problems 94–97, graph each of the following functions using point plotting.

94. $f(x)=5x-3$ **95.** $g(x)=-3x+9$

96. $G(x)=x^2$ **97.** $F(x)=x^3$

Putting the Concepts Together (Sections 9.1–9.6)

We designed these problems so that you can review Sections 9.1 through 9.6 and show your mastery of the concepts. Take time to work these problems before proceeding with the next section. The answers are located at the back of the text on page AN-44.

1. Evaluate: $-25^{\frac{1}{2}}$

2. Evaluate: $(-64)^{-\frac{2}{3}}$

3. Write the expression $\sqrt[4]{3x^3}$ with a rational exponent.

4. Write the expression $7z^{\frac{4}{5}}$ as a radical expression.

5. Simplify using rational exponents: $\sqrt[3]{\sqrt{64x^3}}$

6. Distribute and simplify: $c^{\frac{1}{2}}\left(c^{\frac{3}{2}} + c^{\frac{5}{2}}\right)$

In Problems 7–9, use Laws of Exponents to simplify each expression. Assume all variables in the radicand are greater than or equal to zero. Express answers with positive exponents.

7. $\left(a^{\frac{2}{3}}b^{-\frac{1}{3}}\right)\left(a^{\frac{4}{3}}b^{-\frac{5}{3}}\right)$

8. $\dfrac{x^{\frac{3}{4}}}{x^{\frac{1}{8}}}$

9. $\left(x^{\frac{3}{4}}y^{-\frac{1}{8}}\right)^8$

In Problems 10–19, perform the indicated operation and simplify. Assume all variables in the radicand are greater than or equal to zero.

10. $\sqrt{15a} \cdot \sqrt{2b}$

11. $\sqrt{10m^3n^2} \cdot \sqrt{20mn}$

12. $\sqrt[3]{\dfrac{-32xy^4}{4x^{-2}y}}$

13. $2\sqrt{108} - 3\sqrt{75} + \sqrt{48}$

14. $-5b\sqrt{8b} + 7\sqrt{18b^3}$

15. $\sqrt[3]{16y^4} - y\sqrt[3]{2y}$

16. $(3\sqrt{x})(4\sqrt{x})$

17. $3\sqrt{x} + 4\sqrt{x}$

18. $(2 - 3\sqrt{2})(10 + \sqrt{2})$

19. $(4\sqrt{2} - 3)^2$

In Problems 20 and 21, rationalize the denominator.

20. $\dfrac{3}{2\sqrt{32}}$

21. $\dfrac{4}{\sqrt{3} - 8}$

9.7 Functions Involving Radicals

Objectives

❶ Evaluate Functions Involving Radicals

❷ Find the Domain of a Function Involving a Radical

❸ Graph Functions Involving Square Roots

❹ Graph Functions Involving Cube Roots

Are You Prepared for This Section?

Before getting started, complete the following problems. If you get a problem wrong, go back to the section cited and review the material.

P1. Evaluate: $\sqrt{121}$ [Section 9.1, pp. 616–617]

P2. Simplify: $\sqrt{p^2}$ [Section 9.1, pp. 618–619]

P3. Given $f(x) = x^2 - 4$, find $f(3)$.   [Section 8.3, pp. 542–544]

P4. Solve: $-2x + 3 \geq 0$ [Section 2.8, pp. 150–154]

P5. Graph $f(x) = x^2 + 1$ using point plotting.   [Section 8.4, pp. 549–550]

▶ ❶ **Evaluate Functions Involving Radicals**

Functions whose rule contains a radical can be evaluated by substituting the value of the independent variable into the rule, just as in Section 8.3.

EXAMPLE 1 **Evaluating Functions for Which the Rule Is a Radical Expression**

For the functions $f(x) = \sqrt{x + 2}$ and $g(x) = \sqrt[3]{3x + 1}$, find

(a) $f(7)$ **(b)** $f(10)$ **(c)** $g(-3)$

(continued)

Prepared?...Answers
See page 656.

Solution

(a) $f(x) = \sqrt{x+2}$
$f(7) = \sqrt{7+2}$
$= \sqrt{9}$
$= 3$

(b) $f(x) = \sqrt{x+2}$
$f(10) = \sqrt{10+2}$
$= \sqrt{12}$
$= 2\sqrt{3}$

(c) $g(x) = \sqrt[3]{3x+1}$
$g(-3) = \sqrt[3]{3(-3)+1}$
$= \sqrt[3]{-8}$
$= -2$ ●

Quick ✓

In Problems 1 and 2, find the values for each function.

1. $f(x) = \sqrt{3x+7}$

 (a) $f(3)$ (b) $f(7)$

2. $g(x) = \sqrt[3]{2x+7}$

 (a) $g(-4)$ (b) $g(10)$

▶ ② Find the Domain of a Function Involving a Radical

Recall $\sqrt[n]{a} = b$ means $a = b^n$. This definition showed that for $n \geq 2$ and even, the radicand, a, must be greater than or equal to 0. For $n \geq 3$ and odd, the radicand, a, can be any real number. This leads to a procedure for finding the domain of a function whose rule contains a radical.

> **Finding the Domain of a Function Involving a Radical**
>
> • If the index on a radical is even, then the radicand must be greater than or equal to zero.
>
> • If the index on a radical is odd, then the radicand can be any real number.

EXAMPLE 2 **Finding the Domain of a Radical Function**

Find the domain of each of the following functions:

(a) $f(x) = \sqrt{x-5}$ (b) $G(x) = \sqrt[3]{2x+1}$ (c) $h(t) = \sqrt[4]{5-2t}$

Solution

(a) The function $f(x) = \sqrt{x-5}$ tells us to take the square root of $x - 5$. Square roots can be taken only of numbers greater than or equal to zero, so the radicand, $x - 5$, must be greater than or equal to zero. This requires that

$$x - 5 \geq 0$$

Add 5 to both sides: $x \geq 5$

The domain of f is $\{x \mid x \geq 5\}$, or the interval $[5, \infty)$.

(b) The function $G(x) = \sqrt[3]{2x+1}$ tells us to take the cube root of $2x + 1$. The cube root of any real number can be taken, so the domain of G is $\{x \mid x$ is any real number$\}$, or the interval $(-\infty, \infty)$.

(c) The function $h(t) = \sqrt[4]{5-2t}$ tells us to take the fourth root of $5 - 2t$. Fourth roots can be taken only of numbers greater than or equal to zero, so the radicand, $5 - 2t$, must be greater than or equal to zero:

$$5 - 2t \geq 0$$

Subtract 5 from both sides: $-2t \geq -5$

Divide both sides by -2 (Don't forget to change the direction of the inequality!): $t \leq \dfrac{5}{2}$

The domain of h is $\left\{ t \mid t \leq \dfrac{5}{2} \right\}$, or the interval $\left(-\infty, \dfrac{5}{2} \right]$. ●

Prepared?...Answers

P1. 11 **P2.** $|p|$ **P3.** 5

P4. $\left\{ x \mid x \leq \dfrac{3}{2} \right\}$ or $\left(-\infty, \dfrac{3}{2} \right]$

P5.

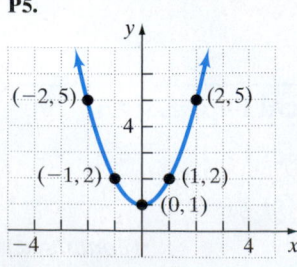

Quick ✓

3. If the index on a radical expression is ____, then the radicand must be greater than or equal to zero. If the index on a radical expression is ____, then the radicand can be any real number.

In Problems 4–6, find the domain of each function.

4. $H(x) = \sqrt{x + 6}$ 5. $g(t) = \sqrt[5]{3t - 1}$ 6. $F(m) = \sqrt[4]{6 - 3m}$

▶ ❸ Graph Functions Involving Square Roots

The **square root function** is given by $f(x) = \sqrt{x}$. The domain of the square root function is $\{x \mid x \geq 0\}$ or, in interval notation, $[0, \infty)$. Graph $f(x) = \sqrt{x}$ by finding some ordered pairs (x, y) such that $y = \sqrt{x}$. Then plot the ordered pairs in the xy-plane and connect the points. To make life easy, choose values of x that are perfect squares $(0, 1, 4, 9,$ and so on). Table 1 shows some points on the graph of $f(x) = \sqrt{x}$. Figure 5 shows the graph of $f(x) = \sqrt{x}$ and illustrates that the range of $f(x) = \sqrt{x}$ is $[0, \infty)$.

Table 1

x	$f(x) = \sqrt{x}$	(x, y) or $(x, f(x))$
0	$f(0) = \sqrt{0} = 0$	$(0, 0)$
1	$f(1) = 1$	$(1, 1)$
4	$f(4) = 2$	$(4, 2)$
9	$f(9) = 3$	$(9, 3)$
16	$f(16) = 4$	$(16, 4)$

Figure 5
$f(x) = \sqrt{x}$

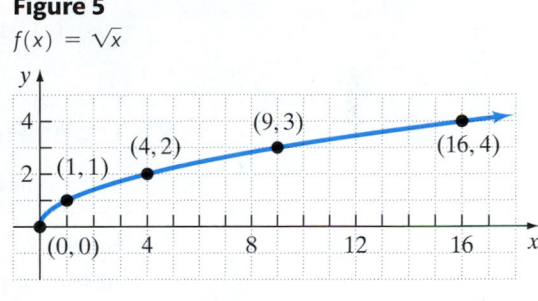

The point-plotting method can be used to graph a variety of functions involving square roots.

EXAMPLE 3   **Graphing a Function Involving a Square Root**

Consider the function $f(x) = \sqrt{x - 2}$.

(a) Find the domain.

(b) Graph the function using point plotting.

(c) Based on the graph, determine the range.

Solution

(a) Because $f(x) = \sqrt{x - 2}$ is a square root function, the radicand, $x - 2$, must be greater than or equal to zero. This requires that

$$x - 2 \geq 0$$

Add 2 to both sides: $x \geq 2$

The domain of f is $\{x \mid x \geq 2\}$, or the interval $[2, \infty)$.

(b) Choose values of x that are greater than or equal to 2 and that will make the radicand a perfect square. See Table 2 on the next page. Figure 6 shows the graph of $f(x) = \sqrt{x - 2}$.

(continued)

Table 2

x	$f(x) = \sqrt{x-2}$	(x, y) or (x, f(x))
2	$f(2) = \sqrt{2-z} = 0$	(2, 0)
3	$f(3) = 1$	(3, 1)
6	$f(6) = 2$	(6, 2)
11	$f(11) = 3$	(11, 3)
18	$f(18) = 4$	(18, 4)

Figure 6

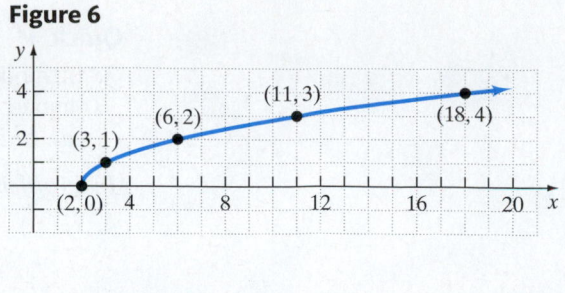

(c) From the graph of $f(x) = \sqrt{x-2}$ given in Figure 6, we can see that the range of $f(x) = \sqrt{x-2}$ is $[0, \infty)$.

> **Quick ✓**
>
> **7.** Consider the function $f(x) = \sqrt{x} + 3$.
>
>   **(a)** Find the domain.
>
>   **(b)** Graph the function using point plotting.
>
>   **(c)** Based on the graph, determine the range.

▶ ❹ Graph Functions Involving Cube Roots

The **cube root function** is given by $f(x) = \sqrt[3]{x}$. The domain of the cube root function is $\{x \mid x \text{ is any real number}\}$ or, in interval notation, $(-\infty, \infty)$. Graph $f(x) = \sqrt[3]{x}$ by finding some ordered pairs (x, y) such that $y = \sqrt[3]{x}$. Then plot the ordered pairs in the xy-plane and connect the points. To make life easy, choose values of x that are perfect cubes $(-8, -1, 0, 1, 8, \text{ and so on})$. See Table 3. Figure 7 shows the graph of $f(x) = \sqrt[3]{x}$.

Table 3

x	$f(x) = \sqrt[3]{x}$	(x, y) or (x, f(x))
-8	$f(-8) = \sqrt[3]{-8} = -2$	(-8, -2)
-1	$f(-1) = -1$	(-1, -1)
$-\dfrac{1}{8}$	$f\left(-\dfrac{1}{8}\right) = -\dfrac{1}{2}$	$\left(-\dfrac{1}{8}, -\dfrac{1}{2}\right)$
0	$f(0) = 0$	(0, 0)
$\dfrac{1}{8}$	$f\left(\dfrac{1}{8}\right) = \dfrac{1}{2}$	$\left(\dfrac{1}{8}, \dfrac{1}{2}\right)$
1	$f(1) = 1$	(1, 1)
8	$f(8) = 2$	(8, 2)

Figure 7

$f(x) = \sqrt[3]{x}$

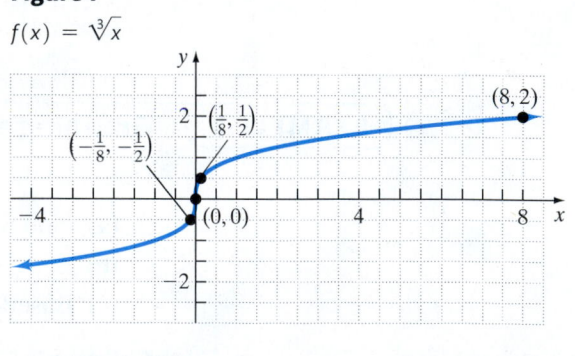

The graph of $f(x) = \sqrt[3]{x}$ in Figure 7 shows that the range of the function is the set of all real numbers or $(-\infty, \infty)$.

The point-plotting method can be used to graph a variety of functions involving cube roots.

EXAMPLE 4 ⟩ **Graphing a Function Involving a Cube Root**

Consider the function $g(x) = \sqrt[3]{x} + 2$.

(a) Find the domain.

(b) Graph the function using point plotting.

(c) Based on the graph, determine the range.

Solution

(a) The function $g(x) = \sqrt[3]{x} + 2$ tells us to take the cube root of x and then add 2. The cube root is defined for all real numbers, so the domain of g is $\{x \mid x \text{ is any real number}\}$, or the interval $(-\infty, \infty)$.

(b) Choose values of x that make the radicand a perfect cube. See Table 4. Figure 8 shows the graph of $g(x) = \sqrt[3]{x} + 2$.

Table 4

x	$g(x) = \sqrt[3]{x} + 2$	(x, y) or $(x, f(x))$
-8	$g(-8) = \sqrt[3]{-8} + 2 = 0$	$(-8, 0)$
-1	$g(-1) = 1$	$(-1, 1)$
0	$g(0) = 2$	$(0, 2)$
1	$g(1) = 3$	$(1, 3)$
8	$g(8) = 4$	$(8, 4)$

Figure 8

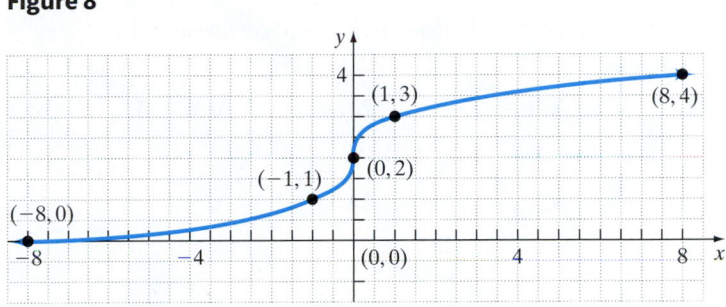

(c) The graph of $g(x) = \sqrt[3]{x} + 2$ shows that the range of $g(x) = \sqrt[3]{x} + 2$ is the set of all real numbers or $(-\infty, \infty)$.

Quick ✓

8. Consider the function $G(x) = \sqrt[3]{x} - 1$.

(a) Find the domain.

(b) Graph the function using point plotting.

(c) Based on the graph, determine the range.

9.7 Exercises MyMathLab® Exercise numbers in green have complete video solutions in MyMathLab or may be accessed using the QR code to the right.

Problems 1–8 are the Quick ✓s that follow the EXAMPLES.

Building Skills

In Problems 9–20, evaluate each radical function at the indicated values. See Objective 1.

9. $f(x) = \sqrt{x + 6}$
   (a) $f(3)$
   (b) $f(8)$
   (c) $f(-2)$

10. $f(x) = \sqrt{x + 10}$
   (a) $f(6)$
   (b) $f(2)$
   (c) $f(-6)$

11. $g(x) = -\sqrt{2x + 3}$
   (a) $g(11)$
   (b) $g(-1)$
   (c) $g\left(\dfrac{1}{8}\right)$

12. $g(x) = -\sqrt{4x + 5}$
   (a) $g(1)$
   (b) $g(10)$
   (c) $g\left(\dfrac{1}{8}\right)$

13. $G(m) = 2\sqrt{5m - 1}$
   (a) $G(1)$
   (b) $G(5)$
   (c) $G\left(\dfrac{1}{2}\right)$

14. $G(p) = 3\sqrt{4p + 1}$
   (a) $G(2)$
   (b) $G(11)$
   (c) $G\left(\dfrac{1}{8}\right)$

15. $H(z) = \sqrt[3]{z + 4}$
   (a) $H(4)$
   (b) $H(-12)$
   (c) $H(-20)$

16. $G(t) = \sqrt[3]{t - 6}$
   (a) $G(7)$
   (b) $G(-21)$
   (c) $G(22)$

17. $f(x) = \sqrt{\dfrac{x - 2}{x + 2}}$
   (a) $f(7)$
   (b) $f(6)$
   (c) $f(10)$

18. $f(x) = \sqrt{\dfrac{x - 4}{x + 4}}$
   (a) $f(5)$
   (b) $f(8)$
   (c) $f(12)$

19. $g(z) = \sqrt[3]{\dfrac{2z}{z - 4}}$
   (a) $g(-4)$
   (b) $g(8)$
   (c) $g(12)$

20. $H(z) = \sqrt[3]{\dfrac{3z}{z + 5}}$
   (a) $H(3)$
   (b) $H(4)$
   (c) $H(-1)$

In Problems 21–36, find the domain of the radical function. See Objective 2.

21. $f(x) = \sqrt{x - 7}$

22. $f(x) = \sqrt{x + 4}$

23. $g(x) = \sqrt{2x + 7}$

24. $g(x) = \sqrt{3x + 7}$

25. $F(x) = \sqrt{4 - 3x}$

26. $G(x) = \sqrt{5 - 2x}$

27. $H(z) = \sqrt[3]{2z + 1}$

28. $G(z) = \sqrt[3]{5z - 3}$

29. $W(p) = \sqrt[4]{7p - 2}$

30. $C(y) = \sqrt[4]{3y - 2}$

31. $g(x) = \sqrt[5]{x - 3}$

32. $g(x) = \sqrt[5]{x + 9}$

33. $f(x) = \sqrt{\dfrac{3}{x + 5}}$

34. $f(x) = \sqrt{\dfrac{3}{x - 3}}$

35. $H(x) = \sqrt[5]{\dfrac{x + 3}{x - 3}}$

36. $H(x) = \sqrt[3]{\dfrac{x - 5}{x + 2}}$

In Problems 37–52, (a) determine the domain of the function; (b) graph the function using point plotting; and (c) based on the graph, determine the range of the function. See Objective 3.

37. $f(x) = \sqrt{x - 4}$

38. $f(x) = \sqrt{x - 1}$

39. $g(x) = \sqrt{x + 2}$

40. $g(x) = \sqrt{x + 5}$

41. $G(x) = \sqrt{2 - x}$

42. $F(x) = \sqrt{4 - x}$

43. $f(x) = \sqrt{x} + 3$

44. $f(x) = \sqrt{x} + 1$

45. $g(x) = \sqrt{x} - 4$

46. $g(x) = \sqrt{x} - 2$

47. $H(x) = 2\sqrt{x}$

48. $h(x) = 3\sqrt{x}$

49. $f(x) = \dfrac{1}{2}\sqrt{x}$

50. $g(x) = \dfrac{1}{4}\sqrt{x}$

51. $G(x) = -\sqrt{x}$

52. $F(x) = \sqrt{-x}$

In Problems 53–58, (a) determine the domain of the function; (b) graph the function using point plotting; and (c) based on the graph, determine the range of the function. See Objective 4.

53. $h(x) = \sqrt[3]{x + 2}$

54. $g(x) = \sqrt[3]{x - 4}$

55. $f(x) = \sqrt[3]{x} - 3$

56. $H(x) = \sqrt[3]{x} + 3$

57. $G(x) = 2\sqrt[3]{x}$

58. $F(x) = 3\sqrt[3]{x}$

Applying the Concepts

59. Distance to a Point on a Graph Suppose that $P = (x, y)$ is a point on the graph of $y = x^2 - 4$. The distance from P to $(0, 1)$ is given by the function

$$d(x) = \sqrt{x^4 - 9x^2 + 25}$$

See the figure.

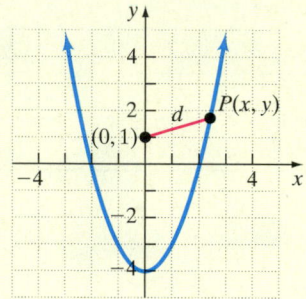

   (a) What is the distance from $P = (0, -4)$ to $(0, 1)$? That is, what is $d(0)$?
   (b) What is the distance from $P = (1, -3)$ to $(0, 1)$? That is, what is $d(1)$?
   (c) What is the distance from $P = (5, 21)$ to $(0, 1)$? That is, what is $d(5)$?

60. Distance to a Point on a Graph Suppose that $P = (x, y)$ is a point on the graph of $y = x^2 - 2$. The distance from P to $(0, 2)$ is given by the function

$$d(x) = \sqrt{x^4 - 7x^2 + 16}$$

See the figure.

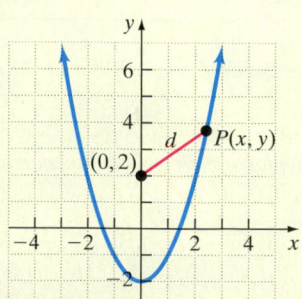

   (a) What is the distance from $P = (0, -2)$ to $(0, 2)$? That is, what is $d(0)$?
   (b) What is the distance from $P = (1, -1)$ to $(0, 2)$? That is, what is $d(1)$?
   (c) What is the distance from $P = (4, 14)$ to $(0, 2)$? That is, what is $d(4)$?

△ **61. Area** A rectangle is inscribed in a semicircle of radius 3, as shown in the figure. Let $P = (x, y)$ be the point in quadrant I that is a vertex of the rectangle and is on the circle.

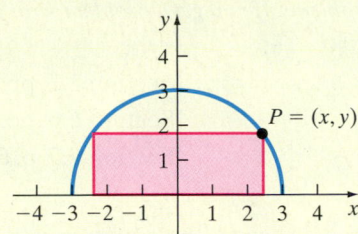

The area A of the rectangle as a function of x is given by

$$A(x) = 2x\sqrt{9 - x^2}$$

(a) What is the area of the rectangle whose vertex is at $(1, 2\sqrt{2})$?

(b) What is the area of the rectangle whose vertex is at $(2, \sqrt{5})$?

(c) What is the area of the rectangle whose vertex is at $(\sqrt{2}, \sqrt{7})$?

△ **62. Area** A rectangle is inscribed in a semicircle of radius 4 as shown in the figure. Let $P = (x, y)$ be the point in quadrant I that is a vertex of the rectangle and is on the circle.

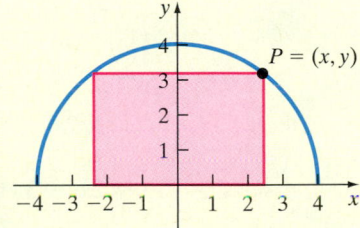

The area A of the rectangle as a function of x is given by

$$A(x) = 2x\sqrt{16 - x^2}$$

(a) What is the area of the rectangle whose vertex is at $(1, \sqrt{15})$?

(b) What is the area of the rectangle whose vertex is at $(2, 2\sqrt{3})$?

(c) What is the area of the rectangle whose vertex is at $(2\sqrt{2}, 2\sqrt{2})$?

Extending the Concepts

63. Use the results of Problems 37–40 to make a generalization about how to obtain the graph of $g(x) = \sqrt{x} + c$ from the graph of $f(x) = \sqrt{x}$.

64. Use the results of Problems 43–46 to make a generalization about how to obtain the graph of $g(x) = \sqrt{x} + c$ from the graph of $f(x) = \sqrt{x}$.

Synthesis Review

In Problems 65–69, add each of the following.

65. $\dfrac{1}{3} + \dfrac{1}{2}$ **66.** $\dfrac{1}{5} + \dfrac{3}{4}$

67. $\dfrac{1}{x} + \dfrac{3}{x + 1}$ **68.** $\dfrac{5}{x - 3} + \dfrac{2}{x + 1}$

69. $\dfrac{4}{x - 1} + \dfrac{3}{x + 1}$

70. Explain how adding rational numbers that do not have denominators with any common factors is similar to adding rational expressions that do not have denominators with any common factors.

Technology Exercises

Technology can graph square root and cube root functions. The figure below shows the graph of $f(x) = \sqrt{x - 2}$ using GeoGebra. Note how the graph shown exists only for $x \geq 2$. Graphing functions is useful in verifying the domain that we find algebraically.

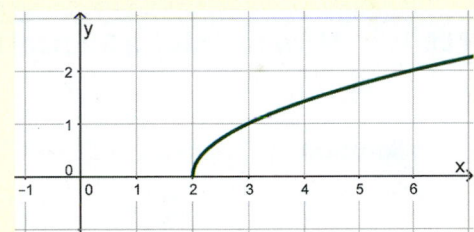

In Problems 71–92, graph the function using technology. Compare the graphs obtained by technology to the hand-drawn graphs in Problems 37–58.

71. $f(x) = \sqrt{x - 4}$ **72.** $f(x) = \sqrt{x - 1}$

73. $g(x) = \sqrt{x + 2}$ **74.** $g(x) = \sqrt{x + 5}$

75. $G(x) = \sqrt{2 - x}$ **76.** $F(x) = \sqrt{4 - x}$

77. $f(x) = \sqrt{x} + 3$ **78.** $f(x) = \sqrt{x} + 1$

79. $g(x) = \sqrt{x} - 4$ **80.** $g(x) = \sqrt{x} - 2$

81. $H(x) = 2\sqrt{x}$ **82.** $h(x) = 3\sqrt{x}$

83. $f(x) = \dfrac{1}{2}\sqrt{x}$ **84.** $g(x) = \dfrac{1}{4}\sqrt{x}$

85. $G(x) = -\sqrt{x}$ **86.** $F(x) = \sqrt{-x}$

87. $h(x) = \sqrt[3]{x + 2}$ **88.** $g(x) = \sqrt[3]{x - 4}$

89. $f(x) = \sqrt[3]{x} - 3$ **90.** $H(x) = \sqrt[3]{x} + 3$

91. $G(x) = 2\sqrt[3]{x}$ **92.** $F(x) = 3\sqrt[3]{x}$

9.8 Radical Equations and Their Applications

Objectives

1 Solve Radical Equations Containing One Radical

2 Solve Radical Equations Containing Two Radicals

3 Solve for a Variable in a Radical Equation

Are You Prepared for This Section?

Before getting started, complete the following problems. If you get a problem wrong, go back to the section cited and review the material.

P1. Solve: $3x - 5 = 0$ [Section 2.2, pp. 91–92]

P2. Solve: $2p^2 + 4p - 6 = 0$ [Section 6.6, pp. 410–415]

P3. Simplify: $\sqrt[3]{(x-5)^3}$ [Section 9.2, p. 622]

When the variable in an equation occurs in a radicand, the equation is called a **radical equation.** Examples of radical equations are

$$\sqrt{3x+1} = 5 \qquad \sqrt[3]{x-5} - 5 = 12 \qquad \sqrt{x-2} - \sqrt{2x+5} = 2$$

This section discusses how to solve radical equations involving one or two radicals.

▶ **1** Solve Radical Equations Containing One Radical

To solve an equation containing a single radical, isolate the radical. Then, as stated in Section 9.3, use the fact that $(\sqrt[n]{a})^n = a$, assuming the radicand is greater than or equal to zero.

EXAMPLE 1   **How to Solve a Radical Equation Containing One Radical**

Solve: $\sqrt{2x-3} - 5 = 0$

Step-by-Step Solution

Step 1: Isolate the radical.

$$\sqrt{2x-3} - 5 = 0$$

Add 5 to both sides: $\sqrt{2x-3} = 5$

Step 2: Raise both sides to the power of the index.

The index is 2, so square both sides: $(\sqrt{2x-3})^2 = 5^2$

$$2x - 3 = 25$$

Step 3: Solve the equation that results.

Add 3 to both sides: $2x - 3 + 3 = 25 + 3$

$$2x = 28$$

Divide both sides by 2: $\dfrac{2x}{2} = \dfrac{28}{2}$

$$x = 14$$

Step 4: Check

$$\sqrt{2x-3} - 5 = 0$$

Let $x = 14$ in the original equation: $\sqrt{2 \cdot 14 - 3} - 5 \overset{?}{=} 0$

$$\sqrt{28 - 3} - 5 \overset{?}{=} 0$$

$$5 - 5 = 0 \quad \text{True}$$

The solution set is $\{14\}$.   ●

It has always been emphasized to check answers whenever solving an equation. This is particularly important when solving equations involving radicals. Why? When radical equations containing an even index are solved, apparent solutions that are not solutions to the original equation can creep in. As seen when rational equations were solved in Chapter 7, these solutions are extraneous.

Prepared?...Answers **P1.** $\left\{\dfrac{5}{3}\right\}$

P2. $\{-3, 1\}$ **P3.** $x - 5$

> **Solving a Radical Equation Containing One Radical**
>
> **Step 1:** Isolate the radical. That is, get the radical by itself on one side of the equation.
>
> **Step 2:** Raise both sides of the equation to the power of the index. This will eliminate the radical from the equation.
>
> **Step 3:** Solve the equation that results.
>
> **Step 4:** Check your answer in the original equation.

Quick ✔

1. When the variable in an equation occurs in a radical, the equation is called a
   _____ _____.

2. When an apparent solution is not a solution of the original equation, the apparent solution is a(n) _____ solution.

3. *True or False* The first step in solving $x + \sqrt{x - 3} = 5$ is to square both sides of the equation.

4. Solve: $\sqrt{3x + 1} - 4 = 0$

EXAMPLE 2   **Solving a Radical Equation Containing One Radical**

Solve:

(a) $\sqrt{3x + 10} + 2 = 4$ **(b)** $\sqrt{5x - 1} + 7 = 5$

Solution

(a)

$$\sqrt{3x + 10} + 2 = 4$$

Subtract 2 from both sides: $\sqrt{3x + 10} + 2 - 2 = 4 - 2$

$$\sqrt{3x + 10} = 2$$

The index is 2, so square both sides: $(\sqrt{3x + 10})^2 = 2^2$

$$3x + 10 = 4$$

Subtract 10 from both sides: $3x + 10 - 10 = 4 - 10$

$$3x = -6$$

Divide both sides by 3: $\dfrac{3x}{3} = \dfrac{-6}{3}$

$$x = -2$$

Work Smart

Just because the apparent solution is negative does not automatically make it extraneous. Determine whether the apparent solution makes the radicand negative by checking the solution.

Check

$$\sqrt{3x + 10} + 2 = 4$$

Let $x = -2$ in the original equation: $\sqrt{3 \cdot (-2) + 10} + 2 \overset{?}{=} 4$

$$\sqrt{-6 + 10} + 2 \overset{?}{=} 4$$

$$\sqrt{4} + 2 \overset{?}{=} 4$$

$$2 + 2 \overset{?}{=} 4$$

$$4 = 4 \quad \text{True}$$

The solution set is $\{-2\}$.

(b)

$$\sqrt{5x - 1} + 7 = 5$$

Subtract 7 from both sides: $\sqrt{5x - 1} + 7 - 7 = 5 - 7$

$$\sqrt{5x - 1} = -2$$

The equation has no real solution because the principal square root of a number cannot be less than 0. Put another way, there is no real number whose principal square root is -2, so the equation has no real solution. The solution set is $\varnothing$ or $\{\ \}$.

⏵ EXAMPLE 3 **Solving a Radical Equation Containing One Radical**

Solve: $\sqrt{x + 5} = x - 1$

Solution

Work Smart

$(x - 1)^2 = x^2 - 2x + 1;$
$(x - 1)^2 \neq x^2 + 1$

$$\sqrt{x + 5} = x - 1$$

The index is 2, so square both sides: $(\sqrt{x + 5})^2 = (x - 1)^2$

$$x + 5 = x^2 - 2x + 1$$

Subtract x and 5 from both sides: $x + 5 - x - 5 = x^2 - 2x + 1 - x - 5$

$$0 = x^2 - 3x - 4$$

Factor: $0 = (x - 4)(x + 1)$

Zero-Product Property: $x - 4 = 0 \text{ or } x + 1 = 0$

$$x = 4 \text{ or } \qquad x = -1$$

Check $\sqrt{x + 5} = x - 1$

$x = 4:$ $\sqrt{4 + 5} \stackrel{?}{=} 4 - 1$ $x = -1:$ $\sqrt{-1 + 5} \stackrel{?}{=} -1 - 1$

$\sqrt{9} \stackrel{?}{=} 3$ $\sqrt{4} \stackrel{?}{=} -2$

$3 = 3$ True $2 = -2$ False

The apparent solution $x = -1$ does not check, so it is an extraneous solution. The solution set is $\{4\}$.

EXAMPLE 4 **Solving a Radical Equation Containing One Radical**

Solve: $\sqrt[3]{3x - 12} + 4 = 1$

Solution

$$\sqrt[3]{3x - 12} + 4 = 1$$

Subtract 4 from both sides: $\sqrt[3]{3x - 12} + 4 - 4 = 1 - 4$

$$\sqrt[3]{3x - 12} = -3$$

The index is 3, so cube both sides: $(\sqrt[3]{3x - 12})^3 = (-3)^3$

$$3x - 12 = -27$$

Add 12 to both sides: $3x - 12 + 12 = -27 + 12$

$$3x = -15$$

Divide both sides by 3: $x = -5$

Check

$$\sqrt[3]{3x - 12} + 4 = 1$$

Let $x = -5$ in the original equation: $\sqrt[3]{3 \cdot (-5) - 12} + 4 \stackrel{?}{=} 1$

$\sqrt[3]{-15 - 12} + 4 \stackrel{?}{=} 1$

$\sqrt[3]{-27} + 4 \stackrel{?}{=} 1$

$-3 + 4 \stackrel{?}{=} 1$

$1 = 1$ True

The solution set is $\{-5\}$.

Sometimes, an equation will contain rational exponents rather than radicals. When solving these problems, rewrite the equation with a radical or use the fact that $(a^r)^s = a^{r \cdot s}$.

EXAMPLE 5 **Solving an Equation Containing a Rational Exponent**

Solve: $(5x - 1)^{\frac{1}{2}} + 3 = 10$

Solution

$$(5x - 1)^{\frac{1}{2}} + 3 = 10$$

Subtract 3 from both sides: $\quad (5x - 1)^{\frac{1}{2}} = 7$

Square both sides: $\quad ((5x - 1)^{\frac{1}{2}})^2 = 7^2$

Use $(a^r)^s = a^{r \cdot s}$: $\quad (5x - 1)^{\frac{1}{2} \cdot 2} = 49$

$$5x - 1 = 49$$

Add 1 to both sides: $\quad 5x = 50$

Divide both sides by 5: $\quad x = 10$

Check

$$(5x - 1)^{\frac{1}{2}} + 3 = 10$$

Let $x = 10$ in the original equation: $\quad (5 \cdot 10 - 1)^{\frac{1}{2}} + 3 \stackrel{?}{=} 10$

$$(49)^{\frac{1}{2}} + 3 \stackrel{?}{=} 10$$

$$7 + 3 = 10$$

$$10 = 10 \quad \text{True}$$

The solution set is $\{10\}$.

Quick ✓

9. Solve: $(2x - 3)^{\frac{1}{3}} - 7 = -4$

⃟ ❷ Solve Radical Equations Containing Two Radicals

EXAMPLE 6 **How to Solve a Radical Equation Containing Two Radicals**

Solve: $\sqrt[3]{p^2 - 4p - 4} = \sqrt[3]{-3p + 2}$

Step-by-Step-Solution

Step 1: Isolate one of the radicals.

The radical on the left side of the equation is isolated: $\quad \sqrt[3]{p^2 - 4p - 4} = \sqrt[3]{-3p + 2}$

Step 2: Raise both sides to the power of the index.

The index is 3, so cube both sides:

$$(\sqrt[3]{p^2 - 4p - 4})^3 = (\sqrt[3]{-3p + 2})^3$$

$$p^2 - 4p - 4 = -3p + 2$$

Step 3: Because there is no radical, solve the equation that results.

Add $3p$ to both sides; subtract 2 from both sides: $\quad p^2 - 4p - 4 + 3p - 2 = -3p + 2 + 3p - 2$

Combine like terms: $\quad p^2 - p - 6 = 0$

Factor: $\quad (p - 3)(p + 2) = 0$

Use the Zero-Product Property: $\quad p - 3 = 0 \quad \text{or} \quad p + 2 = 0$

$$p = 3 \quad \text{or} \quad p = -2$$

Step 4: Check

$$\sqrt[3]{p^2 - 4p - 4} = \sqrt[3]{-3p + 2}$$

$p = -2$:

$$\sqrt[3]{(-2)^2 - 4(-2) - 4} \stackrel{?}{=} \sqrt[3]{-3(-2) + 2}$$

$$\sqrt[3]{4 + 8 - 4} \stackrel{?}{=} \sqrt[3]{6 + 2}$$

$$\sqrt[3]{8} = \sqrt[3]{8} \quad \text{True}$$

$p = 3$:

$$\sqrt[3]{(3)^2 - 4(3) - 4} \stackrel{?}{=} \sqrt[3]{-3(3) + 2}$$

$$\sqrt[3]{9 - 12 - 4} \stackrel{?}{=} \sqrt[3]{-9 + 2}$$

$$\sqrt[3]{-7} = \sqrt[3]{-7} \quad \text{True}$$

Both apparent solutions check. The solution set is $\{-2, 3\}$.

When a radical equation contains two radicals, use the following steps to solve the equation.

> ### Solving a Radical Equation Containing Two Radicals
>
> **Step 1:** Isolate one of the radicals. That is, get one of the radicals by itself on one side of the equation.
>
> **Step 2:** Raise both sides of the equation to the power of the index. This will eliminate one or both radicals from the equation.
>
> **Step 3:** If a radical remains in the equation, then follow the steps for solving a radical equation containing one radical. Otherwise, solve the equation that results.
>
> **Step 4:** Check your answer. Discard any extraneous solutions.

Quick ✓

10. Solve: $\sqrt[3]{m^2 + 4m + 4} = \sqrt[3]{2m + 7}$

EXAMPLE 7 **Solving a Radical Equation—Squaring Twice**

Solve: $\sqrt{3x + 6} - \sqrt{x + 6} = 2$

Solution

$$\sqrt{3x + 6} - \sqrt{x + 6} = 2$$

Add $\sqrt{x + 6}$ to both sides: $\quad \sqrt{3x + 6} = 2 + \sqrt{x + 6}$

Square both sides: $\quad (\sqrt{3x + 6})^2 = (2 + \sqrt{x + 6})^2$

Use $(A + B)^2 = A^2 + 2AB + B^2$: $\quad 3x + 6 = 4 + 4\sqrt{x + 6} + (\sqrt{x + 6})^2$

$(\sqrt{x + 6})^2 = x + 6$: $\quad 3x + 6 = 4 + 4\sqrt{x + 6} + x + 6$

Isolate the radical: $\quad 2x - 4 = 4\sqrt{x + 6}$

Factor out 2: $\quad 2(x - 2) = 4\sqrt{x + 6}$

Divide both sides by 2: $\quad x - 2 = 2\sqrt{x + 6}$

Square both sides: $\quad (x - 2)^2 = (2\sqrt{x + 6})^2$

$$x^2 - 4x + 4 = 4(x + 6)$$

Distribute: $\quad x^2 - 4x + 4 = 4x + 24$

Subtract $4x$ and 24 from both sides: $\quad x^2 - 8x - 20 = 0$

Factor: $\quad (x - 10)(x + 2) = 0$

Use the Zero-Product Property: $\quad x - 10 = 0 \quad \text{or} \quad x + 2 = 0$

$$x = 10 \quad \text{or} \quad x = -2$$

Work Smart

$(2 + \sqrt{x + 6})^2 \neq 2^2 + (\sqrt{x + 6})^2$

Remember,

$(A + B)^2 = (A + B)(A + B)$
$\quad\quad\quad = A^2 + 2AB + B^2$

Work Smart

When there is more than one radical, it is best to isolate the radical with the more complicated radicand.

Check $\quad x = -2$: $\quad \sqrt{3 \cdot (-2) + 6} - \sqrt{-2 + 6} \overset{?}{=} 2 \quad\quad x = 10$: $\quad \sqrt{3 \cdot 10 + 6} - \sqrt{10 + 6} \overset{?}{=} 2$

$$\sqrt{0} - \sqrt{4} \overset{?}{=} 2 \quad\quad\quad\quad\quad\quad \sqrt{36} - \sqrt{16} \overset{?}{=} 2$$

$$0 - 2 \overset{?}{=} 2 \quad\quad\quad\quad\quad\quad\quad\quad 6 - 4 \overset{?}{=} 2$$

$$-2 = 2 \quad \text{False} \quad\quad\quad\quad\quad\quad 2 = 2 \quad \text{True}$$

The apparent solution $x = -2$ does not check, so it is an extraneous solution. The solution set is $\{10\}$. ●

Quick ✓

11. Solve: $\sqrt{2x + 1} - \sqrt{x + 4} = 1$

▶ ❸ Solve for a Variable in a Radical Equation

In many situations, you will need to solve for a variable in a formula. For instance, Example 8 assesses how much error there is in an estimate based on a statistical study. This commonly used formula from statistics contains a radical.

EXAMPLE 8 **Solving for a Variable**

A formula from statistics for finding the margin of error in estimating a population mean is given by

$$E = z \cdot \frac{\sigma}{\sqrt{n}}$$

where E represents the margin of error, σ (sigma) represents a measure of variability, z is a measure of relative position, and n is the sample size.

(a) Solve this equation for n, the sample size.
(b) Find n when $\sigma = 12$, $z = 2$, and $E = 3$.

Solution

(a)
$$E = z \cdot \frac{\sigma}{\sqrt{n}}$$

Multiply both sides by $\sqrt{n}$: $\sqrt{n} \cdot E = z\sigma$

Divide both sides by E: $\sqrt{n} = \dfrac{z\sigma}{E}$

Square both sides: $n = \left(\dfrac{z\sigma}{E}\right)^2$

(b) $n = \left(\dfrac{z\sigma}{E}\right)^2 = \left(\dfrac{2 \cdot 12}{3}\right)^2$

$= 64$

The sample size needs to be $n = 64$. ●

Quick ✓

12. The period of a pendulum is the time it takes to complete one trip back and forth. The period T, in seconds, of a pendulum of length L, in feet, may be approximated using the formula $T = 2\pi\sqrt{\dfrac{L}{32}}$.

(a) Solve the equation for L.
(b) Determine the length of a pendulum whose period is 2π seconds.

9.8 Exercises **MyMathLab®** Exercise numbers in **green** have complete video solutions in MyMathLab or may be accessed using the QR code to the right.

*Problems **1–12** are the **Quick ✓**s that follow the **EXAMPLES**.*

Building Skills

In Problems 13–44, solve each equation. See Objective 1.

13. $\sqrt{x} = 4$

14. $\sqrt{p} = 6$

15. $\sqrt{x - 3} = 2$

16. $\sqrt{y - 5} = 3$

17. $\sqrt{2t + 3} = 5$

18. $\sqrt{3w - 2} = 4$

19. $\sqrt{4x + 3} = -2$

20. $\sqrt{6p - 5} = -5$

21. $\sqrt[3]{4t} = 2$

22. $\sqrt[3]{9w} = 3$

23. $\sqrt[3]{5q + 4} = 4$

24. $\sqrt[3]{7m + 20} = 5$

25. $\sqrt{y} + 3 = 8$

26. $\sqrt{q} - 5 = 2$

27. $\sqrt{x + 5} - 3 = 1$

28. $\sqrt{x - 4} + 4 = 7$

29. $\sqrt{2x + 9} + 5 = 6$

30. $\sqrt{4x + 21} + 2 = 5$

31. $3\sqrt{x} + 5 = 8$

32. $4\sqrt{t} - 2 = 10$

33. $\sqrt{4 - x} - 3 = 0$

34. $\sqrt{6 - w} - 3 = 1$

35. $\sqrt{p} = 2p$

36. $\sqrt{q} = 3q$

37. $\sqrt{x + 6} = x$

38. $\sqrt{2p + 8} = p$

39. $\sqrt{w} = 6 - w$

40. $\sqrt{m} = 12 - m$

41. $\sqrt{17 - 2x} + 1 = x$

42. $\sqrt{1 - 4x} - 5 = x$

43. $\sqrt{w^2 - 11} + 5 = w + 4$

44. $\sqrt{z^2 - z - 7} + 3 = z + 2$

In Problems 45–58, solve each equation. See Objective 2.

45. $\sqrt{x + 9} = \sqrt{2x + 5}$

46. $\sqrt{3x + 1} = \sqrt{2x + 7}$

47. $\sqrt[3]{4x - 3} = \sqrt[3]{2x - 9}$

48. $\sqrt[3]{3y - 2} = \sqrt[3]{5y + 8}$

49. $\sqrt{2w^2 - 3w - 4} = \sqrt{w^2 + 6w + 6}$

50. $\sqrt{2x^2 + 7x - 10} = \sqrt{x^2 + 4x + 8}$

51. $\sqrt{3w + 4} = 2 + \sqrt{w}$

52. $\sqrt{3y - 2} = 2 + \sqrt{y}$

53. $\sqrt{x + 1} - \sqrt{x - 2} = 1$

54. $\sqrt{2x - 1} - \sqrt{x - 1} = 1$

55. $\sqrt{2x + 6} - \sqrt{x - 1} = 2$

56. $\sqrt{2x + 6} - \sqrt{x - 6} = 3$

57. $\sqrt{2x + 5} - \sqrt{x - 1} = 2$

58. $\sqrt{4x + 1} - \sqrt{2x + 1} = 2$

In Problems 59–72, solve each equation.

59. $(2x + 3)^{\frac{1}{2}} = 3$

60. $(4x + 1)^{\frac{1}{2}} = 5$

61. $(6x - 1)^{\frac{1}{4}} = (2x + 15)^{\frac{1}{4}}$ **62.** $(6p + 3)^{\frac{1}{5}} = (4p - 9)^{\frac{1}{5}}$

63. $(x + 3)^{\frac{1}{2}} - (x - 5)^{\frac{1}{2}} = 2$ **64.** $(3x + 1)^{\frac{1}{2}} - (x - 1)^{\frac{1}{2}} = 2$

65. $(x - 1)^{2/3} = 4$

66. $(2x + 3)^{3/5} = 8$

67. $(x - 2)^{3/2} = 64$

68. $(3x - 2)^{3/4} = 8$

69. $(2x^4 - 81)^{1/4} + 3x = 4x$

70. $(17x^4 - 256)^{1/4} + 5x = 7x$

71. $\dfrac{1}{(x - 4)^{-2/3}} - 3 = 6$

72. $\dfrac{1}{(x + 3)^{-2/3}} + 5 = 9$

For Problems 73–78, see Objective 3.

73. Finance Solve $A = P\sqrt{1 + r}$ for r.

74. Centripetal Acceleration Solve $v = \sqrt{ar}$ for a.

75. Volume of a Sphere Solve $r = \sqrt[3]{\dfrac{3V}{4\pi}}$ for V.

76. Surface Area of a Sphere Solve $r = \sqrt{\dfrac{S}{4\pi}}$ for S.

77. Coulomb's Law Solve $r = \sqrt{\dfrac{4F\pi\epsilon_0}{q_1 q_2}}$ for F.

78. Potential Energy Solve $V = \sqrt{\dfrac{2U}{C}}$ for U.

Mixed Practice

In Problems 79–94, solve each equation.

79. $\sqrt{5p - 3} + 7 = 3$

80. $\sqrt{3b - 2} + 8 = 5$

81. $\sqrt{x + 12} = x$

82. $\sqrt{x + 20} = x$

83. $\sqrt{2p + 12} = 4$

84. $\sqrt{3a - 5} = 2$

85. $\sqrt[4]{x + 7} = 2$

86. $\sqrt[5]{x + 23} = 2$

87. $(3x + 1)^{\frac{1}{3}} + 2 = 0$

88. $(5x - 2)^{\frac{1}{3}} + 3 = 0$

89. $\sqrt{10 - x} - x = 10$

90. $\sqrt{9 - x} - x = 11$

91. $\sqrt{2x + 5} = \sqrt{3x - 4}$ **92.** $\sqrt{4c - 5} = \sqrt{3c + 1}$

93. $\sqrt{x - 1} + \sqrt{x + 4} = 5$ **94.** $\sqrt{x - 3} + \sqrt{x + 4} = 7$

95. Solve $\sqrt{x^2} = x + 4$ using the techniques of this section. Now solve $\sqrt{x^2} = x + 4$ using the fact that $\sqrt{x^2} = |x|$. Which approach do you like better?

96. Solve $\sqrt{p^2} = 3p + 4$ using the techniques of this section. Now solve $\sqrt{p^2} = 3p + 4$ using the fact that $\sqrt{p^2} = |p|$. Which approach do you like better?

97. Suppose that $f(x) = \sqrt{x - 2}$.

(a) Solve $f(x) = 0$. What point is on the graph of f?

(b) Solve $f(x) = 1$. What point is on the graph of f?

(c) Solve $f(x) = 2$. What point is on the graph of f?

(d) Use the information obtained in parts (a)–(c) to graph $f(x) = \sqrt{x - 2}$.

(e) Use the graph and the concept of the range of a function to explain why the equation $f(x) = -1$ has no solution.

98. Suppose that $g(x) = \sqrt{x + 3}$.

(a) Solve $g(x) = 0$. What point is on the graph of g?

(b) Solve $g(x) = 1$. What point is on the graph of g?

(c) Solve $g(x) = 2$. What point is on the graph of g?

(d) Use the information obtained in parts (a)–(c) to graph $g(x) = \sqrt{x + 3}$.

(e) Use the graph and the concept of the range of a function to explain why the equation $g(x) = -1$ has no solution.

Applying the Concepts

△ **99. Finding a y-Coordinate** The solutions to the equation

$$\sqrt{4^2 + (y-2)^2} = 5$$

represent the y-coordinates such that the distance from the point $(3, 2)$ to $(-1, y)$ in the Cartesian plane is 5 units.

  (a) Solve the equation for y.
  (b) Plot $(3, 2)$ and the points found in part (a) in the Cartesian plane. Draw and label the lengths of the sides of the figure formed.

△ **100. Finding an x-Coordinate** The solutions to the equation

$$\sqrt{x^2 + 4^2} = 5$$

represent the x-coordinates such that the distance from the point $(0, 3)$ to $(x, -1)$ in the Cartesian plane is 5 units.

  (a) Solve the equation for x.
  (b) Plot $(0, 3)$ and the points found in part (a) in the Cartesian plane. Draw and label the lengths of the sides of the figure formed.

101. Revenue Growth Suppose that the annual revenue R (in millions of dollars) of a company after t years of operating is modeled by the function

$$R(t) = \sqrt[3]{\frac{t}{2}}$$

  (a) After how many years can the company expect to have annual revenue of $1 million?
  (b) After how many years can the company expect to have annual revenue of $2 million?

△ **102. Sphere** The radius r of a sphere whose volume is V is given by

$$r = \sqrt[3]{\frac{3V}{4\pi}}$$

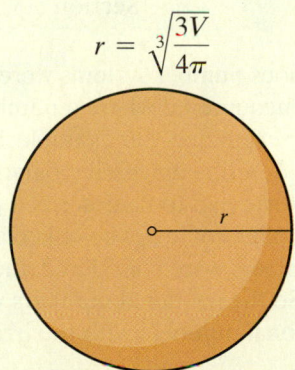

  (a) Find the volume of a sphere whose radius is 3 meters.
  (b) Find the volume of a sphere whose radius is 2 meters.

103. Birth Rates A plural birth is a live birth to twins, triplets, and so forth. The function $R(t) = 26 \cdot \sqrt[10]{t}$ models the plural birth rate R (live births per 1000 live births), where t is the number of years since 1995.

  (a) Use the model to predict the year in which the plural birth rate will be 39. Round to the nearest year.
  (b) Use the model to predict the year in which the plural birth rate will be 36. Round to the nearest year.

104. Money The annual rate of interest r (expressed as a decimal) required to have A dollars after t years from an initial deposit of P dollars is given by

$$r = \sqrt[t]{\frac{A}{P}} - 1$$

  (a) Suppose that you deposit $1000 in an account that pays 5% annual interest so that $r = 0.05$. How much will you have after $t = 2$ years?
  (b) Suppose that you deposit $1000 in an account that pays 5% annual interest so that $r = 0.05$. How much will you have after $t = 3$ years?

Extending the Concepts

105. $\sqrt{\sqrt{x} + 25} = 5$ **106.** $\sqrt{\sqrt{x} + 3} = 2$

107. Solve: $\sqrt{3\sqrt{x} + 1} = \sqrt{2x + 3}$

108. Solve: $\sqrt[3]{2\sqrt{x} - 2} = \sqrt[3]{x} - 1$

Explaining the Concepts

109. Why is it always necessary to check solutions when solving radical equations?

110. How can you tell by inspection that the equation $\sqrt{x} - 2 + 5 = 0$ will have no real solution?

111. Using the concept of domain, explain why radical equations with an even index may have extraneous solutions, but radical equations with an odd index do not have extraneous solutions.

112. Which step in the process of solving a radical equation leads to the possibility of extraneous solutions?

Synthesis Review

In Problems 113–116, consider the set

$$\left\{ 0, -4, 12, \frac{2}{3}, 1.\overline{56}, \sqrt{2^3}, \pi, \sqrt{-5}, \sqrt[3]{-4} \right\}$$

Indicate which of the numbers in the set are ...

113. Integers

114. Rational numbers

115. Irrational numbers

116. Real numbers

117. State the difference between a rational number and an irrational number. Why is $\sqrt{-1}$ not real?

Technology Exercises

Technology can be used to verify solutions obtained algebraically. To solve $\sqrt{2x - 3} = 5$ presented in Example 1, graph $Y_1 = \sqrt{2x - 3}$ and $Y_2 = 5$ and determine the x-coordinate of the point of intersection. The figure below shows the solution to this equation using Desmos.

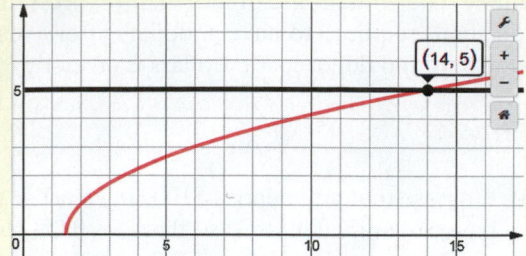

The x-coordinate of the point of intersection is 14, so the solution set is $\{14\}$.

118. Verify your solution to Problem 17 by graphing $Y_1 = \sqrt{2x + 3}$ and $Y_2 = 5$ and then finding the x-coordinate of their point of intersection.

119. Verify your solution to Problem 18 by graphing $Y_1 = \sqrt{3x - 2}$ and $Y_2 = 4$ and then finding the x-coordinate of their point of intersection.

120. When you solved Problem 19 algebraically, you should have determined that the equation has no real solution. Verify this result by graphing $Y_1 = \sqrt{4x + 3}$ and $Y_2 = -2$. Explain what the algebraic solution means graphically.

121. When you solved Problem 20 algebraically, you should have determined that the equation has no real solution. Verify this result by graphing $Y_1 = \sqrt{6x - 5}$ and $Y_2 = -5$. Explain what the algebraic solution means graphically.

9.9 The Complex Number System

Objectives

① Evaluate the Square Root of Negative Real Numbers
② Add or Subtract Complex Numbers
③ Multiply Complex Numbers
④ Divide Complex Numbers
⑤ Evaluate the Powers of i

Are You Prepared for This Section?

Before getting started, complete the following problems. If you get a problem wrong, go back to the section cited and review the material.

P1. List the numbers in the set $\left\{ 8, -\dfrac{1}{3}, -23, 0, \sqrt{2}, 1.\overline{26}, -\dfrac{12}{3}, \sqrt{-5} \right\}$ that are

  (a) Natural numbers   **(b)** Whole numbers
  (c) Integers   **(d)** Rational numbers
  (e) Irrational numbers   **(f)** Real numbers   [Section 1.3, pp. 19–22]

P2. Multiply: $3x(4x - 3)$ [Section 5.3, p. 324]

P3. Multiply: $(z + 4)(3z - 2)$ [Section 5.3, pp. 325–327]

P4. Multiply: $(2y + 5)(2y - 5)$ [Section 5.3, p. 327]

▶ If you look back at Section 1.3, where the various number systems were introduced, you should notice that each time a situation was encountered where a number system could not handle a problem, the number system was expanded. For example, if only the whole numbers were considered, then we could not describe a negative balance in a checking account, so integers were introduced. If the world had to be described only by integers, then parts of a whole, as in ½ a pizza or ¾ of a dollar, could not be talked about, so rational numbers were introduced. If only rational numbers were considered, then it wouldn't be possible to find a number whose square is 2. So the irrational numbers were introduced, such that $(\sqrt{2})^2 = 2$. By combining the rational numbers with the irrational numbers, the real number system was created.

Now, suppose a number whose square is -1 is wanted. Recall that the square of any real number is never negative. This is the *Nonnegativity Property*.

Prepared?...Answers   **P1. (a)** 8
(b) 8, 0   **(c)** 8, −23, 0, −12/3
(d) 8, −1/3, −23, 0, 1.$\overline{26}$, −12/3
(e) $\sqrt{2}$
(f) 8, −$\dfrac{1}{3}$, −23, 0, $\sqrt{2}$, 1.$\overline{26}$, −$\dfrac{12}{3}$
P2. $12x^2 - 9x$ **P3.** $3z^2 + 10z - 8$
P4. $4y^2 - 25$

Nonnegativity Property of Real Numbers

For any real number a, $a^2 \geq 0$.

Because of the Nonnegativity Property, there is no real number solution to the equation

$$x^2 = -1$$

To remedy this situation, a new number is introduced.

Definition

The **imaginary unit,** denoted by *i,* is the number whose square is -1. That is,

$$i^2 = -1$$

Introducing the number *i* establishes a new number system called the **complex number system**.

Definition

Complex numbers are numbers of the form $a + bi$, where *a* and *b* are real numbers. The real number *a* is called the **real part** of the number $a + bi$; the real number *b* is called the **imaginary part** of $a + bi$.

For example, the complex number $6 + 2i$ has the real part 6 and the imaginary part 2. The complex number $4 - 3i = 4 + (-3)i$ has the real part 4 and the imaginary part -3.

When a complex number is written in the form $a + bi$, where *a* and *b* are real numbers, it is in **standard form**. The complex number $a + 0i$ is typically written as *a*. This reminds us that the real number system is a subset of the complex number system. The complex number $0 + bi$ is usually written as *bi*. Any number of the form *bi* is called a **pure imaginary number**. Figure 9 shows the relation among the number systems.

Figure 9
The Complex Number System

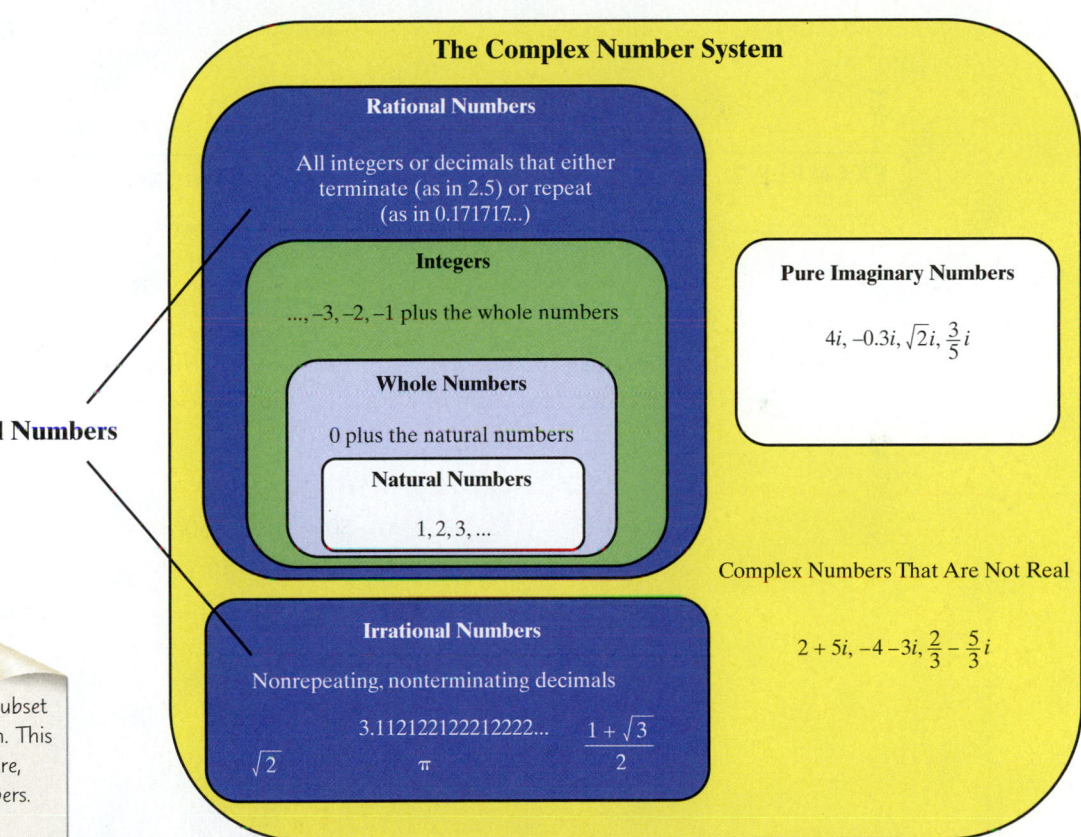

In Other Words
The real number system is a subset of the complex number system. This means that all real numbers are, more generally, complex numbers.

▶ **①** **Evaluate the Square Root of Negative Real Numbers**

First, square roots of negative numbers are defined.

Definition

If *N* is a positive real number, define the **principal square root of** $-N$, denoted by $\sqrt{-N}$, as

$$\sqrt{-N} = \sqrt{N}i$$

where *i* is the number whose square is -1.

EXAMPLE 1 **Evaluating the Square Root of a Negative Number**

Write each of the following as a pure imaginary number.

(a) $\sqrt{-16}$ (b) $\sqrt{-3}$ (c) $\sqrt{-18}$

Solution

(a) $\sqrt{-16} = \sqrt{16}i$ (b) $\sqrt{-3} = \sqrt{3}i$ (c) $\sqrt{-18} = \sqrt{18}i$
$= 4i$ $= 3\sqrt{2}i$ ●

Work Smart

When writing complex numbers whose imaginary part is a radical, as in $\sqrt{3}i$, be sure that the "i" is not written under the radical.

Quick ✓

1. The _____ _____, denoted by i, is the number whose square is -1.

2. Any number of the form bi is called a ____ _____ _____.

3. If N is a positive real number, define the principal square root of $-N$, denoted by $\sqrt{-N}$, as $\sqrt{-N} =$ ____.

4. *True or False* All real numbers are complex numbers.

In Problems 5–7, write each radical as a pure imaginary number.

5. $\sqrt{-36}$ 6. $\sqrt{-5}$ 7. $\sqrt{-12}$

EXAMPLE 2 **Writing Complex Numbers in Standard Form**

Write each of the following in standard form.

(a) $2 - \sqrt{-25}$ (b) $3 + \sqrt{-50}$ (c) $\dfrac{4 - \sqrt{-12}}{2}$

Solution

The standard form of a complex number is $a + bi$.

(a) $2 - \sqrt{-25} = 2 - \sqrt{25}i$
$= 2 - 5i$

(b) $3 + \sqrt{-50} = 3 + \sqrt{50}i$
$\sqrt{50} = \sqrt{25} \cdot \sqrt{2} = 5\sqrt{2}$: $= 3 + 5\sqrt{2}i$

(c) $\dfrac{4 - \sqrt{-12}}{2} = \dfrac{4 - \sqrt{12}i}{2}$

$\sqrt{12} = \sqrt{4} \cdot \sqrt{3} = 2\sqrt{3}$: $= \dfrac{4 - 2\sqrt{3}i}{2}$

Factor out 2: $= \dfrac{2(2 - \sqrt{3}i)}{2}$

Divide out the 2's: $= 2 - \sqrt{3}i$ ●

Quick ✓

In Problems 8–10, write each expression in the standard form of a complex number, $a + bi$.

8. $4 + \sqrt{-100}$ 9. $-2 - \sqrt{-8}$

10. $\dfrac{6 - \sqrt{-72}}{3}$

▶ ❷ Add or Subtract Complex Numbers

You have learned to add, subtract, multiply, and divide using various number systems. Now, we will explore how to perform these operations on complex numbers.

The sum of complex numbers is found by adding the real parts and then adding the imaginary parts.

> **Sum of Complex Numbers**
>
> $$(a + bi) + (c + di) = (a + c) + (b + d)i$$

To subtract two complex numbers, use this rule:

> **Difference of Complex Numbers**
>
> $$(a + bi) - (c + di) = (a - c) + (b - d)i$$

> **In Other Words**
>
> To add two complex numbers, add the real parts, and then add the imaginary parts. To subtract two complex numbers, subtract the real parts, and then subtract the imaginary parts.

EXAMPLE 3 **Adding Complex Numbers**

Add:

(a) $(4 - 3i) + (-2 + 5i)$ **(b)** $(4 + \sqrt{-25}) + (6 - \sqrt{-16})$

Solution

(a) $(4 - 3i) + (-2 + 5i) = [4 + (-2)] + (-3 + 5)i$
$$= 2 + 2i$$

$$\sqrt{-25} = 5i \qquad \sqrt{-16} = 4i$$

(b) $(4 + \sqrt{-25}) + (6 - \sqrt{-16}) = (4 + 5i) + (6 - 4i)$
$$= (4 + 6) + (5 - 4)i$$
$$= 10 + 1i$$
$$= 10 + i$$

EXAMPLE 4 **Subtracting Complex Numbers**

Subtract:

(a) $(-3 + 7i) - (5 - 4i)$ **(b)** $(3 + \sqrt{-12}) - (-2 - \sqrt{-27})$

Solution

(a) $(-3 + 7i) - (5 - 4i) = (-3 - 5) + (7 - (-4))i$
$$= -8 + 11i$$

$$\sqrt{-12} = 2\sqrt{3}i \qquad \sqrt{-27} = 3\sqrt{3}i$$

(b) $(3 + \sqrt{-12}) - (-2 - \sqrt{-27}) = (3 + 2\sqrt{3}i) - (-2 - 3\sqrt{3}i)$
$$= (3 + 2) + [2\sqrt{3} - (-3\sqrt{3})]i$$
$$= 5 + 5\sqrt{3}i$$

> **Work Smart**
>
> Adding or subtracting complex numbers is just like combining like terms. For example,
>
> $(4 - 3x) + (-2 + 5x)$
> $= 4 + (-2) - 3x + 5x$
> $= 2 + 2x$
>
> so
>
> $(4 - 3i) + (-2 + 5i)$
> $= 4 + (-2) - 3i + 5i$
> $= 2 + 2i$

> **Quick ✓**
>
> *In Problems 11–13, add or subtract, as indicated.*
>
> **11.** $(4 + 6i) + (-3 + 5i)$ **12.** $(4 - 2i) - (-2 + 7i)$
>
> **13.** $(4 - \sqrt{-4}) + (-7 + \sqrt{-9})$

▶ ❸ Multiply Complex Numbers

Multiply complex numbers using the same ideas used to multiply polynomials.

EXAMPLE 5

Multiplying Complex Numbers

Multiply:

(a) $4i(3 - 6i)$

(b) $(-2 + 4i)(3 - i)$

Solution

(a) Distribute the $4i$ to each term in the parentheses.

$$4i(3 - 6i) = 4i \cdot 3 - 4i \cdot 6i$$
$$= 12i - 24i^2$$
$$i^2 = -1: \quad = 12i - 24(-1)$$
$$= 24 + 12i$$

(b)
$$\qquad\qquad\qquad\qquad\qquad\text{F}\qquad\text{O}\qquad\text{I}\qquad\text{L}$$
$$(-2 + 4i)(3 - i) = -2 \cdot 3 - 2(-i) + 4i \cdot 3 + 4i(-i)$$
$$= -6 + 2i + 12i - 4i^2$$
$$\text{Combine like terms; } i^2 = -1: \quad = -6 + 14i - 4(-1)$$
$$= -6 + 14i + 4$$
$$= -2 + 14i$$

●

> **Quick ✔**
>
> *In Problems 14 and 15, multiply.*
>
> **14.** $3i(5 - 4i)$ **15.** $(-2 + 5i)(4 - 2i)$

Review the Product Property of Radicals. Notice that the property applies only when $\sqrt[n]{a}$ and $\sqrt[n]{b}$ are real numbers. This means that

$$\sqrt{a} \cdot \sqrt{b} \neq \sqrt{ab} \quad \text{if } a < 0 \text{ or } b < 0$$

How, then, do we perform this multiplication? First write the radical as a complex number using the fact that $\sqrt{-N} = \sqrt{N}i$, and then multiply.

EXAMPLE 6

Multiplying Square Roots of Negative Numbers

Multiply:

(a) $\sqrt{-25} \cdot \sqrt{-4}$

(b) $(2 + \sqrt{-16})(1 - \sqrt{-9})$

Solution

(a) The Product Property of Radicals cannot be used on this product because $\sqrt{-25}$ and $\sqrt{-4}$ are not real numbers. Therefore, express the radicals as pure imaginary numbers and then multiply.

Work Smart

$\sqrt{-25} \cdot \sqrt{-4} \neq \sqrt{(-25)(-4)}$
because the Product Property of
Radicals applies only when the radical
is a real number.

$$\sqrt{-25} \cdot \sqrt{-4} = 5i \cdot 2i$$
$$= 10i^2$$
$$i^2 = -1: \quad = 10(-1)$$
$$= -10$$

(b) First, rewrite each expression as a complex number in standard form.

$$(2 + \sqrt{-16})(1 - \sqrt{-9}) = (2 + 4i)(1 - 3i)$$

$$\text{FOIL:} \quad = 2 \cdot 1 + 2(-3i) + 4i \cdot 1 + 4i(-3i)$$

$$= 2 - 6i + 4i - 12i^2$$

$$\text{Combine like terms; } i^2 = -1: \quad = 2 - 2i - 12(-1)$$

$$= 2 - 2i + 12$$

$$= 14 - 2i$$

> **Quick ✓**
>
> *In Problems 16 and 17, multiply.*
>
> **16.** $\sqrt{-9} \cdot \sqrt{-36}$ **17.** $(2 + \sqrt{-36})(4 - \sqrt{-25})$

▶ Complex Conjugates

A special product that involves the *conjugate* of a complex number is now introduced.

In Other Words
The complex conjugate of $a + bi$ is $a - bi$. The complex conjugate of $a - bi$ is $a + bi$.

> **Complex Conjugate**
>
> If $a + bi$ is a complex number, then its **conjugate** is defined as $a - bi$.

For example,

Complex Number	Conjugate
$4 + 7i$	$4 - 7i$
$-10 - 3i$	$-10 + 3i$

EXAMPLE 7   **Multiplying a Complex Number by Its Conjugate**

Find the product of $4 + 3i$ and its conjugate, $4 - 3i$.

Solution

$$(4 + 3i)(4 - 3i) = 4 \cdot 4 + 4(-3i) + 3i \cdot 4 + 3i(-3i)$$

$$= 16 - 12i + 12i - 9i^2$$

$$= 16 - 9(-1)$$

$$= 16 + 9$$

$$= 25$$

Wow! The product of $4 + 3i$ and its conjugate $4 - 3i$ is 25—a real number! In fact, the result of Example 7 is true in general:

> **Product of a Complex Number and its Conjugate**
>
> The product of a complex number and its conjugate is a nonnegative real number. That is,
>
> $$(a + bi)(a - bi) = a^2 + b^2$$

Notice how multiplying a complex number by its conjugate is different from multiplying $(a + b)(a - b)$:

$$(a + b)(a - b) = a^2 - b^2, \quad \text{whereas} \quad (a + bi)(a - bi) = a^2 + b^2.$$

Do you see why?

▶ **4 Divide Complex Numbers**

Knowing that the product of a complex number and its conjugate is a nonnegative real number makes it possible to divide complex numbers. In Example 8, notice how the approach is similar to rationalizing a denominator. (See Examples 3 and 4 in Section 9.6.)

EXAMPLE 8   **How to Divide Complex Numbers**

Divide: $\dfrac{-3 + i}{5 + 3i}$

Step-by-Step Solution

Step 1: Write the numerator and denominator in standard form, $a + bi$.

The numerator and denominator are already in standard form.

Step 2: Multiply the numerator and denominator by the complex conjugate of the denominator.

Multiply by $1 = \dfrac{5 - 3i}{5 - 3i}$

$$\dfrac{-3 + i}{5 + 3i} = \dfrac{-3 + i}{5 + 3i} \cdot \dfrac{5 - 3i}{5 - 3i}$$

$$= \dfrac{(-3 + i)(5 - 3i)}{(5 + 3i)(5 - 3i)}$$

Step 3: Simplify by writing the quotient in standard form, $a + bi$.

Multiply in the numerator and denominator; $(a + bi)(a - bi) = a^2 + b^2$:

$$= \dfrac{-3 \cdot 5 - 3(-3i) + i \cdot 5 + i(-3i)}{5^2 + 3^2}$$

$$= \dfrac{-15 + 9i + 5i - 3i^2}{25 + 9}$$

Combine like terms; $i^2 = -1$:

$$= \dfrac{-15 + 14i + 3}{34}$$

$$= \dfrac{-12 + 14i}{34}$$

Divide 34 into each term in the numerator to write in standard form:

$$= \dfrac{-12}{34} + \dfrac{14}{34}i$$

Write each fraction in lowest terms:

$$= -\dfrac{6}{17} + \dfrac{7}{17}i$$

●

Below, the steps used in Example 8 are summarized.

Dividing Complex Numbers

Step 1: Write the numerator and denominator in standard form, $a + bi$.

Step 2: Multiply the numerator and denominator by the complex conjugate of the denominator.

Step 3: Simplify by writing the quotient in standard form, $a + bi$.

EXAMPLE 9 **Dividing Complex Numbers**

Divide: $\dfrac{3 + 4i}{2i}$

Solution

The denominator, $2i$, can be written as $0 + 2i$, and the conjugate of $0 + 2i$ is $0 - 2i$. Because $0 - 2i = -2i$, multiply the numerator and denominator by $-2i$.

$$\frac{3 + 4i}{2i} = \frac{3 + 4i}{2i} \cdot \frac{-2i}{-2i}$$

Multiply numerator; multiply denominator: $= \dfrac{(3 + 4i)(-2i)}{(2i)(-2i)}$

Distribute $-2i$: $= \dfrac{-6i - 8i^2}{-4i^2}$

$i^2 = -1$: $= \dfrac{-6i + 8}{4}$

Divide 4 into each term in the numerator to write in standard form: $= \dfrac{8}{4} - \dfrac{6}{4}i$

Write each fraction in lowest terms: $= 2 - \dfrac{3}{2}i$

Work Smart

The numerator and denominator in Example 9 could also have been multiplied by i. Do you know why? Try it yourself! Which approach do you prefer?

In Example 5(b), multiplication was used to find $(-2 + 4i)(3 - i) = -2 + 14i$. Now divide $-2 + 14i$ by $3 - i$. What result do you expect? Verify that $\dfrac{-2 + 14i}{3 - i} = -2 + 4i$. Remember that $\dfrac{A}{B} = C$ can be checked by verifying that $A = B \cdot C$.

> **Quick ✓**
>
> In Problems 21 and 22, divide.
>
> **21.** $\dfrac{-4 + i}{3i}$ **22.** $\dfrac{4 + 3i}{1 - 3i}$

▶ **⑤ Evaluate the Powers of i**

The **powers of i** follow a pattern.

$i^1 = i$ $i^2 = -1$ $i^3 = i^2 \cdot i^1 = -1 \cdot i = -i$ $i^4 = i^2 \cdot i^2 = (-1)(-1) = 1$

$i^5 = i^4 \cdot i = 1 \cdot i = i$ $i^6 = i^4 \cdot i^2 = 1(-1) = -1$ $i^7 = i^4 \cdot i^3 = 1(-i) = -i$ $i^8 = i^4 \cdot i^4 = 1 \cdot 1 = 1$

$i^9 = i^8 \cdot i = 1 \cdot i = i$ $i^{10} = i^8 \cdot i^2 = 1(-1) = -1$ $i^{11} = i^8 \cdot i^3 = 1(-i) = -i$ $i^{12} = (i^4)^3 = (1)^3 = 1$

Do you see the pattern? In the first column, all the expressions simplify to i; in the second column, all the expressions simplify to -1; in the third column, all the expressions simplify to $-i$; in the fourth column, all the expressions simplify to 1. That is, the powers of i repeat with every fourth power. Figure 10 shows the pattern.

Figure 10

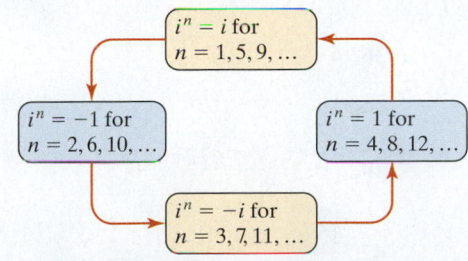

Thus any power of i can be expressed as $1, -1, i,$ or $-i$. Use the steps on the next page to simplify any power of i.

> **Simplifying the Powers of i**
>
> **Step 1:** Divide the exponent of i by 4. Rewrite i^n as $(i^4)^q \cdot i^r$, where q is the quotient and r is the remainder of the division.
>
> **Step 2:** Simplify the product in Step 1 to i^r since $i^4 = 1$.

EXAMPLE 10　**Simplifying Powers of i**

Simplify:

(a) i^{34} 　　　　　　　　　**(b)** i^{101}

Solution

Work Smart

An alternative approach to Example 10:

(a) $i^{34} = (i^2)^{17} = (-1)^{17} = -1$

(b) $i^{101} = i^{100} \cdot i^1 = (i^2)^{50} \cdot i$
$= (-1)^{50} \cdot i$
$= 1 \cdot i$
$= i$

(a) Divide 34 by 4 and obtain a quotient of $q = 8$ and a remainder of $r = 2$, so
$$i^{34} = (i^4)^8 \cdot i^2$$
$$= (1)^8 (-1)$$
$$= -1$$

(b) Divide 101 by 4 and obtain a quotient of $q = 25$ and a remainder of $r = 1$, so
$$i^{101} = (i^4)^{25} \cdot i^1$$
$$= (1)^{25} \cdot i$$
$$= i$$ ●

Quick ✓

In Problems 23 and 24, simplify the power of i.

23. i^{43} 　　　　　　　　**24.** i^{98}

9.9 Exercises　**MyMathLab®**　Exercise numbers in **green** have complete video solutions in MyMathLab or may be accessed using the QR code to the right.

*Problems **1–24** are the Quick ✓ s that follow the **EXAMPLES**.*

Building Skills

In Problems 25–34, write each expression as a pure imaginary number. See Objective 1.

25. $\sqrt{-4}$ 　　**26.** $\sqrt{-25}$ 　　**27.** $-\sqrt{-81}$

28. $-\sqrt{-100}$ 　**29.** $\sqrt{-45}$ 　　**30.** $\sqrt{-48}$

31. $\sqrt{-300}$ 　**32.** $\sqrt{-162}$ 　**33.** $\sqrt{-7}$

34. $\sqrt{-13}$

In Problems 35–42, write each expression as a complex number in standard form. See Objective 1.

35. $5 + \sqrt{-49}$ 　　　　**36.** $4 - \sqrt{-36}$

37. $-2 - \sqrt{-28}$ 　　　**38.** $10 + \sqrt{-32}$

39. $\dfrac{4 + \sqrt{-4}}{2}$ 　　　　　**40.** $\dfrac{10 - \sqrt{-25}}{5}$

41. $\dfrac{4 + \sqrt{-8}}{12}$ 　　　　**42.** $\dfrac{15 - \sqrt{-50}}{5}$

In Problems 43–50, add or subtract, as indicated. See Objective 2.

43. $(4 + 5i) + (2 - 7i)$ 　**44.** $(-6 + 2i) + (3 + 12i)$

45. $(4 + i) - (8 - 5i)$ 　**46.** $(-7 + 3i) - (-3 + 2i)$

47. $(4 - \sqrt{-4}) - (2 + \sqrt{-9})$

48. $(-4 + \sqrt{-25}) + (1 - \sqrt{-16})$

49. $(-2 + \sqrt{-18}) + (5 - \sqrt{-50})$

50. $(-10 + \sqrt{-20}) - (-6 + \sqrt{-45})$

In Problems 51–74, multiply. See Objective 3.

51. $6i(2 - 4i)$ 　　　　**52.** $3i(-2 - 6i)$

53. $-\dfrac{1}{2}i(4 - 10i)$ 　　**54.** $\dfrac{1}{3}i(12 + 15i)$

55. $(2 + i)(4 + 3i)$ 　　**56.** $(3 - i)(1 + 2i)$

57. $(-3 - 5i)(2 + 4i)$ 　**58.** $(5 - 2i)(-1 + 2i)$

59. $(2 - 3i)(4 + 6i)$ 　　**60.** $(6 + 8i)(-3 + 4i)$

61. $(3 - \sqrt{2}i)(-2 + \sqrt{2}i)$ **62.** $(1 + \sqrt{3}i)(-4 - \sqrt{3}i)$

63. $\left(\dfrac{1}{2} - \dfrac{1}{4}i\right)\left(\dfrac{2}{3} + \dfrac{3}{4}i\right)$ **64.** $\left(-\dfrac{2}{3} + \dfrac{4}{3}i\right)\left(\dfrac{1}{2} - \dfrac{3}{2}i\right)$

65. $(3 + 2i)^2$ **66.** $(2 + 5i)^2$

67. $(-4 - 5i)^2$ **68.** $(2 - 7i)^2$

69. $\sqrt{-9} \cdot \sqrt{-4}$ **70.** $\sqrt{-36} \cdot \sqrt{-4}$

71. $\sqrt{-8} \cdot \sqrt{-10}$ **72.** $\sqrt{-12} \cdot \sqrt{-15}$

73. $(2 + \sqrt{-81})(-3 - \sqrt{-100})$

74. $(1 - \sqrt{-64})(-2 + \sqrt{-49})$

In Problems 75–80, (a) find the conjugate of the complex number, and (b) multiply the complex number by its conjugate. See Objective 3.

75. $3 + 5i$ **76.** $5 + 2i$

77. $2 - 7i$ **78.** $9 - i$

79. $-7 + 2i$ **80.** $-1 - 4i$

In Problems 81–94, divide. See Objective 4.

81. $\dfrac{1 + i}{3i}$ **82.** $\dfrac{2 - i}{2i}$

83. $\dfrac{-5 + 2i}{5i}$ **84.** $\dfrac{-4 + 5i}{6i}$

85. $\dfrac{3}{2 + i}$ **86.** $\dfrac{2}{4 + i}$

87. $\dfrac{-2}{-3 - 7i}$ **88.** $\dfrac{-4}{-5 - 3i}$

89. $\dfrac{2 + 3i}{3 - 2i}$ **90.** $\dfrac{2 + 5i}{5 - 2i}$

91. $\dfrac{4 + 2i}{1 - i}$ **92.** $\dfrac{-6 + 2i}{1 + i}$

93. $\dfrac{4 - 2i}{1 + 3i}$ **94.** $\dfrac{5 - 3i}{2 + 4i}$

In Problems 95–102, simplify. See Objective 5.

95. i^{53} **96.** i^{72}

97. i^{43} **98.** i^{110}

99. i^{153} **100.** i^{-45}

101. i^{131} **102.** i^{-26}

Mixed Practice

In Problems 103–116, perform the indicated operation.

103. $(-4 - i)(4 + i)$ **104.** $(-5 + 2i)(5 - 2i)$

105. $(3 + 2i)^2$ **106.** $(-3 + 2i)^2$

107. $\dfrac{-3 + 2i}{3i}$ **108.** $\dfrac{5 - 3i}{4i}$

109. $\dfrac{-4 + i}{-5 - 3i}$ **110.** $\dfrac{-4 + 6i}{-5 - i}$

111. $(10 - 3i) + (2 + 3i)$ **112.** $(-4 + 5i) + (4 - 2i)$

113. $5i^{37}(-4 + 3i)$ **114.** $2i^{57}(3 - 4i)$

115. $\sqrt{-10} \cdot \sqrt{-15}$ **116.** $\sqrt{-8} \cdot \sqrt{-12}$

In Problems 117–122, find the reciprocal of the complex number. Write each number in standard form.

117. $5i$ **118.** $7i$

119. $2 - i$ **120.** $3 - 5i$

121. $-4 + 5i$ **122.** $-6 + 2i$

123. Suppose that $f(x) = x^2$; find **(a)** $f(i)$ **(b)** $f(1 + i)$.

124. Suppose that $f(x) = x^2 + x$; find **(a)** $f(i)$ **(b)** $f(1 + i)$.

125. Suppose that $f(x) = x^2 + 2x + 2$; find **(a)** $f(3i)$ **(b)** $f(1 - i)$.

126. Suppose that $f(x) = x^2 + x - 1$; find **(a)** $f(2i)$ **(b)** $f(2 + i)$.

Applying the Concepts

127. Impedance (Series Circuit) The total impedance, Z, of an ac circuit containing components in series is equivalent to the sum of the individual impedances. Impedance is measured in ohms (Ω) and is expressed as an imaginary number of the form $Z = R + i \cdot X$. Here, R represents resistance and X represents reactance.

  (a) If the impedance in one part of a series circuit is $7 + 3i$ ohms and the impedance of the remainder of the circuit is $3 - 4i$ ohms, find the total impedance of the circuit.

  (b) What is the total resistance of the circuit?

  (c) What is the total reactance of the circuit?

128. Impedance (Parallel Circuit) The total impedance, Z, of an ac circuit consisting of two parallel pathways is given by the formula $\dfrac{1}{Z} = \dfrac{1}{Z_1} + \dfrac{1}{Z_2}$, where Z_1 and Z_2 are the impedances of the pathways. If the impedances of the individual pathways are $Z_1 = 5$ ohms and $Z_2 = 1 - 2i$ ohms, find the total impedance of the circuit.

Extending the Concepts

129. For the function $f(x) = x^2 + 4x + 5$, find **(a)** $f(-2 + i)$ **(b)** $f(-2 - i)$.

130. For the function $f(x) = x^2 - 2x + 2$, find **(a)** $f(1 + i)$ **(b)** $f(1 - i)$.

131. For the function $f(x) = x^3 + 1$, find **(a)** $f(-1)$ **(b)** $f\left(\dfrac{1}{2} + \dfrac{\sqrt{3}}{2}i\right)$ **(c)** $f\left(\dfrac{1}{2} - \dfrac{\sqrt{3}}{2}i\right)$

132. For the function $f(x) = x^3 - 1$, find **(a)** $f(1)$

(b) $f\left(-\dfrac{1}{2} + \dfrac{\sqrt{3}}{2}i\right)$ **(c)** $f\left(-\dfrac{1}{2} - \dfrac{\sqrt{3}}{2}i\right)$.

133. Any complex number z such that $f(z) = 0$ is called a **complex zero** of f. Look at the complex zeros in Problems 129–132. Conjecture a general result regarding the complex zeros of polynomials that have real coefficients.

134. The Complex Plane We are able to plot real numbers on a real number line. We are able to plot ordered pairs (x, y) in a Cartesian plane. Can we plot complex numbers? Yes! A complex number $x + yi$ can be interpreted geometrically as the point (x, y) in the xy-plane. Each point in the plane corresponds to a complex number, and each complex number corresponds to a point in the plane (just as in ordered pairs). In the **complex plane,** the x-axis is called the **real axis** and the y-axis is called the **imaginary axis.**

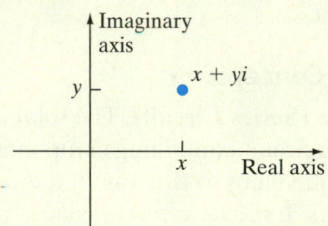

(a) Plot the complex number $4 + 2i$ in the complex plane.

(b) Plot the complex number $-2 + 6i$ in the complex plane.

(c) Plot the complex number $-5i$ in the complex plane.

(d) Plot the complex number -4 in the complex plane.

Explaining the Concepts

135. Explain the relations among the natural numbers, whole numbers, integers, rational numbers, real numbers, and complex number system.

136. Explain why the product of a complex number and its conjugate is a nonnegative real number. That is, explain why $(a + bi)(a - bi) = a^2 + b^2$.

137. How is multiplying two complex numbers related to multiplying two binomials?

138. How is the method used to rationalize denominators related to the method used to write the quotient of two complex numbers in standard form?

Synthesis Review

139. Expand: $(x + 2)^3$ **140.** Expand: $(y - 4)^2$

141. Evaluate: $(3 + i)^3$ **142.** Evaluate: $(4 - 3i)^2$

143. How is raising a complex number to a positive integer power related to raising a binomial to a positive integer power?

Technology Exercises

Technology gives us the ability to add, subtract, multiply, and divide complex numbers. Using a graphing calculator, first put the calculator into complex mode, as shown in Figure 11(a). Figure 11(b) shows the results of Examples 3(a) and 3(b). Figure 11(c) shows the results of Examples 5(b) and 8.

Figure 11

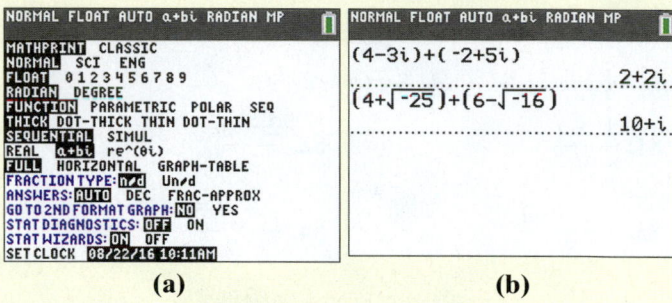

(a)   (b)

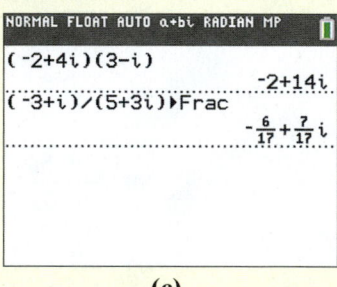

(c)

In Problems 144–151, use technology to simplify each expression. Write your answers in standard form, $a + bi$.

144. $(1.3 - 4.3i) + (-5.3 + 0.7i)$

145. $(-3.4 + 1.9i) - (6.5 - 5.3i)$

146. $(0.3 - 5.2i)(1.2 + 3.9i)$

147. $(-4.3 + 0.2i)(7.2 - 0.5i)$

148. $\dfrac{4}{3 - 8i}$ **149.** $\dfrac{1 - 7i}{-4 + i}$

150. $3i^6 + (5 - 0.3i)(3 + 2i)$

151. $-6i^{14} + 5i^6 - (2 + i)(3 + 11i)$

Chapter 9 Activity: Which One Does Not Belong?

Focus: Simplifying Radicals and Rational Exponents

Time: 15–25 minutes

Group Size: 2–4

5–10 Minutes: Individually, evaluate each row below. Decide which item does not belong and why. Be creative with your reasons!

10–15 Minutes: Discuss each row, and make a list of all members' responses. Is there any example that everyone agrees with?

	A	B	C	D
1	$16^{\frac{3}{2}}$	$-16^{\frac{3}{2}}$	$-16^{\frac{2}{3}}$	$-16^{-\frac{2}{3}}$
2	$\sqrt{(a-b)^3}, a > b$	$\sqrt{a^3 - b^3}$	$\sqrt{(a-b)^3}, a < b$	$\sqrt{a-b}, a > b$
3	The range of $f(x) = x^2$	The range of $f(x) = \sqrt[3]{x}$	The range of $f(x) = |x|$	The range of $f(x) = x^3$
4	i^{212}	i^0	i^{-108}	i

Chapter 9 Review

Section 9.1 Square Roots

KEY CONCEPTS

- The principal square root of any positive number is positive.
- The principal square root of 0 is 0 because $0^2 = 0$. That is, $\sqrt{0} = 0$.
- The square root of a perfect square is a rational number.
- The square root of a positive rational number that is not a perfect square is an irrational number.
- The square root of a negative real number is not a real number.
- For any real number a, $\sqrt{a^2} = |a|$. If $a \geq 0$, $\sqrt{a^2} = a$.

KEY TERMS

Square root
Radical
Principal square root
Radicand
Perfect square

You Should Be Able To...	EXAMPLE	Review Exercises
1 Evaluate square roots of perfect squares (p. 616)	Examples 1 and 2	1–12
2 Determine whether a square root is rational, irrational, or not a real number (p. 618)	Examples 3 and 4	13–16
3 Find square roots of variable expressions (p. 619)	Example 5	17–18

In Problems 1 and 2, find the value of each expression.

1. the square roots of 4

2. the square roots of 81

In Problems 3–12, find the exact value of each expression.

3. $-\sqrt{1}$

4. $-\sqrt{25}$

5. $\sqrt{0.16}$

6. $\sqrt{0.04}$

7. $\dfrac{3}{2}\sqrt{\dfrac{25}{36}}$

8. $\dfrac{4}{3}\sqrt{\dfrac{81}{4}}$

9. $\sqrt{25 - 9}$

10. $\sqrt{169 - 25}$

11. $\sqrt{9^2 - (4)(5)(-18)}$

12. $\sqrt{13^2 - (4)(-3)(-4)}$

In Problems 13–16, determine whether each square root is rational, irrational, or not a real number. Then evaluate the square root. For each square root that is irrational, use a calculator or Appendix D to round your answer to two decimal places.

13. $-\sqrt{9}$

14. $-\frac{1}{2}\sqrt{48}$

15. $\sqrt{14}$

16. $\sqrt{-2}$

In Problems 17 and 18, simplify each square root. Assume that the variable can be any real number.

17. $\sqrt{(4x-9)^2}$

18. $\sqrt{m^2 - 10m + 25}$

Section 9.2   *n*th Roots and Rational Exponents

KEY CONCEPTS

- **Simplifying** $\sqrt[n]{a^n}$
- If $n \geq 2$ is a positive integer and a is a real number, then

$$\sqrt[n]{a^n} = |a| \quad \text{if } n \geq 2 \text{ is even}$$
$$\sqrt[n]{a^n} = a \quad \text{if } n \geq 3 \text{ is odd}$$

- If a is a real number and $n \geq 2$ is an integer, then $a^{\frac{1}{n}} = \sqrt[n]{a}$ provided that $\sqrt[n]{a}$ exists.

- If a is a real number, $\dfrac{m}{n}$ is a rational number in lowest terms with $n \geq 2$, then $a^{\frac{m}{n}} = \sqrt[n]{a^m} = (\sqrt[n]{a})^m$ provided that $\sqrt[n]{a}$ exists.

- If $\dfrac{m}{n}$ is a rational number and if a is a nonzero real number, then

$$a^{-\frac{m}{n}} = \frac{1}{a^{\frac{m}{n}}} \text{ or } \frac{1}{a^{-\frac{m}{n}}} = a^{\frac{m}{n}}, \text{ if } a \neq 0 \text{ and the } n\text{th root of } a \text{ is defined.}$$

KEY TERMS

Principal *n*th root of a
  number *a*
Index
Cube root

You Should Be Able To...	EXAMPLE	Review Exercises
❶ Evaluate *n*th roots (p. 620)	Examples 1 and 2	19–23, 37, 38
❷ Simplify expressions of the form $\sqrt[n]{a^n}$ (p. 622)	Example 3	24–26
❸ Evaluate expressions of the form $a^{\frac{1}{n}}$ (p. 623)	Examples 4 and 5	27–30, 35, 39
❹ Evaluate expressions of the form $a^{\frac{m}{n}}$ (p. 624)	Examples 6 through 9	31–34, 36, 40–42

In Problems 19–26, simplify each radical.

19. $\sqrt[3]{343}$

20. $\sqrt[3]{-125}$

21. $\sqrt[3]{\dfrac{8}{27}}$

22. $\sqrt[4]{81}$

23. $-\sqrt[5]{-243}$

24. $\sqrt[3]{10^3}$

25. $\sqrt[5]{z^5}$

26. $\sqrt[4]{(5p-3)^4}$

In Problems 27–34, evaluate each of the following expressions.

27. $81^{\frac{1}{2}}$

28. $(-256)^{\frac{1}{4}}$

29. $-4^{\frac{1}{2}}$

30. $729^{\frac{1}{3}}$

31. $16^{\frac{7}{4}}$

32. $-(-27)^{\frac{2}{3}}$

33. $-121^{\frac{3}{2}}$

34. $\dfrac{1}{36^{-\frac{1}{2}}}$

In Problems 35–38, use a calculator to write each expression rounded to two decimal places.

35. $(-65)^{\frac{1}{3}}$

36. $4^{\frac{3}{5}}$

37. $\sqrt[3]{100}$

38. $\sqrt[4]{10}$

In Problems 39–42, rewrite each of the following radicals with a rational exponent.

39. $\sqrt[3]{5a}$

40. $\sqrt[5]{p^7}$

41. $(\sqrt[4]{10z})^3$

42. $\sqrt[6]{(2ab)^5}$

Section 9.3   Simplifying Expressions Using the Laws of Exponents

KEY CONCEPTS

KEY TERM

Simplify

- If a and b are real numbers and if r and s are rational numbers, then, assuming the expression is defined, the following rules apply.

Zero-Exponent Rule: $\qquad\qquad a^0 = 1 \qquad\qquad$ if $a \neq 0$

Negative-Exponent Rule: $\qquad\quad a^{-r} = \dfrac{1}{a^r} \qquad$ if $a \neq 0$

Product Rule: $\qquad\qquad\quad a^r \cdot a^s = a^{r+s}$

Quotient Rule: $\qquad\qquad\quad \dfrac{a^r}{a^s} = a^{r-s} = \dfrac{1}{a^{s-r}} \quad$ if $a \neq 0$

Power Rule: $\qquad\qquad\quad (a^r)^s = a^{r \cdot s}$

Product-to-a-Power Rule: $\qquad (a \cdot b)^r = a^r \cdot b^r$

Quotient-to-a-Power Rule: $\qquad \left(\dfrac{a}{b}\right)^r = \dfrac{a^r}{b^r} \qquad$ if $b \neq 0$

Quotient-to-a-Negative-Power Rule: $\quad \left(\dfrac{a}{b}\right)^{-r} = \left(\dfrac{b}{a}\right)^r \quad$ if $a \neq 0, b \neq 0$

You Should Be Able To...	EXAMPLE	Review Exercises
1 Simplify expressions involving rational exponents (p. 628)	Examples 1 through 3	43–48
2 Simplify radical expressions (p. 631)	Example 4	49–52
3 Factor expressions containing rational exponents (p. 632)	Examples 5 and 6	53, 54

In Problems 43–52, simplify the expression. Assume all variables are positive.

43. $4^{\frac{2}{3}} \cdot 4^{\frac{7}{3}}$

44. $\dfrac{k^{\frac{1}{2}}}{k^{\frac{3}{4}}}$

45. $\left(p^{\frac{4}{3}} \cdot q^4\right)^{\frac{3}{2}}$

46. $\left(32a^{-\frac{3}{2}} \cdot b^{\frac{1}{4}}\right)^{\frac{1}{5}}$

47. $5m^{-\frac{2}{3}}\left(2m + m^{-\frac{1}{3}}\right)$

48. $\left(\dfrac{16x^{\frac{1}{3}}}{x^{-\frac{1}{3}}}\right)^{-\frac{1}{2}} + \left(\dfrac{x^{-\frac{3}{2}}}{64x^{-\frac{1}{2}}}\right)^{\frac{1}{3}}$

49. $\sqrt[8]{x^6}$

50. $\sqrt{121x^4y^{10}}$

51. $\sqrt[3]{m^2} \cdot \sqrt{m^3}$

52. $\dfrac{\sqrt[3]{c}}{\sqrt[6]{c^4}}$

In Problems 53 and 54, factor the expression.

53. $2(3m - 1)^{\frac{1}{4}} + (m - 7)(3m - 1)^{\frac{5}{4}}$

54. $3(x^2 - 5)^{\frac{1}{3}} - 4x(x^2 - 5)^{-\frac{2}{3}}$

Section 9.4   Simplifying Radical Expressions Using Properties of Radicals

KEY CONCEPTS

KEY TERM

Simplify

- **Product Property of Radicals**
  If $\sqrt[n]{a}$ and $\sqrt[n]{b}$ are real numbers and $n \geq 2$ is an integer, then $\sqrt[n]{a} \cdot \sqrt[n]{b} = \sqrt[n]{ab}$.

- **Quotient Property of Radicals**
  If $\sqrt[n]{a}$ and $\sqrt[n]{b}$ are real numbers, $b \neq 0$, and $n \geq 2$ is an integer, then $\dfrac{\sqrt[n]{a}}{\sqrt[n]{b}} = \sqrt[n]{\dfrac{a}{b}}$.

You Should Be Able To...	EXAMPLE	Review Exercises
❶ Use the Product Property to multiply radical expressions (p. 634)	Example 1	55, 56, 67–70
❷ Use the Product Property to simplify radical expressions (p. 635)	Examples 2 through 7	57–70
❸ Use the Quotient Property to simplify radical expressions (p. 639)	Examples 8 and 9	71–78
❹ Multiply radicals with unlike indices (p. 640)	Example 10	79, 80

In Problems 55 and 56, use the Product Property to multiply. Assume that all variables can be any real number.

55. $\sqrt{15} \cdot \sqrt{7}$

56. $\sqrt[4]{2ab^2} \cdot \sqrt[4]{6a^2b}$

In Problems 57–62, simplify each radical using the Product Property. Assume that all variables can be any real number.

57. $\sqrt{80}$

58. $\sqrt[3]{-500}$

59. $\sqrt[3]{162m^6n^4}$

60. $\sqrt[4]{50p^8q^4}$

61. $2\sqrt{16x^6y}$

62. $\sqrt{(2x+1)^3}$

In Problems 63–66, simplify each radical using the Product Property. Assume that all variables are greater than or equal to zero.

63. $\sqrt{w^3z^2}$

64. $\sqrt{45x^4yz^3}$

65. $\sqrt[3]{16a^{12}b^5}$

66. $\sqrt{4x^2+8x+4}$

In Problems 67–70, multiply and simplify. Assume that all variables are greater than or equal to zero.

67. $\sqrt{15} \cdot \sqrt{18}$

68. $\sqrt[3]{20} \cdot \sqrt[3]{30}$

69. $\sqrt[3]{-3x^4y^7} \cdot \sqrt[3]{24x^3y^2}$ **70.** $3\sqrt{4xy^2} \cdot 5\sqrt{3x^2y}$

In Problems 71–78, simplify. Assume that all variables are greater than zero.

71. $\sqrt{\dfrac{121}{25}}$

72. $\sqrt{\dfrac{5a^4}{64b^2}}$

73. $\sqrt[3]{\dfrac{54k^2}{9k^5}}$

74. $\sqrt[3]{\dfrac{-160w^{11}}{343w^{-4}}}$

75. $\dfrac{\sqrt{12h^3}}{\sqrt{3h}}$

76. $\dfrac{\sqrt{50a^3b^3}}{\sqrt{8a^5b^{-3}}}$

77. $\dfrac{\sqrt[3]{-8x^7y}}{\sqrt[3]{27xy^4}}$

78. $\dfrac{\sqrt[4]{48m^2n^7}}{\sqrt[4]{3m^6n}}$

In Problems 79 and 80, multiply and simplify.

79. $\sqrt{5} \cdot \sqrt[3]{2}$

80. $\sqrt[4]{8} \cdot \sqrt[6]{4}$

Section 9.5 Adding, Subtracting, and Multiplying Radical Expressions

KEY CONCEPTS

- To add or subtract radicals, the radicals must have the same index and the same radicand.
- To multiply radicals, the index on each radical must be the same. Then multiply the radicands.

KEY TERM

Like radicals

You Should Be Able To...	EXAMPLE	Review Exercises
❶ Add or subtract radical expressions (p. 643)	Examples 1 through 3	81–90
❷ Multiply radical expressions (p. 645)	Examples 4 and 5	91–100

In Problems 81–90, add or subtract as indicated. Assume all variables are greater than or equal to zero.

81. $2\sqrt[4]{x} + 6\sqrt[4]{x}$

82. $7\sqrt[3]{4y} + 2\sqrt[3]{4y} - 3\sqrt[3]{4y}$

83. $5\sqrt{2} - 2\sqrt{12}$

84. $\sqrt{18} + 2\sqrt{50}$

85. $\sqrt[3]{-16z} + \sqrt[3]{54z}$

86. $7\sqrt[3]{8x^2} - \sqrt[3]{-27x^2}$

87. $\sqrt{16a} + \sqrt[6]{729a^3}$

88. $\sqrt{27x^2} - x\sqrt{48} + 2\sqrt{75x^2}$

89. $5\sqrt[3]{4m^5y^2} - \sqrt[6]{16m^{10}y^4}$

90. $\sqrt{y^3-4y^2} - 2\sqrt{y-4} + \sqrt[4]{y^2-8y+16}$

In Problems 91–100, multiply and simplify. Assume all variables are greater than or equal to zero.

91. $\sqrt{3}(\sqrt{5} - \sqrt{15})$

92. $\sqrt[3]{5}(3 + \sqrt[3]{4})$

93. $(3 + \sqrt{5})(4 - \sqrt{5})$

94. $(7 + \sqrt{3})(6 + \sqrt{2})$

95. $(1 - 3\sqrt{5})(1 + 3\sqrt{5})$

96. $(\sqrt[3]{x} + 1)(9\sqrt[3]{x} - 4)$

97. $(\sqrt{x} - \sqrt{5})^2$

98. $(11\sqrt{2} + \sqrt{5})^2$

99. $(\sqrt{2a} - b)(\sqrt{2a} + b)$

100. $(\sqrt[3]{6s} + 2)(\sqrt[3]{6s} - 7)$

Section 9.6 Rationalizing Radical Expressions

KEY CONCEPTS

- To rationalize the denominator containing a single radical with index n, multiply the numerator and denominator by a radical so that the radicand in the denominator is a perfect nth power.
- To rationalize a denominator containing two terms, use the fact that $(A + B)(A - B) = A^2 - B^2$.

KEY TERM

Rationalizing the denominator

You Should Be Able To...	EXAMPLE	Review Exercises
1 Rationalize a denominator containing one term (p. 649)	Examples 1 and 2	101–108, 117
2 Rationalize a denominator containing two terms (p. 652)	Examples 3 and 4	109–116, 118

In Problems 101–116, rationalize the denominator. Assume all variables are greater than or equal to zero.

101. $\dfrac{2}{\sqrt{6}}$

102. $\dfrac{6}{\sqrt{3}}$

103. $\dfrac{\sqrt{48}}{\sqrt{p^3}}$

104. $\dfrac{5}{\sqrt{2a}}$

105. $\dfrac{-2}{\sqrt{6y^3}}$

106. $\dfrac{3}{\sqrt[3]{5}}$

107. $\sqrt[3]{\dfrac{-4}{45}}$

108. $\dfrac{27}{\sqrt[5]{8p^3q^4}}$

109. $\dfrac{6}{7 - \sqrt{6}}$

110. $\dfrac{3}{\sqrt{3} - 9}$

111. $\dfrac{\sqrt{3}}{3 + \sqrt{2}}$

112. $\dfrac{\sqrt{k}}{\sqrt{k} - \sqrt{m}}$

113. $\dfrac{\sqrt{10} + 2}{\sqrt{10} - 2}$

114. $\dfrac{3 - \sqrt{y}}{3 + \sqrt{y}}$

115. $\dfrac{4}{2\sqrt{3} + 5\sqrt{2}}$

116. $\dfrac{\sqrt{5} - \sqrt{6}}{\sqrt{10} + \sqrt{3}}$

117. Simplify: $\dfrac{\sqrt{7}}{3} + \dfrac{6}{\sqrt{7}}$

118. Find the reciprocal: $4 - \sqrt{7}$

Section 9.7 Functions Involving Radicals

KEY CONCEPT

- **Finding the Domain of a Function Involving Radicals**
  1. If the index on a radical expression is even, then the radicand must be greater than or equal to zero.
  2. If the index on a radical expression is odd, then the radicand can be any real number.

KEY TERMS

Square root function
Cube root function

You Should Be Able To...	EXAMPLE	Review Exercises
1 Evaluate functions involving radicals (p. 655)	Example 1	119–122
2 Find the domain of a function involving a radical (p. 656)	Example 2	123–128, 129(a), 130(a), 131(a), 132(a)
3 Graph functions involving square roots (p. 657)	Example 3	129(b)–131(b)
4 Graph functions involving cube roots (p. 658)	Example 4	132(b)

In Problems 119–122, evaluate each radical function at the indicated values.

119. $f(x) = \sqrt{x+4}$
   (a) $f(-3)$
   (b) $f(0)$
   (c) $f(5)$

120. $g(x) = \sqrt{3x-2}$
   (a) $g\left(\dfrac{2}{3}\right)$
   (b) $g(2)$
   (c) $g(6)$

121. $H(t) = \sqrt[3]{t+3}$
   (a) $H(-2)$
   (b) $H(-4)$
   (c) $H(5)$

122. $G(z) = \sqrt{\dfrac{z-1}{z+2}}$
   (a) $G(1)$
   (b) $G(-3)$
   (c) $G(2)$

In Problems 123–128, find the domain of the radical function.

123. $f(x) = \sqrt{3x-5}$

124. $g(x) = \sqrt[3]{2x-7}$

125. $h(x) = \sqrt[4]{6x+1}$

126. $F(x) = \sqrt[5]{2x-9}$

127. $G(x) = \sqrt{\dfrac{4}{x-2}}$

128. $H(x) = \sqrt{\dfrac{x-3}{5}}$

In Problems 129–132, (a) determine the domain of the function; (b) graph the function using point plotting; and (c) based on the graph, determine the range of the function.

129. $f(x) = \dfrac{1}{2}\sqrt{1-x}$

130. $g(x) = \sqrt{x+1} - 2$

131. $h(x) = -\sqrt{x+3}$

132. $F(x) = \sqrt[3]{x+1}$

Section 9.8 Radical Equations and Their Applications

KEY TERMS

Radical equation   Extraneous solution

You Should Be Able To...	EXAMPLE	Review Exercises
❶ Solve radical equations containing one radical (p. 662)	Examples 1 through 5	133–142, 147, 148
❷ Solve radical equations containing two radicals (p. 665)	Examples 6 and 7	143–146
❸ Solve for a variable in a radical equation (p. 667)	Example 8	149, 150

In Problems 133–148, solve each equation.

133. $\sqrt{m} = 13$

134. $\sqrt[3]{3t+1} = -2$

135. $\sqrt[4]{3x-8} = 3$

136. $\sqrt{2x+5} + 4 = 2$

137. $\sqrt{4-k} - 3 = 0$

138. $3\sqrt{t} - 4 = 11$

139. $2\sqrt[3]{m} + 5 = -11$

140. $\sqrt{q+2} = q$

141. $\sqrt{w+11} + 3 = w + 2$

142. $\sqrt{p^2 - 2p + 9} = p + 1$

143. $\sqrt{a+10} = \sqrt{2a-1}$

144. $\sqrt{5x+9} = \sqrt{7x-3}$

145. $\sqrt{c-8} + \sqrt{c} = 4$

146. $\sqrt{x+2} - \sqrt{x+9} = 7$

147. $(4x-3)^{1/3} - 3 = 0$

148. $(x^2 - 9)^{1/4} = 2$

149. Height of a Cone Solve $r = \sqrt{\dfrac{3V}{\pi h}}$ for h.

150. Ball Slide Speed Factor Solve $f_s = \sqrt[3]{\dfrac{30}{v}}$ for v.

Section 9.9 The Complex Number System

KEY CONCEPTS

- **Nonnegativity Property of Real Numbers**
  For any real number a, $a^2 \geq 0$.

- **Imaginary Unit**
  The imaginary unit, denoted by i, is the number whose square is -1.
  That is, $i^2 = -1$.

KEY TERMS

Imaginary Unit
Complex number system
Complex number
Real part
Imaginary part

- **Complex numbers**
  Complex numbers are numbers of the form $a + bi$, where a and b are real numbers. The real number a is called the real part of the number $a + bi$; the real number b is called the imaginary part of $a + bi$.

- **Square Roots of Negative Numbers**
  If N is a positive real number, then the principal square root of $-N$, denoted by $\sqrt{-N}$, is $\sqrt{-N} = \sqrt{N}i$, where i is the number whose square is -1.

- **Sum of Complex Numbers**
  $(a + bi) + (c + di) = (a + c) + (b + d)i$

- **Difference of Complex Numbers**
  $(a + bi) - (c + di) = (a - c) + (b - d)i$

- **Complex Conjugate**
  If $a + bi$ is a complex number, then its conjugate is $a - bi$.

- **Product of a Complex Number and Its Conjugate**
  $(a + bi)(a - bi) = a^2 + b^2$

Standard form
Pure imaginary number
Principal square root
Conjugate
Powers of i

You Should Be Able To...	EXAMPLE	Review Exercises
❶ Evaluate the square root of negative real numbers (p. 671)	Examples 1 and 2	151–154
❷ Add or subtract complex numbers (p. 673)	Examples 3 and 4	155–158
❸ Multiply complex numbers (p. 674)	Examples 5 through 7	159–164
❹ Divide complex numbers (p. 676)	Examples 8 and 9	165–168
❺ Evaluate the powers of i (p. 677)	Example 10	169–170

In Problems 151 and 152, write each expression as a pure imaginary number.

151. $\sqrt{-29}$ **152.** $\sqrt{-54}$

In Problems 153 and 154, write each expression as a complex number in standard form.

153. $14 - \sqrt{-162}$ **154.** $\dfrac{6 + \sqrt{-45}}{3}$

In Problems 155–168, perform the indicated operation.

155. $(3 - 7i) + (-2 + 5i)$

156. $(4 + 2i) - (9 - 8i)$

157. $(8 - \sqrt{-45}) - (3 + \sqrt{-80})$

158. $(1 + \sqrt{-9}) + (-6 + \sqrt{-16})$

159. $(4 - 5i)(3 + 7i)$

160. $\left(\dfrac{1}{2} + \dfrac{2}{3}i\right)(4 - 9i)$

161. $\sqrt{-3} \cdot \sqrt{-27}$

162. $(1 + \sqrt{-36})(-5 - \sqrt{-144})$

163. $(1 + 12i)(1 - 12i)$

164. $(7 + 2i)(5 + 4i)$

165. $\dfrac{4}{3 + 5i}$ **166.** $\dfrac{-3}{7 - 2i}$

167. $\dfrac{2 - 3i}{5 + 2i}$ **168.** $\dfrac{4 + 3i}{1 - i}$

In Problems 169 and 170, simplify.

169. i^{59} **170.** i^{173}

Chapter 9 Test Step-by-step test solutions are found on the Chapter Test Prep Videos available in MyMathLab®, on YouTube™, or may be accessed using the QR code to the right.

1. Evaluate: $49^{-\frac{1}{2}}$

In Problems 2 and 3, simplify using rational exponents.

2. $\sqrt[3]{8x^{\frac{1}{2}}y^3} \cdot \sqrt{9xy^{\frac{1}{2}}}$

3. $\sqrt[5]{(2a^4b^3)^7}$

In Problems 4–9, perform the indicated operation and simplify. Assume all variables are greater than zero.

4. $\sqrt{3m} \cdot \sqrt{13n}$

5. $\sqrt{32x^7y^4}$

6. $\dfrac{\sqrt{9a^3b^{-3}}}{\sqrt{4ab}}$

7. $\sqrt{5x^3} + 2\sqrt{45x}$

8. $\sqrt{9a^2b} - \sqrt[4]{16a^4b^2}$

9. $(11 + 2\sqrt{x})(3 - \sqrt{x})$

In Problems 10 and 11, rationalize the denominator.

10. $\dfrac{-2}{3\sqrt{72}}$

11. $\dfrac{\sqrt{5}}{\sqrt{5} + 2}$

12. For $f(x) = \sqrt{-2x + 3}$, find the following:
    (a) $f(1)$ (b) $f(-3)$

13. Determine the domain of the function
    $g(x) = \sqrt{-3x + 5}$.

14. For $f(x) = \sqrt{x} - 3$, do the following:
    (a) Determine the domain of the function.
    (b) Graph the function using point plotting.
    (c) From the graph, determine the range of the function.

In Problems 15–17, solve the given equations.

15. $\sqrt{x + 3} = 4$

16. $\sqrt{x + 13} - 4 = x - 3$

17. $\sqrt{x - 1} + \sqrt{x + 2} = 3$

In Problems 18–20, perform the indicated operation.

18. $(13 + 2i) + (4 - 15i)$

19. $(4 - 7i)(2 + 3i)$

20. $\dfrac{7 - i}{12 + 11i}$

Cumulative Review Chapters 1–9

1. Evaluate: $6 - 3^2 \div (9 - 3)$

2. Simplify: $(3x + 2y) - (2x - 5y + 3) + 9$

3. Solve: $(3x + 5) - 2 = 7x - 13$

4. Solve and write your answer in interval notation:
   $$6x + \frac{1}{2}(4x - 2) \le 3x + 9$$

5. For $f(x) = 3x^2 - x + 5$, find the following:
   (a) $f(-2)$ (b) $f(3)$

6. Find the domain of $g(x) = \dfrac{x^2 - 9}{x^2 - 2x - 8}$.

7. The annual sales for a certain computer game is given by $n(p) = -50p + 6000$, where n is the number of games sold and p is the price in dollars.
   (a) How many games will be sold in one year if the price were $50?
   (b) At what price will there be no sales during a given year?

8. Find the equation of the line that passes through the points $(-1, 6)$ and $(3, -2)$.

9. Graph the inequality: $6x + 3y > 24$

10. Solve the systemu using substitution:
    $$\begin{cases} 4x - y = 17 \\ 5x + 6y = 14 \end{cases}$$

11. Dried fruit is to be mixed with nuts to create a trail mix blend. The fruit costs $3.45 per pound and the nuts cost $2.10 per pound. How much of each should be used to make 10 pounds of trail mix that will sell for $2.64 per pound if the total revenue should remain the same?

12. If $g(x) = x^3 + 4x - 16$, what is $g(-4)$?

13. Add: $(8x^3 - 4x^2 + 5x + 3) + (2x^2 - 8x + 7)$

14. Multiply: $(2x - 1)(4x^2 + 2x - 9)$

15. Divide: $\dfrac{6x^4 + 13x^3 - 21x^2 - 28x + 37}{2x^2 + 3x - 5}$

16. Factor completely: $8x^2 - 44x - 84$

17. Subtract: $\dfrac{2x}{x - 3} - \dfrac{x + 1}{x + 2}$

18. Solve: $\dfrac{9}{k - 2} = \dfrac{6}{k} + 3$

19. Solve and graph the solution set on a number line:
$3x + 1 < 13 \text{ or } -2x + 3 \le 23$

20. Shawn can paint a certain room in 4 hours. Payton can paint the same room in 6 hours. How long will it take them to paint the room if they work together?

21. Simplify: $\dfrac{\sqrt{50a^3b}}{\sqrt{2a^{-1}b^3}}$

22. Find the domain of the function $f(x) = \sqrt[4]{8 - 3x}$.

23. Solve: $\sqrt{x + 7} - 8 = x - 7$

24. Divide: $\dfrac{3i}{1 - 7i}$

25. Rationalize the denominator: $\dfrac{2}{4 - \sqrt{11}}$

10 Quadratic Equations and Functions

One of the more unusual sports—Punkin Chunkin—holds its World Championship competition every year in Millsboro, Delaware. Participants catapult or fire pumpkins to see who can toss them the farthest (and most accurately). Interestingly, this activity is an application of a quadratic function at work. See Problems 79–82 in Section 10.5.

The Big Picture: Putting It Together

In Chapter 2, we solved linear equations and inequalities in one variable. In Chapter 3, our discussion of "everything linear" concluded by covering linear equations and inequalities in two variables.

This chapter is dedicated to extending the discussion of quadratic equations, $ax^2 + bx + c = 0$, that began in Section 6.6. In Section 6.6, quadratic equations $ax^2 + bx + c = 0$, were solved by factoring and using the Zero-Product Property. Remember, if the expression $ax^2 + bx + c$ was not factorable, there was not a method for solving the equation. The missing piece for solving this type of quadratic equation was the idea of a radical. The familiarity with radicals acquired in Chapter 9 supplies the tools necessary to solve all quadratic equations $ax^2 + bx + c = 0$, regardless of whether the expression $ax^2 + bx + c$ is factorable. In addition to solving quadratic equations, graphing quadratic functions and solving quadratic and polynomial inequalities will be covered.

10.1 Solving Quadratic Equations by Completing the Square

Objectives

❶ Solve Quadratic Equations Using the Square Root Property

❷ Complete the Square in One Variable

❸ Solve Quadratic Equations by Completing the Square

❹ Solve Problems Using the Pythagorean Theorem

Are You Prepared for This Section?

Before getting started, complete the following problems. If you get a problem wrong, go back to the section cited and review the material.

P1. Multiply: $(2p + 3)^2$ [Section 5.3, pp. 329–330]

P2. Factor: $y^2 - 8y + 16$ [Section 6.4, pp. 396–398]

P3. Solve: $x^2 + 5x - 14 = 0$ [Section 6.6, pp. 410–415]

P4. Solve: $x^2 - 16 = 0$ [Section 6.6, pp. 410–415]

P5. Simplify: **(a)** $\sqrt{36}$ **(b)** $\sqrt{45}$ **(c)** $\sqrt{-12}$ [Section 9.1, pp. 616–617; Section 9.4, pp. 635–639; Section 9.9, pp. 671–672]

P6. Find the complex conjugate of $-3 + 2i$. [Section 9.9, p. 675]

P7. Simplify: $\sqrt{x^2}$ [Section 9.1, pp. 618–619]

P8. Define: $|x|$ [Section 1.3, pp. 24–25]

▶ ❶ Solve Quadratic Equations Using the Square Root Property

Suppose we want to solve the quadratic equation

$$x^2 = p$$

where p is any real number. This equation is saying, "Give me all numbers whose square is p." For example, the equation $x^2 = 16$ means "Find all numbers whose square is 16." Because those numbers are -4 and 4, the solution set to the equation $x^2 = 16$ is $\{-4, 4\}$.

In general, to solve an equation of the form $x^2 = p$, take the square root of both sides of the equation.

$$x^2 = p$$

Take the square root of both sides: $\quad \sqrt{x^2} = \sqrt{p}$

$\sqrt{x^2} = |x|: \quad |x| = \sqrt{p}$

$|x| = -x \text{ if } x < 0; |x| = x \text{ if } x \geq 0: \quad -x = \sqrt{p} \quad \text{or} \quad x = \sqrt{p}$

Solve for x: $\quad x = -\sqrt{p} \quad \text{or} \quad x = \sqrt{p}$

This gives the following result.

Work Smart

The Square Root Property is useful for solving equations of the form "some unknown squared equals a real number." To solve this equation, take the square root of both sides of the equation, but do not forget the $\pm$ symbol to obtain the positive and negative square root.

The Square Root Property

If $x^2 = p$, then $x = \sqrt{p}$ or $x = -\sqrt{p}$.

In using the Square Root Property to solve an equation such as $x^2 = p$, the solutions are usually written as $x = \pm\sqrt{p}$, which is read "x equals plus or minus the square root of p." For example, the two solutions of the equation

$$x^2 = 16$$

are

$$x = \pm\sqrt{16}$$
$$= \pm 4$$

Let's use the Square Root Property in an example now.

EXAMPLE 1   **How to Solve a Quadratic Equation Using the Square Root Property**

Solve: $x^2 - 9 = 0$

Step-by-Step Solution

Step 1: Isolate the expression containing the square term.

$$x^2 - 9 = 0$$

Add 9 to both sides: $x^2 = 9$

Step 2: Use the Square Root Property. Don't forget the $\pm$ symbol.

$$x = \pm\sqrt{9}$$

Simplify the radical: $= \pm 3$

Step 3: Isolate the variable, if necessary.

The variable is already isolated.

Step 4: Check Verify your solution(s).

$x = -3$: $(-3)^2 - 9 \stackrel{?}{=} 0$ $x = 3$: $3^2 - 9 \stackrel{?}{=} 0$

$9 - 9 = 0$ True $9 - 9 = 0$ True

The solution set is $\{-3, 3\}$.

The steps that are used to solve a quadratic equation using the Square Root Property are summarized below.

> **Solving a Quadratic Equation Using the Square Root Property**
>
> **Step 1:** Isolate the expression containing the squared term.
> **Step 2:** Use the Square Root Property: If $x^2 = p$, then $x = \pm\sqrt{p}$.
> **Step 3:** Isolate the variable, if necessary.
> **Step 4:** Verify your solution(s).

The equation in Example 1 could also be solved by factoring the difference of two squares:

$$x^2 - 9 = 0$$
$$(x - 3)(x + 3) = 0$$

Zero-Product Property: $x - 3 = 0$ or $x + 3 = 0$

$x = 3$ or $x = -3$

Work Smart

As our mathematical knowledge develops, there is often more than one way to solve a problem.

So there is more than one way to find the solution! However, in solving equations of the form $x^2 = p$, factoring works nicely only when p is a perfect square.

EXAMPLE 2   **Solving a Quadratic Equation Using the Square Root Property**

Solve: $3x^2 - 60 = 0$

Solution

$$3x^2 - 60 = 0$$

Add 60 to both sides of the equation: $3x^2 = 60$

Divide both sides by 3: $x^2 = 20$

Use the Square Root Property: $x = \pm\sqrt{20}$

Simplify the radical: $= \pm 2\sqrt{5}$

Check

$x = -2\sqrt{5}:$ $3(-2\sqrt{5})^2 - 60 \stackrel{?}{=} 0$ $x = 2\sqrt{5}:$ $3(2\sqrt{5})^2 - 60 \stackrel{?}{=} 0$

$(ab)^2 = a^2b^2:$ $3(-2)^2(\sqrt{5})^2 - 60 \stackrel{?}{=} 0$ $3(2)^2(\sqrt{5})^2 - 60 \stackrel{?}{=} 0$

$3 \cdot 4 \cdot 5 - 60 \stackrel{?}{=} 0$ $3 \cdot 4 \cdot 5 - 60 \stackrel{?}{=} 0$

$60 - 60 = 0$ True $60 - 60 = 0$ True

The solution set is $\left\{-2\sqrt{5}, 2\sqrt{5}\right\}$. ●

> **Quick ✓**
>
> **1.** If $x^2 = p$, then $x =$ ___ or $x =$ ____.
>
> *In Problems 2–4, solve the quadratic equation using the Square Root Property.*
>
> **2.** $p^2 = 48$ **3.** $3b^2 = 75$ **4.** $s^2 - 81 = 0$

As the next example illustrates, the solution to a quadratic equation does not have to be real.

EXAMPLE 3 **Solving a Quadratic Equation Using the Square Root Property**

Solve: $y^2 + 14 = 2$

Solution

$$y^2 + 14 = 2$$

Subtract 14 from both sides: $y^2 = -12$

Use the Square Root Property: $y = \pm\sqrt{-12}$

Simplify the radical: $= \pm 2\sqrt{3}i$

Check

$y = -2\sqrt{3}i:$ $(-2\sqrt{3}i)^2 + 14 \stackrel{?}{=} 2$ $y = 2\sqrt{3}i:$ $(2\sqrt{3}i)^2 + 14 \stackrel{?}{=} 2$

$(-2\sqrt{3})^2 i^2 + 14 \stackrel{?}{=} 2$ $(2\sqrt{3})^2 i^2 + 14 \stackrel{?}{=} 2$

$i^2 = -1:$ $4 \cdot 3 \cdot (-1) + 14 \stackrel{?}{=} 2$ $4 \cdot 3 \cdot (-1) + 14 \stackrel{?}{=} 2$

$-12 + 14 \stackrel{?}{=} 2$ $-12 + 14 \stackrel{?}{=} 2$

$2 = 2$ True $2 = 2$ True

The solution set is $\left\{-2\sqrt{3}i, 2\sqrt{3}i\right\}$. ●

> **Quick ✓**
>
> *In Problems 5 and 6, solve the quadratic equation using the Square Root Property.*
>
> **5.** $d^2 = -72$ **6.** $3q^2 + 27 = 0$

EXAMPLE 4 **Solving Quadratic Equations Using the Square Root Property**

Solve:

(a) $(x - 2)^2 = 25$ **(b)** $(y + 5)^2 + 24 = 0$

Solution

(a)

$$(x - 2)^2 = 25$$

Use the Square Root Property: $x - 2 = \pm\sqrt{25}$

Simplify the radical: $x - 2 = \pm 5$

Add 2 to each side: $x = 2 \pm 5$

2 ± 5 means $2 - 5$ or $2 + 5$: $x = 2 - 5$ or $x = 2 + 5$

$= -3$ $= 7$

(continued)

Check

$x = -3$: $(-3 - 2)^2 \overset{?}{=} 25$ $\qquad$ $x = 7$: $(7 - 2)^2 \overset{?}{=} 25$

$\qquad\qquad (-5)^2 \overset{?}{=} 25$ $\qquad\qquad\qquad\qquad 5^2 \overset{?}{=} 25$

$\qquad\qquad\qquad 25 = 25$ True $\qquad\qquad\qquad\quad 25 = 25$ True

The solution set is $\{-3, 7\}$.

(b) $\hfill (y + 5)^2 + 24 = 0$

Subtract 24 from both sides: $\hfill (y + 5)^2 = -24$

Use the Square Root Property: $\hfill y + 5 = \pm\sqrt{-24}$

$\sqrt{-24} = \sqrt{24} \cdot i = \sqrt{4 \cdot 6} i = 2\sqrt{6}i$: $\hfill y + 5 = \pm 2\sqrt{6}i$

Subtract 5 from each side: $\hfill y = -5 \pm 2\sqrt{6}i$

$$y = -5 - 2\sqrt{6}i \quad \text{or} \quad y = -5 + 2\sqrt{6}i$$

Check

$y = -5 - 2\sqrt{6}i$: $\qquad\qquad\qquad\qquad\qquad$ $y = -5 + 2\sqrt{6}i$:

$(-5 - 2\sqrt{6}i + 5)^2 + 24 \overset{?}{=} 0$ $\qquad\qquad$ $(-5 + 2\sqrt{6}i + 5)^2 + 24 \overset{?}{=} 0$

$(-2\sqrt{6}i)^2 + 24 \overset{?}{=} 0$ $\qquad\qquad\qquad$ $(2\sqrt{6}i)^2 + 24 \overset{?}{=} 0$

$4 \cdot 6 \cdot i^2 + 24 \overset{?}{=} 0$ $\qquad\qquad\qquad$ $4 \cdot 6 \cdot i^2 + 24 \overset{?}{=} 0$

$-24 + 24 = 0$ True $\qquad\qquad\qquad$ $-24 + 24 = 0$ True

The solution set is $\{-5 - 2\sqrt{6}i, -5 + 2\sqrt{6}i\}$. $\qquad\qquad\qquad\qquad\qquad$ ●

Quick ✓

In Problems 7 and 8, solve the quadratic equation using the Square Root Property.

7. $(y + 3)^2 = 100$ $\qquad\qquad\qquad$ **8.** $(q - 5)^2 + 20 = 4$

▶ ② **Complete the Square in One Variable**

The idea behind **completing the square** in one variable is to "adjust" the left side of a quadratic equation of the form $x^2 + bx + c = 0$ to make it a perfect square trinomial. Recall that perfect square trinomials have the form

$$A^2 + 2AB + B^2 = (A + B)^2 \quad \text{or} \quad A^2 - 2AB + B^2 = (A - B)^2$$

For example, $x^2 + 6x + 9$ is a perfect square trinomial because $x^2 + 6x + 9 = x^2 + 2 \cdot x \cdot 3 + 3^2 = (x + 3)^2$.

"Adjust" $x^2 + bx + c$ by adding a number to make it a perfect square trinomial. For example, to make $x^2 + 6x$ a perfect square, add 9. Why choose 9? Dividing the coefficient of the first-degree term, 6, by 2 and then squaring the result yields 9. That is $\left(\dfrac{6}{2}\right)^2 = 3^2 = 9$. This approach works in general.

Work Smart

To complete the square, the coefficient of x^2 must be 1.

> **Obtaining a Perfect Square Trinomial**
>
> Multiply the coefficient of the first-degree term by $\dfrac{1}{2}$. Square the result. That is, for $x^2 + bx$, compute $\left(\dfrac{1}{2}b\right)^2$, and add the result to the expression $x^2 + bx$ to obtain $x^2 + bx + \left(\dfrac{1}{2}b\right)^2$.

EXAMPLE 5 **Obtaining a Perfect Square Trinomial**

Determine the number that must be added to each expression to make it a perfect square trinomial. Then factor the expression.

Start	Add	Result	Factored Form
$y^2 + 8y$	$\left(\dfrac{1}{2} \cdot 8\right)^2 = 4^2 = 16$	$y^2 + 8y + 16$	$(y + 4)^2$
$x^2 + 12x$	$\left(\dfrac{1}{2} \cdot 12\right)^2 = 6^2 = 36$	$x^2 + 12x + 36$	$(x + 6)^2$
$a^2 - 20a$	$\left(\dfrac{1}{2}(-20)\right)^2 = (-10)^2 = 100$	$a^2 - 20a + 100$	$(a - 10)^2$
$p^2 - 5p$	$\left(\dfrac{1}{2}(-5)\right)^2 = \left(\dfrac{-5}{2}\right)^2 = \dfrac{25}{4}$	$p^2 - 5p + \dfrac{25}{4}$	$\left(p - \dfrac{5}{2}\right)^2$

Work Smart

It is important to notice that the expression at the "Start" is not equivalent to the "Result" because we are adding a quantity to the expression at the "Start."

Work Smart

It is common to write the value $\left(\dfrac{1}{2}b\right)^2$ as a fraction, not a decimal.

Notice in the factored form that the perfect square trinomial always factors so that

$$x^2 + bx + \left(\frac{b}{2}\right)^2 = \left(x + \frac{b}{2}\right)^2 \quad \text{or} \quad x^2 - bx + \left(\frac{b}{2}\right)^2 = \left(x - \frac{b}{2}\right)^2$$

That is, the perfect square trinomial always factors as $\left(x \pm \dfrac{b}{2}\right)^2$, where the $+$ is used if the coefficient of the first-degree term is positive and the $-$ is used if the coefficient of the first-degree term is negative. The $\dfrac{b}{2}$ represents $\dfrac{1}{2}$ the value of the coefficient of the first-degree term.

Quick ✓

In Problems 9 and 10, determine the number that must be added to the expression to make it a perfect square trinomial. Then factor the expression.

9. $p^2 + 14p$ **10.** $w^2 - 3w$

Figure 1

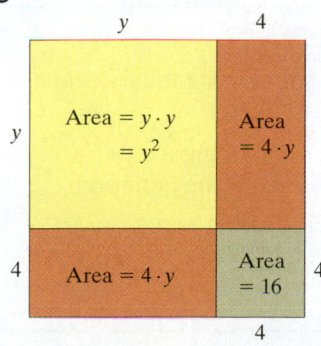

Why is this process called "completing the square"? Consider the expression $y^2 + 8y$ given in Example 5, which is geometrically represented in Figure 1. The yellow area is y^2 and each orange area is $4y$ (for a total area of $8y$). But what is the area of the green region to make the square complete? The dimensions of the green region are 4 by 4, so its area is 16. The area of the entire square region, $(y + 4)^2$, equals the sum of the area of the four regions that make up the square: $y^2 + 4y + 4y + 16 = y^2 + 8y + 16$.

❸ Solve Quadratic Equations by Completing the Square

Until now, we have solved quadratic equations of the form $ax^2 + bx + c = 0$ only when $ax^2 + bx + c$ was factorable. Now we can solve $ax^2 + bx + c = 0$ when $ax^2 + bx + c$ is not factorable. We begin with equations of the form $x^2 + bx + c = 0$. That is, the coefficient of the square term is 1.

▶ **EXAMPLE 6** **How to Solve a Quadratic Equation by Completing the Square**

Solve: $x^2 + 6x + 1 = 0$

Step-by-Step Solution

Step 1: Rewrite $x^2 + bx + c = 0$ as $x^2 + bx = -c$ by subtracting the constant from both sides of the equation.

Subtract 1 from both sides:

$$x^2 + 6x + 1 = 0$$
$$x^2 + 6x = -1$$

Step 2: Complete the square in the expression $x^2 + bx$ by making it a perfect square trinomial. Add the same number to each side of the equation to maintain the equality.

$\left(\dfrac{1}{2} \cdot 6\right)^2 = 3^2 = 9$; add 9 to both sides:

$$x^2 + 6x + 9 = -1 + 9$$
$$x^2 + 6x + 9 = 8$$

Step 3: Factor the perfect square trinomial on the left side of the equation.

$x^2 + 6x + 9 = (x + 3)^2$:

$$(x + 3)^2 = 8$$

Step 4: Solve the equation using the Square Root Property.

$\sqrt{8} = 2\sqrt{2}$:

Subtract 3 from both sides:

$a \pm b$ means $a - b$ or $a + b$:

$$x + 3 = \pm\sqrt{8}$$
$$x + 3 = \pm 2\sqrt{2}$$
$$x = -3 \pm 2\sqrt{2}$$
$$x = -3 - 2\sqrt{2} \quad \text{or} \quad x = -3 + 2\sqrt{2}$$

Step 5: **Check** Verify your solution(s).

$$x^2 + 6x + 1 = 0$$

$x = -3 - 2\sqrt{2}$:

$$(-3 - 2\sqrt{2})^2 + 6(-3 - 2\sqrt{2}) + 1 \overset{?}{=} 0$$
$$9 + 12\sqrt{2} + 8 - 18 - 12\sqrt{2} + 1 \overset{?}{=} 0$$
$$0 = 0 \quad \text{True}$$

$x = -3 + 2\sqrt{2}$:

$$(-3 + 2\sqrt{2})^2 + 6(-3 + 2\sqrt{2}) + 1 \overset{?}{=} 0$$
$$9 - 12\sqrt{2} + 8 - 18 + 12\sqrt{2} + 1 \overset{?}{=} 0$$
$$0 = 0 \quad \text{True}$$

The solution set is $\left\{-3 - 2\sqrt{2}, -3 + 2\sqrt{2}\right\}$. ●

Solving a Quadratic Equation By Completing the Square

Step 1: Rewrite $x^2 + bx + c = 0$ as $x^2 + bx = -c$ by subtracting the constant from both sides of the equation.

Step 2: Complete the square in the expression $x^2 + bx$ by making it a perfect square trinomial. Add the same number to each side of the equation.

Step 3: Factor the perfect square trinomial on the left side of the equation.

Step 4: Solve the equation using the Square Root Property.

Step 5: Verify your solution(s).

Quick ✓

In Problems 11 and 12, solve the equation by completing the square.

11. $b^2 + 2b - 8 = 0$

12. $z^2 - 8z + 9 = 0$

Work Smart

If the coefficient of the square term is not 1, then before using the method of completing the square, divide each side of the equation by the coefficient of the square term so that it becomes 1.

When the coefficient of the square term is not 1, multiply or divide both sides of the equation by a nonzero constant so this coefficient becomes 1. The next example demonstrates this method.

> **EXAMPLE 7** **Solving a Quadratic Equation by Completing the Square When the Coefficient of the Square Term Is Not 1**

Solve: $2x^2 + 4x + 3 = 0$

Solution

To make the coefficient of the square term be 1, divide both sides of the equation by 2.

$$2x^2 + 4x + 3 = 0$$

Divide both sides of the equation by 2: $\dfrac{2x^2 + 4x + 3}{2} = \dfrac{0}{2}$

Simplify: $x^2 + 2x + \dfrac{3}{2} = 0$

Now solve the equation by completing the square.

Subtract $\dfrac{3}{2}$ from both sides: $x^2 + 2x = -\dfrac{3}{2}$

$\left(\dfrac{1}{2} \cdot 2\right)^2 = 1^2 = 1$; add 1 to both sides: $x^2 + 2x + 1 = -\dfrac{3}{2} + 1$

Simplify: $x^2 + 2x + 1 = -\dfrac{1}{2}$

Factor: $(x + 1)^2 = -\dfrac{1}{2}$

Use the Square Root Property: $x + 1 = \pm\sqrt{-\dfrac{1}{2}}$

$\sqrt{-N} = \sqrt{N}\,i$: $x + 1 = \pm\sqrt{\dfrac{1}{2}}\,i$

$\sqrt{\dfrac{1}{2}} = \dfrac{\sqrt{1}}{\sqrt{2}} = \dfrac{\sqrt{1}}{\sqrt{2}} \cdot \dfrac{\sqrt{2}}{\sqrt{2}} = \dfrac{\sqrt{2}}{2}$: $x + 1 = \pm\dfrac{\sqrt{2}}{2}\,i$

Subtract 1 from both sides: $x = -1 \pm \dfrac{\sqrt{2}}{2}\,i$

$a \pm b$ means $a - b$ or $a + b$: $x = -1 - \dfrac{\sqrt{2}}{2}\,i \quad \text{or} \quad x = -1 + \dfrac{\sqrt{2}}{2}\,i$

Work Smart

Notice the solutions in Example 7 are complex conjugates of each other.

The solution set is $\left\{-1 - \dfrac{\sqrt{2}}{2}\,i, -1 + \dfrac{\sqrt{2}}{2}\,i\right\}$. The check is left to you. ●

Quick ✓

In Problems 13 and 14, solve the quadratic equation by completing the square.

13. $2q^2 + 6q - 1 = 0$

14. $3m^2 + 2m + 7 = 0$

Figure 2

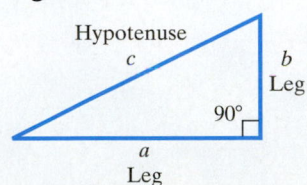

Hypotenuse c

b
Leg

$90°$

a
Leg

> **④ Solve Problems Using the Pythagorean Theorem**

The Pythagorean Theorem is a statement about *right triangles*. A **right triangle** is one that contains a **right angle**—that is, an angle of 90°. The side of the triangle opposite the 90° angle is the **hypotenuse;** the remaining two sides are the **legs.** In Figure 2, c represents the length of the hypotenuse; a and b represent the lengths of the legs. Notice the use of the symbol ⌐ to show the 90° angle.

The Pythagorean Theorem

In a right triangle, the square of the length of the hypotenuse equals the sum of the squares of the lengths of the legs. That is, in the right triangle shown on the previous page in Figure 2,

$$c^2 = a^2 + b^2$$

EXAMPLE 8   **Finding the Hypotenuse of a Right Triangle**

In a right triangle, one leg is of length 5 inches and the other is of length 12 inches. What is the length of the hypotenuse?

Solution

Because this is a right triangle, use the Pythagorean Theorem with $a = 5$ and $b = 12$ to find the length c of the hypotenuse.

$$c^2 = a^2 + b^2$$
$$c^2 = 5^2 + 12^2$$
$$= 25 + 144$$
$$= 169$$

Now use the Square Root Property to find c, the length of the hypotenuse.

$$c = \sqrt{169} = 13$$

The length of the hypotenuse is 13 inches. Because c represents a length, which must be positive, find only the positive square root of 169.   ●

Quick ✓

15. The side of a right triangle opposite the 90° angle is called the _____; the remaining two sides are called ___.

16. *True or False* The Pythagorean Theorem states that for any triangle, the length of the hypotenuse is equal to the sum of the squares of the lengths of the legs.

In Problem 17, the lengths of the legs of a right triangle are given. Find the length of the hypotenuse.

17. $a = 3, b = 4$

EXAMPLE 9   **How Far Can You See?**

The Currituck Lighthouse is located in Corolla, in the Outer Banks of North Carolina. The lighthouse, completed in 1875, stands 162 feet tall, however, a person standing in the observation deck is located 158 feet above the ground. See Figure 3.

Figure 3

The website for the Currituck Lighthouse states that a person standing on the observation deck could see approximately 18 miles. See Figure 4. Assuming that the radius of the Earth is 3960 miles, verify this claim.

Figure 4

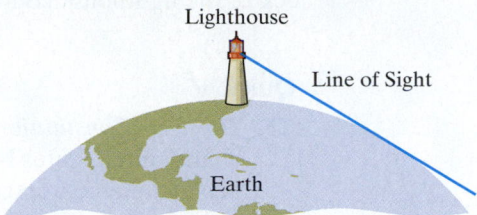

Lighthouse

Line of Sight

Earth

Solution

Step 1: Identify Find how far a person can see from the lighthouse.

Step 2: Name Call this unknown distance d, in miles.

Step 3: Translate To help with the translation, draw a picture. From the center of Earth, draw two lines: one through the lighthouse and the other to the farthest point a person can see from the lighthouse. See Figure 5.

Figure 5

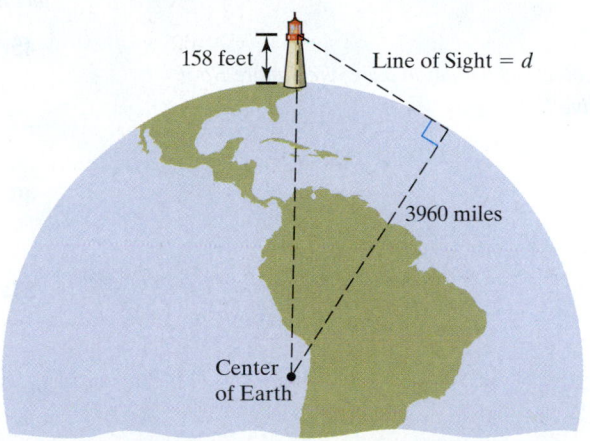

158 feet

Line of Sight = d

3960 miles

Center of Earth

The line of sight and the two lines drawn from the center of Earth form a right triangle, so the angle where the line of sight touches the horizon measures 90°. From the Pythagorean Theorem,

$$\text{Hypotenuse}^2 = \text{Leg}^2 + \text{Leg}^2$$

The length of the hypotenuse is 3960 miles plus 158 feet. Add 3960 miles to 158 feet by converting the height of the tower to miles. Because 158 feet =

$158 \text{ feet} \cdot \dfrac{1 \text{ mile}}{5280 \text{ feet}} = \dfrac{158}{5280} \text{ mile},$ the hypotenuse is $\left(3960 + \dfrac{158}{5280}\right)$ miles. One of the

legs is 3960 miles. The length of the other leg is our unknown, d. Thus

$$3960^2 + d^2 = \left(3960 + \frac{158}{5280}\right)^2 \qquad \textcolor{magenta}{\text{The Model}}$$

Step 4: Solve

$$3960^2 + d^2 = \left(3960 + \frac{158}{5280}\right)^2$$

$\textcolor{magenta}{\text{Subtract } 3960^2 \text{ from both sides:}} \quad d^2 = \left(3960 + \dfrac{158}{5280}\right)^2 - 3960^2$

$\textcolor{magenta}{\text{Use a calculator:}} \quad d^2 \approx 237.000895$

$\textcolor{magenta}{\text{Square Root Property:}} \quad d \approx \sqrt{237.000895}$

$$\approx 15.39 \text{ miles}$$

(continued)

Step 5: Check The solution is less than the distance given on the website.

Step 6: Answer The distance given on the Currituck Lighthouse website appears to overstate the actual distance a person could see. Someone standing on the observation deck of the lighthouse could see about 15.39 miles. ●

Quick ✓

18. The USS *Constitution* (aka *Old Ironsides*) is perhaps the most famous ship from United States Naval history. The mainmast of the Constitution is 220 feet high. Suppose that a sailor climbs the mainmast to a height of 200 feet in order to look for enemy vessels. How far could the sailor see? Assume the radius of the Earth is 3960 miles.

10.1 Exercises · MyMathLab®

Exercise numbers in **green** have complete video solutions in MyMathLab or may be accessed using the QR code to the right.

*Problems 1–18 are the **Quick ✓**s that follow the **EXAMPLES**.*

Building Skills

In Problems 19–44, solve each equation using the Square Root Property. See Objective 1.

19. $y^2 = 100$

20. $x^2 = 81$

21. $p^2 = 50$

22. $z^2 = 48$

23. $m^2 = -25$

24. $n^2 = -49$

25. $w^2 = \dfrac{5}{4}$

26. $z^2 = \dfrac{8}{9}$

27. $x^2 + 5 = 13$

28. $w^2 - 6 = 14$

29. $3z^2 = 48$

30. $4y^2 = 100$

31. $3x^2 = 8$

32. $5y^2 = 32$

33. $2p^2 + 23 = 15$

34. $-3x^2 - 5 = 22$

35. $(d - 1)^2 = -18$

36. $(z + 4)^2 = -24$

37. $3(q + 5)^2 - 1 = 8$

38. $5(x - 3)^2 + 2 = 27$

39. $(3q + 1)^2 = 9$

40. $(2p + 3)^2 = 16$

41. $\left(x - \dfrac{2}{3}\right)^2 = \dfrac{5}{9}$

42. $\left(y + \dfrac{3}{2}\right)^2 = \dfrac{3}{4}$

43. $x^2 + 8x + 16 = 81$

44. $q^2 - 6q + 9 = 16$

In Problems 45–52, complete the square in each expression. Then factor the perfect square trinomial. See Objective 2.

45. $x^2 + 10x$

46. $y^2 + 16y$

47. $z^2 - 18z$

48. $p^2 - 4p$

49. $y^2 + 7y$

50. $x^2 + x$

51. $w^2 + \dfrac{1}{2}w$

52. $z^2 - \dfrac{1}{3}z$

In Problems 53–72, solve each quadratic equation by completing the square. See Objective 3.

53. $x^2 + 4x = 12$

54. $y^2 + 3y = 18$

55. $x^2 - 4x + 1 = 0$

56. $p^2 - 6p + 4 = 0$

57. $a^2 - 4a + 5 = 0$

58. $m^2 - 2m + 5 = 0$

59. $b^2 + 5b - 2 = 0$

60. $q^2 + 7q + 7 = 0$

61. $m^2 = 8m + 3$

62. $n^2 = 10n + 5$

63. $p^2 - p + 3 = 0$

64. $z^2 - 3z + 5 = 0$

65. $2y^2 - 5y - 12 = 0$

66. $3a^2 - 4a - 4 = 0$

67. $3y^2 - 6y + 2 = 0$

68. $2y^2 - 2y - 1 = 0$

69. $2z^2 - 5z + 1 = 0$

70. $2x^2 - 7x + 2 = 0$

71. $2x^2 + 4x + 5 = 0$

72. $2z^2 + 6z + 5 = 0$

In Problems 73–82, the lengths of the legs of a right triangle are given. Find the hypotenuse. Give exact answers and decimal approximations rounded to two decimal places. See Objective 4.

73. $a = 6, b = 8$

74. $a = 7, b = 24$

75. $a = 12, b = 16$

76. $a = 15, b = 8$

77. $a = 5, b = 5$

78. $a = 3, b = 3$

79. $a = 1, b = \sqrt{3}$

80. $a = 2, b = \sqrt{5}$

81. $a = 6, b = 10$

82. $a = 8, b = 10$

In Problems 83–86, use the right triangle shown below and find the missing length. Give exact answers and decimal approximations rounded to two decimal places. See Objective 4.

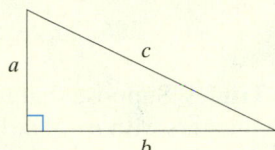

83. $a = 4, c = 8$

84. $a = 4, c = 10$

85. $b = 8, c = 12$

86. $b = 2, c = 10$

Mixed Practice

87. Given that $f(x) = (x - 3)^2$, find all x such that $f(x) = 36$. What points are on the graph of f?

88. Given that $f(x) = (x - 5)^2$, find all x such that $f(x) = 49$. What points are on the graph of f?

89. Given that $g(x) = (x + 2)^2$, find all x such that $g(x) = 18$. What points are on the graph of g?

90. Given that $h(x) = (x + 1)^2$, find all x such that $h(x) = 32$. What points are on the graph of h?

Applying the Concepts

In Problems 91 and 92, find the exact length of the diagonal in each rectangle.

△ **91.**
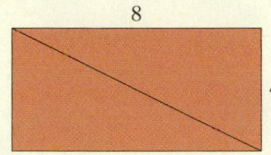

△ **92.**
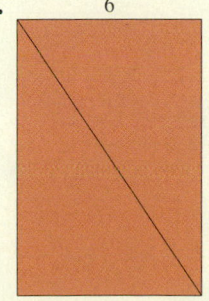

In Problems 93–100, express your answer as a decimal rounded to three decimal places.

93. Golf A golfer hits an errant tee shot that lands in the rough. The golfer finds that the ball is exactly 30 yards to the right of the 100-yard marker, which indicates the distance to the center of the green as shown in the figure. How far is the ball from the center of the green?

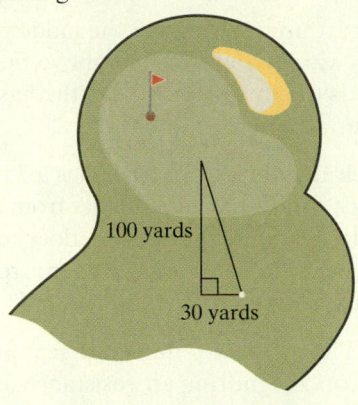

94. Baseball Avisail Garcia plays right field for the Chicago White Sox. He catches a fly ball 40 feet from the right field foul line, as indicated in the figure. How far is it to home plate?

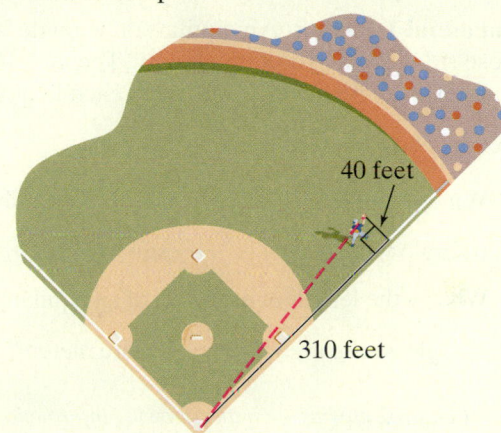

95. Guy Wire A guy wire is a wire used to support telephone poles. Suppose that a guy wire is located 30 feet up a telephone pole and is anchored to the ground 10 feet from the base of the pole. How long is the guy wire?

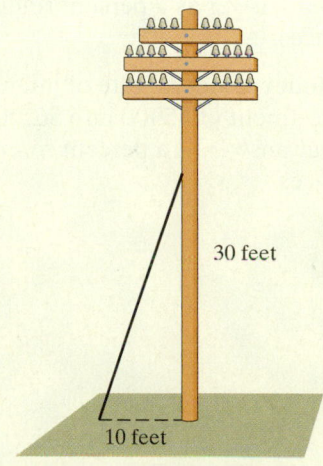

96. Guy Wire A guy wire is used to support an antenna on a rooftop. The wire is located 40 feet up on the antenna and anchored to the roof 8 feet from the base of the antenna. What is the length of the guy wire?

97. Ladder Bob needs to wash the windows on his house. He has a 25-foot ladder and places the base of the ladder 10 feet from the wall on the house.

  (a) How far up the wall will the ladder reach?

  (b) If his windows are 20 feet above the ground, what is the farthest distance the base of the ladder can be from the wall?

98. Fire Truck Ladder A fire truck has a 75-foot ladder. If the truck can safely park 20 feet from a building, how far up the building can the ladder reach, assuming that the base of the ladder is resting on top of the truck and the truck is 10 feet tall?

99. Gravity The distance s that an object falls (in feet) after t seconds, ignoring air resistance, is given by the equation $s = 16t^2$.

  (a) How long does it take an object to fall 16 feet?

  (b) How long does it take an object to fall 48 feet?

  (c) How long does it take an object to fall 64 feet?

△ **100. Equilateral Triangles** An equilateral triangle is one whose sides are all the same length. The area A of an equilateral triangle whose sides are each length x is given by $A = \dfrac{\sqrt{3}}{4}x^2$.

  (a) What is the length of each side of an equilateral triangle whose area is $\dfrac{8\sqrt{3}}{9}$ square feet?

  (b) What is the length of each side of an equilateral triangle whose area is $\dfrac{25\sqrt{3}}{4}$ square meters?

Problems 101 and 102 are based on the following information. If P dollars are invested today at an annual interest rate r compounded once a year, then the value of the account A after 2 years is given by the formula $A = P(1 + r)^2$.

101. Value of Money Find the rate of interest required to turn an investment of $1000 into $1200 after 2 years. Express your answer as a percent rounded to two decimal places.

102. Value of Money Find the rate of interest required to turn an investment of $5000 into $6200 after 2 years. Express your answer as a percent rounded to two decimal places.

Extending the Concepts

If you look carefully at the Pythagorean Theorem, it states, "If we have a right triangle, then $c^2 = a^2 + b^2$, where c is the length of the hypotenuse." In this theorem "a right triangle" represents the hypothesis, and "$c^2 = a^2 + b^2$" represents the conclusion. The **converse** *of a theorem interchanges the hypothesis and conclusion. The converse of the Pythagorean Theorem is true.*

Converse of the Pythagorean Theorem

In a triangle, if the square of the length of one side equals the sum of the squares of the lengths of the other two sides, then the triangle is a right triangle. The $90°$ angle is opposite the longest side.

In Problems 103–106, the lengths of the sides of a triangle are given. Determine whether the triangle is a right triangle. If it is, identify the hypotenuse.

103. 8, 15, 17   **104.** 4, 6, 8

105. 14, 18, 20   **106.** 20, 48, 52

107. Pythagorean Triples Suppose that m and n are positive integers with $m > n$. If $a = m^2 - n^2$, $b = 2mn$, and $c = m^2 + n^2$, show that a, b, and c are the lengths of the sides of a right triangle using the Converse of the Pythagorean Theorem. Any numbers a, b, and c found from the above formulas are called **Pythagorean Triples.**

108. Solve $ax^2 + bx + c = 0$ for x by completing the square.

Synthesis Review

In Problems 109–112, solve each equation.

109. $a^2 - 5a - 36 = 0$ **110.** $p^2 + 4p = 32$

111. $|4q + 1| = 3$ **112.** $\left|\dfrac{3}{4}w - \dfrac{2}{3}\right| = \dfrac{5}{2}$

113. In Problems 109–112, you are asked to solve quadratic and absolute value equations. In solving both types of equations, you must reduce the equation to a simpler equation. What type of equation is the simpler equation?

10.2 Solving Quadratic Equations by the Quadratic Formula

Objectives

1 Solve Quadratic Equations Using the Quadratic Formula

2 Use the Discriminant to Determine the Nature of Solutions of a Quadratic Equation

3 Model and Solve Problems Involving Quadratic Equations

Are You Prepared for This Section?

Before getting started, complete the following problems. If you get a problem wrong, go back to the section cited and review the material.

P1. Simplify: **(a)** $\sqrt{121}$ **(b)** $\sqrt{54}$ [Section 9.1, pp. 616–617; Section 9.4, pp. 635–639]

P2. Simplify: **(a)** $\sqrt{-9}$ **(b)** $\sqrt{-72}$ [Section 9.9, pp. 671–672]

P3. Simplify: $\dfrac{3 + \sqrt{18}}{6}$ [Section 9.4, pp. 635–639]

So far, three methods have been used for solving quadratic equations: (1) factoring, (2) the Square Root Property, and (3) completing the square. Why are three methods needed? Because each method provides the quickest route to the solution when used appropriately. For example, if the quadratic expression is easy to factor, the method of factoring will yield a solution most easily. If the equation is in the form $x^2 = p$ or $(ax \pm b)^2 = p$, using the Square Root Property is fastest. If the quadratic expression is not factorable, completing the square is needed. But the method of completing the square is tedious. Is there an alternative to this method? Yes!

1 Solve Quadratic Equations Using the Quadratic Formula

The method of completing the square leads to a general formula for solving the quadratic equation

$$ax^2 + bx + c = 0 \qquad a \neq 0$$

To complete the square, first get the constant c on the right-hand side of the equation.

$$ax^2 + bx = -c \qquad a \neq 0$$

Divide both sides of the equation by a: $\quad x^2 + \dfrac{b}{a}x = -\dfrac{c}{a}$

Now complete the square on the left side by adding the square of $\dfrac{1}{2}$ of the coefficient of x to both sides of the equation. That is, add

$$\left(\dfrac{1}{2}\cdot\dfrac{b}{a}\right)^2 = \dfrac{b^2}{4a^2}$$

to both sides of the equation. This gives

$$x^2 + \dfrac{b}{a}x + \dfrac{b^2}{4a^2} = -\dfrac{c}{a} + \dfrac{b^2}{4a^2}$$

$$x^2 + \dfrac{b}{a}x + \dfrac{b^2}{4a^2} = \dfrac{b^2}{4a^2} - \dfrac{c}{a}$$

Combine like terms on the right-hand side. The least common denominator on the right-hand side is $4a^2$, so multiply $-\dfrac{c}{a}$ by $\dfrac{4a}{4a}$:

$$x^2 + \dfrac{b}{a}x + \dfrac{b^2}{4a^2} = \dfrac{b^2}{4a^2} - \dfrac{c}{a}\cdot\dfrac{4a}{4a}$$

$$x^2 + \dfrac{b}{a}x + \dfrac{b^2}{4a^2} = \dfrac{b^2}{4a^2} - \dfrac{4ac}{4a^2}$$

$$x^2 + \dfrac{b}{a}x + \dfrac{b^2}{4a^2} = \dfrac{b^2 - 4ac}{4a^2}$$

Prepared?...Answers **P1. (a)** 11 **(b)** $3\sqrt{6}$ **P2. (a)** $3i$ **(b)** $6\sqrt{2}i$ **P3.** $\dfrac{1}{2} + \dfrac{\sqrt{2}}{2}$ or $\dfrac{1 + \sqrt{2}}{2}$

Work Smart
To factor any perfect square trinomial of the form $x^2 + bx + c$, write
$$\left(x + \frac{b}{2}\right)^2.$$

Factor the left-hand side and obtain

$$\left(x + \frac{b}{2a}\right)^2 = \frac{b^2 - 4ac}{4a^2}$$

Assume that $a > 0$ (you'll see why in a little while). This assumption does not affect the results because if $a < 0$, both sides of the equation $ax^2 + bx + c = 0$ could be multiplied by -1 to make it positive. With this assumption, use the Square Root Property and get

$$x + \frac{b}{2a} = \pm\sqrt{\frac{b^2 - 4ac}{4a^2}}$$

$$\sqrt{\frac{a}{b}} = \frac{\sqrt{a}}{\sqrt{b}}: \quad x + \frac{b}{2a} = \pm\frac{\sqrt{b^2 - 4ac}}{\sqrt{4a^2}}$$

$$\sqrt{4a^2} = 2a \text{ since } a > 0: \quad x + \frac{b}{2a} = \pm\frac{\sqrt{b^2 - 4ac}}{2a}$$

Subtract $\frac{b}{2a}$ from both sides: $\quad x = -\frac{b}{2a} \pm \frac{\sqrt{b^2 - 4ac}}{2a}$

Write over a common denominator: $\quad x = \frac{-b \pm \sqrt{b^2 - 4ac}}{2a}$

This is called the *quadratic formula*.

Work Smart: Study Skills
When solving homework problems, always write down the quadratic formula before you begin solving the equation. This will help you memorize the formula.

In Other Words
The quadratic formula says that the solution(s) to the equation $ax^2 + bx + c = 0$ is (are) "the opposite of b plus or minus the square root of b squared minus $4ac$, all over $2a$."

The Quadratic Formula

The solution(s) to the quadratic equation $ax^2 + bx + c = 0, a \neq 0$, is (are) given by the **quadratic formula**

$$x = \frac{-b \pm \sqrt{b^2 - 4ac}}{2a}$$

▶ **EXAMPLE 1** **How to Solve a Quadratic Equation Using the Quadratic Formula**

Solve: $12x^2 + 5x - 3 = 0$

Step-by-Step Solution

$\boxed{a = 12} \quad \boxed{b = 5} \quad \boxed{c = -3}$

Step 1: Write the equation in standard form $ax^2 + bx + c = 0$ and identify the values of a, b, and c.

$12x^2 + 5x - 3 = 0$

Step 2: Substitute the values of a, b, and c into the quadratic formula.

$$x = \frac{-b \pm \sqrt{b^2 - 4ac}}{2a}$$

$$x = \frac{-5 \pm \sqrt{5^2 - 4(12)(-3)}}{2(12)}$$

Step 3: Simplify the expression found in Step 2.

$$= \frac{-5 \pm \sqrt{25 + 144}}{24}$$

$$= \frac{-5 \pm \sqrt{169}}{24}$$

$\sqrt{169} = 13: \quad = \dfrac{-5 \pm 13}{24}$

$$a \pm b \text{ means } a - b \text{ or } a + b: \quad x = \frac{-5 - 13}{24} \quad \text{or} \quad x = \frac{-5 + 13}{24}$$

$$= \frac{-18}{24} \quad \text{or} \quad = \frac{8}{24}$$

$$\text{Simplify:} \quad = -\frac{3}{4} \quad \text{or} \quad = \frac{1}{3}$$

Step 4: Check

$$12x^2 + 5x - 3 = 0$$

$$x = -\frac{3}{4}: \quad 12\left(-\frac{3}{4}\right)^2 + 5\left(-\frac{3}{4}\right) - 3 \overset{?}{=} 0 \qquad x = \frac{1}{3}: \quad 12\left(\frac{1}{3}\right)^2 + 5\left(\frac{1}{3}\right) - 3 \overset{?}{=} 0$$

$$12 \cdot \frac{9}{16} - \frac{15}{4} - 3 \overset{?}{=} 0 \qquad\qquad 12 \cdot \frac{1}{9} + \frac{5}{3} - 3 \overset{?}{=} 0$$

$$\frac{27}{4} - \frac{15}{4} - 3 \overset{?}{=} 0 \qquad\qquad \frac{4}{3} + \frac{5}{3} - 3 \overset{?}{=} 0$$

$$\frac{12}{4} - 3 \overset{?}{=} 0 \qquad\qquad \frac{9}{3} - 3 \overset{?}{=} 0$$

$$0 = 0 \quad \text{True} \qquad\qquad\qquad 0 = 0 \quad \text{True}$$

The solution set is $\left\{-\frac{3}{4}, \frac{1}{3}\right\}$. ●

Work Smart

If $b^2 - 4ac$ is a perfect square, then the quadratic equation can be solved by factoring. In Example 1, $b^2 - 4ac = 169$, a perfect square.

$$12x^2 + 5x - 3 = 0$$
$$(4x + 3)(3x - 1) = 0$$
$$4x + 3 = 0 \text{ or } 3x - 1 = 0$$
$$x = -\frac{3}{4} \quad \text{or} \quad x = \frac{1}{3}$$

Notice that the solutions to the equation in Example 1 are rational numbers and that the expression $b^2 - 4ac$ under the radical in the quadratic formula, 169, is a perfect square. This leads to a generalization. Whenever the expression $b^2 - 4ac$ is a perfect square, the quadratic equation will have rational solutions, and the quadratic equation can be solved by factoring (provided the coefficients of the quadratic equation are rational numbers).

With that said, *any* quadratic equation can be solved using the quadratic formula.

> **Solving a Quadratic Equation Using the Quadratic Formula**
>
> **Step 1:** Write the equation in standard form $ax^2 + bx + c = 0$ and identify the values of a, b, and c.
>
> **Step 2:** Substitute the values of a, b, and c into the quadratic formula.
>
> **Step 3:** Simplify the expression found in Step 2.
>
> **Step 4:** Verify the solution(s).

Quick ✓

1. The solution(s) to the quadratic equation $ax^2 + bx + c = 0, a \neq 0$, are given

   by the quadratic formula $x = \underline{\hspace{3cm}}$.

In Problems 2 and 3, solve each equation using the quadratic formula.

2. $2x^2 - 3x - 9 = 0$ \qquad\qquad 3. $2x^2 + 7x = 4$

▶ **EXAMPLE 2** **Solving a Quadratic Equation Using the Quadratic Formula**

Solve: $3p^2 = 6p - 1$

Solution

First, write the equation in standard form to identify a, b, and c.

$$3p^2 = 6p - 1$$

Subtract $6p$ from both sides;
Add 1 to both sides: $\quad 3p^2 - 6p + 1 = 0$

(continued)

$$\text{Write the quadratic equation using "}p =\text{":} \quad p = \frac{-b \pm \sqrt{b^2 - 4ac}}{2a}$$

$$a = 3, b = -6, c = 1: \quad p = \frac{-(-6) \pm \sqrt{(-6)^2 - 4(3)(1)}}{2(3)}$$

$$= \frac{6 \pm \sqrt{36 - 12}}{6}$$

$$= \frac{6 \pm \sqrt{24}}{6}$$

$$\sqrt{24} = \sqrt{4 \cdot 6} = 2\sqrt{6}: \quad = \frac{6 \pm 2\sqrt{6}}{6}$$

$$\frac{a + b}{c} = \frac{a}{c} + \frac{b}{c}: \quad = \frac{6}{6} \pm \frac{2\sqrt{6}}{6}$$

$$\text{Simplify:} \quad = 1 \pm \frac{\sqrt{6}}{3}$$

$$a \pm b \text{ means } a - b \text{ or } a + b: \quad p = 1 - \frac{\sqrt{6}}{3} \quad \text{or} \quad p = 1 + \frac{\sqrt{6}}{3}$$

Work Smart

$\dfrac{6 \pm 2\sqrt{6}}{6}$ can also be simplified by factoring:

$$\frac{6 \pm 2\sqrt{6}}{6} = \frac{2(3 \pm \sqrt{6})}{6}$$

$$= \frac{3 \pm \sqrt{6}}{3}$$

This is equivalent to $1 \pm \dfrac{\sqrt{6}}{3}$. Ask your instructor which form of the solution is preferred, if any.

We leave it to you to verify the solutions. The solution set is $\left\{ 1 - \dfrac{\sqrt{6}}{3}, 1 + \dfrac{\sqrt{6}}{3} \right\}$. ●

Notice in Example 2 that the value of $b^2 - 4ac$, 24, is positive, but not a perfect square. There are two solutions to the quadratic equation and they are irrational.

> **Quick ✔**
>
> **4.** *True or False* For $3x^2 - 2x = 9$, $a = 3$, $b = -2$, and $c = 9$.
>
> **5.** Solve: $4z^2 + 1 = 8z$

EXAMPLE 3 **Solving a Rational Equation That Leads to a Quadratic Equation**

Solve: $9m + \dfrac{4}{m} = 12$

Solution

First note that m cannot equal 0. To clear the equation of rational expressions, multiply both sides of the equation by the LCD, m.

$$m\left(9m + \frac{4}{m} \right) = 12 \cdot m$$

$$\text{Distribute } m: \quad 9m^2 + 4 = 12m$$

$$\text{Subtract } 12m \text{ from both sides:} \quad 9m^2 - 12m + 4 = 0$$

$$m = \frac{-b \pm \sqrt{b^2 - 4ac}}{2a}$$

$$a = 9, b = -12, c = 4: \quad m = \frac{-(-12) \pm \sqrt{(-12)^2 - 4(9)(4)}}{2(9)}$$

$$= \frac{12 \pm \sqrt{144 - 144}}{18}$$

Work Smart

Notice that the entire expression $-b \pm \sqrt{b^2 - 4ac}$ is in the numerator of the quadratic formula.

$$= \frac{12 \pm \sqrt{0}}{18}$$

$$= \frac{12}{18} = \frac{2}{3}$$

We leave it to you to verify the solution. The solution set is $\left\{\frac{2}{3}\right\}$. ●

In Example 3, there was one solution rather than two (as in Examples 1 and 2). The solution $x = \frac{2}{3}$ is called a **repeated root** because it occurs twice, once for $\frac{12 + 0}{18} = \frac{12}{18} = \frac{2}{3}$ and once for $\frac{12 - 0}{18} = \frac{12}{18} = \frac{2}{3}$. Notice that the value of $b^2 - 4ac$ equals 0, which is the reason for the repeated root. More will be said about this soon.

> **Quick ✓**
>
> *In Problems 6 and 7, solve each equation.*
>
> **6.** $4w + \dfrac{25}{w} = 20$ **7.** $2x = 8 - \dfrac{3}{x}$

▶ **EXAMPLE 4** **Solving a Quadratic Equation Using the Quadratic Formula**

Solve: $y^2 - 4y + 13 = 0$

Solution

$\left\{ -\dfrac{1}{2} - \dfrac{\sqrt{7}}{2}i, -\dfrac{1}{2} + \dfrac{\sqrt{7}}{2}i \right\}$

$$y^2 - 4y + 13 = 0$$

$$1y^2 - 4y + 13 = 0$$

$$y = \frac{-b \pm \sqrt{b^2 - 4ac}}{2a}$$

$a = 1, b = -4, c = 13:$ $$y = \frac{-(-4) \pm \sqrt{(-4)^2 - 4(1)(13)}}{2(1)}$$

$$= \frac{4 \pm \sqrt{16 - 52}}{2}$$

$\sqrt{-36} = 6i:$ $$= \frac{4 \pm \sqrt{-36}}{2} = \frac{4 \pm 6i}{2}$$

$\dfrac{a + b}{c} = \dfrac{a}{c} + \dfrac{b}{c}:$ $$= \frac{4}{2} \pm \frac{6}{2}i = 2 \pm 3i$$

$a \pm b$ means $a - b$ or $a + b:$ $$y = 2 - 3i \quad \text{or} \quad y = 2 + 3i$$

We leave it to you to verify the solution. The solution set is $\{2 - 3i, 2 + 3i\}$. ●

Notice in Example 4 that the value of $b^2 - 4ac$ is negative and that the equation has two complex solutions that are not real.

> **Quick ✓**
>
> **8.** Solve: $z^2 + 2z + 26 = 0$

---

❷ **Use the Discriminant to Determine the Nature of Solutions of a Quadratic Equation**

Work Smart: Study Skills

Think of the *discriminant* as the expression that helps you *discriminate* what types of solutions are possible.

In the quadratic formula $x = \dfrac{-b \pm \sqrt{b^2 - 4ac}}{2a}$, the quantity $b^2 - 4ac$ is called the **discriminant** because its value tells us the number and the type of solutions to expect from the quadratic equation when $a, b,$ and c are rational numbers.

		The Discriminant and the Nature of the Solutions of a Quadratic Equation $ax^2 + bx + c = 0$, Where a, b, and c Are Rational Numbers		
Example	**Value of Discriminant**	**Description of Discriminant**	**Number of Solutions**	**Type of Solution**
1	169	Positive and a perfect square	Two	Rational
2	24	Positive and not a perfect square	Two	Irrational
3	0	Zero	One (repeated root)	Rational
4	−36	Negative	Two	Complex, nonreal

Work Smart

The rules in the box to the right apply only if the coefficients of the quadratic equation are rational numbers.

In Example 4, the solutions are complex conjugates. In general, for any quadratic equation of the form $ax^2 + bx + c = 0$, where a, b, and c are real numbers and $b^2 - 4ac < 0$, the equation has two complex solutions that are not real and are complex conjugates.

This result is a consequence of the quadratic formula. Suppose that $b^2 - 4ac = -N < 0$. Then, by the quadratic formula, the solutions are

$$x = \frac{-b \pm \sqrt{b^2 - 4ac}}{2a} = \frac{-b \pm \sqrt{-N}}{2a}$$

$$= \frac{-b \pm \sqrt{N}i}{2a} = \frac{-b}{2a} \pm \frac{\sqrt{N}}{2a}i$$

which are complex conjugates.

▶ **EXAMPLE 5**   **Determining the Nature of the Solutions of a Quadratic Equation**

For each quadratic equation, determine the discriminant. Use the value of the discriminant to determine whether the quadratic equation has two rational solutions, two irrational solutions, one repeated real solution, or two complex solutions that are not real.

(a) $x^2 - 5x + 2 = 0$ **(b)** $9y^2 + 6y + 1 = 0$ **(c)** $3p^2 - p = -5$

Solution

(a) Compare $x^2 - 5x + 2 = 0$ to the standard form $ax^2 + bx + c = 0$.

$$x^2 - 5x + 2 = 0$$

$$a = 1 \quad b = -5 \quad c = 2$$

Substitute the values for a, b, and c shown above into the discriminant formula $b^2 - 4ac$:

$$b^2 - 4ac = (-5)^2 - 4(1)(2) = 25 - 8 = 17$$

Because $b^2 - 4ac = 17$ is positive but not a perfect square, the quadratic equation has two irrational solutions.

(b) For the quadratic equation $9y^2 + 6y + 1 = 0$, $a = 9$, $b = 6$, and $c = 1$. Substituting these values into $b^2 - 4ac$ gives

$$b^2 - 4ac = 6^2 - 4(9)(1) = 36 - 36 = 0$$

Because $b^2 - 4ac = 0$, the quadratic equation has one repeated rational solution.

(c) Is the quadratic equation $3p^2 - p = -5$ in standard form? No! Add 5 to both sides of the equation and write the equation as $3p^2 - p + 5 = 0$. For $a = 3$, $b = -1$, and $c = 5$, the value of the discriminant, $b^2 - 4ac$, is

$$b^2 - 4ac = (-1)^2 - 4(3)(5) = 1 - 60 = -59$$

Because $-59 < 0$, the quadratic equation has two complex solutions that are not real and are conjugates.

Quick ✓

9. In the quadratic formula, the quantity $b^2 - 4ac$ is called the _____ of the quadratic equation.

10. *True or False* The discriminant of the quadratic equation $4x^2 - 5x + 1 = 0$ is $\sqrt{25 - 4(4)(1)} = \sqrt{9} = 3$.

11. If the discriminant of a quadratic equation is _____, then the quadratic equation has two complex solutions that are not real.

12. *True or False* If the discriminant of a quadratic equation is zero, then the equation has no solution.

13. *True or False* The solutions to a quadratic equation in which the solutions are complex numbers that are not real are complex conjugates.

In Problems 14–16, use the value of the discriminant to determine whether the quadratic equation has two rational solutions, two irrational solutions, one repeated real solution, or two complex solutions that are not real.

14. $2z^2 + 5z + 4 = 0$ 15. $4y^2 + 12y = -9$ 16. $2x^2 - 4x + 1 = 0$

▶ Which Method Should I Use?

We now have four methods for solving quadratic equations:

1. Factoring
2. Square Root Property
3. Completing the Square
4. The Quadratic Formula

You may be asking yourself, "Which method should I use?" and "Does it matter which method I use?" The answer to the second question is that it does not matter which method you use, but one method may be more efficient than the others. Table 1 on the next page contains guidelines to help you solve any quadratic equation. The method that is most efficient for solving a quadratic equation depends on the equation. The value of the discriminant can be used as a guide in choosing the most efficient method.

Notice in Table 1, on the next page, that completing the square is not one of the methods recommended for use in solving a quadratic equation. This is because the quadratic formula is based on completing the square of $ax^2 + bx + c = 0$. Besides, completing the square may be a cumbersome task, whereas the quadratic formula is fairly straightforward. Learning how to complete the square was worth the time, however, because it is used to derive the quadratic formula. Also, completing the square is a skill that will be needed later in this course and in future math courses.

Table 1

Form of the Quadratic Equation	Most Efficient Method	Example
$x^2 = p$, where p is any real number	Square Root Property	$x^2 = 45$ Square Root Property: $x = \pm\sqrt{45}$ $= \pm 3\sqrt{5}$
$ax^2 + c = 0$	Square Root Property	$3p^2 + 12 = 0$ Subtract 12 from both sides: $3p^2 = -12$ Divide both sides by 3: $p^2 = -4$ Square Root Property: $p = \pm\sqrt{-4}$ $= \pm 2i$
$(ax + b)^2 = p$, where p is any real number	Square Root Property	$(2x + 7)^2 = 18$ Square Root Property: $2x + 7 = \pm 3\sqrt{2}$ Subtract 7 from both sides: $2x = -7 \pm 3\sqrt{2}$ Divide by 2: $x = \dfrac{-7 \pm 3\sqrt{2}}{2}$
$ax^2 + bx + c = 0$, where $b^2 - 4ac$ is a perfect square	Factoring or the quadratic formula. Factor if the quadratic expression is easy to factor. Otherwise, use the quadratic formula.	$a = 2, b = 1, c = -10$: $2m^2 + m - 10 = 0$ $b^2 - 4ac = 1^2 - 4(2)(-10) = 1 + 80 = 81$ 81 is a perfect square, so use factoring: $2m^2 + m - 10 = 0$ $(2m + 5)(m - 2) = 0$ $2m + 5 = 0$ or $m - 2 = 0$ $m = -\dfrac{5}{2}$ or $m = 2$
$ax^2 + bx + c = 0$, where $b^2 - 4ac$ is not a perfect square	Quadratic formula	$a = 2, b = 4, c = -1$: $2x^2 + 4x - 1 = 0$ $b^2 - 4ac = 4^2 - 4(2)(-1) = 16 + 8 = 24$ 24 is not a perfect square, so use the quadratic formula (because it's easier than completing the square): $x = \dfrac{-b \pm \sqrt{b^2 - 4ac}}{2a}$ $= \dfrac{-4 \pm \sqrt{24}}{2(2)}$ $= \dfrac{-4 \pm 2\sqrt{6}}{4}$ $= -1 \pm \dfrac{\sqrt{6}}{2}$ $x = -1 - \dfrac{\sqrt{6}}{2}$ or $x = -1 + \dfrac{\sqrt{6}}{2}$
$ax^2 + bx + c = 0$, where $b^2 - 4ac$ is negative	Quadratic formula	$a = 3, b = -4, c = 7$: $3x^2 - 4x + 7 = 0$ $b^2 - 4ac = (-4)^2 - 4(3)(7) = 16 - 84 = -68$ The discriminant is negative, so use the quadratic formula: $x = \dfrac{-b \pm \sqrt{b^2 - 4ac}}{2a}$ $x = \dfrac{-(-4) \pm \sqrt{-68}}{2(3)}$ $= \dfrac{4 \pm \sqrt{-1 \cdot 4 \cdot 17}}{6}$ $= \dfrac{4 \pm 2\sqrt{17}i}{6}$ $= \dfrac{2}{3} \pm \dfrac{1}{3}\sqrt{17}i$ $x = \dfrac{2}{3} - \dfrac{1}{3}\sqrt{17}i$ or $x = \dfrac{2}{3} + \dfrac{1}{3}\sqrt{17}i$

Quick ✓

17. *True or False* If the discriminant of a quadratic equation is a perfect square, then the equation can be solved by factoring.

In Problems 18–20, solve each quadratic equation using any appropriate method.

18. $5n^2 - 45 = 0$ **19.** $-2y^2 + 5y - 6 = 0$ **20.** $3w^2 + 2w = 5$

▶ ❸ Model and Solve Problems Involving Quadratic Equations

Many applied problems, such as the one below, require solving quadratic equations. As always, use the problem-solving strategy from Section 2.5.

EXAMPLE 6   **Revenue**

The revenue R received by a company selling x specialty T-shirts per week is given by the function $R(x) = -0.005x^2 + 30x$.

(a) How many T-shirts must be sold in order for revenue to be $25,000 per week?

(b) How many T-shirts must be sold in order for revenue to be $45,000 per week?

Solution

(a) Step 1: Identify Determine the number x of T-shirts required so that $R = \$25,000$.

Step 2: Name x represents the number of T-shirts sold.

Step 3: Translate The function $R(x) = 25,000$ can be used to solve for the number of T-shirts, x.

$$R(x) = 25{,}000$$
$$-0.005x^2 + 30x = 25{,}000$$
$$-0.005x^2 + 30x - 25{,}000 = 0$$

Step 4: Solve Letting $a = -0.005$, $b = 30$, and $c = -25{,}000$, we have

$$b^2 - 4ac = 30^2 - 4(-0.005)(-25{,}000)$$
$$= 400$$

Because 400 is a perfect square, solve the equation by factoring or using the quadratic formula. It is not obvious how to factor $-0.005x^2 + 30x - 25000$, so use the quadratic formula:

$$\overset{\color{magenta}{b^2 - 4ac = 400}}{\color{magenta}{\downarrow}}$$
$$x = \frac{-30 \pm \sqrt{400}}{2(-0.005)}$$
$$= \frac{-30 \pm 20}{-0.01}$$

$$x = \frac{-30 - 20}{-0.01} = 5000 \quad \text{or} \quad x = \frac{-30 + 20}{-0.01} = 1000$$

Step 5: Check If 1000 T-shirts are sold, then revenue is $R(1000) = -0.005(1000)^2 + 30(1000) = \$25{,}000$. If 5000 T-shirts are sold, then revenue is $R(5000) = -0.005(5000)^2 + 30(5000) = \$25{,}000$.

Step 6: Answer The company needs to sell either 1000 or 5000 T-shirts each week to earn $25,000 in revenue. *(continued)*

(b) **Step 1: Identify** Determine the number x of T-shirts required so that $R = \$45,000$. That is, solve the equation $R(x) = 45,000$.

Step 2: Name x represents the number of T-shirts sold.

Step 3: Translate The function $R(x) = 45,000$ can be used to solve for the number of T-shirts, x.

$$R(x) = 45,000$$
$$-0.005x^2 + 30x = 45,000$$
$$-0.005x^2 + 30x - 45,000 = 0$$

Step 4: Solve Letting $a = -0.005$, $b = 30$, and $c = -45,000$, we have

$$b^2 - 4ac = 30^2 - 4(-0.005)(-45,000) = 0$$

Because the discriminant is 0, the quadratic equation has a single rational solution. Solve the equation by factoring or using the quadratic formula. It is not obvious how to factor $-0.005x^2 + 30x - 45,000$, so use the quadratic formula to solve the equation.

$$x = \frac{-30 \pm \sqrt{0}}{2(-0.005)} \qquad b^2 - 4ac = 0$$
$$= \frac{-30 \pm 0}{-0.01}$$
$$= \frac{-30}{-0.01} = 3000$$

Step 5: Check If 3000 T-shirts are sold, then revenue is $R(3000) = -0.005(3000)^2 + 30(3000) = \$45,000$.

Step 6: Answer The company needs to sell 3000 T-shirts each week to earn $45,000 in revenue. ●

Quick ✓

21. The revenue R received by a Redbox kiosk renting x DVDs per day is given by the function $R(x) = -0.005x^2 + 4x$.

(a) How many DVDs must be rented in order for revenue to be $600 per day?

(b) How many DVDs must be rented in order for revenue to be $800 per day?

EXAMPLE 7 **Designing a Window**

A window designer wishes to design a window so that the diagonal is 20 feet. In addition, the width of the window needs to be 4 feet more than the height. What are the dimensions of the window?

Solution

Step 1: Identify Determine the dimensions of the window. That is, find the width and height of the window.

Figure 6

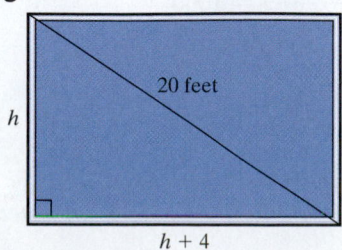

Step 2: Name Let h represent the height of the window in feet. Because the width is 4 feet more than the height, $h + 4$ is the width of the window.

Step 3: Translate Figure 6 illustrates the situation. Notice that the two sides and the diagonal form a right triangle. Express the relationship among the sides using the Pythagorean Theorem.

$$\text{leg}^2 + \text{leg}^2 = \text{hypotenuse}^2: \qquad h^2 + (h+4)^2 = 20^2$$
$$(A+B)^2 = A^2 + 2AB + B^2: \quad h^2 + h^2 + 8h + 16 = 400$$

Combine like terms: $2h^2 + 8h + 16 = 400$

Subtract 400 from both sides: $2h^2 + 8h - 384 = 0$

Divide both sides by 2: $h^2 + 4h - 192 = 0$

Step 4: Solve In the model, $a = 1$, $b = 4$, and $c = -192$. The discriminant is $b^2 - 4ac = 4^2 - 4(1)(-192) = 784 = 28^2$. Thus solve the equation by factoring.

$$h^2 + 4h - 192 = 0$$
$$(h + 16)(h - 12) = 0$$
$$h + 16 = 0 \quad \text{or} \quad h - 12 = 0$$
$$h = -16 \quad \text{or} \quad h = 12$$

Step 5: Check Disregard the solution $h = -16$ because h represents the height of the window. Determine whether a window whose dimensions are 12 feet by $12 + 4 = 16$ feet has a diagonal that is 20 feet by verifying that $12^2 + 16^2 = 20^2$.

$$12^2 + 16^2 \stackrel{?}{=} 20^2$$
$$144 + 256 \stackrel{?}{=} 400$$
$$400 = 400$$

Step 6: Answer The dimensions of the window are 12 feet by 16 feet.   ●

Quick ✓

22. A rectangular plot of land is designed so that its length is 14 meters more than its width. The diagonal of the land is known to be 34 meters. What are the dimensions of the land?

10.2 Exercises   MyMathLab®

*Problems 1–22 are the Quick ✓s that follow the **EXAMPLES**.*

Building Skills

In Problems 23–40, solve each equation using the quadratic formula. See Objective 1.

23. $x^2 - 4x - 12 = 0$

24. $p^2 - 4p - 32 = 0$

25. $6y^2 - y - 15 = 0$

26. $10x^2 + x - 2 = 0$

27. $4m^2 - 8m + 1 = 0$

28. $2q^2 - 4q + 1 = 0$

29. $3w - 6 = \dfrac{1}{w}$

30. $x + \dfrac{1}{x} = 3$

31. $3p^2 = -2p + 4$

32. $5w^2 = -3w + 1$

33. $x^2 - 2x + 7 = 0$

34. $y^2 - 4y + 5 = 0$

35. $2z^2 + 7 = 2z$

36. $2z^2 + 7 = 4z$

37. $4x^2 = 2x + 1$

38. $6p^2 = 4p + 1$

39. $1 = 3q^2 + 4q$

40. $1 = 5w^2 + 6w$

In Problems 41–50, determine the discriminant of each quadratic equation. Use the value of the discriminant to determine whether the quadratic equation has two rational solutions, two irrational solutions, one repeated rational solution, or two complex solutions that are not real. See Objective 2.

41. $x^2 - 5x + 1 = 0$

42. $p^2 + 4p - 2 = 0$

43. $3z^2 + 2z + 5 = 0$

44. $2y^2 - 3y + 5 = 0$

45. $9q^2 - 6q + 1 = 0$

46. $16x^2 + 24x + 9 = 0$

47. $3w^2 = 4w - 2$ **48.** $6x^2 - x = -4$

49. $6x = 2x^2 - 1$ **50.** $10w^2 = 3$

Mixed Practice

In Problems 51–76, solve each equation.

51. $w^2 - 5w + 5 = 0$ **52.** $q^2 - 7q + 7 = 0$

53. $3x^2 + 5x = 8$ **54.** $4p^2 + 5p = 9$

55. $2x^2 = 3x + 35$ **56.** $3x^2 + 5x = 2$

57. $q^2 + 2q + 8 = 0$ **58.** $w^2 + 4w + 9 = 0$

59. $2z^2 = 2(z + 3)^2$ **60.** $3z^2 = 3(z + 1)(z - 2)$

61. $7q - 2 = \dfrac{4}{q}$ **62.** $5m - 4 = \dfrac{5}{m}$

63. $5a^2 - 80 = 0$ **64.** $4p^2 - 100 = 0$

65. $8n^2 + 1 = 4n$ **66.** $4q^2 + 1 = 2q$

67. $27x^2 + 36x + 12 = 0$ **68.** $8p^2 - 40p + 50 = 0$

69. $\dfrac{1}{3}x^2 + \dfrac{2}{9}x - 1 = 0$ **70.** $\dfrac{1}{2}x^2 + \dfrac{3}{4}x - 1 = 0$

71. $(x - 5)(x + 1) = 4$ **72.** $(a - 3)(a + 1) = 2$

73. $\dfrac{x - 2}{x + 2} = x - 3$ **74.** $\dfrac{x - 5}{x + 3} = x - 3$

75. $\dfrac{x - 4}{x^2 + 2} = 2$ **76.** $\dfrac{x - 1}{x^2 + 4} = 1$

77. Suppose that $f(x) = x^2 + 4x - 21$.
    (a) Solve $f(x) = 0$ for x.
    (b) Solve $f(x) = -21$ for x. What points are on the graph of f?

78. Suppose that $f(x) = x^2 + 2x - 8$.
    (a) Solve $f(x) = 0$ for x.
    (b) Solve $f(x) = -8$ for x. What points are on the graph of f?

79. Suppose that $H(x) = -2x^2 - 4x + 1$.
    (a) Solve $H(x) = 0$ for x.
    (b) Solve $H(x) = 2$ for x.

80. Suppose that $g(x) = 3x^2 + x - 1$.
    (a) Solve $g(x) = 0$ for x.
    (b) Solve $g(x) = 4$ for x.

81. What are the zeros of $G(x) = 3x^2 + 2x - 2$?

82. What are the zeros of $F(x) = x^2 + 3x - 3$?

Applying the Concepts

In Problems 83–86, use the Pythagorean Theorem to determine the value of x and the measurements of all sides of the right triangle.

△ **83.**

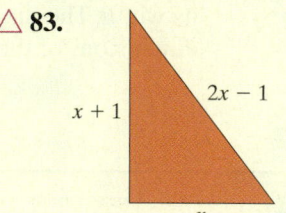

△ **84.**

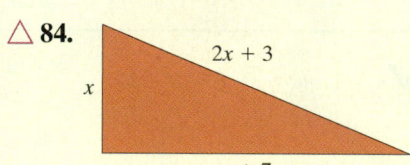

△ **85.**

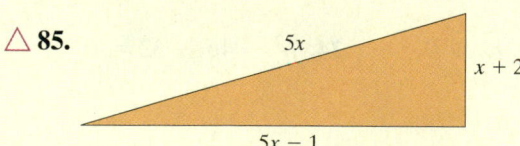

△ **86.**

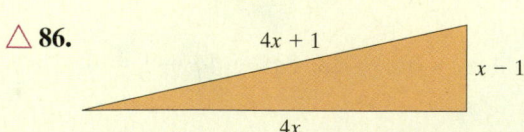

△ **87.** **Area** The area of a rectangle is 40 square inches. The width of the rectangle is 4 inches more than the length. What are the dimensions of the rectangle?

△ **88.** **Area** The area of a rectangle is 60 square inches. The width of the rectangle is 6 inches more than the length. What are the dimensions of the rectangle?

△ **89. Area** The area of a triangle is 25 square inches. The height of the triangle is 3 inches less than the base. What are the base and height of the triangle?

△ **90. Area** The area of a triangle is 35 square inches. The height of the triangle is 2 inches less than the base. What are the base and height of the triangle?

91. Revenue The revenue R received by a company selling x pairs of sunglasses per week is given by the function $R(x) = -0.1x^2 + 70x$.

(a) Find and interpret the values of $R(17)$ and $R(25)$.

(b) How many pairs of sunglasses must be sold in order for revenue to be $10,000 per week?

(c) How many pairs of sunglasses must be sold in order for revenue to be $12,250 per week?

92. Revenue The revenue R received by a company selling x "all-day passes" to a small amusement park per day is given by the function $R(x) = -0.02x^2 + 24x$.

(a) Find and interpret the values of $R(300)$ and $R(800)$.

(b) How many tickets must be sold in order for revenue to be $4000 per day?

(c) How many tickets must be sold in order for revenue to be $7200 per day?

93. Projectile Motion The height s of a ball after t seconds, when thrown straight up with an initial speed of 70 feet per second from an initial height of 5 feet, can be modeled by the function

$$s(t) = -16t^2 + 70t + 5$$

(a) When will the height of the ball be 40 feet? Round your answer to the nearest tenth of a second.

(b) When will the height of the ball be 70 feet? Round your answer to the nearest tenth of a second.

(c) Will the ball ever reach a height of 150 feet? How does the result of the equation tell you this?

94. Projectile Motion The height s of a toy rocket after t seconds, when fired straight up with an initial speed of 150 feet per second from an initial height of 2 feet, can be modeled by the function

$$s(t) = -16t^2 + 150t + 2$$

(a) When will the height of the rocket be 200 feet? Round your answer to the nearest tenth of a second.

(b) When will the height of the rocket be 300 feet? Round your answer to the nearest tenth of a second.

(c) Will the rocket ever reach a height of 500 feet?

△ **95. Similar Triangles** Consult the figure. Suppose that $\triangle ABC$ is similar to $\triangle DEC$. The length of $\overline{BC}$ is 24 inches and the length of $\overline{DE}$ is 6 inches. If the length of $\overline{AB}$ equals the length of $\overline{CE}$, which we call x, find x.

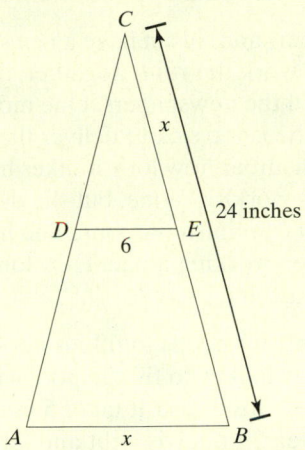

96. Number Sense Three times the square of a number equals the sum of two times the number and 5. Find the number(s).

97. Life Cycle Hypothesis The Life Cycle Hypothesis from economics was presented by Franco Modigliani in 1954. One of its components states that income is a function of age. The function $I(a) = -55a^2 + 5119a - 54,448$ represents the relation between average annual income I and age a.

(a) For what age does average income I equal $40,000? Round your answer to the nearest year.

(b) For what age does average income I equal $50,000? Round your answer to the nearest year.

98. Population The function $P(a) = -0.001a^2 + 0.0573a + 3.6941$ represents the population (in millions) of Americans in 2014, P, that are a years of age. (SOURCE: *United States Census Bureau*)

(a) For what age(s) was the population 4 million in 2014? Round your answer to the nearest year.

(b) For what age(s) was the population 3 million in 2014? Round your answer to the nearest year.

99. Upstream and Back Zene decides to canoe 4 miles upstream on a river to a waterfall and then canoe back. The total trip (excluding the time spent at the waterfall) takes 6 hours. Zene knows she can canoe at an average speed of 5 miles per hour in still water. What is the speed of the current?

100. Round Trip A Cessna aircraft flies 200 miles due west into the jet stream and flies back home on the same route. The total time of the trip (excluding the time on the ground) is 4 hours. The Cessna aircraft can fly 120 miles per hour in still air. What is the net effect of the jet stream on the aircraft?

101. Work Robert and Susan have a newspaper route. When they work the route together, it takes 2 hours to deliver all the newspapers. One morning Robert tells Susan he is too sick to deliver the papers. Susan doesn't remember how long it takes her to deliver the newspapers working alone, but she does remember that Robert can finish the route one hour sooner than she can when working alone. How long will it take Susan to finish the route?

102. Work Demitrius needs to fill his pool. When he rents a water tanker to fill the pool with the help of the hose from his house, it takes 5 hours to fill the pool. This year, money is tight and he can't afford to rent the water tanker to fill the pool. He doesn't remember how long it takes for his house hose to fill the pool, but he does remember that the tanker hose filling the pool alone can finish the job in 8 fewer hours than using his house hose alone. How long will it take Demitrius to fill his pool using only his house hose?

Extending the Concepts

103. Show that the sum of the solutions to a quadratic equation is $-\dfrac{b}{a}$.

104. Show that the product of the solutions to a quadratic equation is $\dfrac{c}{a}$.

105. Show that the real solutions to the equation $ax^2 + bx + c = 0$ are the negatives of the real solutions to the equation $ax^2 - bx + c = 0$. Assume that $b^2 - 4ac \geq 0$.

106. Show that the real solutions to the equation $ax^2 + bx + c = 0$ are the reciprocals of the real solutions to the equation $cx^2 + bx + a = 0$. Assume that $b^2 - 4ac \geq 0$.

Explaining the Concepts

107. Explain the circumstances for which you would use factoring to solve a quadratic equation.

108. Explain the circumstances for which you would use the Square Root Property to solve a quadratic equation.

Synthesis Review

109. (a) Graph $f(x) = x^2 + 3x + 2$ by plotting points.
(b) Solve the equation $x^2 + 3x + 2 = 0$.

(c) Compare the solutions to the equation in part (b) to the x-intercepts of the graph drawn in part (a). What do you notice?

110. (a) Graph $f(x) = x^2 - x - 6$ by plotting points.
(b) Solve the equation $x^2 - x - 6 = 0$.
(c) Compare the solutions to the equation in part (b) to the x-intercepts of the graph drawn in part (a). What do you notice?

111. (a) Graph $g(x) = x^2 - 2x + 1$ by plotting points.
(b) Solve the equation $x^2 - 2x + 1 = 0$.
(c) Compare the solutions to the equation in part (b) to the x-intercepts of the graph drawn in part (a). What do you notice?

112. (a) Graph $g(x) = x^2 + 4x + 4$ by plotting points.
(b) Solve the equation $x^2 + 4x + 4 = 0$.
(c) Compare the solutions to the equation in part (b) to the x-intercepts of the graph drawn in part (a). What do you notice?

Technology Exercises

In Problems 113–116, the graph of the quadratic function f is given. For each function, determine the discriminant of the equation $f(x) = 0$ in order to determine the nature of the solutions of the equation $f(x) = 0$. Compare the nature of solutions based on the discriminant to the graph of the function.

113. $f(x) = x^2 - 7x + 3$ **114.** $f(x) = -x^2 - 3x + 1$

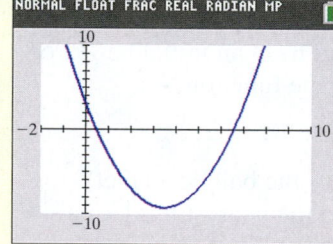 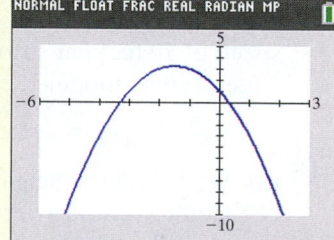

115. $f(x) = -x^2 - 3x - 4$ **116.** $f(x) = x^2 - 6x + 9$

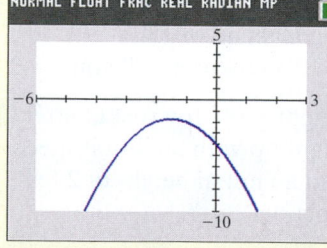 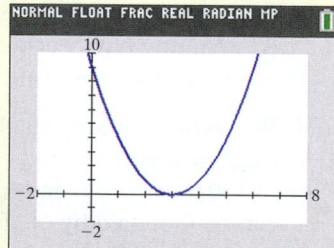

117. (a) Solve the equation $x^2 - 5x - 24 = 0$ algebraically.
(b) Graph $Y_1 = x^2 - 5x - 24$. Compare the x-intercepts of the graph to the solutions found in part (a).

118. (a) Solve the equation $x^2 - 4x - 45 = 0$ algebraically.
(b) Graph $Y_1 = x^2 - 4x - 45$. Compare the x-intercepts of the graph to the solutions found in part (a).

119. (a) Solve the equation $x^2 - 6x + 9 = 0$ algebraically.
 (b) Graph $Y_1 = x^2 - 6x + 9$. Compare the x-intercepts of the graph to the solutions found in part (a).

120. (a) Solve the equation $x^2 + 10x + 25 = 0$ algebraically.
 (b) Graph $Y_1 = x^2 + 10x + 25$. Compare the x-intercepts of the graph to the solutions found in part (a).

121. (a) Solve the equation $x^2 + 5x + 8 = 0$ algebraically.
 (b) Graph $Y_1 = x^2 + 5x + 8$. How is the result of part (a) related to the graph?

122. (a) Solve the equation $x^2 + 2x + 5 = 0$ algebraically.
 (b) Graph $Y_1 = x^2 + 2x + 5$. How is the result of part (a) related to the graph?

10.3 Solving Equations Quadratic in Form

Objective

1 Solve Equations That Are Quadratic in Form

Are You Prepared for This Section?

Before getting started, complete the following problems. If you get a problem wrong, go back to the section cited and review the material.

P1. Factor: $x^4 - 5x^2 - 6$ [Section 6.5, p. 408]

P2. Factor: $2(p + 3)^2 + 3(p + 3) - 5$ [Section 6.5, p. 408]

P3. Simplify: **(a)** $(x^2)^2$ **(b)** $(p^{-1})^2$ [Section 5.2, p. 320]

▶ ❶ Solve Equations That Are Quadratic in Form

Consider the equation $x^4 - 4x^2 - 12 = 0$. Even though this equation is not in the form of a quadratic equation, $ax^2 + bx + c = 0$, the equation can be written as $(x^2)^2 - 4x^2 - 12 = 0$. Let $u = x^2$ in the equation, to get $u^2 - 4u - 12 = 0$, which is of the form $ax^2 + bx + c = 0$. Now solve $u^2 - 4u - 12 = 0$ for u by factoring. Then, using the fact that $u = x^2$, find x, which was the goal in the first place.

In general, if a substitution u transforms an equation into one of the form

$$au^2 + bu + c = 0$$

then the original equation is an **equation quadratic in form.**

It is often hard to tell whether an equation is quadratic in form. Table 2 shows some equations that are quadratic in form and the appropriate substitution.

Prepared? ... Answers
P1. $(x^2 - 6)(x^2 + 1)$
P2. $(2p + 11)(p + 2)$
P3. (a) x^4 **(b)** $p^{-2} = \dfrac{1}{p^2}$

Table 2

Original Equation	Substitution	Equation with Substitution
$2x^4 - 3x^2 + 5 = 0$ $2(x^2)^2 - 3x^2 + 5 = 0$	$u = x^2$	$2u^2 - 3u + 5 = 0$
$3(z - 5)^2 + 4(z - 5) + 1 = 0$	$u = z - 5$	$3u^2 + 4u + 1 = 0$
$-2y + 5\sqrt{y} - 2 = 0$ $-2(\sqrt{y})^2 + 5\sqrt{y} - 2 = 0$	$u = \sqrt{y}$	$-2u^2 + 5u - 2 = 0$

EXAMPLE 1 **How to Solve Equations That Are Quadratic in Form**

Solve: $x^4 + x^2 - 12 = 0$

Step-by-Step Solution

Step 1: Determine the appropriate substitution and write the equation in the form $au^2 + bu + c = 0$.

$$x^4 + x^2 - 12 = 0$$
$$(x^2)^2 + x^2 - 12 = 0$$
Let $u = x^2$: $u^2 + u - 12 = 0$

(continued)

Step 2: Solve the equation $au^2 + bu + c = 0$.

$$(u + 4)(u - 3) = 0$$
$$u + 4 = 0 \quad \text{or} \quad u - 3 = 0$$
$$u = -4 \quad \text{or} \quad u = 3$$

Step 3: Rewrite the solution in terms of the original variable.

We want to know x, so replace u with x^2: $\quad x^2 = -4 \quad$ or $\quad x^2 = 3$

Square Root Property: $\quad x = \pm\sqrt{-4} \quad$ or $\quad x = \pm\sqrt{3}$

$$= \pm 2i$$

Step 4: Check Verify your solutions.

$x = 2i$:
$$(2i)^4 + (2i)^2 - 12 \overset{?}{=} 0$$
$$2^4 i^4 + 2^2 i^2 - 12 \overset{?}{=} 0$$
$$16(1) + 4(-1) - 12 \overset{?}{=} 0$$
$$16 - 4 - 12 \overset{?}{=} 0$$
$$0 = 0 \quad \text{True}$$

$x = -2i$:
$$(-2i)^4 + (-2i)^2 - 12 \overset{?}{=} 0$$
$$(-2)^4 i^4 + (-2)^2 i^2 - 12 \overset{?}{=} 0$$
$$16(1) + 4(-1) - 12 \overset{?}{=} 0$$
$$16 - 4 - 12 \overset{?}{=} 0$$
$$0 = 0 \quad \text{True}$$

$x = \sqrt{3}$:
$$(\sqrt{3})^4 + (\sqrt{3})^2 - 12 \overset{?}{=} 0$$
$$\sqrt{3^4} + \sqrt{3^2} - 12 \overset{?}{=} 0$$
$$\sqrt{81} + 3 - 12 \overset{?}{=} 0$$
$$9 + 3 - 12 \overset{?}{=} 0$$
$$0 = 0 \quad \text{True}$$

$x = -\sqrt{3}$:
$$(-\sqrt{3})^4 + (-\sqrt{3})^2 - 12 \overset{?}{=} 0$$
$$(-1)^4\sqrt{3^4} + (-1)^2\sqrt{3^2} - 12 \overset{?}{=} 0$$
$$\sqrt{81} + 3 - 12 \overset{?}{=} 0$$
$$9 + 3 - 12 \overset{?}{=} 0$$
$$0 = 0 \quad \text{True}$$

The solution set is $\{2i, -2i, \sqrt{3}, -\sqrt{3}\}$. ●

The steps used to solve an equation that is quadratic in form are summarized below.

> **Solving Equations Quadratic in Form**
>
> **Step 1:** Determine the appropriate substitution and write the equation in the form $au^2 + bu + c = 0$.
>
> **Step 2:** Solve the equation $au^2 + bu + c = 0$.
>
> **Step 3:** Rewrite the solution in terms of the original variable.
>
> **Step 4:** Verify your solution(s).

Quick ✓

1. If a substitution u transforms an equation into one of the form $au^2 + bu + c = 0$, then the original equation is called an equation _____ __ ____.

2. For the equation $2(3x + 1)^2 - 5(3x + 1) + 2 = 0$, an appropriate substitution would be $u =$ _____.

3. *True or False* The equation $3\left(\dfrac{x}{x-2}\right)^2 - \dfrac{5x}{x-2} + 3 = 0$ is quadratic in form.

4. What is the appropriate choice for u when solving the equation $2 \cdot \dfrac{1}{x^2} - 6 \cdot \dfrac{1}{x} + 3 = 0$?

In Problems 5 and 6, solve each equation.

5. $x^4 - 13x^2 + 36 = 0$

6. $p^4 - 7p^2 = 18$

Solving Equations That Are Quadratic in Form

Solve: $(z^2 - 5)^2 - 3(z^2 - 5) - 4 = 0$

Solution

$$(z^2 - 5)^2 - 3(z^2 - 5) - 4 = 0$$

Let $u = z^2 - 5$: $u^2 - 3u - 4 = 0$

$$(u - 4)(u + 1) = 0$$

$$u - 4 = 0 \quad \text{or} \quad u + 1 = 0$$

$$u = 4 \quad \text{or} \quad u = -1$$

Replace u with $z^2 - 5$ and solve for z: $z^2 - 5 = 4 \quad \text{or} \quad z^2 - 5 = -1$

$$z^2 = 9 \quad \text{or} \quad z^2 = 4$$

Square Root Property: $z = \pm 3 \quad \text{or} \quad z = \pm 2$

Check

$z = -3$:

$((-3)^2 - 5)^2 - 3((-3)^2 - 5) - 4 \overset{?}{=} 0$

$(9 - 5)^2 - 3(9 - 5) - 4 \overset{?}{=} 0$

$4^2 - 3(4) - 4 \overset{?}{=} 0$

$16 - 12 - 4 \overset{?}{=} 0$

$0 = 0$ True

$z = 3$:

$((3)^2 - 5)^2 - 3((3)^2 - 5) - 4 \overset{?}{=} 0$

$(9 - 5)^2 - 3(9 - 5) - 4 \overset{?}{=} 0$

$4^2 - 3(4) - 4 \overset{?}{=} 0$

$16 - 12 - 4 \overset{?}{=} 0$

$0 = 0$ True

$z = -2$:

$((-2)^2 - 5)^2 - 3((-2)^2 - 5) - 4 \overset{?}{=} 0$

$(4 - 5)^2 - 3(4 - 5) - 4 \overset{?}{=} 0$

$(-1)^2 - 3(-1) - 4 \overset{?}{=} 0$

$1 + 3 - 4 \overset{?}{=} 0$

$0 = 0$ True

$z = 2$:

$((2)^2 - 5)^2 - 3((2)^2 - 5) - 4 \overset{?}{=} 0$

$(4 - 5)^2 - 3(4 - 5) - 4 \overset{?}{=} 0$

$(-1)^2 - 3(-1) - 4 \overset{?}{=} 0$

$1 + 3 - 4 \overset{?}{=} 0$

$0 = 0$ True

The solution set is $\{-3, -2, 2, 3\}$. ●

Quick ✔

In Problems 7 and 8, solve each equation.

7. $(p^2 - 2)^2 - 9(p^2 - 2) + 14 = 0$ **8.** $2(2z^2 - 1)^2 + 5(2z^2 - 1) - 3 = 0$

When both sides of an equation are raised to an even power (such as in squaring both sides of the equation), extraneous solutions may be introduced to the equation. Therefore, it is important that to verify all solutions.

EXAMPLE 3 **Solving Equations That Are Quadratic in Form**

Solve: $3x - 5\sqrt{x} - 2 = 0$

Solution

$$3x - 5\sqrt{x} - 2 = 0$$

$$3(\sqrt{x})^2 - 5\sqrt{x} - 2 = 0$$

Let $u = \sqrt{x}$: $3u^2 - 5u - 2 = 0$

$$(3u + 1)(u - 2) = 0$$

(continued)

$$3u + 1 = 0 \quad \text{or} \quad u - 2 = 0$$
$$3u = -1 \quad \text{or} \quad u = 2$$
$$u = -\frac{1}{3}$$

Replace u with $\sqrt{x}$ and solve for x: $\sqrt{x} = -\dfrac{1}{3}$ or $\sqrt{x} = 2$

Square both sides: $x = \dfrac{1}{9}$ or $x = 4$

Check

$x = \dfrac{1}{9}$: $3 \cdot \dfrac{1}{9} - 5\sqrt{\dfrac{1}{9}} - 2 = 0$ $x = 4$: $3 \cdot 4 - 5\sqrt{4} - 2 = 0$

$\dfrac{1}{3} - 5 \cdot \dfrac{1}{3} - 2 = 0$ $12 - 5 \cdot 2 - 2 = 0$

$\dfrac{1}{3} - \dfrac{5}{3} - \dfrac{6}{3} = 0$ $12 - 10 - 2 = 0$

$0 = 0$ True

$-\dfrac{10}{3} = 0$ False

Work Smart

The equation $\sqrt{x} = -\dfrac{1}{3}$ has no solution because the square root of a positive number is positive, the square root of 0 is 0, and the square root of a negative number is a non-real, complex number.

The solution $x = \dfrac{1}{9}$ is extraneous. The solution set is $\{4\}$.   ●

The equation in Example 3 could also have been solved using the methods introduced in Section 9.8 by isolating the radical and squaring both sides.

> **Quick ✓**
>
> *In Problems 9 and 10, solve the equation.*
>
> **9.** $3w - 14\sqrt{w} + 8 = 0$ **10.** $2q - 9\sqrt{q} - 5 = 0$

EXAMPLE 4 **Solving Equations That Are Quadratic in Form**

Solve: $4x^{-2} + 13x^{-1} - 12 = 0$

Solution

$$4x^{-2} + 13x^{-1} - 12 = 0$$
$$4(x^{-1})^2 + 13x^{-1} - 12 = 0$$

Let $u = x^{-1}$: $4u^2 + 13u - 12 = 0$

Factor: $(4u - 3)(u + 4) = 0$

$$4u - 3 = 0 \quad \text{or} \quad u + 4 = 0$$
$$4u = 3 \quad \text{or} \quad u = -4$$
$$u = \frac{3}{4}$$

Replace u with x^{-1} and solve for x: $x^{-1} = \dfrac{3}{4}$ or $x^{-1} = -4$

$x^{-1} = \dfrac{1}{x}$: $\dfrac{1}{x} = \dfrac{3}{4}$ or $\dfrac{1}{x} = -4$

Take the reciprocal of both sides of the equation: $x = \dfrac{4}{3}$ or $x = \dfrac{1}{-4} = -\dfrac{1}{4}$

Check

$$x = \frac{4}{3}:$$

$$4\left(\frac{4}{3}\right)^{-2} + 13\left(\frac{4}{3}\right)^{-1} - 12 \stackrel{?}{=} 0$$

$$4\left(\frac{3}{4}\right)^2 + 13 \cdot \frac{3}{4} - 12 \stackrel{?}{=} 0$$

$$4 \cdot \frac{9}{16} + \frac{39}{4} - 12 \stackrel{?}{=} 0$$

$$\frac{9}{4} + \frac{39}{4} - \frac{48}{4} \stackrel{?}{=} 0$$

$$0 = 0 \quad \text{True}$$

$$x = -\frac{1}{4}:$$

$$4\left(-\frac{1}{4}\right)^{-2} + 13\left(-\frac{1}{4}\right)^{-1} - 12 \stackrel{?}{=} 0$$

$$4(-4)^2 + 13(-4) - 12 \stackrel{?}{=} 0$$

$$4 \cdot 16 - 52 - 12 \stackrel{?}{=} 0$$

$$64 - 52 - 12 \stackrel{?}{=} 0$$

$$0 = 0 \quad \text{True}$$

The solution set is $\left\{-\frac{1}{4}, \frac{4}{3}\right\}$.

Quick ✓

In Problem 11, solve the equation.

11. $5x^{-2} + 12x^{-1} + 4 = 0$

▶ **EXAMPLE 5** **Solving Equations That Are Quadratic in Form**

Solve: $a^{\frac{2}{3}} + 3a^{\frac{1}{3}} - 28 = 0$

Solution

$$a^{\frac{2}{3}} + 3a^{\frac{1}{3}} - 28 = 0$$

$$\left(a^{\frac{1}{3}}\right)^2 + 3a^{\frac{1}{3}} - 28 = 0$$

Let $u = a^{\frac{1}{3}}$:
$$u^2 + 3u - 28 = 0$$

$$(u + 7)(u - 4) = 0$$

$$u + 7 = 0 \quad \text{or} \quad u - 4 = 0$$

$$u = -7 \quad \text{or} \quad u = 4$$

Replace u with $a^{\frac{1}{3}}$ and solve for a: $\quad a^{\frac{1}{3}} = -7 \quad \text{or} \quad a^{\frac{1}{3}} = 4$

Cube both sides of the equation: $\quad \left(a^{\frac{1}{3}}\right)^3 = (-7)^3 \quad \text{or} \quad \left(a^{\frac{1}{3}}\right)^3 = 4^3$

$$a = -343 \quad \text{or} \quad a = 64$$

Check $a = -343:$

$$(-343)^{\frac{2}{3}} + 3(-343)^{\frac{1}{3}} - 28 = 0$$

$$(\sqrt[3]{-343})^2 + 3 \cdot \sqrt[3]{-343} - 28 \stackrel{?}{=} 0$$

$$(-7)^2 + 3(-7) - 28 \stackrel{?}{=} 0$$

$$49 - 21 - 28 \stackrel{?}{=} 0$$

$$0 = 0 \quad \text{True}$$

$a = 64:$

$$(64)^{\frac{2}{3}} + 3(64)^{\frac{1}{3}} - 28 = 0$$

$$(\sqrt[3]{64})^2 + 3 \cdot \sqrt[3]{64} - 28 \stackrel{?}{=} 0$$

$$4^2 + 3 \cdot 4 - 28 \stackrel{?}{=} 0$$

$$16 + 12 - 28 \stackrel{?}{=} 0$$

$$0 = 0 \quad \text{True}$$

The solution set is $\{-343, 64\}$.

Quick ✓

In Problem 12, solve the equation.

12. $p^{\frac{2}{3}} - 4p^{\frac{1}{3}} - 5 = 0$

*Problems **1–12** are the* **Quick ✔** s *that follow the* **EXAMPLES**.

Building Skills

In Problems 13–48, solve each equation. See Objective 1.

13. $x^4 - 5x^2 + 4 = 0$ **14.** $x^4 - 10x^2 + 9 = 0$

15. $q^4 + 13q^2 + 36 = 0$ **16.** $z^4 + 10z^2 + 9 = 0$

17. $4a^4 - 17a^2 + 4 = 0$ **18.** $4b^4 - 5b^2 + 1 = 0$

19. $p^4 + 6 = 5p^2$ **20.** $q^4 + 15 = 8q^2$

21. $(x - 3)^2 - 6(x - 3) - 7 = 0$

22. $(x + 2)^2 - 3(x + 2) - 10 = 0$

23. $(x^2 - 1)^2 - 11(x^2 - 1) + 24 = 0$

24. $(p^2 - 2)^2 - 8(p^2 - 2) + 12 = 0$

25. $(y^2 + 2)^2 + 7(y^2 + 2) + 10 = 0$

26. $(q^2 + 4)^2 + 3(q^2 + 4) - 4 = 0$

27. $x - 3\sqrt{x} - 4 = 0$ **28.** $x - 5\sqrt{x} - 6 = 0$

29. $w + 5\sqrt{w} + 6 = 0$ **30.** $z + 7\sqrt{z} + 6 = 0$

31. $2x + 5\sqrt{x} = 3$ **32.** $3x = 11\sqrt{x} + 4$

33. $x^{-2} + 3x^{-1} = 28$ **34.** $q^{-2} + 2q^{-1} = 15$

35. $10z^{-2} + 11z^{-1} = 6$ **36.** $10a^{-2} + 23a^{-1} = 5$

37. $x^{\frac{2}{3}} + 3x^{\frac{1}{3}} - 4 = 0$ **38.** $y^{\frac{2}{3}} - 2y^{\frac{1}{3}} - 3 = 0$

39. $z^{\frac{2}{3}} - z^{\frac{1}{3}} = 2$ **40.** $w^{\frac{2}{3}} + 2w^{\frac{1}{3}} = 3$

41. $a + a^{\frac{1}{2}} = 30$ **42.** $b + 3b^{\frac{1}{2}} = 28$

43. $\dfrac{1}{x^2} - \dfrac{5}{x} + 6 = 0$

44. $\dfrac{1}{x^2} - \dfrac{7}{x} + 12 = 0$

45. $\left(\dfrac{1}{x + 2}\right)^2 + \dfrac{4}{x + 2} = 5$

46. $\left(\dfrac{1}{x + 2}\right)^2 + \dfrac{6}{x + 2} = 7$

47. $p^6 - 28p^3 + 27 = 0$

48. $y^6 - 7y^3 - 8 = 0$

Mixed Practice

In Problems 49–62, solve each equation.

49. $8a^{-2} + 2a^{-1} = 1$ **50.** $6b^{-2} - b^{-1} = 1$

51. $z^4 = 4z^2 + 32$ **52.** $x^4 + 3x^2 = 4$

53. $x^{\frac{1}{2}} + x^{\frac{1}{4}} - 6 = 0$ **54.** $c^{\frac{1}{2}} + c^{\frac{1}{4}} - 12 = 0$

55. $w^4 - 5w^2 - 36 = 0$ **56.** $p^4 - 15p^2 - 16 = 0$

57. $\left(\dfrac{1}{x + 3}\right)^2 + \dfrac{2}{x + 3} = 3$

58. $\left(\dfrac{1}{x - 1}\right)^2 + \dfrac{7}{x - 1} = 8$

59. $x - 7\sqrt{x} + 12 = 0$

60. $x - 8\sqrt{x} + 12 = 0$

61. $2(x - 1)^2 - 7(x - 1) = 4$

62. $3(y - 2)^2 - 4(y - 2) = 4$

63. Suppose that $f(x) = x^4 + 7x^2 + 12$. Find the values of x such that
    (a) $f(x) = 12$ **(b)** $f(x) = 6$

64. Suppose that $f(x) = x^4 + 5x^2 + 3$. Find the values of x such that
    (a) $f(x) = 3$ **(b)** $f(x) = 17$

65. Suppose that $g(x) = 2x^4 - 6x^2 - 5$. Find the values of x such that
    (a) $g(x) = -5$ **(b)** $g(x) = 15$

66. Suppose that $h(x) = 3x^4 - 9x^2 - 8$. Find the values of x such that
    (a) $h(x) = -8$ **(b)** $h(x) = 22$

67. Suppose that $F(x) = x^{-2} - 5x^{-1}$. Find the values of x such that
    (a) $F(x) = 6$ **(b)** $F(x) = 14$

68. Suppose that $f(x) = x^{-2} - 3x^{-1}$. Find the values of x such that
    (a) $f(x) = 4$ **(b)** $f(x) = 18$

In Problems 69–74, find the zeros of the function. (Hint: Remember, r is a zero if f(r) = 0.)

69. $f(x) = x^4 + 9x^2 + 14$

70. $f(x) = x^4 - 13x^2 + 42$

71. $g(t) = 6t - 25\sqrt{t} - 9$

72. $h(p) = 8p - 18\sqrt{p} - 35$

73. $s(d) = \dfrac{1}{(d+3)^2} - \dfrac{4}{d+3} + 3$

74. $f(a) = \dfrac{1}{(a-2)^2} + \dfrac{3}{a-2} - 4$

Applying the Concepts

75. (a) Solve $x^2 - 5x + 6 = 0$.
    (b) Solve $(x-3)^2 - 5(x-3) + 6 = 0$. Compare the solutions to part (a).
    (c) Solve $(x+2)^2 - 5(x+2) + 6 = 0$. Compare the solutions to part (a).
    (d) Solve $(x-5)^2 - 5(x-5) + 6 = 0$. Compare the solutions to part (a).
    (e) Conjecture a generalization for the solution of $(x-a)^2 - 5(x-a) + 6 = 0$.

76. (a) Solve $x^2 + 3x - 18 = 0$.
    (b) Solve $(x-1)^2 + 3(x-1) - 18 = 0$. Compare the solutions to part (a).
    (c) Solve $(x+5)^2 + 3(x+5) - 18 = 0$. Compare the solutions to part (a).
    (d) Solve $(x-3)^2 + 3(x-3) - 18 = 0$. Compare the solutions to part (a).
    (e) Conjecture a generalization for the solution of $(x-a)^2 + 3(x-a) - 18 = 0$.

77. Consider the function $f(x) = 2x^2 - 3x + 1$.
    (a) Solve $f(x) = 0$.
    (b) Solve $f(x-2) = 0$. Compare the solutions to part (a).
    (c) Solve $f(x-5) = 0$. Compare the solutions to part (a).
    (d) Conjecture a generalization for the zeros of $f(x-a)$.

78. Consider the function $f(x) = 3x^2 - 5x - 2$.
    (a) Solve $f(x) = 0$.
    (b) Solve $f(x-1) = 0$. Compare the solutions to part (a).
    (c) Solve $f(x-4) = 0$. Compare the solutions to part (a).
    (d) Conjecture a generalization for the zeros of $f(x-a)$.

79. Revenue The function
$$R(x) = \frac{(x-1990)^2}{2} + \frac{3(x-1990)}{2} + 3000$$
models the revenue R (in thousands of dollars) of a start-up computer consulting firm in year x, where $x \geq 1990$.
    (a) Determine and interpret $R(1990)$.
    (b) Solve and interpret $R(x) = 3065$.
    (c) According to the model, in what year can the firm expect to receive \$3350 thousand in revenue?

80. Revenue The function
$$R(x) = \frac{(x-2000)^2}{3} + \frac{5(x-2000)}{3} + 2000$$
models the revenue R (in thousands of dollars) of a start-up computer software firm in year x, where $x \geq 2000$.
    (a) Determine and interpret $R(2000)$.
    (b) Solve and interpret $R(x) = 2250$.
    (c) According to the model, in what year can the firm expect to receive \$2350 thousand in revenue?

Extending the Concepts

All of the problems given in this section resulted in equations quadratic in form that could be factored after the appropriate substitution. However, this is not a necessary requirement to solving equations quadratic in form. In Problems 81–84, determine the appropriate substitution, and then use the quadratic formula to find the value of u. Finally, determine the value of the variable in the equation.

81. $x^4 + 5x^2 + 2 = 0$

82. $x^4 + 7x^2 + 4 = 0$

83. $2(x - 2)^2 + 8(x - 2) - 1 = 0$

84. $3(x + 1)^2 + 6(x + 1) - 1 = 0$

Explaining the Concepts

85. The equation $x - 5\sqrt{x} - 6 = 0$ can be solved either by using the methods of this section or by isolating the radical and squaring both sides. Solve it both ways and explain which approach you prefer.

86. Explain the steps required to solve an equation quadratic in form. Be sure to include an explanation of how to identify the appropriate substitution.

87. Under what circumstances might extraneous solutions occur when one is solving equations quadratic in form?

Synthesis Review

In Problems 88–91, add or subtract the expressions.

88. $(4x^2 - 3x - 1) + (-3x^2 + x + 5)$

89. $(3p^{-2} - 4p^{-1} + 8) - (2p^{-2} - 8p^{-1} - 1)$

90. $3\sqrt{2x} - \sqrt{8x} + \sqrt{50x}$

91. $\sqrt[3]{16a} + \sqrt[3]{54a} - \sqrt[3]{128a^4}$

92. Write a sentence or two that discusses how to add or subtract algebraic expressions, in general.

Technology Exercises

In Problems 93–98, use technology to find the real solutions to the equations. Round your answers to two decimal places, if necessary.

93. $x^4 + 5x^2 - 14 = 0$ **94.** $x^4 - 4x^2 - 12 = 0$

95. $2(x - 2)^2 = 5(x - 2) + 1$

96. $3(x + 3)^2 = 2(x + 3) + 6$

97. $x - 5\sqrt{x} = -3$ **98.** $x + 4\sqrt{x} = 5$

99. (a) Graph $Y = x^2 - 5x - 6$. Find the x-intercepts of the graph.
 (b) Graph $Y = (x + 2)^2 - 5(x + 2) - 6$. Find the x-intercepts of the graph.
 (c) Graph $Y = (x + 5)^2 - 5(x + 5) - 6$. Find the x-intercepts of the graph.
 (d) Make a generalization based upon the results of parts (a), (b), and (c).

100. (a) Graph $Y = x^2 + 4x + 3$. Find the x-intercepts of the graph.
 (b) Graph $Y = (x - 3)^2 + 4(x - 3) + 3$. Find the x-intercepts of the graph.
 (c) Graph $Y = (x - 6)^2 + 4(x - 6) + 3$. Find the x-intercepts of the graph.
 (d) Make a generalization based upon the results of parts (a), (b), and (c).

Putting the Concepts Together (Sections 10.1–10.3)

We designed these problems so that you can review Sections 10.1–10.3 and show your mastery of the concepts. Take time to work these problems before proceeding with the next section. The answers are located at the back of the text on page AN-49.

In Problems 1–3, complete the square in the given expression. Then factor the perfect square trinomial.

1. $z^2 + 10z$

2. $x^2 + 7x$

3. $n^2 - \dfrac{1}{4}n$

In Problems 4–6, solve each quadratic equation using the stated method.

4. $(2x - 3)^2 - 5 = -1$; Square Root Property

5. $x^2 + 8x + 4 = 0$; completing the square

6. $x(x - 6) = -7$; quadratic formula

In Problems 7–10, solve each equation using the method you prefer.

7. $49x^2 - 80 = 0$

8. $p^2 - 8p + 6 = 0$

9. $3y^2 + 6y + 4 = 0$

10. $\dfrac{1}{4}n^2 + n = \dfrac{1}{6}$

In Problems 11–13, determine the discriminant of each quadratic equation. Use the value of the discriminant to determine whether the equation has two rational solutions, two irrational solutions, one repeated rational solution, or two complex solutions that are not real.

11. $9x^2 + 12x + 4 = 0$ **12.** $3x^2 + 6x - 2 = 0$

13. $2x^2 + 6x + 5 = 0$

14. Find the missing length in the right triangle shown below.

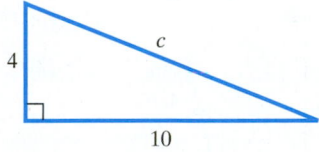

In Problems 15 and 16, solve each equation.

15. $2m + 7\sqrt{m} - 15 = 0$

16. $p^{-2} - 3p^{-1} - 18 = 0$

17. Revenue The revenue R received by a company selling x microwave ovens per day is given by the function $R(x) = -0.4x^2 + 140x$. How many microwave ovens must be sold for revenue to be $12,000 per day?

18. Airplane Ride An airplane flies 300 miles into the wind and then flies home against the wind. The total time of the trip (excluding time on the ground) is 5 hours. If the plane can fly 140 miles per hour in still air, what was the speed of the wind? Round your answer to the nearest tenth.

10.4 Graphing Quadratic Functions Using Transformations

Objectives

1 Graph Quadratic Functions of the Form $f(x) = x^2 + k$

2 Graph Quadratic Functions of the Form $f(x) = (x - h)^2$

3 Graph Quadratic Functions of the Form $f(x) = ax^2$

4 Graph Quadratic Functions of the Form $f(x) = ax^2 + bx + c$

5 Find a Quadratic Function from Its Graph

Prepared?...Answers

P1.

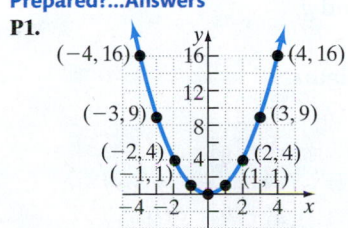

P2.

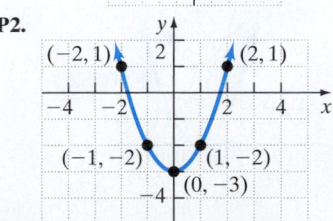

P3. The set of all real numbers

Are You Prepared for This Section?

Before getting started, complete the following problems. If you get a problem wrong, go back to the section cited and review the material.

P1. Graph $y = x^2$ using point plotting. [Section 8.1, pp. 524–526]

P2. Use the point-plotting method to graph $y = x^2 - 3$. [Section 8.1, pp. 524–526]

P3. What is the domain of $f(x) = 2x^2 + 5x + 1$? [Section 8.3, pp. 545–546]

Definition

A **quadratic function** is a function of the form

$$f(x) = ax^2 + bx + c$$

where a, b, and c are real numbers and $a \neq 0$. The domain of a quadratic function consists of all real numbers.

Many situations can be modeled using quadratic functions. For example, Example 10 of Section 8.3 showed that Franco Modigliani used the quadratic function $I(a) = -55a^2 + 5119a - 54{,}448$ to model the relation between average annual income, I, and age, a.

Also, if the effect of air resistance on a projectile is ignored, then the height of the projectile as a function of horizontal distance traveled can be modeled using a quadratic function. See Figure 7 on the next page.

In this section and the next, new techniques for graphing quadratic functions will be explored. In Section 8.1, equations were graphed using point plotting, but this method is inefficient and can lead to incomplete graphs. Remember, a complete graph shows all of its "interesting features," such as its intercepts and high and low points. Two methods for graphing quadratic functions that are superior to the point-plotting

Figure 7

method will be presented here. The first method uses *transformations* and is the subject of this section. The second method, discussed in the next section, uses properties of quadratic functions.

▶ ❶ Graph Quadratic Functions of the Form $f(x) = x^2 + k$

Work Smart

Consider the quadratic function $f(x) = ax^2 + bx + c$, where $a = 1$, $b = 0$, and c is any real number. This is a function of the form $f(x) = x^2 + k$.

We begin by learning how to graph any quadratic function of the form $f(x) = x^2 + k$, such as $g(x) = x^2 + 3$ or $h(x) = x^2 - 4$.

Consider the graph of $y = g(x) = x^2$ in Figure 8, which first appeared in Section 8.1. What effect does adding a real number k to the function $g(x) = x^2$ have on its graph? Let's see!

Figure 8

$y = g(x) = x^2$

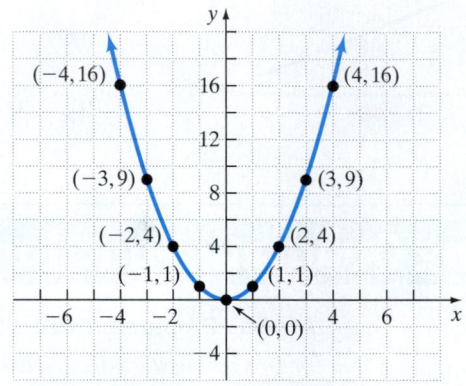

EXAMPLE 1

Graphing a Quadratic Function of the Form $f(x) = x^2 + k$

In the same Cartesian plane, graph $g(x) = x^2$ and $f(x) = x^2 + 3$.

Solution

Begin by obtaining some points on the graphs of g and f. For example, when $x = 0$, then $y = g(0) = 0$ and $y = f(0) = 0^2 + 3 = 3$. When $x = 1$, then $y = g(1) = 1$ and $y = f(1) = 1^2 + 3 = 4$. Table 3 lists these points along with a few others. Notice that while the x-coordinates stay the same, the y-coordinates on the graph of $f(x) = x^2 + 3$ are exactly 3 units greater than the corresponding y-coordinates on the graph of $g(x) = x^2$. Figure 9 shows the graphs of f and g.

Figure 9

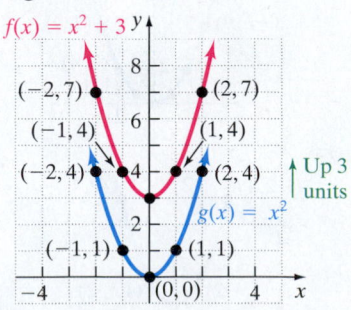

Table 3

x	$g(x) = x^2$	$(x, g(x))$	$f(x) = x^2 + 3$	$(x, f(x))$
-2	$(-2)^2 = 4$	$(-2, 4)$	$(-2)^2 + 3 = 7$	$(-2, 7)$
-1	$(-1)^2 = 1$	$(-1, 1)$	$(-1)^2 + 3 = 4$	$(-1, 4)$
0	0	$(0, 0)$	3	$(0, 3)$
1	1	$(1, 1)$	4	$(1, 4)$
2	4	$(2, 4)$	7	$(2, 7)$

The graph of f is identical to the graph of g, except that it is shifted up 3 units. ●

Let's look at another example.

EXAMPLE 2 **Graphing a Quadratic Function of the Form** $f(x) = x^2 - k$

In the same Cartesian plane, graph $g(x) = x^2$ and $f(x) = x^2 - 4$.

Solution

Table 4 lists some points on the graphs of g and f. Notice that while the x-coordinates stay the same, the y-coordinates on the graph of $f(x) = x^2 - 4$ are exactly 4 units less than the corresponding y-coordinates on the graph of $g(x) = x^2$. Figure 10 shows the graphs of f and g.

Table 4

x	$g(x) = x^2$	$(x, g(x))$	$f(x) = x^2 - 4$	$(x, f(x))$
-2	4	$(-2, 4)$	$(-2)^2 - 4 = 0$	$(-2, 0)$
-1	1	$(-1, 1)$	$(-1)^2 - 4 = -3$	$(-1, -3)$
0	0	$(0, 0)$	-4	$(0, -4)$
1	1	$(1, 1)$	-3	$(1, -3)$
2	4	$(2, 4)$	0	$(2, 0)$

Figure 10

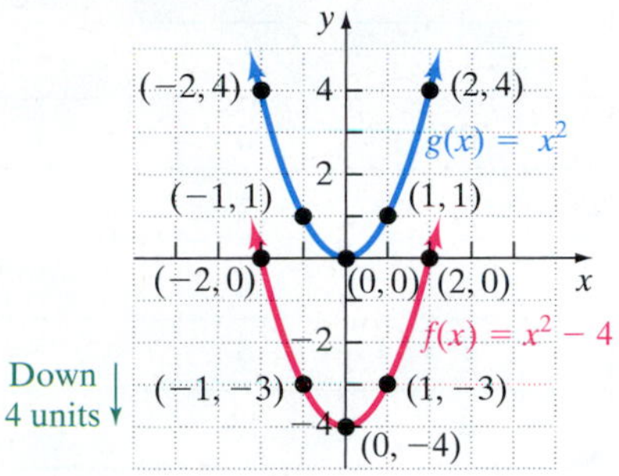

The graph of f is identical to the graph of g except that f is shifted down 4 units. ●

The results of Examples 1 and 2 lead to the following conclusion.

Graphing a Function of the Form $f(x) = x^2 + k$ **or** $f(x) = x^2 - k$

To obtain the graph of $f(x) = x^2 + k$, $k > 0$, from the graph of $y = x^2$, shift the graph of $y = x^2$ up k units. To obtain the graph of $f(x) = x^2 - k$, $k > 0$, from the graph of $y = x^2$, shift the graph of $y = x^2$ down k units.

Quick ✓

1. A _____ _____ is a function of the form $f(x) = ax^2 + bx + c$, where a, b, and c are real numbers and $a \neq 0$.

2. To graph $f(x) = x^2 + k$, $k > 0$, from the graph of $y = x^2$, shift the graph of $y = x^2$ ___ k units. To graph $f(x) = x^2 - k$, $k > 0$, from the graph of $y = x^2$ shift the graph of $y = x^2$ _____ k units.

In Problems 3 and 4, use the graph of $y = x^2$ to graph the quadratic function. Show at least three points on the graph.

3. $f(x) = x^2 + 5$ 4. $f(x) = x^2 - 2$

▶ ❷ **Graph Quadratic Functions of the Form** $f(x) = (x - h)^2$

Now look at the graph of any quadratic function of the form $f(x) = (x - h)^2$ such as $f(x) = (x + 3)^2$ or $f(x) = (x - 2)^2$.

EXAMPLE 3 **Graphing a Quadratic Function of the Form** $f(x) = (x - h)^2$

In the same Cartesian plane, graph $g(x) = x^2$ and $f(x) = (x - 2)^2$.

Solution
Again, use the point-plotting method. Table 5 lists some points on the graphs of g and f. Note that when $g(x) = 0$, $x = 0$, and when $f(x) = 0$, $x = 2$. Also, when $g(x) = 4$, $x = -2$ or 2, and when $f(x) = 4$, $x = 0$ or 4. The graph of f is identical to that of g, except that f is shifted 2 units to the right of g. See Figure 11.

Figure 11

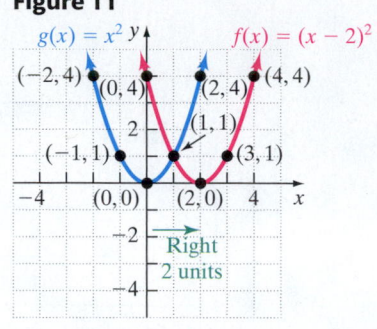

Table 5

x	$g(x) = x^2$	$(x, g(x))$	$f(x) = (x - 2)^2$	$(x, f(x))$
−2	4	(−2, 4)	$(-2 - 2)^2 = 16$	(−2, 16)
−1	1	(−1, 1)	$(-1 - 2)^2 = 9$	(−1, 9)
0	0	(0, 0)	4	(0, 4)
1	1	(1, 1)	1	(1, 1)
2	4	(2, 4)	0	(2, 0)
3	9	(3, 9)	1	(3, 1)
4	16	(4, 16)	4	(4, 4)

Why does this happen? Because 2 is being subtracted from each x-value in the function $f(x) = (x - 2)^2$, the x-values must be greater by 2 in order to obtain the same y-value that was obtained in the graph of $g(x) = x^2$. ●

What if a positive number h is added to x?

EXAMPLE 4 **Graphing a Quadratic Function of the Form** $f(x) = (x + h)^2$

In the same Cartesian plane, graph $g(x) = x^2$ and $f(x) = (x + 3)^2$.

Solution
Table 6 lists some points on the graphs of g and f. Notice that when $g(x) = 0$, then $x = 0$, and when $f(x) = 0$, then $x = -3$. Also, when $g(x) = 4$, then $x = -2$ or 2, and when $f(x) = 4$, then $x = -5$ or -1. The graph of f is identical to that of g, except that f is shifted 3 units to the left of g. See Figure 12.

Figure 12

Table 6

x	$g(x) = x^2$	$(x, g(x))$	$f(x) = (x + 3)^2$	$(x, f(x))$
−5	25	(−5, 25)	$(-5 + 3)^2 = 4$	(−5, 4)
−4	16	(−4, 16)	$(-4 + 3)^2 = 1$	(−4, 1)
−3	9	(−3, 9)	$(-3 + 3)^2 = 0$	(−3, 0)
−2	4	(−2, 4)	1	(−2, 1)
−1	1	(−1, 1)	4	(−1, 4)
0	0	(0, 0)	9	(0, 9)
1	1	(1, 1)	16	(1, 16)
2	4	(2, 4)	25	(2, 25)

Why does this happen? Because 3 is being added to each x-value in the function $f(x) = (x + 3)^2$, the x-values must be smaller by 3 in order to obtain the same y-value that was obtained in the graph of $g(x) = x^2$. ●

Examples 3 and 4 lead to the following conclusion.

Graphing a Function of the Form $f(x) = (x - h)^2$ or $f(x) = (x + h)^2$

To obtain the graph of $f(x) = (x - h)^2, h > 0$, from the graph of $y = x^2$, shift the graph of $y = x^2$ to the right h units. To obtain the graph of $f(x) = (x + h)^2$, $h > 0$, from the graph of $y = x^2$, shift the graph of $y = x^2$ to the left h units.

Quick ✓

5. *True or False* To obtain the graph of $f(x) = (x + 12)^2$ from the graph of $y = x^2$, shift the graph of $y = x^2$ to the right 12 units.

In Problems 6 and 7, use the graph of $y = x^2$ to graph the quadratic function. Show at least three points on the graph.

6. $f(x) = (x + 5)^2$ 7. $f(x) = (x - 1)^2$

Let's do an example where a horizontal shift and a vertical shift are combined.

EXAMPLE 5   **Combining Horizontal and Vertical Shifts**

Graph the function $f(x) = (x + 2)^2 - 3$.

Solution

Graph f in stages. Begin with the graph of $y = x^2$ as shown in Figure 13(a). Shift that graph 2 units to the left to get the graph of $y = (x + 2)^2$ shown in Figure 13(b). Then shift the graph of $y = (x + 2)^2$ down 3 units to get the graph of $y = (x + 2)^2 - 3$ shown in Figure 13(c). At each stage, key points are plotted.

Figure 13

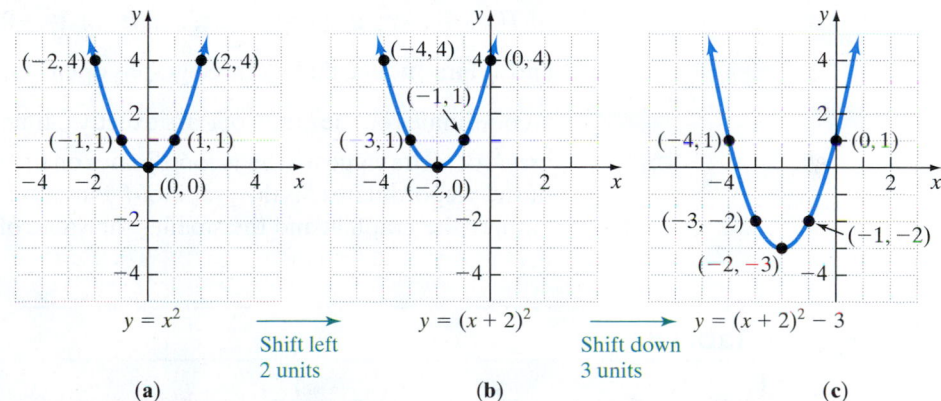

(a) $y = x^2$ → Shift left 2 units → (b) $y = (x + 2)^2$ → Shift down 3 units → (c) $y = (x + 2)^2 - 3$

Note: The order in which the stages are performed to obtain the graph does not matter. In Example 5, the graph could just as easily have shifted down 3 units first and then shifted left 2 units.

Quick ✓

In Problems 8 and 9, graph each quadratic function using horizontal and vertical shifts. Show at least three points on the graph.

8. $f(x) = (x - 3)^2 + 2$ 9. $f(x) = (x + 1)^2 - 4$

▶ ❸ Graph Quadratic Functions of the Form $f(x) = ax^2$

So far, the coefficient of the square term has been equal to 1. What impact does the value of a have on the graph of $f(x) = ax^2 + bx + c$? To answer this question, consider quadratic functions of the form $f(x) = ax^2, a \neq 0$.

First, consider situations in which the value of a is positive. Table 7 shows values on the graphs of $f(x) = x^2$, $g(x) = \frac{1}{2}x^2$, and $h(x) = 2x^2$. Figure 14 shows the graphs of f, g, and h. First, notice that all three graphs open "up." Next, notice that the values of the y-coordinates on the graph of g are exactly $\frac{1}{2}$ of the values of the y-coordinates on the graph of f. The values of the y-coordinates on the graph of h are exactly 2 times the values of the y-coordinates on the graph of f. Put another way, the larger the value of a, the "taller" the graph is, and the smaller the value of a, the "shorter" the graph is.

Work Smart

In a "taller" graph, points with the same x-values have greater y-values. In a "shorter" graph, points with the same x-values have smaller y-values.

Table 7

x	$f(x) = x^2$	$g(x) = \frac{1}{2}x^2$	$h(x) = 2x^2$
-2	$(-2)^2 = 4$	$\frac{1}{2}(-2)^2 = 2$	$2(-2)^2 = 8$
-1	$(-1)^2 = 1$	$\frac{1}{2}(-1)^2 = \frac{1}{2}$	$2(-1)^2 = 2$
0	0	0	0
1	1	$\frac{1}{2}$	2
2	4	2	8

Figure 14

$f(x) = ax^2$. Since $a > 0$, the graphs open up.

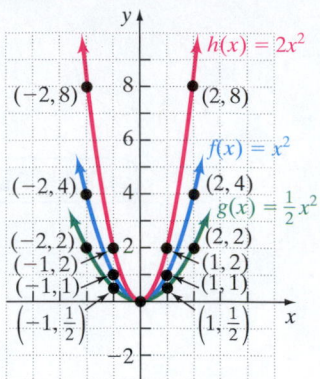

Now consider what happens when a is negative. Table 8 shows values on the graphs of $f(x) = -x^2$, $g(x) = -\frac{1}{2}x^2$, and $h(x) = -2x^2$. Figure 15 shows the graphs of f, g, and h. First, notice that all three graphs open "down." Next, notice that the values of the y-coordinates on the graph of g are exactly $\frac{1}{2}$ times the values of the y-coordinates on the graph of f. The values of the y-coordinates on the graph of h are exactly 2 times the values of the y-coordinates on the graph of f. Put another way, the larger the value of $|a|$, the "taller" the graph is, and the smaller the value of $|a|$, the "shorter" the graph is.

Table 8

x	$f(x) = -x^2$	$g(x) = -\frac{1}{2}x^2$	$h(x) = -2x^2$
-2	$-(-2)^2 = -4$	$-\frac{1}{2}(-2)^2 = -2$	$-2(-2)^2 = -8$
-1	$-(-1)^2 = -1$	$-\frac{1}{2}(-1)^2 = -\frac{1}{2}$	$-2(-1)^2 = -2$
0	0	0	0
1	-1	$-\frac{1}{2}$	-2
2	-4	-2	-8

Figure 15

$f(x) = ax^2$. Since $a < 0$, the graphs open down.

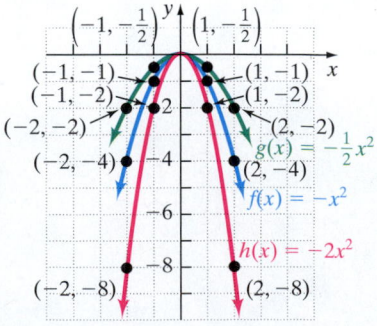

These conclusions are summarized on the following page.

Properties of the Graph of $f(x) = ax^2$

- When $a > 0$, the graph of $f(x) = ax^2$ opens up. If $0 < a < 1$ (a is between 0 and 1), the graph will be "shorter" than that of $y = x^2$. If $a > 1$, the graph will be "taller" than that of $y = x^2$.
- When $a < 0$, the graph of $f(x) = ax^2$ will open down. If $0 < |a| < 1$, the graph will be "shorter" than that of $y = x^2$. If $|a| > 1$, the graph will be "taller" than that of $y = x^2$.
- When $|a| > 1$, we say that the graph is **vertically stretched** by a factor of $|a|$. When $0 < |a| < 1$, we say that the graph is **vertically compressed** by a factor of $|a|$.

Graphing a Function of the Form $f(x) = ax^2$

To obtain the graph of $f(x) = ax^2$ from the graph of $y = x^2$, multiply each y-coordinate on the graph of $y = x^2$ by a.

EXAMPLE 6   **Graphing a Quadratic Function of the Form $f(x) = ax^2$**

Use the graph of $y = x^2$ to obtain the graph of $f(x) = -2x^2$.

Solution
To obtain the graph of $f(x) = -2x^2$ from the graph of $y = x^2$, multiply each y-coordinate on the graph of $y = x^2$ by -2 (the value of a). See Figure 16.

Figure 16

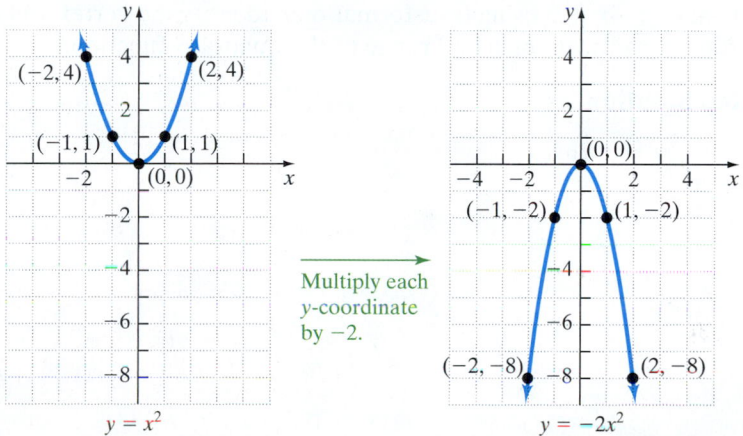

Multiply each
y-coordinate
by -2.

$y = x^2$ $y = -2x^2$

Quick ✓

10. When obtaining the graph of $f(x) = ax^2$ from the graph of $y = x^2$, multiply each _-coordinate on the graph of $y = x^2$ by _. If $|a| > 1$, the graph is _____ _____ by a factor of $|a|$. If $0 < |a| < 1$, the graph is _____ _____ by a factor of $|a|$.

In Problems 11 and 12, use the graph of $y = x^2$ to graph each quadratic function. Label at least 3 points on the graph.

11. $f(x) = 3x^2$ 12. $f(x) = -\dfrac{1}{4}x^2$

---

❹ Graph Quadratic Functions of the Form $f(x) = ax^2 + bx + c$

▶ The graphs in Examples 1–6 are typical of graphs of all quadratic functions. The graph of a quadratic function is called a **parabola** (pronounced puh-răb-uh-luh). Figure 17 on the next page shows two parabolas.

Figure 17

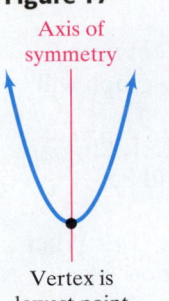
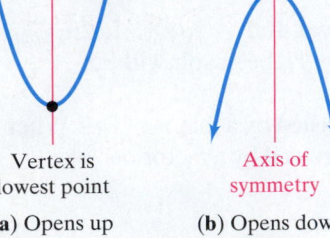

Axis of symmetry

Vertex is highest point

Vertex is lowest point

Axis of symmetry

(a) Opens up
$a > 0$

(b) Opens down
$a < 0$

Work Smart

Although there may be multiple orders that work, it is recommended that, when graphing parabolas using transformations, you obtain the graph that results from the vertical compression or stretch first, followed by the horizontal shift, followed by the vertical shift.

The parabola in Figure 17(a) **opens up** (since $a > 0$) and has a lowest point; the parabola in Figure 17(b) **opens down** (since $a < 0$) and has a highest point. The lowest or highest point of a parabola is the **vertex**. The vertical line passing through the vertex of a parabola is its **axis of symmetry**. If we take the portion of the parabola to the right of the vertex and fold it over the axis of symmetry, it would lie directly on top of the portion of the parabola to the left of the vertex. Therefore, the parabola is symmetric about its axis of symmetry. Note that the axis of symmetry is not part of the graph of the quadratic function, but it will be useful when quadratic functions are graphed using the methods presented in the next section.

The techniques of shifting horizontally, shifting vertically, stretching, and compressing are collectively referred to as **transformations**. To graph any quadratic function of the form $f(x) = ax^2 + bx + c$ using transformations, take the following steps.

Graphing Quadratic Functions Using Transformations

Step 1: Write the function $f(x) = ax^2 + bx + c$ as $f(x) = a(x - h)^2 + k$ by completing the square in x.

Step 2: Graph the function $f(x) = a(x - h)^2 + k$ using transformations.

Notice that the quadratic function $f(x) = ax^2 + bx + c$ must be written as $f(x) = a(x - h)^2 + k$ to determine the horizontal and vertical shifts.

EXAMPLE 7 **How to Graph a Quadratic Function of the Form $f(x) = ax^2 + bx + c$ Using Transformations**

Graph $f(x) = x^2 + 4x + 3$ using transformations. Identify the vertex and axis of symmetry of the parabola. Based on the graph, determine the domain and range of the quadratic function.

Step-by-Step Solution

Step 1: Write the function $f(x) = ax^2 + bx + c$ as $f(x) = a(x - h)^2 + k$ by completing the square in x.

$$f(x) = x^2 + 4x + 3$$

Group the terms involving x: $\qquad = (x^2 + 4x) + 3$

Complete the square in x by taking $\dfrac{1}{2}$ the coefficient of x and squaring the result: $\left(\dfrac{1}{2} \cdot 4\right)^2 = 4$.

Because 4 was added, it must also be subtracted so that the value of the function does not change: $\qquad = (x^2 + 4x + 4) + 3 - 4$

$$= (x^2 + 4x + 4) - 1$$

Factor the perfect square trinomial in parentheses: $\qquad = (x + 2)^2 - 1$

Step 2: Graph the function $f(x) = a(x - h)^2 + k$ using transformations.

The graph of $f(x) = (x + 2)^2 - 1$ is the graph of $y = x^2$ shifted 2 units left and 1 unit down. See Figure 18.

Figure 18

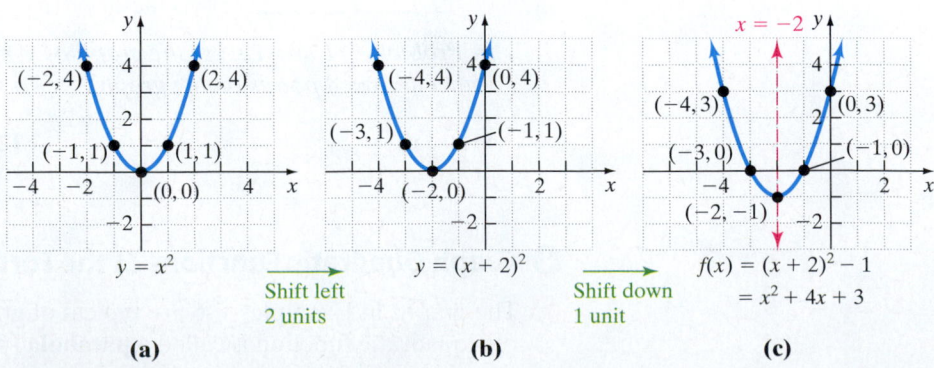

$y = x^2$

$(-2, 4)$ $(2, 4)$ $(-1, 1)$ $(1, 1)$ $(0, 0)$

(a)

Shift left 2 units

$y = (x + 2)^2$

$(-4, 4)$ $(0, 4)$ $(-3, 1)$ $(-1, 1)$ $(-2, 0)$

(b)

Shift down 1 unit

$x = -2$

$f(x) = (x + 2)^2 - 1$
$= x^2 + 4x + 3$

$(-4, 3)$ $(0, 3)$ $(-3, 0)$ $(-1, 0)$ $(-2, -1)$

(c)

From the graph in Figure 18(c), note that the vertex of the parabola is $(-2, -1)$. The axis of symmetry is the line $x = -2$. The domain is the set of all real numbers, or $(-\infty, \infty)$. The range is $\{y | y \geq -1\}$, or $[-1, \infty)$.

EXAMPLE 8

Graphing a Quadratic Function of the Form $f(x) = ax^2 + bx + c$ Using Transformations

Graph $f(x) = -2x^2 + 4x + 1$ using transformations. Identify the vertex and axis of symmetry of the parabola. Based on the graph, determine the domain and range of the quadratic function.

Solution

Write the function $f(x) = ax^2 + bx + c$ as $f(x) = a(x - h)^2 + k$ by completing the square in x.

$$f(x) = -2x^2 + 4x + 1$$

Group the terms involving x: $= (-2x^2 + 4x) + 1$

Factor out the coefficient of x^2, -2, from the parentheses: $= -2(x^2 - 2x) + 1$

Complete the square in x by taking $\dfrac{1}{2}$ the coefficient of x and

squaring the result: $\left(\dfrac{1}{2} \cdot -2\right)^2 = 1$. Add 1 inside the

parentheses. Because 1 is multiplied by -2, we really

subtracted 2, so we must add 2 to offset this: $= -2(x^2 - 2x + 1) + 1 + 2$

Factor the perfect square trinomial in parentheses: $= -2(x - 1)^2 + 3$

Now, graph the function $f(x) = a(x - h)^2 + k$ using transformations. Because $a = -2$, the parabola opens down and is stretched by a factor of 2. Because 1 is subtracted from x, the parabola shifts 1 unit to the right; because $k = 3$, the parabola shifts 3 units up. See Figure 19.

Figure 19

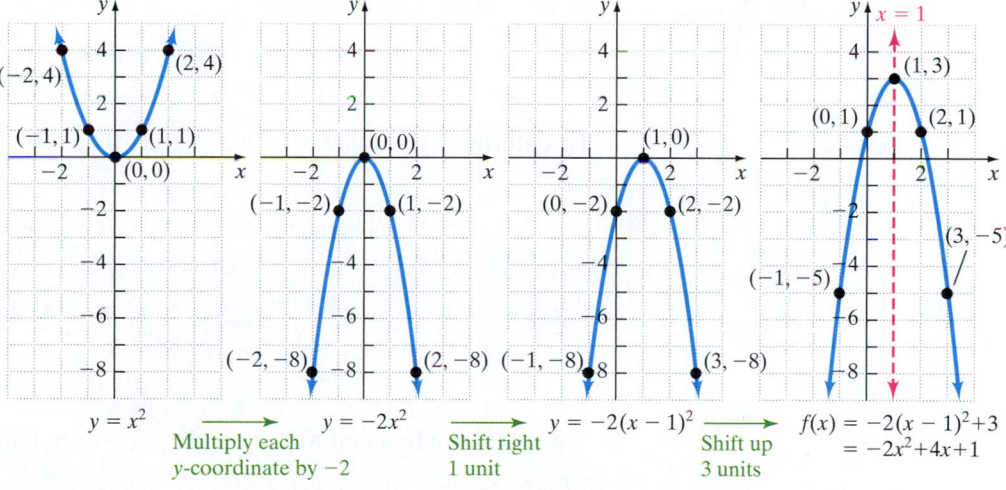

The vertex of the parabola is $(1, 3)$. The axis of symmetry is the line $x = 1$. The domain is the set of all real numbers, or $(-\infty, \infty)$. The range is $\{y \mid y \le 3\}$, or $(-\infty, 3]$. ●

Work Smart: Study Skills

Be sure you know what $y = a(x - h)^2 + k$ represents in reference to the graph of $y = x^2$:

- $|a|$ represents the vertical stretch or compression factor: how tall or short the graph appears.
- The sign of a determines whether the parabola opens up or down.
- h represents the number of units the graph is shifted horizontally.
- k represents the number of units the graph is shifted vertically.
- The vertex of the parabola has coordinates (h, k).

For example, $y = 4(x + 2)^2 - 3$ means that the graph of $y = x^2$ is stretched vertically by a factor of 4, is shifted 2 units to the left, and is shifted down 3 units. The vertex of $y = 4(x + 2)^2 - 3$ is at $(-2, -3)$ and represents the low point of the graph since $a > 0$.

Work Smart

When picking at least three points on the graph, it is best to pick the vertex and one point on each side of the vertex.

Quick ✔

13. *True or False* The graph of $f(x) = -3x^2 + x + 6$ opens down.

In Problems 14 and 15, graph each quadratic function using transformations. Be sure to label at least three points on the graph. Based on the graph, determine the domain and range of each function.

14. $f(x) = -3(x + 2)^2 + 1$ **15.** $f(x) = 2x^2 - 8x + 5$

▶ ⑤ Find a Quadratic Function from Its Graph

In Example 7, the quadratic function $f(x) = x^2 + 4x + 3 = (x + 2)^2 - 1$ was graphed. Notice that the vertex of the parabola is $(-2, -1)$. If f is written as $f(x) = (x - (-2))^2 - 1$ and compared to $f(x) = a(x - h)^2 + k$, then $h = -2$ and $k = -1$. Thus, the vertex of a quadratic function in the form $f(x) = a(x - h)^2 + k$ is (h, k). Now look at Example 8, where $f(x) = -2x^2 + 4x + 1 = -2(x - 1)^2 + 3$. The vertex is $(h, k) = (1, 3)$.

Given the vertex, (h, k), and one additional point on the graph of a quadratic function, the quadratic function $f(x) = a(x - h)^2 + k$ that has this graph may be found. For this reason, $f(x) = a(x - h)^2 + k$ is called the **vertex form** of a quadratic function.

EXAMPLE 9   **Finding the Quadratic Function Given Its Vertex and One Other Point**

Figure 20

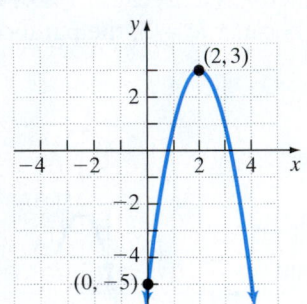

Determine the quadratic function whose graph is given in Figure 20. Write the function in vertex form, $f(x) = a(x - h)^2 + k$.

Solution
The vertex is $(2, 3)$, so $h = 2$ and $k = 3$. Substitute these values into $f(x) = a(x - h)^2 + k$.

$$f(x) = a(x - h)^2 + k$$
$$h = 2, k = 3: \qquad = a(x - 2)^2 + 3$$

To determine the value of a, use the fact that $f(0) = -5$ (the y-intercept).

$$f(x) = a(x - 2)^2 + 3$$
$$x = 0, y = f(0) = -5: \quad -5 = a(0 - 2)^2 + 3$$
$$-5 = a(4) + 3$$
$$-5 = 4a + 3$$
$$\text{Subtract 3 from both sides:} \quad -8 = 4a$$
$$a = -2$$

The quadratic function whose graph is shown in Figure 20 is $f(x) = -2(x - 2)^2 + 3$.

     ●

Quick ✔

In Problem 16, find the quadratic function whose graph is given. Write the function in vertex form, $f(x) = a(x - h)^2 + k$.

16.

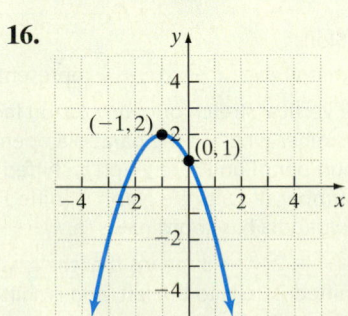

10.4 Exercises MyMathLab®

Exercise numbers in **green** have complete video solutions in MyMathLab or may be accessed using the QR code to the right.

*Problems **1–16** are the Quick ✔s that follow the **EXAMPLES**.*

Building Skills

17. Match each quadratic function to its graph.

(I) $f(x) = x^2 + 3$ (II) $f(x) = (x + 3)^2$

(III) $f(x) = x^2 - 3$ (IV) $f(x) = (x - 3)^2$

(A)   **(B)**

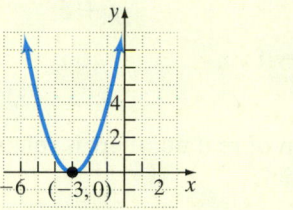 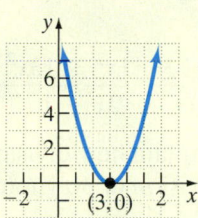

(C)   **(D)**

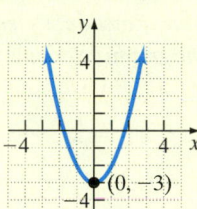 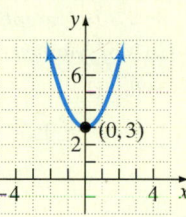

18. Match each quadratic function to its graph.

(I) $f(x) = (x - 2)^2 - 4$ (II) $f(x) = -(x - 2)^2 + 4$

(III) $f(x) = -(x + 2)^2 + 4$ (IV) $f(x) = 2(x - 2)^2 - 4$

(A)   **(B)**

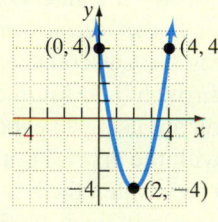 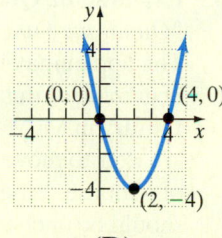

(C)   **(D)**

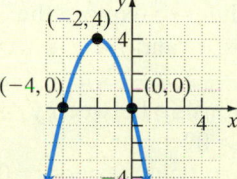 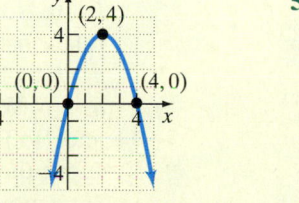

In Problems 19–26, verbally explain how to obtain the graph of the given quadratic function from the graph of $y = x^2$. For example, to obtain the graph of $f(x) = (x - 3)^2 - 6$ from the graph of $y = x^2$, take the graph of $y = x^2$ and shift it 3 units to the right and 6 units down.

19. $f(x) = (x + 10)^2$ **20.** $G(x) = (x - 9)^2$

21. $F(x) = x^2 + 12$ **22.** $g(x) = x^2 - 8$

23. $H(x) = 2(x - 5)^2$ **24.** $h(x) = 4(x + 7)^2$

25. $f(x) = -3(x + 5)^2 + 8$ **26.** $F(x) = -\dfrac{1}{2}(x - 3)^2 - 5$

In Problems 27–30, use the graph of $y = x^2$ to graph the quadratic function. See Objective 1.

27. $f(x) = x^2 + 1$ **28.** $h(x) = x^2 + 6$

29. $f(x) = x^2 - 1$ **30.** $g(x) = x^2 - 7$

In Problems 31–34, use the graph of $y = x^2$ to graph each quadratic function. See Objective 2.

31. $F(x) = (x - 3)^2$ **32.** $F(x) = (x - 2)^2$

33. $h(x) = (x + 2)^2$ **34.** $f(x) = (x + 4)^2$

In Problems 35–40, use the graph of $y = x^2$ to graph each quadratic function. See Objective 3.

35. $g(x) = 4x^2$ **36.** $G(x) = 5x^2$

37. $H(x) = \dfrac{1}{3}x^2$ **38.** $h(x) = \dfrac{3}{2}x^2$

39. $p(x) = -x^2$ **40.** $P(x) = -3x^2$

In Problems 41–54, use the graph of $y = x^2$ to graph each quadratic function. See Objective 4.

41. $f(x) = (x - 1)^2 - 3$ **42.** $g(x) = (x + 2)^2 - 1$

43. $F(x) = (x + 3)^2 + 1$ **44.** $G(x) = (x - 4)^2 + 2$

45. $h(x) = -(x + 3)^2 + 2$ **46.** $H(x) = -(x - 3)^2 + 5$

47. $G(x) = 2(x + 1)^2 - 2$ **48.** $F(x) = 3(x - 2)^2 - 1$

49. $H(x) = -\dfrac{1}{2}(x + 5)^2 + 3$ **50.** $f(x) = -\dfrac{1}{2}(x + 6)^2 + 2$

51. $f(x) = x^2 + 2x - 4$ **52.** $f(x) = x^2 + 4x - 1$

53. $g(x) = x^2 - 4x + 8$ **54.** $G(x) = x^2 - 2x + 7$

In Problems 55–60, determine the quadratic function whose graph is given. Write the function in vertex form, $f(x) = a(x - h)^2 + k$. See Objective 5.

55.   **56.**

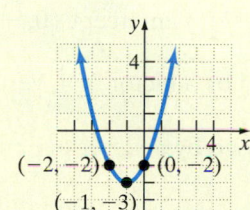 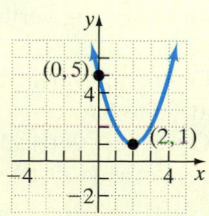

57.   **58.**

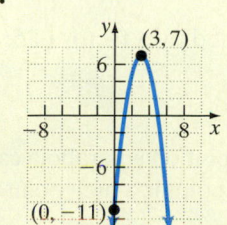 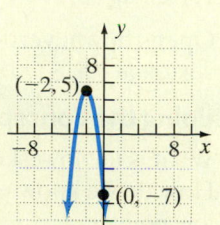

59.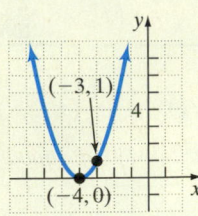

60.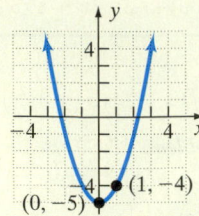

Mixed Practice

In Problems 61–78, write each function in vertex form,
$f(x) = a(x - h)^2 + k$. *Then graph each quadratic function using transformations. Determine the vertex and axis of symmetry. Based on the graph, determine the domain and range of the quadratic function.*

61. $f(x) = x^2 + 6x - 16$ **62.** $f(x) = x^2 + 4x + 5$

63. $F(x) = x^2 + x - 12$ **64.** $h(x) = x^2 - 7x + 10$

65. $H(x) = 2x^2 - 4x - 1$ **66.** $g(x) = 2x^2 + 4x - 3$

67. $P(x) = 3x^2 + 12x + 13$ **68.** $f(x) = 3x^2 + 18x + 25$

69. $F(x) = -x^2 - 10x - 21$ **70.** $g(x) = -x^2 - 8x - 14$

71. $g(x) = -x^2 + 6x - 1$ **72.** $f(x) = -x^2 + 10x - 17$

73. $H(x) = -2x^2 + 8x - 4$ **74.** $h(x) = -2x^2 + 12x - 17$

75. $f(x) = \frac{1}{3}x^2 - 2x + 4$ **76.** $f(x) = \frac{1}{2}x^2 + 2x - 1$

77. $G(x) = -12x^2 - 12x + 1$ **78.** $h(x) = -4x^2 + 4x$

Applying the Concepts

In Problems 79–88, write a quadratic function in the vertex form,
$f(x) = a(x - h)^2 + k$ *with the properties given.*

79. Opens up; vertex and y-intercept at $(3, 0)$

80. Opens up; vertex at $(0, 2)$; contains $(1, 3)$

81. Opens up; vertex at $(-3, 1)$; contains $(-5, 5)$

82. Opens up; vertex at $(4, -2)$; contains $(2, 2)$

83. Opens down; vertex at $(5, -1)$; contains $(6, -2)$

84. Opens down; vertex at $(-4, -7)$; y-intercept $(0, -23)$

85. Opens up; vertically stretched by a factor of 4; vertex at $(9, -6)$.

86. Opens up; vertically compressed by a factor of $\frac{1}{2}$; vertex at $(-5, 0)$.

87. Opens down; vertically compressed by a factor of $\frac{1}{3}$; vertex at $(0, 6)$.

88. Opens down; vertically stretched by a factor of 5; vertex at $(5, 8)$.

Explaining the Concepts

89. What is the lowest or highest point on a parabola called? How do we know whether this point is a high point or a low point?

90. Why does the graph of a quadratic function open up if $a > 0$ and down if $a < 0$?

91. Can a quadratic function have a range of $(-\infty, \infty)$? Justify your answer.

92. Can the graph of a quadratic function have more than one y-intercept? Justify your answer.

Synthesis Review

In Problems 93–95, divide.

93. $\dfrac{349}{12}$ **94.** $\dfrac{4x^2 + 19x - 1}{x + 5}$

95. $\dfrac{2x^4 - 11x^3 + 13x^2 - 8x}{2x - 1}$

96. Explain how division of real numbers is related to division of polynomials.

Technology Exercises

In Problems 97–104, graph each quadratic function. Determine the vertex and axis of symmetry. Based on the graph, determine the range of the function.

97. $f(x) = x^2 + 1.3$ **98.** $f(x) = x^2 - 3.5$

99. $g(x) = (x - 2.5)^2$ **100.** $G(x) = (x + 4.5)^2$

101. $h(x) = 2.3(x - 1.4)^2 + 0.5$

102. $H(x) = 1.2(x + 0.4)^2 - 1.3$

103. $F(x) = -3.4(x - 2.8)^2 + 5.9$

104. $f(x) = 0.3(x + 3.8)^2 - 8.9$

105. **Exploration: Quadratic Functions—a Value** *Open the "Quadratic Functions—Vertex Form" Geogebra applet. The applet may be found using the QR code at the beginning of this section or through the Multimedia Library in MyMathLab. If you are using a keyboard, it is easier to move the sliders with the arrow keys instead of a mouse. When you begin the applet, the boxes should be unchecked. If they are not, uncheck them.*

(a) Move the sliders so that $a = 0$, $h = 0$, and $k = 0$. Now move the a slider to explore the relationship between $y = x^2$ and $f(x) = ax^2$ for $a > 0$ by letting $a = 1$, $a = 2$, $a = 3$, $a = 4$, and $a = 5$. What is the relationship between $y = x^2$ and $f(x) = ax^2$ when $a > 0$?

(b) Move the sliders so that $a = 0$, $h = 0$, and $k = 0$. Now move the a slider to explore the relationship between $y = x^2$ and $f(x) = ax^2$ for $a < 0$ by letting $a = -1$, $a = -2$, $a = -3$, $a = -4$, and $a = -5$. What is the relationship between $y = x^2$ and $f(x) = ax^2$ when $a < 0$?

(c) Now let $a = 0$. What type of function is $f(x)$?

106. Exploration: Quadratic Functions—Vertex Form *Open the "Quadratic Functions—Vertex Form" Geogebra applet.* The applet may be found using the QR code at the beginning of this section or through the Multimedia Library in MyMathLab. Check the "Vertex Form" and "Vertex" boxes. Explore the relationship between $y = x^2$ and $f(x) = a(x - h)^2 + k$. Move the sliders so that $a = 1$, $h = 0$, and $k = 0$. If you are using a keyboard, it is easier to move the sliders with the arrow keys instead of a mouse.

(a) Explore the relationship between h and the x-coordinate of the vertex of the parabola. Move

the slider h to take on the values from -5 to 5. For each value of h, write down the coordinates of the vertex. What relationship exists between h and this point?

(b) Now explore the relationship between k and the y-coordinate of the vertex of the parabola. Reset the slider h back to 0 and let k take on the values from -5 to 5. For each value of k, write down the coordinates of the vertex. What relationship exists between k and this point?

(c) In general, when a quadratic function is written in the form $f(x) = a(x - h)^2 + k$, what are the coordinates of the vertex?

10.5 Graphing Quadratic Functions Using Properties

Objectives

1 Graph Quadratic Functions of the Form $f(x) = ax^2 + bx + c$

2 Find the Maximum or Minimum Value of a Quadratic Function

3 Model and Solve Optimization Problems Involving Quadratic Functions

Are You Prepared for This Section?

Before getting started, complete the following problems. If you get a problem wrong, go back to the section cited and review the material.

P1. Find the intercepts of the graph of $2x + 5y = 20$.   [Section 3.2, pp. 187–190]

P2. Solve: $2x^2 - 3x - 20 = 0$ [Section 6.6, pp. 410–415]

P3. Find the zeros of $f(x) = x^2 - 3x - 4$.   [Section 8.4, p. 553]

In Section 10.4, quadratic functions were graphed using transformations. This section covers graphing a quadratic function by using its intercepts, axis of symmetry, and vertex.

1 Graph Quadratic Functions of the Form *f(x)* = *ax²* + *bx* + *c*

▶ As explained in Section 10.4, a quadratic function $f(x) = ax^2 + bx + c$ can be written in the form $f(x) = a(x - h)^2 + k$ by completing the square in x. The value of a determines whether the graph of the quadratic function (the parabola) opens up or down. In addition, the point with coordinates (h, k) is the vertex of the quadratic function.

A formula that may be used to determine the vertex of a parabola can be found by completing the square in $f(x) = ax^2 + bx + c$, $a \neq 0$, as follows:

$$f(x) = ax^2 + bx + c$$

Group terms involving x: $= (ax^2 + bx) + c$

Factor out a: $= a\left(x^2 + \dfrac{b}{a}x\right) + c$

Complete the square in x by taking $\dfrac{1}{2}$ the coefficient of x and squaring the result: $\left(\dfrac{1}{2} \cdot \dfrac{b}{a}\right)^2 = \dfrac{b^2}{4a^2}$. Because $\dfrac{b^2}{4a^2}$ is added inside the parentheses, subtract $a \cdot \dfrac{b^2}{4a^2} = \dfrac{b^2}{4a}$ outside the parentheses:

Factor the perfect square trinomial; $= a\left(x^2 + \dfrac{b}{a}x + \dfrac{b^2}{4a^2}\right) + c - \dfrac{b^2}{4a}$

multiply c by $\dfrac{4a}{4a}$ to get a common denominator: $= a\left(x + \dfrac{b}{2a}\right)^2 + c \cdot \dfrac{4a}{4a} - \dfrac{b^2}{4a}$

Write expression in the form $f(x) = a(x - h)^2 + k$: $= a\left(x - \left(-\dfrac{b}{2a}\right)\right)^2 + \dfrac{4ac - b^2}{4a}$

Compare $f(x) = a\left(x - \left(-\dfrac{b}{2a}\right)\right)^2 + \dfrac{4ac - b^2}{4a}$ to $f(x) = a(x - h)^2 + k$, and conclude the following:

Prepared?...Answers **P1.** $(0, 4)$, $(10, 0)$ **P2.** $\left\{-\dfrac{5}{2}, 4\right\}$ **P3.** -1 and 4

Work Smart

Another formula for the vertex is

$$\left(-\frac{b}{2a}, \frac{-D}{4a}\right)$$

where $D = b^2 - 4ac$, the discriminant.

The Vertex of a Parabola

Any quadratic function $f(x) = ax^2 + bx + c, a \neq 0$, has vertex

$$\left(-\frac{b}{2a}, \frac{4ac - b^2}{4a}\right)$$

Because the y-coordinate on the graph of any function can be found by evaluating the function at the corresponding x-coordinate, the coordinates of the vertex can be restated as

$$\left(-\frac{b}{2a}, f\left(-\frac{b}{2a}\right)\right)$$

For instance, Example 7 in Section 10.4 showed $f(x) = x^2 + 4x + 3 = (x + 2)^2 - 1$ has a vertex of $(-2, -1)$. In $f(x) = x^2 + 4x + 3, a = 1, b = 4$, and $c = 3$, so the x-coordinate of the vertex is given by $x = -\frac{b}{2a} = -\frac{4}{2(1)} = -2$. The y-coordinate of the vertex is $f(-2) = (-2)^2 + 4(-2) + 3 = -1$.

Because the axis of symmetry intersects the vertex, the axis of symmetry is $x = -\frac{b}{2a}$.

In addition, the parabola will open up if $a > 0$ and down if $a < 0$. By using this information along with the intercepts of the graph, we can obtain a complete graph.

The y-intercept is the value of the quadratic function $f(x) = ax^2 + bx + c$ at $f(0) = c$.

Any x-intercepts are found by solving the quadratic equation

$$f(x) = ax^2 + bx + c = 0$$

This equation has two, one, or no real solutions, depending on the value of the discriminant $b^2 - 4ac$. Use the value of the discriminant to determine how many x-intercepts the graph of the quadratic function has.

The x-Intercepts of the Graph of a Quadratic Function

1. If the discriminant $b^2 - 4ac > 0$, the graph of $f(x) = ax^2 + bx + c$ has two x-intercepts. The graph crosses the x-axis at the solutions to the equation $ax^2 + bx + c = 0$.
2. If the discriminant $b^2 - 4ac = 0$, the graph of $f(x) = ax^2 + bx + c$ has one x-intercept. The graph touches the x-axis at the solution to the equation $ax^2 + bx + c = 0$.
3. If the discriminant $b^2 - 4ac < 0$, the graph of $f(x) = ax^2 + bx + c$ has no x-intercepts. The graph does not cross or touch the x-axis.

Figure 21 illustrates these possibilities for parabolas that open up.

Figure 21

$f(x) = ax^2 + bx + c, a > 0$

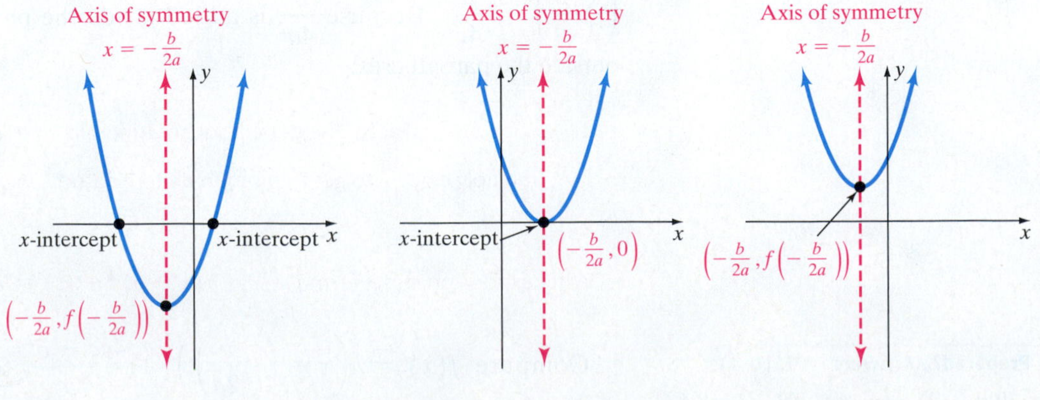

EXAMPLE 1 **How to Graph a Quadratic Function Using Its Properties**

Graph $f(x) = x^2 + 2x - 15$ using its properties.

Step-by-Step Solution

Compare $f(x) = x^2 + 2x - 15$ to $f(x) = ax^2 + bx + c$ and see that $a = 1$, $b = 2$, and $c = -15$.

Step 1: Determine whether the parabola opens up or down.

The parabola opens up because $a = 1 > 0$.

Step 2: Determine the vertex and axis of symmetry.

The x-coordinate of the vertex is

$$x = -\frac{b}{2a} = -\frac{2}{2(1)} = -1$$

The y-coordinate of the vertex is

$$f\left(-\frac{b}{2a}\right) = f(-1)$$
$$= (-1)^2 + 2(-1) - 15$$
$$= 1 - 2 - 15$$
$$= -16$$

The vertex is $(-1, -16)$. The axis of symmetry is the line $x = -\frac{b}{2a} = -1$.

Step 3: Determine the y-intercept, $f(0)$.

$$f(0) = 0^2 + 2(0) - 15$$
$$= -15$$

The y-intercept is $(0, -15)$.

Step 4: Find the discriminant, $b^2 - 4ac$, to determine the number of x-intercepts. Then determine the x-intercepts, if any.

Because $a = 1$, $b = 2$, and $c = -15$, the value of the discriminant is $b^2 - 4ac = (2)^2 - 4(1)(-15) = 64 > 0$, so the parabola has two x-intercepts. Find the x-intercepts by solving

$$f(x) = 0$$
$$x^2 + 2x - 15 = 0$$
$$(x + 5)(x - 3) = 0$$

Factor: $\qquad x + 5 = 0 \quad$ or $\quad x - 3 = 0$

Zero-Product Property: $\qquad x = -5 \quad$ or $\qquad x = 3$

The x-intercepts are $(-5, 0)$ and $(3, 0)$.

Step 5: Plot the vertex, y-intercept, and x-intercepts. Use the axis of symmetry to find an additional point. Draw the graph of the quadratic function.

Use the axis of symmetry to find the additional point $(-2, -15)$ by recognizing that the y-intercept, $(0, -15)$, is 1 unit to the right of the axis of symmetry, so there must be a point 1 unit to the left of the axis of symmetry. Plot the points and draw the graph. See Figure 22.

Figure 22

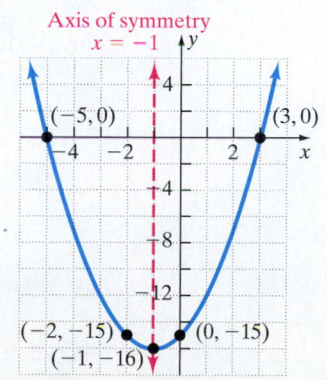

> **Graphing a Quadratic Function Using Its Properties**
>
> To graph any quadratic function of the form $f(x) = ax^2 + bx + c, a \neq 0$, use the following steps:
>
> **Step 1:** Determine whether the parabola opens up or down.
>
> **Step 2:** Determine the vertex and axis of symmetry.
>
> **Step 3:** Determine the y-intercept, $f(0) = c$.
>
> **Step 4:** Determine the discriminant, $b^2 - 4ac$.
> - If $b^2 - 4ac > 0$, then the parabola has two x-intercepts, which are found by solving $f(x) = 0(ax^2 + bx + c = 0)$.
> - If $b^2 - 4ac = 0$, the vertex is the x-intercept.
> - If $b^2 - 4ac < 0$, there are no x-intercepts.
>
> **Step 5:** Plot the vertex, y-intercept, and any x-intercept(s). Use the axis of symmetry to find an additional point. Draw the graph of the quadratic function.

Quick ✓

1. Any quadratic function $f(x) = ax^2 + bx + c, a \neq 0$, will have a vertex whose x-coordinate is $x =$ ____.

2. The graph of $f(x) = ax^2 + bx + c$ will have two x-intercepts if $b^2 - 4ac$ ____ 0.

3. How many x-intercepts does the graph of $f(x) = -2x^2 - 3x + 6$ have?

4. What is the vertex of $f(x) = x^2 + 4x - 3$?

5. Graph $f(x) = x^2 - 4x - 12$ using its properties.

In Example 1, the function was factorable, so the x-intercepts were rational numbers. When the quadratic function cannot be factored, use the quadratic formula to find the x-intercepts. For the purpose of graphing the quadratic function, approximate the x-intercepts rounded to two decimal places.

EXAMPLE 2   **Graphing a Quadratic Function Using Its Properties**

Graph $f(x) = -2x^2 + 12x - 5$ using its properties. Based on the graph, determine the domain and range of the quadratic function.

Solution

In $f(x) = -2x^2 + 12x - 5, a = -2, b = 12$, and $c = -5$. The parabola opens down because $a = -2 < 0$. The x-coordinate of the vertex is

$$x = -\frac{b}{2a} = -\frac{12}{2(-2)} = 3$$

The y-coordinate of the vertex is

$$y = f\left(-\frac{b}{2a}\right)$$
$$= f(3)$$
$$= -2(3)^2 + 12(3) - 5$$
$$= -18 + 36 - 5$$
$$= 13$$

The vertex is $(3, 13)$. The axis of symmetry is the line $x = -\dfrac{b}{2a} = 3$.

The y-intercept is $f(0) = -2(0)^2 + 12(0) - 5 = -5$. Find the discriminant, $b^2 - 4ac$, to determine the number of x-intercepts. Because $a = -2$, $b = 12$, and $c = -5$, $b^2 - 4ac = (12)^2 - 4(-2)(-5) = 104 > 0$. The parabola will have two x-intercepts. Find the x-intercepts by solving

$$f(x) = 0$$
$$-2x^2 + 12x - 5 = 0$$

The equation cannot be solved by factoring (because $b^2 - 4ac$ is not a perfect square), so use the quadratic formula:

Figure 23

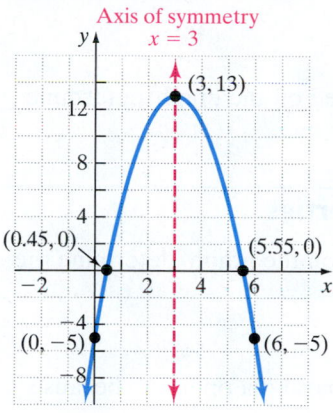

$$x = \frac{-b \pm \sqrt{b^2 - 4ac}}{2a}$$

$a = -2, b = 12, b^2 - 4ac = 104$:
$$= \frac{-12 \pm \sqrt{104}}{2(-2)}$$

$\sqrt{104} = \sqrt{4 \cdot 26} = 2\sqrt{26}$:
$$= \frac{-12 \pm 2\sqrt{26}}{-4}$$

Divide -4 into each term in the numerator and simplify:
$$x = 3 \pm \frac{\sqrt{26}}{2} \qquad \text{Exact solution}$$

Evaluate $3 \pm \dfrac{\sqrt{26}}{2}$ and find the x-intercepts are approximately 0.45 and 5.55. Use the axis of symmetry to find the additional point $(6, -5)$. (The y-intercept, $(0, -5)$, is 3 units to the left of the axis of symmetry, therefore, there must be a point 3 units to the right of the axis of symmetry.) Next, plot the vertex, the intercepts, and the additional point. Draw the graph of the quadratic function. See Figure 23. Based on the graph, the domain is $\{x \mid x \text{ is any real number}\}$ or $(-\infty, \infty)$ and the range is $\{y \mid y \le 13\}$ or $(-\infty, 13]$. ●

Work Smart

Notice that the vertex in Example 2 lies in quadrant I (above the x-axis) and the graph opens down. This tells us the graph must have two x-intercepts.

> **Quick ✓**
>
> **6.** Graph $f(x) = -3x^2 + 12x - 7$ using its properties. Based on the graph, determine the domain and range of the quadratic function.

EXAMPLE 3 | **Graphing a Quadratic Function Using Its Properties**

Graph $g(x) = x^2 - 8x + 16$ using its properties. Based on the graph, determine the domain and range of the quadratic function.

Solution

In $g(x) = x^2 - 8x + 16$, $a = 1$, $b = -8$, and $c = 16$. The parabola opens up because $a = 1 > 0$. The x-coordinate of the vertex is

$$x = -\frac{b}{2a} = -\frac{-8}{2(1)} = 4$$

The y-coordinate of the vertex is

$$y = f\left(-\frac{b}{2a}\right)$$
$$= f(4)$$
$$= 4^2 - 8(4) + 16$$
$$= 16 - 32 + 16$$
$$= 0$$

The vertex is $(4, 0)$. The axis of symmetry is the line $x = -\dfrac{b}{2a} = 4$.

(continued)

Figure 24

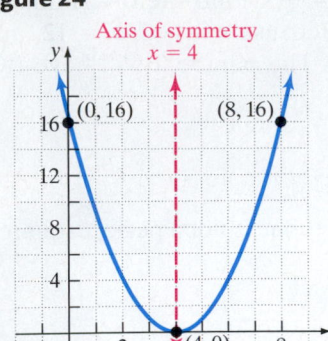

The y-intercept is $f(0) = 0^2 - 8(0) + 16 = 16$. Because $a = 1, b = -8$, and $c = 16, b^2 - 4ac = (-8)^2 - 4(1)(16) = 0$. The vertex $(4, 0)$ is on the x-axis, so the parabola has one x-intercept. Verify this by solving

$$f(x) = 0$$
$$x^2 - 8x + 16 = 0$$
Factor: $\quad (x - 4)^2 = 0$
$$x = 4$$

Use the point $(0, 16)$ and the axis of symmetry to find the additional point $(8, 16)$. The parabola is shown in Figure 24. Based on the graph, the domain is $\{x | x \text{ is any real number}\}$ or $(-\infty, \infty)$ and the range is $\{y | y \geq 0\}$ or $[0, \infty)$. ●

> **Quick ✔**
>
> 7. Graph $f(x) = x^2 + 6x + 9$ using its properties. Based on the graph, determine the domain and range of the quadratic function.

⊙ **EXAMPLE 4** **Graphing a Quadratic Function Using Its Properties**

Graph $F(x) = 2x^2 + 6x + 5$ using its properties. Based on the graph, determine the domain and range of the quadratic function.

Solution
In $F(x) = 2x^2 + 6x + 5, a = 2, b = 6$, and $c = 5$. The parabola opens up because $a = 2 > 0$. The x-coordinate of the vertex is

$$x = -\frac{b}{2a} = -\frac{6}{2(2)} = -\frac{3}{2}$$

The y-coordinate of the vertex is

$$y = f\left(-\frac{b}{2a}\right) = f\left(-\frac{3}{2}\right)$$
$$= 2\left(-\frac{3}{2}\right)^2 + 6\left(-\frac{3}{2}\right) + 5$$
$$= 2\left(\frac{9}{4}\right) - 9 + 5$$
$$= \frac{1}{2}$$

Work Smart

Notice that the vertex lies in quadrant II (above the x-axis) and the graph opens up. This tells us the graph has no x-intercepts.

The vertex is $\left(-\frac{3}{2}, \frac{1}{2}\right)$. The axis of symmetry is the line $x = -\frac{b}{2a} = -\frac{3}{2}$.

The y-intercept is $f(0) = 2(0)^2 + 6(0) + 5 = 5$. Because the parabola opens up and the vertex is in the second quadrant, the parabola has no x-intercepts. Verify this by computing the discriminant with $a = 2, b = 6$, and $c = 5$:

$$b^2 - 4ac = 6^2 - 4(2)(5) = -4 < 0$$

Figure 25

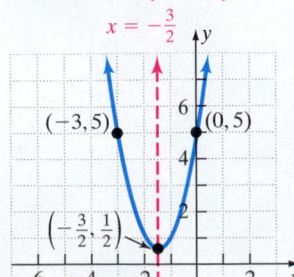

The parabola has no x-intercepts. Use the point $(0, 5)$ and the axis of symmetry to find the additional point $(-3, 5)$. The parabola is shown in Figure 25. Based on the graph, the domain is $\{x | x \text{ is any real number}\}$ or $(-\infty, \infty)$ and the range is $\left\{y | y \geq \frac{1}{2}\right\}$ or $\left[\frac{1}{2}, \infty\right)$. ●

> **Quick ✔**
>
> 8. Graph $G(x) = -3x^2 + 9x - 8$ using its properties. Based on the graph, determine the domain and range of the quadratic function.

Figure 26

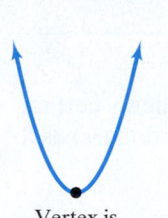

Vertex is
highest point;
y-coordinate is
maximum value

Vertex is
lowest point;
y-coordinate is
minimum value

(a) Opens up
$a > 0$

(b) Opens down
$a < 0$

2 Find the Maximum or Minimum Value of a Quadratic Function

Recall, the graph of a quadratic function $f(x) = ax^2 + bx + c$ is a parabola with vertex at $\left(-\dfrac{b}{2a}, f\left(-\dfrac{b}{2a}\right)\right)$. The vertex is the highest point on the graph if $a < 0$ and the lowest point on the graph if $a > 0$. If the vertex is the highest point then $f\left(-\dfrac{b}{2a}\right)$, the y-coordinate of the vertex, is the **maximum value** of f. If the vertex is the lowest point, then $f\left(-\dfrac{b}{2a}\right)$ is the **minimum value** of f. See Figure 26.

Questions involving **optimization,** the process of finding the maximum or minimum value(s) of a function, can now be answered. In the case of quadratic functions, the maximum $(a < 0)$ or minimum $(a > 0)$ is found at the vertex.

EXAMPLE 5   **Finding the Maximum or Minimum Value of a Quadratic Function**

Determine whether the quadratic function

$$f(x) = 3x^2 + 12x - 7$$

has a maximum or a minimum value. Then find the maximum or minimum value of the function, and indicate when it occurs.

Solution

In $f(x) = 3x^2 + 12x - 7$, $a = 3$, $b = 12$, and $c = -7$. Because $a = 3 > 0$, the graph of the quadratic function opens up, so the function has a minimum value. The minimum value of the function occurs at

$$x = -\frac{b}{2a} = -\frac{12}{2(3)} = -2$$

Work Smart

The vertex is not the maximum or minimum value—the y-coordinate of the vertex is the maximum or minimum value of the function.

The minimum value of the function is

$$f\left(-\frac{b}{2a}\right) = f(-2)$$
$$= 3(-2)^2 + 12(-2) - 7$$
$$= 3(4) - 24 - 7$$
$$= -19$$

The minimum value of the function is -19 and occurs at $x = -2$.   ●

Quick ✓

9. *True or False* For the quadratic function $f(x) = ax^2 + bx + c$, if $a < 0$ then $f\left(-\dfrac{b}{2a}\right)$ is the maximum value of f.

In Problems 10 and 11, determine whether the quadratic function has a maximum or a minimum value. Then find the maximum or minimum value of the function and where it occurs.

10. $f(x) = 2x^2 - 8x + 1$ **11.** $G(x) = -x^2 + 10x + 8$

Properties of quadratic functions are useful to answer questions regarding applications modeled by quadratic functions.

EXAMPLE 6 **Maximizing Revenue**

Suppose the marketing department of Dell Computer has found that, when a certain model of computer is sold at a price of p dollars, the daily revenue R (in dollars) as a function of the price p is the function

$$R(p) = -\frac{1}{4}p^2 + 400p$$

(a) For what price will the revenue be maximized?

(b) What is the maximum daily revenue?

Solution

(a) Notice that the revenue function is a quadratic function whose graph opens down because $a = -\frac{1}{4} < 0$. Therefore, the function has a maximum at

$$p = -\frac{b}{2a} = -\frac{400}{2\left(-\frac{1}{4}\right)} = -\frac{400}{-\frac{1}{2}} = 800$$

Revenue will be maximized when the price is $p = \$800$.

(b) The maximum daily revenue is found by letting $p = \$800$ in the revenue function.

$$R(800) = -\frac{1}{4}(800)^2 + 400 \cdot 800$$

$$= \$160,000$$

The maximum daily revenue is $160,000. See Figure 27 for an illustration.

Figure 27

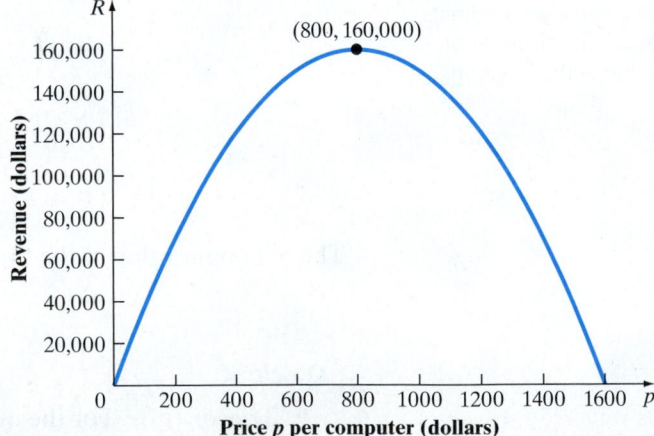

Quick ✔

12. Suppose that the marketing department of Texas Instruments has found that, when a certain model of calculator is sold at a price of p dollars, the daily revenue R (in dollars) as a function of the price p is the function $R(p) = -0.5p^2 + 75p$.

(a) For what price will the daily revenue be maximized?

(b) What is the maximum daily revenue?

▶ ③ Model and Solve Optimization Problems Involving Quadratic Functions

Models that result in quadratic functions will now be discussed. Use the problem-solving strategy introduced in Section 2.5.

EXAMPLE 7   ### Maximizing the Area Enclosed by a Fence

A farmer has 3000 feet of fence with which to enclose a rectangular field. What is the maximum area that can be enclosed by the fence? What are the dimensions of the rectangle that encloses the most area?

Solution

Step 1: Identify Find the dimensions of a rectangle that maximize the area.

Step 2: Name Let w represent the width of the rectangle and let l represent its length, both in feet.

Step 3: Translate See Figure 28.

Figure 28

Because 3000 feet of fence is available, the perimeter of the rectangle will be 3000 feet. Use the perimeter formula of a rectangle to get

$$2l + 2w = 3000$$

The area of the rectangle, $A = lw$, uses two variables, l and w. To express A in terms of only one variable, solve the equation $2l + 2w = 3000$ for l and then substitute for l in the area formula $A = lw$.

Work Smart

The equation $2l + 2w = 3000$ could also be solved for w. Ultimately the same solution would be obtained.

$$2l + 2w = 3000$$
Subtract 2w from both sides: $$2l = 3000 - 2w$$
Divide both sides by 2: $$l = \frac{3000 - 2w}{2}$$
Simplify: $$l = 1500 - w$$

Next let $l = 1500 - w$ in the formula $A = lw$.

$$A = (1500 - w)w$$
$$= -w^2 + 1500w$$

Now, A is a quadratic function of w.

$$A(w) = -w^2 + 1500w \quad \text{The Model}$$

Step 4: Solve Find the dimensions that result in a maximum area enclosed by the fence. The model developed in Step 3 is a quadratic function that opens down (because $a = -1 < 0$), so the vertex is a maximum point on the graph of A. The maximum value occurs at

$$w = -\frac{b}{2a} = -\frac{1500}{2(-1)} = 750$$

(continued)

The maximum value of A is

$$A\left(-\frac{b}{2a}\right) = A(750)$$
$$= -750^2 + 1500(750)$$
$$= 562{,}500 \text{ square feet}$$

Because $l = 1500 - w$, then if the width is 750 feet, the length will be $l = 1500 - 750 = 750$ feet.

Step 5: Check With a length and width of 750 feet, the perimeter is $2(750) + 2(750) = 3000$ feet. The area is $(750 \text{ feet})(750 \text{ feet}) = 562{,}500$ square feet. Everything checks!

Step 6: Answer The largest area that can be enclosed by 3000 feet of fence has an area of 562,500 square feet. Its dimensions are 750 feet by 750 feet.

> ### Quick ✓
>
> **13.** Roberta has 1000 yards of fence to enclose a rectangular field. What is the maximum area that can be enclosed by the fence? What are the dimensions of the rectangle that encloses the most area?

EXAMPLE 8) **Pricing a Charter**

Chicago Tours offers boat charters along the Chicago coastline on Lake Michigan. Normally, a ticket costs $20 per person, but for any group, Chicago Tours will lower the price of a ticket by $0.10 per person for each person in excess of 30. Determine the group size that will maximize revenue. What is the maximum revenue that can be earned from a group sale?

Solution

Step 1: Identify This is a direct translation problem involving revenue. Remember, revenue is price times quantity.

Step 2: Name Let x represent the number in the group in excess of 30.

Step 3: Translate Revenue is price times quantity. If 30 individuals make up a group, the revenue to Chicago Tours will be $20(30)$. If 31 individuals make up a group, revenue will be $19.90(31)$. The revenue for 32 individuals will be $19.80(32)$. In general, for $30 + x$ individuals, revenue will be $(20 - 0.1x)(30 + x)$. Thus the revenue R for a group that has $30 + x$ people in it is given by

$$R(x) = (20 - 0.1x)(30 + x)$$
$$\text{FOIL:} \quad = 600 + 20x - 3x - 0.1x^2$$
$$= -0.1x^2 + 17x + 600 \qquad \text{The Model}$$

Step 4: Solve Find the revenue-maximizing number of individuals in the group. The function R is a quadratic function with a graph that opens down (because $a = -0.1 < 0$), so the vertex is a maximum point. The value of x that results in a maximum is given by

$$x = -\frac{b}{2a} = -\frac{17}{2(-0.1)}$$
$$= 85$$

The maximum revenue is

$$R(85) = -0.1(85)^2 + 17(85) + 600$$
$$= \$1322.50$$

Step 5: Check Because x represents the number of passengers in excess of 30, $30 + 85 = 115$ tickets should be sold to maximize revenue. The cost per ticket is $\$20 - 0.1(85) = \$20 - \$8.50 = \$11.50$. The cost per ticket times the number of passengers is $\$11.50(115) = \$1322.50$. The answer checks!

Step 6: Answer A group sale of 115 passengers will maximize revenue. The maximum revenue is $1322.50.

Quick ✓

14. A flash drive manufacturer charges $100 for each box of flash drives ordered. However, it reduces the price by $1 per box for each box in excess of 30 boxes but less than 90 boxes. Determine the number of boxes of flash drives that should be sold to maximize revenue. What is the maximum revenue?

10.5 Exercises MyMathLab®

Exercise numbers in **green** have complete video solutions in MyMathLab or may be accessed using the QR code to the right.

*Problems **1–14** are the Quick ✓s that follow the **EXAMPLES**.*

Building Skills

In Problems 15–22, (a) find the vertex of each parabola, (b) use the discriminant to determine the number of x-intercepts the graph will have. Then determine the x-intercepts. See Objective 1.

15. $f(x) = x^2 - 6x - 16$ **16.** $g(x) = x^2 + 4x - 12$

17. $G(x) = -2x^2 + 4x - 5$ **18.** $H(x) = x^2 - 4x + 5$

19. $h(x) = 4x^2 + 4x + 1$ **20.** $f(x) = x^2 - 6x + 9$

21. $F(x) = 4x^2 - x - 1$ **22.** $P(x) = -2x^2 + 3x + 1$

In Problems 23–62, graph each quadratic function using its properties by following Steps 1–5 on page 740. Based on the graph, determine the domain and range of the quadratic function. See Objective 1.

23. $f(x) = x^2 - 4x - 5$ **24.** $f(x) = x^2 - 2x - 8$

25. $G(x) = x^2 + 12x + 32$ **26.** $g(x) = x^2 - 12x + 27$

27. $F(x) = -x^2 + 2x + 8$ **28.** $g(x) = -x^2 + 2x + 15$

29. $H(x) = x^2 - 4x + 4$ **30.** $h(x) = x^2 + 6x + 9$

31. $g(x) = x^2 + 2x + 5$ **32.** $f(x) = x^2 - 4x + 7$

33. $h(x) = -x^2 - 10x - 25$ **34.** $P(x) = -x^2 - 12x - 36$

35. $p(x) = -x^2 + 2x - 5$ **36.** $f(x) = -x^2 + 4x - 6$

37. $F(x) = 4x^2 - 4x - 3$ **38.** $f(x) = 4x^2 - 8x - 21$

39. $G(x) = -9x^2 + 18x + 7$ **40.** $g(x) = -9x^2 - 36x - 20$

41. $H(x) = 4x^2 - 4x + 1$ **42.** $h(x) = 9x^2 + 12x + 4$

43. $f(x) = -16x^2 - 24x - 9$ **44.** $F(x) = -4x^2 - 20x - 25$

45. $f(x) = 2x^2 + 8x + 11$ **46.** $F(x) = 3x^2 + 6x + 7$

47. $P(x) = -4x^2 + 6x - 3$ **48.** $p(x) = -2x^2 + 6x + 5$

49. $h(x) = x^2 + 5x + 3$ **50.** $H(x) = x^2 + 3x + 1$

51. $G(x) = -3x^2 + 8x + 2$ **52.** $F(x) = -2x^2 + 6x + 1$

53. $f(x) = 5x^2 - 5x + 2$ **54.** $F(x) = 4x^2 + 4x - 1$

55. $H(x) = -3x^2 + 6x$ **56.** $h(x) = -4x^2 + 8x$

57. $f(x) = x^2 - \dfrac{5}{2}x - \dfrac{3}{2}$ **58.** $g(x) = x^2 + \dfrac{5}{2}x - 6$

59. $G(x) = \dfrac{1}{2}x^2 + 2x - 6$ **60.** $H(x) = \dfrac{1}{4}x^2 + x - 8$

61. $F(x) = -\dfrac{1}{4}x^2 + x + 15$ **62.** $G(x) = -\dfrac{1}{2}x^2 - 8x - 24$

In Problems 63–74, determine whether the quadratic function has a maximum or a minimum value. Then find the maximum or minimum value and where it occurs. See Objective 2.

63. $f(x) = x^2 + 8x + 13$ **64.** $f(x) = x^2 - 6x + 3$

65. $G(x) = -x^2 - 10x + 3$ **66.** $g(x) = -x^2 + 4x + 12$

67. $F(x) = -2x^2 + 12x + 5$ **68.** $H(x) = -3x^2 + 12x - 1$

69. $h(x) = 4x^2 + 16x - 3$ **70.** $G(x) = 5x^2 + 10x - 1$

71. $f(x) = 2x^2 - 5x + 1$ **72.** $F(x) = 3x^2 + 4x - 3$

73. $H(x) = -3x^2 + 4x + 1$ **74.** $h(x) = -4x^2 - 6x + 1$

Applying the Concepts

75. Revenue Function Suppose that the marketing department of Samsung has found that, when a certain model of DVD player is sold at a price of p dollars, the daily revenue R (in dollars) as a function of the price p is $R(p) = -2.5p^2 + 600p$.

(a) For what price will the daily revenue be maximized?

(b) What is the maximum daily revenue?

76. Revenue Function Suppose that the marketing department of Samsung has found that, when a certain model of cell phones is sold at a price of p dollars, the daily revenue R (in dollars) as a function of the price p is $R(p) = -5p^2 + 600p$.

(a) For what price will the daily revenue be maximized?

(b) What is the maximum daily revenue?

77. Marginal Cost The marginal cost of a product can be thought of as the cost of producing one additional unit of output. For example, if the marginal cost of producing the fiftieth unit of a product is $6.30, then it costs $6.30 to increase production from 49 to 50 units of output. Suppose that the marginal cost C (in dollars) to produce x digital cameras is given by $C(x) = 0.05x^2 - 6x + 215$. How many digital cameras should be produced to minimize marginal cost? What is the minimum marginal cost?

78. Marginal Cost (See Problem 77.) The marginal cost C (in dollars) of manufacturing x MP3 players is given by $C(x) = 0.05x^2 - 9x + 435$. How many MP3 players should be manufactured to minimize marginal cost? What is the minimum marginal cost?

79. Punkin Chunkin Suppose that an air cannon in the Punkin Chunkin contest whose muzzle is 10 feet above the ground fires a pumpkin at an angle of 45° to the horizontal with a muzzle velocity of 335 feet per second. The model $s(t) = -16t^2 + 240t + 10$ can be used to estimate the height s of a pumpkin after t seconds.

(a) Determine the time at which the pumpkin is at its maximum height.

(b) Determine the maximum height of the pumpkin.

(c) After how long will the pumpkin strike the ground?

80. Punkin Chunkin Suppose that a catapult in the Punkin Chunkin contest releases a pumpkin 8 feet above the ground at an angle of 45° to the horizontal with an initial speed of 220 feet per second. The model $s(t) = -16t^2 + 155t + 8$ can be used to estimate the height s of a pumpkin after t seconds.

(a) Determine the time at which the pumpkin is at its maximum height.

(b) Determine the maximum height of the pumpkin.

(c) After how long will the pumpkin strike the ground?

81. Punkin Chunkin Suppose that an air cannon in the Punkin Chunkin contest whose muzzle is 10 feet above the ground fires a pumpkin at an angle of 45° to the horizontal with a muzzle velocity of 335 feet per second. The model $h(x) = \dfrac{-32}{335^2}x^2 + x + 10$ can be used to estimate the height h of the pumpkin after it has traveled x feet.

(a) How far from the cannon will the pumpkin reach its maximum height?

(b) What is the maximum height of the pumpkin?

(c) How far will the pumpkin travel before it strikes the ground?

(d) Compare your answer in part (b) of this problem with the answer found in part (b) of Problem 79. Why might the answers differ?

82. Punkin Chunkin Suppose that a catapult in the Punkin Chunkin contest releases a pumpkin 8 feet above the ground at an angle of 45° to the horizontal with an initial speed of 220 feet per second. The model $h(x) = \dfrac{-32}{220^2}x^2 + x + 8$ can be used to estimate the height h of the pumpkin after it has traveled x feet.

(a) How far from the cannon will the pumpkin reach its maximum height?

(b) What is the maximum height of the pumpkin?

(c) How far will the pumpkin travel before it strikes the ground?

(d) Compare your answer in part (b) of this problem with the answer found in part (b) of Problem 80. Why might the answers differ?

83. Life Cycle Hypothesis The Life Cycle Hypothesis from economics was presented by Franco Modigliani in 1954. One of its components states that income is a function of age. The function $I(a) = -55a^2 + 5119a - 54{,}448$ represents the relation between average annual income I and age a.

(a) According to the model, at what age will average income be a maximum?

(b) According to the model, what is the maximum average income?

84. Age and Divorce The percentage of Americans who divorce is a function of age. The polynomial function $D(a) = -0.014a^2 + 1.586a - 28.1889$ models the percent D of Americans of age a who are divorced.

(a) To the nearest year, what is the age for which the highest percentage of Americans are divorced?

(b) To the nearest percentage, what is the percentage of Americans who are divorced?

85. Fun with Numbers The sum of two numbers is 36. Find the numbers such that their product is a maximum.

86. Fun with Numbers The sum of two numbers is 50. Find the numbers such that their product is a maximum.

87. Fun with Numbers The difference of two numbers is 18. Find the numbers such that their product is a minimum.

88. Fun with Numbers The difference of two numbers is 10. Find the numbers such that their product is a minimum.

89. Enclosing a Rectangular Field Maurice has 500 yards of fencing and wishes to enclose a rectangular area. What is the maximum area that can be enclosed by the fence? What are the dimensions of the area enclosed?

90. Enclosing a Rectangular Field Maude has 800 yards of fencing and wishes to enclose a rectangular area. What is the maximum area that can be enclosed by the fence? What are the dimensions of the area enclosed?

91. Maximizing an Enclosed Area A farmer with 2000 meters of fencing wants to enclose a rectangular plot that borders a river. If the farmer does not fence the side along the river, what is the largest area that can be enclosed? What are the dimensions of the enclosed area? See the figure.

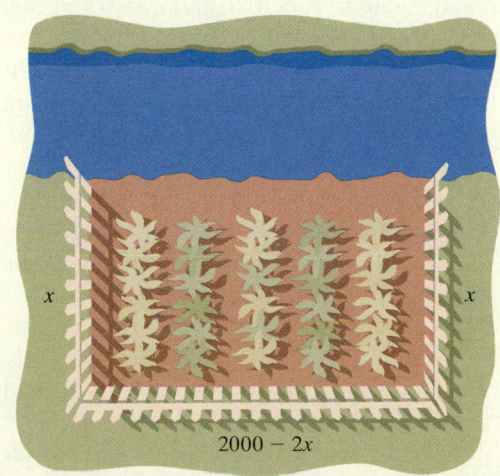

92. Maximizing an Enclosed Area A farmer with 8000 meters of fencing wants to enclose a rectangular plot and then divide it into two plots with a fence parallel to one of the sides. See the figure. What is the largest area that can be enclosed? What are the lengths of the sides of each part of the enclosed area?

93. Constructing Rain Gutters A rain gutter is to be made of aluminum sheets that are 20 inches wide by turning up the edges 90°. What depth will provide maximum cross-sectional area and hence allow the most water to flow?

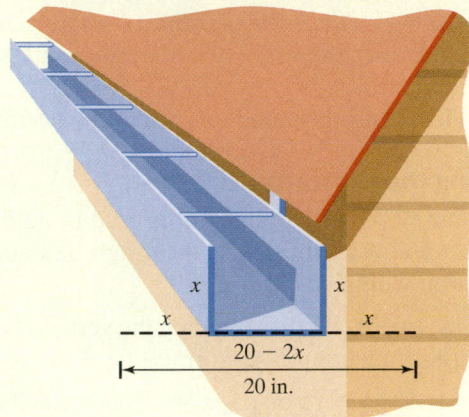

94. Maximizing the Volume of a Box A box with a rectangular base is to be constructed such that the perimeter of the base of the box is to be 40 inches. The height of the box must be 15 inches. Find the dimensions of the box such that the volume is maximized. What is the maximum volume?

95. Revenue Function Weekly demand for jeans at a department store obeys the demand equation

$$x = -p + 110$$

where x is the quantity demanded and p is the price (in dollars).

(a) Express the revenue R as a function of p. (*Hint: $R = xp$*)

(b) What price p maximizes revenue? What is the maximum revenue?

(c) How many pairs of jeans will be sold at the revenue-maximizing price?

96. Revenue Function Demand for hot dogs at a baseball game obeys the demand equation

$$x = -800p + 8000$$

where x is the quantity demanded and p is the price (in dollars).

(a) Express the revenue R as a function of p. (*Hint: $R = xp$*)

(b) What price p maximizes revenue? What is the maximum revenue?

(c) How many hot dogs will be sold at the revenue-maximizing price?

Extending the Concepts

Answer Problems 97 and 98 using the following information: A quadratic function of the form $f(x) = ax^2 + bx + c$ with $b^2 - 4ac > 0$ may also be written in the form $f(x) = a(x - r_1)(x - r_2)$, where r_1 and r_2 are the x-intercepts of the graph of the quadratic function.

97. (a) Find a quadratic function whose x-intercepts are 2 and 6 with $a = 1$, $a = 2$, and $a = -2$.
   (b) How does the value of a affect the intercepts?
   (c) How does the value of a affect the axis of symmetry?
   (d) How does the value of a affect the vertex?

98. (a) Find a quadratic function whose x-intercepts are -1 and 5 with $a = 1$, $a = 2$, and $a = -2$.
   (b) How does the value of a affect the intercepts?
   (c) How does the value of a affect the axis of symmetry?
   (d) How does the value of a affect the vertex?

Explaining the Concepts

99. Explain how the discriminant is used to determine the number of x-intercepts the graph of a quadratic function will have.

100. Provide two methods for finding the vertex of any quadratic function $f(x) = ax^2 + bx + c$.

101. Refer to Example 6 on page 744. Notice that if the price charged for the computer is $0 or $1600, the revenue is $0. It is easy to explain why revenue would be $0 if the price charged were $0, but how can revenue be $0 if the price charged is $1600?

Synthesis Review

In Problems 102–105, graph each function using point plotting.

102. $f(x) = -2x + 12$
103. $G(x) = \dfrac{1}{4}x - 2$
104. $f(x) = x^2 - 5$
105. $f(x) = (x + 2)^2 + 4$

106. For each function in Problems 102–105, explain an alternative method for graphing the function. Which method do you prefer? Why?

Technology Exercises

Technology can be used to determine the coordinates of the vertex of a parabola. In Desmos, once the function is graphed, click on the curve and gray dots will appear at "interesting points," including the vertex. Hover over the point with your mouse to display the ordered pair. See Figure 29.

Figure 29

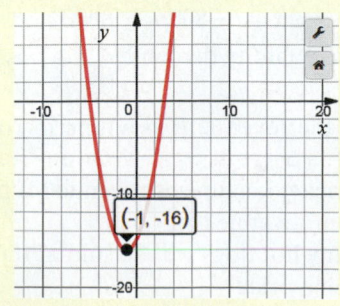

In Problems 107–114, use technology to graph each quadratic function and determine the vertex. If necessary, round your answers to two decimal places.

107. $f(x) = x^2 - 7x + 3$
108. $f(x) = x^2 + 3x + 8$

109. $G(x) = -2x^2 + 14x + 13$

110. $g(x) = -4x^2 - x + 11$

111. $F(x) = 5x^2 + 3x - 20$
112. $F(x) = 3x^2 + 2x - 21$

113. $H(x) = \dfrac{1}{2}x^2 - \dfrac{2}{3}x + 5$
114. $h(x) = \dfrac{3}{4}x^2 + \dfrac{4}{3}x - 1$

115. Exploration: Quadratic Functions—b and c Values *Open the "Quadratic Functions–Standard Form" Geogebra applet. The applet may be found using the QR code at the beginning of this section or through the Multimedia Library in MyMathLab. If you are using a keyboard, it is easier to move the sliders with the arrow keys instead of a mouse. Uncheck the boxes in the applet. Move the sliders so that $a = 1$, $b = 0$, and $c = 0$.*
   (a) Move the b slider to take on the values from -5 to 5. What role does b play in the graph of $y = ax^2 + bx + c$?
   (b) Now reset the sliders so that $a = 1$, $b = 0$, and $c = 0$ and check the y-intercept box. Move the slider c to take on the values from -5 to 5. For each value of c write down the corresponding y-intercept of the parabola. What relationship exists between c and this point?

116. Exploration: Quadratic Functions—Discriminant *Open the "Quadratic Functions—Standard Form" Geogebra applet. The applet may be found using the QR code at the beginning of this section or through the Multimedia Library in MyMathLab. If you are using a keyboard, it is easier to move the sliders with the arrow keys instead of a mouse. Uncheck the boxes in the applet. Move the sliders so that $a = 1$, $b = 0$, and $c = 0$ and check the x-intercept and discriminant boxes.*
   (a) Move the a, b, and c sliders to find two different functions each of which has a discriminant that is a positive number. How many x-intercepts does each of these functions have?
   (b) Now move the a, b, and c sliders to find two different functions each of which has a discriminant that is a negative number. How many x-intercepts does each of these functions have?
   (c) Now move the a, b, and c sliders to find two different functions each of which has a discriminant that is zero. How many x-intercepts does each of these functions have?
   (d) In general, what is the relationship between the discriminant of a quadratic function and the number of x-intercepts it has? In other words, when the discriminant is positive, the function has ___ x-intercepts; when the discriminant is negative, the function has ___ x-intercepts; and when the discriminant is zero, the function has ___ x-intercepts.

10.6 Polynomial Inequalities

Objectives

1. Solve Quadratic Inequalities
2. Solve Polynomial Inequalities

Are You Prepared for This Section?

Before getting started, complete the following problems. If you get a problem wrong, go back to the section cited and review the material.

P1. Write $-4 \leq x < 5$ in interval notation. [Getting Ready, pp. 517–519]

P2. Solve: $3x + 5 > 5x - 3$ [Section 2.8, pp. 150–155]

In Section 2.8, linear inequalities in one variable, such as $2x - 3 > 4x + 5$, were solved using methods similar to solving linear equations. The solution sets of inequalities were represented using either set-builder notation or interval notation.

Solving *polynomial inequalities* is not a simple extension of solving polynomial equations. However, the skills developed in solving polynomial equations are used to solve polynomial inequalities.

> **Definition**
>
> A **polynomial inequality** is an inequality of the form
>
> $$f(x) > 0 \quad f(x) < 0 \quad f(x) \geq 0 \quad f(x) \leq 0$$
>
> where f is a polynomial function.

Some examples of polynomial inequalities are

$$x^2 - 3x - 4 > 0 \quad 2x^2 - 5x + 1 < 0 \quad x^3 - 6x^2 + 7x + 1 \geq 0$$

We begin by looking at quadratic inequalities—a special case of polynomial inequalities.

1 Solve Quadratic Inequalities

> **Definition**
>
> A **quadratic inequality** is an inequality of the form
>
> $$ax^2 + bx + c > 0 \quad \text{or} \quad ax^2 + bx + c < 0 \quad \text{or}$$
> $$ax^2 + bx + c \geq 0 \quad \text{or} \quad ax^2 + bx + c \leq 0$$
>
> where $a \neq 0$.

First a graphical approach to the solution of a quadratic inequality will be presented and then an algebraic method.

To understand the logic behind the graphical approach, consider the following. Suppose we want to solve the inequality $ax^2 + bx + c > 0$. If $f(x) = ax^2 + bx + c$, then we want all x-values such that $f(x) > 0$. Since $f(x)$ represents the y-values of the graph of the function $f(x) = ax^2 + bx + c$, look for all x-values such that the graph of f is *above* the x-axis that is, where y-values are positive. This occurs when the graph lies in either quadrant I or quadrant II of the Cartesian plane. If asked to solve $f(x) < 0$, look for the x-values such that the graph is *below* the x-axis that is, where the y-values are negative. This occurs when the graph lies in either quadrant III or quadrant IV of the Cartesian plane. See Figure 30 on the next page for an illustration of this idea.

Figure 30

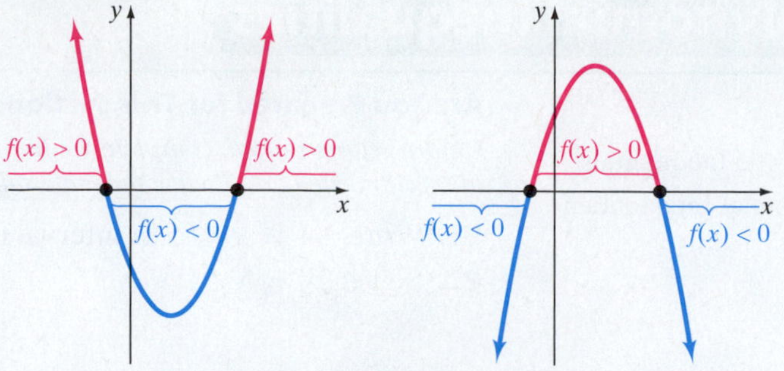

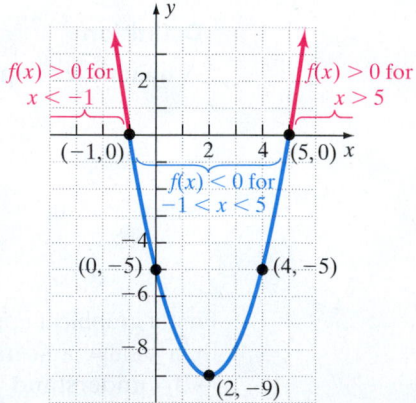

EXAMPLE 1   **How to Solve a Quadratic Inequality Using the Graphical Method**

Solve $x^2 - 4x - 5 \geq 0$ using the graphical method.

Step-by-Step Solution

Step 1: Write the inequality so that $ax^2 + bx + c$ is on one side of the inequality and 0 is on the other.

The inequality is already in the form needed.

$$x^2 - 4x - 5 \geq 0$$

Step 2: Graph the function $f(x) = ax^2 + bx + c$. Label the x-intercepts on the graph.

Graph $f(x) = x^2 - 4x - 5$. Because $a = 1 > 0$, the graph opens up.

x-intercepts: $f(x) = 0$ **y-intercept:** $f(0) = -5$

$$x^2 - 4x - 5 = 0$$ The y-intercept is $(0, -5)$.

$$(x - 5)(x + 1) = 0$$

$$x = 5 \quad \text{or} \quad x = -1$$

The x-intercepts are $(-1, 0)$ and $(5, 0)$.

Vertex: $x = -\dfrac{b}{2a} = -\dfrac{-4}{2(1)} = 2$

$$y = f\left(-\frac{b}{2a}\right) = f(2) = -9$$

The vertex is at $(2, -9)$. Figure 31 shows the graph.

Figure 31

Step 3: From the graph, determine where the function is positive and where it is negative. This information will give the solution set to the inequality.

Figure 31 shows that the graph of $f(x) = x^2 - 4x - 5$ is greater than 0 for $x < -1$ or $x > 5$. Because the inequality is nonstrict ($\geq$), include the x-intercepts in the solution. The solution is $\{x \mid x \leq -1 \text{ or } x \geq 5\}$ using set-builder notation, or $(-\infty, -1] \cup [5, \infty)$ using interval notation. See Figure 32 for a graph of the solution set.

Figure 32

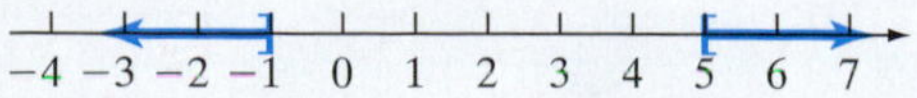

The second method for solving inequalities is algebraic. This method uses the Rules of Signs for multiplying or dividing real numbers. Recall that the product of two positive real numbers is positive, the product of a positive real number and a negative real number is negative, and the product of two negative real numbers is positive.

 Let's solve the inequality from Example 1 using the algebraic method.

EXAMPLE 2   **How to Solve a Quadratic Inequality Using the Algebraic Method**

Solve the inequality from Example 1, $x^2 - 4x - 5 \geq 0$, using the algebraic method.

Step-by-Step Solution

Step 1: Write the inequality so that $ax^2 + bx + c$ is on one side of the inequality and 0 is on the other.

The inequality is already in the form needed.

$$x^2 - 4x - 5 \geq 0$$

Step 2: Determine the solutions to the equation $ax^2 + bx + c = 0$.

$$x^2 - 4x - 5 = 0$$

Factor: $(x - 5)(x + 1) = 0$

Use the Zero-Product Property: $x - 5 = 0$ or $x + 1 = 0$

$x = 5$ or $x = -1$

Step 3: Use the solutions to the equation from Step 2 to separate the real number line into intervals.

Separate the real number line into intervals:

$$(-\infty, -1) \qquad (-1, 5) \qquad (5, \infty)$$

Step 4: Write $x^2 - 4x - 5$ in factored form as $(x - 5)(x + 1)$. Choose a test point within each interval formed in Step 3 to determine the sign of each factor. Then determine the sign of the product. Also write the value of $x^2 - 4x - 5$ at each solution found in Step 2.

The sign of a binomial factor will change only on either side of its corresponding zero. For example, the sign of $x + 1$ changes from negative to positive on either side of $x = -1$. Therefore, when a test point is chosen in each interval, if one number in an interval satisfies the inequality, then all the numbers in that interval will satisfy the inequality.

- In the interval $(-\infty, -1)$, choose a test point of -2. When $x = -2$, the factor $x - 5$ equals $-2 - 5 = -7$, and the factor $x + 1$ equals $-2 + 1 = -1$. Because the product of two negatives is positive, the expression $(x - 5)(x + 1)$ is positive for all x in the interval $(-\infty, -1)$.

- In the interval $(-1, 5)$, let 0 be the test point. When $x = 0$, $x - 5$ is negative and $x + 1$ is positive. Therefore, $(x - 5)(x + 1)$ is negative for all x in the interval $(-1, 5)$.

- In the interval $(5, \infty)$, let 6 be the test point. For $x = 6$, both $x - 5$ and $x + 1$ are positive, so $(x - 5)(x + 1)$ is positive in the interval $(5, \infty)$.

- For $x = -1$ and $x = 5$, the value of $x^2 - 4x - 5 = (x - 5)(x + 1)$ is zero.

(continued)

Table 9 shows the sign of $(x - 5)(x + 1)$ in each interval, as well as the value of $x^2 - 4x - 5$ at $x = -1$ and at $x = 5$.

Table 9

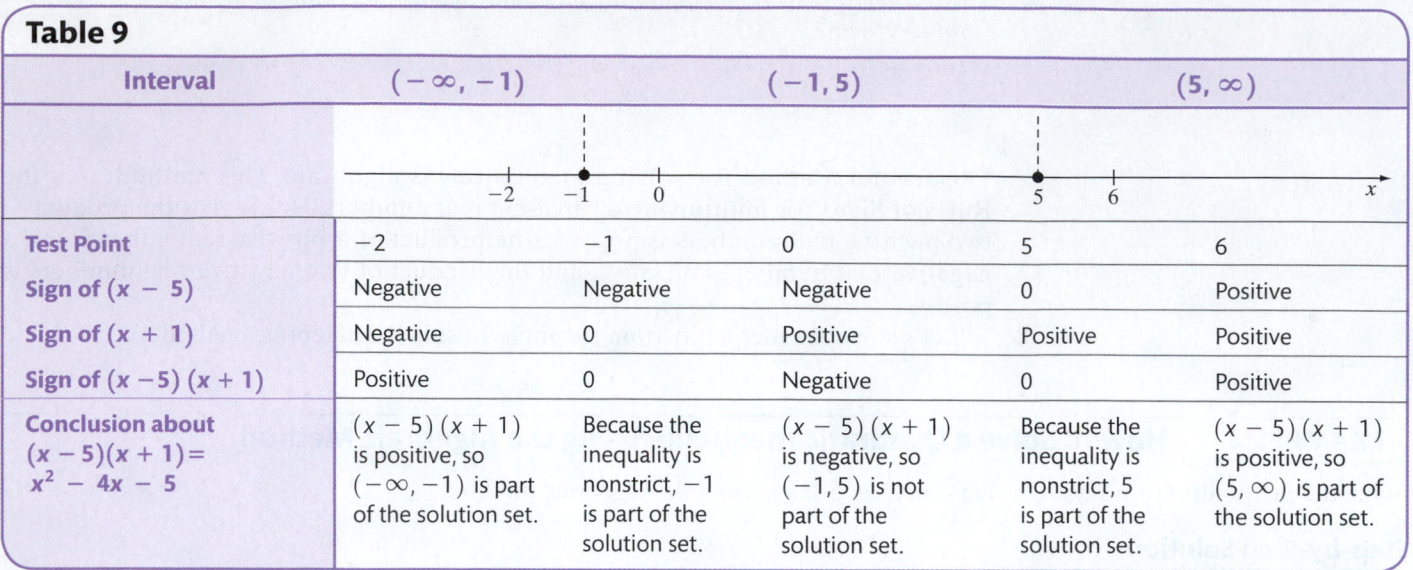

Interval	$(-\infty, -1)$			$(-1, 5)$			$(5, \infty)$
Test Point	-2	-1	0		5		6
Sign of $(x - 5)$	Negative	Negative	Negative		0		Positive
Sign of $(x + 1)$	Negative	0	Positive		Positive		Positive
Sign of $(x - 5)(x + 1)$	Positive	0	Negative		0		Positive
Conclusion about $(x - 5)(x + 1) =$ $x^2 - 4x - 5$	$(x - 5)(x + 1)$ is positive, so $(-\infty, -1)$ is part of the solution set.	Because the inequality is nonstrict, -1 is part of the solution set.	$(x - 5)(x + 1)$ is negative, so $(-1, 5)$ is not part of the solution set.		Because the inequality is nonstrict, 5 is part of the solution set.		$(x - 5)(x + 1)$ is positive, so $(5, \infty)$ is part of the solution set.

Figure 33

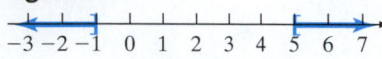

To find where $x^2 - 4x - 5$ is greater than or equal to zero, include -1 and 5 in the solution. The solution set is $\{x \mid x \leq -1 \text{ or } x \geq 5\}$ in set-builder notation, or $(-\infty, -1] \cup [5, \infty)$ in interval notation. See Figure 33 for a graph of the solution set. ●

> **Quick ✓**
>
> In Problem 2, solve the quadratic inequality using the algebraic method.
>
> **2.** $x^2 + 3x - 10 \geq 0$

Solving Quadratic Inequalities

Graphical Method	Algebraic Method
Step 1: Write the inequality so that $ax^2 + bx + c$ is on one side of the inequality and 0 is on the other.	**Step 1:** Write the inequality so that $ax^2 + bx + c$ is on one side of the inequality and 0 is on the other.
Step 2: Graph the function $f(x) = ax^2 + bx + c$. Label the x-intercepts of the graph.	**Step 2:** Determine the solutions to the equation $ax^2 + bx + c = 0$.
Step 3: From the graph, determine where the function is positive and where it is negative. This information will give you the solution set to the inequality.	**Step 3:** Use the solutions to the equation from Step 2 to separate the real number line into intervals.
	Step 4: Write $ax^2 + bx + c$ in factored form. Within each interval formed in Step 3, determine the sign of each factor. Then determine the sign of the product. Also write the value of $ax^2 + bx + c$ at each solution found in Step 2.
	(a) If the product of the factors is positive, then $ax^2 + bx + c > 0$ for all numbers x in the interval.
	(b) If the product of the factors is negative, then $ax^2 + bx + c < 0$ for all numbers x in the interval.

If the inequality is not strict ($\leq$ or $\geq$), include the solutions of $ax^2 + bx + c = 0$ in the solution set.

Now let's do an example where both methods are used.

⏵ **EXAMPLE 3**   **Solving a Quadratic Inequality**

Solve $-x^2 + 10 > 3x$ using both the graphical and the algebraic methods.

Solution

Graphical Method:
To make the computation easier, rearrange the inequality so that the coefficient of x^2 is positive.

$$-x^2 + 10 > 3x$$

Add x^2 to both sides; subtract 10 from both sides: $0 > x^2 + 3x - 10$

$0 > b$ is equivalent to $b < 0$: $x^2 + 3x - 10 < 0$

Graph $f(x) = x^2 + 3x - 10$. Because $a = 1 > 0$, the graph opens up. Find the intercepts and vertex of the function.

x-intercepts: $f(x) = 0$ **y-intercept:** $f(0) = -10$

$$x^2 + 3x - 10 = 0$$
$$(x + 5)(x - 2) = 0$$
$$x + 5 = 0 \quad \text{or} \quad x - 2 = 0$$
$$x = -5 \quad \text{or} \quad x = 2$$

Figure 34

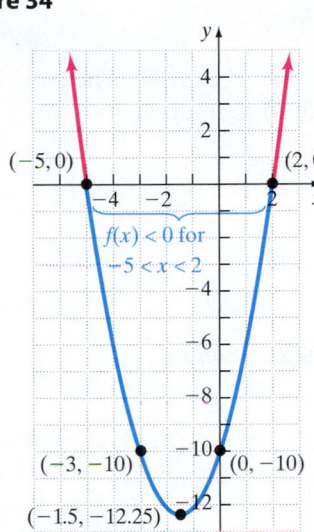

Vertex:

$$x = -\frac{b}{2a} = -\frac{3}{2(1)} = -\frac{3}{2} = -1.5$$

$$y = f\left(-\frac{b}{2a}\right) = f\left(-\frac{3}{2}\right) = -\frac{49}{4} = -12.25$$

The vertex is at $(-1.5, -12.25)$.

Figure 34 shows the graph. The graph of $f(x) = x^2 + 3x - 10$ is below the x-axis (and therefore $x^2 + 3x - 10 < 0$) for $-5 < x < 2$. Because the inequality is strict ($<$) do not include the x-intercepts in the solution. Thus the solution is $\{x \mid -5 < x < 2\}$ in set-builder notation, or $(-5, 2)$ in interval notation.

Algebraic Method:
Rearrange the inequality so that the coefficient of x^2 is positive.

$$-x^2 + 10 > 3x$$

Add x^2 to both sides; subtract 10 from both sides: $0 > x^2 + 3x - 10$

$0 > b$ is equivalent to $b < 0$: $x^2 + 3x - 10 < 0$

Now solve the equation

$$x^2 + 3x - 10 = 0$$
$$(x + 5)(x - 2) = 0$$
$$x + 5 = 0 \quad \text{or} \quad x - 2 = 0$$
$$x = -5 \quad \text{or} \quad x = 2$$

Use the solutions to separate the real number line into the following intervals:

$$(-\infty, -5) \qquad (-5, 2) \qquad (2, \infty)$$

The factored form of $x^2 + 3x - 10$ is $(x + 5)(x - 2)$. Table 10 on the next page shows the sign of each factor and of $(x + 5)(x - 2)$ in each interval. The value of $(x + 5)(x - 2)$ at $x = -5$ and at $x = 2$ is also listed.

(continued)

Table 10

Interval	$(-\infty, -5)$		$(-5, 2)$		$(2, \infty)$

	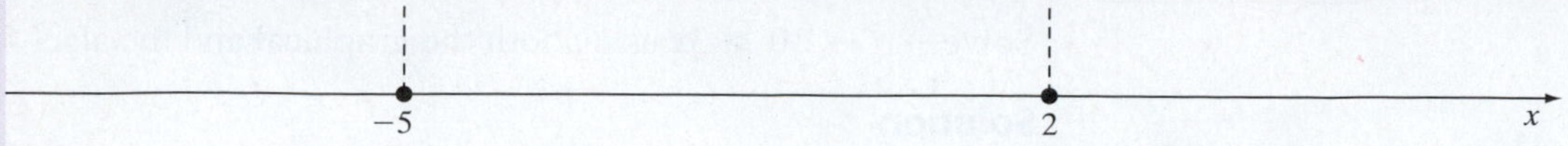				
Test Point	-6	-5	0	2	3
Sign of $(x + 5)$	Negative	0	Positive	Positive	Positive
Sign of $(x - 2)$	Negative	Negative	Negative	0	Positive
Sign of $(x + 5)(x - 2)$	Positive	0	Negative	0	Positive
Conclusion about $(x + 5)(x - 2) < 0$	$(x + 5)(x - 2)$ is positive in the interval $(-\infty, -5)$, so it is not part of the solution set.	The inequality is strict, so -5 is not part of the solution set.	$(x + 5)(x - 2)$ is negative in the interval $(-5, 2)$, so it is part of the solution set.	The inequality is strict, so 2 is not part of the solution set.	$(x + 5)(x - 2)$ is positive in the interval $(2, \infty)$, so it is not part of the solution set.

The inequality is strict, so -5 and 2 are not included in the solution. The solution set is $\{x \mid -5 < x < 2\}$ in set-builder notation, or $(-5, 2)$ in interval notation. Figure 35 shows the graph of the solution set.

Figure 35

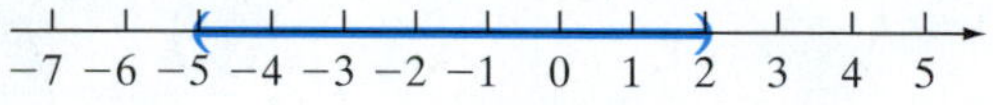

Quick ✓

In Problem 3, solve the quadratic inequality. Graph the solution set.

3. $-x^2 > 2x - 24$

EXAMPLE 4 **Solving a Quadratic Inequality**

Solve $2x^2 > 4x - 1$ using both the graphical and the algebraic methods. Graph the solution set.

Solution

Graphical Method:

$$2x^2 > 4x - 1$$

Subtract $4x$ from both sides; add 1 to both sides: $\quad 2x^2 - 4x + 1 > 0$

Graph $f(x) = 2x^2 - 4x + 1$ to determine where $f(x) > 0$. Because $a = 2 > 0$, the graph opens up.

x-intercepts: $\qquad f(x) = 0$ $\qquad\qquad$ **y-intercept:** $f(0) = 1$

$$2x^2 - 4x + 1 = 0$$

$a = 2; b = -4; c = 1: \quad x = \dfrac{-(-4) \pm \sqrt{(-4)^2 - 4(2)(1)}}{2(2)}$

$$= \frac{4 \pm \sqrt{8}}{4}$$

$$= 1 \pm \frac{\sqrt{2}}{2}$$

$$\approx 0.29 \text{ or } 1.71$$

Vertex:
$$x = -\frac{b}{2a} = -\frac{-4}{2(2)} = 1$$

$$f\left(-\frac{b}{2a}\right) = f(1) = -1$$

The vertex is at $(1, -1)$. Figure 36 shows the graph.

Figure 36

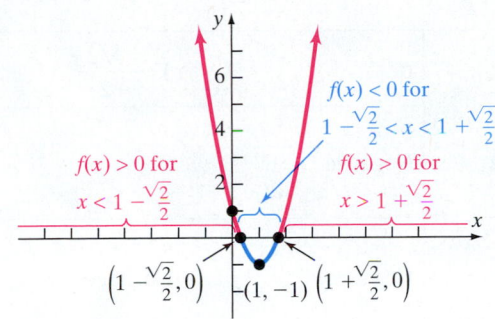

The graph shown in Figure 36 shows that the graph of $f(x) = 2x^2 - 4x + 1$ is greater than 0 for $x < 1 - \dfrac{\sqrt{2}}{2}$ or $x > 1 + \dfrac{\sqrt{2}}{2}$. Because the inequality is strict, do not include the x-intercepts in the solution. Thus the solution is

$$\left\{x \,\middle|\, x < 1 - \frac{\sqrt{2}}{2} \text{ or } x > 1 + \frac{\sqrt{2}}{2}\right\} \text{ in set-builder notation, or}$$

$$\left(-\infty, 1 - \frac{\sqrt{2}}{2}\right) \cup \left(1 + \frac{\sqrt{2}}{2}, \infty\right) \text{ in interval notation.}$$

Algebraic Method:

First, write the inequality in the form $ax^2 + bx + c > 0$.

$$2x^2 > 4x - 1$$

Subtract 4x from both sides; add 1 to both sides: $\quad 2x^2 - 4x + 1 > 0$

Now determine the solutions to the equation $2x^2 - 4x + 1 = 0$.

$$x = \frac{-(-4) \pm \sqrt{(-4)^2 - 4(2)(1)}}{2(2)}$$

$$= \frac{4 \pm \sqrt{8}}{4}$$

$$= 1 \pm \frac{\sqrt{2}}{2}$$

$$\approx 0.29 \text{ or } 1.71$$

Separate the real number line into the following intervals:

$$\left(-\infty, 1 - \frac{\sqrt{2}}{2}\right) \quad \left(1 - \frac{\sqrt{2}}{2}, 1 + \frac{\sqrt{2}}{2}\right) \quad \left(1 + \frac{\sqrt{2}}{2}, \infty\right)$$

Work Smart

The factor $\left(x - \left(1 - \dfrac{\sqrt{2}}{2}\right)\right)$ is close to the factor $x - 0.29$. Similarly, use $x - 1.71$ as an approximation of $\left(x - \left(1 + \dfrac{\sqrt{2}}{2}\right)\right)$.

The solutions to the equation $2x^2 - 4x + 1 = 0$ enable us to factor $2x^2 - 4x + 1$ as $\left(x - \left(1 - \dfrac{\sqrt{2}}{2}\right)\right)\left(x - \left(1 + \dfrac{\sqrt{2}}{2}\right)\right)$. Set up Table 11 on the next page, using 0.29 as an approximation of $1 - \dfrac{\sqrt{2}}{2}$ and 1.71 as an approximation for $1 + \dfrac{\sqrt{2}}{2}$ to help determine the sign of each factor.

(*continued*)

Table 11

Interval	$\left(-\infty, 1-\dfrac{\sqrt{2}}{2}\right)$		$\left(1-\dfrac{\sqrt{2}}{2}, 1+\dfrac{\sqrt{2}}{2}\right)$		$\left(1+\dfrac{\sqrt{2}}{2}, \infty\right)$
		$1-\dfrac{\sqrt{2}}{2} \approx 0.29$		$1+\dfrac{\sqrt{2}}{2} \approx 1.71$	
Test Point	0	$1-\dfrac{\sqrt{2}}{2}$	1	$1+\dfrac{\sqrt{2}}{2}$	2
Sign of $\left(x - \left(1 - \dfrac{\sqrt{2}}{2}\right)\right)$ $\approx (x - 0.29)$	Negative	0	Positive	Positive	Positive
Sign of $\left(x - \left(1 + \dfrac{\sqrt{2}}{2}\right)\right)$ $\approx (x - 1.71)$	Negative	Negative	Negative	0	Positive
Sign of $\left(x - \left(1 - \dfrac{\sqrt{2}}{2}\right)\right)$ $\left(x - \left(1 + \dfrac{\sqrt{2}}{2}\right)\right)$	Positive	0	Negative	0	Positive
Conclusion about $2x^2 - 4x + 1 > 0$	$2x^2 - 4x + 1$ is positive in the interval $\left(-\infty, 1-\dfrac{\sqrt{2}}{2}\right)$, so it is part of the solution set.	Since the inequality is strict, $1-\dfrac{\sqrt{2}}{2}$ is not part of the solution set.	$2x^2 - 4x + 1$ is negative in the interval $\left(1-\dfrac{\sqrt{2}}{2}, 1+\dfrac{\sqrt{2}}{2}\right)$, so it is not part of the solution set.	Since the inequality is strict, $1+\dfrac{\sqrt{2}}{2}$ is not part of the solution set.	$2x^2 - 4x + 1$ is positive in the interval $\left(1+\dfrac{\sqrt{2}}{2}, \infty\right)$, so it is part of the solution set.

Work Smart

Instead of using the signs of the factors, evaluate $2x^2 - 4x + 1$ at each test point. For example, at $x = 0$,
$2x^2 - 4x + 1$
$= 2(0)^2 - 4(0) + 1$
$= 1$
$> 0.$
Therefore, $2x^2 - 4x + 1 > 0$ for all $x < 1 - \dfrac{\sqrt{2}}{2}$.

To find where $2x^2 - 4x + 1$ is positive, do not include $1 - \dfrac{\sqrt{2}}{2}$ or $1 + \dfrac{\sqrt{2}}{2}$ in the solution. The solution set is $\left\{ x \mid x < 1 - \dfrac{\sqrt{2}}{2} \text{ or } x > 1 + \dfrac{\sqrt{2}}{2} \right\}$ in set-builder notation, or $\left(-\infty, 1 - \dfrac{\sqrt{2}}{2}\right) \cup \left(1 + \dfrac{\sqrt{2}}{2}, \infty\right)$ in interval notation.

Figure 37 shows the graph of the solution set.

Figure 37

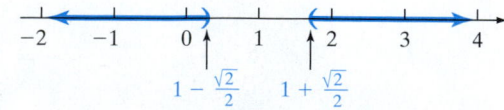

Quick ✔

4. Solve $3x^2 > -x + 5$ using both the graphical and the algebraic methods. Graph the solution set.

▶ ② Solve Polynomial Inequalities

To solve polynomial inequalities, focus on the algebraic method, because we do not yet have experience graphing polynomial functions. To solve polynomial inequalities algebraically, use the same steps used to solve quadratic inequalities algebraically.

EXAMPLE 5 **Solving a Polynomial Inequality of Degree 3**

Solve $x^3 + x^2 - 9x - 9 > 0$ algebraically and graph the solution set.

Solution

The inequality is already written so that the polynomial is on one side of the inequality and 0 is on the other side. Therefore, solve the equation $x^3 + x^2 - 9x - 9 = 0$.

$$x^3 + x^2 - 9x - 9 = 0$$

Factor by grouping: $\quad x^2(x + 1) - 9(x + 1) = 0$

$$(x^2 - 9)(x + 1) = 0$$

$$(x + 3)(x - 3)(x + 1) = 0$$

Set each factor equal to 0: $\quad x + 3 = 0 \quad$ or $\quad x - 3 = 0 \quad$ or $\quad x + 1 = 0$

$$x = -3 \qquad\qquad x = 3 \qquad\qquad x = -1$$

Notice that the solutions of the equation $x^3 + x^2 - 9x - 9 = 0$ separate the number line into *four* intervals: $(-\infty, -3), (-3, -1), (-1, 3)$ and $(3, \infty)$. Determine the sign of each of the factors $(x + 3), (x-3)$, and $(x + 1)$, and then find the sign of the product. Table 12 shows the sign of each factor and the sign of $(x + 3)(x - 3)(x + 1)$ in each interval. The value of $(x + 3)(x - 3)(x + 1)$ at $x = -3, x = -1$, and $x = 3$ is also in the table.

Table 12

Interval	$(-\infty, -3)$		$(-3, -1)$		$(-1, 3)$		$(3, \infty)$
Test Point	-4	-3	-2	-1	1	3	4
Sign of $(x + 3)$	Negative	0	Positive	Positive	Positive	Positive	Positive
Sign of $(x - 3)$	Negative	Negative	Negative	Negative	Negative	0	Positive
Sign of $(x + 1)$	Negative	Negative	Negative	0	Positive	Positive	Positive
Sign of $(x + 3)(x - 3)(x + 1)$	Negative	0	Positive	0	Negative	0	Positive
Conclusion about $(x + 3)(x - 3)(x + 1) > 0$	The product of the factors is negative, so $(-\infty, -3)$ is not part of the solution set.	The inequality is strict, so $x = -3$, is not in the solution set.	The product of the factors is positive, so $(-3, -1)$ is part of the solution set.	The inequality is strict, so $x = -1$, is not in the solution set.	The product is negative, so $(-1, 3)$ is not part of the solution set.	The inequality is strict, so $x = 3$ is not in the solution set.	The product of factors is positive, so $(3, \infty)$ is part of the solution set.

The inequality is strict, so do not include $-3, -1$, or 3 in the solution set. The solution of the inequality $x^3 + x^2 - 9x - 9 > 0$ is $\{x \mid -3 < x < -1 \text{ or } x > 3\}$ in set-builder notation, or $(-3, -1) \cup (3, \infty)$ in interval notation. Figure 38 shows the graph of the solution set.

Figure 38

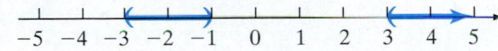

Quick ✓

Solve Problems 5 and 6 algebraically. Graph the solution set.

5. $(2x - 3)(x + 4)(x - 5) \leq 0$

6. $x^3 + 7x^2 - 4x - 28 > 0$

*Problems **1–6** are the Quick ✔s that follow the **EXAMPLES**.*

Building Skills

In Problems 7–10, use the graphs of the quadratic function f to determine the solution.

7. (a) Solve $f(x) > 0$.   **(b)** Solve $f(x) \le 0$.

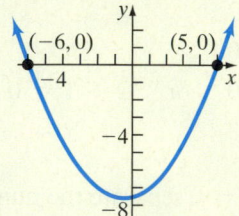

8. (a) Solve $f(x) > 0$.   **(b)** Solve $f(x) \le 0$.

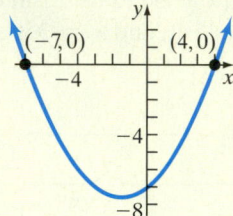

9. (a) $f(x) \ge 0$ **(b)** $f(x) < 0$

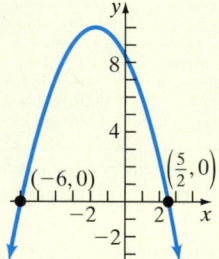

10. (a) $f(x) > 0$ **(b)** $f(x) \le 0$

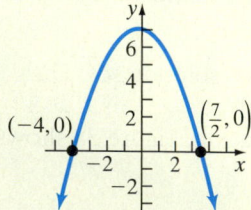

In Problems 11–32, solve each quadratic inequality. Graph the solution set. See Objective 1.

11. $(x - 5)(x + 2) \ge 0$ **12.** $(x - 8)(x + 1) \le 0$

13. $(x + 3)(x + 7) < 0$ **14.** $(x - 4)(x - 10) > 0$

15. $x^2 - 2x - 35 > 0$ **16.** $x^2 + 3x - 18 \ge 0$

17. $n^2 - 6n - 8 \le 0$ **18.** $p^2 + 5p + 4 < 0$

19. $m^2 + 5m \ge 14$ **20.** $z^2 > 7z + 8$

21. $2q^2 \ge q + 15$ **22.** $2b^2 + 5b < 7$

23. $3x + 4 \ge x^2$ **24.** $x + 6 < x^2$

25. $-x^2 + 3x < -10$ **26.** $-x^2 > 4x - 21$

27. $-3x^2 \le -10x - 8$ **28.** $-3m^2 \ge 16m + 5$

29. $x^2 + 4x + 1 < 0$ **30.** $x^2 - 3x - 5 \ge 0$

31. $-2a^2 + 7a \ge -4$ **32.** $-3p^2 < 3p - 5$

In Problems 33–38, solve each quadratic inequality using the graphical method.

33. $z^2 + 2z + 3 > 0$ **34.** $y^2 + y + 5 \ge 0$

35. $2b^2 + 5b \le -6$ **36.** $3w^2 + w < -2$

37. $x^2 - 6x + 9 > 0$ **38.** $p^2 - 8p + 16 \le 0$

In Problems 39–48, solve each polynomial inequality. Graph the solution set. See Objective 2.

39. $(x + 1)(x - 2)(x - 5) > 0$

40. $(x + 3)(x - 1)(x - 3) < 0$

41. $(2x + 1)(x - 4)(x - 9) \le 0$

42. $(3x + 4)(x - 2)(x - 6) \ge 0$

43. $(x + 3)(2x^2 - x - 1) \ge 0$

44. $(x - 2)(3x^2 + x - 2) \le 0$

45. $x^3 + 3x^2 - 4x - 12 \leq 0$

46. $x^3 + 4x^2 - 9x - 36 \geq 0$

47. $4x^3 + 16x^2 - 9x - 36 > 0$

48. $3x^3 + 5x^2 - 12x - 20 < 0$

Mixed Practice

In Problems 49–56, for each function find the values of x that satisfy the given condition.

49. Solve $f(x) < 0$ if $f(x) = x^2 - 5x$.

50. Solve $f(x) > 0$ if $f(x) = x^2 + 4x$.

51. Solve $f(x) \geq 0$ if $f(x) = x^2 - 3x - 28$.

52. Solve $f(x) \leq 0$ if $f(x) = x^2 + 2x - 48$.

53. Solve $g(x) > 0$ if $g(x) = 2x^2 + x - 10$.

54. Solve $F(x) < 0$ if $F(x) = 2x^2 + 7x - 15$.

55. Solve $f(x) < 0$ if $f(x) = 4x^3 - x^2 - 14x$.

56. Solve $g(x) > 0$ if $g(x) = 2x^3 - 7x^2 - 9x$.

In Problems 57–60, find the domain of the given function.

57. $f(x) = \sqrt{x^2 + 8x}$ **58.** $f(x) = \sqrt{x^2 - 5x}$

59. $g(x) = \sqrt{x^2 - x - 30}$ **60.** $G(x) = \sqrt{x^2 + 2x - 63}$

Applying the Concepts

61. Physics A ball is thrown vertically upward with an initial speed of 80 feet per second from a cliff 500 feet above sea level. The height s (in feet) of the ball from the ground after t seconds is $s(t) = -16t^2 + 80t + 500$. For what time t is the ball more than 596 feet above sea level?

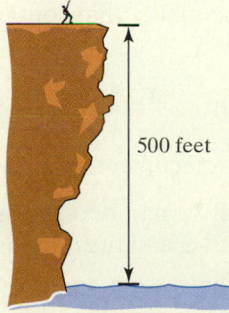

500 feet

62. Physics A water balloon is thrown vertically upward with an initial speed of 64 feet per second from the top of a building 200 feet above the ground. The height s (in feet) of the balloon from the ground after t seconds is $s(t) = -16t^2 + 64t + 200$. For what time t is the balloon more than 248 feet above the ground?

63. Revenue Function Suppose that the marketing department of Panasonic has found that, when a certain model of DVD player is sold at a price of p dollars, the daily revenue R (in dollars) as a function of the price p is $R(p) = -2.5p^2 + 600p$. Determine the prices for which revenue will exceed \$35,750. That is, solve $R(p) > 35,750$.

64. Revenue Function Suppose that the marketing department of Samsung has found that, when a certain model of cell phone is sold at a price of p dollars, the daily revenue R (in dollars) as a function of the price p is $R(p) = -5p^2 + 600p$. Determine the prices for which revenue will exceed \$17,500. That is, solve $R(p) > 17,500$.

Extending the Concepts

In Problems 65–68, solve each inequality by inspection. Then provide a verbal explanation of the solution.

65. $(x + 3)^2 \leq 0$

66. $(x - 4)^2 > 0$

67. $(x + 8)^2 > -2$

68. $(3x + 1)^2 < -2$

69. Write a quadratic inequality that has $[-3, 2]$ as the solution set.

70. Write a quadratic inequality that has $(0, 5)$ as the solution set.

In Problems 71–76, use the techniques presented in this section to solve each inequality algebraically.

71. $(x + 1)(x - 2)(x + 3)(x - 4) > 0$

72. $(x - 4)(x - 1)(x + 5)(x + 6) < 0$

73. $x^4 - 29x^2 + 100 \leq 0$

74. $x^4 - 10x^2 + 9 \geq 0$

75. $\dfrac{x^2 + 5x + 6}{x - 2} > 0$

76. $\dfrac{x^2 - 3x - 10}{x + 1} < 0$

Explaining the Concepts

77. The inequalities $(3x + 2)^2 < 2$ and $(3x + 2)^{-2} > \dfrac{1}{2}$ have the same solution set. Why?

78. The inequality $x^2 + 3 < -2$ has no solution. Explain why.

79. The inequality $x^2 - 1 \geq -1$ has the set of all real numbers as the solution. Explain why.

80. Explain when the endpoints of an interval are included in the solution set of a quadratic inequality.

81. Is the inequality $x^2 + 1 > 1$ true for all real numbers? Explain.

82. Explain how to solve the quadratic inequality $f(x) < 0$ from the graph of $y = f(x)$, where f is a quadratic function.

83. Explain how to solve the quadratic inequality $f(x) > 0$ from the graph of $y = f(x)$, where f is a quadratic function.

Synthesis Review

In Problems 84–87, simplify each expression.

84. $\dfrac{3a^4 b}{12a^{-3} b^5}$

85. $(4mn^{-3})(-2m^4 n)$

86. $\left(\dfrac{3x^4 y}{6x^{-2} y^5} \right)^{\frac{1}{2}}$

87. $\left(\dfrac{9a^{\frac{2}{3}} b^{\frac{1}{2}}}{a^{\frac{-1}{9}} b^{\frac{3}{4}}} \right)^{-1}$

88. Do the Laws of Exponents presented in the "Getting Ready: Laws of Exponents" section also apply to the Laws of Exponents for rational exponents presented in Section 9.3.

Technology Exercises

Technology may be used to solve the quadratic inequality in Example 1. If using a graphing calculator, graph $Y_1 = x^2 - 4x - 5$. Use the ZERO feature of the calculator to find the x-intercepts of the graph. See Figures 39(a) and (b).

Figure 39

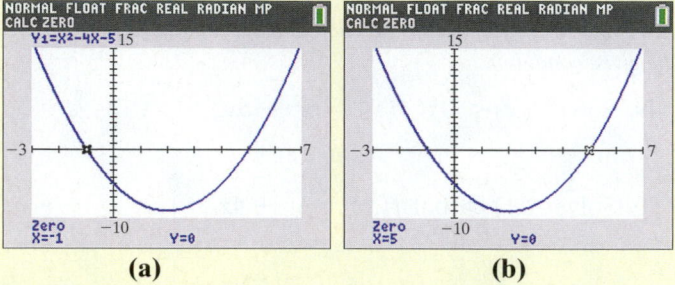

(a)   (b)

From Figures 39(a) and (b), we can see that $x^2 - 4x - 5 \geq 0$ for $x \leq -1$ or $x \geq 5$.

In Problems 89–92, solve each quadratic inequality using technology.

89. $2x^2 + 7x - 49 > 0$

90. $2x^2 + 3x - 27 < 0$

91. $6x^2 + x \leq 40$

92. $8x^2 + 18x \geq 81$

10.7 Rational Inequalities

Objective

1 Solve a Rational Inequality

Are You Prepared for This Section?

Before getting started, complete the following problems. If you get a problem wrong, go back to the section cited and review the material.

P1. Write $-1 < x \leq 8$ in interval notation.   [Getting Ready, pp. 517–519]

P2. Solve: $2x + 3 > 4x - 9$ [Section 2.8, pp. 150–155]

This approach to solving rational inequalities is similar to the algebraic approach to solving quadratic inequalities.

1 Solve a Rational Inequality

▶ A **rational inequality** is an inequality that contains a rational expression. Examples of rational inequalities include

$$\dfrac{1}{x} > 1 \qquad \dfrac{x-1}{x+5} \leq 0 \qquad \dfrac{x^2 + 3x + 2}{x - 5} > 0 \qquad \dfrac{3}{x-5} < \dfrac{4x}{2x-1} + \dfrac{1}{x}$$

To solve rational inequalities, remember the following:

1. The quotient of two positive numbers is positive; the quotient of a positive and a negative number is negative; and the quotient of two negative numbers is positive.

2. The value of a rational expression may be positive, negative, zero, or undefined, depending on the value of its variable.

The value of a rational expression may change signs on either side of a value of the variable that makes the rational expression equal to 0. For example, if $x = 4$, then the value of the rational expression $\dfrac{x-4}{x+5}$ is $\dfrac{4-4}{4+5} = \dfrac{0}{9} = 0$. If $x = 3$, then the value of $\dfrac{x-4}{x+5}$ is $\dfrac{3-4}{3+5} = \dfrac{-1}{8} = -\dfrac{1}{8}$, a negative number. If $x = 5$, then the value of $\dfrac{x-4}{x+5}$ is $\dfrac{5-4}{5+5} = \dfrac{1}{10}$, a positive number.

The value of a rational expression may also change signs on either side of a value of the variable that makes the rational expression undefined. The rational expression $\dfrac{x-4}{x+5}$ is undefined at $x = -5$. If $x = -6$, then the value of $\dfrac{x-4}{x+5}$ is $\dfrac{-6-4}{-6+5} = \dfrac{-10}{-1} = 10$, and if $x = -4$, then $\dfrac{-4-4}{-4+5} = \dfrac{-8}{1} = -8$.

> **In Other Words**
> When $x = 4$, the expression $\dfrac{x-4}{x+5}$ is equal to 0. When $x = 3$, the expression is negative, and when $x = 5$, the expression is positive.

EXAMPLE 1   How to Solve a Rational Inequality

Solve $\dfrac{x+3}{x-4} \geq 0$. Graph the solution set.

Step-by-Step Solution

Step 1: Write the inequality with 0 on one side of the inequality and with a rational expression, written as a single quotient, on the other side.

The inequality is in the form needed.

$$\frac{x+3}{x-4} \geq 0$$

Step 2: Determine the numbers for which the rational expression equals 0 or is undefined.

The rational expression equals 0 when $x = -3$. The rational expression is undefined when $x = 4$.

Step 3: Use the numbers found in Step 2 to separate the real number line into intervals.

Separate the real number line into the following intervals:

Because the rational expression is undefined at $x = 4$, plot an open circle at 4.

Step 4: Choose a test point within each interval formed in Step 3 to determine the sign of the numerator, $x + 3$, and the sign of the denominator, $x - 4$. Then determine the sign of the quotient.

When a test point is chosen in each interval, if one number in an interval satisfies the inequality, then all the numbers in that interval will satisfy the inequality.

- In the interval $(-\infty, -3)$, choose a test point of -4. When $x = -4$, the numerator, $x + 3$, equals -1, and the denominator, $x - 4$, equals -8. Since the quotient of two negatives is positive, the expression $\dfrac{x+3}{x-4} = \dfrac{-1}{-8} = \dfrac{1}{8}$ is positive when $x = -4$. Furthermore, the expression $\dfrac{x+3}{x-4}$ is positive for all x in the interval $(-\infty, -3)$.

- In the interval $(-3, 4)$, let 0 be the test point. When $x = 0$, $x + 3$ is positive, and $x - 4$ is negative. A positive divided by a negative is negative, so $\dfrac{x+3}{x-4} = \dfrac{0+3}{0-4} = \dfrac{3}{-4} = -\dfrac{3}{4}$ is negative when $x = 0$. The expression $\dfrac{x+3}{x-4}$ is negative for all x in the interval $(-3, 4)$.

- In the interval $(4, \infty)$, let 5 be the test point. When $x = 5$, both $x + 3$ and $x - 4$ are positive. A positive divided by a positive is positive, so $\dfrac{x+3}{x-4} = \dfrac{5+3}{5-4} = \dfrac{8}{1} = 8$ is positive. Furthermore, $\dfrac{x+3}{x-4}$ is positive for all x in $(4, \infty)$.

Table 13 on the next page shows these results and the sign of $\dfrac{x+3}{x-4}$ in each interval.

Table 13

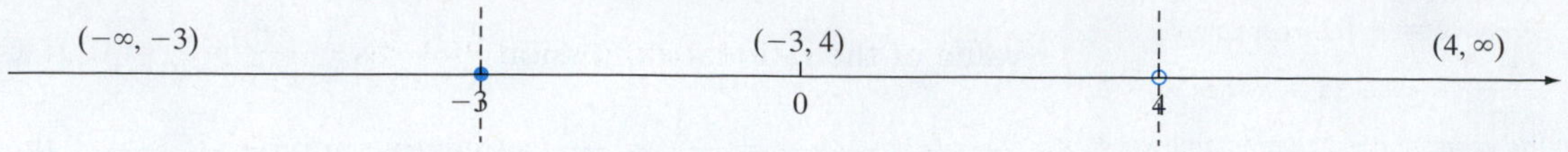

Interval	$(-\infty, -3)$		$(-3, 4)$		$(4, \infty)$
	$x < -3$	$x = -3$	$-3 < x < 4$	$x = 4$	$x > 4$
Test Point	-4	-3	0	4	5
Sign of $x + 3$	Negative	0	Positive	Positive	Positive
Sign of $x - 4$	Negative	Negative	Negative	0	Positive
Sign of $\dfrac{x + 3}{x - 4}$	Positive	0	Negative	Undefined	Positve
Conclusion about $\dfrac{x + 3}{x - 4}$	$\dfrac{x + 3}{x - 4}$ is positive, so $(-\infty, -3)$ is part of the solution set.	Because the inequality is nonstrict, -3 is part of the solution set.	$\dfrac{x + 3}{x - 4}$ is negative, so $(-3, 4)$ is not part of the solution set.	4 cannot be part of the solution set because it causes division by 0.	$\dfrac{x + 3}{x - 4}$ is positive, so $(4, \infty)$ is part of the solution set.

Find where $\dfrac{x + 3}{x - 4}$ is greater than or equal to zero. The solution set is $\{x \mid x \leq -3 \text{ or } x > 4\}$ using set-builder notation, and $(-\infty, -3] \cup (4, \infty)$ using interval notation. Notice that -3 is part of the solution set because $\dfrac{-3 + 3}{-3 - 4} = \dfrac{0}{-7} = 0$, but 4 is not part of the solution set because it is not in the domain of $\dfrac{x + 3}{x - 4}$. Figure 39 shows the graph of the solution set.

Figure 39

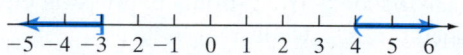

Solving Rational Inequalities

Step 1: Write the inequality so that a rational expression is on one side of the inequality and 0 is on the other. Be sure to write the rational expression as a single quotient in factored form.

Step 2: Determine the values for which the rational expression equals 0 or is undefined.

Step 3: Use the values found in Step 2 to separate the real number line into intervals.

Step 4: Choose a test point within each interval formed in Step 3 to determine the sign of each factor in the numerator and denominator. Then determine the sign of the quotient.

- If the quotient is positive, then the rational expression is positive for all values x in the interval.
- If the quotient is negative, then the rational expression is negative for all values x in the interval.

Also determine the value of the rational expression at each value found in Step 2. If the inequality is not strict ($\leq$ or $\geq$), include the values of the variable for which the rational expression equals 0 in the solution set, but do not include the values for which the rational expression is undefined!

Quick ✓

1. The inequality $\dfrac{2x - 3}{x + 6} > 1$ is an example of a(n) _____ inequality.

2. Solve $\dfrac{x - 7}{x + 3} \geq 0$. Graph the solution set.

3. Solve $\dfrac{1 - x}{x + 5} > 0$. Graph the solution set.

▶ **EXAMPLE 2** **Solving a Rational Inequality**

Solve $\dfrac{x+3}{x-1} > 2$. Graph the solution set.

Solution

First, write the inequality so that a rational expression is on one side of the inequality and 0 is on the other.

$$\frac{x+3}{x-1} > 2$$

Subtract 2 from both sides: $\quad\dfrac{x+3}{x-1} - 2 > 0$

LCD $= x - 1$; multiply -2 by $\dfrac{x-1}{x-1}$: $\quad\dfrac{x+3}{x-1} - 2 \cdot \dfrac{x-1}{x-1} > 0$

Write the rational expression with the LCD: $\quad\dfrac{x+3-2(x-1)}{x-1} > 0$

Distribute -2: $\quad\dfrac{x+3-2x+2}{x-1} > 0$

Combine like terms in the numerator: $\quad\dfrac{-x+5}{x-1} > 0$

The rational expression equals 0 when $x = 5$. The rational expression is undefined when $x = 1$. Separate the real number line into the intervals as shown in Figure 40.

Figure 40

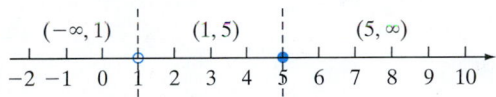

Table 14 shows the sign of $-x + 5$, $x - 1$, and $\dfrac{-x+5}{x-1}$ in each interval. It also shows the value of $\dfrac{-x+5}{x-1}$ at $x = 1$ and $x = 5$.

Table 14

Interval	$(-\infty, 1)$		$(1, 5)$		$(5, \infty)$
Test Point	0	1	3	5	6
Sign of $-x + 5$	Positive	Positive	Positive	0	Negative
Sign of $x - 1$	Negative	0	Positive	Positive	Positive
Sign of $\dfrac{-x+5}{x-1}$	Negative	Undefined	Positive	0	Negative
Conclusion about $\dfrac{-x+5}{x-1}$	$\dfrac{-x+5}{x-1}$ is negative, so $(-\infty, 1)$ is not part of the solution set.	Because $\dfrac{-x+5}{x-1}$ is undefined at $x = 1$, $x = 1$ is not part of the solution set.	$\dfrac{-x+5}{x-1}$ is positive, so $(1, 5)$ is part of the solution set.	Because the inequality is strict, 5 is not part of the solution set.	$\dfrac{-x+5}{x-1}$ is negative, so $(5, \infty)$ is not part of the solution set.

Figure 41

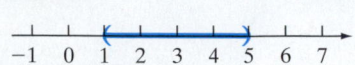

Find the values of x such that $\dfrac{x + 3}{x - 1} > 2$. This is equivalent to determining where $\dfrac{-x + 5}{x - 1} > 0$. Thus the solution set is $\{x \mid 1 < x < 5\}$, or $(1, 5)$ using interval notation. The endpoints of the interval are not in the solution set because the inequality in the original problem is strict. Figure 41 shows the graph of the solution set. ●

Quick ✓

4. Solve $\dfrac{4x + 5}{x + 2} < 3$. Graph the solution set.

10.7 Exercises   MyMathLab®

Problems 1–4 are the Quick ✓s that follow the EXAMPLES.

Building Skills

In Problems 5–20, solve each rational inequality. Graph the solution set. See Objective 1.

5. $\dfrac{x - 4}{x + 1} > 0$

6. $\dfrac{x + 5}{x - 2} > 0$

7. $\dfrac{x + 9}{x - 3} < 0$

8. $\dfrac{x + 8}{x + 2} < 0$

9. $\dfrac{x + 10}{x - 4} \geq 0$

10. $\dfrac{x + 12}{x - 2} \geq 0$

11. $\dfrac{(3x + 5)(x + 8)}{x - 2} \leq 0$ **12.** $\dfrac{(3x - 2)(x - 6)}{x + 1} \geq 0$

13. $\dfrac{x - 5}{x + 1} < 1$

14. $\dfrac{x + 3}{x - 4} > 1$

15. $\dfrac{2x - 9}{x - 3} > 4$

16. $\dfrac{3x + 20}{x + 6} < 5$

17. $\dfrac{3}{x - 4} + \dfrac{1}{x} \geq 0$

18. $\dfrac{2}{x + 3} + \dfrac{2}{x} \leq 0$

19. $\dfrac{3}{x - 2} \leq \dfrac{4}{x + 5}$

20. $\dfrac{1}{x - 4} \geq \dfrac{3}{2x + 1}$

Mixed Practice

In Problems 21–30, solve each inequality. Graph the solution set.

21. $\dfrac{(2x - 1)(x + 3)}{x - 5} > 0$ **22.** $\dfrac{(5x - 2)(x + 4)}{x - 5} < 0$

23. $3 - 4(x + 1) < 11$ **24.** $2x + 3(x - 2) \geq x + 2$

25. $\dfrac{x + 7}{x - 8} \leq 0$ **26.** $\dfrac{x - 10}{x + 5} \leq 0$

27. $(x - 2)(2x + 1) \geq 2(x - 1)^2$

28. $(x + 2)^2 < 3x^2 - 2(x + 1)(x - 2)$

29. $\dfrac{3x - 1}{x + 4} \geq 2$

30. $\dfrac{3x - 7}{x + 2} \leq 2$

In Problems 31–34, for each function, find the values of x that satisfy the given condition.

31. Solve $R(x) \leq 0$ if $R(x) = \dfrac{x - 6}{x + 1}$.

32. Solve $R(x) \geq 0$ if $R(x) = \dfrac{x + 3}{x - 8}$.

33. Solve $R(x) < 0$ if $R(x) = \dfrac{2x - 5}{x + 2}$.

34. Solve $R(x) < 0$ if $R(x) = \dfrac{3x + 2}{x - 4}$.

Applying the Concepts

35. Average Cost Suppose that the daily cost C of manufacturing x bicycles is given by $C(x) = 80x + 5000$. Then the average daily cost $\overline{C}$ is given by $\overline{C}(x) = \dfrac{80x + 5000}{x}$. How many bicycles must be produced each day in order for the average cost to be no more than $130?

36. Average Cost See Problem 35. Suppose that the government imposes a $10 tax on each bicycle manufactured so that the daily cost C of manufacturing x bicycles is now given by $C(x) = 90x + 5000$. Now the average daily cost $\overline{C}$ is given by $\overline{C}(x) = \dfrac{90x + 5000}{x}$. How many bicycles must be produced each day in order for the average cost to be no more than $130?

Extending the Concepts

37. Write a rational inequality that has $(2, \infty)$ as the solution set.

38. Write a rational inequality that has $(-2, 5]$ as the solution set.

Explaining the Concepts

39. In solving the rational inequality $\dfrac{x - 4}{x + 1} \leq 0$, a student determines that the only interval that makes the inequality true is $(-1, 4)$. He states that the solution set is $\{x \mid -1 \leq x \leq 4\}$. What is wrong with this solution?

40. In Step 2 of the steps for solving a rational inequality, the values for which the rational expression equals 0 or is undefined are determined. Then these values are used to form intervals on the real number line. Explain why this guarantees that there is not a change in the sign of the rational expression within any given interval.

Synthesis Review

In Problems 41–46, find the x-intercept(s) of the graph of each function.

41. $F(x) = 6x - 12$

42. $G(x) = 5x + 30$

43. $f(x) = 2x^2 + 3x - 14$

44. $h(x) = -3x^2 - 7x + 20$

45. $R(x) = \dfrac{3x - 2}{x + 4}$

46. $R(x) = \dfrac{x^2 + 5x + 6}{x + 2}$

Technology Exercises

Solutions to rational inequalities can be approximated using the INTERSECT or ZERO (or ROOT) feature of the graphing calculator. Use the ZERO or ROOT feature of the graphing calculator when one side of the inequality is 0. Use the INTERSECT feature when neither side of the inequality is 0. For example, to solve $\dfrac{x - 4}{x + 1} > \dfrac{7}{2}$, graph $Y_1 = \dfrac{x - 4}{x + 1}$ and $Y_2 = \dfrac{7}{2}$. To determine the x-values such that the graph of Y_1 is above that of Y_2, find x-coordinates of the point(s) of intersection. The graph of Y_1 is above that of Y_2 between $x = -3$ and $x = -1$. See Figure 42.

Figure 42

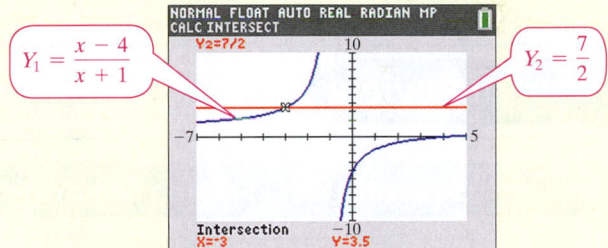

In using a graphing calculator to approximate solutions to inequalities, the solution is typically expressed as a decimal rounded to two decimal places, if exact answers cannot be found. The solution to the inequality $\dfrac{x - 4}{x + 1} > \dfrac{7}{2}$ is $\{x \mid -3 < x < -1\}$, or $(-3, -1)$ using interval notation.

In Problems 47–50, solve each inequality using technology.

47. $\dfrac{x - 5}{x + 1} \leq 3$

48. $\dfrac{x + 2}{x - 5} > -2$

49. $\dfrac{2x + 5}{x - 7} > 3$

50. $\dfrac{2x - 1}{x + 5} \leq 4$

Chapter 10 Activity: Presidential Decision Making

Focus: Developing quadratic equations.

Time: 30–35 minutes

Group size: 2–4

1. Your boss, Huntington Corporation's president, Gerald Cain, is very concerned about his approval rating with his employees. Last year, on January 1, he made some policy changes and saw his approval rating drop. In fact, his approval rating was 48% just before he made the policy changes. One month later, his rating was at 41%, two months later it was at 40%, and at three months it began to climb and was 45%. He discovered that the following function described his approval rating for that year, where x is the month:

   $$R(x) = 3x^2 - 10x + 48$$

   (a) As a group, use the above function to find when President Cain's approval rating will return to the original rating.

   (b) If his approval rating continues to climb, when will he reach a 68% approval rating?

2. On January 1 of this year, President Cain surveyed his employees again and found that 68% of them approved of his leadership skills. At this time, President Cain decided to become very strict with his employees and began a series of new policies. He noticed that his approval rating began steadily to slip and reached an all-time low of 38% on March 30 (3 months later). President Cain is not worried because he knows from last year that this drop in popularity will bottom out and eventually rise. He believes his popularity can be modeled by a quadratic function and needs your help. He has more bad news to deliver but does not want to begin the next round of policy changes until his approval rating is back to approximately 50%.

   (a) As a group, write a quadratic function that would model President Cain's approval rating.

   (b) As a group, develop different ways to advise President Cain on what date to begin his policy changes. Use graphs and computations to prove your point.

Chapter 10 Review

Section 10.1   Solving Quadratic Equations by Completing the Square

KEY CONCEPTS

- **Square Root Property**
  If $x^2 = p$, then $x = \sqrt{p}$ or $x = -\sqrt{p}$.

- **Pythagorean Theorem**
  In a right triangle, the square of the length of the hypotenuse is equal to the sum of the squares of the lengths of the legs. That is, $\text{leg}^2 + \text{leg}^2 = \text{hypotenuse}^2$.

KEY TERMS

Completing the square
Right triangle
Right angle
Hypotenuse
Legs

You Should Be Able To...	EXAMPLE	Review Exercises
❶ Solve quadratic equations using the Square Root Property (p. 691)	Examples 1 through 4	1–10
❷ Complete the square in one variable (p. 694)	Example 5	11–16
❸ Solve quadratic equations by completing the square (p. 695)	Examples 6 and 7	17–26
❹ Solve problems using the Pythagorean Theorem (p. 697)	Examples 8 and 9	27–36

In Problems 1–10, solve each equation using the Square Root Property.

1. $m^2 = 169$

2. $n^2 = 75$

3. $a^2 = -16$

4. $b^2 = \dfrac{8}{9}$

5. $(x - 8)^2 = 81$

6. $(y - 2)^2 - 62 = 88$

7. $(3z + 5)^2 = 100$

8. $7p^2 = 18$

9. $3q^2 + 251 = 11$

10. $\left(x + \dfrac{3}{4}\right)^2 = \dfrac{13}{16}$

In Problems 11–16, complete the square in each expression. Then factor the perfect square trinomial.

11. $a^2 + 30a$

12. $b^2 - 14b$

13. $c^2 - 11c$

14. $d^2 + 9d$

15. $m^2 - \dfrac{1}{4}m$

16. $n^2 + \dfrac{6}{7}n$

In Problems 17–26, solve each quadratic equation by completing the square.

17. $x^2 - 10x + 16 = 0$

18. $y^2 - 3y - 28 = 0$

19. $z^2 - 6z - 3 = 0$

20. $a^2 - 5a - 7 = 0$

21. $b^2 + b + 7 = 0$

22. $c^2 - 6c + 17 = 0$

23. $2d^2 - 7d + 3 = 0$

24. $2w^2 + 2w + 5 = 0$

25. $3x^2 - 9x + 8 = 0$

26. $3x^2 + 4x - 2 = 0$

In Problems 27–32, the lengths of the legs of a right triangle are given. Find the hypotenuse.

27. $a = 9, b = 12$

28. $a = 8, b = 8$

29. $a = 3, b = 6$

30. $a = 10, b = 24$

31. $a = 5, b = \sqrt{11}$

32. $a = 6, b = \sqrt{13}$

In Problems 33–35, use the right triangle shown below to find the missing length.

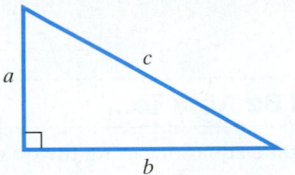

33. $a = 9, c = 12$

34. $b = 5, c = 10$

35. $b = 6, c = 17$

36. Baseball Diamond A baseball diamond is really a square that is 90 feet long on each side. (See the figure.) What is the distance between home plate and second base? Round your answer to the nearest tenth of a foot.

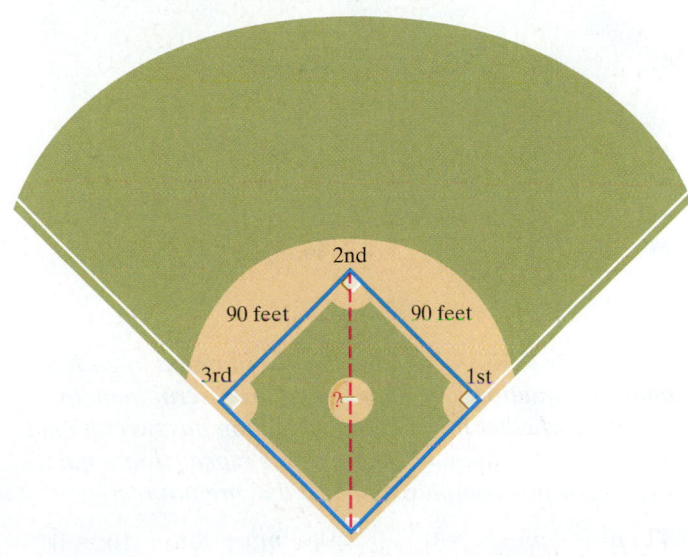

Section 10.2 Solving Quadratic Equations by the Quadratic Formula

KEY CONCEPTS

KEY TERM

Discriminant

- **The Quadratic Formula**

  The solutions to the equation $ax^2 + bx + c = 0, a \neq 0$, are given by

  $$x = \frac{-b \pm \sqrt{b^2 - 4ac}}{2a}$$

- **Discriminant**

  For the quadratic equation $ax^2 + bx + c = 0$, where $a, b,$ and c are rational numbers:

  - If $b^2 - 4ac > 0$, the equation has two unequal real solutions.

  - If $b^2 - 4ac$ is a perfect square, the equation has two rational solutions provided $a, b,$ and c are rational numbers.

- If $b^2 - 4ac$ is not a perfect square, the equation has two irrational solutions provided a, b, and c are rational numbers.
- If $b^2 - 4ac = 0$, the equation has a repeated rational solution.
- If $b^2 - 4ac < 0$, the equation has two complex solutions that are not real.

You Should Be Able To...	EXAMPLE	Review Exercises
1 Solve quadratic equations using the quadratic formula (p. 703)	Examples 1 through 4	37–46
2 Use the discriminant to determine the nature of solutions of a quadratic equation (p. 707)	Example 5	47–52
3 Model and solve problems involving quadratic equations (p. 711)	Examples 6 and 7	63–68

In Problems 37–46, solve each equation using the quadratic formula.

37. $x^2 - x - 20 = 0$

38. $4y^2 = 8y + 21$

39. $3p^2 + 8p = -3$

40. $2q^2 - 3 = 4q$

41. $3w^2 + w = -3$

42. $9z^2 + 16 = 24z$

43. $m^2 - 4m + 2 = 0$

44. $5n^2 + 4n + 1 = 0$

45. $5x + 13 = -x^2$

46. $-2y^2 = 6y + 7$

In Problems 47–52, determine the discriminant of each quadratic equation. Use the value of the discriminant to determine whether the quadratic equation has two rational solutions, two irrational solutions, one repeated rational solution, or two complex solutions that are not real.

47. $p^2 - 5p - 8 = 0$

48. $m^2 + 8m + 16 = 0$

49. $3n^2 + n = -4$

50. $7w^2 + 3 = 8w$

51. $4x^2 + 49 = 28x$

52. $11z - 12 = 2z^2$

In Problems 53–62, solve each equation using any appropriate method.

53. $x^2 + 8x - 9 = 0$

54. $6p^2 + 13p = 5$

55. $n^2 + 13 = -4n$

56. $5y^2 - 60 = 0$

57. $\dfrac{1}{4}q^2 - \dfrac{1}{2}q - \dfrac{3}{8} = 0$

58. $\dfrac{1}{8}m^2 + m + \dfrac{5}{2} = 0$

59. $(w - 8)(w + 6) = -33$

60. $(x - 3)(x + 1) = -2$

61. $9z^2 = 16$

62. $\dfrac{1 - 2x}{x^2 + 5} = 1$

63. Pythagorean Theorem Use the Pythagorean Theorem to determine the value of x and the measurements of the right triangle shown below.

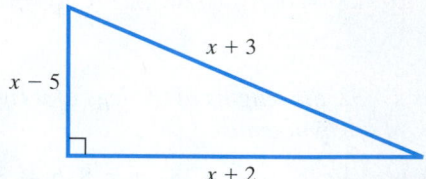

64. Area The area of a rectangle is 108 square centimeters. The width of the rectangle is 3 centimeters less than the length. What are the dimensions of the rectangle?

65. Revenue The revenue R received by a company selling x cell phones per week is given by the function $R(x) = -0.2x^2 + 180x$.

(a) How many cell phones must be sold in order for revenue to be $36,000 per week?

(b) How many cell phones must be sold in order for revenue to be $40,500 per week?

66. Projectile Motion The height s of a ball after t seconds, when it is thrown straight up with an initial speed of 50 feet per second from an initial height of 180 feet, can be modeled by the function $s(t) = -16t^2 + 50t + 180$.

(a) When will the height of the ball be 200 feet? Round your answer to the nearest tenth of a second.

(b) When will the height of the ball be 100 feet? Round your answer to the nearest tenth of a second.

(c) Will the ball ever reach a height of 300 feet? How does the result of the equation tell you this?

67. Pleasure Boat Ride A pleasure boat carries passengers 10 miles upstream and then returns to the starting point. The total time of the trip (excluding the time on land) is 2 hours. If the speed of the current is 3 miles per hour, find the speed of the boat in still water. Round your answer to the nearest tenth of a mile per hour.

68. Work Together, Tom and Beth can wash their car in 30 minutes. By himself, Tom can wash the car in 14 minutes less time than Beth can by herself. How long will it take Beth to wash the car by herself? Round your answer to the nearest tenth of a minute.

Section 10.3 Solving Equations Quadratic in Form

KEY TERM

Equation quadratic in form

You Should Be Able To...	EXAMPLE	Review Exercises
① Solve equations that are quadratic in form (p. 717)	Examples 1 through 5	69–80

In Problems 69–78, solve each equation.

69. $x^4 + 7x^2 - 144 = 0$ **70.** $4w^4 + 5w^2 - 6 = 0$

71. $3(a + 4)^2 - 11(a + 4) + 6 = 0$

72. $(q^2 - 11)^2 - 2(q^2 - 11) - 15 = 0$

73. $y - 13\sqrt{y} + 36 = 0$ **74.** $5z + 2\sqrt{z} - 3 = 0$

75. $p^{-2} - 4p^{-1} - 21 = 0$ **76.** $2b^{\frac{2}{3}} + 13b^{\frac{1}{3}} - 7 = 0$

77. $m^{\frac{1}{2}} + 2m^{\frac{1}{4}} - 8 = 0$ **78.** $\left(\dfrac{1}{x + 5}\right)^2 + \dfrac{3}{x + 5} = 28$

In Problems 79 and 80, find the zeros of the function.

79. $f(x) = 4x - 20\sqrt{x} + 21$ **80.** $g(x) = x^4 - 17x^2 + 60$

Section 10.4 Graphing Quadratic Functions Using Transformations

KEY CONCEPTS

- **Graphing a Function of the Form $f(x) = x^2 + k$ or $f(x) = x^2 - k$**
  To obtain the graph of $f(x) = x^2 + k, k > 0$, from the graph of $y = x^2$, shift the graph of $y = x^2$ vertically up k units. To obtain the graph of $f(x) = x^2 - k, k > 0$, from the graph of $y = x^2$ shift the graph of $y = x^2$ vertically down k units.

- **Graphing a Function of the Form $f(x) = (x - h)^2$ or $f(x) = (x + h)^2$**
  To obtain the graph of $f(x) = (x - h)^2, h > 0$, from the graph of $y = x^2$, shift the graph of $y = x^2$ horizontally to the right h units. To obtain the graph of $f(x) = (x + h)^2, h > 0$, from the graph of $y = x^2$ shift the graph of $y = x^2$ horizontally left h units.

- **Graphing a Function of the Form $f(x) = ax^2$**
  To obtain the graph of $f(x) = ax^2$ from the graph of $y = x^2$, multiply each y-coordinate on the graph of $y = x^2$ by a.

KEY TERMS

Quadratic function
Vertically stretched
Vertically compressed
Parabola
Opens up
Opens down
Vertex
Axis of symmetry
Transformations

You Should Be Able To...	EXAMPLE	Review Exercises
❶ Graph quadratic functions of the form $f(x) = x^2 + k$ (p. 726)	Examples 1 and 2	81–82; 87–90
❷ Graph quadratic functions of the form $f(x) = (x - h)^2$ (p. 727)	Examples 3 through 5	83–84; 87–90
❸ Graph quadratic functions of the form $f(x) = ax^2$ (p. 729)	Example 6	85–86; 89–90
❹ Graph quadratic functions of the form $f(x) = ax^2 + bx + c$ (p. 731)	Examples 7 and 8	91–96
❺ Find a quadratic function from its graph (p. 734)	Example 9	97–100

In Problems 81–90, use the graph of $y = x^2$ to graph the quadratic function.

81. $f(x) = x^2 + 4$

82. $g(x) = x^2 - 5$

83. $h(x) = (x + 1)^2$

84. $F(x) = (x - 4)^2$

85. $G(x) = -4x^2$

86. $H(x) = \frac{1}{5}x^2$

87. $p(x) = (x - 4)^2 - 3$

88. $P(x) = (x + 4)^2 + 2$

89. $f(x) = -(x - 1)^2 + 4$

90. $F(x) = \frac{1}{2}(x + 2)^2 - 1$

In Problems 91–96, graph each quadratic function using transformations. Determine the vertex and axis of symmetry. Based on the graph, determine the domain and range of each function.

91. $g(x) = x^2 - 6x + 10$

92. $G(x) = x^2 + 8x + 11$

93. $h(x) = 2x^2 - 4x - 3$

94. $H(x) = -x^2 - 6x - 10$

95. $p(x) = -3x^2 + 12x - 8$

96. $P(x) = \frac{1}{2}x^2 - 2x + 5$

In Problems 97–100, determine the quadratic function whose graph is given.

97.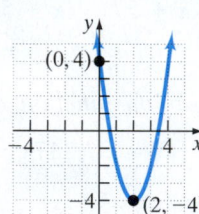

98.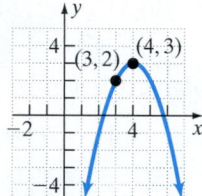

99.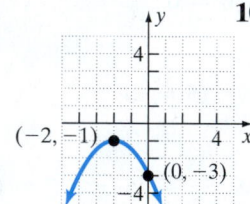

100.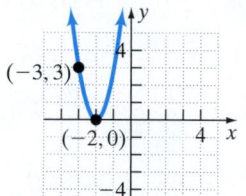

Section 10.5 Graphing Quadratic Functions Using Properties

KEY CONCEPTS

- **Vertex of a Parabola**

  Any quadratic function of the form $f(x) = ax^2 + bx + c, a \neq 0$, will have vertex

  $$\left(-\frac{b}{2a}, f\left(-\frac{b}{2a}\right)\right)$$

- **The x-Intercepts of the Graph of a Quadratic Function**

  1. If $b^2 - 4ac > 0$, the graph of $f(x) = ax^2 + bx + c$ has two x-intercepts.

  2. If $b^2 - 4ac = 0$, the graph of $f(x) = ax^2 + bx + c$ has one x-intercept.

  3. If $b^2 - 4ac < 0$, the graph of $f(x) = ax^2 + bx + c$ has no x-intercepts.

KEY TERMS

Maximum value
Minimum value
Optimization

You Should Be Able To...	EXAMPLE	Review Exercises
❶ Graph quadratic functions of the form $f(x) = ax^2 + bx + c$ (p. 737)	Examples 1 through 4	101–108
❷ Find the maximum or minimum value of a quadratic function (p. 743)	Examples 5 and 6	109–112
❸ Model and solve optimization problems involving quadratic functions (p. 745)	Examples 7 and 8	113–118

In Problems 101–108 graph each quadratic function using its properties by following Steps 1–5 on page 740. Based on the graph, determine the domain and range of each function.

101. $f(x) = x^2 + 2x - 8$ **102.** $F(x) = 2x^2 - 5x + 3$

103. $g(x) = -x^2 + 6x - 7$ **104.** $G(x) = -2x^2 + 4x + 3$

105. $h(x) = 4x^2 - 12x + 9$ **106.** $H(x) = \frac{1}{3}x^2 + 2x + 3$

107. $p(x) = \frac{1}{4}x^2 + 3x + 10$ **108.** $P(x) = -x^2 + 4x - 9$

In Problems 109–112, determine whether the quadratic function has a maximum or a minimum value. Then find the maximum or minimum value and where it occurs.

109. $f(x) = -2x^2 + 16x - 10$

110. $g(x) = 6x^2 - 3x - 1$

111. $h(x) = -4x^2 + 8x + 3$

112. $F(x) = -\frac{1}{3}x^2 + 4x - 7$

113. Revenue Suppose that the marketing department of Zenith has found that, when a certain model of television is sold for a price of p dollars, the daily revenue R (in dollars) as a function of the price p is

$$R(p) = -\frac{1}{3}p^2 + 150p.$$

(a) For what price will the daily revenue be maximized?

(b) What is this maximum daily revenue?

114. Electrical Power In a 120-volt electrical circuit having a resistance of 16 ohms, the available power P (in watts) is given by the function $P(I) = -16I^2 + 120I$, where I represents the current (in amperes).

(a) What current will produce the maximum power in the circuit?

(b) What is this maximum power?

115. Fun with Numbers The sum of two numbers is 24. Find the numbers such that their product is a maximum.

116. Maximizing an Enclosed Area Becky has 15 yards of fencing to make a rectangular kennel for her dog. She will build the kennel next to her garage, so she only needs to enclose three sides. (See the figure.)

(a) What dimensions maximize the area of the kennel?

(b) What is this maximum area?

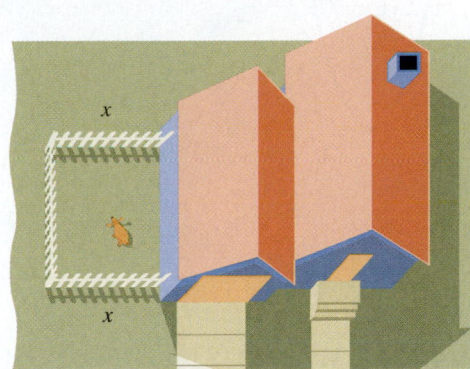

117. Kicking a Football Ted kicks a football at a 45° angle to the horizontal with an initial velocity of 80 feet per second. The model $h(x) = -0.005x^2 + x$ can be used to estimate the height h of the ball after it has traveled x feet.

(a) How far from Ted will the ball reach a maximum height?

(b) What is the maximum height of the ball?

(c) How far will the ball travel before it strikes the ground?

118. Revenue Monthly demand for automobiles at a certain dealership obeys the demand equation $x = -0.002p + 60$, where x is the quantity and p is the price (in dollars).

(a) Express the revenue R as a function of p. (*Hint:* $R = xp$)

(b) What price p maximizes revenue? What is the maximum revenue?

(c) How many automobiles will be sold at the revenue-maximizing price?

Section 10.6   Polynomial Inequalities

KEY CONCEPT

- The steps for solving a quadratic inequality are given on page 754.

KEY TERM

Quadratic inequality
Polynomial inequality

You Should Be Able To	EXAMPLE	Review Exercises
① Solve quadratic inequalities (p. 751)	Examples 1 through 4	119–126
② Solve polynomial inequalities (p. 758)	Example 5	127, 128

In Problems 119 and 120, use the graphs of the quadratic function f to determine the solution.

119. (a) $f(x) > 0$ **(b)** $f(x) < 0$

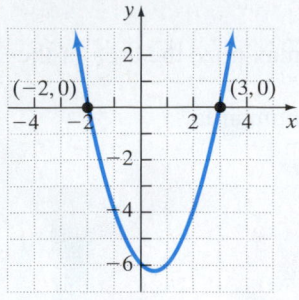

120. (a) $f(x) \geq 0$ **(b)** $f(x) \leq 0$

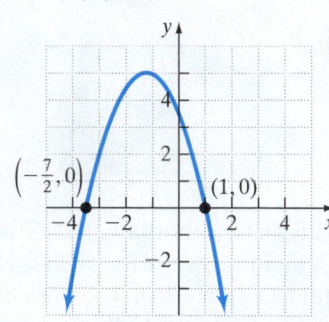

In Problems 121–128, solve each inequality. Graph the solution set.

121. $x^2 - 2x - 24 \leq 0$ **122.** $y^2 + 7y - 8 \geq 0$

123. $3z^2 - 19z + 20 > 0$ **124.** $p^2 + 4p - 2 < 0$

125. $4m^2 - 20m + 25 \geq 0$ **126.** $6w^2 - 19w - 7 \leq 0$

127. $(2x - 3)(x + 1)(x - 2) < 0$

128. $x^3 + 5x^2 - 9x - 45 \geq 0$

Section 10.7 Rational Inequalities

KEY CONCEPT

- The steps for solving any rational inequality are given on page 764.

KEY TERM

Rational inequality

You Should Be Able To...	EXAMPLE	Review Exercises
❶ Solve a rational inequality (p. 762)	Examples 1 and 2	129–138

In Problems 129–136, solve each rational inequality. Graph the solution set.

129. $\dfrac{x - 4}{x + 2} \geq 0$ **130.** $\dfrac{y - 5}{y + 4} < 0$

131. $\dfrac{4}{z^2 - 9} \leq 0$ **132.** $\dfrac{w^2 + 5w - 14}{w - 4} < 0$

133. $\dfrac{m - 5}{m^2 + 3m - 10} \geq 0$ **134.** $\dfrac{4}{n - 2} \leq -2$

135. $\dfrac{a + 1}{a - 2} > 3$ **136.** $\dfrac{4}{c - 2} - \dfrac{3}{c} < 0$

In Problems 137 and 138, for each function, find the values of x that satisfy the given condition.

137. Solve $Q(x) < 0$ if $Q(x) = \dfrac{2x + 3}{x - 4}$.

138. Solve $R(x) \geq 0$ if $R(x) = \dfrac{x + 5}{x + 1}$.

In Problems 1 and 2, complete the square in the given expression. Then factor the perfect square trinomial.

1. $x^2 - 3x$

2. $m^2 + \dfrac{2}{5}m$

In Problems 3–6, solve each equation using any appropriate method you prefer.

3. $9\left(x + \dfrac{4}{3}\right)^2 = 1$

4. $m^2 - 6m + 4 = 0$

5. $2w^2 - 4w + 3 = 0$

6. $\dfrac{1}{2}z^2 - \dfrac{3}{2}z = -\dfrac{7}{6}$

7. Determine the discriminant of $2x^2 + 5x = 4$. Use the value of the discriminant to determine whether the quadratic equation has two rational solutions, two irrational solutions, one repeated real solution, or two complex solutions that are not real.

8. Find the missing length in the right triangle shown below.

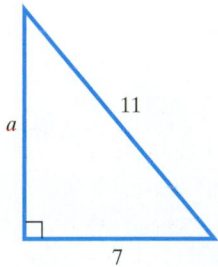

In Problems 9 and 10, solve each equation.

9. $x^4 - 5x^2 - 36 = 0$

10. $6y^{\frac{1}{2}} + 13y^{\frac{1}{4}} - 5 = 0$

In Problems 11 and 12, graph each quadratic function by determining the vertex, intercepts, and axis of symmetry. Based on the graph, determine the domain and range of the quadratic function.

11. $f(x) = (x + 2)^2 - 5$

12. $g(x) = -2x^2 - 8x - 3$

13. Determine the quadratic function whose graph is given.

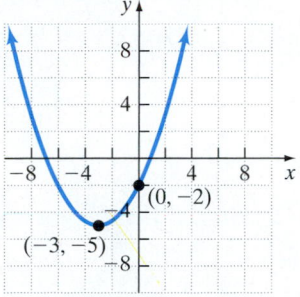

14. Determine whether the quadratic function
$h(x) = -\dfrac{1}{4}x^2 + x + 5$ has a maximum or a
minimum value. Then find the maximum or minimum value.

In Problems 15–17, solve each inequality. Graph the solution set.

15. $2m^2 + m - 15 > 0$

16. $x^3 + 5x^2 - 4x - 20 \le 0$

17. $\dfrac{x + 5}{x - 2} \ge 3 \left\{ x \mid 2 < x \le \dfrac{11}{2} \right\}$ or $\left(2, \dfrac{11}{2} \right]$

18. **Projectile Motion** The height s of a rock after t seconds, when propelled straight up with an initial speed of 80 feet per second from an initial height of 20 feet, can be modeled by the function $s(t) = -16t^2 + 80t + 20$. When will the height of the rock be 50 feet? Round your answer to the nearest tenth of a second.

19. Work Together, Lex and Rupert can roof a house in 16 hours. By himself, Rex can roof the house in 4 hours less time than Rupert can by himself. How long will it take Rupert to roof the house by himself? Round your answer to the nearest tenth of an hour.

20. Revenue A small company has found that, when one of its products is sold for a price of p dollars, the weekly revenue (in dollars) as a function of price p is $R(p) = -0.25p^2 + 170p$.

(a) For what price will the weekly revenue from this product be maximized?

(b) What is this maximum weekly revenue?

21. Maximizing Volume A box with a rectangular base is to be constructed such that the perimeter of the base of the box is 50 inches. The height of the box must be 12 inches.

(a) Find the dimensions that maximize the volume of the box.

(b) What is this maximum volume?

11

Exponential and Logarithmic Functions

We all dream of a long and happy retirement. To make this dream a reality, we must save money during our working years so that our retirement savings can grow through the power of compounding interest. To see how this phenomenon works, see Example 9 in Section 11.2 and Problems 89–92 in Section 11.2.

The Big Picture: Putting It Together

Up to this point, polynomial, rational, and radical expressions and functions have been discussed. We learned how to evaluate, simplify, and solve equations involving these types of algebraic expressions. We also learned how to graph linear functions, quadratic functions, and certain types of radical functions.

This chapter introduces two more types of functions, the *exponential* and *logarithmic functions*. As usual, these functions will be evaluated and graphed, and their properties explored. Equations involving exponential or logarithmic expressions will also be solved.

11.1 Composite Functions and Inverse Functions

Objectives

❶ Form the Composite Function

❷ Determine Whether a Function Is One-to-One

❸ Find the Inverse of a Function Defined by a Map or Set of Ordered Pairs

❹ Obtain the Graph of the Inverse Function from the Graph of a Function

❺ Find the Inverse of a Function Defined by an Equation

Are You Prepared for This Section?

Before getting started, complete the following problems. If you get a problem wrong, go back to the section cited and review the material.

P1. Determine the domain: $R(x) = \dfrac{x^2 - 9}{x^2 + 3x - 28}$ [Section 8.3, pp. 544–545]

P2. If $f(x) = 2x^2 - x + 1$, find **(a)** $f(-2)$ **(b)** $f(a + 1)$. [Section 8.3, pp. 542–544]

P3. The graph of a relation is given. Does the relation represent a function? Why or why not?   [Section 8.3, pp. 541–542]

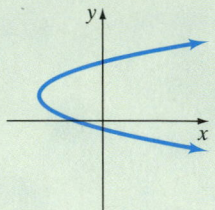

▶ ❶ **Form the Composite Function**

Suppose you decide to buy four shirts at $20 per shirt, and the sales tax on your purchase is 5%. First, you determine the before-tax cost of the shirts to be $20(4) = $80. The after-tax cost of the shirts is $80 + $80(0.05) = $80(1 + 0.05) = 1.05($80) = $84. Thus $C(x) = 20x$ represents the cost C of buying x shirts at $20 per shirt. The after-tax cost S of the x shirts is $1.05(20x) = S(C(x))$.

In general, the total cost can be found as a function of the number of shirts by evaluating $S(C(x))$ and obtaining $S(C(x)) = S(20x) = 1.05(20x) = 21x$. The function $S(C(x))$ is a special type of function called a *composite function*.

Here's another example. Consider the function $y = (x - 3)^4$. If $y = f(u) = u^4$ and $u = g(x) = x - 3$, then, by a substitution process, the original function is: $y = f(u) = f(g(x)) = (x - 3)^4$. This process is called **composition.** To evaluate this function, first evaluate $x - 3$ and then raise the result to the fourth power. Technically, two different functions form the function $y = (x - 3)^4$. Figure 1 illustrates the idea for $x = 5$.

In general, suppose that f and g are two functions and that x is a number in the domain of g. Evaluating g at x yields $g(x)$. If $g(x)$ is in the domain of f, then evaluate f at $g(x)$ and obtain the expression $f(g(x))$. The correspondence from x to $f(g(x))$ is called a *composite function $f \circ g$.*

Figure 1

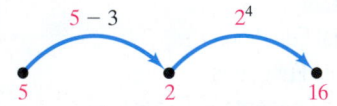

> **In Other Words**
> To find $f(g(x))$, first determine $g(x)$. This is the output of g. Then evaluate f at the output of g. The result is $f(g(x))$.

Definition

Given two functions f and g, the **composite function,** denoted by $f \circ g$ (read as "f composed with g"), is defined by

$$(f \circ g)(x) = f(g(x))$$

The notation $f(g(x))$ is read "f of g of x".

Figure 2 illustrates the definition. Notice that the "inside" function g in $f(g(x))$ is always evaluated first.

Figure 2

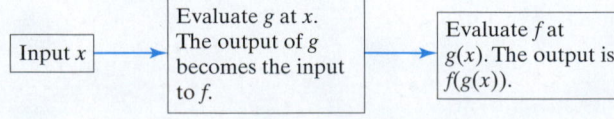

EXAMPLE 1 **Evaluating a Composite Function**

Suppose that $f(x) = x^2 - 3$ and $g(x) = 2x + 1$. Find

(a) $(f \circ g)(3)$ (b) $(g \circ f)(3)$ (c) $(f \circ f)(-2)$

Solution

(a) Using the flowchart in Figure 2, evaluate $(f \circ g)(3)$ as follows:

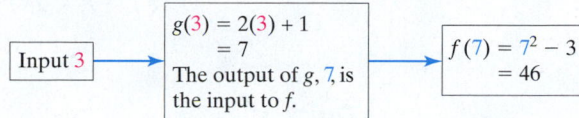

More directly,

$$(f \circ g)(3) = f(g(3)) = f(2(3) + 1) = f(7) = 7^2 - 3 = 49 - 3 = 46$$

(b) $(g \circ f)(3) = g(f(3)) = g(6) = 2(6) + 1 = 13$

$$f(x) = x^2 - 3 \qquad g(x) = 2x + 1$$
$$f(3) = 3^2 - 3$$
$$= 6$$

(c) $(f \circ f)(-2) = f(f(-2)) = f(1) = 1^2 - 3 = -2$

$$f(x) = x^2 - 3 \qquad f(x) = x^2 - 3$$
$$f(-2) = (-2)^2 - 3$$
$$= 1$$

●

> **Quick** ✓
>
> 1. Given two functions f and g, the _____ _____, denoted by $f \circ g$, is defined by $(f \circ g)(x) = f(g(x))$.
> 2. Suppose that $f(x) = 4x - 3$ and $g(x) = x^2 + 1$. Find
>   (a) $(f \circ g)(2)$ (b) $(g \circ f)(2)$ (c) $(f \circ f)(-3)$

Rather than evaluating a composite function at a specific value, a composite function can be written in terms of the independent variable, x.

◉ EXAMPLE 2 **Finding a Composite Function**

Suppose that $f(x) = x^2 + 2x$ and $g(x) = 2x - 1$. Find

(a) $(f \circ g)(x)$ (b) $(g \circ f)(x)$ (c) $(f \circ g)(2)$

Solution

(a)
$$(f \circ g)(x) = f(g(x))$$
$$g(x) = 2x - 1: \quad = f(2x - 1)$$
$$f(x) = x^2 + 2x: \quad = (2x - 1)^2 + 2(2x - 1)$$
FOIL; distribute: $= 4x^2 - 4x + 1 + 4x - 2$
Combine like terms: $= 4x^2 - 1$

Work Smart

$(f \circ g)(x)$ does not mean $(f \cdot g)(x)$.

(continued)

Work Smart

Notice in Example 2(a) and 2(b) that $(f \circ g)(x) \neq (g \circ f)(x)$.

(b)
$$(g \circ f)(x) = g(f(x))$$
$$f(x) = x^2 + 2x: \quad = g(x^2 + 2x)$$
$$g(x) = 2x - 1: \quad = 2(x^2 + 2x) - 1$$
$$\text{Distribute:} \quad = 2x^2 + 4x - 1$$

(c) Instead of finding $(f \circ g)(2)$ using the approach presented in Example 1, find $(f \circ g)(2)$ using the results from part (a).

$$(f \circ g)(2) = f(g(2)) = 4(2)^2 - 1 = 4(4) - 1 = 15 \qquad \bullet$$

Quick ✓

3. *True or False* $(f \circ g)(x) = f(x) \cdot g(x)$
4. If $f(g(x)) = (4x - 3)^5$ and $f(x) = x^5$, what is $g(x)$?
5. Suppose that $f(x) = x^2 - 3x + 1$ and $g(x) = 3x + 2$. Find

    (a) $(f \circ g)(x)$ **(b)** $(g \circ f)(x)$ **(c)** $(f \circ g)(-2)$

▶ ❷ Determine Whether a Function Is One-to-One

Figures 3 and 4 illustrate two different functions represented as mappings. The function in Figure 3 maps states to their population in millions. The function in Figure 4 maps animals to their life expectancy.

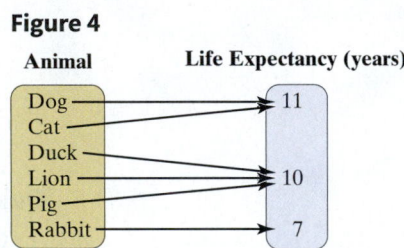

Figure 3

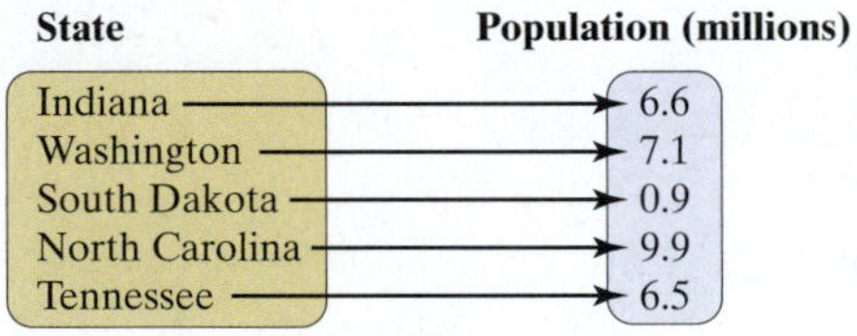

State	Population (millions)
Indiana	6.6
Washington	7.1
South Dakota	0.9
North Carolina	9.9
Tennessee	6.5

Figure 4

Animal	Life Expectancy (years)
Dog	11
Cat	
Duck	
Lion	10
Pig	
Rabbit	7

If you were asked to name the state with a population of 0.9 million based on the function in Figure 3, you would say "South Dakota." If you were asked to name the animal whose life expectancy is 11 years based on the function in Figure 4, you might say "dog" or "cat." What is the difference between the functions in Figures 3 and 4? In Figure 3, one (and only one) element in the domain corresponds to each element in the range. In Figure 4, this is not the case – there is more than one element in the domain that corresponds to an element in the range. For example, "dog" corresponds to "11," but "cat" also corresponds to "11." Functions such as the one in Figure 3 have a special name.

In Other Words

A function is NOT one-to-one if two different inputs correspond to the same output.

Definition

A function is **one-to-one** if any two different inputs in the domain correspond to two different outputs in the range. That is, if x_1 and x_2 are two different inputs to a function f, then $f(x_1) \neq f(x_2)$.

Thus a function is not one-to-one if two different elements in the domain correspond to the same element in the range. The function in Figure 3 is one-to-one because no two different elements in the domain correspond to the same element in the range. The function in Figure 4 is not one-to-one because two different elements in the domain, dog and cat, both correspond to 11.

EXAMPLE 3   **Determining Whether a Function Is One-to-One**

Determine which of the following functions is one-to-one.

(a) For the function to the right, the domain represents the age of five males, and the range represents their HDL (good) cholesterol (mg/dL).

(b) $\{(-2, 6), (-1, 3), (0, 2), (1, 4), (2, 8)\}$

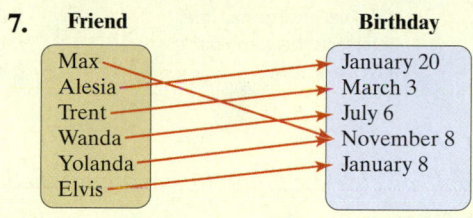

| Age, years | HDL Cholesterol, mg/dL |

Solution

(a) The function is not one-to-one because two different inputs, 55 and 61, correspond to the same output, 38.

(b) The function is one-to-one because no two distinct inputs (the x-coordinates of the ordered pairs) correspond to the same output (the y-coordinates).   ●

Quick ✓

6. A function is _____ if any two different inputs in the domain correspond to two different outputs in the range. That is, if x_1 and x_2 are two different inputs to a function f, then $f(x_1) \neq f(x_2)$.

In Problems 7 and 8, determine whether the function is one-to-one.

7. **Friend**   **Birthday**

Max
Alesia
Trent
Wanda
Yolanda
Elvis

January 20
March 3
July 6
November 8
January 8

8. $\{(-3, 3), (-2, 2), (-1, 1), (0, 0), (1, -1)\}$

▶ Consider the functions $y = 2x - 5$ and $y = x^2 - 4$ shown in Figure 5.

Figure 5

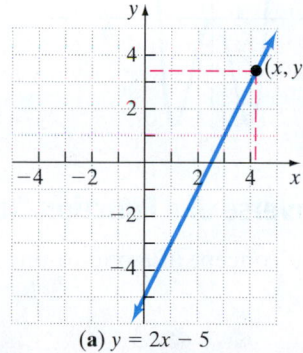

(a) $y = 2x - 5$

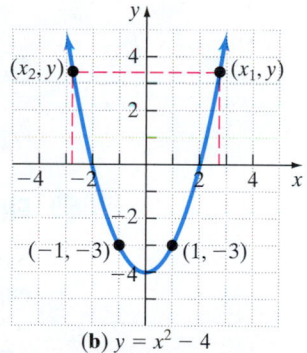

(b) $y = x^2 - 4$

Note, for the function $y = 2x - 5$ shown in Figure 5(a), that any output is the result of only one input. However, for the function $y = x^2 - 4$ shown in Figure 5(b), there are instances where a given output is the result of more than one input. For example, the output -3 is the result of input -1 and of input 1.

If a horizontal line intersects the graph of a function at more than one point, the function cannot be one-to-one because this would mean that two different inputs give the same output. This result is stated formally as the *horizontal line test*.

Horizontal Line Test

If every horizontal line intersects the graph of a function f in at most one point, then f is one-to-one.

EXAMPLE 4

Using the Horizontal Line Test

For each function, use the graph to determine whether the function is one-to-one.

(a) $f(x) = -2x^3 + 1$ **(b)** $g(x) = x^3 - 4x$

Solution

(a) Figure 6(a) illustrates the horizontal line test for $f(x) = -2x^3 + 1$. Because every horizontal line intersects the graph of f exactly once, f must be one-to-one.

(b) Figure 6(b) illustrates the horizontal line test for $g(x) = x^3 - 4x$. The horizontal line $y = 1$ intersects the graph three times, so g is not one-to-one.

Figure 6

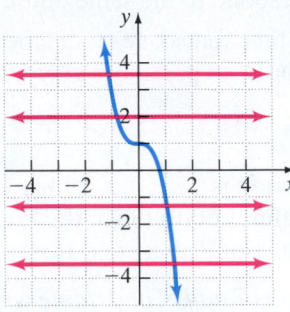

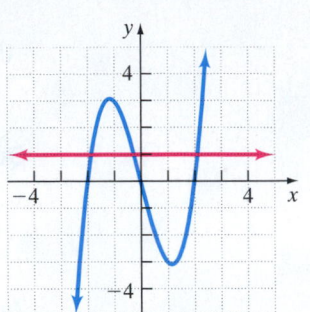

(a) Every horizontal line intersects the graph once; f is one-to-one.

(b) A horizontal line intersects the graph three times; g is not one-to-one.

Work Smart

For a graph NOT to be one-to-one you only need to find one horizontal line that intersects it more than once.

Quick ✓

In Problem 9, use the graph to determine whether the given function is one-to-one.

9. (a) $f(x) = x^4 - 4x^2$ **(b)** $g(x) = x^5 + 4x$

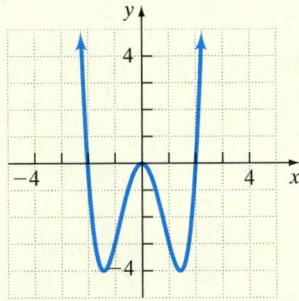

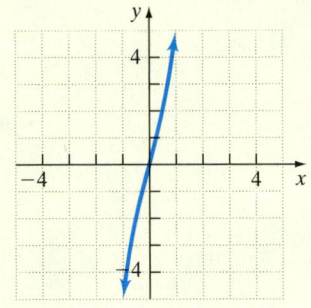

▶ ❸ Find the Inverse of a Function Defined by a Map or Set of Ordered Pairs

Now that the concept of a one-to-one function is understood, *inverse functions* can be explored.

Definition

If f is a one-to-one function with ordered pairs of the form (a, b), then the **inverse function**, denoted f^{-1}, is the set of ordered pairs of the form (b, a).

Work Smart

The symbol f^{-1} represents the inverse function of f. The -1 used in f^{-1} is not an exponent. That is,

$$f^{-1}(x) \neq \frac{1}{f(x)}.$$

In order for the inverse of a function to also be a function, the function must be one-to-one. We begin by finding inverses of functions represented by maps and sets of ordered pairs. To do this, interchange the inputs and outputs. So for each ordered pair (a, b) that is defined in a function f, the ordered pair (b, a) is defined in the inverse function f^{-1}. The inverse undoes what the function does. Suppose a function takes the input 11 and gives the output 3. This can be represented as $(11, 3)$. The inverse would take as input 3 and give the output 11, which can be represented as $(3, 11)$.

EXAMPLE 5   **Finding the Inverse of a Function Defined by a Map**

Find the inverse of the following function. The domain of the function represents the states, and the range represents the states' population in millions. Give the domain and the range of the inverse function.

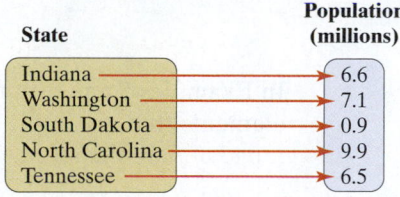

Solution

The elements in the domain represent the inputs to the function. The elements in the range represent the outputs of the function. The function is one-to-one, so the inverse is a function. To find the inverse function, interchange the elements in the domain with the elements in the range. For example, the function receives as input Indiana and outputs 6.6, so the inverse receives as input 6.6 and outputs Indiana. The inverse function is shown below.

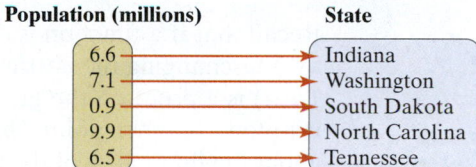

The domain of the inverse function is $\{6.6, 7.1, 0.9, 9.9, 6.5\}$. The range of the inverse function is $\{$Indiana, Washington, South Dakota, North Carolina, Tennessee$\}$.   ●

Quick ✓

10. Find the inverse of the one-to-one function shown below. The domain of the function represents the lengths of the right humerus (in mm), and the range represents the lengths of the right tibia (in mm) of rats sent to space. Give the domain and the range of the inverse function.

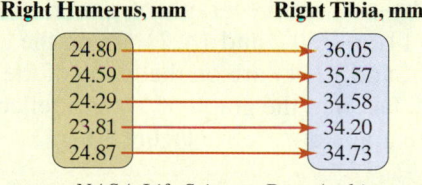

SOURCE: *NASA Life Sciences Data Archive*

EXAMPLE 6   **Finding the Inverse of a Function Defined by a Set of Ordered Pairs**

Find the inverse of the following function. State the domain and the range of the inverse function.

$$\{(-2, 6), (-1, 3), (0, 2), (1, 5), (2, 8)\}$$

Solution

The function is one-to-one, so the inverse is a function. Find the inverse by interchanging the entries in each ordered pair. Thus the inverse function is given by

$$\{(6, -2), (3, -1), (2, 0), (5, 1), (8, 2)\}$$

The domain of the inverse function is $\{6, 3, 2, 5, 8\}$, and the range of the inverse function is $\{-2, -1, 0, 1, 2\}$.   ●

In Examples 5 and 6, notice that the elements that are in the domain of f are the same elements that are in the range of its inverse. In addition, the elements that are in the range of f are the same elements that are in the domain of its inverse.

> **Relation Between the Domain and Range of a Function and Its Inverse**
>
> All elements in the domain of a function are elements in the range of its inverse.
> All elements in the range of a function are elements in the domain of its inverse.

Figure 7

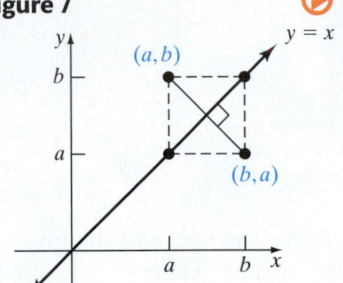

④ Obtain the Graph of the Inverse Function from the Graph of a Function

Recall that if a function is defined by a set of ordered pairs, the inverse can be found by interchanging the entries. Thus if (a, b) is a point on the graph of a function, then (b, a) is a point on the graph of the inverse function. See Figure 7. From the graph, it follows that the point (b, a) on the graph of the inverse function is the reflection about the line $y = x$ of the point (a, b).

The graph of a function f and the graph of its inverse are symmetric with respect to the line $y = x$.

EXAMPLE 7 **Graphing the Inverse Function**

Figure 8(a) shows the graph of a one-to-one function $y = f(x)$. Draw the graph of its inverse.

Solution

Begin by adding the graph of $y = x$ to Figure 8(a). Since the points $(-4, -2)$, $(-3, -1)$, $(-1, 0)$, and $(4, 2)$ are on the graph of f, the points $(-2, -4)$, $(-1, -3)$, $(0, -1)$, and $(2, 4)$ are on the graph of the inverse of f, f^{-1}. Using these points, along with the fact that the graph of f^{-1} is a reflection about the line $y = x$ of the graph of f, draw the graph of f^{-1}. See Figure 8(b).

Figure 8

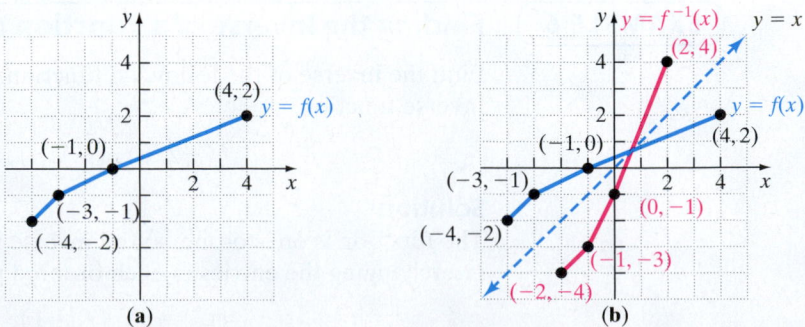

Quick ✓

12. Below is the graph of a one-to-one function $y = f(x)$. Draw the graph of its inverse.

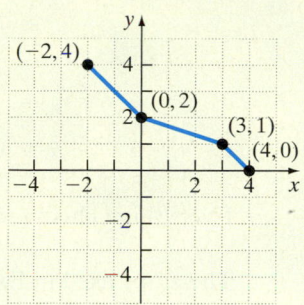

▶ ❺ Find the Inverse of a Function Defined by an Equation

We have learned how to find the inverse of a one-to-one function defined by a map, a set of ordered pairs, or a graph. It is now time to discuss how to find the inverse of a function defined by an equation.

Remember that when a function is defined by an equation, the notation $y = f(x)$ is used. The notation $y = f^{-1}(x)$ can be used to denote the equation whose rule is the inverse function of f.

Because the inverse of a function "undoes" what the original function does, the following relation between a function f and its inverse, f^{-1}, exists.

$$f^{-1}(f(x)) = x \text{ for every } x \text{ in the domain of } f$$
$$f(f^{-1}(x)) = x \text{ for every } x \text{ in the domain of } f^{-1}$$

Let's now find the inverse of a one-to-one function defined by an equation.

Work Smart

If $f(x) = 2x$, then each input is doubled. The inverse function is $f^{-1}(x) = \frac{x}{2}$, so each input is cut in half. If $x = 10$, then $f(10) = 20$. Now substitute the output of f, 20, into f^{-1} to get $f^{-1}(20) = 10$. Notice we are right back where we started!

EXAMPLE 8   **How to Find the Inverse of a One-to-One Function**

Find the inverse of $f(x) = 2x - 3$.

Step-by-Step Solution

First, verify that the function is one-to-one. Because the graph of the function f is a line with y-intercept $(0, -3)$ and slope 2, we know by the horizontal line test that the function is one-to-one and therefore has an inverse function.

Step 1: Replace $f(x)$ with y in the equation for $f(x)$.

$$f(x) = 2x - 3$$
$$y = 2x - 3$$

Step 2: In $y = f(x)$, interchange the variables x and y to obtain $x = f(y)$.

$$x = 2y - 3$$

Step 3: Solve the equation found in Step 2 for y in terms of x.

Add 3 to both sides: $x + 3 = 2y$

Divide both sides by 2: $\dfrac{x + 3}{2} = y$

Step 4: Replace y with $f^{-1}(x)$.

$$f^{-1}(x) = \frac{x + 3}{2}$$

(continued)

Step 5: Check Verify your result by showing that $f^{-1}(f(x)) = x$ and $f(f^{-1}(x)) = x$.

$$f^{-1}(f(x)) = f^{-1}(2x - 3)$$

$$= \frac{2x - 3 + 3}{2}$$

$$= \frac{2x}{2}$$

$$= x$$

$$f(f^{-1}(x)) = f\left(\frac{x + 3}{2}\right)$$

$$= 2\left(\frac{x + 3}{2}\right) - 3$$

$$= x + 3 - 3$$

$$= x$$

Everything checks, so $f^{-1}(x) = \dfrac{x + 3}{2}$. •

The steps used in Example 8 are summarized below.

Finding The Inverse of a One-to-One Function Defined by an Equation

Step 1: Replace $f(x)$ with y in the equation for $f(x)$.

Step 2: In $y = f(x)$, interchange the variables x and y to obtain $x = f(y)$.

Step 3: Solve the equation found in Step 2 for y in terms of x.

Step 4: Replace y with $f^{-1}(x)$.

Step 5: Verify the result by showing that $f^{-1}(f(x)) = x$ and $f(f^{-1}(x)) = x$.

Work Smart

We need to show that *both* $f^{-1}(f(x)) = x$ and $f(f^{-1}(x)) = x$ because function composition is not commutative.

Consider the function $f(x) = 2x - 3$ and its inverse $f^{-1}(x) = \dfrac{x + 3}{2}$ from Example 8. Notice that $f(5) = 2(5) - 3 = 7$. What do you think $f^{-1}(7)$ will equal? Notice that $f^{-1}(7) = \dfrac{7 + 3}{2} = \dfrac{10}{2} = 5$. Thus f^{-1} "undoes" what f did!

Quick ✔

13. *True or False* If f is a one-to-one function so that its inverse is f^{-1}, then $f^{-1}(f(x)) = x$ for every x in the domain of f, and $f(f^{-1}(x)) = x$ for every x in the domain of f^{-1}.

14. *True or False* The notation $f^{-1}(x)$ is equivalent to $\dfrac{1}{f(x)}$.

15. Find the inverse of $g(x) = 5x - 1$.

EXAMPLE 9   **Finding the Inverse of a One-to-One Function**

Find the inverse of $h(x) = x^3 + 4$.

Solution

$$h(x) = x^3 + 4$$

Replace $h(x)$ with y in the equation for $h(x)$: $y = x^3 + 4$

Interchange the variables x and y: $x = y^3 + 4$

Solve the equation for y in terms of x: $x - 4 = y^3$

Take the cube root of both sides: $\sqrt[3]{x - 4} = y$

Replace y with $h^{-1}(x)$: $h^{-1}(x) = \sqrt[3]{x - 4}$

Check Verify your result by showing that $h^{-1}(h(x)) = x$ and $h(h^{-1}(x)) = x$.

$$h^{-1}(h(x)) = h^{-1}(x^3 + 4) \qquad\qquad h(h^{-1}(x)) = h(\sqrt[3]{x - 4})$$
$$= \sqrt[3]{x^3 + 4 - 4} \qquad\qquad\qquad = (\sqrt[3]{x - 4})^3 + 4$$
$$= \sqrt[3]{x^3} \qquad\qquad\qquad\qquad\qquad = x - 4 + 4$$
$$= x \qquad\qquad\qquad\qquad\qquad\qquad = x$$

So $h^{-1}(x) = \sqrt[3]{x - 4}$.

Quick ✓

16. Find the inverse of $f(x) = x^5 + 3$.

11.1 Exercises   MyMathLab®

Exercise numbers in **green** have complete video solutions in MyMathLab or may be accessed using the QR code to the right.

*Problems **1–16** are the **Quick ✓** s that follow the **EXAMPLES**.*

Building Skills

In Problems 17–24, for the given functions f and g, find

(a) $(f \circ g)(3)$ **(b)** $(g \circ f)(-2)$

(c) $(f \circ f)(1)$ **(d)** $(g \circ g)(-4)$

See Objective 1.

17. $f(x) = 2x + 5; g(x) = x - 4$

18. $f(x) = 4x - 3; g(x) = x + 2$

19. $f(x) = x^2 + 4; g(x) = 2x + 3$

20. $f(x) = x^2 - 3; g(x) = 5x + 1$

21. $f(x) = 2x^3; g(x) = -2x^2 + 5$

22. $f(x) = -2x^3; g(x) = x^2 + 1$

23. $f(x) = |x - 10|; g(x) = \dfrac{12}{x + 3}$

24. $f(x) = \sqrt{x + 8}; g(x) = x^2 - 4$

In Problems 25–36, for the given functions f and g, find

(a) $(f \circ g)(x)$ **(b)** $(g \circ f)(x)$

(c) $(f \circ f)(x)$ **(d)** $(g \circ g)(x)$

See Objective 1.

25. $f(x) = x + 1; g(x) = 2x$

26. $f(x) = x - 3; g(x) = 4x$

27. $f(x) = 2x + 7; g(x) = -4x + 5$

28. $f(x) = 3x - 1; g(x) = -2x + 5$

29. $f(x) = x^2; g(x) = x - 3$

30. $f(x) = x^2 + 1; g(x) = x + 1$

31. $f(x) = \sqrt{x}; g(x) = x + 4$

32. $f(x) = \sqrt{x + 2}; g(x) = x - 2$

33. $f(x) = |x + 4|; g(x) = x^2 - 4$

34. $f(x) = |x - 3|; g(x) = x^3 + 3$

35. $f(x) = \dfrac{2}{x + 1}; g(x) = \dfrac{1}{x}$

36. $f(x) = \dfrac{2}{x - 1}; g(x) = \dfrac{4}{x}$

In Problems 37–46, determine which of the following functions is one-to-one. See Objective 2.

37.

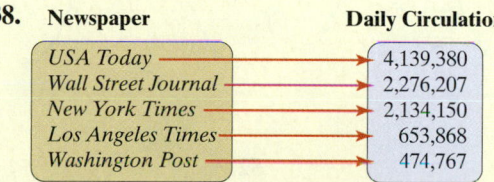

Level of Education	Average Annual Income, 2015
Less than 9th grade	$20,791
9th–12th grade, No diploma	$23,234
High School Graduate	$32,456
Associate's Degree	$42,508
Bachelor's Degree	$62,240

SOURCE: *United States Census Bureau*

38.

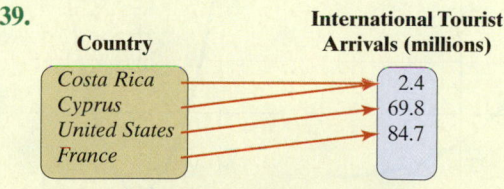

Newspaper	Daily Circulation
USA Today	4,139,380
Wall Street Journal	2,276,207
New York Times	2,134,150
Los Angeles Times	653,868
Washington Post	474,767

SOURCE: *Information Please Almanac*

39.

Country	International Tourist Arrivals (millions)
Costa Rica	2.4
Cyprus	69.8
United States	84.7
France	

SOURCE: *Information Please Almanac*

40.

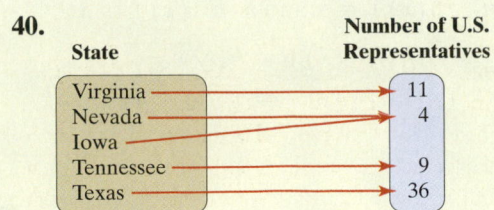

State	Number of U.S. Representatives
Virginia	11
Nevada	4
Iowa	
Tennessee	9
Texas	36

41. $\{(-3, 4), (-2, 6), (-1, 8), (0, 10), (1, 12)\}$

42. $\{(-2, 6), (-1, 3), (0, 0), (1, -3), (2, 6)\}$

43. $\{(-2, 4), (-1, 2), (0, 0), (1, 2), (2, 4)\}$

44. $\{(-2, -8), (-1, -1), (0, 0), (1, 1), (2, 8)\}$

45. $\{(0, -4), (-1, -1), (-2, 0), (1, 1), (2, 4)\}$

46. $\{(-3, 0), (-2, 3), (-1, 0), (0, -3)\}$

In Problems 47–52, use the horizontal line test to determine whether the function whose graph is given is one-to-one. See Objective 2.

47.

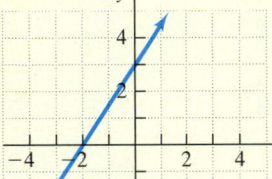

48.

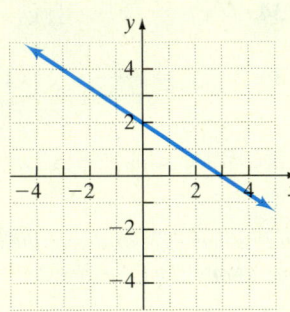

49.

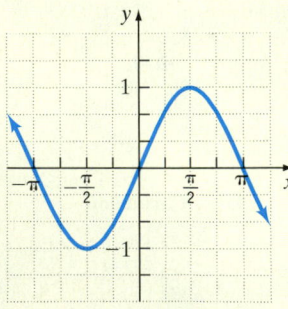

50.

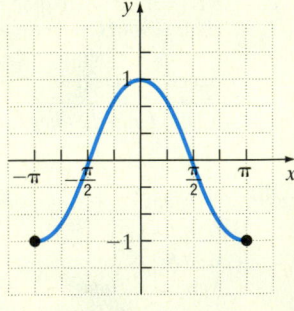

51.

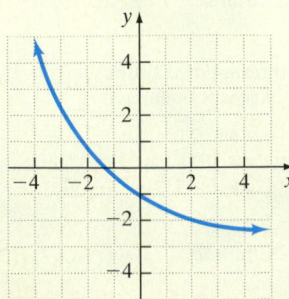

52.

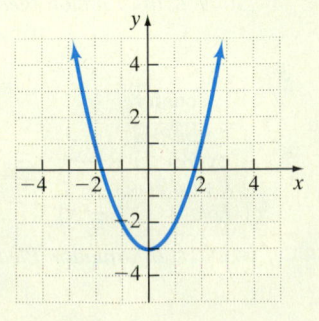

In Problems 53–58, find the inverse of the following one-to-one functions. See Objective 3.

53.

U.S. Coin	Weight (g)
Cent	2.500
Nickel	5.000
Dime	2.268
Quarter	5.670
Half Dollar	11.340
Dollar	8.100

SOURCE: *U.S. Mint website*

54.

Price ($)	Quantity Demanded
2300	152
2000	159
1700	164
1500	171
1300	176

55. $\{(0, 3), (1, 4), (2, 5), (3, 6)\}$

56. $\{(-1, 4), (0, 1), (1, -2), (2, -5)\}$

57. $\{(-2, 3), (-1, 1), (0, -3), (1, 9)\}$

58. $\{(-10, 1), (-5, 4), (0, 3), (5, 2)\}$

In Problems 59–64, the graph of a one-to-one function f is given. Draw the graph of the inverse function f^{-1}. See Objective 4.

59.

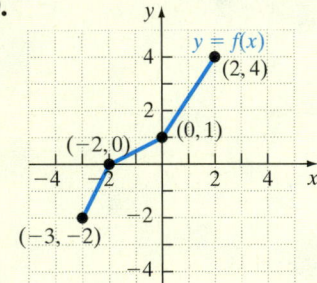

60.

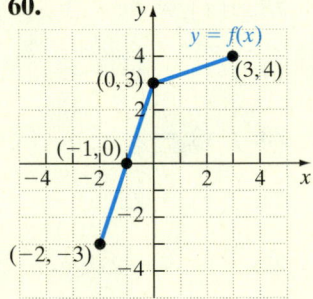

61.

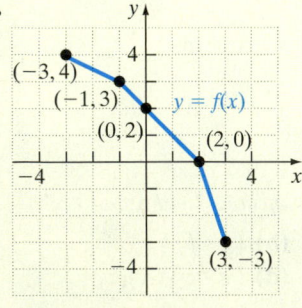

62.

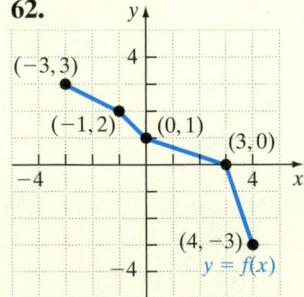

63.

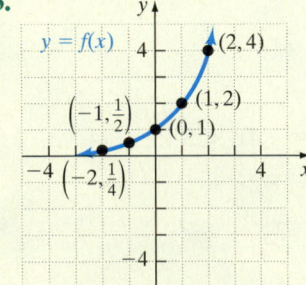

64.

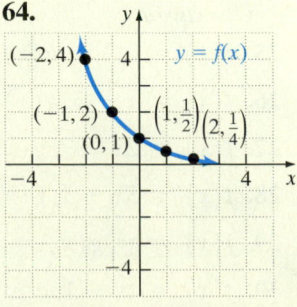

In Problems 65–72, verify that the functions f and g are inverses of each other. See Objective 5.

65. $f(x) = x + 5; g(x) = x - 5$

66. $f(x) = 10x; g(x) = \dfrac{x}{10}$

67. $f(x) = 5x + 7; g(x) = \dfrac{x - 7}{5}$

68. $f(x) = 3x - 5; g(x) = \dfrac{x + 5}{3}$

69. $f(x) = \dfrac{3}{x - 1}; g(x) = \dfrac{3}{x} + 1$

70. $f(x) = \dfrac{2}{x + 4}; g(x) = \dfrac{2}{x} - 4$

71. $f(x) = \sqrt[3]{x + 4}; g(x) = x^3 - 4$

72. $f(x) = \sqrt[3]{2x + 1}; g(x) = \dfrac{x^3 - 1}{2}$

In Problems 73–92, find the inverse function of the given one-to-one function. See Objective 5.

73. $f(x) = 6x$ **74.** $f(x) = 12x$

75. $f(x) = x + 4$ **76.** $g(x) = x + 6$

77. $h(x) = 2x - 7$ **78.** $H(x) = 3x + 8$

79. $G(x) = 2 - 5x$ **80.** $F(x) = 1 - 6x$

81. $g(x) = x^3 + 3$ **82.** $f(x) = x^3 - 2$

83. $p(x) = \dfrac{1}{x + 3}$ **84.** $P(x) = \dfrac{1}{x + 1}$

85. $F(x) = \dfrac{5}{2 - x}$ **86.** $G(x) = \dfrac{2}{3 - x}$

87. $f(x) = \sqrt[3]{x - 2}$ **88.** $f(x) = \sqrt[5]{x + 5}$

89. $R(x) = \dfrac{x}{x + 2}$ **90.** $R(x) = \dfrac{2x}{x + 4}$

91. $f(x) = \sqrt[3]{x - 1} + 4$ **92.** $g(x) = \sqrt[3]{x + 2} - 3$

Applying the Concepts

93. Environmental Disaster An oil tanker hits a rock that rips a hole in the hull of the ship. Oil leaking from the ship forms a circular region around the ship. If the radius r of the circle (in feet) as a function of time t (in hours) is $r(t) = 20t$, express the area A of the circular region contaminated with oil as a function of time. What will be the area of the circular region at 3 hours? (*Hint:* $A(r) = \pi r^2$)

94. Volume of a Balloon The volume V of a hot-air balloon (in cubic meters) as a function of its radius r is given by $V(r) = \dfrac{4}{3}\pi r^3$. If the radius r of the balloon is increasing as a function of time t (in minutes) according to $r(t) = 3\sqrt[3]{t}$, for $t \geq 0$, find the volume of the balloon as a function of time t. What will be the volume of the balloon after 30 minutes?

95. Buying Carpet You want to purchase new carpet for your family room. Carpet is sold by the square yard, but you have measured your family room in square feet. The function $A(x) = \dfrac{x}{9}$ converts the area of a room in square feet to an area in square yards. Suppose that the carpet you have selected is $18 per square yard installed. Then the function $C(A) = 18A$ represents the cost C of installing carpet in a room that measures A square yards.

  (a) Find the cost C as a function of the square footage of the room x.

  (b) If your family room is 15 feet by 21 feet, what will it cost to install the carpet?

96. Tax Time You have a job that pays $20 per hour. Your gross salary G as a function of hours worked h is given by $G(h) = 20h$. Federal tax withholding T on your paycheck is equal to 18% of gross earnings G, so federal tax withholding as a function of gross pay is given by the function $T(G) = 0.18G$.

  (a) Find federal tax withholding T as a function of hours worked h.

  (b) Suppose that you worked 28 hours last week. What will be the federal tax withholding on your paycheck?

97. If $f(4) = 12$ and f is one-to-one, what is $f^{-1}(12)$?

98. If $g(-2) = 7$ and f is one-to-one, what is $g^{-1}(7)$?

99. The domain of a one-to-one function f is $[0, \infty)$, and its range is $[-5, \infty)$. State the domain and the range of f^{-1}.

100. The domain of a one-to-one function f is $[5, \infty)$, and its range is $[0, \infty)$. State the domain and the range of f^{-1}.

101. The domain of a one-to-one function g is $[-4, 10]$, and its range is $(-6, 12)$. State the domain and the range of g^{-1}.

102. The domain of a one-to-one function g is $[0, 15,]$ and its range is $(0, 8)$. State the domain and the range of g^{-1}.

103. Taxes The function $T(x) = 0.15(x - 9275) + 927.50$ represents the tax bill T of a single person whose adjusted gross income is x dollars for income between $9275 and $37,650, inclusive. (SOURCE: *Internal Revenue Service*) Find the inverse function that expresses adjusted gross income x as a function of taxes T. That is, find $x(T)$.

104. Health Costs The annual cost of health insurance H as a function of age a is given by the function $H(a) = 22.8a - 117.5$ for $15 \leq a \leq 90$. (SOURCE: *Statistical Abstract*) Find the inverse function that expresses age a as a function of health insurance cost H. That is, find $a(H)$.

Extending the Concepts

105. If $f(x) = 2x^2 - x + 5$ and $g(x) = x + a$, find a so that the y-intercept of the graph of $(f \circ g)(x)$ is $(0, 20)$.

106. If $f(x) = x^2 - 3x + 1$ and $g(x) = x - a$, find a so that the y-intercept of the graph of $(f \circ g)(x)$ is $(0, -1)$.

Explaining the Concepts

107. Explain what it means for a function to be one-to-one. Why must a function be one-to-one in order for its inverse to be a function?

108. Are all linear functions one-to-one? Explain.

109. Explain why domain of f = range of f^{-1} and range of f = domain of f^{-1}.

110. State the horizontal line test. Why does it work?

Technology Exercises

Many types of technology, such as graphing calculators, have the ability to evaluate composite functions. To obtain the results of Example 1, we would let $Y_1 = f(x) = x^2 - 3$ and $Y_2 = g(x) = 2x + 1$.

Figure 9 shows the results of Example 1 using a graphing calculator.

Figure 9

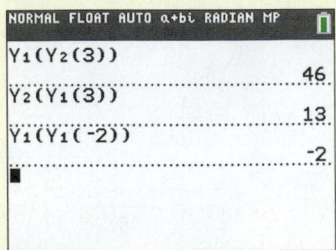

In Problems 111–118, use technology to evaluate the composite functions. Compare your answers with those found in Problems 17–24.

   (a) $(f \circ g)(3)$ **(b)** $(g \circ f)(-2)$
   (c) $(f \circ f)(1)$ **(d)** $(g \circ g)(-4)$

111. $f(x) = 2x + 5; g(x) = x - 4$

112. $f(x) = 4x - 3; g(x) = x + 2$

113. $f(x) = x^2 + 4; g(x) = 2x + 3$

114. $f(x) = x^2 - 3; g(x) = 5x + 1$

115. $f(x) = 2x^3; g(x) = -2x^2 + 5$

116. $f(x) = -2x^3; g(x) = x^2 + 1$

117. $f(x) = |x - 10|; g(x) = \dfrac{12}{x + 3}$

118. $f(x) = \sqrt{x + 8}; g(x) = x^2 - 4$

In Problems 119–122, the functions f and g are inverses. Graph both functions on the same screen, along with the line $y = x$, to see the symmetry of the functions about the line $y = x$.

119. $f(x) = x + 5; g(x) = x - 5$

120. $f(x) = 10x; g(x) = \dfrac{x}{10}$

121. $f(x) = 5x + 7; g(x) = \dfrac{x - 7}{5}$

122. $f(x) = 3x - 5; g(x) = \dfrac{x + 5}{3}$

11.2 Exponential Functions

Objectives

1. Evaluate Exponential Expressions
2. Graph Exponential Functions
3. Define the Number e
4. Solve Exponential Equations
5. Use Exponential Models That Describe Our World

Are You Prepared for This Section?

Before getting started, complete the following problems. If you get a problem wrong, go back to the section cited and review the material.

P1. Evaluate: **(a)** 2^3 **(b)** 2^{-1} **(c)** 3^4 [Section 1.7, pp. 56–57; Section 5.4, pp. 337–340]

P2. Graph: $f(x) = x^2$ [Section 8.4, p. 554]

P3. State the definition of a rational number. [Section 1.3, p. 20]

P4. State the definition of an irrational number. [Section 1.3, p. 21]

P5. Write 3.20349193 as a decimal **(a)** rounded to four decimal places **(b)** truncated to four decimal places. [Section 1.2, pp. 13–14]

P6. Simplify: **(a)** $m^3 \cdot m^5$ **(b)** $\dfrac{a^7}{a^2}$ **(c)** $(z^3)^4$ [Section 5.2, pp. 318–320; Section 5.4, pp. 335–336]

P7. Solve: $x^2 - 5x = 14$ [Section 6.6, pp. 410–414]

Suppose editor Mary Beckwith has just hired you as a proofreader for Pearson Publishing. Mary offers you two options: Option A states that you will be paid $100 for each error you find in the final page proofs of a text. Option B states that you will start with $1 and your payment will double for each error you find in the final page proofs. You know from experience that the final page proofs of a text typically have about 15–20 errors. Which option will you choose?

If there is one error, Option A pays $100, while Option B pays $2. For two errors, Option A pays $2(\$100) = \$200$, while Option B pays $\$2^2 = \$4$. For three errors, Option A pays $3(\$100) = \$300$, while Option B pays $\$2^3 = \$8$. Option A seems to be the way to go. To complete the analysis, set up Table 1, which lists the payment amount as a function of the number of errors in the page proofs. Remember, in Option B, the payment amount doubles each time you find an error.

Prepared?...Answers

P1. (a) 8 **(b)** $\dfrac{1}{2}$ **(c)** 81

P2.
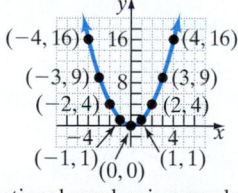

P3. A rational number is a number that can be expressed as a quotient $\dfrac{p}{q}$ of two integers. The integer p is called the numerator, and the integer q, which cannot be 0, is called the denominator. The set of rational numbers are the numbers $\mathbb{Q} = \left\{ x \,\middle|\, x = \dfrac{p}{q}, \text{where } p, q \text{ are integers and } q \neq 0 \right\}$.

P4. An irrational number has a decimal representation that neither repeats nor terminates.

P5. (a) 3.2035 **(b)** 3.2034

P6. (a) m^8 **(b)** a^5 **(c)** z^{12}

P7. $\{-2, 7\}$

Table 1

Number of Errors	Option A Payment	Option B Payment	Number of Errors	Option A Payment	Option B Payment
0	$0	$1	11	$1100	$2048
1	$100	$2	12	$1200	$4096
2	$200	$4	13	$1300	$8192
3	$300	$8	14	$1400	$16,384
4	$400	$16	15	$1500	$32,768
5	$500	$32	16	$1600	$65,536
6	$600	$64	17	$1700	$131,072
7	$700	$128	18	$1800	$262,144
8	$800	$256	19	$1900	$524,288
9	$900	$512	20	$2000	$1,048,576
10	$1000	$1024			

Holy cow! If you find 20 errors, you will get paid over a million dollars! Mary Beckwith better reconsider her offer! If x represents the number of errors, Option A can be expressed as a linear function, $f(x) = 100x$; Option B can be expressed as an *exponential function*, $g(x) = 2^x$.

Definition

An **exponential function** is a function of the form

$$f(x) = a^x$$

where $a \neq 1$ is a positive real number $(a > 0)$. The domain of the exponential function is the set of all real numbers.

The restrictions on the base a will be addressed shortly. The key point is that *the independent variable is in the exponent of the exponential expression.* Contrast this idea with polynomial functions (such as $f(x) = x^2 - 4x$ or $g(x) = 2x^3 + x^2 - 5$), where the independent variable is the base of each exponential expression.

▶ **① Evaluate Exponential Expressions**

In Section 9.2, $a^{\frac{m}{n}}$ was defined, where the base a is a positive real number and the exponent $\frac{m}{n}$ is a rational number.

But what if you want to raise the base a to an irrational number? Although the answer to this question requires advanced mathematics, here is an intuitive explanation.

Suppose you want to find the value of $3^{\sqrt{2}}$. The calculator says that $\sqrt{2} \approx 1.414213562$, so it should seem reasonable that $3^{\sqrt{2}}$ is approximately $3^{1.4}$, where the 1.4 comes from truncating the decimals to the right of the 4 in the tenths position. A better approximation of $3^{\sqrt{2}}$ would be $3^{1.4142}$, where the digits to the right of the ten-thousandths position have been truncated. The more decimals used in approximating $\sqrt{2}$, the better the approximation of $3^{\sqrt{2}}$.

To evaluate expressions of the form a^x using a scientific calculator, enter the base a, press the $\boxed{x^y}$ key, enter the exponent x, and press $\boxed{=}$. To evaluate expressions of the form a^x using a graphing calculator, enter the base a, press the caret $\boxed{\wedge}$ key, enter the exponent x, and press $\boxed{\text{ENTER}}$.

EXAMPLE 1 **Evaluating Exponential Expressions**

Using a calculator, evaluate each of the following expressions. Write as many places as your calculator allows.

(a) $3^{1.4}$ (b) $3^{1.41}$ (c) $3^{1.414}$ (d) $3^{1.4142}$ (e) $3^{\sqrt{2}}$

Solution

(a) $3^{1.4} \approx 4.655536722$ (b) $3^{1.41} \approx 4.706965002$

(c) $3^{1.414} \approx 4.727695035$ (d) $3^{1.4142} \approx 4.72873393$

(e) $3^{\sqrt{2}} \approx 4.728804388$ ●

Quick ✓

1. An exponential function is a function of the form $f(x) = a^x$ where a ___ 0 and a ___ 1.

In Problem 2, use a calculator to evaluate each of the following expressions. Write as many places as your calculator allows.

2. (a) $2^{1.7}$ (b) $2^{1.73}$ (c) $2^{1.732}$ (d) $2^{1.7321}$ (e) $2^{\sqrt{3}}$

Example 1 illustrates that the value of an exponential expression at any real number can be approximated. This is why the domain of any exponential function is the set of all real numbers.

In the definition of an exponential function, $a = 1$ was ruled out, and it was required that a be positive. The base $a = 1$ is excluded because this function is the constant function $f(x) = 1^x = 1$. Exclude bases that are negative to avoid problems for exponents such as $\frac{1}{2}$ or $\frac{3}{4}$. For example, suppose $f(x) = (-2)^x$. In the real number system, $f\left(\frac{1}{2}\right)$ could not be evaluated because $f\left(\frac{1}{2}\right) = (-2)^{\frac{1}{2}} = \sqrt{-2}$, which is not a real number. Finally, exclude a base of 0, because this function is $f(x) = 0^x$, which equals zero for $x > 0$ and is undefined when $x < 0$. When $x = 0$, $f(x) = 0^x$ is *indeterminate* because its value is not precisely determined.

❷ Graph Exponential Functions

▶ Let's learn about properties of exponential functions from their graphs.

EXAMPLE 2 **Graphing an Exponential Function**

Graph the exponential function $f(x) = 2^x$ using point plotting. From the graph, state the domain and the range of the function.

Solution
Begin by locating some points on the graph of $f(x) = 2^x$ as shown in Table 2. Plot the points in Table 2 and connect them in a smooth curve. Figure 10 shows the graph of $f(x) = 2^x$.

Figure 10

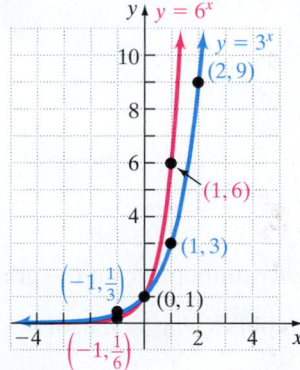

Table 2

x	$f(x) = 2^x$	$(x, f(x))$
-3	$f(-3) = 2^{-3} = \dfrac{1}{2^3} = \dfrac{1}{8}$	$\left(-3, \dfrac{1}{8}\right)$
-2	$f(-2) = 2^{-2} = \dfrac{1}{2^2} = \dfrac{1}{4}$	$\left(-2, \dfrac{1}{4}\right)$
-1	$f(-1) = 2^{-1} = \dfrac{1}{2^1} = \dfrac{1}{2}$	$\left(-1, \dfrac{1}{2}\right)$
0	$f(0) = 2^0 = 1$	$(0, 1)$
1	$f(1) = 2^1 = 2$	$(1, 2)$
2	$f(2) = 2^2 = 4$	$(2, 4)$
3	$f(3) = 2^3 = 8$	$(3, 8)$

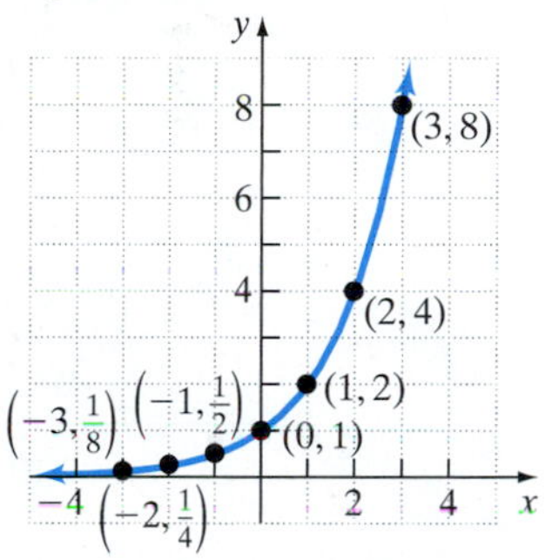

The domain of any exponential function is the set of all real numbers. Notice that no x caused 2^x to be less than or equal to 0. Based on this and the graph, the range of $f(x) = 2^x$ is the set of all positive real numbers $\{y \,|\, y > 0\}$, or $(0, \infty)$ in interval notation. ●

The graph of $f(x) = 2^x$ in Figure 10 is typical of all exponential functions that have a base larger than 1. Figure 11 shows the graphs of two other exponential functions whose bases are larger than 1: $y = 3^x$ and $y = 6^x$. Notice that the larger the base, the steeper the graph is for $x > 0$ and the closer the graph is to the x-axis for $x < 0$.

The information for $f(x) = a^x$, where $a > 1$, is summarized below.

Figure 11

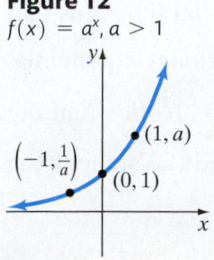

Figure 12
$f(x) = a^x, a > 1$

> **Properties of the Graph of an Exponential Function $f(x) = a^x, a > 1$**
>
> 1. The domain is the set of all real numbers. The range is the set of all positive real numbers.
>
> 2. There are no x-intercepts; the y-intercept is $(0, 1)$.
>
> 3. The graph of f contains the points $\left(-1, \dfrac{1}{a}\right)$, $(0, 1)$, and $(1, a)$.
>
> See Figure 12.

3. Graph the exponential function $f(x) = 4^x$ using point plotting. From the graph, state the domain and the range of the function.

▶ Now consider $f(x) = a^x, 0 < a < 1$.

EXAMPLE 3

Graphing an Exponential Function

Graph the exponential function $f(x) = \left(\dfrac{1}{2}\right)^x$ using point plotting. From the graph, state the domain and the range of the function.

Solution

Begin by locating some points on the graph of $f(x) = \left(\dfrac{1}{2}\right)^x$ as shown in Table 3. Plot the points in Table 3 and connect them in a smooth curve. Figure 13 shows the graph of $f(x) = \left(\dfrac{1}{2}\right)^x$.

Table 3

x	$f(x) = \left(\dfrac{1}{2}\right)^x$	$(x, f(x))$
-3	$f(-3) = \left(\dfrac{1}{2}\right)^{-3} = 2^3 = 8$	$(-3, 8)$
-2	$f(-2) = \left(\dfrac{1}{2}\right)^{-2} = 2^2 = 4$	$(-2, 4)$
-1	$f(-1) = \left(\dfrac{1}{2}\right)^{-1} = 2^1 = 2$	$(-1, 2)$
0	$f(0) = \left(\dfrac{1}{2}\right)^0 = 1$	$(0, 1)$
1	$f(1) = \left(\dfrac{1}{2}\right)^1 = \dfrac{1}{2}$	$\left(1, \dfrac{1}{2}\right)$
2	$f(2) = \left(\dfrac{1}{2}\right)^2 = \dfrac{1}{4}$	$\left(2, \dfrac{1}{4}\right)$
3	$f(3) = \left(\dfrac{1}{2}\right)^3 = \dfrac{1}{8}$	$\left(3, \dfrac{1}{8}\right)$

Figure 13

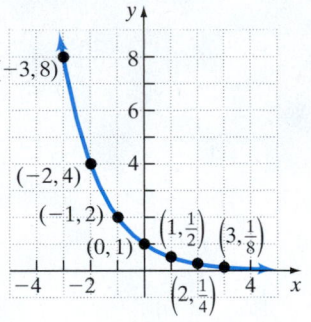

The domain of any exponential function is the set of all real numbers. From the graph, the range of $f(x) = \left(\dfrac{1}{2}\right)^x$ is the set of all positive real numbers $\{y \mid y > 0\}$, or $(0, \infty)$ in interval notation. ●

The graph of $f(x) = \left(\dfrac{1}{2}\right)^x$ in Figure 13 is typical of all exponential functions that have a base between 0 and 1. Figure 14 shows the graphs of two additional exponential functions whose bases are between 0 and 1: $y = \left(\dfrac{1}{3}\right)^x$ and $y = \left(\dfrac{1}{6}\right)^x$. Notice that the smaller the base, the closer the graph is to the x-axis for $x > 0$ and the steeper the graph is for $x < 0$.

Figure 14

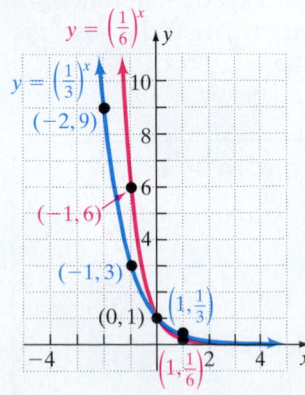

The information about $f(x) = a^x$, where $0 < a < 1$, is now summarized.

Figure 15

$f(x) = a^x, 0 < a < 1$

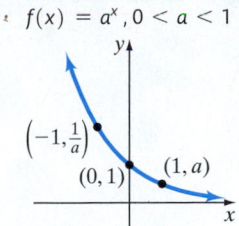

Properties of the Graph of an Exponential Function $f(x) = a^x, 0 < a < 1$

1. The domain is the set of all real numbers. The range is the set of all positive real numbers.

2. There are no x-intercepts; the y-intercept is $(0, 1)$.

3. The graph of f contains the points $\left(-1, \dfrac{1}{a}\right)$, $(0, 1)$, and, $(1, a)$.

   See Figure 15.

Quick ✓

4. The graph of every exponential function $f(x) = a^x$ passes through three points:

   _____, _____, and _____.

5. *True or False* The domain of the exponential function $f(x) = a^x, a > 0, a \neq 1$, is the set of all real numbers.

6. *True or False* The range of the exponential function $f(x) = a^x, a > 0, a \neq 1$, is the set of all real numbers.

7. Graph the exponential function $f(x) = \left(\dfrac{1}{4}\right)^x$ using point plotting. From the graph, state the domain and the range of the function.

EXAMPLE 4 **Graphing an Exponential Function**

Use point plotting to graph $f(x) = 3^{x+1}$. From the graph, state the domain and the range of the function.

Solution

Choose values of x and find the corresponding values of the function. See Table 4. Then plot the ordered pairs and connect them in a smooth curve. See Figure 16.

Table 4

x	$f(x)$	$(x, f(x))$
-3	$f(-3) = 3^{-3+1} = 3^{-2} = \dfrac{1}{3^2} = \dfrac{1}{9}$	$\left(-3, \dfrac{1}{9}\right)$
-2	$f(-2) = 3^{-2+1} = 3^{-1} = \dfrac{1}{3^1} = \dfrac{1}{3}$	$\left(-2, \dfrac{1}{3}\right)$
-1	$f(-1) = 3^{-1+1} = 3^0 = 1$	$(-1, 1)$
0	$f(0) = 3^{0+1} = 3^1 = 3$	$(0, 3)$
1	$f(1) = 3^{1+1} = 3^2 = 9$	$(1, 9)$

Figure 16

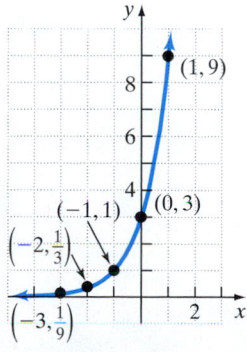

The domain is the set of all real numbers. The range is $\{y \mid y > 0\}$, or $(0, \infty)$ in interval notation.

Quick ✓

In Problems 8 and 9, graph each function using point plotting. From the graph, state the domain and the range of each function.

8. $f(x) = 2^{x-1}$

9. $f(x) = 3^x + 1$

▶ ❸ **Define the Number *e***

Many models use an exponential function whose base is an irrational number symbolized by the letter *e*. The number *e* can be used to model the growth of a stock's price or to estimate the time of death of a carbon-based life form.

> **Definition**
>
> The **number *e*** is defined as the number that the expression
>
> $$\left(1 + \frac{1}{n}\right)^n$$
>
> approaches as *n* increases.

Table 5 shows some values of $\left(1 + \dfrac{1}{n}\right)^n$ as *n* increases. The last number in column 4 is the number *e* correct to nine decimal places.

Table 5

n	$\dfrac{1}{n}$	$1 + \dfrac{1}{n}$	$\left(1 + \dfrac{1}{n}\right)^n$
1	1	2	2
2	0.5	1.5	2.25
5	0.2	1.2	2.48832
10	0.1	1.1	2.59374246
100	0.01	1.01	2.704813829
1,000	0.001	1.001	2.716923932
10,000	0.0001	1.0001	2.718145927
100,000	0.00001	1.00001	2.718268237
1,000,000	0.000001	1.000001	2.718280469
1,000,000,000	10^{-9}	$1 + 10^{-9}$	2.718281827

The exponential function $f(x) = e^x$, whose base is the number *e*, occurs so often in applications that it is sometimes called *the* exponential function. Most calculators have the key $\boxed{e^x}$ or $\boxed{\exp(x)}$, which you can use to evaluate the exponential function $f(x) = e^x$ for a given value of *x*.

Use your calculator to approximate the values of $f(x) = e^x$ for $x = -1, 0,$ and 1 as was done to create Table 6. The graph of the exponential function is shown in Figure 17(a). Because $2 < e < 3$, the graph of $f(x) = e^x$ lies between the graph of $y = 2^x$ and that of $y = 3^x$. See Figure 17(b).

Table 6

x	$f(x) = e^x$
-2	$e^{-2} \approx 0.135$
-1	$e^{-1} \approx 0.368$
0	$e^0 = 1$
1	$e^1 \approx 2.718$
2	$e^2 \approx 7.389$

Figure 17

$f(x) = e^x$

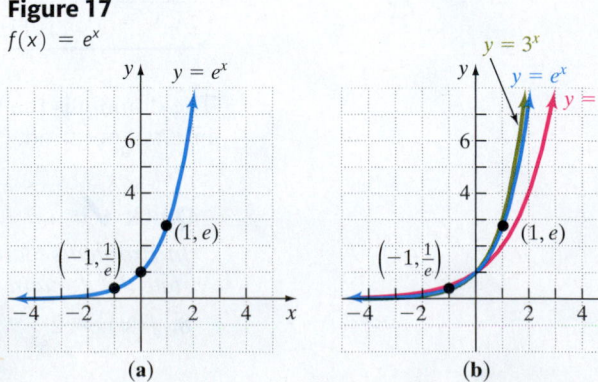

(a)

(b)

Quick ✓

10. What is the value of e rounded to five decimal places?

11. Evaluate each of the following rounded to three decimal places:

    (a) e^4 **(b)** e^{-4}

▶ ❹ Solve Exponential Equations

Equations that involve terms of the form $a^x, a > 0, a \neq 1$, are called **exponential equations.** Some exponential equations can be solved using the Laws of Exponents and the following property.

Work Smart

In order to use the Property for Solving Exponential Equations, both sides of the equation must have the *same* base.

Property For Solving Exponential Equations
$$\text{If } a^u = a^v, \text{ then } u = v.$$

This property results from the fact that exponential functions are one-to-one. In exponential functions, any output is the result of one (and only one) input. That is to say, two different inputs cannot yield the same output. Contrast this with $y = x^2$, which is not one-to-one because two different inputs, -2 and 2, correspond to the same output, 4.

EXAMPLE 5   **How to Solve an Exponential Equation**

Solve: $2^{x-3} = 32$

Step-by-Step Solution

Step 1: Use the Laws of Exponents to write both sides of the equation with the same base.

                        $2^{x-3} = 32$

$32 = 2^5$: $2^{x-3} = 2^5$

Step 2: Set the exponents on each side of the equation equal to each other.

If $a^u = a^v$, then $u = v$: $x - 3 = 5$

Step 3: Solve the equation resulting from Step 2.

Add 3 to both sides: $x - 3 + 3 = 5 + 3$

                          $x = 8$

Step 4: Verify your solution(s).

Work Smart

Did you notice that properties of algebra were used to reduce the exponential equation to a linear equation?

Let $x = 8$:
$$2^{x-3} = 32$$
$$2^{8-3} \stackrel{?}{=} 32$$
$$2^5 \stackrel{?}{=} 32$$
$$32 = 32 \quad \text{True}$$

The solution set is $\{8\}$. ●

Solving Exponential Equations of the Form $a^u = a^v$

Step 1: Use the Laws of Exponents to write both sides of the equation with the same base.

Step 2: Set the exponents on each side of the equation equal to each other.

Step 3: Solve the equation resulting from Step 2.

Step 4: Verify the solution(s).

Quick ✓

In Problems 12 and 13, solve each equation.

12. $5^{x-4} = 5^{-1}$ **13.** $3^{x+2} = 81$

EXAMPLE 6   **Solving Exponential Equations**

Solve the following equations:

(a) $4^{x^2} = 32$ (b) $\dfrac{e^{x^2}}{e^{2x}} = e^8$

Solution

(a)

Rewrite as exponential expressions with a common base:	$4^{x^2} = 32$
	$(2^2)^{x^2} = 2^5$
Use $(a^m)^n = a^{mn}$:	$2^{2x^2} = 2^5$
If $a^u = a^v$, then $u = v$:	$2x^2 = 5$
Divide both sides by 2:	$x^2 = \dfrac{5}{2}$
Take the square root of both sides:	$x = \pm\sqrt{\dfrac{5}{2}}$
Rationalize the denominator:	$x = \pm\dfrac{\sqrt{10}}{2}$

Check

$x = -\dfrac{\sqrt{10}}{2}:$ $4^{\left(-\frac{\sqrt{10}}{2}\right)^2} \overset{?}{=} 32$ $x = \dfrac{\sqrt{10}}{2}:$ $4^{\left(\frac{\sqrt{10}}{2}\right)^2} \overset{?}{=} 32$

$4^{\frac{10}{4}} \overset{?}{=} 32$ $4^{\frac{10}{4}} \overset{?}{=} 32$

$4^{\frac{5}{2}} \overset{?}{=} 32$ $4^{\frac{5}{2}} \overset{?}{=} 32$

$a^{\frac{m}{n}} = (\sqrt[n]{a})^m:$ $(\sqrt{4})^5 \overset{?}{=} 32$ $(\sqrt{4})^5 \overset{?}{=} 32$

$2^5 \overset{?}{=} 32$ $2^5 \overset{?}{=} 32$

$32 = 32$ True $32 = 32$ True

The solution set is $\left\{-\dfrac{\sqrt{10}}{2}, \dfrac{\sqrt{10}}{2}\right\}$.

(b)

	$\dfrac{e^{x^2}}{e^{2x}} = e^8$
Use $\dfrac{a^m}{a^n} = a^{m-n}$:	$e^{x^2 - 2x} = e^8$
If $a^u = a^v$, then $u = v$:	$x^2 - 2x = 8$
Subtract 8 from both sides:	$x^2 - 2x - 8 = 0$
Factor:	$(x - 4)(x + 2) = 0$
Zero-Product Property:	$x = 4$ or $x = -2$

Check

$x = 4:$ $x = -2:$

$\dfrac{e^{4^2}}{e^{2(4)}} \overset{?}{=} e^8$ $\dfrac{e^{(-2)^2}}{e^{2(-2)}} \overset{?}{=} e^8$

$\dfrac{e^{16}}{e^8} \overset{?}{=} e^8$ $\dfrac{e^4}{e^{-4}} \overset{?}{=} e^8$

$e^8 = e^8$ $e^8 = e^8$

The solution set is $\{-2, 4\}$.   ●

Quick ✓

In Problems 14 and 15, solve each equation.

14. $e^{x^2} = e^x \cdot e^{4x}$ **15.** $\dfrac{2^{x^2}}{8} = 2^{2x}$

▶ ❺ Use Exponential Models That Describe Our World

Exponential functions are used in many different disciplines, such as biology (half-life), chemistry (carbon dating), economics (time value of money), and psychology (learning curves). Exponential functions are also used in statistics, as shown in the next example.

EXAMPLE 7 **Exponential Probability**

The manager of a crisis helpline knows that between 3:00 A.M. and 5:00 A.M., 3 calls per hour occur (that's 0.05 call per minute). The following formula from statistics gives the likelihood that a call will occur within t minutes of 3 A.M.

$$F(t) = 1 - e^{-0.05t}$$

Determine the likelihood that a person will call

    (a) within 5 minutes of 3:00 A.M.

    (b) within 20 minutes of 3:00 A.M.

Solution

    (a) The likelihood that a call will occur within 5 minutes of 3:00 A.M. is found by evaluating the function $F(t) = 1 - e^{-0.05t}$ at $t = 5$.

$$F(5) = 1 - e^{-0.05(5)}$$
$$\approx 0.221$$

    The likelihood that a call will occur in this time span is $0.221 = 22.1\%$.

    (b) The likelihood that a call will occur within 20 minutes of 3:00 A.M. is found by evaluating the function $F(t) = 1 - e^{-0.05t}$ at $t = 20$, or $F(20)$.

$$F(20) = 1 - e^{-0.05(20)}$$
$$\approx 0.632$$

    The likelihood that a call will occur in this time span is $0.632 = 63.2\%$.   ●

Quick ✓

16. A bank manager knows that between 3:00 P.M. and 5:00 P.M., 15 people arrive per hour (that's 0.25 people per minute). The following formula from statistics gives the likelihood that a person will arrive within t minutes of 3:00 P.M.

$$F(t) = 1 - e^{-0.25t}$$

Find the likelihood that a person will arrive
(a) within 10 minutes of 3 P.M.   **(b)** within 25 minutes of 3 P.M.

EXAMPLE 8 **Radioactive Decay**

The radioactive **half-life** for a given radioisotope of an element is the time it takes for half the radioactive nuclei in any sample to decay to some other substance. For example, the half-life of plutonium-239 is 24,360 years. Plutonium-239 is particularly dangerous because it emits alpha particles that are absorbed into bone marrow. The maximum amount of plutonium-239 an adult can handle without significant injury is 0.13 microgram (= 0.000000013 gram). Suppose a researcher has a 1-gram sample of plutonium-239. The amount A (in grams) of plutonium-239 after t years is given by

$$A(t) = 1\left(\frac{1}{2}\right)^{\frac{t}{24,360}}$$

How much plutonium-239 is left in the sample after

    (a) 500 years?   **(b)** 24,360 years?   **(c)** 73,080 years?

                                                                  (continued)

Solution

(a) The amount of plutonium-239 left in the sample after 500 years is found by evaluating A at $t = 500$. That is, determine $A(500)$.

$$A(500) = 1\left(\frac{1}{2}\right)^{\frac{500}{24,360}}$$

Use a calculator: ≈ 0.986 gram

After 500 years, approximately 0.986 gram of plutonium-239 will be left in the sample.

(b) The amount of plutonium-239 remaining after 24,360 years is found by evaluating A at $t = 24,360$. That is, determine $A(24,360)$.

$$A(24,360) = 1\left(\frac{1}{2}\right)^{\frac{24,360}{24,360}}$$

$$= 1\left(\frac{1}{2}\right)^{1}$$

$$= 0.5 \text{ gram}$$

After 24,360 years, there will be 0.5 gram of plutonium-239 left.

(c) To find the amount left after 73,080 years, evaluate $A(73,080)$.

$$A(73,080) = 1\left(\frac{1}{2}\right)^{\frac{73,080}{24,360}}$$

$$= 1\left(\frac{1}{2}\right)^{3}$$

$$= \frac{1}{8} \text{ gram}$$

After 73,080 years, there will be $\frac{1}{8} = 0.125$ gram of plutonium-239 left in the sample.

Quick ✔

17. The half-life of thorium-227 is 18.72 days. Suppose that a researcher possesses a 10-gram sample of thorium-227. The amount A (in grams) of thorium-227 left after t days is given by

$$A(t) = 10 \cdot \left(\frac{1}{2}\right)^{\frac{t}{18.72}}$$

How much thorium-227 is left in the sample after

(a) 10 days?   (b) 18.72 days?
(c) 74.88 days?   (d) 100 days?

When money is deposited in a bank, the bank pays interest on the balance in the account. When solving interest problems, we use the term **payment period** as shown in Table 7 to indicate how often the bank pays interest.

Table 7

Payment Period	Number of Times Interest Is Paid
Annually	Once per year
Semiannually	Twice per year
Quarterly	4 times per year
Monthly	12 times per year
Daily	360 times per year

When the interest due at the end of a payment period is added to the principal so that the interest computed at the end of the *next* payment period is based on this new principal amount (old principal + interest), the interest has been **compounded.** **Compound interest** is interest paid on the original principal and all previously earned interest.

The following formula can be used to determine the value of an account after a certain period of time.

Work Smart

When using the compound interest formula, be sure to express the interest rate as a decimal.

> **Compound Interest Formula**
>
> The amount A after t years resulting from a principal P invested at an annual interest rate r compounded n times per year is
>
> $$A = P\left(1 + \frac{r}{n}\right)^{nt}$$

For example, if you deposit $500 into an account paying 3% annual interest compounded monthly, then $P = \$500$, $r = 0.03$, and $n = 12$ (twelve compounding periods per year).

EXAMPLE 9   **Future Value of Money**

Suppose you deposit $3000 into a Roth IRA today. If the deposit earns 8% interest compounded quarterly, determine its future value A after

(a) 1 year   **(b)** 10 years   **(c)** 35 years, when you plan on retiring

Solution

Use the compound interest formula with $P = \$3000$, $r = 0.08$, and $n = 4$ (for quarterly compounding), so

$$A = \$3000\left(1 + \frac{0.08}{4}\right)^{4t}$$
$$= \$3000\left(1 + 0.02\right)^{4t}$$
$$= \$3000\left(1.02\right)^{4t}$$

(a) The value of the account after $t = 1$ year is

$$A = \$3000\left(1.02\right)^{4(1)}$$
$$= \$3000\left(1.02\right)^{4}$$

Use a calculator; see Figure 18: $= \$3000\left(1.08243216\right)$
$$= \$3247.30$$

Figure 18

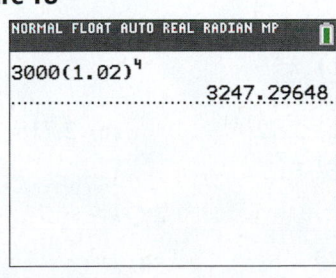

(b) The value after $t = 10$ years is

$$A = \$3000\left(1.02\right)^{4(10)}$$
$$= \$3000\left(1.02\right)^{40}$$

Use a calculator: $= \$3000\left(2.208039664\right)$
$$= \$6624.12$$

(c) The value after $t = 35$ years is

$$A = \$3000\left(1.02\right)^{4(35)}$$
$$= \$3000\left(1.02\right)^{140}$$

Use a calculator: $= \$3000\left(15.99646598\right)$
$$= \$47,989.40$$

Quick ✔

18. Suppose that you deposit $2000 into a Roth IRA today. Determine the future value A of the deposit if it earns 5% interest compounded monthly (twelve times per year) after

(a) 1 year (b) 15 years (c) 30 years, when you plan on retiring

11.2 Exercises MyMathLab®

Exercise numbers in green have complete video solutions in MyMathLab or may be accessed using the QR code to the right.

*Problems **1–18** are the **Quick ✔** s that follow the **EXAMPLES**.*

Building Skills

In Problems 19–22, approximate each number using a calculator. Express your answer rounded to three decimal places. See Objective 1.

19. (a) $3^{2.2}$ (b) $3^{2.23}$ (c) $3^{2.236}$
(d) $3^{2.2361}$ (e) $3^{\sqrt{5}}$

20. (a) $5^{1.4}$ (b) $5^{1.41}$ (c) $5^{1.414}$
(d) $5^{1.4142}$ (e) $5^{\sqrt{2}}$

21. (a) $4^{3.1}$ (b) $4^{3.14}$ (c) $4^{3.142}$
(d) $4^{3.1416}$ (e) 4^{π}

22. (a) $10^{2.7}$ (b) $10^{2.72}$ (c) $10^{2.718}$
(d) $10^{2.7183}$ (e) 10^{e}

In Problems 23–30, the graph of an exponential function is given. Match each graph to one of the following functions. It may prove useful to create a table of values for each function to help you identify the correct graph. See Objective 2.

(a) $f(x) = 2^{x}$ (b) $f(x) = 2^{-x}$ (c) $f(x) = 2^{x+1}$
(d) $f(x) = 2^{x-1}$ (e) $f(x) = -2^{x}$ (f) $f(x) = 2^{x} + 1$
(g) $f(x) = 2^{x} - 1$ (h) $f(x) = -2^{-x}$

23.

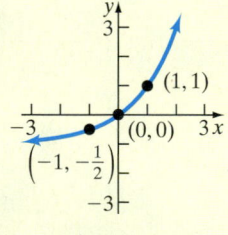

24.

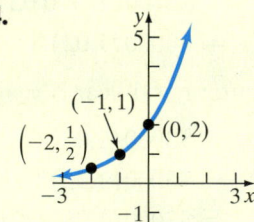

25.

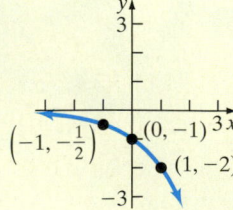

26.

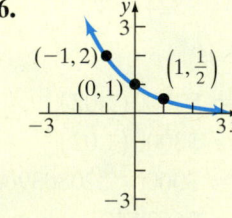

27.

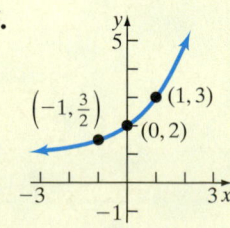

28.

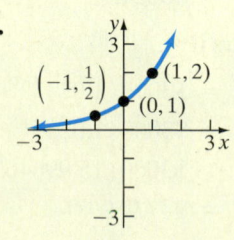

29.

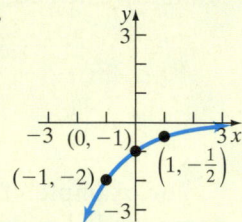

30.

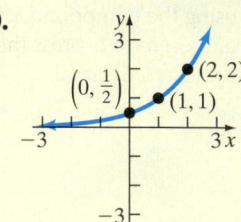

In Problems 31–42, graph each function. State the domain and the range of the function. See Objective 2.

31. $f(x) = 5^{x}$ **32.** $f(x) = 7^{x}$

33. $F(x) = \left(\dfrac{1}{5}\right)^{x}$ **34.** $F(x) = \left(\dfrac{1}{7}\right)^{x}$

35. $h(x) = 2^{x+2}$ **36.** $H(x) = 2^{x-2}$

37. $f(x) = 2^{x} + 3$ **38.** $F(x) = 2^{x} - 3$

39. $F(x) = \left(\dfrac{1}{2}\right)^{x} - 1$ **40.** $G(x) = \left(\dfrac{1}{2}\right)^{x} + 2$

41. $P(x) = \left(\dfrac{1}{3}\right)^{x-2}$ **42.** $p(x) = \left(\dfrac{1}{3}\right)^{x+2}$

In Problems 43–50, approximate each number using a calculator. Express your answer rounded to three decimal places. See Objective 3.

43. (a) $3.1^{2.7}$ (b) $3.14^{2.72}$ (c) $3.142^{2.718}$
(d) $3.1416^{2.7183}$ (e) π^{e}

44. (a) $2.7^{3.1}$ (b) $2.72^{3.14}$ (c) $2.718^{3.142}$
(d) $2.7183^{3.1416}$ (e) e^{π}

45. e^{2} **46.** e^{3} **47.** e^{-2}

48. e^{-3} **49.** $e^{2.3}$ **50.** $e^{1.5}$

In Problems 51–54, graph each function. State the domain and the range of the function. See Objective 3.

51. $g(x) = e^{x-1}$ **52.** $f(x) = e^{x} - 1$

53. $f(x) = -2e^{x}$ **54.** $F(x) = \dfrac{1}{2} e^{x}$

In Problems 55–80, solve each equation. See Objective 4.

55. $2^x = 2^5$ **56.** $3^x = 3^{-2}$

57. $3^{-x} = 81$ **58.** $4^{-x} = 64$

59. $\left(\dfrac{1}{2}\right)^x = \dfrac{1}{32}$ **60.** $\left(\dfrac{1}{3}\right)^x = \dfrac{1}{243}$

61. $5^{x-2} = 125$ **62.** $2^{x+3} = 128$

63. $4^x = 8$ **64.** $9^x = 27$

65. $2^{-x+5} = 16^x$ **66.** $3^{-x+4} = 27^x$

67. $3^{x^2-4} = 27^x$ **68.** $5^{x^2-10} = 125^x$

69. $4^x \cdot 2^{x^2} = 16^2$ **70.** $9^{2x} \cdot 27^{x^2} = 3^{-1}$

71. $2^x \cdot 8 = 4^{x-3}$ **72.** $3^x \cdot 9 = 27^x$

73. $\left(\dfrac{1}{5}\right)^x - 25 = 0$ **74.** $\left(\dfrac{1}{6}\right)^x - 36 = 0$

75. $(2^x)^x = 16$ **76.** $(3^x)^x = 81$

77. $e^x = e^{3x+4}$ **78.** $e^{3x} = e^2$

79. $(e^x)^2 = e^{3x-2}$ **80.** $(e^3)^x = e^2 \cdot e^x$

Mixed Practice

81. Suppose that $f(x) = 2^x$.
  (a) What is $f(3)$? What point is on the graph of f?
  (b) If $f(x) = \dfrac{1}{8}$, what is x? What point is one the graph of f?

82. Suppose that $f(x) = 3^x$.
  (a) What is $f(2)$? What point is on the graph of f?
  (b) If $f(x) = \dfrac{1}{81}$, what is x? What point is on the graph of f?

83. Suppose that $g(x) = 4^x - 1$.
  (a) What is $g(-1)$? What point is on the graph of g?
  (b) If $g(x) = 15$, what is x? What point is on the graph of g?

84. Suppose that $g(x) = 5^x + 1$.
  (a) What is $g(-1)$? What point is on the graph of g?
  (b) If $g(x) = 126$, what is x? What point is on the graph of g?

85. Suppose that $H(x) = 3\left(\dfrac{1}{2}\right)^x$.
  (a) What is $H(-3)$? What point is on the graph of H?
  (b) If $H(x) = \dfrac{3}{4}$, what is x? What point is on the graph of H?

86. Suppose that $F(x) = -2\left(\dfrac{1}{3}\right)^x$.
  (a) What is $F(-1)$? What point is on the graph of F?
  (b) If $F(x) = -18$, what is x? What point is on the graph of F?

Applying the Concepts

87. A Population Model According to the U.S. Census Bureau, the population of the United States in 2016 was 326 million people. In addition, the population of the United States was growing at a rate of 0.99% per year. Assuming that this growth rate continues, the model $P(t) = 326(1.0099)^{t-2016}$ represents the population P (in millions of people) in year t.
  (a) According to this model, what will be the population of the United States in 2021?
  (b) According to this model, what will be the population of the United States in 2050?
  (c) The United States Census Bureau predicts that the United States population will be 439 million in 2050. Compare this estimate to the one obtained in part (b). What might account for any differences?

88. A Population Model According to the U.S. Census Bureau, the population of the world in 2016 was 7563 million people. In addition, the population of the world was growing at a rate of 1.02% per year. Assuming that this growth rate continues, the model $P(t) = 7563(1.0102)^{t-2016}$ represents the population P (in millions of people) in year t.
  (a) According to this model, what will be the population of the world in 2021?
  (b) According to this model, what will be the population of the world in 2025?
  (c) The United States Census Bureau predicts that the world population will be 8000 million (8 billion) in 2025. Compare this estimate to the one obtained in part (b). What might account for any differences?

89. Time Is Money Suppose that you deposit $5000 into a certificate of deposit (CD) today. Determine the future value A of the deposit if it earns 2% interest compounded monthly after
  (a) 1 year.
  (b) 3 years.
  (c) 5 years, when the CD comes due.

90. Time Is Money Suppose that you deposit $8000 into a certificate of deposit (CD) today. Determine the future value A of the deposit if it earns 2% interest compounded quarterly (4 times per year) after

(a) 1 year.   (b) 3 years.
(c) 5 years, when the CD comes due.

91. Do the Compounding Periods Matter? Suppose that you deposit $2000 into an account that pays 3% annual interest. How much will you have after 5 years if interest is compounded

(a) annually?   (b) quarterly?
(c) monthly?   (d) daily?
(e) Based on the results of parts (a)–(d), what impact does increasing the number of compounding periods have on the future value, all other things equal?

92. Do the Compounding Periods Matter? Suppose that you deposit $1000 into an account that pays 6% annual interest. How much will you have after 3 years if interest is compounded

(a) annually?   (b) quarterly?
(c) monthly?   (d) daily?
(e) Based on the results of parts (a)–(d), what impact does increasing the number of compounding periods have on the future value, all other things equal?

93. Depreciation Based on data obtained from the *Kelley Blue Book,* the value V of a Ford Focus that is t years old can be modeled by $V(t) = 19{,}841(0.88)^t$.

(a) According to the model, what is the value of a brand-new Focus?
(b) According to the model, what is the value of a 2-year-old Focus?
(c) According to the model, what is the value of a 5-year-old Focus?

94. Depreciation Based on data obtained from the *Kelley Blue Book*, the value V of a Chevy Malibu that is t years old can be modeled by $V(t) = 25{,}258(0.84)^t$.

(a) According to the model, what is the value of a brand-new Chevy Malibu?
(b) According to the model, what is the value of a 2-year-old Chevy Malibu?
(c) According to the model, what is the value of a 5-year-old Chevy Malibu?

95. Radioactive Decay The half-life of beryllium-11 is 13.81 seconds. Suppose that a researcher possesses a 100-gram sample of beryllium-11. The amount A (in grams) of beryllium-11 after t seconds is given by

$$A(t) = 100\left(\frac{1}{2}\right)^{\frac{t}{13.81}}$$

(a) How much beryllium-11 is left in the sample after 1 second?
(b) How much beryllium-11 is left in the sample after 13.81 seconds?
(c) How much beryllium-11 is left in the sample after 27.62 seconds?
(d) How much beryllium-11 is left in the sample after 100 seconds?

96. Radioactive Decay The half-life of carbon-10 is 19.255 seconds. Suppose that a researcher possesses a 100-gram sample of carbon-10. The amount A (in grams) of carbon-10 after t seconds is given by

$$A(t) = 100\left(\frac{1}{2}\right)^{\frac{t}{19.255}}$$

(a) How much carbon-10 is left in the sample after 1 second?
(b) How much carbon-10 is left in the sample after 19.255 seconds?
(c) How much carbon-10 is left in the sample after 38.51 seconds?
(d) How much carbon-10 is left in the sample after 100 seconds?

For Exercises 97 and 98, use Newton's Law of Cooling, which states that the temperature of a heated object decreases exponentially over time toward the temperature of the surrounding medium.

97. Newton's Law of Cooling Suppose a pizza is removed from a 400°F oven and placed in a room whose temperature is 70°F. The temperature u (in °F) of the pizza at time t (in minutes) can be modeled by $u(t) = 70 + 330e^{-0.072t}$.

(a) According to the model, what will be the temperature of the pizza in 5 minutes?

(b) According to the model, what will be the temperature of the pizza in 10 minutes?

(c) If the pizza can be safely consumed when its temperature is 200°F, will it be ready to eat after cooling for 13 minutes?

98. Newton's Law of Cooling Suppose coffee at a temperature of 170°F is poured into a coffee mug and allowed to cool in a room whose temperature is 70°F. The temperature u (in °F) of the coffee at time t (in minutes) can be modeled by $u(t) = 70 + 100e^{-0.045t}$.

(a) According to the model, what will be the temperature of the coffee in 5 minutes?

(b) According to the model, what will be the temperature of the coffee in 10 minutes?

(c) If the coffee doesn't taste good once its temperature reaches 120°F, will it still be tasty after cooling for 20 minutes?

99. Learning Curve Suppose that a student has 200 vocabulary words to learn. If a student learns 20 words in 30 minutes, the function

$$L(t) = 200(1 - e^{-0.0035t})$$

models the number of words L that the student will learn in t minutes.

(a) How many words will the student learn in 45 minutes?

(b) How many words will the student learn in 60 minutes?

100. Learning Curve Suppose that a student has 50 biology terms to learn. If a student learns 10 terms in 30 minutes, the function

$$L(t) = 50(1 - e^{-0.0223t})$$

models the number of terms L that the student will learn in t minutes.

(a) How many words will the student learn in 45 minutes?

(b) How many words will the student learn in 60 minutes?

101. Current in an RL Circuit The equation governing the amount of current I (in amperes) after time t (in seconds) in a single RL circuit consisting of a resistance R (in ohms), an inductance L (in henrys), and an electromotive force E (in volts) is

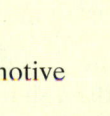

$$I = \frac{E}{R}\left[1 - e^{-(R/L)t}\right]$$

(a) If $E = 120$ volts, $R = 10$ ohms, and $L = 25$ henrys, how much current I is flowing in 0.05 second?

(b) If $E = 240$ volts, $R = 10$ ohms, and $L = 25$ henrys, how much current I is flowing in 0.05 second?

102. Current in an RC Circuit The equation governing the amount of current I (in amperes) after time t (in microseconds) in a single RC circuit consisting of a resistance R (in ohms), a capacitance C (in microfarads), and an electromotive force E (in volts) is

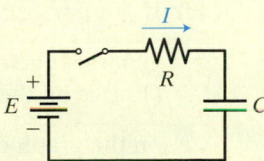

$$I = \frac{E}{R}e^{-t/(RC)}$$

(a) If $E = 120$ volts, $R = 2500$ ohms, and $C = 100$ microfarads, how much current I is flowing initially $(t = 0)$? In 50 microseconds?

(b) If $E = 240$ volts, $R = 2500$ ohms, and $C = 100$ microfarads, how much current I is flowing initially $(t = 0)$? In 50 microseconds?

Extending the Concepts

In Problems 103 and 104, find the exponential function whose graph is given.

103.   **104.**

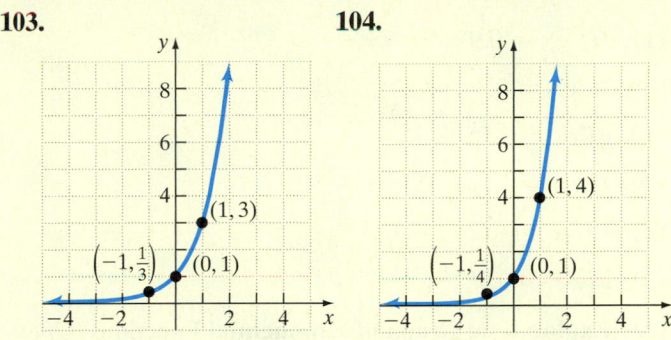

Explaining the Concepts

105. As the base a of an exponential function $f(x) = a^x$ increases (for $a > 1$), what happens to the graph of the exponential function for $x > 0$? What happens to the behavior of the graph for $x < 0$?

106. The graphs of $f(x) = 2^{-x}$ and $g(x) = \left(\frac{1}{2}\right)^x$ are identical. Why?

107. Explain the difference between exponential functions and polynomial functions.

108. Can we solve the equation $2^x = 12$ using the fact that if $a^u = a^v$, then $u = v$? Why or why not?

Synthesis Review

In Problems 109–114, evaluate each expression.

109. Evaluate $x^2 - 5x + 1$ at

(a) $x = -2$ **(b)** $x = 3$

110. Evaluate $x^3 - 5x + 2$ at

(a) $x = -4$ **(b)** $x = 2$

111. Evaluate $\dfrac{4}{x + 2}$ at

(a) $x = -2$ **(b)** $x = 3$

112. Evaluate $\dfrac{2x}{x + 1}$ at

(a) $x = -4$ **(b)** $x = 5$

113. Evaluate $\sqrt{2x+5}$ at

(a) $x = 2$ (b) $x = 11$

114. Evaluate $\sqrt[3]{3x-1}$ at

(a) $x = 3$ (b) $x = 0$

Technology Exercises

In Problems 115–122, graph each function using technology. State the domain and the range of each function.

115. $f(x) = 1.5^x$

116. $G(x) = 3.1^x$

117. $H(x) = 0.9^x$

118. $F(x) = 0.3^x$

119. $g(x) = 2.5^x + 3$

120. $f(x) = 1.7^x - 2$

121. $F(x) = 1.6^{x-3}$

122. $g(x) = 0.3^{x+2}$

123. Exploration: Graph of Exponential Functions *Open the "Graph of Exponential Functions" Geogebra applet.* The applet may be found using the QR code at the beginning of this section or through the Multimedia Library in MyMathLab. In the upper left corner is one "slider" that represents the value for a in the exponential function $f(x) = a^x$. To start, make sure the box is unchecked. If you are using a keyboard, it may be easier to use the arrow keys, instead of the mouse, to move the slider.

(a) Move the a slider to 2. This will produce the graph of $f(x) = 2^x$. State the domain, the range, the intercepts, and the values of the three coordinate points that are labeled on the graph.

(b) Move the a slider to 3. This will produce the graph of $f(x) = 3^x$. State the domain, the range, the intercepts, and the values of the three coordinate points that are labeled on the graph.

(c) Move the a slider to 4. This will produce the graph of $f(x) = 4^x$. State the domain, the range, the intercepts, and the values of the three coordinate points that are labeled on the graph.

(d) Use your results from parts (a) through (c) to generalize the following for *all* exponential functions of the form $f(x) = a^x$:
   i. The domain is _____.
   ii. The range is _____.
   iii. How many x-intercepts are there? What is the y-intercept?
   iv. Conclude the graph of $f(x) = a^x$ contains the points $(-1, \_\_)$ $(0, \_)$, and $(1, \_)$.

124. Exploration: Graph of Exponential Functions *Open the "Graph of Exponential Functions" Geogebra applet.* The applet may be found using the QR code at the beginning of this section or through the Multimedia Library in MyMathLab. In the upper left corner is one "slider" that stands for the value of a in the exponential functions $f(x) = a^x$ and $g(x) = \left(\dfrac{1}{a}\right)^x$. To start, make sure the box for $g(x)$ is checked. If you are using a keyboard, it may be easier to use the arrow keys, instead of the mouse, to move the slider.

(a) Move the a slider to 2. This will produce the graphs of $f(x) = 2^x$ and $g(x) = \left(\dfrac{1}{2}\right)^x$. Notice that the point $(1, 2)$ is on the graph of f and the point $(\_\_, 2)$ is on the graph of g. Also, the point $\left(-1, \dfrac{1}{2}\right)$ is on the graph of f and the point $\left(\_, \dfrac{1}{2}\right)$ is on the graph of g. Both graphs contain the point _____.

(b) Move the a slider to 3. This will produce the graphs of $f(x) = 3^x$ and $g(x) = \left(\dfrac{1}{3}\right)^x$. Notice that the point $(1, 3)$ is on the graph of f and the point $(\_\_, 3)$ is on the graph of g. Also, the point $\left(-1, \dfrac{1}{3}\right)$ is on the graph of f and the point $\left(\_, \dfrac{1}{3}\right)$ is on the graph of g. Both graphs contain the point _____.

(c) If $f(x) = a^x$ and $g(x) = \left(\dfrac{1}{a}\right)^x$, where $a > 1$, in general conclude the following: If the point $(1, a)$ is on the graph of f, the point $(\_\_, a)$ is on the graph of g; if the point $\left(\_, \dfrac{1}{a}\right)$ is on the graph of f, the point $\left(\_, \dfrac{1}{a}\right)$ is on the graph of g; both graphs contain the point _____.

11.3 Logarithmic Functions

Objectives

1. Change Exponential Equations to Logarithmic Equations
2. Change Logarithmic Equations to Exponential Equations
3. Evaluate Logarithmic Functions
4. Determine the Domain of a Logarithmic Function
5. Graph Logarithmic Functions
6. Work with Natural and Common Logarithms
7. Solve Logarithmic Equations
8. Use Logarithmic Models That Describe Our World

Work Smart

It is required that a be positive and not equal to 1 for the same reasons that these restrictions existed for the exponential function.

In Other Words

The logarithm to the base a of x is the number y that we must raise a to in order to obtain x.

Work Smart: Study Skills

In doing problems similar to Examples 1 and 2, it is a good idea to practice saying, "y equals the logarithm to the base a of x is equivalent to a to the y equals x" so that you memorize the definition of a logarithm.

Are You Prepared for This Section?

Before getting started, complete the following problems. If you get a problem wrong, go back to the section cited and review the material.

P1. Solve: $3x + 2 > 0$ [Section 2.8, pp. 150–155]
P2. Solve: $\sqrt{x + 2} = x$ [Section 9.8, pp. 662–665]
P3. Solve: $x^2 = 6x + 7$ [Section 6.6, pp. 410–414]

Consider the "squaring function," $f(x) = x^2$, and the square root function, $g(x) = \sqrt{x}$. How are these two functions related? Well, $f(5) = 25$ and $g(25) = 5$. That is, the input to f is the output of g and the output of f is the input to g, so g "undoes" what f does.

In general, whenever a function is introduced in mathematics, there is a second function that "undoes" it. For example, a square root function undoes the squaring function. A cube root function undoes the cubing function. Is there a function that undoes an exponential function? Yes! This function is the *logarithmic function*.

Definition

The **logarithmic function to the base a**, where $a > 0$ and $a \neq 1$, is denoted by $y = \log_a x$ (read as "y is the logarithm to the base a of x"), and

$$y = \log_a x \qquad \text{is equivalent to} \qquad x = a^y$$

To evaluate logarithmic functions, convert them into their equivalent exponential form. Therefore, go from logarithmic form to exponential form, and back. For example,

$$0 = \log_3 1 \qquad \text{is equivalent to} \qquad 3^0 = 1$$
$$2 = \log_5 25 \qquad \text{is equivalent to} \qquad 5^2 = 25$$
$$-2 = \log_4 \frac{1}{16} \qquad \text{is equivalent to} \qquad 4^{-2} = \frac{1}{16}$$

Notice that the base of the logarithm is the base of the exponential; the argument of the logarithm is what the exponential equals; and the value of the logarithm is the exponent of the exponential expression. A logarithm is just another way of writing an exponential expression.

To see how the logarithmic function undoes the exponential function, consider the function $y = 2^x$. If the input is $x = 3$, then the output is $y = 2^3 = 8$. To undo this function would require that an input of 8 give an output of 3. If $2^3 = 8$, then $3 = \log_2 8$ using the definition of a logarithm. The input of $\log_2 8$ is 8, and its output is 3.

▶ ① Change Exponential Equations to Logarithmic Equations

The definition of a logarithm can be used to rewrite exponential equations as logarithmic equations.

EXAMPLE 1   **Changing Exponential Equations to Logarithmic Equations**

Rewrite each exponential equation as an equivalent logarithmic equation.

    (a) $6^2 = 36$ **(b)** $x^2 = 9$ **(c)** $4^3 = y$

Solution

Use the fact that $y = \log_a x$ is equivalent to $a^y = x$ provided that $a > 0$ and $a \neq 1$.

    (a) If $6^2 = 36$, then $2 = \log_6 36$.   **(b)** If $x^2 = 9$, then $2 = \log_x 9$.
    (c) If $4^3 = y$, then $3 = \log_4 y$.   ●

Prepared?...Answers

P1. $\left\{ x \mid x > -\dfrac{2}{3} \right\}$ or $\left(-\dfrac{2}{3}, \infty \right)$
P2. $\{2\}$ **P3.** $\{-1, 7\}$

Quick ✓

1. The logarithm to the base a of x, denoted $y = \log_a x$, can be expressed in exponential form as _____, where a __ 0 and a __ 1.

In Problems 2 and 3, rewrite each exponential equation as an equivalent equation involving a logarithm.

2. $4^3 = 64$

3. $p^{-2} = 8$

▷ ❷ Change Logarithmic Equations to Exponential Equations

The definition of a logarithm can also be used to rewrite logarithmic equations as exponential equations.

EXAMPLE 2   **Changing Logarithmic Equations to Exponential Equations**

Change each logarithmic equation to an equivalent exponential equation.

(a) $4 = \log_3 81$

(b) $-3 = \log_a \dfrac{1}{27}$

(c) $2 = \log_4 x$

Solution

In each of these problems, use the fact that $y = \log_a x$ is equivalent to $a^y = x$ provided that $a > 0$ and $a \neq 1$.

(a) If $4 = \log_3 81$, then $3^4 = 81$.

(b) If $-3 = \log_a \dfrac{1}{27}$, then $a^{-3} = \dfrac{1}{27}$.

(c) If $2 = \log_4 x$, then $4^2 = x$.   ●

Quick ✓

In Problems 4–6, rewrite each logarithmic equation as an equivalent equation involving an exponent.

4. $4 = \log_2 16$

5. $5 = \log_a 20$

6. $-3 = \log_5 z$

▷ ❸ Evaluate Logarithmic Functions

To find the exact value of a logarithm, write the logarithm in exponential notation and use the fact that if $a^u = a^v$, then $u = v$.

EXAMPLE 3   **Finding the Exact Value of a Logarithmic Expression**

Find the exact value of

(a) $\log_2 32$

(b) $\log_4 \dfrac{1}{16}$

Solution

(a) Let $y = \log_2 32$ and convert this equation into an exponential equation.

$$y = \log_2 32$$

Write the logarithm as an exponent: $\quad 2^y = 32$

$32 = 2^5$: $\quad 2^y = 2^5$

If $a^u = a^v$, then $u = v$: $\quad y = 5$

Therefore, $\log_2 32 = 5$.

(b) Let $y = \log_4 \dfrac{1}{16}$ and convert this equation into an exponential equation.

$$y = \log_4 \frac{1}{16}$$

Write the logarithm as an exponent: $4^y = \dfrac{1}{16}$

$\dfrac{1}{16} = 4^{-2}$: $4^y = 4^{-2}$

If $a^u = a^v$, then $u = v$: $y = -2$

Therefore, $\log_4 \dfrac{1}{16} = -2$.

Quick ✓

In Problems 7 and 8, find the exact value of each logarithmic expression.

7. $\log_5 25$

8. $\log_2 \dfrac{1}{8}$

The equation $y = \log_a x$ could also be written using function notation as $f(x) = \log_a x$. We use this notation in the next example to evaluate a logarithmic function.

EXAMPLE 4   **Evaluating a Logarithmic Function**

Find the value of each of the following, given that $f(x) = \log_2 x$.

(a) $f(2)$

(b) $f\left(\dfrac{1}{4}\right)$

Solution

(a) Finding $f(2)$ means evaluating $\log_2 x$ at $x = 2$, or evaluating $\log_2 2$. To determine this value, follow the approach of Example 3 by letting $y = \log_2 2$ and converting the equation into an exponential equation.

$$y = \log_2 2$$

Write the logarithm as an exponent: $2^y = 2$

$2 = 2^1$: $2^y = 2^1$

If $a^u = a^v$, then $u = v$: $y = 1$

Therefore, $f(2) = 1$.

(b) $f\left(\dfrac{1}{4}\right)$ means evaluating $\log_2 x$ at $x = \dfrac{1}{4}$ or $\log_2\left(\dfrac{1}{4}\right)$. Let $y = \log_2\left(\dfrac{1}{4}\right)$.

$$y = \log_2\left(\frac{1}{4}\right)$$

Write the logarithm as an exponent: $2^y = \dfrac{1}{4}$

$\dfrac{1}{4} = 2^{-2}$: $2^y = 2^{-2}$

If $a^u = a^v$, then $u = v$: $y = -2$

Therefore, $f\left(\dfrac{1}{4}\right) = -2$.

Quick ✓

In Problems 9 and 10, evaluate the function, given that $g(x) = \log_5 x$.

9. $g(25)$

10. $g\left(\dfrac{1}{5}\right)$

▶ ❹ Determine the Domain of a Logarithmic Function

The domain of a function $y = f(x)$ is the set of all x such that the function makes sense, and the range is the set of all outputs of the function. To find the range of the logarithmic function, recognize that $y = f(x) = \log_a x$ is equivalent to $x = a^y$. Because a can be raised to any real number (since $a > 0$ and $a \neq 1$), conclude that y can be any real number in the function $y = f(x) = \log_a x$. In addition, because a^y is positive for any real number, conclude that x must be positive. Since x represents the input of the logarithmic function, the domain of the logarithmic function is the set of all positive real numbers.

Work Smart

Notice that the domain of the logarithmic function is the same as the range of the exponential function. The range of the logarithmic function is the same as the domain of the exponential function.

> **Domain and Range of The Logarithmic Function**
>
> $$\text{Domain of the logarithmic function} = (0, \infty)$$
> $$\text{Range of the logarithmic function} = (-\infty, \infty)$$

Because the domain of the logarithmic function is the set of all positive real numbers, the argument of the logarithmic function must be greater than zero. For example, $f(x) = \log_{10} x$ is defined for $x = 2$, but not for $x = -1$, $x = -8$, or any other $x \leq 0$.

EXAMPLE 5 **Finding the Domain of a Logarithmic Function**

Find the domain of each logarithmic function.

    (a) $f(x) = \log_6 (x - 5)$ **(b)** $G(x) = \log_3 (3x + 1)$

Solution

    (a) The argument of the function $f(x) = \log_6 (x - 5)$ is $x - 5$. The domain of f is the set of all real numbers x such that $x - 5 > 0$. Solve this inequality:

$$x - 5 > 0$$

Add 5 to both sides: $x > 5$

The domain of f is $\{x | x > 5\}$, or $(5, \infty)$ in interval notation.

    (b) The argument of the function $G(x) = \log_3 (3x + 1)$ is $3x + 1$. The domain of G is the set of all real numbers x such that $3x + 1 > 0$.

$$3x + 1 > 0$$

Subtract 1 from both sides: $3x > -1$

Divide both sides by 3: $x > -\dfrac{1}{3}$

The domain of G is $\left\{ x \,\middle|\, x > -\dfrac{1}{3} \right\}$, or $\left(-\dfrac{1}{3}, \infty \right)$ in interval notation.   ●

> **Quick ✓**
>
> *In Problems 11 and 12, find the domain of each logarithmic function.*
>
> **11.** $g(x) = \log_8 (x + 3)$ **12.** $F(x) = \log_2 (5 - 2x)$

▶ ❺ Graph Logarithmic Functions

To graph a logarithmic function $y = \log_a x$, it helps to rewrite the function in exponential form: $x = a^y$. Then choose "nice" values of y and use the expression $x = a^y$ to find the corresponding values of x, as shown in Example 6.

EXAMPLE 6   **Graphing a Logarithmic Function**

Graph $f(x) = \log_2 x$ using point plotting. From the graph, state the domain and the range of the function.

Solution

Rewrite $y = f(x) = \log_2 x$ as $x = 2^y$. Table 8 shows various values of y, the corresponding values of x, and points on the graph of $y = f(x) = \log_2 x$. Plot these ordered pairs, connect them in a smooth curve, and obtain the graph of $f(x) = \log_2 x$ in Figure 19. The domain of f is $\{x \mid x > 0\}$, or $(0, \infty)$ in interval notation. The range of f is the set of all real numbers, or $(-\infty, \infty)$ in interval notation.

Table 8

y	$x = 2^y$	(x, y)
-2	$\dfrac{1}{4}$	$\left(\dfrac{1}{4}, -2\right)$
-1	$\dfrac{1}{2}$	$\left(\dfrac{1}{2}, -1\right)$
0	1	$(1, 0)$
1	2	$(2, 1)$
2	4	$(4, 2)$

Figure 19

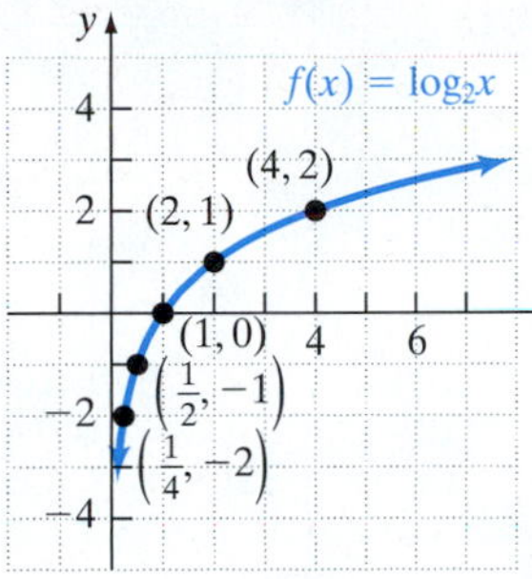

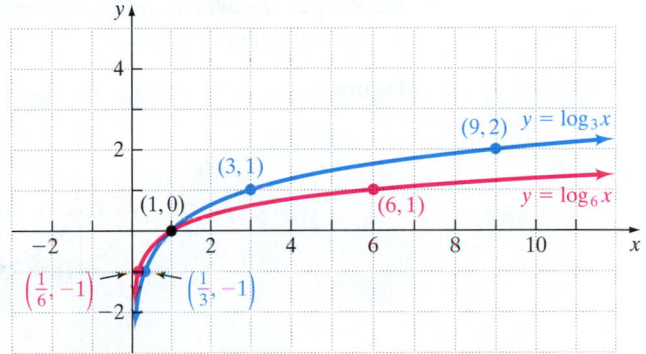

The graph of $f(x) = \log_2 x$ in Figure 19 is typical of all logarithmic functions that have a base larger than 1. Figure 20 shows the graphs of two other logarithmic functions with bases larger than 1, $y = \log_3 x$ and $y = \log_6 x$.

Figure 20

The information about $f(x) = \log_a x$, where $a > 1$, is summarized below.

Figure 21
$f(x) = \log_a x, a > 1$

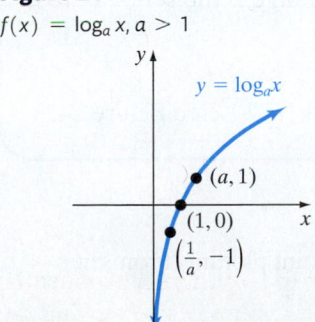

Properties of the Graph of a Logarithmic Function $f(x) = \log_a x, a > 1$

1. The domain is the set of all positive real numbers. The range is the set of all real numbers.
2. There is no y-intercept; the x-intercept is $(1, 0)$.
3. The graph of f contains the points $\left(\dfrac{1}{a}, -1\right)$, $(1, 0)$, and $(a, 1)$. See Figure 21.

Quick ✓

13. Graph the logarithmic function $f(x) = \log_4 x$ using point plotting. From the graph, state the domain and the range of the function.

Now we consider $f(x) = \log_a x, 0 < a < 1$.

Graphing a Logarithmic Function

Graph $f(x) = \log_{1/2} x$ using point plotting. From the graph, state the domain and the range of the function.

Solution

Rewrite $y = f(x) = \log_{1/2} x$ as $x = \left(\dfrac{1}{2}\right)^y$. Table 9 shows various values of y, the corresponding values of x, and points on the graph of $y = f(x) = \log_{1/2} x$. Plot these points, connect them in a smooth curve, and obtain the graph of $f(x) = \log_{1/2} x$ in Figure 22. The domain of f is $\{x \mid x > 0\}$, or $(0, \infty)$ in interval notation. The range of f is the set of all real numbers, or $(-\infty, \infty)$ in interval notation.

Table 9

y	$x = \left(\dfrac{1}{2}\right)^y$	(x, y)
-2	4	$(4, -2)$
-1	2	$(2, -1)$
0	1	$(1, 0)$
1	$\dfrac{1}{2}$	$\left(\dfrac{1}{2}, 1\right)$
2	$\dfrac{1}{4}$	$\left(\dfrac{1}{4}, 2\right)$

Figure 22

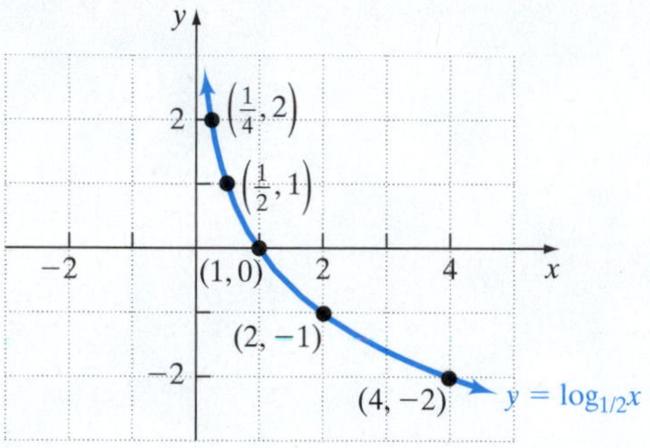

The graph of $f(x) = \log_{1/2} x$ in Figure 22 is typical of all logarithmic functions that have a base between 0 and 1. Figure 23 shows the graph of two more logarithmic functions whose bases are less than 1, $y = \log_{1/3} x$ and $y = \log_{1/6} x$.

Figure 23

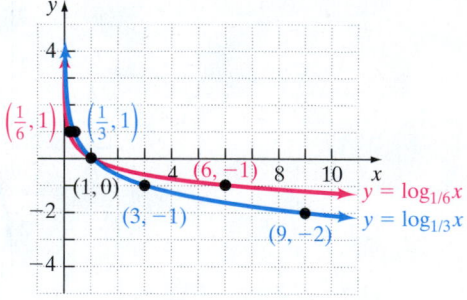

The information about $f(x) = \log_a x$, where the base a is between 0 and 1 ($0 < a < 1$), is summarized below.

Figure 24
$f(x) = \log_a x, 0 < a < 1$

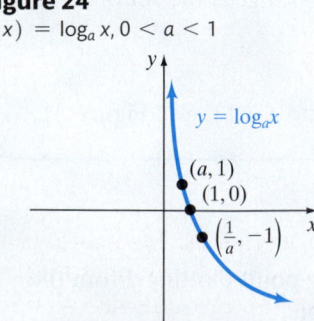

> **Properties of the Graph of a Logarithmic Function $f(x) = \log_a x, 0 < a < 1$**
>
> 1. The domain is the set of all positive real numbers. The range is the set of all real numbers.
> 2. There is no y-intercept; the x-intercept is $(1, 0)$.
> 3. The graph of f contains the points $\left(\dfrac{1}{a}, -1\right)$, $(1, 0)$, and $(a, 1)$. See Figure 24.

Quick ✔

14. Graph the logarithmic function $f(x) = \log_{1/4} x$ using point plotting. From the graph, state the domain and the range of the function.

▶ ⑥ Work with Natural and Common Logarithms

The **natural logarithm function** is a logarithmic function whose base is the number e. This function occurs so frequently in applications that it is given a special symbol, **ln** (from the Latin *logarithmus naturalis*).

> **In Other Words**
> A logarithm to the base e is called the natural logarithm because this function can be used to model many things in nature. In addition, $\log_e x$ is written $\ln x$.

> **Definition**
> The natural logarithm: $y = \ln x$ if and only if $x = e^y$

Figure 25
$y = \ln x$

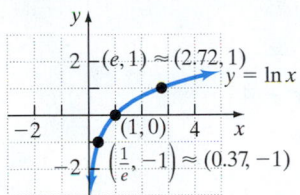

Figure 25 shows the graph of $y = \ln x$.

If the base of a logarithmic function is 10, then we have the **common logarithm function.** If the base a of the logarithmic function is not indicated, it is understood to be 10.

> **Definition**
> The common logarithm: $y = \log x$ if and only if $x = 10^y$

> **In Other Words**
> When there is no base on a logarithm, then the base is understood to be 10. That is, $\log_{10} x$ is written $\log x$.

Figure 26
$y = \log x$

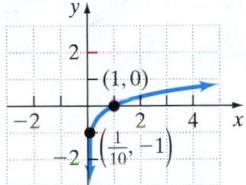

Figure 26 shows the graph of $y = \log x$.

Scientific and graphing calculators have both a natural logarithm button, $\boxed{\ln}$, and a common logarithm button, $\boxed{\log}$. This enables us to approximate the values of logarithms to the base e and the base 10, when the results are not exact.

To evaluate logarithmic expressions using a scientific calculator, enter the argument of the logarithm and then press the $\boxed{\ln}$ or $\boxed{\log}$ button, depending on the base of the logarithm. For example, to evaluate $\log 80$, type in 80 and then press the $\boxed{\log}$ button. The display should show 1.90308999. Try it!

To evaluate logarithmic expressions using a graphing calculator, press the $\boxed{\ln}$ or $\boxed{\log}$ button, depending on the base of the logarithm, and then enter the argument of the logarithm. Finally, press $\boxed{\text{ENTER}}$. For example, to evaluate $\log 80$, press the $\boxed{\log}$ button, type in 80, and then press $\boxed{\text{ENTER}}$. The display should show 1.903089987. Try it!

By the way, it shouldn't be surprising that $\log 80$ is between 1 and 2 since $\log 10 = 1$ (because $10^1 = 10$) and $\log 100 = 2$ (because $10^2 = 100$). It is a good idea to develop number sense for logarithms prior to using your calculator to approximate the value.

EXAMPLE 8 **Evaluating Natural and Common Logarithms on a Calculator**

Using a calculator, evaluate each of the following. Round your answers to three decimal places.

(a) $\ln 20$ (b) $\log 30$ (c) $\ln 0.5$

Solution

(a) $\ln 20 \approx 2.996$ (b) $\log 30 \approx 1.477$ (c) $\ln 0.5 \approx -0.693$ ●

> **Quick ✓**
> In Problems 15–17, evaluate each logarithm using a calculator. Round your answers to three decimal places.
> **15.** $\log 1400$ **16.** $\ln 4.8$ **17.** $\log 0.3$

▶ ⑦ Solve Logarithmic Equations

> **Work Smart**
> Extraneous solutions can occur while solving logarithmic equations. This is why it is important to always check your solutions!

Equations that contain logarithms are called **logarithmic equations.** Take care when solving logarithmic equations, because extraneous solutions (apparent solutions that are not solutions to the original equation) might creep in. To help locate extraneous solutions, remember that in the expression $\log_a M$, both a and M must be positive with $a \neq 1$.

To solve logarithmic equations, rewrite the logarithmic equation as an equivalent exponential equation, using the fact that $y = \log_a x$ is equivalent to $a^y = x$.

EXAMPLE 9 **Solving Logarithmic Equations**

Solve:

(a) $\log_2(3x + 4) = 4$ **(b)** $\log_x 25 = 2$

Solution

(a) Find the solution by writing the logarithmic equation as an exponential equation, using the fact that if $y = \log_a x$, then $a^y = x$.

$$\log_2(3x + 4) = 4$$

If $y = \log_a x$, then $a^y = x$: $2^4 = 3x + 4$

$2^4 = 16$: $16 = 3x + 4$

Subtract 4 from both sides: $12 = 3x$

Divide both sides by 3: $4 = x$

Verify this solution by letting $x = 4$ in the original equation:

$$\log_2(3 \cdot 4 + 4) \stackrel{?}{=} 4$$
$$\log_2(16) \stackrel{?}{=} 4$$

Since $2^4 = 16$, $\log_2 16 = 4$. Therefore, the solution set is $\{4\}$.

(b) Change the logarithmic equation to an exponential equation.

$$\log_x 25 = 2$$

If $y = \log_a x$, then $a^y = x$: $x^2 = 25$

Use the Square Root Property: $x = \pm\sqrt{25} = \pm 5$

Since the base of a logarithm must be positive, $x = -5$ is extraneous. We leave it to you to verify that the solution set is $\{5\}$.   ●

> **Quick ✓**
>
> *In Problems 18 and 19, solve each equation. Be sure to verify your solution.*
>
> **18.** $\log_3(5x + 1) = 4$ **19.** $\log_x 16 = 2$

EXAMPLE 10 **Solving Logarithmic Equations**

Solve each equation and state the exact solution.

(a) $\ln x = 3$ **(b)** $\log(x + 1) = -2$

Solution
Write each logarithmic equation as an exponential equation.

(a) $\ln x = 3$

Write as an exponent; if $y = \ln x$, then $e^y = x$: $e^3 = x$

Verify this solution by letting $x = e^3$ in the original equation:

$$\ln e^3 \stackrel{?}{=} 3$$

The equation $\ln e^3 = 3$ can be written as $\log_e e^3 = 3$, which is equivalent to $e^3 = e^3$, so this is a true statement. The solution set is $\{e^3\}$.

(b) $\log(x + 1) = -2$

Write as an exponent; if $y = \log x$, then $10^y = x$: $10^{-2} = x + 1$

$10^{-2} = 0.01$: $x + 1 = 0.01$

Subtract 1 from both sides: $x = -0.99$

Work Smart

The solution to Example 10(b) illustrates that negative answers may occur in logarithmic equations. Just because an apparent solution is negative does not automatically mean the solution is extraneous.

Verify this solution by letting $x = -0.99$ in the original equation:

$$\log(-0.99 + 1) \overset{?}{=} -2$$
$$\log(0.01) \overset{?}{=} -2$$
$$10^{-2} = 0.01 \quad \text{True}$$

The solution set is $\{-0.99\}$.

> **Quick** ✔
>
> In Problems 20 and 21, solve each equation. Be sure to verify your solution.
>
> **20.** $\ln x = -2$ **21.** $\log(x - 20) = 4$

▶ ❽ Use Logarithmic Models That Describe Our World

Common logarithms often are used when quantities vary from very large to very small. This is because the common logarithm can "scale down" the measurement. For example, if a certain quantity varied from $0.00000001 = 10^{-8}$ to $100{,}000{,}000 = 10^8$, the common logarithm of the same quantity would vary from $\log 10^{-8} = -8$ to $\log 10^8 = 8$.

Physicists define the **intensity of a sound wave** as the amount of energy the sound wave transmits through a given area. The *loudness L* (measured in **decibels** in honor of Alexander Graham Bell) of a sound of intensity x (measured in watts per square meter) is defined as follows.

> **Definition**
>
> The **loudness** L, measured in decibels, of a sound of intensity x, measured in watts per square meter, is
>
> $$L(x) = 10 \log \frac{x}{10^{-12}}$$

The quantity 10^{-12} watt per square meter in the definition is the least intense sound that a human ear can detect. If $x = 10^{-12}$ watt per square meter, then

$$L(10^{-12}) = 10 \log \frac{10^{-12}}{10^{-12}}$$
$$= 10 \log 1$$
$$= 10(0)$$
$$= 0$$

Thus the loudness of the least intense sound a human ear can detect is 0 decibels.

EXAMPLE 11   **Measuring the Loudness of a Sound**

What is the loudness, in decibels, of normal conversation, which has an intensity level of 10^{-6} watt per square meter?

Solution

Evaluate L at $x = 10^{-6}$.

$$L(10^{-6}) = 10 \log \frac{10^{-6}}{10^{-12}}$$

Laws of exponents; $\dfrac{a^m}{a^n} = a^{m-n}$: $= 10 \log 10^{-6-(-12)}$

Simplify: $= 10 \log 10^6$

If $y = \log 10^6$, then $10^y = 10^6$, so $y = 6$: $= 10(6)$

$$= 60 \text{ decibels}$$

The loudness of normal conversation is 60 decibels.

> **Quick** ✔
>
> **22.** An MP3 player has an intensity level of 10^{-2} watt per square meter when set at its maximum level. What is the loudness, in decibels, of the MP3 player on "full blast"?

11.3 Exercises MyMathLab®

Problems **1–22** *are the* **Quick ✔**s *that follow the* **EXAMPLES**.

Building Skills

In Problems 23–30, change each exponential equation to an equivalent equation involving a logarithm. See Objective 1.

23. $64 = 4^3$

24. $16 = 2^4$

25. $\dfrac{1}{8} = 2^{-3}$

26. $\dfrac{1}{9} = 3^{-2}$

27. $a^3 = 19$

28. $b^4 = 23$

29. $5^{-6} = c$

30. $10^{-3} = z$

In Problems 31–40, change each logarithmic equation to an equivalent equation involving an exponent. See Objective 2.

31. $\log_2 16 = 4$

32. $\log_3 81 = 4$

33. $\log_3 \dfrac{1}{9} = -2$

34. $\log_2 \dfrac{1}{32} = -5$

35. $\log_5 a = -3$

36. $\log_6 x = -4$

37. $\log_a 4 = 2$

38. $\log_a 16 = 2$

39. $\log_{1/2} 12 = y$

40. $\log_{1/2} 18 = z$

In Problems 41–48, find the exact value of each logarithm without using a calculator. See Objective 3.

41. $\log_3 1$

42. $\log_5 5$

43. $\log_2 8$

44. $\log_4 16$

45. $\log_4 \left(\dfrac{1}{16}\right)$

46. $\log_5 \left(\dfrac{1}{125}\right)$

47. $\log_{\sqrt{2}} 4$

48. $\log_{\sqrt{3}} 3$

In Problems 49–52, evaluate each function, given that $f(x) = \log_3 x$ and $g(x) = \log_5 x$. See Objective 3.

49. $f(81)$

50. $f(9)$

51. $g\left(\sqrt{5}\right)$

52. $g\left(\sqrt[3]{5}\right)$

In Problems 53–62, find the domain of each function. See Objective 4.

53. $f(x) = \log_2(x - 4)$

54. $f(x) = \log_3(x - 2)$

55. $F(x) = \log_3(2x)$

56. $h(x) = \log_4(5x)$

57. $f(x) = \log_8(3x - 2)$

58. $F(x) = \log_2(4x - 3)$

59. $H(x) = \log_7(2x + 1)$

60. $f(x) = \log_3(5x + 3)$

61. $H(x) = \log_2(1 - 4x)$

62. $G(x) = \log_4(3 - 5x)$

In Problems 63–68, graph each function. From the graph, state the domain and the range of each function. See Objective 5.

63. $f(x) = \log_5 x$

64. $f(x) = \log_7 x$

65. $g(x) = \log_6 x$

66. $G(x) = \log_8 x$

67. $F(x) = \log_{1/5} x$

68. $F(x) = \log_{1/7} x$

In Problems 69–72, change each exponential equation to an equivalent equation involving a logarithm, and change each logarithmic equation to an equivalent equation involving an exponent. See Objective 6.

69. $e^x = 12$

70. $e^4 = M$

71. $\ln x = 4$

72. $\ln(x - 1) = 3$

In Problems 73–76, evaluate each function, given that $H(x) = \log x$ and $P(x) = \ln x$. See Objective 6.

73. $H(0.1)$

74. $H(100{,}000)$

75. $P(e^3)$

76. $P(e^{-3})$

In Problems 77–88, use a calculator to evaluate each expression. Round your answers to three decimal places. See Objective 6.

77. $\log 67$

78. $\log 106$

79. $\ln 5.4$

80. $\ln 10.4$

81. $\log 0.35$

82. $\log 0.78$

83. $\ln 0.2$

84. $\ln 0.4$

85. $\log \dfrac{5}{4}$

86. $\log \dfrac{10}{7}$

87. $\ln \dfrac{3}{8}$

88. $\ln \dfrac{1}{2}$

In Problems 89–108, solve each logarithmic equation. All answers should be exact. See Objective 7.

89. $\log_3(2x + 1) = 2$

90. $\log_3(5x - 3) = 3$

91. $\log_5(20x - 5) = 3$

92. $\log_4(8x + 10) = 3$

93. $\log_a 36 = 2$

94. $\log_a 81 = 2$

95. $\log_a 18 = 2$

96. $\log_a 28 = 2$

97. $\log_a 1000 = 3$

98. $\log_a 243 = 5$

99. $\ln x = 5$

100. $\ln x = 10$

101. $\log(2x - 1) = -1$

102. $\log(2x + 3) = 1$

103. $\ln e^x = -3$

104. $\ln e^{2x} = 8$

105. $\log_3 81 = x$

106. $\log_4 16 = x + 1$

107. $\log_2(x^2 - 1) = 3$

108. $\log_3(x^2 + 1) = 2$

Mixed Practice

109. Suppose that $f(x) = \log_2 x$.

(a) What is $f(16)$? What point is on the graph of f?

(b) If $f(x) = -3$, what is x? What point is on the graph of f?

110. Suppose that $f(x) = \log_5 x$.

(a) What is $f(5)$? What point is on the graph of f?

(b) If $f(x) = -2$, what is x? What point is on the graph of f?

111. Suppose that $G(x) = \log_4 (x + 1)$.

(a) What is $G(7)$? What point is on the graph of G?

(b) If $G(x) = 2$, what is x? What point is on the graph of G?

112. Suppose that $F(x) = \log_2 x - 3$.

(a) What is $F(8)$? What point is on the graph of f?

(b) If $F(x) = -1$, what is x? What point is on the graph of F?

113. Find a so that the graph of $f(x) = \log_a x$ contains the point $(16, 2)$.

114. Find a so that the graph of $f(x) = \log_a x$ contains the point $\left(\dfrac{1}{4}, -2 \right)$.

In Problems 115–120, state the domain of each logarithmic function.

115. $f(x) = \log_3 (x^2 - 4x - 5)$

116. $f(x) = \log_2 (x^2 - 2x - 8)$

117. $f(x) = \log \left(\dfrac{x - 4}{x + 1} \right)$

118. $f(x) = \log \left(\dfrac{x + 3}{x - 2} \right)$

119. $f(x) = \ln |x - 3|$

120. $f(x) = \ln |x + 1|$

Applying the Concepts

121. Loudness of a Whisper A whisper has an intensity level of 10^{-10} watt per square meter. How many decibels is a whisper?

122. Loudness of a Concert If you sit in the front row of a rock concert, you will experience an intensity level of 10^{-1} watt per square meter. How many decibels is a rock concert in the front row? If you move back to the 15th row, you will experience an intensity level of 10^{-2} watt per square meter. How many decibels is a rock concert in the 15th row?

123. Threshold of Pain The threshold of pain has an intensity level of 10^1 watt per square meter. How many decibels is the threshold of pain?

124. Exploding Eardrum Instant perforation of the eardrum occurs at an intensity level of 10^4 watts per square meter. At how many decibels does instant perforation of the eardrum occur?

*Problems 125–128 use the following discussion: The **Richter scale** is one way of converting seismographic readings into numbers that provide an easy reference for measuring the **magnitude M** of an earthquake. All earthquakes are compared to a **zero–level earthquake** whose seismographic reading measures 0.001 millimeter at a distance of 100 kilometers from the epicenter. An earthquake whose seismographic reading measures x millimeters has magnitude M given by*

$$M(x) = \log \left(\frac{x}{10^{-3}} \right)$$

where 10^{-3} is the reading of a zero–level earthquake 100 kilometers from its epicenter.

125. San Francisco, 1906 According to the United States Geological Survey, the San Francisco earthquake of 1906 resulted in a seismographic reading of 63,096 millimeters 100 kilometers from its epicenter. What was the magnitude of this earthquake?

126. Alaska, 1964 According to the United States Geological Survey, an earthquake on March 28, 1964, in Prince William Sound, Alaska, resulted in a seismographic reading of 1,584,893 millimeters 100 kilometers from its epicenter. What was the magnitude of this earthquake? This earthquake was the second largest ever recorded. The largest was the Great Chilean Earthquake of 1960, whose magnitude was 9.5 on the Richter scale.

127. Japan, 2011 According to the United States Geological Survey, an earthquake on March 11, 2011, off the coast of Japan had a magnitude of 8.9. What was the seismographic reading 100 kilometers from its epicenter?

128. Brazil, 2015 According to the United States Geological Survey, an earthquake on November 26, 2015, in Tarauka, Brazil, had a magnitude of 7.6. What was the seismographic reading 100 kilometers from its epicenter?

129. pH The pH of a chemical solution is given by the formula

$$pH = -\log [H^+]$$

where $[H^+]$ is the concentration of hydrogen ions in moles per liter. Values of pH range from 0 to 14. A solution whose pH is 7 is considered neutral. The pH of pure water at 25 degrees Celsius is 7. A solution whose pH is less than 7 is considered acidic, while a solution whose pH is greater than 7 is considered basic.

(a) What is the pH of household ammonia for which $[H^+]$ is 10^{-12} ? Is ammonia basic or acidic?

(b) What is the pH of black coffee for which $[H^+]$ is 10^{-5} ? Is black coffee basic or acidic?

(c) What is the pH of lemon juice for which $[H^+]$ is 10^{-2} ? Is lemon juice basic or acidic?

(d) What is the concentration of hydrogen ions in human blood $(\text{pH} = 7.4)$?

130. Energy of an Earthquake The magnitude and the seismic moment are related to the amount of energy that is given off by an earthquake. The relationship between magnitude and energy is

$$\log E_S = 11.8 + 1.5M$$

where E_S is the energy (in ergs) for an earthquake whose magnitude is M. Note that E_S is the amount of energy given off from the earthquake as seismic waves.

(a) How much energy is given off by an earthquake that measures 5.8 on the Richter scale?

(b) The earthquake on the Rat Islands in Alaska on February 4, 1965, measured 8.7 on the Richter scale. How much energy was given off by this earthquake?

Explaining the Concepts

131. Explain why the base in the logarithmic function $f(x) = \log_a x$ is not allowed to equal 1.

132. Explain why the base in the logarithmic function $f(x) = \log_a x$ must be greater than 0.

133. Explain why the domain of the function $f(x) = \log_a (x^2 + 1)$ is the set of all real numbers.

134. What are the domain and the range of the exponential function $f(x) = a^x$? What are the domain and the range of the logarithmic function $f(x) = \log_a x$? What is the relation between the domain and the range of each function?

Synthesis Review

In Problems 135–140, add or subtract each expression.

135. $(2x^2 - 6x + 1) - (5x^2 + x - 9)$

136. $(x^3 - 2x^2 + 10x + 2) + (3x^3 - 4x - 5)$

137. $\dfrac{3x}{x^2 - 1} + \dfrac{x - 3}{x^2 + 3x + 2}$

138. $\dfrac{x + 5}{x^2 + 5x + 6} - \dfrac{2}{x^2 + 2x - 3}$

139. $\sqrt{8x^3} + x\sqrt{18x}$

140. $\sqrt[3]{27a} + 2\sqrt[3]{a} - \sqrt[3]{8a}$

Technology Exercises

In Problems 141–146, graph each logarithmic function using technology. State the domain and the range of each function.

141. $f(x) = \log(x + 1)$

142. $g(x) = \log(x - 2)$

143. $G(x) = \ln(x) + 1$

144. $F(x) = \ln(x) - 4$

145. $f(x) = 2\log(x - 3) + 1$

146. $G(x) = -\log(x + 1) - 3$

147. Exploration: Graph of Logarithmic Functions *Open the "Graph of Logarithmic Functions" Geogebra applet. The applet may be found using the QR code at the beginning of this section or through the Multimedia Library in MyMathLab. In the upper left corner is one "slider" that represents the value for a in the logarithmic function $f(x) = \log_a x$. If you are using a keyboard, it may be easier to use the arrow keys, instead of the mouse, to move the slider.*

(a) Move the a slider to 2. This will produce the graph of $f(x) = \log_2 x$. State the domain, the range, the intercepts, and the values of the three coordinate points that are labeled on the graph.

(b) Move the a slider to 3. This will produce the graph of $f(x) = \log_3 x$. State the domain, the range, the intercepts, and the values of the three coordinate points that are labeled on the graph.

(c) Move the a slider to 4. This will produce the graph of $f(x) = \log_4 x$. State the domain, the range, the intercepts, and the values of the three coordinate points that are labeled on the graph.

(d) Use your results from parts (a) through (c) to generalize the following for *all* exponential functions of the form $f(x) = \log_a x$:

    i. The domain is _____.

    ii. The range is _____.

    iii. How many y-intercepts are there? What is the x-intercept?

    iv. Conclude the graph of $f(x) = \log_a x$ contains the points $(\underline{\ \ }, -1)$, $(\underline{\ \ }, 0)$, and $(\underline{\ \ }, 1)$.

Putting the Concepts Together (Sections 11.1–11.3)

We designed these problems so that you can review Sections 11.1–11.3 and show your mastery of the concepts. Take time to work these problems before proceeding with the next section. The answers are located at the back of the text on page AN-59.

1. Given the functions $f(x) = 2x + 3$ and $g(x) = 2x^2 - 4x$, find

   (a) $(f \circ g)(x)$ **(b)** $(g \circ f)(x)$
   (c) $(f \circ g)(3)$ **(d)** $(g \circ f)(-2)$
   (e) $(f \circ f)(1)$

2. Find the inverse of the given one-to-one functions.

   (a) $f(x) = 3x + 4$ **(b)** $g(x) = x^3 - 4$

3. Sketch the graph of the inverse of the given one-to-one function.

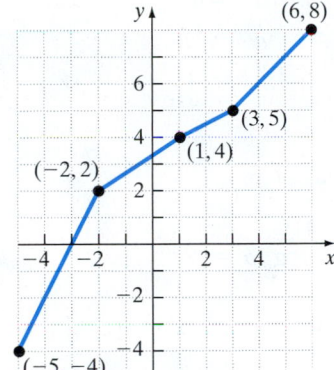

4. Approximate each number using a calculator. Express your answer rounded to three decimal places.

   (a) $2.7^{2.7}$ **(b)** $2.72^{2.72}$ **(c)** $2.718^{2.718}$
   (d) $2.7183^{2.7183}$ **(e)** e^e

5. Change each exponential equation to an equivalent equation involving a logarithm.

   (a) $a^4 = 6.4$ **(b)** $10^x = 278$

6. Change each logarithmic equation to an equivalent equation involving an exponent.

   (a) $\log_2 x = 7$ **(b)** $\ln 16 = M$

7. Find the exact value of each expression without using a calculator.

   (a) $\log_5 625$ **(b)** $\log_{2/3}\left(\dfrac{9}{4}\right)$

8. Determine the domain of the logarithmic function $f(x) = \log_{13}(2x + 12)$.

In Problems 9 and 10, graph each function. State the domain and range of the function.

9. $f(x) = \left(\dfrac{1}{6}\right)^x$ **10.** $g(x) = \log_{3/2} x$

In Problems 11–14, solve each equation. State the exact solution.

11. $3^{-x+2} = 27$

12. $e^x = e^{2x+5}$

13. $\log_2(2x + 5) = 4$

14. $\ln x = 7$

15. Suppose that a student has 150 anatomy and physiology terms to learn. If the student learns 40 terms in 60 minutes, the function $L(t) = 150(1 - e^{-0.0052t})$ models the number of terms L that the student will learn after t minutes. According to the model, how many terms will the student have learned after 90 minutes?

11.4 Properties of Logarithms

Objectives

1 Understand the Properties of Logarithms

2 Write a Logarithmic Expression as a Sum or Difference of Logarithms

3 Write a Logarithmic Expression as a Single Logarithm

4 Evaluate a Logarithm Whose Base Is Neither 10 Nor e

Are You Prepared for This Section?

Before getting started, complete the following problems. If you get a problem wrong, go back to the section cited and review the material.

P1. Round 3.03468 to three decimal points.   [Section 1.3, pp. 13–14]
P2. Evaluate $a^0, a \neq 0$.   [Section 5.4, p. 336]

▶ **1** Understand the Properties of Logarithms

Logarithms have some very useful properties. These properties will be derived directly from the definition of a logarithm and the laws of exponents.

EXAMPLE 1 **Deriving Properties of Logarithms**

Determine the value of the following logarithmic expressions where a is any positive real number, and $a \neq 1$.

(a) $\log_a 1$ (b) $\log_a a$

Solution

(a) Remember, $y = \log_a x$ is equivalent to $x = a^y$, where $a > 0, a \neq 1$.

$$y = \log_a 1$$

Change to an exponent: $a^y = 1$

Because $a \neq 0, a^0 = 1$: $a^y = a^0$

If $a^u = a^v$, then $u = v$: $y = 0$

Thus $\log_a 1 = 0$.

(b) Again, write the logarithm as an exponent in order to evaluate the logarithm. Let $y = \log_a a$, so that

$$y = \log_a a$$

Change to an exponent: $a^y = a$

$a = a^1$: $a^y = a^1$

If $a^u = a^v$, then $u = v$: $y = 1$

Thus $\log_a a = 1$.   ●

The results of Example 1 are summarized below.

$$\log_a 1 = 0 \qquad \log_a a = 1$$

Quick ✓

In Problems 1–4, evaluate each logarithm.

1. $\log_5 1$ **2.** $\ln 1$ **3.** $\log_4 4$ **4.** $\log 10$

How can $3^{\log_3 81}$ be evaluated? The exponent, $\log_3 81$, is 4 (because $y = \log_3 81$ means $3^y = 81$, or $3^y = 3^4$, so $y = 4$). Now $3^{\log_3 81} = 3^4 = 81$, so $3^{\log_3 81} = 81$. This result is true in general.

In Other Words

When the number a is raised to the power $\log_a M$, the result is M.

An Inverse Property of Logarithms

If a and M are positive real numbers, with $a \neq 1$, then

$$a^{\log_a M} = M$$

This is an Inverse Property of Logarithms because if the logarithm to the base a of a positive number M is computed and then a is raised to this power, the result is M.

EXAMPLE 2 **Using an Inverse Property of Logarithms**

(a) $5^{\log_5 20} = 20$ (b) $0.8^{\log_{0.8} \sqrt{23}} = \sqrt{23}$ ●

Quick ✓

In Problems 5 and 6, evaluate each logarithm.

5. $12^{\log_{12} \sqrt{2}}$ **6.** $10^{\log 0.2}$

How can $\log_5 5^6$ be evaluated? If $y = \log_5 5^6$, then $5^y = 5^6$ (since $y = \log_a x$ means $a^y = x$), so $y = 6$. Thus $\log_5 5^6 = 6$. This result is true in general.

In Other Words
The logarithm to the base a of a raised to the power of r is r.

An Inverse Property of Logarithms

If a is a positive real number, with $a \neq 1$, and r is any real number, then

$$\log_a a^r = r$$

This is another Inverse Property of Logarithms because if a is raised to some power r and then the logarithm to the base a of a^r is computed, the result is r.

EXAMPLE 3 **Using an Inverse Property of Logarithms**

    (a) $\log_4 4^3 = 3$ **(b)** $\ln e^{-0.5} = -0.5$ ●

Quick ✓

In Problems 7 and 8, evaluate each logarithm.

7. $\log_8 8^{1.2}$ **8.** $\log 10^{-4}$

▶ ❷ **Write a Logarithmic Expression as a Sum or Difference of Logarithms**

The next two properties deal with logarithms whose arguments are products or quotients.
    Notice that $\log_2 8 = 3$ and that $\log_2 2 + \log_2 4 = 1 + 2 = 3$. Therefore, $\log_2 8 = \log_2 2 + \log_2 4$ and $8 = 4 \cdot 2$. This suggests the following result.

In Other Words
The product rule of logarithms states that the log of a product equals the sum of the logs of the factors.

The Product Rule of Logarithms

If $M, N,$ and a are positive real numbers, with $a \neq 1$, then

$$\log_a (MN) = \log_a M + \log_a N$$

EXAMPLE 4 **Using the Product Rule of Logarithms**

Simplify each of the following logarithms by writing it as the sum of logarithms.

    (a) $\log_2 (5 \cdot 3)$ **(b)** $\ln (6z)$

Solution
Do you notice that each argument contains a product? To simplify, use the Product Rule of Logarithms.

    (a) $\log_2 (5 \cdot 3) = \log_2 5 + \log_2 3$ **(b)** $\ln (6z) = \ln 6 + \ln z$ ●

Work Smart

$\log_a (M + N)$ does *not* equal $\log_a M + \log_a N$.

Quick ✓

9. *True or False* $\log (x + 4) = \log x + \log 4$

In Problems 10 and 11, write each logarithm as the sum of logarithms.

10. $\log_4 (9 \cdot 5)$ **11.** $\log (5w)$

Notice that $\log_2 8 = 3$ and that $\log_2 16 - \log_2 2 = 4 - 1 = 3$. Therefore, $\log_2 8 = \log_2 16 - \log_2 2$ and $8 = \dfrac{16}{2}$. This suggests the result on the next page.

The Quotient Rule of Logarithms

If M, N, and a are positive real numbers, with $a \neq 1$, then

$$\log_a\left(\frac{M}{N}\right) = \log_a M - \log_a N$$

The relationships in the product and quotient rules for logarithms are not coincidental. After all, when exponential expressions with the same base are multiplied, the exponents are added, and when exponential expressions with the same base are divided, the exponents are subtracted. The same thing is going on here!

EXAMPLE 5 | **Using the Quotient Rule of Logarithms**

Simplify each of the following logarithms by writing it as the difference of logarithms.

(a) $\log_2\left(\frac{5}{3}\right)$ (b) $\log\left(\frac{y}{5}\right)$

Solution

Do you notice that each argument contains a quotient? To simplify, use the Quotient Rule of Logarithms.

(a) $\log_2\left(\frac{5}{3}\right) = \log_2 5 - \log_2 3$ (b) $\log\left(\frac{y}{5}\right) = \log y - \log 5$ ●

Work Smart

$\log_a(M - N)$ does *not* equal $\log_a M - \log_a N$.

Quick ✓

In Problems 12 and 13, write each logarithm as the difference of logarithms.

12. $\log_7\left(\frac{9}{5}\right)$ **13.** $\ln\left(\frac{p}{3}\right)$

A single logarithm can be written as the sum or difference of logs when the argument of the logarithm contains both products and quotients.

EXAMPLE 6 | **Writing a Single Logarithm as the Sum or Difference of Logs**

Write $\log_3\left(\frac{4x}{y}\right)$ as the sum or difference of logarithms.

Solution

Notice that the argument of the logarithm contains a product and a quotient. Thus the Product Rule and the Quotient Rule of Logarithms can be used. When applying both rules, it is typically easier to use the Quotient Rule first.

Work Smart

When you need to use both the Product Rule and the Quotient Rule, use the Quotient Rule first.

The Quotient Rule of Logarithms
↓
$$\log_3\left(\frac{4x}{y}\right) = \log_3(4x) - \log_3 y$$

The Product Rule of Logarithms: $= \log_3 4 + \log_3 x - \log_3 y$ ●

Quick ✓

In Problems 14 and 15, write each logarithm as the sum or difference of logarithms.

14. $\log_2\left(\frac{3m}{n}\right)$ **15.** $\ln\left(\frac{q}{3p}\right)$.

Another useful property of logarithms allows us to express powers on the argument of a logarithm as factors.

> **The Power Rule of Logarithms**
>
> If M and a are positive real numbers, with $a \neq 1$, and r is any real number, then
> $$\log_a M^r = r \log_a M$$

EXAMPLE 7   **Using the Power Rule of Logarithms**

Use the Power Rule of Logarithms to express all powers as factors.

    (a) $\log_8 3^5$ **(b)** $\ln x^{\sqrt{3}}$

Solution

    (a) $\log_8 3^5 = 5 \log_8 3$ **(b)** $\ln x^{\sqrt{3}} = \sqrt{3} \ln x$ ●

Work Smart

Whenever you see the word "factor," you should think product (or multiplication).

> **Quick ✔**
>
> *In Problems 16 and 17, write each logarithm so that all powers are factors.*
>
> **16.** $\log_2 5^{1.6}$ **17.** $\log b^5$

Use the direction *expand the logarithm* to mean "Write a logarithm as a sum or difference with all exponents written as factors."

EXAMPLE 8   **Expanding a Logarithm**

Expand the logarithm.

    (a) $\log_2 (x^2 y^3)$ **(b)** $\log\left(\dfrac{100x}{\sqrt{y}}\right)$

Solution

    (a) The argument of the logarithm contains a product, so use the Product Rule of Logarithms to write the single log as the sum of two logs.

$$\text{Product Rule} \downarrow$$
$$\log_2 (x^2 y^3) = \log_2 x^2 + \log_2 y^3$$
$$\text{Write exponents as factors:} \quad = 2 \log_2 x + 3 \log_2 y$$

    (b) The argument of the logarithm contains a quotient and a product.

Work Smart

If the argument of the logarithm contains both a quotient and a product, it is easier to write the quotient as the difference of two logs first.

$$\text{Quotient Rule} \downarrow$$
$$\log\left(\frac{100x}{\sqrt{y}}\right) = \log(100x) - \log\sqrt{y}$$
$$\sqrt{y} = y^{\frac{1}{2}}\text{:} \quad = \log(100x) - \log y^{\frac{1}{2}}$$
$$\text{Product Rule:} \quad = \log 100 + \log x - \log y^{\frac{1}{2}}$$
$$\text{Write exponents as factors:} \quad = \log 100 + \log x - \frac{1}{2}\log y$$
$$\log 100 = 2\text{:} \quad = 2 + \log x - \frac{1}{2}\log y \quad ●$$

> **Quick ✔**
>
> *In Problems 18 and 19, expand each logarithm.*
>
> **18.** $\log_4 (a^2 b)$ **19.** $\log_3\left(\dfrac{9m^4}{\sqrt[3]{n}}\right)$

▶ ❸ Write a Logarithmic Expression as a Single Logarithm

The Product Rule, Quotient Rule, and Power Rule of Logarithms can be used to write the sums and/or differences of logarithms that have the same base as a single logarithm. This skill will be useful when solving certain logarithmic equations in Section 11.5.

EXAMPLE 9   **Writing Expressions as Single Logarithms**

Write each of the following as a single logarithm.

(a) $\log_6 3 + \log_6 12$

(b) $\log(x - 2) - \log x$

Solution

(a) The base of each logarithm is the same, 6. Use the Product Rule to write the sum of the logs as a single log.

$$\overset{\text{Product Rule}}{\underset{\downarrow}{}}$$

$$\log_6 3 + \log_6 12 = \log_6(3 \cdot 12)$$

$$3 \cdot 12 = 36: \quad = \log_6 36$$

$$6^2 = 36, \text{ so } \log_6 36 = 2: \quad = 2$$

(b) The base of each logarithm is the same, 10. Use the Quotient Rule to write the difference of the logs as a single log.

$$\log(x - 2) - \log x = \log\left(\frac{x - 2}{x}\right)$$

●

Quick ✓

In Problems 20 and 21, write each expression as a single logarithm.

20. $\log_8 4 + \log_8 16$ **21.** $\log_3(x + 4) - \log_3(x - 1)$

In order to use the Product Rule or the Quotient Rule to write the sum or difference of logarithms as a single logarithm, the coefficients of the logarithms must be 1. Therefore if logarithms have coefficients, first use the Power Rule to write the coefficient as a power. For example, write $2 \log x$ as $\log x^2$.

EXAMPLE 10   **Writing Expressions as Single Logarithms**

Write each of the following as a single logarithm.

(a) $2 \log_2(x - 1) + \frac{1}{2}\log_2 x$

(b) $\log(x - 1) + \log(x + 1) - 3 \log x$

Solution

(a) Each logarithm has the same base. Because the logarithms have coefficients, use the Power Rule to write the coefficients as exponents.

$$2 \log_2(x - 1) + \frac{1}{2}\log_2 x = \log_2(x - 1)^2 + \log_2 x^{\frac{1}{2}}$$

$$x^{\frac{1}{2}} = \sqrt{x}: \quad = \log_2(x - 1)^2 + \log_2 \sqrt{x}$$

$$\text{Product Rule:} \quad = \log_2\left[(x - 1)^2 \sqrt{x}\right]$$

(b) Work from left to right. Since the first two logs are being added, use the Product Rule to write these logs as a single log.

Product Rule
↓

$$\log(x-1) + \log(x+1) - 3\log x = \log[(x-1)(x+1)] - 3\log x$$

$(x-1)(x+1) = x^2 - 1:\quad = \log(x^2 - 1) - 3\log x$

Write the coefficient as an exponent:$\quad = \log(x^2 - 1) - \log x^3$

Quotient Rule:$\quad = \log\left(\dfrac{x^2 - 1}{x^3}\right)$

Quick ✔

In Problems 22 and 23, write each expression as a single logarithm.

22. $\log_5 x - 3\log_5 2$ $\qquad$ **23.** $\log_2(x+1) + \log_2(x+2) - 2\log_2 x$

▶ ❹ Evaluate a Logarithm Whose Base Is Neither 10 Nor *e*

Section 11.3 discussed how to use a calculator to approximate logarithms whose base was either 10 (the common logarithm) or *e* (the natural logarithm). But what if the base of the logarithm is neither 10 nor *e*? To determine how to evaluate these types of logarithms, the following property is needed. In this property, $M, N,$ and a are positive numbers and $a \neq 1$. The property is a result of the fact that the logarithmic function is a one-to-one function.

One-to-One Property of Logarithms

If $\quad M = N, \quad$ then $\quad \log_a M = \log_a N.$

EXAMPLE 11 **Approximating a Logarithm Whose Base Is Neither 10 Nor *e***

Approximate $\log_2 5$. Round the answer to three decimal places.

Solution
Let $y = \log_2 5$ and then convert the logarithmic equation to an equivalent exponential equation using the fact that if $y = \log_a x$, then $x = a^y$.

$$y = \log_2 5$$
$$2^y = 5$$

Now, use the One-to-One Property and "take the logarithm of both sides." So that a calculator can be used, take either the common log or the natural log of both sides (it doesn't matter which). Let's take the common log of both sides.

In Other Words

"Taking the logarithm" of both sides of an equation is the same type of approach as squaring both sides of an equation.

$$\log 2^y = \log 5$$

Use the Power Rule, $\log a^r = r\log a$:$\quad y\log 2 = \log 5$

Divide both sides by $\log 2$:$\quad y = \dfrac{\log 5}{\log 2}$

Using a calculator, find $\dfrac{\log 5}{\log 2} \approx 2.322$. Thus $\log_2 5 \approx 2.322$.

There is another way to approximate a logarithm whose base is neither 10 nor *e*. It is called the **Change-of-Base Formula.**

Change-of-Base Formula

If $a \neq 1, b \neq 1,$ and M are positive real numbers, then

$$\log_a M = \frac{\log_b M}{\log_b a}$$

Because many calculators have keys for only the common logarithm, $\boxed{\log}$, and the natural logarithm, $\boxed{\ln}$, we use the Change-of-Base Formula with either $b = 10$ or $b = e$:

$$\log_a M = \frac{\log M}{\log a} \quad \text{or} \quad \log_a M = \frac{\ln M}{\ln a}$$

EXAMPLE 12   **Using the Change-of-Base Formula**

Approximate $\log_4 45$. Round your answer to three decimal places.

Solution

Use the Change-of-Base Formula.

Using common logarithms: $\log_4 45 = \dfrac{\log 45}{\log 4}$

≈ 2.746

Using natural logarithms: $\log_4 45 = \dfrac{\ln 45}{\ln 4}$

≈ 2.746

Thus $\log_4 45 \approx 2.746$.

Quick ✓

24. $\log_3 10 = \dfrac{\log \underline{\quad}}{\log \underline{\quad}} = \dfrac{\ln \underline{\quad}}{\ln \underline{\quad}}$

In Problems 25 and 26, approximate each logarithm. Round your answers to three decimal places.

25. $\log_3 32$ **26.** $\log_{\sqrt{2}} \sqrt{7}$

Let's review all the properties of logarithms.

Properties of Logarithms

In the following properties, $M, N, a,$ and b are positive real numbers, $a \neq 1, b \neq 1,$ and r is any real number.

- **Inverse Properties of Logarithms**
  $a^{\log_a M} = M$ and $\log_a a^r = r$

- **The Product Rule of Logarithms**
  $\log_a(MN) = \log_a M + \log_a N$

- **The Quotient Rule of Logarithms**
  $\log_a\left(\dfrac{M}{N}\right) = \log_a M - \log_a N$

- $\log_a 1 = 0$

- **The Power Rule of Logarithms**
  $\log_a M^r = r \log_a M$

- **Change-of-Base Formula**
  $\log_a M = \dfrac{\log_b M}{\log_b a} = \dfrac{\log M}{\log a} = \dfrac{\ln M}{\ln a}$

- **One-to-One Property**
  If $M = N$, then $\log_a M = \log_a N$

- $\log_a a = 1$

11.4 Exercises   MyMathLab®   Exercise numbers in **green** have complete video solutions in MyMathLab or may be accessed using the QR code to the right.

*Problems **1–26** are the Quick ✓ s that follow the **EXAMPLES**.*

Building Skills

In Problems 27–38, use properties of logarithms to find the exact value of each expression. Do not use a calculator. See Objective 1.

27. $\log_2 2^3$ **28.** $\log_5 5^{-3}$

29. $\ln e^{-7}$ **30.** $\ln e^9$

31. $3^{\log_3 5}$ **32.** $5^{\log_5 \sqrt{2}}$

33. $e^{\ln 2}$ **34.** $e^{\ln 10}$

35. $\log_7 7$ **36.** $\log_5 5$

37. $\log 1$ **38.** $\ln 1$

In Problems 39–46, suppose that $\ln 2 = a$ *and* $\ln 3 = b$. *Use properties of logarithms to write each logarithm in terms of a and b. See Objective 2.*

39. $\ln 6$

40. $\ln\dfrac{3}{2}$

41. $\ln 9$

42. $\ln 4$

43. $\ln 12$

44. $\ln 18$

45. $\ln\sqrt{2}$

46. $\ln\sqrt[4]{3}$

In Problems 47–68, write each expression as a sum or difference of logarithms. Express exponents as factors. See Objective 2.

47. $\log(ab)$

48. $\log_4\left(\dfrac{a}{b}\right)$

49. $\log_5 x^4$

50. $\log_3 z^{-2}$

51. $\log_2(xy^2)$

52. $\log_3(a^3 b)$

53. $\log_5(25x)$

54. $\log_2(8z)$

55. $\log_7\left(\dfrac{49}{y}\right)$

56. $\log_2\left(\dfrac{16}{p}\right)$

57. $\ln(e^2 x)$

58. $\ln\left(\dfrac{x}{e^3}\right)$

59. $\log_3(27\sqrt{x})$

60. $\log_2(32\sqrt[4]{z})$

61. $\log_5\left(x^2\sqrt{x^2 + 1}\right)$

62. $\log_3\left(x^3\sqrt{x^2 - 1}\right)$

63. $\log\left(\dfrac{x^4}{\sqrt[3]{x - 1}}\right)$

64. $\ln\left(\dfrac{\sqrt[5]{x}}{(x + 2)^2}\right)$

65. $\log_7\sqrt{\dfrac{x + 1}{x}}$

66. $\log_6\sqrt[3]{\dfrac{x - 2}{x + 1}}$

67. $\log_2\left[\dfrac{x(x - 1)^2}{\sqrt{x + 1}}\right]$

68. $\log_4\left[\dfrac{x^3(x - 3)}{\sqrt[3]{x + 1}}\right]$

In Problems 69–92, write each expression as a single logarithm. See Objective 3.

69. $\log 25 + \log 4$

70. $\log_4 32 + \log_4 2$

71. $\log x + \log 3$

72. $\log_2 6 + \log_2 z$

73. $\log_3 36 - \log_3 4$

74. $\log_2 48 - \log_2 3$

75. $10^{\log 8 - \log 2}$

76. $e^{\ln 24 - \ln 3}$

77. $3\log_3 x$

78. $8\log_2 z$

79. $\log_4(x + 1) - \log_4 x$

80. $\log_5(2y - 1) - \log_5 y$

81. $2\ln x + 3\ln y$

82. $4\log_2 a + 2\log_2 b$

83. $\dfrac{1}{2}\log_3 x + 3\log_3(x - 1)$

84. $\dfrac{1}{3}\log_4 z + 2\log_4(2z + 1)$

85. $\log x^5 - 3\log x$

86. $\log_7 x^4 - 2\log_7 x$

87. $\dfrac{1}{2}\left[3\log x + \log y\right]$

88. $\dfrac{1}{3}\left[\ln(x - 1) + \ln(x + 1)\right]$

89. $\log_8(x^2 - 1) - \log_8(x + 1)$

90. $\log_5(x^2 + 3x + 2) - \log_5(x + 2)$

91. $18\log\sqrt{x} + 9\log\sqrt[3]{x} - \log 10$

92. $10\log_4\sqrt[5]{x} + 4\log_4\sqrt{x} - \log_4 16$

In Problems 93–100, use the Change-of-Base Formula and a calculator to evaluate each logarithm. Round your answer to three decimal places. See Objective 4.

93. $\log_2 10$

94. $\log_3 18$

95. $\log_8 3$

96. $\log_7 5$

97. $\log_{1/3} 19$

98. $\log_{1/4} 3$

99. $\log_{\sqrt{2}} 5$

100. $\log_{\sqrt{3}}\sqrt{6}$

Applying the Concepts

101. Find the value of
$\log_2 3 \cdot \log_3 4 \cdot \log_4 5 \cdot \log_5 6 \cdot \log_6 7 \cdot \log_7 8$.

102. Find the value of $\log_2 4 \cdot \log_4 6 \cdot \log_6 8$.

103. Find the value of
$\log_2 3 \cdot \log_3 4 \cdot \,\cdots\, \cdot \log_n(n + 1) \cdot \log_{n+1} 2$.

104. Find the value of
$\log_3 3 \cdot \log_3 9 \cdot \log_3 27 \cdot \,\cdots\, \cdot \log_3 3^n$.

Extending the Concepts

105. Show that
$$\log_a\left(x + \sqrt{x^2 - 1}\right) + \log_a\left(x - \sqrt{x^2 - 1}\right) = 0.$$

106. Show that
$$\log_a\left(\sqrt{x} + \sqrt{x-1}\right) + \log_a\left(\sqrt{x} - \sqrt{x-1}\right) = 0.$$

107. If $f(x) = \log_a x$, show that $f(AB) = f(A) + f(B)$.

108. Find the domain of $f(x) = \log_a x^2$ and the domain of $g(x) = 2\log_a x$. Since $\log_a x^2 = 2\log_a x$, how can it be that the domains are not equal? Write a brief explanation.

Explaining the Concepts

109. State the Product Rule for Logarithms in your own words.

110. State the Quotient Rule for Logarithms in your own words.

111. Write an example to illustrate
$$\log_2(x + y) \neq \log_2 x + \log_2 y.$$

112. Write an example to illustrate
$$(\log_a x)^r \neq r\log_a x.$$

Synthesis Review

In Problems 113–118 solve each equation.

113. $4x + 3 = 13$

114. $-3x + 10 = 4$

115. $x^2 + 4x + 2 = 0$

116. $3x^2 = 2x + 1$

117. $\sqrt{x + 2} - 3 = 4$

118. $\sqrt[3]{2x} - 2 = -5$

Technology Exercises

We can use the Change-of-Base Formula to graph any logarithmic function using technology. For example, to graph $f(x) = \log_2 x$ on a graphing calculator, graph $Y_1 = \dfrac{\log x}{\log 2}$ or $Y_1 = \dfrac{\ln x}{\ln 2}$.

In Problems 119–122, graph each logarithmic function using technology. State the domain and the range of each function.

119. $f(x) = \log_3 x$

120. $f(x) = \log_5 x$

121. $F(x) = \log_{1/2} x$

122. $G(x) = \log_{1/3} x$

11.5 Exponential and Logarithmic Equations

Objectives

1 Solve Logarithmic Equations Using the Properties of Logarithms

2 Solve Exponential Equations

3 Solve Equations Involving Exponential Models

Are You Prepared for This Section?

Before getting started, complete the following problems. If you get a problem wrong, go back to the section cited and review the material.

P1. Solve: $2x + 5 = 13$ [Section 2.2, pp. 91–92]

P2. Solve: $x^2 - 4x = -3$ [Section 6.6, pp. 410–414]

P3. Solve: $3a^2 = a + 5$ [Section 10.2, pp. 703–707]

P4. Solve: $(x + 3)^2 + 2(x + 3) - 8 = 0$ [Section 10.3, pp. 717–721]

▶ **1** Solve Logarithmic Equations Using the Properties of Logarithms

Section 11.3 showed how to solve logarithmic equations of the form $\log_a x = y$ by changing the logarithmic equation to an equivalent exponential equation. If a logarithmic equation has more than one logarithm in it, however, properties of logarithms need to be used to solve the equation.

The last section introduced the One-to-One Property that if $M = N$, then $\log_a M = \log_a N$. It turns out that the converse of this property is true as well.

> #### One-to-One Property of Logarithms
>
> In the following property, M, N, and a are positive real numbers, with $a \neq 1$.
>
> $$\text{If } \log_a M = \log_a N, \text{ then } M = N.$$

This property is useful for solving equations that contain logarithms with the same base by setting the arguments equal to each other.

EXAMPLE 1

Solving a Logarithmic Equation

Solve: $2 \log_3 x = \log_3 25$

Solution

Both logarithms have the same base, 3, so if the equation is written in the form $\log_a M = \log_a N$, the One-to-One Property can be used.

$$2 \log_3 x = \log_3 25$$

$r \log_a M = \log_a M^r$: $\log_3 x^2 = \log_3 25$

Set the arguments equal to each other: $x^2 = 25$

Square Root Property: $x = -5 \quad \text{or} \quad x = 5$

The apparent solution $x = -5$ is extraneous because the argument of a logarithm must be greater than zero, and -5 causes the argument to be negative. Now check the other apparent solution.

Check $x = 5$: $2 \log_3 5 \stackrel{?}{=} \log_3 25$

$2 \log_3 5 \stackrel{?}{=} \log_3 5^2$

$\log_a m^r = r \log_a m$: $2 \log_3 5 = 2 \log_3 5 \quad$ True

The solution set is $\{5\}$.

●

> **Quick ✔**
>
> **1.** If $\log_a M = \log_a N$, then _____. **2.** Solve: $2 \log_4 x = \log_4 9$

If a logarithmic equation contains more than one logarithm on one side of the equation, then properties of logarithms can be used to rewrite the equation as a single logarithm. Once again, use properties to reduce an equation into a form that is familiar. In this case, use properties of logarithms to express the sum or difference of logarithms as a single logarithm. Then rewrite the equation in exponential form and solve for the unknown.

EXAMPLE 2

Solving a Logarithmic Equation

Solve: $\log_2(x - 2) + \log_2 x = 3$

Solution

Begin by rewriting the left side of the equation as a single logarithm.

$$\log_2(x - 2) + \log_2 x = 3$$

$\log_a M + \log_a N = \log_a(MN)$: $\log_2[x(x - 2)] = 3$

If $y = \log_a M$, then $a^y = M$: $2^3 = x(x - 2)$

Distribute: $8 = x^2 - 2x$

Write in standard form: $x^2 - 2x - 8 = 0$

Factor: $(x - 4)(x + 2) = 0$

Zero-Product Property: $x - 4 = 0 \quad \text{or} \quad x + 2 = 0$

$x = 4 \quad \text{or} \quad x = -2$

Work Smart

The apparent solution $x = -2$ in Example 2 is extraneous because it results in our attempting to find the log of a negative number, not because -2 is negative.

Check $x = 4$:

$\log_2(4 - 2) + \log_2 4 \stackrel{?}{=} 3$

$\log_2 2 + \log_2 4 \stackrel{?}{=} 3$

$1 + 2 = 3 \quad$ True

$x = -2$:

$\log_2(-2 - 2) + \log_2(-2) \stackrel{?}{=} 3$

$x = -2$ is extraneous because it causes the argument to be negative.

The solution set is $\{4\}$.

●

▶ ② Solve Exponential Equations

Section 11.2 explained how to solve exponential equations using the fact that if $a^u = a^v$, then $u = v$. However, in many situations both sides of the equation cannot be written with the same base.

EXAMPLE 3   **Using Logarithms to Solve Exponential Equations**

Solve: $3^x = 5$

Solution
Because 5 cannot be written so that it is 3 raised to some integer power, write the equation $3^x = 5$ as a logarithm.

$$3^x = 5$$

If $a^y = x$, then $y = \log_a x$: $x = \log_3 5$ Exact solution

To find a decimal approximation to the solution, use the Change-of-Base Formula.

$$x = \log_3 5 = \frac{\log 5}{\log 3}$$

$$\approx 1.465 \quad \text{Approximate solution}$$

An alternative approach to solving the equation would be to take either the natural logarithm or the common logarithm of both sides of the equation. If the natural logarithm of both sides of the equation is taken, the following results:

$$3^x = 5$$

If $M = N$, then $\ln M = \ln N$: $\ln 3^x = \ln 5$

$\log_a M^r = r \log_a M$: $x \ln 3 = \ln 5$

Divide both sides by $\ln 3$: $x = \dfrac{\ln 5}{\ln 3}$ Exact solution

$$\approx 1.465 \quad \text{Approximate solution}$$

The solution set is $\left\{ \dfrac{\ln 5}{\ln 3} \right\}$. If the common logarithm of both sides had been taken, the solution set would have been $\left\{ \dfrac{\log 5}{\log 3} \right\}$. ●

EXAMPLE 4   **Using Logarithms to Solve Exponential Equations**

Solve: $4e^{3x} = 10$

Solution
First isolate the exponential expression by dividing both sides of the equation by 4.

$$4e^{3x} = 10$$

$$e^{3x} = \frac{5}{2}$$

Work Smart

You could also take the natural logarithm of both sides:

$$\ln e^{3x} = \ln\left(\frac{5}{2}\right)$$

$$3x = \ln\left(\frac{5}{2}\right)$$

$$x = \frac{\ln\left(\frac{5}{2}\right)}{3}$$

Because $\frac{5}{2}$ cannot be expressed as e raised to an integer power, write the exponential equation as an equivalent logarithmic equation.

$$e^{3x} = \frac{5}{2}$$

If $e^y = x$, then $y = \ln x$: $\ln\left(\frac{5}{2}\right) = 3x$

Divide both sides by 3: $x = \dfrac{\ln\left(\dfrac{5}{2}\right)}{3}$ Exact solution

$x \approx 0.305$ Approximate solution

The check is left to you. The solution set is $\left\{\dfrac{\ln\left(\frac{5}{2}\right)}{3}\right\}$.   ●

Quick ✓

In Problems 7 and 8, solve each equation. Express answers in exact form and as a decimal rounded to three decimal places.

7. $e^{2x} = 5$

8. $3e^{-4x} = 20$

▶ ❸ **Solve Equations Involving Exponential Models**

Section 11.2 examined a variety of models from areas such as statistics, biology, and finance. The following examples, rather than evaluating the models at certain values of the independent variable, solve equations that involve the models.

EXAMPLE 5   **Radioactive Decay**

The half-life of plutonium-239 is 24,360 years. The maximum amount of plutonium-239 that an adult can handle without significant injury is 0.13 microgram ($= 0.000000013$ gram). Suppose a researcher has a 1-gram sample of plutonium-239. The amount A (in grams) of plutonium-239 after t years is given by

$$A(t) = 1\left(\frac{1}{2}\right)^{\frac{t}{24,360}}$$

(a) How long will it take until 0.9 gram of plutonium-239 is left in the sample?

(b) How long will it take until the 1-gram sample is safe–that is, until 0.000000013 gram is left?

Solution

(a) To find the time until $A = 0.9$ gram, solve the equation

$$0.9 = 1\left(\frac{1}{2}\right)^{\frac{t}{24,360}}$$

for t. How can t be removed from the exponent? Use the fact that $\log_a M^r = r \log_a M$ to move the variable.

$$\log 0.9 = \log\left(\frac{1}{2}\right)^{\frac{t}{24,360}}$$

$$\log 0.9 = \frac{t}{24,360}\log\left(\frac{1}{2}\right)$$

(continued)

Multiply both sides by 24,360: $\quad 24{,}360 \log 0.9 = t \log\left(\dfrac{1}{2}\right)$

Divide both sides by $\log\left(\dfrac{1}{2}\right)$: $\quad \dfrac{24{,}360 \log 0.9}{\log\left(\dfrac{1}{2}\right)} = t$

Thus $t = \dfrac{24{,}360 \log 0.9}{\log\left(\dfrac{1}{2}\right)} \approx 3702.8$. After approximately 3703 years, there will be 0.9 gram of plutonium-239 left.

(b) To determine the time until $A = 0.000000013$ gram, solve the equation

$$0.000000013 = 1 \cdot \left(\frac{1}{2}\right)^{\frac{t}{24{,}360}}$$

Take the logarithm of both sides: $\quad \log(0.000000013) = \log\left(\frac{1}{2}\right)^{\frac{t}{24{,}360}}$

$\log_a M^r = r \log_a M$: $\quad \log(0.000000013) = \dfrac{t}{24{,}360} \log\left(\dfrac{1}{2}\right)$

Multiply both sides by 24,360: $\quad 24{,}360 \log(0.000000013) = t \log\left(\dfrac{1}{2}\right)$

Divide both sides by $\log\left(\dfrac{1}{2}\right)$: $\quad \dfrac{24{,}360 \log(0.000000013)}{\log\left(\dfrac{1}{2}\right)} = t$

Thus $t = \dfrac{24{,}360 \log(0.000000013)}{\log\left(\dfrac{1}{2}\right)} \approx 638{,}156.8$. After approximately 638,157 years, the 1-gram sample will be safe! ●

Quick ✔

9. The half-life of thorium-227 is 18.72 days. Suppose a researcher has a 10-gram sample of thorium-227. The amount A (in grams) of thorium-227 after t days is given by

$$A(t) = 10\left(\frac{1}{2}\right)^{\frac{t}{18.72}}$$

(a) How long will it take until 9 grams of thorium-227 is left in the sample?

(b) How long will it take until 3 grams of thorium-227 is left in the sample?

Now let's look at an example involving compound interest. Remember, the compound interest formula states that the future value of P dollars invested in an account paying an annual interest rate r, compounded n times per year for t years, is given by $A = P\left(1 + \dfrac{r}{n}\right)^{nt}$.

EXAMPLE 6 **Future Value of Money**

Suppose you deposit $5000 into a Roth IRA today. If the deposit earns 8% interest compounded quarterly, when will it be worth

(a) $7500?

(b) $10,000? That is, when will your money have doubled?

Solution

With $P = 5000$, $r = 0.08$, and $n = 4$ (compounded quarterly),

$$A = 5000\left(1 + \frac{0.08}{4}\right)^{4t} \quad \text{or} \quad A = 5000(1.02)^{4t}$$

(a) Find the time t when $A = 7500$. That is, solve

$$7500 = 5000(1.02)^{4t}$$

Divide both sides by 5000:	$1.5 = (1.02)^{4t}$
Take the logarithm of both sides:	$\log 1.5 = \log(1.02)^{4t}$
$\log_a M^r = r \log_a M$:	$\log 1.5 = 4t \log(1.02)$
Divide both sides by 4 log(1.02):	$\dfrac{\log 1.5}{4 \log(1.02)} = t$

So $t = \dfrac{\log 1.5}{4 \log(1.02)} \approx 5.12$. After approximately 5.12 years
(5 years, 1.4 months), the account will be worth \$7500.

(b) Find the time t when $A = 10{,}000$. That is, solve

$$10{,}000 = 5000(1.02)^{4t}$$

Divide both sides by 5000:	$2 = (1.02)^{4t}$
Take the logarithm of both sides:	$\log 2 = \log(1.02)^{4t}$
$\log_a M^r = r \log_a M$:	$\log 2 = 4t \log(1.02)$
Divide both sides by 4 log(1.02):	$\dfrac{\log 2}{4 \log(1.02)} = t$

Thus $t = \dfrac{\log 2}{4 \log(1.02)} \approx 8.75$. After approximately 8.75 years (8 years,
9 months), the account will be worth \$10,000.

Quick ✓

10. Suppose that you deposit \$2000 into a Roth IRA today. If the deposit earns 6%
interest compounded monthly, how long will it be before the account is worth
(a) \$3000?
(b) \$4000? That is, when will your money have doubled?

11.5 Exercises MyMathLab®

Exercise numbers in **green** have
complete video solutions in MyMathLab or
may be accessed using the QR code to the right.

*Problems **1–10** are the **Quick ✓**s that follow the **EXAMPLES**.*

Building Skills

In Problems 11–28, solve each equation. See Objective 1.

11. $\log_2 x = \log_2 7$

12. $\log_5 x = \log_5 13$

13. $2 \log_3 x = \log_3 81$

14. $2 \log_3 x = \log_3 4$

15. $\log_6(3x + 1) = \log_6 10$ **16.** $\log(2x - 3) = \log 11$

17. $\dfrac{1}{2} \ln x = 2 \ln 3$

18. $\dfrac{1}{2} \log_2 x = 2 \log_2 2$

19. $\log_2(x + 3) + \log_2 x = 2$

20. $\log_2(x - 7) + \log_2 x = 3$

21. $\log_2(x + 2) + \log_2(x + 5) = \log_2 4$

22. $\log_2(x + 5) + \log_2(x + 4) = \log_2 2$

23. $\log(x + 3) - \log x = 1$

24. $\log_3(x + 5) - \log_3 x = 2$

25. $\log_4(x + 5) - \log_4(x - 1) = 2$

26. $\log_3(x + 2) - \log_3(x - 2) = 4$

27. $\log_4(x + 8) + \log_4(x + 6) = \log_4 3$

28. $\log_3(x + 8) + \log_3(x + 4) = \log_3 5$

In Problems 29–48, solve each equation. Express irrational solutions in exact form and as a decimal rounded to three decimal places. See Objective 2.

29. $2^x = 10$

30. $3^x = 8$

31. $5^x = 20$

32. $4^x = 20$

33. $\left(\dfrac{1}{2}\right)^x = 7$

34. $\left(\dfrac{1}{2}\right)^x = 10$

35. $e^x = 5$

36. $e^x = 3$

37. $10^x = 5$

38. $10^x = 0.2$

39. $3^{2x} = 13$

40. $2^{2x} = 5$

41. $\left(\dfrac{1}{2}\right)^{4x} = 13$

42. $\left(\dfrac{1}{3}\right)^{2x} = 4$

43. $4 \cdot 2^x + 3 = 8$

44. $3 \cdot 4^x - 5 = 10$

45. $-3e^x = -18$

46. $\dfrac{1}{2}e^x = 4$

47. $0.2^{x+1} = 3^x$

48. $0.4^x = 2^{x-3}$

Mixed Practice

In Problems 49–64, solve each equation. Express irrational solutions in exact form and as a decimal rounded to three decimal places.

49. $\log_4 x + \log_4(x - 6) = 2$ **50.** $\log_6 x + \log_6(x + 5) = 2$

51. $5^{3x} = 7$ **52.** $3^{2x} = 4$

53. $3 \log_2 x = \log_2 8$ **54.** $5 \log_4 x = \log_4 32$

55. $\dfrac{1}{3}e^x = 5$ **56.** $-4e^x = -16$

57. $\left(\dfrac{1}{4}\right)^{x+1} = 8^x$ **58.** $9^x = 27^{x-4}$

59. $\log_3 x^2 = \log_3 16$ **60.** $\log_7 x^2 = \log_7 8$

61. $\log_2(x) + \log_2(x - 2) = \log_2(x + 4)$

62. $\log_3 x + \log_3(x - 4) = \log_3(x + 6)$

63. $\log_4 x + \log_4(x - 4) = \log_4 3$

64. $\log_6(x + 1) + \log_6(x - 5) = \log_6 2$

Applying the Concepts

65. A Population Model According to the U.S. Census Bureau, the population of the United States in 2016 was 326 million people. In addition, the population of the United States was growing at a rate of 0.99% per year. Assuming that this growth rate continues, the model $P(t) = 326(1.0099)^{t-2016}$ represents the population P (in millions of people) in year t.

   (a) According to this model, when will the population of the United States be 350 million people?

   (b) According to this model, when will the population of the United States be 471 million people?

66. A Population Model According to the *United States Census Bureau*, the population of the world in 2016 was 7563 million people. In addition, the population of the world was growing at a rate of 1.02% per year. Assuming that this growth rate continues, the model $P(t) = 7563(1.0102)^{t-2016}$ represents the population P (in millions of people) in year t.

(a) According to this model, when will the population of the world be 9.84 billion people?

(b) According to this model, when will the population of the world be 11.58 billion people?

67. Time Is Money Suppose that you deposit $5000 in a certificate of deposit (CD) today. If the deposit earns 2% interest compounded monthly, when will the account be worth

(a) 7000?

(b) 10,000? That is, when will your money have tripled

68. Time Is Money Suppose that you deposit $8000 in a certificate of deposit (CD) today. If the deposit earns 2% interest compounded quarterly, when will it be worth

(a) 10,000?

(b) 24,000? That is, when will your money have tripled?

69. Depreciation Based on data obtained from the *Kelley Blue Book*, the value V of a Ford Focus that is t years old can be modeled by $V(t) = 19,841(0.88)^t$.

(a) According to the model, when will the car be worth $15,000?

(b) According to the model, when will the car be worth $5000?

(c) According to the model, when will the car be worth $1000?

70. Depreciation Based on data obtained from the *Kelley Blue Book*, the value V of a Chevy Malibu that is t years old can be modeled by $V(t) = 25,258(0.84)^t$.

(a) According to the model, when will the car be worth $15,000?

(b) According to the model, when will the car be worth $5000?

(c) According to the model, when will the car be worth $1000?

71. Radioactive Decay The half-life of beryllium-11 is 13.81 seconds. Suppose that a researcher possesses a 100-gram sample of beryllium-11. The amount A (in grams) of beryllium-11 after t seconds is given by

$$A(t) = 100\left(\frac{1}{2}\right)^{\frac{t}{13.81}}$$

(a) When will there be 90 grams of beryllium-11 left in the sample?

(b) When will 25 grams be left?

(c) When will 10 grams of beryllium-11 be left?

72. Radioactive Decay The half-life of carbon-10 is 19.255 seconds. Suppose that a researcher possesses a 100-gram sample of carbon-10. The amount A (in grams) of carbon-10 after t seconds is given by

$$A(t) = 100\left(\frac{1}{2}\right)^{\frac{t}{19.255}}$$

(a) When will there be 90 grams of carbon-10 left in the sample?

(b) When will 25 grams be left?

(c) When will 10 grams be left?

For Exercises 73 and 74, use Newton's Law of Cooling, which states that the temperature of a heated object decreases exponentially over time toward the temperature of the surrounding medium.

73. Newton's Law of Cooling Suppose that a pizza is removed from a 400°F oven and placed in a room where the temperature is 70°F. The temperature u (in °F) of the pizza at time t (in minutes) can be modeled by $u(t) = 70 + 330e^{-0.072t}$.

(a) According to the model, when will the temperature of the pizza be 300°F?

(b) According to the model, when will the temperature of the pizza be 220°F?

74. Newton's Law of Cooling Suppose that coffee that is 170°F is poured into a coffee mug and allowed to cool in a room where the temperature is 70°F. The temperature u (in °F) of the coffee at time t (in minutes) can be modeled by $u(t) = 70 + 100e^{-0.045t}$.

(a) According to the model, when will the temperature of the coffee be 120°F?

(b) According to the model, when will the temperature of the coffee be 100°F?

75. Learning Curve Suppose that a student has 200 vocabulary words to learn. If a student learns 20 words in 30 minutes, the function

$$L(t) = 200(1 - e^{-0.0035t})$$

models the number of words L that the student will learn in t minutes.

(a) How long will it take the student to learn 50 words?

(b) How long will it take the student to learn 150 words?

76. Learning Curve Suppose that a student has 50 biology terms to learn. If a student learns 10 terms in 30 minutes, the function

$$L(t) = 50(1 - e^{-0.0223t})$$

models the number of terms L that the student will learn in t minutes.

(a) How long will it take the student to learn 10 words?

(b) How long will it take the student to learn 40 words?

Extending the Concepts

77. The Rule of 72 The Rule of 72 states that the time for an investment to double in value is approximately given by 72 divided by the annual interest rate. For example, an investment earning 10% annual interest will double in approximately $\frac{72}{10} = 7.2$ years.

(a) According to the Rule of 72, approximately how long will it take an investment to double if it earns 8% annual interest?

(b) Derive a formula that can be used to find the number of years required for an investment to double. (*Hint*: Let $A = 2P$ in the formula $A = P\left(1 + \frac{r}{n}\right)^{nt}$ and solve for t.)

(c) Use the formula derived in part (b) to determine the exact amount of time it takes an investment to double that earns 8% interest compounded monthly. Compare the result to the results given by the Rule of 72.

78. Critical Thinking Suppose you need to open a certificate of deposit (CD). Bank A offers 2% interest compounded daily, and Bank B offers 2.1% interest compounded quarterly. Which bank offers the better deal? Why?

79. Critical Thinking The bacteria in a 2-liter container double every minute. After 30 minutes the container is full. How long did it take to fill half the container?

Synthesis Review

In Problems 80–85, find the following values for each function.

    (a) $f(3)$ **(b)** $f(-2)$ **(c)** $f(0)$

80. $f(x) = 5x + 2$ **81.** $f(x) = -2x + 7$

82. $f(x) = \dfrac{x + 3}{x - 2}$ **83.** $f(x) = \dfrac{x}{x - 5}$

84. $f(x) = \sqrt{x + 5}$ **85.** $f(x) = 2^x$

Technology Exercises

The techniques for solving equations introduced in this chapter apply only to certain types of exponential or logarithmic equations. Solutions for other types of equations are usually studied in calculus using numerical methods. However, technology can be used to approximate solutions. Specifically, with a graphing calculator the INTERSECT feature can be used and in Desmos simply by clicking on the curve the "intersting points" including the intersection points will be displayed. For example, to solve $e^x = 3x + 2$, graph $Y_1 = e^x$ and $Y_2 = 3x + 2$ on the same screen in Desmos. Then click on one of the curves to display the "interesting points" and hover over those points with your mouse to see their coordinates. Each x-coordinate represents an approximate solution, as shown in Figure 27.

Figure 27

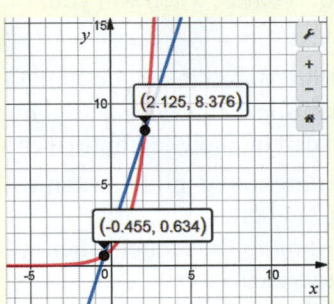

The solution set is $\{-0.46, 2.13\}$ rounded to two decimal places.

In Problems 86–93, solve each equation using technology. Express your answer rounded to two decimal places.

86. $e^x = -3x + 2$ **87.** $e^x = -2x + 5$

88. $e^x = x + 2$ **89.** $e^x = x^2$

90. $e^x + \ln(x) = 2$ **91.** $e^x - \ln(x) = 4$

92. $\ln x = x^2 + 1$ **93.** $\ln x = x^2 - 1$

Chapter 11 Activity: Correct the Quiz

Focus: Solving exponential and logarithmic equations

Time: 10–15 minutes

Group size: 2

In this activity you will work as a team to grade the student quiz shown on the next page. One of you will grade the odd questions, and the other will grade the even questions. If an answer is correct, mark it correct. If an answer is wrong, mark it wrong and show the correct answer.

    Once all of the quiz questions are graded, explain your results to each other and compute the final score for the quiz. Be prepared to discuss your results with the rest of the class.

Student Quiz

Name: *Ima Student*   Quiz Score: _____

Solve the following equations. Express any irrational answers in exact form.

(1) $\log_2 x = 3$	Answer: {9}
(2) $\log_{16} x = \dfrac{3}{4}$	Answer: {8}
(3) $\log(2x) - \log 6 = \log(x - 8)$	Answer: {12}
(4) $3^x = 27$	Answer: {9}
(5) $6^{2x} = 18$	Answer: $\left\{\dfrac{\log 18}{2 \log 6}\right\}$
(6) $4^{x+9} = 7$	Answer: $\left\{\dfrac{\log 7}{\log 4} - 9\right\}$
(7) $9^{3x-1} = 27^{4x}$	Answer: $\left\{-\dfrac{1}{3}\right\}$
(8) $\log_3(2x - 3) = 2$	Answer: $\left\{\dfrac{11}{2}\right\}$

Chapter 11 Review

Section 11.1   Composite Functions and Inverse Functions

KEY CONCEPTS

- $(f \circ g)(x) = f(g(x))$
- **Horizontal Line Test**
  If every horizontal line intersects the graph of a function f in at most one point, then f is one-to-one.
- For the inverse of a function also to be a function, the function must be one-to-one.
- **Relation Between the Domain and Range of a Function and Its Inverse**
  All elements in the domain of a one-to-one function are also elements in the range of its inverse.
  All elements in the range of a one-to-one function are also elements in the domain of its inverse.
- The graph of a one-to-one function f and the graph of its inverse are symmetric with respect to the line $y = x$.
- $f^{-1}(f(x)) = x$ for every x in the domain of f and
  $f(f^{-1}(x)) = x$ for every x in the domain of f^{-1}

KEY TERMS

Composition
Composite function
One-to-one
Inverse function

You Should be Able To...	EXAMPLE	Review Exercises
1 Form the composite function (p. 778)	Examples 1 and 2	1–8
2 Determine whether a function is one-to-one (p. 780)	Examples 3 and 4	9–12
3 Find the inverse of a function defined by a map or set of ordered pairs (p. 782)	Examples 5 and 6	13–16
4 Obtain the graph of the inverse function from the graph of a function (p. 784)	Example 7	17, 18
5 Find the inverse of a function defined by an equation (p. 785)	Examples 8 and 9	19–22

In Problems 1–4, for the given functions f and g, find:

(a) $(f \circ g)(5)$ **(b)** $(g \circ f)(-3)$

(c) $(f \circ f)(-2)$ **(d)** $(g \circ g)(4)$

1. $f(x) = 3x + 5; g(x) = 2x - 1$

2. $f(x) = x - 3; g(x) = 5x + 2$

3. $f(x) = 2x^2 + 1; g(x) = x + 5$

4. $f(x) = x - 3; g(x) = x^2 + 1$

In Problems 5–8, for the given functions f and g, find:

(a) $(f \circ g)(x)$ **(b)** $(g \circ f)(x)$

(c) $(f \circ f)(x)$ **(d)** $(g \circ g)(x)$

5. $f(x) = x + 1; g(x) = 5x$

6. $f(x) = 2x - 3; g(x) = x + 6$

7. $f(x) = x^2 + 1; g(x) = 2x + 1$

8. $f(x) = \dfrac{2}{x + 1}; g(x) = \dfrac{1}{x}$

In Problems 9–12, determine which of the functions are one-to-one.

9. $\{(-5, 8), (-3, 2), (-1, 8), (0, 12), (1, 15)\}$

10. $\{(-4, 2), (-2, 1), (0, 0), (1, -1), (2, 8)\}$

11.
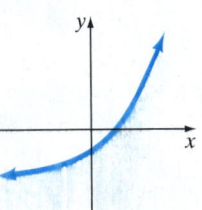

12.
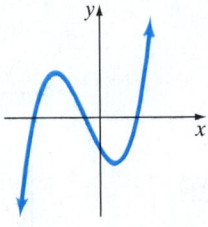

In Problems 13–16, find the inverse of each one-to-one function.

13.
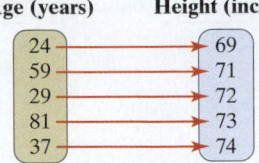

14.
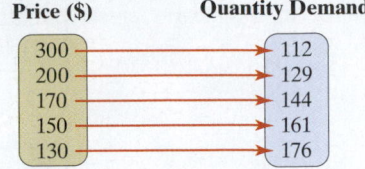

15. $\{(-5, 3), (-3, 1), (1, -3), (2, 9)\}$

16. $\{(-20, 1), (-15, 4), (5, 3), (25, 2)\}$

In Problems 17 and 18, the graph of a one-to-one function f is given. Draw the graph of the inverse function f^{-1}.

17.
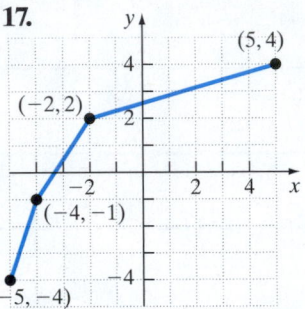

18.
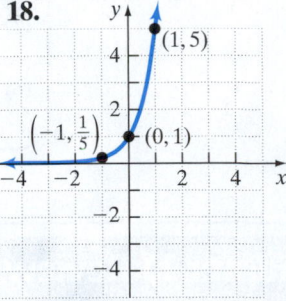

In Problems 19 and 22, the equation of a one-to-one function f is given. Find the equation of the inverse function f^{-1}.

19. $f(x) = 5x$ **20.** $H(x) = 2x + 7$

21. $P(x) = \dfrac{4}{x + 2}$ **22.** $g(x) = 2x^3 - 1$

Section 11.2   Exponential Functions

KEY CONCEPTS

- **Properties of Exponential Functions of the Form $f(x) = a^x$, $a > 0$, $a \neq 1$**

  1. The domain is the set of all real numbers. The range is the set of all positive real numbers.
  2. There are no x-intercepts. The y-intercept is $(0, 1)$.
  3. The graph of an exponential function contains the points $\left(-1, \dfrac{1}{a}\right)$, $(0, 1)$, and $(1, a)$.

- **Property for Solving Exponential Equations of the Form $a^u = a^v$**
  If $a^u = a^v$, then $u = v$.

KEY TERMS

Exponential function
The number e
Exponential equation
Half-life
Payment period
Compound interest

You Should Be Able To...	EXAMPLE	Review Exercises
① Evaluate exponential expressions (p. 792)	Example 1	25–25
② Graph exponential functions (p. 793)	Examples 2 through 4	26–29
③ Define the number e (p. 796)		30
④ Solve exponential equations (p. 797)	Examples 5 and 6	31–36
⑤ Use exponential models that describe our world (p. 799)	Examples 7 through 9	37–40

In Problems 23–25, approximate each number using a calculator. Express your answer rounded to three decimal places.

23. (a) $7^{1.7}$ **(d)** $7^{1.7321}$
(b) $7^{1.73}$ **(e)** $7^{\sqrt{3}}$
(c) $7^{1.732}$

24. (a) $10^{3.1}$ **(d)** $10^{3.1416}$
(b) $10^{3.14}$ **(e)** 10^{π}
(c) $10^{3.142}$

25. (a) $e^{0.5}$ **(d)** $e^{-0.8}$
(b) e^{-1} **(e)** $e^{\sqrt{\pi}}$
(c) $e^{1.5}$

In Problems 26–29, graph each function. State the domain and the range of the function.

26. $f(x) = 9^x$ **27.** $g(x) = \left(\dfrac{1}{9}\right)^x$

28. $H(x) = 4^{x-2}$ **29.** $h(x) = 4^x - 2$

30. State the definition of the number e.

In Problems 31–36, solve each equation.

31. $2^x = 64$ **32.** $25^{x-2} = 125$

33. $27^x \cdot 3^{x^2} = 9^2$ **34.** $\left(\dfrac{1}{4}\right)^x = 16$

35. $(e^2)^{x-1} = e^x \cdot e^7$ **36.** $(2^x)^x = 512$

37. Future Value of Money Suppose that you deposit $2500 into a traditional IRA that pays 4.5% annual interest. How much money will you have after 25 years if interest is compounded

(a) annually?   **(b)** quarterly?
(c) monthly?   **(d)** daily?

38. Radioactive Decay The half-life of the radioactive gas radon is 3.5 days. Suppose a researcher possesses a 100-gram sample of radon gas. The amount A (in grams) of radon after t days is given by $A(t) = 100\left(\dfrac{1}{2}\right)^{\frac{t}{3.5}}$. How much radon gas is left in the sample after

(a) 1 day?   **(b)** 3.5 days?   **(c)** 7 days?   **(d)** 30 days?

39. A Population Model According to the U.S. Census Bureau, the population of Nevada in 2015 was 2.891 million people. In addition, the population of Nevada was growing at a rate of 1.6% per year. Assuming that this growth rate continues, the model $P(t) = 2.891(1.016)^{t-2015}$ represents the population (in millions of people) in year t. According to this model, what will be the population of Nevada

(a) in 2025?   **(b)** in 2030?

40. Newton's Law of Cooling A baker removes a cake from a 350°F oven and places it in a room whose temperature is 72°F. According to Newton's Law of Cooling, the temperature u (in °F) of the cake at time t (in minutes) can be modeled by $u(t) = 72 + 278e^{-0.0835t}$. According to this model, what will be the temperature of the cake

(a) after 15 minutes?   **(b)** after 30 minutes?

Section 11.3 Logarithmic Functions

KEY CONCEPTS

- **The Logarithmic Function to the Base a**
  $y = \log_a x$ is equivalent to $x = a^y$.
- **Properties of the Logarithmic Function of the Form $f(x) = \log_a x, a > 0, a \neq 1$**
  1. The domain is the set of all positive real numbers. The range is the set of all real numbers.
  2. There is no y-intercept. The x-intercept is $(0, 1)$.
  3. The graph of the logarithmic function contains the points $\left(\dfrac{1}{a}, -1\right)$, $(1, 0)$, and $(a, 1)$.

KEY TERMS

Logarithmic function
Natural logarithm function
Common logarithm function
Logarithmic equation
Intensity of a sound wave
Decibels
Loudness

You Should Be Able To...	EXAMPLE	Review Exercises
❶ Change exponential equations to logarithmic equations (p. 807)	Example 1	41–44
❷ Change logarithmic equations to exponential equations (p. 807)	Example 2	45–48
❸ Evaluate logarithmic functions (p. 808)	Examples 3 and 4	49–52
❹ Determine the domain of a logarithmic function (p. 810)	Example 5	53–56
❺ Graph logarithmic functions (p. 810)	Examples 6 and 7	57, 58
❻ Work with natural and common logarithms (p. 813)	Example 8	59–62
❼ Solve logarithmic equations (p. 813)	Examples 9 and 10	63–68
❽ Use logarithmic models that describe our world (p. 815)	Example 11	69, 70

In Problems 41–44, change each exponential equation to an equivalent equation involving a logarithm.

41. $3^4 = 81$

42. $4^{-3} = \dfrac{1}{64}$

43. $b^3 = 5$

44. $10^{3.74} = x$

In Problems 45–48, change each logarithmic equation to an equivalent equation involving an exponent.

45. $\log_8 2 = \dfrac{1}{3}$

46. $\log_5 18 = r$

47. $\ln(x + 3) = 2$

48. $\log x = -4$

In Problems 49–52, find the exact value of each logarithm without using a calculator.

49. $\log_8 128$

50. $\log_6 1$

51. $\log \dfrac{1}{100}$

52. $\log_9 27$

In Problems 53–56, find the domain of each function.

53. $f(x) = \log_2(x + 5)$

54. $g(x) = \log_8(7 - 3x)$

55. $h(x) = \ln(3x)$

56. $F(x) = \log_{1/3}(4x + 10)$

In Problems 57 and 58, graph each function.

57. $f(x) = \log_{5/2} x$

58. $g(x) = \log_{2/5} x$

In Problems 59–62, use a calculator to evaluate each expression. Round your answers to three decimal places.

59. $\ln 24$

60. $\ln \dfrac{5}{6}$

61. $\log 257$

62. $\log 0.124$

In Problems 63–68, solve each logarithmic equation.

63. $\log_7(4x - 19) = 2$

64. $\log_{1/3}(x^2 + 8x) = -2$

65. $\log_a \dfrac{4}{9} = -2$

66. $\ln e^{5x} = 30$

67. $\log(6 - 7x) = 3$

68. $\log_b 75 = 2$

69. Loudness of a Vacuum Cleaner A vacuum cleaner has an intensity level of 10^{-4} watt per square meter. What is the loudness, in decibels, of the vacuum cleaner?

70. The Great New Madrid Earthquake According to the United States Geological Survey, an earthquake on December 16, 1811, in New Madrid, Missouri, had a magnitude of approximately 8.0. What would have been the seismographic reading 100 kilometers from its epicenter?

Section 11.4 Properties of Logarithms

KEY CONCEPTS

- **Properties of Logarithms**
  For the following properties, a, M, and N are positive real numbers, with $a \neq 1$, and r is any real number.
  $\log_a a = 1$ $\log_a(MN) = \log_a M + \log_a N$
  $\log_a 1 = 0$ $\log_a M^r = r \log_a M$
  $a^{\log_a M} = M$ $\log_a\left(\dfrac{M}{N}\right) = \log_a M - \log_a N$
  $\log_a a^r = r$ If $M = N$, then $\log_a M = \log_a N$.

- **Change-of-Base Formula**
  If $a \neq 1$, $b \neq 1$, and M are positive real numbers, then $\log_a M = \dfrac{\log_b M}{\log_b a} = \dfrac{\log M}{\log a} = \dfrac{\ln M}{\ln a}$

You Should Be Able To...	EXAMPLE	Review Exercises
❶ Understand the properties of logarithms (p. 819)	Examples 1 through 3	71–76
❷ Write a logarithmic expression as a sum or difference of logarithms (p. 821)	Examples 4 through 8	77–80
❸ Write a logarithmic expression as a single logarithm (p. 824)	Examples 9 and 10	81–84
❹ Evaluate a logarithm whose base is neither 10 nor e (p. 825)	Examples 11 and 12	85–88

In Problems 71–76, use properties of logarithms to find the exact value of each expression. Do not use a calculator.

71. $\log_4 4^{21}$ **72.** $7^{\log_7 9.34}$

73. $\log_5 5$ **74.** $\log_9 1$

75. $\log_4 12 - \log_4 3$ **76.** $12^{\log_{12} 2 + \log_{12} 8}$

In Problems 77–80, write each expression as a sum and/or difference of logarithms. Write exponents as factors.

77. $\log_7\left(\dfrac{xy}{z}\right)$ **78.** $\log_3\left(\dfrac{81}{x^2}\right)$

79. $\log(1000r^4)$ **80.** $\ln\sqrt{\dfrac{x-1}{x}}$

In Problems 81–84, write each expression as a single logarithm.

81. $4\log_3 x + 2\log_3 y$ **82.** $\dfrac{1}{4}\ln x + \ln 7 - 2\ln 3$

83. $\log_2 3 - \log_2 6$

84. $\log_6(x^2 - 7x + 12) - \log_6(x - 3)$

In Problems 85–88, use the Change-of-Base Formula and a calculator to evaluate each logarithm. Round your answer to three decimal places.

85. $\log_6 50$ **86.** $\log_\pi 2$

87. $\log_{2/3} 6$ **88.** $\log_{\sqrt{5}} 20$

Section 11.5 Exponential and Logarithmic Equations

KEY CONCEPT

- In the following property, M, N, and a are positive real numbers, with $a \neq 1$.

  If $\log_a M = \log_a N$, then $M = N$.

You Should Be Able To...	EXAMPLE	Review Exercises
❶ Solve logarithmic equations using the properties of logarithms (p. 828)	Examples 1 and 2	89–92
❷ Solve exponential equations (p. 830)	Examples 3 and 4	93–96
❸ Solve equations involving exponential models (p. 831)	Examples 5 and 6	97, 98

In Problems 89–96, solve each equation. Express irrational solutions in exact form and as a decimal rounded to three decimal places.

89. $3\log_4 x = \log_4 1000$

90. $\log_3(x + 7) + \log_3(x + 6) = \log_3 2$

91. $\ln(x + 2) - \ln x = \ln(x + 1)$

92. $\dfrac{1}{3}\log_{12} x = 2\log_{12} 2$

93. $2^x = 15$

94. $10^{3x} = 27$

95. $\dfrac{1}{3}e^{7x} = 13$

96. $3^x = 2^{x+1}$

97. Radioactive Decay The half-life of the radioactive gas radon is 3.5 days. Suppose a researcher possesses a 100-gram sample of radon gas. The amount A (in grams) of radon after t days is given by

$$A(t) = 100\left(\frac{1}{2}\right)^{\frac{t}{3.5}}.$$

(a) When will 75 grams of radon gas be left in the sample?

(b) When will 1 gram of radon gas be left in the sample?

98. A Population Model According to the U.S. Census Bureau, the population of Nevada in 2015 was 2.891 million people. In addition, the population of Nevada was growing at a rate of 1.6% per year. Assuming that this growth rate continues, the model $P(t) = 2.891(1.016)^{t-2015}$ represents the population (in millions of people) in year t. According to this model, when will the population of Nevada be

(a) 3.13 million people?

(b) 5.285 million people?

Chapter 11 Test

Step-by-step test solutions are found on the Chapter Test Prep Videos available in MyMathLab®, on You Tube, or may be accessed using the QR code to the right.

1. Determine whether the following function is one-to-one:

   $\{(1, 4), (3, 2), (5, 8), (-1, 4)\}$

2. Find the inverse of $f(x) = 4x - 3$.

3. Approximate each number using a calculator. Express your answer rounded to three decimal places.

   (a) $3.1^{3.1}$　　　　　　　**(d)** $3.1416^{3.1416}$

   (b) $3.14^{3.14}$　　　　　　**(e)** π^{π}

   (c) $3.142^{3.142}$

4. Change $4^x = 19$ to an equivalent equation involving a logarithm.

5. Change $\log_b x = y$ to an equivalent equation involving an exponent.

6. Find the exact value of each expression without using a calculator.

   (a) $\log_3\left(\dfrac{1}{27}\right)$　　　**(b)** $\log 10{,}000$

7. Determine the domain of $f(x) = \log_5(7 - 4x)$.

In Problems 8–9, graph each function. State the domain and the range of the function.

8. $f(x) = 6^x$　　　　　9. $g(x) = \log_{1/9} x$

10. Use the properties of logarithms to find the exact value of each expression. Do not use a calculator.

    (a) $\log_7 7^{10}$　　　　　**(b)** $3^{\log_3 15}$

11. Write the expression $\log_4 \dfrac{\sqrt{x}}{y^3}$ as a sum or difference of logarithms. Express exponents as factors.

12. Write the expression $4 \log M + 3 \log N$ as a single logarithm.

13. Use the Change-of-Base Formula and a calculator to evaluate $\log_{3/4} 10$. Round to three decimal places.

In Problems 14–20, solve each equation. Express irrational solutions in exact form and as a decimal rounded to three places.

14. $4^{x+1} = 2^{3x+1}$　　　　15. $5^{x^2} \cdot 125 = 25^{2x}$

16. $\log_a 64 = 3$　　　　　17. $\log_2(x^2 - 33) = 8$

18. $2 \log_7(x - 3) = \log_7 3 + \log_7 12$

19. $3^{x-1} = 17$

20. $\log(x - 2) + \log(x + 2) = 2$

21. According to the U.S. Census Bureau, International Data Base, the population of Canada in 2016 was 35.9 million people. In addition, the population of Canada was growing at a rate of 0.8% annually. Assuming that this growth rate continues, the model $P(t) = 35.9(1.008)^{t-2016}$ represents the population (in millions of people) in year t.

    (a) According to the model, what will be the population of Canada in 2020?

    (b) According to the model, in what year will the population of Canada be 50 million?

22. Rustling leaves have an intensity of 10^{-11} watt per square meter. How many decibels are rustling leaves?

Cumulative Review Chapters 1–11

1. Solve: $3(5 - 2x) + 8 = 4(x - 7) + 1$

2. Solve: $5 - 3|x - 2| \geq -7$

3. Determine the domain of $f(x) = \dfrac{9 - x^2}{2x^2 - x - 21}$.

4. Graph the linear equation: $4x + 3y = 6$

5. Find the equation of the line that passes through the points $(-10, 17)$ and $(5, -4)$. Write your answer in either slope-intercept or standard form, whichever you prefer.

6. Graph the following system of linear inequalities.

$$\begin{cases} x + 2y \geq 8 \\ 2x - y < 1 \end{cases}$$

In Problems 7 and 8, add, subtract, multiply, or divide as indicated.

7. $(m^2 - 5m + 13) - (6 - 2m - 3m^2)$

8. $(2n + 3)(n^2 - 4n + 6)$

In Problems 9 and 10, factor completely.

9. $16a^2 + 8ab + b^2$

10. $6y^2 - 17y + 7$

In Problems 11 and 12, perform the indicated operations. Be sure to express the final answer in lowest terms.

11. $\dfrac{2x^2 - 9x - 5}{x^2 - 3x - 10} \cdot \dfrac{3x^2 + 2x - 8}{2x^2 - 13x - 7}$

12. $\dfrac{4}{p^2 - 6p + 5} + \dfrac{2}{p^2 - 3p - 10}$

13. Solve: $\dfrac{2}{x - 5} = \dfrac{x - 2}{x + 1} + \dfrac{6x - 12}{x^2 - 4x - 5}$

14. Simplify: $\sqrt{150} + 4\sqrt{6} - \sqrt{24}$

15. Rationalize the denominator: $\dfrac{1 + \sqrt{5}}{3 - \sqrt{5}}$

16. Solve: $\sqrt{x - 8} + \sqrt{x} = 4$

17. Solve: $3x^2 = 4x + 6$

18. Solve: $2a - 7\sqrt{a} + 6 = 0$

19. Graph: $f(x) = -x^2 + 6x - 4$

20. Solve: $3x^2 + 2x - 8 < 0$

21. Graph: $g(x) = 3^x - 4$

22. Evaluate: $\log_9\left(\dfrac{1}{27}\right)$

CHAPTER
12 Conics

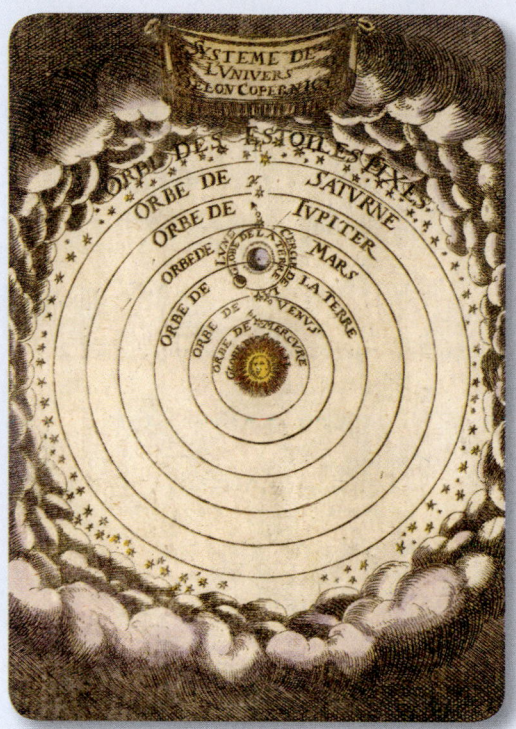

The Big Picture: Putting It Together

In Chapter 3, the Cartesian plane or rectangular coordinate system was introduced. This system lets us make connections between algebra and geometry. In this chapter, that connection will be developed by showing how geometric definitions of certain figures lead to algebraic equations.

This chapter begins by showing how to use algebra to find the distance between any two points in the Cartesian plane. This method is a direct consequence of the Pythagorean Theorem studied in Section 10.1. Knowing the distance formula enables us to present a complete discussion of the *conic sections*.

How could a circle be described to someone? A picture could be drawn to illustrate the verbal description. The Cartesian plane and the distance formula enable us to take the geometric definition of a circle and develop an algebraic equation whose graph represents a circle. This powerful connection between geometry and algebra lets us answer all types of interesting questions. The methods presented in this section form the foundation of an area of mathematics called analytic geometry.

Earth is the center of the universe. Although this statement seems ludicrous to us now, it was commonly believed until the 1500s. A Polish astronomer named Nicolaus Copernicus published a book entitled *De Revolutionibus Orbium Coelestium* (*On the Revolutions of the Celestial Spheres*), which stated that Earth and the other planets orbited the Sun in a circular motion. Copernicus's ideas were not readily accepted by the geocentrists, who held on to the belief that Earth is at the center of the universe. Copernicus's model of planetary motion was later improved upon by the German astronomer Johannes Kepler. In 1609, Kepler published *Astronomia nova* (New Astronomy) in which he proved that the orbit of Mars is an ellipse, with the Sun occupying one of its two foci. See Problems 45–48 in Section 12.4.

Outline

12.1 Distance and Midpoint Formulas

Objectives

① Use the Distance Formula
② Use the Midpoint Formula

Are You Prepared for This Section?

Before getting started, complete the following problems. If you get a problem wrong, go back to the section cited and review the material.

P1. Simplify: **(a)** $\sqrt{64}$ **(b)** $\sqrt{24}$ **(c)** $\sqrt{(x-2)^2}$ [Section 9.1, pp. 616–619; Section 9.4, pp. 635–639]

P2. Find the length of the hypotenuse in a right triangle whose legs are 6 and 8.   [Section 10.1, pp. 697–700]

▶ ① Use the Distance Formula

Section 3.1 discussed how to plot points in the Cartesian plane. The distance between any two points plotted in the Cartesian plane can be algebraically computed, so there is a connection between geometry (literally, "measuring the distance") and algebra. This distance is found using the Pythagorean Theorem.

EXAMPLE 1 **Finding the Distance between Two Points**

Find the distance d between the points $(2, 4)$ and $(5, 8)$.

Solution

First plot the points in the Cartesian plane and connect them with a line, as shown in Figure 1(a). To find the length d, draw a horizontal line through the point $(2, 4)$ and a vertical line through the point $(5, 8)$ and form a right triangle. The right angle of this triangle is at the point $(5, 4)$. The right angle is at this point because in moving horizontally from the point $(2, 4)$, there is no "up or down" movement. For this reason, the y-coordinate of the point at the right angle must be 4. Similarly, moving vertically straight down from the point $(5, 8)$, reveals that the x-coordinate of the point at the right angle must be 5. See Figure 1(b).

Figure 1

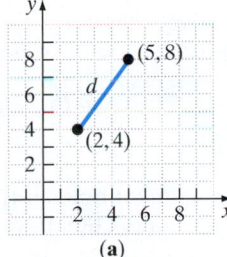   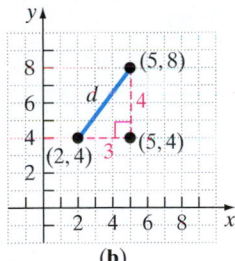

(a)   (b)

One leg of the triangle has length 3 (because $5 - 2 = 3$). The other leg has length 4 (because $8 - 4 = 4$). The Pythagorean Theorem says that

$$d^2 = 3^2 + 4^2$$
$$= 9 + 16$$
$$= 25$$

Square Root Property: $d = \pm 5$

Because d is the length of the hypotenuse, we discard the solution $d = -5$ and find that the hypotenuse is 5. Therefore, the distance between the points $(2, 4)$ and $(5, 8)$ is 5 units. ●

That was quite a bit of work to find the distance between $(2, 4)$ and $(5, 8)$. The *distance formula* lets us easily find the distance between any two points in the Cartesian plane.

Prepared?...Answers P1. (a) 8
(b) $2\sqrt{6}$ **(c)** $|x - 2|$ **P2.** 10

The Distance Formula

The distance between two points $P_1 = (x_1, y_1)$ and $P_2 = (x_2, y_2)$, denoted by $d(P_1, P_2)$, is

$$d(P_1, P_2) = \sqrt{(x_2 - x_1)^2 + (y_2 - y_1)^2}$$

Figure 2

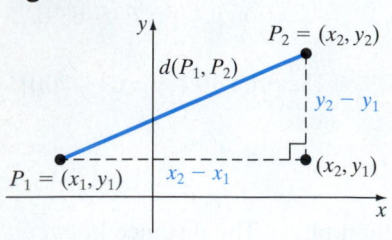

See Figure 2.

Let's justify the formula. Let point P_1 have coordinates (x_1, y_1), and let point P_2 have coordinates (x_2, y_2). If the line joining the points P_1 and P_2 is neither vertical nor horizontal, form a right triangle so that the vertex of the right angle is at the point $P_3 = (x_2, y_1)$, as shown in Figure 3(a).

The vertical distance from P_3 to P_2 is the absolute value of the difference of the y-coordinates, $|y_2 - y_1|$. The horizontal distance from P_1 to P_3 is the absolute value of the difference of the x-coordinates, $|x_2 - x_1|$. See Figure 3(b).

Figure 3

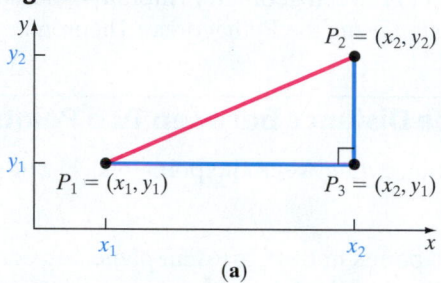 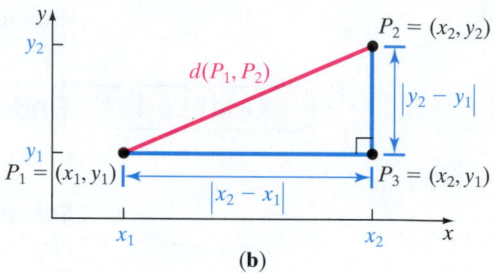

(a)   (b)

The distance $d(P_1, P_2)$ is the length of the hypotenuse of the right triangle, so by the Pythagorean Theorem, it follows that

$$[d(P_1, P_2)]^2 = |x_2 - x_1|^2 + |y_2 - y_1|^2$$
$$= (x_2 - x_1)^2 + (y_2 - y_1)^2$$

Take the square root of both sides: $d(P_1, P_2) = \sqrt{(x_2 - x_1)^2 + (y_2 - y_1)^2}$

Work Smart

Recall that

$\sqrt{a^2 + b^2} \neq \sqrt{a^2} + \sqrt{b^2}$.

If the line joining P_1 and P_2 is horizontal, then the y-coordinate of P_1 equals the y-coordinate of P_2; that is, $y_1 = y_2$. See Figure 4(a). In this case, the distance formula still works, because, for $y_1 = y_2$, it becomes

$$d(P_1, P_2) = \sqrt{(x_2 - x_1)^2 + 0^2}$$
$$= \sqrt{(x_2 - x_1)^2} = |x_2 - x_1|$$

A similar argument holds if the line joining P_1 and P_2 is vertical. See Figure 4(b). The distance formula works in all cases.

Figure 4

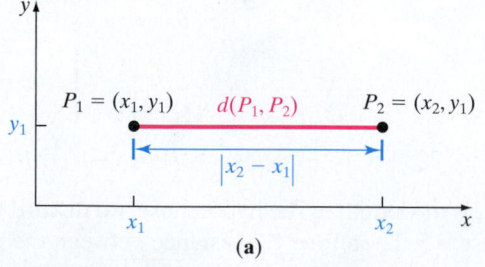 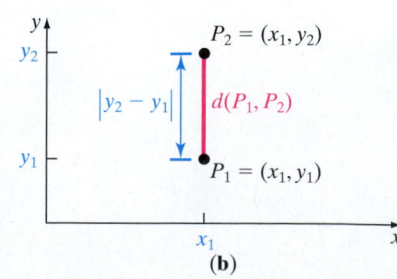

(a)   (b)

The distance between two points $P_1 = (x_1, y_1)$ and $P_2 = (x_2, y_2)$ cannot be negative; it can be 0 only when the two points are identical—that is, when $x_1 = x_2$ and $y_1 = y_2$.

Also, it does not matter whether the distance from P_1 to P_2 or from P_2 to P_1 is computed, because $(x_2 - x_1)^2 = (x_1 - x_2)^2$. This should seem reasonable because the distance from P_1 to P_2 equals the distance from P_2 to P_1.

Figure 5

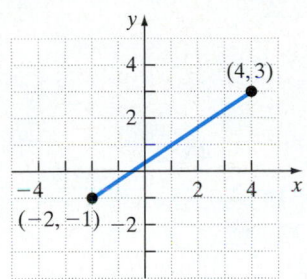

EXAMPLE 2   Finding the Length of a Line Segment

Find the length of the line segment shown in Figure 5.

Solution

The length of the line segment is the distance between the points $(-2, -1)$ and $(4, 3)$. Use the distance formula with $P_1 = (x_1, y_1) = (-2, -1)$ and $P_2 = (x_2, y_2) = (4, 3)$.

$$d = \sqrt{(x_2 - x_1)^2 + (y_2 - y_1)^2}$$

$x_1 = -2, y_1 = -1, x_2 = 4, y_2 = 3:$
$$= \sqrt{(4 - (-2))^2 + (3 - (-1))^2}$$
$$= \sqrt{6^2 + 4^2}$$
$$= \sqrt{36 + 16}$$
$$= \sqrt{52}$$
$$= 2\sqrt{13} \quad \text{Exact}$$
$$\approx 7.21 \quad \text{Approximate}$$

> **Quick ✓**
> 1. The distance between two points $P_1 = (x_1, y_1)$ and $P_2 = (x_2, y_2)$, denoted by $d(P_1, P_2)$, is _____.
> 2. *True or False* The distance between two points can be a negative number.
>
> *In Problems 3 and 4, find the distance between the points.*
> 3. $(3, 8)$ and $(0, 4)$ 4. $(-2, -5)$ and $(4, 7)$

The next example shows how algebra (the distance formula) can be used to solve geometry problems.

EXAMPLE 3   Using Algebra to Solve Geometry Problems

Work Smart

When using the distance formula in applications, make sure that both the x-axis and the y-axis have the same unit of measure, such as inches.

Consider the three points $A = (-1, 1)$, $B = (2, -2)$, and $C = (3, 5)$.

  (a) Plot each point in the Cartesian plane and form the triangle ABC.
  (b) Find the length of each side of the triangle.
  (c) Verify that the triangle is a right triangle.
  (d) Find the area of the triangle.

Solution

  (a) Points A, B, and C are plotted in Figure 6.
  (b) Use the distance formula to find the length of each side of the triangle.

$$d(A, B) = \sqrt{(2 - (-1))^2 + (-2 - 1)^2} = \sqrt{3^2 + (-3)^2}$$
$$= \sqrt{9 + 9} = \sqrt{18} = 3\sqrt{2}$$
$$d(A, C) = \sqrt{(3 - (-1))^2 + (5 - 1)^2} = \sqrt{4^2 + 4^2}$$
$$= \sqrt{16 + 16} = \sqrt{32} = 4\sqrt{2}$$
$$d(B, C) = \sqrt{(3 - 2)^2 + (5 - (-2))^2} = \sqrt{1^2 + 7^2} = \sqrt{1 + 49} = \sqrt{50} = 5\sqrt{2}$$

(continued)

Figure 6

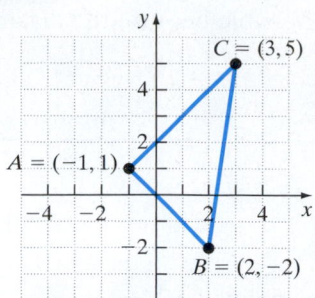

(c) The Pythagorean Theorem says that if you have a right triangle, the sum of squares of the two legs will equal the square of the hypotenuse. The converse of this statement is true as well. That is, if the sum of the squares of two sides of a triangle equals the square of the third side, then the triangle is a right triangle. Thus show that the sum of the squares of two sides of the triangle in Figure 6 equals the square of the third side. In Figure 6, the right angle appears to be at vertex A, which means the side opposite vertex A should be the hypotenuse of the triangle. Show that

$$[d(B, C)]^2 = [d(A, B)]^2 + [d(A, C)]^2$$

The distances from part (b) are known, so

$$(5\sqrt{2})^2 \overset{?}{=} (3\sqrt{2})^2 + (4\sqrt{2})^2$$

$$(ab)^n = a^n \cdot b^n: \quad 25 \cdot 2 \overset{?}{=} 9 \cdot 2 + 16 \cdot 2$$

$$50 \overset{?}{=} 18 + 32$$

$$50 = 50 \quad \text{True}$$

Because $[d(B, C)]^2 = [d(A, B)]^2 + [d(A, C)]^2$, the triangle ABC is a right triangle.

(d) The area of a triangle is $^1/_2$ times the product of the base and the height. From Figure 6, side AB forms the base and side AC forms the height. The length of side AB is $3\sqrt{2}$ and the length of side AC is $4\sqrt{2}$, so the area of triangle ABC is

$$\text{Area} = \frac{1}{2}(\text{Base})(\text{Height}) = \frac{1}{2}(3\sqrt{2})(4\sqrt{2}) = 12 \text{ square units} \quad \bullet$$

Quick ✓

5. Consider the three points $A = (-2, -1)$, $B = (4, 2)$, and $C = (0, 10)$.
   (a) Plot each point in the Cartesian plane and form the triangle ABC.
   (b) Find the length of each side of the triangle.
   (c) Verify that the triangle is a right triangle.
   (d) Find the area of the triangle.

▶ ❷ Use the Midpoint Formula

Consider the two points $P_1 = (x_1, y_1)$ and $P_2 = (x_2, y_2)$ in the Cartesian plane. To find the point $M = (x, y)$ that is on the line segment joining P_1 and P_2 and is the same distance to each of these two points so that $d(P_1, M) = d(M, P_2)$, use the **midpoint formula**.

In Other Words

To find the midpoint of a line segment, average the x-coordinates of the endpoints and average the y-coordinates of the endpoints.

Midpoint Formula

The midpoint $M = (x, y)$ of the line segment from $P_1 = (x_1, y_1)$ to $P_2 = (x_2, y_2)$ is

$$M = \left(\frac{x_1 + x_2}{2}, \frac{y_1 + y_2}{2}\right)$$

EXAMPLE 4   **Finding the Midpoint of a Line Segment**

Find the midpoint of the line segment joining $P_1 = (-2, 3)$ and $P_2 = (4, 7)$. Plot the points P_1, P_2, and their midpoint. Check your answer.

Solution

Substitute $x_1 = -2$, $y_1 = 3$, $x_2 = 4$, and $y_2 = 7$ into the midpoint formula. The coordinates (x, y) of the midpoint M are

$$x = \frac{x_1 + x_2}{2} = \frac{-2 + 4}{2} = \frac{2}{2} = 1$$

and

$$y = \frac{y_1 + y_2}{2} = \frac{3 + 7}{2} = \frac{10}{2} = 5$$

Figure 7

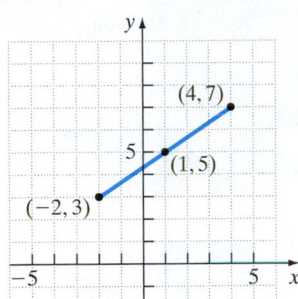

Thus the midpoint of the line segment joining $P_1 = (-2, 3)$ and $P_2 = (4, 7)$ is $M = (1, 5)$. See Figure 7.

Quick ✓

6. The midpoint $M = (x, y)$ of the line segment from $P_1 = (x_1, y_1)$ to $P_2 = (x_2, y_2)$

is _____.

In Problems 7 and 8, find the midpoint of the line segment joining the points.

7. $(3, 8)$ and $(0, 4)$

8. $(-2, -5)$ and $(4, 10)$

12.1 Exercises MyMathLab® Exercise numbers in **green** have complete video solutions in MyMathLab or may be accessed using the QR code to the right.

*Problems **1–8** are the Quick ✓s that follow the EXAMPLES.*

Building Skills

In Problems 9–24, find the distance $d(P_1, P_2)$ between the points P_1 and P_2. See Objective 1.

9.

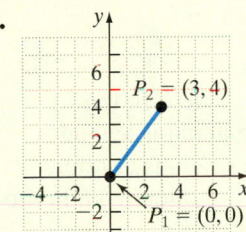

10.

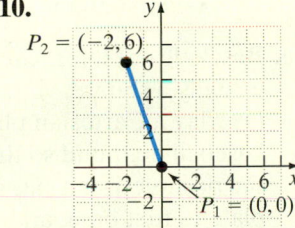

11.

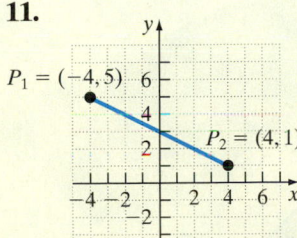

12.

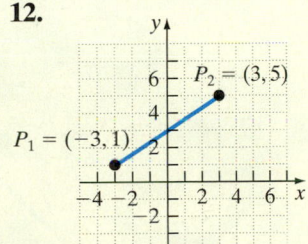

13. $P_1 = (2, 1); P_2 = (6, 4)$

14. $P_1 = (1, 3); P_2 = (4, 7)$

15. $P_1 = (-3, 2); P_2 = (9, -3)$

16. $P_1 = (-10, -3); P_2 = (14, 4)$

17. $P_1 = (-4, 2); P_2 = (2, 2)$

18. $P_1 = (-1, 2); P_2 = (-1, 0)$

19. $P_1 = (0, -3); P_2 = (-3, 3)$

20. $P_1 = (5, 0); P_2 = (-1, -4)$

21. $P_1 = (2\sqrt{2}, \sqrt{5}); P_2 = (5\sqrt{2}, 4\sqrt{5})$

22. $P_1 = (\sqrt{6}, -2\sqrt{2}); P_2 = (3\sqrt{6}, 10\sqrt{2})$

23. $P_1 = (0.3, -3.3); P_2 = (1.3, 0.1)$

24. $P_1 = (-1.7, 1.3); P_2 = (0.3, 2.6)$

In Problems 25–36, find the midpoint of the line segment formed by joining the points P_1 and P_2. See Objective 2.

25. $P_1 = (2, 2); P_2 = (6, 4)$

26. $P_1 = (1, 3); P_2 = (5, 7)$

27. $P_1 = (-3, 2); P_2 = (9, -4)$

28. $P_1 = (-10, -3); P_2 = (14, 7)$

29. $P_1 = (-4, 3); P_2 = (2, 4)$

30. $P_1 = (-1, 2); P_2 = (3, 9)$

31. $P_1 = (0, -3); P_2 = (-3, 3)$

32. $P_1 = (5, 0); P_2 = (-1, -4)$

33. $P_1 = \left(2\sqrt{2}, \sqrt{5}\right); P_2 = \left(5\sqrt{2}, 4\sqrt{5}\right)$

34. $P_1 = \left(\sqrt{6}, -2\sqrt{2}\right); P_2 = \left(3\sqrt{6}, 10\sqrt{2}\right)$

35. $P_1 = (0.3, -3.3); P_2 = (1.3, 0.1)$

36. $P_1 = (-1.7, 1.3); P_2 = (0.3, 2.6)$

Applying the Concepts

37. Consider the three points $A = (0, 3)$, $B = (2, 1)$, and $C = (6, 5)$.

   (a) Plot each point in the Cartesian plane and form the triangle ABC.
   (b) Find the length of each side of the triangle.
   (c) Verify that the triangle is a right triangle.
   (d) Find the area of the triangle.

38. Consider the three points $A = (0, 2)$, $B = (1, 4)$, and $C = (4, 0)$.

   (a) Plot each point in the Cartesian plane and form the triangle ABC.
   (b) Find the length of each side of the triangle.
   (c) Verify that the triangle is a right triangle.
   (d) Find the area of the triangle.

39. Consider the three points $A = (-2, -4)$, $B = (3, 1)$, and $C = (15, -11)$.

   (a) Plot each point in the Cartesian plane and form the triangle ABC.
   (b) Find the length of each side of the triangle.
   (c) Verify that the triangle is a right triangle.
   (d) Find the area of the triangle.

40. Consider the three points $A = (-2, 3)$, $B = (2, 0)$, and $C = (5, 4)$.

   (a) Plot each point in the Cartesian plane and form the triangle ABC.
   (b) Find the length of each side of the triangle.
   (c) Verify that the triangle is a right triangle.
   (d) Find the area of the triangle.

41. Find all points having an x-coordinate of 2 whose distance from the point $(5, 1)$ is 5.

42. Find all points having an x-coordinate of 4 whose distance from the point $(0, 3)$ is 5.

43. Find all points having a y-coordinate of -3 whose distance from the point $(2, 3)$ is 10.

44. Find all points having a y-coordinate of -3 whose distance from the point $(-4, 2)$ is 13.

45. The City of Chicago The city of Chicago's road system is set up like a Cartesian plane, where streets are indicated by the number of blocks they are from Madison Street and State Street. For example, Wrigley Field in Chicago is located at 1060 West Addison, which is 10 blocks west of State Street and 36 blocks north of Madison Street.

City of Chicago, Illinois

Addison Street

1 mile
1 km

Wrigley Field
1060 West Addison

Madison Street

Guaranteed Rate Field
35th and Princeton

35th Street

   (a) Find the distance "as the crow flies" from Madison and State Street to Wrigley Field. Use city blocks as the unit of measurement.
   (b) Guaranteed Rate Field, home of the White Sox, is located at 35th and Princeton, which is 3 blocks west of State Street and 35 blocks south of Madison. Find the distance "as the crow flies" from Madison and State Street to Guaranteed Rate Field.
   (c) Find the distance "as the crow flies" from Wrigley Field to Guaranteed Rate Field.

46. Baseball A major league baseball "diamond" is actually a square, 90 feet on a side (see the figure). Overlay a Cartesian plane on a major league baseball diamond so that the origin is at home plate, the positive x-axis lies in the direction from home plate to first base, and the positive y-axis lies in the direction from home plate to third base.

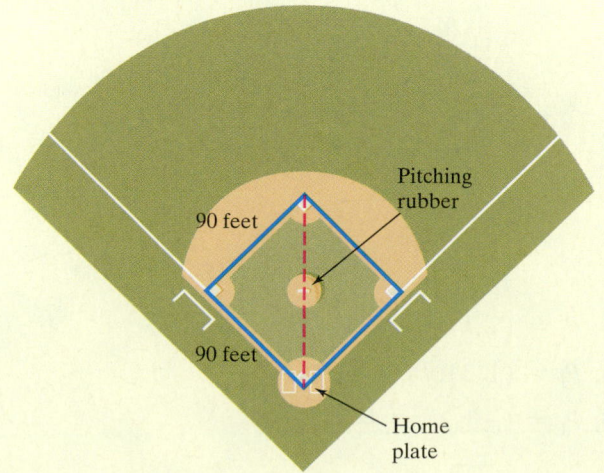

Pitching rubber

90 feet

90 feet

Home plate

(a) What are the coordinates of home plate, first base, second base, and third base? Use feet as the unit of measurement.

(b) Suppose the center fielder is located at $(310, 260)$. How far is he from second base?

(c) Suppose the shortstop is located at $(60, 100)$. How far is he from second base?

Extending the Concepts

47. Baseball Refer to Problem 46.

(a) Suppose the right fielder catches a fly ball at $(320, 20)$. How many seconds will it take him to throw the ball to second base if he can throw 130 feet per second? (*Hint:* Time = distance divided by speed.)

(b) Suppose a runner "tagging up" from first base can run 27 feet per second. Would you "send the runner" as the first base coach if the right fielder requires 0.8 second to catch and throw? Why?

48. Let $P = (x, y)$ be a point on the graph of $y = x^2 - 4$.

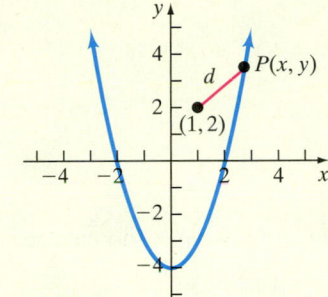

(a) Express the distance d from P to the point $(1, 2)$ as a function of x using the distance formula. See the figure for an illustration of the problem.

(b) What is d if $x = 0$?

(c) What is d if $x = 3$?

Explaining the Concepts

49. How is the distance formula related to the Pythagorean Theorem?

50. How can the distance formula be used to verify that a point is the midpoint of a line segment? What must be done in addition to using the distance formula to verify a midpoint? Why?

Synthesis Review

51. Evaluate 3^2. What is $\sqrt{9}$?

52. Evaluate 8^2. What is $\sqrt{64}$?

53. Evaluate $(-3)^4$. What is $\sqrt[4]{81}$?

54. Evaluate $(-3)^3$. What is $\sqrt[3]{-27}$?

55. Describe the relationship between raising a number to a positive integer power, n, and the nth root of a number.

12.2 Circles

Objectives

1 Write the Standard Form of the Equation of a Circle

2 Graph a Circle

3 Find the Center and Radius of a Circle Given an Equation in General Form

Are You Prepared for This Section?

Before getting started, complete the following problem. If you get the problem wrong, go back to the section cited and review the material.

P1. Complete the square in x: $x^2 - 8x$ [Section 10.1, pp. 695–696]

Conics, an abbreviation for conic sections, are curves that result from the intersection of a right circular cone and a plane. The four conics that will be studied are shown in Figure 8. These conics are *circles* (Figure 8(a)), *ellipses* (Figure 8(b)), *parabolas* (Figure 8(c)), and *hyperbolas* (Figure 8(d)).

Figure 8

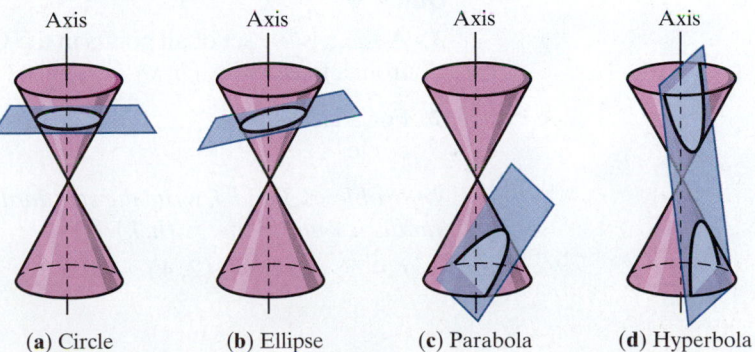

(a) Circle   (b) Ellipse   (c) Parabola   (d) Hyperbola

Prepared?...Answer
P1. $x^2 - 8x + 16$

Circles will be studied in this section, parabolas in Section 12.3, ellipses in Section 12.4, and hyperbolas in Section 12.5.

▶ ❶ Write the Standard Form of the Equation of a Circle

Because of the Cartesian plane, a geometric statement can be translated into an algebraic statement, and vice versa. Consider the following geometric statement that defines a circle.

Figure 9

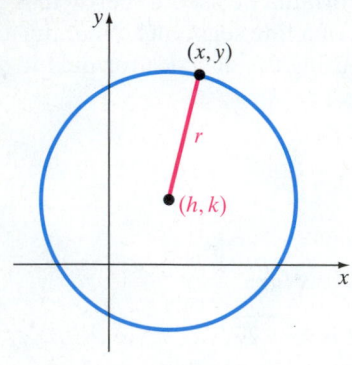

> **Definition**
>
> A **circle** is the set of all points in the Cartesian plane that are a fixed distance r from a fixed point (h, k). The fixed distance r is the **radius,** and the fixed point (h, k) is the **center** of the circle. See Figure 9.

To find an equation that has this graph, let (x, y) represent the coordinates of any point on a circle with radius r and center (h, k). Then the distance between the points (x, y) and (h, k) must always equal r. That is, by the distance formula,

$$\sqrt{(x - h)^2 + (y - k)^2} = r$$

Square both sides of the equation to get

$$(x - h)^2 + (y - k)^2 = r^2$$

and this is the equation of a circle.

Work Smart

Because r represents a distance, and distance is always positive, r is a positive number.

> **Definition**
>
> The **standard form of an equation of a circle** with radius r and center (h, k) is
>
> $$(x - h)^2 + (y - k)^2 = r^2$$

The standard form of a circle of radius r with center at the origin $(0, 0)$ is $x^2 + y^2 = r^2$.

EXAMPLE 1 **Writing the Standard Form of the Equation of a Circle**

Write the standard form of the equation of the circle with radius 4 and center $(2, -3)$.

Solution
Use the equation $(x - h)^2 + (y - k)^2 = r^2$ with $r = 4$, $h = 2$, and $k = -3$.

$$(x - 2)^2 + (y - (-3))^2 = 4^2$$
$$(x - 2)^2 + (y + 3)^2 = 16$$

●

> **Quick ✓**
>
> 1. A _____ is the set of all points in the Cartesian plane that are a fixed distance r from a fixed point (h, k).
>
> 2. For a circle, the _____ is the distance from the center to any point on the circle.
>
> *In Problems 3 and 4, write the standard form of the equation of each circle whose radius is r and center is (h, k).*
>
> 3. $r = 5$; $(h, k) = (2, 4)$ 4. $r = \sqrt{2}$; $(h, k) = (-2, 0)$

⊘ ❷ Graph a Circle

The graph of any equation of the form $(x - h)^2 + (y - k)^2 = r^2$ is a circle with radius r and center (h, k).

EXAMPLE 2 **Graphing a Circle**

Graph the equation: $(x + 2)^2 + (y - 3)^2 = 9$

Solution

The equation is of the form $(x - h)^2 + (y - k)^2 = r^2$, so its graph is a circle. To graph the equation, first identify the center and radius of the circle by comparing the equation to the standard form of the equation of a circle.

$$(x + 2)^2 + (y - 3)^2 = 9$$
$$(x - (-2))^2 + (y - 3)^2 = 3^2$$

$$(x - h)^2 + (y - k)^2 = r^2$$

Therefore, $h = -2$, $k = 3$, and $r = 3$. The circle has center $(-2, 3)$ and radius 3. To graph this circle, first plot the center $(-2, 3)$. Since the radius is 3 units, go 3 units in any direction from the center and find a point on the circle. It is easiest to find the four points left, right, up, and down from the center. These four points are $(-5, 3)$, $(1, 3)$, $(-2, 6)$, and $(-2, 0)$, respectively. Plot these points in Figure 10(a) and use them to draw the graph of the circle in Figure 10(b).

Figure 10

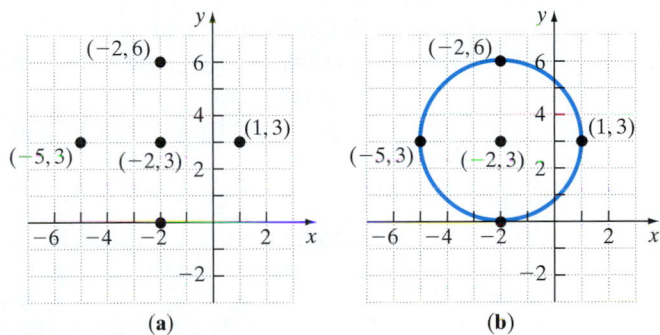

(a)   (b)

Notice that the graph of $(x + 2)^2 + (y - 3)^2 = (x - (-2))^2 + (y - 3)^2 = 9$ is the graph of $x^2 + y^2 = 9$ shifted 2 units left and 3 units up.

Quick ✓

5. *True or False* The center of the circle $(x + 1)^2 + (y - 3)^2 = 25$ is $(1, -3)$.

6. *True or False* The center of the circle $x^2 + y^2 = 9$ is $(0, 0)$; its radius is 3.

In Problems 7 and 8, graph the circle.

7. $(x - 3)^2 + (y - 1)^2 = 4$ **8.** $(x + 5)^2 + y^2 = 16$

⊘ ❸ Find the Center and Radius of a Circle Given an Equation in General Form

Eliminate the parentheses from the standard form of the equation of the circle given in Example 2 to get

$$(x + 2)^2 + (y - 3)^2 = 9$$

FOIL: $x^2 + 4x + 4 + y^2 - 6y + 9 = 9$

Subtract 9 from both sides: $x^2 + y^2 + 4x - 6y + 4 = 0$

Any equation of the form

$$x^2 + y^2 + ax + by + c = 0$$

has a graph that is a circle, a graph that is a point, or no graph at all. For example, the graph of the equation $x^2 + y^2 = 0$ is the single point $(0, 0)$. The equation $x^2 + y^2 + 4 = 0$, or $x^2 + y^2 = -4$, has no graph, because the sum of squares of real numbers is never negative.

In Other Words
The standard form of the equation of a circle is $(x - h)^2 + (y - k)^2 = r^2$. The general form is $x^2 + y^2 + ax + by + c = 0$.

Definition

The **general form of the equation of a circle** is given by the equation

$$x^2 + y^2 + ax + by + c = 0$$

when the graph exists.

If an equation of a circle is in general form, use the method of completing the square to put the equation in standard form, $(x - h)^2 + (y - k)^2 = r^2$, in order to find the center and radius of the circle.

EXAMPLE 3   **Graphing a Circle Whose Equation Is in General Form**

Graph the equation: $x^2 + y^2 + 8x - 2y - 8 = 0$

Solution
To find the center and radius of the circle, put the equation in standard form by completing the square in both x and y (covered in Section 10.1). To do this, group the terms involving x, group the terms involving y, and put the constant on the right side of the equation by adding 8 to both sides.

$$(x^2 + 8x) + (y^2 - 2y) = 8$$

Now complete the square of each expression in parentheses. Remember, any number added to the left side of the equation must also be added to the right.

Figure 11

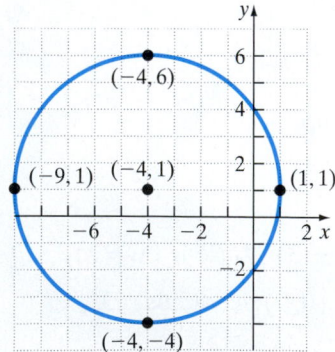

$$(x^2 + 8x + 16) + (y^2 - 2y + 1) = 8 + 16 + 1$$

$$\left(\frac{1}{2} \cdot 8\right)^2 = 16 \qquad \left(\frac{1}{2}(-2)\right)^2 = 1$$

Factor: $(x + 4)^2 + (y - 1)^2 = 25$

The equation is now in the standard form of the equation of a circle. The center is $(-4, 1)$ and the radius is 5. To graph the equation, plot the center and the four points $r = 5$ units to the left, to the right, above, and below the center. Then draw in the circle. See Figure 11.

Notice that the graph of $(x + 4)^2 + (y - 1)^2 = 25$ is the graph of $x^2 + y^2 = 25$ shifted 4 units left and 1 unit up.

Quick ✓

In Problems 9 and 10, graph each circle.

9. $x^2 + y^2 - 6x - 4y + 4 = 0$ **10.** $2x^2 + 2y^2 - 16x + 4y - 38 = 0$

12.2 Exercises   MyMathLab®

Exercise numbers in **green** have complete video solutions in MyMathLab or may be accessed using the QR code to the right.

*Problems **1–10** are the* **Quick ✔** *s that follow the* **EXAMPLES**.

Building Skills

In Problems 11–14, find the center and radius of each circle. Write the standard form of the equation.

11.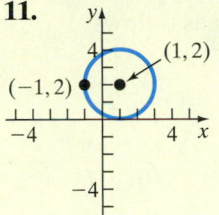

12.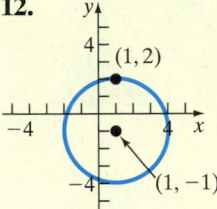

13.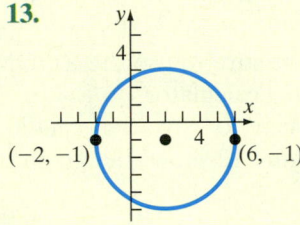

14.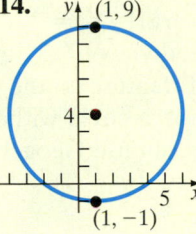

In Problems 15–26, write the standard form of the equation of each circle whose radius is r and center is (h, k). See Objective 1.

15. $r = 3$; $(h, k) = (0, 0)$

16. $r = 5$; $(h, k) = (0, 0)$

17. $r = 2$; $(h, k) = (1, 4)$

18. $r = 4$; $(h, k) = (3, 1)$

19. $r = 6$; $(h, k) = (-2, 4)$

20. $r = 3$; $(h, k) = (1, -4)$

21. $r = 4$; $(h, k) = (0, 3)$

22. $r = 2$; $(h, k) = (1, 0)$

23. $r = 5$; $(h, k) = (5, -5)$

24. $r = 4$; $(h, k) = (-4, 4)$

25. $r = \sqrt{5}$; $(h, k) = (1, 2)$

26. $r = \sqrt{7}$; $(h, k) = (5, 2)$

In Problems 27–36, find the center (h, k) and radius r of each circle. Graph each circle. See Objective 2.

27. $x^2 + y^2 = 36$

28. $x^2 + y^2 = 144$

29. $(x - 4)^2 + (y - 1)^2 = 25$

30. $(x - 2)^2 + (y - 3)^2 = 9$

31. $(x + 3)^2 + (y - 2)^2 = 81$

32. $(x - 5)^2 + (y + 2)^2 = 49$

33. $x^2 + (y - 3)^2 = 64$

34. $(x - 6)^2 + y^2 = 36$

35. $(x - 1)^2 + (y + 1)^2 = \dfrac{1}{4}$

36. $(x - 2)^2 + (y + 2)^2 = \dfrac{1}{4}$

In Problems 37–42, find the center (h, k) and radius r of each circle. Graph each circle. See Objective 3.

37. $x^2 + y^2 - 6x + 2y + 1 = 0$

38. $x^2 + y^2 + 2x - 8y + 8 = 0$

39. $x^2 + y^2 + 10x + 4y + 4 = 0$

40. $x^2 + y^2 + 4x - 12y + 36 = 0$

41. $2x^2 + 2y^2 - 12x + 24y - 72 = 0$

42. $2x^2 + 2y^2 - 28x + 20y + 20 = 0$

Mixed Practice

In Problems 43–48, find the standard form of the equation of each circle.

43. Center at the origin and containing the point $(4, -2)$

44. Center at $(0, 3)$ and containing the point $(3, 7)$

45. Center at $(-3, 2)$ and tangent to the *y*-axis

46. Center at $(2, -3)$ and tangent to the *x*-axis

47. With endpoints of a diameter at $(2, 3)$ and $(-4, -5)$

48. With endpoints of a diameter at $(-5, -3)$ and $(7, 2)$

Applying the Concepts

△ **49.** Find the area and circumference of the circle $(x - 3)^2 + (y - 8)^2 = 64$.

△ **50.** Find the area and circumference of the circle $(x - 1)^2 + (y - 4)^2 = 49$.

△ **51.** Find the area of the square in the figure.

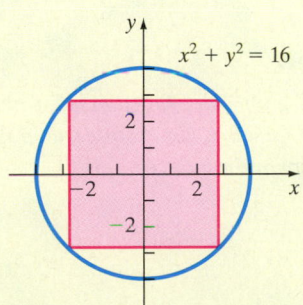

△ **52.** Find the area of the shaded region in the figure.

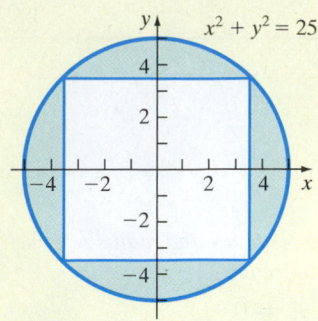

Extending the Concepts

53. Which of the following equation(s) might have the graph shown? (More than one answer may be possible.)

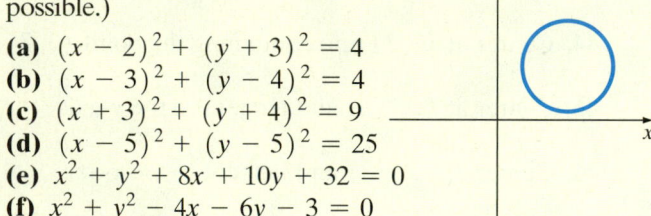

(a) $(x - 2)^2 + y^2 = 1$
(b) $x^2 + (y - 2)^2 = 1$
(c) $(x + 4)^2 + y^2 = 9$
(d) $(x - 5)^2 + y^2 = 25$
(e) $x^2 + y^2 - 8x + 7 = 0$
(f) $x^2 + y^2 + 10x + 18 = 0$

54. Which of the following equation(s) might have the graph shown? (More than one answer may be possible.)

(a) $(x - 2)^2 + (y + 3)^2 = 4$
(b) $(x - 3)^2 + (y - 4)^2 = 4$
(c) $(x + 3)^2 + (y + 4)^2 = 9$
(d) $(x - 5)^2 + (y - 5)^2 = 25$
(e) $x^2 + y^2 + 8x + 10y + 32 = 0$
(f) $x^2 + y^2 - 4x - 6y - 3 = 0$

Explaining the Concepts

55. How is the distance formula related to the definition of a circle?

56. Are circles functions? Why or why not?

57. Is $x^2 = 36 - y^2$ the equation of a circle? Why or why not? If so, what are the center and the radius?

58. Is $3x^2 - 12x + 3y^2 - 15 = 0$ the equation of a circle? Why or why not? If so, what is the center and radius?

Synthesis Review

In Problems 59–63, graph each function using either point plotting or properties of the function. For example, to graph a line, use the slope and y-intercept or the intercepts.

59. $f(x) = 4x - 3$

60. $2x + 5y = 20$

61. $g(x) = x^2 - 4x - 5$

62. $F(x) = -3x^2 + 12x - 12$

63. $G(x) = -2(x + 3)^2 - 5$

64. Present an argument that supports the approach you took to graphing each function in Problems 59–63. For example, in Problem 59, did you use point plotting? Intercepts? Slope? Which method did you use and why?

Technology Exercises

To graph a circle using a graphing calculator, first solve the equation for y. For example, to graph $(x + 2)^2 + (y - 3)^2 = 9$, solve for y as follows:

$$(x + 2)^2 + (y - 3)^2 = 9$$

Subtract $(x + 2)^2$ from both sides: $(y - 3)^2 = 9 - (x + 2)^2$
Take the square root of both sides: $y - 3 = \pm \sqrt{9 - (x + 2)^2}$
Add 3 to both sides: $y = 3 \pm \sqrt{9 - (x + 2)^2}$

Graph the top half $Y_1 = 3 + \sqrt{9 - (x + 2)^2}$ and the bottom half $Y_2 = 3 - \sqrt{9 - (x + 2)^2}$. To get an undistorted view of the graph, be sure to use the ZOOM SQUARE feature on the graphing calculator.

To graph a circle without having to solve the equation for y, you can use Geogebra. Figure 12 shows the graph of $(x + 2)^2 + (y - 3)^2 = 9$.

Figure 12

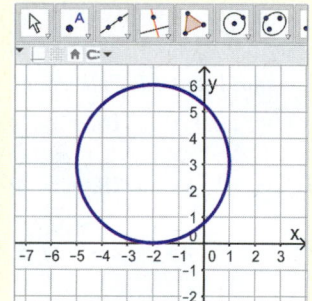

In Problems 65–74, graph each circle using technology. Compare your graphs to the graphs drawn by hand in Problems 27–36.

65. $x^2 + y^2 = 36$

66. $x^2 + y^2 = 144$

67. $(x - 4)^2 + (y - 1)^2 = 25$

68. $(x - 2)^2 + (y - 3)^2 = 9$

69. $(x + 3)^2 + (y - 2)^2 = 81$

70. $(x - 5)^2 + (y + 2)^2 = 49$

71. $x^2 + (y - 3)^2 = 64$

72. $(x - 6)^2 + y^2 = 36$

73. $(x - 1)^2 + (y + 1)^2 = \dfrac{1}{4}$

74. $(x - 2)^2 + (y + 2)^2 = \dfrac{1}{4}$

75. Exploration: Equation of a Circle *Open the "Equation of a Circle—Standard Form" Geogebra applet. The applet may be found using the QR code at the beginning of this section or through the Multimedia Library in MyMathLab. In the upper left corner are three "sliders." Use your mouse to* change the values of the *h, k,* or *r* sliders by clicking on the appropriate slider and dragging the value left or right. As you are moving the sliders, take note of how the position and equation of the circle change. If you are using a keyboard, it may be easier to use the arrow keys, instead of the mouse, to move the slider. First click on the slider with your mouse, and then use the left and right arrow keys to move the slider. Checking on the *Center* or *Radius* box will help you answer the following questions.

(a) Move the slider so that $h = 0$ and $k = 0$. Where is the center of the circle?

(b) Move the "*r*" slider so that *r* increases from 0 to 10. What happens to the radius of the circle as *r* increases? Change the value of *r* from 2 to 3 to 4. As you do this, what happens to the value on the right side of the equation in green?

(c) Move the "*r*" slider to 3. Move the "*h*" slider so that *h* increases from −10 to 10. What role does *h* play in the graph of the circle? In particular, what do you notice when *h* is negative? Zero? Positive?

(d) Move the "*r*" slider to 3. Move the "*k*" slider so that *k* increases from −10 to 10. What role does *k* play in the graph of the circle? In particular, what do you notice when *k* is negative? Zero? Positive?

(e) Check the "baseball" image box on the right-hand side of the applet. This will place a baseball image on the coordinate plane. Move the sliders to find the formula of the circle, in standard form, that best covers the outside edge of the baseball. What is this equation? What is the center of this circle? What is the radius?

(f) Check the "flower" image box on the right-hand side of the applet. This will place a flower image on the coordinate plane. Move the sliders to find the formula of the circle, in standard form, that best covers the outside edge of the center of the flower. What is this equation? What is the center of this circle? What is the radius?

(g) Check the "half dollar" image box on the right-hand side of the applet. This will place a half dollar image on the coordinate plane. Move the sliders to find the formula of the circle, in standard form, that best covers the outside edge of the half dollar. What is this equation? What is the center of this circle? What is the radius?

(h) Check the "lemon" image box on the right-hand side of the applet. This will place a lemon image on the coordinate plane. Move the sliders to find the formula of the circle, in standard form, that best covers the outside edge of the lemon. What is this equation? What is the center of this circle? What is the radius?

12.3 Parabolas

Objectives

1 Graph Parabolas Whose Vertex Is the Origin

2 Find the Equation of a Parabola

3 Graph a Parabola Whose Vertex Is Not the Origin

4 Solve Applied Problems Involving Parabolas

Are You Prepared for This Section?

Before getting started, complete the following problems. If you get a problem wrong, go back to the section cited and review the material.

P1. Identify the vertex and axis of symmetry of $f(x) = -3(x + 4)^2 - 5$. Does the parabola open up or down? Why? [Section 10.4, pp. 731–734]

P2. Identify the vertex and axis of symmetry of the quadratic function $f(x) = 2x^2 - 8x + 1$. Does the parabola open up or down? Why? [Section 10.5, pp. 737–742]

P3. Complete the square of $x^2 - 12x$. [Section 10.1, pp. 694–695]

P4. Solve: $(x - 3)^2 = 25$ [Section 10.1, pp. 691–694]

A discussion of parabolas began in Section 10.4 when quadratic functions were studied. Refresh your memory with the summary on the next page.

Prepared?...Answers **P1.** Vertex: $(-4, -5)$; axis of symmetry: $x = -4$; opens down since $a = -3 < 0$. **P2.** Vertex: $(2, -7)$; axis of symmetry: $x = 2$; opens up since $a = 2 > 0$. **P3.** $x^2 - 12x + 36$ **P4.** $\{-2, 8\}$

Summary Parabolas That Open Up or Down

The graph of $y = a(x - h)^2 + k$ or $y = ax^2 + bx + c$ is a parabola that

1. opens up if $a > 0$ and opens down if $a < 0$.
2. has vertex (h, k).
3. has a vertex whose x-coordinate is $x = -\dfrac{b}{2a}$. The y-coordinate is found by evaluating the equation at the x-coordinate of the vertex.

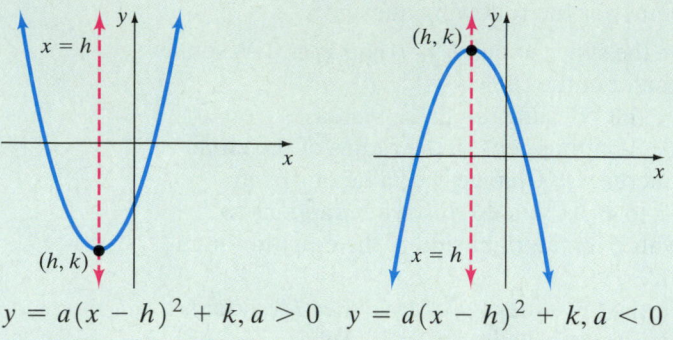

$$y = a(x - h)^2 + k, a > 0 \qquad y = a(x - h)^2 + k, a < 0$$

Figure 13

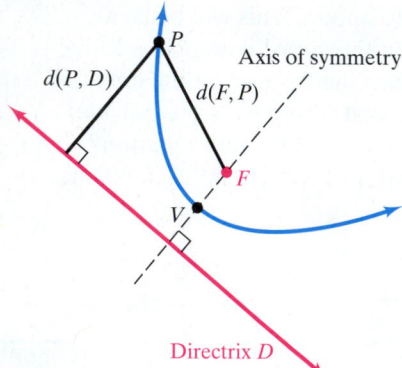

Directrix D

Ideas about parabolas in Sections 10.4 and 10.5 relied on algebra. In this chapter, parabolas (and other conic sections) will be examined from a geometric point of view.

Definition

A **parabola** is the collection of all points P in the plane that are the same distance from a fixed point F as they are from a fixed line D. The point F is called the **focus** of the parabola, and the line D is its **directrix.** In other words, a parabola is the set of points P for which

$$d(F, P) = d(P, D)$$

Figure 13 shows a parabola. The line through the focus F and perpendicular to the directrix D is the **axis of symmetry** of the parabola. The point of intersection of the parabola with its axis of symmetry is the **vertex** V.

Figure 14

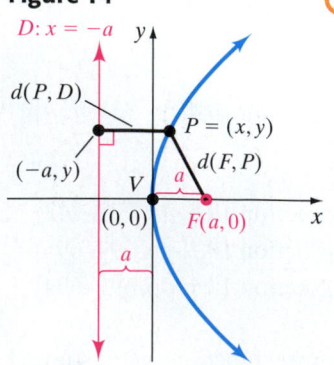

▶ ❶ **Graph Parabolas Whose Vertex Is the Origin**

Develop an equation for the parabola based on the definition just given. For example, the distance from the focus to the vertex (a point on the parabola) must equal the distance from the vertex to the directrix. Let a represent the distance from the focus to the vertex. To develop an equation for a parabola, start by looking at parabolas whose vertex is at the origin. Consider four possibilities—parabolas that open left, parabolas that open right, parabolas that open up, and parabolas that open down.

First obtain the equation of a parabola whose vertex is at the origin and opens right. To do this, position a parabola in a Cartesian plane so that its vertex is at the origin, $(0, 0)$, the focus is on the positive x-axis, and the directrix is a vertical line in quadrants II and III. Because a represents the distance from the vertex to the focus, the focus is located at $(a, 0)$. Also, because the distance from the vertex to the directrix is a, the directrix is the line $x = -a$. See Figure 14. If $P = (x, y)$ is any point on the parabola, then the distance from P to the focus F, $(a, 0)$, must equal the distance from P to the directrix, $x = -a$. Use the point $(-a, y)$ on the directrix in the distance formula. That is,

$$d(F, P) = d(P, D)$$

Use the distance formula: $\sqrt{(x - a)^2 + (y - 0)^2} = \sqrt{(x - (-a))^2 + (y - y)^2}$

Square both sides; simplify: $(x - a)^2 + y^2 = (x + a)^2$

Multiply out binomials: $x^2 - 2ax + a^2 + y^2 = x^2 + 2ax + a^2$

Combine like terms: $y^2 = 4ax$

Thus the equation of the parabola whose vertex is at the origin and opens to the right is $y^2 = 4ax$. Obtain the equations of the three other possibilities (opens left, opens up, or opens down) using similar reasoning. These possibilities are summarized in Table 1.

Table 1 Equations of a Parabola: Vertex at $(0, 0)$; Focus on an Axis; $a > 0$

Vertex	Focus	Directrix	Equation	Axis of Symmetry	Opens
$(0, 0)$	$(a, 0)$	$x = -a$	$y^2 = 4ax$	x-axis	Right
$(0, 0)$	$(-a, 0)$	$x = a$	$y^2 = -4ax$	x-axis	Left
$(0, 0)$	$(0, a)$	$y = -a$	$x^2 = 4ay$	y-axis	Up
$(0, 0)$	$(0, -a)$	$y = a$	$x^2 = -4ay$	y-axis	Down

The graphs of the four parabolas are given in Figure 15.

Figure 15

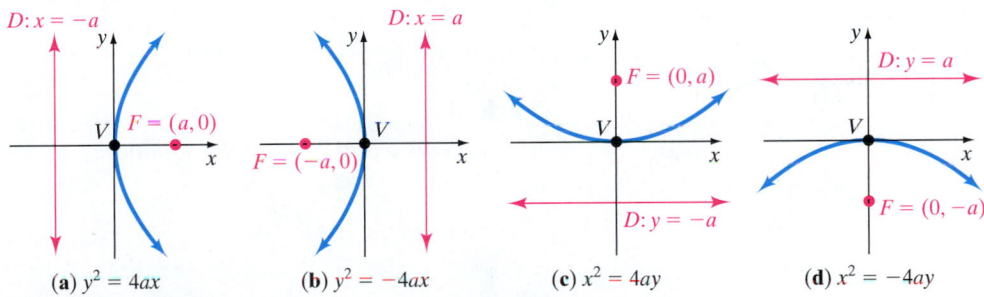

(a) $y^2 = 4ax$ **(b)** $y^2 = -4ax$ **(c)** $x^2 = 4ay$ **(d)** $x^2 = -4ay$

EXAMPLE 1

Graphing a Parabola That Opens Left

Graph the equation $y^2 = -12x$.

Solution

Figure 16

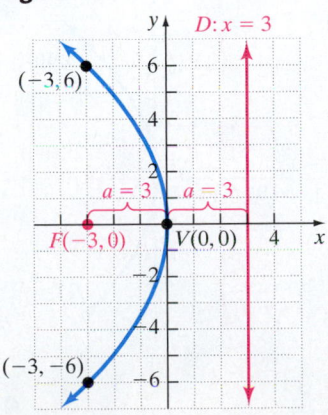

The equation $y^2 = -12x$ is of the form $y^2 = -4ax$, where $-4a = -12$, so $a = 3$. The graph of the equation is a parabola with vertex at $(0, 0)$ and focus at $(-a, 0) = (-3, 0)$, so the parabola opens to the left. The directrix is the line $x = 3$. To graph the parabola, it helps to plot the two points on the graph above and below the focus. Because the points are directly above and below the focus, let $x = -3$ in the equation $y^2 = -12x$ and solve for y.

$$y^2 = -12x$$

Let $x = -3$: $\quad = -12(-3)$

$$= 36$$

Take the square root of both sides: $\quad y = \pm 6$

The points on the parabola above and below the focus are $(-3, 6)$ and $(-3, -6)$. These points help in graphing the parabola because they determine the "opening." See Figure 16.

Quick ✓

1. A _____ is the collection of all points P in the plane that are the same distance from a fixed point F as they are from a fixed line D.

2. The point of intersection of the parabola with its axis of symmetry is called the _____.

3. The line through the focus and perpendicular to the directrix is called the ___ __ _____.

In Problems 4 and 5, graph the equation.

4. $y^2 = 8x$ 5. $y^2 = -20x$

EXAMPLE 2   **Graphing a Parabola That Opens Up**

Graph the equation $x^2 = 8y$.

Solution

The equation $x^2 = 8y$ is of the form $x^2 = 4ay$, where $4a = 8$, so $a = 2$. The graph of the equation is a parabola with vertex at $(0, 0)$ and focus at $(0, a) = (0, 2)$, so the parabola opens up. The directrix is the line $y = -2$. To graph the parabola, plot the two points on the graph to the left and right of the focus. Let $y = 2$ in the equation $x^2 = 8y$ and solve for x.

$$x^2 = 8y = 8(2) = 16$$

Take the square root of both sides: $x = \pm 4$

The points on the parabola to the left and right of the focus are $(-4, 2)$ and $(4, 2)$. See Figure 17.

Figure 17

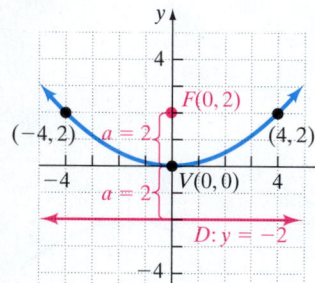

Quick ✓

In Problems 6 and 7, graph the equation.

6. $x^2 = 4y$ 7. $x^2 = -12y$

▶ ❷ Find the Equation of a Parabola

Now use information about a parabola to obtain its equation.

EXAMPLE 3   **Finding an Equation of a Parabola**

Find an equation of the parabola with vertex at $(0, 0)$ and focus at $(5, 0)$. Graph the equation.

Figure 18

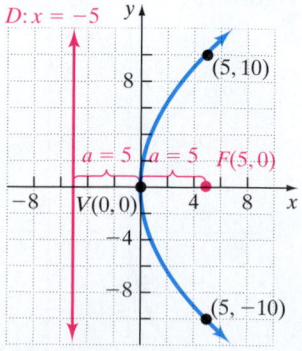

Solution

The distance from the vertex $(0, 0)$ to the focus $(5, 0)$ is $a = 5$. Because the focus lies on the positive x-axis, the parabola opens to the right. This means the equation of the parabola is of the form $y^2 = 4ax$ with $a = 5$:

$$y^2 = 4(5)x$$
$$= 20x$$

Figure 18 shows the graph of $y^2 = 20x$.

EXAMPLE 4   **Finding the Equation of a Parabola**

Work Smart

It is helpful to plot the information given about the parabola before finding its equation. For example, if the vertex and focus of the parabola in Example 3 and 4 are plotted, we can see that it opens to the right.

Find the equation of a parabola with vertex at $(0, 0)$ if its axis of symmetry is the y-axis and its graph contains the point $(4, 3)$. Graph the equation.

Solution

Which of the four equations of parabolas listed in Table 1 should be used? The vertex is at the origin and the axis of symmetry is the y-axis, so the parabola opens either up or down. Because the graph contains the point $(4, 3)$, which is in quadrant I, the parabola must open up. Therefore, the equation of the parabola is of the form $x^2 = 4ay$. Because the point $(4, 3)$ is on the parabola, let $x = 4$ and $y = 3$ in the equation $x^2 = 4ay$ to determine a.

$$x^2 = 4ay$$

$$x = 4, y = 3: \quad 4^2 = 4a(3)$$

$$16 = 12a$$

Divide both sides by 12: $\quad a = \dfrac{16}{12} = \dfrac{4}{3}$

The equation of the parabola is

$$x^2 = 4\left(\dfrac{4}{3}\right)y, \quad \text{or} \quad x^2 = \dfrac{16}{3}y$$

With $a = \dfrac{4}{3}$, the focus is $\left(0, \dfrac{4}{3}\right)$ and the directrix is the line $y = -\dfrac{4}{3}$. Let $y = \dfrac{4}{3}$ to find points left and right of the focus to determine the "opening," so the points $\left(-\dfrac{8}{3}, \dfrac{4}{3}\right)$ and $\left(\dfrac{8}{3}, \dfrac{4}{3}\right)$ are on the graph. Figure 19 shows the graph of the parabola. ●

Figure 19

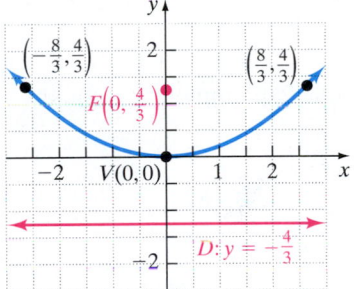

Quick ✓

In Problems 8 and 9, find the equation of the parabola described. Graph the equation.

8. Vertex at $(0, 0)$; focus at $(0, -8)$

9. Vertex at $(0, 0)$; axis of symmetry the x-axis; contains the point $(3, 2)$

▶ ❸ **Graph a Parabola Whose Vertex Is Not the Origin**

Work Smart

Think back to circles. The equation of a circle whose center is (h, k) is $(x - h)^2 + (y - k)^2 = r^2$. The circle shifts horizontally h units and vertically k units.

If a parabola with vertex at the origin and axis of symmetry along a coordinate axis is shifted horizontally h units and then vertically k units, the result is a parabola with vertex at (h, k) and axis of symmetry parallel to either the x- or the y-axis. The equations of these parabolas have the same form as those whose vertex is at the origin, except that x is replaced with $x - h$ (the horizontal shift) and y is replaced with $y - k$ (the vertical shift). Table 2 gives the equations of the four parabolas. Figure 20(a)–(d) on the next page illustrates the graphs for $h > 0$ and $k > 0$.

Table 2 Parabolas with Vertex at (h, k); Axis of Symmetry Parallel to a Coordinate Axis; $a > 0$

Vertex	Focus	Directrix	Equation	Axis of Symmetry	Opens
(h, k)	$(h + a, k)$	$x = h - a$	$(y - k)^2 = 4a(x - h)$	Parallel to x-axis	Right
(h, k)	$(h - a, k)$	$x = h + a$	$(y - k)^2 = -4a(x - h)$	Parallel to x-axis	Left
(h, k)	$(h, k + a)$	$y = k - a$	$(x - h)^2 = 4a(y - k)$	Parallel to y-axis	Up
(h, k)	$(h, k - a)$	$y = k + a$	$(x - h)^2 = -4a(y - k)$	Parallel to y-axis	Down

Figure 20

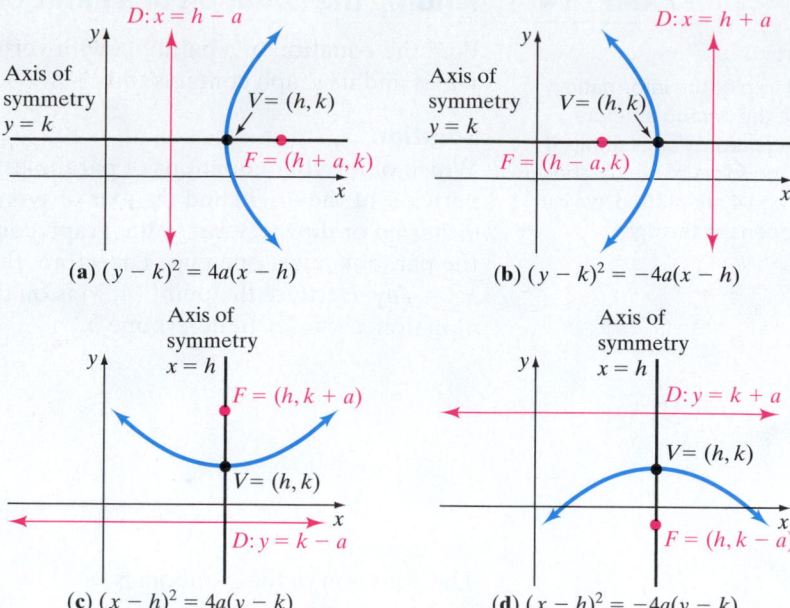

(a) $(y - k)^2 = 4a(x - h)$

(b) $(y - k)^2 = -4a(x - h)$

(c) $(x - h)^2 = 4a(y - k)$

(d) $(x - h)^2 = -4a(y - k)$

EXAMPLE 5

Graphing a Parabola Whose Vertex Is Not the Origin

Graph the parabola $x^2 - 2x + 8y + 25 = 0$.

Solution

Notice the equation is not in any of the forms given in Table 2. Complete the square in x to write the equation in standard form.

$$x^2 - 2x + 8y + 25 = 0$$

Isolate the terms involving x: $\qquad x^2 - 2x = -8y - 25$

Complete the square: $\quad x^2 - 2x + 1 = -8y - 25 + 1$

Simplify: $\quad x^2 - 2x + 1 = -8y - 24$

Factor: $\qquad (x - 1)^2 = -8(y + 3)$

Work Smart

Notice that the graph of $(x - 1)^2 = -8(y + 3)$ is the graph of $x^2 = -8y$ shifted right 1 unit and down 3 units.

The equation is of the form $(x - h)^2 = -4a(y - k)$. This is a parabola that opens down with vertex $(h, k) = (1, -3)$. Since $-4a = -8$, then $a = 2$. Because the parabola opens down, the focus is $a = 2$ units below the vertex at $(1, -5)$. To find two points on the graph to the left and right of the focus, let $y = -5$ in the equation of the parabola.

Let $y = -5$: $\quad (x - 1)^2 = -8(-5 + 3)$

$$(x - 1)^2 = -8(-2)$$

$$(x - 1)^2 = 16$$

Take the square root of both sides: $\quad x - 1 = \pm 4$

Add 1 to both sides: $\quad x = 1 \pm 4$

$$x = 1 - 4 \quad \text{or} \quad x = 1 + 4$$

$$x = -3 \quad \text{or} \quad x = 5$$

Figure 21

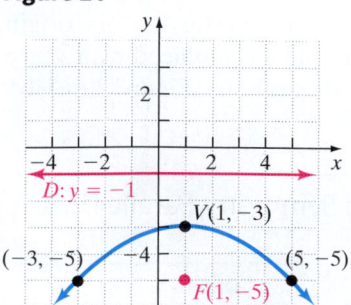

The points $(-3, -5)$ and $(5, -5)$ are on the graph of the parabola. The directrix is $a = 2$ units above the vertex, so $y = -1$ is the directrix. Figure 21 shows the graph. ●

4 Solve Applied Problems Involving Parabolas

The variety of applications of parabolas is astounding. Suppose that a mirror is shaped like a parabola. If a light bulb is placed at the focus of the parabola, then all the rays from the bulb reflect off the mirror in lines parallel to the axis of symmetry. This concept is used in the design of car headlights, flashlights, and searchlights. See Figure 22.

Suppose that rays of light are received by a parabola. When the rays strike the surface of a parabolic mirror whose axis of symmetry is parallel to these rays, they are all reflected to a single point—the focus. This idea is used in the design of some telescopes and satellite dishes. See Figure 23.

Work Smart

Which is the graph of a parabola and which is the graph of a circle?
$x^2 + y^2 - 2y + 3 = 0$
$2y^2 - 12y - x + 22 = 0$
How can you tell the difference between the equation of a circle and the equation of a parabola?
The equation of a parabola has either an x^2-term or a y^2-term but not both. The equation of a circle has the sum of an x^2-term and a y^2-term.

Figure 22

Rays of light

Light at focus

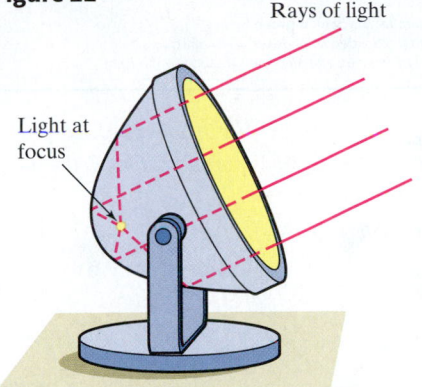

Figure 23

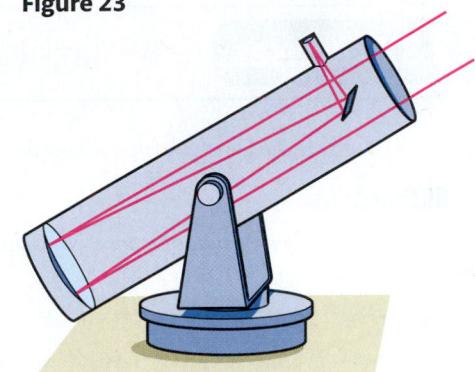

EXAMPLE 6 | **A Satellite Dish**

A satellite dish is shaped like a parabola. See Figure 24(a). Signals strike the surface of the dish and are reflected to a single point, where the receiver of the dish is located. If a satellite dish is 2 feet across at its opening and 6 inches deep (0.5 foot) at its center, at what position should the receiver be placed?

Solution

To find where to locate the receiver of the satellite dish, locate its focus. Draw a parabola in a Cartesian plane so that the vertex is at the origin and the focus is on the positive y-axis. The width of the dish is 2 feet across and its height is 0.5 feet, so two points on the graph are known, as indicated in Figure 24(b).

Figure 24

(a)

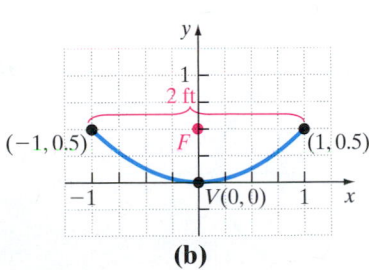

(b)

The parabola is an equation of the form $x^2 = 4ay$. Since $(1, 0.5)$ is a point on the graph, the following can be found:

$$x = 1, y = 0.5: \quad 1^2 = 4a(0.5)$$
$$1 = 2a$$

Divide both sides by 2: $\quad a = \dfrac{1}{2}$

The receiver should be located $\dfrac{1}{2}$ foot from the base of the dish along its axis of symmetry.

Quick ✓

12. A satellite dish is shaped like a parabola. Signals strike the surface of the dish and are reflected to a single point, where the receiver of the dish is located. If the dish is 4 feet across at its opening and 6 inches deep (0.5 foot) at its center, at what position should the receiver be placed?

12.3 Exercises MyMathLab®

Exercise numbers in **green** have complete video solutions in MyMathLab or may be accessed using the QR code to the right.

*Problems **1–12** are the Quick✓s that follow the EXAMPLES.*

Building Skills

In Problems 13–20, the graph of a parabola is given. Match each graph to its equation.

(a) $y^2 = 8x$ **(b)** $y^2 = -8x$

(c) $x^2 = 8y$ **(d)** $x^2 = -8y$

(e) $(y - 2)^2 = 8(x + 1)$ **(f)** $(y - 2)^2 = -8(x + 1)$

(g) $(x + 1)^2 = 8(y - 2)$ **(h)** $(x + 1)^2 = -8(y - 2)$

13.
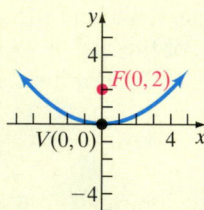

14.
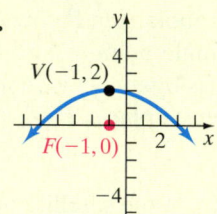

15.
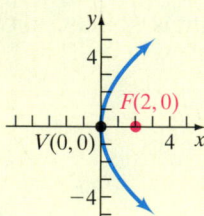

16.
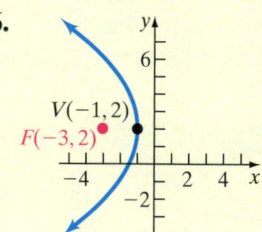

17.
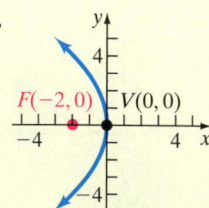

18.
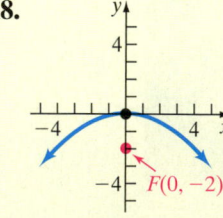

19.
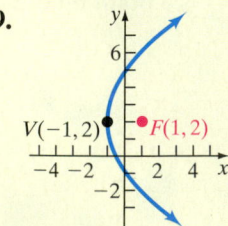

20.
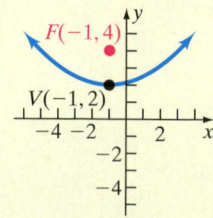

In Problems 21–26, find the vertex, focus, and directrix of each parabola. Graph the parabola. See Objective 1.

21. $x^2 = 24y$ **22.** $x^2 = 28y$

23. $y^2 = -6x$ **24.** $y^2 = 10x$

25. $x^2 = -8y$ **26.** $x^2 = -16y$

In Problems 27–36, find the equation of the parabola described. Graph the parabola. See Objective 2.

27. Vertex at $(0, 0)$; focus at $(5, 0)$

28. Vertex at $(0, 0)$; focus at $(0, 5)$

29. Vertex at $(0, 0)$; focus at $(0, -6)$

30. Vertex at $(0, 0)$; focus at $(-8, 0)$

31. Vertex at $(0, 0)$; contains the point $(6, 6)$; axis of symmetry the y-axis

32. Vertex at $(0, 0)$; contains the point $(2, 2)$; axis of symmetry the x-axis

33. Vertex at $(0, 0)$; directrix the line $y = 3$

34. Vertex at $(0, 0)$; directrix the line $x = -4$

35. Focus at $(-3, 0)$; directrix the line $x = 3$

36. Focus at $(0, -2)$; directrix the line $y = 2$

In Problems 37 and 38, write an equation for each parabola. See Objective 2.

37.
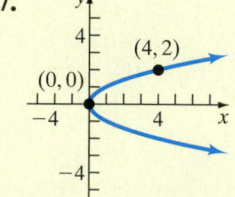

38.
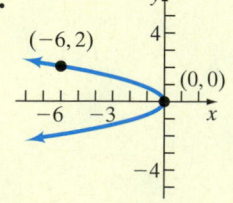

In Problems 39–50, find the vertex, focus, and directrix of each parabola. Graph the parabola. See Objective 3.

39. $(x - 2)^2 = 4(y - 4)$ **40.** $(x + 4)^2 = -4(y - 1)$

41. $(y + 3)^2 = -8(x + 2)$ **42.** $(y - 2)^2 = 12(x + 5)$

43. $(x + 5)^2 = -20(y - 1)$ **44.** $(x - 6)^2 = 2(y - 2)$

45. $x^2 + 4x + 12y + 16 = 0$ **46.** $x^2 + 2x - 8y + 25 = 0$

47. $y^2 - 8y - 4x + 20 = 0$ **48.** $y^2 - 8y + 16x - 16 = 0$

49. $x^2 + 10x + 6y + 13 = 0$ **50.** $x^2 - 4x + 10y + 4 = 0$

Applying the Concepts

51. A Headlight The headlight of a car is in the shape of a parabola. Its diameter is 4 inches and its depth is 1 inch. How far from the vertex should the light bulb be placed so that the rays will be reflected parallel to the axis?

52. A Headlight The headlight of a car is in the shape of a parabola. Suppose the engineers have designed the headlight to be 5 inches wide and want the bulb to be placed at the focus 1 inch from the vertex. What is the depth of the headlight?

---

53. Suspension Bridge The cables of a suspension bridge are in the shape of a parabola, as shown in the figure. The towers supporting the cable are 500 feet apart and 60 feet high. If the cables touch the road surface midway between the towers, what is the height of the cable at a point 150 feet from the center of the bridge?

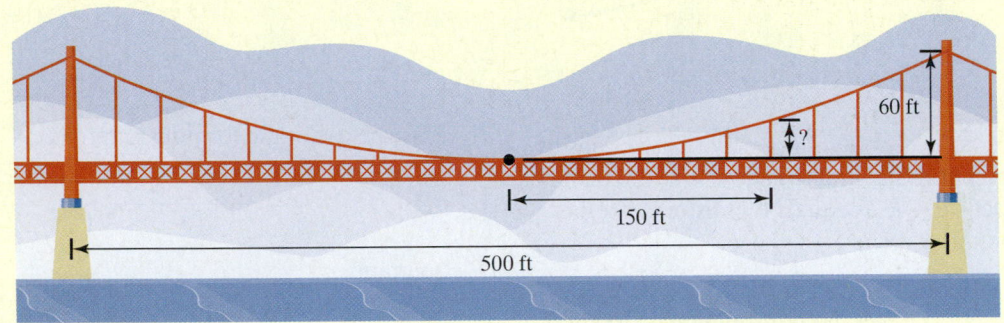

54. Suspension Bridge The cables of a suspension bridge are in the shape of a parabola. The towers supporting the cable are 400 feet apart and 80 feet high. If the cables touch the road surface midway between the towers, what is the height of the cable at a point 100 feet from the center of the bridge?

55. Parabolic Arch Bridge A bridge is built in the shape of a parabolic arch. The arch has a span of 100 feet and a maximum height of 30 feet. See the illustration. Choose a suitable rectangular coordinate system and find the height of the arch at distances of 10, 30, and 50 feet from the center.

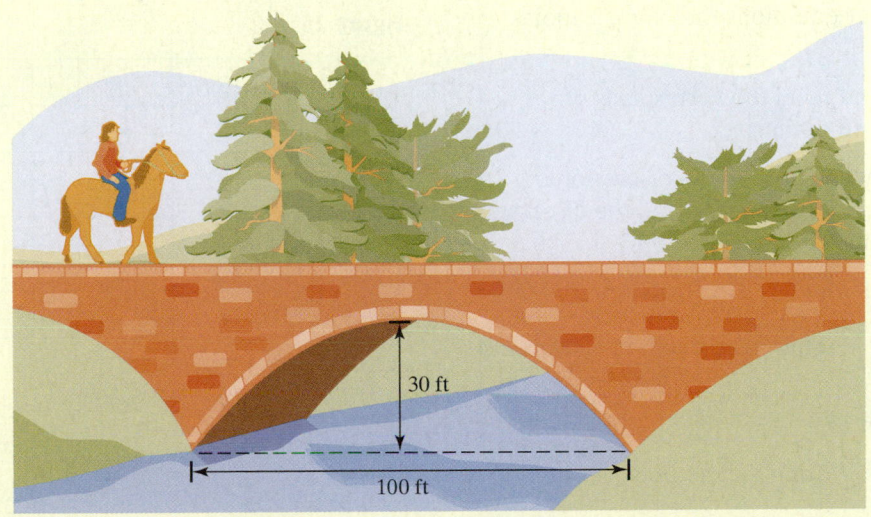

56. Parabolic Arch Bridge A bridge is to be built in the shape of a parabolic arch and is to have a span of 120 feet. The height of the arch at a distance of 30 feet from the center is to be 15 feet. Find the height of the arch at its center.

Extending the Concepts

In Problems 57–60, write an equation for each parabola.

57.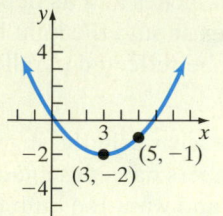

58.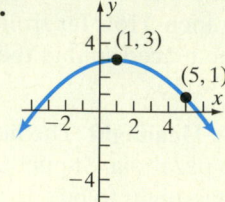

59.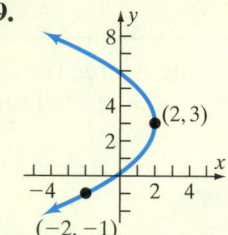

60.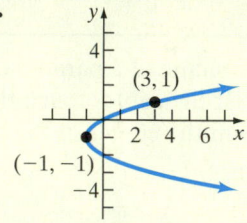

61. For the parabola $x^2 = 8y$, **(a)** verify that $(4, 2)$ is a point on the parabola, and **(b)** show that the distance from the focus to $(4, 2)$ equals the distance from $(4, 2)$ to the directrix.

62. For the parabola $(y - 3)^2 = 12(x + 1)$, **(a)** verify that $(2, 9)$ is a point on the parabola, and **(b)** show that the distance from the focus to $(2, 9)$ equals the distance from $(2, 9)$ to the directrix.

Explaining the Concepts

63. The distance from a point on a parabola to its focus is 8 units. What is the distance from the same point on the parabola to the directrix? Why?

64. Write down the four equations that are parabolas with vertex at (h, k).

65. Draw a parabola and label the vertex, axis of symmetry, focus, and directrix.

66. Explain the difference between the discussion of parabolas presented in this section and the discussion presented in Sections 10.4 and 10.5.

Synthesis Review

67. Graph $y = (x + 3)^2$ using the methods of Section 10.4.

68. Graph $y = (x + 3)^2 = x^2 + 6x + 9$ using the methods of Section 10.5.

69. Graph $y = (x + 3)^2$ using the methods of this section.

70. Graph $4(y + 2) = (x - 2)^2$ using the methods of Section 10.4. (*Hint:* Write the equation in the form $y = a(x - h)^2 + k$.)

71. Graph $4(y + 2) = (x - 2)^2$ using the methods of this section.

72. Compare and contrast the methods of graphing a parabola using the approach in Section 10.4 and the approach in this section. Which do you prefer? Why?

Technology Exercises

To graph a parabola using a graphing calculator, solve the equation for y, just as in graphing circles. This is fairly straightforward if the equation is in the form given in either Table 1 or Table 2. If the equation is not in the form given in Table 1 or Table 2 and it is a parabola that opens up or down, then solving for y is also straightforward. However, if it is a parabola that opens left or right, graph the parabola in two "pieces"—the top half and the bottom half. Use the quadratic formula to accomplish this task. Consider the equation $y^2 + 8y + x - 4 = 0$. This equation is quadratic in y as shown:

$$y^2 + 8y + x - 4 = 0$$

$$\boxed{a = 1} \quad \boxed{b = 8} \quad \boxed{c = x - 4}$$

Use the quadratic formula and obtain

$$y = \frac{-8 \pm \sqrt{8^2 - 4(1)(x - 4)}}{2(1)}$$

so graph

$$Y_1 = \frac{-8 - \sqrt{64 - 4(x - 4)}}{2} \quad \text{and}$$

$$Y_2 = \frac{-8 + \sqrt{64 - 4(x - 4)}}{2}.$$

as shown in Figure 25.

To graph a parabola without having to solve the equation for y, you can use Desmos.

Figure 25

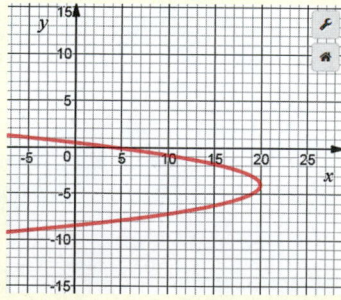

In Problems 73–84, graph each parabola using technology.

73. $x^2 = 24y$ **74.** $x^2 = -8y$

75. $y^2 = -6x$ **76.** $y^2 = 10x$

77. $(x - 2)^2 = 4(y - 4)$ **78.** $(x + 4)^2 = -4(y - 1)$

79. $(y + 3)^2 = -8(x + 2)$ **80.** $(y - 2)^2 = 12(x + 5)$

81. $x^2 + 4x + 12y + 16 = 0$ **82.** $x^2 + 2x - 8y + 25 = 0$

83. $y^2 - 8y - 4x + 20 = 0$ **84.** $y^2 - 8y + 16x - 16 = 0$

12.4 Ellipses

Objectives

1. Graph an Ellipse Whose Center Is the Origin
2. Find the Equation of an Ellipse Whose Center Is the Origin
3. Graph an Ellipse Whose Center Is Not the Origin
4. Solve Applied Problems Involving Ellipses

Are You Prepared for This Section?

Before getting started, complete the following problems. If you get a problem wrong, go back to the section cited and review the material.

P1. Complete the square of $x^2 + 10x$.   [Section 10.1, pp. 694–695]

P2. Graph $f(x) = (x + 2)^2 - 1$ using transformations.   [Section 10.4, p. 732]

▶ 1 Graph an Ellipse Whose Center Is the Origin

An ellipse is a conic section that is obtained through the intersection of a plane and a cone. See Figure 8(b) in Section 12.2.

> **Definition**
>
> An **ellipse** is the collection of points in the plane such that the sum of the distances from two fixed points, called the **foci,** is a constant.

Figure 26

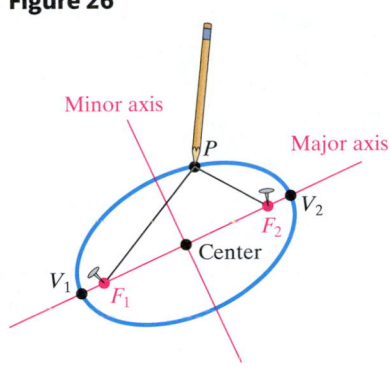

This definition enables us to physically draw an ellipse. To do this, find a piece of string (the length of the string is the constant in the definition). Now stick two thumbtacks on a piece of cardboard so that the distance between them is less than the length of the string. The two thumbtacks represent the foci of the ellipse. Now attach the ends of the string to the thumbtacks and, using the point of a pencil, pull the string taut. Keeping the string taut, rotate the pencil around the two thumbtacks. The pencil traces out an ellipse, as shown in Figure 26.

In Figure 26, the foci are labeled F_1 and F_2. The line containing the foci is called the **major axis.** The midpoint of the line segment joining the foci is called the **center** of the ellipse. The line through the center and perpendicular to the major axis is called the **minor axis.**

The two points of intersection of the ellipse and the major axis are the **vertices,** V_1 and V_2, of the ellipse. The distance from one vertex to the other is called the **length of the major axis.**

In this text, consider only ellipses whose major axis is parallel to (or coincides with) either the x-axis or the y-axis. The equations of an ellipse with center at the origin are given in Table 3.

Table 3	Ellipses with Center at the Origin			
Center	**Major Axis**	**Foci**	**Vertices**	**Equation**
$(0, 0)$	x-axis	$(-c, 0)$ and $(c, 0)$	$(-a, 0)$ and $(a, 0)$	$\dfrac{x^2}{a^2} + \dfrac{y^2}{b^2} = 1$
$(0, 0)$	y-axis	$(0, -c)$ and $(0, c)$	$(0, -a)$ and $(0, a)$	$\dfrac{x^2}{b^2} + \dfrac{y^2}{a^2} = 1$

The graphs of the two ellipses are given in Figure 27.

Did you notice that in the ellipse $\dfrac{x^2}{a^2} + \dfrac{y^2}{b^2} = 1$ in Figure 27(a) and in the ellipse $\dfrac{x^2}{b^2} + \dfrac{y^2}{a^2} = 1$ in Figure 27(b), the focus, the origin, and the endpoint of the minor axis form a right triangle? In both ellipses, $a > b$ and, from the Pythagorean Theorem, $b^2 + c^2 = a^2$, or $b^2 = a^2 - c^2$.

Use the fact that $a > b$ to determine whether the major axis is the x-axis or y-axis. If the larger denominator is associated with the x^2 term, then the major axis is the

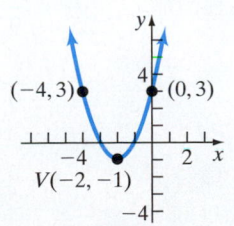

Figure 27

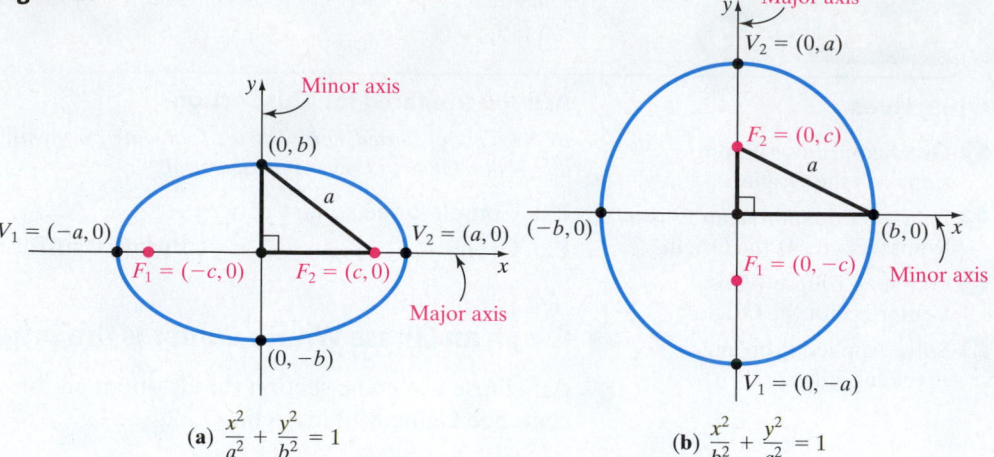

(a) $\dfrac{x^2}{a^2} + \dfrac{y^2}{b^2} = 1$

(b) $\dfrac{x^2}{b^2} + \dfrac{y^2}{a^2} = 1$

Work Smart

The term with the larger denominator tells us which axis is the major axis.

x-axis; if the larger denominator is associated with the y^2 term, then the major axis is the y-axis.

> **Quick ✓**
>
> 1. An _____ is the collection of points in the plane such that the sum of the distances from two fixed points, called the ___, is a constant.
>
> 2. For an ellipse, the line containing the foci is called the _____ ___.
>
> 3. The two points of intersection of the ellipse and the major axis are the _____, V_1 and V_2, of the ellipse.
>
> 4. *True or False* If an ellipse has vertex $(a, 0)$ and focus $(c, 0)$, then $c^2 = a^2 + b^2$.

    Graphing an ellipse whose center is at the origin is fairly straightforward. Find the intercepts by letting $y = 0$ (x-intercepts) and letting $x = 0$ (y-intercepts). When asked to graph an ellipse, be sure to label the foci.

EXAMPLE 1   **Graphing an Ellipse**

Graph the ellipse: $\dfrac{x^2}{25} + \dfrac{y^2}{9} = 1$

Solution

First, notice that the x^2 term has the larger denominator, 25. This means that the major axis is the x-axis and the equation of the ellipse has the form $\dfrac{x^2}{a^2} + \dfrac{y^2}{b^2} = 1$, so $a^2 = 25$ and $b^2 = 9$. The center of the ellipse is the origin, $(0, 0)$. Because $b^2 = a^2 - c^2$, we know that $c^2 = a^2 - b^2 = 25 - 9 = 16$, so $c = 4$. Since the major axis is the x-axis, the foci are $(-4, 0)$ and $(4, 0)$. Now find the intercepts:

x-intercepts: Let $y = 0$: $\dfrac{x^2}{25} + \dfrac{0^2}{9} = 1$ **y-intercepts:**   Let $x = 0$: $\dfrac{0^2}{25} + \dfrac{y^2}{9} = 1$

$$\dfrac{x^2}{25} = 1 \qquad\qquad\qquad\qquad \dfrac{y^2}{9} = 1$$

$$x^2 = 25 \qquad\qquad\qquad\qquad y^2 = 9$$

$$x = \pm 5 \qquad\qquad\qquad\qquad y = \pm 3$$

The intercepts are $(-5, 0)$, $(5, 0)$, $(0, -3)$, and $(0, 3)$. Figure 28 shows the graph of the ellipse.

Figure 28

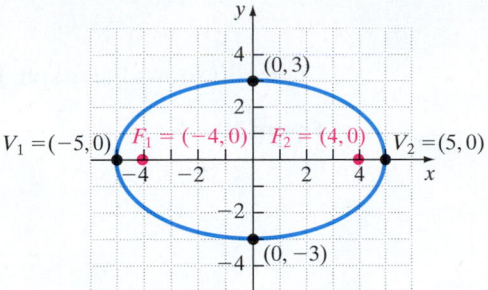

EXAMPLE 2 **Graphing an Ellipse**

Graph the ellipse: $\dfrac{x^2}{4} + \dfrac{y^2}{16} = 1$

Solution

First, notice that the y^2 term has the larger denominator, 16. This means that the major axis is the y-axis and the equation has the form $\dfrac{x^2}{b^2} + \dfrac{y^2}{a^2} = 1$, so $a^2 = 16$ and $b^2 = 4$. The center of the ellipse is the origin, $(0, 0)$. Because $b^2 = a^2 - c^2$, then $c^2 = a^2 - b^2 = 16 - 4 = 12$, so $c = \sqrt{12} = 2\sqrt{3}$. Since the major axis is the y-axis, the foci are $\left(0, -2\sqrt{3}\right)$ and $\left(0, 2\sqrt{3}\right)$. Now find the intercepts:

Figure 29

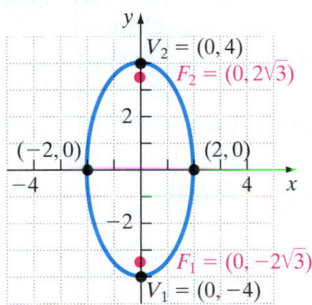

x**-intercepts:** Let $y = 0$: $\quad \dfrac{x^2}{4} + \dfrac{0^2}{16} = 1 \qquad$ y**-intercepts:** Let $x = 0$: $\quad \dfrac{0^2}{4} + \dfrac{y^2}{16} = 1$

$$\dfrac{x^2}{4} = 1 \qquad\qquad\qquad\qquad \dfrac{y^2}{16} = 1$$

$$x^2 = 4 \qquad\qquad\qquad\qquad\quad y^2 = 16$$

$$x = \pm 2 \qquad\qquad\qquad\qquad\quad y = \pm 4$$

The intercepts are $(-2, 0)$, $(2, 0)$, $(0, -4)$, and $(0, 4)$. Figure 29 shows the graph of the ellipse.

Quick ✔

In Problems 5 and 6, graph each ellipse.

5. $\dfrac{x^2}{9} + \dfrac{y^2}{4} = 1$ **6.** $\dfrac{x^2}{16} + \dfrac{y^2}{36} = 1$

▶ ❷ Find the Equation of an Ellipse Whose Center Is the Origin

Just as in working with parabolas, information about an ellipse is used to find its equation.

EXAMPLE 3 **Finding the Equation of an Ellipse**

Find the equation of the ellipse whose center is the origin with a focus at $(-2, 0)$ and vertex at $(-5, 0)$. Graph the ellipse.

Solution

The given focus and vertex lie on the x-axis, so the major axis is the x-axis. The equation of the ellipse must be of the form $\dfrac{x^2}{a^2} + \dfrac{y^2}{b^2} = 1$. The distance from the center of the ellipse to the vertex is $a = 5$ units.

(continued)

The distance from the center of the ellipse to the focus is $c = 2$ units, so $b^2 = a^2 - c^2 = 5^2 - 2^2 = 25 - 4 = 21$. Thus the equation of the ellipse is

$$\frac{x^2}{25} + \frac{y^2}{21} = 1$$

Figure 30 shows the graph. Use $b = \sqrt{21} \approx 4.6$ to graph the y-intercepts.

Figure 30

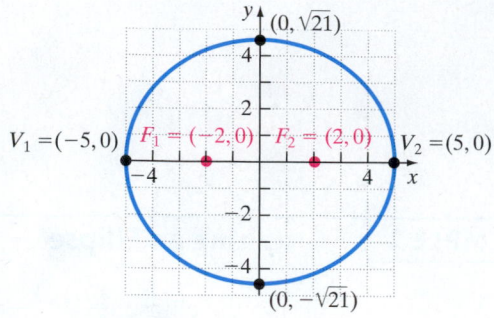

Quick ✓

7. Find the equation of the ellipse whose center is the origin with a focus at $(0, 3)$ and vertex at $(0, 7)$. Graph the ellipse.

▶ ❸ **Graph an Ellipse Whose Center Is Not the Origin**

If an ellipse with center at the origin and major axis coinciding with a coordinate axis is shifted horizontally h units and vertically k units, the result is an ellipse with center at (h, k) and major axis parallel to a coordinate axis. The equations of these ellipses have the same forms as those given for ellipses whose center is the origin, except that x is replaced by $x - h$ (the horizontal shift) and y is replaced by $y - k$ (the vertical shift). Table 4 gives the forms of the equations for these ellipses. Figure 31 shows their graphs.

Work Smart

Do not attempt to memorize Table 4. Instead, understand the roles of a and c in an ellipse—a is the distance from the center to each vertex, and c is the distance from the center to each focus.

Table 4 Ellipses with Center at (h, k) and Major Axis Parallel to a Coordinate Axis

Center	Major Axis	Foci	Vertices	Equation
(h, k)	Parallel to x-axis	$(h + c, k)$	$(h + a, k)$	$\dfrac{(x - h)^2}{a^2} + \dfrac{(y - k)^2}{b^2} = 1$
		$(h - c, k)$	$(h - a, k)$	$a > b$ and $b^2 = a^2 - c^2$
(h, k)	Parallel to y-axis	$(h, k + c)$	$(h, k + a)$	$\dfrac{(x - h)^2}{b^2} + \dfrac{(y - k)^2}{a^2} = 1$
		$(h, k - c)$	$(h, k - a)$	$a > b$ and $b^2 = a^2 - c^2$

Figure 31

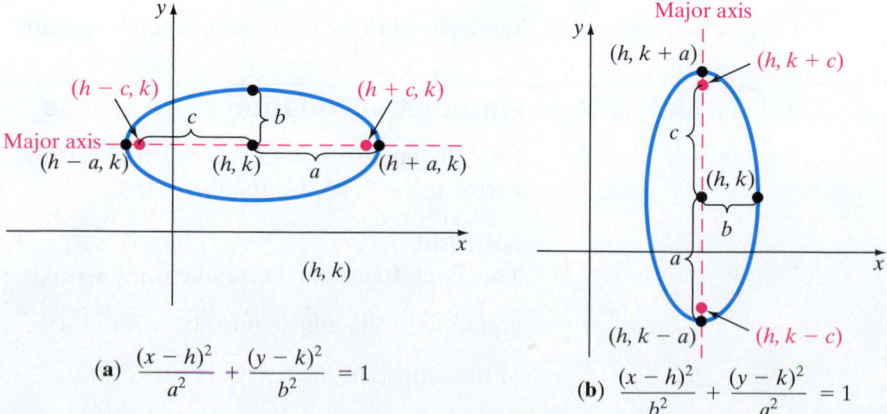

(a) $\dfrac{(x - h)^2}{a^2} + \dfrac{(y - k)^2}{b^2} = 1$

(b) $\dfrac{(x - h)^2}{b^2} + \dfrac{(y - k)^2}{a^2} = 1$

EXAMPLE 4 **Graphing an Ellipse**

Graph the equation: $x^2 + 16y^2 - 4x + 32y + 4 = 0$

Solution

Write the equation in one of the forms given in Table 4. To do this, complete the squares in x and in y.

$$x^2 + 16y^2 - 4x + 32y + 4 = 0$$

Group like variables; place the constant on the right side:
$$(x^2 - 4x) + (16y^2 + 32y) = -4$$

Factor out 16 from $16y^2 + 32y$:
$$(x^2 - 4x) + 16(y^2 + 2y) = -4$$

Complete each square:
$$(x^2 - 4x + 4) + 16(y^2 + 2y + 1) = -4 + 4 + 16$$

Factor:
$$(x - 2)^2 + 16(y + 1)^2 = 16$$

Divide both sides by 16:
$$\frac{(x - 2)^2}{16} + \frac{(y + 1)^2}{1} = 1$$

Work Smart

To complete the square in y, 1 was added, but because it was inside the parentheses with a factor of 16 in front, $16 \cdot 1 = 16$ must be added to the other side.

This is the equation of an ellipse with center at $(2, -1)$. Because the larger value, 16, is the denominator of the x^2 term, the major axis is parallel to the x-axis. Because $a^2 = 16$ and $b^2 = 1$, then $c^2 = a^2 - b^2 = 16 - 1 = 15$. The vertices are $a = 4$ units to the left and right of center at $V_1 = (-2, -1)$ and $V_2 = (6, -1)$. The foci are $c = \sqrt{15}$ units to the left and right of center at $F_1 = (2 - \sqrt{15}, -1)$ and $F_2 = (2 + \sqrt{15}, -1)$. Plot points $b = 1$ unit above and below the center at $(2, 0)$ and $(2, -2)$. Use $2 \pm \sqrt{15} \approx -1.9$ and 5.9 when plotting the foci of the ellipse. See Figure 32.

Work Smart

The graph of $\dfrac{(x - 2)^2}{16} + \dfrac{(y + 1)^2}{1} = 1$

is the graph of $\dfrac{x^2}{16} + \dfrac{y^2}{1} = 1$ shifted

2 units right and 1 unit down.

Figure 32

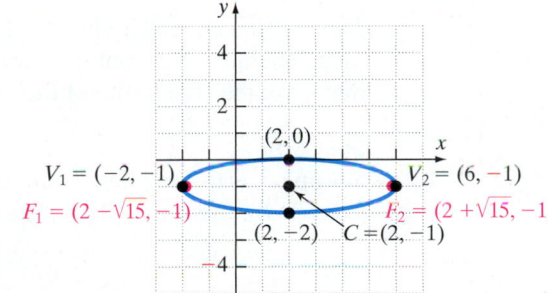

Quick ✓

8. The center of $\dfrac{(x - 3)^2}{25} + \dfrac{(y + 1)^2}{16} = 1$ is _____.

9. Graph the equation: $9x^2 + y^2 + 54x - 2y + 73 = 0$

▶ ❹ Solve Applied Problems Involving Ellipses

Ellipses are found in many applications in science and engineering. For example, the orbits of the planets around the Sun are elliptical, with the Sun's position at a focus. See Figure 33.

Stone and concrete bridges are often the shape of semielliptical arches. Elliptical gears are used in machinery when a variable rate of motion is required.

Ellipses also have an interesting reflection property. If a source of sound (or light) is placed at one focus, the waves transmitted by the source reflect off the ellipse and

Figure 33

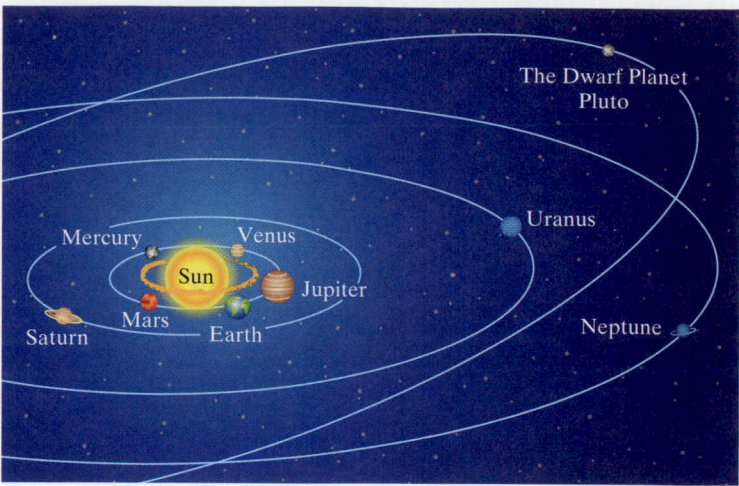

concentrate at the other focus. This is the principle behind whispering galleries, which are rooms with elliptical ceilings. A person standing at one focus of the ellipse can whisper and be heard by a person standing at the other focus, because all the sound waves that reach the ceiling are reflected to the other person. National Statuary Hall in the United States Capital is a whispering gallery.

EXAMPLE 5 **A Whispering Gallery**

The whispering gallery in the Museum of Science and Industry in Chicago is 47.3 feet long. The distance from the center of the room to the foci is 20.3 feet. Find an equation that describes the shape of the room. How high is the room at its center?

Solution
Set up a Cartesian plane so that the center of the ellipse is at the origin and the major axis lies on the x-axis. The equation of the ellipse is

$$\frac{x^2}{a^2} + \frac{y^2}{b^2} = 1$$

Figure 34

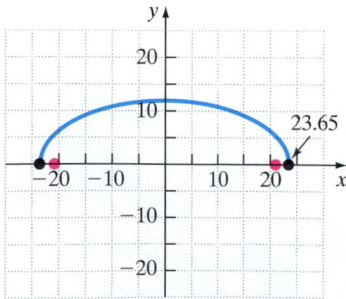

Because the length of the room is 47.3 feet, the distance from the center of the room to each vertex (the end of the room) will be $47.3/2 = 23.65$ feet, so $a = 23.65$ feet. The distance from the center of the room to each focus is $c = 20.3$ feet. See Figure 34. Because $b^2 = a^2 - c^2$, we have $b^2 = 23.65^2 - 20.3^2 = 147.2325$. An equation that describes the shape of the room is given by

$$\frac{x^2}{23.65^2} + \frac{y^2}{147.2325} = 1$$

The height of the room at its center is $b = \sqrt{147.2325} \approx 12.1$ feet.

Quick ✓

10. A hall 100 feet in length is to be designed as a whispering gallery. If the foci are to be located 30 feet from the center, determine an equation that describes the room. What is the height of the room at its center?

12.4 Exercises MyMathLab® Exercise numbers in **green** have complete video solutions in MyMathLab or may be accessed using the QR code to the right.

*Problems **1–10** are the **Quick ✔**s that follow the **EXAMPLES**.*

Building Skills

In Problems 11–14, the graph of an ellipse is given. Match each graph to its equation. See Objective 1.

(a) $x^2 + \dfrac{y^2}{9} = 1$

(b) $\dfrac{x^2}{9} + y^2 = 1$

(c) $\dfrac{x^2}{16} + \dfrac{y^2}{9} = 1$

(d) $\dfrac{x^2}{9} + \dfrac{y^2}{16} = 1$

11.

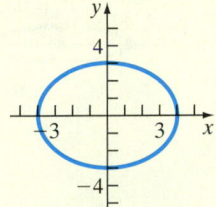

12.

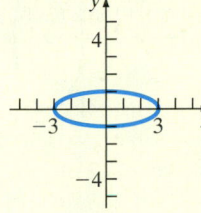

13.

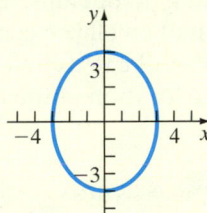

14.

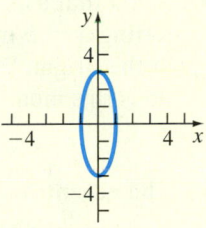

In Problems 15–24, find the vertices and foci of each ellipse. Graph each ellipse. See Objective 1.

15. $\dfrac{x^2}{25} + \dfrac{y^2}{16} = 1$

16. $\dfrac{x^2}{25} + \dfrac{y^2}{4} = 1$

17. $\dfrac{x^2}{36} + \dfrac{y^2}{100} = 1$

18. $\dfrac{x^2}{16} + \dfrac{y^2}{36} = 1$

19. $\dfrac{x^2}{49} + \dfrac{y^2}{4} = 1$

20. $\dfrac{x^2}{121} + \dfrac{y^2}{100} = 1$

21. $x^2 + \dfrac{y^2}{49} = 1$

22. $\dfrac{x^2}{64} + y^2 = 1$

23. $4x^2 + y^2 = 16$

24. $9x^2 + y^2 = 81$

In Problems 25–32, find an equation for each ellipse. Graph each ellipse. See Objective 2.

25. Center at $(0,0)$; focus at $(4,0)$; vertex at $(6,0)$

26. Center at $(0,0)$; focus at $(2,0)$; vertex at $(5,0)$

27. Center at $(0,0)$; focus at $(0,-4)$; vertex at $(0,7)$

28. Center at $(0,0)$; focus at $(0,-1)$; vertex at $(0,5)$

29. Foci at $(\pm6,0)$; vertices at $(\pm10,0)$

30. Foci at $(0,\pm2)$; vertices at $(0,\pm7)$

31. Foci at $(0,\pm5)$; length of the major axis is 16

32. Foci at $(\pm6,0)$; length of the major axis is 20

In Problems 33–42, graph each ellipse. See Objective 3.

33. $\dfrac{(x-3)^2}{9} + \dfrac{(y+2)^2}{25} = 1$

34. $\dfrac{(x-1)^2}{36} + \dfrac{(y+4)^2}{100} = 1$

35. $\dfrac{(x+2)^2}{16} + \dfrac{(y-5)^2}{4} = 1$

36. $\dfrac{(x+5)^2}{64} + \dfrac{(y+1)^2}{16} = 1$

37. $(x-5)^2 + \dfrac{(y+1)^2}{49} = 1$

38. $\dfrac{(x+8)^2}{81} + (y-3)^2 = 1$

39. $4(x+2)^2 + 16(y-1)^2 = 64$

40. $9(x-3)^2 + (y-4)^2 = 81$

41. $4x^2 + y^2 - 24x + 2y - 63 = 0$

42. $16x^2 + 9y^2 - 128x + 54y - 239 = 0$

Applying the Concepts

43. Semielliptical Arch Bridge An arch in the shape of the upper half of an ellipse is used to support a bridge that is to span a river 30 meters wide. The center of the arch is 10 meters above the center of the river. See the figure.

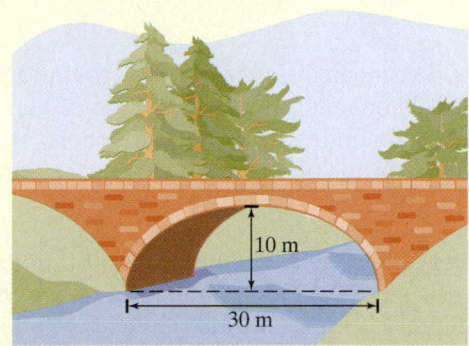

(a) Write the equation for the ellipse in which the x-axis coincides with the water and the y-axis passes through the center of the arch.

(b) Can a rectangular barge that is 18 meters wide and sits 7 meters above the surface of the water fit through the opening of the bridge?

(c) If heavy rains cause the river's level to increase 1.1 meters, will the barge make it through the opening?

44. London Bridge An arch in the shape of the upper half of an ellipse is used to support London Bridge. The main span is 45.6 meters wide. Suppose that the center of the arch is 15 meters above the center of the river.

(a) Write the equation for the ellipse in which the x-axis coincides with the water and the y-axis passes through the center of the arch.

(b) Can a rectangular barge that is 20 meters wide and sits 12 meters above the surface of the water fit through the opening of the bridge?

(c) If heavy rains cause the river's level to increase 1.5 meters, will the barge make it through the opening?

*In Problems 45–48, use the fact that the orbit of a planet about the Sun is an ellipse, with the Sun at one focus. The **aphelion** of a planet is its greatest distance from the Sun, and the **perihelion** is its shortest distance. The **mean distance** of a planet from the Sun is the distance from the center to a vertex of the elliptical orbit. See the illustration.*

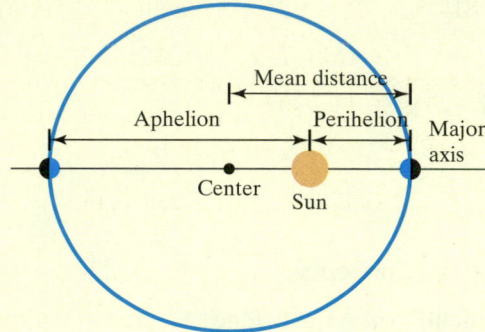

45. Earth The mean distance of Earth from the Sun is 93 million miles. If the aphelion of Earth is 94.5 million miles, what is the perihelion? Write an equation for the orbit of Earth around the Sun.

46. Mars The mean distance of Mars from the Sun is 142 million miles. If the perihelion of Mars is 128.5 million miles, what is the aphelion? Write an equation for the orbit of Mars about the Sun.

47. Jupiter The aphelion of Jupiter is 507 million miles. If the distance from the Sun to the center of its elliptical orbit is 23.2 million miles, what is the perihelion? What is the mean distance? Write an equation for the orbit of Jupiter around the Sun.

48. Dwarf Planet Pluto The perihelion of Pluto is 4551 million miles, and the distance of the Sun from the center of its elliptical orbit is 897.5 million miles. Find the aphelion of Pluto. What is the mean distance of Pluto from the Sun? Write an equation for the orbit of Pluto about the Sun.

Extending the Concepts

In Problems 49–52, write an equation for each ellipse.

49.

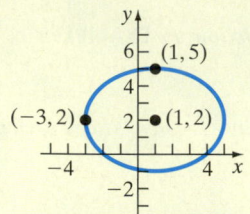

50.

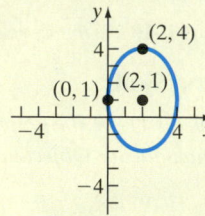

51.

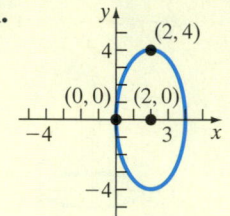

52.

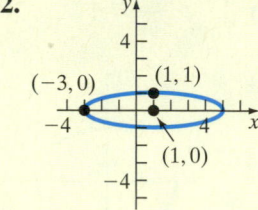

53. Show that a circle is a special kind of ellipse by letting $a = b$ in the equation of an ellipse centered at the origin. What is the value of c in a circle? What does this mean regarding the location of the foci?

54. The **eccentricity** e of an ellipse is defined as the number $\dfrac{c}{a}$, where a is the distance from the center of an ellipse to a vertex and c is the distance from the center of an ellipse to a focus. Because $a > c$, it follows that $e < 1$ for an ellipse. Write a paragraph about the general shape of each of the following ellipses. Be sure to justify your conclusion.

(a) Eccentricity close to 0

(b) Eccentricity $= 0.5$

(c) Eccentricity close to 1

Explaining the Concepts

55. In the ellipse drawn below, the center is at the origin and the major axis is the x-axis. The point F is a focus. In the right triangle (drawn in red), label the lengths of the legs and the hypotenuse using a, b, and c.

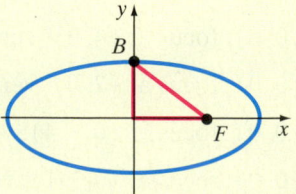

56. How does the ellipse given by the equation $\dfrac{x^2}{25} + \dfrac{y^2}{9} = 1$ differ from the ellipse given by the equation $\dfrac{x^2}{9} + \dfrac{y^2}{25} = 1$? How are they the same?

Synthesis Review

In Problems 57–60, fill in the following table for each function.

x	5	10	100	1000
f(x)				

57. $f(x) = \dfrac{5}{x + 2}$

58. $f(x) = \dfrac{x - 1}{x^2 + 4}$

59. $f(x) = \dfrac{2x + 1}{x - 3}$

60. $f(x) = \dfrac{3x^2 - x + 1}{x^2 + 1}$

In Problems 61 and 62, fill in the following table for each function.

x	5	10	100	1000
f(x)				
g(x)				

61. $f(x) = \dfrac{x^2 + 3x + 1}{x + 1}; g(x) = x + 2$

62. $f(x) = \dfrac{x^2 - 3x + 5}{x + 2}; g(x) = x - 5$

63. For the functions in Problems 57–60, compare the degree of the polynomial in the numerator to the degree of the polynomial in the denominator. Conjecture what happens to a rational function as x increases when the degree of the numerator is less than the degree of the denominator. Conjecture what happens to a rational function as x increases when the degree of the numerator equals the degree of the denominator.

64. For the functions in Problems 61 and 62, write f in the form quotient $+ \dfrac{\text{remainder}}{\text{dividend}}$. That is, perform the division indicated by the rational function. Now compare the values of f to those of g in the table. What does the function g represent?

Technology Exercises

A graphing calculator can be used to graph an ellipse by solving the equation for y. Because ellipses are not functions, graph the upper half and the lower half of an ellipse in two pieces. For example, to graph $\dfrac{x^2}{4} + \dfrac{y^2}{9} = 1$, graph

$$Y_1 = 3\sqrt{1 - \dfrac{x^2}{4}} \text{ and } Y_2 = -3\sqrt{1 - \dfrac{x^2}{4}}.$$

To graph an ellipse without having to solve the equation for y, you can use Desmos. The graph shown in Figure 35 is obtained.

In Problems 65–72, graph each ellipse using technology.

65. $\dfrac{x^2}{25} + \dfrac{y^2}{16} = 1$

66. $\dfrac{x^2}{25} + \dfrac{y^2}{4} = 1$

67. $4x^2 + y^2 = 16$

68. $9x^2 + y^2 = 81$

Figure 35

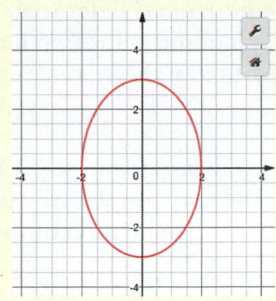

69. $\dfrac{(x - 3)^2}{9} + \dfrac{(y + 2)^2}{25} = 1$

70. $\dfrac{(x - 1)^2}{36} + \dfrac{(y + 4)^2}{100} = 1$

71. $\dfrac{(x + 2)^2}{16} + \dfrac{(y - 5)^2}{4} = 1$

72. $\dfrac{(x + 5)^2}{64} + \dfrac{(y + 1)^2}{16} = 1$

73. Exploration: Equation of an Ellipse *Open the "Equation of an Ellipse" Geogebra applet. The applet may be found using the QR code at the beginning of this section or through the Multimedia Library in MyMathLab. In the upper left corner are four "sliders." Use your mouse to change the values of the h, k, horizontal, or vertical sliders by clicking on the appropriate slider and dragging the value left or right. As you are moving the sliders, take note of how the position and equation of the ellipse change. If you are using a keyboard, it may be easier to use the arrow keys, instead of the mouse, to move the slider. First click on the slider with your mouse, and then use the left and right arrow keys to move the slider. Start with all slider values set to the following: h = 5, k = 2, horizontal = 5, vertical = 3.*

(a) Check the *Segment Length* boxes to display the segment lengths from each focus to the ellipse. Click on the blue point on the ellipse to move it. Notice that as you move the blue point on the ellipse, these segment lengths change. What happens to the sum of the lengths of these segments as you move the point? In other words, what happens to the sum of the distances between the foci and the ellipse?

(b) Check the *Center* box to display the center of the ellipse. What is the center of the current ellipse? Move h to 3. What did this do to the center of the ellipse? Did this move change the length of the major and minor axes?

(c) Move k to 4. What did this do to the center of the ellipse? Did this move change the length of the major and minor axes?

(d) What does the role of h and k play in the graph of the ellipse?

(e) Move the *horizontal* slider to 4. What did this do to the graph of the ellipse? What did this change in the equation of the ellipse?

(f) Now move the *horizontal* slider to 2. What did this do to the graph of the ellipse? What did this change in the equation of the ellipse?

(g) What role does the number shown on the horizontal slider play in the graph and equation of an ellipse?

(h) Move the *vertical* slider to 4. What did this do to the graph of the ellipse? What did this change in the equation of the ellipse?

(i) Now move the *vertical* slider to 1. What did this do to the graph of the ellipse? What did this change in the equation of the ellipse?

(j) What role does the number shown on the vertical slider play in the graph and equation of an ellipse?

12.5 Hyperbolas

Objectives

1 Graph a Hyperbola Whose Center Is the Origin

2 Find the Equation of a Hyperbola Whose Center Is the Origin

3 Find the Asymptotes of a Hyperbola Whose Center Is the Origin

In Other Words

The distance from F_1 to P minus the distance from F_2 to P is a constant value for any point P on a hyperbola.

Figure 36

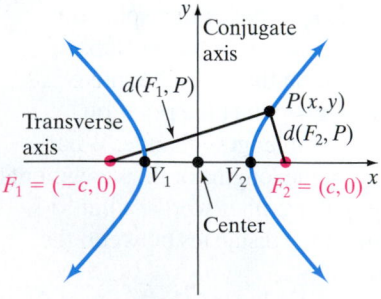

Work Smart

Notice for hyperbolas that $c^2 = a^2 + b^2$, but for ellipses $c^2 = a^2 - b^2$.

Are You Prepared for This Section?

Before getting started, complete the following problems. If you get a problem wrong, go back to the section cited and review the material.

P1. Complete the square: $x^2 - 5x$ [Section 10.1, pp. 694–695]
P2. Solve: $y^2 = 64$ [Section 10.1, pp. 691–694]

▶ **1** Graph a Hyperbola Whose Center Is the Origin

Recall from Section 12.2 that a hyperbola is a conic section that is obtained through the intersection of a plane and two cones. See Figure 8(d) in Section 12.2.

> A **hyperbola** is the collection of all points in the plane the difference of whose distances from two fixed points, called the **foci,** is a positive constant.

Figure 36 illustrates a hyperbola with foci F_1 and F_2. The line containing the foci is the **transverse axis.** The midpoint of the line segment joining the foci is the **center** of the hyperbola. The line through the center and perpendicular to the transverse axis is the **conjugate axis.** The hyperbola consists of two separate curves called **branches.** The two points of intersection of the hyperbola and the transverse axis are the **vertices,** V_1 and V_2, of the hyperbola.

The equation of a hyperbola can be found using the distance formula. However, this information will not be presented here. Instead, the equations of the hyperbolas whose branches open left and right and the equations of hyperbolas whose branches open up and down, with both hyperbolas centered at the origin will be given. See Table 5.

Table 5   Hyperbolas with Center at the Origin

Center	Transverse Axis	Branches Open	Foci	Vertices	Equation
$(0,0)$	x-axis	Left and right	$(-c, 0)$ and $(c, 0)$	$(-a, 0)$ and $(a, 0)$	$\dfrac{x^2}{a^2} - \dfrac{y^2}{b^2} = 1$, where $b^2 = c^2 - a^2$ or $c^2 = a^2 + b^2$
$(0,0)$	y-axis	Up and down	$(0, -c)$ and $(0, c)$	$(0, -a)$ and $(0, a)$	$\dfrac{y^2}{a^2} - \dfrac{x^2}{b^2} = 1$, where $b^2 = c^2 - a^2$ or $c^2 = a^2 + b^2$

The graphs of the two hyperbolas are given in Figure 37.

Figure 37

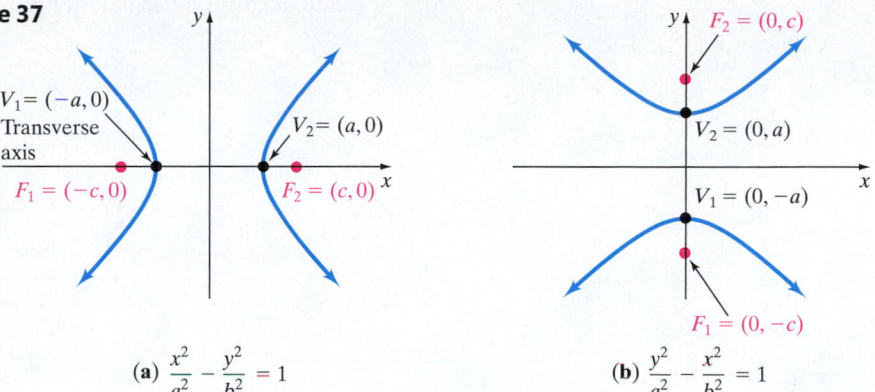

(a) $\dfrac{x^2}{a^2} - \dfrac{y^2}{b^2} = 1$ (b) $\dfrac{y^2}{a^2} - \dfrac{x^2}{b^2} = 1$

Notice the difference between the two equations given in Table 5. When the x^2 term is first, as in Figure 37(a), the transverse axis is the x-axis and the hyperbola opens left and right. When the y^2 term is first, as in Figure 37(b), the transverse axis is the y-axis and the hyperbola opens up and down. In both cases, the value of a^2 is in the denominator of the first term.

> **Quick ✓**
>
> 1. A(n) _____ is the collection of points in the plane the difference of whose distances from two fixed points is a positive constant.
>
> 2. For a hyperbola, the foci lie on a line called the _____ ___.
>
> 3. The line through the center of a hyperbola that is perpendicular to the transverse axis is called the _____ ___.

EXAMPLE 1

Graphing a Hyperbola

Graph the equation: $\dfrac{x^2}{16} - \dfrac{y^2}{4} = 1$

Solution

The equation is of the form $\dfrac{x^2}{a^2} - \dfrac{y^2}{b^2} = 1$. The x^2 term is first, so the hyperbola opens left and right, and the transverse axis is the x-axis. The center of the hyperbola is the origin, $(0,0)$. Because $a^2 = 16$ and $b^2 = 4$, then $c^2 = a^2 + b^2 = 16 + 4 = 20$, so $c = 2\sqrt{5}$. The vertices are at $(\pm a, 0) = (\pm 4, 0)$, and the foci are at $(\pm c, 0) = (\pm 2\sqrt{5}, 0)$.

To obtain the graph, plot the vertices and foci. Then locate points above and below the foci (as in graphing parabolas). Thus let $x = \pm 2\sqrt{5}$ in the equation $\dfrac{x^2}{16} - \dfrac{y^2}{4} = 1$.

$$\frac{(\pm 2\sqrt{5})^2}{16} - \frac{y^2}{4} = 1$$

$$\frac{20}{16} - \frac{y^2}{4} = 1$$

Reduce the fraction: $\dfrac{5}{4} - \dfrac{y^2}{4} = 1$

Subtract $\dfrac{5}{4}$ from both sides: $-\dfrac{y^2}{4} = -\dfrac{1}{4}$

Multiply both sides by -4: $y^2 = 1$

Take the square root of both sides: $y = \pm 1$

(continued)

The points above and below the foci are $\left(\pm 2\sqrt{5}, -1\right)$ and $\left(\pm 2\sqrt{5}, 1\right)$. Use $\pm 2\sqrt{5} \approx \pm 4.5$ to graph the foci. See Figure 38 for the graph of the hyperbola.

Figure 38

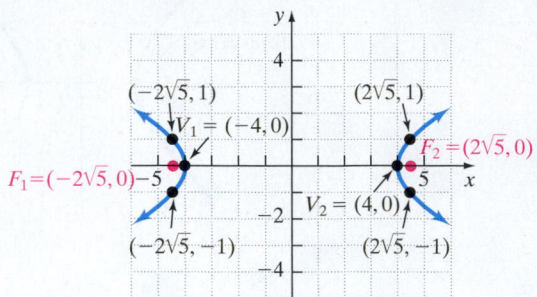

EXAMPLE 2 **Graphing a Hyperbola**

Graph the equation: $4y^2 - 16x^2 = 16$

Solution

Write the equation in one of the forms given in Table 5. To do this, divide both sides of the equation by 16:

$$\frac{y^2}{4} - \frac{x^2}{1} = 1$$

Because the y^2 term is first, the hyperbola opens up and down and the transverse axis is the y-axis. The center of this hyperbola is the origin, $(0, 0)$. Comparing the equation to $\frac{y^2}{a^2} - \frac{x^2}{b^2} = 1$, we find that $a^2 = 4$ and $b^2 = 1$; therefore, $c^2 = a^2 + b^2 = 4 + 1 = 5$. The vertices are at $(0, \pm a) = (0, \pm 2)$. The foci are at $(0, \pm c) = \left(0, \pm \sqrt{5}\right)$. Locate four more points on the hyperbola to the left and right of each focus by letting $y = \pm \sqrt{5}$ in the equation $\frac{y^2}{4} - \frac{x^2}{1} = 1$ and solving for x.

$$\frac{\left(\pm \sqrt{5}\right)^2}{4} - \frac{x^2}{1} = 1$$

$$\frac{5}{4} - \frac{x^2}{1} = 1$$

Subtract $\frac{5}{4}$ from both sides: $-x^2 = -\frac{1}{4}$

Multiply both sides by -1: $x^2 = \frac{1}{4}$

Take the square root of both sides: $x = \pm \frac{1}{2}$

Four additional points on the graph are $\left(-\frac{1}{2}, \pm \sqrt{5}\right)$ and $\left(\frac{1}{2}, \pm \sqrt{5}\right)$. See Figure 39 for the graph of the hyperbola.

Figure 39

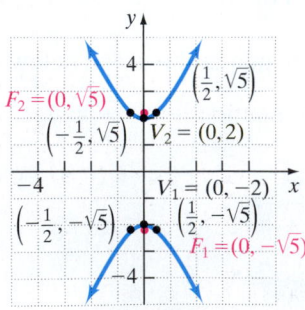

Quick ✓

In Problems 4 and 5, graph each hyperbola.

4. $\frac{x^2}{36} - \frac{y^2}{64} = 1$ **5.** $16y^2 - 9x^2 = 144$

⊙ ❷ Find the Equation of a Hyperbola Whose Center Is the Origin

Just as in working with parabolas and ellipses, we will use information about a hyperbola to find its equation.

EXAMPLE 3 **Finding and Graphing the Equation of a Hyperbola**

Find an equation of the hyperbola with one focus at $(0, -3)$ and vertices at $(0, -2)$ and $(0, 2)$. Graph the equation.

Solution

The center of a hyperbola is located at the midpoint of the vertices (or foci). Because the vertices are at $(0, -2)$ and $(0, 2)$, the center must be at the origin, $(0, 0)$. Notice that the given focus and vertices lie on the y-axis. Therefore, the transverse axis is the y-axis, and the hyperbola opens up and down. A vertex is at $(0, 2)$, so $a = 2$. A focus is at $(0, -3)$, so $c = 3$. With $a = 2$ and $c = 3$, $b^2 = c^2 - a^2 = 3^2 - 2^2 = 9 - 4 = 5$.

The equation must be of the form $\dfrac{y^2}{a^2} - \dfrac{x^2}{b^2} = 1$, so let $a^2 = 4$ and $b^2 = 5$ to get

$$\frac{y^2}{4} - \frac{x^2}{5} = 1$$

To find points to the left and right of each focus, let $y = \pm 3$ in the equation $\dfrac{y^2}{4} - \dfrac{x^2}{5} = 1$.

$$\frac{(\pm 3)^2}{4} - \frac{x^2}{5} = 1$$

$$\frac{9}{4} - \frac{x^2}{5} = 1$$

Subtract $\dfrac{9}{4}$ from both sides: $\quad -\dfrac{x^2}{5} = -\dfrac{5}{4}$

Multiply both sides by -5: $\quad x^2 = \dfrac{25}{4}$

Take the square root of both sides: $\quad x = \pm\dfrac{5}{2}$

Figure 40

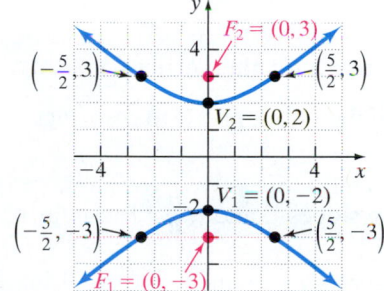

The points to the left and right of the foci are $\left(-\dfrac{5}{2}, \pm 3\right)$ and $\left(\dfrac{5}{2}, \pm 3\right)$. See Figure 40 for the graph of the hyperbola. ●

Look at the equations of the hyperbolas in Examples 1 and 3. For the hyperbola in Example 1, $a^2 = 16$ and $b^2 = 4$, so $a > b$; for the hyperbola in Example 3, $a^2 = 4$ and $b^2 = 5$, so $a < b$. Therefore, for hyperbolas, there are no requirements involving the relative sizes for a and b. For ellipses, however, the relative sizes of a and b determine the major axis.

> **Quick ✓**
>
> **6.** *True or False* In any hyperbola, it must be the case that $a > b$.
>
> **7.** Find the equation of a hyperbola whose vertices are $(-4, 0)$ and $(4, 0)$ and has a focus at $(6, 0)$. Graph the hyperbola.

⊙ ❸ Find the Asymptotes of a Hyperbola Whose Center Is the Origin

As x and y get larger in both the positive and the negative directions, the branches of the hyperbola approach two lines called **asymptotes** of the hyperbola. The asymptotes

Work Smart

Asymptotes provide an alternative method for determining the opening of each branch of the hyperbola, rather than finding and plotting four additional points.

help us graph hyperbolas. Table 6 gives the asymptotes of the two hyperbolas discussed in this section.

Table 6 Asymptotes of a Hyperbola	
Hyperbola	**Asymptotes**
$\dfrac{x^2}{a^2} - \dfrac{y^2}{b^2} = 1$	$y = -\dfrac{b}{a}x$ and $y = \dfrac{b}{a}x$
$\dfrac{y^2}{a^2} - \dfrac{x^2}{b^2} = 1$	$y = -\dfrac{a}{b}x$ and $y = \dfrac{a}{b}x$

Figure 41 illustrates how the asymptotes can be used to help graph a hyperbola. It is important to remember that the asymptotes are not part of the hyperbola—they only serve as guides in graphing the hyperbola.

Work Smart: Study Skills

Which equation is that of an ellipse and which is the equation of a hyperbola?

$$\frac{x^2}{4} + y^2 = 1$$

$$\frac{x^2}{4} - y^2 = 1$$

Summarize how to tell the difference between the equation of an ellipse and the equation of a hyperbola.

Figure 41

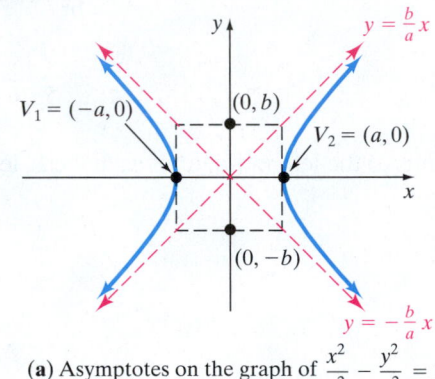 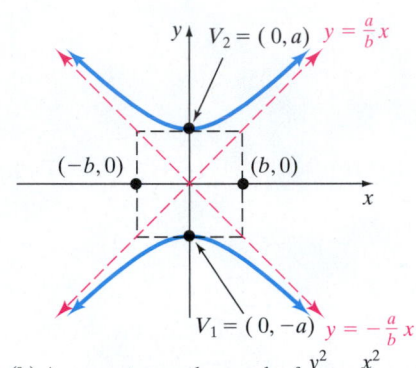

(**a**) Asymptotes on the graph of $\dfrac{x^2}{a^2} - \dfrac{y^2}{b^2} = 1$ (**b**) Asymptotes on the graph of $\dfrac{y^2}{a^2} - \dfrac{x^2}{b^2} = 1$

For example, to graph the equation $\dfrac{x^2}{a^2} - \dfrac{y^2}{b^2} = 1$, first plot the vertices $(-a, 0)$ and $(a, 0)$. Then plot the points $(0, -b)$ and $(0, b)$. Use these four points to construct a rectangle as shown in Figure 41(a).

The lines through the diagonals of this rectangle have slopes $\dfrac{b}{a}$ and $-\dfrac{b}{a}$. These lines are the asymptotes. The equations of these asymptotes are $y = -\dfrac{b}{a}x$ and $y = \dfrac{b}{a}x$. Using this technique enables us to avoid plotting the four additional points plotted earlier.

EXAMPLE 4 **Graphing a Hyperbola and Finding Its Asymptotes**

Graph the equation $9x^2 - 4y^2 = 36$ using the asymptotes as a guide.

Solution

Divide both sides of the equation by 36 to put the equation in the form $\dfrac{x^2}{a^2} - \dfrac{y^2}{b^2} = 1$.

$$\frac{x^2}{4} - \frac{y^2}{9} = 1$$

The center of the hyperbola is the origin, $(0, 0)$. Because the x^2 term is first, the hyperbola opens to the left and right. The transverse axis is along the x-axis. In addition, notice that $a^2 = 4$ and $b^2 = 9$. Because $c^2 = a^2 + b^2$, then $c^2 = 4 + 9 = 13$, so $c = \pm\sqrt{13}$. The vertices are at $(\pm a, 0) = (\pm 2, 0)$; the foci

Figure 42

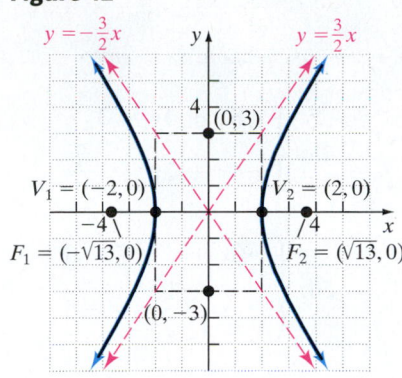

are at $(\pm c, 0) = (\pm\sqrt{13}, 0)$. Since $b^2 = 9$, is then $b = \pm 3$. The equations of the asymptotes are

$$y = \frac{b}{a}x = \frac{3}{2}x \quad \text{and} \quad y = -\frac{b}{a}x = -\frac{3}{2}x$$

To graph the hyperbola, form a rectangle using the points $(\pm a, 0) = (\pm 2, 0)$ and $(0, \pm b) = (0, \pm 3)$. Draw the asymptotes through the vertices of the rectangle. See Figure 42. ●

Quick ✓

8. The asymptotes of the hyperbola $\dfrac{x^2}{a^2} - \dfrac{y^2}{b^2} = 1$ are _____ and _____.

9. Graph the equation $x^2 - 9y^2 = 9$ using the asymptotes as a guide.

10. Graph the equation $\dfrac{y^2}{16} - \dfrac{x^2}{9} = 1$ using the asymptotes as a guide.

12.5 Exercises MyMathLab®

Exercise numbers in **green** have complete video solutions in MyMathLab or may be accessed using the QR code to the right.

Problems 1–10 are the Quick ✓ s that follow the EXAMPLES.

Building Skills

In Problems 11–14, the graph of a hyperbola is given. Match each graph to its equation. See Objective 1.

(a) $\dfrac{x^2}{4} - y^2 = 1$

(b) $x^2 - \dfrac{y^2}{4} = 1$

(c) $\dfrac{y^2}{4} - x^2 = 1$

(d) $y^2 - \dfrac{x^2}{4} = 1$

11.

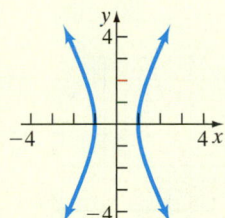

12.

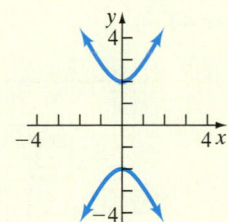

13.

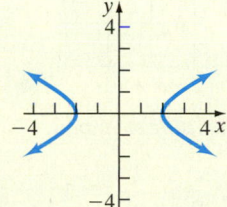

14.

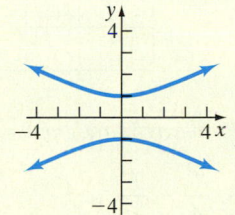

In Problems 15–22, graph each equation. See Objective 1.

15. $\dfrac{x^2}{4} - \dfrac{y^2}{16} = 1$

16. $\dfrac{x^2}{9} - \dfrac{y^2}{16} = 1$

17. $\dfrac{y^2}{25} - \dfrac{x^2}{36} = 1$

18. $\dfrac{y^2}{81} - \dfrac{x^2}{9} = 1$

19. $4x^2 - y^2 = 36$

20. $x^2 - 9y^2 = 36$

21. $25y^2 - x^2 = 100$

22. $4y^2 - 9x^2 = 36$

In Problems 23–28, find the equation for the hyperbola described. Graph the equation. See Objective 2.

23. Center at $(0, 0)$; focus at $(3, 0)$; vertex at $(2, 0)$

24. Center at $(0, 0)$; focus at $(-4, 0)$; vertex at $(-1, 0)$

25. Vertices at $(0, 5)$ and $(0, -5)$; focus at $(0, 7)$

26. Vertices at $(0, 6)$ and $(0, -6)$; focus at $(0, 8)$

27. Foci at $(-10, 0)$ and $(10, 0)$; vertex at $(-7, 0)$

28. Foci at $(-5, 0)$ and $(5, 0)$; vertex at $(-3, 0)$

In Problems 29–34, graph each equation using its asymptotes. See Objective 3.

29. $\dfrac{x^2}{25} - \dfrac{y^2}{9} = 1$

30. $\dfrac{x^2}{16} - \dfrac{y^2}{64} = 1$

31. $\dfrac{y^2}{4} - \dfrac{x^2}{100} = 1$

32. $\dfrac{y^2}{25} - \dfrac{x^2}{100} = 1$

33. $x^2 - y^2 = 4$

34. $y^2 - x^2 = 25$

Mixed Practice

In Problems 35–38, find the equation for the hyperbola described. Graph the equation.

35. Vertices at $(0, -8)$ and $(0, 8)$; asymptote the line $y = 2x$

36. Vertices at $(0, -4)$ and $(0, 4)$; asymptote the line $y = 2x$

37. Foci at $(-3, 0)$ and $(3, 0)$; asymptote the line $y = x$

38. Foci at $(-9, 0)$ and $(9, 0)$; asymptote the line $y = -3x$

Applying the Concepts

In Problems 39–42, write the equation of the hyperbola.

39.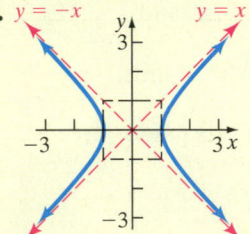

40.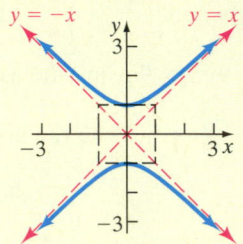

41.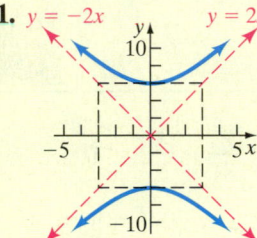

42.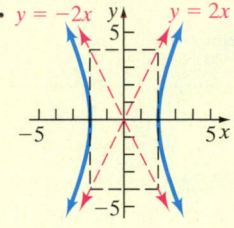

Extending the Concepts

43. Two hyperbolas that have the same set of asymptotes are called **conjugate.** Show that the hyperbolas

$$\frac{x^2}{4} - y^2 = 1 \quad \text{and} \quad y^2 - \frac{x^2}{4} = 1$$

are conjugate. Graph each hyperbola in the same Cartesian plane.

44. The **eccentricity** e of a hyperbola is defined as the number $\frac{c}{a}$. Because $c > a$, it follows that $e > 1$. Describe the general shape of a hyperbola whose eccentricity is close to 1. What is the shape if e is very large?

Explaining the Concepts

45. Explain how the asymptotes of a hyperbola are helpful in obtaining its graph.

46. How can you tell the difference between the equation of a hyperbola and the equation of an ellipse just by looking at the equations?

Synthesis Review

In Problems 47–51, solve each system of equations using either substitution or elimination.

47. $\begin{cases} 2x - 3y = -9 \\ -x + 5y = 8 \end{cases}$

48. $\begin{cases} 3x + 4y = 3 \\ -6x + 2y = -\dfrac{7}{2} \end{cases}$

49. $\begin{cases} 2x - 3y = 6 \\ -6x + 9y = -18 \end{cases}$

50. $\begin{cases} -2x + y = 8 \\ x - \dfrac{1}{2}y = -4 \end{cases}$

51. $\begin{cases} 6x + 3y = 4 \\ -2x - y = -\dfrac{4}{3} \end{cases}$

52. In Problems 47–51, which method did you use more often? Why? Do you think there are situations where substitution is superior to elimination? Are there situations where elimination is superior to substitution? Describe these circumstances.

Technology Exercises

A graphing calculator can be used to graph hyperbolas by solving the equation for y. Because hyperbolas are not functions, graph the upper half and lower half of the hyperbola in two pieces. For example, to graph $\dfrac{x^2}{4} - \dfrac{y^2}{9} = 1$, graph $Y_1 = 3\sqrt{\dfrac{x^2}{4} - 1}$ and $Y_2 = -3\sqrt{\dfrac{x^2}{4} - 1}$.

To graph a hyperbola without having to solve the equation for y, you can use Desmos. The graph shown in Figure 43 is obtained.

Figure 43

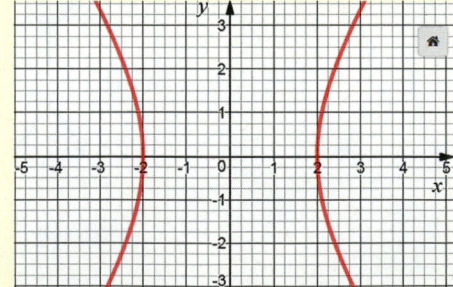

In Problems 53–60, graph each hyperbola using technology.

53. $\dfrac{x^2}{4} - \dfrac{y^2}{16} = 1$

54. $\dfrac{x^2}{9} - \dfrac{y^2}{16} = 1$

55. $\dfrac{y^2}{25} - \dfrac{x^2}{36} = 1$

56. $\dfrac{y^2}{81} - \dfrac{x^2}{9} = 1$

57. $4x^2 - y^2 = 36$

58. $x^2 - 9y^2 = 36$

59. $25y^2 - x^2 = 100$

60. $4y^2 - 9x^2 = 36$

Putting the Concepts Together (Sections 12.1–12.5)

We designed these problems so that you can review Sections 12.1–12.5 and show your mastery of the concepts. Take time to work these problems before proceeding with the next section. The answers are located at the back of the text on page AN-66.

1. Find the exact distance $d(P_1, P_2)$ between points $P_1 = (-6, 4)$ and $P_2 = (3, -2)$.

2. Find the midpoint of the line segment formed by joining the points $P_1 = (-3, 1)$ and $P_2 = (5, -7)$.

In Problems 3 and 4, find the center (h, k) and the radius r of each circle. Graph each circle.

3. $(x + 2)^2 + (y - 8)^2 = 36$

4. $x^2 + y^2 + 6x - 4y - 3 = 0$

In Problems 5 and 6, find the standard form of the equation of each circle.

5. Center $(0, 0)$; contains the point $(-5, 12)$

6. With endpoints of a diameter at $(-1, 5)$ and $(5, -3)$.

In Problems 7 and 8, find the vertex, focus, and directrix of each parabola. Graph each parabola.

7. $(x + 2)^2 = -4(y - 4)$

8. $y^2 + 2y - 8x + 25 = 0$

In Problems 9 and 10, find an equation for each parabola described.

9. Vertex at $(-1, -2)$; focus at $(-1, -5)$

10. Vertex at $(-3, 3)$; contains the point $(-1, 7)$; axis of symmetry parallel to the *x*-axis

In Problems 11 and 12, find the center, vertices, and foci of each ellipse. Graph each ellipse.

11. $x^2 + 9y^2 = 81$

12. $\dfrac{(x + 1)^2}{36} + \dfrac{(y - 2)^2}{49} = 1$

In Problems 13 and 14, find an equation for each ellipse described.

13. Foci at $(0, \pm 6)$; vertices at $(0, \pm 9)$

14. Center at $(3, -4)$; vertex at $(7, -4)$; focus at $(6, -4)$

In Problems 15 and 16, find the vertices, foci, and asymptotes for each hyperbola. Graph each hyperbola using the asymptotes as a guide.

15. $\dfrac{y^2}{81} - \dfrac{x^2}{9} = 1$

16. $25x^2 - y^2 = 25$

17. Find an equation for a hyperbola with center at $(0, 0)$, focus at $(0, -5)$, and vertex at $(0, -2)$.

18. A large floodlight is in the shape of a parabola with diameter 36 inches and depth 12 inches. How far from the vertex should the light bulb be placed so that the rays will be reflected parallel to the axis?

12.6 Systems of Nonlinear Equations

Objectives

1. Solve a System of Nonlinear Equations Using Substitution

2. Solve a System of Nonlinear Equations Using Elimination

Prepared?...Answers **P1.** $(2, -1)$
P2. $\left(-\dfrac{1}{2}, \dfrac{5}{2}\right)$
P3. $\{(x, y) \,|\, 3x - 5y = 4\}$

Are You Prepared for This Section?

Before getting started, complete the following problems. If you get a problem wrong, go back to the section cited and review the material.

P1. Solve the system using substitution: $\begin{cases} y = 2x - 5 \\ 2x - 3y = 7 \end{cases}$ [Section 4.2, pp. 260–263]

P2. Solve the system using elimination: $\begin{cases} 2x - 4y = -11 \\ -x + 5y = 13 \end{cases}$ [Section 4.3, pp. 268–271]

P3. Solve the system: $\begin{cases} 3x - 5y = 4 \\ -6x + 10y = -8 \end{cases}$ [Section 4.3, p. 272]

Recall from Section 4.1 that a system of equations is a grouping of two or more equations, each containing one or more variables. Systems of linear equations were solved in Chapter 4. Solving systems of equations where the equations are not linear will now be examined. Systems containing only two equations with two unknowns will be addressed. In a **system of nonlinear equations** in two variables, at least one of the equations is not linear. That is, at least one of the equations cannot be written in the form $Ax + By = C$. Some examples of nonlinear systems of equations containing two unknowns follow.

$$\begin{cases} x + y^2 = 5 & \text{(1) A parabola} \\ 2x + y = 4 & \text{(2) A line} \end{cases} \qquad \begin{cases} x^2 + y^2 = 9 & \text{(1) A circle} \\ -x^2 + y = 9 & \text{(2) A parabola} \end{cases}$$

In Section 4.1, the solution of a system of linear equations were represented geometrically by showing the point of intersection of the equations in the system. The same could be done for nonlinear systems—the point(s) of intersection represent the solution(s) to the system.

As you worked through Chapter 4, you may have developed a sense of when substitution or when elimination was the best approach for solving a system. The same is true of nonlinear systems—sometimes substitution is best, sometimes elimination is best. Experience and a degree of imagination are your friends when solving these problems.

▶ ❶ Solve a System of Nonlinear Equations Using Substitution

The steps for solving a system of nonlinear equations using substitution are identical to the steps for solving a system of linear equations using substitution.

EXAMPLE 1 **How to Solve a System of Nonlinear Equations Using Substitution**

Solve the following system of equations using substitution: $\begin{cases} 3x - y = -2 & \text{(1) A line} \\ 2x^2 - y = 0 & \text{(2) A parabola} \end{cases}$

Step-by-Step Solution

Step 1: Solve equation (1) for y.

Equation (1): $\quad 3x - y = -2$

Add y to both sides; add 2 to both sides: $\quad y = 3x + 2$

Step 2: Substitute $3x + 2$ for y in equation (2).

Equation (2): $\qquad\qquad 2x^2 - y = 0$

$$2x^2 - (3x + 2) = 0$$

Distribute: $\qquad 2x^2 - 3x - 2 = 0$

Step 3: Solve for x.

Factor: $\qquad (2x + 1)(x - 2) = 0$

Zero-Product Property: $\quad 2x + 1 = 0 \quad \text{or} \quad x - 2 = 0$

$$x = -\frac{1}{2} \quad \text{or} \qquad x = 2$$

Step 4: Let $x = -\dfrac{1}{2}$ and $x = 2$ in equation (1) to determine y.

Equation (1): $\qquad 3x - y = -2$

$x = -\frac{1}{2}: \quad 3\left(-\frac{1}{2}\right) - y = -2$

$$-\frac{3}{2} - y = -2$$

Add $\frac{3}{2}$ to both sides: $\qquad -y = -\frac{1}{2}$

Multiply both sides by -1: $\qquad y = \frac{1}{2}$

$$x = 2: \quad 3(2) - y = -2$$
$$6 - y = -2$$

Subtract 6 from both sides: $\quad -y = -8$

Multiply both sides by -1: $\quad y = 8$

The apparent solutions are $\left(-\dfrac{1}{2}, \dfrac{1}{2}\right)$ and $(2, 8)$.

Step 5: Check

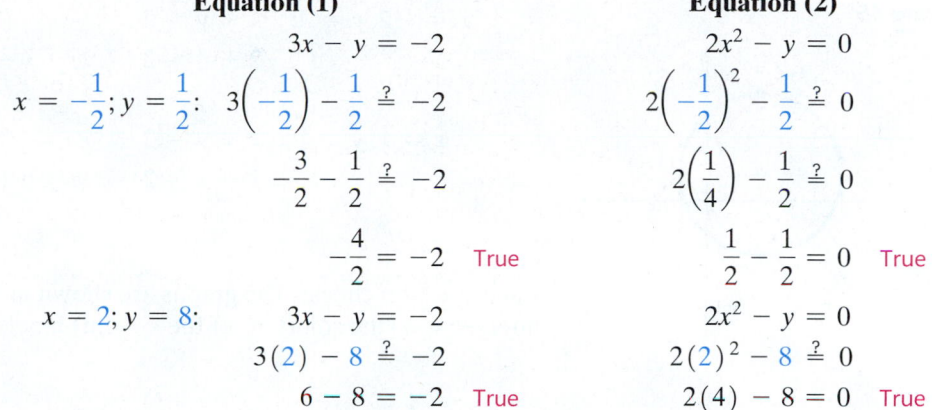

	Equation (1)	Equation (2)
	$3x - y = -2$	$2x^2 - y = 0$

$x = -\dfrac{1}{2}; y = \dfrac{1}{2}:$ $\quad 3\left(-\dfrac{1}{2}\right) - \dfrac{1}{2} \overset{?}{=} -2$ $\qquad 2\left(-\dfrac{1}{2}\right)^2 - \dfrac{1}{2} \overset{?}{=} 0$

$-\dfrac{3}{2} - \dfrac{1}{2} \overset{?}{=} -2$ $\qquad 2\left(\dfrac{1}{4}\right) - \dfrac{1}{2} \overset{?}{=} 0$

$-\dfrac{4}{2} = -2$ True $\qquad \dfrac{1}{2} - \dfrac{1}{2} = 0$ True

$x = 2; y = 8:$ $\qquad 3x - y = -2$ $\qquad 2x^2 - y = 0$

$3(2) - 8 \overset{?}{=} -2$ $\qquad 2(2)^2 - 8 \overset{?}{=} 0$

$6 - 8 = -2$ True $\qquad 2(4) - 8 = 0$ True

Figure 44

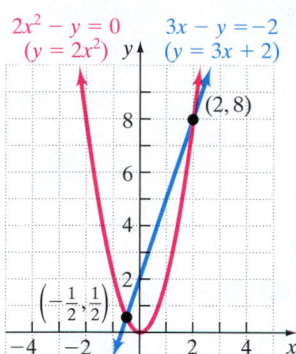

Each solution checks. Figure 44 shows the graphs of each equation in the system with the points of intersection (the solutions of the system) labeled at $\left(-\dfrac{1}{2}, \dfrac{1}{2}\right)$ and $(2, 8)$. ●

> **Quick ✓**
>
> 1. Solve the following system of equations using substitution:
> $$\begin{cases} 2x + y = -1 \\ x^2 - y = 4 \end{cases}$$

EXAMPLE 2 **Solving a System of Nonlinear Equations Using Substitution**

Solve the following system of equations using substitution:

$$\begin{cases} x + y = 2 & \text{(1) A line} \\ (x + 2)^2 + (y - 1)^2 = 9 & \text{(2) A circle} \end{cases}$$

Solution

Solve equation (1) for y.

$$x + y = 2$$

Subtract x from both sides: $\quad y = -x + 2$

Substitute this expression for y in equation (2), $(x + 2)^2 + (y - 1)^2 = 9$, and then solve for x.

Substitute $-x + 2$ for y in equation (2): $\quad (x + 2)^2 + (-x + 2 - 1)^2 = 9$

Combine like terms: $\quad (x + 2)^2 + (1 - x)^2 = 9$

Use FOIL: $\quad x^2 + 4x + 4 + 1 - 2x + x^2 = 9$

Combine like terms: $\quad 2x^2 + 2x + 5 = 9$

Put in standard form: $\quad 2x^2 + 2x - 4 = 0$

Factor: $\quad 2(x^2 + x - 2) = 0$

Divide both sides by 2: $\quad x^2 + x - 2 = 0$

Factor: $\quad (x + 2)(x - 1) = 0$

Zero-Product Property: $\quad x + 2 = 0 \quad \text{or} \quad x - 1 = 0$

Solve for x: $\quad x = -2 \quad \text{or} \quad x = 1$

(continued)

Let $x = -2$ and $x = 1$ in equation (1) to determine y.

$$x + y = 2$$

$x = -2$: $-2 + y = 2$ $x = 1$: $1 + y = 2$

Add 2 to both sides: $y = 4$ Subtract 1 from both sides: $y = 1$

The apparent solutions are $(-2, 4)$ and $(1, 1)$.

Figure 45

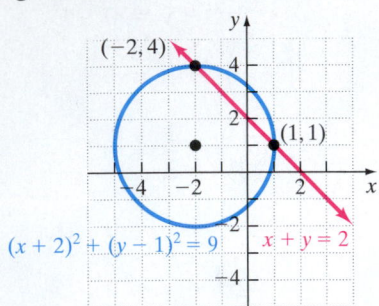

$(x + 2)^2 + (y - 1)^2 = 9$ $x + y = 2$

Check	Equation (1)	Equation (2)

$x = -2; y = 4$: $x + y = 2$ $(x + 2)^2 + (y - 1)^2 = 9$

$\qquad\qquad -2 + 4 = 2$ True $(-2 + 2)^2 + (4 - 1)^2 \overset{?}{=} 9$

$\qquad\qquad\qquad\qquad\qquad\qquad\qquad 3^2 = 9$ True

$x = 1; y = 1$: $x + y = 2$ $(x + 2)^2 + (y - 1)^2 = 9$

$\qquad\qquad 1 + 1 = 2$ True $(1 + 2)^2 + (1 - 1)^2 \overset{?}{=} 9$

$\qquad\qquad\qquad\qquad\qquad\qquad\qquad 3^2 = 9$ True

Each solution checks. The graphs are shown in Figure 45 with the points of intersection (the solutions of the system) labeled at $(-2, 4)$ and $(1, 1)$.

Quick ✓

2. Solve the following system of equations using substitution:

$$\begin{cases} 2x + y = 0 \\ (x - 4)^2 + (y + 2)^2 = 9 \end{cases}$$

▶ ❷ Solve a System of Nonlinear Equations Using Elimination

Now let's discuss the method of elimination. Use elimination to solve a system of nonlinear equations in the same way it was used to solve systems of linear equations.

Remember, in using elimination, to get the coefficients of one variable to be additive inverses.

EXAMPLE 3   **How to Solve a System of Nonlinear Equations by Elimination**

Solve the following system of equations using elimination: $\begin{cases} x^2 + y^2 = 13 & \text{(1) A circle} \\ x^2 - y = 7 & \text{(2) A parabola} \end{cases}$

Step-by-Step Solution

Step 1: Multiply equation (2) by -1 so the coefficients of x^2 become additive inverses.

$$\begin{cases} x^2 + y^2 = 13 & \text{(1)} \\ -x^2 + y = -7 & \text{(2)} \end{cases}$$

Step 2: Add equations (1) and (2) to eliminate x^2. Solve the resulting equation for y.

Add: $\begin{cases} x^2 + y^2 = 13 & \text{(1)} \\ -x^2 + y = -7 & \text{(2)} \end{cases}$

$$y^2 + y = 6$$

Put in standard form: $y^2 + y - 6 = 0$

Factor: $(y + 3)(y - 2) = 0$

Zero-Product Property: $y + 3 = 0$ or $y - 2 = 0$

$$y = -3 \quad \text{or} \quad y = 2$$

Step 3: Solve for x using equation (2).

$$x^2 - y = 7 \quad (2)$$

Using $y = -3$: $\quad x^2 - (-3) = 7$ $\qquad\qquad$ Using $y = 2$: $\quad x^2 - 2 = 7$

$$x^2 + 3 = 7 \qquad\qquad\qquad\qquad x^2 = 9$$

$$x^2 = 4 \qquad\qquad\qquad\qquad\qquad x = \pm 3$$

$$x = \pm 2$$

Work Smart

Use equation (2) in Step 3 because it's easier to work with.

The apparent solutions are $(-2, -3)$, $(2, -3)$, $(-3, 2)$, and $(3, 2)$.

Step 4: Check

The check is left to you to verify that all four of the apparent solutions are solutions to the system. The four points $(-2, -3)$, $(2, -3)$, $(-3, 2)$, and $(3, 2)$ are the points of intersection of the graphs. See Figure 46.

Figure 46

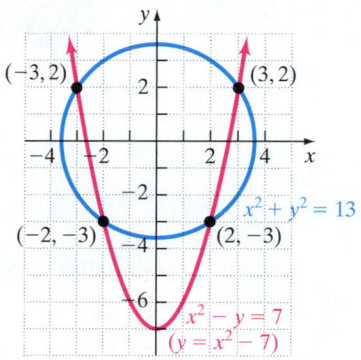

EXAMPLE 4

Solving a System of Nonlinear Equations by Elimination

Solve the following system of equations using elimination:

$$\begin{cases} x^2 - y^2 = 4 & \text{(1) A hyperbola} \\ x^2 - y \ = 0 & \text{(2) A parabola} \end{cases}$$

Solution

Multiply equation (2) by -1 to get the coefficients of x^2 to be additive inverses.

$$\begin{cases} x^2 - y^2 = 4 & \text{(1)} \\ -x^2 + y \ = 0 & \text{(2)} \end{cases}$$

Add equations (1) and (2) to eliminate x^2. Solve the resulting equation for y.

Figure 47

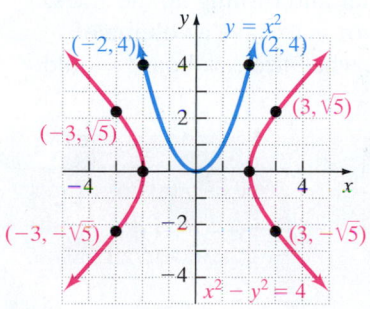

$$\begin{cases} \quad x^2 - y^2 = 4 & \text{(1)} \\ \underline{-x^2 + y \ = 0} & \text{(2)} \\ \qquad -y^2 + y = 4 \end{cases}$$

Add y^2 to both sides; subtract y from both sides: $\quad y^2 - y + 4 = 0$

This is a quadratic equation whose discriminant is $b^2 - 4ac = (-1)^2 - 4(1)(4) = 1 - 16 = -15$. The equation has no real solution. Therefore, the system of equations is inconsistent. The solution set is $\varnothing$ or $\{\ \}$. Figure 47 shows the graphs of $x^2 - y^2 = 4$ and $x^2 - y = 0$ ($y = x^2$). Figure 47 confirms the algebraic result.

Quick ✔

Solve each system of nonlinear equations using elimination.

3. $\begin{cases} x^2 + y^2 = 16 \\ x^2 - 2y = 8 \end{cases}$ $\qquad\qquad\qquad\qquad$ **4.** $\begin{cases} x^2 - y \ = -4 \\ x^2 + y^2 = 9 \end{cases}$

12.6 Exercises · MyMathLab®

Exercise numbers in green have complete video solutions in MyMathLab or may be accessed using the QR code to the right.

*Problems **1–4** are the Quick ✔ s that follow the EXAMPLES.*

Building Skills

In Problems 5–12, solve the system of nonlinear equations using the method of substitution. See Objective 1.

5. $\begin{cases} y = x^2 + 4 \\ y = x + 4 \end{cases}$ **6.** $\begin{cases} y = x^3 + 2 \\ y = x + 2 \end{cases}$

7. $\begin{cases} y = \sqrt{25 - x^2} \\ x + y = 7 \end{cases}$ **8.** $\begin{cases} y = \sqrt{100 - x^2} \\ x + y = 14 \end{cases}$

9. $\begin{cases} x^2 + y^2 = 4 \\ y = x^2 - 2 \end{cases}$ **10.** $\begin{cases} x^2 + y^2 = 16 \\ y = x^2 - 4 \end{cases}$

11. $\begin{cases} xy = 4 \\ x^2 + y^2 = 8 \end{cases}$ **12.** $\begin{cases} xy = 1 \\ x^2 - y = 0 \end{cases}$

In Problems 13–20, solve the system of nonlinear equations using the method of elimination. See Objective 2.

13. $\begin{cases} x^2 + y^2 = 4 \\ y^2 - x = 4 \end{cases}$ **14.** $\begin{cases} x^2 + y^2 = 8 \\ x^2 + y^2 + 4y = 0 \end{cases}$

15. $\begin{cases} x^2 + y^2 = 7 \\ x^2 - y^2 = 25 \end{cases}$ **16.** $\begin{cases} 4x^2 + 16y^2 = 16 \\ 2x^2 - 2y^2 = 8 \end{cases}$

17. $\begin{cases} x^2 + y^2 = 6y \\ x^2 = 3y \end{cases}$ **18.** $\begin{cases} 2x^2 + y^2 = 18 \\ x^2 - y^2 = 9 \end{cases}$

19. $\begin{cases} x^2 - 2x - y = 8 \\ 6x + 2y = -4 \end{cases}$ **20.** $\begin{cases} 2x^2 - 5x + y = 12 \\ 14x - 2y = -16 \end{cases}$

Mixed Practice

In Problems 21–36, solve the system of nonlinear equations using any method you wish.

21. $\begin{cases} y = x^2 - 6x + 4 \\ 5x + y = 6 \end{cases}$ **22.** $\begin{cases} y = x^2 + 4x + 5 \\ x - y = 9 \end{cases}$

23. $\begin{cases} x^2 + y^2 = 16 \\ x^2 - y^2 = 16 \end{cases}$ **24.** $\begin{cases} x^2 + y^2 = 25 \\ x^2 - y^2 = 25 \end{cases}$

25. $\begin{cases} (x - 4)^2 + y^2 = 25 \\ x - y = -3 \end{cases}$

26. $\begin{cases} (x + 5)^2 + (y - 2)^2 = 100 \\ 8x + y = 18 \end{cases}$

27. $\begin{cases} (x - 1)^2 + (y + 2)^2 = 4 \\ y^2 + 4y - x = -1 \end{cases}$

28. $\begin{cases} (x + 2)^2 + (y - 1)^2 = 4 \\ y^2 - 2y - x = 5 \end{cases}$

29. $\begin{cases} (x + 3)^2 + 4y^2 = 4 \\ x^2 + 6x - y = 13 \end{cases}$ **30.** $\begin{cases} 9x^2 + 4y^2 = 36 \\ x^2 + (y - 7)^2 = 4 \end{cases}$

31. $\begin{cases} x^2 - y^2 = 21 \\ x + y = 7 \end{cases}$ **32.** $\begin{cases} y - 2x = 1 \\ 2x^2 + y^2 = 1 \end{cases}$

33. $\begin{cases} x^2 + 2y^2 = 16 \\ 4x^2 - y^2 = 24 \end{cases}$ **34.** $\begin{cases} 4x^2 + 3y^2 = 4 \\ 6y^2 - 2x^2 = 3 \end{cases}$

35. $\begin{cases} x^2 + y^2 = 25 \\ y = -x^2 + 6x - 5 \end{cases}$ **36.** $\begin{cases} x^2 + y^2 = 65 \\ y = -x^2 + 9 \end{cases}$

Applying the Concepts

37. Fun with Numbers The difference of two numbers is 2. The sum of their squares is 34. Find the numbers.

38. Fun with Numbers The sum of two numbers is 8. The sum of their squares is 160. Find the numbers.

△ **39. Perimeter and Area of a Rectangle** The perimeter of a rectangle is 48 feet. The area of the rectangle is 140 square feet. Find the dimensions of the rectangle.

△ **40. Perimeter and Area of a Rectangle** The perimeter of a rectangle is 64 meters. The area of the rectangle is 240 square meters. Find the dimensions of the rectangle.

41. Constructing a Box A rectangular piece of cardboard, whose area is 190 square centimeters, is made into an open box by cutting a 2-centimeter square from each corner and turning up the sides. See the figure. If the box is to have a volume of 180 cubic centimeters, what size cardboard should you start with?

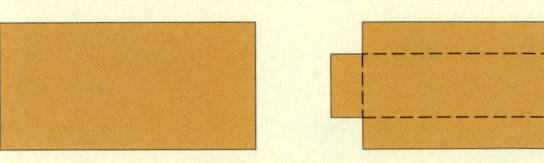

42. Fencing A farmer has 132 yards of fencing available to enclose a 900-square-yard region in the shape of adjoining squares with sides of length x and y. See the figure. Find x and y.

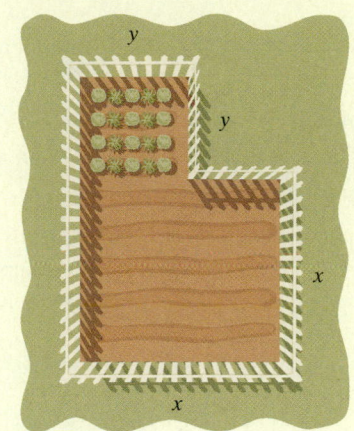

Extending the Concepts

In Problems 43–46, solve the system of nonlinear equations. Do not attempt to graph the equations in the system.

43. $\begin{cases} y^2 + y + x^2 - x - 2 = 0 \\ y + 1 + \dfrac{x-2}{y} = 0 \end{cases}$

44. $\begin{cases} x^3 - 2x^2 + y^2 + 3y - 4 = 0 \\ x - 2 + \dfrac{y^2 - y}{x^2} = 0 \end{cases}$

45. $\begin{cases} \ln x = 4 \ln y \\ \log_3 x = 2 + 2 \log_3 y \end{cases}$

46. $\begin{cases} \ln x = 5 \ln y \\ \log_2 x = 3 + 2 \log_2 y \end{cases}$

47. If r_1 and r_2 are two solutions of a quadratic equation $ax^2 + bx + c = 0$, then it can be shown that

$$r_1 + r_2 = -\frac{b}{a} \quad \text{and} \quad r_1 r_2 = \frac{c}{a}$$

Solve this system of equations for r_1 and r_2.

48. A circle and a line intersect at most twice. A circle and a parabola intersect at most four times. How many times do you think a circle and the graph of a polynomial of degree 3 can intersect? What about a circle and the graph of a polynomial of degree 4? What about a circle and the graph of a polynomial of degree n? Explain your conclusions using an algebraic argument.

Synthesis Review

In Problems 49–53, evaluate the functions given that $f(x) = 3x + 4$ and $g(x) = 2^x$.

49. (a) $f(1)$ **(b)** $g(1)$ **50. (a)** $f(2)$ **(b)** $g(2)$

51. (a) $f(3)$ **(b)** $g(3)$ **52. (a)** $f(4)$ **(b)** $g(4)$

53. (a) $f(5)$ **(b)** $g(5)$

54. Use the results of Problems 49–53 to compute

(a) $f(2) - f(1)$ **(b)** $f(3) - f(2)$

(c) $f(4) - f(3)$ **(d)** $f(5) - f(4)$

(e) $\dfrac{g(2)}{g(1)}$ **(f)** $\dfrac{g(3)}{g(2)}$

(g) $\dfrac{g(4)}{g(3)}$ **(h)** $\dfrac{g(5)}{g(4)}$

(i) Make a generalization about $f(n+1) - f(n)$ for $n \ge 1$ an integer. Make a generalization about $\dfrac{g(n+1)}{g(n)}$ for $n \ge 1$ an integer.

Technology Exercises

In Problems 55–64, use technology to solve the following systems of nonlinear equations. If necessary, round answers to three decimal places.

55. $\begin{cases} y = x^2 - 6x + 4 \\ 5x + y = 6 \end{cases}$ **56.** $\begin{cases} y = x^2 + 4x + 5 \\ x - y = 9 \end{cases}$

57. $\begin{cases} x^2 + y^2 = 16 \\ x^2 - y^2 = 16 \end{cases}$ **58.** $\begin{cases} x^2 + y^2 = 25 \\ x^2 - y^2 = 25 \end{cases}$

59. $\begin{cases} (x-4)^2 + y^2 = 25 \\ x - y = -3 \end{cases}$

60. $\begin{cases} (x+5)^2 + (y-2)^2 = 100 \\ 8x + y = 18 \end{cases}$

61. $\begin{cases} (x-1)^2 + (y+2)^2 = 4 \\ x^2 + 4y - x = -1 \end{cases}$

62. $\begin{cases} (x+2)^2 + (y-1)^2 = 4 \\ y^2 - 2y - x = 5 \end{cases}$

63. $\begin{cases} x^2 + 4y^2 = 4 \\ x^2 + 6x - y = -13 \end{cases}$ **64.** $\begin{cases} 9x^2 + 4y^2 = 36 \\ x^2 + (y-7)^2 = 4 \end{cases}$

Chapter 12 Activity: How Do You Know That...?

Focus: A sharing of ideas to identify topics contained in the study of conics

Time: 30–35 minutes

Group size: 2–4

For each question, every member of the group should spend 1–2 minutes individually listing "How they know ...." At the end of the allotted time, the group should convene and conduct a 2–3-minute discussion of the different responses.

How Do You Know ...

...that a triangle with vertices at $(2, 6)$, $(0, -2)$ and $(5, 1)$ is an isosceles triangle?

...the coordinates for the midpoint between two given ordered pairs?

...that $x^2 + y^2 + 4x - 8y = 16$ is an equation of a circle and not a parabola?

...which way a parabola opens?

...whether an ellipse's center is at the origin or another point?

...that a hyperbola will not intersect its asymptotes?

...that a circle and a parabola can have 0, 1, 2, 3, or 4 points of intersection? (If necessary, use a sketch to support your response.)

Chapter 12 Review

Section 12.1   Distance and Midpoint Formulas

KEY CONCEPTS

- **The Distance Formula**
  The distance between two points $P_1 = (x_1, y_1)$ and $P_2 = (x_2, y_2)$, denoted by $d(P_1, P_2)$, is
  $d(P_1, P_2) = \sqrt{(x_2 - x_1)^2 + (y_2 - y_1)^2}$.
- **The Midpoint Formula**
  The midpoint $M = (x, y)$ of the line segment from $P_1 = (x_1, y_1)$ to $P_2 = (x_2, y_2)$ is $M = \left(\dfrac{x_1 + x_2}{2}, \dfrac{y_1 + y_2}{2}\right)$.

You Should Be Able To...	EXAMPLE	Review Exercises
❶ Use the Distance Formula (p. 845)	Examples 1 through 3	1–6, 12
❷ Use the Midpoint Formula (p. 848)	Example 4	7–11

In Problems 1–6, find the distance $d(P_1, P_2)$ between points P_1 and P_2.

1. $P_1 = (0, 0)$ and $P_2 = (-4, -3)$

2. $P_1 = (-3, 2)$ and $P_2 = (5, -4)$

3. $P_1 = (-1, 1)$ and $P_2 = (5, 3)$

4. $P_1 = (6, -7)$ and $P_2 = (6, -1)$

5. $P_1 = \left(\sqrt{7}, -\sqrt{3}\right)$ and $P_2 = \left(4\sqrt{7}, 5\sqrt{3}\right)$

6. $P_1 = (-0.2, 1.7)$ and $P_2 = (1.3, 3.7)$

In Problems 7–11, find the midpoint of the line segment formed by joining the points P_1 and P_2.

7. $P_1 = (-1, 6)$ and $P_2 = (-3, 4)$

8. $P_1 = (7, 0)$ and $P_2 = (5, -4)$

9. $P_1 = \left(-\sqrt{3}, 2\sqrt{6}\right)$ and $P_2 = \left(-7\sqrt{3}, -8\sqrt{6}\right)$

10. $P_1 = (5, -2)$ and $P_2 = (0, 3)$

11. $P_1 = \left(\dfrac{1}{4}, \dfrac{2}{3}\right)$ and $P_2 = \left(\dfrac{5}{4}, \dfrac{1}{3}\right)$

12. Consider the three points
  $A = (-2, 2), B = (1, -1), C = (-1, -3)$.

  (a) Plot each point in the Cartesian plane and form the triangle ABC.

  (b) Find the length of each side of the triangle.

  (c) Verify that the triangle is a right triangle.

  (d) Find the area of the triangle.

Section 12.2 Circles

KEY CONCEPTS

- **Standard Form of an Equation of a Circle**

  The standard form of the equation of a circle with radius r and center (h, k) is $(x - h)^2 + (y - k)^2 = r^2$.

- **General Form of the Equation of a Circle**

  The general form of the equation of a circle is $x^2 + y^2 + ax + by + c = 0$ when the graph exists.

KEY TERMS

Conic sections
Circle
Radius
Center

You Should Be Able To...	EXAMPLE	Review Exercises
1 Write the standard form of the equation of a circle (p. 852)	Example 1	13–20
2 Graph a circle (p. 853)	Example 2	15–18; 21–30
3 Find the center and radius of a circle given an equation in general form (p. 853)	Example 3	27–30

In Problems 13 and 14, find the center and radius of each circle. Write the standard form of the equation.

13.
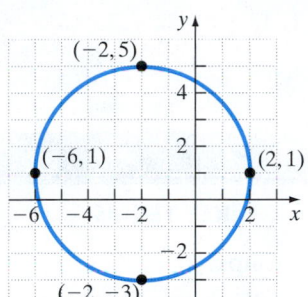

14.
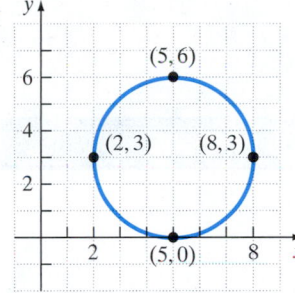

In Problems 15–18, write the standard form of the equation of each circle whose radius is r and center is (h, k). Graph each circle.

15. $r = 4$; $(h, k) = (0, 0)$ **16.** $r = 3$; $(h, k) = (-3, 1)$

17. $r = 1$; $(h, k) = (5, -2)$ **18.** $r = \sqrt{7}$; $(h, k) = (4, 0)$

In Problems 19 and 20, find the standard form of the equation of each circle.

19. Center at $(2, -1)$; contains the point $(5, 3)$.

20. Endpoints of a diameter at $(1, 7)$ and $(-3, -1)$.

In Problems 21–26, find the center (h, k) and the radius r of each circle. Graph each circle.

21. $x^2 + y^2 = 25$

22. $(x - 1)^2 + (y - 2)^2 = 4$

23. $x^2 + (y - 4)^2 = 16$

24. $(x + 1)^2 + (y + 6)^2 = 49$

25. $(x + 2)^2 + \left(y - \dfrac{3}{2}\right)^2 = \dfrac{1}{4}$

26. $(x + 3)^2 + (y + 3)^2 = 4$

In Problems 27–30, find the center (h, k) and the radius r of each circle. Graph each circle.

27. $x^2 + y^2 + 6x + 10y - 2 = 0$

28. $x^2 + y^2 - 8x + 4y + 16 = 0$

29. $x^2 + y^2 + 2x - 4y - 4 = 0$

30. $x^2 + y^2 - 10x - 2y + 17 = 0$

Section 12.3 Parabolas

KEY CONCEPTS

- **Equations of a Parabola: Vertex at $(0, 0)$; $a > 0$**

  See Table 1 on page 859.

- **Equations of a Parabola: Vertex at (h, k); $a > 0$**

  See Table 2 on page 861.

KEY TERMS

Parabola
Focus
Directrix
Axis of symmetry
Vertex

You Should Be Able To...	EXAMPLE	Review Exercises
❶ Graph parabolas whose vertex is the origin (p. 858)	Examples 1 and 2	31–36
❷ Find the equation of a parabola (p. 860)	Examples 3 and 4	31–34
❸ Graph a parabola whose vertex is not the origin (p. 861)	Example 5	37–39
❹ Solve applied problems involving parabolas (p. 863)	Example 6	40

In Problems 31–34, find the equation of the parabola described. Graph each parabola.

31. Vertex at $(0,0)$; focus at $(0, -3)$

32. Focus at $(-4, 0)$; directrix the line $x = 4$

33. Vertex at $(0,0)$; contains the point $(8, -2)$; axis of symmetry the x-axis

34. Vertex at $(0,0)$; directrix the line $y = -2$; axis of symmetry the y-axis

In Problems 35–39, find the vertex, focus, and directrix of each parabola. Graph each parabola.

35. $x^2 = 2y$ **36.** $y^2 = 16x$

37. $(x + 1)^2 = 8(y - 3)$

38. $(y - 4)^2 = -2(x + 3)$

39. $x^2 - 10x + 3y + 19 = 0$

40. Radio Telescope The U.S. Naval Research Laboratory has a giant radio telescope with a dish that is shaped like a parabola. The signals that are received by the dish strike the surface of the dish and are reflected to a single point, where the receiver is located. If the giant dish is 300 feet across and 44 feet deep at its center, at what position should the receiver be placed?

Section 12.4 Ellipses

KEY CONCEPTS

- **Ellipses with Center at $(0, 0)$**
  See Table 3 on page 867.

- **Ellipses with Center at (h, k)**
  See Table 4 on page 870.

KEY TERMS

Ellipse
Foci
Major axis
Center
Minor axis
Vertices

You Should Be Able To...	EXAMPLE	Review Exercises
❶ Graph an ellipse whose center is the origin (p. 867)	Examples 1 and 2	41, 42
❷ Find the equation of an ellipse whose center is the origin (p. 869)	Example 3	43–45
❸ Graph an ellipse whose center is not the origin (p. 870)	Example 4	46–47
❹ Solve applied problems involving ellipses (p. 871)	Example 5	48

In Problems 41 and 42, find the vertices and foci of each ellipse. Graph each ellipse.

41. $\dfrac{x^2}{9} + y^2 = 1$ **42.** $9x^2 + 4y^2 = 36$

In Problems 43–45, find an equation for each ellipse. Graph each ellipse.

43. Center at $(0,0)$; focus at $(0,3)$; vertex at $(0,5)$

44. Center at $(0,0)$; focus at $(-2, 0)$; vertex at $(-6, 0)$

45. Foci at $(\pm 8, 0)$; vertices at $(\pm 10, 0)$

In Problems 46 and 47, find the center, vertices, and foci of each ellipse. Graph each ellipse.

46. $\dfrac{(x - 1)^2}{49} + \dfrac{(y + 2)^2}{25} = 1$

47. $25(x + 3)^2 + 9(y - 4)^2 = 225$

48. Semielliptical Arch Bridge An arch in the shape of the upper half of an ellipse is used to support a bridge that spans a river 60 feet wide. The center of the arch is 16 feet above the center of the river.

(a) Write the equation for the ellipse in which the x-axis coincides with the water and the y-axis passes through the center of the arch.

(b) Can a rectangular barge that is 25 feet wide and sits 12 feet above the surface of the water fit through the opening of the bridge?

Section 12.5   Hyperbolas

KEY CONCEPTS

- **Hyperbolas with Center at the Origin**
  See Table 5 on page 876.

- **Asymptotes of a Hyperbola**
  See Table 6 on page 880.

KEY TERMS

Hyperbola
Foci
Transverse axis
Center
Conjugate axis
Branches
Vertices

You Should Be Able To...	EXAMPLE	Review Exercises
❶ Graph a hyperbola whose center is the origin (p. 876)	Examples 1 and 2	49–53
❷ Find the equation of a hyperbola whose center is the origin (p. 879)	Example 3	54–56
❸ Find the asymptotes of a hyperbola whose center is the origin (p. 879)	Example 4	52, 53

In Problems 49–51, find the vertices and foci of each hyperbola. Graph each hyperbola.

49. $\dfrac{x^2}{4} - \dfrac{y^2}{9} = 1$

50. $\dfrac{y^2}{25} - \dfrac{x^2}{49} = 1$

51. $16y^2 - 25x^2 = 400$

In Problems 52 and 53, graph each hyperbola using the asymptotes as a guide.

52. $\dfrac{x^2}{36} - \dfrac{y^2}{36} = 1$

53. $\dfrac{y^2}{25} - \dfrac{x^2}{4} = 1$

In Problems 54–56, find an equation for each hyperbola described. Graph each hyperbola.

54. Center at $(0,0)$; focus at $(-4, 0)$; vertex at $(-3, 0)$

55. Vertices at $(0, -3)$ and $(0, 3)$; focus at $(0, 5)$

56. Vertices at $(0, \pm 4)$; asymptote the line $y = \dfrac{4}{3}x$

Section 12.6   Systems of Nonlinear Equations

KEY TERM

System of nonlinear equations

You Should Be Able To...	EXAMPLE	Review Exercises
❶ Solve a system of nonlinear equations using substitution (p. 884)	Examples 1 and 2	57–60; 65–76
❷ Solve a system of nonlinear equations using elimination (p. 886)	Examples 3 and 4	61–64; 65–76

In Problems 57–60, solve the system of nonlinear equations using the method of substitution.

57. $\begin{cases} 4x^2 + y^2 = 10 \\ y = x \end{cases}$

58. $\begin{cases} y = 2x^2 + 1 \\ y = x + 2 \end{cases}$

59. $\begin{cases} 6x - y = 5 \\ xy = 1 \end{cases}$

60. $\begin{cases} x^2 + y^2 = 26 \\ x^2 - 2y^2 = 23 \end{cases}$

In Problems 61–64, solve the system of nonlinear equations using the method of elimination.

61. $\begin{cases} 4x - y^2 = 0 \\ 2x^2 + y^2 = 16 \end{cases}$

62. $\begin{cases} x^2 - y = -2 \\ x^2 + y = 4 \end{cases}$

63. $\begin{cases} 4x^2 - 2y^2 = 2 \\ -x^2 + y^2 = 2 \end{cases}$

64. $\begin{cases} x^2 + y^2 = 8x \\ y^2 = 3x \end{cases}$

In Problems 65–72, solve the system of nonlinear equations using the method you prefer.

65. $\begin{cases} y = x + 2 \\ y = x^2 \end{cases}$

66. $\begin{cases} x^2 + 2y = 9 \\ 5x - 2y = 5 \end{cases}$

67. $\begin{cases} x^2 + y^2 = 36 \\ x - y = -6 \end{cases}$

68. $\begin{cases} y = 2x - 4 \\ y^2 = 4x \end{cases}$

69. $\begin{cases} x^2 + y^2 = 9 \\ x + y = 7 \end{cases}$

70. $\begin{cases} 2x^2 + 3y^2 = 14 \\ x^2 - y^2 = -3 \end{cases}$

71. $\begin{cases} x^2 + y^2 = 16 \\ x^2 + 4y = 16 \end{cases}$

72. $\begin{cases} x = 4 - y^2 \\ x = 2y + 4 \end{cases}$

73. Fun with Numbers The sum of two numbers is 12. The difference of their squares is 24. Find the two numbers.

△**74. Perimeter and Area of a Rectangle** The perimeter of a rectangle is 34 centimeters. The area of the rectangle is 60 square centimeters. Find the dimensions of the rectangle.

△**75. Dimensions of a Rectangle** The area of a rectangle is 2160 square inches. The diagonal of the rectangle is 78 inches. Find the dimensions of the rectangle.

△**76. Dimensions of a Triangle** A right triangle has a perimeter of 36 feet and a hypotenuse of 15 feet. Find the lengths of the legs of the right triangle.

Chapter 12 Test

Step-by-step test solutions are found on the Chapter Test Prep Videos available in MyMathLab®, on YouTube*, or may be accessed using the QR code to the right.*

1. Find the distance $d(P_1, P_2)$ between points $P_1 = (-1, 3)$ and $P_2 = (3, -5)$.

2. Find the midpoint of the line segment formed by joining the points $P_1 = (-7, 6)$ and $P_2 = (5, -2)$.

In Problems 3 and 4, find the center (h, k) and the radius r of each circle. Graph each circle.

3. $(x - 4)^2 + (y + 1)^2 = 9$

4. $x^2 + y^2 + 10x - 4y + 13 = 0$

In Problems 5 and 6, find the standard form of the equation of each circle.

5. Radius $r = 6$ and center $(h, k) = (-3, 7)$

6. Center at $(-5, 8)$; contains the point $(3, 2)$

In Problems 7 and 8, find the vertex, focus, and directrix of each parabola. Graph each parabola.

7. $(y + 2)^2 = 4(x - 1)$

8. $x^2 - 4x + 3y - 8 = 0$

In Problems 9 and 10, find an equation for each parabola described.

9. Vertex at $(0, 0)$; focus at $(0, -4)$

10. Focus at $(3, 4)$; directrix the line $x = -1$

In Problems 11 and 12, find the vertices and foci of each ellipse. Graph each ellipse.

11. $9x^2 + 25y^2 = 225$

12. $\dfrac{(x - 2)^2}{9} + \dfrac{(y + 4)^2}{16} = 1$

In Problems 13 and 14, find an equation for each ellipse described.

13. Center at $(0, 0)$; focus at $(0, -4)$; vertex at $(0, -5)$

14. Vertices at $(-1, 7)$ and $(-1, -3)$; focus at $(-1, -1)$

In Problems 15 and 16, find the vertices, foci, and asymptotes for each hyperbola. Graph each hyperbola using the asymptotes as a guide.

15. $x^2 - \dfrac{y^2}{4} = 1$

16. $16y^2 - 25x^2 = 1600$

17. Find an equation for a hyperbola with foci at $(\pm 8, 0)$ and vertex at $(-3, 0)$.

In Problems 18 and 19, solve the system of nonlinear equations using the method you prefer.

18. $\begin{cases} x^2 + y^2 = 17 \\ x + y = -3 \end{cases}$

19. $\begin{cases} x^2 + y^2 = 9 \\ 4x^2 - y^2 = 16 \end{cases}$

20. An arch in the shape of the upper half of an ellipse is used to support a bridge spanning a creek 30 feet wide. The center of the arch is 10 feet above the center of the creek.

(a) Write the equation for the ellipse in which the *x*-axis coincides with the creek and the *y*-axis passes through the center of the arch.

(b) What is the height of the arch at a distance 12 feet from the center of the creek?

13 Sequences, Series, and the Binomial Theorem

Population growth has been a topic of debate among scientists for a long time. In fact, over 200 years ago, the English economist and mathematician Thomas Robert Malthus anonymously published a paper predicting that the world's population would overwhelm the Earth's capacity to sustain it—he claimed that food supplies increase arithmetically while population grows geometrically. See Problems 55 and 56 in Section 13.1 and Problem 99 in Section 13.3.

The Big Picture: Putting It Together

In Chapter 8, the domain of a function (inputs) was defined as the set of real numbers so that the outputs of the function make sense. In Sections 13.1–13.3, a special type of function called a *sequence* is introduced. The domain of a sequence is the set of natural numbers (that is, the positive integers). Functions whose domain is the set of all real numbers enable us to model situations in which the value of an independent variable, such as a population of bacteria, changes continuously. In contrast, sequences make it possible to model situations in which the value of a *dependent* variable changes at discrete intervals of time, such as weekly changes in the value of a deposit at a bank.

In Chapter 5, a formula for expanding $(x + a)^2$ was introduced. Does any formula exist for expanding $(x + a)^n$, where n is an integer greater than 2? Yes! This formula is discussed in Section 13.4.

Outline

13.1 Sequences

Objectives

1. Write the First Few Terms of a Sequence
2. Find a Formula for the nth Term of a Sequence
3. Use Summation Notation

Are You Prepared for This Section?

Before getting started, complete the following problems. If you get a problem wrong, go back to the section cited and review the material.

P1. Evaluate $f(x) = x^2 - 4$ at **(a)** $x = 3$ **(b)** $x = -7$. [Section 8.3, pp. 538–541]

P2. If $g(x) = 2x - 3$, find $g(1) + g(2) + g(3)$. [Section 8.3, pp. 538–541]

P3. In the function $f(n) = n^2 - 4$, what is the independent variable? [Section 8.3, pp. 538–541]

Definition

A **sequence** is a function whose domain is the set of positive integers.

Because a sequence is a function, it has a graph. The function $f(x) = \dfrac{1}{x}$ for $x > 0$ is graphed in Figure 1(a). If all the points on this graph were removed except those whose x-coordinates are positive integers—that is, if all points were removed except $(1, 1)$, $\left(2, \dfrac{1}{2}\right), \left(3, \dfrac{1}{3}\right)$, and so on, the remaining points would be the graph of the sequence $f(n) = \dfrac{1}{n}$, as shown in Figure 1(b). Notice that n is used to represent the independent variable in a sequence. This serves as a reminder that n is a positive integer, or natural number.

Figure 1

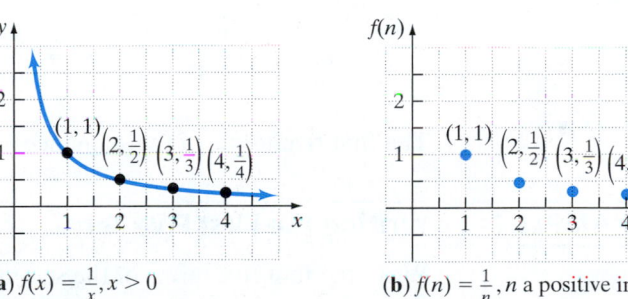

(a) $f(x) = \dfrac{1}{x}, x > 0$ **(b)** $f(n) = \dfrac{1}{n}, n$ a positive integer

A sequence may be represented by listing its values in order. For example, the sequence whose graph is given in Figure 1(b) might be represented as

$$f(1), f(2), f(3), f(4), \ldots \quad \text{or} \quad 1, \frac{1}{2}, \frac{1}{3}, \frac{1}{4}, \ldots$$

The numbers in the ordered list are the **terms** of the sequence. Notice that the list never ends, as the **ellipsis** (the three dots) indicates. A sequence that does not end is an **infinite sequence.** In contrast, **finite sequences** have a domain that is the first n positive integers. For example,

$$40, 44, 48, 52, 56, 60, 64$$

is a finite sequence because it contains $n = 7$ terms.

In Other Words
The word *infinite* means "without bound," so infinite sequences have no bound. The word *finite* means "bounded in number," so a finite sequence has a certain number of terms.

▶ 1 Write the First Few Terms of a Sequence

In a sequence, the traditional notation $f(n)$ is not used. This is done to distinguish sequences from functions whose domain is the set of all real numbers. Instead, the notation a_n is used to indicate that the function is a sequence. The value of the subscript represents the value of the independent variable, so a_4 means to evaluate the function defined by the sequence a_n at $n = 4$. For the sequence $f(n) = \dfrac{1}{n}$, write the function as $a_n = \dfrac{1}{n}$ and evaluate the function as follows:

$$a_1 = 1, \quad a_2 = \frac{1}{2}, \quad a_3 = \frac{1}{3}, \quad a_4 = \frac{1}{4}, \quad \ldots, \quad a_n = \frac{1}{n}$$

In addition, just as functions are named f, g, F, G, and so on, sequences can be named. Usually, lowercase letters from the beginning of the alphabet are used to name a sequence, as in a_n, b_n, c_n, and so on although this naming scheme is not necessary.

When a formula for the nth term (sometimes called the **general term**) of a sequence is known, the sequence is represented by placing braces around the formula for the nth term, rather than by writing out the terms of the sequence. For example, the sequence whose nth term is $b_n = \left(\dfrac{1}{3}\right)^n$ can be written as

$$\{b_n\} = \left\{ \left(\frac{1}{3}\right)^n \right\}$$

or as

$$b_1 = \frac{1}{3}, \quad b_2 = \frac{1}{9}, \quad b_3 = \frac{1}{27}, \quad \ldots, \quad b_n = \left(\frac{1}{3}\right)^n$$

EXAMPLE 1 **Writing the First Five Terms of a Sequence**

Write the first five terms of the sequence $\{a_n\} = \left\{ \dfrac{n}{n+1} \right\}$.

Solution
To find the first five terms of the sequence, evaluate the function at $n = 1, 2, 3, 4$, and 5.

$$a_1 = \frac{1}{1+1} = \frac{1}{2}, \quad a_2 = \frac{2}{2+1} = \frac{2}{3}, \quad a_3 = \frac{3}{3+1} = \frac{3}{4}, \quad a_4 = \frac{4}{4+1} = \frac{4}{5}, \quad a_5 = \frac{5}{5+1} = \frac{5}{6}$$

The first five terms of the sequence are $\dfrac{1}{2}, \dfrac{2}{3}, \dfrac{3}{4}, \dfrac{4}{5}$, and $\dfrac{5}{6}$. ●

EXAMPLE 2 **Writing the First Five Terms of a Sequence**

Write the first five terms of the sequence $\{b_n\} = \{(-1)^n n^2\}$.

Solution
The first five terms of the sequence are

$$b_1 = (-1)^1 \cdot 1^2 = -1, \quad b_2 = (-1)^2 \cdot 2^2 = 4, \quad b_3 = (-1)^3 \cdot 3^2 = -9,$$
$$b_4 = (-1)^4 \cdot 4^2 = 16, \quad b_5 = (-1)^5 \cdot 5^2 = -25$$

The first five terms of the sequence are $-1, 4, -9, 16$, and -25. ●

Notice in Example 2 that the signs of the terms in the sequence **alternate**. This occurs when there are factors such as $(-1)^{n+1}$, which equals 1 if n is odd and -1 if n is even, or $(-1)^n$, which equals -1 if n is odd and 1 if n is even.

Quick ✓

1. A(n) _____ is a function whose domain is the set of positive integers.

2. A sequence that does not end is said to be a(n) _____ sequence. A(n) _____ sequence has a domain that is the first n positive integers.

3. *True or False* A sequence is a function.

In Problems 4 and 5, write the first five terms of the given sequence.

4. $\{a_n\} = \{2n - 3\}$

5. $\{b_n\} = \{(-1)^n \cdot 4n\}$

▶ ❷ Find a Formula for the *n*th Term of a Sequence

Sometimes a sequence is indicated by an observed pattern in the first few terms that makes it possible to determine the formula for the *n*th term. In the example that follows, a sufficient number of terms of the sequence are given so that a natural choice for the *n*th term is suggested.

EXAMPLE 3 **Determining a Sequence from a Pattern**

Find a formula for the *n*th term of each sequence.

(a) $3, 7, 11, 15, \ldots$

(b) $2, 4, 8, 16, \ldots$

(c) $1, -8, 27, -64, 125, \ldots$

Solution

The goal in all these problems is to find a formula in terms of *n* such that $n = 1$ is the first term, $n = 2$ is the second term, and so on.

(a) When $n = 1$, then $a_1 = 3$; when $n = 2$, then $a_2 = 7$. Notice that each subsequent term increases by 4. A formula for the *n*th term is given by
$$a_n = 4n - 1.$$

(b) The terms of the sequence are all powers of 2 with the first term equaling 2^1, the second term equaling 2^2, and so on. A formula for the *n*th term is given by
$$b_n = 2^n.$$

(c) Notice that the sign of the terms alternate with the first term being positive. Thus $(-1)^{n+1}$ must be part of the formula. Ignoring the sign, notice that the terms are all perfect cubes. A formula for the *n*th term is given by
$$c_n = (-1)^{n+1}n^3.$$ ●

Quick ✓

In Problems 6 and 7, find a formula for the nth term of each sequence.

6. $5, 7, 9, 11, \ldots$

7. $\dfrac{1}{2}, -\dfrac{1}{3}, \dfrac{1}{4}, -\dfrac{1}{5}, \ldots$

▶ ❸ Use Summation Notation

In other mathematics courses, such as statistics and calculus, it is important to find the sum of the first *n* terms of a sequence $\{a_n\}$ — that is,

$$a_1 + a_2 + a_3 + \cdots + a_n$$

Mathematicians express the sum using **summation notation.** With this notation, write $a_1 + a_2 + a_3 + \cdots + a_n$ as

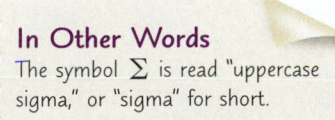

In Other Words
The symbol Σ is read "uppercase sigma," or "sigma" for short.

$$\sum_{i=1}^{n} a_i = a_1 + a_2 + a_3 + \cdots + a_n$$

The symbol Σ is an instruction to sum, or add up, the terms. The integer *i* is called the **index** of the sum; it tells you where to start the sum and where to end it. The expression

$$\sum_{i=1}^{n} a_i$$

means to add the terms of the sequence $\{a_i\}$ from $i = 1$ through $i = n$. Read the expression $\sum_{i=1}^{n} a_i$ as "the sum of *a* sub *i* from $i = 1$ to $i = n$." When a finite number of terms are to be added, the sum is called a **partial sum.**

EXAMPLE 4 **Finding a Partial Sum**

Write out the sum and determine its value.

$$\text{(a)} \ \sum_{i=1}^{4} (2i + 5) \qquad\qquad \text{(b)} \ \sum_{i=1}^{5} (i^2 - 5)$$

Solution

(a) $\displaystyle\sum_{i=1}^{4} (2i + 5) = \underbrace{(2 \cdot 1 + 5)}_{i\,=\,1} + \underbrace{(2 \cdot 2 + 5)}_{i\,=\,2} + \underbrace{(2 \cdot 3 + 5)}_{i\,=\,3} + \underbrace{(2 \cdot 4 + 5)}_{i\,=\,4}$

$\qquad\qquad\quad = 7 + 9 + 11 + 13$

$\qquad\qquad\quad = 40$

(b) $\displaystyle\sum_{i=1}^{5} (i^2 - 5) = \underbrace{(1^2 - 5)}_{i\,=\,1} + \underbrace{(2^2 - 5)}_{i\,=\,2} + \underbrace{(3^2 - 5)}_{i\,=\,3} + \underbrace{(4^2 - 5)}_{i\,=\,4} + \underbrace{(5^2 - 5)}_{i\,=\,5}$

$\qquad\qquad\quad = -4 + (-1) + 4 + 11 + 20$

$\qquad\qquad\quad = 30$

●

> **Quick ✓**
>
> **8.** When there is a finite number of terms to be added, the sum is called a _____ ___.
>
> *In Problems 9 and 10, write out the sum and determine its value.*
>
> **9.** $\displaystyle\sum_{i=1}^{3} (4i - 1)$ $\qquad\qquad$ **10.** $\displaystyle\sum_{i=1}^{5} (i^3 + 1)$

The index of summation does not have to begin with 1. In addition, the index of summation does not have to be i. For example,

$$\sum_{k=3}^{6} (2k + 1) = \underbrace{(2 \cdot 3 + 1)}_{k\,=\,3} + \underbrace{(2 \cdot 4 + 1)}_{k\,=\,4} + \underbrace{(2 \cdot 5 + 1)}_{k\,=\,5} + \underbrace{(2 \cdot 6 + 1)}_{k\,=\,6}$$

$$= 7 + 9 + 11 + 13$$

$$= 40$$

Notice that the terms of this partial sum are identical to those in Example 4(a). What is the moral of the story? The same sum can be made to look entirely different by changing the starting point of the index of summation and changing the variable that represents the index.

Now reverse the process. Rather than writing out a sum, express a sum using summation notation.

EXAMPLE 5 **Writing a Sum in Summation Notation**

Express each sum using summation notation.

(a) $1^2 + 2^2 + 3^2 + \cdots + 10^2$

(b) $2 + 1 + \dfrac{2}{3} + \dfrac{1}{2} + \dfrac{2}{5} + \cdots + \dfrac{1}{6}$

Solution

(a) The sum $1^2 + 2^2 + 3^2 + \cdots + 10^2$ has 10 terms, each of the form i^2, and starts at $i = 1$ and ends at $i = 10$:

$$1^2 + 2^2 + 3^2 + \cdots + 10^2 = \sum_{i=1}^{10} i^2$$

(b) First, figure out the pattern. With a little investigation,

$$2 + 1 + \frac{2}{3} + \frac{1}{2} + \frac{2}{5} + \cdots + \frac{1}{6} \text{ can be written as}$$

$$\frac{2}{1} + \frac{2}{2} + \frac{2}{3} + \frac{2}{4} + \frac{2}{5} + \cdots + \frac{2}{12}, \text{ so that the } n\text{th term of the sum is } \frac{2}{n}. \text{ There}$$

are $n = 12$ terms, so

$$2 + 1 + \frac{2}{3} + \frac{1}{2} + \frac{2}{5} + \cdots + \frac{1}{6} = \sum_{n=1}^{12} \left(\frac{2}{n}\right)$$

●

Work Smart

The index of summation can be any variable and can start at any value. Keep this in mind when checking your answers.

> **Quick ✓**
>
> *In Problems 11 and 12, write each sum using summation notation.*
>
> **11.** $1 + 4 + 9 + \cdots + 144$
>
> **12.** $1 + \frac{1}{2} + \frac{1}{4} + \cdots + \frac{1}{32}$

13.1 Exercises MyMathLab®

Exercise numbers in **green** have complete video solutions in MyMathLab or may be accessed using the QR code to the right.

*Problems **1–12** are the Quick ✓s that follow the EXAMPLES.*

Building Skills

In Problems 13–24, write the first five terms of each sequence. See Objective 1.

13. $\{3n + 5\}$

14. $\{n - 4\}$

15. $\left\{\dfrac{n}{n + 2}\right\}$

16. $\left\{\dfrac{n + 4}{n}\right\}$

17. $\{(-1)^n n\}$

18. $\{(-1)^{n+1} n\}$

19. $\{2^n + 1\}$

20. $\{3^n - 1\}$

21. $\left\{\dfrac{2n}{2^n}\right\}$

22. $\left\{\dfrac{3n}{3^n}\right\}$

23. $\left\{\dfrac{n}{e^n}\right\}$

24. $\left\{\dfrac{n^2}{2}\right\}$

In Problems 25–32, the given pattern continues. Write the nth term of each sequence suggested by the pattern. See Objective 2.

25. $2, 4, 6, 8, \ldots$

26. $5, 10, 15, 20, \ldots$

27. $\dfrac{1}{2}, \dfrac{2}{3}, \dfrac{3}{4}, \dfrac{4}{5}, \ldots$

28. $\dfrac{1}{2}, 1, \dfrac{3}{2}, 2, \dfrac{5}{2}, \ldots$

29. $3, 6, 11, 18, \ldots$

30. $0, 7, 26, 63, \ldots$

31. $-1, 4, -9, 16, \ldots$

32. $1, -\dfrac{1}{2}, \dfrac{1}{4}, -\dfrac{1}{8}, \ldots$

In Problems 33–44, write out each sum and determine its value. See Objective 3.

33. $\displaystyle\sum_{i=1}^{4} (5i + 1)$

34. $\displaystyle\sum_{i=1}^{5} (3i + 2)$

35. $\displaystyle\sum_{i=1}^{5} \frac{i^2}{2}$

36. $\displaystyle\sum_{i=1}^{4} \frac{i^3}{2}$

37. $\displaystyle\sum_{k=1}^{3} 2^k$

38. $\displaystyle\sum_{k=1}^{4} 3^k$

39. $\displaystyle\sum_{k=1}^{5} [(-1)^{k+1} \cdot 2k]$

40. $\displaystyle\sum_{k=1}^{8} [(-1)^k \cdot k]$

41. $\displaystyle\sum_{j=1}^{10} 5$

42. $\displaystyle\sum_{j=1}^{8} 2$

43. $\displaystyle\sum_{k=3}^{7} (2k - 1)$

44. $\displaystyle\sum_{j=5}^{10} (k + 4)$

In Problems 45–52, express each sum using summation notation. See Objective 3.

45. $1 + 2 + 3 + \cdots + 15$

46. $1 + 3 + 5 + \cdots + 17$

47. $1 + \dfrac{1}{2} + \dfrac{1}{3} + \cdots + \dfrac{1}{12}$

48. $1 + \dfrac{1}{2} + \dfrac{1}{4} + \cdots + \dfrac{1}{2^{15}}$

49. $1 - \dfrac{1}{3} + \dfrac{1}{9} - \dfrac{1}{27} + \cdots + (-1)^{9+1}\left(\dfrac{1}{3^{9-1}}\right)$

50. $\dfrac{2}{3} - \dfrac{4}{9} + \dfrac{8}{27} + \cdots + (-1)^{15+1}\left(\dfrac{2}{3}\right)^{15}$

51. $5 + (5 + 2 \cdot 1) + (5 + 2 \cdot 2) + (5 + 2 \cdot 3) + \cdots$
$+ (5 + 2 \cdot 10)$

52. $3 + 3 \cdot \dfrac{1}{2} + 3 \cdot \dfrac{1}{4} + \cdots + 3 \cdot \left(\dfrac{1}{2}\right)^{11}$

Applying the Concepts

53. The Future Value of Money Suppose that you place $12,000 in your company 401(k) plan, which pays 6% interest compounded quarterly. The balance in the account after n quarters is given by

$$a_n = 12{,}000\left(1 + \frac{0.06}{4}\right)^n$$

(a) Find the value of the account after 1 quarter.
(b) Find the value of the account after 1 year.
(c) Find the value of the account after 10 years.

54. The Future Value of Money Suppose that you place $5000 into a company 401(k) plan that pays 8% interest compounded monthly. The balance in the account after n months is given by

$$a_n = 5000\left(1 + \frac{0.08}{12}\right)^n$$

(a) Find the value of the account after 1 month.
(b) Find the value of the account after 1 year.
(c) Find the value of the account after 10 years.

55. Population Growth According  to the U.S. Census Bureau, the population of the United States in 2016 was 326 million people. In addition, the population of the United States was growing at a rate of 0.99% per year. A model for the population of the United States is given by

$$p_n = 326(1.0099)^n$$

where n is the number of years after 2016.

(a) Use this model to predict the United States population in 2020 to the nearest million.
(b) Use this model to predict the United States population in 2050 to the nearest million.

56. Population Growth According to the United States Census Bureau, the population of the world in 2016 was 7563 million people. In addition, the population of the world was growing at a rate of 1.02% per year. A model for the population of the world is given by

$$p_n = 7563(1.0102)^n$$

where n is the number of years after 2016.

(a) Use this model to predict the world population in 2020 to the nearest million.
(b) Use this model to predict the world population in 2050 to the nearest million.

57. Fibonacci Sequence Let

$$u_n = \frac{\left(1 + \sqrt{5}\right)^n - \left(1 - \sqrt{5}\right)^n}{2^n \cdot \sqrt{5}}$$

define the nth term of a sequence. Find the first 10 terms of the sequence. This sequence is called the

Fibonacci sequence. The terms of the sequence are called **Fibonacci numbers.**

58. Pascal's Triangle The triangular array of numbers shown below is called Pascal's Triangle. Each number in the triangle is found by adding the entries directly above to the left and right of the number. For example, the 4 in the fourth row is found by adding the 1 and 3 above the 4.

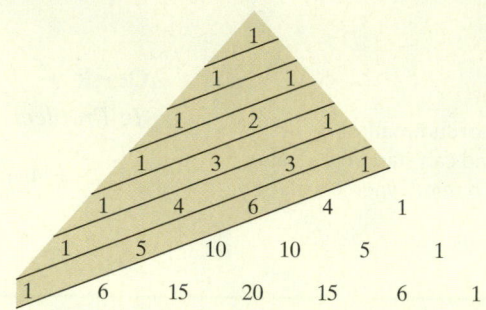

Divide the triangular array using diagonal lines (as shown). Now find the sum of the numbers in each of the highlighted diagonal rows. Do you recognize the sequence? (*Hint:* See Problem 57.)

Extending the Concepts

*A second way of defining a sequence is to assign a value to the first (or first few) term(s) and specify the nth term by a formula or equation that involves one or more of the terms preceding it. Sequences defined this way are said to be defined **recursively**, and the rule or formula is called a **recursive formula**. For example, $s_1 = 3$ and $s_n = 2s_{n-1}$ is a recursively defined sequence where $s_1 = 3$, $s_2 = 2s_1 = 2(3) = 6$, $s_3 = 2s_2 = 2(6) = 12$, $s_4 = 2s_3 = 2(12) = 24$, and so on.*

In Problems 59–62, a sequence is defined recursively. Write the first five terms of the sequence.

59. $a_1 = 10, a_n = 1.05a_{n-1}$ **60.** $b_1 = 20, b_n = 3b_{n-1}$

61. $b_1 = 8, b_n = n + b_{n-1}$ **62.** $c_1 = 1000,$
$c_n = 1.01c_{n-1} + 100$

63. Fibonacci Sequence Use the result of Problem 57 to do the following problems:

(a) Compute the ratio $\dfrac{u_{n+1}}{u_n}$ for the first 10 terms.

(b) As n gets large, what number does this ratio approach? This number is referred to as the **golden ratio.** Rectangles whose sides are in this ratio were considered pleasing to the eye by the Greeks. For example, the façade of the Parthenon was constructed using the golden ratio.

(*continued*)

(c) Compute the ratio $\dfrac{u_n}{u_{n+1}}$ for the first 10 terms.

(d) As n gets large, what number does this ratio approach? This number is also referred to as the **golden ratio.** This ratio is believed to have been used in the construction of the Great Pyramid in Egypt. The ratio equals the sum of the areas of the four face triangles divided by the total surface area of the Great Pyramid.

64. Investigate various applications that lead to a Fibonacci sequence, such as art, architecture, or financial markets. Write an essay on these applications.

Explaining the Concepts

65. Explain how a sequence and a function differ.

66. What does the graph of a sequence look like when compared to the graph of a function? Use the function $f(x) = 3x + 1$ and the sequence $\{a_n\} = \{3n + 1\}$ when doing the comparison.

67. Write a sentence that explains the meaning of the symbol Σ.

68. What does it mean when a sequence alternates?

Synthesis Review

In Problems 69–71, (a) determine the slope of the linear function, and (b) compute f(1), f(2), f(3), and f(4).

69. $f(x) = 4x - 6$

70. $f(x) = 2x - 10$

71. $f(x) = -5x + 8$

72. Compute $f(2) - f(1), f(3) - f(2)$, and $f(4) - f(3)$ for each of the functions in Problems 69–71. How is the difference in the value of the function for consecutive values of the independent variable related to the slope?

Technology Exercises

Technology can be used to write the terms of a sequence. For example, Figure 2 shows the first five terms of the sequence $\{b_n\} = \{(-1)^n n^2\}$ that we studied in Example 2. Figure 3 shows how the terms of the sequence can be listed in a table using SEQuence mode in a graphing calculator.

Figure 2

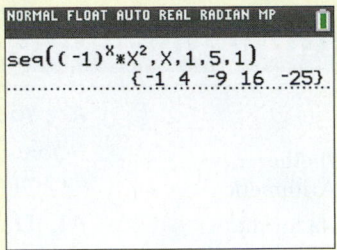

Figure 3

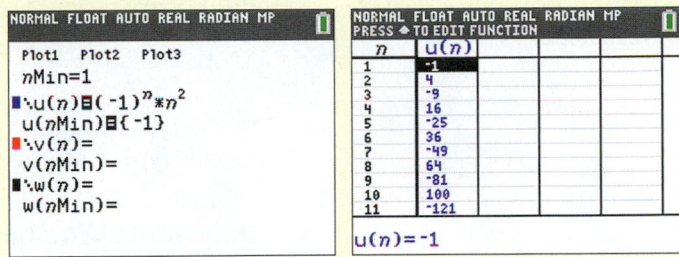

In Problems 73–80, use technology to find the first five terms of the sequence.

73. $\{3n + 5\}$

74. $\{n - 4\}$

75. $\left\{\dfrac{n}{n + 2}\right\}$

76. $\left\{\dfrac{n + 4}{n}\right\}$

77. $\{(-1)^n n\}$

78. $\{(-1)^{n+1} n\}$

79. $\{2^n + 1\}$

80. $\{3^n - 1\}$

Technology can also be used to find the sum of a sequence. For example, Figure 4 shows the result of the sum $\displaystyle\sum_{i=1}^{5} (i^2 - 5)$ that was investigated in Example 4(b) using a graphing calculator.

Figure 4

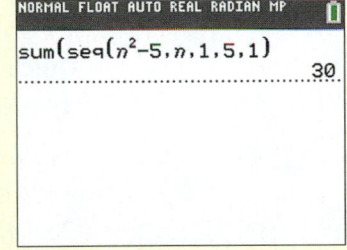

In Problems 81–88, use technology to find the sum.

81. $\displaystyle\sum_{i=1}^{4} (5i + 1)$

82. $\displaystyle\sum_{i=1}^{5} (3i + 2)$

83. $\displaystyle\sum_{i=1}^{5} \dfrac{i^2}{2}$

84. $\displaystyle\sum_{i=1}^{4} \dfrac{i^3}{2}$

85. $\displaystyle\sum_{k=1}^{3} 2^k$

86. $\displaystyle\sum_{k=1}^{4} 3^k$

87. $\displaystyle\sum_{k=1}^{5} [(-1)^{k+1} 2k]$

88. $\displaystyle\sum_{k=1}^{8} [(-1)^k k]$

13.2 Arithmetic Sequences

Objectives

1 Determine Whether a Sequence Is Arithmetic

2 Find a Formula for the nth Term of an Arithmetic Sequence

3 Find the Sum of an Arithmetic Sequence

Are You Prepared for This Section?

Before getting started, complete the following problems. If you get a problem wrong, go back to the section cited and review the material.

P1. Determine the slope of $y = -3x + 1$.   [Section 3.4 pp. 205–206]

P2. If $g(x) = 5x + 2$, find $g(3)$.   [Section 8.3, pp. 538–541]

P3. Solve: $\begin{cases} x - 3y = -17 \\ 2x + y = 1 \end{cases}$ [Section 4.2, pp. 260–263; Section 4.3, pp. 268–271]

In the last section, sequences in general were discussed. Now it's time to consider a specific type of sequence called an *arithmetic sequence.*

▶ **1** Determine Whether a Sequence Is Arithmetic

When the difference between successive terms of a sequence is always the same number, the sequence is called an **arithmetic sequence** (sometimes called an **arithmetic progression**). For example, the sequence

$$2, 6, 10, 14, \ldots$$

is arithmetic because the constant difference between consecutive terms is 4. If the first term is called a_1 and the **common difference** between consecutive terms is called d, then the terms of an arithmetic sequence follow the pattern

$$a_1, a_1 + d, a_1 + 2d, a_1 + 3d, \ldots$$

EXAMPLE 1   Determining Whether a Sequence Is Arithmetic

Determine whether the sequence $3, 9, 15, 21, \ldots$ is arithmetic. If it is, find the first term a_1 and the common difference d.

Solution

If the difference between consecutive terms is constant, then the sequence is arithmetic. The sequence $3, 9, 15, 21, \ldots$ is arithmetic because this difference is 6 ($= 9 - 3$ or $15 - 9$ or $21 - 15$). The first term is $a_1 = 3$, and the common difference is $d = 6$. ●

> ### Quick ✔
>
> **1.** In a(n) _____ sequence, the difference between consecutive terms is constant.
>
> *In Problems 2 and 3, determine which of the following sequences is arithmetic. If the sequence is arithmetic, determine the first term a_1 and common difference d.*
>
> **2.** $-3, -1, 1, 3, 5, \ldots$ **3.** $3, 9, 27, 81, \ldots$

EXAMPLE 2   Determining Whether a Sequence Is Arithmetic

Show that the sequence $\{s_n\} = \{2n + 7\}$ is arithmetic. Find the first term and the common difference.

Solution

Listing the first few terms and showing that the difference between consecutive terms is the same would be more of a demonstration than a proof. To prove the sequence is arithmetic, show that for *any* consecutive terms, the difference is the same number. This is done by evaluating the sequence at $(n - 1)$ and at n.

Then compute the difference between these values. If it is constant, the sequence is arithmetic.

$$s_{n-1} = 2(n-1) + 7 = 2n - 2 + 7 = 2n + 5 \quad \text{and} \quad s_n = 2n + 7$$

Now compute $s_n - s_{n-1}$.

$$s_n - s_{n-1} = (2n + 7) - (2n + 5)$$

Distribute: $= 2n + 7 - 2n - 5$

Combine like terms: $= 2$

The difference between *any* consecutive terms is 2, so the sequence is arithmetic with common difference $d = 2$. To find the first term, evaluate s_1 and find $s_1 = 2(1) + 7 = 9$, so $a_1 = 9$. ●

EXAMPLE 3 **Determining Whether a Sequence Defined by a Function Is Arithmetic**

Show that the sequence $\{b_n\} = \{n^2\}$ is not arithmetic.

Solution
As in Example 2, evaluate the sequence at $(n - 1)$ and n. Then show that the difference between consecutive terms is not constant.

$$b_{n-1} = (n - 1)^2 = n^2 - 2n + 1 \quad \text{and} \quad b_n = n^2$$

Now compute $b_n - b_{n-1}$.

$$b_n - b_{n-1} = n^2 - (n^2 - 2n + 1)$$

Distribute: $= n^2 - n^2 + 2n - 1$

Combine like terms: $= 2n - 1$

The difference between consecutive terms is not constant—its value depends on n. Therefore, the sequence is not arithmetic. ●

> **Quick ✓**
>
> *In Problems 4–6, determine whether the sequence is arithmetic. If it is, state the first term a_1 and the common difference d.*
>
> **4.** $\{a_n\} = \{3n - 8\}$ **5.** $\{b_n\} = \{n^2 - 1\}$ **6.** $\{c_n\} = \{5 - 2n\}$

▶ ❷ Find a Formula for the *n*th Term of an Arithmetic Sequence

Suppose a_1 is the first term of an arithmetic sequence whose common difference is d. Find a formula for a_n, the nth term of the sequence. To do this, list the first few terms of the sequence.

$$a_1 = a$$
$$a_2 = a_1 + d = a_1 + 1 \cdot d$$
$$a_3 = a_2 + d = (a_1 + d) + d = a_1 + 2d$$
$$a_4 = a_3 + d = (a_1 + 2d) + d = a_1 + 3d$$
$$a_5 = a_4 + d = (a_1 + 3d) + d = a_1 + 4d$$
$$\vdots$$
$$a_n = a_{n-1} + d = [a_1 + (n - 2)d] + d = a_1 + (n - 1)d$$

The *n*th Term of an Arithmetic Sequence

For an arithmetic sequence $\{a_n\}$ whose first term is a_1 and whose common difference is d, the nth term is determined by the formula

$$a_n = a_1 + (n - 1)d$$

EXAMPLE 4 **Finding a Formula for the *n*th Term of an Arithmetic Sequence**

(a) Write a formula for the *n*th term of an arithmetic sequence whose 4th term is 8 and whose common difference is -3.

(b) Find the 14th term of the sequence.

Solution

(a) Find a formula for the *n*th term of an arithmetic sequence. It is known that $d = -3$ and that $a_4 = 8$.

$$\text{nth term:} \quad a_n = a_1 + (n-1)d$$
$$d = -3; a_4 = 8: \quad a_4 = 8 = a_1 + (4-1)(-3)$$

Solve this equation for a_1, the first term of the sequence.

$$8 = a_1 - 9$$
$$\text{Add 9 to both sides:} \quad 17 = a_1$$

The formula for the *n*th term is

$$a_n = a_1 + (n-1)d: \quad a_n = 17 + (n-1)(-3)$$
$$\text{Distribute:} \quad = 17 - 3n + 3$$
$$\text{Combine like terms:} \quad = -3n + 20$$

(b) To find the 14th term, let $n = 14$ in $a_n = -3n + 20$.

$$a_{14} = -3(14) + 20$$
$$= -42 + 20$$
$$= -22$$

The 14th term in the sequence is -22. ●

Quick ✓

7. For an arithmetic sequence $\{a_n\}$ whose first term is a_1 and whose common difference is d, the *n*th term is determined by the formula _____.

8. (a) Write a formula for the *n*th term of an arithmetic sequence whose 5th term is 25 and whose common difference is 6.
   (b) Find the 14th term of the sequence.

EXAMPLE 5 **Finding a Formula for the *n*th Term of an Arithmetic Sequence**

The 4th term of an arithmetic sequence is 7, and the 10th term is 31.

(a) Find the first term and the common difference.

(b) Give a formula for the *n*th term of the sequence.

Solution

(a) The *n*th term of an arithmetic sequence is $a_n = a_1 + (n-1)d$, where a_1 is the first term and d is the common difference. Since $a_4 = 7$ and $a_{10} = 31$:

$$\begin{cases} a_4 = a_1 + (4-1)d \\ a_{10} = a_1 + (10-1)d \end{cases} \quad \text{or} \quad \begin{cases} 7 = a_1 + 3d \quad (1) \\ 31 = a_1 + 9d \quad (2) \end{cases}$$

This is a system of two linear equations with two variables, a and d. This system can be solved by elimination. Subtracting equation (2) from equation (1) yields

$$-24 = -6d$$
$$\text{Divide both sides by } -6: \quad 4 = d$$

Let $d = 4$ in equation (1) to find a_1.

$$7 = a_1 + 3(4)$$
$$7 = a_1 + 12$$

Subtract 12 from both sides: $\quad -5 = a_1$

The first term is $a_1 = -5$ and the common difference is $d = 4$.

(b) A formula for the nth term is

$$a_n = a_1 + (n-1)d$$
$a_1 = -5; d = 4: \quad = -5 + (n-1)(4)$

Distribute: $\quad = -5 + 4n - 4$

Combine like terms: $\quad = 4n - 9$ ●

> ### Quick ✓
>
> **9.** The 5th term of an arithmetic sequence is 7, and the 13th term is 31.
>
> **(a)** Find the first term and the common difference.
>
> **(b)** Give a formula for the nth term of the sequence.

▶ ❸ Find the Sum of an Arithmetic Sequence

The next result gives a formula for finding the sum of the first n terms of an arithmetic sequence.

Work Smart

There are two formulas for finding the sum of the first n terms of an arithmetic sequence. Use Formula (1) if you know n, the first term a_1, and the common difference d; use Formula (2) if you know n, the first term a_1, and the last term a_n.

> **Sum of the First n Terms of an Arithmetic Sequence**
>
> Let $\{a_n\}$ be an arithmetic sequence with first term a_1 and common difference d. The sum S_n of the first n terms of $\{a_n\}$ is
>
> $$S_n = \frac{n}{2}[2a_1 + (n-1)d] \quad (1) \quad \text{or} \quad S_n = \frac{n}{2}[a_1 + a_n] \quad (2)$$

The following shows where these results come from.

$$S_n = a_1 + a_2 + a_3 + \cdots + a_n$$
$$= \underbrace{a_1}_{a_1} + \underbrace{(a_1 + d)}_{a_2} + \underbrace{(a_1 + 2d)}_{a_3} + \cdots + \underbrace{(a_1 + (n-1)d)}_{a_n}$$

The sequence S_n can also be represented by reversing the order in which the terms are added, so that

$$S_n = a_n + a_{n-1} + \cdots + a_1$$
$$= \underbrace{(a_1 + (n-1)d)}_{a_n} + \underbrace{(a_1 + (n-2)d)}_{a_{n-1}} + \cdots + \underbrace{a_1}_{a_1}$$

Add these two different representations of S_n as follows:

$$
\begin{array}{lllllll}
S_n = & a_1 & + (a_1 + d) & + (a_1 + 2d) & + \cdots + (a_1 + (n-2)d) & + (a_1 + (n-1)d) \\
S_n = & (a_1 + (n-1)d) & + (a_1 + (n-2)d) & + (a_1 + (n-3)d) & + \cdots + \quad (a_1 + d) & + \quad a_1 \\
\hline
2S_n = & 2a_1 + (n-1)d & + 2a_1 + (n-1)d & + 2a_1 + (n-1)d & + \cdots + 2a_1 + (n-1)d & + 2a_1 + (n-1)d
\end{array}
$$

Thus $2S_n$ is $2a_1 + (n-1)d$ added to itself n times, or $n[2a_1 + (n-1)d]$. Therefore,

$$2S_n = n[2a_1 + (n-1)d]$$

Divide both sides by 2: $\quad S_n = \dfrac{n}{2}[2a_1 + (n-1)d] \quad$ Formula (1)

This is the first formula in the box. If the expression $2a_1 + (n - 1)d$ is rewritten as $a_1 + a_1 + (n - 1)d$, notice that $a_1 + (n - 1)d$ is a_n.

$$S_n = \frac{n}{2}[a_1 + a_n] \quad \text{Formula (2)}$$

There are two ways to find the sum of the first n terms of an arithmetic sequence. Notice that Formula (1) involves the first term a_1 and the common difference d, whereas Formula (2) involves the first term a_1 and the last term a_n. Use whichever is easier.

EXAMPLE 6   Finding the Sum of the First n Terms of an Arithmetic Sequence

Find the sum S_n of the first 50 terms of the arithmetic sequence 2, 5, 8, 11, ...

Solution

Because the first term is $a_1 = 2$ and the common difference is $d = 5 - 2 = 3$, use the formula $S_n = \frac{n}{2}[2a_1 + (n - 1)d]$ to find the sum.

$$S_n = \frac{n}{2}[2a_1 + (n - 1)d]$$

$$n = 50, a_1 = 2, d = 3: \quad S_{50} = \frac{50}{2}[2(2) + (50 - 1)(3)]$$

$$= 25[4 + 49(3)]$$

$$= 25[151]$$

$$= 3775$$

The sum of the first 50 terms, S_{50}, of the arithmetic sequence 2, 5, 8, 11, . . . is 3775. ●

Quick ✓

10. Find the sum S_n of the first 100 terms of the arithmetic sequence whose first term is 5 and whose common difference is 2.

11. Find the sum S_n of the first 70 terms of the arithmetic sequence 1, 5, 9, 13, ...

EXAMPLE 7   Finding the Sum of the First n Terms of an Arithmetic Sequence

Find the sum S_n of the first 40 terms of the arithmetic sequence $\{-2n + 50\}$.

Solution

The first term is $a_1 = -2(1) + 50 = 48$ and the 40th term is $a_{40} = -2(40) + 50 = -30$. Because the first term and the last term of the sequence are known, use the formula $S_n = \frac{n}{2}[a_1 + a_n]$ to find the sum.

$$S_n = \frac{n}{2}[a_1 + a_n]$$

$$n = 40, a_1 = 48, a_{40} = -30: \quad S_{40} = \frac{40}{2}[48 + (-30)]$$

$$= 360$$

The sum of the first 40 terms, S_{40}, of the arithmetic sequence $\{-2n + 50\}$ is 360. ●

Quick ✓

12. Find the sum S_n of the first 50 terms of the arithmetic sequence whose first term is 4 and 50th term 298.

13. Find the sum S_n of the first 75 terms of the arithmetic sequence $\{-3n + 100\}$.

EXAMPLE 8 **Creating a Floor Design**

A ceramic tile floor is designed in the shape of a trapezoid 10 feet wide at the base and 5 feet wide at the top. See Figure 5. The tiles, which measure 6 inches by 6 inches, are to be placed so that each successive row contains one fewer tile than the preceding row. How many tiles are required?

Figure 5

Solution
Each tile is 6 inches (0.5 foot) wide, and the bottom row is 10 feet wide, so the bottom row requires 20 tiles. Similar reasoning tells us the top row requires 10 tiles. Because each successive row has one fewer tile, the total number of tiles required is

$$S = 20 + 19 + 18 + \cdots + 11 + 10$$

This is the sum of an arithmetic sequence with common difference -1. The number of terms to be added is $n = 11$, the first term is $a_1 = 20$, and the last term is $a_{11} = 10$. The sum S is

$$S_n = \frac{n}{2}[a_1 + a_n] \qquad S_{11} = \frac{11}{2}(20 + 10) = 165$$

In all, 165 tiles will be required. ●

Quick ✔

14. In the corner section of a theater, the first row has 20 seats. Each subsequent row has 2 more seats, and there are a total of 30 rows. How many seats are in this section?

13.2 Exercises MyMathLab® Exercise numbers in **green** have complete video solutions in MyMathLab or may be accessed using the QR code to the right.

Problems 1–14 are the **Quick ✔** *s that follow the* **EXAMPLES**.

Building Skills

In Problems 15–22, verify that the given sequence is arithmetic. Then find the first term and common difference. See Objective 1.

15. $\{n + 5\}$ 16. $\{n - 1\}$ 17. $\{7n + 2\}$

18. $\{10n + 1\}$ 19. $\{7 - 3n\}$ 20. $\{5 - 2n\}$

21. $\left\{\frac{1}{2}n + 5\right\}$ 22. $\left\{\frac{1}{4}n + \frac{3}{4}\right\}$

In Problems 23–30, find a formula for the nth term of the arithmetic sequence whose first term a_1 and common difference d are given. What is the 5th term? See Objective 2.

23. $a_1 = 4; d = 3$ 24. $a_1 = 8; d = 3$

25. $a_1 = 10; d = -5$ 26. $a_1 = 12; d = -3$

27. $a_1 = 2; d = \frac{1}{3}$ 28. $a_1 = -3; d = \frac{1}{2}$

29. $a_1 = 5; d = -\frac{1}{5}$ 30. $a_1 = -\frac{4}{3}; d = -\frac{2}{3}$

In Problems 31–36, write a formula for the nth term of each arithmetic sequence. Use the formula to find the 20th term in each arithmetic sequence. See Objective 2.

31. 2, 7, 12, 17, . . .   **32.** −5, −1, 3, 7, . . .

33. 12, 9, 6, 3, . . .   **34.** 20, 14, 8, 2, . . .

35. $1, \dfrac{5}{4}, \dfrac{3}{2}, \dfrac{7}{4}, \ldots$ **36.** $10, \dfrac{19}{2}, 9, \dfrac{17}{2}, \ldots$

In Problems 37–44, find the first term and the common difference of the arithmetic sequence described. Give a formula for the nth term of the sequence. See Objective 2.

37. 3rd term is 17; 7th term is 37

38. 5th term is 7; 9th term is 19

39. 4th term is −2; 8th term is 26

40. 2nd term is −9; 8th term is 15

41. 5th term is −1; 12th term is −22

42. 6th term is −8; 12th term is −38

43. 3rd term is 3; 9th term is 0

44. 5th term is 5; 13th term is 7

For Problems 45–56, see Objective 3.

45. Find the sum of the first 30 terms of the sequence 2, 8, 14, 20, ...

46. Find the sum of the first 40 terms of the sequence 1, 8, 15, 22, ...

47. Find the sum of the first 25 terms of the sequence −8, −5, −2, 1, ...

48. Find the sum of the first 75 terms of the sequence −9, −5, −1, 3, ...

49. Find the sum of the first 40 terms of the sequence 10, 3, −4, −11, ...

50. Find the sum of the first 50 terms of the sequence 12, 4, −4, −12, ...

51. Find the sum of the first 40 terms of the arithmetic sequence $\{4n - 3\}$.

52. Find the sum of the first 80 terms of the arithmetic sequence $\{2n - 13\}$.

53. Find the sum of the first 75 terms of the arithmetic sequence $\{-5n + 70\}$.

54. Find the sum of the first 35 terms of the arithmetic sequence $\{-6n + 25\}$.

55. Find the sum of the first 30 terms of the arithmetic sequence $\left\{5 + \dfrac{2}{3}n\right\}$.

56. Find the sum of the first 28 terms of the arithmetic sequence $\left\{7 - \dfrac{3}{2}n\right\}$.

Applying the Concepts

57. Find x so that $x + 3, 2x + 1$, and $5x + 2$ are consecutive terms of an arithmetic sequence.

58. Find x so that $2x, 3x + 2$, and $5x + 3$ are consecutive terms of an arithmetic sequence.

59. A Stack of Cans Suppose that the bottom row in a stack of cans contains 35 cans. Each layer contains one can fewer than the layer below it. The top row has 1 can. How many cans are in the stack?

60. A Pile of Bricks Suppose that the bottom row in a pile of bricks contains 46 bricks. Each layer contains 2 bricks fewer than the layer below it. The top row has 2 bricks. How many bricks are in the stack?

61. The Theater An auditorium has 40 seats in the first row and 25 rows in all. Each successive row contains 2 additional seats. How many seats are in the auditorium?

62. Mosaic A mosaic is designed in the shape of an equilateral triangle, 20 feet on each side. Each tile in the mosaic is in the shape of an equilateral triangle, 12 inches to a side. The tiles are to alternate in color as shown in the illustration. How many tiles of each color will be required?

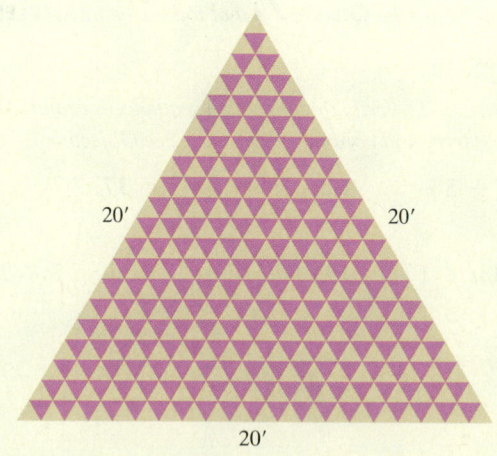

Extending the Concepts

In Problems 63–66, determine the number of terms that are in each arithmetic sequence.

63. $-5, -2, 1, \ldots, 244$

64. $-9, -5, -1, \ldots, 219$

65. $108, 101, 94, \ldots, -326$

66. $99, 93, 87, \ldots, -339$

67. **Salary** Suppose you have just been hired at the starting salary of $32,000 per year. Your contract guarantees you a $2500 raise each year. How many years will it take before your aggregate salary is $757,500? (*Hint:* Your aggregate salary is $32,000 + (\$32,000 + \$2500) + \cdots.$)

68. **Stadium Construction** How many rows are in the corner section of a stadium containing 2030 seats if the first row has 14 seats and each successive row has 4 additional seats?

Explaining the Concepts

69. Explain how you can determine whether a sequence is arithmetic.

70. Provide an explanation that justifies the formula for the sum of the first n terms of an arithmetic sequence.

Synthesis Review

In Problems 71–73, (a) determine the base a of the exponential function, and (b) compute $f(1), f(2), f(3),$ and $f(4)$.

71. $f(x) = 3^x$

72. $f(x) = 4^x$

73. $f(x) = 10\left(\dfrac{1}{2}\right)^x$

74. Compute $\dfrac{f(2)}{f(1)}, \dfrac{f(3)}{f(2)},$ and $\dfrac{f(4)}{f(3)}$ for each of the functions in Problems 71–73. How is the ratio in the value of the function for consecutive values of the independent variable related to the base?

Technology Exercises

In Problems 75–78, use technology to find the sum of each sequence.

75. $\{3.45n + 4.12\}; n = 20$

76. $\{2.67n - 1.23\}; n = 25$

77. $85.9 + 83.5 + 81.1 + \cdots; n = 25$

78. $-11.8 + (-8.2) + (-4.6) + \cdots; n = 30$

13.3 Geometric Sequences and Series

Objectives

1. Determine Whether a Sequence Is Geometric
2. Find a Formula for the nth Term of a Geometric Sequence
3. Find the Sum of a Geometric Sequence
4. Find the Sum of a Geometric Series
5. Solve Annuity Problems

Are You Prepared for This Section?

Before getting started, complete the following problems. If you get a problem wrong, go back to the section cited and review the material.

P1. If $g(x) = 4^x$, evaluate $g(1), g(2),$ and $g(3)$. [Section 11.2, pp. 792–795]

P2. Simplify: $\dfrac{x^4}{x^3}$ [Section 5.4, pp. 334–335]

1 Determine Whether a Sequence Is Geometric

The *difference* between consecutive terms in an arithmetic sequence is constant. If the *ratio* of consecutive terms in a sequence is constant, then the sequence is a **geometric sequence.** For example, the sequence

$$2, 4, 8, 16, \ldots$$

is geometric because the ratio of consecutive terms is $2 \left(= \dfrac{4}{2} = \dfrac{8}{4} = \dfrac{16}{8} \right)$. The first term is called a_1 and the **common ratio** of consecutive terms is called r. The terms of a geometric sequence follow the pattern

$$a_1, \ a_1 r, \ a_1 r^2, \ a_1 r^3, \ldots$$

Prepared?...Answers

P1. $g(1) = 4, g(2) = 16, g(3) = 64$

P2. x

EXAMPLE 1 **Determining Whether a Sequence Is Geometric**

Determine whether the sequence $2, 6, 18, 54, 162, \ldots$ is geometric. If it is, state the first term and the common ratio.

Solution

The sequence is geometric if the ratio of consecutive terms is a constant. Because the ratio of consecutive terms is 3 $\left(= \dfrac{6}{2} = \dfrac{18}{6} = \dfrac{54}{18} = \dfrac{162}{54} \right)$, the sequence is geometric.

The first term is $a_1 = 2$ and the common ratio is $r = 3$. ●

Quick ✓

1. In a(n) _____ sequence the ratio of consecutive terms is constant.

In Problems 2–4, determine whether the sequence is geometric. If it is, state the first term and common ratio.

2. $4, 8, 16, 32, 64, \ldots$ **3.** $5, 10, 16, 23, 31, \ldots$ **4.** $9, 3, 1, \dfrac{1}{3}, \dfrac{1}{9}, \ldots$

EXAMPLE 2 **Determining Whether a Sequence Is Geometric**

Determine whether the sequence $\{b_n\} = \{3^{-n}\}$ is geometric. If it is, find the common ratio.

Solution

Show that for *any* consecutive terms, the ratio is the same number. Do this by evaluating the sequence at $(n - 1)$ and n. Then compute the ratio of these two values. If the ratio is constant, the sequence is geometric.

$$b_{n-1} = 3^{-(n-1)} = 3^{-n+1} \quad \text{and} \quad b_n = 3^{-n}$$

Work Smart

Notice that the technique used in Example 2 is similar to the approach used in Section 13.2 to show a sequence is arithmetic.

Now compute $\dfrac{b_n}{b_{n-1}}$.

$$\frac{b_n}{b_{n-1}} = \frac{3^{-n}}{3^{-n+1}}$$

$a^{m+n} = a^m \cdot a^n: \quad = \dfrac{3^{-n}}{3^{-n} \cdot 3^1}$

Divide out like factors: $\quad = \dfrac{1}{3}$

The ratio of *any* two consecutive terms is the constant $\dfrac{1}{3}$, so the sequence is geometric with $r = \dfrac{1}{3}$. ●

EXAMPLE 3 **Determining Whether a Sequence Is Geometric**

Determine whether the sequence $\{a_n\} = \{n^2 + 1\}$ is geometric. If it is, find the common ratio.

Solution

If the ratio $\dfrac{a_n}{a_{n-1}}$ is constant, then the sequence is geometric.

$$\frac{a_n}{a_{n-1}} = \frac{n^2 + 1}{(n-1)^2 + 1}$$

$$\text{Use FOIL:} \quad = \frac{n^2 + 1}{n^2 - 2n + 1 + 1}$$

$$\text{Combine like terms:} \quad = \frac{n^2 + 1}{n^2 - 2n + 2}$$

The rational expression cannot be simplified any further. Because the ratio $\dfrac{a_n}{a_{n-1}}$ depends on the value of n, it is not constant. Therefore, the sequence is not geometric. ●

Quick ✓

In Problems 5–7, determine whether the sequence is geometric. If it is geometric, find the common ratio.

5. $\{a_n\} = \{5^n\}$ **6.** $\{b_n\} = \{n^2\}$

7. $\{c_n\} = \left\{ 5\left(\dfrac{2}{3}\right)^n \right\}$

▶ ❷ Find a Formula for the *n*th Term of a Geometric Sequence

If a_1 is the first term of a geometric sequence with common ratio $r \neq 0$, what is a formula for the *n*th term of a_n? Let's list the first few terms to find out.

$$a_1 = 1a_1 = a_1 r^0$$
$$a_2 = ra_1 = a_1 r^1$$
$$a_3 = ra_2 = a_1 r^2$$
$$a_4 = ra_3 = a_1 r^3$$
$$a_5 = ra_4 = a_1 r^4$$
$$\vdots$$
$$a_n = ra_{n-1} = r(a_1 r^{n-2}) = a_1 r^{n-1}$$

This leads to the following result.

> **The *n*th Term of a Geometric Sequence**
>
> If a geometric sequence $\{a_n\}$ has first term a_1 and common ratio r, then the *n*th term is determined by the formula
>
> $$a_n = a_1 r^{n-1} \qquad r \neq 0$$

EXAMPLE 4   ### Finding a Particular Term of a Geometric Sequence

Consider the geometric sequence $8, 6, \dfrac{9}{2}, \dfrac{27}{8}, \ldots$.

(a) Find a formula for the *n*th term.

(b) Find the 11th term of the sequence.

(continued)

Solution

(a) The sequence is geometric. The first term is $a_1 = 8$. The common ratio, r, is the ratio of any two consecutive terms. So

$$r = \frac{6}{8} = \frac{\frac{9}{2}}{6} = \frac{\frac{27}{8}}{\frac{9}{2}} = \frac{3}{4}.$$ Now substitute these values in the formula for

the nth term of a geometric sequence.

$$a_n = a_1 r^{n-1}$$

$$a_1 = 8; r = \frac{3}{4}: \quad = 8\left(\frac{3}{4}\right)^{n-1}$$

(b) Find the eleventh term by letting $n = 11$ in the formula found in part (a).

$$a_n = 8\left(\frac{3}{4}\right)^{n-1}$$

$$a_{11} = 8\left(\frac{3}{4}\right)^{11-1}$$

$$= 8\left(\frac{3}{4}\right)^{10}$$

$$\approx 0.45051$$

Quick ✓

In Problems 8 and 9, find a formula for the nth term of each geometric sequence. Use this result to find the 9th term of the sequence.

8. $a_1 = 5, r = 2$ **9.** $\{50, 25, 12.5, 6.25, \ldots\}$

▶ ❸ Find the Sum of a Geometric Sequence

The next result gives us a formula for finding the sum of the first n terms of a geometric sequence.

Sum of the First n Terms of a Geometric Sequence

Let $\{a_n\}$ be a geometric sequence with first term a_1 and common ratio r, where $r \neq 0, r \neq 1$. The sum S_n of the first n terms of $\{a_n\}$ is

$$S_n = a_1 \cdot \frac{1 - r^n}{1 - r} \quad r \neq 0, r \neq 1$$

Where does this formula come from? The sum S_n of the first n terms of $\{a_n\} = \{a_1 r^{n-1}\}$ is

$$S_n = a_1 + a_1 r + a_1 r^2 + \cdots + a_1 r^{n-1}$$

Multiply both sides by r to obtain

$$rS_n = a_1 r + a_1 r^2 + a_1 r^3 + \cdots + a_1 r^n$$

Now subtract rS_n from S_n:

$$S_n = a_1 + a_1 r + a_1 r^2 + \cdots + a_1 r^{n-1}$$
$$\underline{rS_n = \qquad a_1 r + a_1 r^2 + a_1 r^3 + \cdots + a_1 r^n}$$
$$S_n - rS_n = a_1 - a_1 r^n$$

Factor S_n from the expression on the left and factor a_1 from the expression on the right.

$$S_n(1 - r) = a_1(1 - r^n)$$

Divide both sides by $1 - r$ (since $r \neq 1$) and solve for S_n.

$$S_n = a_1 \cdot \frac{1 - r^n}{1 - r}$$

EXAMPLE 5 **Finding the Sum of the First *n* Terms of a Geometric Sequence**

Find the sum of the first 10 terms of the sequence $2, 6, 18, 54, \ldots$.

Solution

Find the sum of the first 10 terms of a geometric sequence with $a_1 = 2$ and common ratio $r = 3$.

$$S_n = a_1 \cdot \frac{1 - r^n}{1 - r}$$

$$a_1 = 2, r = 3, n = 10: \quad S_{10} = 2 \cdot \frac{1 - 3^{10}}{1 - 3}$$

$$= 2 \cdot \frac{-59{,}048}{-2}$$

$$= 59{,}048 \qquad \bullet$$

EXAMPLE 6 **Finding the Sum of the First *n* Terms of a Geometric Sequence in Summation Notation**

Find the sum: $\displaystyle\sum_{n=1}^{8}\left[5\left(\frac{1}{2}\right)^n\right]$

Express your answer to as many decimal places as your calculator allows.

Solution

Expand this sum as follows:

$$\sum_{n=1}^{8}\left[5\left(\frac{1}{2}\right)^n\right] = \frac{5}{2} + \frac{5}{4} + \cdots + \frac{5}{256}$$

Work Smart

$$\sum_{n=1}^{8} 5\left(\frac{1}{2}\right)^n$$

$$= 5\left(\frac{1}{2}\right)^1 + 5\left(\frac{1}{2}\right)^2 + 5\left(\frac{1}{2}\right)^3 + 5\left(\frac{1}{2}\right)^4$$

$$+ \cdots + 5\left(\frac{1}{2}\right)^8$$

Find the sum of the first $n = 8$ terms of a geometric sequence with $a_1 = \dfrac{5}{2}$ and common ratio $r = \dfrac{1}{2}$.

$$S_n = a_1 \cdot \frac{1 - r^n}{1 - r}$$

$$a_1 = \frac{5}{2}, r = \frac{1}{2}, n = 8: \quad S_8 = \frac{5}{2} \cdot \frac{1 - \left(\dfrac{1}{2}\right)^8}{1 - \dfrac{1}{2}}$$

$$= \frac{5}{2} \cdot \frac{0.99609375}{\dfrac{1}{2}}$$

$$= 4.98046875 \qquad \bullet$$

Quick ✓

10. For a geometric sequence with first term a_1 and common ratio r, where $r \neq 0, r \neq 1$, the sum of the first n terms is _____.

In Problems 11 and 12, find the sum.

11. $3 + 6 + 12 + 24 + \cdots + 3 \cdot 2^{12}$

12. $\displaystyle\sum_{n=1}^{10} \left[8 \left(\frac{1}{2} \right)^n \right]$

▶ ❹ Find the Sum of a Geometric Series

An infinite sum of the form

$$a_1 + a_1 r + a_1 r^2 + \cdots + a_1 r^{n-1} + \cdots$$

whose first term is a_1 and whose common ratio is r, is called an **infinite geometric series** and is denoted by

$$\sum_{n=1}^{\infty} a_1 r^{n-1}$$

> **In Other Words**
> A sequence is a list of terms. A series is the sum of an infinite number of terms.

The sum of the first n terms of a geometric sequence is given by the formula

$$S_n = a_1 \cdot \frac{1 - r^n}{1 - r}$$

Distribute the a_1:
$$= \frac{a_1 - a_1 r^n}{1 - r}$$

Divide each term by $1 - r$:
$$= \frac{a_1}{1 - r} - \frac{a_1 r^n}{1 - r}$$

As n gets larger and larger, the expression S_n approaches the value $\dfrac{a_1}{1 - r}$ because r^n approaches 0 provided that $-1 < r < 1$.

> **Sum of a Geometric Series**
>
> If $-1 < r < 1$, the sum of the terms of an infinite geometric series with first term a_1 and common ratio r is
>
> $$\sum_{n=1}^{\infty} a_1 r^{n-1} = \frac{a_1}{1 - r}$$

EXAMPLE 7 **Finding the Sum of a Geometric Series**

Find the sum of the geometric series: $4 + 2 + 1 + \dfrac{1}{2} + \cdots$

Solution

This is a geometric series with $a_1 = 4$ and common ratio $r = \dfrac{1}{2}$ $\left(= \dfrac{2}{4} = \dfrac{1}{2} = \dfrac{\frac{1}{2}}{1} \right)$.

Since the common ratio r is between -1 and 1, use the formula for the sum of a geometric series to find that

$$4 + 2 + 1 + \frac{1}{2} + \cdots = \frac{a_1}{1 - r} = \frac{4}{1 - \dfrac{1}{2}} = 8$$

●

Quick ✔

13. If $-1 < r < 1$, then the sum of the geometric series $\displaystyle\sum_{n=1}^{\infty} a_1 r^{n-1} =$ _____.

In Problems 14 and 15, find the sum of the geometric series.

14. $10 + \dfrac{5}{2} + \dfrac{5}{8} + \dfrac{5}{32} + \cdots$

15. $\displaystyle\sum_{n=1}^{\infty} \left(\dfrac{1}{3}\right)^n$

Work Smart: Study Skills

Let's summarize the formulas for arithmetic and geometric sequences where a_1 is the first term of the sequence, d is the common difference, and r is the common ratio.

	Arithmetic	Geometric
nth term	$a_n = a_1 + (n-1)d$	$a_n = a_1 r^{n-1}, r \neq 0$
Sum of the first n terms	$S_n = \dfrac{n}{2}[2a_1 + (n-1)d]$ $= \dfrac{n}{2}(a_1 + a_n)$	$S_n = a_1 \cdot \dfrac{1 - r^n}{1 - r}, r \neq 0, r \neq 1$

If you were asked to find the 5th term of the sequence $5, 20, 80, 320, \dots$, which formula would you use?

If you were asked to find the sum of the first 5 terms of the sequence $5, 9, 13, 17, \dots$, which formula would you use?

EXAMPLE 8 **Writing a Repeating Decimal as a Fraction**

Express $0.\overline{1}$ as a fraction in lowest terms.

Solution

The line over the 1 indicates that the 1 repeats indefinitely. That is,

$$0.\overline{1} = 0.11111\dots$$
$$= 0.1 + 0.01 + 0.001 + \dots$$

This is a geometric series with $a_1 = 0.1$ and common ratio $r = 0.1 \left(= \dfrac{0.01}{0.1} = \dfrac{0.001}{0.01}\right)$.

Since $-1 < r < 1$, use the formula for the sum of a geometric series to find that

$$0.\overline{1} = 0.1 + 0.01 + 0.001 + \cdots$$

$a_1 = 0.1, r = 0.1$ in $\dfrac{a_1}{1 - r}$:
$$= \dfrac{0.1}{1 - 0.1}$$
$$= \dfrac{0.1}{0.9}$$

Multiply numerator and denominator by 10:
$$= \dfrac{1}{9}$$

Thus $0.\overline{1} = \dfrac{1}{9}$.

Quick ✔

16. Express $0.\overline{2}$ as a fraction in lowest terms.

EXAMPLE 9 **The Multiplier**

Suppose that Americans spend 90% of every additional dollar they earn. Economists would say that an individual's **marginal propensity to consume** is 0.90. For example, if Roberta earns a dollar, she will spend $0.9(\$1) = \$0.90$ of it and save \$0.10. Whoever earns \$0.90 (from Roberta) will spend 90% of it, or $0.9(\$0.90) = \$0.81$. This process of spending continues and results in a geometric series as follows:

$$\$1 + \$0.90 + \$0.81 + \$0.729 + \cdots$$

The sum of this geometric series is called the **multiplier.** The multiplier is used to determine the total impact of spending on the U.S. economy. Suppose the government gives a child-tax rebate of \$500 to Roberta. Determine the impact of the rebate on the U.S. economy if Americans spend 90% of every dollar they earn.

Solution

The total impact of the \$500 tax rebate on the U.S. economy is

$$\$500 + \$500(0.9) + \$500(0.9)^2 + \$500(0.9)^3 + \cdots$$

This is a geometric series with first term $a_1 = 500$ and common ratio $r = 0.9$. The sum of this series is

$$\$500 + \$500(0.9) + \$500(0.9)^2 + \$500(0.9)^3 + \cdots = \frac{\$500}{1 - 0.9}$$

$$= \$5000$$

The United States economy will grow by \$5000 because of the child-tax credit to Roberta.

●

Quick ✓

17. Redo Example 9 if the marginal propensity to consume is 95%.

▶ ⑤ Solve Annuity Problems

Section 11.2 introduced the compound interest formula, which lets us calculate the future value of a lump sum of money that is deposited in an account that pays interest compounded periodically. Often, though, money is invested at periodic intervals of time. An **annuity** is a sequence of equal periodic deposits that may be made annually, quarterly, monthly, or daily.

When deposits are made at the same time the interest is credited, the annuity is called **ordinary.** Only ordinary annuities will be discussed here. The **amount of an annuity** is the sum of all deposits made plus all interest paid.

Suppose the interest an account earns is i percent per payment period (expressed as a decimal). For example, if an account pays 6% compounded monthly (12 times a year), then $i = \dfrac{0.06}{12} = 0.005$. Let's develop a formula for the amount of an annuity.

Suppose $\$P$ is deposited each payment period for n payment periods in an account that earns i% per payment period. When the last deposit is made at the nth payment period, the first deposit has earned interest compounded for $n - 1$ payment periods, the second deposit of $\$P$ has earned interest compounded for $n - 2$ payment periods, and so on. Table 1 shows the value of each deposit after n deposits have been made.

Table 1

Deposit	1	2	3	$n - 1$	n
Future Value of Deposit of P	$P(1 + i)^{n-1}$	$P(1 + i)^{n-2}$	$P(1 + i)^{n-3}$	$P(1 + i)$	P

Work Smart

The common ratio is $1+i$ because

$$\frac{(1 + i)^{n-2}}{(1 + i)^{n-3}} = (1 + i)^{n-2-(n-3)}$$

$$= (1 + i)^{n-2-n+3}$$

$$= (1 + i)^1$$

$$= 1 + i$$

The amount A of the annuity is the sum of the amounts shown in Table 1, namely,

$$A = P(1 + i)^{n-1} + P(1 + i)^{n-2} + P(1 + i)^{n-3} + \cdots + P(1 + i) + P$$

$$= P[(1 + i)^{n-1} + (1 + i)^{n-2} + (1 + i)^{n-3} + \cdots + (1 + i) + 1]$$

$$= P[1 + (1 + i) + \cdots + (1 + i)^{n-3} + (1 + i)^{n-2} + (1 + i)^{n-1}]$$

The expression is the sum of a geometric sequence with n terms, first term $a_1 = P$, and a common ratio $r = (1 + i)$. Using $S_n = a_1 \dfrac{1 - r^n}{1 - r}$, gives

$$A = P[1 + (1 + i) + \cdots + (1 + i)^{n-3} + (1 + i)^{n-2} + (1 + i)^{n-1}]$$

$$= P \cdot \frac{1 - (1 + i)^n}{1 - (1 + i)} = P \cdot \frac{1 - (1 + i)^n}{-i} = P \cdot \frac{(1 + i)^n - 1}{i}$$

We have the following result.

> **Amount of an Annuity**
>
> If P represents the deposit in dollars made at each payment period for an annuity at i percent interest per payment period, then the amount A of the annuity after n payment periods is
>
> $$A = P \cdot \frac{(1 + i)^n - 1}{i}$$

EXAMPLE 10 **Determining the Amount of an Annuity**

To save for retirement, Alejandro decides to put $100 into a Roth Individual Retirement Account (IRA) every month for the next 30 years. What will be the value of the IRA after 30 years if his account earns 6% interest compounded monthly?

Solution

This is an ordinary annuity with $n = 12 \cdot 30 = 360$ payments and deposits of $P = \$100$. The rate of interest per payment period is $i = \dfrac{0.06}{12} = 0.005$. The amount after 30 years (360 deposits) is

$$A = \$100 \left[\frac{(1 + 0.005)^{360} - 1}{0.005} \right]$$

$$= \$100 [1004.515042]$$

$$= \$100{,}451.50$$ ●

Quick ✓

18. To save for retirement, Magglio decides to place $500 into a Roth Individual Retirement Account (IRA) every quarter (every three months) for the next 30 years. What will be the value of the IRA after 30 years if his account earns 8% interest compounded quarterly?

13.3 Exercises MyMathLab® Exercise numbers in green have complete video solutions in MyMathLab or may be accessed using the QR code to the right.

*Problems **1–18** are the Quick ✓s that follow the **EXAMPLES**.*

Building Skills

In Problems 19–26, show that the given sequence is a geometric sequence. Then find the first term and common ratio. See Objective 1.

19. $\{4^n\}$

20. $\{(-2)^n\}$

21. $\left\{\left(\dfrac{2}{3}\right)^n\right\}$

22. $\left\{\dfrac{2^n}{3}\right\}$

23. $\{3 \cdot 2^{-n}\}$

24. $\left\{-10\left(\dfrac{1}{2}\right)^n\right\}$

25. $\left\{\dfrac{5^{n-1}}{2^n}\right\}$

26. $\left\{\dfrac{3^{-n}}{2^{n-1}}\right\}$

In Problems 27–34, (a) find a formula for the nth term of the geometric sequence whose first term and common ratio are given, and (b) use the formula to find the 8th term. See Objective 2.

27. $a_1 = 10, r = 2$

28. $a_1 = 2, r = 3$

29. $a_1 = 100, r = \dfrac{1}{2}$

30. $a_1 = 30, r = \dfrac{1}{3}$

31. $a_1 = 1, r = -3$

32. $a_1 = 1, r = -4$

33. $a_1 = 100, r = 1.05$

34. $a_1 = 500, r = 1.04$

In Problems 35–40, find the indicated term of each geometric sequence. See Objective 2.

35. 10th term of $3, 6, 12, 24, \ldots$

36. 12th term of $1, 3, 9, 27, \ldots$

37. 15th term of $4, -2, 1, -\dfrac{1}{2}, \ldots$

38. 8th term of $10, -20, 40, -80, \ldots$

39. 9th term of $0.5, 0.05, 0.005, 0.0005, \ldots$

40. 10th term of $0.4, 0.04, 0.004, 0.0004, \ldots$

In Problems 41–48, find the sum. If necessary, express your answer to as many decimal places as your calculator allows. See Objective 3.

41. $2 + 4 + 8 + \cdots + 2^{12}$

42. $3 + 9 + 27 + \cdots + 3^{10}$

43. $50 + 20 + 8 + \dfrac{16}{5} + \cdots + 50\left(\dfrac{2}{5}\right)^{10-1}$

44. $10 + 5 + \dfrac{5}{2} + \cdots + 10\left(\dfrac{1}{2}\right)^{12-1}$

45. $\displaystyle\sum_{n=1}^{10} [3 \cdot 2^n]$

46. $\displaystyle\sum_{n=1}^{12} [5 \cdot 2^n]$

47. $\displaystyle\sum_{n=1}^{8} \left[\dfrac{4}{2^{n-1}}\right]$

48. $\displaystyle\sum_{n=1}^{14} \left[10\left(\dfrac{1}{2}\right)^{n-1}\right]$

In Problems 49–58, find the sum of each geometric series. If necessary, express your answer to as many decimal places as your calculator allows. See Objective 4.

49. $1 + \dfrac{1}{2} + \dfrac{1}{4} + \cdots$

50. $1 + \dfrac{1}{3} + \dfrac{1}{9} + \cdots$

51. $10 + \dfrac{10}{3} + \dfrac{10}{9} + \cdots$

52. $20 + 5 + \dfrac{5}{4} + \cdots$

53. $6 - 2 + \dfrac{2}{3} - \dfrac{2}{9} + \cdots$

54. $12 - 3 + \dfrac{3}{4} - \dfrac{3}{16} + \cdots$

55. $\displaystyle\sum_{n=1}^{\infty} \left(5\left(\dfrac{1}{5}\right)^n\right)$

56. $\displaystyle\sum_{n=1}^{\infty} \left(10\left(\dfrac{1}{3}\right)^n\right)$

57. $\displaystyle\sum_{n=1}^{\infty} \left(12\left(-\dfrac{1}{3}\right)^{n-1}\right)$

58. $\displaystyle\sum_{n=1}^{\infty} \left(100\left(-\dfrac{1}{2}\right)^{n-1}\right)$

In Problems 59–62, express each repeating decimal as a fraction in lowest terms. See Objective 4.

59. $0.\overline{5}$

60. $0.\overline{3}$

61. $0.\overline{89}$

62. $0.\overline{45}$

Mixed Practice

In Problems 63–74, determine whether the given sequence is arithmetic, geometric, or neither. If the sequence is arithmetic, find the common difference; if it is geometric, find the common ratio.

63. $\{5n + 1\}$

64. $\{8 - 3n\}$

65. $\{2n^2\}$

66. $\{n^2 - 2\}$

67. $\left\{\dfrac{2^{-n}}{5}\right\}$

68. $\left\{\dfrac{2}{3^n}\right\}$

69. $54, 36, 24, 16, \ldots$

70. $100, 20, 4, \dfrac{4}{5}, \ldots$

71. $2, 6, 10, 14, \ldots$

72. $15, 12, 9, 6, \ldots$

73. $1, 2, 3, 5, 8, \ldots$

74. $5, -2, 3, -1, 2, \ldots$

Applying the Concepts

75. Find x so that $x, x + 2,$ and $x + 3$ are consecutive terms of a geometric sequence.

76. Find x so that $x - 1, x,$ and $x + 2$ are consecutive terms of a geometric sequence.

77. Salary Increases Suppose that you have been hired at an annual salary of $40,000 per year. You have been promised a raise of 5% for each of the next 10 years.
   (a) What will be your salary at the beginning of your 2nd year?
   (b) What will be your salary at the beginning of your 10th year?
   (c) How much will you have earned cumulatively once you have finished your 10th year?

78. Salary Increases Suppose that you have been hired at an annual salary of $45,000 per year. Historically, the typical raise is 4% each year. You expect to be at the company for the next 10 years.
   (a) What will be your salary at the beginning of your 2nd year?
   (b) What will be your salary at the beginning of your 10th year?
   (c) How much will you have earned cumulatively once you have finished your 10th year?

79. Depreciation of a Car Suppose that you have just purchased a Honda Accord for $20,000. Historically, the car depreciates by 8% each year, so that next year the car is worth $20,000(0.92). What will the value of the car be after you have owned it for 5 years?

80. Depreciation of a Car Suppose that you have just purchased a Chevy Impala for $16,000. Historically, the car depreciates by 10% each year, so that next year the car is worth $16,000(0.9). What will the value of the car be after you have owned it for 4 years?

81. Pendulum Swings A pendulum makes its first swing through an arc of 3 feet. On each successive swing, the length of the arc is 0.95 of the previous length.
   (a) What is the length of the arc after 10 swings?
   (b) On which swing is the length of the arc less than 1 foot for the first time?
   (c) After 10 swings, what total length will the pendulum have swung?
   (d) When it stops, what total length will the pendulum have swung?

82. Bouncing Balls A ball is dropped from a height of 30 feet. Each time it strikes the ground, it bounces up to 0.8 of the previous height.

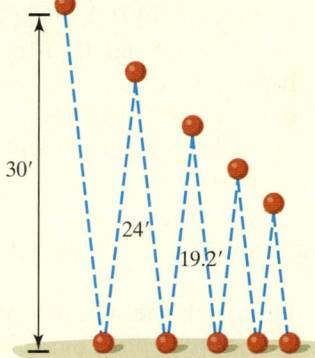

   (a) What height will the ball bounce up to after it strikes the ground for the 4th time?
   (b) How high will it bounce after it strikes the ground for the 5th time?
   (c) How many times does the ball need to strike the ground before its bounce is less than 6 inches?
   (d) What total distance does the ball travel before it stops bouncing?

83. A Job Offer You are interviewing for a job and receive two offers:

A: $30,000 to start, with guaranteed annual increases of 5% for the first 5 years

B: $31,000 to start with guaranteed annual increases of 4% for the first 5 years.

Which offer is best if your goal is to be making as much money as possible after the 5th year? Which is best if your goal is to make as much money as possible over the entire 5 years of the contract?

84. Be an Agent Suppose that you are an agent for a professional baseball player. Management has just offered your client a 5-year contract with a first-year

(continued)

salary of $1,500,000. Beyond that, your client has three choices:

A: A bonus of $75,000 each year (including the first year)

B: An annual increase of 4% per year beginning after the first year

C: An annual increase of $80,000 per year beginning after the first year

Which option provides the most money over the 5-year period? Which the least? Which would you choose? Why?

85. The Multiplier Suppose the marginal propensity to consume throughout the U.S. economy is 0.98. What is the multiplier for the U.S. economy?

86. The Multiplier Suppose the marginal propensity to consume throughout the U.S. economy is 0.96. What is the multiplier for the U.S. economy?

87. Stock Price One method of pricing a stock is based on the stream of future dividends of the stock. Suppose that a stock pays $P per year in dividends and, historically, the dividend has been increased by *i*% per year. If you desire an annual rate of return of *r*%, this method of pricing a stock states that the price you should pay is the present value of an infinite stream of payments:

$$\text{Price} = P + P \cdot \frac{1+i}{1+r} + P \cdot \left(\frac{1+i}{1+r}\right)^2 + P \cdot \left(\frac{1+i}{1+r}\right)^3 + \cdots$$

The price of the stock is the sum of a geometric series. Suppose that a stock pays an annual dividend of $2.00 and, historically, the dividend has been increased by 2% per year. You desire an annual rate of return of 9%. What is the most you should pay for the stock?

88. Stock Price Refer to Problem 87. Suppose that a stock pays an annual dividend of $3.00 and, historically, the dividend has been increased by 3% per year. You desire an annual rate of return of 10%. What is the most you should pay for the stock?

89. 401(k) Christine contributes $100 each month into her 401(k) retirement plan. What will be the value of Christine's 401(k) in 30 years if the per annum rate of return is assumed to be 8% compounded monthly?

90. Saving for a Home Jolene wants to purchase a new home. Suppose she invests $400 a month into a money market fund. If the per annum interest rate of return on the money market fund is 1.2% compounded monthly, how much will Jolene have for a down payment in 4 years?

91. Roth IRA Jackson contributes $500 each quarter into his Roth IRA. What will be the value of Jackson's IRA in 25 years if the per annum rate of return is assumed to be 6% compounded quarterly?

92. Retirement Raymont is planning on retiring in 15 years, so he contributes $1500 into his IRA every 6 months (semiannually). What will be the value of the IRA when Raymont retires if the per annum interest rate is 10% compounded semiannually?

93. What's My Payment? Suppose that Aaliyah wants to have $1,500,000 in her 401(k) retirement account in 35 years. How much does she need to contribute each month if the account earns 10% interest compounded monthly?

94. What's My Payment? Suppose that Sophia wants to have $2,000,000 in her 401(k) retirement account in 25 years. How much does she need to contribute each quarter if the account earns 12% interest compounded quarterly?

Extending the Concepts

95. Express $0.4\overline{9}$ as a fraction in lowest terms.

96. Express $0.85\overline{9}$ as a fraction in lowest terms.

97. Find the sum: $2 + 4 + 8 + \cdots + 1,073,741,824$

98. Can a sequence be both arithmetic and geometric? Give reasons for your answer.

99. Which yields faster growth in the terms of a sequence— an arithmetic sequence or a geometric sequence with $r > 1$? Why? Explain why Thomas Robert Malthus's conjecture  that food supplies grow arithmetically while population grows geometrically provides a recipe for disaster unless population growth is curbed.

Explaining the Concepts

100. How do you determine whether a sequence is geometric?

101. How do you determine whether a geometric series has a sum?

Synthesis Review

102. Express $\dfrac{1}{3}$ as a repeating decimal. Express $\dfrac{2}{3}$ as a repeating decimal.

103. What is $\dfrac{1}{3} + \dfrac{2}{3}$? Now add the repeating decimals found in Problem 102. Conjecture the value of $0.9999999\ldots$

104. Prove that $0.9999\ldots = 0.\overline{9}$ equals 1.

Technology Exercises

In Problems 105–108, use technology to find the sum of each sequence.

105. $4 + 4.8 + 5.76 + \cdots + 4(1.2)^{15-1}$

106. $3 + 4.8 + 7.68 + \cdots + 3(1.6)^{20-1}$

107. $\sum_{n=1}^{20} [1.2(1.05)^n]$

108. $\sum_{n=1}^{25} [1.3(0.55)^n]$

Putting the Concepts Together (Sections 13.1–13.3)

We designed these problems from Sections 13.1 through 13.3 so that you can review the chapter so far and show your mastery of the concepts. Take time to work these problems before proceeding with the next section. The answers are located at the back of the text on page AN-70.

In Problems 1–6, determine if the sequence is arithmetic, geometric, or neither. If arithmetic or geometric, determine the first term and the common difference or common ratio.

1. $\dfrac{3}{4}, \dfrac{3}{16}, \dfrac{3}{64}, \dfrac{3}{256}, \cdots$

2. $\{2(n + 3)\}$

3. $\left\{\dfrac{7n + 2}{9}\right\}$

4. $1, -4, 9, -16, \ldots$

5. $\{3 \cdot 2^{n+1}\}$

6. $\{n^2 - 5\}$

7. Write out the sum and evaluate: $\sum_{k=1}^{6} [3k + 4]$

8. Express the sum using summation notation:

$$\frac{1}{2(6 + 1)} + \frac{1}{2(6 + 2)} + \frac{1}{2(6 + 3)} + \cdots + \frac{1}{2(6 + 12)}$$

In Problems 9–12, write a formula for the nth term of the indicated sequence. Write the first five terms of each sequence.

9. arithmetic: $a_1 = 25, d = -2$

10. arithmetic: $a_4 = 9, d = 11$

11. geometric: $a_4 = \dfrac{9}{25}, r = \dfrac{1}{5}$

12. geometric: $a_1 = 150, r = 1.04$

In Problems 13–15, find the indicated sum.

13. $2 + 6 + 18 + \cdots + 2 \cdot (3)^{11-1}$

14. $2 + 7 + 12 + 17 + \cdots + [2 + (20 - 1) \cdot 5]$

15. $1000 + 100 + 10 + \cdots$

16. Table Seating A restaurant uses square tables in the main dining area that seat four people. For larger parties, tables can be placed together. Two tables will seat 6 people, three tables will seat 8 people, and so on (see diagram). How many tables are needed to seat a party of 24 people?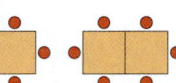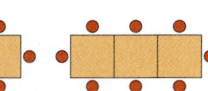

13.4 The Binomial Theorem

Objectives

1 Compute Factorials

2 Evaluate a Binomial Coefficient

3 Expand a Binomial

Are You Prepared for This Section?

Before getting started, complete the following problems. If you get a problem wrong, go back to the section cited and review the material.

P1. Multiply: $(x - 5)^2$ [Section 5.3, pp. 328–329]

P2. Multiply: $(2x + 3)^2$ [Section 5.3, pp. 328–329]

▶ **1 Compute Factorials**

Suppose you want to find the product of the first 12 positive integers, or

$$12 \cdot 11 \cdot 10 \cdot 9 \cdot 8 \cdot 7 \cdot 6 \cdot 5 \cdot 4 \cdot 3 \cdot 2 \cdot 1$$

Prepared?...Answers

P1. $x^2 - 10x + 25$ **P2.** $4x^2 + 12x + 9$

This product equals 479,001,600. Not only is this a big number, but writing all the factors is time-consuming. A shorthand method for writing this product is *factorial notation*.

> **Definition**
>
> If $n \geq 0$ is an integer, the **factorial symbol $n!$** (read "n factorial") is defined as
>
> $$0! = 1 \qquad 1! = 1$$
> $$n! = n(n-1)(n-2) \cdots \cdot 3 \cdot 2 \cdot 1 \quad \text{if } n \geq 2$$

For example, $2! = 2 \cdot 1 = 2, 3! = 3 \cdot 2 \cdot 1 = 6, 4! = 4 \cdot 3 \cdot 2 \cdot 1 = 24$, and so on. Table 2 lists the values of $n!$ for $0 \leq n \leq 7$.

Table 2

n	0	1	2	3	4	5	6	7
$n!$	1	1	2	6	24	120	720	5040

Because

$$n! = n\underbrace{(n-1)(n-2) \cdots \cdot 3 \cdot 2 \cdot 1}_{(n-1)!}$$

use the formula

$$n! = n(n-1)!$$

to find successive factorials. For example, because $7! = 5040$,

$$8! = 8 \cdot 7! = 8(5040) = 40,320$$

Use your calculator's factorial key to see how fast factorials increase in value. Find the value of $69!$. What happens when you try to find $70!$? In fact, $70!$ is larger than 10^{100} (a **googol**), the largest number that most calculators can display.

EXAMPLE 1 **Computing Factorials**

Compute the value of $\dfrac{12!}{9!}$.

Solution

You could directly compute $12!$ and then $9!$, but this would be inefficient. Use the properties of factorials instead.

$$\frac{12!}{9!} = \frac{12 \cdot 11 \cdot 10 \cdot 9!}{9!}$$

Divide out $9!$: $\quad = 12 \cdot 11 \cdot 10$

$$= 1320$$

See Figure 6 for this calculation on a graphing calculator. ●

Figure 6

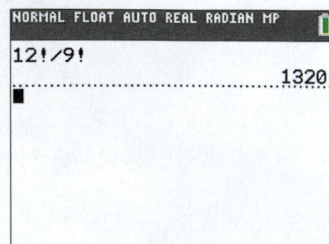

NORMAL FLOAT AUTO REAL RADIAN MP

12!/9!
                              1320

> **Quick ✔**
>
> **1.** If $n \geq 2$ is an integer, then $n! = $ _____.
>
> **2.** $0! = $ __; $1! = $ __.
>
> *In Problems 3 and 4, find the value of each factorial.*
>
> **3.** $5!$
>
> **4.** $\dfrac{7!}{3!}$

▶ ❷ Evaluate a Binomial Coefficient

The formula for expanding $(x + a)^n$ for $n = 2$ is known. The *Binomial Theorem* is a formula for expanding $(x + a)^n$ for any positive integer n. If $n = 1, 2, 3,$ or 4, the expansion of $(x + a)^n$ is straightforward:

$(x + a)^1$	$= x + a$	Two terms, beginning with x^1 and ending with a^1
$(x + a)^2$	$= x^2 + 2ax + a^2$	Three terms, beginning with x^2 and ending with a^2
$(x + a)^3$	$= x^3 + 3ax^2 + 3a^2x + a^3$	Four terms, beginning with x^3 and ending with a^3
$(x + a)^4$	$= x^4 + 4ax^3 + 6a^2x^2 + 4a^3x + a^4$	Five terms, beginning with x^4 and ending with a^4

Each expansion of $(x + a)^n$ begins with x^n and ends with a^n. As you read from left to right, the powers of x decrease by 1, while the powers of a increase by 1. The number of terms is $n + 1$. Notice that the degree of each monomial in the expansion equals n. For example, in the expansion of $(x + a)^3$, each monomial $(x^3, 3ax^2, 3a^2x, a^3)$ is of degree 3. As a result, a conjecture might be that the expansion of $(x + a)^n$ would look like this:

$$(x + a)^n = x^n + \underline{} ax^{n-1} + \underline{} a^2 x^{n-2} + \cdots + \underline{} a^{n-1} x + a^n$$

where the blanks are numbers to be found. This is correct, as will be seen shortly.

Before we go any further, we need to introduce the symbol $\binom{n}{j}$, which is read "n taken j at a time" or "n choose j":

Work Smart

Do not write $\binom{n}{j}$ as $\left(\dfrac{n}{j}\right)$.

Definition

If j and n are integers with $0 \leq j \leq n$, the symbol $\binom{n}{j}$ is defined as

$$\binom{n}{j} = \frac{n!}{j!(n-j)!}$$

EXAMPLE 2

Evaluating $\binom{n}{j}$

Find:

(a) $\binom{4}{1}$ **(b)** $\binom{6}{2}$ **(c)** $\binom{5}{4}$

Solution

(a) Here, $n = 4$ and $j = 1$, so

$$\binom{4}{1} = \frac{4!}{1!(4-1)!} = \frac{4!}{1! \cdot 3!} = \frac{4 \cdot 3!}{1! \cdot 3!} = \frac{4 \cdot 3!}{1 \cdot 3!} = 4$$

$$6! = 6 \cdot 5 \cdot 4!$$

(b) $\binom{6}{2} = \dfrac{6!}{2!(6-2)!} = \dfrac{6!}{2! \cdot 4!} = \dfrac{6 \cdot 5 \cdot 4!}{2 \cdot 1 \cdot 4!} = \dfrac{6 \cdot 5 \cdot 4!}{2 \cdot 1 \cdot 4!} = \dfrac{30}{2} = 15$

(c) $\binom{5}{4} = \dfrac{5!}{4!(5-4)!} = \dfrac{5!}{4! \cdot 1!} = \dfrac{5 \cdot 4!}{4! \cdot 1} = \dfrac{5 \cdot 4!}{4! \cdot 1} = 5$ ●

Quick ✓

5. *True or False* $\binom{n}{j} = \dfrac{j!}{n!(n-j)!}$ **6.** *True or False* $\binom{5}{1} = 5$

In Problems 7 and 8, evaluate each expression.

7. $\binom{7}{1}$ **8.** $\binom{6}{3}$

Four useful formulas involving the symbol $\dbinom{n}{j}$ are

$$\dbinom{n}{0} = 1 \qquad \dbinom{n}{1} = n \qquad \dbinom{n}{n-1} = n \qquad \dbinom{n}{n} = 1$$

Suppose that the various values of the symbol $\dbinom{n}{j}$ are arranged in a triangular display, as shown next and in Figure 7.

$$\dbinom{0}{0}$$

$$\dbinom{1}{0} \quad \dbinom{1}{1}$$

$$\dbinom{2}{0} \quad \dbinom{2}{1} \quad \dbinom{2}{2}$$

$$\dbinom{3}{0} \quad \dbinom{3}{1} \quad \dbinom{3}{2} \quad \dbinom{3}{3}$$

$$\dbinom{4}{0} \quad \dbinom{4}{1} \quad \dbinom{4}{2} \quad \dbinom{4}{3} \quad \dbinom{4}{4}$$

$$\dbinom{5}{0} \quad \dbinom{5}{1} \quad \dbinom{5}{2} \quad \dbinom{5}{3} \quad \dbinom{5}{4} \quad \dbinom{5}{5}$$

Figure 7
Pascal's Triangle

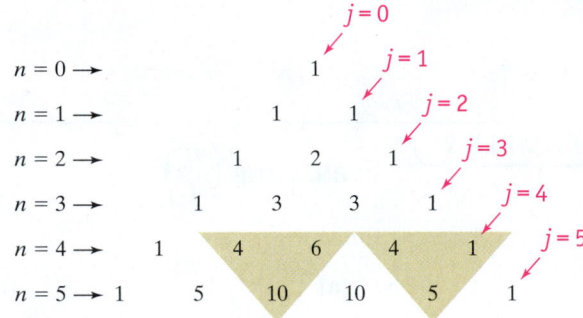

This display is called **Pascal's triangle,** named after Blaise Pascal (1623–1662), a French mathematician.

Pascal's triangle has 1s down the sides. To get any other entry, add the two nearest entries in the row above it. The shaded triangles in Figure 7 illustrate this feature. Using this feature, find the row corresponding to $n = 6$.

$$n = 5 \rightarrow \qquad 1 \quad 5 \quad 10 \quad 10 \quad 5 \quad 1$$

$$n = 6 \rightarrow \quad 1 \quad 6 \quad 15 \quad 20 \quad 15 \quad 6 \quad 1$$

Although Pascal's triangle is an interesting and organized display of the symbol $\dbinom{n}{j}$, in practice it is not all that helpful. For example, to find the value of $\dbinom{12}{8}$, you would need to produce 12 rows of the triangle before seeing the answer. It is much faster to use the definition of $\dbinom{n}{j}$.

▶ ❸ Expand a Binomial

The formula for the **Binomial Theorem** is presented on the next page.

> **The Binomial Theorem**
>
> Let x and a be real numbers. For any positive integer n,
>
> $$(x + a)^n = \binom{n}{0}x^n + \binom{n}{1}ax^{n-1} + \binom{n}{2}a^2x^{n-2} + \cdots + \binom{n}{j}a^jx^{n-j} + \cdots + \binom{n}{n}a^n$$

You should now see why there is a need to discuss the symbol $\binom{n}{j}$; these symbols are the numerical coefficients that appear in the expansion of $(x + a)^n$. Because of this, the symbol $\binom{n}{j}$ is called a **binomial coefficient.**

EXAMPLE 3 Expanding a Binomial

Use the Binomial Theorem to expand $(x + 3)^4$.

Solution
In the Binomial Theorem, let $a = 3$ and $n = 4$. Then

$$(x + 3)^4 = \binom{4}{0}x^4 + \binom{4}{1}3^1 \cdot x^{4-1} + \binom{4}{2}3^2 \cdot x^{4-2} + \binom{4}{3}3^3 \cdot x^{4-3} + \binom{4}{4}3^4$$

Work Smart

In the expansion, notice that the exponent on x **decreases** by 1 for each term as you move to the right, while the exponent on 3 **increases** by 1.

Use Pascal's triangle in Figure 7 or use $\binom{n}{j} = \dfrac{n!}{j!\,(n-j)!}$ to evaluate the binomial coefficients.

$$= 1 \cdot x^4 + 4 \cdot 3 \cdot x^3 + 6 \cdot 9 \cdot x^2 + 4 \cdot 27 \cdot x + 1 \cdot 81$$
$$= x^4 + 12x^3 + 54x^2 + 108x + 81$$

EXAMPLE 4 Expanding a Binomial

Expand $(2y - 3)^5$ using the Binomial Theorem.

Solution
First, rewrite the expression $(2y - 3)^5$ as $[2y + (-3)]^5$. Then use the Binomial Theorem with $n = 5$, $x = 2y$, and $a = -3$.

$$[2y+(-3)]^5 = \binom{5}{0}(2y)^5 + \binom{5}{1}(-3)^1(2y)^{5-1} + \binom{5}{2}(-3)^2(2y)^{5-2} + \binom{5}{3}(-3)^3(2y)^{5-3} + \binom{5}{4}(-3)^4(2y)^{5-4} + \binom{5}{5}(-3)^5$$

Use Pascal's triangle or use the formula $\binom{n}{j} = \dfrac{n!}{j!\,(n-j)!}$ to evaluate binomial coefficients.

$$= 1(2y)^5 + 5(-3)^1(2y)^4 + 10(-3)^2(2y)^3 + 10(-3)^3(2y)^2 + 5(-3)^4(2y)^1 + 1(-3)^5$$
$$= 32y^5 + 5 \cdot (-3) \cdot 16y^4 + 10 \cdot 9 \cdot 8y^3 + 10 \cdot (-27) \cdot 4y^2 + 5 \cdot 81 \cdot 2y - 243$$
$$= 32y^5 - 240y^4 + 720y^3 - 1080y^2 + 810y - 243$$

Quick ✓

In Problems 9 and 10, expand each binomial using the Binomial Theorem.

9. $(x + 2)^4$ 　　　　　　　　　　　　**10.** $(2p - 1)^5$

13.4 Exercises   MyMathLab®

Exercise numbers in green have
complete video solutions in MyMathLab or
may be accessed using the QR code to the right.

*Problems **1–10** are the* **Quick ✔** *s that follow the* **EXAMPLES**.

Building Skills

In Problems 11–18, evaluate each factorial expression. See Objective 1.

11. $3!$ **12.** $4!$ **13.** $8!$ **14.** $9!$

15. $\dfrac{10!}{8!}$ **16.** $\dfrac{12!}{10!}$ **17.** $\dfrac{8!}{5!}$ **18.** $\dfrac{9!}{6!}$

In Problems 19–22, evaluate each expression. See Objective 2.

19. $\binom{7}{2}$ **20.** $\binom{8}{5}$ **21.** $\binom{10}{4}$ **22.** $\binom{12}{9}$

In Problems 23–38, expand each expression using the Binomial Theorem. See Objective 3.

23. $(x + 1)^5$ **24.** $(x - 1)^4$

25. $(x - 4)^4$ **26.** $(x + 5)^5$

27. $(3p + 2)^4$ **28.** $(2q + 3)^4$

29. $(2z - 3)^5$ **30.** $(3w - 4)^4$

31. $(x^2 + 2)^4$ **32.** $(y^2 - 3)^4$

33. $(2p^3 + 1)^5$ **34.** $(3b^2 + 2)^5$

35. $(x + 2)^6$ **36.** $(p - 3)^6$

37. $(2p^2 - q^2)^4$ **38.** $(3x^2 + y^3)^4$

Applying the Concepts

39. Use the Binomial Theorem to find the numerical value of $(1.001)^4$ correct to five decimal places. (*Hint:* $(1.001)^4 = (1 + 10^{-3})^4$.)

40. Use the Binomial Theorem to find the numerical value of $(1.001)^5$ correct to five decimal places. (*Hint:* $(1.001)^5 = (1 + 10^{-3})^5$.)

41. Use the Binomial Theorem to find the numerical value of $(0.998)^5$ correct to five decimal places.

42. Use the Binomial Theorem to find the numerical value of $(0.997)^5$ correct to five decimal places.

Extending the Concepts

Notice that in the formula for expanding a binomial $(x + a)^n$, the first term is $\binom{n}{0}a^0 \cdot x^n$, the second term is $\binom{n}{1}a^1 \cdot x^{n-1}$, the third term is $\binom{n}{2}a^2 \cdot x^{n-2}$, and so on. In general, the jth term in a binomial expansion of $(x + a)^n$ is $\binom{n}{j-1}a^{j-1}x^{n-j+1}$. Use this result to find the indicated term in Problems 43–46.

43. The 3rd term in the expansion of $(x + 2)^7$

44. The 4th term in the expansion of $(x - 1)^{10}$

45. The 6th term in the expansion of $(2p - 3)^8$

46. The 7th term in the expansion of $(3p + 1)^9$

47. Show that $\binom{n}{n-1} = n$ and $\binom{n}{n} = 1$.

48. Show that if n and j are integers with $0 \le j \le n$, then $\binom{n}{j} = \binom{n}{n-j}$.

Explaining the Concepts

49. Write the first four rows of Pascal's triangle.

50. Describe the pattern of exponents of x in the expansion of $(x + a)^n$. Describe the pattern of exponents of a in the expansion of $(x + a)^n$.

51. What is true about the degree of each monomial in the expansion of $(x + y)^n$?

52. Explain how you might find a particular term in a binomial expansion. For example, how might you find the 5th term in the expansion of $(x + 3)^8$?

Synthesis Review

53. If $f(x) = x^4$, find $f(a - 2)$ with the aid of the Binomial Formula.

54. If $g(x) = x^5 + 3$, find $g(z + 1)$ with the aid of the Binomial Formula.

55. If $H(x) = x^5 - 4x^4$, find $H(p + 1)$ with the aid of the Binomial Formula.

56. If $h(x) = 2x^5 + 5x^4$, find $h(a + 3)$ with the aid of the Binomial Formula.

Chapter 13 Activity: Discover the Relation

Focus: Review of objectives for arithmetic and geometric sequences and the Binomial Theorem

Time: 30–35 minutes

Group size: 2–4

As a group, discover the relationship between the results found from Column A and Column B (that is, $<$, $>$, or $=$). Be sure to discuss any differences in outcomes.

	Column A	Column B
1	The sum of the first five terms of $$a_n = \frac{3}{5}n + 1 \quad A < B$$	The sum of the first five terms of $$a_n = -\frac{1}{4}(n - 1) + 4$$
2	$$\sum_{i=1}^{4}(i^2 + 2i) \quad A = B$$	$$\sum_{i=1}^{4} i^2 + \sum_{i=1}^{4} 2i$$
3	The 8th term of an arithmetic sequence when $a_1 = 3$ and $d = \dfrac{3}{2}$ $\quad A > B$	The 27th term of an arithmetic sequence when $a_1 = \dfrac{5}{3}$ and $d = \dfrac{1}{3}$
4	The 10th term of a geometric sequence with $a_1 = 3$ and $r = \sqrt{2}$ $\quad A < B$	The 9th term of a geometric sequence with $a_1 = 5$ and $r = \sqrt{3}$
5	The coefficient of the 4th term of $(x + y)^{10}$ $\quad A < B$	The coefficient of the 3rd term of $(8x - y)^4$

Chapter 13 Review

Section 13.1   Sequences

KEY TERMS

Sequence

Terms

Ellipsis

Infinite sequence

Finite sequence

General term

Alternate

Summation notation

Index

Partial sum

You Should Be Able To...	EXAMPLE	Review Exercises
1 Write the first few terms of a sequence (p. 897)	Examples 1 and 2	1–6
2 Find a formula for the nth term of a sequence (p. 899)	Example 3	7–12
3 Use summation notation (p. 899)	Examples 4 and 5	13–20

In Problems 1–6, write the first five terms of each sequence.

1. $\{-3n + 2\}$

2. $\left\{\dfrac{n - 2}{n + 4}\right\}$

3. $\{5^n + 1\}$

4. $\{(-1)^{n-1} \cdot 3n\}$

5. $\left\{\dfrac{n^2}{n + 1}\right\}$

6. $\left\{\dfrac{\pi^n}{n}\right\}$

In Problems 7–12, the given pattern continues. Write the nth term of each sequence suggested by the pattern.

7. $-3, -6, -9, -12, -15, \ldots$

8. $\dfrac{1}{3}, \dfrac{2}{3}, 1, \dfrac{4}{3}, \dfrac{5}{3}, \ldots$

9. $5, 10, 20, 40, 80, \ldots$

10. $-\dfrac{1}{2}, 1, -\dfrac{3}{2}, 2, \ldots$

11. $6, 9, 14, 21, 30, \ldots$

12. $0, \dfrac{1}{3}, \dfrac{1}{2}, \dfrac{3}{5}, \ldots$

In Problems 13–16, write out each sum and determine its value.

13. $\displaystyle\sum_{k=1}^{5}(5k-2)$

14. $\displaystyle\sum_{k=1}^{6}\left(\dfrac{k+2}{2}\right)$

15. $\displaystyle\sum_{i=1}^{5}(-2i)$

16. $\displaystyle\sum_{i=1}^{4}\dfrac{i^2-1}{3}$

In Problems 17–20, express each sum using summation notation.

17. $(4+3\cdot1)+(4+3\cdot2)+(4+3\cdot3)+\cdots+(4+3\cdot15)$

18. $\dfrac{1}{3^1}+\dfrac{1}{3^2}+\dfrac{1}{3^3}+\cdots+\dfrac{1}{3^8}$

19. $\dfrac{1^3+1}{1+1}+\dfrac{2^3+1}{2+1}+\dfrac{3^3+1}{3+1}+\cdots+\dfrac{10^3+1}{10+1}$

20. $(-1)^{1-1}\cdot1^2+(-1)^{2-1}\cdot2^2+(-1)^{3-1}\cdot3^2+\cdots+(-1)^{7-1}\cdot7^2$

Section 13.2   Arithmetic Sequences

KEY CONCEPTS

- **The *n*th Term of an Arithmetic Sequence**
  For an arithmetic sequence whose first term is a_1 and whose common difference is d, the nth term is $a_n = a_1 + (n-1)d$.

- **Sum of the First *n* Terms of an Arithmetic Sequence**
  For an arithmetic sequence whose first term is a_1 and whose common difference is d, the sum of the first n terms is

  $$S_n = \dfrac{n}{2}[2a_1 + (n-1)d] \quad \text{or} \quad S_n = \dfrac{n}{2}[a_1 + a_n].$$

KEY TERMS

Arithmetic sequence
Common difference

You Should Be Able To...	EXAMPLE	Review Exercises
1 Determine whether a sequence is arithmetic (p. 904)	Examples 1 through 3	21–26
2 Find a formula for the *n*th term of an arithmetic sequence (p. 905)	Examples 4 and 5	27–32, 37
3 Find the sum of an arithmetic sequence (p. 907)	Examples 6 through 8	33–36, 38

In Problems 21–26, determine whether the sequence is arithmetic. If so, find the common difference.

21. $4, 10, 16, 22, \ldots$

22. $-1, \dfrac{1}{2}, 2, \dfrac{7}{2}, \ldots$

23. $-2, -5, -9, -14, \ldots$

24. $-1, 3, -5, 7, \ldots$

25. $\{4n+7\}$

26. $\left\{\dfrac{n+1}{2n}\right\}$

In Problems 27–32, find a formula for the nth term of the arithmetic sequence. Use the formula to find the 25th term of the sequence.

27. $a_1 = 3; d = 8$

28. $a_1 = -4; d = -3$

29. $7, \dfrac{20}{3}, \dfrac{19}{3}, 6, \ldots$

30. $11, 17, 23, 29, \ldots$

31. The 3rd term is 7 and the 8th term is 25.

32. The 4th term is -20 and the 7th term is -32.

33. Find the sum of the first 30 terms of the sequence $-1, 9, 19, 29, \ldots$.

34. Find the sum of the first 40 terms of the sequence $5, 2, -1, -4, \ldots$.

35. Find the sum of the first 60 terms of the sequence $\{-2n - 7\}$.

36. Find the sum of the first 50 terms of the sequence $\left\{\dfrac{1}{4}n + 3\right\}$.

37. Cicadas Seventeen-year cicadas emerge every 17 years to mate, lay eggs, and start the next 17-year cycle. In 2004, the Brood X cicada (the largest brood of the 17-year cicada) emerged in Maryland, Kentucky, Tennessee, and parts of surrounding states. Determine when the Brood X cicada will first appear in the twenty-second century.

38. Wind Sprints At a certain football practice, players would run wind sprints for exercise. Starting at the goal line, players would sprint to the 10-yard line and back to the goal line. The players would then sprint to the 20-yard line and back to the goal line. This would continue for the 30-yard line, 40-yard line, and 50-yard line. Determine the total distance a player would run during the wind sprints.

Section 13.3   Geometric Sequences and Series

KEY CONCEPTS

- **The _n_th Term of a Geometric Sequence**
  For a geometric sequence whose first term is a_1 and whose common ratio is r, the nth term is $a_n = a_1 r^{n-1}$.

- **Sum of the First _n_ Terms of a Geometric Sequence**
  For a geometric sequence whose first term is a_1 and whose common ratio is r, the sum of the first n terms is $S_n = a_1 \cdot \dfrac{1 - r^n}{1 - r}, r \neq 0, r \neq 1$.

- **Sum of a Geometric Series**
  If $-1 < r < 1$, then the sum of the geometric series $\displaystyle\sum_{n=1}^{\infty} a_1 r^{n-1} = \dfrac{a_1}{1 - r}$.

KEY TERMS

Geometric sequence
Common ratio
Geometric series
Marginal propensity
  to consume
Multiplier

You Should Be Able To...	EXAMPLE	Review Exercises
1 Determine whether a sequence is geometric (p. 911)	Examples 1 through 3	39–44
2 Find a formula for the nth term of a geometric sequence (p. 913)	Example 4	45–48, 57
3 Find the sum of a geometric sequence (p. 914)	Examples 5 and 6	49–52, 58
4 Find the sum of a geometric series (p. 916)	Examples 7 through 9	53–56
5 Solve annuity problems (p. 918)	Example 10	59–62

In Problems 39–44, determine whether the given sequence is geometric. If so, determine the common ratio.

39. $\dfrac{1}{3}, 2, 12, 72, \ldots$

40. $-1, 3, -9, 27, \ldots$

41. $1, 1, 2, 6, \ldots$

42. $6, 4, \dfrac{8}{3}, \dfrac{16}{9}, \ldots$

43. $\{5(-2)^n\}$

44. $\{3n - 14\}$

In Problems 45–48, find a formula for the nth term of the geometric sequence. Use the formula to find the 10th term of the sequence.

45. $a_1 = 4, r = 3$

46. $a_1 = 8, r = \dfrac{1}{4}$

47. $a_1 = 5, r = -2$

48. $a_1 = 1000, r = 1.08$

In Problems 49–52, find the sum. If necessary, express your answer to as many decimal places as your calculator allows.

49. $2 + 4 + 8 + \cdots + 2^{15}$

50. $40 + 5 + \dfrac{5}{8} + \cdots + 40\left(\dfrac{1}{8}\right)^{13-1}$

51. $\displaystyle\sum_{n=1}^{12}\left[\dfrac{3}{4}(2)^{n-1}\right]$

52. $\displaystyle\sum_{n=1}^{16}\left[-4(3^n)\right]$

In Problems 53–56, find the sum of each geometric series. If necessary, express your answer to as many decimal places as your calculator allows.

53. $\displaystyle\sum_{n=1}^{\infty}\left[20\left(\dfrac{1}{4}\right)^n\right]$

54. $\displaystyle\sum_{n=1}^{\infty}\left[50\left(-\dfrac{1}{2}\right)^{n-1}\right]$

55. $1 + \dfrac{1}{5} + \dfrac{1}{25} + \cdots$

56. $0.8 + 0.08 + 0.008 + 0.0008 + \cdots$

57. Radioactive Decay The radioactive isotope tritium has a half-life of about 12 years. If there were 200 grams of the isotope initially, use a geometric sequence to determine how much would remain after 72 years.

58. Computer Virus In January 2004, the Mydoom e-mail worm was declared the worst e-mail worm incident in virus history, accounting for roughly 20–30% of worldwide e-mail traffic. Suppose the virus was initially sent to 5 e-mail addresses and that, upon receipt, it sends itself out to 5 e-mail addresses from the address book of the infected computer. If each cycle of e-mails, including the initial sending, takes 1 minute to complete, how many total e-mails will have been sent after 15 minutes?

59. 403(b) Each quarter, Scott contributes $900 to a 403(b) plan at work. His employer agrees to match half of employee contributions up to $600 per quarter. What will be the value of Scott's 403(b) in 25 years if the per annum rate of return is assumed to be 7% compounded quarterly?

60. Lottery Payment The winner of a state lottery has the option of receiving about $2 million per year for 26 years (after taxes) or a lump sum payment of about $28 million (after taxes). Assuming all winnings will be invested at a per annum rate of return of 6.5% compounded annually, which option yields the most money after 26 years?

61. What's My Payment? Sheri starts her career when she is 22 years old and wants to have $2,500,000 in her 401(k) retirement account when she retires in 40 years. How much does she need to contribute each month if the account earns 9% interest compounded monthly?

62. College Savings Plan On Samantha's 8th birthday, her parents open a 529 college savings plan for her. They plan to contribute $400 per month until she turns 18. The per annum rate of return is assumed to be 5.25% compounded monthly, and the cost per credit hour at a private 4-year university is locked in at a rate of $340 per hour. What will be the value of the plan when Samantha turns 18, and how many credit hours will the plan cover?

Section 13.4   The Binomial Theorem

KEY CONCEPTS

- **Factorial symbol $n!$**
  If $n \geq 0$ is an integer, the factorial symbol $n!$ is defined as
  $0! = 1 \qquad 1! = 1 \qquad n! = n(n-1)(n-2)\cdots\cdot 3\cdot 2\cdot 1 \quad$ if $n \geq 2$.

- **The symbol $\dbinom{n}{j}$**
  If j and n are integers with $0 \leq j \leq n$, then $\dbinom{n}{j} = \dfrac{n!}{j!\,(n-j)!}$.

- **The Binomial Theorem**
  Let x and a be real numbers. For any positive integer n,
  $$(x+a)^n = \binom{n}{0}x^n + \binom{n}{1}ax^{n-1} + \binom{n}{2}a^2 x^{n-2} + \cdots + \binom{n}{j}a^j x^{n-j} + \cdots + \binom{n}{n}a^n.$$

KEY TERMS

Factorial symbol
Googol
Pascal's triangle
Binomial coefficient

You Should Be Able To...	EXAMPLE	Review Exercises
1 Compute factorials (p. 923)	Example 1	63–66
2 Evaluate a binomial coefficient (p. 925)	Example 2	67–70
3 Expand a binomial (p. 926)	Examples 3 and 4	71–78

In Problems 63–66, evaluate the expression.

63. $5!$

64. $\dfrac{11!}{7!}$

65. $\dfrac{10!}{6!}$

66. $\dfrac{13!}{6!7!}$

In Problems 67–70, evaluate each binomial coefficient.

67. $\dbinom{7}{3}$

68. $\dbinom{10}{5}$

69. $\dbinom{8}{8}$

70. $\dbinom{6}{0}$

In Problems 71–76, expand each expression using the Binomial Theorem.

71. $(z + 1)^4$

72. $(y - 3)^5$

73. $(3y + 4)^6$

74. $(2x^2 - 3)^4$

75. $(3p - 2q)^4$

76. $(a^3 + 3b)^5$

77. Find the 4th term in the expansion of $(x - 2)^8$.

78. Find the 7th term in the expansion of $(2x + 1)^{11}$.

Chapter 13 Test   *Step-by-step test solutions are found on the Chapter Test Prep Videos available in* MyMathLab®, *on* YouTube, *or may be accessed using the QR code to the right.*

In Problems 1–6, determine whether the sequence is arithmetic, geometric, or neither. If arithmetic or geometric, determine the first term and the common difference or common ratio.

1. $-15, -7, 1, 9, \ldots$

2. $\{(-4)^n\}$

3. $\left\{\dfrac{4}{n!}\right\}$

4. $\left\{\dfrac{2n - 3}{5}\right\}$

5. $-3, 2, 0, 5, 3, \ldots$

6. $\{7 \cdot 3^n\}$

7. Write out the sum and evaluate: $\displaystyle\sum_{i=1}^{5}\left[\dfrac{3}{i^2} + 2\right]$

8. Express the sum using summation notation:
$$\dfrac{3}{5} + \dfrac{2}{3} + \dfrac{5}{7} + \dfrac{3}{4} + \cdots + \dfrac{5}{6}$$

In Problems 9–12, write a formula for the nth term of the indicated sequence. Write the first five terms of each sequence.

9. arithmetic: $a_1 = 6, d = 10$

10. arithmetic: $a_1 = 0, d = -4$

11. geometric: $a_1 = 10, r = 2$

12. geometric: $a_3 = 9, r = -3$

In Problems 13–15, find the indicated sum.

13. $-2 + 2 + 6 + \cdots + [4(20 - 1) - 2]$

14. $\dfrac{1}{9} - \dfrac{1}{3} + 1 - 3 + \cdots + \dfrac{1}{9}(-3)^{12-1}$

15. $216 + 72 + 24 + 8 + \cdots$

16. Evaluate $\dfrac{15!}{8!7!}$.

17. Evaluate $\dbinom{12}{5}$.

18. Expand $(5m - 2)^4$ using the Binomial Theorem.

19. A new car sold for \$31,000. If the vehicle loses 15% of its value each year, how much will it be worth after 10 years?

20. A weightlifter begins his routine by benching 100 pounds and increases the weight by 30 pounds for each set. If he does 10 repetitions in each set, what is the total weight lifted after 5 sets?

Cumulative Review Chapters 1–13

1. Solve $\dfrac{1}{2}(x + 2) = \dfrac{5}{4}(x - 3y)$ for y.

2. Evaluate $f(x) = x^2 - x + 7$ for $x = 2$ and $x = -3$.

In Problems 3–8, find all solutions to the indicated equation.

3. $\dfrac{1}{2}x - 2 = \dfrac{1}{3}(x + 1) + 3$

4. $5x^2 - 3x = 2$

5. $3x^2 + 7x - 2 = 0$

6. $\sqrt{2x + 1} - 3 = 8$

7. $4^{x+1} = 8^{2x-3}$

8. $x^2(2x + 1) + 40 = (x^2 - 8)(x - 5)$

In Problems 9 and 10, solve the indicated inequality. Write your answer in interval notation.

9. $\dfrac{2}{3}x + 1 > \dfrac{1}{4}x - \dfrac{3}{2}$

10. $3x^2 - 2x \le 3 - 10x$

In Problems 11 and 12, factor the expression completely.

11. $2x^2 - 5x - 18$

12. $6x^3 - 3x^2 + 4x - 2$

In Problems 13–15, perform the indicated operation and simplify. Write complex numbers in standard form.

13. $(5x - 3)(4x^2 - 2x + 1)$

14. $\dfrac{x}{x + 4} - \dfrac{3}{x - 1}$

15. $\dfrac{3 - i}{2 + i}$

16. Find the domain of the function $f(x) = \sqrt{x - 15} + \sqrt{2x - 5}$.

17. Find the equation of the line that passes through the points $(2, -3)$ and $(1, 4)$.

18. Solve the system: $\begin{cases} 2x + 3y = 5 \\ x - 2y = 6 \end{cases}$

19. Graph the quadratic function $f(x) = 2x^2 - 8x - 3$. Label the vertex and axis of symmetry.

20. Write the standard form of the equation of the circle whose center is $(4, -3)$ and whose radius is $r = 6$ units. Graph the circle.

21. Sketch the graph of the ellipse given by the equation $4x^2 + y^2 = 64$.

22. Find the sum of the first 20 terms of the arithmetic sequence whose first term is $a = -47$ and whose common difference is $d = 12$.

23. Find the sum of the infinite geometric series:
$$2, \dfrac{3}{2}, \dfrac{9}{8}, \dfrac{27}{32}, \ldots .$$

24. Mowing Lawns The Robomower® automatic lawnmower can mow a 7500-square-foot lot in 5 hours. It takes the Mowbot® automatic lawnmower 6 hours to mow the same lot. How long would it take both machines to mow the lot if they work together?

25. Aluminum Alloy The most commonly used aluminum alloy is aluminum 3003, which is often used to make rain gutters. A manufacturer of rain gutters has 100 metric tons of an aluminum alloy that is 2.5% manganese, but this percent is too high. How much pure aluminum must be added to the 100 metric tons in order to obtain a desired alloy that is 1.2% manganese?

Appendix A Synthetic Division

Objectives

1. Divide Polynomials Using Synthetic Division
2. Use the Remainder and Factor Theorems

Are You Prepared for This Section?

Before getting started, complete the following problems. If you get a problem wrong, go back to the section cited and review the material.

P1. Simplify: $\dfrac{15x^5}{12x^3}$ [Section 5.4, pp. 335–336]

P2. Divide: $\dfrac{2x^3 + 7x^2 - 17x - 10}{2x + 1}$ [Section 5.5, pp. 348–351]

When polynomials are added, subtracted, and multiplied, the result is also a polynomial. A polynomial divided by a polynomial, however, may not be a polynomial.

▶ 1 Divide Polynomials Using Synthetic Division

To find the quotient and remainder when dividing a polynomial of degree 1 or higher by $x - c$, a version of long division called **synthetic division** can be used.

To see how synthetic division works, use long division to divide the polynomial $2x^3 - 5x^2 - 7x + 20$ by $x - 3$. Synthetic division is just long division written in a more compact form. For example, in the long division below, the terms in red ink are not really necessary because they are identical to the terms directly above them. The subtraction signs are not necessary, because subtraction is understood. With these items removed, the division shown on the right results.

Work Smart

Synthetic division can be used only when the divisor is of the form $x - c$ or $x + c$. Remember that $x + b$ can be rewritten in the form $x - (-b)$.

$$
\begin{array}{r}
2x^2 + \ x \ - 4 \quad \leftarrow \text{Quotient}\\
x - 3\overline{)2x^3 - 5x^2 - 7x + 20} \\
\underline{-(2x^3 - 6x^2)} \\
x^2 - 7x \\
\underline{-(x^2 - 3x)} \\
-4x + 20 \\
\underline{-(-4x + 12)} \\
8 \quad \leftarrow \text{Remainder}
\end{array}
$$

$$
\begin{array}{r}
2x^2 + x - 4 \\
x - 3\overline{)2x^3 - 5x^2 - 7x + 20} \\
\underline{-6x^2} \\
x^2 \\
\underline{-3x} \\
-4x \\
\underline{12} \\
8
\end{array}
$$

The x's that appear in the division on the right are not necessary if the positioning of each coefficient is done carefully. As long as the rightmost number under the division symbol is the constant, the number to its left is the coefficient of x, and so on, the x's can be removed.

$$
\begin{array}{r}
2x^2 + x - 4x \\
x - 3\overline{)2 -5 - 7 \ \ 20} \\
\boxed{-6} \\
1 \\
\boxed{-3} \\
4 \\
\boxed{12} \\
8
\end{array}
$$

This display can be made more compact by moving the rows up so that the boxed numbers align horizontally.

$$
\begin{array}{r}
2x^2 + x - 4 \\
x - 3\overline{)2 \ \ -5 \ \ -7 \ \ \ 20} \\
\underline{-6 \ \ -3 \ \ \ 12} \\
\Box \ \ \ 1 \ \ -4 \ \ \ \ 8
\end{array}
$$

Because the leading coefficient of the divisor is always 1, the leading coefficient of the dividend will always be the leading coefficient of the quotient. Thus the leading coefficient of the quotient, 2, is placed in the boxed position.

$$
\begin{array}{r}
2x^2 + x - 4 \\
x - 3\overline{)2 \quad -5 \quad -7 \quad 20} \\
\underline{-6 \quad -3 \quad 12} \\
2 \quad 1 \quad -4 \quad 8
\end{array}
$$

Notice that the first three numbers in Row 3 are the coefficients of the quotient. The last number in Row 3 is the remainder. Now, the top row above is not needed.

$$
\begin{array}{r}
x - 3\overline{)2 \quad -5 \quad -7 \quad 20} \quad \text{Row 1} \\
\underline{-6 \quad -3 \quad 12} \quad \text{Row 2 (subtract)} \\
2 \quad 1 \quad -4 \quad 8 \quad \text{Row 3}
\end{array}
$$

Remember, the entries in Row 3 are obtained by subtracting the entries in Row 2 from the entries in Row 1. Rather than subtracting the entries in Row 2, change the sign of each entry and then add. With this modification, the display becomes

$$
\begin{array}{r}
x - 3\overline{)2 \quad -5 \quad -7 \quad 20} \quad \text{Row 1} \\
\underline{6 \quad 3 \quad -12} \quad \text{Row 2 (add)} \\
2 \quad 1 \quad -4 \quad 8 \quad \text{Row 3}
\end{array}
$$

Notice that each entry in Row 2 is 3 times the entry that is one column to the left in Row 3 (for example, the 6 in Row 2 is 3 times 2; the 3 in Row 2 is 3 times 1, and so on). Remove the $x - 3$ and replace it with 3. The entries in Row 3 give us the quotient and remainder.

$$
\begin{array}{r}
3\overline{)2 \qquad -5 \qquad -7 \qquad 20} \quad \text{Row 1} \\
6 \qquad 3 \qquad -12 \quad \text{Row 2 (add)} \\
\underline{2^{3(2)\nearrow} \quad 1^{3(1)\nearrow} \quad -4^{3(-4)\nearrow} \quad 8} \quad \text{Row 3} \\
2x^2 \quad + x \quad - \quad 4 \qquad 8 \quad \boxed{\text{Remainder}} \\
\underbrace{\qquad\qquad\qquad\qquad}_{\text{Quotient}}
\end{array}
$$

Work Smart

If there are any missing powers of x in the dividend, insert a coefficient of 0 for the missing term when doing synthetic division.

(**EXAMPLE 1**)   **How to Use Synthetic Division to Divide Polynomials**

Use synthetic division to find the quotient and remainder when $3x^3 + 11x^2 + 14$ is divided by $x + 4$.

Step-by-Step Solution

Step 1: Write the dividend in standard form. Then copy the coefficients of the dividend. Remember to insert a 0 for any missing power of x.

$$
\begin{array}{l}
3x^3 + 11x^2 + 14 = 3x^3 + 11x^2 + 0x + 14 \\
\qquad\qquad\qquad 3 \qquad 11 \qquad 0 \qquad 14 \quad \text{Row 1}
\end{array}
$$

Step 2: Insert the division symbol. Rewrite the divisor in the form $x - c$ and insert the value of c to the left of the division symbol.

$$
\begin{array}{r}
x + 4 = x - (-4) \\
-4\overline{)3 \quad 11 \quad 0 \quad 14} \quad \text{Row 1}
\end{array}
$$

Step 3: Bring the 3 down two rows and enter it in Row 3.

$$
\begin{array}{r}
-4\overline{)3 \quad 11 \quad 0 \quad 14} \quad \text{Row 1} \\
\downarrow \qquad\qquad\qquad \text{Row 2} \\
\underline{} \\
3 \qquad\qquad\qquad \text{Row 3}
\end{array}
$$

Step 4: Multiply the latest entry in Row 3 by -4 and place the result in Row 2, one column over to the right.

$$
\begin{array}{r|rrrr}
-4)\!\!\! & 3 & 11 & 0 & 14 \ \ \text{Row 1}\\
& \downarrow & -12 & & \quad\ \ \text{Row 2}\\
\hline
& 3^{\,-4(3)\nearrow} & & & \quad\ \ \text{Row 3}
\end{array}
$$

Step 5: Add the entry in Row 2 to the entry above it in Row 1. Enter the sum in Row 3.

$$
\begin{array}{r|rrrr}
-4)\!\!\! & 3 & 11 & 0 & 14 \ \ \text{Row 1}\\
& \downarrow & -12 & & \quad\ \ \text{Row 2}\\
\hline
& 3^{\,-4(3)\nearrow} & -1 & & \quad\ \ \text{Row 3}
\end{array}
$$

Step 6: Repeat Steps 4 and 5 until no more entries are available in Row 1.

$$
\begin{array}{r|rrrr}
-4)\!\!\! & 3 & 11 & 0 & 14 \ \ \text{Row 1}\\
& \downarrow & -12 & 4 & -16 \ \ \text{Row 2}\\
\hline
& 3^{\,-4(3)\nearrow} & -1^{\,-4(-1)\nearrow} & 4^{\,-4(4)\nearrow} & -2 \ \ \text{Row 3}
\end{array}
$$

Step 7: The final entry in Row 3, -2, is the remainder; the other entries in Row 3 (3, -1 and 4) are the coefficients of the quotient, in descending order of degree. The quotient is a polynomial whose degree is one less than the degree of the dividend.

Quotient: $3x^2 - x + 4$
Remainder: -2

Step 8: Check
(Quotient)(Divisor) + Remainder = Dividend

$$
\begin{aligned}
(3x^2 - x + 4)&(x + 4) + (-2)\\
&= 3x^3 + 12x^2 - x^2 - 4x + 4x + 16 - 2\\
&= 3x^3 + 11x^2 + 14
\end{aligned}
$$

Therefore, $\dfrac{3x^3 + 11x^2 + 14}{x + 4} = 3x^2 - x + 4 - \dfrac{2}{x + 4}.$ ●

Let's do another example, where we consolidate all the steps given in Example 1.

EXAMPLE 2 ## Dividing Two Polynomials Using Synthetic Division

Use synthetic division to find the quotient and remainder when $x^4 - 5x^3 - 6x^2 + 33x - 15$ is divided by $x - 5$.

Solution
The divisor is $x - 5$, so $c = 5$.

$$
\begin{array}{r|rrrrr}
5)\!\!\! & 1 & -5 & -6 & 33 & -15\\
& \downarrow & 5 & 0 & -30 & 15\\
\hline
& 1 & 0 & -6 & 3 & 0
\end{array}
$$

The dividend is a fourth-degree polynomial, so the quotient is a third-degree polynomial, $x^3 + 0x^2 - 6x + 3 = x^3 - 6x + 3$, and the remainder is 0. Thus

$$
\frac{x^4 - 5x^3 - 6x^2 + 33x - 15}{x - 5} = x^3 - 6x + 3
$$ ●

In Example 2, because $\dfrac{x^4 - 5x^3 - 6x^2 + 33x - 15}{x - 5} = x^3 - 6x + 3$, we know that $x - 5$ and $x^3 - 6x + 3$ are factors of $x^4 - 5x^3 - 6x^2 + 33x - 15$. Therefore, we can write

$$
x^4 - 5x^3 - 6x^2 + 33x - 15 = (x - 5)(x^3 - 6x + 3)
$$

Quick ✓

1. *True or False* We can divide $-4x^3 + 5x^2 + 10x - 3$ by $x^2 - 2$ using synthetic division.

2. *True or False* We can divide $-4x^3 + 5x^2 + 10x - 3$ by $2x + 1$ using synthetic division.

3. To divide a polynomial by $x + 7$ using synthetic division, use $c =$ ___

In Problems 4–6, use synthetic division to find the quotient.

4. $\dfrac{2x^3 + x^2 - 7x - 13}{x - 2}$ 5. $\dfrac{x + 1 - 3x^2 + 4x^3}{x + 2}$ 6. $\dfrac{x^4 + 8x^3 + 15x^2 - 2x - 6}{x + 3}$

❷ Use the Remainder and Factor Theorems

In Example 1, synthetic division was used to find the quotient and remainder when $3x^3 + 11x^2 + 14$ is divided by $x + 4$. Letting $f(x) = 3x^3 + 11x^2 + 14$ yields $f(-4) = -2$, which is equal to the remainder when $3x^3 + 11x^2 + 14$ is divided by $x + 4$. So the value of the function f at $x = -4$ is the same as the remainder when f is divided by $x + 4 = x - (-4)$. This result is true in general! It is called the *Remainder Theorem*.

In Other Words

A theorem is a big idea that can be shown to be true in general. The word comes from a Greek verb meaning "to view."

The Remainder Theorem

Let f be a polynomial function. When $f(x)$ is divided by $x - c$, the remainder is $f(c)$.

EXAMPLE 3 **Using the Remainder Theorem**

Use the Remainder Theorem to find the remainder when $f(x) = 2x^3 - 3x + 8$ is divided by $x + 3$.

Solution

The divisor is $x + 3 = x - (-3)$, so the Remainder Theorem says that the remainder is $f(-3)$:

$$f(x) = 2x^3 - 3x + 8$$
$$f(-3) = 2(-3)^3 - 3(-3) + 8$$
$$= 2(-27) + 9 + 8$$
$$= -54 + 9 + 8$$
$$= -37$$

When $f(x) = 2x^3 - 3x + 8$ is divided by $x + 3$, the remainder is -37.

Check Using synthetic division, we find that the remainder is, in fact, -37.

$$
\begin{array}{r|rrrr}
-3) & 2 & 0 & -3 & 8 \\
    &   & -6 & 18 & -45 \\
\hline
    & 2 & -6 & 15 & -37 \leftarrow \text{Remainder}
\end{array}
$$

Quick ✓

7. Use the Remainder Theorem to find the remainder when
$f(x) = 3x^3 + 10x^2 - 9x - 4$ is divided by

(a) $x - 2$ (b) $x + 4$

APPENDIX A

Work Smart

"If and only if" statements are used to compress two statements into one. The Factor Theorem is two statements:

1. If $f(c) = 0$, then $x - c$ is a factor of f.

2. If $x - c$ is a factor of f, then $f(c) = 0$.

Example 2 showed that when the remainder is 0, the quotient and divisor are factors of the dividend. The Remainder Theorem can be used to determine whether an expression of the form $x - c$ is a factor of the dividend. This result is called the *Factor Theorem*.

> **The Factor Theorem**
>
> Let f be a polynomial function. Then $x - c$ is a factor of $f(x)$ if and only if $f(c) = 0$.

The Factor Theorem can be used to determine whether a polynomial has a particular factor.

EXAMPLE 4 **Using the Factor Theorem**

Use the Factor Theorem to determine whether the function $f(x) = 2x^3 - 3x^2 - 18x - 8$ has the factor

(a) $x - 3$ **(b)** $x + 2$

Solution

The Factor Theorem states that if $f(c) = 0$, then $x - c$ is a factor of f.

(a) Because $x - 3$ is of the form $x - c$ with $c = 3$, find the value of $f(3)$.

$$f(3) = 2(3)^3 - 3(3)^2 - 18(3) - 8 = -35 \neq 0$$

Since $f(3) \neq 0$, $x - 3$ is not a factor of f.

(b) Because $x + 2 = x - (-2)$ is of the form $x - c$ with $c = -2$, find the value of $f(-2)$. Rather than evaluating the function using substitution, use synthetic division.

$$
\begin{array}{r|rrrr}
-2 & 2 & -3 & -18 & -8 \\
   &   & -4 & 14 & 8 \\
\hline
   & 2 & -7 & -4 & 0 \leftarrow \text{Remainder}
\end{array}
$$

The remainder is 0, so $f(-2) = 0$, and therefore $x + 2$ is a factor of f.

This means the dividend can be written as the product of the quotient and divisor. The quotient is $2x^2 - 7x - 4$ and the divisor is $x + 2$, so

$$2x^3 - 3x^2 - 18x - 8 = (x + 2)(2x^2 - 7x - 4)$$

●

In Other Words

If $f(c) = 0$, then $f(x)$ can be written in factored form as $f(x) = (x - c)(\text{quotient})$

> **Quick ✓**
>
> **8.** Use the Factor Theorem to determine whether $x - c$ is a factor of $f(x) = 2x^3 - 9x^2 - 6x + 5$ for the given values of c. If $x - c$ is a factor, then write f in factored form. That is, write $f(x) = (x - c)(\text{quotient})$.
>
> **(a)** $c = -2$ **(b)** $c = 5$

A.1 Exercises MyMathLab® Exercise numbers in **green** have complete video solutions in MyMathLab or may be accessed using the QR code to the right.

Problems 1–8 are the Quick ✓s that follow the EXAMPLES.

Building Skills

In Problems 9–22, divide using synthetic division. See Objective 1.

9. $\dfrac{x^2 - 3x - 10}{x - 5}$

10. $\dfrac{x^2 + 4x - 12}{x - 2}$

11. $\dfrac{2x^2 + 11x + 12}{x + 4}$

12. $\dfrac{3x^2 + 19x - 40}{x + 8}$

13. $\dfrac{x^2 - 3x - 14}{x - 6}$

14. $\dfrac{x^2 + 2x - 17}{x - 4}$

15. $\dfrac{x^3 - 19x - 15}{x - 5}$

16. $\dfrac{x^3 - 13x - 17}{x + 3}$

17. $\dfrac{3x^4 - 5x^3 - 21x^2 + 17x + 25}{x - 3}$

18. $\dfrac{2x^4 - x^3 - 38x^2 + 16x + 103}{x + 4}$

19. $\dfrac{x^4 - 40x^2 + 109}{x + 6}$

20. $\dfrac{a^4 - 65a^2 + 55}{a - 8}$

21. $\dfrac{2x^3 + 3x^2 - 14x - 15}{x - \dfrac{5}{2}}$

22. $\dfrac{3x^3 + 13x^2 + 8x - 12}{x - \dfrac{2}{3}}$

In Problems 23–30, use the Remainder Theorem to find the remainder. See Objective 2.

23. $f(x) = x^2 - 5x + 1$ is divided by $x - 2$

24. $f(x) = x^2 + 4x - 5$ is divided by $x + 2$

25. $f(x) = x^3 - 2x^2 + 5x - 3$ is divided by $x + 4$

26. $f(x) = x^3 + 3x^2 - x + 1$ is divided by $x - 3$

27. $f(x) = 2x^3 - 4x + 1$ is divided by $x - 5$

28. $f(x) = 3x^3 + 2x^2 - 5$ is divided by $x + 3$

29. $f(x) = x^4 + 1$ is divided by $x - 1$

30. $f(x) = x^4 - 1$ is divided by $x - 1$

In Problems 31–38, use the Factor Theorem to determine whether $x - c$ is a factor of the given function for the given value of c. If $x - c$ is a factor, then write f in factored form. That is, write $f(x) = (x - c)(quotient)$. See Objective 2.

31. $f(x) = x^2 - 3x + 2; c = 2$

32. $f(x) = x^2 + 5x + 6; c = 3$

33. $f(x) = 2x^2 + 5x + 2; c = -2$

34. $f(x) = 3x^2 + x - 2; c = -1$

35. $f(x) = 4x^3 - 9x^2 - 49x - 30; c = 3$

36. $f(x) = 2x^3 - 9x^2 - 2x + 24; c = 1$

37. $f(x) = 4x^3 - 7x^2 - 5x + 6; c = -1$

38. $f(x) = 5x^3 + 8x^2 - 7x - 6; c = -2$

39. If $\dfrac{f(x)}{x - 5} = 3x + 5$, find $f(x)$.

40. If $\dfrac{f(x)}{x + 3} = 2x + 7$, find $f(x)$.

41. If $\dfrac{f(x)}{x - 3} = x + 8 + \dfrac{4}{x - 3}$, find $f(x)$.

42. If $\dfrac{f(x)}{x - 3} = x^2 + 2 + \dfrac{7}{x - 3}$, find $f(x)$.

Extending the Concepts

43. Find the sum of a, b, c, and d if
$$\dfrac{2x^3 - 3x^2 - 26x - 37}{x + 2} = ax^2 + bx + c + \dfrac{d}{x + 2}.$$

44. What is the remainder when
$f(x) = 2x^{30} - 3x^{20} + 4x^{10} - 2$ is divided by $x - 1$?

Explaining the Concepts

45. If f is a polynomial of degree n and it is divided by $x + 4$, the quotient will be a polynomial of degree $n - 1$. Explain why.

46. Explain the Remainder Theorem in your own words. Explain the Factor Theorem in your own words.

47. Suppose that you were asked to divide $8x^3 - 3x + 1$ by $x + 3$. Would you use long division or synthetic division? Why?

Appendix B Geometry Review

B.1 Lines and Angles

Objectives

1. Understand the Terms Point, Line, and Plane
2. Work with Angles
3. Find the Measures of Angles Formed by Parallel Lines

The word *geometry* comes from the Greek words *geo*, meaning "earth" and *metra*, meaning "measure." The Greek scholar Euclid collected and organized the geometry known in his day into a logical system more than two thousand years ago. Euclid's system forms the basis of the geometry we study today.

❶ Understand the Terms Point, Line, and Plane

A **point** has no size, only position, and is usually designated by a capital letter as shown below.

•
P

Figure 1

A **line** is a set of points extending infinitely far in opposite directions. A line has no width or height, just length, and is uniquely determined by two points. For example, the line in Figure 1 passes through the points *A* and *B*. The notation for the line shown is $\overleftrightarrow{AB}$.

Figure 2

A **ray** is a half-line with one **endpoint,** and extends infinitely far in one direction. See Figure 2. The notation for the ray shown is $\overrightarrow{AB}$.

A **line segment** is a portion of a line that has a beginning and an end. See Figure 3. If two line segments have the *same length*, they are said to be *congruent*. The notation for the congruent line segments shown are $\overline{AB}$ and $\overline{CD}$.

Figure 3

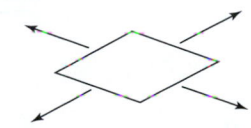

A **plane** is the set of points that forms a flat surface that extends indefinitely. A plane has no thickness. See Figure 4. The arrows indicate that the plane extends indefinitely in each direction.

Figure 4

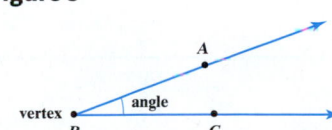

❷ Work with Angles

Suppose two rays with a common endpoint are drawn as in Figure 5. The amount of rotation from one ray to the second ray is called the **angle** between the rays. The common endpoint is called the **vertex.** In Figure 5, the name of the angle is $\angle ABC$, $\angle CBA$, or $\angle B$. Angles are measured in **degrees,** which is symbolized °. One full rotation represents 360°. The notation $m\angle A = 60°$ means "the measure of angle *A* is 60 degrees." Because 60° is $\frac{1}{6}$ of 360°, an angle whose measure is 60° is $\frac{1}{6}$ of a full rotation. Two angles with the same measure are congruent.

Some angles are classified by their measure.

Figure 5

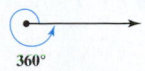

vertex
B   angle   *C*
A

> **In Other Words**
> In geometry, **congruent** means two figures are exactly the same size and shape.

Work Smart

One full rotation is represented below.

360°

Definitions

An angle that measures 90° is a **right angle.** The symbol ⌐ denotes a right angle. A right angle is $\frac{1}{4}$ of a full rotation. See Figure 6(a).

An angle whose measure is between 0° and 90° is an **acute angle.** See Figure 6(b).

An angle whose measure is between 90° and 180° is an **obtuse angle.** See Figure 6(c).

An angle whose measure is 180° is a **straight angle.** A straight angle is $\frac{1}{2}$ of a full rotation. See Figure 6(d).

Figure 6

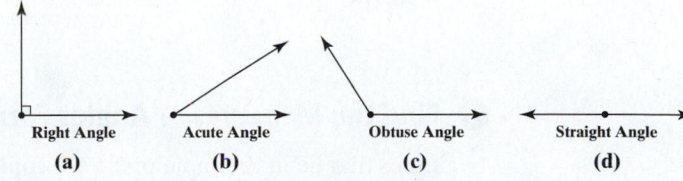

Right Angle	Acute Angle	Obtuse Angle	Straight Angle
(a)	(b)	(c)	(d)

Definitions

Two angles whose measures sum to 90° are **complementary** angles. Each angle is the **complement** of the other. Two angles whose measures sum to 180° are **supplementary** angles. Each angle is the **supplement** of the other. See Figure 7.

Figure 7

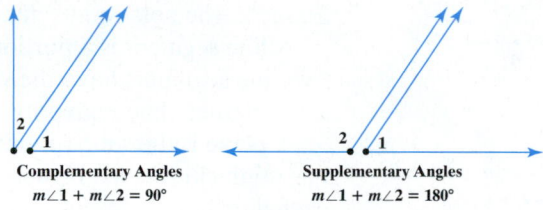

Complementary Angles
$m\angle 1 + m\angle 2 = 90°$

Supplementary Angles
$m\angle 1 + m\angle 2 = 180°$

EXAMPLE 1   **Finding the Complement of an Angle**

Find the complement of an angle whose measure is 18°.

Solution
Two angles are complementary if their sum is 90°. An angle that is complementary to an 18° angle measures $90° - 18° = 72°$. ●

EXAMPLE 2   **Finding the Supplement of an Angle**

Find the supplement of an angle whose measure is 97°.

Solution
Two angles are supplementary if their sum is 180°. An angle that is supplementary to a 97° angle measures $180° - 97° = 83°$. ●

❸ Find the Measures of Angles Formed by Parallel Lines

Lines that lie in the same plane are **coplanar.**

Definitions

- **Parallel lines** are lines in the same plane that never meet, as shown in Figure 8(a).
- **Intersecting lines** meet or cross in one point. See Figure 8(b).
- **Perpendicular lines** are two lines that intersect to form right angles (90°). See Figure 8(c).

Figure 8

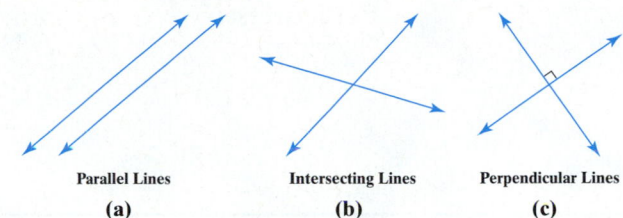

Parallel Lines   Intersecting Lines   Perpendicular Lines
(a)   (b)   (c)

Figure 9
$m\angle 1 = m\angle 3$
$m\angle 2 = m\angle 4$

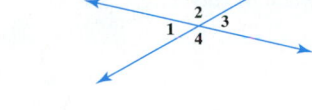

Two lines that intersect form four angles. Two angles that are opposite each other are called **vertical angles.** Vertical angles have equal measures. **Adjacent angles** have the same vertex and share a side. In Figure 9, angles 1 and 3 are vertical angles, and angles 1 and 2 are adjacent angles. Angles 1 and 2 are also supplementary angles. Other pairs of adjacent angles are angles 2 and 3, angles 3 and 4, and angles 1 and 4.

A line that cuts two parallel lines is called a **transversal.** In Figure 10, lines m and n are parallel, and the transversal is labeled t. In this figure, certain angles have special names.

Figure 10
m is parallel to n

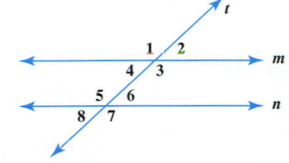

- There are four pairs of **corresponding angles:**

  $\angle 1$ and $\angle 5$ $\angle 2$ and $\angle 6$ $\angle 3$ and $\angle 7$ $\angle 4$ and $\angle 8$

- There are two pairs of **alternate interior angles:**

  $\angle 3$ and $\angle 5$ $\angle 4$ and $\angle 6$

- There are two pairs of **alternate exterior angles:**

  $\angle 1$ and $\angle 7$ $\angle 2$ and $\angle 8$

Angles formed by parallel lines are related in the following way.

Parallel Lines Cut by a Transversal

If two parallel lines are cut by a transversal, then

- Corresponding angles are equal in measure.
- Alternate interior angles are equal in measure.
- Alternate exterior angles are equal in measure.

EXAMPLE 3   **Finding the Measure of Corresponding and Alternate Interior Angles**

Given that lines m and n are parallel, t is a transversal, and the measure of angle 1 is 85°, find the measures of angles 2, 3, 4, 5, 6, and 7.

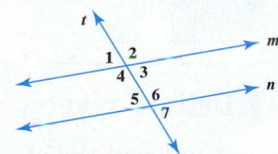

Solution
$m\angle 2 = 180° - 85° = 95°$ because $\angle 1$ and $\angle 2$ are supplementary angles.
$m\angle 3 = 85°$ because $\angle 1$ and $\angle 3$ are vertical angles.
$m\angle 4 = 180° - 85° = 95°$ because $\angle 3$ and $\angle 4$ are supplementary angles. $\angle 1$ and $\angle 4$ are also supplementary angles.
$m\angle 5 = 85°$ because $\angle 1$ and $\angle 5$ are corresponding angles (or because $\angle 3$ and $\angle 5$ are alternate interior angles).
$m\angle 6 = 95°$ because $\angle 2$ and $\angle 6$ are corresponding angles.
$m\angle 7 = 85°$ because $\angle 1$ and $\angle 7$ are alternate exterior angles.

Quick ✔

11. _____ lines are lines in the same plane that never meet.

12. Suppose two lines intersect. The two angles that are opposite each other are called _____ angles.

13. *True or False* If two parallel lines are cut by a transversal, the corresponding angles are equal in measure.

14. Find the measure of angles 1–7, given that lines *m* and *n* are parallel and *t* is a transversal.

B.1 Exercises MyMathLab®

Problems 1–14 are the Quick ✔s that follow the EXAMPLES.

Building Skills

In Problems 15–22, classify each angle as right, acute, obtuse, or straight. See Objective 2.

15.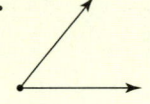

16.

17.

18.

19.

20.

21.

22.

In Problems 23–26, find the complement of each angle. See Objective 2.

23. 32°   24. 19°

25. 73°   26. 51°

In Problems 27–30, find the supplement of each angle. See Objective 2.

27. 67°   28. 145°

29. 8°   30. 106°

In Problems 31 and 32, find the measure of angles 1–7, given that lines m and n are parallel and t is a transversal. See Objective 3.

31.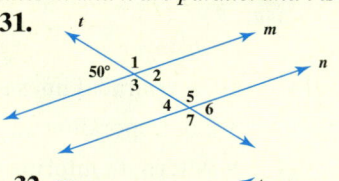

32.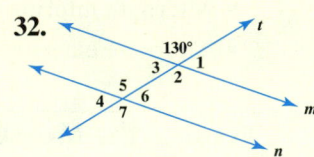

B.2 Polygons

Objectives

❶ Define Polygon
❷ Work with Triangles
❸ Identify Quadrilaterals
❹ Work with Circles

❶ Define Polygon

A **polygon** is a closed figure in a plane consisting of line segments that meet at the **vertices.** A **regular polygon** is a polygon in which the sides are congruent and the angles are congruent. Figure 11 shows four regular polygons.

Figure 11
Regular polygons: All the sides are the same length; all the angles have the same measure.

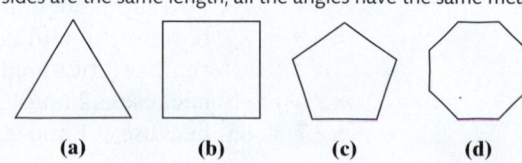

(a)   (b)   (c)   (d)

In Other Words

Vertices is the plural form of the word *vertex.* The word *polygon* comes from the Greek words *poly,* which means "many," and *gon,* which means "angle." See Figure 11 on the previous page.

A polygon is named according to the number of sides. Table 1 summarizes the names of the polygons with 3 to 10 sides. A **triangle** is a polygon with three sides. Figure 11(a) on the previous page is a regular triangle. A **quadrilateral** is a polygon with four sides. Figure 11(b) is a regular quadrilateral (a *square*). A **pentagon** is a polygon with five sides. Figure 11(c) is a regular pentagon. An **octagon** is a polygon with eight sides. Figure 11(d) shows a regular octagon.

Table 1

Polygons	
Number of Sides	**Name of Polygon**
3	Triangle
4	Quadrilateral
5	Pentagon
6	Hexagon
7	Heptagon
8	Octagon
9	Nonagon
10	Decagon

Figure 12

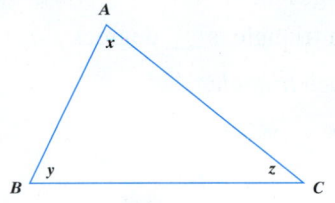

❷ Work with Triangles

In triangle ABC, shown in Figure 12, angles A, B, and C are called **interior angles.** The sum of the measures of the interior angles of a triangle is $180°$. If x, y, and z represent the measures of angles A, B, and C, respectively, then

$$x + y + z = 180°$$

EXAMPLE 1   **Finding the Measure of an Interior Angle of a Triangle**

Find the measure of angle C in the triangle shown in Figure 13.

Figure 13

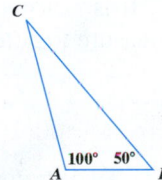

Solution

The measure of angle A is $100°$ and the measure of angle B is $50°$. Because the sum of the three interior angles of a triangle is $180°$, the measure of angle C can be found as follows:

$$m\angle C = 180° - 50° - 100° = 30°$$ ●

We classify triangles by the lengths of their sides. A triangle with three congruent sides is an **equilateral** triangle. A triangle with two congruent sides is an **isosceles** triangle, and a triangle with no congruent sides is a **scalene** triangle. See Figure 14.

Figure 14

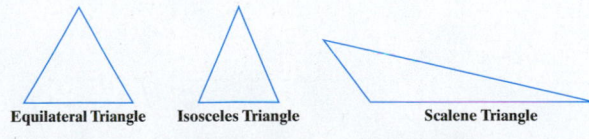

Equilateral Triangle   Isosceles Triangle   Scalene Triangle

Figure 15

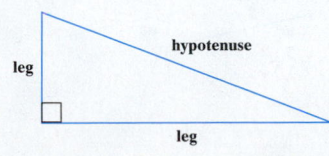

A **right triangle** is a triangle that contains a right ($90°$) angle. In a right triangle, the longest side is the **hypotenuse**, and the remaining two sides are the **legs**. See Figure 15.

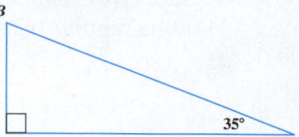

EXAMPLE 2   **Finding the Measure of an Angle of a Right Triangle**

Find the measure of angle B in the right triangle shown in Figure 16.

Figure 16

Solution

The sum of the interior angles of a triangle is 180° degrees and the right angle measures 90°, so

$$m\angle B = 180° - 90° - 35° = 55°$$

Quick ✓

1. A triangle in which two sides are congruent is a(n) _____ triangle.

2. A(n) ____ triangle is a triangle that contains a 90° angle.

3. The sum of the measures of the interior angles in a triangle is ___ degrees.

In Problems 4 and 5, find the measure of angle B in each triangle.

4.

5.

Congruent Triangles

Figure 17

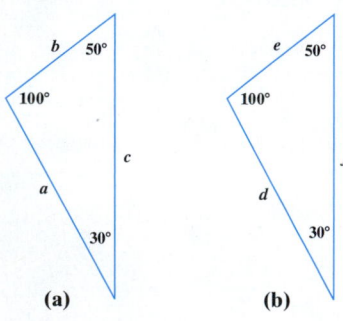

(a)   (b)

Work Smart

In geometry, "corresponding" refers to sides or angles that sit in the same position in two different figures.

In Other Words

A *postulate* is a statement that is taken to be true without proof.

Two triangles are **congruent** if the corresponding angles have the same measure and the corresponding sides have the same length. See Figures 17(a) and 17(b). The corresponding angles in triangles (a) and (b) are equal. Also, the lengths of the corresponding sides are equal: $a = d, b = e,$ and $c = f$.

It is not necessary to verify that all three angles and all three sides are the same measure to determine whether two triangles are congruent.

Determining Congruent Triangles

1. Two triangles are congruent if two pairs of corresponding angles and the length of the corresponding side between the two angles are equal. This is called the Angle-Side-Angle (ASA) postulate. See Figure 18(a).

2. Two triangles are congruent if the lengths of the corresponding sides of the triangles are equal. This is called the Side-Side-Side (SSS) postulate. See Figure 18(b).

3. Two triangles are congruent if the lengths of two pairs of corresponding sides and the measure of the angle between the two sides are equal. This is called the Side-Angle-Side (SAS) postulate. See Figure 18(c).

Figure 18

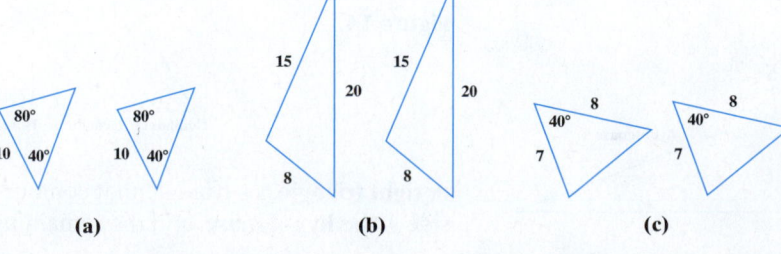

(a)   (b)   (c)

Figure 19

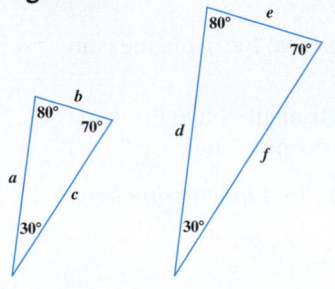

Work Smart

Congruent figures are exactly the same size and shape. Similar figures are the same shape but different sizes.

Similar Triangles

Two triangles are **similar** if the corresponding angles are equal and the lengths of the corresponding sides are proportional. That is, triangles are similar if they have the same shape. In Figure 19, the triangles are similar because the corresponding angles are equal and the corresponding sides are proportional: $\frac{a}{d} = \frac{b}{e} = \frac{c}{f}$. It is not necessary to verify that all the angles are congruent and all the sides are proportional to determine whether two triangles are similar.

> **Determining Similar Triangles**
>
> 1. Two triangles are similar if two pairs of corresponding angles are equal. See Figure 20(a).
> 2. Two triangles are similar if the lengths of all the corresponding sides of the triangles are proportional. See Figure 20(b).
> 3. Two triangles are similar if two pairs of corresponding sides are proportional and the angles between the corresponding sides are congruent. See Figure 20(c).

Figure 20

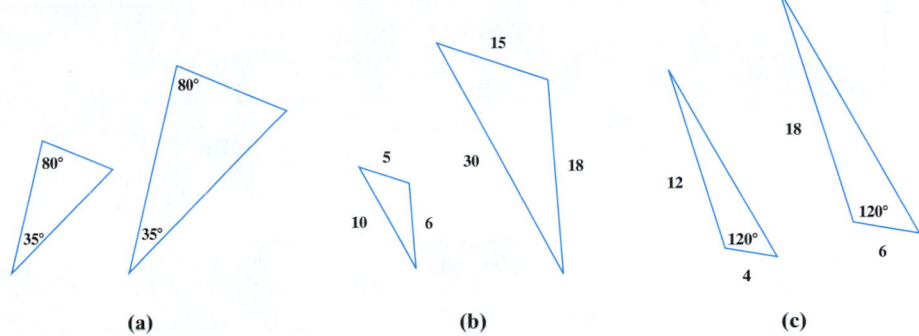

(a)   (b)   (c)

EXAMPLE 3 **Finding the Missing Length in Similar Triangles**

Given that the triangles in Figure 21 are similar, find the missing length.

Figure 21

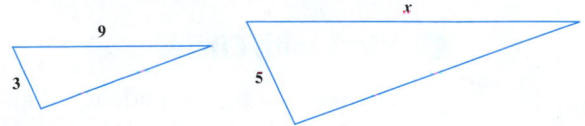

Solution

Because the triangles are similar, the corresponding sides are proportional. That is, $\frac{3}{5} = \frac{9}{x}$. Solve this equation for x.

$$\frac{3}{5} = \frac{9}{x}$$

Multiply both sides by the LCD, 5x: $5x\left(\frac{3}{5}\right) = 5x\left(\frac{9}{x}\right)$

Simplify: $3x = 45$

Divide both sides by 3: $x = 15$

The missing length is 15 units.

Quick ✓

6. Two triangles are _____ if the corresponding angles have the same measure and the corresponding sides have the same length.

7. Two triangles are _____ if corresponding angles of the triangles have the same measure and the lengths of the corresponding sides are proportional.

In Problem 8, given that the following triangles are similar, find the missing length.

8.

❸ Identify Quadrilaterals

A **quadrilateral** is a polygon with four sides. A **parallelogram** is a quadrilateral in which both pairs of opposite sides are parallel. A **rectangle** is a parallelogram that contains a right angle. A **square** is a rectangle with all sides of equal length. A **rhombus** is a parallelogram that has all sides equal in length. A **trapezoid** is a quadrilateral with exactly one pair of opposite sides that are parallel. See Table 2.

Table 2

Figure	Sketch	Figure	Sketch
Parallelogram		Rhombus	
Rectangle		Trapezoid	
Square			

❹ Work with Circles

Figure 22

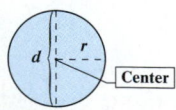

A **circle** is a figure made up of all points in the plane that are a fixed distance from a point called the **center.** A **radius** of the circle is any line segment drawn from the center of the circle to any point on the circle. A **diameter** of the circle is any line segment that has endpoints on the circle and passes through the center of the circle. See Figure 22. Notice that the length of the diameter d of a circle is twice the length of the radius r. That is, $d = 2r$.

EXAMPLE 4 **Finding the Length of a Diameter of a Circle**

Find the length of a diameter of a circle with radius 4 centimeters (cm).

Solution

The length of a diameter is twice the length of a radius.

$$d = 2r$$
$$d = 2 \cdot 4 \text{ cm}$$
$$d = 8 \text{ cm}$$

The diameter of the circle is 8 cm.

EXAMPLE 5   **Finding the Length of a Radius of a Circle**

Find the length of a radius of a circle with diameter 18 yards.

Solution

The length of a radius is one-half the length of a diameter.

$$r = \frac{1}{2}d$$

$$r = \frac{1}{2} \cdot 18 \text{ yards}$$

$$r = 9 \text{ yards}$$

The radius of the circle is 9 yards.

Quick ✓

9. A _____ of a circle is any line segment drawn from the center of the circle to any point on the circle.

10. *True or False* The length of a diameter of a circle is exactly twice the length of a radius.

In Problems 11–14, find the length of a radius or diameter of each circle.

11. $d = 15$ inches, find r.

12. $d = 24$ feet, find r.

13. $r = 3.6$ yards, find d.

14. $r = 9$ cm, find d.

B.2 Exercises   MyMathLab®

*Problems **1–14** are the **Quick ✓** s that follow the **EXAMPLES**.*

Building Skills

In Problems 15–18, find the measure of the missing angle of the triangle. See Objective 2.

15.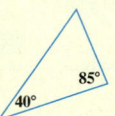

16.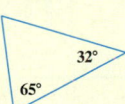

17.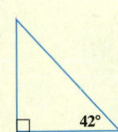

18.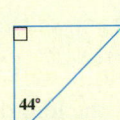

In Problems 19–22, determine the length of the missing side of the similar triangle. See Objective 2.

19.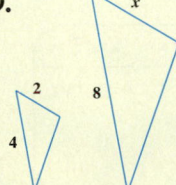

20.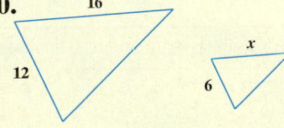

21.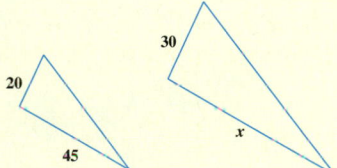

22.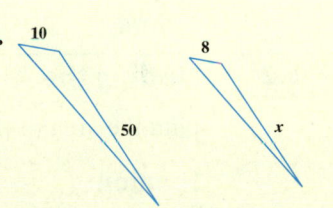

In Problems 23–26, find the length of a diameter of the circle. See Objective 4.

23. $r = 5$ in.

24. $r = 16$ feet

25. $r = 2.5$ cm

26. $r = 5.9$ in.

In Problems 27–30, find the length of a radius of the circle. See Objective 4.

27. $d = 14$ cm

28. $d = 58$ inches

29. $d = 11$ yards

30. $d = 27$ feet

B.3 Perimeter and Area of Polygons and Circles

Objectives

1 Find the Perimeter and Area of a Rectangle and a Square

2 Find the Perimeter and Area of a Parallelogram and a Trapezoid

3 Find the Perimeter and Area of a Triangle

4 Find the Circumference and Area of a Circle

The **perimeter** of a polygon is the distance around the polygon. Put another way, the perimeter of a polygon is the sum of the lengths of the sides.

1 Find the Perimeter and Area of a Rectangle and a Square

A rectangle is a polygon, so the perimeter of a rectangle is the sum of the lengths of the sides.

EXAMPLE 1 **Finding the Perimeter of a Rectangle**

Find the perimeter of the rectangle in Figure 23.

Solution

The perimeter of the rectangle is the sum of the lengths of the sides, so

$$\text{Perimeter} = 11 \text{ feet} + 8 \text{ feet} + 11 \text{ feet} + 8 \text{ feet}$$
$$= 38 \text{ feet}$$

●

Figure 23

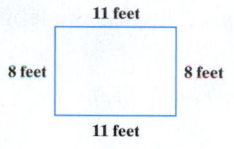

Did you notice that the perimeter from Example 1 can also be written as follows?

$$\text{Perimeter} = 2 \cdot 11 \text{ feet} + 2 \cdot 8 \text{ feet}$$

In general, the perimeter of a rectangle is $P = 2l + 2w$, where l is the length and w is the width of the rectangle.

A different measure of a polygon is its *area*. The **area** of a polygon is the amount of surface the polygon covers. Consider the rectangle shown in Figure 24. The number of 1-unit by 1-unit squares within the rectangle shows that the area of this rectangle is 6 square units. The area can also be found by multiplying the number of units of length by the number of units of width. In other words, the area of a rectangle is the product of its length and width.

Figure 24

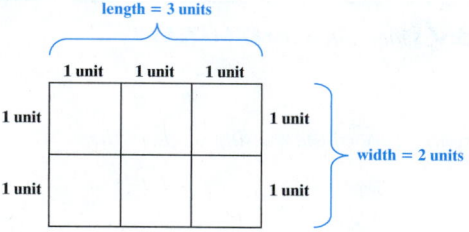

EXAMPLE 2 **Finding the Area of a Rectangle**

Find the area of the rectangle in Figure 25.

Figure 25

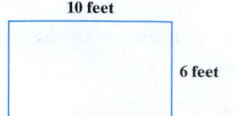

Solution

The area of the rectangle is the product of the length and the width, so

$$\text{Area} = 6 \text{ feet} \cdot 10 \text{ feet}$$
$$= 60 \text{ square feet}$$

●

Below are the formulas for the perimeter and area of a rectangle.

Figure	Sketch	Perimeter	Area
Rectangle		$P = 2l + 2w$	$A = lw$

EXAMPLE 3 **Finding the Perimeter and Area of a Rectangle**

Find (a) the perimeter and (b) the area of the rectangle shown in Figure 26.

Solution

Figure 26

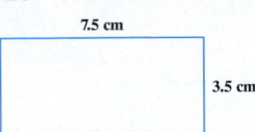

7.5 cm

3.5 cm

(a) The perimeter of the rectangle is

$$P = 2l + 2w$$
$$= 2 \cdot 7.5 \text{ cm} + 2 \cdot 3.5 \text{ cm}$$
$$= 15 \text{ cm} + 7 \text{ cm}$$
$$= 22 \text{ cm}$$

(b) The area of the rectangle is

$$A = lw$$
$$= 7.5 \text{ cm} \cdot 3.5 \text{ cm}$$
$$= 26.25 \text{ square cm}$$ ●

Quick ✔

1. The _____ of a polygon is the distance around the polygon.

2. The ____ of a polygon is the amount of surface the polygon covers.

In Problems 3 and 4, find the perimeter and area of each rectangle.

3. 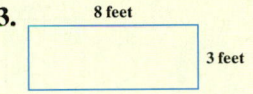 8 feet

3 feet

4. 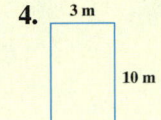 3 m

10 m

A square is a rectangle that has four congruent sides, so the perimeter of a square is

$$\text{Perimeter} = \text{side} + \text{side} + \text{side} + \text{side} = 4 \cdot \text{side} = 4s$$

where s is the length of each side.

EXAMPLE 4 **Finding the Perimeter of a Square**

Find the perimeter of the square in Figure 27.

Figure 27

3 cm

Solution

The perimeter of the square is

$$\text{Perimeter} = 4s$$
$$= 4 \cdot 3 \text{ cm}$$
$$= 12 \text{ cm}$$ ●

The area of a rectangle is the product of the length and the width. In a square, the sides are congruent, so

$$\text{Area} = \text{side} \cdot \text{side} = \text{side}^2 = s^2$$

where s is the length of each side of the square.

EXAMPLE 5 **Finding the Area of a Square**

Figure 28

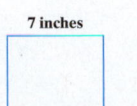

7 inches

Find the area of the square in Figure 28.

Solution

The area of the square is

$$\text{Area} = s^2$$
$$= (7 \text{ inches})^2$$
$$= 49 \text{ square inches}$$ ●

Below are the formulas for the perimeter and area of a square.

Figure	Sketch	Perimeter	Area
Square		$P = 4s$	$A = s^2$

Quick ✓

5. *True or False* The area of a square is the sum of the lengths of the sides.

In Problems 6 and 7, find the perimeter and area of each square.

6. 4 cm

7. 1.5 yards

(**EXAMPLE 6**) **Finding the Perimeter and Area of a Geometric Figure**

Find (a) the perimeter and (b) the area of the region shown in Figure 29.

Figure 29

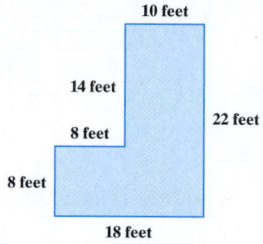

Solution

(a) The perimeter is the distance around the polygon, so

$$\text{Perimeter} = 8 \text{ feet} + 8 \text{ feet} + 14 \text{ feet} + 10 \text{ feet} + 22 \text{ feet} + 18 \text{ feet}$$
$$= 80 \text{ feet}$$

Figure 30

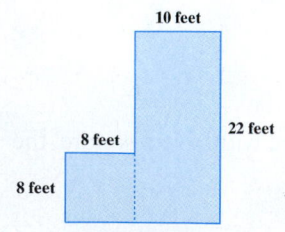

(b) The region can be divided into an 8-foot by 8-foot square plus a 10-foot by 22-foot rectangle. See Figure 30. The total area is the area of the square plus the area of the rectangle.

$$\text{Area} = \text{area of square} + \text{area of rectangle}$$
$$= (8 \text{ feet})^2 + (10 \text{ feet})(22 \text{ feet})$$
$$= 64 \text{ square feet} + 220 \text{ square feet}$$
$$= 284 \text{ square feet}$$

Quick ✓

In Problem 8, find the perimeter and area of the figure.

8.

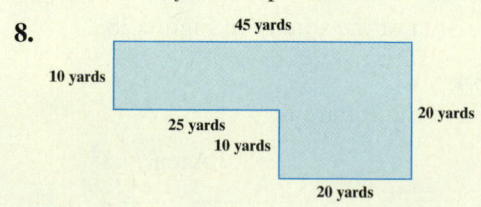

EXAMPLE 7 | **Painting a Room**

You've decided to paint your rectangular bedroom. Two walls are 14 feet long and 8 feet high, and the other two walls are 10 feet long and 8 feet high.

(a) Ignoring the window and door openings in the bedroom, what is the area of the walls in your bedroom?

(b) You know that 1 gallon of paint will cover 300 square feet. How many 1-gallon cans of paint must you purchase to paint your bedroom?

Solution

(a) The bedroom has two walls that are 14 feet long and 8 feet high. The area of these two walls is

$$\text{Area} = 2lw$$
$$= 2 \cdot 14 \text{ feet} \cdot 8 \text{ feet}$$
$$= 224 \text{ square feet}$$

The area of the other two walls is

$$\text{Area} = 2lw$$
$$= 2 \cdot 10 \text{ feet} \cdot 8 \text{ feet}$$
$$= 160 \text{ square feet}$$

The total area to be painted is

$$\text{Area} = 224 \text{ square feet} + 160 \text{ square feet}$$
$$= 384 \text{ square feet}$$

(b) A 1-gallon can of paint covers 300 square feet. You have 384 square feet to paint, so you will need to purchase 2 gallons of paint. ●

❷ Find the Perimeter and Area of a Parallelogram and a Trapezoid

Recall that a parallelogram is a quadrilateral with parallel opposite sides. A trapezoid is a quadrilateral with exactly one pair of opposite sides that are parallel. The following table gives the formulas for the perimeter and area of parallelograms and trapezoids.

Figure	Sketch	Perimeter	Area
Parallelogram		$P = 2a + 2b$	$A = bh$
Trapezoid		$P = a + b + c + B$	$A = \dfrac{1}{2}h(b + B)$

EXAMPLE 8 | **Finding the Perimeter and Area of a Parallelogram**

Find (a) the perimeter and (b) the area of parallelogram shown in Figure 31.

Solution

(a) The perimeter of the parallelogram is

$$P = 2a + 2b$$
$$= 2 \cdot 12 \text{ feet} + 2 \cdot 5 \text{ feet}$$
$$= 24 \text{ feet} + 10 \text{ feet}$$
$$= 34 \text{ feet}$$

Figure 31

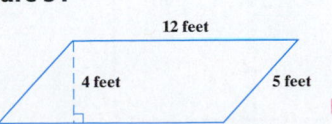

(b) The area of the parallelogram is

$$A = bh$$
$$= 12 \text{ feet} \cdot 4 \text{ feet}$$
$$= 48 \text{ square feet}$$

●

EXAMPLE 9 **Finding the Perimeter and Area of a Trapezoid**

Find (a) the perimeter and (b) the area of the trapezoid shown in Figure 32.

Solution

Figure 32

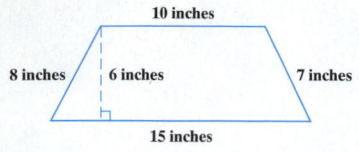

(a) The perimeter of the trapezoid is

$$\text{Perimeter} = 7 \text{ inches} + 8 \text{ inches} + 10 \text{ inches} + 15 \text{ inches}$$
$$= 40 \text{ inches}$$

(b) The area of the trapezoid is

$$A = \frac{1}{2} h (b + B)$$

$$= \frac{1}{2} \cdot 6 \text{ inches} \cdot (10 \text{ inches} + 15 \text{ inches})$$

$$= \frac{1}{2} \cdot 6 \text{ inches} \cdot 25 \text{ inches}$$

$$= 75 \text{ square inches}$$

●

Quick ✓

9. To find the area of a trapezoid, use the formula $A =$ _____ , where _ is the height of the trapezoid and the bases have lengths _ and _ .

In Problems 10 and 11, find the perimeter and area of each figure.

10.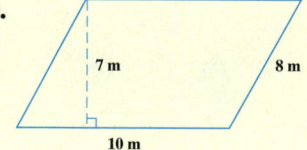

11.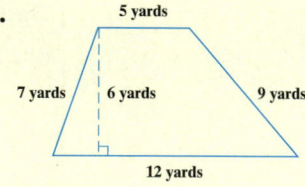

---

❸ Find the Perimeter and Area of a Triangle

Recall that a triangle is a polygon with three sides. The following table gives the formulas for the perimeter and area of a triangle.

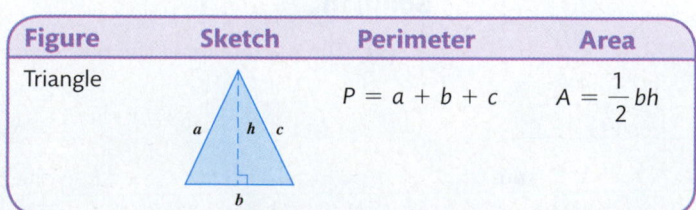

Figure	Sketch	Perimeter	Area
Triangle		$P = a + b + c$	$A = \frac{1}{2} bh$

EXAMPLE 10 **Finding the Perimeter and Area of a Triangle**

Find (a) the perimeter and (b) the area of the triangle shown in Figure 33.

Figure 33

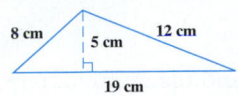

Solution

(a) To find the perimeter of the triangle, add the lengths of the three sides of the triangle.

$$\text{Perimeter} = a + b + c$$
$$= 8 \text{ cm} + 12 \text{ cm} + 19 \text{ cm}$$
$$= 39 \text{ cm}$$

(b) Use $A = \dfrac{1}{2} bh$ with base $b = 19$ cm and height $h = 5$ cm.

$$A = \frac{1}{2} bh$$
$$= \frac{1}{2} \cdot 19 \text{ cm} \cdot 5 \text{ cm}$$
$$= 47.5 \text{ square cm}$$ ●

Quick ✓

12. *True or False* The area of a triangle is given by the formula $A = \dfrac{1}{2} bh$, where b is the base and h is the height.

In Problems 13 and 14, find the perimeter and area of each triangle.

13.

6 mm 5 mm 3 mm

8 mm

14.

5 feet 13 feet

12 feet

4 Find the Circumference and Area of a Circle

The **circumference** of a circle is the distance around a circle. A diameter or a radius of a circle can be used to find the circumference of a circle according to the formulas given below. The formula for the area of a circle is also given.

Figure	Sketch	Perimeter	Area
Circle	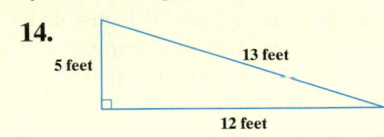	$C = \pi d$ where d is the length of a diameter $C = 2\pi r$ where r is the length of a radius	$A = \pi r^2$

EXAMPLE 11 **Finding the Circumference of a Circle**

Figure 34

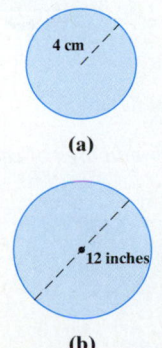

(a)

12 inches

(b)

Find the exact and approximate circumferences of the circles in Figure 34. Use 3.1415 for π and round to the nearest hundredth.

Solution

(a) The length of the radius is 4 cm, so use the formula $C = 2\pi r$.

$$C = 2\pi r$$
$r = 4$ cm: $\quad = 2 \cdot \pi \cdot 4 \text{ cm}$
$$= 8 \cdot \pi \text{ cm}$$
Use a calculator: $\quad \approx 25.13 \text{ cm}$

The exact circumference of the circle is 8π cm, and the approximate circumference is 25.13 cm, rounded to the nearest hundredth of a centimeter.

(b) The length of the diameter is 12 inches, so use $C = \pi d$.

$$C = \pi d$$

$d = 12$ inches: $= \pi \cdot 12$ inches

$= 12\pi$ inches

Use a calculator: ≈ 37.70 inches

The exact circumference of the circle is 12π inches, and the approximate circumference is 37.70 inches, rounded to the nearest hundredth of an inch. ●

EXAMPLE 12 **Finding the Area of a Circle**

Find the exact and approximate area of the circle in Figure 35. Use 3.1415 for π and round to the nearest hundredth.

Figure 35

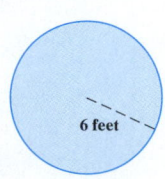
6 feet

Solution

The circle in Figure 35 has a radius whose length is 6 feet, so substitute $r = 6$ feet in the equation $A = \pi r^2$.

$$A = \pi r^2$$

$$= \pi \cdot (6 \text{ feet})^2$$

$$= \pi (36) \text{ square feet}$$

$$= 36\pi \text{ square feet}$$

Use a calculator: ≈ 113.10 square feet

The area of the circle is exactly 36π square feet, or approximately 113.10 square feet. ●

Work Smart

Units of measure are important and should always be included in the answer. Perimeter is measured with linear units such as feet (ft), inches (in.), or centimeters (cm). Area is measured with squares (square feet or ft², square inches or in.², square centimeters or cm²).

Work Smart

For calculations involving circles, to find the exact value means leave π in the answer. For approximate values, it is best to use the π key on your calculator. Because not everyone has access to technology, we used 3.1415 for π. Answers may vary slightly.

Quick ✓

15. The _____ of a circle is the distance around the circle.

16. *True or False* The area of a circle is given by the formula $A = \pi d^2$.

In Problems 17 and 18, find the exact and approximate circumference and area of each circle. Use 3.1415 for π and round to the nearest hundredth.

17.

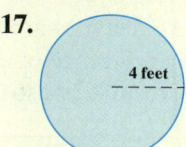
4 feet

18.

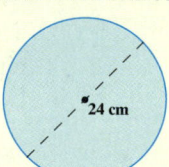
24 cm

B.3 Exercises MyMathLab®

*Problems **1–18** are the Quick ✓s that follow the EXAMPLES.*

Building Skills

In Problems 19–22, find the perimeter and area of each rectangle. See Objective 1.

19.

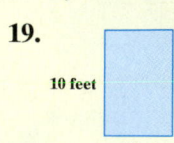
10 feet
4 feet

20.

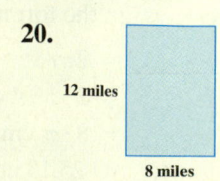
12 miles
8 miles

21.

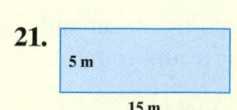
5 m
15 m

22.

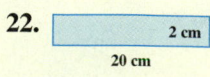
2 cm
20 cm

In Problems 23 and 24, find the perimeter and area of each square. See Objective 1.

23.   **24.**

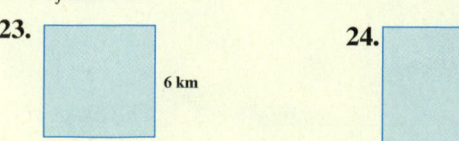
6 km
13 yards

In Problems 25–28, find the perimeter and area of each figure. See Objective 1.

25.

8 feet
15 feet
6 feet
7 feet
14 feet
22 feet

26.

8 m
3 m
5 m
5 m
2 m
3 m

27.

13 m
2 m
8 m
2 m
8 m
2 m

28.

5 yards
20 yards
5 yards
10 yards
10 yards

In Problems 29–36, find the perimeter and area of each quadrilateral. See Objective 2.

29.

6 feet
5 feet
9 feet

30.

4 cm
2 cm
14 cm

31.

9 mm
10 mm
4 mm

32.

19 in.
20 in.
8 in.

33.

8 in.
8 in.
7 in.
8 in.
16 in.

34.

10 m
6 m
5 m
6 m
14 m

35.

8 cm
8 cm
7 cm
10 cm
19 cm

36.

11 feet
8 feet
5 feet
6 feet
20 feet

In Problems 37–40, find the perimeter and area of each triangle. See Objective 3.

37.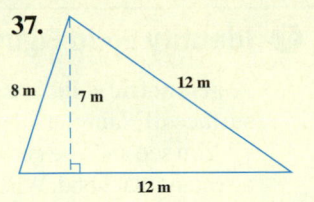

8 m
7 m
12 m
12 m

38.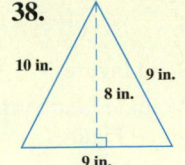

10 in.
8 in.
9 in.
9 in.

39.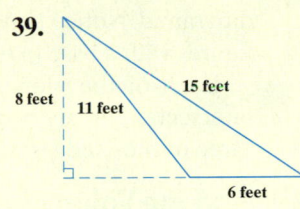

8 feet
11 feet
15 feet
6 feet

40.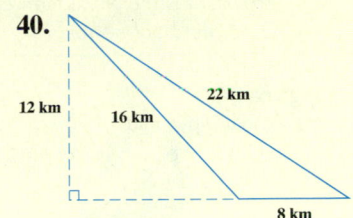

12 km
16 km
22 km
8 km

In Problems 41–44, find (a) the circumference and (b) the area of each circle. For both the circumference and the area, provide exact answers and approximate answers rounded to the nearest hundredth. Use 3.1415 for π. See Objective 4.

41.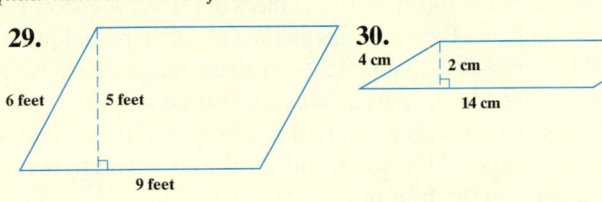

16 in.

42.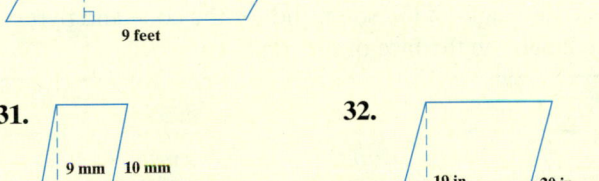

9 mm

43.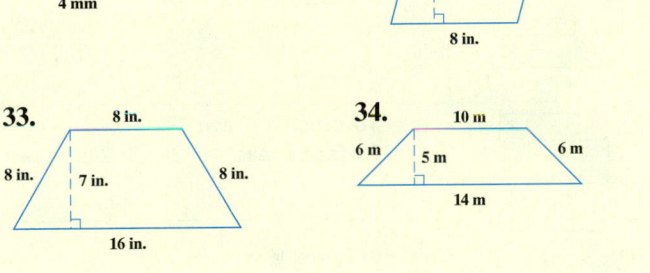

20 cm

44.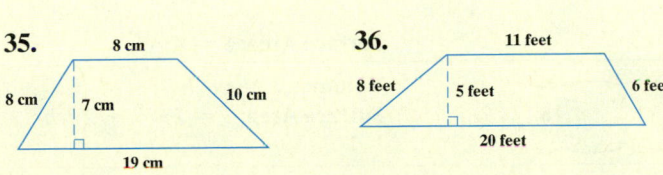

3 miles

Applying the Concepts

In Problems 45 and 46, find the exact area of the shaded region.

45.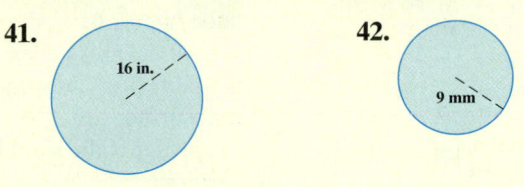

2
2

46.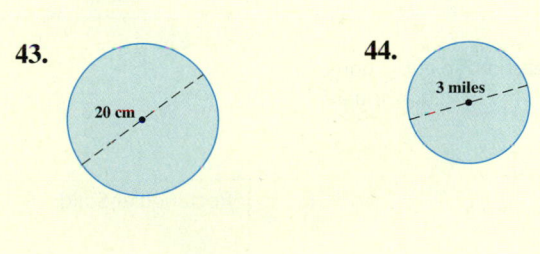

2
2

47. How many feet, to the nearest hundredth, will a wheel with a diameter of 20 inches have traveled after five revolutions?

48. How many feet, to the nearest hundredth, will a wheel with a diameter of 18 inches have traveled after three revolutions?

B.4 Volume and Surface Area

Objectives

❶ Identify Solid Figures
❷ Find the Volumes and Surface Areas of Solid Figures

> **In Other Words**
> *Polyhedra* is the plural form of the Greek word *polyhedron*.

Figure 36
Hexagonal Dipyramid

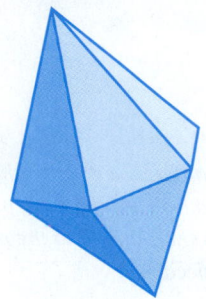

Figure 37

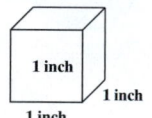

1 inch
1 inch
1 inch

Work Smart

Volume is measured in cubic units. Surface area is measured in square units.

❶ Identify Solid Figures

A **geometric solid** is a three-dimensional region of space enclosed by planes and curved surfaces. Examples of geometric solids are cubes, pyramids, spheres, cylinders, and cones.

You see and use geometric solids every day. When you grab a box of cereal, you are holding a rectangular solid. When you reach for a can of soup, you are reaching for a circular cylinder.

We first discuss *polyhedra*. A **polyhedron** is a three-dimensional solid formed by connecting polygons. Figure 36 shows an example of a polyhedron called a hexagonal dipyramid. Notice that the top and bottom are hexagonal pyramids, or three-dimensional figures with a base (a hexagon) and *faces* that are triangles (one for each side of the base).

Each of the planes of a polyhedron is called a **face.** A line segment that is the intersection of any two faces of a polyhedron is called an **edge.** A point of intersection of three or more edges is called a **vertex.**

❷ Find the Volumes and Surface Areas of Solid Figures

The **volume** of a polyhedron is the measure of the number of units of space contained in the solid. Volume can be used to describe the amount of soda in a can or the amount of cereal in a box. Volume is measured in cubic units. For example, the cube in Figure 37 represents 1 cubic inch.

The **surface area** of a polyhedron is the sum of the areas of the faces of the polyhedron. For example, because each face in the cube shown in Figure 37 has an area of 1 square inch, and a cube has six faces, the surface area of the cube is 6 square inches. Because surface area is the sum of the areas of each polygon in the polyhedron, surface area is measured in square units.

Table 3 shows some common geometric solids, along with formulas to find their volume and surface area. Note h represents the height of the solid, and for the cone and pyramid, s represents the slant height, or the height on the face of the solid.

Table 3

Solids		Formulas
Cube		**Volume:** $V = s^3$ **Surface Area:** $S = 6s^2$
Rectangular Solid		**Volume:** $V = lwh$ **Surface Area:** $S = 2lw + 2lh + 2wh$
Sphere		**Volume:** $V = \dfrac{4}{3}\pi r^3$ **Surface Area:** $S = 4\pi r^2$
Right Circular Cylinder		**Volume:** $V = \pi r^2 h$ **Surface Area:** $S = 2\pi r^2 + 2\pi rh$
Cone		**Volume:** $V = \dfrac{1}{3}\pi r^2 h$ **Surface Area:** $S = \pi r^2 + \pi rs$
Square Pyramid		**Volume:** $V = \dfrac{1}{3}b^2 h$ **Surface Area:** $S = b^2 + 2bs$

Figure 38

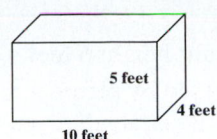

EXAMPLE 1 · **Finding the Volume and Surface Area of a Rectangular Solid**

Find the volume and surface area of the rectangular solid shown in Figure 38.

Solution
From Figure 38, $l = 10$ feet, $h = 5$ feet, and $w = 4$ feet. The volume of the rectangular solid is

$$V = lwh$$
$$= (10 \text{ feet})(4 \text{ feet})(5 \text{ feet})$$
$$= 200 \text{ cubic feet}$$

The surface area of the rectangular solid is

$$S = 2lw + 2lh + 2wh$$
$$= 2(10 \text{ feet})(4 \text{ feet}) + 2(10 \text{ feet})(5 \text{ feet}) + 2(4 \text{ feet})(5 \text{ feet})$$
$$= 220 \text{ square feet}$$

●

EXAMPLE 2 · **Finding the Volume of a Right Circular Cylinder**

Figure 39

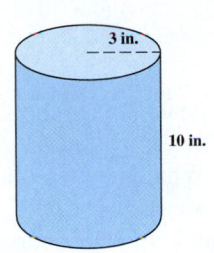

Find the exact and approximate volume and surface area of the right circular cylinder shown in Figure 39. Use 3.1415 for π and round to the nearest hundredth.

Solution
From Figure 39, $h = 10$ inches and $r = 3$ inches. The volume of the right circular cylinder is
$$V = \pi r^2 h$$
$$= \pi (3 \text{ in.})^2 (10 \text{ in.})$$
$$= 90\pi \text{ in.}^3$$

Use a calculator: $\approx 282.74 \text{ in.}^3$

The volume of the right circular cylinder is exactly 90π cubic inches and approximately 282.74 cubic inches.

The surface area of the right circular cylinder is
$$V = 2\pi r^2 + 2\pi rh$$
$$= 2\pi (3 \text{ in.})^2 + 2\pi (3 \text{ in.})(10 \text{ in.})$$
$$= 18\pi \text{ in.}^2 + 60\pi \text{ in.}^2$$
$$= 78\pi \text{ in.}^2$$

Use a calculator: $\approx 245.04 \text{ in.}^2$

The surface area of the right circular cylinder is exactly 78π square inches and approximately 245.04 square inches.

●

Quick ✓

1. A _____ is a three-dimensional solid formed by connecting polygons.

2. The _____ of a polyhedron is the sum of the areas of the faces of the polyhedron.

3. *True or False* The volume of a right circular cylinder is measured in square units.

4. *True or False* The volume of a rectangular solid is the product of its length, width, and height.

In Problems 5 and 6, find the volume and surface area of the solid. For Problem 6, find the exact and approximate values. Use 3.1415 for π and round to the nearest hundredth.

5.

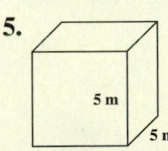

6.

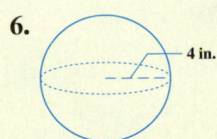

B.4 Exercises MyMathLab®

*Problems **1–6** are the Quick ✔ s that follow the* **EXAMPLES**.

Building Skills

See Objective 2.

7. Find the volume *V* and surface area *S* of a rectangular box with length 10 feet, width 5 feet, and height 12 feet.

8. Find the volume *V* and surface area *S* of a rectangular box with length 2 meters, width 6 meters, and height 9 meters.

9. Find the volume *V* and surface area *S* of a square pyramid with height 10 feet, slant height 12 feet, and base 8 feet.

10. Find the volume *V* and surface area *S* of a square pyramid with height 5 m, slant height 8 m, and base 10 m.

In Problems 11–16, find the exact and approximate volume, V, and surface area, S. Use 3.1415 for π and round to the nearest hundredth, if necessary.

11. A sphere with radius 6 centimeters.

12. A sphere with radius 10 inches.

13. A right circular cylinder with radius 2 inches and height 8 inches.

14. A right circular cylinder with radius 3 inches and height 6 inches.

15. A cone with radius 6 mm, slant height 10 mm, and height 8 mm.

16. A cone with radius 4 feet, slant height 5 feet, and height 3 feet.

Applying the Concepts

17. Rain Gutter A rain gutter is in the shape of a rectangular solid. How much water, in cubic inches, can the gutter hold if it is 4 inches in height, 3 inches wide, and 12 feet long?

18. Water for the Horses A trough for horses in the shape of a rectangular solid is 10 feet long, 2 feet wide, and 3 feet deep. How much water can the trough hold?

19. A Can of Peaches A can of peaches is in the shape of a right circular cylinder. The can has a 4-inch diameter and is 6 inches tall. What is the volume of the can? What is the surface area of the can? Express your answers as decimals rounded to the nearest hundredth.

20. Coffee Can A coffee can is in the shape of a right circular cylinder. The can has an 8-inch diameter and is 10 inches tall. What is the volume of the can? What is the surface area of the can? Express your answers as decimals rounded to the nearest hundredth.

21. Ice Cream Cone A waffle cone for ice cream has a diameter of 8 cm and a height of 16 cm. How much ice cream can the cone hold if the ice cream is flush with the top of the cone? Express your answer as a decimal rounded to the nearest hundredth.

22. Water Cooler The cups at a water cooler are cone-shaped. How much water can a cup hold if it has a 5-inch diameter and is 8 inches in height? Express your answer as a decimal rounded to the nearest hundredth.

Appendix C More on Systems of Linear Equations

C.1 A Review of Systems of Linear Equations in Two Variables

Objectives

1. Determine Whether an Ordered Pair Is a Solution of a System of Linear Equations
2. Solve a System of Two Linear Equations by Graphing
3. Solve a System of Two Linear Equations by Substitution
4. Solve a System of Two Linear Equations by Elimination
5. Identify Inconsistent Systems
6. Write the Solution of a System with Dependent Equations

Are You Prepared for This Section?

Before getting started, complete the following problems. If you get a problem wrong, go back to the section cited and review the material.

P1. Evaluate $2x - 3y$ for $x = 5, y = 4$. [Section 1.8, pp. 64–65]

P2. Determine whether the point $(4, -1)$ is on the graph of the equation $2x - 3y = 11$. [Section 3.1, pp. 173–174]

P3. Graph: $y = 3x - 7$ [Section 3.4, pp. 206–208]

P4. Find the equation of the line parallel to $y = -3x + 1$ containing the point $(2, 3)$. [Section 3.6, pp. 223–224]

P5. Determine the slope and y-intercept of $4x - 3y = 15$. [Section 3.4, pp. 205–206]

P6. What is the additive inverse of 4? [Section 1.4, pp. 30–31]

P7. Solve: $2x - 3(-3x + 1) = -36$ [Section 2.2, pp. 92–94]

Recall from Section 3.2 that an equation in two variables is linear if it can be written in the form $Ax + By = C$, where $A, B,$ and C are real numbers and A and B are not both zero.

A **system of linear equations** is a grouping of two or more linear equations, each of which contains one or more variables.

EXAMPLE 1

Examples of Systems of Linear Equations

(a) $\begin{cases} 2x + y = 5 \\ x - 5y = -10 \end{cases}$ Two linear equations containing two variables, x and y

(b) $\begin{cases} x + 3y + z = 8 \\ 3x - y + 6z = 12 \\ -4x - y + 2z = -1 \end{cases}$ Three linear equations containing three variables, $x, y,$ and z

Use a brace, as shown in the systems in Example 1, to indicate that a *system* of equations is involved. In this section, concentrate on systems of two linear equations containing two variables, such as the system in Example 1(a).

1 Determine Whether an Ordered Pair Is a Solution of a System of Linear Equations

A **solution** of a system of equations consists of values for the variables that are solutions of each equation of the system. Represent the solution of a system of two linear equations containing two unknowns as an ordered pair, (x, y).

EXAMPLE 2

Determining Whether an Ordered Pair Is a Solution of a System of Linear Equations

Determine whether the given ordered pairs are solutions of the system of equations.

$$\begin{cases} 2x + 3y = 9 \\ -5x - 3y = 0 \end{cases}$$

(a) $(6, -1)$ **(b)** $(-3, 5)$

Solution

To make it easier to describe this process, call $2x + 3y = 9$ equation (1) and $-5x - 3y = 0$ equation (2).

$$\begin{cases} 2x + 3y = 9 & \text{(1)} \\ -5x - 3y = 0 & \text{(2)} \end{cases}$$

(continued)

Prepared?...Answers **P1.** -2 **P2.** Yes

P3.

P4. $y = -3x + 9$ **P5.** slope $= \dfrac{4}{3}$; y-intercept $= -5$ **P6.** 4 **P7.** $\{-3\}$

(a) Let $x = 6$ and $y = -1$ in both equations (1) and (2). If both equations are true, then $(6, -1)$ is a solution.

Equation (1):	$2x + 3y = 9$	Equation (2):	$-5x - 3y = 0$
$x = 6, y = -1$:	$2(6) + 3(-1) \overset{?}{=} 9$	$x = 6, y = -1$:	$-5(6) - 3(-1) \overset{?}{=} 0$
	$12 - 3 \overset{?}{=} 9$		$-30 + 3 \overset{?}{=} 0$
	$9 = 9$ True		$-27 = 0$ False

> **Work Smart**
>
> A solution of a system must satisfy *all* of the equations in the system.

Although $x = 6$, and $y = -1$ satisfy equation (1), they do not satisfy equation (2); therefore, $(6, -1)$ is not a solution of the system of equations.

(b) Let $x = -3$ and $y = 5$ in both equations (1) and (2). If both equations are true, then $(-3, 5)$ is a solution.

Equation (1):	$2x + 3y = 9$	Equation (2):	$-5x - 3y = 0$
$x = -3, y = 5$:	$2(-3) + 3(5) \overset{?}{=} 9$	$x = -3, y = 5$:	$-5(-3) - 3(5) \overset{?}{=} 0$
	$-6 + 15 \overset{?}{=} 9$		$15 - 15 \overset{?}{=} 0$
	$9 = 9$ True		$0 = 0$ True

Because $x = -3$, and $y = 5$ satisfy both equations (1) and (2), the ordered pair $(-3, 5)$ is a solution of the system of equations.   ●

> **Work Smart**
>
> It is a good idea to number the equations in a system so that it is easier to keep track of your work.

When solving homework problems, number each equation as in Example 1.

Quick ✔

1. A _____ __ ____ _____ is a grouping of two or more linear equations, each of which contains one or more variables.

2. A _____ of a system of equations consists of values of the variables that satisfy each equation of the system.

3. Which of the following points is a solution of the system of equations?

$$\begin{cases} 2x + 3y = 7 \\ 3x + y = -7 \end{cases}$$

  (a) $(2, 1)$ **(b)** $(-4, 5)$ **(c)** $(-2, -1)$

Visualizing the Solutions in a System of Two Linear Equations

The graph of each equation in a system of linear equations is a line, so a system of two linear equations containing two variables represents a pair of lines. The graphs of the two lines can appear in one of three ways:

1. **Intersect:** If the lines intersect, then the system of equations has one solution given by the point of intersection. The system is **consistent** and the equations are **independent**. See Figure 1(a).

2. **Parallel:** If the lines are parallel, then the system of equations has no solution because the lines never intersect. The system is **inconsistent** and the equations are **independent**. See Figure 1(b).

3. **Coincident:** If the lines lie on top of each other (are coincident), then the system of equations has infinitely many solutions. The solution set is the set of all points on the line. The system is **consistent** and the equations are **dependent**. See Figure 1(c).

Figure 1

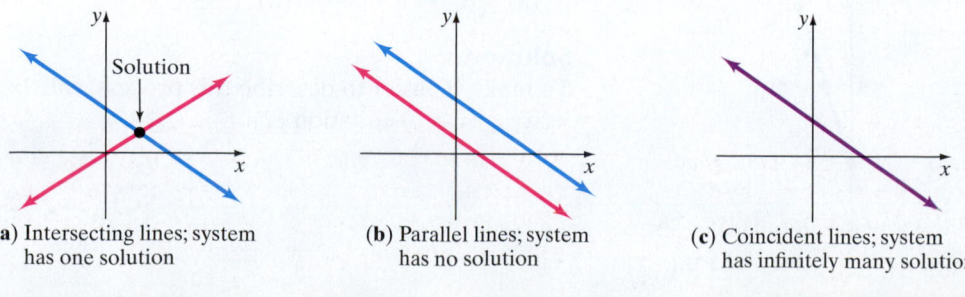

(a) Intersecting lines; system has one solution

(b) Parallel lines; system has no solution

(c) Coincident lines; system has infinitely many solutions

For now, concentrate on solving systems that have a single solution.

▶ ❷ Solve a System of Two Linear Equations by Graphing

The solution of a system of equations is the point or points of intersection, if any, of the two graphs, because both equations are satisfied at the point(s) of intersection.

EXAMPLE 3 **Solving a System of Two Linear Equations Using Graphing**

Solve the following system by graphing: $\begin{cases} x + y = -1 \\ -2x + y = -7 \end{cases}$

Solution

First, name $x + y = -1$ equation (1) and name $-2x + y = -7$ equation (2).

$$\begin{cases} x + y = -1 & (1) \\ -2x + y = -7 & (2) \end{cases}$$

Figure 2

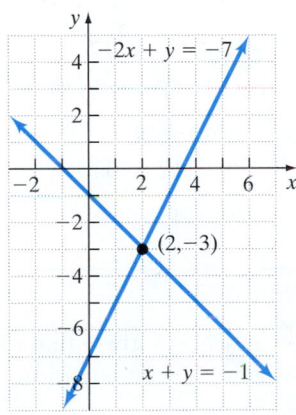

In order to graph each equation, put them in slope-intercept form. Equation (1) in slope-intercept form is $y = -x - 1$, which has slope -1 and y-intercept $(0, -1)$. Equation (2) in slope-intercept form is $y = 2x - 7$, which has slope 2 and y-intercept $(0, -7)$. Figure 2 shows their graphs. Note that the lines could have been graphed using the intercepts. The lines appear to intersect at $(2, -3)$, so the ordered pair $(2, -3)$ is the solution to the system.

Check Let $x = 2$ and $y = -3$ in both equations in the system:

Equation (1):	$x + y = -1$	Equation (2):	$-2x + y = -7$
$x = 2, y = -3$:	$2 + (-3) \overset{?}{=} -1$	$x = 2, y = -3$:	$-2(2) + (-3) \overset{?}{=} -7$
	$2 - 3 \overset{?}{=} -1$		$-4 - 3 \overset{?}{=} -7$
	$-1 = -1$ True		$-7 = -7$ True

Both equations are satisfied, so the solution is the ordered pair $(2, -3)$. ●

▶ ❸ Solve a System of Two Linear Equations by Substitution

Work Smart

Obtaining exact solutions using graphical methods can be difficult. Therefore, algebraic methods are usually preferred.

Finding the exact point of intersection of two lines can be difficult if the x- and y-coordinates of that point are not integers. Therefore, algebraic methods, rather than graphical methods, for solving systems of equations may be preferred. One algebraic method is *substitution*. The goal of the substitution method is to obtain a single linear equation involving a single unknown.

EXAMPLE 4   **How to Solve a System of Two Equations by Substitution**

Solve the following system by substitution: $\begin{cases} 3x + y = -9 & (1) \\ -2x + 3y = 17 & (2) \end{cases}$

Step-by-Step Solution

Step 1: Solve one of the equations for one of the unknowns.

It is easier to solve equation (1) for y since the coefficient of y is 1.

$$3x + y = -9$$

Subtract 3x from both sides: $\qquad y = -3x - 9$

Step 2: Substitute $-3x - 9$ for y in equation (2).

Equation (2): $\qquad -2x + 3y = 17$

$$-2x + 3(-3x - 9) = 17$$

Step 3: Solve the equation for x.

Distribute the 3: $\qquad -2x - 9x - 27 = 17$

Combine like terms: $\qquad -11x - 27 = 17$

Add 27 to both sides: $\qquad -11x = 44$

Divide both sides by -11: $\qquad x = -4$

Step 4: Substitute -4 for x in the equation from Step 1.

$$y = -3x + 9$$
$$y = -3(-4) - 9$$
$$y = 12 - 9$$
$$y = 3$$

Step 5: Check Verify that $x = -4$ and $y = 3$ satisfy each equation in the original system.

Equation (1): $\qquad 3x + y = -9$
$$3(-4) + 3 \overset{?}{=} -9$$
$$-12 + 3 \overset{?}{=} -9$$
$$-9 = -9 \quad \text{True}$$

Equation (2): $\qquad -2x + 3y = 17$
$$-2(-4) + 3(3) \overset{?}{=} 17$$
$$8 + 9 \overset{?}{=} 17$$
$$17 = 17 \quad \text{True}$$

Both equations are satisfied, so the solution is the ordered pair $(-4, 3)$.

Work Smart

When using substitution, if possible, solve for the variable whose coefficient is 1 or -1 in order to simplify the algebra.

> **Solving a System of Two Linear Equations by Substitution**
>
> **Step 1:** Solve one of the equations for one of the unknowns. Choose the equation that is easier to solve for a variable. Typically, this would be an equation that has a variable whose coefficient is 1 or -1.
>
> **Step 2:** Substitute the expression found in Step 1 into the *other* equation. The result will be a single linear equation in one unknown.
>
> **Step 3:** Solve the linear equation in one unknown found in Step 2.
>
> **Step 4:** Substitute the value of the variable found in Step 3 into one of the original equations to find the value of the other variable.
>
> **Step 5:** Check your answer by substituting the ordered pair into both of the original equations.

EXAMPLE 5   **Solving a System of Two Equations by Substitution**

Solve the following system by substitution: $\begin{cases} 2x - 3y = -6 & (1) \\ -8x + 3y = 3 & (2) \end{cases}$

Solution

Solve equation (1) for x because it seems easiest.

Equation (1): $\quad 2x - 3y = -6$

Add 3y to both sides: $\qquad 2x = 3y - 6$

Divide both sides by 2: $\qquad x = \dfrac{3y - 6}{2}$

Divide both terms in the numerator by 2: $\qquad x = \dfrac{3}{2}y - 3$

Now substitute $\dfrac{3}{2}y - 3$ for x in equation (2) and then solve for y.

Equation (2): $\qquad -8x + 3y = 3$

$$-8\left(\frac{3}{2}y - 3\right) + 3y = 3$$

Distribute the -8: $\quad -8 \cdot \dfrac{3}{2}y - (-8) \cdot 3 + 3y = 3$

Multiply: $\qquad -12y + 24 + 3y = 3$

Combine like terms: $\qquad -9y + 24 = 3$

Subtract 24 from both sides: $\qquad -9y = -21$

Divide both sides by -9: $\qquad y = \dfrac{-21}{-9}$

$$y = \frac{7}{3}$$

Now substitute $\dfrac{7}{3}$ for y in $x = \dfrac{3}{2}y - 3$ to find the value of x.

$$x = \frac{3}{2}\left(\frac{7}{3}\right) - 3$$

Multiply: $\quad x = \dfrac{7}{2} - 3$

$$x = \frac{1}{2}$$

Check Let $x = \dfrac{1}{2}$ and $y = \dfrac{7}{3}$ in both equations in the system:

Equation (1): $\qquad 2x - 3y = -6$ $\qquad\qquad$ Equation (2): $\qquad -8x + 3y = 3$

$$2\left(\frac{1}{2}\right) - 3\left(\frac{7}{3}\right) \overset{?}{=} -6 \qquad\qquad -8\left(\frac{1}{2}\right) + 3\left(\frac{7}{3}\right) \overset{?}{=} 3$$

$$1 - 7 \overset{?}{=} -6 \qquad\qquad\qquad -4 + 7 \overset{?}{=} 3$$

$$-6 = -6 \quad \text{True} \qquad\qquad\qquad 3 = 3 \quad \text{True}$$

Both equations are satisfied, so the solution is the ordered pair $\left(\dfrac{1}{2}, \dfrac{7}{3}\right)$. $\qquad$ ●

Quick ✓

In Problems 11 and 12, solve the system using substitution.

11. $\begin{cases} y = -3x - 5 \\ 5x + 3y = 1 \end{cases}$ $\qquad\qquad$ **12.** $\begin{cases} 2x + y = -2 \\ -3x - 2y = -2 \end{cases}$

▶ ❹ Solve a System of Two Linear Equations by Elimination

Using substitution in Example 5 led to equations containing fractions. A second algebraic method for solving a system of linear equations is the *elimination method*. This method is usually preferred over the substitution method if substitution leads to fractions.

The basic idea in using elimination is to get the coefficients of one of the variables to be additive inverses, or opposites, such as 5 and -5, so that the equations can be added together to yield a single linear equation involving one unknown.

Let's go over an example that shows how to solve a system of linear equations by elimination.

EXAMPLE 6 **How to Solve a System of Linear Equations by Elimination**

Solve: $\begin{cases} 5x + 2y = -5 & (1) \\ -2x - 4y = -14 & (2) \end{cases}$

Step-by-Step Solution

Step 1: The first goal is to get the coefficients on one of the variables to be additive inverses.

Make the coefficients of y additive inverses by multiplying equation (1) by 2.

$\begin{cases} 5x + 2y = -5 & (1) \\ -2x - 4y = -14 & (2) \end{cases}$

Multiply both sides of (1) by 2: $\begin{cases} 2(5x + 2y) = 2(-5) & (1) \\ -2x - 4y = -14 & (2) \end{cases}$

Use the Distributive Property: $\begin{cases} 10x + 4y = -10 & (1) \\ -2x - 4y = -14 & (2) \end{cases}$

Step 2: Now add equations (1) and (2) to eliminate the variable y and then solve for x.

$\begin{cases} 10x + 4y = -10 & (1) \\ -2x - 4y = -14 & (2) \end{cases}$

Add (1) and (2): $\qquad 8x \qquad = -24$

Divide both sides by 8: $\qquad x = -3$

The x-coordinate of the solution is -3.

Step 3: Substitute   for x in either equation (1) or (2) and solve for y. We will use equation (1).

Equation (1): $\qquad 5x + 2y = -5$

$x = -3$: $\quad 5(-3) + 2y = -5$

$-15 + 2y = -5$

Add 15 to both sides: $\qquad 2y = 10$

Divide both sides by 2: $\qquad y = 5$

The y-coordinate of the solution is 5. The solution appears to be $(-3, 5)$.

Step 4: Check Verify that $x = -3$ and $y = 5$ satisfy each equation in the original system.

Equation (1): $\qquad 5x + 2y = -5$

$5(-3) + 2(5) \stackrel{?}{=} -5$

$-15 + 10 \stackrel{?}{=} -5$

$-5 = -5$ **True**

Equation (2): $\qquad -2x - 4y = -14$

$-2(-3) - 4(5) \stackrel{?}{=} -14$

$6 - 20 \stackrel{?}{=} -14$

$-14 = -14$ **True**

The solution is the ordered pair $(-3, 5)$. ●

The following steps can be used to solve a system of linear equations by elimination.

> **Solving a System of Linear Equations by Elimination**
>
> **Step 1:** Write both equations of the system in standard form, $Ax + By = C$. If necessary, multiply both sides of one equation or both equations by a nonzero constant so that the coefficients of one of the variables are additive inverses.
>
> **Step 2:** Add equations (1) and (2) to eliminate the variable whose coefficients are now additive inverses. Solve the resulting equation for the unknown.
>
> **Step 3:** Substitute the value of the variable found in Step 2 into one of the original equations to find the value of the remaining variable.
>
> **Step 4:** Check the answer by substituting the ordered pair in both of the original equations.

What allows us to add two equations and use the result to replace an equation? Remember, an equation is a statement that the left side equals the right side. When equation (2) is added to equation (1), the same quantity is added to both sides of equation (1).

EXAMPLE 7 **Solving a System of Linear Equations by Elimination**

Solve: $\begin{cases} \dfrac{5}{2}x + 2y = 5 & (1) \\ \dfrac{3}{2}x + \dfrac{3}{2}y = \dfrac{9}{4} & (2) \end{cases}$

Solution

Because both equations have fractions, the first goal is to clear the fractions by multiplying both sides of equation (1) by 2 and both sides of equation (2) by 4.

$$\begin{cases} \dfrac{5}{2}x + 2y = 5 & (1) \\ \dfrac{3}{2}x + \dfrac{3}{2}y = \dfrac{9}{4} & (2) \end{cases}$$

Work Smart

If you are not afraid of fractions, you can eliminate x by multiplying both sides of equation (1) by -3 and both sides of equation (2) by 5. Try it!

Multiply both sides of (1) by 2:

Multiply both sides of (2) by 4:

$$\begin{cases} 2\left(\dfrac{5}{2}x + 2y\right) = 2 \cdot 5 & (1) \\ 4\left(\dfrac{3}{2}x + \dfrac{3}{2}y\right) = 4 \cdot \dfrac{9}{4} & (2) \end{cases}$$

Use the Distributive Property:

$$\begin{cases} 5x + 4y = 10 & (1) \\ 6x + 6y = 9 & (2) \end{cases}$$

Work Smart

Although it is not necessary to divide both sides of equation (2) by 3, it makes solving the problem easier. Do you see why?

Divide both sides of equation (2) by 3:

$$\begin{cases} 5x + 4y = 10 & (1) \\ 2x + 2y = 3 & (2) \end{cases}$$

Multiply both sides of equation (2) by -2:

Add (1) and (2):

$$\begin{cases} 5x + 4y = 10 & (1) \\ -4x - 4y = -6 & (2) \\ \hline \quad x \quad\quad = 4 \end{cases}$$

Now substitute 4 for x into either original equation. In this case, substitute into equation (1).

Equation (1): $\dfrac{5}{2}x + 2y = 5$

$x = 4$: $\dfrac{5}{2}(4) + 2y = 5$

$10 + 2y = 5$

Subtract 10 from both sides: $2y = -5$

Divide both sides by 2: $y = -\dfrac{5}{2}$

It appears that $x = 4$ and $y = -\dfrac{5}{2}$.

The check is left to you. The solution is the ordered pair $\left(4, -\dfrac{5}{2}\right)$. ●

Quick ✓

13. The basic idea in using the elimination method is to get the coefficients of one of the variables to be _____ _____, such as 5 and -5.

In Problems 14–16, solve the system using elimination. **Note:** *Problem 14 is the system that we solved by using substitution in Example 5.*

14. $\begin{cases} 2x - 3y = -6 \\ -8x + 3y = 3 \end{cases}$

15. $\begin{cases} -2x + y = 4 \\ -5x + 3y = 7 \end{cases}$

16. $\begin{cases} -3x + 2y = 3 \\ 4x - 3y = -6 \end{cases}$

Work Smart

The figure below shows a consistent and independent system.

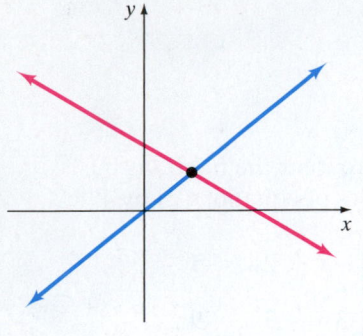

Which Method Should I Use?

When should each of the three methods be used for solving systems of two linear equations containing two unknowns? Suggestions are given below.

Method	Advantages/Disadvantages	When Should I Use It?
Graphical	This method allows the answer to be seen, but if the solutions are not integers, it can be difficult to determine the solution.	When a visual solution is required
Substitution	This method gives exact solutions. The algebra can be easy if one of the variables has a coefficient of 1.	If one of the coefficients of the variables is 1 or one of the variables is already isolated (as in $x =$ or $y =$)
Elimination	This method gives exact solutions. It is easy to use when none of the variables has a coefficient of 1.	If both equations are in standard form ($Ax + By = C$)

▶ ⑤ Identify Inconsistent Systems

Examples 2 through 7 dealt only with consistent and independent systems of equations—that is, systems with a single solution. Let's now consider systems that are inconsistent, which means the lines representing their equations are parallel.

EXAMPLE 8

An Inconsistent System

Solve: $\begin{cases} 3x + 2y = 2 & (1) \\ -6x - 4y = 8 & (2) \end{cases}$

Solution

Use the elimination method to solve this system because none of the variables has a coefficient of 1. Notice that the coefficients of x can be made additive inverses by multiplying equation (1) by 2.

$$\begin{cases} 3x + 2y = 2 & (1) \\ -6x - 4y = 8 & (2) \end{cases}$$

Multiply (1) by 2: $\begin{cases} 2(3x + 2y) = 2(2) & (1) \\ -6x - 4y = 8 & (2) \end{cases}$

Use the Distributive Property: $\begin{cases} 6x + 4y = 4 & (1) \\ -6x - 4y = 8 & (2) \end{cases}$

Add (1) and (2): $\qquad\qquad 0 = 12$

Work Smart

When solving a system of equations, if you end up with a statement "0 = some nonzero constant," or a false statement with no variables, the system is inconsistent. Graphically, the lines representing the system are parallel.

The equation $0 = 12$ is false. Conclude that the system has no solution, so the solution set is $\varnothing$ or $\{\ \}$. The system is inconsistent. ●

Figure 3 shows the graph of the lines from Example 8. Notice that the two lines both have slope $-\dfrac{3}{2}$. Equation (1) has a y-intercept of $(0, 1)$, and equation (2) has a y-intercept of $(0, -2)$. Therefore, the lines are parallel and do not intersect. This geometric statement is equivalent to the algebraic statement that the system has no solution.

Figure 3

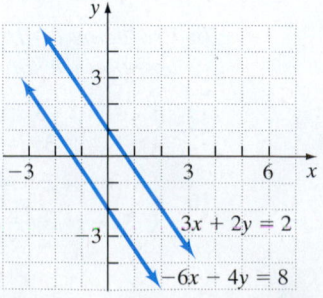

▶ ❻ Write the Solution of a System with Dependent Equations

Let's now consider systems that are consistent but have dependent equations, which means the lines representing their equations are coinciding (in other words, the same line).

EXAMPLE 9   **Solving a System with Dependent Equations**

Solve: $\begin{cases} 3x + y = 1 & (1) \\ -6x - 2y = -2 & (2) \end{cases}$

Solution

Use the substitution method because solving equation (1) for y is straightforward.

Equation (1):	$3x + y = 1$
Subtract 3x from both sides:	$y = -3x + 1$

Substitute $-3x + 1$ for y in equation (2).

Equation (2):	$-6x - 2y = -2$
	$-6x - 2(-3x + 1) = -2$
Distribute the -2:	$-6x + 6x - 2 = -2$
	$-2 = -2$

The equation $-2 = -2$ is true. This means that any values of x and y that satisfy $3x + y = 1$ or $-6x - 2y = -2$ are solutions. Some ordered pairs that satisfy both equations are $(0, 1)$, $(1, -2)$, and $(2, -5)$.

The system is consistent and its equations are dependent (the value of y that makes the equation true depends on the value of x), so there are infinitely many solutions. Write the solution in either of two equivalent ways:

$$\{(x, y) \,|\, 3x + y = 1\} \qquad \text{or} \qquad \{(x, y) \,|\, -6x - 2y = -2\} \qquad \bullet$$

Figure 4 illustrates the system in Example 9. The graphs of the two equations are lines, and both have slope -3 and y-intercept $(0, 1)$. The lines are coincident.

Look back at the equations in Example 9. Notice that the terms in equation (2) are -2 times the terms in equation (1). This is another way to identify dependent equations when given two equations with two unknowns.

Figure 4

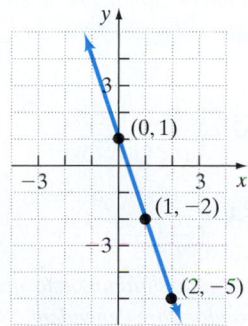

In Other Words

The solution to a consistent system with dependent equations is "the set of all ordered pairs such that one of the equations in the system is true."

C.1 Exercises MyMathLab®

Exercise numbers in **green** have complete video solutions in MyMathLab or may be accessed using the QR code to the right.

Problems 1–22 are the Quick ✔s that follow the EXAMPLES.

Building Skills

In Problems 23–26, determine whether the ordered pairs listed are solutions of the system of linear equations. See Objective 1.

23. $\begin{cases} 2x + y = 13 \\ -5x + 3y = 6 \end{cases}$
  (a) $(5, 3)$
  (b) $(3, 7)$

24. $\begin{cases} x - 2y = -11 \\ 3x + 2y = -1 \end{cases}$
  (a) $(-5, 3)$
  (b) $(-3, 4)$

25. $\begin{cases} 5x + 2y = 9 \\ -10x - 4y = -18 \end{cases}$
  (a) $(1, 2)$
  (b) $\left(2, -\dfrac{1}{2}\right)$

26. $\begin{cases} -3x + y = 5 \\ 6x - 2y = 6 \end{cases}$
  (a) $(-2, -1)$
  (b) $(2, 0)$

In Problems 27–30, use the graph of the system to determine whether the system is consistent or inconsistent. If consistent, indicate whether the equations are independent or dependent. See Objective 1.

27. $\begin{cases} x + y = 1 \\ x + 2y = 0 \end{cases}$

28. $\begin{cases} -2x + y = 4 \\ 2x + y = 0 \end{cases}$

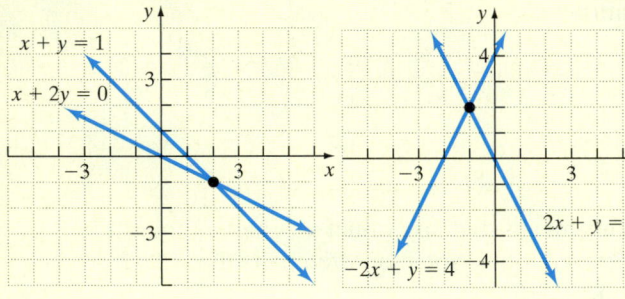

29. $\begin{cases} x - 2y = -2 \\ x - 2y = 2 \end{cases}$

30. $\begin{cases} 3x + y = 1 \\ -6x - 2y = -2 \end{cases}$

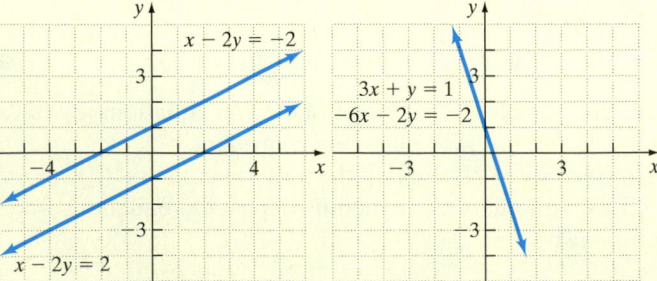

In Problems 31–34, solve the system of equations by graphing. See Objective 2.

31. $\begin{cases} y = 3x \\ y = -2x + 5 \end{cases}$

32. $\begin{cases} y = -2x + 4 \\ y = 2x - 4 \end{cases}$

33. $\begin{cases} 2x + y = 2 \\ x + 3y = -9 \end{cases}$

34. $\begin{cases} -x + 2y = -9 \\ 2x + y = -2 \end{cases}$

In Problems 35–42, solve the system of equations using substitution. See Objective 3.

35. $\begin{cases} y = -\dfrac{1}{2}x + 1 \\ 2x + y = 10 \end{cases}$

36. $\begin{cases} 3x + y = -4 \\ y = 4x + 17 \end{cases}$

37. $\begin{cases} x = \dfrac{2}{3}y \\ 3x - y = -3 \end{cases}$

38. $\begin{cases} y = \dfrac{1}{2}x \\ x - 4y = -4 \end{cases}$

39. $\begin{cases} 2x - 4y = 2 \\ x + 2y = 0 \end{cases}$

40. $\begin{cases} 3x + 2y = 0 \\ 6x + 2y = 5 \end{cases}$

41. $\begin{cases} x + y = 10{,}000 \\ 0.05x + 0.07y = 650 \end{cases}$

42. $\begin{cases} x + y = 5000 \\ 0.04x + 0.08y = 340 \end{cases}$

In Problems 43–50, solve the system of equations using elimination. See Objective 4.

43. $\begin{cases} x + y = -5 \\ -x + 2y = 14 \end{cases}$

44. $\begin{cases} x + y = -6 \\ -2x - y = 0 \end{cases}$

45. $\begin{cases} x + 2y = -5 \\ 3x + 3y = 9 \end{cases}$

46. $\begin{cases} -3x + 2y = -5 \\ 2x - y = 10 \end{cases}$

47. $\begin{cases} 2x + 5y = -3 \\ x + \dfrac{5}{4}y = -\dfrac{1}{2} \end{cases}$

48. $\begin{cases} x + 2y = -\dfrac{8}{3} \\ 3x - 3y = 5 \end{cases}$

49. $\begin{cases} 0.05x + 0.1y = 5.25 \\ 0.08x - 0.02y = 1.2 \end{cases}$

50. $\begin{cases} 0.04x + 0.06y = 2.1 \\ 0.06x - 0.03y = 0.15 \end{cases}$

In Problems 51–54, use either substitution or elimination to show that the system is inconsistent. Draw a graph to support your result. See Objective 5.

51. $\begin{cases} 3x + y = 1 \\ -6x - 2y = -4 \end{cases}$

52. $\begin{cases} -2x + 4y = 9 \\ x - 2y = -3 \end{cases}$

53. $\begin{cases} 5x - 2y = 2 \\ -10x + 4y = 3 \end{cases}$

54. $\begin{cases} 6x - 4y = 6 \\ -3x + 2y = 3 \end{cases}$

In Problems 55–60, use either substitution or elimination to show that the system is consistent and its equations are dependent. Solve the system and draw a graph to support your solution. See Objective 6.

55. $\begin{cases} y = \dfrac{1}{2}x + 1 \\ 2x - 4y = -4 \end{cases}$

56. $\begin{cases} y = -\dfrac{2}{3}x + 3 \\ 2x + 3y = 9 \end{cases}$

57. $\begin{cases} x + 3y = 6 \\ -\dfrac{x}{3} - y = -2 \end{cases}$ **58.** $\begin{cases} -4x + y = 8 \\ x - \dfrac{y}{4} = -2 \end{cases}$

59. $\begin{cases} \dfrac{1}{3}x - 2y = 6 \\ -\dfrac{1}{2}x + 3y = -9 \end{cases}$ **60.** $\begin{cases} \dfrac{5}{4}x - \dfrac{1}{2}y = 6 \\ -\dfrac{5}{3}x + \dfrac{2}{3}y = -8 \end{cases}$

Mixed Practice

In Problems 61–68, solve the system of equations using either substitution or elimination.

61. $\begin{cases} x + 3y = 0 \\ -2x + 4y = 30 \end{cases}$ **62.** $\begin{cases} 2x + y = -1 \\ -3x - 2y = 7 \end{cases}$

63. $\begin{cases} x = 5y - 3 \\ -3x + 15y = 9 \end{cases}$ **64.** $\begin{cases} y = \dfrac{1}{2}x + 2 \\ x - 2y = -4 \end{cases}$

65. $\begin{cases} 2x - 4y = 18 \\ 3x + 5y = -3 \end{cases}$ **66.** $\begin{cases} 12x + 45y = 0 \\ 8x + 6y = 24 \end{cases}$

67. $\begin{cases} \dfrac{5}{6}x - \dfrac{1}{3}y = -5 \\ -x + \dfrac{2}{5}y = 1 \end{cases}$ **68.** $\begin{cases} \dfrac{1}{3}x - \dfrac{1}{2}y = -5 \\ -\dfrac{4}{5}x + \dfrac{6}{5}y = 1 \end{cases}$

Applying the Concepts

In Problems 69–72, write each equation in the system of equations in slope-intercept form. Use the slope-intercept form to determine the number of solutions the system has.

69. $\begin{cases} 2x + y = -5 \\ 5x + 3y = 1 \end{cases}$ **70.** $\begin{cases} 4x - 2y = 8 \\ -10x + 5y = 5 \end{cases}$

71. $\begin{cases} 3x - 2y = -2 \\ -6x + 4y = 4 \end{cases}$ **72.** $\begin{cases} 2x - y = -5 \\ -4x + 3y = 9 \end{cases}$

△ **73. Parallelogram** Use the parallelogram shown to answer parts (a) and (b).

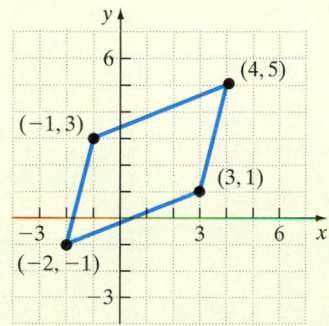

(a) Find the equation of the line for the diagonal through the points $(-1, 3)$ and $(3, 1)$. Find the equation of the line for the diagonal through the points $(-2, -1)$ and $(4, 5)$.
(b) Find the point of intersection of the diagonals.

△ **74. Rhombus** A rhombus is a parallelogram whose adjacent sides are congruent. Use the rhombus to answer parts (a), (b), and (c).

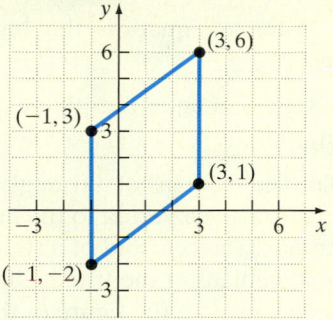

(a) Find the equation of the line for the diagonal through the points $(-1, 3)$ and $(3, 1)$. Find the equation of the line for the diagonal through the points $(-1, -2)$ and $(3, 6)$.
(b) Find the point of intersection of the diagonals.
(c) Compare the slopes of the diagonals. What can be said about the diagonals of a rhombus?

Extending the Concepts

75. Which of the following ordered pairs could be a solution to the system graphed below?

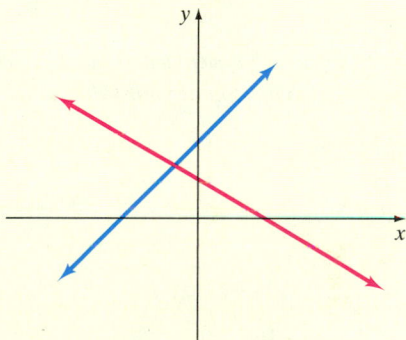

(a) $(2, 4)$ (b) $(-2, 0)$ (c) $(-3, 1)$
(d) $(5, -2)$ (e) $(-1, -3)$ (f) $(-1, 3)$

76. Which of the following systems of equations could have the graph below?

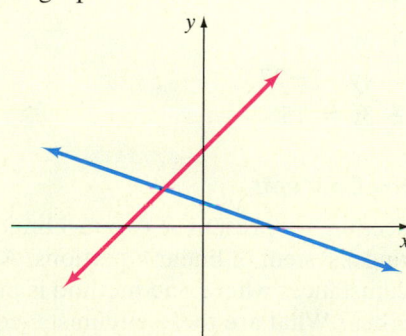

(a) $\begin{cases} 2x + 3y = 12 \\ 2x + y = -2 \end{cases}$ (b) $\begin{cases} 2x + 3y = 3 \\ -2x + y = 2 \end{cases}$

(c) $\begin{cases} 2x - 3y = 12 \\ x + 2y = 2 \end{cases}$

77. For the system $\begin{cases} Ax + 3By = 2 \\ -3Ax + By = -11 \end{cases}$, find A and B such that $x = 3, y = 1$ is a solution.

78. Write a system of equations that has $(3, 5)$ as a solution.

79. Write a system of equations that has $(-1, 4)$ as a solution.

△ **80. Centroid** The medians of a triangle are the line segments from each vertex to the midpoint of the opposite side. The centroid of a triangle is the point where the medians of the triangle intersect. Use the information given in the figure of the triangle to find its centroid.

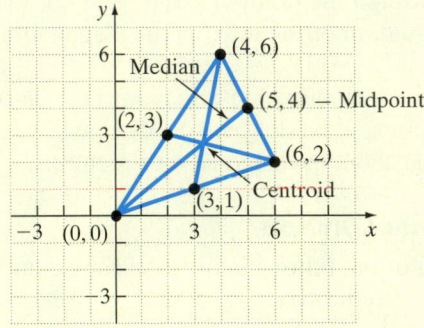

In Problems 81–84, solve each system using either substitution or elimination. Draw a graph to support your solution.

81. $\begin{cases} 3x + y = 5 \\ x + y = 3 \\ x + 3y = 7 \end{cases}$

82. $\begin{cases} x + y = -2 \\ -3x + 2y = 16 \\ 2x - 4y = -16 \end{cases}$

83. $\begin{cases} y = \dfrac{2}{3}x - 5 \\ 4x - 6y = 30 \\ x - 5y = 11 \end{cases}$

84. $\begin{cases} -4x + 3y = 33 \\ 3x - 4y = -37 \\ 2x - 3y = 15 \end{cases}$

Explaining the Concepts

85. In this section, we presented two algebraic methods for solving a system of linear equations. Are there any circumstances where one method is preferable to the other? What are these circumstances?

86. Describe geometrically the three possibilities for a solution of a system of two linear equations containing two variables.

87. The solution of a system of two linear equations in two unknowns is $x = 3, y = -2$. Where do the lines in the system intersect? Why?

88. In the process of solving a system of linear equations, what tips you off that the system is consistent and the equations are dependent? What tips you off that the system is inconsistent?

Technology Exercises

Technology can be used to approximate the point of intersection between two equations. We illustrate this feature by doing Example 3 using Desmos. Start by graphing each equation in the system. Then click on one of the lines. This will bring up the points of interest. Hover your mouse over the intersection point and it will be displayed. See Figure 5. The solution is the ordered pair $(2, -3)$.

Figure 5

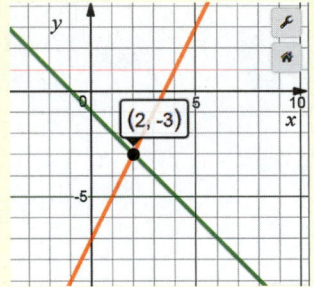

In Problems 89–96, use technology to solve each system of equations. If necessary, express your solution rounded to two decimal places.

89. $\begin{cases} y = 3x - 1 \\ y = -2x + 5 \end{cases}$

90. $\begin{cases} y = \dfrac{3}{2}x - 4 \\ y = -\dfrac{1}{4}x + 3 \end{cases}$

91. $\begin{cases} 3x - y = -1 \\ -4x + y = -3 \end{cases}$

92. $\begin{cases} -6x - 2y = 4 \\ 5x + 3y = -2 \end{cases}$

93. $\begin{cases} 4x - 3y = 1 \\ -8x + 6y = -2 \end{cases}$

94. $\begin{cases} -2x + 5y = -2 \\ 4x - 10y = 1 \end{cases}$

95. $\begin{cases} 2x - 3y = 12 \\ 5x + y = -2 \end{cases}$

96. $\begin{cases} x - 3y = 21 \\ x + 6y = -2 \end{cases}$

C.2 Systems of Linear Equations in Three Variables

Objectives

1. Solve Systems of Three Linear Equations
2. Identify Inconsistent Systems
3. Write the Solution of a System with Dependent Equations
4. Model and Solve Problems Involving Three Linear Equations

Are You Prepared for This Section?

Before getting started, complete the following problem. If you get the problem wrong, go back to the section cited and review the material.

P1. Evaluate the expression $3x - 2y + 4z$
for $x = 1, y = -2$, and $z = 3$. [Section 1.8, pp. 64–65]

Prepared?...Answer **P1.** 19

1 Solve Systems of Three Linear Equations

An example of a linear equation in three variables is $2x - y + z = 8$. An example of a system of three linear equations containing three variables, x, y, and z is

$$\begin{cases} x + 3y + z = 8 \\ 3x - y + 6z = 12 \\ -4x - y + 2z = -1 \end{cases}$$

Systems of three linear equations with three variables have the same possible numbers of solutions as a system of two linear equations containing two variables:

1. **Exactly one solution**—A consistent system with independent equations
2. **No solution**—An inconsistent system
3. **Infinitely many solutions**—A consistent system with dependent equations

Think of a system of three linear equations containing three variables as a geometry problem. The graph of each equation in a system of linear equations containing three variables is a plane in space. A system of three linear equations containing three variables represents three planes in space. Figure 6 illustrates some of the possibilities.

Recall that a **solution** to a system of equations consists of values for the variables that are solutions of each equation of the system. Write the solution to a system of three equations containing three unknowns as an **ordered triple** (x, y, z).

Figure 6

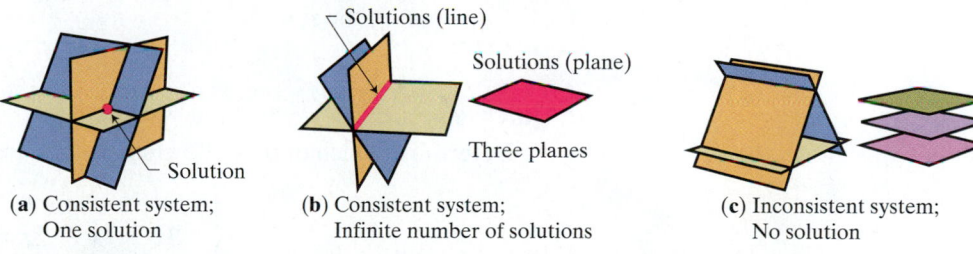

(**a**) Consistent system;
    One solution

(**b**) Consistent system;
    Infinite number of solutions

(**c**) Inconsistent system;
    No solution

Quick ✔

1. If a system of equations has no solution, it is said to be _____. If a system of equations has infinitely many solutions, the system is said to be _____ and the equations are _____.

2. A _____ to a system of equations consists of values for the variables that are solutions of each equation of the system.

3. *True or False* A system of three linear equations containing three variables always has at least one solution.

4. *True or False* When the planes in a system of equations are parallel, the system is inconsistent and has no solution.

EXAMPLE 1 **Determining Whether Values Are a Solution of a System**

Determine which of the following ordered triples are solutions of the system.

$$\begin{cases} x + y + z = 0 \\ 2x - y + 3z = 17 \\ -3x + 2y - z = -21 \end{cases}$$

(a) $(1, 3, -4)$ **(b)** $(3, -5, 2)$

Solution

First, name the system's equations (1), (2), and (3).

$$\begin{cases} x + y + z = 0 & \text{(1)} \\ 2x - y + 3z = 17 & \text{(2)} \\ -3x + 2y - z = -21 & \text{(3)} \end{cases}$$

(a) Let $x = 1$, $y = 3$, and $z = -4$ in equations (1), (2), and (3). If all three equations are true, then $(1, 3, -4)$ is a solution.

Equation (1): $x + y + z = 0$ Equation (2): $2x - y + 3z = 17$ Equation (3): $-3x + 2y - z = -21$

$1 + 3 + (-4) \stackrel{?}{=} 0$ $2(1) - 3 + 3(-4) \stackrel{?}{=} 17$ $-3(1) + 2(3) - (-4) \stackrel{?}{=} -21$

$0 = 0$ True $2 - 3 - 12 \stackrel{?}{=} 17$ $-3 + 6 + 4 \stackrel{?}{=} -21$

$-13 = 17$ False $7 = -21$ False

Although this ordered triple satisfies (1), it does not satisfy (2) or (3). Therefore, the ordered triple $(1, 3, -4)$ is not a solution.

(b) Let $x = 3$, $y = -5$, and $z = 2$ in equations (1), (2), and (3).

Equation (1): $x + y + z = 0$ Equation (2): $2x - y + 3z = 17$ Equation (3): $-3x + 2y - z = -21$

$3 + (-5) + (2) \stackrel{?}{=} 0$ $2(3) - (-5) + 3(2) \stackrel{?}{=} 17$ $-3(3) + 2(-5) - (2) \stackrel{?}{=} -21$

$0 = 0$ True $6 + 5 + 6 \stackrel{?}{=} 17$ $-9 - 10 - 2 \stackrel{?}{=} -21$

$17 = 17$ True $-21 = -21$ True

Because the ordered triple $(3, -5, 2)$ satisfies all three equations, it is a solution to the system. ●

Quick ✔

5. Determine which of the following ordered triples are solutions of the system.

$$\begin{cases} x + y + z = 3 \\ 3x + y - 2z = -23 \\ -2x - 3y + 2z = 17 \end{cases}$$

(a) $(3, 2, -2)$ **(b)** $(-4, 1, 6)$

Typically, to solve a system of three linear equations containing three variables, the *elimination method* is used. The first step in solving a system of three linear equations with three unknowns is to reduce the system to one with two linear equations with two unknowns. Then solve the smaller system using the methods of Section C.1.

To reduce the system of three linear equations to a system of two linear equations, various methods may be used. Eliminate one variable by multiplying equations by nonzero constants to get the coefficients of the variables to be additive inverses. Add these equations to remove that variable. Also interchange any two equations or multiply (or divide) each side of an equation by the same nonzero constant. Another approach is to replace an equation by the sum of that equation and a multiple of a second equation.

▶ EXAMPLE 2   How to Solve a System of Three Linear Equations

Use the method of elimination to solve the system:
$$\begin{cases} x + y - z = -1 & (1) \\ 2x - y + 2z = 8 & (2) \\ -3x + 2y + z = -9 & (3) \end{cases}$$

Step-by-Step Solution

Step 1: The first step is to eliminate the same variable from two of the equations. Use equation (1) to eliminate the variable x from equations (2) and (3). Do this by multiplying equation (1) by -2 and adding the result to equation (2). The resulting equation is equation (4). Multiply equation (1) by 3 and add the result to equation (3). The resulting equation is equation (5).

$$\begin{aligned} x + y - z &= -1 \quad (1) \\ 2x - y + 2z &= 8 \quad (2) \end{aligned}$$

Multiply (1) by -2:
$$\begin{aligned} -2x - 2y + 2z &= 2 \quad (1) \\ 2x - y + 2z &= 8 \quad (2) \\ \hline \text{Add:} \quad -3y + 4z &= 10 \end{aligned}$$

$$\begin{aligned} x + y - z &= -1 \quad (1) \\ -3x + 2y + z &= -9 \quad (3) \end{aligned}$$

Multiply (1) by 3:
$$\begin{aligned} 3x + 3y - 3z &= -3 \quad (1) \\ -3x + 2y + z &= -9 \quad (3) \\ \hline \text{Add:} \quad 5y - 2z &= -12 \end{aligned}$$

$$\begin{cases} x + y - z = -1 & (1) \\ -3y + 4z = 10 & (4) \\ 5y - 2z = -12 & (5) \end{cases}$$

Step 2: Treat equations (4) and (5) as a system of two equations with two variables. Eliminate z by multiplying equation (5) by 2 and then adding equations (4) and (5). The result is equation (6).

$$\begin{cases} -3y + 4z = 10 & (4) \\ 5y - 2z = -12 & (5) \end{cases}$$

$$\begin{aligned} -3y + 4z &= 10 \quad (4) \\ \text{Multiply (5) by 2:} \quad 10y - 4z &= -24 \quad (5) \\ \hline \text{Add:} \quad 7y &= -14 \quad (6) \end{aligned}$$

Step 3: Solve equation (6) for y.

$$7y = -14 \quad (6)$$

Divide both sides by 7:
$$y = -2$$

Step 4: Substitute -2 for y in equation (4) and solve for z.

$$\begin{aligned} -3y + 4z &= 10 \quad (4) \\ -3(-2) + 4z &= 10 \\ 6 + 4z &= 10 \\ 4z &= 4 \\ z &= 1 \end{aligned}$$

Step 5: Substitute -2 for y and   for z in equation (1) and solve for x.

$$\begin{aligned} x + y - z &= -1 \quad (1) \\ x + (-2) - 1 &= -1 \\ x - 3 &= -1 \\ x &= 2 \end{aligned}$$

The solution appears to be $(2, -2, 1)$.

Step 6: Verify that $(2, -2, 1)$ is the solution.

Equation (1): $x + y - z = -1$

$x = 2, y = -2, z = 1$: $\quad 2 + (-2) - 1 \overset{?}{=} -1$

$$-1 = -1 \quad \text{True}$$

Equation (2): $\quad 2x - y + 2z = 8$

$$2(2) - (-2) + 2(1) \overset{?}{=} 8$$
$$4 + 2 + 2 \overset{?}{=} 8$$
$$8 = 8 \quad \text{True}$$

Equation (3): $-3x + 2y + z = -9$

$$-3(2) + 2(-2) + 1 \overset{?}{=} -9$$
$$-6 - 4 + 1 \overset{?}{=} -9$$
$$-9 = -9 \quad \text{True}$$

The solution, $(2, -2, 1)$, checks.

Work Smart

In Example 2, we eliminated x from equations (2) and (3). We could also have eliminated y from equation (2) by adding equations (1) and (2). We could then eliminate y from equation (3) by multiplying equation (1) by -2 and adding the result to equation (3). Had we done this, we would have ended up with the following system of two equations with two unknowns:

$$\begin{cases} 3x + z = 7 \\ -5x + 3z = -7 \end{cases}$$

The solution of this system is $x = 2$ and $z = 1$. Substituting these values into equation (1) yields $y = -2$, so the solution to the system is $(2, -2, 1)$, which agrees with the solution we found in Example 2. There is more than one way to solve a problem!

Solving a System of Three Linear Equations Containing Three Unknowns by Elimination

Step 1: Select two of the equations and eliminate one of the variables from one of the equations. Select any two other equations and eliminate the *same variable* from one of the equations.

Step 2: This will leave two equations with two unknowns. Solve this system using the techniques from Section C.1.

Step 3: Use the two known values of the variables found in Step 2 to find the value of the third variable.

Step 4: Check the answer.

Work Smart

By eliminating a variable in two of the equations in Step 1, we are creating a system of two equations with two unknowns we can solve. Remember, in mathematics it is helpful to relate a problem to one we already know how to solve.

Quick ✓

6. Use the elimination method to solve the system: $\begin{cases} x + y + z = -3 \\ 2x - 2y - z = -7 \\ -3x + y + 5z = 5 \end{cases}$

EXAMPLE 3 **Solving a System of Three Linear Equations**

Use the elimination method to solve the system: $\begin{cases} 4x \qquad + z = 4 & (1) \\ 2x + 3y \qquad = -4 & (2) \\ 2y - 4z = -15 & (3) \end{cases}$

Solution
Eliminate z from equation (3) by multiplying equation (1) by 4 and adding the result to equation (3). The resulting equation is equation (4).

$$\begin{array}{l} 4x \quad + z = 4 \quad (1) \\ 2y - 4z = -15 \quad (3) \end{array}$$

Multiply (1) by 4:
$$\begin{array}{l} 16x \qquad + 4z = 16 \quad (1) \\ \underline{\qquad 2y - 4z = -15} \quad (3) \\ \text{Add:} \quad 16x + 2y \qquad = 1 \end{array}$$

$$\begin{cases} 4x \qquad + z = 4 & (1) \\ 2x + 3y \qquad = -4 & (2) \\ 16x + 2y \qquad = 1 & (4) \end{cases}$$

Now focus on the system containing equations (2) and (4). To eliminate the variable x, multiply equation (2) by -8 and then add equations (2) and (4). The result is equation (5).

$$\begin{cases} 2x + 3y = -4 & (2) \\ 16x + 2y = 1 & (4) \end{cases}$$

Multiply (2) by -8:
$$\begin{array}{l} -16x - 24y = 32 \quad (2) \\ \underline{\quad 16x + 2y = 1} \quad (4) \\ \text{Add:} \qquad -22y = 33 \quad (5) \end{array}$$

Solve equation (5) for y by dividing both sides of the equation by -22.

$$\text{Equation (5):} \quad -22y = 33$$

$$y = \frac{33}{-22} = -\frac{3}{2}$$

Now substitute $-\dfrac{3}{2}$ for y in equation (2) and solve for x.

$$2x + 3y = -4 \quad (2)$$

Let $y = -\dfrac{3}{2}$ in Equation (2): $\quad 2x + 3\left(-\dfrac{3}{2}\right) = -4$

$$2x - \dfrac{9}{2} = -4$$

$$2x = \dfrac{1}{2}$$

$$x = \dfrac{1}{4}$$

Now substitute $\dfrac{1}{4}$ for x in equation (1) and solve for z.

$$4x + z = 4 \quad (1)$$

$$4\left(\dfrac{1}{4}\right) + z = 4$$

$$1 + z = 4$$

$$z = 3$$

The check is left to you. The solution is the ordered triple $\left(\dfrac{1}{4}, -\dfrac{3}{2}, 3\right)$. ●

> **Quick** ✔
> **7.** Use the elimination method to solve the system: $\begin{cases} 2x \quad\quad - 4z = -7 \\ x + 6y \quad\quad = 5 \\ 2y - z = 2 \end{cases}$

❷ Identify Inconsistent Systems

Examples 2 and 3 were consistent systems with independent equations resulting in a single solution. Now let's look at an inconsistent system.

EXAMPLE 4 **An Inconsistent System of Linear Equations**

Use the elimination method to solve the system: $\begin{cases} x + 2y - z = 4 & (1) \\ -2x + 3y + z = -4 & (2) \\ x + 9y - 2z = 1 & (3) \end{cases}$

Solution
Notice that equation (1) can be used to eliminate x from equations (2) and (3). This can be done by multiplying equation (1) by 2 and adding the result to equation (2). The resulting equation is equation (4).

Work Smart

These figures illustrate two different inconsistent systems geometrically.

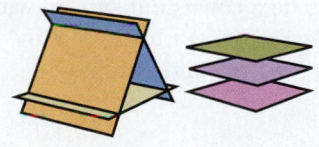

$x + 2y - z = 4$ (1) Multiply by 2: $\quad 2x + 4y - 2z = 8$ (1)

$-2x + 3y + z = -4$ (2) $\quad\quad\quad\quad\quad\quad -2x + 3y + z = -4$ (2)

Add: $\quad\quad\quad\quad\quad\quad\quad\quad 7y - z = 4$ (4)

Next, multiply equation (1) by -1 and add the result to equation (3). The resulting equation is equation (5).

$x + 2y - z = 4$ (1) Multiply by -1: $\quad -x - 2y + z = -4$ (1)

$x + 9y - 2z = 1$ (3) $\quad\quad\quad\quad\quad\quad\quad x + 9y - 2z = 1$ (3)

Add: $\quad\quad\quad\quad\quad\quad\quad\quad\quad 7y - z = -3$ (5)

Work Smart

Whenever you end up with a false statement such as $0 = -7$, you have an inconsistent system.

Now focus on the system containing equations (4) and (5). Multiply equation (4) by -1 and then add equations (4) and (5).

$$7y - z = 4 \quad (4) \quad \text{Multiply by } -1: \quad -7y + z = -4 \quad (4)$$
$$7y - z = -3 \quad (5) \qquad\qquad\qquad\qquad\quad \underline{7y - z = -3} \quad (5)$$
$$\text{Add:} \qquad\qquad\qquad 0 = -7 \quad (6) \quad \text{False}$$

Equation (6) now states that $0 = -7$, which is a false statement. Therefore, the system is inconsistent and has no solution. The solution set is $\varnothing$ or $\{\ \}$.

> **Quick ✓**
>
> **8.** Use the elimination method to solve the system: $\begin{cases} x - y + 2z = -7 \\ -2x + y - 3z = 5 \\ x - 2y + 3z = 2 \end{cases}$

▶ ③ Write the Solution of a System with Dependent Equations

Now let's look at a system of dependent equations.

(**EXAMPLE 5**) **Solving a System with Dependent Equations**

Work Smart

These figures illustrate two different dependent systems.

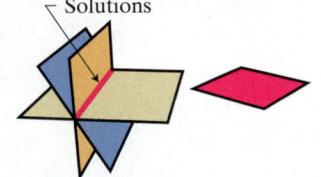

Solutions

Use the elimination method to solve the system: $\begin{cases} x - 3y - z = 4 & (1) \\ x - 2y + 2z = 5 & (2) \\ 2x - 5y + z = 9 & (3) \end{cases}$

Solution

Use equation (1) to eliminate x from equations (2) and (3). Do this by multiplying equation (1) by -1 and adding the result to equation (2). The resulting equation is equation (4).

$$x - 3y - z = 4 \quad (1) \quad \text{Multiply by } -1: \quad -x + 3y + z = -4 \quad (1)$$
$$x - 2y + 2z = 5 \quad (2) \qquad\qquad\qquad\qquad \underline{x - 2y + 2z = 5} \quad (2)$$
$$\text{Add:} \qquad\qquad\qquad y + 3z = 1 \quad (4)$$

Next, multiply equation (1) by -2 and add the result to equation (3). The resulting equation is equation (5).

$$x - 3y - z = 4 \quad (1) \quad \text{Multiply by } -2: \quad -2x + 6y + 2z = -8 \quad (1)$$
$$2x - 5y + z = 9 \quad (3) \qquad\qquad\qquad\qquad \underline{2x - 5y + z = 9} \quad (3) \qquad \begin{cases} y + 3z = 1 & (4) \\ y + 3z = 1 & (5) \end{cases}$$
$$\text{Add:} \qquad\qquad\qquad y + 3z = 1 \longrightarrow$$

Now concentrate on the system containing equations (4) and (5).

$$y + 3z = 1 \quad (4) \qquad \text{Multiply by } -1: \quad -y - 3z = -1 \quad (4)$$
$$y + 3z = 1 \quad (5) \qquad\qquad\qquad\qquad\qquad \underline{y + 3z = 1} \quad (5)$$
$$\text{Add:} \qquad\qquad\qquad\qquad 0 = 0 \quad (6)$$

The statement $0 = 0$ in equation (6) indicates that it may be a consistent system with dependent equations. Show how the values for x and y *depend* on the value of z by letting z represent any real number. Then solve equation (4) for y and obtain y in terms of z.

$$\text{Equation (4):} \quad y + 3z = 1$$
$$\text{Subtract } 3z \text{ from both sides:} \qquad y = -3z + 1$$

Let $y = -3z + 1$ in equation (1) and solve for x in terms of z.

Equation (1):	$x - 3y - z = 4$
Let $y = -3z + 1$ in (1):	$x - 3(-3z + 1) - z = 4$
Distribute:	$x + 9z - 3 - z = 4$
Combine like terms:	$x + 8z - 3 = 4$
Subtract $8z$ from both sides; add 3 to both sides:	$x = -8z + 7$

The solution to the system is $\{(x, y, z) \mid x = -8z + 7, y = -3z + 1, z \text{ is any real}$ number$\}$. To find specific solutions to the system, choose any value of z and use the equations $x = -8z + 7$ and $y = -3z + 1$ to determine x and y. Some specific solutions to the system are $(7, 1, 0)$, $(-1, -2, 1)$, and $(15, 4, -1)$. ●

Quick ✔

9. Use the elimination method to solve the system: $\begin{cases} x - y + 3z = 2 \\ -x + 2y - 5z = -3 \\ 2x - y + 4z = 3 \end{cases}$

Notice that in Examples 2 through 5, we did not always eliminate x first and y second. The order in which variables are eliminated from a system does not matter, and different approaches to solving the problem will lead to the right answer if done correctly. For example, in Example 4 we could have chosen to eliminate z from equations (1) and (3).

▶ ❹ Model and Solve Problems Involving Three Linear Equations

Now let's look at problems involving systems of three equations containing three variables.

EXAMPLE 6

Production

A manufacturer makes three different models of swing sets. The Monkey takes 2 hours to cut the wood, 2 hours to stain, and 3 hours to assemble. The Gorilla takes 3 hours to cut the wood, 4 hours to stain, and 4 hours to assemble. The King Kong takes 4 hours to cut the wood, 5 hours to stain, and 5 hours to assemble. The company has 61 hours available to cut the wood, 73 hours available to stain, and 83 hours available to assemble each day. How many of each type of swing set can be manufactured each day?

Solution

Step 1: Identify Determine the number of Monkey, Gorilla, and King Kong swing sets that can be manufactured each day.

Step 2: Name Let m represent the number of Monkey swing sets, g represent the number of Gorilla swing sets, and k represent the number of King Kong swing sets.

Step 3: Translate Organize the given information in Table 4.

Table 4

	Monkey	Gorilla	King Kong	Total Hours Available
Cut Wood	2	3	4	61
Stain	2	4	5	73
Assemble	3	4	5	83

Making m Monkey swing sets takes $2m$ hours to cut the wood. Similarly, it takes $3g$ and $4k$ hours to cut the wood for g Gorilla and k King Kong sets. Since 61 hours are available,

$$2m + 3g + 4k = 61 \quad \text{Equation (1)}$$

(continued)

It takes $2m$ hours to stain m Monkey sets, $4g$ hours to stain g Gorilla sets, and $5k$ hours to stain k King Kong sets. Since 73 hours are available,

$$2m + 4g + 5k = 73 \quad \text{Equation (2)}$$

$3m$ hours are needed to assemble m Monkey sets, $4g$ hours to assemble g Gorilla sets, and $5k$ hours to assemble k King Kong sets. Since 83 hours are available,

$$3m + 4g + 5k = 83 \quad \text{Equation (3)}$$

Combine equations (1), (2), and (3) to form the following system:

$$\begin{cases} 2m + 3g + 4k = 61 & \text{(1)} \\ 2m + 4g + 5k = 73 & \text{(2)} \quad \text{The Model} \\ 3m + 4g + 5k = 83 & \text{(3)} \end{cases}$$

Step 4: Solve The solution of this system of equations is $m = 10$, $g = 7$, and $k = 5$.

Step 5: Check Manufacturing 10 Monkeys, 7 Gorillas, and 5 King Kongs requires $2(10) + 3(7) + 4(5) = 61$ hours to cut the wood; $2(10) + 4(7) + 5(5) = 73$ hours to stain; and $3(10) + 4(7) + 5(5) = 83$ hours to assemble.

Step 6: Answer The company can manufacture 10 Monkey swing sets, 7 Gorilla swing sets, and 5 King Kong swing sets each day. ●

Quick ✔

10. The Mowing 'Em Down lawn mower company makes three styles of lawn mowers. The 21-inch model requires 2 hours to mold, 3 hours for engine manufacturing, and 1 hour to assemble. The 24-inch model requires 3 hours to mold, 3 hours for engine manufacturing, and 1 hour to assemble. The 40-inch model requires 4 hours to mold, 4 hours for engine manufacturing, and 2 hours to assemble. The company has 81 hours available to mold, 95 hours available for engine manufacturing, and 35 hours available to assemble each day. How many of each type of mower can be manufactured each day?

C.2 Exercises MyMathLab®

Exercise numbers in **green** have complete video solutions in MyMathLab or may be accessed using the QR code to the right.

Problems **1–10** *are the* **Quick ✔** *s that follow the* **EXAMPLES.**

Building Skills

In Problems 11 and 12, determine whether the ordered triples listed are solutions of the system of linear equations. See Objective 1.

11. $\begin{cases} x + y + 2z = 6 \\ -2x - 3y + 5z = 1 \\ 2x + y + 3z = 5 \end{cases}$

    (a) $(6, 2, -1)$ **(b)** $(-3, 5, 2)$

12. $\begin{cases} 2x + y - 2z = 6 \\ -2x + y + 5z = 1 \\ 2x + 3y + z = 13 \end{cases}$

    (a) $(3, 2, 1)$
    (b) $(10, -4, 5)$

In Problems 13–20, solve each system of three linear equations containing three unknowns. See Objective 1.

13. $\begin{cases} x + y + z = 5 \\ -2x - 3y + 2z = 8 \\ 3x - y - 2z = 3 \end{cases}$

14. $\begin{cases} x + 2y - z = 4 \\ 2x - y + 3z = 8 \\ -2x + 3y - 2z = 10 \end{cases}$

15. $\begin{cases} x - 3y + z = 13 \\ 3x + y - 4z = 13 \\ -4x - 4y + 2z = 0 \end{cases}$

16. $\begin{cases} x + 2y - 3z = -19 \\ 3x + 2y - z = -9 \\ -2x - y + 3z = 26 \end{cases}$

17. $\begin{cases} x - 4y + z = 5 \\ 4x + 2y + z = 2 \\ -4x + y - 3z = -8 \end{cases}$

18. $\begin{cases} 2x + 2y - z = -7 \\ x + 2y - 3z = -8 \\ 4x - 2y + z = -11 \end{cases}$

19. $\begin{cases} x - 3y = 12 \\ 2y - 3z = -9 \\ 2x + z = 7 \end{cases}$

20. $\begin{cases} 2x + z = -7 \\ 3y - 2z = 17 \\ -4x - y = 7 \end{cases}$

In Problems 21 and 22, show that each system of equations has no solution. See Objective 2.

21. $\begin{cases} x + y - 2z = 6 \\ -2x - 3y + z = 12 \\ -3x - 4y + 3z = 2 \end{cases}$

22. $\begin{cases} -x + 4y - z = 8 \\ 4x - y + 3z = 9 \\ 2x + 7y + z = 0 \end{cases}$

In Problems 23–26, solve each system with dependent equations. See Objective 3.

23. $\begin{cases} x + y + z = 4 \\ -2x - y + 2z = 6 \\ x + 2y + 5z = 18 \end{cases}$

24. $\begin{cases} x + 2y + z = 4 \\ -3x + y + 4z = -2 \\ -x + 5y + 6z = 6 \end{cases}$

25. $\begin{cases} x + 3z = 5 \\ -2x + y = 1 \\ y + 6z = 11 \end{cases}$

26. $\begin{cases} 2x - y = 2 \\ -x + 5z = 3 \\ -y + 10z = 8 \end{cases}$

Mixed Practice

In Problems 27–40, solve each system of equations.

27. $\begin{cases} 2x - y + 2z = 1 \\ -2x + 3y - 2z = 3 \\ 4x - y + 6z = 7 \end{cases}$

28. $\begin{cases} x - y + 3z = 2 \\ -2x + 3y - 8z = -1 \\ 2x - 2y + 4z = 7 \end{cases}$

29. $\begin{cases} x - y + z = 5 \\ -2x + y - z = 2 \\ x - 2y + 2z = 1 \end{cases}$

30. $\begin{cases} x - y + 2z = 3 \\ 2x + y - 2z = 1 \\ 4x - y + 2z = 0 \end{cases}$

31. $\begin{cases} 2y - z = -3 \\ -2x + 3y = 10 \\ 4x + 3z = -11 \end{cases}$

32. $\begin{cases} x - 3z = -3 \\ 3y + 4z = -5 \\ 3x - 2y = 6 \end{cases}$

33. $\begin{cases} x - 2y + z = 5 \\ -2x + y - z = 2 \\ x - 5y - 4z = 8 \end{cases}$

34. $\begin{cases} x + 2y - z = -4 \\ -2x + 4y - z = 6 \\ 2x + 2y + 3z = 1 \end{cases}$

35. $\begin{cases} x + 2y - z = 1 \\ 2x + 7y + 4z = 11 \\ x + 3y + z = 4 \end{cases}$

36. $\begin{cases} x + y - 2z = 3 \\ -2x - 3y + z = -7 \\ x + 2y + z = 4 \end{cases}$

37. $\begin{cases} x + y + z = 5 \\ 3x + 4y + z = 16 \\ -x - 4y + z = -6 \end{cases}$

38. $\begin{cases} x + y + z = 4 \\ 2x + 3y - z = 8 \\ x + y - z = 3 \end{cases}$

39. $\begin{cases} x + y + z = 3 \\ -x + \dfrac{1}{2}y + z = \dfrac{1}{2} \\ -x + 2y + 3z = 4 \end{cases}$

40. $\begin{cases} x + \dfrac{1}{2}y + \dfrac{1}{2}z = \dfrac{3}{2} \\ -x + 2y + 3z = 1 \\ 3x + 4y + 5z = 7 \end{cases}$

Applying the Concepts

41. Role Reversal Write a system of three linear equations containing three unknowns that has the solution $(2, -1, 3)$.

42. Role Reversal Write a system of three linear equations containing three unknowns that has the solution $(-4, 1, -3)$.

43. Curve Fitting The function $f(x) = ax^2 + bx + c$ is a quadratic function, where a, b, and c are constants.

(a) If $f(1) = 4$, then $4 = a(1)^2 + b(1) + c$ or $a + b + c = 4$. Find two additional linear equations if $f(-1) = -6$ and $f(2) = 3$.

(b) Use the three linear equations found in part (a) to determine a, b, and c. What is the quadratic function that contains the points $(-1, -6)$, $(1, 4)$, and $(2, 3)$?

44. Curve Fitting The function $f(x) = ax^2 + bx + c$ is a quadratic function, where a, b, and c are constants.

(a) If $f(-1) = 6$, then $6 = a(-1)^2 + b(-1) + c$ or $a - b + c = 6$. Find two additional linear equations if $f(1) = 2$ and $f(2) = 9$.

(b) Use the three linear equations found in part (a) to determine a, b, and c. What is the quadratic function that contains the points $(-1, 6)$, $(1, 2)$, and $(2, 9)$?

45. Electricity: Kirchhoff's Rules An application of Kirchhoff's Rule to the circuit shown results in the following system of equations:

$$\begin{cases} i_1 + i_3 = i_2 \\ -3 - 3i_1 + 2i_3 = 0 \\ -22 + 4i_2 + 2i_3 = 0 \end{cases}$$

In the system circuit, V is the voltage, Ω is resistance, and i is the current. Find the currents i_1, i_2, and i_3.

46. Electricity: Kirchhoff's Rules An application of Kirchhoff's Rule to the circuit shown results in the following system of equations:

$$\begin{cases} i_1 + i_3 = i_2 \\ -8 - 5i_1 + 8i_3 = 0 \\ -48 + 6i_2 + 8i_3 = 0 \end{cases}$$

In the system circuit, V is the voltage, Ω is resistance, and i is the current. Find the currents i_1, i_2, and i_3.

47. Minor League Baseball In the Joliet Slammers baseball stadium, there are three types of seats available. Club seats are $14, reserved seats are $12, and lawn seats are $7. The stadium capacity is 6000. If all the seats are sold, the total revenue to the club is $64,125. If all of the club seats are sold, $\frac{4}{5}$ of the reserved seats are sold, and $\frac{1}{5}$ of the lawn seats

are sold, the total revenue is $45,189. How many are there of each kind of seat?

48. Theater Revenues A theater has 600 seats, divided into orchestra, main floor, and balcony seating. Orchestra seats sell for $80, main floor seats for $60, and balcony seats for $25. If all the seats are sold, the total revenue to the theater is $33,500. One evening, all the orchestra seats were sold, $\frac{3}{5}$ of the main seats were sold, and $\frac{4}{5}$ of the balcony seats were sold. The total revenue collected was $24,640. How many are there of each kind of seat?

49. Nutrition Nancy's dietitian wants her to consume 470 mg of sodium, 89 g of carbohydrates, and 20 g of protein for breakfast. This morning, Nancy wants to have Chex® cereal, 2% milk, and orange juice for breakfast. Each serving of Chex® cereal contains 220 mg of sodium, 26 g of carbohydrates, and 1 g of protein. Each serving of 2% milk contains 125 mg of sodium, 12 g of carbohydrates, and 8 g of protein. Each serving of orange juice contains 0 mg of sodium, 26 g of carbohydrates, and 2 g of protein. How many servings of each does Nancy need?

50. Nutrition Antonio is on a special diet that requires he consume 1325 calories, 172 grams of carbohydrates, and 63 grams of protein for lunch. He wishes to have a Broccoli and Cheese Baked Potato, Chicken BLT Salad, and a medium Coke. Each Broccoli and Cheese Baked Potato has 480 calories, 80 g of carbohydrates, and 9 g of protein. Each Chicken BLT Salad has 310 calories, 10 g of carbohydrates, and 33 g of protein. Each Coke has 140 calories, 37 g of carbohydrates, and 0 g of protein. How many servings of each does Antonio need?

51. Finance Sachi has $25,000 to invest. Her financial planner suggests that she diversify her investment into three investment categories: Treasury bills that yield 1% simple interest annually, municipal bonds that yield 2% simple interest annually, and corporate bonds that yield 6% simple interest annually. Sachi would like to earn $580 per year in income. In addition, Sachi wants her investment in Treasury bills to be $7000 more than her investment in corporate bonds. How much should Sachi invest in each investment category?

52. Finance Delu has $15,000 to invest. She decides to place some of the money into a savings account paying 0.50% annual interest, some in Treasury bonds paying 3% annual interest, and some in a mutual fund paying 10% annual interest. Delu would like to earn $500 per year in income. In addition, Delu wants her investment in the savings account to be twice the amount in the mutual fund. How much should Delu invest in each investment category?

△ **53. Geometry** A circle is inscribed in $\triangle ABC$ as shown in the figure. Suppose that $AB = 6$, $AC = 14$, and $BC = 12$. Find the length of $\overline{AM}$, $\overline{BN}$, and $\overline{CO}$. (*Hint:* $\overline{AM} \cong \overline{AO}$; $\overline{BM} \cong \overline{BN}$; $\overline{NC} \cong \overline{OC}$)

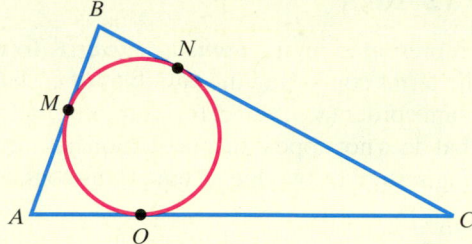

55. $\begin{cases} \dfrac{2}{5}x + \dfrac{1}{2}y - \dfrac{1}{3}z = 0 \\[2mm] \dfrac{3}{5}x - \dfrac{1}{4}y + \dfrac{1}{2}z = 10 \\[2mm] -\dfrac{1}{5}x + \dfrac{1}{4}y - \dfrac{1}{6}z = -4 \end{cases}$

56. $\begin{cases} x + y + z + w = 0 \\ 2x - 3y - z + w = -17 \\ 3x + y + 2z - w = 8 \\ -x + 2y - 3z + 2w = -7 \end{cases}$

57. $\begin{cases} x + y + z + w = 3 \\ -2x - y + 3z - w = -1 \\ 2x + 2y - 2z + w = 2 \\ -x + 2y - 3z + 2w = 12 \end{cases}$

Extending the Concepts

In Problems 54–57, solve each system of equations.

54. $\begin{cases} \dfrac{1}{4}x + \dfrac{1}{4}y + \dfrac{1}{2}z = 6 \\[2mm] -\dfrac{1}{8}x + \dfrac{1}{2}y - \dfrac{1}{5}z = -5 \\[2mm] \dfrac{1}{2}x + \dfrac{1}{2}y - \dfrac{1}{2}z = -3 \end{cases}$

Explaining the Concepts

58. Suppose that $(3, 2, 5)$ is the only solution of a system of three linear equations containing three variables. What does this mean geometrically?

59. Why is it necessary to eliminate the same variable in Step 1 of the box "Solving a System of Three Linear Equations Containing Three Unknowns by Elimination" (page 976)?

C.3 Using Matrices to Solve Systems

Objectives

❶ Write the Augmented Matrix of a System

❷ Write the System from the Augmented Matrix

❸ Perform Row Operations on a Matrix

❹ Solve Systems Using Matrices

❺ Solve Consistent Systems with Dependent Equations and Solve Inconsistent Systems

Are You Prepared for This Section?

Before getting started, complete the following problems. If you get a problem wrong, go back to the section cited and review the material.

P1. Determine the coefficients of the expression
$4x - 2y + z$. [Section 1.8, p. 66]

P2. Solve $x - 4y = 3$ for x. [Section 2.4, pp. 113–115]

P3. Evaluate $3x - 2y + z$ when $x = 1$, $y = -3$, and $z = 2$. [Section 1.8, pp. 64–65]

A way to solve systems of linear equations that streamlines the notation and makes working with systems more manageable is to use a *matrix*.

Prepared?...Answers **P1.** $4, -2, 1$
P2. $x = 4y + 3$ **P3.** 11

Definition

A **matrix** is a rectangular array of numbers.

A matrix has rows and columns. The number of rows and columns determines the size of the matrix. For example, a matrix with 2 rows and 3 columns is called a "2 by 3 matrix," which is denoted "2 × 3 matrix." Below are some examples of matrices.

$$
\underset{\text{2 rows}}{}\; \overset{\textbf{2 × 3 matrix}}{\underset{\text{3 columns}}{\begin{bmatrix} 3 & -1 & 4 \\ 8 & 0 & -5 \end{bmatrix}}} \qquad \underset{\text{3 rows}}{}\; \overset{\textbf{3 × 3 matrix}}{\underset{\text{3 columns}}{\begin{bmatrix} 2 & -8 & 12 \\ 0 & 7 & -2 \\ 5 & -2 & 1 \end{bmatrix}}}
$$

Work Smart

A spreadsheet such as Microsoft Excel is a matrix!

▶ ❶ Write the Augmented Matrix of a System

A system of linear equations may be represented in an *augmented matrix*. To write the **augmented matrix** of a system, write the terms containing the variables of each equation on the left side of the equal sign in the same order (x, y, and z, for example), and write the constants on the right side. A variable that does not appear in an equation has a coefficient of 0. For example, consider the following system of two linear equations containing two unknowns:

$$
\begin{cases} a_1 x + b_1 y = c_1 \\ a_2 x + b_2 y = c_2 \end{cases} \text{ is written as an augmented matrix as } \begin{bmatrix} a_1 & b_1 & c_1 \\ a_2 & b_2 & c_2 \end{bmatrix}
$$

coefficients of x — coefficients of y — constants

Notice that the first column represents the coefficients of the variable x, the second column represents the coefficients of the variable y, the vertical bar represents the equal signs, and the constants are to the right of the vertical bar. The first row in the matrix is from the first equation, and the second row in the matrix is from the second equation.

EXAMPLE 1

Writing the Augmented Matrix of a System of Linear Equations

Write each system of linear equations as an augmented matrix.

(a) $\begin{cases} x - 3y = 2 \\ -2x + 5y = 7 \end{cases}$

(b) $\begin{cases} 2x + 3y - z = 1 \\ x - 2z + 1 = 0 \\ -4x - y + 3z = 5 \end{cases}$

Solution

(a) In the augmented matrix, the first column represents the coefficients of the variable x. The second column represents the coefficients of the variable y. The vertical line signifies the equal signs. The third column represents the constants to the right of the equal sign.

$$
\overset{x \quad\quad y}{\begin{bmatrix} 1 & -3 & 2 \\ -2 & 5 & 7 \end{bmatrix}} \quad \begin{array}{l} x - 3y = 2 \\ -2x + 5y = 7 \end{array}
$$

(b) Write the system with all the variables on the left side of the equal sign and the constants on the right. Write 0 as the coefficient of any missing variable. To this end, rewrite the system

$$
\begin{cases} 2x + 3y - z = 1 \\ x - 2z + 1 = 0 \\ -4x - y + 3z = 5 \end{cases} \text{ as } \begin{cases} 2x + 3y - z = 1 \\ x + 0y - 2z = -1 \\ -4x - y + 3z = 5 \end{cases}
$$

The augmented matrix is

$$
\begin{bmatrix} 2 & 3 & -1 & 1 \\ 1 & 0 & -2 & -1 \\ -4 & -1 & 3 & 5 \end{bmatrix}
$$

▶ ❷ Write the System from the Augmented Matrix

EXAMPLE 2 **Writing the System from the Augmented Matrix**

Write the system of linear equations corresponding to each augmented matrix.

(a) $\left[\begin{array}{cc|c} 2 & 1 & -5 \\ -1 & 3 & 2 \end{array}\right]$

(b) $\left[\begin{array}{ccc|c} 1 & -3 & 2 & 5 \\ -2 & 0 & 4 & -3 \\ -1 & 4 & 1 & 0 \end{array}\right]$

Solution

(a) The augmented matrix has two rows, so it represents a system of two equations. There are two columns to the left of the vertical bar, so the system has two variables, which we can call *x* and *y*. The system of equations is

$$\begin{cases} 2x + y = -5 \\ -x + 3y = 2 \end{cases}$$

(b) The augmented matrix has three rows, so it represents a system of three equations. The three columns to the left of the vertical bar represent three variables, which we can call *x*, *y*, and *z*. The system of equations is

$$\begin{cases} x - 3y + 2z = 5 \\ -2x \qquad + 4z = -3 \\ -x + 4y + z = 0 \end{cases}$$ ●

▶ ❸ Perform Row Operations on a Matrix

Use row operations on an augmented matrix to solve the corresponding system of equations.

In Other Words

The first row operation is like reversing two equations in a system. The second row operation is like multiplying both sides of an equation by a nonzero constant. The third row operation is like adding two equations and replacing an equation with the sum.

Row Operations

1. Interchange any two rows.
2. Multiply *all* entries in a row by a nonzero constant.
3. Replace a row by the sum of that row and a nonzero multiple of some other row.

These are the same operations that were performed on systems of equations in Section C.2. The main reason for using a matrix to solve a system of equations is that its notation is more efficient and helps us organize the mathematics. To see how row operations work, consider the following augmented matrix.

$$\begin{bmatrix} 1 & 3 & | & -1 \\ -2 & 1 & | & 3 \end{bmatrix}$$

Suppose we want to apply a row operation that results in a matrix whose entry in row 2, column 1 is a 0. The current entry here is -2. The row operation to use is

Multiply each entry in row 1 by 2, and then add the result to the corresponding entry in row 2. Have this result replace the current row 2.

To streamline this a little, notation is needed. Use R_2 to represent the new entries in row 2, and use r_1 and r_2 to represent the original entries in rows 1 and 2, respectively, and then express the row operation given above by

The "new" row 2 is obtained by…   Multiplying each entry in row 1 by 2 and adding the result to the corresponding entry in row 2

$$R_2 = 2r_1 + r_2$$

This row operation is demonstrated below.

$$\begin{bmatrix} 1 & 3 & | & -1 \\ -2 & 1 & | & 3 \end{bmatrix} \xrightarrow{R_2 = 2r_1 + r_2} \begin{bmatrix} 1 & 3 & | & -1 \\ 2(1) + (-2) & 2(3) + 1 & | & 2(-1) + 3 \end{bmatrix} = \begin{bmatrix} 1 & 3 & | & -1 \\ 0 & 7 & | & 1 \end{bmatrix}$$

We now have a 0 in row 2, column 1, as desired. Why would this be useful? Notice the second row represents the equation $7y = 1$, which is easy to solve.

EXAMPLE 3 **Applying a Row Operation to an Augmented Matrix**

Apply the row operation $R_2 = -3r_1 + r_2$ to the augmented matrix

$$\begin{bmatrix} 1 & 2 & | & -3 \\ 3 & 4 & | & 5 \end{bmatrix}$$

Solution

The row operation $R_2 = -3r_1 + r_2$ tells us to multiply the entries in row 1 by -3 and then add the result to the entries in row 2. The result should replace the current row 2.

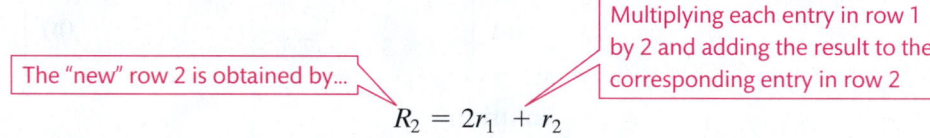

$$\begin{bmatrix} 1 & 2 & | & -3 \\ 3 & 4 & | & 5 \end{bmatrix} \xrightarrow{R_2 = -3r_1 + r_2} \begin{bmatrix} 1 & 2 & | & -3 \\ -3(1) + 3 & -3(2) + 4 & | & -3(-3) + 5 \end{bmatrix} = \begin{bmatrix} 1 & 2 & | & -3 \\ 0 & -2 & | & 14 \end{bmatrix}$$

●

Quick ✓

9. Apply the row operation $R_2 = 4r_1 + r_2$ to the augmented matrix

$$\begin{bmatrix} 1 & -2 & | & 5 \\ -4 & 5 & | & -11 \end{bmatrix}$$

EXAMPLE 4   **Finding a Particular Row Operation**

For the augmented matrix

$$\left[\begin{array}{cc|c} 1 & -3 & -12 \\ 0 & 1 & 5 \end{array}\right]$$

find and perform a row operation that will result in the entry in row 1, column 2 becoming a 0.

Solution
To get a 0 in row 1, column 2, multiply row 2 by 3 and add the result to row 1. That is, apply the row operation $R_1 = 3r_2 + r_1$.

$$\left[\begin{array}{cc|c} 1 & -3 & -12 \\ 0 & 1 & 5 \end{array}\right] \xrightarrow{R_1 = 3r_2 + r_1} \left[\begin{array}{cc|c} 3(0)+1 & 3(1)+(-3) & 3(5)+(-12) \\ 0 & 1 & 5 \end{array}\right] = \left[\begin{array}{cc|c} 1 & 0 & 3 \\ 0 & 1 & 5 \end{array}\right]$$ ●

Work Smart

If you want to change the entries in row 1, then you should multiply the entries in some other row by a "well-chosen" constant and then add this result to the entries in row 1.

Quick ✔

10. For the augmented matrix $\left[\begin{array}{cc|c} 1 & 5 & 13 \\ 0 & 1 & 2 \end{array}\right]$, find and perform a row operation that will result in the entry in row 1, column 2 becoming a 0.

❹ Solve Systems Using Matrices

▶ To solve a system of linear equations using matrices, use row operations on the augmented matrix of the system to obtain a matrix that is in *row echelon form*.

Definition

A matrix is in **row echelon form** when

1. The entry in row 1, column 1 is a 1, and 0s appear below it.

2. The first nonzero entry in each row below the first row is a 1, 0s appear below it, and it appears to the right of the first nonzero entry in any row above.

3. Any rows that contain all 0s to the left of the vertical bar appear at the bottom.

For a system of two equations containing two variables with a single solution, the row echelon form of the augmented matrix is

$$\left[\begin{array}{cc|c} 1 & a & b \\ 0 & 1 & c \end{array}\right]$$

where a, b, and c are real numbers. For a system of three equations containing three variables with a single solution, the row echelon form is

$$\left[\begin{array}{ccc|c} 1 & a & b & d \\ 0 & 1 & c & e \\ 0 & 0 & 1 & f \end{array}\right]$$

where $a, b, c, d, e,$ and f are real numbers.

The augmented matrix $\left[\begin{array}{cc|c} 1 & a & b \\ 0 & 1 & c \end{array}\right]$ is equivalent to the system $\begin{cases} x + ay = b \ (1) \\ y = c \ (2) \end{cases}$.

Because c is a known number, we know y. Then substitute to find x. For example, the augmented matrix $\left[\begin{array}{cc|c} 1 & 3 & -5 \\ 0 & 1 & -3 \end{array}\right]$ is in row echelon form and is equivalent to the system $\begin{cases} x + 3y = -5 \ (1) \\ y = -3 \ (2) \end{cases}$. Equation (2) states that $y = -3$. Using substitution reveals that $x = 4$, so the solution is $(4, -3)$.

EXAMPLE 5 **How to Solve a System of Two Linear Equations Containing Two Variables Using Matrices**

Solve: $\begin{cases} 3x - 2y = -19 \\ x + 2y = 7 \end{cases}$

Step-by-Step Solution

Step 1: Write the augmented matrix of the system.

$$\begin{bmatrix} 3 & -2 & | & -19 \\ 1 & 2 & | & 7 \end{bmatrix}$$

Step 2: The entry in row 1, column 1 should be 1, so interchange rows 1 and 2.

$$\begin{bmatrix} 1 & 2 & | & 7 \\ 3 & -2 & | & -19 \end{bmatrix}$$

Step 3: The entry in row 2, column 1 should be 0, so use the row operation $R_2 = -3r_1 + r_2$. The entries in row 1 remain unchanged.

$R_2 = -3r_1 + r_2$: $\begin{bmatrix} 1 & 2 & | & 7 \\ 0 & -8 & | & -40 \end{bmatrix}$

Step 4: Now the entry in row 2, column 2 should be 1. This is accomplished by multiplying row 2 by $-\dfrac{1}{8}$. Use the row operation $R_2 = -\dfrac{1}{8}r_2$.

$R_2 = -\dfrac{1}{8}r_2$: $\begin{bmatrix} 1 & 2 & | & 7 \\ 0 & 1 & | & 5 \end{bmatrix}$

Step 5: Row 2 says that $y = 5$. Substitute this value into row 1, which is $x + 2y = 7$. Solve for x.

$$x + 2y = 7$$
Let $y = 5$: $\quad x + 2(5) = 7$
$$x + 10 = 7$$
$$x = -3$$

Step 6: Check: Let $x = -3$ and $y = 5$ in equations (1) and (2) to verify the solution.

Equation (1): $\quad 3x - 2y = -19$
$x = -3, y = 5$: $\quad 3(-3) - 2(5) \overset{?}{=} -19$
$$-9 - 10 \overset{?}{=} -19$$
$$-19 = -19 \quad \text{True}$$

Equation (2): $\quad x + 2y = 7$
$x = -3, y = 5$: $\quad -3 + 2(5) \overset{?}{=} 7$
$$-3 + 10 \overset{?}{=} 7$$
$$7 = 7 \quad \text{True}$$

The solution is the ordered pair $(-3, 5)$.

Quick ✓

11. *True or False* The matrix $\begin{bmatrix} 1 & 3 & | & 2 \\ 0 & 1 & | & -5 \end{bmatrix}$ is in row echelon form.

12. Solve the following system using matrices: $\begin{cases} 2x - 4y = 20 \\ 3x + y = 16 \end{cases}$

EXAMPLE 6 **How to Solve a System of Three Linear Equations Containing Three Variables Using Matrices**

Solve: $\begin{cases} x + y + z = 3 \\ -2x - 3y + 2z = 13 \\ 4x + 5y + z = -3 \end{cases}$

Step-by-Step Solution

Step 1: Write the augmented matrix of the system.

$$\begin{bmatrix} 1 & 1 & 1 & | & 3 \\ -2 & -3 & 2 & | & 13 \\ 4 & 5 & 1 & | & -3 \end{bmatrix}$$

Step 2: The entry in row 1, column 1 should be 1. This is already done.

$$\begin{bmatrix} 1 & 1 & 1 & | & 3 \\ -2 & -3 & 2 & | & 13 \\ 4 & 5 & 1 & | & -3 \end{bmatrix}$$

Step 3: The entry in row 2, column 1 should be 0, so use the row operation $R_2 = 2r_1 + r_2$. The entry in row 3, column 1 should also be 0, so use the row operation $R_3 = -4r_1 + r_3$. The entries in row 1 remain unchanged.

$$\begin{array}{c} \\ R_2 = 2r_1 + r_2: \\ R_3 = -4r_1 + r_3: \end{array} \begin{bmatrix} 1 & 1 & 1 & | & 3 \\ 0 & -1 & 4 & | & 19 \\ 0 & 1 & -3 & | & -15 \end{bmatrix}$$

Step 4: Now the entry in row 2, column 2 should be 1. This is accomplished by interchanging rows 2 and 3.

$$\begin{bmatrix} 1 & 1 & 1 & | & 3 \\ 0 & 1 & -3 & | & -15 \\ 0 & -1 & 4 & | & 19 \end{bmatrix}$$

Step 5: The entry in row 3, column 2 should be 0. Use the row operation $R_3 = r_2 + r_3$.

$$\begin{array}{c} \\ \\ R_3 = r_2 + r_3: \end{array} \begin{bmatrix} 1 & 1 & 1 & | & 3 \\ 0 & 1 & -3 & | & -15 \\ 0 & 0 & 1 & | & 4 \end{bmatrix}$$

Step 6: The entry in row 3, column 3 should be 1. This is already done.

$$\begin{bmatrix} 1 & 1 & 1 & | & 3 \\ 0 & 1 & -3 & | & -15 \\ 0 & 0 & 1 & | & 4 \end{bmatrix}$$

Step 7: The augmented matrix is in row echelon form. Write the corresponding system of equations and solve.

$$\begin{cases} x + y + z = 3 & (1) \\ y - 3z = -15 & (2) \\ z = 4 & (3) \end{cases}$$

Equation (3) says $z = 4$. Let $z = 4$ in equation (2) to find y.

$$\begin{array}{rl} \text{Equation (2):} & y - 3z = -15 \\ z = 4: & y - 3(4) = -15 \\ & y - 12 = -15 \\ & y = -3 \end{array}$$

Let $y = -3$ and $z = 4$ in equation (1) to find that $x = 2$.

Step 8: Check: Let $x = 2$, $y = -3$, and $z = 4$ in each equation in the original system to verify the solution.

We leave it to you to verify the solution, $(2, -3, 4)$.

The solution is the ordered triple $(2, -3, 4)$. ●

Work Smart

Look at Step 4 in Example 6. Other ways to obtain a 1 in row 2, column 2 are to use the row operation $R_2 = 2r_3 + r_2$ or $R_2 = -r_2$.

Quick ✔

In Problems 13 and 14, solve each system of equations using matrices.

13. $\begin{cases} x - y + 2z = 7 \\ 2x - 2y + z = 11 \\ -3x + y - 3z = -14 \end{cases}$

14. $\begin{cases} 2x + 3y - 2z = 1 \\ 2x - 4z = -9 \\ 4x + 6y - z = 14 \end{cases}$

Solving a System of Linear Equations Using Matrices

Step 1: Write the augmented matrix of the system.

Step 2: Perform row operations so the entry in row 1, column 1 is 1.

Step 3: Perform row operations so all the entries below the 1 in row 1, column 1 are 0s.

Step 4: Perform row operations so the entry in row 2, column 2 is 1. Be sure the entries in column 1 remain unchanged.

Step 5: If solving a system of three linear equations, continue to perform row operations so that the entry below the 1 in row 2, column 2, is 0.

Step 6: Perform row operations so that the entry in row 3, column 3, is 1.

Step 7: With the augmented matrix in row echelon form, write the corresponding system of equations. Substitute to find the values of the variables.

Step 8: Check your answer.

▶ ❺ Solve Consistent Systems with Dependent Equations and Solve Inconsistent Systems

The matrix method for solving a system of linear equations also identifies systems that have infinitely many solutions (dependent equations) and systems with no solution (inconsistent systems).

EXAMPLE 7

Solving a Consistent System with Dependent Equations Using Matrices

Solve: $\begin{cases} x + y + 3z = 3 \\ -2x - y - 8z = -5 \\ 3x + 2y + 11z = 8 \end{cases}$

Solution

Write the augmented matrix of the system and perform row operations to get the matrix into row echelon form.

$$\begin{bmatrix} 1 & 1 & 3 & | & 3 \\ -2 & -1 & -8 & | & -5 \\ 3 & 2 & 11 & | & 8 \end{bmatrix} \xrightarrow[\substack{R_2 = 2r_1 + r_2 \\ R_3 = -3r_1 + r_3}]{} \begin{bmatrix} 1 & 1 & 3 & | & 3 \\ 0 & 1 & -2 & | & 1 \\ 0 & -1 & 2 & | & -1 \end{bmatrix} \xrightarrow[R_3 = r_2 + r_3]{} \begin{bmatrix} 1 & 1 & 3 & | & 3 \\ 0 & 1 & -2 & | & 1 \\ 0 & 0 & 0 & | & 0 \end{bmatrix}$$

Notice that the last row is all 0s. The augmented matrix is in row echelon form. The corresponding system of equations is

$$\begin{cases} x + y + 3z = 3 & \text{(1)} \\ y - 2z = 1 & \text{(2)} \\ 0 = 0 & \text{(3)} \end{cases}$$

The statement $0 = 0$ in equation (3) indicates that the equations are dependent. Let z represent any real number, and solve equation (2) for y in terms of z.

$$\text{Equation (2):} \quad y - 2z = 1$$
$$\text{Add } 2z \text{ to both sides:} \quad y = 2z + 1$$

Let $y = 2z + 1$ in equation (1) and solve for x in terms of z.

Equation (1): $x + y + 3z = 3$

Let $y = 2z + 1$: $x + (2z + 1) + 3z = 3$

Combine like terms: $x + 5z + 1 = 3$

Subtract $5z + 1$ from both sides: $x = -5z + 2$

The solution to the system is $\{(x, y, z) \mid x = -5z + 2, y = 2z + 1, z$ is any real number$\}$. To find specific solutions to the system, choose any value of z and use the equations $x = -5z + 2$ and $y = 2z + 1$ to determine x and y. Some specific solutions to the system are $(2, 1, 0)$, $(-3, 3, 1)$, and $(7, -1, -1)$.

Evaluate the system at some of the specific solutions to see that all of them satisfy the original system. This indicates that our solution is correct. ●

Quick ✓

In Problems 15 and 16, solve each system of equations using matrices.

15. $\begin{cases} 2x + 5y = -6 \\ -6x - 15y = 18 \end{cases}$

16. $\begin{cases} x + y - 3z = 8 \\ 2x + 3y - 10z = 19 \\ -x - 2y + 7z = -11 \end{cases}$

EXAMPLE 8   **Solving an Inconsistent System Using Matrices**

Solve: $\begin{cases} 2x + y - z = 5 \\ -x + 2y + z = 1 \\ 3x + 4y - z = -2 \end{cases}$

Solution

Write the augmented matrix of the system and perform row operations to get the matrix into row echelon form.

$$\left[\begin{array}{ccc|c} 2 & 1 & -1 & 5 \\ -1 & 2 & 1 & 1 \\ 3 & 4 & -1 & -2 \end{array} \right] \xrightarrow{R_1 = r_2 + r_1} \left[\begin{array}{ccc|c} 1 & 3 & 0 & 6 \\ -1 & 2 & 1 & 1 \\ 3 & 4 & -1 & -2 \end{array} \right] \xrightarrow[\substack{R_2 = r_1 + r_2 \\ R_3 = -3r_1 + r_3}]{} \left[\begin{array}{ccc|c} 1 & 3 & 0 & 6 \\ 0 & 5 & 1 & 7 \\ 0 & -5 & -1 & -20 \end{array} \right]$$

$$\xrightarrow{R_2 = \frac{1}{5} r_2} \left[\begin{array}{ccc|c} 1 & 3 & 0 & 6 \\ 0 & 1 & \frac{1}{5} & \frac{7}{5} \\ 0 & -5 & -1 & -20 \end{array} \right] \xrightarrow{R_3 = 5r_2 + r_3} \left[\begin{array}{ccc|c} 1 & 3 & 0 & 6 \\ 0 & 1 & \frac{1}{5} & \frac{7}{5} \\ 0 & 0 & 0 & -13 \end{array} \right]$$

The entry in row 3, column 3 needs to be 1. This cannot be done (without changing the entries in row 3, column 1 or 2). The corresponding system of equations is

$$\begin{cases} x + 3y \phantom{{}+ \frac{1}{5}z} = 6 & (1) \\ y + \frac{1}{5}z = \frac{7}{5} & (2) \\ \phantom{x + 3y + \frac{1}{5}z ={}} 0 = -13 & (3) \end{cases}$$

The statement $0 = -13$ in equation (3) indicates that the system is inconsistent and therefore has no solution. The solution is $\varnothing$ or $\{\}$. ●

Quick ✓

In Problems 17 and 18, solve each system of equations using matrices.

17. $\begin{cases} -2x + 3y = 4 \\ 10x - 15y = 2 \end{cases}$

18. $\begin{cases} -x + 2y - z = 5 \\ 2x + y + 4z = 3 \\ 3x - y + 5z = 0 \end{cases}$

C.3 Exercises · MyMathLab®

Exercise numbers in **green** have complete video solutions in MyMathLab or may be accessed using the QR code to the right.

*Problems **1–18** are the **Quick ✓**s that follow the **EXAMPLES**.*

Building Skills

In Problems 19–26, write the augmented matrix of the given system of equations. See Objective 1.

19. $\begin{cases} x - 3y = 2 \\ 2x + 5y = 1 \end{cases}$ **20.** $\begin{cases} -x + y = 6 \\ 5x - y = -3 \end{cases}$

21. $\begin{cases} x + y + z = 3 \\ 2x - y + 3z = 1 \\ -4x + 2y - 5z = -3 \end{cases}$

22. $\begin{cases} x + y - z = 2 \\ -2x + y - 4z = 13 \\ 3x - y - 2z = -4 \end{cases}$

23. $\begin{cases} -x + y - 2 = 0 \\ 5x + y + 5 = 0 \end{cases}$

24. $\begin{cases} 6x + 4y + 2 = 0 \\ -x - y + 1 = 0 \end{cases}$

25. $\begin{cases} x + z = 2 \\ 2x + y = 13 \\ x - y + 4z + 4 = 0 \end{cases}$

26. $\begin{cases} 2x + 7y = 1 \\ -x - 6z = 5 \\ 5x + 2y - 4z + 1 = 0 \end{cases}$

In Problems 27–32, write the system of linear equations corresponding to each augmented matrix. Use x, y; or x, y, z as variables. See Objective 2.

27. $\begin{bmatrix} 2 & 5 & | & 3 \\ -4 & 1 & | & 10 \end{bmatrix}$

28. $\begin{bmatrix} -1 & 4 & | & 0 \\ 2 & -7 & | & -9 \end{bmatrix}$

29. $\begin{bmatrix} 1 & 5 & -3 & | & 2 \\ 0 & 3 & -1 & | & -5 \\ 4 & 0 & 8 & | & 6 \end{bmatrix}$

30. $\begin{bmatrix} -2 & 3 & 0 & | & 3 \\ 4 & 1 & -8 & | & -6 \\ 7 & 2 & -5 & | & 10 \end{bmatrix}$

31. $\begin{bmatrix} 1 & -2 & 9 & | & 2 \\ 0 & 1 & -5 & | & 8 \\ 0 & 0 & 1 & | & \frac{4}{3} \end{bmatrix}$

32. $\begin{bmatrix} 1 & 4 & -1 & | & \frac{2}{3} \\ 0 & 1 & 3 & | & 0 \\ 0 & 0 & 1 & | & 10 \end{bmatrix}$

In Problems 33–38, perform each row operation on the given augmented matrix. See Objective 3.

33. $\begin{bmatrix} 1 & -3 & | & 2 \\ -2 & 5 & | & 1 \end{bmatrix}$

(a) $R_2 = 2r_1 + r_2$ followed by
(b) $R_2 = -r_2$

34. $\begin{bmatrix} 1 & 5 & | & 7 \\ 3 & 11 & | & 13 \end{bmatrix}$

(a) $R_2 = -3r_1 + r_2$ followed by
(b) $R_2 = -\dfrac{1}{4}r_2$

35. $\begin{bmatrix} 1 & 1 & -1 & | & 4 \\ 2 & 5 & 3 & | & -3 \\ -1 & -3 & 2 & | & 1 \end{bmatrix}$

(a) $R_2 = -2r_1 + r_2$ followed by
(b) $R_3 = r_1 + r_3$

36. $\begin{bmatrix} 1 & -1 & 1 & | & 6 \\ -2 & 1 & -3 & | & 3 \\ 3 & 2 & -2 & | & -5 \end{bmatrix}$

(a) $R_2 = 2r_1 + r_2$ followed by
(b) $R_3 = -3r_1 + r_3$

37. $\begin{bmatrix} 1 & 1 & 1 & | & 4 \\ 0 & 5 & 3 & | & -3 \\ 0 & -4 & 2 & | & 8 \end{bmatrix}$

(a) $R_2 = r_3 + r_2$ followed by
(b) $R_3 = \dfrac{1}{2}r_3$

38. $\begin{bmatrix} 1 & -3 & 4 & | & 11 \\ 0 & 3 & 6 & | & -12 \\ 0 & -2 & -3 & | & 8 \end{bmatrix}$

(a) $R_2 = \dfrac{1}{3}r_2$ followed by
(b) $R_3 = 2r_2 + r_3$

In Problems 39–44, solve each system of equations using matrices. See Objective 4.

39. $\begin{cases} 2x + 3y = 1 \\ -x + 4y = -28 \end{cases}$

40. $\begin{cases} 3x + 5y = -1 \\ -2x - 3y = 2 \end{cases}$

41. $\begin{cases} x + 5y - 2z = 23 \\ -2x - 3y + 5z = -11 \\ 3x + 2y + z = 4 \end{cases}$

42. $\begin{cases} x - 4y + 2z = -2 \\ -3x + y - 3z = 12 \\ 2x + 3y - 4z = -20 \end{cases}$

43. $\begin{cases} 4x + 5y - 2z = 0 \\ -8x + 10y - 2z = -13 \\ 15y + 4z = -7 \end{cases}$

44. $\begin{cases} 2x + y - 3z = 4 \\ 2y + 6z = 18 \\ -4x - y = -14 \end{cases}$

In Problems 45–50, solve each system of equations using matrices. See Objective 5.

45. $\begin{cases} x - 3y = 3 \\ -2x + 6y = -6 \end{cases}$

46. $\begin{cases} -4x + 2y = 4 \\ 6x - 3y = -6 \end{cases}$

47. $\begin{cases} x - 2y + z = 3 \\ -2x + 4y - z = -5 \\ -8x + 16y + z = -21 \end{cases}$

48. $\begin{cases} x - 3y + 2z = 4 \\ 3x + y + 4z = -5 \\ -x - 7y = 5 \end{cases}$

49. $\begin{cases} x + y - 2z = 4 \\ -4x + 3z = -4 \\ -2x + 2y - z = 4 \end{cases}$

50. $\begin{cases} -x + 3y + 3z = 0 \\ x + 5y = -2 \\ 8y + 3z = -2 \end{cases}$

Mixed Practice

In Problems 51–56, an augmented matrix of a system of linear equations is given. Write the system of equations corresponding to the given matrix. Use x, y; or x, y, z as variables. Determine whether the system is consistent with independent equations, consistent with dependent equations, or inconsistent. If it is consistent with independent equations, give the solution.

51. $\begin{bmatrix} 1 & 4 & | & -5 \\ 0 & 1 & | & -2 \end{bmatrix}$

52. $\begin{bmatrix} 1 & -2 & | & 3 \\ 0 & 1 & | & -5 \end{bmatrix}$

53. $\begin{bmatrix} 1 & 3 & -2 & | & 6 \\ 0 & 1 & 5 & | & -2 \\ 0 & 0 & 0 & | & 4 \end{bmatrix}$

54. $\begin{bmatrix} 1 & -2 & -4 & | & 6 \\ 0 & 1 & -5 & | & -3 \\ 0 & 0 & 0 & | & 0 \end{bmatrix}$

55. $\begin{bmatrix} 1 & -2 & -1 & | & 3 \\ 0 & 1 & -2 & | & -8 \\ 0 & 0 & 1 & | & 5 \end{bmatrix}$

56. $\begin{bmatrix} 1 & 2 & -1 & | & -7 \\ 0 & 1 & 2 & | & -4 \\ 0 & 0 & 1 & | & -3 \end{bmatrix}$

In Problems 57–78, solve each system of equations using matrices.

57. $\begin{cases} x - 3y = 18 \\ 2x + y = 1 \end{cases}$

58. $\begin{cases} x + 5y = 2 \\ -2x + 3y = 9 \end{cases}$

59. $\begin{cases} 2x + 4y = 10 \\ x + 2y = 3 \end{cases}$

60. $\begin{cases} 5x - 2y = 3 \\ -15x + 6y = -9 \end{cases}$

61. $\begin{cases} x - 6y = 8 \\ 2x + 8y = -9 \end{cases}$

62. $\begin{cases} 3x + 3y = -1 \\ 2x + y = 1 \end{cases}$

63. $\begin{cases} 4x - y = 8 \\ 2x - \dfrac{1}{2}y = 4 \end{cases}$

64. $\begin{cases} 5x - 2y = 10 \\ 2x - \dfrac{4}{5}y = 4 \end{cases}$

65. $\begin{cases} x + y + z = 0 \\ 2x - 3y + z = 19 \\ -3x + y - 2z = -15 \end{cases}$

66. $\begin{cases} x + y + z = 5 \\ 2x - y + 3z = -3 \\ -x + 2y - z = 10 \end{cases}$

67. $\begin{cases} 2x + y - z = 13 \\ -x - 3y + 2z = -14 \\ -3x + 2y - 3z = 3 \end{cases}$

68. $\begin{cases} 2x - y + 2z = 13 \\ -x + 2y - z = -14 \\ 3x + y - 2z = -13 \end{cases}$

69. $\begin{cases} 2x - y + 3z = 1 \\ -x + 3y + z = -4 \\ 3x + y + 7z = -2 \end{cases}$

70. $\begin{cases} -x + 2y + z = 1 \\ 2x - y + 3z = -3 \\ -x + 5y + 6z = 2 \end{cases}$

71. $\begin{cases} 3x + y - 4z = 0 \\ -2x - 3y + z = 5 \\ -x - 5y - 2z = 3 \end{cases}$

72. $\begin{cases} -x + 4y - 3z = 1 \\ 3x + y - z = -3 \\ x + 9y - 7z = -1 \end{cases}$

73. $\begin{cases} 2x - y + 3z = -1 \\ 3x + y - 4z = 3 \\ x + 7y - 2z = 2 \end{cases}$

74. $\begin{cases} x + y + z = 8 \\ 2x + 3y + z = 19 \\ 2x + 2y + 4z = 21 \end{cases}$

75. $\begin{cases} 3x + 5y + 2z = 6 \\ 10y - 2z = 5 \\ 6x + 4z = 8 \end{cases}$

76. $\begin{cases} 2x + y + 3z = 3 \\ 2x - 3y = 7 \\ 4y + 6z = -2 \end{cases}$

77. $\begin{cases} x - z = 3 \\ 2x + y = -3 \\ 2y - z = 7 \end{cases}$

78. $\begin{cases} x + 2y - z = 3 \\ y + z = 1 \\ x - 3z = 2 \end{cases}$

Applying the Concepts

79. Curve Fitting The function $f(x) = ax^2 + bx + c$ is a quadratic function, where $a, b,$ and c are constants.

   (a) If $f(-1) = 6$, then $6 = a(-1)^2 + b(-1) + c$ or $a - b + c = 6$. Find two additional linear equations if $f(1) = 0$ and $f(2) = 3$.

   (b) Use the three linear equations found in part (a) to determine $a, b,$ and c. What is the quadratic function that contains the points $(-1, 6), (1, 0),$ and $(2, 3)$?

80. Curve Fitting The function $f(x) = ax^2 + bx + c$ is a quadratic function, where $a, b,$ and c are constants.

   (a) If $f(-1) = -6$, then $-6 = a(-1)^2 + b(-1) + c$ or $a - b + c = -6$. Find two additional linear equations if $f(1) = 0$ and $f(2) = -3$.

   (b) Use the three linear equations found in part (a) to determine $a, b,$ and c. What is the quadratic function that contains the points $(-1, -6), (1, 0),$ and $(2, -3)$?

81. Finance Carissa has $20,000 to invest. Her financial planner suggests that she diversify her investment into three investment categories: Treasury bills that yield 3% simple interest, municipal bonds that yield 4% simple interest, and corporate bonds that yield 6% simple interest. Carissa would like to earn $820 per year in income. In addition, she wants her investment in Treasury bills to be $3000 more than her investment in corporate bonds. How much should Carissa invest in each investment category?

82. Finance Marlon has $12,000 to invest. He decides to place some of the money into a savings account paying 0.50% interest, some in Treasury bonds paying 3% interest, and some in a mutual fund paying 10% interest. Marlon would like to earn $325 per year in income. In addition, Marlon wants his investment in the savings account to be $4000 more than the amount in Treasury bonds. How much should Marlon invest in each investment category?

Extending the Concepts

*Sometimes it is advantageous to write a matrix in **reduced row echelon form**. In this form, row operations are used to obtain entries that are 0 above and below the leading 1 in a row. Augmented matrices for systems in reduced row echelon form with two equations containing two variables and three equations containing three variables, with a single solution, are shown below.*

$$\begin{bmatrix} 1 & 0 & | & a \\ 0 & 1 & | & b \end{bmatrix} \qquad \begin{bmatrix} 1 & 0 & 0 & | & a \\ 0 & 1 & 0 & | & b \\ 0 & 0 & 1 & | & c \end{bmatrix}$$

The obvious advantage to writing an augmented matrix in reduced row echelon form is that the solution to the system is readily seen. In the system with two equations containing two variables, the solution is $x = a$ and $y = b$. In the system with three equations containing three variables, the solution is $x = a, y = b,$ and $z = c$.

   In Problems 83–86, solve the system of equations by writing the augmented matrix in reduced row echelon form.

83. $\begin{cases} 2x + y = 1 \\ -3x - 2y = -5 \end{cases}$

84. $\begin{cases} 2x + y = -1 \\ -3x - 2y = -3 \end{cases}$

85. $\begin{cases} x + y + z = 3 \\ 2x + y - 4z = 25 \\ -3x + 2y + z = 0 \end{cases}$

86. $\begin{cases} x + y + z = 3 \\ 3x - 2y + 2z = 38 \\ -2x - 3z = -19 \end{cases}$

Explaining the Concepts

87. Write a paragraph that outlines the strategy for putting an augmented matrix in row echelon form.

88. When solving a system of linear equations using matrices, how do you know when the system is inconsistent?

89. What would be the next row operation on the given augmented matrix?

$$\begin{bmatrix} 1 & 3 & | & 8 \\ 0 & 5 & | & 10 \end{bmatrix}$$

90. What would you recommend as the next row operation on the given augmented matrix? Why?

$$\begin{bmatrix} 1 & 3 & -2 & | & 4 \\ 0 & 5 & 3 & | & 2 \\ 0 & -4 & 6 & | & -5 \end{bmatrix}$$

Technology Exercises

Graphing calculators have the ability to solve systems of linear equations by writing the augmented matrix in row echelon form. We enter the augmented matrix into the graphing calculator and name it A. See Figure 7(a). Figure 7(b) shows the results of Example 6 using a graphing calculator.

Figure 7

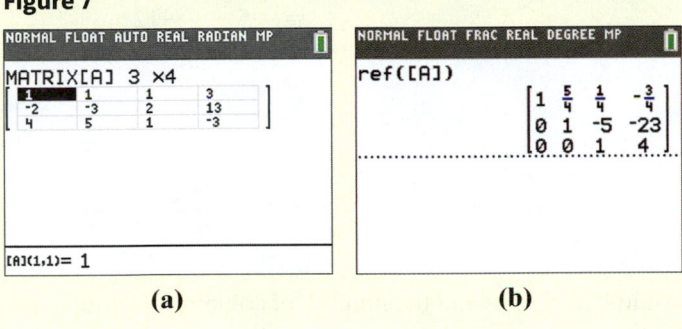

(a)   (b)

Notice that the row echelon form of the augmented matrix found using the graphing calculator differs from the row echelon form in the algebraic solution presented in Example 6, yet both matrices provide the same solution! This is because the two solutions use different row operations to obtain the row echelon form.

In Problems 91–94, solve each system of equations using technology.

91. $\begin{cases} 2x + 3y = 1 \\ -3x - 4y = -3 \end{cases}$

92. $\begin{cases} 3x + 2y = 4 \\ -5x - 3y = -4 \end{cases}$

93. $\begin{cases} 2x + 3y - 2z = -12 \\ -3x + y + 2z = 0 \\ 4x + 3y - z = 3 \end{cases}$

94. $\begin{cases} 2x - 3y - 4z = 16 \\ -3x + y + 2z = -23 \\ 4x + 3y - z = 13 \end{cases}$

C.4 Determinants and Cramer's Rule

Objectives

1 Evaluate the Determinant of a 2 × 2 Matrix

2 Use Cramer's Rule to Solve a System of Two Equations

3 Evaluate the Determinant of a 3 × 3 Matrix

4 Use Cramer's Rule to Solve a System of Three Equations

Are You Prepared for This Section?

Before getting started, complete the following problems. If you get a problem wrong, go back to the section cited and review the material.

P1. Evaluate: $4 \cdot 2 - 3(-3)$ [Section 1.7, pp. 57–58]

P2. Simplify: $\dfrac{18}{6}$ [Section 1.2, pp. 12–13]

At this point, it is known how to solve systems of linear equations using substitution, elimination, and row operations on augmented matrices. In this section, another method is introduced. This method, which requires that the number of equations equal the number of variables, is called *Cramer's Rule*. It is based on the concept of a *determinant*.

▶ **1** Evaluate the Determinant of a 2 × 2 Matrix

A matrix is **square** if the number of rows equals the number of columns. It is possible to computer the *determinant* of any square matrix.

Work Smart

The methods for solving systems presented in this section work *only* when the number of equations equals the number of variables.

Definition

Suppose a, b, c, and d are four real numbers. The **determinant** of a 2×2 matrix $\begin{bmatrix} a & b \\ c & d \end{bmatrix}$, denoted $\begin{vmatrix} a & b \\ c & d \end{vmatrix}$, is

$$\begin{vmatrix} a & b \\ c & d \end{vmatrix} = ad - bc$$

Notice that the determinant of a matrix is a number, not a matrix.

EXAMPLE 1 **Evaluating a 2 × 2 Determinant**

Evaluate each determinant:

(a) $\begin{vmatrix} 5 & 2 \\ 3 & 4 \end{vmatrix}$

(b) $\begin{vmatrix} -1 & 3 \\ -4 & 5 \end{vmatrix}$

Solution

(a) $\begin{vmatrix} 5 & 2 \\ 3 & 4 \end{vmatrix} = 5(4) - 3(2)$

$= 20 - 6$

$= 14$

(b) $\begin{vmatrix} -1 & 3 \\ -4 & 5 \end{vmatrix} = -1(5) - (-4)(3)$

$= -5 - (-12)$

$= -5 + 12$

$= 7$

Quick ✓

1. $D = \begin{vmatrix} a & b \\ c & d \end{vmatrix} = $ _____.

2. A matrix is _____ if the number of rows and the number of columns are equal.

In Problems 3 and 4, evaluate each determinant.

3. $\begin{vmatrix} 5 & 3 \\ 4 & 6 \end{vmatrix}$

4. $\begin{vmatrix} -2 & -5 \\ 1 & 7 \end{vmatrix}$

② Use Cramer's Rule to Solve a System of Two Equations

Use 2×2 determinants to solve a system of two equations containing two variables.

Cramer's Rule for Two Equations Containing Two Variables

The solution to the system of equations

$$\begin{cases} ax + by = s & (1) \\ cx + dy = t & (2) \end{cases}$$

is given by

$$x = \frac{\begin{vmatrix} s & b \\ t & d \end{vmatrix}}{\begin{vmatrix} a & b \\ c & d \end{vmatrix}} = \frac{D_x}{D} \qquad y = \frac{\begin{vmatrix} a & s \\ c & t \end{vmatrix}}{\begin{vmatrix} a & b \\ c & d \end{vmatrix}} = \frac{D_y}{D}$$

where

$$D = \begin{vmatrix} a & b \\ c & d \end{vmatrix} = ad - bc \neq 0$$

Look very carefully at the pattern in Cramer's Rule. The denominator in the solution is the determinant of the coefficients of the variables.

$$\begin{cases} ax + by = s \\ cx + dy = t \end{cases} \qquad D = \begin{vmatrix} a & b \\ c & d \end{vmatrix}$$

In the solution for x, the numerator is the determinant formed by replacing the entries in the first column (the coefficients of x) in D by the system's constants. That is,

$$D_x = \begin{vmatrix} s & b \\ t & d \end{vmatrix}$$

In the solution for y, the numerator is the determinant formed by replacing the entries in the second column (the coefficients of y) in D by the constants. That is,

$$D_y = \begin{vmatrix} a & s \\ c & t \end{vmatrix}$$

EXAMPLE 2 **How to Solve a System of Two Linear Equations Containing Two Variables Using Cramer's Rule**

Use Cramer's Rule to solve the system: $\begin{cases} x + 2y = 0 \\ -2x - 8y = -9 \end{cases}$

Step-by-Step Solution

Step 1: Find the determinant of the coefficients of the variables, D.

$$D = \begin{vmatrix} 1 & 2 \\ -2 & -8 \end{vmatrix} = 1(-8) - (-2)(2) = -8 - (-4) = -8 + 4 = -4$$

Step 2: Since $D \neq 0$, find D_x by replacing the first column in D with the system's constants. Find D_y by replacing the second column in D with the constants.

$$D_x = \begin{vmatrix} 0 & 2 \\ -9 & -8 \end{vmatrix} = 0(-8) - (-9)(2) = 0 - (-18) = 18$$

$$D_y = \begin{vmatrix} 1 & 0 \\ -2 & -9 \end{vmatrix} = 1(-9) - (-2)(0) = -9 - 0 = -9$$

Step 3: Find $x = \dfrac{D_x}{D}$ and $y = \dfrac{D_y}{D}$.

$$x = \frac{D_x}{D} = \frac{18}{-4} = -\frac{9}{2} \qquad y = \frac{D_y}{D} = \frac{-9}{-4} = \frac{9}{4}$$

Step 4: Check: $x = -\dfrac{9}{2}$ and $y = \dfrac{9}{4}$

The check of the solution $\left(-\dfrac{9}{2}, \dfrac{9}{4}\right)$ is left to you. ●

If the determinant of the coefficients of the variables, D, is found to be 0, then the system of equations is either consistent with dependent equations, or inconsistent. To determine if the system has no solution or infinitely many solutions, solve the system using the methods of Section C.1, C.2, or C.3.

Quick ✓

In Problems 5 and 6, use Cramer's Rule to solve the system, if possible.

5. $\begin{cases} 3x + 2y = 1 \\ -2x - y = 1 \end{cases}$ **6.** $\begin{cases} 4x - 2y = 8 \\ -6x + 3y = 3 \end{cases}$

▶ ❸ **Evaluate the Determinant of a 3 × 3 Matrix**

To use Cramer's Rule to solve a system of three equations containing three variables, define a 3 × 3 determinant.

The **determinant of a 3 × 3 matrix** is symbolized by

$$\begin{vmatrix} a_{1,1} & a_{1,2} & a_{1,3} \\ a_{2,1} & a_{2,2} & a_{2,3} \\ a_{3,1} & a_{3,2} & a_{3,3} \end{vmatrix}$$

where $a_{1,1}, a_{1,2}, \ldots, a_{3,3}$ are real numbers.

The subscript is used to identify the row and column of an entry. For example, $a_{2,3}$ is the entry in row 2, column 3.

> **Definition**
>
> The **value of the determinant of a 3 × 3 matrix** may be defined in terms of 2 × 2 determinants as follows:
>
>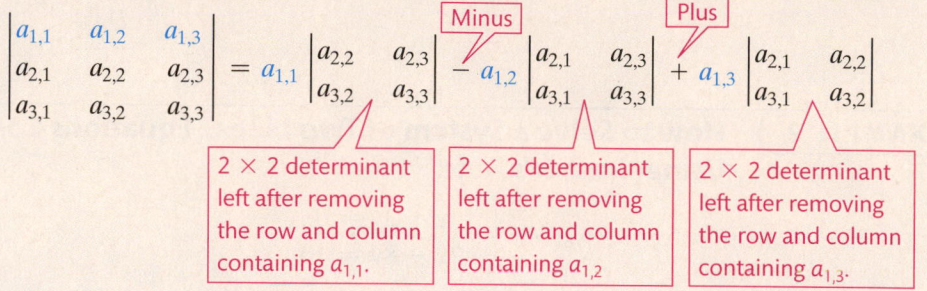
>
> | 2 × 2 determinant left after removing the row and column containing $a_{1,1}$. | 2 × 2 determinant left after removing the row and column containing $a_{1,2}$ | 2 × 2 determinant left after removing the row and column containing $a_{1,3}$. |

The 2 × 2 determinants shown in the definition above are called **minors** of the 3 × 3 determinant. Notice that once again a problem has been reduced to something we already know how to do—here the 3 × 3 determinant was reduced to three different 2 × 2 determinants.

The formula given in the definition above is easiest to remember by noting that each entry in row 1 is multiplied by the 2 × 2 determinant that remains after the row and column containing the entry have been removed.

EXAMPLE 3   **Evaluating a 3 × 3 Determinant**

Evaluate: $\begin{vmatrix} 3 & 2 & -4 \\ 1 & 7 & -3 \\ 0 & 2 & -5 \end{vmatrix}$

Solution

$$\begin{vmatrix} 3 & 2 & -4 \\ 1 & 7 & -3 \\ 0 & 2 & -5 \end{vmatrix} = 3\begin{vmatrix} 7 & -3 \\ 2 & -5 \end{vmatrix} - 2\begin{vmatrix} 1 & -3 \\ 0 & -5 \end{vmatrix} + (-4)\begin{vmatrix} 1 & 7 \\ 0 & 2 \end{vmatrix}$$

$$= 3(-35 - (-6)) - 2(-5 - 0) - 4(2 - 0)$$
$$= 3(-29) - 2(-5) - 4(2)$$
$$= -87 + 10 - 8$$
$$= -85$$

The definition shows one way to evaluate a 3 × 3 determinant—by *expanding* across row 1. In fact, the expansion can take place across any row or down any column. The terms to be added or subtracted consist of the row (or column) entry times the value of the 2 × 2 determinant that remains after removing the row and column containing the entry. There is only one glitch—the signs of the terms in the expansion change depending on the row or column that is expanded on. The signs of the terms obey the following scheme:

$$\begin{matrix} + & - & + \\ - & + & - \\ + & - & + \end{matrix}$$

For example, if we choose to expand down column 2, we obtain

$$\begin{vmatrix} a_{1,1} & a_{1,2} & a_{1,3} \\ a_{2,1} & a_{2,2} & a_{2,3} \\ a_{3,1} & a_{3,2} & a_{3,3} \end{vmatrix} = \overbrace{-a_{1,2}}^{\text{Minus}} \begin{vmatrix} a_{2,1} & a_{2,3} \\ a_{3,1} & a_{3,3} \end{vmatrix} + \overbrace{a_{2,2}}^{\text{Plus}} \begin{vmatrix} a_{1,1} & a_{1,3} \\ a_{3,1} & a_{3,3} \end{vmatrix} - \overbrace{a_{3,2}}^{\text{Minus}} \begin{vmatrix} a_{1,1} & a_{1,3} \\ a_{2,1} & a_{2,3} \end{vmatrix}$$

EXAMPLE 4

Evaluating a 3 × 3 Determinant

Evaluate: $\begin{vmatrix} 3 & 2 & -4 \\ 1 & 7 & -3 \\ 0 & 2 & -5 \end{vmatrix}$ (from Example 3), by expanding down column 1.

Solution

Expand down column 1 because it has a 0 in it.

$$\begin{vmatrix} 3 & 2 & -4 \\ 1 & 7 & -3 \\ 0 & 2 & -5 \end{vmatrix} = 3\begin{vmatrix} 7 & -3 \\ 2 & -5 \end{vmatrix} - 1\begin{vmatrix} 2 & -4 \\ 2 & -5 \end{vmatrix} + 0\begin{vmatrix} 2 & -4 \\ 7 & -3 \end{vmatrix}$$

$$= 3(-35 - (-6)) - 1(-10 - (-8)) + 0(-6 - (-28))$$

$$= 3(-29) - 1(-2) + 0$$

$$= -87 + 2$$

$$= -85 \qquad \bullet$$

Notice that the results of Example 3 and 4 are the same! However, the computation in Example 4 was a little easier because we expanded down the column that contains a 0.

To make the computation easier, expand across the row that contains the most 0s or down the column that contains the most 0s.

Work Smart

Expand across the row, or down the column, with the most 0s.

> **Quick ✓**
>
> 7. Evaluate: $\begin{vmatrix} 2 & -3 & 5 \\ 0 & 4 & -1 \\ 3 & 8 & -7 \end{vmatrix}$
>
> 8. Evaluate: $\begin{vmatrix} 3 & 2 & 1 \\ 1 & 1 & -3 \\ -5 & -1 & -5 \end{vmatrix}$

4 ## Use Cramer's Rule to Solve a System of Three Equations

Cramer's Rule can also be applied to a system of three linear equations.

> ### Cramer's Rule for Three Equations Containing Three Variables
>
> For the system of three equations containing three variables
>
> $$\begin{cases} a_1 x + b_1 y + c_1 z = d_1 \\ a_2 x + b_2 y + c_2 z = d_2 \\ a_3 x + b_3 y + c_3 z = d_3 \end{cases}$$
>
> with
>
> $$D = \begin{vmatrix} a_1 & b_1 & c_1 \\ a_2 & b_2 & c_2 \\ a_3 & b_3 & c_3 \end{vmatrix} \neq 0 \quad D_x = \begin{vmatrix} d_1 & b_1 & c_1 \\ d_2 & b_2 & c_2 \\ d_3 & b_3 & c_3 \end{vmatrix} \quad D_y = \begin{vmatrix} a_1 & d_1 & c_1 \\ a_2 & d_2 & c_2 \\ a_3 & d_3 & c_3 \end{vmatrix} \quad D_z = \begin{vmatrix} a_1 & b_1 & d_1 \\ a_2 & b_2 & d_2 \\ a_3 & b_3 & d_3 \end{vmatrix}$$
>
> then
>
> $$x = \frac{D_x}{D} \qquad y = \frac{D_y}{D} \qquad z = \frac{D_z}{D}$$

EXAMPLE 5 **How to Use Cramer's Rule**

Use Cramer's Rule to solve the following system:

$$\begin{cases} x - 2y + z = -9 \\ -3x + y - 2z = 5 \\ 4x + 3z = 1 \end{cases}$$

Step-by-Step Solution

Step 1: Find the determinant of the coefficients of the variables, D. Choose to expand down column 2. Do you see why?

$$D = \begin{vmatrix} 1 & -2 & 1 \\ -3 & 1 & -2 \\ 4 & 0 & 3 \end{vmatrix} = -(-2)\begin{vmatrix} -3 & -2 \\ 4 & 3 \end{vmatrix} + 1\begin{vmatrix} 1 & 1 \\ 4 & 3 \end{vmatrix} - 0\begin{vmatrix} 1 & 1 \\ -3 & -2 \end{vmatrix}$$

$$= 2[-3(3) - 4(-2)] + 1[1(3) - 4(1)] - 0$$
$$= 2(-1) + 1(-1) - 0$$
$$= -2 - 1$$
$$= -3$$

Step 2: Since $D \neq 0$, continue by determining D_x by replacing column 1 in D with the system's constants.

$$D_x = \begin{vmatrix} -9 & -2 & 1 \\ 5 & 1 & -2 \\ 1 & 0 & 3 \end{vmatrix} \overset{\text{Expand down column 2}}{=} -(-2)\begin{vmatrix} 5 & -2 \\ 1 & 3 \end{vmatrix} + 1\begin{vmatrix} -9 & 1 \\ 1 & 3 \end{vmatrix} - 0\begin{vmatrix} -9 & 1 \\ 5 & -2 \end{vmatrix}$$

$$= 2[15 - (-2)] + 1[-27 - 1] - 0$$
$$= 2(17) - 28$$
$$= 6$$

Find D_y by replacing column 2 in D with the system's constants.

$$D_y = \begin{vmatrix} 1 & -9 & 1 \\ -3 & 5 & -2 \\ 4 & 1 & 3 \end{vmatrix} \overset{\text{Expand down column 1}}{=} 1\begin{vmatrix} 5 & -2 \\ 1 & 3 \end{vmatrix} - (-3)\begin{vmatrix} -9 & 1 \\ 1 & 3 \end{vmatrix} + 4\begin{vmatrix} -9 & 1 \\ 5 & -2 \end{vmatrix}$$

$$= -15$$

Find D_z by replacing column 3 in D with the system's constants.

$$D_z = \begin{vmatrix} 1 & -2 & -9 \\ -3 & 1 & 5 \\ 4 & 0 & 1 \end{vmatrix} \overset{\text{Expand down column 2}}{=} -(-2)\begin{vmatrix} -3 & 5 \\ 4 & 1 \end{vmatrix} + 1\begin{vmatrix} 1 & -9 \\ 4 & 1 \end{vmatrix} - 0\begin{vmatrix} 1 & -9 \\ -3 & 5 \end{vmatrix}$$

$$= -9$$

Step 3: Find $x = \dfrac{D_x}{D}$, $y = \dfrac{D_y}{D}$, and $z = \dfrac{D_z}{D}$.

$$x = \frac{D_x}{D} = \frac{6}{-3} \qquad y = \frac{D_y}{D} = \frac{-15}{-3} \qquad z = \frac{D_z}{D} = \frac{-9}{-3}$$
$$= -2 \qquad\qquad\qquad = 5 \qquad\qquad\qquad = 3$$

Step 4: Check.

$$\begin{array}{ccc}
x - 2y + z = -9 & -3x + y - 2z = 5 & 4x + 3z = 1 \\
-2 - 2(5) + 3 \overset{?}{=} -9 & -3(-2) + 5 - 2(3) \overset{?}{=} 5 & 4(-2) + 3(3) \overset{?}{=} 1 \\
-2 - 10 + 3 \overset{?}{=} -9 & 6 + 5 - 6 \overset{?}{=} 5 & -8 + 9 \overset{?}{=} 1 \\
-9 = -9 \ \text{True} & 5 = 5 \ \text{True} & 1 = 1 \ \text{True}
\end{array}$$

The solution is the ordered triple $(-2, 5, 3)$. ●

Work Smart: Study Skills

At the end of Section C.1, we summarized the methods that can be used to solve a system of linear equations in two variables. Make up a summary table of your own for solving systems of linear equations in three variables. When is elimination appropriate? When would you use matrices? When is Cramer's Rule best?

Quick ✔

9. Suppose that you wish to solve a system of equations using Cramer's Rule and find that $D = 4, D_x = 2, D_y = -8,$ and $D_z = -4.$ What is the solution of the system?

10. Use Cramer's Rule to solve the system
$$\begin{cases} x - y + 3z = -2 \\ 4x + 3y + z = 9 \\ -2x + 5z = 7 \end{cases}$$

In the case where $D = 0,$ Cramer's Rule does not apply. To determine whether the system has no solution or infinitely many solutions, it is necessary to solve the system using the methods of Section C.1, C.2, or C.3.

C.4 Exercises MyMathLab®

Exercise numbers in **green** have complete video solutions in MyMathLab or may be accessed using the QR code to the right.

*Problems **1–10** are the **Quick ✔** s that follow the **EXAMPLES**.*

Building Skills

In Problems 11–14, find the value of each determinant. See Objective 1.

11. $\begin{vmatrix} 4 & 2 \\ 1 & 3 \end{vmatrix}$

12. $\begin{vmatrix} 5 & 3 \\ 2 & 4 \end{vmatrix}$

13. $\begin{vmatrix} -2 & -4 \\ 1 & 3 \end{vmatrix}$

14. $\begin{vmatrix} -8 & 5 \\ -4 & 3 \end{vmatrix}$

In Problems 15–22, solve each system of equations using Cramer's Rule, if applicable. See Objective 2.

15. $\begin{cases} x + y = -4 \\ x - y = -12 \end{cases}$

16. $\begin{cases} x + y = 6 \\ x - y = 4 \end{cases}$

17. $\begin{cases} 2x + 3y = 3 \\ -3x + y = -10 \end{cases}$

18. $\begin{cases} 2x + 4y = -6 \\ 3x + 2y = 7 \end{cases}$

19. $\begin{cases} 3x + 4y = 1 \\ -6x + 8y = 4 \end{cases}$

20. $\begin{cases} 4x - 2y = 9 \\ -8x - 2y = -11 \end{cases}$

21. $\begin{cases} 2x - 6y - 12 = 0 \\ 3x - 5y - 11 = 0 \end{cases}$

22. $\begin{cases} 3x - 6y - 2 = 0 \\ x + 2y - 4 = 0 \end{cases}$

In Problems 23–28, find the value of each determinant. See Objective 3.

23. $\begin{vmatrix} 2 & 0 & -1 \\ 3 & 8 & -3 \\ 1 & 5 & -2 \end{vmatrix}$

24. $\begin{vmatrix} -2 & 1 & 6 \\ -3 & 2 & 5 \\ 1 & 0 & -2 \end{vmatrix}$

25. $\begin{vmatrix} -3 & 2 & 3 \\ 0 & 5 & -2 \\ 1 & 4 & 8 \end{vmatrix}$

26. $\begin{vmatrix} 8 & 4 & -1 \\ 2 & -7 & 1 \\ 0 & 5 & -3 \end{vmatrix}$

27. $\begin{vmatrix} 0 & 2 & 1 \\ 1 & -6 & -4 \\ -3 & 4 & 5 \end{vmatrix}$

28. $\begin{vmatrix} -3 & 4 & -2 \\ 1 & -2 & 0 \\ 0 & 6 & 6 \end{vmatrix}$

In Problems 29–42, solve each system of equations using Cramer's Rule, if applicable. See Objective 4.

29. $\begin{cases} x - y + z = -4 \\ x + 2y - z = 1 \\ 2x + y + 2z = -5 \end{cases}$

30. $\begin{cases} x + y - z = 6 \\ x + 2y + z = 6 \\ -x - y + 2z = -7 \end{cases}$

31. $\begin{cases} x + y + z = 4 \\ 5x + 2y - 3z = 7 \\ 2x - y - z = 5 \end{cases}$

32. $\begin{cases} x + y + z = -3 \\ -2x - 3y - z = 1 \\ 2x - y - 3z = -5 \end{cases}$

33. $\begin{cases} 2x + y - z = 4 \\ -x + 2y + 2z = -6 \\ 5x + 5y - z = 6 \end{cases}$

34. $\begin{cases} -x + 2y - z = 2 \\ 2x + y + 2z = -6 \\ -x + 7y - z = 0 \end{cases}$

35. $\begin{cases} 3x + y + z = 5 \\ x + y - 3z = 9 \\ -5x - y - 5z = -10 \end{cases}$

36. $\begin{cases} x - 2y - z = 1 \\ 2x + 2y + z = 3 \\ 6x + 6y + 3z = 6 \end{cases}$

37. $\begin{cases} 2x + z = 27 \\ -x - 3y = 6 \\ x - 2y + z = 27 \end{cases}$

38. $\begin{cases} x + 2y + z = 0 \\ -x - 3y - 2z = 3 \\ 2x - 3z = 7 \end{cases}$

39. $\begin{cases} 5x + 3y = 2 \\ -10x + 3z = -3 \\ y - 2z = -9 \end{cases}$

40. $\begin{cases} 2x + 4y = 0 \\ -2x + z = -5 \\ -4y - 3z = 1 \end{cases}$

41. $\begin{cases} x + y + z = 3 \\ -y + 2z = 1 \\ -x + z = 0 \end{cases}$

42. $\begin{cases} 3y - z = -1 \\ x + 5y - z = -4 \\ -3x + 6y + 2z = 11 \end{cases}$

Mixed Practice

In Problems 43–46, solve for x.

43. $\begin{vmatrix} x & 3 \\ 1 & 2 \end{vmatrix} = 7$

44. $\begin{vmatrix} -2 & x \\ 3 & 4 \end{vmatrix} = 1$

45. $\begin{vmatrix} x & -1 & -2 \\ 1 & 0 & 4 \\ 3 & 2 & 5 \end{vmatrix} = 5$

46. $\begin{vmatrix} 2 & x & -1 \\ 3 & 5 & 0 \\ -4 & 1 & 2 \end{vmatrix} = 0$

Applying the Concepts

Problems 47–50, use the following result. Determinants can be used to find the area of a triangle. If (a_1, b_1), (a_2, b_2), and (a_3, b_3) are the vertices of a triangle, the area of the triangle is $|D|$, where

$$D = \frac{1}{2} \begin{vmatrix} a_1 & a_2 & a_3 \\ b_1 & b_2 & b_3 \\ 1 & 1 & 1 \end{vmatrix}$$

△ **47. Geometry: Area of a Triangle** Given the points $A = (1, 1)$, $B = (5, 1)$, and $C = (5, 6)$,

    (a) Plot the points in the Cartesian plane and form triangle ABC.

    (b) Find the area of the triangle ABC.

△ **48. Geometry: Area of a Triangle** Given the points $A = (-1, -1)$, $B = (3, 2)$, and $C = (0, 6)$,

    (a) Plot the points in the Cartesian plane and form triangle ABC.

    (b) Find the area of the triangle ABC.

△ **49. Geometry: Area of a Parallelogram** Find the area of a parallelogram by doing the following:

    (a) Plot the points $A = (2, 1)$, $B = (7, 2)$, $C = (8, 4)$, and $D = (3, 3)$ in the Cartesian plane.

    (b) Form triangle ABC and find the area of triangle ABC.

    (c) Find the area of the triangle ADC.

    (d) Conclude that the diagonal of the parallelogram forms two triangles of equal area. Use this result to find the area of the parallelogram.

△ **50. Geometry: Area of a Parallelogram** Find the area of a parallelogram by doing the following:

(a) Plot the points $A = (-3, -2)$, $B = (3, 1)$, $C = (4, 4)$, and $D = (-2, 1)$ in the Cartesian plane.

(b) Form triangle ABC and find the area of triangle ABC.

(c) Find the area of the triangle ADC.

(d) Conclude that the diagonal of the parallelogram forms two triangles of equal area. Use this result to find the area of the parallelogram.

△ **51. Geometry: Equation of a Line** An equation of the line containing the two points (x_1, y_1) and (x_2, y_2) may be expressed as the determinant

$$\begin{vmatrix} x & y & 1 \\ x_1 & y_1 & 1 \\ x_2 & y_2 & 1 \end{vmatrix} = 0$$

(a) Find the equation of the line containing $(3, 2)$ and $(5, 1)$ using the determinant.

(b) Verify your result by using the slope formula and the point-slope formula.

△ **52. Geometry: Collinear Points** The distinct points (x_1, y_1), (x_2, y_2), and (x_3, y_3) are collinear if and only if

$$\begin{vmatrix} x_1 & y_1 & 1 \\ x_2 & y_2 & 1 \\ x_3 & y_3 & 1 \end{vmatrix} = 0$$

(a) Plot the points $(-3, -2)$, $(1, 2)$, and $(7, 8)$ in the Cartesian plane.

(b) Show that the points are collinear using the determinant.

(c) Show that the points are collinear using the idea of slopes.

Extending the Concepts

53. Evaluate the determinant $\begin{vmatrix} 3 & -2 \\ 1 & 4 \end{vmatrix}$. Now interchange rows 1 and 2 and recompute the determinant. What do you notice? Do you think that this result is true in general?

54. Evaluate the determinant $\begin{vmatrix} -3 & 1 \\ 6 & 5 \end{vmatrix}$. Multiply the entries in column 2 by 3 and recompute the determinant. What do you notice? Do you think that this result is true in general?

Technology Exercises

Graphing calculators have the ability to evaluate 2×2 determinants. First, enter the matrix into the graphing calculator and name it A. Then compute the determinant of matrix A. Figure 8 shows the results of Example 1(a) using a graphing calculator.

Figure 8

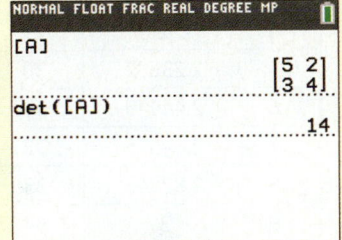

Because graphing calculators have the ability to compute determinants of matrices, they can be used to solve systems of equations using Cramer's Rule. Figure 9 shows the results of solving Example 2 using Cramer's Rule, where $A = D$, $B = D_x$, and $C = D_y$.

Figure 9

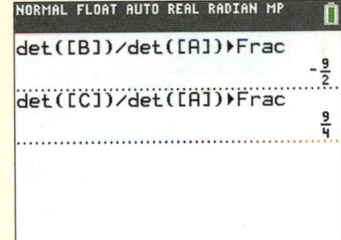

In Problems 55–60, use technology to solve the system of equations.

55. $\begin{cases} x + y = -4 \\ x - y = -12 \end{cases}$

56. $\begin{cases} x + y = 6 \\ x - y = 4 \end{cases}$

57. $\begin{cases} 2x + 3y = 3 \\ -3x + y = -10 \end{cases}$

58. $\begin{cases} 2x + 4y = -6 \\ 3x + 2y = 7 \end{cases}$

59. $\begin{cases} x - y + z = -4 \\ x + 2y - z = 1 \\ 2x + y + 2z = -5 \end{cases}$

60. $\begin{cases} x + y - z = 6 \\ x + 2y + z = 6 \\ -x - y + 2z = -7 \end{cases}$

Appendix D Table of Square Roots

n	$\sqrt{n}$	n	$\sqrt{n}$	n	$\sqrt{n}$	n	$\sqrt{n}$
1	1	26	5.09902	51	7.14143	76	8.71780
2	1.41421	27	5.19615	52	7.21110	77	8.77496
3	1.73205	28	5.29150	53	7.28011	78	8.83176
4	2	29	5.38516	54	7.34847	79	8.88819
5	2.23607	30	5.47723	55	7.41620	80	8.94427
6	2.44949	31	5.56776	56	7.48331	81	9
7	2.64575	32	5.65685	57	7.54983	82	9.05539
8	2.82843	33	5.74456	58	7.61577	83	9.11043
9	3	34	5.83095	59	7.68115	84	9.16515
10	3.16228	35	5.91608	60	7.74597	85	9.21954
11	3.31662	36	6	61	7.81025	86	9.27362
12	3.46410	37	6.08276	62	7.87401	87	9.32738
13	3.60555	38	6.16441	63	7.93725	88	9.38083
14	3.74166	39	6.24500	64	8	89	9.43398
15	3.87298	40	6.32456	65	8.06226	90	9.48683
16	4	41	6.40312	66	8.12404	91	9.53939
17	4.12311	42	6.48074	67	8.18535	92	9.59166
18	4.24264	43	6.55744	68	8.24621	93	9.64365
19	4.35890	44	6.63325	69	8.30662	94	9.69536
20	4.47214	45	6.70820	70	8.36660	95	9.74679
21	4.58258	46	6.78233	71	8.42615	96	9.79796
22	4.69042	47	6.85565	72	8.48528	97	9.84886
23	4.79583	48	6.92820	73	8.54400	98	9.89949
24	4.89898	49	7	74	8.60233	99	9.94987
25	5	50	7.07107	75	8.66025	100	10

Appendix E Transformations

Objectives

1. Graph Functions Using Vertical and Horizontal Shifts
2. Graph Functions Using Compressions and Stretches
3. Graph Functions Using Reflections about the x-Axis and the y-Axis

At this stage, if you were asked to graph any of the functions defined by $y = x$, $y = x^2$, $y = \sqrt{x}$, $y = \sqrt[3]{x}$, or $y = |x|$, your response should be "Yes, I recognize these functions and know the general shapes of their graphs." (If this is not your answer, review Section 8.4, Table 7.)

Sometimes we are asked to graph a function that is "almost" like one that we already know how to graph. This section will develop techniques for graphing such functions. Collectively, these techniques are referred to as **transformations**.

▶ 1 Graph Functions Using Vertical and Horizontal Shifts

EXAMPLE 1 Vertical Shift Up

Use the graph of $f(x) = |x|$ to obtain the graph of $g(x) = |x| + 4$.

Solution

Begin by obtaining some points on the graphs of f and g. For example, when $x = 0$, then $y = f(0) = 0$ and $y = g(0) = 4$. Table 1 lists these points as well as a few others. Notice that the y-coordinates on the graph of $g(x) = |x| + 4$ are exactly 4 units greater than the corresponding y-coordinates on the graph of $f(x) = |x|$. Figure 1 shows the graphs of f and g.

Table 1

x	$f(x) = \|x\|$	$(x, f(x))$	$g(x) = \|x\| + 4$	$(x, g(x))$
-2	$\|-2\| = 2$	$(-2, 2)$	$\|-2\| + 4 = 6$	$(-2, 6)$
-1	$\|-1\| = 1$	$(-1, 1)$	$\|-1\| + 4 = 5$	$(-1, 5)$
0	0	$(0, 0)$	4	$(0, 4)$
1	1	$(1, 1)$	5	$(1, 5)$
2	2	$(2, 2)$	6	$(2, 6)$

Figure 1

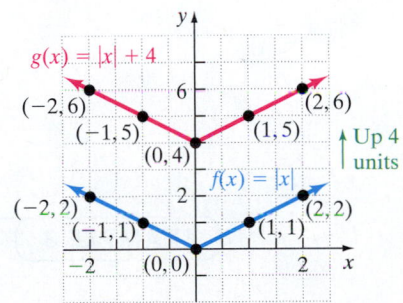

The graph of g is identical to the graph of f, except that it is shifted up 4 units.

EXAMPLE 2 **Vertical Shift Down**

Use the graph of $f(x) = |x|$ to obtain the graph of $g(x) = |x| - 5$.

Solution

Table 2 lists some points on the graphs of f and g. Notice that the y-coordinates on the graph of g are 5 units less than the corresponding y-coordinates on the graph of f. Figure 2 shows the graphs of f and g.

Table 2

x	$f(x) = \lvert x \rvert$	$(x, f(x))$	$g(x) = \lvert x \rvert - 5$	$(x, g(x))$
-2	2	$(-2, 2)$	-3	$(-2, -3)$
-1	1	$(-1, 1)$	-4	$(-1, -4)$
0	0	$(0, 0)$	-5	$(0, -5)$
1	1	$(1, 1)$	-4	$(1, -4)$
2	2	$(2, 2)$	-3	$(2, -3)$

Figure 2

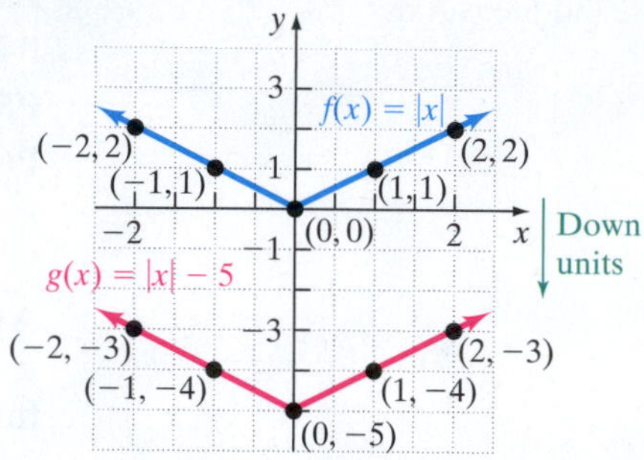

The graph of g is identical to the graph of f, except that it is shifted down 5 units. ●

The results of Examples 1 and 2 lead to the following conclusion.

> If a positive real number k is added to the output of a function, $y = f(x)$, the graph of the new function, $y = f(x) + k$, is the graph of f **shifted vertically up** k units.
>
> If a positive real number k is subtracted from the output of a function, $y = f(x)$, the graph of the new function, $y = f(x) - k$, is the graph of f **shifted vertically down** k units.

Quick ✓

1. If k is a positive real number, to obtain the graph of $y = f(x) + k$ from the graph of $y = f(x)$, shift the graph of $y = f(x)$ _____ k units. To obtain the graph of $y = f(x) - k$ from the graph of $y = f(x)$, shift the graph of $y = f(x)$ _____ k units.

2. Use the graph of $y = |x|$ to graph the function $f(x) = |x| + 5$. Show at least three points.

3. Use the graph of $y = x^2$ to graph the function $f(x) = x^2 - 4$. Show at least three points.

EXAMPLE 3 **Horizontal Shift to the Right**

In the same Cartesian plane, graph $f(x) = x^3$ and $g(x) = (x - 2)^3$.

Solution

Use the point-plotting method. Table 3 on the next page lists some points on the graphs of f and g. Note that when $f(x) = 0$, $x = 0$ and that when $g(x) = 0$, $x = 2$. Also, when $f(x) = -1$, $x = -1$, and when $g(x) = -1$, $x = 1$. The graph of g is identical to that of f, except that g is shifted 2 units to the right. See Figure 3.

Figure 3

Table 3

x	$f(x) = x^3$	$(x, f(x))$	$g(x) = (x - 2)^3$	$(x, g(x))$
-2	-8	$(-2, -8)$	$(-2 - 2)^3 = -64$	$(-2, -64)$
-1	-1	$(-1, 1)$	$(-1 - 2)^3 = -27$	$(-1, -27)$
0	0	$(0, 0)$	-8	$(0, -8)$
1	1	$(1, 1)$	-1	$(1, -1)$
2	8	$(2, 8)$	0	$(2, 0)$

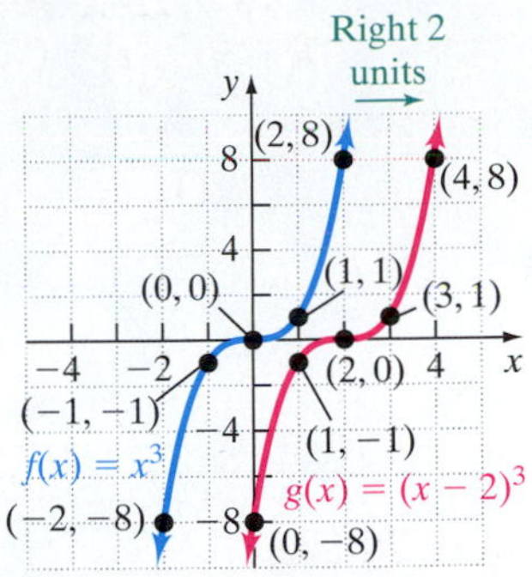

Why does this happen? Because 2 is getting subtracted from each x-value in the function $g(x) = (x - 2)^3$, the x-values must be greater by 2 in order to obtain the same y-value that was obtained in the graph of $f(x) = x^3$.

EXAMPLE 4 **Horizontal Shift to the Right**

In the same Cartesian plane, graph $f(x) = x^3$ and $g(x) = (x + 2)^3$.

Solution
Again, use the point-plotting method. Table 4 lists some points on the graphs of f and g. Notice that when $f(x) = 0, x = 0$, and when $g(x) = 0, x = -2$. Also, when $f(x) = 1, x = 1$, and when $g(x) = 1, x = -1$. The graph of g is identical to that of f, except that g is shifted 2 units to the left. See Figure 4.

Figure 4

Table 4

x	$f(x) = x^3$	$(x, f(x))$	$g(x) = (x + 2)^3$	$(x, g(x))$
-2	-8	$(-2, -8)$	0	$(-2, 0)$
-1	-1	$(-1, -1)$	1	$(-1, 1)$
0	0	$(0, 0)$	8	$(0, 8)$
1	1	$(1, 1)$	27	$(1, 27)$
2	8	$(2, 8)$	64	$(2, 64)$

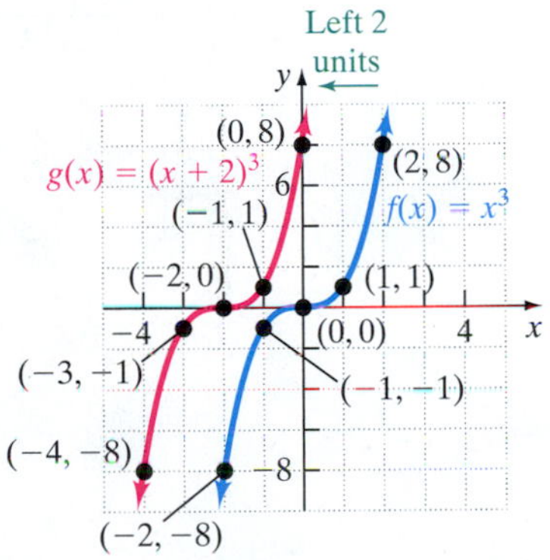

Why does this happen? Because 2 is being added to each x-value in the function $g(x) = (x + 2)^3$, the x-values must be smaller by 2 in order to obtain the same y-value that was obtained in the graph of $f(x) = x^3$.

The results of Examples 3 and 4 lead to the following conclusion.

In Other Words
For $y = f(x - h)$, add h to each x-coordinate on the graph of $y = f(x)$. For $y = f(x + h)$, subtract h from each x-coordinate on the graph of $y = f(x)$.

If the argument x of a function is replaced by $x - h, h > 0$, the graph of the new function, $y = f(x - h)$, is the graph of f **shifted horizontally right** h units.

If the argument x of a function is replaced by $x + h, h > 0$, the graph of the new function, $y = f(x + h)$, is the graph of f **shifted horizontally left** h units.

Quick ✓

4. *True or False* To obtain the graph of $f(x) = |x - 3|$ from the graph of $y = |x|$, shift the graph of $y = |x|$ to the right 3 units.

5. Use the graph of $y = x^2$ to graph the function $f(x) = (x + 3)^2$. Show at least three points.

6. Use the graph of $y = \sqrt{x}$ to graph the function $f(x) = \sqrt{x - 1}$. Show at least three points.

EXAMPLE 5 **Combining Vertical and Horizontal Shifts**

Use the graph of $f(x) = \sqrt{x}$ to obtain the graphs of $g(x) = \sqrt{x + 2} - 3$.

Solution

Graph g in stages. First, start with the graph of $f(x)$ as shown in Figure 5(a). Next, to get the graph of $y = \sqrt{x + 2}$, shift the graph of f, 2 units to the left. See Figure 5(b). Finally, to get the graph of $y = \sqrt{x + 2} - 3$, shift the graph of $y = \sqrt{x + 2}$ down 3 units. See Figure 5(c). Note the points plotted on each graph. Using key points can be helpful in keeping track of the transformation that has taken place.

Figure 5

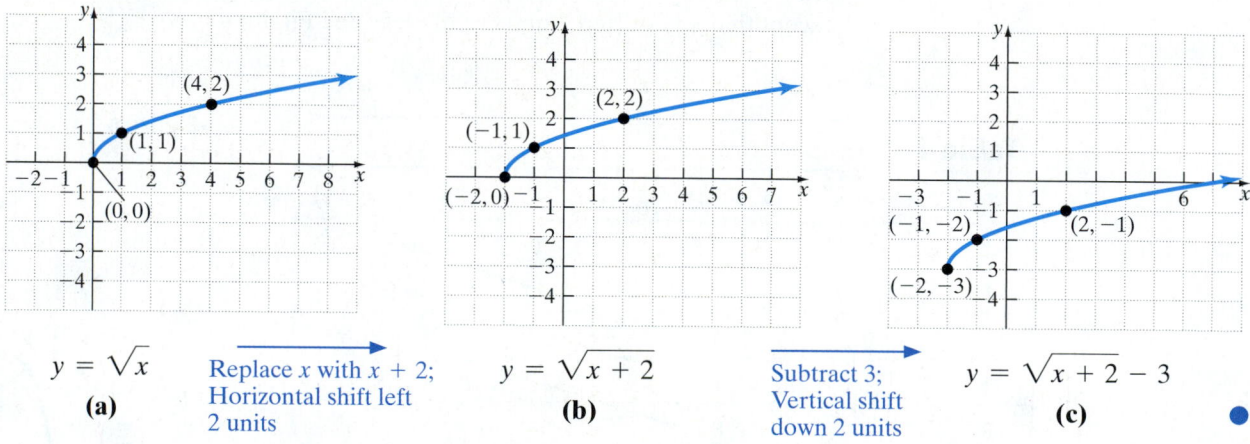

In Example 5, if the vertical shift had been done first, followed by the horizontal shift, the final graph would have been the same. Try it for yourself.

▶ ❷ Graph Functions Using Compressions and Stretches

EXAMPLE 6 **Vertical Stretch**

Use the graph of $f(x) = \sqrt[3]{x}$ to obtain the graph of $g(x) = 3\sqrt[3]{x}$.

Solution

To see the relationship between the graphs of f and g, form Table 5 on the next page, listing points on each graph. For each x-value, the y-coordinate of a point on the graph of g is 3 times as large as the corresponding y-coordinate on the graph of f. To obtain the graph of $g(x) = 3\sqrt[3]{x}$, vertically stretch the graph of $f(x) = \sqrt[3]{x}$ by a factor of 3. For example, $(1, 1)$ is on the graph of f, but $(1, 3)$ is on the graph of g. See Figure 6.

Table 5

x	$y = f(x) = \sqrt[3]{x}$	$y = g(x) = 3\sqrt[3]{x}$
-8	$\sqrt[3]{-8} = -2$	$3\sqrt[3]{-8} = 3(-2) = -6$
-1	$\sqrt[3]{-1} = -1$	$3\sqrt[3]{-1} = 3(-1) = -3$
0	0	0
1	1	3
8	2	6

Figure 6

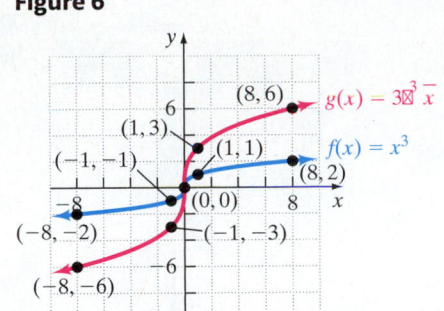

EXAMPLE 7   **Vertical Compression**

Use the graph of $f(x) = x^2$ to obtain the graph of $g(x) = \dfrac{1}{2}x^2$.

Solution

To see the relationship between the graphs of f and g, form Table 6, listing points on each graph. For each x-value, the y-coordinate of a point on the graph of g is $\dfrac{1}{2}$ as large as the corresponding y-coordinate on the graph of f. The graph of $f(x) = x^2$ is vertically compressed by a factor of $\dfrac{1}{2}$ to obtain the graph of $g(x) = \dfrac{1}{2}x^2$. For example, $(2, 4)$ is on the graph of f, but $(2, 2)$ is on the graph of g. See Figure 7.

Table 6

x	$y = f(x) = x^2$	$y = g(x) = \dfrac{1}{2}x^2$
-2	$(-2)^2 = 4$	$\dfrac{1}{2}(-2)^2 = \dfrac{1}{2}(4) = 2$
-1	$(-1)^2 = 1$	$\dfrac{1}{2}(-1)^2 = \dfrac{1}{2}(1) = \dfrac{1}{2}$
0	0	0
1	1	$\dfrac{1}{2}$
2	4	2

Figure 7

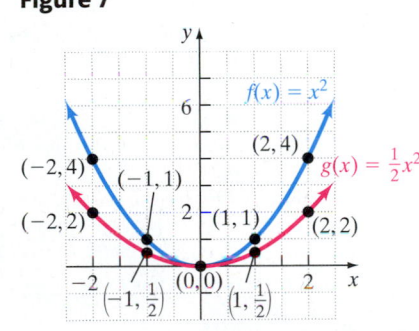

Examples 6 and 7 lead to the following conclusion.

When the right side of a function $y = f(x)$ is multiplied by a positive number a, the graph of the new function, $y = af(x)$ is obtained by multiplying each y-coordinate on the graph of $y = f(x)$ by a. The new graph is a **vertically compressed** (if $0 < a < 1$) or a **vertically stretched** (if $a > 0$) version of the graph $y = f(x)$.

In Other Words

For $y = af(x)$, the factor a is "outside" the function, so it affects the y-coordinates. Multiply each y-coordinate on the graph of $y = f(x)$ by a.

Quick ✓

7. To obtain the graph of $y = af(x)$ from the graph of $y = f(x)$, multiply each __-coordinate on the graph of $y = f(x)$ by __. If $0 < a < 1$, say that the graph is _____ _____ by a factor of a. If $a > 0$, say that the graph is _____ _____ by a factor of a.

8. Use the graph of $y = |x|$ to graph the function $f(x) = 2|x|$. Show at least three points.

9. Use the graph of $y = \sqrt{x}$ to graph the function $f(x) = \frac{1}{2}\sqrt{x}$. Show at least three points.

What happens if the argument x of a function $y = f(x)$ is multiplied by a positive number a, creating a new function $y = f(ax)$? To find the answer, look at the following exploration.

EXAMPLE 8 **Horizontal Stretch and Compression Technology Exploration**

Use technology to graph the following functions on the same screen:

$$Y_1 = f(x) = x^2 \qquad Y_2 = f(2x) = (2x)^2 \qquad Y_3 = f\left(\frac{1}{2}x\right) = \left(\frac{1}{2}x\right)^2$$

Create a table of values and explore the relation between the x-and y-coordinates of each function.

Solution

You should have obtained the graphs in Figure 8. Look at Table 7. Notice that $(-4, 16)$, $(-2, 4)$ and, $(2, 4)$ are points on the graph of $Y_1 = x^2$ and the points $(-2, 16)$, $(-1, 4)$ and, $(1, 4)$ are points on the graph of $Y_2 = (2x)^2$. For a given y-coordinate, the x-coordinate on the graph of Y_2 is $\frac{1}{2}$ of the x-coordinate on Y_1. It can be concluded that the graph of Y_2 is obtained by multiplying the x-coordinate of each point on the graph of Y_1 by $\frac{1}{2}$. The graph of $Y_2 = (2x)^2$ is the graph of $Y_1 = x^2$ *compressed* horizontally.

Figure 8

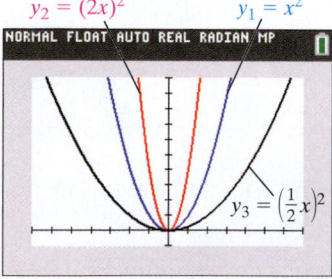

Table 7

X	Y1	Y2	Y3
-5	25	100	6.25
-4	16	64	4
-3	9	36	2.25
-2	4	16	1
-1	1	4	.25
0	0	0	0
1	1	4	.25
2	4	16	1
3	9	36	2.25
4	16	64	4
5	25	100	6.25

X= -5

Now, look at Table 7 again. Notice that $(-1, 1)$, $(1, 1)$, and $(2, 4)$ are points on the graph of $Y_1 = x^2$ and the points $(-2, 1)$, $(2, 1)$, and $(4, 4)$ are points on the graph of $Y_3 = \left(\frac{1}{2}x\right)^2$. For a given y-coordinate, the x-coordinate on the graph of Y_2 is 2 times the x-coordinate on Y_1. It can be concluded that the graph of Y_2 is obtained by multiplying the x-coordinate of each point on the graph of Y_1 by 2. The graph of $Y_3 = \left(\frac{1}{2}x\right)^2$ is the graph of $Y_1 = x^2$ *stretched* horizontally.

Based on this exploration, the following results.

> If the argument x of a function $y = f(x)$ is multiplied by a positive number a, the graph of the new function, $y = f(ax)$ is obtained by multiplying each x-coordinate of by $\dfrac{1}{a}$. A **horizontally compression** results if $a > 0$ and a **horizontal stretch** results if $0 < a < 1$.

Quick ✓

10. To obtain the graph of $y = f(ax)$ from the graph of $y = f(x)$, multiply each ___-coordinate on the graph of $y = f(x)$ by ___. If $0 < a < 1$, say that the graph is _____ _____ by a factor of $\dfrac{1}{a}$. If $a > 0$, say that the graph is _____ _____ by a factor of $\dfrac{1}{a}$.

11. Use the graph of $y = x^3$ to graph the function $f(x) = \left(\dfrac{1}{3}x\right)^3$. Show at least three points.

12. Use the graph of $y = \sqrt[3]{x}$ to graph the function $f(x) = \sqrt[3]{2x}$. Show at least three points.

▶ ❸ Graph Functions Using Reflections about the x-Axis and the y-Axis

EXAMPLE 9 **Reflection about the x-Axis**

In the same Cartesian plane, graph $f(x) = |x|$ and $g(x) = -|x|$.

Solution

Use the point-plotting method. Table 8 on the next page lists some points on the graphs of f and g. Notice that when $x = 1$, $f(x) = 1$ and when $x = 1$, $g(x) = -1$. Also, when $x = 2$, $f(x) = 2$, and when $x = 2$, $g(x) = -2$. The graph of g is identical to that of f, except that g is a reflection of f about the x-axis. See Figure 9.

Table 8

x	$f(x) = |x|$	$(x, f(x))$	$g(x) = -|x|$	$(x, g(x))$
-2	2	$(-2, 2)$	-2	$(-2, -2)$
-1	1	$(-1, 1)$	-1	$(-1, -1)$
0	0	$(0, 0)$	0	$(0, 0)$
1	1	$(1, 1)$	-1	$(1, -1)$
2	2	$(2, 2)$	-2	$(2, -2)$

Figure 9

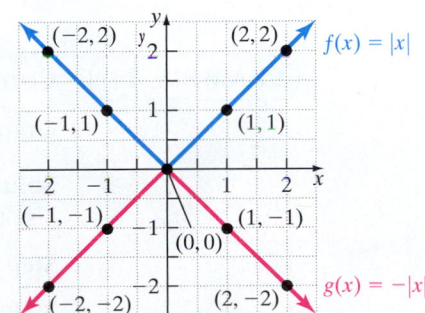

EXAMPLE 10 **Reflection about the y-Axis**

In the same Cartesian plane, graph $f(x) = \sqrt[3]{x}$ and $g(x) = \sqrt[3]{-x}$.

Solution

Again, use the point-plotting method. Table 9 lists some points on the graphs of f and g. Notice when $f(x) = -2, x = -8$ and when $g(x) = -2, x = 8$. Also, when $f(x) = -1, x = -1$, and when $g(x) = -1, x = 1$. The graph of g is identical to that of f, except that g is a reflection of f about the y-axis. See Figure 10.

Figure 10

Table 9

x	$f(x) = \sqrt[3]{x}$	$(x, f(x))$	$g(x) = -\lvert x \rvert$	$(x, g(x))$
-8	$\sqrt[3]{-8} = -2$	$(-8, -2)$	$\sqrt[3]{-(-8)} = \sqrt[3]{8} = 2$	$(-8, 2)$
-1	$\sqrt[3]{-1} = -1$	$(-1, -1)$	$\sqrt[3]{-(-1)} = \sqrt[3]{1} = 1$	$(-1, 1)$
0	0	$(0, 0)$	0	$(0, 0)$
1	1	$(1, 1)$	-1	$(1, -1)$
8	2	$(8, 2)$	-2	$(8, -2)$

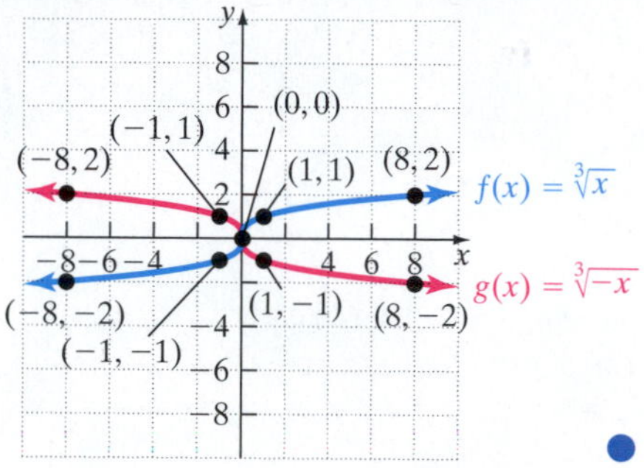

When the right side of the function $y = f(x)$ is multiplied by -1, the graph of the new function $y = -f(x)$ is the **reflection about the x-axis** of the graph of the function $y = f(x)$.

When the graph of the function $y = f(x)$ is known, the graph of the new function $y = f(-x)$ is the **reflection about the y-axis** of the graph of the function $y = f(x)$.

In Other Words

For $y = -f(x)$, multiply each y-coordinate on the graph of $y = f(x)$ by -1. For $y = f(-x)$, multiply each x-coordinate on the graph of $y = f(x)$ by -1.

Quick ✓

13. The graph of $y = -f(x)$ is a reflection of the graph of $y = f(x)$ about the ___-axis. The graph of $y = f(-x)$ is a reflection of the graph of $y = f(x)$ about the ___-axis.

14. Use the graph of $y = \sqrt{x}$ to graph the function $f(x) = -\sqrt{x}$. Show at least three points.

15. Use the graph of $y = x^3$ to graph the function $f(x) = (-x)^3$. Show at least three points.

EXAMPLE 11 **Combining Graphing Procedures**

Graph the function $f(x) = 2\sqrt{x + 1} - 3$. Find the domain and the range of f.

Solution

Use the following steps to obtain the graph of f:

Step 1: $y = \sqrt{x}$ Square root function

Step 2: $y = 2\sqrt{x}$ Multiply by 2; vertical stretch of the graph of $y = \sqrt{x}$ by a factor of 2.

Step 3: $y = 2\sqrt{x + 1}$ Replace x by $x + 1$; horizontal shift to the left 1 unit.

Step 4: $y = 2\sqrt{x + 1} - 3$ Subtract 3; vertical shift down 3 units.

See Figure 11.

Figure 11

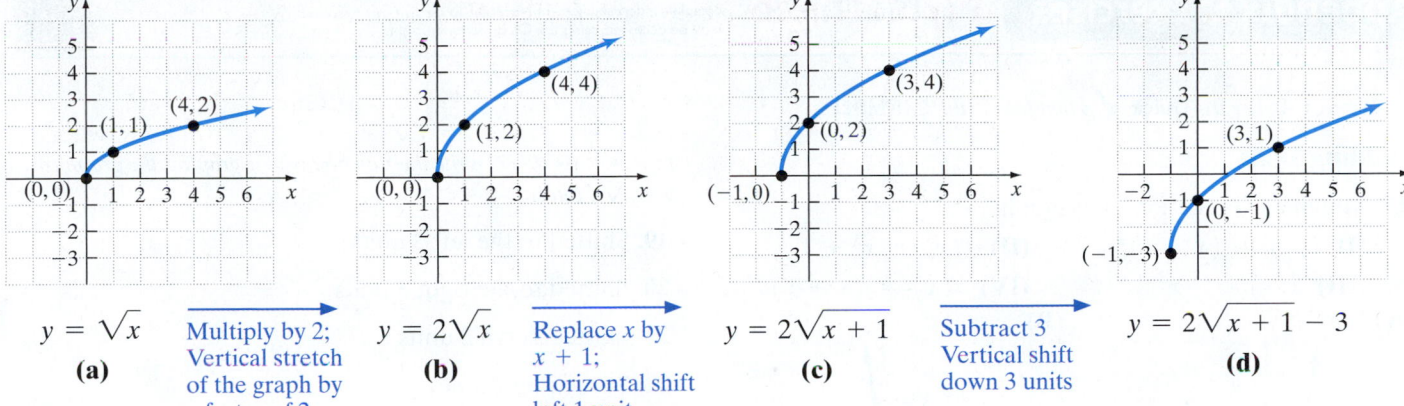

$y = \sqrt{x}$ Multiply by 2; Vertical stretch of the graph by a factor of 2
(a)

$y = 2\sqrt{x}$ Replace x by $x + 1$; Horizontal shift left 1 unit
(b)

$y = 2\sqrt{x + 1}$ Subtract 3 Vertical shift down 3 units
(c)

$y = 2\sqrt{x + 1} - 3$
(d)

The domain of f is $[-1, \infty)$ and the range is $[-3, \infty)$

Work Smart

Although the order in which transformations are performed can be altered, you may consider using the following order for consistency:

1. Reflections
2. Compressions and stretches
3. Shifts

Other orderings of the steps shown in Example 11 would also result in the graph of f. For example, try this one:

Step 1: $y = \sqrt{x}$ Square root function

Step 2: $y = \sqrt{x + 1}$ Replace x by $x + 1$; horizontal shift to the left 1 unit.

Step 3: $y = 2\sqrt{x + 1}$ Multiply by 2; vertical stretch of the graph of $y = \sqrt{x}$ by a factor of 2.

Step 4: $y = 2\sqrt{x + 1} - 3$ Subtract 3; vertical shift down 3 units.   ●

Quick ✔

16. Graph $f(x) = -2(x - 1)^2$ using the techniques of shifting, compressing, stretching, and/or reflecting. Start with the graph of the basic function and show at least three points.

▶ Summary of Graphing Techniques

To Graph:	Draw the Graph of f and :	Functional Change to $f(x)$:
Vertical Shifts		
$y = f(x) + k, k > 0$	Raise the graph of f by k units.	Add k to $f(x)$.
$y = f(x) - k, k > 0$	Lower the graph of f by k units.	Subtract k from $f(x)$.
Horizontal Shifts		
$y = f(x + h), h > 0$	Shift the graph of f to the left h units.	Replace x by $x + h$.
$y = f(x - h), h > 0$	Shift the graph of f to the right h units.	Replace x with $x - h$.
Compressing or Stretching		
$y = af(x), a > 0$	Multiply each y-coordinate by a. Stretch the graph of f vertically if $a > 1$. Compress the graph of f vertically if $0 < a < 1$.	Multiply $f(x)$ by a.
$y = f(ax), a > 0$	Multiply each x-coordinate by $\frac{1}{a}$. Stretch the graph of f horizontally if $0 < a < 1$. Compress the graph of f horizontally if $a > 1$.	Replace x by ax.
Reflection about the x-Axis		
$y = -f(x)$	Reflect the graph of f about the x-axis.	Multiply $f(x)$ by -1.
Reflection about the y-Axis		
$y = f(-x)$	Reflect the graph of f about the y-axis.	Replace x by $-x$.

Appendix E: Exercises MyMathLab®

Exercise numbers in **green** have complete video solutions in MyMathLab or may be accessed using the QR code to the right.

Problems 1–16 are the Quick ✔s that follow the EXAMPLES.

Building Skills

17. Match each function to its graph:

(**I**) $f(x) = (x - 1)^3$ (**II**) $f(x) = x^3 + 1$

(**III**) $f(x) = -x^3$ (**IV**) $f(x) = (x + 1)^3$

(**A**)

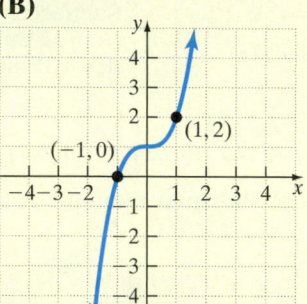

(**B**)

(**C**)

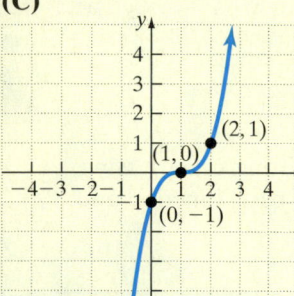

(**D**)

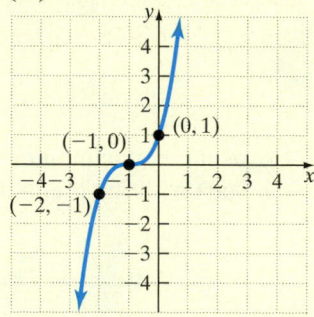

18. Match each function to its graph:

(**I**) $f(x) = -|2x|$ (**II**) $f(x) = |x + 1| - 2$

(**III**) $f(x) = |x + 1| + 2$ (**IV**) $f(x) = -|x + 1| + 2$

(**A**)

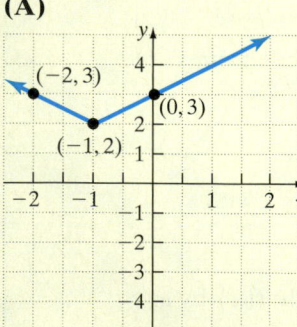

(**B**)

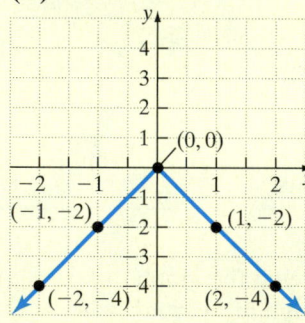

(**C**)

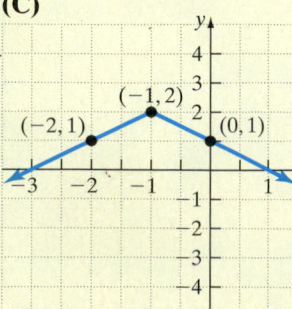

(**D**)

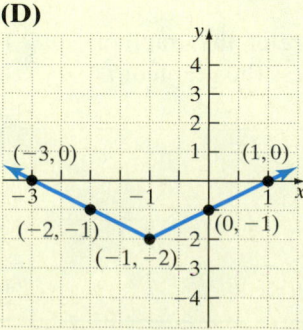

In Problems 19–26, write the function whose graph is the graph of $y = \sqrt{x}$ but is:

19. Shifted to the left 3 units

20. Shifted to the right 3 units

21. Shifted down 2 units

22. Shifted up 2 units

23. Reflected about the *x*-axis

24. Reflected about the *y*-axis

25. Horizontally stretched by a factor of 2

26. Vertically stretched by a factor of 2

In Problems 27–42, graph each function using the techniques of shifting, compressing, stretching, and/or reflecting. Start with the graph of the basic function (for example, $y = x^2$) and show all stages. Be sure to show at least three key points. Find the domain and the range of each function.

27. $f(x) = x^3 + 1$ **28.** $f(x) = x^3 - 1$

29. $g(x) = (x - 1)^2$ **30.** $g(x) = (x + 1)^2$

31. $h(x) = \sqrt{2x}$ **32.** $h(x) = \sqrt{\frac{1}{2}x}$

33. $f(x) = \frac{1}{2}|x|$ **34.** $f(x) = 3|x|$

35. $g(x) = \sqrt[3]{x - 1} + 2$ **36.** $g(x) = \sqrt[3]{x + 1} - 2$

37. $h(x) = \sqrt{-x}$ **38.** $h(x) = -\sqrt{x}$

39. $f(x) = 2|x - 1| + 2$ **40.** $f(x) = \frac{1}{2}|x - 1| + 3$

41. $f(x) = -\sqrt{x + 2} + 3$ **42.** $f(x) = 2\sqrt{-x} + 1$

43. Use the graph of *f* illustrated below as the first step toward graphing each of the following functions.

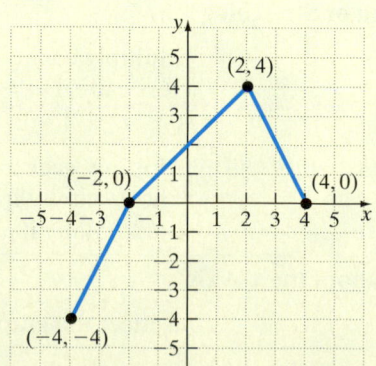

(a) $h(x) = f(x) + 2$

(b) $g(x) = f(x - 1)$

(c) $F(x) = f(2x)$

(d) $k(x) = f\left(\dfrac{1}{2}x\right)$

(e) $G(x) = f(-x) - 2$

(f) $p(x) = f\left(\dfrac{1}{2}x\right) - 1$

(g) $r(x) = 2f(x - 1) + 2$

(h) $j(x) = -f(2x)$

44. Use the graph of f illustrated below as the first step toward graphing each of the following functions.

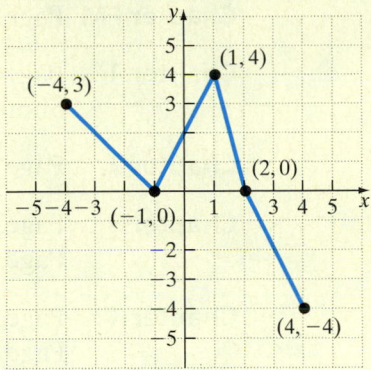

(a) $h(x) = f(x) - 2$

(b) $g(x) = f(x + 1)$

(c) $F(x) = 2f(x)$

(d) $k(x) = \dfrac{1}{2}f(x)$

(e) $G(x) = -f(x) + 2$

(f) $p(x) = \dfrac{1}{2}f(x) + 1$

(g) $r(x) = 2f(x + 1) - 2$

(h) $j(x) = 2f(-x)$

Photo Credits

Answers to Selected Exercises

Chapter 1 Operations on Real Numbers and Algebraic Expressions

Section 1.1 Success in Mathematics Answers will vary.

Section 1.2 Fractions, Decimals, and Percents **1.** prime **2.** factors; product **3.** $2 \cdot 2 \cdot 3$ **4.** $2 \cdot 2 \cdot 2 \cdot 3 \cdot 5$ **5.** prime **6.** $3 \cdot 3 \cdot 13$
7. least common multiple **8.** 24 **9.** 360 **10.** 126 **11.** 180 **12.** numerator; denominator **13.** equivalent fractions **14.** $\frac{5}{10}$ **15.** $\frac{30}{48}$ **16.** least common denominator
17. $\frac{1}{4} = \frac{3}{12}; \frac{5}{6} = \frac{10}{12}$ **18.** $\frac{9}{20} = \frac{36}{80}; \frac{11}{16} = \frac{55}{80}$ **19.** lowest terms **20.** $\frac{9}{16}$ **21.** in lowest terms **22.** $\frac{2}{7}$ **23.** hundredths **24.** tenths **25.** thousands **26.** thousandths
27. 0.2 **28.** 0.93 **29.** 1.40 **30.** 690.00 **31.** 60.0 **32.** 0.4 **33.** 0.833... or $0.8\overline{3}$ **34.** 1.375 **35.** $0.\overline{428571}$ **36.** $\frac{3}{5}$ **37.** $\frac{17}{100}$ **38.** $\frac{5}{8}$ **39.** hundred; 35; 35 **40.** 0.23
41. 0.01 **42.** 0.724 **43.** 1.27 **44.** 100% **45.** 15% **46.** 80% **47.** 137.2% **48.** 0.4% **49.** $5 \cdot 5$ **51.** $2 \cdot 2 \cdot 7$ **53.** $3 \cdot 7$ **55.** $2 \cdot 2 \cdot 3 \cdot 3$ **57.** $2 \cdot 5 \cdot 5$ **59.** 53 is prime.
61. $2 \cdot 2 \cdot 3 \cdot 3 \cdot 7$ **63.** 42 **65.** 126 **67.** 210 **69.** 90 **71.** 60 **73.** 72 **75.** $\frac{8}{12}$ **77.** $\frac{18}{24}$ **79.** $\frac{21}{3}$ **81.** $\frac{4}{8}$ and $\frac{3}{8}$ **83.** $\frac{9}{15}$ and $\frac{10}{15}$ **85.** $\frac{3}{36}$ and $\frac{10}{36}$ **87.** $\frac{20}{90}$ and $\frac{35}{90}$ and $\frac{21}{90}$ **89.** $\frac{2}{3}$
91. $\frac{19}{9}$ **93.** $\frac{1}{3}$ **95.** 6 **97.** hundredths place **99.** tens place **101.** thousandths place **103.** 578.2 **105.** 354.68 **107.** 3682.010 **109.** 30 **111.** 0.625 **113.** $0.\overline{285714}$
115. 0.3125 **117.** $0.\overline{230769}$ **119.** 1.16 **121.** $\frac{3}{4}$ **123.** $\frac{1}{2}$ **125.** $\frac{491}{500}$ **127.** 0.37 **129.** 0.0602 **131.** 0.001 **133.** 20% **135.** 27.5% **137.** 200% **139.** 2.2 **141.** 2.67
143. 0.519 **145.** 84 months **147.** every 20 days **149.** $\frac{13}{20}$ **151.** 70% **153.** 77.27% **155. (a)** 33.33% **(b)** 16.67% **(c)** 25% **157.** 21.43% **159.** 2, 3, 5, 7, 11, 13, 17,
19, 23, 29, 31, 37, 41, 43, 47, 53, 59, 61, 67, 71, 73, 79, 83, 89, 97

Section 1.3 The Number Systems and the Real Number Line **1.** {1, 3, 5, 7} **2.** {Alabama, Alaska, Arkansas, Arizona} **3.** $\varnothing$ or { } **4.** True **5.** rational
6. 12 **7.** 12, 0 **8.** $-5, 12, 0$ **9.** $\frac{11}{5}, -5, 12, 2.\overline{76}, 0, \frac{18}{4}$ **10.** 2.737737773... **11.** All numbers listed **12.** origin **13.**
14. inequality **15.** $<$ **16.** $<$ **17.** $>$ **18.** $>$ **19.** $=$ **20.** $<$ **21.** absolute value **22.** 15 **23.** $\frac{3}{4}$ **24.** -4 **25.** $A = \{0, 1, 2, 3, 4\}$ **27.** $D = \{1, 2, 3, 4\}$
29. $E = \{ \}$ or $\varnothing$ **31.** 3 **33.** $-4, 3, 0$ **35.** 2.303003000... **37.** All numbers listed **39.** π **41.** $\frac{5}{5} = 1$ **43.** **45.** True
47. True **49.** True **51.** False **53.** $<$ **55.** $>$ **57.** $>$ **59.** $=$ **61.** 12 **63.** 4 **65.** $\frac{3}{8}$ **67.** 2.1 **69. (a)**
(b) $-4.5, -1, -\frac{1}{2}, \frac{3}{5}, 1, 3.5, |-7| = 7$ **(c) (i)** $-1, 1, |-7| = 7$ **(ii)** All numbers listed
71. -100, Integer, Rational, Real **73.** -10.5, Rational, Real **75.** $\frac{75}{25}$, Natural, Whole, Integer, Rational, Real **77.** 7.56556555..., Irrational, Real
79. True **81.** False **83.** True **85.** True **87.** True **89.** Irrational numbers **91.** Real numbers **93.** 0 **95.** True **97.** True **99.** {7, 8, 9, 10, 11, 12, 13, 14, 15}
101. {10, 11, 12, 13, 14, 15} **103.** {11, 12} **105.** {0, 2, 4, 6, 8, 10} **107. (a)** {1}, {2}, {3}, {4}, {1, 2}, {1, 3}, {1, 4}, {2, 3}, {2, 4}, {3, 4}, {1, 2, 3}, {1, 2, 4}, {1, 3, 4},
{2, 3, 4}, {1, 2, 3, 4}, $\varnothing$ **(b)** 16 **109.** A rational number is any number that can be written as the quotient of two integers, denominator not equal to zero.
Natural numbers, whole numbers, and integers are rational numbers. Terminating and repeating decimals are also rational numbers.

Section 1.4 Adding, Subtracting, Multiplying, and Dividing Integers **1.** sum **2.** 14 **3.** -8 **4.** 3 **5.** -1 **6.** -4 **7.** 14 **8.** negative **9.** -4 **10.** -3
11. -24 **12.** 56 **13.** additive inverse; opposite; 0 **14.** -7 **15.** 21 **16.** $\frac{8}{5}$ **17.** -5.75 **18.** difference **19.** (-10) **20.** 80 **21.** -178 **22.** 21 **23.** 0 **24.** -58
25. 12 **26.** 550 **27.** positive **28.** -21 **29.** -52 **30.** 80 **31.** 108 **32.** 325 **33.** True **34.** 108 **35.** -360 **36.** $\frac{1}{6}$ **37.** $-\frac{1}{2}$ **38.** dividend; divisor; quotient **39.** True
40. -5 **41.** $-\frac{18}{5}$ **42.** 9 **43.** 15 **45.** 4 **47.** -4 **49.** -19 **51.** 21 **53.** -328 **55.** -11 **57.** 43 **59.** 325 **61.** -125 **63.** 11 **65.** -8 **67.** -28 **69.** 54 **71.** 0 **73.** -41
75. -31 **77.** 172 **79.** 40 **81.** -56 **83.** 0 **85.** 144 **87.** -126 **89.** -90 **91.** 210 **93.** 120 **95.** $\frac{1}{8}$ **97.** $-\frac{1}{4}$ **99.** 1 **101.** 5 **103.** 7 **105.** -15 **107.** $\frac{7}{2}$ **109.** $-\frac{10}{7}$
111. $\frac{35}{4}$ **113.** -72 **115.** 60 **117.** 171 **119.** -15 **121.** -42 **123.** $\frac{15}{4}$ **125.** 40 **127.** -238 **129.** $28 + (-21) = 7$ **131.** $-21 - 47 = -68$ **133.** $-12 \cdot 18 = -216$
135. $-36 \div (-108)$ or $\frac{-36}{-108} = \frac{1}{3}$ **137.** -47 **139.** -25 **141.** -3.25 dollars **143.** -6 yards **145.** $-\$48$ **147.** 14 miles **149.** $655 **151.** No; -125 cases
153. 25,725 feet **155.** $-3, -5$ **157.** $-12, 2$ **159. (a)** 1, 2, 1.5, $1.\overline{6}$, 1.6, 1.625, 1.615, 1.619, 1.618,... **(b)** 1.618 **(c)** Answers may vary
161. The problem $42 \div 4$ may be written equivalently as $42 \cdot \frac{1}{4}$.

Section 1.5 Adding, Subtracting, Multiplying, and Dividing Rational Numbers **1.** $\frac{27}{32}$ **2.** $-\frac{8}{3}$ **3.** $-\frac{6}{25}$ **4.** $\frac{3}{4}$ **5.** $-\frac{3}{22}$ **6.** 1 **7.** $\frac{1}{12}$ **8.** $\frac{5}{7}$ **9.** -4 **10.** $-\frac{20}{31}$
11. $\frac{50}{49}$ **12.** $-\frac{3}{8}$ **13.** $-\frac{16}{7}$ **14.** $\frac{5}{27}$ **15.** $-5; 7$ **16.** $\frac{10}{11}$ **17.** $-\frac{3}{7}$ **18.** $\frac{1}{7}$ **19.** $-\frac{6}{5}$ **20.** $\frac{29}{42}$ **21.** $\frac{5}{36}$ **22.** $-\frac{13}{4}$ **23.** $-\frac{3}{20}$ **24.** $-\frac{25}{16}$ **25.** $\frac{15}{4}$ **26.** 21.014 **27.** 64.57
28. -59.448 **29.** -71.412 **30.** 4.78 **31.** -899.5 **32.** -0.1035 **33.** 0.0135 **34.** 0.25 **35.** 8.36 **36.** -0.094 **37.** $\frac{2}{3}$ **39.** $-\frac{19}{9}$ **41.** $-\frac{1}{2}$ **43.** $\frac{12}{25}$ **45.** -25 **47.** $-\frac{2}{3}$
49. 8 **51.** $\frac{1}{6}$ **53.** $\frac{5}{3}$ **55.** $-\frac{1}{5}$ **57.** $\frac{5}{6}$ **59.** $-\frac{1}{9}$ **61.** $-\frac{2}{3}$ **63.** $\frac{1}{11}$ **65.** 32 **67.** $\frac{3}{2}$ **69.** $\frac{1}{2}$ **71.** 2 **73.** $\frac{1}{3}$ **75.** $\frac{5}{2}$ **77.** $-\frac{13}{12}$ **79.** $\frac{1}{4}$ **81.** $-\frac{1}{6}$ **83.** $\frac{33}{40}$ **85.** -6.5 **87.** 8.4 **89.** 55.92

91. 1.49 **93.** 42.55 **95.** 24.94 **97.** -9 **99.** 24.3 **101.** -490 **103.** $-\dfrac{11}{30}$ **105.** $-\dfrac{4}{3}$ **107.** $-\dfrac{1}{12}$ **109.** $-\dfrac{1}{4}$ **111.** $-\dfrac{129}{35}$ **113.** 1.6 **115.** $-\dfrac{2}{21}$ **117.** -50.526 **119.** -16

121. -15 **123.** -6.58 **125.** -7.9 **127.** $-\dfrac{25}{14}$ **129.** 58.39 **131.** -30.96 **133.** $\dfrac{1}{8}$ **135.** 21 hours **137.** 18 students **139.** \$18 **141.** \$2.40 **143.** 13.2 **145.** $\dfrac{86}{15}$

147. The expression $6 \div 2$ means to take 6 and divide it into groups of 2. How many times can we divide 6 into groups of 2? The answer is 3, so $6 \div 2 = 3$. The expression $6 \div \dfrac{1}{2}$ means to take 6 and divide it into groups of $\dfrac{1}{2}$. There are twelve $\dfrac{1}{2}$'s in 6, as shown in the figure, so $6 \div \dfrac{1}{2} = 12$.

Putting the Concepts Together (Sections 1.2–1.5) **1.** $\dfrac{7}{8} = \dfrac{35}{40}; \dfrac{9}{20} = \dfrac{18}{40}$ **2.** $\dfrac{1}{3}$ **3.** $0.\overline{285714}$ **4.** $\dfrac{3}{8}$ **5.** 0.123 **6.** 6.25% **7. (a)** $-12, -\dfrac{14}{7} = -2, 0, 3$

(b) $-12, -\dfrac{14}{7} = -2, -1.25, 0, 3, 11.2$ **(c)** $\sqrt{2}$ **(d)** All the numbers in the set are real numbers. **8.** $<$ **9.** -11 **10.** -65 **11.** -27 **12.** 27 **13.** -5.5 **14.** -10

15. -100 **16.** 72 **17.** -5 **18.** 16 **19.** -9 **20.** -3 **21.** $\dfrac{31}{5}$ **22.** $\dfrac{31}{36}$ **23.** $-\dfrac{17}{36}$ **24.** $\dfrac{9}{5}$ **25.** $-\dfrac{1}{28}$ **26.** 0 **27.** 10.76 **28.** 7.646 **29.** 1.46 **30.** 22.232

Section 1.6 Properties of Real Numbers **1.** multiplicative identity **2.** 8 feet **3.** $8\dfrac{1}{3}$ hours **4.** $5\dfrac{1}{2}$ pounds **5.** $b; a$ **6.** $a; b; b; a$ **7.** 22

8. $\dfrac{3}{20}$ **9.** 11.98 **10.** -13 **11.** $-\dfrac{36}{331}$ **12.** 349 **13.** 14 **14.** 14 **15.** -24.2 **16.** $\dfrac{50}{13}$ **17.** 0 **18.** undefined **19.** 156 inches **20.** 45 meters **23.** $10\dfrac{1}{2}$ gallons

25. $11\dfrac{1}{4}$ pounds **27.** 27 minutes **29.** $21\dfrac{1}{2}$ feet **31.** 480 minutes **33.** 25 quarts **35.** Additive Inverse Property **37.** Identity Property of Multiplication

39. Multiplicative Inverse Property **41.** Additive Inverse Property **43.** $\dfrac{0}{a} = 0$ **45.** Commutative Property of Multiplication **47.** Commutative Property

of Addition **49.** $\dfrac{a}{0}$ is undefined **51.** 29 **53.** 18 **55.** 0 **57.** -65 **59.** 347 **61.** -90 **63.** undefined **65.** -34 **67.** 0 **69.** 0 **71.** $-\dfrac{20}{3}$ **73.** \$203.16 **75.** $-3 - (4 - 10)$

77. $-15 + 10 - (4 - 8)$ **79.** $4\dfrac{1}{2}$ hours **81.** 44 feet per second **83.** 16 **85.** There is no real number that equals 1 when multiplied by 0.

87. The quotient $\dfrac{0}{4} = 0$ because $4 \cdot 0 = 0$. The quotient $\dfrac{4}{0}$ is undefined because we should be able to determine a real number $\square$ such that $0 \cdot \square = 4$.

But because the product of 0 and every real number is 0, there is no replacement value for $\square$. **89.** The product of a nonzero real number and its

multiplicative inverse (reciprocal) equals 1, the multiplicative identity.

Section 1.7 Exponents and the Order of Operations **1.** base; exponent; power **2.** 11^5 **3.** $(-7)^4$ **4.** 16 **5.** 49 **6.** $-\dfrac{1}{216}$ **7.** 0.81 **8.** -16 **9.** 16 **10.** 15 **11.** 24

12. 23 **13.** -19 **14.** 40 **15.** -63 **16.** 4 **17.** $-\dfrac{8}{7}$ **18.** $\dfrac{4}{7}$ **19.** $\dfrac{5}{3}$ **20.** 20 **21.** 40 **22.** -9 **23.** 48 **24.** $-\dfrac{1}{6}$ **25.** 12 **26.** -108 **27.** 10 **28.** 4 **29.** $5^2 \cdot$ **31.** $\left(-\dfrac{3}{5}\right)^3$ **33.** 64

35. 64 **37.** 1000 **39.** -1000 **41.** -1000 **43.** 2.25 **45.** -64 **47.** -1 **49.** 0 **51.** $\dfrac{1}{64}$ **53.** $-\dfrac{1}{27}$ **55.** 14 **57.** -3 **59.** 2500 **61.** 160 **63.** 20 **65.** 4 **67.** $\dfrac{3}{5}$ **69.** -1

71. 42 **73.** 5 **75.** -4 **77.** 115 **79.** -5 **81.** $\dfrac{169}{4}$ **83.** -24 **85.** $-\dfrac{13}{12}$ **87.** 11 **89.** 12 **91.** $-\dfrac{1}{2}$ **93.** 36 **95.** 1 **97.** 24 **99.** 5 **101.** $\dfrac{2}{5}$ **103.** 3 **105.** $\dfrac{3}{2}$ **107.** $\dfrac{64}{27}$ **109.** $-\dfrac{1}{6}$

111. $2^3 \cdot 3^2$ **113.** $2^4 \cdot 3$ **115.** $(4 \cdot 3 + 6) \cdot 2$ **117.** $(4 + 3) \cdot (4 + 2)$ **119.** $(6 - 4) + (3 - 1)$ **121.** \$514.93 **123.** 603.19 in.^{2} **125.** \$1060.90 **127.** 115.75°

129. The expression -3^2 means "take the opposite of three squared": $-(3 \cdot 3) = -9$. The base is the number 3. The expression $(-3)^2$ means use -3 as a

base twice: $-3 \cdot -3 = 9$.

Section 1.8 Simplifying Algebraic Expressions **1.** variable **2.** evaluate **3.** -7 **4.** 2 **5.** \$216 **6.** True **7.** $5x^2; 3xy$ **8.** $9ab; -3bc; 5ac; -ac^2$

9. $\dfrac{2mn}{5}; -\dfrac{3n}{7}$ **10.** 2 **11.** 1 **12.** -1 **13.** 5 **14.** $-\dfrac{2}{3}$ **15.** False **16.** like **17.** like **18.** unlike **19.** unlike **20.** like **21.** $b; c$ **22.** $6x + 12$ **23.** $-5x - 10$

24. $-2k + 14$ **25.** $6x + 9$ **26.** $4; 9$ **27.** $-5x$ **28.** $-4x^2$ **29.** $-8x + 3$ **30.** $-2a + 9b - 4$ **31.** $12ac - 5a + b$ **32.** $8ab^2 - a^2b$ **33.** $2rs - \dfrac{3}{2}r^2 - 5$

34. Remove all parentheses and combine like terms. **35.** $-2x - 1$ **36.** $-2m - n - 7$ **37.** $a - 11b$ **38.** $6x - 2$ **39.** 13 **41.** 17 **43.** -21 **45.** $\dfrac{1}{4}$ **47.** 81

49. 4 **51.** $2x^3, 3x^2; -x, 6; 2, 3, -1, 6$ **53.** $z^2, \dfrac{2y}{3}; 1, \dfrac{2}{3}$ **55.** unlike **57.** like **59.** like **61.** unlike **63.** $3m + 6$ **65.** $18n^2 + 12n - 6$ **67.** $-x + y$ **69.** $-4x + 3y$

71. $3x$ **73.** $6z$ **75.** $10m + 10n$ **77.** $2.2x^7$ **79.** $10y^6$ **81.** $-6w - 12y + 13z$ **83.** $-3k + 15$ **85.** $4n - 8$ **87.** $-3x + 3$ **89.** $4n - 2$ **91.** $-4n + 20$ **93.** $\dfrac{5}{6}x$

95. $-\dfrac{11}{2}$ **97.** $-3.5x - 6$ **99.** 32 **101.** 27 **103.** 0 **105.** -13 **107.** -7 **109.** -12 **111.** 29 **113.** -70 **115.** 44 **117.** $-\dfrac{3}{2}$ **119.** 36 **121.** \$78.70 **123.** \$4819 **125. (a)** $8w - 8$

(b) 32 yards **127.** \$210.43 **129.** $-3x^2 + 7x - 3$ **131.** The sum $2x^2 + 4x^2$ is not equal to $6x^4$ because when we combine like terms, we add the coefficients

of the like terms and keep the variables and exponents the same. Put another way, $2x^2 + 4x^2 = (2 + 4)x^2 = 6x^2$.

Chapter 1 Review **1.** $3 \cdot 5 \cdot 5$ **2.** $3 \cdot 29$ **3.** $3 \cdot 3 \cdot 3 \cdot 3$ **4.** prime **5.** 72 **6.** 72 **7.** $\dfrac{14}{30}$ **8.** $\dfrac{12}{4}$ **9.** $\dfrac{1}{6} = \dfrac{4}{24}; \dfrac{3}{8} = \dfrac{9}{24}$ **10.** $\dfrac{9}{16}$ **11.** $\dfrac{27}{48}; \dfrac{7}{24} = \dfrac{14}{48}$ **11.** $\dfrac{5}{12}$ **12.** $\dfrac{1}{2}$

13. $\dfrac{4}{5}$ **14.** 21.76 **15.** 15 **16.** $0.88\ldots = 0.\overline{8}$ **17.** 0.28125 **18.** 1.83 **19.** 2.4 **20.** $\dfrac{3}{5}$ **21.** $\dfrac{3}{8}$ **22.** $\dfrac{108}{125}$ **23.** 0.41 **24.** 7.60 **25.** 0.0903 **26.** 0.0035 **27.** 23% **28.** 117%

29. 4.5% **30.** 300% **31. (a)** $\dfrac{3}{5}$ **(b)** 60% **32.** $A = \{0, 1, 2, 3, 4, 5, 6\}$ **33.** $B = \{1, 2, 3\}$ **34.** $C = \{-2, -1, 0, 1, 2, 3, 4, 5\}$ **35.** $D = \{\ \}$ or $\varnothing$

36. $\dfrac{9}{3} = 3, 11$ **37.** $0, \dfrac{9}{3} = 3, 11$ **38.** $-6, 0, \dfrac{9}{3} = 3, 11$ **39.** $-6, -3.25, 0, \dfrac{9}{3} = 3, 11, \dfrac{5}{7}$ **40.** 5.030030003… **41.** All numbers listed

42.

43. False **44.** True **45.** True **46.** True **47.** False **48.** $-\dfrac{1}{2}$ **49.** 7 **50.** -6 **51.** $=$ **52.** $<$ **53.** $>$ **54.** $=$ **55.** $>$

56. $<$ **57.** Rational numbers are numbers that can be written as the quotient of two integers, provided the denominator does not equal 0. Rational numbers can also be written as decimals that either terminate, or do not terminate but repeat a block of decimals. Irrational numbers can be written as decimals that neither terminate nor repeat. **58.** natural numbers **59.** 7 **60.** -4 **61.** -34 **62.** -95 **63.** -4 **64.** 47 **65.** -53 **66.** -115 **67.** -22 **68.** -7 **69.** 21 **70.** 67

71. 26 **72.** -36 **73.** 12 **74.** -40 **75.** -1118 **76.** -8037 **77.** -715 **78.** $-11{,}130$ **79.** 5 **80.** -12 **81.** 5 **82.** -25 **83.** -8 **84.** $-\dfrac{16}{5}$ **85.** $-\dfrac{10}{3}$ **86.** $-\dfrac{30}{7}$ **87.** -13

88. 45 **89.** $-43 + 101 = 58$ **90.** $45 + (-28) = 17$ **91.** $-10 - (-116) = 106$ **92.** $74 - 56 = 18$ **93.** $13 + (-8) = 5$ **94.** $-60 - (-10) = -50$

95. $-21 \cdot (-3) = 63$ **96.** $54 \cdot (-18) = -972$ **97.** $-34 \div (-2)$ or $\dfrac{-34}{-2} = 17$ **98.** $-49 \div 14$ or $\dfrac{-49}{14} = -\dfrac{7}{2}$ **99.** 26 yards **100.** $-3°F$ **101.** $24°F$

102. 87 points **103.** $\dfrac{5}{4}$ **104.** $-\dfrac{5}{28}$ **105.** $-\dfrac{1}{20}$ **106.** $-\dfrac{3}{2}$ **107.** $\dfrac{4}{17}$ **108.** $-\dfrac{2}{3}$ **109.** $-\dfrac{3}{10}$ **110.** -32 **111.** $\dfrac{1}{3}$ **112.** $-\dfrac{2}{5}$ **113.** $\dfrac{3}{7}$ **114.** 3 **115.** $\dfrac{7}{20}$ **116.** $\dfrac{31}{36}$ **117.** $-\dfrac{59}{245}$

118. $\dfrac{13}{12}$ **119.** $-\dfrac{19}{12}$ **120.** $-\dfrac{11}{4}$ **121.** 0 **122.** $-\dfrac{23}{24}$ **123.** 48.5 **124.** -24.66 **125.** 82.98 **126.** -53.74 **127.** 0.0804 **128.** -260.154 **129.** 18.4 **130.** -25.79 **131.** 2.3

132. -22.9 **133.** -5.418 **134.** -0.732 **135.** $-\$98.93$; yes **136.** 24 friends **137.** $\dfrac{23}{2}$ or $11\dfrac{1}{2}$ inches **138.** $\$186.81$ **139.** Associative Property of Multiplication

140. Multiplicative Inverse Property **141.** Multiplicative Inverse Property **142.** Commutative Property of Multiplication **143.** Commutative Property of Multiplication **144.** Additive Inverse Property **145.** Identity Property of Addition **146.** Identity Property of Addition **147.** Commutative Property of Addition **148.** Multiplicative Identity Property **149.** Multiplication Property of Zero **150.** Associative Property of Addition **151.** 29 **152.** 99 **153.** 18

154. 121 **155.** 3.4 **156.** 5.3 **157.** -33 **158.** 6 **159.** undefined **160.** 0 **161.** -334 **162.** 2 **163.** 0 **164.** 1 **165.** 0 **166.** 130 **167.** $-\dfrac{5}{3}$ **168.** $-\dfrac{150}{13}$ **169.** 3^4

170. $\left(\dfrac{2}{3}\right)^3$ **171.** $(-4)^2$ **172.** $(-3)^3$ **173.** 125 **174.** -125 **175.** 81 **176.** -125 **177.** -81 **178.** $\dfrac{1}{64}$ **179.** -4 **180.** -76 **181.** 206 **182.** 32 **183.** 2 **184.** $\dfrac{1}{2}$

185. $\dfrac{6}{5}$ **186.** $\dfrac{7}{5}$ **187.** 21 **188.** -18 **189.** -729 **190.** -3 **191.** $3x^2, -x, 6$; 3, -1, 6 **192.** $2x^2y^3, -\dfrac{y}{5}, 2, -\dfrac{1}{5}$ **193.** like **194.** unlike **195.** unlike **196.** like **197.** $-3x$

198. $-4x - 15$ **199.** $-4.1x^4 + 0.3x^3$ **200.** $-3x^4 + 6x^2 + 12$ **201.** $18 - x$ **202.** $4x - 18$ **203.** $9x - 4$ **204.** -1 **205.** $\$98.70$

Chapter 1 Test   **1.** 42 **2.** $\dfrac{7}{22}$ **3.** 1.44 **4.** $\dfrac{17}{40}$ **5.** 0.006 **6.** 18.3% **7.** $\dfrac{1}{3}$ **8.** $\dfrac{9}{4}$ **9.** $-\dfrac{320}{3}$ **10.** -102 **11.** -12.16 **12.** -20 **13.** undefined **14.** -14 **15.** 55

16. (a) 6 **(b)** 0, 6 **(c)** $-2, 0, 6$ **(d)** $-2, -\dfrac{1}{2}, 0, 2.5, 6$ **(e)** none **(f)** All those listed. **17.** $<$ **18.** $=$ **19.** -7 **20.** 15 **21.** -102 **22.** -343 **23.** $-16x - 28$

24. $-4x^2 + 5x + 4$ **25.** $\$531.85$ **26.** $4x + 10$

Chapter 2 Equations and Inequalities in One Variable

Section 2.1 Linear Equations: The Addition and Multiplication Properties of Equality   **1.** solution **2.** yes **3.** no **4.** yes **5.** no **6.** True

7. $\{32\}$ **8.** $\{14\}$ **9.** $\{12\}$ **10.** $\{-15\}$ **11.** $\left\{\dfrac{7}{3}\right\}$ **12.** $\left\{-\dfrac{13}{12}\right\}$ **13.** $\$12{,}455$ **14.** Multiplication **15.** $\{2\}$ **16.** $\{-2\}$ **17.** $\left\{\dfrac{5}{2}\right\}$ **18.** $\left\{-\dfrac{7}{3}\right\}$ **19.** $\{9\}$

20. $\{-9\}$ **21.** $\{-30\}$ **22.** $\{6\}$ **23.** $\left\{\dfrac{8}{3}\right\}$ **24.** $\left\{-\dfrac{5}{14}\right\}$ **25.** Yes **27.** No **29.** Yes **31.** Yes **33.** $\{20\}$ **35.** $\{-12\}$ **37.** $\{19\}$ **39.** $\{-13\}$ **41.** $\{2\}$ **43.** $\left\{\dfrac{1}{4}\right\}$

45. $\left\{\dfrac{19}{24}\right\}$ **47.** $\{-6.1\}$ **49.** $\{5\}$ **51.** $\{-4\}$ **53.** $\left\{\dfrac{7}{2}\right\}$ **55.** $\left\{-\dfrac{5}{2}\right\}$ **57.** $\{21\}$ **59.** $\{121\}$ **61.** $\{40\}$ **63.** $\left\{\dfrac{3}{5}\right\}$ **65.** $\left\{-\dfrac{1}{3}\right\}$ **67.** $\{9\}$ **69.** $\left\{-\dfrac{4}{9}\right\}$ **71.** $\left\{-\dfrac{5}{3}\right\}$

73. $\{2\}$ **75.** $\{-3\}$ **77.** $\left\{\dfrac{2}{3}\right\}$ **79.** $\{-6\}$ **81.** $\{19\}$ **83.** $\{283\}$ **85.** $\{50\}$ **87.** $\{41.1\}$ **89.** $\left\{\dfrac{20}{3}\right\}$ **91.** $\{-4\}$ **93.** $\{-12\}$ **95.** $\left\{\dfrac{1}{2}\right\}$ **97.** $\left\{\dfrac{3}{16}\right\}$ **99.** $\left\{-\dfrac{5}{4}\right\}$

101. $\$24{,}963.27$ **103.** $\$68$ **105.** 12 boxes **107.** $r = 0.18$ or 18% **109.** $x = 48 - \lambda$ **111.** $x = \dfrac{14}{\theta}$ **113.** $\lambda = \dfrac{50}{9}$ **115.** $\theta = -\dfrac{6}{7}$ **117.** To find the solution of an equation means to find all values of the variable that satisfy the equation. **119.** An algebraic expression differs from an equation in that the expression does not contain an equal sign, and an equation does. An equation using the expression $x - 10$ is $x - 10 = 22$. Solving for x, $x = 32$. **121.** The Addition Property states that we may add (or subtract) any real number to (or from) both sides of an equation and maintain the equality. It is used to isolate the variable. To solve $x - 5 = 12$, use the Addition Property of Equality to add 5 to each side.

Section 2.2 Linear Equations: Using the Properties Together   **1.** True **2.** $\{3\}$ **3.** $\{18\}$ **4.** $\{2\}$ **5.** $\left\{\dfrac{3}{2}\right\}$ **6.** $\{2\}$ **7.** $\{-18\}$ **8.** $\left\{\dfrac{5}{2}\right\}$ **9.** $\{-8\}$

10. $\{2\}$ **11.** $\left\{\dfrac{4}{3}\right\}$ **12.** $\{13\}$ **13.** $\{6\}$ **14.** $\{2\}$ **15.** $\{6\}$ **16.** $\left\{-\dfrac{7}{2}\right\}$ **17.** False **18.** $\{1\}$ **19.** $\left\{-\dfrac{5}{3}\right\}$ **20.** 52 hours **21.** $\{1\}$ **23.** $\{-2\}$ **25.** $\{-3\}$ **27.** $\left\{\dfrac{3}{2}\right\}$

29. $\{-3\}$ **31.** $\{12\}$ **33.** $\{4\}$ **35.** $\{-2\}$ **37.** $\{-5\}$ **39.** $\{-8\}$ **41.** $\{-5\}$ **43.** $\{-21\}$ **45.** $\{-6\}$ **47.** $\{-8\}$ **49.** $\{3\}$ **51.** $\left\{-\dfrac{7}{2}\right\}$ **53.** $\{7\}$ **55.** $\{-18\}$

57. $\left\{\dfrac{5}{3}\right\}$ **59.** $\left\{-\dfrac{27}{16}\right\}$ **61.** $\{2\}$ **63.** $\left\{-\dfrac{3}{4}\right\}$ **65.** $\left\{\dfrac{1}{3}\right\}$ **67.** $\{0\}$ **69.** $\left\{\dfrac{15}{2}\right\}$ **71.** $\left\{\dfrac{1}{2}\right\}$ **73.** $\{-7\}$ **75.** $\left\{\dfrac{7}{2}\right\}$ **77.** McDonald's: 16 grams; Burger King: 22 grams

79. width: $\dfrac{13}{3} = 4\dfrac{1}{3}$ feet; length: $\dfrac{32}{3} = 10\dfrac{2}{3}$ feet **81.** $\$16$ **83.** 8 feet by 13 feet **85.** $\left\{\dfrac{51}{7}\right\}$ **87.** $\{2.47\}$ **89.** $\{-5.6\}$ **91.** $\dfrac{20}{3}$ **93.** $-\dfrac{5}{4}$ **95.** The first is an expression because it does not have an equal sign. The second is an equation because it does have an equal sign. In general, an expression contains the sum, difference, product, and/or quotient of terms. An equation may be thought of as the equality of two algebraic expressions. **97.** Adding $2x$ to both sides will lead to the correct solution, but it may be easier to combine like terms on the left side of linear equation instead of adding $2x$ to each side. A series of steps leading to the solution would be (1) combining like terms on the left side of the equation; (2) isolating x on the right side by subtracting $5x$ from both sides; (3) isolating the constants by adding 5 to each side; (4) dividing both sides by 4 to obtain $x = 2$.

Section 2.3 Solving Linear Equations Involving Fractions and Decimals; Classifying Equations   1. least common denominator 2. {10}

3. $\left\{-\dfrac{3}{7}\right\}$ 4. 70 5. {−11} 6. $\left\{-\dfrac{10}{3}\right\}$ 7. 100 8. {100} 9. {50} 10. 1.25 11. {50} 12. {160} 13. {4} 14. {3000} 15. conditional equation

16. contradiction; identity 17. True 18. $\varnothing$ or { } 19. all real numbers 20. all real numbers 21. $\varnothing$ or { } 22. all real numbers; identity

23. {4}; conditional 24. $\varnothing$ or { }; contradiction 25. $\varnothing$ or { }; contradiction 26. $500 27. {−6} 29. {6} 31. {−2} 33. $\left\{\dfrac{9}{2}\right\}$ 35. $\left\{\dfrac{2}{3}\right\}$ 37. $\left\{-\dfrac{2}{5}\right\}$

39. {−20} 41. {1} 43. {30} 45. {−4} 47. {50} 49. {4.8} 51. {150} 53. {−12} 55. {60} 57. {5} 59. {2} 61. {75} 63. $\varnothing$ or { }; contradiction

65. all real numbers; identity 67. $\left\{-\dfrac{1}{2}\right\}$; conditional equation 69. $\varnothing$ or { }; contradiction 71. $\varnothing$ or { }; contradiction 73. all real numbers; identity

75. $\left\{-\dfrac{3}{4}\right\}$ 77. all real numbers 79. $\left\{-\dfrac{1}{3}\right\}$ 81. {−20} 83. $\left\{\dfrac{7}{2}\right\}$ 85. $\varnothing$ or { } 87. {−2} 89. {3} 91. $\varnothing$ or { } 93. {−4} 95. {39} 97. {0} 99. $\left\{\dfrac{2}{3}\right\}$

101. {−1.70} 103. {−13.2} 105. $50 107. $18,000 109. $14.50 111. $80 113. 15 quarters 115. 6 units 117. $12,050

119. Answers will vary. A linear equation with one solution is $2x + 5 = 11$. A linear equation with no solution is $2x + 5 = 6 + 2x - 9$. A linear equation that is an identity is $2(x + 5) - 3 = 4x - (2x - 7)$. To form an identity or a contradiction, the variable expressions must be eliminated, leaving either a true (identity) or a false (contradiction) statement. 121. The student didn't multiply each term of the equation by 6. Solving using that (incorrect) method gives the second step $4x - 5 = 3x$, and solving for x gives $x = 5$. The correct method to solve the equation is to multiply ALL terms by 6. The correct second step is $4x - 30 = 3x$, producing the correct solution $x = 30$.

Section 2.4 Evaluating Formulas and Solving Formulas for a Variable   1. formula 2. 59° F 3. size 40 4. $312.50 5. 5936 persons

6. principal; Interest 7. $25 interest; $2525 total 8. perimeter 9. area 10. volume 11. radius 12. True 13. 36 square inches

14. (a) 187 ft^2 (b) $46.75 15. 28.27 ft^2 16. The extra large pizza is the better buy. extra large $\approx$ $0.07/in.^{2}; small $\approx$ $0.14/in.^{2}

17. $t = \dfrac{d}{r}$ 18. $l = \dfrac{V}{wh}$ 19. $C = \dfrac{5}{9}(F - 32)$ 20. $h = \dfrac{S - 2\pi r^2}{2\pi r}$ 21. $y = 2x - 6$ 22. $y = \dfrac{15 - 4x}{6}$ or $y = -\dfrac{2}{3}x + \dfrac{5}{2}$ 23. 186 miles 25. $104.00 27. $650

29. 20° C 31. $20 33. (a) 50 units (b) 144 square units 35. (a) 36.2 meters (b) 70 square meters 37. (a) 36 units (b) 81 square units

39. (a) 10π cm $\approx$ 31.4 cm (b) 25π cm^2 $\approx$ 78.5 cm^2 41. $\dfrac{616}{9} = 68\dfrac{4}{9}$ in.^{2} 43. $r = \dfrac{d}{t}$ 45. $d = \dfrac{C}{\pi}$ 47. $t = \dfrac{I}{Pr}$ 49. $b = \dfrac{2A}{h}$ 51. $a = P - b - c$

53. $r = \dfrac{A - P}{Pt}$ 55. $b = \dfrac{2A}{h} - B$ 57. $y = -3x + 12$ 59. $y = 2x - 5$ 61. $y = \dfrac{-4x + 13}{3}$ or $y = -\dfrac{4}{3}x + \dfrac{13}{3}$ 63. $y = 3x - 12$

65. (a) $C = R - P$ (b) $450 67. (a) $r = \dfrac{I}{Pt}$ (b) 0.03 or 3% 69. (a) $x = Z\sigma + \mu$ (b) 130 71. (a) $m = \dfrac{y - 5}{x}$ (b) −2 73. (a) $r = \dfrac{A - P}{Pt}$ (b) 0.04 or 4%

75. (a) $h = \dfrac{V}{\pi r^2}$ (b) 5 mm 77. (a) $b = \dfrac{2A}{h}$ (b) 18 ft 79. 1834.05 calories 81. (a) $h = \dfrac{V}{\pi r^2}$ (b) 10 inches 83. medium (12″) 85. (a) 12 hours (b) $336

87. 50.5 square inches 89. 96π cm^3 $\approx$ 301.59 cm^3 91. (a) $I = 4T + 16{,}575$ (b) $75,003 93. (a) 62 tiles (b) $372 (c) Yes 95. (a) 4948 ft^2 (b) $1237

97. (a) 1080 in.^{2} (b) 7.5 ft^2 99. Multiply by $\dfrac{1\,\text{ft}^2}{144\,\text{in}^2}$ 101. Both answers are correct. When solving for y, the first student left the entire numerator over 2. The second student simplified the expression to $y = \dfrac{-x + 6}{2} = \dfrac{-x}{2} + \dfrac{6}{2} = -\dfrac{1}{2}x + 3$.

Putting the Concepts Together (Sections 2.1–2.4)   1. (a) Yes (b) No 2. (a) No (b) Yes 3. $\left\{-\dfrac{2}{3}\right\}$ 4. {−40} 5. {−6} 6. {3} 7. {−6}

8. $\left\{-\dfrac{7}{3}\right\}$ 9. $\left\{-\dfrac{57}{2}\right\}$ 10. {25} 11. {6} 12. $\left\{\dfrac{74}{5}\right\}$ or {14.8} 13. $\varnothing$ or { } 14. all real numbers 15. $5000 16. (a) $b = \dfrac{2A}{h} - B$ (b) 6 in.

17. (a) $h = \dfrac{V}{\pi r^2}$ (b) 13 in. 18. $y = -\dfrac{3}{2}x + 7$

Section 2.5 Problem Solving: Direct Translation   1. $5 + 17$ 2. $7 - 4$ 3. $\dfrac{25}{3}$ 4. $-2 \cdot 6$ 5. $2a - 2$ 6. $5(m - 6)$ 7. $z + 50$ 8. $x - 15$

9. $75 - d$ 10. $3l - 2$ 11. $2q + 3$ 12. $3b - 5$ 13. equations 14. $3y = 21$ 15. $x - 10 = \dfrac{x}{2}$ 16. $3(n + 2) = 15$ 17. $3n + 2 = 15$

18. mathematical modeling 19. $s + \dfrac{2}{3}s = 15$, where s represents the amount Sean pays; Sean pays $9 and Connor pays $6. 20. False

21. $n + (n + 2) + (n + 4) = 270$; 88, 90, 92 22. $n + (n + 2) + (n + 4) + (n + 6) = 72$; 15, 17, 19, 21 23. $x + (x + 24) + \dfrac{1}{2}(x + 24) = 76$,

where x = length of smallest piece of ribbon; 16 inches, 40 inches, 20 inches 24. $x + 2x = 18{,}000$, where x = amount invested in stocks; $6000 in

stocks; $12,000 in bonds 25. 150 miles 26. 1.5 hours 27. $5 + x$ or $x + 5$ 29. $x\left(\dfrac{2}{3}\right)$ or $\dfrac{2}{3}x$ 31. $\dfrac{1}{2}x$ 33. $x - (-25)$ 35. $\dfrac{x}{3}$ 37. $x + \dfrac{1}{2}$ 39. $6x + 9$

41. $2(13.7 + x)$ 43. $2x + 31$ 45. Knights: r; Clippers: $r + 5$ 47. Bill's amount: b; Jan's amount: $b + 0.55$ 49. Janet's share: j; Kathy's share: $200 - j$

51. number adults: a; number children: $1433 - a$ 53. $x + 15 = -34$ 55. $35 = 3x - 7$ 57. $\dfrac{x}{-4} + 5 = 36$ 59. $2(x + 6) = x + 3$

61. x = number in millions of users in 35–44 age category; $x + 2.6$ = number in millions of users in 25–34 age category; $x + (x + 2.6) = 20$

63. w = number of apps on Wendy's phone; $w - 15$ = number of apps on Jessica's phone; $w + (w - 15) = 31$

65. x = length of middle piece; $x - 4$ = length of shortest piece; $x + 2$ = length of longest piece; $x + (x - 4) + (x + 2) = 46$

67. 83 **69.** 54, 55, 56 **71.** Verrazano-Narrows Bridge; 4260 ft; Golden Gate Bridge: 4200 ft **73.** $11,215 **75.** CDs: $8500; bonds: $11,500
77. stocks: $20,000; bonds: $12,000 **79.** Smart Start: 43 g; Go Lean Crunch: 38 g **81.** $29,140 **83.** 150 miles **85.** 2000 pages **87.** Jensen: $36,221; Maureen: $35,972
89. 94 **91.** Answers may vary. **93.** Answers may vary. **95.** 20°; 40°; 120° **97.** The process of taking a verbal description of a problem and developing a mathematical equation that can be used to solve the problem is mathematical modeling. We often make assumptions to make the mathematics more manageable in the mathematical models. **99.** Answers may vary. Both students are correct; however, the value for n will not be the same for both students.

Section 2.6 Problem Solving: Problems Involving Percent
1. 100 **2.** True **3.** 54 **4.** 2.4 **5.** 3.2 **6.** 3.5 **7.** 40% **8.** 37.5% **9.** 110% **10.** 50

11. 80 **12.** 75 **13.** 40,660,000 **14.** $7300 **15.** $60,475 **16.** $2.40 **17.** $659 **18.** $151,020 **19.** 80 **21.** 14 **23.** 4.8 **25.** 120 **27.** 72

29. 20 **31.** 40% **33.** 7.5% **35.** 200% **37.** $54 **39.** $102,000 **41.** $25,000 **43.** $2240 **45.** $490 **47.** $400 **49.** winner: 530; loser: 318 **51.** $285,700

53. 282,606 therapists **55.** 27 million males **57.** 17.2%; 20.5% **59.** $24 **61.** 28.6% **63. (a)** $34,268.80 **(b)** 36% **65.** 2.1% **67.** The equation should be $x + 0.05x = 12.81$. Jack's new hourly wage is a percentage of his current hourly wage, so multiplying 5% by his original wage gives his hourly raise. His current hourly wage is $12.20.

Section 2.7 Problem Solving: Geometry and Uniform Motion
1. 90 **2.** 39° and 51° **3.** 70° and 110° **4.** 180 **5.** 35°, 40°, and 105°

6. True **7.** False **8.** width is 1.5 feet; length is 3 feet **9.** 5 feet **10.** False **11.** Mariko's speed is 13 mph and Luis's speed is 8 mph

12. It takes $\frac{1}{2}$ hour to catch up to Tanya. Each of you has traveled 20 miles. **13.** 45° and 135° **15.** 44°, 46° **17.** 42°, 44°, 94 **19.** 18°, 72°, 90°,

21. length = 26 feet; width = 18 feet **23.** 42 yards by 84 yards **25.** shorter base: 15 m; longer base: 30 m **27. (a)** $62t$ **(b)** $68t$ **(c)** $62t + 68t$
(d) $62t + 68t = 585$ **29.** $528(t + 10) = 704t$ **31.** 26.5 inches **33.** 40 ft **35.** 40 ft; 32 ft **37. (a)** length = 11 yards; width = 19 yards **(b)** 209 yd²
39. 13 hours **41.** fast car: 60 mph; slow car: 48 mph **43.** freeway: 4.5 hours; 2-lane: 1.5 hours **45.** 6 mph **47.** 49°, 49°, 82° **49.** $x = 3$ **51.** $x = 15$
53. $x = 12$ **55.** Complementary angles are those whose measures sum to 90° and supplementary angles have measures that sum to 180°.
57. Answers may vary. The equation $65t + 40t = 115$ is describing the sum of distances, whereas the equation $65t - 40t = 115$ represents the difference of distances.

Section 2.8 Solving Linear Inequalities in One Variable
1. True **2.** 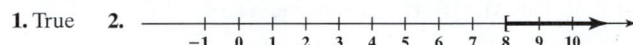 **3.**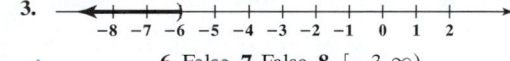

4. 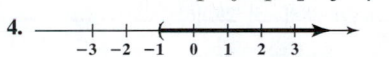 **5.** 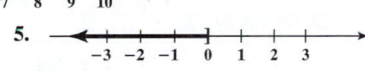 **6.** False **7.** False **8.** $[-3, \infty)$

9. $(-\infty, 12)$ **10.** $(-\infty, 2.5]$ **11.** $(125, \infty)$ **12.** solve **13.** $10 < 15$; Addition Property

14. $\{n \mid n > 3\}; (3, \infty)$ 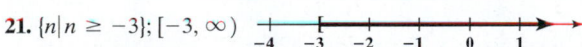 **15.** $\{x \mid x < 4\}; (-\infty, 4)$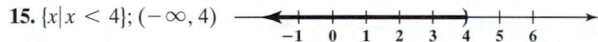

16. $\{n \mid n \leq -4\}; (-\infty, -4]$ 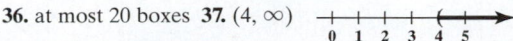 **17.** $\{x \mid x > -1\}; (-1, \infty)$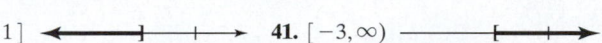

18. $1 < 4$; Multiplication Property of Inequality **19.** $-3 > -5$; Multiplication Property of Inequality **20.** $\{k \mid k < 12\}; (-\infty, 12)$

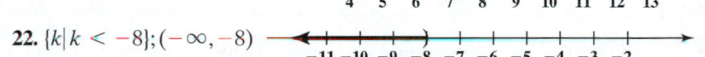

21. $\{n \mid n \geq -3\}; [-3, \infty)$ **22.** $\{k \mid k < -8\}; (-\infty, -8)$

23. $\left\{p \mid p \geq \frac{3}{5}\right\}; \left[\frac{3}{5}, \infty\right)$ **24.** $\{x \mid x > 7\}; (7, \infty)$

25. $\{n \mid n > -3\}; (-3, \infty)$ **26.** $\{x \mid x > -4\}; (-4, \infty)$

27. $\{x \mid x \leq -6\}; (-\infty, -6]$ **28.** $\{x \mid x > 8\}; (8, \infty)$

29. $\left\{x \mid x \leq \frac{19}{8}\right\}; \left(-\infty, \frac{19}{8}\right]$ **30.** True **31.** True

32. $\{x \mid x \geq 3\}; [3, \infty)$ **33.** $\varnothing$ or $\{\ \}$

34. $\{x \mid x > 4\}; (4, \infty)$ **35.** $\{x \mid x \text{ is any real number}\}; (-\infty, \infty)$

36. at most 20 boxes **37.** $(4, \infty)$ **39.** $(-\infty, -1]$ **41.** $[-3, \infty)$

43. $(-\infty, 4)$ **45.** $(-\infty, 3)$ **47.** $\varnothing$ or $\{\ \}$ **49.** $(-\infty, \infty)$ **51.** $<$; Addition Property of Inequality
53. $>$; Multiplication Property of Inequality **55.** $\leq$; Addition Property of Inequality **57.** $\leq$; Multiplication Property of Inequality

59. $\{x \mid x < 4\}; (-\infty, 4)$  **61.** $\{x \mid x \geq 2\}; [2, \infty)$ **63.** $\{x \mid x \leq 5\}; (-\infty, 5]$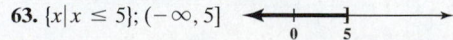

65. $\{x|x > -7\}; (-7, \infty)$ **67.** $\{x|x > 3\}; (3, \infty)$ **69.** $\{x|x \geq 2\}; [2, \infty)$

71. $\{x|x \geq -1\}; [-1, \infty)$ **73.** $\{x|x > -7\}; (-7, \infty)$ **75.** $\{x|x \leq 0\}; (-\infty, 0]$

77. $\{x|x < -20\}; (-\infty, -20)$ **79.** $\varnothing$ or $\{ \}$

81. $\{n|n$ is any real number$\}; (-\infty, \infty)$ **83.** $\{n|n > 5\}; (5, \infty)$

85. $\{w|w$ is any real number$\}; (-\infty, \infty)$ **87.** $\left\{y\middle|y < -\dfrac{3}{2}\right\}; \left(-\infty, -\dfrac{3}{2}\right)$

89. $x \geq 16,000$ **91.** $x \leq 20,000$ **93.** $x > 12,000$ **95.** $x > 0$ **97.** $x \leq 0$ **99.** $\{x|x > 4\}; (4, \infty)$

101. $\left\{x\middle|x < \dfrac{3}{4}\right\}; \left(-\infty, \dfrac{3}{4}\right)$ **103.** $\{x|x$ is any real number$\}; (-\infty, \infty)$

105. $\{a|a < -1\}; (-\infty, -1)$ **107.** $\{n|n$ is any real number$\}; (-\infty, \infty)$

109. $\left\{x\middle|x \geq \dfrac{4}{3}\right\}; \left[\dfrac{4}{3}, \infty\right)$ **111.** $\{x|x < 25\}; (-\infty, 25)$ **113.** $\varnothing$ or $\{ \}$

115. $\{x|x > 5.9375\}; (5.9375, \infty)$ **117.** at most 1250 miles **119.** at least 32 points **121.** more than 1.7 cubic yards

123. greater than \$50,361.11 **125.** at least 74 **127.** $\{x|-33 < x < -14\}$ **129.** $\{x|-4 \leq x \leq 6\}$ **131.** $\{x|-2 \leq x < 9\}$ **133.** $\{x|-8 \leq x < 4\}$

135. A left parenthesis is used to indicate that the solution is greater than a number. A left bracket is used to show that the solution is greater than or equal to a given number. **137.** In solving an inequality, when the variables are eliminated and a true statement results, the solution is all real numbers. In solving an inequality, when the variables are eliminated and a false statement results, the solution is the empty set.

Chapter 2 Review 1. No **2.** No **3.** No **4.** Yes **5.** $\{16\}$ **6.** $\{20\}$ **7.** $\{-16\}$ **8.** $\{-7\}$ **9.** $\{-95\}$ **10.** $\{50\}$ **11.** $\{24\}$ **12.** $\{80\}$ **13.** $\{-6\}$ **14.** $\{5\}$

15. $\left\{-\dfrac{1}{3}\right\}$ **16.** $\left\{\dfrac{3}{8}\right\}$ **17.** $\{4\}$ **18.** $\{5\}$ **19.** \$23,475 **20.** \$2.55 **21.** $\{-4\}$ **22.** $\{4\}$ **23.** $\{9\}$ **24.** $\{-21\}$ **25.** $\{-4\}$ **26.** $\{-6\}$ **27.** $\{4\}$ **28.** $\{-5\}$ **29.** $\{6\}$

30. $\{-6\}$ **31.** $\left\{\dfrac{4}{3}\right\}$ **32.** $\{5\}$ **33.** $\{2\}$ **34.** $\{7\}$ **35.** 14 **36.** width $= 19$ yards; length $= 29$ yards **37.** $\left\{-\dfrac{35}{12}\right\}$ **38.** $\left\{-\dfrac{62}{3}\right\}$ **39.** $\{-2\}$ **40.** $\left\{\dfrac{6}{5}\right\}$

41. $\left\{\dfrac{3}{2}\right\}$ **42.** $\{3\}$ **43.** $\{4\}$ **44.** $\{-1\}$ **45.** $\left\{-\dfrac{7}{2}\right\}$ **46.** $\{-3\}$ **47.** $\{58\}$ **48.** $\{-10.5\}$ **49.** $\varnothing$ or $\{ \}$; contradiction **50.** $\varnothing$ or $\{ \}$; contradiction

51. $\{0\}$; conditional equation **52.** $\{0\}$; conditional equation **53.** all real numbers; identity **54.** all real numbers; identity **55.** \$15.75

56. 3 dimes **57.** 48 in.^{2} **58.** 64 cm **59.** $\dfrac{3}{2}$ yards **60.** 15 mm **61.** $H = \dfrac{V}{LW}$ **62.** $P = \dfrac{I}{rt}$ **63.** $W = \dfrac{S - 2LH}{2L + 2H}$ **64.** $M = \dfrac{\rho - mv}{V}$

65. $y = \dfrac{-2x + 10}{3}$ or $y = -\dfrac{2}{3}x + \dfrac{10}{3}$ **66.** $y = 2x - 5$ **67. (a)** $P = \dfrac{A}{(1 + r)^t}$ **(b)** \$2238.65 **68. (a)** $h = \dfrac{A - 2\pi r^2}{2\pi r}$ **(b)** 5 cm **69.** \$7.50

70. $\dfrac{9}{4}\pi$ ft$^2 \approx 7.1$ ft^2 **71.** $x - 6$ **72.** $x - 8$ **73.** $-8x$ **74.** $\dfrac{x}{10}$ **75.** $2(6 + x)$ **76.** $4(5 - x)$ **77.** $6 + x = 2x + 5$ **78.** $6x - 10 = 2x + 1$

79. $x - 8 = \dfrac{1}{2}x$ **80.** $\dfrac{6}{x} = 10 + x$ **81.** $4(2x + 8) = 16$ **82.** $5(2x - 8) = -24$ **83.** Sarah's age: s; Jacob's age: $s + 7$

84. Consuelo's speed: c; Maria's speed: $2c$ **85.** Max's amount: m; Irene's amount: $m - 6$ **86.** Victor's amount: v; Larry's amount: $350 - v$

87. 153 pounds **88.** 12, 13, 14 **89.** Juan: \$11,000; Roberto: \$9000 **90.** 100 miles **91.** 5.2 **92.** 60 **93.** 13% **94.** 50 **95.** \$18.50 **96.** \$30 **97.** \$40

98. \$200 **99.** \$125,000 **100.** winner: 500 votes: loser: 400 votes **101.** 80°, 10° **102.** 100°, 80° **103.** 30°, 60°, 90° **104.** 50°, 45°, 85°

105. length $= 31$ in.; width $= 8$ in. **106.** length $= 28$ cm; width $= 7$ cm **107. (a)** length $= 20$ ft; width $= 40$ ft **(b)** 800 ft^2

108. 50 ft; 40 ft **109.** 5 hours **110.** 65 mph

111. **112.** **113.** **114.**

115. **116.** **117.** $(-\infty, -4)$ **118.** $[7, \infty)$ **119.** $[2, \infty)$ **120.** $(-\infty, 3)$

121. $\left\{x\middle|x < -\dfrac{13}{2}\right\}; \left(-\infty, -\dfrac{13}{2}\right)$ **122.** $\left\{x\middle|x \geq -\dfrac{7}{3}\right\}; \left[-\dfrac{7}{3}, \infty\right)$

123. $\left\{x\middle|x \geq -\dfrac{4}{5}\right\}; \left[-\dfrac{4}{5}, \infty\right)$ **124.** $\{x|x > -12\}; (-12, \infty)$

125. $\varnothing$ or $\{ \}$ **126.** $\{x|x$ is any real number$\}; (-\infty, \infty)$

127. $\{x \mid x > 3\}$; $(3, \infty)$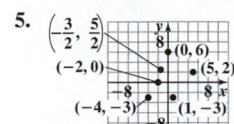

128. $\left\{x \mid x < -\dfrac{38}{5}\right\}$; $\left(-\infty, -\dfrac{38}{5}\right)$

129. at most 65 miles **130.** more than 150

Chapter 2 Test **1.** $\{-17\}$ **2.** $\left\{-\dfrac{4}{9}\right\}$ **3.** $\{4\}$ **4.** $\left\{\dfrac{10}{13}\right\}$ **5.** $\left\{\dfrac{5}{8}\right\}$ **6.** $\{5\}$ **7.** $\varnothing$ or $\{\,\}$ **8.** all real numbers **9. (a)** $l = \dfrac{V}{wh}$ **(b)** 9 in.

10. (a) $y = -\dfrac{2}{3}x + 4$ **(b)** $y = -\dfrac{4}{3}$ **11.** $6(x-8) = 2x - 5$ **12.** 60 **13.** 15, 16, 17 **14.** 10 in., 24 in., 26 in. **15.** 3.5 hours

16. shorter piece is 5 feet; longer is 16 feet **17.** $36 **18.** $\{x \mid x \le 6\}$; $(-\infty, 6]$ **19.** $\left\{x \mid x > \dfrac{5}{4}\right\}$; $\left(\dfrac{5}{4}, \infty\right)$

20. at most 400 miles

Chapter 3 Introduction to Graphing and Equations of Lines

Section 3.1 The Rectangular Coordinate System and Equations in Two Variables

1. x-axis; y-axis; origin **2.** x-coordinate; y-coordinate **3.** False **4.** False

5. 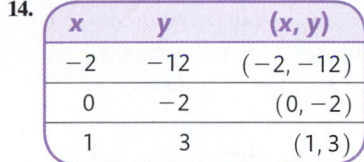 $\left(-\dfrac{3}{2}, \dfrac{5}{2}\right)$ **(a)** I **(b)** III **(c)** IV **(d)** x-axis **(e)** y-axis **(f)** II

6. 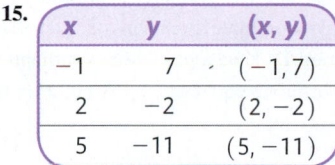 **(a)** II **(b)** I **(c)** III **(d)** x-axis **(e)** y-axis **(f)** IV

7. (a) $(2,3)$ **(b)** $(1,-3)$ **(c)** $(-3,0)$ **(d)** $(-2,-1)$ **(e)** $(0,2)$ **8.** True **9. (a)** Yes **(b)** No **(c)** No

10. (a) No **(b)** Yes **(c)** Yes **11.** $(3,4)$ **12.** $(-3,1)$ **13.** $\left(\dfrac{1}{2}, -\dfrac{2}{3}\right)$

14.

x	y	(x, y)
−2	−12	(−2, −12)
0	−2	(0, −2)
1	3	(1, 3)

15.

x	y	(x, y)
−1	7	(−1, 7)
2	−2	(2, −2)
5	−11	(5, −11)

16.

x	y	(x, y)
−5	2	(−5, 2)
−2	−4	(−2, −4)
2	−12	(2, −12)

17.

x	y	(x, y)
−6	−6	(−6, −6)
−1	−4	(−1, −4)
2	$-\dfrac{14}{5}$	$\left(2, -\dfrac{14}{5}\right)$

18.(a)

x (kWh)	C ($)	(x, C)
50	51.85	(50, 51.85)
100	93.69	(100, 93.69)
1500	135.54	(150, 135.54)

$(50, 51.85)$; $(100, 93.69)$; $(150, 135.54)$

(b)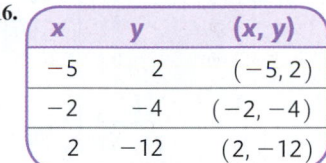

19.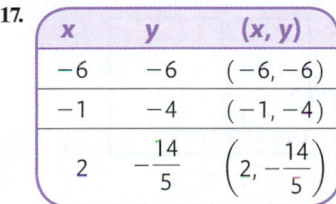

Quadrant I: B
Quadrant II: A, E
Quadrant III: C
Quadrant IV: D, F

21.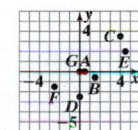

Quadrant I: C, E
Quadrant III: F
Quadrant IV: B
x-axis: A, G
y-axis: D, G

23.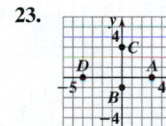

Positive x-axis: A
Negative x-axis: D
Positive y-axis: C
Negative y-axis: B

25. $A(4,0)$: x-axis; $B(-3,2)$: quadrant II; $C(1,-4)$: quadrant IV; $D(-2,-4)$: quadrant III; $E(3,5)$: quadrant I; $F(0,-3)$: y-axis

27. A No
B Yes
C Yes

29. A Yes
B No
C Yes

31. A Yes
B No
C Yes

33. $(4,1)$ **35.** $(5,-1)$ **37.** $(-3,3)$

39.

x	y	(x, y)
−3	3	(−3, 3)
0	0	(0, 0)
1	−1	(1, −1)

41.

x	y	(x, y)
−2	7	(−2, 7)
−1	4	(−1, 4)
4	−11	(4, −11)

43.

x	y	(x, y)
−1	8	(−1, 8)
2	2	(2, 2)
3	0	(3, 0)

45.

x	y	(x, y)
−4	6	(−4, 6)
1	6	(1, 6)
12	6	(12, 6)

47.

x	y	(x, y)
1	$\dfrac{7}{2}$	$\left(1, \dfrac{7}{2}\right)$
−4	1	(−4, 1)
−2	2	(−2, 2)

49.

x	y	(x, y)
4	7	(4, 7)
−4	3	(−4, 3)
−6	2	(−6, 2)

51.

x	y	(x, y)
0	−3	(0, −3)
−2	0	(−2, 0)
2	−6	(2, −6)

53. $A(2,-16)$
$B(-3,-1)$
$C\left(-\dfrac{1}{3}, -9\right)$

55. $A(2,-6)$
$B(0,0)$
$C\left(\dfrac{1}{6}, -\dfrac{1}{2}\right)$

57. $A(4,-8)$
$B(4,-19)$
$C(4,5)$

59. $A(3,4)$
$B(-6,-2)$
$C\left(\dfrac{1}{2}, \dfrac{7}{3}\right)$

61. $A\left(-4, -\dfrac{4}{3}\right)$
$B(-2, -1)$
$C\left(-\dfrac{2}{3}, -\dfrac{7}{9}\right)$

63. $A(20, 23)$
$B(-4, -17)$
$C(2.6, -6)$

65. (a) \$265,000 **(b)** \$235,000
(c) After 8 years
(d) After 3 years, the book value is \$255,000.

(e)

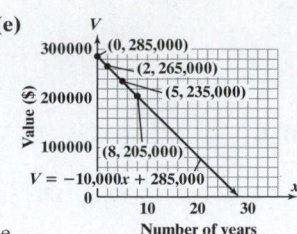

67. (a) 30.5% **(b)** 32.78% **(c)** 34.68% **(d)** 2060 **(e)** Answers may vary. The result is not reasonable since it is unlikely the trend will continue over this time frame.

69.

a	b	(a, b)
2	−8	(2, −8)
0	−4	(0, −4)
−5	6	(−5, 6)

71.

p	q	(p, q)
0	$\dfrac{10}{3}$	$\left(0, \dfrac{10}{3}\right)$
$\dfrac{5}{2}$	0	$\left(\dfrac{5}{2}, 0\right)$
−10	$\dfrac{50}{3}$	$\left(-10, \dfrac{50}{3}\right)$

73. $k = 4$ **75.** $k = 2$
77. $k = \dfrac{1}{2}$
79. Points may vary; line

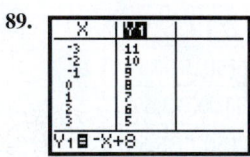

81.

x	y	(x, y)
−2	0	(−2, 0)
−1	−3	(−1, −3)
0	−4	(0, −4)
1	−3	(1, −3)
2	0	(2, 0)

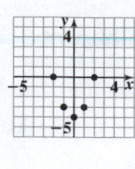

83.

x	y	(x, y)
−2	10	(−2, 10)
−1	3	(−1, 3)
0	2	(0, 2)
1	1	(1, 1)
2	−6	(2, −6)

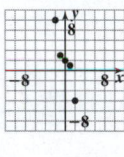

85. The first quadrant is the upper right-hand corner of the rectangular coordinate system. The quadrants are then II, III, and IV going in a counterclockwise direction. Points in quadrant I have both the *x*- and *y*-coordinates positive; points in quadrant II have a negative *x*-coordinate and a positive *y*-coordinate; points in quadrant III have both the *x*- and *y*-coordinates negative; points in quadrant IV have a positive *x*-coordinate and a negative *y*-coordinate. A point on the *x*-axis has *y*-coordinate equal to zero. A point on the *y*-axis has *x*-coordinate that is equal to zero.

87.

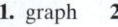

89. **91.** **93.**

Section 3.2 Graphing Equations in Two Variables

1. graph **2.** **3.** **4.** linear; standard form **5.** Linear **6.** Not linear **7.** Linear **8.** line **9.**

10. **11. (a)** (0, 3000), (10,000, 3800), (25,000, 5000) **(b)**

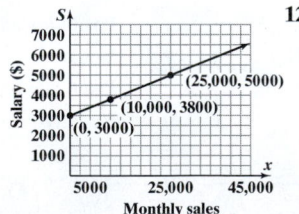

12. intercepts

13. Intercepts: (0, 3), (4, 0); *x*-intercept: (4, 0); *y*-intercept: (0, 3) **14.** Intercept: (0, −2); *y*-intercept: (0, −2); no *x*-intercept **15.** False

16. **17.** **18.** **19.** **20.** **21.** vertical; (a, 0) **23.**
22. horizontal; (0, b)

24. **25.** Linear **27.** Not linear **29.** Not linear **31.** Linear **33.** $y = 2x$ **35.** $y = 4x - 2$

37. $y = -2x + 5$
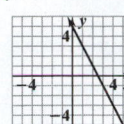

39. $x + y = 5$
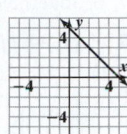

41. $-2x + y = 6$
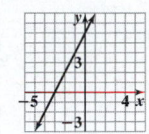

43. $4x - 2y = -8$
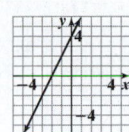

45. $x = -4y$
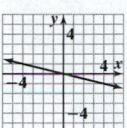

47. $y + 7 = 0$
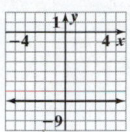

49. $y - 2 = 3(x + 1)$
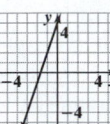

51. $(0, -5), (5, 0)$

53. $(0, 4), (2, 0)$

55. $(0, -3)$

57. $(-5, 0)$

59. $(0, -4), (-6, 0)$

61. $(0, 0)$

63. $(0, -5), (5, 0)$

65. $(0, 8), (6, 0)$

67. $(4, 0)$

69. $(0, -2)$

71. $3x + 6y = 18$
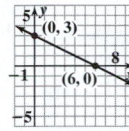

73. $-x + 5y = 15$
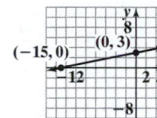

75. $\frac{1}{2}x - y = 3$
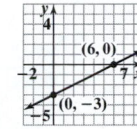

77. $9x - 2y = 0$
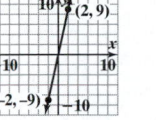

79. $y = -\frac{1}{2}x + 3$
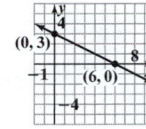

81. $\frac{1}{3}y + 2 = 2x$
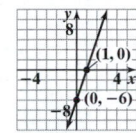

83. $\frac{x}{2} + \frac{y}{3} = 1$
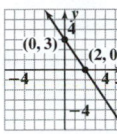

85. $4y - 2x + 1 = 0$
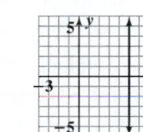

87. $x = 5$
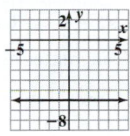

89. $y = -6$

91. $y - 12 = 0$
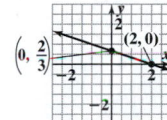

93. $3x - 5 = 0$

95. $y = 2x - 5$
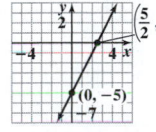

97. $y = -5$
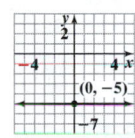

99. $2x + 5y = -20$
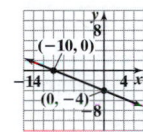

101. $2x = -6y + 4$
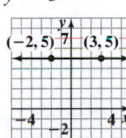

103. $x - 3 = 0$
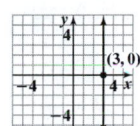

105. $3y - 12 = 0$
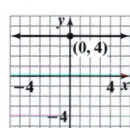

107. $y = 2$

109. $x = 7$

111. $y = 5$
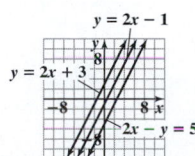

113. $x = -2$
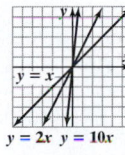

115. $y = 4$

117. $x = -9$

119. $x = 2y$

121. $y = x + 2$

123. (a) $(0, 500), (4, 900), (10, 1500)$

(b)
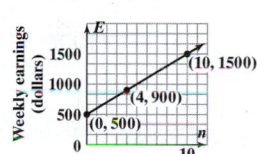

(c) If she sells 0 cars, her weekly earnings are $500.

125. The "steepness" of the lines is the same.
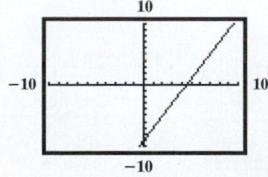

127. The lines get more steep as the coefficient of x gets larger.
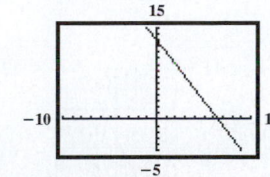

129. $(0, -6), (-2, 0), (3, 0)$

131. $(0, 14), (-3, 0), (2, 0), (5, 0)$

133. The graph of an equation is the set of all ordered pairs (x, y) that make the equation a true statement.

135. Two points are needed to graph a line. A third point is used to verify your results.

137. $y = 2x - 9$
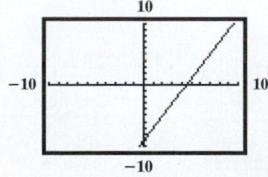

139. $y + 2x = 13$ or $y = -2x + 13$
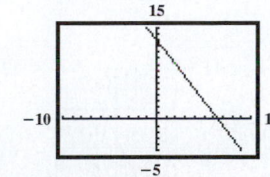

141. $y = -6x^2 + 1$
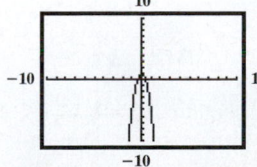

Section 3.3 Slope

1. $\dfrac{3}{5}$

2. False

3. True

4. positive

5.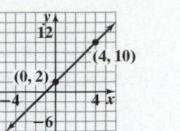

$m = 2$; y increases by 2 when x increases by 1

6.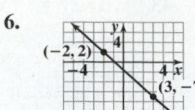

$m = -\dfrac{9}{5}$; y decreases by 9 when x increases by 5, or y increases by 9 when x decreases by 5.

7. 0; undefined

8. Slope undefined; when y increases by 1, there is no change in x.

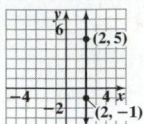

9. $m = 0$; there is no change in y when x increases by 1 unit.

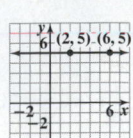

10. (a) 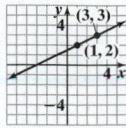   **(b)** 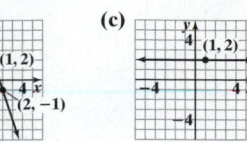   **(c)**

(c) $m = \dfrac{1}{2}$; for every 2-unit increase in x, there is a 1-unit increase in y.

11. 8%

12. $m = 0.12$; between 10,000 and 14,000 miles driven, the average annual cost of operating a Chevy Cruze is $0.12 per mile.

13. $m = -\dfrac{3}{2}$ **15.** $m = \dfrac{1}{2}$ **17.** $m = -\dfrac{2}{3}$

19. (a), (b)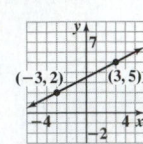

21. (a), (b)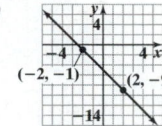

(c) $m = -2$; the value of y decreases by 2 when x increases by 1.

23. $m = -2$; y decreases by 2 when x increases by 1.   **25.** $m = -1$; y decreases by 1 when x increases by 1.   **27.** $m = -\dfrac{5}{3}$; y decreases by 5 when x increases by 3.   **29.** $m = \dfrac{3}{2}$; y increases by 3 when x increases by 2.   **31.** $m = \dfrac{2}{3}$; y increases by 2 when x increases by 3.   **33.** $m = \dfrac{1}{3}$; y increases by 1 when x increases by 3.   **35.** $m = 2$; y increases by 2 when x increases by 1.   **37.** m is undefined.   **39.** $m = 0$ **41.** $m = 0$; the line is horizontal, so there is no change in y when x increases by 1.   **43.** slope is undefined; the line is vertical, so there is no change in x when y increases by 1.

45. 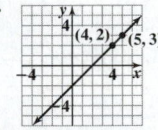   **47.** 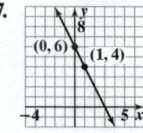   **49.** 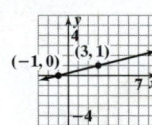   **51.** 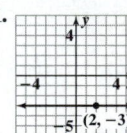   **53.** 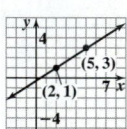   **55.**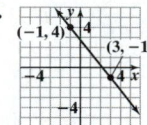

57. 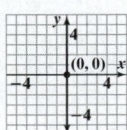   **59.** 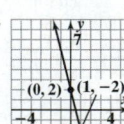   **61.** 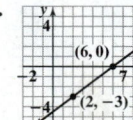   **63.** 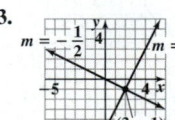   **65.** 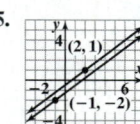   **67.**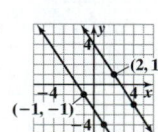

69. $\dfrac{1}{3}$ **71.** 12 in. or 1 ft   **73.** 16%   **75.** $m = 2.239$ million; the population was increasing at an average rate of about 2.239 million people per year.

77. Points may vary.
$(-2, 1)$, $(0, -5)$; $m = -3$

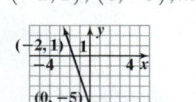

79. Points may vary.
$(-2, -2)$, $(0, 4)$; $m = 3$

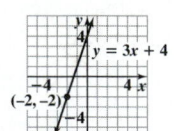

81. $m = -2$ **83.** $m = \dfrac{q}{p}$ **85.** $m = \dfrac{6}{a-6}$ **87.** $MR = 2$; For every hot dog sold, revenue increases by $2.   **89.** The line is a vertical line. Answers will vary, but the points should be of the form (a, y_1) and (a, y_2), where a is a specific value and y_1, y_2 are any two different values. Because the line is a vertical line, the slope is undefined.

91. (a) rise = 3; run = 4; slope = $\dfrac{3}{4}$ **(b)** rise = 4; run = 6; slope = $\dfrac{2}{3}$ **(c)** rise = -4; run = 3; slope = $\dfrac{-4}{3} = -\dfrac{4}{3}$

(d) rise = 2; run = -4; slope = $\dfrac{2}{-4} = -\dfrac{1}{2}$ **(e)** rise = 0; run = 5; slope = 0   **(f)** rise = 7; run = 0; slope is undefined   **(g)** 1   **(h)** -1

Section 3.4 Slope-Intercept Form of a Line

1. $y = mx + b$ **2.** slope: 4, y-intercept: $(0, -3)$ **3.** slope: -3, y-intercept: $(0, 7)$ **4.** slope: $-\dfrac{2}{5}$, y-intercept: $(0, 3)$ **5.** slope: 0; y-intercept: $(0, 8)$

6. slope: undefined; y-intercept: none

7. 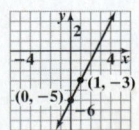   **8.** 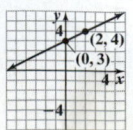   **9.** 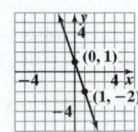   **10.** 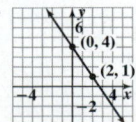   **11.** 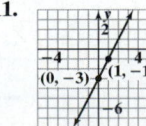   **12.** 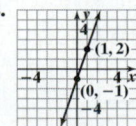   **13.**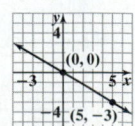

14. point-plotting, using intercepts, using slope and a point

15. $y = 3x - 2$ **16.** $y = -\dfrac{1}{4}x + 3$ **17.** $y = -1$ **18.** $y = x$

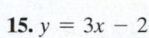

19. (a) 2075 grams   **(b)** 2933 grams   **(c)** slope = 143; Birth weight increases by 143 grams for each additional week of pregnancy.   **(d)** A gestation period of 0 weeks does not make sense.   **(e)**

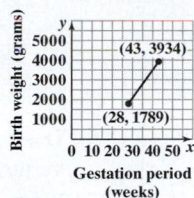

20. (a) $y = 0.38x + 50$ **(b)** \$78.50   **(c)** 90 miles   **(d)**

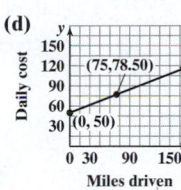

21. $m = 5$; y-intercept $= (0, 2)$ **23.** $m = 1$; y-intercept $= (0, -9)$

25. $m = -10$; y-intercept $= (0, 7)$ **27.** $m = -1$; y-intercept $= (0, -9)$

29. $m = -2$; y-intercept $= (0, 4)$ **31.** $m = -\dfrac{2}{3}$; y-intercept $= (0, 8)$

33. $m = \dfrac{5}{3}$; y-intercept $= (0, -3)$ **35.** $m = \dfrac{1}{2}$; y-intercept $= \left(0, -\dfrac{5}{2}\right)$ **37.** $m = 0$; y-intercept $= (0, -5)$ **39.** m is undefined; no y-intercept

41. 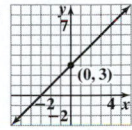   **43.** 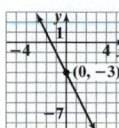   **45.** 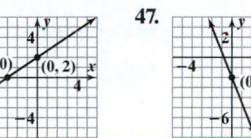   **47.** 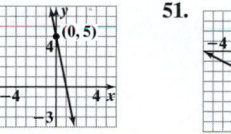   **49.**   **51.** 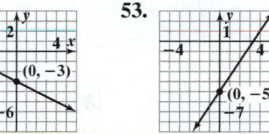   **53.**

55. 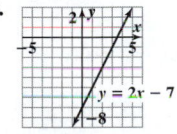   **57.** $y = -x + 8$ **59.** $y = \dfrac{6}{7}x - 6$ **61.** $y = -\dfrac{1}{3}x + \dfrac{2}{3}$ **63.** $x = -5$ **65.** $y = 3$ **67.** $y = 5x$

69. 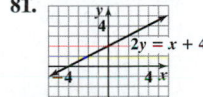   **71.** 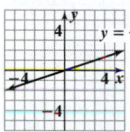   **73.** 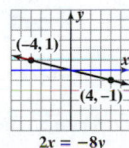   **75.** 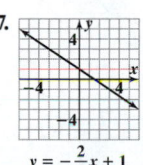   **77.** 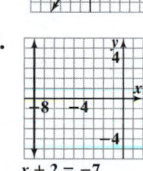   **79.**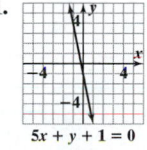

81.   **83.**   **85.**   **87.**   **89.**   **91.**

93. (a) 62 minutes   **(b)** 2012   **(c)** Each year, Americans spent 21.7 additional minutes on their mobile devices.   **(d)** No, because the number of daily minutes isn't infinite.

(e)

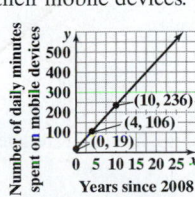

95. (a) $y = 0.08x + 400$ **(b)** \$496   **(c)**

97. $B = -4$ **99.** $A = -4$ **101.** $B = -3$

103. (a) $y = 40x + 4000$ **(b)** \$24,000   **(c)** 375 calculators   **(d)**

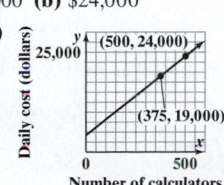

105. The line has a positive slope with a negative y-intercept. Possible equations are (a) and (e).

107. (a) Answers will vary.   **(b)** Answers will vary.   **(c)** Answers will vary.   **(d)** $\dfrac{1}{4}, \dfrac{1}{2}, 2, 4$; As the value of the slope increases, the line gets steeper as it rises from left to right.   **(e)** $-\dfrac{1}{4}, -\dfrac{1}{2}, -2, -4$; As the value of the slope decreases, the line gets steeper as it falls from left to right.   **(f)** The line appears to be horizontal.   **(g)** The line appears to be vertical.

Section 3.5 Point-Slope Form of a Line

1. $y - y_1 = m(x - x_1)$ **2.** True

3. $y = 3x - 5$
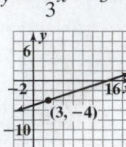

4. $y = \frac{1}{3}x - 5$
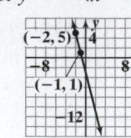

5. $y = -4x - 3$
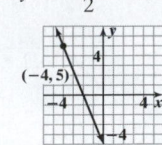

6. $y = -\frac{5}{2}x - 5$
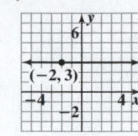

7. $y = 3$
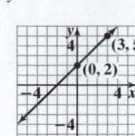

8. $y = x + 2$
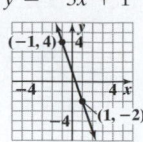

9. $y = -3x + 1$

10. $x = 3$
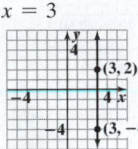

11. Horizontal line: $y = b$;
Vertical line: $x = a$;
Point-slope: $y - y_1 = m(x - x_1)$;
Slope-intercept: $y = mx + b$;
Standard form: $Ax + By = C$

12. (a) $y = -100x + 675$
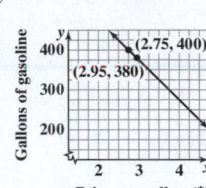

(b) 275 gallons

(c) The number of gallons of
gasoline sold will decrease
by 100 if the price per gallon
increases by \$1.

13. $y = 3x - 1$
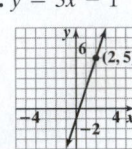

15. $y = -2x$
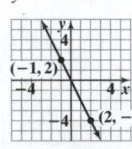

17. $y = \frac{1}{4}x - 3$
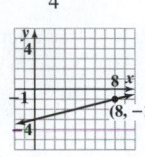

19. $y = -6x + 13$
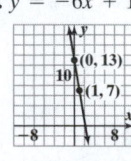

21. $y = -7$
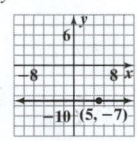

23. $x = -4$
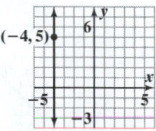

25. $y = \frac{2}{3}x + 2$
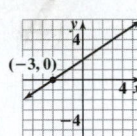

27. $y = -\frac{3}{4}x$
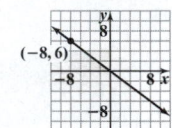

29. $x = -3$ **31.** $y = -5$ **33.** $y = -4.3$ **35.** $x = \frac{1}{2}$ **37.** $y = 2x + 4$ **39.** $y = -4x + 6$

41. $y = -\frac{3}{2}x - \frac{5}{2}$ **43.** $y = 2x - 5$ **45.** $y = -3$ **47.** $x = 2$ **49.** $y = 0.25x + 0.575$

51. $y = x - \frac{11}{4}$

53. $y = 5x - 22$
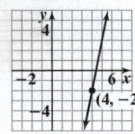

55. $y = 5$
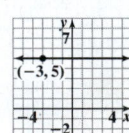

57. $y = x + 2$
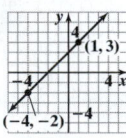

59. $y = \frac{1}{2}x + 4$
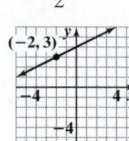

61. $x = 5$
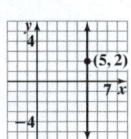

63. $y = -7x + 2$
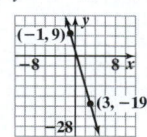

65. $y = -\frac{2}{3}x + 7$
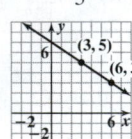

67. $y = -\frac{3}{2}x$

69. $y = \frac{2}{5}x - 2$

71. (a) When 60 packages are shipped, the expenses are \$1635.

(b)
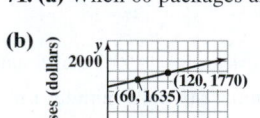

(c) $y = 2.25x + 1500$
(d) \$1950
(e) Expenses increase by \$2.25 for each
additional package.

73. (a) $(600, 15)$; $(750, 6)$ **(b)**
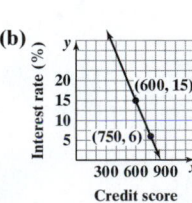

(c) $y = -0.06x + 51$ **(d)** 9% **(e)** For each 1-point increase in credit score, interest rate goes
down 0.06%. **75.** $y = 3x + 14$ **77.** $y = -2x - 2$ **79.** $y = 6x - \frac{7}{2}$

81. $y = -x - 7$ **83.** $y = \frac{1}{5}x - 2$ **85.** $y = -\frac{3}{2}x + 2$

87. Yes. It makes no difference whether $(x_1, y_1) = (3, 1)$ or $(x_1, y_1) = (4, 7)$ is chosen, the equation of the line will be the same.

Section 3.6 Parallel and Perpendicular Lines

1. slopes; y-intercepts; x-intercepts

2. Not parallel 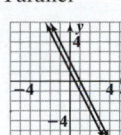   **3.** Parallel 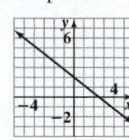   **4.** Not parallel 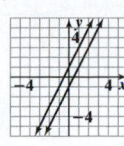   **5.** $y = 2x - 1$ 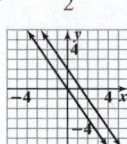   **6.** $y = -\dfrac{3}{2}x$ 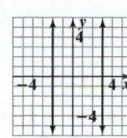   **7.** $x = 3$ 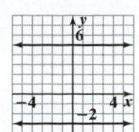   **8.** $y = 5$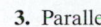

9. perpendicular **10.** False **11.** $\dfrac{1}{4}$ **12.** $-\dfrac{4}{5}$ **13.** 8

14. Perpendicular 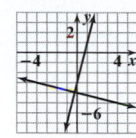   **15.** Not perpendicular 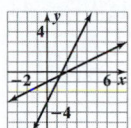   **16.** Perpendicular 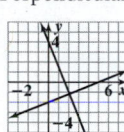   **17.** $y = -\dfrac{1}{2}x$ 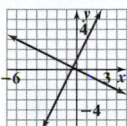   **18.** $y = \dfrac{3}{2}x + 2$ 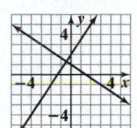   **19.** $y = -5$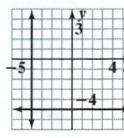

20. $x = 3$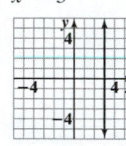

	Slope of the Given Line	Slope of a Line Parallel to the Given Line	Slope of a Line Perpendicular to the Given Line
21.	$m = -3$	-3	$\dfrac{1}{3}$
23.	$m = \dfrac{1}{2}$	$\dfrac{1}{2}$	-2
25.	$m = -\dfrac{4}{9}$	$-\dfrac{4}{9}$	$\dfrac{9}{4}$
27.	$m = 0$	0	undefined

29. perpendicular **31.** parallel **33.** perpendicular **35.** perpendicular **37.** neither **39.** parallel **41.** parallel

43. $y = 3x - 14$ **45.** $y = -4x - 4$ **47.** $y = -7$ **49.** $x = -1$ **51.** $y = \dfrac{3}{2}x - 13$ **53.** $y = -\dfrac{1}{2}x - \dfrac{21}{2}$ **55.** $y = -2x + 11$

57. $y = \dfrac{1}{4}x$ **59.** $x = -2$ **61.** $y = 5$ **63.** $y = \dfrac{5}{2}x$ **65.** $y = -\dfrac{3}{5}x - 9$

67. $y = 7x - 26$ 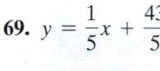   **69.** $y = \dfrac{1}{5}x + \dfrac{43}{5}$ 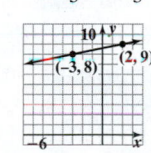   **71.** $y = -7x + 41$ 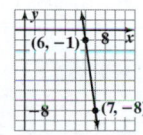   **73.** $y = -2x - 10$ 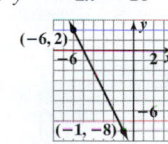   **75.** $y = 3x - 2$ 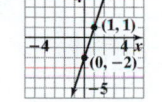   **77.** $x = 5$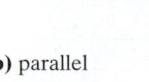

79. $y = -\dfrac{4}{3}x + 2$ 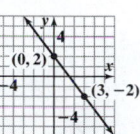   **81.** $y = -2x - 5$ 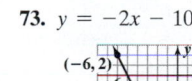   **83. (a)** $m_1 = 3$, $m_2 = -3$ **(b)** neither   **85. (a)** $m_1 = 2$, $m_2 = 2$ **(b)** parallel

87. (a) $m_1 = \dfrac{1}{2}$, $m_2 = \dfrac{1}{2}$ **(b)** parallel   **89. (a)** $m_1 = \dfrac{5}{3}$, $m_2 = -\dfrac{3}{5}$ **(b)** perpendicular

91.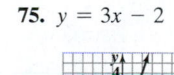
parallelogram

93.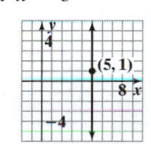
rectangle

95. right triangle

97. right triangle

99. $B = 2$ **101.** $A = 4$ **103.**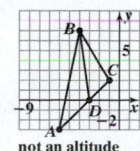
not an altitude

105. Pamels is incorrect. If the products of the slopes is -1 (or the slopes are negative reciprocals), the lines are perpendicular. Because the product of the slopes is $m = \dfrac{3}{2} \cdot -\dfrac{3}{2} = -\dfrac{9}{4} \neq -1$, the lines are not perpendicular.

107. (d) The slopes of the lines are equal; the y-intercepts are different.

Putting the Concepts Together (Sections 3.1–3.6)

1. Yes

2.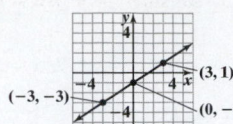
(3, 1)
(−3, −3)
(0, −1)

3.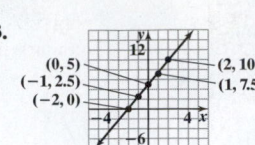
(0, 5)
(−1, 2.5)
(−2, 0)
(2, 10)
(1, 7.5)

4. (a) x-intercept is $\left(-\frac{3}{4}, 0\right)$ **(b)** y-intercept is $(0, 3)$

5. x-intercept is $\left(\frac{3}{2}, 0\right)$; y-intercept is $(0, 2)$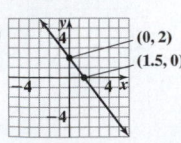
(0, 2)
(1.5, 0)

6. (a) $m = -\frac{2}{3}$ **(b)** y-intercept $= \left(0, -\frac{4}{3}\right)$ **7.** $m = -\frac{1}{3}$

8. (a) $m = \frac{2}{5}$ **(b)** $m = -\frac{5}{2}$ **9.** The lines are parallel. Answers may vary. **10.** $y = 3x + 1$ **11.** $y = -6x - 2$ **12.** $y = -2x + 7$

13. $y = -\frac{5}{2}x - 20$ **14.** $y = \frac{1}{4}x + 5$ **15.** $y = -8$ **16.** $x = 2$ **17.** $m = 11$. Every package increases expenses by $11.

18. (a) $y = 8350x - 2302$, where x is the weight in carats, and y is the price in dollars. **(b)** For every 1-carat increase in weight, the cost increases by $8350.
(c) $4044

Section 3.7 Linear Inequalities in Two Variables

1. (a) No **(b)** Yes **(c)** Yes **2. (a)** Yes **(b)** Yes **(c)** No **3.** solid; dashed **4.** half-planes

5. 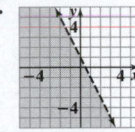 **6.** 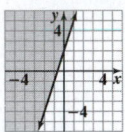 **7.** 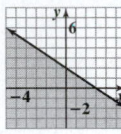 **8.** 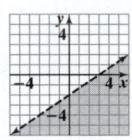 **9.** 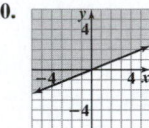 **10.** 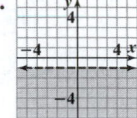 **11.**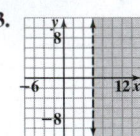

12. 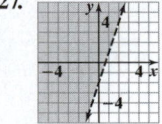 **13.** 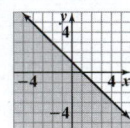 **14. (a)** $0.2s + 0.25t \le 2$ **(b)** Yes **(c)** No **15.** A is a solution. **17.** C is a solution.

19. A and C are solutions. **21.** B is a solution. **23.** B is a solution. **25.** A is a solution.

27. 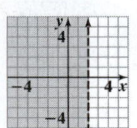 **29.** 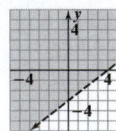 **31.** 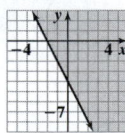 **33.** 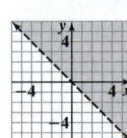 **35.** 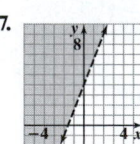 **37.**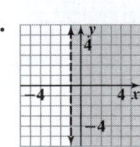

39. 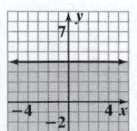 **41.** 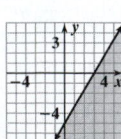 **43.** 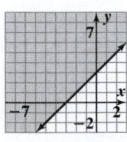 **45.** 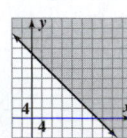 **47.** 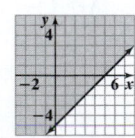 **49.**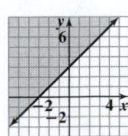

51. 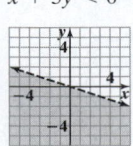 **53.** 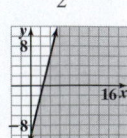 **55.** 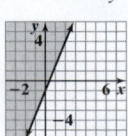 **57.** $x + y \ge 26$ **59.** $x - y \le 5$ **61.** $x \le y - 3$

63. $x + 3y < 0$ **65.** $2x - \frac{1}{2}y \ge 5$ **67.** $5x \ge 2 + 2y$ **69. (a)** $3s + 5a \le 120$ **(b)** No **(c)** Yes

71. (a) $50x + 40y \ge 4000$ **(b)** Yes **(c)** No

73. 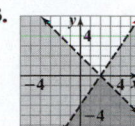 **75.** 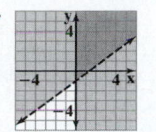 **77.**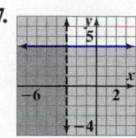

79. If (a, b) satisfies the inequality, then (a, b) is a solution to the linear inequality. If (a, b) satisfies $Ax + By = C$, then (a, b) lies on the boundary line.

81.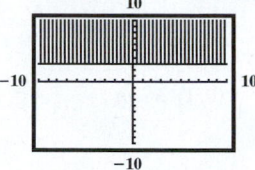

83.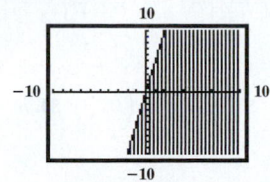

85.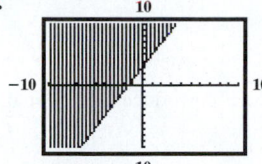

87.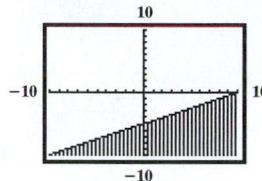

89.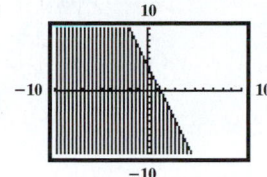

91.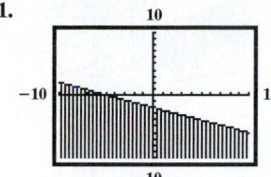

Chapter 3 Review

1–4.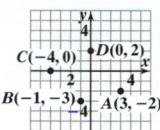

$A(3, -2)$ quadrant IV;
$B(-1, -3)$ quadrant III;
$C(-4, 0)$ x-axis; $D(0, 2)$
y-axis

5. $A(1, 4)$; quadrant I; $B(-3, 0)$; x-axis **6.** $A(0, 1)$; y-axis; $B(-2, 2)$; quadrant II **7.** A: Yes; B: No

8. A: Yes; B: No **9. (a)** $\left(4, -\dfrac{4}{3}\right)$ **(b)** $(-6, 2)$ **10. (a)** $(4, 4)$ **(b)** $(-2, -2)$

11.

x	y	(x, y)
-2	-8	(-2, -8)
0	-5	(0, -5)
4	1	(4, 1)

12.

x	y	(x, y)
-3	5	(-3, 5)
2	0	(2, 0)
4	-2	(4, -2)

13.

x	y	(x, y)
-18	2	(-18, 2)
-6	-2	(-6, -2)
24	-12	(24, -12)

14.

x	y	(x, y)
-3	-8	(-3, -8)
3	1	(3, 1)
5	4	(5, 4)

15.

x	C	(x, C)
1	7	(1, 7)
2	9	(2, 9)
3	11	(3, 11)

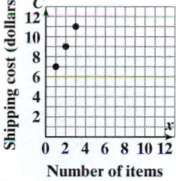

16.

x	E	(x, E)
500	1050	(500, 1050)
1000	1100	(1000, 1100)
2000	1200	(2000, 1200)

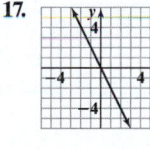

17.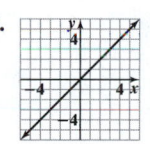

18.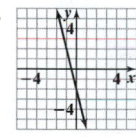

19.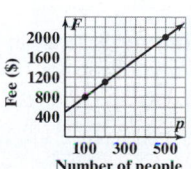

20.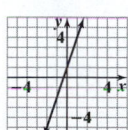

21.

p	C	(p, C)
20	80	(20, 80)
50	140	(50, 140)
80	200	(80, 200)

22.

p	F	(p, F)
100	800	(100, 800)
200	1100	(200, 1100)
500	2000	(500, 2000)

23. $(-2, 0)$, $(0, -4)$ **24.** $(0, 1)$ **25.** $(0, 9)$, $(-3, 0)$ **26.** $(0, -6)$, $(3, 0)$ **27.** $(3, 0)$ **28.** $\left(0, -\dfrac{2}{5}\right)$, $(1, 0)$

29. 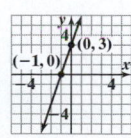 **30.** 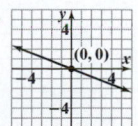 **31.** 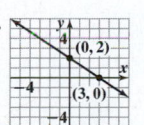 **32.** 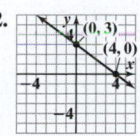 **33.** 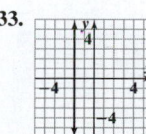 **34.**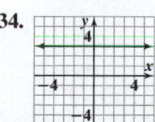

35. 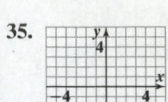 **36.** 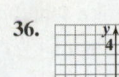 **37.** $m = \dfrac{4}{3}$ **38.** $m = -2$ **39.** $m = -8$ **40.** $m = 2$ **41.** $m = \dfrac{5}{2}$ **42.** $m = -\dfrac{1}{6}$ **43.** undefined **44.** $m = 0$ **45.** $m = 0$

46. undefined **47.** 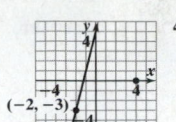 **48.** 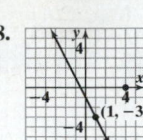 **49.** 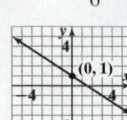 **50.**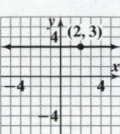

51. $m = 45$; the cost to produce 1 additional bicycle is \$45. **52.** 5% **53.** $m = -1; \left(0, \dfrac{1}{2}\right)$ **54.** $m = 1; \left(0, -\dfrac{3}{2}\right)$ **55.** $m = \dfrac{3}{4}; (0, 1)$ **56.** $m = -\dfrac{2}{5}; \left(0, \dfrac{8}{5}\right)$

57. 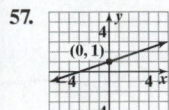 **58.** 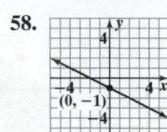 **59.** 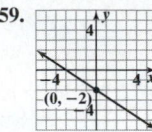 **60.** 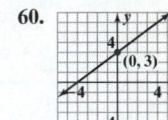 **61.** 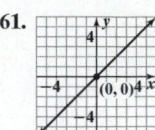 **62.**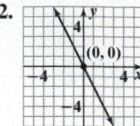

63. 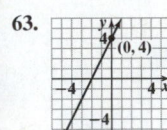 **64.** 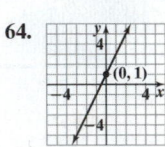 **65.** $y = -3x + 5$ **66.** $y = \dfrac{1}{5}x + 10$ **67.** $x = -12$ **68.** $y = -4$ **69.** $y = x - 20$

70. $y = -x - 8$ **71. (a)** \$360 **(b)** 7 days **(c)** 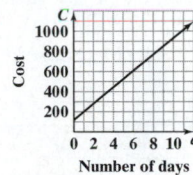 **72. (a)** $(22, 418), (35, 665), m = 19$ **(b)** It costs \$19 more for each additional day. **(c)** $C = 19d$ **(d)** \$152 **73.** $y = 6x - 3$ **74.** $y = -2x + 8$

75. $y = -\dfrac{1}{2}x + \dfrac{1}{2}$ **76.** $y = \dfrac{2}{3}x - \dfrac{7}{3}$ **77.** $y = -\dfrac{1}{2}$ **78.** $x = -\dfrac{4}{7}$

79. $y = \dfrac{8}{7}x + 8$ **80.** $y = \dfrac{3}{2}x - 6$ **81.** $y = 3x - 4$ **82.** $y = -\dfrac{2}{5}x - 5$ **83.** $y = 7$ **84.** $x = 4$ **85.** $A = -4d + 24$ **86.** $F = -\dfrac{1}{3}m + 15$ **87.** No

88. Yes **89.** $y = -x + 2$ **90.** $y = 2x + 8$ **91.** $y = -3x + 7$ **92.** $y = -\dfrac{3}{2}x + 1$ **93.** $x = 5$ **94.** $y = -12$ **95.** -2 **96.** $-\dfrac{9}{4}$ **97.** Yes

98. No **99.** $y = \dfrac{1}{3}x + 5$ **100.** $y = -\dfrac{1}{2}x + 1$ **101.** $y = -\dfrac{3}{2}x - \dfrac{3}{2}$ **102.** $y = x + 1$ **103.** A: Yes; B: Yes; C: No **104.** A: No; B: No; C: Yes

105. 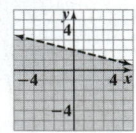 **106.** 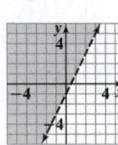 **107.** 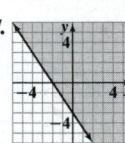 **108.** 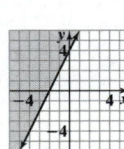 **109.** 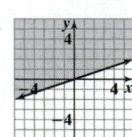 **110.**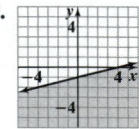

111. 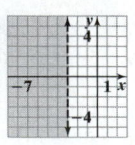 **112.** 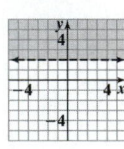 **113.** $0.25x + 0.01y \geq 12$ **114.** $2x - \dfrac{1}{2}y \leq 10$

Chapter 3 Test **1.** No **2. (a)** $(4, 0)$ **(b)** $\left(0, -\dfrac{4}{3}\right)$ **3. (a)** $m = \dfrac{4}{3}$ **(b)** $(0, 8)$

4. 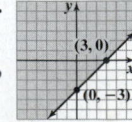 **5.** 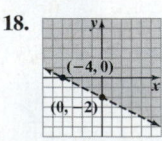 **6.** $m = -\dfrac{1}{6}$ **7. (a)** $-\dfrac{3}{2}$ **(b)** $\dfrac{2}{3}$ **8.** neither; Answers may vary. **9.** $y = -4x - 15$

10. $y = 2x + 14$ **11.** $y = -3x - 11$ **12.** $y = \dfrac{1}{2}x - 2$ **13.** $y = -\dfrac{3}{2}x + 8$

14. $y = 5$ **15.** $x = -2$ **16.** 2.5 women per year

17. 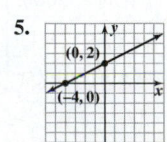 **18.** 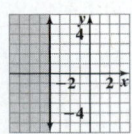 **19.**

Cumulative Reviews Chapters 1–3

1. -16 **2.** $\frac{1}{4}$ **3.** -4 **4.** 8 **5.** $4m^2 + 5m - 9$ **6.** $\{-7\}$ **7.** $\left\{-\frac{25}{12}\right\}$ **8.** $B = \frac{2A - hb}{h}$ or $B = \frac{2A}{h} - b$

9. $\{x \mid x > -4\}$ or $(-4, \infty)$ **10.** $\{x \mid x \le 5\}$ or $(-\infty, 5]$ **11.** 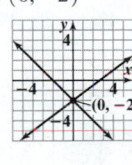 **12.** 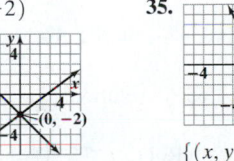 **13.**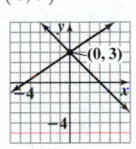

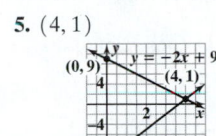 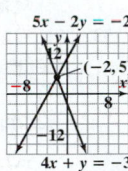

14. $y = -\frac{4}{3}x + 2$ **15.** $y = -3x - 8$ **16.** 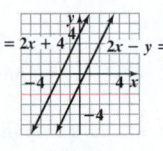 **17.** Shawn needs to score at least 91 on the final exam to earn an A.

18. A person 62 inches tall would be considered obese if she or he weighed 160 pounds or more. **19.** The angles measure 55° and 125°. **20.** The cylinder should be about 5.96 inches tall. **21.** The three consecutive even integers are 24, 26, and 28.

Chapter 4 Systems of Linear Equations and Inequalities in Two Variables

Section 4.1 Solving Systems of Linear Equations by Graphing
1. system of linear equations **2.** solution **3. (a)** No **(b)** Yes **(c)** No

4. (a) Yes **(b)** Yes **(c)** Yes

5. $(4, 1)$ 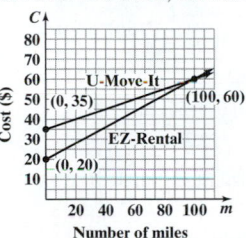 **6.** $(-2, 5)$ 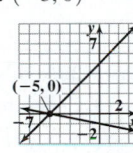 **7.** 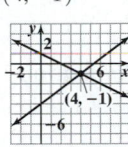 no solution; $\{\ \}$ or $\varnothing$ **8.** 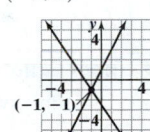 $\{(x, y) \mid 3x - y = -2\}$

9. consistent **10.** independent **11.** inconsistent **12.** True **13.** One solution; consistent; independent **14.** infinitely many solutions; consistent; dependent **15.** no solution; inconsistent

16. After 100 miles; the cost is $60.   **17. (a)** No **(b)** Yes **(c)** No **19. (a)** No **(b)** Yes **(c)** Yes **21. (a)** No **(b)** No **(c)** No

23. $(-5, 0)$ 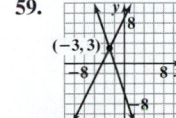 **25.** $(4, -1)$ 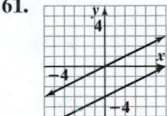 **27.** $(-1, 1)$ 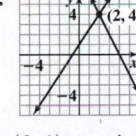 **29.** $(4, -2)$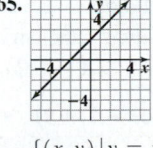

31. 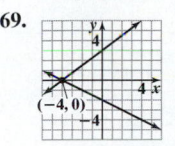 no solution; $\{\ \}$ or $\varnothing$ **33.** $(0, -2)$ 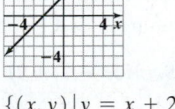 **35.** 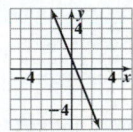 $\{(x, y) \mid 4x + 2y = 6\}$ **37.** $(0, 3)$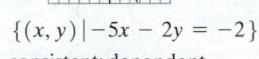

39. one solution; consistent; independent; $(3, 2)$ **41.** no solution; inconsistent; $\{\ \}$ or $\varnothing$ **43.** infinitely many solutions; consistent; dependent; $\{(x, y) \mid 2x + y = 0\}$ **45.** one solution; consistent; independent; $(-1, -2)$ **47.** one solution; consistent; independent **49.** no solution; inconsistent **51.** infinitely many solutions; consistent; dependent **53.** one solution; consistent; independent

55. no solution; inconsistent **57.** infinitely many solutions; consistent; dependent

59. 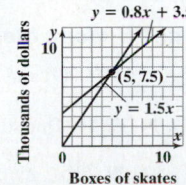 $(-3, 3)$; consistent; independent **61.** no solution $\{\ \}$ or $\varnothing$; inconsistent **63.** $(2, 4)$; consistent; independent **65.** $\{(x, y) \mid y = x + 2\}$; consistent; dependent **67.** $\{(x, y) \mid -5x - 2y = -2\}$; consistent; dependent

69. 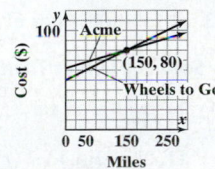 $(-4, 0)$; consistent; independent **71.** The break-even point is $(100, 2500)$. San Diego T-shirt Company must sell 100 T-shirts to break even at a cost/revenue of $2500. **73.** The break-even point is $(5, 7.5)$. The company needs to sell 5 boxes of skates to break even at a cost/revenue of $7500. **75.** $\begin{cases} y = 62 + 0.12x \\ y = 50 + 0.2x \end{cases}$ The cost for driving 150 miles is the same for both companies ($80). Choose Acme to drive 200 miles.

77.

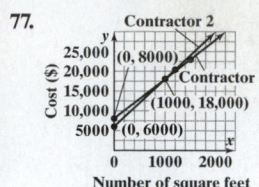
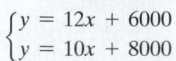

$$\begin{cases} y = 12x + 6000 \\ y = 10x + 8000 \end{cases}$$

When the cabin contains 1000 square feet, both contractors will charge the same amount. That cost is $18,000.

79. $c = 3$ **81.** $c = 1$

83.

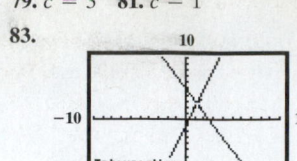

85.

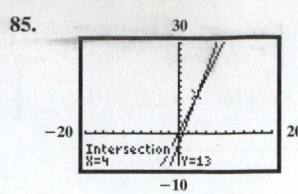

87. $\{(x, y) \mid 4x - 3y = 1\}$

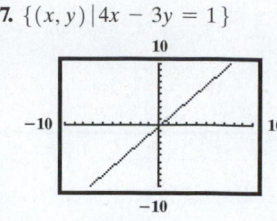

89. The three possibilities for a solution of a system of two linear equations containing two variables are (1) lines intersecting at a single point (one solution); (2) the lines are coincident (a single line in the plane) so an infinite number of solutions; (3) parallel lines (no solution).

Section 4.2 Solving Systems of Linear Equations Using Substitution
1. (2) **2.** $(2, 4)$ **3.** $(-3, 5)$ **4.** $\left(\frac{3}{2}, 7\right)$ **5.** $\left(-\frac{4}{3}, \frac{1}{2}\right)$ **6.** $\varnothing$ or $\{\,\}$

7. infinitely many solutions **8.** False **9.** $\{(x, y) \mid y = 5x + 2\}$ **10.** no solution; $\{\,\}$ or $\varnothing$ **11.** no solution; $\{\,\}$ or $\varnothing$ **12.** Equilibrium price $= \$2$;

Number of sodas $= 1200$ **13.** $(4, -1)$ **15.** $(-1, 1)$ **17.** $(-3, -4)$ **19.** $(-4, -7)$ **21.** $\left(-\frac{2}{3}, 2\right)$ **23.** $\left(2, -\frac{1}{3}\right)$ **25.** $\left(\frac{1}{4}, \frac{1}{6}\right)$ **27.** no solution; $\{\,\}$ or $\varnothing$;

inconsistent **29.** $\{(x, y) \mid y = 4x - 1\}$; consistent and dependent **31.** no solution; $\{\,\}$ or $\varnothing$; inconsistent **33.** $\{(x, y) \mid y = -5x - 4\}$; consistent and dependent **35.** $(-2, -1)$ **37.** no solution; $\{\,\}$ or $\varnothing$ **39.** $\{(x, y) \mid y = 2x - 8\}$ **41.** $(-3, 5)$ **43.** no solution; $\{\,\}$ or $\varnothing$ **45.** $\{(x, y) \mid 3x - y = 1\}$

47. $\left(\frac{5}{4}, -\frac{7}{4}\right)$ **49.** $\{(x, y) \mid 3x - 2y = 2\}$ **51.** $\left(-\frac{5}{2}, 4\right)$ **53.** length 10 ft; width 7 ft **55.** $8000 in money market; $4000 in international fund

57. $1,000,000; $35,000 **59.** 5 and 12 **61.** $A = \frac{7}{6}$; $B = -\frac{1}{2}$ **63.** Answers may vary. **65.** Answers may vary. **67.** A reasonable first step is to multiply the first

equation by 6 to clear fractions and to multiply the second equation by 4 to clear fractions. Then the system would be $\begin{cases} 2x - y = 4 \\ 6x + 2y = -5 \end{cases}$. Solve for y, because its coefficient is -1. Answers may vary.

Section 4.3 Solving Systems of Linear Equations Using Elimination
1. additive inverses **2.** $(-4, -2)$ **3.** $(8, 3)$ **4.** $(6, -5)$ **5.** $(5, 0)$
6. dependent; infinitely **7.** no solution **8.** infinitely many solutions **9.** infinitely many solutions **10.** Cheeseburger: $1.75; shake: $2.25

11. $(2, -1)$ **13.** $(6, 4)$ **15.** $\left(-\frac{11}{4}, \frac{1}{2}\right)$ **17.** $(-1, -3)$ **19.** no solution; $\{\,\}$ or $\varnothing$; inconsistent **21.** $\{(x, y) \mid 3x + y = -1\}$; consistent, dependent

23. $\{(x, y) \mid 2x - 3y = 10\}$; consistent, dependent **25.** no solution; $\{\,\}$ or $\varnothing$; inconsistent **27.** $(-5, 8)$ **29.** $\left(\frac{3}{2}, -\frac{3}{4}\right)$ **31.** no solution; $\{\,\}$ or $\varnothing$

33. $(12, -9)$ **35.** $\{(x, y) \mid 2x - 5y = 1\}$ **37.** $\left(\frac{6}{29}, -\frac{8}{29}\right)$ **39.** $\left(\frac{1}{2}, 4\right)$ **41.** $(-1, 2.1)$ **43.** $(-2, -6)$ **45.** $(50, 30)$ **47.** $(1, 5)$ **49.** $(5, 2)$ **51.** $\left(-\frac{9}{17}, \frac{31}{17}\right)$

53. $\left(\frac{8}{3}, -\frac{4}{7}\right)$ **55.** $\{(x, y) \mid 3x - 2y = 6\}$ **57.** $(1, -3)$ **59.** $(-4, 0)$ **61.** $\left(-\frac{8}{5}, \frac{27}{10}\right)$ **63.** $(-4, -5)$ **65.** no solution; $\{\,\}$ or $\varnothing$ **67.** $\left(\frac{1}{2}, \frac{1}{3}\right)$

69. hamburger 280 cal; Coke 210 cal **71.** tomatoes: 10 hr; zucchini: 15 hr; no **73.** 60 lb Arabica; 40 lb Robusta **75.** 70° and 20° **77.** $\left(-\frac{3}{a}, 1\right)$

79. $\left(-3a - b, \frac{3}{2}a - \frac{3}{2}b\right)$ **81.** It is easier to eliminate y because the signs on y are opposites. Multiply the second equation by 3, to form $6x + 3y = 15$.

Then add the equations, obtaining $7x = 21$. So $x = 3$. Substitute $x = 3$ into either the first equation or the second equation and solve for y, $y = -1$.
83. Both of you will obtain the correct solution. Other strategies are: multiplying equation (1) by one-third and adding; solving equation (2) for y and substituting this expression into equation (1); and graphing the equations and finding the point of intersection.

Putting the Concepts Together (Section 4.1–4.3)
1. (a) No (b) Yes (c) No **2.** (a) infinitely many (b) consistent (c) dependent **3.** (a) one

(b) consistent (c) independent **4.** $(1, 1)$ **5.** $(5, -1)$ **6.** $(1, -5)$ **7.** $(-1, 0)$ **8.** $(-4, 5)$ **9.** $(-3, -1)$ **10.** $\left(-\frac{1}{3}, 3\right)$ **11.** $\left\{(x, y) \mid x = \frac{y}{4} + 1\right\}$

12. $(2.5, 3)$ **13.** no solution; $\{\,\}$ or $\varnothing$

Section 4.4 Solving Direct Translation, Geometry, and Uniform Motion Problems Using Systems of Linear Equations
1. 43 and 61

2. length: 150 yards width: 50 yards; **3.** True **4.** 180° **5.** 36° and 54° **6.** 49° and 131° **7.** True **8.** airspeed: 350 mph; wind resistance: 50 mph

9. $2x$; $12 + \frac{1}{2}y$ **11.** $2l + 2w$ **13.** $8(r - c)$; 16 **15.** 33, 49 **17.** 14, 37 **19.** Thursday 35,000; Friday 42,000 **21.** $12,000 in stocks; $9000 in bonds

23. length 12 ft; width 9 ft **25.** length 25 m; width 10 m **27.** 40°, 50° **29.** 22.5°, 157.5° **31.** bike 11 mph; wind 1 mph **33.** current 0.4 mph; still water 3.9 mph
35. northbound 60 mph; southbound 72 mph **37.** 2.5 hr **39.** wind 25 mph; Piper 175 mph **41.** 24

Section 4.5 Solving Mixture Problems Using Systems of Linear Equations

2.

	Number · Cost per ticket	= Amount
Single-game ticket	s	50	50s
Single-game ticket with concession credit	c	66	66c
Total	12		728

9.

	Number · Cost	= Total
Adult	a	4	4a
Student	s	1.5	1.5s
Total	215		580

$$\begin{cases} a + s = 215 \\ 4a + 1.5s = 580 \end{cases}$$

11.

	Principal · Rate	= Interest
CD	c	0.015	0.015c
Treasury note	t	0.02	0.025
Total	1600		27

$$\begin{cases} c + t = 1600 \\ 0.015c + 0.02t = 27 \end{cases}$$

13.

	lb · Price	= Total
Mild	m	7.5	7.5m
Robust	r	10	10r
Total	12	8.75	8.75(12)

$$\begin{cases} m + r = 12 \\ 7.5m + 10r = 105 \end{cases}$$

1. rate; amount
3. 45 dimes; 40 quarters **4.** principal; rate; time **5.** True **6.** $40,000 in AA-rated bond; $50,000 in B-rated bond **7.** 5 pounds of Brazilian coffee and 15 pounds of Colombian coffee **8.** Mix 120 gallons of the ice cream with 5% butterfat and 80 gallons of the ice cream with 15% butterfat.

15. 32a; 24c **17.** 0.05A; 0.07B **19.** 5.85r; 4.20y; 128.85 **21.** 315 student and 75 nonstudent tickets **23.** 400 flats **25.** 60 nickels, 90 dimes **27.** first-class $0.49; postcard $0.34 **29.** $7500 in 5% account; $2500 in 8% account **31.** $3200 in risky plan; $1800 in safer plan **33.** 2 lb arbequina; 3 lb green **35.** 48.9 lb of the $2.75 per pound coffee; 51.1 lb of the $5 per pound coffee **37.** 80 lb rye; 100 lb bluegrass **39.** 20 ml 30% saline solution **41.** 210 liters **43.** 1.2 gal **45.** $6200 at 5%; $3800 at 7.5% loss **47.** The percentage of ethanol in the final solution is greater than the percentage of ethanol in either of the two original solutions.

Section 4.6 Systems of Linear Inequalities

1. solution **2. (a)** solution **(b)** not a solution **3.** dashed; solid

4. 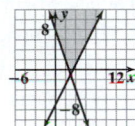 **5.** 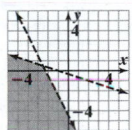 **6.** 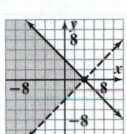 **7.** 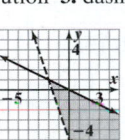 **8. (a)**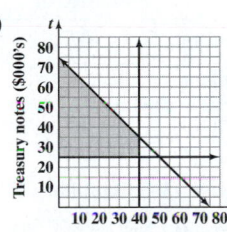

9. (a) Yes **(b)** No **(c)** Yes **11. (a)** Yes **(b)** No **(c)** No

13. 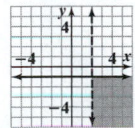 **15.** 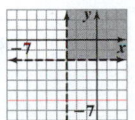 **17.** 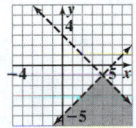 **19.** 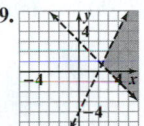 **21.**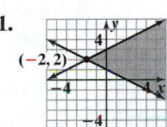

(b) Yes **(c)** No

23. 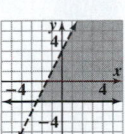 **25.** 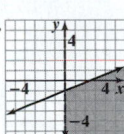 **27.** 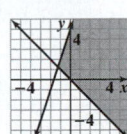 **29.** 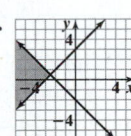 **31.** 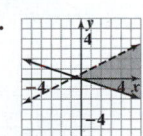 **33.**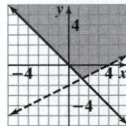

35. 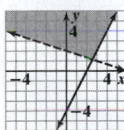 **37.** 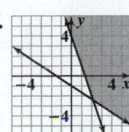 **39. (a)**   **(b)** No **(c)** Yes **41. (a)** 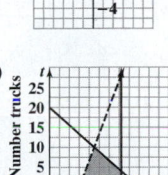 **(b)** No **(c)** No

43.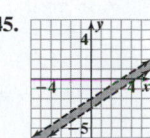

no solution; { } or ∅

45.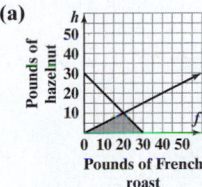

47. $$\begin{cases} y \geq 0 \\ y \geq x \\ y \leq \frac{1}{2}x + 3 \end{cases}$$

49. Answers may vary. Graph each inequality in the system. The overlapping shaded region represents the solution of the system.

51. (a) $$\begin{cases} y \geq x \\ y \geq -2x \end{cases}$$ **(b)** $$\begin{cases} y \leq x \\ y \geq -2x \end{cases}$$ **(c)** $$\begin{cases} y \geq x \\ y \leq -2x \end{cases}$$ **(d)** $$\begin{cases} y \leq x \\ y \leq -2x \end{cases}$$ Answers may vary.

Chapter 4 Review **1. (a)** No **(b)** Yes **(c)** No **2. (a)** No **(b)** No **(c)** Yes **3. (a)** Yes **(b)** Yes **(c)** Yes **4. (a)** No **(b)** No **(c)** No

5. 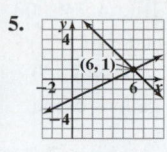 **6.** 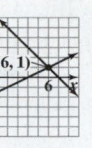 **7.** 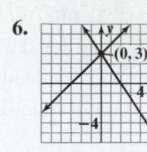 **8.** 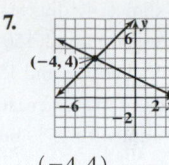 **9.** 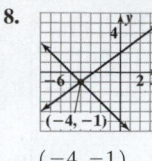 **10.** 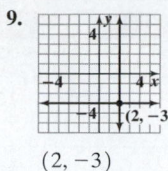 **11.**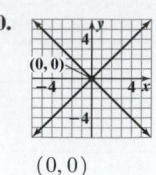

$(6, 1)$ $(0, 3)$ $(-4, 4)$ $(-4, -1)$ $(2, -3)$ $(0, 0)$ no solution

12. 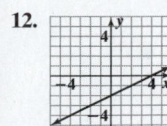 **13.** none; inconsistent **14.** none; inconsistent **15.** infinitely many; consistent; dependent **16.** infinitely many; consistent; dependent **17.** one; consistent; independent **18.** one; consistent; independent

$\{(x, y) \mid 3x - 6y = 12\}$

19. (a) $\begin{cases} y = 70 + 0.1x \\ y = 100 + 0.04x \end{cases}$ **(b)** 500 flier

20. (a) $\begin{cases} y = 50 + 40x \\ y = 350 + 10x \end{cases}$ **(b)** 10 sq yd

21. $(2, 1)$ **22.** $(1, -1)$ **23.** $(7, -2)$ **24.** $(4, -2)$ **25.** $(18, 11)$ **26.** $(-6, 8)$ **27.** $(2, -2)$ **28.** $(3, -1)$

29. $\{(x, y) \mid y = -2x + 4\}$ **30.** no solution; $\{\}$ or $\varnothing$ **31.** no solution; $\{\}$ or $\varnothing$ **32.** $\{(x, y) \mid 8x + 6y = 6\}$

33. $\left(\dfrac{3}{2}, 1\right)$ **34.** $\left(\dfrac{1}{3}, 1\right)$ **35.** width 125 m; length 200 m

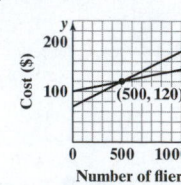

(c) Printer A

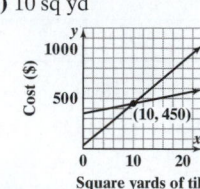

(c) tile

36. -3 **37.** $(0, -12)$ **38.** $(-3, 7)$ **39.** $(-3, 4)$ **40.** $(-3, 0)$ **41.** no solution; $\{\}$ or $\varnothing$ **42.** $\{(x, y) \mid 2x + 5y = 3\}$ **43.** $(-2, 2)$

44. $(3, -2.5)$ **45.** $\left(\dfrac{1}{2}, -2\right)$ **46.** $\left(-\dfrac{3}{2}, 1\right)$ **47.** $\left\{(x, y) \mid \dfrac{1}{3}x - \dfrac{1}{2}y = 1\right\}$ **48.** $\left(\dfrac{6}{7}, \dfrac{5}{2}\right)$ **49.** $(-20, -13)$ **50.** $\left(\dfrac{1}{3}, \dfrac{2}{3}\right)$ **51.** no solution; $\{\}$ or $\varnothing$

52. $\left(-\dfrac{1}{7}, 0\right)$ **53.** 2.4 lb cookies; 1.6 lb chocolates **54.** 400 individual tickets; 325 block tickets **55.** $\dfrac{3}{8}, \dfrac{1}{3}$ **56.** 26, 32

57. $32,000 in stocks, $18,000 in bonds **58.** 14 notebooks, 10 calculators **59.** 77.5°, 102.5° **60.** 40°, 50° **61.** 15 m by 22 m **62.** 110°, 40°

63. plane 450 mph, wind 50 mph **64.** current 2 mph, paddling 6 mph **65.** cyclist 10 mph, wind 2 mph **66.** faster 8 mph; slower 6 mph

67.

Number	·	Value	=	Total Value
Dimes	d	0.10		$0.10d$
Nickels	n	0.05		$0.05n$
Total	35			2.25

68.

	P	·	r	=	I
Savings	s		0.065		$0.065s$
Mutual Fund	m		0.08		$0.08m$
Total	15,000				1050

69. 676 **70.** 4 quarters **71.** $7000 at 5%, $12,000 at 3% **72.** $10,000 bonds, $15,000 stocks **73.** 3 quarts 30% sugar solution

74. 60 pints **75.** 3 lb peanuts; 2 lb almonds **76.** 4 liters 35% acid; 16 liters 60% acid **77. (a)** Yes **(b)** Yes **(c)** No **78. (a)** No **(b)** Yes **(c)** Yes

79. (a) Yes **(b)** No **(c)** No **80. (a)** No **(b)** Yes **(c)** Yes **81. (a)** No **(b)** Yes **(c)** No **82. (a)** No **(b)** Yes **(c)** No

83. 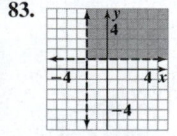 **84.** **85.** **86.** **87.**

88. 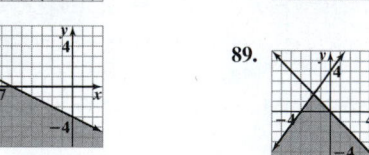 **89.** **90.** **91.** **92.**

93. 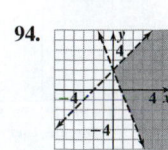 **94.** **95.**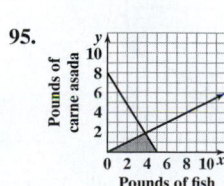

Pounds of carne asada / Pounds of fish

Chapter 4 Test **1. (a)** No **(b)** No **(c)** Yes **2. (a)** Yes **(b)** No **(c)** No **3. (a)** one **(b)** consistent **(c)** independent
4. (a) none **(b)** inconsistent **(c)** not applicable **5.**

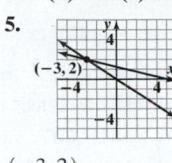

6.

7. $(0, -3)$ **8.** $(-12, -1)$ **9.** $(-3, 3)$

10. $(-2, -4)$ **11.** $\left(3, -\dfrac{1}{2}\right)$ **12.** $(2.5, 3)$

$(-3, 2)$ $(4, 2)$

13. $\left\{(x, y) \,\middle|\, x + \dfrac{3y}{2} = \dfrac{3}{2}\right\}$
14. airplane 200 mph; wind 25 mph
15. 7 containers vanilla; 4 containers peach
16. 52.5°, 127.5°
17. 15 basketballs, 25 volleyballs

18. 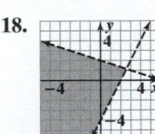 **19.** 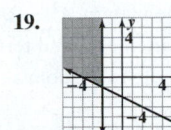 **20.**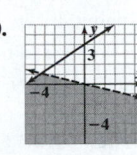

Chapter 5 Exponents and Polynomials

Section 5.1 Adding and Subtracting Polynomials
1. monomial **2.** 1 **3.** Yes; coefficient 12; degree 6 **4.** No **5.** Yes; coefficient 10; degree 0
6. No **7.** $m + n$ **8.** Yes; coefficient 3; degree 7 **9.** No **10.** Yes; coefficient -1; degree 3 **11.** True **12.** standard form **13.** Yes; degree 3 **14.** No

15. No **16.** Yes; degree 4 **17.** False **18.** $12x^2 + 3x + 3$ **19.** $2z^4 + 4z^3 - z^2 + z - 1$ **20.** $5x^2y + 6x^2y^2 - xy^2$ **21.** $5x^3 - 15x^2 + 3x + 7$

22. $8y^3 - 3y^2 - 3y - 6$ **23.** $11x^2y + 2x^2y^2 - 9xy^2$ **24.** $a^2b + a^2b^3 + 2ab^2 - 5ab$ **25.** $7a^2 - 4ab - 12b^2$ **26. (a)** 1 **(b)** -214 **(c)** 101 **27. (a)** 40 **(b)** -20

28. \$1875 **29.** Yes; coefficient 8; degree 3 **31.** Yes; coefficient $\dfrac{1}{7}$; degree 2 **33.** No **35.** Yes; coefficient 12; degree 5 **37.** No **39.** Yes; coefficient 4; degree 0

41. Yes; $6x^2 - 10$; degree 2; binomial **43.** No **45.** No **47.** Yes; $\dfrac{1}{8}$; degree 0; monomial **49.** Yes; $5z^3 - 10z^2 + z + 12$; degree 3; polynomial **51.** No

53. Yes; $-\dfrac{1}{2}t^4 + 3t^2 + 6t$; degree 4; trinomial **55.** Yes; $2xy^4 + 3x^2y^2 + 4$; degree 5; trinomial **57.** $7x - 10$ **59.** $-2m^2 + 5$ **61.** $-2p^2 + p + 8$

63. $-3y^2 - 2y - 4$ **65.** $\dfrac{5}{4}p^2 + \dfrac{1}{6}p - 3$ **67.** $6m^2 - 4mn - n^2$ **69.** $4n^2 - 8n + 5$ **71.** $3x - 16$ **73.** $14x^2 - 3x - 5$ **75.** $4y^3 - 3y - 4$ **77.** $2y^3 - y^2 - 4y$

79. $\dfrac{14}{9}q^2 - \dfrac{23}{8}q + 2$ **81.** $-8m^2n^2 - 4mn - 7$ **83.** $-4x - 5$ **85. (a)** 3 **(b)** 48 **(c)** 13 **87. (a)** -2 **(b)** $\dfrac{3}{4}$ **(c)** 4.75 **89.** 45 **91.** -328 **93.** $-3t + 3$

95. $3x^2 - 3x - 3$ **97.** $9xy^2 + 1$ **99.** $-y^2 + 15y + 1$ **101.** $\dfrac{7}{3}q^2 + \dfrac{5}{3}$ **103.** $13d^2 + d + 10$ **105.** $6a^2 + 5a - 3$ **107.** $10p$ **109.** $-2x^2 - 2xy + y^2$

111. $-5x + 12$ **113.** $3x^2 + 4x - 3$ **115.** $-14x^2 + 4x - 13$ **117.** $-3x + 5$ **119.** 26 ft **121. (a)** 135 feet **(b)** 30 feet **123. (a)** $-2.125x^2 + 105x$
(b) \$1250 **125. (a)** 10x **(b)** \$500 **(b)** \$15,000 **127.** $16x - 12$ **129.** $7x + 18$ **131.** $2x - 5$ **133.** Answers may vary. **135.** To find the degree of a polynomial in one variable, use the largest exponent on any term in the polynomial. To find the degree of a polynomial in more than one variable, find the degree of each term in the polynomial. That is, the degree of the term ax^my^n is the sum of the exponents, $m + n$. The highest degree of all of the terms is the degree of the polynomial. **137.** 1

Section 5.2 Multiplying Monomials: The Product and Power Rules
1. exponential **2.** $m + n$ **3.** False **4.** 27 **5.** -3125 **6.** c^8 **7.** y^9 **8.** a^9b^6 **9.** $m \cdot n$
10. 2^8 **11.** $(-3)^6$ **12.** b^{10} **13.** $(-y)^8$ **14.** False **15.** $8n^3$ **16.** $-125x^{12}$ **17.** $49a^6b^2$ **18.** $8a^{11}$ **19.** $-18m^3n^9$ **20.** $-16x^4y^4$ **21.** 1024 **23.** -128 **25.** m^9 **27.** b^{20}

29. x^8 **31.** p^9 **33.** $(-n)^7$ **35.** 64 **37.** 64 **39.** m^{14} **41.** $(-b)^{20}$ **43.** $27x^6$ **45.** $25z^4$ **47.** $\dfrac{1}{4}a^2$ **49.** $81p^{28}q^8$ **51.** $-8m^6n^3$ **53.** $25x^2y^4z^6$ **55.** $12x^5$ **57.** $-40a^{10}$

59. $-7m^4$ **61.** $6x^7$ **63.** $6x^6y^4$ **65.** $-5m^2n^4$ **67.** $-40x^5y^6$ **69.** $4x^3$ **71.** x^4 **73.** $28b^6$ **75.** $81b^4 + 16b^2$ **77.** $9p^5$ **79.** $27x^6y^2$ **81.** $-30cd^2$ **83.** $25x^8$ **85.** $-2x^7y^6$

87. $72x^{14} + 9x^{24}$ **89.** $4q^6$ **91.** $16x^{14}$ **93.** $-\dfrac{24}{5}m^{17}n^9$ **95.** x^6 **97.** $48x^2$ **99.** The product rule for exponential expressions is a "shortcut" for writing out each exponential expression in expanded form. For example, $x^2 \cdot x^3 = (x \cdot x) \cdot (x \cdot x \cdot x) = x^5$. **101.** The product to a power rule generalizes the result of using a product as a factor several times. That is, $(a \cdot b)^n = (a \cdot b) \cdot (a \cdot b) \cdot (a \cdot b) \ldots = a \cdot a \cdot a \cdot \ldots \cdot b \cdot b \cdot b \cdot \ldots = a^n \cdot b^n$. **103.** The expression $4x^2 + 3x^2$ is a sum of terms: Because they are like terms, add the coefficients and retain the common base. The expression $(4x^2)(3x^2)$ is a product: Multiply the constant terms and, since the bases are the same, add the exponents on the variable bases.

Section 5.3 Multiplying Polynomials
1. $3x^3 - 6x^2 + 12x$ **2.** $-2a^4b^3 + 4a^3b^3 - 3a^3b^2$ **3.** $3n^5 - \dfrac{9}{8}n^4 - \dfrac{6}{7}n^3$ **4.** $x^2 + 9x + 14$ **5.** $2n^2 - 5n - 3$
6. $6p^2 - 7p + 2$ **7.** First, Outer, Inner, Last **8.** $x^2 + 7x + 12$ **9.** $y^2 + 2y - 15$ **10.** $a^2 - 6a + 5$ **11.** $2x^2 + 11x + 12$ **12.** $6y^2 + y - 15$
13. $6a^2 - 11a + 4$ **14.** $15x^2 - 14xy - 8y^2$ **15.** $A^2 - B^2$ **16.** $a^2 - 16$ **17.** $9w^2 - 49$ **18.** $x^2 - 4y^6$ **19.** $A^2 - 2AB + B^2$; $A^2 + 2AB + B^2$
20. perfect square trinomial **21.** False **22.** $z^2 - 18z + 81$ **23.** $p^2 + 2p + 1$ **24.** $16 - 8a + a^2$ **25.** $9z^2 - 24z + 16$ **26.** $25p^2 + 10p + 1$
27. $4w^2 + 28wy + 49y^2$ **28.** False **29.** $x^3 - 8$ **30.** $3y^3 + 4y^2 + 8y - 8$ **31.** $x^3 - 8$ **32.** $3y^3 + 5y^2 - 17y + 5$ **33.** $-24a^3 - 34a^2 + 10a$
34. $x^3 + 4x^2 + x - 6$ **35.** $6x^2 - 10x$ **37.** $2n^2 - 3n$ **39.** $12n^4 + 6n^3 - 15n^2$ **41.** $4x^4y^2 - 3x^3y^3$ **43.** $x^2 + 12x + 35$ **45.** $y^2 + 2y - 35$
47. $6m^2 + 11my - 10y^2$ **49.** $x^2 + 5x + 6$ **51.** $q^2 - 13q + 42$ **53.** $6x^2 + 7x - 3$ **55.** $x^4 + 4x^2 + 3$ **57.** $42 - 13x + x^2$ **59.** $10u^2 + 17uv + 6v^2$
61. $10a^2 - ab - 2b^2$ **63.** $x^2 - 9$ **65.** $4z^2 - 25$ **67.** $16x^4 - 1$ **69.** $4x^2 - 9y^2$ **71.** $x^2 - \dfrac{1}{9}$ **73.** $x^2 - 4x + 4$ **75.** $25k^2 - 30k + 9$ **77.** $x^2 + 4xy + 4y^2$
79. $4a^2 - 12ab + 9b^2$ **81.** $x^2 + x + \dfrac{1}{4}$ **83.** $x^3 + x^2 - 5x - 2$ **85.** $2y^3 - 12y^2 + 19y - 3$ **87.** $2x^3 - 7x^2 + 4x + 3$ **89.** $2b^3 + 2b^2 - 24b$ **91.** $-x^3 + 9x$
93. $10y^5 + 3y^4 + 4y^3 + 3y^2 + 2y + 2$ **95.** $b^3 + 2b^2 - 5b - 6$ **97.** $x^4 - 25$ **99.** $z^4 - 81$ **101.** $-2x^2 - 37x$ **103.** $4x^6 + 12x^3 + 9$

105. $x^6 + x^4 - 3x^2 - 3$ **107.** $2x^3 + x^2 - 7x - 2$ **109.** $-5x^7 + 4x^6 - 6x^5$ **111.** $5x^3 + 21x^2 + 2x$ **113.** $-3w^3 + 3w^2 + 36w$ **115.** $3a^3 + 24a^2 + 48a$

117. $2n^2 + 6n$ **119.** $24ab$ **121.** $-x^2 + 3x + 2$ **123.** $4x^2 + 4x + 1$ **125.** $x^3 - 3x^2 + 3x - 1$ **127.** $2x^2 - 9$ **129.** $2x^2 + 7x - 15$ **131.** $2x^2 + x - 1$

133. $4x^3 + x^2 - 3x$ **135.** $x^2 + 12x + 20$. The sum of the areas of the four interior rectangles is $x^2 + 2x + 10x + 20 = x^2 + 12x + 20$. The product of the width and length is $(x + 10)(x + 2) = x^2 + 12x + 20$. Multiplying two binomials produces the same result as finding the area of the rectangle using the sum of the four interior rectangles. **137.** $x^2 + 3x + 2$ **139.** $(x^2 - x)$ square inches **141.** $(\pi x^2 + 4\pi x + 4\pi)$ square feet **143. (a)** a^2, ab, ab, b^2 **(b)** $a^2 + 2ab + b^2$ **(c)** length: $a + b$, width: $a + b$, area: $(a + b)^2 = a^2 + 2ab + b^2$ **145.** $x^3 + 2x^2 - 4x - 8$ **147.** $9 - x^2 - 2xy - y^2$

149. $x^3 + 6x^2 + 12x + 8$ **151.** $z^4 + 12z^3 + 54z^2 + 108z + 81$ **153.** To square the binomial $(3a - 5)^2$, use the pattern $A^2 - 2AB + B^2$. This means square the first term, take the product of -2 times the first term times the second term, and square the second term. So $(3a - 5)^2 = (3a)^2 - 2(3a)(5) + 5^2 = 9a^2 - 30a + 25$. **155.** The FOIL method can only be used to multiply two binomials.

Section 5.4 Dividing Monomials: The Quotient Rule and Integer Exponents

1. a^{m-n} **2.** False **3.** $3^2 = 9$ **4.** $\dfrac{7c}{5}$ **5.** $-\dfrac{3wz^7}{2}$ **6.** $\dfrac{a^n}{b^n}; b \neq 0$ **7.** $\dfrac{p^4}{16}$

8. $-\dfrac{8a^6}{b^{12}}$ **9.** $\dfrac{81m^4}{n^{12}}$ **10.** 1 **11. (a)** 1 **(b)** -1 **(c)** 1 **12. (a)** 1 **(b)** 2 **(c)** 1 **13.** $\dfrac{1}{a^n}$ **14. (a)** $\dfrac{1}{16}$ **(b)** $\dfrac{1}{16}$ **(c)** $-\dfrac{1}{16}$ **15.** $\dfrac{1}{8}$ **16. (a)** $\dfrac{1}{y^8}$ **(b)** $\dfrac{1}{y^8}$ **(c)** $-\dfrac{1}{y^8}$ **17. (a)** $\dfrac{2}{m^5}$

(b) $\dfrac{1}{32m^5}$ **(c)** $-\dfrac{1}{32m^5}$ **18.** 9 **19.** -100 **20.** $\dfrac{5z^2}{2}$ **21.** $\left(\dfrac{b}{a}\right)^n$ **22.** $\dfrac{8}{7}$ **23.** $\dfrac{125}{27a^3}$ **24.** $\dfrac{9n^8}{4}$ **25.** $-\dfrac{20}{a^2}$ **26.** $\dfrac{3}{mn}$ **27.** $-\dfrac{4a^3b^3}{3}$ **28.** $\dfrac{9x^2}{7y^3}$ **29.** $\dfrac{z^6}{9y^4}$ **30.** $\dfrac{4w^4}{49z^6}$ **31.** $-\dfrac{2}{p^{11}}$ **32.** $-\dfrac{4}{k}$

33. 16 **35.** x^9 **37.** $4y^3$ **39.** $-\dfrac{2m^7}{3}$ **41.** $2m^8n^2$ **43.** $\dfrac{27}{8}$ **45.** $\dfrac{x^{15}}{27}$ **47.** $\dfrac{x^{20}}{y^{28}}$ **49.** $\dfrac{49a^4b^2}{c^6}$ **51.** 1 **53.** -1 **55.** 18 **57.** 1 **59.** 1 **61.** $24a$ **63.** $\dfrac{1}{1000}$ **65.** $\dfrac{1}{m^2}$ **67.** $-\dfrac{1}{a^2}$

69. $-\dfrac{4}{y^3}$ **71.** $\dfrac{11}{18}$ **73.** $\dfrac{25}{4}$ **75.** $\dfrac{z^2}{3}$ **77.** $\dfrac{m^6}{8n^3}$ **79.** 16 **81.** $6x^4$ **83.** $\dfrac{5m^3}{2}$ **85.** $40m^3$ **87.** $\dfrac{5z^5}{6y^4}$ **89.** $-\dfrac{7m^{10}n^6}{3}$ **91.** $\dfrac{7y^4}{z^2}$ **93.** $\dfrac{y^4}{16x^4}$ **95.** $\dfrac{48}{b^3}$ **97.** 4 **99.** $\dfrac{1}{8}$ **101.** 81 **103.** $\dfrac{1}{x^9}$

105. $7x^3 + 5x$ **107.** $-\dfrac{3z^3}{2x^3}$ **109.** $\dfrac{1}{4x^3y^9}$ **111.** $-\dfrac{1}{a^{12}}$ **113.** $\dfrac{27}{m^6}$ **115.** $\dfrac{x^4y^6}{9}$ **117.** $\dfrac{6}{p^7}$ **119.** 12 **121.** 0 **123.** $\dfrac{16y}{27x^{18}}$ **125.** y **127.** $216x^3y^6$ **129.** $\dfrac{5ab^7}{3}$

131. $3x^3$ cubic meters **133.** $V = \dfrac{\pi d^2 h}{4}$ **135.** $9\pi x^3$ square units **137.** $\dfrac{1}{x^n}$ **139.** $x^{6n-2}y^2$ **141.** $\dfrac{x^{6a^2}y^{3ab}}{z^{3ac}}$ **143.** $\dfrac{9}{8a^{10n}}$ **145.** The quotient $\dfrac{11^6}{11^4} \neq 1^2$

because $\dfrac{11^6}{11^4} = \dfrac{11 \cdot 11 \cdot 11 \cdot 11 \cdot 11 \cdot 11}{11 \cdot 11 \cdot 11 \cdot 11} = \dfrac{\cancel{11} \cdot \cancel{11} \cdot \cancel{11} \cdot \cancel{11} \cdot 11 \cdot 11}{\cancel{11} \cdot \cancel{11} \cdot \cancel{11} \cdot \cancel{11}} = 11^2$. **147.** When we simplify the expression $-12x^0$, the exponent 0 applies only to the base x, so $-12x^0 = -12 \cdot 1 = -12$. However, $(-12x)^0 = 1$ because the entire expression is raised to the 0 power. **149.** His answer is incorrect. To raise a factor to a power, multiply exponents. Your friend added exponents.

Putting the Concepts Together (Sections 5.1–5.4)

1. Yes; degree 6; trinomial **2. (a)** 0 **(b)** -4 **(c)** 2 **3.** $8x^4 - 4x^2 + 14$ **4.** $5x^2y + 3xy + 2y^2$

5. $-6m^3n^2 + 2m^2n^4$ **6.** $5x^2 - 17x - 12$ **7.** $8x^2 - 26xy + 21y^2$ **8.** $25x^2 - 64$ **9.** $4x^2 + 12xy + 9y^2$ **10.** $6a^3 + 22a^2 - 40a$

11. $16m^4 + 4m^3 + 10m^2 - 20m - 24$ **12.** $-8x^8z^9$ **13.** $-\dfrac{15}{mn}$ **14.** $-3a^3b^2$ **15.** $\dfrac{1}{2x^2y^2}$ **16.** $\dfrac{7b^2}{2a}$ **17.** $\dfrac{r^5}{q^7t^2}$ **18.** $\dfrac{8}{27r^6}$ **19.** $\dfrac{y^4}{16z^6}$ **20.** $\dfrac{64y^8}{81x^{24}}$

Section 5.5 Dividing Polynomials

1. $4x^4$; $8x^2$ **2.** $2n^2 - 4n + 1$ **3.** $6k^2 - 9 + \dfrac{5}{2k^2}$ **4.** $\dfrac{xy^3}{4} + \dfrac{2y}{x} - \dfrac{1}{x^2}$ **5.** standard **6.** quotient; remainder; dividend

7. $x - 8$ **8.** $x - 4$ **9.** $4x + 5 + \dfrac{6}{x + 3}$ **10.** $4x^2 - 11x + 23 - \dfrac{45}{x + 2}$ **11.** $x^2 + 4x + 10 + \dfrac{60}{2x - 5}$ **12.** $4x - 3 + \dfrac{-7x + 7}{x^2 + 2}$ **13.** $2x - 1$

15. $3a + 9 - \dfrac{1}{a^2}$ **17.** $\dfrac{n^4}{5} - \dfrac{2n^2}{5} - 1$ **19.** $\dfrac{5r^2}{3} - 3$ **21.** $-x^2 + 3x - \dfrac{9}{x^2}$ **23.** $\dfrac{3}{8} + \dfrac{z^2}{2} - \dfrac{z}{4}$ **25.** $-7x + 5$ **27.** $-\dfrac{6y}{x} + 15$ **29.** $-\dfrac{5}{a} - \dfrac{2c^2}{a^2b}$ **31.** $x - 7$

33. $x - 5$ **35.** $x^2 + 6x - 3$ **37.** $x^3 - 3x^2 + 6x - 2$ **39.** $x^2 - 2x + 3 + \dfrac{5}{x + 1}$ **41.** $2x + 3$ **43.** $x^2 + 6x + 7 + \dfrac{16}{x - 2}$ **45.** $2x^3 + 5x^2 + 9x - 4 - \dfrac{17}{x - 4}$

47. $x^2 + 4x - 3 + \dfrac{2}{2x - 1}$ **49.** $x - 4 - \dfrac{4}{5 + x}$ **51.** $2x - 1 + \dfrac{6}{1 + 2x}$ **53.** $x^2 + 4$ **55.** $a^2 + a - 30$ **57.** $-x^2 - x - 6$ **59.** $-2ab + 2b^2$

61. $\dfrac{2}{x^2} + \dfrac{7}{2} - \dfrac{3x^2}{2} + 3x$ **63.** $\dfrac{2x^7y^8z^{12}}{3}$ **65.** $2x - 6$ **67.** $n^2 - 6n + 9$ **69.** $-10x^5 - 10x^3 + 40x^6$ **71.** $6pq - 2q^2 - p^2$ **73.** $x^3 + 6x^2 + 4x - 5$

75. $15rs^2 - 4r^2s$ **77.** $-3x^5 + 6x^3 - 3x^2$ **79.** $4x$ **81.** $-x + \dfrac{2}{x^2} - \dfrac{4}{x}$ **83.** $-\dfrac{1}{2x} - \dfrac{2}{x^2} + \dfrac{1}{x^3}$ **85.** x ft **87.** $(z + 3)$ in. **89.** $(4x + 4)$ ft

91. (a) $0.004x^2 - 0.8x + 180 + \dfrac{5000}{x}$ **(b)** \$182.11 per computer **93.** 6 **95.** When the divisor is a monomial, divide each term in the numerator by the denominator. When dividing by a binomial, use long division. Answers may vary.

Section 5.6 Applying Exponent Rules: Scientific Notation

1. scientific **2.** positive **3.** False **4.** 4.32×10^2 **5.** 1.0302×10^4 **6.** 5.432×10^6

7. 9.3×10^{-2} **8.** 4.59×10^{-5} **9.** 8×10^{-8} **10.** True **11.** True **12.** 310 **13.** 0.901 **14.** 170,000 **15.** 7 **16.** 0.00089 **17.** 6×10^7 **18.** 8×10^{-3}

19. 1.5×10^4 **20.** 2.8×10^{-5} **21.** 4×10^5 **22.** 2×10^{-4} **23.** 5×10^3 **24.** 6.25×10^{-5} **25.** $1.125 \times 10^{10} = 11,125,000,000$ gallons **26.** $1.9345 \times 10^7 = $ 19,345,000 visitors **27.** 3×10^5 **29.** 6.4×10^7 **31.** 5.1×10^{-4} **33.** 1×10^{-9} **35.** 8.007×10^9 **37.** 3.09×10^{-5} **39.** 6.2×10^2 **41.** 4×10^0 **43.** 7.303×10^9

45. 1.89×10^{13} **47.** 9.3×10^7 miles **49.** 3×10^{-5} mm **51.** 2.5×10^{-7} m **53.** 3.11×10^{-3} kg **55.** 420,000 **57.** 100,000,000 **59.** 0.0039 **61.** 0.4 **63.** 3760

65. 0.0082 **67.** 0.00006 **69.** 7,050,000 **71.** 0.000000000000001 **73.** 0.00225 **75.** 194,500,000 **77.** 3×10^9 **79.** 8.4×10^{-3} **81.** 3×10^8 **83.** 2.5×10^1

85. 8×10^8 **87.** 9×10^{15} **89.** 1.116×10^7 miles **91.** 0.000001 mile; 0.00000168 mile **93. (a)** 4.1×10^9 **(b)** 3.19×10^8 **(c)** 12.9 pounds **95.** 2.5×10^{-4} m

97. 8×10^{-10} m **99.** 7.15×10^{-8} m **101.** $1.2348\pi \times 10^{-14}$ m^3 **103.** $2.88\pi \times 10^{-25}$ m^3 **105.** To convert a number written in decimal notation to scientific notation, count the number of decimal places, N, that the decimal point must be moved to arrive at a number x such that $1 \le x < 10$. If the original number is greater than or equal to 1, move the decimal point to the left that many places and write the number in the form $x \times 10^N$. If the original number is between 0 and 1, move the decimal point to the right that many places and write the number in the form $x \times 10^{-N}$. **107.** The number 34.5×10^4 is incorrect because the number 34.5 is not a number between 1 (inclusive) and 10. The correct answer is 3.45×10^5.

Chapter 5 Review Exercises **1.** Yes; coefficient 4; degree 3 **2.** No **3.** No **4.** Yes; coefficient 1; degree 3 **5.** No **6.** No **7.** Yes; degree 0; monomial **8.** Yes; degree 5; binomial **9.** Yes; degree 6; trinomial **10.** Yes; degree 10; trinomial **11.** $9x^2 + 8x - 2$ **12.** $m^3 + mn - 5m$ **13.** $-x^2y + 12x$ **14.** $-7y^2 - 6yz + 9z^2$ **15.** $-2x^2 - 2$ **16.** $-20y^2 - 8y + 5$ **17. (a)** 0 **(b)** 8 **(c)** 2 **18. (a)** 3 **(b)** 2 **(c)** $\dfrac{11}{4}$ **19.** 0 **20.** 29 **21.** 279,936 **22.** $-\dfrac{1}{243}$ **23.** 16,777,216 **24.** 1 **25.** x^{13} **26.** m^6 **27.** r^{12} **28.** m^{24} **29.** $1024x^5$ **30.** $64n^6$ **31.** $81x^8y^4$ **32.** $4x^6y^8$ **33.** $15x^6$ **34.** $-36a^4$ **35.** $16y^5$ **36.** $-12p^6$ **37.** $12w^4$ **38.** $-\dfrac{3}{4}z^3$ **39.** $108x^8$ **40.** $80a^6$ **41.** $-8x^5 + 6x^4 - 2x^3$ **42.** $2x^7 + 4x^6 - x^4$ **43.** $6x^2 - 7x - 5$ **44.** $4x^2 - 5x - 6$ **45.** $x^2 - 3x - 40$ **46.** $w^2 + 9w - 10$ **47.** $24m^3 - 22m^2 + 7m - 3$ **48.** $8y^5 + 12y^4 + 4y^3 + 6y^2 - 6y - 9$ **49.** $x^2 + 8x + 15$ **50.** $2x^2 - 17x + 8$ **51.** $6m^2 + 17m - 14$ **52.** $48m^2 - 26m - 4$ **53.** $21x^2 + 5xy - 6y^2$ **54.** $20x^2 + 7xy - 3y^2$ **55.** $x^2 - 16$ **56.** $4x^2 - 25$ **57.** $4x^2 + 12x + 9$ **58.** $49x^2 - 28x + 4$ **59.** $9x^2 - 16y^2$ **60.** $64m^2 - 36n^2$ **61.** $25x^2 - 20xy + 4y^2$ **62.** $4a^2 + 12ab + 9b^2$ **63.** $x^2 - 0.25$ **64.** $r^2 - 0.0625$ **65.** $y^2 + \dfrac{4}{3}y + \dfrac{4}{9}$ **66.** $x^3 - 4x^2 - 22x$ **67.** 36 **68.** $\dfrac{1}{343}$ **69.** x^4 **70.** $\dfrac{1}{x^8}$ **71.** 1 **72.** -1 **73.** 1 **74.** -1 **75.** $\dfrac{5x^2}{2y^3}$ **76.** $\dfrac{x^2}{3y^8}$ **77.** $\dfrac{x^{15}}{y^{10}}$ **78.** $\dfrac{343}{x^6}$ **79.** $\dfrac{8m^6n^3}{p^{12}}$ **80.** $\dfrac{81m^4n^8}{p^{20}}$ **81.** $-\dfrac{1}{25}$ **82.** 64 **83.** $\dfrac{81}{16}$ **84.** 27 **85.** $\dfrac{7}{12}$ **86.** $\dfrac{1}{8}$ **87.** $\dfrac{2x^3y^5}{3}$ **88.** $\dfrac{3}{7xy^{10}}$ **89.** $\dfrac{9m^4}{16}$ **90.** $\dfrac{64}{9m^6n^6}$ **91.** $\dfrac{8r^4s^6}{9}$ **92.** $7a^3 + 7$ **93.** $6x^5 - 4x^4 + 5$ **94.** $3x^4 + 5x^2 - 6x$ **95.** $4n^3 + 1 - \dfrac{5}{2n^4}$ **96.** $6n - 4 - \dfrac{16}{5n^2}$ **97.** $-\dfrac{p^3}{8} - \dfrac{1}{4} + \dfrac{1}{2p^2}$ **98.** $-\dfrac{p^2}{2} + 1 - \dfrac{3}{2p^2}$ **99.** $4x - 7$ **100.** $x + 6$ **101.** $x^2 + 7x + 5 + \dfrac{6}{x - 1}$ **102.** $2x^2 - 3x - 12 - \dfrac{16}{x - 2}$ **103.** $x^2 - 2x + 4$ **104.** $3x^2 + 15x + 77 + \dfrac{378}{x - 5}$ **105.** 2.7×10^7 **106.** 1.23×10^9 **107.** 6×10^{-5} **108.** 3.05×10^{-6} **109.** 3×10^0 **110.** 8×10^0 **111.** 0.0006 **112.** 0.00125 **113.** 613,000 **114.** 80,000 **115.** 0.37 **116.** 54,000,000 **117.** 6×10^3 **118.** 4.2×10^{-8} **119.** 2×10^2 **120.** 2×10^8 **121.** 4×10^{15} **122.** 2×10^2

Chapter 5 Test **1.** Yes; degree 5; binomial **2. (a)** 5 **(b)** 21 **(c)** 26 **3.** $7x^2y^2 - 6x - 3y$ **4.** $10m^3 + m^2 - 6$ **5.** $-6x^5 + 18x^4 - 15x^3$ **6.** $2x^2 - 3x - 35$ **7.** $4x^2 - 28x + 49$ **8.** $16x^2 - 9y^2$ **9.** $6x^3 + x^2 - 25x + 8$ **10.** $2x - \dfrac{8}{3} + \dfrac{3}{x^3}$ **11.** $3x^2 - 11x + 33 - \dfrac{94}{x + 3}$ **12.** $-12x^4y^6$ **13.** $\dfrac{2m^3}{3n^5}$ **14.** $\dfrac{n^{24}}{m^{30}}$ **15.** $\dfrac{x^{22}}{y^{14}}$ **16.** $\dfrac{8m^7}{n^7}$ **17.** 1.2×10^{-5} **18.** 210,100 **19.** 3.57×10^4 **20.** 2×10^{-7}

Chapter 5 Cumulative Review Chapters 1–5 **1. (a)** $\left\{\sqrt{25}\right\}$ **(b)** $\left\{0, \sqrt{25}\right\}$ **(c)** $\left\{-6, -\dfrac{4}{2}, 0, \sqrt{25}\right\}$ **(d)** $\left\{-6, -\dfrac{4}{2}, 0, 1.4, \sqrt{25}\right\}$ **(e)** $\left\{\sqrt{7}\right\}$ **(f)** $\left\{-6, -\dfrac{4}{2}, 0, 1.4, \sqrt{7}, \sqrt{25}\right\}$ **2.** $-\dfrac{4}{9}$ **3.** -19 **4.** $6x^3 + 5x^2 - 3x$ **5.** $-18x + 8$ **6.** $\{1\}$ **7.** $4(x - 5) = 2x + 10$ **8.** $640 **9.** 45 mph **10.** $\{x \mid x < -3\}$ or $(-\infty, -3)$; **11.** $-\dfrac{1}{2}$ **12.** $-\dfrac{3}{2}$ **13.** 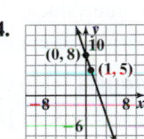 **14.** **15.** $(3, -7)$ **16.** No solution **17.** $6x^3 - 3x^2 + 7x$ **18.** $28m^2 - 13m - 6$ **19.** $9m^2 - 4n^2$ **20.** $49x^2 + 14xy + y^2$ **21.** $4m^3 - 19m + 15$ **22.** $\dfrac{2}{x} + \dfrac{1}{y}$ **23.** $x^2 - 3x + 9$ **24.** $-24n^4$ **25.** $-\dfrac{5n^8}{2m^2}$ **26.** $\dfrac{1}{64x^6y^{24}z^{12}}$ **27.** $\dfrac{1}{2x^{12}y^3}$ **28.** 6.05×10^{-5} **29.** 2,175,000 **30.** 7.14×10^5

Chapter 6 Factoring Polynomials

Section 6.1 Greatest Common Factor and Factoring by Grouping **1.** greatest common factor **2.** factors **3.** 8 **4.** 3 **5.** 7 **6.** $7y^2$ **7.** $2z$ **8.** $4xy^3$ **9.** $2x + 3$ **10.** $3(k + 8)$ **11.** factor **12.** Distributive **13.** $5z(z + 6)$ **14.** $12p(p - 1)$ **15.** $4y(4y^2 - 3y + 1)$ **16.** $2m^2n^2(3m^2 + 9mn^2 - 11n^3)$ **17.** $-4y(y - 2)$ **18.** $-3a(2a^2 - 4a + 1)$ **19.** False **20.** $(a - 5)(2a + 3)$ **21.** $(z + 5)(7z - 4)$ **22.** $(x + y)(4 + b)$ **23.** $(3z - 1)(2a - 3b)$ **24.** $(2b + 1)(4 - 5a)$ **25.** $3(z + 4)(z^2 + 2)$ **26.** $2n(n + 1)(n^2 - 2)$ **27.** 4 **29.** 1 **31.** 6 **33.** x^2 **35.** $7x$ **37.** $15ab^2$ **39.** $2abc^2$ **41.** $x - 1$ **43.** $2(x - 4)^2$ **45.** $6(x - 3)^2$ **47.** $6(2x - 3)$ **49.** $x(x + 12)$ **51.** $5x^2y(1 - 3xy)$ **53.** $3x(x^2 + 2x - 1)$ **55.** $-5x(x^2 - 2x + 3)$ **57.** $3(3m^5 - 6m^3 - 4m^2 + 27)$ **59.** $-4z(3z^2 - 4z + 2)$ **61.** $-5(3b^3 + b - 2)$ **63.** $15ab^2(ab^2 - 4b + 3a^2)$ **65.** $(x - 3)(x - 5)$ **67.** $(x - 1)(x^2 + y^2)$ **69.** $(4x + 1)(x^2 + 2x + 5)$ **71.** $(x + 3)(y + 4)$ **73.** $(y + 1)(z - 1)$ **75.** $(x - 1)(x^2 + 2)$ **77.** $(2t - 1)(t^2 - 2)$ **79.** $(2t - 1)(t^3 - 3)$ **81.** $4(y - 5)$ **83.** $7m(4m^2 + m + 9)$ **85.** $6m^2n(2mnp - 3)$ **87.** $3(p + 3)(3p + 1)$ **89.** $3(2x - y)(3a - 2b)$ **91.** $(x - 2)(x - 2)$ or $(x - 2)^2$ **93.** $3x(5x - 2)(x^2 + 2)$ **95.** $-3x(x^2 - 2x + 3)$ **97.** $4b(4b - 3)$ **99.** $(4y + 3)(3x - 2)$ **101.** $\dfrac{1}{3}x^2\left(x - \dfrac{2}{3}\right)$ **103.** $-2t(8t - 75)$ **105.** $150(140 - p)$ **107.** $4x^3(2x^2 - 7)$ **109.** $2\pi r(r + 4)$ **111.** $x - 2$ **113.** $4x^{2n} + 5$ **115.** $3x^3 - 4x^2 + 2$ **117.** $x^4 - 3x + 2$

119. $\frac{3}{5}x^3 - \frac{1}{2}x + 1$ **121.** To find the greatest common factor: (1) Determine the GCF of the coefficients of the variable factors (the largest number that divides evenly into a set of numbers). (2) For each variable factor common to all the terms, determine the smallest exponent that the variable factor is raised to. (3) Find the product of these common factors. This is the GCF. To factor the GCF from a polynomial: (1) Identify the GCF of the terms that make up the polynomial. (2) Rewrite each term as the product of the GCF and the remaining factor. (3) Use the Distributive Property "in reverse" to factor out the GCF. (4) Check using the Distributive Property. **123.** $3a(x + y) - 4b(x + y) \neq (3a - 4b)(x + y)^2$ because when checking, the product $(3a - 4b)(x + y)^2$ is equal to $(3a - 4b)(x + y)^2 = (3a - 4b)(x^2 + 2xy + y^2)$, not to $3a(x + y) - 4b(x + y) = 3ax + 3ay - 4bx - 4by$.

Section 6.2 Factoring Trinomials of the Form $x^2 + bx + c$

1. quadratic trinomial **2.** 24; -10 **3.** False **4.** $(y + 5)(y + 4)$ **5.** $(z - 7)(z - 2)$ **6.** True **7.** $(y - 5)(y + 3)$ **8.** $(w - 3)(w + 4)$ **9.** prime **10.** prime **11.** $(q + 9)(q - 5)$ **12.** $(x + 4y)(x + 5y)$ **13.** $(m + 7n)(m - 6n)$ **14. (a)** $x^2 - x - 12$ **(b)** $n^2 + 9n - 10$ **15.** $(n + 8)(n - 7)$ **16.** $(y - 7)(y - 5)$ **17.** False **18.** $4(m + 3)(m - 7)$ **19.** $3z(z - 1)(z + 5)$ **20.** $3b^2(a - 6)(a - 2)$ **21.** $-1(w + 5)(w - 2)$ **22.** $-2a(a + 6)(a - 2)$ **23.** $(x + 2)(x + 3)$ **25.** $(m + 3)(m + 6)$ **27.** $(x - 3)(x - 12)$ **29.** $(p - 2)(p - 6)$ **31.** $(x - 4)(x + 3)$ **33.** prime **35.** $(z - 3)(z + 15)$ **37.** $(x - 2y)(x - 3y)$ **39.** $(r - 2s)(r + 3s)$ **41.** $(x - 4y)(x + y)$ **43.** prime **45.** $3(x + 2)(x - 1)$ **47.** $3n(n - 3)(n - 5)$ **49.** $5z(x - 8)(x + 4)$ **51.** $x(x - 6)(x + 3)$ **53.** $-2(y - 2)^2$ **55.** $x^2(x + 8)(x - 4)$ **57.** $x(x + 5)(x - 3)$ **59.** $(x + 4y)(x - 7y)$ **61.** prime **63.** $(k - 5)(k + 4)$ **65.** $2w(w^2 - 4w - 6)$ **67.** $(st - 3)(st - 5)$ **69.** $-3p(p - 2)(p + 1)$ **71.** $-(x + 3)^2$ **73.** prime **75.** $n^2(n - 6)(n + 5)$ **77.** $2(x - 5)(x^2 + 3)$ **79.** $(s + 5)(s + 7)$ **81.** prime **83.** $(mn - 6)(mn - 2)$ **85.** $-x(x - 14)(x + 2)$ **87.** $(y - 6)^2$ **89.** $-2x(x^2 - 10x + 18)$ **91.** $-7xy(3x^2 + 2y)$ **93.** $-16(t + 1)(t - 5)$ **95.** $(x + 6)$ meters and $(x + 3)$ meters **97.** $(x + 5)$ inches and $(x - 3)$ inches **99.** $-7, -5, 5, 7$ **101.** $-20, -4, 4, 20$ **103.** 1 **105.** 6, 10, 12 **107.** Yes, $(5 + x)(2 - x) = (x + 5)(-x + 2) = -1(x + 5)(x - 2)$ **109.** In the trinomial $x^2 + bx + c$, where both b and c are negative, find two numbers m and n so that $m \cdot n = c$ and $m + n = b$. m and n will have opposite signs, and the factor with the larger absolute value is negative. For example, $x^2 - 3x - 28 = (x - 7)(x + 4)$.

Section 6.3 Factoring Trinomials of the Form $ax^2 + bx + c, a \neq 1$

1. $-6; 1$ **2.** $12; -13$ **3.** $(3x + 4)(x - 2)$ **4.** $(5z + 3)(2z + 3)$ **5.** $3(4x + 3)(2x - 1)$ **6.** $-(5n - 1)(2n - 3)$ **7.** $(3x + 7y)(2x + 3y)$ **8.** $-3(4a - 5b)(3a + 2b)$ **9.** $(3x + 2)(x + 1)$ **10.** $(7y + 1)(y + 3)$ **11.** $(3x - 4)(x - 3)$ **12.** $(5p - 1)(p - 4)$ **13.** $(2n + 1)(n - 9)$ **14.** $(4w + 3)(w - 2)$ **15.** $(3x + 2)(4x + 3)$ **16.** $(6y - 5)(2y + 7)$ **17.** Look for a common factor **18.** False **19.** $-4(2x + 3y)(x - 5y)$ **20.** $3(6x - y)(5x + 2y)$ **21.** $-(6y + 1)(y - 4)$ **22.** $-3(2x - 5)(x + 3)$ **23.** $(2x + 3)(x + 5)$ **25.** $(5w - 2)(w + 3)$ **27.** $(2w + 1)(2w - 5)$ **29.** $(9z - 2)(3z + 1)$ **31.** $(3y + 2)(2y - 3)$ **33.** $(2m - 1)(2m + 5)$ **35.** $(3n + 1)(4n + 5)$ **37.** $(6x - 5)(3x + 1)$ **39.** $2(3x + 2y)(2x - y)$ **41.** $4(2x + 1)(x + 3)$ **43.** $(3x + 5)(2x - 1)$ **45.** $-(4p + 3)(2p - 3)$ **47.** $(2x + 3)(x + 1)$ **48.** $(5n + 2)(n + 1)$ **51.** $(5y - 3)(y + 1)$ **53.** $-(4p + 1)(p - 3)$ **55.** $(5w - 2)(w + 3)$ **57.** $(7t + 2)(t + 5)$ **59.** $(6n - 5)(n - 2)$ **61.** $(x - 2)(5x - 1)$ **63.** $(2x + y)(x + y)$ **65.** $(2m + n)(m - 2n)$ **67.** $2(3x - 2y)(x + y)$ **69.** $-(2x - 3)(x + 5)$ **71.** $(3x - 4)(5x - 1)$ **73.** $-(y + 4)(4y - 3)$ **75.** $2(x - 2y)(5x + 6y)$ **77.** prime **79.** $-3(8x^2 + 4x - 3)$ **81.** $4x^2y^2(x - 2y - 1)$ **83.** $(m + 3n)(4m + n)$ **85.** prime **87.** $6(2x + 5y)(2x - y)$ **89.** $3(3m + 8n)(2m - n)$ **91.** $-2x^2(2x + 5)(x - 1)$ **93.** $-2x(4x + 3)(3x - 5)$ **95.** $(x^2 + 1)(2x - 7)(3x - 2)$ **97.** $(2x + 7)$ m and $(3x - 4)$ m **99.** $2x - 5$ **101.** $(9z^2 + 8)(3z^2 + 2)$ **103.** $(3x^n + 1)(x^n + 6)$ **105.** ± 2 or ± 14 **107.** Trial and error may be used when the value of a and/or the value of c are prime numbers or numbers with few factors. Grouping can be used when the product of a and c is not a very large number. Examples may vary.

Section 6.4 Factoring Special Products

1. perfect square trinomial **2.** $(A - B)^2$ **3.** $(x - 6)^2$ **4.** $(4x + 5)^2$ **5.** $(3a - 10b)^2$ **6.** $(z - 4)^2$ **7.** $(2n + 3)^2$ **8.** prime **9.** $4(z + 3)^2$ **10.** $2a(5a + 4)^2$ **11.** difference; two squares **12.** $(P - Q)(P + Q)$ **13.** $(z - 5)(z + 5)$ **14.** $(9m - 4n)(9m + 4n)$ **15.** $\left(4a - \frac{2}{3}b\right)\left(4a + \frac{2}{3}b\right)$ **16.** True **17.** $(10k^2 - 9w)(10k^2 + 9w)$ **18.** $(x - 2)(x + 2)(x^2 + 4)$ **19.** False **20.** $3(7x - 4)(7x + 4)$ **21.** $-2y(x^2 - 2y^3)(x^2 + 2y^3)$ **22.** sum; two cubes **23.** $(A - B)(A^2 + AB + B^2)$ **24.** False **25.** $(z + 5)(z^2 - 5z + 25)$ **26.** $(2p - 3q^2)(4p^2 - 6pq^2 + 9q^4)$ **27.** $-2a(2a + 3)(4a^2 - 6a + 9)$ **28.** $-3(5b - 1)(25b^2 + 5b + 1)$ **29.** $(x + 5)^2$ **31.** $(2p - 1)^2$ **33.** $(4x + 3)^2$ **35.** $(x - 2y)^2$ **37.** $(2z - 3)^2$ **39.** $(x - 7)(x + 7)$ **41.** $(2x - 5)(2x + 5)$ **43.** $(10n^4 - 9p^2)(10n^4 + 9p^2)$ **45.** $(k - 2)(k + 2)(k^2 + 4)(k^4 + 16)$ **47.** $(5p^2 - 7q)(5p^2 + 7q)$ **49.** $(3 + x)(9 - 3x + x^2)$ **51.** $(2x - 3y)(4x^2 + 6xy + 9y^2)$ **53.** $(x^2 - 2y)(x^4 + 2x^2y + 4y^2)$ **55.** $(3c + 4d^3)(9c^2 - 12cd^3 + 16d^6)$ **57.** $2(2x + 5y)(4x^2 - 10xy + 25y^2)$ **59.** $(4m + 5n)^2$ **61.** $2(x - 3)^2$ **63.** $-(2a - 5)(3a + 8)$ **65.** $2x(x^2 + 5x + 8)$ **67.** $x^2y^2(x - y)(x + y)$ **69.** $(x^4 - 5y^5)(x^4 + 5y^5)$ **71.** $(b + 9)(b + 4)$ **73.** $3s(s^2 + 2)(s^4 - 2s^2 + 4)$ **75.** $2x(x - 3)(x + 3)(x^2 + 9)$ **77.** $(2x + 5)(3x + 2)$ **79.** $4(x - 6)(x + 3)$ **81.** prime **83.** $2(x - 2)^2$ **85.** prime **87.** $xy^3(x + 6)(x^2 - 6x + 36)$ **89.** $3n^2(4n - 1)^2$ **91.** $(x - 2)(x + 2)(2x + 5)$ **93.** $(2y + 5)(y - 4)(y + 4)$ **95.** $(2x + 5)$ meters **97.** $-3(2x - 1)$ **99.** $x(x^2 - 3xy + 3y^2)$ **101.** $(x - 2)(x + 4)$ **103.** $(x + 1)(2a - 5)(a - 6)$ **105.** $x(5x + 13)$ **107.** First identify the problem as a sum or a difference of cubes. Identify A and B, and rewrite the problem as $A^3 + B^3$ or $A^3 - B^3$. The sum of cubes can be factored into the product of a binomial and a trinomial, where the binomial factor is $A + B$ and the trinomial factor is A^2 minus the product AB plus B^2. The difference of cubes also can be factored into the product of a binomial and a trinomial, but here the binomial factor is $A - B$ and the trinomial factor is the sum of A^2, the product AB, and B^2. **109.** The correct factorization of $x^3 + y^3$ is $(x + y)(x^2 - xy + y^2)$. The trinomial factor $(x^2 - xy + y^2)$ is not a perfect square trinomial. A perfect square trinomial is $(x - y)^2 = (x^2 - 2xy + y^2)$.

Section 6.5 Summary of Factoring Techniques   **1.** greatest common factor **2.** $2(p-5)(p+9)$ **3.** $-3x(3x+1)(5x-2)$

4. $(4x+1)(2x+5)$ **5.** $(6m+7)(2m-5)$ **6.** $(p-6q)^2$ **7.** $3(5x+3)^2$ **8.** $(5y-4)(25y^2+20y+16)$ **9.** $-3(2a-b)(4a^2+2ab+b^2)$

10. False **11.** $(x-1)(x+1)(x^2+1)$ **12.** $-4y(3x-2)(3x+2)$ **13.** $(2x+3)(x^2+2)$ **14.** $3(2x+3)(x-1)(x+1)$ **15.** $-3(z^2-3z+7)$

16. $3x(2y^2+5x^2)$ **17.** $(x-10)(x+10)$ **19.** $(t+3)(t-2)$ **21.** $(x+y)(1+2a)$ **23.** $(a-2)(a^2+2a+4)$ **25.** $(a-3b)(a+2b)$

27. $(2x-7)(x+1)$ **29.** $2(x-5y)(x+2y)$ **31.** prime **33.** $(3-a)(3+a)$ **35.** $(u-11)(u-3)$ **37.** $(x-a)(y-b)$ **39.** $(w+2)(w+4)$

41. $(6a-7b^2)(6a+7b^2)$ **43.** $(x+4m)(x-2m)$ **45.** $(3xy-2)(2xy-3)$ **47.** $(x+1)(x^2+1)$ **49.** $6(2z^2+2z+3)$ **51.** $(7c-1)(2c+3)$

53. $(3m+4n^2)(9m^2-12mn^2+16n^4)$ **55.** $2j^2(j-1)(j+1)(j^2+1)$ **57.** $(4a-b)(2a+5b)$ **59.** $2a(a^2+4)$ **61.** $3(2z-1)(2z+1)$

63. prime **65.** $2a(2a+b)(4a^2-2ab+b^2)$ **67.** $(pq+7)(pq-1)$ **69.** $(s+2)^2(s-2)$ **71.** $-2x(3x+1)(2x-1)$ **73.** $(5v+2)(2v-1)$

75. $-n^2(n-4)(n+1)$ **77.** $-2ab(2a^2-a+1)$ **79.** $-p(p-4)(p+3)$ **81.** $-8x(2x-3y)(2x+3y)$ **83.** $2(n-5)(n^2-3)$

85. $4(4x-3)(x+1)$ **87.** $(3x^2+2)(x^2+4)$ **89.** $-xy(2x-3y)(x+y)$ **91.** $x(x-5)(x-3)$ **93.** $2x$ ft., $(2x+1)$ ft., $(x-3)$ ft.

95. $(y^2+4)(y^2-6)$ **97.** $(z^3-2)(4z^3-5)$ **99.** $3(r-2)(3r-5)$ **101.** $(y+2)(2y-3)$ **103.** $2(3x+4)(9x+13)$ **105.** $2x(3x+5)$

107. $3(4x-3)(4x-1)$ **109.** $(2m-n+p)(2m-n-p)$ **111.** $(x+y-z)(x-y+z)$ **113.** The student initially wrote the

sum of terms as a product of terms by enclosing the binomial $-3x-12$ in parentheses. Then, in the second step, the student wrote the

factor $-3(x-4)$ and concluded that the factorization of $x^2+4x-3x-12$ is $(x-3)(x+4)$, even though both terms do not have

the common factor of $(x+4)$. The student obtained the correct answer, but the first two steps are incorrect; the correct factorization is

$x^2+4x-3x-12 = x(x+4)-3(x+4) = (x+4)(x-3)$.

Putting the Concepts Together (Sections 6.1–6.5)   **1.** $5x^2y$ **2.** $(x-4)(x+1)$ **3.** $(x^2-3)(x^4+3x^2+9)$ **4.** $(2x+1)(6x+5z)$

5. $(x+6y)(x-y)$ **6.** $(x+4)(x^2-4x+16)$ **7.** prime **8.** $3(x+6y)(x-2y)$ **9.** $4z^2(3z^3-11z-6)$ **10.** prime **11.** $m^2(m+2)(4m-3)$

12. $(5p-2)(p-3)$ **13.** $(2m+5)(5m-3)$ **14.** $6(3m-1)(2m+1)$ **15.** $(2m-5)^2$ **16.** $(5x+4y)(x-y)$ **17.** $S = 2\pi r(h+r)$

18. $h = 16t(3-t)$ or $h = -16t(t-3)$

Section 6.6 Solving Polynomial Equations by Factoring   **1.** 2 **2.** $a = 0; b = 0$ **3.** $\{-3, 0\}$ **4.** $\left\{-\dfrac{5}{4}, 2\right\}$ **5.** quadratic **6.** second

7. False **8.** $\{2, 4\}$ **9.** $\left\{-\dfrac{1}{2}, 3\right\}$ **10.** $\left\{-\dfrac{5}{2}, 1\right\}$ **11.** $\{-5, -4\}$ **12.** $\left\{-2, \dfrac{2}{5}\right\}$ **13.** $\left\{-4, \dfrac{1}{3}\right\}$ **14.** False **15.** $\{-6, 4\}$ **16.** $\left\{-\dfrac{3}{5}, 1\right\}$ **17.** $\left\{\dfrac{4}{3}\right\}$

18. $\{-5\}$ **19.** $\{-6, 3\}$ **20.** $\{-2, 3\}$ **21.** After 1 second and after 4 seconds **22.** True **23.** $\{-2, -1, 0\}$ **24.** $\left\{-1, 1, \dfrac{4}{3}\right\}$ **25.** $\{-4, 0\}$

27. $\{-3, 9\}$ **29.** $\left\{-\dfrac{1}{3}, 5\right\}$ **31.** linear **33.** quadratic **35.** $\{-1, 4\}$ **37.** $\{-7, -2\}$ **39.** $\left\{-\dfrac{1}{2}, 0\right\}$ **41.** $\left\{-\dfrac{1}{2}, 2\right\}$ **43.** $\{3\}$ **45.** $\{0, 6\}$

47. $\{-2, 3\}$ **49.** $\{-2\}$ **51.** $\left\{\dfrac{1}{4}, 1\right\}$ **53.** $\{-4, 6\}$ **55.** $\{-5, 10\}$ **57.** $\{-5, 1\}$ **59.** $\{-3, 0, 2\}$ **61.** $\{-3, -2, 2\}$ **63.** $\left\{-\dfrac{3}{2}, 2, -2\right\}$

65. $\left\{-\dfrac{3}{5}, 4\right\}$ **67.** $\{-4, 5\}$ **69.** $\{-5\}$ **71.** $\left\{-\dfrac{3}{4}, 7\right\}$ **73.** $\{-4, 2\}$ **75.** $\left\{-4, -\dfrac{1}{2}, 4\right\}$ **77.** $\{20\}$ **79.** $\{-10, 10\}$ **81.** $\{-6, 2\}$ **83.** $\left\{-\dfrac{1}{2}, \dfrac{3}{2}, 5\right\}$

85. $\{0, 8\}$ **87.** $\left\{-4, \dfrac{3}{2}\right\}$ **89.** $\left\{-6, \dfrac{1}{2}\right\}$ **91.** 1 sec, 3 sec **93.** 9 **95.** -4 and -3 or 3 and 4 **97.** $-12, -10,$ and $-8,$ or 8, 10, and 12 **99.** 15 by 17 units

101. 8 teams **103.** $(x-3)(x+5) = 0$; at least degree 2 **105.** $(z-6)^2 = 0$; at least degree 2 **107.** $(x+3)(x-1)(x-5) = 0$; at least degree 3

109. $\{a, -b\}$ **111.** $\{0, 2a\}$

113. The student divided by the variable x instead of writing
the quadratic equation in standard form, factoring, and using
the Zero-Product Property. The correct solution is

$$15x^2 = 5x$$
$$15x^2 - 5x = 0$$
$$5x(3x-1) = 0$$
$$5x = 0 \quad \text{or} \quad 3x - 1 = 0$$
$$x = 0 \quad \text{or} \qquad 3x = 1$$
$$x = \dfrac{1}{3}$$

115. We write a quadratic equation in standard
form, $ax^2 + bx + c = 0$, so that when the quadratic
polynomial on the left side of the equation is
factored, we can apply the Zero-Product Property,
which says that if $a \cdot b = 0$, then $a = 0$ or $b = 0$.

Section 6.7 Modeling and Solving Problems with Quadratic Equations   **1. (a)** After 4 seconds and after 6 seconds **(b)** After 10 seconds

2. 8 km by 13 km **3.** height = 8 yards, base = 12 yards **4.** $x = 12, x - 3 = 9$ **5.** 7 inches and 24 inches **6.** 6 inches high, 8 inches wide **7.** 4, 14

9. 2, 9 **11.** base = 26; height = 8 **13.** base = 18; height = 16 **15.** base = 13; height = 11 **17.** $B = 53; b = 43$ **19.** 12, 5 **21.** 9, 12, 15 **23.** 256, 252, 240,

220, 192, 156, 112, 60, 0 feet **25.** 6 sec **27.** width = 4 m; length = 12 m **29.** base = 6 ft; height = 18 ft **31.** 30 inches

33. width = 12 mm; length = 25 mm **35.** width = 4 cm; length = 25 cm **37. (a)** length = 45 in.; width = 28 in. **(b)** 48 in. by 31 in. **39.** 6 ft

41. width = 7 ft; length = 14 feet **43.** If you have two positive numbers that satisfy the projectile motion word problem, one answer is the height of

the object on its upward path, and the other represents the height of the object on its downward path. There is only one positive number that satisfies

the motion problem if the object is dropped from a height above the ground or if the answer represents the maximum height of the projectile.

Chapter 6 Review   **1.** 12 **2.** 27 **3.** 10 **4.** 4 **5.** x^2 **6.** m **7.** $15ab^2$ **8.** $6x^3y^2z$ **9.** $2(2a+1)^2$ **10.** $9(x-y)$ **11.** $-6a^2(3a+4)$ **12.** $-3x(3x-4)$

13. $5y^2z(3+y^5+4y)$ **14.** $7xy(x^2-3xy+2y^2)$ **15.** $(5-y)(x+2)$ **16.** $(a+b)(z+y)$ **17.** $(5m+2n)(m+3n)$ **18.** $(2x+y)(y+x)$

19. $(x + 2)(8 - y)$ **20.** $(y^2 + 1)(x - 3)$ **21.** $(x + 3)(x + 2)$ **22.** $(x + 2)(x + 4)$ **23.** $(x - 7)(x + 3)$ **24.** $(x + 5)(x - 2)$ **25.** prime

26. prime **27.** $(x - 3y)(x - 5y)$ **28.** $(m + 5n)(m - n)$ **29.** $-(p + 6)(p + 5)$ **30.** $-(y - 5)(y + 3)$ **31.** $3x(x^2 + 11x + 12)$

32. $4(x + 1)(x + 8)$ **33.** $2(x - 7y)(x + 6y)$ **34.** $4y(y + 5)(y - 2)$ **35.** $(5y - 6)(y + 4)$ **36.** $(y - 7)(6y + 1)$ **37.** $(2x - 3)(x - 1)$

38. $(2x + 7)(3x + 1)$ **39.** prime **40.** $(4m + 3)(2m + 3)$ **41.** $3m(m + n)(3m + 7n)$ **42.** $2(7m + n)(m + n)$ **43.** $-x(5x + 2)(3x - 1)$

44. $-2p^2(3p - 1)(2p + 1)$ **45.** $(2x - 3)^2$ **46.** $(x - 5)^2$ **47.** $(x + 3y)^2$ **48.** prime **49.** $2(2m + 1)^2$ **50.** $2(m - 6)^2$ **51.** $(2x - 5y)(2x + 5y)$

52. $(7x - 6y)(7x + 6y)$ **53.** prime **54.** prime **55.** $(x - 3)(x + 3)(x^2 + 9)$ **56.** $(x - 5)(x + 5)(x^2 + 25)$ **57.** $(m + 3)(m^2 - 3m + 9)$

58. $(m + 5)(m^2 - 5m + 25)$ **59.** $(3p - 2)(9p^2 + 6p + 4)$ **60.** $(4p - 1)(16p^2 + 4p + 1)$ **61.** $(y^3 + 4z^2)(y^6 - 4y^3z^2 + 16z^4)$

62. $(2y + 3z^2)(4y^2 - 6yz^2 + 9z^4)$ **63.** $(5a - 2b)(3a^2 - 5b^2)$ **64.** $(4a - 3b)(3a + b)$ **65.** prime **66.** $(x - 12y)(x + 2y)$ **67.** $x(x - 7)(x + 6)$

68. $3x^4(x - 7)(x - 3)$ **69.** $(3x + 1)(2x + 3)$ **70.** $(2z + 3)(5z - 3)$ **71.** $(3x + 2)(9x^2 - 6x + 4)$ **72.** $(2z - 1)(4z^2 + 2z + 1)$

73. $2(2y - 1)(y + 5)$ **74.** $xy^2(5x - 3)(x - 1)$ **75.** $(5k - 9m)(5k + 9m)$ **76.** $(x^2 - 3)(x^2 + 3)$ **77.** prime **78.** prime **79.** $\left\{\dfrac{3}{2}, 4\right\}$

80. $\left\{-7, -\dfrac{1}{2}\right\}$ **81.** $\{-3, 15\}$ **82.** $\{2, 5\}$ **83.** $\{0, -2\}$ **84.** $\left\{-\dfrac{9}{2}, 0\right\}$ **85.** $\{-1, 3\}$ **86.** $\{-3, -1\}$ **87.** $\{-4, -2\}$ **88.** $\left\{-\dfrac{1}{2}, \dfrac{3}{2}\right\}$

89. $\{-14, 0, 3\}$ **90.** $\{-3, 0, 6\}$ **91.** $\left\{-3, \dfrac{4}{3}, 3\right\}$ **92.** $\left\{-\dfrac{5}{2}, -2, 2\right\}$ **93.** 5 sec **94.** 2 sec and 3 sec **95.** 9 ft by 6 ft **96.** 3 yd by 5 yd

97. 8 ft, 6 ft **98.** 24 ft, 10 ft

Chapter 6 Test 1. $4x^4y^2$ **2.** $(x - 4)(x + 4)(x^2 + 16)$ **3.** $9x(2x - 3)(x + 1)$ **4.** $(x - 7)(y - 4)$ **5.** $(3x + 5)(9x^2 - 15x + 25)$

6. $(y - 12)(y + 4)$ **7.** $(6m + 5)(m - 1)$ **8.** prime **9.** $(x - 5)(4 + y)$ **10.** $3y(x - 7)(x + 2)$ **11.** prime

12. $2(x^2 - 3y)(x^4 + 3x^2y + 9y^2)$ **13.** $3x(3x + 1)(x + 4)$ **14.** $(3m + 2)(2m + 1)$ **15.** $2(2m^2 - 3mn + 2)$ **16.** $(5x + 7y)^2$

17. $\left\{-\dfrac{1}{5}, -3\right\}$ **18.** $\{0, 2\}$ **19.** length = 7 in.; width = 5 in. **20.** legs: 12 in., 5 in., hypotenuse: 13 in.

Getting Ready for Intermediate Algebra

1. (a) 1 **(b)** 0, 1, -6 **(c)** 0, 1, -6, $\dfrac{2}{5}$, -0.83, $0.5454\ldots$ **(d)** $1.010010001\ldots$ **(e)** 0, 1, -6, $\dfrac{2}{5}$, -0.83, $0.5454\ldots$, $0.010010001\ldots$ **2.** -20 **3.** $-\dfrac{6}{5}$ **4.** $-\dfrac{21}{2}$

5. $\dfrac{4}{3}$ **6.** $\dfrac{3}{2}$ **7.** 74 **8.** $-\dfrac{3}{2}$ **9.** $\{3\}$ **10.** $\left\{\dfrac{5}{4}\right\}$ **11.** The angles measure 55° and 125°. **12.** $\{x \mid x \geq 7\}$; $[7, \infty)$; See Graphing Answer Section

13. A person 62 inches tall would be considered obese if they weighed 160 pounds or more. **14.** 5 **15.** 1 **16.** $5y^3 + 5y^2 - y - 6$ **17.** $9x^8$

18. $-4x^3 + 20x^2 - 12x$ **19.** $12x^2 - 17x - 5$ **20.** $4x^2 - 28x + 49$ **21.** $25x^2 - 9y^2$ **22.** $\dfrac{y}{4z^3}$ **23.** $2a^2 - 6a + 1$ **24.** $3x^2 - 5x + 2 - \dfrac{9}{x + 5}$

25. $-4x(x^3 - 4x + 5)$ **26.** $(2z - 1)(4z^2 + 3)$ **27.** $(w + 5)(w - 7)$ **28.** $-3c(c - 8)(c + 3)$ **29.** prime **30.** $(3y - 4)(y + 4)$

31. $(p + 2)(p - 2)(p^2 + 4)$ **32.** $(2a + 5b)^2$ **33.** $\left\{-\dfrac{5}{2}, 3\right\}$ **34.** $\{4, 5\}$ **35.** $\left\{-\dfrac{7}{2}, \dfrac{2}{3}\right\}$

Chapter 7 Rational Expressions and Equations

Section 7.1 Simplifying Rational Expressions **1.** rational expression **2. (a)** $-\dfrac{1}{8}$ **(b)** $\dfrac{3}{16}$ **3. (a)** -1 **(b)** 9 **4.** 4 **5.** $\dfrac{11}{2}$ **6.** undefined **7.** $x = -7$

8. $n = -\dfrac{5}{3}$ **9.** $-4, 10$ **10.** simplify **11.** False **12.** $\dfrac{n + 4}{2}$ **13.** $-\dfrac{z + 1}{2}$ **14.** $\dfrac{a + 7}{2a + 7}$ **15.** $\dfrac{z^2 - 3}{z - 2}$ **16.** $\dfrac{2k + 1}{2k - 1}$ **17.** False **18.** -7 **19.** $\dfrac{-4}{4x - 1}$

20. $-\dfrac{5z + 1}{3}$ **21. (a)** 2 **(b)** $\dfrac{1}{2}$ **(c)** 0 **23. (a)** 1 **(b)** 3 **(c)** $\dfrac{5}{3}$ **25. (a)** -3 **(b)** 0 **(c)** $-\dfrac{15}{7}$ **27. (a)** 0 **(b)** $-\dfrac{7}{2}$ **(c)** $-\dfrac{3}{4}$ **29.** 0 **31.** 5 **33.** $\dfrac{3}{2}$ **35.** $-6, 6$ **37.** $2, 5$

39. $-2, 0, 3$ **41.** $\dfrac{x - 2}{3}$ **43.** $z + 2$ **45.** $\dfrac{1}{p + 2}$ **47.** -1 **49.** $-2k$ **51.** $\dfrac{x - 1}{2x + 3}$ **53.** $\dfrac{6}{x - 5}$ **55.** -3 **57.** $\dfrac{b - 5}{4}$ **59.** $\dfrac{x + 3}{x - 3}$ **61.** $\dfrac{3x - 4}{x - 5}$ **63.** $-(x + y)$

65. $\dfrac{x^2 + 1}{x + 1}$ **67.** $-\dfrac{4 + c}{c - 4}$ **69.** $\dfrac{2(x - 2)}{3(x - 5)}$ **71.** $-\dfrac{x - 3}{6x - 5}$ **73.** $\dfrac{2}{t^2 + 9}$ **75.** $\dfrac{-3}{w - 1}$ **77. (a)** 5 mg/mL **(b)** 4 mg/mL **79.** Yes; BMI = 27.5 **81.** \$15,600

83. $\dfrac{c^4 + 1}{c^6 + 1}$ **85.** $\dfrac{(x - 2)(x^2 + 3)}{x^3 - 9}$ **87.** $\dfrac{(t^2 + 4)(t + 2)}{t^2 - 2t + 4}$ **89. (a)** The x^2s cannot be divided out because they are not factors. **(b)** When the factors $x - 2$ are

divided out, the result is $\dfrac{1}{3}$, not 3. **91.** A rational expression is undefined if the denominator is equal to zero. Division by zero is not allowed in the real

number system. The expression $\dfrac{x^2 - 9}{3x - 6}$ is undefined for $x = 2$ because when 2 is substituted for x, we get $\dfrac{2^2 - 9}{3(2) - 6} = \dfrac{4 - 9}{6 - 6} = \dfrac{-5}{0}$, which is undefined.

93. $\dfrac{x - 7}{7 - x} = -1$ because $x - 7$ and $7 - x$ are opposites: $\dfrac{x - 7}{7 - x} = \dfrac{x - 7}{-1(-7 + x)} = \dfrac{x - 7}{-1(x - 7)} = \dfrac{\cancel{x - 7}}{-1(\cancel{x - 7})} = \dfrac{1}{-1} = -1$. The expression $\dfrac{x - 7}{x + 7}$ is in

simplest form because $x - 7$ and $x + 7$ are not opposites and they have no common factor.

Section 7.2 Multiplying and Dividing Rational Expressions **1.** False **2.** False **3.** $\dfrac{5p(p + 3)}{2}$ **4.** $\dfrac{7}{3(x + 5)}$ **5.** $-\dfrac{1}{x - 3}$ **6.** $-\dfrac{3}{7}$ **7.** $\dfrac{1}{3a}$ **8.** $\dfrac{3}{2}; \dfrac{4}{3}$

9. $\dfrac{6(x + 1)}{x(2x - 1)}$ **10.** $\dfrac{(x - 3)^2}{(x - 4)^2}$ **11.** $\dfrac{q + 1}{(q - 5)(q + 5)}$ **12.** $\dfrac{x - 1}{4}$ **13.** $\dfrac{7x}{4(x - 2)}$ **14.** $\dfrac{6m}{m + 2n}$ **15.** $\dfrac{2x}{x - 5}$ **17.** $\dfrac{x}{(x + 1)^2}$ **19.** $p - 1$ **21.** 1 **23.** $\dfrac{1}{3x}$ **25.** $\dfrac{2(m - 4)}{m + 4}$

27. x **29.** 1 **31.** $\dfrac{10}{3}$ **33.** 1 **35.** $-\dfrac{4}{11}$ **37.** $\dfrac{4}{3y(y-3)}$ **39.** $\dfrac{1}{3}$ **41.** $\dfrac{3x}{x-4}$ **43.** $\dfrac{2(4w+1)}{4w-1}$ **45.** $-\dfrac{(x-y)^2}{x}$ **47.** $-\dfrac{1}{2}$ **49.** $\dfrac{3(4n+3)(n-3)}{2(n+4)}$ **51.** $\dfrac{x}{4y^2(x-2y)}$

53. $\dfrac{2a-b}{2a(a+b)}$ **55.** $\dfrac{2x+5}{2(x+3)}$ **57.** -3 **59.** $-\dfrac{x(x+3)}{(x+1)^2}$ **61.** $\dfrac{a+b}{2}$ **63.** $\dfrac{1}{3}$ **65.** $\dfrac{(x-a)^2}{2(x+a)}$ **67.** $\dfrac{1}{3(x-y)}$ **69.** $\dfrac{9}{8}$ **71.** $\dfrac{x-y}{(x+y)^3}$ **73.** $\dfrac{4x}{(x-2)(x-3)}$ **75.** $\dfrac{1}{x}$ ft^2

77. $\dfrac{2(x+2)}{x-3}$ in.^{2} **79.** $\dfrac{1-x}{4}$ **81.** $\dfrac{2a^2}{3}$ **83.** $\dfrac{(x+y)^2}{(x^2-xy+y^2)(x+2)}$ **85.** $3p^5q^2(p^2+3pq+9q^2)(p-q)$ **87.** $6x+12$ **89.** To multiply two rational expressions,

(1) factor the polynomials in the numerator and denominator; (2) multiply the numerators and denominators using $\dfrac{a}{b}\cdot\dfrac{c}{d}=\dfrac{a\cdot c}{b\cdot d}$; (3) divide out common

factors in the numerators and denominators. Leave the result in factored form. **91.** The error is in incorrectly dividing out the factors $(n-6)$ and

$(6-n)$. The correct method is $\dfrac{n^2-2n-24}{6n-n^2}=\dfrac{^{-1}\cancel{(n-6)}(n+4)}{n\cancel{(6-n)}}=\dfrac{-(n+4)}{n}$.

Section 7.3 Adding and Subtracting Rational Expressions with a Common Denominator **1.** $\dfrac{a+b}{c}$ **2.** False **3.** $\dfrac{4}{x-2}$ **4.** $x+1$ **5.** $\dfrac{11x-3}{6x-5}$

6. $\dfrac{1}{x-5}$ **7.** $\dfrac{2(4y-3)}{2y-5}$ **8.** $\dfrac{z+1}{2z}$ **9.** $x+3$ **10.** $\dfrac{x+8}{3}$ **11.** $\dfrac{x+2}{x-7}$ **12.** True **13.** $\dfrac{3x-1}{x-5}$ **14.** 2 **15.** $\dfrac{2}{n-3}$ **16.** $\dfrac{2k+3}{2(k-1)}$ **17.** $\dfrac{7p}{4}$ **19.** $2n$ **21.** $\dfrac{2a-1}{a}$

23. $\dfrac{2(5c-2)}{c-1}$ **25.** 2 **27.** $7x$ **29.** 2 **31.** $\dfrac{1}{a-5}$ **33.** $\dfrac{x}{x+3}$ **35.** $-\dfrac{x}{2}$ **37.** $-\dfrac{c+3}{c}$ **39.** 3 **41.** -1 **43.** $x-1$ **45.** 1 **47.** $\dfrac{2(2p+3q)}{p-q}$ **49.** $\dfrac{2p^2+2p+1}{p-1}$ **51.** $\dfrac{4b}{a-b}$

53. -1 **55.** $\dfrac{2}{x-y}$ **57.** $-\dfrac{1}{x}$ **59.** -1 **61.** 0 **63.** $n-1$ **65.** $\dfrac{3}{v+3}$ **67.** $\dfrac{x+1}{x-1}$ **69.** $\dfrac{x^2+3}{x(x-3)}$ **71.** $2(x-1)$ **73.** $\dfrac{3x+3y-2}{x-y}$ **75.** $\dfrac{p+2q}{p^2-6q^2}$ **77.** $12a$

79. $\dfrac{n-3}{n+1}$ **81.** $\dfrac{4}{n}$ **83.** $\dfrac{x}{x+3}$ **85.** $\dfrac{4x-2}{x+2}$ **87.** $\dfrac{2(6x-1)}{x}$ cm **89.** $\dfrac{6x+7y}{(x+y)(x-y)}$ **91.** $-\dfrac{x(x-1)}{(3x-1)(x+2)}$ **93.** $\dfrac{-5x+12}{x-2}$ **95.** $-4n-6$

97. The expression is incorrect because the subtraction symbol was not applied to both terms of the second numerator. The correct method is

$\dfrac{x-2}{x}-\dfrac{x+4}{x}=\dfrac{x-2-x-4}{x}=\dfrac{-6}{x}$. **99.** The correct answer is (c). The answer (a) $\dfrac{7}{2x}$ is incorrect because the denominators were added; and the

answer (b) $\dfrac{7}{x^2}$ is incorrect because the denominators were multiplied.

Section 7.4 Finding the Least Common Denominator and Forming Equivalent Rational Expressions **1.** least common denominator

2. False **3.** $24x^2y^3$ **4.** $42a^3b^3$ **5.** True **6.** $15z(z+1)$ **7.** $(x-1)(x+5)^2$ **8.** $-3(x-7)(x+7)$ **9.** $\dfrac{x+3}{x+3}$ **10.** $\dfrac{12p^2}{16p^3(p-2)}$

11. LCD $=60a^2b^3$; $\dfrac{3}{6a^2b}=\dfrac{30b^2}{60a^2b^3}$, $\dfrac{-5}{20ab^3}=\dfrac{-15a}{60a^2b^3}$ **12.** LCD $=(x-5)(x-2)(x+1)$; $\dfrac{5}{x^2-4x-5}=\dfrac{5x-10}{(x-5)(x-2)(x+1)}$;

$\dfrac{-3}{x^2-7x+10}=\dfrac{-3x-3}{(x-5)(x-2)(x+1)}$ **13.** $25x^2$ **15.** $60x^3y^2$ **17.** $x(x+1)$ **19.** $2x+1$ **21.** $4(b-3)$ **23.** $p(p+1)(p-2)$ **25.** $(r+2)^2(r+1)(r-2)$

27. $-(x-4)$ **29.** $-2(x+3)(x-3)$ **31.** $-(x-1)(x+2)$ **33.** $\dfrac{12x^2}{3x^3}$ **35.** $\dfrac{3c+c^2}{a^2b^2c^2}$ **37.** $\dfrac{x^2-2x-8}{(x-4)(x+4)}$ **39.** $\dfrac{9n^2-9n}{6(n-1)(n+1)}$ **41.** $\dfrac{4t^2-4t}{t-1}$ **43.** $\dfrac{6xy}{9y^2}$, $\dfrac{4}{9y^2}$

45. $\dfrac{6a+2}{4a^3}$, $\dfrac{4a^3-a^2}{4a^3}$ **47.** $\dfrac{2m+2}{m(m+1)}$, $\dfrac{3m}{m(m+1)}$ **49.** $\dfrac{2y^2-5y+2}{4y(2y-1)}$, $\dfrac{y^2}{4y(2y-1)}$ **51.** $\dfrac{x+1}{(x-1)(x+1)}$, $\dfrac{-2x}{(x-1)(x+1)}$

53. $\dfrac{4x}{(x+2)(x-2)}$, $\dfrac{2x-4}{(x+2)(x-2)}$ **55.** $\dfrac{3x^2+9x}{(x-7)(x+3)(x+1)}$, $\dfrac{5x+5}{(x-7)(x+3)(x+1)}$ **57.** $\dfrac{x^2-2x-3}{(x+3)(x-3)^2}$, $\dfrac{x^2+5x+6}{(x+3)(x-3)^2}$

59. $\dfrac{6x-3}{(x+4)(2x-1)}$, $\dfrac{2x-1}{(x+4)(2x-1)}$ **61.** $3x$; $\dfrac{3}{3x}$, $\dfrac{1}{3x}$ **63.** $(r-4)(r+4)$; $\dfrac{12r+48}{(r-4)(r+4)}$, $\dfrac{12r-48}{(r-4)(r+4)}$ **65.** $(x+2)(x-2)(x^2-2x+4)(x^2+2x+4)$

67. $p^2(p-1)(p-3)$ **69.** $4a^2b^2(a+b)(a-b)(a^2+ab+b^2)$ **71.** To write an equivalent rational expression with a given denominator: (1) Write each

denominator in factored form. (2) Determine the "missing factor(s)." (Answer the question "What factor(s) does the new denominator have that is

missing from the original denominator?") (3) Multiply the original expression by $1=\dfrac{\text{missing factor}}{\text{missing factor}}$. (4) Find the product and leave the denominator

in factored form.

Section 7.5 Adding and Subtracting Rational Expressions with Unlike Denominators **1.** 72 **2.** $\dfrac{25}{36}$ **3.** $\dfrac{7}{30}$ **4.** least common denominator

5. True **6.** $\dfrac{3b+10a^2}{24a^3b^2}$ **7.** $\dfrac{6x^2+35y}{90x^3y^2}$ **8.** $\dfrac{8x-2}{(x-4)(x+2)}$ **9.** $\dfrac{3n-13}{(n-3)(n+1)}$ **10.** $\dfrac{1}{z+2}$ **11.** $\dfrac{2}{(x-5)(x+5)}$ **12.** $\dfrac{-16ab-15}{20a^2b^3}$ **13.** $\dfrac{2(x-10)}{x(x-4)}$

14. $\dfrac{x-11}{(x-5)(x+4)(x-1)}$ **15.** $\dfrac{(x+8)(x-1)}{4x(x-3)}$ **16.** True **17.** $\dfrac{p-5}{p-1}$ **18.** $\dfrac{y+4}{y+1}$ **19.** $\dfrac{2x+1}{x-1}$ **20.** $\dfrac{-3(x+1)}{x+3}$ **21.** $\dfrac{4x}{(x-1)(x+1)}$ **22.** $\dfrac{2}{x-2}$ **23.** $-\dfrac{5}{6}$

25. $\dfrac{5}{3x}$ **27.** $\dfrac{2a^2+7a-3}{(2a-1)(2a+1)}$ **29.** $-\dfrac{2}{x-4}$ **31.** $\dfrac{a+4}{4a}$ **33.** $\dfrac{5}{x}$ **35.** $\dfrac{8}{75}$ **37.** $\dfrac{(m-4)(m+4)}{m}$ **39.** $\dfrac{2(4x-3)}{(x-3)(x+3)}$ **41.** $\dfrac{2}{x+3}$ **43.** $\dfrac{x}{x-2}$ **45.** $\dfrac{20-9y}{12y^2}$

47. $\dfrac{6}{5x}$ **49.** $-\dfrac{3}{2x+3}$ **51.** $\dfrac{2(n^2-2)}{n(n-2)}$ **53.** $\dfrac{2x^2-5x+15}{x(x-3)}$ **55.** $\dfrac{y-2}{y-3}$ **57.** $\dfrac{2n+5}{2n+1}$ **59.** $\dfrac{4x-5}{x-2}$ **61.** $\dfrac{3x-28}{x-9}$ **63.** $\dfrac{1}{(x+1)(x-1)}$ **65.** $-\dfrac{13}{2a(a-2)}$

67. $-\dfrac{n^2 - n - 3}{(n + 2)(n - 2)(n + 3)}$ **69.** $\dfrac{-1}{(m - 2)(m - 4)}$ **71.** $\dfrac{-3n + 5}{(n + 3)(n - 2)}$ **73.** $\dfrac{x}{x - 2}$ **75.** $\dfrac{x - 4}{x}$ **77.** $\dfrac{6}{m(m - 2)}$ **79.** $\dfrac{x^2 + 4}{(x + 2)(x + 3)}$ **81.** $\dfrac{x + 3}{(x + 1)(x - 1)}$

83. $\dfrac{2x^2 - x + 10}{(x + 3)(x - 2)(x + 2)}$ **85.** $\dfrac{2x^2 - 5x + 3}{(x + 6)(x - 7)}$ **87.** $\dfrac{7x - 5}{4}$ units **89.** $\dfrac{a + 2}{a(a - 1)^2}$ **91.** $\dfrac{x^3 - 2x^2 + 4x + 3}{x^2(x + 1)(x - 1)}$ **93.** $\dfrac{1}{a + b}$ **95.** $\dfrac{x^2 + 2x + 5}{(x - 4)(x + 1)}$

97. $\dfrac{-x^3 + 6x^2 - 11x + 10}{x - 2}$ **99.** The steps to add or subtract rational expressions with unlike denominators are (1) find the least common denominator; (2) rewrite each rational expression with the common denominator; (3) add or subtract the rational expressions from Step 2; (4) simplify the result.

101. To complete the subtraction problem $\dfrac{3x + 1}{(2x + 5)(x - 1)} - \dfrac{x - 4}{(2x + 5)(x - 1)}$, distribute the subtraction sign to both the x and the -4 in the numerator of the second term to form $\dfrac{3x + 1 - x + 4}{(2x + 5)(x - 1)} = \dfrac{2x + 5}{(2x + 5)(x - 1)}$. Then divide out the common factor $(2x + 5)$ to arrive at the result $\dfrac{1}{x - 1}$.

Section 7.6 Complex Rational Expressions
1. complex rational expression **2.** simplify **3.** $\dfrac{k + 1}{2}$ **4.** $\dfrac{1}{2n}$ **5.** False **6.** $\dfrac{3}{2}$ **7.** $\dfrac{1}{3 + y}$ **8.** $\dfrac{3}{p + 1}$ **9.** 13

10. $\dfrac{y + 2x}{2y - x}$ **11.** $\dfrac{2}{17}$ **13.** $\dfrac{3(x - 2)}{2}$ **15.** $\dfrac{9}{x - 3}$ **17.** $\dfrac{n(m + 2n)}{2m}$ **19.** $\dfrac{5b^2 + 4a}{a(5b + 4)}$ **21.** $-\dfrac{2}{y + 4}$ **23.** $\dfrac{3(y - 24)}{2(5 - 6y^2)}$ **25.** $\dfrac{15}{16}$ **27.** $\dfrac{1}{2}$ **29.** $\dfrac{b - 7}{b}$ **31.** $\dfrac{x + 4}{x}$ **33.** $-\dfrac{x + y}{x^2y^2}$

35. $\dfrac{ab + 1}{ab + 2}$ **37.** $\dfrac{18n}{5}$ **39.** $\dfrac{x^2 + 2xy - y^2}{xy}$ **41.** $-\dfrac{1}{2x + 7}$ **43.** $-b$ **45.** $\dfrac{x + 1}{x + 3}$ **47.** $\dfrac{2b - a}{2b}$ **49.** $\dfrac{-4x + 17}{2x - 1}$ **51.** $\dfrac{n + 1}{5n - 9}$ **53.** $\dfrac{11(2n + 3)}{72}$ **55.** $-\dfrac{x^3}{x + 3}$

57. (a) $R = \dfrac{R_1 R_2}{R_2 + R_1}$ **(b)** $\dfrac{15}{4}$ ohms **59.** $\dfrac{x}{1 + 3x}$ **61.** $\dfrac{x}{2 - 5x}$ **63.** $\dfrac{2x + 1}{x + 1}$ **65.** $\dfrac{5 - 2x}{2 - x}$ **67.** The expression $\dfrac{\dfrac{2}{x + 1}}{\dfrac{1}{x - 5}}$ is not in simplified form. Simplified form is the form $\dfrac{p}{q}$, where p and q are polynomials that have no common factors. **69.** Answers may vary.

Putting the Concepts Together (Sections 7.1–7.6)
1. -35 **2. (a)** The expression is undefined for $a = 6$. **(b)** The expression is undefined for $y = 0$ and $y = -4$. **3. (a)** $\dfrac{a - 4b}{2}$ **(b)** $-\dfrac{x + 2}{x + 1}$ **4.** $8(a + 2b)(a - 4b)$ **5.** $\dfrac{35x}{5x^2(3x - 1)}$ **6.** $\dfrac{-2y^2}{y + 1}$ **7.** $\dfrac{2(m - 1)^2}{m^2}$ **8.** $\dfrac{2(y - 1)}{5(y - 3)}$

9. $4x + 1$ **10.** $\dfrac{1}{x + 1}$ **11.** $\dfrac{1}{2x - 1}$ **12.** $\dfrac{15m + 17}{(m + 3)(3m + 2)}$ **13.** $\dfrac{m}{(m + 5)(m + 4)}$ **14.** $-\dfrac{1}{m - 1}$ **15.** $\dfrac{a(a + 24)}{3(a^2 + 6)}$

Section 7.7 Rational Equations
1. rational equation **2.** $\left\{\dfrac{2}{3}\right\}$ **3.** $\{-18\}$ **4.** $\{-6\}$ **5.** $\{4\}$ **6.** $\{-5, 6\}$ **7.** extraneous solutions

8. $\{5\}$ **9.** True **10.** $\{\ \}$ or $\varnothing$ **11.** After 2.5 hours and after 10 hours **12.** $x = \dfrac{4g}{R}$ **13.** $r = 1 - \dfrac{a}{S}$ or $r = \dfrac{S - a}{S}$ **14.** $p = \dfrac{fq}{q - f}$

15. $\{2\}$ **17.** $\left\{-\dfrac{3}{4}\right\}$ **19.** $\{-7\}$ **21.** $\left\{\dfrac{1}{2}\right\}$ **23.** $\{7\}$ **25.** $\{\ \}$ or $\varnothing$

27. $\{-3\}$ **29.** $\left\{\dfrac{3}{4}\right\}$ **31.** $\left\{-\dfrac{5}{6}\right\}$ **33.** $\{-4\}$ **35.** $\{-3, -2\}$

37. $\{-1\}$ **39.** $\left\{-\dfrac{2}{3}, \dfrac{1}{2}\right\}$ **41.** $\{\ \}$ or $\varnothing$ **43.** $\{-3, 4\}$ **45.** $\left\{-\dfrac{1}{2}\right\}$

47. $y = \dfrac{2}{x}$ **49.** $R = \dfrac{E}{I}$ **51.** $b = \dfrac{2A}{h} - B$ or $b = \dfrac{2A - Bh}{h}$

53. $y = \dfrac{x}{z} - 3$ or $y = \dfrac{x - 3z}{z}$ **55.** $S = \dfrac{RT}{T - R}$ **57.** $y = \dfrac{an - p}{am}$

59. $x = \dfrac{Ay}{y - A}$ **61.** $y = \dfrac{xz}{2z - 6x}$ **63.** $x = \dfrac{cy}{y - 1}$ **65.** $\dfrac{4x + 5}{x(x + 5)}$

67. $\{-2, 3\}$ **69.** $\dfrac{x + 7}{2}$ **71.** $\left\{-\dfrac{1}{6}, 1\right\}$ **73.** $\{-1, 5\}$ **75.** $\dfrac{x^3 - x^2 + 4}{x(x + 2)(x - 2)}$

77. $\dfrac{(x - 3)(x - 1)}{3x}$ **79.** $\dfrac{5}{a}$ **81.** $\{\ \}$ or $\varnothing$ **83.** 2 **85.** $\{2\}$ **87.** $\{-3\}$

89. After 1 hour and after 9 hours **91.** 80% **93.** $k = 33$ **95.** The error occurred when the student incorrectly multiplied $-2[(x - 1)(x + 1)]$ on the right side of the equation. The correct solution is

$$\dfrac{2}{x - 1} - \dfrac{4}{x^2 - 1} = -2$$

$$\dfrac{2}{x - 1} - \dfrac{4}{(x - 1)(x + 1)} = -2$$

$$[(x - 1)(x + 1)]\left(\dfrac{2}{x - 1} - \dfrac{4}{(x - 1)(x + 1)}\right) = -2[(x - 1)(x + 1)]$$

$$2(x + 1) - 4 = -2(x - 1)(x + 1)$$

$$2x + 2 - 4 = -2(x^2 - 1)$$

$$2x - 2 = -2x^2 + 2$$

$$2x^2 + 2x - 4 = 0$$

$$x^2 + x - 2 = 0$$

$$(x + 2)(x - 1) = 0$$

$$x + 2 = 0 \text{ or } x - 1 = 0$$

$$x = -2 \qquad x = 1$$

$x = 1$ is extraneous, so the solution set is $\{-2\}$.

97. To *simplify* a rational expression means to add, subtract, multiply, or divide the rational expressions and express the result in a form in which there are no common factors in the numerator or denominator. To *solve* an equation means to find the value(s) of the variable that satisfy the equation. Answers will vary.

Section 7.8 Models Involving Rational Equations 1. proportion 2. $d;c$ 3. $\left\{-\dfrac{2}{3}\right\}$ 4. $\{3\}$ 5. \$233.38 6. 210 miles 7. similar 8. $XY = 8$ 9. 60 feet

10. constant; $\dfrac{1}{4}$ 11. 2 hours 12. $3\dfrac{1}{3}$ hours or 3 hours, 20 minutes 13. 20 mph 14. 6 mph 15. $\{12\}$ 17. $\left\{\dfrac{18}{7}\right\}$ 19. $\{16\}$ 21. $\left\{\dfrac{24}{5}\right\}$ 23. $\left\{-\dfrac{4}{5}\right\}$ 25. $\{-2\}$

27. $\left\{\dfrac{5}{7}\right\}$ 29. $\{-3, 2\}$ 31. $\{2, 5\}$ 33. $\{11\}$ 35. $XZ = 12$ 37. $n = 3; ZY = 5$ 39. $x = 30; AB = 28$ 41. $2x$ 43. $t - 3$ 45. $\dfrac{1}{5} + \dfrac{1}{3} = \dfrac{1}{t}$ 47. $\dfrac{1}{b + 4} + \dfrac{1}{b} = \dfrac{1}{7}$

49. $r - 2$ 51. $r - 48$ 53. $\dfrac{4}{14 - c} = \dfrac{7}{14 + c}$ 55. $\dfrac{12}{r + 3} = \dfrac{8}{r - 3}$ 57. 145 miles 59. $8\dfrac{1}{3}$ lb 61. 84, 696 rubles 63. 24 ft 65. 1.875 hr 67. $\dfrac{24}{7} \approx 3.4$ minutes

69. $\dfrac{36}{7} \approx 5.1$ hr 71. 15 hr 73. 24 hr 75. 4 km per hour 77. $\dfrac{5}{3}$ hour or 1 hour, 40 minutes 79. 40 mph 81. 35 mph 83. $x = 6$ 85. $x = \dfrac{98}{9}$ 87. $x = \dfrac{25}{6}$

89. The time component is not correctly placed. The times in the table should be $\dfrac{10}{r + 1}$ and $\dfrac{10}{r - 1}$. The equation will incorporate the 5 hours: time traveled downstream + time traveled upstream = 5 hours. 91. One equation to solve the problem is $\dfrac{1}{8} + \dfrac{1}{6} = \dfrac{1}{t}$, where t is the number of hours required to complete the job together. Answers may vary.

Chapter 7 Review 1. (a) $\dfrac{1}{2}$ (b) -4 (c) 1 2. (a) 0 (b) -1 (c) 1 3. (a) 0 (b) 9 (c) 49 4. (a) -1 (b) $-\dfrac{2}{3}$ (c) 42 5. $\dfrac{7}{3}$ 6. $\dfrac{1}{2}$ 7. -5 or 5 8. none

9. -10 or -2 10. -1 or 4 11. $\dfrac{y + 2}{5}$ 12. $\dfrac{x(x - 4)}{5}$ 13. $\dfrac{3}{k + 2}$ 14. $-\dfrac{2}{x - 3}$ 15. $\dfrac{x + 5}{2x - 1}$ 16. $\dfrac{3x - 1}{2(x^2 - 2x + 4)}$ 17. $\dfrac{2m^2}{n^2}$ 18. $\dfrac{y^3}{6x}$ 19. $\dfrac{14}{9n^2}$ 20. $\dfrac{1}{ab^7}$

21. $-\dfrac{5(x - 2)}{x + 3}$ 22. $-\dfrac{4(x - 3)(x + 3)}{(x - 9)^2}$ 23. $\dfrac{x - 2}{4(x - 3)}$ 24. $15x^3$ 25. $\dfrac{x + 5}{x + 6}$ 26. $\dfrac{y + 1}{y - 3}$ 27. $\dfrac{x^2}{(x + 1)(x + 9)}$ 28. $\dfrac{y - 2}{3y + 1}$ 29. $\dfrac{9}{x - 3}$ 30. $\dfrac{10}{x + 4}$ 31. m 32. $\dfrac{1}{m}$

33. $\dfrac{1}{m - 2}$ 34. $2m + 3$ 35. $\dfrac{1}{3m}$ 36. $\dfrac{2}{b}$ 37. $y + 7$ 38. $\dfrac{1}{m - 6}$ 39. $-\dfrac{7}{x - 4}$ 40. $\dfrac{4}{a - b}$ 41. $\dfrac{3x - 5}{(x - 5)(x + 5)}$ 42. $\dfrac{x + 4}{x - 3}$ 43. $12x^4y^7$ 44. $60a^3b^4c^7$ 45. $8a(a + 2)$

46. $5a(a + 6)$ 47. $4(x - 3)(x + 1)$ 48. $x(x - 7)(x + 7)$ 49. $\dfrac{6xy^6}{x^4y^7}$ 50. $\dfrac{11a^4b^3}{a^7b^5}$ 51. $\dfrac{x^2 + x - 2}{(x - 2)(x + 2)}$ 52. $\dfrac{m^2 - 4}{(m + 7)(m - 2)}$ 53. $\dfrac{20x^2y}{24x^5}; \dfrac{21}{24x^5}$ 54. $\dfrac{12}{10a^3}; \dfrac{11a^2b}{10a^3}$

55. $\dfrac{4x^2 + 8x}{(x - 2)(x + 2)}; \dfrac{-6}{(x - 2)(x + 2)}$ 56. $\dfrac{3m + 9}{(m - 5)(m + 3)}; \dfrac{2m}{(m - 5)(m + 3)}$ 57. $\dfrac{2m + 4}{(m + 7)(m - 2)(m + 2)}; \dfrac{m^2 - m - 2}{(m + 7)(m - 2)(m + 2)}$

58. $\dfrac{n^2 + 3n - 10}{n(n - 5)(n + 5)}; \dfrac{n^2}{n(n - 5)(n + 5)}$ 59. $\dfrac{12 + 8x}{9x^2}$ 60. $\dfrac{5y + x}{10x^2y^2}$ 61. $\dfrac{x^2 - 5x + 14}{(x + 7)(x - 7)}$ 62. $\dfrac{2x^2 + 13x - 9}{(2x + 3)(2x - 3)}$ 63. $\dfrac{-2(2x + 5)}{x(x - 2)}$ 64. $\dfrac{p - 7}{p - 4}$ 65. $\dfrac{3x - 1}{4x + 1}$

66. $\dfrac{3x - 4}{(x - 1)(x - 2)}$ 67. $\dfrac{m - 4}{m - 3}$ 68. $-\dfrac{1}{(y + 1)(y + 2)}$ 69. $\dfrac{3}{m + 2}$ 70. $\dfrac{m^2 + n^2}{(m - n)(m + n)}$ 71. $\dfrac{5x + 12}{x + 3}$ 72. $\dfrac{7x + 13}{x + 2}$ 73. $-\dfrac{3}{23}$ 74. $\dfrac{18}{11}$ 75. $\dfrac{2m^2 - 10m}{m^2 + 10}$

76. $\dfrac{6 + 4m^2}{m(6 - 5m)}$ 77. $\dfrac{1}{6}$ 78. $-\dfrac{y + 8}{y - 2}$ 79. $\dfrac{2(m + 1)}{m + 4}$ 80. $-a$ 81. $x = -5$ or $x = 0$ 82. $x = -6$ or $x = 6$ 83. $\{-4\}$ 84. $\left\{\dfrac{2}{3}\right\}$ 85. $\{5\}$ 86. $\left\{-4, -\dfrac{1}{3}\right\}$ 87. $\{38\}$

88. $\left\{\dfrac{9}{7}\right\}$ 89. $\{ \}$ or $\varnothing$ 90. $\left\{-\dfrac{5}{2}\right\}$ 91. $\{6\}$ 92. $\{0\}$ 93. $k = \dfrac{4}{y}$ 94. $y = \dfrac{x}{6}$ 95. $y = \dfrac{xz}{x - z}$ 96. $z = \dfrac{xy}{x + y}$ 97. $\{4\}$ 98. $\{5\}$ 99. $\{1\}$ 100. $\{3\}$ 101. $x = 15$.

102. $x = 3.5$ 103. 8 tanks 104. \$26.25 105. 1.2 hours 106. $\dfrac{15}{8} \approx 1.9$ hours 107. 15 min 108. $\dfrac{28}{3} \approx 9.3$ hours 109. 9 mph 110. 29 mph 111. 70 mph

112. 45 mph

Chapter 7 Test 1. 2 2. $x = 5$ or $x = -2$ 3. $-\dfrac{x + 3}{2}$ 4. $\dfrac{14}{5x^2}$ 5. $\dfrac{25}{9}$ 6. $\dfrac{x - y}{2x - 3y}$ 7. y 8. $\dfrac{x - 5}{x + 3}$ 9. $-\dfrac{1}{y - z}$ or $\dfrac{1}{z - y}$ 10. $\dfrac{2(x + 3)(x - 1)}{(x - 2)(2x + 1)}$

11. $\dfrac{x^2 - 5x - 2}{(x + 2)(x + 3)(x - 1)}$ 12. $\dfrac{y - 3}{3y}$ 13. $\{-20, 5\}$ 14. $\{2\}$ 15. $y = \dfrac{xz}{x - z}$ 16. $\{5\}$ 17. $x = \dfrac{24}{5}$ 18. \$33 19. 36 min 20. 4 mph

Cumulative Review Chapters 1-7 1. -144 2. $9x - 11$ 3. $\left\{\dfrac{19}{6}\right\}$ 4. all real numbers 5. $\{4\}$ 6. 7 feet 7. 44, 46, 48 8. $x \geq 1$;

9. $9x^2 - 12xy + 4y^2$ 10. $\dfrac{16}{a^{16}b^{12}}$ 11. $\dfrac{27z^9}{y^{12}}$ 12. $2b(4a - 7)(a + 6)$ 13. $\left\{0, \dfrac{1}{6}, 5\right\}$ 14. 10 cm 15. 0 16. $-\dfrac{3(x + 1)}{5x(2x - 3)}$ 17. $\dfrac{(x - 2)(x + 1)}{4(x + 2)}$ 18. $\dfrac{1}{x - 6}$

19. $\dfrac{17x + 30}{2(x - 2)(x + 2)}$ 20. $\dfrac{4(10 - 3x)}{85x}$ 21. $\{ \}$ or $\varnothing$ 22. $\{-16\}$ 23. 10 inches 24. 9 hr 25. 7 mph 26. $\dfrac{4}{3}$ 27. $y = -\dfrac{3}{2}x + 5$ or $3x + 2y = 10$

28. 29. $-\dfrac{3}{5}; \dfrac{5}{3}$ 30. $(-4, 3)$

Getting Ready for Chapter 8: Interval Notation

1. closed interval   **2.** left endpoint; right endpoint   **3.** $[-3, 2]$ **4.** $[3, 6)$

5. $(-\infty, 3]$ **6.** $\left(\dfrac{1}{2}, \dfrac{7}{2}\right)$ **7.** $0 < x \le 5$

8. $-6 < x < 0$ **9.** $x > 5$ **10.** $x \le \dfrac{8}{3}$

11. $[2, 10]$;   **13.** $[-4, 0)$;   **15.** $[6, \infty)$;

17. $\left(-\infty, \dfrac{3}{2}\right)$;   **19.** $1 < x < 8$;

21. $-5 < x \le 1$;   **23.** $x < 5$;   **25.** $x \ge 3$;

Chapter 8 Graphs, Relations, and Functions

Section 8.1 Graphs of Equations   **1.** origin   **2.** True

3. A: Quadrant I;
B: Quadrant IV;
C: y-axis;
D: Quadrant III

4. A: Quadrant II;
B: x-axis;
C: Quadrant IV;
D: Quadrant I

5. True   **6.** (a) No (b) Yes (c) Yes

7. (a) Yes (b) No (c) Yes

8. **9.** **10.** **11.** **12.**

13. intercepts   **14.** False
15. Intercepts: $(-5, 0)$, $(0, -0.9)$, $(1, 0)$, $(6.7, 0)$; x-intercepts: $(-5, 0)$, $(1, 0)$, $(6.7, 0)$ y-intercept: $(0, -0.9)$

16. (a) \$200 thousand **(b)** \$350 thousand **(c)** The capacity of the refinery is 700 thousand gallons of gasoline per hour. **(d)** The intercept is $(0, 100)$. The cost of \$100 thousand for refining 0 gallons of gasoline can be thought of as fixed costs. **17.** A: $(2, 3)$ I; B: $(-5, 2)$ II; C: $(0, -2)$ y-axis; D: $(-4, -3)$ III; E: $(3, -4)$ IV; F: $(4, 0)$ x-axis

19. A: quadrant I;
B: quadrant III;
C: x-axis;
D: quadrant IV;
E: y-axis;
F: quadrant II

21. (a) yes **(b)** no **(c)** yes **(d)** yes **23. (a)** yes **(b)** no **(c)** no **(d)** yes

25. (a) no **(b)** yes **(c)** yes **(d)** yes

27. $y = 4x$ **29.** $y = -\dfrac{1}{2}x$ **31.** $y = x + 3$ **33.** $y = -3x + 1$ **35.** $y = \dfrac{1}{2}x - 4$

37. $2x + y = 7$ **39.** $y = -x^2$ **41.** $y = 2x^2 - 8$ **43.** $y = |x|$ **45.** $y = |x - 1|$ **47.** $y = x^3$ **49.** $y = x^3 + 1$

51. $x^2 - y = 4$ **53.** $x = y^2 - 1$ **55.** $(-2, 0)$ and $(0, 3)$
57. $(-2, 0)$, $(1, 0)$, and $(0, -4)$
59. $a = \dfrac{7}{4}$ **61.** $b = 4$ **63. (a)** 400 ft^2 **(b)** 25 feet; 625 ft^2 **(c)** The x-intercepts are $x = 0$ and $x = 50$. These values form the bounds for the width of the opening. The y-intercept is $y = 0$. The area of the opening will be 0 ft^2 when the width is 0 feet. **65. (a)** \$39.99; \$39.99 **(b)** \$3437.49 **(c)** $(0, 39.99)$; The monthly cost will be \$39.99 if no minutes are used.

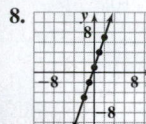
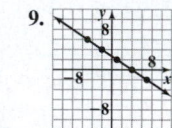
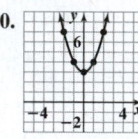
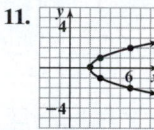
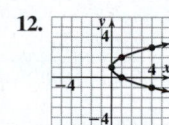
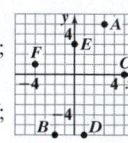
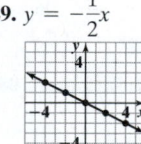
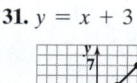
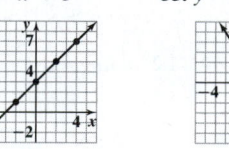
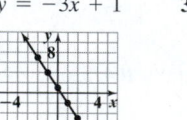
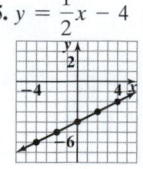
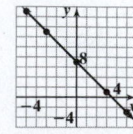
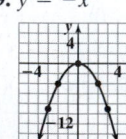
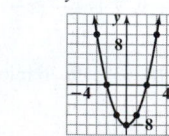
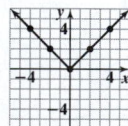
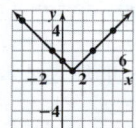
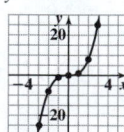
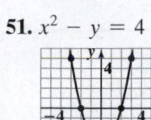

67. Vertical line with an x-intercept of 4.

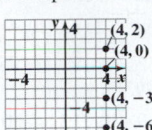

69. Answers will vary. One possible graph is shown.

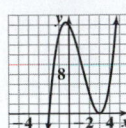

71. Answers will vary. One possibility: $y = 0$

73. A complete graph is one that shows enough of the graph so that anyone who is looking at it will visualize the rest of it as an obvious continuation of what is shown. A complete graph should show all the interesting features of the graph, such as intercepts and high/low points.

75. The point-plotting method of graphing an equation requires one to choose certain values of one variable and use the equation to find the corresponding values of the other variable. These points are then plotted and connected in a smooth curve.

77. $y = 3x - 9$

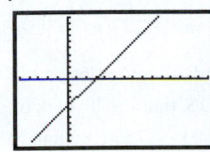

79. $y = -x^2 + 8$

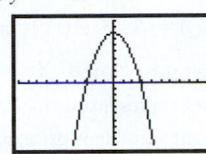

81. $y + 2x^2 = 13$

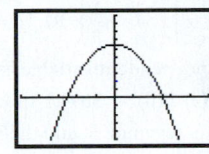

83. $y = x^3 - 6x + 1$

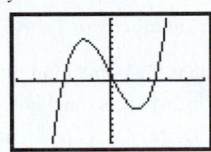

Section 8.2 Relations

1. corresponds; depends **2.** {(Max, November 8), (Alesia, January 20), (Trent, March 3), (Yolanda, November 8), (Wanda, July 6), (Elvis, January 8)} **3.**

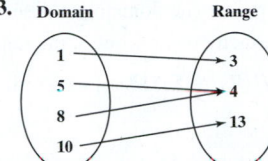

4. domain; range **5.** Domain: {Max, Alesia, Trent, Yolanda, Wanda, Elvis}; Range: {January 20, March 3, July 6, November 8, January 8} **6.** Domain: {1, 5, 8, 10}; Range: {3, 4, 13} **7.** Domain: {−2, −1, 2, 3, 4}; Range: {−3, −2, 0, 2, 3} **8.** True **9.** False **10.** Domain: {$x | -2 \le x \le 4$} or [−2, 4]; Range: {$y | -2 \le y \le 2$} or [−2, 2] **11.** Domain: {$x | x$ is a real number} or $(-\infty, \infty)$; Range: {$y | y$ is a real number} or $(-\infty, \infty)$

12.

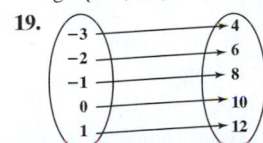

Domain: {$x | x$ is a real number} or $(-\infty, \infty)$; Range: {$y | y$ is a real number} or $(-\infty, \infty)$

13.

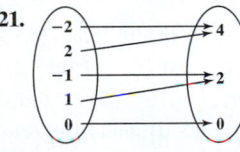

Domain: {$x | x$ is a real number} or $(-\infty, \infty)$; Range: {$y | y \ge -8$} or $[-8, \infty)$

14.

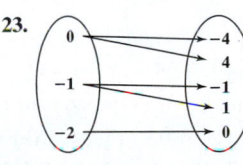

Domain: {$x | x \ge 1$} or $[1, \infty)$; Range: {$y | y$ is a real number} or $(-\infty, \infty)$

15. {(USA Today, 2.28.), (Wall Street Journal, 2.06), (New York Times, 1.1), (Los Angeles Times, 0.82), (Washington Post, 0.70)}; Domain: {USA Today, Wall Street Journal, New York Times, Los Angeles Times, Washington Post}; Range: {2.28, 2.06, 1.1, 0.82, 0.70} **17.** {(Less than 9th Grade, $20,791), (9th–12th Grade – no diploma, $23,234), (High School Graduate, $32,456), (Associate's Degree, $42,508), (Bachelor's Degree, $62,240)}; Domain: {Less than 9th Grade, 9th–12th Grade – no diploma, High School Graduate, Associate's Degree, Bachelor's Degree}; Range: {$20,791, $23,234, $32,456, $42,508, $62,240}

19. Domain: {−3, −2, −1, 0, 1}; Range: {4, 6, 8, 10, 12}

21. Domain: {−2, −1, 0, 1, 2}; Range: {0, 2, 4}

23. Domain: {−2, −1, 0}; Range: {−4, −1, 0, 1, 4}

25. Domain: {−3, −2, 0, 2, 3}; Range: {−3, −1, 2, 3}
27. Domain: {$x | -4 \le x \le 4$} or [−4, 4]; Range: {$y | -2 \le y \le 2$} or [−2, 2]
29. Domain: {$x | -1 \le x \le 3$} or [−1, 3]; Range: {$y | 0 \le y \le 4$} or [0, 4]
31. Domain: {$x | x$ is a real number} or $(-\infty, \infty)$; Range: {$y | y \ge -3$} or $[-3, \infty)$

33. Domain: {$x | x$ is a real number} or $(-\infty, \infty)$; Range: {$y | y$ is a real number} or $(-\infty, \infty)$ **35.** Domain: {$x | x$ is a real number} or $(-\infty, \infty)$; Range: {$y | y$ is a real number} or $(-\infty, \infty)$ **37.** Domain: {$x | x$ is a real number} or $(-\infty, \infty)$; Range: {$y | y$ is a real number} or $(-\infty, \infty)$ **39.** Domain: {$x | x$ is a real number} or $(-\infty, \infty)$; Range: {$y | y \le 0$} or $(-\infty, 0]$ **41.** Domain: {$x | x$ is a real number} or $(-\infty, \infty)$; Range: {$y | y \ge -8$} or $[-8, \infty)$ **43.** Domain: {$x | x$ is a real number} or $(-\infty, \infty)$; Range: {$y | y \ge 0$} or $[0, \infty)$ **45.** Domain: {$x | x$ is a real number} or $(-\infty, \infty)$; Range: {$y | y \ge 0$} or $[0, \infty)$ **47.** Domain: {$x | x$ is a real number} or $(-\infty, \infty)$; Range: {$y | y$ is a real number} or $(-\infty, \infty)$ **49.** Domain: {$x | x$ is a real number} or $(-\infty, \infty)$; Range: {$y | y$ is a real number} or $(-\infty, \infty)$ **51.** Domain: {$x | x$ is a real number} or $(-\infty, \infty)$; Range: {$y | y \ge -4$} or $[-4, \infty)$ **53.** Domain: {$x | x \ge -1$} or $[-1, \infty)$; Range: {$y | y$ is a real number} or $(-\infty, \infty)$

55. **(a)** Domain: {$x | 0 < x < 50$} or (0, 50); Range: {$y | 0 < y \le 625$} or (0, 625] **(b)** Width must be less than $\frac{1}{2}$ the perimeter.

57. **(a)** Domain: {$m | 0 \le m \le 15{,}120$} or [0, 15,120]; Range: {$C | 100 \le C \le 3380$} or [100, 3380] **(b)** $21 \cdot 12 \cdot 60 = 15{,}120$ minutes

59. Actual graphs will vary but all should be horizontal lines. **61.** A relation is a correspondence between two sets called the domain and range. The domain is the set of all inputs, and the range is the set of all outputs.

Section 8.3 An Introduction to Functions

1. function **2.** false **3.** Function; Domain: {Max, Alesia, Trent, Yolanda, Wanda, Elvis}; Range: {January 20, March 3, July 6, November 8, January 8} **4.** Not a function **5.** Function; Domain: {−3, −2, −1, 0, 1}; Range: {0, 1, 2, 3} **6.** Not a function **7.** Function **8.** Not a function **9.** Function **10.** True **11.** Function **12.** Not a function **13.** 14 **14.** −4 **15.** −18 **16.** 2 **17.** dependent; independent; argument **18.** $2x - 9$

19. $2x - 4$ **20.** domain **21.** $(-\infty, \infty)$ **22.** $\{x \mid x \neq 3\}$ **23.** $\{r \mid r > 0\}$ or $(0, \infty)$ **24. (a)** Independent variable: t; dependent variable: A
(b) $A(30) \approx 706.86$ square miles. After 30 days, the area contaminated with oil will be a circle covering about 706.86 square miles. **25.** Function;
Domain: $\{$ Virginia, Nevada, Arkansas, Tennessee, Texas $\}$; Range: $\{4, 9, 11, 36\}$ **27.** Not a function; Domain: $\{150, 174, 180\}$; Range: $\{118, 130, 140\}$
29. Function; Domain: $\{0, 1, 2, 3\}$; Range: $\{3, 4, 5, 6\}$ **31.** Function; Domain: $\{-3, 1, 4, 7\}$; Range: $\{5\}$ **33.** Not a function; Domain: $\{-10, -5, 0\}$;
Range: $\{1, 2, 3, 4\}$ **35.** Function **37.** Function **39.** Not a function **41.** Function **43.** Not a function **45.** Function **47.** Not a function **49.** Function
51. Function **53. (a)** $f(0) = 3$; **(b)** $f(3) = 9$ **(c)** $f(-2) = -1$ **55. (a)** $f(0) = 2$ **(b)** $f(3) = -13$ **(c)** $f(-2) = 12$ **57. (a)** $f(0) = 0$ **(b)** $f(3) = 0$
(c) $f(-2) = 10$ **59. (a)** $f(0) = 3$ **(b)** $f(3) = -3$ **(c)** $f(-2) = -3$ **61. (a)** $f(-x) = -2x - 5$ **(b)** $f(x + 2) = 2x - 1$ **(c)** $f(2x) = 4x - 5$
(d) $-f(x) = -2x + 5$ **(e)** $f(x + h) = 2x + 2h - 5$ **63. (a)** $f(-x) = 7 + 5x$ **(b)** $f(x + 2) = -3 - 5x$ **(c)** $f(2x) = 7 - 10x$ **(d)** $-f(x) = -7 + 5x$
(e) $f(x + h) = 7 - 5x - 5h$ **65.** $f(2) = 7$ **67.** $s(-2) = 16$ **69.** $F(-3) = 5$ **71.** $F(4) = -6$ **73.** $\{x \mid x$ is any real number $\}$ or $(-\infty, \infty)$

75. $\{z \mid z \neq 5\}$ **77.** $\{x \mid x$ is any real number $\}$ or $(-\infty, \infty)$ **79.** $\left\{x \mid x \neq -\dfrac{1}{3}\right\}$ **81.** $C = -6$ **83.** $A = 5$ **85.** $A(r) = \pi r^2$; 50.27 in.^{2} **87.** $G(h) = 15h$;

$375 **89. (a)** The dependent variable is the population, P, and the independent variable is the age, a. **(b)** $P(20) = 4433.3$ thousand; The population of U.S.
residents who were 20 years of age in 2014 was roughly 4,433,000. **(c)** $P(0) = 3694.1$ thousand; $P(0)$ represents residents of the U.S. that are 0 years of
age in 2014 was roughly 3,694,000. **91. (a)** The dependent variable is revenue, R, and the independent variable is price, p. **(b)** $R(50) = 7500$; Selling MP3
players for $50 will yield a daily revenue of $7500 for the company. **(c)** $R(120) = 9600$; Selling MP3 players for $120 will yield a daily revenue of $9600
for the company. **93.** $\{r \mid r > 0\}$ or $(0, \infty)$ **95.** $\{h \mid 0 \leq h \leq 60\}$ or $[0, 60]$ **97.** $\{p \mid 0 \leq p \leq 120\}$ or $[0, 120]$ **99. (a) (i)** -5 **(ii)** 1 **(iii)** 1 **(b) (i)** 13
(ii) 4 **(iii)** 4 **101.** Answers will vary. **103.** A function is a relation between two sets, the domain and the range. The domain is the set of all inputs to the
function, and the range is the set of all outputs. In a function, each input in the domain corresponds to exactly one output in the range. **105.** The four

forms of a function are map, ordered pairs, equation, and graph. **107.** $f(2) = 7$ **109.** $F(-3) = 5$ **111.** $H(7) = 5$ **113.** $F(4) = -0.65$ or $-\dfrac{13}{20}$

Section 8.4 Functions and Their Graphs **1.** graph; function **2.** 4; -7

3. 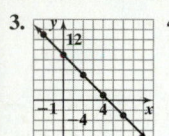 **4.** 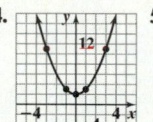 **5.** 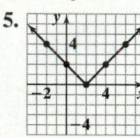 **6. (a)** Domain: $\{x \mid x$ is a real number $\}$; $(-\infty, \infty)$; Range: $\{y \mid y \leq 1\}$; $(-\infty, 1]$ **(b)** x-intercepts:
$(-2, 0)$ and $(2, 0)$; y-intercept: $(0, 1)$ **7.** $f(3) = 8$; $(-2, 4)$ **8. (a)** $f(-3) = -15$; $f(1) = -3$
(b) Domain: $\{x \mid x$ is a real number $\}$ or $(-\infty, \infty)$ **(c)** Range: $\{y \mid y$ is a real number $\}$ or
$(-\infty, \infty)$ **(d)** x-intercepts: $(-2, 0)$, $(0, 0)$ and $(2, 0)$; y-intercept: $(0, 0)$ **(e)** $\{3\}$

9. (a) No **(b)** $f(3) = -2$; $(3, -2)$ is on the graph **(c)** $x = 5$; $(5, -8)$ is on the graph **10.** yes **11.** no **12.** yes **13.** -2 and 2

14. 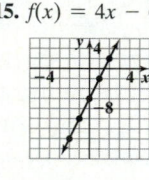 **15.** $f(x) = 4x - 6$ **17.** $h(x) = x^2 - 2$ **19.** $G(x) = |x - 1|$ **21.** $g(x) = x^3$

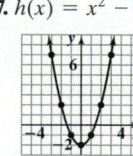 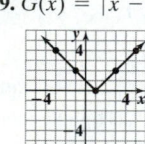 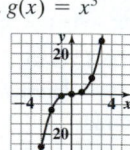

23. (a) Domain: $\{x \mid x$ is a real number $\}$ or $(-\infty, \infty)$; Range: $\{y \mid y$ is a real number $\}$ or $(-\infty, \infty)$ **(b)** $(0, 2)$ and $(1, 0)$ **(c)** 1 **25. (a)** Domain:
$\{x \mid x$ is a real number $\}$ or $(-\infty, \infty)$; Range: $\{y \mid y \geq -2.25\}$ or $[-2.25, \infty)$ **(b)** $(-2, 0)$, $(4, 0)$, and $(0, -2)$ **(c)** -2, 4 **27. (a)** Domain:
$\{x \mid x$ is a real number $\}$ or $(-\infty, \infty)$; Range: $\{y \mid y$ is a real number $\}$ or $(-\infty, \infty)$ **(b)** $(-3, 0)$, $(-1, 0)$, $(2, 0)$, and $(0, -3)$ **(c)** -3, -1, 2
29. (a) Domain: $\{x \mid x$ is a real number $\}$ or $(-\infty, \infty)$; Range: $\{y \mid y \geq 0\}$ or $[0, \infty)$ **(b)** $(-3, 0)$, $(3, 0)$, and $(0, 9)$ **(c)** -3, 3 **31. (a)** Domain:
$\{x \mid x \leq 4\}$ or $(-\infty, 4]$; Range: $\{y \mid y \leq 3\}$ or $(-\infty, 3]$ **(b)** $(-2, 0)$ and $(0, 2)$ **(c)** -2 **33. (a)** $f(-7) = -2$ **(b)** $f(-3) = 3$ **(c)** $f(6) = 2$
(d) negative **(e)** $\{-6, -1, 4\}$ **(f)** $\{x \mid -7 \leq x \leq 6\}$ or $[-7, 6]$ **(g)** $\{y \mid -2 \leq y \leq 3\}$ or $[-2, 3]$ **(h)** $(-6, 0)$, $(-1, 0)$ and $(4, 0)$ **(i)** $(0, -1)$
(j) $\{-7, 2\}$ **(k)** $x = -3$ **(l)** -6, -1, 4 **35. (a)** $F(-2) = 3$ **(b)** $F(3) = -6$ **(c)** $x = -1$ **(d)** $(-4, 0)$ **(e)** $(0, 2)$ **37. (a)** no **(b)** $f(3) = 3$; $(3, 3)$ **(c)** 4; $(4, 7)$
(d) no **39. (a)** yes **(b)** $g(6) = 1$; $(6, 1)$ **(c)** -12; $(-12, 10)$ **(d)** yes **41. (c)** **43. (e)** **45. (d)**

47. 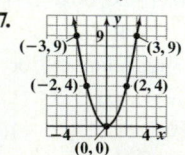 **49.** 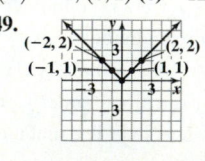 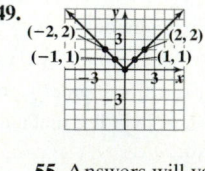 **51. (a)** III **(b)** I **(c)** IV **(d)** V **(e)** II

53. 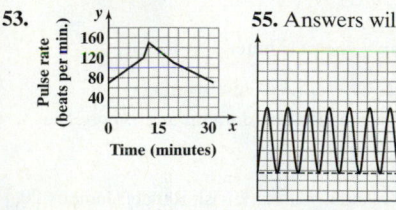 **55.** Answers will vary. **57.** The person's weight increases until age 30, then oscillates back and forth between
158 pounds and 178 pounds, then slowly levels off at about 150 pounds.
59. One possibility: Answers will vary.

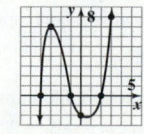 **61.** A function cannot have more than one output for a given input.
So, there cannot be two outputs for the input 0.
63. The range of a function is the set of all outputs of the elements
in the domain. **65.** The zeros of a function are the input values that
make the function equal to zero.

Putting the Concepts Together (Sections 8.1–8.4)

1. A: x-axis; B: quadrant II; **2. (1)** yes **(b)** yes **(c)** no **3.** $y = |x| + 3$ **4.** $y = \frac{1}{2}x^2 - 1$ **5.** $(-3, 0)(-1, 0), (0, -3), (2, 0)$;
C: quadrant I; D: y-axis;
E: quadrant III; F: quadrant IV.

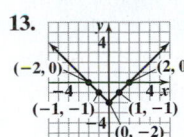

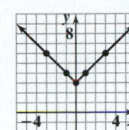

x-intercepts: $(-3, 0)(-1, 0), (2, 0)$;
y-intercept: $(0, -3)$

6. The relation is a function because each element in the domain corresponds to only one element in the range. $\{(-2, -1), (-1, 0), (0, 1), (1, 2), (2, 3)\}$

7. (a) Function **(b)** Not a function **8.** Yes; Domain: $\{-4, -1, 0, 3, 6\}$; Range: $\{-3, -2, 2, 6\}$ **9.** The graph passes the vertical line test. $f(5) = -6$

10. 4 **11. (a)** -17 **(b)** -34 **(c)** $-5x + 20$ **(d)** $-5x + 23$ **12. (a)** $\{h | h$ is any real number$\}$ or $(-\infty, \infty)$ **(b)** $\left\{w | w \neq -\frac{1}{3}\right\}$

13.

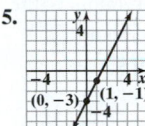

Domain: $\{x | x$ is any real number$\}$ or $(-\infty, \infty)$. Range: $\{y | y \geq -2\}$ or $[-2, \infty)$ **14. (a)** $h(2.5) = 80$; After 2.5 seconds, the
height of the ball is 80 feet. **(b)** $\{t | 0 \leq t \leq 3.8\}$ or $[0, 3.8]$ **(c)** $\{h | 0 \leq h \leq 105\}$ or $[0, 105]$ **(d)** 1.25 seconds
15. (a) no **(b)** -12; $(-2, -12)$ **(c)** -4; $(-4, -22)$ **(d)** yes

Section 8.5 Linear Functions and Models **1.** slope; y-intercept **2.** line **3.** False **4.** -2; $(0, 3)$

5.

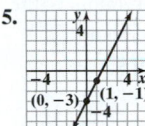

6.

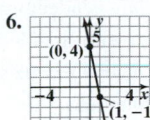

7.

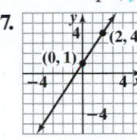

8.

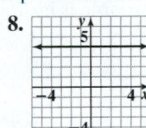

9. 5 **10.** -8 **11.** 12

12. (a) $\{x | x \geq 0\}$; $[0, \infty)$ **(b)** $40 **13. (a)** $C(x) = 81x + 2000$ **14. (a)** $C(x) = 0.18x + 250$ **15. (a)**
(c) $68 **(d)** 130 miles **(b)** $2405 **(c)** 10 bicycles **(b)** $[0, \infty)$ **(c)** $307.60
(e) **(d)** **(d)** 180 miles

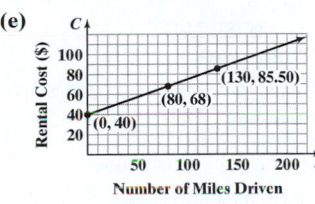

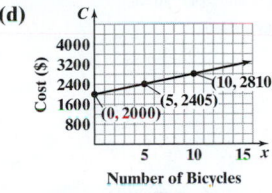

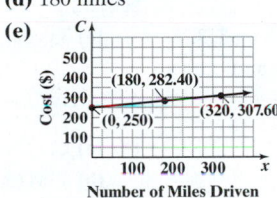

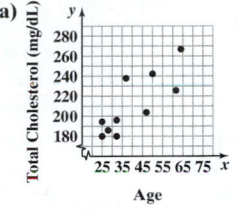

(e)

(f) You may drive between 0 and
250 miles.

(b) As age increases, total
cholesterol also increases.

16. nonlinear **17.** linear, positive slope **18. (a)** Answers will vary. Using $(25, 180)$ and $(65, 269)$: $y = f(x) = 2.225x + 124.375$

(b)

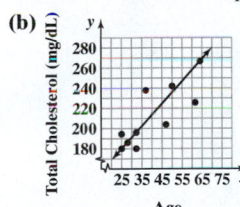

(c) 211 mg/dL **(d)** Each year, a male's total cholesterol increases by 2.225 mg/dL; No

19. $F(x) = 5x - 2$ **21.** $G(x) = -3x + 7$ **23.** $H(x) = -2$ **25.** $f(x) = \frac{1}{2}x - 4$

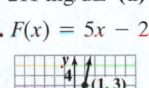

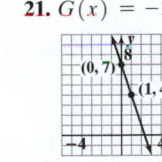

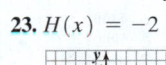

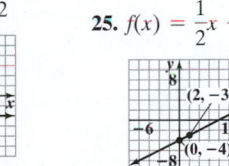

27. $F(x) = -\frac{5}{2}x + 5$ **29.** $G(x) = -\frac{3}{2}x$ **31.** -5 **33.** 8 **35.** 6 **37.** 9 **39.** nonlinear **41.** linear; positive slope

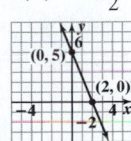

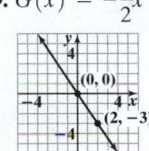

43. (a)

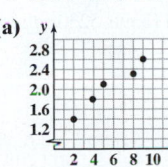

(b) Answers will vary. Using the
points $(4, 1.8)$ and $(9, 2.6)$, the
equation is $y = 0.16x + 1.16$.

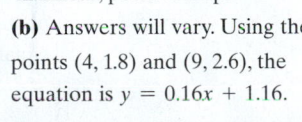

45. (a)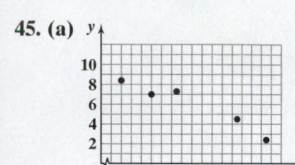

(b) Answers will vary. Using the points $(1.2, 8.4)$ and $(4.1, 2.4)$, the equation is $y = -2.07x + 10.88$.

(c)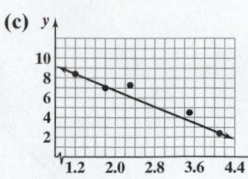

47. (a) 3 **(b)** $(0, 2)$ **(c)** $-\dfrac{2}{3}$ **(d)** 1; $(1, 5)$

(e) $\{x \mid x \le -1\}$; $(-\infty, -1]$

(f)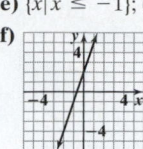

49. (a) $\{3\}$; -2; $(3, -2)$; $(3, -2)$

(b) $\{x \mid x > 3\}$; $(3, \infty)$

(c)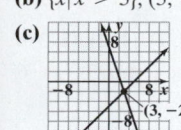

51. $f(x) = 2x + 2$; $f(-2) = -2$

53. $h(x) = -\dfrac{7}{4}x + \dfrac{49}{4}$; $h\left(\dfrac{1}{2}\right) = \dfrac{91}{8}$

55. (a) 3 **(b)** -1 **(c)** 2

(d) x-intercept: $(2, 0)$;

y-intercept: $(0, -2)$

(e) $f(x) = x - 2$

57. (a) $\{x \mid 9926 \le x \le 37{,}450\}$; $[9926, 37{,}450]$ **(b)** \$2538.60

(c) Independent variable: adjusted gross income; dependent variable: tax bill

(d) 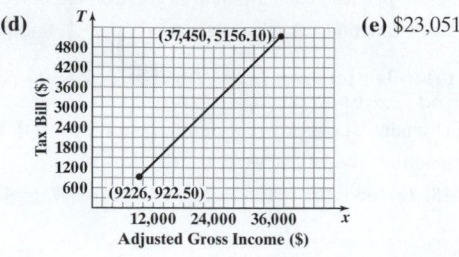 **(e)** \$23,051

59. (a) $\{m \mid m \ge 0\}$ or $[0, \infty)$

(b) 2; The base fare is \$2.00 before any distance is driven.

(c) \$9.50

(d)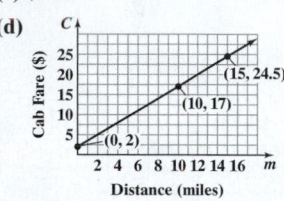

(e) A person can travel 7.5 miles in a cab for \$13.25.

(f) A person can ride between 0 miles and 25 miles.

61. (a) The independent variable is age; the dependent variable is insurance cost.

(b) $\{a \mid 15 \le a \le 90\}$ or $[15, 90]$

(c) \$566.50

(d)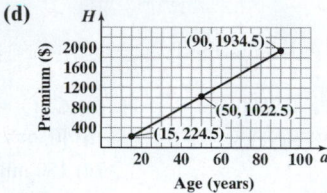

(e) 48 years

63. (a) $B(m) = 0.05m + 5.95$

(b) The independent variable is minutes; the dependent variable is bill amount.

(c) $\{m \mid m \ge 0\}$ or $[0, \infty)$ **(d)** \$20.95

(e) 240 minutes

(f)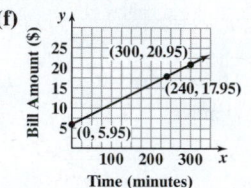

(g) You can talk between 0 minutes and 250 minutes.

65. (a) $V(x) = -900x + 2700$

(b) $\{x \mid 0 \le x \le 3\}$ or $[0, 3]$

(c) \$1800

(d) The V-intercept is $(0, 2700)$, and the x-intercept is $(3, 0)$.

(e) After two years

(f)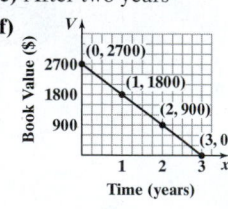

67. (a) $C(x) = 7250x - 1735$

(b) \$3847.50 **(c)** The cost of diamonds increases at a rate of \$7250 per carat.

(d) 0.97 carat

71. (a)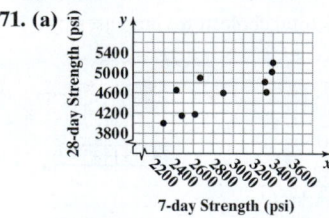

(b) Linear **(c)** Answers will vary. Using the points $(2300, 4070)$ and $(3390, 5220)$, the equation is $y = 1.06x + 1632$.

69. (a) $C(x) = 0.774x + 1543.978$

(b) \$12,551 billion

(c) If personal disposable income increase by \$1, personal consumption increase by 0.77.

(d) \$14.952 billion

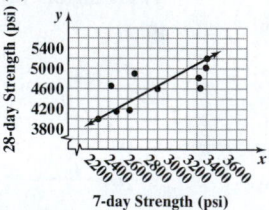

(e) 4823 psi

(f) If the 7-day strength is increased by 1 psi, then the 28-day strength will increase by 1.06 psi.

73. (a) No **(b)**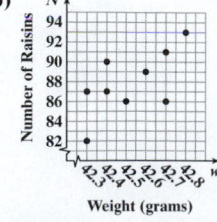

(c) Answers will vary. Using the points $(42.3, 82)$ and $(42.8, 93)$, the equation is $N = 22w - 848.6$.

(d)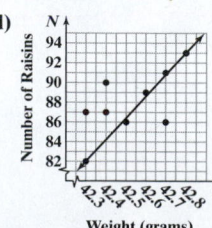

(e) $N(w) = 22w - 848.6$ **(f)** approximately 86 raisins

(g) If the weight increases by 1 gram, then the number of raisins increases by 22 raisins.

75. (a) 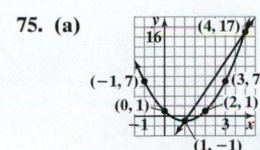 **(b)** 6 **(c)** $y = 6x - 7$

(d) 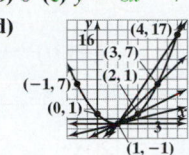 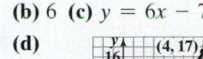 $x = 3$: slope = 4; $y = 4x - 5$; $x = 2$: slope = 2; $y = 2x - 3$; $x = 1.5$: slope = 1;

$y = x - 2$; $x = 1.1$: slope = 0.2; $y = 0.2x - 1.2$ **(e)** As x approaches 1, the slope decreases and gets closer to 0.

77. (a)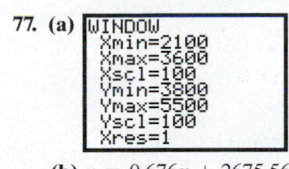

(b) $y = 0.676x + 2675.562$

79. (a)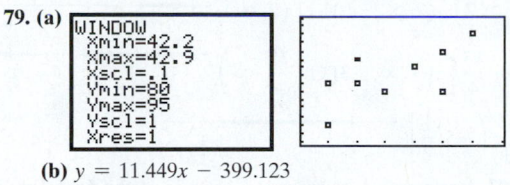

(b) $y = 11.449x - 399.123$

Section 8.6 Compound Inequalities

1. intersection **2.** and; or **3.** True **4.** False **5.** $\{1, 3, 5\}$ **6.** $\{2, 4, 6\}$ **7.** $\{1, 2, 3, 4, 5, 6, 7\}$

8. $\{1, 2, 3, 4, 5, 6, 8\}$ **9.** $\varnothing$ or $\{\ \}$ **10.** $\{1, 2, 3, 4, 5, 6, 7, 8\}$ **11.** $\{x \mid 2 < x < 7\}$; $(2, 7)$

12. $\{x \mid x \leq -3 \text{ or } x > 2\}$; $(-\infty, -3] \cup (2, \infty)$ **13.** $\{x \mid x \geq 2\}$; $[2, \infty)$

14. $\{x \mid -3 < x < 3\}$; $(-3, 3)$ **15.** $\{x \mid 1 < x < 3\}$; $(1, 3)$ **16.** $\{\ \}$ or $\varnothing$

17. $\{1\}$ **18.** $\{x \mid -1 < x < 3\}$; $(-1, 3)$

19. $\left\{ x \mid \dfrac{5}{4} < x \leq 2 \right\}$; $\left(\dfrac{5}{4}, 2 \right]$ **20.** $\{x \mid -6 \leq x \leq -2\}$; $[-6, -2]$

21. $\{x \mid x < -2 \text{ or } x > 5\}$; $(-\infty, -2) \cup (5, \infty)$

22. $\{x \mid x \leq 2 \text{ or } x > 6\}$; $(-\infty, 2] \cup (6, \infty)$ **23.** $\{x \mid x \geq 1\}$; $[1, \infty)$

24. $\{x \mid x < 4 \text{ or } x > 9\}$; $(-\infty, 4) \cup (9, \infty)$

25. $\{x \mid x \text{ is any real number}\}$; $(-\infty, \infty)$ **26.** $\{x \mid x \geq -1\}$; $[-1, \infty)$

27. Income between $37,450 and $90,750.

28. $[800, 1500]$. **29.** $\{1, 4, 5, 6, 7, 8, 9\}$ **31.** $\{5, 7, 9\}$ **33.** $\varnothing$ or $\{\ \}$

35. (a) $A \cap B = \{x \mid -2 < x \leq 5\}$; $(-2, 5]$

(b) $A \cup B = \{x \mid x \text{ is any real number}\}$; $(-\infty, \infty)$

37. (a) $E \cap F = \varnothing$ or $\{\ \}$ **(b)** $E \cup F = \{x \mid x < -1 \text{ or } x > 3\}$; $(-\infty, -1) \cup (3, \infty)$

39. (a) $\{x \mid -2 \leq x \leq 2\}$; $[-2, 2]$ **(b)** $(-\infty, -2) \cup (2, \infty)$

41. (a) $\{x \mid -3 < x < 3\}$; $(-3, 3)$ **(b)** $(-\infty, -3] \cup [3, \infty)$

43. $\{x \mid -2 \leq x < 3\}$; $[-2, 3)$ **45.** $\varnothing$ or $\{\ \}$ **47.** $\{x \mid x < -2\}$; $(-\infty, -2)$

49. $\{1\}$ **51.** $\{x \mid -1 \leq x < 3\}$; $[-1, 3)$

53. $\left\{ x \mid -\dfrac{2}{3} \leq x \leq \dfrac{3}{2} \right\}$; $\left[-\dfrac{2}{3}, \dfrac{3}{2} \right]$ **55.** $\left\{ x \mid -1 < x \leq \dfrac{4}{5} \right\}$; $\left(-1, \dfrac{4}{5} \right]$

57. $\{x \mid 0 \leq x \leq 8\}$; $[0, 8]$ **59.** $\{x \mid -6 \leq x \leq -2\}$; $[-6, -2]$

61. $\varnothing$ or $\{\ \}$ **63.** $\left\{ x \mid -\dfrac{5}{3} < x \leq 5 \right\}$; $\left(-\dfrac{5}{3}, 5 \right]$ **65.** $\{x \mid -4 < x \leq 3\}$; $(-4, 3]$

67. $\{x \mid x < -2 \text{ or } x > 3\}$; $(-\infty, -2) \cup (3, \infty)$

69. $\{x \mid x < -2 \text{ or } x > 5\}$; $(-\infty, -2) \cup (5, \infty)$

71. $\{x \mid x \text{ is any real number}\}$; $(-\infty, \infty)$

73. $\{x \mid x < -1 \text{ or } x > 4\}$; $(-\infty, -1) \cup (4, \infty)$

75. $\{x \mid x \leq -3 \text{ or } x > 6\}$; $(-\infty, -3] \cup (6, \infty)$

77. $\{x | x \text{ is any real number}\}$; $(-\infty, \infty)$

79. $\left\{x \middle| x < 0 \text{ or } x > \dfrac{5}{2}\right\}$; $(-\infty, 0) \cup \left(\dfrac{5}{2}, \infty\right)$

81. $\{a | -3 \le a < 0\}$; $[-3, 0)$

83. $\{x | x \text{ is any real number}\}$; $(-\infty, \infty)$

85. $\left\{x \middle| -2 \le x \le \dfrac{8}{3}\right\}$; $\left[-2, \dfrac{8}{3}\right]$

87. $\{x | x < -10 \text{ or } x > 2\}$; $(-\infty, -10) \cup (2, \infty)$

89. $\{x | 2 < x < 5\}$; $(2, 5)$

91. $\left\{x \middle| x \le -3 \text{ or } x > \dfrac{15}{4}\right\}$; $\left(-\infty, -3\right] \cup \left(\dfrac{15}{4}, \infty\right)$

93. $\left\{x \middle| -5 < x \le \dfrac{1}{2}\right\}$; $\left(-5, \dfrac{1}{2}\right]$

95. $a = 1$ and $b = 8$ **97.** $a = 12$ and $b = 30$ **99.** $a = -1$ and $b = 23$

101. $90 < x < 140$ **103.** Joanna needs to score at least a 77 on the final. That is, $77 \le x \le 100$ (assuming 100 is the max score, otherwise $77 \le x \le 104$).

105. The amount withheld ranges between \$89.50 and \$104.50, inclusive. **107.** Total sales between \$50,000 and \$250,000

109. The electrical usage ranged from 850 kwh and 1370 kwh.

111. **Step 1:**

$a < b$

$a + a < a + b$

$2a < a + b$

$\dfrac{2a}{2} < \dfrac{a + b}{2}$

$a < \dfrac{a + b}{2}$

Step 2:

$a < b$

$a + b < b + b$

$a + b < 2b$

$\dfrac{a + b}{2} < \dfrac{2b}{2}$

$\dfrac{a + b}{2} < b$

Step 3:

Since $a < \dfrac{a + b}{2}$ and $\dfrac{a + b}{2} < b$, it follows that $a < \dfrac{a + b}{2} < b$.

113. $\{ \ \}$ or $\varnothing$ **115.** This is a contradiction. There is no solution. If, during simplification, the variable terms all cancel out and a contradiction results, then there is no solution to the inequality. **117.** If $x < 2$ then $x - 2 < 2 - 2 \Rightarrow x - 2 < 0$. When multiplying both sides of the inequality by $(x - 2)$ in the second step, the direction of the inequality must switch.

Section 8.7 Absolute Value Equations and Inequalities

1. $\{-7, 7\}$ **2.** $\{-1, 1\}$ **3.** $a; -a$ **4.** $2x + 3 = -5$ **5.** $\{-2, 5\}$ **6.** $\left\{-\dfrac{5}{3}, 3\right\}$

7. $\left\{-1, \dfrac{9}{5}\right\}$ **8.** $\{-5, 1\}$ **9.** True **10.** $\varnothing$ or $\{ \ \}$ **11.** $\varnothing$ or $\{ \ \}$ **12.** $\{-1\}$ **13.** $u; v; u; -v$ **14.** $\left\{-8, -\dfrac{2}{3}\right\}$ **15.** $\{-2, 3\}$ **16.** $\{-3, 0\}$

17. $\{2\}$ **18.** $-a < u < a$ **19.** $-10; 10$ **20.** $\{x | -5 \le x \le 5\}$; $[-5, 5]$

21. $\left\{x \middle| -\dfrac{3}{2} < x < \dfrac{3}{2}\right\}$; $\left(-\dfrac{3}{2}, \dfrac{3}{2}\right)$ **22.** $\{x | -8 < x < 2\}$; $(-8, 2)$

23. $\{x | -2 \le x \le 5\}$; $[-2, 5]$ **24.** $\varnothing$ or $\{ \ \}$

25. False **26.** $\{x | -2 < x < 2\}$; $(-2, 2)$ **27.** $\{x | -1 \le x \le 7\}$; $[-1, 7]$

28. $\{x | -2 \le x \le 1\}$; $[-2, 1]$ **29.** $\left\{x \middle| -\dfrac{7}{3} < x < 3\right\}$; $\left(-\dfrac{7}{3}, 3\right)$

30. $u < -a; u > a$ **31.** $-7; 7$ **32.** $\{x | x \le -6 \text{ or } x \ge 6\}$; $(-\infty, -6] \cup [6, \infty)$

33. $\left\{x \middle| x < -\dfrac{5}{2} \text{ or } x > \dfrac{5}{2}\right\}$; $\left(-\infty, -\dfrac{5}{2}\right) \cup \left(\dfrac{5}{2}, \infty\right)$ **34.** $<$

35. $\{x | x < -7 \text{ or } x > 1\}$; $(-\infty, -7) \cup (1, \infty)$

36. $\left\{x \middle| x \le -\dfrac{1}{2} \text{ or } x \ge 2\right\}$; $\left(-\infty, -\dfrac{1}{2}\right] \cup [2, \infty)$

37. $\left\{x \middle| x < -\dfrac{5}{3} \text{ or } x > 3\right\}$; $\left(-\infty, -\dfrac{5}{3}\right) \cup (3, \infty)$

38. $\left\{x \middle| x \ne -\dfrac{5}{2}\right\}$; $\left(-\infty, -\dfrac{5}{2}\right) \cup \left(-\dfrac{5}{2}, \infty\right)$

39. $\{x | x \text{ is any real number}\}$; $(-\infty, \infty)$

40. $\{x | x \text{ is any real number}\}$; $(-\infty, \infty)$ **41.** The acceptable belt width is between $\dfrac{127}{32}$ inches and $\dfrac{129}{32}$ inches.

42. The percentage of Americans that have been shot at is between 7.3% and 10.7%, inclusive. **43.** $\{-10, 10\}$ **45.** $\varnothing$ or $\{ \ \}$ **47.** $\{-1, 7\}$ **49.** $\left\{-1, \dfrac{13}{3}\right\}$

51. $\{-5, 5\}$ **53.** $\left\{-\dfrac{11}{2}, \dfrac{5}{2}\right\}$ **55.** $\{-4, 10\}$ **57.** $\{0\}$ **59.** $\left\{-\dfrac{7}{3}, 3\right\}$ **61.** $\left\{-7, \dfrac{3}{5}\right\}$ **63.** $\{1, 3\}$ **65.** $\{2\}$ **67.** $\varnothing$ or $\{\ \}$

69. $\{x \mid -9 < x < 9\}; (-9, 9)$ 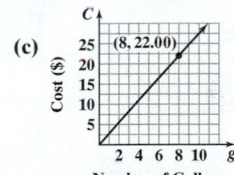 **71.** $\{x \mid -3 \le x \le 11\}; [-3, 11]$

73. $\left\{x \mid -3 < x < \dfrac{7}{3}\right\}; \left(-3, \dfrac{7}{3}\right)$ **75.** $\varnothing$ or $\{\ \}$

77. $\{x \mid 0 < x < 6\}; (0, 6)$ **79.** $\left\{x \mid -1 < x < \dfrac{9}{5}\right\}; \left(-1, \dfrac{9}{5}\right)$

81. $\{x \mid 1.995 < x < 2.005\}; (1.995, 2.005)$

83. $\{y \mid y < 3 \text{ or } y > 7\}; (-\infty, 3) \cup (7, \infty)$

85. $\left\{x \mid x \le -2 \text{ or } x \ge \dfrac{1}{2}\right\}; (-\infty, -2] \cup \left[\dfrac{1}{2}, \infty\right)$

87. $\{y \mid y \text{ is any real number}\}; (-\infty, \infty)$

89. $\left\{x \mid x < -2 \text{ or } x > \dfrac{4}{5}\right\}; (-\infty, -2) \cup \left(\dfrac{4}{5}, \infty\right)$

91. $\{x \mid x < 0 \text{ or } x > 1\}; (-\infty, 0) \cup (1, \infty)$

93. $\{x \mid x \le -2 \text{ or } x \ge 3\}; (-\infty, -2] \cup [3, \infty)$
95. (a) $\{-5, 5\}$ (b) $\{x \mid -5 \le x \le 5\}; [-5, 5]$ (c) $\{x \mid x < -5 \text{ or } x > 5\}; (-\infty, -5) \cup (5, \infty)$
97. (a) $\{-5, 1\}$ (b) $\{x \mid -5 < x < 1\}; (-5, 1)$ (c) $\{x \mid x \le -5 \text{ or } x \ge 1\}; (-\infty, -5] \cup [1, \infty)$
99. $\{x \mid x < -5 \text{ or } x > 5\}; (-\infty, -5) \cup (5, \infty)$

101. $\{-4, -1\}$ **103.** $\{-5, 5\}$ **105.** $\{-4, 3\}$ **107.** $\varnothing$ or $\{\ \}$ **109.** $\varnothing$ or $\{\ \}$

111. $\left\{x \mid x \le -\dfrac{7}{3} \text{ or } x \ge 1\right\}; \left(-\infty, -\dfrac{7}{3}\right] \cup [1, \infty)$

113. $\left\{x \mid x < 0 \text{ or } x > \dfrac{4}{3}\right\}; (-\infty, 0) \cup \left(\dfrac{4}{3}, \infty\right)$ **115.** $\{-1, 1\}$ **117.** $\varnothing$ or $\{\ \}$ **119.** $\left\{-8, \dfrac{4}{7}\right\}$

121. $|x - 5| < 3; \{x \mid 2 < x < 8\}; (2, 8)$ **123.** $|2x - (-6)| > 3; \left\{x \mid x < -\dfrac{9}{2} \text{ or } x > -\dfrac{3}{2}\right\}; \left(-\infty, -\dfrac{9}{2}\right) \cup \left(-\dfrac{3}{2}, \infty\right)$

125. The acceptable rod lengths are between 5.6995 inches and 5.7005 inches, inclusive. **127.** An unusual IQ score would be less than 70.6 or greater

than 129.4. **129.** $\left\{-\dfrac{5}{2}\right\}$ **131.** $\{2\}$ **133.** $\varnothing$ or $\{\ \}$ **135.** $\{x \mid x \le -5\}; (-\infty, -5]$ **137.** The absolute value, when isolated, is equal to a negative number,

which is not possible. **139.** The absolute value, when isolated, is less than -3. Since absolute values are always nonnegative, this is not possible.

Section 8.8 Variation **1.** Variation **2.** $y = kx$

3. (a) $C(g) = 2.75g$
(b) \$12.65

4. $y = \dfrac{k}{x}$ **5.** (a) $y = \dfrac{6}{x}$ (b) 1.5 **6.** (a) $V(l) = \dfrac{15{,}000}{l}$ (b) 300 oscillations per second **7.** joint **8.** 1750 joules

9. combined variation **10.** approximately 1.44 ohms **11.** (a) $k = 6$ (b) $y = 6x$ (c) $y = 42$ **13.** (a) $k = \dfrac{3}{7}$ (b) $y = \dfrac{3}{7}x$

(c)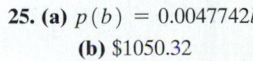

(c) $y = 12$ **15.** (a) $k = \dfrac{1}{2}$ (b) $y = \dfrac{1}{2}x$ (c) $y = 15$ **17.** (a) $k = 20$ (b) $y = \dfrac{20}{x}$ (c) $y = 4$ **19.** (a) $k = 21$

(b) $y = \dfrac{21}{x}$ (c) $y = \dfrac{3}{4}$ **21.** (a) $k = \dfrac{1}{4}$ (b) $y = \dfrac{1}{4}xz$ (c) $y = 27$ **23.** (a) $k = \dfrac{13}{10}$ (b) $Q = \dfrac{13x}{10y}$ (c) $Q = \dfrac{117}{40}$

25. (a) $p(b) = 0.0047742b$
(b) \$1050.32

27. (a) $C(w) = 5.6w$
(b) \$19.60

29. 96 feet per second **31.** (a) $D(p) = \dfrac{375}{p}$ (b) 125 bags of candy **33.** 450 cc

35. approximately 119.8 pounds **37.** 2250 newtons **39.** 1.4007×10^{-7} newtons

41. 360 pounds **43.** (a) 314.16 (b) approximately 942.48 meters per minute;
approximately 15.708 meters per second (c) $k \approx 0.056$ (d) approximately
7.86 newtons

(c)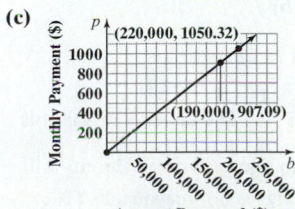

(c)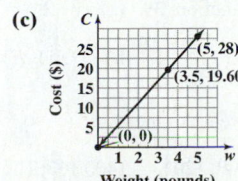

Chapter 8 Review

1. A: quadrant IV;
B: quadrant III;
C: y-axis; D: quadrant II;
E: x-axis; F: quadrant I

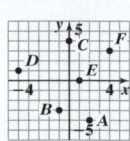

2. A: x-axis;
B: quadrant I;
C: quadrant III; D: quadrant II;
E: quadrant IV;
F: y-axis

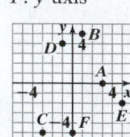

3. **(a)** yes **(b)** no
(c) no **(d)** yes
4. **(a)** no **(b)** yes
(c) yes **(d)** no
5. $y = x + 2$

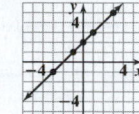

6. $2x + y = 3$

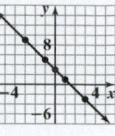

7. $y = -x^2 + 4$

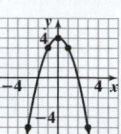

8. $y = |x + 2| - 1$

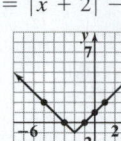

9. $y = x^3 + 2$

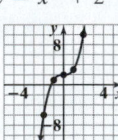

10. $x = y^2 + 1$

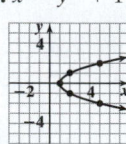

11. $(-3, 0), (0, -1), (0, 3)$
x-intercept: $(-3, 0)$;
y-intercepts: $(0, -1), (0, 3)$
12. **(a)** \$599.99
(b) \$3499.99

13. {(Penny, 2.500), (Nickel, 5.000), (Dime, 2.268), (Quarter, 5.670), (Half Dollar, 11.340), (Dollar, 8.100)}
Domain: { Penny, Nickel, Dime, Quarter, Half Dollar, Dollar }; Range: { 2.268, 2.500, 5.000, 5.670, 8.100, 11.340 }
14. { (16, \$12.99), (28, \$14.99) (30, \$14.99) (59, \$24.99) (85, \$29.99) }; Domain: {16, 28, 30, 59, 85}; Range: {\$12.99, \$14.99, \$24.99, \$29.99}

15. Domain: $\{-4, -2, 2, 3, 6\}$;
Range: $\{-9, -1, 5, 7, 8\}$

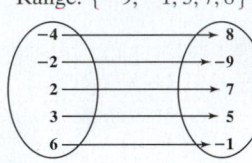

16. Domain: $\{-2, 1, 3, 5\}$;
Range: $\{1, 4, 7, 8\}$

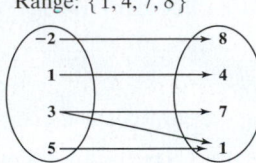

17. Domain: $\{x \,|\, x$ is a real number $\}$ or $(-\infty, \infty)$
Range: $\{y \,|\, y$ is a real number $\}$ or $(-\infty, \infty)$
18. Domain: $\{x \,|\, -6 \le x \le 4\}$ or $[-6, 4]$
Range: $\{y \,|\, -4 \le y \le 6\}$ or $[-4, 6]$
19. Domain: $\{2\}$; Range: $\{y \,|\, y$ is a real number $\}$ or $(-\infty, \infty)$
20. Domain: $\{x \,|\, x \ge -1\}$ or $[-1, \infty)$; Range: $\{y \,|\, y \ge -2\}$ or $[-2, \infty)$

21. $y = x + 2$
Domain: $\{x \,|\, x$ is a real number $\}$ or $(-\infty, \infty)$
Range: $\{y \,|\, y$ is a real number $\}$ or $(-\infty, \infty)$

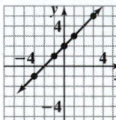

22. $2x + y = 3$
Domain: $\{x \,|\, x$ is a real number $\}$ or $(-\infty, \infty)$
Range: $\{y \,|\, y$ is a real number $\}$ or $(-\infty, \infty)$

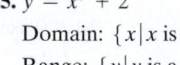

23. $y = -x^2 + 4$
Domain: $\{x \,|\, x$ is a real number $\}$ or $(-\infty, \infty)$
Range: $\{y \,|\, y \le 4\}$ or $(-\infty, 4]$

24. $y = |x + 2| - 1$
Domain: $\{x \,|\, x$ is a real number $\}$ or $(-\infty, \infty)$
Range: $\{y \,|\, y \ge -1\}$ or $[-1, \infty)$

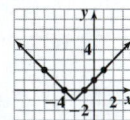

25. $y = x^3 + 2$
Domain: $\{x \,|\, x$ is a real number $\}$ or $(-\infty, \infty)$
Range: $\{y \,|\, y$ is a real number $\}$ or $(-\infty, \infty)$

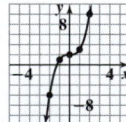

26. $x = y^2 + 1$
Domain: $\{x \,|\, x \ge 1\}$ or $[1, \infty)$
Range: $\{y \,|\, y$ is a real number $\}$ or $(-\infty, \infty)$

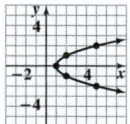

27. **(a)** Domain: $\{x \,|\, 0 \le x \le 44{,}640\}$ or $[0, 44{,}640]$; Range: $\{y \,|\, 40 \le y \le 2122\}$ or $[40, 2122]$ **(b)** Answers may vary. **28.** Domain: $\{t \,|\, 0 \le t \le 4\}$ or $[0, 4]$; Range: $\{y \,|\, 0 \le y \le 121\}$ or $[0, 121]$ **29.** **(a)** Not a function. Domain: $\{-1, 5, 7, 9\}$; Range: $\{-2, 0, 2, 3, 4\}$ **(b)** Function. Domain: { Camel, Macaw, Deer, Fox, Tiger, Crocodile }; Range: $\{14, 22, 35, 45, 50\}$ **30.** **(a)** Function; Domain: $\{-3, -2, 2, 4, 5\}$; Range: $\{-1, 3, 4, 7\}$ **(b)** Not a function; Domain: { Red, Blue, Green, Black }; Range: { Camry, Taurus, Windstar, Durango } **31.** Function **32.** Not a function **33.** Not a function **34.** Function **35.** Not a function **36.** Function **37.** Function **38.** Not a function **39.** **(a)** $f(-2) = -5$ **(b)** $f(3) = 10$

40. **(a)** $g(0) = -\dfrac{1}{3}$ **(b)** $g(2) = -5$ **41.** **(a)** $F(5) = -3$ **(b)** $F(-x) = 2x + 7$ **42.** **(a)** $G(7) = 15$ **(b)** $G(x + h) = 2x + 2h + 1$

43. $\{x \,|\, x$ is a real number $\}$ or $(-\infty, \infty)$ **44.** $\left\{ w \,\middle|\, w \ne -\dfrac{5}{2} \right\}$ **45.** $\{t \,|\, t \ne 5\}$ **46.** $\{t \,|\, t$ is a real number $\}$ or $(-\infty, \infty)$ **47.** **(a)** The dependent variable
is the population, P, and the independent variable is the number of years after 1900, t. **(b)** $P(120) = 1400.167$; The population of Orange County will be roughly 1,400,167 in 2020. **(c)** $P(-70) = 1066.337$; The population of Orange County was roughly 1,066,337 in 1830. This is not reasonable. (The population of the entire Florida territory was roughly 35,000 in 1830.) **48.** **(a)** The dependent variable is percent of the population with an advanced degree, P, and the independent variable is age, a. **(b)** $P(30) = 7.9$; According to the model, 7.9% of 30-year-olds have an advanced degree.

49. $f(x) = 2x - 5$ **50.** $g(x) = x^2 - 3x + 2$ **51.** $h(x) = (x - 1)^3 - 3$ **52.** $f(x) = |x + 1| - 4$ **53. (a)** Domain: $\{x | x$ is a real number $\}$ or $(-\infty, \infty)$; Range: $\{y | y$ is a real number $\}$ or $(-\infty, \infty)$ **(b)** $(0, 2)$ and $(4, 0)$

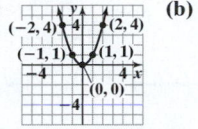
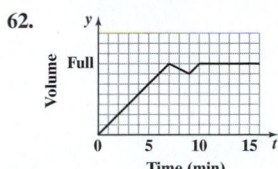
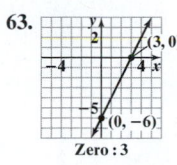
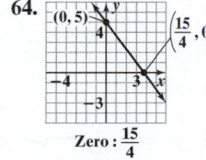
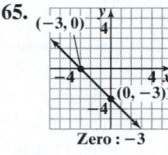

54. (a) Domain: $\{x | x$ is a real number $\}$ or $(-\infty, \infty)$; Range: $\{y | y \geq -3\}$ or $[-3, \infty)$ **(b)** $(-2, 0)$, $(2, 0)$, $(0, -3)$

55. (a) Domain: $\{x | x$ is a real number $\}$ or $(-\infty, \infty)$; Range: $\{y | y$ is a real number $\}$ or $(-\infty, \infty)$ **(b)** $(0, 0)$ and $(2, 0)$

56. (a) Domain: $\{x | x \geq -3\}$ or $[-3, \infty)$; Range: $\{y | y \geq 1\}$ or $[1, \infty)$ **(b)** $(0, 3)$ **57. (a)** 4 **(b)** 1 **(c)** -1 and 3

58. (a)   **(b)**

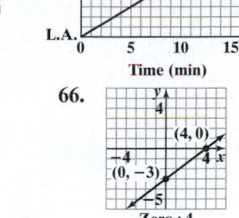

59. (a) yes
(b) $h(-2) = -11$; $(-2, -11)$
(c) $x = \dfrac{11}{2}$; $\left(\dfrac{11}{2}, 4\right)$

60. (a) no
(b) $g(3) = \dfrac{29}{5}$; $\left(3, \dfrac{29}{5}\right)$
(c) $x = -10$; $(-10, -2)$

61.

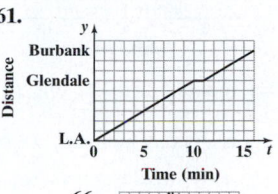

62.

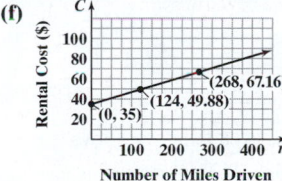

63.

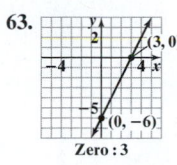

Zero: 3

64.

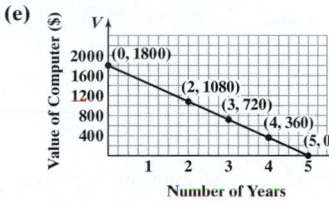

Zero: $\dfrac{15}{4}$

65.

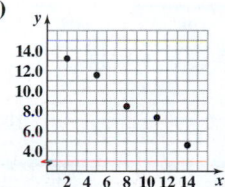

Zero: -3

66.

Zero: 4

67. (a) $[0, \infty)$
(b) \$260; This is the monthly payment for the car.
(c) \$460

(d)

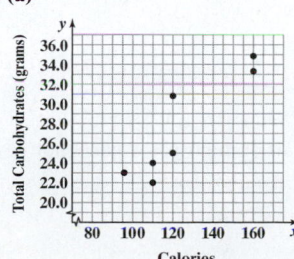

(e) $[0, 1450]$

68. (a) The independent variable is years; the dependent variable is value.
(b) $\{x | 0 \leq x \leq 5\}$ or $[0, 5]$
(c) \$1800 **(d)** \$1080

(e)

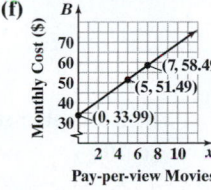

(f) After 5 years

69. (a) $L(x) = -\dfrac{4}{75}x + 45$ **(b)** 7% **(c)** The auto loan rate decreases approximately by 0.05% for every 1-unit increase in FICO score. **(d)** 722

70. (a) $H(x) = -x + 220$ **(b)** 175 beats per minute
(c) The maximum recommended heart rate for men under stress decreases at a rate of 1 beat per minute per year.
(d) 52 years

71. (a) $C(m) = 0.12m + 35$
(b) The independent variable is m; the dependent variable is C.
(c) $\{m | m \geq 0\}$ or $[0, \infty)$
(d) \$49.88
(e) 268 miles were driven
(f)

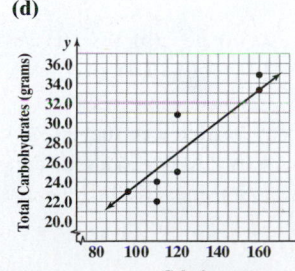

72. (a) $B(x) = 3.50x + 33.99$
(b) The independent variable is x; the dependent variable is B.
(c) $\{x | x \geq 0\}$ or $[0, \infty)$
(d) \$51.49
(e) 7 pay-per-view movies
(f)

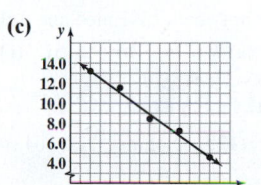

73. (a)

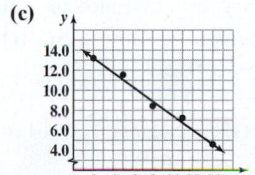

(b) Answers will vary. Using the points $(2, 13.3)$ and $(14, 4.6)$, the equation is $y = -0.725x + 14.75$.

(c)

74. (a)

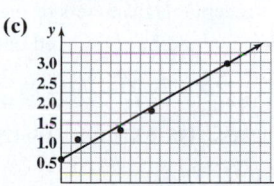

(b) Answers will vary. Using the points $(0, 0.6)$ and $(4.2, 3.0)$, the equation is $y = \dfrac{4}{7}x + 0.6$.

(c)

75. (a)

(b) Approximately linear
(c) Answers will vary. Using the points $(96, 23.2)$ and $(160, 33.3)$, the equation is $y = 0.158x + 8.032$.

(d)

(e) 30.2 grams
(f) In a one-cup serving of cereal, total carbohydrates will increase by 0.158 gram for each 1-calorie increase.

76. (a)

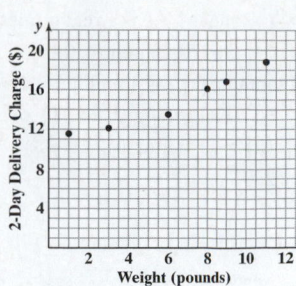

(b) Linear with positive slope
(c) Answers will vary. Using (3, 12.2) and (9, 16.9), $y = 0.78x + 9.85$

(d)

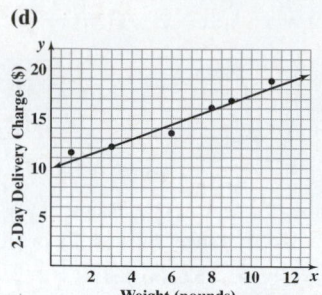

(e) $13.75
(f) If the weight increases by 1 pound, the shipping charge increases $0.78.

77. $\{-1, 0, 1, 2, 3, 4, 6, 8\}$ **78.** $\{2, 4\}$ **79.** $\{1, 2, 3, 4\}$ **80.** $\{1, 2, 3, 4, 6, 8\}$ **81. (a)** $\{x \mid 2 < x \le 4\}; (2, 4]$

(b) $\{x \mid x \text{ is any real number}\}; (-\infty, \infty)$ **82. (a)** $\{\ \}$ or $\varnothing$

(b) $\{x \mid x < -2 \text{ or } x \ge 3\}; (-\infty, -2) \cup [3, \infty)$

83. $\{x \mid -1 < x < 4\}; (-1, 4)$ Graph:

84. $\{x \mid -5 < x < -1\}; (-5, -1)$ Graph:

85. $\{x \mid x < -2 \text{ or } x > 2\}; (-\infty, -2) \cup (2, \infty)$ Graph:

86. $\{x \mid x \le 0 \text{ or } x \ge 4\}; (-\infty, 0] \cup [4, \infty)$ Graph:

87. $\{\ \}$ or $\varnothing$ **88.** $\{x \mid -2 \le x < 4\}; [-2, 4)$ Graph:

89. $\{x \mid x \le -2 \text{ or } x > 3\}; (-\infty, -2] \cup (3, \infty)$ Graph:

90. $\{x \mid x \text{ is any real number}\}; (-\infty, \infty)$ Graph:

91. $\{x \mid x < -10 \text{ or } x > 6\}; (-\infty, -10) \cup (6, \infty)$ Graph:

92. $\left\{x \mid -\dfrac{3}{2} \le x < \dfrac{5}{8}\right\}; \left[-\dfrac{3}{2}, \dfrac{5}{8}\right)$ Graph: **93.** $70 \le x \le 75$

94. The electric usage varied from roughly 1033.8 kilowatt hours up to roughly 1952.5 kilowatt hours. (recall, x is the number *above* 800).

95. $\{-4, 4\}$ **96.** $\left\{\dfrac{1}{3}, 3\right\}$ **97.** $\{-5, 13\}$ **98.** $\{-3, -1\}$ **99.** $\{\ \}$ or $\varnothing$ **100.** $\left\{-\dfrac{1}{2}, 2\right\}$

101. $\{x \mid -2 < x < 2\}; (-2, 2)$

102. $\left\{x \mid x \le -\dfrac{7}{2} \text{ or } x \ge \dfrac{7}{2}\right\}; \left(-\infty, -\dfrac{7}{2}\right] \cup \left[\dfrac{7}{2}, \infty\right)$

103. $\{x \mid -5 \le x \le 1\}; [-5, 1]$ **104.** $\left\{x \mid x \le \dfrac{1}{2} \text{ or } x \ge 1\right\}; \left(-\infty, \dfrac{1}{2}\right] \cup [1, \infty)$

105. $\{x \mid x \text{ is a real number}\}; (-\infty, \infty)$ **106.** $\{\ \}$ or $\varnothing$

107. $\{x \mid 4.99 \le x \le 5.01\}; [4.99, 5.01]$

108. $\left\{x \mid x < -\dfrac{1}{2} \text{ or } x > \dfrac{7}{2}\right\}; \left(-\infty, -\dfrac{1}{2}\right) \cup \left(\dfrac{7}{2}, \infty\right)$

109. The acceptable diameters of the bearing are between 0.502 inch and 0.504 inch, inclusive. **110.** Tensile strengths below 36.08 lb/in.² or above 43.92 lb/in.² would be considered unusual. **111. (a)** $k = 5$ **(b)** $y = 5x$ **(c)** 50 **112. (a)** $k = -6$ **(b)** $y = -6x$ **(c)** -48 **113. (a)** $k = 60$

(b) $y = \dfrac{60}{x}$ **(c)** 12 **114. (a)** $k = \dfrac{3}{4}$ **(b)** $y = \dfrac{3}{4}xz$ **(c)** 42 **115. (a)** $k = 72$ **(b)** $s = \dfrac{72}{t^2}$ **(c)** 8 **116. (a)** $k = \dfrac{8}{5}$ **(b)** $w = \dfrac{8x}{5z}$ **(c)** $\dfrac{9}{10}$

117. 6 inches **118.** $352.19 **119.** 1200 kilohertz **120.** 12 ohms **121.** 704 cubic cm **122.** 375 cubic inches

Chapter 8 Test

1. *A:* quadrant IV;
B: *y*-axis;
C: *x*-axis;
D: quadrant I;
E: quadrant III;
F: quadrant II

2. (a) no **(b)** yes **(c)** yes
3. $(-3, 0), (0, 1), (0, 3)$
x-intercept: $(-3, 0)$;
y-intercepts: $(0, 1)$ $(0, 3)$
4. (a) At 6 seconds the car is traveling 30 miles per hour. **(b)** $(0, 0)$: At the start of the trip, the car is not moving. $(32, 0)$: After 32 seconds, the speed of the car is 0 miles per hour.

5. Domain: $\{-4, 2, 5, 7\}$
   Range: $\{-7, -2, -1, 3, 8, 12\}$

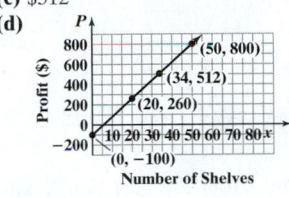

6. Domain: $\left\{x \left| -\dfrac{5\pi}{2} \le x \le \dfrac{5\pi}{2}\right.\right\}$ or $\left[-\dfrac{5\pi}{2}, \dfrac{5\pi}{2}\right]$
   Range: $\{y | 1 \le y \le 5\}$ or $[1, 5]$

7. $y = x^2 - 3$

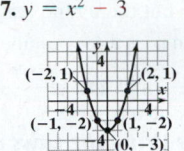

Domain: $\{x | x \text{ is a real number}\}$ or $(-\infty, \infty)$; Range: $\{y | y \ge -3\}$ or $[-3, \infty)$

8. Function. Domain: $\{-5, -3, 0, 2\}$; Range: $\{3, 7\}$ **9.** Not a function. Domain: $\{x | x \le 3\}$ or $(-\infty, 3]$; Range: $\{y | y \text{ is a real number}\}$ or $(-\infty, \infty)$
10. No   **11.** $f(x + h) = -3x - 3h + 11$ **12. (a)** $g(-2) = 5$ **(b)** $g(0) = -1$ **(c)** $g(3) = 20$

13. $f(x) = x^2 + 3$

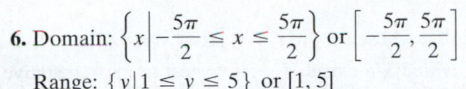

14. (a) The dependent variable is the ticket price, P, and the independent variable is the number of years after 1996, x.
   (b) $P(25) = 9.94$; According to the model, the average ticket price in 2021 $(x = 25)$ was \$9.94.   **(c)** 2030

15. $\{x | x \ne -2\}$ **16. (a)** yes   **(b)** $h(3) = -3; (3, -3)$ **(c)** $x = -3; (-3, 27)$ **(d)** $\dfrac{12}{5}$

17. (a) The car stops accelerating when the speed stops increasing. Thus, the car stops accelerating after 6 seconds.
   (b) The car has a constant speed when the graph is horizontal. Thus, the car maintains a constant speed for 18 seconds.

18. (a) $P(x) = 18x - 100$
(b) $\{x | x \ge 0\}$ or $[0, \infty)$
(c) \$512
(d)

(e) 48 shelves

19. (a)

(b) Approximately linear.
(c) Answers will vary. Using the points $(6, 95)$ and $(18, 170)$, the equation is $y = 6.25x + 57.5$.

(d)

(e) 113.75 kilograms
(f) A Shetland pony's weight will increase by 6.25 kilograms for each one-month increase in age.

20. $\{-4, -1\}$ **21.** $\{x | -2 \le x < 6\}$; $[-2, 6)$

22. $\left\{x \left| x < -\dfrac{3}{2} \text{ or } x > 4\right.\right\}$; $\left(-\infty, -\dfrac{3}{2}\right) \cup (4, \infty)$ **23.** $\{x | 2 < x < 8\}$; $(2, 8)$

24. $\{x | x \le -2 \text{ or } x \ge 3\}$; $(-\infty, -2] \cup [3, \infty)$ **25.** 20 pounds   **26.** 792 cm^2

Chapter 9 Radicals and Rational Exponents

Section 9.1 Square Roots   **1.** radical sign   **2.** principal square root   **3.** $-4, 4$ **4.** 9   **5.** 30   **6.** $\dfrac{3}{2}$ **7.** 0.4   **8.** 13   **9.** 15
10. 10   **11.** 14   **12.** 7   **13.** True   **14.** rational; 20   **15.** irrational; ≈ 6.32 **16.** not a real number   **17.** rational; -14 **18.** $|a|$ **19.** 14   **20.** $|z|$
21. $|2x + 3|$ **22.** $|p - 6|$ **23.** 1   **25.** -10 **27.** $\dfrac{1}{2}$ **29.** 0.6   **31.** 1.6   **33.** not a real number   **35.** rational; 8   **37.** rational; $\dfrac{1}{4}$ **39.** irrational; 6.63
41. irrational; 7.07   **43.** not a real number   **45.** 8   **47.** 19   **49.** $|r|$ **51.** $|x + 4|$ **53.** $|4x - 3|$ **55.** $|2y + 3|$ **57.** 13   **59.** 17   **61.** not a real number
63. 15   **65.** -8 **67.** 6   **69.** not a real number   **71.** -4 **73.** 13   **75.** The square roots of 36 are -6 and 6; $\sqrt{36} = 6$ **77.** $\dfrac{4\sqrt{13}}{3}$; 4.81   **79.** Answers
will vary. One reasonable explanation follows. Because $a^2 \ge 0$, the radicand of $\sqrt{a^2}$ is greater than or equal to 0 (nonnegative). The principal square
root of a nonnegative number is nonnegative. The absolute value ensures this result.

Section 9.2 *nth* Roots and Rational Exponents   **1.** index   **2.** 4   **3.** 3   **4.** -6 **5.** not a real number   **6.** $\dfrac{1}{2}$ **7.** 3.68   **8.** 2.99   **9.** 5   **10.** $|z|$

11. $3x - 2$ **12.** 2   **13.** $-\dfrac{2}{3}$ **14.** $\sqrt[n]{a}$ **15.** 5   **16.** -3 **17.** -8 **18.** not a real number   **19.** $\sqrt{b}$ **20.** $(8b)^{\frac{1}{5}}$ **21.** $\left(\dfrac{mn^5}{3}\right)^{\frac{1}{8}}$ **22.** $4y^{\frac{1}{3}}$ **23.** $\sqrt[n]{a^m}$; $\left(\sqrt[n]{a}\right)^m$

24. 64   **25.** 9   **26.** -8 **27.** 16   **28.** not a real number   **29.** 13.57   **30.** 1.74   **31.** $a^{\frac{3}{8}}$ **32.** t^3 **33.** $(12ab^3)^{\frac{9}{4}}$ **34.** $\dfrac{1}{9}$ **35.** 4   **36.** $\dfrac{1}{(13x)^{\frac{3}{2}}}$ **37.** 5   **39.** -3

41. -5 **43.** $-\dfrac{1}{2}$ **45.** 3   **47.** 2.92   **49.** 1.86   **51.** 5   **53.** $|m|$ **55.** $x - 3$ **57.** $-|3p + 1|$ **59.** 2   **61.** -6 **63.** 2   **65.** -2 **67.** $\dfrac{2}{5}$ **69.** -5 **71.** not a real number

73. $(3x)^{\frac{1}{5}}$ **75.** $\left(\dfrac{x}{3}\right)^{\frac{1}{4}}$ **77.** 32   **79.** -64 **81.** 16   **83.** 16   **85.** 8   **87.** $\dfrac{1}{12}$ **89.** 125   **91.** 32   **93.** $x^{\frac{3}{4}}$ **95.** $(3x)^{\frac{2}{5}}$ **97.** $\left(\dfrac{5x}{y}\right)^{\frac{3}{2}}$ **99.** $(9ab)^{\frac{4}{3}}$ **101.** 4.47
103. 10.08   **105.** 1.26   **107.** $5x$ **109.** -10 **111.** 8   **113.** 243   **115.** not a real number   **117.** $\dfrac{3}{4}$ **119.** 0.2   **121.** 127   **123.** not a real number   **125.** $3p - 5$

127. -18 **129.** 2   **131.** 8   **133.** 16   **135.** 10; 10   **137. (a)** about 21.25°F   **(b)** about 17.36°F   **(c)** about -15.93°F   **139. (a)** $\sqrt{\dfrac{8r\rho_w g}{3C\rho}}$ m/s

(b) about 7.38 m/s   **141.** $(-9)^{\frac{1}{2}} = \sqrt{-9}$, but there is no real number whose square is -9. However, $-9^{\frac{1}{2}} = -1 \cdot 9^{\frac{1}{2}} = -1\sqrt{9} = -3$.

143. If $\dfrac{m}{n}$, in lowest terms, is positive, then $a^{\frac{m}{n}}$ is a real number provided $\sqrt[n]{a}$ exists. If $\dfrac{m}{n}$, in lowest terms, is negative, then $a^{\frac{m}{n}}$ is a real number provided $a \neq 0$ and $\sqrt[n]{a}$ exists.   **145.** $(x + 2)(x - 1)^3$ **147.** $\dfrac{z - 6}{z - 1}$

Section 9.3 Simplifying Expressions Using the Laws of Exponents

1. $a^r b^r$ **2.** a^{r+s} **3.** $5^{\frac{11}{12}}$ **4.** 8   **5.** 10   **6.** $ab^{\frac{5}{6}}$ **7.** $x^{\frac{2}{3}}$ **8.** $\dfrac{4x^{\frac{1}{2}}}{y^{\frac{2}{3}}}$ **9.** $\dfrac{5x^{\frac{5}{8}}}{y^{\frac{1}{8}}}$ **10.** $\dfrac{200a^{\frac{1}{2}}}{b^{\frac{2}{3}}}$

11. 6   **12.** $2a^2b^3$ **13.** $\sqrt[12]{x^5}$ **14.** $\sqrt[6]{a}$ **15.** $x^{\frac{1}{2}}(20x + 9)$ **16.** $\dfrac{12x + 1}{x^{\frac{2}{3}}}$ **17.** 25   **19.** 8   **21.** $\dfrac{1}{2^{\frac{7}{6}}}$ **23.** $\dfrac{1}{x^{\frac{7}{12}}}$ **25.** 2   **27.** $\dfrac{125}{8}$ **29.** $x^{\frac{1}{2}}y^{\frac{2}{9}}$ **31.** $\dfrac{x^{\frac{1}{6}}}{y^{\frac{1}{3}}}$ **33.** $\dfrac{2a}{b^{\frac{3}{4}}}$

35. $\dfrac{x^{\frac{1}{18}}}{2y^{\frac{4}{9}}}$ **37.** $8x^{\frac{1}{8}}y^{\frac{1}{2}}$ **39.** x^4 **41.** 2   **43.** $-2ab^4$ **45.** $\sqrt[4]{x}$ **47.** $\sqrt[6]{x^5}$ **49.** $\sqrt[8]{x^3}$ **51.** $\sqrt[6]{3^7}$ **53.** 1   **55.** $5x^{\frac{1}{2}}(8x + 15)$ **57.** $8(x + 2)^{\frac{2}{3}}(3x + 1)$ **59.** $\dfrac{6x + 5}{x^{\frac{1}{2}}}$

61. $\dfrac{2(10x - 27)}{(x - 4)^{\frac{1}{3}}}$ **63.** $5(x^2 + 4)^{\frac{1}{2}}(x^2 + 3x + 4)$ **65.** 2   **67.** 4   **69.** 10   **71.** 5   **73.** 0   **75.** $\dfrac{1}{48}$ **77.** $x^2 - 2x^{\frac{1}{2}}$ **79.** $\dfrac{2}{y^{\frac{1}{3}}} + 6y^{\frac{2}{3}}$

81. $4z^3 - 32 = 4(z - 2)(z^2 + 2z + 4)$ **83.** 5   **85.** 3   **87.** $\sqrt[24]{x}$ **89.** 36   **91.** $\{x \mid x > -1\}$ or $(-1, \infty)$ **93.** $4a^2 - 8a - 6$ **95.** -6

Section 9.4 Simplifying Radical Expressions Using Properties of Radicals

1. $\sqrt[n]{ab}$ **2.** $\sqrt{77}$ **3.** $\sqrt[4]{42}$ **4.** $\sqrt{x^2 - 25}$ **5.** $\sqrt[5]{20p^4}$ **6.** 1, 4, 9, 16, 25, 36   **7.** 1, 8, 27, 64, 125, 216   **8.** $4\sqrt{3}$ **9.** $12\sqrt[3]{2}$ **10.** $10|a|\sqrt{2}$ **11.** Fully simplified   **12.** $2 + \sqrt{5}$ **13.** $\dfrac{-1 + 2\sqrt{2}}{2}$ or $-\dfrac{1}{2} + \sqrt{2}$

14. $5a^3\sqrt{3}$ **15.** $3a^2\sqrt{2a}$ **16.** $-4x^2y^3\sqrt[3]{2y}$ **17.** $2ab^2\sqrt[4]{ab^3}$ **18.** $4\sqrt{3}$ **19.** $2a^2\sqrt[3]{15}$ **20.** $8ab^3\sqrt[3]{6a}$ **21.** $\dfrac{\sqrt{13}}{7}$ **22.** $\dfrac{3p}{2}$ **23.** $\dfrac{q\sqrt[4]{3}}{2}$ **24.** $2a^2$ **25.** $-2x$

26. $\dfrac{5a\sqrt[3]{a}}{b}$ **27.** $\sqrt[12]{10,125}$ **28.** $2\sqrt[6]{2250}$ **29.** $\sqrt[3]{60}$ **30.** $\sqrt{60}$ **31.** $\sqrt{15ab}$ if $a, b \geq 0$ **33.** $\sqrt{x^2 - 49}$ **35.** $\sqrt{5}$ **37.** $5\sqrt{2}$ **39.** $3\sqrt[3]{2}$ **41.** $4|x|\sqrt{3}$ **43.** $-3x$

45. $2|m|\sqrt[4]{2}$ **47.** $2|p|\sqrt{3q}$ **49.** $9m^2\sqrt{2}$ **51.** $y^6\sqrt{y}$ **53.** $c^2\sqrt[3]{c^2}$ **55.** $5pq^2\sqrt{5p}$ **57.** $-2x^3\sqrt[3]{2}$ **59.** $-m\sqrt[5]{16m^3n^2}$ **61.** $(x - y)\sqrt[4]{x - y}$

63. $2\sqrt[3]{x^3 - y^3}$ **65.** 5   **67.** $3 + \sqrt{2}$ **69.** $\dfrac{1 - \sqrt{2}}{2}$ **71.** 5   **73.** 4   **75.** 2   **77.** $5x\sqrt{3}$ **79.** $2b\sqrt[3]{3b}$ **81.** $18ab^2\sqrt{10}$ **83.** $3pq\sqrt[4]{4p}$ **85.** $-2ab\sqrt[5]{3a}$

87. $2(x - y)\sqrt[4]{3(x - y)}$ **89.** $\dfrac{\sqrt{3}}{4}$ **91.** $\dfrac{x\sqrt[4]{5}}{2}$ **93.** $\dfrac{3y}{5x}$ **95.** $-\dfrac{3x^3}{4y^4}$ **97.** 2   **99.** 4   **101.** $2a\sqrt{2}$ **103.** $\dfrac{2a^2\sqrt{2}}{b}$ **105.** $\dfrac{16a^3}{3b}$ **107.** $a^2\sqrt[3]{26}$ **109.** $\dfrac{3x^3\sqrt{5}}{y}$

111. $\sqrt[6]{432}$ **113.** $\sqrt[6]{12}$ **115.** $3\sqrt[6]{12}$ **117.** $\sqrt[3]{18}$ **119.** $\dfrac{\sqrt[3]{5x}}{2}$ **121.** $\sqrt[3]{45a^2}$ **123.** $6a^2\sqrt{2}$ **125.** $3a\sqrt[3]{2b^2}$ **127.** $\dfrac{-2\sqrt[3]{2}}{a}$ **129.** $-10m\sqrt[3]{4}$ **131.** $3ab^2\sqrt[3]{3ab}$

133. 6   **135. (a)**

(b) $3\sqrt{5}$ units   **137. (a)** roughly $1,587,000   **(b)** roughly $2,381,000   **139. (a)** $4x^2$ **(b)** 36

141. $x = -3 - \sqrt{6}$ or $x = -3 + \sqrt{6}$ **143.** $x = \dfrac{-2 - \sqrt{7}}{3}$ or $x = \dfrac{-2 + \sqrt{7}}{3}$ **145.** The index of each radical must be the same. We can use the Laws of Exponents to rewrite radicals so that they have common indices.   **147.** $\left\{\dfrac{5}{2}\right\}$ **149.** $\left\{x \mid x \leq \dfrac{130}{3}\right\}$ **151.** Answers will vary.

Section 9.5 Adding, Subtracting, and Multiplying Radical Expressions

1. like radicals   **2.** False   **3.** $13\sqrt{13y}$ **4.** $7\sqrt[4]{5}$ **5.** $6\sqrt{2}$ **6.** $-8x\sqrt[3]{2x}$ **7.** Fully simplified   **8.** $(8z + 5)\sqrt[3]{z}$ **9.** $2\sqrt{m}$ **10.** $3\sqrt{6} - 30$ **11.** $3\sqrt[3]{12} - 2\sqrt[3]{3}$ **12.** $-74 - 27\sqrt{3}$ **13.** False   **14.** conjugates   **15.** False

16. $9 - 4\sqrt{5}$ **17.** $25 - 6\sqrt{14}$ **18.** 1   **19.** $10\sqrt{2}$ **21.** $2\sqrt[3]{x}$ **23.** $14\sqrt{5x}$ **25.** $11\sqrt[3]{5} - 11\sqrt{5}$ **27.** $8\sqrt{2}$ **29.** $-2\sqrt[3]{3}$ **31.** $-25\sqrt[3]{2}$ **33.** $9\sqrt{6x}$

35. $4\sqrt{2} + 3\sqrt{10}$ **37.** $32x\sqrt{3x}$ **39.** $2x\sqrt{3} - 11x\sqrt{2}$ **41.** $(3x - 8)\sqrt[3]{2}$ **43.** $5\sqrt{x - 1}$ **45.** $3\sqrt{x}$ **47.** $5\sqrt[3]{x}$ **49.** $2\sqrt{3} - 3\sqrt{6}$ **51.** $\sqrt{6} + 3\sqrt{2}$

53. $\sqrt[3]{12} - 2\sqrt[3]{3}$ **55.** $3\sqrt{2x} - 2x\sqrt{5}$ **57.** $12 + 3\sqrt{3} + 4\sqrt{2} + \sqrt{6}$ **59.** $12 - 6\sqrt{7} + 2\sqrt{3} - \sqrt{21}$ **61.** $6\sqrt{7} - 30$ **63.** $\sqrt{6} - 12\sqrt{3} + 9\sqrt{2} - 4$

65. $31 - 2\sqrt{15}$ **67.** $4 + 2\sqrt{3}$ **69.** $7 - 2\sqrt{10}$ **71.** $x - 2\sqrt{2x} + 2$ **73.** 1   **75.** -11 **77.** $2x - 3y$ **79.** $\sqrt[3]{x^2} + \sqrt[3]{x} - 12$ **81.** $\sqrt[3]{4a^2} - 25$

83. $\sqrt{15} + 5\sqrt{2}$ **85.** $26x^2\sqrt{7x}$ **87.** -13 **89.** $2\sqrt[3]{7} + \sqrt[3]{28}$ **91.** $-4 + 12\sqrt{2}$ **93.** $20\sqrt{2}$ **95.** $8 - 4\sqrt{15}$ **97.** $(3x + 2)\sqrt[3]{5y}$ **99.** $-7y$

101. $5 + x + 4\sqrt{x + 1}$ **103.** $27 + x - 10\sqrt{x + 2}$ **105.** $\dfrac{11\sqrt{5}}{25}$ **107. (a)** $3\sqrt{3x}$ **(b)** $6\sqrt{3}$ **(c)** $6x$

109. Check $x = -2 + \sqrt{5}$:

$0 = x^2 + 4x - 1$

$0 \overset{?}{=} (-2 + \sqrt{5})^2 + 4(-2 + \sqrt{5}) - 1$

$0 \overset{?}{=} (-2)^2 + 2(-2)(\sqrt{5}) + (\sqrt{5})^2 - 8 + 4\sqrt{5} - 1$

$0 \overset{?}{=} 4 - 4\sqrt{5} + \sqrt{25} - 8 + 4\sqrt{5} - 1$

$0 \overset{?}{=} 9 - 9$

$0 = 0$ true

The value is a solution.

Check $x = -2 - \sqrt{5}$:

$0 = x^2 + 4x - 1$

$0 \overset{?}{=} (-2 - \sqrt{5})^2 + 4(-2 - \sqrt{5}) - 1$

$0 \overset{?}{=} (-2)^2 - 2(-2)(\sqrt{5}) + (\sqrt{5})^2 - 8 - 4\sqrt{5} - 1$

$0 \overset{?}{=} 4 + 4\sqrt{5} + \sqrt{25} - 8 - 4\sqrt{5} - 1$

$0 \overset{?}{=} 9 - 9$

$0 = 0$ true

The value is a solution.

111. The perimeter is $30\sqrt{2}$ units. The area is 108 square units.   **113.** $12\sqrt{6}$ square units   **115.** To add or subtract radicals, the indices and radicands must be the same. Then use the Distributive Property "in reverse" to add the "coefficients" of the radicals.   **117.** $12a^5b^5$ **119.** $6y^2 + y - 2$ **121.** $25w^2 - 4$

Section 9.6 Rationalizing Radical Expressions

1. rationalizing the denominator   **2.** $\sqrt{11}$ **3.** $\dfrac{\sqrt{3}}{3}$ **4.** $\dfrac{\sqrt{10}}{4}$ **5.** $\dfrac{\sqrt{10x}}{2x}$ **6.** $\dfrac{4\sqrt[3]{9}}{3}$ **7.** $\dfrac{\sqrt[3]{150}}{10}$

8. $\dfrac{3\sqrt[4]{p^3}}{p}$ **9.** $-2 - \sqrt{7}$ **10.** $2(\sqrt{3} - 1)$ **11.** $\dfrac{\sqrt{3} + 1}{2}$ **12.** $\dfrac{5 + \sqrt{10} + 4\sqrt{5} + 4\sqrt{2}}{3}$ **13.** $\dfrac{\sqrt{2}}{2}$ **15.** $-\dfrac{2\sqrt{3}}{5}$ **17.** $\dfrac{\sqrt{3}}{2}$ **19.** $\dfrac{\sqrt{3}}{3}$ **21.** $\dfrac{\sqrt{2p}}{p}$ **23.** $\dfrac{2\sqrt{2y}}{y^2}$

25. $\sqrt[3]{4}$ **27.** $\dfrac{\sqrt[3]{7q^2}}{q}$ **29.** $-\dfrac{\sqrt[3]{60}}{10}$ **31.** $\dfrac{\sqrt[3]{50y^2}}{5y}$ **33.** $-\dfrac{4\sqrt[4]{27x}}{3x}$ **35.** $\dfrac{12\sqrt[5]{m^2n^3}}{mn}$ **37.** $2(\sqrt{6} + 2)$ **39.** $5(\sqrt{5} - 2)$ **41.** $2(\sqrt{7} + \sqrt{3})$ **43.** $\dfrac{\sqrt{5} + \sqrt{3}}{2}$

45. $\dfrac{p - \sqrt{pq}}{p - q}$ **47.** $-3(2\sqrt{3} - 3\sqrt{2})$ or $3(3\sqrt{2} - 2\sqrt{3})$ **49.** $-8 - 3\sqrt{7}$ or $-(8 + 3\sqrt{7})$ **51.** $\dfrac{13\sqrt{6} - 46}{38}$ **53.** $\dfrac{p + 4\sqrt{p} + 4}{p - 4}$ **55.** $\dfrac{2 - 3\sqrt{2}}{2}$

57. $\dfrac{4\sqrt{3}}{3}$ **59.** $\dfrac{\sqrt{10} - \sqrt{2}}{2}$ **61.** $\dfrac{22\sqrt{3}}{3}$ **63.** 0 **65.** $\dfrac{1}{2}$ **67.** $\dfrac{\sqrt{2}}{4}$ **69.** $\dfrac{2\sqrt{3}}{3}$ **71.** $\sqrt{3} - 2$ **73.** $2(\sqrt{5} - 2)$ **75.** 2 **77.** $\dfrac{\sqrt{3}}{3}$ **79.** $\dfrac{\sqrt[3]{18}}{6}$ **81.** $\dfrac{5 - \sqrt{3}}{22}$

83. $\dfrac{\sqrt{6} - \sqrt{2}}{4}$ **85.** $\dfrac{1}{3(\sqrt{2} - 1)}$ **87.** $\dfrac{x - h}{x + \sqrt{xh}}$

89. $\dfrac{(\sqrt{6})^2 + 2 \cdot \sqrt{6} \cdot \sqrt{2} + (\sqrt{2})^2}{4^2} \stackrel{?}{=} \left(\dfrac{\sqrt{2} + \sqrt{3}}{2}\right)^2$

$\dfrac{6 + 2\sqrt{12} + 2}{16} \stackrel{?}{=} \dfrac{2 + \sqrt{3}}{4}$

$\dfrac{8 + 2 \cdot 2\sqrt{3}}{16} \stackrel{?}{=} \dfrac{2 + \sqrt{3}}{4}$

$\dfrac{8 + 4\sqrt{3}}{16} \stackrel{?}{=} \dfrac{2 + \sqrt{3}}{4}$

$\dfrac{2 + \sqrt{3}}{4} = \dfrac{2 + \sqrt{3}}{4}$

91. (a) $\dfrac{1}{\sqrt{x + h} + \sqrt{x}}$ (b) $\dfrac{1}{2\sqrt{x}}$ (c) $\dfrac{1}{4}$ (d) $2; (4, 2)$ (e) $y = \dfrac{1}{4}x + 1$

(f)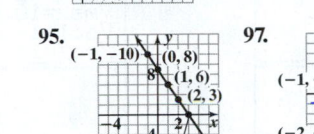

93. It is called rationalization because we are rewriting the expression so that the denominator is a rational number.

95. 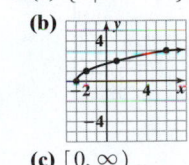   **97.**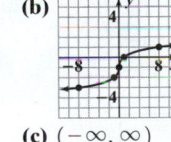

Putting the Concepts Together (Sections 9.1–9.6)

1. -5 **2.** $\dfrac{1}{16}$ **3.** $(3x^3)^{\frac{1}{4}}$ **4.** $7\sqrt[5]{z^4}$ **5.** $2x^{\frac{1}{2}}$ or $2\sqrt{x}$ **6.** $c^2 + c^3$ **7.** $\dfrac{a^2}{b^2}$ **8.** $x^{\frac{5}{8}}$ or $\sqrt[8]{x^5}$ **9.** $\dfrac{x^6}{y}$

10. $\sqrt{30ab}$ **11.** $10m^2n\sqrt{2n}$ **12.** $-2xy$ **13.** $\sqrt{3}$ **14.** $11b\sqrt{2b}$ **15.** $y\sqrt[3]{2y}$ **16.** $12x$ **17.** $7\sqrt{x}$ **18.** $14 - 28\sqrt{2} = 14(1 - 2\sqrt{2})$ **19.** $41 - 24\sqrt{2}$

20. $\dfrac{3\sqrt{2}}{16}$ **21.** $-\dfrac{4(\sqrt{3} + 8)}{61}$

Section 9.7 Functions Involving Radicals

1. (a) 4 (b) $2\sqrt{7}$ **2.** (a) -1 (b) 3 **3.** even; odd   **4.** $\{x \mid x \geq -6\}$ or $[-6, \infty)$

5. $\{t \mid t$ is any real number$\}$ or $(-\infty, \infty)$ **6.** $\{m \mid m \leq 2\}$ or $(-\infty, 2]$

7. (a) $\{x \mid x \geq -3\}$ or $[-3, \infty)$ **8.** (a) $\{x \mid x$ is any real number$\}$ or $(-\infty, \infty)$ **9.** (a) 3 (b) $\sqrt{14}$ (c) 2 **11.** (a) -5 (b) -1 (c) $-\dfrac{\sqrt{13}}{2}$

(b) 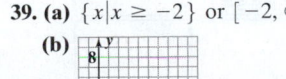   (b) 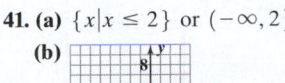   **13.** (a) 4 (b) $4\sqrt{6}$ (c) $\sqrt{6}$ **15.** (a) 2 (b) -2 (c) $-2\sqrt[3]{2}$

17. (a) $\dfrac{\sqrt{5}}{3}$ (b) $\dfrac{\sqrt{2}}{2}$ (c) $\dfrac{\sqrt{6}}{3}$ **19.** (a) 1 (b) $\sqrt[3]{4}$ (c) $\sqrt[3]{3}$

(c) $[0, \infty)$ (c) $(-\infty, \infty)$ **21.** $\{x \mid x \geq 7\}$ or $[7, \infty)$ **23.** $\left\{x \mid x \geq -\dfrac{7}{2}\right\}$ or $\left[-\dfrac{7}{2}, \infty\right)$

25. $\left\{x \mid x \leq \dfrac{4}{3}\right\}$ or $\left(-\infty, \dfrac{4}{3}\right]$ **27.** $\{z \mid z$ is any real number$\}$ or $(-\infty, \infty)$ **29.** $\left\{p \mid p \geq \dfrac{2}{7}\right\}$ or $\left[\dfrac{2}{7}, \infty\right)$ **31.** $\{x \mid x$ is any real number$\}$ or $(-\infty, \infty)$

33. $\{x \mid x > -5\}$ or $(-5, \infty)$ **35.** $\{x \mid x \neq 3\}$ or $(-\infty, 3) \cup (3, \infty)$

37. (a) $\{x \mid x \geq 4\}$ or $[4, \infty)$ **39.** (a) $\{x \mid x \geq -2\}$ or $[-2, \infty)$ **41.** (a) $\{x \mid x \leq 2\}$ or $(-\infty, 2]$ **43.** (a) $\{x \mid x \geq 0\}$ or $[0, \infty)$

(b)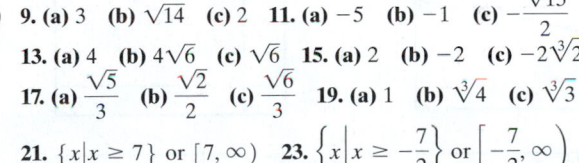

(c) $[0, \infty)$ (c) $[0, \infty)$ (c) $[0, \infty)$ (c) $[3, \infty)$

45. (a) $\{x \mid x \geq 0\}$ or $[0, \infty)$

(b)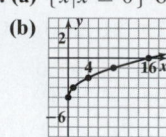

(c) $[-4, \infty)$

47. (a) $\{x \mid x \geq 0\}$ or $[0, \infty)$

(b)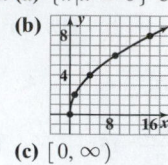

(c) $[0, \infty)$

49. (a) $\{x \mid x \geq 0\}$ or $[0, \infty)$

(b)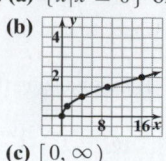

(c) $[0, \infty)$

51. (a) $\{x \mid x \geq 0\}$ or $[0, \infty)$

(b)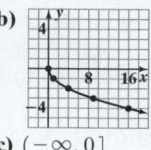

(c) $(-\infty, 0]$

53. (a) $\{x \mid x \text{ is any real number}\}$ or $(-\infty, \infty)$

(b)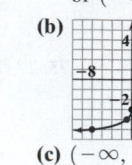

(c) $(-\infty, \infty)$

55. (a) $\{x \mid x \text{ is any real number}\}$ or $(-\infty, \infty)$

(b)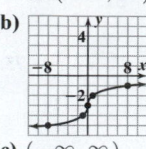

(c) $(-\infty, \infty)$

57. (a) $\{x \mid x \text{ is any real number}\}$ or $(-\infty, \infty)$

(b)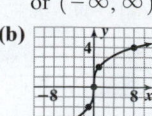

(c) $(-\infty, \infty)$

59. (a) 5 units

(b) $\sqrt{17} \approx 4.123$ units

(c) $5\sqrt{17} \approx 20.616$ units

61. (a) $4\sqrt{2} \approx 5.657$ square units

(b) $4\sqrt{5} \approx 8.944$ square units

(c) $2\sqrt{14} \approx 7.483$ square units

63. Shift the graph of $f(x) \mid c \mid$ units to the right if $c < 0$ or c units to the left if $c > 0$. **65.** $\dfrac{5}{6}$ **67.** $\dfrac{4x + 1}{x(x + 1)}$ **69.** $\dfrac{7x + 1}{(x - 1)(x + 1)}$ or $\dfrac{7x + 1}{x^2 - 1}$

71.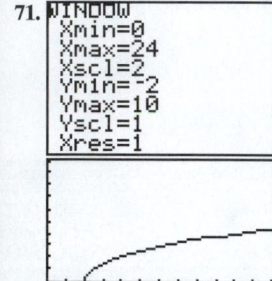

73.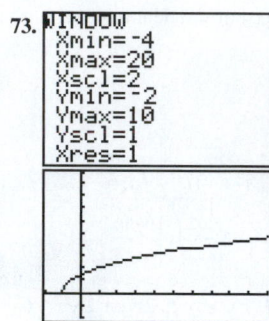

75.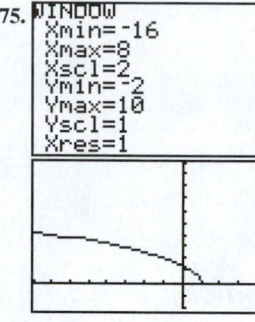

77.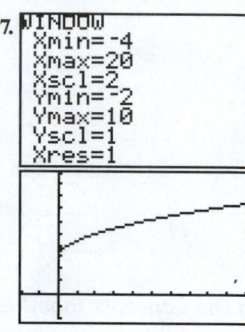

79.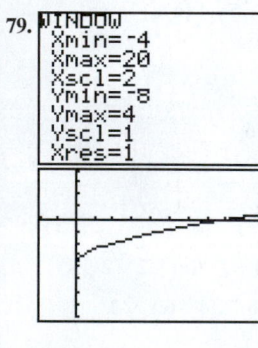

81.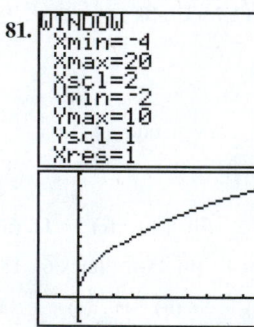

83.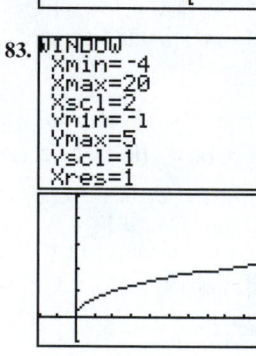

85.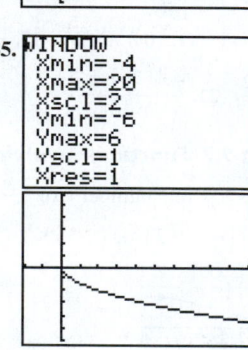

87.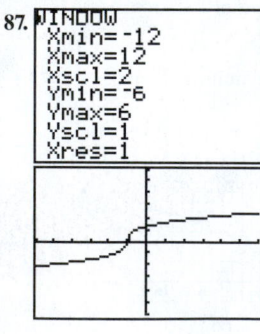

89.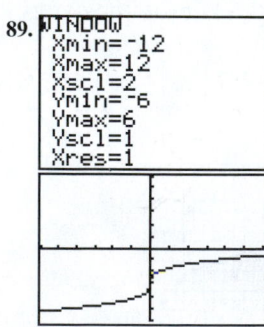

91.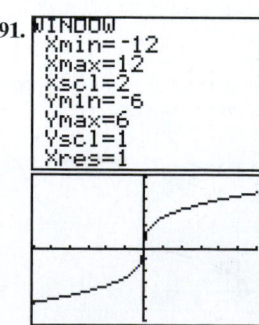

Section 9.8 Radical Equations and Their Applications **1.** radical equation **2.** extraneous **3.** False **4.** $\{5\}$ **5.** $\{-5\}$ **6.** $\varnothing$ or $\{\ \}$ **7.** $\{4\}$

8. $\{-3\}$ **9.** $\{15\}$ **10.** $\{-3, 1\}$ **11.** $\{12\}$ **12. (a)** $L = \dfrac{8T^2}{\pi^2}$ **(b)** 32 feet **13.** $\{16\}$ **15.** $\{7\}$ **17.** $\{11\}$ **19.** $\varnothing$ or $\{\ \}$ **21.** $\{2\}$ **23.** $\{12\}$

25. $\{25\}$ **27.** $\{11\}$ **29.** $\{-4\}$ **31.** $\{1\}$ **33.** $\{-5\}$ **35.** $\left\{0, \dfrac{1}{4}\right\}$ **37.** $\{3\}$ **39.** $\{4\}$ **41.** $\{4\}$ **43.** $\{6\}$ **45.** $\{4\}$ **47.** $\{-3\}$ **49.** $\{-1, 10\}$

51. $\{0, 4\}$ **53.** $\{3\}$ **55.** $\{5\}$ **57.** $\{2, 10\}$ **59.** $\{3\}$ **61.** $\{4\}$ **63.** $\{6\}$ **65.** $\{-7, 9\}$ **67.** $\{18\}$ **69.** $\{3\}$ **71.** $\{-23, 31\}$ **73.** $r = \dfrac{A^2 - P^2}{P^2}$

75. $V = \dfrac{4}{3}\pi r^3$ **77.** $F = \dfrac{q_1 q_2 r^2}{4\pi\varepsilon_0}$ **79.** $\varnothing$ or $\{\ \}$ **81.** $\{4\}$ **83.** $\{2\}$ **85.** $\{9\}$ **87.** $\{-3\}$ **89.** $\{-6\}$ **91.** $\{9\}$ **93.** $\{5\}$ **95.** $\{-2\}$

97. (a) $\{2\}$; $(2, 0)$ **(b)** $\{3\}$; $(3, 1)$ **(c)** $\{6\}$; $(6, 2)$

(d)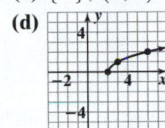

(e) The equation $f(x) = -1$ has no solution because the graph of the function does not go below the x-axis.

99. (a) $y = 5$ or $y = -1$

(b)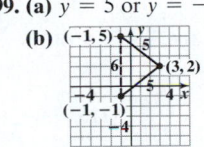

101. (a) after 2 years **(b)** after 16 years

103. (a) in the year 2053 **(b)** in the year 2021

105. $\{0\}$ **107.** $\left\{-\dfrac{3}{4}, 0\right\}$

109. If the index is even, it is important so that you can identify extraneous solutions. Regardless of the index, it is important to make sure that your answer is correct. **111.** In radical expressions with an even index, the radicand must be greater than or equal to zero. After raising both sides of the equation to an even index, it is possible to obtain solutions that result in a negative radicand. We do not have any restrictions on the value of the radicand when the index is odd, so we don't need to be worried about extraneous solutions. **113.** $0, -4, 12$ **115.** $\sqrt{2^3}, \pi, \sqrt[3]{-4}$ **117.** A rational number is a number that can be written as the quotient of two integers, where the denominator is not zero; a rational number is also a number where the decimal either terminates, or does not terminate but repeats. An irrational number has a decimal form that neither terminates nor repeats. The square root of -1 is not real because there is no real number whose square is -1.

119. 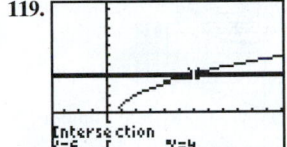 **121.** 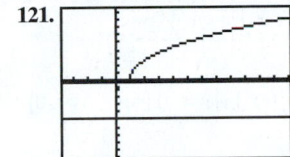 The two graphs do not intersect. Therefore, the equation has no real solution.

Section 9.9 The Complex Number System **1.** imaginary unit **2.** pure imaginary number **3.** $\sqrt{N}i$ **4.** True **5.** $6i$ **6.** $\sqrt{5}i$ **7.** $2\sqrt{3}i$ **8.** $4 + 10i$

9. $-2 - 2\sqrt{2}i$ **10.** $2 - 2\sqrt{2}i$ **11.** $1 + 11i$ **12.** $6 - 9i$ **13.** $-3 + i$ **14.** $12 + 15i$ **15.** $2 + 24i$ **16.** -18 **17.** $38 + 14i$ **18.** $-3 - 5i$ **19.** 73 **20.** 29

21. $\dfrac{1}{3} + \dfrac{4}{3}i$ **22.** $-\dfrac{1}{2} + \dfrac{3}{2}i$ **23.** $-i$ **24.** -1 **25.** $2i$ **27.** $-9i$ **29.** $3\sqrt{5}i$ **31.** $10\sqrt{3}i$ **33.** $\sqrt{7}i$ **35.** $5 + 7i$ **37.** $-2 - 2\sqrt{7}i$ **39.** $2 + i$ **41.** $\dfrac{1}{3} + \dfrac{\sqrt{2}}{6}i$

43. $6 - 2i$ **45.** $-4 + 6i$ **47.** $2 - 5i$ **49.** $3 - 2\sqrt{2}i$ **51.** $24 + 12i$ **53.** $-5 - 2i$ **55.** $5 + 10i$ **57.** $14 - 22i$ **59.** 26 **61.** $-4 + 5\sqrt{2}i$ **63.** $\dfrac{25}{48} + \dfrac{5}{24}i$

65. $5 + 12i$ **67.** $-9 + 40i$ **69.** -6 **71.** $-4\sqrt{5}$ **73.** $84 - 47i$ **75. (a)** $3 - 5i$ **(b)** 34 **77. (a)** $2 + 7i$ **(b)** 53 **79. (a)** $-7 - 2i$ **(b)** 53 **81.** $\dfrac{1}{3} - \dfrac{1}{3}i$

83. $\dfrac{2}{5} + i$ **85.** $\dfrac{6}{5} - \dfrac{3}{5}i$ **87.** $\dfrac{3}{29} - \dfrac{7}{29}i$ **89.** i **91.** $1 + 3i$ **93.** $-\dfrac{1}{5} - \dfrac{7}{5}i$ **95.** i **97.** $-i$ **99.** i **101.** $-i$ **103.** $-15 - 8i$ **105.** $5 + 12i$ **107.** $\dfrac{2}{3} + i$

109. $\dfrac{1}{2} - \dfrac{1}{2}i$ **111.** 12 **113.** $-15 - 20i$ **115.** $-5\sqrt{6}$ **117.** $-\dfrac{1}{5}i$ **119.** $\dfrac{2}{5} + \dfrac{1}{5}i$ **121.** $-\dfrac{4}{41} - \dfrac{5}{41}i$ **123. (a)** -1 **(b)** $2i$

125. (a) $-7 + 6i$ **(b)** $4 - 4i$ **127. (a)** $10 - i$ ohms **(b)** 10 ohms **(c)** -1 ohm **129. (a)** 0 **(b)** 0 **131. (a)** 0 **(b)** 0 **(c)** 0 **133.** For a polynomial with real coefficients, the zeros will be real numbers or will occur in conjugate pairs. If the complex number $a + bi$ is a complex zero of the polynomial, then its conjugate $a - bi$ is also a complex zero. **135.** The set of natural numbers is a subset of the set of whole numbers, which is a subset of the set of integers, which is a subset of the set of rational numbers, which is a subset of the set of real numbers, which is a subset of the set of complex numbers. Thus all real numbers are, more generally, complex numbers. **137.** Both methods can rely on the "FOIL" method or the Distributive Property.

139. $x^3 + 6x^2 + 12x + 8$ **141.** $18 + 26i$ **143.** Both methods rely on special product formulas to obtain the results. **145.** $-9.9 + 7.2i$

147. $-30.86 + 3.59i$ **149.** $-\dfrac{11}{17} + \dfrac{27}{17}i$ **151.** $6 - 25i$

Chapter 9 Review **1.** $-2, 2$ **2.** $-9, 9$ **3.** -1 **4.** -5 **5.** 0.4 **6.** 0.2 **7.** $\dfrac{5}{4}$ **8.** 6 **9.** 4 **10.** 12 **11.** 21 **12.** 11 **13.** rational; -3

14. irrational; -3.46 **15.** irrational; 3.74 **16.** not a real number **17.** $|4x - 9|$ **18.** $|m - 5|$ **19.** 7 **20.** -5 **21.** $\dfrac{2}{3}$ **22.** 3 **23.** 3 **24.** 10

25. z **26.** $|5p - 3|$ **27.** 9 **28.** not a real number **29.** -2 **30.** 9 **31.** 128 **32.** -9 **33.** -1331 **34.** 6 **35.** -4.02 **36.** 2.30 **37.** 4.64

38. 1.78 **39.** $(5a)^{1/3}$ **40.** $p^{7/5}$ **41.** $(10z)^{3/4}$ **42.** $(2ab)^{5/6}$ **43.** 64 **44.** $\dfrac{1}{k^{1/4}}$ **45.** $p^2 \cdot q^6$ or $(p \cdot q^3)^2$ **46.** $\dfrac{2b^{1/20}}{a^{3/10}}$ or $2\left(\dfrac{b}{a^6}\right)^{1/20}$ **47.** $10m^{1/3} + \dfrac{5}{m}$

48. $\dfrac{1}{2x^{1/3}}$ **49.** $\sqrt[4]{x^3}$ **50.** $11x^2y^5$ **51.** $m^2\sqrt[6]{m}$ **52.** $\dfrac{1}{\sqrt[3]{c}}$ **53.** $(3m-1)^{1/4}(3m^2-22m+9)$ **54.** $\dfrac{(3x+5)(x-3)}{(x^2-5)^{2/3}}$ **55.** $\sqrt{105}$ **56.** $\sqrt[4]{12a^3b^3}$

57. $4\sqrt{5}$ **58.** $-5\sqrt[3]{4}$ **59.** $3m^2n\sqrt[3]{6n}$ **60.** $p^2|q|\sqrt[4]{50}$ **61.** $8|x^3|\sqrt{y}$ as long as $y \geq 0$ **62.** $(2x+1)\sqrt{2x+1}$ as long as $2x+1 \geq 0$ **63.** $wz\sqrt{w}$

64. $3x^2z\sqrt{5yz}$ **65.** $2a^4b\sqrt[3]{2b^2}$ **66.** $2(x+1)$ **67.** $3\sqrt{30}$ **68.** $2\sqrt[3]{75}$ **69.** $-2x^2y^3\sqrt[3]{9x}$ **70.** $30xy\sqrt{3xy}$ **71.** $\dfrac{11}{5}$ **72.** $\dfrac{a^2\sqrt{5}}{8b}$ **73.** $\dfrac{\sqrt[3]{6}}{k}$

74. $\dfrac{-2w^5\sqrt[3]{20}}{7}$ **75.** $2h$ **76.** $\dfrac{5b^3}{2a}$ **77.** $-\dfrac{2x^2}{3y}$ **78.** $\dfrac{2n\sqrt{n}}{m}$ **79.** $\sqrt[6]{500}$ **80.** $2\sqrt[12]{2}$ **81.** $8\sqrt[4]{x}$ **82.** $6\sqrt[3]{4y}$ **83.** $5\sqrt{2}-4\sqrt{3}$ **84.** $13\sqrt{2}$

85. $\sqrt[3]{2z}$ **86.** $17\sqrt[3]{x^2}$ **87.** $7\sqrt{a}$ **88.** $9x\sqrt{3}$ **89.** $4m\sqrt[3]{4m^2y^2}$ **90.** $(y-1)\sqrt{y-4}$ **91.** $\sqrt{15}-3\sqrt{5}$ **92.** $3\sqrt[3]{5}+\sqrt[3]{20}$ **93.** $7+\sqrt{5}$

94. $42+7\sqrt{2}+6\sqrt{3}+\sqrt{6}$ **95.** -44 **96.** $9\sqrt[3]{x^2}+5\sqrt[3]{x}-4$ **97.** $x-2\sqrt{5x}+5$ **98.** $247+22\sqrt{10}$ **99.** $2a-b^2$ **100.** $\sqrt[3]{36s^2}-5\sqrt[3]{6s}-14$

101. $\dfrac{\sqrt{6}}{3}$ **102.** $2\sqrt{3}$ **103.** $\dfrac{4\sqrt{3p}}{p^2}$ **104.** $\dfrac{5\sqrt{2a}}{2a}$ **105.** $-\dfrac{\sqrt{6y}}{3y^2}$ **106.** $\dfrac{3\sqrt[3]{25}}{5}$ **107.** $-\dfrac{\sqrt[3]{300}}{15}$ **108.** $\dfrac{27\sqrt[5]{4p^2q}}{2pq}$ **109.** $\dfrac{42+6\sqrt{6}}{43}$

110. $-\dfrac{\sqrt{3}+9}{26}$ **111.** $\dfrac{\sqrt{3}(3-\sqrt{2})}{7}$ or $\dfrac{3\sqrt{3}-\sqrt{6}}{7}$ **112.** $\dfrac{k+\sqrt{km}}{k-m}$ **113.** $\dfrac{7+2\sqrt{10}}{3}$ **114.** $\dfrac{9-6\sqrt{y}+y}{9-y}$ or $\dfrac{y-6\sqrt{y}+9}{9-y}$

115. $\dfrac{10\sqrt{2}-4\sqrt{3}}{19}$ **116.** $\dfrac{8\sqrt{2}-3\sqrt{15}}{7}$ **117.** $\dfrac{25\sqrt{7}}{21}$ **118.** $\dfrac{4+\sqrt{7}}{9}$ **119. (a)** 1 **(b)** 2 **(c)** 3 **120. (a)** 0 **(b)** 2 **(c)** 4 **121. (a)** 1

(b) -1 **(c)** 2 **122. (a)** 0 **(b)** 2 **(c)** $\dfrac{1}{2}$ **123.** $\left\{x \mid x \geq \dfrac{5}{3}\right\}$ or $\left[\dfrac{5}{3}, \infty\right)$ **124.** $\{x \mid x$ is any real number$\}$ or $(-\infty, \infty)$ **125.** $\left\{x \mid x \geq -\dfrac{1}{6}\right\}$ or $\left[-\dfrac{1}{6}, \infty\right)$

126. $\{x \mid x$ is any real number$\}$ or $(-\infty, \infty)$ **127.** $\{x \mid x > 2\}$ or $(2, \infty)$ **128.** $\{x \mid x \geq 3\}$ or $[3, \infty)$

129. (a) $\{x \mid x \leq 1\}$ or $(-\infty, 1]$ **130. (a)** $\{x \mid x \geq -1\}$ or $[-1, \infty)$ **131. (a)** $\{x \mid x \geq -3\}$ or $[-3, \infty)$ **132. (a)** $\{x \mid x$ is any real number$\}$ or $(-\infty, \infty)$

(b) 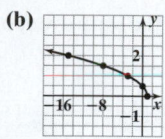 **(b)** 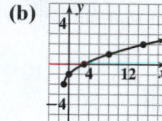 **(b)** 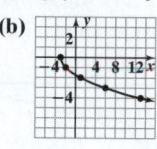 **(b)**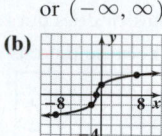

(c) $\{y \mid y \geq 0\}$ or $[0, \infty)$ **(c)** $\{y \mid y \geq -2\}$ or $[-2, \infty)$ **(c)** $\{y \mid y \leq 0\}$ or $(-\infty, 0]$ **(c)** $\{y \mid y$ is any real number$\}$ or $(-\infty, \infty)$

133. $\{169\}$ **134.** $\{-3\}$ **135.** $\left\{\dfrac{89}{3}\right\}$ **136.** $\varnothing$ or $\{\ \}$ **137.** $\{-5\}$ **138.** $\{25\}$ **139.** $\{-512\}$ **140.** $\{2\}$ **141.** $\{5\}$ **142.** $\{2\}$

143. $\{11\}$ **144.** $\{6\}$ **145.** $\{9\}$ **146.** $\varnothing$ or $\{\ \}$ **147.** $\left\{\dfrac{15}{2}\right\}$ **148.** $\{-5, 5\}$ **149.** $h = \dfrac{3V}{\pi r^2}$ **150.** $v = \dfrac{30}{f_s^3}$ **151.** $\sqrt{29}i$ **152.** $3\sqrt{6}i$

153. $14 - 9\sqrt{2}i$ **154.** $2 + \sqrt{5}i$ **155.** $1 - 2i$ **156.** $-5 + 10i$ **157.** $5 - 7\sqrt{5}i$ **158.** $-5 + 7i$ **159.** $47 + 13i$ **160.** $8 - \dfrac{11}{6}i$ **161.** -9

162. $67 - 42i$ **163.** 145 **164.** $27 + 38i$ **165.** $\dfrac{6}{17} - \dfrac{10}{17}i$ **166.** $-\dfrac{21}{53} - \dfrac{6}{53}i$ **167.** $\dfrac{4}{29} - \dfrac{19}{29}i$ **168.** $\dfrac{1}{2} + \dfrac{7}{2}i$ **169.** $-i$ **170.** i

Chapter 9 Test **1.** $\dfrac{1}{7}$ **2.** $6y\sqrt[12]{x^8y^3}$ **3.** $2a^5b^4\sqrt[5]{4a^3b}$ **4.** $\sqrt{39mn}$ **5.** $4x^3y^2\sqrt{2x}$ **6.** $\dfrac{3a}{2b^2}, a > 0, b > 0$ **7.** $(x+6)\sqrt{5x}$ **8.** $a\sqrt{b}$

9. $33 - 5\sqrt{x} - 2x$ **10.** $-\dfrac{\sqrt{2}}{18}$ **11.** $5 - 2\sqrt{5}$ **12. (a)** 1 **(b)** 3 **13.** $\left\{x \mid x \leq \dfrac{5}{3}\right\}$ or $\left(-\infty, \dfrac{5}{3}\right]$ **14. (a)** $\{x \mid x \geq 0\}$ or $[0, \infty)$

(b) 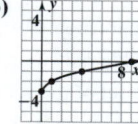 **(c)** $\{y \mid y \geq -3\}$ or $[-3, \infty)$ **15.** $\{13\}$ **16.** $\{3\}$ **17.** $\{2\}$ **18.** $17 - 13i$ **19.** $29 - 2i$ **20.** $\dfrac{73}{265} - \dfrac{89}{265}i$

Cumulative Review Chapters 1–9 **1.** $\dfrac{9}{2}$ **2.** $x + 7y + 6$ **3.** $\{4\}$ **4.** $(-\infty, 2]$ **5. (a)** 19 **(b)** 29 **6.** $\{x \mid x \neq 4, -2\}$

7. (a) 3500 games **(b)** \$120 **8.** $y = -2x + 4$ **9.** 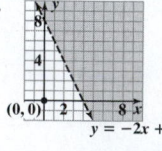 **10.** $(4, -1)$ **11.** 4 pounds of dried fruit and 6 pounds of nuts

12. -96 **13.** $8x^3 - 2x^2 - 3x + 10$ **14.** $8x^3 - 20x + 9$ **15.** $3x^2 + 2x - 6 + \dfrac{7}{2x^2 + 3x - 5}$ **16.** $4(2x+3)(x-7)$

17. $\dfrac{x^2 + 6x + 3}{(x-3)(x+2)}$ **18.** $\{-1, 4\}$ **19.** $\{x \mid x < 4$ or $x \geq 10\}$, or $(-\infty, 4) \cup [10, \infty)$ [number line from -4 to 12]

20. 2.4 hours **21.** $\dfrac{5a^2}{|b|}$ **22.** $\left\{x \mid x \leq \dfrac{8}{3}\right\}$ or $\left(-\infty, \dfrac{8}{3}\right]$ **23.** $\{2\}$ **24.** $-\dfrac{21}{50} + \dfrac{3}{50}i$ **25.** $\dfrac{8 + 2\sqrt{11}}{5}$

Chapter 10 Quadratic Equations and Functions

Section 10.1 Solving Quadratic Equations by Completing the Square 1. $\sqrt{p}; -\sqrt{p}$ 2. $\{-4\sqrt{3}, 4\sqrt{3}\}$ 3. $\{-5, 5\}$ 4. $\{-9, 9\}$ 5. $\{-6\sqrt{2}i, 6\sqrt{2}i\}$
6. $\{-3i, 3i\}$ 7. $\{-13, 7\}$ 8. $\{5 - 4i, 5 + 4i\}$ 9. $49; (p + 7)^2$ 10. $\dfrac{9}{4}; \left(w - \dfrac{3}{2}\right)^2$ 11. $\{-4, 2\}$ 12. $\{4 - \sqrt{7}, 4 + \sqrt{7}\}$ 13. $\left\{\dfrac{-3 - \sqrt{11}}{2}, \dfrac{-3 + \sqrt{11}}{2}\right\}$
14. $\left\{-\dfrac{1}{3} - \dfrac{2\sqrt{5}}{3}i, -\dfrac{1}{3} + \dfrac{2\sqrt{5}}{3}i\right\}$ 15. hypotenuse; legs 16. False 17. $c = 5$ 18. Approximately 17.32 miles 19. $\{-10, 10\}$ 21. $\{-5\sqrt{2}, 5\sqrt{2}\}$
23. $\{-5i, 5i\}$ 25. $\left\{-\dfrac{\sqrt{5}}{2}, \dfrac{\sqrt{5}}{2}\right\}$ 27. $\{-2\sqrt{2}, 2\sqrt{2}\}$ 29. $\{-4, 4\}$ 31. $\left\{-\dfrac{2\sqrt{6}}{3}, \dfrac{2\sqrt{6}}{3}\right\}$ 33. $\{-2i, 2i\}$ 35. $\{1 - 3\sqrt{2}i, 1 + 3\sqrt{2}i\}$
37. $\{-5 - \sqrt{3}, -5 + \sqrt{3}\}$ 39. $\left\{-\dfrac{4}{3}, \dfrac{2}{3}\right\}$ 41. $\left\{\dfrac{2}{3} - \dfrac{\sqrt{5}}{3}, \dfrac{2}{3} + \dfrac{\sqrt{5}}{3}\right\}$ 43. $\{-13, 5\}$ 45. $x^2 + 10x + 25; (x + 5)^2$ 47. $z^2 - 18z + 81; (z - 9)^2$
49. $y^2 + 7y + \dfrac{49}{4}; \left(y + \dfrac{7}{2}\right)^2$ 51. $w^2 + \dfrac{1}{2}w + \dfrac{1}{16}; \left(w + \dfrac{1}{4}\right)^2$ 53. $\{-6, 2\}$ 55. $\{2 - \sqrt{3}, 2 + \sqrt{3}\}$ 57. $\{2 - i, 2 + i\}$ 59. $\left\{-\dfrac{5}{2} - \dfrac{\sqrt{33}}{2}, -\dfrac{5}{2} + \dfrac{\sqrt{33}}{2}\right\}$
61. $\{4 - \sqrt{19}, 4 + \sqrt{19}\}$ 63. $\left\{\dfrac{1}{2} - \dfrac{\sqrt{11}}{2}i, \dfrac{1}{2} + \dfrac{\sqrt{11}}{2}i\right\}$ 65. $\left\{-\dfrac{3}{2}, 4\right\}$ 67. $\left\{1 - \dfrac{\sqrt{3}}{3}, 1 + \dfrac{\sqrt{3}}{3}\right\}$ 69. $\left\{\dfrac{5}{4} - \dfrac{\sqrt{17}}{4}, \dfrac{5}{4} + \dfrac{\sqrt{17}}{4}\right\}$
71. $\left\{-1 - \dfrac{\sqrt{6}}{2}i, -1 + \dfrac{\sqrt{6}}{2}i\right\}$ 73. 10 75. 20 77. $5\sqrt{2}; 7.07$ 79. 2 81. $2\sqrt{34}; 11.66$ 83. $b = 4\sqrt{3} \approx 6.93$ 85. $a = 4\sqrt{5} \approx 8.94$
87. $\{-3, 9\}; (-3, 36), (9, 36)$ 89. $\{-2 - 3\sqrt{2}, -2 + 3\sqrt{2}\}; (-2 - 3\sqrt{2}, 18), (-2 + 3\sqrt{2}, 18)$ 91. $4\sqrt{5}$ units 93. approximately 104.403 yards
95. approximately 31.623 feet 97. (a) approximately 22.913 feet (b) 15 feet 99. (a) 1 second (b) approximately 1.732 seconds (c) 2 seconds
101. approximately 9.54% 103. The triangle is a right triangle; the hypotenuse is 17. 105. The triangle is not a right triangle.
107. $c^2 = (m^2 + n^2)^2 = m^4 + 2m^2n^2 + n^4$ 109. $\{-4, 9\}$ 111. $\left\{-1, \dfrac{1}{2}\right\}$ 113. In both cases, the simpler equations are linear.
$a^2 + b^2 = (m^2 - n^2)^2 + (2mn)^2$
$= m^4 - 2m^2n^2 + n^4 + 4m^2n^2$
$= m^4 + 2m^2n^2 + n^4$
Because c^2 and $a^2 + b^2$ result in the same
expression, $a, b,$ and c are the lengths of the
sides of a right triangle.

Section 10.2 Solving Quadratic Equations by the Quadratic Formula 1. $\dfrac{-b \pm \sqrt{b^2 - 4ac}}{2a}$ 2. $\left\{-\dfrac{3}{2}, 3\right\}$ 3. $\left\{-4, \dfrac{1}{2}\right\}$ 4. False

5. $\left\{1 - \dfrac{\sqrt{3}}{2}, 1 + \dfrac{\sqrt{3}}{2}\right\}$ 6. $\left\{\dfrac{5}{2}\right\}$ 7. $\left\{2 - \dfrac{\sqrt{10}}{2}, 2 + \dfrac{\sqrt{10}}{2}\right\}$ 8. $\{-1 - 5i, -1 + 5i\}$ 9. discriminant 10. False 11. negative 12. False 13. True
14. Two complex solutions that are not real 15. One repeated rational solution 16. Two irrational solutions 17. True 18. $\{-3, 3\}$
19. $\left\{\dfrac{5}{4} - \dfrac{\sqrt{23}}{4}i, \dfrac{5}{4} + \dfrac{\sqrt{23}}{4}i\right\}$ 20. $\left\{-\dfrac{5}{3}, 1\right\}$ 21. (a) 200 or 600 DVDs (b) 400 DVDs 22. 16 meters by 30 meters 23. $\{-2, 6\}$ 25. $\left\{-\dfrac{3}{2}, \dfrac{5}{3}\right\}$
27. $\left\{1 - \dfrac{\sqrt{3}}{2}, 1 + \dfrac{\sqrt{3}}{2}\right\}$ 29. $\left\{1 - \dfrac{2\sqrt{3}}{3}, 1 + \dfrac{2\sqrt{3}}{3}\right\}$ 31. $\left\{-\dfrac{1}{3} - \dfrac{\sqrt{13}}{3}, -\dfrac{1}{3} + \dfrac{\sqrt{13}}{3}\right\}$ 33. $\{1 - \sqrt{6}i, 1 + \sqrt{6}i\}$ 35. $\left\{\dfrac{1}{2} - \dfrac{\sqrt{13}}{2}i, \dfrac{1}{2} + \dfrac{\sqrt{13}}{2}i\right\}$
37. $\left\{\dfrac{1}{4} - \dfrac{\sqrt{5}}{4}, \dfrac{1}{4} + \dfrac{\sqrt{5}}{4}\right\}$ 39. $\left\{-\dfrac{2}{3} - \dfrac{\sqrt{7}}{3}, -\dfrac{2}{3} + \dfrac{\sqrt{7}}{3}\right\}$ 41. 21; two irrational solutions 43. -56; two complex solutions that are not real
45. 0; one repeated rational solution 47. -8; two complex solutions that are not real 49. 44; two irrational solutions 51. $\left\{\dfrac{5}{2} - \dfrac{\sqrt{5}}{2}, \dfrac{5}{2} + \dfrac{\sqrt{5}}{2}\right\}$
53. $\left\{-\dfrac{8}{3}, 1\right\}$ 55. $\left\{-\dfrac{7}{2}, 5\right\}$ 57. $\{-1 - \sqrt{7}i, -1 + \sqrt{7}i\}$ 59. $\left\{-\dfrac{3}{2}\right\}$ 61. $\left\{\dfrac{1}{7} - \dfrac{\sqrt{29}}{7}, \dfrac{1}{7} + \dfrac{\sqrt{29}}{7}\right\}$ 63. $\{-4, 4\}$ 65. $\left\{\dfrac{1}{4} - \dfrac{1}{4}i, \dfrac{1}{4} + \dfrac{1}{4}i\right\}$ 67. $\left\{-\dfrac{2}{3}\right\}$
69. $\left\{-\dfrac{1}{3} - \dfrac{2\sqrt{7}}{3}, -\dfrac{1}{3} + \dfrac{2\sqrt{7}}{3}\right\}$ 71. $\{2 - \sqrt{13}, 2 + \sqrt{13}\}$ 73. $\{1 - \sqrt{5}, 1 + \sqrt{5}\}$ 75. $\left\{\dfrac{1}{4} - \dfrac{3\sqrt{7}}{4}i, \dfrac{1}{4} + \dfrac{3\sqrt{7}}{4}i\right\}$ 77. (a) $\{-7, 3\}$
(b) $\{-4, 0\}; (-4, -21), (0, -21)$ 79. (a) $\left\{-1 - \dfrac{\sqrt{6}}{2}, -1 + \dfrac{\sqrt{6}}{2}\right\}$ (b) $\left\{-1 - \dfrac{\sqrt{2}}{2}, -1 + \dfrac{\sqrt{2}}{2}\right\}$ 81. $\dfrac{-1 - \sqrt{7}}{3}, \dfrac{-1 + \sqrt{7}}{3}$ 83. $x = 3$; the three
sides measure 3, 4, and 5 units 85. Either $x = 1$ and the three sides measure 3, 4, and 5 units, or $x = 5$ and the three sides measure 7, 24, and 25 units.
87. $-2 + 2\sqrt{11}$ inches by $2 + 2\sqrt{11}$ inches, which is approximately 4.633 inches by 8.633 inches. 89. The base is $\dfrac{3}{2} + \dfrac{\sqrt{209}}{2}$ inches, which is
approximately 8.728 inches; the height is $-\dfrac{3}{2} + \dfrac{\sqrt{209}}{2}$ inches, which is approximately 5.728 inches. 91. (a) $R(17) = 1161.1$; if 17 pairs of sunglasses are
sold per week, then the company's revenue will be $1161.10. $R(25) = 1687.5$; if 25 pairs of sunglasses are sold per week, then the company's revenue
will be $1687.50. (b) either 200 or 500 pairs of sunglasses (c) 350 pairs of sunglasses 93. (a) after approximately 0.6 second and after approximately
3.8 seconds (b) after approximately 1.3 seconds and after approximately 3.0 seconds (c) No; the solutions to the equation are
complex solutions that are not real. 95. 12 inches 97. (a) ages 25 and 68 (b) ages 30 and 63 99. approximately 4.3 miles per hour

101. approximately 4.6 hours

103. By the quadratic formula, the solutions of the equation

$ax^2 + bx + c = 0$ are $x = \dfrac{-b - \sqrt{b^2 - 4ac}}{2a}$ and $x = \dfrac{-b + \sqrt{b^2 - 4ac}}{2a}$.

The sum of these two solutions is

$\dfrac{-b - \sqrt{b^2 - 4ac}}{2a} + \dfrac{-b + \sqrt{b^2 - 4ac}}{2a} = \dfrac{-2b}{2a} = -\dfrac{b}{a}$.

105. The solutions of $ax^2 + bx + c = 0$ are $x = \dfrac{-b \pm \sqrt{b^2 - 4ac}}{2a}$.

The solutions of $ax^2 - bx + c = 0$ are

$x = \dfrac{-(-b) \pm \sqrt{(-b)^2 - 4ac}}{2a} = \dfrac{b \pm \sqrt{b^2 - 4ac}}{2a}$.

Now, the negatives of the solutions to $ax^2 - bx + c = 0$

are $-\left(\dfrac{b \pm \sqrt{b^2 - 4ac}}{2a} \right) = \dfrac{-b \mp \sqrt{b^2 - 4ac}}{2a} = \dfrac{-b \pm \sqrt{b^2 - 4ac}}{2a}$,

which are the solutions to $ax^2 + bx + c = 0$.

107. Use factoring if the discriminant is a perfect square.

109. (a) 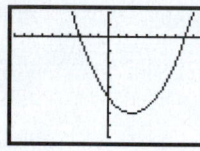 **(b)** $x = -1$ or $x = -2$

(c) The x-intercepts of the function $f(x) = x^2 + 3x + 2$ are -2 and -1, which are the same as the solutions of the equation $x^2 + 3x + 2 = 0$.

111. (a) 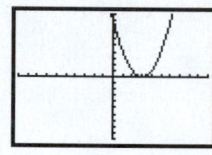 **(b)** $x = 1$

(c) The x-intercept of the function $g(x) = x^2 - 2x + 1$ is 1, which is the same as the solution of the equation $x^2 - 2x + 1 = 0$.

113. The discriminant is 37; the equation has two irrational solutions. This conclusion based on the discriminant is apparent in the graph because the graph has two x-intercepts. **115.** The discriminant is -7; the equation has two complex solutions that are not real. This conclusion based on the discriminant is apparent in the graph because the graph has no x-intercepts.

117. (a) $x = -3$ or $x = 8$

(b) The x-intercepts are -3 and 8.

119. (a) $x = 3$

(b) The x-intercept is 3.

121. (a) $x = -\dfrac{5}{2} \pm \dfrac{\sqrt{7}}{2} i$

(b) The graph has no x-intercepts.

The x-intercepts of $y = x^2 - 5x - 24$ are the same as the solutions of $x^2 - 5x - 24 = 0$.

The x-intercept of $y = x^2 - 6x + 9$ is the same as the solution of $x^2 - 6x + 9 = 0$.

$y = x^2 + 5x + 8$ has no x-intercepts, and the solutions of $x^2 + 5x + 8 = 0$ are not real.

Section 10.3 Solving Equations Quadratic in Form **1.** quadratic in form **2.** $3x + 1$ **3.** True **4.** $u = \dfrac{1}{x}$ **5.** $\{-3, -2, 2, 3\}$ **6.** $\{-3, 3, \sqrt{2}i, -\sqrt{2}i\}$

7. $\{-3, -2, 2, 3\}$ **8.** $\left\{ -\dfrac{\sqrt{3}}{2}, \dfrac{\sqrt{3}}{2}, -i, i \right\}$ **9.** $\left\{ \dfrac{4}{9}, 16 \right\}$ **10.** $\{25\}$ **11.** $\left\{ -\dfrac{5}{2}, -\dfrac{1}{2} \right\}$ **12.** $\{-1, 125\}$ **13.** $\{-2, -1, 1, 2\}$ **15.** $\{-3i, 3i, -2i, 2i\}$

17. $\left\{ -2, -\dfrac{1}{2}, \dfrac{1}{2}, 2 \right\}$ **19.** $\{-\sqrt{3}, -\sqrt{2}, \sqrt{2}, \sqrt{3}\}$ **21.** $\{2, 10\}$ **23.** $\{-3, -2, 2, 3\}$ **25.** $\{-2i, 2i, -\sqrt{7}i, \sqrt{7}i\}$ **27.** $\{16\}$ **29.** $\varnothing$ or $\{ \ \}$ **31.** $\left\{ \dfrac{1}{4} \right\}$ **33.** $\left\{ -\dfrac{1}{7}, \dfrac{1}{4} \right\}$

35. $\left\{ -\dfrac{2}{3}, \dfrac{5}{2} \right\}$ **37.** $\{-64, 1\}$ **39.** $\{-1, 8\}$ **41.** $\{25\}$ **43.** $\left\{ \dfrac{1}{3}, \dfrac{1}{2} \right\}$ **45.** $\left\{ -\dfrac{11}{5}, -1 \right\}$ **47.** $\left\{ 1, 3, -\dfrac{1}{2} - \dfrac{\sqrt{3}}{2}i, -\dfrac{1}{2} + \dfrac{\sqrt{3}}{2}i, -\dfrac{3}{2} - \dfrac{3\sqrt{3}}{2}i, -\dfrac{3}{2} + \dfrac{3\sqrt{3}}{2}i \right\}$

49. $\{-2, 4\}$ **51.** $\{-2i, 2i, -2\sqrt{2}, 2\sqrt{2}\}$ **53.** $\{16\}$ **55.** $\{-3, 3, -2i, 2i\}$ **57.** $\left\{ -\dfrac{10}{3}, -2 \right\}$ **59.** $\{9, 16\}$ **61.** $\left\{ \dfrac{1}{2}, 5 \right\}$ **63. (a)** $0, -\sqrt{7}i, \sqrt{7}i$

(b) $-\sqrt{6}i, \sqrt{6}i, -i, i$ **65. (a)** $0, -\sqrt{3}, \sqrt{3}$ **(b)** $-\sqrt{2}i, \sqrt{2}i, -\sqrt{5}, \sqrt{5}$ **67. (a)** $-1, \dfrac{1}{6}$ **(b)** $-\dfrac{1}{2}, \dfrac{1}{7}$ **69.** $-\sqrt{7}i, \sqrt{7}i, -\sqrt{2}i, \sqrt{2}i$ **71.** $\dfrac{81}{4}$ **73.** $-\dfrac{8}{3}, -2$

75. (a) $x = 2$ or $x = 3$ **(b)** $x = 5$ or $x = 6$; comparing these solutions to those in part (a), we note that $5 = 2 + 3$ and $6 = 3 + 3$. **(c)** $x = 0$ or $x = 1$; comparing these solutions to those in part (a), we note that $0 = 2 - 2$ and $1 = 3 - 2$. **(d)** $x = 7$ or $x = 8$; comparing these solutions to those in part (a), we note that $7 = 2 + 5$ and $8 = 3 + 5$. **(e)** The solution set of the equation $(x - a)^2 - 5(x - a) + 6 = 0$ is $\{2 + a, 3 + a\}$.

77. (a) $x = \dfrac{1}{2}$ or $x = 1$ **(b)** $x = \dfrac{5}{2}$ or $x = 3$; comparing these solutions to those in part (a), we note that $\dfrac{5}{2} = \dfrac{1}{2} + 2$ and $3 = 1 + 2$.

(c) $x = \dfrac{11}{2}$ or $x = 6$; comparing these solutions to those in part (a), we note that $\dfrac{11}{2} = \dfrac{1}{2} + 5$ and $6 = 1 + 5$. **(d)** For $f(x) = 2x^2 - 3x + 1$, the zeros

of $f(x - a)$ are $\dfrac{1}{2} + a$ and $1 + a$. **79. (a)** $R(1990) = 3000$; the revenue in 1990 was $3000 thousand (or \$3,000,000) **(b)** $x = 2000$; in the year 2000,

revenue was \$3065 thousand (or \$3,065,000) **(c)** 2015 **81.** $x = \pm\dfrac{\sqrt{10 + 2\sqrt{17}}}{2} i$ or $x = \pm\dfrac{\sqrt{10 - 2\sqrt{17}}}{2} i$ **83.** $x = \pm\dfrac{3\sqrt{2}}{2} i$ **85.** $\{36\}$; Answers will

vary. **87.** Extraneous solutions may result after squaring both sides of the equation. **89.** $p^{-2} + 4p^{-1} + 9$ **91.** $5\sqrt[3]{2a} - 4a\sqrt[3]{2a} = (5 - 4a)\sqrt[3]{2a}$

93. Let $Y_1 = x^4 + 5x^2 - 14$.   **95.** Let $Y_1 = 2(x-2)^2$ and $Y_2 = 5(x-2) + 1$.   **97.** Let $Y_1 = x - 5\sqrt{x}$ and $Y_2 = -3$.

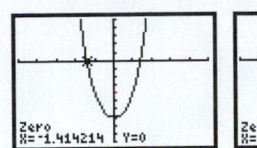 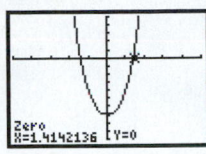   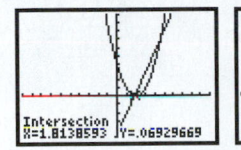 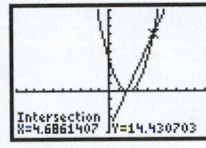   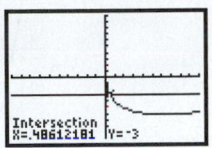

The solution set is approximately $\{-1.41, 1.41\}$.   The solution set is approximately $\{1.81, 4.69\}$.   The solution set is approximately $\{0.49\}$.

99. (a) $Y_1 = x^2 - 5x - 6$ **(b)** $Y_1 = (x+2)^2 - 5(x+2) - 6$ **(c)** $Y_1 = (x+5)^2 - 5(x+5) - 6$

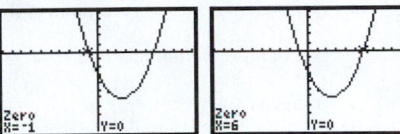   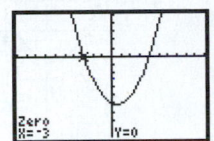 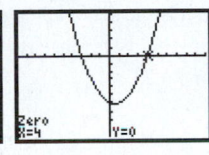   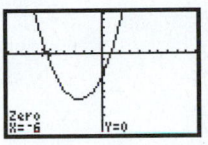 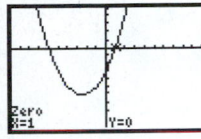

The x-intercepts are -1 and 6.   The x-intercepts are -3 and 4.   The x-intercepts are -6 and 1.

(d) The x-intercepts of the graph of $y = f(x) = x^2 - 5x - 6$ are -1 and 6. The x-intercepts of the graph of
$y = f(x+a) = (x+a)^2 - 5(x+a) - 6$ are $-1 - a$ and $6 - a$.

Putting the Concepts Together (Sections 10.1–10.3)   **1.** $z^2 + 10z + 25 = (z+5)^2$ **2.** $x^2 + 7x + \dfrac{49}{4} = \left(x + \dfrac{7}{2}\right)^2$ **3.** $n^2 - \dfrac{1}{4}n + \dfrac{1}{64} = \left(n - \dfrac{1}{8}\right)^2$

4. $\left\{\dfrac{1}{2}, \dfrac{5}{2}\right\}$ **5.** $\{-4 - 2\sqrt{3}, -4 + 2\sqrt{3}\}$ **6.** $\{3 - \sqrt{2}, 3 + \sqrt{2}\}$ **7.** $\left\{-\dfrac{4\sqrt{5}}{7}, \dfrac{4\sqrt{5}}{7}\right\}$ **8.** $\{4 - \sqrt{10}, 4 + \sqrt{10}\}$ **9.** $\left\{-1 - \dfrac{\sqrt{3}}{3}i, -1 + \dfrac{\sqrt{3}}{3}i\right\}$

10. $\left\{-2 - \dfrac{\sqrt{42}}{3}, -2 + \dfrac{\sqrt{42}}{3}\right\}$ **11.** $b^2 - 4ac = 0$; the quadratic equation has one repeated rational solution. **12.** $b^2 - 4ac = 60$; the quadratic equation has

two irrational solutions. **13.** $b^2 - 4ac = -4$; the quadratic equation has two complex solutions that are not real. **14.** $c = \sqrt{116} = 2\sqrt{29}$

15. $\left\{\dfrac{9}{4}\right\}$ **16.** $\left\{\dfrac{1}{6}, -\dfrac{1}{3}\right\}$ **17.** Revenue will be \$12,000 when either 150 microwaves or 200 microwaves are sold. **18.** The speed of the wind was

approximately 52.9 miles per hour.

Section 10.4 Graphing Quadratic Functions Using Transformations   **1.** quadratic function **2.** up; down

3. 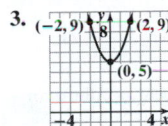   **4.** 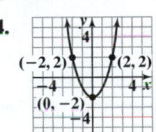   **5.** False   **6.** 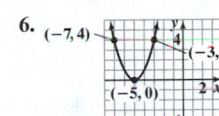   **7.** 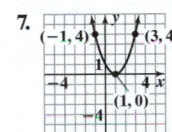   **8.**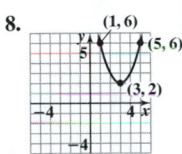

9. 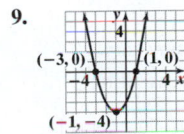   **10.** y; a; vertically stretched; vertically compressed   **11.** 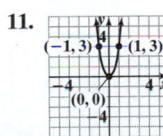   **12.** 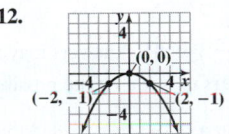   **13.** True   **14.** 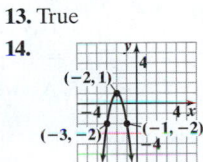   Domain: $\{x \mid x$ is any real number$\}$ or $(-\infty, \infty)$; Range: $\{y \mid y \le 1\}$ or $(-\infty, 1]$

15. 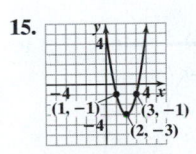   Domain: $\{x \mid x$ is any real number$\}$ or $(-\infty, \infty)$ Range: $\{y \mid y \ge -3\}$ or $[-3, \infty)$ **16.** $f(x) = -(x+1)^2 + 2$ **17.** (I) (D); (II) (A); (III) (C); (IV) (B) **19.** shift 10 units to the left **21.** shift 12 units up **23.** vertically stretch by a factor of 2 (multiply the y-coordinates by 2) and shift 5 units to the right **25.** multiply the y-coordinates by -3 (which means it opens down and is stretched vertically by a factor of 3), shift 5 units to the left, and shift up 8 units

27. 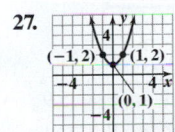   **29.** 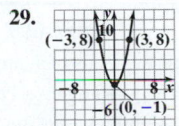   **31.** 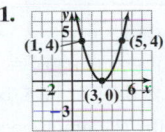   **33.** 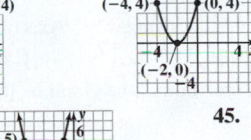   **35.** 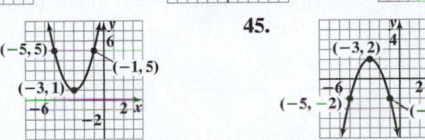   **37.**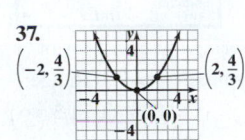

39. 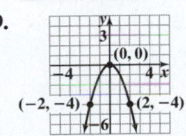   **41.** 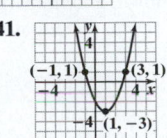   **43.** 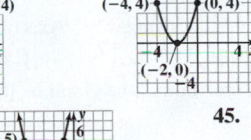   **45.** 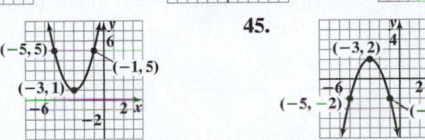   **47.**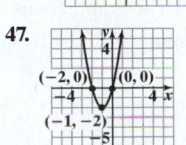

49.

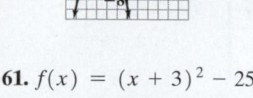

51. $f(x) = (x + 1)^2 - 5$

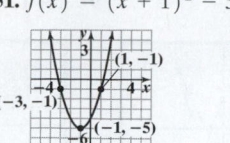

vertex is $(-1, -5)$;

53. $g(x) = (x - 2)^2 + 4$

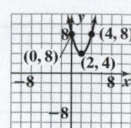

55. $f(x) = (x + 1)^2 - 3$

57. $f(x) = -2(x - 3)^2 + 7$

59. $f(x) = (x + 4)^2$

61. $f(x) = (x + 3)^2 - 25$

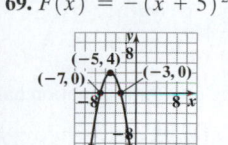

vertex is $(-3, -25)$; axis of symmetry is $x = -3$; domain is the set of all real numbers or $(-\infty, \infty)$; range is $\{y \mid y \geq -25\}$ or $[-25, \infty)$

63. $F(x) = \left(x + \dfrac{1}{2}\right)^2 - \dfrac{49}{4}$

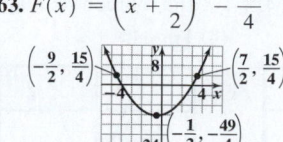

vertex is $\left(-\dfrac{1}{2}, -\dfrac{49}{4}\right)$; axis of symmetry is $x = -\dfrac{1}{2}$; domain is the set of all real numbers or $(-\infty, \infty)$; range is $\left\{y \mid y \geq -\dfrac{49}{4}\right\}$ or $\left[-\dfrac{49}{4}, \infty\right)$

65. $H(x) = 2(x - 1)^2 - 3$

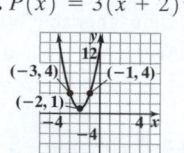

vertex is $(1, -3)$; axis of symmetry is $x = 1$; domain is the set of all real numbers or $(-\infty, \infty)$; range is $\{y \mid y \geq -3\}$ or $[-3, \infty)$

67. $P(x) = 3(x + 2)^2 + 1$

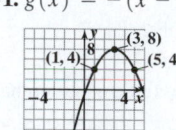

vertex is $(-2, 1)$; axis of symmetry is $x = -2$; domain is the set of all real numbers or $(-\infty, \infty)$; range is $\{y \mid y \geq 1\}$ or $[1, \infty)$

69. $F(x) = -(x + 5)^2 + 4$

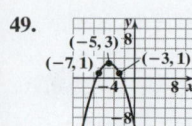

vertex is $(-5, 4)$; axis of symmetry is $x = -5$; domain is the set of all real numbers or $(-\infty, \infty)$; range is $\{y \mid y \leq 4\}$ or $(-\infty, 4]$

71. $g(x) = -(x - 3)^2 + 8$

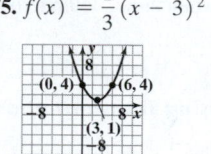

vertex is $(3, 8)$; axis of symmetry is $x = 3$; domain is the set of all real numbers or $(-\infty, \infty)$; range is $\{y \mid y \leq 8\}$ or $(-\infty, 8]$

73. $H(x) = -2(x - 2)^2 + 4$

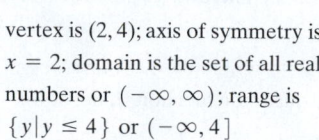

vertex is $(2, 4)$; axis of symmetry is $x = 2$; domain is the set of all real numbers or $(-\infty, \infty)$; range is $\{y \mid y \leq 4\}$ or $(-\infty, 4]$

75. $f(x) = \dfrac{1}{3}(x - 3)^2 + 1$

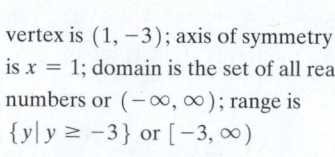

vertex is $(3, 1)$; axis of symmetry is $x = 3$; domain is the set of all real numbers or $(-\infty, \infty)$; range is $\{y \mid y \geq 1\}$ or $[1, \infty)$

77. $G(x) = -12\left(x + \dfrac{1}{2}\right)^2 + 4$

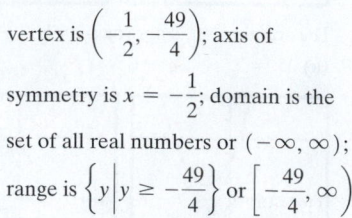

vertex is $\left(-\dfrac{1}{2}, 4\right)$; axis of symmetry is $x = -\dfrac{1}{2}$; domain is the set of all real numbers or $(-\infty, \infty)$; range is $\{y \mid y \leq 4\}$ or $(-\infty, 4]$

79. Answers may vary. One possibility: $f(x) = (x - 3)^2$ **81.** Answers may vary. One possibility: $f(x) = (x + 3)^2 + 1$ **83.** Answers may vary. One possibility: $f(x) = -(x - 5)^2 - 1$ **85.** Answers may vary. One possibility: $f(x) = 4(x - 9)^2 - 6$ **87.** Answers may vary. One possibility: $f(x) = -\dfrac{1}{3}x^2 + 6$ **89.** The highest or lowest point on a parabola is called the vertex. If $a < 0$, the vertex is the highest point; if $a > 0$, the vertex is the lowest point. **91.** No. Explanations may vary. **93.** $29 + \dfrac{1}{12}$ **95.** $x^3 - 5x^2 + 4x - 2 + \dfrac{-2}{2x - 1}$

97.

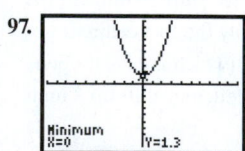

Vertex: $(0, 1.3)$ Axis of symmetry: $x = 0$ Range: $\{y \mid y \geq 1.3\}$ or $[1.3, \infty)$

99.

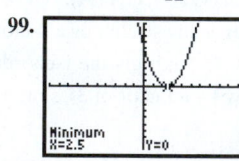

Vertex: $(2.5, 0)$ Axis of symmetry: $x = 2.5$ Range: $\{y \mid y \geq 0\}$ or $[0, \infty)$

101.

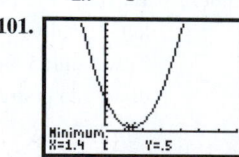

Vertex: $(1.4, 0.5)$ Axis of symmetry: $x = 1.4$ Range: $\{y \mid y \geq 0.5\}$ or $[0.5, \infty)$

103.

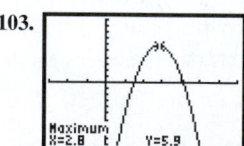

Vertex: $(2.8, 5.9)$ Axis of symmetry: $x = 2.8$ Range: $\{y \mid y \leq 5.9\}$ or $(-\infty, 5.9]$

105. (a) The graph opens up, and the larger the value of a, the "taller" the graph is. **(b)** The graph opens down, and the larger the value of $|a|$, the "taller" the graph is. **(c)** The function is linear.

Section 10.5 Graphing Quadratic Functions Using Properties

1. $-\dfrac{b}{2a}$ 2. $>$ 3. 2 4. $(-2, -7)$

5.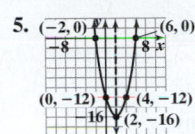

6.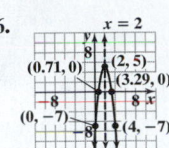

Domain: $\{x \mid x$ is any real number$\}$ or $(-\infty, \infty)$; Range: $\{y \mid y \le 5\}$ or $(-\infty, 5]$

7.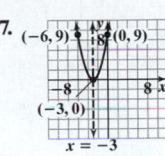

Domain: $\{x \mid x$ is any real number$\}$ or $(-\infty, \infty)$; Range: $\{y \mid y \ge 0\}$ or $[0, \infty)$

8.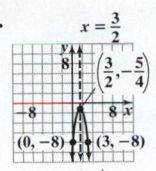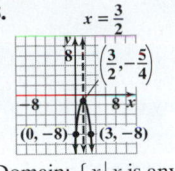

Domain: $\{x \mid x$ is any real number$\}$ or $(-\infty, \infty)$; Range: $\left\{y \mid y \le -\dfrac{5}{4}\right\}$ or $\left(-\infty, -\dfrac{5}{4}\right]$

9. True 10. minimum; -7 at $x = 2$
11. maximum; 33 at $x = 5$ 12. (a) The revenue will be maximized at a price of \$75. (b) The maximum revenue is \$2812.50. 13. The maximum area that can be enclosed is 62,500 square feet. The dimensions are 250 feet by 250 feet.

14. The number of boxes that should be sold to maximize revenue is 65, and the maximum revenue is \$4225. 15. (a) $(3, -25)$ (b) The discriminant is positive; there are two x-intercepts: $(-2, 0)$ and $(8, 0)$. 17. (a) $(1, -3)$ (b) The discriminant is negative; there are no x-intercepts. 19. (a) $\left(-\dfrac{1}{2}, 0\right)$

(b) The discriminant is zero; there is one x-intercept: $\left(-\dfrac{1}{2}, 0\right)$. 21. (a) $\left(\dfrac{1}{8}, -\dfrac{17}{16}\right)$ (b) The discriminant is positive; there are two x-intercepts: $(-0.39, 0)$ and $(0.64, 0)$.

23.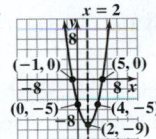

Domain: $\{x \mid x$ is any real number$\}$ or $(-\infty, \infty)$
Range: $\{y \mid y \ge -9\}$ or $[-9, \infty)$

25.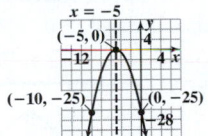

Domain: $\{x \mid x$ is any real number$\}$ or $(-\infty, \infty)$
Range: $\{y \mid y \ge -4\}$ or $[-4, \infty)$

27.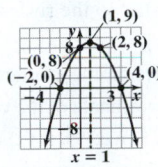

Domain: $\{x \mid x$ is any real number$\}$ or $(-\infty, \infty)$
Range: $\{y \mid y \le 9\}$ or $(-\infty, 9]$

29.

Domain: $\{x \mid x$ is any real number$\}$ or $(-\infty, \infty)$
Range: $\{y \mid y \ge 0\}$ or $[0, \infty)$

31.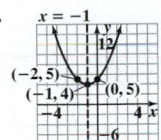

Domain: $\{x \mid x$ is any real number$\}$ or $(-\infty, \infty)$
Range: $\{y \mid y \ge 4\}$ or $[4, \infty)$

33.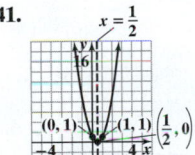

Domain: $\{x \mid x$ is any real number$\}$ or $(-\infty, \infty)$
Range: $\{y \mid y \le 0\}$ or $(-\infty, 0]$

35.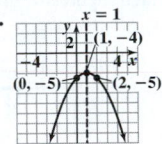

Domain: $\{x \mid x$ is any real number$\}$ or $(-\infty, \infty)$
Range: $\{y \mid y \le -4\}$ or $(-\infty, -4]$

37.

Domain: $\{x \mid x$ is any real number$\}$ or $(-\infty, \infty)$
Range: $\{y \mid y \ge -4\}$ or $[-4, \infty)$

39.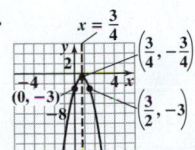

Domain: $\{x \mid x$ is any real number$\}$ or $(-\infty, \infty)$
Range: $\{y \mid y \le 16\}$ or $(-\infty, 16]$

41.

Domain: $\{x \mid x$ is any real number$\}$ or $(-\infty, \infty)$
Range: $\{y \mid y \ge 0\}$ or $[0, \infty)$

43.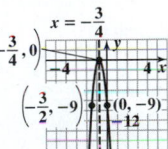

Domain: $\{x \mid x$ is any real number$\}$ or $(-\infty, \infty)$
Range: $\{y \mid y \le 0\}$ or $(-\infty, 0]$

45.

Domain: $\{x \mid x$ is any real number$\}$ or $(-\infty, \infty)$
Range: $\{y \mid y \ge 3\}$ or $[3, \infty)$

47.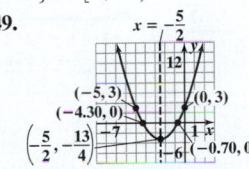

Domain: $\{x \mid x$ is any real number$\}$ or $(-\infty, \infty)$
Range: $\left\{y \mid y \le -\dfrac{3}{4}\right\}$ or $\left(-\infty, -\dfrac{3}{4}\right]$

49.

Domain: $\{x \mid x$ is any real number$\}$ or $(-\infty, \infty)$
Range: $\left\{y \mid y \ge -\dfrac{13}{4}\right\}$ or $\left[-\dfrac{13}{4}, \infty\right)$

51.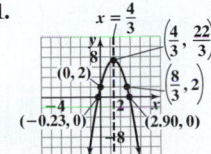

Domain: $\{x \mid x$ is any real number$\}$ or $(-\infty, \infty)$
Range: $\left\{y \mid y \le \dfrac{22}{3}\right\}$ or $\left(-\infty, \dfrac{22}{3}\right]$

53.

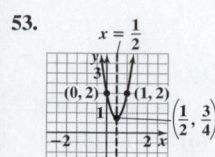

Domain: $\{x \mid x$ is any real number$\}$ or $(-\infty, \infty)$

Range: $\left\{y \mid y \geq \dfrac{3}{4}\right\}$ or $\left[\dfrac{3}{4}, \infty\right)$

55.

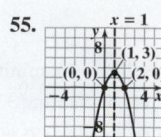

Domain: $\{x \mid x$ is any real number$\}$ or $(-\infty, \infty)$

Range: $\{y \mid y \leq 3\}$ or $(-\infty, 3]$

57.

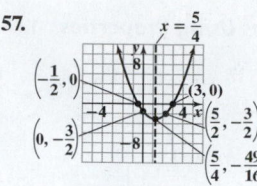
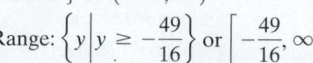

Domain: $\{x \mid x$ is any real number$\}$ or $(-\infty, \infty)$

Range: $\left\{y \mid y \geq -\dfrac{49}{16}\right\}$ or $\left[-\dfrac{49}{16}, \infty\right)$

59.

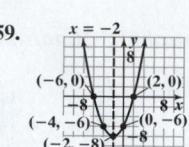

Domain: $\{x \mid x$ is any real number$\}$ or $(-\infty, \infty)$

Range: $\{y \mid y \geq -8\}$ or $[-8, \infty)$

61.

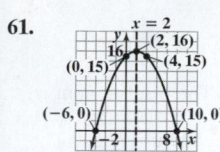

Domain: $\{x \mid x$ is any real number$\}$ or $(-\infty, \infty)$;

Range: $\{y \mid y \leq 16\}$ or $(-\infty, 16]$

63. minimum; -3 at $x = -4$ **65.** maximum; 28 at $x = -5$ **67.** maximum; 23 at $x = 3$ **69.** minimum; -19 at $x = -2$ **71.** minimum; $-\dfrac{17}{8}$ at $x = \dfrac{5}{4}$ **73.** maximum; $\dfrac{7}{3}$ at $x = \dfrac{2}{3}$ **75. (a)** \$120 **(b)** \$36,000

77. 60; \$35 **79. (a)** after 7.5 seconds **(b)** 910 feet **(c)** about 15.042 seconds **81. (a)** about 1753.52 feet from the cannon **(b)** about 886.76 feet **(c)** about 3517 feet from the cannon **(d)** The two answers are close. Explanations may vary. **83. (a)** about 46.5 years **(b)** \$64,661.75 **85.** 18 and 18 **87.** -9 and 9 **89.** 15,625 square yards; 125 yards $\times$ 125 yards **91.** 500,000 square meters; 500 m $\times$ 1000 m and the long side is parallel to the river **93.** 5 inches **95. (a)** $R = -p^2 + 110p$ **(b)** \$55; \$3025 **(c)** 55 pairs **97. (a)** $f(x) = x^2 - 8x + 12$; $f(x) = 2x^2 - 16x + 24$; $f(x) = -2x^2 + 16x - 24$ **(b)** The value of a has no effect on the x-intercepts. **(c)** The value of a has no effect on the axis of symmetry.

(d) The x-coordinate of the vertex is 4, which does not depend on a. However, the y-coordinate is $-4a$, which does depend on a. **99.** If the discriminant is positive, the equation $ax^2 + bx + c = 0$ will have two distinct real solutions, which means the graph of $f(x) = ax^2 + bx + c$ will have two x-intercepts. If the discriminant is zero, the equation $ax^2 + bx + c = 0$ will have one real solution, which means the graph of $f(x) = ax^2 + bx + c$ will have one x-intercept. If the discriminant is negative, the equation $ax^2 + bx + c = 0$ will have no real solutions, which means the graph of $f(x) = ax^2 + bx + c$ will have no x-intercepts.

101. If the price is too high, the quantity demanded will be 0.

103.

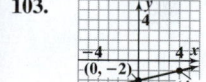
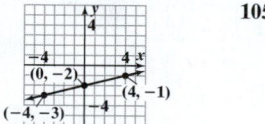

105.

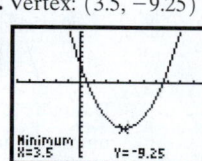

107. Vertex: $(3.5, -9.25)$

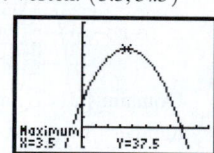

109. Vertex: $(3.5, 37.5)$

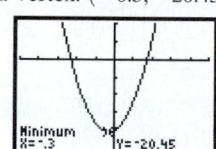

111. Vertex: $(-0.3, -20.45)$

113. Vertex: $(0.67, 4.78)$

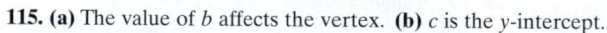

115. (a) The value of b affects the vertex. **(b)** c is the y-intercept.

Section 10.6 Polynomial Inequalities **1.** $\{x \mid x \leq -5 \text{ or } x \geq 2\}; (-\infty, -5] \cup [2, \infty)$

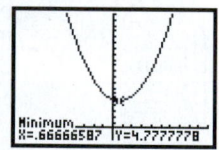

2. $\{x \mid x \leq -5 \text{ or } x \geq 2\}; (-\infty, -5] \cup [2, \infty)$ **3.** $\{x \mid -6 < x < 4\}; (-6, 4)$

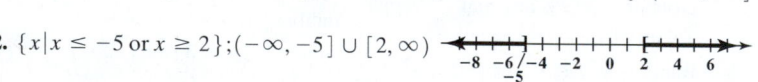

4. $\left\{x \mid x < \dfrac{-1 - \sqrt{61}}{6} \text{ or } x > \dfrac{-1 + \sqrt{61}}{6}\right\}; \left(-\infty, \dfrac{-1 - \sqrt{61}}{6}\right) \cup \left(\dfrac{-1 + \sqrt{61}}{6}, \infty\right)$

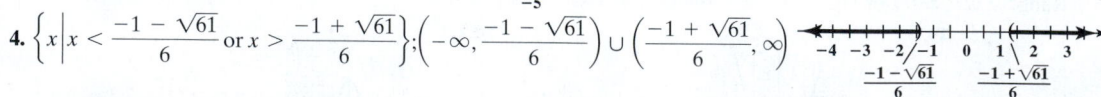

5. $\left\{x \mid x \leq -4 \text{ or } \dfrac{3}{2} \leq x \leq 5\right\}$ or $(-\infty, -4] \cup \left[\dfrac{3}{2}, 5\right]$

6. $\{x \mid -7 < x < -2 \text{ or } x > 2\}$ or $(-7, -2) \cup (2, \infty)$

7. (a) $\{x \mid x < -6 \text{ or } x > 5\}$ or $(-\infty, -6) \cup (5, \infty)$ **(b)** $\{x \mid -6 \leq x \leq 5\}$ or $[-6, 5]$

9. (a) $\left\{x \mid -6 \leq x \leq \dfrac{5}{2}\right\}$ or $\left[-6, \dfrac{5}{2}\right]$ **(b)** $\left\{x \mid x < -6 \text{ or } x > \dfrac{5}{2}\right\}$ or $(-\infty, -6) \cup \left(\dfrac{5}{2}, \infty\right)$

11. $\{x \mid x \leq -2 \text{ or } x \geq 5\}$ or $(-\infty, -2] \cup [5, \infty)$

13. $\{x|-7 < x < -3\}$ or $(-7, -3)$

15. $\{x|x < -5 \text{ or } x > 7\}$ or $(-\infty, -5) \cup (7, \infty)$

17. $\{n|3 - \sqrt{17} \le n \le 3 + \sqrt{17}\}$ or $\left[3 - \sqrt{17}, 3 + \sqrt{17}\right]$

19. $\{m|m \le -7 \text{ or } m \ge 2\}$ or $(-\infty, -7] \cup [2, \infty)$

21. $\left\{q\middle|q \le -\dfrac{5}{2} \text{ or } q \ge 3\right\}$ or $\left(-\infty, -\dfrac{5}{2}\right] \cup [3, \infty)$

23. $\{x|-1 \le x \le 4\}$ or $[-1, 4]$

25. $\{x|x < -2 \text{ or } x > 5\}$ or $(-\infty, -2) \cup (5, \infty)$

27. $\left\{x\middle|x \le -\dfrac{2}{3} \text{ or } x \ge 4\right\}$ or $\left(-\infty, -\dfrac{2}{3}\right] \cup [4, \infty)$

29. $\{x|-2 - \sqrt{3} < x < -2 + \sqrt{3}\}$ or $\left(-2 - \sqrt{3}, -2 + \sqrt{3}\right)$

31. $\left\{a\middle|-\dfrac{1}{2} \le a \le 4\right\}$ or $\left[-\dfrac{1}{2}, 4\right]$

33. $\{z|z \text{ is any real number}\}$ or $(-\infty, \infty)$

35. $\varnothing$ or $\{ \}$

37. $\{x|x \ne 3\}$ or $(-\infty, 3) \cup (3, \infty)$

39. $\{x|-1 < x < 2 \text{ or } x > 5\}$ or $(-1, 2) \cup (5, \infty)$

41. $\left\{x\middle|x \le -\dfrac{1}{2} \text{ or } 4 \le x \le 9\right\}$ or $\left(-\infty, -\dfrac{1}{2}\right] \cup [4, 9]$

43. $\left\{x\middle|-3 \le x \le -\dfrac{1}{2} \text{ or } x \ge 1\right\}$ or $\left[-3, -\dfrac{1}{2}\right] \cup [1, \infty)$

45. $\{x|x \le -3 \text{ or } -2 \le x \le 2\}$ or $(-\infty, -3] \cup [-2, 2]$

47. $\left\{x\middle|-4 < x < -\dfrac{3}{2} \text{ or } x > \dfrac{3}{2}\right\}$ or $\left(-4, -\dfrac{3}{2}\right) \cup \left(\dfrac{3}{2}, \infty\right)$

49. $\{x|0 < x < 5\}$ or $(0, 5)$ **51.** $\{x|x \le -4 \text{ or } x \ge 7\}$ or $(-\infty, -4] \cup [7, \infty)$ **53.** $\left\{x\middle|x < -\dfrac{5}{2} \text{ or } x > 2\right\}$ or $\left(-\infty, -\dfrac{5}{2}\right) \cup (2, \infty)$

55. $\left\{x\middle|x < -\dfrac{7}{4} \text{ or } 0 < x < 2\right\}$ or $\left(-\infty, -\dfrac{7}{4}\right) \cup (0, 2)$ **57.** $\{x|x \le -8 \text{ or } x \ge 0\}$ or $(-\infty, -8] \cup [0, \infty)$ **59.** $\{x|x \le -5 \text{ or } x \ge 6\}$ or $(-\infty, -5] \cup [6, \infty)$ **61.** between 2 and 3 seconds after the ball is thrown **63.** between \$110 and \$130 **65.** $x = -3$; a perfect square cannot be negative. Therefore, the only solution will be where the perfect square expression equals zero, which is -3. **67.** all real numbers; a perfect square must always be zero or greater. Therefore, it must always be larger than -2. **69.** Answers may vary. One possibility follows: $x^2 + x - 6 \le 0$ **71.** $(-\infty, -3) \cup (-1, 2) \cup (4, \infty)$ **73.** $[-5, -2] \cup [2, 5]$ **75.** $\{x|-3 < x < -2 \text{ or } x > 2\}$ or $(-3, -2) \cup (2, \infty)$ **77.** Answers will vary. One possibility follows: The inequalities have the same solution set because they are equivalent. **79.** The square of a real number is greater than or equal to zero. Therefore, $x^2 - 1$ is greater than or equal to -1 for all real numbers. **81.** No. The inequality $x^2 + 1 > 1$ is true for all real numbers except $x = 0$. **83.** To solve the quadratic inequality $f(x) > 0$ from the graph of $y = f(x)$, where x is a quadratic function, determine where the graph lies above the x-axis. This information will give you the solution set. **85.** $-\dfrac{8m^5}{n^2}$ **87.** $\dfrac{b^{\frac{1}{4}}}{9a^{\frac{7}{9}}}$

89.

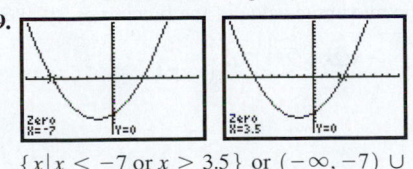

$\{x|x < -7 \text{ or } x > 3.5\}$ or $(-\infty, -7) \cup (3.5, \infty)$

91.

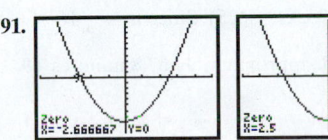

$\{x|-2.67 \le x \le 2.5\}$ or $[-2.67, 2.5]$

Section 10.7 Rational Inequalities   **1.** rational   **2.** $\{x|x < -3 \text{ or } x \geq 7\}$; $(-\infty, -3) \cup [7, \infty)$

3. $\{x|-5 < x < 1\}$; $(-5, 1)$ **4.** $\{x|-2 < x < 1\}$; $(-2, 1)$

5. $\{x|x < -1 \text{ or } x > 4\}$ or $(-\infty, -1) \cup (4, \infty)$

7. $\{x|-9 < x < 3\}$ or $(-9, 3)$

9. $\{x|x \leq -10 \text{ or } x > 4\}$ or $(-\infty, -10] \cup (4, \infty)$

11. $\left\{x \middle| x \leq -8 \text{ or } -\dfrac{5}{3} \leq x < 2\right\}$ or $(-\infty, -8] \cup \left[-\dfrac{5}{3}, 2\right)$

13. $\{x|x > -1\}$ or $(-1, \infty)$ **15.** $\left\{x \middle| \dfrac{3}{2} < x < 3\right\}$ or $\left(\dfrac{3}{2}, 3\right)$

17. $\{x|0 < x \leq 1 \text{ or } x > 4\}$ or $(0, 1] \cup (4, \infty)$

19. $\{x|-5 < x < 2 \text{ or } x \geq 23\}$ or $(-5, 2) \cup [23, \infty)$

21. $\left\{x \middle| -3 < x < \dfrac{1}{2} \text{ or } x > 5\right\}$ or $\left(-3, \dfrac{1}{2}\right) \cup (5, \infty)$

23. $\{x|x > -3\}$; $(-3, \infty)$ **25.** $\{x|-7 \leq x < 8\}$ or $[-7, 8)$

27. $\{x|x \geq 4\}$; $[4, \infty)$ **29.** $\{x|x < -4 \text{ or } x \geq 9\}$ or $(-\infty, -4) \cup [9, \infty)$

31. $\{x|-1 < x \leq 6\}$ or $(-1, 6]$ **33.** $\left\{x \middle| -2 < x < \dfrac{5}{2}\right\}$ or $\left(-2, \dfrac{5}{2}\right)$ **35.** 100 or more bicycles   **37.** Answers will vary. One possibility: $\dfrac{1}{x-2} > 0$

39. Because -1 is not in the domain of the variable x, the solution set is $\{x|-1 < x \leq 4\}$.   **41.** $(2, 0)$ **43.** $\left(-\dfrac{7}{2}, 0\right), (2, 0)$ **45.** $\left(\dfrac{2}{3}, 0\right)$

47. $\{x|x \leq -4 \text{ or } x > -1\}$ or $(-\infty, -4] \cup (-1, \infty)$ **49.** $\{x|7 < x < 26\}$ or $(7, 26)$

Chapter 10 Review   **1.** $\{-13, 13\}$ **2.** $\{-5\sqrt{3}, 5\sqrt{3}\}$ **3.** $\{-4i, 4i\}$ **4.** $\left\{-\dfrac{2\sqrt{2}}{3}, \dfrac{2\sqrt{2}}{3}\right\}$ **5.** $\{-1, 17\}$ **6.** $\{2 - 5\sqrt{6}, 2 + 5\sqrt{6}\}$ **7.** $\left\{-5, \dfrac{5}{3}\right\}$

8. $\left\{-\dfrac{3\sqrt{14}}{7}, \dfrac{3\sqrt{14}}{7}\right\}$ **9.** $\{-4\sqrt{5}i, 4\sqrt{5}i\}$ **10.** $\left\{-\dfrac{3}{4} - \dfrac{\sqrt{13}}{4}, -\dfrac{3}{4} + \dfrac{\sqrt{13}}{4}\right\}$ **11.** $a^2 + 30a + 225; (a + 15)^2$ **12.** $b^2 - 14b + 49; (b - 7)^2$

13. $c^2 - 11c + \dfrac{121}{4}; \left(c - \dfrac{11}{2}\right)^2$ **14.** $d^2 + 9d + \dfrac{81}{4}; \left(d + \dfrac{9}{2}\right)^2$ **15.** $m^2 - \dfrac{1}{4}m + \dfrac{1}{64}; \left(m - \dfrac{1}{8}\right)^2$ **16.** $n^2 + \dfrac{6}{7}n + \dfrac{9}{49}; \left(n + \dfrac{3}{7}\right)^2$ **17.** $\{2, 8\}$ **18.** $\{-4, 7\}$

19. $\{3 - 2\sqrt{3}, 3 + 2\sqrt{3}\}$ **20.** $\left\{\dfrac{5}{2} - \dfrac{\sqrt{53}}{2}, \dfrac{5}{2} + \dfrac{\sqrt{53}}{2}\right\}$ **21.** $\left\{-\dfrac{1}{2} - \dfrac{3\sqrt{3}}{2}i, -\dfrac{1}{2} + \dfrac{3\sqrt{3}}{2}i\right\}$ **22.** $\{3 - 2\sqrt{2}i, 3 + 2\sqrt{2}i\}$ **23.** $\left\{\dfrac{1}{2}, 3\right\}$

24. $\left\{-\dfrac{1}{2} - \dfrac{3}{2}i, -\dfrac{1}{2} + \dfrac{3}{2}i\right\}$ **25.** $\left\{\dfrac{3}{2} - \dfrac{\sqrt{15}}{6}i, \dfrac{3}{2} + \dfrac{\sqrt{15}}{6}i\right\}$ **26.** $\left\{-\dfrac{2}{3} - \dfrac{\sqrt{10}}{3}, -\dfrac{2}{3} + \dfrac{\sqrt{10}}{3}\right\}$ **27.** $c = 15$ **28.** $c = 8\sqrt{2}$ **29.** $c = 3\sqrt{5}$ **30.** $c = 26$

31. $c = 6$ **32.** $c = 7$ **33.** $b = 3\sqrt{7}$ **34.** $a = 5\sqrt{3}$ **35.** $a = \sqrt{253}$ **36.** approximately 127.3 feet   **37.** $\{-4, 5\}$ **38.** $\left\{-\dfrac{3}{2}, \dfrac{7}{2}\right\}$

39. $\left\{-\dfrac{4}{3} - \dfrac{\sqrt{7}}{3}, -\dfrac{4}{3} + \dfrac{\sqrt{7}}{3}\right\}$ **40.** $\left\{1 - \dfrac{\sqrt{10}}{2}, 1 + \dfrac{\sqrt{10}}{2}\right\}$ **41.** $\left\{-\dfrac{1}{6} - \dfrac{\sqrt{35}}{6}i, -\dfrac{1}{6} + \dfrac{\sqrt{35}}{6}i\right\}$ **42.** $\left\{\dfrac{4}{3}\right\}$ **43.** $\{2 - \sqrt{2}, 2 + \sqrt{2}\}$

44. $\left\{-\dfrac{2}{5} - \dfrac{1}{5}i, -\dfrac{2}{5} + \dfrac{1}{5}i\right\}$ **45.** $\left\{-\dfrac{5}{2} - \dfrac{3\sqrt{3}}{2}i, -\dfrac{5}{2} + \dfrac{3\sqrt{3}}{2}i\right\}$ **46.** $\left\{-\dfrac{3}{2} - \dfrac{\sqrt{5}}{2}i, -\dfrac{3}{2} + \dfrac{\sqrt{5}}{2}i\right\}$ **47.** 57; two irrational solutions   **48.** 0; one repeated rational

solution   **49.** -47; two complex solutions that are not real   **50.** -20; two complex solutions that are not real   **51.** 0; one repeated rational solution

52. 25; two rational solutions   **53.** $\{-9, 1\}$ **54.** $\left\{-\dfrac{5}{2}, \dfrac{1}{3}\right\}$ **55.** $\{-2 - 3i, -2 + 3i\}$ **56.** $\{-2\sqrt{3}, 2\sqrt{3}\}$ **57.** $\left\{1 - \dfrac{\sqrt{10}}{2}, 1 + \dfrac{\sqrt{10}}{2}\right\}$

58. $\{-4 - 2i, -4 + 2i\}$ **59.** $\{-3, 5\}$ **60.** $\{1 - \sqrt{2}, 1 + \sqrt{2}\}$ **61.** $\left\{-\dfrac{4}{3}, \dfrac{4}{3}\right\}$ **62.** $\{-1 - \sqrt{3}i, -1 + \sqrt{3}i\}$ **63.** $x = 10$; the three sides measure

5, 12, and 13   **64.** 12 centimeters by 9 centimeters   **65. (a)** either 300 or 600 cell phones   **(b)** 450 cell phones   **66. (a)** after approximately 0.5 second and after approximately 2.7 seconds   **(b)** after approximately 4.3 seconds   **(c)** No; the solutions to the equation are complex solutions that are not real. Another explanation: The vertex is $(1.5625, 219.0625)$. Since $a = -16 < 0$, the parabola opens down, so the maximum value of the function is 219.0625.

67. approximately 10.8 miles per hour   **68.** approximately 67.8 minutes   **69.** $\{-4i, 4i, -3, 3\}$ **70.** $\left\{-\dfrac{\sqrt{3}}{2}, \dfrac{\sqrt{3}}{2}, -\sqrt{2}i, \sqrt{2}i\right\}$ **71.** $\left\{-\dfrac{10}{3}, -1\right\}$

72. $\{-4, 4, -2\sqrt{2}, 2\sqrt{2}\}$ **73.** $\{16, 81\}$ **74.** $\left\{\dfrac{9}{25}\right\}$ **75.** $\left\{-\dfrac{1}{3}, \dfrac{1}{7}\right\}$ **76.** $\left\{-343, \dfrac{1}{8}\right\}$ **77.** $\{16\}$ **78.** $\left\{-\dfrac{36}{7}, -\dfrac{19}{4}\right\}$ **79.** $\left\{\dfrac{9}{4}, \dfrac{49}{4}\right\}$ **80.** $\{-2\sqrt{3}, -\sqrt{5}, \sqrt{5}, 2\sqrt{3}\}$

81. 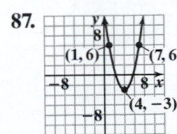 **82.** 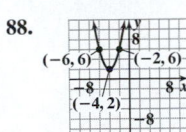 **83.** 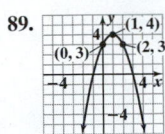 **84.** 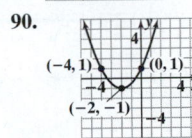 **85.** 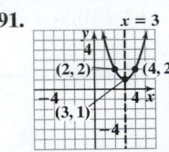 **86.**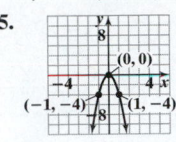

87. 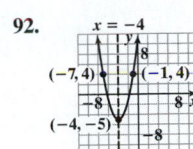 **88.** 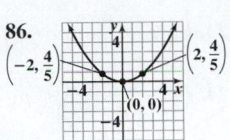 **89.** 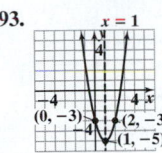 **90.** 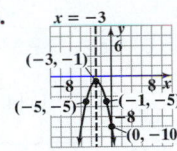 **91.** 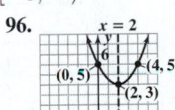 vertex is $(3, 1)$; axis of symmetry is $x = 3$; Domain: $\{x | x \text{ is any real number}\}$ or $(-\infty, \infty)$; Range: $\{y | y \geq 1\}$ or $[1, \infty)$

92. 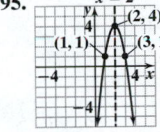 vertex is $(-4, -5)$; axis of symmetry is $x = -4$ **93.** 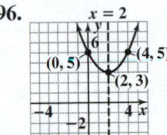 vertex is $(1, -5)$; axis of symmetry is $x = 1$ **94.** vertex is $(-3, -1)$; axis of symmetry is $x = -3$

Domain: $\{x | x \text{ is any real number}\}$ or $(-\infty, \infty)$ Domain: $\{x | x \text{ is any real number}\}$ or $(-\infty, \infty)$ Domain: $\{x | x \text{ is any real number}\}$ or $(-\infty, \infty)$
Range: $\{y | y \geq -5\}$ or $[-5, \infty)$ Range: $\{y | y \geq -5\}$ or $[-5, \infty)$ Range: $\{y | y \leq -1\}$ or $(-\infty, -1]$

95. 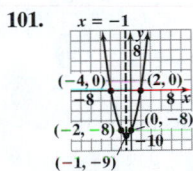 vertex is $(2, 4)$; axis of symmetry is $x = 2$
Domain: $\{x | x \text{ is any real number}\}$ or $(-\infty, \infty)$
Range: $\{y | y \leq 4\}$ or $(-\infty, 4]$
96. 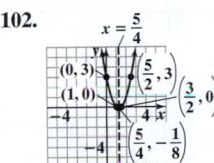 vertex is $(2, 3)$; axis of symmetry is $x = 2$
Domain: $\{x | x \text{ is any real number}\}$ or $(-\infty, \infty)$
Range: $\{y | y \geq 3\}$ or $[3, \infty)$

97. $f(x) = 2(x - 2)^2 - 4$ or $f(x) = 2x^2 - 8x + 4$ **98.** $f(x) = -(x - 4)^2 + 3$ or $f(x) = -x^2 + 8x - 13$

99. $f(x) = -\dfrac{1}{2}(x + 2)^2 - 1$ or $f(x) = -\dfrac{1}{2}x^2 - 2x - 3$ **100.** $f(x) = 3(x + 2)^2$ or $f(x) = 3x^2 + 12x + 12$

101.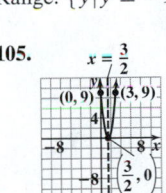
Domain: $\{x | x \text{ is any real number}\}$ or $(-\infty, \infty)$
Range: $\{y | y \geq -9\}$ or $[-9, \infty)$
102.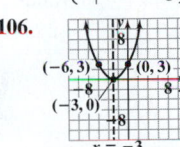
Domain: $\{x | x \text{ is any real number}\}$ or $(-\infty, \infty)$
Range: $\left\{y \middle| y \geq -\dfrac{1}{8}\right\}$ or $\left[-\dfrac{1}{8}, \infty\right)$
103.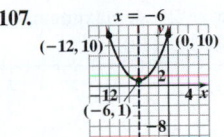
Domain: $\{x | x \text{ is any real number}\}$ or $(-\infty, \infty)$
Range: $\{y | y \leq 2\}$ or $(-\infty, 2]$
104.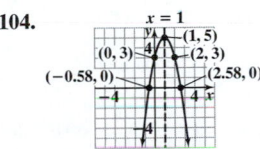
Domain: $\{x | x \text{ is any real number}\}$ or $(-\infty, \infty)$
Range: $\{y | y \leq 5\}$ or $(-\infty, 5]$

105.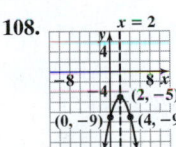
Domain: $\{x | x \text{ is any real number}\}$ or $(-\infty, \infty)$
Range: $\{y | y \geq 0\}$ or $[0, \infty)$
106.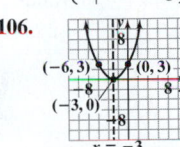
Domain: $\{x | x \text{ is any real number}\}$ or $(-\infty, \infty)$
Range: $\{y | y \geq 0\}$ or $[0, \infty)$
107.
Domain: $\{x | x \text{ is any real number}\}$ or $(-\infty, \infty)$
Range: $\{y | y \geq 1\}$ or $[1, \infty)$
108.
Domain: $\{x | x \text{ is any real number}\}$ or $(-\infty, \infty)$
Range: $\{y | y \leq -5\}$ or $(-\infty, -5]$

109. maximum; 22 at $x = 4$ **110.** minimum; $-\dfrac{11}{8}$ at $x = \dfrac{1}{4}$ **111.** maximum; 7 at $x = 1$ **112.** maximum; 5 at $x = 6$ **113.** (a) \$225 (b) \$16,875

114. (a) 3.75 amperes (b) 225 watts

115. both numbers are 12 **116.** (a) 3.75 yards by 7.5 yards (b) 28.125 square yards **117.** (a) 100 feet (b) 50 feet (c) 200 feet

118. (a) $R(p) = -0.002p^2 + 60p$ (b) \$15,000; \$450,000 (c) 30 automobiles per month **119.** (a) $\{x | x < -2 \text{ or } x > 3\}$ or $(-\infty, -2) \cup (3, \infty)$

(b) $\{x | -2 < x < 3\}$ or $(-2, 3)$ **120.** (a) $\left\{x \middle| -\dfrac{7}{2} \leq x \leq 1\right\}$ or $\left[-\dfrac{7}{2}, 1\right]$ (b) $\left\{x \middle| x \leq -\dfrac{7}{2} \text{ or } x \geq 1\right\}$ or $\left(-\infty, -\dfrac{7}{2}\right] \cup [1, \infty)$

121. $\{x | -4 \leq x \leq 6\}$ or $[-4, 6]$

122. $\{y \mid y \le -8 \text{ or } y \ge 1\}$ or, $(-\infty, -8] \cup [1, \infty)$

123. $\left\{z \mid z < \dfrac{4}{3} \text{ or } z > 5\right\}$ or $\left(-\infty, \dfrac{4}{3}\right) \cup (5, \infty)$

124. $\{p \mid -2 - \sqrt{6} < p < -2 + \sqrt{6}\}$ or $(-2 - \sqrt{6}, -2 + \sqrt{6})$

125. $\{m \mid m \text{ is any real number}\}$ or $(-\infty, \infty)$

126. $\left\{w \mid -\dfrac{1}{3} \le w \le \dfrac{7}{2}\right\}$ or $\left[-\dfrac{1}{3}, \dfrac{7}{2}\right]$

127. $\left\{x \mid x < -1 \text{ or } \dfrac{3}{2} < x < 2\right\}$ or $(-\infty, -1) \cup \left(\dfrac{3}{2}, 2\right)$

128. $\{x \mid -5 \le x \le -3 \text{ or } x \ge 3\}$ or $[-5, -3] \cup [3, \infty)$

129. $\{x \mid x < -2 \text{ or } x \ge 4\}$ or $(-\infty, -2) \cup [4, \infty)$

130. $\{y \mid -4 < y < 5\}$ or $(-4, 5)$ **131.** $\{z \mid -3 < z < 3\}$ or $(-3, 3)$

132. $\{w \mid w < -7 \text{ or } 2 < w < 4\}$ or $(-\infty, -7) \cup (2, 4)$

133. $\{m \mid -5 < m < 2 \text{ or } m \ge 5\}$ or $(-5, 2) \cup [5, \infty)$

134. $\{n \mid 0 \le n < 2\}$ or $[0, 2)$ **135.** $\left\{a \mid 2 < a < \dfrac{7}{2}\right\}$ or $\left(2, \dfrac{7}{2}\right)$

136. $\{c \mid c < -6 \text{ or } 0 < c < 2\}$ or $(-\infty, -6) \cup (0, 2)$ **137.** $\left\{x \mid -\dfrac{3}{2} < x < 4\right\}$ or $\left(-\dfrac{3}{2}, 4\right)$
138. $\{x \mid x \le -5 \text{ or } x > -1\}$ or $(-\infty, -5] \cup (-1, \infty)$

Chapter 10 Test   **1.** $x^2 - 3x + \dfrac{9}{4}; \left(x - \dfrac{3}{2}\right)^2$ **2.** $m^2 + \dfrac{2}{5}m + \dfrac{1}{25}; \left(m + \dfrac{1}{5}\right)^2$ **3.** $\left\{-\dfrac{5}{3}, -1\right\}$ **4.** $\{3 - \sqrt{5}, 3 + \sqrt{5}\}$ **5.** $\left\{1 - \dfrac{\sqrt{2}}{2}i, 1 + \dfrac{\sqrt{2}}{2}i\right\}$

6. $\left\{\dfrac{3}{2} - \dfrac{\sqrt{3}}{6}i, \dfrac{3}{2} + \dfrac{\sqrt{3}}{6}i\right\}$ **7.** 57; two irrational solutions   **8.** $a = 6\sqrt{2}$ **9.** $\{-3, 3, -2i, 2i\}$ **10.** $\left\{\dfrac{1}{81}\right\}$

11.
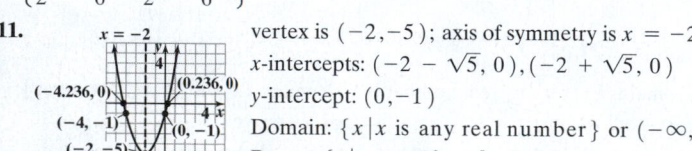
vertex is $(-2, -5)$; axis of symmetry is $x = -2$
x-intercepts: $(-2 - \sqrt{5}, 0), (-2 + \sqrt{5}, 0)$
y-intercept: $(0, -1)$
Domain: $\{x \mid x \text{ is any real number}\}$ or $(-\infty, \infty)$
Range: $\{y \mid y \ge -5\}$ or $[-5, \infty)$

12.
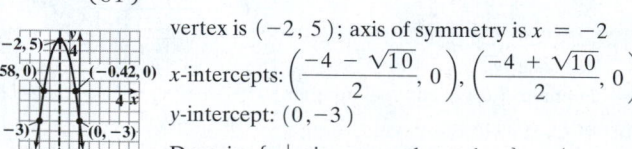
vertex is $(-2, 5)$; axis of symmetry is $x = -2$
x-intercepts: $\left(\dfrac{-4 - \sqrt{10}}{2}, 0\right), \left(\dfrac{-4 + \sqrt{10}}{2}, 0\right)$
y-intercept: $(0, -3)$
Domain: $\{x \mid x \text{ is any real number}\}$ or $(-\infty, \infty)$
Range: $\{y \mid y \le 5\}$ or $(-\infty, 5]$

13. $f(x) = \dfrac{1}{3}(x + 3)^2 - 5$ or $f(x) = \dfrac{1}{3}x^2 + 2x - 2$ **14.** maximum; 6

15. $\left\{m \mid m < -3 \text{ or } m > \dfrac{5}{2}\right\}$ or $(-\infty, -3) \cup \left(\dfrac{5}{2}, \infty\right)$

16. $\{x \mid x \le -5 \text{ or } -2 \le x \le 2\}$ or $(-\infty, -5] \cup [-2, 2]$

17. $\left\{x \mid 2 < x \le \dfrac{11}{2}\right\}$ or $\left(2, \dfrac{11}{2}\right]$

18. 0.4 second and 4.6 seconds   **19.** 34.1 hours   **20. (a)** $340   **(b)** $28,900   **21. (a)** 12.5 in. by 12.5 in. by 12 in.   **(b)** 1875 cubic inches

Chapter 11 Exponential and Logarithmic Functions

Section 11.1 Composite Functions and Inverse Functions

1. composite function   **2. (a)** 17   **(b)** 26   **(c)** -63 **3.** False   **4.** $g(x) = 4x - 3$ **5. (a)** $9x^2 + 3x - 1$ **(b)** $3x^2 - 9x + 5$ **(c)** 29

6. one-to-one   **7.** not one-to-one   **8.** one-to-one   **9. (a)** not one-to-one   **(b)** one-to-one

10.
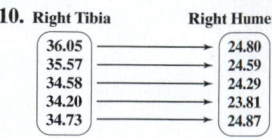
The domain of the inverse function is $\{36.05, 35.57, 34.58, 34.20, 34.73\}$. The range of the inverse function is $\{24.80, 24.59, 24.29, 23.81, 24.87\}$.

11. $\{(3, -3), (2, -2), (1, -1), (0, 0), (-1, 1)\}$. The domain of the inverse function is $\{3, 2, 1, 0, -1\}$. The range of the inverse function is $\{-3, -2, -1, 0, 1\}$.

12. 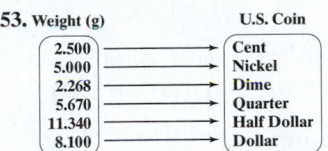   **13.** True   **14.** False   **15.** $g^{-1}(x) = \dfrac{x+1}{5}$ **16.** $f^{-1}(x) = \sqrt[5]{x} - 3$ **17. (a)** 3 **(b)** -3 **(c)** 19 **(d)** -12 **19. (a)** 85 **(b)** 19

(c) 29 **(d)** -7 **21. (a)** -4394 **(b)** -507 **(c)** 16 **(d)** -1453 **23. (a)** 8 **(b)** $\dfrac{4}{5}$ **(c)** 1 **(d)** $-\dfrac{4}{3}$ **25. (a)** $(f \circ g)(x) = 2x + 1$

(b) $(g \circ f)(x) = 2x + 2$ **(c)** $(f \circ f)(x) = x + 2$ **(d)** $(g \circ g)(x) = 4x$

27. (a) $(f \circ g)(x) = -8x + 17$ **(b)** $(g \circ f)(x) = -8x - 23$ **(c)** $(f \circ f)(x) = 4x + 21$ **(d)** $(g \circ g)(x) = 16x - 15$ **29. (a)** $(f \circ g)(x) = x^2 - 6x + 9$
(b) $(g \circ f)(x) = x^2 - 3$ **(c)** $(f \circ f)(x) = x^4$ **(d)** $(g \circ g)(x) = x - 6$ **31. (a)** $(f \circ g)(x) = \sqrt{x+4}$ **(b)** $(g \circ f)(x) = \sqrt{x} + 4$ **(c)** $(f \circ f)(x) = \sqrt[4]{x}$
(d) $(g \circ g)(x) = x + 8$ **33. (a)** $(f \circ g)(x) = x^2$ **(b)** $(g \circ f)(x) = x^2 + 8x + 12$ **(c)** $(f \circ f)(x) = ||x + 4| + 4|$ **(d)** $(g \circ g)(x) = x^4 - 8x^2 + 12$

35. (a) $(f \circ g)(x) = \dfrac{2x}{x+1}$, where $x \neq -1, 0$ **(b)** $(g \circ f)(x) = \dfrac{x+1}{2}$, where $x \neq -1$ **(c)** $(f \circ f)(x) = \dfrac{2(x+1)}{x+3}$, where $x \neq -1, -3$

(d) $(g \circ g)(x) = x$, where $x \neq 0$ **37.** one-to-one   **39.** not one-to-one   **41.** one-to-one   **43.** not one-to-one   **45.** one-to-one   **47.** one-to-one

49. not one-to-one   **51.** one-to-one

53.

Weight (g)	U.S. Coin
2.500	Cent
5.000	Nickel
2.268	Dime
5.670	Quarter
11.340	Half Dollar
8.100	Dollar

55. $\{(3,0),(4,1),(5,2),(6,3)\}$ **57.** $\{(3,-2),(1,-1),(-3,0),(9,1)\}$

59. 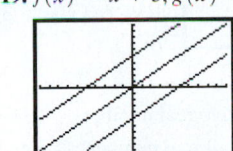   **61.** 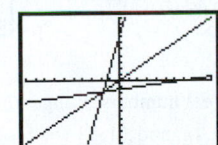   **63.** 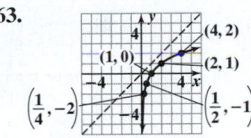   **65.** $f(g(x)) = (x - 5) + 5 = x$
$g(f(x)) = (x + 5) - 5 = x$

67. $f(g(x)) = 5\left(\dfrac{x-7}{5}\right) + 7 = x - 7 + 7 = x$ **69.** $f(g(x)) = \dfrac{3}{\left(\dfrac{3}{x} + 1\right) - 1} = \dfrac{3}{\dfrac{3}{x}} = 3 \cdot \dfrac{x}{3} = x$

$g(f(x)) = \dfrac{(5x+7) - 7}{5} = \dfrac{5x}{5} = x$ $g(f(x)) = \dfrac{3}{\dfrac{3}{x-1}} + 1 = 3 \cdot \dfrac{x-1}{3} + 1 = x - 1 + 1 = x$

71. $f(g(x)) = \sqrt[3]{(x^3 - 4) + 4} = \sqrt[3]{x^3} = x$ **73.** $f^{-1}(x) = \dfrac{x}{6}$ **75.** $f^{-1}(x) = x - 4$ **77.** $h^{-1}(x) = \dfrac{x+7}{2}$ **79.** $G^{-1}(x) = \dfrac{2-x}{5}$
$g(f(x)) = (\sqrt[3]{x} + 4)^3 - 4 = x + 4 - 4 = x$

81. $g^{-1}(x) = \sqrt[3]{x} - 3$ **83.** $p^{-1}(x) = \dfrac{1}{x} - 3$ **85.** $F^{-1}(x) = 2 - \dfrac{5}{x}$ **87.** $f^{-1}(x) = x^3 + 2$

89. $R^{-1}(x) = \dfrac{2x}{1-x}$ **91.** $f^{-1}(x) = (x - 4)^3 + 1$ **93.** $A(t) = 400\pi t^2; 3600\pi \approx 11{,}309.73$ sq ft   **95. (a)** $C(x) = 2x$ **(b)** \$630   **97.** $f^{-1}(12) = 4$

99. Domain of $f^{-1}: [-5, \infty)$; Range of $f^{-1}: [0, \infty)$ **101.** Domain of $g^{-1}: (-6, 12)$; Range of $g^{-1}: [-4, 10]$ **103.** $x(T) = \dfrac{T + 463.75}{0.15}$ for $927.50 \leq T \leq 5183.75$

105. $\left\{-\dfrac{5}{2}, 3\right\}$ **107.** A function is one-to-one provided no two different inputs correspond to the same output. A function must be one-to-one in order for
the inverse to be a function because we interchange the inputs and outputs to obtain the inverse. Thus, if a function is not one-to-one, the inverse will have
a single input corresponding to two different outputs.   **109.** The domain of f equals the range of f^{-1} because we interchange the roles of the inputs and
outputs to obtain the inverse of a function. The same logic explains why the range of f equals the domain of f^{-1}.   **111. (a)** 3 **(b)** -3 **(c)** 19 **(d)** -12

113. (a) 85 **(b)** 19 **(c)** 29 **(d)** -7 **115. (a)** -4394 **(b)** -507 **(c)** 16 **(d)** -1453 **117. (a)** 8 **(b)** $\dfrac{4}{5}$ **(c)** 1 **(d)** $-\dfrac{4}{3}$

119. $f(x) = x + 5; g(x) = x - 5$ **121.** $f(x) = 5x + 7; g(x) = \dfrac{x-7}{5}$

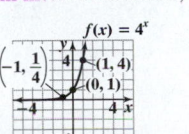   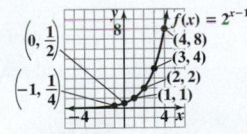

Section 11.2 Exponential Functions

1. $>; \neq$ **2. (a)** 3.249009585 **(b)** 3.317278183 **(c)** 3.321880096 **(d)** 3.32211036 **(e)** 3.321997085

3. The domain of f is all real numbers or, in
interval notation, $(-\infty, \infty)$. The range of f is
$\{y | y > 0\}$ or, in interval notation, $(0, \infty)$.

4. $\left(-1, \dfrac{1}{a}\right); (0, 1); (1, a)$ **5.** True   **6.** False

7. The domain of f is all real numbers or, in interval
notation, $(-\infty, \infty)$. The range of f is $\{y | y > 0\}$
or, in interval notation, $(0, \infty)$.

8. The domain of f is all real numbers or,
in interval notation, $(-\infty, \infty)$. The range
of f is $\{y | y > 0\}$ or, in interval notation,
$(0, \infty)$.

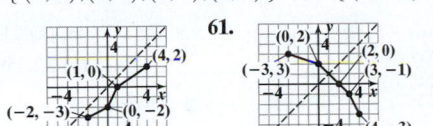

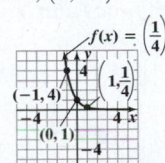

9. The domain of f is all real numbers or, in interval notation, $(-\infty, \infty)$. The range of f is $\{y \mid y > 1\}$ or, in interval notation, $(1, \infty)$.

10. 2.71828   **11. (a)** 54.598 **(b)** 0.018   **12.** $\{3\}$ **13.** $\{2\}$ **14.** $\{0, 5\}$ **15.** $\{-1, 3\}$ **16. (a)** 0.918 or 91.8% **(b)** 0.998 or 99.8%   **17. (a)** approximately 6.91 grams **(b)** 5 grams **(c)** 0.625 gram **(d)** approximately 0.247 gram   **18. (a)** \$2102.32 **(b)** \$4227.41 **(c)** \$8935.49   **19. (a)** 11.212 **(b)** 11.587 **(c)** 11.664 **(d)** 11.665 **(e)** 11.665   **21. (a)** 73.517 **(b)** 77.708 **(c)** 77.924 **(d)** 77.881 **(e)** 77.880   **23.** g   **25.** e   **27.** f   **29.** h

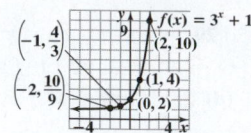

31. 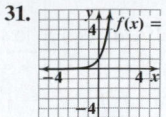 Domain: all real numbers or $(-\infty, \infty)$; Range: $\{y \mid y > 0\}$ or $(0, \infty)$

33. 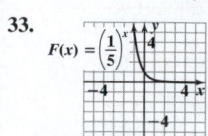 Domain: all real numbers or $(-\infty, \infty)$; Range: $\{y \mid y > 0\}$ or $(0, \infty)$

35. 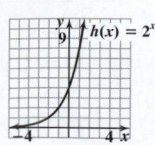 Domain: all real numbers or $(-\infty, \infty)$; Range: $\{y \mid y > 0\}$ or $(0, \infty)$

37. 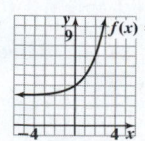 Domain: all real numbers or $(-\infty, \infty)$; Range: $\{y \mid y > 3\}$ or $(3, \infty)$

39. 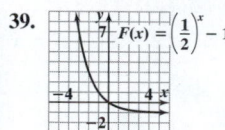 Domain: all real numbers or $(-\infty, \infty)$; Range: $\{y \mid y > -1\}$ or $(-1, \infty)$

41. 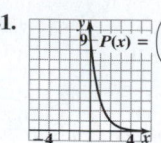 Domain: all real numbers or $(-\infty, \infty)$; Range: $\{y \mid y > 0\}$ or $(0, \infty)$

43. (a) 21.217 **(b)** 22.472 **(c)** 22.460 **(d)** 22.460 **(e)** 22.459   **45.** 7.389   **47.** 0.135   **49.** 9.974

51. 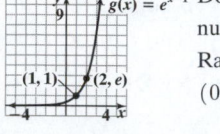 Domain: all real numbers or $(-\infty, \infty)$; Range: $\{y \mid y > 0\}$ or $(0, \infty)$

53. 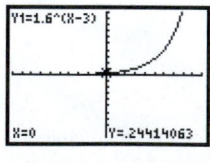 Domain: all real numbers or $(-\infty, \infty)$; Range: $\{y \mid y < 0\}$ or $(-\infty, 0)$

55. $\{5\}$ **57.** $\{-4\}$ **59.** $\{5\}$ **61.** $\{5\}$ **63.** $\left\{\dfrac{3}{2}\right\}$ **65.** $\{1\}$ **67.** $\{-1, 4\}$ **69.** $\{-4, 2\}$ **71.** $\{9\}$ **73.** $\{-2\}$ **75.** $\{-2, 2\}$ **77.** $\{-2\}$ **79.** $\{2\}$ **81. (a)** $f(3) = 8; (3, 8)$ **(b)** $x = -3; \left(-3, \dfrac{1}{8}\right)$

83. (a) $g(-1) = -\dfrac{3}{4}; \left(-1, -\dfrac{3}{4}\right)$ **(b)** $x = 2; (2, 15)$ **85. (a)** $H(-3) = 24; (-3, 24)$ **(b)** $x = 2; \left(2, \dfrac{3}{4}\right)$

87. (a) approximately 342.5 million people **(b)** approximately 455.7 million people **(c)** Answers may vary. One possibility is that the population is not growing exponentially.   **89. (a)** \$5100. 92 **(b)** \$5308.92 **(c)** \$5525.29   **91. (a)** \$2318.55 **(b)** \$2322.37 **(c)** \$2323.23 **(d)** \$2323.65 **(e)** The future value is higher with more compounding periods.   **93. (a)** \$19,841 **(b)** \$15,365 **(c)** \$10,471   **95. (a)** approximately 95.105 grams **(b)** 50 grams **(c)** 25 grams **(d)** approximately 0.661 gram   **97. (a)** approximately 300.233°F **(b)** approximately 230.628°F **(c)** yes   **99. (a)** approximately 29 words **(b)** approximately 38 words   **101. (a)** approximately 0.238 ampere **(b)** approximately 0.475 ampere   **103.** $y = 3^x$

105. As x increases, the graph increases very rapidly. As x decreases, the graph approaches the x-axis.   **107.** Answers may vary. The big difference is that exponential functions are of the form $f(x) = a^x$ (the variable is in the exponent), while polynomial functions are of the form $f(x) = a_n x^n + a_{n-1} x^{n-1} + \cdots + a_1 x + a_0$ (the variable is a base).   **109. (a)** 15 **(b)** -5 **111. (a)** undefined **(b)** $\dfrac{4}{5}$ **113. (a)** 3 **(b)** $3\sqrt{3}$

115. $f(x) = 1.5^x$

Domain: all real numbers or $(-\infty, \infty)$; Range: $\{y \mid y > 0\}$ or $(0, \infty)$

117. $H(x) = 0.9^x$

Domain: all real numbers or $(-\infty, \infty)$; Range: $\{y \mid y > 0\}$ or $(0, \infty)$

119. $g(x) = 2.5^x + 3$

Domain: all real numbers or $(-\infty, \infty)$; Range: $\{y \mid y > 3\}$ or $(3, \infty)$

121. $F(x) = 1.6^{x-3}$

Domain: all real numbers or $(-\infty, \infty)$; Range: $\{y \mid y > 0\}$ or $(0, \infty)$

123. (a) Domain: the set of all real numbers; Range: the set of all positive real numbers; intercepts: $(0, 1); (-1, 1/2), (0, 1),$ and $(1, 2)$ **(b)** Domain: the set of all real numbers; Range: the set of all positive real numbers; intercepts: $(0, 1); (-1, 1/3), (0, 1), (1, 3)$ **(c)** Domain: the set of all real numbers; Range: the set of all positive real numbers; intercepts: $(0, 1); (-1, 1/4), (0, 1),$ and $(1, 4)$ **(d) (i)** the set of all real numbers; **(ii)** the set of all positive real numbers; **(iii)** Zero; $(0, 1)$ **(iv)** $(-1, 1/a), (0, 1),$ and $(1, a)$.

Section 11.3 Logarithmic Functions

1. $x = a^y; >; \neq$ **2.** $3 = \log_4 64$ **3.** $-2 = \log_p 8$ **4.** $2^4 = 16$ **5.** $a^5 = 20$ **6.** $5^{-3} = z$ **7.** 2   **8.** -3

9. 2   **10.** -1 **11.** $\{x \mid x > -3\}$ or $(-3, \infty)$ **12.** $\left\{x \mid x < \dfrac{5}{2}\right\}$ or $\left(-\infty, \dfrac{5}{2}\right)$

13. The domain of f is $\{x \mid x > 0\}$ or, in interval notation, $(0, \infty)$. The range of f is all real numbers or, in interval notation, $(-\infty, \infty)$.

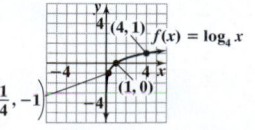

14. The domain of f is $\{x \mid x > 0\}$ or, in interval notation, $(0, \infty)$. The range of f is all real numbers or, in interval notation, $(-\infty, \infty)$.

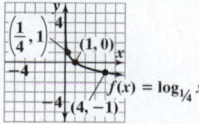

15. 3.146   **16.** 1.569   **17.** −0.523   **18.** {16}   **19.** {4}   **20.** {e^{-2}}   **21.** {10,020}   **22.** 100 decibels   **23.** $3 = \log_4 64$ **25.** $-3 = \log_2\left(\dfrac{1}{8}\right)$

27. $\log_a 19 = 3$ **29.** $\log_5 c = -6$ **31.** $2^4 = 16$ **33.** $3^{-2} = \dfrac{1}{9}$ **35.** $5^{-3} = a$ **37.** $a^2 = 4$ **39.** $\left(\dfrac{1}{2}\right)^y = 12$ **41.** 0   **43.** 3   **45.** −2   **47.** 4   **49.** 4   **51.** $\dfrac{1}{2}$

53. $\{x\,|\,x > 4\}$ or $(4, \infty)$ **55.** $\{x\,|\,x > 0\}$ or $(0, \infty)$ **57.** $\left\{x\,\middle|\,x > \dfrac{2}{3}\right\}$ or $\left(\dfrac{2}{3}, \infty\right)$ **59.** $\left\{x\,\middle|\,x > -\dfrac{1}{2}\right\}$ or $\left(-\dfrac{1}{2}, \infty\right)$ **61.** $\left\{x\,\middle|\,x < \dfrac{1}{4}\right\}$ or $\left(-\infty, \dfrac{1}{4}\right)$

63. 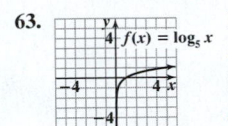 Domain: $\{x\,|\,x > 0\}$ or $(0, \infty)$ Range: all real numbers or $(-\infty, \infty)$

65. 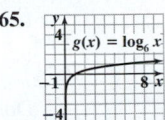 Domain: $\{x\,|\,x > 0\}$ or $(0, \infty)$ Range: all real numbers or $(-\infty, \infty)$

67. 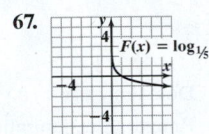 Domain: $\{x\,|\,x > 0\}$ or $(0, \infty)$ Range: all real numbers or $(-\infty, \infty)$

69. $\ln 12 = x$ **71.** $e^4 = x$ **73.** −1   **75.** 3   **77.** 1.826   **79.** 1.686   **81.** −0.456   **83.** −1.609   **85.** 0.097   **87.** −0.981   **89.** {4}   **91.** $\left\{\dfrac{13}{2}\right\}$ **93.** {6}

95. $\{3\sqrt{2}\}$ **97.** {10}   **99.** {e^5}   **101.** $\left\{\dfrac{11}{20}\right\}$ **103.** {−3}   **105.** {4}   **107.** {−3, 3}   **109.** (a) $f(16) = 4$; $(16, 4)$ (b) $x = \dfrac{1}{8}$; $\left(\dfrac{1}{8}, -3\right)$

111. (a) $G(7) = \dfrac{3}{2}$; $\left(7, \dfrac{3}{2}\right)$ (b) $x = 15$; $(15, 2)$ **113.** $a = 4$ **115.** $\{x\,|\,x < -1\,\text{or}\,x > 5\}$ or $(-\infty, -1) \cup (5, \infty)$ **117.** $\{x\,|\,x < -1\,\text{or}\,x > 4\}$ or $(-\infty, -1) \cup (4, \infty)$ **119.** $\{x\,|\,x \neq 3\}$ or $(-\infty, 3) \cup (3, \infty)$ **121.** 20 decibels   **123.** 130 decibels   **125.** approximately 7.8 on the Richter scale

127. approximately 794,328   **129.** (a) 12; basic (b) 5; acidic (c) 2; acidic (d) $10^{-7.4}$ mole per liter   **131.** The base of $f(x) = \log_a x$ cannot equal 1 because $y = \log_a x$ is equivalent to $x = a^y$ and a does not equal 1 in the exponential function. In addition, the graph would be a vertical line ($x = 1$), which is not a function.   **133.** The domain of $f(x) = \log_a(x^2 + 1)$ is the set of all real numbers because $x^2 + 1 > 0$ for all x.

135. $-3x^2 - 7x + 10$ **137.** $\dfrac{4x^2 + 2x + 3}{(x + 1)(x - 1)(x + 2)}$ **139.** $5x\sqrt{2x}$

141. $f(x) = \log(x + 1)$ 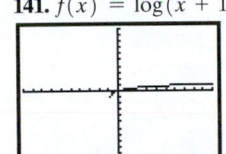 Domain: $\{x\,|\,x > -1\}$ or $(-1, \infty)$; Range: all real number or $(-\infty, \infty)$

143. $G(x) = \ln(x) + 1$ 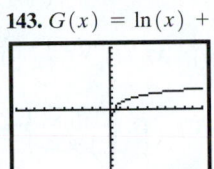 Domain: $\{x\,|\,x > 0\}$ or $(0, \infty)$; Range: all real number or $(-\infty, \infty)$

145. $f(x) = 2\log(x - 3) + 1$ 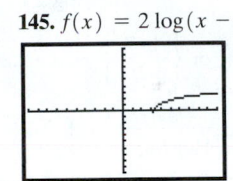 Domain: $\{x\,|\,x > 3\}$ or $(3, \infty)$; Range: all real number or $(-\infty, \infty)$

147. (a) Domain: the set of all positive real numbers; Range: the set of all real numbers; intercepts: $(1, 0)$; $(1/2, -1)$, $(1, 0)$, and $(2, 1)$ (b) Domain: the set of all positive real numbers; Range: the set of all real numbers; intercepts: $(1, 0)$; $(1/3, -1)$, $(1, 0)$, and $(3, 1)$ (c) Domain: the set of all positive real numbers; Range: the set of all real numbers; intercepts: $(1, 0)$; $(1/4, -1)$, $(1, 0)$, and $(4, 1)$ (d) (i) the set of all positive real numbers; (ii) the set of all real numbers; (iii) Zero; $(1, 0)$; (iv) $(1/a, -1)$, $(1, 0)$, and $(a, 1)$.

Putting the Concepts Together (Sections 11.1–11.3)

1. (a) $(f \circ g)(x) = 4x^2 - 8x + 3$ (b) $(g \circ f)(x) = 8x^2 + 16x + 6$ (c) $(f \circ g)(3) = 15$ (d) $(g \circ f)(-2) = 6$ (e) $(f \circ f)(1) = 13$ **2.** (a) $f^{-1}(x) = \dfrac{x - 4}{3}$ (b) $g^{-1}(x) = \sqrt[3]{x + 4}$

3.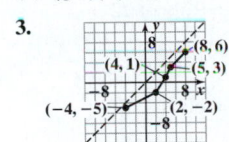

4. (a) 14.611 (b) 15.206 (c) 15.146 (d) 15.155 (e) 15.154   **5.** (a) $\log_a 6.4 = 4$ (b) $\log 278 = x$ **6.** (a) $2^7 = x$ (b) $e^M = 16$ **7.** (a) 4 (b) −2   **8.** $\{x\,|\,x > -6\}$ or $(-6, \infty)$

9. 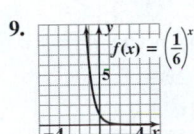 Domain: all real numbers or $(-\infty, \infty)$ Range: $\{y\,|\,y > 0\}$ or $(0, \infty)$

10. 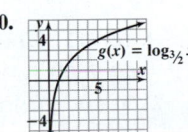 Domain: $\{x\,|\,x > 0\}$ or $(0, \infty)$ Range: all real numbers or $(-\infty, \infty)$

11. {−1}   **12.** {−5}   **13.** $\left\{\dfrac{11}{2}\right\}$ **14.** {e^7}   **15.** approximately 56 terms

Section 11.4 Properties of Logarithms

1. 0   **2.** 0   **3.** 1   **4.** 1   **5.** $\sqrt{2}$ **6.** 0.2   **7.** 1.2   **8.** −4   **9.** False   **10.** $\log_4 9 + \log_4 5$ **11.** $\log 5 + \log w$ **12.** $\log_7 9 - \log_7 5$ **13.** $\ln p - \ln 3$ **14.** $\log_2 3 + \log_2 m - \log_2 n$ **15.** $\ln q - \ln 3 - \ln p$ **16.** $1.6\log_2 5$ **17.** $5\log b$ **18.** $2\log_4 a + \log_4 b$

19. $2 + 4\log_3 m - \dfrac{1}{3}\log_3 n$ **20.** 2   **21.** $\log_3\left(\dfrac{x + 4}{x - 1}\right)$ **22.** $\log_5\dfrac{x}{8}$ **23.** $\log_2\dfrac{x^2 + 3x + 2}{x^2}$ **24.** 10; 3; 10; 3   **25.** 3.155   **26.** 2.807   **27.** 3   **29.** −7

31. 5   **33.** 2   **35.** 1   **37.** 0   **39.** $a + b$ **41.** $2b$ **43.** $2a + b$ **45.** $\dfrac{1}{2}a$ **47.** $\log a + \log b$ **49.** $4\log_5 x$ **51.** $\log_2 x + 2\log_2 y$ **53.** $2 + \log_5 x$

55. $2 - \log_7 y$ **57.** $2 + \ln x$ **59.** $3 + \dfrac{1}{2}\log_3 x$ **61.** $2\log_5 x + \dfrac{1}{2}\log_5(x^2 + 1)$ **63.** $4\log x - \dfrac{1}{3}\log(x - 1)$ **65.** $\dfrac{1}{2}\log_7(x + 1) - \dfrac{1}{2}\log_7 x$

67. $\log_2 x + 2\log_2(x - 1) - \dfrac{1}{2}\log_2(x + 1)$ **69.** 2   **71.** $\log(3x)$ **73.** 2   **75.** 4   **77.** $\log_3 x^3$ **79.** $\log_4\left(\dfrac{x + 1}{x}\right)$ **81.** $\ln(x^2 y^3)$

83. $\log_3\left[\sqrt{x}(x - 1)^3\right]$ **85.** $\log(x^2)$ **87.** $\log(x\sqrt{xy})$ **89.** $\log_8(x - 1)$ **91.** $\log\left(\dfrac{x^{12}}{10}\right)$ **3.** 3.322   **95.** 0.528   **97.** −2.680   **99.** 4.644   **101.** 3   **103.** 1

105. $\log_a(x + \sqrt{x^2 - 1}) + \log_a(x - \sqrt{x^2 - 1})$
$= \log_a\left[(x + \sqrt{x^2 - 1})(x - \sqrt{x^2 - 1})\right]$
$= \log_a\left[x^2 - x\sqrt{x^2 - 1} + x\sqrt{x^2 - 1} - (x^2 - 1)\right]$
$= \log_a(x^2 - x^2 + 1)$
$= \log_a 1$
$= 0$

107. If $f(x) = \log_a x$, then
$f(AB) = \log_a(AB)$
$= \log_a A + \log_a B$
$= f(A) + f(B)$

109. Answers will vary. One possibility: The logarithm of the product of two expressions equals the sum of the logarithms of the two expressions.

111. Answers will vary. One possibility:
$\log_2(2 + 4) \neq \log_2 2 + \log_2 4$

113. $\left\{\dfrac{5}{2}\right\}$ **115.** $\left\{-2 - \sqrt{2}, -2 + \sqrt{2}\right\}$ **117.** $\{47\}$

119. $f(x) = \log_3 x = \dfrac{\log x}{\log 3}$

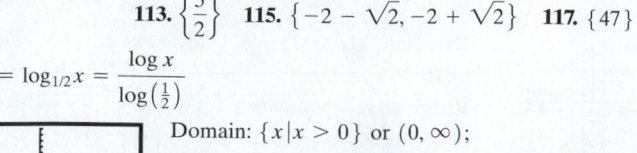

Domain: $\{x \mid x > 0\}$ or $(0, \infty)$;
Range: all real numbers or $(-\infty, \infty)$

121. $F(x) = \log_{1/2} x = \dfrac{\log x}{\log\left(\frac{1}{2}\right)}$

Domain: $\{x \mid x > 0\}$ or $(0, \infty)$;
Range: all real numbers or $(-\infty, \infty)$

Section 11.5 Exponential and Logarithmic Equations   **1.** $M = N$ **2.** $\{3\}$ **3.** $\{8\}$ **4.** $\{-2\}$ **5.** $\left\{\dfrac{\ln 11}{\ln 2}\right\}$ or $\left\{\dfrac{\log 11}{\log 2}\right\}$; $\{3.459\}$

6. $\left\{\dfrac{\ln 3}{2\ln 5}\right\}$ or $\left\{\dfrac{\log 3}{2\log 5}\right\}$; $\{0.341\}$ **7.** $\left\{\dfrac{\ln 5}{2}\right\}$; $\{0.805\}$ **8.** $\left\{-\dfrac{\ln\left(\frac{20}{3}\right)}{4}\right\}$; $\{-0.474\}$ **9. (a)** approximately 2.85 days **(b)** approximately 32.52 days

10. (a) approximately 6.77 years **(b)** approximately 11.58 years   **11.** $\{7\}$ **13.** $\{9\}$ **15.** $\{3\}$ **17.** $\{81\}$ **19.** $\{1\}$ **21.** $\{-1\}$

23. $\left\{\dfrac{1}{3}\right\}$ **25.** $\left\{\dfrac{7}{5}\right\}$ **27.** $\{-5\}$ **29.** $\left\{\dfrac{1}{\log 2}\right\} \approx \{3.322\}$ or $\left\{\dfrac{\ln 10}{\ln 2}\right\} \approx \{3.322\}$ **31.** $\left\{\dfrac{\log 20}{\log 5}\right\} \approx \{1.861\}$ or $\left\{\dfrac{\ln 20}{\ln 5}\right\} \approx \{1.861\}$

33. $\left\{\dfrac{\log 7}{\log\left(\frac{1}{2}\right)}\right\} \approx \{-2.807\}$ or $\left\{\dfrac{\ln 7}{\ln\left(\frac{1}{2}\right)}\right\} \approx \{-2.807\}$ **35.** $\{\ln 5\} \approx \{1.609\}$ **37.** $\{\log 5\} \approx \{0.699\}$ **39.** $\left\{\dfrac{\log 13}{2\log 3}\right\} \approx \{1.167\}$ or

$\left\{\dfrac{\ln 13}{2\ln 3}\right\} \approx \{1.167\}$ **41.** $\left\{\dfrac{\log 3}{4\log\left(\frac{1}{2}\right)}\right\} \approx \{-0.396\}$ or $\left\{\dfrac{\ln 3}{4\ln\left(\frac{1}{2}\right)}\right\} \approx \{-0.396\}$ **43.** $\left\{\dfrac{\log\left(\frac{5}{4}\right)}{\log 2}\right\} \approx \{0.322\}$ or $\left\{\dfrac{\ln\left(\frac{5}{4}\right)}{\ln 2}\right\} \approx \{0.322\}$

45. $\{\ln 6\} \approx \{1.792\}$ **47.** $\left\{\dfrac{\log 0.2}{\log 3 - \log 0.2}\right\} \approx \{-0.594\}$ or $\left\{\dfrac{\ln 0.2}{\ln 3 - \ln 0.2}\right\} \approx \{-0.594\}$ **49.** $\{8\}$ **51.** $\left\{\dfrac{\log 7}{3\log 5}\right\} \approx \{0.403\}$ or

$\left\{\dfrac{\ln 7}{3\ln 5}\right\} \approx \{0.403\}$ **53.** $\{2\}$ **55.** $\{\ln 15\} \approx \{2.708\}$ **57.** $\left\{-\dfrac{2}{5}\right\}$ **59.** $\{-4, 4\}$ **61.** $\{4\}$ **63.** $\{2 + \sqrt{7}\} \approx \{4.646\}$

65. (a) 2023 **(b)** 2053   **67. (a)** approximately 16.8 years **(b)** approximately 34.7 years   **69. (a)** approximately 2.188 years **(b)** approximately 10.782 years **(c)** approximately 23.372 years   **71. (a)** approximately 2.099 seconds **(b)** 27.62 seconds **(c)** approximately 45.876 seconds

73. (a) approximately 5 minutes **(b)** approximately 11 minutes   **75. (a)** approximately 82 minutes **(b)** approximately 396 minutes (or 6.6 hours)

77. (a) approximately 9 years **(b)** $t = \dfrac{\log 2}{n\log\left(1 + \dfrac{r}{n}\right)}$ **(c)** approximately 8.693 years, which is about the same as the result from the Rule of 72

79. 29 minutes   **81. (a)** 1 **(b)** 11 **(c)** 7   **83. (a)** $-\dfrac{3}{2}$ **(b)** $\dfrac{2}{7}$ **(c)** 0   **85. (a)** 8 **(b)** $\dfrac{1}{4}$ **(c)** 1   **87.** approximately $\{1.06\}$ **89.** approximately $\{-0.70\}$

91. approximately $\{0.05, 1.48\}$ **93.** approximately $\{0.45, 1\}$

Chapter 11 Review   **1. (a)** 32 **(b)** -9 **(c)** 2 **(d)** 13   **2. (a)** 24 **(b)** -28 **(c)** -8 **(d)** 112   **3. (a)** 201 **(b)** 24 **(c)** 163 **(d)** 14   **4. (a)** 23 **(b)** 37 **(c)** -8 **(d)** 290   **5. (a)** $(f \circ g)(x) = 5x + 1$ **(b)** $(g \circ f)(x) = 5x + 5$ **(c)** $(f \circ f)(x) = x + 2$ **(d)** $(g \circ g)(x) = 25x$ **6. (a)** $(f \circ g)(x) = 2x + 9$ **(b)** $(g \circ f)(x) = 2x + 3$ **(c)** $(f \circ f)(x) = 4x - 9$ **(d)** $(g \circ g)(x) = x + 12$ **7. (a)** $(f \circ g)(x) = 4x^2 + 4x + 2$ **(b)** $(g \circ f)(x) = 2x^2 + 3$ **(c)** $(f \circ f)(x) = x^4 + 2x^2 + 2$ **(d)** $(g \circ g)(x) = 4x + 3$ **8. (a)** $(f \circ g)(x) = \dfrac{2x}{x + 1}$, where $x \neq -1, 0$ **(b)** $(g \circ f)(x) = \dfrac{x + 1}{2}$, where $x \neq -1$ **(c)** $(f \circ f)(x) = \dfrac{2(x + 1)}{x + 3}$, where $x \neq -1, -3$ **(d)** $(g \circ g)(x) = x$, where $x \neq 0$ **9.** not one-to-one   **10.** one-to-one   **11.** one-to-one   **12.** not one-to-one

13.

Height (inches)	Age
69	24
71	59
72	29
73	81
74	37

14.

Quantity Demanded	Price ($)
112	300
129	200
144	170
161	150
176	130

15. $\{(3, -5), (1, -3), (-3, 1), (9, 2)\}$ **16.** $\{(1, -20), (4, -15), (3, 5), (2, 25)\}$

17.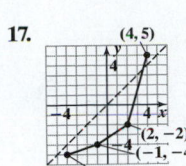

18.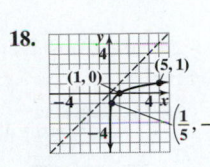

19. $f^{-1}(x) = \dfrac{x}{5}$ **20.** $H^{-1}(x) = \dfrac{x-7}{2}$ **21.** $P^{-1}(x) = \dfrac{4}{x} - 2$ **22.** $g^{-1}(x) = \sqrt[3]{\dfrac{x+1}{2}}$

23. (a) 27.332 (b) 28.975 (c) 29.088 (d) 29.093 (e) 29.091   **24.** (a) 1258.925 (b) 1380.384
(c) 1386.756 (d) 1385.479 (e) 1385.456   **25.** (a) 1.649 (b) 0.368 (c) 4.482 (d) 0.449 (e) 5.885

26. 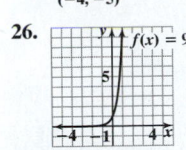 Domain: all real numbers
or $(-\infty, \infty)$;
Range: $\{y \mid y > 0\}$ or
$(0, \infty)$

27. 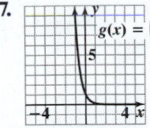 Domain: all real numbers
or $(-\infty, \infty)$;
Range: $\{y \mid y > 0\}$ or
$(0, \infty)$

28. 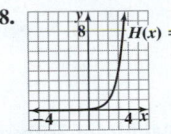 Domain: all real numbers
or $(-\infty, \infty)$;
Range: $\{y \mid y > 0\}$ or
$(0, \infty)$

29. 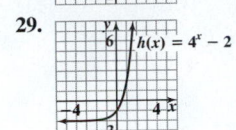 Domain: all real numbers or
$(-\infty, \infty)$;
Range: $\{y \mid y > -2\}$ or
$(-2, \infty)$

30. The number e is defined as the number that the expression $\left(1 + \dfrac{1}{n}\right)^{n}$ approaches as n
increases.   **31.** $\{6\}$ **32.** $\left\{\dfrac{7}{2}\right\}$ **33.** $\{-4, 1\}$ **34.** $\{-2\}$ **35.** $\{9\}$ **36.** $\{-3, 3\}$

37. (a) $7513.59 (b) $7652.33 (c) $7684.36 (d) $7700.00

38. (a) approximately 82.034 grams (b) 50 grams (c) 25 grams (d) approximately 0.263 gram   **39.** (a) approximately 3.388 million people
(b) approximately 3.668 million people   **40.** (a) approximately 151.449 °F (b) approximately 94.706 °F   **41.** $\log_3 81 = 4$ **42.** $\log_4\left(\dfrac{1}{64}\right) = -3$

43. $\log_b 5 = 3$ **44.** $\log x = 3.74$ **45.** $8^{1/3} = 2$ **46.** $5^r = 18$ **47.** $e^2 = x + 3$ **48.** $10^{-4} = x$ **49.** $\dfrac{7}{3}$ **50.** 0 **51.** -2 **52.** $\dfrac{3}{2}$ **53.** $\{x \mid x > -5\}$ or

$(-5, \infty)$ **54.** $\left\{x \mid x < \dfrac{7}{3}\right\}$ or $\left(-\infty, \dfrac{7}{3}\right)$ **55.** $\{x \mid x > 0\}$ or $(0, \infty)$ **56.** $\left\{x \mid x > -\dfrac{5}{2}\right\}$ or $\left(-\dfrac{5}{2}, \infty\right)$

57.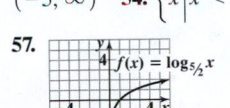

58.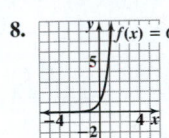

59. 3.178   **60.** -0.182 **61.** 2.410   **62.** -0.907 **63.** $\{17\}$ **64.** $\{-9, 1\}$ **65.** $\left\{\dfrac{3}{2}\right\}$ **66.** $\{6\}$ **67.** $\{-142\}$

68. $\{5\sqrt{3}\}$ **69.** 80 decibels   **70.** 100,000 millimeters   **71.** 21   **72.** 9.34   **73.** 1   **74.** 0   **75.** 1   **76.** 16

77. $\log_7 x + \log_7 y - \log_7 z$ **78.** $4 - 2\log_3 x$ **79.** $3 + 4\log r$ **80.** $\dfrac{1}{2}\ln(x-1) - \dfrac{1}{2}\ln x$

81. $\log_3(x^4 y^2)$ **82.** $\ln\left(\dfrac{7\sqrt[4]{x}}{9}\right)$ **83.** -1 **84.** $\log_6(x-4)$ **85.** 2.183   **86.** 0.606   **87.** -4.419 **88.** 3.723   **89.** $\{10\}$ **90.** $\{-5\}$ **91.** $\{\sqrt{2}\} \approx \{1.414\}$

92. $\{64\}$ **93.** $\left\{\dfrac{\log 15}{\log 2}\right\} \approx \{3.907\}$ or $\left\{\dfrac{\ln 15}{\ln 2}\right\} \approx \{3.907\}$ **94.** $\left\{\dfrac{\log 27}{3}\right\} \approx \{0.477\}$ **95.** $\left\{\dfrac{\ln 39}{7}\right\} \approx \{0.523\}$

96. $\left\{\dfrac{\log 2}{\log 3 - \log 2}\right\} \approx \{1.710\}$ or $\left\{\dfrac{\ln 2}{\ln 3 - \ln 2}\right\} \approx \{1.710\}$ **97.** (a) after approximately 1.453 days (b) after approximately 23.253 days

98. (a) about 2020 (b) about 2053

Chapter 11 Test   **1.** not one-to-one   **2.** $f^{-1}(x) = \dfrac{x+3}{4}$ **3.** (a) 33.360 (b) 36.338 (c) 36.494 (d) 36.463 (e) 36.462   **4.** $\log_4 19 = x$ **5.** $b^y = x$

6. (a) -3 (b) 4   **7.** $\left\{x \mid x < \dfrac{7}{4}\right\}$ or $\left(-\infty, \dfrac{7}{4}\right)$

8. 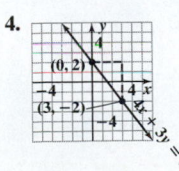 Domain: all real numbers or $(-\infty, \infty)$;
Range: $\{y \mid y > 0\}$ or $(0, \infty)$

9. 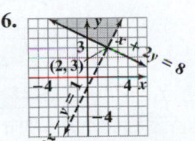 Domain: $\{x \mid x > 0\}$ or $(0, \infty)$;
Range: all real numbers or $(-\infty, \infty)$

10. (a) 10 (b) 15   **11.** $\dfrac{1}{2}\log_4 x - 3\log_4 y$ **12.** $\log(M^4 N^3)$ **13.** -8.004 **14.** $\{1\}$ **15.** $\{1, 3\}$ **16.** $\{4\}$ **17.** $\{-17, 17\}$ **18.** $\{9\}$

19. $\left\{\dfrac{\log 17 + \log 3}{\log 3}\right\} \approx \{3.579\}$ or $\left\{\dfrac{\ln 17 + \ln 3}{\ln 3}\right\} \approx \{3.579\}$ **20.** $\{2\sqrt{26}\} \approx \{10.198\}$ **21.** (a) approximately 37.1 million people
(b) about 2058   **22.** 10 decibels

Cumulative Review Chapters 1–11   **1.** $\{5\}$ **2.** $\{x \mid -2 \le x \le 6\}$ or $[-2, 6]$ **3.** $\left\{x \mid x \ne -3 \text{ and } x \ne \dfrac{7}{2}\right\}$

4.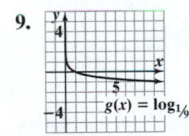

5. $y = -\dfrac{7}{5}x + 3$ or $7x + 5y = 15$ **6.**   **7.** $4m^2 - 3m + 7$ **8.** $2n^3 - 5n^2 + 18$ **9.** $(4a + b)^2$

10. $(3y - 7)(2y - 1)$ **11.** $\dfrac{3x - 4}{x - 7}$

12. $\dfrac{6(p + 1)}{(p - 5)(p - 1)(p + 2)}$ **13.** $\{4\}$ **14.** $7\sqrt{6}$

15. $2 + \sqrt{5}$ **16.** $\{9\}$ **17.** $\left\{\dfrac{2 - \sqrt{22}}{3}, \dfrac{2 + \sqrt{22}}{3}\right\}$ **18.** $\left\{\dfrac{9}{4}, 4\right\}$

19.

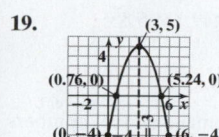

20. $\left\{x \mid -2 < x < \dfrac{4}{3}\right\}$ or $\left(-2, \dfrac{4}{3}\right)$

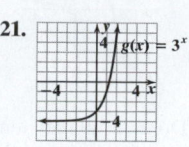

21.

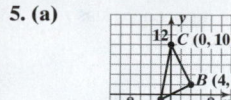

22. $-\dfrac{3}{2}$

Chapter 12 Conics

Section 12.1 Distance and Midpoint Formulas
1. $d = \sqrt{(x_2 - x_1)^2 + (y_2 - y_1)^2}$ **2.** False **3.** 5 **4.** $6\sqrt{5} \approx 13.42$

5. (a)

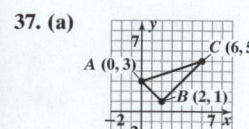

(b) $d(A, B) = 3\sqrt{5}; d(A, C) = 5\sqrt{5}; d(B, C) = 4\sqrt{5}$ **(c)** $[d(A, C)]^2 = [d(A, B)]^2 + [d(B, C)]^2$

(d) 30 square units **6.** $M = \left(\dfrac{x_1 + x_2}{2}, \dfrac{y_1 + y_2}{2}\right)$ **7.** $\left(\dfrac{3}{2}, 6\right)$ **8.** $\left(1, \dfrac{5}{2}\right)$ **9.** 5 **11.** $4\sqrt{5} \approx 8.94$

13. 5 **15.** 13 **17.** 6 **19.** $3\sqrt{5} \approx 6.71$ **21.** $3\sqrt{7} \approx 7.94$ **23.** $\sqrt{12.56} \approx 3.54$ **25.** $(4, 3)$ **27.** $(3, -1)$

29. $\left(-1, \dfrac{7}{2}\right)$ **31.** $\left(-\dfrac{3}{2}, 0\right)$ **33.** $\left(\dfrac{7\sqrt{2}}{2}, \dfrac{5\sqrt{5}}{2}\right)$ **35.** $(0.8, -1.6)$

37. (a)

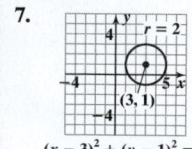

39. (a)

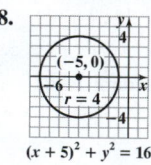

(b) $d(A, B) = 2\sqrt{2} \approx 2.83; d(B, C) = 4\sqrt{2} \approx 5.66;$
$d(A, C) = 2\sqrt{10} \approx 6.32$

(c) $[d(A, B)]^2 + [d(B, C)]^2 \overset{?}{=} [d(A, C)]^2$
$\left(2\sqrt{2}\right)^2 + \left(4\sqrt{2}\right)^2 \overset{?}{=} \left(2\sqrt{10}\right)^2$
$4 \cdot 2 + 16 \cdot 2 \overset{?}{=} 4 \cdot 10$
$8 + 32 \overset{?}{=} 40$
$40 = 40 \leftarrow$ True

Therefore, triangle ABC is a right triangle.

(d) 8 square units

(b) $d(A, B) = 5\sqrt{2} \approx 7.07; d(B, C) = 12\sqrt{2} \approx 16.97;$
$d(A, C) = 13\sqrt{2} \approx 18.38$

(c) $[d(A, B)]^2 + [d(B, C)]^2 \overset{?}{=} [d(A, C)]^2$
$\left(5\sqrt{2}\right)^2 + \left(12\sqrt{2}\right)^2 \overset{?}{=} \left(13\sqrt{2}\right)^2$
$25 \cdot 2 + 144 \cdot 2 \overset{?}{=} 169 \cdot 2$
$50 + 288 \overset{?}{=} 338$
$338 = 338 \leftarrow$ True

Therefore, triangle ABC is a right triangle.

(d) 60 square units

41. $(2, -3), (2, 5)$ **43.** $(-6, -3), (10, -3)$ **45. (a)** approximately 37.36 blocks **(b)** approximately 35.13 blocks **(c)** approximately 71.34 blocks
47. (a) approximately 1.85 seconds **(b)** No. The ball will reach second base (2.65 seconds) before the runner (3.33 seconds). **49.** Answers will vary.
Essentially, the Pythagorean Theorem is a theorem that relates the hypotenuse to the lengths of the legs in a right triangle. The hypotenuse in the
right triangle is the distance between the two points in the Cartesian plane. **51.** 9; 3 **53.** 81; 3 **55.** If n is a positive integer and $a^n = b$, then
$\sqrt[n]{b} = \begin{cases} |a| & \text{if } n \text{ is even} \\ a & \text{if } n \text{ is odd.} \end{cases}$

Section 12.2 Circles
1. circle **2.** radius **3.** $(x - 2)^2 + (y - 4)^2 = 25$ **4.** $(x + 2)^2 + y^2 = 2$ **5.** False **6.** True

7.

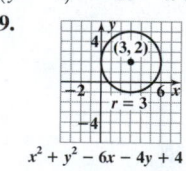

$(x - 3)^2 + (y - 1)^2 = 4$

8.

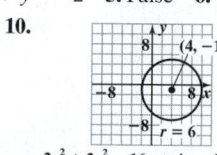

$(x + 5)^2 + y^2 = 16$

9.

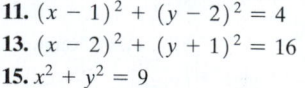

$x^2 + y^2 - 6x - 4y + 4 = 0$

10.

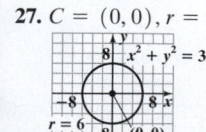

$2x^2 + 2y^2 - 16x + 4y - 38 = 0$

11. $(x - 1)^2 + (y - 2)^2 = 4$
13. $(x - 2)^2 + (y + 1)^2 = 16$
15. $x^2 + y^2 = 9$
17. $(x - 1)^2 + (y - 4)^2 = 4$

19. $(x + 2)^2 + (y - 4)^2 = 36$ **21.** $x^2 + (y - 3)^2 = 16$ **23.** $(x - 5)^2 + (y + 5)^2 = 25$ **25.** $(x - 1)^2 + (y - 2)^2 = 5$
27. $C = (0, 0), r = 6$

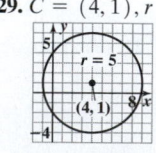

$x^2 + y^2 = 36$

29. $C = (4, 1), r = 5$

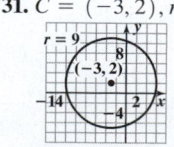

31. $C = (-3, 2), r = 9$

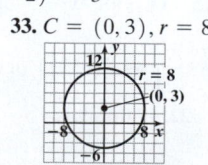

33. $C = (0, 3), r = 8$

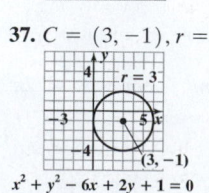

35. $C = (1, -1), r = \dfrac{1}{2}$

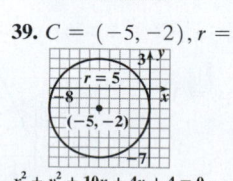

37. $C = (3, -1), r = 3$

$x^2 + y^2 - 6x + 2y + 1 = 0$

39. $C = (-5, -2), r = 5$

$x^2 + y^2 + 10x + 4y + 4 = 0$

41. $C = (3, -6), r = 9$

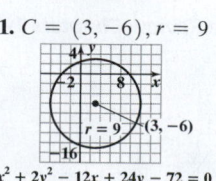

$2x^2 + 2y^2 - 12x + 24y - 72 = 0$

43. $x^2 + y^2 = 20$ **45.** $(x + 3)^2 + (y - 2)^2 = 9$ **47.** $(x + 1)^2 + (y + 1)^2 = 25$ **49.** $A = 64\pi$ square units; $C = 16\pi$ units
51. 32 square units **53.** (a), (e) **55.** The distance formula is used along with the definition of a circle to obtain the equation of the circle.
57. Yes, since it can be written as $x^2 + y^2 = 36$. Center: $(0, 0)$; radius $= 6$.

59. $f(x) = 4x - 3$

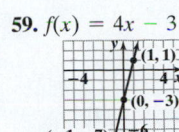
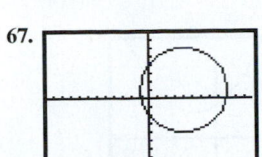

61. $g(x) = x^2 - 4x - 5$

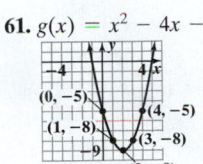
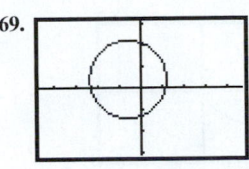

63. $G(x) = -2(x + 3)^2 - 5$

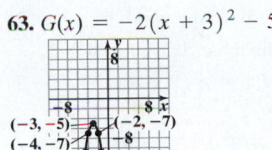
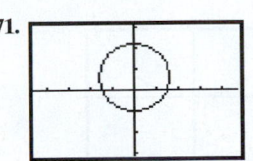

65.

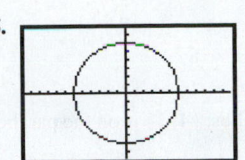

The graph here agrees with that in Problem 27.

67.

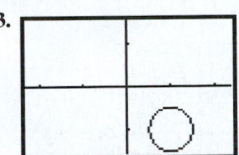

The graph here agrees with that in Problem 29.

69.

The graph here agrees with that in Problem 31.

71.

The graph here agrees with that in Problem 33.

73.

The graph here agrees with that in Problem 35.

75. (a) $(0, 0)$ **(e)** $(x - 4.5)^2 + (y - 4.5)^2 = 6.76$; $(4.5, 4.5)$; 2.6 **(f)** $(x + 4)^2 + (y + 4)^2 = 1.44$; $(-4, 4)$; 1.2
(g) $(x + 3.5)^2 + (y + 3.5)^2 = 5.76$; $(-3.5, -3.5)$; 2.4 **(h)** $(x - 3)^2 + (y + 3)^2 = 5.76$; $(3, -3)$; 2.4

Section 12.3 Parabolas **1.** parabola **2.** vertex **3.** axis of symmetry

4. $D: x = -2$

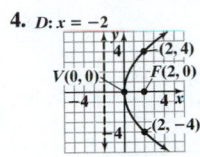

5.

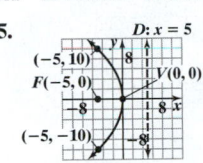

6.

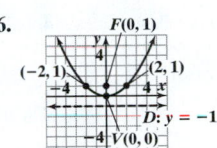

7.

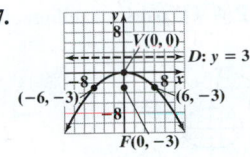

8. $x^2 = -32y$

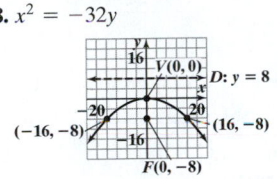

9. $y^2 = \dfrac{4}{3}x$

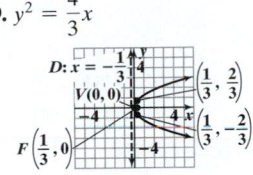

10. $(-3, 2)$

11. $D: x = -6$

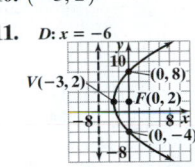

12. The receiver should be located 2 feet from the base of the dish, along its axis of symmetry.

13. c **15.** a **17.** b **19.** e

21. Vertex: $(0, 0)$; focus: $(0, 6)$; directrix: $y = -6$

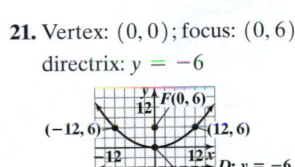

23. Vertex: $(0, 0)$; focus: $\left(-\dfrac{3}{2}, 0\right)$; directrix: $x = \dfrac{3}{2}$

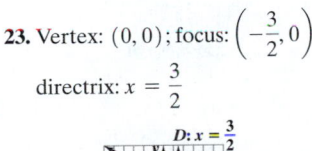

25. Vertex: $(0, 0)$; focus: $(0, -2)$; directrix: $y = 2$

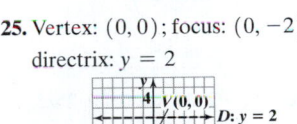
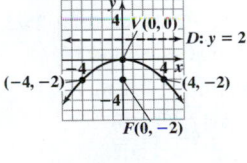

27. $y^2 = 20x$

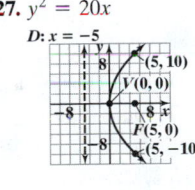

29. $x^2 = -24y$

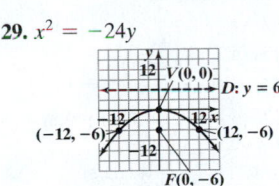

31. $x^2 = 6y$

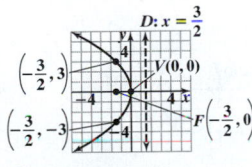

33. $x^2 = -12y$

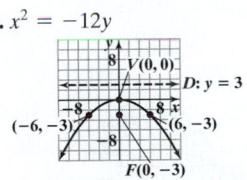
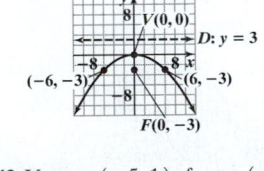

35. $y^2 = -12x$

37. $y^2 = x$

39. Vertex: $(2, 4)$; focus: $(2, 5)$; directrix: $y = 3$

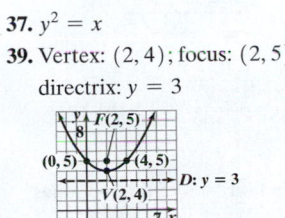

41. Vertex: $(-2, -3)$; focus: $(-4, -3)$; directrix: $x = 0$

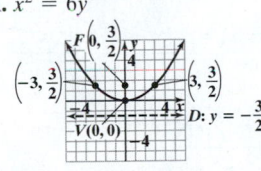

43. Vertex: $(-5, 1)$; focus: $(-5, -4)$; directrix: $y = 6$

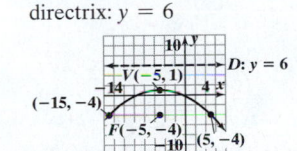

45. Vertex: $(-2, -1)$; focus: $(-2, -4)$; directrix: $y = 2$

47. Vertex: $(1, 4)$; focus: $(2, 4)$; directrix: $x = 0$

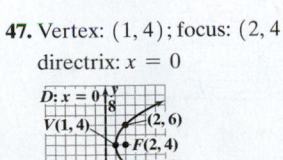

49. Vertex: $(-5, 2)$; focus: $\left(-5, \dfrac{1}{2}\right)$; directrix: $y = \dfrac{7}{2}$

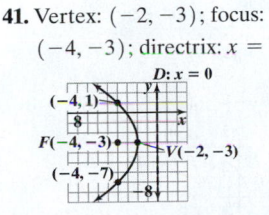
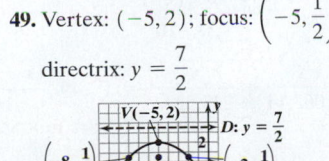

51. 1 inch from the vertex, along the axis of symmetry **53.** 21.6 feet
55. The height of the bridge is 28.8 feet at a distance of 10 feet from the center, 19.2 feet at a distance of 30 feet from the center, and 0 feet (i.e., ground level) at a distance of 50 feet from the center.
57. $(x - 3)^2 = 4(y + 2)$ **59.** $(y - 3)^2 = -4(x - 2)$

61. (a) Let $x = 4$ and $y = 2$:
$$4^2 \stackrel{?}{=} 8 \cdot 2$$
$16 = 16 \leftarrow$ True
Thus, $(4, 2)$ is on the parabola.

(b) The focus of the parabola is $F(0, 2)$, and the directrix is $D: y = -2$.
$d(F, P) = \sqrt{(0 - 4)^2 + (2 - 2)^2} = \sqrt{16} = 4$
$d(P, D) = 2 - (-2) = 4$
Thus, $d(F, P) = d(P, D) = 4$.

63. 8 units **65.** See Figure 13.

67. $y = (x + 3)^2$

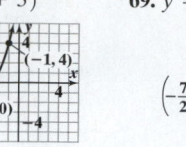

69. $y = (x + 3)^2$

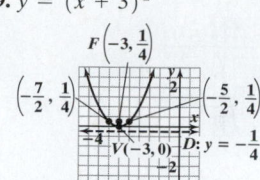

71. $4(y + 2) = (x - 2)^2$

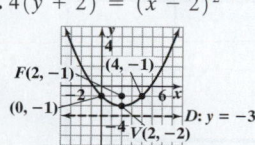

73.

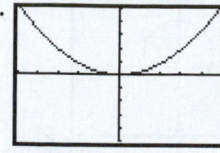

75.

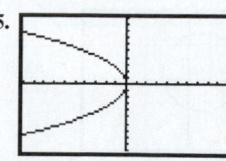

77.

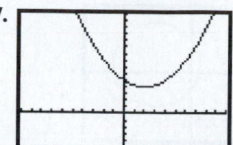

79.

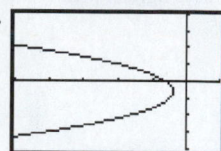

81.

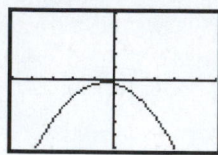

83.

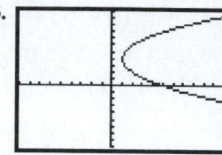

Section 12.4 Ellipses

1. ellipse; foci **2.** major axis **3.** vertices **4.** False

5.

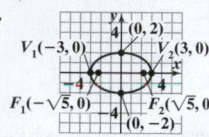

6.

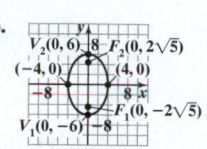

7. $\dfrac{x^2}{40} + \dfrac{y^2}{49} = 1$

8. $(3, -1)$

9.

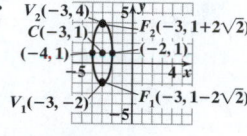

10. $\dfrac{x^2}{2500} + \dfrac{y^2}{1600} = 1$; 40 feet

11. c **13.** d

15. Foci: $(-3, 0)$ and $(3, 0)$; vertices: $(-5, 0)$ and $(5, 0)$

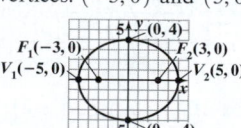

17. Foci: $(0, -8)$ and $(0, 8)$; vertices: $(0, -10)$ and $(0, 10)$

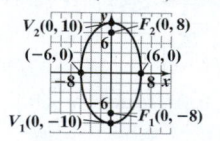

19. Foci: $\left(-3\sqrt{5}, 0\right)$ and $\left(3\sqrt{5}, 0\right)$; vertices: $(-7, 0)$ and $(7, 0)$

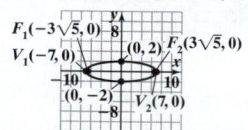

21. Foci: $\left(0, -4\sqrt{3}\right)$ and $\left(0, 4\sqrt{3}\right)$; vertices: $(0, -7)$ and $(0, 7)$

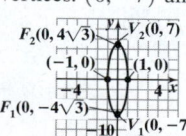

23. Foci: $\left(0, -2\sqrt{3}\right)$ and $\left(0, 2\sqrt{3}\right)$; vertices: $(0, -4)$ and $(0, 4)$

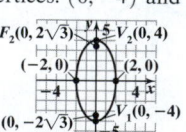

25. $\dfrac{x^2}{36} + \dfrac{y^2}{20} = 1$

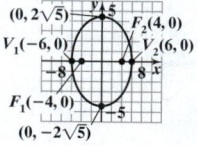

27. $\dfrac{x^2}{33} + \dfrac{y^2}{49} = 1$

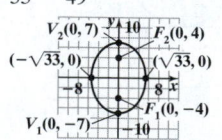

29. $\dfrac{x^2}{100} + \dfrac{y^2}{64} = 1$

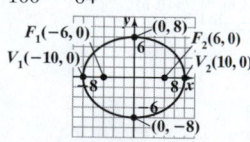

31. $\dfrac{x^2}{39} + \dfrac{y^2}{64} = 1$

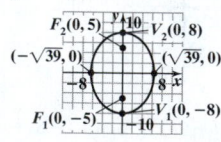

33. $\dfrac{(x - 3)^2}{9} + \dfrac{(y + 2)^2}{25} = 1$

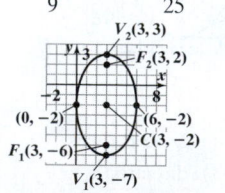

35. $\dfrac{(x + 2)^2}{16} + \dfrac{(y - 5)^2}{4} = 1$

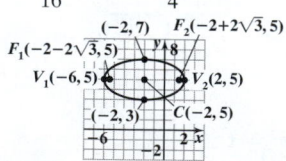

37. $(x - 5)^2 + \dfrac{(y + 1)^2}{49} = 1$

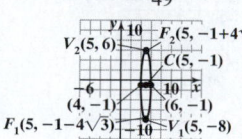

39. $4(x + 2)^2 + 16(y - 1)^2 = 64$

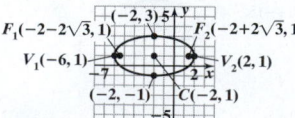

41. $4x^2 + y^2 - 24x + 2y - 63 = 0$

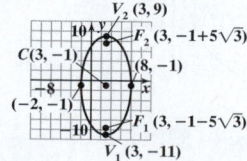

43. (a) $\dfrac{x^2}{225} + \dfrac{y^2}{100} = 1$ **(b)** Yes

(c) No

45. Perihelion = 91.5 million miles;
$$\dfrac{x^2}{8649} + \dfrac{y^2}{8646.75} = 1$$

47. Perihelion = 460.6 million miles; mean distance = 483.8 million miles; $\dfrac{x^2}{234{,}062.44} + \dfrac{y^2}{233{,}524.2} = 1$ **49.** $\dfrac{(x - 1)^2}{16} + \dfrac{(y - 2)^2}{9} = 1$

51. $\dfrac{(x-2)^2}{4} + \dfrac{y^2}{16} = 1$

55.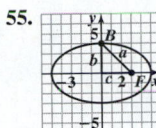

53. Let $a = b$, then

$$\dfrac{x^2}{a^2} + \dfrac{y^2}{b^2} = 1$$

$$\dfrac{x^2}{a^2} + \dfrac{y^2}{a^2} = 1$$

$$a^2\left(\dfrac{x^2}{a^2} + \dfrac{y^2}{a^2}\right) = a^2(1)$$

$$x^2 + y^2 = a^2$$

which is the equation of a circle with center $(0,0)$ and radius a. $c = 0$; The foci are located at the center point.

57.

x	5	10	100	1000
$f(x)$	0.71429	0.41667	0.04902	0.00499

59.

x	5	10	100	1000
$f(x)$	5.5	3	2.07216	2.00702

61.

x	5	10	100	1000
$f(x)$	6.83333	11.90909	101.99010	1001.99900
$g(x)$	7	12	102	1002

63. In Problems 57 and 58, the degree of the numerator is less than the degree of the denominator. In Problems 59 and 60, the degree of the numerator and the degree of the denominator are the same.

Conjecture 1: If the degree of the numerator of a rational function is less than the degree of the denominator, then as x increases, the value of the function will approach zero (0).

Conjecture 2: If the degree of the numerator of a rational function equals the degree of the denominator, then as x increases, the value of the function will approach the ratio of the leading coefficients of the numerator and denominator.

65. 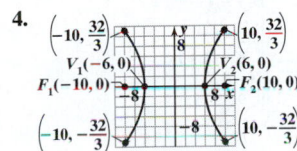 **67.** 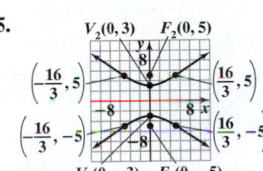 **69.** 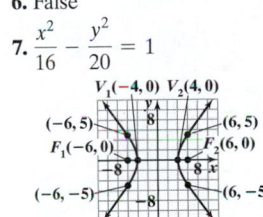 **71.**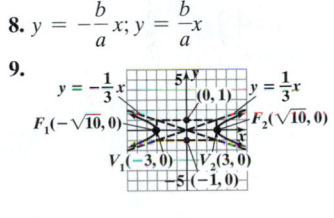

73. (a) The sum stays the same. **(b)** The center is at $(5,2)$ and changes to $(3,2)$. This move did not change the length of the major or minor axis. **(c)** The center changed to $(3,4)$, This move did not change the length of the major or minor axis. **(d)** h and k represent the center (h,k) of the ellipse. **(e)** This changed the length of the horizontal (major) axis and changed the denominator of the first term. **(f)** This changed the length of the horizontal (minor) axis and changed the denominator of the first term. **(g)** The number on the horizontal slider changes the length of the horizontal aixs. **(h)** This changed the length of the vertical (major) axis and changed the denominator of the second term. **(i)** This changed the length of the vertical (minor) axis and changed the denominator of the second term. **(j)** The number on the vertical slider changes the length of the vertical axis.

Section 12.5 Hyperbolas **1.** hyperbola **2.** transverse axis **3.** conjugate axis

4.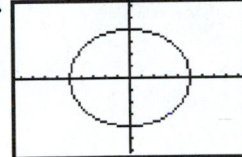

5.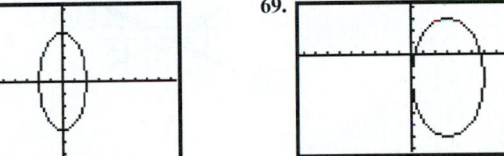

6. False

7. $\dfrac{x^2}{16} - \dfrac{y^2}{20} = 1$

8. $y = -\dfrac{b}{a}x;\ y = \dfrac{b}{a}x$

9.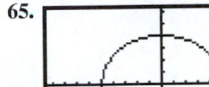

10.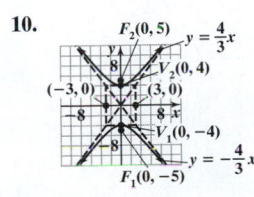

11. (b) **13. (a)**

15. $\dfrac{x^2}{4} - \dfrac{y^2}{16} = 1$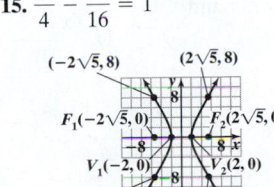

17. $\dfrac{y^2}{25} - \dfrac{x^2}{36} = 1$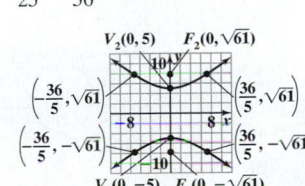

19. $4x^2 - y^2 = 36$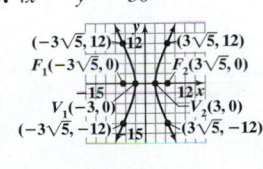

21. $25y^2 - x^2 = 100$

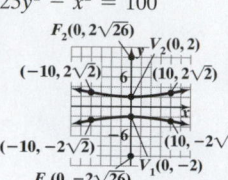

23. $\dfrac{x^2}{4} - \dfrac{y^2}{5} = 1$

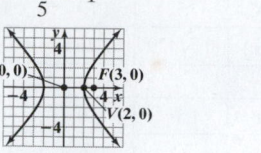

25. $\dfrac{y^2}{25} - \dfrac{x^2}{24} = 1$

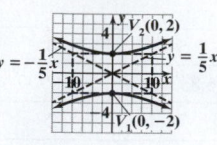

27. $\dfrac{x^2}{49} - \dfrac{y^2}{51} = 1$

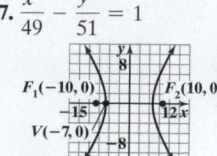

29.

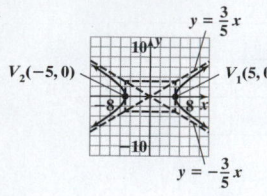

31.

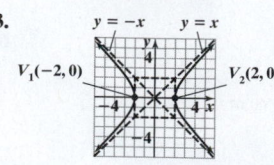

33.

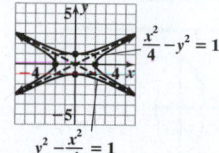

35. $\dfrac{y^2}{64} - \dfrac{x^2}{16} = 1$

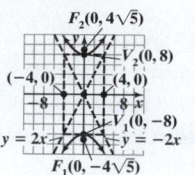

37. $\dfrac{x^2}{4.5} - \dfrac{y^2}{4.5} = 1$

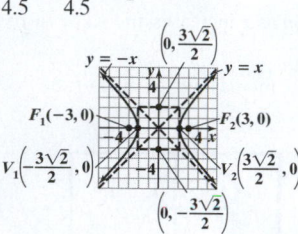

39. $x^2 - y^2 = 1$

41. $\dfrac{y^2}{36} - \dfrac{x^2}{9} = 1$

43. The asymptotes of both hyperbolas are $y = -\dfrac{1}{2}x$ and $y = \dfrac{1}{2}x$. Thus, they are conjugates.

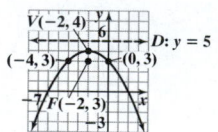

45. Answers will vary. **47.** $(-3, 1)$

49. $\{(x, y) \mid 2x - 3y = 6\}$

51. $\{(x, y) \mid 6x + 3y = 4\}$

53.

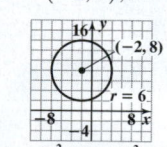

55.

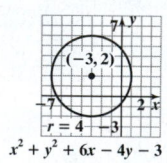

57.

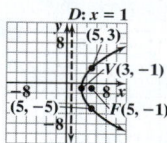

59.

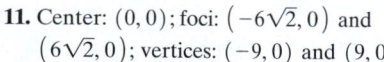

Putting the Concepts Together (Sections 12.1–12.5)   **1.** $3\sqrt{13}$ **2.** $(1, -3)$

3. $C = (-2, 8), r = 6$

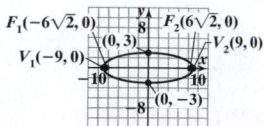

$(x + 2)^2 + (y - 8)^2 = 36$

4. $C = (-3, 2), r = 4$

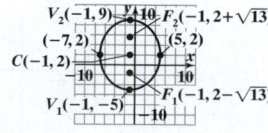

$x^2 + y^2 + 6x - 4y - 3 = 0$

5. $x^2 + y^2 = 169$

6. $(x - 2)^2 + (y - 1)^2 = 25$

7. Vertex: $(-2, 4)$; focus: $(-2, 3)$; directrix: $y = 5$

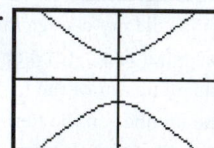

8. Vertex: $(3, -1)$; focus: $(5, -1)$; directrix: $x = 1$

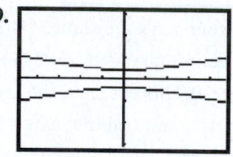

9. $(x + 1)^2 = -12(y + 2)$ **10.** $(y - 3)^2 = 8(x + 3)$

11. Center: $(0, 0)$; foci: $\left(-6\sqrt{2}, 0\right)$ and $\left(6\sqrt{2}, 0\right)$; vertices: $(-9, 0)$ and $(9, 0)$

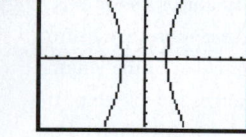

12. Center: $(-1, 2)$; vertices: $(-1, -5)$ and $(-1, 9)$; foci: $\left(-1, 2 - \sqrt{13}\right)$ and $\left(-1, 2 + \sqrt{13}\right)$

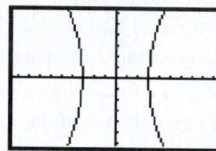

13. $\dfrac{x^2}{45} + \dfrac{y^2}{81} = 1$

14. $\dfrac{(x - 3)^2}{16} + \dfrac{(y + 4)^2}{7} = 1$

15. Vertices: $(0, -9)$ and $(0, 9)$;
foci: $\left(0, -3\sqrt{10}\right)$ and $\left(0, 3\sqrt{10}\right)$;
asymptotes: $y = 3x$ and $y = -3x$

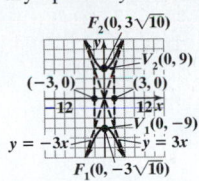

16. Vertices: $(-1, 0)$ and $(1, 0)$;
foci: $\left(-\sqrt{26}, 0\right)$ and $\left(\sqrt{26}, 0\right)$;
asymptotes: $y = -5x$ and $y = 5x$

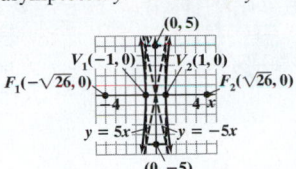

17. $\dfrac{y^2}{4} - \dfrac{x^2}{21} = 1$

18. 6.75 inches from the vertex, along its axis of symmetry

Section 12.6 Systems of Nonlinear Equations

1. $(-3, 5)$ and $(1, -3)$ **2.** $\left(\dfrac{11}{5}, -\dfrac{22}{5}\right)$ and $(1, -2)$ **3.** $(0, -4), (-2\sqrt{3}, 2), (2\sqrt{3}, 2)$

4. $\varnothing$ or $\{\}$ **5.** $(0, 4)$ and $(1, 5)$ **7.** $(3, 4)$ and $(4, 3)$ **9.** $(0, -2), \left(-\sqrt{3}, 1\right),$ and $\left(\sqrt{3}, 1\right)$ **11.** $(-2, -2)$ and $(2, 2)$

13. $(0, -2), (0, 2), \left(-1, -\sqrt{3}\right),$ and $\left(-1, \sqrt{3}\right)$ **15.** $\varnothing$ **17.** $(0, 0), (-3, 3),$ and $(3, 3)$ **19.** $(-3, 7)$ and $(2, -8)$ **21.** $(-1, 11)$ and $(2, -4)$

23. $(-4, 0)$ and $(4, 0)$ **25.** $(0, 3)$ and $(1, 4)$ **27.** $\left(0, -2 - \sqrt{3}\right), \left(0, -2 + \sqrt{3}\right), (1, -4),$ and $(1, 0)$ **29.** $\varnothing$

31. $(5, 2)$ **33.** $\left(-\dfrac{8}{3}, -\dfrac{2\sqrt{10}}{3}\right), \left(-\dfrac{8}{3}, \dfrac{2\sqrt{10}}{3}\right), \left(\dfrac{8}{3}, -\dfrac{2\sqrt{10}}{3}\right),$ and $\left(\dfrac{8}{3}, \dfrac{2\sqrt{10}}{3}\right)$ **35.** $(0, -5), (3, 4), (4, 3)$ and $(5, 0)$

37. Either -5 and -3, or 3 and 5 **39.** 14 feet by 10 feet   **41.** 19 cm by 10 cm   **43.** $(0, -2), (0, 1),$ and $(2, -1)$ **45.** $(81, 3)$

47. If $r_1 = \dfrac{-b + \sqrt{b^2 - 4ac}}{2a}$, then $r_2 = \dfrac{-b - \sqrt{b^2 - 4ac}}{2a}$; if $r_1 = \dfrac{-b - \sqrt{b^2 - 4ac}}{2a}$, then $r_2 = \dfrac{-b + \sqrt{b^2 - 4ac}}{2a}$.

49. (a) $f(1) = 7$ (b) $g(1) = 2$ **51.** (a) $f(3) = 13$ (b) $g(3) = 8$ **53.** (a) $f(5) = 19$ (b) $g(5) = 32$

55. $(-1, 11)$ and $(2, -4)$ **57.** $(-4, 0)$ and $(4, 0)$ **59.** $(0, 3)$ and $(1, 4)$ **61.** $(0.056, -0.237)$ and $(2.981, -1.727)$ **63.** $\varnothing$

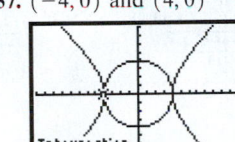
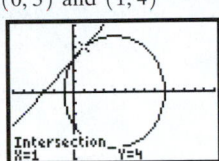
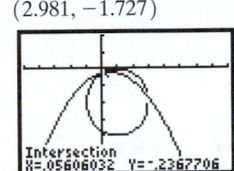
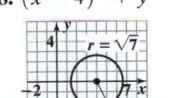
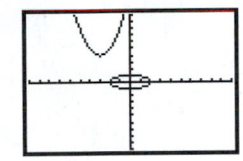

Chapter 12 Review

1. 5   **2.** 10   **3.** $2\sqrt{10} \approx 6.32$ **4.** 6   **5.** $3\sqrt{19} \approx 13.08$ **6.** 2.5   **7.** $(-2, 5)$ **8.** $(6, -2)$ **9.** $\left(-4\sqrt{3}, -3\sqrt{6}\right)$ **10.** $\left(\dfrac{5}{2}, \dfrac{1}{2}\right)$ **11.** $\left(\dfrac{3}{4}, \dfrac{1}{2}\right)$

12. (a)

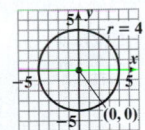

(b) $d(A, B) = 3\sqrt{2} \approx 4.24$;
$d(B, C) = 2\sqrt{2} \approx 2.83$;
$d(A, C) = \sqrt{26} \approx 5.10$

(c) $[d(A, B)]^2 + [d(B, C)]^2 \overset{?}{=} [d(A, C)]^2$
$(3\sqrt{2})^2 + (2\sqrt{2})^2 \overset{?}{=} \left(\sqrt{26}\right)^2$
$9 \cdot 2 + 4 \cdot 2 \overset{?}{=} 26$
$18 + 8 \overset{?}{=} 26$
$26 = 26 \leftarrow$ True
Therefore, triangle ABC is a right triangle.

(d) 6 square units

13. $C = (-2, 1); r = 4;$
$(x + 2)^2 + (y - 1)^2 = 16$

14. $C = (5, 3); r = 3;$
$(x - 5)^2 + (y - 3)^2 = 9$

15. $x^2 + y^2 = 16$

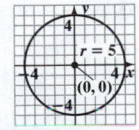

16. $(x + 3)^2 + (y - 1)^2 = 9$

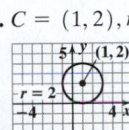

17. $(x - 5)^2 + (y + 2)^2 = 1$

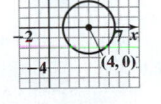

18. $(x - 4)^2 + y^2 = 7$

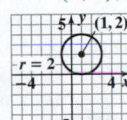

19. $(x - 2)^2 + (y + 1)^2 = 25$ **21.** $C = (0, 0), r = 5$

20. $(x + 1)^2 + (y - 3)^2 = 20$

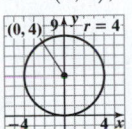

22. $C = (1, 2), r = 2$

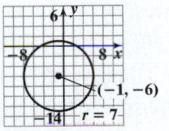

23. $C = (0, 4), r = 4$

24. $C = (-1, -6), r = 7$

25. $C = \left(-2, \dfrac{3}{2}\right), r = \dfrac{1}{2}$

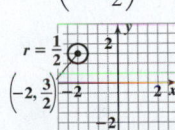

26. $C = (-3, -3), r = 2$

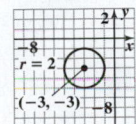

27. $C = (-3, -5), r = 6$

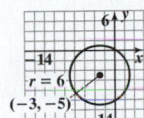

28. $C = (4, -2), r = 2$

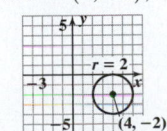

29. $C = (-1, 2), r = 3$

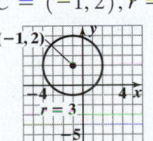

30. $C = (5, 1), r = 3$

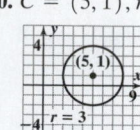

31. $x^2 = -12y$

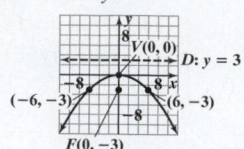

32. $y^2 = -16x$

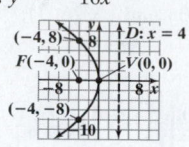

33. $y^2 = \dfrac{1}{2}x$

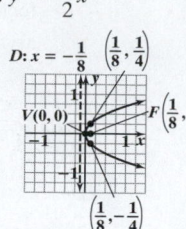

34. $x^2 = 8y$

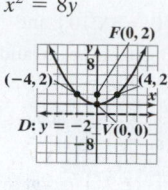

35. Vertex: $(0, 0)$; focus $\left(0, \dfrac{1}{2}\right)$; directrix: $y = -\dfrac{1}{2}$

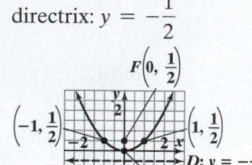

36. Vertex: $(0, 0)$; focus: $(4, 0)$; directrix $x = -4$

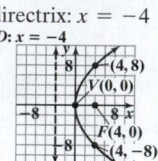

37. Vertex: $(-1, 3)$; focus: $(-1, 5)$; directrix: $y = 1$

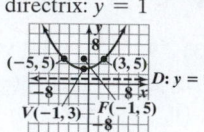

38. Vertex: $(-3, 4)$; focus: $\left(-\dfrac{7}{2}, 4\right)$; directrix: $x = -\dfrac{5}{2}$

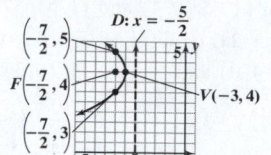

39. Vertex: $(5, 2)$; focus: $\left(5, \dfrac{5}{4}\right)$; directrix: $y = \dfrac{11}{4}$

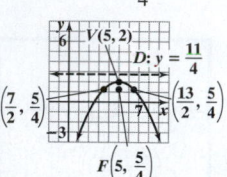

40. approximately 127.84 feet from the center of the dish, along its axis of symmetry

41. Foci: $\left(-2\sqrt{2}, 0\right)$ and $\left(2\sqrt{2}, 0\right)$; vertices: $(-3, 0)$ and $(3, 0)$

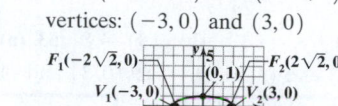

42. Foci: $\left(0, -\sqrt{5}\right)$ and $\left(0, \sqrt{5}\right)$; vertices: $(0, -3)$, and $(0, 3)$

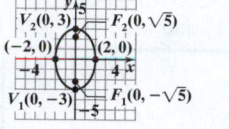

43. $\dfrac{x^2}{16} + \dfrac{y^2}{25} = 1$

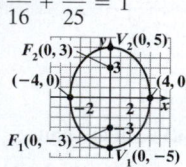

44. $\dfrac{x^2}{36} + \dfrac{y^2}{32} = 1$

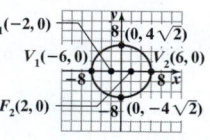

45. $\dfrac{x^2}{100} + \dfrac{y^2}{36} = 1$

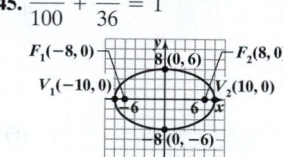

46. Center: $(1, -2)$; vertices: $(-6, -2)$ and $(8, -2)$; foci: $\left(1 - 2\sqrt{6}, -2\right)$ and $\left(1 + 2\sqrt{6}, -2\right)$

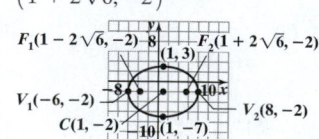

47. Center: $(-3, 4)$; vertices: $(-3, -1)$ and $(-3, 9)$; foci: $(-3, 0)$ and $(-3, 8)$

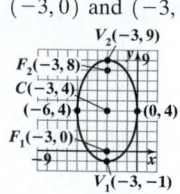

48. (a) $\dfrac{x^2}{900} + \dfrac{y}{256} = 1$

(b) Yes

49. Vertices: $(-2, 0)$ and $(2, 0)$; foci: $\left(-\sqrt{13}, 0\right)$ and $\left(\sqrt{13}, 0\right)$

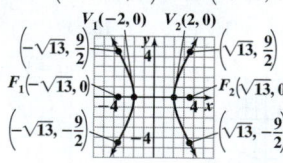

50. Vertices: $(0, -5)$ and $(0, 5)$; foci: $\left(0, -\sqrt{74}\right)$ and $\left(0, \sqrt{74}\right)$

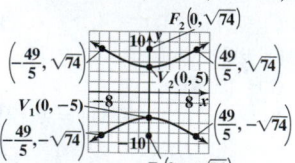

51. Vertices: $(0, -5)$ and $(0, 5)$; foci: $\left(0, -\sqrt{41}\right)$ and $\left(0, \sqrt{41}\right)$

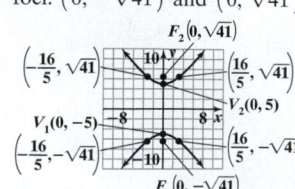

52. Asymptotes: $y = x$ and $y = -x$

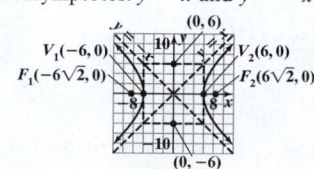

53. Asymptotes: $y = \dfrac{5}{2}x$ and $y = -\dfrac{5}{2}x$

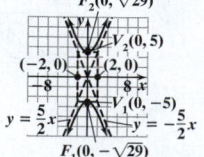

54. $\dfrac{x^2}{9} - \dfrac{y^2}{7} = 1$

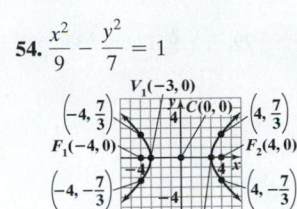

55. $\dfrac{y^2}{9} - \dfrac{x^2}{16} = 1$

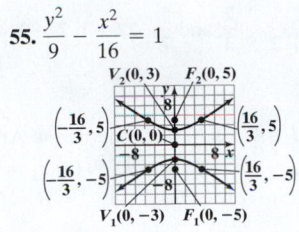

56. $\dfrac{y^2}{16} - \dfrac{x^2}{9} = 1$

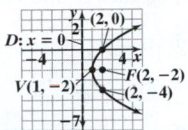

57. $\left(\sqrt{2}, \sqrt{2}\right)$ and $\left(-\sqrt{2}, -\sqrt{2}\right)$

58. $\left(-\dfrac{1}{2}, \dfrac{3}{2}\right)$ and $(1, 3)$

59. $\left(-\dfrac{1}{6}, -6\right)$ and $(1, 1)$

60. $(-5, -1), (-5, 1), (5, -1),$ and $(5, 1)$

61. $\left(2, -2\sqrt{2}\right)$ and $\left(2, 2\sqrt{2}\right)$

62. $(-1, 3)$ and $(1, 3)$

63. $\left(-\sqrt{3}, -\sqrt{5}\right), \left(-\sqrt{3}, \sqrt{5}\right), \left(\sqrt{3}, -\sqrt{5}\right),$ and $\left(\sqrt{3}, \sqrt{5}\right)$ **64.** $(0, 0), \left(5, -\sqrt{15}\right),$ and $\left(5, \sqrt{15}\right)$ **65.** $(2, 4)$ and $(-1, 1)$

66. $(-7, -20)$ and $\left(2, \dfrac{5}{2}\right)$ **67.** $(0, 6)$ and $(-6, 0)$ **68.** $(4, 4)$ and $(1, -2)$ **69.** $\varnothing$ **70.** $(-1, -2), (-1, 2), (1, -2),$ and $(1, 2)$

71. $(-4, 0), (4, 0),$ and $(0, 4)$ **72.** $(4, 0)$ and $(0, -2)$ **73.** 7 and 5 **74.** 12 cm by 5 cm **75.** 72 inches by 30 inches **76.** 12 feet and 9 feet

Chapter 12 Test **1.** $4\sqrt{5}$ **2.** $(-1, 2)$
3. $C = (4, -1), r = 3$

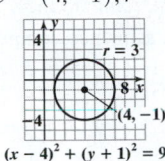

$(x-4)^2 + (y+1)^2 = 9$

4. $C = (-5, 2), r = 4$

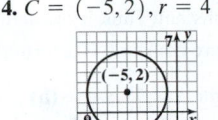

$x^2 + y^2 + 10x - 4y + 13 = 0$

5. $(x+3)^2 + (y-7)^2 = 36$
6. $(x+5)^2 + (y-8)^2 = 100$

7. Vertex $(1, -2)$; focus $(2, -2)$; directrix $x = 0$

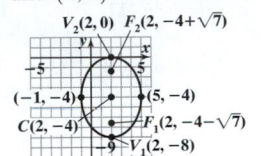

8. Vertex $(2, 4)$; focus $\left(2, \dfrac{13}{4}\right)$; directrix $y = \dfrac{19}{4}$

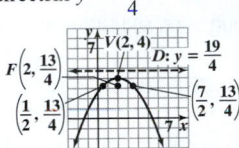

9. $x^2 = -16y$
10. $(y-4)^2 = 8(x-1)$

11. Foci: $(-4, 0)$ and $(4, 0)$; vertices: $(-5, 0)$ and $(5, 0)$

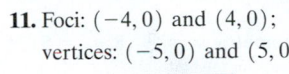

12. Foci: $\left(2, -4 - \sqrt{7}\right)$ and $\left(2, -4 + \sqrt{7}\right)$ vertices: $(2, -8)$ and $(2, 0)$

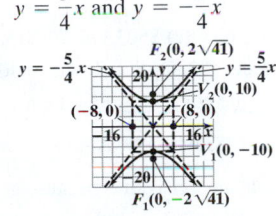

13. $\dfrac{x^2}{9} + \dfrac{y^2}{25} = 1$

14. $\dfrac{(x+1)^2}{16} + \dfrac{(y-2)^2}{25} = 1$

15. Vertices: $(-1, 0)$ and $(1, 0)$; foci: $\left(-\sqrt{5}, 0\right)$ and $\left(\sqrt{5}, 0\right)$; asymptotes: $y = 2x$ and $y = -2x$

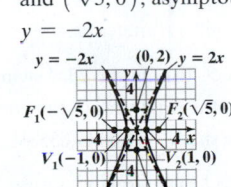

16. Vertices: $(0, -10)$ and $(0, 10)$; foci: $\left(0, -2\sqrt{41}\right)$ and $\left(0, 2\sqrt{41}\right)$; asymptotes: $y = \dfrac{5}{4}x$ and $y = -\dfrac{5}{4}x$

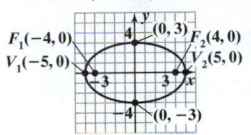

17. $\dfrac{x^2}{9} - \dfrac{y^2}{55} = 1$ **18.** $(-4, 1)$ and $(1, -4)$ **19.** $\left(-\sqrt{5}, -2\right), \left(-\sqrt{5}, 2\right), \left(\sqrt{5}, -2\right),$ and $\left(\sqrt{5}, 2\right)$ **20. (a)** $\dfrac{x^2}{225} + \dfrac{y^2}{100} = 1$ **(b)** 6 feet

Chapter 13 Sequences, Series, and the Binomial Theorem

Section 13.1 Sequences **1.** sequence **2.** infinite; finite **3.** True **4.** $-1, 1, 3, 5, 7$ **5.** $-4, 8, -12, 16, -20$ **6.** $a_n = 2n + 3$ **7.** $b_n = \dfrac{(-1)^{n+1}}{n+1}$

8. partial sum **9.** $3 + 7 + 11 = 21$ **10.** $2 + 9 + 28 + 65 + 126 = 230$ **11.** $\displaystyle\sum_{k=1}^{12} k^2$ **12.** $\displaystyle\sum_{n=1}^{6} \left(\dfrac{1}{2}\right)^{n-1}$ **13.** $8, 11, 14, 17, 20$ **15.** $\dfrac{1}{3}, \dfrac{1}{2}, \dfrac{3}{5}, \dfrac{2}{3}, \dfrac{5}{7}$

17. $-1, 2, -3, 4, -5$ **19.** $3, 5, 9, 17, 33$ **21.** $1, 1, \dfrac{3}{4}, \dfrac{1}{2}, \dfrac{5}{16}$ **23.** $\dfrac{1}{e}, \dfrac{2}{e^2}, \dfrac{3}{e^3}, \dfrac{4}{e^4}, \dfrac{5}{e^5}$ **25.** $a_n = 2n$ **27.** $a_n = \dfrac{n}{n+1}$ **29.** $a_n = n^2 + 2$ **31.** $a_n = (-1)^n n^2$

33. 54 **35.** $\dfrac{55}{2}$ **37.** 14 **39.** 6 **41.** 50 **43.** 45 **45.** $\displaystyle\sum_{k=1}^{15} k$ **47.** $\displaystyle\sum_{i=1}^{12} \dfrac{1}{i}$ **49.** $\displaystyle\sum_{i=1}^{9} (-1)^{i+1}\left(\dfrac{1}{3^{i-1}}\right)$ **51.** $\displaystyle\sum_{k=1}^{11} (2k + 3)$ **53. (a)** \$12,180 **(b)** \$12,736.36

(c) \$21,768.22 **55. (a)** 339 million **(b)** 456 million **57.** $1, 1, 2, 3, 5, 8, 13, 21, 34, 55$ **59.** $10, 10.5, 11.025, 11.57625,$ and 12.1550625 **61.** $8, 10, 13, 17,$ and 22

63. (a) $1; 2; 1.5; 1.\overline{6}; 1.6; 1.625; \dfrac{21}{13} \approx 1.615385; \dfrac{34}{21} \approx 1.619048; \dfrac{55}{34} \approx 1.617647; \dfrac{89}{55} \approx 1.618182$ **(b)** around 1.618 **(c)** $1; 0.5; \dfrac{2}{3} = 0.\overline{6}; 0.6; 0.625;$

$\dfrac{8}{13} \approx 0.615385; \dfrac{13}{21} \approx 0.619048; \dfrac{21}{34} \approx 0.617647; \dfrac{34}{55} \approx 0.618182; \dfrac{55}{89} \approx 0.617978$ **(d)** around 0.618 **65.** The main difference between a function and a sequence is that the domain of a function is based on the real number system, whereas the domain of a sequence is the positive integers. **67.** The symbol $\sum$ means to add up the terms of a sequence. **69. (a)** $m = 4$ **(b)** $f(1) = -2; f(2) = 2; f(3) = 6; f(4) = 10$ **71. (a)** $m = -5$

(b) $f(1) = 3; f(2) = -2; f(3) = -7; f(4) = -12$ **73.** 8, 11, 14, 17, and 20   **75.** $\frac{1}{3}, \frac{1}{2}, \frac{3}{5}, \frac{2}{3}$, and $\frac{5}{7}$ **77.** $-1, 2, -3, 4$, and -5 **79.** 3, 5, 9, 17, and 33

81. 54   **83.** $\frac{55}{2}$ **85.** 14   **87.** 6

Section 13.2 Arithmetic Sequences   1. arithmetic   **2.** Arithmetic; $a_1 = -3$; $d = 2$ **3.** Not arithmetic   **4.** Arithmetic $a_1 = -5$; $d = 3$
5. Not arithmetic   **6.** Arithmetic $a_1 = 3$; $d = -2$ **7.** $a_n = a_1 + (n-1)d$ **8. (a)** $a_n = 6n - 5$ **(b)** 79   **9. (a)** $a_1 = -5$; $d = 3$ **(b)** $a_n = 3n - 8$
10. 10,400   **11.** 9730   **12.** 7550   **13.** -1050 **14.** 1470 seats   **15.** $a_n - a_{n-1} = d = 1$; $a_1 = 6$ **17.** $a_n - a_{n-1} = d = 7$; $a_1 = 9$

19. $a_n - a_{n-1} = d = -3$; $a_1 = 4$ **21.** $a_n - a_{n-1} = d = \frac{1}{2}$; $a_1 = \frac{11}{2}$ **23.** $a_n = 3n + 1$; $a_5 = 16$ **25.** $a_n = -5n + 15$; $a_5 = -10$ **27.** $a_n = \frac{1}{3}n + \frac{5}{3}$;
$a_5 = \frac{10}{3}$ **29.** $a_n = -\frac{1}{5}n + \frac{26}{5}$; $a_5 = \frac{21}{5}$ **31.** $a_n = 5n - 3$; $a_{20} = 97$ **33.** $a_n = -3n + 15$; $a_{20} = -45$ **35.** $a_n = \frac{1}{4}n + \frac{3}{4}$; $a_{20} = \frac{23}{4}$

37. $a_1 = 7$; $d = 5$; $a_n = 5n + 2$ **39.** $a_1 = -23$; $d = 7$; $a_n = 7n - 30$ **41.** $a_1 = 11$; $d = -3$; $a_n = -3n + 14$ **43.** $a_1 = 4$; $d = -\frac{1}{2}$; $a_n = -\frac{1}{2}n + \frac{9}{2}$

45. $S_{30} = 2670$ **47.** $S_{25} = 700$ **49.** $S_{40} = -5060$ **51.** $S_{40} = 3160$ **53.** $S_{75} = -9000$ **55.** $S_{30} = 460$ **57.** $x = -\frac{3}{2}$ **59.** There are 630 cans in the
stack.   **61.** There are 1600 seats in the auditorium.   **63.** There are 84 terms in the sequence.   **65.** There are 63 terms in the sequence.   **67.** It will take
about 15.22 years.   **69.** A sequence is arithmetic if the difference in consecutive terms is constant. Thus, if the actual terms are given, compute the
difference between each pair of consecutive terms and determine whether the differences are constant. If a formula is given for the sequence, compute
$a_n - a_{n-1}$ and determine whether the result is a constant–if it is, then the sequence is arithmetic.   **71. (a)** 3   **(b)** 3; 9; 27; 81   **73. (a)** $\frac{1}{2}$ **(b)** 5; $\frac{5}{2}, \frac{5}{4}, \frac{5}{8}$
75. 806.9   **77.** 1427.5

Section 13.3 Geometric Sequences and Series   1. geometric   **2.** Geometric; $a_1 = 4$, $r = 2$ **3.** Not geometric   **4.** Geometric; $a_1 = 9$, $r = \frac{1}{3}$

5. Geometric; $r = 5$ **6.** Not geometric   **7.** Geometric; $r = \frac{2}{3}$ **8.** $a_n = 5 \cdot 2^{n-1}$; $a_9 = 1280$ **9.** $a_n = 50\left(\frac{1}{2}\right)^{n-1}$; $a_9 = 0.1953125$ or $a_9 = \frac{25}{128}$

10. $S_n = a_1 \cdot \frac{1 - r^n}{1 - r}$ **11.** 24,573   **12.** 7.9921875   **13.** $\frac{a_1}{1 - r}$ **14.** $\frac{40}{3}$ **15.** $\frac{1}{2}$ **16.** $\frac{2}{9}$ **17.** The U.S. economy will grow by \$10,000.   **18.** \$244,129.08

19. $\frac{a_n}{a_{n-1}} = r = 4$; $a_1 = 4$ **21.** $\frac{a_n}{a_{n-1}} = r = \frac{2}{3}$; $a_1 = \frac{2}{3}$ **23.** $\frac{a_n}{a_{n-1}} = r = \frac{1}{2}$; $a_1 = \frac{3}{2}$ **25.** $\frac{a_n}{a_{n-1}} = r = \frac{5}{2}$; $a_1 = \frac{1}{2}$ **27. (a)** $a_n = 10 \cdot 2^{n-1}$ **(b)** $a_8 = 1280$

29. (a) $a_n = 100 \cdot \left(\frac{1}{2}\right)^{n-1}$ **(b)** $a_8 = \frac{25}{32}$ **31. (a)** $a_n = (-3)^{n-1}$ **(b)** $a_8 = -2187$ **33. (a)** $a_n = 100 \cdot (1.05)^{n-1}$ **(b)** $a_8 = 100 \cdot (1.05)^7 \approx 140.71$

35. $a_{10} = 1536$ **37.** $a_{15} = \frac{1}{4096}$ **39.** $a_9 = 0.000000005$ **41.** 8190   **43.** 83.3245952   **45.** 6138   **47.** 7.96875   **49.** 2   **51.** 15   **53.** $\frac{9}{2}$ **55.** $\frac{5}{4}$ **57.** 9

59. $\frac{5}{9}$ **61.** $\frac{89}{99}$ **63.** arithmetic; $d = 5$ **65.** Neither   **67.** geometric; $r = \frac{1}{2}$ **69.** geometric; $r = \frac{2}{3}$ **71.** arithmetic; $d = 4$ **73.** Neither   **75.** $x = -4$
77. (a) \$42,000   **(b)** \$62,053   **(c)** \$503,116   **79.** \$13,182   **81. (a)** About 1.891 feet   **(b)** On the 23rd swing   **(c)** About 24.08 feet   **(d)** The pendulum will have
swung a total of 60 feet.   **83.** Option A will yield the larger annual salary in the final year of the contract, and option B will yield the larger cumulative
salary over the life of the contract.   **85.** The multiplier is 50.   **87.** \$31.14 per share   **89.** \$149,035.94   **91.** \$114,401.52   **93.** \$395.09, or about \$395
95. $0.4\overline{9} = \frac{1}{2}$ **97.** 2,147,483,646   **99.** A geometric sequence with $r > 1$ yields faster growth because of the effect of compounding. Answers will vary.
However, a reasonable answer is that because geometric growth is faster than arithmetic growth, the population will grow beyond the ability of food
supplies to sustain the population, and hunger will ensue.   **101.** A geometric series has a sum provided that the common ratio r is between -1 and 1.
103. 1   **105.** approximately 288.1404315   **107.** approximately 41.66310217

Putting the Concepts Together (Sections 13.1–13.3)   1. Geometric with $a_1 = \frac{3}{4}$ and common ratio $r = \frac{1}{4}$ **2.** Arithmetic with $a_1 = 8$ and

$d = 2$ **3.** Arithmetic with $a_1 = 1$ and $d = \frac{7}{9}$ **4.** Neither arithmetic nor geometric   **5.** Geometric with $a_1 = 12$ and $r = 2$ **6.** Neither arithmetic nor

geometric   **7.** 87   **8.** $\sum_{i=1}^{12} \frac{1}{2(6 + i)}$ **9.** $a_n = 27 - 2n$; 25, 23, 21, 19, and 17   **10.** $a_n = 11n - 35$; $-24, -13, -2, 9$, and 20   **11.** $a_n = 45 \cdot \left(\frac{1}{5}\right)^{n-1}$; 45, 9,

$\frac{9}{5}, \frac{9}{25}$, and $\frac{9}{125}$ **12.** $a_n = 150 \cdot (1.04)^{n-1}$; 150, 156, 162.24, 168.7296, and 175.478784   **13.** $S_{11} = 177,146$ **14.** $S_{20} = 990$ **15.** $\frac{10,000}{9}$ **16.** A party of
24 people would require 11 tables.

Section 13.4 The Binomial Theorem   1. $n(n-1)(n-2) \cdot \cdots \cdot 3 \cdot 2 \cdot 1$ **2.** 1; 1   **3.** 120   **4.** 840   **5.** False   **6.** True   **7.** 7   **8.** 20
9. $x^4 + 8x^3 + 24x^2 + 32x + 16$ **10.** $32p^5 - 80p^4 + 80p^3 - 40p^2 + 10p - 1$ **11.** 6   **13.** 40,320   **15.** 90   **17.** 336   **19.** 21   **21.** 210
23. $x^5 + 5x^4 + 10x^3 + 10x^2 + 5x + 1$ **25.** $x^4 - 16x^3 + 96x^2 - 256x + 256$ **27.** $81p^4 + 216p^3 + 216p^2 + 96p + 16$
29. $32z^5 - 240z^4 + 720z^3 - 1080z^2 + 810z - 243$ **31.** $x^8 + 8x^6 + 24x^4 + 32x^2 + 16$ **33.** $32p^{15} + 80p^{12} + 80p^9 + 40p^6 + 10p^3 + 1$
35. $x^6 + 12x^5 + 60x^4 + 160x^3 + 240x^2 + 192x + 64$ **37.** $16p^8 - 32p^6q^2 + 24p^4q^4 - 8p^2q^6 + q^8$ **39.** 1.00401   **41.** 0.99004   **43.** $84x^5$

45. $-108,864p^3$ **47.** $\binom{n}{n-1} = \dfrac{n!}{(n-1)!(n-(n-1))!} = \dfrac{n!}{(n-1)!1!} = \dfrac{n\cdot(n-1)!}{(n-1)!} = n;\ \binom{n}{n} = \dfrac{n!}{n!(n-n)!} = \dfrac{n!}{n!0!} = \dfrac{n!}{n!} = 1$

49.
$$
\begin{matrix}
& & 1 & & \\
& 1 & & 1 & \\
1 & & 2 & & 1 \\
1 & & 3 & 3 & 1
\end{matrix}
$$

51. The degree of each monomial equals n. **53.** $a^4 - 8a^3 + 24a^2 - 32a + 16$ **55.** $p^5 + p^4 - 6p^3 - 14p^2 - 11p - 3$

Chapter 13 Review **1.** $-1, -4, -7, -10,$ and -13 **2.** $-\dfrac{1}{5}, 0, \dfrac{1}{7}, \dfrac{1}{4},$ and $\dfrac{1}{3}$ **3.** $6, 26, 126, 626,$ and 3126 **4.** $3, -6, 9, -12,$ and 15 **5.** $\dfrac{1}{2}, \dfrac{4}{3}, \dfrac{9}{4}, \dfrac{16}{5},$

and $\dfrac{25}{6}$ **6.** $\pi, \dfrac{\pi^2}{2}, \dfrac{\pi^3}{3}, \dfrac{\pi^4}{4},$ and $\dfrac{\pi^5}{5}$ **7.** $a_n = -3n$ **8.** $a_n = \dfrac{n}{3}$ **9.** $a_n = 5\cdot 2^{n-1}$ **10.** $a_n = (-1)^n\cdot\dfrac{n}{2}$ **11.** $a_n = n^2 + 5$ **12.** $a_n = \dfrac{n-1}{n+1}$.

13. 65 **14.** $\dfrac{33}{2}$ **15.** -30 **16.** $\dfrac{26}{3}$ **17.** $\displaystyle\sum_{i=1}^{15}(4+3i)$ **18.** $\displaystyle\sum_{i=1}^{8}\dfrac{1}{3^i}$ **19.** $\displaystyle\sum_{i=1}^{10}\dfrac{i^3+1}{i+1}$ **20.** $\displaystyle\sum_{i=1}^{7}[(-1)^{i-1}\cdot i^2]$ **21.** arithmetic with $d = 6$ **22.** arithmetic

with $d = \dfrac{3}{2}$ **23.** not arithmetic **24.** not arithmetic **25.** arithmetic with $d = 4$ **26.** not arithmetic **27.** $a_n = 8n - 5; a_{25} = 195$

28. $a_n = -3n - 1; a_{25} = -76$ **29.** $a_n = -\dfrac{1}{3}n + \dfrac{22}{3}; a_{25} = -1$ **30.** $a_n = 6n + 5; a_{25} = 155$ **31.** $a_n = \dfrac{18}{5}n - \dfrac{19}{5}; a_{25} = \dfrac{431}{5}$

32. $a_n = -4n - 4; a_{25} = -104$ **33.** 4320 **34.** -2140 **35.** -4080 **36.** $\dfrac{1875}{4}$, or 468.75 **37.** 2106 **38.** 300 yards **39.** geometric with $r = 6$

40. geometric with $r = -3$ **41.** not geometric **42.** geometric with $r = \dfrac{2}{3}$ **43.** geometric with $r = -2$ **44.** not geometric **45.** $a_n = 4\cdot 3^{n-1}$;

$a_{10} = 78,732$ **46.** $a_n = 8\cdot\left(\dfrac{1}{4}\right)^{n-1}; a_{10} = \dfrac{1}{32,768}$ **47.** $a_n = 5\cdot(-2)^{n-1}; a_{10} = -2560$ **48.** $a_n = 1000\cdot(1.08)^{n-1}; a_{10} \approx 1999.005$

49. $65,534$ **50.** ≈ 45.71428571 **51.** $\dfrac{12,285}{4}$ or 3071.25 **52.** $-258,280,320$ **53.** $\dfrac{20}{3}$ **54.** $\dfrac{100}{3}$ **55.** $\dfrac{5}{4}$ **56.** $\dfrac{8}{9}$ **57.** After 72 years, there will be

3.125 grams of the tritium remaining. **58.** After 15 minutes, about 38.15 billion e-mails will have been sent. **59.** After 25 years, Scott's 403(b) will be
worth $360,114.89. **60.** The lump sum option would yield more money after 26 years. **61.** Sheri would need to contribute $534.04, or about $534,
each month to reach her goal. **62.** When Samantha turns 18, the plan will be worth $62,950.79 and will cover about 185 credit hours. **63.** 120
64. 7920 **65.** 5040 **66.** 1716 **67.** 35 **68.** 252 **69.** 1 **70.** 1 **71.** $z^4 + 4z^3 + 6z^2 + 4z + 1$ **72.** $y^5 - 15y^4 + 90y^3 - 270y^2 + 405y - 243$
73. $729y^6 + 5832y^5 + 19,440y^4 + 34,560y^3 + 34,560y^2 + 18,432y + 4096$ **74.** $16x^8 - 96x^6 + 216x^4 - 216x^2 + 81$
75. $81p^4 - 216p^3q + 216p^2q^2 - 96pq^3 + 16q^4$ **76.** $a^{15} + 15a^{12}b + 90a^9b^2 + 270a^6b^3 + 405a^3b^4 + 243b^5$ **77.** $-448x^5$ **78.** $14,784x^5$

Chapter 13 Test **1.** arithmetic with $a_1 = -15$ and $d = 8$ **2.** geometric with $a_1 = -4$ and $r = -4$ **3.** neither arithmetic nor geometric

4. arithmetic with $a_1 = -\dfrac{1}{5}$ and $d = \dfrac{2}{5}$ **5.** neither arithmetic nor geometric **6.** geometric with $a_1 = 21$ and $r = 3$ **7.** $\dfrac{17,269}{1200}$ **8.** $\displaystyle\sum_{i=1}^{8}\dfrac{i+2}{i+4}$

9. $a_n = 10n - 4; 6, 16, 26, 36,$ and 46 **10.** $a_n = 4 - 4n; 0, -4, -8, -12,$ and -16 **11.** $a_n = 10\cdot 2^{n-1}; 10, 20, 40, 80,$ and 160

12. $a_n = (-3)^{n-1}; 1, -3, 9, -27,$ and 81 **13.** 720 **14.** $-\dfrac{132,860}{9}$ **15.** 324 **16.** 6435 **17.** 792 **18.** $625m^4 - 1000m^3 + 600m^2 - 160m + 16$

19. $6103.11 **20.** 8000 lb

Cumulative Review Chapters 1–13 **1.** $y = \dfrac{3x-4}{15}$ or $y = \dfrac{1}{5}x - \dfrac{4}{15}$ **2.** $f(2) = 9; f(-3) = 19$ **3.** $\{32\}$ **4.** $\left\{-\dfrac{2}{5}, 1\right\}$

5. $\left\{\dfrac{-7-\sqrt{73}}{6}, \dfrac{-7+\sqrt{73}}{6}\right\}$ **6.** $\{60\}$ **7.** $\left\{\dfrac{11}{4}\right\}$ **8.** $\{-4, -2, 0\}$ **9.** $(-6, \infty)$ **10.** $\left[-3, \dfrac{1}{3}\right]$ **11.** $(x+2)(2x-9)$ **12.** $(2x-1)(3x^2+2)$

13. $20x^3 - 22x^2 + 11x - 3$ **14.** $\dfrac{(x-6)(x+2)}{(x+4)(x-1)}$ **15.** $1 - i$ **16.** $\{x\,|\,x \ge 15\}$ or $[15, \infty)$

17. $y = -7x + 11$ **18.** $(4, -1)$ **19.** 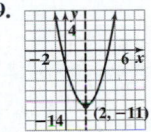 **20.** $(x-4)^2 + (y+3)^2 = 36$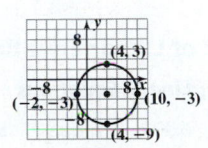

21.   **22.** 1340 **23.** 8 **24.** It would take about 2.73 hours to mow the lot if both machines worked together.

25. $\dfrac{325}{3}$ metric tons of pure aluminum must be added.

Appendix A Synthetic Division

1. False **2.** False **3.** -7 **4.** $2x^2 + 5x + 3 - \dfrac{7}{x-2}$ **5.** $4x^2 - 11x + 23 - \dfrac{45}{x+2}$ **6.** $x^3 + 5x^2 - 2$

7. (a) 42 **(b)** 0 **8. (a)** $f(-2) = -35$; $x + 2$ is not a factor **(b)** $f(5) = 0$; $f(x) = (x-5)(2x^2 + x - 1)$ **9.** $x + 2$ **11.** $2x + 3$ **13.** $x + 3 + \dfrac{4}{x-6}$

15. $x^2 + 5x + 6 + \dfrac{15}{x-5}$ **17.** $3x^3 + 4x^2 - 9x - 10 - \dfrac{5}{x-3}$ **19.** $x^3 - 6x^2 - 4x + 24 - \dfrac{35}{x+6}$ **21.** $2x^2 + 8x + 6$

23. -5 **25.** -119 **27.** 231 **29.** 2 **31.** $x - 2$ is a factor; $f(x) = (x-2)(x-1)$ **33.** $x + 2$ is a factor; $f(x) = (x+2)(2x+1)$

35. $x - 3$ is not a factor **37.** $x + 1$ is a factor; $f(x) = (x+1)(4x^2 - 11x + 6)$ **39.** $f(x) = 3x^2 - 10x - 25$ **41.** $f(x) = x^2 + 5x - 20$

43. $a = 2, b = -7, c = -12$, and $d = -13$, thus, $a + b + c + d = -30$ **45.** The dividend is the polynomial f and has degree n. Remember, (Divisor)(Quotient) + Remainder = Dividend. Since the remainder must be 0 or a polynomial that has lower degree than f, the degree n must be obtained from the product of the divisor and the quotient. The divisor is $x + 4$, which is of degree 1, so the quotient must be of degree $n - 1$. **47.** Answers may vary. Because $x + 3 = x - (-3)$, synthetic division is more likely efficient.

Appendix B Geometry Review

Section B.1 Lines and Angles **1.** congruent **2.** angle **3.** right **4.** acute **5.** obtuse **6.** straight **7.** right **8.** False **9.** $75°$; $165°$

10. $30°$; $120°$ **11.** Parallel **12.** False **13.** True **14.** $m\angle 1 = 140°$; $m\angle 2 = 40°$; $m\angle 3 = 140°$; $m\angle 4 = 40°$; $m\angle 5 = 140°$; $m\angle 6 = 40°$; $m\angle 7 = 140°$

15. acute **17.** right **19.** straight **21.** obtuse **23.** $58°$ **25.** $17°$ **27.** $113°$ **29.** $172°$

31. $m\angle 1 = 130°$; $m\angle 2 = 50°$; $m\angle 3 = 130°$; $m\angle 4 = 50°$; $m\angle 5 = 130°$; $m\angle 6 = 50°$; $m\angle 7 = 130°$

Section B.2 Polygons **1.** isosceles **2.** right **3.** 180 **4.** $70°$ **5.** $42°$ **6.** Congruent **7.** Similar **8.** 14 units **9.** radius **10.** True

11. $\dfrac{15}{2}$ inches or 7.5 inches **12.** 12 feet **13.** 7.2 yards **14.** 18 cm **15.** $55°$ **17.** $48°$ **19.** 4 units **21.** 67.5 units **23.** 10 inches **25.** 5 cm

27. 7 cm **29.** $\dfrac{11}{2}$ yards or 5.5 yards

Section B.3 Perimeter and Areas of Polygons and Circles **1.** perimeter **2.** area **3.** Perimeter: 22 feet; Area: 24 square feet

4. Perimeter: 26 m; Area: 30 square m **5.** False **6.** Perimeter: 16 cm; Area 16 square cm **7.** Perimeter: 6 yards; Area: 2.25 square yards

8. Perimeter: 130 yards: Area: 650 square yards **9.** $\dfrac{1}{2}h(b + B)$; h; b; B **10.** Perimeter: 36 m; Area: 70 square m

11. Perimeter: 33 yards; Area: 51 square yards **12.** True **13.** Perimeter: 19 mm: Area: 12 square mm **14.** Perimeter: 30 feet; Area: 30 square feet

15. Circumference **16.** False **17.** Circumference: 8π feet $\approx$ 25.13 feet; Area: 16π square feet $\approx$ 50.27 square feet

18. Circumference: 24π cm $\approx$ 75.40 cm; Area: 144π square cm $\approx$ 452.39 square cm **19.** Perimeter: 28 feet; Area: 40 square feet

21. Perimeter: 40 m; Area: 75 square m **23.** Perimeter; 24 km; Area: 36 square km **25.** Perimeter: 72 feet: Area: 218 square feet

27. Perimeter: 54 m; Area: 62 square m **29.** Perimeter: 30 feet: Area: 45 square feet **31.** Perimeter: 28 mm; Area: 36 square mm

33. Perimeter: 40 in; Area: 84 square in. **35.** Perimeter: 45 cm; Area: 94.5 square cm **37.** Perimeter: 32 m; Area: 42 square m

39. Perimeter: 32 feet; Area: 24 square feet **41.** Circumference; 32π in. $\approx$ 100.53 in; Area: 256π square in. $\approx$ 804.25 square in.

43. Circumference: 20π cm $\approx$ 62.83 cm; Area 100π square cm $\approx$ 314.16 square cm **45.** π square units **47.** about 26.18 feet

Section B. 4 Volume and Surface Area **1.** polyhedron **2.** surface area **3.** False **4.** True **5.** Volume: 125 cubic m; Surface area: 150 square m

6. Volume: $\dfrac{256}{3}\pi$ cubic in. $\approx$ 268.08 cubic in.; Surface area: 64π square in. $\approx$ 201.06 square in. **7.** Volume: 600 cubic feet; Surface Area: 460 square feet

9. Volume: 288π cubic centimeters $\approx$ 904.78 cubic centimeters; Surface Area: 144π square centimeters $\approx$ 452.39 square centimeters

11. Volume: 32π cubic inches $\approx$ 100.53 cubic inches: Surface Area: 40π square inches $\approx$ 125.66 square inches

13. Volume: $\dfrac{800}{3}\pi$ cubic mm $\approx$ 837.76 cubic mm **15.** Volume $\dfrac{640}{3}$ cubic feet; Surface Area: 256 square feet **17.** 1728 cubic inches

19. 75.40 cubic inches; 100.53 square inches **21.** approximately 268.08 cubic cm

Appendix C More on Systems of Linear Equations

Section C.1 A Review of Systems of Linear Equations in Two Variables **1.** system of linear equations **2.** solution **3. (a)** no **(b)** yes **(c)** no

4. inconsistent **5.** consistent; dependent **6.** False **7.** True **8.** True **9.** $(3, 1)$ **10.** $(-2, 3)$ **11.** $(-4, 7)$ **12.** $(-6, 10)$ **13.** additive inverses **14.** $\left(\dfrac{1}{2}, \dfrac{7}{3}\right)$

15. $(-5, -6)$ **16.** $(3, 6)$

17. $\varnothing$ or $\{\ \}$ **18.** $\varnothing$ or $\{\ \}$ **19.** dependent **20.** $\{(x, y) \mid -3x + 2y = 8\}$ **21.** $(-5, 2)$ **22.** $(12, 15)$

23. (a) no **(b)** yes **25. (a)** yes **(b)** yes **27.** consistent; independent **29.** inconsistent

31. $(1, 3)$

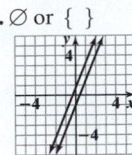

33. $(3, -4)$

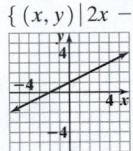

35. $(6, -2)$ **37.** $(-2, -3)$ **39.** $\left(\frac{1}{2}, -\frac{1}{4}\right)$ **41.** $(2500, 7500)$ **51.** $\varnothing$ or $\{\ \}$

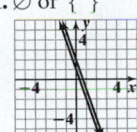

43. $(-8, 3)$ **45.** $(11, -8)$ **47.** $\left(\frac{1}{2}, -\frac{4}{5}\right)$ **49.** $(25, 40)$

53. $\varnothing$ or $\{\ \}$ **55.** $\{(x, y) \mid 2x - 4y = -4\}$ **57.** $\{(x, y) \mid x + 3y = 6\}$ **59.** $\left\{(x, y) \mid \frac{1}{3}x - 2y = 6\right\}$ **61.** $(-9, 3)$ **63.** $\{(x, y) \mid x = 5y - 3\}$

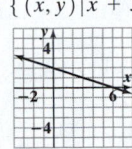

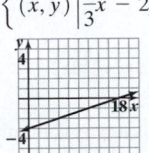

65. $\left(\frac{39}{11}, -\frac{30}{11}\right)$ **67.** $\varnothing$ or $\{\ \}$

69. $y = -2x - 5$; $y = -\frac{5}{3}x + \frac{1}{3}$; exactly one solution

71. $y = \frac{3}{2}x + 1$; $y = \frac{3}{2}x + 1$; infinite number of solutions

73. (a) $y = -\frac{1}{2}x + \frac{5}{2}$; $y = x + 1$ **(b)** $(1, 2)$ **75. (c)** and **(f)** **77.** $A = \frac{7}{6}$ and $B = -\frac{1}{2}$ **79.** Answers will vary. One possibility follows: $\begin{cases} x + y = 3 \\ x - y = -5 \end{cases}$

81. $(1, 2)$ **83.** $(6, -1)$

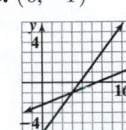

85. Yes, typically, we use substitution when one of the equations is solved for one of the variables or when the coefficient on one of the variables is 1 (which makes it easy to solve for that variable). Otherwise, we use elimination.

87. $(3, -2)$

89. $(1.2, 2.6)$

91. $(4, 13)$

93. $\{(x, y) \mid 4x - 3y = 1\}$

95. approximately $(0.35, -3.76)$

Section C.2 Systems of Linear Equations in Three Variables
1. inconsistent; consistent; dependent **2.** solution **3.** False **4.** True

5. (a) no **(b)** yes **6.** $(-3, 1, -1)$ **7.** $\left(-\frac{5}{2}, \frac{5}{4}, \frac{1}{2}\right)$ **8.** $\varnothing$ or $\{\ \}$; the system is inconsistent **9.** $\{(x, y, z) \mid x = -z + 1, y = 2z - 1, z$ is any real number$\}$ **10.** The company can manufacture fourteen 21-inch, eleven 24-inch, and five 40-inch **11. (a)** no **(b)** yes **13.** $(3, -2, 4)$

15. $(3, -4, -2)$ **17.** $\left(\frac{1}{3}, -\frac{2}{3}, 2\right)$ **19.** $(3, -3, 1)$ **21.** $\varnothing$ or $\{\ \}$ **23.** $\{(x, y, z) \mid x = 3z - 10, y = -4z + 14, z$ is any real number$\}$

25. $\{(x, y, z) \mid x = -3z + 5, y = -6z + 11, z$ is any real number$\}$ **27.** $\left(0, 2, \frac{3}{2}\right)$ **29.** $\varnothing$ or $\{\ \}$ **31.** $(-5, 0, 3)$ **33.** $\left(-\frac{7}{2}, -\frac{7}{2}, \frac{3}{2}\right)$

35. $\{(x, y, z) \mid x = 5z - 5, y = -2z + 3, z$ is any real number$\}$ **37.** $\left(\frac{11}{2}, 0, -\frac{1}{2}\right)$ **39.** $\left\{(x, y, z) \mid x = \frac{1}{3}z + \frac{2}{3}, y = -\frac{4}{3}z + \frac{7}{3}, z$ is any real number$\right\}$

41. Answers will vary. One possibility follows. $\begin{cases} x + y + z = 4 \\ x - y + z = 6 \\ x + y - z = -2 \end{cases}$ **43. (a)** $a - b + c = -6$; $4a + 2b + c = 3$ **(b)** $a = -2, b = 5, c = 1$; $f(x) = -2x^2 + 5x + 1$ **45.** $i_1 = 1, i_2 = 4,$ and $i_3 = 3$ **47.** There are 450 club seats, 3795 reserved seats, and 1755 lawn seats in the stadium.

49. Nancy needs 1 serving of Chex® cereal, 2 servings of 2% milk, and 1.5 servings of orange juice. **51.** \$12,000 in Treasury bills, \$8000 in municipal bonds, and \$5000 in corporate bonds. **53.** $\overline{AM} = 4, \overline{BN} = 2,$ and $\overline{OC} = 10$ **55.** $(10, -4, 6)$ **57.** $(-2, 1, 0, 4)$

59. To create a system of two equations and two unknowns, something we already know how to solve.

Section C.3 Using Matrices to Solve Systems
1. matrix **2.** augmented **3.** 4; 3 **4.** False **5.** $\begin{bmatrix} 3 & -1 & | & -10 \\ -5 & 2 & | & 0 \end{bmatrix}$

6. $\begin{bmatrix} 1 & 2 & -2 & | & 11 \\ -1 & 0 & -2 & | & 4 \\ 4 & -1 & 1 & | & 3 \end{bmatrix}$ **7.** $\begin{cases} x - 3y = 7 \\ -2x + 5y = -3 \end{cases}$ **8.** $\begin{cases} x - 3y + 2z = 4 \\ 3x \quad - z = -1 \\ -x + 4y \quad = 0 \end{cases}$ **9.** $\begin{bmatrix} 1 & -2 & | & 5 \\ 0 & -3 & | & 9 \end{bmatrix}$ **10.** $R_1 = -5r_2 + r_1$; $\begin{bmatrix} 1 & 0 & | & 3 \\ 0 & 1 & | & 2 \end{bmatrix}$

11. True **12.** $(6, -2)$ **13.** $(3, -2, 1)$ **14.** $\left(\frac{7}{2}, \frac{2}{3}, 4\right)$ **15.** $\{(x, y) \mid 2x + 5y = -6\}$

16. $\{(x, y, z) \mid x = -z + 5, y = 4z + 3, z$ is any real number$\}$ **17.** $\varnothing$ or $\{\ \}$ **18.** $\varnothing$ or $\{\ \}$ **19.** $\begin{bmatrix} 1 & -3 & | & 2 \\ 2 & 5 & | & 1 \end{bmatrix}$ **21.** $\begin{bmatrix} 1 & 1 & 1 & | & 3 \\ 2 & -1 & 3 & | & 1 \\ -4 & 2 & -5 & | & -3 \end{bmatrix}$

23. $\begin{bmatrix} -1 & 1 & | & 2 \\ 5 & 1 & | & -5 \end{bmatrix}$ **25.** $\begin{bmatrix} 1 & 0 & 1 & | & 2 \\ 2 & 1 & 0 & | & 13 \\ 1 & -1 & 4 & | & -4 \end{bmatrix}$ **27.** $\begin{cases} 2x + 5y = 3 \\ -4x + y = 10 \end{cases}$ **29.** $\begin{cases} x + 5y - 3z = 2 \\ 3y - z = -5 \\ 4x \quad + 8z = 6 \end{cases}$ **31.** $\begin{cases} x - 2y + 9z = 2 \\ y - 5z = 8 \\ z = \frac{4}{3} \end{cases}$

33. (a) $\begin{bmatrix} 1 & -3 & | & 2 \\ 0 & -1 & | & 5 \end{bmatrix}$ **(b)** $\begin{bmatrix} 1 & -3 & | & 2 \\ 0 & 1 & | & -5 \end{bmatrix}$ **35. (a)** $\begin{bmatrix} 1 & 1 & -1 & | & 4 \\ 0 & 3 & 5 & | & -11 \\ -1 & -3 & 2 & | & 1 \end{bmatrix}$ **(b)** $\begin{bmatrix} 1 & 1 & -1 & | & 4 \\ 0 & 3 & 5 & | & -11 \\ 0 & -2 & 1 & | & 5 \end{bmatrix}$

37. (a) $\begin{bmatrix} 1 & 1 & 1 & | & 4 \\ 0 & 1 & 5 & | & 5 \\ 0 & -4 & 2 & | & 8 \end{bmatrix}$ **(b)** $\begin{bmatrix} 1 & 1 & 1 & | & 4 \\ 0 & 1 & 5 & | & 5 \\ 0 & -2 & 1 & | & 4 \end{bmatrix}$ **39.** $(8, -5)$ **41.** $(-2, 5, 0)$ **43.** $\left(\dfrac{29}{34}, -\dfrac{47}{85}, \dfrac{11}{34}\right)$ **45.** $\{(x, y) \mid x - 3y = 3\}$ **47.** $\varnothing$ or $\{\ \}$

49. $\left\{(x, y, z) \mid x = \dfrac{3}{4}z + 1, y = \dfrac{5}{4}z + 3, z \text{ is any real number}\right\}$ **51.** $\begin{cases} x + 4y = -5 & (1) \\ \quad\quad y = -2 & (2) \end{cases}$ consistent and independent; $(3, -2)$

53. $\begin{cases} x + 3y - 2z = 6 & (1) \\ \quad\quad y + 5z = -2 & (2) \\ \quad\quad\quad\quad 0 = 4 & (3) \end{cases}$ inconsistent; $\varnothing$ or $\{\ \}$ **55.** $\begin{cases} x - 2y - z = 3 & (1) \\ \quad\quad y - 2z = -8 & (2) \\ \quad\quad\quad\quad z = 5 & (3) \end{cases}$ consistent and independent; $(12, 2, 5)$

57. $(3, -5)$ **59.** $\varnothing$ or $\{\ \}$ **61.** $\left(\dfrac{1}{2}, -\dfrac{5}{4}\right)$ **63.** $\{(x, y) \mid 4x - y = 8\}$ **65.** $(3, -4, 1)$ **67.** $(4, 0, -5)$

69. $\{(x, y, z) \mid x = -2z - 0.2, y = -z - 1.4, z \text{ is any real number}\}$ **71.** $\varnothing$ or $\{\ \}$ **73.** $\left(\dfrac{3}{10}, \dfrac{1}{10}, -\dfrac{1}{2}\right)$ **75.** $\left(\dfrac{5}{3}, \dfrac{2}{5}, -\dfrac{1}{2}\right)$

77. $(-2, 1, -5)$ **79. (a)** $a + b + c = 0; 4a + 2b + c = 3$ **(b)** $a = 2, b = -3, c = 1; f(x) = 2x^2 - 3x + 1$
81. \$8000 in Treasury bills, \$7000 in municipal bonds, and \$5000 in corporate bonds **83.** $(-3, 7)$ **85.** $(2, 5, -4)$ **87.** Answers will vary.

89. Multiply each entry in row 2 by $\dfrac{1}{5}$: $R_2 = \dfrac{1}{5}r_2$. **91.** $(5, -3)$ **93.** $(4, -2, 7)$

Section C.4 Determinants and Cramer's Rule 1. $ad - bc$ **2.** square **3.** 18 **4.** -9 **5.** $(-3, 5)$ **6.** Cramer's Rule does not apply because $D = 0$.

7. -91 **8.** 20 **9.** $\left(\dfrac{1}{2}, -2, -1\right)$ **10.** $(-1, 4, 1)$ **11.** 10 **13.** -2 **15.** $(-8, 4)$ **17.** $(3, -1)$ **19.** $\left(-\dfrac{1}{6}, \dfrac{3}{8}\right)$ **21.** $\left(\dfrac{3}{4}, -\dfrac{7}{4}\right)$ **23.** -9 **25.** -163 **27.** 0

29. $(-2, 1, -1)$ **31.** $(3, -1, 2)$ **33.** Cramer's Rule does not apply. **35.** Cramer's Rule does not apply. **37.** $(12, -6, 3)$

39. $\left(\dfrac{7}{5}, -\dfrac{5}{3}, \dfrac{11}{3}\right)$ **41.** $(1, 1, 1)$ **43.** $x = 5$ **45.** $x = -2$

47. (a)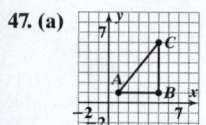
49. (a) 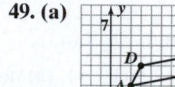 **(b)**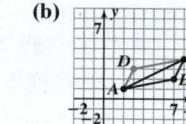

51. (a) $x + 2y = 7$ **(b)** $x + 2y = 7$
53. $14: -14:$ answers may vary.
55. $(-8, 4)$
57. $(3, -1)$
59. $(-2, 1, -1)$

(b) The area of triangle ABC is 10 square units.

The area of triangle ABC is 4.5 square units.

(c) The area of triangle ADC is 4.5 square units.
(d) 9 square units

Appendix E Transformations

1. up; down **2.** 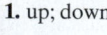 **3.** 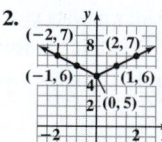 **4.** True **5.**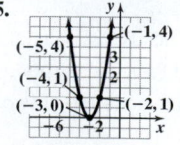

6. 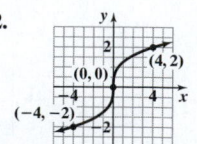 **7.** y; a; vertically compressed; vertically stretched **8.** 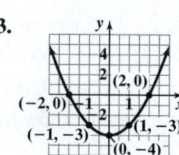 **9.** 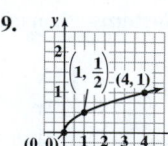 **10.** x; $\dfrac{1}{a}$; horizontally stretched; horizontally compressed

11. 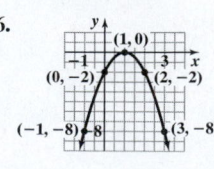 **12.**  **13.** x; y **14.** 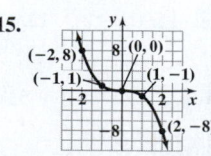 **15.**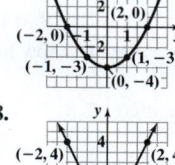

16. 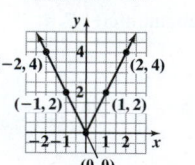 **17.** (I) (C); (II) (B); (III) (A); (IV) (D) **19.** $y = \sqrt{x} + 3$ **21.** $y = \sqrt{x} - 2$ **23.** $y = -\sqrt{x}$

25. $y = \sqrt{\dfrac{1}{2}x}$

27.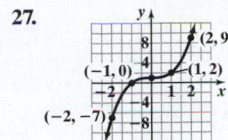
Domain: All real numbers;
Range: All real numbers

29.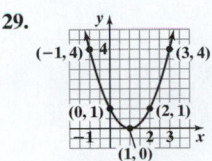
Domain: All real numbers;
Range: $[0, \infty)$

31.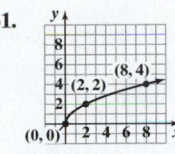
Domain: $[0, \infty)$;
Range: $[0, \infty)$

33.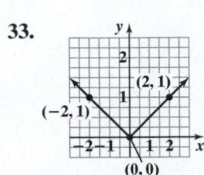
Domain: All real numbers;
Range: $[0, \infty)$

35.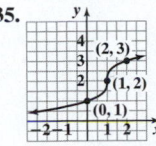
Range: All real numbers;
Domain: All real numbers

37.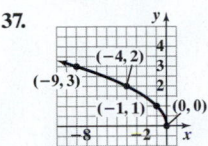
Domain: $(-\infty, 0]$;
Range: $[0, \infty)$

39.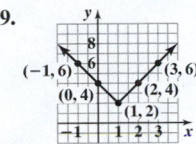
Domain: All real numbers;
Range: $[2, \infty)$

41.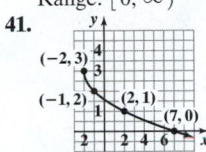
Domain: $[-2, \infty)$;
Range: $(-\infty, 3]$

43. (a)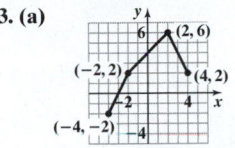
(b)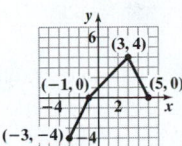
(c)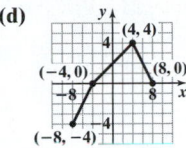
(d)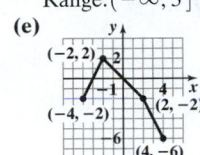

(e)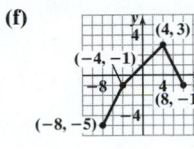
(f)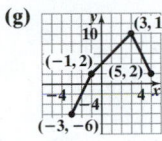
(g)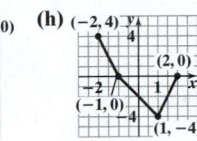
(h)

Applications Index

Subject Index

Properties of the Real Number System (Chapter 1)

	Addition	Multiplication
Identity Properties	$a + 0 = 0 + a = a$	$1 \cdot a = a \cdot 1 = a$
Inverse Properties	$a + (-a) = (-a) + a = 0$	$a \cdot \dfrac{1}{a} = \dfrac{1}{a} \cdot a = 1 (a \neq 0)$
Commutative Property	$a + b = b + a$	$a \cdot b = b \cdot a$
Associative Property	$(a + b) + c = a + (b + c)$	$(ab)c = a(bc)$
Distributive Property	$a(b + c) = ab + ac$	
Division Properties	$\dfrac{0}{a} = 0, a \neq 0; \dfrac{a}{a} = 1, a \neq 0; \dfrac{a}{0}$ is undefined, $a \neq 0$	

Order of Operations (Chapter 1)

How to Evaluate Expressions

1. Perform all operations with grouping symbols first. If multiple grouping symbols, begin with the innermost grouping symbol and work outward.

2. Then evaluate the exponents.

3. Next multiplication and division, working from left to right.

4. Perform addition and subtraction, working from left to right.

Geometry Formulas (Chapter 2)

Figure	Formulas	Figure	Formulas
Square	**Area:** $A = s^2$ **Perimeter:** $P = 4s$	Cube	**Volume:** $V = s^3$ **Surface Area:** $S = 6s^2$
Rectangle	**Area:** $A = lw$ **Perimeter:** $P = 2l + 2w$	Rectangular Solid	**Volume:** $V = lwh$ **Surface Area:** $S = 2lw + 2lh + 2wh$
Circle	**Area:** $A = \pi r^2$ **Circumference:** $C = 2\pi r = \pi d$	Sphere	**Volume:** $V = \dfrac{4}{3}\pi r^3$ **Surface Area:** $S = 4\pi r^2$
Triangle	**Area:** $A = \dfrac{1}{2}bh$ **Perimeter:** $P = a + b + c$	Right Circular Cylinder	**Volume:** $V = \pi r^2 h$ **Surface Area:** $S = 2\pi r^2 + 2\pi rh$
Trapezoid	**Area:** $A = \dfrac{1}{2}h(B + b)$ **Perimeter:** $P = a + b + c + B$	Cone	**Volume:** $V = \dfrac{1}{3}\pi r^2 h$
Parallelogram	**Area:** $A = ah$ **Perimeter:** $P = 2a + 2b$		

Formulas for Lines and Slope (Chapter 3)

Standard form of a line	$Ax + By = C$
Equation of a vertical line	$x = a$ where $(a, 0)$ is the x-intercept
Equation of a horizontal line	$y = b$ where $(0, b)$ is the y-intercept
Slope of a line	$m = \dfrac{y_2 - y_1}{x_2 - x_1}, x_1 \neq x_2$ Slope undefined if $x_1 = x_2$
Point-slope form of a line	$y - y_1 = m(x - x_1)$
Slope-intercept form of a line	$y = mx + b$

Functions (Chapter 8)

- A **function** is a special type of relation where any given input, x, corresponds to only one output y. Functions can be represented through maps, sets of ordered pairs, equations, or graphs.
- **Vertical Line Test:** A set of points in the xy-plane is the graph of a function if and only if every vertical line intersects the graph in at most one point.
- The graph of a function, f, is the set of all ordered pairs $(x, f(x))$.
- When only an equation of a function is given, the **domain** of the function is the largest set of real numbers for which $f(x)$ is a real number.
- The **range** of a function is the set of all outputs of the function.

Steps for Factoring (Chapter 6)

Step 1: Factor out the Greatest Common Factor (GCF), if any exists.

Step 2: Count the number of terms.

Step 3: (a) 2 terms
- Is it the difference of two squares? If so,
  $A^2 - B^2 = (A - B)(A + B)$
- Is it the sum of two squares? if so, stop! The expression is prime.
- Is it the difference of two cubes? If so,
  $A^3 - B^3 = (A - B)(A^2 + AB + B^2)$
- Is it the sum of two cubes? If so,
  $A^3 + B^3 = (A + B)(A^2 - AB + B^2)$

(b) 3 terms
- Is it a perfect square trinomial? If so,
  $A^2 + 2AB + B^2 = (A + B)^2$ or
  $A^2 - 2AB + B^2 = (A - B)^2$
- Is the coefficient of the square term 1? If so, $x^2 + bx + c = (x + m)(x + n)$, where $mn = c$ and $m + n = b$
- Is the coefficient of the square term different from 1? If so,
  (a) Use factoring by grouping
  (b) Use trial and error

(c) 4 terms
- Use factoring by grouping

Step 4: Check your work by multiplying out the factored form.

The Rules of Exponents (Chapter 5, Chapter 8)

If a and b are real numbers and if r and s are rational numbers, then assuming the expression is defined,

Zero-Exponent Rule:	$a^0 = 1$	if $a \neq 0$
Negative-Exponent Rule:	$a^{-r} = \dfrac{1}{a^r}$	if $a \neq 0$.
Product Rule:	$a^r \cdot a^s = a^{r+s}$	
Quotient Rule:	$\dfrac{a^r}{a^s} = a^{r-s} = \dfrac{1}{a^{s-r}}$	if $a \neq 0$
Power Rule:	$(a^r)^s = a^{r \cdot s}$	
Product-to-a-Power Rule:	$(a \cdot b)^r = a^r \cdot b^r$	
Quotient-to-a-Power Rule:	$\left(\dfrac{a}{b}\right)^r = \dfrac{a^r}{b^r}$	if $b \neq 0$
Quotient-to-a-Negative-Power Rule:	$\left(\dfrac{a}{b}\right)^{-r} = \left(\dfrac{b}{a}\right)^r$	if $a \neq 0, b \neq 0$

Working with Rational Expressions (Chapter 7)

Multiplying Rational Expressions	$\dfrac{a}{b} \cdot \dfrac{c}{d} = \dfrac{ac}{bd}$	$b \neq 0, d \neq 0$
Adding Rational Expressions	$\dfrac{a}{c} + \dfrac{b}{c} = \dfrac{a + b}{c}$	$c \neq 0$
Subtracting Rational Expressions	$\dfrac{a}{c} - \dfrac{b}{c} = \dfrac{a - b}{c}$	$c \neq 0$
Dividing Rational Expressions	$\dfrac{a}{b} \div \dfrac{c}{d} = \dfrac{\frac{a}{b}}{\frac{c}{d}} = \dfrac{a}{b} \cdot \dfrac{d}{c} = \dfrac{ad}{bc}$	$b \neq 0, c \neq 0, d \neq 0$